Periodic Table of the Elements

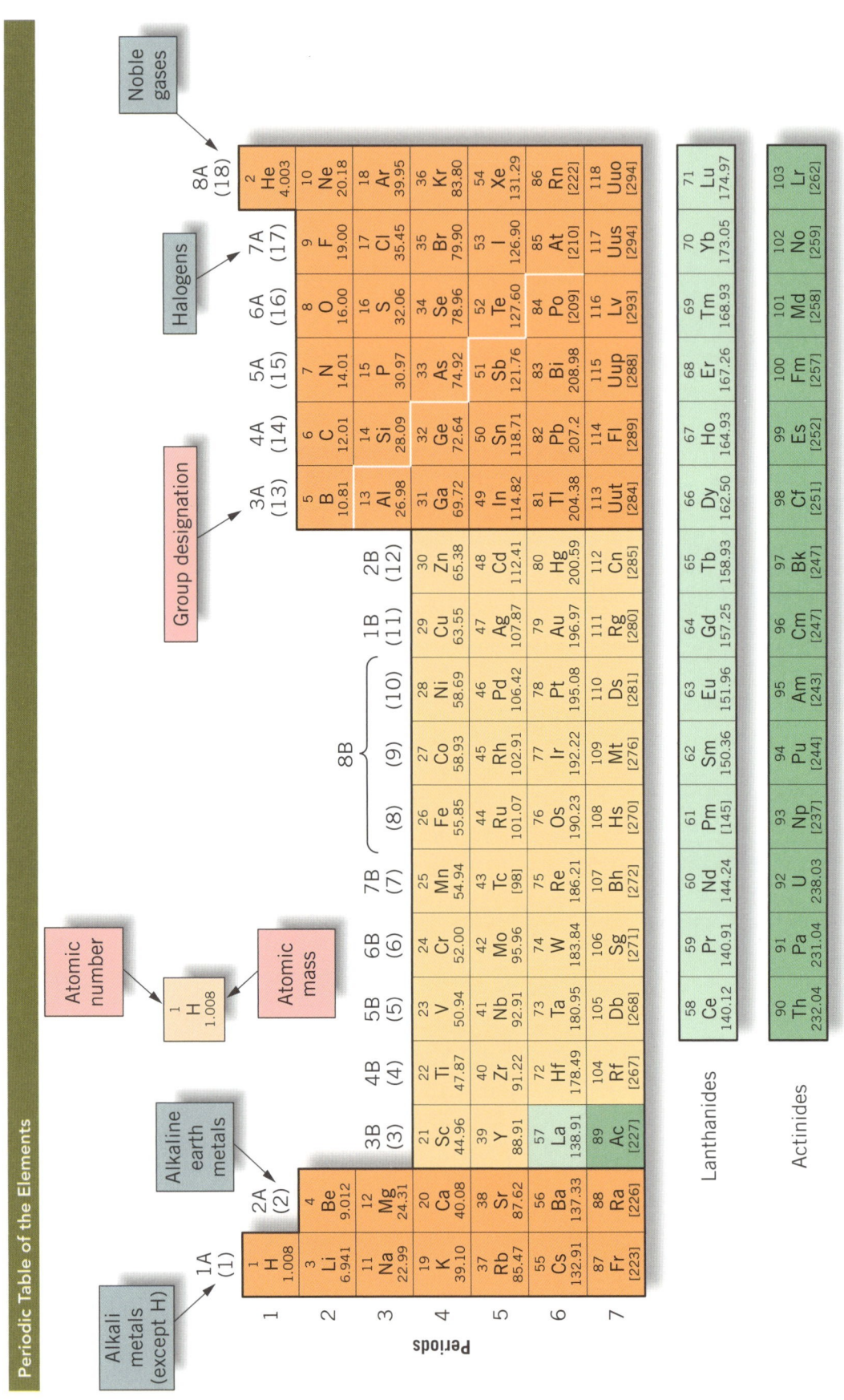

Alkali metals (except H)

Alkaline earth metals

Atomic number → 1 H 1.008 ← **Atomic mass**

Group designation

Halogens

Noble gases

Periods

	1A (1)	2A (2)	3B (3)	4B (4)	5B (5)	6B (6)	7B (7)	8B (8)	8B (9)	8B (10)	1B (11)	2B (12)	3A (13)	4A (14)	5A (15)	6A (16)	7A (17)	8A (18)
1	1 H 1.008																	2 He 4.003
2	3 Li 6.941	4 Be 9.012											5 B 10.81	6 C 12.01	7 N 14.01	8 O 16.00	9 F 19.00	10 Ne 20.18
3	11 Na 22.99	12 Mg 24.31											13 Al 26.98	14 Si 28.09	15 P 30.97	16 S 32.06	17 Cl 35.45	18 Ar 39.95
4	19 K 39.10	20 Ca 40.08	21 Sc 44.96	22 Ti 47.87	23 V 50.94	24 Cr 52.00	25 Mn 54.94	26 Fe 55.85	27 Co 58.93	28 Ni 58.69	29 Cu 63.55	30 Zn 65.38	31 Ga 69.72	32 Ge 72.64	33 As 74.92	34 Se 78.96	35 Br 79.90	36 Kr 83.80
5	37 Rb 85.47	38 Sr 87.62	39 Y 88.91	40 Zr 91.22	41 Nb 92.91	42 Mo 95.96	43 Tc [98]	44 Ru 101.07	45 Rh 102.91	46 Pd 106.42	47 Ag 107.87	48 Cd 112.41	49 In 114.82	50 Sn 118.71	51 Sb 121.76	52 Te 127.60	53 I 126.90	54 Xe 131.29
6	55 Cs 132.91	56 Ba 137.33	57 La 138.91	72 Hf 178.49	73 Ta 180.95	74 W 183.84	75 Re 186.21	76 Os 190.23	77 Ir 192.22	78 Pt 195.08	79 Au 196.97	80 Hg 200.59	81 Tl 204.38	82 Pb 207.2	83 Bi 208.98	84 Po [209]	85 At [210]	86 Rn [222]
7	87 Fr [223]	88 Ra [226]	89 Ac [227]	104 Rf [267]	105 Db [268]	106 Sg [271]	107 Bh [272]	108 Hs [270]	109 Mt [276]	110 Ds [281]	111 Rg [280]	112 Cn [285]	113 Uut [284]	114 Fl [289]	115 Uup [288]	116 Lv [293]	117 Uus [294]	118 Uuo [294]

Lanthanides

58 Ce 140.12	59 Pr 140.91	60 Nd 144.24	61 Pm [145]	62 Sm 150.36	63 Eu 151.96	64 Gd 157.25	65 Tb 158.93	66 Dy 162.50	67 Ho 164.93	68 Er 167.26	69 Tm 168.93	70 Yb 173.05	71 Lu 174.97

Actinides

90 Th 232.04	91 Pa 231.04	92 U 238.03	93 Np [237]	94 Pu [244]	95 Am [243]	96 Cm [247]	97 Bk [247]	98 Cf [251]	99 Es [252]	100 Fm [257]	101 Md [258]	102 No [259]	103 Lr [262]

ATOMIC MASSES OF THE ELEMENTS

This table is based on the 2007 table at *Pure Appl. Chem.,* **81**, 2131–2156 (2009) with changes to the values for lutetium, molybdenum, nickel, ⬤rbium and zinc from the 2005 table, and additions from IUPAC 2011 Periodic Table of the Elements for flerovium and livermorium. Mass ⬤ber of the longest-lived isotope of hassium from *Phys. Rev. Lett.,* **97** 242501 (2006). The number in parentheses following the atomic mass is the estimated uncertainty in the last digit.

At No	Symbol	Name	Atomic Mass	Notes	At No	Symbol	Name	Atomic Mass	Notes
89	Ac	Actinium	[227]	5	101	Md	Mendelevium	[258]	5
13	Al	Aluminium	26.9815386(8)		80	Hg	Mercury	200.59(2)	
95	Am	Americium	[243]	5	42	Mo	Molybdenum	95.96(2)	1
51	Sb	Antimony	121.760(1)	1	60	Nd	Neodymium	144.242(3)	1
18	Ar	Argon	39.948(1)	1, 2	10	Ne	Neon	20.1797(6)	1, 3
33	As	Arsenic	74.92160(2)		93	Np	Neptunium	[237]	5
85	At	Astatine	[210]	5	28	Ni	Nickel	58.6934(4)	
56	Ba	Barium	137.327(7)		41	Nb	Niobium	92.90638(2)	
97	Bk	Berkelium	[247]	5	7	N	Nitrogen	14.0067(2)	1, 2, 4
4	Be	Beryllium	9.012182(3)		102	No	Nobelium	[259]	5
83	Bi	Bismuth	208.98040(1)		76	Os	Osmium	190.23(3)	1
107	Bh	Bohrium	[272]	5	8	O	Oxygen	15.9994(3)	1, 2, 4
5	B	Boron	10.811(7)	1, 2, 3, 4	46	Pd	Palladium	106.42(1)	1
35	Br	Bromine	79.904(1)	4	15	P	Phosphorus	30.973762(2)	
48	Cd	Cadmium	112.411(8)	1	78	Pt	Platinum	195.084(9)	
55	Cs	Cesium	132.9054519(2)		94	Pu	Plutonium	[244]	5
20	Ca	Calcium	40.078(4)	1	84	Po	Polonium	[209]	5
98	Cf	Californium	[251]	5	19	K	Potassium	39.0983(1)	1
6	C	Carbon	12.0107(8)	1, 2, 4	59	Pr	Praseodymium	140.90765(2)	
58	Ce	Cerium	140.116(1)	1	61	Pm	Promethium	[145]	5
17	Cl	Chlorine	35.453(2)	3, 4	91	Pa	Protactinium	231.03588(2)	5
24	Cr	Chromium	51.9961(6)		88	Ra	Radium	[226]	5
27	Co	Cobalt	58.933195(5)		86	Rn	Radon	[222]	5
112	Cn	Copernicium	[285]	5	75	Re	Rhenium	186.207(1)	
29	Cu	Copper	63.546(3)	2	45	Rh	Rhodium	102.90550(2)	
96	Cm	Curium	[247]	5	111	Rg	Roentgenium	[280]	5
110	Ds	Darmstadtium	[281]	5	37	Rb	Rubidium	85.4678(3)	1
105	Db	Dubnium	[268]	5	44	Ru	Ruthenium	101.07(2)	1
66	Dy	Dysprosium	162.500(1)	1	104	Rf	Rutherfordium	[265]	5
99	Es	Einsteinium	[252]	5	62	Sm	Samarium	150.36(2)	1
68	Er	Erbium	167.259(3)	1	21	Sc	Scandium	44.955912(6)	
63	Eu	Europium	151.964(1)	1	106	Sg	Seaborgium	[271]	5
100	Fm	Fermium	[257]	5	34	Se	Selenium	78.96(3)	
114	Fl	Flerovium	[289]	5	14	Si	Silicon	28.0855(3)	2, 4
9	F	Fluorine	18.9984032(5)		47	Ag	Silver	107.8682(2)	1
87	Fr	Francium	[223]	5	11	Na	Sodium	22.98976928(2)	
64	Gd	Gadolinium	157.25(3)	1	38	Sr	Strontium	87.62(1)	1, 2
31	Ga	Gallium	69.723(1)		16	S	Sulfur	32.065(5)	1, 2, 4
32	Ge	Germanium	72.64(1)		73	Ta	Tantalum	180.94788(2)	
79	Au	Gold	196.966569(4)		43	Tc	Technetium	[98]	5
72	Hf	Hafnium	178.49(2)		52	Te	Tellurium	127.60(3)	1
108	Hs	Hassium	[270]	5	65	Tb	Terbium	158.92535(2)	
2	He	Helium	4.002602(2)	1, 2	81	Tl	Thallium	204.3833(2)	4
67	Ho	Holmium	164.93032(2)		90	Th	Thorium	232.03806(2)	1, 5
1	H	Hydrogen	1.00794(7)	1, 2, 3, 4	69	Tm	Thulium	168.93421(2)	
49	In	Indium	114.818(3)		50	Sn	Tin	118.710(7)	1
53	I	Iodine	126.90447(3)		22	Ti	Titanium	47.867(1)	
77	Ir	Iridium	192.217(3)		74	W	Tungsten	183.84(1)	
26	Fe	Iron	55.845(2)		118	Uuo	Ununoctium	[294]	5
36	Kr	Krypton	83.798(2)	1, 3	117	Uus	Ununseptium	[294]	5
57	La	Lanthanum	138.90547(7)	1	115	Uup	Ununpentium	[288]	5
103	Lr	Lawrencium	[262]	5	113	Uut	Ununtrium	[284]	5
82	Pb	Lead	207.2(1)	1, 2	92	U	Uranium	238.02891(3)	1, 3, 5
3	Li	Lithium	6.941(2)	1, 2, 3, 4	23	V	Vanadium	50.9415(1)	
116	Lv	Livermorium	[293]	5	54	Xe	Xenon	131.293(6)	1, 3
71	Lu	Lutetium	174.9668(1)	1	70	Yb	Ytterbium	173.054(5)	1
12	Mg	Magnesium	24.3050(6)	4	39	Y	Yttrium	88.90585(2)	
25	Mn	Manganese	54.938045(5)		30	Zn	Zinc	65.38(2)	
109	Mt	Meitnerium	[276]	5	40	Zr	Zirconium	91.224(2)	1

1. Geological specimens are known in which the element has an isotopic composition outside the limits for normal material. The difference between the atomic mass of the element in such specimens and that given in the Table may exceed the stated uncertainty.

2. Range in isotopic composition of normal terrestrial material prevents a more precise ⬤ being given; the tabulated value should be applicable to any normal material.

⬤odified isotopic compositions may be found in commercially available material because it has been subject to an undisclosed or inadvertant isotopic fractionation. Substantial deviations in atomic mass of the element from that given in the Table can occur.

4. IUPAC recommends a range of masses for H, Li, B, C, N, O, Mg, Si, S, Cl, Br, Tl. For simplicity we have decided to use the single masses. In On the Cutting Edge 0.3, these masses and their ranges are discussed further.

5. Element has no stable nuclides. The value enclosed in brackets, e.g. [209], indicates the mass number of the longest-lived isotope of the element. However three such elements (Th, Pa, and U) do have a characteristic terrestrial isotopic composition, and for these an atomic mass is tabulated.

CHEMISTRY

The Molecular Nature
of Matter

7th Edition

CHEMISTRY
The Molecular Nature of Matter

Neil D. Jespersen
St. John's University, New York

Alison Hyslop
St. John's University, New York

With significant contributions by
James E. Brady
St. John's University, New York

WILEY

VICE PRESIDENT and PUBLISHER Petra Recter
ACQUISITION EDITOR Nicholas Ferrari
SENIOR PROJECT EDITOR Jennifer Yee
SENIOR MARKETING MANAGER Kristine Ruff
INTERIOR DESIGNER Thomas Nery
COVER DESIGNER Thomas Nery
SENIOR PHOTO EDITOR Mary Ann Price
PHOTO RESEARCHER Lisa Passmore
SENIOR PRODUCT DESIGNER Geraldine Osnato
MEDIA SPECIALIST Daniela DiMaggio
SENIOR CONTENT MANAGER Kevin Holm
SENIOR PRODUCTION EDITOR Elizabeth Swain
COVER PHOTO © Isak55/Shutterstock

This book was set in 10.5 Adobe Garamond by Prepare and printed and bound by Courier Kendallville.
The cover was printed by Courier Kendallville.

This book is printed on acid free paper.

Founded in 1807, John Wiley & Sons, Inc. has been a valued source of knowledge and understanding for more than 200 years, helping people around the world meet their needs and fulfill their aspirations. Our company is built on a foundation of principles that include responsibility to the communities we serve and where we live and work. In 2008, we launched a Corporate Citizenship Initiative, a global effort to address the environmental, social, economic, and ethical challenges we face in our business. Among the issues we are addressing are carbon impact, paper specifications and procurement, ethical conduct within our business and among our vendors, and community and charitable support. For more information, please visit our website: www.wiley.com/go/citizenship.

Evaluation copies are provided to qualified academics and professionals for review purposes only, for use in their courses during the next academic year. These copies are licensed and may not be sold or transferred to a third party. Upon completion of the review period, please return the evaluation copy to Wiley. Return instructions and a free of charge return shipping label are available at www.wiley.com/go/returnlabel. Outside of the United States, please contact your local representative.

Main ISBN 978-1-118-51646-1
Binder version ISBN 978-1-118-41392-0

Printed in the United States of America
10 9 8 7 6 5 4 3 2 1

About the Authors

Neil D. Jespersen is a Professor of Chemistry at St. John's University in New York. He earned a B.S. with Special Attainments in Chemistry at Washington and Lee University (VA) and his Ph.D. in Analytical Chemistry with Joseph Jordan at The Pennsylvania State University. He has received awards for excellence in teaching and research from St. John's University and the E. Emmit Reid Award in college teaching from the American Chemical Society's Middle Atlantic Region. He chaired the Department of Chemistry for 6 years and has mentored the St. John's student ACS club for over 30 years while continuing to enjoy teaching Quantitative and Instrumental Analysis courses, along with General Chemistry. He has been an active contributor to the Eastern Analytical Symposium, chairing it in 1991. Neil authors the Barrons AP Chemistry Study Guide; has edited 2 books on Instrumental Analysis and Thermal Analysis; and has 4 chapters in research monographs, 50 refereed publications, and 150 abstracts and presentations. He is active at the local, regional and national levels of the American Chemical Society, and served on the ACS Board of Directors and was named a Fellow of the ACS in 2013. When there is free time you can find him playing tennis, baseball, and soccer with four grandchildren, or traveling with his wife Marilyn.

Alison Hyslop received her BA degree from Macalester College in 1986 and her Ph.D. from the University of Pennsylvania under the direction of Michael J. Therien in 1998. Alison currently chairs the Department of Chemistry at St. John's University, New York where she is an Associate Professor. She has been teaching graduate and undergraduate courses since 2000. She was a visiting Assistant Professor at Trinity College (CT) from 1998 to 1999. She was a visiting scholar at Columbia University (NY) in 2005 and in 2007 and at Brooklyn College in 2009, where she worked on research projects in the laboratory of Brian Gibney. Her research focuses on the synthesis and study of porphyrin-based light harvesting compounds. When not in the laboratory, she likes to hike in upstate New York, and practice tae kwon do.

James E. Brady received his BA degree from Hofstra College in 1959 and his Ph.D. from Penn State University under the direction of C. David Schmulbach in 1963. He is Professor Emeritus at St. John's University, New York, where he taught graduate and undergraduate courses for 35 years. His first textbook, *General Chemistry: Principles and Structure*, coauthored with Gerard Humiston, was published in 1975. An innovative feature of the text was 3D illustrations of molecules and crystal structures that could be studied with a stereo viewer that came tucked into a pocket inside the rear cover of the book. The popularity of his approach to teaching general chemistry is evident in the way his books have shaped the evolution of textbooks over the last 35 years. He has been the principal coauthor of various versions of this text, along with John Holum, Joel Russell, Fred Senese, Neil Jespersen, and Alison Hyslop. In 1999, Jim retired from St. John's University to devote more time to writing, and since then he has coauthored four editions of this text. He and his wife, June, enjoy their current home in Jacksonville, Florida where Jim is also an avid photographer.

Brief Table of Contents

Table of Contents

Special Topics

Preface

The seventh edition of our textbook continues to emphasize the molecular nature of matter, strong problem solving, and clarity of writing that was the basis of the sixth edition of *Chemistry: The Molecular Nature of Matter* by Neil D. Jespersen and James E. Brady. The relationship between the molecular level and the observable macroscopic properties of matter is presented in increased detail to reinforce and expand this fundamental concept.

Neil Jespersen continues his role as lead author as this text evolves in the electronic age. Neil is an analytical chemist, respected educator, and award-winning teacher who spearheaded the emphasis on the connection between the microscopic view and the macroscopic properties we experience in everyday life. **Alison Hyslop** has more than proven herself as a contributing author on the previous edition, and will continue to contribute to future editions. Alison is an inorganic chemist with extensive experience teaching graduate and undergraduate inorganic chemistry as well as general chemistry. She currently chairs her department and works collaboratively to enhance the chemistry degree programs. **James Brady** has taken an advisory role in this edition. His vision and guidance formed the philosophy and organization of the book. From completely introducing all topics before they are used to never skipping steps in solving problems, his leadership has made this book accessible for all chemistry students.

Philosophy and Goals

The philosophy of the text is based on our conviction that a general chemistry course serves a variety of goals in the education of a student. First, of course, it must provide a foundation in the basic facts and concepts of chemistry upon which theoretical models can be constructed. The general chemistry course should also give the student an appreciation of the central role that chemistry plays among the sciences, as well as the importance of chemistry in society and day-to-day living. In addition, it should enable the student to develop skills in analytical thinking and problem solving. With these thoughts in mind, our aim in structuring the text was to provide a logical progression of topics arranged to provide the maximum flexibility for the teacher in organizing his or her course. In this text, we were guided by **three principal goals**. The **first** was to strengthen the connection between observations on the macroscopic scale and the behavior of atoms, molecules, and ions at the atomic level while introducing all concepts in a logical and understandable manner. The **second** was to further enhance and streamline our approach to teaching effective problem-solving skills. This includes emphasis on estimation and answer checking. The **third** goal was to provide a seamless, total solution to the General Chemistry course by fully integrating the textbook content with online assessment, answer-specific responses, and resources delivered within *WileyPLUS*.

Emphasizing the Molecular View of Nature

The value of the molecular approach in teaching chemistry is well accepted and has always been a cornerstone in the approach taken by Jim Brady and his co-authors in presenting chemistry for many years. From his first text, in which novel three-dimensional computer-drawn representations of molecules and crystal structures were presented and observed using stereoscopic viewers, up through the 6th edition of this text, the atomic/molecular view has dominated the pedagogy. This new edition builds on that tradition by employing the "molecular basis of chemistry" as a powerful central theme of the text. Through this approach, the student will gain a sound appreciation of the nature of matter and how structure determines properties. Some actions we have taken to accomplish this are as follows:

Chapter Zero: A Very Brief History of Chemistry This new edition of the textbook begins with the formation of atoms from the origin of the universe. By discussing how atoms were initially formed and then moving on to the structure of the atoms through discoveries of the subatomic particles, we lay the groundwork for the atomic and molecular view of matter and outline how these concepts are used throughout the text. We provide a brief introduction to the distribution of elements throughout the earth and introduce students to the way we visualize molecules and chemical reactions.

Macro-to-Micro Illustrations To help students make the connection between the macroscopic world we see and events that take place at the molecular level, we have a substantial number of illustrations that combine both views. A photograph, for example, will show a chemical reaction as well as an artist's rendition of the chemical interpretation of what is taking place between the atoms, molecules, or ions involved.

The goal is to show how models of nature enable chemists to better understand their observations and to get students to visualize and describe events at the molecular level.

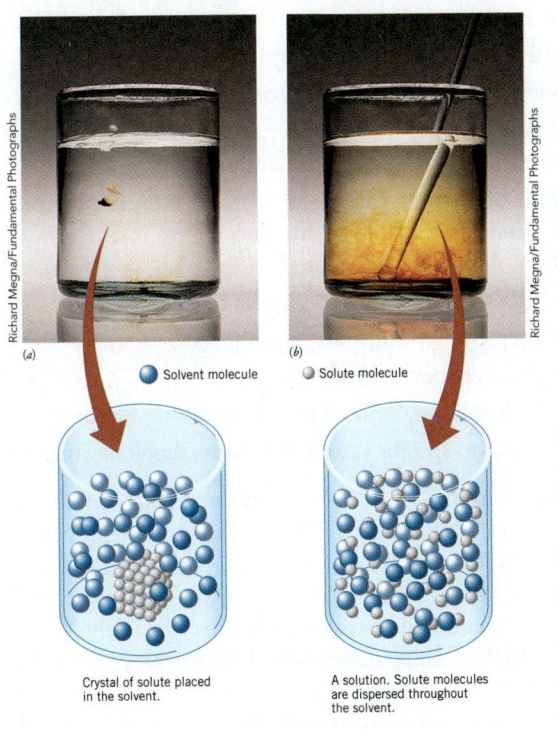

Figure 4.1 | Formation of a solution of iodine molecules in alcohol. (*a*) A crystal of iodine, I₂, on its way to the bottom of the beaker is already beginning to dissolve, the purplish iodine crystal forming a reddish brown solution. In the hugely enlarged view beneath the photo, we see the iodine molecules still bound in a crystal. For simplicity, the solute and solvent particles are shown as spheres. (*b*) Stirring the mixture helps the iodine molecules to disperse in the solvent, as illustrated in the molecular view below the photo.

Problem Solving and the Connection between Textbook and *WileyPLUS* When students solve the end-of-chapter problems in *WileyPLUS*, the feedback to the answers submitted will guide the students to the correct answer. For the answers that are incorrect, the responses will give an explanation as to why the answer is wrong. In addition, we have included question and answer-specific hints and answer-specific feedback to give the students more assistance in solving problems.

Learning Objectives The learning objectives for each chapter have been explicitly stated at the beginning of each chapter. These learning objectives give the students guidance as to what they will learn after they have mastered each section. In addition, all of the end-of-chapter Questions and Problems are organized by the learning objectives and labeled by the section headers.

Developing Problem-solving Skills

We strongly believe that problem solving reinforces the learning of concepts and that assisting students in improving their skills in this area is one of the critical aspects of teaching

chemistry. We also believe that it is possible to accommodate students who come into the course with a wide range of problem-solving abilities so that they will finish the course with skill sets that will make them successful in later chemistry courses.

We continue to use a **"chemical tools"** model and approach to aid in teaching problem analysis. This approach encourages students to think of basic skills, such as converting from grams to moles, as tools that can be combined in various ways to solve more complex problems. Students and instructors have responded positively to this concept in earlier editions and we continue to employ this strategy in problem analysis. Tools are identified by an icon in the margin when they are introduced in a chapter and the tools are summarized at the end of each chapter.

TOOLS

Tools for Problem Solving The following tools were introduced in this chapter. Study them carefully so you can select the appropriate tool when needed.

Criteria for a balanced ionic equation (Section 4.3)
To be balanced, an equation that includes the formulas of ions must satisfy two criteria: (1) the number of atoms of each kind must be the same on both sides of the equation, and (2) the net electrical charge shown on each side of the equation must be the same.

List of strong acids (Section 4.4)
The common strong acids are perchloric acid, $HClO_4$, chloric acid, $HClO_3$, hydrochloric acid, HCl, hydrobromic acid, HBr, hydroiodic acid, HI, nitric acid, HNO_3, and sulfuric acid, H_2SO_4.

A significant strength of previous editions was the four-step problem-solving process of *Analysis, Assembling the Tools, Solution,* and asking *Is the Answer Reasonable?*, which was applied to all worked examples. Like a mechanic we perform an Analysis to understand and plan how the problem can be solved. Then the Tools needed to do the work are assembled and used to provide the Solution. This reinforces the notion that the Tools can be combined in various ways to solve complex problems. The complete solution showing every step in a logical sequence is presented. Finally, as a mechanic always tests the repair job, we show how scientists test their answers while asking "Is the Answer Reasonable?"

Example 4.3
Writing the Equation for the Ionization of a Molecular Base

Dimethylamine, $(CH_3)_2NH$, is a base that is soluble in water. It attracts boll weevils (an agricultural pest) so they can be destroyed, since this insect has caused billions in losses to cotton crops in the United States. Write an equation for the ionization of $(CH_3)_2NH$ in water.

Analysis: We've been told that $(CH_3)_2NH$ is a base, so it's going to react with water to form hydroxide ion. This gives us two reactants and one product. We need to determine the formula for the second product to write and balance the equation.

Assembling the Tools: The tool is the general equation for the ionization of a weak base with water, Equation 4.4, which we use as a template for writing the formulas of reactants and products.

Solution: The reactants in the equation are $(CH_3)_2NH$ and H_2O. According to Equation 4.4, when the base reacts with water it takes an H^+ from H_2O, becoming $(CH_3)_2NH_2^+$ and leaving OH^- behind. The equation for the reaction is

$$(CH_3)_2NH(aq) + H_2O \rightleftharpoons (CH_3)_2NH_2^+(aq) + OH^-(aq)$$

Is the Answer Reasonable? Compare the equation we've written with the general equation for reaction of a base with water. Notice that the formula for the product has one more H and a positive charge, and that the H^+ has been added to the nitrogen. Also, notice that the water has become OH^- when it loses H^+. The equation is therefore correct.

We continue to provide at least two **Practice Exercises** following the worked examples that give the student an opportunity to apply the principles used to solve the preceding example. These have been thoroughly reviewed and in some cases expanded. The answers to all of the Practice Exercises are available to the student in Appendix B at the back of the book.

The **end-of-chapter Questions and Problems** have undergone a reworking to ensure that they provide an increasing range of difficulty, from routine drill-type problems to significantly more difficult ones, and have been organized by the learning objectives. Many problems require students to draw on knowledge acquired in earlier chapters. For example, in many of the problems in Chapter 4 and beyond, the chemical name of a compound in question is given rather than the formula, so students must apply (and review if necessary) the rules of nomenclature presented in Chapter 2.

One of the main goals of chemistry instruction is to help students develop the ability to solve problems that are more thought-provoking than typical review problems. Recognizing that students often have difficulty with solving problems that require application of several different concepts, we continue to use the Analyzing and Solving Multi-Concept Problems feature. These problems are more difficult than those in a typical worked example and require the use of concepts presented in more than one chapter. Students must combine two or more concepts before reaching a solution, and they must reduce a complex problem into a sum of simpler parts. Problems of this type first appear in Chapter 4 after students have had a chance to work on basic problem skills and after sufficient concepts have been introduced in earlier chapters to make such problems meaningful. Analyzing and Solving Multi-Concept Problems addresses instructor frustration and students' deficiencies in problem solving by teaching students how to deconstruct problems and emphasize the actual thinking that goes into solving problems.

Available in *WileyPLUS*, we include problem sets titled **Bringing it Together** that consist mostly of problems that require students to apply concepts developed in two or more of the preceding chapters. These problem sets are available for groups of four to five chapters. Problems have been selected to provide a range of difficulties so as to challenge students of varying levels of achievement.

The WileyPLUS Advantage

WileyPLUS is a research-based online environment for effective teaching and learning. *WileyPLUS* is packed with interactive study tools and resources–including the complete online textbook.

WileyPLUS addresses the needs of students, empowering them to be successful.

The 7th Edition *WileyPLUS* course that accompanies *Chemistry: The Molecular Nature of Matter* includes:

- QuickStart assignments and presentations that are pre-loaded for every chapter.
- Lecture Note PowerPoint presentation slides

- Image Gallery that includes all line art, and tables
- Test Bank questions
- Classroom Response System (Clicker) questions
- Solutions Manuals
- A database of 3D molecules (available in *WileyPLUS*)
- All content mapped to learning objectives
- New visualizations of key concepts
- End-of-chapter questions are available to be used for assessment, assignable and automatically graded
- End-of-chapter questions that have multiple forms of assistance, available to students at the instructor's discretion. Assistance includes:
  - Question and answer-specific hints
  - Step-by-step tutorials (Go tutorials)
  - Answer-specific feedback
  - Office Hour, worked problem-solving videos
  - Links to specific sections of the textbook or other media

In addition, *WileyPLUS* is now equipped with an adaptive learning module called ORION. Based on cognitive science, *WileyPLUS* with ORION provides students with a personal, adaptive learning experience so they can build their proficiency on concepts and use their study time effectively. *WileyPLUS* with ORION helps students learn by learning about them.

WileyPLUS with ORION is great as:

- An adaptive **pre-lecture tool** that assesses your students' conceptual knowledge so they come to class better prepared,

Begin

A **personalized study guide** that helps students understand both strengths and areas where they need to invest more time, especially in preparation for quizzes and exams.

Unique to ORION, students begin by taking a quick diagnostic for any chapter. This will determine each student's baseline proficiency on each topic in the chapter. Students see their individual diagnostic report to help them decide what to do next with the help of ORION's recommendations.

Practice

For each topic, students can either Study or Practice. **Study** directs the student to the specific topic they choose in *WileyPLUS*, where they can read from the e-textbook, or use the variety of relevant resources available there. Students can also **practice**, using questions and feedback powered by ORION's adaptive learning engine. Based on the results of their diagnostic and ongoing practice, ORION will present students with questions appropriate for their current level of understanding and will continuously adapt to each student, helping them build their proficiency.

ORION includes a number of reports and ongoing recommendations for students to help them maintain their

Maintain

proficiency over time for each topic. Students can easily access ORION from multiple places within *WileyPLUS*. It does not require any additional registration, and there is no additional charge for students using this adaptive learning system.

About the Adaptive Engine

ORION includes a powerful algorithm that feeds questions to students based on their responses to the diagnostic and to the practice questions. Students who answer questions correctly at one difficulty level will soon be given questions at the next difficulty level. If students start to answer some of those questions incorrectly, the system will present questions of lower difficulty. The adaptive engine also takes into account other factors such as reported confidence levels, time spent on each question, and changes in response options before submitting answers.

The questions used for the adaptive practice are numerous and are not found in the *WileyPLUS* assignment area. This ensures that students will not be encountering questions in ORION that they may also encounter in their *WileyPLUS* assessments.

ORION also offers a number of reporting options available for instructors so that instructors can easily monitor student usage and performance.

Significant Changes in the 7th Edition

As noted earlier, our mission in developing this revision was to sharpen the focus of the text as it relates to the relationship between behavior at the molecular level and properties observed at the macroscopic level.

As much as possible, chapters are written to stand alone as instructional units, enabling instructors to modify the chapter sequence to suit the specific needs of their students. For example, if instructors wish to cover the chapter dealing with the properties of gases (Chapter 10) early in the course, they can easily do so. While we believe this chapter fits best in sequence with the chapters dealing with the other states of matter, we realize that there are other valid organizational preferences and the chapter has been written to accommodate them.

Some of the more significant changes to the organization are the following:

- Short essays addressing special topics are spread throughout the book. Those titled *Chemistry Outside the Classroom* and *Chemistry and Current Affairs* provide descriptions of real-world, practical applications of chemistry to industry, medicine, and the environment. Essays titled *On the Cutting Edge* serve to highlight chemical phenomena that are of current research interest and that have potential practical applications in the future. A list of these special topics appears at the end of the Table of Contents. In these essays we have included discussions on the IUPAC recommendations for using a range of atomic masses for

some elements and the use of these ranges in forensic science.

- **Chapter 0** is entirely new and sets the tone for the rest of the text. It provides an introduction to the important topics that we will address in this book: atomic theory, macroscopic properties rely on the microscopic properties, energy changes, and the geometric shapes of molecules. The atomic theory is introduced after a discussion of the origins of the elements from the start of the universe, through multiple supernova. A clear connection is made between observations at the macroscopic level and their interpretation at the molecular level.

- **Chapter 1** is devoted to measurements and their units. In this edition, we start with the scientific method and the classification of matter, then we move on to scientific measurements. The importance of quantitative measurements with respect to physical properties is introduced along with the concepts of intensive and extensive properties. The uncertainty of measurements is described. Significant figures are developed to provide the student with a logical method for assessing data. Finally, the method of dimensional analysis is discussed and applied to familiar calculations to develop confidence at an early stage.

- **Chapter 2** continues the discussion begun in Chapter 0 on the structure of the atom. We introduce the periodic table in this chapter as well as molecules, chemical formulas, and chemical reactions. The concepts of chemical reactions and chemical equations are presented and described by drawings of molecules and through the use of chemical symbols.

- **Chapter 3** covers the mole concept and stoichiometry. We have separated the discussion on the mole and Avogadro's number to emphasize the importance of these concepts.

- **Chapter 5** deals with redox reactions and includes a revised section on redox titrations to connect this procedure to the one introduced in Chapter 4. Redox reactions are presented in this chapter because many concomitant laboratory experiments use redox reactions.

- **Chapter 7** is a logical extension of Chapter 0 in our discussion of how our understanding of the atom has developed. The fundamentals of the quantum mechanical atom are introduced to the extent that the material is relevant to the remainder of the text. The discussion concerning orbitals has been expanded to include *f* orbitals.

- **Chapter 8** is the first of two chapters dealing with chemical bonding. We have moved the section devoted to some common kinds of organic compounds to the end of the chapter to allow for a more logical flow of concepts within the chapter. The section also serves as a brief introduction to organic chemistry for students whose major requires only one semester of chemistry. For instructors who do not wish to discuss organic compounds at this point in the course, the section is easily skipped and may be covered with Chapter 22.

- Chapter 12 discusses the physical properties of solutions. The discussion on concentration units has been rewritten to integrate the temperature-independent concentration units with the temperature-dependent concentration units.
- Chapter 13 covers the kinetics of chemical reactions, including mechanisms, and catalysis with the section on integrated rate laws is expanded in this new edition.
- Chapter 20 discusses nuclear reactions and their applications. In this chapter, we have utilized the masses of the subatomic particles as defined by the National Institutes of Standards and Technology. Additional Review Problems were added that address the utilization of radioactive elements in quantitative analysis.
- Chapter 22 provides an expanded discussion of organic chemistry with an emphasis on organic structures and functional groups. The number of Practice Exercises has been increased to give the student the opportunity to work through some examples while reading the chapter.

TEACHING AND LEARNING RESOURCES

A comprehensive package of supplements has been created to assist both the teacher and the student and includes the following:

For Students

Study Guide by Neil Jespersen of St. John's University. This guide has been written to further enhance the understanding of concepts. It is an invaluable tool for students and contains chapter overviews, additional worked-out problems giving detailed steps involved in solving them, alternate problem-solving approaches, as well as extensive review exercises. (ISBN: 978-1-118-70508-7)

Student Solutions Manual by Alison Hyslop, of St. John's University, with contributions by Duane Swank, of Pacific Lutheran University. The manual contains worked-out solutions for text problems whose answers appear in Appendix B. (ISBN: 978-1-118-70494-3)

Laboratory Manual for Principles of General Chemistry, 10th Edition, by Jo Beran of Texas A&M University, Kingsville. This comprehensive laboratory manual is for use in the general chemistry course. This manual is known for its broad selection of topics and experiments, and for its clear, user-friendly layout and design. Containing enough material for two or three terms, this lab manual emphasizes techniques, helping students learn the appropriate time and situation for their correct use. The accompanying Instructor's Manual presents the details of each experiment, including overviews, an instructor's lecture outline, teaching hints, and answers to pre-lab and laboratory questions. The Instructor's Manual also contains answers to the pre-laboratory assignment and laboratory questions. (ISBN: 978-1-118-62151-6)

For Instructors

Instructor's Manual by Scott Kirkby of East Tennessee State University. In addition to lecture outlines, alternate syllabi, and chapter overviews, this manual contains suggestions for small group active-learning projects, class discussions, tips for first-time instructors, class demonstrations, short writing projects, and relevant web links for each chapter.

Test Bank by Justin Meyer of South Dakota School of Mines and Technology. The test bank contains over 2,300 questions including: multiple-choice, true-false, short answer questions, fill in the blank questions, and critical thinking problems. A computerized version of the entire test bank is available with full editing features to help instructors customize tests.

Instructor's Solutions Manual by Alison Hyslop, of St. John's University, with contributions by Duane Swank, of Pacific Lutheran University, contains worked-out solutions to all end-of-chapter problems.

Digital Image Archive—The text web site includes downloadable files of text images in JPEG format. Instructors may use these images to customize their presentations and to provide additional visual support for quizzes and exams.

PowerPoint Lecture Slides by Mark Vitha, of Drake University and Nicholas Kingsley, of the University of Michigan – Flint, highlight key chapter concepts, contain numerous clicker questions, and include examples and illustrations that help reinforce and test students' grasp of essential topics. The slides feature images from the text that are customizable to fit your course.

PowerPoint Slides with Text Images—PPT slides containing images, tables, and figures from the text.

Personal Response Systems/"Clicker" Questions—A bank of questions is available for anyone using personal response systems technology in their classroom.

All instructor supplements can be requested from your local Wiley sales representative.

ACKNOWLEDGMENTS

In this edition it is a pleasure to welcome Alison Hyslop who contributed significantly to the sixth edition and now is a co-author for the seventh edition. She has important insights that will continually improve this text and we look forward to an on-going productive collaboration. At the same time we celebrate the tradition of excellence in chemistry teaching and lucid writing set forth by Jim Brady for many years. In his role as consultant, mentor, and friend he contributes greatly with his support and encouragement.

We express our fond thanks to our spouses, June Brady, Marilyn Jespersen, and Peter de Rege, and our children, Mark and Karen Brady, Lisa Fico and Kristen Pierce, and Nora, Alexander, and Joseph de Rege, for their constant support, understanding, and patience.

They have been, and continue to be, a constant source of inspiration for us all.

We deeply appreciate the contributions of others who have helped in preparing materials for this edition. In particular, Conrad Bergo of East Stroudsburg University, for reviewing the answers and solutions for accuracy. We would also like to thank the following colleagues at St. John's University for helpful discussions: Gina Florio, Steven Graham, Renu Jain, Elise Megehee, Jack Preses, Richard Rosso, Joseph Serafin, and Enju Wang.

It is with particular pleasure that we thank the staff at Wiley for their careful work, encouragement, and sense of humor, particularly our editors, Nicholas Ferrari and Jennifer Yee.

We are also grateful for the efforts of Senior Marketing Manager Kristine Ruff, Senior Product Designer Geraldine Osnato, Media Specialist Daniela DiMaggio, our Photo Editor, Mary Ann Price, our Designer, Thomas Nery, the entire production team, and especially Elizabeth Swain for her tireless attention to getting things right. Our thanks also go to Rebecca Dunn at Preparé (the compositor) for their unflagging efforts toward changing a manuscript into a book.

We express gratitude to the colleagues whose careful reviews, helpful suggestions, and thoughtful criticism of previous editions as well as the current edition manuscript have been so important in the development of this book. Additional thanks go to those who participated in the media development by creating content and reviewing extensively. Our thanks go out to the reviewers of previous editions, your comments and suggestions have been invaluable to us over the years. Thank you to the reviewers of the current edition, and to the authors and reviewers of the supporting media package:

Ahmed Ahmed, *Cornell University*

Georgia Arbuckle-Keil, *Rutgers University*

Pamela Auburn, *Lonestar College*

Stewart Bachan, *Hunter College, CUNY*

Suzanne Bart, *Purdue University*

Susan Bates, *Ohio Northern University*

Peter Bastos, *Hunter College, CUNY*

Shay Bean, *Chattanooga State Community College*

Tom Berke, *Brookdale Community College*

Thomas Bertolini, *University of Southern California*

Chris Bowers, *Ohio Northern University*

William Boyke, *Brookdale Community College*

Rebecca Broyer, *University of Southern California*

Robert Carr, *Francis Marion University*

Mary Carroll, *Union College*

Jennifer Cecile, *Appalachian State University*

Nathan Crawford, *Northeast Mississippi Community College*

Patrick Crawford, *Augustana College*

Mapi Cuevas, *Santa Fe College*

Ashley Curtis, *Auburn University*

Mark Cybulski, *Miami University, Ohio*

Michael Danahy, *Bowdoin College*

Scott Davis, *Mansfield University*

Donovan Dixon, *University of Central Florida*

Doris Espiritu, *City Colleges of Chicago- Wright College*

Theodore Fickel, *Los Angeles Valley College*

Andrew Frazer, *University of Central Florida*

Eric Goll, *Brookdale Community College*

Eric J. Hawrelak, *Bloomsburg University of Pennsylvania*

Paul Horton, *Indian River State College*

Christine Hrycyna, *Purdue University*

Dell Jensen, *Augustana College*

Nicholas Kingsley, *University of Michigan-Flint*

Jesudoss Kingston, *Iowa State University*

Gerald Korenowski, *Rensselaer Polytechnic Institute*

William Lavell, *Camden County College*

Chuck Leland, *Black Hawk College*

Lauren Levine, *Kutztown University*

Harpreet Malhotra, *Florida State College at Jacksonville*

Ruhullah Massoudi, *South Carolina State University*

Scott McIndoe, *University of Victoria*

Justin Meyer, *South Dakota School of Mines and Technology*

John Milligan, *Los Angeles Valley College*

Troy Milliken, *Jackson State University*

Alexander Nazarenko, *SUNY College at Buffalo*

Anne-Marie Nickel, *Milwaukee School of Engineering*

Fotis Nifiatis, *SUNY-Plattsburgh*

Mya Norman, *University of Arkansas*

Jodi O'Donnell, *Siena College*

Ngozi Onyia, *Rockland Community College*

Ethel Owus, *Santa Fe College*

Maria Pacheco, *Buffalo State College*

Manoj Patil, *Western Iowa Tech Community College*

Cynthia Peck, *Delta College*

John Pollard, *University of Arizona*

Rodney Powell, *Central Carolina Community College*

Daniel Rabinovich, *University of North Carolina- Charlotte*

Lydia Martinez Rivera, *University of Texas at San Antonio*

Brandy Russell, *Gustavus Adolphus College*

Aislinn Sirk, *University of Victoria*

Christine Snyder, *Ocean County College*

Bryan Spiegelberg, *Rider University*

John Stankus, *University of the Incarnate Word*

John Stubbs, *The University of New England*

Luyi Sun, *Texas State University-San Marcos*

Mark Tapsak, *Bloomsburg University of Pennsylvania*

Loretta Vogel, *Ocean County College*

Daniel Wacks, *University of Redlands*

Crystal Yau, *Community College of Baltimore County*

Curtis Zaleski, *Shippensburg University*

Mu Zheng, *Tennessee State University*

Greg Zimmerman, *Bloomsburg University*

A Very Brief History of Chemistry

Chapter Outline

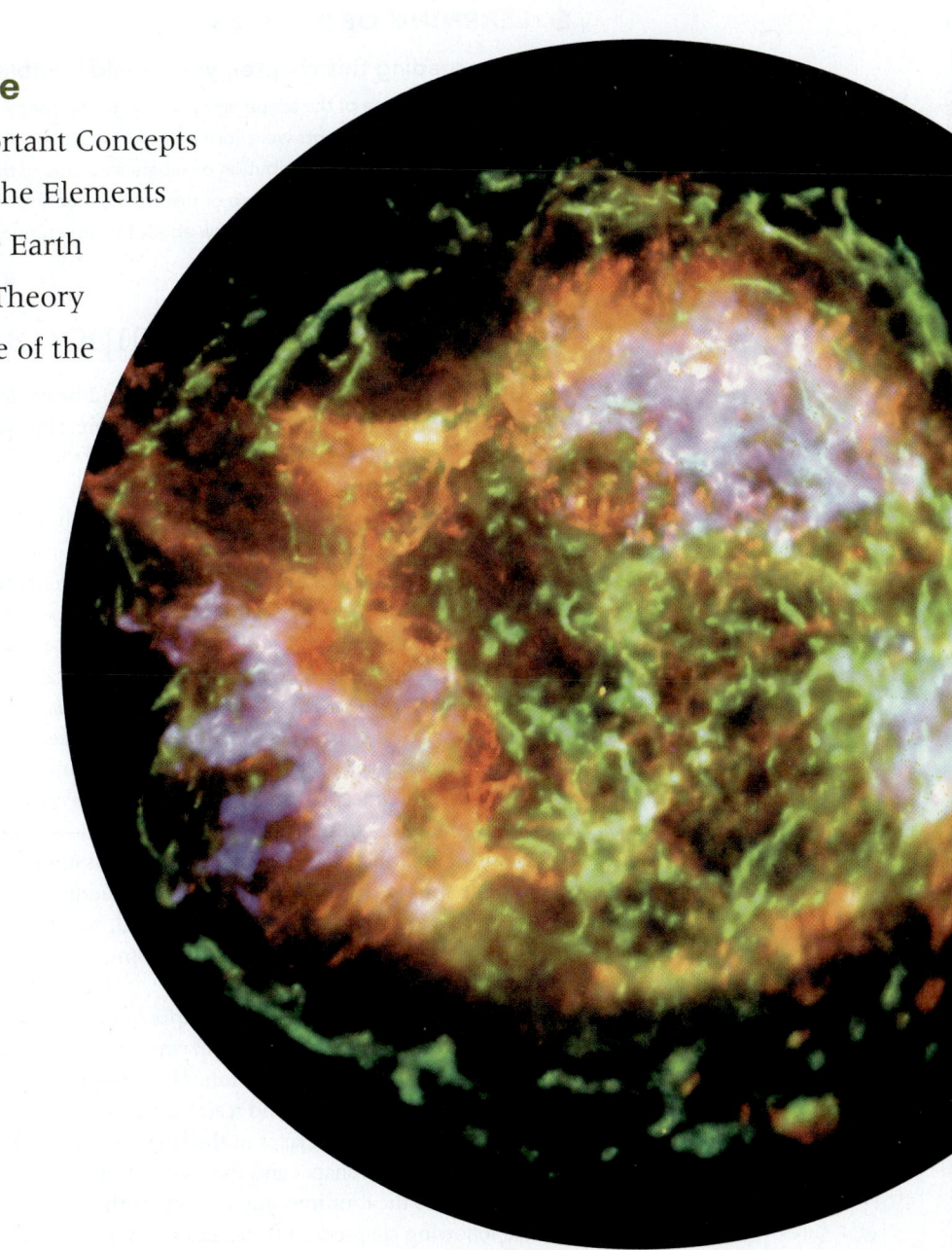

NASA/CXC/SAO/P. Slane et al.

This Chapter in Context

In this introductory chapter we attempt to answer the large questions: "Where did we come from?" and "Where are we going?" In suggesting where we came from, we draw upon cosmology's current theories about the start of the universe and the sequential synthesis of the elements. To the question of where we are going, this chapter suggests some of the "Important Concepts" that the science of chemistry uses to entice us toward the future.

These goals also set the theme on how scientists ply their trade. All of the information in this text is the result of a scientist asking a question, and then through scientific observation and research finding an answer. The same person who asks a question may not find the answer, and the time between asking and answering a question can be minutes or hundreds of years. In the end, we have an explanation of how chemists describe the physical world around us.

Advances in modern chemistry, physics, and mathematics allow us to explain our chemical surroundings with more clarity than ever. There will be more involved details to master, especially if your career plans include a significant amount of chemical work. However, you should be aware of the big picture and modern ideas. If you do that, it will certainly make the study of chemistry more meaningful for you. So, sit back and enjoy this chapter.

● LEARNING OBJECTIVES

After reading this chapter, you should be able to:

- develop a sense of the scope and purpose of the chemical sciences
- learn how the elements were formed
- understand that the distribution of substances around the world is not accidental
- appreciate the powerful nature of the atomic theory
- understand how we came to know about the structure of the atom

0.1 | Chemistry's Important Concepts

Although this seems to be a rather large and heavy textbook, and it must contain a lot of information that needs to be learned, there are a few guiding ideas that bring it all together. The intent of this section is to give an overall view of the main concepts of chemistry, and then we will fill in the details as we go along.

The *atomic theory* as explained by John Dalton in 1813 is the first of these important concepts. This theory describes atoms, the basic building blocks of our world. Dalton, in the most fundamental way, described the nature of atoms and how they interact with each other. Since then, chemists and physicists have been working out the fine details of atomic structure and chemical interactions. Many of these details are described in later chapters.

The second important concept is that we can tell a lot about what happens on the atomic scale with *careful observations on the laboratory or macroscopic scale*. In fact, until recently when instruments were developed to see, really detect, individual atoms and molecules, this was the only way that scientists could deduce what was happening.

Our third concept is that knowledge of *energy changes* and the probability of different arrangements of atoms help scientists predict how atoms interact. All energy of atoms can be classified as either *kinetic energy* (energy of motion) or *potential energy* (energy of position) and the sum of the two cannot change. In addition, atoms and molecules will tend toward the most probable arrangement. In general, we find that chemical reactions occur when the energy, potential and kinetic, of the atoms decreases and/or the atoms achieve their most probable arrangement.

The significance of geometric shapes of molecules is the fourth important concept. Large molecules such as DNA, RNA, enzymes, and antibodies have a three-dimensional structure that is important to their function. The three-dimensional shapes of much smaller molecules also affect their properties and reactivity. Indeed, the three-dimensional shape of these smaller structures dictates the shapes of the larger molecules. In this book we develop understanding three-dimensional shapes and the relationship between structure, properties, and reactivity.

These are the four important concepts that are developed throughout this book. Each of the following chapters adds increasing layers of details and depth. Recalling these concepts throughout your chemistry course will help keep you from being overwhelmed by the amount of material.

Using the chapter titles alone, assign one or two of chemistry's big ideas to each chapter and explain why you made your choice.[1]

0.2 | Supernovas and the Elements

In the Beginning

We turn to physical cosmology for one of our most important ideas, the **"big-bang" theory**, to begin the story of chemistry. The big-bang theory postulates that the universe, as we know it, experienced a tremendous explosion of energy and subatomic particles approximately 14 billion years ago and that it has been expanding ever since.

Perhaps the first experimental data that suggested that the universe is expanding were observations by Edwin Hubble and others that the majority of stars and galaxies seem to shine with light that is shifted toward the red end of the visible spectrum. The well-known Doppler effect that explains why the whistle of a train has a higher pitch, or frequency, when the train is approaching and a lower pitch as it moves away, was used to give meaning to the red-shift observations. "Hubble's law" proposes that the size of the red shift is proportional to the distance and speed of the star moving away from the earth. Cosmologists concluded that the only way to explain these data was to propose a universe that was expanding in all directions.

Working backwards, it was not difficult to imagine that the entire universe started from a single point that physicists call a singularity. Over time, the observations made by astronomers have all been explained by this theory. Interestingly, one of the supporting experiments was the serendipitous discovery in 1964 by two astronomers, Penzias and Wilson, who were trying to make very accurate measurements with a radio telescope. A persistent static was present no matter where they pointed the telescope. They expended great effort to clean the telescope in an attempt to remove the static. They even scrubbed off the "white dielectric material," also known as pigeon droppings, from the telescope, to no avail. In the end, they questioned whether the static was more significant than just being some random noise. After careful calculations they concluded that the static was indeed microwave radiation characteristic of a temperature that matched the predicted temperature of the universe after cooling for 14 billion years. Today this is recognized as evidence that strongly supports the big-bang theory.

The First Elements

Using the big-bang theory, quantum mechanics, and some complex mathematics, physicists and cosmologists are able to provide us with some ideas on how the universe might have developed. The extreme temperature, density, and pressure of the singularity at the start of the universe allowed only the most basic particles such as quarks to exist. Within one second after the big bang, the universe expanded and cooled to about ten billion degrees, allowing the basic units of matter, quarks, in groups of three, to form protons and neutrons. Within three minutes, the temperature dropped to about a billion degrees, allowing **nucleosynthesis**, creation of atomic nuclei, to occur. In nucleosynthesis collisions between protons and neutrons resulted in the formation of deuterium, helium, and lithium nuclei.

When the universe became cool enough that nucleosynthesis could no longer occur, 91% of all atoms were hydrogen atoms, 8% were helium atoms, and all the rest comprised less than 1% of all the atoms as shown in Table 0.1. As the universe cooled further, electrons combined with these nuclei to form neutral atoms.

TABLE 0.1	Estimates of the Most Abundant Isotopes in the Solar System
Isotope	Solar System Atom Percent
Hydrogen-1	90.886
Helium-4	8.029
Oxygen-16	0.457
Carbon-12	0.316
Nitrogen-14	0.102
Neon-20	0.100

[1]Answers to the Practice Exercises are found in Appendix B at the back of the book.

If the initial expansion of the universe had an even distribution of atoms, it would have remained as a dark, uniform, sea of atoms. Instead there were small disturbances in the dispersion of matter that grew with time. This led to the formation of the first stars as the matter coalesced.

Practice Exercise 0.2

Suggest conditions that are favorable for nucleosynthesis.

Practice Exercise 0.3

Suggest why only the lightest elements were formed during the big bang.

Elements Formed in Stars

As the stars grew in size, the temperature and pressure within each star increased to the point where nuclear fusion of hydrogen nuclei into helium started and stars began to shine. Heat generated from the fusion of hydrogen to helium maintained the volume and pressures within a star for millions of years. During that time the helium, being heavier than hydrogen, concentrated in the core, or center, of the star. In the core, the helium interfered with the collisions of the hydrogen nuclei and the rate of these nuclear reactions decreased. The star then cooled and contracted under gravitational forces. As the size decreased, the temperature and pressure of the core rose again and at about 100 million degrees the fusion of helium nuclei into carbon began. After a while the carbon was concentrated in the core and the helium core became a layer surrounding the carbon core. The hydrogen was still mostly found in the outer layer of the star.

Continuing, the carbon nuclei entered into nuclear reactions that produced argon. As the amount of argon increased, it migrated inward and became the core, now surrounded by a layer rich in carbon, then a helium layer, and finally the outer hydrogen layer.

We now have a pattern: Each successively heavier element becomes concentrated in the core of the star and when there are enough nuclei, nuclear reactions begin to produce an even heavier *nucleus*, which then concentrates in the core and repeats the process. In that way, oxygen and silicon cores are formed and then are forced out into layers by heavier elements. These layers and their nuclear reactions produce large quantities of heat to fuel and expand the star. A rapidly expanding star cannot generate enough heat to keep the hydrogen layer white hot, and as it cools the color becomes red. Stars like this are called **red giants**. Figure 0.1 illustrates the layered structure of a red giant star.

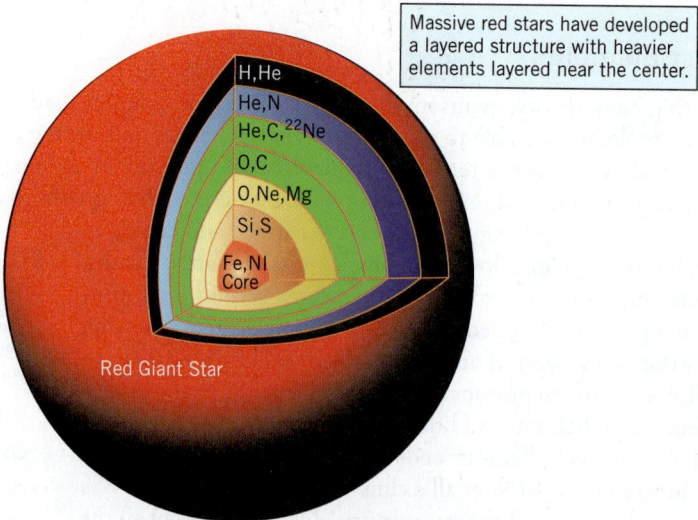

Massive red stars have developed a layered structure with heavier elements layered near the center.

H,He
He,N
He,C, ^{22}Ne
O,C
O,Ne,Mg
Si,S
Fe,NI Core

Red Giant Star

Figure 0.1 | Artist's rendering of the layered structure of a red giant star. Layering increases the density of specific nuclei that can fuse into larger elements, forming new layers.

Suggest why a core and enriched layers of nuclei are needed for nucleosynthesis in stars.

Suggest why, after some 14 billion years, hydrogen still makes up the overwhelming majority of the atoms in the universe.

Practice Exercise 0.4

Practice Exercise 0.5

Elements Formed in Supernovas

Finally, the silicon started fusing in nuclear reactions to form iron. The iron-forming reaction actually consumes heat and starts to cool the core. This cooling causes a cataclysmic collapse of the star, and as the nuclei rush toward the core the increase in pressure and density does two things. First, the speeding nuclei destroy many of the iron nuclei, creating a rich mixture of smaller particles such as helium nuclei and neutrons. In addition, the temperature of the collapsing star reaches levels that cannot be achieved even in the most massive stars. At its culmination, the collapsing star disintegrates, spewing all of its matter into interstellar space. This is called a **supernova**; in it exists a mix of nuclei that have very high energies and an atom density that has sufficient numbers of collisions to create even the heaviest elements. These conditions for nucleosynthesis last for less than a minute, perhaps for just seconds, when the expansion and cooling then make these reactions improbable.

The remnants of a supernova are eventually brought together to form a new star to repeat the process. In some instances, the formation of the new star leaves a ring of debris around it. This debris eventually accretes (clumps together) to form planets, moons, and asteroids.

Why don't elements heavier than iron form in stars?

What conditions do supernovas provide for synthesis of heavier elements?

Practice Exercises 0.6

Practice Exercises 0.7

0.3 | Elements and the Earth

As the stars formed, planets also formed from the debris surrounding the stars. The formation of the planets and the composition of the planets depended upon the matter that was available.

Planet Building

Nebula is the word that describes the debris left after the formation of a star forms a disk that can accrete into planets, moons, and asteroids. Depending on the debris, the planets can be rocky like the earth, Mars, and Venus or gaseous as Jupiter and Saturn. The final chemical makeup of a planet depends on the materials that accreted at the start and the elements that were retained by the gravitational forces of the planet itself. You can find a list of all the known elements inside the front cover of this book.

Table 0.2 lists the atom abundance in the whole earth, the crust, the oceans, and the atmosphere. We might expect that the distribution of the elements will be uniform on earth because the nebula that the earth condensed from had a relatively uniform distribution. Taking a quick look around us, we see that the earth does not have a uniform distribution of elements either on or below the surface, while the atmosphere and oceans tend to have more uniform compositions.

Distribution of the Elements

This uneven distribution can often be understood based on the properties of the elements such as their melting points, densities, and solubilities. When the earth formed some 4.5 billion years ago, the solid dust and gas particles in the nebula were slowly attracted to

TABLE 0.2	Estimates of the atom percentages in the earth as a whole, the crust, the atmosphere, and the oceans.*			
Element	Earth	Crust	Atmosphere	Oceans
Oxygen	48.2	59.0	20.9	33.02
Iron	14.8	2.1		
Silicon	15.0	20.4		
Magnesium	16.4	2.0		0.03
Hydrogen	0.67	2.9		66.06
Chlorine				0.34
Sodium	0.20	2.1		0.17
Sulfur	0.52			0.02
Calcium	1.1	2.2		0.006
Potassium	0.01	1.1		0.006
Aluminum	1.5	0.57		
Bromine				0.0067
Carbon	0.16			0.0028
Nitrogen			78.1	
Argon			0.96	

*Blank entries indicate that an atom is present to a negligible extent.

each other by gravitational and electrostatic forces. Once the earth formed, it began heating due to the radioactive elements releasing heat as they decayed to stable isotopes. In addition, bombardment by meteorites also heated the earth's surface while continued gravitational contraction also added more heat.

Eventually a large proportion of the earth melted and iron and nickel migrated to the inner core. Based on measurements of seismic waves (vibrations due to earthquakes) the actual inner core of the earth is composed of solid iron and nickel that is surrounded by a liquid layer of these metals. The outer core is superheated lava. Surrounding the core is the mantle of superheated rock that comprises about 85% of the earth's mass. The outer layer, comprising the lighter substances that we observe as solid rock and soil, is a ten-mile-thick crust. Figure 0.2 illustrates these features of the inner workings of the earth.

The outer core, mantle, and crust of the earth are not very fluid, and so different materials did not have the opportunity to separate on a massive scale as the core did. However you may have seen the exotic patterns of crystals in a granite counter-top. Minerals do separate, but only in small areas. That is also why the surface of the earth is not uniform. Minerals or elements in the crust will separate to a small extent due to similarities in their composition and structure as well as by melting points. For instance, gold atoms tend to aggregate with other gold atoms, rather than silicate minerals (silicon-based rocks) because they have distinctly different

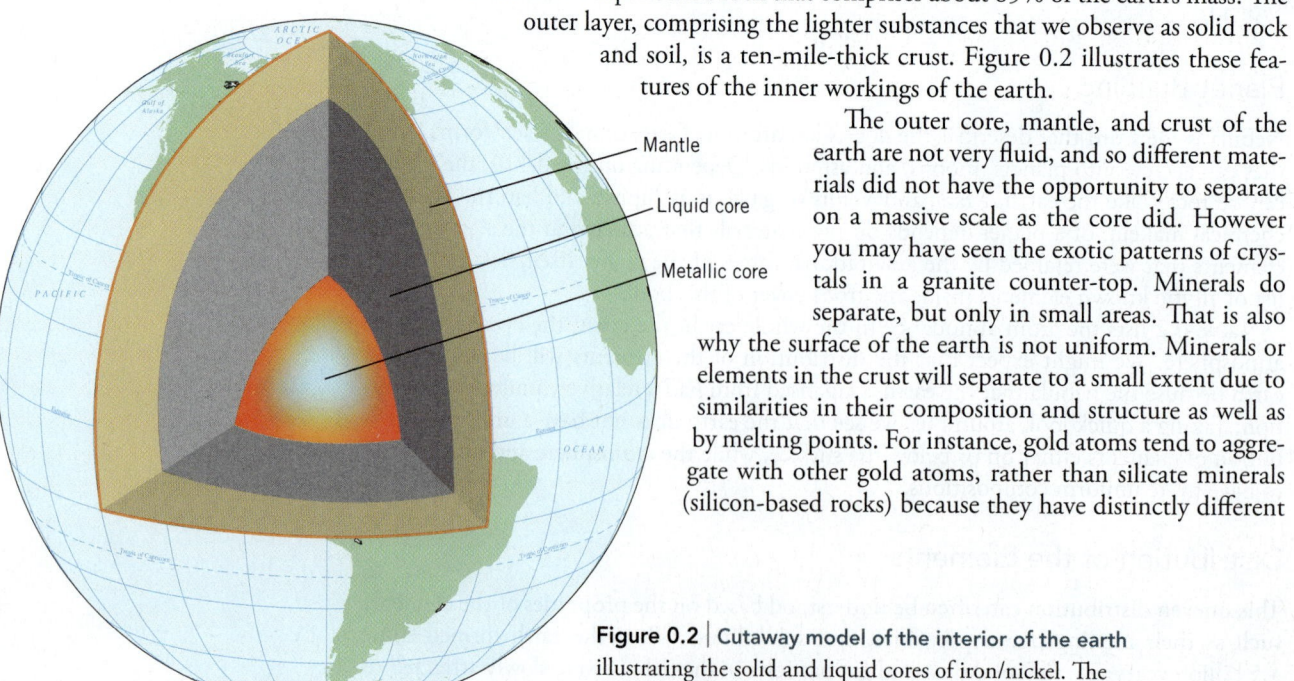

Mantle

Liquid core

Metallic core

Figure 0.2 | Cutaway model of the interior of the earth illustrating the solid and liquid cores of iron/nickel. The mantle and the thin layer called the crust are shown.

crystal structures, densities, and melting points. When the crust cooled the gold separated from the silicate rocks as they solidified. This process is called differential crystallization. Due to the tremendous reservoir of heat in the earth's core, these processes continue to this day and are seen frequently in the form of earthquakes and volcanic eruptions.

0.4 | Dalton's Atomic Theory

Today we define an atom as the smallest representative sample of an element and a compound as a substance that contains two or more elements, always in a fixed ratio by mass. Originally, the concept of atoms began nearly 2500 years ago when certain Greek philosophers expressed the belief that matter is ultimately composed of tiny indivisible particles. The Greek word meaning "not cut," is the source of our modern word "atom." The philosophers had no experimental evidence for their concept of atoms and many argued against this idea. Scientific support for the existence of atoms awaited the discovery of the *law of definite proportions* and the *law of conservation of mass*. Amazingly, these two important general observations about the nature of compounds and chemical reactions became apparent through the work of many early chemists (or alchemists) whose lab equipment only measured mass and volume.

Laws of Chemical Combination

Prior to the 19th century, progress in science was slow because there was little understanding of the need for accurate measurements. As we will discuss in Chapter 1, accurate, precise, and reproducible measurements are necessary in all the sciences. Despite the lack of accuracy and precision of early scientific work, over the course of time data accumulated that revealed some principles that apply to all chemical compounds and chemical reactions.

The first principle is that, when a compound is formed, elements always combine in the same proportion by mass. For example, when hydrogen and oxygen combine to form water the mass of oxygen is always eight times the mass of hydrogen—never more and never less. Similar observations apply to every compound we study. Such observations led to a generalization, known as the law of definite proportions (or law of definite composition). That law states that *in any chemical compound the elements are always combined in a definite proportion by mass.*

The second observation is that when a reaction is carried out in a sealed vessel, so that nothing can escape or enter, the total mass after the reaction is over is exactly the same as at the start. For instance, if we place hydrogen and oxygen into a sealed container and initiate the reaction to form water, the mass of water and whatever hydrogen or oxygen is left over is the same as the mass of the hydrogen and oxygen we started with. Such observations, repeated over and over for large numbers of chemical reactions, led to the generalization known as the law of conservation of mass. This law states that *mass is neither lost nor created during a chemical reaction.*

Law of Definite Proportions
In a given chemical compound, the elements are always combined in the *same proportions by mass.*

Law of Conservation of Mass
No detectable gain or loss of mass occurs in chemical reactions. Mass is *conserved.*

In Chapter 3 you will see how these laws can be used to perform calculations related to chemical composition.

The Atomic Theory

The laws of definite proportions and conservation of mass served as the *experimental foundation* for the atomic theory. They prompted the question: "What must be true about the nature of matter, given the truth of these laws?"

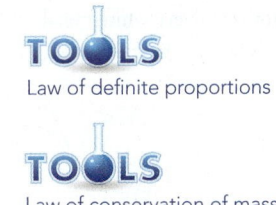

TOOLS
Law of definite proportions

TOOLS
Law of conservation of mass

At the beginning of the 19th century, John Dalton (1766–1844), an English scientist, used the Greek concept of atoms to make sense out of the laws of definite proportions and conservation of mass. Dalton reasoned that if atoms really exist, they must have certain properties to account for these laws. He postulated what those properties must be, and the list of postulates constitutes what we now call **Dalton's atomic theory**.

Dalton's Atomic Theory

1. Matter consists of tiny particles called *atoms*.
2. In any sample of a *pure element, all the atoms are identical* in mass and other properties.
3. The atoms of *different elements differ in mass and other properties*.
4. When atoms of *different elements combine to form compounds*, new and more complex particles form. However, in a given compound the constituent atoms are always present in the *same fixed numerical ratio*.
5. *Atoms are indestructible*. In chemical reactions, the atoms rearrange but they do not themselves break apart.

Modern Experimental Evidence for Atoms

In the early years of the 19th century scientists and alchemists had little more than rudimentary balances and graduated cylinders to make measurements. Modern chemical instrumentation now provides additional proof that atoms actually exist. Although atoms and most molecules are so incredibly tiny that even the most powerful optical microscopes are unable to detect them, experiments have been performed that provide pretty convincing evidence that atoms are real.

Scientists have developed very sensitive instruments that are able to map the surfaces of solids with remarkable resolution. One such instrument is called a **scanning tunneling microscope.** It was invented in the early 1980s by Gird Binnig and Heinrich Rohrer and earned them the 1986 Nobel Prize in physics. With this instrument, the tip of a sharp metal probe is brought very close to an electrically conducting surface and an electric current bridging the gap is begun. The flow of current is extremely sensitive to the distance between the tip of the probe and the sample. As the tip is moved across the surface, the height of the tip is continually adjusted to keep the current flow constant. By accurately recording the height fluctuations of the tip, a map of the hills and valleys on the surface is obtained. The data are processed using a computer to reveal an image of atoms on the surface as illustrated in Figure 0.3. Other, more complex instruments are also used to "observe" individual atoms and molecules, thus increasing our confidence that atoms exist and the atomic theory is correct. A newer instrument called the **atomic force microscope** is described in On the Cutting Edge 0.1.

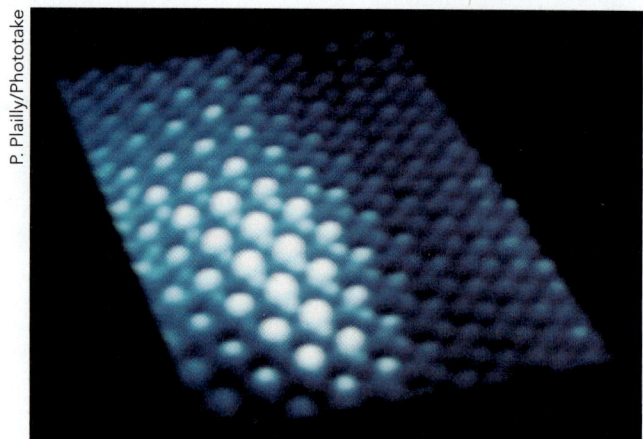

P. Plailly/Phototake

Figure 0.3 | **Individual atoms can be imaged using a scanning tunneling microscope.** This STM micrograph reveals the pattern of individual atoms of palladium deposited on a graphite surface. Palladium is a silvery white metal used in alloys such as white gold and dental crowns.

0.5 | Internal Structure of the Atom

The earliest models of atoms imagined them to be indestructible and totally unable to be broken into smaller pieces. However, as you probably know, atoms are not quite as indestructible as Dalton and other early philosophers had thought. During the late 1800s and early 1900s, experiments were performed that demonstrated that atoms are composed of **subatomic particles**. From this work the current theoretical model of atomic structure evolved. We will examine it in general terms in this chapter. A more detailed discussion of the electronic structure of the atom will follow in Chapter 7.

From M.F. Crommie, C.P. Lutz, D.M. Eigler. Confinement of electrons to quantum corrals on a metal surface. *Science 262, 218-220 (1993)* Reprinted with permission from AAAS. Image originally created by IBM.

ON THE CUTTING EDGE 0.1

Seeing and Manipulating Atoms and Molecules

Atoms and small molecules are incredibly tiny. Experiments have shown that they have diameters of the order of several billionths of an inch. For example, the diameter of a carbon atom is about 6 billionths of an inch (6×10^{-9} in.). As you will learn in Chapter 1, the prefix nano implies 10^{-9}, so when we examine matter at the nanoscale level, we are looking at very small structures, usually with dimensions of perhaps tens to hundreds of atoms. Nanotechnology deals with using such small-scale objects and the special properties that

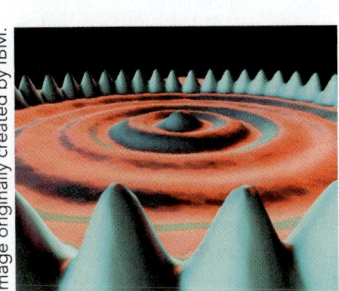

Figure 1 Scientists have control of not only the atomic landscape, but the electronic landscape also. Here they have positioned 48 iron atoms into a circular ring in order to "corral" some surface state electrons and force them into this circular structure. The ripples in the ring of atoms are predicted by quantum theory that is introduced in Chapter 8.

accompany them to develop useful applications. Ultimately, the goal of nanotechnology (also sometimes called molecular nanotechnology) is to be able to build materials from the atom up. Such technology doesn't quite exist yet, but scientists are beginning to make progress in that direction. This discussion, therefore, is kind of a progress report that will give you some feeling of where science is now and where it's heading—sort of a glimpse at the future.

There are several reasons why there is so much interest in nanotechnology. For one, the properties of materials are related to their structures. By controlling structures at the atomic and molecular level, we can (in principle) tailor materials to have specific properties. Driving much of the research in this area are the continuing efforts by computer and electronics designers to produce ever-smaller circuits. The reductions in size achieved through traditional methods are near their limit, so new ways to achieve smaller circuits and smaller electrical devices are being sought.

Molecular Self-Assembly

An area of research that is of great interest today is the field of molecular self-assembly, in which certain molecules, when brought together, spontaneously arrange themselves into desirable structures. Biological systems employ this strategy in constructing structures such as cellular membranes. The goal of scientists is to mimic biology by designing molecules that will self-assemble into specific arrangements.

Visualizing and Manipulating Very Tiny Structures

What has enabled scientists to begin the exploration of the nanoworld is the development of tools that allow them to see and sometimes manipulate individual atoms and molecules. We've already discussed one of these important devices, the scanning tunneling microscope (STM), when we discussed experimental evidence for atoms (see Section 0.4). This instrument, which can only be used with electrically conducting samples, makes it possible to image individual atoms. What is very interesting is that it can also be used to move atoms around on a surface. To illustrate this, scientists have arranged atoms to corral electrons (Figure 1). Although this experiment may not have much practical use, it demonstrates that one of the required capabilities for working with substances at the molecular level is achievable.

To study nonconducting samples, a device called an atomic force microscope (AFM) can be used. Figure 2 illustrates its basic principles. A very sharp stylus (sort of like the needle used in a DJ's vinyl record player) is moved across the surface of the sample under study. Forces between the tip of the probe and the surface molecules cause the probe to flex as it follows the ups and downs of the bumps that are the individual molecules and atoms. A mirrored surface attached to the probe reflects a laser beam at angles proportional to the amount of deflection of the probe. A sensor picks up the signal from the laser and translates it into data that can be analyzed by a computer to give three-dimensional images of the sample's surface. A typical image produced by an AFM is shown in Figure 3.

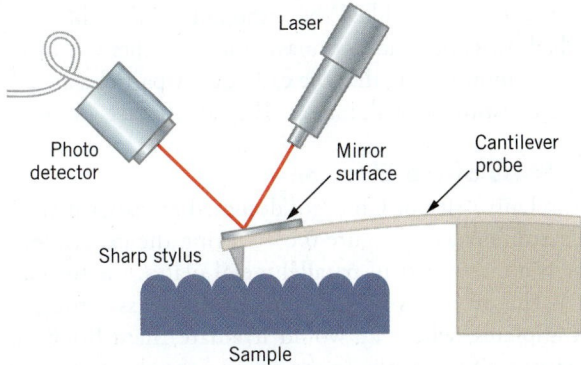

Figure 2 An atomic force microscope (AFM). A sharp stylus attached to the end of a cantilever probe rides up and down over the surface features of the sample. A laser beam, reflected off a mirrored surface at the end of the probe, changes angle as the probe moves up and down. A photo detector reads these changes and sends the information to a computer, which translates the data into an image.

Labels in Figure 2: Laser; Photo detector; Mirror surface; Cantilever probe; Sharp stylus; Sample

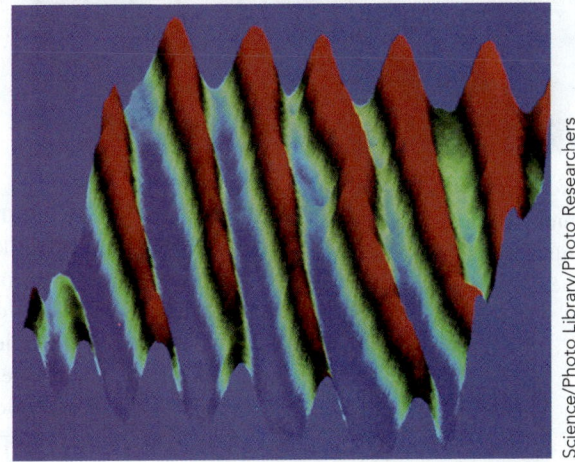

Science/Photo Library/Photo Researchers

Figure 3 Silver nanowires imaged using an atomic force microscope. This colored micrograph shows thin silver nanowires spaced about 0.8 millionth of an inch apart on a calcium fluoride crystal surface. Such wires could be used for miniature electronics.

Discovery of the Electron, Proton, and Neutron

Our current knowledge of atomic structure was pieced together from facts obtained from experiments by scientists that began in the 19th century. In 1834, Michael Faraday discovered that the passage of electricity through aqueous solutions could cause chemical changes. This was the first hint that matter was electrical in nature. Later in that century, scientists began to experiment with *gas discharge tubes* in which a high-voltage electric current was passed through a gas at low pressure in a glass tube (Figure 0.4). Such a tube is fitted with a pair of metal *electrodes*, and when the electricity begins to flow between them, the gas in the tube glows. This flow of electricity is called an *electric discharge*, which is how the tubes got their name.

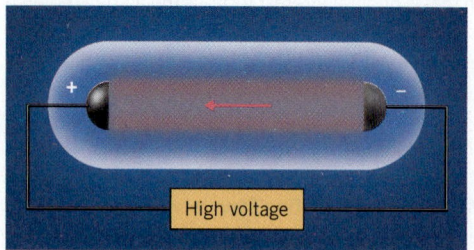

Figure 0.4 | **A gas discharge tube.** Cathode rays flow from the negatively charged cathode to the positively charged anode.

The physicists who first studied this phenomenon did not know what caused the tube to glow, but tests soon revealed that negatively charged particles were moving from the negative electrode (the **cathode**) to the positive electrode (the **anode**). The physicists called these emissions **cathode rays**, since they appeared to originate from the cathode.

Measuring the Charge-to-Mass Ratio of the Electron

In 1897, the British physicist J. J. Thomson modified a cathode ray tube, a special gas discharge tube, to make quantitative measurements of the properties of cathode rays, Figure 0.5. In Thomson's tube, a beam of cathode rays was focused on a glass surface coated with a phosphor, a substance that glows when the cathode rays strike it (point 1). The cathode ray beam passed between the poles of a magnet and between a pair of metal electrodes that could be given electrical charges. The magnetic field tends to bend the beam in one direction (toward point 2), while the charged electrodes bend the beam in the opposite direction (toward point 3). By adjusting the charge on the electrodes, the two effects can be made to cancel, and from the amount of charge on the electrodes required to balance the effect of the magnetic field, Thomson was able to calculate the first bit of quantitative information about a cathode ray particle—the ratio of its charge to its mass (often expressed as e/m, where e stands for charge and m stands for mass). The charge-to-mass ratio has a value of -1.76×10^8 coulombs/gram, where the coulomb (C) is a standard unit of electrical charge and the negative sign reflects the negative charge on the particle.

No matter what gas filled the cathode ray tube, the cathode rays all had the same charge-to-mass ratio and otherwise behaved the same, demonstrating that the cathode ray particles are a fundamental constituent of all matter. They are, in fact, *electrons*.

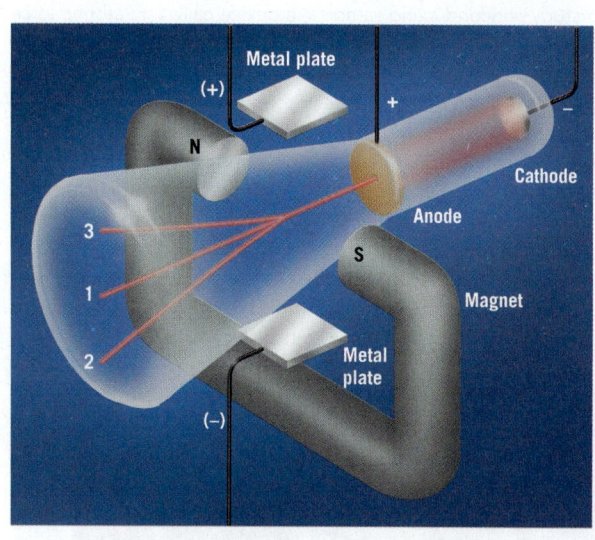

Figure 0.5 | **Thomson's cathode ray tube,** which was used to measure the charge-to-mass ratio for the electron.

Measuring the Charge and Mass of the Electron

In 1909, Robert Millikan, at the University of Chicago, designed an experiment that enabled him to measure the electron's charge (Figure 0.6). During the experiment he sprayed a fine mist of oil droplets above a pair of parallel metal plates, the top one of which had a small hole in it. As the oil drops settled, some would pass through this hole into the space between the plates, where he would irradiate them briefly with X rays. The X rays knocked electrons off molecules in the air, and the electrons became attached to the oil drops, giving them an electrical charge. By observing the rate of fall of the charged drops both when the metal plates were electrically charged and when they were not, Millikan was able to calculate the amount of charge carried by each drop. When he examined his results, he found that all the values he obtained were whole-number multiples of -1.60×10^{-19} C. He reasoned that since a drop could only pick up whole numbers of electrons, this value must be the charge carried by each individual electron.

Once Millikan had measured the electron's charge, its mass could then be calculated from Thomson's charge-to-mass ratio. This mass was calculated to be 9.09×10^{-28} g. More precise measurements have since been made, and the mass of the electron is currently reported to be $9.10938291 \times 10^{-28}$ g. Thomson's early measurements are in good agreement with today's more precise measurements.

Discovery of the Proton

The removal of electrons from an atom gives a positively charged particle (called an *ion*). To study these particles, holes were drilled in the cathode, and rays moving in the opposite direction of the cathode rays were observed. These were called canal rays and their behavior depended on the gas that filled the tube. Soon a modification was made in the construction of the cathode ray tube to produce a new device called a *mass spectrometer* to make better measurements on these new rays. This apparatus is described in On the Cutting Edge 0.2 and was used to measure the charge-to-mass ratios of positive ions. These ratios were found to vary, depending on the chemical nature of the gas in the discharge tube, showing that their masses also varied. The lightest positive particle observed was produced when hydrogen was in the tube, and its mass

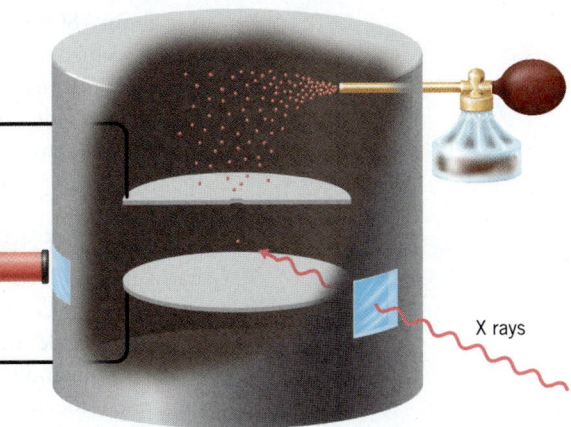

Figure 0.6 | Millikan's oil drop experiment. Electrons, which are ejected from molecules in the air by the X rays, are picked up by very small drops of oil falling through the tiny hole in the upper metal plate. By observing the rate of fall of the charged oil drops, with and without electrical charges on the metal plates, Millikan was able to calculate the charge carried by an electron.

ON THE CUTTING EDGE (0.2

The Mass Spectrometer and the Experimental Measurement of Atomic Masses

When a spark is passed through a gas, electrons are knocked off the gas molecules. Because electrons are negatively charged, the particles left behind carry positive charges; they are called *positive ions*. These positive ions have different masses, depending on the masses of the molecules from which they are formed. Thus, some molecules have large masses and give heavy ions, while others have small masses and give light ions.

The device that is used to study the positive ions produced from gas molecules is called a *mass spectrometer* (illustrated in the figure). In a mass spectrometer, positive ions are created by passing an electrical spark (called an *electric discharge*) through a sample of the particular gas being studied. As the positive ions are formed, they are attracted to a negatively charged metal plate that has a small hole in its center. Some of the positive ions pass through this hole and travel onward through a tube that passes between the poles of a powerful magnet, as shown in the accompanying figure.

One of the properties of charged particles, both positive and negative, is that their paths curve as they pass through a magnetic field. This is exactly what happens to the positive ions in the mass spectrometer as they pass between the poles of the magnet. However, the extent to which their paths bend depends on the masses and velocities of the ions. This is because the path of a heavy ion, like that of a speeding cement truck, is difficult to change, but the path of a light ion, like that of a motorcycle, is influenced more easily. As a result, heavy ions emerge from between the magnet's poles along different lines than the lighter ions. In effect, an entering beam containing ions of different masses is sorted by the magnet into a number of beams, each containing ions of the same mass. This spreading out of the ion beam thus produces an array of different beams called a *mass spectrum*.

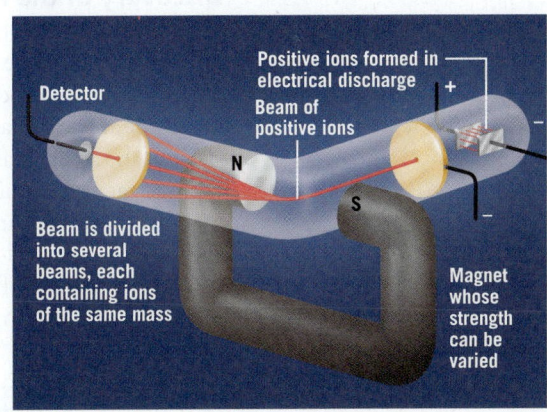

There are many types of mass spectrometers. In one type, as the one shown here, the strength of the magnetic field is gradually changed, which sweeps the beams of ions across a detector located at the end of the tube. As a beam of ions strikes the detector, its intensity is measured and the masses of the particles in the beam are computed based on the strength of the magnetic field, the speed of the particles, along with the geometry of the apparatus.

Among the benefits derived from measurements using the mass spectrometer are very accurate isotopic masses and relative isotopic abundances. These serve as the basis for the very precise values of the atomic masses that you find in the table inside the front cover. (Isotopes are atoms of the same element with slightly different masses. They are discussed in Section 0.5.)

was about 1800 times as heavy as an electron. When other gases were used, their masses always seemed to be whole-number multiples of the mass observed for hydrogen ions. This suggested the possibility that clusters of the positively charged particles made from hydrogen atoms made up the positively charged particles of other gases. The hydrogen atom, minus an electron, thus seemed to be a fundamental particle in all matter and was named the proton, after the Greek word *protos*, meaning "first."

Discovery of the Atomic Nucleus

Early in the 20th century, Hans Geiger and Ernest Marsden, working under Ernest Rutherford at Great Britain's Manchester University, studied what happened when *alpha rays* hit thin gold foils. Alpha rays are composed of particles having masses four times those of the proton and bearing two positive charges; they are emitted by certain unstable atoms in a phenomenon called *radioactive decay*. Most of the alpha particles passed right on through the foils to the phosphorescent screen as if they were virtually empty space (Figure 0.7). A significant number of alpha particles, however, were deflected at very large angles. Some were even deflected backward, as if they had hit a stone wall. Part of the genius of this experiment was to arrange the phosphorescent screen completely around the gold foil so that the alpha particles deflected at large angles would be observed. Rutherford was so astounded that he compared the result to that of firing a 15-inch artillery shell at a piece of tissue paper and having it come back and hit the gunner! From studying the angles of deflection of the particles, Rutherford reasoned that only something extraordinarily massive and positively charged could cause such an occurrence. Since most of the alpha particles went straight through, he further reasoned that the gold atoms in the foils must be mostly empty space. Rutherford's ultimate conclusion was that virtually all of the mass of an atom must be concentrated in a particle having a very small volume located in the center of the atom. He called this massive particle the atom's nucleus.

Discovery of the Neutron

From the way alpha particles were scattered by a metal foil, Rutherford and his students were able to estimate the number of positive charges on the nucleus of an atom of the metal. This had to be equal to the number of protons in the nucleus. When they computed the nuclear mass based on this number of protons, however, the value always fell short of the actual mass. In fact, Rutherford found that for many atoms only about half of the nuclear mass could be accounted for by protons. This led him to suggest that there were other particles in the nucleus that had a mass close to or equal to that of a proton, but with no electrical charge. This suggestion initiated a search that finally ended in 1932 with the discovery of the neutron by Sir James Chadwick, a British physicist who was awarded the Nobel Prize in 1935 for this discovery.

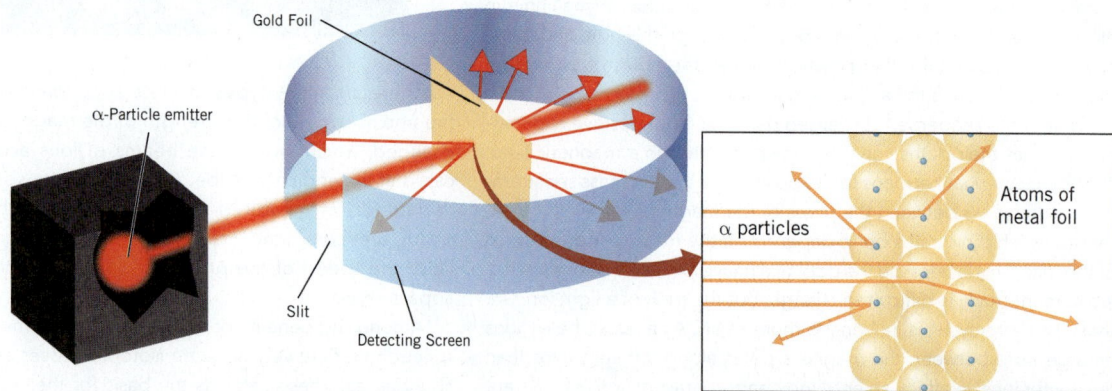

Figure 0.7 | **Some alpha particles are deflected by a thin gold foil.** Some hit something very massive head-on and are deflected backward. Most sail through. Some, making near misses with the massive "cores" (nuclei), are still deflected, because alpha particles have the same kind of charge (+) as these cores.

Subatomic Particles

The experiments we just described showed that atoms are composed of three principal kinds of subatomic particles: **protons**, **neutrons**, and **electrons**. These experiments also revealed that the center of an atom, or nucleus, is a very tiny, extremely dense core, which is where an atom's protons and neutrons are found. Because they are found in nuclei, protons and neutrons are sometimes called **nucleons**. The electrons in an atom surround the nucleus and fill the remaining volume of the atom. (*How* the electrons are distributed around the nucleus is the subject of Chapter 7.) The properties of the subatomic particles are summarized in Table 0.3, and the general structure of the atom is illustrated in Figure 0.8.

■ Physicists have found many subatomic particles. However, protons, neutrons, and electrons are the most important for the chemist.

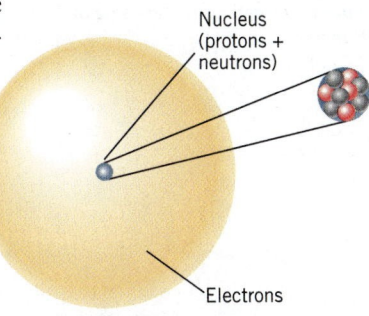

TABLE 0.3	Properties of the Subatomic Particles		
Particle	Mass (g)	Electrical Charge	Symbol
Electron	$9.10938291 \times 10^{-28}$	1−	$_{-1}^{0}e$
Proton	$1.672621777 \times 10^{-24}$	1+	$_{1}^{1}H^{+}, _{1}^{1}p$
Neutron	$1.674927351 \times 10^{-24}$	0	$_{0}^{1}n$

Figure 0.8 | The internal structure of an atom. An atom is composed of a tiny nucleus that holds all of the protons (red) and neutrons (grey). The electrons are in the space outside the nucleus.

Subatomic Particles in Atoms

As noted above, two of the subatomic particles carry electrical charges. Protons carry a single unit of **positive charge**, and electrons carry a single unit of **negative charge**. Two particles that have the same electrical charge repel each other, and two particles that have opposite charges attract each other. In an atom the negatively charged electrons are attracted to the positively charged protons. In fact, it is this attraction that holds the electrons around the nucleus. Neutrons have no charge and neither attract nor repel protons and electrons.

Because of their negative charges, electrons repel each other. The repulsions between the electrons keep them spread out throughout the volume of the atom, and it is the balance between the attractions the electrons feel toward the nucleus and the repulsions they feel toward each other that controls the size of atoms. Protons also repel each other, but they are able to stay together in the small volume of the nucleus because their repulsions are offset by powerful nuclear binding forces that involve other subatomic particles that are studied in particle physics.

Matter as we generally find it in nature appears to be electrically neutral, which means that it contains equal numbers of positive and negative charges. Therefore, *in a neutral atom, the number of electrons must equal the number of protons.*

The proton and neutron are much more massive than the electron, about 1800 times heavier, so in any atom almost all of the atomic mass is contributed by the particles found in the nucleus. It is also interesting to note, however, that the diameter of the atom is approximately 10,000 times the diameter of its nucleus, so almost all of the *volume* of an atom is occupied by its electrons, which fill the space around the nucleus. (To place this on a more meaningful scale, if the nucleus had a diameter the size of the period at the end of this sentence, it would be at the center of an atom with a diameter of approximately ten feet.)

Atomic Numbers and Mass Numbers

What distinguishes one element from another is the number of protons in their nuclei. In fact, this allows us to revise Dalton's definition of an **element** to a *substance whose atoms all contain the same number of protons.* Thus, we can assign each element a unique number, called its **atomic number** (Z), that is equal to the number of protons in its nucleus. The sum of the protons and neutrons within the nucleus of an atom is called its **mass number** and is given the symbol A.

$$\text{Atomic number } (Z) = \text{number of protons}$$

$$\text{Mass number } (A) = \text{number of protons} + \text{number of neutrons}$$

TOOLS

Numbers of subatomic particles in atoms

■ Atomic symbols are one- or two-letter abbreviations of the name of an element.

■ In neutral atoms the number of protons (A) equals the number of electrons.

Currently we have found or synthesized 118 different elements that are listed along with their names and symbols inside the front cover of this book. Each element is assigned its own unique **chemical symbol**, or **atomic symbol**, which can be used as shorthand for the name of the element.

Most elements exist in nature as mixtures of atoms called **isotopes** that are virtually identical in their chemical and physical properties. However, *the isotopes of a given element have atoms with the same number of protons but different numbers of neutrons.* Therefore, every isotope is fully defined by two numbers, its atomic number and its mass number. Sometimes these numbers are added to the left of the chemical symbol as a subscript and a superscript, respectively. Thus, if X stands for the chemical symbol for the element, an isotope of X is represented as

$$^A_Z X$$

The isotope of uranium used in nuclear reactors, uranium-235, has a mass number of 235 and can be symbolized as follows:

Mass number (number of protons and number of neutrons) $^{235}_{92}U$

Atomic number (number of protons)

As indicated, the name of this isotope is uranium-235 or U-235, and an atom contains 92 protons and $(235 - 92) = 143$ neutrons. If we refer to a neutral atom of uranium it must also contain 92 electrons. In writing the symbol for the isotope, the atomic number is often omitted because it is redundant. Every atom of uranium has 92 protons, and every atom that has 92 protons is an atom of uranium. Therefore, this uranium isotope can be represented simply as ^{235}U.

In naturally occurring uranium, a more abundant isotope is ^{238}U. Atoms of this isotope also have 92 protons, but the number of neutrons is 146. Thus, atoms of ^{235}U and ^{238}U have the identical number of protons but differ in the number of neutrons.

An Important Word about Problem Solving

Learning chemistry is far more than memorizing facts and formulas. To a large degree, your success in this course will be tied to your ability to solve problems, both numerical and nonnumerical. Because this is so important, one of the principal goals of this textbook is to help you develop and improve your problem-solving skills. Toward this end we include a large number of detailed worked examples throughout the book. You should study these and then work the practice exercises that follow. The answers to all of the practice exercises are in Appendix B at the back of the book, so you can check your work as you proceed.

You will find each of the examples divided into four steps intended to guide you through the thought processes followed by good problem solvers. We encourage you to practice this approach when you tackle problems on your own. As you gain experience, these steps will merge into a seamless, effective way to solve problems, not only in chemistry but in other subjects as well.

Analysis: If you intend to drive from your home to someplace you haven't been before, you don't just jump in the car and start driving. You first give some thought to where you're going and the route you have to take. Solving chemistry problems begins the same way. In the *Analysis* section, we identify exactly what is being asked and then plan what we have to do arrive at the solution. Sometimes the analysis will be relatively simple, but with more complex problems, you may have to spend some time thinking about how to solve the problem. You may find it necessary to review a concept to be certain you're on the right track.

Assembling the Tools: After we've determined how to proceed, the next step is to assemble the various chemical tools that apply to the particular problem at hand. These are the tools that have been developed in the text and are identified by the "Tool" icon in the margin. Each tool is chosen because it can be utilized to solve the problem.

Solution: At last we're ready to work out the answer by applying the appropriate tool to each step of the plan we developed in the *Analysis* section. This is really the simplest part of the problem, because we've already figured out what we have to do and which tools we need to do the job.

Is the Answer Reasonable? The preceding step has given us an answer, but is it the right answer? Does the answer make sense? It is always prudent to conclude your problem-solving experience with a quick check to see whether the answer is reasonable. As we progress, you will learn some techniques to help you check your answer.

In solving a problem, you have to be flexible. There may be times when you're able to figure out part of a problem in the *Analysis* step, but the full solution isn't apparent. If this happens, proceed with the next two steps and then come back to the *Analysis* again to try to plan the rest of the solution.

The most important thing to remember in working problems in this book is that all of the concepts and tools necessary to solve them have been given to you. If you're struggling, don't despair. Take a break and come back with a fresh perspective. You can be successful.

Example 0.1
Counting Protons, Neutrons, and Electrons

How many electrons, protons, and neutrons does the neutral atom (isotope) Cr-52 have?

Analysis: This problem asks for all three of the major subatomic particles in the Cr-52 isotope that has a mass number of 52. We will need to find the atomic number for Cr to use our tools.

Assembling the Tools: We have tools that define atomic number (Z) and mass number (A) in terms of the numbers of neutrons and protons. Those same tools specify that neutral elements have the same numbers of protons and electrons.

Solution: From the front cover we find that Cr is the symbol for chromium, so $Z = 24$ and $A = 52$, and we conclude

$$\text{Protons} = 24 \qquad \text{Electrons} = 24 \qquad \text{Neutrons} = 52 - 24 = 28$$

Is the Answer Reasonable? One check is to be sure that the sum of the number of protons and neutrons is the isotope mass number. A second check is that the number of any of the particles is not larger than the isotope mass number (the largest number given in the problem), and in most cases the number of electrons, protons, or neutrons is usually close to half of the mass number. A final check is that the number of protons equals the number of electrons because it is a neutral atom. Our answers fulfill these conditions.

Write the symbol for the isotope of plutonium (Pu) that contains 146 neutrons. How many electrons does it have? (*Hint:* Review the tools for writing isotope symbols and counting electrons.)

Practice Exercise 0.8

How many protons, neutrons, and electrons are in each atom of $^{35}_{17}\text{Cl}$? Can we discard the 35 or the 17 or both from this symbol without losing the ability to solve the problem? Explain your reasoning.

Practice Exercise 0.9

Relative Atomic Masses of Elements

Before subatomic particles were discovered, a significant body of data had already been developed that showed that atoms of different elements had different distinctive masses. In fact, one of the most useful concepts to come from Dalton's atomic theory is that atoms of an element have a constant, characteristic **atomic mass** (or **atomic weight**). This concept opened the door to the determination of chemical formulas and ultimately to one of the most useful devices chemists have for organizing chemical information, the **periodic table** of the elements. But how could the masses of atoms be measured without a knowledge of atomic structure?

Individual atoms are much too small to weigh in the traditional manner. However, the *relative masses* of the atoms of elements can be determined *provided we know the ratio in which the atoms occur in a compound*. Let's look at an example to see how this could work.

Early chemists such as Gay-Lussac noticed that when gases react, their volumes are always ratios of small whole numbers. Avogadro proposed that a given volume of a gas will always contain the same number of molecules as long as the conditions are held constant. Although many scientists, including Dalton, refused to accept Avogadro's hypothesis, it turned out to be the answer that allowed the ratio of atoms in many compounds to be determined. For instance, if one liter of hydrogen reacts with one liter of fluorine to make two liters of hydrogen fluoride, the simplest explanation is that hydrogen and fluorine must be diatomic (H_2 and F_2) and hydrogen fluoride contains one atom each of hydrogen and one of fluorine (HF).

Once the ratio of hydrogen atoms to fluorine atoms in hydrogen fluoride is established it means that in any sample of this substance the fluorine-to-hydrogen atom ratio is always 1 to 1. It is also found that when a sample of hydrogen fluoride is decomposed, the mass of fluorine obtained is always 19.0 times larger than the mass of hydrogen, so the fluorine-to-hydrogen mass ratio is always 19.0 to 1.00.

- F-to-H atom ratio: 1 to 1
- F-to-H mass ratio: 19.0 to 1.00

The only way we can explain a 1-to-1 atom ratio and a 19.0-to-1.00 mass ratio is *if each fluorine atom is 19.0 times heavier than each H atom.*

Notice that even though we haven't found the actual masses of F and H atoms, we now know how their masses compare (i.e., we know their *relative masses*). Similar procedures, with other elements in other compounds, are able to establish relative mass relationships among the other elements as well. What we need next is a way to place all of these masses on the same mass scale.

Carbon-12: Standard for the Atomic Mass Scale

To establish a uniform mass scale for atoms it is necessary to select a standard against which the relative masses can be compared. Currently, the agreed-upon reference uses the most abundant isotope of carbon, carbon-12, ^{12}C. From this reference, one atom of this isotope is exactly 12 units of mass, which are called **atomic mass units**. Some prefer to use the symbol **amu** for the atomic mass unit. The internationally accepted symbol is **u**, which is the symbol we will use throughout the rest of the book. By assigning 12 u to the mass of one atom of ^{12}C, the size of the atomic mass unit is established to be $\frac{1}{12}$ of the mass of a single carbon-12 atom:

- 1 atom of ^{12}C has a mass of 12 u (exactly)
- 1 u equals the mass of $\frac{1}{12}$ atom of ^{12}C (exactly)

In modern terms, the atomic mass of an element is the average mass of the element's atoms (as they occur in nature) relative to an atom of carbon-12, which is assigned a mass of 12 units. Thus, if an average atom of an element has a mass twice that of a ^{12}C atom, its atomic mass would be 24 u.

In general, the mass number of an isotope differs slightly from the atomic mass of the isotope. For instance, the isotope ^{35}Cl has an atomic mass of 34.968852 u. In fact, the only isotope that has an atomic mass equal to its mass number is ^{12}C, since *by definition* the mass of this atom is exactly 12 u.

Aluminum atoms have a mass that is 2.24845 times that of an atom of ^{12}C. What is the atomic mass of aluminum? (*Hint:* Recall that we have a tool that gives the relationship between the atomic mass unit and ^{12}C.)

Practice Exercise 0.10

How much heavier is the average atom of naturally occurring copper than an atom of ^{12}C? Refer to the table inside the front cover of the book for the necessary data.

Practice Exercise 0.11

The definition of the size of the atomic mass unit is really quite arbitrary. Could another atom have been chosen, and could the atomic mass unit have been another fraction of the mass of that atom? Give an example. Provide some reasons why scientists would choose ½ of a C-12 atom as the definition of the atomic mass unit.

Practice Exercise 0.12

Chemists generally work with the mixture of isotopes that occur naturally for a given element. Because the composition of this isotopic mixture is very nearly constant regardless of the source of the element, we can speak of an *average mass of an atom* of the element—average in terms of mass. For example, naturally occurring hydrogen is almost entirely a mixture of two isotopes, with a trace of a third, in the relative proportions given in Table 0.4. The "average mass of an atom" of the element hydrogen, as it occurs in nature, has a mass that is 0.083992 times that of a ^{12}C atom. Since 0.083992 × 12.000 u = 1.0079 u, the average atomic mass of hydrogen is 1.0079 u. Notice that this average value is only a little larger than the atomic mass of ^{1}H because naturally occurring hydrogen contains mostly ^{1}H, only a little ^{2}H, and a trace of ^{3}H as shown in Table 0.4.

Scientists still measure isotope masses and calculate average atomic masses to increase the accuracy of their values. The International Union of Pure and Applied Chemistry, **IUPAC**, an international body of scientists responsible for setting standards in chemistry, when necessary, revises the average atomic masses to reflect this research. On the Cutting Edge 0.3 discusses the most recent findings by IUPAC that some elements cannot be assigned a single average atomic mass.

■ Hydrogen has a third isotope called tritium with two neutrons in its nucleus. The amount of tritium in hydrogen samples is vanishingly small.

TABLE 0.4   Abundance of Hydrogen Isotopes

Hydrogen Isotope	Mass	Percentage Abundance
^{1}H	1.007825 u	99.985
^{2}H	2.0140 u	0.015
^{3}H	3.01605 u	trace

Average Atomic Masses from Isotopic Abundances

Originally, the relative atomic masses of the elements were determined in a way similar to that described for hydrogen and fluorine in our earlier discussion. A sample of a compound was analyzed for the mass of each element in its formula, and from the ratio of the elements in the formula the relative atomic masses were calculated. These relative masses were then adjusted, if necessary, to relate them to ^{12}C. Today, methods, such as high-resolution mass spectrometry discussed in On the Cutting Edge 0.2, are used to precisely measure both the relative abundances of the isotopes of the elements and their atomic masses. The **relative abundance** (also called the atom fraction) of an isotope is defined as the fraction of all the atoms of an element that exist as a given isotope.

$$\text{Relative abundance} = \frac{\text{atoms of one isotope}}{\text{total atoms of all isotopes}} = \frac{\text{atom percentage}}{100\%}$$

This kind of information has made it possible to calculate more precise values of the average atomic masses. The average atomic mass for any element can be calculated by multiplying the relative abundance of each isotope by its mass and adding the values together. Example 0.2 illustrates how this calculation is done.

ON THE CUTTING EDGE (0.3

Atomic Masses Are Changing (Again)

We all like to think that there are some solid, unchangeable constants in science, such as the speed of light and the atomic mass unit. As we saw in this chapter, the first measurements of quantities such as the electron's *e/m* or its mass are rather imprecise. Continued measurements and improved instruments refine these quantities to better and better values. The International Union of Pure and Applied Chemistry, IUPAC, and the National Institute of Standards and Technology, NIST, in the United States are two organizations that evaluate new data and announce changes, if needed, to the scientific community. Periodically IUPAC publishes reports to inform scientists that recent research requires that atomic masses need to be adjusted, and scientists have become quite accustomed to this practice.

Recently IUPAC published a report concerning changes of atomic masses. It said that they could not determine a single value for the atomic masses of ten elements listed in Table 1. Apparently isotopes of these elements are dispersed in different ratios around the world. For example, the atomic mass of hydrogen varies from 1.00785 to 1.00811, and if the sample comes from a reagent chemical it will have a higher mass than if it came from an auto exhaust. Similar bimodal situations occur with other elements. We will see later in this chapter that this information has already been used in crime scene investigations.

The end result is that IUPAC recommends that selected elements have a range of atomic masses rather than a single value. As we will see in the next chapter, all experimental measurements have some uncertainty attached and this report emphasizes that point dramatically. Except for lithium, the range of values in the table are less than one-tenth of one percent. The calculations in this text will not be affected by the uncertainties presented by the IUPAC.

TABLE 1	Ranges of Atomic Masses Proposed by IUPAC	
Elements whose atomic weights are now presented as intervals are shown below		
Element name	From	To the interval
Hydrogen	1.00794(7)	[1.00784; 1.00811]
Lithium	6.941(2)	[6.938; 6997]
Boron	10.811(7)	[10.806; 10.821]
Carbon	12.0107(8)	[12.0096; 12.0116]
Nitrogen	14.0067(2)	[14.00643; 14.00728]
Oxygen	15.9994(3)	[15.99903; 15.99977]
Silicon	28.0855(3)	[28.084; 28.086]
Sulfur	32.065(5)	[32.059; 32.076]
Chlorine	35.453(2)	[35.446; 35.457]
Thallium	204.3833(2)	[204.382; 204.385]

Example 0.2
Calculating Average Atomic Masses from Isotopic Abundances

Naturally occurring chlorine is a mixture of two isotopes. In every sample of this element, 75.77% of the atoms are ^{35}Cl and 24.23% are atoms of ^{37}Cl. The accurately measured atomic mass of ^{35}Cl is 34.9689 u and that of ^{37}Cl is 36.9659 u. From these data, calculate the average atomic mass of chlorine.

Analysis: We will need to determine the relative abundance of each isotope from their atom percentages. Then we can multiply the exact mass of each isotope by their relative abundances to determine how much of the total mass is contributed by each isotope. Adding all the contributions together will give the average mass.

Assembling the Tools: The tool for calculating the relative abundances from the atom percentages is needed. If a sample is made up of more than one isotope, then the mass contribution of one of the isotopes, X, in the sample is calculated using the equation

$$\text{mass contribution of } X = (\text{exact mass of } X) \times (\text{relative abundance of } X)$$

Solution: We will calculate the mass contributions of the two isotopes as

$$\text{mass contribution of } ^{35}Cl = 34.9689 \text{ u} \times \frac{75.77\% \ ^{35}Cl}{100\%} = 26.496 \text{ u}$$

$$\text{mass contribution of } ^{37}Cl = 36.9659 \text{ u} \times \frac{24.23\% \ ^{37}Cl}{100\%} = 8.957 \text{ u}$$

Now we add these contributions to give us the total mass of the "average atom."

$$26.496 \text{ u} + 8.957 \text{ u} = 35.453 \text{ u rounded to } 35.45 \text{ u}$$

Notice that in this two-step problem we kept one extra significant figure until the final rounding to four significant figures.

Is the Answer Reasonable? For a quick check: We know that the answer must lie between the highest mass and the lowest mass; it does and that should satisfy us. To analyze this further, the answer should be closer to the mass of ^{35}Cl because ^{35}Cl is more abundant than ^{37}Cl. Our result is closer to 35 than 37; therefore, we can feel pretty confident our answer is correct.

Naturally occurring boron is composed of 19.9% of ^{10}B atoms and 80.1% of ^{11}B atoms. Atoms of ^{10}B have a mass of 10.0129 u and those of ^{11}B have a mass of 11.0093 u. Calculate the average atomic mass of boron. (*Hint:* Convert percent abundance to relative abundance.)

Practice Exercise 0.13

Neon, the gas used in neon lamps, is composed of 90.483% of ^{20}Ne atoms, 0.271% of ^{21}Ne atoms, and 9.253% of ^{22}Ne atoms. The ^{20}Ne atoms have a mass of 19.992 u, ^{21}Ne atoms have a mass of 20.994, and those of ^{22}Ne have a mass of 21.991 u. Calculate the average atomic mass of neon.

Practice Exercise 0.14

Recent investigations have shown that measuring isotope abundances may help solve crime cases. In this case, slight variations of isotopes from one location to another were able to provide information in a possible murder case, as discussed in On the Cutting Edge 0.4.

ON THE CUTTING EDGE (0.4)

Isotope Ratios Help Solve Crime

The variation in the distribution of isotopes throughout the world has even made it into forensic science. In February of 1971, a woman's body was found under a highway overpass at the southern end of Lake Panasoffkee in central Florida. She had been murdered. The detectives in the case could not identify her or her murderer, and she was buried as Jane Doe.

In 1987 the body was exhumed for further investigation, but forensic science at the time did not lead to any more conclusions. However, in 2012, the skeleton was brought to the Tampa Bay Cold Case Project at the University of South Florida, shavings were taken of her tooth enamel and bones, and samples of hair were collected. These samples were then analyzed for isotopes of lead, carbon, and oxygen, which gave clues as to her history.

The ratio of isotopes of different elements vary in the rocks and soil depending on their geographical location. Using a mass spectrometer (On the Cutting Edge 0.2) chemists can determine these ratios and match the ratios to locations throughout the world.

The first element to give an indication of where she was from was lead. Tetraethyl lead shown in the Figure was used in gasoline until the 1970s, and this lead contaminated the air, soil, food, and bodies of growing children. The lead used in Europe came from Australia, which has a different isotopic signature compared to the lead used in the United States. This indicated that the woman was from Europe.

The isotopes of the second element, oxygen, pointed toward southern Europe. When water evaporates from the sea, the water with the heavier isotopes of oxygen, ^{18}O and ^{17}O, precipitate closer to the

A model of tetraethyl lead

coast, and the oxygen in her bones had a higher concentration of the heavier isotopes. The combination of the isotopic patterns of the lead and oxygen pointed to a geographic origin of Greece with a 60–70% probability.

The last element analyzed for isotopes was carbon. Using the woman's hair, and examining the ratios of carbon isotopes along the length of hair, the scientists were able to conclude that she was a recent arrival in the United States. The United States has a corn-based diet while Europeans have a wheat-based diet, and the corn in the United States has a higher ratio of the heavier carbon isotopes.

| Summary

Organized by Learning Objective

Develop a sense of the scope and purpose of the chemical sciences

Chemistry has some main ideas to keep in mind. These include the atomic theory developed by Dalton. Next is the idea that the atomic scale is reflected in the macroscopic world. Another is that energy changes and probability allow us to understand why chemicals react. Last is that the three-dimensional structure of molecules often dictates their function.

Learn how the elements were formed

Hydrogen, deuterium, and helium, with trace amounts of lithium were formed in the **big bang**. The elements up to iron were produced in stars, mainly red giants. The remaining natural elements were formed in **supernovas**.

Understand that the distribution of substances around the world is not accidental

As the matter from supernovas accreted, hydrogen tended to form new stars while the heavier elements formed planets. As the molten earth cooled, elements and compounds with different physical properties separated, resulting in their nonuniform distribution as we observe.

Appreciate the powerful nature of the atomic theory

Dalton based his **atomic theory** on the **law of definite proportions** and the **law of conservation of mass**. Dalton's theory proposed that matter consists of indestructible atoms with masses that do not change during chemical reactions. During a chemical reaction, atoms may change partners, but they are neither created nor destroyed. Today, using modern instruments scientists are able to "see" atoms on the surfaces of solids.

Understand how we came to know about the structure of the atom

Atoms can be split into **subatomic particles**, such as **electrons** (with one negative charge, $1-$), **protons** (with one positive charge, $(1+)$, and **neutrons** (with zero charge). The number of protons is called the **atomic number** (Z) of the element. Each element has a different atomic number. The electrons are found outside the **nucleus**; their number equals the atomic number in a neutral atom. **Isotopes** of an element have identical atomic numbers but different numbers of neutrons. An element's **atomic mass** (**atomic weight**) is the relative mass of its atoms on a scale in which atoms of carbon-12 have a mass of exactly 12 u (atomic mass units). Most elements occur in nature as uniform mixtures of a small number of isotopes, whose masses differ slightly. However, all isotopes of an element have very nearly identical chemical properties.

TOOLS

Tools for Problem Solving The following tools were introduced in this chapter. Study them carefully so you can select the appropriate tool when needed.

Law of definite proportions (Section 0.4)

A given compound will always have the same atomic composition and the elements will always be present in the same proportions by mass.

Law of conservation of mass (Section 0.4)

Mass is not lost or gained in a chemical reaction.

Numbers of subatomic particles in atoms (Section 0.5)

Number of electrons = number of protons (in neutral atoms only)
Atomic number (Z) = number of protons
Mass number (A) = number of protons + number of neutrons

Symbols for the elements (Section 0.5)

Each element has a one- or two-letter abbreviation with the first letter always capitalized and the second letter, if present, is in lower case.

Atomic symbols for isotopes (Section 0.5)

The mass number (A) comes before the element symbol as a superscript and the atomic number (Z) also comes before the element as a subscript.

WileyPLUS = *WileyPLUS*, an online teaching and learning solution. *Note to instructors:* Many of the end-of-chapter problems are available for assignment via the WileyPLUS system. **www.wileyplus.com.** Review Problems are presented in pairs separated by blue rules. Answers to problems whose numbers appear in blue are given in Appendix B. More challenging problems are marked with an asterisk *.

|Review Questions

Chemistry's Important Concepts

0.1 Suggest two other important concepts of chemistry. Are these ideas appropriate for an introductory chemistry course?

0.2 Knowing what we know today, which of the postulates of the atomic theory are wrong?

0.3 In what ways does the atomic theory affect the discovery of new drugs?

0.4 Shake a few crystals of salt into your hand and look at the crystals. What does the shape of the salt crystal tell us?

0.5 Are there any examples of crystals in nature that suggest the underlying atomic arrangement?

0.6 Heat is one form of energy. If a reaction gives off heat, will the substances that remain have more or less energy than when the reaction started?

0.7 Can you think of a chemical reaction that starts as soon as the chemicals are mixed? Can you deduce how the atoms lose energy in that reaction?

0.8 Some large biological molecules rely on molecular shape for their function. Can you describe one of these?

Supernovas and the Elements

0.9 What elements were formed **(a)** during the first moments of the big-bang, **(b)** in red giant stars, and **(c)** during a supernova?

0.10 What effect does the formation of iron have on a red giant star?

0.11 What is the predominant element in the solar system today?

Elements and the Earth

0.12 Describe the earth's core. Why isn't the core a different element such as lead or aluminum?

0.13 What physical factors caused elements and minerals to separate in the earth?

0.14 Is this separation still occurring on the earth?

Atomic Theory

0.15 What measuring devices did early chemists have available for establishing the laws of conservation of mass and definite proportions?

0.16 In your own words, describe how Dalton's atomic theory explains the law of conservation of mass and the law of definite proportions.

0.17 Which of the laws of chemical combination is used to define the term compound?

Internal Structure of the Atom

0.18 What are the names, symbols, and electrical charges of the three subatomic particles introduced in this chapter?

0.19 Where in an atom is nearly all of its mass concentrated? Explain your answer in terms of the particles that contribute to this mass.

0.20 What is a nucleon? Which ones have we studied?

0.21 How was the charge-to-mass ratio of the electron determined?

0.22 How did Robert Millikan determine the charge of an electron, and how did this allow the mass of the electron to be determined?

0.23 How was the proton discovered?

0.24 What experiment did Rutherford carry out to determine the existence of the nucleus?

0.25 Why don't we count the electrons when determining the mass of an element?

0.26 Define the terms atomic number and mass number. What symbols are used to designate these terms?

0.27 Consider the symbol $^a_b X$, where X stands for the chemical symbol for an element. What information is given in locations **(a)** a and **(b)** b?

0.28 Write the symbols of the isotopes that contain the following. (Use the table of atomic masses and numbers printed inside the front cover for additional information, as needed.)

(a) An isotope of iodine whose atoms have 78 neutrons

(b) An isotope of strontium whose atoms have 52 neutrons

(c) An isotope of cesium whose atoms have 82 neutrons

(d) An isotope of fluorine whose atoms have 9 neutrons

0.29 What is wrong with the following statement? "The atomic mass of an atom of chlorine is 35.453 u."

0.30 The atomic number of silver, Ag, is 47 and it has an average atomic mass of 107.87. Why is it impossible to determine the number of neutrons in the nucleus of a silver atom?

Review Problems

Internal Structure of the Atom

0.31 A certain element X forms a compound with oxygen in which there are two atoms of X for every three atoms of O. In this compound, 1.125 g of X are combined with 1.000 g of oxygen. Use the average atomic mass of oxygen to calculate the average atomic mass of X. Use your calculated atomic mass to identify the element X.

0.32 Nitrogen reacts with a metal to form a compound in which there are three atoms of the metal for each atom of nitrogen. If 1.486 g of the metal reacts with 1.000 g of nitrogen, what is the calculated atomic mass of the metal? Use your calculated atomic mass to identify the metal.

0.33 Give the numbers of neutrons, protons, and electrons in the atoms of each of the following isotopes: **(a)** radium-226, **(b)** ^{206}Pb, **(c)** carbon-14, **(d)** ^{23}Na. (Use the table of atomic masses and numbers printed inside the front cover for additional information, as needed.)

0.34 Give the numbers of electrons, protons, and neutrons in the atoms of each of the following isotopes: **(a)** cesium-137, **(b)** ^{238}U, **(c)** iodine-131, **(d)** ^{285}Cn. (As necessary, consult the table of atomic masses and numbers printed inside the front cover.)

0.35 Iodine-131 is used to treat overactive thyroids; it has a mass of 130.9061 u. Give the number of protons, neutrons, and electrons in the atom.

0.36 In the polyatomic ion TcO_4^-, Tc has a hypothetical charge of 7+. The ^{99}Tc isotope is used in medicine for imaging purposes. It has a mass of 98.90625 u. Give the number of protons, neutrons, and electrons for a Tc^{7+} ion.

0.37 One chemical substance in natural gas is a compound called methane. Its molecules are composed of carbon and hydrogen, and each molecule contains four atoms of hydrogen and one atom of carbon. In this compound, 0.33597 g of hydrogen is combined with 1.0000 g of carbon-12. Use this information to calculate the atomic mass of the element hydrogen.

0.38 Carbon tetrachloride contains one carbon and four chlorine atoms. For this compound 11.818 g of chlorine combine with 1.000 g of C-12. Using this information, calculate the atomic mass of chlorine.

0.39 If an atom of carbon-12 had been assigned a relative mass of 24.0000 u, what would be the average atomic mass of hydrogen relative to this mass?

0.40 One atom of ^{109}Ag has a mass that is 9.0754 times that of a ^{12}C atom. What is the atomic mass of this isotope of silver expressed in atomic mass units?

0.41 Naturally occurring copper is composed of 69.17% of ^{63}Cu, with an atomic mass of 62.9396 u, and 30.83% of ^{65}Cu, with an atomic mass of 64.9278 u. Use these data to calculate the average atomic mass of copper.

0.42 Naturally occurring magnesium (one of the elements in milk of magnesia) is composed of 78.99% of ^{24}Mg (atomic mass = 23.9850 u), 10.00% of ^{25}Mg (atomic mass = 24.9858 u), and 11.01% of ^{26}Mg (atomic mass = 25.9826 u). Use these data to calculate the average atomic mass of magnesium.

Additional Exercises

0.43 N_2O is often called nitrous oxide or "laughing gas," and NO_2 is a well-known air pollutant. How would you explain that these molecules follow the law of definite proportions?

0.44 An element has 24 protons in its nucleus.

   (a) Identify this element.

   (b) Write the symbol for the element's isotope. That will have a mass closest to its average atomic mass.

   (c) How many neutrons are in the isotope you described in part (b)?

   (d) How many electrons are in neutral atoms of this element?

   (e) How many times heavier than ^{12}C is the average atom of this element?

0.45 Iron is composed of four isotopes with the percentage abundances and atomic masses given in the following table. Calculate the average atomic mass of iron.

Isotope	Percentage Abundance	Atomic Mass
^{54}Fe	5.80%	53.9396
^{56}Fe	91.72%	55.9349
^{57}Fe	2.20%	56.9354
^{58}Fe	0.28%	57.9333

0.46 What color are the protons, neutrons, and electrons in the following figure? What element is it? What is the mass number of the atom? Write the symbol for the atom using the $^A_Z X$ notation.

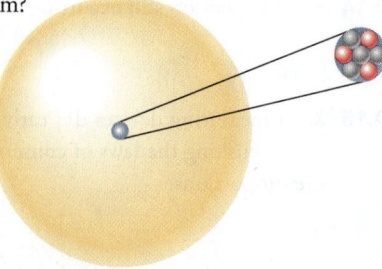

Exercises in Critical Thinking

0.47 Immediately after the big bang, helium and other very light elements up to about lithium were formed. It was not until stars were formed that heavier atoms up to iron were created. The heaviest naturally occurring elements did not form until the explosion of massive stars in supernovas. Look at the periodic table and divide it up by these masses. Is there any trend or information about these elements that you can determine from this information?

0.48 Avogadro proposed that, at the same temperature and pressure, containers of equal volumes would contain the same numbers of molecules of a gas. How did this allow scientists to discover that elements such as hydrogen, oxygen, and chlorine were diatomic. Why was this important?

0.49 Scientists often validate measurements, such as measuring the circumference of the earth, by using two independent methods to measure the same value. Describe two independent methods for determining the atomic mass of an element. Explain how these methods are truly independent.

0.50 Sir James Chadwick discovered the neutron. The web site for the Nobel Prize committee contains his acceptance speech for the Nobel Prize he won in physics in 1935. Read the speech and draw a diagram that summarizes the experiment.

Scientific Measurements

Chapter Outline

Tom Zagwodzki/Goddard Space Flight Center

In the previous chapter we saw that chemistry is an atomic and molecular science. If we understand the nature of our atoms and molecules on the microscopic scale, we will be able to understand the properties of chemicals on the macroscopic scale. We obtain this understanding by making precise observations and numerical measurements. For example in the opening picture, a laser beam is shot to the moon to measure the distance between the earth and moon in the Lunar Laser Ranging Experiment. By measuring the time it takes the light to return to the earth, the distance can be accurately determined. In order to study chemistry, scientists systematically make and report measurements that allow them to explain the microscopic world.

This Chapter in Context

LEARNING OBJECTIVES

After reading this chapter, you should be able to:

- explain the scientific method
- classify matter
- develop an understanding of which physical and chemical properties are important to measure
- understand measurements and the use of SI units
- understand how errors in measurements occur and are reported using significant figures
- learn how to effectively use dimensional analysis
- learn to determine the density of substances and to use density as a conversion factor

1.1 | Laws and Theories: The Scientific Method

Scientists are curious creatures who want to know what makes the world "tick" (Figure 1.1). The approach they take to their work is generally known as the **scientific method** (Figure 1.2), which basically boils down to an iterative process of gathering information and formulating explanations.

In the sciences, we usually gather information by performing experiments in laboratories under controlled conditions so the **observations** we make are **reproducible**, or can be repeated and the same results obtained. An observation is a statement that accurately describes something we see, hear, taste, feel, or smell. The observations we make while performing experiments are referred to as **data**. In the opening photo of the lunar laser ranging experiment the data collected are the sending and receiving times of the laser pulse. In order for the observations to be considered valid, they have to be reproducible.

Data gathered during an experiment often lead us to draw conclusions. A **conclusion** is a statement that's based on what we think about a series of observations. For example, consider the following statements about the fermentation of grape juice to make wine:

1. Before fermentation, grape juice is very sweet and contains no alcohol.
2. After fermentation, the grape juice is no longer as sweet and it contains a great deal of alcohol.
3. In fermentation, sugar is converted into alcohol.

Statements 1 and 2 are observations because they describe properties of the grape juice that can be tasted and smelled. Statement 3 is a conclusion because it *interprets* the observations that are available.

Experimental Observations and Scientific Laws

One of the goals of science is to organize facts so that relationships or generalizations among the data can be established. For example, if we study the behavior of gases, such as

Figure 1.1 | A research chemist working in a modern laboratory. Reproducible conditions in a laboratory permit experiments to yield reliable results.

©AP/Wide World Photos

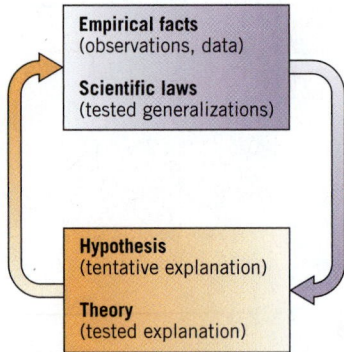

Figure 1.2 | **The scientific method is iterative** where observations suggest explanations, which suggest new experiments, which suggest new explanations, and so on.

■ The pressure of the gas is inversely proportional to its volume, so the smaller the volume, the larger the pressure.

air, we discover that the volume of a gas depends on a number of physical properties, including the amount of the gas, its temperature, and its pressure. The observations relating these physical properties are our data.

By studying data obtained from many different experiments performed at different temperatures and pressures we observe that when the temperature of the gas is held constant, compressing the gas into half of its original volume causes the pressure of the gas to double. If we were to repeat our experiments with many different gases, we would find that this generalization is applicable to all of them. Such a broad generalization, based on the results of many experiments, is called a **law** or **scientific law**, which is often expressed in the form of mathematical equations. For example, if we represent the pressure of a gas by the symbol P and its volume by V, the inverse relationship between pressure and volume can be written as

$$P = \frac{C}{V}$$

where C is a proportionality constant. (We will discuss gases and the laws relating to them in greater detail in Chapter 10.)

Hypotheses and Theories: Models of Nature

As useful as they may be, laws only state what happens; they do not provide explanations. *Why, for example, are gases so easily compressed to a smaller volume?* Or, *what must gases be like at the most basic, elementary, level for them to behave as they do?* Answering such questions when they first arise is no simple task and requires much speculation. But over time scientists build mental pictures, called **theoretical models**, that enable them to explain observed laws.

In the development of a theoretical model, researchers form tentative explanations called **hypotheses**. They then perform experiments that test predictions derived from the model. Sometimes the results show that the model is wrong. When this happens, the model must be abandoned or modified to account for the new data. Eventually, if the model survives repeated testing, it achieves the status of a theory. A **theory** is a *well-tested explanation of the behavior of nature*—for example, the atomic theory explains the composition of matter and the changes that matter can undergo. Keep in mind, however, that it is impossible to perform every test that might show a theory to be wrong, so we can never prove *absolutely* that a theory is correct.

Luck and ingenuity sometimes play an important role in the process of science. For example, in 2004, Andre Geim and Konstantin Novoselov used Scotch tape and a lump of graphite, a form of carbon, to extract a single layer of carbon atoms from the graphite, or *graphene*, Figure 1.3. This new extraction technique allowed others to discover

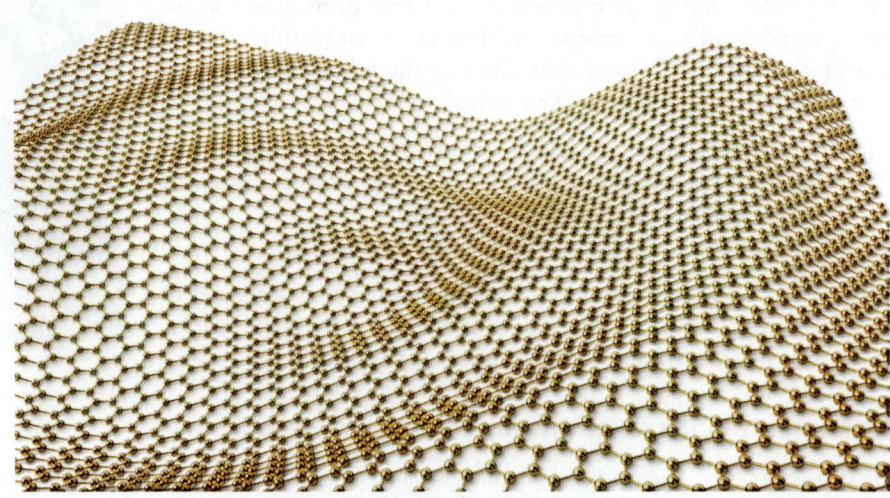

Figure 1.3 | **Graphene.** One layer of carbon atoms forms an incredibly strong and flexible substance.

graphene's properties, such as being incredibly strong for being only one atom thick and being one of the best conductors of electricity known. Yet, had it not been for Geim and Novoselov's creativity, application of the scientific method to these further discoveries might not have been made.

As a final note, some of the most spectacular and dramatic changes in science occur when major theories are proved to be wrong. While this happens only rarely, when it does occur, scientists are sent scrambling to develop new theories, and exciting new frontiers are opened.

What is the process of the scientific method? (*Hint:* Use Figure 1.2.)[1]

Practice Exercise 1.1

The Atomic Theory as a Model of Nature

Virtually every scientist would agree that one of the most significant theoretical models of nature ever formulated is the atomic theory, which was discussed in some detail in Section 0.3. According to this theory, all chemical substances are composed of tiny submicroscopic particles called **atoms**, which combine in various ways to form all the complex materials we find in the **macroscopic**, visible world around us. This theory forms the foundation for the way scientists think about nature. The atomic theory and how it enables us to explain chemical facts forms the central theme of this book as well.

■ *Macroscopic* commonly refers to physical objects that are measurable and observable by the naked eye.

1.2 | Matter and Its Classifications

We mentioned above that one of our goals is to be able to relate things we observe around us and in the laboratory to the way individual atoms and their combinations behave at the submicroscopic level. To begin this discussion we need to study how chemistry views the macroscopic world.

Matter Defined

Chemistry is concerned with the properties and transformations of matter. **Matter** is defined as anything that occupies space and has mass. It is the stuff our universe is made of. All of the objects around us, from rocks to pizza to people, are examples of matter.

Notice that our definition of matter uses the term *mass* rather than *weight*. The words mass and weight are often used interchangeably even though they refer to different things. **Mass** *refers to how much matter there is in a given object,*[2] *whereas* **weight** *refers to the force with which the object is attracted by gravity.* For example, a golf ball contains a certain amount of matter and has a certain mass, which is the same regardless of the golf ball's location. However, the *weight* of a golf ball on earth is about six times greater than it would be on the moon because the gravitational attraction of the earth is six times that of the moon. Because mass does not vary from place to place, we use mass rather than weight when we specify the amount of matter in an object. Mass is measured with an instrument called a balance, which we will discuss in Section 1.4.

---

[1]Answers to the Practice Exercises are found in Appendix B at the back of the book.
[2]Mass is a measure of an object's momentum, or resistance to a change in motion. Something with a large mass, such as a truck, contains a lot of matter and is difficult to stop once it's moving. An object with less mass, such as a baseball, is much easier to stop.

(a)   (b)   (c)

Figure 1.4 | **Some elements that occur naturally.** (a) Sulfur. (b) Gold. (c) Carbon in the form of diamonds. (Coal is also made up mostly of carbon.)

Elements

Chemistry is especially concerned with **chemical reactions**, which are transformations that alter the chemical compositions of substances. An important type of chemical reaction is **decomposition**, in which one substance is changed into two or more other substances. For example, if we pass electricity through molten (melted) sodium chloride (salt), the silvery metal sodium and the pale green gas chlorine are formed. This change has decomposed sodium chloride into two simpler substances. No matter how we try, however, sodium and chlorine cannot be decomposed further by chemical reactions into still simpler substances that can be stored and studied.

In chemistry, *substances that cannot be decomposed into simpler materials by chemical reactions are called* **elements**. Sodium and chlorine are two examples. Others you may be familiar with include iron, aluminum, sulfur, and carbon (as in graphite, diamonds, and graphene). Some elements are gases at room temperature. Examples include oxygen, nitrogen, hydrogen, chlorine, and helium. Elements are the simplest forms of matter that chemists work with directly. All more complex substances are composed of elements in various combinations. Figure 1.4 shows some samples of elements that occur uncombined with other elements in nature.

Chemical Symbols for Elements

So far, scientists have discovered 90 naturally occurring elements and have made 28 more, for a total of 118. Each element is assigned a unique **chemical symbol**, which is used as an abbreviation for the name of the element. The names and chemical symbols of the elements are given on the inside front cover of this book. In most cases, an element's chemical symbol is formed from one or two letters of its English name. For instance, the symbol for carbon is C, for bromine it is Br, and for silicon it is Si. For some elements, the symbols are derived from their non-English names. Those that come from their Latin names are listed in Table 1.1. The symbol for tungsten (W) comes from *wolfram*, the German name of the element. Regardless of the origin of the symbol, the first letter is *always* capitalized and the second letter, if there is one, is *always* lowercase.

Chemical symbols are also used to stand for atoms of elements when we write *chemical formulas* such as H_2O (water) and CO_2 (carbon dioxide). We will have a lot more to say about formulas later.

TABLE 1.1	Elements That Have Symbols Derived from Their Latin Names	
Element	Symbol	Latin Name
Sodium	Na	Natrium
Potassium	K	Kalium
Iron	Fe	Ferrum
Copper	Cu	Cuprum
Silver	Ag	Argentum
Gold	Au	Aurum
Mercury	Hg	Hydrargyrum
Antimony	Sb	Stibium
Tin	Sn	Stannum
Lead	Pb	Plumbum

Compounds

By means of chemical reactions, elements combine in various specific proportions to give all of the more complex substances in

(a)   (b)   (c)

Figure 1.5 | Mixtures can have variable compositions. Orange juice, pancake syrup, and Coca-Cola are mixtures that contain sugar. The amount of sugar varies from one to another because the composition can vary from one mixture to another.

nature. Thus, hydrogen and oxygen combine to form water (H_2O), and sodium and chlorine combine to form sodium chloride (NaCl, common table salt). Water and sodium chloride are examples of compounds. A compound is a substance formed from two or more different elements in which the elements are always combined in the same, fixed (i.e., constant) proportions by mass. For example, if any sample of pure water is decomposed, the mass of oxygen obtained is always eight times the mass of hydrogen.

Mixtures

Elements and compounds are examples of **pure substances**.[3] The **composition** of a pure substance is always the same, regardless of its source. Pure substances are rare, however. Usually, we encounter mixtures of compounds or elements. Unlike elements and compounds, **mixtures** *can have variable compositions.* For example, Figure 1.5 shows three mixtures that contain sugar. They have different degrees of sweetness because the amount of sugar in a given amount varies from one to the other.

Mixtures can be either homogeneous or heterogeneous. A **homogeneous mixture** has the same properties throughout the sample. An example is a thoroughly stirred mixture of sugar in water. We call such a homogeneous mixture a **solution**. Solutions need not be liquids, just homogeneous. For example, the alloy used in the U.S. 5, 10, 25 and 50 cent coins are solid solutions of copper and nickel, and clean air is a gaseous solution of oxygen, nitrogen, and a number of other gases.

A **heterogeneous mixture** consists of two or more regions, called **phases**, that differ in properties. A mixture of olive oil and vinegar in a salad dressing, for example, is a two-phase mixture in which the oil floats on the vinegar as a separate layer (Figure 1.6). The U.S. one cent coin (the penny) is mostly zinc coated with copper. The phases in a mixture don't have to be chemically different substances like oil and vinegar, however. A mixture of ice and liquid water is a two-phase heterogeneous mixture in which the phases have the same chemical composition but occur in different *physical states* (a term we discuss further in Section 1.3).

Physical and Chemical Changes

The process we use to create a mixture involves a **physical change**, because no new chemical substances form. Shown in Figure 1.7, powdered samples of the elements iron and

Figure 1.6 | A heterogeneous mixture. The salad dressing contains vinegar and vegetable oil (plus assorted other flavorings). Vinegar and oil do not dissolve in each other, and they form two layers. The mixture is heterogeneous because each of the separate phases (oil, vinegar, and other solids) has its own set of properties that differ from the properties of the other phases.

---

[3]We have used the term *substance* rather loosely until now. Strictly speaking, **substance** really means *pure substance*. Each unique chemical element and compound is a *substance*; a *mixture* consists of two or more substances.

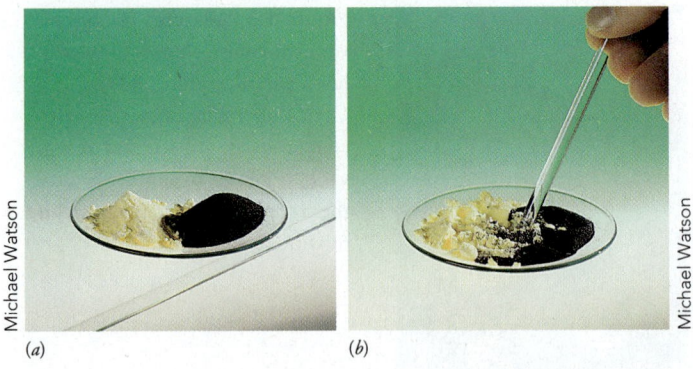

Michael Watson

(a)   (b)

Michael Watson

Figure 1.7 | Formation of a mixture of iron and sulfur. (*a*) Samples of powdered sulfur and powdered iron. (*b*) A mixture of sulfur and iron is made by stirring the two powders together.

Michael Watson

Figure 1.8 | Formation of a mixture does not change chemical composition. Here we see that forming the mixture has not changed the iron and sulfur into a compound of these two elements. The mixture can be separated by pulling the iron out with a magnet. Making a mixture involves a physical change.

Charles D. Winters/Getty Images

Figure 1.9 | "Fools gold." The mineral pyrite (also called iron pyrite) is a compound of iron and sulfur. When the compound forms, the properties of iron and sulfur disappear and are replaced by the properties of the compound. Pyrite has an appearance that caused some miners to mistake it for real gold.

sulfur are simply dumped together and stirred. The mixture forms, but both elements retain their original properties. To separate the mixture, we could similarly use just physical changes. For example, we could remove the iron by stirring the mixture with a magnet—a physical operation. The iron powder sticks to the magnet as we pull it out, leaving the sulfur behind (Figure 1.8). The mixture also could be separated by treating it with a liquid called carbon disulfide, which is able to dissolve the sulfur but not the iron. Filtering the sulfur solution from the solid iron, followed by evaporation of the liquid carbon disulfide from the sulfur solution, gives the original components, iron and sulfur, separated from each other.

The formation of a compound involves a **chemical change** (a chemical reaction) because the chemical makeup of the substances involved are changed. Iron and sulfur, for example, combine to form a compound often called "fool's gold" because it glitters like real gold (Figure 1.9). In this compound the elements no longer have the same properties they had before they combined, and they cannot be separated by physical means. The decomposition of fool's gold into iron and sulfur is also a chemical reaction.

The relationships among elements, compounds, and mixtures are shown in Figure 1.10.

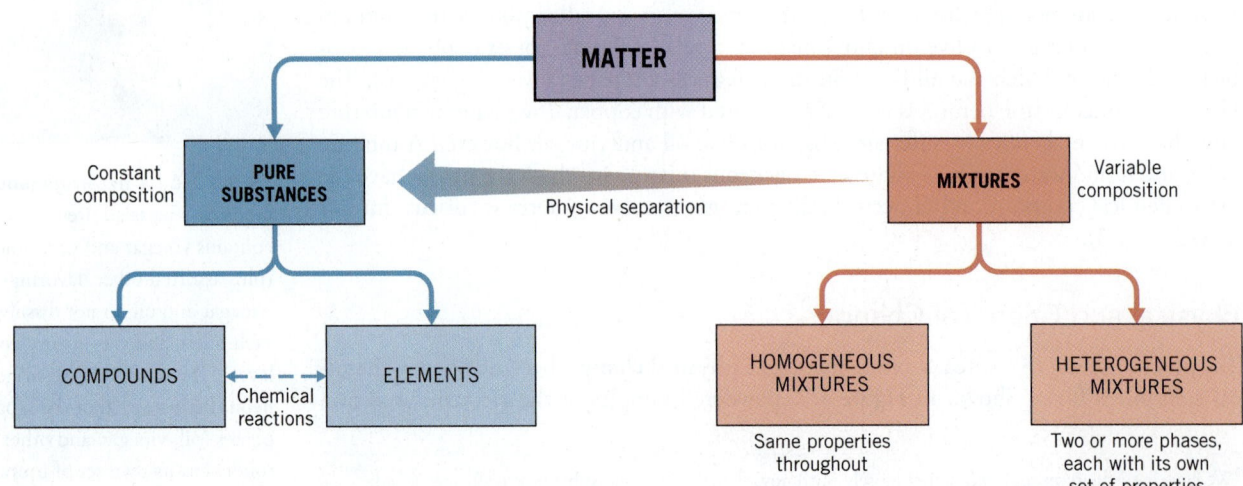

Figure 1.10 | Classification of matter.

Classify the following as an element, compound, or mixture. If it is a mixture, decide if it is a homogeneous or heterogeneous mixture: (a) lead, (b) black coffee, (c) sugar, (d) lasagna, (e) phosphorus. (*Hint:* What are elements, compounds, and mixtures?)

Practice Exercise 1.2

1.3 | Physical and Chemical Properties

In chemistry we use the **properties** (characteristics) of substances to identify them and to distinguish one from another. To organize our thinking, we can classify properties into two different types, physical and chemical, which separate properties by whether the act of observing the property changes the substance, or we categorize properties into extensive or intensive properties, which separates properties by how the property relates to the amount of substance that is being observed.

Physical Properties

The first way to classify properties is whether or not the chemical composition of an object is changed by the act of observing the property. A **physical property** is *one that can be observed without changing the chemical makeup of a substance*. For example, a physical property of gold is that it is yellow. The act of observing this property (color) doesn't change the chemical makeup of the gold. Neither does observing that gold conducts electricity, so color and electrical conductivity are physical properties.

Sometimes, observing a physical property does lead to a physical change. To measure the melting point of ice, for example, we observe the temperature at which the solid begins to melt (Figure 1.11). This is a physical change because it does not lead to a change in chemical composition; both ice and liquid water are composed of water molecules.

States of Matter

When we observe ice melting, we see two states of matter, liquid and solid. Water can also exist as a gas that we call water vapor or steam. Although ice, liquid water, and steam have quite different appearances and physical properties, they are just different forms of water. **Solid**, **liquid**, and **gas** are the most common states of matter. As with water, most substances are able to exist in all three of these states, and the state we observe generally depends on the temperature. The properties of solids, liquids, and gases relate to the different ways the particles on the atomic scale are organized (Figure 1.12), and a change from one state to another is a physical change.

Daniel Smith/Corbis

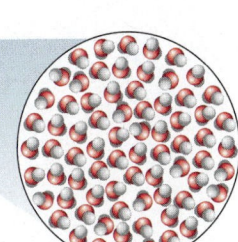

Figure 1.11 | **Liquid water and ice are both composed of water molecules.** Melting the ice cube does not change the chemical composition of the molecules.

Chemical Properties

A **chemical property** *describes how a substance undergoes a chemical change (chemical reaction)*. When a chemical reaction takes place, the chemicals interact to form entirely *different* substances with different chemical and physical properties. As previously described, pyrite is formed by the chemical reaction of sulfur and iron, and the product does not resemble sulfur or iron. It's a golden solid that is not attracted by a magnet.

The ability of iron to react with sulfur to form pyrite is a chemical property of iron. When we observe this property, we conduct a reaction, so after we've made the

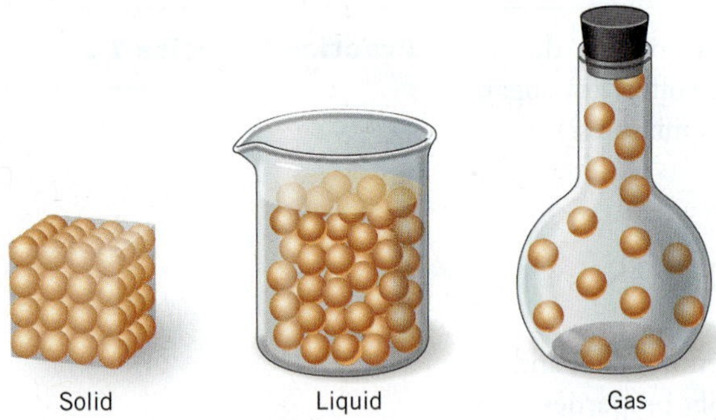

Figure 1.12 | **Solid, liquid, and gaseous states viewed using the atomic model of matter.** (*a*) In a solid, the particles are tightly packed and cannot move easily. (*b*) In a liquid, the particles are still close together but can readily move past one another. (*c*) In a gas, particles are far apart with much empty space between them.

observation we no longer have the same substances as before. In describing a chemical property, we usually refer to a chemical reaction.

Intensive and Extensive Properties

Another way of classifying a property is according to whether or not it depends on the size of the sample. For example, two different pieces of gold can have different volumes, but both have the same characteristic shiny yellow color and both will begin to melt at the same temperature. Color and melting point (and boiling point, too) are examples of **intensive properties**—*properties that are independent of sample size.* Volume, on the other hand, is an **extensive property**—a *property that depends on sample size.* Mass is another extensive property.

A job chemists often perform is *chemical analysis.* They're asked, "What is a particular sample composed of?" To answer such a question, the chemist relies on the properties of the chemicals that make up the sample. For identification purposes, intensive properties are more useful than extensive ones because every sample of a given substance exhibits the same set of intensive properties.

Color, freezing point, and boiling point are examples of intensive physical properties that can help us identify substances. Chemical properties are also intensive properties and also can be used for identification. For example, gold miners were able to distinguish between real gold and fool's gold, or pyrite, (Figure 1.9) by heating the material in a flame. Nothing happens to the gold, but the pyrite sputters, smokes, and releases bad-smelling fumes because of its ability, when heated, to react chemically with oxygen in the air.

Practice Exercise 1.3

Each of the following can be classified as a physical or chemical change. Identify which type of change is given and explain your choice.

(a) Sugar caramelizes.
(b) Iron wire glows red when an electric current is passed through it.
(c) Liquid nitrogen boils at $-196\,°C$.
(d) Tea is made sweeter by adding sugar.

Practice Exercise 1.4

Each of these properties can be classified as intensive or extensive. Identify which type of property is given and explain your choice.

(a) The color of a blue paint chip.
(b) The amount of paint needed to paint a room.
(c) The thickness of the paint when it is well stirred.
(d) The weight of the paint left in the can after the room is painted.

1.4 | Measurement of Physical and Chemical Properties

Qualitative and Quantitative Observations

Earlier you learned that an important step in the scientific method is observation. In general, observations fall into two categories, qualitative and quantitative. **Qualitative observations**, such as the color of a chemical or that a mixture becomes hot when a reaction occurs, do not involve numerical information. **Quantitative observations** are those **measurements** that do yield numerical data. You make such observations in everyday life, for example, when you glance at your watch or step onto a bathroom scale. In chemistry, we make various measurements that aid us in describing both chemical and physical properties.

Measurements Include Units

Measurements involve numbers, but they differ from the numbers used in mathematics in two crucial ways.

First, measurements always involve a comparison. When you say that a person is six feet tall, you're really saying that the person is six times taller than a reference object that is 1 foot in length, where foot is an example of a **unit of measurement**. *Both the number and the unit are essential parts of the measurement*, because the unit gives a sense of size. For example, if you were told that the distance between two points is 25, you would naturally ask "25 what?" The distance could be 25 inches, 25 feet, 25 miles, or 25 of any other unit that's used to express distance. A number without a unit is really meaningless.

The second important difference is that measurements always involve uncertainty; they are *inexact*. The act of measurement involves an estimation of one sort or another, and both the observer and the instruments used to make the measurement have inherent physical limitations. As a result, measurements always include some uncertainty, which can be minimized but never entirely eliminated. We will say more about this topic in Section 1.5.

International System of Units (SI Units)

A standard system of units is essential if measurements are to be made consistently. In the sciences, metric-based units are used. The advantage of working with metric units is that converting to larger or smaller values can be done simply by moving a decimal point, because metric units are related to each other by simple multiples of ten.

The original metric system has been simplified and now is called the **International System of Units**, abbreviated **SI** from the French name, *Le Système International d'Unités*. The SI is the dominant system of units in science and engineering, although there is still some usage of older metric units.

The SI has as its foundation a set of **base units** for seven measured quantities (Table 1.2). For now, we will focus on the base units for length, mass, time, and temperature. We will discuss the unit for amount of substance, the mole, at length in Chapter 3. The unit for electrical current, the ampere, will be discussed briefly when we study electrochemistry in Chapter 19. The unit for luminous intensity, the candela, will not be discussed in this book.

Most of the base units are defined in terms of reproducible physical phenomena. For instance, the meter is defined as exactly the distance light travels in a vacuum in 1/299,792,458 of a second. Only the base unit for mass is defined by an object made by human hands, a carefully preserved platinum–iridium alloy cylinder stored at the International Bureau of Weights and Measures in France (Figure 1.13). This cylinder serves indirectly as the calibrating standard for all "weights" used for scales and balances throughout the world.[4]

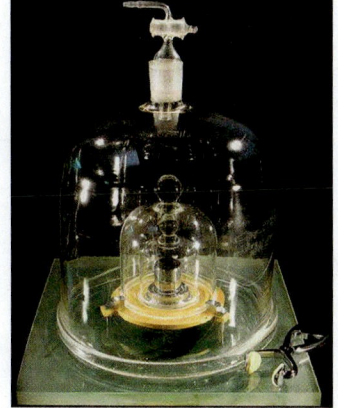

Figure 1.13 | **The international standard kilogram.** The SI standard for mass is made of a platinum–iridium alloy and is protected under two bell jars, as shown, at the International Bureau of Weights and Measures in France. Precise copies, called secondary standards, are maintained by many countries.

TOOLS

Base SI Units

TABLE 1.2	The SI Base Units	
Measurement	**Unit**	**Abbreviation**
Length	meter	m
Mass	kilogram	kg
Time	second	s
Electric current	ampere	A
Temperature	kelvin	K
Amount of substance	mole	mol
Luminous intensity	candela	cd

---

[4] Scientists are working on a method of accurately counting atoms whose masses are accurately known. Their goal is to develop a new definition of the kilogram that doesn't depend on an object that can be stolen, lost, or destroyed.

In scientific measurements, *all* physical quantities will have units that are combinations of the seven base SI units. For example, there is no SI base unit for area, but to calculate area we multiply length by width. Therefore, the *unit* for area is derived by multiplying the *unit* for length by the *unit* for width. Length and width are measurements that have the SI base unit of the **meter (m)**.

$$\text{length} \times \text{width} = \text{area}$$
$$(\text{meter}) \times (\text{meter}) = (\text{meter})^2$$
$$\text{m} \times \text{m} = \text{m}^2$$

The SI **derived unit** for area is therefore m^2 (read as *meters squared*, or *square meter*).

In deriving SI units, we employ a very important concept that we will use repeatedly throughout this book when we perform calculations: *Units undergo the same kinds of mathematical operations that numbers do.* We will see how this fact can be used to convert from one unit to another in Section 1.6.

EXAMPLE 1.1
Deriving SI Units

Linear momentum is a measure of the "push" a moving object has, equal to the object's mass times its velocity. What is the SI derived unit for linear momentum?

Analysis: To derive a unit for a quantity first we express it in terms of simpler quantities. We're told that linear momentum is mass times velocity. Therefore, the SI unit for linear momentum will be the SI unit for mass times the SI unit for velocity. Since velocity is not one of the base SI units, we will need to find the SI units for velocity.

Assembling the Tools: The tool we need is the list of SI base units in Table 1.2. We will also have to recall that velocity is distance traveled (length) per unit time.

Solution: We start by expressing the given information as an equation:

$$\text{mass} \times \text{velocity} = \text{linear momentum}$$

Next we write the same equation with units, kg for mass and m/s for velocity, because units undergo the same mathematical operations as the numbers.

$$\text{kg} \times \text{m/s} = \text{kg m/s or kg m s}^{-1}$$

Our result is that linear momentum has units of kg m/s or kg m s^{-1}.

■ Notice that for derived units, scientists usually omit the multiplication sign but always leave a space between units.

Is the Answer Reasonable? *Before leaving a problem, it is always wise to examine the answer to see whether it makes sense.* In this problem, the check is simple. The derived unit for linear momentum should be the product of units for mass and velocity, and this is obviously true.

Practice Exercise 1.5

The volume of a sphere is given by the formula $V = \frac{4}{3}\pi r^3$, where r is the radius of the sphere. From this equation, determine the SI units for volume. (*Hint:* r is a length, so it must have a length unit.)

Practice Exercise 1.6

When you "step hard on the gas" in a car you feel an invisible force pushing you back in your seat. This force, F, equals the product of your mass, m, times the acceleration, a, of the car. In equation form, this is $F = ma$. Acceleration is the change in velocity, v, with time, t:

$$a = \frac{\text{change in } v}{\text{change in } t}$$

Therefore, the units of acceleration are those of velocity divided by time. What is the SI derived unit for force expressed in SI base units?

TABLE 1.3	Some Non-SI Metric Units Commonly Used in Chemistry		
Measurement	Unit	Abbreviation	Value in SI Units
Length	angstrom	Å	$1\,Å = 0.1\,nm = 10^{-10}\,m$
Mass	atomic mass unit	u (amu)	$1\,u = 1.66054 \times 10^{-27}\,kg$ (rounded to six digits)
	metric ton	t	$1\,t = 10^3\,kg$
Time	minute	min.	$1\,min. = 50\,s$
	hour	h	$1\,h = 60\,min. = 3600\,s$
Temperature	degree Celsius	°C	$T_K = t_C + 273.15$
Volume	liter	L	$1\,L = 1000\,cm^3$

355 mL   454 g   1 liter
            454 g

Michael Watson

Figure 1.14 | Metric units are commonplace on consumer products.

Other Unit Systems

Some older metric units that are not part of the SI system are still used in the laboratory and in the scientific literature. Some of these units are listed in Table 1.3; others will be introduced as needed in upcoming chapters.

The United States is one of the only nations still using the **English system** of units, which measures distance in inches, feet, and miles; volume in ounces, quarts, and gallons; and mass in ounces and pounds. However, many of the English units are defined with reference to base SI units. Beverages, food packages, tools, and machine parts are often labeled in metric units (Figure 1.14). Common conversions between the English system and the SI are given in Table 1.4 and inside the back cover of this book.[5]

Decimal Multipliers

Sometimes the basic units are either too large or too small to be used conveniently. For example, the meter is inconvenient for expressing the size of very small things such as bacteria. The SI solves this problem by forming larger or smaller units by applying **decimal multipliers** to the base units. Table 1.5 lists the most commonly used decimal multipliers and the prefixes used to identify them. Those listed in boldface type are the ones most commonly encountered in chemistry.

When the name of a unit is preceded by one of these prefixes, the size of the unit is modified by the corresponding decimal multiplier. For instance, the prefix *kilo* indicates a multiplying factor of 10^3, or 1000. Therefore, a *kilo*meter is a unit of length equal to

■ A modified base unit can be conveniently converted to the base unit by removing the prefix and inserting $\times 10^x$.

$$25.2\,pm = 25.2 \times 10^{-12}\,m$$

A prefix can be inserted by adjusting the exponential part of a number to match the definition of a prefix. Then the ten raised to an exponent is removed and replaced by the prefix.

$$2.34 \times 10^{10}\,g = 23.4 \times 10^9\,g = 23.4\,Gg$$

TABLE 1.4	Some Useful Conversions	
Measurement	English Unit	English/SI Equality[a]
Length	inch	1 in. = 2.54 cm
	yard	1 yd = 0.9144 m
	mile	1 mi = 1.609 km
Mass	pound	1 lb = 453.6 g
	ounce (mass)	1 oz = 28.35 g
Volume	gallon	1 gal = 3.785 L
	quart	1 qt = 946.4 mL
	ounce (fluid)	1 oz = 29.6 mL

[a]These equalities allow us to convert English to metric or metric to English units.

[5]Originally, these conversions were established by measurement. For example, if a metric ruler is used to measure the length of an inch, it is found that 1 in. equals 2.54 cm. Later, to avoid confusion about the accuracy of such measurements, it was agreed that these relationships would be taken to be exact. For instance, 1 in. is now defined as *exactly* 2.54 cm. Exact relationships also exist for the other quantities, but for simplicity many have been rounded off. For example, 1 lb = 453.59237 g, *exactly*.

SI prefixes

TABLE 1.5	SI Prefixes—Their Meanings and Values[a]			
Prefix	Meaning	Symbol	Prefix Value[b] (numerical)	Prefix Value[b] (power of ten)
exa		E		10^{18}
peta		P		10^{15}
tera		T		10^{12}
giga	billions of	G	1000000000	10^{9}
mega	millions of	M	1000000	10^{6}
kilo	thousands of	k	1000	10^{3}
hecto		h		10^{2}
deka		da		10^{1}
deci	tenths of	d	0.1	10^{-1}
centi	hundredths of	c	0.01	10^{-2}
milli	thousandths of	m	0.001	10^{-3}
micro	millionths of	μ	0.000001	10^{-6}
nano	billionths of	n	0.000000001	10^{-9}
pico	trillionths of	p	0.000000000001	10^{-12}
femto		f		10^{-15}
atto		a		10^{-18}

[a]Prefixes in red type are used most often.

[b]Numbers in these columns can be interchanged with the corresponding prefix.

1000 meters.[6] The symbol for kilometer (km) is formed by applying the symbol meaning kilo (k) as a prefix to the symbol for meter (m). Thus 1 km = 1000 m (or 1 km = 10^3 m). Similarly a decimeter (dm) is 1/10th of a meter, so 1 dm = 0.1 m (1 dm = 10^{-1} m).

Units in laboratory measurements

Laboratory Measurements

Length, volume, mass, and temperature are the most common measurements made in the laboratory.

Length

The SI base unit for length, the meter (m), is too large for most laboratory purposes. More convenient units are the **centimeter (cm)** and the **millimeter (mm)**. Using Table 1.5 we see that they are related to the meter as follows.

$$1 \text{ cm} = 10^{-2} \text{ m} = 0.01 \text{ m}$$

$$1 \text{ mm} = 10^{-3} \text{ m} = 0.001 \text{ m}$$

It is also useful to know the relationships

$$1 \text{ m} = 100 \text{ cm} = 1000 \text{ mm}$$

$$1 \text{ cm} = 10 \text{ mm}$$

■ An older, non-SI unit called the angstrom (Å) is often used to describe dimensions of atomic- and molecular-sized particles.

1 Å = 0.1 nm = 10^{-10} m

---

[6]In the sciences, powers of 10 are often used to express large and small numbers. The quantity 10^3 means $10 \times 10 \times 10 = 1000$. Similarly, the quantity $6.5 \times 10^2 = 6.5 \times 10 \times 10 = 650$. Numbers less than 1 have negative exponents when expressed as powers of 10. Thus, the fraction 1⁄10 is expressed as 10^{-1}, so the quantity 10^{-3} means 1⁄10 × 1⁄10 × 1⁄10 = 1⁄1000. A value of $6.5 \times 10^{-3} = 6.5 \times$ 1⁄10 × 1⁄10 × 1⁄10 = 0.0065. Numbers written as 6.5×10^2 and 6.5×10^{-3}, with the decimal point between the first and second digit, are said to be expressed in scientific notation.

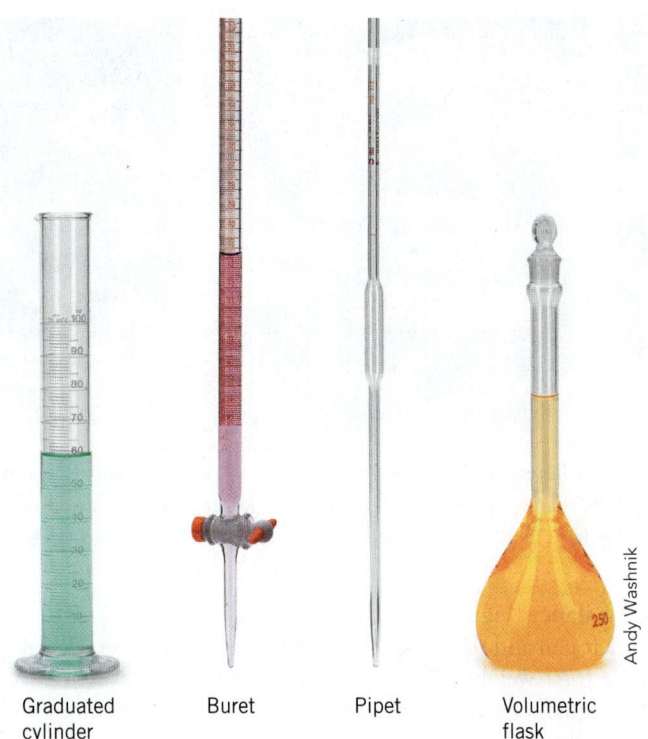

Andy Washnik

| Graduated cylinder | Buret | Pipet | Volumetric flask |

Figure 1.15 | **Common laboratory glassware used for measuring volumes.** Graduated cylinders are used for measuring volumes to the nearest milliliter. Precise measurements of volumes are made using burets, pipets, and volumetric flasks.

Volume

Volume is a derived unit with dimensions of (length)^3. With these dimensions expressed in meters, the derived SI unit for volume is the **cubic meter, m^3**.

In chemistry, measurements of volume usually arise when we measure amounts of liquids. The traditional metric unit of volume used for this is the **liter (L)**. In SI terms, a liter is defined as exactly 1 cubic decimeter.

$$1 \text{ L} = 1 \text{ dm}^3 \quad \text{(exactly)} \tag{1.1}$$

However, even the liter is too large to conveniently express most volumes measured in the lab. The glassware we normally use, as illustrated in Figure 1.15, is marked in **milliliters (mL)**.[7]

$$1 \text{ L} = 1000 \text{ mL}$$

Because 1 dm = 10 cm, then 1 dm^3 = 1000 cm^3. Therefore, 1 mL is exactly the same as 1 cm^3.

$$1 \text{ cm}^3 = 1 \text{ mL}$$

$$1 \text{ L} = 1000 \text{ cm}^3 = 1000 \text{ mL}$$

Sometimes you may see cm^3 abbreviated cc for cubic centimeter (especially in medical applications), although the SI frowns on this symbol. Figure 1.16 compares the cubic meter, liter, and milliliter.

Mass

In the SI, the base unit for mass is the **kilogram (kg)**, although the **gram (g)** is a more conveniently sized unit for most laboratory measurements. One gram, of course, is $\frac{1}{1000}$ of a kilogram (1 kilogram = 1000 g, so 1 g must equal 0.001 kg).

Mass is measured by comparing the weight of a sample with the weights of known standard masses. (Recall that mass and weight are not the same thing.) The instrument

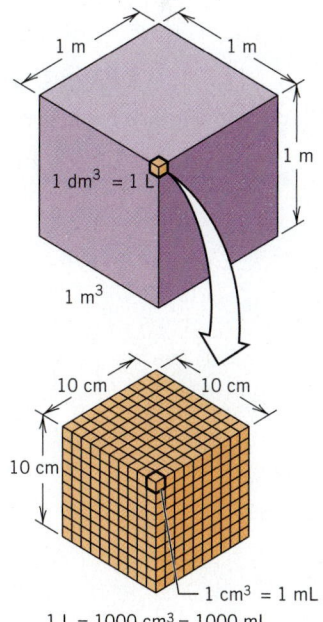

1 dm^3 = 1 L

1 m^3

1 cm^3 = 1 mL
1 L = 1000 cm^3 = 1000 mL

Figure 1.16 | **Comparing volume units.** A cubic meter (m^3) is approximately equal to a cubic yard; 1000 cm^3 is approximately a quart, and approximately 30 cm^3 is equal to one fluid ounce.

---

[7]Use of the abbreviations L for liter and mL for milliliter is rather recent. Confusion between the printed letter l and the number 1 prompted the change from l to L and ml to mL. You may encounter the abbreviation ml in other books or on older laboratory glassware.

(a)

Michael Watson

(b)

© 2005 Richard Megna/Fundamental Photographs

(c)

Charles D. Winters/Photo Researchers

Figure 1.17 | **Typical laboratory balances.** (*a*) A traditional two-pan analytical balance capable of measurements to the nearest 0.0001 g. (*b*) A modern top-loading balance capable of mass measurements to the nearest 0.1 g. (*c*) A modern analytical balance capable of measurements to the nearest 0.0001 g.

used is called a **balance** (Figure 1.17). For the balance in Figure 1.17*a*, we would place our sample on the left pan and then add standard masses to the other. When the weight of the sample and the total weight of the standards are in balance (when they match), their masses are then equal. Figure 1.18 shows the masses of some common objects in SI units.

Temperature

Temperature is usually measured with a thermometer (Figure 1.19). Thermometers are graduated in *degrees* according to one of two temperature scales. On the **Fahrenheit scale** water freezes at 32 °F and boils at 212 °F. (See Figure 1.20.) On the **Celsius scale** water freezes at 0 °C and boils at 100 °C, which means there are 100 degree units between the freezing and boiling points of water, while on the Fahrenheit scale this same temperature range is spanned by 180 degree units. Consequently, five Celsius degrees are the same as nine Fahrenheit degrees. We can use the following equation as a tool to convert between these temperature scales.

TOOLS

Celsius-to-Fahrenheit conversions

$$t_F = \left(\frac{9\ ^\circ F}{5\ ^\circ C}\right) t_C + 32\ ^\circ F \qquad (1.2)$$

In this equation, t_F is the Fahrenheit temperature and t_C is the Celsius temperature. As noted earlier, units behave like numbers in calculations, and we see in Equation 1.2 that °C "cancels out" to leave only °F. The 32 °F is added to account for the fact that the freezing point of water (0 °C) occurs at 32 °F on the Fahrenheit scale. Equation 1.2 can easily be rearranged to permit calculating the Celsius temperature from the Fahrenheit temperature.

$$(t_F - 32\ ^\circ F)\left(\frac{5\ ^\circ C}{9\ ^\circ F}\right) = t_C$$

Paper clip
0.4 g

Penny   2.5 g

A 220 lb football player   100 kg

Jim Cummins/Taxi/Getty Images, Inc.
paper clip: iStockphoto; penny: iStockphoto;
water: Washnik

One cup of water (8 fluid ounces)   About 250 g of water

Figure 1.18 | **Masses of several common objects.**

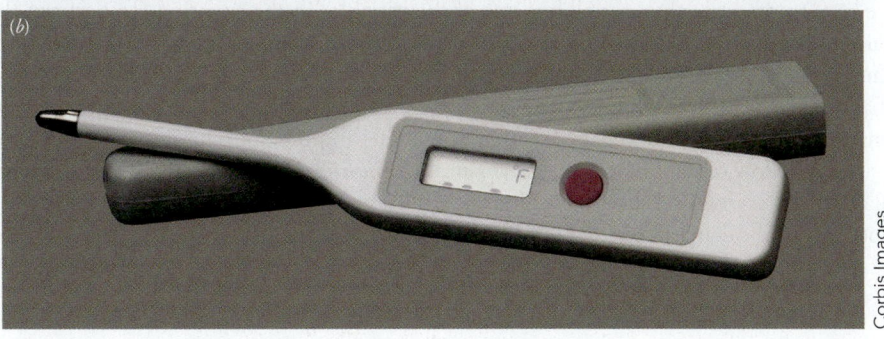

Figure 1.19 | **Typical laboratory thermometers.** (*a*) A traditional mercury thermometer. (*b*) An electronic thermometer.

The SI unit of temperature is the **kelvin (K)**, which is the degree unit on the **Kelvin scale**. Notice that the temperature unit is K, not °K (the degree symbol, °, is omitted). Also notice that the name of the unit, kelvin, is not capitalized.

Figure 1.20 shows how the Kelvin, Celsius, and Fahrenheit temperature scales relate to each other. Notice that the kelvin is *exactly* the same size as the Celsius degree. *The only difference between these two temperature scales is the zero point.* The zero point on the Kelvin scale is called **absolute zero** and corresponds to nature's lowest temperature. It is 273.15 degree units below the zero point on the Celsius scale, which means that 0 °C equals 273.15 K, and 0 K equals −273.15 °C. To convert from Celsius to Kelvin temperatures the following equation applies.

$$T_K = (t_C + 273.15)\ °C\ \frac{1\ K}{1\ °C} \qquad (1.3)$$

This amounts to simply adding 273.15 to the Celsius temperature to obtain the Kelvin temperature. Often we are given Celsius temperatures rounded to the nearest degree, in which case we round 273.15 to 273. Thus, 25 °C equals (25 + 273) K or 298 K.

■ The name of the temperature scale, the Kelvin scale, is capitalized, but the name of the unit, the kelvin, is not. However, the symbol for the kelvin is a capital K.

TOOLS

Celsius-to-Kelvin conversions

■ We will use a capital *T* to stand for the Kelvin temperature and a lowercase *t* (as in t_C) to stand for the Celsius temperature. This conforms to the usage described by the International Bureau of Weights and Measures in Sevres, France, and the National Institute of Standards and Technology in Gaithersburg, Maryland.

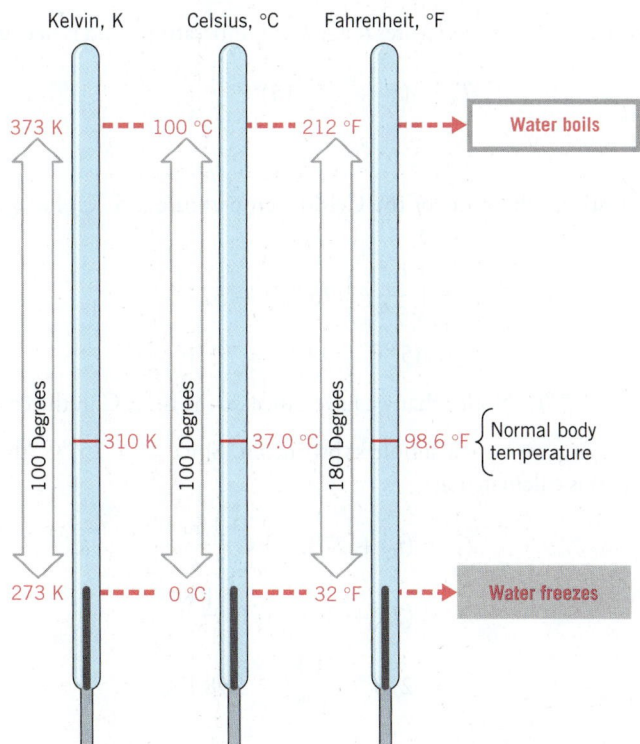

Figure 1.20 | **Comparison among Kelvin, Celsius, and Fahrenheit temperature scales.**

Types of Mathematical Calculations

We will discuss many problems throughout the book that require mathematical calculations. In approaching them you will see that they can be divided into two general types. The type described in Example 1.2 involves applying a mathematical equation in which we have numerical values for all but one of the variables in the equation. To perform the calculation, we solve the equation by substituting known values for the variables until there is only one unknown variable that can be calculated. The second type of calculation that we will encounter, which is described in detail in Section 1.6, is one where we convert one set of units into another set of units using a method called dimensional analysis.

If you follow our approach of *Analysis* and *Assembling the Tools*, the nature of the types of calculations involved should become apparent. As you will see, some problems involve both kinds of calculations.

Example 1.2
Converting among Temperature Scales

Thermal pollution, the release of large amounts of heat into rivers and other bodies of water, is a serious problem near power plants and can affect the survival of some species of fish. For example, trout will die if the temperature of the water rises above approximately 25 °C. (a) What is this temperature in °F? (b) Rounded to the nearest whole degree unit, what is this temperature in kelvins?

Analysis: Both parts of the problem here deal with temperature conversions. Therefore, we ask, "What tools do we have that relate the temperature scales to each other?"

Assembling the Tools: The first tool, Equation 1.2, relates Fahrenheit temperatures to Celsius temperatures, and is needed to answer part (a).

$$t_F = \left(\frac{9\ °F}{5\ °C}\right)t_C + 32\ °F$$

Equation 1.3 relates Kelvin temperatures to Celsius temperatures, and is needed for part (b).

$$T_K = (t_C + 273.15)°C\,\frac{1\ K}{1\ °C}$$

Solution:

Part (a): We substitute the value of the Celsius temperature (25 °C) for t_C and calculate the answer

$$t_F = \left(\frac{9\ °F}{5\ °C}\right)(25\ °C) + 32\ °F$$

$$= 45\ °F + 32\ °F = 77\ °F$$

Therefore, 25 °C = 77 °F. (Notice that we have canceled the unit °C in the equation above.)

Part (b): Once again, we have a simple substitution. Since $t_C = 25\ °C$, the Kelvin temperature (rounded) is calculated as

$$T_K = (t_C + 273.15)°C\,\frac{1\ K}{1\ °C}$$

$$= (25 + 273.15)°C\,\frac{1\ K}{1\ °C}$$

$$= 298\ °C\,\frac{1\ K}{1\ °C} = 298\ K$$

Thus, 25 °C = 298 K.

● **Are the Answers Reasonable?** For numerical calculations, ask yourself, "Is the answer too large or too small?" Judging the answers to such questions serves as a check on the arithmetic as well as on the method of obtaining the answer. For part (a), we know that a Fahrenheit degree is about half the size of a Celsius degree, so 25 Celsius degrees should be about 50 Fahrenheit degrees, then we add 32 °F, so the Fahrenheit temperature should be approximately 32 °F + 50 °F = 82 °F. The answer of 77 °F is quite close.

For part (b), we recall that 0 °C = 273 K. A temperature above 0 °C must be higher than 273 K. Our calculation, therefore, appears to be correct.

What is 355 °C in °F? (*Hint:* What tool relates these two temperature scales?)

Convert 55 °F to its Celsius temperature. What Kelvin temperature corresponds to 68 °F (expressed to the nearest whole kelvin unit)?

Practice Exercise 1.7

Practice Exercise 1.8

1.5 | The Uncertainty of Measurements

Measurements always have some **uncertainty** or **error**. One type of error is due to uncalibrated instruments, poor lab technique or perhaps an inappropriate method. These affect the accuracy of a result and are called determinate errors. Such errors will not average out to the true value even after many tries and these errors have a sign to indicate if they are above or below the true value. The good news is that determinate errors can often be discovered and eliminated.

Random errors are another source of uncertainty. These arise from the limitations in our ability to read the scales of measuring instruments or the noise in electrical equipment. Statistically, measurements that have only random error will cluster around a central value which we assume is close to the true value. This central value is easily determined by taking the **mean** or **average** of the measurements.

Uncertainties in Measurements

One kind of error that cannot be eliminated arises when we attempt to obtain a measurement by reading the scale on an instrument. Consider, for example, reading the same temperature from each of the two thermometers in Figure 1.21.

The marks on the left thermometer are one degree apart, and we can see that the temperature lies between 24 °C and 25 °C. When reading any scale, we always record the last digit to the nearest tenth of the smallest scale division. Looking closely, therefore, we might estimate that the fluid column falls about 3/10 of the way between the marks for 24 and 25 degrees, so we can report the temperature to be 24.3 °C. However, it would be foolish to say that the temperature is *exactly* 24.3 °C. The last digit is only an estimate, and this thermometer might be read as 24.2 °C by one observer or 24.4 °C by another. Because of this, there is an uncertainty of ±0.1 °C in the measured temperature. We can express this by writing the temperature as 24.3 ± 0.1 °C.

The thermometer on the right has marks that are 1/10 of a degree apart, which allows us to estimate the temperature as 24.32 °C. In this case, we are estimating the hundredths place and the uncertainty is ±0.01 °C, therefore the temperature is reported as 24.32 ± 0.01 °C. Notice that because the thermometer on the right is more finely graduated, we are able to obtain measurements with smaller uncertainties. We would have more confidence in temperatures read from the thermometer on the right in Figure 1.21 because it has more digits and a smaller amount of uncertainty. *The reliability of a measurement is indicated by the number of digits used to represent it.*

By convention in science, *all digits in a measurement up to and including the first estimated digit are recorded.* If a reading measured with the thermometer on the right seemed exactly on the 24 °C mark, we would record the temperature as 24.00 °C, not 24 °C, to show that the thermometer can be read to the nearest 1/100 of a degree.

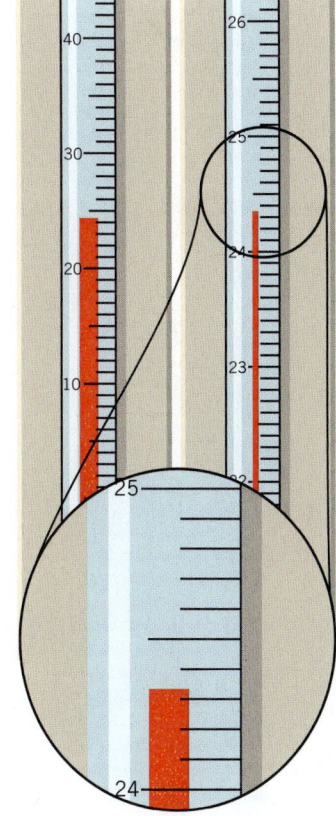

Figure 1.21 | **Thermometers with different scales give readings with different precision.** The thermometer on the left has marks that are one degree apart, allowing the temperature to be estimated to the nearest tenth of a degree. The thermometer on the right has marks every 0.1 °C. This scale permits estimation of the hundredths place.

Modern laboratory instruments such as balances and meters often have digital displays. If you read the mass of a beaker on a digital balance as 65.23 grams, everyone else observing the same beaker will see the same display and obtain exactly the same reading. There seems to be no uncertainty, or estimation, in reading this type of scale. However, scientists agree that the uncertainty is ±½ of the last readable digit. Using this definition our digital reading may be written as 65.230 ± 0.005 grams.[8]

Significant Figures

Counting significant figures

The concepts discussed above are so important that we have special terminology to describe numbers that come from measurements.

Digits that result from a measurement such that only the digit farthest to the right is not known with certainty are called **significant figures** (*or* **significant digits**).

The number of significant figures in a measurement is equal to the number of digits known for sure *plus* one that is estimated. Let's look at our two temperature measurements:

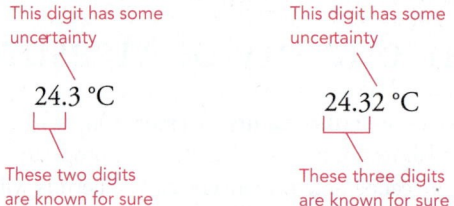

The first measurement, 24.3 °C, has three significant figures; the second, 24.32 °C, has four significant figures.

When Are Zeros Significant?

Usually, it is simple to determine the number of significant figures in a measurement; we just count the digits. Thus, the number 3.25 has three significant figures and 56.215 has five of them. When zeros are part of a number they can sometimes cause confusion.

We will use the following rules when zeros are involved:

Zeros to the left of the first nonzero digit, called leading zeros, are never counted as significant. For instance, a length of 2.3 mm is the same as 0.0023 m. Since we are dealing with the same measured value, its number of significant figures cannot change when we change the units. Both quantities have two significant figures.

Zeros imbedded inside a number are always significant. Thus, 2047 m and 3.096 g have four significant figures each because the zeros are imbedded.

Trailing zeros are (a) always counted as significant if the number has a decimal point and (b) not counted as significant if the number does not have a decimal point. Thus, 4.500 m and 630.0 g have four significant figures each because the zeros would not be written unless those digits were known to be zeros. The numbers 3200 and 20 have two and one significant figures, respectively.

There are times when a measurement or calculation requires the trailing zeros to be significant. The best way to handle this is to use **scientific notation** to write the number. For example, suppose you were told that a football game was attended by 45,000 people. If we want to report the number of sports fans as 45,000 give or take a thousand, we can write the rough estimate as 4.5×10^4. The 4.5 shows the number of significant figures and the 10^4 tells us the location of the decimal point. On the other hand, suppose the sports fans were carefully counted using an aerial photograph, so that the count could be reported to be 45,000—give or take about 100 people. In this case, the value represents $45,000 \pm 100$ attendees and contains three significant figures, with the uncertain digit being the zero in the hundreds place. In scientific notation, this can be expressed as 4.50×10^4. This time the 4.50 shows three significant figures and an uncertainty of $\pm 0.01 \times 10^4$ or ± 100 people.

---

[8]Some instructors may wish to maintain a uniform policy of assigning an uncertainty of ±1 in the last readable digit for both analog and digital scale readings.

Thus, a simple statement such as "there were 45,000 people attending the game" is ambiguous. We can't tell how many significant digits the number has from the number alone. The best we can do is to say "45,000 has *at least* two significant figures."

Accuracy and Precision

Two words often used in reference to measurements are accuracy and precision. **Accuracy** *refers to how close a measurement is to the true or the accepted true value.* **Precision** *refers to how close repeated measurements come to their average.* Notice that the two terms are not synonyms, because accuracy is related to determinate errors and precision is related to random errors. A practical example of how accuracy and precision differ is illustrated in Figure 1.22. For measurements to be accurate, the measuring device must be carefully calibrated (adjusted) with a standard reference so it gives correct readings. For example, to calibrate an electronic balance, a known reference mass is placed on the balance and a calibration routine within the balance is initiated. Once calibrated, the balance will give accurate readings, the accuracy of which is determined by the quality of the standard mass used. Standard reference masses, often "traceable" to the international prototype kilogram in Paris (shown in Figure 1.13), can be purchased from scientific supply companies.

Precision refers to how closely repeated measurements of the same quantity come to each other. In general, the smaller the uncertainty (i.e., the "plus or minus" part of the measurement), the more precise the measurement. This translates as: *The more significant figures in a measured quantity, the more precise the measurement.*

We usually assume that a very precise measurement is also of high accuracy. We can be wrong, however, if our instruments are improperly calibrated. For example, the improperly marked ruler in Figure 1.23 might yield measurements that vary by a hundredth of a centimeter (± 0.01 cm), but all the measurements would be too large by 1 cm—a case of good precision but poor accuracy.

Significant Figures in Calculations

When several measurements are obtained in an experiment they are usually combined in some way to calculate a desired quantity. For example, to determine the area of a rectangular carpet we require two measurements, length and width, which are then multiplied to give the answer we want. To get some idea of how precise the area really is, we need a way to take into account the precision of the various values used in the calculation. To make sure this happens, we follow certain rules according to the kinds of arithmetic being performed.

Multiplication and Division

For multiplication and division, the number of significant figures in the answer should be equal to the number of significant figures in the least precise measurement. The least precise measurement is the number with the fewest significant figures. Let's look at a typical problem involving some measured quantities.

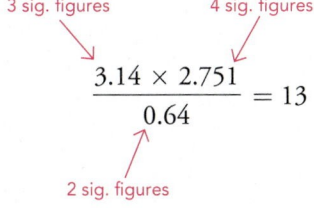

$$\overset{\text{3 sig. figures}}{\overbrace{3.14} \times \overset{\text{4 sig. figures}}{2.751}} \over \underset{\text{2 sig. figures}}{0.64}} = 13$$

Figure 1.22 | **The difference between precision and accuracy in the game of golf.** Golfer 1 hits shots that are precise (because they are tightly grouped), but the accuracy is poor because the balls are not near the target (the "true" value). Golfer 2 needs help. The shots are neither precise nor accurate. Golfer 3 wins the prize with shots that are precise (tightly grouped) and accurate (in the hole).

Golfer 1: Precise inaccurate

Golfer 2: Imprecise inaccurate

Golfer 3: Accurate precise

TOOLS

Significant figures: multiplication and division

How accurate would measurements be with this ruler?

Figure 1.23 | **An improperly marked ruler.** This improperly marked ruler will yield measurements that are each wrong by one whole unit. The measurements might be precise, but the accuracy would be very poor.

The result displayed on a calculator[9] is 13.49709375. However, the least precise factor, 0.64, has only two significant figures, so the answer should have only two. The correct answer, 13, is obtained by rounding off the calculator answer.[10] When we multiply and divide measurements, the units of those measurements are multiplied and divided in the same way as the numbers.

Addition and Subtraction

TOOLS

Significant figures: addition and subtraction

For addition and subtraction, the answer should have the same number of decimal places as the quantity with the fewest number of decimal places. As an example, consider the following addition of measured quantities.

$$
\begin{array}{r}
3.247 \\
41.36 \\
+125.2 \quad \longleftarrow \text{(This number has only 1 decimal place.)} \\
\hline
169.8 \quad \longleftarrow \text{(This answer has been rounded to 1 decimal place.)}
\end{array}
$$

In this calculation, the digits beneath the 6 and the 7 are unknown; they could be anything. (They're not necessarily zeros because if we *knew* they were zeros, then zeros would have been written there.) Adding an unknown digit to the 6 or 7 will give an answer that's also unknown, so for this sum we are not justified in writing digits in the second and third places after the decimal point. Therefore, we round the answer to the nearest tenth. We must also recall that we can only add and subtract numbers that have identical units, and the answer will have the same units. When a calculation has both addition (or subtraction) and multiplication (or division), do the addition/subtraction first to the correct number of significant figures, and then perform the multiplication and division operations to the correct number of significant figures.

Exact Numbers

TOOLS

Significant figures: exact numbers

Numbers that come from definitions, such as 12 in. = 1 ft, and those that come from a direct count, such as the number of people in a small room, have no uncertainty, and we can assume that they have an infinite number of significant figures. Therefore, exact numbers do not affect the number of significant figures in multiplication or division calculations.

Practice Exercise 1.9

Perform the following calculations involving measurements and round the results so they have the correct number of significant figures and proper units. (*Hint:* Apply the rules for significant figures described in this section, and keep in mind that units behave as numbers do in calculations.)

(a)   21.0233 g + 21.0 g

(b)   10.0324 g ÷ 11.7 mL

(c) $\dfrac{14.25 \text{ cm} \times 12.334 \text{ cm}}{(2.223 \text{ cm} - 1.04 \text{ cm})}$

---

[9]Calculators usually give too many significant figures. An exception is when the answer has zeros at the right that are significant figures. For example, an answer of 1.200 is displayed on most calculators as 1.2. If the zeros belong in the answer, be sure to write them down.

[10]When we wish to round off a number at a certain point, we simply drop the digits that follow if the first of them is less than 5. Thus, 8.1634 rounds to 8.16 if we wish to have only two decimal places. If the first digit after the point of round off is larger than 5, or if it is 5 followed by other nonzero digits, then we add 1 to the preceding digit. Thus 8.167 and 8.1653 both round to 8.17. Finally, when the digit after the point of round off is a 5 and no nonzero digits follow the 5, then we drop the 5 if the preceding digit is even and add 1 if it is odd. Thus, 8.165 rounds to 8.16 and 8.17500 rounds to 8.18.

Perform the following calculations involving measurements and round the results so that they are written to the correct number of significant figures and have the correct units.

(a) 54.183 g − 0.0278 g

(b) 10.0 g + 1.03 g + 0.243 g

(c) 43.4 in. $\times \dfrac{1 \text{ ft}}{12 \text{ in.}}$

(d) $\dfrac{1.03 \text{ m} \times 2.074 \text{ m} \times 2.9 \text{ m}}{12.46 \text{ m} + 4.778 \text{ m}}$

1.6 | Dimensional Analysis

Earlier we mentioned that for numerical problems we often do not have a specific equation to solve; instead, all we need to do is convert one set of units to another. After analyzing the problem and assembling the necessary information to solve it, scientists usually use a technique commonly called **dimensional analysis** to help them perform the correct arithmetic. As you will see, this method also helps in analyzing the problem and selecting the tools needed to solve it.

Conversion Factors

In dimensional analysis we treat a numerical problem as one involving a conversion of units (the dimensions) from one kind to another. To do this we use one or more *conversion factors* to change the units of the given quantity to the units of the answer:

$$(\text{given quantity}) \times (\text{conversion factor}) = (\text{desired quantity})$$

A **conversion factor** *is a fraction formed from a* valid *equality or equivalence between units and is used to switch from one system of measurement and units to another.* To illustrate, suppose we want to express a person's height of 72.0 inches in centimeters. To do this we need the relationship between the inch and the centimeter. We can obtain this from Table 1.4:

$$2.54 \text{ cm} = 1 \text{ in.} \quad \text{(exactly)} \tag{1.4}$$

If we divide both sides of this equation by 1 in., we obtain a conversion factor.

$$\frac{2.54 \text{ cm}}{1 \text{ in.}} = \frac{1 \text{ in.}}{1 \text{ in.}} = 1$$

Notice that we have canceled the units from both the numerator and denominator of the center fraction, leaving the first fraction equaling 1. As mentioned earlier, *units behave just as numbers do in mathematical operations*; this is a key part of dimensional analysis. Let's see what happens if we multiply 72.0 inches, the height that we mentioned, by this fraction.

$$72.0 \text{ in.} \times \frac{2.54 \text{ cm}}{1 \text{ in.}} = 183 \text{ cm}$$

$$(\text{given quantity}) \times (\text{conversion factor}) = (\text{desired quantity})$$

Because we have multiplied 72.0 in. by something that is equal to 1, we haven't changed the person's height. We have, however, changed the units. Notice that we have canceled the unit inches. The only unit left is centimeters, which is the unit we want for the answer. The result, therefore, is the person's height in centimeters.

One of the benefits of dimensional analysis is that it often lets you know when you have done the *wrong* arithmetic. From the relationship in Equation 1.4, we can actually construct two conversion factors:

$$\frac{2.54 \text{ cm}}{1 \text{ in.}} \quad \text{and} \quad \frac{1 \text{ in.}}{2.54 \text{ cm}}$$

■ To construct a valid conversion factor, the relationship between the units must be true. For example, the statement 3 ft = 41 in. is false. Although you might make a conversion factor out of it, any answers you would calculate are sure to be wrong. Correct answers require correct relationships between units.

■ The relationship between the inch and the centimeter is exact, so that the numbers in 1 in. = 2.54 cm have an infinite number of significant figures.

We used the first one correctly, but what would have happened if we had used the second by mistake?

$$72.0 \text{ in.} \times \frac{1 \text{ in.}}{2.54 \text{ cm}} = 28.3 \text{ in.}^2/\text{cm}$$

In this case, none of the units cancel. We get units of in.^{2}/cm because inches times inches is inches squared. Even though our calculator may be very good at arithmetic, we've got the wrong answer. *Dimensional analysis lets us know we have the wrong answer because the units are wrong!*

There is a general strategy for all problems that involve a conversion. We always start with a given piece of data along with its units. Then we write down the desired units that we want the answer to have. This gives us the start and end points of our calculation. Then, all we need to do is find the correct conversion factors that will lead us from one set of units to the next. Sometimes this will require one conversion factor, other times it may take two or more to complete the conversion. The next examples illustrate this process.

Example 1.3
Dimensional Analysis Applied to Metric Prefixes

The world's longest documented hair in 2004 belonged to Xie Qiuping of China and was measured at 5.627 m. Convert 5.627 m to millimeters (mm).

Analysis: We are asked to convert a number with meter units to another number that has millimeter units, a problem that involves conversion of metric prefixes.

Assembling the Tools: To solve this problem, our tool will be a conversion factor that relates the unit meter to the unit millimeter. From Table 1.5, the table of decimal multipliers, the prefix "milli" means "$\times 10^{-3}$," so we can write

$$1 \text{ mm} = 10^{-3} \text{ m}$$

Notice that this relationship connects the units given to the units desired.

From this relationship, we can form two conversion factors.

$$\frac{1 \text{ mm}}{10^{-3} \text{ m}} \quad \text{and} \quad \frac{10^{-3} \text{ m}}{1 \text{ mm}}$$

We now have all the information we need to solve the problem.

Solution: We start with the given quantity and its units on the left and the units of the desired answer on the right.

$$5.627 \text{ m} = ? \text{ mm}$$

We have to cancel the unit meter, so we need to multiply by the conversion factor with meter in the denominator. Therefore, we select the one on the left as our tool. This gives

$$5.627 \text{ m} \times \frac{1 \text{ mm}}{10^{-3} \text{ m}} = 5.627 \times 10^3 \text{ mm}$$

Notice that we have expressed the answer to four significant figures because that is how many there are in the given quantity, 5.627 m. The equality that relates meters and millimeters involves exact numbers because it is a definition.

Is the Answer Reasonable? We know that millimeters are much smaller than meters, so 5.627 m must represent a lot of millimeters. We also know that conversions between prefixes will not change the 5.627 part of our number, just the power of 10. Our answer, therefore, makes sense.

Example 1.4
Using Dimensional Analysis

A liter, which is slightly larger than a quart, is defined as 1 cubic decimeter (1 dm³). How many liters are there in exactly 1 cubic meter (1 m³)?

Analysis: Let's begin once again by stating the problem in equation form.

$$1 \text{ m}^3 = ? \text{ L}$$

We don't have any direct conversions between cubic meters and liters. There was, however, a definition of the liter in terms of base metric units that we can combine with metric prefixes to do the job.

Assembling the Tools: The relationship between liters and cubic decimeters was given in Equation 1.1,

$$1 \text{ L} = 1 \text{ dm}^3$$

From the table of decimal multipliers, we also know the relationship between decimeters and meters,

$$1 \text{ dm} = 0.1 \text{ m}$$

but we need a relationship between cubic units. Since units undergo the same kinds of operations numbers do, we simply cube each side of this equation (being careful to cube both the numbers and the units).

$$(1 \text{ dm})^3 = (0.1 \text{ m})^3$$
$$1 \text{ dm}^3 = 0.001 \text{ m}^3 \tag{1.5}$$

Notice how Equations 1.5 and 1.1 provide a path from the given units to those we seek. In the pathway below, the equalities are shown inside the colored arrows, but we still need to make the appropriate ratio using the equalities.

$$\text{m}^3 \xrightarrow{\; 1 \text{ dm}^3 = 0.001 \text{ m}^3 \;} \text{dm}^3 \xrightarrow{\; 1 \text{ dm}^3 = 1 \text{ L} \;} \text{L}$$

Now we are ready to solve the problem.

Solution: The first step is to eliminate the units m³, using Equation 1.5.

$$1 \text{ m}^3 \times \frac{1 \text{ dm}^3}{0.001 \text{ m}^3} = 1000 \text{ dm}^3$$

Then we use Equation 1.1 to take us from dm³ to L.

$$1000 \text{ dm}^3 \times \frac{1 \text{ L}}{1 \text{ dm}^3} = 1000 \text{ L}$$

Thus, 1 m³ = 1000 L.

When a problem involves the use of two or more conversion factors, often they can be "strung together" in a "chain calculation." For example, this problem can be set up as follows.

$$1 \text{ m}^3 \times \frac{1 \text{ dm}^3}{0.001 \text{ m}^3} \times \frac{1 \text{ L}}{1 \text{ dm}^3} = 1000 \text{ L}$$

Since all of our conversion factors involve exact numbers and we were given exactly one cubic meter, our answer is an exact number also.

Is the Answer Reasonable? One liter is about a quart. A cubic meter is about a cubic yard. Therefore, we expect a large number of liters in a cubic meter, so our answer seems reasonable. (Notice here that in our analysis we have approximated the quantities in the calculation in units of quarts and cubic yards, which may be more familiar than liters and m³. We get a feel for the approximate magnitude of the answer using our familiar units and then relate this to the actual units of the problem.)

Example 1.5
Applying Dimensional Analysis to Non-SI Units

Some mountain climbers are susceptible to high-altitude pulmonary edema (HAPE), a life-threatening condition that causes fluid retention in the lungs. It can develop when a person climbs rapidly to heights greater than 2.5×10^3 meters (2500 m). What is this distance expressed in feet?

Analysis: The problem can be stated as

$$2.5 \times 10^3 \text{ m} = ? \text{ ft}$$

in which we convert meters to feet. Since Table 1.4 does not have a relationship between feet and meters we need to develop a sequence of conversions by looking at the units we can convert. One sequence may be:

$$2.5 \times 10^3 \text{ m} \xrightarrow{\text{1 m=100 cm}} \text{centimeters} \xrightarrow{\text{2.54 cm=1 in.}} \text{inches} \xrightarrow{\text{12 in.=1 ft}} \text{feet}$$

or if we consult the inside back cover of this book we find the conversion factor 1 yard = 0.9144 meters. Another multistep sequence could be

$$2.5 \times 10^3 \text{ m} \xrightarrow{\text{0.9144 m=1 yd}} \text{yards} \xrightarrow{\text{1 yd=3 ft}} \text{feet}$$

We will choose the first sequence for this example.

Assembling the Tools: The tools we will need can be found by looking in the appropriate tables as shown.

$$1 \text{ cm} = 10^{-2} \text{ m} \quad \text{(from Table 1.5)}$$
$$1 \text{ in.} = 2.54 \text{ cm} \quad \text{(from Table 1.4)}$$
$$1 \text{ ft} = 12 \text{ in.}$$

Solution: Now we apply dimensional analysis by following our planned sequence of eliminating unwanted units to bring us to the units of the answer.

$$2.5 \times 10^3 \text{ m} \times \frac{1 \text{ cm}}{10^{-2} \text{ m}} \times \frac{1 \text{ in.}}{2.54 \text{ cm}} \times \frac{1 \text{ ft}}{12 \text{ in.}} = 8.2 \times 10^3 \text{ ft}$$

This time the answer has been rounded to two significant figures because that's how many there were in the measured distance. Notice that the numbers 12 and 2.54 do not affect the number of significant figures in the answer because they are exact numbers derived from definitions.

As suggested in the analysis, this is not the only way we could have solved this problem. Other sets of conversion factors could have been chosen. For example, we could have followed the second sequence of conversions and used: 1 yd = 0.9144 m and 3 ft = 1 yd. Then the problem would have been set up as follows:

$$2500 \text{ m} \times \frac{1 \text{ yd}}{0.9144 \text{ m}} \times \frac{3 \text{ ft}}{1 \text{ yd}} = 8200 \text{ ft} \qquad \text{(Remember, this number has two significant figures.)}$$

Many problems that you meet, just like this one, have more than one path to the answer. *The important thing is for you to be able to reason your way through a problem and find a set of relationships that can take you from the given information to the answer.* Dimensional analysis can help you search for these relationships if you keep in mind the units that must be eliminated by cancellation.

● **Is the Answer Reasonable?** Let's do some approximate arithmetic to get a feel for the size of the answer. A meter is slightly longer than a yard, so let's approximate the given distance, 2500 m, as 2500 yd. In 2500 yd, there are $3 \times 2500 = 7500$ ft. Since the meter is a bit longer than a yard, our answer should be a bit longer than 7500 ft, so the answer of 8200 ft seems to be reasonable.

Use dimensional analysis to convert an area of 124 square feet to square meters. (*Hint:* What relationships would be required to convert feet to meters?)

Practice Exercise 1.11

Use dimensional analysis to perform the following conversions: (a) 3.00 yd^3 to $inches^3$, (b) 1.25 km to centimeters, (c) 3.27 ounces to grams, (d) 20.2 miles/gallon to kilometers/liter.

Practice Exercise 1.12

Equivalencies

Up to now we've constructed conversion factors from relationships that are literally equalities. We can also make conversion factors from expressions that show how one thing is *equivalent* to another. For instance, if you buy a pair of shoes for $85, we can say you converted $85 into a pair of shoes or $85 is equivalent to a pair of shoes. We would write this as

$$\$85 \Leftrightarrow 1 \text{ pair of shoes}$$

where the symbol $\Leftrightarrow$ is read as "is equivalent to." Mathematically, this equivalence sign works the same as an equal sign and we can construct two conversion factors:

■ The equivalence symbol, $\Leftrightarrow$, acts just like an equal sign.

$$\frac{\$85}{1 \text{ pair of shoes}} \quad \text{or} \quad \frac{1 \text{ pair of shoes}}{\$85}$$

that allow us to convert from shoes to dollars or from dollars to shoes. We can use the first conversion factor to determine how much it would cost to buy three pairs of shoes:

$$3 \text{ pairs of shoes} \left(\frac{\$85}{1 \text{ pair of shoes}} \right) = \$255$$

The "units" pairs of shoes cancel, leaving just the cost of the shoes, $255.

1.7 | Density and Specific Gravity

In our earlier discussion of properties we noted that intensive properties are useful for identifying substances. One of the interesting things about extensive properties is that if you take the ratio of two of them, the resulting quantity is usually independent of sample size. In effect, the sample size cancels out and the calculated quantity becomes an intensive property. A useful property obtained this way is **density**, *which is defined as the ratio of an object's mass to its volume.* Using the symbols d for density, m for mass, and V for volume, we can express this mathematically as

$$d = \frac{m}{V}$$

(1.6)

TOOLS

Density

Notice that to determine an object's density we make two measurements, mass and volume.

Example 1.6
Calculating Density

A sample of blood completely fills an 8.20 cm³ vial. The empty vial has a mass of 10.30 g and the filled vial has a mass of 18.91 g. What is the density of blood in units of g/cm³?

Analysis: This problem asks you to connect the mass and volume of blood with its density. We are given the volume of the blood and its mass can be calculated. Once we have the mass and volume, we can use the definition of density to solve the problem.

Assembling the Tools: The law of conservation of mass is one tool we need. It can be stated as follows:

mass of full vial = mass of empty vial + mass of blood

The other tool needed is the definition of density, given by Equation 1.6,

$$d = \frac{m}{V}$$

Solution: The mass of the blood is the difference between the masses of the full and empty vials:

mass of blood = 18.91 g − 10.30 g = 8.61 g

To determine the density we simply take the ratio of mass to volume:

$$\text{density} = \frac{m}{V} = \frac{8.61 \text{ g}}{8.20 \text{ cm}^3} = 1.05 \text{ g cm}^{-1}$$

This could also be written as density = 1.05 g/mL, because 1 cm³ = 1 mL.

Is the Answer Reasonable? First, the answer has the correct units, so that's encouraging. In the calculation we are dividing 8.61 by a number that is slightly smaller, 8.20. The answer should be slightly larger than one, which it is, so a density of 1.05 g/cm³ seems reasonable.

Practice Exercise 1.13

A 15.0 mL sample of polystyrene used in insulated foam cups weighs 0.244 g. What is the density of this polystyrene? (*Hint:* What is the relationship between mass and volume in density calculations?)

Practice Exercise 1.14

A crystal of salt was grown and found to have a mass of 0.547 g. When the crystal was placed in a graduated cylinder that contained 5.70 mL oil, the volume increased to 5.95 mL. What is the density of salt? Give your answer in g cm⁻³ since the density of solids are reported in g cm⁻³.

■ In chemistry, reference data are commonly tabulated at 25 °C, which is close to room temperature. Biologists often carry out their experiments at 37 °C because that is our normal body temperature.

■ Although the density of water varies slightly with temperature, it is very close to 1.00 g/cm³ when the temperature is close to room temperature, 25 °C.

Each pure substance has its own characteristic density (Table 1.6). Gold, for instance, is much more dense than iron. Each cubic centimeter of gold has a mass of 19.3 g, so its density is 19.3 g/cm³, while iron has a density of 7.86 g/cm³. By comparison, the density of water is 1.00 g/cm³, and the density of air at room temperature is about 0.0012 g/cm³.

Most substances, such as the fluid in the bulb of a thermometer, expand slightly when they are heated, so the amount of matter packed into each cubic centimeter is less. Therefore, density usually decreases slightly with increasing temperature.[11] For solids and liquids the size of this change is small, as you can see from the data for water in Table 1.7. When only two or three significant figures are required, we can often ignore the variation of density with small changes in temperature.

---

[11]Liquid water behaves oddly. Its maximum density is at 4 °C, so when water at 0 °C is warmed, its density increases until the temperature reaches 4 °C. As the temperature is increased further, the density of water gradually decreases.

TABLE 1.6	Densities of Some Common Substances in g/cm³ at Room Temperature[a]
Water	1.00
Aluminum	2.70
Iron	7.87
Silver	10.5
Gold	19.3
Glass	2.2
Air	0.0012

[a]P. J. Linstrom and W. G. Mallard, Eds., *NIST Chemistry WebBook, NIST Standard Reference Database Number 69*, National Institute of Standards and Technology, Gaithersburg MD, 20899, http://webbook.nist.gov (retrieved September 15, 2013).

TABLE 1.7	Density of Water as a Function of Temperature[a]	
Temperature (°C)		Density (g/cm³)
10		0.999702
15		0.999103
20		0.998207
25		0.997048
30		0.995649
50		0.988035
100		0.958367

[a]P. J. Linstrom and W. G. Mallard, Eds., *NIST Chemistry WebBook, NIST Standard Reference Database Number 69*, National Institute of Standards and Technology, Gaithersburg MD, 20899, http://webbook.nist.gov (retrieved September 15, 2013).

Density as a Conversion Factor

A useful property of density is that it provides a way to convert between the mass and volume of a substance. It defines the relationship, or equivalence, between the substance's mass and its volume. For instance, the density of gold (19.3 g/cm³) tells us that 19.3 g of the metal is equivalent to a volume of 1.00 cm³,

$$19.3 \text{ g gold} \Leftrightarrow 1.00 \text{ cm}^3 \text{ gold}$$

In setting up calculations for dimensional analysis, this equivalence can be used to form the two conversion factors:

$$\frac{19.3 \text{ g gold}}{1.00 \text{ cm}^3 \text{ gold}} \quad \text{and} \quad \frac{1.00 \text{ cm}^3 \text{ gold}}{19.3 \text{ g gold}}$$

The following example illustrates how we use density in calculations.

Example 1.7
Calculations Using Density

Seawater has a density of about 1.03 g/mL. (a) What mass of seawater would fill a sampling vessel to a volume of 225 mL? (b) What is the volume, in milliliters, of 45.0 g of seawater?

Analysis: For both parts of this problem, we are relating the mass of a material to its volume. Density provides a direct relationship between mass and volume, so this should be a one-step conversion.

Assembling the Tools: Density, Equation 1.6, is the only tool that we need to convert between these two quantities. We write the equivalence for this problem as

$$1.03 \text{ g seawater} \Leftrightarrow 1.00 \text{ mL seawater}$$

From this relationship we can construct two conversion factors. These will be the tools we use to obtain the answers.

$$\frac{1.03 \text{ g seawater}}{1.00 \text{ mL seawater}} \quad \text{and} \quad \frac{1.00 \text{ mL seawater}}{1.03 \text{ g seawater}}$$

Solution to (a): The question can be restated as:

$$225 \text{ mL seawater} \Leftrightarrow ? \text{ g seawater}$$

We need to eliminate the unit *mL seawater*, so we choose the conversion factor on the left as our tool.

$$225 \text{ mL seawater} \times \frac{1.03 \text{ g seawater}}{1.00 \text{ mL seawater}} = 232 \text{ g seawater}$$

Thus, 225 mL of seawater has a mass of 232 g.

Solution to (b): The question is:

$$45.0 \text{ g seawater} \Leftrightarrow ? \text{ mL seawater}$$

This time we need to eliminate the unit *g seawater*, so we use the conversion factor on the right as our tool.

$$45.0 \text{ g seawater} \times \frac{1.00 \text{ mL seawater}}{1.03 \text{ g seawater}} = 43.7 \text{ mL seawater}$$

Thus, 45.0 g of seawater has a volume of 43.7 mL.

Are the Answers Reasonable? Notice that the density tells us that 1 mL of seawater has a mass of slightly more than 1 g. Thus, for part (a), we might expect that 225 mL of seawater should have a mass slightly more than 225 g. Our answer, 232 g, is reasonable. For part (b), 45 g of seawater should have a volume not too far from 45 mL, so our answer of 43.7 mL is the right size.

Practice Exercise 1.15 A gold-colored metal object has a mass of 365 g and a volume of 22.12 cm^3. Is the object composed of pure gold? (*Hint:* How does the density of the object compare with that of pure gold?)

Practice Exercise 1.16 A certain metal alloy has a density of 12.6 g/cm^3. How many pounds would 0.822 ft^3 of this alloy weigh? (*Hint:* What is the density of this alloy in units of lb/ft^3?)

Specific Gravity

Density is the ratio of mass to volume of a substance. We used units of grams for the mass and cubic centimeters for the volume in our examples. This is entirely reasonable since most liquids and solids have densities between 0.5 and 20 g/cm^3. There are many different units for mass. Kilograms, grams, micrograms, pounds, and ounces come to mind, and there are undoubtedly more. Volume also has many possible units: cm^3, liters, ounces, gallons, and so on. From those listed, there are 20 different possible ratios of mass and volume units for density. If each set of units required a separate table, this would result in each substance having 20 different densities listed in tables.

The concept of *specific gravity solves this problem*. The **specific gravity** for a substance is simply the density of that substance divided by the density of water. The units for the two densities must be the same so that specific gravity will be a *dimensionless number*. In addition other experimental conditions, such as temperature, for determining the two densities must be the same.

TOOLS
Specific gravity

$$\text{Specific gravity} = \frac{\text{density of substance}}{\text{density of water}} \qquad (1.7)$$

To use the specific gravity, the scientist simply selects the specific gravity of a substance and then multiplies it by the density of water that has the units desired. Now we can have a relatively compact table that lists the specific gravity for our chemical substances and then a second short table of the density of water, perhaps in the 20 different units suggested above.

CHEMISTRY OUTSIDE THE CLASSROOM (1.1)

Density and Wine

Density, or specific gravity, is one of the basic measurements in the wine-making process. The essential chemical reaction is that yeast cells feed on the sugars in the grape juice, and one of the products is ethanol (ethyl alcohol) and the other product is carbon dioxide. This is the fermentation process. The two main sugars in grapes are glucose and fructose. The fermentation reaction for these sugars is

Steve Hyde/Flicker/Getty Images, Inc.

$$C_6H_{12}O_6(aq) \longrightarrow 2CH_3CH_2OH(aq) + 2CO_2(g)$$

If there is too little sugar the amount of alcohol in the product will be low. Too much sugar will result in a maximum amount of alcohol, about 13%, but there will be leftover sugar. Wines that are sweet with leftover sugars or low in alcohol are not highly regarded.

To make high-quality wines the natural sugar content of the grapes is monitored carefully, toward the end of the growing season. When the sugar content reaches the optimum level the grapes are harvested, crushed, and pressed to start the fermentation process.

A very quick way to determine the sugar content of the grapes is to determine the density of the juice. The higher the density, the higher the sugar content. Vintners use a simple device called a hydrometer. The **hydrometer** is a weighted glass bulb with a graduated stem. Placed in a liquid and given a slight spin the hydrometer will stay centered in the graduated cylinder and sink to a level that is proportional to the density. After the density is read from the hydrometer scale, the vintner consults a calibration chart to find the corresponding sugar content. Once the fermentation ends, the alcohol content must also be determined. Again, this is done by measuring the density. To do this, a sample of wine is distilled by heating it to boiling and condensing the alcohol and water that are vaporized. When the required amount of liquid has been condensed, its density is determined using another hydrometer. Using calibration tables, the density reading is converted into the percentage alcohol.

Warren McConnaughie/Alamy

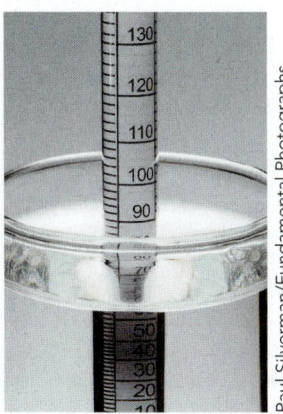

Paul Silverman/Fundamental Photographs

Hydrometer used for measuring sugar content in crushed grapes before fermentation.

Hydrometer used for measuring alcohol content in a distilled wine sample after fermentation.

Example 1.8
Using Specific Gravity

Concentrated sulfuric acid is sold in bottles with a label that states that the specific gravity at 25 °C is 1.84. How many cubic feet of sulfuric acid will weigh 55.5 pounds?

Analysis: We rewrite the question as an equation,

$$55.5 \text{ lb sulfuric acid} = ? \text{ ft}^3 \text{ sulfuric acid}$$

We saw how to use density to convert mass to volume. The specific gravity needs to be converted to density units, preferably pounds and cubic feet.

Assembling the Tools: Using Equation 1.7 which defines specific gravity,

$$\text{specific gravity} = \frac{\text{density of substance}}{\text{density of water}}$$

will allow us to calculate the density of sulfuric acid. Table 1.8 provides a list of densities. We will then use the density as a conversion factor as we did in Example 1.7.

Solution: Rearranging the specific gravity equation we get

$$\text{density sulfuric acid} = (\text{specific gravity}) \times (\text{density of water}) = 1.84 \times 62.4 \text{ lb/ft}^3$$
$$\text{density sulfuric acid} = 114.8 \text{ lb/ft}^3$$

TABLE 1.8
Density of Water Expressed with Different Units
1.00 g/mL
1000 kg/m³
62.4 lb/ft³
8.34 lb/gallon

We now perform the conversion with the conversion factor from the density,

$$55.5 \text{ lb sulfuric acid} \times \frac{1 \text{ ft}^3}{114.8 \text{ lb}} = 0.483 \text{ ft}^3$$

Is the Answer Reasonable? We can do some quick estimates. The density seems correct: since the specific gravity is approximately 2 and the density of water is approximately 60, an answer near 120 is expected. For the second part, we see that 114 is about twice the size of 55.5 so we expect an answer of about 0.5. Our answer is very close to that, suggesting that our answers are reasonable.

Practice Exercise 1.17

Table wines have a minimum specific gravity of 1.090. If a barrel of wine contains 79 gallons, what is the mass of the wine in the barrel in pounds? (*Hint:* What is the relationship between specific gravity and density?)

Practice Exercise 1.18

Diabetes can cause a decrease in the specific gravity of urine. If normal urine has a specific gravity between 1.002 and 1.030, and a sample of urine has a density of 1.008 oz/liq. oz., does the urine sample have a low specific gravity?

Importance of Reliable Measurements

We saw earlier that substances can be identified by their properties. If we are to rely on properties such as density for identification of substances, it is very important that our measurements be reliable. We must have some idea of what the measurements' accuracy and precision are.

The importance of accuracy is obvious. If we have no confidence that our measured values are close to the true values, we certainly cannot trust any conclusions that are based on the data we have collected (Table 1.9).

Precision of measurements can be equally important. For example, suppose we have a gold ring and we want to determine its purity—that is, 24 carat gold, 22 carat gold, or 18 carat gold. We could determine the mass of the ring and then its volume, compute the density of the ring, and then compare our experimental density with the density of 24 carat gold (19.3 g/mL), 22 carat gold (17.7 g/mL), or 18 carat gold (17 g/mL). If we used a graduated cup measure and a kitchen scale, we would obtain a volume of 1.0 mL and a mass of 18 g, and the density of the ring would then be 18 g/mL. On the other hand, if we measured the mass of the ring with a laboratory balance capable of measurements to the nearest ±0.001 g and used volumetric glassware to measure the volume, we would find that the mass of the ring is 18.153 g and the volume is 1.03 mL. The density would be 17.6 g/mL, to the correct number of significant figures.

■ The carat system for gold states that pure gold is referred to as 24 carat gold. Gold that is 50% gold by mass will be 12 carat. Gold that is less than 24 carats is usually alloyed with cheaper metals such as silver or copper.

TABLE 1.9		
	Graduated Cup/ Kitchen Scale	Laboratory Balance/ Volumetric Glassware
Mass (g)	18	18.153
Volume (mL)	1.0	1.03
Density (g/mL)	18	17.6
Number of significant figures	2	3
Error (g/mL)	±1	±0.1
Range of possible values (g/mL)	17 – 19	17.5 – 17.7

If we compare this density to the densities of gold, rounded to two significant figures, 24 carat gold has a density of 19 g/mL and 18 carat gold has a density of 17 g/mL; then the ring could be anywhere from 18 to 24 carat gold. Instead, using the more precise measurements, the density of the ring is 17.6 g/mL, which matches the density of 22 carat gold, which has a density of 17.7 g/mL. This is one example of why precise and accurate measurements are important. Every minute of every day there are multi-million-dollar transactions made in the business world that depend on accurate and precise measurements such as this.

Summary

Organized by Learning Objective

Explain the scientific method
Chemistry is a science that studies the properties and composition of **matter**, which is defined as anything that has **mass** and occupies space. Chemistry employs the scientific method in which **observations** are used to collect **empirical facts**, or **data**, that can be summarized in **scientific laws**. **Models** of nature begin as **hypotheses** that mature into **theories** when they survive repeated testing.

Classify matter
An **element**, which is identified by its **chemical symbol**, cannot be decomposed into something simpler by a **chemical reaction**. Elements combine in fixed proportions to form **compounds**. Elements and compounds are **pure substances** that may be combined in varying proportions to give **mixtures**. If a mixture has two or more **phases**, it is **heterogeneous**. A one-phase **homogeneous** mixture is called a **solution**. Formation or separation of a mixture into its components can be accomplished by a **physical change**. Formation or decomposition of a compound takes place by a **chemical change** that changes the chemical makeup of the substances involved.

Develop an understanding of which physical and chemical properties are important to measure
Physical properties are measured without changing the chemical composition of a sample. **Solid, liquid,** and **gas** are the most common **states of matter.** The properties of the states of matter can be related to the different ways the individual atomic-size particles are organized. A **chemical property** describes a chemical reaction a substance undergoes. **Intensive properties** are independent of sample size; **extensive properties** depend on sample size.

Understand measurements and the use of SI units
Qualitative observations lack numerical information, whereas **quantitative observations** require numerical measurements. The units used for scientific measurements are based on the set of seven **SI base units**, which can be combined to give various **derived units**. These all can be scaled to larger or smaller sized units by applying **decimal multiplying factors**. Convenient units for length and volume are, respectively, **centimeters** or **millimeters**, and **liters** or **milliliters**. **Mass** is a measure of the amount of matter in an object and is measured with a **balance** and is expressed in units of **kilograms** or **grams**. Temperature is measured in units of **degrees Celsius** (or **Fahrenheit**) using a thermometer. For many calculations, temperature must be expressed in **kelvins (K)**. The zero point on the **Kelvin temperature scale** is called **absolute zero**.

Understand how error in measurements occur and are reported using significant figures
The **precision** of a measured quantity is expressed by the number of **significant figures** that it contains, which equals the number of digits known for sure plus the first one that possesses some uncertainty. Measured values are **precise** if they contain many significant figures and differ from each other by small amounts. A measurement is **accurate** if its value lies very close to the true value. When measurements are combined in calculations, rules help us determine the correct number of significant figures in the answer. **Exact numbers** are considered to have an infinite number of significant figures.

Learn how to effectively use dimensional analysis
Dimensional analysis is based on the ability of units to undergo the same mathematical operations as numbers. **Conversion factors** are constructed from *valid relationships* between units. These relationships can be either equalities or **equivalencies** (indicated by the symbol ⇔) between units.

Learn to determine the density of substances and to use density as a conversion factor
Density is an intensive property equal to the ratio of a sample's mass to its volume. Besides serving as a means for identifying substances, density provides a conversion factor that relates mass to volume. **Specific gravity** is the ratio of the density of a substance to the density of water and is a **dimensionless** quantity. This serves as a convenient way to have density data available in a large number of mass and volume units.

TOOLS

Tools for Problem Solving The following tools were introduced in this chapter. Study them carefully so you can select the appropriate tool when needed.

Base SI units (Section 1.4)
The eight basic units of the SI system are used to derive the units for all scientific measurements.

SI prefixes (Section 1.4)
The prefixes are used to create larger and smaller units.

Units in laboratory measurements (Section 1.4)

Conversion for commonly used laboratory measurements.

Length: 1 m = 100 cm = 1000 mm

Volume: 1 L = 1000 mL = 1000 cm³

Temperature conversions: Celsius to Fahrenheit and Celsius to Kelvin (Section 1.4)

$$t_F = \left(\frac{9\,°F}{5\,°C}\right)t_C + 32\,°F$$

$$T_K = (t_C + 273.15)°C\left(\frac{1\,K}{1\,°C}\right)$$

Counting significant figures (Section 1.5)

- All nonzero digits are significant.
- Zeros to the *left* of the first nonzero digit are never significant.
- Zeros between significant digits, imbedded zeros, are significant.
- Zeros on the end of a number (a) with a decimal point are significant; (b) those without a decimal point are assumed not to be significant. (To avoid confusion, scientific notation should be used.)

Significant figures: multiplication and division (Section 1.5)

Round the answer to the same number of significant figures as the factor with the fewest significant figures.

Significant figures: addition and subtraction (Section 1.5)

Round the answer to the same number of decimal places as the quantity with the fewest decimal places.

Significant figures: exact numbers (Section 1.5)

Numbers from definitions, coefficients, and subscripts are exact and do not affect the number of significant figures of a calculation.

Density (Section 1.7)

$$d = \frac{m}{V}$$

Specific gravity (Section 1.7)

$$\text{Specific gravity} = \frac{d_{substance}}{d_{water}}$$

WileyPLUS = *WileyPLUS*, an online teaching and learning solution. *Note to instructors:* Many of the end-of-chapter problems are available for assignment via the WileyPLUS system. **www.wileyplus.com.** Review Problems are presented in pairs separated by blue rules. Answers to problems whose numbers appear in blue are given in Appendix B. More challenging problems are marked with an asterisk ∗.

| Review Questions

Laws and Theories: The Scientific Method

1.1 After some thought, give two reasons why a course in chemistry will benefit you in the pursuit of your particular major.

1.2 What steps are involved in the scientific method?

1.3 What is the difference between (a) a law and a theory, (b) an observation and a conclusion, (c) an observation and data?

1.4 Can a theory be proved to be correct? Can a theory be proved to be wrong?

Matter and Its Classifications

1.5 Define matter. Which are examples of matter? (a) air, (b) an idea, (c) a bowl of soup, (d) a squirrel, (e) sodium chloride, (f) the sound of an explosion

1.6 Define (a) element, (b) compound, (c) mixture, (d) homogeneous, (e) heterogeneous, (f) phase, (g) solution.

1.7 Which kind of change, chemical or physical, is needed to change a compound into its elements?

1.8 What is the chemical symbol for each of the following elements? **(a)** fluorine, **(b)** selenium, **(c)** nickel, **(d)** argon, **(e)** lithium, **(f)** phosphorus, **(g)** iodine, **(h)** gallium, **(i)** mercury, **(j)** manganese

1.9 What is the name of each of the following elements? **(a)** Na, **(b)** Zn, **(c)** Si, **(d)** Sn, **(e)** Mg, **(f)** W, **(g)** Co, **(h)** Al, **(i)** O, **(j)** N

1.10 For each of the following molecular pictures, state whether it represents a pure substance or a mixture. For pure substances, state whether it represents an element or a compound. For mixtures, state whether it is homogeneous or heterogeneous.

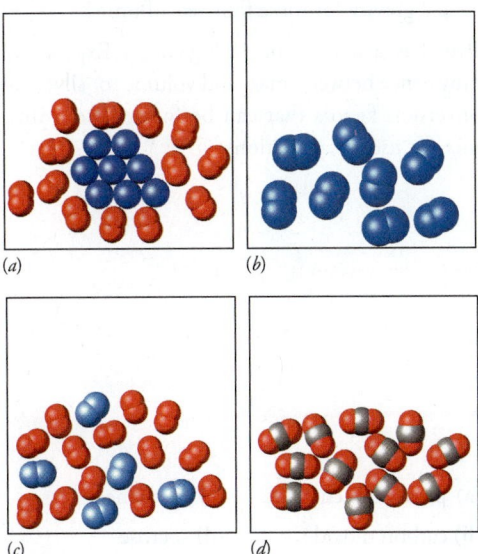

(a)　　　*(b)*

(c)　　　*(d)*

1.11 Consider the following four samples of matter.

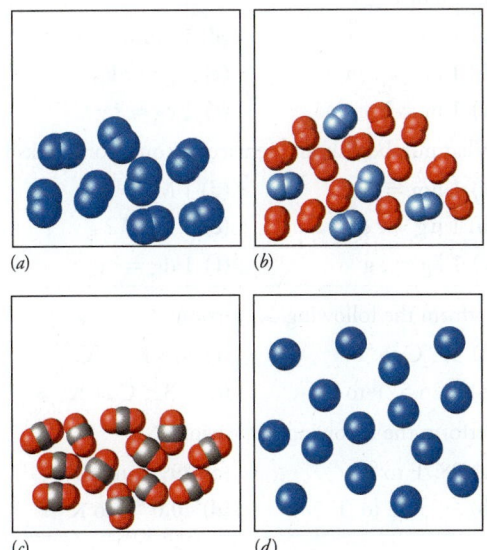

(a)　　　*(b)*

(c)　　　*(d)*

(a) Which sample(s) consist of only one element?
(b) Which sample(s) consist of only one compound?
(c) Which sample(s) consist of diatomic molecules?

Physical and Chemical Properties

1.12 What is a physical change? What is a chemical change? What is the chief distinction between physical and chemical changes?

1.13 "A sample of calcium (an electrically conducting white metal that is shiny, relatively soft, melts at 850 °C, and boils at 1440 °C) was placed into liquid water that was at 25 °C. The calcium reacted slowly with the water to give bubbles of gaseous hydrogen and a solution of the substance calcium hydroxide." Based on this description, what physical changes and what chemical changes occurred?

1.14 In places like Saudi Arabia, freshwater is scarce and is recovered from seawater. When seawater is boiled, the water evaporates and the steam can be condensed to give pure water that people can drink. If all the water is evaporated, solid salt is left behind. Are the changes described here chemical or physical?

1.15 How does a chemical property differ from a physical property?

1.16 Distinguish between an extensive and an intensive property.

1.17 Determine whether each of the following is an intensive or extensive property, and justify your reasoning:
　(a) mass　　　　　　　**(e)** physical state
　(b) boiling point　　　**(f)** density
　(c) color　　　　　　 **(g)** volume
　(d) surface area

1.18 Describe one or more physical properties of each state of matter that distinguishes it from the other states of matter:
　(a) gas　　　　**(b)** liquid　　　　**(c)** solid

Measurement of Physical and Chemical Properties

1.19 Why must measurements always be written with a unit?

1.20 What is the only SI base unit that includes a prefix?

1.21 Which SI units are mainly used in chemistry?

1.22 What are derived units? Give two examples of derived units.

1.23 What is the meaning of each of the following prefixes? What abbreviation is used for each prefix?
　(a) centi　　　　**(e)** nano
　(b) milli　　　　**(f)** pico
　(c) kilo　　　　 **(g)** mega
　(d) micro

1.24 What reference points do we use in calibrating the scale of a thermometer? What temperature on the Celsius scale do we assign to each of these reference points?

1.25 In each pair, which is larger:
　(a) A Fahrenheit degree or a Celsius degree?
　(b) A Celsius degree or a kelvin?
　(c) A Fahrenheit degree or a kelvin?

The Uncertainty of Measurements

1.26 Define the term *significant figures.*

1.27 Explain how to round numbers.

1.28 What is the difference between *accuracy* and *precision*?

1.29 Suppose a length had been reported to be 31.24 cm. What is the minimum uncertainty implied in this measurement?

1.30 What is the difference between the treatment of significant figures in addition and multiplication?

Dimensional Analysis

1.31 When constructing a conversion factor between two units, which unit becomes the numerator and which one becomes the denominator?

1.32 Suppose someone suggested using the fraction 3 yd/1 ft as a conversion factor to change a length expressed in feet to its equivalent in yards. What is wrong with this conversion factor? Can we construct a valid conversion factor relating centimeters to meters from the equation 1 cm = 1000 m? Explain your answer.

1.33 In 1 hour there are 3600 seconds. By what conversion factor would you multiply 250 seconds to convert it to hours? By what conversion factor would you multiply 3.84 hours to convert it to seconds?

Review Problems

Physical and Chemical Properties

1.39 Determine whether each of the following is a physical or chemical change, and explain your reasoning.
   (a) Copper conducts electricity.
   (b) Gallium metal melts in your hand.
   (c) Bread turns brown in a toaster.
   (d) Wine turns to vinegar.
   (e) Cement hardens.

1.40 Determine whether each of the following is a physical or chemical change, and explain your reasoning.
   (a) Iron rusts.
   (b) Kernels of corn are heated to make popcorn.
   (c) Molten copper is mixed with molten gold to make an alloy.
   (d) Heavy cream is churned to make butter.
   (e) Water vapor cools into water.

1.41 At room temperature, what is the state of each of the following? If necessary, look up the information in a reference source.
   (a) hydrogen   **(c)** nitrogen
   (b) aluminum   **(d)** mercury

1.42 At room temperature, determine the appropriate phase for each of the following substances. (Look up the substance in data tables if needed.)

1.34 If you were to convert the measured length 4.165 ft to yards by multiplying by the conversion factor (1 yd/3 ft), how many significant figures should the answer contain? Why?

Density and Specific Gravity

1.35 Write the equation that defines density. Identify the symbols in the equation.

1.36 Compare density and specific gravity. What is the difference between the two? When would specific gravity be used?

1.37 Give four sets of units for density. What mathematical operation must be carried out to convert the density into specific gravity for these four sets of units?

1.38 Silver has a density of 10.5 g cm^{-3}. Express this as an equivalence between mass and volume for silver. Write two conversion factors that can be formed from this equivalence for use in calculations.

   (a) potassium chloride   **(c)** methane
   (b) carbon dioxide   **(d)** sucrose

Measurement of Physical and Chemical Properties

1.43 What number should replace the question mark in each of the following?
   (a) 1 cm = ? m   **(d)** 1 dm = ? m
   (b) 1 km = ? m   **(e)** 1 g = ? kg
   (c) 1 m = ? pm   **(f)** 1 cg = ? g

1.44 What numbers should replace the question marks below?
   (a) 1 nm = ? m   **(d)** 1 Mg = ? g
   (b) 1 μg = ? g   **(e)** 1 mg = ? g
   (c) 1 kg = ? g   **(f)** 1 dg = ? g

1.45 Perform the following conversions.
   (a) 57 °C to °F   **(c)** 378 K to °C
   (b) −25.5 °F to °C   **(d)** −31 °C to K

1.46 Perform the following conversions.
   (a) 98 °F to °C   **(c)** 299 K to °C
   (b) −55 °C to °F   **(d)** 40.0 °C to K

1.47 The temperature of the core of the sun is estimated to be about 15.7 million K. What is this temperature in °C and °F?

1.48 Natural gas is mostly methane, a substance that boils at a temperature of 109 K. What is its boiling point in °C and °F?

1.49 A healthy dog has a temperature ranging from 37.8 °C to 39.2 °C. Is a dog with a temperature of 103.5 °F within the normal range?

1.50 In South America, the warmest recorded temperature of 120.0 °F was in Rivadavia, Argentina. Was Death Valley's record temperature of 56.7 °C warmer or colder than the temperature in South America?

Uncertainty of Measurements

1.51 The length of a wire was measured using two different rulers. What are the lengths of the wire and how many significant figures are in each measurement?

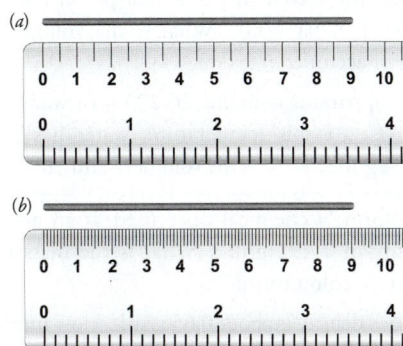

1.52 What are the temperatures being measured in the two thermometers, and how many significant figures are in each measurement?

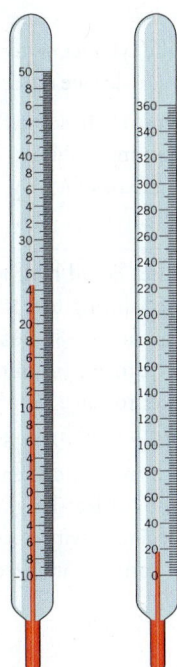

1.53 How many significant figures do the following measured quantities have?

(a) 37.53 cm   (d) 0.00024 kg

(b) 37.240 cm   (e) 2400 mL

(c) 202.0 g

1.54 How many significant figures do the following measured quantities have?

(a) 0.0230 g   (d) 614.00 mg

(b) 105.303 m   (e) 10 L

(c) 0.007 kg

1.55 Perform the following arithmetic and round off the answers to the correct number of significant figures. Include the correct units with the answers.

(a) 0.0023 m × 315 m

(b) 84.25 kg − 0.01075 kg

(c) (184.45 g − 94.45 g)/(31.4 mL − 9.9 mL)

(d) (23.4 g + 102.4 g + 0.003 g)/(6.478 mL)

(e) (313.44 cm − 209.1 cm) × 8.2234 cm

1.56 Perform the following arithmetic and round off the answers to the correct number of significant figures. Include the correct units with the answers.

(a) 3.58 g/1.739 mL

(b) 4.02 mL + 0.001 mL

(c) (22.4 g − 8.3 g)/(1.142 mL − 0.002 mL)

(d) (1.345 g + 0.022 g)/(13.36 mL − 8.4115 mL)

(e) (74.335 m − 74.332 m)/(4.75 s × 1.114 s)

1.57 Which are exact numbers and which ones have a finite number of significant figures?

(a) the volume of piece of aluminum

(b) the number of centimeters in an inch

(c) today's temperature

(d) the number of seconds in a year

1.58 Which are exact numbers and which ones have a finite number of significant figures?

(a) the mass of 454 g of sugar

(b) the distance from the earth to the moon

(c) the width of your chemistry book

(d) the number of milliliters in a gallon

Dimensional Analysis

1.59 Perform the following conversions.

(a) 32.0 dm/s to km/hr   (d) 137.5 mL to L

(b) 8.2 mg/mL to µg/L   (e) 0.025 L to mL

(c) 75.3 mg to kg   (f) 342 pm^2 to dm^2

1.60 Perform the following conversions.

(a) 92 dL to µm^3 (d) 230 km^3 to m^3

(b) 22 ng to µg   (e) 87.3 cm s^{-2} to km hr^{-2}

(c) 83 pL to nL   (f) 238 mm^2 to nm^2

1.61 Perform the following conversions. If necessary, refer to Tables 1.3 and 1.4.

(a) 36 in. to cm   (d) 1 cup (8 oz) to mL

(b) 5.0 lb to kg   (e) 55 mi/hr to km/hr

(c) 3.0 qt to mL   (f) 50.0 mi to km

1.62 Perform the following conversions. If necessary, refer to Tables 1.3 and 1.4.

(a) 250 mL to qt   (d) 1.75 L to fluid oz

(b) 3.0 ft to m   (e) 35 km/hr to mi/hr

(c) 1.62 kg to lb   (f) 80.0 km to mi

1.63 Perform the following conversions.

(a) 8.4 ft^2 to cm^2 (c) 231 ft^3 to cm^3

(b) 223 mi^2 to km^2

1.64 Perform the following conversions.

(a) 2.4 yd^2 to m^2 (c) 9.1 ft^3 to L

(b) 8.3 in.^{2} to mm^2

1.65 The human stomach can expand to hold up to 4.2 quarts of food. A pistachio nut has a volume of about 0.9 mL. Use this information to estimate the maximum number of pistachios that can be eaten in one sitting.

1.66 In the movie *Cool Hand Luke* (1967), Luke wagers that he can eat 50 eggs in one hour. The prisoners and guards bet against him, saying, "Fifty eggs gotta weigh a good six pounds. A man's gut can't hold that." A peeled, chewed chicken egg has a volume of approximately 53 mL. If Luke's stomach has a volume of 4.2 quarts, does he have any chance of winning the bet?

1.67 The winds in a hurricane can reach almost 200 miles per hour. What is this speed in meters per second? (Assume three significant figures.)

1.68 A bullet is fired at a speed of 2435 ft/s. What is this speed expressed in kilometers per hour?

1.69 A bullet leaving the muzzle of a pistol was traveling at a speed of 2230 feet per second. What is this speed in miles per hour?

1.70 On average, water flows over Niagara Falls at a rate of 2.05×10^5 cubic feet per second. One cubic foot of water weighs 62.4 lb. Calculate the rate of water flow in tons of water per day. (1 ton = 2000 lb)

1.71 The brightest star in the night sky in the northern hemisphere is Sirius. Its distance from earth is estimated to be 8.7 light years. A light year is the distance light travels in one year. Light travels at a speed of 3.00×10^8 m/s. Calculate the distance from earth to Sirius in miles. (1 mi = 5280 ft)

*⁕**1.72** One degree of latitude on the earth's surface equals 60.0 nautical miles. One nautical mile equals 1.151 statute miles. (A *statute mile* is the distance over land that we normally associate with the unit mile.) Calculate the circumference of the earth in statute miles.

Density and Specific Gravity

1.73 Calculate the density of kerosene, in g/mL, if its mass is 36.4 g and its volume is 45.6 mL.

1.74 Calculate the density of magnesium, in g/cm^3, if its mass is 14.3 g and its volume is 8.46 cm^3.

1.75 Acetone, the solvent in some nail polish removers, has a density of 0.791 g/mL. What is the volume, in mL, of 25.0 g of acetone?

1.76 A glass apparatus contains 26.223 g of water when filled at 25 °C. At this temperature, water has a density of 0.99704 g/mL. What is the volume, in mL, of the apparatus?

1.77 Chloroform, a chemical once used as an anesthetic, has a density of 1.492 g/mL. What is the mass in grams of 185 mL of chloroform?

1.78 Gasoline's density is about 0.65 g/mL. How much does 34 L (approximately 18 gallons) weigh in kilograms? In pounds?

1.79 A graduated cylinder was filled with water to the 15.0 mL mark and weighed on a balance. Its mass was 27.35 g. An object made of silver was placed in the cylinder and completely submerged in the water. The water level rose to 18.3 mL. When reweighed, the cylinder, water, and silver object had a total mass of 62.00 g. Calculate the density of silver in g cm^{-3}.

1.80 Titanium is a metal used to make golf clubs. A rectangular bar of this metal measuring 1.84 cm × 2.24 cm × 2.44 cm was found to have a mass of 45.7 g. What is the density of titanium in g cm^{-3}?

1.81 The space shuttle uses liquid hydrogen as its fuel. The external fuel tank used during takeoff carries 227,641 lb of hydrogen with a volume of 385,265 gallons. Calculate the density of liquid hydrogen in units of lb/gal and g/mL. (Express your answer to three significant figures.) What is the specific gravity of liquid hydrogen?

1.82 You are planning to make a cement driveway that has to be 10.1 feet wide, 32.3 feet long, and 4.00 inches deep on average. The specific gravity of concrete is 0.686. How many kilograms of concrete are needed for the job?

Additional Exercises

1.83 Some time ago, a U.S. citizen traveling in Canada observed that the price of regular gasoline was 1.299 Canadian dollars per liter. The exchange rate at the time was 1.001 Canadian dollars per one U.S. dollar. Calculate the price of the Canadian gasoline in units of U.S. dollars per gallon. (Just the week before, the traveler had paid $3.759 per gallon in the United States.)

1.84 Driving to work one day, one of the authors of this book had a gas mileage of 28.5 km/L. Another author got 31 mi/gal. Which author had the more efficient car? Why?

1.85 You are the science reporter for a daily newspaper, and your editor has asked you to write a story based on a report in the scientific literature. The report states that analysis of the sediments in Hausberg Tarn (elevation 4350 m) on the side of Mount Kenya (elevation 4600–4700 m) shows that the average temperature of the water rose by 4.0 °C between 350 BC and 450 AD. Your editor wants all the data expressed in the English system of units. Make the appropriate conversions.

1.86 An astronomy web site states that neutron stars have a density of 1.00×10^8 tons per cubic centimeter. The site does not specify whether "tons" means metric tons (1 metric ton = 1000 kg) or English tons (1 English ton = 2000 pounds). How many grams would one teaspoon of a neutron star weigh if the density were in metric tons per cm^3? How many grams would the teaspoon weigh if the density were in English tons per cm^3? (One teaspoon is defined as 5.00 mL.)

1.87 The star Arcturus is 3.50×10^{14} km from the earth. How many days does it take for light to travel from Arcturus to earth? What is the distance to Arcturus in light years? One light year is the distance light travels in one year (365 days); light travels at a speed of 3.00×10^8 m/s.

1.88 A pycnometer is a glass apparatus used for accurately determining the density of a liquid. When dry and empty, a certain pycnometer had a mass of 27.314 g. When filled with distilled water at 25.0 °C, it weighed 36.842 g. When filled with chloroform (a liquid once used as an anesthetic before its toxic properties were known), the apparatus weighed 41.428 g. At 25.0 °C, the density of water is 0.99704 g/mL. **(a)** What is the volume of the pycnometer? **(b)** What is the density of chloroform?

1.89 Radio waves travel at the speed of light, 3.00×10^8 m/s. If you were to broadcast a question to the Mars Rover, Spirit, on Mars, which averages 225,000,000 miles from earth, what is the minimum time that you would have to wait to receive a reply?

1.90 Suppose you have a job in which you earn $7.35 for each 30 minutes that you work.

(a) Express this information in the form of an equivalence between dollars earned and minutes worked.

(b) Use the equivalence defined in (a) to calculate the number of dollars earned in 1 hr 45 min.

(c) Use the equivalence defined in (a) to calculate the number of minutes you would have to work to earn $333.50.

1.91 When an object floats in water, it displaces a volume of water that has a weight equal to the weight of the object. If a ship has a weight of 4255 tons, how many cubic feet of seawater will it displace? Seawater has a density of 1.025 g cm^{-3}; 1 ton = 2000 lb, exactly.

1.92 Aerogel or "solid smoke" is a novel material that is made of silicon dioxide, like glass, but is a thousand times less dense than glass because it is extremely porous. Material scientists at NASA's Jet Propulsion Laboratory created the lightest aerogel ever in 2002, with a density of 0.00011 pounds per cubic inch. The material was used for thermal insulation in the 2003 Mars Exploration Rover. If the maximum space for insulation in the spacecraft's hull was 2510 cm^3, what mass (in grams) did the aerogel insulation add to the spacecraft?

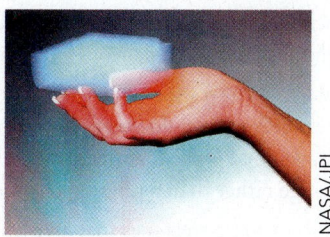

NASA/JPL

1.93 A liquid known to be either ethanol (ethyl alcohol) or methanol (methyl alcohol) was found to have a density of 0.798 ± 0.001 g/mL. Consult tabulated data to determine which liquid it is. What other measurements could help to confirm the identity of the liquid?

1.94 An unknown liquid was found to have a density of 69.22 lb/ft^3. The density of ethylene glycol (the liquid used in antifreeze) is 1.1088 g/mL. Could the unknown liquid be ethylene glycol?

1.95 There exists a single temperature at which the value reported in °F is numerically the same as the value reported in °C. What is this temperature?

1.96 In the text, the Kelvin scale of temperature is defined as an absolute scale in which one Kelvin degree unit is the same size as one Celsius degree unit. A second absolute temperature scale exists called the Rankine scale. On this scale, one Rankine degree unit (°R) is the same size as one Fahrenheit degree unit. **(a)** What is the only temperature at which the Kelvin and Rankine scales possess the same numerical value? Explain your answer. **(b)** What is the boiling point of water expressed in °R?

*1.97 Density measurements can be used to analyze mixtures. For example, the density of solid sand (without air spaces) is about 2.84 g/mL. The density of gold is 19.3 g/mL. If a 1.00 kg sample of sand containing some gold has a density of 3.10 g/mL (without air spaces), what is the percentage of gold in the sample?

*1.98 An artist's statue has a surface area of 14.6 ft^2. The artist plans to apply gold leaf to the statue and wants the coating to be 2.50 μm thick. If the price of gold were $1,774.10 per troy ounce, how much would it cost to give the statue its gold coating? (1 troy ounce = 31.1035 g; the density of gold is 19.3 g/mL.)

1.99 What is the volume in cubic millimeters of a 3.54 carat diamond, given that the density of diamond is 3.51 g/mL? (1 carat = 200 mg) Assuming that the diamond is generally spherical, what is the approximate diameter of the diamond in mm?

Exercises in Critical Thinking

1.100 A solution is defined as a uniform mixture consisting of a single phase. With our vastly improved abilities to "see" smaller and smaller particles, down to the atomic level, present an argument for the proposition that all mixtures are heterogeneous. Present the argument that the ability to observe objects as small as an atom has no effect on the definitions of heterogeneous and homogeneous.

1.101 How do you know that Coca-Cola is not a compound? What experiments could you perform to prove it? What would you do to prove that the rusting of iron is a chemical change rather than a physical change?

1.102 Find two or more web sites that give the values for each of the seven base SI units. Keeping in mind that not all web sites provide reliable information, which web site do you believe provides the most reliable values? Justify your answer.

1.103 Reference books such as the *Handbook of Chemistry and Physics* report the specific gravities of substances instead of their densities. Using the definition of specific gravity, discuss the relative merits of specific gravity and density in terms of their usefulness as a physical property.

*1.104 A student used a 250 mL graduated cylinder having volume markings every 2 mL to carefully measure 100 mL of water for an experiment. A fellow student said that by reporting the volume as "100 mL" in her lab notebook, she was only entitled to one significant figure. The first student disagreed. Why did her fellow student say the reported volume had only one significant figure?

*1.105 Because of the serious consequences of lead poisoning, the Federal Centers for Disease Control in Atlanta has set a threshold of concern for lead levels in children's blood. This threshold was based on a study that suggested that lead levels in blood as low as *10 micrograms of lead per deciliter of blood* can result in subtle effects of lead toxicity. Suppose a child had a lead level in her blood of 2.5×10^{-4} grams of lead per liter of blood. Is this person in danger of exhibiting the effects of lead poisoning?

*1.106 Gold has a density of 19.31 g cm^{-3}. How many grams of gold are required to provide a gold coating 0.500 mm thick on a ball bearing having a diameter of 2.000 mm?

*1.107 A Boeing 747 jet airliner carrying 568 people burns about 5.0 gallons of jet fuel per mile. What is the rate of fuel consumption in units of gallons per person per mile? Is this better or worse than the rate of fuel consumption in an automobile carrying two people that gets 21.5 miles per gallon? If the airliner were making the 3470 mile trip from New York to London, how many pounds of jet fuel would be consumed? (Jet fuel has a density of 0.817 g/mL.) Considering the circumstances, how many significant figures should be used in reporting the volume of jet fuel consumed? Justify your answer.

*1.108 Download a table of data for the density of water between its freezing and boiling points. Use a spreadsheet program to plot **(a)** the density of water versus temperature and **(b)** the volume of a kilogram of water versus temperature. Interpret the significance of these plots.

1.109 List the physical and chemical properties mentioned in this chapter. What additional physical and chemical properties can you think of to extend this list?

*1.110 A barge pulled by a tugboat can be described as a rectangular box that is open on the top. A certain barge is 30.2 feet wide and 12.50 feet high, with the base of each side being 116 feet long. If the barge itself weighs 6.58×10^4 pounds, what will be the draft (depth in the water) if the cargo weighs 1.12×10^6 pounds in **(a)** seawater with a density of 1.025 g/mL and **(b)** freshwater with a density of 0.997 g/mL?

2

Elements, Compounds, and the Periodic Table

Chapter Outline

2.1 | The Periodic Table

2.2 | Metals, Nonmetals, and Metalloids

2.3 | Molecules and Chemical Formulas

2.4 | Chemical Reactions and Chemical Equations

2.5 | Ionic Compounds

2.6 | Nomenclature of Ionic Compounds

2.7 | Molecular Compounds

2.8 | Nomenclature of Molecular Compounds

© SKrow/iStockphoto

© SKrow/iStockphoto

© skegbydave/iStockphoto

This Chapter in Context

In this chapter we consider one of the most recognizable icons in all of science, the *periodic table*. Using two simple observations, the relative masses of the elements and the reactions they enter into, much of the known data about the elements was organized into logical rows and columns. Our chapter-opening photo has several smart phones, each displaying a type of list that you might like to keep. An app list arranges the apps in the order that you set, so they can be called up quickly. Your to-do list may be arranged by the importance of each task, and the playlist may be arranged based on the music you like. We will find that the chemists' periodic table is another useful way to arrange information. From there, this chapter discusses the types of elements by classifying them in various ways. We finish the chapter by describing the use of chemical equations to explain how chemicals react with each other and the systematic naming of simple chemical compounds.

LEARNING OBJECTIVES

After reading this chapter, you should be able to:

- describe the information in and the organization of the periodic table
- understand the distribution of metals, nonmetals, and metalloids within the periodic table
- explain the information embodied in a chemical formula
- understand the nature of a balanced chemical equation and how it relates to the atomic theory
- use the periodic table and ion charges to write chemical formulas of ionic compounds
- name ionic compounds and write chemical formulas from chemical names
- understand the difference between ionic and molecular compounds
- name molecular compounds

2.1 | The Periodic Table

When we study different kinds of substances, we find that some are elements and others are compounds. Among the compounds, some are composed of discrete molecules while others are *ionic compounds*, made up of atoms that have acquired electrical charges. Some elements, such as iron and chromium, have properties we associate with metals, whereas others, such as carbon and sulfur, do not have metallic properties and are said to be nonmetallic. If we were to continue on this way, without attempting to build our subject around some central organizing structure, it would not be long before we became buried beneath a mountain of information of seemingly unconnected facts. Chemists have organized the information into the *periodic table*, a section of which is on the cover of the book.

Mendeleev's Periodic Table

The need for organization was recognized by many early chemists, and there were numerous attempts to discover relationships among the chemical and physical properties of the elements. The **periodic table** we use today is based primarily on the efforts of a Russian chemist, Dmitri Ivanovich Mendeleev (1834–1907) and a German physicist, Julius Lothar Meyer (1830–1895). Working independently, these scientists developed similar periodic tables only a few months apart in 1869. Mendeleev is usually given the credit, however, because his work was published first.

Mendeleev was preparing a chemistry textbook for his students at the University of St. Petersburg. Looking for some pattern among the properties of the elements, he found that when he arranged them in order of increasing atomic mass, similar chemical properties were repeated over and over again at regular intervals. For instance, the elements lithium (Li), sodium (Na), potassium (K), rubidium (Rb), and cesium (Cs) are soft metals that are very reactive toward water. They form compounds with chlorine that have a 1-to-1 ratio of metal to chlorine. Similarly, the elements that immediately follow each of these also constitute a set with similar chemical properties. Thus, beryllium (Be) follows lithium, magnesium (Mg) follows sodium, calcium (Ca) follows potassium, strontium (Sr) follows rubidium, and barium (Ba) follows cesium. All of these elements form a water-soluble chlorine compound with a 1-to-2 metal to chlorine atom ratio. Mendeleev used such observations to construct his periodic table.

The elements in Mendeleev's table are arranged in order of increasing atomic mass. When the sequence is broken at the right places, the elements fall naturally into columns in which the elements in a given column have similar chemical properties. Mendeleev's genius rested on his placing elements with similar properties in the same column even when this made him assume that some atomic masses were wrong or left occasional gaps in the table. Mendeleev reasoned, correctly, that the elements that belonged in these gaps had simply not yet been discovered. In fact, on the basis of the location of these gaps, Mendeleev was able to predict with remarkable accuracy the properties of these yet-to-be-found substances. His predictions also helped serve as a guide in the search for those missing elements.

■ Periodic refers to a repeating **property**—in this case, chemical properties.

Arrangement of the Modern Periodic Table

When the concept of atomic numbers was developed, it was soon realized that the elements in Mendeleev's table were arranged in precisely the order of increasing atomic number. That it is the atomic number—the number of protons in the nucleus of an atom—that determines the order of elements in the table is very significant. We will see later that this has important implications with regard to the relationship between the number of electrons in an atom and the atom's chemical properties.

The modern periodic table is shown in Figure 2.1 and also appears on the inside front cover of the book. We will refer to the table frequently, so it is important for you to become familiar with it and the terminology applied to it.

TOOLS
Periodic table

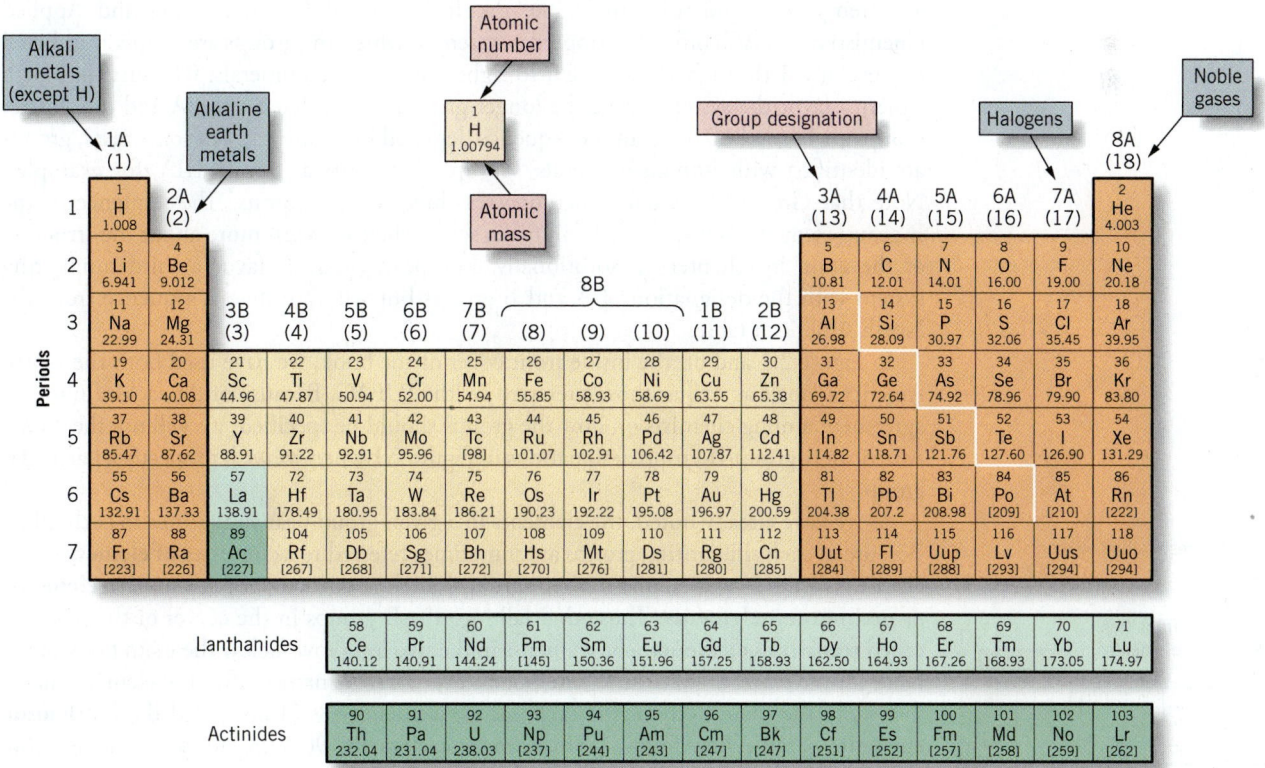

Figure 2.1 | **The modern periodic table.** At room temperature, mercury and bromine are liquids. Eleven elements are gases, including the noble gases and the diatomic gases of hydrogen, oxygen, nitrogen, fluorine, and chlorine. The remaining elements are solids.

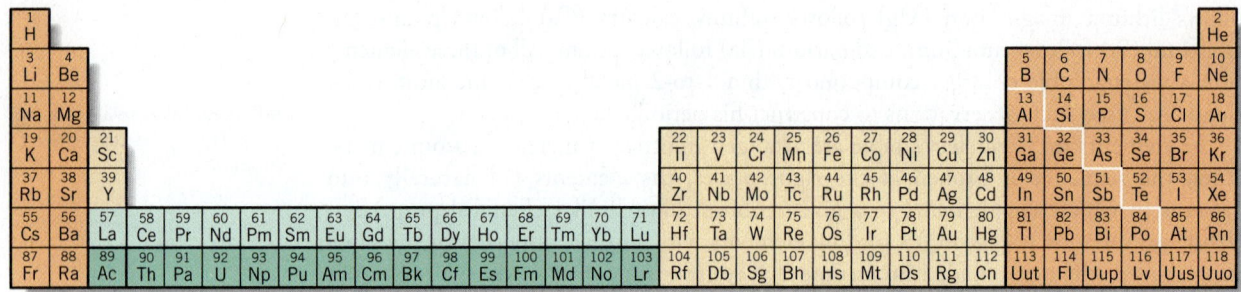

Figure 2.2 | **Extended form of the periodic table.** The two long rows below the main body of the table in Figure 2.1 are placed in their proper places in the table.

Special Terminology of the Periodic Table

In the modern periodic table the elements are arranged in order of increasing atomic number. The horizontal rows in the table are called **periods**, and for identification purposes the periods are numbered. Below the main body of the table are two long rows of 14 elements each. These actually belong in the main body of the table following lanthanum (La: $Z = 57$) and actinium (Ac: $Z = 89$), as shown in Figure 2.2. They are almost always placed below the table simply to conserve space. If the fully spread-out table is printed on one page, the type is so small that it's difficult to read. Notice also that in the fully extended form of the table, there is a great deal of empty space. An important requirement of a detailed atomic theory, which we will get to in Chapter 7, is that it must explain not only the repetition of properties, but also why there is so much empty space in the table.

■ Recall that the symbol *Z* stands for the atomic number.

The vertical columns in the periodic table are called **groups**, also identified by numbers. However, there is not uniform agreement among chemists on the numbering system. In an attempt to standardize the table, the International Union of Pure and Applied Chemistry (IUPAC) officially adopted a system in which the groups are simply numbered sequentially, 1 through 18, from left to right using Arabic numerals. Chemists in North America favor the system where the longer groups are labeled 1A to 8A and the shorter groups are labeled 1B to 8B in the sequence depicted in Figure 2.1. (In some texts, groups are identified with Roman numerals; Group 3A appears as Group IIIA, for example.) Note that Group 8B actually encompasses three short columns. The sequence of the B-group elements is unique and will make sense when we learn more about the structure of the atom in Chapter 7. Additionally, European chemists favor a third numbering system with the designation of A and B groups but with a different sequence from the North American table.

In Figure 2.1 and on the inside front cover of the book, we have used both the North American labels as well as those preferred by the IUPAC. Because of the lack of uniform agreement among chemists on how the groups should be specified, we will use the North American A-group/B-group designations in Figure 2.1 when we wish to specify a particular group.

As we have already noted, the elements in a given group bear similarities to each other. Because of such similarities, groups are sometimes referred to as **families of elements**. The elements in the longer columns (the A groups) are known as the **representative elements** or **main group elements**. Those that fall into the B groups in the center of the table are called **transition elements**. The elements in the two long rows below the main body of the table are the **inner transition elements**, and each row is named after the element that it follows in the main body of the table. Thus, elements 58–71 are called the **lanthanide elements** because they follow lanthanum, and elements 90–103 are called the **actinide elements** because they follow actinium.

Some of the groups have acquired common names. For example, except for hydrogen, the Group 1A elements are metals. They form compounds with oxygen that dissolve in water to give solutions that are strongly alkaline, or caustic. As a result, they are called the **alkali metals**

Representative elements

Transition elements

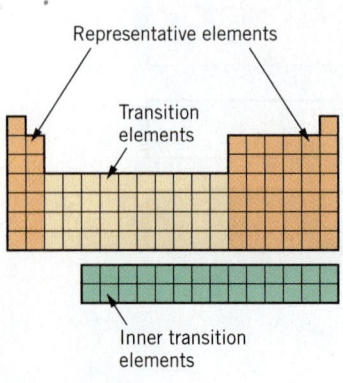

Inner transition elements

or simply the alkalis. The Group 2A elements are also metals. Their oxygen compounds are alkaline, too, but many compounds of the Group 2A elements are unable to dissolve in water and are found in deposits in the ground. Because of their properties and where they occur in nature, the Group 2A elements became known as the **alkaline earth metals**.

On the right side of the table, in Group 8A, are the **noble gases**. They used to be called the inert gases until it was discovered that the heavier members of the group show a small degree of chemical reactivity. The term noble is used when we wish to suggest a very limited degree of chemical reactivity. Gold, for instance, is often referred to as a noble metal because so few chemicals are capable of reacting with it.

Finally, the elements of Group 7A are called the **halogens**, derived from the Greek word meaning "sea" or "salt." Chlorine (Cl), for example, is found in familiar table salt, a compound that accounts in large measure for the salty taste of seawater. The other groups of the representative elements have less frequently used names, and we will name those groups based on the first element in the family. For example, Group 5A is the **nitrogen family**.

■ The Group 6A elements are also called the *chalcogens*, and the Group 5A elements are called the *pnictogens*.

2.2 | Metals, Nonmetals, and Metalloids

The periodic table organizes all sorts of chemical and physical information about the elements and their compounds. It allows us to study systematically the way properties vary with an element's position within the table and, in turn, makes the similarities and differences among the elements easier to understand and remember.

Even a casual inspection of the periodic table reveals that some elements are familiar metals and that others, equally well known, are not metals. Most of us recognize **metals** such as lead, iron, or gold and **nonmetals** such as oxygen or nitrogen. A closer look at the nonmetallic elements, though, reveals that some of them, silicon and arsenic to name two, have properties that lie between those of true metals and true nonmetals. These elements are called **metalloids**. The elements are not evenly divided into the categories of metals, nonmetals, and metalloids. (See Figure 2.3.) About 75% of the elements are metals, approximately 20 are nonmetals, and only a handful are metalloids.

TOOLS

Periodic table: Metals, nonmetals, and metalloids

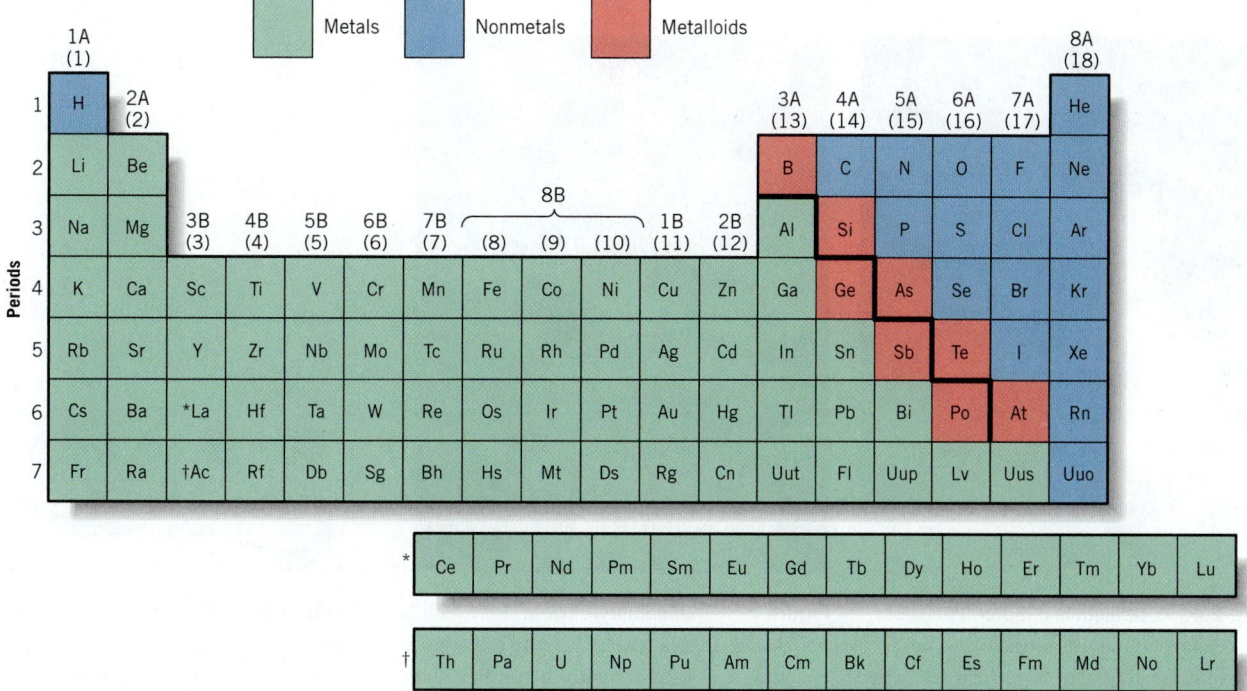

Figure 2.3 | **Distribution of metals, nonmetals, and metalloids** among the elements in the periodic table.

Metals

You probably know a metal when you see one, and you are familiar with their physical properties. Metals tend to have a shine so unique that it's called a metallic luster. For example, the silvery sheen of the surface of potassium in Figure 2.4 would most likely lead you to identify potassium as a metal even if you had never seen or heard of it before. We also know that metals conduct electricity. Few of us would hold an iron nail in our hand and poke it into an electrical outlet. In addition, we know that metals conduct heat very well. On a cool day, metals always feel colder to the touch than do neighboring nonmetallic objects because metals conduct heat away from your hand very rapidly. Nonmetals seem less cold because they can't conduct heat away as quickly and so their surfaces warm up faster.

Other properties that metals possess, to varying degrees, are **malleability**—the ability to be hammered or rolled into thin sheets—and **ductility**—the ability to be drawn into wire. The ability of gold to be hammered into foils a few atoms thick depends on the malleability of gold (Figure 2.5), and the manufacture of electrical wire is based on the ductility of copper.

Hardness is another physical property that can be used to descibe metals. Some, such as chromium or iron, are indeed quite hard; but others, including copper and lead, are rather soft. The alkali metals such as potassium (Figure 2.4) are so soft they can be cut with a knife, but they are also so chemically reactive that we rarely get to see them as free elements.

All the metallic elements, except mercury, are solids at room temperature (Figure 2.6). Mercury's low freezing point and fairly high boiling point make it useful as a fluid in thermometers. Most of the other metals have much higher melting points. Tungsten, for example, has the highest melting point of any metal, which explains its use as a filament that glows white-hot in an electric light bulb.

The chemical properties of metals vary tremendously. Some, such as gold and platinum, are very unreactive toward almost all chemical agents. This property, plus their natural beauty and rarity, makes them highly prized for use in jewelry. Other metals, however, are so reactive that few people except chemists and chemistry students

Figure 2.4 | **Potassium is a metal.** Potassium reacts quickly with moisture and oxygen to form a white coating. Due to its high reactivity, it is stored under oil to prevent water and oxygen from reacting with it.

Figure 2.5 | **Malleability of gold.** Pure gold is not usually used in jewelry because it is too malleable. It is used decoratively to cover domes since it can be hammered into very thin sheets called gold leaf.

Figure 2.6 | **Mercury droplet.** The metal mercury (once known as quicksilver) is a liquid at room temperature, unlike other metals, which are solids.

ever get to see them in their "free" states. For instance, the metal sodium reacts very quickly with oxygen or moisture in the air, and its bright metallic surface tarnishes almost immediately.

Nonmetals

Substances such as plastics, wood, and glass that lack the properties of metals are said to be *nonmetallic*, and an element that has nonmetallic properties is called a nonmetal. Most often, we encounter the nonmetals in the form of compounds or mixtures of compounds. There are some nonmetals, however, that are very important to us in their elemental forms. The air we breathe, for instance, contains mostly nitrogen and oxygen. Both are gaseous, colorless, and odorless nonmetals. Since we can't see, taste, or smell them, however, it's difficult to experience their existence. (Although the gentle summer breeze or a frigid winter gale quickly lets us know they are all around us!) Probably the most commonly observed nonmetallic element is carbon. We find it as the graphite in pencils, as coal, and as the charcoal used for barbecues. It also occurs in a more valuable form as diamond (Figure 2.7). Although diamond and graphite differ in appearance, each is a form of elemental carbon.

Slightly less than half of the nonmetals are solids at room temperature and atmospheric pressure, while a little more than half are gases. Photographs of some of the nonmetallic elements appear in Figure 2.8. Their properties are almost completely opposite those of metals. Nonmetals lack the characteristic appearance of a metal; solids are dull, not lustrous. They are poor conductors of heat and, with the exception of the graphite form of carbon, are also poor conductors of electricity.

The nonmetallic elements lack the malleability and ductility of metals. A lump of sulfur crumbles when hammered and breaks apart when pulled on. Diamond cutters rely on the brittle nature of carbon when they split a gem-quality stone by carefully striking a quick blow with a sharp blade.

Figure 2.7 | **Diamonds.** Gems such as this are simply another form of the element carbon.

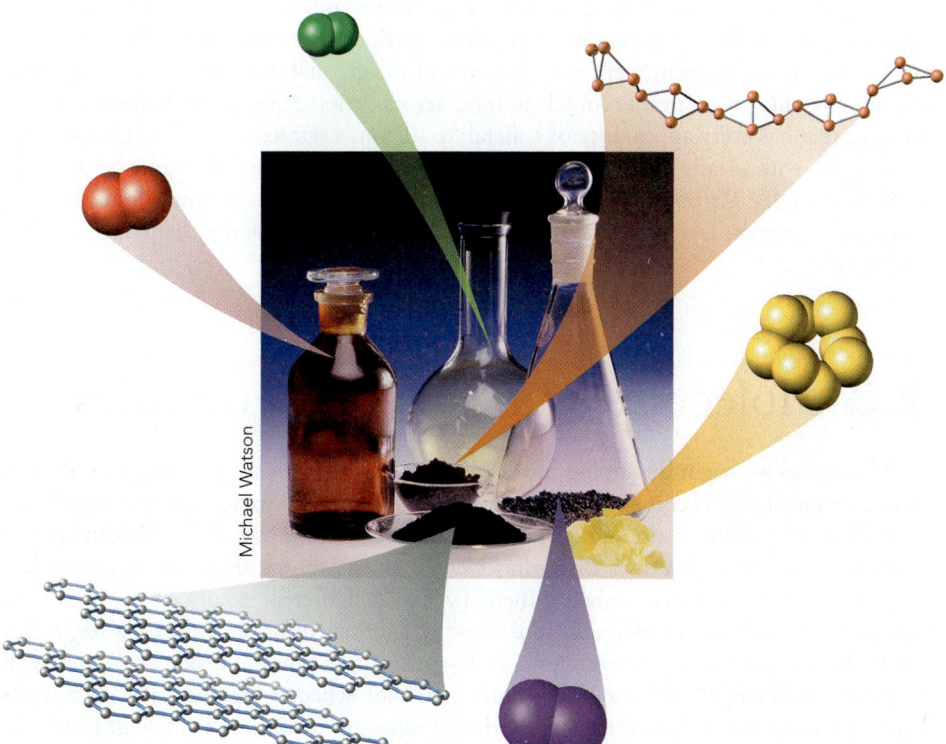

Figure 2.8 | **Some nonmetallic elements.** In the bottle on the left is dark-red liquid bromine, which vaporizes easily to give a deeply colored orange vapor. Pale green chlorine fills the round flask in the center. Solid iodine lines the bottom of the flask on the right and gives off a violet vapor. Powdered red phosphorus occupies the dish in front of the flask of chlorine, and black powdered graphite is in the watch glass. Also shown are lumps of yellow sulfur.

Figure 2.9 | **Modern electronic circuits rely on the semiconductor properties of silicon.** The silicon wafer shown here contains more electronic components (10 billion) than there are people on our entire planet (about 7 billion)!

As with metals, nonmetals exhibit a broad range of chemical reactivities. Fluorine, for instance, is extremely reactive. It reacts readily with almost all of the other elements. At the other extreme is helium, the gas used to inflate children's balloons and the blimps seen at major sporting events. This element does not react with anything. Another unreactive element is argon, which chemists use when they want to provide a blanket of inert (unreactive) gas surrounding instruments or reaction vessels that need to be protected from atmospheric gases such as oxygen, carbon dioxide, and water.

Metalloids

The properties of metalloids lie between those of metals and nonmetals. This shouldn't surprise us since the metalloids are located between the metals and the nonmetals in the periodic table. In most respects, metalloids behave as nonmetals, both chemically and physically. However, in their most important physical property, electrical conductivity, they somewhat resemble metals. Metalloids tend to be **semiconductors**; they conduct electricity, but not nearly as well as metals. This property, particularly as found in silicon and germanium, is responsible for the remarkable progress made during the last six decades in the field of solid-state electronics. The operation of every computer, smart phone, LCD TV, tablet display, CD player, and AM-FM radio relies on transistors and integrated circuits made from semiconductors. Perhaps the most amazing advance of all has been the fantastic reduction in the size of electronic components that semiconductors have allowed (Figure 2.9). To it, we owe the development of small and versatile cell phones, cameras, flash drives, MP3 players, calculators, and computers. The heart of these devices is an integrated circuit that begins as a wafer of extremely pure silicon (or germanium) that is etched and chemically modified into specialized arrays of thousands of transistors.

Metallic and Nonmetallic Character

The occurrence of the metalloids between the metals and the nonmetals is our first example of trends in properties within the periodic table. We will frequently see that as we move from position to position across a period or down a group in the table, chemical and physical properties change in a gradual way. There are few abrupt changes in the characteristics of the elements as we scan across a period or down a group. The location of the metalloids can be seen, then, as an example of the gradual transition between metallic and nonmetallic properties. From left to right across Period 3, we go from aluminum, an element that has every appearance of a metal; to silicon, a semiconductor; to phosphorus, an element with clearly nonmetallic properties. A similar gradual change is seen going down Group 4A. Carbon is a nonmetal, silicon and germanium are metalloids, and tin and lead are metals. Trends such as these are useful to spot because they help us remember properties.

2.3 | Molecules and Chemical Formulas

In this section we begin describing how the atoms discussed in the previous sections can combine into chemical compounds. When describing these substances at the atomic level, it is frequently useful to use mental or graphic images of the way the individual atoms combine. Because we live in a three-dimensional world, atoms are often represented as spheres (or circles if you're drawing them by hand). Different colors are used to help differentiate atoms of one element from those of another. In most of the artwork in this book, the standard color scheme shown in Figure 2.10 and on the inside back cover will be followed. Scientists are not locked into this color scheme, however, and sometimes colors are chosen to emphasize some particular aspect of a substance. In such cases, the legend for the figure will identify which colors are associated with which elements.

Atoms of different elements also have different sizes, so often the size of the sphere used to represent an atom will be large for larger atoms and small for smaller atoms. This is done when we particularly wish to emphasize size difference among atoms. The sizes of the different spheres in Figure 2.10 roughly indicate the relative sizes of the atoms. As suggested by the following drawings, a chlorine atom is larger than an oxygen atom.

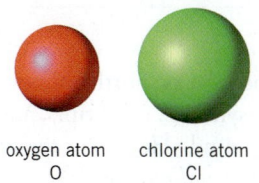

oxygen atom
O

chlorine atom
Cl

Molecules

Atoms combine in a variety of ways to form all of the more complex substances we find in nature and synthesize in the laboratory. One type of substance consists of discrete particles called **molecules**, each of which is made up of two or more atoms. Enormous numbers of different compounds as well as many elements exist in nature as molecules. Another type of compound (called an ionic compound) consists of electrically charged atoms and will be discussed in Section 2.5.

Chemical Formulas

To describe chemical substances, both elements and compounds, we commonly use **chemical formulas,** in which chemical symbols are used to represent atoms of the elements. For a **free element** (one that is not combined with another element in a compound) we often simply use the chemical symbol. Thus, the element sodium is represented by its symbol, Na, which is interpreted to mean one atom of sodium. Chemists often will use a symbol such as Na as another way of writing the word sodium.

Many of the elements we encounter frequently are found in nature as **diatomic molecules** (molecules composed of two atoms each). Among them are the gases hydrogen, oxygen, nitrogen, and chlorine. A **subscript** following the chemical symbol is used to indicate the number of atoms of an element in a molecule. Thus, the formula for molecular hydrogen is H_2, and those for oxygen, nitrogen, and chlorine are O_2, N_2, and Cl_2, respectively. Drawings of these molecules are shown in Figure 2.11. A more complete list of such elements is found in Table 2.1. This would be a good time to learn the formulas of these elements because you will come upon them often throughout the course.

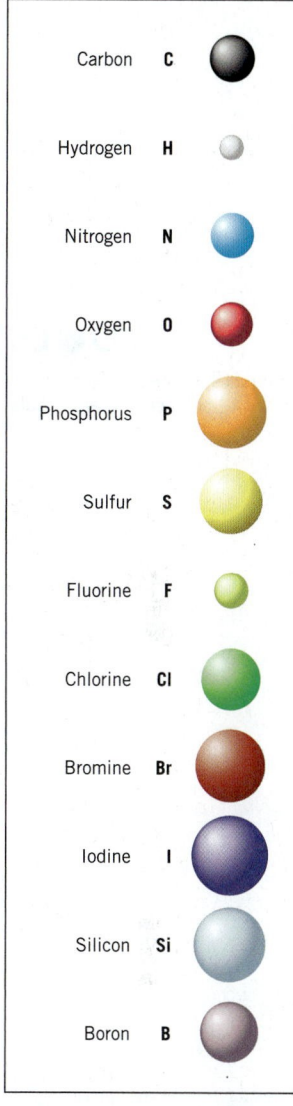

Carbon	C
Hydrogen	H
Nitrogen	N
Oxygen	O
Phosphorus	P
Sulfur	S
Fluorine	F
Chlorine	Cl
Bromine	Br
Iodine	I
Silicon	Si
Boron	B

Figure 2.10 | Colors used to represent atoms of different elements. The sizes of the spheres roughly illustrate the relative sizes of the different atoms. The atomic symbols of the elements are shown next to their names.

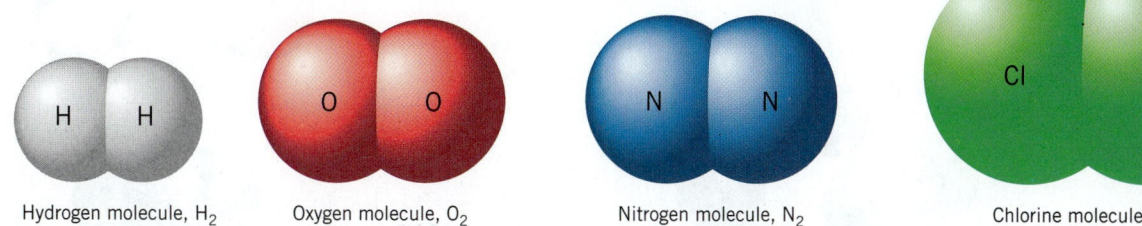

Hydrogen molecule, H_2

Oxygen molecule, O_2

Nitrogen molecule, N_2

Chlorine molecule, Cl_2

Figure 2.11 | Models that depict the diatomic molecules of hydrogen, oxygen, nitrogen, and chlorine. Each contains two atoms per molecule; their different sizes reflect differences in the sizes of the atoms that make up the molecules.

TABLE 2.1	Elements That Occur Naturally as Diatomic Molecules		
Hydrogen	H_2	Fluorine	F_2
Nitrogen	N_2	Chlorine	Cl_2
Oxygen	O_2	Bromine	Br_2
		Iodine	I_2

TOOLS

Chemical symbols and subscripts in a chemical formula

Just as chemical symbols can be used as shorthand notations for the names of elements, a chemical formula including the symbols, subscripts, and parentheses is a shorthand way of writing the name for a compound. However, *the most important characteristic of a formula is that it specifies the identity and number of each element in that substance.*

In the formula of a compound, each element is identified by its chemical symbol. Water, H_2O, consists of two atoms of hydrogen and one of oxygen. The lack of a subscript after the symbol O means that the formula specifies just one atom of oxygen. Another example is methane, CH_4 one of the combustible compounds found in "**natural gas**," which is used for cooking in the kitchen and Bunsen burners in the lab. The formula tells us that methane molecules are composed of one atom of carbon and the subscript indicates four atoms of hydrogen.

When sufficient information is available it is possible to represent how the atoms in a molecule are connected to each other and even the three-dimensional shape of the molecule. To show how the atoms are connected, we can use chemical symbols to represent the atoms and dashes to indicate the chemical bonds that bind the atoms to each other; this kind of figure is often called a structural formula.

$$H-O-H$$

$$H_2O$$

water

$$H-\overset{\displaystyle H}{\underset{\displaystyle H}{\overset{|}{\underset{|}{C}}}}-H$$

$$CH_4$$

methane

■ Molecular model kits can be purchased to construct your own models. These kits have color-coded "balls" with holes drilled so that when the "sticks" are added the molecules will have the correct geometry.

To understand the three-dimensional structure of molecules we can construct "ball-and-stick" models. For a **ball-and-stick model**, spheres representing atoms are connected by sticks that indicate the connections between the atoms. In a space-filling model, the relative sizes of the atoms and how they take up space in the molecule is shown. Figure 2.12 gives the ball-and-stick models of methane and chloroform. Figure 2.13 shows the more realistic **space-filling models** of water, methane, and chloroform.

For more complicated compounds, we sometimes find formulas that contain parentheses. An example is the formula for urea, $CO(NH_2)_2$, which tells us that the group

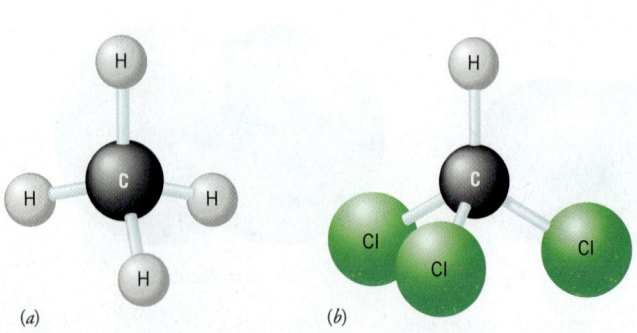

Figure 2.12 | **Ball-and-stick models** of (*a*) methane and (*b*) chloroform.

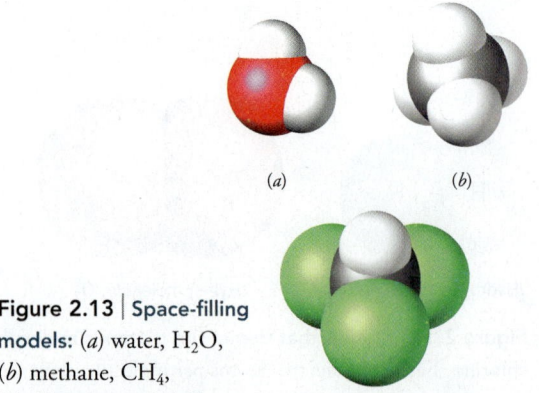

Figure 2.13 | **Space-filling models:** (*a*) water, H_2O, (*b*) methane, CH_4, (*c*) chloroform, $CHCl_3$.

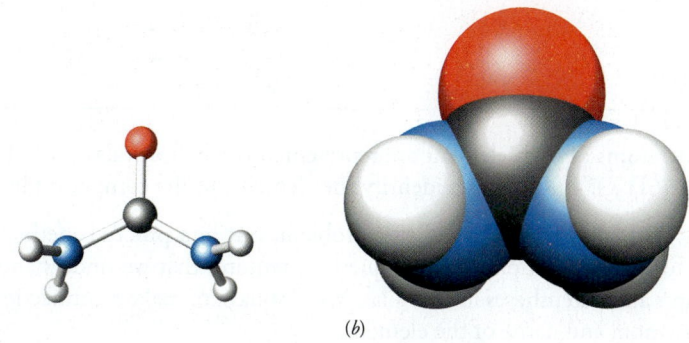

Figure 2.14 | Models of the urea molecule, $CO(NH_2)_2$:
(*a*) ball-and-stick model;
(*b*) space-filling model.

of atoms within the parentheses, NH_2, occurs twice. (The formula for urea could also be written as CON_2H_4, but there are good reasons for writing certain formulas with parentheses, as you will see later.) A ball-and-stick model and a space-filling model of urea are shown in Figure 2.14.

Hydrates: Crystals That Contain Water in Fixed Proportions

Certain compounds form crystals that contain water molecules. An example is ordinary plaster of Paris—the material often used to coat the interior walls of buildings. Plaster of Paris consists of crystals of a compound called calcium sulfate, $CaSO_4$, which contain two molecules of water for each $CaSO_4$. These water molecules are not held very tightly and can be driven off by heating the crystals. The dried crystals absorb water again if exposed to moisture, and the amount of water absorbed always gives crystals in which the H_2O-to-$CaSO_4$ ratio is 2-to-1. Compounds whose crystals contain water molecules in fixed ratios are called **hydrates**, and they are quite common. The formula for this hydrate of calcium sulfate is written $CaSO_4 \cdot 2H_2O$ to show that there are two molecules of water per $CaSO_4$. The raised dot is used to indicate that the water molecules are not bound very tightly in the crystal and can be removed.

Sometimes the dehydration (removal of water) of hydrate crystals produces changes in color. For example, copper sulfate forms bright blue crystals with the formula $CuSO_4 \cdot 5H_2O$ in which there are five water molecules for each $CuSO_4$. When the blue crystals are heated, most of the water is driven off and the solid that remains, now nearly pure $CuSO_4$, is almost white (Figure 2.15). If left exposed to the air, the $CuSO_4$ will absorb moisture and form blue $CuSO_4 \cdot 5H_2O$ again.

■ When all of the water is removed from a hydrate, it is said to be "anhydrous" or without water.

Counting Atoms in Formulas: A Necessary Skill

Counting the number of atoms of the elements in a chemical formula is a very important skill you will have to use correctly many times in the coming chapters. To count atoms we need to interpret the meaning of subscripts, parentheses, and perhaps coefficients if the compound is a hydrate. Let's look at how these principles are applied in the following example.

Richard Megna/Fundamental Photographs

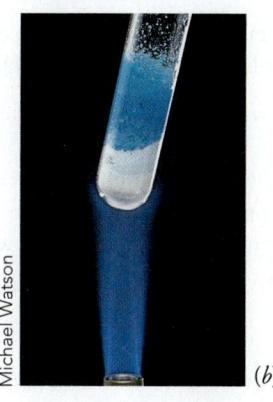
Michael Watson

Figure 2.15 | **Water can be driven from hydrates by heating.** (*a*) Blue crystals of copper sulfate pentahydrate, $CuSO_4 \cdot 5H_2O$, about to be heated. (*b*) The hydrate readily loses water when heated. The light-colored solid observed in the lower half of the test tube is pure $CuSO_4$.

Example 2.1
Counting Atoms in Formulas

How many atoms of each element are represented by the formulas (a) $(CH_3)_3COH$ and (b) $CoCl_2 \cdot 6H_2O$? In each case, identify the elements in the compound by name.

Analysis: This may not be a difficult problem, but let's proceed methodically just for practice. To answer both parts of this question requires that we understand the meaning of subscripts and parentheses in formulas. We also have to make a connection between the chemical symbol and name of the element.

Assembling the Tools: Here are the three tools we will use: (1) The subscript following a symbol indicates how many of that element are part of the formula; a subscript of 1 is implied if there is no subscript. (2) A quantity within parentheses is repeated a number of times equal to the subscript that follows. (3) A raised dot in a formula indicates the substance is a hydrate in which the number preceding H_2O specifies how many water molecules are present.

Solution: (a) For $(CH_3)_3COH$ we must recognize that all the atoms within the parentheses occur three times.

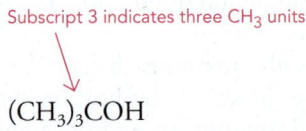

Subscript 3 indicates three CH_3 units

$$(CH_3)_3COH$$

Each CH_3 contains one C and three H atoms, so three of them contain three C and nine H atoms. In the COH unit, there is one additional C and one additional H, which gives a total of four C atoms and ten H atoms. The molecule also contains one O atom. Therefore, the formula $(CH_3)_3COH$ shows

$$4\,C \qquad 10\,H \qquad 1\,O$$

The elements in the compound are carbon (C), hydrogen (H), and oxygen (O).

(b) $CoCl_2 \cdot 6H_2O$ is the formula for a hydrate, as indicated by the raised dot. It contains six water molecules, each with two H and one O, for every $CoCl_2$.

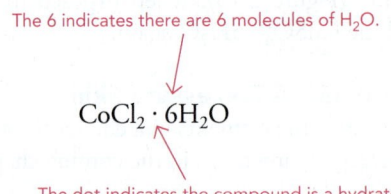

The 6 indicates there are 6 molecules of H_2O.

$$CoCl_2 \cdot 6H_2O$$

The dot indicates the compound is a hydrate.

Therefore, the formula $CoCl_2 \cdot 6H_2O$ represents

$$1\,Co \qquad 2\,Cl \qquad 12\,H \qquad 6\,O$$

Checking the table inside the front cover of the book, we see that the elements here are cobalt (Co), chlorine (Cl), hydrogen (H), and oxygen (O).

Are the Answers Reasonable? The only way to check the answer here is to perform a recount.

Practice Exercise 2.1[1]

How many atoms of each element are expressed by the following formulas? (*Hint:* Pay special attention to counting elements within parentheses.)

(a) SF_6 (b) $(C_2H_5)_2N_2H_2$ (c) $Ca_3(PO_4)_2$ (d) $Co(NO_3)_2 \cdot 6H_2O$

---

[1]Answers to the Practice Exercises are found in Appendix B at the back of the book.

What are the names and how many atoms of each element are present in each of the formulas?

(a) NH_4NO_3 (b) $FeNH_4(SO_4)_2$ (c) $Mo(NO_3)_2 \cdot 5H_2O$ (d) $C_6H_4ClNO_2$

Atoms, Molecules, and the Law of Definite Proportions

Let's now use our methods of representing atoms and molecules to understand how Dalton's atomic theory accounts for the law of definite proportions. According to the theory, all of the molecules of a compound are alike and contain atoms in the same numerical ratio. Thus all water molecules have the formula H_2O and contain two atoms of hydrogen and one of oxygen.

Today we know that oxygen atoms are much heavier than hydrogen atoms. In fact, one oxygen atom weighs 16 times as much as a hydrogen atom. We haven't said how much mass this is, so let's say that one hydrogen atom has a mass of one unit. On this scale an oxygen atom would then weigh 16 times as much, or 16 mass units.

1 H atom   1 mass unit
1 O atom   16 mass units

In Figure 2.16*a* we see one water molecule with two hydrogen atoms and one oxygen atom. The mass of oxygen in that molecule (16 mass units) is eight times the total mass of hydrogen (2 mass units). In Figure 2.16*b* we have five water molecules containing a total of 10 hydrogen atoms and 5 oxygen atoms. Notice that the total mass of oxygen in the five molecules is, once again, eight times the total mass of hydrogen. Regardless of the number of water molecules in the sample, the total mass of oxygen is eight times the total mass of hydrogen.

Recall from Section 0.4 that in any sample of pure water, the mass of oxygen present is always found experimentally to be eight times the mass of hydrogen—an example of how the law of definite proportions applies to water. Figure 2.16 explains *why* the law works in terms of the atomic theory.

The Law of Multiple Proportions

One of the real successes of Dalton's atomic theory was that it predicted another chemical law—one that had not been discovered yet. This law is called the law of multiple proportions and applies to atoms that are able to form two or more different compounds with each other.

Law of Multiple Proportions

Whenever two elements form more than one compound, the different masses of one element that combine with the same mass of the other element are in the ratio of small whole numbers.

2 H atoms 2 mass units	10 H atoms 10 mass units
1 O atom 16 mass units	5 O atoms 80 mass units
(a)	*(b)*

Figure 2.16 | Law of Definite Proportions. Regardless of the number of water molecules, the mass of oxygen is always eight times the mass of hydrogen.

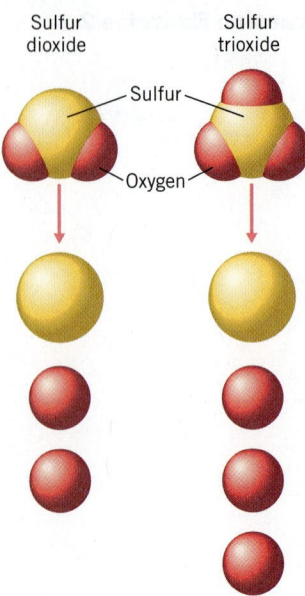

Sulfur dioxide · Sulfur trioxide

Sulfur

Oxygen

Figure 2.17 | **Oxygen compounds of sulfur** demonstrate the law of multiple proportions. Illustrated here are molecules of sulfur trioxide and sulfur dioxide. Each has one sulfur atom, and therefore the same mass of sulfur. The oxygen ratio is 2-to-3, both by atoms and by mass.

To see what this means, consider the two compounds sulfur dioxide, SO_2, and sulfur trioxide, SO_3, which are illustrated in Figure 2.17. In one molecule of each of these compounds there is one atom of sulfur, so both molecules must have the same mass of sulfur. Now let's focus on the oxygen. The SO_2 molecule has two O atoms; the SO_3 molecule has three O atoms. This means that the ratio of the atoms of O in the two compounds is 2-to-3.

$$\frac{\text{atoms of O in } SO_2}{\text{atoms of O in } SO_3} = \frac{2 \text{ atoms of O}}{3 \text{ atoms of O}} = \frac{2}{3}$$

Because all O atoms have the same mass, the *ratio* of the masses of O in the two molecules *must* be the same as the ratio of the atoms, and this ratio (⅔) is a ratio of small whole numbers.

Molecules Small and Large

So far we have discussed small molecules that have just a few atoms in each of them. Indeed, most molecules you will encounter in this book are considered to be small. However, nature presents us with some very large molecules as well, particularly in living organisms. For example, DNA is the molecule that contains the genetic code that differentiates one form of life from another. To do this a DNA molecule has millions of atoms woven into a very complex structure. A short segment of a DNA molecule is illustrated in Figure 2.18. We will say more about DNA in Chapter 22.

The Relationship between Atoms, Molecules, and the World We See

At this point you may begin to think that the formulas and shapes of molecules come to chemists mysteriously "out of the blue." This is hardly the case. When a chemical is first prepared or isolated from nature, its formula needs to be confirmed. The compound, for example, might be produced from an experiment in the form of a white powder, and there's nothing about its formula or the arrangement of the atoms within it that comes from the outward appearance of the substance. To acquire such knowledge chemists perform experiments, some of which will be described later in this book, that enable them to calculate what the formula of the substance is. Once a formula is known, we might speculate on the shape of the molecule, but a lot of work and very expensive and sophisticated instruments are required to know for sure. (If you take advanced courses in chemistry, it is likely you will get hands-on experience using such instruments.) It is important for you to understand that when we describe the formulas of compounds and the structures of molecules, such information is the culmination of the work of many scientists over many years.

Figure 2.18 | **Some molecules are extremely large.** Shown here is a short segment of a DNA molecule, the structure of which is responsible for the differences between the various species of living things on earth. An entire DNA molecule contains millions of atoms.

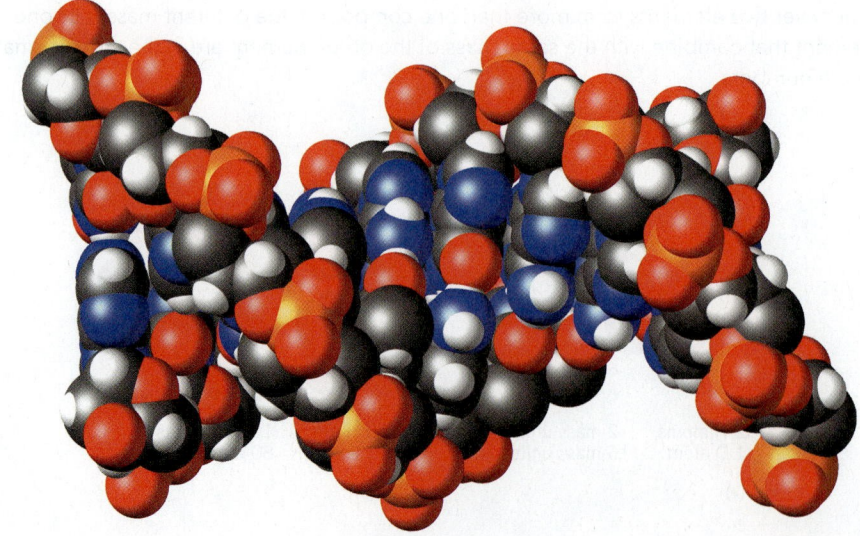

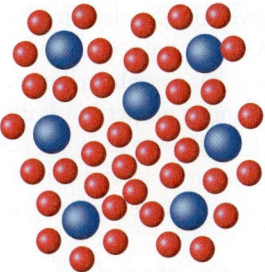

Figure 2.19 | A portion of a homogeneous mixture viewed at the atomic/molecular level. Red and blue spheres represent two different substances (not two different elements). One substance is uniformly distributed throughout the other.

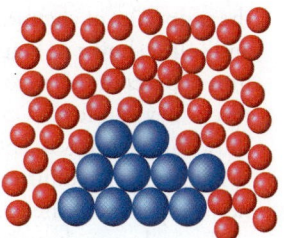

Figure 2.20 | A portion of a heterogeneous mixture viewed at the atomic/molecular level. Two substances exist in separate phases in a heterogeneous mixture.

Mixtures at the Atomic/Molecular Level

Earlier we noted that mixtures differ from elements and compounds in that mixtures can have variable compositions. Figure 2.19 illustrates this at the atomic/molecular level, where we have used different-color spheres to stand for two substances in a homogeneous mixture (also called a solution). Notice that the two substances are uniformly mixed. The difference between homogeneous and heterogeneous mixtures on the molecular level can be seen by comparing Figures 2.19 and 2.20.

2.4 | Chemical Reactions and Chemical Equations

Chemical reactions are at the heart of chemistry. When they occur, dramatic changes often occur among the chemicals involved. While chemical reactions are interesting to observe in the laboratory, they have an enormous number of applications in industry and ordinary everyday living. In fact, the chemical industry is a major part of the world economy. Some reactions occur rapidly and violently, such as the explosions used to demolish this large building (Figure 2.21). Other reactions are less violent. For example, you may use Clorox® as a bleach because the active ingredients react with stains in clothing and also destroy bacteria. These are just two examples; many others lie in the pages ahead.

Gamma-Rapho/Getty Images, Inc.

Figure 2.21 | Hundreds of strategically placed explosives cause a large building to collapse, in a controlled manner, in a matter of seconds.

Figure 2.22 | **The combustion of methane.** Gas-burning stoves that use natural gas (methane) as a fuel are common in many parts of the United States. The reaction consumes oxygen and produces carbon dioxide and water vapor.

TOOLS

Coefficients in an equation

To understand chemical reactions, we need to observe how they lead to changes among the properties of chemical substances. Consider, for example, the mixture of elements iron and sulfur shown in Figure 1.8 in Section 1.2. The sulfur has a bright yellow color, and the iron appears in this mixture as a black powder with magnetic properties. If these elements react, they can form a compound called iron sulfide (also known as "fools gold"), and as shown in Figure was 1.9, it doesn't look like either sulfur or iron, and it is not magnetic. When iron and sulfur combine chemically, the properties of the elements give way to the new properties of the compound.

To see how a chemical change occurs at the atomic level, let's study the combustion of methane, CH_4 (Figure 2.22). The reaction consumes CH_4 and oxygen (O_2, which is how oxygen occurs in nature) and forms carbon dioxide, CO_2, and water, H_2O. Figure 2.23 illustrates the reaction at the atomic/molecular level. For this reaction, CH_4 and O_2 are the **reactants**, the substances present before the reaction begins, and are shown on the left in Figure 2.23. On the right we see the **products** of the reaction, which are the molecules present after the reaction occurs. The arrow indicates that the reactants undergo the change to form the products.

Instead of drawing pictures to describe a chemical change, chemists use chemical symbols to describe reactions by writing chemical equations. A **chemical equation** uses chemical formulas to describe what happens when a chemical reaction occurs. As also shown in Figure 2.23, it describes the before-and-after picture of the chemical substances involved.

$$CH_4 + 2O_2 \longrightarrow CO_2 + 2H_2O \qquad (2.1)$$

The symbols stand for atoms of the elements involved. The numbers that precede O_2 and H_2O are called **coefficients**. In this equation, the coefficients tell us how many CH_4 and O_2 molecules react and how many CO_2 and H_2O molecules are formed. Note that when no coefficient is written in front of a formula, it is assumed to be equal to 1. The arrow in a chemical equation is read as "reacts to yield." Thus, this chemical equation would be read as *methane and oxygen react to yield carbon dioxide and water*.

Chemical Reactions and Conservation of Mass

The law of conservation of mass says that mass is neither created nor destroyed in a chemical reaction. In Figure 2.23, observe that before the reaction begins there are four H atoms, four O atoms, and one C atom among the reactants. They are found in one CH_4 molecule and two O_2 molecules. After the reaction is over, we still have the same number of atoms of each kind, but they have become rearranged into one CO_2 molecule and two H_2O molecules. Because atoms are neither lost nor created during the reaction, the total mass must remain the same. Thus, by applying the concept of atoms and the postulate of Dalton's atomic theory that says atoms simply rearrange during a chemical reaction, we've accounted for the law of conservation of mass.

Coefficients in Equations and the Law of Conservation of Mass

All chemical reactions obey the law of conservation of mass, which means that in any reaction the total numbers of atoms of each kind before and after a reaction are the same. When we write chemical equations we adjust the numbers of molecules on each side of the

Figure 2.23 | **The reaction of methane,** CH_4, with oxygen, O_2 to give carbon dioxide, CO_2, and water, viewed at the atomic-molecular level. On the left are methane and oxygen molecules before reaction, and on the right are the carbon dioxide and water molecules that are present after the reaction is complete. Below the drawings is the chemical equation for the reaction.

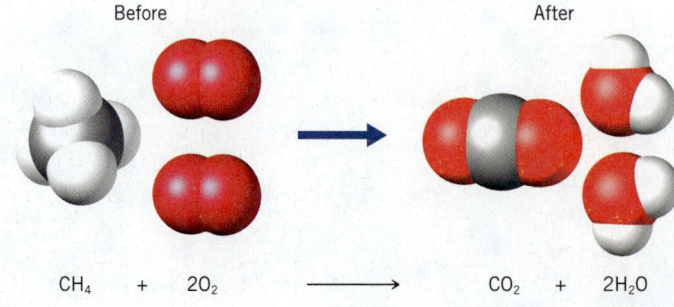

arrow to make the equation conform to this principle. We say we balance the equation, and we accomplish this by adjusting the coefficients in front of reactant and product molecules. In the equation for the combustion of methane, the coefficients in front of O_2 on the left and H_2O on the right make this a **balanced equation**.

Let's look at another example, the combustion of butane, C_4H_{10}, the fuel in disposable lighters (Figure 2.24).

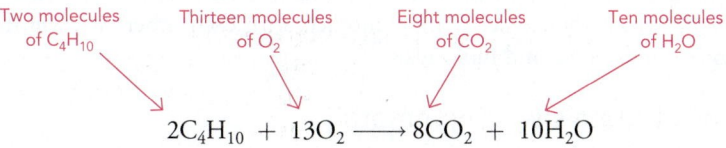

Two molecules of C_4H_{10} Thirteen molecules of O_2 Eight molecules of CO_2 Ten molecules of H_2O

$$2C_4H_{10} + 13O_2 \longrightarrow 8CO_2 + 10H_2O$$

The 2 before the C_4H_{10} tells us that two molecules of butane react. This involves a total of 8 carbon atoms and 20 hydrogen atoms, as we see in Figure 2.25. Notice that we have multiplied the numbers of atoms of C and H in one molecule of C_4H_{10} by the coefficient 2. On the right we find 8 molecules of CO_2, which contain a total of 8 carbon atoms. Similarly, 10 water molecules contain 20 hydrogen atoms. Finally, there are 26 oxygen atoms on both sides of the equation. You will learn to balance equations such as this in Chapter 3. For now, however, you should be able to determine when an equation is balanced and when it is not.

Example 2.2
Determining Whether an Equation Is Balanced

Determine whether or not the following chemical equations are balanced. Support your conclusions by writing how many of each element is on either side of the arrow.

(a) $Fe(OH)_3 + 2HNO_3 \longrightarrow Fe(NO_3)_3 + 2H_2O$

(b) $BaCl_2 + H_2SO_4 \longrightarrow BaSO_4 + 2HCl$

(c) $C_6H_{12}O_6 + 6O_2 \longrightarrow 6CO_2 + 6H_2O$

Analysis: The statement of the problem asks whether the equations are balanced. You can check that an equation is balanced if each element has the same number of atoms on either side of the arrow.

Assembling the Tools: The meaning of subscripts and parentheses in formulas as well as the meaning of coefficients in front of formulas are the tools we will use to count atoms. Note that in two of the equations, oxygen appears in both reactants and both products. We must be sure to count all of the atoms.

Solution: (a) Reactants: 1 Fe, 9 O, 5 H, 2 N; Products: 1 Fe, 11 O, 4 H, 3 N. Only Fe has the same number of atoms on either side of the arrow. This equation *is not* balanced.

(b) Reactants: 1 Ba, 2 Cl, 2 H, 1 S, 4 O; Products: 1 Ba, 2 Cl, 2 H, 1 S, 4 O. This equation *is* balanced.

(c) Reactants: 6 C, 12 H, 18 O; Products: 6 C, 12 H, 18 O. This equation *is* balanced.

Is the Answer Reasonable? The appropriate way to check this is to recount the atoms. Try counting the atoms in the reverse direction this time.

How many atoms of each element appear on each side of the arrow in the following equation? Is this reaction balanced? (*Hint:* Recall that coefficients multiply the elements in the entire formula.)

$$4NH_3 + 3O_2 \longrightarrow 2N_2 + 6H_2O$$

Figure 2.24 | The combustion of butane, C_4H_{10}. Several commercial products, such as the cigarette lighter and the BBQ lighter shown here, burn butane, producing water and carbon dioxide.

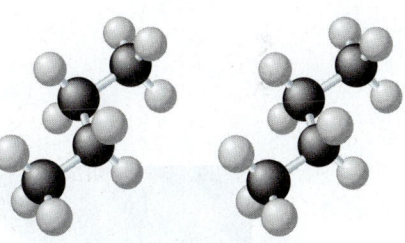

Two molecules of butane contain 8 atoms of C and 20 atoms of H.

Figure 2.25 | Understanding coefficients in an equation. The expression $2C_4H_{10}$ describes two molecules of butane, each of which contains 4 carbon and 10 hydrogen atoms. This gives a total of 8 carbon and 20 hydrogen atoms.

Practice Exercise 2.3

Practice Exercise 2.4

Is this equation balanced?

$$2(NH_4)_3PO_4 + 3Ba(C_2H_3O_2)_2 \longrightarrow Ba_3(PO_4)_2 + 6NH_4C_2H_3O_2$$

2.5 | Ionic Compounds

Most of the substances that we encounter on a daily basis are not free elements but are compounds in which the elements are combined with each other. We will discuss two types of compounds: ionic and molecular.

Reactions of Metals with Nonmetals

Under appropriate conditions, atoms are able to transfer electrons between one another when they react to yield electrically charged particles called **ions**. This is what happens, for example, when the metal sodium combines with the nonmetal chlorine. As shown in Figure 2.26, when sodium, a typical shiny metal, and chlorine, a pale green gas, are mixed, a vigorous reaction takes place yielding a white powder, sodium chloride. The equation for the reaction is

$$2Na(s) + Cl_2(g) \longrightarrow 2NaCl(s)$$

The changes that take place at the atomic level are also illustrated in Figure 2.26.

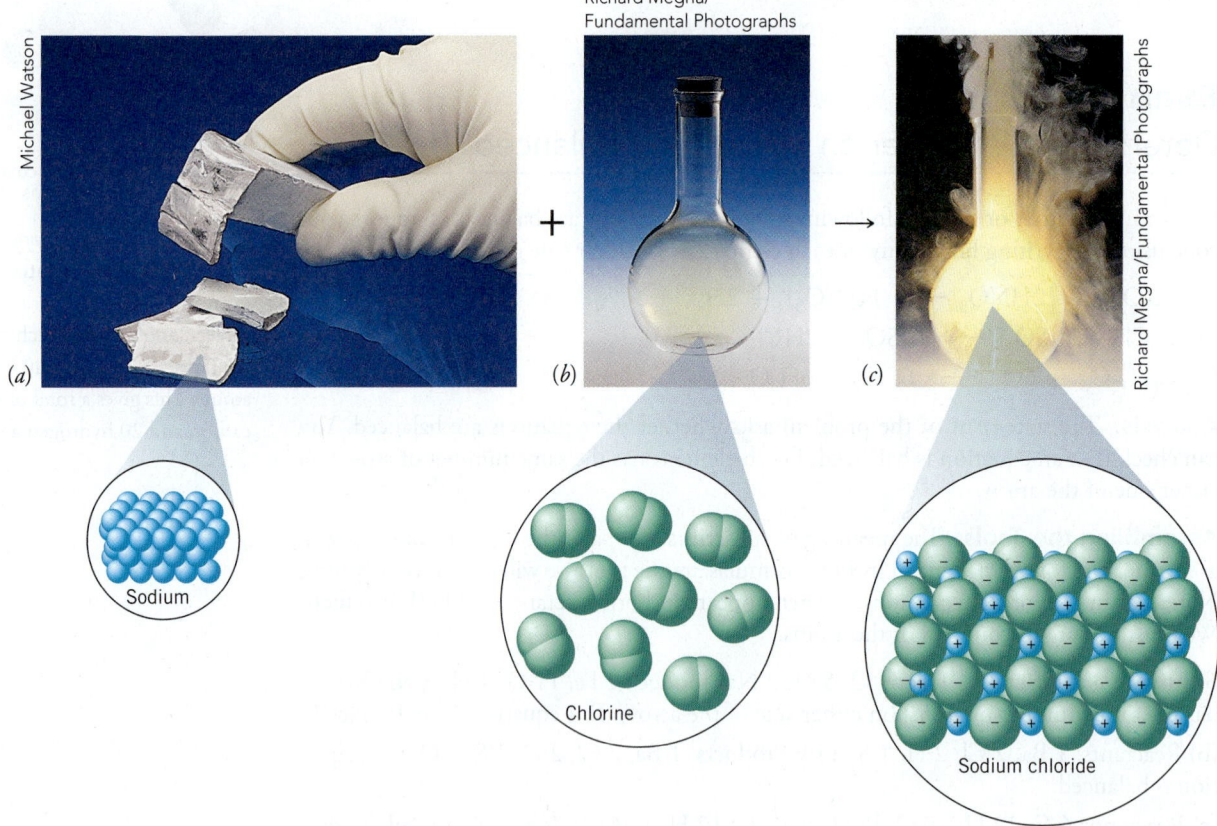

Richard Megna/
Fundamental Photographs

Michael Watson

Richard Megna/Fundamental Photographs

(*a*)   +   (*b*)   →   (*c*)

Sodium

Chlorine

Sodium chloride

Figure 2.26 | **Sodium reacts with chlorine** to give the ionic compound sodium chloride, with the reaction viewed at the atomic level. (*a*) Freshly cut sodium has a shiny metallic surface. The metal reacts with oxygen and moisture, so it cannot be touched with bare fingers. (*b*) Chlorine is a pale green gas. (*c*) When a small piece of sodium is melted in a metal spoon and thrust into the flask of chlorine, it burns brightly as the two elements react to form sodium chloride. The smoke coming from the flask is composed of fine crystals of salt. The electrically neutral atoms and molecules react to yield positive and negative ions, which are held to each other by electrostatic attractions (attractions between opposite electrical charges).

The formation of the ions in sodium chloride results from the transfer of electrons between the reacting atoms. Specifically, each sodium atom gives up one electron to a chlorine atom. We can diagram the changes in equation form by using the symbol e^- to stand for an electron.

$$\overset{\overset{\displaystyle e^-}{\frown}}{Na + Cl} \longrightarrow Na^+ + Cl^-$$

The electrically charged particles formed in this reaction are a sodium ion (Na^+) and a chloride ion (Cl^-). The sodium ion has a positive 1+ charge, indicated by the superscript plus sign, because the loss of an electron leaves it with one more proton in its nucleus than there are electrons outside. Similarly, by gaining one electron the chlorine atom has added one more negative charge, so the chloride ion has a single negative charge indicated by the minus sign. Solid sodium chloride is composed of these charged sodium and chloride ions and is said to be an **ionic compound**.

As a general rule, *ionic compounds are formed when metals react with nonmetals.* In the electron transfer, however, not all atoms gain or lose just one electron; some gain or lose more. For example, when calcium atoms react they lose two electrons to form Ca^{2+} ions, and when oxygen atoms form ions they each gain two electrons to give O^{2-} ions. Notice that in writing the formulas for ions, the number of positive or negative charges is indicated by a superscript before the positive or negative charge. (We will have to wait until a later chapter to study the reasons why certain atoms gain or lose one electron each, whereas other atoms gain or lose two or more electrons.)

For each of the following atoms or ions, give the number of protons and the number of electrons in one particle: (a) an Fe atom, (b) an Fe^{3+} ion, (c) an N^{3-} ion, (d) an N atom. (*Hint:* Recall that electrons have a negative charge, and ions that have a negative charge must have gained electrons.)

For each of the following atoms or ions, give the number of protons and the number of electrons in one particle: (a) an O atom, (b) an O^{2-} ion, (c) an Al^{3+} ion, (d) an Al atom.

Looking at the structure of sodium chloride in Figure 2.26, it is impossible to say that a particular Na^+ ion belongs to a particular Cl^- ion. The ions in a crystal of NaCl are simply packed in the most efficient way, so that positive ions and negative ions can be as close to each other as possible. In this way, the attractions between oppositely charged ions, which are responsible for holding the compound together, can be as strong as possible.

Since discrete units don't exist in ionic compounds, the subscripts in their formulas are always chosen to specify the smallest whole-number ratio of the ions. This is why the formula of sodium chloride is given as NaCl rather than Na_2Cl_2 or Na_3Cl_3. The idea of a "smallest unit" of an ionic compound is still quite often useful. Therefore, we take the smallest unit of an ionic compound to be whatever is represented in its formula and call this unit a **formula unit**. Thus, one formula unit of NaCl consists of one Na^+ and one Cl^-, whereas one formula unit of the ionic compound $CaCl_2$ consists of one Ca^{2+} and two Cl^- ions. (In a broader sense, we can use the term formula unit to refer to whatever is represented by a formula. Sometimes the formula specifies a set of ions, as in NaCl; sometimes it is a molecule, as in O_2 or H_2O; sometimes it can be just an ion, as in Cl^- or Ca^{2+}; and sometimes it might be just an atom, as in Na.)

Experimental Evidence Exists for Ions in Compounds

We know that metals conduct electricity because electrons can move from one atom to the next in a wire when connected to a battery. Solid ionic compounds are poor conductors of electricity as are other substances such as water. However, if an ionic compound is dissolved in water or is heated to a high temperature so that it melts, the resulting liquids are able to conduct electricity easily. These observations suggest that ionic compounds are composed

■ Here we are concentrating on what happens to the individual atoms, so we have not shown chlorine as diatomic Cl_2 molecules.

Ionic compounds

Practice Exercise 2.5

Practice Exercise 2.6

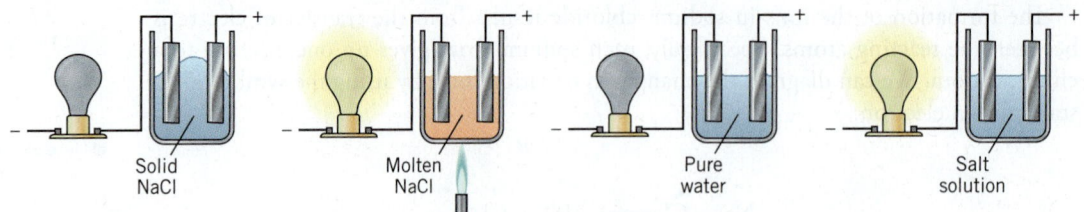

Figure 2.27 | **An apparatus to test for electrical conductivity.** The electrodes are dipped into the substance to be tested. If the light bulb glows when electricity is applied, the sample is an electrical conductor. Here we see that solid sodium chloride does not conduct electricity, but when the solid is melted it does conduct. Liquid water, a molecular compound, is not a conductor of electricity because it does not contain electrically charged particles. However, when the salt is dissolved in water, the solution conducts electricity.

of charged ions rather than neutral molecules and that these ions when made mobile by dissolving or melting can conduct electricity. Figure 2.27 illustrates how the electrical conductivity can be tested.

Formulas of Ionic Compounds

We have noted that metals combine with nonmetals to form ionic compounds. In such reactions, metal atoms lose one or more electrons to become positively charged ions and nonmetal atoms gain one or more electrons to become negatively charged ions. In referring to these particles, a positively charged ion is called a **cation** (pronounced CAT-I-on) and a negatively charged ion is called an **anion** (pronounced AN-I-on). Thus, solid NaCl is composed of sodium cations and chloride anions.

Ions of Representative Metals and Nonmetals

The periodic table can help us remember the kinds of ions formed by many of the representative elements (elements in the A groups of the periodic table). For example, except for hydrogen, the neutral atoms of the Group 1A elements always lose one electron each when they react, thereby becoming ions with a charge of 1+. Similarly, atoms of the Group 2A elements always lose two electrons when they react, so these elements always form ions with a charge of 2+. In Group 3A, the only important positive ion we need consider now is that of aluminum, Al^{3+}; an aluminum atom loses three electrons when it reacts to form this ion.

All these ions are listed in Table 2.2. *Notice that the number of positive charges on each of the cations is the same as the group number when we use the North American numbering of the groups in the periodic table.* Thus, sodium is in Group 1A and forms an ion with a 1+ charge, barium (Ba) is in Group 2A and forms an ion with a 2+ charge, and aluminum is in Group 3A and forms an ion with a 3+ charge. Although this generalization doesn't work for all the metallic elements (for example, the transition elements), it does help us remember what happens to the metallic elements of Groups 1A and 2A and aluminum when they react.

■ While hydrogen is a nonmetal, it usually loses an electron to form a positive ion. However, in certain compounds it can gain an electron.

TO**O**LS

Predicting cation charge

TABLE 2.2	Ions Formed from the Representative Elements					
Group Number						
1A	**2A**	**3A**	**4A**	**5A**	**6A**	**7A**
H^+						
Li^+	Be^{2+}		C^{4-}	N^{3-}	O^{2-}	F^-
Na^+	Mg^{2+}	Al^{3+}	Si^{4-}	P^{3-}	S^{2-}	Cl^-
K^+	Ca^{2+}				Se^{2-}	Br^-
Rb^+	Sr^{2+}				Te^{2-}	I^-
Cs^+	Ba^{2+}					

Among the nonmetals on the right side of the periodic table we also find some useful generalizations. For example, when they combine with metals, the halogens (Group 7A) form ions with one negative charge (written as 1−) and the nonmetals in Group 6A form ions with two negative charges (written as 2−). *Notice that the number of negative charges on the anion is equal to the number of spaces to the right that we have to move in the periodic table to get to a noble gas.*

TOOLS
Prediction anion charge

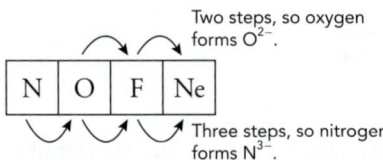

Writing Formulas for Ionic Compounds

All chemical compounds are electrically neutral, so the ions in an ionic compound always occur in a ratio such that the total positive charge is equal to the total negative charge. This is why the formula for sodium chloride is NaCl; the l-to-l ratio of Na^+ to Cl^- gives electrical neutrality. In addition, as we've already mentioned, discrete molecules do not exist in ionic compounds, so we always use the smallest set of subscripts that specify the correct ratio of the ions. The following, therefore, are the rules we use in writing the formulas of ionic compounds.

■ Electrical neutrality means that the net charge is zero. If the total positive charges equal the total negative charges the compound will be neutral.

Rules for Writing Formulas of Ionic Compounds

1. The positive ion is given first in the formula. (This isn't required by nature, but it is a custom we always follow.)
2. The subscripts in the formula must produce an electrically neutral formula unit. (Nature does require electrical neutrality.)
3. The subscripts should be the smallest set of whole numbers possible. For instance, if all subscripts are even, divide them by 2. (You may have to repeat this simplification step several times.)
4. The charges on the ions are not included in the finished formula for the substance.
5. When a subscript is 1 it is left off; no subscript implies a subscript of 1.

TOOLS
Formulas for ionic compounds

Example 2.3
Writing Formulas for Ionic Compounds

Write the formulas for the ionic compounds formed from (a) Ba and S, (b) Al and Cl, and (c) Al and O.

Analysis: We need to determine the charges on the cation and the anion and then find how many of each will give a net charge of zero. When writing the formula the cation will always be written first.

Assembling the Tools: First, we need the tool to figure out the charges of the ions from the periodic table. Then we apply the tool that summarizes the rules for writing the formula of ionic compounds.

Solution: (a) The element Ba is in Group 2A, so the charge on its ion is 2+. Sulfur is in Group 6A, so its ion has a charge of 2−. Therefore, the ions are Ba^{2+} and S^{2-}. Since the charges are equal but opposite, a 1-to-1 ratio will give a neutral formula unit. Therefore, writing the Ba first, the formula is BaS. Notice that we have *not* included the charges on the ions in the finished formula.

(b) By using the periodic table, the ions of these elements are Al^{3+} and Cl^-. We can obtain a neutral formula unit by combining one Al^{3+} with three Cl^-. (The charge on Cl is 1−; the 1 is understood.)

$$1(3+) + 3(1-) = 0$$

The cation comes first so the formula is $AlCl_3$.

(c) For these elements, the ions are Al^{3+} and O^{2-}. In the formula we seek there must be the same number of positive charges as negative charges. This number must be a whole-number multiple of both 3 and 2. The smallest number that satisfies this condition is 6, so there must be two Al^{3+} and three O^{2-} in the formula.

$$
\begin{array}{ll}
2Al^{3+} & 2(3+) = 6+ \\
3O^{2-} & \underline{3(2-) = 6-} \\
& sum = 0
\end{array}
$$

The formula is Al_2O_3.

A method that you may have seen before uses the *number* of positive charges for the subscript of the anion and the *number* of negative charges as the subscript for the cation, as shown in this diagram.

$$Al\,\text{③}+ \quad \times \quad O\,\text{②}-$$

In the end, always be sure to check that the subscripts cannot be reduced to smaller numbers.

Practice Exercise 2.7

Write formulas for ionic compounds formed from (a) Na and F, (b) Na and O, (c) Mg and F, and (d) Al and C. (*Hint:* One element must form a cation, and the other will form an anion based on its position in the periodic table.)

Practice Exercise 2.8

Write the formulas for the compounds made from (a) Ca and N, (b) Al and Br, (c) K and S, (d) Cs and Cl.

Many of our most important chemicals are ionic compounds. We have mentioned NaCl, common table salt, and $CaCl_2$, which is a substance often used to melt ice on walkways in the winter. Other examples are sodium fluoride, NaF, used by dentists to give fluoride treatments to teeth, and calcium oxide, CaO, an important ingredient in cement.

Cations of Transition and Post-transition Metals

The transition elements are located in the center of the periodic table, from Group 3B on the left to Group 2B on the right (Groups 3 to 12 if using the IUPAC system). All of them lie to the left of the metalloids, and they all are metals. Included here are some of our most familiar metals, including iron, chromium, copper, silver, and gold.

Most of the transition metals are much less reactive than the metals of Groups 1A and 2A, but when they do react they also transfer electrons to nonmetal atoms to form ionic compounds. However, the charges on the ions of the transition metals do not follow as straightforward a pattern as do those of the alkali and alkaline earth metals. One of the characteristic features of the transition metals is the ability of many of them to form more than one positive ion. Iron, for example, can form two different ions, Fe^{2+} and Fe^{3+}. This means that iron can form more than one compound with a given nonmetal. For example, iron forms two compounds containing chloride ions with the formulas $FeCl_2$ and $FeCl_3$. With oxygen, we find the compounds FeO and Fe_2O_3. As usual, we see that the formulas contain the ions in a ratio that gives electrical neutrality. Some of the most common ions of the transition metals are given in Table 2.3. Notice that one of the ions of mercury is diatomic Hg_2^{2+}. It consists of two Hg ions joined by the same kind of bond found in molecular substances. The simple Hg^+ ion does not exist.

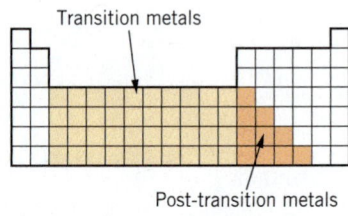
Transition metals

Post-transition metals

Distribution of transition and post-transition metals in the periodic table.

Practice Exercise 2.9

Write all the formulas for the chlorides and oxides formed by (a) chromium and (b) copper. (*Hint:* There are more than one chloride and one oxide for each of these transition metals.)

Practice Exercise 2.10

Write the formulas for the sulfides and nitrides of (a) gold and (b) titanium.

The **post-transition metals** are those metals that occur in the periodic table immediately following the transition metals. The two most common and important ones are tin (Sn) and lead (Pb). Except for bismuth, post-transition metals have the ability to form two different ions and therefore two different compounds with a given nonmetal. For example,

TABLE 2.3 Ions of Some Transition Metals and Post-transition Metals

Transition Metals		Transition Metals	
Titanium	Ti^{2+}, Ti^{3+}, Ti^{4+}	Silver	Ag^+
Chromium	Cr^{2+}, Cr^{3+}	Cadmium	Cd^{2+}
Manganese	Mn^{2+}, Mn^{3+}	Gold	Au^+, Au^{3+}
Iron	Fe^{2+}, Fe^{3+}	Mercury	Hg_2^{2+}, Hg^{2+}
Cobalt	Co^{2+}, Co^{3+}	**Post-transition Metals**	
Nickel	Ni^{2+}	Tin	Sn^{2+}, Sn^{4+}
Copper	Cu^+, Cu^{2+}	Lead	Pb^{2+}, Pb^{4+}
Zinc	Zn^{2+}	Bismuth	Bi^{3+}

tin forms two oxides, SnO and SnO_2. Lead also forms two oxides that have similar formulas (PbO and PbO_2). The ions that these metals form are also included in Table 2.3.

Compounds Containing Polyatomic Ions

The ionic compounds that we have discussed so far have been **binary compounds**—compounds formed from just two different elements. There are many other ionic compounds that contain more than two elements. These substances usually contain **polyatomic ions**, which are ions that are themselves composed of two or more atoms linked by the same kinds of bonds that hold molecules together. Polyatomic ions differ from molecules, however, in that they contain either too many or too few electrons to make them electrically neutral. Table 2.4 lists some important polyatomic ions. These ions will be used often in the coming chapters so it is important to learn their formulas, charges, and names.

The formulas of compounds formed from polyatomic ions are determined in the same way as are those of binary ionic compounds: the ratio of the ions must be such that the formula unit is electrically neutral, and the smallest set of whole-number subscripts is used. One difference in writing formulas with polyatomic ions is that parentheses are needed around the polyatomic ion if a subscript is required.

TOOLS
Polyatomic ions

TABLE 2.4 Formulas and Names of Some Polyatomic Ions

Ion	Name (Alternate name in parentheses)	Ion	Name (Alternate name in parentheses)
NH_4^+	Ammonium ion	CO_3^{2-}	Carbonate ion
Hg_2^{2+}	Mercury(I) ion	HCO_3^-	Hydrogen carbonate ion (bicarbonate ion)[b]
H_3O^+	Hydronium ion[a]	SO_3^{2-}	Sulfite ion
OH^-	Hydroxide ion	HSO_3^-	Hydrogen sulfite ion (bisulfite ion)[b]
CN^-	Cyanide ion	SO_4^{2-}	Sulfate ion
NO_2^-	Nitrite ion	HSO_4^-	Hydrogen sulfate ion (bisulfate ion)[b]
NO_3^-	Nitrate ion	SCN^-	Thiocyanate ion
ClO^- or OCl^-	Hypochlorite ion	$S_2O_3^{2-}$	Thiosulfate ion
ClO_2^-	Chlorite ion	CrO_4^{2-}	Chromate ion
ClO_3^-	Chlorate ion	$Cr_2O_7^{2-}$	Dichromate ion
ClO_4^-	Perchlorate ion	PO_4^{3-}	Phosphate ion
MnO_4^-	Permanganate ion	HPO_4^{2-}	Monohydrogen phosphate ion
$C_2H_3O_2^-$	Acetate ion	$H_2PO_4^-$	Dihydrogen phosphate ion
$C_2O_4^{2-}$	Oxalate ion		

[a]You will only encounter this ion in aqueous solutions.

[b]You will often see and hear the alternate names for these ions.

Example 2.4
Formulas That Contain Polyatomic Ions

One of the minerals responsible for the strength of bones is the ionic compound calcium phosphate, which is formed from Ca^{2+} and PO_4^{3-}. Write the formula for this compound.

Analysis: The cation and anion are given along with their charges, and the formula needs to be electrically neutral. We need to be prepared to use parentheses if more than one polyatomic ion is needed.

Assembling the Tools: The essential tool for solving this problem is to follow the rules for determining the subscripts needed to write the formula.

Solution: It may be immediately obvious that we need three calcium ions and two phosphate ions to attain electrical neutrality. If it isn't obvious, we can use the method mentioned at the end of Example 2.3 to determine these subscripts.

$$Ca^{②+} \diagdown PO_4^{③-}$$

Since the phosphate ion contains two elements, parentheses are needed to include the entire ion and show that the PO_4^{3-} ion occurs two times in the formula unit.

$$Ca_3(PO_4)_2$$

Is the Answer Reasonable? We double-check to see that electrical neutrality is achieved for the compound. We have six positive charges from the three Ca^{2+} ions and six negative ions. The sum is zero, and our compound is electrically neutral as required.

Practice Exercise 2.11 Write the formula for the ionic compound formed from (a) Na^+ and CO_3^{2-} and (b) NH_4^+ and SO_4^{2-}. (*Hint:* Compounds containing polyatomic ions must also be electrically neutral.)

Practice Exercise 2.12 Write the formula for the ionic compound formed from (a) potassium and acetate ions, (b) strontium and nitrate ions, and (c) Fe^{3+} and acetate ions.

Polyatomic ions are found in a large number of very important compounds. Examples include $CaSO_4$ (calcium sulfate, found in plaster of Paris or gypsum), $NaHCO_3$ (sodium bicarbonate, also called baking soda), $NaOCl$ (sodium hypochlorite, in liquid laundry bleach), $NaNO_2$ (sodium nitrite, a meat preservative), $MgSO_4$ (magnesium sulfate, also known as Epsom salts), and $NH_4H_2PO_4$ (ammonium dihydrogen phosphate, a fertilizer).

2.6 | Nomenclature of Ionic Compounds

In conversation, chemists rarely use formulas to describe compounds. Instead, names are used. For example, you already know that water is the name for the compound having the formula H_2O and that sodium chloride is the name of $NaCl$.

At one time there was no uniform procedure for assigning names to compounds, and those who discovered compounds used whatever method they wished. Today, we know of more than 50 million different chemical compounds,[2] so it is necessary to have a

---

[2]CAS Registry, the most comprehensive and high-quality compendium of publicly disclosed chemical information, now includes 70 million compounds!

TABLE 2.5	Monatomic Negative Ions								
H^-	Hydride	C^{4-}	Carbide	N^{3-}	Nitride	O^{2-}	Oxide	F^-	Fluoride
		Si^{4-}	Silicide	P^{3-}	Phosphide	S^{2-}	Sulfide	Cl^-	Chloride
				As^{3-}	Arsenide	Se^{2-}	Selenide	Br^-	Bromide
				Te^{2-}	Telluride	I^-	Iodide		

TOOLS

Monatomic anion names

logical system for naming them. Chemists around the world now agree on a systematic method for naming substances that is overseen by the IUPAC. By using basic methods we are able to write the correct formula given the name for the many compounds we will encounter. Additionally, we will be able to take a formula and correctly name it, since up to this point we have used common names for substances. In addition, when we first name a compound in this book, we will give the IUPAC name first, followed by the common name, if there is one, in parentheses. We will subsequently use the common name.

Naming Ionic Compounds of Representative Elements

In this section we discuss the **nomenclature** (naming) of simple inorganic ionic compounds. In general, **inorganic compounds** are substances that are not derived from hydrocarbons such as methane (CH_4), ethane (C_2H_6), and other carbon–hydrogen compounds. In naming ionic compounds, our goal is a name that someone else could use to write the formula.

For ionic compounds, the name of the cation is given first, followed by the name of the anion. This is the same as the sequence in which the ions appear in the formula. If the metal in the compound forms only one cation, such as Na^+ or Ca^{2+}, the cation takes the English name of the metal. The anion in a binary compound is formed from a nonmetal and its name is created by adding the suffix *-ide* to the stem of the name for the nonmetal. An example is KBr, potassium bromide. Table 2.5 lists some common **monatomic** (one-atom) negative ions and their names. It is also useful to know that the *-ide* suffix is usually used only for monatomic ions, with just two common exceptions—the *hydroxide ion* (OH^-) and the *cyanide ion* (CN^-).

To form the name of an ionic compound, we simply specify the names of the cation and anion. We do not need to state how many cations or anions are present, since once we know what the ions are we can assemble the correct formula as we just did in the previous section.

TOOLS

Naming ionic compounds

Example 2.5
Naming Compounds and Writing Formulas

(a) What is the name of $SrBr_2$? (b) What is the formula for aluminum selenide?

Analysis: Both compounds are ionic, and we will name the first one using the names of the elements with the appropriate ending for the anion. For the second compound, we will write the formula using the concept of electrical neutrality.

Assembling the Tools: We will use the tool for naming ionic compounds and the concept that ionic compounds must be electrically neutral.

Solution: (a) The compound $SrBr_2$ is composed of the elements Sr and Br. Sr is a metal from Group 2A, and Br is a nonmetal from Group 7A. Compounds of a metal and nonmetal are ionic, so we use the rules for naming ionic compounds. The cation simply takes the name of the metal, which is strontium. The anion's name is derived from bromine by replacing *-ine* with *-ide*; it is the bromide ion. The name of the compound is strontium bromide.

(b) Aluminum is a metal from Group 3A and forms the cation Al^{3+}. The *-ide* ending of selenide suggests the anion is composed of a single atom of a nonmetal. The only one that begins with the letters "selen-" is selenium, Se. (See the table inside the front cover.) The anion that is formed from selenium (Group 6A) is Se^{2-}. Since the correct formula must represent an electrically neutral formula unit, we use the number of charges on one ion as the subscript of the other—the formula is Al_2Se_3.

Are the Answers Reasonable? The easiest check is to reverse the process to see if the names you came up with will result in the formulas you started with and vice versa. Using our rules for strontium bromide we get $SrBr_2$. Similarly, our rules for Al_2Se_3 result in the name aluminum selenide.

Practice Exercise 2.13

Give the correct formulas for (a) potassium sulfide, (b) barium bromide, (c) sodium cyanide, (d) aluminum hydroxide, and (e) calcium phosphide. (*Hint:* Recall what the ending *-ide* means.)

Practice Exercise 2.14

Give the correct names for (a) $AlCl_3$, (b) $Ba(OH)_2$, (c) NaBr, (d) CaF_2, and (e) K_3P.

Naming Cations of Transition Metals

Earlier we learned that many of the transition metals and post-transition metals are able to form more than one positive ion. Our naming process must be able to clearly identify which of the possible cations is present in a compound.

TOOLS

Using the Stock system

The currently preferred method for naming ions of metals, which can have more than one charge in compounds, is called the **Stock system**. Here we use the English name followed, *without a space*, by the numerical value of the charge written as a Roman numeral in parentheses.[3] Examples using the Stock system are shown below.

Fe^{2+}	iron(II)		$FeCl_2$	iron(II) chloride
Fe^{3+}	iron(III)		$FeCl_3$	iron(III) chloride
Cr^{2+}	chromium(II)		CrS	chromium(II) sulfide
Cr^{3+}	chromium(III)		Cr_2S_3	chromium(III) sulfide

Remember that the *Roman numeral equals the positive charge on the metal ion*; it is not necessarily a subscript in the formula. For example, copper forms two oxides, one containing the Cu^+ ion and the other containing the Cu^{2+} ion. Their formulas are Cu_2O and CuO and their names are as follows:[4]

Cu^+	copper(I)		Cu_2O	copper(I) oxide
Cu^{2+}	copper(II)		CuO	copper(II) oxide

These copper compounds illustrate that in deriving the formula from the name, you must figure out the formula from the ionic charges, as previously discussed in this section and illustrated in Example 2.7.

---

[3]Silver and nickel are almost always found in compounds as Ag^+ and Ni^{2+}, respectively. Therefore, AgCl and $NiCl_2$ are almost always called simply silver chloride and nickel chloride.

[4]For some metals, such as copper and lead, one of their ions is much more commonly found in compounds than any of their others. For example, most common copper compounds contain Cu^{2+} and most common lead compounds contain Pb^{2+}. For compounds of these metals, if the charge is not indicated by a Roman numeral, we assume the ion present has a 2+ charge. Thus, it is not unusual to find $PbCl_2$ called lead chloride or for $CuCl_2$ to be called copper chloride.

Example 2.6
Naming Compounds and Writing Formulas

The compound $MnCl_2$ has a number of commercial uses, including the manufacture of batteries, disinfectants, and natural gas purifiers. What is the name of the compound?

Analysis: We can see that the compound is an ionic compound since it is made up of a metal and a nonmetal, so we use the rules for naming ionic compounds.

Manganese (Mn) is a transition element, and transition elements often form more than one cation, so we apply the Stock method. We also need to determine the charge on the manganese cation. We can figure this out because the sum of the charges on the manganese and chlorine ions must equal zero, and because the only monatomic ion of chlorine is Cl^-.

Assembling the Tools: The tool we will use to find the charge on the metal ion is the requirement that the compound be electrically neutral. Also, because the cation is that of a transition metal that may have more than one charge, the Stock system must be used to specify the charge on the cation.

Solution: The anion of chlorine (the chloride ion) is Cl^-, so a total of two negative charges are supplied by the two Cl^- ions. Therefore, for $MnCl_2$ to be electrically neutral, the Mn ion must carry two positive charges, 2+. The cation is named as manganese(II), and the name of the compound is manganese(II) chloride.

Is the Answer Reasonable? Performing a quick check of the arithmetic assures us we've got the correct charges on the ions. Everything appears to be okay.

Example 2.7
Naming Compounds and Writing Formulas

What is the formula for cobalt(III) fluoride?

Analysis: To answer this question, we first need to determine the charges on the two ions. Then we assemble the ions into a chemical formula being sure to achieve an electrically neutral formula unit.

Assembling the Tools: We will use the same tools as in Example 2.6.

Solution: Cobalt(III) corresponds to Co^{3+}. The fluoride ion is F^-. To obtain an electrically neutral substance, we must have three F^- ions for each Co^{3+} ion, so the formula is CoF_3.

Is the Answer Reasonable? We can check to see that we have the correct formulas of the ions and that we've combined them to achieve an electrically neutral formula unit. This will tell us we've obtained the correct answer.

Name the compounds Li_2S, Mg_3P_2, $NiCl_2$, $TiCl_2$, and Fe_2O_3. Use the Stock system where appropriate. (*Hint:* Determine which metals can have more than one charge.)

Practice Exercise 2.15

Write formulas for (a) aluminum sulfide, (b) strontium fluoride, (c) titanium(IV) oxide, (d) cobalt(II) oxide, and (e) gold(III) oxide.

Practice Exercise 2.16

Naming Ionic Compounds Containing Polyatomic Ions

The extension of the nomenclature system to include ionic compounds containing polyatomic ions is straightforward. Most of the polyatomic ions listed in Table 2.4 are anions and their names are used without modification as the second word in the name of the compound. For example, Na_2SO_4 contains the sulfate ion, SO_4^{2-}, and is called sodium

TOOLS
Naming with polyatomic ions

sulfate. Similarly, $Cr(NO_3)_3$ contains the nitrate ion, NO_3^-. Chromium is a transition element, and in this compound its charge must be 3+ to balance the negative charges of three NO_3 ions. Therefore, $Cr(NO_3)_3$ is called chromium(III) nitrate.

Among the ions in Table 2.4, the only cations that form compounds that can be isolated are the ammonium ion, NH_4^+, and the mercury(I) ion Hg_2^{2+}. They form ionic compounds such as NH_4Cl (ammonium chloride) and Hg_2Cl_2 (mercury(I) chloride). Ammonium salts, including $(NH_4)_2SO_4$ (ammonium sulfate), are written with the NH_4^+ first even though it is not a metal cation. Notice that $(NH_4)_2SO_4$ is composed of two polyatomic ions.

Example 2.8
Naming Compounds and Writing Formulas

What is the name of $Mg(ClO_4)_2$, a compound used commercially for removing moisture from gases?

Analysis: We recognize that this ionic compound contains a polyatomic ion since ClO_4 is in parentheses. If you've learned the contents of Table 2.4, you know that "ClO_4" is the formula (without the charge) of the perchlorate ion. With the charge, the ion's formula is ClO_4^-. We need to decide if the Stock system is required to name the metal Mg.

Assembling the Tools: We will use the names of polyatomic ions in Table 2.4 and the tools about deciding whether or not to use the Stock system.

Solution: Magnesium is in Group 2A, and the only ion it forms is Mg^{2+}. Therefore, we don't need to use the Stock system; it is named simply as "magnesium." The anion is perchlorate, so the name of $Mg(ClO_4)_2$ is *magnesium perchlorate*.

Is the Answer Reasonable? We can check to be sure we've named the anion correctly, and we have. The metal is magnesium, which only forms Mg^{2+}. Therefore, the answer seems to be correct.

Practice Exercise 2.17

What are the names of (a) Li_2CO_3, (b) $KMnO_4$, and (c) $Fe(OH)_3$? (*Hint:* Recall the names of the polyatomic ions and the positions of Li, K, and Fe in the periodic table.)

Practice Exercise 2.18

Write the formulas for (a) potassium chlorate, (b) sodium hypochlorite, and (c) nickel(II) phosphate.

Naming Hydrates

In Section 2.3 we discussed compounds called hydrates, such as $CuSO_4 \cdot 5H_2O$. Usually, hydrates are ionic compounds whose crystals contain water molecules in fixed proportions relative to the ionic substance. To name them, we provide two pieces of information: the name of the ionic compound and the number of water molecules in the formula. The number of water molecules is specified using the following Greek prefixes:

TOOLS

Greek prefixes

mono	= 1	penta	= 5	octa	= 8
di	= 2	hexa	= 6	nona	= 9
tri	= 3	hepta	= 7	deca	= 10
tetra	= 4				

These prefixes precede the word "hydrate." Thus, $CuSO_4 \cdot 5H_2O$ is named as "copper sulfate *pentahydrate*." Similarly, $CaSO_4 \cdot 2H_2O$ is named calcium sulfate dihydrate, and $FeCl_3 \cdot 6H_2O$ is iron(III) chloride hexahydrate.[5]

---

[5]Chemical suppliers (who do not always follow current rules of nomenclature) sometimes indicate the number of water molecules using a number and a dash. For example, one supplier lists $Ca(NO_3)_2 \cdot 4H_2O$ as "Calcium nitrate, 4-hydrate."

2.7 | Molecular Compounds

The concept of molecules dates to the time of Dalton's atomic theory, where a part of his theory was that atoms of elements combine in fixed numerical ratios to form "molecules" of a compound. By our modern definition a molecule is an *electrically neutral particle consisting of two or more atoms*. Accordingly, the term molecule applies to many elements such as H_2 and O_2 as well as to compounds.

Experimental Evidence for Molecules

One phenomenon that points to the existence of molecules is called **Brownian motion**, named after Robert Brown (1773–1858), the Scottish botanist who first observed it. When very small particles such as tiny grains of pollen are suspended in a liquid and observed under a microscope, the tiny particles are seen to be constantly jumping and jiggling about. It appears as though they are continually being knocked back and forth by collisions with something. One explanation is that this "something" is molecules of the liquid. Molecules of the liquid constantly bombard the microscopic particles, but because the suspended particles are so small, the collisions are not occurring equally on all sides. The unequal numbers of collisions cause the lightweight particles to jerk about. There is additional evidence for molecules, and today scientists accept the existence of molecules as fact.

Looking more closely, within molecules the atoms are held to each other by attractions called **chemical bonds**, which are electrical in nature. In **molecular compounds** chemical bonds arise from the sharing of electrons between one atom and another. We will discuss such bonds at considerable length in Chapters 9 and 10. What is important to know about molecules now is that *the group of atoms that make up a molecule move about together and behave as a single particle*, just as the various parts that make up a car move about as one unit. The chemical formulas that we write to describe the compositions of molecules are called molecular formulas, which specify the actual numbers of atoms of each kind that make up a single molecule.

Compare the structures of water and sodium chloride in Figure 2.28. Water is a discrete unit with the two hydrogen atoms bonded to the oxygen atom. In contrast, in sodium chloride, the ions are packed as close as possible to each other; each cation has six anions next to it, and each anion has six cations next to it. It cannot be said that one sodium ion "belongs" to one chloride ion. Instead, there is an attraction between each ion and its nearest neighbors of the opposite charge.

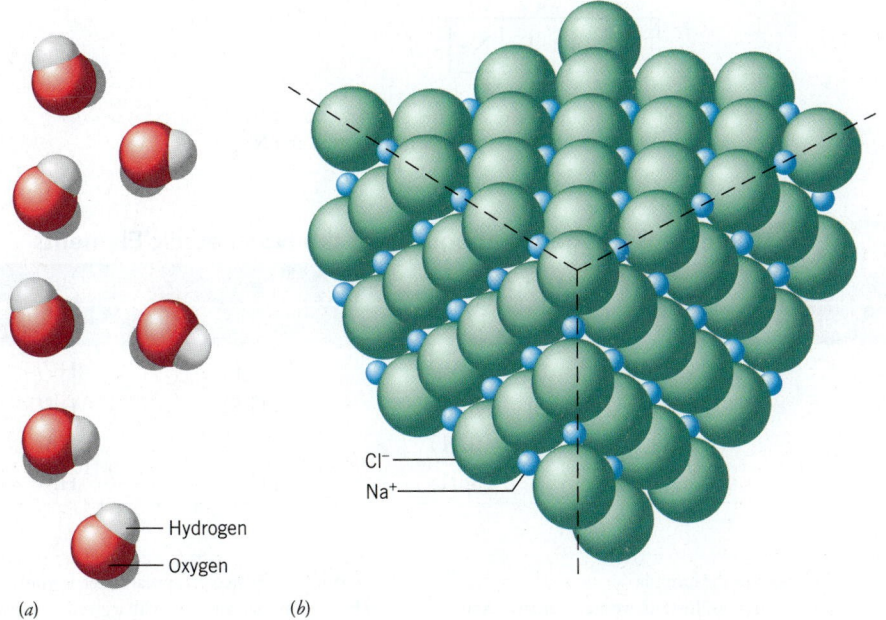

Cl⁻
Na⁺

— Hydrogen
— Oxygen

(a)   *(b)*

Figure 2.28 | Molecular and ionic substances. (*a*) In water there are discrete molecules that each consist of one atom of oxygen and two atoms of hydrogen. Each particle has the formula H_2O. (*b*) In sodium chloride, ions are packed in the most efficient way. Each Na^+ is surrounded by six Cl^-, and each Cl^- is surrounded by six Na^+. Because individual molecules do not exist, we simply specify the ratio of ions as NaCl.

Molecular Compounds Made from Nonmetals

As a general rule, *molecular compounds are formed when nonmetallic elements combine.* For example, you learned that H_2 and O_2 combine to form molecules of water. Similarly, carbon and oxygen combine to form either carbon monoxide, CO, or carbon dioxide, CO_2. (Both are gases that are formed in various amounts as products in the combustion of fuels such as wood, gasoline, and charcoal.) Although molecular compounds can be formed by the direct combination of elements, often they are the products of reactions between compounds. You will encounter many such reactions in your study of chemistry.

Although there are relatively few nonmetals, the number of molecular substances formed by them is huge. This is because of the variety of ways in which nonmetals combine as well as the varying degrees of complexity of their molecules. Variety and complexity reach a maximum with compounds in which carbon is combined with a handful of other elements such as hydrogen, oxygen, and nitrogen. There are so many of these compounds, in fact, that their study encompasses the largest chemical specialties called **organic chemistry** and **biochemistry**.

Molecules vary in size from small to very large. Some contain as few as two atoms. Molecules of water (H_2O), for example, have 3 atoms and those of ordinary table sugar ($C_{12}H_{22}O_{11}$) have 45. There also are molecules that are very large, such as those that occur in plastics and in living organisms, some of which contain millions of atoms.

At this early stage we can only begin to look for signs of order among the vast number of nonmetal–nonmetal compounds. To give you a taste of the subject, we will look briefly at some simple compounds that the nonmetals form with hydrogen, as well as some simple compounds of carbon.

Hydrogen-Containing Compounds

■ Many of the nonmetals also form more complex compounds with hydrogen, but we will not discuss them here.

TOOLS

Predicting formulas of nonmetal hydrogen compounds

Hydrogen forms a variety of compounds with other elements, and the formulas of the simple hydrogen compounds of the nonmetals (called **nonmetal hydrides**) are given in Table 2.6.[6] These compounds provide an opportunity to observe how we can use the periodic table as an aid in remembering factual information—in this case, the formulas of the hydrogen compounds of the nonmetals. Notice that the number of hydrogen atoms combined with the nonmetal atom equals *the number of spaces to the right that we have to move in the periodic table to get to a noble gas.* (You will learn why this is so in Chapter 9, but for now we can just use the periodic table as diagrammed below to help us remember the formulas.)

Two steps, so oxygen combines with two hydrogens to give H_2O.

| N | O | F | Ne |

Three steps, so nitrogen combines with three hydrogens to give NH_3.

TABLE 2.6   **Simple Hydrogen Compounds of the Nonmetallic Elements**

Period	Group			
	4A	5A	6A	7A
2	CH_4	NH_3	H_2O	HF
3	SiH_4	PH_3	H_2S	HCl
4	GeH_4	AsH_3	H_2Se	HBr
5		SbH_3	H_2Te	HI

[6] Table 2.6 shows how the formulas are normally written. The order in which the hydrogens appear in the formula is not of concern to us now. Instead, we are interested in the number of hydrogens that combine with a given nonmetal.

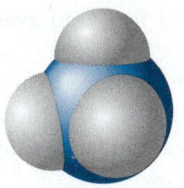
Ammonia, NH_3

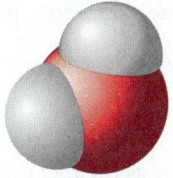
Water, H_2O

Hydrogen fluoride, HF

Figure 2.29 | Nonmetal hydrides of nitrogen, oxygen, and fluorine.

Also note in Table 2.6 that the formulas of the hydrogen compounds within a given group of the periodic table are similar for nonmetals. If you know the formula for the hydrogen compound of the top member of the group, then you know the formulas of all of the members in that group.

We live in a three-dimensional world, and this is reflected in the three-dimensional shapes of molecules. The shapes of the simple nonmetal hydrogen containing compounds of nitrogen, oxygen, and fluorine are illustrated as space-filling models in Figure 2.29. The geometric shapes of molecules will be described further in Chapter 10.

Carbon Compounds: The Basis of Organic Chemistry

Among all the elements, carbon is unique in the variety of compounds it forms with elements such as hydrogen, oxygen, and nitrogen, and their study constitutes the major specialty, organic chemistry. The term "organic" here comes from an early belief that these compounds could only be made by living organisms. We now know this isn't true, but the name organic chemistry persists nonetheless.

Organic compounds are around us everywhere and we will frequently use such substances as examples in our discussions. Therefore, it will be helpful if you can begin to learn some of them now.

The study of organic chemistry begins with **hydrocarbons** (compounds of hydrogen and carbon). The simplest hydrocarbon is methane, CH_4, which is a member of a series of hydrocarbons with the general formula C_nH_{2n+2}, where n is an integer (i.e., a whole number). The first six members of this series, called the **alkane** series, are listed in Table 2.7 along with their boiling points. Notice that as the molecules become larger, their boiling points increase. Molecules of methane, ethane, and propane are illustrated as space-filling models in Figure 2.30.

The alkanes are common substances. They are the principal constituents of petroleum from which most of our useful fuels are produced. Methane, as we have mentioned, is often used for heating and cooking. Gas-fired barbecues and some homes use propane as a fuel, and butane is the fuel in inexpensive cigarette lighters.[7] Hydrocarbons with higher

■ Our goal here is to introduce some basic organic substances so we can use them in later discussions.

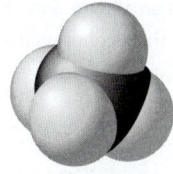
Methane, CH_4

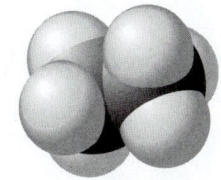
Ethane, C_2H_6

Propane, C_3H_8

Figure 2.30 | The first three members of the alkane series of hydrocarbons.

TABLE 2.7	Hydrocarbons Belonging to the Alkane Series		
Compound	Name	Boiling Point	Structural Formula
CH_4	methane[a]	−161.5	CH_4
C_2H_6	ethane[a]	−88.6	CH_3CH_3
C_3H_8	propane[a]	−42.1	$CH_3CH_2CH_3$
C_4H_{10}	butane[a]	−0.5	$CH_3CH_2CH_2CH_3$
C_5H_{12}	pentane	36.1	$CH_3CH_2CH_2CH_2CH_3$
C_6H_{14}	hexane	68.7	$CH_3CH_2CH_2CH_2CH_2CH_3$

[a]Gases at room temperature (25 °C) and atmospheric pressure

---

[7] Propane and butane are gases when they're at the pressure of the air around us, but become liquids when compressed. When you purchase these substances, they are liquids with pressurized gas above them. The gas can be drawn off and used by opening a valve to the container.

boiling points are found in gasoline, kerosene, paint thinners, diesel fuel, and even candle wax.

Alkanes are not the only class of hydrocarbons. For example, there are three hydrocarbons with two carbon atoms each. They are, in addition to ethane, C_2H_6, ethene (ethylene), C_2H_4, and ethyne (*acetylene*), C_2H_2. Ethylene is a common reactant for making plastics, while acetylene is known best as the fuel used in welding torches.

The hydrocarbons serve as the foundation for organic chemistry. Derived from them are various other classes of organic compounds. An example is the class of compounds called **alcohols**, in which the atoms OH replace a hydrogen atom. Thus, *methanol*, CH_3OH (also called *methyl alcohol*), is related to methane, CH_4, by removing one H and replacing it with OH (Figure 2.31). Methanol is used as a fuel and as a raw material for making more complex organic chemicals. Another familiar alcohol is ethanol (also called *ethyl alcohol*), C_2H_5OH. Ethanol, known as grain alcohol because it is often obtained from the fermentation of grains, is in alcoholic beverages. It is also mixed with gasoline to reduce petroleum consumption. A 10% ethanol/90% gasoline mixture is known as gasohol, and an 85% mixture of ethanol and gasoline is called E85.

Alcohols constitute just one class of compound derived from hydrocarbons. We will discuss some others after you've learned more about how atoms bond to each other and about the structures of molecules.

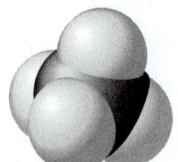

Methane

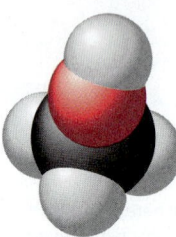

Methanol

Figure 2.31 | Relationship between an alkane and an alcohol. The alcohol methanol is derived from methane by replacing one H by OH.

Writing Formulas for Organic Compounds

Organic formulas can be written in different ways, depending on what information is needed. A molecular formula such as C_2H_6 for ethane or C_3H_8 for propane simply indicates the number of each type of atom in the molecule. The order of the elements in the molecular formula starts with carbon since it is the basis for organic compounds. This is followed by hydrogen, and then the rest of the elements are written in alphabetical order. For example, sucrose is $C_{12}H_{22}O_{11}$. There are exceptions; for example, if we wish to emphasize an alcohol, the OH is separated and listed last, as in ethanol, C_2H_5OH. Condensed structural formulas indicate how the carbon atoms are connected. Ethane is written as CH_3CH_3, and propane is $CH_3CH_2CH_3$ in the condensed structural formula format.

Practice Exercise 2.19

Gasoline used in modern cars is a complex mixture of hundreds of different organic compounds. Less than 1% of gasoline is actually octane. Write the formula for octane using the molecular and condensed structural formats. (*Hint:* To figure out the number of carbon atoms, recall the meaning of the Greek prefix "octa.")

Practice Exercise 2.20

On the basis of the discussions in this section, what are the formulas of (a) propanol and (b) butanol? Write both the molecular formula and the condensed structural formula.

2.8 | Nomenclature of Molecular Compounds

Binary Molecular Compounds

Just as in naming the ionic compounds, we want to be able to translate a chemical formula into a name that anyone with a background in chemistry can understand. For a binary molecular compound, therefore, we must indicate which two elements are present and the number of atoms of each in a molecule of the substance.

To identify the first element in a formula, we just specify its English name. Thus, for HCl the first word in the name is "hydrogen" and for PCl$_5$ the first word is "phosphorus." To identify the second element, we append the suffix *-ide* to the stem of the element's English name just as we did for the monatomic anions in ionic compounds.

To form the name of the compound, we place the two parts of the name one after another. Therefore, the name of HCl is hydrogen chloride. However, to name PCl$_5$, we need a way to specify the number of Cl atoms bound to the phosphorus in the molecule. This is done using the same Greek prefixes listed in the hydrates part of Section 2.6; the main difference is that the prefix mono- is often omitted.

To name PCl$_5$, therefore, we add the prefix penta- to chloride to give the name phosphorus pentachloride. Notice how easily this allows us to translate the name back into the formula.

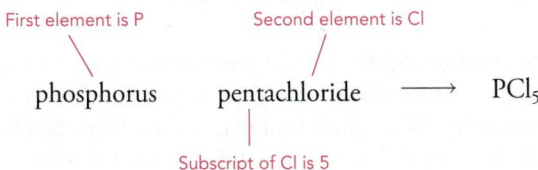

The prefix mono- is used when we want to emphasize that only one atom of a particular element is present. For instance, carbon forms two compounds with oxygen, CO and CO$_2$. To clearly distinguish between them, the first is called carbon monoxide (one of the "o"s is omitted to make the name easier to pronounce) and the second is carbon dioxide.

As indicated above, the prefix mono- is often omitted from a name. Therefore, in general, if there is no prefix before the name of an element, we take it to mean there is only one atom of that element in the molecule. An exception to this is in the names of binary compounds of nonmetals with hydrogen. An example is hydrogen sulfide. The name tells us the compound contains the two elements hydrogen and sulfur. We don't have to be told how many hydrogens are in the molecule because, as you learned earlier, we can use the periodic table to determine the number of hydrogen atoms in molecules of the simple nonmetal, hydrogen-containing compounds. Sulfur is in Group 6A, so to get to the noble gas column (Group 8A) we have to move two steps to the right; the number of hydrogens combined with the atom of sulfur is two. The formula for hydrogen sulfide is therefore H$_2$S.

Example 2.9
Naming Compounds and Writing Formulas

(a) What is the name of AsCl$_3$? (b) What is the formula for dinitrogen tetraoxide?

Analysis: (a) In naming compounds, the first step is to determine the type of compound involved. Looking at the periodic table, we see that AsCl$_3$ is made up of two nonmetals, so we conclude that it is a molecular compound. (b) To write the formula from the name, we convert the prefixes to numbers.

Assembling the Tools: (a) We apply the tool for naming molecular compounds from the molecule's formula described previously. (b) The meanings of the Greek prefixes will be used to convert the prefixes to numbers that we apply as subscripts for the chemical symbols in the formula.

Solution: (a) In $AsCl_3$, As is the symbol for arsenic and Cl is the symbol for chlorine. The first word in the name is just arsenic and the second will contain chloride with an appropriate prefix to indicate number. There are three Cl atoms, so the prefix is tri. Therefore, the name of the compound is arsenic trichloride.

(b) As we did earlier for phosphorus pentachloride, we convert the prefixes to numbers and apply them as subscripts.

■ Sometimes we drop the *a* before an *o* for ease of pronunciation. N_2O_4 would then be named dinitrogen tetroxide.

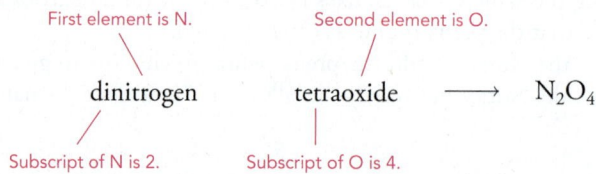

First element is N. Second element is O.

dinitrogen   tetraoxide $\longrightarrow$ N_2O_4

Subscript of N is 2.   Subscript of O is 4.

Are the Answers Reasonable? To feel comfortable with the answers, be sure to double-check for careless errors. Next, take your answers and reverse the process. Does arsenic trichloride result in the original formula, $AsCl_3$? Does N_2O_4 have the name of dinitrogen tetraoxide? We can say yes to both and have confidence in our work.

Practice Exercise 2.21   Name the following compounds using Greek prefixes when needed: (a) PCl_3, (b) SO_2, (c) Cl_2O_7, and (d) H_2S. (*Hint:* See the list of prefixes above.)

Practice Exercise 2.22   Write formulas for the following compounds: (a) arsenic pentachloride, (b) sulfur hexachloride, (c) disulfur dichloride, and (d) hydrogen telluride.

Common Names for Molecular Compounds

Not every compound is named according to the systematic procedure we have described so far. Many familiar substances were discovered long before a systematic method for naming them had been developed, and they acquired common names that are so well known that no attempt has been made to rename them. For example, following the scheme described above we might expect that H_2O would have the name hydrogen oxide (or even dihydrogen monoxide). Although this isn't wrong, the common name, water, is so well known that it is always used. Another example is ammonia, NH_3, whose odor you have no doubt experienced while using household ammonia solutions or the glass cleaner Windex®. Common names are used for the other hydrogen-containing compounds of the nonmetals in Group 5A as well. The compound PH_3 is called phosphine and AsH_3 is called arsine; both are powerful poisons.

Common names are also used for very complex substances. An example is sucrose, which is the chemical name for table sugar, $C_{12}H_{22}O_{11}$. The structure of this compound is pretty complex, and its name assigned following the systematic method is equally complex. It is much easier to say the simple name sucrose, and be understood, than to struggle with the cumbersome systematic name for this common compound.

Naming Molecular and Ionic Compounds

In this chapter we've discussed how to name two classes of compounds, molecular and ionic, and you saw that slightly different rules apply to each. To name chemical compounds successfully we need to make a series of decisions based on the rules we just covered. We can summarize this decision process in a flowchart such as the one shown in Figure 2.32. The next example illustrates how to use the flowchart, and when working on the Review Questions, you may want to refer to Figure 2.32 until you are able to develop the skills that will enable you to work without it.

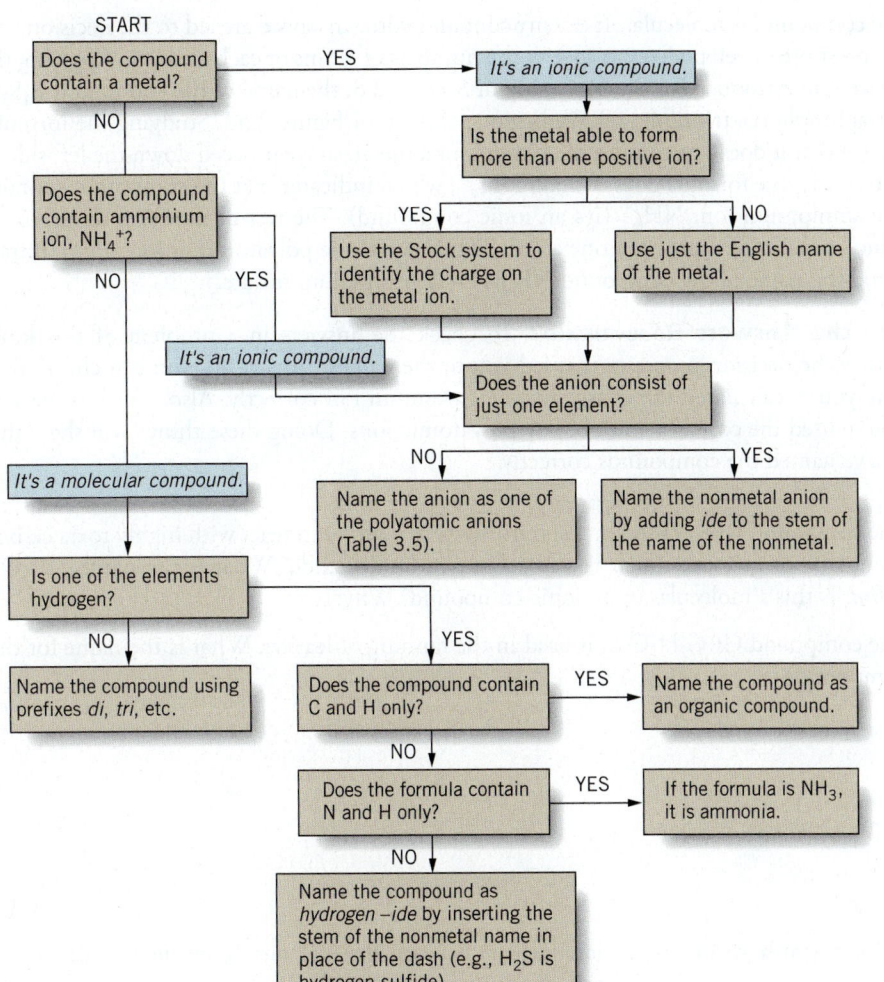

START

Does the compound contain a metal? — YES → It's an ionic compound.

NO

Is the metal able to form more than one positive ion?

YES → Use the Stock system to identify the charge on the metal ion.

NO → Use just the English name of the metal.

Does the compound contain ammonium ion, NH_4^+?

NO   YES

It's an ionic compound.

Does the anion consist of just one element?

NO → Name the anion as one of the polyatomic anions (Table 3.5).

YES → Name the nonmetal anion by adding *ide* to the stem of the name of the nonmetal.

It's a molecular compound.

Is one of the elements hydrogen?

NO → Name the compound using prefixes *di*, *tri*, etc.

YES → Does the compound contain C and H only? — YES → Name the compound as an organic compound.

NO

Does the formula contain N and H only? — YES → If the formula is NH_3, it is ammonia.

NO

Name the compound as *hydrogen –ide* by inserting the stem of the nonmetal name in place of the dash (e.g., H_2S is hydrogen sulfide).

Figure 2.32 | Flowchart for naming molecular and ionic compounds.

TO⚗LS
Flowchart for naming compounds

Example 2.10
Applying the Rules for Naming Compounds

What is the name of (a) $CrCl_3$, (b) P_4S_3, and (c) NH_4NO_3?

Analysis: We need to start by determining whether or not the compounds are ionic or molecular, and then follow the rules given for naming the appropriate type of compound.

Assembling the Tools: For each compound, we use the tools summarized in Figure 2.32 and proceed through the decision processes to arrive at the name of the compound.

Solution: (a) Starting at the top of Figure 2.32, we first determine that the compound contains a metal (Cr), so it's an ionic compound. Next, we see that the metal is a transition element, and chromium is one of those that forms more than one cation, so we have to apply the Stock method. To do this, we need to know the charge on the metal ion. We can figure this out using the charge on the anion and the fact that the compound must be electrically neutral overall. The anion is formed from chlorine, so its charge is 1− (the anion is Cl^-). Since there are three chlorine ions, the metal ion must be Cr^{3+}; we name the metal chromium(III). Next, we see that there is only one nonmetallic element in the compound, Cl, so the name of the anion ends in *-ide*; it's the chloride ion. The compound $CrCl_3$ is therefore named chromium(III) chloride. (b) Once again, we start at the top of Figure 2.32. First, we determine that the compound doesn't contain a metal or NH_4^+, so the compound is molecular. It doesn't contain hydrogen, so we are led to the decision that we must use Greek prefixes to specify the numbers of atoms of each element. Applying the procedure introduced at the beginning of Section 2.8, the name of the compound P_4S_3 is tetraphosphorus trisulfide. (c) We begin at the top of Figure 2.32. Studying the formula, we see that it does not contain the symbol for a metal, so we proceed down the left side of the figure. The formula does contain NH_4^+, which indicates that the compound contains the ammonium ion, NH_4^+ (it's an ionic compound). The rest of the formula is NO^{3-}, which consists of more than one atom. This suggests the polyatomic anion, NO_3 (nitrate ion). The name of the compound NH_4NO_3 is ammonium nitrate.

Are the Answers Reasonable? To check the answers in a problem of this kind, review the decision processes that led you to the names. In part (a), you can check to be sure you've calculated the charge on the chromium ion correctly. Also, check to be sure you've used the correct names of any polyatomic ions. Doing these things will show that you've named the compounds correctly.

Practice Exercise 2.23   The compound I_2O_5 is used in respirators where it serves to react with highly toxic carbon monoxide to give the much less toxic gas, carbon dioxide. What is the name of I_2O_5? (*Hint:* Is this a molecular or an ionic compound? Why?)

Practice Exercise 2.24   The compound $Cr(C_2H_3O_2)_3$ is used in the tanning of leather. What is the name for this compound?

| Summary

Organized by Learning Objective

Describe the information in and the organization of the periodic table

In the modern **periodic table** the elements are arranged in rows, called **periods**, in order of increasing atomic number. The rows are stacked so that elements in the columns, called **groups** or **families**, have similar chemical and physical properties. The A-group elements (IUPAC Groups 1, 2, and 13–18) are called **representative elements**; the B-group elements (IUPAC Groups 3–12) are called **transition elements**. The **inner transition elements** consist of the lanthanides, which follow La ($Z = 57$), and the actinides, which follow Ac ($Z = 89$). Certain groups are given family names: Group 1A (Group 1), except for hydrogen, are the **alkali metals** (the alkalis); Group 2A (Group 2) are the **alkaline earth metals**; Group 7A (Group 17) are the **halogens**; and Group 8A (Group 18) are the **noble gases**.

Understand the distribution of the metals, nonmetals, and metalloids within the periodic table

Most elements are **metals**; they occupy the lower left-hand region of the periodic table (to the left of a line drawn approximately from boron, B, to astatine, At). **Nonmetals** are found in the upper right-hand region of the table. **Metalloids** occupy a narrow band between the metals and nonmetals.

Metals exhibit a **metallic luster**, tend to be **ductile** and **malleable**, and conduct heat and electricity. Nonmetals tend to be brittle, lack metallic luster, and are nonconductors of electricity. Many nonmetals are gases. **Metalloids** have properties intermediate between those of metals and nonmetals and are **semiconductors** of electricity.

Explain the information embodied in a chemical formula

Chemical symbols can be used in drawings to indicate how atoms are attached to each other in compounds. Three-dimensional representations of molecules can be **ball-and-stick** or **space-filling models**. Some compounds form solids called **hydrates**, which contain water molecules in definite proportions.

Using atoms we can account for the law of definite proportions. The atomic theory also led to the discovery of the law of multiple proportions.

Understand the nature of a balanced chemical equation and how it relates to the atomic theory

Significant changes in the properties of substances are observed when **chemical reactions** occur, and chemical equations are used to show how the reactants change to products. **Coefficients** in front of formulas indicate the number of molecules that react or are formed and are used to balance an equation, making the total numbers of atoms of each kind the same on both sides of the arrow. A **balanced equation** conforms to the law of conservation of mass.

Use the periodic table and ion charges to write chemical formulas of ionic compounds

Binary ionic compounds are formed when metals react with nonmetals; electrons are transferred from a metal to a nonmetal.

The metal atom becomes a positive ion (a **cation**); the nonmetal atom becomes a negative ion (an **anion**). The formula of an ionic compound specifies the smallest whole-number ratio of the ions. The smallest unit of an ionic compound is called a **formula unit**, which specifies the smallest whole-number ratio of the ions that is electrically neutral. Many ionic compounds also contain **polyatomic** ions—ions that are composed of two or more atoms.

Name ionic compounds and write chemical formulas from chemical names

In naming an ionic compound, the cation is specified first, followed by the anion. Metal cations take the English name of the element, and when more than one positive ion can be formed by the metal, the **Stock system** is used to identify the amount of positive charge on the cation. This is done by placing a Roman numeral equal to the positive charge in parentheses following the name of the metal. Simple **monatomic** anions are formed by nonmetals, and their names are formed by adding the suffix *-ide* to the stem of the nonmetal's name. Only two common polyatomic anions (cyanide and hydroxide) end in the suffix *-ide*.

Understand the difference between ionic and molecular compounds

Molecules are electrically neutral particles consisting of two or more atoms. Molecules are held together by **chemical bonds** that arise from the sharing of electrons between atoms. The formulas we write for molecules are **molecular formulas**. Molecular compounds are formed when nonmetals combine with each other. The simple **nonmetal hydrides** have formulas that can be remembered by the position of the nonmetal in the periodic table. **Organic compounds** are **hydrocarbons**, or compounds considered to be derived from hydrocarbons by replacing H atoms with other atoms.

We found that it is often possible to distinguish ionic compounds from molecular compounds by their ability to **conduct electricity**. Molecular compounds are generally poor electrical conductors, whereas ionic compounds when melted into the liquid state or dissolved in water will conduct electricity readily.

Name molecular compounds

The system of **nomenclature** for **binary** molecular **inorganic compounds** uses a set of Greek prefixes to specify the numbers of atoms of each kind in the formula of the compound. The first element in the formula is specified by its English name; the second element takes the suffix *-ide*, which is added to the stem of the English name. For simple nonmetal hydrogen compounds, it is not necessary to specify the number of hydrogens in the formula. Many familiar substances, as well as very complex molecules, are usually identified by **common names**. The decision tree in Figure 2.32 is a helpful tool in naming ionic and molecular compounds.

Tools for Problem Solving
The following tools were introduced in this chapter. Study them carefully so you can select the appropriate tool when needed.

Periodic table (Section 2.1)
The periodic table lists the atoms by atomic number and organizes them by their properties in periods and groups. We can obtain atomic numbers and average masses of the elements from the periodic table.

Periodic table: Metals, nonmetals, or metalloids (Section 2.2)
We can tell whether an element is a metal, nonmetal, or metalloid from its position in the periodic table.

Chemical symbols and subscripts in a chemical formula (Section 2.3)
In a chemical formula, the chemical symbol stands for an atom of an element. Subscripts show the number of atoms of each kind, and when a subscript follows parentheses, it multiplies everything within the parentheses.

Coefficients in an equation (Section 2.4)
Coefficients are only used to balance an equation and indicate the number of chemical units of each type that are present in reactants and products.

Ionic compounds (Section 2.5)
Ionic compounds are formed when metals react with nonmetals and the ions formed combine into a compound.

Predicting cation charge (Section 2.5)
The metals in Group 1A have a +1 charge and metals in Group 2A have a +2 charge.

Predicting anion charge (Section 2.5)
From a nonmetal's position in the periodic table, we can determine the charge of the monatomic anions.

Formulas for ionic compounds (Section 2.5)
Following the rules gives us the correct formulas for ionic compounds with electrically neutral formula units and with subscripts in the smallest set of whole numbers.

Polyatomic ions (Section 2.5)
The names, formulas, and charges of these ions are listed in Table 2.4.

Monatomic anion names (Section 2.6)
This list in Table 2.5 gives the common names of anions that must be remembered so you can use them to name ionic compounds.

Naming ionic compounds (Section 2.6)
These rules give us a systematic method for naming ionic compounds.

Using the Stock system (Section 2.6)
For metals that can form ions with more than one possible charge, the Stock system specifies the charge of a cation by placing a Roman numeral in parentheses just after the name of the cation.

Naming with polyatomic ions (Section 2.6)
Naming compounds that contain polyatomic anions is done by specifying the cation name, using the Stock system if needed, and then specifying the polyatomic anion name as given in Table 2.4.

Greek prefixes (Section 2.6)
This is a list of the Greek prefixes from one to ten. They are used in naming hydrates and molecular compounds.

Predicting formulas of nonmetal hydrogen compounds (Section 2.7)
From a nonmetal's position in the periodic table we can write the formula of its simple hydrogen compound. These are given in Table 2.6.

Naming binary molecular compounds (Section 2.8)
These rules give us a logical system for naming binary molecular compounds by specifying the number of each type of atom using Greek prefixes.

Flowchart for naming compounds (Section 2.8)
This flowchart in Figure 2.32 can be followed for naming both ionic and molecular compounds.

WileyPLUS=*WileyPLUS*, an online teaching and learning solution. *Note to instructors*: Many of the end-of-chapter problems are available for assignment via the WileyPLUS system. **www.wileyplus.com.** Review Problems are presented in pairs separated by blue rules. Answers to problems whose numbers appear in blue are given in Appendix B. More challenging problems are marked with an asterisk ★.

| Review Questions

The Periodic Table

2.1 In the compounds formed by Li, Na, K, Rb, and Cs with chlorine, how many atoms of Cl are there per atom of each metal? In the compounds formed by Be, Mg, Ca, Sr, and Ba with chlorine, how many atoms of Cl are there per atom of each metal? How did this kind of information lead Mendeleev to develop his periodic table?

2.2 On what basis did Mendeleev construct his periodic table? On what basis are the elements arranged in the modern periodic table?

2.3 Using their positions in the periodic table, explain why is it not surprising that strontium-90, a radioactive isotope, replaces calcium in newly formed bones.

2.4 In the refining of copper, sizable amounts of silver and gold are recovered. Why is this not surprising?

2.5 Why would you reasonably expect cadmium to be a contaminant in zinc but not in silver?

2.6 Using the symbol for nitrogen, $^{14}_{7}N^{0}_{2}$, indicate what information is conveyed by the two superscripts, and what information is conveyed by the two subscripts.

2.7 Make a rough sketch of the periodic table and mark off those areas where you would find **(a)** the representative elements, **(b)** the transition elements, and **(c)** the inner transition elements.

2.8 Which of the following is

(a) an alkali metal? Ca, Cu, In, Li, S

(b) a halogen? Ce, Hg, Si, O, I

(c) a transition element? Pb, W, Ca, Cs, P

(d) a noble gas? Xe, Se, H, Sr, Zr

(e) a lanthanide element? Th, Sm, Ba, F, Sb

(f) an actinide element? Ho, Mn, Pu, At, Na

(g) an alkaline earth metal? Mg, Fe, K, Cl, Ni

Metals, Nonmetals, and Metalloids

2.9 Name five physical properties that we usually observe for metals.

2.10 Why is mercury used in thermometers? Why is tungsten used in light bulbs?

2.11 Which nonmetals occur as monatomic gases (i.e., gases whose particles consist of single atoms)?

2.12 Which two elements exist as liquids at room temperature and pressure?

2.13 Which physical property of metalloids distinguishes them from metals and nonmetals?

2.14 Sketch the shape of the periodic table and mark off those areas where we find **(a)** metals, **(b)** nonmetals, and **(c)** metalloids.

2.15 Most periodic tables have a heavy line that looks like a staircase starting from boron down to polonium. What information does this line convey?

2.16 Which metals can you think of that are commonly used to make jewelry? Why isn't iron used to make jewelry? Why isn't potassium used?

2.17 What trends—regular changes in physical or chemical properties—in the periodic table have been mentioned in this chapter?

2.18 Find a periodic table on the internet that lists physical properties of the elements. Can you distinguish trends in the periodic table based on **(a)** melting point, **(b)** boiling point, or **(c)** density?

Molecules and Chemical Formulas

2.19 What are two ways to interpret a chemical symbol?

2.20 What is the difference between an atom and a molecule?

2.21 Write the formulas and names of the elements that exist in nature as diatomic molecules.

2.22 Atoms of which elements are usually represented by the following drawings? Give their names and chemical symbols.

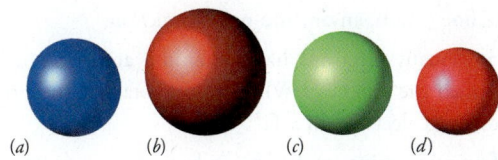

(a)   (b)   (c)   (d)

2.23 Atoms of which elements are usually represented by the following drawings? Give their names and chemical symbols.

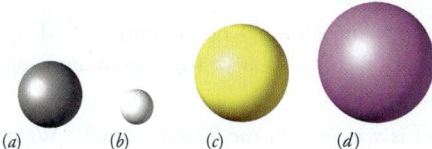

(a)   (b)   (c)   (d)

2.24 A DNA molecule is small in actual size but contains an enormous number of atoms. Surprisingly, the molecule is made up of atoms of just five different elements. Study Figure 2.18 and identify which elements those are.

Chemical Reactions and Chemical Equations

2.25 What do we mean when we say a chemical equation is balanced? Why do we balance chemical equations?

2.26 For a chemical reaction, what do we mean by the term reactants? What do we mean by the term products?

2.27 The combustion of a thin wire of magnesium metal (Mg) in an atmosphere of pure oxygen produces the brilliant light of a flashbulb, once commonly used in photography. After the reaction, a thin film of magnesium oxide is seen on the inside of the bulb. The equation for the reaction is

$$2Mg + O_2 \longrightarrow 2MgO$$

(a) State in words how this equation is read.

(b) Give the formula(s) of the reactants.

(c) Give the formula(s) of the products.

Ionic Compounds

2.28 Describe what kind of event must occur (involving electrons) if the atoms of two different elements are to react to form an ionic compound.

2.29 With what kind of elements do metals react?

2.30 What is an ion? How does it differ from an atom or a molecule?

2.31 Why do we use the term formula unit for ionic compounds instead of the term molecule? Can we use that term for molecules, atoms, and ions? Explain why.

2.32 Consider the sodium atom and the sodium ion.

(a) Write the chemical symbol of each.

(b) Do these particles have the same number of nuclei?

(c) Do they have the same number of protons?

(d) Could they have different numbers of neutrons?

(e) Do they have the same number of electrons?

2.33 Define cation, anion, and polyatomic ion.

2.34 How many electrons has a titanium atom lost if it has formed the ion Ti^{4+}? What are the total numbers of protons and electrons in a Ti^{4+} ion?

2.35 If an atom gains an electron to become an ion, what kind of electrical charge does the ion have?

2.36 How many electrons has a nitrogen atom gained if it has formed the ion N^{3-}? How many protons and electrons are in an N^{3-} ion?

2.37 What are the formulas of the ions formed by **(a)** iron, **(b)** cobalt, **(c)** mercury, **(d)** chromium, **(e)** tin, and **(f)** manganese?

2.38 What is wrong with the formula $RbCl_3$? What is wrong with the formula SNa_2? Rewrite the formulas so they are correct.

2.39 A student wrote the formula for an ionic compound of titanium as Ti_2O_4. What is wrong with this formula? What should the formula be?

2.40 Which of the following formulas are incorrect? Write the formulas for the compounds correctly. **(a)** NaO_2, **(b)** $RbCl$, **(c)** K_2S, **(d)** Al_2Cl_3, **(e)** MgP_2

2.41 What are the formulas (including charges) for **(a)** cyanide ion, **(b)** ammonium ion, **(c)** nitrate ion, **(d)** sulfite ion, **(e)** chlorate ion, and **(f)** sulfate ion?

2.42 What are the formulas (including charges) for **(a)** hypochlorite ion, **(b)** bisulfate ion, **(c)** phosphate ion, **(d)** dihydrogen phosphate ion, **(e)** permanganate ion, and **(f)** oxalate ion?

2.43 What are the names of the following ions? **(a)** $Cr_2O_7^{2-}$, **(b)** OH^-, **(c)** $C_2H_3O_2^-$, **(d)** CO_3^{2-}, **(e)** CN^-, **(f)** ClO_4^-

2.44 Write the correct formulas for the compounds formed when the following react **(a)** calcium and chlorine, **(b)** magnesium and oxygen, **(c)** aluminum and oxygen, and **(d)** sodium and sulfur.

2.45 Write the unbalanced equations for the following reactions:

(a) Iron(III) hydroxide reacts with hydrogen chloride forming water and iron(III) chloride.

(b) Silver nitrate is reacted with barium chloride to form silver chloride and barium nitrate.

2.46 Write the unbalanced equations for the following reactions:

(a) Propane reacts with oxygen to form carbon dioxide and water.

(b) Sodium metal is added to water and the products are sodium hydroxide and hydrogen gas.

Molecular Compounds

2.47 With what kind of elements do nonmetals react?

2.48 Which are the only elements that exist as free, individual atoms when not chemically combined with other elements?

2.49 Write chemical formulas for the elements that normally exist in nature as diatomic molecules.

2.50 Which kind of elements normally combine to form molecular compounds?

2.51 Why are nonmetals found in more compounds than are metals, even though there are fewer nonmetals than metals?

2.52 Most compounds of aluminum are ionic, but a few are molecular. How do we know that Al_2Cl_6 is molecular?

2.53 Without referring to Table 2.6 but using the periodic table, write chemical formulas for the simplest hydrogen compounds of **(a)** carbon, **(b)** nitrogen, **(c)** tellurium, and **(d)** iodine.

2.54 The simplest hydrogen compound of phosphorus is phosphine, a highly flammable and poisonous compound with an odor of decaying fish. What is the formula for phosphine?

2.55 Astatine, a member of the halogen family, forms a compound with hydrogen. Predict its chemical formula.

2.56 Under appropriate conditions, tin can be made to form a simple molecular compound with hydrogen. Predict its formula.

2.57 Write the chemical formulas for **(a)** methane, **(b)** ethane, **(c)** propane, and **(d)** butane. Give one practical use for each of these hydrocarbons.

2.58 What are the formulas for **(a)** methanol and **(b)** ethanol?

2.59 What is the formula for the alkane decane, which has 10 carbon atoms?

2.60 Candle wax is a mixture of hydrocarbons, one of which is an alkane with 23 carbon atoms. What is the formula for this hydrocarbon?

2.61 The formula for a compound is correctly given as $C_6H_{12}O_6$. State two reasons why we expect this to be a molecular compound, rather than an ionic compound.

2.62 Explore the internet and find a reliable source of structures for molecular compounds. For Problems 2.57 to 2.61, print out the ball-and-stick and space-filling models of the compounds mentioned.

Nomenclature of Ionic and Molecular Compounds

2.63 What is the difference between a binary compound and one that is diatomic? Give examples that illustrate this difference.

2.64 In naming the compounds discussed in this chapter, why is it important to know whether a compound is molecular or ionic?

2.65 In naming ionic compounds of the transition elements, why is it essential to give the charge on the anion?

2.66 Describe **(a)** the three situations in which Greek prefixes are used and **(b)** when Roman numerals are used.

|Review Problems

Molecules and Chemical Formulas

2.67 The compound $Cr(C_2H_3O_2)_3$ is used in the tanning of leather. How many atoms of each element are given in this formula?

2.68 Asbestos, a known cancer-causing agent, has a typical formula of $Ca_3Mg_5(Si_4O_{11})_2(OH)_2$. How many atoms of each element are given in this formula?

2.69 Epsom salts is a hydrate of magnesium sulfate, $MgSO_4 \cdot 7H_2O$. What is the formula of the substance that remains when Epsom salts is completely dehydrated?

2.70 Rochelle salt is the tetrahydrate of $KNaC_4H_4O_6$, which means there are four molecules of water per $KNaC_4H_4O_6$. Write the formula for Rochelle salt.

2.71 A molecule of acetic acid, formed when wine spoils and becomes sour, is shown below. Write the chemical formula for the molecule.

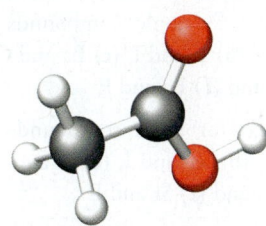

2.72 A molecule of dimethyl sulfide is shown below. The compound is a sulfurous gas with a disagreeable odor produced from breakdown products of phytoplankton through biological interactions. Write the chemical formula for the molecule. Is there a way you can write this formula using parentheses?

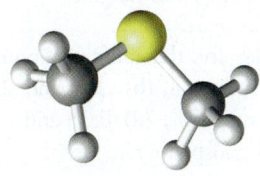

2.73 Write the chemical formula for the molecule illustrated below.

2.74 Write the chemical formula for the molecule illustrated below.

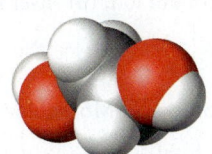

2.75 For the molecule in Problem 2.73, which of the following structural formulas is correct?

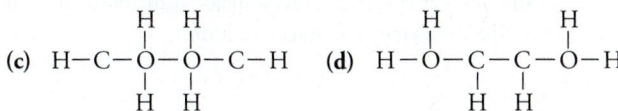

2.76 For the molecule in Problem 2.74, which of the following structural formulas is correct?

(a) $H-O-\overset{\displaystyle H}{\underset{\displaystyle H}{C}}-\overset{\displaystyle H}{\underset{\displaystyle H}{C}}-O-H$ (b) $H-\overset{\displaystyle H}{\underset{\displaystyle H}{C}}-O-\overset{\displaystyle H}{\underset{\displaystyle H}{C}}-O-H$

(c) $H-C-\overset{\displaystyle H}{\underset{\displaystyle H}{O}}-\overset{\displaystyle H}{\underset{\displaystyle H}{O}}-C-H$ (d) $H-O-\overset{\displaystyle H}{\underset{\displaystyle H}{C}}-\overset{\displaystyle H}{\underset{\displaystyle H}{C}}-O-H$

2.77 How many atoms of each element are represented in each of the following formulas? For each, name the elements present. (a) $K_2C_2O_4$, (b) H_2SO_3, (c) $C_{12}H_{26}$, (d) $HC_2H_3O_2$, (e) $(NH_4)_2HPO_4$

2.78 How many atoms of each kind are represented in the following formulas? For each, name the elements present. (a) H_3PO_4, (b) $Ca(H_2PO_4)_2$, (c) C_4H_9Br, (d) $Fe_3(AsO_4)_2$, (e) $C_3H_5(OH)_3$

2.79 How many atoms of each kind are represented in the following formulas? For each, name the elements present. (a) $Ni(ClO_4)_2$, (b) $COCl_2$, (c) $K_2Cr_2O_7$, (d) CH_3CO_2H, (e) $(NH_4)_2HPO_4$

2.80 How many atoms of each kind are represented in the following formulas? For each, name the elements present. (a) $CH_3CH_2CO_2C_3H_7$, (b) $MgSO_4 \cdot 7H_2O$, (c) $KAl(SO_4)_2 \cdot 12H_2O$, (d) $Cu(NO_3)_2$, (e) $(CH_3)_3COH$

2.81 How many atoms of each element are represented in each of the following expressions? (a) $7CH_3CO_2H$, (b) $2(NH_2)_2CO$, (c) $5C_3H_5(OH)_3$

2.82 How many atoms of each element are represented in each of the following expressions? (a) $3N_2O$, (b) $4(CH_3)_2S$, (c) $2CuSO_4 \cdot 5H_2O$

Chemical Reactions and Chemical Equations

2.83 Consider the balanced equation

$$2Fe(NO_3)_3 + 3Na_2CO_3 \longrightarrow Fe_2(CO_3)_3 + 6NaNO_3$$

(a) How many atoms of Na are on each side of the equation? **(b)** How many atoms of C are on each side of the equation? **(c)** How many atoms of O are on each side of the equation? **(d)** How many atoms of Fe are on each side of the equation?

2.84 Consider the balanced equation for the combustion of hexane, a component of gasoline:

$$2C_6H_{14} + 19O_2 \longrightarrow 12CO_2 + 14H_2O$$

(a) How many atoms of C are on each side of the equation? **(b)** How many atoms of H are on each side of the equation? **(c)** How many atoms of O are on each side of the equation?

2.85 When sulfur impurities in fuels burn, they produce pollutants such as sulfur dioxide, a major contributor to acid rain. The following is a typical reaction.

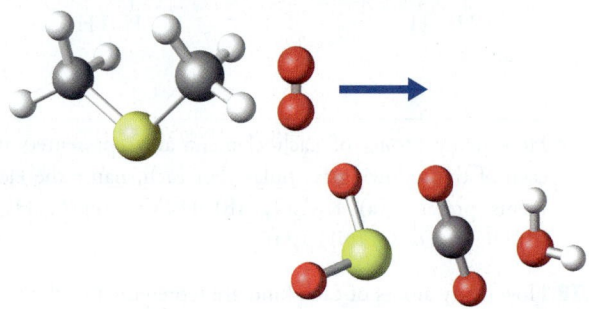

On the left are reactant molecules and on the right are product molecules in a chemical reaction. Is this equation balanced? Write the formulas for the reactants and products.

2.86 Race car drivers can get extra power by burning methyl alcohol with nitrous oxide. Below on the left are the reactant molecules and on the right are product molecules of the chemical reaction. Is this equation balanced? Write the formulas for the reactants and products.

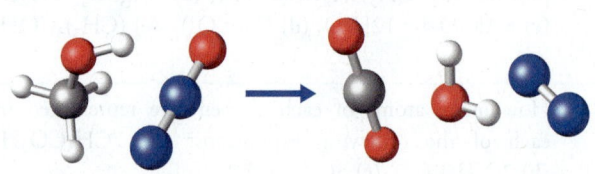

2.87 Is the following chemical equation for the combustion of octane (C_8H_{18}), a component of gasoline, balanced?

$$C_8H_{18} + 12O_2 \longrightarrow 8CO_2 + 9H_2O$$

2.88 Is the following chemical equation balanced? This reaction is used for the production of nitric acid, HNO_3, and is one of the reactions responsible for acid rain.

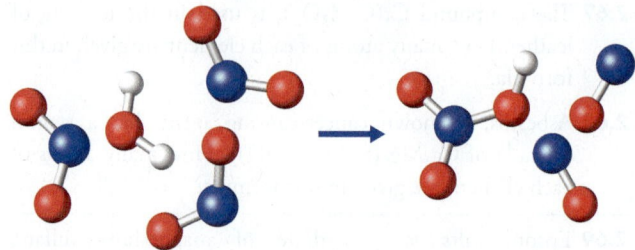

Ionic Compounds

2.89 Use the periodic table, but not Table 2.2, to write the symbols for the ions of **(a)** K, **(b)** Br, **(c)** Mg, **(d)** S, and **(e)** Al.

2.90 Use the periodic table, but not Table 2.2, to write the symbols for ions of **(a)** barium, **(b)** oxygen, **(c)** fluorine, **(d)** strontium, and **(e)** rubidium.

2.91 Write formulas for ionic compounds formed between **(a)** Na and Br, **(b)** K and I, **(c)** Ba and O, **(d)** Mg and Br, **(e)** Al and S, and **(f)** Ba and F.

2.92 Write formulas for ionic compounds formed between **(a)** Al and Br, **(b)** Rb and I, **(c)** Sr and S, **(d)** Mg and N, **(e)** Ga and O, and **(f)** Sr and F.

2.93 Write the formulas for the ionic compounds formed by the following transition metals with the chloride ion, Cl^-: **(a)** cobalt, **(b)** titanium, and **(c)** manganese.

2.94 Write the formulas for the ionic compounds formed by the following transition metals with the bromide ion, Br^-: **(a)** copper, **(b)** zinc, and **(c)** gold.

2.95 Write formulas for the ionic compounds formed from **(a)** K^+ and nitrate ion, **(b)** Ca^{2+} and acetate ion, **(c)** ammonium ion and Cl^-, **(d)** Fe^{3+} and carbonate ion, and **(e)** Mg^{2+} and phosphate ion.

2.96 Write formulas for the ionic compounds formed from **(a)** Zn^{2+} and hydroxide ion, **(b)** Ag^+ and chromate ion, **(c)** Ba^{2+} and sulfite ion, **(d)** Rb^+ and sulfate ion, and **(e)** Li^+ and bicarbonate ion.

2.97 Each of the following metals can form two compounds with oxygen. Write their formulas. **(a)** lead, **(b)** tin, **(c)** manganese, **(d)** iron, and **(e)** copper

2.98 Write formulas for the ionic compounds formed from Cl^- and **(a)** cadmium ion, **(b)** silver ion, **(c)** zinc ion, and **(d)** nickel ion.

Nomenclature of Ionic and Molecular Compounds

2.99 Name the following ionic compounds: **(a)** CaS, **(b)** $AlBr_3$, **(c)** Na_3P, **(d)** Ba_3As_2, **(e)** Rb_2S.

2.100 Name the following ionic compounds: **(a)** NaF, **(b)** Mg_2C, **(c)** Li_3N, **(d)** Al_2O_3, **(e)** K_2Se.

2.101 Name the following molecular compounds: **(a)** SiO_2, **(b)** XeF_4, **(c)** P_4O_{10}, **(d)** Cl_2O_7.

2.102 Name the following molecular compounds: **(a)** ClF_3, **(b)** S_2Cl_2, **(c)** N_2O_5, **(d)** $AsCl_5$.

2.103 Name the following ionic compounds using the Stock system: **(a)** FeS, **(b)** CuO, **(c)** SnO_2, **(d)** $CoCl_2 \cdot 6H_2O$.

2.104 Name the following ionic compounds using the Stock system: **(a)** Mn_2O_3, **(b)** Hg_2Cl_2, **(c)** PbS, **(d)** $CrCl_3 \cdot 4H_2O$.

2.105 Name the following. If necessary, refer to Table 2.4 in Section 2.5. **(a)** $NaNO_2$, **(b)** $KMnO_4$, **(c)** $MgSO_4 \cdot 7H_2O$, **(d)** KSCN

2.106 Name the following. If necessary, refer to Table 2.4 in Section 2.5. **(a)** K_3PO_4, **(b)** $NH_4C_2H_3O_2$, **(c)** $Fe_2(CO_3)_3$, **(d)** $Na_2S_2O_3 \cdot 5H_2O$

2.107 Identify each of the following as molecular or ionic, and give its name:

(a) $CrCl_2$	**(e)** KIO_3	**(h)** AgCN
(b) S_2Cl_2	**(f)** P_4O_6	**(i)** $ZnBr_2$
(c) SO_3	**(g)** $CaSO_3$	**(j)** H_2Se
(d) $NH_4C_2H_3O_2$		

2.108 Identify each of the following as molecular or ionic, and give its name:

(a) $V(NO_3)_3$	**(e)** $GeBr_4$	**(h)** I_2O_4
(b) Au_2S	**(f)** K_2CrO_4	**(i)** I_4O_9
(c) Au_2S_3	**(g)** $Fe(OH)_2$	**(j)** P_4Se_3
(d) $Co(C_2H_3O_2)_2$		

2.109 Write formulas for the following.
(a) sodium monohydrogen phosphate
(b) lithium selenide
(c) chromium(III) acetate
(d) disulfur decafluoride
(e) nickel(II) cyanide
(f) iron(III) oxide
(g) antimony pentafluoride

2.110 Write formulas for the following.
(a) dialuminum hexachloride
(b) tetraarsenic decaoxide
(c) magnesium hydroxide
(d) copper(II) bisulfate
(e) ammonium thiocyanate
(f) potassium thiosulfate
(g) diiodine pentaoxide

2.111 Write formulas for the following.
(a) ammonium sulfide
(b) chromium(III) sulfate hexahydrate
(c) silicon tetrafluoride
(d) molybdenum(IV) sulfide
(e) tin(IV) chloride
(f) hydrogen selenide
(g) tetraphosphorus heptasulfide

2.112 Write formulas for the following.
(a) mercury(II) acetate
(b) barium hydrogen sulfite
(c) boron trichloride
(d) calcium phosphide
(e) magnesium dihydrogen phosphate
(f) calcium oxalate
(g) xenon tetrafluoride

2.113 Which of the following formulas are incorrect? Write the correct formula if possible.

(a) HgCl	**(e)** MnO_4K
(b) $MgSO_3$	**(f)** CaF
(c) FeS	**(g)** NaO_2
(d) $AlCO_3$	

2.114 Which of the following formulas are incorrect? Write the correct formula if possible.

(a) CrO_4Ca	**(e)** $CsCl_2$
(b) $FeSO_3$	**(f)** $BaBr_3$
(c) $NH_4H_2PO_4$	**(g)** $Ca_2Cr_2O_7$
(d) TiO_3	

2.115 Which of the following names are incorrect? Write the correct name if possible; if there are several possibilities; write them all. If the name is correct, write the formula.
(a) iron sulfide
(b) sodium(II) chloride
(c) calcium dibromide
(d) aluminum carbonate
(e) calcium hypochlorite
(f) magnesium(III) phosphate
(g) potassium sulfide(II)

2.116 Which of the following names are incorrect? Write the correct name if possible; if there are several possibilities, write them all. If the name is correct, write the formula.
(a) sulfate of barium
(b) potassium bromide
(c) iron(III) carbonate
(d) aluminum(III) phosphate

(e) calcium selenide

(f) trimagnesium diphosphate

(g) dilithium difluoride

Additional Exercises

2.119 The following are models of molecules of two compounds composed of carbon, hydrogen, and oxygen. The one on the left is ethanol, the alcohol that's added to gasoline with the hope of reducing our dependence on foreign oil supplies. The one on the right is dimethyl ether, a compound that is a gas at room temperature and is used as an aerosol propellant. How many atoms of carbon, hydrogen, and oxygen are in each of these molecules? Find a way to write the formula of one of them using parentheses so that the two compounds can be distinguished one from the other.

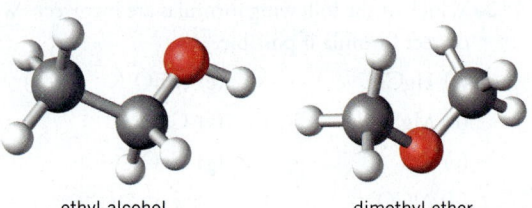

ethyl alcohol   dimethyl ether

2.120 A student obtained a sample from an experiment that had the following composition. If the experiment was repeated and another sample was obtained, would it have to have the exact same composition? Would the results of these experiments illustrate the law of definite proportions? Explain your answer.

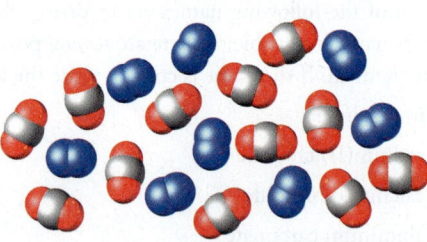

2.121 Suppose you wanted to separate the sample illustrated in the preceding question into its constituent chemical elements. **(a)** Would a chemical change be sufficient? **(b)** If not, what could be accomplished by a chemical change? **(c)** Would the elements isolated by these changes necessarily be composed of individual atoms, or would at least one be composed of molecules?

2.122 The elements in Group 1A and Group 7A of the periodic table are some of the most reactive elements. What is the difference in reactivity between these two groups?

2.117 The compounds Se_2S_6 and Se_2S_4 have been shown to be antidandruff agents. What are their names?

2.118 The compound P_2S_5 is used to manufacture safety matches. What is the name of this compound?

2.123 What are the formulas for mercury(I) nitrate dihydrate and mercury(II) nitrate monohydrate?

2.124 Consider the following substances: Cl_2, CaO, HBr, $CuCl_2$, AsH_3, $NaNO_3$, and NO_2.

(a) Which are binary substances?

(b) Which is a triatomic molecule?

(c) In which do we find only electron sharing?

(d) Which are diatomic?

(e) In which do we find only attractions between ions?

(f) Which are molecular?

(g) Which are ionic?

2.125 Write the balanced chemical equation for the reaction between elements with atomic numbers of **(a)** 20 and 35, **(b)** 6 and 17, **(c)** 13 and 16. For each of these, determine the ratio of the mass of the heavier element to the lighter element in the compound.

2.126 Write the balanced gas phase chemical equation for the reaction of dinitrogen pentoxide with sulfur dioxide to form sulfur trioxide and nitrogen oxide. What small, whole-number ratios are expected for oxygen in the nitrogen oxides and the sulfur oxides?

2.127 Bromine is a diatomic molecule, and it has two isotopes, ^{79}Br and ^{81}Br, with a natural abundance of about 50% for each isotope. How many different Br_2 molecules can be made? What instrument would you use to distinguish between the different molecules?

2.128 Write the symbols, with charges, of all atoms and ions with just ten electrons. Write all of the compounds that can be made by combining the substances with ten electrons.

2.129 Using the figures below, match the image with the correct molecular formula: CH_3CH_2OH, NaCl, $SnCl_4$, and H_2O. Write the names of the compounds.

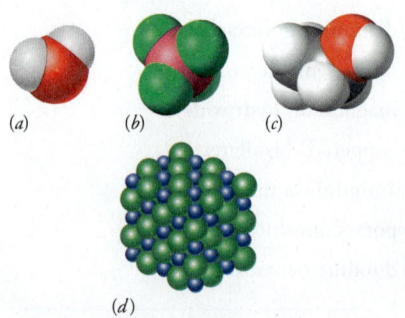

(a)   (b)   (c)

(d)

Exercises in Critical Thinking

2.130 Imagine a world where, for some reason, hydrogen and helium have not been discovered. Would Mendeleev have had enough information to predict their existence?

2.131 Around 1750 Benjamin Franklin knew of two opposite types of electric charge, produced by rubbing a glass rod or amber rod with fur. He decided that the charge developed on the glass rod should be the "positive" charge, and from there on charges were defined. What would have changed if Franklin had decided the amber rod was the positive charge?

2.132 Explore the internet and find a reliable source of physical properties of elements and compounds. Justify how you decided the site was reliable.

2.133 Spreadsheet applications such as Microsoft Excel can display data in a variety of ways, some of which are shown throughout this book. What method of displaying periodic trends (for example, line graphs, tables, bar graphs, 3-D views, etc.) is most effective for your learning style? Explain your answer by stating why your chosen display is better than the others.

3

The Mole and Stoichiometry

Chapter Outline

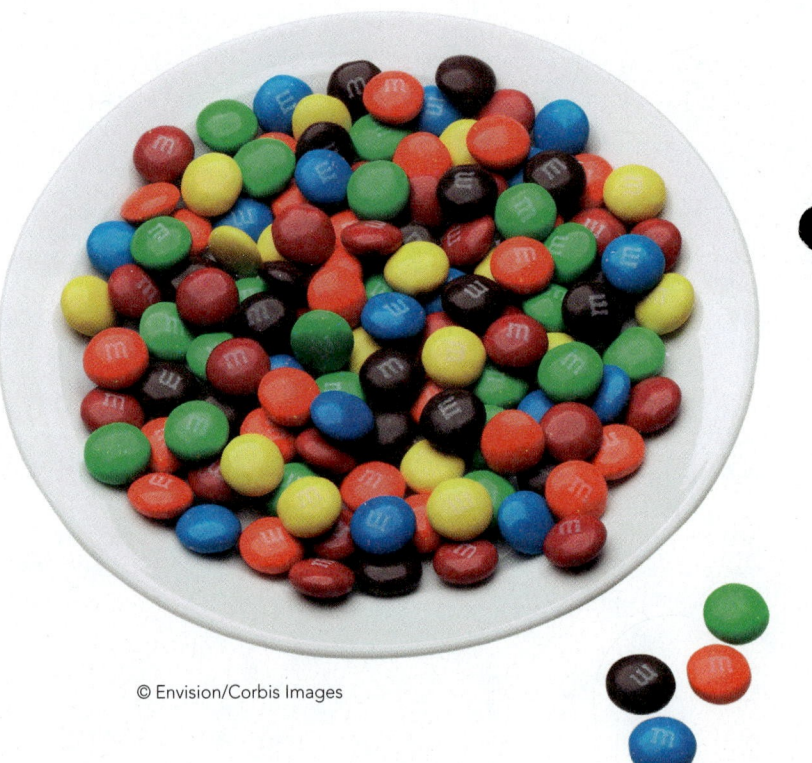

© Envision/Corbis Images

In this chapter we will learn the fundamentals of chemical calculations called **stoichiometry** (stoy-kee-AH-meh-tree), which loosely translates as "the measure of the elements." These calculations are key for success in the chemistry laboratory. You will also find this chapter to be important for future courses in organic chemistry, biochemistry, and almost any other advanced laboratory course in the sciences.

Stoichiometry involves converting chemical formulas and equations that represent individual atoms, molecules, and formula units to the laboratory scale that uses milligrams, grams, and even kilograms of these substances, just as the number of M&Ms in the bowl can be measured by weighing the M&Ms. To do this, we introduce the concept of the *mole*. The mole allows the chemist to scale up from the atomic/molecular level to the laboratory scale, much as a fast-food restaurant scales up the amount of ingredients from a single hamburger to mass-production as in Figure 3.1. Our stoichiometric calculations are usually conversions from one set of units to another using dimensional analysis. To be successful using dimensional analysis calculations we need two things: a knowledge of the equalities that can be made into conversion factors and a logical sequence of steps to guide us from the starting set of units to the desired units. Figure 3.6, at the end of this chapter, is a flowchart that organizes the sequence of conversion steps and the conversion factors used in stoichiometric calculations in this chapter. In the later chapters, as we learn new concepts, they will be added to this flowchart to illustrate how many of our chemical ideas are interrelated.

This Chapter in Context

LEARNING OBJECTIVES

After reading this chapter, you should be able to:

- explain how the mole and Avogadro's number serve as conversion factors between the molecular and laboratory scales of matter
- perform calculations involving moles and amount of substance
- calculate percentage composition of a compound
- describe how to determine the empirical and molecular formulas of a given compound
- perform calculations involving moles of reactants and products in a chemical reaction
- explain how to determine the limiting reactant in a given reaction and the amount of substance remaining unreacted
- calculate the percentage yield of a given reaction

3.1 | The Mole and Avogadro's Number

We can tell from the fundamental measurements of the mass of the proton, neutron, and electron in Chapter 0 that even the largest of the atoms must have extremely small masses and correspondingly small sizes. Consequently, any sample of matter that is observable by the naked eye must have very large numbers of atoms or molecules. The methods of

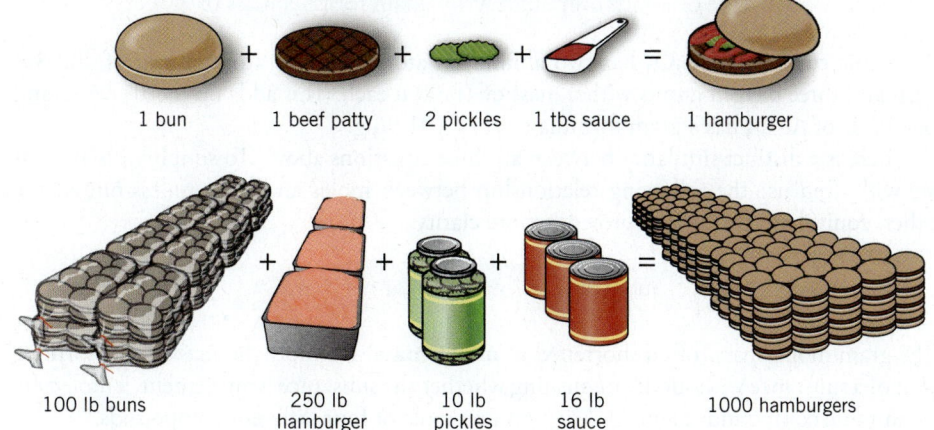

1 bun   +   1 beef patty   +   2 pickles   +   1 tbs sauce   =   1 hamburger

100 lb buns   +   250 lb hamburger   +   10 lb pickles   +   16 lb sauce   =   1000 hamburgers

Figure 3.1 | **Hamburgers on the small and large scale.** Making a single hamburger requires one bun, one hamburger patty, two pickles, and a tablespoon of sauce. To make a thousand hamburgers we need 100 lb of rolls, 250 lb of hamburger meat, 10 lb of pickles, and 16 lb of sauce.

Figure 3.2 | Moles of elements.
Each sample of these elements—
iron, mercury, copper, and
sulfur—contains the same
number of atoms.

TOOLS
Atomic mass

■ Many chemists still use terms like
molecular weight and atomic weight
for molecular mass and atomic mass.

TOOLS
Molecular mass and formula mass

TOOLS
Molar mass

calculation developed in Chapter 1, along with the mole concept, allow us to count by weighing and then use that information to solve some very interesting problems.

Counting by weighing is familiar to everyone even if we are not aware of it. The number of M&Ms in a bag are measured by weighing the bag, and the example of mass-produced hamburgers illustrated another use of counting by weighing. A quarter pound of rice counts out the correct number of rice grains to accompany a family meal. Weighing a bag of dimes, knowing that each dime weighs 2.27 grams, will allow you to calculate the number of coins. Similarly, the mass of a chemical substance can be used to determine the number of atoms or molecules in the sample. This last conversion is possible because of the mole concept.

Defining the Mole

In Chapter 1 we noted that the mole is the SI unit for the amount of substance. The amount of substance does not refer to the mass or volume of your sample, but it does refer to the number of atoms, molecules, or formula units, etc. in your sample. Exactly one **mole (mol)** *is defined as a number equal to the number of atoms in exactly 12 grams of ^{12}C atoms*. From this definition and the fact that the average atomic masses in the periodic table are relative values, we can deduce that we will have a mole of atoms of any element if we weigh an amount equal to the atomic mass in grams (this is often called the gram atomic mass).

> 1 mole of element X = gram atomic mass of X

For example, the atomic mass of sulfur is 32.06 u, and one mole of sulfur will weigh 32.06 g and it will have as many atoms as exactly 12 g of carbon-12. Figure 3.2 shows one mole of some common elements—iron, mercury, copper, and sulfur.

The Mole Concept Applied to Compounds

Molecules and ionic compounds discussed in Chapter 2 have definite formulas. For molecular compounds and elements, adding the atomic masses of all atoms in the formula results in the **molecular mass** (sometimes called the **molecular weight**). The gram molecular mass of a molecular substance (the mass in grams equal to the molecular mass) is also equal to one mole of those molecules.

> 1 mole of molecule X = gram molecular mass of X

For example, the molecular mass of water is 18.02 u, the sum of the atomic masses of two H atoms and one O atom. Similarly, the gram **formula mass** of an ionic compound is the sum of the atomic masses of all the atoms in the formula of an ionic compound expressed with units of grams.

> 1 mole of ionic compound X = gram formula mass of X

The ionic compound Al_2O_3 has two aluminum atoms with an atomic mass of 26.98 u each and three oxygen atoms with a mass of 16.00 u each. This adds up to 101.96 u, and one mole of Al_2O_3 has a gram formula mass of 101.96 g.

There is a distinct similarity between all three equations above. To simplify discussions we will often use the following relationship between moles and mass unless one of the other, equivalent, definitions provides more clarity.

> 1 mole of X = gram molar mass of X

The gram molar mass (often shortened to **molar mass**) is simply the mass of the formula unit of a substance without distinguishing whether the substance is an element, a molecule, or an ionic compound. Figure 3.3 depicts one mole of four different compounds.

Figure 3.3 | **Moles of compounds.** One mole of four different compounds (clockwise from the upper left) copper sulfate pentahydrate, water, sodium chromate, and sodium chloride. Each sample contains the same number of formula units or molecules.

Michael Watson

Converting between Mass and Moles

At this point, we recognize the above relationships or equalities as the necessary information for conversion problems similar to those in Chapter 1. Now, however, the problems will be couched in chemical terms.

Example 3.1
Converting from Grams to Moles

Titanium(IV) oxide is one of the best sunscreens because it completely blocks ultraviolet radiation from reaching the skin. In a preparation of TiO_2, we start with a 23.5 g sample of titanium. How many moles of Ti do we have?

Analysis: The problem starts with a certain mass of titanium and asks us to convert it to moles.

Assembling the Tools: We have a tool for converting the mass into moles that states 1 mol X is equal to the atomic mass of X in grams. Now to make it specific for titanium, replace X with the symbol for titanium and use the atomic mass of Ti to get

$$1 \text{ mol Ti} = 47.867 \text{ g Ti}$$

(When we work problems that include tabulated data, such as the atomic mass of Ti, we will, if possible, *keep at least one more significant figure* than the given information.)

Solution: Start by setting up the problem in the form of an equation showing the number and its units that we start with and the units we want when we're finished.

$$23.5 \text{ g Ti} = ? \text{ mol Ti}$$

start   end

Now use the equality between mass and moles to set up the ratio that will cancel the grams of titanium:

$$23.5 \text{ g Ti} \times \left(\frac{1 \text{ mol Ti}}{47.867 \text{ g Ti}} \right) = 0.491 \text{ mol Ti}$$

Is the Answer Reasonable? First, be sure that the units cancel properly. Second, round all numbers to one or two significant figures and calculate an estimated answer. One way to round the numbers gives us $^{25}/_{50}$, and that is equal to ½. Our answer is close enough (not close would be off by a factor of 10 or more) to ½, so our answer is reasonable.

■ Solving a stoichiometry problem is much like giving directions from your house to your college. You need both the starting and ending points. Stating the problem as an equation gives you these reference points, and the conversion factors get us from the start to the end.

Example 3.2
Converting from Moles to Grams

We need 0.254 moles of $FeCl_3$ for a certain experiment. How many grams do we need to weigh?

Analysis: In the last problem the conversion was between moles and mass for an element; now we are converting between moles and mass, but we have a compound, so we need an additional step to calculate the molar mass of $FeCl_3$.

Assembling the Tools: We need the tool for calculating the molar mass of $FeCl_3$. That is the sum of masses of one mole of iron atoms and three moles of chlorine atoms.

$$\text{Molar mass } FeCl_3 = 55.845 \text{ g/mol} + (3 \times 35.453 \text{ g/mol}) = 162.204 \text{ g/mol}$$

Now the tool that represents the equality between mass and moles can be written as

$$1 \text{ mol } FeCl_3 = 162.204 \text{ g } FeCl_3$$

Solution: The problem is set up by stating the question as an equation:

$$0.254 \text{ mol } FeCl_3 = ? \text{ g } FeCl_3$$

Then we construct the conversion factor from the equality between mass and moles to perform the conversion:

$$0.254 \text{ mol } FeCl_3 \times \left(\frac{162.204 \text{ g } FeCl_3}{1 \text{ mol } FeCl_3} \right) = 41.2 \text{ g } FeCl_3$$

Is the Answer Reasonable? First, we verify that the units cancel properly. Next, we can do an approximate calculation by rounding 0.254 to 0.25 and 162.204 to 160. The arithmetic becomes $0.25 \times 160 = 40$, which gives a result that is very close to the calculated value. We can also round to one significant figure, which gives $0.3 \times 200 = 60$, and we still conclude that the answer is reasonable. Remember that these estimates tell us that our more precise answer, 41.2 g $FeCl_3$, is in the right ballpark.

Practice Exercise 3.1

How many moles of aluminum are there in a 3.47 gram sheet of aluminum foil used to wrap your sandwich for lunch today? (*Hint:* Recall the tool that relates the mass of an element to moles of that element.)

Practice Exercise 3.2

In a reaction to synthesize nitrogen triiodide, 0.023 mol of I_2 are needed. How many grams need to be weighed to measure 0.023 mol of I_2?

Practice Exercise 3.3

Your laboratory balance can weigh samples to three decimal places. If the uncertainty in your weighing is ± 0.002 g, what is the uncertainty expressed in terms of moles, if the sample being weighed is pure potassium sulfate?

Avogadro's Number

■ Avogadro's number was named for Amedeo Avogadro (1776–1856), an Italian chemist who was one of the pioneers of stoichiometry.

The SI definition of the mole refers to a number equal to the number of atoms in exactly 12 g of ^{12}C. Just what is that number? After much experimentation the scientific community agrees that the value, to four significant figures, is 6.022×10^{23}. This value has been named **Avogadro's number**. Now we can write a very important relationship between the atomic scale and the laboratory scale as

TOOLS
Avogadro's number

$$1 \text{ mole of } X = 6.022 \times 10^{23} \text{ units of } X$$

The units of our chemicals can be atoms, molecules, formula units, and so on. This means that one mole of xenon atoms is the same as 6.022×10^{23} atoms of Xe. Similarly, 6.022×10^{23} molecules of NO_2 represents one mole of nitrogen dioxide molecules.

Using Avogadro's Number

The relationships developed above allow us to connect the laboratory scale with the atomic scale using our standard dimensional analysis calculations as shown in the next two examples.

■ Avogadro's number is the link between the moles of a substance and its elementary units. If a problem is on the laboratory scale (atoms or molecules are not mentioned) then Avogadro's number is not needed in the calculation.

Example 3.3
Converting from the Laboratory Scale to the Atomic Scale

Tungsten wire is the filament inside most old-style incandescent light bulbs. In a typical light bulb, the tungsten filament weighs 0.635 grams. How many atoms of tungsten are there in such a light bulb filament?

Analysis: Let's first translate the question into an equation:

$$0.635 \text{ g tungsten} = ? \text{ atoms tungsten}$$

Here we do not have any tool that directly relates grams of tungsten to atoms of tungsten. However, we can start with the grams of tungsten and make the following *sequence of conversions*.

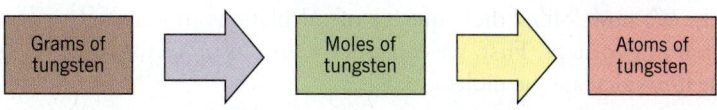

Assembling the Tools: The first tool we need is the mass-to-moles tool,

$$183.84 \text{ g W} = 1 \text{ mol W}$$

to construct conversion factors between grams of tungsten and moles of tungsten. Next, we need Avogadro's number that allows us to construct conversion factors between the moles of tungsten atoms and the number of tungsten atoms,

$$1 \text{ mol W} = 6.022 \times 10^{23} \text{ atoms W}$$

Solution: Having expressed the question as an equation, we can use it along with the two tools for the conversions to construct the two conversion factors. The first is

$$\frac{1 \text{ mol W}}{183.84 \text{ g W}}$$

and the second conversion factor is

$$\frac{6.022 \times 10^{23} \text{ atoms W}}{1 \text{ mol W}}$$

We multiply the original 0.635 grams of tungsten by the conversion factors to get

$$0.635 \text{ g W} \times \left(\frac{1 \text{ mol W}}{183.84 \text{ g W}} \right) \times \left(\frac{6.022 \times 10^{23} \text{ atoms W}}{1 \text{ mol W}} \right) = 2.08 \times 10^{21} \text{ atoms W}$$

Is the Answer Reasonable? The most important check in this type of question involves the magnitude of the numbers. We know that even the smallest measurable sample of a chemical must contain a very large number of atoms. Our answer is a very large number, and therefore the answer seems reasonable.

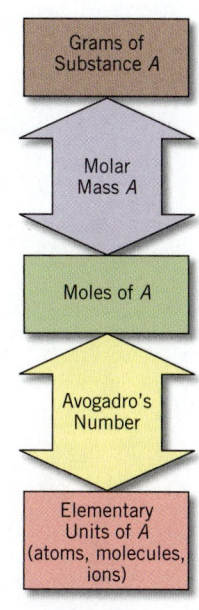

General sequence of calculations to convert between mass and elementary units of a substance. Arrows indicate which tools apply to each conversion.

Example 3.4
Calculating the Mass of a Molecule

Carbon tetrachloride was used as a dry-cleaning fluid until it was found to be carcinogenic. What is the average mass of one molecule of carbon tetrachloride?[1]

Analysis: First, we can set up the question as an equation:

$$1 \text{ molecule carbon tetrachloride} = ? \text{ g carbon tetrachloride}$$

We can see that this problem has many steps that need to be combined to obtain the complete solution. First, we need to identify the chemical formula from its name; then we need to use that to calculate the molar mass of the compound. Finally, we need to construct the appropriate conversion factors to first convert one molecule to moles and then convert those moles into the mass of the carbon tetrachloride molecule.

Assembling the Tools: We first use the nomenclature tools in Chapter 2 to determine the formula for carbon tetrachloride. Next, we need the tool that uses Avogadro's number,

$$6.022 \times 10^{23} \text{ molecules carbon tetrachloride} = 1 \text{ mol carbon tetrachloride}$$

After that we need to calculate the molar mass of carbon tetrachloride and use it to calculate the mass of one molecule of carbon tetrachloride.

Solution: Using nomenclature tools, we determine that the formula for carbon tetrachloride is CCl_4. We now follow the sequence of calculations in our analysis, and will do the calculation in two steps. First, we construct a conversion factor using Avogadro's number and use it to calculate the moles of CCl_4.

$$1 \text{ molecule } CCl_4 \times \left(\frac{1 \text{ mol } CCl_4}{6.022 \times 10^{23} \text{ molecules } CCl_4} \right) = 1.661 \times 10^{-24} \text{ mol } CCl_4$$

For the molar mass of CCl_4,

$$\text{Molar mass } CCl_4 = 12.011 \text{ g/mol} + (4 \times 35.453 \text{ g/mol}) = 153.823 \text{ g } CCl_4/\text{mol}$$

We added the masses of one mole of carbon and four moles of chlorine atoms, getting 153.823 g/mol. We then use this molar mass to make a second conversion factor to convert mol CCl_4 to g CCl_4.

$$1.661 \times 10^{-24} \text{ mol } CCl_4 \times \left(\frac{153.823 \text{ g } CCl_4}{1 \text{ mol } CCl_4} \right) = 2.555 \times 10^{-22} \text{ g } CCl_4$$

One molecule can be considered to be an exact number. Therefore, the number of significant figures to keep in the answer depends upon the number of significant figures taken from the tabulated data. Four significant figures were used for Avogadro's number and four significant figures were kept in the answer.

Is the Answer Reasonable? We expect that a single molecule, even a very large molecule, would have a very small mass. Since our answer is very small, it seems reasonable.

Practice Exercise 3.4

How many atoms of gold are there in one ounce of gold? (*Hint:* How many grams are in an ounce?)

Practice Exercise 3.5

What is the average mass of one molecule of table sugar, $C_{12}H_{22}O_{11}$?

---

[1]Since atomic masses given in the periodic table are weighted averages of naturally occurring isotopes, we cannot determine the exact mass of a molecule unless we know the isotope of each atom. We can, however, calculate an average mass as requested in this question.

Practice Exercise 3.6

Most chemistry laboratories have balances that can weigh to the nearest milligram. Would it be possible to weigh 5.64×10^{18} molecules of octadecane, $C_{18}H_{38}$, on such a balance?

3.2 | The Mole, Formula Mass, and Stoichiometry

The mole ratio concept that we develop here is a very powerful and important chemical tool. It allows the chemist to start with the amount of one substance and then find the chemically equivalent amount of another substance without doing the actual experiment.

Mole-to-Mole Conversion Factors

The chemical formula gives us the ratio of the substances in the formula, and using these ratios, we can determine the amounts of the other substances in the formula.

Consider the chemical formula for water, H_2O:

- One molecule of water contains 2 H atoms and 1 O atom.
- Two molecules of water contain 4 H atoms and 2 O atoms.
- A dozen molecules of water contain 2 dozen H atoms and 1 dozen O atoms.
- A mole of molecules of water contains 2 moles of H atoms and 1 mole of O atoms.

Whether we're dealing with atoms, dozens of atoms, or moles of atoms, the chemical formula tells us that the ratio of H atoms to O atoms is always 2-to-1. In addition, we can write the following equivalencies concerning water molecules and moles of water molecules:

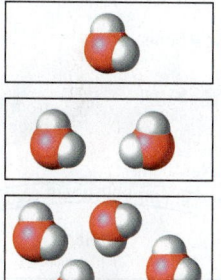

One water molecule	One mole of water molecules
1 molecule $H_2O \Leftrightarrow 2$ atoms H	1 mol $H_2O \Leftrightarrow 2$ mol H
1 molecule $H_2O \Leftrightarrow 1$ atom O	1 mol $H_2O \Leftrightarrow 1$ mol O
1 atom O $\Leftrightarrow$ 2 atoms H	1 mol O $\Leftrightarrow$ 2 mol H

Recall that the symbol $\Leftrightarrow$ means "is chemically equivalent to" and that it is treated mathematically as an equal sign (see Section 1.6).

Within chemical compounds, moles of atoms always combine in the same ratio as the individual atoms themselves.

TOOLS
Mole ratios

This tool allows us to use the atom-to-atom ratios in a chemical formula to easily prepare mole-to-mole conversion factors for calculations on the laboratory scale. For example, in the formula P_4O_{10}, the subscripts mean that there are 4 atoms of P for every 10 atoms of O in the molecule. On the laboratory scale, this also means that there are 4 moles of P for every 10 moles of O in 1 mole of this compound. We can relate P and O within the compound using the following conversion factors.

$$4 \text{ mol P} \Leftrightarrow 10 \text{ mol O} \quad \text{from which we write} \quad \frac{4 \text{ mol P}}{10 \text{ mol O}} \quad \text{or} \quad \frac{10 \text{ mol O}}{4 \text{ mol P}}$$

The formula P_4O_{10} also implies other equivalencies, each with its two associated conversion factors.

$$1 \text{ mol } P_4O_{10} \Leftrightarrow 4 \text{ mol P} \quad \text{or} \quad \frac{1 \text{ mol } P_4O_{10}}{4 \text{ mol P}} \quad \text{or} \quad \frac{4 \text{ mol P}}{1 \text{ mol } P_4O_{10}}$$

$$1 \text{ mol } P_4O_{10} \Leftrightarrow 10 \text{ mol O} \quad \text{or} \quad \frac{1 \text{ mol } P_4O_{10}}{10 \text{ mol O}} \quad \text{or} \quad \frac{10 \text{ mol O}}{1 \text{ mol } P_4O_{10}}$$

Example 3.5
Calculating the Amount of a Compound by Analyzing One Element

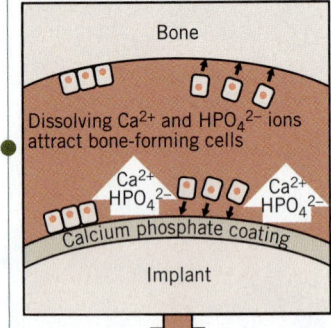

Dissolving Ca^{2+} and HPO_4^{2-} ions attract bone-forming cells

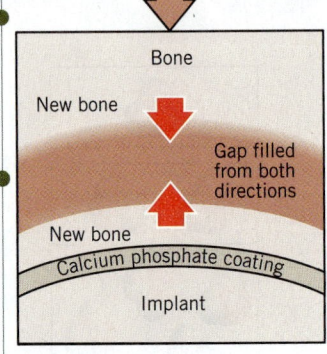

Some surfaces on bone implants are coated with calcium phosphate to permit the bone to actually bond to the surface.

Calcium phosphate is widely found in nature in the form of natural minerals. It is also found in bones and some kidney stones. In one case a sample is found to contain 0.864 moles of phosphorus. How many moles of $Ca_3(PO_4)_2$ are in that sample?

Analysis: Let's state the question in equation form first:

$$0.864 \text{ mol P} = ? \text{ mol } Ca_3(PO_4)_2$$

Since we start with moles of one substance and end with moles of a second, this is the appropriate place to use the mole ratio conversion factor.

Assembling the Tools: All we need is the mole ratio tool that relates P to the chemical formula $Ca_3(PO_4)_2$. We write it as

$$2 \text{ mol P} \Leftrightarrow 1 \text{ mol } Ca_3(PO_4)_2$$

Solution: Starting with the equivalence we expressed above, our tool is rearranged into a conversion factor so that the mol P cancels and we are left with the mol $Ca_3(PO_4)_2$. Applying that ratio we get

$$0.864 \text{ mol P} \left(\frac{1 \text{ mol } Ca_3(PO_4)_2}{2 \text{ mol P}} \right) = 0.432 \text{ mol } Ca_3(PO_4)_2$$

Is the Answer Reasonable? For a quick check, you can round 0.864 to 1 and divide by 2 to get 0.5. There is little difference between our estimate 0.5 and the calculated answer 0.432, and we conclude our answer is reasonable.

Practice Exercise 3.7

Aluminum sulfate is analyzed, and the sample contains 0.0774 moles of sulfate ions. How many moles of aluminum are in the sample? (*Hint:* Recall the tool for writing the correct formula for aluminum sulfate.)

Practice Exercise 3.8

How many moles of nitrogen atoms are combined with 8.60 mol of oxygen atoms in dinitrogen pentoxide?

Practice Exercise 3.9

Sodium bicarbonate is used to soften water. How many moles of oxygen atoms are found in 1.29 moles of sodium bicarbonate?

Mass-to-Mass Calculations

One common use of stoichiometry in the lab occurs when we need to determine the mass of one reactant, *B*, needed to combine with a given mass of second reactant, *A*, to make a compound. These calculations are summarized by the following sequence of steps to convert the given mass of compound *A* to the mass of compound *B*.

Mass-to-mass conversions using formulas

In the following example we see how this is applied.

Example 3.6
Calculating the Amount of One Element from the Amount of Another in a Compound

Chlorophyll, the green pigment in leaves, has the formula $C_{55}H_{72}MgN_4O_5$. If 0.0011 g of Mg is available to a plant for chlorophyll synthesis, how many grams of carbon will be required to completely use up the magnesium?

Analysis: Let's begin, as usual, by restating the problem as an equivalency

$$0.0011 \text{ g Mg} \Leftrightarrow ? \text{ g C}$$

We will use the sequence of steps in the preceding example to relate the mass of one substance to the mass of another. Our first step is to convert the mass of Mg to moles of Mg. Once we know the moles of Mg, we can convert that to the moles of C using the formula of the compound. Finally, we can calculate the mass of the second substance, C, from the moles using its molar mass.

Assembling the Tools: For the first step, we will need the mass-to-moles tool to convert the mass of Mg to moles of Mg,

$$24.3050 \text{ g Mg} = 1 \text{ mol Mg}$$

Next we need a mole ratio to convert moles of Mg to moles of C,

$$1 \text{ mol Mg} \Leftrightarrow 55 \text{ mol C} \quad \text{(these are exact numbers)}$$

Finally, the mass-to-moles tool for carbon is

$$1 \text{ mol C} = 12.011 \text{ g C}$$

Our complete sequence for the problem, with numbers rounded to three significant figures, is

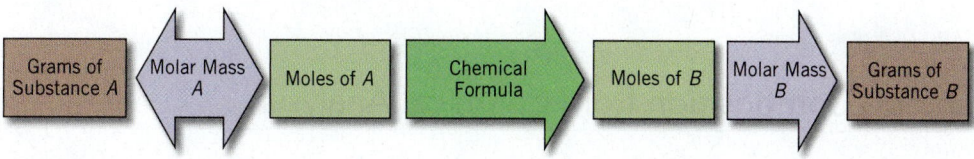

| 1 mol Mg ⇔ 24.3 g Mg | 1 mol Mg ⇔ 55 mol C | 1 mol C ⇔ 12.0 g C |

$$0.0011 \text{ g Mg} \longrightarrow \text{mol Mg} \longrightarrow \text{mol C} \longrightarrow \text{g C}$$

The general sequence for dealing with problems like the one we're solving here is illustrated below. The arrows indicate which tools are used for each conversion.

Grams of Substance *A* ⟷ Molar Mass *A* → Moles of *A* → Chemical Formula → Moles of *B* → Molar Mass *B* → Grams of Substance *B*

Solution: We now set up the solution by forming conversion factors so the units cancel.

$$0.0011 \text{ g Mg} \left(\frac{1 \text{ mol Mg}}{24.3 \text{ g Mg}} \right) \left(\frac{55 \text{ mol C}}{1 \text{ mol Mg}} \right) \left(\frac{12.0 \text{ g C}}{1 \text{ mol C}} \right) = 0.030 \text{ g C}$$

A plant cell must supply 0.030 g C for every 0.0011 g Mg to completely use up the magnesium in the synthesis of chlorophyll.

> ■ In a problem that asks to convert the amount (grams, moles, or atomic scale units) of one substance into the amount of a different substance, the mole-to-mole relationship between the two substances must be used.

Is the Answer Reasonable? After checking that our units cancel properly, a quick estimate can be made by rounding all numbers to one significant figure, giving the following expression (without units):

$$0.001\left(\frac{1 \times 50 \times 10}{20 \times 1 \times 1}\right) = \left(\frac{0.5}{20}\right) = \frac{0.05}{2} = 0.025$$

This value is close to the answer we got and gives us confidence that it is reasonable. (Note that if we rounded the 55 up to 60 our estimate would have been 0.030, which would still confirm our conclusion.)

Practice Exercise 3.10   How many grams of iron are needed to combine with 25.6 g of O to make Fe_2O_3? (*Hint*: Recall the mole ratios that the formula represents.)

Practice Exercise 3.11   Bromine trifluoride explodes upon contact with water. If a sample of bromine trifluoride has 0.163 g of bromine, how many grams of fluorine are in the sample?

3.3 | Chemical Formula and Percentage Composition

The formula of the compound gives the mole ratio of elements in a compound. Another way to describe a compound's makeup is to use the relative the masses of the elements in the compound as a list of percentages by mass, called the compound's **percentage composition**.

Percentage Composition

The **percentage by mass** of an element in a sample is the number of grams of the element present in 100 g of the sample. In general, a percentage by mass is found by using the following equation.

Percentage composition

$$\text{Percentage by mass of elements} = \frac{\text{mass of element}}{\text{mass of whole sample}} \times 100\% \qquad (3.1)$$

We can determine the percentage composition based on chemical analysis of a substance as shown in the next example.

Example 3.7
Calculating a Percentage Composition from Chemical Analysis

A sample of a liquid with a mass of 8.657 g was decomposed into its elements and gave 5.217 g of carbon, 0.9620 g of hydrogen, and 2.478 g of oxygen. What is the percentage composition of this compound?

Analysis: Solving problems often requires that we know the meaning of key terms, in this case percentage composition, which is the list of the percentage by mass of the elements in the compound.

Assembling the Tools: The tool we need is the percentage by mass equation, Equation 3.1. We are given the mass of each element in the sample, and the total mass, which is also the sum of the masses of the elements.

Solution: Using Equation 3.1 for each element gives three equations that we use to compute the needed percentages:

$$\text{For C:} \quad \frac{5.217 \text{ g}}{8.657 \text{ g}} \times 100\% = 60.26\% \text{ C}$$

$$\text{For H:} \quad \frac{0.9620 \text{ g}}{8.657 \text{ g}} \times 100\% = 11.11\% \text{ H}$$

$$\text{For O:} \quad \frac{2.478 \text{ g}}{8.657 \text{ g}} \times 100\% = 28.62\% \text{ O}$$

Sum of percentages: 99.99%

Is the Answer Reasonable? The "check" is that the percentages must add up to 100%, allowing for small differences caused by rounding. We can also check the individual results by rounding all the numbers to one significant figure to estimate the results. For example, the percentage C would be estimated as $5/9 \times 100$, which is a little over 50% and is close to our answer.

When 0.5462 g of a compound was decomposed, 0.2012 g of nitrogen and 0.3450 g of oxygen were isolated. What is the percentage composition of this compound? Explain how you can determine if there are any other elements present in this compound. (*Hint:* What formula gives the percentage by mass of an element?)

Practice Exercise 3.12

An organic compound weighing 0.6672 g is decomposed, giving 0.3481 g carbon and 0.0870 g hydrogen. What are the percentages of hydrogen and carbon in this compound? Is it likely that this compound contains another element?

Practice Exercise 3.13

We can also determine the percentage composition of a compound from its chemical formula. If we consider one mole of a substance, its molar mass will be the mass of the whole sample in Equation 3.1. The numerator in Equation 3.1 will be the mass associated with one of the elements in the formula. In the next section we see how this calculated percentage is useful for identifying a substance.

Percentage Composition and Chemical Identity

We can use Equation 3.1 to determine the percentage composition of any chemical compound from its formula. Nitrogen and oxygen, for example, form all of the following compounds: N_2O, NO, NO_2, N_2O_3, N_2O_4, and N_2O_5. To identify an unknown sample of a compound of nitrogen and oxygen, one might compare the percentage composition found by experiment with the calculated, or theoretical, percentages for each possible formula. A strategy for matching formulas with mass percentages is outlined in the following example.

Example 3.8
Identifying a Compound Based on Percentage Composition

Do the mass percentages of 25.94% N and 74.06% O match the formula N_2O_5?

Analysis: To calculate the mass percentages we need to use Equation 3.1. Looking at the equation, we will need the masses of N, O, and N_2O_5 in a sample of the compound. *If we choose 1 mole of the given compound to be this sample, it will be easy to determine the mass of the oxygen, nitrogen, and dinitrogen pentoxide that make up one mole of N_2O_5.*

Assembling the Tools: Our tool will be Equation 3.1, the equation to calculate the percentage by mass of an element in a compound. We will also need the tool for calculating the molar mass of N_2O_5 and the mole ratios we can obtain from that formula. We need to perform a mass-to-moles conversion for nitrogen and oxygen atoms.

The required relationships are:

$$2 \text{ mol N atoms} \Leftrightarrow 1 \text{ mol } N_2O_5$$

$$5 \text{ mol O atoms} \Leftrightarrow 1 \text{ mol } N_2O_5$$

$$1 \text{ mol N atoms} = 14.01 \text{ g N atoms}$$

$$1 \text{ mol O atoms} = 16.00 \text{ g O atoms}$$

Solution: We know that 1 mol of N_2O_5 must contain 2 mol N and 5 mol O from the mol ratio tool. The corresponding number of grams of N and O are found as follows.

$$1 \text{ mol } N_2O_5\left(\frac{2 \text{ mol N}}{1 \text{ mol } N_2O_5}\right)\left(\frac{14.01 \text{ g N}}{1 \text{ mol N}}\right) = 28.02 \text{ g N}$$

$$1 \text{ mol } N_2O_5\left(\frac{5 \text{ mol O}}{1 \text{ mol } N_2O_5}\right)\left(\frac{16.00 \text{ g O}}{1 \text{ mol O}}\right) = 80.00 \text{ g O}$$

$$1 \text{ mole } N_2O_5 = 108.02 \text{ g } N_2O_5$$

Now we can calculate the percentages.

For % N: $\qquad \dfrac{28.02 \text{ g}}{108.02 \text{ g}} \times 100\% = 25.94\% \text{ N in } N_2O_5$

For % O: $\qquad \dfrac{80.00 \text{ g}}{108.02 \text{ g}} \times 100\% = 74.06\% \text{ O in } N_2O_5$

These experimental values do match the theoretical percentages for the formula N_2O_5.

Is the Answer Reasonable? The easiest check for this problem is to be sure that all percentages add up to 100%. They do add to 100%, and so the calculated result is reasonable.

Practice Exercise 3.14

Calculate the theoretical percentage composition of N_2O. (*Hint:* Recall the definition of percentage composition.)

Practice Exercise 3.15

Calculate the theoretical percentage compositions for NO, NO_2, N_2O_3, and N_2O_4. Which of these compounds produced the data in Practice Exercise 3.12?

In working Practice Exercise 3.15, you may have noticed that the percentage composition of NO_2 was the same as that of N_2O_4. This is because both compounds have the same mole ratio. All compounds with the same mole ratios will have the same percentage compositions.

3.4 | Determining Empirical and Molecular Formulas

One of the major activities of chemists is to synthesize compounds that have never existed before. In pharmaceutical research, chemists often synthesize entirely new compounds, or isolate new compounds from plant and animal tissues. They must then determine the formula and structure of the new compound. Modern chemists use mass spectroscopy and other modern instruments for structure analysis. However, they still rely upon elemental analysis, where the compound is decomposed chemically to find the masses of elements within a given amount of compound to determine *empirical formulas*. Let's see how such experimental mass measurements, expressed in a variety of ways, can be used to determine these formulas.

A form of the element phosphorus known as "white phosphorus" is *pyrophoric*—that is, it spontaneously burns when exposed to air. The compound that forms when white phosphorus burns in oxygen consists of molecules with the formula P_4O_{10}. When a formula gives the composition of one *molecule*, it is called a **molecular formula**. Notice, however, that both the subscripts 4 and 10 are divisible by 2, so the *smallest* numbers that tell us the *ratio* of P to O are 2 and 5. We can write a simpler (but less informative) formula that expresses this ratio, P_2O_5. This is called the **empirical formula** because it can also be obtained from an experimental analysis of the compound.

The empirical formula expresses the simplest ratio of the atoms of each element in a compound. We already know that the ratio of atoms in a compound is the same as a ratio of the moles of those atoms in the compound. We will determine the simplest ratio of moles from experimental data. The experimental data we need is any information that allows us to determine the moles of each element in a sample of the compound. We will investigate three types of data that can be used to determine empirical formulas. They are (a) masses of the elements, (b) percentage composition, and (c) combustion data, also known as indirect analysis. In all three, the goal is to obtain the simplest ratio of moles of each element in the formula.

TOOLS

Empirical formulas

Empirical Formulas from Mass Data

If we determine the mass of each element in a pure sample of a compound we can calculate the moles of each element. From that we can find the simplest ratio of moles, which by definition is the empirical formula. In many instances, we can analyze a sample for all but one element and then use the law of conservation of mass to calculate the missing mass, as shown in the next example.

Example 3.9
Calculating an Empirical Formula from Mass Data

A 2.57 g sample of a compound composed of only tin and chlorine was found to contain 1.17 g of tin. What is the compound's empirical formula?

Analysis: The subscripts in an empirical formula can be interpreted as the relative number of moles of the elements in a compound. If we can find the *mole* ratio of Sn to Cl, we will have the empirical formula. The first step, therefore, is to convert the numbers of grams of Sn and Cl to the numbers of moles of Sn and Cl. Then we convert these numbers into their simplest positive *whole-number*, or *integer*, ratio.

The problem did not give the mass of chlorine in the 2.57 g sample. However, we can calculate it using the tool that expresses the law of conservation of mass.

Assembling the Tools: Our first tool will be the law of conservation of mass from Chapter 0, which requires that the mass of compound equal the sum of the mass of Cl and the mass of Sn. The molar mass is the tool that allows us to calculate the moles of each element.

$$118.7 \text{ g Sn} = 1 \text{ mol Sn}$$

$$35.45 \text{ g Cl} = 1 \text{ mol Cl}$$

Solution: First, we find the mass of Cl in 2.57 g of compound:

$$\text{Mass of Cl} = 2.57 \text{ g compound} - 1.17 \text{ g Sn} = 1.40 \text{ g Cl}$$

Now we use the molar mass equalities to create ratios to convert the mass data for tin and chlorine into moles.

$$1.17 \text{ g Sn} \left(\frac{1 \text{ mol Sn}}{118.7 \text{ g Sn}} \right) = 0.00986 \text{ mol Sn}$$

$$1.40 \text{ g Cl} \left(\frac{1 \text{ mol Cl}}{35.45 \text{ g Cl}} \right) = 0.0395 \text{ mol Cl}$$

We could now write a formula: $Sn_{0.00986}Cl_{0.0395}$, which does express the mole ratio, but subscripts also represent atom ratios and need to be integers. To convert the decimal subscripts to integers, we begin by dividing each by the smallest number in the set. *This is always the way to begin the search for whole-number subscripts; pick the smallest number of the set as the divisor.* It's guaranteed to make at least one subscript a whole number—namely, 1. Here, we divide both numbers by 0.00986.

$$Sn_{\frac{0.00986}{0.00986}}Cl_{\frac{0.0395}{0.00986}} = Sn_{1.00}Cl_{4.01}$$

In most cases if, after this step, a calculated subscript differs from a whole number by less than 0.1, we can safely round to the nearest whole number. We may round 4.01 to 4, so the empirical formula is $SnCl_4$.

Is the Answer Reasonable? In addition to the fact that whole-number subscripts were easily found, you should also recall from Chapter 2 that tin forms either the Sn^{2+} or the Sn^{4+} ion and that chlorine forms only the Cl^- ion. Therefore either $SnCl_2$ or $SnCl_4$ is a reasonable compound, and $SnCl_4$ was our answer.

Practice Exercise 3.16

A 1.525 g sample of a compound between nitrogen and hydrogen contains 1.333 g of nitrogen. Calculate its empirical formula. (*Hint:* How many grams of hydrogen are there?)

Practice Exercise 3.17

Sulfur forms two different compounds with oxygen, SO_2 and SO_3. A 1.525 g sample of a compound between sulfur and oxygen was prepared by burning 0.7625 g of sulfur in air and collecting the product. From the empirical formula for the compound formed, which compound is it?

Sometimes our strategy of using the lowest common divisor does not give whole numbers. If a decimal value corresponding to a rational fraction results, we can obtain integers by multiplying by the denominator of the rational fraction.

Let's go through an empirical formula calculation for the analysis of a compound of iron and oxygen called "black iron oxide," which is found in the mineral magnetite and is used in magnetic ink. When a 2.448 g sample is analyzed it is found to have 1.771 g of Fe and 0.677 g of O. We can determine the moles of iron and oxygen

$$1.771 \text{ g Fe} \left(\frac{1 \text{ mol Fe}}{55.845 \text{ g Fe}} \right) = 0.03171 \text{ mol Fe}$$

$$0.677 \text{ g O} \left(\frac{1 \text{ mol O}}{16.00 \text{ g O}} \right) = 0.0423 \text{ mol O}$$

The mineral magnetite, like any magnet, is able to affect the orientation of a compass needle.

These results let us write the formula as $Fe_{0.03171}O_{0.0423}$. We know that we need whole numbers for the subscripts since the formula represents atoms as well as moles, so let's divide by the smaller number, 0.03171.

$$Fe_{\frac{0.03171}{0.03171}}O_{\frac{0.0423}{0.03171}} = Fe_{1.000}O_{1.33}$$

This time we cannot round 1.33 to 1.0 because the decimal we want to round, 0.33, is larger than the 0.1 criterion established previously. In a mole sense, the ratio of 1 to 1.33 is correct; we just need a way to re-state this ratio in whole numbers. To do this, we need to recognize that the decimal 0.33 represents ⅓ (one-third). The denominator is 3, and if we multiply both subscripts by 3 we will get

$$Fe_{(1.000 \times 3)}O_{(1.33 \times 3)} = Fe_{3.000}O_{3.99}$$

or a formula of Fe_3O_4. Table 3.1 lists some common decimals and their related fractions.

A less mathematical method that also works is trial and error. First, multiply all subscripts by 2; if that does not give integer subscripts, go back and try 3, then 4, 5, and so on. In this example you would have found integer subscripts after multiplying by 3.

TABLE 3.1	Decimal and Rational Fractions
Decimal	Fraction[a]
0.20	⅕
0.25	¼
0.33	⅓
0.40	⅖
0.50	½
0.60	⅗
0.66	⅔
0.75	¾
0.80	⅘

[a]Use the denominator of the fractions as a multiplier to increase whole-number subscripts in empirical formulas.

TOOLS

Determination of integer subscripts

Practice Exercise 3.18

A compound of nitrogen and iodine is an extremely sensitive contact explosive that snaps and gives off a purple gas when a small crystal is touched lightly. A sample of this compound was analyzed to give 0.259 g nitrogen and 7.04 g iodine. What is the empirical formula of this compound? (*Hint:* What are the atomic masses of nitrogen and iodine?)

Practice Exercise 3.19

When aluminum is produced by electrolysis we get 5.68 tons of aluminum and 5.04 tons of oxygen. What is the empirical formula of the compound that is being electrolyzed?

Empirical Formulas from Experimental Mass Percentages

It is not always possible to obtain the masses of every element in a compound by the use of just one weighed sample. Two or more analyses carried out on different samples are often needed. For example, suppose a compound that needs to be analyzed is known to consist exclusively of calcium, chlorine, and oxygen. The mass of calcium in one weighed sample and the mass of chlorine in another sample would be determined in separate experiments. Then the mass data for calcium and chlorine would be converted to percentages by mass *so that the data from different samples relate to the same sample size.* The percentage of oxygen would be calculated by difference because % Ca + % Cl + % O = 100%. If we choose our sample size to be a 100 g sample of our compound, each mass percentage of an element would represent the number of grams of that element in the sample. From here, the masses are next converted into the corresponding number of moles of each element. Finally, mole proportions are converted to integers in the way we just studied, giving us the subscripts for the empirical formula. Let's see how this works.

TOOLS

Empirical formulas from percentage composition

Example 3.10
Calculating an Empirical Formula from Percentage Composition

A white powder used in paints, enamels, and ceramics has the following percentage composition: Ba, 69.6%; C, 6.09%; and O, 24.3%. What is its empirical formula? What is the name of this compound?

■ We see a general principle that by assuming a 100 g sample, all percent signs can be easily changed to gram units for our calculations.

Analysis: If we do the calculation starting with 100 grams of this compound, then the percentages of the elements given in the problem are numerically the same as the masses of these elements in a 100 g sample.

Now that we have the masses of the elements we can calculate the moles of each element and then the integer values for the subscripts as we did in the previous examples.

Assembling the Tools: We will need the molar mass as our tool to determine the moles of each element. We will also need the procedures for finding integer subscripts.

Solution: Assuming a 100 g sample of the compound, we convert 69.6% Ba to 69.6 g Ba, 6.09% C to 6.09 g C, and 24.3% O to 24.3 g O which are then converted to moles.

$$\text{Ba:} \quad 69.6 \text{ g Ba} \left(\frac{1 \text{ mol Ba}}{137.3 \text{ g Ba}} \right) = 0.507 \text{ mol Ba}$$

$$\text{C:} \quad 6.09 \text{ g C} \left(\frac{1 \text{ mol C}}{12.01 \text{ g C}} \right) = 0.507 \text{ mol C}$$

$$\text{O:} \quad 24.3 \text{ g O} \left(\frac{1 \text{ mol O}}{16.00 \text{ g O}} \right) = 1.52 \text{ mol O}$$

Our preliminary empirical formula is then

$$\text{Ba}_{0.507}\text{C}_{0.507}\text{O}_{1.52}$$

We next divide each subscript by the smallest value, 0.507.

$$\text{Ba}_{\frac{0.507}{0.507}}\text{C}_{\frac{0.507}{0.507}}\text{O}_{\frac{1.52}{0.507}} = \text{Ba}_{1.00}\text{C}_{1.00}\text{O}_{3.00}$$

The subscripts are whole numbers, so the empirical formula is $BaCO_3$, representing barium carbonate.

Is the Answer Reasonable: First of all, we found simple, whole-number subscripts, which strongly suggest that the answer is right. In addition, our knowledge of ionic compounds and the polyatomic ions from Chapter 2 tells us that the combination of a barium ion, Ba^{2+}, and the carbonate ion, CO_3^{2-}, yields the same formula, $BaCO_3$. We can conclude that our answer is reasonable.

Practice Exercise 3.20

A white solid used to whiten paper has the following percentage composition: Na, 32.4%; S, 22.6%. The unanalyzed element is oxygen. What is the compound's empirical formula? (*Hint:* What law allows you to calculate the % oxygen?)

Practice Exercise 3.21

Cinnamon gets some of its flavor from cinnamaldehyde that is 81.79% C, 6.10% H, and the rest is oxygen. Determine the empirical formula for this compound.

The law of conservation of mass and its use with percentage composition is important because it allowed us to determine the amount of three substances using only two experiments. This in itself is a considerable saving in time and effort. Additionally, it is often difficult to analyze a sample for certain elements such as oxygen, and using percentage measurements helps avoid this problem.

Empirical Formulas from Indirect Analysis

In practice, a compound is seldom broken down completely to its *elements* in a quantitative analysis. Instead, the compound is changed into other *compounds*. The reactions separate the elements by capturing each one entirely (quantitatively) in a *separate* compound *whose formula is known*.

ON THE CUTTING EDGE | 3.1

Combustion Analysis

Determining the mass of carbon and hydrogen in a compound can be done by burning the compound in pure oxygen in the presence of substances called catalysts that speed up the reaction to ensure complete conversion of carbon to CO_2 and hydrogen to H_2O. The stream of gases passes through a pre-weighed tube containing anhydrous calcium sulfate and then through a second pre-weighed tube containing sodium hydroxide. The first tube absorbs the water and the second absorbs the carbon dioxide, as seen in Figure 1.

The increase in masses of these tubes represents the masses of CO_2 and H_2O from which the masses of carbon and hydrogen can be calculated. If a compound contains oxygen, it will be determined by subtracting the mass of hydrogen and carbon from the total mass burned. Compounds containing nitrogen, sulfur, or a halogen are more difficult to analyze and often require additional time-consuming experiments.

Modern instruments automate the process of analyzing the combustion products using a technique called gas chromatography. After burning a weighed sample, advanced catalysts ensure that all the carbon, nitrogen, and sulfur atoms are converted to CO_2, N_2, and SO_2. One instrument automatically takes a small sample of the gases and injects it into a gas chromatograph. Within the instrument the gases travel through a tubular column packed with absorbents that cause each gas to travel at a different rate. At the end of the column a thermal conductivity detector, which measures the ability of a gas to dissipate heat, senses each component of the mixture. The result is a chromatogram with a peak for each separate substance, as shown in Figure 2. The area of each peak is proportional to the amount of each gas, and the internal computer computes the percentage composition.

All instruments require standard samples of high purity and known composition to calibrate the readout. In this case, very pure compounds, called standards, with known percentages of each element, are burned and then analyzed to develop conversion factors relating peak areas to the amounts of each substance. Typically, the classical method takes about 30 minutes per sample, and the instruments, once calibrated, take between 2 and 5 minutes per sample.

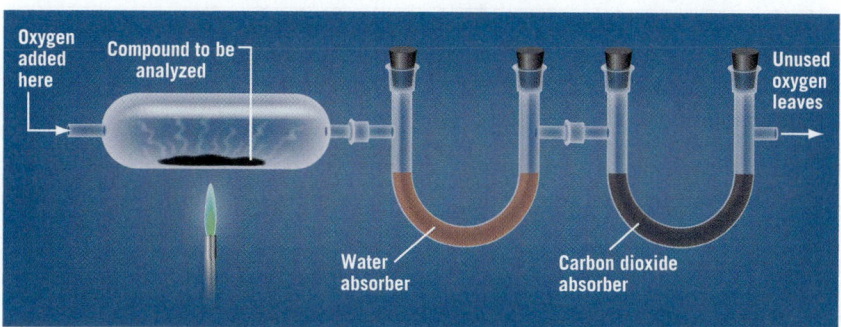

Figure 1 Classic CH analysis. This experimental setup shows pure oxygen added to the compound. Combustion products are absorbed by reaction with chemicals in the U-tubes. Calcium sulfate absorbs water and sodium hydroxide absorbs carbon dioxide.

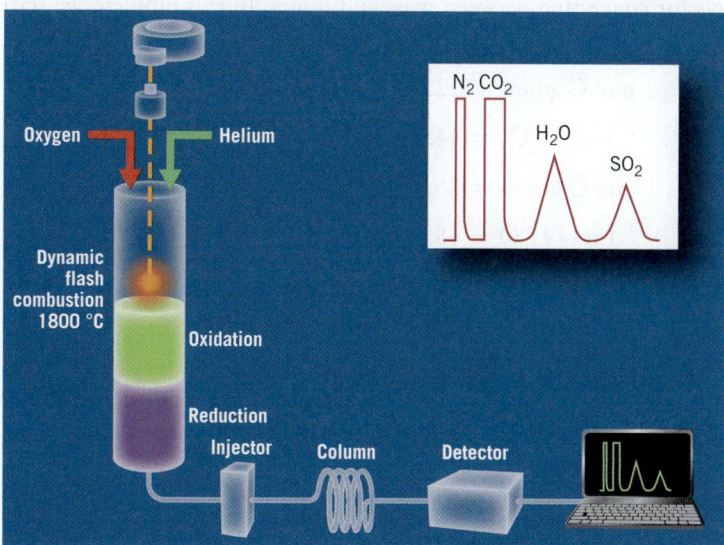

Figure 2 Diagram of an automated CHNS elemental analysis system. The diagram illustrates the combustion chamber, the chromatography column, a thermal conductivity detector, and a sample of a chromatogram with peaks for N_2, CO_2, H_2O, and SO_2.

In Example 3.11, we illustrate the indirect analysis of a compound made entirely of carbon, hydrogen, and oxygen. Such compounds burn completely in pure oxygen—it is called a *combustion reaction*—and the sole products are carbon dioxide and water. (This particular kind of indirect analysis is sometimes called a **combustion analysis**.) The complete combustion of methyl alcohol (CH_3OH), for example, occurs according to the following equation.

$$2CH_3OH + 3O_2 \longrightarrow 2CO_2 + 4H_2O$$

The carbon dioxide and water can be separated and are individually weighed. Notice that all of the carbon atoms in the original compound end up among the CO_2 molecules, and all of the hydrogen atoms are in H_2O molecules. In this way at least two of the original elements, C and H, are quantitatively measured.

We will calculate the mass of carbon in the CO_2 collected, which equals the mass of carbon in the original sample. Similarly, we will calculate the mass of hydrogen in the H_2O collected, which equals the mass of hydrogen in the original sample. When added together, the mass of C and mass of H are less than the total mass of the sample because the sample also contains oxygen. The law of conservation of mass allows us to subtract the sum of the C and H masses from the original sample mass to obtain the mass of oxygen in the sample of the compound.

Example 3.11
Empirical Formula from Indirect Analysis

A 0.5438 g sample of a pure liquid consisting of only C, H, and O was burned in 100% oxygen, and 1.039 g of CO_2 and 0.6369 g of H_2O were obtained. What is the empirical formula of the compound?

Analysis: There are several steps to this problem. First, we need to calculate the mass of the elements, C and H, by determining the number of grams of C in the CO_2 and the number of grams of H in the H_2O. Then the mass of oxygen is determined by difference. Next, we use the masses of C, H, and O to calculate the moles of each. Finally, we use our procedures to convert the moles of each element into integer subscripts in the empirical formula.

Assembling the Tools: To convert the masses of CO_2 and H_2O to grams of C and H we will need our tools for converting between mass and moles. In this problem we start with the equalities:

$$1 \text{ mol C atoms} = 12.011 \text{ g C atoms}$$

$$1 \text{ mol CO}_2 = 44.010 \text{ g CO}_2$$

$$1 \text{ mol O atoms} = 15.999 \text{ g O atoms}$$

$$1 \text{ mol H}_2\text{O} = 18.015 \text{ g H}_2\text{O}$$

Next we need the mole ratio equivalencies that we derive from the formulas.

$$1 \text{ mol C atoms} \Leftrightarrow 1 \text{ mol CO}_2$$

$$2 \text{ mol H atoms} \Leftrightarrow 1 \text{ mol H}_2\text{O}$$

This is similar to the calculations in Example 3.6, where we use the normal conversion sequence from grams of compound to grams of an element in the compound.

We then need the law of conservation of mass,

$$\text{mass of compound} = \text{mass of C} + \text{mass of H} + \text{mass of O}$$

to determine the oxygen content.

After we have the mass of each element, we will convert all the masses to moles. The final step is to convert the calculated moles of each element into an integer subscript for our formula.

Solution: First, we find the number of grams of C in the CO_2 as

$$1.039 \text{ g CO}_2 \left(\frac{1 \text{ mol CO}_2}{44.009 \text{ g CO}_2} \right) \left(\frac{1 \text{ mol C}}{1 \text{ mol CO}_2} \right) \left(\frac{12.011 \text{ g C}}{1 \text{ mol C}} \right) = 0.2836 \text{ g C}$$

For the number of grams of H in 0.6369 g of H_2O we calculate

$$0.6369 \text{ g H}_2\text{O} \left(\frac{1 \text{ mol H}_2\text{O}}{18.015 \text{ g H}_2\text{O}} \right) \left(\frac{2 \text{ mol H}}{1 \text{ mol H}_2\text{O}} \right) \left(\frac{1.0079 \text{ g H}}{1 \text{ mol H}} \right) = 0.07127 \text{ g H}$$

The total mass of C and H is therefore the sum of these two quantities.

$$\text{Total mass of C and H} = 0.2836 \text{ g C} + 0.07127 \text{ g H} = 0.3549 \text{ g}$$

The difference between this total and the 0.5438 g mass of the original sample is the mass of oxygen (the only other element present).

$$\text{Mass of O} = 0.5438 \text{ g} - 0.3549 \text{ g} = 0.1889 \text{ g O}$$

Now we can convert the masses of the elements to an empirical formula.

For C: $\qquad 0.2836 \text{ g C} \left(\frac{1 \text{ mol C}}{12.011 \text{ g C}} \right) = 0.02361 \text{ mol C}$

For H: $\qquad 0.07127 \text{ g H} \left(\frac{1 \text{ mol H}}{1.0079 \text{ g H}} \right) = 0.07071 \text{ mol H}$

For O: $\qquad 0.1889 \text{ g O} \left(\frac{1 \text{ mol O}}{15.999 \text{ g O}} \right) = 0.01181 \text{ mol O}$

Our preliminary empirical formula is thus $C_{0.02361}H_{0.070701}O_{0.01181}$. We divide all of these subscripts by the smallest number, 0.01181.

$$C_{\frac{0.02361}{0.01181}} H_{\frac{0.07071}{0.01181}} O_{\frac{0.01181}{0.01181}} = C_{1.999}H_{5.987}O_1$$

The results are acceptably close to integers, to conclude that the empirical formula is C_2H_6O.

Is the Answer Reasonable: Our checks on problems need to be quick and efficient. The fact that the integer subscripts were found easily suggests that the answer is correct and we can usually stop here. If more confirmation is needed, we can estimate the answers to the individual steps.

A sample containing only sulfur and carbon is completely burned in air to produce 0.640 g of SO_2 and 0.220 g of CO_2. What is the empirical formula? (*Hint:* Use the tools for relating grams of a compound to grams of an element.)

Practice Exercise 3.22

Calculate the empirical formula of a compound in which the combustion of a 5.048 g sample of the compound of C, H, and O gave 7.406 g of CO_2 and 3.027 g of H_2O.

Practice Exercise 3.23

Molecular Formulas from Empirical Formulas and Molecular Masses

The empirical formula is the accepted formula unit for ionic compounds. For molecular compounds, however, chemists prefer *molecular* formulas because they give the number of atoms of each type in a molecule, rather than just the simplest ratio of moles of elements in a compound as the empirical formula does.

Sometimes an empirical and molecular formula are the same. Two examples are H_2O and NH_3. When they are different, the subscripts of the molecular formula are integer multiples of those in the empirical formula. The subscripts of the molecular formula P_4O_{10}, for example, are each two times those in the empirical formula, P_2O_5, as you saw earlier. The molecular mass of P_4O_{10} is likewise two times the empirical formula mass of P_2O_5. This observation provides us with a way to find out the molecular formula for a compound provided we have a way of determining experimentally the molecular mass of the compound. If the experimental molecular mass equals the calculated empirical formula mass, the empirical formula is the same as the molecular formula. Otherwise, the molecular mass will be a whole-number multiple of the empirical formula mass. Whatever the integer is, it's a common multiplier for the subscripts of the empirical formula.

■ There are many methods for determining molecular masses. They are discussed in Chapters 10 and 12. Instruments such as the mass spectrometers described in Chapter 2 can also be used.

Example 3.12
Determining a Molecular Formula from an Empirical Formula and a Molecular Mass

Styrene, the raw material for polystyrene foam plastics, has an empirical formula of CH. Its molecular mass is 104. What is its molecular formula?

Analysis: Since we know the empirical formula and the molar mass of the compound, styrene, we need to find out how many empirical formula units make up one molecule. Then the molecular formula will have subscripts that are an integer multiple of the empirical formula subscripts.

Assembling the Tools: The relationship between empirical and molecular formulas tells us that the molecular mass of styrene divided by the formula mass of the empirical formula will result in an integer that represents the number of empirical formula units in the molecule itself.

$$\frac{\text{Molecular mass of styrene}}{\text{Empirical formula mass of CH}}$$

To obtain the molecular formula, all subscripts of the empirical formula are multiplied by that integer.

Solution: For the empirical formula, CH, the formula mass is

$$12.01 + 1.008 = 13.02$$

To find how many CH units weighing 13.02 are in a mass of 104, we divide.

$$\frac{104}{13.02} = 7.99$$

Rounding this to 8, we find that eight CH units make up the molecular formula of styrene, and styrene must have subscripts 8 times those in CH. Styrene, therefore, is C_8H_8.

Is the Answer Reasonable? The molecular mass of C_8H_8 is approximately $(8 \times 12) + (8 \times 1) = 104$, which is consistent with the molecular mass we started with.

Practice Exercise 3.24

The empirical formula of hydrazine is NH_2 and its molecular mass is 32.0. What is its molecular formula? (*Hint:* Recall the relationship between the molecular and empirical formula.)

Practice Exercise 3.25

After determining that the empirical formulas of two different compounds were CH_2Cl and $CHCl$, a student mixed up the data for the molecular masses. However, the student knew that one compound had a molecular mass of 100 and the other had a molecular mass of 289. What are the likely molecular formulas of the two compounds?

3.5 | Stoichiometry and Chemical Equations

Writing and Balancing Chemical Equations

Here we will see that a *balanced* chemical equation is a very useful tool for problem solving. We learned in Chapter 2 that a *chemical equation* is a shorthand, quantitative description of a chemical reaction. An equation is balanced when all atoms present among the reactants (written to the left of the arrow) are also present among the products (written to the right of the arrow). Coefficients, numbers placed in front of formulas, are multiplier numbers for their respective formulas that are used to balance an equation.

Always approach the balancing of an equation as a two-step process.

Step 1. *Write the unbalanced "equation."* Organize the formulas in the pattern of an equation with plus signs and an arrow (think of the arrow as an equal sign because we need to end up with the same number of each atom on both sides). Use *correct* formulas. (You learned to write many of them in Chapter 2, so keep on practicing your formula-writing skills.)

Step 2. *Adjust the coefficients to get equal numbers of each kind of atom on both sides of the arrow.* When doing Step 2, make no changes in the formulas, either in the atomic symbols or their subscripts. If you do, the equation will involve different substances from those intended. It often helps to start the process with the most complex formula, leaving elements and simple compounds to the end.

We'll begin with simple equations that can be balanced easily by inspection. An example is the reaction of zinc metal with hydrochloric acid (see the margin photo). First, we need the correct formulas, and this time we'll include the physical states because they may not be obvious. The reactants are zinc, $Zn(s)$, and hydrochloric acid, $HCl(aq)$, an aqueous solution of the gas hydrogen chloride. We also need formulas for the products. Zn changes to a water-soluble compound, zinc chloride, $ZnCl_2(aq)$, and hydrogen gas, $H_2(g)$, bubbles out as the other product. (Recall that hydrogen is one of the elements that occurs naturally as a *diatomic molecule*, not as an atom.)

Step 1. Write an unbalanced equation.

$$Zn(s) + HCl(aq) \longrightarrow ZnCl_2(aq) + H_2(g) \qquad \text{(unbalanced)}$$

Step 2. Adjust the coefficients to get equal numbers of each kind of atom on both sides of the arrow.

There is not one simple set of rules for adjusting coefficients—as you may have discovered, experience is the greatest help. Experience has taught chemists that the following guidelines often get to the solution most directly when they are applied in the order given.

Some Guidelines for Balancing Equations

1. Start balancing with the most complicated formula first. Elements, particularly H_2 and O_2, should be left until the end.
2. Balance atoms that appear in only two formulas: one as a reactant and the other as a product. (Leave elements that appear in three or more formulas until later.)
3. Balance as a group those polyatomic ions that appear unchanged on both sides of the arrow.

Using the guidelines given here, we'll leave the $Zn(s)$ and $H_2(g)$ until later. The two remaining formulas have chlorine in common. Because there are two Cl to the right of the arrow but only one to the left, we put a 2 in front of the HCl on the left side. Remember that this also sets the coefficient of $ZnCl_2$ as 1, which by custom is not written. The result is

$$Zn(s) + 2HCl(aq) \longrightarrow ZnCl_2(aq) + H_2(g)$$

Zinc metal reacts with hydrochloric acid.

Richard Megna/Fundamental Photographs

■ If there is a chance of confusion, we refer to H_2 as molecular hydrogen and H as atomic hydrogen. You may also see H_2 referred to as dihydrogen.

TOOLS

Balancing chemical equations

We then balance the hydrogen and zinc and find that no additional coefficient changes are needed. Everything is now balanced. On each side we find 1 Zn, 2 H, and 2 Cl. One complication is that an infinite number of *balanced* equations can be written for any given reaction! We might, for example, have adjusted the coefficients so that our equation came out as follows.

$$2Zn(s) + 4HCl(aq) \longrightarrow 2ZnCl_2(aq) + 2H_2(g)$$

This equation is also balanced. For simplicity, chemists prefer the *smallest* whole-number coefficients when writing balanced equations.

■ Although the smallest whole-number coefficients are preferred, stoichiometric calculations still work as long as the equation is balanced.

Example 3.13
Writing a Balanced Equation

Sodium hydroxide and phosphoric acid, H_3PO_4, react as aqueous solutions to give sodium phosphate and water. The sodium phosphate remains in solution. Write the balanced equation for this reaction.

Analysis: First, we need to write an unbalanced equation that includes the reactant formulas on the left-hand side and the product formulas on the right. Then we need to use the above procedures for balancing the equation.

Assembling the Tools: First, we need the formula for the reactants and products to write the unbalanced equation. We are given the formula for phosphoric acid, and from the nomenclature tools in Chapter 2, the formula of sodium hydroxide is NaOH and sodium phosphate is Na_3PO_4. The formula for water is H_2O. Finally, we use our guidelines for balancing chemical equations.

■ When a reaction occurs in aqueous solution, water is not pure water, so the (*l*) is not appropriate. Also, the term (*aq*) is meaningless since we would be saying "an aqueous solution of water." In solution chemistry, therefore, the state of water is not specified.

Solution: We write the unbalanced equation by placing the reactants to the left of the arrow and all products to the right. The designation (*aq*) is added for all substances dissolved in water (H_2O itself does not have the (*aq*); we'll not give it any designation when it is in its liquid state).

$$NaOH(aq) + H_3PO_4(aq) \longrightarrow Na_3PO_4(aq) + H_2O \quad \text{(unbalanced)}$$

We will focus on balancing the Na and P atoms first, since H and O atoms appear in more than two formulas. Focusing on the Na first, there are 3 Na on the right side, so we put a 3 in front of NaOH on the left, as a trial. Remember that this step also places an unwritten 1 in front of the Na_3PO_4 formula.

$$3NaOH(aq) + H_3PO_4(aq) \longrightarrow Na_3PO_4(aq) + H_2O \quad \text{(unbalanced)}$$

Now that Na is in balance we can focus on the P. Since all of the phosphorus is found in the PO_4 units, we will balance the phosphate units as a group of atoms rather than individually. We see that the PO_4 unit is already balanced and that means that the H_3PO_4 has a coefficient of 1. Only the coefficient for water has not been assigned. We can see that the reactant side has 6 H and 3 O atoms (notice that we don't count the oxygen atoms in the PO_4 units since they have already been balanced), which will produce 3 H_2O molecules. Thus the coefficient of water should be three.

$$3NaOH(aq) + H_3PO_4(aq) \longrightarrow Na_3PO_4(aq) + 3H_2O \quad \text{(balanced)}$$

We now have a balanced equation.

Is the Answer Reasonable? To check if we have a balanced chemical reaction, we need to count the atoms on both sides of the arrow. On each side we have 3 Na, 1 PO_4, 6 H, and 3 O besides those in PO_4, and since the coefficients for $H_3PO_4(aq)$ and $Na_3PO_4(aq)$ are 1 our coefficients cannot be reduced to smaller whole numbers.

When aqueous solutions of barium nitrate, $Ba(NO_3)_2$, and ammonium phosphate, $(NH_4)_3PO_4$, are mixed, a reaction occurs in which solid barium phosphate, $Ba_3(PO_4)_2$, separates from the solution. The other product is $NH_4NO_3(aq)$. Write the balanced equation. (*Hint:* What steps are needed to balance a chemical equation?)

Practice Exercise 3.26

Write the balanced chemical equation that describes what happens when a solution containing calcium chloride is mixed with a solution containing potassium carbonate and the product of the reaction is solid calcium carbonate and a solution of potassium chloride.

Practice Exercise 3.27

The strategy of balancing whole units of polyatomic ions, like $PO_4{}^{3-}$, as a group is extremely useful. Using this method, we have less atom counting to do and balancing equations is often easier.

Calculations That Use Balanced Chemical Equations

So far we have focused on mole ratios between elements within a single compound. We have seen that the essential conversion factor between substances within a compound is the mole ratio obtained from the compound's formula. Now, we will extend this mole ratio concept to relate substances involved in a chemical reaction.

Establishing Mole-to-Mole Ratios

To see how chemical equations can be used to obtain mole-to-mole relationships, consider the equation that describes the burning of octane (C_8H_{18}) in oxygen (O_2) to give carbon dioxide and water vapor:

$$2C_8H_{18}(l) + 25\,O_2(g) \longrightarrow 16CO_2(g) + 18H_2O(g)$$

■ The chemical equation gives the relative amounts of each of the molecules that participate in a reaction. It does not mean that 2 octane molecules actually collide with 25 O_2 molecules. The actual reaction occurs in many steps that the chemical equation does not show.

When two molecules of liquid octane react with twenty-five molecules of oxygen gas, sixteen molecules of carbon dioxide gas and eighteen molecules of water vapor are produced.

This statement immediately suggests many equivalence relationships that can be used to build conversion factors for stoichiometry problems:

$$2 \text{ molecules } C_8H_{18} \Leftrightarrow 25 \text{ molecules } O_2$$

$$2 \text{ molecules } C_8H_{18} \Leftrightarrow 16 \text{ molecules } CO_2$$

$$2 \text{ molecules } C_8H_{18} \Leftrightarrow 18 \text{ molecules } H_2O$$

$$25 \text{ molecules } O_2 \Leftrightarrow 16 \text{ molecules } CO_2$$

$$25 \text{ molecules } O_2 \Leftrightarrow 18 \text{ molecules } H_2O$$

$$16 \text{ molecules } CO_2 \Leftrightarrow 18 \text{ molecules } H_2O$$

Any of these *microscopic* relationships can be scaled up to the *macroscopic* laboratory scale by multiplying both sides of the equivalency by Avogadro's number, which effectively allows us to replace "molecules" with "moles" or "mol":

$$2 \text{ mol } C_8H_{18} \Leftrightarrow 25 \text{ mol } O_2$$

$$2 \text{ mol } C_8H_{18} \Leftrightarrow 16 \text{ mol } CO_2$$

$$2 \text{ mol } C_8H_{18} \Leftrightarrow 18 \text{ mol } H_2O$$

$$25 \text{ mol } O_2 \Leftrightarrow 16 \text{ mol } CO_2$$

$$25 \text{ mol } O_2 \Leftrightarrow 18 \text{ mol } H_2O$$

$$16 \text{ mol } CO_2 \Leftrightarrow 18 \text{ mol } H_2O$$

We can interpret the equation on a macroscopic (mole) scale as follows:

Two moles of liquid octane react with twenty-five moles of oxygen gas to produce sixteen moles of carbon dioxide gas and eighteen moles of water vapor.

TOOLS

Equivalences obtained from balanced chemical equations

To generalize this statement, we can say that the coefficients in balanced chemical reactions gives us the mole-to-mole relationships that we can use in equivalence and conversion factors. To use these equivalencies in a stoichiometry problem, the equation must be a **balanced equation**. This means that every atom found in the reactants must also be found somewhere in the products. You must always check to see whether this is true for a given equation before you can use the coefficients to build equivalencies and conversion factors.

First, let's see how mole-to-mole relationships obtained from a chemical equation can be used to convert moles of one substance to moles of another when both substances are involved in a chemical reaction.

Example 3.14
Stoichiometry of Chemical Reactions

How many moles of sodium phosphate can be made from 0.240 mol of sodium hydroxide by the following unbalanced equation?

$$NaOH(aq) + H_3PO_4(aq) \longrightarrow Na_3PO_4(aq) + H_2O$$

Analysis: We are given an unbalanced equation and will have to balance it first before we can find an appropriate mole ratio. Then the question asks us to relate amounts, in moles, of two different substances, for which we will need a mole ratio.

Assembling the Tools: The first tool we will use is the tool for balancing chemical equations, and the second will be the one that allows us to use the equivalences derived from the coefficients of the balanced chemical equation.

Solution: We start by balancing the equation as we did previously in Example 3.13,

$$3NaOH(aq) + H_3PO_4(aq) \longrightarrow Na_3PO_4(aq) + 3H_2O$$

Then we look at the question to see what we want to equate:

$$0.240 \text{ mol NaOH} \Leftrightarrow \text{? mol Na}_3PO_4$$

From the coefficients of the balanced chemical equation, we can write the equivalence between NaOH and Na_3PO_4 as

$$3 \text{ mol NaOH} \Leftrightarrow 1 \text{ mol Na}_3PO_4$$

This enables us to prepare the mole ratio conversion factor that we need,

$$\frac{1 \text{ mol Na}_3PO_4}{3 \text{ mol NaOH}}$$

We can now convert 0.240 mol NaOH to the number of moles of Na_3PO_4:

$$0.240 \text{ mol NaOH} \left(\frac{1 \text{ mol Na}_3PO_4}{3 \text{ mol NaOH}} \right) = 0.0800 \text{ mol Na}_3PO_4$$

The result states that we can make 0.0800 mol Na_3PO_4 from 0.240 mol NaOH. Recall that the coefficients, 1 and 3, in the conversion factor are exact numbers, so our answer has three significant figures.

● **Is the Answer Reasonable?** The balanced equation tells us that 3 mol NaOH ⇔ 1 mol Na_3PO_4, so the actual number of moles of Na_3PO_4 (0.0800 mol) should be one-third the actual number of moles of NaOH (0.240 mol), and it is. We can also check that the units cancel correctly.

In the reaction $2SO_2(g) + O_2(g) \longrightarrow 2SO_3(g)$, how many moles of O_2 are needed to produce 6.76 moles of SO_3? (*Hint:* Write the equality that relates O_2 to SO_3.)

Practice Exercise 3.28

How many moles of sulfuric acid, H_2SO_4, are needed to react with 0.366 mol of NaOH by the following balanced chemical equation?

Practice Exercise 3.29

$$2NaOH(aq) + H_2SO_4(aq) \longrightarrow Na_2SO_4(aq) + 2H_2O$$

Mass-to-Mass Calculations

The most common stoichiometric calculation the chemist does is to relate grams of one substance with grams of another in a chemical reaction. For example, consider glucose, $C_6H_{12}O_6$, one of the body's primary energy sources. The body combines glucose and oxygen, in a process called metabolism, to give carbon dioxide and water. The balanced equation for the overall reaction is

$$C_6H_{12}O_6(aq) + 6O_2(aq) \longrightarrow 6CO_2(aq) + 6H_2O$$

How many grams of oxygen must the body take in to completely process 1.00 g of glucose? The problem can be expressed as

$$1.00 \text{ g } C_6H_{12}O_6 \Leftrightarrow ? \text{ g } O_2$$

● The first thing we should notice about this problem is that we're relating two *different substances* in a reaction. The equality that relates the substances is the mole-to-mole relationship between glucose and O_2 given by the chemical equation. In this case, the equation tells us that 1 mol $C_6H_{12}O_6$ ⇔ 6 mol O_2. It is very important to realize that there is no direct conversion between the mass of $C_6H_{12}O_6$ and the mass of O_2. We need to convert the mass of glucose to moles of glucose, then we use the mole ratio to convert moles of glucose to moles of oxygen, and finally we convert moles of oxygen to mass of oxygen. This sequence, indicating where we use the mole-to-mole equivalence, is shown below

$$\boxed{1 \text{ mol } C_6H_{12}O_6 \Leftrightarrow 6 \text{ mol } O_2}$$

$$1.00 \text{ g } C_6H_{12}O_6 \longrightarrow \text{mol } C_6H_{12}O_6 \overset{\diagup}{\longrightarrow} \text{mol } O_2 \longrightarrow \text{g } O_2$$

Two molar masses are used, once for converting 1.00 g of glucose to moles and again for converting moles of O_2 to grams of O_2.

Figure 3.4 outlines this flow for *any* stoichiometry problem that relates reactant or product masses. If we know the *balanced equation* for a reaction and the *mass* of any reactant or product, we can calculate the required or expected mass of *any* other substance in the equation. Example 3.15 shows how it works.

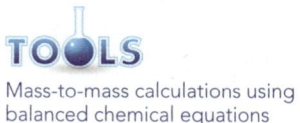

TOOLS

Mass-to-mass calculations using balanced chemical equations

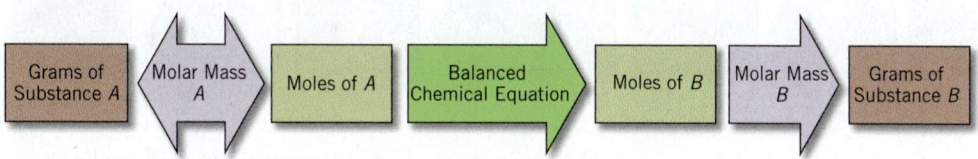

Figure 3.4 | The sequence of calculations for solving stoichiometry problems. This sequence applies to all calculations that start with the mass of one substance *A* and require the mass of a second substance *B* as the answer. Each box represents a measured or calculated quantity. Each arrow represents one of our chemical tools.

Example 3.15
Stoichiometric Mass Calculations

Portland cement is a mixture of the oxides of calcium, aluminum, and silicon. The raw material for calcium oxide is calcium carbonate, which occurs as the chief component of limestone. When calcium carbonate is strongly heated it decomposes. One product, carbon dioxide, is driven off to leave the desired calcium oxide as the only other product.

A chemistry student needs to prepare 1.50×10^2 g of calcium oxide in order to test a particular "recipe" for Portland cement. How many grams of calcium carbonate should be used, assuming 100% conversion to product?

Analysis: As usual, this is a multi-part problem. First, we need to determine the formulas of the compounds and then write a balanced chemical reaction. Then we will perform the conversions from the given 1.50×10^2 grams of calcium oxide to moles of calcium oxide, then to moles of calcium carbonate, and then to grams of calcium carbonate.

Assembling the Tools: We start with the nomenclature tools in Chapter 2 to get the formulas of the compounds and then write the balanced chemical equation, which is

■ Special reaction conditions are often shown with words or symbols above the arrow. In this reaction, temperatures above 2000 °C are needed and this is indicated with the word, *heat*, above the arrow.

$$CaCO_3(s) \xrightarrow{\text{heat}} CaO(s) + CO_2(g)$$

Now we can write our sequence of conversions using formulas,

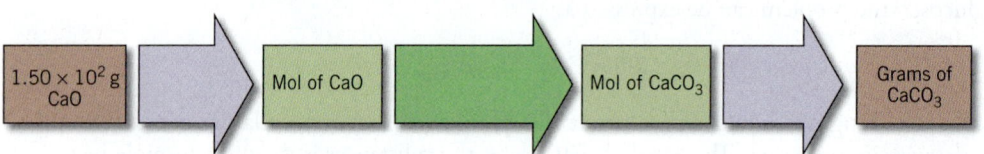

We need tools for converting the mass to moles for CaO in the first step and the mass to moles for $CaCO_3$ in the last step. They are

$$56.08 \text{ g CaO} = 1 \text{ mol CaO} \quad \text{and} \quad 1 \text{ mol CaCO}_3 = 100.09 \text{ g CaCO}_3$$

Finally, we need a tool for the mole-to-mole conversion in a balanced chemical equation, which gives us the equivalence

$$1 \text{ mol CaO} \Leftrightarrow 1 \text{ mol CaCO}_3$$

Solution: We start by writing the question as an equation.

$$1.50 \times 10^2 \text{ g CaO} \Leftrightarrow ? \text{ g CaCO}_3$$

Reviewing our sequence of calculation we can write the sequence of steps indicating where we use each of our tools.

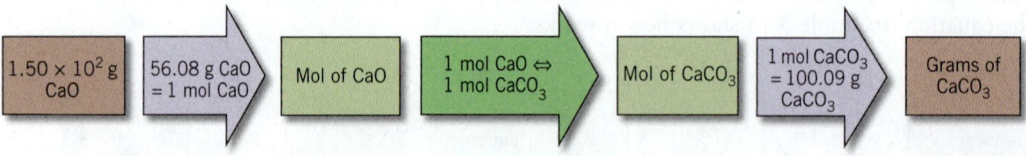

Starting from the left, we assemble conversion factors so the units cancel correctly:

$$1.50 \times 10^2 \text{ g } \cancel{\text{CaO}} \left(\frac{1 \text{ mol } \cancel{\text{CaO}}}{56.08 \text{ g } \cancel{\text{CaO}}} \right) \left(\frac{1 \text{ mol } \cancel{\text{CaCO}_3}}{1 \text{ mol } \cancel{\text{CaO}}} \right) \left(\frac{100.09 \text{ g CaCO}_3}{1 \text{ mol } \cancel{\text{CaCO}_3}} \right) = 268 \text{ g CaCO}_3$$

Notice how the calculation flows from grams of CaO to moles of CaO, then to moles of CaCO₃ (using the balanced chemical equation), and finally to grams of CaCO₃. We cannot emphasize too much that *the key step in all calculations of reaction stoichiometry is the use of the balanced equation.*

Is the Answer Reasonable? In a mass-to-mass calculation like this, the first check is the magnitude of the answer compared to the starting mass. In the majority of reactions the calculated mass is not less than ⅕ of the starting mass nor is it larger than five times the starting mass. Our result is reasonable based on this criterion. We can make a more detailed check by first checking that the units cancel properly. We can also round to one or two significant figures and estimate the answer. We would estimate $\frac{150 \times 100}{50} = 300$, and this value is close to our answer of 268, giving us confidence that the answer is reasonable.

■ This 5 to ⅕ range for conversions of the mass of *A* to the mass of *B* holds true for most (maybe 90–95%) calculations. If your answer is outside this range, do a very careful check for errors.

Example 3.16
Stoichiometric Mass Calculations

The thermite reaction is one of the most spectacular reactions with flames, sparks, and glowing molten iron. Here aluminum reacts with iron(III) oxide to produce aluminum oxide and metallic iron. So much heat is generated that the iron forms in the liquid state (Figure 3.5).

A certain welding operation requires at least 86.0 g of iron each time a weld is made. What are the minimum masses, in grams, of iron(III) oxide and aluminum that must be used for each weld?

Analysis: First, we need to determine the formulas of the compounds so we can write a balanced chemical equation. Next, we need to convert the 86.0 g Fe to the needed mass of iron(III) oxide. Remember that all problems in reaction stoichiometry must be solved at the mole level because an equation's coefficients disclose *mole* ratios, not mass ratios. The sequence of conversions to use is

Figure 3.5 | The thermite reaction. Pictured here is a device for making white-hot iron by the reaction of aluminum with iron oxide and letting the molten iron run down into a mold between ends of two steel railroad rails, welding the rails together.

courtesy Skatebiker http://en.wikipedia.org/wiki/

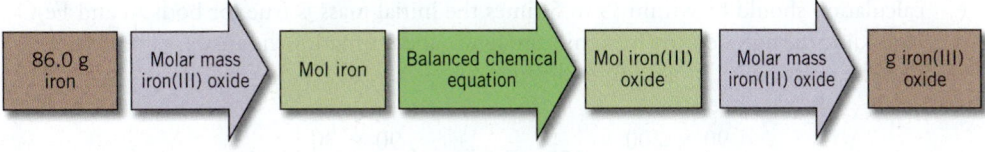

For the second calculation we calculate the number of grams of Al needed, but we know that we must first find the number of *moles* of Al required. For this second calculation the sequence is

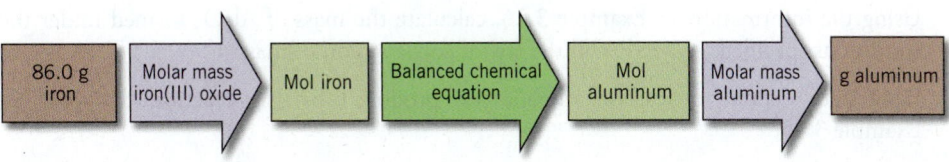

Assembling the Tools: We use our nomenclature tools to determine that Al and Fe_2O_3 are the reactants and that Fe and Al_2O_3 are the products. The balanced chemical equation is:

$$2Al(s) + Fe_2O_3(s) \longrightarrow Al_2O_3(s) + 2Fe(l)$$

The following conversion equalities are needed to calculate the mass of Fe_2O_3

$$55.85 \text{ g Fe} = 1 \text{ mol Fe}$$
$$1 \text{ mol Fe}_2O_3 \Leftrightarrow 2 \text{ mol Fe}$$
$$159.70 \text{ g Fe}_2O_3 = 1 \text{ mol Fe}_2O_3$$

Next, we need the following conversion equalities to calculate the mass of aluminum:

$$55.85 \text{ g Fe} = 1 \text{ mol Fe}$$
$$2 \text{ mol Al} \Leftrightarrow 2 \text{ mol Fe}$$
$$26.98 \text{ g Al} = 1 \text{ mol Al}$$

Solution: Now we state the first part of the problem in mathematical form:

$$86.0 \text{ g Fe} \Leftrightarrow ? \text{ g Fe}_2O_3$$

We'll set up the first calculation as a chain. The steps are summarized below each conversion factor.

$$86.0 \text{ g Fe} \left(\frac{1 \text{ mol Fe}}{55.85 \text{ g Fe}} \right) \left(\frac{1 \text{ mol Fe}_2O_3}{2 \text{ mol Fe}} \right) \left(\frac{159.7 \text{ g Fe}_2O_3}{1 \text{ mol Fe}_2O_3} \right) = 123 \text{ g Fe}_2O_3$$

$$\text{grams Fe} \longrightarrow \text{moles Fe} \longrightarrow \text{moles Fe}_2O_3 \longrightarrow \text{grams Fe}_2O_3$$

A minimum of 123 g of Fe_2O_3 is required to make 86.0 g of Fe.

To calculate the mass of aluminum, we follow a similar sequence, using the conversion factors in order so that the units cancel as shown. We employ another chain calculation to find the mass of Al needed to make 86.0 g of Fe.

$$86.0 \text{ g Fe} \left(\frac{1 \text{ mol Fe}}{55.85 \text{ g Fe}} \right) \left(\frac{2 \text{ mol Al}}{2 \text{ mol Fe}} \right) \left(\frac{26.98 \text{ g Al}}{1 \text{ mol Al}} \right) = 41.5 \text{ g Al}$$

$$\text{grams Fe} \longrightarrow \text{moles Fe} \longrightarrow \text{moles Al} \longrightarrow \text{grams Al}$$

■ We could simplify the mole ratio to 1 mol Al ⇔ 1 mol Fe. Leaving the 2-to-2 ratio maintains the connection to the coefficients in the balanced equation.

Are the Answers Reasonable? The estimate that our answers in a mass-to-mass calculation should be within ⅕ to 5 times the initial mass is true for both Al and Fe_2O_3. Rounding the numbers to one significant figure and estimating the answer (after rechecking that the units cancel properly) results in

$$\frac{90 \times 200}{60 \times 2} = 150 \text{ g Fe}_2O_3 \text{ and } \frac{90 \times 30}{60} = 45 \text{ g Al}$$

Both estimates are close to our calculated values and give confidence that our answers are reasonable.

Practice Exercise 3.30

Using the information in Example 3.16, calculate the mass of Al_2O_3 formed under the conditions specified. (*Hint:* Recall the law of conservation of mass.)

Practice Exercise 3.31

How many grams of carbon dioxide are produced by the reaction described in Example 3.15?

3.6 | Limiting Reactants

Limiting Reactants Viewed at the Molecular Level

We've seen that balanced chemical equations can tell us how to mix reactants together in just the right proportions to get a certain amount of product. For example, ethanol, C_2H_5OH, is prepared industrially as follows:

$$C_2H_4 + H_2O \longrightarrow C_2H_5OH$$
$$\text{ethylene} \qquad\qquad\qquad \text{ethanol}$$

We can interpret the equation on a laboratory scale using moles. Every mole of ethylene that reacts requires one mole of water to produce one mole of ethanol. Let's look at this reaction at the molecular level. Now the equation tells us that one molecule of ethylene will react with one molecule of water to give one molecule of ethanol.

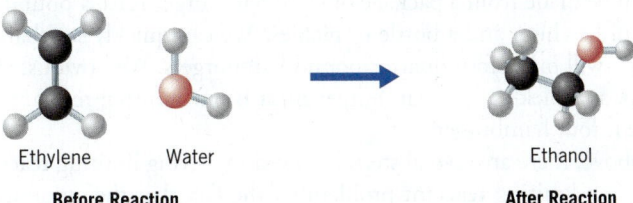

Ethylene   Water   Ethanol

Before Reaction   **After Reaction**

If we have three molecules of ethylene reacting with three molecules of water, then three ethanol molecules are produced:

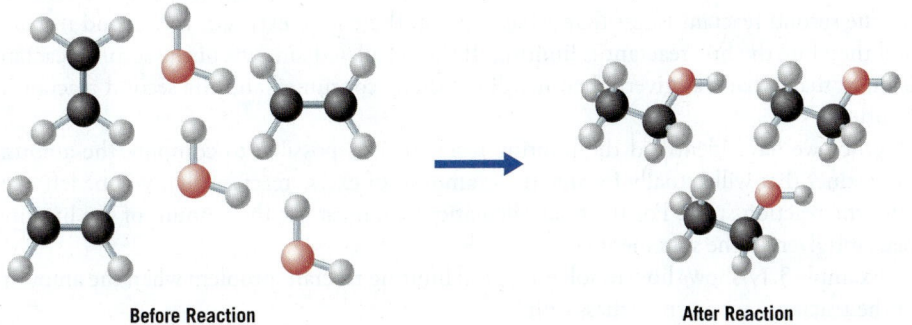

Before Reaction   **After Reaction**

■ Notice that in both the "before" and "after" views of the reactions, the numbers of carbon, hydrogen, and oxygen atoms are the same.

What happens if we mix three molecules of ethylene with five molecules of water? The ethylene will be completely used up before all the water is, and the product will contain two unreacted water molecules:

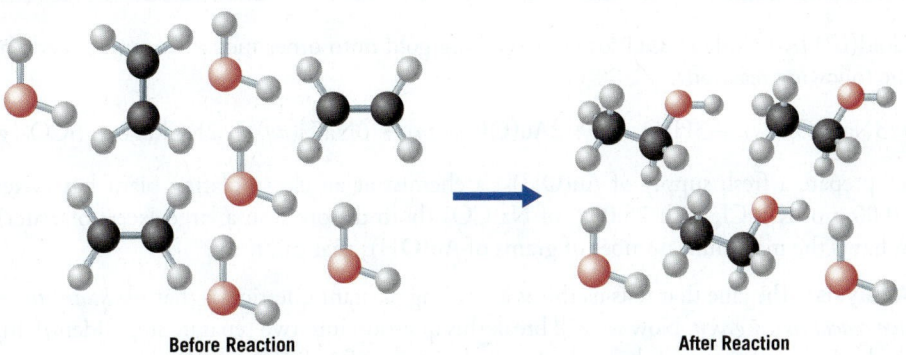

Before Reaction   **After Reaction**

We don't have enough ethylene to use up all the water. The excess water remains after the reaction stops. This situation can be a problem in the manufacture of chemicals because not only do we waste one of our reactants (water, in this case), but we also obtain a product that is contaminated with unused reactant.

In this reaction mixture, ethylene is called the **limiting reactant** because it limits the amount of product (ethanol) that forms. The water is called an **excess reactant**, because we have more of it than is needed to completely consume all the ethylene.

To predict the amount of product we'll actually obtain in a reaction, we need to know which of the reactants is the limiting reactant. In the example above, we saw that we needed only 3 H_2O molecules to react with 3 C_2H_4 molecules, but we had 5 H_2O molecules, so H_2O is present in excess and C_2H_4 is the limiting reactant. We could also have reasoned that 5 molecules of H_2O would require 5 molecules of C_2H_4, and since we have only 3 molecules of C_2H_4, it must be the limiting reactant.

A Strategy for Solving Limiting Reactant Problems

At the start of this chapter we referred to the billions of hamburger buns produced each year to reflect the large number of atoms and molecules we work with. Construction of hamburgers can also be a limiting reactant problem. For instance, how many quarterpound hamburgers can be made from a package of eight hamburger rolls, a pound of hamburger meat, a bottle of ketchup, and a bottle of pickles? We can quickly see that one pound of hamburger meat will make four quarter-pound hamburgers. We obviously have an excess of buns, catsup, and pickles. The hamburger meat is our limiting reactant, and the most we can prepare is four hamburgers.

TOOLS
Limiting reactant calculations

As we saw above, there are several steps involved in solving limiting reactant problems. First, we identify a limiting reactant problem by the fact that the amount of more than one reactant is given. Next, we need to identify which reactant is the limiting reactant, and finally we solve the problem based on the amount of limiting reactant at hand. The first step is easy. Next, when finding the limiting reactant, we arbitrarily pick one of the reactants and calculate how much of the second reactant is needed. If the amount calculated for the second reactant is less than what is given, there is an excess of the second reactant and therefore the first reactant is limiting. If the calculated amount of the second reactant is more than what was given, then it will be totally consumed and the second reactant is limiting.

Once we have identified the limiting reactant, it is possible to compute the amount of product that will actually form and the amount of excess reactant that will be left over after the reaction stops. For the final calculations we must use the amount of the limiting reactant given in the statement of the problem.

Example 3.17 shows how to solve a typical limiting reactant problem when the amounts of the reactants are given in mass units.

Example 3.17
Limiting Reactant Calculation

Gold(III) hydroxide is used for electroplating gold onto other metals. It can be made by the following reaction.

$$2KAuCl_4(aq) + 3Na_2CO_3(aq) + 3H_2O \longrightarrow 2Au(OH)_3(aq) + 6NaCl(aq) + 2KCl(aq) + 3CO_2(g)$$

To prepare a fresh supply of $Au(OH)_3$, a chemist at an electroplating plant has mixed 20.00 g of $KAuCl_4$ with 25.00 g of Na_2CO_3 (both dissolved in a large excess of water). What is the maximum number of grams of $Au(OH)_3$ that can form?

Analysis: The clue that tells us this is a limiting reactant question is that *the quantities of two reactants are given*. Now we will break this question into two separate steps, identifying the limiting reactant and then calculating the grams of $Au(OH)_3$ produced.

Step 1: To identify the limiting reactant, we arbitrarily pick one of the reactants ($KAuCl_4$ or Na_2CO_3; we were told that water is in excess, so we know that it is not a limiting reactant), and calculate how much of the second reactant is needed. Based on the result we will be able to decide which reactant is limiting. We will need to use a combination of our stoichiometry tools to solve this problem, using the following sequence:

Step 2: Once we've identified the limiting reactant we can use it to calculate the amount of $Au(OH)_3$ produced using the following sequence of conversions. Notice that this sequence is essentially the same as the one we used in Figure 3.4.

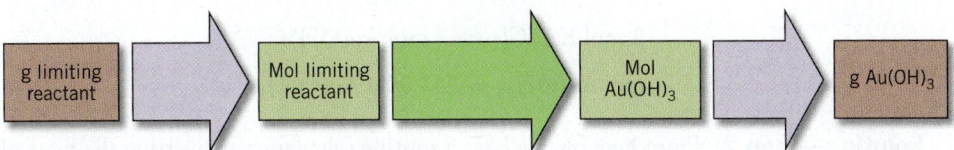

Assembling the Tools—Step 1: Our tool for limiting reactant calculations outlines the process for identifying the limiting reactant. First, we calculate the amount of reactant 1 that will react with reactant 2.

This is our typical mass-to-mass calculation that requires the following relationships:

$$1 \text{ mol KAuCl}_4 = 377.88 \text{ g KAuCl}_4$$

$$2 \text{ mol KAuCl}_3 \Leftrightarrow 3 \text{ mol Na}_2CO_3$$

$$1 \text{ mol Na}_2CO_3 = 105.99 \text{ g Na}_2CO_3$$

Solution—Step 1: In the following two calculations, we will show the process used to identify the limiting reactant. In solving a limiting reactant problem you will need to do only one of these calculations.

We start with $KAuCl_4$ as the reactant to work with and calculate how many grams of Na_2CO_3 *are needed* to react with 20.00 g of $KAuCl_4$. We'll set up a chain calculation as follows.

$$20.00 \text{ g KAuCl}_4 \left(\frac{1 \text{ mol KAuCl}_4}{377.88 \text{ g KAuCl}_4} \right) \left(\frac{3 \text{ mol Na}_2CO_3}{2 \text{ mol KAuCl}_4} \right) \left(\frac{105.99 \text{ g Na}_2CO_3}{1 \text{ mol Na}_2CO_3} \right) = 8.415 \text{ g Na}_2CO_3$$

$$\text{grams KAuCl}_4 \longrightarrow \text{moles KAuCl}_4 \longrightarrow \text{moles Na}_2CO_3 \longrightarrow \text{grams Na}_2CO_3$$

We find that 20.00 g of $KAuCl_4$ needs 8.415 g of Na_2CO_3. The given amount of 25.00 g of Na_2CO_3 is more than enough to let the $KAuCl_4$ react completely. *We conclude that $KAuCl_4$ is the limiting reactant and that Na_2CO_3 is present in excess.*

Although we already know the limiting reactant, let's demonstrate what would have happened if Na_2CO_3 was chosen instead of $KAuCl_4$. This time we start with Na_2CO_3 as the reactant to work with and calculate how many grams of $KAuCl_4$ are needed to react with 25.00 g of Na_2CO_3. Again we perform a chain calculation using the appropriate conversion factors

$$25.00 \text{ g Na}_2CO_3 \left(\frac{1 \text{ mol Na}_2CO_3}{105.99 \text{ g Na}_2CO_3} \right) \left(\frac{2 \text{ mol KAuCl}_4}{3 \text{ mol Na}_2CO_3} \right) \left(\frac{377.88 \text{ g KAuCl}_4}{1 \text{ mol KAuCl}_4} \right) = 59.42 \text{ g KAuCl}_4$$

$$\text{grams Na}_2CO_3 \longrightarrow \text{moles Na}_2CO_3 \longrightarrow \text{moles KAuCl}_4 \longrightarrow \text{grams KAuCl}_4$$

We find that 25.00 g Na_2CO_3 would require almost three times the mass of $KAuCl_4$ provided, so *we again conclude that $KAuCl_4$ is the limiting reactant.*

The result from either calculation above is sufficient to designate $KAuCl_4$ as the limiting reactant. Now we proceed to Step 2.

■ After identifying the limiting reactant, return to the statement of the problem and use the amount of the limiting reactant given in the problem to perform further calculations.

Assembling the Tools—Step 2: Since $KAuCl_4$ is the limiting reactant, we can calculate the mass of $Au(OH)_3$ using the sequence of steps

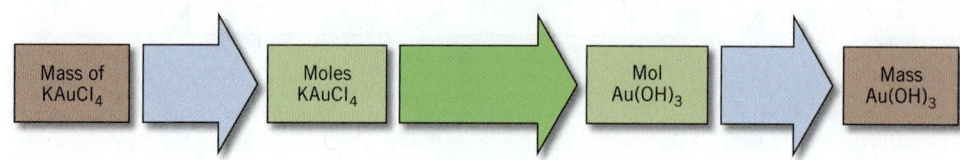

Now we write the equalities we need to finish the problem:

$$1 \text{ mol } KAuCl_4 = 377.88 \text{ g } KAuCl_4$$

$$2 \text{ mol } KAuCl_4 \Leftrightarrow 2 \text{ mol } Au(OH)_3$$

$$1 \text{ mol } Au(OH)_3 = 247.99 \text{ g } Au(OH)_3$$

Solution—Step 2: From here on, we have a routine calculation converting the mass of the limiting reactant, $KAuCl_4$, to the mass of product, $Au(OH)_3$. We set up the following chain calculation.

$$20.00 \text{ g } KAuCl_4 \left(\frac{1 \text{ mol } KAuCl_4}{377.88 \text{ g } KAuCl_4}\right)\left(\frac{2 \text{ mol } Au(OH)_3}{2 \text{ mol } KAuCl_4}\right)\left(\frac{247.99 \text{ g } Au(OH)_3}{1 \text{ mol } Au(OH)_3}\right) = 13.13 \text{ g } Au(OH)_3$$

$$\text{grams } KAuCl_4 \longrightarrow \text{moles } KAuCl_4 \longrightarrow \text{moles } Au(OH)_3 \longrightarrow \text{grams } Au(OH)_3$$

Thus from 20.00 g of $KAuCl_4$ we can make a maximum of 13.13 g of $Au(OH)_3$.

In this synthesis, some of the initial 25.00 g of Na_2CO_3 is left over. Since one of our calculations showed that 20.00 g of $KAuCl_4$ requires only 8.415 g of Na_2CO_3 out of 25.00 g Na_2CO_3, the difference, (25.00 g − 8.415 g) = 16.58 g of Na_2CO_3, remains unreacted. It is possible that the chemist used an excess to ensure that every last bit of the very expensive $KAuCl_4$ would be changed to $Au(OH)_3$.

Are the Answers Reasonable? First, the resulting masses are within the range of ⅕ to 5 times the starting mass and are not unreasonable. Again, we check that our units cancel properly and then we estimate the answer as,

$$\frac{20 \times 200}{400} = 10 \text{ g } Au(OH)_3,$$

which is close enough to our answer to give confidence that the calculation was done correctly.

Practice Exercise 3.32

The reaction between the limestone and hydrochloric acid produces carbon dioxide as shown in the reaction

$$CaCO_3(s) + 2HCl(aq) \longrightarrow CO_2(g) + CaCl_2(aq) + H_2O$$

How many grams of CO_2 can be made by reacting 125 g of $CaCO_3$ with 125 g of HCl? How many grams of which reactant are left over? (*Hint:* Find the limiting reactant.)

Practice Exercise 3.33

In an industrial process for making nitric acid, the first step is the reaction of ammonia with oxygen at high temperature in the presence of platinum. Nitrogen monoxide forms as follows.

$$4NH_3 + 5O_2 \longrightarrow 4NO + 6H_2O$$

How many grams of nitrogen monoxide can form if a mixture initially contains 30.00 g of NH_3 and 40.00 g of O_2?

3.7 | Theoretical Yield and Percentage Yield

In most experiments designed for chemical synthesis, the amount of a product actually isolated falls short of the calculated maximum amount. Losses occur for several reasons. Some are mechanical, such as materials sticking to glassware. In some reactions, losses occur by the evaporation of a volatile product. In others, a product is a solid that separates from the solution as it forms because it is largely insoluble. The solid is removed by filtration. What stays in solution, although relatively small, contributes to some loss of product.

One of the common causes of obtaining less than the stoichiometric amount of a product is the occurrence of a **competing reaction**. It produces a **by-product**, a substance made by a reaction that competes with the **main reaction**. The synthesis of phosphorus trichloride, for example, gives some phosphorus pentachloride as well, because PCl_3 can react further with Cl_2.

Main reaction: $$2P(s) + 3Cl_2(g) \longrightarrow 2PCl_3(l)$$

Competing reaction: $$PCl_3(l) + Cl_2(g) \longrightarrow PCl_5(s)$$

The competition is between newly formed PCl_3 and unreacted phosphorus for unreacted chlorine.

The **actual yield** of desired product is simply how much is isolated, stated in either mass units or moles. The **theoretical yield** of the product is what must be obtained if no losses occur. When less than the theoretical yield of product is obtained, chemists generally calculate the *percentage yield* of product to describe how well the preparation went. The **percentage yield** is the actual yield calculated as a percentage of the theoretical yield.

■ When determining the percentage yield, you must be given the actual yield of the experiment since it cannot be calculated.

$$\text{Percentage yield} = \frac{\text{actual yield}}{\text{theoretical yield}} \times 100\% \qquad (3.2)$$

TOOLS

Theoretical, actual, and percentage yields

Both the actual and theoretical yields must be in the same units.

It is important to realize that the actual yield is an experimentally determined quantity. It cannot be calculated. The theoretical yield is always a calculated quantity based on a chemical equation and the amounts of the reactants available.

Let's now work an example that combines the determination of the limiting reactant with a calculation of percentage yield.

Example 3.18
Calculating a Percentage Yield

A chemist set up a synthesis of phosphorus trichloride by mixing 12.0 g of phosphorus with 35.0 g chlorine gas and obtained 41.7 g of solid phosphorus trichloride. Calculate the percentage yield of this compound.

Analysis: We start by determining the formulas for the reactants and products and then balancing the chemical equation. Phosphorus is $P(s)$ and chlorine, a diatomic gas, is $Cl_2(g)$ and the product is $PCl_3(s)$. The balanced equation is

$$2P(s) + 3Cl_2(g) \longrightarrow 2PCl_3(s)$$

Now we notice that the masses of *both* reactants are given, so this must be a limiting reactant problem. The first step is to figure out which reactant, P or Cl_2, is the limiting reactant, because we must base all calculations on the limiting reactant. Next we can calculate the theoretical yield of product, $PCl_3(s)$. Finally we calculate the percentage yield. The three main steps are summarized as

Determine the limiting reactant	$\longrightarrow$	Determine the theoretical yield	$\longrightarrow$	Calculate the percentage yield

Assembling the Tools: To solve the first two steps, our basic tools are the relationships

$$1 \text{ mol P} = 30.97 \text{ g P}$$

$$1 \text{ mol Cl}_2 = 70.90 \text{ g Cl}_2$$

$$3 \text{ mol Cl}_2 \Leftrightarrow 2 \text{ mol P}$$

For the percentage yield our tool is Equation 3.2.

Solution: In any limiting reactant problem, we can arbitrarily pick one reactant and do a calculation to see whether it can be entirely used up. We'll choose phosphorus and see whether there is enough to react with 35.0 g of chlorine. The following calculation gives us the answer.

$$12.0 \text{ g P} \left(\frac{1 \text{ mol P}}{30.97 \text{ g P}} \right) \left(\frac{3 \text{ mol Cl}_2}{2 \text{ mol P}} \right) \left(\frac{70.90 \text{ g Cl}_2}{1 \text{ mol Cl}_2} \right) = 41.2 \text{ g Cl}_2$$

Thus, 41.2 g of Cl_2 are needed but only 35.0 g of Cl_2 are provided, so there is not enough Cl_2 to react with 12.0 g of P. The Cl_2 will be all used up before the P is used up, so Cl_2 is the limiting reactant. We therefore base the calculation of the theoretical yield of PCl_3 on Cl_2. (We must be careful to use the 35.0 g of Cl_2 given in the problem, *not* the 41.2 g calculated while we determined the limiting reactant.)

To find the theoretical yield of PCl_3, we calculate how many grams of PCl_3 could be made from 35.0 g of Cl_2 if everything went perfectly according to the equation given.

$$35.0 \text{ g Cl}_2 \left(\frac{1 \text{ mol Cl}_2}{70.90 \text{ g Cl}_2} \right) \left(\frac{2 \text{ mol PCl}_3}{3 \text{ mol Cl}_2} \right) \left(\frac{137.32 \text{ g PCl}_3}{1 \text{ mol PCl}_3} \right) = 45.2 \text{ g PCl}_3$$

grams $Cl_2 \longrightarrow$ moles $Cl_2 \longrightarrow$ moles $PCl_3 \longrightarrow$ grams PCl_3

The actual yield was 41.7 g of PCl_3, not 45.2 g, so the percentage yield is calculated as follows

$$\text{Percentage yield} = \frac{41.7 \text{ g PCl}_3}{45.2 \text{ g PCl}_3} \times 100\% = 92.3\%$$

Thus 92.3% of the theoretical yield of PCl_3 was obtained.

● **Is the Answer Reasonable?** The obvious check is that the calculated or theoretical yield can never be *less* than the actual yield. Second, our theoretical yield is within the range of $\frac{1}{5}$ to 5 times the starting amount. Finally, one way to estimate the theoretical yield, after checking that all units cancel properly, is

$$\frac{40 \times 2 \times 140}{70 \times 3} = \frac{80 \times 140}{210} \approx \frac{110}{2} \approx 50$$

which is close to the 45.2 g PCl_3 we calculated.

In the synthesis of aspirin we react salicylic acid with acetic anhydride. The balanced chemical equation is:

$$2HOOCC_6H_4OH + C_4H_6O_3 \longrightarrow 2HOOCC_6H_4O_2C_2H_3 + H_2O$$

salicylic acid   acetic anhydride   acetyl salicylic acid   water

Practice Exercise 3.34

If we mix together 28.2 grams of salicylic acid with 15.6 grams of acetic anhydride in this reaction, we obtain 30.7 grams of aspirin. What are the theoretical and percentage yields of our experiment? (*Hint:* What is the limiting reactant?)

Ethanol, C_2H_5OH, can be converted to acetic acid (the acid in vinegar), $HC_2H_3O_2$, by the action of sodium dichromate in aqueous sulfuric acid according to the following equation.

Practice Exercise 3.35

$$3C_2H_5OH(aq) + 2Na_2Cr_2O_7(aq) + 8H_2SO_4(aq) \longrightarrow$$
$$3HC_2H_3O_2(aq) + 2Cr_2(SO_4)_3(aq) + 2Na_2SO_4(aq) + 11H_2O$$

In one experiment, 24.0 g of C_2H_5OH, 90.0 g of $Na_2Cr_2O_7$, and an excess of sulfuric acid were mixed, and 26.6 g of acetic acid ($HC_2H_3O_2$) was isolated. Calculate the theoretical and percentage yields of $HC_2H_3O_2$.

Multi-Step Reactions

Many chemical synthesis reactions involve more than one step to produce a product. Sometimes there can be more than ten separate reactions from the initial reactants to the final products. In such reactions, the product of one reaction is the reactant for the next. As a result, the overall percentage yield is the product of the percentage yields for all the steps along the way. The following equation is used to calculate percent yields for multi-step syntheses.

$$\text{Overall \% yield} = \left(\frac{\% \text{ yield}_1}{100} \times \frac{\% \text{ yield}_2}{100} \times ... \right)100\%$$

TOOLS
Multi-step percentage yield

In producing a certain drug, one synthetic route involves three steps with percentage yields of 87.2%, 91.1%, and 86.3%. An alternate synthesis uses two steps with percentage yields of 85.5% and 84.3%. Based on the overall percent yield, which is the preferred synthesis?

Practice Exercise 3.36

|Summary

Organized by Learning Objective

Explain how the mole and Avogadro's number serve as conversion factors between the molecular and laboratory scales of matter

In the SI definition, one **mole** of any substance is an amount with the same number, 6.022×10^{23} (**Avogadro's number**), of atoms, molecules, or formula units as there are atoms in 12 g (exactly) of carbon-12. Chemical equations show how atoms and molecules react on the molecular scale or the mole scale. Moles of substances have masses that are convenient for **laboratory scale** experiments.

Perform calculations involving moles and amount of substance

Atomic masses listed on the periodic table are relative to an atomic mass of 12 (exactly) for C-12. The atomic mass of one mole of a monatomic element is often called the **gram atomic mass**. The mass of a mole of molecules is equal to the sum of the atomic masses of all the atoms in the formula and is called the **gram molecular mass**. Similarly, the mass of a mole of an ionic substance is equal to the masses of all the atoms in the ionic formula and is called the **gram formula mass**.

Calculate percentage composition of a compound

Molar mass is a general term that may be used in place of the gram atomic mass, gram molecular mass, or gram formula mass and used for grams-to-moles or moles-to-grams conversions.

Describe how to determine the empirical and molecular formulas of a given compound

An **empirical formula** gives the smallest whole-number ratio of the atoms in a substance. The actual composition of a molecule is given by its **molecular formula**. An empirical formula is generally the *only* formula we write for ionic compounds.

Perform calculations involving moles of reactants and products in a chemical reaction

Stoichiometric calculations are generally problems in which units are converted in a logical sequence of steps (see Figure 3.6). Conversion factors used in these calculations are found in the

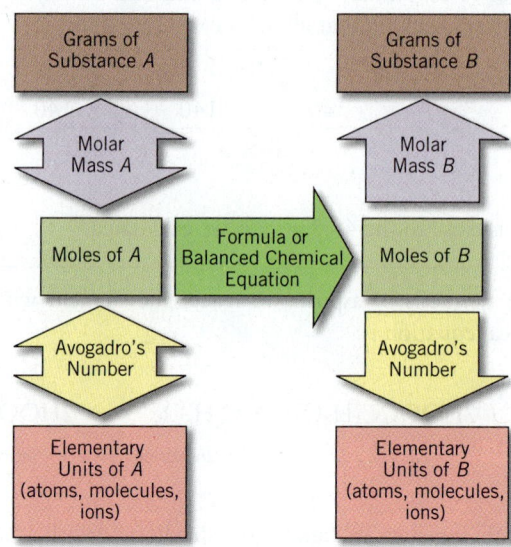

Figure 3.6 | **Stoichiometry pathways.** This diagram summarizes all of the possible stoichiometric calculations encountered in this chapter. The boxes represent the units that we start with and want to end at. Arrows between the boxes indicate the tools that provide the needed conversion factors.

molar mass, Avogadro's number, the chemical formula, or the balanced chemical reaction.

Explain how to determine the limiting reactant in a given reaction and the amount of substance remaining unreacted

A reactant taken in a quantity less than required by another reactant, as determined by the reaction's stoichiometry, is called the **limiting reactant**.

Calculate the percentage yield of a given reaction

The **theoretical yield** of a product can be no more than permitted by the limiting reactant. Sometimes **competing reactions** (side reactions) producing by-products reduce the **actual yield**. The ratio of the actual to the theoretical yields, expressed as a percentage, is the **percentage yield**.

 **TOOLS**

Tools for Problem Solving The following tools were introduced in this chapter. Study them carefully so you can select the appropriate tool when needed.

Atomic mass (Section 3.1)

$$\text{Gram atomic mass of } X = \text{molar mass of } X = 1 \text{ mole } X$$

Molecular mass and formula mass (Section 3.1)

$$\text{Gram molecular mass of } X = \text{molar mass of } X = 1 \text{ mole } X$$
$$\text{Gram formula mass of } X = \text{molar mass of } X = 1 \text{ mole } X$$

Molar mass (Section 3.1)
The sum of the masses of the elements in any chemical formula.

$$\text{Molar mass of } X = 1 \text{ mole of } X$$

Avogadro's number (Section 3.1)

$$1 \text{ mole } X = 6.022 \times 10^{23} \text{ particles of } X$$

Mole ratios (Section 3.2)
Subscripts in a formula establish atom ratios and mole ratios between the elements in the substance.

Mass-to-mass conversions using formulas (Section 3.2)
These steps are required for a mass-to-mass conversion problem using a chemical formula; also see Figure 3.6.

Percentage composition (Section 3.3)
This describes the composition of a compound and can be the basis for computing the empirical formula.

$$\text{Percent of } X = \frac{\text{mass of } X \text{ in the sample}}{\text{mass of the entire sample}} \times 100\%$$

Empirical formula (Section 3.4)
The empirical formula expresses the simplest ratio of the atoms of each element in a compound.

Determination of integer subscripts (Section 3.4)
When determining an empirical formula, dividing all molar amounts by the smallest value often normalizes subscripts to integers. If decimals remain, multiply the subscript by a small whole number until integer subscripts result.

Empirical formulas from percentage composition (Section 3.4)
Percentage composition correlates information from different experiments, to determine empirical formulas.

Balancing chemical equations (Section 3.5)
Balancing equations involves writing the unbalanced equation and then adjusting the coefficients to get equal numbers of each kind of atom on both sides of the arrow.

Equivalencies obtained from balanced equations (Section 3.5)
The coefficients in balanced chemical equations give us relationships between all reactants and products that can be used in factor-label calculations.

Mass-to-mass calculations using balanced chemical equations (Section 3.5)
A logical sequence of conversions allows calculation of all components of a chemical reaction. See Figure 3.6.

Limiting reactant calculations (Section 3.6)
When the amounts of at least two reactants are given, one will be used up before the other and dictate the amont of products formed and the amount of excess reactant left over.

Theoretical, actual, and percentage yields (Section 3.7)
The theoretical yield is based on the limiting reactant whether stated, implied, or calculated. The actual yield must be determined by experiment, and the percentage yield relates the magnitude of the actual yield to the percentage yield.

$$\text{Percentage yield} = \frac{\text{actual mass by experiment}}{\text{theoretical mass by calculation}} \times 100\%$$

Multi-step percentage yield (Section 3.7)
Modern chemical synthesis often involves more than one distinct reaction or step. The overall percentage yield of a multi-step synthesis is

$$\text{Overall \% yield} = \left(\frac{\text{\% yield}_1}{100} \times \frac{\text{\% yield}_2}{100} \times ...\right)100\%$$

Review Questions

The Mole and Avogadro's Number

3.1   What is the definition of the mole?

3.2   Why are moles used, when all stoichiometry problems could be done using only atomic mass units?

3.3   Which contains more molecules: 2.5 mol of H_2O or 2.5 mol of H_2? Which contains more atoms? Which weighs more?

✱3.4   How would Avogadro's number change if the atomic mass unit were to be redefined as 2×10^{-27} kg, exactly?

3.5   What information is required to convert grams of a substance into molecules of that same substance?

3.6   Using atomic mass units, how would you estimate the number of atoms in a gram of iron?

The Mole, Formula Mass, and Stoichiometry

3.7   How many moles of iron atoms are in one mole of Fe_2O_3? How many iron atoms are in one mole of Fe_2O_3?

3.8   Write all the mole-to-mole conversion factors that can be written based on the following chemical formulas: **(a)** SO_2, **(b)** As_2O_3, **(c)** K_2SO_4, **(d)** Na_2HPO_4.

3.9   What information is required to convert grams of a substance into moles of that same substance?

3.10   Why is the expression "1.0 mol of oxygen" ambiguous? Why doesn't a similar ambiguity exist in the expression "64 g of oxygen?"

3.11   The atomic mass of aluminum is 26.98. What specific conversion factors does this value make available for relating a mass of aluminum (in grams) and a quantity of aluminum given in moles?

Chemical Formula and Percentage Composition

3.12   If you are given the mass of two elements and the mass of the compound for a sample, how can you determine if there is another element in the compound?

3.13   How can percentage composition be used to identify a compound?

Determining Empirical and Molecular Formulas

3.14   In general, what fundamental information, obtained from experimental measurements, is required to calculate the empirical formula of a compound?

3.15   Why can percentage composition be used to determine empirical formula, but not molecular formula?

3.16   Why are empirical formulas always used for ionic compounds?

3.17   Under what circumstances can we change, or assign, subscripts in a chemical formula?

3.18   How many distinct empirical formulas are shown by the following models for compounds formed between elements *A* and *B*? Explain. (Element *A* is represented by a black sphere and element *B* by a light gray sphere.)

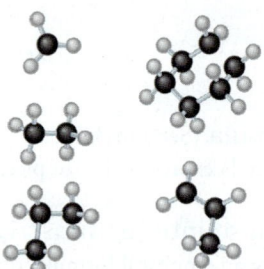

Stoichiometry and Chemical Equations

3.19   When balancing a chemical reaction, what numbers can be changed? When balancing chemical equations which scientific law is being used?

3.20   When given the *unbalanced* equation

$$Na(s) + Cl_2(g) \longrightarrow NaCl(s)$$

and asked to balance it, student A wrote

$$Na(s) + Cl_2(g) \longrightarrow NaCl_2(s)$$

and student B wrote

$$2Na(s) + Cl_2(g) \longrightarrow 2NaCl(s)$$

Both equations are balanced, but which student is correct? Explain why the other student's answer is incorrect.

3.21   Give a step-by-step procedure for estimating the number of grams of *A* required to completely react with 10 moles of *B*, given the following information:

A and *B* react to form A_5B_2.

A has a molecular mass of 100.0. *B* has a molecular mass of 200.0.

There are 6.02×10^{23} molecules of *A* in a mole of *A*.

Which of these pieces of information weren't needed?

3.22   If two substances react completely in a 1-to-1 ratio *both* by mass and by moles, what must be true about these substances?

3.23   What information is required to determine how many grams of sulfur would react with a gram of arsenic?

3.24 A mixture of 0.020 mol of Mg and 0.020 mol of Cl_2 reacted completely to form $MgCl_2$ according to the equation

$$Mg + Cl_2 \longrightarrow MgCl_2$$

What information describes the *stoichiometry* of this reaction? What information gives the *scale* of the reaction?

3.25 In a report to a supervisor, a chemist described an experiment in the following way: "0.0800 mol of H_2O_2 decomposed into 0.0800 mol of H_2O and 0.0400 mol of O_2." Express the chemistry and stoichiometry of this reaction by a conventional chemical equation.

3.26 On April 16, 1947, in Texas City, Texas, two cargo ships, the *Grandcamp* and the *High Flier*, were each loaded with approximately 2000 tons of ammonium nitrate fertilizer. The *Grandcamp* caught fire and exploded, followed by the *High Flier*. More than 600 people were killed and one-third of the city was destroyed. Considering a much smaller mass, how would you count the number of N_2 molecules that could be produced after the explosion of 1.00 kg of NH_4NO_3?

Limiting Reactants

3.27 What is a limiting reactant? And why does the limiting reactant determine how much product is formed?

3.28 Molecules containing A and B react to form AB as shown below. Based on the equations and the contents of the boxes labeled "Initial," sketch for each reaction the molecular models of what is present after the reaction is over. (In both cases, the species B exists as B_2. In reaction 1, A is monatomic; in reaction 2, A exists as diatomic molecules A_2.)

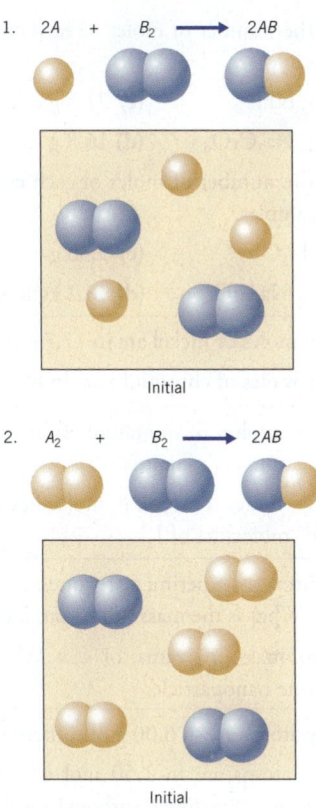

1. $2A + B_2 \longrightarrow 2AB$

Initial

2. $A_2 + B_2 \longrightarrow 2AB$

Initial

Theoretical Yield and Percentage Yield

3.29 Explain why the actual yield is always less than the theoretical yield.

3.30 Is a 45% yield ever a good percentage yield? Why?

Review Problems

The Mole and Avogadro's Number

3.31 Determine the mass in grams of each of the following:

(a) 1.35 mol Fe   (c) 0.876 mol Ca

(b) 24.5 mol O

3.32 Determine the mass in grams of the following:

(a) 0.546 mol S   (c) 8.11 mol Al

(b) 3.29 mol N

3.33 Calculate the formula mass of each of the following to the maximum number of significant figures possible using the table of atomic masses inside the front cover.

(a) $NaHCO_3$ (d) potassium dichromate

(b) $(NH_4)_2CO_3$ (e) aluminum sulfate

(c) $CuSO_4 \cdot 5H_2O$

3.34 Calculate the formula mass of each of the following to the maximum number of significant figures possible using (a)

the periodic table and (b) the table of atomic masses. Both tables are found inside the front cover.

(a) $Pb(C_2H_5)_4$ (d) calcium nitrate

(b) $Na_2SO_4 \cdot 10H_2O$ (e) magnesium phosphate

(c) $Fe_4[Fe(CN)_6]_3$

3.35 Calculate the mass in grams of the following.

(a) 1.25 mol $Ca_3(PO_4)_2$

(b) 0.600 µmol C_4H_{10}

(c) 0.625 mmol iron(III) nitrate

(d) 1.45 mol ammonium carbonate

3.36 What is the mass in grams of the following?

(a) 0.322 mmol $POCl_3$

(b) 4.31×10^{-3} mol $(NH_4)_2HPO_4$

(c) 0.754 mol zinc chloride

(d) 0.194 µmol potassium chlorate

3.37 Calculate the number of moles of each compound in the following samples.

(a) 1.56 ng NH_3 (c) 21.5 g calcium carbonate

(b) 6.98 μg Na_2CrO_4 (d) 16.8 g strontium nitrate

3.38 Calculate the number of moles of each compound in the following samples.

(a) 4.29 g H_2O_2 (c) 9.36 g calcium hydroxide

(b) 4.65 mg $NaAuCl_4$ (d) 38.2 kg lead(II)sulfate

3.39 How many moles of nickel are in 17.7 g of Ni?

3.40 How many moles of chromium are in 85.7 g of Cr?

3.41 How many moles of tantalum atoms correspond to 1.56×10^{21} atoms of tantalum?

3.42 How many moles of iodine molecules correspond to 1.80×10^{24} molecules of I_2?

3.43 A nanotechnology experiment requires 2×10^{12} atoms of potassium. What is the mass of this sample?

3.44 What is the mass, in grams, of 4×10^{17} atoms of gold present in one nanoparticle?

3.45 How many atoms are in 6.00 g of carbon-12?

3.46 How many atoms are in 1.50 mol of carbon-12? How many grams does this much carbon-12 weigh?

The Mole, Formula Mass, and Stoichiometry

3.47 Sucrose (table sugar) has the formula $C_{12}H_{22}O_{11}$. In this compound, what is the

(a) atom ratio of C to H?   (c) atom ratio of H to O?

(b) mole ratio of C to O?   (d) mole ratio of H to O?

3.48 Nail polish remover is usually the volatile liquid ethyl acetate, $CH_3COOC_2H_5$. In this compound, what is the

(a) atom ratio of C to O?   (c) atom ratio of C to H?

(b) mole ratio of C to O?   (d) mole ratio of C to H?

3.49 How many moles of Bi atoms are needed to combine with 1.58 mol of O atoms to make Bi_2O_3?

3.50 How many moles of vanadium atoms, V, are needed to combine with 0.565 mol of O atoms to make vanadium pentoxide, V_2O_5?

3.51 How many moles of Cr are in 2.16 mol of Cr_2O_3?

3.52 How many moles of O atoms are in 4.25 mol of calcium carbonate, $CaCO_3$, the chief constituent of seashells?

3.53 Aluminum sulfate, $Al_2(SO_4)_3$, is a compound used in sewage treatment plants to remove pollutants by causing them to precipitate.

(a) Construct a pair of conversion factors that relate moles of aluminum to moles of sulfur for this compound.

(b) Construct a pair of conversion factors that relate moles of sulfur to moles of $Al_2(SO_4)_3$.

(c) How many moles of Al are in a sample of this compound if the sample also contains 0.900 mol S?

(d) How many moles of S are in 1.16 mol $Al_2(SO_4)_3$?

3.54 Magnetite is a magnetic iron ore. Its formula is Fe_3O_4.

(a) Construct a pair of conversion factors that relate moles of Fe to moles of Fe_3O_4.

(b) Construct a pair of conversion factors that relate moles of Fe to moles of O in Fe_3O_4.

(c) How many moles of Fe are in 2.75 mol of Fe_3O_4?

(d) If this compound could be prepared from Fe_2O_3 and O_2, how many moles of Fe_2O_3 would be needed to prepare 4.50 mol Fe_3O_4?

3.55 How many moles of H_2 and N_2 can be formed by the decomposition of 0.145 mol of ammonia, NH_3?

3.56 How many moles of S are needed to combine with 0.225 mol Al to give Al_2S_3?

3.57 How many moles of UF_6 would have to be decomposed to provide enough fluorine to prepare 1.25 mol of CF_4? (Assume sufficient carbon is available.)

3.58 How many moles of Fe_3O_4 are required to supply enough iron to prepare 0.260 mol Fe_2O_3? (Assume sufficient oxygen is available.)

3.59 How many atoms of carbon are combined with 4.13 moles of hydrogen atoms in a sample of the compound propane, C_3H_8? (Propane is used as the fuel in gas barbecues.)

3.60 How many atoms of hydrogen are found in 2.31 mol of propane, C_3H_8?

3.61 What is the total number of C, H, and O atoms in 0.260 moles of glucose, $C_6H_{12}O_6$?

3.62 What is the total number of N, H, and O atoms in 0.356 mol of ammonium nitrate, NH_4NO_3, an important fertilizer?

3.63 Calcium carbide, CaC_2, was once used to make signal flares for ships. Water dripped onto CaC_2 reacts to give acetylene (C_2H_2), which burns brightly. One sample of CaC_2 contains 0.150 mol of carbon. How many moles and how many grams of calcium are also in the sample?

3.64 Iodized salt contains a trace amount of calcium iodate, $Ca(IO_3)_2$, to help prevent a thyroid condition called goiter. How many moles of iodine atoms are in 0.500 mol of $Ca(IO_3)_2$? How many grams of calcium iodate are needed to supply this much iodine?

3.65 How many moles of nitrogen, N, are in 0.650 mol of ammonium carbonate? How many grams of this compound supply this much nitrogen?

3.66 How many moles of nitrogen, N, are in 0.556 mol of ammonium nitrate? How many grams of this compound supply this much nitrogen?

3.67 How many kilograms of a fertilizer made of pure $(NH_4)_2CO_3$ would be required to supply 1 kilogram of nitrogen to the soil?

3.68 How many kilograms of a fertilizer made of pure P_2O_5 would be required to supply 1 kilogram of phosphorus to the soil?

Chemical Formula and Percentage Composition

3.69 Calculate the percentage composition by mass for each element in the following:

(a) $NH_4H_2PO_4$ (c) sodium dihydrogen phosphate

(b) $(CH_3)_2CO$ (d) calcium sulfate dihydrate

3.70 Calculate the percentage composition by mass for each element in the following:

(a) $(CH_3)_2N_2H_2$ (c) iron(III) nitrate

(b) C_3H_8 (d) aluminum sulfate

3.71 Which has a higher percentage of oxygen: morphine $(C_{17}H_{19}NO_3)$ or heroin $(C_{21}H_{23}NO_5)$?

3.72 Which has a higher percentage of nitrogen: carbamazepine, an anticonvulsant, $(C_{15}H_{12}N_2O)$ or carbetapentane, a cough suppressant, $(C_{20}H_{31}NO_3)$?

3.73 Freon is a trade name for a group of gaseous compounds once used as propellants in aerosol cans. Which has a higher percentage of chlorine: Freon-12 (CCl_2F_2) or Freon-141b $(C_2H_3Cl_2F)$?

3.74 Which has a higher percentage of fluorine: Freon-12 (CCl_2F_2) or Freon 113 $(C_2Cl_3F_3)$?

3.75 It was found that 2.35 g of a compound of phosphorus and chlorine contained 0.539 g of phosphorus. What are the percentages by mass of phosphorus and chlorine in this compound?

3.76 An analysis revealed that 5.67 g of a compound of nitrogen and oxygen contained 1.47 g of nitrogen. What are the percentages by mass of nitrogen and oxygen in this compound?

3.77 Phencyclidine ("angel dust") is $C_{17}H_{25}N$. A sample suspected of being this illicit drug was found to have a percentage composition of 84.71% C, 10.42% H, and 5.61% N. Do these data acceptably match the theoretical data for phencyclidine?

3.78 The hallucinogenic drug LSD has the molecular formula $C_{20}H_{25}N_3O$. One suspected sample contained 74.07% C, 7.95% H, and 9.99% N.

(a) What is the percentage of O in the sample?

(b) Are these data consistent for LSD?

3.79 How many grams of O are combined with 7.14×10^{21} atoms of N in the compound dinitrogen pentoxide?

3.80 How many grams of C are combined with 4.25×10^{23} atoms of H in the compound C_5H_{12}?

Determining Empirical Formulas and Molecular Formulas

3.81 Write empirical formulas for the following compounds.

(a) S_2Cl_2 (d) As_2O_6

(b) $C_6H_{12}O_6$ (e) H_2O_2

(c) NH_3

3.82 What are the empirical formulas of the following compounds?

(a) $C_2H_4(OH)_2$ (d) B_2H_6

(b) $H_2S_2O_8$ (e) C_2H_5OH

(c) C_4H_{10}

3.83 Radioactive sodium pertechnetate is used as a brain-scanning agent in medicine. Quantitative analysis of a sample of sodium pertechnetate with a mass of 0.896 g found 0.111 g of sodium and 0.477 g of technetium. The remainder was oxygen. Calculate the empirical formula of sodium pertechnetate.

3.84 A sample of Freon was found to contain 0.423 g of C, 2.50 g of Cl, and 1.34 g of F. What is the empirical formula of this compound?

3.85 Cinnamic acid, a compound related to the flavor component of cinnamon, is 72.96% carbon, 5.40% hydrogen, and the rest is oxygen. What is the empirical formula of this acid?

3.86 Vanillin, a compound used as a flavoring agent in food products, has the following percentage composition: 63.2% C, 5.29% H, and 31.6% O. What is the empirical formula of vanillin?

3.87 A dry-cleaning fluid composed of only carbon and chlorine was found to be composed of 14.5% C and 85.5% Cl (by mass). What is the empirical formula of this compound?

3.88 One compound of mercury with a molar mass of 519 contains 77.26% Hg, 9.25% C, and 1.17% H (with the balance being O). Calculate the empirical formula.

3.89 When 0.684 g of an organic compound containing only carbon, hydrogen, and oxygen was burned in oxygen, 1.312 g of CO_2 and 0.805 g of H_2O were obtained. What is the empirical formula of the compound?

3.90 Methyl ethyl ketone (often abbreviated MEK) is a powerful solvent with many commercial uses. A sample of this compound (which contains only C, H, and O) weighing 0.822 g was burned in oxygen to give 2.01 g of CO_2 and 0.827 g of H_2O. What is the empirical formula for MEK?

3.91 When 6.853 mg of a sex hormone was burned in a combustion analysis, 19.73 mg of CO_2 and 6.391 mg of H_2O were obtained. What is the empirical formula of the compound?

3.92 When a sample of a compound in the vitamin D family was burned in a combustion analysis, 5.983 mg of the compound gave 18.490 mg of CO_2 and 6.232 mg of H_2O. What is the empirical formula of the compound?

3.93 The following are empirical formulas and the masses per mole for three compounds. What are their molecular formulas?

(a) NaS_2O_3; 270.4 g/mol   (c) C_2HCl; 181.4 g/mol

(b) C_3H_2Cl; 147.0 g/mol

3.94 The following are empirical formulas and the masses per mole for three compounds. What are their molecular formulas?

(a) Na_2SiO_3; 732.6 g/mol   (c) CH_3O; 62.1 g/mol

(b) $NaPO_3$; 305.9 g/mol

3.95 The compound described in Problem 3.91 was found to have a molecular mass of 290. What is its molecular formula?

3.96 The compound described in Problem 3.92 was found to have a molecular mass of 399. What is the molecular formula of this compound?

3.97 A sample of a compound of mercury and bromine with a mass of 0.389 g was found to contain 0.111 g bromine. Its molecular mass was found to be 561. What are its empirical and molecular formulas?

3.98 A 0.6662 g sample of "antimonal saffron," which is a red pigment used in painting, was found to contain 0.4017 g of antimony. The remainder was sulfur. The formula mass of this compound is 404. What are the empirical and molecular formulas of this pigment?

3.99 A sample of a compound of C, H, N, and O, was found to have a percentage composition of 27.91% C, 2.341% H, and 32.56% N. Its formula mass is 129. Calculate its empirical and molecular formulas.

3.100 Strychnine, a deadly poison, has a formula mass of 334 and a percentage composition of 75.42% C, 6.63% H, 8.38% N, and the balance oxygen. Calculate the empirical and molecular formulas of strychnine.

Stoichiometry and Chemical Equations

3.101 Balance the following equations.

(a) $Mg(OH)_2 + HBr \longrightarrow MgBr_2 + H_2O$

(b) $HCl + Ca(OH)_2 \longrightarrow CaCl_2 + H_2O$

(c) $Al_2O_3 + H_2SO_4 \longrightarrow Al_2(SO_4)_3 + H_2O$

(d) $KHCO_3 + H_3PO_4 \longrightarrow$
$$K_2HPO_4 + H_2O + CO_2$$

(e) $C_9H_{20} + O_2 \longrightarrow CO_2 + H_2O$

3.102 Balance the following equations.

(a) $CaO + HNO_3 \longrightarrow Ca(NO_3)_2 + H_2O$

(b) $Na_2CO_3 + Mg(NO_3)_2 \longrightarrow MgCO_3 + NaNO_3$

(c) $(NH_4)_3PO_4 + NaOH \longrightarrow$
$$Na_3PO_4 + NH_3 + H_2O$$

(d) $LiHCO_3 + H_2SO_4 \longrightarrow Li_2SO_4 + H_2O + CO_2$

(e) $C_4H_{10}O + O_2 \longrightarrow CO_2 + H_2O$

3.103 Write the following as balanced chemical equations.

(a) Calcium hydroxide reacts with hydrogen chloride to form calcium chloride and water.

(b) Silver nitrate and calcium chloride react to form calcium nitrate and silver chloride.

(c) Lead nitrate reacts with sodium sulfate to form lead sulfate and sodium nitrate.

(d) Iron(III) oxide and carbon react to form iron and carbon dioxide.

3.104 Write the following as balanced chemical equations.

(a) Sulfur dioxide reacts with oxygen to form sulfur trioxide.

(b) Sodium bicarbonate and sulfuric acid react to form sodium sulfate, water, and carbon dioxide.

(c) Tetraphosphorous decaoxide and water react together and form phosphoric acid.

(d) Butane reacts with oxygen to form carbon dioxide and water.

3.105 Using the diagram below, write the balanced chemical equation.

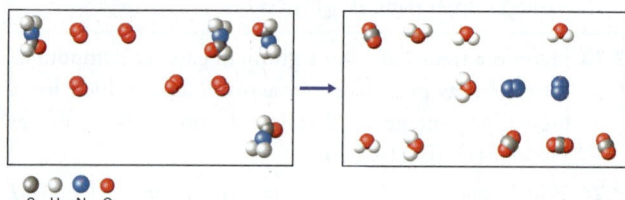

C H N O

3.106 Using the diagram below, write the balanced chemical equation.

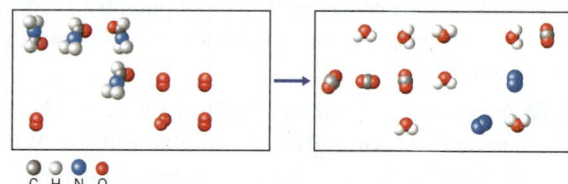

C H N O

3.107 The balanced chemical equation for the combustion of propane, a common heating fuel, is

$$C_3H_8 + 5O_2 \longrightarrow 3CO_2 + 4H_2O$$

Draw a molecular-level diagram of this chemical reaction. Use the color chart in Section 2.3 to distinguish the atoms from each other.

3.108 Draw a molecular-level diagram for the reaction in which sulfur dioxide reacts with molecular oxygen to form sulfur trioxide. Be sure to write and balance the equation first. Use the color chart in Section 2.3 to distinguish the atoms from each other.

3.109 Write the equation that expresses in acceptable chemical shorthand the following statement: "Iron can be made to react with molecular oxygen to give iron(III) oxide."

3.110 The conversion of one air pollutant, nitrogen monoxide, produced in vehicle engines, into another pollutant, nitrogen dioxide, occurs when nitrogen monoxide reacts with molecular oxygen in the air. Write the balanced equation for this reaction.

3.111 A balanced chemical equation contains the term "$2Ba(OH)_2 \cdot 8H_2O$." How many atoms of each element does this represent in the molecular view? How many moles of each element does this represent on the laboratory scale?

3.112 A balanced chemical equation contains the term "$3Ca_3(PO_4)_2$." How many atoms of each element does this represent in the molecular view? How many moles of each element does this represent on the laboratory scale?

3.113 Chlorine is used by textile manufacturers to bleach cloth. Excess chlorine is destroyed by its reaction with sodium thiosulfate, $Na_2S_2O_3$, as follows.

$$Na_2S_2O_3(aq) + 4Cl_2(g) + 5H_2O \longrightarrow$$
$$2NaHSO_4(aq) + 8HCl(aq)$$

(a) How many moles of $Na_2S_2O_3$ are needed to react with 0.12 mol of Cl_2?

(b) How many moles of HCl can form from 0.12 mol of Cl_2?

(c) How many moles of H_2O are required for the reaction of 0.12 mol of Cl_2?

(d) How many moles of H_2O react if 0.24 mol HCl is formed?

3.114 The octane in gasoline burns according to the following equation.

$$2C_8H_{18} + 25O_2 \longrightarrow 16CO_2 + 18H_2O$$

(a) How many moles of O_2 are needed to react fully with 6.84 mol of octane?

(b) How many moles of CO_2 can be formed from 0.511 mol of octane?

(c) How many moles of water are produced by the combustion of 8.20 mol of octane?

(d) If this reaction is used to synthesize 6.00 mol of CO_2, how many moles of oxygen are needed? How many moles of octane?

3.115 Propane burns according to the following equation:

$$C_3H_8 + 5O_2 \longrightarrow 3CO_2 + 4H_2O$$

(a) How many grams of O_2 are needed to react fully with 3.45 mol of propane?

(b) How many grams of CO_2 can form from 0.177 mol of propane?

(c) How many grams of water are produced by the combustion of 4.86 mol of propane?

3.116 The following reaction is used to extract gold from pre-treated gold ore:

$$2Au(CN)_2^-(aq) + Zn(s) \longrightarrow 2Au(s) + Zn(CN)_4^-(aq)$$

(a) How many grams of Zn are needed to react with 0.11 mol of $Au(CN)_2^-$?

(b) How many grams of Au can form from 0.11 mol of $Au(CN)_2^-$?

(c) How many grams of $Au(CN)_2^-$ are required for the reaction of 0.11 mol of Zn?

3.117 The incandescent white of a fireworks display is caused by the reaction of phosphorus with O_2 to give P_4O_{10}.

(a) Write the balanced chemical equation for the reaction.

(b) How many grams of O_2 are needed to combine with 6.85 g of P?

(c) How many grams of P_4O_{10} can be made from 8.00 g of O_2?

(d) How many grams of P are needed to make 7.46 g of P_4O_{10}?

3.118 The combustion of butane, C_4H_{10}, produces carbon dioxide and water. When one sample of C_4H_{10} was burned, 4.46 g of water was formed.

(a) Write the balanced chemical equation for the reaction.

(b) How many grams of butane were burned?

(c) How many grams of O_2 were consumed?

(d) How many grams of CO_2 were formed?

3.119 In *dilute* nitric acid, HNO_3, copper metal dissolves according to the following equation.

$$3Cu(s) + 8HNO_3(aq) \longrightarrow$$
$$3Cu(NO_3)_2(aq) + 2NO(g) + 4H_2O$$

How many grams of HNO_3 are needed to dissolve 11.45 g of Cu according to this equation?

3.120 The reaction of hydrazine, N_2H_4, with hydrogen peroxide, H_2O_2, has been used in rocket engines. One way these compounds react is described by the equation

$$N_2H_4 + 7H_2O_2 \longrightarrow 2HNO_3 + 8H_2O$$

According to this equation, how many grams of H_2O_2 are needed to react completely with 852 g of N_2H_4?

3.121 Oxygen gas can be produced in the laboratory by decomposition of hydrogen peroxide (H_2O_2):

$$2H_2O_2(l) \longrightarrow 2H_2O + O_2(g)$$

How many kg of O_2 can be produced from 1.0 kg of H_2O_2?

3.122 Oxygen gas can be produced in the laboratory by decomposition of potassium chlorate ($KClO_3$):

$$KClO_3(s) \longrightarrow 2KCl(s) + 3O_2(g)$$

How many kg of O_2 can be produced from 1.0 kg of $KClO_3$?

Limiting Reactants

3.123 Using the balanced equation

$$2C_2H_6S(g) + 9O_2(g) \longrightarrow 4CO_2(g) + 2SO_2(g) + 6H_2O(g)$$

determine the number of SO_2 units formed in the following molecular representation.

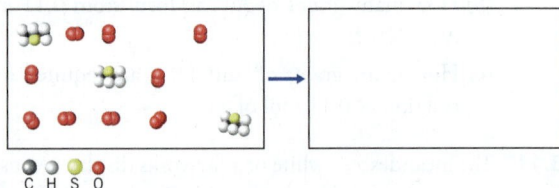

C H S O

3.124 Using the balanced chemical equation in Problem 3.123 and the diagram below, determine what the limiting reactant was.

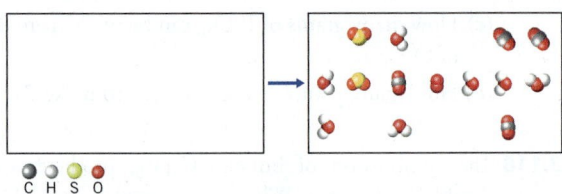

C H S O

3.125 The thermite reaction between powdered aluminum and iron(III) oxide,

$$2Al + Fe_2O_3 \longrightarrow Al_2O_3 + 2Fe$$

produces so much heat the iron that forms is molten, as described in Example 3.16. Suppose that in one batch of reactants 4.20 mol of Al was mixed with 1.75 mol of Fe_2O_3.

(a) Which reactant, if either, was the limiting reactant?

(b) Calculate the number of grams of iron that can be formed from this mixture of reactants.

3.126 Ethanol (C_2H_5OH) is synthesized for industrial use by the following reaction, carried out at very high pressure:

$$C_2H_4(g) + H_2O(g) \longrightarrow C_2H_5OH(l)$$

What is the maximum amount, in kg, of ethanol that can be produced when 1.62 kg of ethylene (C_2H_4) and 0.0148 kg of steam are placed into the reaction vessel?

3.127 Silver nitrate, $AgNO_3$, reacts with iron(III) chloride, $FeCl_3$, to give silver chloride, $AgCl$, and iron(III) nitrate, $Fe(NO_3)_3$. A solution containing 18.0 g of $AgNO_3$ was mixed with a solution containing 32.4 g of $FeCl_3$. How many grams of which reactant remains after the reaction is over?

3.128 Chlorine dioxide, ClO_2, has been used as a disinfectant in air-conditioning systems. It reacts with water according to the equation

$$6ClO_2 + 3H_2O \longrightarrow 5HClO_3 + HCl$$

If 142.0 g of ClO_2 is mixed with 38.0 g of H_2O, how many grams of which reactant remain if the reaction is complete?

3.129 Some of the acid in acid rain is produced by the following reaction:

$$3NO_2(g) + H_2O(l) \longrightarrow 2HNO_3(aq) + NO(g)$$

If a falling raindrop weighing 0.050 g comes into contact with 1.0 mg of $NO_2(g)$, how many milligrams of HNO_3 can be produced?

3.130 Phosphorus pentachloride reacts with water to give phosphoric acid and hydrogen chloride according to the following equation.

$$PCl_5 + 4H_2O \longrightarrow H_3PO_4 + 5HCl$$

In one experiment, 0.360 mol of PCl_5 was slowly added to 2.88 mol of water.

(a) Which reactant, if either, was the limiting reactant?

(b) How many grams of HCl were formed in the reaction?

Theoretical Yield and Percentage Yield

3.131 Barium sulfate, $BaSO_4$, is made by the following reaction.

$$Ba(NO_3)_2(aq) + Na_2SO_4(aq) \longrightarrow BaSO_4(s) + 2NaNO_3(aq)$$

An experiment was begun with 75.00 g of $Ba(NO_3)_2$ and an excess of Na_2SO_4. After collecting and drying the product, 64.45 g of $BaSO_4$ was obtained. Calculate the theoretical yield and percentage yield of $BaSO_4$.

3.132 The Solvay process for the manufacture of sodium carbonate begins by passing ammonia and carbon dioxide through a solution of sodium chloride to make sodium bicarbonate and ammonium chloride. The equation for the overall reaction is

$$H_2O + NaCl + NH_3 + CO_2 \longrightarrow NH_4Cl + NaHCO_3$$

In the next step, sodium bicarbonate is heated to give sodium carbonate and two gases, carbon dioxide and steam.

$$2NaHCO_3 \longrightarrow Na_2CO_3 + CO_2 + H_2O$$

What is the theoretical yield of sodium carbonate, expressed in grams, if 120 g of NaCl was used in the first reaction? If 85.4 g of Na_2CO_3 was obtained, what was the percentage yield?

3.133 Aluminum sulfate can be made by the following reaction.

$$2AlCl_3(aq) + 3H_2SO_4(aq) \longrightarrow Al_2(SO_4)_3(aq) + 6HCl(aq)$$

It is quite soluble in water, so to isolate it the solution has to be evaporated to dryness. This drives off the volatile HCl, but the residual solid has to be heated to a little over 200 °C to drive off all of the water. In one experiment, 25.0 g of $AlCl_3$ was mixed with 30.0 g of H_2SO_4. Eventually, 28.46 g of pure $Al_2(SO_4)_3$ was isolated. Calculate the percentage yield.

3.134 The combustion of methyl alcohol in an abundant excess of oxygen follows the equation

$$2CH_3OH + 3O_2 \longrightarrow 2CO_2 + 4H_2O$$

When 6.40 g of CH_3OH was mixed with 10.2 g of O_2 and ignited, 6.12 g of CO_2 was obtained. What was the percentage yield of CO_2?

3.135 Manganese(III) fluoride, MnF_3, can be prepared by the following reaction.

$$2MnI_2(s) + 13F_2(g) \longrightarrow 2MnF_3(s) + 4IF_5(l)$$

If the percentage yield of MnF_3 is always approximately 56%, how many grams of MnF_3 can be expected if 10.0 grams of each reactant is used in an experiment?

3.136 The potassium salt of benzoic acid, potassium benzoate $(KC_7H_5O_2)$, can be made by the action of potassium permanganate on toluene (C_7H_8) as follows.

$$C_7H_8 + 2KMnO_4 \longrightarrow$$
$$KC_7H_5O_2 + 2MnO_2 + KOH + H_2O$$

If the yield of potassium benzoate cannot realistically be expected to be more than 71%, what is the minimum number of grams of toluene needed to produce 11.5 g of potassium benzoate?

Additional Exercises

*__3.137__ Mercury is an environmental pollutant because it can be converted by certain bacteria into the extremely poisonous substance dimethyl mercury, $(CH_3)_2Hg$. This compound ends up in the food chain and accumulates in the tissues of aquatic organisms, particularly fish, which renders them unsafe to eat. It is estimated that in the United States 263 tons of mercury are released into the atmosphere each year. If only 1.0% of this mercury is eventually changed to $(CH_3)_2Hg$, how many pounds of this compound are formed annually?

*__3.138__ Lead compounds are often highly colored and are toxic to mold, mildew, and bacteria, properties that in the past were useful for paints used before 1960. Today we know lead is very hazardous and it is not used in paint; however, old paint is still a problem. If a certain lead-based paint contains 14.5% $PbCr_2O_7$ and 73% of the paint evaporates as it dries, what mass of lead will be in a paint chip that weighs 0.15 g?

*__3.139__ A superconductor is a substance that is able to conduct electricity without resistance, a property that is very desirable in the construction of large electromagnets. Metals have this property if cooled to temperatures a few degrees above absolute zero, but this requires the use of expensive liquid helium (boiling point 4 K). Scientists have discovered materials that become superconductors at higher temperatures, but they are ceramics. Their brittle nature has so far prevented them from being made into long wires. A recently discovered compound of magnesium and boron, which consists of 52.9% Mg and 47.1% B, shows special promise as a high-temperature superconductor because it is inexpensive to make and can be fabricated into wire relatively easily. What is the formula of this compound?

*__3.140__ A compound of Ca, C, N, and S was subjected to quantitative analysis and formula mass determination, and the following data were obtained. A 0.250 g sample was mixed with Na_2CO_3 to convert all of the Ca to 0.160 g of $CaCO_3$. A 0.115 g sample of the compound was carried through a series of reactions until all of its S was changed to 0.344 g of $BaSO_4$. A 0.712 g sample was processed to liberate all of its N as NH_3, and 0.155 g of NH_3

was obtained. The formula mass was found to be 156. Determine the empirical and molecular formulas of this compound.

3.141 Ammonium nitrate will detonate if ignited in the presence of certain impurities. The equation for this reaction at a high temperature is

$$2NH_4NO_3(s) \xrightarrow{>300\,°C} 2N_2(g) + O_2(g) + 4H_2O(g)$$

Notice that all of the products are gases, so they occupy a vastly greater volume than the solid reactant. How many moles of *all* gases are produced from 1 mol of NH_4NO_3? If 1.00 ton of NH_4NO_3 exploded according to this equation, how many moles of *all* gases would be produced? (1 ton = 2000 lb.)

*__3.142__ A lawn fertilizer is rated as 6.00% nitrogen, meaning 6.00 g of N in 100 g of fertilizer. The nitrogen is present in the form of urea, $(NH_2)_2CO$. How many grams of urea are present in 100 g of the fertilizer to supply the rated amount of nitrogen?

*__3.143__ Nitrogen is the "active ingredient" in many quick-acting fertilizers. You are operating a farm of 1500 acres to produce soybeans. Which of the following fertilizers will you choose as the most economical for your farm?

(a) NH_4NO_3 at $625 for 25 kg

(b) $(NH_4)_2HPO_4$ at $55 for 1 kg

(c) urea, CH_4ON_2, at $60 for 5 kg

(d) ammonia, NH_3 at $128 for 50 kg

*__3.144__ Based solely on the amount of available carbon, how many grams of sodium oxalate, $Na_2C_2O_4$, could be obtained from 125 g of C_6H_6? (Assume that no loss of carbon occurs in any of the reactions needed to produce the $Na_2C_2O_4$.)

*__3.145__ According to NASA, the space shuttle's external fuel tank for the main propulsion system carries 1,361,936 lb of liquid oxygen and 227,641 lb of liquid hydrogen. During takeoff, these chemicals are consumed as they react to form water. If the reaction is continued until all of one reactant is gone, how many pounds of which reactant are left over?

*3.146 For a research project, a student decided to test the effect of the lead(II) ion (Pb^{2+}) on the ability of salmon eggs to hatch. This ion was obtainable from the water-soluble salt, lead(II) nitrate, $Pb(NO_3)_2$, which the student decided to make by the following reaction. (The desired product was to be isolated by the slow evaporation of the water.)

$$PbO(s) + 2HNO_3(aq) \longrightarrow Pb(NO_3)_2(aq) + H_2O$$

Losses of product for various reasons were expected, and a yield of 86.0% was expected. In order to have 5.00 g of product at this yield, how many grams of PbO should be taken? (Assume that sufficient nitric acid, HNO_3, would be used.)

3.147 Chlorine atoms cause chain reactions in the stratosphere that destroy ozone that protects the earth's surface from ultraviolet radiation. The chlorine atoms come from chlorofluorocarbons, compounds that contain carbon, fluorine, and chlorine, which were used for many years as refrigerants. One of these compounds is Freon-12, CF_2Cl_2. If a sample contains 1.0×10^{-9} g of Cl, how many grams of F should be present if all of the F and Cl atoms in the sample came from CF_2Cl_2 molecules?

3.148 Lime, CaO, can be produced in two steps as shown in the equations below. If the percentage yield of the first step is 83.5% and the percentage yield of the second step is 71.4%, what is the expected overall percentage yield for producing CaO from $CaCl_2$?

$$CaCl_2(aq) + CO_2(g) + H_2O \longrightarrow CaCO_3(s) + 2HCl(aq)$$

$$CaCO_3(s) \xrightarrow{\text{heat}} CaO(s) + CO_2(g)$$

Exercises in Critical Thinking

*3.149 A newspaper story describing the local celebration of Mole Day on October 23 (selected for Avogadro's number, 6.02×10^{23}) attempted to give the readers a sense of the size of the number by stating that a mole of M&Ms would be equal to 18 tractor trailers full. Assuming that an M&M occupies a volume of about 0.5 cm^3, calculate the dimensions of a cube required to hold one mole of M&Ms. Would 18 tractor trailers be sufficient?

*3.150 Suppose you had one mole of pennies and you were going to spend 500 million dollars each and every second until you spent your entire fortune. How many years would it take you to spend all this cash? (Assume 1 year = 365 days.)

*3.151 Using the above two questions as examples, devise a creative way to demonstrate the size of the mole, or Avogadro's number.

3.152 List the different ways in which a chemist could use the information used to determine empirical formulas.

4

Molecular View of Reactions in Aqueous Solutions

Chapter Outline

© Cristian Baitg/iStockphoto

155

This Chapter in Context

Water is an amazing substance. Composed of just three atoms, it is one of the most common compounds on earth, and its ability to dissolve so many different kinds of materials such as the oxygen for the fish to breathe or the salts that allow seawater to support sea life makes it responsible for the evolution of life as we know it.

A property of water that sets it apart from other liquids is its ability to dissolve many ionic compounds, enabling their particles to mix at the molecular level where they are able to easily react.

LEARNING OBJECTIVES

After reading this chapter, you should be able to:

- describe solutions qualitatively and quantitatively
- understand that electrolytes are soluble ionic compounds and nonelectrolytes do not produce ions in solutions
- write balanced molecular, ionic, and net ionic equations
- identify acids and bases based on their names and formulas and write balanced equations showing hydronium and hydroxide ions
- learn how to translate formulas to names and names to formulas for common acids and bases
- use the principles of metathesis to predict reaction products and to plan a chemical synthesis
- define and calculate molarity, and use it as a conversion factor
- learn to use molarity in stoichiometric calculations
- understand the methods and calculations used in titrations and chemical analyses

4.1 | Describing Solutions

To begin this subject we have to define some terms. A **solution** is a homogeneous mixture in which molecules or ions move freely (Figure 4.1). When a solution is prepared, at least two substances are needed. One is the **solvent**, and all of the others are **solutes**. The solutes are mixed or dissolved in the solvent, and in this chapter only aqueous solutions are used, so the solvent will be liquid water.[1] A solute is any substance dissolved in the solvent. It might be a gas, like the oxygen dissolved in the water that allows the fish shown in the photo on the preceding page to live. Some solutes are liquids, like ethylene glycol, which is dissolved in water to make antifreeze, which protects a car's radiator from freezing in the winter. Solids, of course, can be solutes, like the sugar dissolved in lemonade or the salts dissolved in seawater. One way to describe a solution is to specify the solvent and solutes.

Another way to describe a solution is to specify the **concentration** of its solutes. A concentration is always expressed as a ratio of the amount of solute to the amount of solvent (or to the amount of solution). A **percentage concentration**, for example, is a "solute-to-solution" ratio that we calculate as the number of grams of solute per gram of solution multiplied by exactly 100%. Thus, the concentration of sodium chloride in seawater is often given as 3% NaCl, which means 0.03 g NaCl/g seawater.

The relative ratios of solute to solvent are often described qualitatively, without numbers. In a **dilute solution** the solute to solvent ratio is small—such as when a few crystals of sugar are dissolved in a glass of water. In a **concentrated solution**, the solute to solvent ratio is large (Figure 4.2). Pancake syrup, for example, is a concentrated solution of sugar in water. Concentrated and dilute are relative terms. For example, a solution of 100 g of sugar in 100 mL of water is concentrated only when compared to one with just 10 g of sugar in 100 mL of water, but the latter solution is concentrated when compared to one that has just 1 g of sugar in 100 mL of water.

Usually there is a limit to how much solute will dissolve in a given amount of solution. When a solution contains as much solute as it will hold at a given temperature, it

■ When water is a component of a solution, it is usually considered to be the solvent even when it is present in small amounts.

---

[1] Liquid water is a typical and very common solvent, but the solvent can actually be in any physical state: solid, liquid, or gas.

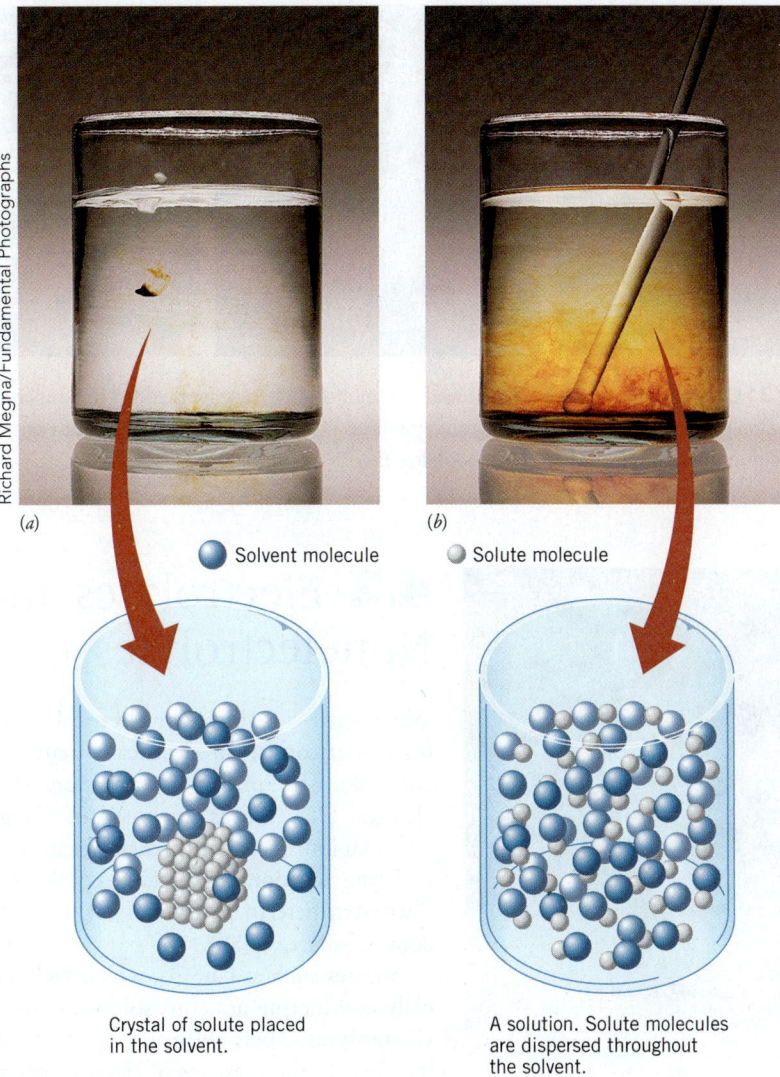

Solvent molecule   Solute molecule

Crystal of solute placed
in the solvent.

A solution. Solute molecules
are dispersed throughout
the solvent.

Richard Megna/Fundamental Photographs

Figure 4.1 | Formation of a solution of iodine molecules in alcohol. (*a*) A crystal of iodine, I_2, on its way to the bottom of the beaker is already beginning to dissolve, the purplish iodine crystal forming a reddish brown solution. In the hugely enlarged view beneath the photo, we see the iodine molecules still bound in a crystal. For simplicity, the solute and solvent particles are shown as spheres. (*b*) Stirring the mixture helps the iodine molecules to disperse in the solvent, as illustrated in the molecular view below the photo.

is a **saturated solution**. If more solute is added it simply does not dissolve. A solid that is not dissolved is called a **precipitate**, and a chemical reaction that produces a precipitate is called a precipitation reaction. The **solubility** of a solute is the amount required to make a saturated solution, usually expressed as grams dissolved in 100 g of solvent at a given temperature. The temperature must be specified because solubility varies with temperature. An **unsaturated solution** has less solute than required for saturation, and more solute can dissolve.

Solubility of a solute usually increases if the temperature increases, which means that more solute can be dissolved by heating a saturated solution. If the temperature of a warm saturated solution is lowered, the additional solute should precipitate from the solution, and indeed, this tends to happen. However, sometimes the solute doesn't precipitate, leaving us with a **supersaturated solution**, a solution that actually contains more solute than required for saturation. Supersaturated solutions are unstable and can only be prepared if there are no solids present. If even a tiny crystal or dust particle is present, or is added, the extra solute precipitates (Figure 4.3).

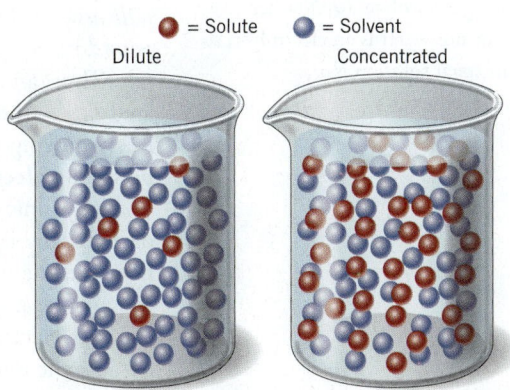

= Solute   = Solvent
Dilute   Concentrated

Figure 4.2 | Dilute and concentrated solutions. The dilute solution on the left has fewer solute molecules per unit volume than the more concentrated solution on the right.

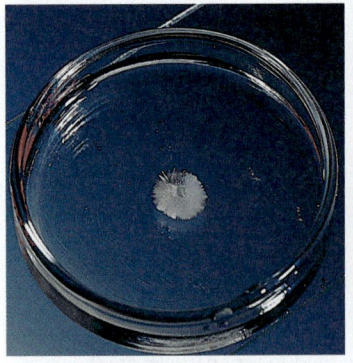  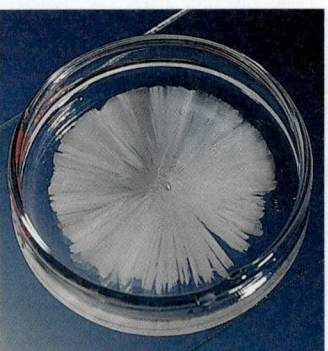 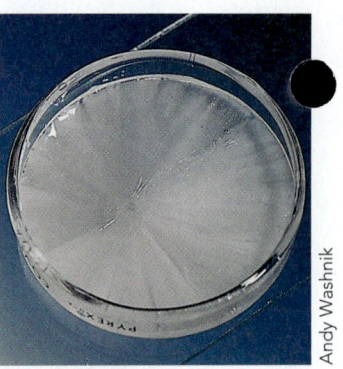

Andy Washnik

Figure 4.3 | Crystallization. When a small seed crystal of sodium acetate is added to a supersaturated solution of the compound, excess solute crystallizes rapidly until the solution is just saturated. The crystallization shown in this sequence took less than 10 seconds!

Michael Watson

(*a*)

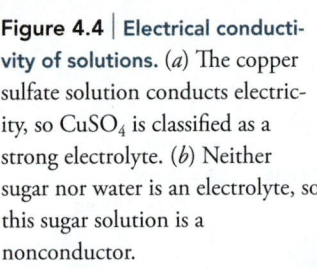

Figure 4.4 | **Electrical conductivity of solutions.** (*a*) The copper sulfate solution conducts electricity, so $CuSO_4$ is classified as a strong electrolyte. (*b*) Neither sugar nor water is an electrolyte, so this sugar solution is a nonconductor.

Michael Watson

(*b*)

4.2 | Electrolytes and Nonelectrolytes

Water itself is a very poor electrical conductor because it consists of electrically neutral molecules that are unable to transport electrical charges. However, as we noted in Chapter 2, when an ionic compound dissolves in water the resulting solution conducts electricity well, as illustrated here in Figure 4.4 for a solution of copper sulfate, $CuSO_4$.

Solutes such as $CuSO_4$, which yield electrically conducting aqueous solutions, are called **electrolytes**. Their ability to conduct electricity suggests the presence of electrically charged particles that move freely within the solution. The generally accepted reason is that when an ionic compound dissolves in water, the ions separate from each other and enter the solution as more or less independent particles surrounded by molecules of the solvent. This change is called the **dissociation** (*breaking apart*) of the ionic compound, and is illustrated in Figure 4.5. In general, *we will assume that in water the dissociation of any salt (a term that applies to ionic compounds) is complete* and that the solution contains no undissociated formula units of the salt. Thus, an aqueous solution of $CuSO_4$ is really a solution that contains the dissociated Cu^{2+} and SO_4^{2-} ions. Because solutions of ionic compounds contain so many freely moving ions, they are strong conductors of electricity. Therefore, salts are said to be **strong electrolytes**.

Many ionic compounds have low solubilities in water. For example, only a tiny amount of AgBr, the light-sensitive compound in light-responsive sunglasses, dissolves in water; however, the very small amount of it that does dissolve is completely dissociated. Even though the number of ions in the solution is extremely small and the solution doesn't conduct electricity well, it is still convenient to think of AgBr as a strong electrolyte because the ions are completely dissociated in aqueous solution.

Aqueous solutions of most molecular, covalently bonded compounds do not conduct electricity, and such solutes are called **nonelectrolytes**. Examples are sugar (Figure 4.4*b*), ethylene glycol (the solute in antifreeze solutions), and alcohol. All three consist of uncharged molecules that stay intact and simply intermingle with water molecules when they dissolve (Figure 4.6).

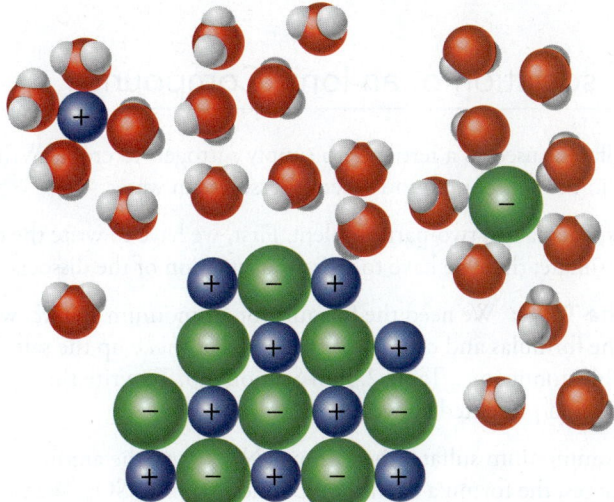

Figure 4.5 | Dissociation of an ionic compound as it dissolves in water. Ions separate from the solid and become surrounded by water molecules, or hydrated. In the solution the ions move freely, which enables the solution to conduct electricity.

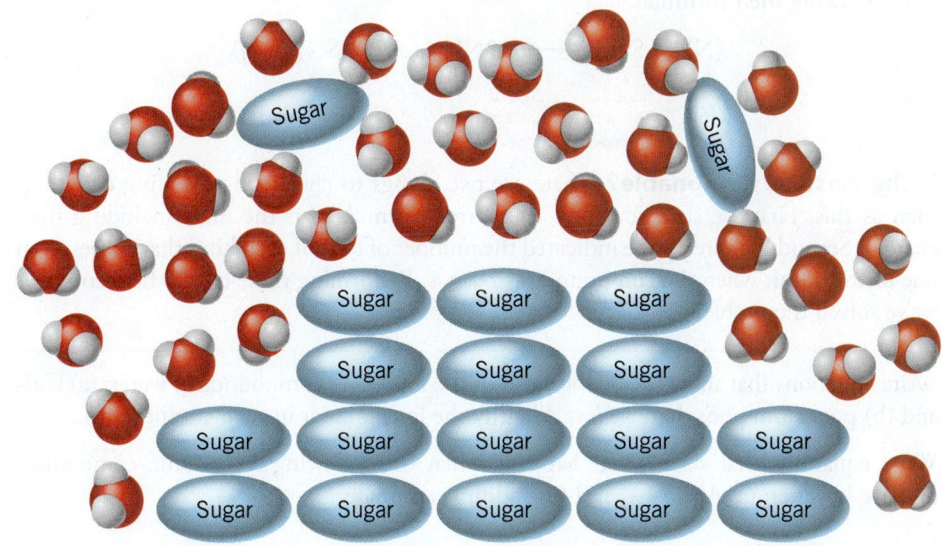

Figure 4.6 | Formation of an aqueous solution of a nonelectrolyte. When a nonelectrolyte dissolves in water, the molecules of solute separate from each other, stay intact, do not dissociate into smaller particles, and mingle with the water molecules. (For clarity the sugar is rendered as an oblate sphere.)

Dissociation Reactions

We can describe the dissociation of an ionic compound with a chemical equation. Thus, for the dissociation of calcium chloride in water we write

$$CaCl_2(s) \longrightarrow Ca^{2+}(aq) + 2Cl^-(aq)$$

We use the symbol (*aq*) after a particle to mean that it is a **hydrated ion** (*surrounded by water molecules in the solution*). By writing the formulas of the ions separately, we mean that they are essentially independent of each other in the solution. Notice that each formula unit of $CaCl_2(s)$ releases three ions, one $Ca^{2+}(aq)$ and two $Cl^-(aq)$.

Often, when the context is clear that the system is aqueous, the symbols (*s*) and (*aq*) are omitted. They are "understood." You might, therefore, see an equation such as

$$CaCl_2 \longrightarrow Ca^{2+} + 2Cl^-$$

Polyatomic ions generally remain intact as dissociation occurs. When copper sulfate dissolves, for example, both Cu^{2+} and SO_4^{2-} ions are released.

$$CuSO_4(s) \longrightarrow Cu^{2+}(aq) + SO_4^{2-}(aq)$$

Example 4.1
Writing the Equation for the Dissociation of an Ionic Compound

Ammonium sulfate is used as a fertilizer to supply nitrogen to crops. Write the equation for the dissociation of this compound when it dissolves in water.

Analysis: This is actually a two-part problem. First, we have to write the correct formula for ammonium sulfate; then we have to write the equation of the dissociation.

Assembling the Tools: We need the formula for ammonium sulfate, which means we need to know the formulas and charges of the ions that make up the salt. The tool to use is the table of polyatomic ions, Table 2.4, in Section 2.5. To write the equation, we need to follow the method presented above.

Solution: For ammonium sulfate, the cation is NH_4^+ and the anion is SO_4^{2-}. Keeping in mind the charges, the formula of the compound is $(NH_4)_2SO_4$. We write the formula for the solid on the left of the equation and indicate its state by (s). The ions are written on the right side of the equation and are shown to be in aqueous solution by the symbol (aq) following their formulas.

$$(NH_4)_2SO_4(s) \longrightarrow 2NH_4^+(aq) + SO_4^{2-}(aq)$$

The subscript 2 becomes the coefficient for NH_4^+.

Is the Answer Reasonable? There are two things to check when writing equations such as this. First, be sure you have the correct formulas for the ions, including their charges. Second, be sure you've indicated the number of ions of each kind that comes from one formula unit when the compound dissociates. Performing these checks here confirms we've solved the problem correctly.

Practice Exercise 4.1

Write equations that show the dissociation of the following compounds in water: (a) CaI_2 and (b) potassium phosphate. (*Hint:* Identify the ions present in each compound.)

Practice Exercise 4.2

Write equations that show what happens when the following solid ionic compounds dissolve in water: (a) $Al(NO_3)_3$ and (b) sodium carbonate.

4.3 | Equations for Ionic Reactions

Another concept we will cover before we explore the many different reactions that can occur in aqueous solutions is writing equations. Often, ionic compounds react with each other when their aqueous solutions are combined. For example, when solutions of lead(II) nitrate, $Pb(NO_3)_2$, and potassium iodide, KI, are mixed, a bright yellow precipitate of lead(II) iodide, PbI_2, forms (Figure 4.7). The chemical equation for the reaction is

$$Pb(NO_3)_2(aq) + 2KI(aq) \longrightarrow PbI_2(s) + 2KNO_3(aq)$$

where we write (s) following PbI_2 to note that it is a precipitate. This is called a **molecular equation** because all the formulas are written with the ions together, as if the substances in solution consist of neutral "molecules." Let's look at other ways we might write the chemical equation.

Soluble ionic compounds are fully dissociated in solution, so $Pb(NO_3)_2$, KI, and KNO_3 are not present in the solution as intact units or "molecules." To show this, we can write the formulas of all soluble strong electrolytes in "dissociated" form to give the **ionic equation** for the reaction.

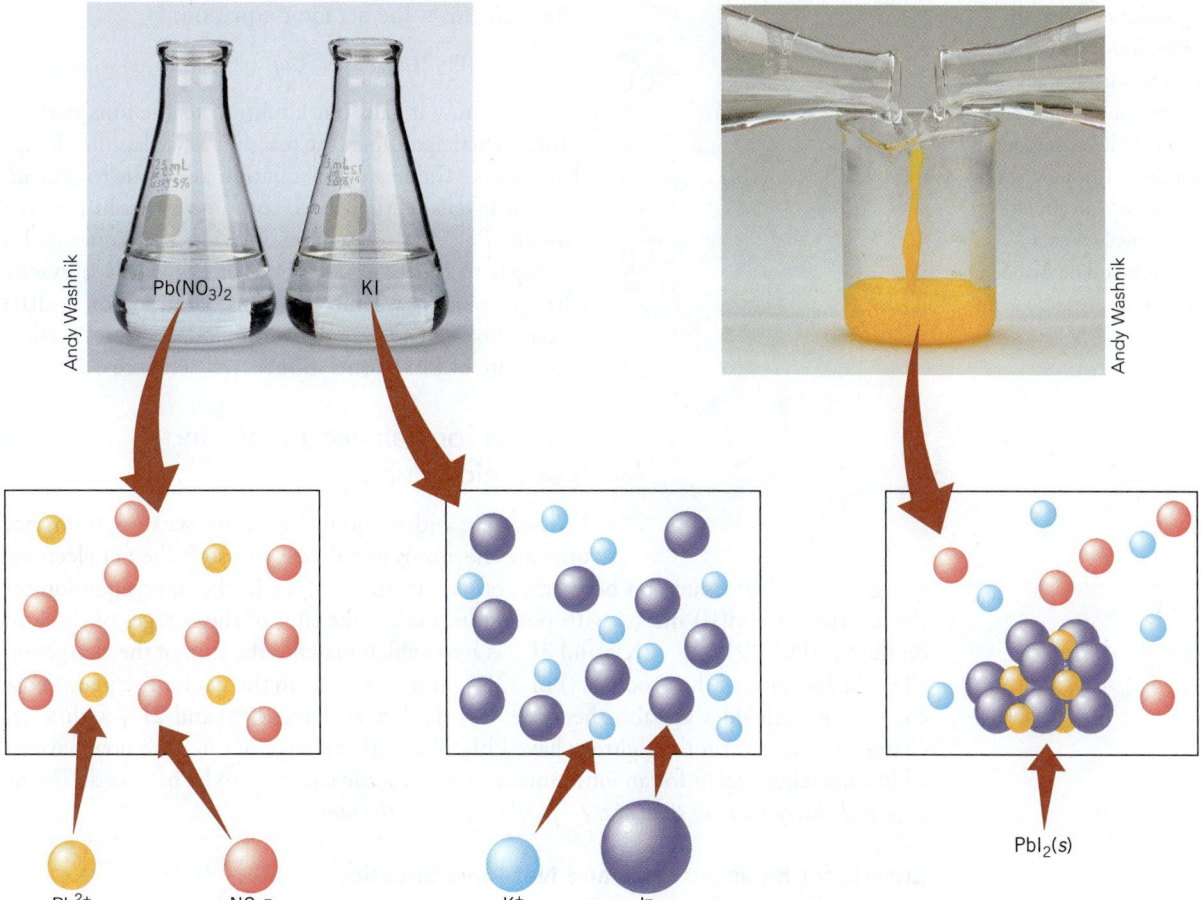

Figure 4.7 | **The reaction of Pb(NO₃)₂ with KI.** On the left are flasks containing solutions of lead nitrate and potassium iodide. These solutes exist as separated ions (shown as spheres for simplicity) in their respective solutions. On the right, we observe that when the solutions of the ions are combined, there is an immediate reaction as the Pb^{2+} ions join with the I^- ions to give a precipitate of solid, yellow PbI_2. If the $Pb(NO_3)_2$ and KI were combined in a 1-to-2 mole ratio, the solution surrounding the precipitate would now contain only K^+ and NO_3^- ions (the ions of KNO_3).

$$Pb(NO_3)_2(aq) \quad + \quad 2KI(aq) \quad \longrightarrow \quad PbI_2(s) \quad + \quad 2KNO_3(aq)$$

$$Pb^{2+}(aq) + 2NO_3^-(aq) + 2K^+(aq) + 2I^-(aq) \longrightarrow PbI_2(s) + 2K^+(aq) + 2NO_3^-(aq)$$

Notice that we have *not* separated PbI_2 into its ions in this equation. This is because PbI_2 has an extremely low solubility in water; it is essentially insoluble. When the Pb^{2+} and I^- ions meet in the solution, insoluble PbI_2 forms and separates as a precipitate. Therefore, after the reaction is over, the Pb^{2+} and I^- ions are no longer able to move independently. They are trapped in the insoluble product.

The ionic equation gives a clearer picture of what is actually going on at the molecular level in the solution during the reaction. The Pb^{2+} and I^- ions come together to form the product. The other ions, K^+ and NO_3^-, are unchanged by the reaction. Ions that do not actually take part in a reaction are called **spectator ions**; in a sense, they just "stand by and watch the action." To emphasize the actual reaction that occurs, we can write the **net ionic equation**, which is obtained by eliminating spectator ions from the ionic equation.

$$Pb^{2+}(aq) + 2NO_3^-(aq) + 2K^+(aq) + 2I^-(aq) \longrightarrow PbI_2(s) + 2K^+(aq) + 2NO_3^-(aq)$$

Figure 4.8 | Another reaction that forms lead iodide. The net ionic equation tells us that any soluble lead compound will react with any soluble iodide compound to give lead iodide. This prediction is borne out here as a precipitate of lead iodide is formed when a solution of sodium iodide is added to a solution of lead acetate.

Andy Washnik

What remains is the net ionic equation,

$$Pb^{2+}(aq) + 2I^-(aq) \longrightarrow PbI_2(s)$$

Notice how it calls our attention to the ions that are actually participating in the reaction as well as the change that occurs. The net ionic equation allows us to *generalize*, and it tells us that if we combine *any* solution that contains Pb^{2+} with *any* other solution that contains I^-, we ought to expect a precipitate of PbI_2. This is exactly what happens if we mix aqueous solutions of lead(II) acetate, $Pb(C_2H_3O_2)_2$, and sodium iodide, NaI. A yellow precipitate of PbI_2 forms immediately (Figure 4.8).

Criteria for Balanced Ionic and Net Ionic Equations

In the ionic and net ionic equations we've written, not only are the atoms in balance, but so is the net electrical charge, which is the same on both sides of the equation. Thus, in the ionic equation for the reaction of lead(II) nitrate with potassium iodide, the sum of the charges of the ions on the left (Pb^{2+}, $2NO_3^-$, $2K^+$, and $2I^-$) is zero, which matches the sum of the charges on all of the formulas of the products (PbI_2, $2K^+$, and $2NO_3^-$).[2] In the net ionic equation the charges on both sides are also the same: On the left we have Pb^{2+} and $2I^-$, with a net charge of zero, and on the right we have PbI_2, also with a charge of zero. We now have an additional requirement for an ionic equation or net ionic equation to be balanced: *The net electrical charge on both sides of the equation must be the same.*

Criteria for a balanced ionic equation

Criteria for Balanced Ionic and Net Ionic Equations

1. *Material balance:* There must be the same number of atoms of each kind on both sides of the arrow.
2. *Electrical balance:* The *net* electrical charge on the left must equal the *net* electrical charge on the right (the charge does not necessarily have to be zero).

Let's go through an example using these criteria to construct the molecular, ionic, and net ionic equations for a reaction.

Example 4.2
Writing Molecular, Ionic, and Net Ionic Equations

Write the molecular, ionic, and net ionic equations for the reaction of aqueous solutions of lead(II) chlorate and sodium iodide, which yields a precipitate of lead(II) iodide and leaves the compound sodium chlorate in solution.

Analysis: Once again, this is a multi-part question. To write the equations, we need to know the formulas of the reactants and products. Then we can proceed to construct the three equations to answer the questions.

Assembling the Tools: The first tool we need is the set of rules of nomenclature you learned in Chapter 2. (If necessary, review them.) The requirement for electrical neutrality is the next tool we use to obtain the correct formulas of the reactants and products. Then we can use the methods for writing, balancing, and constructing the three kinds of equations asked for in the question.

---

[2]There is no charge written for the formula of a compound such as PbI_2, so as we add up charges, we take the charge on PbI_2 to be zero.

Solution: Following the nomenclature rules in Chapter 2, the ions involved are: Pb^{2+}, ClO_3^-, Na^+, and I^-, and the formulas are:

Reactants		Products	
lead(II) chlorate	$Pb(ClO_3)_2$	lead(II) iodide	PbI_2
sodium iodide	NaI	sodium chlorate	$NaClO_3$

The Molecular Equation: We assemble the chemical formulas into the molecular equation and balance it.

$$Pb(ClO_3)_2(aq) + 2NaI(aq) \longrightarrow PbI_2(s) + 2NaClO_3(aq)$$

Notice that we've indicated which substances are in solution and which substance is a precipitate. This is the *balanced molecular equation*.

The Ionic Equation: To write the ionic equation, we write the formulas of all soluble salts in dissociated form and the formulas of precipitates in "molecular" form. We are careful to use the subscripts and coefficients in the molecular equation to obtain the coefficients of the ions in the ionic equation.

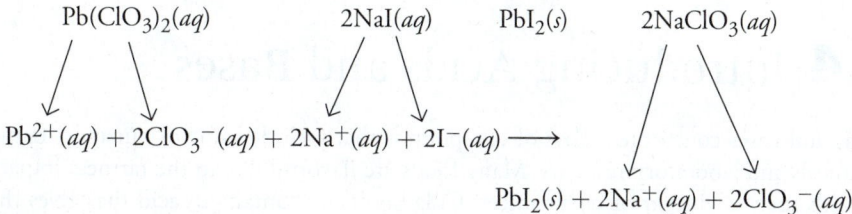

This is the *balanced ionic equation*. Notice that to write the ionic equation it is necessary to know both the formulas and charges of the ions.

The Net Ionic Equation: We obtain the net ionic equation from the ionic equation by eliminating spectator ions, which are Na^+ and ClO_3^- (they're exactly the same on both sides of the arrow).

$$Pb^{2+}(aq) + 2\cancel{ClO_3^-(aq)} + \cancel{2Na^+(aq)} + 2I^-(aq) \longrightarrow PbI_2(s) + \cancel{2Na^+(aq)} + \cancel{2ClO_3^-(aq)}$$

What's left is the *net ionic equation*.

$$Pb^{2+}(aq) + 2I^-(aq) \longrightarrow PbI_2(s)$$

Notice that this is the same net ionic equation as in the reaction of lead(II) nitrate with potassium iodide.

Are the Answers Reasonable? When you look back over a problem such as this, things to ask yourself are (1) "Have I written the correct formulas for the reactants and products?" (2) "Is the molecular equation balanced correctly?" (3) "Have I divided the soluble ionic compounds into their ions correctly, being careful to properly apply the subscripts of the ions and the coefficients in the molecular equation?" and (4) "Have I identified and eliminated the correct ions from the ionic equation to obtain the net ionic equation?" If each of these questions can be answered in the affirmative, as they can here, you have solved the problem correctly.

When solutions of $(NH_4)_2SO_4$ and $Ba(NO_3)_2$ are mixed, a precipitate of $BaSO_4$ forms, leaving soluble NH_4NO_3 in the solution. Write the molecular, ionic, and net ionic equations for the reaction. (*Hint:* Remember that polyatomic ions do not break apart when ionic compounds dissolve in water.)

Practice Exercise 4.3

Write molecular, ionic, and net ionic equations for the reaction of aqueous solutions of cadmium chloride and sodium sulfide to give a precipitate of cadmium sulfide and a solution of sodium chloride.

Practice Exercise 4.4

CHEMISTRY OUTSIDE THE CLASSROOM (4.1

Painful Precipitates—Kidney Stones

Each year, more than a million people in the United States are treated at hospitals because of very painful kidney stone attacks. A kidney stone is a hard mass developed from crystals that separate from the urine and build up on the inner surfaces of the kidney. The formation of the stones is caused primarily by the build-up of Ca^{2+}, $C_2O_4^{2-}$, and PO_4^{3-} ions in the urine. When the concentrations of these ions become large enough, the urine becomes supersaturated with respect to calcium oxalate and/or calcium phosphate and precipitates begin to form (70% to 80% of all kidney stones are made up of calcium oxalate and phosphate). If the crystals remain tiny enough, they can travel through the urinary tract and pass out of the body in the urine without being noticed. Sometimes, however, they continue to grow without being passed and can cause

intense pain if they become stuck in the urinary tract.

Kidney stones don't all look alike. Their color depends on what substances are mixed with the inorganic precipitates (e.g., proteins or blood). Most are yellow or brown, as seen in the accompanying photo, but they can be tan, gold, or even black. Stones can be round, jagged, or even have branches. They vary in size from mere specks to pebbles to stones as big as golf balls!

Custom Medical Stock Photo, Inc.

A calcium oxalate kidney stone. Kidney stones such as this can be extremely painful.

4.4 | Introducing Acids and Bases

Acids and bases constitute a class of compounds that include some of our most familiar chemicals and laboratory reagents. Many foods are flavorful due to the tartness imparted by acids such as vinegar or citrus juices. Cola beverages contain an acid that gives them their unique taste. More powerful acids are used in automobile batteries, and to clean rust and other deposits from metals. Bases, on the other hand, can be found in the white crystals of lye in some drain cleaners, in the white substance that makes milk of magnesia opaque, and in household ammonia.

There are some general properties that are common to aqueous solutions of acids and bases. As noted above, foods that contain acids generally have a tart (sour) taste; on the other hand, bases have a somewhat bitter taste and a "slippery feel." (CAUTION: Taste and feel are **never** used as laboratory tests for acids or bases because some are extremely destructive to living tissues and others are quite poisonous. ***Never taste, feel, or smell any laboratory chemicals!***)

For most purposes, we find that the following modified versions of Swedish chemist Svante Arrhenius'[3] definitions work satisfactorily when we deal with aqueous solutions of acids or bases.

■ The Arrhenius definition of acids and bases is just one definition; other definitions will be discussed in Chapter 16.

The Arrhenius Definition of Acids and Bases:
An **acid** is a substance that reacts with water to produce hydronium ions, H_3O^+.
A **base** is a substance that produces hydroxide ion, OH^-, in water.

■ Even the formula H_3O^+ is something of a simplification. In water the H^+ ion is associated with more than one molecule of water, but we use the formula H_3O^+ as a simple representation.

Originally Arrhenius suggested that acids create H^+ ions. Today we know that in aqueous solutions hydrogen ions, H^+, attach themselves to water molecules to form **hydronium ions**, H_3O^+. When H_3O^+ reacts with something, it gives up a hydrogen ion, so we can think of H^+ as the active ingredient in H_3O^+. Therefore, we often use the term *hydrogen ion* as a substitute for *hydronium ion*, and in many equations, we use $H^+(aq)$ to stand for $H_3O^+(aq)$. Whenever you see the symbol $H^+(aq)$, we are actually referring to $H_3O^+(aq)$.

Formation of H_3O^+ by Acids and OH^- by Bases

Almost all **acids** are molecular substances that react with water to produce ions, one of which is H_3O^+. For example, when gaseous molecular HCl dissolves in water, a hydrogen ion, H^+, transfers from the HCl molecule to a water molecule. The reaction at the

---

[3]Arrhenius proposed his theory of acids and bases in 1884 in his Ph.D. thesis. He won the Nobel Prize for his work in 1903.

HCl   H₂O   Cl⁻   H₃O⁺

Figure 4.9 | Ionization of HCl in water. Collisions between HCl molecules and water molecules lead to a transfer of H⁺ from HCl to H₂O, giving Cl⁻ and H₃O⁺ as products.

molecular level is depicted in Figure 4.9 using space-filling models,[4] and is represented by the chemical equation

$$HCl(g) + H_2O(l) \longrightarrow H_3O^+(aq) + Cl^-(aq)$$

This is an **ionization reaction** because ions form where none existed before. Because the solution contains ions, it conducts electricity, so acids are also classified as electrolytes. The important point is that the transfer of hydrogen ions to water molecules is characteristic of all Arrhenius acids in aqueous solution.

Bases fall into two categories: ionic compounds that contain OH⁻ or O²⁻, and molecular compounds that react with water to give hydroxide ions. Because solutions of bases also contain ions, they conduct electricity and are electrolytes.

Ionic bases include metal hydroxides, such as NaOH and Ca(OH)₂. When dissolved in water, they dissociate just like other soluble ionic compounds.

$$NaOH(s) \longrightarrow Na^+(aq) + OH^-(aq)$$

$$Ca(OH)_2(s) \longrightarrow Ca^{2+}(aq) + 2OH^-(aq)$$

Strong Acids and Bases

Ionic compounds such as NaCl and CaCl₂ dissociate essentially 100% into ions in water. No "molecules" or formula units of either NaCl or CaCl₂ are detectable in their aqueous solutions. Because these solutions contain so many ions, they are good conductors of electricity, so ionic compounds are said to be *strong electrolytes*.

Hydrochloric acid is also a strong electrolyte. Its ionization in water is essentially complete; its solutions are strongly acidic, and it is said to be a *strong acid*. In general, acids that are strong electrolytes are called **strong acids**. There are relatively few strong acids; the most common ones are as follows:

List of strong acids

Strong Acids

HClO₄(aq)	perchloric acid
HClO₃(aq)	chloric acid
HCl(aq)	hydrochloric acid
HBr(aq)	hydrobromic acid
HI(aq)	hydroiodic acid[5]
HNO₃(aq)	nitric acid
H₂SO₄(aq)	sulfuric acid

Metal hydroxides are ionic compounds, so they are also strong electrolytes. Those that are soluble are the hydroxides of Group 1A and the hydroxides of calcium, strontium, and barium in Group 2A. Solutions of these compounds are strongly basic, so these substances are considered to be **strong bases.** The hydroxides of other metals have very low solubilities in water yet they are strong electrolytes in the sense that the small amounts that do dissolve are completely dissociated. However, because of their low solubility in water, their solutions do not have a high concentration of hydroxide ions.

---

[4]To emphasize the transfer of the H⁺, we have shown the positive charge to be on just one of the H atoms in H₃O⁺. Actually the charge is distributed evenly over all three H atoms. In effect, the entire H₃O unit carries a single positive charge.

[5]Sometimes the first "o" in the name of HI(aq) is dropped for ease of pronunciation to give *hydriodic acid.*

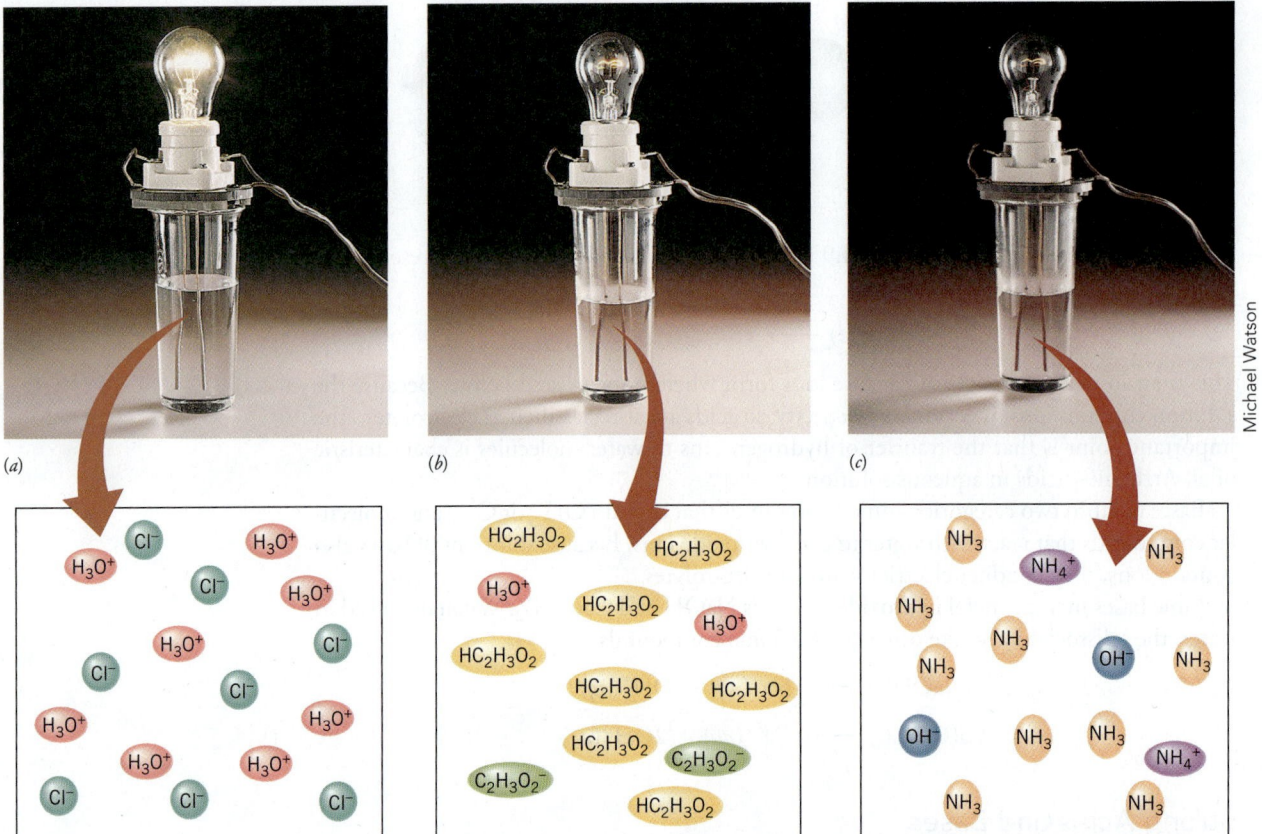

All the HCl is ionized in the solution, so there are many ions present.

Only a small fraction of the acetic acid is ionized, so there are few ions to conduct electricity. Most of the acetic acid is present as neutral molecules of $HC_2H_3O_2$.

Only a small fraction of the ammonia is ionized, so few ions are present to conduct electricity. Most of the ammonia is present as neutral molecules of NH_3.

Michael Watson

Figure 4.10 | **Electrical conductivity of solutions of strong and weak acids and bases at equal concentrations.** (*a*) HCl is 100% ionized and is a strong conductor, enabling the light to glow brightly. (*b*) $HC_2H_3O_2$ is a weaker conductor than HCl because the extent of its ionization is far less, so the light is dimmer. (*c*) NH_3 also is a weaker conductor than HCl because the extent of its ionization is low, and the light remains dim.

Weak Acids and Bases

Most acids are not completely ionized in water. For instance, a solution of ethanoic acid (commonly called acetic acid, the acid that gives vinegar its sour taste), $HC_2H_3O_2$, is a relatively poor conductor of electricity compared to a solution of HCl with the same concentration (Figure 4.10). An acid, such as acetic acid, that is not fully ionized in water is classified as a **weak electrolyte** and is a **weak acid.**

An acetic acid solution is a poor conductor because in solution only a small fraction of the dissolved acid exists as H_3O^+ and $C_2H_3O_2^-$ ions. The rest is present as molecules of $HC_2H_3O_2$ since the $C_2H_3O_2^-$ ions have a strong tendency to react with H_3O^+ when the ions meet in the solution. As a result, there are two opposing reactions occurring simultaneously at the same rate in what is called a **chemical equilibrium** (Figure 4.11). One reaction forms the ions,

$$HC_2H_3O_2(aq) + H_2O \longrightarrow H_3O^+(aq) + C_2H_3O_2^-(aq)$$

and the other removes ions,

$$H_3O^+(aq) + C_2H_3O_2^-(aq) \longrightarrow HC_2H_3O_2(aq) + H_2O$$

■ For water in an equation, we use $H_2O(l)$ when it is a pure liquid; however, in many cases it is not a pure liquid, and using $H_2O(aq)$ is redundant, so we do not use a label.

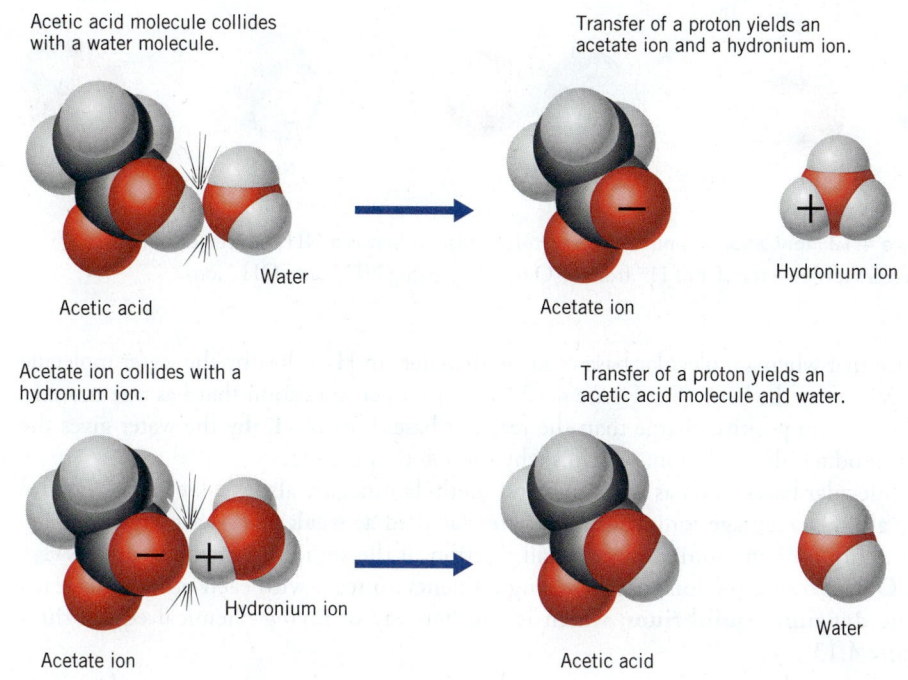

Acetic acid molecule collides with a water molecule.

Transfer of a proton yields an acetate ion and a hydronium ion.

Water

Acetic acid

Acetate ion

Hydronium ion

Acetate ion collides with a hydronium ion.

Transfer of a proton yields an acetic acid molecule and water.

Hydronium ion

Water

Acetate ion

Acetic acid

Figure 4.11 | **Equilibrium in a solution of acetic acid.** Two opposing reactions take place simultaneously in a solution of acetic acid. Molecules of acid collide with molecules of water and form acetate ions and H_3O^+ ions. Meanwhile, acetate ions collide with H_3O^+ ions to give acetic acid molecules and water molecules.

To represent the forward reaction and the reverse reaction at the same time, a double arrow is inserted in the forward reaction

$$HC_2H_3O_2(aq) + H_2O \rightleftharpoons H_3O^+(aq) + C_2H_3O_2^-(aq)$$

The **forward reaction** (read from left to right) forms the ions; the **reverse reaction** (from right to left) removes them from the solution. The following practice exercises review the process of writing ionization equations for acids.

Propanoic acid, $HC_3H_7O_2$, is only slightly ionized when dissolved in water. Write the appropriate chemical equation to describe this statement. (*Hint:* Use acetic acid as a guide.)

Practice Exercise 4.5

Nitric acid, HNO_3, is a strong acid that can be used to make explosives. Nitrous acid, HNO_2, is a weak acid thought to be responsible for certain cancers of the intestinal system. Write the chemical equations that show the ionization of these two acids in water.

Practice Exercise 4.6

Molecular Bases

The simplest and most common molecular base is the gas ammonia, NH_3, which dissolves in water and reacts to give a basic solution by the ionization reaction

$$NH_3(aq) + H_2O \rightleftharpoons NH_4^+(aq) + OH^-(aq)$$

Organic compounds called amines, in which fragments of hydrocarbons replace hydrogen atoms in ammonia, are similar to ammonia in their behavior toward water. An example is methylamine, CH_3NH_2, in which a *methyl group*, CH_3, replaces a hydrogen atom in ammonia. When methylamine, or any soluble amine, is dissolved in water this reaction occurs

$$CH_3NH_2(aq) + H_2O \rightleftharpoons CH_3NH_3^+(aq) + OH^-(aq)$$

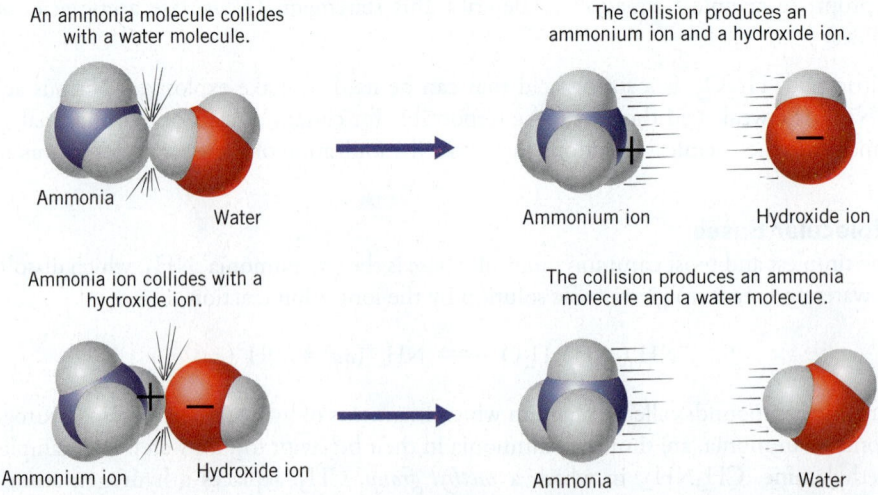

NH₃   H₂O   NH₄⁺   OH⁻

Figure 4.12 | **Ionization of ammonia in water.** Collisions between NH_3 molecules and water molecules lead to a transfer of H^+ from H_2O to NH_3, giving NH_4^+ and OH^- ions.

Notice that when a molecular base reacts with water, an H^+ is lost by the water molecule and gained by the base. (See Figure 4.12.) One product is a cation that has one more H and one more positive charge than the reactant base. Loss of H^+ by the water gives the other product, the OH^- ion, which is why the solution is basic.

Molecular bases, such as ammonia and methylamine, are also weak electrolytes and have a low percentage ionization; they are classified as **weak bases**. See Figure 4.10*c*. In a solution of ammonia, only a small fraction of the solute is ionized to give NH_4^+ and OH^- because the ions have a strong tendency to react with each other. This leads to the **dynamic equilibrium**, which is another way of saying chemical equilibrium (Figure 4.13),

$$NH_3(aq) + H_2O \rightleftharpoons NH_4^+(aq) + OH^-(aq)$$

in which most of the base is present as NH_3 molecules. Writing ionization reactions for weak bases is often more difficult than writing ionization reactions for weak acids, so a worked example is given below.

General Ionization Equations

Many times we may wish to discuss how *any* acid or base reacts with water. This is done using a "*general*" equation. In the case of a strong acid dissolved in water it is

$$HX(aq) + H_2O \longrightarrow H_3O^+(aq) + X^-(aq) \qquad (4.1)$$

TOOLS

Ionization of acids and bases in water

An ammonia molecule collides with a water molecule.

The collision produces an ammonium ion and a hydroxide ion.

Ammonia   Water   Ammonium ion   Hydroxide ion

Ammonia ion collides with a hydroxide ion.

The collision produces an ammonia molecule and a water molecule.

Ammonium ion   Hydroxide ion   Ammonia   Water

Figure 4.13 | **Equilibrium in a solution of the weak base ammonia.** Collisions between water and ammonia molecules produce ammonium and hydroxide ions. The reverse process, which involves collisions between ammonium ions and hydroxide ions, removes ions from the solution and forms ammonia and water molecules.

where $X^-(aq)$ represents the anion of the strong acid. For any strong base, $M(OH)_n$, the reaction is

$$M(OH)_n \longrightarrow M^{n+}(aq) + nOH^-(aq) \qquad (4.2)$$

and M^{n+} represents the cation of a strong base.

The general equations for any weak acid, HA, (where A^- represents the anion of a weak acid) and any weak base, B, are

$$HA(aq) + H_2O \rightleftharpoons H_3O^+(aq) + A^-(aq) \qquad (4.3)$$

and

$$B(aq) + H_2O \rightleftharpoons HB^+(aq) + OH^-(aq) \qquad (4.4)$$

Example 4.3
Writing the Equation for the Ionization of a Molecular Base

Dimethylamine, $(CH_3)_2NH$, is a base that is soluble in water. It attracts boll weevils (an agricultural pest) so they can be destroyed, since this insect has caused billions in losses to cotton crops in the United States. Write an equation for the ionization of $(CH_3)_2NH$ in water.

Analysis: We've been told that $(CH_3)_2NH$ is a base, so it's going to react with water to form hydroxide ion. This gives us two reactants and one product. We need to determine the formula for the second product to write and balance the equation.

Assembling the Tools: The tool is the general equation for the ionization of a weak base with water, Equation 4.4, which we use as a template for writing the formulas of reactants and products.

Solution: The reactants in the equation are $(CH_3)_2NH$ and H_2O. According to Equation 4.4, when the base reacts with water it takes an H^+ from H_2O, becoming $(CH_3)_2NH_2^+$ and leaving OH^- behind. The equation for the reaction is

$$(CH_3)_2NH(aq) + H_2O \rightleftharpoons (CH_3)_2NH_2^+(aq) + OH^-(aq)$$

Is the Answer Reasonable? Compare the equation we've written with the general equation for reaction of a base with water. Notice that the formula for the product has one more H and a positive charge, and that the H^+ has been added to the nitrogen. Also, notice that the water has become OH^- when it loses H^+. The equation is therefore correct.

Triethylamine, $(C_2H_5)_3N$, is a base in water. Write an equation for its reaction with the solvent. (*Hint:* How do nitrogen-containing bases react toward water?)

Practice Exercise 4.7

Ethylamine, a base in water, has the following structure:

Practice Exercise 4.8

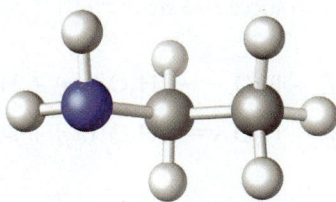

It is used in the manufacture of many herbicides. Sketch the structure of the nitrogen-containing product that is formed when it reacts with water. Write the equation for the ionization of ethylamine in water.

Let's briefly summarize the results of our discussion.

Strong acids and bases are strong electrolytes.
Weak acids and bases are weak electrolytes.

In describing dynamic equilibria such as those above, we will often talk about the position of equilibrium, which is the extent to which the forward reaction proceeds toward completion. If very little of the products are present at equilibrium, the forward reaction has not gone far toward completion and we say, "the position of equilibrium lies to the left side of the arrow," toward the reactants. On the other hand, if large amounts of the products are present at equilibrium, we say, "the position of equilibrium lies to the right side of the arrow."

For any weak electrolyte, when equilibrium is reached only a small percentage of the solute is actually ionized, so the position of equilibrium lies to the left. To call acetic acid a weak acid, for example, is just another way of saying that the forward reaction in the equilibrium is far from completion. It is also important to understand that calling acetic acid a weak acid does not imply a small concentration of $HC_2H_3O_2$. Rather, the concentration of H_3O^+ is small because only a small fraction of the solute is ionized. Even a concentrated solution of $HC_2H_3O_2$ is weakly acidic.

With strong electrolytes such as HCl, the tendency of the forward ionization reaction to occur is very large, while the tendency of the reverse reaction to occur is extremely small. As a result, all of the HCl molecules dissolved in water become converted to H_3O^+ and Cl^- ions; the acid becomes 100% ionized. For this reason, *we do not use double arrows in describing what happens when HCl(g) or any other strong electrolyte undergoes ionization or dissociation.*

Polyprotic Acids

At this point you may be asking why only one of the four hydrogen atoms in acetic acid, $HC_2H_3O_2$, that we discussed previously ionizes and is transferred to H_2O.

$$HC_2H_3O_2(aq) + H_2O \rightleftharpoons H_3O^+(aq) + C_2H_3O_2^-(aq)$$

The structures of the acetic acid molecule and the acetate ion are shown in Figure 4.14, with the hydrogen that can be lost by the acetic acid molecule indicated in the drawing. We see that the hydrogen that ionizes is attached to acetic acid differently from the other three. (We will discuss the meaning of this in detail in Chapter 16.) To make things a bit easier at this point, we will write the formulas of all acids with hydrogen as the first atom. Only this hydrogen will ionize, while all the others are **nonionizing hydrogens**.

So far we have discussed acids that are capable of furnishing only one H^+ per molecule and are said to be **monoprotic acids. Polyprotic acids** can furnish more than one H^+ per molecule. They undergo reactions similar to those of HCl and $HC_2H_3O_2$, except that the loss of H^+ by the acid occurs in two or more steps. Thus, the ionization of sulfuric acid, a **diprotic acid**, takes place by two successive steps.

$$H_2SO_4(aq) + H_2O \longrightarrow H_3O^+(aq) + HSO_4^-(aq)$$
$$HSO_4^-(aq) + H_2O \rightleftharpoons H_3O^+(aq) + SO_4^{2-}(aq)$$

For sulfuric acid the single arrow indicates that the first ionization step is characteristic of a strong acid, while the second ionization step is characteristic of a weak acid.

Acetic acid molecule
$HC_2H_3O_2$

Only this H comes off as H^+.

Acetate ion
$C_2H_3O_2^-$

Figure 4.14 | **Acetic acid and acetate ion.** The structures of acetic acid and acetate ion are illustrated here. In acetic acid, only the hydrogen attached to an oxygen can come off as H^+.

Example 4.4
Writing Equations for Ionization Reactions of Polyprotic Acids

Phosphoric acid, H_3PO_4, is a weak triprotic acid found in cola soft drinks such as Coca-Cola, where it adds a touch of tartness to the beverage. Write equations for its stepwise ionization in water.

Analysis: We are told that H_3PO_4 is a triprotic acid, which is also indicated by the three hydrogens at the beginning of the formula. Because three hydrogens come off the molecule, there are three steps in the ionization, with each step removing one H^+:

$$H_3PO_4 \xrightarrow{-H^+} H_2PO_4^- \xrightarrow{-H^+} HPO_4^{2-} \xrightarrow{-H^+} PO_4^{3-}$$

Notice that the loss of H^+ decreases the number of hydrogens by one and increases the negative charge by one unit. Also, the product of one step serves as the reactant in the next step.

Assembling the Tools: We'll use Equation 4.3 for the ionization of a weak acid as a tool in writing the chemical equation for each step.

Solution: The first step is the reaction of H_3PO_4 with water to give H_3O^+ and $H_2PO_4^-$.

$$H_3PO_4(aq) + H_2O \rightleftharpoons H_3O^+(aq) + H_2PO_4^-(aq)$$

The second and third steps are similar to the first.

$$H_2PO_4^-(aq) + H_2O \rightleftharpoons H_3O^+(aq) + HPO_4^{2-}(aq)$$

$$HPO_4^{2-}(aq) + H_2O \rightleftharpoons H_3O^+(aq) + PO_4^{3-}(aq)$$

Is the Answer Reasonable? Check to see whether the equations are balanced in terms of atoms and charge. If any mistakes were made, something would be out of balance and we would discover the error. In this case, all of the equations are balanced, so we can feel confident we've written them correctly.

Citric acid is the acid in citrus fruits such as lemons, limes, and oranges. Write equations for the stepwise ionization of citric acid, $H_3C_6H_5O_7$, in water. (*Hint:* Determine how many of the hydrogen atoms will ionize.)

Practice Exercise 4.9

Hydrogen sulfide is responsible for the smell of rotten eggs, and dissolves in water to form hydrosulfuric acid. Write the stepwise ionization equations for aqueous hydrosulfuric acid, $H_2S(aq)$.

Practice Exercise 4.10

Acid and Basic Oxides

In addition to the simple acids and bases that we have already discussed, there are some compounds that *become* acids (or bases) when dissolved in water. For the most part, soluble oxides of the nonmetals will react with water to form acids while the soluble oxides of the metals form bases.

Acidity of Nonmetal Oxides

The acids we've discussed so far have been molecules containing hydrogen atoms that can be transferred to water molecules. Nonmetal oxides form another class of compounds that yield acidic solutions in water. Examples are SO_3, CO_2, and N_2O_5, whose aqueous solutions contain H_3O^+. These oxides are called **acid anhydrides**, where *anhydride* means

Figure 4.15 | **Oxide ion reacts with water.** When a soluble metal oxide dissolves, the oxide ion takes an H^+ from a water molecule. The result is two hydroxide ions.

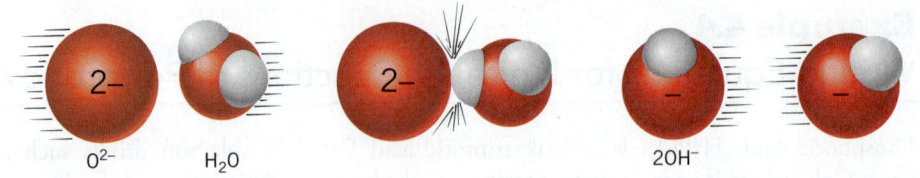

"without water." They react with water to form molecular acids containing hydrogen, which are then able to undergo reaction with water to yield H_3O^+.

$$SO_3(g) + H_2O \longrightarrow H_2SO_4(aq) \quad \text{sulfuric acid}$$

$$N_2O_5(g) + H_2O \longrightarrow 2HNO_3(aq) \quad \text{nitric acid}$$

$$CO_2(g) + H_2O \longrightarrow H_2CO_3(aq) \quad \text{carbonic acid}$$

Although carbonic acid is too unstable to be isolated as a pure compound, its solutions in water are quite common. Carbon dioxide from the atmosphere dissolves in rainwater and the waters of lakes and streams, where it exists partly as carbonic acid and its ions (i.e., H_3O^+, HCO_3^-, and CO_3^{2-}). This makes these waters naturally slightly acidic. Carbonic acid is also present in carbonated beverages.

Not all nonmetal oxides are acidic anhydrides, just those that are able to react with water. For example, carbon monoxide doesn't react with water, so its solutions in water are not acidic; carbon monoxide, therefore, is not an acidic anhydride.

Basicity of Metal Oxides

Soluble metal oxides are **base anhydrides** because they react with water to form the hydroxide ion as one of the products. Calcium oxide is typical.

$$CaO(s) + H_2O \longrightarrow Ca(OH)_2(aq)$$

This reaction occurs when water is added to dry cement or concrete because calcium oxide or "quicklime" is an ingredient in these materials. In this case it is the oxide ion, O^{2-}, that actually forms the OH^-. See Figure 4.15.

$$O^{2-} + H_2O \longrightarrow 2OH^-$$

Even insoluble metal hydroxides and oxides are bases because they are able to react with acids. We will study these reactions in Section 4.6.

4.5 | Acid–Base Nomenclature

In the naming of acids, there are patterns that help organize names of acids and the anions that come from them when the acids are reacted with bases. Since we use the names of the acids and bases as often as the formulas, we need to understand and know how they are named.

Hydrogen Compounds of Nonmetals

Many of the binary compounds of hydrogen with nonmetals are acidic, and in their aqueous solutions they are referred to as **binary acids**. Some examples are HCl, HBr, and H_2S. In naming these substances as acids, we add the prefix *hydro-* and the suffix *-ic* to the stem of the nonmetal name, followed by the word *acid*. For example, aqueous solutions of hydrogen chloride and hydrogen sulfide are named as follows:

Name of the molecular compound		Name of the binary acid in water	
$HCl(g)$	hydrogen chloride	$HCl(aq)$	*hydrochloric acid*
$H_2S(g)$	hydrogen sulfide	$H_2S(aq)$	*hydrosulfuric acid*

Notice that the gaseous molecular substances are named in the usual way as binary compounds. *It is their aqueous solutions that are named as acids.*

When an acid reacts with a base, as will be discussed in Section 4.6, a salt is produced that contains the anion of the acid and the cation of the base. Thus, HBr yields salts containing the bromide ion, Br^-, and NaOH gives salts containing the sodium ion, Na^+. The salt produced is sodium bromide, NaBr.

Naming Oxoacids

Acids that contain hydrogen, oxygen, plus another element are called **oxoacids**. Examples are H_2SO_4 and HNO_3. The names of these acids are derived from the polyatomic ions, which we learned in Section 2.5.

Just like the binary acids, oxoacids react with bases to form a salt that contains the anion of the acid and the cation of the base. We will use our knowledge of the names of these polyatomic ions to form the names of oxoacids. The suffix of the name of the polyatomic ion is related to that of its parent acid in the following manner.

Relationship	Example
(1) -*ate* anions give -*ic* acids:	NO_3^- (nit*rate* ion) $\longrightarrow$ HNO_3 (nit*ric* acid)
(2) -*ite* anions give -*ous* acids:	NO_2^- (nit*rite* ion) $\longrightarrow$ HNO_2 (nit*rous* acid)

TOOLS

Acid and polyatomic anion names

This relationship between the suffix of the anion and suffix of the acid carries over to other anions that end in -*ate* and -*ite*. Examples include the thiocyanate ion, SCN^-, which has a name that ends in -*ate*, so SCN^- becomes thiocyan*ic acid*, HSCN; the acetate anion gives acetic acid, and the citrate anion gives citric acid. Knowing the names and formulas of the polyatomic ions in Table 2.4 will help you in naming the acids.

The halogens can form as many as four different polyatomic anions and oxoacids. In naming oxoacids, the prefixes *per-* and *hypo-* carry over from the name of the parent anion. Thus perchlorate ion, ClO_4^-, gives perchloric acid, $HClO_4$, and, hypochlorite ion, ClO^- gives hypochlorous acid, HClO.

HClO	*hypochlorous acid* (usually written HOCl)	$HClO_3$	*chloric acid*
$HClO_2$	*chlorous acid*	$HClO_4$	*perchloric acid*

Example 4.5
Naming Acids

Bromine forms four oxoacids, similar to those of chlorine. What is the name of the acid $HBrO_2$?

Analysis: In this problem we will reason by analogy and by analyzing the names of the oxoacids of the halogens.

Assembling the Tools: The tools are the names of the polyatomic anions and how they relate to the names of the acids.

Solution: The oxoacids of chlorine are

HClO	hypochlorous acid	$HClO_3$	chloric acid
$HClO_2$	chlorous acid	$HClO_4$	perchloric acid

The formula of the acid $HBrO_2$ is similar to chlorous acid, so to name it we will use the stem of the element name bromine (brom-) in place of chlor-. Therefore, the name of $HBrO_2$ is *bromous acid.*

Are the Answers Reasonable? There's really not much we can do to check the answers here. If $HClO_2$ is chlorous acid, then it seems reasonable that $HBrO_2$ would be bromous acid.

Practice Exercise 4.11

Practice Exercise 4.12

Name the aqueous solutions of the following acids: HF, HI. (*Hint:* These are binary acids.)

Using the colors of the atoms to identify the elements, name the acid having the structure below.

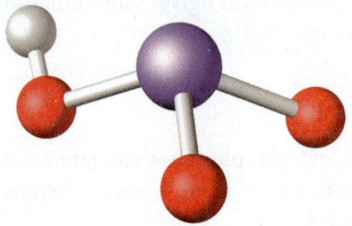

Naming Bases

Metal compounds that contain the ions OH^- or O^{2-}, such as NaOH and Na_2O, are ionic and are named just like any other ionic compound. Thus, NaOH is sodium hydroxide and Na_2O is sodium oxide.

Molecular bases such as NH_3 (ammonia)[6] and CH_3NH_2 (methylamine) are specified by just giving the name of the molecule. The naming of molecular bases is complicated, but you can often identify them by finding the word "amine" in them.

4.6 | Double Replacement (Metathesis) Reactions

In our discussion of the reaction of KI with $Pb(NO_3)_2$, you saw that the net ionic equation reveals a change in the number of ions in solution when the reaction takes place. Such changes characterize ionic reactions in general. In this section you will learn how we can use the existence or nonexistence of a net ionic equation as a guide to determine whether or not an ionic reaction occurs in a solution of mixed solutes.

A net ionic equation will exist (and a reaction will occur) if one of the following occurs:

Predicting net ionic equations

A precipitate is formed from a mixture of soluble reactants.
An acid and a base react together to form a salt and usually water.
A weak electrolyte is formed from a mixture of strong electrolytes.
A gas is formed from a mixture of reactants.

It is also important to note that there will be no net reaction if we cannot write a net ionic equation! In addition, net ionic equations meeting these criteria generally go to completion and we will use a single arrow in their equations.

Predicting Precipitation Reactions

The reaction between lead(II) nitrate and potassium iodide,

$$Pb(NO_3)_2(aq) + 2KI(aq) \longrightarrow PbI_2(s) + 2KNO_3(aq)$$

is just one example of a large class of ionic reactions, called **metathesis reactions**, in which cations and anions change partners. Metathesis reactions are also sometimes called **double replacement reactions**. (In the formation of the products, PbI_2 and KNO_3, the I^- re-

---

[6]Solutions of ammonia in water are sometimes called *ammonium hydroxide*, although there is no evidence that the species NH_4OH actually exists.

TABLE 4.1	Solubility Rules for Ionic Compounds in Water

TOOLS
Solubility rules

Soluble Compounds

1. All compounds of the alkali metals (Group 1A) are soluble.
2. All salts containing NH_4^+, NO_3^-, ClO_4^-, ClO_3^-, and $C_2H_3O_2^-$ are soluble.
3. All chlorides, bromides, and iodides (salts containing Cl^-, Br^-, or I^-) are soluble except when combined with Ag^+, Pb^{2+}, and Hg_2^{2+} (note the subscript "2").
4. All sulfates (salts containing SO_4^{2-}) are soluble except those of Pb^{2+}, Ca^{2+}, Sr^{2+}, Hg_2^{2+}, and Ba^{2+}.

Insoluble Compounds

5. All metal hydroxides (ionic compounds containing OH^-) and all metal oxides (ionic compounds containing O^{2-}) are insoluble except those of Group 1A and of Ca^{2+}, Sr^{2+}, and Ba^{2+}.

   When metal oxides do dissolve, they react with water to form hydroxides. The oxide ion, O^{2-}, does not exist in water. For example,

   $$Na_2O(s) + H_2O \longrightarrow 2NaOH(aq)$$

6. All salts that contain PO_4^{3-}, CO_3^{2-}, SO_3^{2-}, and S^{2-} are insoluble, except those of Group 1A and NH_4^+.

places NO_3^- in the lead compound and NO_3^- replaces I^- in the potassium compound.) Metathesis reactions in which a precipitate forms are also called **precipitation reactions**.

Lead(II) nitrate and potassium iodide react because one of the products is insoluble. Such reactions can be predicted if we know which substances are soluble and which are insoluble. To help us, we can apply a set of **solubility rules** (Table 4.1) that tell us, in many cases, whether an ionic compound is soluble or insoluble. (We will investigate why substances dissolve in Chapter 12.) To make the rules easier to remember, they are divided into two categories. The first includes compounds that are soluble, with some exceptions. The second describes compounds that are generally insoluble, with some exceptions. Some examples will help clarify their use.

Rule 1 states that all compounds of the alkali metals are soluble in water. This means that you can expect any compound containing Na^+ or K^+, or any of the other Group 1A metal ions, regardless of the anion, to be soluble. If one of the reactants in a metathesis reaction is Na_3PO_4, you now know from Rule 1 that it is soluble. Therefore, you would write it in dissociated form ($3Na^+(aq) + PO_4^{3-}(aq)$) in the ionic equation. Similarly, Rule 6 states, in part, that all carbonate compounds are insoluble except those of the alkali metals and the ammonium ion. If one of the products in a metathesis reaction is $CaCO_3$, you'd expect it to be insoluble because the cation is not an alkali metal or NH_4^+. Therefore, you would write its formula in the undissociated form as $CaCO_3(s)$ in the ionic equation.

Example 4.6 illustrates how we can use the rules to predict the outcome of a reaction.

Example 4.6
Predicting Precipitation Reactions and Writing Their Equations

Predict whether a reaction will occur when aqueous solutions of $Pb(NO_3)_2$ and $Fe_2(SO_4)_3$ are mixed. Write the molecular, ionic, and net ionic equations.

Analysis: For the molecular equation, let's start by writing the reactants on the left.

$$Pb(NO_3)_2 + Fe_2(SO_4)_3 \longrightarrow$$

To complete the equation we have to determine the makeup of the products. Next, we write the ionic equation by writing soluble strong electrolytes in ionic form. Finally, the spectator ions are crossed out leaving the net ionic equation. If a net ionic equation exists at this point, it tells us that a reaction does indeed take place.

Assembling the Tools: The tools you've learned that apply here are (1) rules for writing the formula of a salt from the formulas of its ions, (2) the rules for writing ionic and net ionic equations, and (3) the solubility rules.

Solution: The reactants, $Pb(NO_3)_2$ and $Fe_2(SO_4)_3$, contain the ions Pb^{2+} and NO_3^-, and Fe^{3+} and SO_4^{2-}, respectively. To write the formulas of the products, we interchange anions, being careful that the formula units are electrically neutral. This gives $PbSO_4$ as one possible product and $Fe(NO_3)_3$ as the other. The unbalanced molecular equation at this point is

$$Pb(NO_3)_2 + Fe_2(SO_4)_3 \longrightarrow PbSO_4 + Fe(NO_3)_3 \quad \text{(unbalanced)}$$

Next, let's determine solubilities. The reactants are ionic compounds and we are told that they are in solution, so we know they are water soluble (solubility Rules 2 and 4 tell us this also). For the products, we find that Rule 2 says that all nitrates are soluble, so $Fe(NO_3)_3$ is soluble; Rule 4 tells us that the sulfate of Pb^{2+} is *insoluble*. This means that a precipitate of $PbSO_4$ will form. Writing (*aq*) and (*s*) following appropriate formulas, the unbalanced molecular equation is

$$Fe_2(SO_4)_3(aq) + Pb(NO_3)_2(aq) \longrightarrow Fe(NO_3)_3(aq) + PbSO_4(s) \quad \text{(unbalanced)}$$

When balanced, we obtain the *molecular equation*.

$$Fe_2(SO_4)_3(aq) + 3Pb(NO_3)_2(aq) \longrightarrow 2Fe(NO_3)_3(aq) + 3PbSO_4(s) \quad \text{(balanced)}$$

Next, we expand this to give the *ionic equation* in which soluble compounds are written in dissociated (separated) form as ions, and insoluble compounds are written in "molecular" form. Once again, we are careful to apply the subscripts of the ions and the coefficients.

$$2Fe^{3+}(aq) + 3SO_4^{2-}(aq) + 3Pb^{2+}(aq) + 6NO_3^-(aq) \longrightarrow 2Fe^{3+}(aq) + 6NO_3^-(aq) + 3PbSO_4(s)$$

By removing spectator ions (Fe^{3+} and NO_3^-), we obtain

$$3Pb^{2+}(aq) + 3SO_4^{2-}(aq) \longrightarrow 3PbSO_4(s)$$

Finally, we reduce the coefficients to give us the correct *net ionic equation*.

$$Pb^{2+}(aq) + SO_4^{2-}(aq) \longrightarrow PbSO_4(s)$$

The existence of the net ionic equation confirms that a reaction does take place between lead(II) nitrate and iron(III) sulfate.

Is the Answer Reasonable? One of the main things we have to check in solving a problem such as this is that we've written the correct formulas of the products. Carefully check that all charges for the ions are correct and that subscripts in formulas are those needed for neutral formula units. For example, in this problem some students might be tempted (without thinking) to write Fe^{2+}, SO_4^-, or $Pb(SO_4)_2$ and $Fe_2(NO_3)_3$, or even $Pb(SO_4)_3$ and $Fe_2(NO_3)_2$. *These are common errors.* Once we're sure the formulas of the products are right, we check that we've applied the solubility rules correctly.

Practice Exercise 4.13

Show that, in aqueous solutions, there is a net reaction between $Zn(NO_3)_2$ and $Ca(OH)_2$. (*Hint:* Write molecular, ionic, and net ionic equations.)

Practice Exercise 4.14

Predict what occurs on mixing the following solutions, by writing the molecular, ionic, and net ionic equations for the reactions that take place: (a) $AgNO_3$ and NH_4Cl, (b) sodium sulfide and lead(II) acetate.

TOOLS

Neutralization of an
acid and a base

Predicting Acid–Base Reactions

One of the most important properties of acids and bases is their reaction with each other, a reaction referred to as **acid–base neutralization**. For example, when solutions of hydrobromic acid, HBr(aq), and the base potassium hydroxide, KOH(aq), are mixed the following reaction occurs.

$$HBr(aq) + KOH(aq) \longrightarrow KBr(aq) + H_2O$$

When the reactants are combined in a 1-to-1 mole ratio, the acidic and basic properties of the solutes disappear and the resulting solution is neither acidic nor basic. We say that the acid and base neutralize each other; in other words, an *acid–base neutralization* has occurred. Arrhenius was the first to suggest that an acid–base neutralization is simply the combination of a hydrogen ion with a hydroxide ion to produce a water molecule, thus making H^+ ions and OH^- ions disappear.

The reaction of an acid with a base also produces an ionic compound, a salt, as one of the products. The reaction of HBr(aq) with KOH(aq) forms potassium bromide. Most salts can be formed by reacting the appropriate acid and base.

In general, neutralization reactions can be viewed as metathesis reactions. Using the reaction above,

$$HBr(aq) + KOH(aq) \longrightarrow KBr(aq) + H_2O$$

Writing this as an ionic equation gives

$$H^+(aq) + Br^-(aq) + K^+(aq) + OH^-(aq) \longrightarrow K^+(aq) + Br^-(aq) + H_2O$$

where we have used H^+ as shorthand for H_3O^+. The net ionic equation is obtained by removing spectator ions.

$$H^+(aq) + OH^-(aq) \longrightarrow H_2O$$

In this case, a net ionic equation exists because of the formation of a very weak electrolyte, H_2O, instead of a precipitate. In fact, we find this same net ionic equation for any reaction between a strong acid and any soluble strong base—that is, one with OH^-.

The formation of water in a neutralization reaction is such a strong driving force for reaction that it will form even if the acid is weak or if the base is insoluble, or both. Here are some examples.

Reaction of a Weak Acid with a Strong Base
Molecular equation:

$$\underset{\text{weak acid}}{HC_2H_3O_2(aq)} + \underset{\text{strong base}}{NaOH(aq)} \longrightarrow NaC_2H_3O_2(aq) + H_2O$$

Net ionic equation:

$$HC_2H_3O_2(aq) + OH^-(aq) \longrightarrow C_2H_3O_2^-(aq) + H_2O$$

This reaction is illustrated in Figure 4.16.

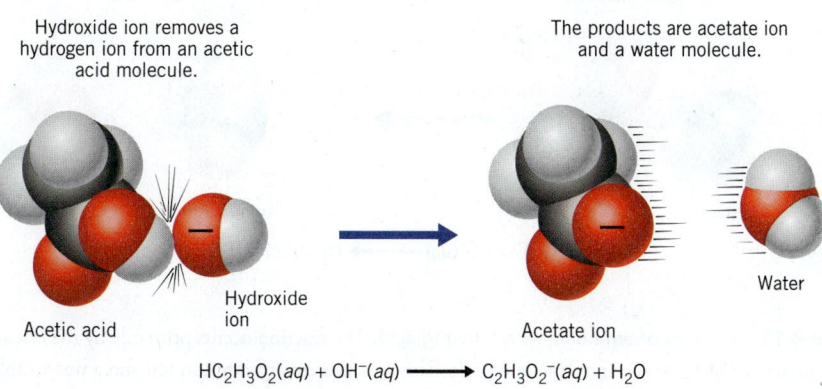

Hydroxide ion removes a hydrogen ion from an acetic acid molecule.

The products are acetate ion and a water molecule.

Acetic acid

Hydroxide ion

Acetate ion

Water

$$HC_2H_3O_2(aq) + OH^-(aq) \longrightarrow C_2H_3O_2^-(aq) + H_2O$$

Figure 4.16 | Net reaction of acetic acid with a strong base. The neutralization of acetic acid by hydroxide ion occurs primarily by the removal of H^+ from acetic acid molecules by OH^- ions.

Figure 4.17 | Hydrochloric acid is neutralized by milk of magnesia. A solution of hydrochloric acid is added to a beaker containing milk of magnesia. The thick white solid in milk of magnesia is magnesium hydroxide, $Mg(OH)_2$, which is able to neutralize the acid. The mixture is clear where some of the solid $Mg(OH)_2$ has already reacted and dissolved.

Andy Washnik

Reaction of a Strong Acid with an Insoluble Base

Molecular equation:

$$2HCl(aq) + Mg(OH)_2(s) \longrightarrow MgCl_2(aq) + 2H_2O$$
$$\text{strong acid} \qquad \text{insoluble base}$$

Net ionic equation:

$$2H^+(aq) + Mg(OH)_2(s) \longrightarrow Mg^{2+}(aq) + 2H_2O$$

Figure 4.17 shows the reaction of hydrochloric acid with milk of magnesia, which contains $Mg(OH)_2$.

Reaction of a Weak Acid with an Insoluble Base

Molecular equation:

$$2HC_2H_3O_2(aq) + Mg(OH)_2(s) \longrightarrow Mg(C_2H_3O_2)_2(aq) + 2H_2O$$
$$\text{weak acid} \qquad \text{insoluble base}$$

Net ionic equation:

$$2HC_2H_3O_2(aq) + Mg(OH)_2(s) \longrightarrow Mg^{2+}(aq) + 2C_2H_3O_2^-(aq) + 2H_2O$$

Notice that in the last two examples, the formation of water drives the reaction, even though one of the reactants is insoluble. To correctly write the ionic and net ionic equations, it is important to know both the solubility rules and which acids are strong and weak. If you've learned the list of strong acids, you can expect that any acid *not* on the list will be a weak acid. (Unless specifically told otherwise, you should assume weak acids to be water soluble.)

Reaction of an Acid with a Weak Base

Acid–base neutralization doesn't always involve the formation of water. We see this in the reaction of an acid with a weak base such as NH_3, as illustrated in Figure 4.18. For a strong acid such as HCl, we have

Molecular equation:

$$HCl(aq) + NH_3(aq) \longrightarrow NH_4Cl(aq)$$

Net ionic equation (using H^+ as shorthand for H_3O^+):

$$H^+(aq) + NH_3(aq) \longrightarrow NH_4^+(aq)$$

With a weak acid such as $HC_2H_3O_2$, we have

Molecular equation:

$$HC_2H_3O_2(aq) + NH_3(aq) \longrightarrow NH_4C_2H_3O_2(aq)$$

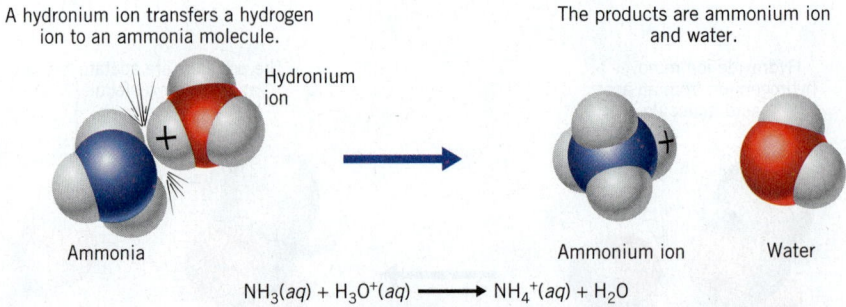

A hydronium ion transfers a hydrogen ion to an ammonia molecule.

Hydronium ion

Ammonia

The products are ammonium ion and water.

Ammonium ion   Water

$$NH_3(aq) + H_3O^+(aq) \longrightarrow NH_4^+(aq) + H_2O$$

Figure 4.18 | Reaction of ammonia with a strong acid. The reaction occurs primarily by the transfer of a proton, from H_3O^+, to the ammonia molecule. This produces an ammonium ion and a water molecule.

Ionic and net ionic equation:

$$HC_2H_3O_2(aq) + NH_3(aq) \longrightarrow NH_4^+(aq) + C_2H_3O_2^-(aq)$$

Even though solutions of $HC_2H_3O_2$ contain some H^+, and solutions of NH_3 contain some OH^-, when these solutions are mixed the predominant reaction is between molecules of acid and base. This is illustrated in Figure 4.19.

Write the molecular, ionic, and net ionic equations for the neutralization of $HNO_3(aq)$ by $Ca(OH)_2(aq)$. (*Hint:* First determine whether the acid and base are strong or weak.)

Write molecular, ionic, and net ionic equations for the reaction of (a) HCl with KOH and (b) barium hydroxide with hydrochloric acid.

Write molecular, ionic, and net ionic equations for the reaction of the weak base methylamine, CH_3NH_2, with formic acid, $HCHO_2$ (a weak acid).

Practice Exercise 4.15

Practice Exercise 4.16

Practice Exercise 4.17

Acid Salts

If a polyprotic acid is neutralized with a base in a stepwise manner, the neutralization can be halted after each hydrogen has been removed. For example, the reaction of 1.0 mol H_2SO_4 with 1.0 mol of NaOH gives the HSO_4^- ion, which, if the solution is evaporated, yields the salt $NaHSO_4$. This compound is called an **acid salt** because its anion, HSO_4^-, is capable of furnishing additional H^+. The *complete neutralization* of a polyprotic acid means that all of the ionizable hydrogens have reacted with a base.

We have already encountered the names of the acid salt anions in Chapter 2 in the list of polyatomic anions in Table 2.4. However, as a reminder, for ions such as HSO_4^-, we specify the number of hydrogens that can still be reacted if the anion were to be treated with additional base. Thus, HSO_4^- is called the hydrogen sulfate ion, and $H_2PO_4^-$ is the dihydrogen phosphate ion. These two anions give the following salts with Na^+:

$NaHSO_4$	sodium hydrogen sulfate
NaH_2PO_4	sodium dihydrogen phosphate

For acid salts of diprotic acids, the prefix *bi-* is still often used.

$NaHCO_3$	sodium bicarbonate or sodium hydrogen carbonate

The prefix bi- does *not* mean "two"; instead, it means that there is an acidic sulfate.

Ammonia molecule collides with an
acetic acid molecule and extracts
a hydrogen ion from the acid.

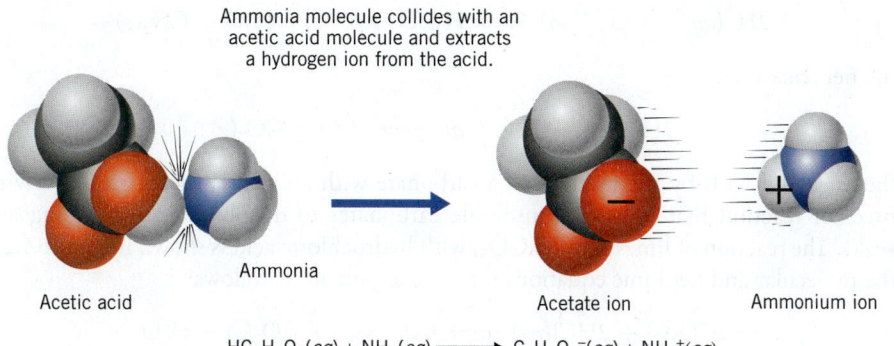

Acetic acid   Ammonia   Acetate ion   Ammonium ion

$$HC_2H_3O_2(aq) + NH_3(aq) \longrightarrow C_2H_3O_2^-(aq) + NH_4^+(aq)$$

Figure 4.19 | Reaction of acetic acid with ammonia. The collision of an ammonia molecule with an acetic acid molecule leads to a transfer of H^+ from the acetic acid to ammonia and the formation of the acetate and hydroxide ions.

Practice Exercise 4.18

What is the formula for the product of the reaction of one mole of NaOH with one mole of sulfurous acid? What is the name for this compound? (*Hint:* What information do we get from the prefix *bi-* and the suffix *-ite*?)

Practice Exercise 4.19

Write molecular equations for the stepwise neutralization of arsenic acid by sodium hydroxide. What are the names of the salts that are formed? Identify the step in which the acid is completely neutralized.

Figure 4.20 | **The reaction of sodium bicarbonate with hydrochloric acid.** The bubbles contain the gas carbon dioxide.

Michael Watson

Predicting Reactions in Which a Gas is Formed

Sometimes a product of a metathesis reaction is a substance that normally is a gas at room temperature and is not very soluble in water. The most common example is carbon dioxide. This product forms when an acid reacts with either a bicarbonate or carbonate salt. For example, as a sodium bicarbonate solution is added to hydrochloric acid, bubbles of carbon dioxide are released (Figure 4.20). This is the same reaction that occurs if you take sodium bicarbonate to soothe an upset stomach. Stomach acid is HCl and its reaction with the $NaHCO_3$ both neutralizes the acid and produces CO_2 gas (burp!). The molecular equation for the metathesis reaction is

$$HCl(aq) + NaHCO_3(aq) \longrightarrow NaCl(aq) + H_2CO_3(aq)$$

Carbonic acid, H_2CO_3, is too unstable to be isolated in pure form. When it forms in appreciable amounts as a product in a metathesis reaction, it decomposes into its anhydride, CO_2, and water. Carbon dioxide is only slightly soluble in water, so most of the CO_2 bubbles out of the solution. The decomposition reaction is

$$H_2CO_3(aq) \rightleftharpoons H_2O + CO_2(g)$$

Therefore, the overall molecular equation for the reaction is

$$HCl(aq) + NaHCO_3(aq) \longrightarrow NaCl(aq) + H_2O + CO_2(g)$$

The ionic equation is

$$H^+(aq) + Cl^-(aq) + Na^+(aq) + HCO_3^-(aq) \longrightarrow Na^+(aq) + Cl^-(aq) + H_2O + CO_2(g)$$

and the net ionic equation is

$$H^+(aq) + HCO_3^-(aq) \longrightarrow H_2O + CO_2(g)$$

Similar results are obtained if we begin with a carbonate instead of a bicarbonate. In this case, two hydrogen ions combine with carbonate ions to give H_2CO_3, which subsequently decomposes to water and CO_2.

$$2H^+(aq) + CO_3^{2-}(aq) \longrightarrow H_2CO_3(aq) \longrightarrow H_2O + CO_2(g)$$

The net reaction is

$$2H^+(aq) + CO_3^{2-}(aq) \longrightarrow H_2O + CO_2(g)$$

The release of CO_2 by the reaction of a carbonate with an acid is such a strong driving force for reaction that it enables insoluble carbonates to dissolve in acids (strong and weak). The reaction of limestone, $CaCO_3$, with hydrochloric acid is shown in Figure 4.21. The molecular and net ionic equations for the reaction are as follows:

$$CaCO_3(s) + 2HCl(aq) \longrightarrow CaCl_2(aq) + CO_2(g) + H_2O$$

$$CaCO_3(s) + 2H^+(aq) \longrightarrow Ca^{2+}(aq) + CO_2(g) + H_2O$$

Figure 4.21 | **Limestone reacts with acid.** Bubbles of CO_2 are formed in the reaction of limestone ($CaCO_3$) with hydrochloric acid.

Andy Washnik

Carbon dioxide is not the only gas formed in metathesis reactions. Table 4.2 lists others and the reactions that form them.

TABLE 4.2	Gases Formed in Metathesis Reactions	
Gases Formed in Reactions with Acids		
Gas	**Reactant Type**	**Equation for Formation[a]**
H_2S	Sulfides	$2H^+ + S^{2-} \longrightarrow H_2S$
HCN	Cyanides	$H^+ + CN^- \longrightarrow HCN$
CO_2	Carbonates	$2H^+ + CO_3^{2-} \longrightarrow (H_2CO_3) \longrightarrow H_2O + CO_2$
	Bicarbonates (hydrogen carbonates)	$H^+ + HCO_3^- \longrightarrow (H_2CO_3) \longrightarrow H_2O + CO_2$
SO_2	Sulfites	$2H^+ + SO_3^{2-} \longrightarrow (H_2SO_3) \longrightarrow H_2O + SO_2$
	Bisulfites (hydrogen sulfites)	$H^+ + HSO_3^- \longrightarrow (H_2SO_3) \longrightarrow H_2O + SO_2$
Gas Formed in Reaction with Bases		
Gas	**Reactant Type**	**Equation for Formation**
NH_3	Ammonium salts[b]	$NH_4^+ + OH^- \longrightarrow NH_3 + H_2O$

[a]Formulas in parentheses are of unstable compounds that break down according to the continuation of the sequence.

[b]In writing a metathesis reaction, you may be tempted sometimes to write NH_4OH as a formula for "ammonium hydroxide." This compound does not exist. In water, it is nothing more than a solution of NH_3.

TOOLS
Gases formed in metathesis reactions

CHEMISTRY OUTSIDE THE CLASSROOM 4.2

Hard Water and Its Problems

Precipitation reactions occur around us all the time and we hardly ever take notice until they cause a problem. One common problem is caused by **hard water**—ground water that contains the "hardness ions," Ca^{2+}, Mg^{2+}, Fe^{2+}, or Fe^{3+}, in concentrations high enough to form precipitates with ordinary soap. Soap normally consists of the sodium salts of organic acids derived from animal fats or oils (so-called *fatty acids*). An example is sodium stearate, $NaC_{17}H_{35}O_2$. The negative ion of the soap forms an insoluble "scum" with hardness ions, which reduces the effectiveness of the soap for removing dirt and grease.

Hardness ions can be removed from water in a number of ways. One way is to add hydrated sodium carbonate, $Na_2CO_3 \cdot 10H_2O$, often called washing soda, to the water. The carbonate ion forms insoluble precipitates with the hardness ions; an example is $CaCO_3$.

$$Ca^{2+}(aq) + CO_3^{2-}(aq) \longrightarrow CaCO_3(s)$$

Once precipitated, the hardness ions are not available to interfere with the soap.

In localities where the earth contains limestone, composed mainly of $CaCO_3$, a particularly bothersome type of hard water is often found. Rain and other natural waters contain dissolved CO_2, which the water picks up from the atmosphere. Carbon dioxide is the anhydride of carbonic acid, H_2CO_3, a weak diprotic acid. This causes the water to be slightly acidic, primarily from the first step in the ionization of the acid.

$$CO_2(g) + H_2O \rightleftharpoons H_2CO_3(aq)$$

$$H_2CO_3(aq) + H_2O \rightleftharpoons H_3O^+(aq) + HCO_3^-(aq)$$

As the slightly acidic water seeps through the limestone, some of the insoluble $CaCO_3$ dissolves to give soluble calcium bicarbonate.

$$CaCO_3(s) + H_2CO_3(aq) \longrightarrow Ca^{2+}(aq) + 2HCO_3^-(aq)$$

When this ground water is pumped and distributed in a community water supply, the presence of Ca^{2+} gives the kinds of problems discussed above.

A more serious problem occurs when this hard water is heated in hot water heaters or spattered onto glassware and the walls of shower stalls. When solutions containing Ca^{2+} and HCO_3^- are heated or when the water evaporates, the reverse reactions occur and insoluble $CaCO_3$ precipitates as follows.

$$Ca^{2+}(aq) + 2HCO_3^-(aq) \longrightarrow H_2O + CO_2(g) + CaCO_3(s)$$

Evaporation of the hard water leaves spots of $CaCO_3$ on glass and other surfaces that are difficult to remove. In hot water heaters, the problem is more troublesome. The $CaCO_3$ precipitate sticks to the inner walls of pipes and hot water boilers and is called *boiler scale*. In locations that have high concentrations of Ca^{2+} and HCO_3^- in the water supply, boiler scale is a very serious problem, as illustrated in the accompanying photograph.

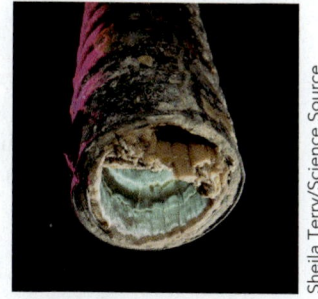

Sheila Terry/Science Source

Boiler scale built up on the inside of a water pipe.

Example 4.7
Predicting Reactions in Which a Gas Is Formed and Writing Their Equations

What reaction (if any) occurs when solutions of ammonium carbonate, $(NH_4)_2CO_3$, and propanoic acid (also called propionic acid), $HC_3H_5O_2$, are mixed?

Analysis: To determine whether a reaction occurs, we need to know whether a net ionic equation exists after we drop spectator ions from an ionic equation. We begin by writing a potential metathesis equation in molecular form. Then we examine the reactants and products to see whether any are solids, weak electrolytes, or substances that give gases. Then we form the ionic equation and search for spectator ions, which we eliminate to obtain the net ionic equation.

Assembling the Tools: Our tools for working a problem such as this are the list of strong acids, the solubility rules (Table 4.1), and the list of gases formed in metathesis reactions (Table 4.2).

Solution: First we construct a molecular equation, treating the reaction as a metathesis. For the acid, we take the cation to be H^+ and the anion to be $C_3H_5O_2^-$. Therefore, after exchanging cations between the two anions we can obtain the following balanced molecular equation:

$$(NH_4)_2CO_3 + 2HC_3H_5O_2 \longrightarrow 2NH_4C_3H_5O_2 + H_2CO_3$$

In the statement of the problem we are told that we are working with a *solution* of $HC_3H_5O_2$, so we know it's soluble. Also, it is not on the list of strong acids (the first tool mentioned above), so we expect it to be a weak acid; we will write it in molecular form in the ionic equation.

Next, we recognize that H_2CO_3 decomposes into $CO_2(g)$ and H_2O. (This information is also found in Table 4.2.) Let's rewrite the molecular equation taking this into account.

$$(NH_4)_2CO_3 + 2HC_3H_5O_2 \longrightarrow 2NH_4C_3H_5O_2 + CO_2(g) + H_2O$$

Now we need to determine which of the ionic substances are soluble. According to the solubility rules, all ammonium salts are soluble, and we know that all salts are strong electrolytes. Therefore, we will write $(NH_4)_2CO_3$ and $NH_4C_3H_5O_2$ in dissociated form.

$$2NH_4^+(aq) + CO_3^{2-}(aq) + 2HC_3H_5O_2(aq) \longrightarrow 2NH_4^+(aq) + 2C_3H_5O_2^-(aq) + CO_2(g) + H_2O$$

The only spectator ion is NH_4^+. Cancelling the $2NH_4^+$ ions on both sides of the equation gives

$$CO_3^{2-}(aq) + 2HC_3H_5O_2(aq) \longrightarrow 2C_3H_5O_2^-(aq) + CO_2(g) + H_2O$$

A net reaction *does* occur, as indicated by the existence of a net ionic equation.

Is the Answer Reasonable? There are some common errors that people make in working problems of this kind, so it is important to double check. Be sure you've written the formulas of the products correctly. (If you need review, you might look at Example 4.6.) Look for weak acids. (You need to know the list of strong ones; if an acid isn't on the list, it's a weak acid.) Look for gases, or substances that decompose into gases. (Be sure you've studied Table 4.2.) Check for insoluble compounds. (You need to know the solubility rules in Table 4.1.) If you've checked each step, it is likely your result is correct.

Example 4.8
Predicting Metathesis Reactions and Writing Their Equations

What reaction (if any) occurs in water between potassium nitrate and ammonium chloride?

Analysis: Once again, we need to know whether a net ionic equation exists. We follow the same path as in Example 4.7, but first we have to convert the names of the compounds into chemical formulas.

Assembling the Tools: The tools we need are (1) nomenclature rules from Chapter 2, (2) the solubility rules (Table 4.1), and (3) the list of gases formed in a metathesis reaction (Table 4.2).

Solution: First we apply the nomenclature rules. The ions in potassium nitrate are K^+ and NO_3^-, so the salt has the formula KNO_3. In ammonium chloride, the ions are NH_4^+ and Cl^-, so the salt is NH_4Cl.

Next we write the molecular equation, being sure to construct correct formulas for the products.

$$KNO_3 + NH_4Cl \longrightarrow KCl + NH_4NO_3$$

Looking over the substances in the equation, we don't find any that are weak acids or that decompose to give gases. Next, we check solubilities. According to solubility Rule 2, both KNO_3 and NH_4Cl are soluble. By solubility Rules 1 and 2, both products are also soluble in water. The anticipated molecular equation is therefore

$$KNO_3(aq) + NH_4Cl(aq) \longrightarrow KCl(aq) + NH_4NO_3(aq)$$

and the ionic equation is

$$K^+(aq) + NO_3^-(aq) + NH_4^+(aq) + Cl^-(aq) \longrightarrow K^+(aq) + Cl^-(aq) + NH_4^+(aq) + NO_3^-(aq)$$

Notice that the right side of the equation is the same as the left side except for the order in which the ions are written. When we eliminate spectator ions, everything goes. There is no net ionic equation, which means there is no net reaction.

Is the Answer Reasonable? Once again, we perform the same checks here as in Example 4.7, and they tell us our answer is right.

Knowing that salts of the formate ion, CHO_2^-, are water soluble, predict the reaction between $Co(OH)_2$ and formic acid, $HCHO_2$. Write molecular, ionic, and net ionic equations. (*Hint:* Apply the tools we used in Example 4.7.)

Practice Exercise 4.20

Show that in aqueous solutions there is no net reaction between $MgCl_2$ and $(NH_4)_2SO_4$.

Practice Exercise 4.21

Using Metathesis Reactions to Synthesize Salts

Chemists often need to synthesize their own chemicals, and one practical use for metathesis reactions is chemical synthesis. For example, in the manufacture of photographic film used by dentists and physicians for X rays, the X ray–sensitive component is a mixture of silver bromide and silver iodide formed by the reaction of silver nitrate with potassium bromide and iodide. Both reactants are water soluble, but the silver bromide and iodide are insoluble.

In planning a synthesis, the most important criterion is that the desired pure compound should be easily separated from the reaction mixture. This leads to three principal approaches.

1. If the desired compound is insoluble in water, then we can start with two soluble reactants. After the reaction is over, the compound can be separated from the mixture by filtration.

TOOLS
Synthesis by metathesis

2. If the desired compound is soluble, a neutralization reaction, where stoichiometric amounts of acid and base are mixed, will result in a solution of the pure compound.
3. Another way to prepare a soluble substance is to add an acid, which supplies the anion, to an excess of a metal carbonate, sulfide, or sulfite, which supplies the cation. (Metal sulfides or sulfites work, but are usually avoided since H_2S and SO_2 are both poisonous gases.)

In cases 2 and 3 the resulting solution can be evaporated to recover the pure solid.

Example 4.9
Synthesizing a Compound by Metathesis

What reaction might we use to synthesize nickel sulfate, $NiSO_4$?

Analysis: Questions of this kind are somewhat open ended because there is often more than one reaction we can use to make the same compound. First we need to know whether $NiSO_4$ is soluble or not. Then we can select our reactants.

Assembling the Tools: One tool we are definitely going to need is the solubility rules. The better you know them, the easier answering the question will be. Next, we use the decision-making tool for synthesizing compounds by metathesis reactions to determine our path. We may also need to know how to write equations for acid–base neutralization or reactions of acids with carbonates.

Solution: Since $NiSO_4$ is soluble in water, our best path is the reaction of an acid with a carbonate or hydroxide. The acid supplies the anion, so we would use H_2SO_4. The base would be $Ni(OH)_2$, and if we use a carbonate it would be $NiCO_3$. The solubility rules tell us both of these nickel-containing reactants are insoluble. Here are the two possible reactions written in molecular form.

$$H_2SO_4(aq) + Ni(OH)_2(s) \longrightarrow NiSO_4(aq) + 2H_2O$$
$$H_2SO_4(aq) + NiCO_3(s) \longrightarrow NiSO_4(aq) + CO_2(g) + H_2O$$

Is the Answer Reasonable? All we can do here is check the reasoning, which seems sound. The two equations yield the desired product, and if the reaction is carried out with an excess of the insoluble reactant, filtration will give a solution containing only $NiSO_4$.

Practice Exercise 4.22 Write an equation for the preparation of copper(II) nitrate using an acid–base neutralization reaction.

Practice Exercise 4.23 Suppose you wished to prepare cobalt(II) sulfide using a metathesis reaction. What criteria would the reactants have to satisfy *in this particular case*? Write an equation for a reaction utilizing a set of reactants that meet these criteria.

4.7 | Molarity

Solutions are very convenient for carrying out many reactions, and to deal with their stoichiometry we need some quantitative tools—for example, a way to express *concentrations*. Earlier we used percentage concentration (grams of solute per gram of solution multiplied by 100%) as an example of a concentration unit. To deal with the stoichiometry of reactions in solution, we want to use moles, not mass, so we express the amount of solute in moles and the amount of solution in liters.

The **molar concentration**, or **molarity** (abbreviated *M*), of a solution is defined as *the number of moles of solute per liter of solution*.

TOOLS
Molarity

$$\text{Molarity} = \frac{\text{moles of solute}}{\text{liters of solution}} \qquad (4.5)$$

Thus, a solution that contains 0.100 mol of NaCl in 1.00 L has a molarity of 0.100 M, and we would refer to this solution as 0.100 *molar* NaCl or as 0.100 M NaCl. The same concentration would result if we dissolved 0.0100 mol of NaCl in 0.100 L (100 mL) of solution, because the *ratio* of moles of solute to volume of solution is the same.

$$\frac{0.100 \text{ mol NaCl}}{1.00 \text{ L NaCl solution}} = \frac{0.0100 \text{ mol NaCl}}{0.100 \text{ L NaCl solution}} = 0.100 \ M \text{ NaCl}$$

Using Molarity as a Conversion Factor

Whenever we have to deal with a stoichiometry problem that involves an amount of a chemical and a volume of a solution of that substance, you can expect that solving the problem will involve molarity.

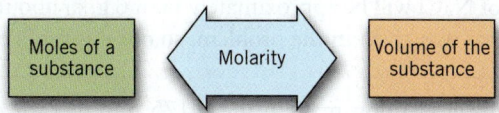

Molarity is a tool that provides the conversion factors we need to convert between moles and volume (either in liters or milliliters). Consider, for example, a solution labeled 0.100 M NaCl. The unit M always translates to mean "moles per liter," so we can write

$$0.100 \ M \text{ NaCl} = \frac{0.100 \text{ mol NaCl}}{1.00 \text{ L NaCl solution}}$$

This gives us an equivalence relationship between "mol NaCl" and "L soln" that we can use to form two conversion factors.[7]

$$0.100 \text{ mol NaCl} \Leftrightarrow 1.00 \text{ L soln}$$

$$\frac{0.100 \text{ mol NaCl}}{1.00 \text{ L NaCl soln}} \quad \text{and} \quad \frac{1.00 \text{ L NaCl soln}}{0.100 \text{ mol NaCl}}$$

Example 4.10
Calculating the Molarity of a Solution

To study the effect of dissolved salt on the rusting of an iron sample, a student prepared a solution of NaCl by dissolving 1.461 g of NaCl in enough water to make 250.0 mL of solution. What is the molarity of this solution?

Analysis: When given the amount of solute and the volume of a solution, we go back to the definition of molarity to calculate the concentration.

Assembling the Tools: The tool we'll use to solve this problem is Equation 4.5, which defines molarity as the ratio of *moles of solute* to *liters of solution*,

$$\text{molarity} = \frac{? \text{ mol NaCl}}{? \text{ L soln}}$$

Therefore, we have to convert 1.461 g of NaCl to moles of NaCl using the molar mass. We also have to change 250.0 mL to liters. The tool for this is 1000 mL = 1 L.

Solution: The number of moles of NaCl is found using the formula mass of NaCl, 58.443 g mol^{-1}.

$$1.461 \text{ g NaCl} \times \frac{1 \text{ mol NaCl}}{58.443 \text{ g NaCl}} = 0.024999 \text{ mol NaCl}$$

---

[7]Sometimes it is easier to translate the "1.00 L" part of these factors into the equivalent 1000 mL.

$$\frac{0.100 \text{ mol NaCl}}{1000 \text{ mL NaCl soln}} \quad \text{and} \quad \frac{1000 \text{ mL NaCl soln}}{0.100 \text{ mol NaCl}}$$

To find the volume of the solution in liters,

$$250.0 \text{ mL} \times \frac{1 \text{ L}}{1000 \text{ mL}} = 0.2500 \text{ L}$$

The ratio of moles to liters, therefore, is

$$\frac{0.024999 \text{ mol NaCl}}{0.2500 \text{ L}} = 0.1000 \text{ } M \text{ NaCl}$$

Is the Answer Reasonable? Let's use our answer to do a rough calculation of the amount of NaCl in the solution. If our answer is right, we should find a value not too far from the amount given in the problem (1.461 g). If we round the formula mass of NaCl to 60 and use 0.1 M as an approximate concentration, then one liter of the solution contains 0.1 mol of NaCl, or approximately 6 g of NaCl (one-tenth of 60 g). But 250 mL is only ¼ of a liter, so the mass of NaCl will be approximately ¼ of 6 g, or about 1.5 g. This is pretty close to the amount that was given in the problem, so our answer is probably correct.

Practice Exercise 4.24

What is the molarity of a solution made using 0.175 mol of LiBr dissolved in enough water to make 0.350 L of solution? (*Hint:* What equation is used to determine the molarity of a solution?)

Practice Exercise 4.25

What is the molarity of a solution made using 2.75 g of KI dissolved in enough water to make 125 mL of solution?

Example 4.11
Using Molar Concentrations

How many milliliters of 0.250 M NaCl solution must be measured to obtain 0.100 mol of NaCl?

Analysis: We can restate the problem as follows:

$$0.100 \text{ mol NaCl} \Leftrightarrow ? \text{ mL soln}$$

To find the answer, we need the conversion factor to take us from moles to volume, and that is molarity.

Assembling the Tools: Our tool will be the molarity, which translates to

$$0.250 \text{ } M \text{ NaCl} = \frac{0.250 \text{ mol NaCl}}{1 \text{ L NaCl soln}}$$

The fraction on the right relates moles of NaCl to liters of the solution and we can derive two conversion factors.

$$\frac{0.250 \text{ mol NaCl}}{1 \text{ L NaCl soln}} \quad \text{and} \quad \frac{1 \text{ L NaCl soln}}{0.250 \text{ mol NaCl}}$$

To obtain the answer, we select the one on the right, which will allow us to cancel the unit "mol NaCl."

Solution: Using the second factor with 0.100 mol NaCl, we get

$$0.100 \text{ mol NaCl} \times \frac{1.00 \text{ L NaCl}}{0.250 \text{ mol NaCl}} = 0.400 \text{ L of } 0.250 \text{ } M \text{ NaCl}$$

Because 0.400 L corresponds to 400 mL, our answer is 400 mL (which is better written as 4.00×10^2 mL to show the correct significant figures) of 0.250 M NaCl provides 0.100 mol of NaCl.

Is the Answer Reasonable? The molarity tells us that one liter contains 0.250 mol NaCl, so we need somewhat less than half of a liter to obtain just 0.100 mol. The answer, 400 mL, is reasonable.

If a reaction requires 0.142 moles of $NaNO_3$, how many milliliters of 0.500 *M* $NaNO_3$ are needed? (*Hint:* Molarity gives the equivalence between moles of solute and volume of solution in liters.)

A student measured 175 mL of 0.250 *M* HCl solution into a beaker. How many moles of HCl were in the beaker? How many grams of HCl are in the beaker?

Obtaining Moles of Solute from Molarity and Volume

If we must prepare a volume of a solution having a desired molarity (e.g., 250.0 mL of 0.0800 *M* Na_2CrO_4), we have to calculate the amount of solute that will be in the solution after it is made. To do this, we can rearrange Equation 4.5 and solve for moles.

Thus, in 250.0 mL of 0.0800 *M* Na_2CrO_4 there are

$$0.2500 \text{ L soln} \times \frac{0.0800 \text{ mol Na}_2\text{CrO}_4}{1 \text{ L soln}} = 0.0200 \text{ mol Na}_2\text{CrO}_4$$

■ Equation 4.5 can be rearranged as molarity × volume (L) = moles of solute.

Figure 4.22 shows how we would use a 250 mL volumetric flask to prepare such a solution.[8] (A **volumetric flask** is a narrow-necked flask having an etched mark high on its

(a) (b) (c)

(d) (e) (f)

Figure 4.22 | The preparation of a solution having a known molarity. (*a*) A 250 mL beaker contains a weighed amount of solute. (*b*) The solid is completely dissolved in distilled water. (*c*) The solution is transferred quantitatively to the 250 mL volumetric flask by rinsing the beaker several times with distilled water and adding that to the volumetric flask. (*d*) Water is being added. (*e*) More water is added to bring the level of the solution to the etched line. When filled to the line etched around its neck, the flask contains exactly 250 mL of solution. (*f*) The flask is stoppered and then inverted several times to mix its contents thoroughly.

[8]Note that in preparing the solution described in Figure 4.22, we do not simply add 250 mL of water to the solute. If we did, the final volume would be slightly larger than 250 mL, which would make the molarity slightly less than desired. To obtain an accurate molarity, we add enough water to give a total volume of 250 mL. The actual volume of water added is not important, only that the *total final volume* is 250 mL.

neck. When filled to the mark, the flask contains precisely the volume given by the flask's label.)

What we can conclude from this is that *any time you know both the molarity and volume of a solution, you can easily calculate the number of moles of solute in it.*

Example 4.12
Preparing a Solution with a Known Molarity

Strontium nitrate, $Sr(NO_3)_2$, can help in easing skin irritations. Suppose we need to prepare 250.0 mL of 0.100 M $Sr(NO_3)_2$ solution. How many grams of strontium nitrate are required?

Analysis: The essence of this problem is that we know the volume and molarity of the final solution, from which we can calculate the number of moles of $Sr(NO_3)_2$ in it. Once we know the moles of $Sr(NO_3)_2$, its mass is calculated using the molar mass of the salt.

Assembling the Tools: The first tool we will use is Equation 4.5 re-written to find the moles of $Sr(NO_3)_2$.

Molarity of $Sr(NO_3)_2$ × volume of solution in liters = moles $Sr(NO_3)_2$

Then we will use the molar mass of $Sr(NO_3)_2$

1 mol $Sr(NO_3)_2$ = 211.64 g $Sr(NO_3)_2$

from which we will construct the necessary conversion factor.

Solution: We apply Equation 4.5, first converting volume 250.0 mL to 0.2500 L. Multiplying the molarity by the volume in liters takes the following form.

$$\frac{0.100 \text{ mol } Sr(NO_3)_2}{1.00 \text{ L } Sr(NO_3)_2} \times 0.2500 \text{ L } Sr(NO_3)_2 = 0.02500 \text{ mol } Sr(NO_3)_2$$

$\uparrow$ molarity $\qquad\qquad$ $\uparrow$ volume (L) $\qquad\qquad$ $\uparrow$ moles of solute

Finally, we convert from moles to grams using the molar mass of $Sr(NO_3)_2$.

$$0.02500 \text{ mol } Sr(NO_3)_2 \times \frac{211.64 \text{ g } Sr(NO_3)_2}{1 \text{ mol } Sr(NO_3)_2} = 5.29 \text{ g } Sr(NO_3)_2$$

Thus, to prepare the solution we need to dissolve 5.29 g of $Sr(NO_3)_2$ by adding water to obtain a total volume of 250.0 mL.

We could also have set this up as a chain calculation as follows, with the conversion factors strung together.

$$0.2500 \text{ L } Sr(NO_3)_2 \text{ soln} \times \frac{0.100 \text{ mol } Sr(NO_3)_2}{1.00 \text{ L } Sr(NO_3)_2 \text{ soln}} \times \frac{211.62 \text{ g } Sr(NO_3)_2}{1 \text{ mol } Sr(NO_3)_2} = 5.29 \text{ g } Sr(NO_3)_2$$

Is the Answer Reasonable? We can check the answer using estimations. If we needed a full liter of this solution, it would contain 0.1 mol of $Sr(NO_3)_2$. The molar mass of the salt is 211.64 g mol^{-1}, so 0.1 mol is slightly more than 20 g. However, we are working with just a quarter of a liter (250 mL), so the amount of $Sr(NO_3)_2$ needed is slightly more than a quarter of 20 g, or 5 g. The answer, 5.29 g, is close to this, so it makes sense.

Practice Exercise 4.28

Suppose you wanted to prepare 50 mL of 0.2 M $Sr(NO_3)_2$ solution. Using the kind of approximate arithmetic we employed in the *Is the Answer Reasonable* step in Example 4.12, estimate the number of grams of $Sr(NO_3)_2$ required. (*Hint:* How many moles of $Sr(NO_3)_2$ would be in one liter of the solution?)

Practice Exercise 4.29

How many grams of $AgNO_3$ are needed to prepare 250.0 mL of 0.0125 M $AgNO_3$ solution?

Practice Exercise 4.30

How many grams of sodium sulfate are needed to prepare 500.0 mL of a 0.0150 M sodium sulfate solution?

Diluting Solutions

Laboratory chemicals are usually purchased in concentrated form and must be *diluted* (made less concentrated) before being used. This is accomplished by adding more solvent to the solution, which spreads the solute through a larger volume and causes the concentration (the amount per unit volume) to decrease.

During **dilution**, the number of moles of solute remains constant. This means that the product of molarity and volume, which equals the moles of solute, must be the same for both the concentrated and diluted solution.

$$\left(\begin{array}{c} \text{Volume of} \\ \text{dilute solution} \\ \textit{to be prepared} \end{array} \right) \times M_{\text{dilute}} = \left(\begin{array}{c} \text{Volume of} \\ \text{concentrated solution} \\ \textit{to be used} \end{array} \right) \times M_{\text{conc}}$$

<u>Moles of solute in the dilute solution</u>   <u>Moles of solute in the concentrated solution</u>

Or,

$$V_{\text{dil}} \times M_{\text{dil}} = V_{\text{conc}} \times M_{\text{conc}} \qquad (4.6)$$

TOOLS
Dilution equation

Any units can be used for volume in Equation 4.6 provided that they are the same on both sides of the equation. This means we can solve dilution problems using *milliliters* directly in Equation 4.6.

Example 4.13
Preparing a Solution of Known Molarity by Dilution

How can we prepare 100.0 mL of 0.0400 M $K_2Cr_2O_7$ from 0.200 M $K_2Cr_2O_7$?

Analysis: This is the way such a question comes up in the lab, but what it is really asking is, "How many milliliters of 0.200 M $K_2Cr_2O_7$ (the more concentrated solution) must be diluted to give a solution with a final volume of 100.0 mL and a final molarity of 0.0400 M?" Once we see the question this way, what we have to do becomes clear.

Assembling the Tools: Because it's a dilution problem, the tool we need to solve the problem is Equation 4.6.

$$V_{\text{dil}} \times M_{\text{dil}} = V_{\text{conc}} \times M_{\text{conc}}$$

Solution: It's a good idea to assemble the data first, noting what is missing (and therefore what has to be calculated).

$$V_{dil} = 100.0 \text{ mL} \qquad\qquad M_{dil} = 0.0400 \ M$$
$$V_{conc} = ? \qquad\qquad M_{conc} = 0.200 \ M$$

Next, we substitute into Equation 4.6.

$$100.0 \text{ mL} \times 0.0400 \ M = V_{conc} \times 0.200 \ M$$

Solving for V_{conc} gives

$$V_{conc} = \frac{100.0 \text{ mL} \times 0.0400 \ M}{0.200 \ M}$$

$$= 20.0 \text{ mL}$$

Therefore, the answer to the question as asked is: We would withdraw 20.0 mL of 0.200 M $K_2Cr_2O_7$, place it in a 100 mL volumetric flask, and then add water until the final volume is exactly 100 mL. (See Figure 4.23.)

Is the Answer Reasonable? Notice that the concentrated solution is 5 times more concentrated than the dilute solution ($5 \times 0.04 = 0.2$). To reduce the concentration by a factor of 5 requires that we increase the volume by a factor of 5, and we see that 100 mL is 5 times 20 mL. The answer appears to be correct.

Practice Exercise 4.31

To what final volume must 100.0 mL of 0.125 M H_2SO_4 solution be diluted to give a 0.0500 M H_2SO_4 solution? (*Hint:* Write the equation we used for dilution problems.)

Practice Exercise 4.32

How many milliliters of water have to be *added* to 150 mL of 0.50 M HCl to reduce the concentration to 0.10 M HCl?

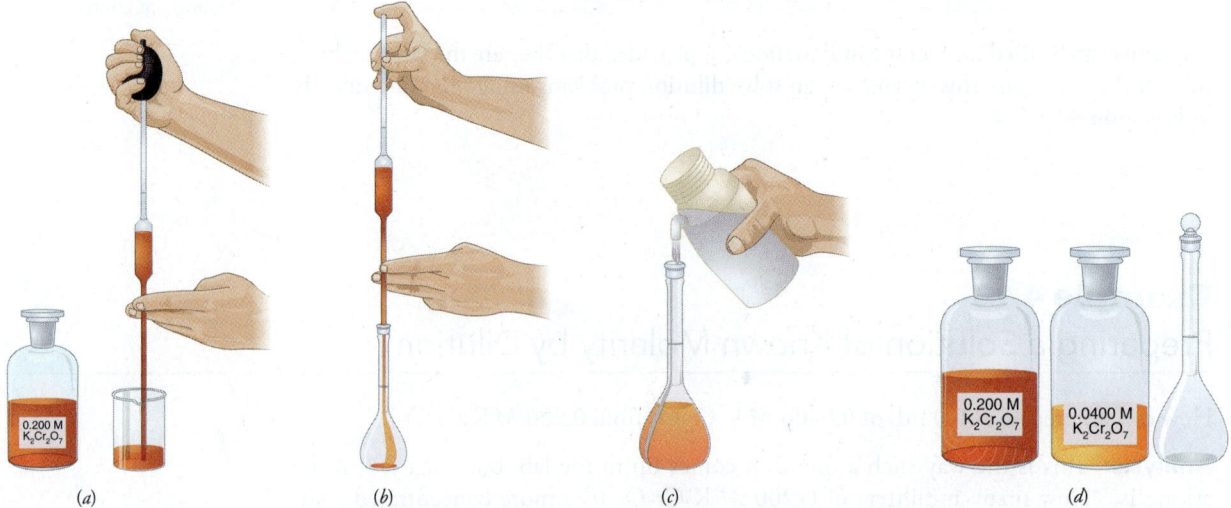

(*a*) (*b*) (*c*) (*d*)

Figure 4.23 | **Preparing a solution by dilution.** (*a*) The calculated volume of the more concentrated solution is withdrawn from the stock solution by means of a volumetric pipet. (*b*) The solution is allowed to drain entirely from the pipet into the volumetric flask and then the pipet tip is gently touched to the liquid. (*c*) Water is added to the flask, the contents are mixed, and the final volume is brought up to the etch mark on the narrow neck of the flask. (*d*) The new solution is put into a labeled container.

4.8 | Solution Stoichiometry

Stoichiometry calculations are simply conversions from one set of chemical units to an-other, as we saw in Chapter 3. To be successful at these calculations it is essential to understand the sequence of conversions between units and where we find the conversion factors needed to move from one unit to the next. The diagram in Figure 3.6 illustrated how we could start with mass, elementary units (atoms, ions, formula units or molecules), or moles and then logically use conversion factors to reach the desired units. Importantly, all stoichiometric calculations require that the starting units be converted to moles on the way to the final answer. Solution stoichiometry simply adds one more starting point to the flowchart—the volume, in liters, of solution. It also uses the molarity of the solution as the conversion factor between volume and moles. The updated flowchart is shown in Figure 4.24. Finally, it is important to note that, unlike the other conversion factors, the molarity cannot be looked up in any tables. The best way to become familiar with solution stoichiometry is to see how some common problems are solved.

TOOLS
Stoichiometry pathways

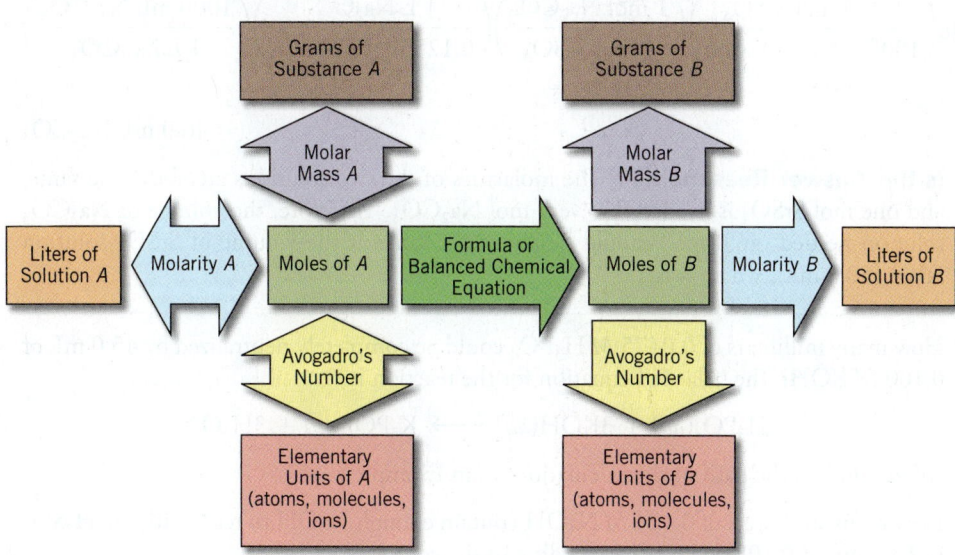

Figure 4.24 | **Paths for working stoichiometry problems involving chemical reactions.** This diagram summarizes the sequences used for efficient stoichiometric conversions from one set of units to another. The use of molarity as a conversion factor has been added compared to Figure 3.6.

Example 4.14
Stoichiometry Involving Reactions in Solution

Strontium carbonate is used to produce the red colors in fireworks. Since strontium is often found naturally as strontium sulfate, how many milliliters of 0.125 M Na_2CO_3 are needed to react with 50.0 mL of 0.115 M $SrSO_4$ solution?

$$Na_2CO_3(aq) + SrSO_4(aq) \longrightarrow SrCO_3(s) + Na_2SO_4(aq)$$

Analysis: The sequence of conversions needed are diagrammed below.

Strontium produces the red color in fireworks.

© drbimages/iStockphoto

Each arrow represents a conversion factor needed to perform the conversion, and we can enter them as

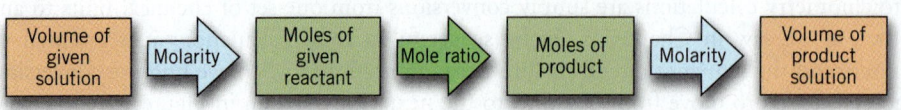

We just need the volume of the $SrSO_4$, its molarity, the balanced equation for the mole ratio and the molarity of the Na_2CO_3 to determine the answer.

Assembling the Tools: Our tools are the sequence of operations described in the analysis section and then the conversion factors associated with the molarity of the appropriate solutions and the mole ratio that we obtain from the balanced chemical equation.

Solution: We need to calculate how many milliliters of Na_2CO_3 are equivalent to 50 mL of $SrSO_4$.

$$50.0 \text{ mL } SrSO_4 \text{ soln} \left(\frac{0.115 \text{ mol } SrSO_4}{1000 \text{ mL } SrSO_4 \text{ soln}} \right) \left(\frac{1 \text{ mol } Na_2CO_3}{1 \text{ mol } SrSO_4} \right) \left(\frac{1 \text{ L } Na_2CO_3}{0.125 \text{ mol } Na_2CO_3} \right) \left(\frac{1000 \text{ mL } Na_2CO_3}{1 \text{ L } Na_2CO_3} \right)$$

M SrSO₄   mole ratio   M Na₂CO₃   volume conversion

$$= 46.0 \text{ mL } Na_2CO_3$$

Is the Answer Reasonable? The molarities of the two solutions are about the same, and one mol $SrSO_4$ is needed for every mol Na_2CO_3. Therefore, the volume of Na_2CO_3 solution needed (46.0 mL) should be about the same as the volume of $SrSO_4$ solution taken (50.0 mL), which it is.

Practice Exercise 4.33

How many milliliters of 0.0475 M H_3PO_4 could be completely neutralized by 45.0 mL of 0.100 M KOH? The balanced equation for the reaction is

$$H_3PO_4(aq) + 3KOH(aq) \longrightarrow K_3PO_4(aq) + 3H_2O$$

(*Hint:* Outline the path of the calculations as in Example 4.14.)

Practice Exercise 4.34

How many milliliters of 0.124 M NaOH contain enough NaOH to react with the H_2SO_4 in 15.4 mL of 0.108 M H_2SO_4 according to the equation

$$2NaOH(aq) + H_2SO_4(aq) \longrightarrow Na_2SO_4(aq) + 2H_2O$$

Using Net Ionic Equations in Calculations

In the preceding problem, we worked with a molecular equation in solving a stoichiometry problem. Ionic and net ionic equations can also be used, but this requires that we work with the concentrations of the ions in solution.

Calculating Concentrations of Ions in Solutions of Electrolytes

The concentrations of the ions in a solution of a strong electrolyte are obtained from the formula and molar concentration of the solute. For example, suppose we are working with a solution labeled "0.20 M $CaCl_2$." In 1.0 L of this solution there is 0.20 mol of $CaCl_2$, which is fully dissociated into Ca^{2+} and Cl^- ions.

$$CaCl_2 \longrightarrow Ca^{2+} + 2Cl^-$$

Based on the stoichiometry of the dissociation, 1 mol Ca^{2+} and 2 mol Cl^- are formed from each 1 mol of $CaCl_2$. We can find the ion concentrations in the solution using dimensional analysis. With a bit of experience you will be able to do this by inspection.

For example, we can find the concentrations of the Ca^{2+} and the Cl^- ions as follows.

$$\frac{0.20 \ \text{mol CaCl}_2}{1 \ \text{L solution}} \left(\frac{1 \ \text{mol Ca}^{2+}}{1 \ \text{mol CaCl}_2} \right) = 0.20 \ M \ Ca^{2+}$$

The result is that the molarity of Ca^{2+} is 0.20 M. We will do a similar calculation for the chloride ions

$$\frac{0.20 \ \text{mol CaCl}_2}{1 \ \text{L solution}} \left(\frac{2 \ \text{mol Cl}^-}{1 \ \text{mol CaCl}_2} \right) = 0.40 \ M \ Cl^-$$

The molarity of Cl^- is 0.40 M.

In the example just given, notice that the ion concentrations are equal to the concentration of $CaCl_2$ multiplied by the number of ions of each kind that are released when a formula unit of $CaCl_2$ dissociates. This provides a very simple way to find the concentrations of the ions.

The concentration of a particular ion equals the concentration of the salt multiplied by the number of ions of that kind in one formula unit of the salt.

TOOLS

Molarity of ions in salt solution

Example 4.15
Calculating the Concentrations of Ions in a Solution

What are the molar concentrations of the ions in 0.20 M aluminum sulfate?

Analysis: We just learned that the molarity of the salt times the number of an ion in the formula unit will give us the molarity of the ion. Since dimensional analysis always works, we will use that approach to be sure.

Assembling the Tools: Several tools are required: (1) the formulas of the ions and the rules for writing chemical formulas, both from Chapter 2, (2) the equation for the dissociation of the salt in water, and (3) the method we just developed for finding molar concentrations of the ions using the coefficients in the dissociation equation.

Solution: The ions are Al^{3+} and SO_4^{2-}, so the salt must be $Al_2(SO_4)_3$. When $Al_2(SO_4)_3$ dissolves, it dissociates as follows:

$$Al_2(SO_4)_3(s) \longrightarrow 2Al^{3+}(aq) + 3SO_4^{2-}(aq)$$

For the first step we are going to rewrite the 0.20 M $Al_2(SO_4)_3$ as

$$\frac{0.20 \ \text{mol Al}_2(SO_4)_3}{1 \ \text{L of solution}}$$

and this will be the starting point for the calculation. The dimensional analysis equations are set up as

$$\frac{0.20 \ \text{mol Al}_2(SO_4)_3}{1 \ \text{L solution}} \times \frac{2 \ \text{mol Al}^{3+}}{1 \ \text{mol Al}_2(SO_4)_3} = \frac{0.40 \ \text{mol Al}^{3+}}{1 \ \text{L solution}} = 0.40 \ M \ Al^{3+}$$

$$\frac{0.20 \ \text{mol Al}_2(SO_4)_3}{1 \ \text{L solution}} \times \frac{3 \ \text{mol SO}_4^{2-}}{1 \ \text{mol Al}_2(SO_4)_3} = \frac{0.60 \ \text{mol SO}_4^{2-}}{1 \ \text{L solution}} = 0.60 \ M \ SO_4^{2-}$$

We've calculated that the solution contains 0.40 M Al^{3+} and 0.60 M SO_4^{2-}.

Is the Answer Reasonable? Each formula unit of $Al_2(SO_4)_3$ yields two Al^{3+} ions and three SO_4^{2-} ions. Therefore, 0.20 mol $Al_2(SO_4)_3$ yields 0.40 mol Al^{3+} and 0.60 mol SO_4^{2-}, and we conclude that our calculations were correct.

Example 4.16
Calculating the Concentration of a Salt from the Concentration of One of Its Ions

A student, using a common laboratory experiment, found that the sulfate ion concentration in a solution of $Al_2(SO_4)_3$ was 0.90 M. What was the concentration of $Al_2(SO_4)_3$ in the solution?

Analysis: Once again, we use the formula of the salt to determine the number of ions released when it dissociates. This time we use the information to work backward to find the salt concentration.

Assembling the Tools: The tools are the same as in Example 4.15.

Solution: Let's set up the problem using dimensional analysis to be sure of our procedure.

$$\frac{0.90 \text{ mol } SO_4^{2-}}{1 \text{ L solution}} \times \frac{1 \text{ mol } Al_2(SO_4)_3}{3 \text{ mol } SO_4^{2-}} = \frac{0.30 \text{ mol } Al_2(SO_4)_3}{1 \text{ L solution}} = 0.30 \text{ } M \text{ } Al_2(SO_4)_3$$

Our result is 0.30 M $Al_2(SO_4)_3$.

Is the Answer Reasonable? We'll use the reasoning approach to check our answer. We know that one mol $Al_2(SO_4)_3$ yields three mol SO_4^{2-} in solution. Therefore, the number of moles of $Al_2(SO_4)_3$ is only one-third the number of moles of SO_4^{2-}. So the concentration of $Al_2(SO_4)_3$ must be one-third of 0.90 M, or 0.30 M.

Practice Exercise 4.35

What are the molar concentrations of the ions in 0.40 M $FeCl_3$? (*Hint:* How many ions of each kind are formed when $FeCl_3$ dissociates?)

Practice Exercise 4.36

In a solution of Na_3PO_4, the PO_4^{3-} concentration was determined to be 0.250 M. What was the sodium ion concentration in the solution?

Stoichiometry Calculations

You have seen that a net ionic equation is convenient for describing the net chemical change in an ionic reaction. Example 4.17 illustrates how such equations can be used in stoichiometric calculations.

Example 4.17
Stoichiometric Calculations Using a Net Ionic Equation

A solution of $AgNO_3$ is added to a solution of $CaBr_2$, producing a precipitate of AgBr.

How many milliliters of 0.100 M Ag^+ solution are needed to react completely with 25.0 mL of 0.400 M $CaBr_2$ solution? The net ionic equation for the reaction is:

$$Ag^+(aq) + Br^-(aq) \longrightarrow AgBr(s)$$

Analysis: In many ways, this problem is similar to Example 4.14. However, to use the net ionic equation, we will need to work with the concentrations of the ions. Therefore, to solve the problem, the following steps, which can be extracted from Figure 4.24, will be used

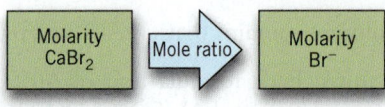

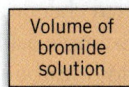 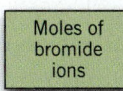 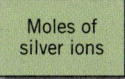 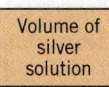

The first sequence converts the molarity of $CaBr_2$ to molarity of Br^- ions and the second allows calculation of the volume of the silver solution.

Assembling the Tools: In addition to the information in Figure 4.24, we need the conversion factors indicated in the arrows above. Molarities are never tabulated so we must find them in the statement of the question; the mole ratios come from formulas or balanced equations.

Solution: We will write the molarity of calcium bromide as the ratio:

$$\frac{0.400 \text{ mol } CaBr_2}{1 \text{ L solution}}$$

and the bromide ions concentration is calculated as:

$$\frac{0.400 \text{ mol } CaBr_2}{1 \text{ L solution}} \times \frac{2 \text{ mol } Br^-}{1 \text{ mol } CaBr_2} = 0.800 \text{ } M \text{ } Br^-$$

Now we can find the volume of the Ag^+ solution that is needed using the three steps shown in the flowchart.

$$25.0 \text{ mL } Br^- \times \frac{0.800 \text{ mol } Br^-}{1000 \text{ mL } Br^-} \times \frac{1 \text{ mol } Ag^+}{1 \text{ mol } Br^-} \times \frac{1000 \text{ mL } Ag^+}{0.100 \text{ mol } Ag^+} = 2.00 \times 10^2 \text{ mL } Ag^+$$

$$= 2.00 \times 10^2 \text{ mL } Ag^+ \text{ solution is needed}$$

Is the Answer Reasonable? The silver ion concentration is one-eighth as large as the bromide ion concentration. Since the ions react one-for-one, we will need eight times as much silver ion solution as bromide solution. Eight times the amount of bromide solution, 25 mL, is 200 mL, which is the answer we obtained. Therefore, the answer appears to be correct.

Suppose 18.4 mL of 0.100 M $AgNO_3$ solution was needed to react completely with 20.5 mL of $CaCl_2$ solution. What is the molarity of the $CaCl_2$ solution? (*Hint:* How can you calculate molarity from moles and volume, and how can you calculate the molarity of the $CaCl_2$ solution from the molarity of Cl^-?)

Practice Exercise 4.37

How many milliliters of 0.500 M KOH are needed to react completely with 60.0 mL of 0.250 M $FeCl_2$ solution to precipitate $Fe(OH)_2$? The net ionic equation is:

Practice Exercise 4.38

$$Fe^{2+}(aq) + 2OH^-(aq) \longrightarrow Fe(OH)_2(s)$$

4.9 | Titrations and Chemical Analysis

Chemical analyses fall into two categories. In a **qualitative analysis** we simply determine which substances are present in a sample without measuring their amounts. In a **quantitative analysis**, our goal is to measure the amounts of the various substances in a sample.

When chemical reactions are used in a quantitative analysis, a useful strategy is to capture *all* of a desired chemical species in a compound with a known formula. From the amount of this compound obtained, we can determine how much of the desired chemical species was present in the original sample. Such an analysis was described in Example 3.11 in Section 3.4 in which we described the combustion analysis of a compound of carbon, hydrogen, and oxygen.

Example 4.18
Calculation Involving a Quantitative Analysis

A chemist was asked to analyze a sample of chlordane, $C_{10}H_6Cl_8$, that was discovered during demolition of an old work shed. The EPA banned the sale of this insecticide in the United States in 1988 because of its potential for causing cancer. Reactions were carried out on a 1.446 g sample converting all of the chlorine to chloride ions. This aqueous solution required 91.22 mL of 0.1400 M AgNO$_3$ to precipitate all of the chloride ion as AgCl. What was the percentage of chlordane in the original solution? The precipitation reaction was

$$Ag^+(aq) + Cl^-(aq) \longrightarrow AgCl(s)$$

Analysis: The percentage of chlordane in the sample will be calculated as follows:

$$\% \ C_{10}H_6Cl_8 \text{ by mass} = \frac{\text{mass of } C_{10}H_6Cl_8 \text{ in sample}}{\text{mass of sample}} \times 100\%$$

We already have the mass of the sample, so we need to use the data to calculate the mass of $C_{10}H_6Cl_8$.

From the molarity and volume of the AgNO$_3$ solution, we can calculate the moles of Ag^+ that reacts with Cl^-, which also equals the number of moles of Cl in the chlordane sample. We will use the moles of Cl to calculate the moles of $C_{10}H_6Cl_8$ and then the grams of $C_{10}H_6Cl_8$. Once we have the mass of $C_{10}H_6Cl_8$ we can calculate the percentage of $C_{10}H_6Cl_8$.

Assembling the Tools: Equation 4.5, rearranged, is the tool with which we calculate moles of Ag^+ that react.

$$\text{Molarity of } Ag^+ \text{ soln} \times \text{volume of } Ag^+ \text{ soln} = \text{moles of } Ag^+$$

As noted, the net ionic equation tells us that moles of Ag^+ equals moles of Cl^-, which equals moles Cl.

The formula for chlordane gives us the stoichiometric ratio of moles of $C_{10}H_6Cl_8$ to moles of Cl

$$1 \text{ mol } C_{10}H_6Cl_8 \Leftrightarrow 8 \text{ mol Cl}$$

To find grams of $C_{10}H_6Cl_8$, use the molar mass of $C_{10}H_6Cl_8$, which is 409.778 g mol^{-1}.

$$1 \text{ mol } C_{10}H_6Cl_8 \Leftrightarrow 409.778 \text{ g}$$

Solution: We will solve this problem stepwise. In 0.1400 M AgNO$_3$, the molarity of Ag^+ is 0.1400 M. Applying Equation 4.5,

$$0.09122 \text{ L Ag}^+ \text{ soln} \times \frac{0.1400 \text{ mol Ag}^+}{1.000 \text{ L Ag}^+ \text{ soln}} = 0.012771 \text{ mol Ag}^+$$

From the stoichiometry of the equation, 1 mol Ag$^+$ $\Leftrightarrow$ 1 mol Cl$^-$,

$$0.012771 \; \text{mol Ag}^+ \times \frac{1 \; \text{mol Cl}^-}{1 \; \text{mol Ag}^+} = 0.012771 \; \text{mol Cl}^-$$

This is the amount of Cl$^-$ that came from the sample, so the chlordane in the sample must have contained 0.012771 mol Cl. Now we use the formula for chlordane to calculate moles of $C_{10}H_6Cl_8$ in the sample.

$$0.012771 \; \text{mol Cl}^- \times \frac{1 \; \text{mol } C_{10}H_6Cl_8}{8 \; \text{mol Cl}^-} = 1.5964 \times 10^{-3} \; \text{mol } C_{10}H_6Cl_8$$

Using the formula mass of chlordane, we calculate the mass of the insecticide in the sample.

$$(1.5964 \times 10^{-3} \; \text{mol } C_{10}H_6Cl_8) \times \frac{409.778 \; \text{g } C_{10}H_6Cl_8}{1 \; \text{mol } C_{10}H_6Cl_8} = 0.65417 \; \text{g } C_{10}H_6Cl_8$$

The percentage of chlordane in the sample was

$$\frac{0.65417 \; \text{g } C_{10}H_6Cl_8}{1.446 \; \text{g sample}} \times 100\% = 45.24\% \; \text{(rounded correctly)}$$

The solution of insecticide contained 45.24% chlordane by mass.

Is the Answer Reasonable? We have used approximately 0.1 L of 0.14 M Ag$^+$ solution, which contains 0.014 mol of Ag$^+$ to react with 0.014 mol of Cl$^-$, which is the amount of Cl in the chlordane. The moles of chlordane are 1/8th of 0.014. For simplicity, suppose it was one-tenth. The moles of chlordane would then be 0.0014 mol. With a molar mass of about 400 g mol^{-1}, 0.001 mol of $C_{10}H_6Cl_8$ would weigh 0.4 g and 0.0014 mol would weigh around 0.6 g. Our answer, 0.654 g, is not far from this, so the calculations seem to be correct.

A solution containing Na_2SO_4 was treated with 0.150 M $BaCl_2$ solution until all the sulfate ion had reacted to form $BaSO_4$. The net reaction

$$Ba^{2+}(aq) + SO_4^{2-}(aq) \longrightarrow BaSO_4(s)$$

required 28.40 mL of the $BaCl_2$ solution. How many grams of Na_2SO_4 were in the solution? (*Hint:* How do moles of SO_4^{2-} relate to moles of Na_2SO_4?)

Practice Exercise 4.39

When 35.00 mL of Na_2CO_3 was reacted with 29.06 mL of a 0.125 M $CaCl_2$ solution, all of the carbonate ion precipitated as $CaCO_3$. How many grams of $NaCO_3$ were in the solution?

Practice Exercise 4.40

Acid–Base Titrations

Titration is an important laboratory procedure used in performing chemical analyses. The apparatus is shown in Figure 4.25. The long tube is called a **buret**, which is marked for volumes, usually in increments of 0.10 mL. The valve at the bottom of the buret is called a **stopcock**, and it permits controlled addition of the **titrant** (*the solution in the buret*) to the receiving vessel (the beaker shown in the drawing).

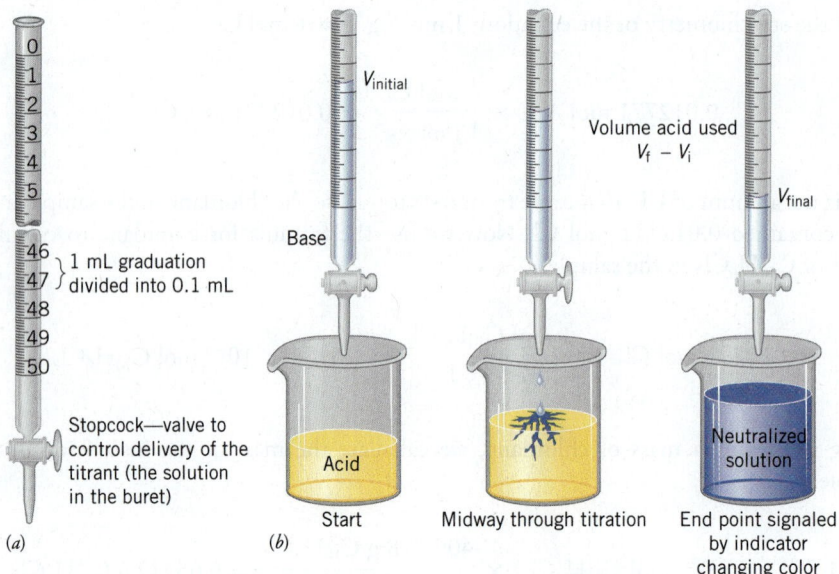

Figure 4.25 | Titration. (*a*) A buret. (*b*) The titration of an acid by a base in which an acid–base indicator, bromothymol blue, is used to signal the end point, which is the point at which all of the acid has been neutralized and addition of the base is halted.

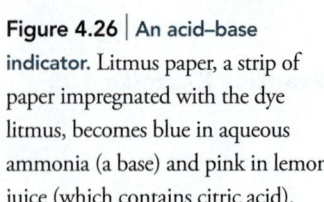

Figure 4.26 | An acid–base indicator. Litmus paper, a strip of paper impregnated with the dye litmus, becomes blue in aqueous ammonia (a base) and pink in lemon juice (which contains citric acid).

In a typical titration, a solution containing one reactant is placed in the receiving vessel. A solution of the other reactant is then added gradually from the buret. (One of the two solutions has a precisely known concentration and is called a **standard solution**.) This addition is continued until something (usually a visual effect, like a color change) signals that the two reactants have been combined in just the right proportions to give a complete reaction.

In acid–base titrations, an **acid–base indicator** is used to detect the completion of the reaction by a change in color. Indicators are dyes that have one color in an acidic solution and a different color in a basic solution. Litmus is one indicator that is red when acidic and blue when basic, as show in Figure 4.26. Phenolphthalein is a common indicator for titrations; it changes from colorless to pink when a solution changes from acidic to basic. This color change is very abrupt, and occurs with the addition of only one final drop of the titrant just as the end of the reaction is reached. When we observe the color change, the **end point** has been reached and the addition of titrant is stopped. We then record the total volume of the titrant that's been added to the receiving flask.

Example 4.19
Calculation Involving an Acid–Base Titration

A student prepares a solution of hydrochloric acid that is approximately 0.1 *M* and wants to determine its precise concentration. A 25.00 mL portion of the HCl solution is transferred to a flask, and after a few drops of indicator are added, the HCl solution is titrated with 0.0775 *M* standard NaOH solution. The titration requires exactly 37.46 mL of the NaOH solution to reach the end point. What is the molarity of the HCl solution?

Analysis: The first step is to write the balanced equation, which will give us the mole ratio between HCl and NaOH. The reaction is an acid–base neutralization, so the product is a salt plus water. Following the procedures developed earlier, the reaction is

$$HCl(aq) + NaOH(aq) \longrightarrow NaCl(aq) + H_2O$$

We have to go from molarity and volume of the NaOH solution to the molarity of the HCl solution. Solving the problem will follow essentially the same route as in Example 4.14 in Section 4.8.

Assembling the Tools: Our tools will be the relationship between molarity and volume of the NaOH solution to calculate moles of NaOH that react, the stoichiometry of the balanced equation to obtain the moles of HCl, and the definition of molarity (to calculate the molarity of the HCl solution from moles of HCl and the volume of the HCl sample taken).

Solution: From the molarity and volume of the NaOH solution, we calculate the number of moles of NaOH consumed in the titration.

$$0.03746 \text{ L NaOH soln} \times \frac{0.0775 \text{ mol NaOH}}{1.000 \text{ L NaOH soln}} = 2.903 \times 10^{-3} \text{ mol HCl}$$

The coefficients in the equation tell us that NaOH and HCl react in a one-to-one mole ratio,

$$2.903 \times 10^{-3} \text{ mol NaOH} \times \frac{1 \text{ mol HCl}}{1 \text{ mol NaOH}} = 2.903 \times 10^{-3} \text{ mol HCl}$$

so in this titration, we have 2.903×10^{-3} mol HCl in the flask. To calculate the molarity of the HCl, we simply apply the definition of molarity using the ratio of the number of moles of HCl that reacted to the volume (in liters) of the HCl solution used.

$$\text{Molarity of HCl soln} = \frac{2.903 \times 10^{-3} \text{ mol HCl}}{0.02500 \text{ L HCl soln}}$$

$$= 0.116 \, M \text{ HCl} \quad (\text{rounded})$$

The molarity of the hydrochloric acid is 0.116 M. (Note that we've rounded the answer to three significant figures to match the data given in the problem.)

Is the Answer Reasonable? If the concentrations of the NaOH and HCl were the same, the volumes used would have been equal. However, the volume of the NaOH solution used is larger than the volume of HCl solution. This must mean that the HCl solution is more concentrated than the NaOH solution. The value we obtained, 0.116 M, is larger than 0.0775 M, so our answer makes sense.

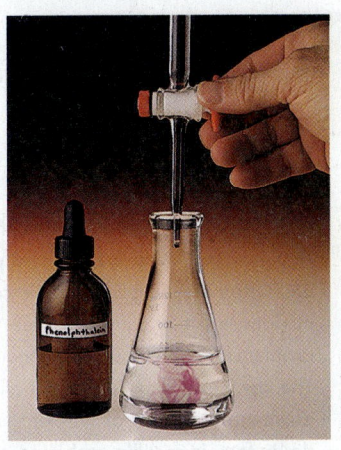

A solution of HCl is titrated with a solution of NaOH using phenolphthalein as the acid–base indicator. The pink color that phenolphthalein has in a basic solution can be seen where a drop of the NaOH solution has entered the HCl solution, to which a few drops of the indicator had been added.

Michael Watson

In a titration, a sample of H_2SO_4 solution having a volume of 15.00 mL required 36.42 mL of 0.147 M NaOH solution for complete neutralization. What is the molarity of the H_2SO_4 solution? (*Hint*: Check to be sure your chemical equation is correctly written and balanced.)

Practice Exercise 4.41

"Stomach acid" is a hydrochloric acid solution. A sample of stomach acid having a volume of 5.00 mL required 11.00 mL of 0.0100 M KOH solution for neutralization in a titration. What was the molar concentration of HCl in this fluid? If we assume a density of 1.00 g mL^{-1} for the fluid, what was the percentage by weight of HCl?

Practice Exercise 4.42

Multi-Concept Problems

The worked examples we've provided so far have been relatively simple, and each has focused on one topic or concept. However, in life and in chemistry, problems are not always that simple. Often they involve multiple concepts that are not immediately obvious, so it is difficult to see how to proceed. Although each problem is unique, there is a general strategy you can learn to apply. The key is realizing that almost all difficult problems can be broken down into some set of simpler tasks that you already know how to do. In fact, in real-life problems, the analysis will sometimes reveal tasks that you have not yet learned how to accomplish and point the way to learning new concepts or tools.

In the *Analyzing and Solving Multi-Concept Problems* section in this and subsequent chapters, our approach will be to look at the big picture first and deal with the details later. The goal will be to break the problem down into manageable parts. There is no simple "formula" for doing this, so don't be discouraged if solving these kinds of problems takes some considerable thought. You should also realize that with complex problems there is often more than one path to the solution. As you study the examples we provide, you may see an alternative way to arrive at the answer. As long as the reasoning is sound and leads to the same answer as ours, you are to be applauded!

Finally, by way of assurance, keep in mind that none of the multi-concept problems you encounter in this text will involve concepts that have not been previously discussed.

Analyzing and Solving Multi-Concept Problems

Magnesium hydroxide, $Mg(OH)_2(s)$, suspended in water, is used to relieve acid indigestion in several over-the-counter preparations. It can be prepared by adding a base to a solution containing Mg^{2+} ions. When 40.0 mL of 0.200 M NaOH solution is added to 25.0 mL of 0.300 M $MgCl_2$ solution, what mass of $Mg(OH)_2$ is formed, and what are the concentrations of Na^+ and Mg^{2+} left in solution?

Richard Megna/Fundamental Photographs

Creamy milk of magnesia is an aqueous suspension of magnesium hydroxide.

Analysis: Overall, we can view this as a complex stoichiometry problem. For all stoichiometry problems we need to describe the reaction with a balanced chemical equation (or formula). Then, we will need to determine the initial concentrations of the four ions, Na^+, OH^-, Mg^{2+}, and Cl^-, so we will have to write ionic and net ionic equations to take into account the dissociation of each solute.

We have a limiting reactant problem (notice that the moles of both reactants is known since we know their molarities and volumes). You solved this kind of problem in Chapter 3.

In this ionic reaction, two of the four ions will react. One will be used up, but some of the other will be left over, and we need to calculate how much. The other two ions are spectator ions and their mole amounts will not change, but their concentrations will when the solutions are mixed.

Let's summarize how we have broken this down to four simpler parts.

Part 1: Write the balanced molecular, ionic, and net ionic equations to identify which ions react.

Part 2: Calculate the moles of each ion before the reaction.

Part 3: Perform a limiting reactant calculation to determine the amount of product and the amount of excess reactant left in solution.

Part 4: Use the total volume to determine the molarities of all of the ions in solution after the reaction.

PART 1

Assembling the Tools: We need to set up a metathesis and ionic equation, and balance them using the solubility rules and our methods for writing formulas and dissociating compounds.

Solution:

The balanced molecular equation for the reaction is

$$MgCl_2(aq) + 2NaOH(aq) \longrightarrow Mg(OH)_2(s) + 2NaCl(aq)$$

from which we construct the ionic and net ionic equations.

$$Mg^{2+}(aq) + 2Cl^-(aq) + 2Na^+(aq) + 2OH^-(aq) \longrightarrow Mg(OH)_2(s) + 2Cl^-(aq) + 2Na^+(aq)$$

$$Mg^{2+}(aq) + 2OH^-(aq) \longrightarrow Mg(OH)_2(s)$$

PART 2

Assembling the Tools: To determine the moles of each ion in solution, we use the relationship between molarity, volume, and moles of substance ($M \times L = mol$, Equation 4.5, rearranged).

Solution: Let's begin by determining the number of moles of sodium hydroxide and magnesium chloride.

$$0.0400 \text{ L NaOH soln} \times \frac{0.200 \text{ mol NaOH}}{1.00 \text{ L NaOH soln}} = 8.00 \times 10^{-3} \text{ mol NaOH}$$

$$0.0250 \text{ L MgCl}_2 \text{ soln} \times \frac{0.300 \text{ mol MgCl}_2}{1.00 \text{ L MgCl}_2 \text{ soln}} = 7.50 \times 10^{-3} \text{ mol MgCl}_2$$

Now, using the moles of each compound we can calculate of the moles of each ion, keeping in mind that 1 mol of $MgCl_2$ gives 2 mol Cl^-:

Moles of Ions before Reaction

Mg^{2+}	7.50×10^{-3} mol	Cl^-	15.0×10^{-3} mol
Na^+	8.00×10^{-3} mol	OH^-	8.00×10^{-3} mol

PART 3

Assembling the Tools: We will solve the limiting reactant problem using the tools from Chapter 3 to determine the mass of product, $Mg(OH)_2$, and the moles of the excess reactant.

Solution: From the coefficients of the equation, 1 mol $Mg^{2+} \Leftrightarrow 2$ mol OH^-. This means that 7.50×10^{-3} mol Mg^{2+} would require 15.0×10^{-3} mol OH^-. But we have only 8.00×10^{-3} mol OH^-. Therefore, OH^- must be the limiting reactant. The amount of Mg^{2+} that does react to form $Mg(OH)_2$ can be found as follows.

$$8.00 \times 10^{-3} \text{ mol OH}^- \times \frac{1 \text{ mol Mg}^{2+}}{2 \text{ mol OH}^-} = 4.00 \times 10^{-3} \text{ mol Mg}^{2+}$$

(This amount reacts.)

Since we started with 7.50×10^{-3} mol Mg^{2+}, we can determine 3.50×10^{-3} mol Mg^{2+} will remain in solution. The mass of $Mg(OH)_2$ that forms is

$$4.00 \times 10^{-3} \text{ mol Mg}^{2+} \times \frac{1 \text{ mol Mg(OH)}_2}{1 \text{ mol Mg}^{2+}} \times \frac{58.32 \text{ g Mg(OH)}_2}{1 \text{ mol Mg(OH)}_2} = 0.233 \text{ g Mg(OH)}_2$$

PART 4

Assembling the Tools: We can calculate the final concentrations of Na^+ and Mg^{2+} using the definition of molarity, Equation 4.5, along with the information from Parts 2 and 3.

Solution: Let's begin by tabulating the number of moles of each ion left in solution after the reaction is complete.

Ion	Moles Present before Reaction	Moles That React	Moles Present after Reaction
Mg^{2+}	7.50×10^{-3} mol	4.00×10^{-3} mol	3.50×10^{-3} mol
Na^+	8.00×10^{-3} mol	0 mol	8.00×10^{-3} mol

The problem asks for the concentrations of the ions in the final reaction mixture, so we must now divide each quantity in the last column by the total volume of the final solution (40.0 mL + 25.0 mL = 65.0 mL or 0.0650 L). For example, for Mg^{2+}, its concentration is

$$\frac{3.50 \times 10^{-3} \text{ mol } Mg^{2+}}{0.0650 \text{ L soln}} = 0.0538 \ M \ Mg^{2+}$$

Performing similar calculations for the other ions gives the following:

Concentrations of Ions after Reaction

Mg^{2+} 0.0538 M	Na^+ 0.123 M

Are the Answers Reasonable? All the reasoning we've done seems to be correct, which is reassuring. For the amount of $Mg(OH)_2$ that forms, 0.001 mol would weigh approximately 0.06 g, so 0.004 mol would weigh 0.24 g. Our answer, 0.233 g $Mg(OH)_2$, is reasonable. The final concentrations of the spectator ions are lower than the initial concentrations, so that makes sense, too, because of the dilution effect.

Practice Exercise 4.43

A sample of a mixture containing $CaCl_2$ and $MgCl_2$ weighed 2.000 g. The sample was dissolved in water, and H_2SO_4 was added until the precipitation of $CaSO_4$ was complete.

$$CaCl_2(aq) + H_2SO_4(aq) \longrightarrow CaSO_4(s) + 2HCl(aq)$$

The $CaSO_4$ was filtered, dried completely, and weighed. A total of 0.736 g of $CaSO_4$ was obtained. What were the percentages by mass of $CaCl_2$ and $MgCl_2$ in the original sample?

| Summary

Organized by Learning Objective

Describe solutions qualitatively and quantitatively

A **solution** is a homogeneous mixture in which one or more **solutes** are dissolved in a **solvent**. A solution may be **dilute** or **concentrated**, depending on the amount of solute dissolved in a given amount of solvent. **Concentration** (e.g., **percentage concentration**) is a ratio of the amount of solute to either the amount of solvent or the amount of solution. The amount of solute required to give a **saturated** solution at a given temperature and volume is called the solute's **solubility**. Additional solute can be dissolved in an **unsaturated** solution, but **supersaturated** solutions are unstable and tend to give a **precipitate**.

Understand that electrolytes are soluble ionic compounds and nonelectrolytes do not produce ions in solutions

Substances that **dissociate** or **ionize** in water to produce cations and anions are **electrolytes**; those that do not are called **nonelectrolytes**. In water, ionic compounds are completely dissociated into ions and are **strong electrolytes**.

Write balanced molecular, ionic, and net ionic equations

Reactions that occur in solution between ions are called **ionic reactions**. Equations for these reactions can be written in three different ways. In **molecular equations**, complete formulas for all reactants and products are used. In an **ionic equation**, soluble strong electrolytes are written in dissociated (ionized) form; "molecular" formulas are used for solids, gases, and weak electrolytes. A **net ionic equation** is obtained by eliminating **spectator ions** from the ionic equation, and such an equation allows us to identify other combinations of reactants that give the same net reaction. An ionic or net ionic equation is balanced only if both atoms *and* charges are balanced.

Identify acids and bases based on their names and formulas and write balanced equations showing hydronium and hydroxide ions

An **acid** is a substance that produces hydronium ions, H_3O^+, when dissolved in water, and a **base** produces hydroxide ions, OH^-, when dissolved in water. The oxides of nonmetals are generally **acidic anhydrides** and react with water to give acids. Metal oxides are usually **basic anhydrides** because they tend to react with water to give metal hydroxides or bases.

Strong acids and **strong bases** are also strong electrolytes. **Weak acids** and **weak bases** are **weak electrolytes**, which are incompletely ionized in water. In a solution of a weak electrolyte there is a **chemical equilibrium (dynamic equilibrium)** between the non-ionized molecules of the solute and the ions formed by the reaction of the solute with water.

Learn how to translate formulas to names and names to formulas for common acids and bases

Binary acids are named with the prefix *hydro-* and the suffix *–ic* added to the stem of the nonmetal name, followed by the word *acid*. The names of the **oxoacids** are derived from the polyatomic ions: *-ate* anions give *-ic* acids, and *-ite* anions give *-ous* acids. The metal oxides are ionic compounds and named as ionic compounds. The molecular bases are named using the names of the molecules.

Use the principles of metathesis to predict reaction products and to plan a chemical synthesis

Metathesis or **double replacement** reactions take place when anions and cations of two salts change partners. A metathesis reaction will occur if there is a net ionic equation. This happens if (1) a **precipitate** forms from soluble reactants, (2) an **acid–base neutralization** occurs, (3) a gas is formed, or (4) a weak electrolyte forms from soluble strong electrolytes. You should learn the **solubility rules** (Table 4.1). Strong acids react with strong bases in neutralization reactions to produce a salt and water. Acids react with insoluble oxides and hydroxides to form water and the corresponding salt. Many acid–base neutralization reactions can be viewed as a type of metathesis reaction in which one product is water. Be sure to learn the reactions that produce gases in metathesis reactions, which are listed in Table 4.2.

Define and calculate molarity, and use it as a conversion factor

Molarity is the ratio of moles of solute to liters of solution. Molarity provides two conversion factors relating moles of solute and the volume of a solution.

$$\frac{\text{mol solute}}{1 \text{ L soln}} \quad \text{and} \quad \frac{1 \text{ L soln}}{\text{mol solute}}$$

Concentrated solutions of known molarity can be diluted quantitatively using volumetric glassware such as pipets and volumetric flasks. When a solution is diluted by adding solvent, the amount of solute doesn't change but the concentration decreases.

Learn to use molarity in stoichiometric calculations

Molarity can be used as a conversion factor for solutions to convert the volume of the solution into moles. In ionic reactions, the concentrations of the ions in a solution of a salt can be derived from the molar concentration of the salt, taking into account the number of ions formed per formula unit of the salt.

Understand the methods and calculations used in titrations and chemical analyses

Titration is a technique used to make quantitative measurements of the amounts of solutions needed to obtain a complete reaction. The apparatus is a long graduated tube called a **buret** that has a **stopcock** at the bottom, which is used to control the flow of **titrant**. In an acid–base titration, the **end point** is normally detected visually using an **acid–base indicator**. A color change indicates complete reaction, at which time the addition of titrant is stopped and the volume added is recorded.

TOOLS

Tools for Problem Solving
The following tools were introduced in this chapter. Study them carefully so you can select the appropriate tool when needed.

Criteria for a balanced ionic equation (Section 4.3)
To be balanced, an equation that includes the formulas of ions must satisfy two criteria: (1) the number of atoms of each kind must be the same on both sides of the equation, and (2) the net electrical charge shown on each side of the equation must be the same.

List of strong acids (Section 4.4)
The common strong acids are percholoric acid, $HClO_4$, chloric acid, $HClO_3$, hydrochloric acid, HCl, hydrobromic acid, HBr, hydroiodic acid, HI, nitric acid, HNO_3, and sulfuric acid, H_2SO_4.

Ionization of acids and bases in water (Section 4.4)
The anion of a strong acid is often given the symbol X^- while the symbol for the anion of a weak acid is A^-.

$$HX + H_2O \longrightarrow H_3O^+ + X^- \quad \text{strong acids}$$
$$HA + H_2O \rightleftharpoons H_3O^+ + A^- \quad \text{weak acids}$$

The cation of an ionic strong base is usually given the symbol M^{n+} and a molecular weak base is given the symbol B. Their reactions are

$$M(OH)_n \longrightarrow M^{n+} + n OH^- \quad \text{strong bases}$$
$$B + H_2O \rightleftharpoons BH^+ + OH^- \quad \text{weak bases}$$

Acid and polyatomic anion names (Section 4.5)

We use the names of the polyatomic anions to derive the names of acids. When the anion ends in *-ate* the acid ends in *-ic*. When the anion ends in *-ite* the acid ends in *-ous*.

Predicting net ionic equations (Section 4.6)

A net ionic equation will exist and a reaction will occur when:

- A precipitate is formed from a mixture of soluble reactants.
- An acid reacts with a base. *This includes strong or weak acids reacting with strong or weak bases or insoluble metal hydroxides or oxides.*
- A weak electrolyte is formed from a mixture of strong electrolytes.
- A gas is formed from a mixture of reactants.

Solubility rules (Section 4.6)

The rules in Table 4.1 serve as the tool we use to determine whether a particular salt is soluble in water.

Neutralization of an acid and a base (Section 4.6)

The reaction of an acid and a base that forms water and a salt is a neutralization reaction.

Gases formed in metathesis reactions (Section 4.6)

Hydrogen sulfide, hydrogen cyanide, carbon dioxide, sulfur dioxide, and ammonia can be formed in metathesis reactions.

Synthesis by metathesis (Section 4.6)

To synthesize a salt by a metathesis reaction, there are three approaches: form an insoluble salt, conduct a neutralization reaction to form the salt and water, and add an acid to a metal carbonate, sulfide or sulfite to form a gas and the salt.

Molarity (Section 4.7)

$$\text{Molarity}(M) = \frac{\text{moles of solute}}{\text{liters of solution}}$$

Dilution equation (Section 4.7)

$$V_{dil} \times M_{dil} = V_{conc} \times M_{conc}$$

Stoichiometry pathways (Section 4.8)

Figure 4.24 gives a summary of the various paths and conversion factors used in stoichiometry calculations to convert from one set of chemical units to another. This diagram adds molarity as a conversion factor to Figure 3.6.

Molarity of ions in a salt solution (Section 4.8)

When using a net ionic equation to work stoichiometry problems, we need the concentrations of the ions in the solutions. The concentration of a particular ion equals the concentration of the salt multiplied by the number of ions of that kind in one formula unit of the salt.

WileyPLUS = *WileyPLUS*, an online teaching and learning solution. *Note to instructors:* Many of the end-of-chapter problems are available for assignment via the WileyPLUS system. **www.wileyplus.com.** Review Problems are presented in pairs separated by blue rules. Answers to problems whose numbers appear in blue are given in Appendix B. More challenging problems are marked with an asterisk ✱.

| Review Questions

Describing Solutions

4.1 Define: **(a)** solvent, **(b)** solute, **(c)** concentration.

4.2 Describe: **(a)** concentrated, **(b)** dilute, **(c)** saturated, **(d)** unsaturated, **(e)** supersaturated, **(f)** solubility.

4.3 Why are chemical reactions often carried out using solutions?

4.4 Describe what will happen if a crystal of sugar is added to **(a)** a saturated sugar solution, **(b)** a supersaturated solution of sugar, and **(c)** an unsaturated solution of sugar.

4.5 What is the meaning of the term *precipitate*? What condition must exist for a precipitate to form spontaneously in a solution?

4.6 Explain how a solution can be called concentrated and dilute at the same time.

Electrolytes and Nonelectrolytes

4.7 Why is an electrolyte able to conduct electricity while a nonelectrolyte cannot?

4.8 Which compounds are likely to be electrolytes and which are likely to be nonelectrolytes? $CuBr_2$, $C_{12}H_{22}O_{11}$, CH_3OH, iron(II) chloride, $(NH_4)_2SO_4$, ethanol.

4.9 What does it mean when we say that an ion is "hydrated?"

4.10 Define "dissociation" as it applies to ionic compounds that dissolve in water. Why don't strong bases "ionize"?

Equations for Ionic Reactions

4.11 How can you tell that the following is a net ionic equation?

$$Al^{3+}(aq) + 3OH^-(aq) \longrightarrow Al(OH)_3(s)$$

4.12 What two conditions must be fulfilled by a balanced ionic equation? The following equation is not balanced. How do we know? Find the errors and fix them.

$$3Sc^{3+}(aq) + 2HPO_4^{2-}(aq) \longrightarrow Sc_3(PO_4)_2(s) + 2H^+(aq)$$

Introducing Acids and Bases

4.13 Give two general properties of an acid. Give two general properties of a base.

4.14 If you believed a solution was basic, which color litmus paper (blue or pink) would you use to test the solution to see whether you were correct? What would you observe if you've selected correctly? Why would the other color litmus paper not lead to a conclusive result?

4.15 How did Arrhenius define an acid and a base?

4.16 How does ionization differ from dissociation?

4.17 Which of the following undergo dissociation in water? Which undergo ionization? **(a)** NaOH, **(b)** HNO_3, **(c)** H_2SO_4

4.18 Which of the following oxides would yield an acidic solution when they react with water? Which would give a basic solution? **(a)** P_4O_{10}, **(b)** K_2O, **(c)** SeO_3, **(d)** Cl_2O_7

4.19 What is a *dynamic equilibrium*? Using acetic acid as an example, describe why all the $HC_2H_3O_2$ molecules are not ionized in water.

4.20 Why don't we use double arrows in the equation for the reaction of a strong acid with water?

4.21 Which of the following are strong acids? **(a)** HCN, **(b)** HNO_3, **(c)** H_2SO_3, **(d)** HCl, **(e)** $HCHO_2$, **(f)** HNO_2

4.22 Which are classified as strong bases when dissolved in water? **(a)** C_5H_5N, **(b)** $Ba(OH)_2$, **(c)** KOH, **(d)** $C_6H_5NH_2$, **(e)** Cs_2O, **(f)** N_2O_5

4.23 Methylamine, CH_3NH_2, reacts with hydronium ions in very much the same manner as ammonia.

$$CH_3NH_2(aq) + H_3O^+(aq) \longrightarrow CH_3NH_3^+(aq) + H_2O$$

On the basis of what you have learned so far in this course, sketch the molecular structures of CH_3NH_2 and $CH_3NH_3^+$ (the methylammonium ion).

4.24 A student was asked to draw the structure of the ion formed when propanoic acid (an organic acid similar to acetic acid) underwent ionization in water. Below is the structure the student drew. What is wrong with the structure, and what would you do to correct it?

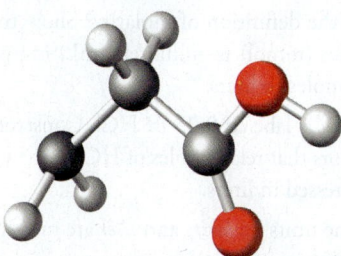

4.25 Would the molecule shown below be acidic or basic in water? What would you do to the structure to show what happens when the substance reacts with water? Write an equation for the ionization of this compound in water. (The compound is a weak electrolyte.)

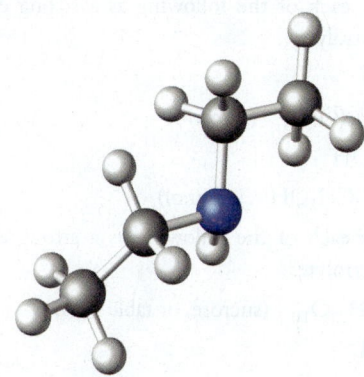

Acid–Base Nomenclature

4.26 Explain the difference between the names of $H_2Se(g)$ and $H_2Se(aq)$.

4.27 Iodine, like chlorine, forms four oxoacids and one binary acid. What are the formulas and the names of the acids? Explain how the names of the oxoacids vary as the number of oxygen increases.

4.28 For the acids in the preceding question, **(a)** write the formulas and, **(b)** name the ions formed by removing a hydrogen ion (H^+) from each acid.

4.29 Explain how the two acid salts of phosphoric acids are formed from the reaction of H_3PO_4 with NaOH, and name them.

Double Replacement (Metathesis) Reactions

4.30 What factors lead to the existence of a net ionic equation in a reaction between ions?

4.31 Explain the three processes that can drive an ionic reaction.

4.32 Silver bromide is "insoluble." What does this mean about the concentrations of Ag^+ and Br^- in a saturated solution of AgBr? What makes it possible for a precipitate of AgBr to form when solutions of the soluble salts $AgNO_3$ and NaBr are mixed?

4.33 What gas is formed if HCl is added to **(a)** $NaHCO_3$, **(b)** Na_2S, and **(c)** potassium sulfite?

Molarity

4.34 What is the definition of molarity? Show that the ratio of millimoles (mmol) to milliliters (mL) is equivalent to the ratio of moles to liters.

4.35 A solution is labeled 0.25 M HCl. Construct two conversion factors that relate moles of HCl to the volume of solution expressed in liters.

4.36 When the units *molarity* and *liter* are multiplied, what are the resulting units?

Review Problems

Electrolytes and Nonelectrolytes

4.43 Classify each of the following as a strong electrolyte or nonelectrolyte.
(a) KCl
(b) $C_3H_5(OH)_3$ (glycerin)
(c) NaOH
(d) CH_3CH_2OH (ethanol)

4.44 Classify each of the following as a strong electrolyte or nonelectrolyte.
(a) $C_{12}H_{22}O_{11}$ (sucrose, or table sugar)
(b) NH_3
(c) CH_3OH (methyl alcohol)
(d) H_2SO_4

4.45 Write equations for the dissociation of the following in water: **(a)** LiF, **(b)** $KMnO_4$, **(c)** potassium carbonate, **(d)** cobalt(II) nitrate.

4.46 Write equations for the dissociation of the following in water: **(a)** $CaCl_2$, **(b)** $(NH_4)_2SO_4$, **(c)** sodium acetate, **(d)** copper(II) perchlorate.

Equations for Ionic Reactions

4.47 Write balanced ionic and net ionic equations for these reactions.
(a) $(NH_4)_2CO_3(aq) + BaCl_2(aq) \longrightarrow$
$$NH_4Cl(aq) + BaCO_3(s)$$
(b) $CuCl_2(aq) + NaOH(aq) \longrightarrow$
$$Cu(OH)_2(s) + NaCl(aq)$$

4.37 When a solution labeled 0.50 M HNO_3 is diluted with water to give 0.25 M HNO_3, what happens to the number of moles of HNO_3 in the solution?

4.38 Two bottles, A and B, are labeled "0.100 M $CaCl_2$" and "0.200 M $CaCl_2$," respectively. Both solutions in the bottles contain the same number of moles of $CaCl_2$. If solution A has a volume of 50.0 mL, what is the volume of solution B?

4.39 Describe the steps to take in diluting a solution of 0.500 M HCl to make 250 mL of 0.100 M HCl.

Titrations and Chemical Analyses

4.40 What is the difference between a qualitative analysis and a quantitative analysis?

4.41 Describe each of the following: **(a)** buret, **(b)** titration, **(c)** titrant, and **(d)** end point.

4.42 What is the function of an indicator in a titration? What color is phenolphthalein in **(a)** an acidic solution and **(b)** a basic solution?

(c) $FeSO_4(aq) + Na_3PO_4(aq) \longrightarrow$
$$Fe_3(PO_4)_2(s) + Na_2SO_4(aq)$$
(d) $AgC_2H_3O_2(aq) + NiCl_2(aq) \longrightarrow$
$$AgCl(s) + Ni(C_2H_3O_2)_2(aq)$$

4.48 Write balanced ionic and net ionic equations for these reactions.
(a) $CuSO_4(aq) + BaCl_2(aq) \longrightarrow$
$$CuCl_2(aq) + BaSO_4(s)$$
(b) $Fe(NO_3)_3(aq) + LiOH(aq) \longrightarrow$
$$LiNO_3(aq) + Fe(OH)_3(s)$$
(c) $Na_3PO_4(aq) + CaCl_2(aq) \longrightarrow$
$$Ca_3(PO_4)_2(s) + NaCl(aq)$$
(d) $Na_2S(aq) + AgC_2H_3O_2(aq) \longrightarrow$
$$NaC_2H_3O_2(aq) + Ag_2S(s)$$

4.49 The following equation shows the formation of cobalt(II) hydroxide, a compound used to improve the drying properties of lithographic inks.

$Co^{2+}(aq) + 2Cl^-(aq) + 2Na^+(aq) + 2OH^-(aq) \longrightarrow$
$$Co(OH)_2(s) + 2Na^+(aq) + 2Cl^-(aq)$$

Which are the spectator ions? Write the net ionic equation.

4.50 Microscopic imperfections in glass are usually harmless, but imperfections of nickel(II) sulfide can cause the glass to crack spontaneously. To study this process, nickel(II) sulfide can be synthesized from sodium sulfide and nickel(II) sulfate.

$Ni^{2+}(aq) + SO_4^{2-}(aq) + 2Na^+(aq) + S^{2-}(aq) \longrightarrow$
$$NiS(s) + 2Na^+(aq) + SO_4^{2-}(aq)$$

Which are the spectator ions? Write the net ionic equation.

Introducing Acids and Bases

4.51 Pure $HClO_4$ is a molecular substance. In water it is a strong acid. Write an equation for its ionization in water.

4.52 Write an equation for the ionization of hydrogen bromide, a molecular substance and a strong acid, in water.

4.53 Pure HI is a gas at room temperature and reacts with water as a strong acid. Write an equation for its reaction with water.

4.54 When chloric acid reacts with water, it reacts as a strong acid. Write an equation for this reaction.

4.55 Hydrazine is a toxic substance that can form when household ammonia is mixed with a bleach such as Clorox. Its formula is N_2H_4, and it is a weak base. Write a chemical equation showing its reaction with water.

4.56 Pyridine, C_5H_5N, is a fishy-smelling compound used as an intermediate in making insecticides. It is a weak base. Write a chemical equation showing its reaction with water.

4.57 Nitrous acid, HNO_2, is a weak acid that can form when sodium nitrite, a meat preservative, reacts with stomach acid (HCl). Write an equation showing the ionization of HNO_2 in water.

4.58 Pentanoic acid, $HC_5H_9O_2$, is found in a plant called valerian, which cats seem to like almost as much as catnip, and is a weak acid. Write an equation showing its reaction with water.

4.59 Atmospheric carbon dioxide dissolves in raindrops, resulting in a solution of carbonic acid that makes rain slightly acidic. Since carbonic acid is a diprotic acid, write the chemical equations that describe its stepwise ionization.

4.60 Arsenic acid, H_3AsO_4, is a very toxic weak acid. It undergoes ionization in three steps. Write chemical equations for the equilibria involved in each of these reactions.

Acid–Base Nomenclature

4.61 Name these acids: **(a)** $H_2S(g)$, **(b)** $H_2S(aq)$.

4.62 Name these acids: **(a)** $HF(g)$, **(b)** $HF(aq)$.

4.63 Name these acids that bromine forms. **(a)** $HBrO_4$, **(b)** $HBrO_3$, **(c)** $HBrO_2$, **(d)** $HBrO$, **(e)** HBr.

4.64 Selenium forms a number of acids, name them: **(a)** H_2Se, **(b)** $HSeO_3$, **(c)** $HSeO_2$.

4.65 For the acids in Problem 4.63, name the ions formed by removing a hydrogen ion (H^+) from each acid.

4.66 For the acids in Problem 4.64, name the ions formed by removing a hydrogen ion (H^+) from each acid.

4.67 Write the formula for **(a)** chromic acid, **(b)** carbonic acid, and **(c)** oxalic acid.

4.68 Write the formula for **(a)** permanganic acid, **(b)** sulfurous acid, and **(c)** dichromic acid.

4.69 Name the following acid salts: **(a)** $NaHCO_3$, **(b)** KH_2PO_4, **(c)** $(NH_4)_2HPO_4$.

4.70 Name the following acid salts: **(a)** $KHSO_4$, **(b)** $LiHSO_3$, **(c)** $PbHAsO_4$.

4.71 Name the following oxoacids and give the names and formulas of the salts formed from them by neutralization with NaOH: **(a)** $HOCl$, **(b)** HIO_2, **(c)** $HClO_4$.

4.72 Name the following oxoacids and give the names and formulas of the salts formed from them by neutralization with KOH: **(a)** $HClO_2$, **(b)** HNO_2, **(c)** $HBrO_4$.

4.73 The formula for the sulfite ion is SO_3^{2-}. What is the formula for sulfuric acid?

4.74 The formula for the arsenate ion is AsO_4^{3-}. What is the formula for arsenous acid?

4.75 Butanoic acid (also called butyric acid), $HC_4H_7O_2$, gives rancid butter its bad odor. What is the name of the salt $NaC_4H_7O_2$?

4.76 Oxalic acid, $H_2C_2O_4$, is the poison in rhubarb leaves. Name its potassium salt.

4.77 Sodium formate, $NaCHO_2$, is used in fabric dyeing. The acid form is found in the venom of ant bites. What is the name of the acid that this salt can be made from?

4.78 Potassium stearate, $KC_{18}H_{36}O_2$, is an effective soap for washing your hands or clothes. What acid is reacted with KOH to make this compound?

Precipitation Reactions

4.79 Use the solubility rules to decide which compounds are *soluble* in water.

  (a) $Ca(NO_3)_2$ **(d)** silver nitrate
  (b) $FeCl_2$ **(e)** barium sulfate
  (c) $Ni(OH)_2$ **(f)** copper(II) carbonate

4.80 Predict which compounds are *soluble* in water.

  (a) $HgBr_2$ **(d)** ammonium phosphate
  (b) $Sr(NO_3)_2$ **(e)** lead(II) iodide
  (c) Hg_2Br_2 **(f)** lead(II) acetate

4.81 Complete and balance the following molecular equations, being careful to apply the solubility rules. Write balanced ionic and net ionic equations for the reactions.

  (a) $FeSO_4(aq) + K_3PO_4(aq) \longrightarrow$
  (b) $AgC_2H_3O_2(aq) + AlCl_3(aq) \longrightarrow$
  (c) chromium(III) chloride + barium hydroxide $\longrightarrow$

4.82 Complete and balance the following molecular equations, being careful to apply the solubility rules. Write balanced ionic and net ionic equations for the reactions.

  (a) $Fe(NO_3)_3(aq) + KOH(aq) \longrightarrow$
  (b) $Na_3PO_4(aq) + SrCl_2(aq) \longrightarrow$
  (c) lead(II) acetate + ammonium sulfate $\longrightarrow$

4.83 Write the molecular, ionic, and net ionic equations for the reaction between NaCl and $AgNO_3$ solutions that forms solid AgCl leaving $NaNO_3$ in water.

4.84 Write the molecular, ionic, and net ionic equations for the aqueous reaction between $NaCO_3$ and $CaCl_2$ that forms solid $CaCO_3$ and a sodium chloride solution.

4.85 Aqueous solutions of sodium sulfide and copper(II) nitrate are mixed. A precipitate of copper(II) sulfide forms at once. The solution that remains contains sodium nitrate. Write the molecular, ionic, and net ionic equations for this reaction.

4.86 If an aqueous solution of iron(III) sulfate (a compound used in dyeing textiles and also for etching aluminum) is mixed with a solution of barium chloride, a precipitate of barium sulfate forms and the solution that remains contains iron(III) chloride. Write the molecular, ionic, and net ionic equations for this reaction.

Acid–Base Reactions

4.87 Complete and balance the following equations. For each, write the molecular, ionic, and net ionic equations. (All of the products are soluble in water.)

(a) $Ca(OH)_2(aq) + HNO_3(aq) \longrightarrow$

(b) $Al_2O_3(s) + HCl(aq) \longrightarrow$

(c) $Zn(OH)_2(s) + H_2SO_4(aq) \longrightarrow$

4.88 Complete and balance the following equations. For each, write the molecular, ionic, and net ionic equations. (All of the products are soluble in water.)

(a) $HC_2H_3O_2(aq) + Mg(OH)_2(s) \longrightarrow$

(b) $HClO_4(aq) + NH_3(aq) \longrightarrow$

(c) $H_2CO_3(aq) + NH_3(aq) \longrightarrow$

***4.89** How would the electrical conductivity of a solution of barium hydroxide change as a solution of sulfuric acid is added slowly to it? Use a net ionic equation to justify your answer.

4.90 How would the electrical conductivity of a solution of acetic acid change as a solution of ammonia is added slowly to it? Use a net ionic equation to justify your answer.

Ionic Reactions That Produce Gases

4.91 Write balanced net ionic equations for the reaction between:

(a) $HNO_3(aq) + K_2CO_3(aq)$

(b) $Ca(OH)_2(aq) + NH_4NO_3(aq)$

4.92 Write balanced net ionic equations for the reaction between:

(a) $H_2SO_4(aq) + NaHSO_3(aq)$

(b) $HNO_3(aq) + (NH_4)_2CO_3(aq)$

4.93 Sodium sulfide and hydrochloric acid react to form a gas. Write the balanced molecular and net ionic equation for the reaction.

4.94 Write the balanced molecular and net ionic equation for the reaction between sulfuric acid and potassium cyanide.

Double Replacement (Metathesis) Reactions

4.95 Explain why the following reactions take place.

(a) $CrCl_3 + 3NaOH \longrightarrow Cr(OH)_3 + 3NaCl$

(b) $ZnO + 2HBr \longrightarrow ZnBr_2 + H_2O$

4.96 Explain why the following reactions take place.

(a) $MnCO_3 + H_2SO_4 \longrightarrow MnSO_4 + H_2O + CO_2$

(b) $Na_2C_2O_4 + 2HNO_3 \longrightarrow 2NaNO_3 + H_2C_2O_4$

4.97 Complete and balance the molecular, ionic, and net ionic equations for the following reactions.

(a) $HNO_3 + Cr(OH)_3 \longrightarrow$

(b) $HClO_4 + NaOH \longrightarrow$

(c) $Cu(OH)_2 + HC_2H_3O_2 \longrightarrow$

(d) $ZnO + H_2SO_4 \longrightarrow$

4.98 Complete and balance the molecular, ionic, and net ionic equations for the following reactions.

(a) $NaHSO_3 + HBr \longrightarrow$

(b) $(NH_4)_2CO_3 + NaOH \longrightarrow$

(c) $(NH_4)_2CO_3 + Ba(OH)_2 \longrightarrow$

(d) $FeS + HCl \longrightarrow$

4.99 Write balanced molecular, ionic, and net ionic equations for the following pairs of reactants. If all ions cancel, indicate that no reaction (N.R.) takes place.

(a) sodium sulfite and barium nitrate

(b) formic acid ($HCHO_2$) and potassium carbonate

(c) ammonium bromide and lead(II) acetate

(d) ammonium perchlorate and copper(II) nitrate

4.100 Write balanced molecular, ionic, and net ionic equations for the following pairs of reactants. If all ions cancel, indicate that no reaction (N.R.) takes place.

(a) ammonium sulfide and sodium hydroxide

(b) chromium(III) sulfate and potassium carbonate

(c) silver nitrate and chromium(III) acetate

(d) strontium hydroxide and magnesium chloride

4.101 Choose reactants that would yield the following net ionic equations. Write molecular equations for each.

(a) $HCO_3^-(aq) + H^+(aq) \longrightarrow H_2O + CO_2(g)$

(b) $Fe^{2+}(aq) + 2OH^-(aq) \longrightarrow Fe(OH)_2(s)$

(c) $Ba^{2+}(aq) + SO_3^{2-}(aq) \longrightarrow BaSO_3(s)$

(d) $2Ag^+(aq) + S^{2-}(aq) \longrightarrow Ag_2S(s)$

(e) $ZnO(s) + 2H^+(aq) \longrightarrow Zn^{2+}(aq) + H_2O$

4.102 Suppose that you wanted to prepare copper(II) carbonate by a precipitation reaction involving Cu^{2+} and CO_3^{2-}. Which of the following pairs of reactants could you use as solutes? For the pairs you did not use, explain why they are not appropriate for synthesizing $CuCO_3$.

(a) $Cu(OH)_2 + Na_2CO_3$ (b) $CuSO_4 + (NH_4)_2CO_3$

(c) $Cu(NO_3)_2 + CaCO_3$ (d) $CuCl_2 + K_2CO_3$

(e) $CuS + NiCO_3$

Molarity

4.103 Calculate the molarity of a solution prepared by dissolving

(a) 0.17 mol sulfuric acid in 85 mL of solution.

(b) 7.5×10^{-3} mol of sodium nitrate in a total volume of 13.5 mL of solution.

4.104 Calculate the molarity of a solution that contains

(a) 2.00×10^{-3} mol cobalt(II) sulfate in 35.0 mL of solution.

(b) 2.75 mole potassium hydroxide in a total volume of 135 mL of solution.

4.105 Calculate the molarity of a solution prepared by dissolving

(a) 4.00 g of sodium hydroxide in a total volume of 100.0 mL of solution.

(b) 16.0 g of calcium chloride in a total volume of 250.0 mL of solution.

4.106 Calculate the molarity of a solution that contains

(a) 3.60 g of sulfuric acid in 450.0 mL of solution.

(b) 0.001 g of iron(II) nitrate in 12.0 mL of solution.

4.107 How many milliliters of $0.265\ M\ NaC_2H_3O_2$ are needed to supply 14.3 g $NaC_2H_3O_2$?

4.108 How many milliliters of $0.615\ M\ HNO_3$ contain 1.67 g HNO_3?

4.109 Calculate the number of grams of each solute that has to be taken to make each of the following solutions.

(a) 125 mL of $0.200\ M$ NaCl

(b) 250.0 mL of $0.360\ M\ C_6H_{12}O_6$

(c) 250.0 mL of $0.250\ M\ H_2SO_4$

4.110 How many grams of solute are needed to make each of the following solutions?

(a) 250.0 mL of $0.100\ M$ potassium sulfate

(b) 100.0 mL of $0.250\ M$ iron(III) chloride

(c) 500.0 mL of $0.400\ M$ barium acetate

4.111 If 25.0 mL of $0.56\ M\ H_2SO_4$ is diluted to a volume of 125 mL, what is the molarity of the resulting solution?

4.112 A 150 mL sample of $0.450\ M\ HNO_3$ is diluted to 450 mL. What is the molarity of the resulting solution?

4.113 To what volume must 25.0 mL of $18.0\ M\ H_2SO_4$ be diluted to produce $1.50\ M\ H_2SO_4$?

4.114 To what volume must 50.0 mL of $1.50\ M$ HCl be diluted to produce $0.200\ M$ HCl?

4.115 How many milliliters of water must be added to 150.0 mL of $2.50\ M$ KOH to give a $1.00\ M$ solution? (Assume the volumes are additive.)

4.116 How many milliliters of water must be added to 120.0 mL of $1.50\ M$ HCl to give $1.00\ M$ HCl? (Assume the volumes are additive.)

Solution Stoichiometry

4.117 Calculate the number of moles of each of the ions in the following solutions.

(a) 32.3 mL of $0.455\ M\ CaCl_2$

(b) 50.0 mL of $0.408\ M\ AlCl_3$

4.118 Calculate the number of moles of each of the ions in the following solutions.

(a) 18.5 mL of $0.402\ M\ (NH_4)_2CO_3$

(b) 30.0 mL of $0.359\ M\ Al_2(SO_4)_3$

4.119 Calculate the concentrations of each of the ions in (a) $0.25\ M\ Cr(NO_3)_2$, (b) $0.10\ M\ CuSO_4$, (c) $0.16\ M\ Na_3PO_4$, (d) $0.075\ M\ Al_2(SO_4)_3$.

4.120 Calculate the concentrations of each of the ions in (a) $0.060\ M\ Ca(OH)_2$, (b) $0.15\ M\ FeCl_3$, (c) $0.22\ M\ Cr_2(SO_4)_3$, (d) $0.60\ M\ (NH_4)_2SO_4$.

4.121 In a solution of $Al_2(SO_4)_3$ the Al^{3+} concentration is $0.125\ M$. How many grams of $Al_2(SO_4)_3$ are in 50.0 mL of this solution?

4.122 In a solution of $NiCl_2$, the Cl^- concentration is $0.0556\ M$. How many grams of $NiCl_2$ are in 225 mL of this solution?

4.123 How many milliliters of $0.258\ M\ NiCl_2$ solution are needed to react completely with 20.0 mL of $0.153\ M\ Na_2CO_3$ solution? How many grams of $NiCO_3$ will be formed? The reaction is

$$Na_2CO_3(aq) + NiCl_2(aq) \longrightarrow NiCO_3(s) + 2NaCl(aq)$$

4.124 How many milliliters of $0.100\ M$ NaOH are needed to completely neutralize 25.0 mL of $0.250\ M\ H_2C_4H_4O_6$? The reaction is

$$2NaOH(aq) + H_2C_4H_4O_6(aq) \longrightarrow$$
$$Na_2C_4H_4O_6(aq) + 2H_2O$$

4.125 What is the molarity of an aqueous solution of potassium hydroxide if 21.34 mL is exactly neutralized by 20.78 mL of $0.116\ M$ HCl? Write and balance the molecular equation for the reaction.

4.126 What is the molarity of an aqueous phosphoric acid solution if 12.88 mL is completely neutralized by 26.04 mL of $0.1024\ M$ NaOH? Write and balance the molecular equation for the reaction.

4.127 Aluminum sulfate, $Al_2(SO_4)_3$, is used in water treatment to remove fine particles suspended in the water. When made basic, a gel-like precipitate forms that removes the fine particles as it settles. In an experiment, a student planned to react $Al_2(SO_4)_3$ with $Ba(OH)_2$. How many grams of $Al_2(SO_4)_3$ are needed to react with 85.0 mL of $0.0500\ M\ Ba(OH)_2$?

4.128 How many grams of baking soda, $NaHCO_3$, are needed to react with 162 mL of stomach acid having an HCl concentration of $0.052\ M$?

4.129 How many milliliters of $0.150\ M\ FeCl_3$ solution are needed to react completely with 20.0 mL of $0.0450\ M$

AgNO$_3$ solution? How many grams of AgCl will be formed? The net ionic equation for the reaction is

$$Ag^+(aq) + Cl^-(aq) \longrightarrow AgCl(s)$$

4.130 How many grams of cobalt(II) chloride are needed to react completely with 60.0 mL of 0.200 M KOH solution? The net ionic equation for the reaction is

$$Co^{2+}(aq) + 2OH^-(aq) \longrightarrow Co(OH)_2(s)$$

4.131 Consider the reaction of aluminum chloride with silver acetate. How many milliliters of 0.250 M aluminum chloride would be needed to react completely with 20.0 mL of 0.500 M silver acetate solution? The net ionic equation for the reaction is

$$Ag^+(aq) + Cl^-(aq) \longrightarrow AgCl(s)$$

4.132 How many milliliters of ammonium sulfate solution having a concentration of 0.250 M are needed to react completely with 50.0 mL of 1.00 M sodium hydroxide solution? The net ionic equation for the reaction is

$$NH_4^+(aq) + OH^-(aq) \longrightarrow NH_3(g) + H_2O$$

4.133 Suppose that 4.00 g of solid Fe$_2$O$_3$ is added to 25.0 mL of 0.500 M HCl solution. What will the concentration of the Fe^{3+} be when all of the HCl has reacted? What mass of Fe$_2$O$_3$ will not have reacted?

4.134 Suppose 3.50 g of solid Mg(OH)$_2$ is added to 30.0 mL of 0.500 M H$_2$SO$_4$ solution. What will the concentration of Mg^{2+} be when all of the acid has been neutralized? How many grams of Mg(OH)$_2$ will not have dissolved?

Titrations and Chemical Analyses

4.135 In a titration, 23.25 mL of 0.105 M NaOH was needed to react with 21.45 mL of HCl solution. What is the molarity of the acid?

4.136 A 12.5 mL sample of vinegar, containing acetic acid, was titrated using 0.504 M NaOH solution. The titration re-

quired 20.65 mL of the base. What was the molar concentration of acetic acid in the vinegar?

4.137 Lactic acid, HC$_3$H$_5$O$_3$, is a monoprotic acid that forms when milk sours. An 18.5 mL sample of a solution of lactic acid required 17.25 mL of 0.155 M NaOH to reach an end point in a titration. **(a)** How many moles of lactic acid were in the sample? **(b)** How many grams of lactic acid were in the sample?

4.138 Oxalic acid, a diprotic acid having the formula H$_2$C$_2$O$_4$, is used to clean the rust out of radiators in cars. A sample of an oxalic acid mixture was analyzed by titrating a 0.1000 g sample dissolved in water with 0.0200 M NaOH. A volume of 15.20 mL of the base was required to completely neutralize the oxalic acid. What was the percentage by mass of oxalic acid in the sample?

*4.139 A certain lead ore contains the compound PbCO$_3$. A sample of the ore weighing 1.526 g was treated with nitric acid, which dissolved the PbCO$_3$. The resulting solution was filtered from undissolved rock and required 29.22 mL of 0.122 M Na$_2$SO$_4$ to completely precipitate all of the lead as PbSO$_4$. **(a)** How many moles of lead were in the ore sample? **(b)** How many grams of lead were in the ore sample? **(c)** What is the percentage by mass of lead in the ore? **(d)** Would you expect the same results if the solid PbSO$_4$ was collected, washed, dried, and weighed and the final mass was used to answer this question?

*4.140 An ore of barium contains BaCO$_3$. A 1.542 g sample of the ore was treated with HCl to dissolve the BaCO$_3$. The resulting solution was filtered to remove insoluble material and then treated with H$_2$SO$_4$ to precipitate BaSO$_4$. The precipitate was filtered, dried, and found to weigh 1.159 g. **(a)** How many moles of barium were in the ore sample? **(b)** How many grams of barium were in the ore sample? **(c)** What is the percentage by mass of barium in the ore? **(d)** If the molarity and volume of the sulfuric acid needed for the precipitation was noted, could you get the same answers?

| Additional Exercises

4.141 If a solution of sodium phosphate (also known as trisodium phosphate, or TSP), Na$_3$PO$_4$, is poured into seawater, precipitates of calcium phosphate and magnesium phosphate are formed. (Magnesium and calcium ions are among the principal ions found in seawater.) Write net ionic equations for these reactions.

4.142 With which of the following will the weak acid HCHO$_2$ react? For those with which there is a reaction, write the formulas of the products. **(a)** KOH, **(b)** MgO, **(c)** NH$_3$

*4.143 Suppose that 25.0 mL of 0.440 M NaCl is added to 25.0 mL of 0.320 M AgNO$_3$.
(a) How many moles of AgCl would precipitate?
(b) What would be the concentrations of each of the ions in the reaction mixture after the reaction?

*4.144 A mixture is prepared by adding 25.0 mL of 0.185 M Na$_3$PO$_4$ to 34.0 mL of 0.140 M Ca(NO$_3$)$_2$.
(a) What mass of Ca$_3$(PO$_4$)$_2$ will be formed?
(b) What will be the concentrations of each of the ions in the mixture after reaction?

4.145 Classify each of the following as a strong electrolyte, weak electrolyte, or nonelectrolyte.
(a) LiBr   **(b)** glucose (a sugar)
(c) Ca(OH)$_2$ **(d)** NH$_3$

*4.146 Aspirin is a monoprotic acid called acetylsalicylic acid. Its formula is HC$_9$H$_7$O$_4$. A certain pain reliever was analyzed for aspirin by dissolving 0.118 g of it in water and titrating it with 0.0300 M KOH solution. The titration required 14.76 mL of base. What is the percentage by weight of aspirin in the drug?

*4.147 In an experiment, 40.0 mL of 0.270 *M* barium hydroxide was mixed with 25.0 mL of 0.330 *M* aluminum sulfate.

(a) Write the net ionic equation for the reaction that takes place.

(b) What is the total mass of precipitate that forms?

(c) What are the molar concentrations of the ions that remain in the solution after the reaction is complete?

*4.148 How many milliliters of 0.10 *M* HCl must be added to 50.0 mL of 0.40 *M* HCl to give a final solution that has a molarity of 0.25 *M*?

Multi-Concept Problems

*4.150 Magnesium sulfate forms a hydrate known as *Epsom salts*. A student dissolved 1.24 g of this hydrate in water and added a barium chloride solution until the precipitation reaction was complete. The precipitate was filtered, dried, and found to weigh 1.174 g. Determine the formula for Epsom salts.

*4.151 Qualitative analysis of an unknown acid found only carbon, hydrogen, and oxygen. In a quantitative analysis, a 10.46 mg sample was burned in oxygen and gave 22.17 mg CO_2 and 3.40 mg H_2O. The molecular mass was determined to be 166 g mol^{-1}. When a 0.1680 g sample of the acid was titrated with 0.1250 *M* NaOH, the end point was reached after 16.18 mL of the base had been added. (a) What is the molecular formula for the acid? (b) Is the acid mono-, di-, or triprotic?

Exercises in Critical Thinking

4.154 Is a 0.250 *M* KCl solution concentrated or dilute? Explain your answer.

4.155 Compare the advantages and disadvantages of performing a titration using the mass of the sample and titrant rather than the volume.

4.156 What kinds of experiments could you perform to measure the solubility of a substance in water? Describe the procedure you would use and the measurements you would make. What factors would limit the precision of your measurements?

*4.157 Describe experiments, both qualitative and quantitative, that you could perform to show that lead chloride is more soluble in water than lead iodide.

4.158 How could you check the accuracy of a 100 mL volumetric flask?

4.159 A white substance was known to be either magnesium hydroxide, calcium hydroxide, or zinc oxide. A 1.25 g sample of the substance was mixed with 50.0 mL of water and 100.0 mL of 0.500 *M* HCl was added. The mixture was stirred, causing all of the sample to dissolve. The excess HCl in the solution required 36.0 mL of 0.200 *M*

4.149 Write an equation for the reaction of sodium hydroxide with the chloride salt of the cation with the following structure. How many grams of sodium hydroxide are needed to completely react with 12.4 g of the salt?

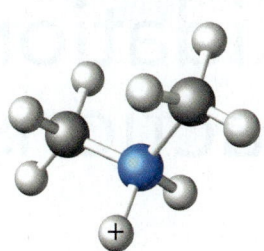

*4.152 A mixture was known to contain both KNO_3 and K_2SO_3. To 0.486 g of the mixture, dissolved in enough water to give 50.00 mL of solution, was added 50.00 mL of 0.150 *M* HCl (an excess of HCl). The reaction mixture was heated to drive off all of the SO_2, and then 25.00 mL of the reaction mixture was titrated with 0.100 *M* KOH. The titration required 13.11 mL of the KOH solution to reach an end point. What was the percentage by mass of K_2SO_3 in the original mixture of KNO_3 and K_2SO_3?

*4.153 A 25.0 mL sample of vinegar with a density of 1.01 g mL^{-1}, containing acetic acid, was titrated using 0.504 *M* NaOH solution. The titration required 42.54 mL of the base. What was the molar concentration and the percent by mass of acetic acid in the vinegar?

NaOH for complete neutralization in a titration. What was the identity of the white substance?

*4.160 A steel cylinder with a diameter of 10.0 cm and a length of 30.0 cm was coated with a thin layer of zinc. The cylinder was dipped briefly in hydrochloric acid, which completely dissolved the coating of zinc, giving Zn^{2+} in the solution. When made basic, the mixture yielded 0.226 g of a $Zn(OH)_2$ precipitate. If a zinc atom occupies an average area of 6.31×10^4 pm^2 on the surface of the cylinder, how many layers of atoms thick was the zinc coating on the cylinder?

*4.161 Suppose a classmate doubted that an equilibrium really exists between acetic acid and its ions in an aqueous solution. What argument would you use to convince the person that such an equilibrium does exist?

*4.162 When Arrhenius originally proposed that ions exist in solution, his idea was not well received. Propose another explanation for the conduction of electricity in molten salts and aqueous salt solutions.

*4.163 Carbon dioxide is one obvious contributor to excessive global warming. What is your plan for controlling CO_2 emissions? What are the advantages and disadvantages of your plan?

5

Oxidation–Reduction Reactions

Chapter Outline

James A Isbell/Shutterstock

This footprint glows dramatically due to traces of blood reacting with a luminol solution, perhaps left near a crime scene. The blue glow is a chemiluminescent process (a light emitting chemical reaction) that is enhanced by the iron in hemoglobin. Popular TV programs based on crime scene investigations have made this sight common knowledge. We will see how this reaction really works since it is one of many oxidation–reduction reactions covered in this chapter. You already learned about some important reactions that take place in aqueous solutions in Chapter 4. This chapter expands on that knowledge with a discussion of a class of reactions that involve the transfer of one or more electrons from one reactant to another. In the luminol reaction, oxygen ultimately reacts with luminol in a very energetic oxidation–reduction reaction that produces light. In biological systems oxidation–reduction reactions are involved in the metabolism of food, and many other reactions that are necessary for life. In addition to biochemical reactions, there are many other oxidation–reduction reactions that we encounter every day. These include many practical examples that range from combustion of fuels for cooking food and moving lawnmowers to launching space shuttles. Oxidation–reduction reactions are found in batteries, solar cells, disinfectants, cleaners, and water purification. Although we will introduce you to some of them in this chapter, we will wait until Chapter 19 to discuss some of the others.

This Chapter in Context

LEARNING OBJECTIVES

After reading this chapter, you should be able to:

- define oxidation, reduction, oxidizing agents, reducing agents, and oxidation numbers
- balance equations for oxidation–reduction reactions in acidic or basic solutions
- illustrate, at the molecular level, how a metal reacts with an acid
- explain the activity series of metals and use it to predict the product of a redox reaction involving a metal
- describe the reaction with oxygen of organic compounds, metals, and nonmetals
- perform calculations involving stoichiometry of a redox reaction

5.1 | Oxidation–Reduction Reactions

Among the first reactions studied by early scientists were those that involved oxygen. The combustion of fuels and the reactions of metals with oxygen to give oxides were described by the word oxidation. The removal of oxygen from metal oxides to give the metals in their elemental forms was described as reduction.

In 1789, the French chemist Antoine Lavoisier discovered that combustion involves the reaction of chemicals in various fuels, like wood and coal, not just with the air but more specifically with the oxygen in air. Over time, scientists came to realize that such reactions were actually special cases of a much more general phenomenon, one in which electrons are transferred from one substance to another. Collectively, electron transfer reactions came to be called **oxidation–reduction reactions**, or simply **redox reactions**. The term **oxidation** was used to describe the loss of electrons by one reactant, and **reduction** to describe the gain of electrons by another. For example, the reaction between sodium and chlorine to yield sodium chloride involves a loss of electrons by sodium (*oxidation* of sodium) and a gain of electrons by chlorine (*reduction* of chlorine). We can write these changes in equation form including the electrons, which we represent by the symbol e^-.

$$Na \longrightarrow Na^+ + e^- \quad \text{(oxidation)}$$
$$Cl_2 + 2e^- \longrightarrow 2Cl^- \quad \text{(reduction)}$$

We say that sodium, Na, is oxidized and chlorine, Cl_2, is reduced.

Oxidation and reduction always occur together. No substance is ever oxidized unless something else is reduced, and the total number of electrons lost by one substance is always the

■ Oxygen had just been discovered by Joseph Priestley in 1774.

■ The term redox was coined to emphasize that reduction and oxidation must always occur together in chemical reactions.

■ Notice that when we write equations of this type, the electron appears as a "product" if the process is oxidation and as a "reactant" if the process is reduction.

Figure 5.1 | "Chlorine" bleach. This common household product is a dilute aqueous solution of sodium hypochlorite, NaOCl, which destroys fabric stains by oxidizing them to colorless products.

Bloomberg/Getty Images, Inc.

■ Electron loss must equal electron gain or else the law of conservation of mass would be violated.

TOOLS
Oxidizing and reducing agents

same as the total number gained by the other. If this were not true, electrons would appear as a product or a reactant of the overall reaction, and this is never observed. In the reaction of sodium with chlorine, for example, the overall reaction is

$$2Na + Cl_2 \longrightarrow 2NaCl$$

When two sodium atoms are oxidized, two electrons are lost, which is exactly the number of electrons gained when one Cl_2 molecule is reduced.

For a redox reaction to occur, one substance must accept electrons from the other. The substance that accepts the electrons is called the **oxidizing agent**; it is the agent that allows the other substance to lose electrons and be oxidized. Similarly, the substance that supplies the electrons is called the **reducing agent** because it helps something else to be reduced. In our example above, sodium serves as a reducing agent when it supplies electrons to chlorine. In the process, sodium is oxidized. Chlorine is an oxidizing agent when it accepts electrons from the sodium, and when that happens, chlorine is reduced to chloride ion. One way to remember this is by the following summary:

The reducing agent is the substance that is oxidized.
The oxidizing agent is the substance that is reduced.

Redox reactions are very common as mentioned in the introduction. They occur in batteries, which are built so that the electrons transferred can pass through some external circuit where they are, for example, able to light an LED flashlight or power an iPod. The metabolism of foods, which supplies our bodies with energy, also occurs by a series of redox reactions. Natural and manmade air pollutants are often the products of redox reactions. And ordinary household bleach works by oxidizing substances that stain fabrics, making them colorless or easier to remove from the fabric (see Figure 5.1).

Example 5.1
Identifying Oxidation–Reduction

Jung-Pang Wu/Getty Images

Fireworks display.

The bright light in fireworks displays is often produced by the reaction between magnesium and molecular oxygen, and the product is magnesium oxide, an ionic compound. Which element is oxidized and which is reduced? What are the oxidizing and reducing agents?

Analysis: This question asks us to apply the definitions presented above, so we need to know how the electrons are transferred in the reaction by determining which ions are involved.

Assembling the Tools: As stated, magnesium oxide is an ionic compound. Using the tools you learned in Chapter 2, the locations of the elements in the periodic table will tell what ions are involved and the formula of the product of this reaction.

Solution: From the periodic table, magnesium in Group 2A forms Mg^{2+} ions by losing two electrons.

$$Mg \longrightarrow Mg^{2+} + 2e^-$$

At the same time, oxygen in Group 6A forms O^{2-} ions by gaining two electrons. Since the reactant is molecular oxygen, O_2, the equation for the reaction forming oxide ions is:

$$O_2 + 4e^- \longrightarrow 2O^{2-}$$

We use the definitions of oxidizing and reducing agent to arrive at our answers.

By losing electrons, *magnesium is oxidized, so it is the reducing agent.* By gaining electrons, O_2 *is reduced and must be the oxidizing agent.*

Is the Answer Reasonable? First, as with ionic equations, *the number of atoms of each kind and the net charge must be the same on both sides.* We see that this is true for both equations. Also, we must always have one reactant gaining electrons and the other losing them, and we do.

When sodium reacts with molecular oxygen, O_2, the product is sodium peroxide, Na_2O_2, which contains the peroxide ion, O_2^{2-}. In this reaction, is O_2 oxidized or reduced? (*Hint:* Which reactant gains electrons, and which loses electrons?)

Practice Exercise 5.1

Identify the substances oxidized and reduced, and the oxidizing and reducing agents in the reaction of aluminum and chlorine to form aluminum chloride.

Practice Exercise 5.2

Using Oxidation Numbers to Follow Redox Reactions

Unlike the reaction of magnesium with oxygen, not all reactions with oxygen produce ionic products. For example, sulfur reacts with oxygen to give sulfur dioxide, SO_2, which is a molecular compound. Nevertheless, it is convenient to also view this as a redox reaction, but this requires that we change the way we define oxidation and reduction. To do this, chemists developed a bookkeeping system called **oxidation numbers**, which provides a way to keep tabs on electron transfers.

The oxidation number of an element in a particular compound is assigned according to a set of rules, which are described below. For simple, monatomic ions in a compound such as NaCl, the oxidation numbers are the same as the charges on the ions, so in NaCl the oxidation number of Na^+ is +1 and the oxidation number of Cl^- is −1. The real value of using oxidation numbers is that they can also be assigned to atoms in molecular compounds. In such cases, it is important to realize that the oxidation number does not actually equal a charge on an atom. To be sure to differentiate oxidation numbers from actual electrical charges, we will specify the sign *before* the number when writing oxidation numbers, and *after* the number when writing electrical charges. Thus, a sodium ion has a charge of 1+ and an oxidation number of +1.

A term that is frequently used interchangeably with oxidation number is **oxidation state**. In NaCl, sodium has an oxidation number of +1 and is said to be "in the +1 oxidation state." Similarly, the chlorine in NaCl is said to be in the −1 oxidation state. When there are several possible oxidation states, the oxidation state of an element is given by writing the oxidation number as a Roman numeral in parentheses after the name of the element. For example, "iron(III)" means iron in the +3 oxidation state. A summary of the terms used in oxidation and reduction reactions is given in Table 5.1.

We can use these new terms now to redefine a redox reaction:

A redox reaction is a chemical reaction in which changes in oxidation numbers occur.

We can follow redox reactions by taking note of the changes in oxidation numbers. To do this, however, we must be able to assign oxidation numbers to atoms in a quick and simple way.

TABLE 5.1	Summary of Oxidation and Reduction Using $2Na + Cl_2 \longrightarrow 2NaCl$
Sodium	**Chlorine**
Oxidized	Reduced
Reducing agent	Oxidizing agent
Oxidation number increases	Oxidation number decreases
Loses electrons	Gains electrons

Rules for the Assignment of Oxidation Numbers

We can use some basic knowledge learned earlier plus the set of rules below to determine the oxidation numbers of the atoms in almost any compound. These rules are a *hierarchy* where a rule can supersede a rule below it if necessary. This allows us to deal with conflicts between two rules.

Assigning oxidation numbers

■ Rule 2 means that O in O_2, P in P_4, and S in S_8 *all* have oxidation numbers of zero.

Rules for Assigning Oxidation Numbers

Note: When there is a conflict between two of these rules or an ambiguity in assigning an oxidation number, apply the rule that comes first and ignore the conflicting rule.

1. Oxidation numbers must add up to the charge on the molecule, formula unit, or ion.
2. All the atoms of free elements have oxidation numbers of zero.
3. Metals in Groups 1A, 2A, and Al have +1, +2, and +3 oxidation numbers, respectively.
4. H and F, in compounds, have +1 and −1 oxidation numbers, respectively.
5. Oxygen has a −2 oxidation number.
6. Group 7A elements have a −1 oxidation number.
7. Group 6A elements have a −2 oxidation number.
8. Group 5A elements have a −3 oxidation number.

These rules can be applied to molecular and ionic compounds as well as to elements and ions. For example, oxygen, O_2, is a free element and has a zero charge; the oxidation number for each oxygen atom in O_2 must be zero, following Rules 1 and 2. Or take the simple ionic compound, NaCl: the sodium is in Group 1A and has an oxidation number of +1, from Rule 3; and from Rule 6, chlorine has an oxidation number of −1. These add up to zero, which is the overall charge of the formula unit which, necessarily, also obeys Rule 1.

The numbered rules and their order in the list given above come into play when an element is capable of having more than one oxidation state. For example, you learned that transition metals can form more than one ion. Iron, for example, forms Fe^{2+} and Fe^{3+} ions, so in an iron compound we have to use the rules to figure out which iron ion is present. Similarly, when nonmetals are combined with hydrogen and oxygen in compounds or polyatomic ions, their oxidation numbers can vary and must be calculated using the rules. For example, in formaldehyde, CH_2O, a preservative for biological specimens, the oxidation of the carbon atom can be calculated using the numbered rules. The two hydrogens together are $2 \times (+1)$ and the oxygen has an oxidation number of −2. Together they add up to zero, and since the molecule is neutral, the carbon must have an oxidation number of 0, according to Rule 1.

With this as background, let's look at some more examples that illustrate how we apply the rules.

Example 5.2
Assigning Oxidation Numbers

Titanium dioxide, TiO_2, is a white pigment used in making paint. A now outmoded process of making TiO_2 from its ore involved $Ti(SO_4)_2$ in one step of the process. What is the oxidation number of titanium in $Ti(SO_4)_2$?

Analysis: The key here is recognizing SO_4 as the sulfate ion, SO_4^{2-}. Because we're only interested in the oxidation number of titanium, we use the charge on SO_4^{2-} to equal the net oxidation number of the polyatomic ion.

Assembling the Tools: We will use the rules for assigning oxidation numbers and, most important, Rule 1 to sum the oxidation numbers to equal the overall charge of the formula unit. The charge on a polyatomic ion is used as the net oxidation number of the ion.

Solution: The sulfate ion has a charge of 2−, which we can take to be its net oxidation number. Then, we apply the summation rule (Rule 1) to find the oxidation number of titanium, which we will represent by x.

$$(\text{charge of all Ti atoms}) + (\text{charge of all } SO_4^{2-} \text{ ions}) = 0 \quad (\text{Rule 1})$$

$$(1 \text{ atom Ti}) \times (x) + (2 \, SO_4^{2-} \text{ ions}) \times (-2) = 0$$

$$x \quad + \quad -4 \quad = 0$$

Solving this, $x = +4$, and the oxidation number of titanium is +4.

Is the Answer Reasonable? We have the sum of oxidation numbers adding up to zero, the charge on $Ti(SO_4)_2$, so our answer is reasonable.

As we have already discussed, transition metals can have more than one oxidation state. In addition, the oxidation states of the nonmetals can vary depending on the compound they are in. For example, sulfur can be found in H_2S, S_8 (elemental sulfur), SO_2, SO_3, and H_2SO_4. The oxidation states for sulfur in these compounds are −2, 0, +4, +6 and +6, respectively.

Hydrogen and oxygen very often have oxidation numbers of +1 and −2, respectively. However, in compounds called **hydrides**, hydrogen has an oxidation number of −1. An example of this is calcium hydride, CaH_2, where the calcium has a +2 oxidation number so the two hydrogen atoms must be −1 each. Compounds that have oxygen with a −1 oxidation number are called **peroxides**. Sodium peroxide has the formula Na_2O_2 and we see that since sodium always has a +1 oxidation number, the oxygen atoms must be −1 each. In addition we can tell from the formula that sodium peroxide is a molecular, not ionic substance. Using our rules for assigning oxidation numbers we can determine if the compound is a peroxide or hydride.

Example 5.3
Assigning Oxidation Numbers

The air bags used as safety devices in modern autos are inflated by the explosive decomposition of the ionic compound sodium azide, NaN_3. The reaction gives elemental sodium and gaseous nitrogen. What is the average oxidation number of a nitrogen atom in sodium azide?

Analysis: We're told that NaN_3 is ionic, so we know that there are a cation and an anion. The cation is the sodium ion, Na^+, which means that we can use Rule 1 to calculate the oxidation number of N.

Assembling the Tools: We will use the same tool as in Example 5.2—namely, the assignment of oxidation numbers, and focus on Rules 1 and 3.

Solution: We tackle the problem just using the rules for assigning oxidation numbers. We know that sodium exists as the ion Na^+ and that the compound is neutral overall. Therefore,

Na	$(1 \text{ atom}) \times (+1) = +1$	(Rule 3)
N	$(3 \text{ atoms}) \times (x) = 3x$	
	Sum = 0	(Rule 1)

© Richard Olivier/CORBIS

Detonation of sodium azide releases nitrogen gas, which rapidly inflates an air bag during a crash.

We can take the above statements and write them algebraically as

$$(1 \text{ atom}) \times (+1) + (3 \text{ atoms}) \times (x) = 0$$

$$3x = -1$$

$$x = -\tfrac{1}{3}$$

so each nitrogen must have an oxidation number of $-\tfrac{1}{3}$.

Is the Answer Reasonable? We can see that the math works out and the total of the oxidation numbers is zero.

While our answer was correct for the question that was asked in Example 5.3, we should also do a reality check. Importantly, there could not be three separate nitrogen anions each with a *charge* of $\tfrac{1}{3}-$, because it would mean that each nitrogen atom had acquired one-third of an electron. *Only whole numbers of electrons are involved in electron transfers.* The best explanation is that the anion must be a single particle with a negative one charge—namely, N_3^-, which is called the azide ion.

Practice Exercise 5.3

The chlorite ion, ClO_2^-, is a potent disinfectant, and solutions of it are sometimes used to disinfect air-conditioning systems in cars. What is the oxidation number of chlorine in this ion? (*Hint:* The sum of the oxidation numbers is not zero.)

Practice Exercise 5.4

Assign oxidation numbers to each atom in (a) $NiCl_2$, (b) Mg_2TiO_4, (c) $K_2Cr_2O_7$, (d) H_3PO_4, (e) $V(C_2H_3O_2)_3$, and (f) NH_4^+.

Identifying Oxidation and Reduction Processes

Oxidation numbers can be used in several ways. One is to define oxidation and reduction in the most comprehensive manner, as follows.

Oxidation involves an increase in oxidation number.
Reduction involves a decrease in oxidation number.

Let's see how these definitions apply to the reaction of hydrogen with chlorine. To avoid confusing oxidation numbers with actual electrical charges, we will write oxidation numbers directly above the chemical symbols of the elements

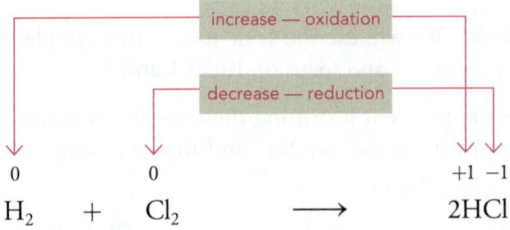

Notice that we have assigned the atoms in H_2 and Cl_2 oxidation numbers of zero, since they are free elements. The changes in oxidation number tell us that hydrogen is oxidized and chlorine is reduced.

Example 5.4
Using Oxidation Numbers to Follow Redox Reaction

Is the following a redox reaction?

$$2KCl + MnO_2 + 2H_2SO_4 \longrightarrow K_2SO_4 + MnSO_4 + Cl_2 + 2H_2O$$

If so, identify the substance oxidized and the substance reduced, as well as the oxidizing and reducing agents.

Analysis: To determine whether the reaction is a redox reaction and to give the specific answers requested, we will need to determine the oxidation numbers and then whether oxidation numbers are changing and how they are changing in the reaction.

Assembling the Tools: Our tools will be the rules for assigning oxidation numbers and the revised definitions of oxidation and reduction. Finally, we use the definitions of oxidizing and reducing agents.

Solution: To determine whether redox is occurring here, we first assign oxidation numbers to each atom on both sides of the equation. Following the rules, we get

$$\overset{+1\,-1}{2KCl} + \overset{+4\,-2}{MnO_2} + \overset{+1\,+6\,-2}{H_2SO_4} \longrightarrow \overset{+1\,+6\,-2}{K_2SO_4} + \overset{+2\,+6\,-2}{MnSO_4} + \overset{0}{Cl_2} + \overset{+1\,-2}{2H_2O}$$

Next, we look for changes, keeping in mind that an increase in oxidation number is oxidation and a decrease is reduction.

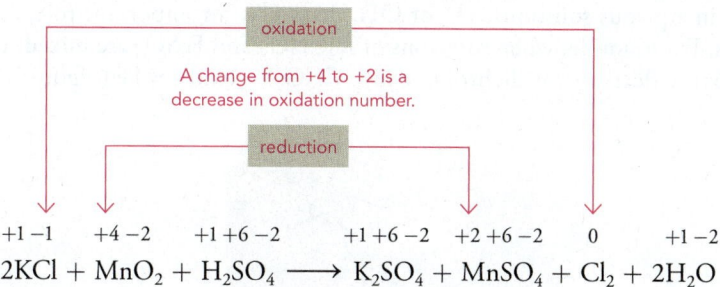

Thus, the Cl in KCl is oxidized to Cl_2 and the Mn in MnO_2 is reduced to Mn^{2+} in $MnSO_4$. The reducing agent is KCl and the oxidizing agent is MnO_2. (Notice that when we identify the oxidizing and reducing agents, we give the entire formula units that contain the atoms with oxidation numbers that change.)

Is the Answer Reasonable? There are lots of things we could check here. We have found changes that lead us to conclude that redox is happening. We can also identify one change as oxidation and the other as reduction, which gives us confidence that we've done the rest of the work correctly.

Consider the following reactions:

$$N_2O_5 + 2NaHCO_3 \longrightarrow 2NaNO_3 + 2CO_2 + H_2O$$

$$KClO_3 + 3HNO_2 \longrightarrow KCl + 3HNO_3$$

Which one is a redox reaction? For the redox reaction, which compound is oxidized and which is reduced? (*Hint:* How is redox defined using oxidation numbers?)

Practice Exercise 5.5

Practice Exercise 5.6

Chlorine dioxide, ClO_2, is used to kill bacteria in the dairy industry, meat industry, and other food and beverage industry applications. It is unstable, but can be made by the following reaction:

$$Cl_2 + 2NaClO_2 \longrightarrow 2ClO_2 + 2NaCl$$

Identify the substances oxidized and reduced, as well as the oxidizing and reducing agents in the reaction.

5.2 | Balancing Redox Equations

Many redox reactions take place in aqueous solution and many of these involve ions. An example is the reaction of laundry bleach with substances in the wash water. The active ingredient in the bleach is hypochlorite ion, OCl^-, which is the oxidizing agent in these reactions. To study redox reactions, it is often helpful to write the ionic and net ionic equations, just as we did in our analysis of metathesis reactions earlier in Chapter 4. In addition, one of the easiest ways to balance equations for redox reactions is to follow a procedure called the ion–electron method.

The Ion–Electron Method: A Divide and Conquer Approach

In the **ion–electron method**, we divide the oxidation and reduction processes into individual equations called **half-reactions** that are balanced separately. In each half-reaction both the atoms and charge have to balance. Then, we combine the balanced half-reactions to obtain the fully balanced net ionic equation.

In balancing the half-reactions, we must take into account that for many redox reactions in aqueous solutions, H^+ or OH^- ions play an important role, as do water molecules. For example, when solutions of $K_2Cr_2O_7$ and $FeSO_4$ are mixed, the acidity of the mixture decreases as dichromate ion, $Cr_2O_7^{2-}$, oxidizes Fe (Figure 5.2). This is

■ For simplicity, we are using H^+ to stand for H_3O^+ throughout our discussions in this chapter.

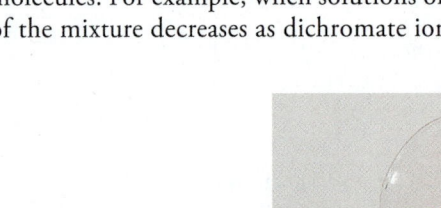

Peter Lerman

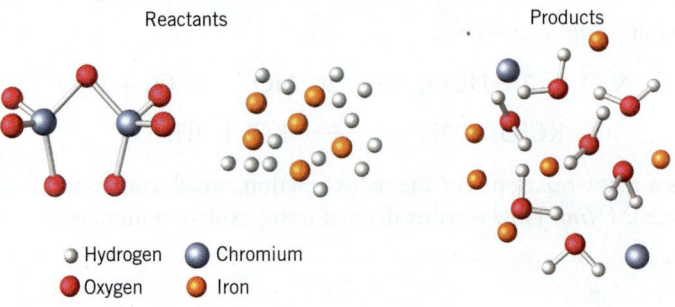

Reactants   Products

○ Hydrogen   ○ Chromium
● Oxygen   ● Iron

Figure 5.2 | **A redox reaction taking place in an acidic solution.** A solution of $K_2Cr_2O_7$ oxidizes Fe^{2+} to Fe^{3+} in an acidic solution, while the orange dichromate ion is reduced to Cr^{3+}.

because the reaction uses up H^+ as a reactant and produces H_2O as a product. In other reactions, OH^- is consumed, while in still others, H_2O is a reactant. Another fact is that in many cases the products (or even the reactants) of a redox reaction will differ depending on the acidity of the solution. For example, in an acidic solution, MnO_4^- is reduced to the Mn^{2+} ion, but in a neutral or slightly basic solution, the reduction product is the insoluble MnO_2.

Because of these factors, redox reactions are generally carried out in solutions containing a substantial excess of either acid or base, so before we can apply the ion–electron method, we have to know whether the reaction occurs in an acidic or a basic solution. (In this book, this information will always be given.)

Balancing Redox Equations in Acidic Solutions

As you just learned, $Cr_2O_7^{2-}$ reacts with Fe^{2+} in an acidic solution to give Cr^{3+} and Fe^{3+} as products. This information permits us to write the **skeleton equation**, which shows only the ions (or molecules) involved in the redox changes. We will leave out the spectator ions, K^+ and SO_4^{2-}, since they do not react in this redox reaction.

$$Cr_2O_7^{2-} + Fe^{2+} \longrightarrow Cr^{3+} + Fe^{3+}$$

We then proceed through the steps described below to find the balanced equation. As you will see, the ion–electron method will tell us how H^+ and H_2O are involved in the reaction; we don't need to know this in advance.

Step 1. **Divide the skeleton equation into half-reactions.** We choose one of the reactants, $Cr_2O_7^{2-}$, and write it at the left of an arrow. On the right, we write what $Cr_2O_7^{2-}$ changes to, which is Cr^{3+}. This gives us the beginnings of one half-reaction. For the second half-reaction, we write the other reactant, Fe^{2+}, on the left and the other product, Fe^{3+}, on the right. Notice that we are careful to use the complete formulas for the ions that appear in the skeleton equation. Except for hydrogen and oxygen, the same elements must appear on both sides of a given half-reaction.

$$Cr_2O_7^{2-} \longrightarrow Cr^{3+}$$
$$Fe^{2+} \longrightarrow Fe^{3+}$$

Step 2. **Balance atoms other than H and O.** There are two Cr atoms on the left and only one on the right, so we place a coefficient of 2 in front of Cr^{3+}. The second half-reaction is already balanced in terms of atoms, so we leave it as is:

$$Cr_2O_7^{2-} \longrightarrow 2Cr^{3+}$$
$$Fe^{2+} \longrightarrow Fe^{3+}$$

Step 3. **Balance oxygen by adding H_2O to the side that needs O.** There are seven oxygen atoms on the left of the first half-reaction and none on the right. Therefore, we add $7H_2O$ to the right side of the first half-reaction to balance the oxygens. (There is no oxygen imbalance in the second half-reaction, so there's nothing to do there.)

$$Cr_2O_7^{2-} \longrightarrow 2Cr^{3+} + 7H_2O$$
$$Fe^{2+} \longrightarrow Fe^{3+}$$

■ We use H_2O, not O or O_2, to balance oxygen atoms, because H_2O is what is actually reacting or being produced in the solution.

The oxygen atoms now balance, but we've created an imbalance in hydrogen. We address that issue next.

Step 4. **Balance hydrogen by adding H^+ to the side that needs H.** After adding the water, we see that the first half-reaction now has 14 hydrogens on the right and none on the left. To balance hydrogen, we add $14H^+$ to the left side of the half-reaction. When you do this step (or others) be careful to write the charges on the ions. If they are omitted, you will not obtain a balanced equation in the end.

$$14H^+ + Cr_2O_7^{2-} \longrightarrow 2Cr^{3+} + 7H_2O$$
$$Fe^{2+} \longrightarrow Fe^{3+}$$

Now each half-reaction is balanced in terms of atoms. Next, we will balance the charge.

Step 5. Balance the charge by adding electrons. First, we compute the net electrical charge on each side. For the first half-reaction we have

$$14H^+ + Cr_2O_7^{2-} \longrightarrow 2Cr^{3+} + 7H_2O$$

Net charge $= (14+) + (2-) = 12+$ Net charge $= 2(3+) + 0 = 6+$

The algebraic difference between the net charges on the two sides equals the number of electrons that must be added to the more positive (or less negative) side. In this instance, we must add $6e^-$ to the left side of the half-reaction.

$$6e^- + 14H^+ + Cr_2O_7^{2-} \longrightarrow 2Cr^{3+} + 7H_2O$$

This half-reaction is now complete; it is balanced in terms of both atoms and charge. (We can check this by recalculating the charge on both sides.)

To balance the other half-reaction, we add one electron to the right.

$$Fe^{2+} \longrightarrow Fe^{3+} + e^-$$

Now this reaction is balanced for both mass and charge.

Step 6. Make the number of electrons gained equal to the number lost and then add the two half-reactions. At this point we have the two balanced half-reactions:

$$6e^- + 14H^+ + Cr_2O_7^{2-} \longrightarrow 2Cr^{3+} + 7H_2O$$
$$Fe^{2+} \longrightarrow Fe^{3+} + e^-$$

Six electrons are gained in the first, but only one is lost in the second Therefore, before combining the two equations we multiply all of the coefficients of the second half-reaction by 6.

$$6e^- + 14H^+ + Cr_2O_7^{2-} \longrightarrow 2Cr^{3+} + 7H_2O$$
$$6(Fe^{2+} \longrightarrow Fe^{3+} + e^-)$$

(sum) $6e^- + 14H^+ + Cr_2O_7^{2-} + 6Fe^{2+} \longrightarrow 2Cr^{3+} + 7H_2O + 6Fe^{3+} + 6e^-$

Step 7. Cancel anything that is the same on both sides. This is the final step. Six electrons cancel from both sides to give the final balanced equation.

$$14H^+ + Cr_2O_7^{2-} + 6Fe^{2+} \longrightarrow 2Cr^{3+} + 7H_2O + 6Fe^{3+}$$

Notice that both the charge and the atoms balance.

In some reactions, after adding the two half-reactions you may have H_2O or H^+ on both sides—for example, $6H_2O$ on the left and $2H_2O$ on the right. Cancel as many as you can. Thus,

$$\ldots + 6H_2O \ldots \longrightarrow \ldots + 2H_2O \ldots$$

reduces to

$$\ldots + 4H_2O \ldots \longrightarrow \ldots$$

The following is a summary of the steps we've followed for balancing an equation for a redox reaction in an acidic solution. If you don't skip any steps and you perform them in the order given, you will always obtain a properly balanced equation.

Ion–Electron Method—Acidic Solution
Step 1. Divide the equation into two half-reactions.
Step 2. Balance atoms other than H and O.
Step 3. Balance O by adding H_2O.
Step 4. Balance H by adding H^+.
Step 5. Balance net charge by adding e^-.
Step 6. Make e^- gain equal e^- loss; then add half-reactions.
Step 7. Cancel anything that's the same on both sides.

TOOLS

Ion–electron method for acidic solutions

Example 5.5
Using the Ion–Electron Method

Balance the following equation:

$$MnO_4^- + H_2SO_3 \longrightarrow SO_4^{2-} + Mn^{2+}$$

The reaction occurs in an acidic solution.

Analysis: In using the ion–electron method, there's not much to analyze. It's necessary to know the steps and to follow them in sequence.

Assembling the Tools: Follow the seven steps of the ion–electron method for acidic solutions to balance the equation.

Solution: We follow the steps given above.

Step 1. Divide the skeleton equation into two half-reactions.

$$MnO_4^- \longrightarrow Mn^{2+}$$
$$H_2SO_3 \longrightarrow SO_4^{2-}$$

Step 2. Balance atoms other than H and O. There is nothing to do for this step since all the atoms except H and O are already in balance.

Step 3. Add H_2O to balance oxygens.

$$MnO_4^- \longrightarrow Mn^{2+} + 4H_2O$$
$$H_2O + H_2SO_3 \longrightarrow SO_4^{2-}$$

Step 4. Add H^+ to balance H.

$$8H^+ + MnO_4^- \longrightarrow Mn^{2+} + 4H_2O$$
$$H_2O + H_2SO_3 \longrightarrow SO_4^{2-} + 4H^+$$

Step 5. Balance the charge by adding electrons to the more positive sides.

$$5e^- + 8H^+ + MnO_4^- \longrightarrow Mn^{2+} + 4H_2O$$
$$H_2O + H_2SO_3 \longrightarrow SO_4^{2-} + 4H^+ + 2e^-$$

Step 6. Make electron loss equal to electron gain, then add the half-reactions.

$$2 \times (5e^- + 8H^+ + MnO_4^- \longrightarrow Mn^{2+} + 4H_2O)$$

$$5 \times (H_2O + H_2SO_3 \longrightarrow SO_4^{2-} + 4H^+ + 2e^-)$$

$$10e^- + 16H^+ + 2MnO_4^- + 5H_2O + 5H_2SO_3 \longrightarrow 2Mn^{2+} + 8H_2O + 5SO_4^{2-} + 20H^+ + 10e^-$$

Step 7. Cancel $10e^-$, $16H^+$, and $5H_2O$ from both sides. The final equation is

$$2MnO_4^- + 5H_2SO_3 \longrightarrow 2Mn^{2+} + 3H_2O + 5SO_4^{2-} + 4H^+$$

Is the Answer Reasonable? The check involves two steps. First, we check that each side of the equation has the same number of atoms of each element, which it does. Second, we check to be sure that the net charge is the same on both sides. On the left we have $2MnO_4^-$ with a net charge of 2−. On the right we have $2Mn^{2+}$ and $4H^+$ (total charge = 8+) along with $5SO_4^{2-}$ (total charge = 10−), so the net charge is also 2−. Having both atoms and charge in balance makes it a balanced equation and confirms that we've worked the problem correctly.

Practice Exercise 5.7

The following equation is not balanced. Explain why the equation is not balanced, and balance it using the ion–electron method. (*Hint:* What are the criteria for a balanced ionic equation?)

$$Al + Cu^{2+} \longrightarrow Al^{3+} + Cu$$

Practice Exercise 5.8

The element technetium (atomic number 43) is radioactive and one of its isotopes, ^{99}Tc, is used in medicine for diagnostic imaging. The isotope is usually obtained in the form of the pertechnetate anion, TcO_4^-, but its use sometimes requires the technetium to be in a lower oxidation state. Reduction can be carried out using Sn^{2+} in an acidic solution. The skeleton equation is

$$TcO_4^- + Sn^{2+} \longrightarrow Tc^{4+} + Sn^{4+} \quad \text{(acidic solution)}$$

Balance the equation using the ion–electron method.

Balancing Redox Equations for Basic Solutions

In a basic solution, the concentration of H^+ is very small; the dominant species are H_2O and OH^-. Strictly speaking, these should be used to balance the half-reactions. However, the simplest way to obtain a balanced equation for a basic solution is to first pretend that the solution is acidic. We balance the equation using the seven steps just described, and then we use a simple three-step procedure described below to convert the equation to the correct form for a basic solution. The conversion uses the fact that H^+ and OH^- react in a 1-to-1 ratio to give H_2O.

TOOLS

Ion–electron method for basic solutions

Additional Steps in the Ion–Electron Method for Basic Solutions
Step 8. Add to both sides of the equation the same number of OH^- as there are H^+.
Step 9. Combine H^+ and OH^- (on the same side of the arrow) to form H_2O.
Step 10. Cancel any H_2O that you can.

As an example, suppose we want to balance the following equation in a basic solution:

$$SO_3^{2-} + MnO_4^- \longrightarrow SO_4^{2-} + MnO_2$$

Following Steps 1 through 7 for acidic solutions gives

$$2H^+ + 3SO_3^{2-} + 2MnO_4^- \longrightarrow 3SO_4^{2-} + 2MnO_2 + H_2O$$

Conversion of this equation to one appropriate for a basic solution proceeds as follows.

Step 8. **Add to both sides of the equation the same number of OH⁻ as there are H⁺.** The equation for acidic solution has $2H^+$ on the left, so we add $2OH^-$ to *both sides*. This gives

$$2OH^- + 2H^+ + 3SO_3^{2-} + 2MnO_4^- \longrightarrow 3SO_4^{2-} + 2MnO_2 + H_2O + 2OH^-$$

Step 9. **Combine H⁺ and OH⁻ to form H₂O.** The left side has $2OH^-$ and $2H^+$, which become $2H_2O$. So in place of $2OH^- + 2H^+$ we write $2H_2O$.

$$2OH^- + 2H^+ + 3SO_3^{2-} + 2MnO_4^- \longrightarrow 3SO_4^{2-} + 2MnO_2 + H_2O + 2OH^-$$

$$2H_2O + 3SO_3^{2-} + 2MnO_4^- \longrightarrow 3SO_4^{2-} + 2MnO_2 + H_2O + 2OH^-$$

Step 10. **Cancel any H₂O that you can.** In this equation, one H_2O can be eliminated from both sides. The final equation, balanced for basic solution, is

$$H_2O + 3SO_3^{2-} + 2MnO_4^- \longrightarrow 3SO_4^{2-} + 2MnO_2 + 2OH^-$$

Consider the following half-reaction balanced for an acidic solution:

$$2H_2O + SO_2 \longrightarrow SO_4^{2-} + 4H^+ + 2e^-$$

What is the balanced half-reaction for a basic solution? (*Hint:* Half-reactions are treated like balanced equations when adjusting to basic solution.)

Balance the following equation for a basic solution.

$$MnO_4^- + C_2O_2^{2-} \longrightarrow MnO_2 + CO_3^{2-}$$

Practice Exercise 5.9

Practice Exercise 5.10

5.3 | Acids as Oxidizing Agents

Earlier you learned that one of the properties of acids is that they react with bases. Another important property is their ability to react with certain metals. These are redox reactions in which the metal is oxidized and the acid is reduced. However, in these reactions, the part of the acid that's reduced depends on the composition of the acid itself as well as on the metal.

When a piece of zinc is placed into a solution of hydrochloric acid, bubbling is observed and the zinc gradually dissolves (Figure 5.3). The chemical reaction is

$$Zn(s) + 2HCl(aq) \longrightarrow ZnCl_2(aq) + H_2(g)$$

for which the net ionic equation is

$$Zn(s) + 2H^+(aq) \longrightarrow Zn^{2+}(aq) + H_2(g)$$

In this reaction, zinc is oxidized and hydrogen ions are reduced. Stated another way, the H^+ of the acid is the oxidizing agent.

Many metals react with acids just as zinc does—by being oxidized by hydrogen ions. In these reactions a metal salt and gaseous hydrogen are the products.

Metals that are able to react with acids such as HCl and H_2SO_4 to give hydrogen gas are said to be more active than hydrogen (H_2). For other metals, however, hydrogen ions

Figure 5.3 | **Zinc reacts with hydrochloric acid.** Bubbles of hydrogen are formed when a solution of hydrochloric acid comes in contact with metallic zinc. In the molecular view (not balanced), the hydronium ions react with the zinc metal atoms (dark gray spheres). Two hydrogen ions gain two electrons from the metal atom and form a hydrogen molecule, and the zinc is oxidized to a cation with a 2+ charge.

are not powerful enough to cause their oxidation. Copper, for example, is significantly less reactive than zinc or iron, and H^+ cannot oxidize it. Copper is an example of a metal that is less active than H_2.

The Anion Determines the Oxidizing Power of an Acid

Hydrochloric acid contains hydronium ions, H_3O^+ ions (which is sometimes abbreviated as H^+, and referred to as the hydrogen ion) and Cl^- ions. The hydronium ion in hydrochloric acid can be an oxidizing agent because it can be reduced to H_2. However, the Cl^- ion in the solution has no tendency at all to be an oxidizing agent, so in a solution of HCl the only oxidizing agent is H_3O^+. The same applies to dilute solutions of sulfuric acid, H_2SO_4.

The hydrogen ion in water is actually a rather poor oxidizing agent, so hydrochloric acid and sulfuric acid have rather poor oxidizing abilities. They are often called **nonoxidizing acids**, even though their hydronium ions are able to oxidize certain metals. Therefore, in a nonoxidizing acid, the anion of the acid is a weaker oxidizing agent than H_3O^+ and the anion of the acid is more difficult to reduce than H_3O^+. Some acids contain anions that are stronger oxidizing agents than H_3O^+ and are called **oxidizing acids**. (See Table 5.2.)

Nitrate Ion as an Oxidizing Agent

Nitric acid, HNO_3, ionizes in water to give H^+ and NO_3^- ions. In this solution, the nitrate ion is a more powerful oxidizing agent than the hydrogen ion. This makes it able to oxidize metals that H^+ cannot, such as copper and silver. For example, the molecular equation for the reaction of concentrated HNO_3 with copper, shown in Figure 5.4, is

$$Cu(s) + 4HNO_3(aq) \longrightarrow Cu(NO_3)_2(aq) + 2NO_2(g) + 2H_2O$$

TABLE 5.2	Nonoxidizing and Oxidizing Acids	

Nonoxidizing Acids

$HCl(g)$

$H_2SO_4(aq)^a$

$H_3PO_4(aq)$

Most organic acids (e.g., $HC_2H_3O_2$)

Oxidizing Acids	Conditions	Reduction Reactions
HNO_3	Concentrated	$NO_3^- + 2H^+ + e^- \longrightarrow NO_2(g) + H_2O$
	Dilute	$NO_3^- + 4H^+ + 3e^- \longrightarrow NO(g) + 2H_2O$
	Very dilute, with strong reducing agent	$NO_3^- + 10H^+ + 8e^- \longrightarrow NH_4^+(aq) + 3H_2O$
H_2SO_4	Hot, concentrated, with strong reducing agent	$SO_4^{2-} + 10H^+ + 8e^- \longrightarrow H_2S(g) + 4H_2O$
	Hot, concentrated	$SO_4^{2-} + 4H^+ + 2e^- \longrightarrow SO_2(g) + 2H_2O$

$^a H_2SO_4$ is a nonoxidizing acid when cold and dilute.

TOOLS

Oxidizing and nonoxidizing acids

$$Cu + 4HNO_3 \longrightarrow Cu(NO_3)_2 + 2NO_2 + 2H_2O$$

Michael Watson

Figure 5.4 | The reaction of copper with concentrated nitric acid. A copper penny reacts vigorously with concentrated nitric acid, as this sequence of photographs shows. The dark red-brown vapors are nitrogen dioxide, the same gas that gives smog its characteristic color. The atomic view shows the copper in the solid, the addition of nitric acid, and finally, the formation of the copper nitrate, nitrogen dioxide as a gas, and water molecules.

If we convert this to a net ionic equation and then assign oxidation numbers, we can see that NO_3^- is the oxidizing agent and Cu is the reducing agent.

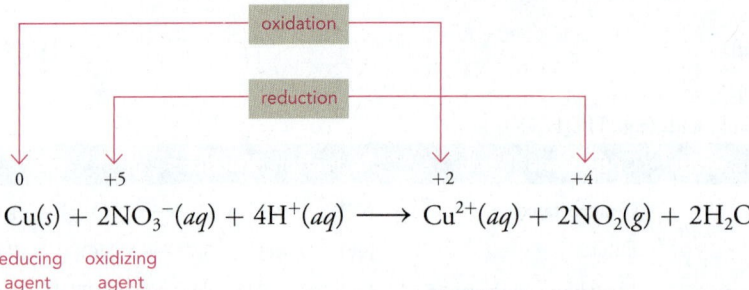

$$Cu(s) + 2NO_3^-(aq) + 4H^+(aq) \longrightarrow Cu^{2+}(aq) + 2NO_2(g) + 2H_2O$$

Notice that in this reaction, no hydrogen gas is formed. The H^+ ions of the HNO_3 are an essential part of the reaction, but they just become part of the water molecules without a change in oxidation number.

The nitrogen-containing product formed in the reduction of nitric acid depends on the concentration of the acid and the reducing power of the metal. With concentrated nitric acid, nitrogen dioxide, NO_2, is often the reduction product. With dilute nitric acid, the product is often nitrogen monoxide (also called nitric oxide), NO, instead. Copper, for example, reacts as follows:

Concentrated HNO₃

$$Cu(s) + 4H^+(aq) + 2NO_3(aq) \longrightarrow Cu(aq) + 2NO_2(g) + 2H_2O$$

Dilute HNO₃

$$3Cu(s) + 8H^+(aq) + 2NO_3(aq) \longrightarrow 3Cu^{2+}(aq) + 2NO(g) + 4H_2O$$

Nitric acid is a very effective oxidizing acid. All metals except the most unreactive ones, such as platinum and gold, are oxidized by it. Nitric acid also does a good job of oxidizing organic compounds, so it is wise to be especially careful when working with this acid in the laboratory. Very serious accidents have occurred when inexperienced people have used concentrated nitric acid around organic substances.

Hot Concentrated Sulfuric Acid: Another Oxidizing Acid

In a dilute solution, the sulfate ion of sulfuric acid has little tendency to serve as an oxidizing agent. However, if the sulfuric acid is both concentrated and hot, it becomes a fairly potent oxidizer. For example, copper is not bothered by cool dilute H_2SO_4, but it is attacked by hot concentrated H_2SO_4 according to the following equation:

$$Cu + 2H_2SO_4(hot, conc.) \longrightarrow CuSO_4 + SO_2 + 2H_2O$$

Because of this oxidizing ability, hot concentrated sulfuric acid can be very dangerous. The liquid is viscous and can stick to the skin, causing severe burn.

Practice Exercise 5.11 Write the balanced half-reactions for the reaction of zinc with hydrogen ions. (*Hint:* Be sure to place the electrons on the correct sides of the half-reactions.)

Practice Exercise 5.12 Write balanced molecular, ionic, and net ionic equations for the reaction of hydrochloric acid with (a) magnesium and (b) aluminum. (Both are oxidized by hydrogen ions.)

5.4 | Redox Reactions of Metals

The formation of hydrogen gas in the reaction of a metal with an acid is a special case of a more general phenomenon—one element displacing (pushing out) another element from a compound by means of a redox reaction. In the case of a metal–acid reaction, it is the metal that displaces hydrogen from the acid, changing $2H^+$ to H_2.

Another reaction of this same general type occurs when one metal displaces another metal from its compounds, and is illustrated by the experiment shown in Figure 5.5. Here we see a brightly polished strip of metallic zinc that is dipped into a solution of copper sulfate. After the zinc is in the solution for a while, a reddish brown deposit of metallic copper forms on the zinc, and if the solution were analyzed, we would find that it now contains zinc ions, as well as some remaining unreacted copper ions.

The results of this experiment can be summarized by the following net ionic equation:

$$Zn(s) + Cu^{2+}(aq) \longrightarrow Cu(s) + Zn^{2+}(aq)$$

Metallic zinc is oxidized as copper ion is reduced. In the process, Zn^{2+} ions have taken the place of the Cu^{2+} ions, so a solution of copper sulfate is changed to a solution of zinc sulfate. An atomic-level view of what's happening at the surface of the zinc during the reaction is depicted in Figure 5.6. A reaction such as this, in which one element replaces another in a compound, is sometimes called a **single replacement reaction**.

Michael Watson

Figure 5.5 | The reaction of zinc with copper ion. (*Left*) A piece of shiny zinc next to a beaker containing a copper sulfate solution. (*Center*) When the zinc is placed in the solution, copper ions are reduced to the free metal while the zinc dissolves. (*Right*) After a while the zinc becomes coated with a red-brown layer of copper. Notice that the solution is a lighter blue than before, showing that some of the copper ions have left the solution.

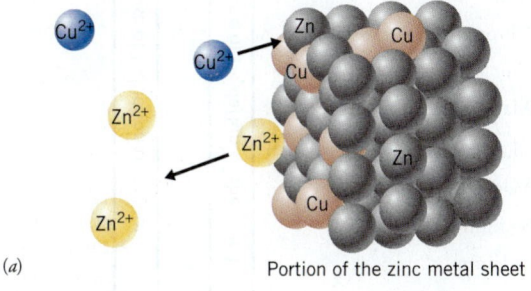

(*a*)

Portion of the zinc metal sheet

Two electrons are transferred from the zinc atom to the copper ion.

The result is a zinc ion and a copper atom.

(*b*)

Figure 5.6 | An atomic-level view of the reaction of copper ions with zinc. (*a*) Copper ions (blue) collide with the zinc surface, where they pick up electrons from zinc atoms (gray). The zinc atoms become zinc ions (yellow) and enter the solution. The copper ions become copper atoms (red-brown) and stick to the surface of the zinc. (For clarity, the water molecules of the solution and the sulfate ions are not shown.) (*b*) A close-up view of the exchange of electrons that leads to the reaction.

Activity Series of Metals

In the reaction of zinc with copper ion, the more "active" zinc displaces the less "active" copper in a compound, where we have used the word active to mean "easily oxidized." This is actually a general phenomenon: *an element that is more easily oxidized will displace one that is less easily oxidized from its compounds.* By comparing the relative ease of oxidation of various metals using experiments like the one pictured in Figure 5.6, we can arrange metals in order of their ease of oxidation. This yields the **activity series** shown in Table 5.3. In this table, metals at the bottom are more easily oxidized (are more active) than those at the top. This means that *a given ion will be displaced from its compounds by any element below it in the table.*

■ The strongest oxidizing agent in a solution of a "nonoxidizing" acid is H⁺.

Notice that we have included hydrogen in the activity series. Metals below hydrogen in the series can displace hydrogen from solutions containing H^+. These are the metals that are capable of reacting with nonoxidizing acids. On the other hand, metals above hydrogen in the table do not react with acids having H^+ as the strongest oxidizing agent.

Metals at the very bottom of the table are very easily oxidized and are extremely strong reducing agents. They are so reactive, in fact, that they are able to reduce the hydrogen in water molecules. Sodium, for example, reacts vigorously (see Figure 5.7).

$$2Na(s) + 2H_2O \longrightarrow H_2(g) + 2NaOH(aq)$$

For metals below hydrogen in the activity series, there's a parallel between the ease of oxidation of the metal and the speed with which it reacts with H^+. For example, in

TOOLS

Activity series of metal

TABLE 5.3	Activity Series for Some Metals (and Hydrogen)

		Element		Ion
Least Active	Do not react with nonoxidizing acids	Gold		Au³⁺
		Mercury		Hg²⁺
		Silver		Ag⁺
		Copper		Cu²⁺
		HYDROGEN		H⁺
		Lead		Pb²⁺
		Tin		Sn²⁺
		Cobalt		Co²⁺
		Cadmium		Cd²⁺
React with nonoxidizing acids to produce hydrogen		Iron		Fe²⁺
		Chromium		Cr³⁺
		Zinc		Zn²⁺
		Manganese		Mn²⁺
		Aluminum		Al³⁺
		Magnesium		Mg²⁺
		Sodium	Increasing ease of oxidation of the metal / Increasing ease of reduction of the ion	Na⁺
		Calcium		Ca²⁺
	React with water to form hydrogen	Strontium		Sr²⁺
		Barium		Ba²⁺
		Cesium		Cs⁺
		Potassium		K⁺
Most Active		Rubidium		Rb⁺

$$2HCl(aq) + Fe(s) \longrightarrow$$
$$FeCl_2(aq) + H_2(g)$$

$$2HCl(aq) + Zn(s) \longrightarrow$$
$$ZnCl_2(aq) + H_2(g)$$

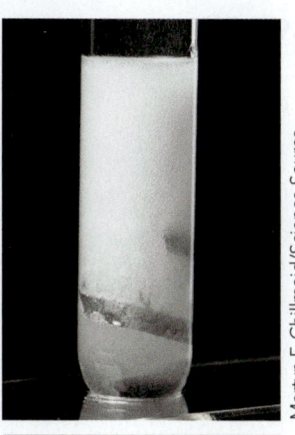

$$2HCl(aq) + Mg(s) \longrightarrow$$
$$MgCl_2(aq) + H_2(g)$$

Figure 5.7 | **Metallic sodium reacts violently with water.** In the reaction, sodium is oxidized to Na^+ and water molecules are reduced to give hydrogen gas and hydroxide ions. The heat of the reaction ignites the hydrogen. The heat melts the sodium, which can be seen glowing and sending flames from the surface of the water. When the reaction is over, the solution contains sodium hydroxide.

Figure 5.8 | **The relative ease of oxidation of metals as demonstrated by their reaction with hydrogen ions of an acid.** The products are hydrogen gas and the metal ion in solution. All three test tubes contain $HCl(aq)$ at the same concentration and temperature. The first contains pieces of iron, the second has bits of zinc, and the third some magnesium. Among these three metals we conclude that more bubbles of hydrogen formed is related to the ease of oxidation.

Figure 5.8, we see samples of iron, zinc, and magnesium reacting with solutions of hydrochloric acid. In each test tube the initial HCl concentration is the same, but we see that the magnesium reacts more rapidly than zinc, which reacts more rapidly than iron. You can see that this matches the order of reactivity in Table 5.3—namely, magnesium is more easily oxidized than zinc, which is more easily oxidized than iron.

Using the Activity Series to Predict Reactions

The activity series in Table 5.3 makes it possible to predict the outcome of single replacement redox reactions, as illustrated in Example 5.6.

Example 5.6
Using the Activity Series

What will happen if an iron nail is dipped into a solution containing copper(II) sulfate? If a reaction occurs, write its molecular equation.

Analysis: Before deciding what *will* happen, we have to consider, what *could* happen? If a chemical reaction were to occur, iron would have to react with the copper sulfate. Is this possibly a *metal* reacting with the *salt of another metal*? This suggests the possibility of a single replacement reaction.

Assembling the Tools: The tool we use to predict such reactions is the activity series of the metals in Table 5.3. We also will need our nomenclature rules for ionic compounds.

Solution: Examining Table 5.3, we see that iron is below copper. This means iron is more easily oxidized than copper, so we expect metallic iron to displace copper ions from the solution. *A reaction will occur.* (We've answered one part of the question.) The formula for copper(II) sulfate is $CuSO_4$. To write an equation for the reaction, we have to know the final oxidation state of the iron. In the table, this is indicated as $+2$, so the Fe atoms change to Fe^{2+} ions. To write the formula of the salt in the solution we pair Fe^{2+} with SO_4^{2-} to give $FeSO_4$. Copper(II) ions are reduced to copper atoms.

Our analysis told us that a reaction *will* occur and it also gave us the products, so the equation is

$$Fe(s) + CuSO_4(aq) \longrightarrow Cu(s) + FeSO_4(aq)$$

Is the Answer Reasonable? We can check the activity series again to be sure we've reached the correct conclusion, and we can check to be sure the equation we've written has the correct formulas and is balanced correctly. Doing this confirms that we've got the right answers.

CHEMISTRY OUTSIDE THE CLASSROOM 5.1

Polishing Silver — The Easy Way

If you have ever tried it, polishing silverware with abrasives and creams is a messy and dirty affair often ending with unsatisfactory results. Since we now know the principle that an active metal will reduce a compound of a less active metal, let's use it to solve a practical problem.

Generally, silver tarnishes by a gradual reaction with hydrogen sulfide, present in trace amounts in the air. The product of the reaction is silver sulfide, Ag_2S, which is black and forms a dull film over the bright metal. Polishing a silver object using a mild abrasive cream restores the shine, but it also gradually removes silver from the object as the silver sulfide is rubbed away. Usually the toxic silver tarnish is washed down the drain, adding a pollutant to the water.

An alternative method, and one that involves little effort, involves lining the bottom of a sink with aluminum foil, adding warm water and detergent (which acts as an electrolyte), and then submerging the tarnished silver object in the detergent–water mixture while keeping it in contact with the aluminum as shown in Figure 1. In a short time, the silver sulfide is reduced, restoring the shine and depositing the freed silver metal on the object. As this happens, a small amount of the aluminum foil is oxidized and dissolves. Besides requiring little physical effort, this silver polishing technique *doesn't remove silver from the object being polished.* Also, the *silver, considered a heavy metal, is being kept out of the water* and there is *no mess to dirty your hands.*

(a)

(b)

(c)

Andy Washnik

Figure 1 Using aluminum to remove tarnish from silver. (a) A badly tarnished silver vase stands next to a container of detergent (Soilax) dissolved in water. The bottom of the container is lined with aluminum foil. (b) The vase, shown partly immersed in the detergent, rests on the aluminum foil. (c) After a short time, the vase is removed, rinsed with water, and wiped with a soft cloth. Where the vase was immersed in the liquid, much of the tarnish has been reduced to metallic silver.

Now let's think about what happens if an iron nail is dipped into a solution of aluminum sulfate. As in Example 5.6 we compare the positions of the element iron and the ion Al^{3+} in the activity series. We see that the ion Al^{3+} is below the iron, and this tells us that the elemental form of aluminum is more active than iron. We must conclude that *elemental iron cannot reduce Al^{3+} and so no reaction occurs.*

Suppose an aqueous mixture is prepared containing magnesium sulfate ($MgSO_4$), copper(II) sulfate ($CuSO_4$), metallic magnesium, and metallic copper. What reaction, if any, will occur? (*Hint:* What reactions could occur?)

Write a chemical equation for the reaction that will occur, if any, when (a) aluminum metal is added to a solution of copper(II) chloride and (b) silver metal is added to a solution of magnesium sulfate. If no reaction will occur, write "no reaction" in place of the products.

Practice Exercise 5.13

Practice Exercise 5.14

5.5 | Molecular Oxygen as an Oxidizing Agent

Oxygen is a plentiful chemical; it's in the air and available to anyone who wants to use it, a chemist or not. Furthermore, O_2 is a very reactive oxidizing agent, so its reactions have been well studied. When these reactions are rapid, with the evolution of light and heat, we call them **combustion**. The products of reactions with oxygen are generally oxides—molecular oxides when oxygen reacts with nonmetals and ionic oxides when oxygen reacts with metals.

Reaction of Nonmetals with Oxygen

Most nonmetals combine as readily with oxygen as do the metals, and their reactions usually occur rapidly enough to be described as combustion. To most people, the most important nonmetal combustion reaction is that of carbon because the reaction is a source of heat. Coal and charcoal, for example, are common carbon fuels. Coal is used worldwide in large amounts to generate electricity, and charcoal is a popular fuel for grilling hamburgers. If plenty of oxygen is available, the combustion of carbon gives CO_2, but when the supply of O_2 is limited, some CO forms as well. Manufacturers that package charcoal briquettes, therefore, print a warning on the bag that the charcoal shouldn't be used indoors for cooking or heating because of the danger of carbon monoxide poisoning.

Sulfur is another nonmetal that burns readily in oxygen to produce sulfur dioxide. One important use for this reaction is in the manufacture of sulfuric acid. The first step in the manufacturing process is the combustion of sulfur to produce sulfur dioxide. In addition, sulfur dioxide also forms when sulfur compounds burn, and the presence of sulfur and sulfur compounds in coal and petroleum is a major source of air pollution. Power plants that burn coal are making strides to remove the SO_2 from their exhausts.

Nitrogen also reacts with oxygen to form nitrogen oxides: NO, NO_2, N_2O, N_2O_3, N_2O_4, and N_2O_5. Nitrogen oxide and nitrogen dioxide are formed when nitrogen and oxygen are heated to high enough temperatures and pressures such as the temperature of a car engine. Dinitrogen oxide, also known as nitrous oxide or laughing gas, is used as an anesthetic by dentists and as the propellant in canned whipped cream, Reddi-wip.

The label on a bag of charcoal displays a warning about carbon monoxide.

Oxidation of Organic Compounds

Experience has taught you that certain materials burn. For example, if you had to build a fire to keep warm, you no doubt would look for combustible materials like twigs, logs, or other pieces of wood to use as fuel. The car you drive is probably powered, at least in part, by the combustion of gasoline. Wood and gasoline are examples of substances or mixtures of substances that chemists call organic compounds—compounds whose structures are determined primarily by the linking together of carbon atoms. When organic compounds burn, the products of the reactions are usually easy to predict.

Hydrocarbons as Fuels

■ All reactions of hydrocarbons with oxygen produce CO_2, CO, and C in different proportions. However, we will assume that just one product is formed: for a plentiful supply of oxygen, carbon dioxide; for a limited supply of oxygen, carbon monoxide; and for an extremely limited supply of oxygen, carbon.

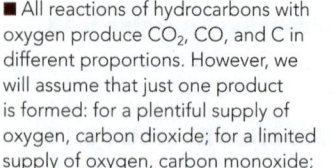

TOOLS

Hydrocarbon combustion with a plentiful supply of O_2

Fuels such as natural gas, gasoline, kerosene, heating oil, and diesel fuel are examples of hydrocarbons—compounds containing only the elements carbon and hydrogen. Natural gas is composed principally of methane, CH_4. Gasoline is a mixture of hydrocarbons; the most familiar, in terms of name recognition, is octane, C_8H_{18}. Kerosene, heating oil, and diesel fuel are mixtures of hydrocarbons in which the molecules contain even more atoms of carbon and hydrogen. When hydrocarbons are burned in air, besides water, a mixture of carbon-containing products are formed, such as carbon dioxide, carbon monoxide, and carbon as soot. If the conditions are carefully managed, the products can be controlled.

When hydrocarbons burn in a plentiful supply of oxygen, the products of combustion are mainly carbon dioxide and water. Thus, methane and octane combine with oxygen according to the equations

$$CH_4 + 2O_2 \longrightarrow CO_2 + 2H_2O$$

$$2C_8H_{18} + 25O_2 \longrightarrow 16CO_2 + 18H_2O$$

Many people don't realize that water is one of the products of the combustion of hydrocarbons, even though they have seen evidence for it. Perhaps you've seen clouds of condensed water vapor coming from the exhaust pipes of automobiles on cold winter days, or you may have noticed that shortly after you first start a car, drops of water fall from the exhaust pipe. This is water that has been formed during the combustion of the gasoline. Similarly, the "smokestacks" of power stations release clouds of condensed water vapor (Figure 5.9), which is often mistaken for smoke from fires used to generate power to make electricity. Actually, many of today's power stations produce very little smoke because they burn natural gas instead of coal.

■ You can tell whether smoke or water vapor is coming from a smokestack. If the cloud is steam, there will be a clear region just above the smokestack. If the emissions are smoke, no clear region will exist.

TOOLS

Hydrocarbon combustion with a limited supply of O_2

When the supply of oxygen is somewhat restricted during the combustion of a hydrocarbon, some of the carbon is converted to carbon monoxide. The formation of CO is a pollution problem associated with the use of gasoline engines, as you may know

$$2CH_4 + 3O_2 \longrightarrow 2CO + 4H_2O \qquad \text{(in a limited oxygen supply)}$$

TOOLS

Hydrocarbon combustion with an extremely limited supply of O_2

When the oxygen supply is extremely limited, only the hydrogen of a hydrocarbon mixture is converted to the oxide (water) and the carbon atoms emerge as elemental carbon. For example, when a candle burns, the fuel is a high-molecular-weight hydrocarbon (e.g., $C_{20}H_{42}$) and incomplete combustion forms tiny particles of carbon that glow brightly, creating the

Figure 5.9 | **Water is a product of the combustion of hydrocarbons.** Here we see clouds of condensed water vapor coming from the stacks of an oil-fired electric generating plant during the winter.

Picture Net/©Corbis

Figure 5.10 | **Incomplete combustion of a hydrocarbon.** The bright yellow color of a candle flame is caused by glowing particles of elemental carbon. Here we see that a black deposit of carbon is formed when the flame contacts a cold porcelain surface.

Andy Washnik

candle's light. If a cold surface is held in the flame, the unburned carbon results in a black deposit, as seen in Figure 5.10.

An important commercial reaction is the incomplete combustion of methane in a very limited oxygen supply, which follows the equation

$$CH_4 + O_2 \longrightarrow C + 2H_2O \quad \text{(in a very limited oxygen supply)}$$

The carbon that forms is very finely divided and would be called soot by almost anyone observing the reaction. Nevertheless, such soot has considerable commercial value when collected and marketed under the name **lampblack**. This sooty form of carbon is used to manufacture inks, and tons of carbon is used in the production of rubber tires, where it serves as a binder and a filler. When soot from incomplete combustion is released into the air, its tiny particles constitute a component of air pollution referred to as **particulates**, which contribute to the haziness of smog.

■ This finely divided form of carbon is also called *carbon black*.

Combustion of Organic Compounds That Contain Oxygen

Earlier we mentioned that you might choose wood to build a fire. The chief combustible ingredient in wood is cellulose, a fibrous material that gives plants their structural strength. Cellulose is composed of the elements carbon, hydrogen, and oxygen. Each cellulose molecule consists of many small, identical groups of atoms that are linked together to form a very long molecule, although the lengths of the molecules differ. For this reason we cannot specify a molecular formula for cellulose. Instead, we use the empirical formula, $C_6H_{10}O_5$, which represents the small, repeating "building block" units in large cellulose molecules. When cellulose burns, the products are mainly carbon dioxide and water. The only difference between its reaction and the reaction of a hydrocarbon with oxygen is that some of the oxygen in the products comes from the cellulose.

■ The formula for cellulose can be expressed as $(C_6H_{10}O_5)n$, which indicates that the molecule contains the $C_6H_{10}O_5$ unit repeated some large number n times.

$$C_6H_{10}O_5 + 6O_2 \longrightarrow 6CO_2 + 5H_2O$$

The complete combustion of all other organic compounds containing only carbon, hydrogen, and oxygen produces the same products, CO_2 and H_2O, and follows similar equations.

Burning Organic Compounds That Contain Sulfur

A major pollution problem is caused by the release into the atmosphere of sulfur dioxide formed by the combustion of fuels that contain sulfur or its compounds. *The products of the combustion of organic compounds of sulfur are carbon dioxide, water, and sulfur dioxide.* A typical reaction is

TOOLS

Combustion of an organic compound containing C, H, and O

TOOLS

Combustion of an organic compound containing sulfur

$$2C_2H_5SH + 9O_2 \longrightarrow 4CO_2 + 6H_2O + 2SO_2$$

A solution of sulfur dioxide in water is acidic, and when rain falls through polluted air it picks up SO_2 and becomes "acid rain." Some SO_2 is also oxidized to SO_3, which reacts with moisture to give H_2SO_4, making the acid rain even more acidic.

Reactions with Oxygen Are Not Always Combustions

The luminol reaction that gave us the striking photo at the start of this chapter is an example of a reaction with molecular oxygen that is not a combustion reaction. The first step in the reaction is the decomposition of hydrogen peroxide that is enhanced or catalyzed by the iron in hemoglobin.

$$2H_2O_2 \longrightarrow 2H_2O + O_2$$

In this reaction half of the oxygen is reduced and half is oxidized. Next, the oxygen reacts with the luminol, which is an anion in basic solution,

$$C_8H_5N_3O_2^{2-} + O_2 \longrightarrow C_8H_5NO_4^{2-} + N_2$$

In the process, some of the energy of the reaction is converted to the blue light, and this is an example of chemiluminesence.

Practice Exercise 5.15

Write a balanced chemical equation for the combustion of thiophene, C_4H_4S, which is a precursor in synthesizing pharmaceuticals and agrochemicals. (*Hint:* What happens to sulfur under these conditions?)

Practice Exercise 5.16

Write a balanced equation for the combustion of propylene glycol, $C_3H_8O_2$, a solvent used for injectable drugs. Isoprene is emitted by trees and is used in the production of synthetic rubber.

Reactions of Metals with Oxygen

We don't often think of metals as undergoing combustion, but have you ever lit a sparkler on the Fourth of July? The source of light is the reaction of the metal magnesium with oxygen. In fact, the light emitted by a magnesium flare, shown in Figure 5.11, is enough to light up large areas such as a battlefield. The equation for the reaction is

$$2Mg + O_2 \longrightarrow 2MgO$$

Figure 5.11 | **Magnesium flares.** Magnesium burns so brightly that flares of burning magnesium easily illuminate a nighttime tank firing range.

Koichi Kamoshida/Getty Images

Most metals react directly with oxygen, although not so spectacularly, and usually we refer to the reaction as **corrosion** or **tarnishing** because the oxidation products dull the shiny metal surface. Iron, for example, is oxidized fairly easily, especially in the presence of moisture. As you know, under these conditions the iron corrodes—it rusts. **Rust** is a form of iron(III) oxide, Fe_2O_3, that also contains an appreciable amount of absorbed water. The formula for rust is therefore normally given as $Fe_2O \cdot xH_2O$ to indicate its somewhat variable composition. Although the rusting of iron is a slow reaction, the combination of iron with oxygen can be speeded up if the metal is heated to a very high temperature under a stream of pure O_2 (see Figure 5.12).

An aluminum surface, unlike that of iron, is not noticeably dulled by the reaction of aluminum with oxygen. Aluminum is found around the home in uses ranging from aluminum foil to aluminum window frames, and it definitely appears shiny. However, aluminum is a rather easily oxidized metal, as can be seen from its position in the activity series (Table 5.3). A freshly exposed surface of the metal does react very quickly with oxygen and becomes coated with a very thin film of aluminum oxide, Al_2O_3, so thin that it doesn't obscure the shininess of the metal beneath. This oxide coating adheres very tightly to the surface of the metal and makes it very difficult for additional oxygen to combine with the aluminum. Therefore, further oxidation of aluminum occurs very slowly. On the other hand, powdered aluminum, which has a large surface area, reacts violently when ignited and is part of the propellant in the booster rockets for the space shuttle.

Figure 5.12 | Hot steel wool burns in oxygen. The reaction of hot steel wool with oxygen speeds up the oxidation process, and the sparks and flames result.

Write a balanced chemical equation for the reaction of molecular oxygen with strontium metal to form the oxide. (*Hint:* Strontium, Sr, is in the same group in the periodic table as calcium.)

Practice Exercise 5.17

The oxide formed in the reaction shown in Figure 5.12 is iron(III) oxide. Write a balanced chemical equation for the reaction.

Practice Exercise 5.18

5.6 | Stoichiometry of Redox Reactions

In general, working stoichiometry problems involving redox reactions follows the same principles we've applied to other reactions. The principal difference is that the chemical equations are more complex. Nevertheless, once we have a balanced equation, the moles of substances involved in the reaction are related by the coefficients in the balanced equation.

Because so many reactions involve oxidation and reduction, it should not be surprising that they have useful applications in the lab. Some redox reactions are especially useful in chemical analyses, particularly in titrations. In acid–base titrations, we often need a specific indicator for each reaction to signal the end point. However, there are not many indicators that can be used to conveniently signal the end points in redox titrations. In practice, there are a few oxidizing agents that are highly colored and can act as their own indicator, so we rely on color changes of the reactants themselves.

One of the most useful reactants for redox titrations is potassium permanganate, $KMnO_4$, especially when the reaction can be carried out in an acidic solution. Permanganate ion is a powerful oxidizing agent, so it oxidizes most substances that are capable of being oxidized. That's one reason why it is used. Especially important, though, is the fact that the ion has an intense purple color and its reduction product in acidic solution is the almost colorless Mn^{2+} ion. Therefore, when a solution of $KMnO_4$ is added from a buret to a solution of a reducing agent, the chemical reaction that occurs forms a nearly colorless

Figure 5.13 | **Reduction of MnO$_4^-$ by Fe^{2+}.** A solution of KMnO is added to a stirred acidic solution containing Fe^{2+}. The reaction oxidizes the pale blue-green Fe^{2+} to Fe^{3+}, while the MnO$_4^-$ is reduced to the almost colorless Mn^{2+} ion. The purple color of the permanganate will continue to be destroyed until all of the Fe^{2+} has reacted. Only then will the iron-containing solution take on a pink or purple color. This ability of MnO$_4^-$ to signal the completion of the reaction makes it especially useful in redox titrations, where it serves as its own indicator.

product. This is illustrated in Figure 5.13, where we see a solution of KMnO$_4$ being poured into an acidic solution containing Fe^{2+}. As the KMnO$_4$ solution is added, the purple color continues to be destroyed as long as there is any reducing agent left. In a titration, after the last trace of the reducing agent has been consumed, the MnO$_4^-$ ion in the next drop of titrant has nothing to react with, so it colors the solution pink. This signals the end of the titration. In this way, permanganate ion serves as its own indicator in redox titrations. The next multi-concept problem illustrates a typical analysis using KMnO$_4$ in a redox titration.

■ Concentrated solutions of MnO$_4^-$ are purple colored, but dilute solutions look pink.

Analyzing and Solving Multi-Concept Problems

All the iron in a 2.00 g sample of an iron ore was dissolved in an acidic solution and converted to Fe^{2+}, which was then titrated with 0.100 M KMnO$_4$ solution. In the titration the iron was oxidized to Fe^{3+}. The titration required 27.45 mL of the KMnO$_4$ solution to reach the end point. (a) How many grams of iron were in the sample? (b) What was the percentage of iron in the sample? (c) If the iron was present in the sample as Fe$_2$O$_3$, what was the percentage by mass of Fe$_2$O$_3$ in the sample?

Analysis: This problem has several parts to it, so to solve it we will divide it into three parts. The first will be to construct a balanced equation to get the correct stoichiometry, the second will be to find the mass of iron and the percentage of iron in the sample, and the third will be to find the percentage by mass of the Fe$_2$O$_3$. In the final two parts, we will use tools we learned in Chapters 3 and 4, as well as tools from this chapter.

PART 1

Assembling the Tools: To start with, let's use the method for balancing redox reactions in acidic solutions as a tool, even though the question does not explicitly state that it is needed.

Solution: The compound KMnO$_4$ provides MnO$_4^-$, and the K$^+$ is a spectator ion and is not included in balancing the equation. The skeleton equation for the reaction is

$$Fe^{2+} + MnO_4^- \longrightarrow Fe^{3+} + Mn^{2+}$$

Balancing it by the ion–electron method for acidic solutions gives

$$5Fe^{2+} + MnO_4^- + 8H^+ \longrightarrow 5Fe^{3+} + Mn^{2+} + 4H_2O$$

PART 2

Assembling the Tools: With this information, let's look over the tools we'll use to answer the first two questions. We'll tackle the calculation of percent Fe_2O_3 afterward.

The first tool relates the molarity and volume of the $KMnO_4$; this will give us moles of $KMnO_4$ used in the titration. We will use the coefficients of the equation, the molar mass, and the equation for calculating the percentage of iron as the tools in this section.

Solution: The number of moles of MnO_4^- consumed in the reaction is calculated from the volume of the solution used in the titration and its concentration.

$$0.02745 \text{ L MnO}_4^- \text{ soln}\left(\frac{0.100 \text{ mol MnO}_4^-}{1 \text{ L MnO}_4^- \text{ soln}}\right) = 0.002745 \text{ mol MnO}_4^-$$

Next, we use the coefficients of the equation to calculate the number of moles of Fe^{2+} that reacted. The balanced chemical equation tells us five moles of Fe^{2+} react per mole of MnO_4^- consumed.

$$0.002745 \text{ mol MnO}_4^-\left(\frac{5 \text{ mol Fe}^{2+}}{1 \text{ mol MnO}_4^-}\right) = 0.01372 \text{ mol Fe}^{2+}$$

This is the number of moles of iron in the ore sample, so the mass of iron in the sample is

$$0.01372 \text{ mol Fe}\left(\frac{55.845 \text{ g Fe}}{1 \text{ mol Fe}}\right) = 0.766 \text{ g Fe}$$

Next, we calculate the percentage of iron in the sample, which is the mass of iron divided by the mass of the sample, all multiplied by 100%.

$$\%\text{Fe} = \frac{\text{g Fe}}{\text{g sample}} \times 100\%$$

$$\%\text{Fe} = \frac{0.766 \text{ g Fe}}{2.00 \text{ g sample}} \times 100\% = 38.3\%$$

The answer to part (b) is that the sample is 38.3% iron.

PART 3

Assembling the Tools: In Part 2 we determined that 0.0137 mol Fe had reacted. We need to convert this to moles and then mass of Fe_2O_3. We will use the chemical formula, Fe_2O_3, and the molar mass of Fe_2O_3 to calculate the mass of Fe_2O_3 and also percentage composition equation analogous to the one in Part 2.

Solution: (c) The chemical formula for the iron oxide gives us

$$1 \text{ mol Fe}_2O_3 \Leftrightarrow 2 \text{ mol Fe}$$

This provides the conversion factor we need to determine how many moles of Fe_2O_3 were present in the sample. Working with the number of moles of Fe,

$$0.0137 \text{ mol Fe}\left(\frac{1 \text{ mol Fe}_2O_3}{2 \text{ mol Fe}}\right) = 0.00685 \text{ mol Fe}_2O_3$$

This is the number of moles of Fe_2O_3 in the sample. Since the formula mass of Fe_2O_3 is 159.69 g mol^{-1}, the mass of Fe_2O_3 in the sample was

$$0.00685 \text{ mol } Fe_2O_3\left(\frac{159.69 \text{ g } Fe_2O_3}{1 \text{ mol } Fe_2O_3}\right) = 1.094 \text{ g } Fe_2O_3$$

Finally, the percentage of Fe_2O_3 in the sample was

$$\%Fe_2O_3 = \frac{1.094 \text{ g } Fe_2O_3}{2.00 \text{ g sample}} \times 100\% = 54.7\% \ Fe_3O_3$$

The ore sample contained 54.8% Fe_2O_3.

Are the Answers Reasonable? For the first part, we have to make sure each side of the equation has the same number of atoms of each element, and for the charge we need to check that the net charge is the same on both sides.

For the second part, we can use some approximate arithmetic to estimate the answer. In the titration we used approximately 30 mL, or 0.030 L, of the $KMnO_4$ solution, which is 0.10 M. Multiplying these tells us we've used approximately 0.003 mol of $KMnO_4$. From the coefficients of the equation, five times as many moles of Fe^{2+} react, so the amount of Fe in the sample is approximately $5 \times 0.003 = 0.015$ mol. The atomic mass of Fe is about 55 g/mol, so the mass of Fe in the sample is approximately $0.015 \times 55 = 0.8$ g. Our answer, 0.766 g Fe, is reasonable.

For the last part, we've noted that the amount of Fe in the sample is approximately 0.015 mol. The amount of Fe_2O_3 that contains this much Fe is 0.0075 mol. The formula mass of Fe_2O_3 is about 160, so the mass of Fe_2O_3 in the sample was approximately $0.0075 \times 160 = 1.2$ g, which isn't too far from the mass we obtained (1.095 g Fe_2O_3). Since 1.09 g is about half of the total sample mass of 2.00 g, the sample was approximately 50% Fe_2O_3, in agreement with the answer we obtained.

| Summary

Organized by Learning Objective

Define oxidation, reduction, oxidizing agents, reducing agents, and oxidation numbers

Oxidation is the loss of electrons or an algebraic increase in oxidation number; **reduction** is the gain of electrons or an algebraic decrease in oxidation number. Both always occur together in **redox** reactions. The substance oxidized is the **reducing agent**; the substance reduced is the **oxidizing agent**. **Oxidation numbers** are a bookkeeping device that we use to follow changes in redox reactions. They are assigned according to the rules in Section 5.1. The term **oxidation state** is equivalent to oxidation number. These terms are summarized in Table 5.1.

Balance equations for oxidation–reduction reactions in acidic or basic solutions

For equations that cannot be balanced by inspection, the **ion–electron method** provides a systematic method for balancing a redox reaction in aqueous solution. According to this method, the *skeleton* net ionic equation is divided into two **half-reactions**, which are balanced separately before being recombined to give the final balanced net ionic equation. For reactions in basic solutions the equation is balanced as if it occurred in an acidic

solution, and it is converted to its proper form for basic solutions by adding an appropriate number of OH^- to both sides and canceling H_2O if possible.

Illustrate, at the molecular level, how a metal reacts with an acid

In **nonoxidizing acids**, the strongest oxidizing agent is H^+ (Table 5.2). The reaction of a metal with a nonoxidizing acid gives hydrogen gas and a salt of the acid. Only metals more active than hydrogen react this way. These are metals that are located below hydrogen in the **activity series** (Table 5.3). **Oxidizing acids**, like HNO_3, contain an anion that is a stronger oxidizing agent than H^+, and they are able to oxidize many metals that nonoxidizing acids cannot.

Explain the activity series of metals and use it to predict the product of a redox reaction involving a metal

If one metal is more easily oxidized than another, it can displace the other metal from its compounds by a redox reaction. Such reactions are sometimes called **single replacement reactions**. Atoms of the more active metal become ions; ions of the less active metal generally become atoms. In this manner, any metal in the **activity series** can displace any of the others above it in the series from their compounds.

Describe the reaction with oxygen of organic compounds, metals, and nonmetals

Combustion is the rapid reaction of a substance with oxygen accompanied by the evolution of heat and light. Combustion of a hydrocarbon in the presence of excess oxygen gives mostly CO_2 and H_2O. When the supply of oxygen is limited, mainly CO forms, and in a very limited supply of oxygen the products are H_2O and mostly very finely divided, elemental carbon. When hydrocarbons are burned in air, a mixture of H_2O, CO_2, CO, and C is formed. The combustion of organic compounds containing only carbon, hydrogen, and oxygen also gives the same products. Most nonmetals also burn in oxygen to give

molecular oxides. Sulfur burns to give SO_2, which also forms when sulfur-containing fuels burn. Nitrogen reacts with oxygen to give a variety of nitrogen oxides, including NO and NO_2. Many metals combine with oxygen to produce metal oxides in a process often called **corrosion**, but only sometimes is the reaction rapid enough to be considered combustion.

Perform calculations involving stoichiometry of a redox reaction

Titrations involving redox reactions can be carried out, just be sure the chemical equation is properly balanced. Potassium permanganate is often used in redox titrations because it is a powerful oxidizing agent and serves as its own indicator. In acidic solutions, the purple MnO_4^- ion is reduced to the nearly colorless Mn^{2+} ion.

Tools for Problem Solving

The following tools were introduced in this chapter. Study them carefully so that you can select the appropriate tool when needed.

Oxidizing and reducing agents (Section 5.1)
The substance reduced is the oxidizing agent; the substance oxidized is the reducing agent.

Assigning oxidation numbers (Section 5.1)
The rules permit us to assign oxidation numbers to elements in compounds and ions.

Oxidation and reduction reactions (Section 5.1)
In an oxidation reaction, the substance loses electrons; in a reduction reaction, the reactant gains electrons. Oxidation and reduction reactions always occur together.

Ion–electron method for acidic solutions (Section 5.2)
Use this method when you need to obtain a balanced net ionic equation for a redox reaction in an acidic solution.

Ion–electron method for basic solutions (Section 5.2)
Use this method when you need to obtain a balanced net ionic equation for a redox reaction in a basic solution.

Oxidizing and nonoxidizing acids (Table 5.2, Section 5.3)
Nonoxidizing acids will react with metals below hydrogen in Table 5.3 to give H_2 and the metal ion.

Activity series of metals (Table 5.3, Section 5.4)
A metal in the table will reduce the ion of any metal above it in the table, leading to a single replacement reaction.

Combustion with oxygen (Section 5.5)
Hydrocarbon combustion with plentiful supply of oxygen

$$\text{hydrocarbon} + O_2 \longrightarrow CO_2 + H_2O \quad \text{(plentiful supply of } O_2\text{)}$$

Hydrocarbon combustion with limited supply of oxygen

$$\text{hydrocarbon} + O_2 \longrightarrow CO + H_2O \quad \text{(limited supply of } O_2\text{)}$$

Hydrocarbon combustion with extremely limited supply of oxygen

$$\text{hydrocarbon} + O_2 \longrightarrow C + H_2O \quad \text{(very limited supply of } O_2\text{)}$$

Combustion of compounds containing C, H, and O
Complete combustion gives CO_2 and H_2O,

$$(\text{C, H, O compound}) + O_2 \longrightarrow CO_2 + H_2O \quad \text{(complete combustion)}$$

Combustion of organic compounds containing sulfur
If an organic compound contains sulfur, SO_2 is formed in addition to CO_2 and H_2O when the compound is burned.

WileyPLUS = *WileyPLUS,* an online teaching and learning solution. *Note to instructors*: Many of the end-of-chapter problems are available for assignment via the WileyPLUS system. **www.wileyplus.com.** Review Problems are presented in pairs separated by blue rules. Answers to problems whose numbers appear in blue are given in Appendix B. More challenging problems are marked with an asterisk *.

Review Questions

Oxidation–Reduction Reactions

5.1 Define oxidation and reduction in terms of **(a)** electron transfer and **(b)** oxidation numbers.

5.2 Why must both oxidation and reduction occur simultaneously during a redox reaction? What is an oxidizing agent and what happens to it in a redox reaction? What is a reducing agent and what happens to it in a redox reaction?

5.3 In the compound As_4O_6, arsenic has an *oxidation number* of +3. What is the *oxidation state* of arsenic in this compound?

5.4 Are the following redox reactions? Explain.

$$2NO_2 \longrightarrow N_2O_4$$

$$2CrO_4^{2-} + 2H^+ \longrightarrow 2Cr_2O_7^{2-} + H_2O$$

5.5 If the oxidation number of nitrogen in a certain molecule changes from +3 to −2 during a reaction, is the nitrogen oxidized or reduced? How many electrons are gained or lost by the nitrogen atom?

Balancing Redox Reactions

5.6 When balancing redox reactions, which side of a half-reaction gets the electrons?

5.7 Why is water used to balance oxygen atoms in a redox reaction, and not O_2?

5.8 What are the net charges on the left and right sides of the following equations? Add electrons as necessary to make each of them a balanced half-reaction.

(a) $NO_3^- + 10H^+ \longrightarrow NH_4^+ + 3H_2O^+$

(b) $Cl_2 + 4H_2O \longrightarrow 2ClO_2^- + 8H^+$

5.9 In Question 5.8, which half-reaction represents oxidation? Which represents reduction?

5.10 The following equation is not balanced.

$$Ag + Fe^{2+} \longrightarrow Ag^+ + Fe$$

Why? Use the ion–electron method to balance it. Can you balance this reaction by inspection?

Acids as Oxidizing Agents

5.11 What is a nonoxidizing acid? Give two examples. What is the oxidizing agent in a nonoxidizing acid?

5.12 What is the strongest oxidizing agent in an aqueous solution of nitric acid?

5.13 What are the possible products of the reduction of nitric acid?

5.14 What are the possible products of the reduction of sulfuric acid?

Redox Reactions of Metals

5.15 What is a *single replacement reaction*?

5.16 If a metal is able to react with a solution of HCl, where must the metal stand relative to hydrogen in the activity series?

5.17 Which metals in Table 5.3 will not react with nonoxidizing acids?

5.18 Which metals in Table 5.3 will react with water? Write chemical equations for each of these reactions.

5.19 Where in the activity series do we find the best reducing agents? Where do we find the best oxidizing agents?

5.20 When manganese reacts with silver ions, is manganese oxidized or reduced? Is it an oxidizing agent or a reducing agent?

Molecular Oxygen as an Oxidizing Agent

5.21 Define combustion, rusting, and tarnishing.

5.22 Why is "loss of electrons" described as oxidation?

5.23 What are the major products produced in the combustion of $C_{10}H_{22}$ under the following conditions: **(a)** an excess of oxygen, **(b)** a slightly limited oxygen supply, **(c)** a very limited supply of oxygen, **(d)** the compound is burned in air?

5.24 If one of the impurities in diesel fuel has the formula C_2H_6S, what products will be formed when it burns? Write a balanced chemical equation for the reaction.

5.25 Burning ammonia in an atmosphere of oxygen produces stable N_2 molecules as one of the products. What is the other product? Write the balanced chemical equation for the reaction.

5.26 What does burning sulfur in air produce? What is the limiting reactant in this reaction?

Review Problems

Oxidation–Reduction Reactions

5.29 Assign oxidation numbers to the atoms in the following: **(a)** ClO_4^-, **(b)** Cl^-, **(c)** SF_6, and **(d)** $Au(NO_3)_3$.

5.30 Assign oxidation numbers to the atoms in the following: **(a)** S^{2-}, **(b)** SO_2, **(c)** P_4, and **(d)** PH_3.

5.31 Assign oxidation numbers to each atom in the following: **(a)** NaOBr, **(b)** $NaBrO_2$, **(c)** $NaBrO_3$, and **(d)** $NaBrO_4$.

5.32 Assign oxidation numbers to the elements in the following: **(a)** $MnCl_2$, **(b)** MnO_4^-, **(c)** MnO_4^{2-}, and **(d)** MnO_2.

5.33 Assign oxidation numbers to the elements in the following: **(a)** Bi_2S_3, **(b)** $CeCl_4$, **(c)** CsO_2, and **(d)** O_2F_2.

5.34 Assign oxidation numbers to the elements in the following: **(a)** $Sr(BrO_3)_2$, **(b)** Cr_2S_3, **(c)** OF_2, and **(d)** HOF.

5.35 Titanium burns in pure nitrogen to form TiN. What are the oxidation states of titanium and nitrogen in TiN?

5.36 Zirconia, which is ZrO_2, is used to make ceramic knives. What are the oxidation states of zirconium and oxygen in zirconia?

5.37 Ozone, O_3, is an allotrope of oxygen and is one of the oxidants responsible for the haze over the Smoky Mountains. What is the oxidation number of the oxygen atoms in ozone?

5.38 The other major air pollutant is NO_2. What are the oxidation numbers of the atoms in NO_2?

5.39 For the following reactions, identify the substance oxidized, the substance reduced, the oxidizing agent, and the reducing agent.

(a) $2HNO_3 + 3H_3AsO_3 \longrightarrow$
$$2NO + 3H_3AsO_4 + H_2O$$
(b) $NaI + 3HOCl \longrightarrow NaIO_3 + 3HCl$
(c) $2KMnO_4 + 5H_2C_2O_4 + 3H_2SO_4 \longrightarrow$
$$10CO_2 + K_2SO_4 + 2MnSO_4 + 8H_2O$$
(d) $6H_2SO_4 + 2Al \longrightarrow Al_2(SO_4)_3 + 3SO_2 + 6H_2O$

5.40 For the following reactions, identify the substance oxidized, the substance reduced, the oxidizing agent, and the reducing agent.

(a) $Cu + 2H_2SO_4 \longrightarrow CuSO_4 + SO_2 + 2H_2O$
(b) $3SO_2 + 2HNO_3 + 2H_2O \longrightarrow 3H_2SO_4 + 2NO$

Stoichiometry of Redox Reactions

5.27 How does a titration of a redox reaction differ from the titration of an acid with a base?

5.28 Potassium permanganate is often used for redox titrations. Why?

(c) $5H_2SO_4 + 4Zn \longrightarrow 4ZnSO_4 + H_2S + 4H_2O$
(d) $I_2 + 10HNO_3 \longrightarrow 2HIO_3 + 10NO_2 + 4H_2O$

5.41 When chlorine is added to drinking water to kill bacteria, some of the chlorine is changed into ions by the following equilibrium:

$$Cl_2(aq) + H_2O \rightleftharpoons H^+(aq) + Cl^-(aq) + HOCl(aq)$$

In the forward reaction (the reaction going from left to right), which substance is oxidized and which is reduced? In the reverse reaction, which is the oxidizing agent and which is the reducing agent?

5.42 One pollutant in smog is nitrogen dioxide, NO_2. The gas has a reddish brown color and is responsible for the red-brown color associated with this type of air pollution. Nitrogen dioxide is also a contributor to acid rain because when rain passes through air contaminated with NO_2, it dissolves and undergoes the following reaction:

$$3NO_2(g) + H_2O \longrightarrow NO(g) + 2H^+(aq) + 2NO_3^-(aq)$$

In this reaction, which element is reduced and which is oxidized? Which is the oxidizing agent and which is the reducing agent?

Balancing Redox Equations

5.43 Balance the following equations for reactions occurring in an acidic solution.

(a) $S_2O_3^{2-} + OCl^- \longrightarrow Cl^- + S_4O_6^{2-}$
(b) $NO_3^- + Cu \longrightarrow NO_2 + Cu^{2+}$
(c) $IO_3^- + H_3AsO_3 \longrightarrow I^- + H_3AsO_4$
(d) $SO_4^{2-} + Zn \longrightarrow Zn^{2+} + SO_2$

5.44 Balance the following equations for reactions occurring in an acidic solution.

(a) $NO_3^- + Zn \longrightarrow NH_4^+ + Zn^{2+}$
(b) $Cr^{3+} + BiO_3^- \longrightarrow Cr_2O_7^{2-} + Bi^{3+}$
(c) $I_2 + OCl^- \longrightarrow IO_3^- + Cl^-$
(d) $Mn^{2+} + BiO_3^- \longrightarrow MnO_4^- + Bi^{3+}$

5.45 Balance the following equations for reactions occurring in an acidic solution.

(a) $Sn + NO_3^- \longrightarrow SnO_2 + NO$
(b) $PbO_2 + Cl^- \longrightarrow PbCl_2 + Cl_2$

(c) $Ag + NO_3^- \longrightarrow NO_2 + Ag^+$

(d) $Fe^{3+} + NH_3OH^+ \longrightarrow Fe^{2+} + N_2O$

5.46 Balance the following equations for reactions occurring in an acidic solution.

(a) $H_2C_2O_4 + HNO_2 \longrightarrow CO_2 + NO$

(b) $HNO_2 + MnO_4^- \longrightarrow Mn^{2+} + NO_3^-$

(c) $H_3PO_2 + Cr_2O_7^{2-} \longrightarrow H_3PO_4 + Cr^{3+}$

(d) $XeF_2 + Cl^- \longrightarrow Xe + F^- + Cl_2$

5.47 Balance the equations for the following reactions occurring in a basic solution.

(a) $2CrO_4^{2-} + S^{2-} \longrightarrow S + CrO_4^{2-}$

(b) $MnO_4^- + C_2O_4^{2-} \longrightarrow CO_2 + MnO_2$

(c) $ClO_3^- + N_2H_4 \longrightarrow NO + Cl^-$

(d) $SO_3^{2-} + MnO_4^- \longrightarrow SO_4^{2-} + MnO_2$

5.48 Balance the equations for the following reactions occurring in a basic solution.

(a) $CrO_2^- + S_2O_8^{2-} \longrightarrow CrO_4^{2-} + SO_4^{2-}$

(b) $SO_3^{2-} + 2CrO_4^{2-} \longrightarrow SO_4^- + CrO_2^-$

(c) $O_2 + N_2H_4 \longrightarrow H_2O_2 + N_2$

(d) $Au + CN^- + O_2 \longrightarrow Au(CN)_4^- + OH^-$

5.49 When very dilute nitric acid reacts with a strong reducing agent such as magnesium, the nitrate ion is reduced to ammonium ion. Write a balanced net ionic equation for the reaction.

5.50 Hydroiodic acid reduces chlorine to hydrochloric acid and iodine. Write a balanced net ionic equation for the reaction.

5.51 Syn-gas, once known as water-gas, is a mixture of hydrogen and carbon monoxide. Syn-gas can be produced by passing steam over red-hot charcoal. Write the balanced chemical equation for producing syn-gas.

5.52 Corn is grown for human consumption, feeding livestock, and producing ethanol for fuel. The starches in corn are broken down by enzymes into dextrose, $C_6H_{12}O_6$, and a yeast fermentation process produces carbon dioxide and ethanol, CH_3CH_2OH, from the dextrose. Write the balanced chemical equation for the production of ethanol.

5.53 Calcium oxalate is one of the minerals found in kidney stones. If a strong acid is added to calcium oxalate, the compound will dissolve and the oxalate ion will be changed to oxalic acid (a weak acid). Oxalate ion is a moderately strong reducing agent. Write a balanced net ionic equation for the oxidation of $H_2C_2O_4$ by $K_2Cr_2O_7$ in an acidic solution. The reaction yields Cr^{3+} and CO_2 among the products.

5.54 Laundry bleach such as Clorox is a dilute solution of sodium hypochlorite, NaOCl. Write a balanced net ionic equation for the reaction of NaOCl with $Na_2S_2O_3$. The OCl^- is reduced to chloride ion and the $S_2O_3^{2-}$ is oxidized to sulfate ion.

5.55 Ozone, O_3, is a very powerful oxidizing agent, and in some places ozone is used to treat water to kill bacteria and make it safe to drink. One of the problems with this method of purifying water is that if there is any bromide ion in the water, it becomes oxidized to bromate ion, which has shown evidence of causing cancer in test animals. Assuming that ozone is reduced to water, write a balanced chemical equation for the reaction. (Assume an acidic solution.)

5.56 Chlorine is a good bleaching agent because it is able to oxidize substances that are colored to give colorless reaction products. It is used in the pulp and paper industry as a bleach, but after it has done its work, residual chlorine must be removed. This is accomplished using sodium thiosulfate, $Na_2S_2O_3$, which reacts with the chlorine, reducing it to chloride ion. The thiosulfate ion is changed to sulfate ion, which is easily removed by washing with water. Write a balanced chemical equation for the reaction of chlorine with thiosulfate ion, assuming an acidic solution.

Acids as Oxidizing Agents

5.57 Write balanced molecular, ionic, and net ionic equations for the reactions of the following metals with hydrochloric acid to give hydrogen plus the metal ion in solution.

(a) Manganese (gives Mn^{2+})

(b) Cadmium (gives Cd^{2+})

(c) Tin (gives Sn^{2+})

5.58 Write balanced molecular, ionic, and net ionic equations for the reactions of the following metals with hydrochloric acid to give hydrogen plus the metal ion in solution.

(a) Cobalt (gives Co^{2+})

(b) Cesium (gives Cs^+)

(c) Zinc (gives Zn^{2+})

5.59 Write balanced molecular, ionic, and net ionic equations for the reaction of each of the following metals with dilute sulfuric acid.

(a) Nickel (gives Ni^{2+})

(b) Chromium (gives Cr^{3+})

5.60 Write balanced molecular, ionic, and net ionic equations for the reaction of each of the following metals with dilute sulfuric acid.

(a) Barium (gives Ba^{2+})

(b) Aluminum (gives Al^{3+})

5.61 On the basis of the discussions in this chapter, suggest balanced chemical equations for the oxidation of metallic silver to Ag^+ ion with (a) dilute HNO_3 and (b) concentrated HNO_3.

5.62 On the basis of the discussions in this chapter, suggest balanced chemical equations for the oxidation of metallic mercury to Hg^{2+} ion with (a) dilute HNO_3 and (b) concentrated HNO_3.

5.63 Hot, concentrated sulfuric acid is a fairly strong oxidizing agent. Write a balanced net ionic equation for the oxidation of metallic copper to copper(II) ion by hot, concentrated H_2SO_4, in which the sulfur is reduced to SO_2. Write a balanced molecular equation for the reaction.

5.64 Hot, concentrated sulfuric acid is a fairly strong oxidizing agent. Write a balanced net ionic equation for the oxidation of carbon, graphite, to carbon dioxide by hot, concentrated H_2SO_4, in which the sulfur is reduced to SO_2. Write a balanced molecular equation for the reaction.

Redox Reactions of Metals

5.65 Use Table 5.3 to predict the outcome of the following reactions. If no reaction occurs, write N.R. If a reaction occurs, write a balanced chemical equation for it.

(a) $Mn + Fe^{2+} \longrightarrow$ (c) $Mg + Co^{2+} \longrightarrow$

(b) $Cd + Zn^{2+} \longrightarrow$ (d) $Cr + Sn^{2+} \longrightarrow$

5.66 Use Table 5.3 to predict the outcome of the following reactions. If no reaction occurs, write N.R. If a reaction occurs, write a balanced chemical equation for it.

(a) $Fe + Mg^{2+} \longrightarrow$ (c) $Ag^+ + Fe \longrightarrow$

(b) $Cr + Pb^{2+} \longrightarrow$ (d) $Ag + Au^{3+} \longrightarrow$

5.67 The following reactions occur spontaneously.

$$Pu + 3Tl^+ \longrightarrow Pu^{3+} + 3Tl$$
$$Ru + Pt^{2+} \longrightarrow Ru^{2+} + Pt$$
$$2Tl + Ru^{2+} \longrightarrow 2Tl^+ + Ru$$

List the metals Pu, Pt, and Tl in order of increasing ease of oxidation.

5.68 The following reactions occur spontaneously.

$$2Y + 3Ni^{2+} \longrightarrow 2Y^{3+} + 3Ni$$
$$2Mo + 3Ni^{2+} \longrightarrow 2Mo^{3+} + 3Ni$$
$$Y^{3+} + Mo \longrightarrow Y + Mo^{3+}$$

List the metals Y, Ni, and Mo in order of increasing ease of oxidation.

5.69 The following reaction occurs spontaneously.

$$Ru^{2+}(aq) + Cd(s) \longrightarrow Ru(s) + Cd^{2+}(aq)$$

What reaction will occur if a mixture is prepared containing the following: $Cd(s)$, $Cd(NO_3)_2(aq)$, $Tl(s)$, and $TlCl(aq)$? (Refer to the information in Problem 5.67.)

5.70 When chromium metal is dipped into a solution of nickel(II) chloride, some Cr dissolves and Ni metal deposits on the surface of the chromium. When chromium metal is dipped into a solution of ytterrium(III) chloride, no reaction occurs. Referring to Problem 5.68 which of these reactions will occur spontaneously? Explain the reason for your choice.

(a) $2Mo^{3+} + 3Cr \longrightarrow 3Cr^{2+} + 2Mo$

(b) $2Mo + 3Cr^{2+} \longrightarrow 2Mo^{3+} + 3Cr$

5.71 Write a balanced equation for the reaction depicted in the following visualization:

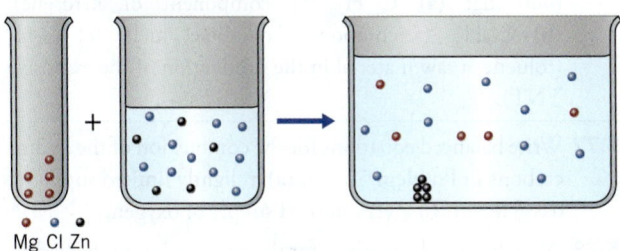

Mg Cl Zn

5.72 Write a balanced equation for the reaction depicted in the following visualization:

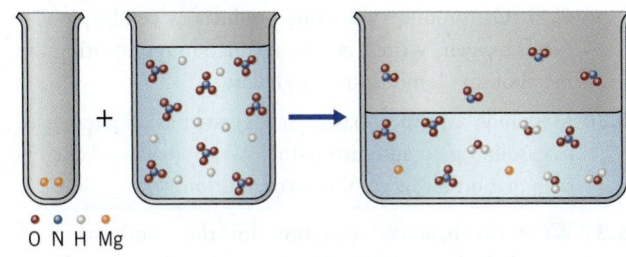

O N H Mg

5.73 Write the balanced chemical equation for the following reaction.

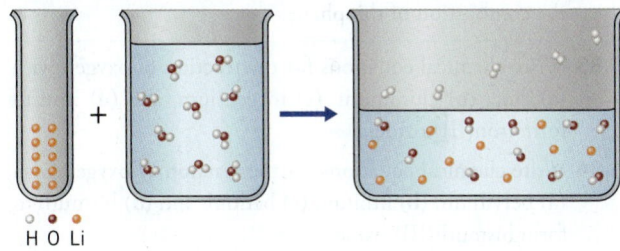

H O Li

5.74 Write the balanced chemical equation for the following reaction.

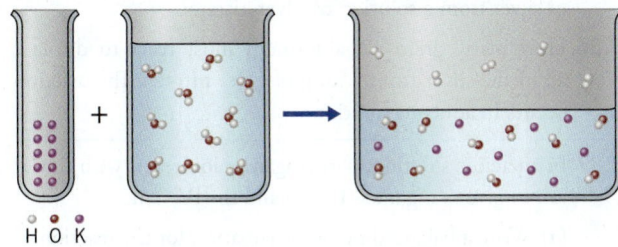

H O K

Molecular Oxygen as an Oxidizing Agent

5.75 Write balanced chemical equations for the complete combustion (in the presence of excess oxygen) of the following: (a) C_6H_6 (benzene, an important industrial chemical and solvent), (b) C_4H_{10} (butane, a fuel used in cigarette lighters), (c) $C_{21}H_{44}$ (a component of paraffin wax used in candles).

5.76 Write balanced chemical equations for the complete combustion (in the presence of excess oxygen) of the following: **(a)** $C_{12}H_{26}$ (a component of kerosene), **(b)** $C_{18}H_{36}$ (a component of diesel fuel), **(c)** C_7H_8 (toluene, a raw material in the production of the explosive TNT).

5.77 Write balanced equations for the combustion of the hydrocarbons in Problem 5.75 in **(a)** a slightly limited supply of oxygen and **(b)** a very limited supply of oxygen.

5.78 Write balanced equations for the combustion of the hydrocarbons in Problem 5.76 in **(a)** a slightly limited supply of oxygen and **(b)** a very limited supply of oxygen.

5.79 The metabolism of carbohydrates such as glucose, $C_6H_{12}O_6$, produces the same products as combustion in excess oxygen. Write a chemical equation representing the metabolism (combustion) of glucose.

5.80 Methanol, CH_3OH, has been suggested as an alternative to gasoline as an automotive fuel. Write a balanced chemical equation for its complete combustion.

5.81 Write the balanced equation for the combustion of dimethylsulfide, $(CH_3)_2S$, in an abundant supply of oxygen.

5.82 Thiophene, C_4H_4S, is an impurity in crude oil and is a source of pollution if not removed. Write an equation for the combustion of thiophene.

5.83 Write chemical equations for the reaction of oxygen with **(a)** zinc, **(b)** aluminum, **(c)** magnesium, and **(d)** iron to form iron(III) oxide.

5.84 Write chemical equations for the reaction of oxygen with **(a)** beryllium, **(b)** lithium, **(c)** barium, and **(d)** bismuth to form bismuth(III) oxide.

Stoichiometry of Redox Reactions

5.85 How many grams of copper must react to displace 12.0 g of silver from a solution of silver nitrate?

5.86 How many grams of aluminum must react to displace all of the silver from 25.0 g of silver nitrate? The reaction occurs in aqueous solution.

5.87 In an acidic solution, permanganate ion reacts with tin(II) ion to give manganese(II) ion and tin(IV) ion.

(a) Write a balanced net ionic equation for the reaction.

(b) How many milliliters of 0.230 M potassium permanganate solution are needed to react completely with 40.0 mL of 0.250 M tin(II) chloride solution?

5.88 In an acidic solution, bisulfite ion reacts with chlorate ion to give sulfate ion and chloride ion.

(a) Write a balanced net ionic equation for the reaction.

(b) How many milliliters of 0.150 M sodium chlorate solution are needed to react completely with 30.0 mL of 0.450 M sodium bisulfite solution?

5.89 Iodate ion reacts with sulfite ion to give sulfate ion and iodide ion.

(a) Write a balanced net ionic equation for the reaction.

(b) How many grams of sodium sulfite are needed to react with 5.00 g of sodium iodate?

5.90 Potable water (drinking water) should not have manganese concentrations in excess of 0.05 mg/mL. If the manganese concentration is greater than 0.1 mg/mL, it imparts a foul taste to the water and discolors laundry and porcelain surfaces. Manganese(II) ion is oxidized to permanganate ion by bismuthate ion, BiO_3^-, in an acidic solution. In the reaction, BiO_3^- is reduced to Bi^{3+}.

(a) Write a balanced net ionic equation for the reaction.

(b) How many milligrams of $NaBiO_3$ are needed to oxidize the manganese in 18.5 mg of manganese(II) sulfate?

5.91 Sulfites are used worldwide in the wine industry as antioxidant and antimicrobial agents. However, sulfites have also been identified as causing certain allergic reactions suffered by asthmatics, and the FDA mandates that sulfites be identified on the label if they are present at levels of 10 ppm (parts per million) or higher. The analysis of sulfites in wine uses the "Ripper method" in which a standard iodine solution, prepared by the reaction of iodate and iodide ions, is used to titrate a sample of the wine. The iodine is formed in the reaction

$$IO_3^- + 5I^- + 6H^+ \longrightarrow 3I_2 + 3H_2O$$

The iodine is held in solution by adding an excess of I^-, which combines with I_2 to give I_3^-. In the titration, the SO_3^{2-} is converted to SO_2 by acidification, and the reaction during the titration is

$$SO_2 + I_3^- + 2H_2O \longrightarrow SO_4^{2-} + 3I^- + 4H^+$$

Starch is added to the wine sample to detect the end point, which is signaled by the formation of a dark blue color when excess iodine binds to the starch molecules. In a certain analysis, 0.0421 g of $NaIO_3$ was dissolved in dilute acid and excess NaI was added to the solution, which was then diluted to a total volume of 100.0 mL. A 50.0 mL sample of wine was then acidified and titrated with the iodine-containing solution. The volume of iodine solution required was 2.47 mL. **(a)** What was the molarity of the iodine (actually, I_3^-) in the standard solution? **(b)** How many grams of SO_2 were in the wine sample? **(c)** If the density of the wine was 0.96 g/mL, what was the percentage of SO_2 in the wine? **(d)** Parts per million (ppm) is calculated in a manner similar to percent (which is equivalent to *parts per hundred*).

$$ppm = \frac{\text{grams of component}}{\text{grams of sample}} \times 10^6 \, ppm$$

What was the concentration of sulfite in the wine, expressed as parts per million SO_2?

5.92 Methylbromide, CH_3Br, is used in agriculture to fumigate soil to rid it of pests such as nematodes. It is injected directly into the soil, but over time it has a tendency to escape before it can undergo natural degradation to innocuous products. Soil chemists have found that ammonium thiosulfate, $(NH_4)_2S_2O_3$, a nitrogen and sulfur fertilizer, drastically reduces methylbromide emissions by causing it to degrade. In a chemical analysis to determine the purity of a batch of commercial ammonium thiosulfate, a chemist first prepared a standard solution of iodine following the procedure in Problem 5.91. First, 0.462 g KIO_3 was dissolved in water to make 125 mL of solution. The solution was made acidic and treated with excess potassium iodide, which caused the following reaction to take place:

$$IO_3^- + 8I^- + 6H^+ \longrightarrow 3I_3^- + 3H_2O$$

The solution containing the I_3^- ion was then diluted to exactly 250.0 mL in a volumetric flask. Next, the chemist dissolved 0.2180 g of the fertilizer in water, added starch indicator, and titrated it with the standard I_3^- solution. The reaction was

$$2S_2O_3^{2-} + I_3^- \longrightarrow S_4O_6^{2-} + 3I^-$$

The titration required 27.99 mL of the I_3^- solution.

(a) What was the molarity of the I_3^- solution used in the titration?

(b) How many grams of $(NH_4)_2S_2O_3$ were in the fertilizer sample?

(c) What was the percentage by mass of $(NH_4)_2S_2O_3$ in the fertilizer?

5.93 A sample of a copper ore with a mass of 0.4225 g was dissolved in acid. A solution of potassium iodide was added, which caused the reaction

$$2Cu^{2+}(aq) + 5I^-(aq) \longrightarrow I_3^-(aq) + 2CuI(s)$$

The I_3^- that formed reacted quantitatively with exactly 29.96 mL of 0.02100 M $Na_2S_2O_3$ according to the following equation.

$$I_3^-(aq) + 2S_2O_3^{2-}(aq) \longrightarrow 3I^-(aq) + S_4O_6^{2-}(aq)$$

(a) What was the percentage by mass of copper in the ore?

(b) If the ore contained $CuCO_3$, what was the percentage by mass of $CuCO_3$ in the ore?

5.94 A 1.362 g sample of an iron ore that contained Fe_3O_4 was dissolved in acid and all of the iron was reduced to Fe^{2+}. The solution was then acidified with H_2SO_4 and titrated with 39.42 mL of 0.0281 M $KMnO_4$, which oxidized the iron to Fe^{3+}. The net ionic equation for the reaction is

$$5Fe^{2+} + MnO_4^- + 8H^+ \longrightarrow 5Fe^{3+} + Mn^{2+} + 4H_2O$$

(a) What was the percentage by mass of iron in the ore?

(b) What was the percentage by mass of Fe_3O_4 in the ore?

5.95 Hydrogen peroxide, H_2O_2, solution can be purchased in drug stores for use as an antiseptic. A sample of such a solution weighing 1.000 g was acidified with H_2SO_4 and titrated with a 0.02000 M solution of $KMnO_4$. The net ionic equation for the reaction is

$$6H^+ + 5H_2O_2 + 2MnO_4^- \longrightarrow 5O_2 + 2Mn^{2+} + 8H_2O$$

The titration required 17.60 mL of $KMnO_4$ solution.

(a) How many grams of H_2O_2 reacted?

(b) What is the percentage by mass of the H_2O_2 in the original antiseptic solution?

5.96 Sodium nitrite, $NaNO_2$, is used as a preservative in meat products such as frankfurters and bologna. In an acidic solution, nitrite ion is converted to nitrous acid, HNO_2, which reacts with permanganate ion according to the equation

$$H^+ + 5HNO_2 + 2MnO_4^- \longrightarrow 5NO_3 + 2Mn^{2+} + 3H_2O$$

A 1.000 g sample of a water-soluble solid containing $NaNO_2$ was dissolved in dilute H_2SO_4 and titrated with 0.01000 M $KMnO_4$ solution. The titration required 12.15 mL of the $KMnO_4$ solution. What was the percentage by mass of $NaNO_2$ in the original 1.000 g sample?

5.97 A sample of a chromium-containing alloy weighing 3.450 g was dissolved in acid, and all the chromium in the sample was oxidized to $2CrO_4^{2-}$. It was then found that 3.18 g of Na_2SO_3 was required to reduce the $2CrO_4^{2-}$ to CrO_2^- in a basic solution, with the SO_3^{2-} being oxidized to SO_4^{2-}.

(a) Write a balanced equation for the reaction of $2CrO_4^{2-}$ with SO_3^{2-} in a basic solution.

(b) How many grams of chromium were in the alloy sample?

(c) What was the percentage by mass of chromium in the alloy?

5.98 Solder is an alloy containing the metals tin and lead. A particular sample of this alloy weighing 1.50 g was dissolved in acid. All of the tin was then converted to the +2 oxidation state. Next, it was found that 0.368 g of $Na_2Cr_2O_7$ was required to oxidize the Sn^{2+} to Sn^{4+} in an acidic solution. In the reaction the chromium was reduced to Cr^{3+} ion.

(a) Write a balanced net ionic equation for the reaction between Sn^{2+} and $Cr_2O_7^{2-}$ in an acidic solution.

(b) Calculate the number of grams of tin that were in the sample of solder.

(c) What was the percentage by mass of tin in the solder?

5.99 Both calcium chloride and sodium chloride are used to melt ice and snow on roads in the winter. A certain company was marketing a mixture of these two compounds for this purpose. A chemist, wanting to analyze the mixture, dissolved 2.463 g of it in water and precipitated calcium oxalate by adding sodium oxalate, $Na_2C_2O_4$. The calcium oxalate was carefully filtered from the solution, dissolved in sulfuric acid, and titrated with 0.1000 M $KMnO_4$ solution. The reaction that occurred was

$$6H^+ + 5H_2C_2O_4 + 2MnO_4^- \longrightarrow$$
$$10CO_2 + 2Mn^{2+} + 8H_2O$$

The titration required 21.62 mL of the $KMnO_4$ solution.

(a) How many moles of $C_2O_4^{2-}$ were present in the calcium oxalate precipitate?

(b) How many grams of calcium chloride were in the original 2.463 g sample?

(c) What was the percentage by mass of calcium chloride in the sample?

| Additional Exercises

*__5.101__ Use oxidation numbers to show that the fermentation of glucose, $C_6H_{12}O_6$, to carbon dioxide and ethanol, C_2H_5OH, is a redox reaction.

5.102 Waste water treatment often has at least one oxidation–reduction step. In the collection of wastewater, chlorine can be added to control corrosion by hydrogen sulfide to give sulfur and chloride ions. What is the balanced equation for the reaction that occurs in this step?

5.103 Bromine gas is used to make fire-retardant chemicals used on children's pajamas. However, to dispose of excess bromine, it is passed through a solution of sodium hydroxide. The products are NaBr and NaOBr. What is being reduced and what is being oxidized? Write a balanced equation for the reaction.

5.104 What is the oxidation number of sulfur in the tetrathionate ion, $S_4O_6^{2-}$?

5.105 Balance the following equations using the information in the skeleton reaction.

(a) $H_3AsO_3 + Cr_2O_7^{2-} \longrightarrow H_3AsO_4 + Cr^{3+}$

(b) $HNO_2 + I^- \longrightarrow I_2 + NO$

(c) $NiO_2 + Mn(OH)_2 \longrightarrow Mn_2O_3 + Ni(OH)_2$

(d) $Fe(OH)_2 + O_2 \longrightarrow Fe(OH)_3 + OH^-$

5.106 In Practice Exercise 5.6 (Section 5.1), some of the uses of chlorine dioxide were described along with a reaction that could be used to make ClO_2. Another reaction that is used to make this substance is

$$HCl + NaOCl + 2NaClO_2 \longrightarrow 2ClO_2 + 2NaCl + NaOH$$

5.100 One way to analyze a sample for nitrite ion is to acidify a solution containing NO_2^- and then allow the HNO_2 that is formed to react with iodide ion in the presence of excess I^-. The reaction is

$$2HNO_2 + 2H^+ + 3I^- \longrightarrow 2NO + 2H_2O + I_3^-$$

Then the I_3^- is titrated with $Na_2S_2O_3$ solution using starch as an indicator.

$$I_3^- + 2S_2O_3^{2-} \longrightarrow 3I^- + S_4O_6^{2-}$$

In a typical analysis, a 1.104 g sample that was known to contain $NaNO_2$ was treated as described above. The titration required 29.25 mL of 0.3000 M $Na_2S_2O_3$ solution to reach the end point.

(a) How many moles of I_3^- had been produced in the first reaction?

(b) How many moles of NO_2^- had been in the original 1.104 g sample?

(c) What was the percentage by mass of $NaNO_2$ in the original sample?

Which element is oxidized? Which element is reduced? Which substance is the oxidizing agent and which is the reducing agent?

5.107 What is the average oxidation number of carbon in (a) C_2H_5OH (grain alcohol), (b) $C_{12}H_{22}O_{11}$ (sucrose, table sugar), (c) $CaCO_3$ (limestone), and (d) $NaHCO_3$ (baking soda)?

5.108 The following chemical reactions are observed to occur in aqueous solution.

$$2Al + 3Cu^{2+} \longrightarrow 2Al^{3+} + 3Cu$$
$$2Al + 3Fe^{2+} \longrightarrow 3Fe + 2Al^{3+}$$
$$Pb^{2+} + Fe \longrightarrow Pb + Fe^{2+}$$
$$Fe + Cu^{2+} \longrightarrow Fe^{2+} + Cu$$
$$2Al + 3Pb^{2+} \longrightarrow 3Pb + 2Al^{3+}$$
$$Pb + Cu^{2+} \longrightarrow Pb^{2+} + Cu$$

Arrange the metals Al, Pb, Fe, and Cu in order of increasing ease of oxidation without referring to Table 5.3.

5.109 In Problem 5.108, were all of the experiments described actually necessary to establish the order?

5.110 According to the activity series in Table 5.3, which of the following metals react with nonoxidizing acids: (a) silver, (b) gold, (c) zinc, (d) magnesium?

5.111 In each of the following pairs, choose the metal that would most likely react more rapidly with a nonoxidizing acid such as HCl: (a) aluminum or iron, (b) zinc or nickel, and (c) cadmium or magnesium.

5.112 Use Table 5.3 to predict whether the following displacement reactions should occur. If no reaction occurs, write N.R. If a reaction does occur, write a balanced chemical equation for it.

(a) $Zn + Sn^{2+} \longrightarrow$ (d) $Zn + Co^{2+} \longrightarrow$

(b) $Cr + H^+ \longrightarrow$ (e) $Mn + Pb^{2+} \longrightarrow$

(c) $Pb + Cd^{2+} \longrightarrow$

5.113 Sucrose, $C_{12}H_{22}O_{11}$, is ordinary table sugar. Write a balanced chemical equation representing the metabolism of sucrose. (See Review Problem 5.79.)

5.114 Balance the following equations by the ion–electron method.

(a) $NBr_3 \longrightarrow N_2 + Br^- + HOBr$ (basic solution)

(b) $Cl_2 \longrightarrow Cl^- + ClO_3^-$ (basic solution)

(c) $H_2SeO_3 + H_2S \longrightarrow S + Se$ (acidic solution)

(d) $MnO_2 + SO_3^{2-} \longrightarrow$
$$Mn^{2+} + S_2O_3^{2-} \text{ (acidic solution)}$$

(e) $XeO_3 + I^- \longrightarrow Xe + I_2$ (acidic solution)

(f) $(CN)_2 \longrightarrow CN^- + OCN^-$ (basic solution)

5.115 Lead(IV) oxide reacts with hydrochloric acid to give chlorine. The unbalanced equation for the reaction is $PbO_2 + Cl^- \longrightarrow PbCl_2 + Cl_2$. How many grams of PbO_2 must react to give 15.0 g of Cl_2?

5.116 A solution contains $Ce(SO_4)_3^{2-}$ at a concentration of 0.0150 M. It was found that in a titration, 25.00 mL of this solution reacted completely with 23.44 mL of 0.032 M $FeSO_4$ solution. The reaction gave Fe^{3+} as a product in the solution. In this reaction, what is the final oxidation state of the Ce?

*5.117 A copper bar with a mass of 12.340 g is dipped into 255 mL of 0.125 M $AgNO_3$ solution. When the reaction that occurs has finally ceased, what will be the mass of unreacted copper in the bar? If all the silver that forms adheres to the copper bar, what will be the total mass of the bar after the reaction?

*5.118 A bar of copper weighing 32.00 g was dipped into 50.0 mL of 0.250 M $AgNO_3$ solution. If all of the silver that deposits adheres to the copper bar, how much will the bar weigh after the reaction is complete? Write and balance any necessary chemical equations.

5.119 As described in the *Chemistry Outside the Classroom 5.1*, silver tarnishes in the presence of hydrogen sulfide to form Ag_2S, which is how silver spoons tarnish. The silver can be polished and made lustrous again by placing the piece of tarnished silver in a pan with aluminum foil and detergent solution. The silver must touch the aluminum foil. Write balanced equations for both the tarnishing and the polishing of silver.

5.120 Titanium(IV) can be reduced to titanium(III) by the addition of zinc metal. Sulfur dioxide can be reduced to elemental sulfur by hydrogen. For both of these reactions, write the balanced equations. Why could we not include these compounds in the activity series table?

5.121 A researcher planned to use chlorine gas in an experiment and wished to trap excess chlorine to prevent it from escaping into the atmosphere. To accomplish this, the reaction of sodium thiosulfate ($Na_2S_2O_3$) with chlorine gas in an acidic aqueous solution to give sulfate ion and chloride ion would be used. How many grams of $Na_2S_2O_3$ are needed to trap 4.25 g of chlorine?

5.122 A sample of a tin ore weighing 0.3000 g was dissolved in an acid solution and all the tin in the sample was changed to tin(II). In a titration, 8.08 mL of 0.0500 M $KMnO_4$ solution was required to oxidize the tin(II) to tin(IV).

(a) What is the balanced equation for the reaction in the titration?

(b) How many grams of tin were in the sample?

(c) What was the percentage by mass of tin in the sample?

(d) If the tin in the sample had been present in the compound SnO_2, what would have been the percentage by mass of SnO_2 in the sample?

Multi-Concept Problems

*5.123 Biodiesel is formed from the reaction of oils with methanol or ethanol. One of the products is methyl octanoate, $C_9H_{18}O_2$, which burns completely in a diesel engine. Give a balanced equation for this reaction. If the density of methyl octanoate is 0.877 g/mL, how many grams of CO_2 will be formed from the burning of 1 gallon of methyl octanoate? Assuming that you drive a diesel-powered car that gets 32 miles per gallon of biodiesel fuel and you drive 14,525 miles per year, what is the carbon footprint of this activity in terms of pounds of CO_2 per year?

*5.124 In June 2002, the Department of Health and Children in Ireland began a program to distribute tablets of potassium iodate to households as part of Ireland's National Emergency Plan for Nuclear Accidents. Potassium iodate provides iodine, which when taken during a nuclear emergency, works by "topping off " the thyroid gland to prevent the uptake of radioactive iodine that might be released into the environment by a nuclear accident. To test the potency of the tablets, a chemist dissolved one in 100.0 mL of water, made the solution acidic, and then added excess potassium iodide, which caused the following reaction to occur.

$$IO_3^- + 8I^- + 6H^+ \longrightarrow 3I_3^- + 3H_2O$$

The resulting solution containing I_3^- was titrated with 0.0500 M $Na_2S_2O_3$ solution, using starch indicator to detect the end point. (In the presence of iodine, starch turns dark blue. When the $S_2O_3^{2-}$ has consumed all of the iodine, the solution becomes colorless.) The titration required 22.61 mL of the thiosulfate solution to reach the end point. The reaction during the titration was

$$I_3^- + 2S_2O_3^{2-} \longrightarrow 3I^- + S_4O_6^{2-}$$

How many milligrams of KIO_3 were in the tablet?

*5.125 An organic compound contains carbon, hydrogen, and sulfur. A sample of it with a mass of 1.045 g was burned in oxygen to give gaseous CO_2, H_2O, and SO_2. These gases were passed through 500.0 mL of an acidified 0.0200 M $KMnO_4$ solution, which caused the SO_2 to be oxidized to SO_4^{2-}. Only part of the available $KMnO_4$ was reduced to Mn^{2+}. Next, 50.00 mL of 0.0300 M $SnCl_2$ was added to a 50.00 mL portion of this solution, which still contained unreduced $KMnO_4$. There was more than enough added $SnCl_2$ to cause all of the remaining MnO_4^- in the 50 mL portion to be reduced to Mn^{2+}. The excess Sn^{2+} that still remained after the reaction was then titrated with 0.0100 M $KMnO_4$, requiring 27.28 mL of the $KMnO_4$ solution to reach the end point. What was the percentage of sulfur in the original sample of the organic compound that had been burned?

*5.126 A mixture is made by combining 325 mL of 0.0200 M $Na_2Cr_2O_7$ with 425 mL of 0.060 M $Fe(NO_3)_2$. Initially, the H^+ concentration in the mixture is 0.400 M. Dichromate ion oxidizes Fe^{2+} to Fe^{3+} and is reduced to Cr^{3+}. After the reaction in the mixture has ceased, how many milliliters of 0.0100 M NaOH will be required to neutralize the remaining H^+?

*5.127 A solution containing 0.1244 g of $K_2C_2O_4$ was acidified, changing the $C_2O_4^{2-}$ ions to $H_2C_2O_4$. The solution was then titrated with 13.93 mL of a $KMnO_4$ solution to reach a faint pink end point. In the reaction, $H_2C_2O_4$ was oxidized to CO_2 and MnO_4^- was reduced to Mn^{2+}. What was the molarity of the $KMnO_4$ solution used in the titration?

*5.128 It was found that a 20.0 mL portion of a solution of oxalic acid, $H_2C_2O_4$, requires 6.25 mL of 0.200 M $K_2Cr_2O_7$ for complete reaction in an acidic solution. In the reaction, the oxidation product is CO_2 and the reduction product is Cr^{3+}. How many milliliters of 0.450 M NaOH are required to completely neutralize the $H_2C_2O_4$ in a separate 20.00 mL sample of the same oxalic acid solution?

*5.129 A 15.00 mL sample of a solution containing oxalic acid, $H_2C_2O_4$, was titrated with 0.02000 M $KMnO_4$. The titration required 18.30 mL of the $KMnO_4$ solution. If 19.69 mL of a solution of NaOH is needed to react completely with 25.00 mL of this same oxalic acid solution, what is the concentration of the NaOH solution? In the reaction, oxalate ion ($C_2O_4^{2-}$) is oxidized to CO_2 and MnO_4^- is reduced to Mn^{2+}.

*5.130 A solution with a volume of 500.0 mL contained a mixture of SO_3^{2-} and $S_2O_3^{2-}$. A 100.0 mL portion of the solution was found to react with 80.00 mL of 0.0500 M $2CrO_4^{2-}$ in a basic solution to give CrO_2^-. The only sulfur-containing product was SO_4^{2-}. After the reaction, the solution was treated with excess 0.200 M $BaCl_2$ solution, which precipitated $BaSO_4$. This solid was filtered from the solution, dried, and found to weigh 0.9336 g. Explain in detail how you can determine the molar concentrations of SO_3^{2-} and $S_2O_3^{2-}$ in the original solution.

|Exercises in Critical Thinking

*5.131 The ion $OSCN^-$ is found in human saliva. Discuss the problems in assigning oxidation numbers to the atoms in this ion. Suggest a reasonable set of oxidation numbers for the atoms in $OSCN^-$.

*5.132 We described the ion–electron method for balancing redox equations in Section 5.2. Can you devise an alternate method using oxidation numbers?

*5.133 Assuming that a chemical reaction with DNA could lead to damage causing cancer, would a very strong or a weak oxidizing agent have a better chance of being a carcinogen? Justify your answer.

*5.134 Would you expect atomic oxygen and chlorine to be better or worse oxidizing agents than molecular oxygen and molecular chlorine? Justify your answer.

*5.135 Do we live in an oxidizing or reducing environment? What effect might our environment have on chemistry we do in the laboratory? What effect might the environment have on the nature of the chemicals (minerals, etc.) we find on earth?

*5.136 Antioxidants are touted as helping people live longer. What are antioxidants? What role do they play in longevity? In terms of this chapter, how else could antioxidants be defined? What from this chapter would make a good antioxidant?

Energy and Chemical Change

6

© Chris Hepburn/iStockphoto

A mid-afternoon snack of chocolate chip cookies and a glass of milk gives a student energy to get through their studying. The amount of energy in the snack can be calculated and related back to the bonds that are broken and formed in the digestion of the food.

251

This Chapter in Context

Energy is a term we often see in the news. We hear reports on the cost of energy and what's happening to the world's energy supplies. One might come away with the notion that energy is something you can hold in your hands and place in a bottle. Energy is not like matter, however. Instead, it is something that matter can possess—something that enables objects to move or cause other objects to move.

Nearly every chemical and physical change is accompanied by a *change* in energy. The energy changes associated with the evaporation and condensation of water drive global weather systems. The combustion of fuels produces energy changes we use to power cars and generate electricity, and our bodies use energy released in the metabolism of foods, such as chocolate chip cookies and a glass of milk, to drive biochemical processes. This chapter introduces **thermochemistry**, the branch of chemistry that deals specifically with the energy absorbed or released by chemical reactions. Thermochemistry has many practical applications, but it is also of great theoretical importance, because it provides an essential link between laboratory measurements (such as temperature changes) and events on the molecular level that occur when molecules form or break apart.

You will probably perform a thermochemistry experiment in the lab portion of your chemistry course. All it takes is a thermometer and a foam cup to make the measurements. Research-grade instruments may be much more elaborate and expensive because they incorporate more sophisticated temperature measurement and control systems. However, the basic measurement, temperature change due to a chemical or physical change, is the same in both cases.

Thermochemistry is part of the science of **thermodynamics**, the study of energy transfer and energy transformation. Thermodynamics allows scientists to predict whether a proposed physical change or chemical reaction can occur under a given set of conditions. It is an essential part of chemistry (and all of the natural sciences). We'll continue our study of thermodynamics in Chapter 19.

LEARNING OBJECTIVES

After reading this chapter, you should be able to:

- explain the difference between potential and kinetic energy and the law of conservation of energy
- explain the connection between temperature, energy, and the concept of a state function
- determine an amount of heat exchanged from the temperature change of an object
- describe the energy changes in exothermic and endothermic reactions
- state the first law of thermodynamics and explain how it applies to chemistry
- explain the difference between a heat of reaction obtained at constant pressure and the heat of reaction at constant volume
- describe the assumptions and utility of thermochemical equations
- use Hess's law to determine the enthalpy of a reaction
- determine and use standard heats of formation to solve problems

6.1 | Energy: The Ability to Do Work

■ Work is a result of a force causing an object to move. A moving car has energy because it can move another car in a collision.

TOOLS

Kinetic energy

As mentioned in the preceding introduction, energy is intangible; you can't hold it in your hand to study it and you can't put it in a bottle. **Energy** is something an object has if the object is able to do work. It can be possessed by the object in two different ways, as kinetic energy and as potential energy.

Kinetic energy (KE) is the energy an object has when it is moving. It depends on the object's mass and velocity by the equation:

$$KE = \tfrac{1}{2}\,mv^2 \qquad\qquad (6.1)$$

where m is the mass and v is the velocity. The larger the object's mass and the greater its velocity, the more kinetic energy it has.

Potential energy (PE) is energy an object has that can be changed to kinetic energy; it can be thought of as **stored energy**. For example, when you wind an old style alarm clock, you transfer energy to a spring. The spring holds this stored energy (potential energy) and gradually releases it, in the form of kinetic energy, to make the clock's mechanism work.

Chemicals also possess potential energy, which is sometimes called **chemical energy**. When chemical reactions occur, the changes in chemical energy possessed by the substances lead to either the absorption or release of energy (as heat or light, for instance). For example, the explosive reaction between hydrogen and oxygen in the main engines of the space shuttle, shown in Figure 6.1, produces light, heat, and the expanding gases that help lift the vehicle from its launch pad.

Corbis Images

Figure 6.1 | **Liquid hydrogen and oxygen serve as fuel for the space shuttle.** The three almost invisible points of blue flame come from the main engines of the space shuttle, which consume hydrogen and oxygen in the formation of water.

Potential Energy

An important aspect of potential energy is the way it depends on the positions of objects that experience attractions or repulsions toward other objects. For example, a book has potential energy because it experiences a gravitational attraction toward the earth. Lifting the book, which changes its position, increases the potential energy. This energy is supplied by the person doing the lifting. Letting the book fall allows the potential energy to decrease. The lost potential energy is changed to kinetic energy as the book gains speed during its descent.

How potential energy varies with position for objects that attract or repel each other can be illustrated by magnets depicted in Figure 6.2. When the magnets are

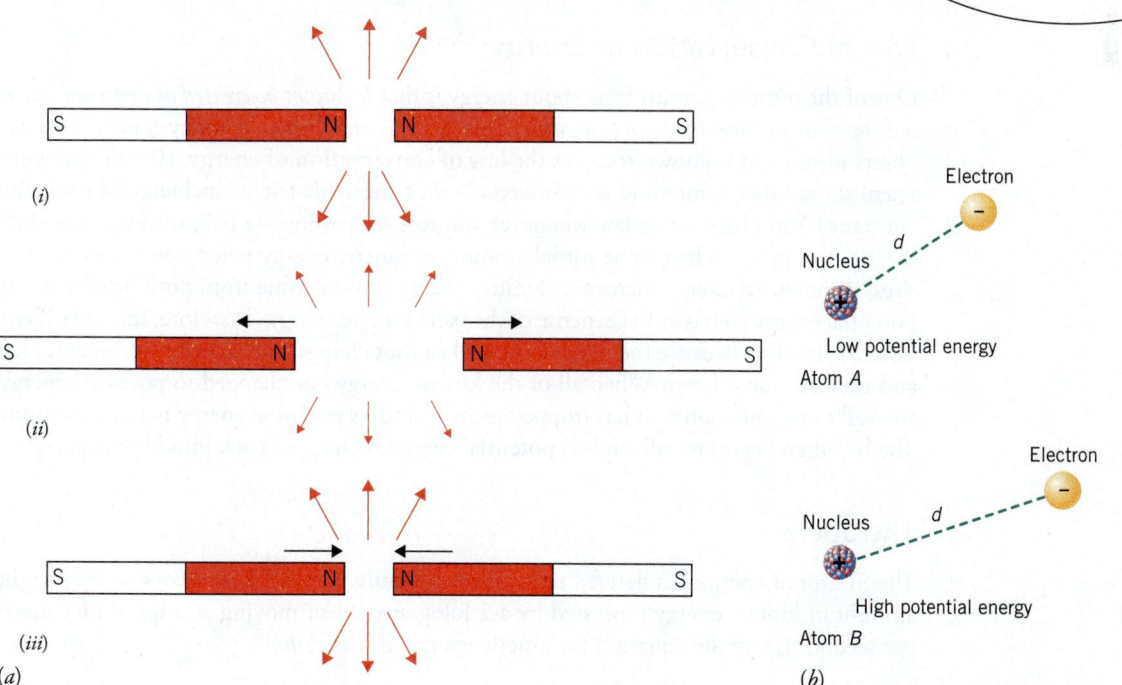

Figure 6.2 | **The potential energy of two magnets with the same poles facing depends on their distance apart.** (*a*) The red arrows represent the amount of repulsion between the two magnets. (*i*) The north poles of two magnets repel when placed near each other. (*ii*) When the magnets are moved apart, their potential energy decreases. (*iii*) When they are moved together, their potential energy increases. (*b*) The positively charged nucleus attracts the negatively charged electron. As the electron is brought closer to the nucleus, the potential energy of the electron decreases. It takes work to separate the electron and the nucleus, so the electron in atom *B* has more potential energy than the electron in atom *A*.

held in a certain way, the end of one magnet repels the end of another; we say the "north pole" of one repels the "north pole" of the other. Because the repulsions push the magnets apart, we have to do some work to move them closer together. This causes their potential energy to increase; the energy we expended pushing them together is now stored by the magnets. If we release them, they will push each other apart, and this "push" could also be made to do work for us. Likewise, if we turn one magnet around, the ends of the two magnets will attract each other—that is, the "south pole" of one magnet attracts the "north pole" of the other. Now, to pull the magnets apart, we have to do work, and this will increase their potential energy.

■ Unlike kinetic energy, there is no single, simple equation that can be used for the amount of potential energy of an object.

Potential energy changes

Factors That Affect Potential Energy
• Potential energy increases when objects that *attract* each other move apart, and decreases when they move toward each other.
• Potential energy increases when objects that *repel* each other move toward each other, and decreases when they move apart.

If you ever have trouble remembering this, think back to the magnets and how the poles attract or repel each other.

In chemical systems, the amount of chemical energy is determined by the type and arrangement of the atoms. There aren't north and south poles, but there are attractions and repulsions between electrical charges. The attraction between opposite charges is similar to the attraction between two magnets in which the opposite poles are next to each other. Electrons are attracted to protons because of their opposite electrical charges. Electrons repel electrons and nuclei repel other nuclei because they have the same kind of electrical charge. Changes in the relative positions of these particles, as atoms join to form molecules or break apart when molecules decompose, lead to changes in potential energy. These are the kinds of potential energy changes that lead to the release or absorption of energy by chemical systems during chemical reactions.

Law of Conservation of Energy

One of the most important facts about energy is that *it cannot be created or destroyed; it can only be changed from one form to another.* This fact was established by many experiments and observations, and is known today as the **law of conservation of energy**. (Recall that when scientists say that something is "conserved," they mean that it is unchanged or remains constant.) You observe this law whenever you toss something—a ball, for instance—into the air. You give the ball some initial amount of kinetic energy when you throw it. As it rises, its potential energy increases. Because energy cannot come from nothing, the rise in potential energy comes at the expense of the ball's kinetic energy. Therefore, the ball's $\frac{1}{2}mv^2$ becomes smaller. Because the mass of the ball cannot change, the velocity (v) becomes less and the ball slows down. When all of the kinetic energy has changed to potential energy, the ball can go no higher; it has stopped moving and its potential energy is at a maximum. The ball then begins to fall, and its potential energy is changed back into kinetic energy.

The Joule

The SI unit of energy is a derived unit called the **joule** (symbol **J**) and corresponds to the amount of kinetic energy possessed by a 2 kilogram object moving at a speed of 1 meter per second. Using the equation for kinetic energy, $KE = \frac{1}{2} mv^2$.

$$1 \text{ J} = \tfrac{1}{2}(2 \text{ kg}) \left(\frac{1 \text{ m}}{1 \text{ s}}\right)^2$$

$$1 \text{ J} = 1 \text{ kg m}^2\text{s}^{-2}$$

The joule is actually a rather small amount of energy and in most cases we will use the larger unit, the **kilojoule (kJ)**; $1 \text{ kJ} = 1000 \text{ J} = 10^3 \text{ J}$.

Another energy unit you may be familiar with is called the **calorie (cal)**. Originally, it was defined as the energy needed to raise the temperature of 1 gram of water by 1 degree Celsius. With the introduction of the SI, the calorie has been redefined as follows:

$$1 \text{ cal} = 4.184 \text{ J} \quad \text{(exactly)} \tag{6.2}$$

The larger unit **kilocalorie (kcal)**, which equals 1000 calories, can also be related to the kilojoule.

$$1 \text{ kcal} = 4.184 \text{ kJ}$$

The nutritional or dietary Calorie (note the capital C), Cal, is actually one kilocalorie.

$$1 \text{ Cal} = 1 \text{ kcal} = 4.184 \text{ kJ}$$

While joules and kilojoules are the standard units of energy, calories and kilocalories are still in common use, so you may need to convert joules into calories and vice versa.

Chocolate pudding, made with 2% milk, claims to have 140 Calories per serving. How many joules of energy is this? Use the appropriate metric prefix to avoid using exponential notation in your answer. (*Hint:* Review the metric prefixes in Table 1.5.)

The label on a bag of Oreo cookies states that one serving of three cookies provides 160 Calories. How many kilocalories, calories, kilojoules, and joules of energy are equivalent to one serving of Oreo cookies?

Practice Exercise 6.1

Practice Exercise 6.2

6.2 | Heat, Temperature, and Internal Energy

If we think about a chunk of matter, it is reasonable to wonder about the energy it contains and how to measure it. Before we discuss this energy, we first need to understand the nature of heat and temperature, which will be important in how we approach how this energy is measured.

Heat and Temperature

Up to now we have used the word "temperature" many times, and now we need to define it precisely. If we think at the molecular level, atoms, molecules, or ions are constantly moving and colliding with each other, which means that any object, even if we cannot see it moving, will have some kinetic energy. Since it is not at all probable that all particles have exactly the same kinetic energy, there must be some distribution of energies that we will see shortly. However, this group of kinetic energies must have an average. At this point we define *the **temperature** of an object as being proportional to the **average kinetic energy** of its particles. The higher the average kinetic energy, the higher the temperature.* Recalling that $KE = \frac{1}{2}mv^2$, increasing the average kinetic energy cannot measurably change the masses of the atoms, so it must increase their speeds. As a result the number and energy of particles colliding with a thermometer must increase. Likewise, when the temperature is reduced, the particles move slower and the average kinetic energy of the molecules is lower, so the number of collisions and the force of the collisions of the particles is lower.

Heat is the kinetic energy (also called **thermal energy**) that is *transferred* between objects caused by differences in their temperatures. On the molecular level, when a hot object is placed in contact with a cold one, the faster atoms of the hot object collide with, and lose some kinetic energy to, the slower atoms of the cold object (Figure 6.3). This decreases the average kinetic energy of the particles of the

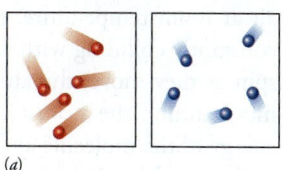
(a)

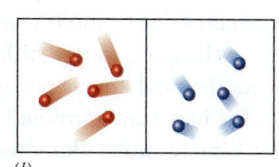
(b)

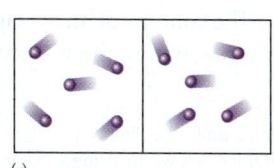
(c)

Figure 6.3 | **Energy transfer from a warmer to a cooler object.** (*a*) The longer trails on the left denote higher kinetic energies of a hot object like hot water just before it's poured into a cold coffee cup, indicated with shorter trails. (*b*) Collisions between the hot water molecules and the cooler molecules in the cup's material cause the water molecules to slow down and the cup's molecules to speed up, as kinetic energy is transferred from the water to the cup. (*c*) Thermal equilibrium is established: the temperatures of the water and the cup wall are now equal.

hot object, causing its temperature to drop. At the same time, the average kinetic energy of the particles in the cold object is raised, causing the temperature of the cold object to rise. Eventually, the average kinetic energies of the atoms in both objects become the same and the objects reach the same temperature, or **thermal equilibrium**. Thus, the transfer of heat is interpreted as a transfer of kinetic energy between two objects. For example, heat will flow from a hot cup of coffee into the cooler surroundings. Eventually the coffee and surroundings come to thermal equilibrium (and now the coffee is considered to be cold.) Overall, the temperature of the coffee has dropped and the temperature of the surroundings has increased a bit.

Energy that is transferred as heat comes from an object's internal energy. **Internal energy, E_{internal}**, is the sum of energies for all of the individual particles in a sample of matter.[1] We can summarize this in equation form as

$$E_{\text{internal}} = \text{PE} + \text{KE}$$

where PE represents *all of the forces* in the universe acting on the particles in an object and KE represents the kinetic energy due to *all of the motions* of these particles. Since it is impossible to quantify all of the attractive and repulsive forces acting on an object we cannot determine the total PE of an object. Similarly, we cannot quantify all of the motions of the particles in an object as we move through the universe and cannot determine the total KE of an object. However, even if we cannot determine the value of E, we can determine *changes* in E.

In studying both chemical and physical changes, we will be interested in the *change* in internal energy that accompanies the process. This is defined as ΔE, where the symbol Δ (Greek letter delta) signifies a change.

$$\Delta E = E_{\text{final}} - E_{\text{initial}}$$

For a chemical reaction, E_{final} corresponds to the internal energy of the products, so we'll write it as E_{products}. Similarly, we'll use the symbol $E_{\text{reactants}}$ for E_{initial}. So for a chemical reaction the change in internal energy is given by

$$\Delta E = E_{\text{products}} - E_{\text{reactants}} \qquad \text{(6.3)}$$

Notice an important convention illustrated by this equation—namely, changes in something like temperature, Δt_{C},[2] or in internal energy, ΔE, are always figured by taking "final minus initial" or "products minus reactants." This means that if a system *absorbs* energy from its surroundings during a change, its final energy is greater than its initial energy and ΔE is positive. This is what happens, for example, when photosynthesis occurs or when a battery is charged. As the system (the battery) absorbs the energy, its internal energy increases and is then available for later use elsewhere.

Temperature and Average Molecular Kinetic Energy

All of the particles within any object are in constant motion. For example, in a sample of air at room temperature, oxygen and nitrogen molecules travel faster than rifle bullets, constantly colliding with each other and with the walls of their container. The molecules spin as they move, the atoms within the molecules jiggle and vibrate, and the electrons move around the atoms; these internal molecular motions also contribute to the kinetic energy of the molecule and, thus, to the internal energy of the sample. We'll use the term **molecular kinetic energy** for the energy associated with such motions. Each particle has a certain value of molecular kinetic energy at any given moment. Molecules are continually exchanging energy with each other during collisions, but as long as the sample is isolated, the total kinetic energy of all the molecules remains constant.

The idea that atoms and molecules are in constant random motion forms the basis for the **kinetic molecular theory**. It is this theory that tells us in part that *temperature* is directly proportional to the *average kinetic energy* of the atoms and molecules of an object.

The concept of an average kinetic energy implies that there is a distribution of kinetic energies among the molecules in an object. Let's examine this further. At any particular

■ Recall that kinetic energy is the energy of motion and is given by $\text{KE} = \frac{1}{2}mv^2$, where m is the mass of an object and v is its velocity.

■ Temperature is related to the *average* molecular kinetic energy; internal energy is the *total* of all molecular kinetic energies.

---

[1] Sometimes the symbol U is used for internal energy.
[2] Remember, we are using t_{C} for Celsius temperature and T_{K}, or T, for temperature in Kelvin.

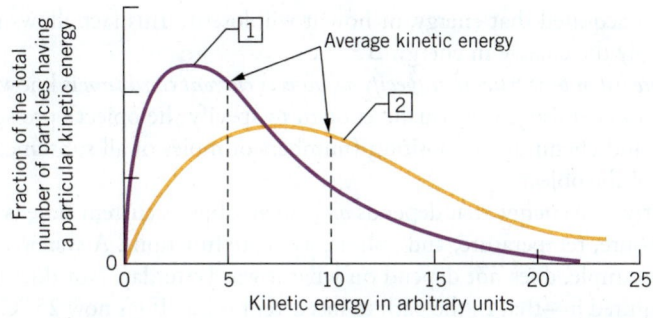

Figure 6.4 | The distribution of kinetic energies among gas particles. The distribution of individual kinetic energies changes in going from a lower temperature, curve 1, to a higher temperature, curve 2. The highest point on each curve is the most probable value of the kinetic energy for that temperature—that is, the one we would most frequently find if we could observe and measure the kinetic energy of each molecule. At the lower temperature, this peak is at lower values for the kinetic energy. At the higher temperature, more molecules have high speeds and fewer molecules have low speeds, so the maximum shifts to the right and the curve flattens.

moment, the individual particles in an object are moving at different velocities, which gives them a broad range of kinetic energies. An extremely small number of particles will be standing still, so their kinetic energies will be zero. Another small number will have very large velocities; these will have very large kinetic energies. Between these extremes there will be many molecules with intermediate amounts of kinetic energy.

The graphs in Figure 6.4 show the kinetic energy distributions among molecules in a sample at two different temperatures. The vertical axis represents the *fraction* of molecules with a given kinetic energy (i.e., the number of molecules with a given KE divided by the total number of molecules in the sample). Each curve in Figure 6.4 describes how the fraction of molecules with a given KE varies with KE.

The curves start out with a fraction equal to zero when KE equals zero. This is because the fraction of molecules with zero KE (corresponding to molecules that are motionless) is essentially zero, regardless of the temperature. As we move along the KE axis we find that the fraction of the molecules increases, reaches a peak, and then declines toward zero again. For very large values of KE, the fraction drops off again because very few molecules have very large velocities.

Each curve in Figure 6.4 has a characteristic peak or maximum corresponding to the most frequently experienced values for molecular KE. Since the curves are not symmetrical, the *average* values of molecular KE lie slightly to the right of the maxima. Notice that when the temperature increases (going from curve 1 to 2), the curve flattens and the maximum shifts to a higher value, as does the average molecular KE. In fact, if *we double the kelvin temperature, the average* KE *also doubles, so the kelvin temperature is directly proportional to the average* KE. The reason the curve flattens is because the area under each curve corresponds to the *sum* of all the fractions, and when all the fractions are added, they must equal one. (No matter how you cut up a pie, if you add all the fractions, you must end up with *one* pie).

We will find the graphs illustrated in Figure 6.4 very useful later when we discuss the effect of temperature on such properties as the rates of evaporation of liquids and the rates of chemical reactions.

State Functions

Equation 6.3 defines what we mean by a change in the internal energy of a chemical system during a reaction. This energy change can be made to appear entirely as heat, and as you will learn soon, we can measure heat. However, there is no way to actually measure either $E_{products}$ or $E_{reactants}$, since this would require measuring all of the molecular motions, attractions, and repulsions. Fortunately, we are more interested in ΔE than we are in the absolute amounts of energy that the reactants or products have.

Even though we cannot measure internal energy, it is important to discuss it. As it turns out, *the energy of an object depends only on the object's current condition*. It doesn't matter

■ "State," as used in thermochemistry, doesn't have the same meaning as when it's used in terms such as "solid state" or "liquid state."

■ Remember, the symbol Δ denotes a change between some initial and final state.

how the object acquired that energy, or how it will lose it. This fact allows us to be concerned with only the change in energy, ΔE.

The complete list of properties that specify an object's current condition is known as the **state** of the object. In chemistry, it is usually enough to specify the object's pressure, temperature, volume, and chemical composition (numbers of moles of all substances present) to give the state of the object.

Any property, like energy, that depends *only* on an object's current state is called a **state function**. Pressure, temperature, and volume are state functions. A system's current temperature, for example, does not depend on what it was yesterday. Nor does it matter *how* the system acquired it—that is, the path to its current value. If it's now 25 °C, we know all we can or need to know about its temperature. Also, if the temperature were to increase—say, to 35 °C—the change in temperature, Δt_C, is simply the difference between the final and the initial temperatures:

$$\Delta t = (t_C)_{\text{final}} - (t_C)_{\text{initial}} \tag{6.4}$$

To make the calculation of Δt, we do not have to know what caused the temperature change; all we need are the initial and final values. *This independence from the method or mechanism by which a change occurs is the important feature of all state functions. It also means that the difference between state functions must be independent of how we got from one state to another.* As we'll see later, the advantage of recognizing that some property is a state function is that many calculations are then much easier.

Practice Exercise 6.3

When monitoring a reaction, the initial temperature of the reaction is 25 °C, it rose to 67 °C, and then fell to a final temperature of 32 °C. What is the overall change in temperature? (*Hint:* Temperature is a state function.)

6.3 | Measuring Heat

By measuring the amount of heat that is absorbed or released by an object, we are able to quantitatively study nearly any type of energy transfer. For example, if we want to measure the energy transferred by an electrical current, we can force the electric current through something with high electrical resistance, like the heating element that glows orange inside a toaster, and the energy transferred by the current becomes heat.

In our study of energy transfer, it is very important to specify the **boundary** across which the heat flows. The boundary might be visible (as the walls of a beaker) or invisible (as the boundary that separates warm air from cold air along a weather front). The boundary encloses the **system**, which is the object we are interested in studying. Everything outside the system is called the **surroundings**. The system and the surroundings together comprise the **universe**.

Three types of systems are possible, depending on whether matter or energy can cross the boundary as shown in Figure 6.5.

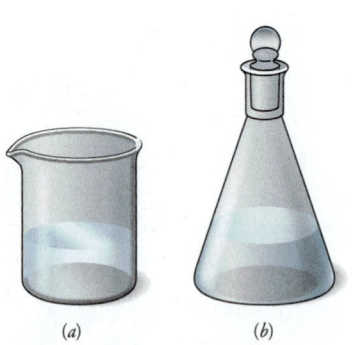

(a)　　　　(b)　　　　(c)

Figure 6.5 | Open, closed, and isolated systems. The open system (*a*) allows the exchange of both energy and mass; the closed system (*b*) only allows the exchange of energy; and the isolated system (*c*) is sealed and insulated (flask inside a flask) so no heat or mass can be exchanged with the surroundings.

- **Open systems** can gain or lose mass and energy across their boundaries. The human body is an example of an open system.
- **Closed systems** can absorb or release energy, but not mass, across the boundary. The mass of a closed system is constant, no matter what happens inside. A light bulb is an example of a closed system.
- **Isolated systems** cannot exchange matter or energy with their surroundings. Because energy cannot be created or destroyed, the energy of an isolated system is constant, no matter what happens inside. Processes that occur within an isolated system are called *adiabatic*, from the Greek *a + diabatos*, meaning not passable. A stoppered Thermos bottle is a good approximation of an isolated system.

Heat and Temperature Change

If the only type of energy that is transferred between two objects is heat energy, then all of the heat lost by one object must be gained by the second object since we have to obey the law of conservation of energy. Unfortunately, there is no instrument available that directly measures heat. Instead, we measure temperature changes, and then use them to calculate changes in heat energy. Experience tells you that the more heat you add to an object, the more its temperature will rise. In fact, experiments show that the temperature change, Δt, is *directly proportional* to the amount of heat absorbed, which we will identify by the symbol q. This can be expressed in the form of an equation as

$$q = C\,\Delta t \qquad (6.5)$$

TOOLS

Heat capacity and q

where C is a proportionality constant called the **heat capacity** of the object. The units of heat capacity are usually $J\,°C^{-1}$ (or $J/°C$), which expresses the amount of energy needed to raise the temperature of an object by 1 °C.

Heat capacity depends on two factors. One is the size of the sample; if we double the size of the sample, we need twice as much heat to cause the same temperature increase. Heat capacity also depends on the sample's composition. For example, it's found that it takes more heat to raise the temperature of one gram of water by 1 °C than to cause the same temperature change in one gram of iron.

Specific Heat

From the preceding discussion, we see that different samples of a given substance will have different heat capacities, and this is what is observed. For example, if we have 10.0 g of water, it must absorb 41.8 J of heat energy to raise its temperature by 1.00 °C. The heat capacity of the sample is found by solving Equation 6.5 for C.

$$C = \frac{q}{\Delta t} = \frac{41.8\ J}{1.00\ °C} = 41.8\ J/°C$$

On the other hand, a 100 g sample of water must absorb 418 J of heat to have its temperature raised by 1.00 °C. The heat capacity of this sample is

$$C = \frac{q}{\Delta t} = \frac{418\ J}{1.00\ °C} = 418\ J/°C$$

Thus, a 10-fold increase in sample size has produced a 10-fold increase in the heat capacity. This indicates that the heat capacity is directly proportional to the mass of the sample.

$$C = m \times s \qquad (6.6)$$

where m represents the mass of the sample and s is the symbol for a constant called the **specific heat capacity** (or simply the **specific heat**). If the heat capacity has units of $J\,°C^{-1}$ and if the mass is in grams, the specific heat capacity has units of $J\,g^{-1}\,°C^{-1}$. The units tell us that the specific heat capacity is the amount of heat required to raise the temperature of 1 gram of a substance by 1 °C.

Because C depends on the size of the sample, *heat capacity is an extensive property*. When we discussed density in Section 1.7, you learned that if we take the ratio of two extensive properties, we can obtain an intensive property—one that is independent of the size of the sample. Because heat capacity and mass both depend on sample size, their ratio, obtained by solving Equation 6.6 for s, yields an intensive property that is the same for any sample of a substance.

If we substitute Equation 6.6 into Equation 6.5, we can obtain an equation for q in terms of specific heat.

$$q = ms\,\Delta t \qquad (6.7)$$

TOOLS

Specific heat capacity and q

We can use this equation to calculate the specific heat of water from the information given above, which states that 10.0 g of water requires 41.8 J to have its temperature raised by 1.00 °C. Solving for s and substituting gives

$$s = \frac{q}{m\,\Delta t} = \frac{41.8 \text{ J}}{10.0 \text{ g} \times 1.00 \text{ °C}}$$

$$= 4.18 \text{ J g}^{-1}\text{°C}^{-1}$$

Recall that the older energy unit calorie (cal) is currently defined as 1 cal = 4.184 J. Therefore, the specific heat of water expressed in calories is

$$s = 1.000 \text{ cal g}^{-1}\text{°C}^{-1} \quad \text{(for H}_2\text{O)}$$

In measuring heat, we often use an apparatus containing water, so the specific heat of water is a quantity we will use in working many problems. Every substance has its own characteristic specific heat, and some are listed in Table 6.1.

When comparing or working with mole-sized quantities of substances, we can use the **molar heat capacity**, which is the amount of heat needed to raise the temperature of 1 mole of a substance by 1 °C. Molar heat capacity equals the specific heat times the molar mass and has units of J mol^{-1}°C^{-1}.

TABLE 6.1	Specific Heats
Substance	Specific Heat (J g^{-1}°C^{-1})
Carbon (graphite)	0.711
Copper	0.387
Ethyl alcohol	2.45
Gold	0.129
Granite	0.803
Iron	0.4498
Lead	0.128
Olive oil	2.0
Silver	0.235
Water (liquid)	4.184

CHEMISTRY OUTSIDE THE CLASSROOM 6.1

Water, Climate, and the Body's "Thermal Cushion"

Compared with most substances, *water has a very high specific heat. A body of water can therefore gain or lose a substantial amount of heat without undergoing a large change in temperature.* Because of this, the oceans of the world have a very significant moderating effect on climate. This is particularly apparent when we compare temperature extremes of locales near the oceans with those inland away from the sea and large lakes (such as the Great Lakes in the upper United States). Places near the sea tend to have cooler summers and milder winters than places located inland because the ocean serves as a thermal "cushion," absorbing heat in the summer and giving some of it back during the winter.

Warm and cool ocean currents also have global effects on climate. For example, the warm waters of the Gulf of Mexico are carried by the Gulf Stream across the Atlantic Ocean and keep winter relatively mild for Ireland, England, and Scotland. By comparison, northeastern Canada, which is at the same latitude as the British Isles, has much colder winters.

Water also serves as a thermal cushion for the human body. The adult body is about 60% water by mass, so it has a high heat capacity. In other words, the body can exchange considerable energy with the environment but experience only a small change in temperature. This makes it relatively easy for the body to maintain a steady temperature of 37 °C, which is vital to survival. With a substantial thermal cushion, the body adjusts to large and sudden changes in outside temperature while experiencing very small fluctuations of its core temperature.

Direction of Heat Flow

Heat is energy that is transferred from one object to another. This means that the heat lost by one object is the same as the heat gained by the other. To indicate the direction of heat flow, we assign a positive sign to q if the heat is gained and a negative sign if the heat is lost. For example, if a piece of warm iron is placed into a beaker of cool water and the iron loses 10.0 J of heat, the water gains 10.0 J of heat. For the iron, $q = -10.0$ J and for the water, $q = +10.0$ J.

The relationship between the algebraic signs of q in a transfer of heat can be stated in a general way by the equation

$$q_1 = -q_2 \tag{6.8}$$

where 1 and 2 refer to the objects between which the heat is transferred.

TOOLS

Heat transfer

Example 6.1
Determining the Heat Capacity of an Object

If you dry your hair with a blow drier and wear earrings at the same time, sometimes you feel the earrings get very warm—almost enough to burn your ear. Suppose that a set of earrings at 85.40 °C is dropped into an insulated coffee cup containing 25.0 g of water at 25.00 °C. The temperature of the water rises to 25.67 °C. What is the heat capacity of the earrings in J/°C?

Analysis: In this question, heat is being transferred from a warm object, the earrings, to the cooler water. We are being asked for the heat capacity, C, of the warm object. We can calculate the amount of heat gained by the water, which is the same as the amount of heat lost by the earrings. Using this amount of heat, we can then calculate the heat capacity of the earrings.

Assembling the Tools: First, for the amount of heat gained by the water, q_{H_2O}, we will apply Equation 6.7,

$$q = ms\,\Delta t$$

as our tool, which uses the specific heat of water ($4.184\ \text{J g}^{-1}\,°\text{C}^{-1}$), the mass of water, and the temperature change for water to calculate specific heat.

Then, we can use Equation 6.8, the tool for calculating the transfer of heat from the earrings to the water, and write:

$$q_{\text{earrings}} = -q_{H_2O}$$

We have to be very careful about the algebraic signs. Because the water is gaining heat, q_{H_2O} will be a positive quantity, and because the earrings lose heat, q_{earrings} will be a negative quantity. The negative sign on the right assures us that when we substitute the positive value for q_{H_2O}, the sign of q_{earrings} will be negative.

Once we've found q_{earrings}, we will utilize Equation 6.5,

$$q = C\,\Delta t$$

the tool that relates heat capacity to temperature change and amount of heat exchanged. To find C, we divide both sides by the temperature change, Δt:

$$C = q/\Delta t$$

Solution: The temperature of the water rises from 25.00 °C to 25.67 °C, so for the water,

$$\Delta t_{H_2O} = t_{\text{final}} - t_{\text{initial}} = 25.67\ °\text{C} - 25.00\ °\text{C} = +0.67\ °\text{C}$$

The specific heat of water is 4.184 J g^{-1} °C^{-1} and its mass is 25.0 g. Therefore, for water, the heat absorbed is

$$q_{H_2O} = ms\,\Delta t$$

$$= 25.0\ g \times 4.184\ J\,g^{-1}°C^{-1} \times 0.67\ °C$$

$$= +7.0 \times 10^1\ J$$

Therefore, $q_{earrings} = -7.0 \times 10^1$ J.

The temperature of the earrings decreases from 85.40 °C to 25.67 °C, so

$$\Delta t_{earrings} = t_{final} - t_{initial} = 25.67\ °C - 85.40\ °C = -59.73\ °C$$

The heat capacity of the earrings is then

$$C = \frac{q_{earrings}}{\Delta t_{earrings}} = \frac{-7.0 \times 10^1\ J}{-59.73\ °C} = 1.2\ J/°C$$

Is the Answer Reasonable? In any calculation involving energy transfer, we first check to see that all quantities have the correct signs. Heat capacities are positive for common objects, so the fact that we've obtained a positive value for C tells us we've handled the signs correctly.

For the transfer of a given amount of heat, the larger the heat capacity, the smaller the temperature change. The heat capacity of the water is $C = ms = 25.0\ g \times 4.18\ J\,g^{-1}°C^{-1} = 105\ J\,°C^{-1}$ and water changes temperature by 0.66 °C. The size of the temperature change for the earrings, 59.74 °C, is almost 100 times as large as that for the water, so the heat capacity should be about ¹⁄₁₀₀ that of the water. Dividing 105 J °C^{-1} by 100 gives 1.05 J °C^{-1}, which is not too far from our answer, so our calculations seem to be reasonable.

Example 6.2
Calculating Heat from a Temperature Change, Mass, and Specific Heat

If a piece of copper wire with a mass of 20.9 g changes in temperature from 25.00 to 28.00 °C, how much heat has it absorbed?

Analysis: The question asks us to connect the heat absorbed by the wire with its temperature change, Δt. We need to know either the heat capacity of the wire or the specific heat of copper and its mass. We don't know the heat capacity of the wire. However, we do know the mass of the wire and the fact that it is made of copper, so we can look up the specific heat and calculate the amount of heat absorbed.

Assembling the Tools: We can start with Equation 6.7,

$$q = ms\,\Delta t$$

as our tool to solve the problem. The specific heat of copper is listed in Table 6.1: 0.387 J g^{-1} °C^{-1}.

Solution: The mass, m, of the wire is 20.9 g, the specific heat, s, is 0.387 J g^{-1} °C^{-1}, and the temperature increases from 25.00 to 28.00 °C, so Δt is 3.00 °C. Using these values in Equation 6.7 gives

$$q = ms\,\Delta t$$

$$= (20.9\ g) \times (0.387\ J\,g^{-1}°C^{-1}) \times (3.00\ °C)$$

$$= 24.3\ J$$

Thus, only 24.3 J raises the temperature of 20.9 g of copper by 3.00 °C. Because Δt is positive, so is q, 24.3 J. Thus, the sign of the energy change is in agreement with the fact that the wire absorbs heat.

Is the Answer Reasonable? If the wire had a mass of only 1 g and its temperature increased by 1 °C, we'd know from the specific heat of copper (let's round it to 0.39 J g^{-1} °C^{-1}) that the ring would absorb 0.39 J. For a 3 °C increase, the answer would be three times as much, or 1.2 J. For a piece of copper a little heavier than 20 g, the heat absorbed would be 20 times as much, or about 24 J. So our answer is reasonable.

Example 6.3
Calculating a Final Temperature from the Specific Heat, Heat Lost or Gained, and the Mass

If a 25.2 g piece of silver absorbs 365 J of heat, what will the final temperature of the silver be if the initial temperature is 22.2 °C? The specific heat of silver is 0.235 J g^{-1} °C^{-1}.

Analysis: We are being asked to find the final temperature of the silver after it absorbs 365 J of heat. This question is easier to solve in two steps. First, we will rearrange Equation 6.7 to determine the change in temperature, and then we will use the initial temperature and the change in temperature to find the final temperature.

Assembling the Tools: Just as in Example 6.2, we will use Equation 6.7, $q = ms\,\Delta t$, as our tool to solve the problem. Then, we will rearrange Equation 6.4, $\Delta t = t_{final} - t_{initial}$, and use it to calculate the final temperature.

Solution: We are looking for the final temperature, so let's start by rearranging Equation 6.7 by dividing both sides by m and s to get

$$\frac{q}{ms} = \Delta t$$

The amount of heat gained is 365 J, the mass of the silver is 25.2 g, and the specific heat for silver is 0.235 J g^{-1} °C^{-1}. Use these numbers in the equation and solve:

$$\frac{365\ \text{J}}{25.2\ \text{g} \times 0.235\ \text{J g}^{-1}\,°\text{C}^{-1}} = \Delta t$$

$$\Delta t = 61.6\ °\text{C}$$

This just gives us the change in temperature. We are really interested in the final temperature, so we will use this change of temperature and the initial temperature to determine the final temperature.

$$\Delta t = t_{final} - t_{initial}$$

$$61.6\ °\text{C} = t_{final} - 22.2\ °\text{C}$$

$$t_{final} = 61.6\ °\text{C} + 22.2\ °\text{C}$$

$$t_{final} = 83.8\ °\text{C}$$

Is the Answer Reasonable? Heat is being added to the system, so for a quick check we would expect the temperature of the silver to increase, and it does. For a more detailed check we can estimate how much it increases since it depends on the amount of silver and the specific heat of silver. If we round off the numbers, we can see whether we are on the right track: the amount of heat is about 350 J, the mass is about 25 g, and the specific heat is about $0.2 \text{ J g}^{-1} \, ^{\circ}\text{C}^{-1}$.

$$\frac{350 \text{ J}}{25 \text{ g} \times 0.2 \text{ J g}^{-1} \, ^{\circ}\text{C}^{-1}} = 70 \, ^{\circ}\text{C}$$

The change in temperature of 70 °C is very close to 61.6 °C, so it is reasonable to think that we are correct.

Practice Exercise 6.4

A ball bearing at 220.0 °C is dropped into a cup containing 250.0 g of water at 20.0 °C. The water and ball bearing come to a temperature of 30.0 °C. What is the heat capacity of the ball bearing in J/°C? (*Hint:* How are the algebraic signs of q related for the ball bearing and the water?)

Practice Exercise 6.5

The temperature of 255 g of water is changed from 25.0 to 30.0 °C. How much energy was transferred into the water? Calculate your answer in joules, kilojoules, calories, and kilocalories.

Practice Exercise 6.6

Silicon, used in computer chips, has a heat capacity of $0.705 \text{ J g}^{-1} \, ^{\circ}\text{C}^{-1}$. If 549 J of heat is absorbed by 7.54 g of silicon, what will the final temperature be if the initial temperature is 25.0 °C?

6.4 | Energy of Chemical Reactions

Almost every chemical reaction involves the absorption or release of energy. As the reaction occurs, the potential energy (also called chemical energy) changes. To understand the origin of this energy change we need to explore the origin of potential energy in chemical systems.

In Chapter 2 we introduced you to the concept of *chemical bonds*, which are the attractive forces that bind atoms to each other in molecules, or ions to each other in ionic compounds. In this chapter we presented the idea that since particles experience attractions or repulsions, potential energy changes occur when the particles come together or move apart. We can now bring these concepts together to understand the origin of energy changes in chemical reactions.

Exothermic and Endothermic Reactions

Chemical reactions generally involve *both* the breaking and making of chemical bonds. In most reactions, when bonds *form*, things that attract each other move closer together, decreasing the potential energy of the reacting system. When bonds break, on the other hand, things that are normally attracted to each other are forced apart, which increases the potential energy of the reacting system. Every reaction, therefore, has a certain net overall potential energy change, the difference between the "cost" of breaking bonds and the "profit" from making new bonds.

In many reactions, the products have *less* potential energy (chemical energy) than the reactants. When the gas methane burns in a Bunsen burner, for example, molecules of CH_4 and O_2 with large amounts of chemical energy but relatively low amounts of molecular kinetic energy change into products (CO_2 and H_2O) that have less chemical energy but much more molecular kinetic energy. Thus, some chemical energy changes into molecular *kinetic* energy, and this increase leads to a temperature increase in the reaction mixture. If the reaction is occurring in an uninsulated system, some of this energy can be transferred to the surroundings as heat. The net result, then, is that the drop in chemical energy appears as heat that's transferred to the surroundings. In the chemical equation, therefore, we can write heat as a product.

■ *exo* means "out"
endo means "in"
therm means "heat"

$$CH_4(g) + 2O_2(g) \longrightarrow CO_2(g) + 2H_2O(g) + \text{heat}$$

Any reaction in which heat is a product, or energy is released, is said to be **exothermic**.

In some reactions the products have more chemical energy than the reactants. For example, green plants make energy-rich molecules of glucose ($C_6H_{12}O_6$) and oxygen from carbon dioxide and water by a multistep process called photosynthesis. It requires a continuous supply of energy, which is supplied by the sun. The plant's green pigment, chlorophyll, is the solar energy absorber for photosynthesis.

© Michele Molinari/Alamy

$$6CO_2 + 6H_2O + \text{solar energy} \xrightarrow[\text{many steps}]{\text{chlorophyll}} C_6H_{12}O_6 + 6O_2$$

Reactions that consume energy, or energy is a reactant, are said to be **endothermic**. Usually such reactions change kinetic energy into potential energy (chemical energy), so the temperature of the system tends to drop as the reaction proceeds. If the reaction is occurring in an uninsulated vessel, the temperature of the surroundings will drop as heat energy flows into the system.

Through an endothermic reaction, photosynthesis converts solar energy, trapped by the green pigment chlorophyll, into plant compounds that are rich in chemical energy.

Energy of Breaking Bonds and Forming Bonds

The strength of a chemical bond is measured by how much energy is needed to break it, or by how much energy is released when the bond forms. When a bond forms, the larger the amount of energy lost, the stronger the bond.

Breaking weak bonds requires relatively little energy compared to the amount of energy released when strong bonds form. This is the key to understanding why burning fuels such as CH_4 produce heat. Weaker bonds between carbon and hydrogen within the fuel break so that stronger bonds in the water and carbon dioxide molecules can form, as in Figure 6.6. Burning a hydrocarbon fuel produces one molecule of carbon dioxide for each carbon atom and one molecule of water for each pair of hydrogen atoms in the fuel. Generally, the more carbon and hydrogen atoms a molecule of the fuel contains, the more strong bonds can be formed when it burns—and the more heat it will produce, per molecule.

We will have a lot more to say about chemical bonds and their strengths in Chapters 8 and 9.

Would the explosive reaction of hydrogen and oxygen to form water be exothermic or endothermic? (*Hint:* What is the definition of exothermic and endothermic?)

When ammonium nitrate is dissolved in water, the container holding the solution becomes cold. Is the reaction exothermic or endothermic?

Practice Exercise 6.7

Practice Exercise 6.8

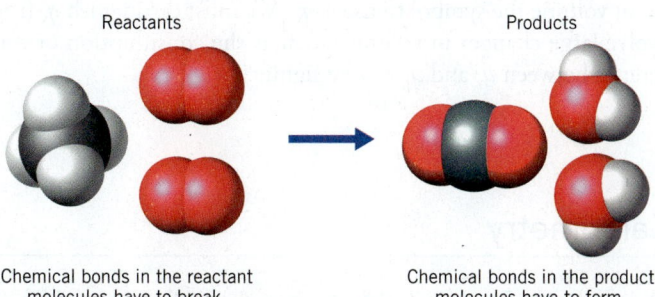

Reactants   Products

Chemical bonds in the reactant   Chemical bonds in the product
molecules have to break.   molecules have to form.

Figure 6.6 | **Energy changes are associated with bond breaking and bond forming.** In the combustion of methane, the C—H bonds in CH_4 molecules and the O—O bonds in O_2 molecules have to break so that the C—O bonds in CO_2 and the O—H bonds in H_2O can form. Bond breaking absorbs energy, whereas bond forming releases energy. In this reaction, more energy is released in bond formation than in bond breaking and the reaction is exothermic. Another way of saying this is that this reaction converts chemical potential energy into heat energy.

6.5 | Heat, Work, and the First Law of Thermodynamics

■ There is no instrument that directly measures energy. A calorimeter does not do this; its thermometer is the only part that provides raw data, the temperature change. We use the change in temperature to calculate the energy change.

The amount of heat absorbed or released in a chemical reaction is called the **heat of reaction**. Heats of reaction are determined by measuring the change in temperature they cause in their surroundings, using an apparatus called a **calorimeter**. The calorimeter is often just a container with a known heat capacity in which the reaction is carried out. We can calculate the heat of reaction by measuring the temperature change the reaction causes in the calorimeter. The science of using a calorimeter for determining heats of reaction is called **calorimetry**.

Calorimeter design is not standard; it varies according to the kind of reaction and the precision desired. Calorimeters are usually designed to measure heats of reaction under conditions of either constant volume or constant pressure. We have constant volume conditions if we run the reaction in a closed, rigid container. Running the reaction in an open container imposes constant-pressure conditions.

Pressure is an important variable in calorimetry, as we'll see in a moment. If you have ever blown up a balloon or worked with an auto or bicycle tire, you have already learned something about *pressure*. You know that when a tire holds air at a relatively high pressure, it is hard to dent or push in its sides because "something" is pushing back. That something is the air pressure inside the tire.

Pressure is the amount of *force* acting on a unit of area; it's the ratio of force to area:

$$\text{Pressure} = \frac{\text{force}}{\text{area}}$$

You can increase the pressure in a tire by pushing or forcing air into it. Its pressure, when you stop, is whatever pressure you have added *above the initial pressure*—namely, the atmospheric pressure. We describe the air in the tire as being *compressed*, meaning that it's at a higher pressure than the atmospheric pressure.

Atmospheric pressure is the pressure exerted by the mixture of gases in our atmosphere. In English units this ratio of force to area is approximately 14.7 lb in.$^{-2}$ at sea level. The ratio does vary a bit with temperature and weather, and it varies considerably with altitude. The value of 14.7 lb in.$^{-2}$ is very close to two other common pressure units, the **standard atmosphere** (abbreviated **atm**) and the **bar**. A container that is open to the atmosphere at sea level is under a constant pressure of about 14.7 lb in.$^{-2}$, which is approximately 1 bar or 1 atm.

■ More precisely, 14.696 lb in.$^{-2}$ = 1.0000 atm = 1.0133 bar.

Up to now we've used the symbol q for heat. Since the heat can be measured when we keep the system at a constant pressure we will use the symbol q_p. When the heat is measured at constant volume the symbol to use is q_v. We must distinguish q_v from q_p. In reactions that involve large changes in volume (such as the consumption or production of a gas), the difference between q_v and q_p can be significant.

Example 6.4
Computing q_v and q_p Using Calorimetry

A gas-phase chemical reaction occurs inside an apparatus similar to that shown in Figure 6.7. The reaction vessel is a cylinder topped by a piston. The piston can be locked in place with a pin. The cylinder is immersed in an insulated bucket containing a precisely weighed amount of water. A separate experiment determined that the calorimeter (which includes the piston, cylinder, bucket, and water) has a heat capacity of 8.101 kJ/°C. The reaction was run twice with identical amounts of reactants each time. The following data were collected.

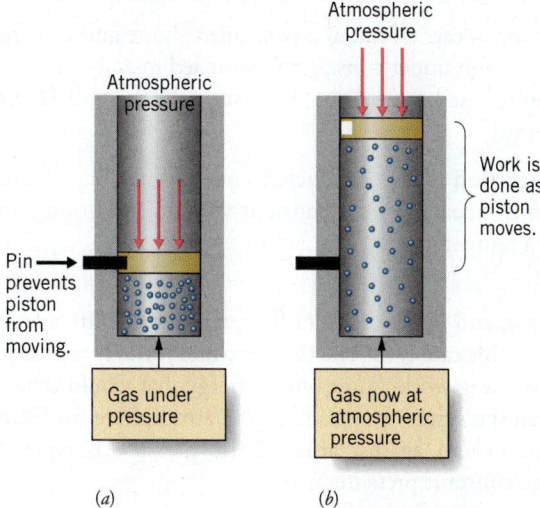

Figure 6.7 | **Expansion of a gas.** (*a*) A gas is confined under pressure in a cylinder fitted with a piston that is held in place by a sliding pin. (*b*) When the piston is released, the gas inside the cylinder expands and pushes the piston upward against the opposing pressure of the atmosphere. As it does so, the gas does some work on the surroundings.

Run	Pin Position	Initial Bucket Temperature (°C)	Final Bucket Temperature (°C)
1	*a* (piston locked)	24.00	28.91
2	*b* (piston unlocked)	27.32	31.54

Determine q_v and q_p for this reaction.

Analysis: The key to solving this problem is realizing that the heat capacity and temperature change can be used to calculate the heat absorbed by the calorimeter, which is equal in magnitude to the amount of heat lost by the reaction.

Assembling the Tools: The tools are Equation 6.5, $q = C\,\Delta t$, which makes it possible to measure the heat gained by the calorimeter and Equation 6.8, $q_{calorimeter} = -q_{reaction}$, which relates the heat lost by the reaction to the heat gained by the calorimeter. Run 1 will give the heat at constant volume, q_v, since the piston cannot move. Run 2 will give the heat at constant pressure, q_p, because with the piston unlocked the entire reaction will run under atmospheric pressure.

Solution: For Run 1, Equation 6.5 gives the heat absorbed by the calorimeter as

$$q = C\,\Delta t = (8.101 \text{ kJ/°C}) \times (28.91\text{ °C} - 24.00\text{ °C}) = 39.8 \text{ kJ}$$

Because the calorimeter gains heat, this amount of heat is released by the reaction, so q_v *for the reaction* must be negative. Therefore, $q_v = -39.8$ kJ.
For Run 2, conducted at constant pressure,

$$q = C\,\Delta t = (8.101 \text{ kJ/°C}) \times (31.54\text{ °C} - 27.32\text{ °C}) = 34.2 \text{ kJ}$$

so $q_p = -34.2$ kJ.

Is the Answer Reasonable? The arithmetic is straightforward, but in any calculation involving heat, always check to see that the signs of the heats in the problem make sense. The calorimeter absorbs heat, so its heats are positive. The reaction releases this heat, so its heats must be negative.

Practice Exercise 6.9

An exothermic reaction is carried out at a constant volume and then repeated under constant pressure. The reaction under constant pressure led to an increase in the volume of the reaction. Which run released more heat, the first or the second? (*Hint:* Apply the law of conservation of energy.)

Practice Exercise 6.10

For an exothermic reaction that is conducted under conditions of constant pressure and expanded, and then is repeated under constant volume conditions, which run will have the larger increase in temperature?

Let's explore why q_v and q_p are different for reactions that involve a significant volume change. The system in this case is the reacting mixture. If the system expands against atmospheric pressure, it is doing work. Some of the energy that would otherwise appear as heat will be used up when the system pushes back the atmosphere. In Example 6.4, the work done to expand the system against atmospheric pressure is equal to the amount of "missing" heat in the constant pressure case:

$$\text{work} = (-39.8 \text{ kJ}) - (-34.2 \text{ kJ}) = -5.6 \text{ kJ}$$

The minus sign indicates that energy is leaving the system. This is called **expansion work** (or more precisely, **pressure–volume work**). A common example of pressure–volume work is the work done by the expanding gases in a cylinder of a car engine as it moves a piston.

Another example is the work done by expanding gases to lift a rocket from the ground. The amount of expansion work, w, done can be computed from atmospheric pressure and the volume change that the system undergoes:[3]

$$w = -P\,\Delta V \tag{6.9}$$

■ The sign of *w* confirms that the system loses energy by doing work pushing back the atmosphere.

P is the *opposing pressure* against which the piston pushes, and ΔV is the change in the volume of the system (the gas) during the expansion; that is, $\Delta V = V_{\text{final}} - V_{\text{initial}}$. Because V_{final} is greater than V_{initial}, ΔV must be positive. This makes the expansion work negative.

First Law of Thermodynamics

The **first law of thermodynamics** is actually the *same as the law of conservation of energy*: "energy cannot be created or destroyed." Energy may be converted from one form to another in many different ways. Some examples are: Kinetic energy can be converted into potential energy by storing water behind a dam, and potential energy may be converted into kinetic or electrical energy when a battery is used to propel a toy car or light an LED flashlight. In light of our discussion of heat and work, let's investigate how this concept applies.

---

[3]$P\,\Delta V$ must have units of energy if it is referred to as work. Work is accomplished when an opposing force, F, is pushed through some distance or length, L. The amount of work done is equal to the strength of the opposing force multiplied by the distance the force is moved:

$$\text{Work} = F \times L$$

Because pressure is force (F) per unit area, and area is simply length squared, L^2, we can write the following equation for pressure.

$$P = \frac{F}{L^2}$$

Volume (or a volume change) has dimensions of length cubed, L^3, so pressure times the volume change is

$$P\,\Delta V = \frac{F}{L^2} \times L^3 = F \times L$$

but, as explained previously, $F \times L$ also equals work.

In chemistry, *a minus sign on an energy transfer always means that the system loses energy.* Consider what happens whenever work can be done or heat can flow in the system shown in Figure 6.7*b*. If work is negative (as in an expansion) the system loses energy and the surroundings gain it. We say work is done *by* the system. If heat is negative (as in an exothermic reaction) the system loses energy and the surroundings gain it. Either event causes a drop in internal energy. On the other hand, if work is positive (as in a compression), the system gains energy and the surroundings lose it. We say that work is done *on* the system. If heat is positive (as in an endothermic reaction), again, the system will gain energy and the surroundings will lose it. Either positive work or positive heat causes a positive change or increase in internal energy.

Work and heat are simply alternative ways to transfer energy. Using this sign convention, we can relate the work w and the heat q that go into the system to the internal energy change (ΔE) the system undergoes:

$$\Delta E = q + w \qquad (6.10)$$

TOOLS

First law of thermodynamics

In Section 6.2 we pointed out that the internal energy depends only on the current state of the system; we said that *internal energy, E, is a state function.* This statement (together with Equation 6.10, which is a definition of internal energy change) is a statement of the first law of thermodynamics.

ΔE is independent of how a change takes place; it depends only on the state of the system at the beginning and the end of the change. The values of q and w depend, however, on what happens *between* the initial and final states. Thus, neither q nor w is a state function. Their values depend on the *path* of the change. For example, consider the discharge of an automobile battery by two different paths (see Figure 6.8). Both paths take us between the same two states, one being the fully charged state and the other the fully discharged state. Because E is a state function and because both paths have the same initial and final states, ΔE *must be the same for both paths.* But lets take a closer look at q and w.

In path 1, we short-circuit the battery by placing a heavy wrench across the terminals. Sparks fly and the wrench becomes very hot as the battery quickly discharges. Heat is given off, but *the system does no work* ($w = 0$). All of ΔE appears as heat.

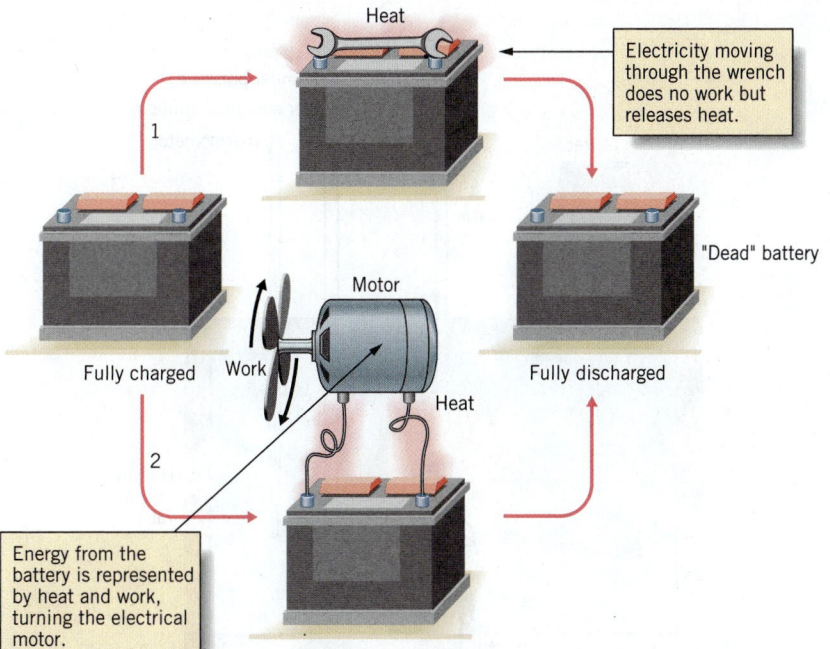

Figure 6.8 | **Energy, heat, and work.** The complete discharge of a battery along two different paths yields the same total amount of energy, ΔE. However, if the battery is simply short-circuited with a heavy wrench, as shown in path 1, this energy appears entirely as heat. Path 2 gives part of the total energy as heat, but much of the energy appears as work done by the motor.

In path 2, we discharge the battery by using it to operate a motor. Along this path, some of the energy represented by ΔE appears as work (running the motor) and the rest appears as heat (from the friction within the motor and the electrical resistance of the wires).

There are two vital lessons here. The first is that neither q nor w is a state function. Their values depend *entirely* on the path between the initial and final states. The second lesson is that the sum of q and w, ΔE, is the same, regardless of the path taken, as long as we start and stop at the same places.

6.6 | Heats of Reaction

As we have shown in Section 6.5, the change of energy in a system can be determined by measuring the amount of heat and work that is exchanged in a system. Since the amount of work that is either done by a system or on a system depends on both pressure and volume, the heat of a reaction is measured under conditions of either constant volume or constant pressure.

ΔE, Constant-Volume Calorimetry

The heats of specific reactions are sometimes given labels—for example, the heat produced by a combustion reaction is called the **heat of combustion**. Because combustion reactions require oxygen and produce gaseous products, we have to measure heats of combustion in a sealed container. Figure 6.9 shows the apparatus that is usually used to determine heats of combustion. The instrument is called a **bomb calorimeter** because the vessel holding the reaction itself resembles a small bomb. The "bomb" has rigid walls, so the change in volume, ΔV, is zero when the reaction occurs. This means, of course, that $P \Delta V$ must also be zero, and no expansion work is done, so w in Equation 6.10 is zero. Therefore, the heat of reaction measured in a bomb calorimeter is the **heat of reaction at constant volume, q_v**, and corresponds to ΔE.

$$\Delta E = q_v$$

Food scientists determine dietary Calories in foods and food ingredients by burning them in a bomb calorimeter. The reactions that break down foods in the body are complex, but they have the same initial and final states as the combustion reaction for the food.

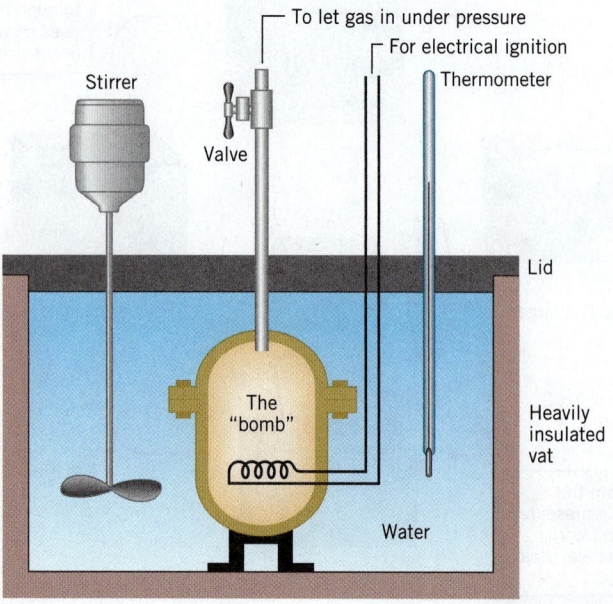

Figure 6.9 | **A bomb calorimeter.**
The water bath is usually equipped with devices for adding or removing heat from the water, thus keeping its temperature constant up to the moment when the reaction occurs in the bomb. The reaction chamber is of fixed volume, so $P \Delta V$ must equal zero for reactions in this apparatus.

Example 6.5
Bomb Calorimetry

When 1.000 g of olive oil was completely burned in pure oxygen in a bomb calorimeter such as the one shown in Figure 6.9, the temperature of the water bath increased from 22.000 °C to 26.049 °C. (a) How many dietary Calories are in olive oil, per gram? The heat capacity of the calorimeter is 9.032 kJ/°C. (b) Olive oil is almost pure glyceryl trioleate, $C_{57}H_{104}O_6$. The equation for its combustion is

$$C_{57}H_{104}O_6(l) + 80O_2(g) \longrightarrow 57CO_2(g) + 52H_2O(l)$$

Assuming that the olive oil burned in part (a) was pure glyceryl trioleate, what is the change in internal energy, ΔE, in units of kilojoules, for the combustion of one mole of glyceryl trioleate?

Analysis: Bomb calorimetry measures q_v, which is equal to the internal energy change for the reaction. In part (a), the question is asking for the *amount of heat* released in the combustion reaction of olive oil in units of dietary Calories or kcal. All of the heat released by the reaction is absorbed by the calorimeter. By calculating the heat absorbed by the calorimeter we will be able to then calculate the amount of heat released by the reaction. Also, since only one gram was burned, we will be calculating the amount of heat per gram of olive oil.

For part (b), the question is asking for the change in *energy released per mole* of glyceryl trioleate (olive oil) burned. We can use the information from part (a), in which we calculated the change in heat energy per gram, and the molar mass of the glyceryl trioleate to calculate the ΔE on a per mole basis.

Assembling the Tools: In part (a), Equation 6.5, $q_{calorimeter} = C \Delta t$, is the tool to compute the heat absorbed by the calorimeter. Then we'll use the tool for heat transfer, $q_{calorimeter} = -q_{reaction}$, to get the heat released by the combustion reaction.

Since the given heat capacity has units of kJ/°C, $q_{reaction}$ will have units of kJ. We'll need to convert kJ to dietary Calories, which are actually kilocalories (kcal), using the relationship

$$1 \text{ kcal} = 4.184 \text{ kJ}$$

Part (b) asks for the change in internal energy, ΔE, per mole. The heat calculated in part (a) is equal to ΔE for the combustion of 1.000 g of glyceryl trioleate. The molar mass of glyceryl trioleate is the tool to convert ΔE per gram to ΔE per mole.

Solution: First, we compute the heat absorbed by the calorimeter when 1.000 g of olive oil is burned, using Equation 6.5:

$$q_{calorimeter} = C \Delta t = (9.032 \text{ kJ/°C}) \times (26.049 \text{ °C} - 22.000 \text{ °C}) = 36.57 \text{ kJ}$$

Changing the algebraic sign gives the heat of combustion. Dividing by the 1.000 g of olive oil used gives us $q_v = -36.57$ kJ/g. Part (a) asks for the heat in dietary Calories, which are equivalent to kilocalories (kcal).

$$\frac{-36.57 \text{ kJ}}{1.000 \text{ g oil}} \times \frac{1 \text{ kcal}}{4.184 \text{ kJ}} \times \frac{1 \text{ Cal}}{1 \text{ kcal}} = -8.740 \text{ Cal/g oil}$$

or we can say "8.740 dietary Calories are released when one gram of olive oil is burned." For part (b), we'll convert the heat produced per gram to the heat produced per mole, using the molar mass of $C_{57}H_{104}O_6$, 885.4 g/mol:

$$\frac{-36.57 \text{ kJ}}{1.000 \text{ g } C_{57}H_{104}O_6} \times \frac{885.4 \text{ g } C_{57}H_{104}O_6}{1 \text{ mol } C_{57}H_{104}O_6} = -3.238 \times 10^4 \text{ kJ/mol } C_{57}H_{104}O_6$$

$$= -3.238 \times 10^4 \text{ kJ/mol}$$

Since this is heat at constant volume, we have $\Delta E = q_v = -3.238 \times 10^4$ kJ for combustion of 1 mol of $C_{57}H_{104}O_6$.

Is the Answer Reasonable? Always check the signs of calculated heats first in calorimetry calculations since the signs follow the heat flow. If heat is given off, the sign for q is negative, and in combustion reactions, heat is released, so the sign of q must be negative. If you look at a bottle of olive oil at home or in a store you will see that the label states that a tablespoon has about 100 dietary Calories (with a tablespoon being 15 mL). Even though olive oil is less dense than water, 15 mL would weigh about 15 g. Therefore, 1 g of olive oil would have about 120 Cal/15 g ≈ 9 Cal/g. So, it would seem reasonable that a gram of olive oil would have about 9 dietary Calories per gram.

Practice Exercise 6.11

Since it can be obtained in very high purity, benzoic acid is one of the classic materials used for determining the heat capacity of bomb calorimeters. The heat of combustion of benzoic acid, $HC_7H_5O_2$ is, -3227 kJ mol^{-1}. When 2.85 g of $HC_7H_5O_2$ (molar mass = 122.12 g/mol) was burned in a bomb calorimeter, the temperature of the calorimeter changed from 24.05 °C to 29.19 °C. What is the heat capacity of the calorimeter in units of kJ/°C? (*Hint:* Which equation relates temperature change, energy, and heat capacity?)

Practice Exercise 6.12

A 1.50 g sample of pure sucrose is burned in a bomb calorimeter that has a heat capacity of 8.930 kJ/°C. The temperature of the water jacket rises from 24.00 °C to 26.77 °C. What is ΔE for the combustion of 1 mole of sucrose, $C_{12}H_{22}O_{11}$? How many dietary Calories per gram does sucrose provide when metabolized?

ΔH, Constant-Pressure Calorimetry

Most reactions that are of interest to us do not occur at constant volume. Instead, they run in open containers such as test tubes, beakers, and flasks, where they experience the constant pressure of the atmosphere, and we can measure the **heat of reaction at constant pressure, q_p**. When reactions are run under constant pressure, they may transfer energy as heat, q_p, and as expansion work, w, so to calculate ΔE we need Equation 6.11:

■ Biochemical reactions also occur at constant pressure.

$$\Delta E = q_p + w \qquad (6.11)$$

This is inconvenient. If we want to calculate the internal energy change for the reaction, we'll have to measure its volume change and then use Equation 6.9. To avoid this problem, scientists have defined a "corrected" internal energy called **enthalpy**, or **H**. Enthalpy is defined by the equation

■ From the Greek *en + thalpein*, meaning to heat or to warm

$$H = E + PV$$

At constant pressure,

$$\Delta H = \Delta E + P\,\Delta V$$

$$\Delta H = (q_p + w) + P\,\Delta V$$

From Equation 6.9, $P\,\Delta V = -w$, so

$$\Delta H = (q_p + w) + (-w)$$

$$\Delta H = q_p \qquad (6.12)$$

Like E, H is a state function.

As with internal energy, an **enthalpy change, ΔH**, is defined by the equation

$$\Delta H = H_{\text{final}} - H_{\text{initial}}$$

For a chemical reaction, this can be rewritten as follows:

$$\Delta H = H_{\text{products}} - H_{\text{reactants}} \qquad (6.13)$$

Positive and negative values of ΔH have the same interpretation as positive and negative values of ΔE.

Significance of the Sign of ΔH:
For an endothermic change, ΔH is positive.
For an exothermic change, ΔH is negative.

Algebraic sign of ΔH

The difference between ΔH and ΔE for a reaction equals $P\,\Delta V$. This difference can be fairly large for reactions that produce or consume gases, because these reactions can have very large volume changes. For reactions that involve only solids and liquids, though, the values of ΔV are tiny, so ΔE and ΔH for these reactions are nearly identical.

A very simple constant-pressure calorimeter, dubbed the coffee-cup calorimeter, is made of two nested and capped cups made of foamed polystyrene, a very good insulator (Figure 6.10). A reaction occurring between aqueous solutions in such a calorimeter exchanges very little heat with the surroundings, particularly if the reaction is fast. The temperature change is rapid and easily measured. We can use Equation 6.5, $q = C\,\Delta t$, to find the heat of reaction, if we have determined the heat capacity of the calorimeter and its contents. The coffee cup and the thermometer absorb only a tiny amount of heat, and we can usually ignore them in our calculations, and just assume that all of the heat from the reaction is transferred to the reaction mixture.

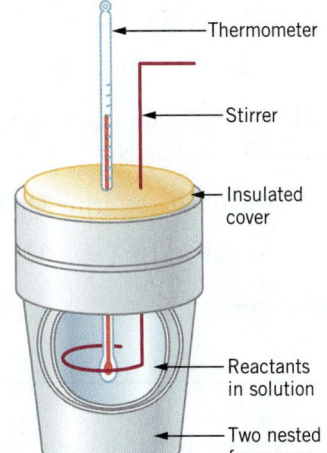

Thermometer

Stirrer

Insulated cover

Reactants in solution

Two nested foam cups

Figure 6.10 | A coffee-cup calorimeter used to measure heats of reaction at constant pressure. Typically, aqueous solutions of the reactants, whose temperatures have been measured, are combined in the calorimeter and the temperature change of the mixture is noted as the reaction proceeds to completion. The temperature change and the heat capacities of the solutions containing the reactants are used to calculate the heat released or absorbed in the reaction.

Analyzing and Solving Multi-Concept Problems

The reaction of hydrochloric acid and sodium hydroxide solutions is very rapid, exothermic, and follows the equation:

$$HCl(aq) + NaOH(aq) \longrightarrow NaCl(aq) + H_2O$$

In one experiment, a student placed 50.0 mL of 1.00 M HCl at 25.5 °C in a coffee-cup calorimeter. To this was added 50.0 mL of 1.50 M NaOH solution also at 25.5 °C. (The density of 1.00 M HCl is 1.02 g mL^{-1} and that of 1.50 M NaOH is 1.06 g mL^{-1}.) The mixture was stirred, and the temperature quickly increased to 32.2 °C. What is ΔH expressed in kJ per mole of the limiting reactant? What actually reacts in this experiment, and what chemical reaction do the thermochemical results actually represent?

Analysis: Sometimes complex questions are best approached from the final result desired, uncovering information we need as we go. In this case we need to know the enthalpy of reaction, which requires a heat capacity and temperature

change using equation 6.7, $q = ms\,\Delta t$. Since we are given volumes of solutions we will need to convert volumes to mass using the given densities. If we want the enthalpy in J mol^{-1}, we need to determine the moles that react. Since we were given the molarity and volume of both the NaOH and the HCl, this appears to be a limiting reactant problem similar to those we solved in Example 3.17 in Section 3.6. The final question suggests that we need to look more deeply at the molecular reaction to understand the chemistry of this process.

Strategy: While it does not matter which step we do first, our strategy will include (1) determining the heat produced, $q_{reaction}$, from the thermochemical data, which includes conversion of volumes to masses; (2) calculating the moles of the limiting reactant as done in Chapter 3; (3) dividing the $q_{reaction}$ by the moles of limiting reactant to obtain the $\Delta H°$ for one mole of reactant; and (4) addressing the last question of what we are really measuring by looking at the net ionic equation.

PART 1

Assembling the Tools: We need to solve the equation $q = ms\,\Delta t$. Because the solutions are relatively dilute, we can assume that their specific heats, s, are close to that of water, which is 4.184 J g^{-1} °C^{-1}. We need to convert volume to mass using the density as a conversion factor as done in Example 1.7 in Section 1.7. Last, we need equation 6.4, $\Delta t = t_{final} - t_{initial}$.

Solution: The density of 1.00 M HCl is 1.02 g mL^{-1} and that of 1.50 M NaOH is 1.06 g mL^{-1}. For the HCl solution, the mass is:

$$\text{mass of HCl} = 50.0 \text{ mL} \times \frac{1.02 \text{ g}}{1.00 \text{ mL}} = 51.0 \text{ g}$$

The mass of the NaOH solution is:

$$\text{mass of NaOH} = 50.0 \text{ mL} \times \frac{1.06 \text{ g}}{1.00 \text{ mL}} = 53.0 \text{ g}$$

Adding these we get the total mass of the solution: $m = 51.0 + 53.0 = 104.0$ g. The reaction changes the system's temperature by ($t_{\text{final}} - t_{\text{initial}}$), so

$$\Delta t = 32.2 \,^{\circ}\text{C} - 25.5 \,^{\circ}\text{C} = +6.7 \,^{\circ}\text{C}$$

(We will neglect the heat lost to the coffee cup itself, to the thermometer, or to the surrounding air.) So the heat produced in the reaction is

$$q = 104.0 \text{ g} \times 4.184 \text{ J g}^{-1}\,^{\circ}\text{C}^{-1} \times 6.7 \,^{\circ}\text{C}$$

$$= 2.9 \times 10^3 \text{ J} = 2.9 \text{ kJ}$$

Based on the energy transfer where $q_{\text{calorimeter}} = -q_{\text{reaction}}$ we say that the heat of the reaction is -2.9 kJ.

PART 2

Assembling the Tools: Now we need to use our tools for limiting reactant problems (see Section 3.6) to determine the moles that reacted to give the heat we just calculated.

Solution: We ask how many mL of NaOH solution is needed to react with the 50.0 mL of HCl that were given.

$$? \text{ mL of NaOH} = 50.0 \text{ mL of HCl}$$

We can use the two molarities, written as moles per 1000 mL, as conversion factors and the balanced equation given with the problem for the mol ratio conversion factor.

$$50.0 \text{ mL HCl} \times \frac{1.00 \text{ mol HCl}}{1000 \text{ mL HCl}} \times \frac{1 \text{ mol NaOH}}{1 \text{ mol HCl}} \times \frac{1000 \text{ mL NaOH}}{1.50 \text{ mol NaOH}} = 33.3 \text{ mL NaOH}$$

Since we are using 50.0 mL NaOH, this tells us that all of the HCl is used up and is the limiting reactant. The moles of HCl is calculated starting with the volume of 50.0 mL,

$$50.0 \text{ mL HCl} \times \frac{1.00 \text{ mol HCl}}{1000 \text{ mL HCl}} = 5.00 \times 10^{-2} \text{ mol HCl}$$

PART 3

Assembling the Tools: Our tool is the definition of a heat of reaction, which is q_{reaction} divided by the number of moles of limiting reactant.

Solution: Finally, we can calculate the $\Delta H°$ of reaction:

$$\Delta H° = \frac{2.9 \text{ kJ}}{5.00 \times 10^{-2} \text{ mol HCl}} = 58 \text{ kJ mol}^{-1} \text{ HCl}$$

PART 4

Assembling the Tools: We use the tool for net ionic equations for neutralization reactions developed in Section 4.6.

Solution: The ionic equation for the reaction given is:

$$Na^+(aq) + OH^-(aq) + H^+(aq) + Cl^-(aq) \longrightarrow H_2O + Na^+(aq) + Cl^-(aq)$$

And the net ionic equation is:

$$OH^-(aq) + H^+(aq) \longrightarrow H_2O$$

From this we can conclude that the spectator ions do not contribute to the thermochemistry of this reaction. We might postulate that the $\Delta H°$ of reaction for all strong acid − strong base reactions will be the same.

Are the Answers Reasonable? For the main calculation of the heat of reaction, we follow the heat flow: since the temperature of the calorimeter increases, $q_{calorimeter}$ is positive. Therefore $q_{reaction}$ must be negative, which we have indicated with the negative sign on the $\Delta H°$. Rechecking the math and unit cancellations shows that the numbers are correct. The use of a net ionic equation for a thermochemical result is also justified on the basis that only the H^+ (or H_3O^+) and OH^- actually react.

Practice Exercise 6.13

For the *Analyzing and Solving Multi-Concept Problems* example we just solved, calculate ΔH per mole of NaOH. (*Hint:* How many moles of NaOH were neutralized in the reaction?)

Practice Exercise 6.14

The exact same procedure as described in the *Analyzing and Solving Multi-Concept Problems* example we just solved was performed, except that the acid neutralized was 1.00 *M* acetic acid, $HC_2H_3O_2$, which has a density of 1.00 g mL^{-1}, and the temperature change was +6.6 °C. What is the net ionic reaction for this experiment, and what is its $\Delta H°$?

6.7 | Thermochemical Equations

As we have seen, the amount of heat a reaction produces or absorbs depends on the number of moles of reactants we combine. It makes sense that if we burn two moles of carbon, we're going to get twice as much heat as we would if we had burned one mole. For heats of reaction to have meaning, we must describe the **system** completely. Our description must include amounts and concentrations of reactants, amounts and concentrations of products, temperature, and pressure, because all of these things can influence heats of reaction.

■ Unless we specify otherwise, whenever we write ΔH we mean ΔH for the system, not the surroundings.

Chemists have agreed to a set of **standard states** to make it easier to report and compare heats of reaction. Most thermochemical data are reported for a pressure of 1 bar, or (for substances in aqueous solution) a concentration of 1 *M*. A temperature of 25 °C (298 K) is often specified as well, although temperature is not part of the definition of standard states in thermochemistry.

■ IUPAC recommends the use of the bar, instead of atmospheres, for standard states.

$\Delta H°$, Enthalpy Change for a Reaction at Standard State

The **standard heat of reaction,** $\Delta H°$, is the value of ΔH for a reaction occurring under standard conditions and involving the actual numbers of *moles* specified by the coefficients of the equation. To show that ΔH is for *standard* conditions, a degree sign is added to ΔH to make $\Delta H°$ (pronounced "delta H naught" or "delta H zero"). The units of $\Delta H°$ are really kilojoules/reaction.

To illustrate clearly what we mean by $\Delta H°$, let us use the reaction between gaseous nitrogen and hydrogen that produces gaseous ammonia.

$$N_2(g) + 3H_2(g) \longrightarrow 2NH_3(g)$$

When 1.000 mol of N_2 and 3.000 mol of H_2 react to form 2.000 mol of NH_3 at 25 °C and 1 bar, the reaction releases 92.38 kJ. Hence, for the reaction *as given by the preceding chemical equation,* $\Delta H° = -92.38$ kJ. Often the enthalpy change is given immediately after the equation; for example,

$$N_2(g) + 3H_2(g) \longrightarrow 2NH_3(g) \qquad \Delta H° = -92.38 \text{ kJ}$$

TOOLS

Thermochemical equations

■ Sometimes $\Delta H°_{rxn}$ is used for the heat of a thermochemical reaction.

An equation that also shows the value of $\Delta H°$ is called a **thermochemical equation.** It always gives the physical states of the reactants and products, and *its $\Delta H°$ value is true only when the coefficients of the reactants and products mean moles of the corresponding substances.* The above equation, for example, shows a release of 92.38 kJ if two moles of NH_3 form. If we were to make twice as much or 4.000 mol of NH_3 (from 2.000 mol of N_2 and 6.000 mol of H_2), then twice as much heat (184.8 kJ) would be released. On the other hand, if only 0.5000 mol of N_2 and 1.500 mol of H_2 were to react to form 1.000 mole of NH_3, then only half as much heat (46.19 kJ) would be released. To describe the various sizes of the reactions just described, we write the following thermochemical equations

$$N_2(g) + 3H_2(g) \longrightarrow 2NH_3(g) \qquad \Delta H° = -92.38 \text{ kJ}$$
$$2N_2(g) + 6H_2(g) \longrightarrow 4NH_3(g) \qquad \Delta H° = -184.8 \text{ kJ}$$
$$\tfrac{1}{2}N_2(g) + \tfrac{3}{2}H_2(g) \longrightarrow NH_3(g) \qquad \Delta H° = -46.19 \text{ kJ}$$

Because the coefficients of a thermochemical equation always mean *moles,* not molecules, we may use fractional coefficients. (In the kinds of equations you've seen up until now, fractional coefficients were not allowed because we cannot have fractions of individual molecules, but we can have fractions of moles in a thermochemical equation.)

It is important to specify physical states for all reactants and products in thermochemical equations. The combustion of 1 mol of methane, for example, has different values of $\Delta H°$ if the water produced is in its liquid or its gaseous state.

$$CH_4(g) + 2O_2(g) \longrightarrow CO_2(g) + 2H_2O(l) \qquad \Delta H° = -890.5 \text{ kJ}$$
$$CH_4(g) + 2O_2(g) \longrightarrow CO_2(g) + 2H_2O(g) \qquad \Delta H° = -802.3 \text{ kJ}$$

The difference in $\Delta H°$ values for these two reactions is the amount of energy that would be released by the physical change of 2 mol of water vapor at 25 °C to 2 mol of liquid water at 25 °C.

Example 6.6
Writing a Thermochemical Equation

The following thermochemical equation is for the exothermic reaction of hydrogen and oxygen that produces water.

$$2H_2(g) + O_2(g) \longrightarrow 2H_2O(l) \qquad \Delta H° = -571.8 \text{ kJ}$$

What is the thermochemical equation for this reaction when 1.000 mol of H_2O is produced?

Analysis: The given equation is for 2.000 mol of H_2O, and any changes in the coefficient for water must be made identically to all other coefficients, as well as to the value of $\Delta H°$.

Assembling the Tools: We will use the tools for the definition of a thermochemical equation, the relationship between coefficients and ΔH, and the fact that when we change the amount of reactants, we change the amount of heat for the reaction.

Solution: Since we are dividing the number of moles of water in half, we must divide the rest of the coefficients by 2 and the heat of the reaction by 2 to obtain

$$H_2(g) + \tfrac{1}{2}O_2(g) \longrightarrow H_2O(l) \qquad \Delta H° = -258.9 \text{ kJ}$$

Is the Answer Reasonable? Compare the equation just found with the initial equation to see that the coefficients and the value of $\Delta H°$ are all divided by 2.

The combustion of methane can be represented by the following thermochemical equation:

Practice Exercise 6.15

$$CH_4(g) + 2O_2(g) \longrightarrow CO_2(g) + 2H_2O(l) \qquad \Delta H° = -890.5 \text{ kJ}$$

Write the thermochemical equation for the combustion of methane when 3 moles of $CO_2(g)$ are formed. (*Hint:* What do you have to do to the coefficient of $CH_4(g)$ to give $3CO_2(g)$?)

What is the thermochemical equation for the formation of 2.50 mol of $NH_3(g)$ from $H_2(g)$ and $N_2(g)$?

Practice Exercise 6.16

6.8 | Hess's Law

Recall from Section 6.6 that enthalpy is a state function. This important fact permits us to calculate heats of reaction for reactions that we cannot actually carry out in the laboratory. You will see soon that to accomplish this, we will combine known thermochemical equations, which usually requires that we manipulate them in some way.

Manipulating Thermochemical Equations

In the last section you learned that if we change the size of a reaction by multiplying or dividing the coefficients of a thermochemical equation by some factor, the value of $\Delta H°$ is multiplied by the same factor. Another way to manipulate a thermochemical equation is to change its direction. For example, the thermochemical equation for the combustion of carbon in oxygen to give carbon dioxide is

$$C(s) + O_2(g) \longrightarrow CO_2(g) \qquad \Delta H° = -393.5 \text{ kJ}$$

The reverse reaction, which is extremely difficult to carry out, would be the decomposition of carbon dioxide to carbon and oxygen. The law of conservation of energy requires that its value of $\Delta H°$ equals +393.5 kJ.

$$CO_2(g) \longrightarrow C(s) + O_2(g) \qquad \Delta H° = +393.5 \text{ kJ}$$

In effect, these two thermochemical equations tell us that the combustion of carbon is exothermic (as indicated by the negative sign of $\Delta H°$) and that the reverse reaction is endothermic. The same amount of energy is involved in both reactions—it is just the direction of energy flow that is different. The lesson to learn here is that we can *reverse any thermochemical equation as long as we change the sign of its $\Delta H°$.*

Enthalpy of Reactions

What we're leading up to is a method for combining known thermochemical equations in a way that will allow us to calculate an unknown $\Delta H°$ for some other reaction. Let's revisit the combustion of carbon to see how this works.

We can imagine two paths leading from 1 mol each of carbon and oxygen to 1 mol of carbon dioxide.

One-Step Path. Let C and O_2 react to give CO_2 directly.

$$C(s) + O_2(g) \longrightarrow CO_2(g) \qquad \Delta H° = -393.5 \text{ kJ}$$

Two-Step Path. Let C and O_2 react to give CO, and then let CO react with more O_2 to give CO_2.

Step 1. $C(s) + \frac{1}{2}O(g) \longrightarrow CO(g)$ $\qquad \Delta H° = -110.5$ kJ

Step 2. $CO(g) + \frac{1}{2}O(g) \longrightarrow CO(g)$ $\qquad \Delta H° = -283.0$ kJ

Overall, the two-step path consumes 1 mol each of C and O_2 to make 1 mol of CO_2, just like the one-step path. The initial and final states for the two routes to CO_2, in other words, are identical.

Because $\Delta H°$ depends only on the initial and final states and is independent of path, the values of $\Delta H°$ for both routes must be identical. We can see that this is exactly true simply by adding the equations for the two-step path and comparing the result with the equation for the single step.

Step 1. $\qquad\qquad C(s) + \frac{1}{2}O(g) \longrightarrow CO(g)$ $\qquad\qquad \Delta H° = -110.5$ kJ

Step 2. $\qquad\qquad CO(g) + \frac{1}{2}O(g) \longrightarrow CO_2(g)$ $\qquad\qquad \Delta H° = -283.0$ kJ

$\overline{\qquad\qquad\qquad \cancel{CO(g)} + C(s) + O_2(g) \longrightarrow CO_2(g) + \cancel{CO(g)} \qquad \Delta H° = -393.5 \text{ kJ}}$

The equation resulting from adding Steps 1 and 2 has $CO(g)$ appearing identically on opposite sides of the arrow. We can cancel them to obtain the net equation. *Such a cancellation is permitted only when both the formula and the physical state of a species are identical on opposite sides of the arrow.* The net thermochemical equation for the two-step process, therefore, is

$$C(s) + O_2(g) \longrightarrow CO_2(g) \qquad \Delta H° = -393.5 \text{ kJ}$$

The results, chemically and thermochemically, are identical for both routes to CO_2.

Enthalpy Diagrams

TOOLS

Enthalpy diagram

The energy relationships among the alternative pathways for the same overall reaction can be depicted using a graphical construction called an **enthalpy diagram**. Figure 6.11 is an enthalpy diagram for the formation of CO_2 from C and O_2. The vertical axis is an energy scale for enthalpy. Each horizontal line corresponds to a certain total amount of enthalpy, which we can't actually measure, for all the substances listed on the line in the physical states specified. We can measure differences in enthalpy, however, and that's what we use the diagram for. Lines higher up the enthalpy scale represent larger amounts of enthalpy, so going from a lower line to a higher line corresponds to an increase in enthalpy and a positive value for $\Delta H°$ (an endothermic change). The size of $\Delta H°$ is represented by the vertical distance between the two lines. Likewise, going from a higher line to a lower one represents a decrease in enthalpy and a negative value for $\Delta H°$ (an exothermic change).

Notice in Figure 6.11 that the line for the absolute enthalpy of $C(s) + O_2(g)$, taken as the sum, is above the line for the final product, CO_2. On the left side of Figure 6.11, we see a long downward arrow that connects the enthalpy level for the reactants, $C(s) + O_2(g)$, to that of the final product, $CO_2(g)$. This is for the one-step path, the direct path.

Figure 6.11 | **An enthalpy diagram for the formation of $CO_2(g)$ from its elements by two different paths.** On the left is path 1, the direct conversion of $C(s)$ and $O_2(g)$ to $CO_2(g)$. On the right, path 2 shows two shorter, downward-pointing arrows. The first step of path 2 takes the elements to $CO(g)$, and the second step takes $CO(g)$ to $CO_2(g)$. The overall enthalpy change is identical for both paths since enthalpy is a state function.

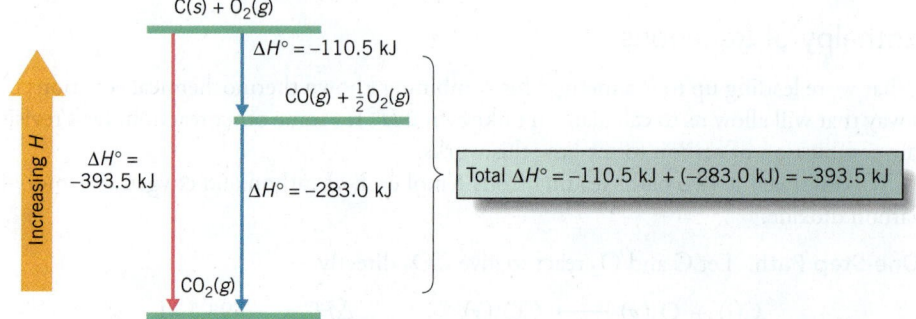

On the right we have the two-step path. Here, Step 1 brings us to an intermediate enthalpy level corresponding to the intermediate products, $CO(g) + \frac{1}{2}O_2(g)$. These include the mole of $CO(g)$ made in the first step plus 1 mol of O_2 that has not reacted. Then, Step 2 occurs to give the final product. What the enthalpy diagram shows is that the total decrease in enthalpy is the same regardless of the path.

Example 6.7
Preparing an Enthalpy Diagram

Hydrogen peroxide, H_2O_2, decomposes into water and oxygen by the following equation:

$$H_2O_2(l) \longrightarrow H_2O(l) + \frac{1}{2}O_2(g)$$

Construct an enthalpy diagram for the following two reactions of hydrogen and oxygen, and use the diagram to determine the value of $\Delta H°$ for the decomposition of hydrogen peroxide,

$$H_2(g) + O_2(g) \longrightarrow H_2O_2(l) \qquad \Delta H° = -188 \text{ kJ}$$

$$H_2(g) + \frac{1}{2}O_2(g) \longrightarrow H_2O(l) \qquad \Delta H° = -286 \text{ kJ}$$

Analysis: In an enthalpy diagram, the reactions are written so as to show the amount of heat gained or lost by the reactants to form the products. In this question, we will be using this diagram to determine the enthalpy of one overall reaction.

Assembling the Tools: Three tools will be necessary to solve this problem. The first will be the tool for drawing enthalpy diagrams in which we will place the reactants and products and their states on the energy scale. The second will be the tool for deciding the algebraic sign of $\Delta H°$, that is, which way the arrows point: up for endothermic reactions or down for exothermic reactions. Finally, for the third tool, we will need to use the given thermochemical equations and the fact that enthalpies of reactions are state functions to calculate the enthalpy for the decomposition of $H_2O_2(l)$.

Solution: The two given reactions are exothermic, so their values of $\Delta H°$ will be associated with downward-pointing arrows. The highest enthalpy level, therefore, must be for the elements themselves, $H_2(g)$ and $O_2(g)$. Notice that the first reaction requires an entire mole of O_2, whereas the second needs only 1 mol of O_2. On the top line we'll include enough O_2 for the formation of H_2O_2 by the first reaction. The lowest level must be for the product of the reaction with the largest negative $\Delta H°$, the formation of water by the second equation. Keep in mind that the second reaction only needs a half a mole of O_2, so this will leave $\frac{1}{2}O_2(g)$ unreacted when we form $H_2O(l)$, as indicated on the bottom line. The enthalpy level for the H_2O_2, formed by a reaction with the less negative $\Delta H°$, must be in between.

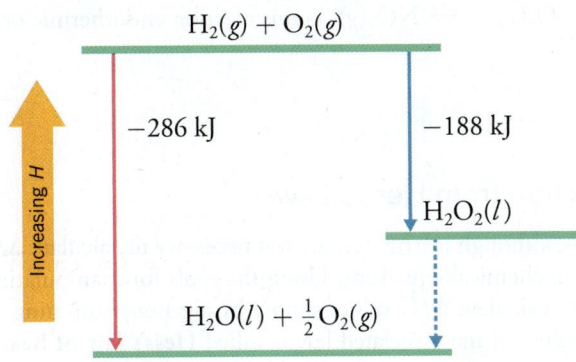

Notice that the gap on the right side (represented by the dashed arrow) corresponds *exactly* to the change of $H_2O_2(l)$ into $H_2O(l) + \frac{1}{2}O_2(g)$, the reaction for which we seek $\Delta H°$. This enthalpy separation corresponds to the difference between the enthalpy changes -286 kJ and -188 kJ. Thus for the decomposition of $H_2O_2(l)$ into $H_2O(l)$ and $\frac{1}{2}O_2(g)$,

$$\Delta H° = [-286 \text{ kJ } (-188 \text{ kJ})] = -98 \text{ kJ}$$

Our completed enthalpy diagram is as follows, indicating that $\Delta H°$ for the decomposition of 1 mol of $H_2O_2(l)$ is -98 kJ.

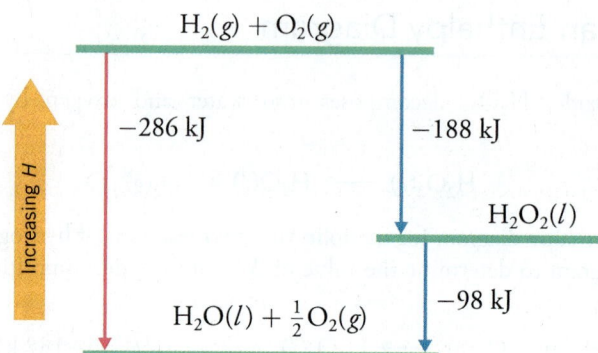

Is the Answer Reasonable? First, be sure that the chemical formulas are arrayed properly on the horizontal lines in the enthalpy diagram. The arrows pointing downward should indicate negative values of $\Delta H°$. These arrows must be *consistent with the given equations* (in other words, they must point from reactants to products). Finally, notice that the total amount of energy associated with the two-step process—namely, $(-188$ kJ$)$ plus $(-98$ kJ$)$—equals -286 kJ, the energy for the one-step process.

Practice Exercise 6.17

Two oxides of copper can be made from copper by the following reactions.

$$2Cu(s) + O_2(g) \longrightarrow 2CuO(s) \qquad \Delta H° = -310 \text{ kJ}$$
$$2Cu(s) + \tfrac{1}{2}O_2(g) \longrightarrow Cu_2O(s) \qquad \Delta H° = -169 \text{ kJ}$$

Using these data, construct an enthalpy diagram that can be used to find $\Delta H°$ for the reaction $Cu_2O(s) + \tfrac{1}{2}O_2(g) \longrightarrow 2CuO(s)$. Is the reaction endothermic or exothermic? (*Hint:* Remember, an arrow points down for a negative $\Delta H°$ and up for a positive $\Delta H°$.)

Practice Exercise 6.18

Consider the following thermochemical equations:

$$\tfrac{1}{2}N_2(g) + \tfrac{1}{2}O_2(g) \longrightarrow NO(g) \qquad \Delta H° = -90.4 \text{ kJ}$$
$$NO(g) + \tfrac{1}{2}O_2(g) \longrightarrow NO_2(g) \qquad \Delta H° = -56.6 \text{ kJ}$$

Using these data, construct an enthalpy diagram that can be used to find $\Delta H°$ for the reaction $\tfrac{1}{2}N_2(g) + O_2(g) \longrightarrow NO_2(g)$. Is this reaction endothermic or exothermic?

Heats of Reaction from Hess's Law

Enthalpy diagrams, although instructive, are not necessary to calculate $\Delta H°$ for a reaction from known thermochemical equations. Using the tools for manipulating equations, we ought to be able to calculate $\Delta H°$ values simply by algebraic summing. G. H. Hess was the first to realize this, so the associated law is called **Hess's law of heat summation** or, simply, **Hess's law**.

■ Germain Henri Hess (1802–1850) anticipated the law of conservation of energy in the law named after him.

Hess's Law

The value of $\Delta H°$ for any reaction that can be written in steps equals the sum of the values of $\Delta H°$ of each of the individual steps.

For example, suppose we add the two equations in Practice Exercise 6.18 and then cancel anything that's the same on both sides; in this case the $NO(g)$ on both sides cancel.

Hess's Law

$$\tfrac{1}{2}N_2(g) + \tfrac{1}{2}O_2(g) \longrightarrow NO(g)$$

$$NO(g) + \tfrac{1}{2}O(g) \longrightarrow NO_2(g)$$

$$\tfrac{1}{2}N_2(g) + \cancel{NO(g)} + \tfrac{1}{2}O_2(g) + \tfrac{1}{2}O_2(g) \longrightarrow \cancel{NO(g)} + NO_2(g)$$

$$O_2(g)$$

Notice that we've combined the two $\tfrac{1}{2}O_2(g)$ to give $O_2(g)$. Rewriting the equation gives

$$\tfrac{1}{2}N_2(g) + O_2(g) \longrightarrow NO_2(g)$$

This is the equation for which we were asked to determine $\Delta H°$. According to Hess's law, we should be able to obtain $\Delta H°$ for this reaction by simply adding the $\Delta H°$ values for the equations we've added.

$$\Delta H° = (+90.4 \text{ kJ}) + (-56.6 \text{ kJ}) = +33.8 \text{ kJ}$$

This was the answer to Practice Exercise 6.18. Thus, we've added the two given equations to obtain the desired equation, and we've added their $\Delta H°$ values to obtain the desired $\Delta H°$.

The chief use of Hess's law is to calculate the enthalpy change for a reaction for which such data cannot be determined experimentally or are otherwise unavailable. Often this requires that we manipulate equations, so let's recapitulate the few rules that govern these operations.

Rules for Manipulating Thermochemical Equations

1. When an equation is reversed—written in the opposite direction—the sign of $\Delta H°$ must also be reversed.[4]
2. Formulas canceled from both sides of an equation must be for the substance in identical physical states.
3. If all the coefficients of an equation are multiplied or divided by the same factor, the value of $\Delta H°$ must likewise be multiplied or divided by that factor.

Manipulating thermochemical equations

Example 6.8
Using Hess's Law

Carbon monoxide is often used in metallurgy to remove oxygen from metal oxides and thereby give the free metal. The thermochemical equation for the reaction of CO with iron(III) oxide, Fe_2O_3, is

$$Fe_2O_3(s) + 3CO(g) \longrightarrow 2Fe(s) + 3CO_2(g) \qquad \Delta H° = -26.7 \text{ kJ}$$

---

[4]To illustrate, the reverse of the equation,

$$C(s) + O_2(g) \longrightarrow CO_2(g) \qquad \Delta H° = -394 \text{ kJ}$$

is the following equation.

$$CO_2(g) \longrightarrow C(s) + O_2(g) \qquad \Delta H° = +394 \text{ kJ}$$

Use this equation and the equation for the combustion of CO,

$$CO(g) + \tfrac{1}{2}O(g) \longrightarrow CO_2(g) \qquad \Delta H° = -283.0 \text{ kJ}$$

to calculate the value of $\Delta H°$ for the following reaction:

$$2Fe(s) + \tfrac{3}{2}O_2(g) \longrightarrow Fe_2O_3(s)$$

Analysis: We are given two thermochemical equations and are asked to find $\Delta H°$ for the third equation. We will need to manipulate the first two equations in order for them to give the third equation when they are added together.

Assembling the Tools: The tools will be Hess's law and the rules for manipulating thermochemical equations.

Solution: In this problem, we cannot simply add the two given equations, because this will not produce the equation we want. We first have to manipulate these equations so that when we add them we will get the target equation. In performing these adjustments, we have to keep our eye on our target—the final desired equation. When we've done this, we can then add the adjusted $\Delta H°$ values to obtain the desired $\Delta H°$. We can manipulate the two given equations as follows.

Step 1. We begin by trying to get the iron atoms to come out right. The target equation must have 2Fe on the left, but the first equation above has 2Fe to the right of the arrow. To move it to the left, we reverse the entire equation, remembering also to reverse the sign of $\Delta H°$. This also puts Fe$_2$O$_3$ to the right of the arrow, which is where it needs to be. Now we have

$$2Fe(s) + 3CO_2(g) \longrightarrow Fe_2O_3(s) + 3CO(g) \qquad \Delta H° = +26.7 \text{ kJ}$$

Step 2. There must be $\tfrac{3}{2}O_2$ on the left, and we must be able to cancel three CO and three CO$_2$ when the equations are added. If we multiply the second of the equations given above by 3, including the value for $\Delta H°$, we will obtain the necessary coefficients. When we have done this, we have

$$3CO(g) + \tfrac{3}{2}O_2(g) \longrightarrow 3CO_2(g) \qquad \Delta H° = 3 \times (-283.0 \text{ kJ}) = -849.0 \text{ kJ}$$

Now let's put our two equations together and find the answer:

$$2Fe(s) + 3\cancel{CO_2(g)} \longrightarrow Fe_2O_3(s) + 3\cancel{CO(g)} \qquad \Delta H° = +26.7 \text{ kJ}$$
$$3\cancel{CO(g)} + \tfrac{3}{2}O_2(g) \longrightarrow 3\cancel{CO_2(g)} \qquad \Delta H° = -849.0 \text{ kJ}$$
$$\text{Sum: } 2Fe(s) + \tfrac{3}{2}O_2(g) \longrightarrow Fe_2O_3(s) \qquad \Delta H° = -822.3 \text{ kJ}$$

Thus, the value of $\Delta H°$ for the oxidation of 2 mol Fe(s) to 1 mol Fe$_2$O$_3$(s) is -822.3 kJ. (The reaction is very exothermic.)

Is the Answer Reasonable? There is no quick check, but for each step, double-check that you have heeded the rules for manipulating thermochemical equations. The most common error is forgetting to perform the same mathematical operation on the $\Delta H°$ as was done to the equation. Also, be careful with the algebraic signs when adding the $\Delta H°$ values.

Practice Exercise 6.19

Given the following thermochemical equations,

$$2NO(g) + O_2(g) \longrightarrow 2NO_2(g) \qquad \Delta H° = -113.2 \text{ kJ}$$
$$2N_2O(g) + 3O_2(g) \longrightarrow 4NO_2(g) \qquad \Delta H° = -28.0 \text{ kJ}$$

calculate $\Delta H°$ for the reaction $N_2O(g) + \tfrac{1}{2}O(g) \longrightarrow 2NO(g)$ (*Hint:* What needs to be done to have the product, NO, on the correct side of the equation?)

Practice Exercise 6.20

Ethanol, C_2H_5OH, is made industrially by the reaction of water with ethylene, C_2H_4. Calculate the value of $\Delta H°$ for the reaction

$$C_2H_4(g) + H_2O(l) \longrightarrow C_2H_5OH(l)$$

given the following thermochemical equations:

$$C_2H_4(g) + 3O_2(g) \longrightarrow 2CO_2(g) + 2H_2O(l) \qquad \Delta H° = -1411.1 \text{ kJ}$$
$$C_2H_5OH(l) + 3O_2(g) \longrightarrow 2CO_2(g) + 3H_2O(l) \qquad \Delta H° = -1367.1 \text{ kJ}$$

6.9 | Standard Heats of Reaction

Enormous databases of thermochemical equations have been compiled to make it possible to calculate almost any heat of reaction using Hess's law. The most frequently tabulated reactions are combustion reactions, phase changes, and formation reactions. We'll discuss the enthalpy changes that accompany phase changes in Chapter 11.

The **standard heat of combustion, $\Delta H_c°$,** of a substance is the amount of heat released when one mole of that substance is completely burned in pure oxygen gas, with all reactants and products brought to 25 °C and 1 bar of pressure. All carbon in the **fuel** becomes carbon dioxide gas, and all of the fuel's hydrogen becomes liquid water. *Combustion reactions are always exothermic, so $\Delta H_c°$ is always negative.*

Example 6.9
Using the Standard Heat of Combustion to Estimate CO_2 Production

How many moles of carbon dioxide gas are produced by a gas-fired power plant for every 1.00 MJ (megajoule) of energy it produces? The plant burns methane, $CH_4(g)$, for which $\Delta H_c°$ is −890 kJ/mol.

Analysis: We can restate the problem as an equivalency relation,

$$? \text{ mol } CO_2(g) \Leftrightarrow 1.00 \text{ MJ released}$$

We need to link moles of carbon dioxide with megajoules of heat produced.

Assembling the Tools: The tool we will use is the balanced thermochemical equation. Remembering that, when a hydrocarbon burns, the products are CO_2 and H_2O, the first step in the solution will be to write a balanced chemical equation for the reaction of CH_4 with O_2. Because $\Delta H_c°$ gives the heat evolved *per mole* of CH_4, we will have to be sure the coefficient of CH_4 is 1. In the equation CH_4, O_2, and CO_2 will be gases and H_2O will be a liquid (because that's the standard form of H_2O at 25 °C.)

Once we have the thermochemical equation, the coefficients will let us relate moles of CO_2 to kilojoules of heat released. To relate megajoules with kilojoules, our tools will be the SI prefixes *kilo* and *mega*.

$$1 \text{ MJ} = 10^6 \text{ J} \quad \text{and} \quad 1 \text{ kJ} = 10^3 \text{ J}$$

Solution: First, we write and balance the thermochemical equation, remembering that $\Delta H°$ is $\Delta H_c°$.

$$CH_4(g) + 2O_2(g) \longrightarrow CO_2(g) + 2H_2O(l) \qquad \Delta H_c° = -890 \text{ kJ}$$

Thus, 1 mol CO_2 is formed for every 890 kJ of heat that is released. The number of moles of CO_2 released for the production of 1.00 MJ of energy is

$$1.00 \text{ MJ} \times \frac{10^6 \text{ J}}{1 \text{ MJ}} \times \frac{1 \text{ kJ}}{10^3 \text{ J}} \times \frac{1 \text{ mol } CO_2}{890 \text{ kJ}} = 1.12 \text{ mol } CO_2$$

Is the Answer Reasonable? When 1 mol CH_4 burns, 890 kJ are released. This is just a little less than 1000 kJ (or 1 MJ), so the amount of CH_4 needed to release 1 MJ should be a bit larger than 1 mol. Our answer, 1.12 mol CO_2, is reasonable.

Practice Exercise 6.21

The heat of combustion, ΔH_c°, of acetone, C_3H_6O, is 1790.4 kJ/mol. How many kilojoules of heat are evolved in the combustion of 12.5 g of acetone? (*Hint:* Calculate the molar mass of acetone.)

Practice Exercise 6.22

n-Octane, $C_8H_{18}(l)$, has a standard heat of combustion of 5450.5 kJ/mol. A 15 gallon automobile fuel tank could hold about 480 moles of *n*-octane. How much heat could be produced by burning a full tank of *n*-octane?

Standard Heats of Formation and Hess's Law Equation

The **standard enthalpy of formation, ΔH_f°**, of a substance, also called its **standard heat of formation**, is the amount of heat absorbed or evolved when *one mole* of the substance is formed at 25 °C and 1 bar from its *elements in their standard states*. An element is in its **standard state** when it is at 25 °C and 1 bar and in its most stable form and physical state (solid, liquid, or gas). Oxygen, for example, is in its standard state only as a gas at 25 °C and 1 bar and only as O_2 molecules, not as O atoms or O_3 (ozone) molecules. Carbon must be in the form of graphite, not diamond, to be in its standard state, because the graphite form of carbon is the more stable form under standard conditions.

Standard enthalpies of formation for a variety of substances are given in Table 6.2, and a more extensive table of standard heats of formation can be found in Appendix C, Table C.2. Notice in particular that *all values of ΔH_f° for the elements in their standard states are zero.* (Forming an element *from itself* would yield no change in enthalpy.) In most tables, values of ΔH_f° for the elements are not included for this reason.

It is important to remember the meaning of the subscript f in the symbol ΔH_f°. It is applied to a value of ΔH° only when *one mole* of the substance is *formed from its elements in their standard states*. Consider, for example, the following four thermochemical equations and their corresponding values of ΔH°.

■ Older thermochemical data used a standard pressure of 1 atm, not 1 bar. Because 1 atm = 1.01325 bar, the new definition made little difference in tabulated heats of reaction.

$$H_2(g) + \tfrac{1}{2}O_2(g) \longrightarrow H_2O(l) \qquad \Delta H_f^\circ = -285.9 \text{ kJ/mol}$$

$$2H_2(g) + O_2(g) \longrightarrow 2H_2O(l) \qquad \Delta H^\circ = -571.8 \text{ kJ}$$

$$CO(g) + \tfrac{1}{2}O(g) \longrightarrow CO_2(g) \qquad \Delta H^\circ = -283.0 \text{ kJ}$$

$$2H(g) + O(g) \longrightarrow H_2O(l) \qquad \Delta H^\circ = -971.1 \text{ kJ}$$

Only in the first equation is ΔH° given the subscript f. It is the only reaction that satisfies both of the conditions specified previously for standard enthalpies of formation. The second equation shows the formation of *two* moles of water, not one. The third involves a *compound* as one of the reactants. The fourth involves the elements as *atoms*, which are *not standard states* for these elements. Also notice that the units of ΔH_f° are *kilojoules per mole of product*, not just kilojoules, because the value is for the formation of 1 mol of the

TABLE 6.2	Standard Enthalpies of Formation of Typical Substances		
Substance	$\Delta H_f°$ (kJ mol^{-1})	Substance	$\Delta H_f°$ (kJ mol^{-1})
Ag(s)	0	$H_2O_2(l)$	−187.6
AgBr(s)	−100.4	HBr(g)	−36
AgCl(s)	−127.0	HCl(g)	−92.30
Al(s)	0	HI(g)	26.6
$Al_2O_3(s)$	−1669.8	$HNO_3(l)$	−173.2
C(s) (graphite)	0	$H_2SO_4(l)$	−811.32
CO(g)	−110.5	$HC_2H_3O_2(l)$	−487.0
$CO_2(g)$	−393.5	Hg(l)	0
$CH_4(g)$	−74.848	Hg(g)	60.84
$CH_3Cl(g)$	−82.0	$I_2(s)$	0
$CH_3I(g)$	14.2	K(s)	0
$CH_3OH(l)$	−238.6	KCl(s)	−435.89
$CO(NH_2)_2(s)$ (urea)	−333.19	$K_2SO_4(s)$	−1433.7
$CO(NH_2)_2(aq)$	−391.2	$N_2(g)$	0
$C_2H_2(g)$	226.75	$NH_3(g)$	−46.19
$C_2H_4(g)$	52.284	$NH_4Cl(s)$	−315.4
$C_2H_6(g)$	−84.667	NO(g)	90.37
$C_2H_5OH(l)$	−277.63	$NO_2(g)$	33.8
Ca(s)	0	$N_2O(g)$	81.57
$CaBr_2(s)$	−682.8	$N_2O_4(g)$	9.67
$CaCO_3(s)$	−1207	$N_2O_5(g)$	11
$CaCl_2(s)$	−795.0	Na(s)	0
CaO(s)	−635.5	$NaHCO_3(s)$	−947.7
$Ca(OH)_2(s)$	−986.59	$Na_2CO_3(s)$	−1131
$CaSO_4(s)$	−1432.7	NaCl(s)	−411.0
$CaSO_4 \cdot \frac{1}{2}H_2O(s)$	−1575.2	NaOH(s)	−426.8
$CaSO_4 \cdot 2H_2O(s)$	−2021.1	$Na_2SO_4(s)$	−1384.5
$Cl_2(g)$	0	$O_2(g)$	0
Fe(s)	0	Pb(s)	0
$Fe_2O_3(s)$	−822.2	PbO(s)	−219.2
$H_2(g)$	0	S(s)	0
$H_2O(g)$	−241.8	$SO_2(g)$	−296.9
$H_2O(l)$	−285.9	$SO_3(g)$	−395.2

compound (from its elements). We can obtain the enthalpy of formation of two moles of water ($\Delta H°$ for the second equation) simply by multiplying the $\Delta H_f°$ value for 1 mol of H_2O by the factor 2 mol $H_2O(l)$.

$$\frac{-285.9 \text{ kJ}}{1 \text{ mol } H_2O(l)} \times 2 \text{ mol } H_2O(l) = -571.8 \text{ kJ}$$

Example 6.10
Writing an Equation for a Standard Heat of Formation

What equation for the formation of nitric acid, $HNO_3(l)$, is used along with the symbol ΔH_f°?

Analysis: Answering the question requires that we know the definition of ΔH_f°, which demands that the equation must show *only one mole* of the product and elements as reactants. We must also be careful to include the physical states the substances will have at 25 °C and 1.0 bar of pressure.

Assembling the Tools: The first tool we will use is the one for writing thermochemical equations and writing equations for the standard heat of formation of a compound. We begin with its formula and take whatever fractions of moles of the elements are required to make it. The other tool we will use is Table 6.2, where the value of ΔH_f° for $HNO_3(l)$ is listed as -173.2 kJ mol^{-1}.

Solution: The three elements, H, N, and O, all occur as diatomic molecules in the gaseous state, so the following fractions of moles supply exactly enough of each to make one mole of HNO_3.

$$\tfrac{1}{2}H_2(g) + \tfrac{1}{2}N_2(g) + \tfrac{3}{2}O_2(g) \longrightarrow HNO_3(l) \qquad \Delta H_f^\circ = -173.2 \text{ kJ/mol}$$

Is the Answer Reasonable? The answer correctly shows only 1 mol of HNO_3, and this governs the coefficients for the reactants. So, simply check to make sure everything is balanced.

Practice Exercise 6.23

Write the thermochemical equation that would be used to represent the standard heat of formation of $KBr(s)$. (*Hint:* Be sure you show substances in their standard states.)

Practice Exercise 6.24

Write the thermochemical equation that would be used to represent the standard heat of formation of sodium bicarbonate, $NaHCO_3(s)$.

Standard enthalpies of formation are useful because they provide a convenient method for applying Hess's law without having to manipulate thermochemical equations. To see how this works, let's look again at Example 6.7, where we calculated ΔH_f° for the decomposition of H_2O_2.

$$H_2O_2(l) \longrightarrow H_2O(l) + \tfrac{1}{2}O_2(g)$$

The two equations we used were

(1) $H_2(g) + O_2(g) \longrightarrow H_2O_2(l) \qquad \Delta H_f^\circ \ (1) = -188 \text{ kJ}$

(2) $H_2(g) + \tfrac{1}{2}O_2(g) \longrightarrow H_2O(l) \qquad \Delta H_f^\circ \ (2) = -286 \text{ kJ}$

This time we have numbered the equations and identified the enthalpy changes as ΔH_f° because in each case we're forming one mole of the compound from elements in their standard states. To combine these equations to find ΔH° for the decomposition of H_2O_2, we have to reverse equation (1), which means we change the sign of its ΔH.

(1 reversed) $\quad H_2O_2(l) \longrightarrow H_2(g) + O_2(g) \qquad \Delta H^\circ = -\Delta H_f^\circ \ (1) = +188 \text{ kJ}$

(2) $\quad H_2(g) + \tfrac{1}{2}O_2(g) \longrightarrow H_2O(l) \qquad\qquad\qquad \Delta H^\circ \ (2) = -286 \text{ kJ}$

$\quad H_2O_2(l) \longrightarrow H_2O(l) + \tfrac{1}{2}O_2(g) \qquad\qquad\qquad \Delta H^\circ = -\ 98 \text{ kJ}$

Thus, the desired $\Delta H°$ was obtained by subtracting the $\Delta H_f°$ for the reactant (H_2O_2) from the $\Delta H_f°$ for the product (H_2O). In fact, this works for *any reaction* where we know the heats of formation of all of the reactants and products. So, if we're working with heats of formation, Hess's law can be restated as follows: The net $\Delta H°$ reaction equals the sum of the heats of formation of the products minus the sum of the heats of formation of the reactants, each $\Delta H_f°$ value multiplied by the appropriate coefficient given by the thermo-chemical equation. In other words, we can express Hess's law in the form of the **Hess's law equation:**

$$\Delta H°_{reaction} = \left[\text{sum of } \Delta H_f° \text{ of all of the products} \right] -$$

$$\left[\text{sum of } \Delta H_f° \text{ of all of the reactants} \right] \qquad \textbf{(6.14)}$$

As long as we have access to a table of standard heats of formation, using Equation 6.14 avoids having to manipulate a series of thermochemical equations. Finding the $\Delta H°$ reaction reduces to performing arithmetic, as we see in the next example

■ We can also use heats of combustion in Hess's law calculations. In doing so, remember that the heats of combustion of elements are not zero and that simple products such as CO_2 and H_2O are obtained from the combustion of the elements C and H_2.

TOOLS

Hess's law equation

Example 6.11
Using Hess's Law and Standard Enthalpies of Formation

Some chefs keep baking soda, $NaHCO_3$, handy to put out grease fires. When thrown on the fire, baking soda partly smothers the fire and the heat decomposes it to give CO_2, which further smothers the flame. The equation for the decomposition of $NaHCO_3$ is

$$2NaHCO_3(s) \longrightarrow Na_2CO_3(s) + H_2O(l) + CO_2(g)$$

Use the standard heats of formation to calculate $\Delta H°$ for this reaction in kilojoules.

Analysis: We are being asked to calculate $\Delta H°$ for a reaction. We are not given a series of reactions to manipulate, but we are asked to use the $\Delta H_f°$ for the reactants and the products. Therefore, we will use Hess's law equation to calculate $\Delta H°_{reaction}$.

Assembling the Tools: Hess's law equation (Equation 6.14)

$$\Delta H°_{reaction} = \left[\text{sum of } \Delta H_f° \text{ of all of the products} \right] - \left[\text{sum of } \Delta H_f° \text{ of all of the reactants} \right]$$

is our basic tool for computing values of $\Delta H°$. We will subtract the sum of the $\Delta H°$ for the reactants from the sum of the $\Delta H°$ for the products to calculate the $\Delta H°$ reaction. For each of the reactants and products, we compute its $\Delta H°$ by multiplying its $\Delta H_f°$ (from Table 6.2) by its coefficient in the balanced chemical equation.

Solution: First, let's set up the calculation using symbols for the quantities we will use.

$$\Delta H° = \left[1 \text{ mol } Na_2CO_3(s) \times \Delta H_f° \, Na_2CO_3(s) + 1 \text{ mol } H_2O(l) \times \Delta H_f° \, H_2O(l) + \right.$$

$$\left. 1 \text{ mol } CO_2(g) \times \Delta H_f° \, CO_2(g) \right] - \left[2 \text{ mol } NaHCO_3(s) \times \Delta H_f° \, NaHCO(s) \right]$$

Now we go to Table 6.2 to find the values of $\Delta H_f°$ for each substance in its proper physical state:

$$\left[(1 \text{ mol } Na_2CO_3) \times \left(\frac{-1131 \text{ kJ}}{1 \text{ mol } Na_2CO_3} \right) + (1 \text{ mol } H_2O) \times \left(\frac{-285.9 \text{ kJ}}{1 \text{ mol } H_2O} \right) \right.$$

$$\left. + (1 \text{ mol } CO_2) \times \left(\frac{-393.5 \text{ kJ}}{1 \text{ mol } CO_2} \right) \right] - \left[(2 \text{ mol } NaHCO_3) \times \left(\frac{-947.7 \text{ kJ}}{1 \text{ mol } NaHCO_3} \right) \right]$$

This reduces to

$$\Delta H° = (-1810 \text{ kJ}) - (-1895 \text{ kJ}) = +85 \text{ kJ}$$

Thus, under standard conditions, the reaction is endothermic by 85 kJ. (Notice that we did not have to manipulate any equations.)

Is the Answer Reasonable? Double-check that all of the coefficients found in the chemical equation are correctly applied as multipliers on the $\Delta H_f°$ quantities. Be sure that the signs of the values of $\Delta H_f°$ have all been carefully used. Keeping track of the algebraic signs requires particular care and is the greatest source of error in these calculations. Also check to be sure you've followed the correct order of subtraction specified in Equation 6.14; in other words, that you have subtracted the $\Delta H°$ of the *reactants* from the $\Delta H°$ of the *products*.

Practice Exercise 6.25

Use heats of formation data from Table 6.2 to calculate $\Delta H°$ for the reaction $CaCl_2(s) + H_2SO_4(l) \longrightarrow CaSO_4(s) + 2HCl(g)$. (*Hint:* Be sure to follow the correct order of subtraction.)

Practice Exercise 6.26

Write thermochemical equations corresponding to $\Delta H_f°$ for $SO_3(g)$ and $SO_2(g)$ and show how they can be manipulated to calculate $\Delta H°$ for the following reaction:

$$SO_3(g) \longrightarrow SO_2(g) + \tfrac{1}{2}O_2(g) \qquad \Delta H° = ?$$

Use Equation 6.14 to calculate $\Delta H°$ for the reaction using the heats of formation of $SO_3(g)$ and $SO_2(g)$. How do the answers compare?

Practice Exercise 6.27

Calculate $\Delta H°$ for the following reactions:

(a) $2NO(g) + O_2(g) \longrightarrow 2NO_2(g)$

(b) $NaOH(s) + HCl(g) \longrightarrow NaCl(s) + H_2O(l)$

CHEMISTRY AND CURRENT AFFAIRS 6.2

Runaway Reactions: The Importance of Thermodynamics

In 1998, a runaway reaction occurred in a reaction vessel in which a dye was being made at a Morton International Plant in Paterson, New Jersey. The reaction overheated, causing pressure to build up inside the vessel. The vessel ruptured, ejecting the contents through the roof. The resultant vapor cloud ignited and produced an explosion and fire that injured nine employees and destroyed the facility (Figure 1).

Most chemical reactions are exothermic. The heat given off by the reaction will raise the temperature of the reaction mixture unless it is removed. The heat may be removed through normal losses to the surroundings, by deliberate cooling, such as pumping a cold fluid around the outside (jacket) of the reaction vessel, or by other techniques. When the rate at which the reaction generates heat is faster than the rate at which the equipment can remove the heat, a point of no return is reached (Figure 2). The reaction mixture will heat up out of control. As the temperature rises, the reaction rate increases exponentially, thus increasing the rate at which heat is released and causing the temperature to rise even faster. As the temperature rises, the solvent and any other volatile components turn into gases, which increases the pressure inside the reactor. In a properly designed reactor, a safety vent will open to relieve the pressure. However, if this vent is not large enough, the pressure will eventually cause the vessel to rupture,

releasing a mixture of liquid and vapor. If the vapor cloud is flammable and it finds an ignition source, an explosion and fireball can result.

Runaway reactions can occur in the laboratory as well as in manufacturing. To prevent runaway reactions and their potentially disastrous results, it is important to know how much heat a reaction will release, how fast the heat is released, and how fast the heat can be removed from the reaction equipment. Hazards can be anticipated by calculating the potential adiabatic temperature rise, pressure increases, and vapor volumes generated.

In the Morton International incident, *ortho*-nitrochlorobenzene (*o*-NCB) was reacted with 2-ethylhexylamine (2-EHA) to form Automate Yellow 96 dye.

The heat of reaction was known from calorimetry experiments, but not included in the incident report. However, the heat of reaction can be estimated using standard heats of formation for the exact compounds, or for simpler compounds as long as the same types of bonds are being broken and formed. Using compounds called chlorobenzene and methylamine as model compounds for the reaction, a heat of reaction of −176 kJ/mol *o*-NCB is estimated. At 195 °C, the Yellow 96 starts to decompose exothermically and generates gaseous by-products.

Figure 1

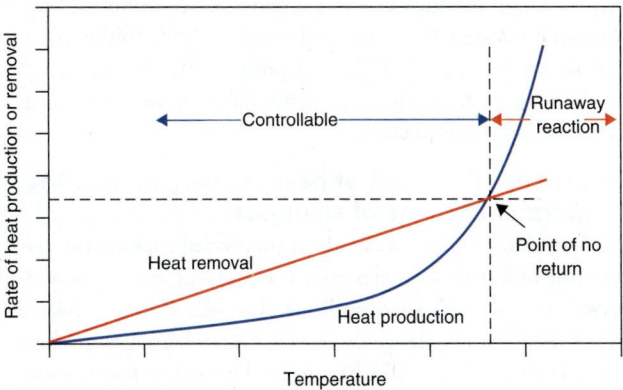

Heat Generation versus Heat Removal

Figure 2 If the reaction temperature exceeds the point of no return, heat generation will exceed the ability of the surroundings to remove the heat and the reaction will run away uncontrolled.

To avoid having to handle the o-NCB several times, since it is a poison, all of the reactants were added at once, and the mixture was slowly heated up in stages. At low temperatures, the reaction was slow and the heat could be removed by passing cold water through the reactor jacket. As the reaction progressed, it slowed down and the reactor was heated. This process continued until the reaction was completed at 150 °C. This staged heating procedure is inherently more hazardous: if the reaction is heated too quickly, it could get out of control.

The size of the process was increased from a 1000 gallon spherical reactor to a 2000 gallon spherical reactor. When switching to larger reactors, the surface area available for heat removal increases approximately with the square of the diameter. The volume, which corresponds to reaction mass and total heat generated, increases approximately with the cube of the diameter. Thus, doubling the reactor volume corresponds to only about 1.6 times the surface area available for heat removal. At this larger scale, 50% of the batches had small runaway reactions as compared to 20% for the smaller-scale reactions.

On the day of the accident, the operators heated the reaction vessel to 90 °C. They heated the reactor too quickly, so very little of

the reaction had occurred at the lower temperatures at which the heat could be safely removed. Now, most of the reactants remained and were reacting rapidly. The temperature overshot the point of no return and continued to rise rapidly despite the application of cooling water. The temperature rose past the 150 °C maximum operating temperature. At about 185 °C, the pressure had increased enough to cause the relief vent to open. The temperature continued to rise and went from 190 °C to 260 °C in less than 30 seconds. With the onset of gaseous decomposition of the product, the vent was not large enough to accommodate the excess gases and pressure, and the reactor ruptured.

Contributed by Gary Buske
Source: U. S. Chemical Safety and Hazard Investigation Board
Office of External Relations
2175 K Street, N. W., Suite 400
Washington, DC 20037
National Technical Information Service Report Number: PB2000-107721
[www.cbs.gov/ (completed investigations August 16, 2000)]

Summary

Organized by Learning Objective

Explain the difference between potential and kinetic energy and the law of conservation of energy

Thermochemistry, which deals with the energy absorbed or released in chemical reactions, is a part of **thermodynamics**, the study of energy transfer and energy transformation. **Heat** is energy (**thermal energy**) that transfers between objects having different temperatures. Heat continues to flow until the two objects come to the same temperature and **thermal equilibrium** is established.

An object has **kinetic energy (KE)** if it is moving, $KE = \frac{1}{2}mv^2$, and **potential energy (PE)** when it experiences attractions or repulsions toward other objects. When objects that attract are moved apart, PE increases. Similarly, when objects that repel are pushed together, PE increases. Because electrical attractions and repulsions occur within atoms, molecules, and

ions, substances have **chemical energy**, a form of **potential energy**. The particles that make up matter are in constant random motion, so they also possess kinetic energy—specifically, **molecular kinetic energy**. According to the kinetic molecular theory, the **average molecular kinetic energy** is directly proportional to the kelvin temperature. According to the law of conservation of energy, energy can be neither created nor destroyed; instead, it can only be changed from one form to another. The SI unit of energy is the **joule**; 4.184 J = 1 cal (**calorie**).

Explain the connection between temperature, energy, and the concept of a state function

The **state** of a system is the list of properties that describe its current condition. The **internal energy** of a system, E, is a **state function**, which is a property that depends only on the current state of the system. E equals the sum of the system's molecular kinetic and potential energies. A change in E is defined as $\Delta E = E_{final} - E_{initial}$, although absolute amounts of E cannot be measured or calculated. For a chemical reaction, the definition translates to $\Delta E = E_{products} - E_{reactants}$. A positive value for ΔE, which can be measured, means that a system absorbs energy from its surroundings during a change.

Determine an amount of heat exchanged from the temperature change of an object

The **boundary** across which heat is transferred encloses the **system** (the object we're interested in). Everything else in the **universe** is the system's **surroundings**. The heat flow q is related to the temperature change Δt by $q = C\Delta t$, where C is the **heat capacity** of the system (the heat needed to change the temperature of the system by one degree Celsius). The heat capacity for a pure substance can be computed from its mass, m, using the equation $C = ms$, where s is the specific heat of the material (the heat needed to change the temperature of 1 g of a substance by 1 °C). Water has an unusually high **specific heat**. We can compute a heat flow when we know the mass and specific heat of an object using the equation $q = ms\,\Delta t$. The heat, q, is given a positive sign when it flows into a system and a negative sign when it flows out.

Describe the energy changes in exothermic and endothermic reactions

Bond breaking increases potential energy (chemical energy); bond formation decreases potential energy (chemical energy). In an **exothermic** reaction, chemical energy is changed to molecular kinetic energy. If the system is **adiabatic** (no heat leaves it), the internal temperature increases. Otherwise, the heat has a tendency to leave the system. In **endothermic** reactions, molecular kinetic energy of the reactants is converted into potential energy of the products. This tends to lower the system's temperature and lead to a flow of heat into the system.

State the first law of thermodynamics and explain how it applies to chemistry

The change in chemical potential energy in a reaction is the **heat of reaction, q,** which can be measured at constant volume or constant pressure. Pressure is the ratio of force to the area over which the force is applied. **Atmospheric pressure** is the pressure exerted by the mixture of gases in our atmosphere. When the volume change, ΔV, occurs at constant opposing pressure, P, the associated **pressure–volume work (expansion work)** is given by $w = -P\,\Delta V$. The energy expended in doing this pressure–volume work causes heats of reaction measured at constant volume (q_v) to differ numerically from heats measured at constant pressure (q_p).

The **first law of thermodynamics**, a specific example of the law of conservation of energy, says that no matter how the change in energy accompanying a reaction may be allocated between q and w, their sum, ΔE, is the same: $\Delta E = q + w$. The algebraic sign for q and w is negative when the system gives off heat to or does work on the surroundings. The sign is positive when the system absorbs heat or receives work energy done to it.

Explain the difference between a heat of reaction obtained at constant pressure and the heat of reaction at constant volume

When the volume of a system cannot change, as in a **bomb calorimeter**, w is zero, and q_v is the **heat of reaction at constant volume, ΔE**. When the system is under conditions of constant pressure, the energy of the system is called its **enthalpy, H**. At constant pressure, $\Delta H = \Delta E + P\,\Delta V$. The **heat of reaction at constant pressure, q_p,** is the **enthalpy change, ΔH**. Its value differs from that of ΔE by the work involved in interacting with the atmosphere when the change occurs at constant atmospheric pressure. In general, the difference between ΔE and ΔH is quite small. Exothermic reactions have negative values of ΔH; endothermic changes have positive values.

Describe the assumptions and utility of thermochemical equations

A balanced chemical equation that includes both the enthalpy change and the physical states of the substances is called a **thermochemical equation**. Coefficients in a thermochemical equation represent mole quantities of reactants and products, so it is reasonable to have fractional coefficients such as $\frac{1}{2}$. Such equations can be added, reversed (reversing also the sign of ΔH), or multiplied by a constant multiplier (doing the same to ΔH). If formulas are canceled or added, they must be of substances in identical physical states.

The reference conditions for thermochemistry, called **standard conditions**, are 25 °C and 1 bar of pressure. An enthalpy change measured under these conditions is called the **standard enthalpy of reaction** or the **standard heat of reaction**, given the symbol $\Delta H°$.

Use Hess's law to determine the enthalpy of a reaction

Hess's law of heat summation is possible because enthalpy is a state function. Values of $\Delta H°$ can be determined by the manipulation of any combination of thermochemical equations that add up to the final net equation. The units for $\Delta H°$ are generally joules or kilojoules. An **enthalpy diagram** provides a graphical description of the enthalpy changes for alternative paths from reactants to products.

Determine and use standard heats of formation to solve problems

When the enthalpy change is for the complete combustion of one mole of a pure substance under standard conditions in pure oxygen, $\Delta H°$ is called the **standard heat of combustion** of the compound and is symbolized as $\Delta H_c°$. When the enthalpy change is for the formation of one mole of a substance under standard conditions from its elements in their standard states, $\Delta H°$ is called the **standard heat of formation** of the compound and is symbolized as $\Delta H_f°$, (usually in units of kilojoules per mole, kJ mol^{-1}). The value of $\Delta H°$ for a reaction can be calculated from tabulated values of $\Delta H_f°$ using the **Hess's law equation**.

TOOLS

Tools for Problem Solving The following tools were introduced in this chapter. Study them carefully so you can select the appropriate tool when needed.

Kinetic energy (Section 6.1)

$$\mathrm{KE} = \tfrac{1}{2}\,mv^2$$

Potential energy changes (Section 6.1)
You should know how potential energy varies when the distance changes between objects that attract or repel.

Heat capacity and q (Section 6.3)

$$q = C\,\Delta t$$

Specific heat capacity and q (Section 6.3)

$$q = ms\,\Delta t$$

Heat transfer (Section 6.3)
When heat is transferred between two objects, the size of q is identical for both objects, but the algebraic signs of q are opposite.

$$q_1 = -q_2$$

First law of thermodynamics (Section 6.5)

$$\Delta E = q + w$$

Algebraic sign of ΔH (Section 6.6)
The sign of ΔH indicates the direction of energy flow.

> For an endothermic change, ΔH is positive.
> For an exothermic change, ΔH is negative.

Thermochemical equations (Section 6.7)
These equations show enthalpy changes ($\Delta H°$) for reactions where the amounts of reactants and products in moles are represented by the coefficients.

Enthalpy diagram (Section 6.8)
We use enthalpy diagrams to provide a graphical picture of the enthalpy changes associated with a set of thermochemical equations that are combined to give some net reaction.

Hess's law (Section 6.8)
Hess's law allows us to combine thermochemical equations to give a final desired equation and its associated $\Delta H°$.

Manipulating thermochemical equations (Section 6.8)
Changing the direction of a reaction changes the sign of $\Delta H°$. When the coefficients are multiplied by a factor, the value of $\Delta H°$ is multiplied by the same factor. Formulas can be canceled only when the substances are in the same physical state. These rules are used in applying Hess's law.

Hess's law equation (Section 6.9)

$$\Delta H° = \left[\text{sum of } \Delta H_f° \text{ of all of the products}\right] \times \left[\text{sum of } \Delta H_f° \text{ of all of the reactants}\right]$$

|Review Questions

Energy: The Ability to Do Work

6.1 Give definitions for **(a)** energy, **(b)** kinetic energy, and **(c)** potential energy.

6.2 How are the terms in Review Question 6.1 related to each other?

6.3 State the equation used to calculate an object's kinetic energy. Define the symbols used in the equation. Which variable has a larger effect on kinetic energy when it is doubled?

6.4 State the law of conservation of energy. Describe how it explains the motion of a child on a swing.

6.5 A pendulum such as a swinging chandelier continuously converts kinetic energy to potential energy and back again. Describe how these energies vary during a single swing of the pendulum.

6.6 A slinky toy, really a large spring, is held in a fixed position at the top and stretched vertically from the bottom. When released the free end rises and falls. Explain how the total, kinetic, and potential energies change as the spring moves from its lowest point to its highest point. How does this differ from the pendulum in the previous question?

6.7 What is meant by the term chemical energy?

6.8 How does the potential energy change (increase, decrease, or no change) for each of the following?

(a) Two electrons come closer together.

(b) An electron and a proton become farther apart.

(c) Two atomic nuclei approach each other.

(d) A ball rolls downhill.

6.9 What is the SI unit of energy? How much energy (in joules and in calories) does a 75 kg object have if it's moving at 45 m s⁻¹?

6.10 Why is heat considered a waste product in a car engine?

Heat, Temperature, and Internal Energy

6.11 Define heat. How do heat and temperature differ?

6.12 How is internal energy related to molecular kinetic and potential energy? How is a change in internal energy defined for a chemical reaction?

6.13 On a molecular level, how is thermal equilibrium achieved **(a)** when a hot object is placed in contact with a cold object, and **(b)** when two liquids are mixed together under

the same conditions of mass and temperature as part **(a)**? **(c)** Which of these two experiments will achieve thermal equilibrium quicker?

6.14 Consider the distribution of molecular kinetic energies shown in the following diagram for a gas at 25 °C:

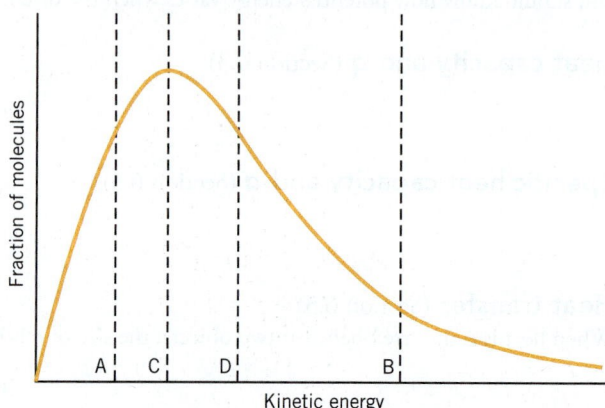

(a) Which point corresponds to the most frequently occurring (also called the most probable) molecular kinetic energy?

(b) Which point corresponds to the average molecular kinetic energy?

(c) If the temperature of the gas is raised to 50 °C, how will the height of the curve at B change?

(d) If the temperature of the gas is raised to 50 °C, how will the height of the curve at A change? How will the maximum height of the curve change?

6.15 Suppose the temperature of an object is raised from 100 °C to 200 °C by heating it with a Bunsen burner. Which of the following will be true?

(a) The average molecular kinetic energy will increase.

(b) The total kinetic energy of all the molecules will increase.

(c) The number of fast-moving molecules will increase.

(d) The number of slow-moving molecules will increase.

(e) The chemical potential energy will decrease.

6.16 A quart of boiling water will cause a more severe burn if it's spilled on you than will just a drop of boiling water. Explain this in terms of the amounts of kinetic energy possessed by the water molecules in the two samples.

6.17 Consider the three systems depicted in the following drawing:

Label them as isolated, closed, and open. What are the differences between the three types of systems?

6.18 What is a state function? Give four examples that meet your definition.

6.19 How would you determine whether an experimental parameter is a state function?

6.20 How can the state of a system be specified?

Measuring Heat

6.21 What do the terms system and surroundings mean? What is the difference between an isolated system and a closed system? What is the universe in terms of thermodynamics?

6.22 What are the names of the thermal properties whose values can have the following units?

(a) $J\,g^{-1}\,°C^{-1}$ (b) $J\,mol^{-1}\,°C^{-1}$ (c) $J\,°C^{-1}$ (d) J

6.23 For samples with the same mass, which kind of substance needs more energy to undergo an increase of 5 °C, something with a large specific heat or something with a small specific heat? Explain.

6.24 How do heat capacity and specific heat differ?

6.25 What is the meaning of a negative value for q?

6.26 Suppose object A has twice the specific heat and twice the mass of object B. If the same amount of heat is applied to both objects, how will the temperature change of A be related to the temperature change in B?

Energy of Chemical Reactions

6.27 In a certain chemical reaction, there is a decrease in the potential energy (chemical energy) as the reaction proceeds.

(a) How does the total kinetic energy of the particles change?

(b) How does the temperature of the reaction mixture change?

6.28 What term do we use to describe a reaction that liberates heat to its surroundings? How does the chemical energy change during such a reaction? What is the algebraic sign of q for such a reaction?

6.29 What term is used to describe a reaction that absorbs heat from the surroundings? How does the chemical energy change during such a reaction? What is the algebraic sign of q for such a reaction?

6.30 When gasoline burns, it reacts with oxygen in the air and forms hot gases consisting of carbon dioxide and water vapor. How does the potential energy of the gasoline and oxygen compare with the potential energy of the carbon dioxide and water vapor?

Heat, Work, and the First Law of Thermodynamics

6.31 Write the equation that states the first law of thermodynamics. In your own words, what does this statement mean in terms of energy exchanges between a system and its surroundings?

6.32 How are heat and work defined?

6.33 Devise an example, similar to the one described in Figure 6.8 to distinguish between heat and work and how they obey the first law of thermodynamics.

6.34 Why are heat and work not state functions?

Heats of Reaction

6.35 When we measure the heat of combustion of glucose, $C_6H_{12}O_6$, in a bomb calorimeter, what products are formed in the reaction? Is the heat we measure equal to ΔE or ΔH?

6.36 Consider the reaction

$$C_{12}H_{22}O_{11}(s) + 12O_2(g) \longrightarrow 12CO_2(g) + 11H_2O(l)$$

Are the values of ΔE and ΔH expected to be appreciably different?

6.37 How is enthalpy defined?

6.38 What is the sign of ΔH for an endothermic change?

6.39 If the enthalpy of a system increases by 100 kJ, what must be true about the enthalpy of the surroundings? Why?

6.40 If a system containing gases expands and pushes back a piston against a constant opposing pressure, what equation describes the work done on the system?

6.41 Why do standard reference values for temperature and pressure have to be selected when we consider and compare heats of reaction for various reactions? What are the values for the standard temperature and standard pressure?

Thermochemical Equations

6.42 What distinguishes a thermochemical equation from an ordinary chemical equation?

6.43 Why are fractional coefficients permitted in a balanced thermochemical equation? If a formula in a thermochemical equation has a coefficient of $\frac{1}{2}$, what does it signify?

Hess's Law

6.44 What fundamental fact about ΔH makes Hess's law possible?

6.45 What two conditions must be met by a thermochemical equation so that its standard enthalpy change can be given the symbol $\Delta H_f°$?

6.46 Describe what must be done with the standard enthalpy of reaction if

(a) a reaction is reversed (i.e., products written as reactants and reactants written as products).

(b) a reaction is multiplied by three.

(c) a reaction is divided by two.

Review Problems

Energy, the Ability to Do Work

6.49 If a car increases its speed from 30 mph to 60 mph, by what factor does the kinetic energy of the car increase? By what factor will the kinetic energy change if the speed decreases to 10 mph?

6.50 If the mass of a truck is doubled—for example, when it is loaded—by what factor does the kinetic energy of the truck increase? By what factor does the kinetic energy change if the mass is one-tenth of the original mass?

6.51 What is the kinetic energy, in joules, of a mole of oxygen moving at a speed of 255 m s^{-1} (use the appropriate prefix to avoid exponential notation)?

6.52 What is the kinetic energy, in joules, of a molecule of carbon tetrachloride moving at a speed of 255 m s^{-1} (use the appropriate prefix to avoid exponential notation)?

Measuring Heat

6.53 How much heat, in joules and in calories, must be removed from 1.75 mol of water to lower its temperature from 25.0 °C to 15.0 °C?

6.54 How much heat, in joules and calories, is needed to increase the temperature of 215 mL of water from 25.0 °C to 99.0 °C?

6.55 How many grams of water can be heated from 25.0 °C to 35.0 °C by the heat released from 85.0 g of iron that cools from 85.0 °C to 35.0 °C?

6.56 How many grams of copper can be cooled from 67.0 °C to 32.3 °C by the heat gained by 100.0 g of water that has an increase in temperature from 27.0 °C to 32.3 °C?

6.57 A 50.0 g piece of a metal at 100.0 °C was plunged into 100.0 g of water at 24.0 °C. The temperature of the resulting mixture became 28.0 °C.

(a) How many joules did the water absorb?

(b) How many joules did the metal lose?

(c) What is the heat capacity of the metal sample?

(d) What is the specific heat of the metal?

Standard Heats of Reaction

6.47 What two additional thermochemical equations are needed to determine the standard heat of formation of a hydrocarbon from its heat of combustion?

6.48 Peptides, small parts of proteins, contain nitrogen and sulfur atoms in addition to carbon and hydrogen in their molecules. What thermochemical equations will be needed to determine the heat of formation of a peptide if the heat of combustion is determined?

6.58 A sample of copper was heated to 120.00 °C and then thrust into 200.0 g of water at 25.00 °C. The temperature of the mixture became 26.50 °C.

(a) How much heat in joules was absorbed by the water?

(b) How many joules did the copper sample lose?

(c) What is the heat capacity of the copper sample?

(d) What was the mass in grams of the copper sample?

6.59 Calculate the molar heat capacity of iron in J mol^{-1} °C^{-1}. Its specific heat is 0.4498 J g^{-1} °C^{-1}.

6.60 What is the molar heat capacity of ethyl alcohol, C_2H_5OH, in units of J mol^{-1} °C^{-1}, if its specific heat is 0.586 cal g^{-1} °C^{-1}?

6.61 A vat of 4.54 kg of water underwent a decrease in temperature from 60.25 °C to 58.65 °C. How much energy in kilojoules left the water? (For this range of temperature, use a value of 4.18 J g^{-1} °C^{-1} for the specific heat of water.)

6.62 A container filled with 2.46 kg of water underwent a temperature change from 25.24 °C to 27.31 °C. How much heat, measured in kilojoules, did the water absorb?

Energy of Chemical Reactions

6.63 Nitric acid neutralizes potassium hydroxide. To determine the heat of reaction, a student placed 55.0 mL of 1.3 M HNO$_3$ in a coffee-cup calorimeter, noted that the temperature was 23.5 °C, and added 55.0 mL of 1.3 M KOH, also at 23.5 °C. The mixture was stirred quickly with a thermometer, and its temperature rose to 31.8 °C. Write the balanced equation for the reaction. Calculate the heat of reaction in joules. Assume that the specific heats of all solutions are 4.18 J g^{-1} °C^{-1} and that all densities are 1.00 g mL^{-1}. Calculate the heat of reaction per mole of acid (in units of kJ mol^{-1}).

6.64 In the reaction between formic acid (HCHO$_2$) and sodium hydroxide, water and sodium formate (NaCHO$_2$) are formed. To determine the heat of reaction, 75.0 mL of 1.07 M HCHO$_2$ was placed in a coffee-cup calorimeter at a temperature of 22.4 °C, and 45.0 mL of 1.78 M NaOH, also at 22.4 °C, was added. The mixture was stirred quickly with a thermometer, and its temperature rose to 23.4 °C. Write the balanced chemical equation for the reaction. Calculate the heat of reaction in joules. Assume that the specific heats of all solutions are 4.18 J g^{-1} °C^{-1} and that all densities are 1.00 g mL^{-1}. Calculate the heat of reaction per mole of acid (in units of kJ mol^{-1}).

6.65 A 1.000 mol sample of propane, a gas used for cooking in many rural areas, was placed in a bomb calorimeter with excess oxygen and ignited. The initial temperature of the calorimeter was 25.000 °C and its total heat capacity was 97.13 kJ °C^{-1}. The reaction raised the temperature of the calorimeter to 27.282 °C.

(a) Write the balanced chemical equation for the reaction in the calorimeter.

(b) How many joules were liberated in this reaction?

(c) What is the heat of reaction of propane with oxygen expressed in kilojoules per mole of C$_3$H$_8$ burned?

6.66 Toluene, C$_7$H$_8$, is used in the manufacture of explosives such as TNT (trinitrotoluene). A 1.500 g sample of liquid toluene was placed in a bomb calorimeter along with excess oxygen. When the combustion of the toluene was initiated, the temperature of the calorimeter rose from 25.000 °C to 26.413 °C. The products of the combustion were CO$_2$(g) and H$_2$O(l), and the heat capacity of the calorimeter was 45.06 kJ °C^{-1}.

(a) Write the balanced chemical equation for the reaction in the calorimeter.

(b) How many joules were liberated by the reaction?

(c) How many joules would be liberated under similar conditions if 1.000 mol of toluene was burned?

Heat, Work, and the First Law of Thermodynamics

6.67 If a system does 45 J of work and receives 28 J of heat, what is the value of ΔE for this change?

6.68 If a system has 48 J of work done on it and absorbs 22 J of heat, what is the value of ΔE for this change?

6.69 An automobile engine converts heat into work via a cycle. The cycle must finish exactly where it started, so the energy at the start of the cycle must be exactly the same as the energy at the end of the cycle. If the engine is to do 100 J of work per cycle, how much heat must it absorb?

6.70 Chargers for cell phones get warm while they are being used. Some of the energy that they are using is being used to power the cell phone and the rest is wasted as heat. If a cell phone battery needs 235 J of energy and 345 J are wasted as heat, how many joules are required to charge the cell phone?

6.71 If the engine in Problem 6.69 absorbs 250 joules of heat per cycle, how much work can it do per cycle?

6.72 If a battery can release 535 J of energy and 455 J are used for work, how much energy is released as heat?

Thermochemical Equations

6.73 Ammonia reacts with oxygen as follows:

$$4NH_3(g) + 7O_2(g) \longrightarrow 4NO_2(g) + 6H_2O(g)$$
$$\Delta H = -1132 \text{ kJ}$$

(a) Calculate the enthalpy change for the combustion of 1.00 mol of NH$_3$.

(b) Write the thermochemical equation for the reaction in which one mole of H$_2$O is formed.

6.74 One thermochemical equation for the reaction of carbon monoxide with oxygen is

$$3CO(g) + \tfrac{3}{2}O_2(g) \longrightarrow 3CO_2(g) \qquad \Delta H° = -849 \text{ kJ}$$

(a) Write the thermochemical equation for the reaction of 2.00 mol of CO.

(b) What is the $\Delta H°$ for the reaction that produces 1.00 mol of CO$_2$?

6.75 Magnesium burns in air to produce a bright light and is often used in fireworks displays. The combustion of magnesium can be described by the following thermochemical equation:

$$2Mg(s) + O_2(g) \longrightarrow 2MgO(s) \qquad \Delta H° = -1203 \text{ kJ}$$

How much heat (in kilojoules) is liberated by the combustion of 6.54 g of magnesium?

6.76 Methanol is the fuel in "canned heat" containers (e.g., Sterno) that are used to heat foods at cocktail parties. The combustion of methanol can be described by the following thermochemical equation:

$$2CH_3OH(l) + 3O_2(g) \longrightarrow 2CO_2(g) + 4H_2O(g)$$
$$\Delta H° = -1199 \text{ kJ}$$

How many kilojoules are liberated by the combustion of 46.0 g of methanol?

6.77 Methane burns with oxygen to produce carbon dioxide and water as a gas. The balanced thermochemical equation is

$$CH_4(g) + 2O_2(g) \longrightarrow CO_2(g) + 2H_2O(g)$$
$$\Delta H° = -802 \text{ kJ}$$

How much methane, in grams, must be burned to release 432 kJ of heat?

6.78 Methanol, as described in Problem 6.76, is used to heat food. The combustion of methanol can be described by the following thermochemical equation:

$$2CH_3OH(l) + 3O_2(g) \longrightarrow 2CO_2(g) + 4H_2O(g)$$
$$\Delta H° = -1199 \text{ kJ}$$

How much methanol, in grams, is needed to release 625 kJ of heat?

Hess's Law

*6.79 Construct an enthalpy diagram that shows the enthalpy changes for a one-step conversion of germanium, $Ge(s)$, into $GeO_2(s)$, the dioxide. On the same diagram, show the two-step process, first to the monoxide, $GeO(s)$, and then its conversion to the dioxide. The relevant thermochemical equations are the following:

$$Ge(s) + \tfrac{1}{2}O_2(g) \longrightarrow GeO(s) \qquad \Delta H° = -255 \text{ kJ}$$

$$Ge(s) + O_2(g) \longrightarrow GeO_2(s) \qquad \Delta H° = -534.7 \text{ kJ}$$

Using this diagram, determine $\Delta H°$ for the following reaction:

$$GeO(s) + \tfrac{1}{2}O_2(g) \longrightarrow GeO_2(s)$$

*6.80 Construct an enthalpy diagram for the formation of $NO_2(g)$ from its elements by two pathways: first, from its elements and, second, by a two-step process, also from the elements. The relevant thermochemical equations are

$$\tfrac{1}{2}N_2(g) + O_2(g) \longrightarrow NO_2(g) \qquad \Delta H° = +33.8 \text{ kJ}$$

$$\tfrac{1}{2}N_2(g) + \tfrac{1}{2}O_2(g) \longrightarrow NO(g) \qquad \Delta H° = +90.37 \text{ kJ}$$

$$NO(g) + \tfrac{1}{2}O_2(g) \longrightarrow NO_2(g) \qquad \Delta H° = ?$$

Be sure to note the signs of the values of $\Delta H°$ associated with arrows pointing up or down. Using the diagram, determine the value of $\Delta H°$ for the third equation.

*6.81 Show how the equations

$$N_2O_4(g) \longrightarrow 2NO_2(g) \qquad \Delta H° = +57.93 \text{ kJ}$$

$$2NO(g) + O_2(g) \longrightarrow 2NO_2(g) \qquad \Delta H° = -113.14 \text{ kJ}$$

can be manipulated to give $\Delta H°$ for the following reaction:

$$2NO(g) + O_2(g) \longrightarrow N_2O_4(g)$$

6.82 We can generate hydrogen chloride by heating a mixture of sulfuric acid and potassium chloride according to the equation

$$2KCl(s) + H_2SO_4(l) \longrightarrow 2HCl(g) + K_2SO_4(s)$$

Calculate $\Delta H°$ in kilojoules for this reaction from the following thermochemical equations:

$$HCl(g) + KOH(s) \longrightarrow KCl(s) + H_2O(l) \qquad \Delta H° = -203.6 \text{ kJ}$$

$$H_2SO_4(l) + 2KOH(s) \longrightarrow K_2SO_4(s) + 2H_2O(l) \qquad \Delta H° = -342.4 \text{ kJ}$$

6.83 Calculate $\Delta H°$ in kilojoules for the following reaction, the preparation of the unstable acid nitrous acid, HNO_2.

$$HCl(g) + NaNO_2(s) \longrightarrow HNO_2(l) + NaCl(s)$$

Use the following thermochemical equations:

$$2NaCl(s) + H_2O(l) \longrightarrow 2HCl(g) + Na_2O(s) \qquad \Delta H° = +507.31 \text{ kJ}$$

$$NO(g) + NO_2(g) + Na_2O(s) \longrightarrow 2NaNO_2(s) \qquad \Delta H° = -427.14 \text{ kJ}$$

$$NO(g) + NO_2(g) \longrightarrow N_2O(g) + O_2(g) \qquad \Delta H° = -42.68 \text{ kJ}$$

$$2HNO_2(l) \longrightarrow N_2O(g) + O_2(g) + H_2O(l) \qquad \Delta H° = +34.35 \text{ kJ}$$

6.84 Calcium hydroxide reacts with hydrochloric acid by the following equation:

$$Ca(OH)_2(aq) + 2HCl(aq) \longrightarrow CaCl_2(aq) + 2H_2O(l)$$

Calculate $\Delta H°$ in kilojoules for this reaction, using the following equations as needed:

$$CaO(s) + 2HCl(aq) \longrightarrow CaCl_2(aq) + H_2O(l) \qquad \Delta H° = -186 \text{ kJ}$$

$$CaO(s) + H_2O(l) \longrightarrow Ca(OH)_2(s) \qquad \Delta H° = -65.1 \text{ kJ}$$

$$Ca(OH)_2(s) \xrightarrow{\text{dissolving in water}} Ca(OH)_2(aq) \qquad \Delta H° = -12.6 \text{ kJ}$$

6.85 Given the following thermochemical equations,

$$CaO(s) + Cl_2(g) \longrightarrow CaOCl_2(s) \qquad \Delta H° = -110.9 \text{ kJ}$$

$$Ca(OH)_2(s) \longrightarrow CaO(s) + H_2O(l) \qquad \Delta H° = +65.1 \text{ kJ}$$

$$H_2O(l) + CaOCl_2(s) + 2NaBr(s) \longrightarrow$$
$$2NaCl(s) + Ca(OH)_2(s) + Br_2(l) \qquad \Delta H° = -60.2 \text{ kJ}$$

calculate the value of $\Delta H°$ (in kilojoules) for the reaction

$$\tfrac{1}{2}Cl_2(g) + NaBr(s) \longrightarrow \tfrac{1}{2}Br_2(l) + NaCl(s)$$

6.86 Given the following thermochemical equations,

$$2Cu(s) + S(s) \longrightarrow Cu_2S(s) \qquad \Delta H° = -79.5 \text{ kJ}$$

$$S(s) + O_2(g) \longrightarrow SO_2(g) \qquad \Delta H° = -297 \text{ kJ}$$

$$Cu_2S(s) + 2O_2(g) \longrightarrow 2CuO(s) + SO_2(g) \qquad \Delta H° = -527.5 \text{ kJ}$$

calculate the standard enthalpy of formation (in kJ mole^{-1}) of $CuO(s)$.

6.87 Given the following thermochemical equations,

$$3Mg(s) + 2NH_3(g) \longrightarrow Mg_3N_2(s) + 3H_2(g) \qquad \Delta H° = -371 \text{ kJ}$$

$$\tfrac{1}{2}N_2(g) + \tfrac{3}{2}H_2(g) \longrightarrow NH_3(g) \qquad \Delta H° = -46 \text{ kJ}$$

calculate $\Delta H°$ (in kilojoules) for the following reaction:

$$3Mg(s) + N_2(g) \longrightarrow Mg_3N_2(s)$$

6.88 Given the following thermochemical equations,

$$4NH_3(g) + 7O_2(g) \longrightarrow 4NO_2(g) + 6H_2O(g)$$
$$\Delta H° = -1132 \text{ kJ}$$

$$6NO_2(g) + 8NH_3(g) \longrightarrow 7N_2(g) + 12H_2O(g)$$
$$\Delta H° = -2740 \text{ kJ}$$

calculate the value of $\Delta H°$ (in kilojoules) for the reaction

$$4NH_3(g) + 3O_2(g) \longrightarrow 2N_2(g) + 6H_2O(g)$$

Standard Heats of Reaction

6.89 Which of the following thermochemical equations can have $\Delta H_f°$ for the heat of the reaction? If it cannot, then why not?

(a) $CO(NH_2)_2(aq) \longrightarrow CO(NH_2)_2(s)$

(b) $C + O + 2N + 4H \longrightarrow CO(NH_2)_2(aq)$

(c) $C(s, \text{graphite}) + \frac{1}{2}O_2(g) + N_2(g) + 4H_2(g) \longrightarrow$
$$CO(NH_2)_2(aq)$$

(d) $2C(s, \text{graphite}) + O_2(g) + 2N_2(g) + 8H_2(g) \longrightarrow$
$$2CO(NH_2)_2(aq)$$

6.90 Which of the following thermochemical equations can have $\Delta H_f°$ for the heat of the reaction? If it cannot, then why not?

(a) $Na_2SO_4(s) + HCl(g) \longrightarrow H_2SO_4(l) + 2NaCl(s)$

(b) $H_2(g) + S(s) + 2O_2(g) \longrightarrow H_2SO_4(l)$

(c) $2H + S + 4O \longrightarrow H_2SO_4(l)$

(d) $\frac{1}{2}H_2(g) + \frac{1}{2}S(s) + O_2(g) \longrightarrow \frac{1}{2}H_2SO_4(l)$

6.91 Write the thermochemical equations, including values of $\Delta H_f°$ in kilojoules per mole (from Table 6.2), for the formation of each of the following compounds from their elements, with everything in standard states.

(a) $HC_2H_3O_2(l)$, acetic acid

(b) $C_2H_5OH(l)$, ethyl alcohol

(c) $CaSO_4 \cdot 2H_2O(s)$, gypsum

(d) $Na_2SO_4(s)$, sodium sulfate

6.92 Write the thermochemical equations, including values of $\Delta H_f°$ in kilojoules per mole (from Appendix C), for the formation of each of the following compounds from their elements, with everything in standard states.

(a) $MgCl \cdot 2H_2O(s)$ (c) $POCl_3(g)$

(b) $(NH_4)_2Cr_2O_7(s)$ (d) $CO(NH_2)_2(s)$

6.93 Using data in Table 6.2, calculate $\Delta H°$ in kilojoules for the following reactions.

(a) $2H_2O_2(l) \longrightarrow 2H_2O(l) + O_2(g)$

(b) $HCl(g) + NaOH(s) \longrightarrow NaCl(s) + H_2O(l)$

6.94 Using data in Table 6.2, calculate $\Delta H°$ in kilojoules for the following reactions.

(a) $CH_4(g) + Cl_2(g) \longrightarrow CH_3Cl(g) + HCl(g)$

(b) $2NH_3(g) + CO_2(g) \longrightarrow CO(NH_2)_2(s) + H_2O(l)$

6.95 The value for the standard heat of combustion, $\Delta H°$ combustion, for sucrose, $C_{12}H_{22}O_{11}$, is -5.65×10^3 kJ mol^{-1}. Write the thermochemical equation for the combustion of 1 mol of sucrose and calculate the value of $\Delta H_f°$ for this compound. The sole products of combustion are $CO_2(g)$ and $H_2O(l)$. Use data in Table 6.2 as necessary.

6.96 The thermochemical equation for the combustion of acetylene gas, $C_2H_2(g)$, is

$$2C_2H_2(g) + 5O_2(g) \longrightarrow 4CO_2(g) + 2H_2O(l)$$
$$\Delta H° = -2599.3 \text{ kJ}$$

Using data in Table 6.2, determine the value of $\Delta H_f°$ for acetylene gas.

| Additional Problems

6.97 Look at the list of substances in Table 6.1. Which one will require the most energy to raise the temperature of a 50 g sample from 20 °C to 30 °C? Which substance will have the largest temperature increase if 40 kJ of heat was absorbed by 30 g of the substance?

*6.98 A dilute solution of hydrochloric acid with a mass of 610.29 g and containing 0.33183 mol of HCl was exactly neutralized in a calorimeter by the sodium hydroxide in 615.31 g of a comparably dilute solution. The temperature increased from 16.784 to 20.610 °C. The specific heat of the HCl solution was 4.031 J g^{-1} °C^{-1}; that of the NaOH solution was 4.046 J g^{-1} °C^{-1}. The heat capacity of the calorimeter was 77.99 J °C^{-1}. Write the balanced equation for the reaction. Use the

data above to calculate the heat evolved. What is the heat of neutralization per mole of HCl? Assume that the original solutions made independent contributions to the total heat capacity of the system following their mixing.

*6.99 A 2.00 kg piece of granite with a specific heat of 0.803 J g^{-1} °C^{-1} and a temperature of 95.0 °C is placed into 2.00 L of water at 22.0 °C. When the granite and water come to the same temperature, what will that temperature be?

6.100 In the recovery of iron from iron ore, the reduction of the ore is actually accomplished by reactions involving carbon monoxide. Use the following thermochemical equations,

$$Fe_2O_3(s) + 3CO(g) \longrightarrow 2Fe(s) + 3CO_2(g)$$
$$\Delta H° = -28 \text{ kJ}$$

$$3Fe_2O_3(s) + CO(g) \longrightarrow 2Fe_3O_4(s) + CO_2(g)$$
$$\Delta H° = -59 \text{ kJ}$$

$$Fe_3O_4(s) + CO(g) \longrightarrow 3FeO(s) + CO_2(g)$$
$$\Delta H° = +38 \text{ kJ}$$

to calculate $\Delta H°$ for the reaction

$$FeO(s) + CO(g) \longrightarrow Fe(s) + CO_2(g)$$

6.101 Use the results of Problem 6.100 and the data in Table 6.2 to calculate the value of $\Delta H_f°$ for FeO. Express the answer in units of kilojoules per mole.

6.102 The amino acid glycine, $C_2H_5NO_2$, is one of the compounds used by the body to make proteins. The equation for its combustion is

$$4C_2H_5NO_2(s) + 9O_2(g) \longrightarrow$$
$$8CO_2(g) + 10H_2O(l) + 2N_2(g)$$

For each mole of glycine that burns, 973.49 kJ of heat is liberated. Use this information, plus values of $\Delta H_f°$ for the products of combustion, to calculate $\Delta H_f°$ for glycine.

6.103 The value of $\Delta H_f°$ for HBr(g) was first evaluated using the following standard enthalpy values obtained experimentally. Use these data to calculate the value of $\Delta H_f°$ for HBr(g).

$$Cl_2(g) + 2KBr(aq) \longrightarrow Br_2(aq) + 2KCl(aq)$$
$$\Delta H° = -96.2 \text{ kJ}$$

$$H_2(g) + Cl_2(g) \longrightarrow 2HCl(g) \qquad \Delta H° = -184\text{kJ}$$

$$HCl(aq) + KOH(aq) \longrightarrow KCl(aq) + H_2O(l)$$
$$\Delta H° = -57.3\text{kJ}$$

$$HBr(aq) + KOH(aq) \longrightarrow KBr(aq) + H_2O(l)$$
$$\Delta H° = -57.3\text{kJ}$$

$$HCl(g) \xrightarrow{\text{dissolving in water}} HCl(aq) \qquad \Delta H° = -77.0 \text{ kJ}$$

$$Br_2(g) \xrightarrow{\text{dissolving in water}} Br_2(aq) \qquad \Delta H° = -4.2\text{kJ}$$

$$HBr(g) \xrightarrow{\text{dissolving in water}} HBr(aq) \qquad \Delta H° = -79.9 \text{ kJ}$$

6.104 Acetylene, C_2H_2, is a gas commonly burned in welding because of its hot flame. Small amounts can be formed in the reaction of calcium carbide, CaC_2, with water. Given the following thermochemical equations, calculate the value of $\Delta H_f°$ for acetylene in units of kilojoules per mole.

$$CaO(s) + H_2O(l) \longrightarrow Ca(OH)_2(s) \qquad \Delta H° = -65.3 \text{ kJ}$$

$$CaO(s) + 3C(s) \longrightarrow CaC_2(s) + CO(g) \quad \Delta H° = +462.3 \text{ kJ}$$

$$CaCO_3(s) \longrightarrow CaO(s) + CO_2(g) \qquad \Delta H° = +178 \text{ kJ}$$

$$CaC_2(s) + 2H_2O(l) \longrightarrow Ca(OH)_2(s) + C_2H_2(g)$$
$$\Delta H° = -126 \text{ kJ}$$

$$2C(s) + O_2(g) \longrightarrow CO_2(g) \qquad \Delta H° = -220 \text{ kJ}$$

$$2H_2O(l) \longrightarrow 2H_2(g) + O_2(g) \qquad \Delta H° = +572 \text{ kJ}$$

6.105 The reaction for the metabolism of sucrose, $C_{12}H_{22}O_{11}$, is the same as for its combustion in oxygen to yield $CO_2(g)$ and $H_2O(l)$. The standard heat of formation of sucrose is $-2230 \text{ kJ mol}^{-1}$. Use the data in Table 6.2 to compute the amount of energy (in kJ) released by metabolizing 1 oz (28.4 g) of sucrose.

6.106 Consider the following thermochemical equations:

$$CH_3OH(l) + O_2(g) \longrightarrow HCHO_2(l) + H_2O(l)$$
$$\Delta H° = -411 \text{ kJ}$$

$$CO(g) + 2H_2(g) \longrightarrow CH_3OH(l) \qquad \Delta H° = -128 \text{ kJ}$$

$$HCHO_2(l) \longrightarrow CO(g) + H_2O(l) \qquad \Delta H° = -33 \text{ kJ}$$

Suppose the first equation is reversed and divided by 2, The second and third equations are multiplied by $\frac{1}{2}$, and then the three adjusted equations are added. What is the net reaction, and what is the value of $\Delta H°$ for the net reaction?

6.107 Chlorofluoromethanes (CFMs) are carbon compounds of chlorine and fluorine and are also known as Freons. Examples are Freon-11 ($CFCl_3$) and Freon-12 (CF_2Cl_2), which were used as aerosol propellants. Freons have also been used in refrigeration and air-conditioning systems. In 1995 Mario Molina, F. Sherwood Rowland, and Paul Crutzen were awarded the Nobel Prize mainly for demonstrating how these and other CFMs contribute to the "ozone hole" that develops at the end of the Antarctic winter. In other parts of the world, reactions such as those shown below occur in the upper atmosphere where ozone protects the earth's inhabitants from harmful ultraviolet radiation. In the stratosphere CFMs absorb high-energy radiation from the sun and split off chlorine atoms that hasten the decomposition of ozone, O_3. Possible reactions are

$$O_3(g) + Cl(g) \longrightarrow O_2(g) + ClO(g) \qquad \Delta H° = -126 \text{ kJ}$$
$$ClO(g) + O(g) \longrightarrow Cl(g) + O_2(g) \qquad \Delta H° = -268 \text{ kJ}$$
$$O_3(g) + O(g) \longrightarrow 2O_2(g) \qquad \Delta H° = ?$$

The O atoms in the second equation come from the breaking apart of O_2 molecules caused by ultraviolet radiation from the sun. Use the first two equations to calculate the value of $\Delta H°$ (in kilojoules) for the last equation, the net reaction for the removal of O_3 from the atmosphere.

6.108 Use the information imbedded in Problem 6.107 to explain why both molecular oxygen and ozone protect us from harmful ultraviolet radiation.

6.109 Suppose a truck with a mass of 14.0 tons (1 ton = 2000 lb) is traveling at a speed of 45.0 mi/hr. If the truck driver slams on the brakes, the kinetic energy of the truck is changed to heat as the brakes slow the truck to a stop. How much would the temperature of 5.00 gallons of water increase if all of this heat could be absorbed by the water?

6.110 How much work must be done to form one mole of CH_4 and enough oxygen from CO_2 and H_2O? Why is it so difficult to carry out this reaction?

6.111 A cold $-15\,°C$ piece of copper metal weighing 7.38 g was placed in a beaker as illustrated, holding 55 mL of water at 32.4 °C.

Multi-Concept Problems

*6.115 For ethanol, C_2H_5OH, which is mixed with gasoline to make the fuel gasohol, $\Delta H_f^° = -277.63$ kJ/mol. Calculate the number of kilojoules released when 1.25 gallons of ethanol is burned completely. The density of ethanol is 0.787 g cm^{-3}. Use the data in Table 6.2 to help in the computation.

6.116 Both calcium and potassium react with water to form hydrogen and the corresponding hydroxide. The potassium hydroxide is soluble in water but the calcium hydroxide is not. Calculate the heats of reaction for both the reaction of potassium with water and that of calcium with water. Considering the metal, on a per mole basis and a per gram basis, which reaction gives off more heat? How many electrons are transferred in each reaction? On a per mole of electron basis, which reaction gives off more heat?

Exercises in Critical Thinking

6.119 Growing wheat and converting it into bread requires energy. In the process, the wheat absorbs sunlight and through photosynthesis converts carbon dioxide and water into carbohydrates. The plants are harvested and the wheat is ground into flour. The flour is transported to a bakery where it is baked into bread. The bread is sold and then consumed by the customer. After eating a piece of bread, the customer goes for a walk. What are the different forms of energy, and how is the energy converted from one form to another in these processes?

6.120 Suppose we compress a spring, tie it up tightly, and then dissolve the spring in acid. What happens to the potential energy contained in the compressed spring? How could you measure it?

6.121 Carefully and precisely describe the difference between H, ΔH, and $\Delta H^°$.

(a) Draw an arrow showing the flow of heat.

(b) What is the sign of q for the water? What is the sign of q for the copper?

(c) What is the final temperature of the system?

6.112 Both Na_2CO_3 and $NaHCO_3$ can be used to neutralize acids. Using the information from Appendix C, which base gives off the most heat in neutralizing 1.0 mole of hydrochloric acid?

*6.113 When 4.56 g of a solid mixture composed of NH_4Cl and $CaCl_2$ was dissolved in 100.0 mL water, the temperature of the water rose by 5.33 °C. How many grams of each substance was in the mixture?

*6.114 Using the results from Analyzing and Solving Multi-Concept Problems in Section 6.6 and the Practice Exercise 6.14 just after it, determine the heat of reaction for the ionization of one mole of acetic acid in water.

6.117 As a routine safety procedure, acids and bases should be stored separately in a laboratory. Concentrated hydrochloric acid is 12.0 M and the molarity of 25% ammonium hydroxide is 13.4 M. Both are sold in 2.5 L bottles. If there were an accident and the two bottles broke and the acid and base mixed completely, how much heat would evolve?

$$NH_4OH(aq) + HCl(aq) \longrightarrow NH_4Cl(aq) + H_2O(l)$$
$$\Delta H^° = -50.49 \text{ kJ}$$

6.118 In an experiment, 95.0 mL of 0.225 M silver nitrate was mixed rapidly with 47.5 mL of 0.225 M calcium chloride in a coffee-cup calorimeter. If a reaction occurred, was it exothermic or endothermic? If the reaction started at 23.7 °C, what is the final temperature of the solution?

6.122 Why do we usually use $\Delta H^°$ rather than $\Delta E^°$ when we discuss energies of reaction? Develop an argument for using $\Delta E^°$ instead.

6.123 Explain why we can never determine values for E or H. Make a list of the factors we would need to know to determine E or H.

6.124 Find the heats of formation of some compounds that are explosives. What do they have in common? Compare them to the $\Delta H_f^°$ values of stable compounds such as H_2O, NaCl, and $CaCO_3$. Formulate a generalization to summarize your observations.

6.125 A 100 g piece of granite and a 100 g piece of lead are placed in an oven that is at 100 °C. The initial temperatures of the pieces are 25 °C. After a couple of minutes, neither one is at 100 °C, but which one is warmer? Which one will reach 100 °C first? How much energy will it take for both of the objects to reach 100 °C?

The Quantum Mechanical Atom

© Radius Images/Corbis

Sparks fly when the metal is cut by a carbon dioxide laser. The light given off by these sparks is a form of energy that is released when electrons release their excess energy. In this chapter, we will begin our discussion of electrons and the role they play in atoms and molecules.

Why are sodium ions cations and chlorine ions anions? Why do nonmetals combine to form molecules, while molecules made of metals are very rare? Why are the noble gases virtually inert? Why does the periodic table have the arrangement and shape it has? Many more questions could also be posed, but they are minor compared to major problems that were becoming apparent in the description of the atom that developed in the late 1800s and early 1900s, a description that relied on classical physics.

First, classical physics, the physics of the late 1800s, predicted that atoms simply could not exist. It predicted that electrons must quickly lose energy and collapse into the nucleus—*the collapsing atom paradox*. Classical physics also predicted that as objects are heated to higher and higher temperatures, they should emit more and more light of higher and higher energy. That they emit very little ultraviolet radiation led physicists to call this the *ultraviolet catastrophe*. Finally, particles, such as electrons, passing through very small openings acted as waves, a fact that could be explained only if we regarded particles as waves leading to the concept of the wave/particle duality of matter. These problems in describing the fundamental nature of matter made it clear that an entirely new set of theoretical concepts was needed. These concepts are commonly embodied in quantum mechanics (also called wave mechanics or quantum theory), which is now the cornerstone of modern chemistry.

In this chapter, we will discuss quantum mechanics, how electrons behave in atoms, and how this affects the properties of the elements.

This Chapter in Context

■ When experiment and theory do not agree, the theory must be modified or discarded. Classical physics could not resolve the collapsing atom paradox. It also could not correctly predict the crucial characteristics of the radiation emitted by hot samples of the elements.

LEARNING OBJECTIVES

After reading this chapter, you should be able to:

- describe light as both a particle and a wave, and use the equations describing the wave and particle nature of light
- calculate the wavelength of a line in the spectrum of hydrogen and relate it to the energy levels of an electron
- using the Bohr model of hydrogen, explain how a line spectrum of hydrogen is generated
- describe the main features of the wave mechanical model of an atom
- define and use the three major quantum numbers and their possible values
- describe electron spin and apply it to explain paramagnetism and diamagnetism along with the Pauli exclusion principle
- write the ground state electron configuration of an atom
- explain the connection between the periodic table and electron configurations of atoms
- draw the shapes and orientations of *s*, *p*, and *d* orbitals in an atom
- describe how the electron configurations of the elements and the periodic table help predict the physical and chemical properties of the elements

7.1 | Electromagnetic Radiation

The Nature of Light

You've learned that objects can have energy in only two ways, as kinetic energy and as potential energy. You also learned that energy can be transferred between things, and in Chapter 6 our principal focus was on the transfer of heat. Energy can also be transferred in the form of light or electromagnetic radiation. As we will see in this chapter, light is a very important form of energy in chemistry since it allows us to delve into the structure of atoms and molecules and, in some cases, their reactions. For example, some chemical systems emit visible light as they react often by a process called chemiluminescence (see Figure 7.1*b* and *c*).

Many experiments have shown that radiation carries energy through space by means of **waves**. Waves are an oscillation that moves outward from a disturbance (think of ripples moving away from a pebble dropped into a pond). In the case of **electromagnetic radiation**, the disturbance can be a pulsing, or oscillating, electric charge that creates an

■ Like shaking a birthday present to try to guess its contents, scientists can "shake up" an atom by adding energy. Instead of hearing the rattle of a present, light is emitted. The energy and intensity of the light emitted helps to describe the inner workings of the atom.

■ Electricity and magnetism are closely related to each other. A moving charge creates an electric current, which in turn creates a magnetic field around it. This is the fundamental idea behind electric motors. A moving magnetic field creates an electric field or current. This is also the idea behind electric generators and turbines.

Figure 7.1 | **Light is given off in a variety of chemical reactions.** (*a*) Combustion. (*b*) Light sticks. (*c*) A lightning bug.

oscillating electric field. The oscillating electric field creates an oscillating magnetic field, and the two together are known as **electromagnetic waves**.

Wavelength and Frequency

An electromagnetic wave is often depicted as a sine wave that has an amplitude, wavelength, and frequency (for simplicity we do not show the magnetic component of the wave). Figure 7.2 shows how the **amplitude** or intensity of the wave varies with time and with distance as the wave travels through space. The amplitude of the wave is related to the intensity or brightness of the radiation. In Figure 7.2*a*, we see two complete oscillations or cycles of the wave during a one-second interval. The number of cycles per second is called the frequency of the electromagnetic radiation, and its symbol is ν, the Greek letter nu. In the SI, so frequency is given the unit "per second," which is $1/second$, or second^{-1}, another name for this unit is **hertz (Hz)**.

As electromagnetic radiation moves away from its source, the positions of maximum and minimum amplitude (peaks and troughs) are regularly spaced. The peak-to-peak distance is called the radiation's **wavelength**, symbolized by λ (the Greek letter lambda). See Figure 7.2*b*. Because wavelength is a distance, it has units of length (for example, meters).

We can see that if we multiply the wavelength by frequency, the result is the velocity of the wave:

$$\text{meters} \times \frac{1}{\text{second}} = \frac{\text{meters}}{\text{seconds}} = \text{velocity}$$

$$\text{m} \times \frac{1}{\text{s}} = \frac{\text{m}}{\text{s}} = \text{m s}^{-1}$$

The velocity of electromagnetic radiation in a vacuum is a constant and is called the *speed of light* with the symbol *c*. Its value to three significant figures is 3.00×10^8 m s^{-1}.

$$c = 3.00 \times 10^8 \text{ m s}^{-1}$$

■ Electromagnetic waves don't need a medium to travel through as do water and sound waves. They can cross empty space. The speed of the electromagnetic wave in a vacuum is the same, about 3.00×10^8 m/s, no matter how the radiation is created. The symbol for frequency, ν, is the Greek letter nu, pronounced "new."

■ The SI symbol for the second is s.

$$\text{s}^{-1} = \frac{1}{\text{s}}$$

■ The speed of light is one of our most carefully measured constants and is used to define the meter. The precise value of the speed of light in a vacuum is 2.99792458×10^8 m/s, and a meter is defined as exactly the distance traveled by light in 1/299,792,458 of a second.

Figure 7.2 | **Two views of electromagnetic radiation.** (*a*) The frequency, ν, of a light wave is the number of complete oscillations each second. Here two cycles span a one-second time interval, so the frequency is 2 cycles per second, or 2 Hz. (*b*) Electromagnetic radiation frozen in time. This curve shows how the amplitude varies along the direction of travel. The distance between two peaks is the wavelength, λ, of the electromagnetic radiation.

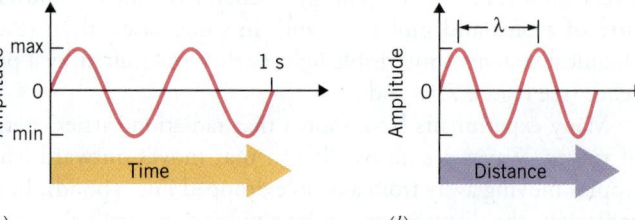

From the preceding discussion we obtain a very important relationship that allows us to convert between λ and ν.

$$\lambda \times \nu = c = 3.00 \times 10^8 \text{ m s}^{-1} \tag{7.1}$$

TOOLS
Wavelength–frequency relationship

Example 7.1
Calculating Frequency from Wavelength

Mycobacterium tuberculosis, the organism that causes tuberculosis, can be completely destroyed by irradiation with ultraviolet light with a wavelength of 254 nm. What is the frequency of this radiation?

Analysis: We need to realize that this question is not a conversion as our stoichiometry problems were, but will require an equation to solve.

Assembling the Tools: The tool needed is the relationship between wavelength and frequency, Equation 7.1. When rearranged it becomes

$$\nu = \frac{c}{\lambda}$$

However, we must be careful about the units, so another tool is the table of SI prefixes in Chapter 1.

Solution: We substitute $3.00 \times 10^8 \text{ m s}^{-1}$ for c and 254 nm for λ, the wavelength. However, to cancel units correctly, we must convert the wavelength from nm, nanometers to meters. Recall that nm means nanometer and the prefix nano implies the factor "$\times 10^{-9}$", so 1 nm = 10^{-9} m. Substituting gives

$$\nu = \frac{3.00 \times 10^{-8} \text{ m s}^{-1}}{254 \text{ nm}} \times \frac{1 \text{ nm}}{1 \times 10^{-9} \text{ m}}$$

$$= 1.18 \times 10^{15} \text{ s}^{-1}$$

$$= 1.18 \times 10^{15} \text{ Hz}$$

Is the Answer Reasonable? One way to test whether our answer is correct is to see if the given wavelength, multiplied by our calculated frequency, gives us the speed of light. To estimate the answer and do the calculation without a calculator, we round the numbers to one or two significant figures. Rounding 254 nm to 250×10^{-9} m and our answer to $1 \times 10^{15} \text{ s}^{-1}$ we calculate

$$(250 \times 10^{-9} \text{ m}) \times (1 \times 10^{15} \text{ s}^{-1}) = 250 \times 10^6 \text{ m s}^{-1} = 2.5 \times 10^8 \text{ m s}^{-1}$$

which is close to the actual speed of light, within one significant figure. Using the actual data and a calculator gives us the exact value for the speed of light.

Example 7.2
Calculating Frequency from Wavelength

Radio station WKXR is an AM radio station broadcasting from Asheboro, North Carolina, at a frequency of 1260 kHz. What is the wavelength of these radio waves expressed in meters?

Analysis: We use the same analysis as in Example 7.1.

Assembling the Tools: This question will use the same equation as Example 7.1. This time we solve for wavelength in meters.

$$\lambda = \frac{c}{\nu}$$

Solution: Solving Equation 7.1 for the wavelength, we first have to consider the units. The frequency in kilohertz has two parts: kilo stands for 10^3 and hertz stands for s^{-1}. Making those substitutions, we write

$$\lambda = \frac{c}{\nu} = \frac{3.00 \times 10^8 \text{ m s}^{-1}}{1260 \text{ kHz}} = \frac{3.00 \times 10^8 \text{ m s}^{-1}}{1260 \times 10^3 \text{ s}^{-1}} = 238 \text{ m}$$

Canceling the units and working out the mathematics, we find that the wavelength is 238 meters.

Is the Answer Reasonable? If this wavelength is correct, we should be able to multiply it by the original frequency (in Hz) and get the speed of light, as we did in the previous example. Another approach is to test our answer by dividing the speed of light by the wavelength to get back the frequency. Rounding to one significant figure, we get

$$\frac{3 \times 10^8 \text{ m s}^{-1}}{2 \times 10^3 \text{ m}} = 1.5 \times 10^6 \text{ Hz} = 1500 \text{ kHz}$$

which is reasonably close to the original frequency of 1260 kHz.

Practice Exercise 7.1

Helium derives its name from the Latin name for the sun. Helium was discovered in 1868 when spectroscopists found that the 588 nm wavelength (among others) was missing from the sun's spectrum. What is the frequency of this radiation? (*Hint:* Recall the metric prefixes so units will cancel correctly.)

Practice Exercise 7.2

Radio station KRED in Eureka, California, broadcasts electromagnetic radiation at a frequency of 92.3 MHz (megahertz). What is the wavelength of these radio waves, expressed in meters?

Electromagnetic Spectrum

Electromagnetic radiation comes in a broad range of frequencies called the **electromagnetic spectrum**, illustrated in Figure 7.3. Some portions of the spectrum have popular names. For example, radio waves are electromagnetic radiations having very low frequencies (and therefore very long wavelengths). Microwaves, which also have low frequencies, are emitted by radar instruments such as those the police use to monitor the speeds of cars. In microwave ovens, similar radiation is used to heat water in foods, causing the food to cook quickly. Infrared radiation is emitted by hot objects and consists of the range of frequencies that can make molecules of most substances vibrate internally. You can't see infrared radiation, but you can feel how your body absorbs it by holding your hand near a hot radiator; the absorbed radiation makes your hand warm. Gamma rays are at the high-frequency end of the electromagnetic spectrum. They are produced by certain elements that are radioactive. X rays are very much like gamma rays, but they are usually made by special equipment. Both X rays and gamma rays penetrate living things easily.

Most of the time, you are bombarded with electromagnetic radiation from all portions of the electromagnetic spectrum, Figure 7.3*a*. Radio and TV signals pass through you; you feel infrared radiation when you sense the warmth of a radiator; X rays and gamma rays fall on you from space; and light from a lamp reflects into your eyes from the page you're reading. Of all these radiations, your eyes are able to sense only a very narrow band of wavelengths ranging from about 400 to 700 nm, Figure 7.3*b*. This band is called the **visible spectrum** and consists of all the colors you can see, from red through orange, yellow, green, blue, and violet. White light is composed of all these colors and can be separated into them by focusing a beam of white light through a prism, which spreads the various

■ There is an inverse relationship between wavelength and frequency. The lower the frequency, the longer the wavelength.

■ Radar stands for radio detection and ranging.

■ The speed of electromagnetic radiation was computed to be about 3×10^8 m/s. The fact that the same speed was determined experimentally for light supported the hypothesis that light was a form of electromagnetic radiation.

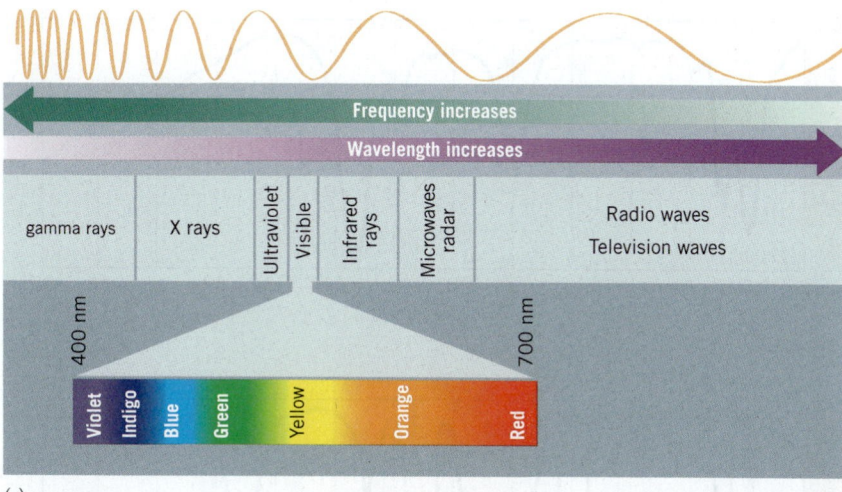

(a)

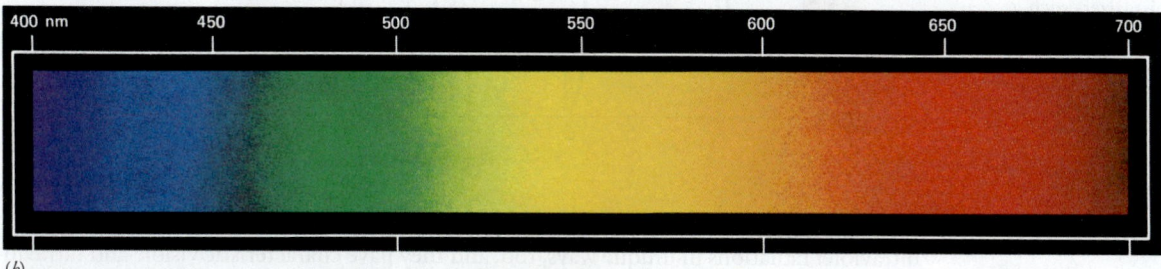

(b)

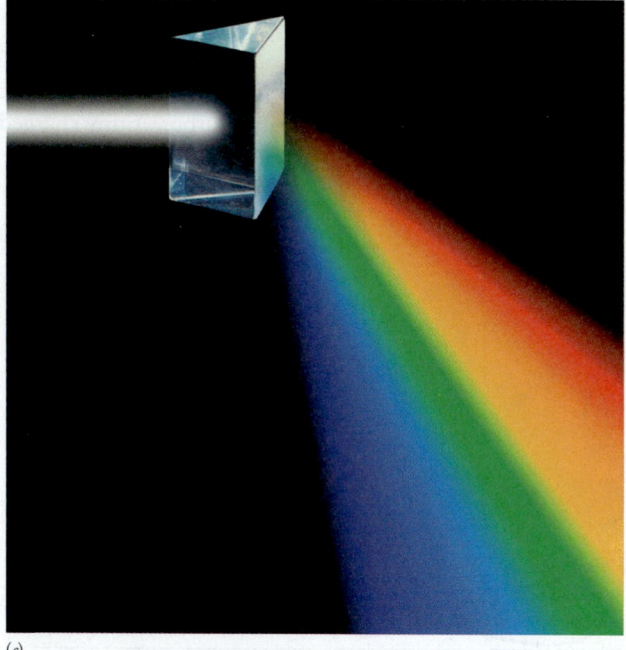

(c)

Figure 7.3 | **The electromagnetic spectrum.** (*a*) The electromagnetic spectrum is divided into regions according to the wavelengths of the radiation. (*b*) The visible spectrum is composed of wavelengths that range from about 400 to 700 nm. (*c*) This shows the production of a visible spectrum by splitting white light into its rainbow of colors.

wavelengths apart. The visible spectrum is shown in Figure 7.3*b*. An illustration showing the production of a visible spectrum is given in Figure 7.3*c*.

The way substances absorb electromagnetic radiation often can help us characterize them. For example, each substance absorbs a uniquely different set of infrared frequencies. A plot of the intensities of absorption versus the wavelength is called an infrared absorption spectrum. It can be used to identify a compound, because each infrared spectrum is

Figure 7.4 | **Absorption of infrared radiation by an organic compound.** (*a*) Infrared absorption spectrum of methanol. In an infrared spectrum, the usual practice is to show the amount of light absorbed increasing from top to bottom in the graph. Thus, for example, there is a peak in the percentage of light absorbed at about 3 mm. (*Spectrum courtesy Sadtler Research Laboratories, Inc., Philadelphia, Pa.*). (*b*) Infrared spectrum of acetone, which is found in nail polish remover mixed with water, fragrance, dyes, and sometimes moisturizers such as lanolin.

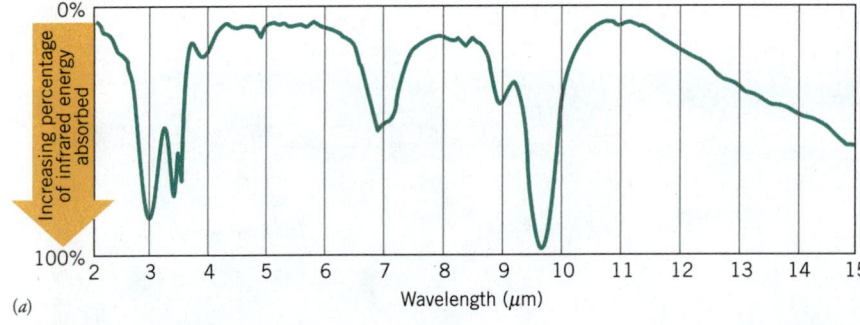
(*a*)

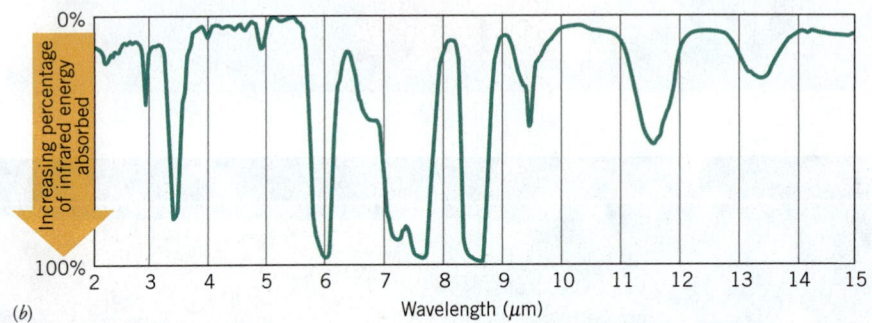
(*b*)

as unique as a set of fingerprints. (See Figure 7.4.) Many substances absorb visible and ultraviolet radiations in unique ways, too, and they have characteristic visible and ultraviolet absorption spectra (Figure 7.5).

Light as a Stream of Photons

When an electromagnetic wave passes an object, the oscillating electric and magnetic fields may interact with it, as ocean waves interact with a buoy in a harbor. A tiny charged

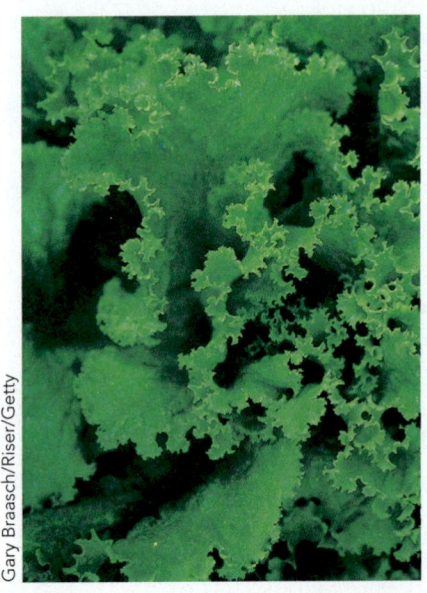
(*a*)

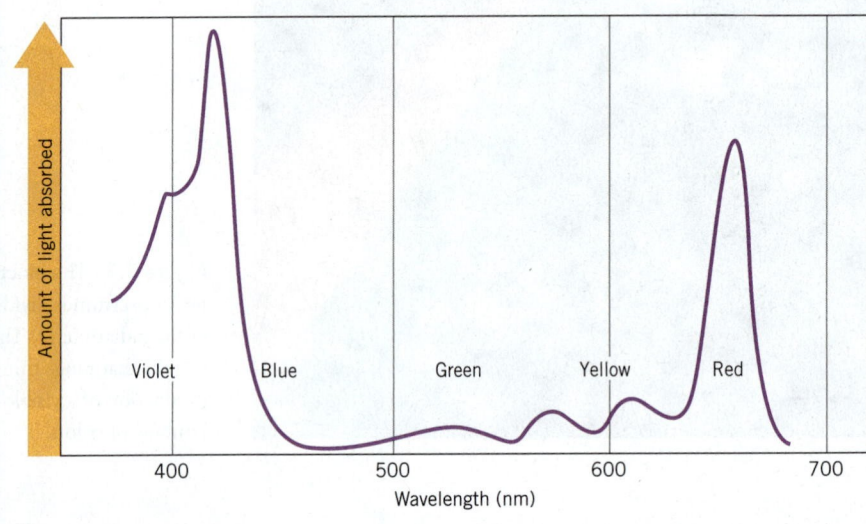
(*b*)

Figure 7.5 | **Absorption of light by chlorophyll.** (*a*) Chlorophyll is the green pigment plants use to harvest solar energy for photosynthesis. (*b*) In this visible absorption spectrum of chlorophyll, the percentage of light absorbed increases from bottom to top. Thus, there is a peak in the light absorbed at about 420 nm and another at about 660 nm. This means the pigment strongly absorbs blue-violet and red light. The green color we see is the light that is not absorbed, but is reflected. (Our eyes are most sensitive to green, so we don't notice the yellow components of the reflected light.)

particle placed in the path of the wave will be yanked back and forth by the changing electric and magnetic fields. For example, when a radio wave strikes an antenna, electrons within the antenna begin to bounce up and down, creating an alternating current that can be detected and decoded electronically. Because the wave exerts a force on the antenna's electrons and moves them through a distance, work is done. Thus, as energy is lost by the source of the wave (the radio transmitter), energy is gained by the electrons in the antenna.

Energy of a Photon

A series of groundbreaking experiments showed that classical physics does not correctly describe energy transfer by electromagnetic radiation. In 1900 a German physicist named Max Planck (1858–1947) proposed that electromagnetic radiation can be viewed as a stream of tiny packets or **quanta** of energy that were later called photons. Each photon travels at the speed of light. Planck proposed, and Albert Einstein (1879–1955) confirmed, that the *energy of a photon of electromagnetic radiation is proportional to the radiation's frequency*, not to its intensity or brightness as had been believed up to that time.

■ The energy of one photon is called one **quantum** of energy.

$$\text{Energy of a photon:} \quad E = h\nu \tag{7.2}$$

TOOLS

Energy of a photon

■ The value of Planck's constant is 6.626068×10^{-34} J s. It has units of energy (joules) multiplied by time (seconds).

In this expression, h is a proportionality constant that we now call **Planck's constant**. Note that Equation 7.2 relates two representations of electromagnetic radiation. The left-hand side of the equation deals with a property of particles (energy per photon); the right-hand side deals with a property of waves (the frequency). Quantum theory unites the two representations, so we can use whichever representation of electromagnetic radiation is convenient for describing experimental results.

Photoelectric Effect

One of the earliest clues to the relationship between the frequency of light and its energy was the discovery of the **photoelectric effect**. In the latter part of the 19th century, it was found that certain metals acquired a positive charge when they were illuminated by light. Apparently, light is capable of kicking electrons out of the surface of the metal.

As this phenomenon was studied, it was discovered that electrons would only leave a metal's surface if the frequency of the incident radiation was above some minimum value, the threshold frequency. This threshold frequency differs for different metals, depending on how easily the metal atom releases the electrons. In addition, above the threshold frequency, the kinetic energy of the emitted electron increases with increasing frequency of the light. However, this kinetic energy does not depend on the intensity of the light. In fact, if the frequency of the light is below the minimum frequency, no electrons are observed at all, no matter how bright the light is. To physicists of that time, this was very perplexing because they believed the energy of light was related to its brightness. The explanation of the phenomenon was finally given by Albert Einstein in the form of a very simple equation:

■ Brighter light delivers more photons; higher frequency light delivers higher energy photons.

$$\text{KE} = h\nu - \text{BE}$$

where KE is the kinetic energy of the electron that is emitted, $h\nu$ is the energy of the photon that strikes the metal surface, and BE is the minimum energy needed to eject the electron from the metal's surface (this can be considered the *binding energy*). Stated another way, part of the energy of the photon is needed just to get the electron off the surface of the metal, the BE. Any energy left over ($h\nu - \text{BE}$) becomes the electron's kinetic energy.

Analyzing and Solving Multi-Concept Problems

Using your microwave oven, how many moles of photons are needed to heat your cup of tea? Consider a cup of tea to be 237 mL of water starting at 25.0 °C and the drinking temperature to be 85.5 °C. Also, most microwave ovens have Klystron tubes tuned to emit microwaves that are absorbed by water molecules at a wavelength of 12.24 cm.

Analysis: This problem requires that we calculate the energy needed to heat a cup of tea. We can calculate the total energy needed using the tools presented in Chapter 6. Once we know the total energy, we can determine the number of photons from the tools presented in this section. Then we can use the information from Chapter 3 to calculate the moles of photons.

Strategy: We have delineated three parts for this calculation: (1) calculate the total energy needed, (2) calculate the number of photons needed, and (3) convert the number of photons to moles of photons.

PART 1

Assembling the Tools: To determine the energy needed to heat a cup of water, we need the calorimetry equation, Equation 6.7 in Section 6.3.

$$\text{Heat energy (J)} = (\text{mass})(\text{specific heat})(\text{temperature change}) = ms\,\Delta T$$

Solution: We know that the specific heat of water is $4.184\ \text{J g}^{-1}\ {}^{\circ}\text{C}^{-1}$, and the mass is determined by converting 237 mL of water to 237 g of water using the density of water, 1.00 g/mL. The change in temperature is $85.5\ {}^{\circ}\text{C} - 25.0\ {}^{\circ}\text{C}$ or $60.5\ {}^{\circ}\text{C}$. Putting this together we get

$$\text{Heat energy (J)} = (237\ \text{g})(4.184\ \text{J g}^{-1}\ {}^{\circ}\text{C}^{-1})(60.5\ {}^{\circ}\text{C}) = 5.999 \times 10^4\ \text{J}$$

Notice that we will keep one extra significant figure until the final result is calculated.

PART 2

Assembling the Tools: We can calculate the energy of one photon using Equation 7.2, $E = h\nu$, combined with Equation 7.1, $\lambda \times \nu = c$, to get

$$E = \frac{hc}{\lambda}$$

Then we divide that into the total energy from Part 1 to determine the number of photons.

Solution: The energy of one photon is

$$E = \frac{(6.626 \times 10^{-34}\ \text{J s})(2.998 \times 10^8\ \text{m s}^{-1})}{(12.24 \times 10^{-2}\ \text{m})} = 1.623 \times 10^{-24}\ \text{J photon}^{-1}$$

The number of photons is

$$\text{number of photons} = \frac{5.999 \times 10^4\ \text{J}}{1.623 \times 10^{-24}\ \text{J photon}^{-1}} = 3.696 \times 10^{28}\ \text{photons}$$

PART 3

Assembling the Tools: We use the tool for converting between elementary particles to moles, more commonly known as Avogadro's number.

$$1\ \text{mole photons} = 6.022 \times 10^{23}\ \text{photons}$$

Solution: We use our conversion factor to calculate moles from elementary units:

$$3.696 \times 10^{28}\ \text{photons}\left(\frac{1\ \text{mol photons}}{6.022 \times 10^{23}\ \text{photons}}\right) = 6.138 \times 10^4\ \text{mol photons}$$

Correctly rounded the answer is 6.14×10^4 mol photons.

The idea that electromagnetic radiation can be represented as either a stream of photons or a wave is a cornerstone of the quantum theory. Physicists were able to use this concept to understand many experimental results that classical physics simply could not explain. The success of the quantum theory in describing radiation paved the way for a second startling realization: that electrons, like radiation, could be represented as both waves and particles. We now turn our attention to the first experimental evidence that led to our modern quantum mechanical model of atomic structure: the existence of discrete lines in atomic spectra.

7.2 | Line Spectra and the Rydberg Equation

The spectrum described in Figure 7.3 is called a **continuous spectrum** because it contains a continuous unbroken distribution of light of *all* colors. It is formed when the light from an object that's been heated to a very high temperature (such as the filament in an electric light bulb), is split by a prism and displayed on a screen. A rainbow after a summer shower appears as a continuous spectrum that most people have seen. In this case, tiny water droplets in the air spread out the colors contained in sunlight.

A rather different kind of spectrum is observed if we examine the light that is given off when an *electric discharge*, or spark, passes through a gas such as hydrogen. The electric discharge is an electric current that excites, or energizes, the electrons in the atoms of the gas. When this occurs we say that an atom is in an **excited state.** The atoms then emit the absorbed energy in the form of light as the electrons return to a lower energy state. When a narrow beam of this light is passed through a prism, as shown in Figure 7.6, we do *not* see a continuous spectrum. Instead, only a few colors are observed, displayed as a series of individual lines. This series of lines is called the element's **atomic spectrum** or **emission spectrum.** Figure 7.7 shows the visible portions of the atomic spectra of two common elements, sodium and hydrogen, and how they compare with a continuous spectrum. Since each spectrum of these elements is quite different, each element has a unique atomic spectrum that is as characteristic as a fingerprint.

■ Atoms of an element can often be excited thermally by adding them to the flame of a Bunsen burner.

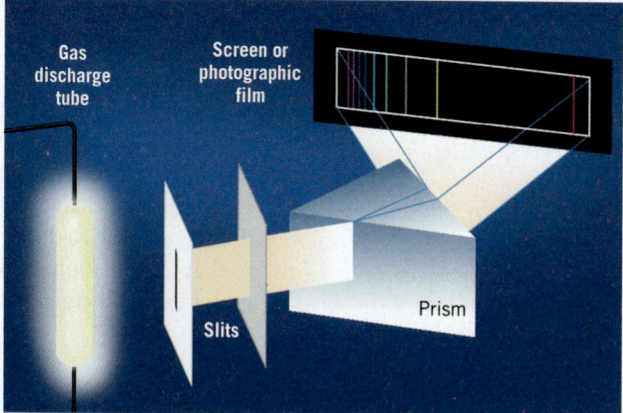

Figure 7.6 | Production and observation of an atomic spectrum. Light emitted by excited atoms is formed into a narrow beam by the slits. It then passes through a prism, which divides the light into narrow beams with frequencies that are characteristic of the particular element that's emitting the light. When these beams fall on a screen, a series of lines is observed, which is why the spectrum is also called a line spectrum.

The Spectrum of Hydrogen

The first success in explaining atomic spectra quantitatively came with the study of the spectrum of hydrogen. This is the simplest element, because its atoms have only one electron, and it produces a simple spectrum with few lines.

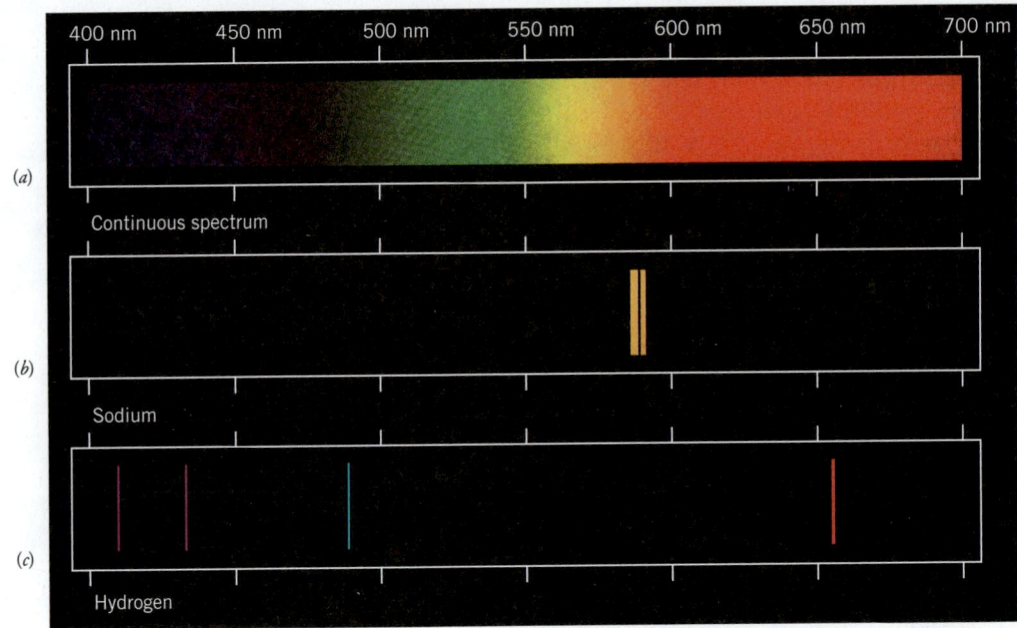

400 nm   450 nm   500 nm   550 nm   600 nm   650 nm   700 nm

(a)

Continuous spectrum

(b)

Sodium

(c)

Hydrogen

Figure 7.7 | **Continuous and atomic emission spectra.** (*a*) The continuous visible spectrum produced by the sun or an incandescent lamp. (*b*) The atomic emission spectrum produced by sodium. The emission spectrum of sodium actually contains more than 90 lines in the visible region. The two brightest lines are shown here. All the others are less than 1% as bright as these. (*c*) The atomic spectrum (line spectrum) produced by hydrogen. There are only four lines in this visible spectrum. They vary in brightness by only a factor of five, so they are all shown.

The Rydberg Equation

The atomic spectrum of hydrogen actually consists of several series of lines. One series is in the visible region of the electromagnetic spectrum and is shown in Figure 7.7. A second series is in the ultraviolet region, and another is in the infrared. In 1885, J. J. Balmer found an equation that was able to give the wavelengths of the lines in the visible portion of the spectrum. This was extended to a more general equation, the **Rydberg equation**, that could be used to calculate the wavelengths of all the spectral lines of hydrogen.

TOOLS

Rydberg equation

$$\frac{1}{\lambda} = R_{\text{H}}\left(\frac{1}{n_1^2} - \frac{1}{n_2^2}\right)$$

The symbol λ stands for the wavelength of the emitted light, R_{H} is a constant (109,678 cm^{-1}), and n_1 and n_2 are variables whose values are positive integers: 1, 2, 3 . . . The only restriction is that the value of n_2 must be larger than n_1. (This ensures that the calculated wavelength has a positive value.) Thus, if $n_1 = 1$, acceptable values of n_2 are 2, 3, 4 . . . . The Rydberg constant, R_{H}, is an empirical constant, which means it was chosen so that the equation gives values for λ that match the ones determined experimentally. The use of the Rydberg equation is straightforward, as illustrated in the following example.

Example 7.3
Calculating the Wavelength of a Line in the Hydrogen Spectrum

The lines in the visible portion of the hydrogen spectrum are called the Balmer series, for which $n_1 = 2$ in the Rydberg equation. Calculate, to four significant figures, the wavelength in nanometers of the spectral line in this series for which $n_2 = 4$.

Analysis: The question states that the Rydberg equation will be used for this calculation.

● **Assembling the Tools:** The Rydberg equation is given above. The values for n_1 and n_2 are given and the unknown is the wavelength, λ. We will also need the SI prefixes from Chapter 1.

● **Solution:** To solve this problem, we substitute values into the Rydberg equation, which will give us $\frac{1}{\lambda}$. Substituting $n_1 = 2$ and $n_2 = 4$ into the Rydberg equation gives

$$\frac{1}{\lambda} = 109{,}678 \text{ cm}^{-1}\left(\frac{1}{2^2} - \frac{1}{4^2}\right)$$

$$\frac{1}{\lambda} = 109{,}678 \text{ cm}^{-1}\left(\frac{1}{4} - \frac{1}{16}\right)$$

$$= 109{,}678 \text{ cm}^{-1}(0.25000 - 0.06250)$$

$$= 109{,}678 \text{ cm}^{-1}(0.18750)$$

$$= 2.05646 \times 10^4 \text{ cm}^{-1}$$

Now that we have $\frac{1}{\lambda}$, taking the reciprocal gives the wavelength in centimeters.

$$\lambda = \frac{1}{2.05646 \times 10^{-1} \text{ cm}^{-1}}$$

$$= 4.86272 \times 10^{-5} \text{ cm}$$

Finally, we convert centimeters to nanometers:

$$\lambda = 4.86272 \times 10^{-5} \text{ cm} \times \frac{10^{-2} \text{ m}}{1 \text{ cm}} \times \frac{1 \text{ nm}}{10^{-9} \text{ m}}$$

$$= 486.272 \text{ nm}$$

The answer rounded to four significant figures is 486.3 nm. Notice that the values for n are exact numbers, and we could have kept all six significant figures.

● **Is the Answer Reasonable?** Besides double-checking the arithmetic, we can note that the wavelength falls within the visible region of the spectrum, 400 to 700 nm. We can also check the answer against the experimental spectrum of hydrogen in Figure 7.7. This wavelength corresponds to the turquoise line in the hydrogen spectrum.

Calculate the wavelength in micrometers, μm, of radiation expected when $n_1 = 4$ and $n_2 = 6$. Report your result to three significant figures. Where does this light come in the electromagnetic spectrum? (*Hint:* Use n_1 and n_2 to calculate the value of the term in parentheses first.)

Practice Exercise 7.3

Calculate the wavelength in nanometers of the spectral line in the visible spectrum of hydrogen for which $n_1 = 2$ and $n_2 = 3$. What color is this line?

Practice Exercise 7.4

The discovery of the Rydberg equation was both exciting and perplexing. The fact that the wavelength of any line in the hydrogen spectrum can be calculated by a simple equation involving just one constant and the reciprocals of the squares of two whole numbers is remarkable. Physicists had to study the behavior of the electron in the atom to account for such simplicity.

7.3 | The Bohr Theory

Quantized Energies of Electrons in Atoms

Earlier you saw that there is a simple relationship between the frequency of light and its energy, $E = h\nu$. Because excited atoms emit light of only certain frequencies, it must be true that only specific energy changes are able to take place within the atoms. For instance,

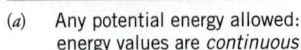

(a) Any potential energy allowed: (b) Potential energy restricted:
     energy values are *continuous*   energy values are *discrete*

Figure 7.8 | **An analogy of quantized states.** (*a*) The rabbit on a slope is free to move to any height less than the height of the hill. Its potential energy can take on any value between the maximum (at the hill top) and the minimum (at the bottom). Similarly, the energy of a free electron can take on any value. (*b*) The rabbit trapped on some steps is found only at one of the three heights: at the bottom (the lowest, or ground state), middle, or top step. Similarly, the energy of the electron trapped inside an atom is restricted to certain values, which correspond to the various energy levels in an atom. Some of the energy levels may be unoccupied.

in the spectrum of hydrogen there is a red line (Figure 7.7) that has a wavelength of 656.4 nm, which is a frequency of 4.567×10^{14} Hz. The energy of a photon of this light is 3.026×10^{-19} J. Whenever a hydrogen atom emits red light, the frequency of the light is always precisely 4.567×10^{14} Hz, and the energy of the atom decreases by exactly 3.026×10^{-19} J, never more and never less. Atomic spectra, then, tell us that *when an excited atom loses energy, not just any arbitrary amount can be lost.* The same is true if the atom gains energy.

The most successful explanation of the emission and absorption of energy is that, in an atom, an electron can have only certain definite amounts of energy and no others. We say that the electron is restricted to certain **energy levels**, and that the energy of the electron is **quantized**.

The energy of an electron in an atom might be compared to the energy of the rabbit shown in Figure 7.8*b*. The rabbit on the steps can be "stable" on only one of the ledges, so it has certain specific amounts of potential energy as determined by the "energy levels" of the various ledges. If the rabbit jumps to a higher ledge, its potential energy is increased. When it hops to a lower ledge, its potential energy decreases. If the rabbit tries to occupy heights between the ledges, it immediately falls to the lower ledge. Therefore, the energy changes for the rabbit are restricted to the differences in potential energy between the ledges.

Similarly, the electron can only have energies corresponding to the set of electron energy levels in the atom. When the atom is supplied with energy (by an electric discharge, for example), an electron is raised from a low-energy level to a higher one. When the electron drops back, energy equal to the difference between the two levels is released and emitted as a photon. Because only certain energy jumps can occur, only certain frequencies of light can appear in the spectrum.

The existence of specific energy levels in atoms, as implied by atomic spectra, forms the foundation of all theories about electronic structure. Any model of the atom that attempts to describe the positions or motions of electrons must also account for atomic spectra.

The Bohr Model of Hydrogen

The first theoretical model of the hydrogen atom that successfully accounted for the Rydberg equation was proposed in 1913 by Niels Bohr (1885–1962), a Danish physicist. In his model, Bohr likened the electron moving around the nucleus to a planet circling the sun. He suggested that the electron moves around the nucleus along fixed paths, or orbits. This model broke with the classical laws of physics by placing restrictions on the sizes of the orbits and the energies that electrons could have in given orbits. This ultimately led Bohr to propose an equation that described the energy of the electron in the atom. The equation includes a number of physical constants such as the mass of the electron, its charge, and Planck's constant. It also contains an integer, n, that Bohr called a **quantum**

■ $E = h\nu$
$h = 6.626 \times 10^{-34}$ J s
$E = (6.626 \times 10^{-34} \text{ J s}) \times$
   $(4.567 \times 10^{14} \text{ s}^{-1}) =$
   3.026×10^{-19} J

■ In a certain sense we can consider that the potential energy of the rabbit on a stair is quantized.

■ Niels Bohr won the 1922 Nobel Prize in physics for his work on his model of the hydrogen atom.

■ Classical physical laws, such as those discovered by Isaac Newton, place no restrictions on the sizes or energies of orbits.

number. Each of the orbits is identified by its value of n. When all the constants are combined, Bohr's equation becomes

$$E = \frac{-b}{n^2} \qquad (7.3)$$

where E is the energy of the electron and b is the combined constant (its value is 2.18×10^{-18} J). The allowed values of n are whole numbers that range from 1 to ∞. From this equation the energy of the electron in any particular orbit could be calculated.

Because of the negative sign in Equation 7.3, the lowest (most negative) energy value occurs when $n = 1$, which corresponds to the *first Bohr orbit*. The lowest energy state of an atom is the most stable one and is called the **ground state**. For hydrogen, the ground state occurs when its electron has $n = 1$. According to Bohr's theory, this orbit brings the electron closest to the nucleus. Conversely, an atom with $n = \infty$ would correspond to an "unbound" electron that had escaped from the nucleus. Such an electron has an energy of zero in Bohr's theory. The negative sign in Equation 7.3 ensures that any electron with a finite value of n has a lower energy than an unbound electron. Thus, energy is released when a free electron is bound to a proton to form a hydrogen atom.

When a hydrogen atom absorbs energy, as it does when an electric discharge passes through it, the electron is raised from the orbit having $n = 1$ to a higher orbit: $n = 2$ or $n = 3$ or even higher. The hydrogen atom is now in an excited state. These higher orbits are less stable than the lower ones, so the electron quickly drops to a lower orbit. When this happens, energy is emitted in the form of light (see Figure 7.9). Since the energy of the electron in a given orbit is fixed, a drop from one particular orbit to another, say, from $n = 2$ to $n = 1$, always releases the same amount of energy, and the frequency of the light emitted is always precisely the same.

The success of Bohr's theory was in its ability to account for the Rydberg equation. When the atom emits a photon, an electron drops from a higher initial energy E_{high} to a lower final energy E_{low}. If the initial quantum number of the electron is n high and the final quantum number is n_{low}, then the energy change, calculated as a positive quantity, is

$$\Delta E = E_{\text{high}} - E_{\text{low}}$$

$$= \left(\frac{-b}{n_{\text{high}}^2}\right) - \left(\frac{-b}{n_{\text{low}}^2}\right)$$

This can be rearranged to give

$$\Delta E = b\left(\frac{1}{n_{\text{high}}^2} - \frac{1}{n_{\text{low}}^2}\right) \qquad \text{with } n_{\text{high}} > n_{\text{low}}$$

By combining Equations 7.1 and 7.2, the relationship between the energy "ΔE" of a photon and its wavelength λ is

$$\Delta E = \frac{hc}{\lambda} = hc\left(\frac{1}{\lambda}\right)$$

Substituting and solving for $1/\lambda$ gives

$$\frac{1}{\lambda} = \frac{b}{hc}\left(\frac{1}{n_{\text{low}}^2} - \frac{1}{n_{\text{high}}^2}\right) \qquad \text{with } n_{\text{high}} > n_{\text{low}}$$

Notice how closely this equation derived from Bohr's theory matches the Rydberg equation, which was obtained solely from the experimentally measured atomic spectrum of hydrogen. Equally satisfying is that the combination of constants, b/hc, has a value of 109,730 cm^{-1}, which differs by only 0.05% from the experimentally derived value of R_{H} in the Rydberg equation.

■ Bohr's equation for the energy actually is $E = \dfrac{2\pi^2 m e^4}{n^2 h^2}$, where m is the mass of the electron, e is the charge on the electron, n is the quantum number, and h is Planck's constant. Therefore, in Equation 7.3, $b = \dfrac{2\pi^2 m e^4}{h^c} = 2.18 \times 10^{-18}$ J.

■ We say $n = 0, 1, 2, 3, \ldots, \infty$.

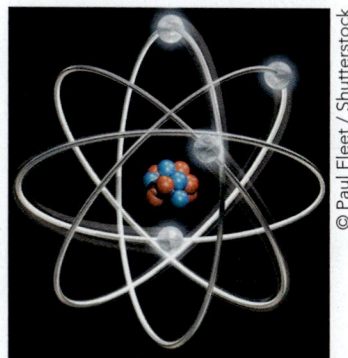

The solar system model of the atom was the way that Niels Bohr's model was presented to the public. Today this simple picture of the atom makes a nice corporate logo, but the idea of an atom with electrons orbiting a nucleus as planets orbit a sun has been replaced by the wave mechanical model.

© Paul Fleet / Shutterstock

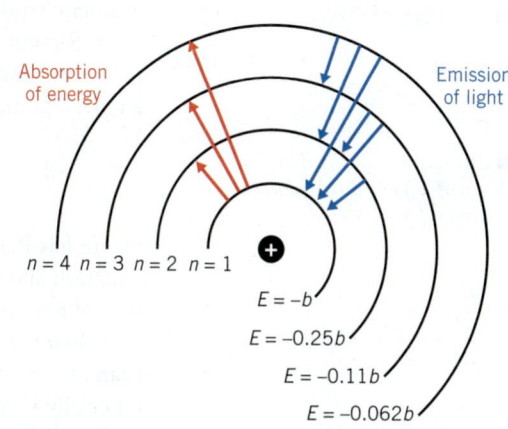

Figure 7.9 | **Absorption of energy and emission of light by the hydrogen atom.** When the atom absorbs energy, the electron is raised to a higher energy level. When the electron falls to a lower energy level, light of a particular energy and frequency is emitted.

Because Bohr envisioned the electron as being in orbit around the nucleus, the quantized energy changes also implied that there were fixed, or quantized, orbits. Angular momentum, a quantity from classical physics, describes circular motion, and Bohr derived equations in which the angular momentum would be multiples of the quantum number n. Using these equations Bohr was able to calculate the radius of the electron orbits in his model. The smallest orbit had a size of 53 pm (picometers). This distance is now known as the **Bohr radius**.

Practice Exercise 7.5

What is the energy of the radiation emitted when an electron in a hydrogen drops from an energy level of $n_{high} = 6$ to $n_{low} = 4$? Report your result to three significant figures. (*Hint:* What equation will allow you to calculate the energy of a light emitted from an excited electron in a hydrogen atom?)

Practice Exercise 7.6

If the energy of light emitted from a hydrogen atom as an electron moves from an orbit of $n = 4$ to another orbit is 2.04×10^{-18} J, what energy level is the electron moving to?

Failure of the Bohr Model

Bohr's model of the atom was both a success and a failure. By calculating the energy changes that occur between energy levels, Bohr was able to account for the Rydberg equation and, therefore, for the atomic spectrum of hydrogen. However, the theory was not able to explain quantitatively the spectra of atoms with more than one electron, and all attempts to modify the theory to make it work failed. It became clear that Bohr's picture of the atom was flawed and that another theory would have to be found. Nevertheless, the concepts of quantum numbers and fixed energy levels were important steps forward.

7.4 | The Wave Mechanical Model

Bohr's efforts to develop a theory of electronic structure were doomed from the very beginning because the classical laws of physics—those known in his day—simply do not apply to objects as small as the electron. Classical physics fails for atomic particles because matter at the atomic level is below the limit that our physical senses can perceive. When physicists were proposing that photons (particles) and the waves of electromagnetic radiation were the same thing, it became apparent that particles such as electrons may behave as waves. This idea was proposed in 1924 by a young French graduate student, Louis de Broglie.

In Section 7.1 we discussed that light waves are characterized by their wavelengths and their frequencies. The same is true of matter waves. De Broglie suggested that the wavelength of a matter wave, λ, is given by the equation

$$\lambda = \frac{h}{mv} \tag{7.4}$$

■ All the objects that had been studied by scientists up until the time of Bohr's model of the hydrogen atom were large and massive in comparison with the electron, so no one had detected the limits of classical physics.

■ De Broglie was awarded the Nobel Prize in physics in 1929.

■ De Broglie originally proposed that $mv^2 = hc/\lambda$. This becomes Equation 7.4.

where h is Planck's constant, m is the particle's mass, and v is its velocity. Notice that this equation allows us to connect a wave property, wavelength, with mass, which is characteristic of a particle, allowing us to describe the electron as either a particle or a wave.

When first encountered, the concept of a particle of matter behaving as a wave rather than as a solid object is difficult to comprehend. This book certainly seems solid enough, especially if you drop it on your toe! The reason for the book's apparent solidity is that in de Broglie's equation (Equation 7.4) the mass appears in the denominator. This means that heavy objects have extremely short wavelengths. The peaks of the matter waves for heavy objects are so close together that the wave properties go unnoticed and can't even be measured experimentally. But tiny particles with very small masses have much longer wavelengths, so their wave properties become an important part of their overall behavior.

Electron Diffraction and Wave Properties of Electrons

Perhaps by now you've begun to wonder whether there is any way to prove that matter has wave properties. Actually, these properties can be demonstrated by a phenomenon that you have probably witnessed. When raindrops fall on a quiet pond, ripples spread out from where the drops strike the water, as shown in Figure 7.10. When two sets of ripples cross, there are places where the waves are *in phase*, which means that the peak of one wave coincides with the peak of the other. At these points the amplitudes of the waves add and the height of the water is equal to the sum of the heights of the two crossing waves. At other places the crossing waves are out of phase, which means the peak of one wave occurs at the trough of the other. In these places the amplitudes of the waves cancel. This reinforcement and cancellation of wave amplitudes, referred to, respectively, as *constructive* and *destructive interference*, is a phenomenon called **diffraction**. It is examined more closely in Figure 7.11. Note how diffraction creates characteristic **interference fringes** when waves pass through adjacent pinholes or reflect off closely spaced grooves. You have seen interference fringes yourself if you've ever noticed the rainbow of colors that shine from the surface of a DVD (Figure 7.12). When white light, which contains all the visible wavelengths, is reflected from the closely spaced "bumps" on the disk, it is divided into many individual light beams. For a given angle between the incoming and reflected light, the light waves experience interference with each other for all wavelengths (colors) except for one wavelength, which is reinforced. We see the wavelength of light that is reinforced, and as the angle changes the wavelengths that are reinforced change. The result is a rainbow of colors reflected from the DVD.

Diffraction is a phenomenon that can be explained only as a property of waves, and we have seen how it can be demonstrated with water waves and light waves. Experiments can also be done to show that electrons, protons, and neutrons experience diffraction, which demonstrates their wave nature (see Figure 7.13). In fact, electron diffraction is the principle on which the electron microscope is based. (See Chemistry and Current Affairs 7.1.)

Figure 7.10 | **Diffraction of water waves on the surface of a pond.** As the waves cross, the amplitudes increase where the waves are in phase and cancel where they are out of phase.

■ Gigantic waves, called rogue waves, with heights up to 100 ft have been observed in the ocean and are believed to be formed when a number of wave sets moving across the sea become in-phase simultaneously.

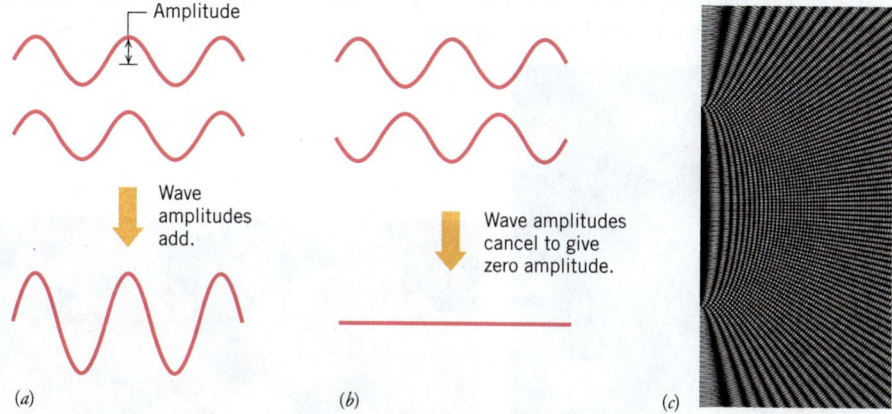

(a) Waves in phase produce constructive interference and an increase in intensity. *(b)* Waves out of phase produce destructive interference that results in the cancellation of intensity. *(c)* Waves passing through two pinholes fan out and interfere with each other, producing an interference pattern characteristic of waves.

Figure 7.11 | **Constructive and destructive interference.**

Figure 7.12 | **Diffraction of light from a DVD.** Colored interference fringes are produced by the diffraction of reflected light from the closely spaced spiral of "bumps" on the thin inner aluminum surface of a compact disc.

Figure 7.13 | Experimental evidence of wave behavior in electrons. Electrons passing one at a time through a double slit. Each spot shows an electron impact on a detector. As more and more electrons are passed through the slits, interference fringes are observed.

Electrons passing through double slit:

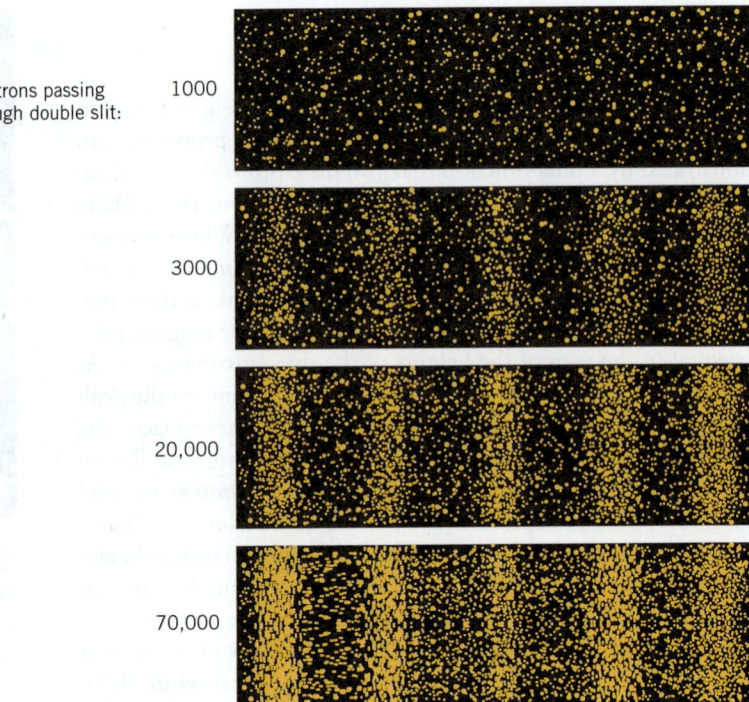

1000

3000

20,000

70,000

CHEMISTRY AND CURRENT AFFAIRS 7.1

The Electron Microscope

The usefulness of a microscope in studying small specimens is limited by its ability to distinguish between closely spaced objects. We call this ability the resolving power of the microscope. Through optics, it is possible to increase the magnification and thereby increase the resolving power, but only within limits. These limits depend on the wavelength of the light that is used. Objects with diameters less than the wavelength of the light cannot be seen in detail. Since the smallest wavelength of visible light is about 400 nm, objects smaller than this can't be seen clearly with a microscope that uses visible light.

The electron microscope uses electron waves to "see" very small objects. De Broglie's equation, $\lambda = h/mv$, suggests that if an electron, proton, or neutron has a very high velocity, its wavelength will be very small. In the electron microscope, high-voltage electrodes accelerate electrons. This gives electron waves with typical wavelengths of about 2 pm to 12 pm that strike the sample and are then focused magnetically (using "magnetic lenses") onto a fluorescent screen where they form a visible image. Because of certain difficulties, the actual resolving power of the instrument is quite a bit less than the wavelength of the electron waves—generally on the order of 1 to 6 nm. Some high-resolution electron microscopes, however, are able to reveal the shadows of individual atoms in very thin specimens through which the electron beam passes.

Brand X/Superstock

(a)

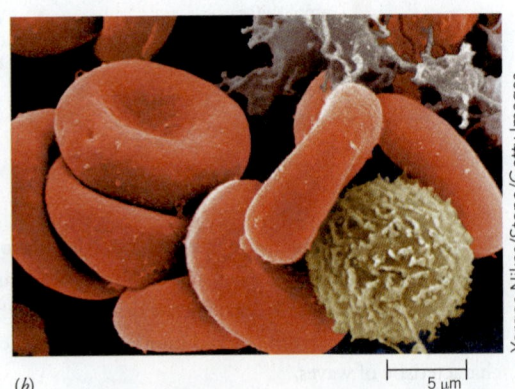

Yorgos Nikas/Stone/Getty Images

(b)

5 µm

Figure 1 A modern electron microscope (a) operated by a trained technician is used to obtain (b) electron micrographs such as the one shown here depicting red and white blood cells.

Quantized Energy of Bound Electrons

Standing Waves and Quantum Numbers

Before we can discuss how electron waves behave in atoms, we need to know a little more about waves in general. There are basically two kinds of waves, traveling waves and standing waves. On a lake or ocean the wind produces waves whose crests and troughs move across the water's surface, as shown in Figure 7.14. The water moves up and down while the crests and troughs travel horizontally in the direction of the wind. These are examples of **traveling waves**.

A more important kind of wave for us is the standing wave. An example is the vibrating string of a guitar. When the string is plucked, its center vibrates up and down while the ends, of course, remain fixed. The crest, or point of maximum amplitude of the wave, occurs at one position. At the ends of the string are points of zero amplitude, called **nodes**, and their positions are also fixed. A **standing wave**, then, is one in which the crests and nodes do not change position. One of the interesting things about standing waves is that they lead naturally to "quantum numbers." Let's see how this works using the guitar as an example.

As you know, many notes can be played on a guitar string by shortening its effective length with a finger placed between frets along the neck of the instrument. But even without shortening the string, we can play a variety of notes. For instance, if the string is touched momentarily at its midpoint at the same time it is plucked, the string vibrates as shown in Figure 7.15 and produces a tone an octave higher. The wave that produces this higher tone has a wavelength exactly half of that formed when the untouched string is plucked. In Figure 7.15 we see that other wavelengths are possible, too, and each gives a different note.

If you examine Figure 7.15, you will see that there are some restrictions on the wavelengths that can exist. Not just any wavelength is possible, because the nodes at either end of the string are in fixed positions. The only waves that can occur are those for which a half-wavelength is repeated *exactly* a whole number of times. Expressed another way, the length of the string is a whole-number multiple of half-wavelengths. In a mathematical form we could write this as

$$L = n\left(\frac{\lambda}{2}\right) \tag{7.5}$$

where L is the length of the string, λ is a wavelength (therefore, $\lambda/2$ is half the wavelength), and n is an integer. Rearranging this to solve for the wavelength gives

$$\lambda = \frac{2L}{n}$$

We see that the waves that are possible are determined quite naturally by a set of whole numbers (similar to quantum numbers).

We are now in a position to demonstrate how quantum theory unites wave and particle descriptions to build a simple but accurate model of a bound electron. Let's look at an electron that is confined to a wire of length L. To start, we will assume that the wire is infinitely thin, so that the electron can only move in straight lines along the wire. The wire is clamped in place at either end, and its ends cannot move up or down.

Figure 7.14 | **Traveling waves.** Waves breaking on this beach are the result of traveling waves from out in the ocean.

© Gallo Images/Alamy

■ The wavelength of this wave is actually twice the length of the string.

■ Notes played this way are called harmonics.

SuperStock/Age Fotostock America, Inc.

Notes played on a guitar rely on standing waves. The ends of the strings correspond to nodes of the standing waves. Different notes can be played by shortening the effective lengths of the strings with fingers placed along the neck of the instrument.

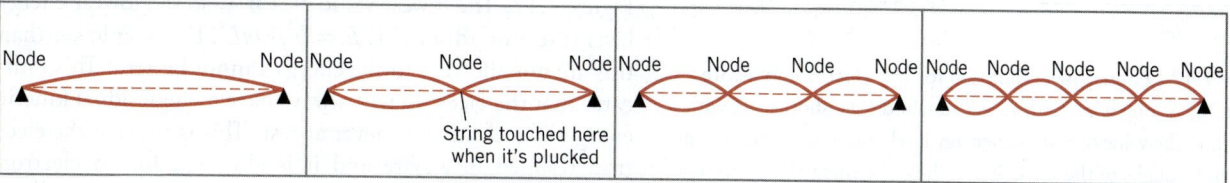

Figure 7.15 | **Standing waves on a guitar string.** Panes show 1, 2, 3, and 4 half-waves, respectively. The number of nodes is one more than the number of half-waves.

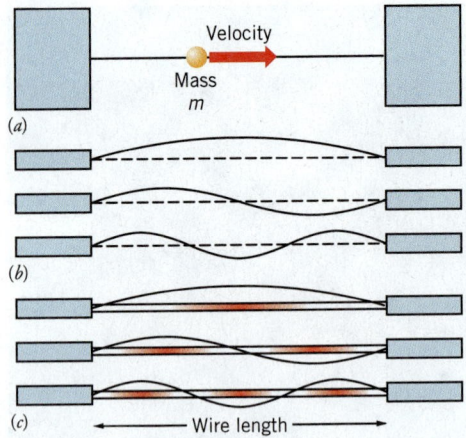

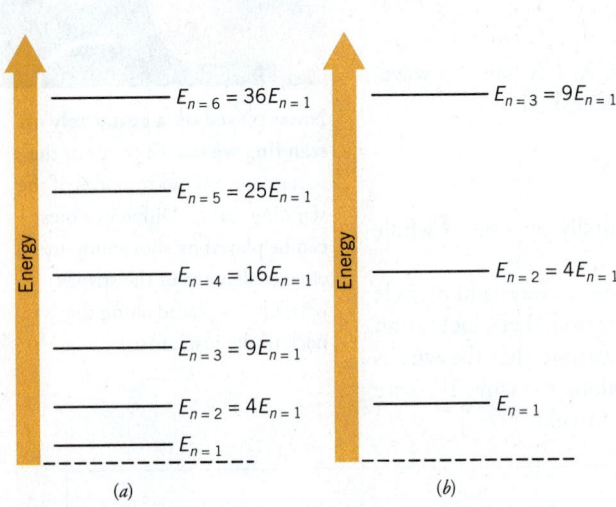

Figure 7.16 | Three models of an electron on an infinitely thin wire of length L. (*a*) In the classical model, the bead can have any velocity or position. (*b*) Classical waves like a guitar string. (*c*) Quantum mechanical model, where wave amplitude is related to the probability of the bead's position.

First, let's consider a classical particle model: the "bead on a wire" model shown in Figure 7.16*a*. The bead can slide in either direction along the wire, like a bead on an abacus. If the bead's mass is *m* and its velocity is *v*, its kinetic energy is given by

$$E = \frac{1}{2}mv^2$$

The bead can have any velocity, even zero, so the energy *E* can have any value, even zero. No position on the wire is any more favorable than any other, and the bead is equally likely to be found anywhere on the wire. There is no reason why the bead's position and velocity cannot be known simultaneously.

Now consider a classical wave along the wire, Figure 7.16*b*. It is exactly like the guitar string we looked at in Figure 7.15. The ends of the wire are clamped in place, so there must be a whole number of peaks and troughs along the wire. The wavelength is restricted to values calculated by Equation 7.5. We can see that the quantum number, *n*, is just the number of peaks and troughs along the wave. It has integer values (1, 2, 3, . . .) because you can't have half a peak or half a trough.

Now let's use the de Broglie relation, Equation 7.4, to unite these two classical models. Our goal is to derive an expression for the energy of an electron trapped on the wire (Figure 7.16*c*). The de Broglie relation lets us relate particle velocities with wavelengths. Rearranging Equation 7.4, we have

$$v = \left(\frac{h}{m\lambda}\right)$$

and substituting this equation for velocity in the equation for kinetic energy gives

$$E = \frac{1}{2}mv^2 = \frac{1}{2}m\left(\frac{h}{m\lambda}\right)^2$$

$$= \frac{h^2}{2m\lambda^2}$$

This equation will give us the energy of the electron from its wavelength. We can replace λ with its equivalent values shown in Equation 7.5, which gives the wavelength of the standing wave in terms of the wire length *L* and the quantum number *n*. The result is

$$E = \left(\frac{h^2}{2m\left(\frac{2L}{n}\right)^2}\right)$$

$$= \frac{n^2h^2}{8mL^2} \qquad (7.6)$$

This equation has a number of profound implications. The fact that the electron's energy depends on an integer, *n*, means that *only certain energy states are allowed*. The allowed states are plotted on the energy level diagram shown in Figure 7.17. The lowest value of *n* is 1, so the lowest energy level (the ground state) is $E = h^2/8mL^2$. Energies lower than this are not allowed, so the energy cannot be zero! This indicates that the electron will always have some residual kinetic energy. The electron is never at rest. This is true for the electron trapped in a wire and it is also true for an electron trapped in an atom. Thus, *quantum theory resolves the collapsing atom paradox.*

Figure 7.17 | Energy level diagram for the electron-on-a-wire model. (*a*) A long wire, with *L* = 2 nm, and (*b*) a short wire, with *L* = 1 nm. The value of *n* at each energy level indicates the number of peaks and troughs of the wave. Notice how the energy levels become more closely spaced when the electron has more room to move. Notice also that the energy in the ground state (E_n = 1) is not zero.

Note that the spacing between energy levels is proportional to $1/L^2$. This means that when the wire is made longer, the energy levels become more closely spaced. In general, *the more room an electron has to move in, the smaller the spacings between its energy levels.* Different molecules have different ways to confine their electrons. This means that the different molecules will absorb light at different wavelengths, thus different substances can have different colors.

Calculate the energy of an electron in its lowest energy state trapped on a one-dimensional wire with a length of 1 nm. Repeat the calculation using a length of 2 nm. (*Hint:* What is the value of n for a ground state?)

Practice Exercise 7.7

If an electron moves from a one-dimensional wire with a length of 1 nm at $n = 3$ to $n = 2$, what wavelength of light could be emitted by this energy change?

Practice Exercise 7.8

Also note that as the energy increases, the spacing between the levels are closer together. At a high enough energy, the energy differences become negligible, and then electrons have enough energy and are actually removed from the atom. This relates to the ultraviolet catastrophe in that the energy of light released when an electron drops from a high energy state to a lower energy state reaches a limit, the limit of removing the electron from the atom. The ultraviolet catastrophe is now resolved.

Wave Function

A **wave function** (symbolized by the symbol ψ, the Greek letter psi) is the quantum mechanical description of an electron. This mathematical construct, based on the quantum numbers, can be used to describe the shape of the electron wave and its energy. The wave function is not an oscillation of the wire, like a guitar wave, nor is it an electromagnetic wave. The wave's amplitude at any given point can be related to the probability of finding the electron there.

The electron waves shown in Figure 7.16c show that unlike the bead-on-a-wire model, the electron is more likely to be found at some places on the wire than others. For the ground state, with $n = 1$, the electron is most likely to be found in the center of the wire. Where the amplitude is zero, such as at the ends of the wire or the center of the wire in the $n = 2$ state, there is a zero probability of finding the electron! Points where the amplitude of the electron wave is zero are called nodes. Notice that the higher the quantum number n, the more nodes the electron wave has, and from Equation 7.6, the more energy the electron has. It is generally true that *the more nodes an electron wave has, the higher its energy.*

The model of the atom has changed over time, from the Greeks with the indivisible particle to the structure of the nucleus developed in the late 19th and early 20th centuries. Now, the theory has been modified to describe the electrons as waves as well as particles.

■ The probability of finding an electron at a given point is proportional to the amplitude of the electron wave squared. Thus, peaks and troughs in the electron wave indicate places where there is the greatest buildup of negative charge.

Models of the Atom

Model	Approx. Date
Indivisible particle	ca 400 BC
Billiard ball elements	1803
Plum pudding model	1897
Nuclear model	1910
Solar system model	1913
Wave mechanical model	1926

7.5 | Quantum Numbers of Electrons in Atoms

In 1926 Erwin Schrödinger (1887–1961), an Austrian physicist, became the first scientist to successfully apply the concept of the wave nature of matter to an explanation of electronic structure. His work and the theory that developed from it are highly mathematical. In this book, we will use a qualitative description of electronic structure, and the main points of the theory can be understood without all the math.

■ Schrödinger won the Nobel Prize in physics in 1933 for his work. The equation he developed that gives electronic wave functions and energies is known as *Schrödinger's equation.*

Electron Waves in Atoms Are Called Orbitals

Schrödinger developed an equation that can be solved to give wave functions and energy levels for electrons trapped inside atoms. Wave functions for electrons in atoms are called **orbitals**. Not all of the energies of the waves are different, but most are. *Energy changes within an atom are simply the result of an electron changing from a wave pattern with one energy to a wave pattern with a different energy.*

We will be interested in two properties of orbitals, their energies and their shapes. Their energies are important because when an atom is in its most stable state, or *ground state*, the atom's electrons have waveforms with the lowest possible energies. The shapes of the wave patterns (i.e., where their amplitudes are large and where they are small) are important because the theory tells us that the amplitude of a wave at any particular place is related to the likelihood of finding the electron there. This will be important when we study how and why atoms form chemical bonds to each other.

In much the same way that the characteristics of a wave on a one-dimensional guitar string can be related to a single integer, wave mechanics tells us that the three-dimensional electron waves (orbitals) can be characterized by a set of *three* integer quantum numbers, n, ℓ, and m_ℓ. In discussing the energies of the orbitals, it is usually most convenient to sort the orbitals into groups according to these quantum numbers.

The Principal Quantum Number, n

The quantum number n is called the **principal quantum number**, and all orbitals that have the same value of n are said to be in the same **shell.** The values of n can range from $n = 1$ to $n = \infty$. The shell with $n = 1$ is called the *first shell*, the shell with $n = 2$ is the *second shell*, and so forth. The various shells are also sometimes identified alphabetically, beginning (for no significant reason) with K for the first shell ($n = 1$).

The principal quantum number is related to the size of the electron wave (i.e., how far the wave effectively extends from the nucleus). The higher the value of n, the larger is the electron's average distance from the nucleus. This quantum number is also related to the energy of the orbital. As n increases, the energies of the orbitals also increase.

Bohr's theory took into account only the principal quantum number n. His theory worked fine for hydrogen because hydrogen just happens to be the one element in which all orbitals having the same value of n also have the same energy. Bohr's theory failed for atoms other than hydrogen, however, because when the atom has more than one electron, orbitals with the same value of n can have different energies.

The Secondary Quantum Number, ℓ

The **secondary quantum number,** ℓ, divides the shells into smaller groups of orbitals called **subshells**. The value of n determines which values of ℓ are allowed. For a given n, ℓ can range from $\ell = 0$ to $\ell = (n - 1)$. Thus, when $n = 1$, $(n - 1) = 0$, so the only value of ℓ that's allowed is zero. This means that when $n = 1$, there is only one subshell (the shell and subshell are really identical). When $n = 2$, ℓ can have a value of 0 or 1. (The maximum value of $\ell = n - 1 = 2 - 1 = 1$.) That is, when $n = 2$, there are two subshells: $n = 2$ and $\ell = 0$, and $n = 2$ and $\ell = 1$. The relationship between n and the allowed values of ℓ are summarized in the table in the margin.

Subshells could be identified by the numerical value of ℓ. However, to avoid confusing numerical values of n with those of ℓ, a letter code is normally used to specify the value of ℓ.

Value of ℓ	0	1	2	3	4	5	. . .
Letter designations	*s*	*p*	*d*	*f*	*g*	*h*	. . .

To designate a particular subshell, we write the value of its principal quantum number followed by the letter code for the subshell. For example, the subshell with $n = 2$ and $\ell = 1$ is the 2*p* subshell; the subshell with $n = 4$ and $\ell = 0$ is the 4*s* subshell. Notice that because of the relationship between n and ℓ, every shell has an *s* subshell (1*s*, 2*s*, 3*s*, etc.). All the

■ Electron waves are described by the term *orbital* to differentiate them from the notion of *orbits*, which was part of the Bohr model of the atom.

■ "Most stable" almost always means "lowest energy."

■ The term shell comes from an early notion that electrons are arranged in atoms in a manner similar to the layers, or shells, found within an onion.

■ ℓ is also called the **azimuthal quantum number** and the **orbital angular momentum number**.

■ The number of subshells in a given shell equals the value of n for that shell. For example, when $n = 3$, there are three subshells.

Relationship between n and ℓ

Value of n	Value of ℓ
1	0
2	0, 1
3	0, 1, 2
4	0, 1, 2, 3
5	0, 1, 2, 3, 4
n	0, 1, 2, . . . $(n - 1)$

shells except the first have a *p* subshell (2*p*, 3*p*, 4*p*, etc.). All but the first and second shells have a *d* subshell (3*d*, 4*d*, etc.); and so forth.

What are the values of *n* and ℓ for the following subshells? (a) 4*d*, (b) 5*f*, and (c) 7*s* (*Hint:* How are the subshell designations related to the value of ℓ?)

Practice Exercise 7.9

What subshells would be found in the shells with *n* = 2 and *n* = 5? How is the number of subshells related to *n*?

Practice Exercise 7.10

The secondary quantum number determines the shape of the orbital, which we will examine more closely later. Except for the special case of hydrogen, which has only one electron, the value of ℓ also indicates the relative energies of the orbitals in a shell. This means that in atoms with two or more electrons, the subshells within a given shell differ slightly in energy, with the energy of the subshell increasing with increasing ℓ. Therefore, within a given shell, the *s* subshell is lowest in energy, *p* is the next lowest, followed by *d*, then *f*, and so on. For example,

$$4s < 4p < 4d < 4f$$

— increasing energy →

The Magnetic Quantum Number, m_ℓ

The third quantum number, m_ℓ, is known as the **magnetic quantum number.** It divides the subshells into individual orbitals, and its values are related to the way the individual orbitals are oriented relative to each other in space. As with ℓ, there are restrictions as to the possible values of m_ℓ, which can range from $+\ell$ to $-\ell$. When $\ell = 0$, m_ℓ can have only the value 0 because +0 and −0 are the same. An *s* subshell, then, has just a single orbital. When $\ell = 1$, the possible values of m_ℓ are +1, 0, and −1. A *p* subshell therefore has three orbitals: one with $\ell = 1$ and $m_\ell = 1$, another with $\ell = 1$ and $m_\ell = 0$, and a third with $\ell = 1$ and $m_\ell = -1$. Similarly, we find that a *d* subshell has five orbitals and an *f* subshell has seven orbitals. The numbers of orbitals in the subshells are easy to remember because they follow a simple arithmetic progression.

$$\begin{array}{cccc} s & p & d & f & \cdots \\ 1 & 3 & 5 & 7 & \cdots \end{array}$$

■ Spectroscopists used m_ℓ to explain additional lines that appear in atomic spectra when atoms emit light while in a magnetic field. This explains how this quantum number got its name.

The Whole Picture

The relationships among all three quantum numbers are summarized in Table 7.1. In addition, the relative energies of the subshells in an atom containing two or more electrons

TABLE 7.1	Summary of Relationships among the Quantum Numbers *n*, ℓ, m_ℓ			
Value of *n*	**Value of ℓ**	**Value of m_ℓ**	**Subshell**	**Number of Orbitals**
1	0	0	1*s*	1
2	0	0	2*s*	1
	1	−1, 0, 1	2*p*	3
3	0	0	3*s*	1
	1	−1, 0, 1	3*p*	3
	2	−2, −1, 0, 1, 2	3*d*	5
4	0	0	4*s*	1
	1	−1, 0, 1	4*p*	3
	2	−2, −1, 0, 1, 2	4*d*	5
	3	−3, −2, −1, 0, 1, 2, 3,	4*f*	7

■ The origins of these letters come from spectroscopists description of the lines they observed.
  s = sharp
  p = principal
  d = diffuse

Figure 7.18 | Approximate energy level diagram for atoms with two or more electrons. The quantum numbers associated with the orbitals in the first two shells are also shown.

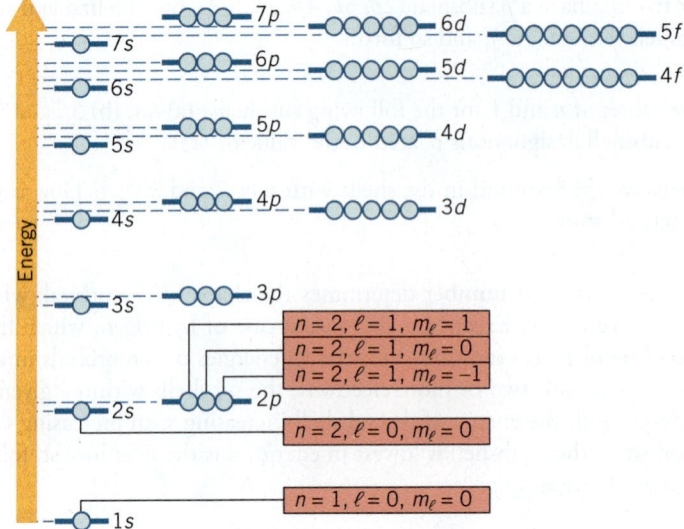

■ In an atom, an electron can take on many different energies and wave shapes, each of which is called an orbital that is identified by a set of values for n, ℓ, and m_ℓ. When the electron wave possesses a given set of n, ℓ, and m_ℓ, we say the electron "occupies the orbital" with that set of quantum numbers.

are depicted in Figure 7.18. Several important features should be noted. First, observe that each orbital on this energy diagram is indicated by a separate circle—one for an s subshell, three for a p subshell, and so forth. Second, notice that all the orbitals of a given subshell have the *same* energy. Third, note that, in going upward on the energy scale, the spacing between successive shells decreases as the number of subshells increases. This leads to the overlapping of shells having different values of n. For instance, the $4s$ subshell is lower in energy than the $3d$ subshell, $5s$ is lower than $4d$, and $6s$ is lower than $5d$.

We will see shortly that Figure 7.18 is very useful for predicting the electronic structures of atoms. Before discussing this, however, we must study another very important property of the electron, a property called spin. Electron spin gives rise to a fourth quantum number.

7.6 | Electron Spin

Recall from Section 7.3 that an atom is in its most stable state (its ground state) when its electrons have the lowest possible energies. This occurs when the electrons "occupy" the lowest energy orbitals that are available. But what determines how the electrons "fill" these orbitals? Fortunately, there are some simple rules that govern both the maximum number of electrons that can be in a particular orbital and how orbitals with the same energy become filled. One important factor that influences the distribution of electrons is the phenomenon known as *electron spin*.

Spin Quantum Number

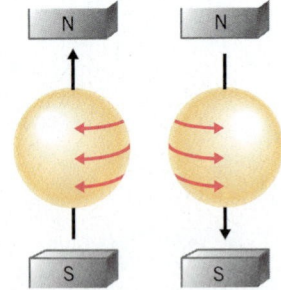

The electron can spin in either of two directions in the presence of an external magnetic field. Without a magnetic field the electrons and their spins are usually oriented in random directions.

■ Electrons don't actually spin, but it is useful to picture the electron as spinning.

When a beam of atoms with an odd number of electrons is passed through an uneven magnetic field, the beam is split in two, as shown in Figure 7.19. The splitting occurs because the electrons within the atoms interact with the magnetic field in two different ways. The electrons behave like tiny magnets, and they are attracted to one or the other of the poles depending on their orientation. This can be explained by imagining that an electron spins around its axis, like a toy top. A moving charge creates a moving electric field, which in turn creates a magnetic field. The spinning electrical charge of the electron creates its own magnetic field. This **electron spin** could occur in two possible directions, which accounts for the two beams.

Electron spin gives us a fourth quantum number for the electron, called the **spin quantum number,** m_s, which can take on two possible values: $m_s = +\frac{1}{2}$ or $m_s = -\frac{1}{2}$, corresponding to the two beams in Figure 7.19. The actual values of m_s and the reason

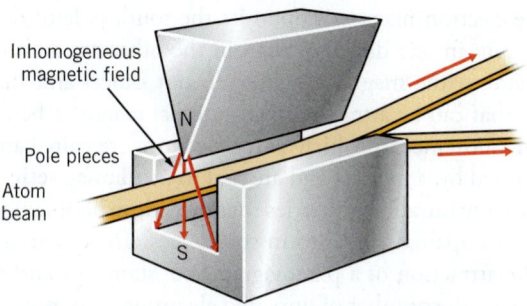

Inhomogeneous
magnetic field

Pole pieces

Atom
beam

N

S

Figure 7.19 | **The discovery of electron spin.** In this classic experiment by Stern and Gerlach, a beam of atoms with an odd number of electrons is passed through an uneven magnetic field, created by magnet pole faces of different shapes. The beam splits in two, indicating that the electrons in the atoms behave as tiny magnets, which are attracted to one or the other of the poles depending on their orientation. Electron spin was proposed to account for the two possible orientations for the electron's magnetic field.

they are not integers aren't very important to us, but the fact that there are *only* two values is very significant.

Pauli Exclusion Principle

In 1925 an Austrian physicist, Wolfgang Pauli (1900–1958), expressed the importance of electron spin in determining electronic structure. The **Pauli exclusion principle** states that *no two electrons in the same atom can have identical values for all four of their quantum numbers.* Suppose two electrons were to occupy the $1s$ orbital of an atom. Each electron would have $n = 1$, $\ell = 0$, and $m_\ell = 0$. Since these three quantum numbers are the same for both electrons, the exclusion principle requires that their fourth quantum numbers (the spin quantum numbers) be different; one electron must have $m_s = +\frac{1}{2}$ and the other, $m_s = -\frac{1}{2}$. No more than two electrons can occupy the $1s$ orbital of the atom simultaneously because there are only two possible values of m_s. Thus, the Pauli exclusion principle is really telling us that when *the maximum number of electrons, two, are in the same orbital, they must have opposite spins.*

The limit of two electrons per orbital also limits the maximum electron populations of the shells and subshells. For the subshells we have

Subshell	Number of Orbitals	Maximum Number of Electrons
s	1	2
p	3	6
d	5	10
f	7	14

■ Pauli received the 1945 Nobel Prize in physics for his discovery of the exclusion principle.

The maximum electron population per shell is $2n^2$:

Shell	Subshells	Maximum Shell Population	
1	$1s$	2	
2	$2s2p$	8	$(2 + 6)$
3	$3s3p3d$	18	$(2 + 6 + 10)$
4	$4s4p4d4f$	32	$(2 + 6 + 10 + 14)$

■ Remember that a shell is a group of orbitals with the same value of n. A subshell is a group of orbitals with the same values of n and ℓ.

Without looking at the tables in the text, how many electrons are there in a d subshell? (*Hint:* How many orbitals are in the d subshell?)

Practice Exercise 7.11

Determine the maximum number of electrons that are in the $n = 5$ shell.

Practice Exercise 7.12

Paramagnetism and Diamagnetism

We have seen that when two electrons occupy the same orbital they must have different values of m_s, and we say that the spins of the electrons are *paired,* or simply that the electrons are *paired.* Such pairing leads to the cancellation of the magnetic effects of the

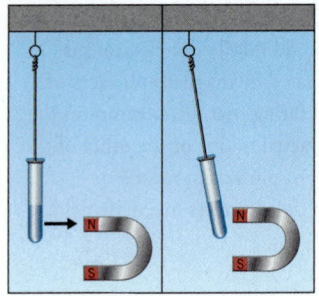

A paramagnetic substance is attracted to a magnetic field.

electrons because the north pole of one electron magnet is opposite the south pole of the other. Atoms with more electrons that spin in one direction than in the other are said to contain *unpaired* electrons. For these atoms, the magnetic effects do not cancel and the atoms themselves become tiny magnets that can be attracted to an external magnetic field. This attraction is called **paramagnetism**. Substances in which all the electrons are paired are not attracted to, but actually weakly repelled by, a magnet and are said to be **diamagnetic**.

Paramagnetism and diamagnetism are measurable properties that provide experimental verification of the presence or absence of unpaired electrons in substances. The quantitative measurement of the strength of the attraction of a paramagnetic substance toward a magnetic field makes it possible to calculate the number of unpaired electrons in its atoms, molecules, or ions.

7.7 | Energy Levels and Ground State Electron Configurations

The distribution of electrons among the orbitals of an atom is called the atom's **electronic structure** or **electron configuration**. This is very useful information about an element because the arrangement of electrons in the outer parts of an atom, which is determined by its electron configuration, controls the chemical properties of the element.

We are interested in the ground state electron configurations of the elements. This is the configuration that yields the lowest energy for an atom and can be predicted for many of the elements by the use of the energy level diagram in Figure 7.18 and application of the Pauli exclusion principle. To see how we go about this, let's begin with the simplest atom of all, hydrogen.

Hydrogen has an atomic number, Z, equal to 1, so a neutral hydrogen atom has one electron. In its ground state this electron occupies the lowest energy orbital that's available, which is the $1s$ orbital. To indicate symbolically the electron configuration we list the subshells that contain electrons and indicate their electron populations by appropriate superscripts. Thus, the electron configuration of hydrogen is written as

$$\text{H} \qquad 1s^1$$

Another way of expressing electron configurations that we will sometimes find useful is the **orbital diagram**. In it, a circle will represent each orbital and half-headed arrows will be used to indicate the individual electrons, head up for spin in one direction and head down for spin in the other. The orbital diagram for hydrogen is simply

$$\text{H} \qquad \underset{1s}{\textcircled{\uparrow}}$$

■ Sometimes instead of a circle, a line is used to represent an orbital:

$$\text{H} \quad \underset{1s}{\underline{\uparrow}}$$

Ground State Electron Configurations

To arrive at the electron configuration of an atom of another element, we imagine that we begin with a hydrogen atom and then add one proton after another (plus whatever neutrons are also needed) until we obtain the nucleus of the atom of interest. We also add electrons, one at a time to the lowest available orbital, until we have added enough electrons to give the neutral atom of the element. This process for obtaining the electronic structure of an atom is known as the **aufbau principle**.

■ The word *aufbau* is German for *building up*.

How s Orbitals Fill

Let's look at the way this works for helium, which has $Z = 2$. This atom has two electrons, both of which occupy the $1s$ orbital. The electron configuration of helium can therefore be written as

$$\text{He} \quad 1s^2 \qquad \text{or} \qquad \text{He} \quad \underset{1s}{\textcircled{\uparrow\downarrow}}$$

Notice that the orbital diagram shows that both electrons in the $1s$ orbital are paired.

We can proceed in the same fashion to predict the electron configurations of most of the elements in the periodic table. For example, the next two elements in the table are lithium, Li ($Z = 3$), and beryllium, Be ($Z = 4$), which have three and four electrons, respectively. For each of these, the first two electrons enter the $1s$ orbital with their spins paired. The Pauli exclusion principle tells us that the $1s$ subshell is filled with two electrons, and Figure 7.18 shows that the orbital of next lowest energy is the $2s$, which can also hold up to two electrons. Therefore, the third electron of lithium and the third and fourth electrons of beryllium enter the $2s$. We can represent the electronic structures of lithium and beryllium as

Li $1s^2 2s^1$ or   Li   ⑪ ①
                                          $1s$ $2s$

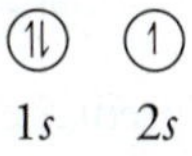

Be $1s^2 2s^2$ or   Be   ⑪ ⑪
                                          $1s$ $2s$

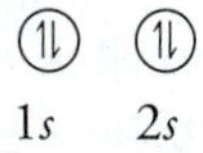

Filling *p* and *d* Orbitals

After beryllium comes boron, B ($Z = 5$). Referring to Figure 7.18, we see that the first four electrons of this atom complete the $1s$ and $2s$ subshells, so the fifth electron must be placed into the $2p$ subshell.

B $1s^2 2s^2 2p^1$

In the orbital diagram for boron, the fifth electron can be put into any one of the $2p$ orbitals—which one doesn't matter because they are all of equal energy.

B   ⑪   ⑪   ①○○
      $1s$ $2s$ $2p$

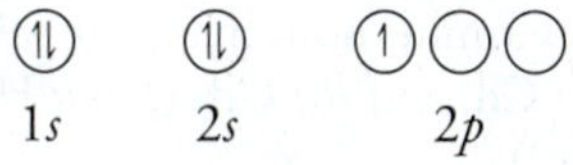

Notice, however, that when we give this orbital diagram we show *all* of the orbitals of the $2p$ subshell even though two of them are empty.

Next we come to carbon, which has six electrons. As before, the first four electrons complete the $1s$ and $2s$ orbitals. The remaining two electrons go in the $2p$ subshell to give

C $1s^2 2s^2 2p^2$

Now, however, to write the orbital diagram we have to make a decision as to where to put the two *p* electrons. (At this point you may have an unprintable suggestion! Bear up. It's really not all that bad.) To make this decision, we apply **Hund's rule**: *when electrons are placed in a set of orbitals of equal energy, they are spread out as much as possible to give as few paired electrons as possible.* Both theory and experiment have shown that if we follow this rule, we obtain the electron configuration with the lowest energy. For carbon, it means that the two *p* electrons are in separate orbitals and their spins are in the same direction.[1]

C   ⑪   ⑪   ①①○
      $1s$ $2s$ $2p$

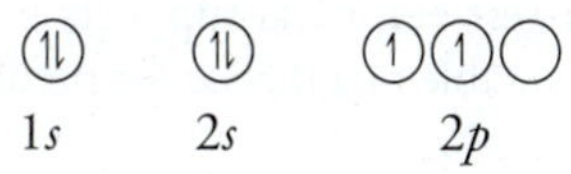

[1] Hund's rule gives us the *lowest* energy (ground state) distribution of electrons among the orbitals. However, configurations such as

C   ⑪ ⑪ ①↓○
C   ⑪ ⑪ ⑪○○
      $1s$ $2s$ $2p$

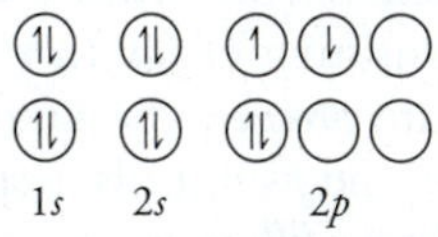

are not impossible—it is just that neither of them corresponds to the lowest energy distribution of electrons in the carbon atom.

Applying the Pauli exclusion principle and Hund's rule, we can now complete the electron configurations and orbital diagrams for the rest of the elements of the second period.

		1s	2s	2p
N	$1s^2 2s^2 2p^3$	⇅	⇅	↑ ↑ ↑
O	$1s^2 2s^2 2p^4$	⇅	⇅	⇅ ↑ ↑
F	$1s^2 2s^2 2p^5$	⇅	⇅	⇅ ⇅ ↑
Ne	$1s^2 2s^2 2p^6$	⇅	⇅	⇅ ⇅ ⇅

We can continue to predict electron configurations in this way, using Figure 7.18 as a guide to tell us which subshells become occupied and in what order. For instance, after completing the 2p subshell at neon, Figure 7.18 predicts that the next two electrons enter the 3s, followed by the filling of the 3p. Then we find the 4s lower in energy than the 3d, so it is filled first. Next, the 3d is completed before we go on to fill the 4p, and so forth.

Practice Exercise 7.13	Draw orbital diagrams for (a) Na, (b) S, and (c) Ar. (*Hint:* Recall how we indicate paired electron spins.)
Practice Exercise 7.14	Use Figure 7.18 to predict the electron configurations of (a) Si, (b) Ti, and (c) Se.
Practice Exercise 7.15	Can an element with an even atomic number be paramagnetic? (*Hint:* Try writing the orbital diagrams of a few of the transition elements in Period 4.)
Practice Exercise 7.16	Use orbital diagrams to determine how many unpaired electrons are in each of the elements (a) Mg, (b) Ge, (c) Cd, and (d) Gd. (*Hint:* How are electrons distributed in an atom's electron configuration?)

You might be thinking that you'll have to consult Figure 7.18 whenever you need to write down the ground state electron configuration of an element. However, as you will see in the next section, all the information contained in this figure is also contained in the periodic table.

7.8 | Periodic Table and Ground State Electron Configurations

In Chapter 3 you learned that when Mendeleev constructed his periodic table, elements with similar chemical properties were arranged in vertical columns called groups. Later work led to the expanded version of the periodic table we use today. The basic structure of this table is one of the strongest empirical supports for the quantum theory, and it also permits us to use the periodic table as a device for predicting electron configurations.

Predicting Ground State Electron Configurations

Consider, for example, the way the periodic table is laid out (Figure 7.20). On the left there is a block of *two* columns shown in blue; on the right there is a block of *six* columns shown in pink; in the center there is a block of *ten* columns shown in yellow, and below the table there are two rows consisting of *fourteen* elements each shown in gray. These numbers—2, 6, 10, and 14—are *precisely* the numbers of electrons that the quantum theory tells us can occupy s, p, d, and f subshells, respectively. In fact, *we can use the structure of the periodic table to predict the filling order of the subshells when we write the electron configuration of an element.*

TOOLS

Electron configurations using the periodic table

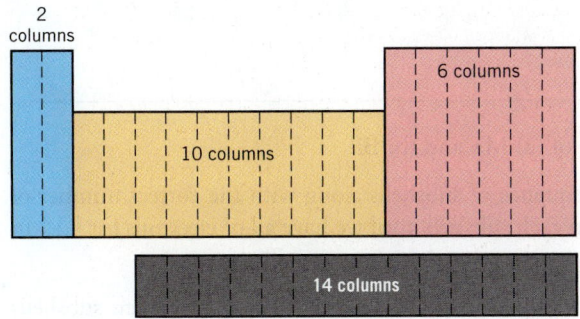

Figure 7.20 | The overall column structure of the periodic table. The table is naturally divided into regions of 2, 6, 10, and 14 columns, which are the numbers of electrons that can occupy *s*, *p*, *d*, and *f* subshells.

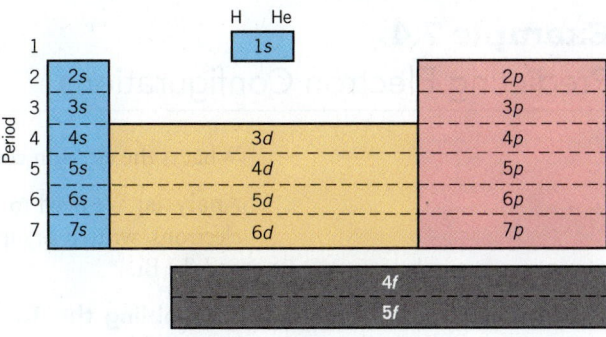

Figure 7.21 | Arrangement of subshells in the periodic table. This format illustrates the sequence for filling subshells.

To use the periodic table to predict electron configurations, we follow the aufbau principle as before. We start with hydrogen and then move through the table row after row until we reach the element of interest, noting as we go along which regions of the table we pass through. For example, consider the element calcium. Using Figure 7.21, we obtain the electron configuration

$$\text{Ca} \qquad 1s^2 2s^2 2p^6 3s^2 3p^6 4s^2$$

Refer to a periodic table and Figure 7.21 as we go through the process of writing an electron configuration.

Filling Periods 1, 2, and 3

The first period has only two elements, H and He, and the first subshell is $1s$. The electrons for H and He are placed in the $1s$ orbital, and as we have seen, for He we have $1s^2$ as the electron configuration. Next we move across the second period, where the first two elements, Li and Be, are in the s block. Two electrons added here enter the $2s$ subshell and are written as $2s^2$. Then we move to the block of six columns, the p block, and as we move across this region in the second row we fill the $2p$ subshell with six electrons, writing $2p^6$. Now we go on to the third period, filling the $3s$ subshell with two electrons, $3s^2$, and then filling the $3p$ subshell with six electrons, $3p^6$.

Filling Periods 4 and 5

Next, we move to the fourth period. To get to calcium in the example above we step through two elements in the two-column block, filling the $4s$ subshell with two electrons. Up to calcium we have only filled s and p subshells. Figure 7.21 shows that after calcium the $3d$ subshell is the next to be filled. Ten electrons are added for the ten elements in the first row of the transition elements. After the $3d$ subshell we complete the fourth period by filling the $4p$ subshell with six electrons. The fifth period fills the $5s$, $4d$, and $5p$ subshells in sequence with 2, 10, and 6 electrons, respectively.

Filling Periods 6 and 7

The inner transition elements (f subshells) begin to fill in the sixth and seventh periods. Period six fills the $6s$, $4f$, $5d$, and $6p$ subshells with 2, 14, 10, and 6 electrons, respectively. There are many irregularities in this sequence, and Appendix C should be consulted for the correct electron configurations. The seventh period fills the $7s$, $5f$, $6d$, and $7p$ subshells. Once again, the sequence is based on the energies of the subshells, and there are many irregularities that make it advisable to consult Appendix C for the correct electron configurations.

Notice that in the above patterns the s and p subshells always have the same number as the period they are in. The d subshells always have a number that is one less than the period they are in. Finally, the f subshells are always two less than the period in which they reside.

Example 7.4
Predicting Electron Configurations

What is the electron configuration of (a) Mn and (b) Bi?

Analysis: We need to write the sequence of subshells along with the correct number of electrons, written as superscripts, in each subshell until we have all of electrons for (a) Mn and (b) Bi.

Assembling the Tools: We use the periodic table as a tool to tell us which subshells become filled. Try not to refer to Figures 7.18, 7.20, or 7.21.

Solution: (a) To get to manganese, find Mn on the periodic table and then start at the top and work your way down:

Period 1 Fill the $1s$ subshell with 2 electrons
Period 2 Fill the $2s$ and $2p$ subshells
Period 3 Fill the $3s$ and $3p$ subshells
Period 4 Fill the $4s$ and then move 5 places in the $3d$ region to Mn

The electron configuration of Mn is therefore

$$\text{Mn} \qquad 1s^2 2s^2 2p^6 3s^2 3p^6 4s^2 3d^5$$

This configuration is correct as is. However, in elements with electrons in d and f orbitals, we can also group all subshells of the same shell together. For manganese, this gives

$$\text{Mn} \qquad 1s^2 2s^2 2p^6 3s^2 3p^6 3d^5 4s^2$$

We'll see later that writing the configuration in shell number order is convenient when working out ground state electronic configurations for ions.

(b) To get to bismuth, we fill the following subshells:

Period 1 Fill the $1s$
Period 2 Fill the $2s$ and $2p$
Period 3 Fill the $3s$ and $3p$
Period 4 Fill the $4s$, $3d$, and $4p$
Period 5 Fill the $5s$, $4d$, and $5p$
Period 6 Fill the $6s$, $4f$, $5d$, and then add 3 electrons to the $6p$

This gives

$$\text{Bi} \qquad 1s^2 2s^2 2p^6 3s^2 3p^6 4s^2 3d^{10} 4p^6 5s^2 4d^{10} 5p^6 6s^2 4f^{14} 5d^{10} 6p^3$$

Grouping subshells with the same value of n gives

$$\text{Bi} \qquad 1s^2 2s^2 2p^6 3s^2 3p^6 3d^{10} 4s^2 4p^6 4d^{10} 4f^{14} 5s^2 5p^6 5d^{10} 6s^2 6p^3$$

Again, either of these configurations is correct.

Is the Answer Reasonable? We can count the number of electrons to be sure we have 25 for Mn and 83 for Bi and none have been left out or added. Also, we can look at Appendix C to check our configurations.

Practice Exercise 7.17 | Use the periodic table to predict the electron configurations of (a) Mg, (b) Ge, (c) Cd, and (d) Gd. Group subshells of the same shell together. (*Hint:* Recall which areas of the periodic table represent s, p, d, and f electrons.)

Practice Exercise 7.18 | Use the periodic table to predict the electron configurations of (a) O, S, Se and (b) P, N, Sb. What is the same about all the elements in (a), and all the elements in (b)?

Abbreviated Electron Configurations

The electron configurations for larger elements can become unwieldy and difficult to work with. Chemists often write **abbreviated electron configurations** (also called shorthand configurations) to make things simpler.

When writing an abbreviated electron configuration for any element, Pb, for example, the complete electron configuration is divided into two groups as shown below. The first group is represented by the symbol for the noble gas immediately preceding Pb in the periodic table—that is, [Xe]. The second group consists of the remaining electron configuration.

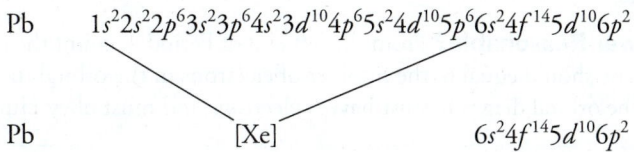

The abbreviated electron configuration is often rearranged in the order of shell numbers.

$$\text{Pb} \qquad [\text{Xe}]4f^{14}5d^{10}6s^26p^2$$

Abbreviated electron configurations can be written for the representative elements. Examples of the complete electron configuration and the abbreviated configuration for sodium and magnesium. are shown below.

$$\text{Na} \quad 1s^22s^22p^63s^1 \qquad \text{Na} \quad [\text{Ne}]3s^1$$
$$\text{Mg} \quad 1s^22s^22p^63s^2 \qquad \text{Mg} \quad [\text{Ne}]3s^2$$

The abbreviated electron configurations for Mn and Bi that we worked with in Example 7.4 are

$$\text{Mn} \quad [\text{Ar}]4s^23d^5 \qquad \text{and} \qquad \text{Bi} \quad [\text{Xe}]6s^24f^{14}5d^{10}6p^3$$

We can also write the configurations in order of shell numbers, n, as

$$\text{Mn} \quad [\text{Ar}]3d^54s^2 \qquad \text{and} \qquad \text{Bi} \quad [\text{Xe}]4f^{14}5d^{10}6s^26p^3$$

Example 7.5
Writing Shorthand Electron Configurations

What is the shorthand electron configuration of chromium? Draw the orbital diagram for chromium based on this shorthand configuration. Is chromium paramagnetic or diamagnetic?

Analysis: This is the same as the previous example except that we list only the electrons in Period 4, while the remaining electrons are represented by the symbol for the appropriate noble gas in square brackets. Then we need to write out the abbreviated electron configuration in an orbital diagram to determine how many unpaired electrons there are to see if chromium is diamagnetic or paramagnetic.

Assembling the Tools: We will need the periodic table, the methods for drawing orbital diagrams, and the conditions that make an element paramagnetic.

Solution: Chromium is in Period 4. The preceding noble gas is argon, Ar, in Period 3. The abbreviated configuration starts with the symbol for argon in brackets followed by the electron configuration of the remaining electrons. After Ar, we move to Period 4 and cross the "s region" by adding two electrons to the $4s$ subshell, and then we add four electrons to the $3d$ subshell. Therefore, the shorthand electron configuration for Cr is

$$\text{Cr} \qquad [\text{Ar}]4s^23d^4$$

Placing the electrons that are in the highest shell farthest to the right gives

$$Cr \qquad [Ar]3d^44s^2$$

To draw the orbital diagram, we simply distribute the electrons in the $3d$ and $4s$ orbitals following Hund's rule. This gives

Cr [Ar] ⟨↑⟩⟨↑⟩⟨↑⟩⟨↑⟩⟨ ⟩ ⟨↑↓⟩
                    $3d$ $4s$

Four of the $3d$ orbitals are half-filled. The atom contains four unpaired electrons and is paramagnetic.

Is the Answer Reasonable? From the left end of Period 4, count the groups needed to reach Cr. These should equal 6, the number of electrons in the orbitals beyond Ar as written above. The orbital diagram must have 6 electrons and must obey Hund's rule.

Practice Exercise 7.19

Write shorthand configurations and abbreviated orbital diagrams for (a) Zr and (b) Po. Where appropriate, place the electrons that are in the highest shell farthest to the right. (*Hint:* What noble gas is right before the given atom?)

Practice Exercise 7.20

Write shorthand configurations and abbreviated orbital diagrams for (a) P and (b) Sn. Where appropriate, place the electrons that are in the highest shell farthest to the right. How many unpaired electrons does each of these atoms have?

Valence Shell Electron Configurations

The abbreviated electron configuration, arranged in shell number order, helps emphasize the electrons that are of most interest with respect to chemical properties. When two atoms come together to form new chemical bonds, it must be due to the interactions of the electrons furthest from the nucleus. It also makes sense that electrons buried deep within the atomic structure most likely do not contribute significantly to the chemical properties of an atom. These electrons furthest from the nucleus are the electrons that we are most interested in.

Each group within the periodic table has elements with similar chemical and physical properties that vary regularly within the group. We are now ready to understand the reason for these similarities in terms of the electronic structures of atoms.

Chemical Properties and Valence Shell Configurations

Periodic table and chemical properties

When we consider the chemical reactions of an atom (particularly one of the representative elements), our attention is usually focused on the distribution of electrons in the **outer shell** of the atom (those electrons where n is the largest). This is because the **outer electrons**, those in the outer shell, are the ones that are exposed to other atoms when the atoms react. (Later we will see that d electrons are also very important in reactions involving the transition elements.) It seems reasonable, therefore, that elements with similar properties should have similar outer shell electron configurations. For example, let's look at the alkali metals of Group 1A. Going by our rules, we obtain the following abbreviated electron configurations and complete electron configurations.

Li	[He] $2s^1$	or	$1s^22s^1$
Na	[Ne] $3s^1$	or	$1s^22s^22p^63s^1$
K	[Ar] $4s^1$	or	$1s^22s^22p^63s^23p^64s^1$
Rb	[Kr] $5s^1$	or	$1s^22s^22p^63s^23p^63d^{10}4s^24p^65s^1$
Cs	[Xe] $6s^1$	or	$1s^22s^22p^63s^23p^63d^{10}4s^24p^64d^{10}5s^25p^66s^1$

The abbreviated electron configuration clearly illustrates the similarities of the outermost electron configurations in these elements. Each of these elements has only one outer shell electron that is in an s subshell (shown in bold, red, type). We know that when they react, the alkali metals each lose one electron to form ions with a charge of 1+. For each, the electron that is lost is this outer s electron, and the electron configuration of the ion that is formed is the same as that of the preceding noble gas.

Li^+	$1s^2$		He	$1s^2$
Na^+	$1s^2 2s^2 2p^6$		Ne	$1s^2 2s^2 2p^6$
K^+	$1s^2 2s^2 2p^6 3s^2 3p^6$		Ar	$1s^2 2s^2 2p^6 3s^2 3p^6$

etc.

If you write the abbreviated electron configurations of the members of any of the groups in the periodic table, you will find the same kind of similarity among the configurations of the outer shell electrons. The differences are in the value of the principal quantum number of these outer electrons.

For the representative elements (those in the longer columns), the only electrons that are normally important in controlling chemical properties are the ones in the outer shell. This outer shell is known as the **valence shell**, and it is always the occupied shell with the largest value of n. The electrons in the valence shell are called **valence electrons**. (The term *valence* comes from the study of chemical bonding and relates to the combining capacity of an element, but that's not important here.)

For the representative elements it is very easy to determine the electron configuration of the valence shell by using the periodic table. *The valence shell always consists of just the s and p subshells that we encounter crossing the period that contains the element in question.* Thus, to determine the valence shell configuration of sulfur, a Period 3 element, we note that to reach sulfur in Period 3 we need to place two electrons into the $3s$ and four electrons into the $3p$ subshells. The valence shell configuration of sulfur is therefore

$$S \qquad 3s^2 3p^4$$

and we can say that sulfur has six valence electrons.

Example 7.6
Writing Valence Shell Configurations

Predict the electron configuration of the valence shell of arsenic ($Z = 33$).

Analysis: To determine the number of s and p electrons in the highest shell, we write the electron configuration grouping the subshells in each shell together.

Assembling the Tools: Our tools are the periodic table and the application of the definition of valence shell electrons.

Solution: To reach arsenic in Period 4, we add electrons to the $4s$, $3d$, and $4p$ subshells. However, the $3d$ is not part of the fourth shell and therefore not part of the valence shell, so all we need be concerned with are the electrons in the $4s$ and $4p$ subshells. This gives us the valence shell configuration of arsenic,

$$As \qquad 4s^2 4p^3$$

Is the Answer Reasonable? We can check that only s and p electrons are counted and that there are five valence electrons. The answer does seem reasonable.

Give an example of a valence shell with more than eight electrons. If that is not possible, explain why. (*Hint:* Are there any elements in which the highest numbered d subshell electrons are equal to the highest numbered s or p shell electrons?)

Practice Exercise 7.21

What is the valence shell electron configuration of (a) Se, (b) Sn, and (c) I? How many valence electrons do each of these elements have?

Practice Exercise 7.22

Some Unexpected Electron Configurations

The rules you've learned for predicting electron configurations work most of the time, but not always. Appendix C gives the electron configurations of all of the elements as determined experimentally. Close examination reveals that there are quite a few exceptions to the rules. Some of these exceptions are important to us because they occur with common elements.

Two important exceptions are for chromium and copper. Following the rules, we would expect the configurations of chromium and copper to be

$$\text{Cr} \qquad [\text{Ar}]\, 3d^4 4s^2$$
$$\text{Cu} \qquad [\text{Ar}]\, 3d^9 4s^2$$

However, the actual electron configurations, determined experimentally, are

$$\text{Cr} \qquad [\text{Ar}]\, 3d^5 4s^1$$
$$\text{Cu} \qquad [\text{Ar}]\, 3d^{10} 4s^1$$

The corresponding orbital diagrams are

For chromium, an electron is moved from the $4s$ subshell to give a $3d$ subshell that is exactly half-filled. For copper the $4s$ electron is borrowed to give a completely filled $3d$ subshell. A similar thing happens with silver and gold, which have filled $4d$ and $5d$ subshells, respectively.

$$\text{Ag} \qquad [\text{Kr}]\, 4d^{10} 5s^1$$
$$\text{Au} \qquad [\text{Xe}]\, 4f^{14} 5d^{10} 6s^1$$

Half-filled and filled d subshells (particularly the latter) have some special stability that makes such borrowing energetically favorable. This subtle but important phenomenon affects not only the ground state configurations of atoms but also the relative stabilities of some of the ions formed by the transition elements. Similar irregularities occur among the lanthanide and actinide elements.

7.9 | Atomic Orbitals: Shapes and Orientations

To picture what electrons are doing within the atom, we are faced with imagining an object that behaves like a particle in some experiments and like a wave in others. There is nothing in our worldly experience that is comparable. Fortunately we can still think of the electron as a particle in the usual sense by speaking in terms of the statistical probability of the electron being found at a particular place. We can then use quantum mechanics to mathematically connect the particle and wave representations of the electron. Even though we may have trouble imagining an object that can be represented both ways, mathematics describes its behavior very accurately.

Describing the electron's position in terms of statistical probability is based on more than simple convenience. The German physicist Werner Heisenberg showed mathematically that it is impossible to measure with complete precision both a particle's velocity and position at the same instant. To measure an electron's position or velocity, we have to bounce another particle, such as a photon, off it. Thus, the very act of making the measurement alters the electron's position and velocity. We cannot determine both exact position and exact velocity simultaneously, no matter how cleverly we make the measurements. This is the Heisenberg **uncertainty principle**, which is often stated mathematically as

■ Werner Heisenberg received the Nobel Prize in physics in 1932 for the creation of quantum mechanics.

$$\Delta x = \frac{h}{4\pi m}\left(\frac{1}{\Delta v}\right)$$

where Δx is the minimum uncertainty in the particle's location, h is Planck's constant, m is the mass of the particle, and Δv is the minimum uncertainty in the particle's velocity.

The theoretical limitations on measuring speed and position are not significant for large objects. However, for small particles such as the electron, these limitations prevent us from ever knowing or predicting where in an atom an electron will be at a particular instant, so we speak of probabilities instead.

Wave mechanics views the probability of finding an electron at a given point in space as equal to the square of the amplitude of the electron wave (given by the square of the wave function, ψ^2) at that point. It seems quite reasonable to relate probability to amplitude, or intensity, because where a wave is intense its presence is strongly felt. The amplitude is squared because, mathematically, the amplitude can be either positive or negative, but probability only makes sense if it is positive. Squaring the amplitude assures us that the probabilities will be positive. We need not be very concerned about this point, however.

The notion of electron probability leads to three very important concepts. One is that an electron behaves as if it were spread out around the nucleus in a sort of **electron cloud.** Figure 7.22a is a *dot-density diagram* that illustrates the way the probability of finding the electron varies in space for a 1s orbital. In those places where the dot density is large (i.e., where there are large numbers of dots per unit volume), the amplitude of the wave is large and the probability of finding the electron is also large.

The second important concept that stems from the notion that the electron probability varies from place to place is **electron density**, which relates to how much of the electron's charge is packed into a given region of space. In regions of high probability there is a high concentration of electrical charge (and mass) and the electron density is large; in regions of low probability, the electron density is small. Figure 7.22b shows how the electron probability for a 1s orbital varies as we move away from the nucleus. As you might expect, the probability of finding the electron close to the nucleus is large and decreases with increasing distance from the nucleus.

In a third view we ask ourselves, "What is the probability that an electron will occupy a thin spherical shell from r to $r + x$, where x is very small, away from the nucleus?" If we plot the probabilities for these shells, the graph that we get looks like Figure 7.22c. This is called a radial distribution plot. Interestingly, the maximum for such a plot for the hydrogen atom falls precisely at 53 pm, the Bohr radius.

Remember that an electron confined to a tiny space no longer behaves much like a particle. It's more like a cloud of negative charge. Like clouds made of water vapor, the density of the cloud varies from place to place. In some places the cloud is dense; in others the cloud is thinner and may be entirely absent. This is a useful picture to keep in mind as you try to visualize the shapes of atomic orbitals.

■ Notice that from the Heisenberg Uncertainty Principle, we can generally measure the particle's location more precisely if the particle is heavier. Notice also that the greater the uncertainty in the velocity, the smaller the uncertainty in the particle's location.

■ The amplitude of an electron wave is described by a **wave function**, which is usually given the symbol ψ (the Greek letter psi). The probability of finding the electron in a given location is given by ψ^2. The sign of the wave function is important when they are combined, as we will discuss in Chapter 9.

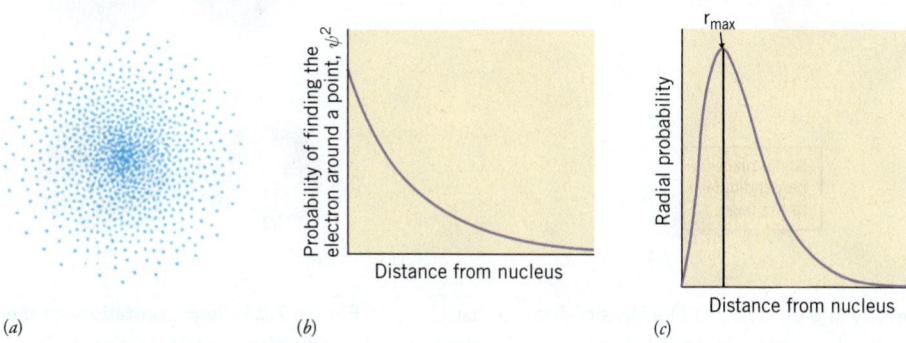

(a)   *(b)*   *(c)*

Figure 7.22 │ **Electron distribution in the 1s orbital of a hydrogen atom.** (*a*) A dot-density diagram that illustrates the electron probability distribution for a 1s electron. (*b*) A graph that shows how the probability of finding the 1s electron around a given point, ψ^2, decreases as the distance from the nucleus increases. (*c*) A graph of the radial probability distribution of a 1s electron that shows the probability of finding the 1s electron in a volume between r and $r + x$ from the nucleus. The maximum is the same as the Bohr radius.

Shapes and Sizes of *s* and *p* Orbitals

In looking at the way the electron density distributes itself in atomic orbitals, we are interested in three things—the *shape* of the orbital, its *size,* and its *orientation* in space relative to other orbitals.

The electron density in an orbital doesn't end abruptly at some particular distance from the nucleus. It gradually fades away. Therefore, to define the size and shape of an orbital, it is useful to picture some imaginary surface enclosing, say, 90% of the electron density of the orbital, and the probability of finding the electron is everywhere the same on that surface. For the 1*s* orbital in Figure 7.22, we find that if we go out a given distance from the nucleus in *any* direction, the probability of finding the electron is the same. This means that all the points of equal probability lie on the surface of a sphere, so we can say that the shape of the orbital is spherical. In fact, all *s* orbitals are spherical. As suggested earlier, their sizes increase with increasing *n*. This is illustrated in Figure 7.23. Notice that beginning with the 2*s* orbital, there are certain places where the electron density drops to zero. These are spherically shaped nodes of the *s* orbital electron waves. It is interesting that electron waves have nodes just like the waves on a guitar string. For *s* orbital electron waves, however, the nodes consist of imaginary spherical *surfaces* on which the electron density is zero.

The *p* orbitals are quite different from *s* orbitals, as shown in Figure 7.24. Notice that the electron density is equally distributed in two regions on opposite sides of the nucleus. Figure 7.24*a* illustrates the two "lobes" of *a single* 2*p* orbital. Between the lobes is a **nodal plane**—an imaginary flat surface on which every point has an electron density of zero. The size of the *p* orbitals also increases with increasing *n* as illustrated by the cross section of a 3*p* orbital in Figure 7.24*b*. The 3*p* and higher *p* orbitals have additional nodes besides the nodal plane that passes through the nucleus.

Figure 7.25*a* illustrates the shape of a surface of constant probability for a 2*p* orbital. Often chemists will simplify this shape by drawing two "balloons" connected at the nucleus and pointing in opposite directions as shown in Figure 7.25*b*. Both representations emphasize the point that a *p* orbital has two equal-sized lobes that extend in opposite directions along a line that passes through the nucleus.

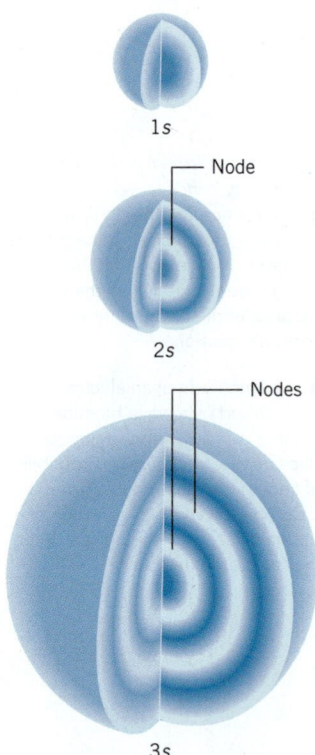

Figure 7.23 | Size variations among *s* orbitals. The orbitals become larger as the principal quantum number, *n*, becomes larger.

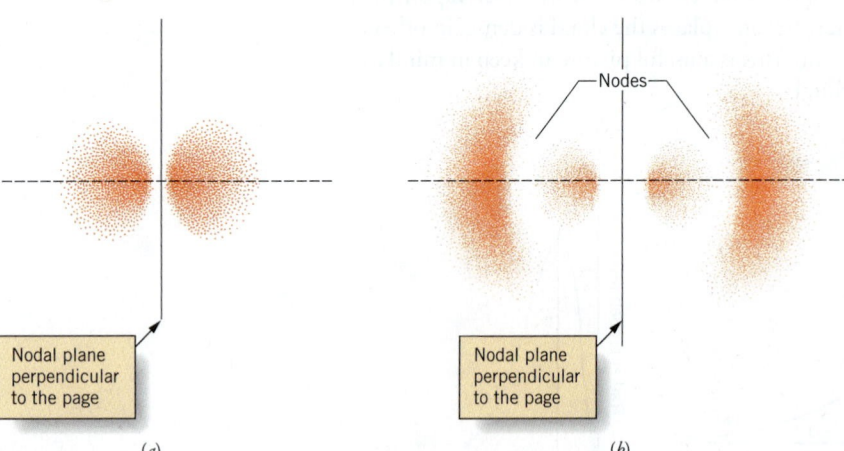

Figure 7.24 | Distribution of electron density in *p* orbitals. (*a*) Dot-density diagram that represents a cross section of the probability distribution in a 2*p* orbital. There is a nodal plane between the two lobes of the orbital. (*b*) Cross section of a 3*p* orbital. Note the nodes in the electron density that are in addition to the nodal plane passing through the nucleus.

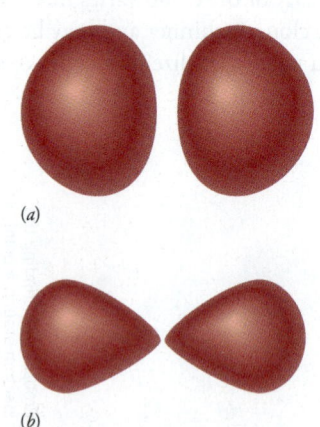

Figure 7.25 | Representations of the shapes of *p* orbitals. (*a*) Shape of a surface of constant probability for a 2*p* orbital. (*b*) A simplified representation of a *p* orbital that emphasizes the directional nature of the orbital.

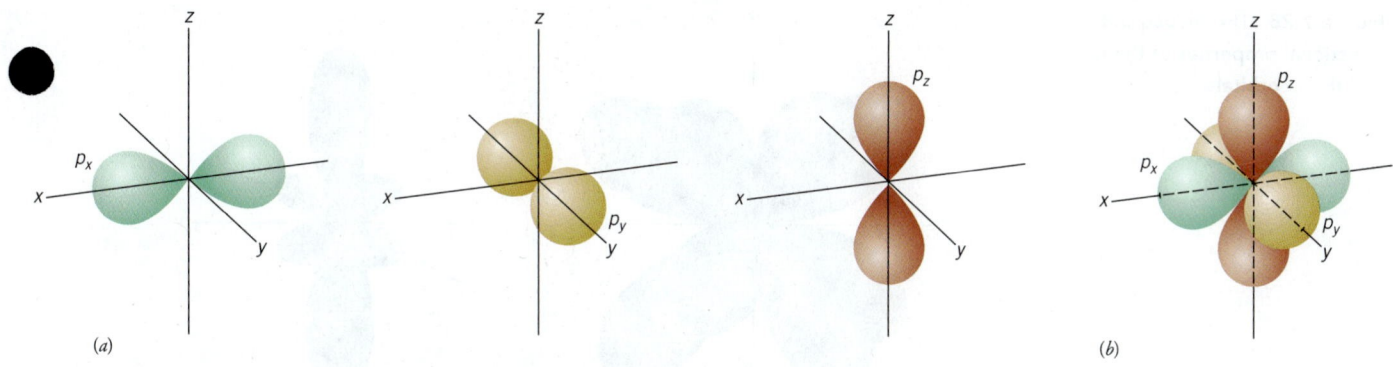

Figure 7.26 | **The orientations of the three _p_ orbitals in a _p_ subshell.** (_a_) Because the direction of maximum electron density lie along lines that are mutually perpendicular like the axes of an _xyz_ coordinate system, the orbitals are labeled p_x, p_y, and p_z. (_b_) The shapes and directions of the three orbitals of a _p_ subshell are shown separately.

Orientations of _p_ Orbitals

As you've learned, a _p_ subshell consists of three orbitals of equal energy. Wave mechanics tells us that the lines along which the orbitals have their maximum electron densities are oriented at 90° angles to each other, corresponding to the axes of an imaginary _xyz_ coordinate system, shown both separately and together in Figure 7.26. For convenience in referring to the individual _p_ orbitals they are often labeled according to the axis along which they lie. The _p_ orbital concentrated along the _x_ axis is labeled p_x, and so forth.

Shapes and Orientations of _d_ Orbitals in a _d_ Subshell

The shapes of the _d_ orbitals, illustrated in Figure 7.27, are a bit more complex than are those of the _p_ orbitals. Notice that four of the five _d_ orbitals have the same shape and consist of four lobes of electron density. These four _d_ orbitals each have two perpendicular nodal planes that intersect at the nucleus. These orbitals differ only in their orientations around the nucleus (their labels come from the mathematics of wave mechanics). The fifth _d_ orbital, labeled d_{z^2}, has two lobes that point in opposite directions along the _z_-axis plus a doughnut-shaped ring of electron density around the center that lies in the _x–y_ plane. The two nodes for the d_{z^2} orbital are conic surfaces whose peaks meet at the nucleus. We will see that the _d_ orbitals are important in the formation of chemical bonds in certain molecules, and that their shapes and orientations are important in understanding the properties of the transition metals, which we will discuss in Chapter 21.

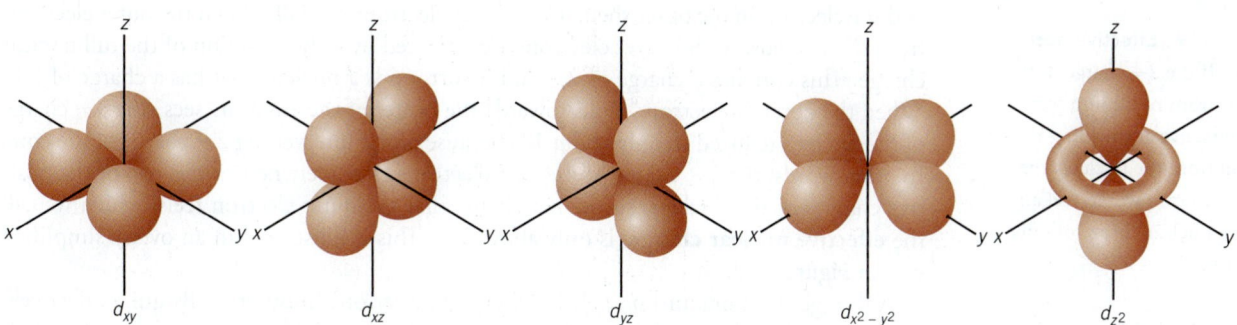

Figure 7.27 | **The shapes and directional properties of the five orbitals of a _d_ subshell.**

Figure 7.28 | The shapes and directional properties of the f_{xyz} and the f_{z^3} orbitals.

The shapes of the seven f orbitals are more complex than the d orbitals, having more lobes, nodes, and a variety of shapes, as shown in Figure 7.28 for two of the f orbitals, f_{xyz} and f_{z^3}. The other five are similar. Use of f orbitals for bonding will not be discussed in this book. However, we should note that each f orbital has three nodal planes.

7.10 | Periodic Table and Properties of the Elements

There are many chemical and physical properties that vary in a more or less systematic way according to an element's position in the periodic table. For example, in Chapter 2 we noted that the metallic character of the elements increases from top to bottom in a group and decreases from left to right across a period. In this section we discuss several physical properties of the elements that have an important influence on chemical properties. We will see how these properties, atomic radii, *ionization energy*, and *electron affinity*, correlate with an atom's electron configuration, and because electron configuration is also related to the location of an element in the periodic table, we will study their periodic variations as well.

Effective Nuclear Charge

Many of an atom's properties are determined by the amount of positive charge felt by the atom's outer electrons. Except for hydrogen, this positive charge is always *less* than the full nuclear charge, because the negative charge of the electrons in inner shells partially offsets, or "neutralizes," the positive charge of the nucleus.

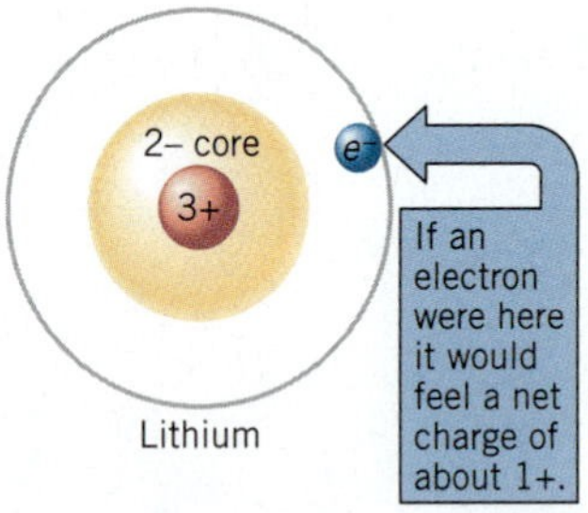

To gain a better understanding of this, consider the element lithium, which has the electron configuration $1s^2 2s^1$. The **core electrons** ($1s^2$), which lie beneath the valence shell ($2s^1$), are tightly packed around the nucleus and for the most part lie between the nucleus and the electron in the outer shell. These inner electrons partially shield the outer electrons from the nucleus, so the outer electrons are attracted by only a fraction of the full nuclear charge. This core has a charge of $2-$ and it surrounds a nucleus that has a charge of $3+$. When the outer $2s$ electron "looks toward" the center of the atom, it "sees" the $3+$ charge of the nucleus reduced to only about $1+$ because of the intervening $2-$ charge of the core. In other words, the $2-$ charge of the core electrons effectively neutralizes two of the positive charges of the nucleus, so the net charge that the outer electron feels, which we call the **effective nuclear charge**, is only about $1+$. This is illustrated in an overly simplified way in Figure 7.29.

Although electrons in inner shells shield the electrons in outer shells quite effectively from the nuclear charge, electrons in the *same* shell are much less effective at shielding each other. For example, in the element beryllium ($1s^2 2s^2$) each of the electrons in the outer $2s$ orbital is shielded quite well from the nuclear charge by the inner $1s^2$ core, but one $2s$

Figure 7.29 | **Effective nuclear charge.** If the $2-$ charge of the $1s^2$ core electrons of lithium were 100% effective at shielding the $2s$ electron from the nucleus, the valence electron would feel an effective nuclear charge of only about $1+$.

■ An electron spends very little time between the nucleus and another electron in the same shell, so it shields that other electron poorly.

electron doesn't shield the other $2s$ electron very well at all. This is because electrons in the same shell are at about the same average distance from the nucleus, and in attempting to stay away from each other they only spend a very small amount of time one below the other, which is what's needed to provide shielding. Since electrons in the same shell hardly shield each other at all from the nuclear charge, *the effective nuclear charge felt by the outer electrons is determined primarily by the difference between the charge on the nucleus and the charge on the core.* With this as background, let's examine some properties controlled by the effective nuclear charge.

Atomic and Ionic Sizes

The wave nature of the electron makes it difficult to define exactly what we mean by the "size" of an atom or ion. As we've seen, the electron cloud doesn't simply stop at some particular distance from the nucleus; instead it gradually fades away. Nevertheless, atoms and ions do behave in many ways as though they have characteristic sizes. For example, in a whole host of hydrocarbons, ranging from methane (CH_4, natural gas) to octane (C_8H_{18}, in gasoline) to many others, the distance between the nuclei of carbon and hydrogen atoms is virtually the same. This would suggest that carbon and hydrogen have the same relative sizes in each of these compounds.

■ The C—H distance in most hydrocarbons is about 110 pm (110×10^{-12} m).

Experimental measurements reveal that the diameters of atoms range from about 1.4×10^{-10} to 5.7×10^{-10} m. Their radii, which is the usual way that size is specified, range from about 7.0×10^{-11} to 2.9×10^{-10} m. Such small numbers are difficult to comprehend. A million carbon atoms placed side by side in a line would extend a little less than 0.2 mm, or about the diameter of the period at the end of this sentence.

The sizes of atoms and ions are rarely expressed in meters because the numbers are so cumbersome. Instead, a unit is chosen that makes the values easier to comprehend. A unit that scientists have traditionally used is called the **angstrom** (symbolized **Å**), which is defined as

■ The angstrom is named after Anders Jonas Ångström (1814–1874), a Swedish physicist who was the first to measure the wavelengths of the four most prominent lines of the hydrogen spectrum.

$$1 \text{ Å} = 1 \times 10^{-10} \text{ m}$$

However, the angstrom is not an SI unit, and in many current scientific journals, atomic dimensions are given in picometers, or sometimes in nanometers. In this book, we will normally express atomic dimensions in picometers, but because much of the scientific literature has these quantities in angstroms, it may be useful to remember the conversions:

■ 1 pm = 10^{-12} m and 1 nm = 10^{-9} m.

$$1 \text{ Å} = 100 \text{ pm}$$
$$1 \text{ Å} = 0.1 \text{ nm}$$

Periodic Variations

The variations in atomic radii within the periodic table are illustrated in Figure 7.30. There we see that atoms generally become larger going from top to bottom in a group, and they become smaller going from left to right across a period. To understand these variations we must consider two factors: (1) the value of the principal quantum number of the valence electrons, and (2) the effective nuclear charge felt by the valence electrons.

Going from top to bottom within a group, the effective nuclear charge felt by the outer electrons remains nearly constant, while the principal quantum number of the valence shell increases. For example, consider the elements of Group 1A. For lithium, the valence shell configuration is $2s^1$; for sodium, it is $3s^1$; for potassium, it is $4s^1$; and so forth. For each of these elements, the core has a negative charge that is one less than the nuclear charge, so the valence electron of each experiences a nearly constant effective nuclear charge of about 1+. However, as we descend the group, the value of n for the valence shell increases, and as you learned in Section 7.5, the larger the value of n, the larger the orbital. Therefore, the atoms become larger as we go down a group simply because the orbitals containing the valence electrons become larger. This same argument applies whether the valence shell orbitals are s or p.

Periodic trends in atomic size

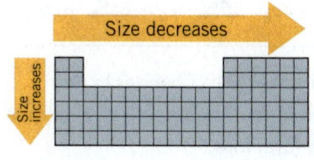

General variation of atomic size within the periodic table.

■ Large atoms are found in the lower left of the periodic table, and small atoms are found in the upper right.

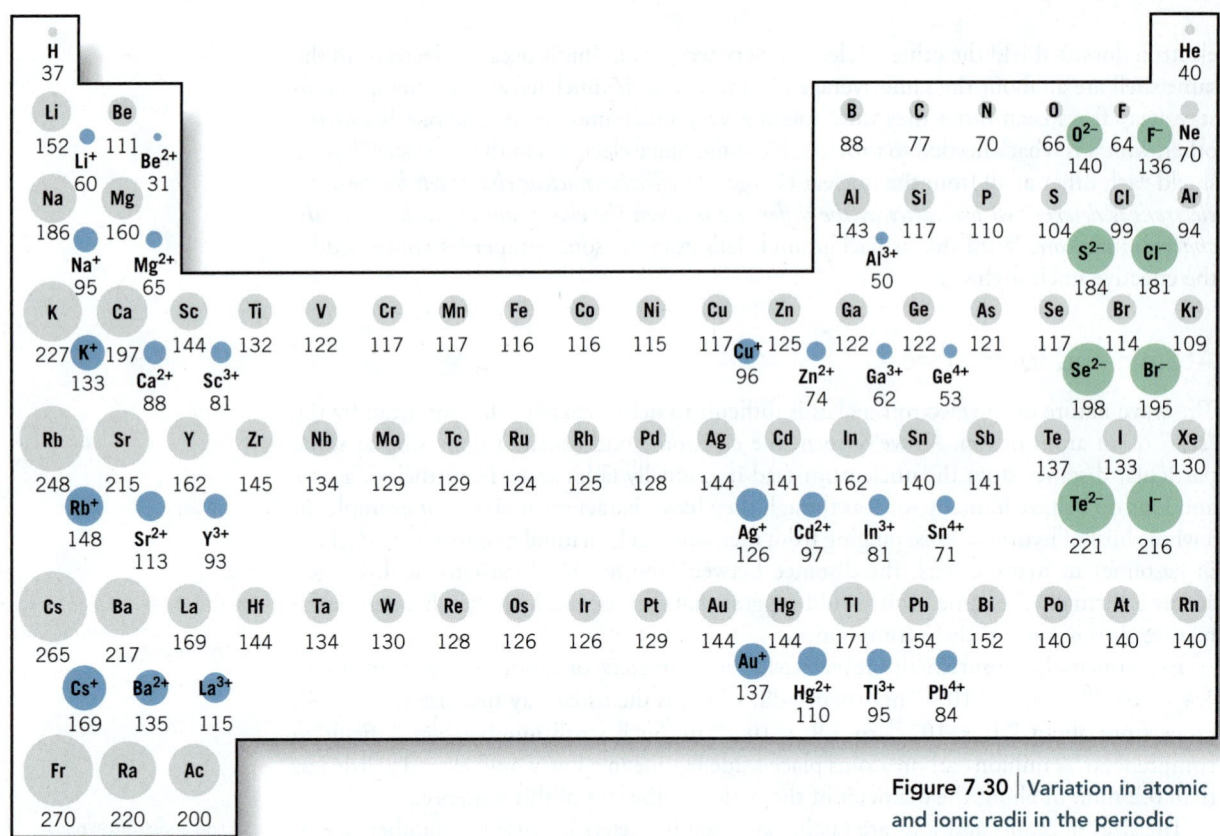

Figure 7.30 | Variation in atomic and ionic radii in the periodic table. Values are in picometers.

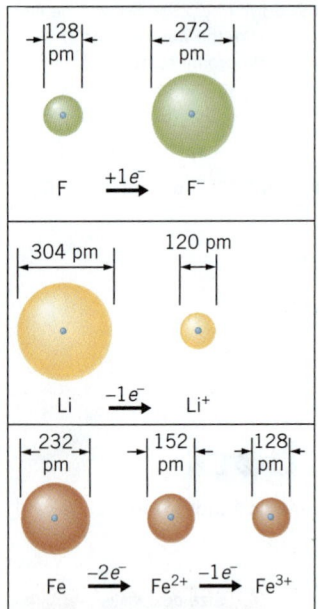

Figure 7.31 | Changes in size when atoms gain or lose electrons to form ions.

Periodic trends in ionic size

Moving from left to right across a period, electrons are added to the same shell. The orbitals holding the valence electrons all have the *same* value of *n*. In this case we have to examine the variation in the effective nuclear charge felt by the valence electrons.

As we move from left to right across a period, the nuclear charge increases, and the outer shells of the atoms become more populated but the inner core remains the same. For example, from lithium to fluorine the nuclear charge increases from 3+ to 9+. The core ($1s^2$) stays the same, however. As a result, the outer electrons feel an increase in positive charge (i.e., effective nuclear charge), which causes them to be drawn inward and thereby causes the sizes of the atoms to decrease.

Across a row of transition elements or inner transition elements, the size variations are less pronounced than among the representative elements. This is because the outer shell configuration remains essentially the same while an inner shell is filled. From atomic numbers 21 to 30, for example, the outer electrons occupy the $4s$ subshell while the underlying $3d$ subshell is gradually completed. The amount of shielding provided by the addition of electrons to this inner $3d$ level is greater than the amount of shielding that would occur if the electrons were added to the outer shell, so the effective nuclear charge felt by the outer electrons increases more gradually. As a result, the decrease in size with increasing atomic number is also more gradual.

Ion Size Compared to Atomic Size

Figure 7.30 also illustrates how sizes of the ions compare with those of the neutral atoms. As you can see, when atoms gain or lose electrons to form ions, rather significant size changes take place. The reasons are easy to understand and remember.

When electrons are added to an atom, the mutual repulsions between them increase. This causes the electrons to push apart and occupy a larger volume. Therefore, *negative ions are often about 1.5 to 2 times larger than the atoms from which they are formed* (Figure 7.31).

When electrons are removed from an atom, the electron–electron repulsions decrease, which allows the remaining electrons to be pulled closer together around the nucleus. Therefore, *positive ions are always smaller than the atoms from which they are formed*. As

Figure 7.30 shows, cations often are only ½ to ⅔ the size of their parent atom. This is also illustrated in Figure 7.31 for the elements lithium and iron. For lithium, removal of the outer 2s electron completely empties the valence shell and exposes the smaller $1s^2$ core. When a metal is able to form more than one positive ion, the sizes of the ions decrease as the amount of positive charge on the ion increases. To form the Fe^{2+} ion, an iron atom loses its outer 4s electrons. To form the Fe^{3+} ion, an additional electron is lost from the 3d subshell that lies beneath the 4s. Comparing sizes, we see that the radius of an iron atom is 116 pm whereas the radius of the Fe^{2+} ion is 76 pm. Removing yet another electron to give Fe^{3+} decreases electron–electron repulsions in the d subshell and gives the Fe^{3+} ion a radius of 64 pm.

■ Adding electrons creates an ion that is larger than the neutral atom; removing electrons produces an ion that is smaller than the neutral atom.

■ Li $1s^2 2s^1$
   Li^+ $1s^2$

■ Fe   [Ar] $3d^6 4s^2$
   Fe^{2+} [Ar] $3d^6$
   Fe^{3+} [Ar] $3d^5$

Practice Exercise 7.23

Use the periodic table to choose the largest atom or ion in each set.

(a) Ge, Te, Se, Sn (b) C, F, Br, Ga (c) Cr, Cr^{2+}, Cr^{3+} (d) O, O^{2-}, S, S^{2-}

(*Hint:* Recall that electrons repel each other.)

Practice Exercise 7.24

Use the periodic table to determine the smallest atom or ion in each group.

(a) Si, Ge, As, P (b) Fe^{2+}, Fe^{3+}, Fe (c) Db, W, Tc, Fe (d) Br^-, I^-, Cl^-

Ionization Energy

The **ionization energy** (abbreviated **IE**) *is the energy required to remove an electron from an isolated, gaseous atom or ion in its ground state.* For an atom of an element X, it is the increase in potential energy associated with the change

$$X(g) \longrightarrow X^+(g) + e^-$$

In effect, the ionization energy is a measure of how much work is required to pull an electron from an atom, so it reflects how tightly the electron is held by the atom. Usually, the ionization energy is expressed in units of kilojoules per mole (kJ/mol), so we can also view it as the energy needed to remove one mole of electrons from one mole of gaseous atoms.

Table 7.2 lists the ionization energies of the first 12 elements. As you can see, atoms with more than one electron have more than one ionization energy. These correspond to the stepwise removal of electrons, one after the other. Lithium, for example, has three ionization energies because it has three electrons. Removing the outer 2s electrons from one mole of isolated lithium atoms to give one mole of gaseous lithium ions, Li^+, requires

TABLE 7.2	Successive Ionization Energies in kJ/mol for Hydrogen through Magnesium							
	1st	**2nd**	**3rd**	**4th**	**5th**	**6th**	**7th**	**8th**
H	1312							
He	2372	5250						
Li	520	7297	11,810					
Be	899	1757	14,845	21,000				
B	800	2426	3659	25,020	32,820			
C	1086	2352	4619	6221	37,820	47,260		
N	1402	2855	4576	7473	9442	53,250	64,340	
O	1314	3388	5296	7467	10,987	13,320	71,320	84,070
F	1680	3375	6045	8408	11,020	15,160	17,860	92,010
Ne	2080	3963	6130	9361	12,180	15,240	–	–
Na	495	4563	6913	9541	13,350	16,600	20,113	25,666
Mg	737	1450	7731	10,545	13,627	17,995	21,700	25,662

Note the sharp increase in ionization energy when crossing the "staircase," indicating that the last of the valence electrons has been removed.

ON THE CUTTING EDGE (7.2

Photoelectron Spectroscopy

The photoelectric effect, where electrons can be ejected from a surface by a beam of light, was described earlier in this chapter. In 1921 Albert Einstein was awarded the Nobel Prize in physics for explaining this phenomenon. To do so he viewed electromagnetic radiation as particles, or photons, with each photon having a quantum of energy equal to its frequency multiplied by Planck's constant, $h\nu$. Einstein then postulated that electrons are attracted to the nucleus with a certain amount of energy that he called the binding energy, BE. He then reasoned that if a photon has more energy than an electron's binding energy, the photon would be able to eject an electron out of the atom. In simpler terms, he showed that it is the energy of the photon that ejects an electron rather than the number of photons.

In addition to the atomic view of the photoelectric process, we know that energy must be conserved. The total energy entering the system is the energy of the photon, $h\nu$. That must exceed the binding energy, BE, and the excess energy appears as the kinetic energy, KE, of the ejected electron.

$$h\nu = BE + KE$$

Researchers can measure the energy of the incoming photon from its frequency and can measure the kinetic energies of the ejected electrons, to determine the binding energies of the ejected electrons.

The binding energy is closely related to the ionization energy for a given electron. In practice, photoelectron spectroscopy, PES, can be divided into two closely related methods. Ultraviolet photoelectron spectroscopy (UPS) uses UV radiation to eject electrons from a sample; X-ray photoelectron spectroscopy, XPS, that uses X rays. In 1981, Kai Siegbahn was awarded the Nobel Prize in physics for the use of XPS in chemical analysis. Current research using PES involves chemical analysis, the study of surfaces and molecules that bind to them, and fine details about molecular structure. Under the appropriate conditions, the oxidation state of an atom can be determined based on the observed binding energy.

Instruments used for PES must provide a monochromatic beam (close to a single wavelength) of UV light or X rays. To ensure that the ejected electrons get to the detector before colliding with another molecule, the sample chamber must have an ultra-high vacuum. The ejected electrons then enter a detector that monitors the number and the kinetic energy of the electrons. This information is collected by a computer and may be shown graphically on the computer monitor. See Figure 1.

The typical result of a PES experiment is a plot of the number of electrons versus the kinetic energy of the ejected electrons. Figure 2 illustrates an X-ray PES spectrum used for chemical analysis.

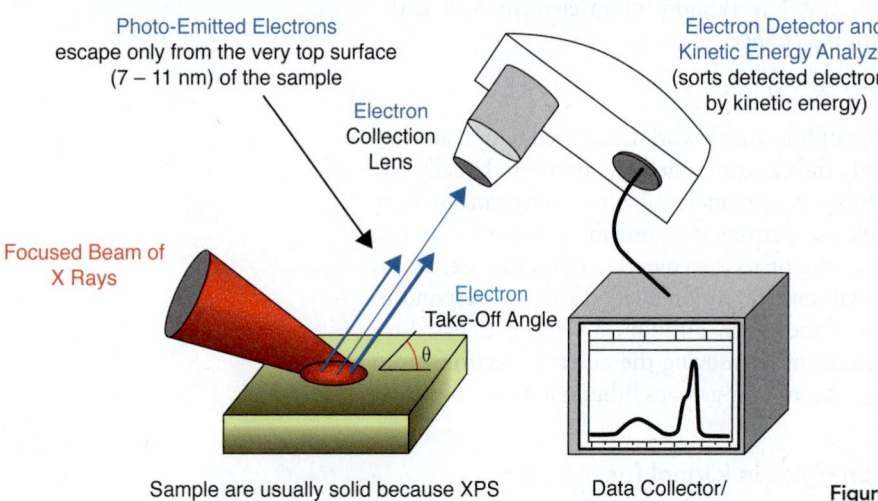

Figure 1 A conceptualized drawing of a photoelectron spectrometer.

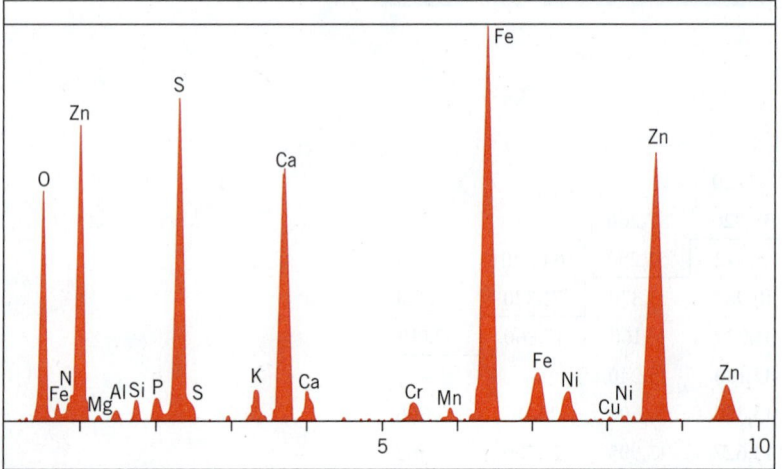

Figure 2 A conceptual graph of intensity versus kinetic energy generated by a PES experiment.

520 kJ; so the *first ionization energy* of lithium is 520 kJ/mol. The second IE of lithium is 7297 kJ/mol, and corresponds to the process

$$Li^+(g) \longrightarrow Li^{2+}(g) + e^-$$

This involves the removal of an electron from the now-exposed 1*s* core of lithium and requires more than thirteen times the energy used to remove the first electron. Removal of the third (and last) electron requires the third IE, which is 11,810 kJ/mol. In general, successive ionization energies always increase because each subsequent electron is being pulled away from an increasingly more positive ion, and that requires more work.

Earlier we saw that Einstein's explanation of the photoelectric effect, which is an ionization process, led to the concept of light quanta, or photons. This phenomena is used in a powerful research technique called photoelectron spectroscopy that is described in On the Cutting Edge 7.2.

Periodic Trends in IE

Within the periodic table there are trends in the way IE varies that are useful to know and to which we will refer in later discussions. We can see these by examining Figure 7.32, which shows a graph of how the first ionization energy varies with an element's position in the periodic table. Notice that the elements with the largest ionization energies are the nonmetals in the upper right of the periodic table, and that those with the smallest ionization energies are the metals in the lower left of the table. In general, then, the following trends are observed.

Ionization energy generally increases from bottom to top within a group and increases from left to right within a period. Overall the ionization energy increases from the lower left corner of the periodic table to the upper right corner. This is usually referred to as a diagonal trend.

■ Ionization energies are additive. For example,

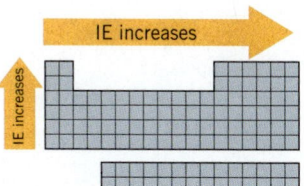

$$Li(g) \longrightarrow Li^+(g) + e^-$$
$$IE_1 = 520 \text{ kJ}$$
$$Li^+(g) \longrightarrow Li^{2+}(g) + e^-$$
$$IE_2 = 7297 \text{ kJ}$$
_____
$$Li(g) \longrightarrow Li^{2+}(g) + 2e^-$$
$$IE_{total} = IE_1 + IE_2 = 7817 \text{ kJ}$$

General variation of ionization energy (as an exothermic quantity) within the periodic table.

TOOLS

Periodic trends in ionization energy

Figure 7.32 | **Variation in first ionization energy with location in the periodic table.** Elements with the largest ionization energies are in the upper right of the periodic table. Those with the smallest ionization energies are at the lower left.

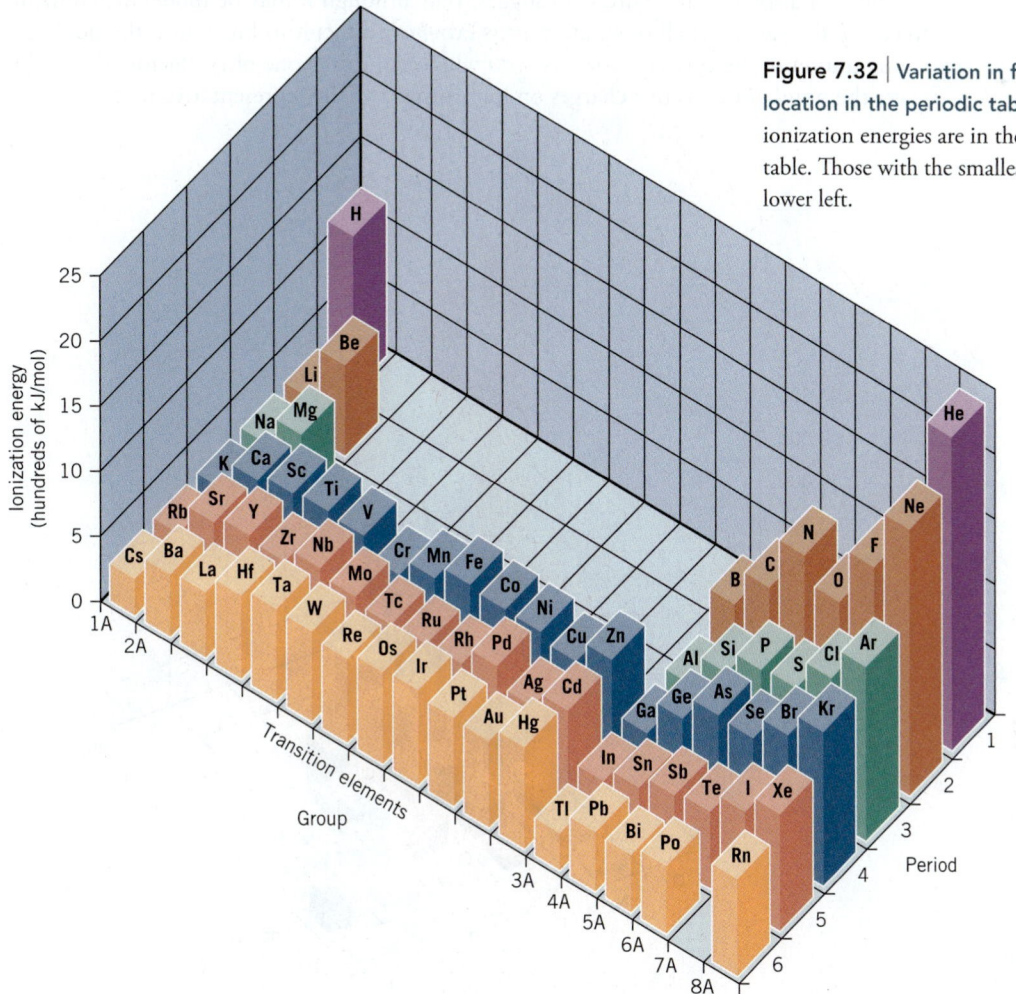

The same factors that affect atomic size also affect ionization energy. As the value of n increases going down a group, the orbitals become larger and the outer electrons are farther from the nucleus. Electrons farther from the nucleus are bound less tightly, so IE decreases from top to bottom.

As you can see, there is a gradual overall increase in IE as we move from left to right across a period, although the horizontal variation of IE is somewhat irregular. The reason for the overall trend is the increase in effective nuclear charge felt by the valence electrons as we move across a period. As we've seen, this draws the valence electrons closer to the nucleus and leads to a decrease in atomic size as we move from left to right. However, the increasing effective nuclear charge also causes the valence electrons to be held more tightly, which makes it more difficult to remove them.

The results of these trends place elements with the largest IE in the upper right-hand corner of the periodic table. It is very difficult to cause these atoms to lose electrons. In the lower left-hand corner of the table are elements that have loosely held valence electrons. These elements form positive ions relatively easily, as you learned in Chapter 2.

■ It is often helpful to remember that the trends in IE are just the opposite of the trends in atomic size within the periodic table; when size increases, IE decreases.

Stability of Noble Gas Configurations

Table 7.2 shows that, for a given representative element, successive ionization energies increase gradually until the valence shell is emptied. Then a very much larger increase in IE occurs as the core is broken into. This is illustrated graphically in Figure 7.33 for the Period 2 elements lithium through fluorine. For lithium, we see that the first electron (the 2s electron) is removed rather easily, but the second and third electrons, which come from the 1s core, are much more difficult to dislodge. For beryllium, the large jump in IE occurs after two electrons (the two 2s electrons) are removed. In fact, for all of these elements, the large increase in IE happens when the core is broken into.

The data displayed in Figure 7.33 suggest that although it may be moderately difficult to empty the valence shell of an atom, it is *extremely* difficult to break into the noble gas configuration of the core electrons. As you will learn, this is one of the factors that influences the number of positive charges on ions formed by the representative metals.

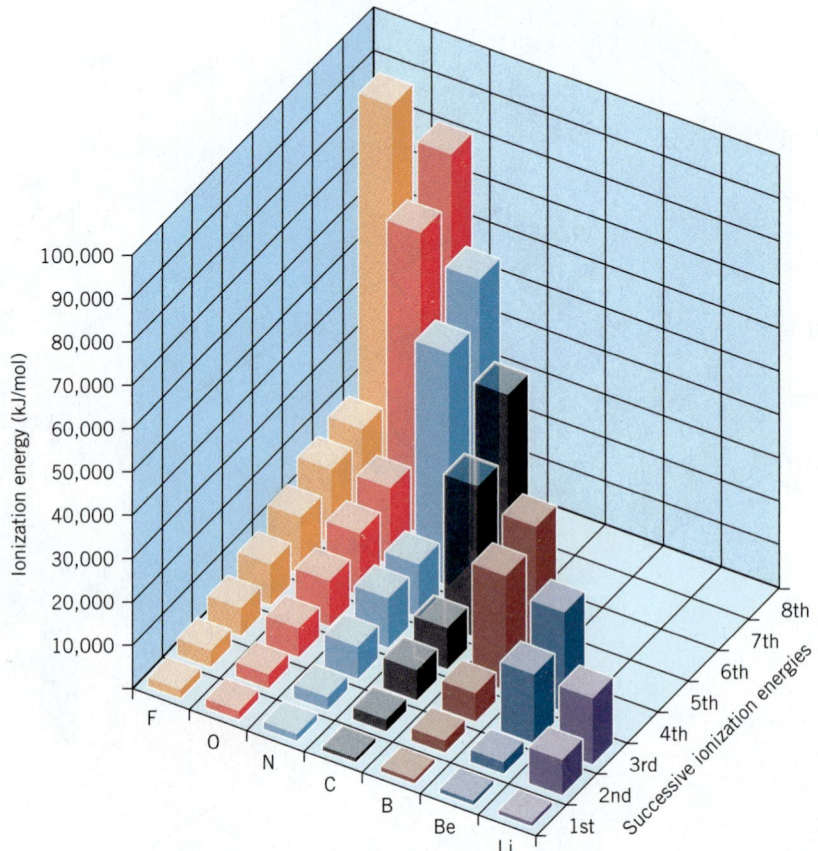

Figure 7.33 | Variations in successive ionization energies for the elements lithium through fluorine.

Practice Exercise 7.25

Use the periodic table to select the atom with the most positive value for its first ionization energy, IE: (a) Na, Sr, Be, Rb; (b) B, Al, C, Si. (*Hint:* Use the diagonal relationships of the IE values within the periodic table.)

Practice Exercise 7.26

Use Table 7.2 to determine which of the following elements is expected to have the most positive ionization energy, IE: (a) Na^+, Mg^+, H, C^{2+}; (b) Ne, F, Mg^{2+}, Li^+.

Electron Affinity

The **electron affinity** (abbreviated **EA**) *is the potential energy change associated with the addition of an electron to a gaseous atom or ion in its ground state.* For an element X, it is the change in potential energy associated with the process

$$X(g) + e^- \longrightarrow X^-(g)$$

As with ionization energy, electron affinities are usually expressed in units of kilojoules per mole, so we can also view the EA as the energy change associated with adding one mole of electrons to one mole of gaseous atoms or ions.

For nearly all the elements, the addition of one electron to the neutral atom is exothermic, and the EA is given as a negative value. This is because the incoming electron experiences an attraction to the nucleus that causes the potential energy to be lowered as the electron approaches the atom. However, when a second electron must be added, as in the formation of the oxide ion, O^{2-}, work must be done to force the electron into an already negative ion. This is an endothermic process where energy must be added and the EA has a positive value.

Notice that more energy is absorbed adding an electron to the O^- ion than is released by adding an electron to the O atom. Overall, the formation of an isolated oxide ion leads to a net increase in potential energy (so we say its formation is *endothermic*). The same applies to the formation of *any* negative ion with a charge larger than $1-$.

Change	EA (kJ/mol)
$O(g) + e^- \longrightarrow O^-(g)$	-141
$O^-(g) + e^- \longrightarrow O^{2-}(g)$	$+844$
$O(g) + 2e^- \longrightarrow O^{2-}(g)$	$+703$ (net)

Periodic Variations

The electron affinities of the representative elements are listed in Table 7.3, and we see that periodic trends in electron affinity roughly parallel those for ionization energy.

Although there are some irregularities, overall the electron affinities of the elements become more *exothermic* going from left to right across a period and from bottom to top in a group.

This shouldn't be surprising, because a valence shell that loses electrons easily (low IE) will have little attraction for additional electrons (small EA). On the other hand, a valence

TOOLS

Periodic trends in electron affinity

TABLE 7.3	Electron Affinities of the Representative Elements (kJ/mol)						
1A	2A	3A	4A	5A	6A	7A
H						
-73						
Li	Be	B	C	N	O	F
-60	0	-27	-122	$\sim+9$	-141	-328
Na	Mg	Al	Si	P	S	Cl
-53	0	-44	-134	-72	-200	-349
K	Ca	Ga	Ge	As	Se	Br
-48	-2.37	-30	-120	-77	-195	-325
Rb	Sr	In	Sn	Sb	Te	I
-47	-5	-30	-107	-101	-190	-295
Cs	Ba	Tl	Pb	Bi	Po	At
-45	-14	-30	-35	-110	-183	-270

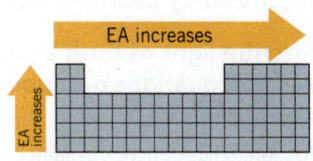

General variation of electron affinity (as an exothermic quantity) within the periodic table.

shell that holds its electrons tightly will also tend to bind an additional electron tightly.

Irregularities in Periodic Trends

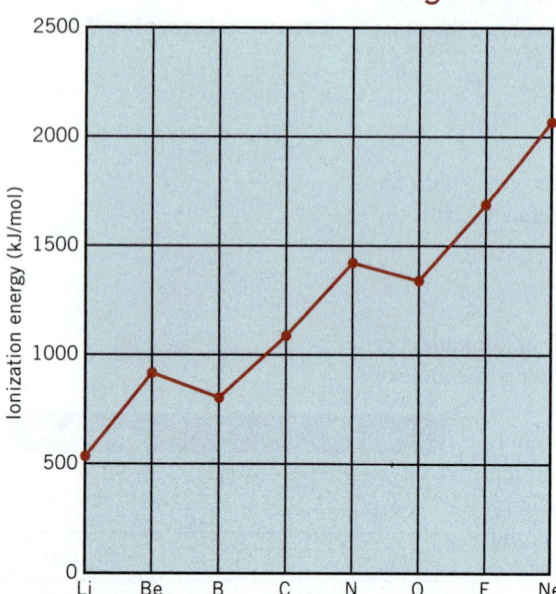

Figure 7.34 | A closer look at the variation in ionization energy for the Period 2 elements Li through Ne.

As mentioned previously, the variation in first ionization energy across a period is not a smooth one, as seen in the graph in Figure 7.34 for the elements in Period 2. The first irregularity occurs between Be and B, where the IE increases from Li to Be but then decreases from Be to B. This happens because there is a change in the nature of the subshell from which the electron is being removed. For Li and Be, the electron is removed from the 2s subshell, but at boron the first electron comes from the higher energy 2p subshell where it is not bound so tightly.

Another irregularity occurs between nitrogen and oxygen. For nitrogen, the electron that's removed comes from a singly occupied orbital. For oxygen, the electron is taken from an orbital that already contains an electron. We can diagram this as follows:

For oxygen, repulsions between the two electrons in the p orbital that's about to lose an electron help the electron leave. This "help" is absent for the electron that's about to leave the p orbital of nitrogen. As a result, it is not as difficult to remove one electron from an oxygen atom as it is to remove one electron from a nitrogen atom.

As with ionization energy, there are irregularities in the periodic trends for electron affinity. For example, the Group 2A elements have little tendency to acquire electrons because their outer shell s orbitals are filled. The incoming electron must enter a higher energy p orbital. We also see that the EA for elements in Group 5A are either endothermic or only slightly exothermic. This is because the incoming electron must enter an orbital already occupied by an electron.

One of the most interesting irregularities occurs between Periods 2 and 3 among the nonmetals. In any group, the element in Period 2 has a less exothermic electron affinity than the element below it. The reason seems to be the small size of the nonmetal atoms of Period 2, which are among the smallest elements in the periodic table. Repulsions between the many electrons in the small valence shells of these atoms leads to a lower than expected attraction for an incoming electron and a less exothermic electron affinity than the element below in Period 3.

| Summary

Organized by Learning Objective

Describe light as both a particle and a wave, and use the equations describing the wave and particle nature of light

Electromagnetic energy, or light energy, travels through space at a constant speed of 3.00×10^8 m s^{-1} in the form of waves. The **wavelength**, λ, and **frequency**, ν, of the wave are related by the equation $\lambda\nu = c$, where c is the **speed of light**. The SI unit for frequency is the **hertz** (**Hz**, 1 Hz = 1 s^{-1}). Light also behaves as if it consists of small packets of energy called **photons** or **quanta**, based on the **photoelectric effect**. The energy delivered by a photon is proportional to the frequency of the light, and is given by the equation $E = h\nu$, where h is **Planck's constant**.

White light is composed of all the frequencies visible to the eye and can be split into a **continuous spectrum**. Visible light represents only a small portion of the entire **electromagnetic spectrum**, which also includes **X rays**, **ultraviolet**, **infrared**, **microwaves**, and **radio** and **TV** waves.

Calculate the wavelength of a line in the spectrum of hydrogen and relate it to the energy levels of an electron

The occurrence of **line spectra** tells us that atoms can emit energy only in discrete amounts and suggests that the energy of the electron is **quantized**; that is, the electron is restricted to certain specific **energy levels** in an atom.

Using the Bohr model of hydrogen, explain how a line spectrum of hydrogen is generated

Niels Bohr recognized this and, although his theory was later shown to be incorrect, he was the first to propose a model that was able to account for the **Rydberg equation**. Bohr was the first to introduce the idea of **quantum numbers**.

Describe the main features of the wave mechanical model of an atom

The wave behavior of electrons and other tiny particles, which can be demonstrated by **diffraction** experiments, was suggested by de Broglie. Schrödinger applied wave theory to the atom and launched the theory we call **wave mechanics** or **quantum mechanics**. This theory tells us that electron waves in atoms are **standing waves** whose crests and **nodes** are stationary.

Define and use the three major quantum numbers and their possible values

Each standing wave, or **orbital**, is characterized by three quantum numbers, n, ℓ, and m_ℓ (**principal**, **secondary**, and **magnetic quantum numbers**, respectively). **Shells** are designated by n (which can range from 1 to ∞), **subshells** by ℓ (which can range from 0 to $n - 1$), and orbitals within subshells by m_ℓ (which can range from $-\ell$ to $+\ell$).

Describe electron spin and apply it to explain paramagnetism and diamagnetism along with the Pauli exclusion principle

The electron has magnetic properties that are explained in terms of spin. The **spin quantum number**, m_s, can have values of $+\frac{1}{2}$ or $-\frac{1}{2}$. The **Pauli exclusion** principle limits orbitals to a maximum population of two electrons with **paired spins**. Substances with unpaired electrons are **paramagnetic** and are weakly attracted to a magnetic field. Substances with only paired electrons are **diamagnetic** and are slightly repelled by a magnetic field.

Write the ground state electron configuration of an atom

The **electron configuration** of an element in its ground state is obtained by filling orbitals beginning with the 1s subshell and following the Pauli exclusion principle and **Hund's rule** (which states that electrons spread out as much as possible in orbitals of equal energy).

Explain the connection between the periodic table and electron configurations of atoms

The periodic table serves as a guide in predicting electron configurations. **Abbreviated configurations** show subshell populations outside a noble gas **core**. **Valence shell configurations** show the populations of subshells in the **outer shell** of an atom of the representative elements. Sometimes we represent electron configurations using **orbital diagrams**. The extra stability of half-filled and filled subshells changes the configurations for some transition elements such as chromium and copper.

Draw the shapes and orientations of s, p, and d orbitals in an atom

The **Heisenberg uncertainty principle** says we cannot know exactly the position and velocity of an electron both at the same instant. Consequently, wave mechanics describes the probable locations of electrons in atoms. In each orbital the electron is conveniently viewed as an **electron cloud** with a varying **electron density**. All s orbitals are spherical; each p orbital consists of two lobes with a **nodal plane** between them. A p subshell has three p orbitals whose axes are mutually perpendicular and point along the x, y, and z axes of an imaginary coordinate system centered at the nucleus. Four of the five d orbitals in a d subshell have the same shape, with four lobes of electron density each. The fifth has two lobes of electron density pointing in opposite directions along the z axis and a ring of electron density in the x–y plane.

Describe how the electron configurations of the elements and the periodic table help predict the physical and chemical properties of the elements

The amount of positive charge felt by the valence electrons of an atom is the **effective nuclear charge**. This is less than the actual nuclear charge because core electrons partially shield the valence electrons from the full positive charge of the nucleus. **Atomic radii** depend on the value of n of the valence shell orbitals and the effective nuclear charge experienced by the valence electrons. Atomic radii decrease from left to right in a period and from bottom to top in a group in the periodic table. Negative ions are larger than the atoms from which they are formed; positive ions are smaller than the atoms from which they are formed.

Ionization energy (IE) is the energy needed to remove an electron from an isolated gaseous atom, molecule, or ion in its ground state; it is endothermic. The first ionization energies of the elements increase from left to right in a period and from bottom to top in a group. (Irregularities occur in a period when the nature of the orbital from which the electron is removed changes and when the electron removed is first taken from a doubly occupied p orbital.) Successive ionization energies become larger, but there is a very large jump when the next electron must come from the noble gas core beneath the valence shell.

Electron affinity (EA) is the potential energy change associated with the addition of an electron to a gaseous atom or ion in its ground state. For atoms, the first EA is usually exothermic. When more than one electron is added to an atom, the overall potential energy change is endothermic. In general, electron affinity becomes more exothermic from left to right in a period and from bottom to top in a group. (However, the EA of second period nonmetals is less exothermic than for the nonmetals of the third period. Irregularities across a period occur when the electron being added must enter the next higher-energy subshell and when it must enter a half-filled p subshell.)

Periodic trends are regular changes of properties within the periods and groups of the periodic table. These include the variation of atomic size, ionic size, ionization energy, electron affinity, and effective nuclear charge. Periodic and group trends in properties can often be combined into a **diagonal trend** from the lower left corner (Fr) to the upper right corner (F). Atomic size, ionization energy, and electron affinity can be viewed as diagonal trends.

TOOLS

Tools for Problem Solving The following tools were introduced in this chapter. Study them carefully so that you can select the appropriate tool when needed.

Wavelength–frequency relationship (Section 7.1)

$$\lambda \times \nu = c$$

Energy of a photon (Section 7.1)

$$E = h\nu = h\frac{c}{\lambda}$$

Rydberg equation (Section 7.2)

$$\frac{1}{\lambda} = R_{\text{H}}\left(\frac{1}{n_1^2} - \frac{1}{n_2^2}\right)$$

Electron configurations using the periodic table (Section 7.8)

The periodic table can be used as an aid in writing electron configurations of the elements.

Periodic table and chemical properties (Section 7.8)

Valence electrons define chemical properties of the elements. Elements within a group (family) have similarities in their valence electrons and thus similar chemical properties.

Periodic trends in atomic and ionic size (Section 7.10)

The relative sizes of atoms and ions increase moving down the periodic table and decrease moving across the periodic table.

Periodic trends in ionization energy (Section 7.10)

The ionization energy of an atom increases moving up the periodic table and across the periodic table.

Periodic trends in electron affinity (Section 7.10)

The electron affinity of an atom is more exothermic moving up and across the periodic table.

WileyPLUS = *WileyPLUS*, an online teaching and learning solution. *Note to instructors*: Many of the end-of-chapter problems are available for assignment via the WileyPLUS system. **www.wileyplus.com.** Review Problems are presented in pairs separated by blue rules. Answers to problems whose numbers appear in blue are given in Appendix B. More challenging problems are marked with an asterisk ✱.

Review Questions

Electromagnetic Radiation

7.1 In general terms, why do we call light electromagnetic radiation?

7.2 In general, what does the term frequency imply? What is meant by the term frequency of light? What symbol is used for it, and what is the SI unit (and symbol) for frequency?

7.3 What is meant by the term wavelength of light? What symbol is used for it?

7.4 Sketch a picture of a wave and label its wavelength and its amplitude. Indicate how frequency could be measured.

7.5 Which property of light waves is a measure of the brightness of the light? Which specifies the color of the light? Which is related to the energy of the light?

7.6 Arrange the following regions of the electromagnetic spectrum in order of increasing wavelength (i.e., shortest wavelength ⟶ longest wavelength): microwave, TV, ultraviolet, visible, X rays, infrared, gamma rays. What are the general wavelengths associated with each region? Arrange them in terms of increasing frequency and in terms of increasing energy.

7.7 What wavelength range is covered by the *visible spectrum*?

7.8 Arrange the following colors of visible light in order of increasing wavelength: orange, green, blue, yellow, violet, red.

7.9 What is the equation that relates the wavelength and frequency of a light wave? (Define all symbols used.)

7.10 How is the frequency of a particular type of radiation related to the energy associated with it? (Give an equation, defining all symbols.)

7.11 What is a photon?

7.12 Show that the energy of a photon is given by the equation

$$E = \frac{hc}{\lambda}$$

7.13 Examine each of the following pairs and state which of the two has the higher energy: **(a)** microwaves and infrared, **(b)** visible light and infrared, **(c)** ultraviolet light and X rays, **(d)** visible light and ultraviolet light.

7.14 What is a quantum of energy?

Line Spectra and the Rydberg Equation

7.15 What is an atomic spectrum? How does it differ from a continuous spectrum?

7.16 What fundamental fact is implied by the existence of atomic spectra?

The Bohr Theory

7.17 Describe Niels Bohr's model of the structure of the hydrogen atom.

7.18 In qualitative terms, how did Bohr's model account for the atomic spectrum of hydrogen?

7.19 Why does the equation for the energy of an electron in a Bohr atom have a negative sign?

7.20 In what way was Bohr's theory a success? How was it a failure?

The Wave Mechanical Model

7.21 How does the behavior of very small particles differ from that of the larger, more massive objects that we encounter in everyday life? Why don't we notice this same behavior for the larger, more massive objects?

7.22 Describe the phenomenon called diffraction. How can this be used to demonstrate that de Broglie's theory was correct?

7.23 What experiment could you perform to determine whether a beam was behaving as a wave or as a stream of particles?

7.24 What is wave/particle duality?

7.25 What is the difference between a traveling wave and a standing wave?

7.26 What is the collapsing atom paradox?

7.27 How does quantum mechanics resolve the collapsing atom paradox?

7.28 What are the names used to refer to the theories that apply the matter–wave concept to electrons in atoms?

7.29 What is the term used to describe a particular waveform of a standing wave for an electron?

7.30 What are the three properties of orbitals in which we are most interested? Why?

Quantum Numbers of Electrons in Atoms

7.31 What are the allowed values of the principal quantum number? Of the secondary quantum numbers? Of the magnetic quantum numbers?

7.32 What information does each type of quantum number give for an atomic orbital?

7.33 How does the value of n in the Rydberg equation relate to the principle quantum number, n?

***7.34** Why is **(a)** the d subshell in the Period 4 designated as $3d$ and **(b)** the f subshell in Period 7 designated as $5f$?

7.35 Why does every shell contain an s subshell?

7.36 How many orbitals are found in **(a)** an s subshell, **(b)** a p subshell, **(c)** a d subshell, and **(d)** an f subshell?

7.37 If the value of m_ℓ for an electron in an atom is 2, could another electron in the same subshell have $m_\ell = -3$?

Electron Spin

7.38 What physical property of electrons leads us to propose that they spin like a toy top?

7.39 Explain the two magnetic properties that are affected by the number of unpaired electrons.

7.40 What is the Pauli exclusion principle? What effect does it have on the populating of orbitals by electrons?

7.41 What are the possible values of the spin quantum number?

7.42 Suppose an electron in an atom has the following set of quantum numbers: $n = 2$, $\ell = -1$, $m_\ell = -1$, $m_s = +\frac{1}{2}$. What set of quantum numbers is impossible for another electron in this same atom?

Energy Levels and Ground State Electron Configuration

7.43 What do we mean by the term *electronic structure*?

7.44 What is the "ground state" of an atom?

7.45 Within any given shell, how do the energies of the s, p, d, and f subshells compare? How do the energies of the orbitals belonging to a given subshell compare?

7.46 What fact about the energies of subshells was responsible for the apparent success of Bohr's theory about electronic structure?

7.47 Give the electron configurations of the elements in Period 2 of the periodic table.

Periodic Table and Ground State Electron Configuration

7.48 Using your own words, describe how to use the periodic table to write the electron configuration of an element.

7.49 Give the correct electron configurations of **(a)** Cr and **(b)** Cu. Explain why they do not have the expected electron configurations.

7.50 What is the correct electron configuration of silver? Why is this electron configuration different from the expected electron configuration.

7.51 How are the electron configurations of the elements in a given group similar? Illustrate your answer by writing shorthand configurations for the elements in Group 6A.

7.52 Define the terms *valence shell* and *valence electrons*. Define *core electrons*.

Atomic Orbitals: Shapes and Orientations

7.53 Why do we use probabilities when we discuss the position of an electron in the space surrounding the nucleus of an atom?

7.54 Sketch the approximate shape of **(a)** a 1*s* orbital and **(b)** a 2*p* orbital.

7.55 How does the size of a given type of orbital vary with *n*?

7.56 How are the *p* orbitals of a given *p* subshell oriented relative to each other?

7.57 What is a *nodal plane*? How are the number of nodal planes relate to the value of ℓ?

7.58 What is a radial node? How are the number of radial nodes related to the value of *n* and ℓ?

7.59 How many nodal planes does a *p* orbital have? How many does a *d* orbital have?

7.60 On appropriate coordinate axes, sketch the shape of the following *d* orbitals: **(a)** d_{xy}, **(b)** $d_{x^2-y^2}$, **(c)** d_{z^2}.

Periodic Table and the Properties of the Elements

7.61 What is the meaning of *effective nuclear charge?* How does the effective nuclear charge felt by the outer electrons vary going down a group? How does it change as we go from left to right across a period?

****7.62** Explain why a 3*s* electron in Al experiences a greater effective nuclear charge than a 3*p* electron.

Atomic Size

7.63 Explain why the atomic size varies across the periodic table and down the periodic table.

7.64 Explain why cations are smaller than their uncharged atoms and anions are larger than their uncharged atoms.

7.65 In what region of the periodic table are the largest atoms found? Where are the smallest atoms found?

7.66 Going from left to right in the periodic table, why are the size changes among the transition elements more gradual than those among the representative elements?

Ionization Energy

7.67 Define ionization energy. Why are ionization energies of atoms and positive ions endothermic quantities?

7.68 For oxygen, write an equation for the change associated with **(a)** its first ionization energy and **(b)** its third ionization energy.

7.69 Explain why ionization energy increases from left to right in a period and decreases from top to bottom in a group.

7.70 Why is an atom's second ionization energy always larger than its first ionization energy?

7.71 Why is the fifth ionization energy of carbon so much larger than its fourth?

7.72 Why is the first ionization energy of aluminum less than the first ionization energy of magnesium?

7.73 Why does phosphorus have a larger first ionization energy than sulfur?

Electron Affinity

7.74 Define *electron affinity*. Why are electron affinities usually exothermic?

7.75 For sulfur, write an equation for the change associated with **(a)** its first electron affinity and **(b)** its second electron affinity. How should they compare?

7.76 Why does Cl have a more exothermic electron affinity than F? Why does Br have a less exothermic electron affinity than Cl?

7.77 Why is the second electron affinity of an atom always endothermic?

7.78 How is electron affinity related to effective nuclear charge? On this basis, explain the relative magnitudes of the electron affinities of oxygen and fluorine.

| Review Problems

Electromagnetic Radiation

7.79 What is the frequency in hertz of blue light having a wavelength of 436 nm?

7.80 Ultraviolet light with a wavelength of more than 295 nm has little germicidal value. What is the frequency that corresponds to this wavelength?

7.81 Ozone protects the earth's inhabitants from the harmful effects of ultraviolet light arriving from the sun. This shielding is a maximum for UV light having a wavelength of 295 nm. What is the frequency in hertz of this light?

7.82 The meter is defined as the length of the path light travels in a vacuum during the time interval of 1/299,792,458 of a second. It is recommended that a helium–neon laser is used for defining the meter. The light from the laser has a wavelength of 632.99139822 nm. What is the frequency of this light, in hertz?

7.83 In New York City, radio station WCBS broadcasts its FM signal at a frequency of 101.1 megahertz (MHz). What is the wavelength of this signal in meters?

7.84 Sodium vapor lamps are often used in residential street lighting. Sodium gives off two yellow lines of light having frequencies of 5.09×10^{14} Hz and 5.08×10^{14} Hz as shown in Figure 7.7. What are the wavelengths of light in nanometers?

7.85 Calculate the energy in joules of a photon of red light having a frequency of 4.0×10^{14} Hz. What is the energy of one mole of these photons?

7.86 Calculate the energy in joules of a photon of green light having a wavelength of 563 nm.

Line Spectra and the Rydberg Equation

7.87 In the spectrum of hydrogen, there is a line with a wavelength of 410.3 nm. **(a)** What color is this line? **(b)** What is its frequency? **(c)** What is the energy of each of its photons?

7.88 In the spectrum of sodium, there is a line with a wavelength of 589 nm. **(a)** What color is this line? **(b)** What is its frequency? **(c)** What is the energy of each of its photons?

7.89 Use the Rydberg equation to calculate the wavelength in nanometers of the spectral line of hydrogen for which $n_2 = 6$ and $n_1 = 3$. (Report your answer using three significant figures.) In what region of the electromagnetic spectrum would this line be found? Would we be expected to see the light corresponding to this spectral line? Explain your answer.

7.90 Use the Rydberg equation to calculate the wavelength in nanometers of the spectral line of hydrogen for which $n_2 = 5$ and $n_1 = 2$. (Report your answer using three significant figures.) In what region of the electromagnetic spectrum would this line be found? Would we be expected to see the light corresponding to this spectral line? Explain your answer.

7.91 Calculate the wavelength and energy in joules of the spectral line produced in the hydrogen spectrum when an electron falls from the tenth Bohr orbit to the fourth. (Report your answer using three significant figures.) In which region of the electromagnetic spectrum (UV, visible, or infrared) is the line?

*7.92 Calculate the energy in joules and the wavelength in nanometers of the spectral line produced in the hydrogen spectrum when an electron falls from the fourth Bohr orbit to the first. (Report your answer using three significant figures.) In which region of the electromagnetic spectrum (UV, visible, or infrared) is the line?

Quantum Numbers of Electrons in Atoms

7.93 What is the letter code for a subshell with **(a)** $\ell = 1$ and **(b)** $\ell = 3$?

7.94 What is the letter code for a subshell with **(a)** $\ell = 2$ and **(b)** $\ell = 4$?

7.95 What is the value of ℓ for **(a)** an f orbital and **(b)** a d orbital?

7.96 What is the value of ℓ for **(a)** a p orbital and **(b)** a g orbital?

7.97 What are the values of n and ℓ for the subshells: **(a)** $3s$, **(b)** $5d$?

7.98 Give the values of n and ℓ for the subshells: **(a)** $4p$, **(b)** $6f$.

7.99 For the shell with $n = 6$, what are the possible values of ℓ?

7.100 What values of ℓ are possible for a shell with $n = 4$?

7.101 In a particular shell, the largest value of ℓ is 7. What is the value of n for this shell?

7.102 What is the value of n for a shell if the largest value of ℓ is 5?

7.103 What are the possible values of m_ℓ for a subshell with **(a)** $\ell = 1$ and **(b)** $\ell = 3$?

7.104 If the value of ℓ for an electron in an atom is 5, what are the possible values of m_ℓ that this electron could have?

7.105 If the value of m_ℓ for an electron in an atom is -4, what is the smallest value of ℓ that the electron could have? What is the smallest value of n that the electron could have?

7.106 How many orbitals are there in an h subshell ($\ell = 5$)? What are the possible values of m_ℓ?

7.107 Give the complete set of quantum numbers for all of the electrons that could populate the $2p$ subshell of an atom.

7.108 Give the complete set of quantum numbers for all of the electrons that could populate the $3d$ subshell of an atom.

*7.109 In an antimony atom, how many electrons have **(a)** $\ell = 1$? **(b)** $m_\ell = 2$?

*7.110 In an atom of barium, how many electrons have **(a)** $\ell = 0$ and **(b)** $m_\ell = 1$?

Energy Levels and Ground State Electron Configurations

7.111 Give the electron configurations of **(a)** S, **(b)** K, **(c)** Ti, and **(d)** Sn.

7.112 Write the electron configurations of **(a)** As, **(b)** Cl, **(c)** Ni, and **(d)** Si.

7.113 Which of the following atoms in their ground states are expected to be paramagnetic: **(a)** Mn, **(b)** As, **(c)** S, **(d)** Sr, **(e)** Ar?

7.114 Which of the following atoms in their ground states are expected to be diamagnetic: **(a)** Ba, **(b)** Se, **(c)** Zn, **(d)** Si?

7.115 How many unpaired electrons would be found in the ground state of **(a)** Mg, **(b)** P, and **(c)** V?

7.116 How many unpaired electrons would be found in the ground state of (a) Cs, (b) S, and (c) Ni?

7.117 Write the abbreviated electron configurations for (a) Ni, (b) Cs, (c) Ge, (d) Br, and (e) Bi.

7.118 Write the abbreviated electron configurations for (a) Al, (b) Se, (c) Ba, (d) Sb, and (e) Gd.

7.119 Draw complete orbital diagrams for (a) Mg and (b) Ti.

7.120 Draw complete orbital diagrams for (a) As and (b) Ni.

7.121 Draw orbital diagrams for the abbreviated configurations of (a) Ni, (b) Cs, (c) Ge, and (d) Br.

7.122 Draw orbital diagrams for the abbreviated configurations of (a) Al, (b) Se, (c) Ba, and (d) Sb.

7.123 What is the value of n for the valence shells of (a) Sn, (b) K, (c) Br, and (d) Bi?

7.124 What is the value of n for the valence shells of (a) Al, (b) Se, (c) Ba, and (d) Sb?

7.125 Give the configuration of the valence shell for (a) Na, (b) Al, (c) Ge, and (d) P.

7.126 Give the configuration of the valence shell for (a) Mg, (b) Br, (c) Ga, and (d) Pb.

7.127 Draw the orbital diagram for the valence shell of (a) Na, (b) Al, (c) Ge, and (d) P.

7.128 Draw the orbital diagram for the valence shell of (a) Mg, (b) Br, (c) Ga, and (d) Pb.

Periodic Table and Properties of the Elements

7.129 If the core electrons were 100% effective at shielding the valence electrons from the nuclear charge and the valence electrons provided no shielding for each other, what would be the effective nuclear charge felt by a valence electron in (a) Na, (b) S, (c) Cl?

7.130 If the core electrons were 100% effective at shielding the valence electrons from the nuclear charge and the valence electrons provided no shielding for each other, what would be the effective nuclear charge felt by a valence electron in (a) Mg, (b) Si, (c) Br?

7.131 Choose the larger atom in each pair: (a) Mg or S; (b) As or Bi.

7.132 Choose the larger atom in each pair: (a) Al or Ar; (b) Tl or In.

7.133 Place the following in order of expected increasing size: Ge, As, Sn, Sb.

7.134 Place the following in order of increasing size: N^{3-}, Mg^{2+}, Na^+, Ne, F^-, O^{2-}.

7.135 Choose the larger particle in each pair: (a) Na or Na^+; (b) Co^{3+} or Co^{2+}; (c) Cl or Cl^-.

7.136 Choose the larger particle in each pair: (a) S or S^{2-}; (b) Al^{3+} or Al; (c) Au^+ or Au^{3+}.

7.137 Choose the atom with the larger ionization energy in each pair: (a) B or N; (b) Se or S; (c) Cl or Ge.

7.138 Choose the atom with the larger ionization energy in each pair: (a) Li or Rb; (b) Al or F; (c) F or C.

7.139 Choose the atom with the more exothermic electron affinity in each pair: (a) I or Br; (b) Ga or As.

7.140 Choose the atom with the more exothermic electron affinity in each pair: (a) S or As; (b) Si or N.

7.141 Use the periodic table to select the element in the following list for which there is the largest difference between the second and third ionization energies: Na, Mg, Al, Si, P, Se, Cl.

*__7.142__ Use the periodic table to select the element in the following list for which there is the largest difference between the fourth and fifth ionization energies: Na, Mg, Al, Si, P, Se, Cl.

Additional Exercises

7.143 The human ear is sensitive to sound ranging from 20.0 to 2.00×10^4 Hz. The speed of sound is 330 m/s in air, and 1500 m/s under water. What is the longest and the shortest wavelength that can be heard (a) in air and (b) under water?

*__7.144__ Microwaves are used to heat food in microwave ovens. The microwave radiation is absorbed by moisture in the food. This heats the water, and as the water becomes hot, so does the food. How many photons having a wavelength of 3.00 mm would have to be absorbed by 1.00 g of water to raise its temperature by 1.00 °C?

*__7.145__ In the spectrum of hydrogen, there is a line with a wavelength of 410.3 nm. Use the Rydberg equation to calculate the value of n for the higher energy Bohr orbit involved in the emission of this light. Assume the value of n for the lower energy orbit equals 2.

*__7.146__ Calculate the wavelength in nanometers of the shortest wavelength of light emitted by a hydrogen atom. In what region of the electromagnetic spectrum is this light?

*__7.147__ Which of the following electronic transitions could lead to the emission of light from an atom?

$$1s \longrightarrow 4p \longrightarrow 3d \longrightarrow 5f \longrightarrow 4d \longrightarrow 2p$$

7.148 Calculate the wavelength of an electron moving at (a) the speed of light, (b) one-tenth the speed of light, and (c) 3×10^3 m/s. Calculate the wavelength of a baseball weighing 145 g traveling at (d) 85 mi/hr and (e) one-tenth the speed of light.

7.149 What, if anything, is wrong with the following electron configurations for atoms in their ground states? (a) $1s^2 2s^1 2p^3$, (b) $[Kr]3d^7 4s^2$, (c) $1s^2 2s^2 2p^4$, (d) $[Xe]4f^{14} 5d^8 6s^1$

7.150 Suppose students gave the following orbital diagrams for the $2s$ and $2p$ subshell in the ground state of an atom. What, if anything, is wrong with them? Are any of these electron distributions impossible?

(a) ⃝ ⇂⃝ ⇂⇂⃝ ⇂⇂⃝   **(b)** ↑⃝ ⇂⃝⃝⃝

(c) ↑⃝ ↑⃝↑⃝↑⃝   **(d)** ↑↑⃝ ↑⃝↑⃝⃝

7.151 How many electrons are in p orbitals in an atom of gallium?

7.152 What are the quantum numbers of the electrons that are lost by an atom of cobalt when it forms the ion Co^{2+}?

7.153 The removal of an electron from the hydrogen atom corresponds to raising the electron to the Bohr orbit that has $n = \infty$. On the basis of this statement, calculate the ionization energy of hydrogen in units of **(a)** joules per atom and **(b)** kilojoules per mole. Compare this number to the value given in Table 7.2.

7.154 Use orbital diagrams to illustrate what happens when an oxygen atom gains two electrons. On the basis of what you have learned about electron affinities and electron configurations, why is it extremely difficult to place a third electron on the oxygen atom?

*7.155 From the data available in this chapter, determine the ionization energy of **(a)** F^-, **(b)** O^-, and **(c)** O^{2-}. Are any of these energies exothermic?

7.156 For an oxygen atom, which requires more energy, the addition of two electrons or the removal of one electron?

*7.157 Write out the orbital diagram of N in the ground state. Without adding any more orbitals, write out all the different excited state electron configurations that can be written for N.

*7.158 The ions He^+ and Li^{2+} have line spectra that can be calculated using a modified Bohr equation. The difference is that the nuclear charge, Z, must be incorporated into the equation

$$\frac{1}{\lambda} = \frac{b}{hc}\left(\frac{Z^2}{n_{low}^2} - \frac{Z^2}{n_{high}^2}\right)$$

What would be the wavelengths of light emitted for an electron dropping from $n = 5$ to $n = 1$ for Li^{2+} and He^+ and H? What are the corresponding energies?

|Multi-Concept Problems

*7.159 A neon sign is a gas discharge tube in which electrons traveling from the cathode to the anode collide with neon atoms in the tube and knock electrons off of them. As electrons return to the neon ions and drop to lower energy levels, light is given off. How fast would an electron have to be moving to eject an electron from an atom of neon, which has a first ionization energy equal to 2080 kJ mol^{-1}?

7.160 How many grams of water could have its temperature raised by 5.0 °C by a mole of photons that have a wavelength of **(a)** 600 nm and **(b)** 300 nm?

7.161 It has been found that when the chemical bond between chlorine atoms in Cl_2 is formed, 328 kJ is released per

mole of Cl_2 formed. What is the wavelength of light that would be required to break chemical bonds between chlorine atoms?

*7.162 Using the ionization energy for sodium, would a photon with a wavelength of 23.7 nm be able to transfer enough energy to an electron in a sodium atom to cause it to ionize if all of the energy of the photon is transferred to the $3s^1$ electron of sodium? What is the maximum kinetic energy of the ejected electron?

*7.163 Using photons with a wavelength of 23.7 nm, determine the kinetic energies of all of the electrons that can be observed from a sample of boron.

|Exercises in Critical Thinking

7.164 Our understanding of the quantum mechanical atom has been developing since the early 1900s. Has quantum mechanics had any effect on the daily lives of people?

7.165 When a copper atom loses an electron to become a Cu^+ ion, what are the possible quantum numbers of the electron that was lost?

7.166 Paired electrons cancel each other's magnetic fields. Why can't unpaired electrons have opposite spins and cancel each other's magnetic fields also?

7.167 Placing a small piece of an element from Group 1A in water results in increasingly rapid and violently spectacular reactions as we progress from lithium down to cesium. What information in this chapter makes this behavior understandable?

8

The Basics of Chemical Bonding

Chapter Outline

The Empire State Building, an iconic structure of the New York City skyline, is a steel frame structure with an outer skin of stone. Just as the steel structure holds the 102-story building up, so do chemical bonds hold together the molecules that we will study in this chapter.

© Jim Zuckerman/Corbis

In Chapter 2 we classified substances into two broad categories, ionic compounds and molecular compounds. Ionic compounds, such as ordinary table salt, consist of anions and cations that bind to each other by electrostatic forces of attraction. We also said that in molecular substances, such as water, the atoms are held to each other by the sharing of electrons. Now that you've learned about the electronic structures of atoms, we can explore the attractions between atoms or ions, called **chemical bonds**, in greater depth. Our goal is to gain some insight into the reasons certain combinations of atoms prefer electron transfer and the formation of ions (leading to *ionic bonding*), while other combinations bind by electron sharing (leading to *covalent bonding*).

As with electronic structure, models of chemical bonding have also evolved, and in this chapter we introduce you to relatively simple theories. Although more complex theories exist (some of which we describe in Chapters 9, 21, and 22), the basic concepts you will study in this chapter still find many useful applications in modern chemical thought.

This Chapter in Context

LEARNING OBJECTIVES

After reading this chapter, you should be able to:

- describe the necessary condition for bond formation that produces a stable compound
- explain the factors involved in ionic bonding
- write electron configurations of ions
- write Lewis symbols for atoms and ions
- describe covalent bonds through the energetics of bond formation, octet rule, and multiple bonds
- explain how electronegativity affects the polarity of covalent bonds and the reactivity of the elements
- draw Lewis structures for covalent molecules and calculate formal charges for individual atoms in a molecule
- draw and explain resonance structures
- classify organic compounds and identify functional groups

8.1 | Energy Requirements for Bond Formation

For a *stable* compound to be formed from its atoms, the reaction must be exothermic. In other words, ΔH_f° must be negative. Two examples of such reactions are the formation of H_2O by the reaction of H_2 with O_2 in Figure 8.1a and the formation of NaCl by the reaction of Na with Cl_2, shown in Figure 8.1b. Both of these reactions release a large

(a)

(b)

Courtesy NASA

Richard Megna/Fundamental Photographs

Figure 8.1 | (*a*) **Reaction of hydrogen with oxygen.** This photo shows the three main engines of the space shuttle at full power an instant before the booster rockets fire and lift the space craft from its launch pad. The violent reaction of hydrogen with oxygen provides the thrust produced by the main engines. (*b*) **Exothermic reaction of sodium with chlorine.** A small piece of melted sodium immediately ignites when dipped into a flask containing chlorine gas, producing light and releasing a lot of heat. The smoke coming from the flask is composed of fine crystals of sodium chloride.

amount of energy when they occur and produce very stable compounds. Typically, compounds with positive heats of formation tend to be unstable and can decompose violently, such as nitroglycerine, a powerful explosive.

When a reaction is exothermic, the potential energy of the particles in the system decreases. Therefore, to understand the formation of chemical bonds, we need to understand how bond formation can lead to a lowering of the potential energy of the atoms involved. In the pages ahead we will examine this for both ionic and molecular substances.

■ Endothermic processes increase PE
Exothermic processes decrease PE

8.2 | Ionic Bonding

When sodium chloride is formed from its elements, each sodium atom loses one electron to form a sodium ion, Na^+, and each chlorine atom gains one electron to become a chloride ion, Cl^-.

$$Na \longrightarrow Na^+ + e^-$$

$$Cl + e^- \longrightarrow Cl^-$$

Once formed, these ions become tightly packed together, as illustrated in Figure 8.2, because their opposite charges attract. This attraction between positive and negative ions in an ionic compound is what we call an **ionic bond**.

The attraction between Na^+ and Cl^- ions may seem reasonable, but *why* are electrons transferred between these and other atoms? Why does sodium form Na^+ and not Na^- or Na^{2+}? And why does chlorine form Cl^- instead of Cl^+ or Cl^{2-}? To answer such questions we have to consider factors that are related to the potential energy of the system of reactants and products. If we consider the formation of a compound from its elements, a stable compound must be lower in potential energy as compared to its elements and the reaction must be exothermic.

Figure 8.2 | **The packing of ions in NaCl.** Electrostatic forces hold the ions in place in the solid. These forces constitute ionic bonds.

Importance of the Lattice Energy

In the formation of an ionic compound such as NaCl, it is not the electron transfer itself that leads to a stable substance. We can see this if we examine the ionization energy of Na and the electron affinity of Cl. On the scale of a mole of gaseous atoms, we have

$$Na(g) \longrightarrow Na^+(g) + e^- \qquad +495 \text{ kJ mol}^{-1} \quad \text{(IE of sodium)}$$

$$Cl(g) + e^- \longrightarrow Cl^-(g) \qquad \underline{-349 \text{ kJ mol}^{-1}} \quad \text{(EA of chlorine)}$$

$$\text{Net} \quad +146 \text{ kJ mol}^{-1}$$

Notice that forming gaseous sodium and chloride ions from gaseous sodium and chlorine atoms *increases* the potential energy of the system. This tells us that if the IE and EA were the only energy changes involved, ionic sodium chloride would not form from atoms of sodium and chlorine. So where does the stability of the compound come from?

In the calculation above, we looked at the formation of gaseous ions, but salt is not a gas; it's a solid. Therefore, we need to see how the energy would change if the gaseous ions are condensed to give the solid, and to do this we must examine a quantity called the lattice energy.

The **lattice energy** is the energy released when ions are brought together from an infinite separation (i.e., a cloud of gaseous ions) to form one mole of the solid compound.[1]

■ The name lattice energy comes from the word *lattice*, which is used to describe the regular pattern of ions or atoms in a crystal. Lattice energies are exothermic and so are given negative signs.

---

[1] In some books, the lattice energy is defined as *the energy needed to separate the ions in one mole of a solid to give a cloud of gaseous ions*. This would correspond to the change

$$NaCl(s) \longrightarrow Na^+(g) + Cl^-(g)$$

The *magnitude* of the energy change is the same as for the process described above in the body of the text, but the algebraic sign of the energy change would be positive instead of negative (endothermic instead of exothermic).

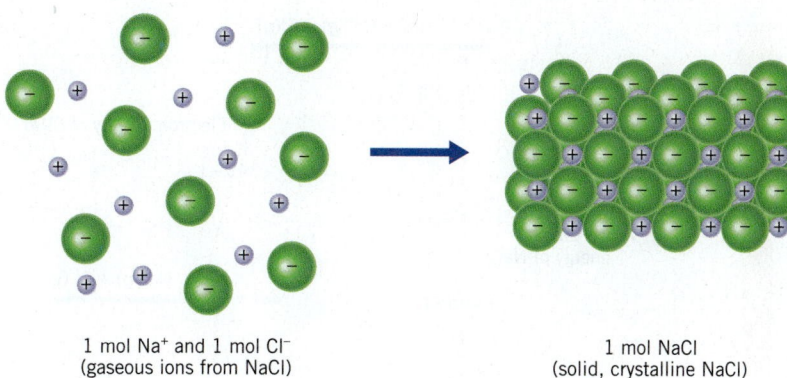

Figure 8.3 | **The lattice energy of NaCl.** The energy released when one mole of gaseous Na^+ and one mole of gaseous Cl^- condense to form one mole of solid NaCl is the lattice energy. For NaCl, this amounts to -786 kJ.

1 mol Na^+ and 1 mol Cl^-
(gaseous ions from NaCl)

1 mol NaCl
(solid, crystalline NaCl)

For sodium chloride, the change associated with the lattice energy is pictured in Figure 8.3, and in equation form can be represented as

$$Na^+(g) + Cl^-(g) \longrightarrow NaCl(s)$$

Because Na^+ and Cl^- ions attract each other, their potential energy decreases as they come together, so the process releases energy and is exothermic. In fact, the lattice energy for sodium chloride is very exothermic and is equal to -786 kJ mol^{-1}. If we include the lattice energy along with the ionization energy of Na and electron affinity of Cl, we have

$$Na(g) \longrightarrow Na^+(g) + e^- \qquad +495 \text{ kJ} \quad \text{(IE of sodium)}$$

$$Cl(g) + e^- \longrightarrow Cl^-(g) \qquad -349 \text{ kJ} \quad \text{(EA of chlorine)}$$

$$Na^+(g) + Cl^-(g) \longrightarrow NaCl(s) \qquad \underline{-786 \text{ kJ}} \quad \text{(lattice energy)}$$

$$\text{Net} \quad -640 \text{ kJ}$$

Thus, the release of energy equal to the lattice energy provides a large net lowering of the potential energy as solid NaCl is formed. We can also say that it is the lattice energy that provides the stabilization necessary for the formation of NaCl. Without it, the compound could not exist.

Determining Lattice Energies

You may be wondering about our starting point in these energy calculations—namely, gaseous sodium atoms and chlorine atoms. In nature, sodium is a solid metal and chlorine consists of gaseous Cl_2 molecules. Let's look at a complete analysis of the energy changes taking this into account by using an enthalpy diagram similar to the type discussed in Chapter 6. It is called a **Born–Haber cycle** after the scientists who were the first to use it to calculate lattice energies.

In Chapter 6 you learned that the enthalpy change for a process is the same regardless of the path we follow from start to finish. With this in mind, we will construct two different paths from the free elements (solid sodium and gaseous chlorine molecules) to the solid ionic compound, sodium chloride, as shown in Figure 8.4.

The starting point for both paths is the same: the free elements sodium and chlorine. The direct path starts at the lower left takes us to the product, NaCl(s), and has as its enthalpy change the heat of formation of NaCl, ΔH_f°.

$$Na(s) + \tfrac{1}{2}Cl(g) \longrightarrow NaCl(s) \qquad \Delta H_f^\circ = -411 \text{ kJ}$$

The alternative path is divided into a number of steps. The first two, labeled 1 and 2, have ΔH° values that can be measured experimentally and both are endothermic. They change Na(s) and $Cl_2(g)$ into gaseous atoms, Na(g) and Cl(g). The next two steps (Steps 3 and 4) change the atoms to ions, first by the endothermic ionization energy (IE) of Na followed by the exothermic electron affinity (EA) of Cl. This brings us to the gaseous ions,

Figure 8.4 | The Born–Haber cycle for the formation of sodium chloride. One path leads directly from the elements Na(s) and Cl₂(g) to NaCl(s). The other, upper, path follows a series of changes that include the lattice energy and take us to the same final product.

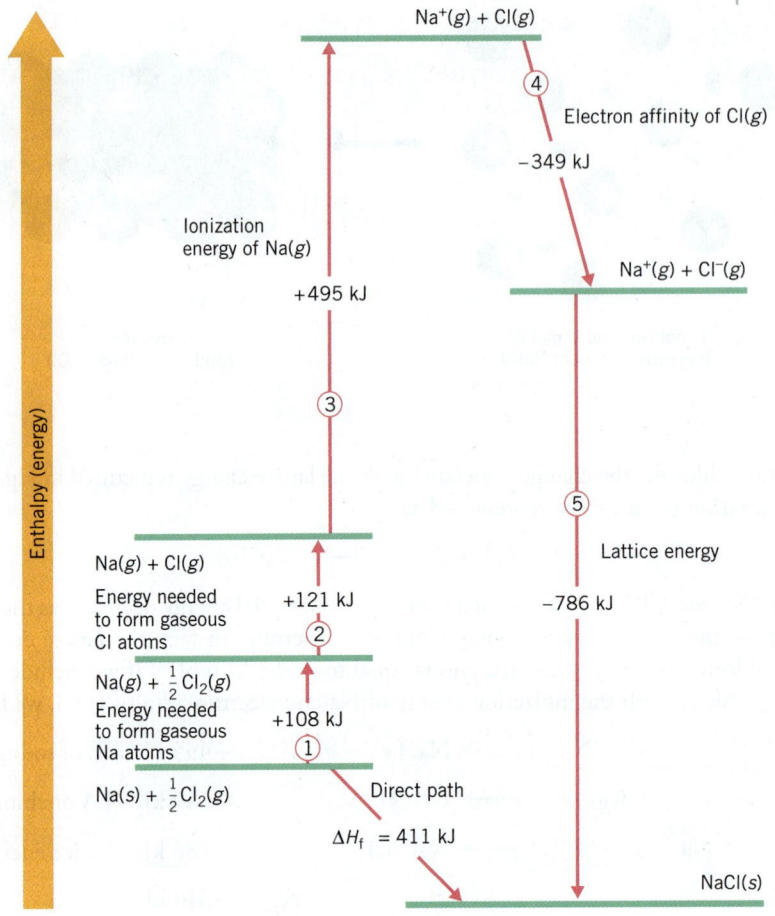

$Na^+(g) + Cl^-(g)$, depicted on the right in Figure 8.4. Notice that at this point, if we add all the energy changes, the ions are at a considerably higher energy than the reactants. If these were the only energy terms involved in the formation of NaCl, the heat of formation would be endothermic and the compound would be unstable; it could not be formed by the direct combination of the elements.

The last step on the right in Figure 8.4 (Step 5) finally brings us to solid NaCl and corresponds to the lattice energy. To make the net energy changes the same along both paths, the energy released when the ions condense to form the solid must equal −786 kJ. Therefore, the calculated lattice energy of NaCl must be −786 kJ mol⁻¹.

What we see in this analysis is that it is the lattice energy that permits NaCl to be formed from its elements. In fact, for any ionic compound, the chief stabilizing influence is the lattice energy, which when released is large enough to overcome the net energy input required to form the ions from the elements.

Besides affecting the ability of ionic compounds to form, lattice energies are also important in determining the solubilities of ionic compounds in water and other solvents. We will explore this topic in Chapter 12.

Effect of Ionic Size and Charge on the Lattice Energy

The lattice energies of some ionic compounds are given in Table 8.1. As you can see, they are all very large exothermic quantities. Their magnitudes depend on a number of factors, including the charges on the ions and their sizes.

For two ions with charges q_1 and q_2 separated by a distance r, the potential energy can be calculated from **Coulomb's law:**

$$E = k\frac{q_1 q_2}{r}$$

(8.1)

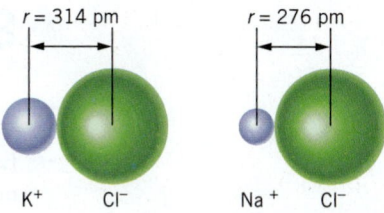

Figure 8.5 | The effect of ionic size on the lattice energy. Because Na^+ is smaller than K^+, the distance between ions is smaller in NaCl than in KCl, which in turn causes the lattice energy of NaCl to be larger than that of KCl.

TABLE 8.1	Lattice Energies of Some Ionic Compounds	
Compound	Ions	Lattice Energy (kJ mol^{-1})
LiCl	Li^+ and Cl^-	−853
NaCl	Na^+ and Cl^-	−786
KCl	K^+ and Cl^-	−715
LiF	Li^+ and F^-	−1036
CaCl$_2$	Ca^{2+} and Cl^-	−2258
AlCl$_3$	Al^{3+} and Cl^-	−5492
CaO	Ca^{2+} and O^{2-}	−3401
Al$_2$O$_3$	Al^{3+} and O^{2-}	−15,916

where k is a proportionality constant.[2] In an ionic solid, q_1 and q_2 have opposite signs, so E is calculated to be a negative quantity. That's why the lattice energy has a negative sign.

When the charges on the ions become larger (i.e., when q_1 and q_2 become larger), E becomes more negative, which means the potential energy becomes lower (more negative). This explains why salts of Ca^{2+} have larger lattice energies than comparable salts of Na^+, and those containing Al^{3+} have even larger lattice energies.

Because the distance between the ions, r, appears in the denominator in Equation 8.1, E becomes a larger negative quantity as r becomes smaller. As a result, compounds formed from small ions (which can get close together) have larger lattice energies than those formed from large ions. For example, a Na^+ ion is smaller than a K^+ ion. Figure 8.5 shows that the cation–anion distances observed in solid NaCl are smaller than in KCl, which causes NaCl to have a larger lattice energy than KCl.

Choose the ionic compound that would have the larger lattice energy: (a) LiBr or CaBr$_2$, (b) LiF or KF, (c) CaO or BaS. (*Hint:* What two factors affect lattice energy?)

Practice Exercise 8.1

Construct an energy diagram similar to the one in Figure 8.4 to determine the heat of formation of LiF for LiF using data from Chapter 7 and

Practice Exercise 8.2

$$Li(s) \longrightarrow Li(g) \qquad \Delta H_f^\circ = 155 \text{ kJ/mol}$$

$$\tfrac{1}{2}F_2(g) \longrightarrow F(g) \qquad \Delta H_f^\circ = 78.5 \text{ kJ/mol}$$

Factors That Determine the Formation of Cations and Anions

We've shown that for an ionic compound to form, the energy lowering produced by release of the lattice energy must exceed the energy rise associated with forming the ions from the neutral atoms. This requires that the cation be formed from an atom of relatively low ionization energy; such atoms are found among the metals. Nonmetals, at the upper right of the periodic table, have large ionization energies but generally exothermic electron affinities. A nonmetal atom has little tendency to lose electrons, but forming an anion from a nonmetal atom can actually help to lower the energy. As a result, cations are formed from metals and anions are formed from nonmetals because this leads to the greatest lowering of the energy. In fact, metals combine with nonmetals to form ionic compounds simply because ionic bonding is favored energetically over other types whenever atoms with small ionization energies combine with atoms that have large exothermic electron affinities.

---

[2] In this case, we're using the symbol q to mean electric charge, not heat as we did in Chapter 6. Because the number of letters in the alphabet is limited, it's not uncommon in science for the same letter to be used to stand for different quantities. This usually doesn't present a problem as long as the symbol is defined in the context in which it is used.

8.3 | Octet Rule and Electron Configurations of Ions

Earlier we raised the question about why sodium forms Na^+ and chlorine forms Cl^-, and not Na^{2+} or Cl^{2-}. To answer this question, we will have to look at the electron configurations of the atoms and the ions.

Stability of the Noble Gas Configuration

Let's begin by examining what happens when a sodium atom loses an electron. The electron configuration of Na is

$$Na \qquad 1s^2 2s^2 2p^6 3s^1$$

The electron that is lost is the one least tightly held, which is the single outer $3s$ electron. The electronic structure of the Na^+ ion, then, is

$$Na^+ \qquad 1s^2 2s^2 2p^6$$

Notice that this is identical to the electron configuration of the noble gas neon. We say the Na^+ ion has achieved a noble gas configuration.

The energy needed to remove the first electron from Na is relatively small and can easily be recovered by the release of the lattice energy when an ionic compound containing Na^+ is formed. However, removal of a second electron from sodium is very difficult because it involves breaking into the $2s^2 2p^6$ core. Forming the Na^{2+} ion is therefore very endothermic, as we can see by adding the first and second ionization energies (Table 7.2).

$$Na(g) \longrightarrow Na^+(g) + e^- \qquad \text{1st IE} = 496 \text{ kJ mol}^{-1}$$

$$\underline{Na^+(g) \longrightarrow Na^{2+}(g) + e^- \qquad \text{2nd IE} = 4563 \text{ kJ mol}^{-1}}$$

$$\text{Total} \quad 5059 \text{ kJ mol}^{-1}$$

This value is so large because the second electron is removed from the noble gas core.

Even though the Na^{2+} ion would give a compound such as $NaCl_2$ a larger lattice energy than NaCl (e.g., see the lattice energy for $CaCl_2$), it would not be large enough to make the formation of the compound exothermic. As a result, $NaCl_2$ cannot form. The same applies to other compounds of sodium, so when sodium forms a cation, electron loss stops once the Na^+ ion is formed and a noble gas electron configuration is reached.

Similar situations exist for other metals, too. For example, the first two electrons to be removed from a calcium atom come from the $4s$ valence shell (outer shell).

$$Ca \qquad 1s^2 2s^2 2p^6 3s^2 3p^6 4s^2$$

$$Ca^{2+} \qquad 1s^2 2s^2 2p^6 3s^2 3p^6$$

■ For calcium:

1st IE = 590 kJ/mol
2nd IE = 1145 kJ/mol
3rd IE = 4912 kJ/mol

The energy needed to accomplish this can be recovered by the release of the lattice energy when a compound containing Ca^{2+} forms. Electron loss ceases at this point, however, because of the huge amount of energy needed to break into the noble gas core. As a result, a calcium atom loses just two electrons when it reacts.

For sodium and calcium, as well as other metals of Groups 1A and 2A and aluminum, the large ionization energy of the noble gas core just below their outer shells limits the number of electrons they lose, so the ions that are formed have noble gas electron configurations.

Nonmetals also tend to achieve noble gas configurations when they form anions. For example, when a chlorine atom reacts, it gains one electron.

$$Cl \qquad 1s^2 2s^2 2p^6 3s^2 3p^5$$

$$Cl^- \qquad 1s^2 2s^2 2p^6 3s^2 3p^6$$

At this point, we have a noble gas configuration (that of argon). Electron gain ceases, because if another electron were to be added, it would have to enter an orbital in the next higher shell, which is very energetically unfavorable. Similar arguments apply to the other nonmetals as well.

The Octet Rule

In the preceding discussion you learned that a balance of energy factors causes many atoms to form ions that have a noble gas electron configuration. Historically, this is expressed in the form of a generalization:

When they form ions, atoms of most of the representative elements tend to gain or lose electrons until they have obtained an electron configuration identical to that of the nearest noble gas.

Because all the noble gases except helium have outer shells with eight electrons, this rule has become known as the **octet rule**, which can be stated as follows: *Atoms tend to gain or lose electrons until they have achieved an outer shell that contains an **octet of electrons** (eight electrons)*.

Cations That Do Not Obey the Octet Rule

The octet rule, as applied to ionic compounds, really works well only for the cations of the Group 1A and 2A metals and aluminum, and for the anions of the nonmetals. It does not work well for the transition metals and post-transition metals (the metals that follow a row of transition metals).

To obtain the correct electron configurations of the cations of these metals, we apply the following rules:

Obtaining the Electron Configuration of a Cation
1. The first electrons to be lost by an atom or ion are those from the shell with the largest value of n (i.e., the outer s and p electrons).
2. As electrons are removed from a given shell, they come from the highest-energy occupied subshell first, before any are removed from a lower-energy subshell. This means that the p electrons are removed before the s electrons.
3. Electrons in d orbitals are removed after the ns electrons since they are in an $n - 1$ shell.

Let's look at two examples.

Tin (a post-transition metal) forms two ions, Sn^{2+} and Sn^{4+}. The electron configurations are

$$Sn \qquad [Kr]\, 4d^{10}5s^{2}5p^{2}$$

$$Sn^{2+} \qquad [Kr]\, 4d^{10}5s^{2}$$

$$Sn^{4+} \qquad [Kr]\, 4d^{10}$$

Notice that the Sn^{2+} ion is formed by the loss of the higher-energy $5p$ electrons first. Then, further loss of the two $5s$ electrons gives the Sn^{4+} ion. However, neither of these ions has a noble gas configuration.

For the transition elements, the first electrons lost are the s electrons of the outer shell. Then, if additional electrons are lost, they come from the underlying d subshell. An example is iron, which forms the ions Fe^{2+} and Fe^{3+}. The element iron has the electron configuration

$$Fe \qquad [Ar]\, 3d^{6}4s^{2}$$

Iron loses its $4s$ electrons fairly easily to give Fe^{2+}, which has the electron configuration

$$Fe^{2+} \qquad [Ar]\, 3d^{6}$$

TOOLS

Octet rule

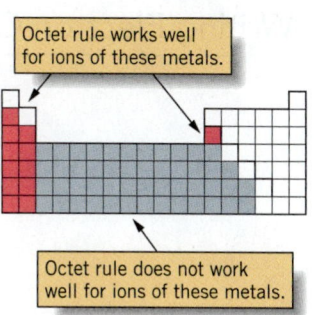

Octet rule works well for ions of these metals.

Octet rule does not work well for ions of these metals.

TOOLS

Order in which electrons are lost from an atom

■ These rules also give the correct electron configurations for the metals of Groups 1A and 2A.

The Fe^{3+} ion results when another electron is removed, this time from the $3d$ subshell.

$$Fe^{3+} \qquad [Ar]\, 3d^5$$

Iron is able to form Fe^{3+} because the $3d$ subshell is close in energy to the $4s$, so it is not very difficult to remove the third electron. Notice once again that the first electrons to be removed come from the shell with the largest value of n (the $4s$ subshell). Then, after this shell is emptied, the next electrons are removed from the shell below.

Because so many of the transition elements are able to form ions in a way similar to that of iron, the ability to form more than one positive ion is usually cited as one of the characteristic properties of the transition elements. Frequently, one of the ions formed has a 2+ charge, which arises from the loss of the two outer s electrons. Ions with larger positive charges result when additional d electrons are lost. Unfortunately, it is not easy to predict exactly which ions can form for a given transition metal, nor is it simple to predict their relative stabilities with respect to being oxidized or reduced.

Example 8.1
Writing Electron Configurations of Ions

How do the electron configurations change (a) when a nitrogen atom forms the N^{3-} ion and (b) when an antimony atom forms the Sb^{3+} ion?

Analysis: For the nonmetals the octet rule works, so the ion that is formed by nitrogen will have a noble gas configuration.

Antimony is a post-transition element, so we don't expect its cation to obey the octet rule. We will have to examine the electron configuration of the neutral atoms and determine which electrons are lost.

Assembling the Tools: The octet rule serves as the tool to determine the electron configuration of N^{3-}. To determine the electron configuration of Sb^{3+}, our tools will be the procedure for writing abbreviated electron configurations for atoms as well as the rules describing the order in which electrons are lost from an atom.

Solution: (a) Following the method for writing abbreviated configurations, the electron configuration for nitrogen is

$$N \qquad [He]\, 2s^2 2p^3$$

To form N^{3-}, three electrons are gained. These enter the $2p$ subshell because it is the lowest available energy level and will complete the octet; the configuration for the ion is therefore

$$N^{3-} \qquad [He]\, 2s^2 2p^6$$

(b) The rules tell us that when a cation is formed, electrons are removed first from the outer shell of the atom (the shell with the largest value of the principal quantum number, n). Within a given shell, electrons are always removed first from the subshell highest in energy. Let's begin with the ground state electron configuration for antimony:

$$Sb \qquad [Kr]\, 4d^{10} 5s^2 5p^3$$

To form the Sb^{3+} ion, three electrons must be removed. These will come from the outer shell, which has $n = 5$. Within this shell, the energies of the subshells increase in the order $s < p < d < f$. Therefore, the $5p$ subshell is higher in energy than the $5s$, so all three electrons are removed from the $5p$. This gives

$$Sb^{3+} \qquad [Kr]\, 4d^{10} 5s^2$$

Are the Answers Reasonable? In Chapter 2 we used the periodic table to figure out the charges on the anions of the nonmetals. For nitrogen, we would take three steps to the right to get to the nearest noble gas, neon. The electron configuration we obtained for N^{3-} is that of neon, so our answer should be correct.

For antimony, we had to remove three electrons, which completely emptied the $5p$ subshell. That's also good news, because ions do not tend to have partially filled s or p subshells (although partially filled d subshells are not uncommon for the transition metals). If we had taken the electrons from any other subshells, the Sb^{3+} ion would have had a partially filled $5p$ subshell.

Example 8.2
Writing Electron Configurations of Ions

What is the electron configuration of the V^{3+} ion? Give the orbital diagram for the ion.

Analysis: To obtain the electron configuration of a cation, begin with the electron configuration of the neutral atom. In this case we will remove three electrons to obtain the electron configuration of the ion.

Assembling the Tools: Once again, our tools will be the method for deriving electron configurations of atoms and the rules that tell us the order in which electrons are lost by an atom. We will also use Hund's rule, which tells us how electrons populate orbitals of a given subshell.

Solution: Following the usual method, the electron configuration of vanadium is

$$V \qquad \underbrace{1s^2 2s^2 2p^6 3s^2 3p^6}_{\text{Argon core}} 3d^3 4s^2 \qquad \text{or} \qquad [Ar]3d^3 4s^2$$

Notice that we've written the configuration showing the outer shell $4s$ electrons farthest to the right. To form the V^{3+} cation, three electrons must be removed from the neutral atom. We have to keep in mind that the electrons are lost first from the occupied shell with highest n. Therefore, the first two come from the $4s$ subshell and the third comes from the $3d$. This means we won't have to take any from the $3s$ or $3p$ subshells, so the argon core will remain intact.

$$V^{3+} \qquad [Ar]\, 3d^2$$

To form the orbital diagram, we show all five orbitals of the $3d$ subshell and then spread out the two electrons with spins unpaired (Hund's rule). This gives

$$V^{3+} \qquad [Ar] \;\; \textcircled{\scriptsize ↑}\textcircled{\scriptsize ↑}\bigcirc\bigcirc\bigcirc$$
$$3d$$

Is the Answer Reasonable? First, we check that we've written the correct electron configuration of vanadium, which we have. (A quick count of the electrons gives 23, which is the atomic number of vanadium.) We've also taken electrons away from the atom following the rules, so the electron configuration of the ion seems okay. Finally, we remembered to show all five orbitals of the $3d$ subshell, even though only two of them are occupied.

What is wrong with the following electron configuration of the In^+ ion? What should the electron configuration be?

$$In^+ \qquad 1s^2 2s^2 2p^6 3s^2 3p^6 3d^{10} 4s^2 4p^6 4d^{10} 5s^1 5p^1$$

(*Hint:* Check the rules that tell us the order in which electrons are lost by an atom or ion.)

Practice Exercise 8.3

How do the electron configurations change when a chromium atom forms the ions: (a) Cr^{2+}, (b) Cr^{3+}, (c) Cr^{6+}?

Practice Exercise 8.4

How are the electron configurations of S^{2-}, Cl^-, and K^+ related?

Practice Exercise 8.5

Bettmann/©Corbis

Gilbert N. Lewis, chemistry professor at the University of California, helped develop theories of chemical bonding. In 1916, he proposed that atoms form bonds by sharing pairs of electrons.

Lewis symbols

8.4 | Lewis Symbols: Keeping Track of Valence Electrons

We have seen how the valence shells of atoms change when electrons are transferred to form ions. We will soon see the way many atoms share their valence electrons with each other when they form covalent bonds. In these discussions it will help to keep track of valence electrons. We can do this using a simple bookkeeping device called Lewis symbols, named after their inventor, the American chemist, G. N. Lewis (1875–1946).

To draw the **Lewis symbol** for an element, we write its chemical symbol surrounded by dots (or some other similar mark), each of which represents a valence electron of the atom. For example, the element lithium has the Lewis symbol

$$\text{Li}\cdot$$

in which the single dot stands for lithium's single valence electron. In fact, each element in Group 1A has a similar Lewis symbol, because each has only one valence electron. The Lewis symbols for all of the Group 1A metals are

$$\text{Li}\cdot \quad \text{Na}\cdot \quad \text{K}\cdot \quad \text{Rb}\cdot \quad \text{Cs}\cdot$$

The Lewis symbols for the eight A-group elements of Period 2 are[3]

Group	1A	2A	3A	4A	5A	6A	7A	8A
Symbol	Li·	·Be·	·Ḃ·	·Ċ·	·N̈·	:Ö·	:F̈·	:N̈e·

The elements below each of these in their respective groups have the same electron structure in their Lewis symbols. Notice that when an atom has more than four valence electrons, the additional electrons are shown to be paired with others. Also notice that *for the representative elements, the group number is equal to the number of valence electrons* when the North American convention for numbering groups is followed.

Example 8.3
Writing Lewis Symbols

What is the Lewis symbol for arsenic?

Analysis: We start with the number of valence electrons, which we can obtain from the group number. Then we distribute the electrons (dots) around the chemical symbol.

Assembling the Tools: Our tool is the method described above for constructing the Lewis symbol.

Solution: The symbol for arsenic is As and we find it in Group 5A. The element therefore has five valence electrons. The first four are placed around the symbol for arsenic as follows:

$$\cdot\overset{\displaystyle\cdot}{\text{As}}\cdot$$

The fifth electron is paired with one of the first four. This gives

$$\cdot\overset{\displaystyle\cdot}{\text{As}}:$$

---

[3] For beryllium, boron, and carbon, the number of unpaired electrons in the Lewis symbol does not agree with the number predicted from the atom's electron configuration. Boron, for example, has two electrons paired in its $2s$ orbital and a third electron in one of its $2p$ orbitals; therefore, there is actually only one unpaired electron in a boron atom. The Lewis symbols are drawn as shown, however, because when beryllium, boron, and carbon form bonds, they *behave* as if they have two, three, and four unpaired electrons, respectively.

The location of the fifth electron doesn't really matter, so equally valid Lewis symbols are

$$:\ddot{As}\cdot \quad \text{or} \quad \cdot\ddot{As}\cdot \quad \text{or} \quad \cdot\ddot{As}\cdot$$

Is the Answer Reasonable? There's not much to check here. Have we got the correct chemical symbol? Yes. Do we have the right number of dots? Yes.

Draw the Lewis structures for lead and tellurium. (*Hint:* Which groups are lead and tellurium in?)

Practice Exercise 8.6

Using Lewis Symbols to Represent Ionic Compounds

Although we will use Lewis symbols mostly to follow the fate of valence electrons in covalent bonds, they can also be used to describe what happens during the formation of ions. For example, when a sodium atom reacts with a chlorine atom, the sodium loses an electron to the chlorine, which we might depict as

$$Na \overset{\curvearrowright}{} \cdot \ddot{\underset{..}{Cl}}: \longrightarrow Na^+ + \left[:\ddot{\underset{..}{Cl}}:\right]^-$$

The valence shell of the sodium atom is emptied, so no dots remain. The outer shell of chlorine, which formerly had seven electrons, gains one to give a total of eight. The brackets are drawn around the chloride ion to show that all eight electrons are the exclusive property of the Cl⁻ ion.

We can diagram a similar reaction between calcium and chlorine atoms.

$$:\ddot{\underset{..}{Cl}} \overset{\curvearrowleft}{} \cdot Ca \overset{\curvearrowright}{} \cdot \ddot{\underset{..}{Cl}}: \longrightarrow Ca^{2+} + 2\left[:\ddot{\underset{..}{Cl}}:\right]^-$$

Example 8.4
Using Lewis Symbols

Use Lewis symbols to diagram the reaction that occurs between sodium and oxygen atoms to give Na^+ and O^{2-} ions.

Analysis: For electrical neutrality, the formula will be Na_2O, so we will use two sodium atoms and one oxygen atom. Each sodium atom will lose one electron to give Na^+ and the oxygen will gain two electrons to give O^{2-}.

Assembling the Tools: The principal tool is the method for constructing the Lewis symbol for an element and its ions.

Solution: Our first task is to draw the Lewis symbols for Na and O.

$$Na\cdot \quad \cdot\ddot{O}\cdot$$

It takes two electrons to complete the octet around oxygen. Each Na supplies one. Therefore,

$$Na \overset{\curvearrowright}{} \cdot\ddot{\underset{..}{O}}\cdot \overset{\curvearrowleft}{} Na \longrightarrow 2Na^+ + \left[:\ddot{\underset{..}{O}}:\right]^{2-}$$

Notice that we have put brackets around the oxide ion.

Is the Answer Reasonable? We have accounted for all the valence electrons (an important check), the net charge is the same on both sides of the arrow (the equation is balanced), and we've placed the brackets around the oxide ion to emphasize that the octet belongs exclusively to that ion.

Use Lewis symbols to diagram the formation of CaI_2 from Ca and I atoms. (*Hint:* Begin by determining how many electrons are gained or lost by each atom.)

Practice Exercise 8.7

Diagram the reaction between magnesium and oxygen atoms to give Mg^{2+} and O^{2-} ions.

Practice Exercise 8.8

8.5 | Covalent Bonds

Most of the substances we encounter in our daily lives are not ionic. Instead, they are composed of electrically neutral *molecules*. The chemical bonds that bind the atoms to each other in such molecules are electrical in nature, but arise from the sharing of electrons rather than by electron transfer.

Energy Changes on Bond Formation

In Section 8.2 we saw that for ionic bonding to occur, the energy-lowering effect of the lattice energy must be greater than the combined net energy-raising effects of the formation of gaseous atoms, ionization energy (IE), and electron affinity (EA). Many times this is not possible. For example, nonmetals, which often have large ionization energies, combine together to form molecules. In such cases, nature uses a different way to lower the energy—electron sharing.

Let's look at what happens when two hydrogen atoms join to form a hydrogen molecule (Figure 8.6). As the two atoms approach each other, the electron of each atom begins to feel the attraction of both nuclei. This causes the electron density around each nucleus to shift toward the region between the two atoms. Therefore, as the distance between the nuclei decreases, there is an increase in the probability of finding either electron near either nucleus. In effect, as the molecule is formed, each of the hydrogen atoms in the H_2 molecule shares the two electrons.

In the H_2 molecule, while the buildup of electron density between the two atoms attracts both nuclei and pulls them together, the two nuclei also repel each other, as they are the same charge. In the molecule that forms, therefore, the atoms are held at a distance at which all these attractions and repulsions are balanced. Overall, the nuclei are kept from separating, and the net force of attraction produced by sharing the pair of electrons is called a **covalent bond**.

Bond Energy and Bond Length

Every covalent bond is characterized by two physical quantities—namely, the average distance between the nuclei held together by the bond and the amount of energy needed to separate the two atoms to produce neutral atoms again. In the hydrogen molecule, the nuclei are at an average distance, or **bond length** (or **bond distance**), of 75 pm. Because a covalent bond holds atoms together, work must be done (energy must be supplied) to separate them. The amount of energy needed to "break" the bond (or the energy released when the bond is formed) is called the **bond energy**.

■ As the distance between the nuclei and the electron cloud that lies between them decreases, the potential energy decreases until the repulsions between the two nuclei cause the potential energy to increase.

Figure 8.7 shows how the potential energy changes when two hydrogen atoms come together to form H_2. We see that the minimum potential energy occurs at a bond length of 75 pm, and that 1 mol of hydrogen molecules is more stable than 2 mol of hydrogen atoms by 435 kJ. In other words, the bond energy of H_2 is 435 kJ/mol.

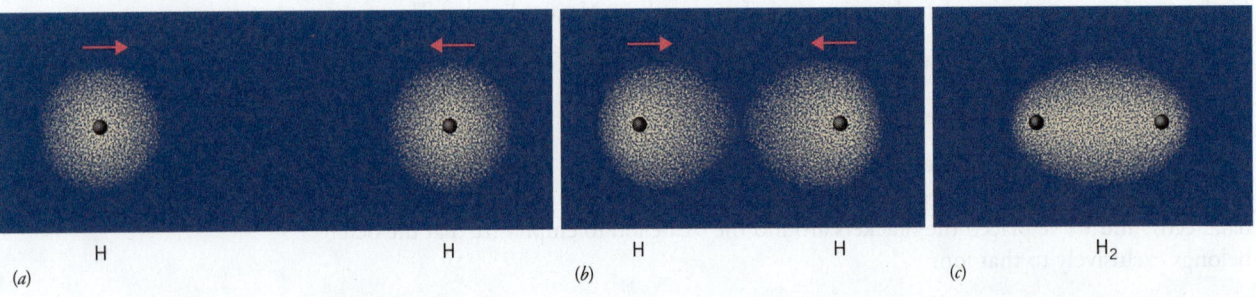

Figure 8.6 | Formation of a covalent bond between two hydrogen atoms. (*a*) Two H atoms separated by a large distance. (*b*) As the atoms approach each other, their electron densities are pulled into the region between the two nuclei. (*c*) In the H_2 molecule, the electron density is concentrated between the nuclei. Both electrons in the bond are shared between both nuclei.

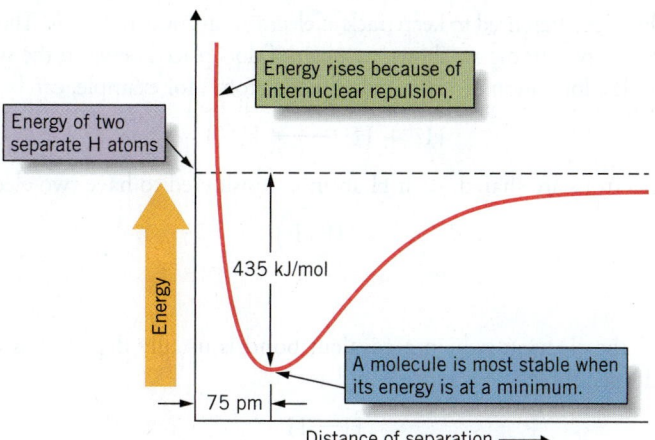

Figure 8.7 | Changes in the total potential energy of two hydrogen atoms as they form H$_2$. The energy of the molecule reaches a minimum when there is a balance between the attractions and repulsions.

In the figure: Energy rises because of internuclear repulsion. Energy of two separate H atoms. 435 kJ/mol. A molecule is most stable when its energy is at a minimum. 75 pm. Energy. Distance of separation →

In general, forming any covalent bond leads to a lowering of the energy and breaking covalent bonds leads to an increase in energy. As noted in Chapter 6, the net energy change we observe in a chemical reaction is the result of energies associated with the breaking and making of bonds.

CHEMISTRY AND CURRENT AFFAIRS (8.1)

Sunlight and Skin Cancer

The ability of light to provide the energy for chemical reactions enables life to exist on our planet. Green plants absorb sunlight and, with the help of chlorophyll, convert carbon dioxide and water into carbohydrates (e.g., sugars and cellulose), which are essential constituents of the food chain. However, not all effects of sunlight are so beneficial.

© Mick Roessler/©Corbis

Dawn of a new day brings the risk of skin cancer to those particularly susceptible. Fortunately, understanding the risk allows us to protect ourselves with clothing and sunblock creams.

As you know, light packs energy that's proportional to its frequency, and if the photons that are absorbed by a substance have enough energy, they can rupture chemical bonds and initiate chemical reactions. Light that is able to do this has frequencies in the ultraviolet (UV) region of the electromagnetic spectrum, and the sunlight bombarding the earth contains substantial amounts of UV radiation. Fortunately, about 45 to 55 km above the Earth's surface, in the stratosphere, is a layer of ozone (O_3) that absorbs most of the incoming UV rays, protecting life on the planet. However, some UV radiation does get through, and the part of the spectrum of most concern is called "UV-B" with wavelengths between 280 and 320 nm.

What makes UV-B so dangerous is its ability to affect the DNA in our cells. (The structure of DNA and its replication is discussed in Chapter 22.) Absorption of UV radiation causes constituents of the DNA, called pyrimidine bases, to undergo reactions that form bonds between them. This causes transcription errors when DNA replicates during cell division, giving rise to genetic mutations that can lead to skin cancers. These skin cancers fall into three classes—basal cell carcinomas, squamous cell carcinomas, and melanomas (the last being the most dangerous). Each year there are more than 1 million cases of skin cancer diagnosed. It is estimated that more than 90% of skin cancers are due to absorption of UV-B radiation.

Pairing of Electrons in Covalent Bonds

Before joining to form H$_2$, each of the separate hydrogen atoms has one electron in a $1s$ orbital. When these electrons are shared, the $1s$ orbital of each atom is, in a sense, filled. Because the electrons now share the same space, they become paired as required by the Pauli exclusion principle; that is, m_s is $+\frac{1}{2}$ for one of the electrons and $-\frac{1}{2}$ for the other. Since electrons involved almost always become paired when atoms form covalent bonds, a covalent bond is sometimes referred to as an **electron pair bond**.

■ In Chapter 7 you learned that when two electrons occupy the same orbital and therefore share the same space, their spins must be paired. The pairing of electrons is an important part of the formation of a covalent bond.

Lewis symbols are often used to keep track of electrons in covalent bonds. The electrons that are shared between two atoms are shown as a pair of dots placed between the symbols for the bonded atoms. The formation of H_2 from hydrogen atoms, for example, can be depicted as

$$H\cdot \; + \; H\cdot \; \longrightarrow \; H\!:\!H$$

Because the electrons are shared, each H atom is considered to have two electrons.

(Colored circles emphasize that each of the H atoms act as if the two electrons are theirs.)

For simplicity, the electron pair in a covalent bond is usually depicted as a single dash. Thus, the hydrogen molecule is represented as

$$H\!-\!H$$

A formula drawn this way is called a **Lewis formula** or **Lewis structure**. It is also called a **structural formula** because it shows which atoms are present in the molecule *and* how they are attached to each other.

The Octet Rule and Covalent Bonding

You have seen that when a nonmetal atom forms an anion, electrons are gained until the *s* and *p* subshells of its valence shell are completed. The tendency of a nonmetal atom to finish with a completed valence shell, usually consisting of eight electrons, also influences the number of electrons the atom tends to acquire by sharing, and it thereby affects the number of covalent bonds the atom forms.

Hydrogen, with just one electron in its 1*s* orbital, completes its valence shell by obtaining a share of just one electron from another atom, so a hydrogen atom forms just one covalent bond. When this other atom is hydrogen, the H_2 molecule is formed.

Many atoms form covalent bonds by sharing enough electrons to give them complete *s* and *p* subshells in their outer shells. This is the noble gas configuration mentioned earlier and is the basis of the octet rule described in Section 8.3. As applied to covalent bonding, the octet rule can be stated as follows:

Octet rule and covalent bonding

When atoms form covalent bonds, they tend to share sufficient electrons so as to achieve an outer shell having eight electrons.

Often, the octet rule can be used to explain the number of covalent bonds an atom forms. This number normally equals the number of electrons the atom must acquire to have a total of eight (an octet) in its outer shell. For instance, the halogens (Group 7A) all have seven valence electrons. The Lewis symbol for a typical member of this group, chlorine, is

$$\cdot \ddot{\underset{\cdot\cdot}{C}}l\!:$$

We can see that only one electron is needed to complete its octet. Of course, chlorine can actually gain this electron and become a chloride ion. This is what it does when it forms an ionic compound such as sodium chloride (NaCl). When chlorine combines with another nonmetal, however, the complete transfer of an electron is not energetically favorable. Therefore, in forming such molecules as HCl or Cl_2, chlorine gets the one electron it needs by forming a covalent bond.

$$H\cdot \; + \; \cdot \ddot{\underset{\cdot\cdot}{C}}l\!: \; \longrightarrow \; H\!:\!\ddot{\underset{\cdot\cdot}{C}}l\!: \quad \text{or} \quad H\!-\!\ddot{\underset{\cdot\cdot}{C}}l\!:$$

$$:\!\ddot{\underset{\cdot\cdot}{C}}l\cdot \; + \; \cdot \ddot{\underset{\cdot\cdot}{C}}l\!: \; \longrightarrow \; :\!\ddot{\underset{\cdot\cdot}{C}}l\!:\!\ddot{\underset{\cdot\cdot}{C}}l\!: \quad \text{or} \quad :\!\ddot{\underset{\cdot\cdot}{C}}l\!-\!\ddot{\underset{\cdot\cdot}{C}}l\!:$$

There are many nonmetals that form more than one covalent bond. For example, the three most important elements in biochemical systems are carbon, nitrogen, and oxygen.

$$\cdot \dot{C}\cdot \qquad \cdot \dot{\ddot{N}}\cdot \qquad \cdot \ddot{\underset{\cdot\cdot}{O}}\cdot$$

You've already encountered the simplest hydrogen compounds of these elements: methane, CH_4, ammonia, NH_3, and water, H_2O. Their Lewis structures are

$$\begin{array}{ccc} \text{H} & \text{H} & \text{H} \\ \text{H} \!:\! \ddot{\text{C}} \!:\! \text{H} & \text{H} \!:\! \ddot{\text{N}} \!:\! \text{H} & \text{H} \!:\! \ddot{\ddot{\text{O}}} \!: \\ \text{H} & & \end{array}$$

$$\text{or} \qquad\qquad \text{or} \qquad\qquad \text{or}$$

$$\begin{array}{ccc} \text{H} & \text{H} & \text{H} \\ | & | & | \\ \text{H} \!-\! \text{C} \!-\! \text{H} & \text{H} \!-\! \ddot{\text{N}} \!-\! \text{H} & \text{H} \!-\! \ddot{\ddot{\text{O}}} \!: \\ | & & \\ \text{H} & & \end{array}$$

| Methane | Ammonia | Water |

In the ball-and-stick drawings of the molecules, the "sticks" represent the covalent bonds between the atoms.

Multiple Bonds

The bond produced by the sharing of *one* pair of electrons between two atoms is called a **single bond**. There are, however, many molecules in which more than one pair of electrons are shared between two atoms. For example, we can diagram the formation of the bonds in CO_2 as follows.

$$:\!\ddot{\text{O}} \cdot \rightleftarrows \cdot \text{C} \cdot \quad \cdot \ddot{\text{O}}\!: \longrightarrow :\!\ddot{\text{O}} :: \text{C} :: \ddot{\text{O}}\!:$$

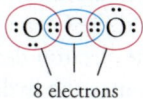

The carbon atom shares two of its valence electrons with one oxygen atom and two with the other. At the same time, each oxygen atom shares two electrons with carbon. The result is the formation of two **double bonds**. Notice that in the Lewis formula, both of the shared electron pairs are placed between the symbols for the two atoms joined by the double bond. Once again, if we circle the valence shell electrons that "belong" to each atom, we see that each has an octet.

$$:\!\ddot{\text{O}} :: \text{C} :: \ddot{\text{O}}\!:$$

8 electrons

The Lewis structure for CO_2, using dashes, is

$$:\!\ddot{\text{O}} \!=\! \text{C} \!=\! \ddot{\text{O}}\!:$$

Sometimes three pairs of electrons are shared between two atoms. The most abundant gas in the atmosphere, nitrogen, occurs in the form of diatomic molecules, N_2. As we've seen, the Lewis symbol for nitrogen is

$$\cdot \dot{\ddot{\text{N}}} \!:$$

and each nitrogen atom needs three electrons to complete its octet. When the N_2 molecule is formed, each of the nitrogen atoms shares three electrons with the other.

$$:\!\dot{\text{N}} \cdot \rightleftarrows \cdot \dot{\text{N}}\!: \longrightarrow :\!\text{N} ::: \text{N}\!:$$

■ The arrows here simply indicate how the electrons can combine to form the electron pair bonds in the molecule.

■ How we place the unshared pairs of electrons around the oxygen is unimportant. Two equally valid Lewis structure for CO_2 are

$$:\!\ddot{\text{O}} \!=\! \text{C} \!=\! \ddot{\text{O}}\!: \quad \text{and} \quad \ddot{\text{O}} \!=\! \text{C} \!-\! \ddot{\ddot{\text{O}}}$$

The result is called a **triple bond**. Again, notice that we place all three electron pairs of the bond between the two atoms. We count all of these electrons as though they belong to both of the atoms. Each nitrogen atom therefore has an octet.

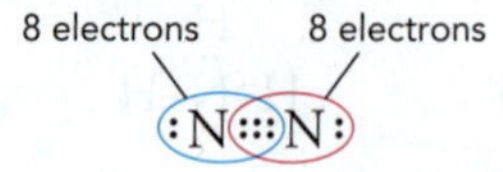

8 electrons   8 electrons

:N:::N:

The triple bond is usually represented by three dashes, so the bonding in the N_2 molecule is normally shown as

:N≡N:

Practice Exercise 8.9

In order to complete an octet, how many bonds do the following atoms need to make with (*a*) sulfur, (*b*) phosphorus, and (*c*) silicon? (*Hint:* How many valence electrons does each atom have?)

Practice Exercise 8.10

For each atom that does not have an octet, how many electrons does it need or how many should it lose?

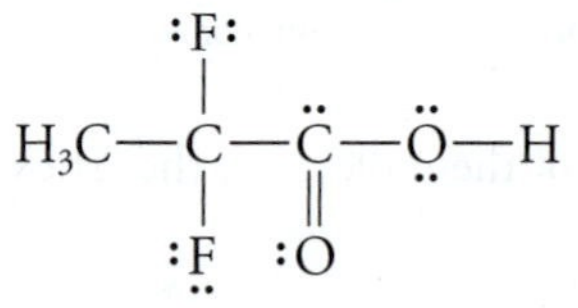

$$H_3C—\overset{\overset{\displaystyle :\ddot{F}:}{|}}{\underset{\underset{\displaystyle :\ddot{F}:}{|}}{C}}—\overset{\overset{\displaystyle }{\ddot{C}}}{\underset{\underset{\displaystyle :O}{||}}{}}—\ddot{\ddot{O}}—H$$

8.6 | Bond Polarity and Electronegativity

When two identical atoms form a covalent bond, as in H_2 or Cl_2, each atom has an equal share of the bond's electron pair. The electron density at both ends of the bond is the same, because the electrons are equally attracted to both nuclei. However, when different kinds of atoms combine, as in HCl, one nucleus usually attracts the electrons in the bond more strongly than the other.

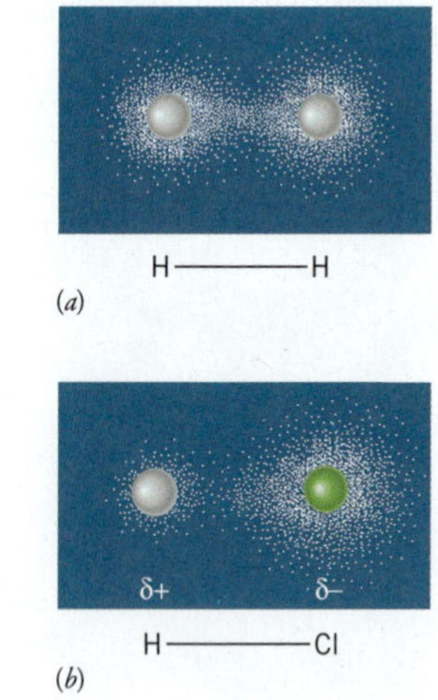

Polar and Nonpolar Bonds

The result of unequal attractions for the bonding electrons is an unbalanced distribution of electron density within the bond. For example, chlorine atoms have a greater attraction for electrons in a bond than do hydrogen atoms. In the HCl molecule, therefore, the electron cloud is pulled more tightly around the Cl, and that end of the molecule experiences a slight buildup of negative charge. The electron density that shifts toward the chlorine is removed from the hydrogen, which causes the hydrogen end to acquire a slight positive charge. These charges are less than full $1+$ and $1-$ charges and are called **partial charges**, which are usually indicated by the lowercase Greek letter delta, δ (see Figure 8.8). Partial charges can also be indicated on Lewis structures. For example,

$$H—\ddot{\underset{\cdot\cdot}{Cl}}:$$
$$\delta^+ \quad \delta^-$$

A bond that carries partial positive and negative charges on opposite ends is called a **polar covalent bond**, or often simply a **polar bond** (the word *covalent* is understood). The term *polar* comes from the notion of *poles* of equal but opposite charge at either end of the bond. Because *two poles* of electric charge are involved, the bond is said to be an **electric dipole**.

The polar bond in HCl causes the molecule as a whole to have opposite charges on either end, so the HCl molecule as a whole is an **electric dipole**. We say that HCl is a

(*a*)

H———H

$\delta+$ $\delta-$

H———Cl

(*b*)

Figure 8.8 | Equal and unequal sharing of electrons in a covalent bond. Each of the diagrams illustrates the distribution of electron density of the shared electron pair in a bond. (*a*) In H_2, the electron density in the bond is spread equally over both atoms. (*b*) In HCl, more than half of the electron density of the bond is concentrated around chlorine, causing opposite ends of the bond to carry partial electrical charges.

polar molecule. The magnitude of its polarity is expressed quantitatively by its **dipole moment** (symbol μ), which is equal to the amount of charge on either end of the molecule, q, multiplied by the distance between the charges, r.

Dipole moment

$$\mu = q \times r$$

$(8.2)^4$

Table 8.2 lists the dipole moments and bond lengths for some diatomic molecules. The dipole moments are reported in **debye** units (symbol D), where $1\ D = 3.34 \times 10^{-30}$ C m (coulomb × meter).

By separate experiments, it is possible to measure both μ and r (which corresponds to the bond length in a diatomic molecule such as HCl). Knowledge of μ and r makes it possible to calculate the amount of charge on opposite ends of the dipole. For HCl, such calculations show that q equals 0.17 electronic charge units, which means the hydrogen carries a charge of $+0.17e^-$ and the chlorine a charge of $-0.17e^-$.

One of the main reasons we are concerned about whether a molecule is polar or not is because many physical properties, such as melting point and boiling point, are affected by it. This is because polar molecules attract each other more strongly than do nonpolar molecules. The positive end of one polar molecule attracts the negative end of another. The strength of the attraction depends on both the amount of charge on either end of the molecule and the distance between the charges; in other words, it depends on the molecule's dipole moment.

TABLE 8.2	Dipole Moments and Bond Lengths for Some Diatomic Molecules	
Compound	Dipole Moment (D)	Bond Length (pm)
HF	1.83	91.7
HCl	1.11	127
HBr	0.83	141
HI	0.45	161
CO	0.11	113
NO	0.16	115

Source: National Institutes of Standards and Technology. P. J. Linstrom and W. G. Mallard, Eds., *NIST Chemistry WebBook*, NIST Standard Reference Database Number 69, National Institute of Standards and Technology, Gaithersburg, MD 20899, http://webbook.nist.gov (retrieved September 15, 2013).

Example 8.5
Calculating the Charge on the End of a Polar Molecule

The HF molecule has a dipole moment of 1.83 D and a bond length of 91.7 pm. What is the amount of charge, in electronic charge units, on either end of the bond?

Analysis: This is going to involve substituting values into Equation 8.2. We will have to be especially careful of the units since the debye has charge expressed in coulombs, but the answer is in electronic charge units.

Assembling the Tools: The primary tool for solving the problem is Equation 8.2.

$$\mu = q \times r$$

Since we are also going to have to convert between coulombs (C) and electronic charge units, we can find the charge on an electron from inside the rear cover of the book (i.e., an electronic charge unit) which equals 1.602×10^{-19} C

$$1\ e^- \Leftrightarrow 1.602 \times 10^{-19}\ C$$

Solution: We will solve Equation 8.2 for q.

$$q = \frac{\mu}{r}$$

The debye unit, D, $= 3.34 \times 10^{-30}$ C m, so the dipole moment of HF is

$$\mu = (1.83\ D)\,(3.34 \times 10^{-30}\ C\ m/D) = 6.11 \times 10^{-30}\ C\ m$$

---

[4] Once again, we're using the symbol q to mean electric charge, not heat as in Chapter 6.

The SI prefix p (pico) means $\times\ 10^{-12}$, so the bond length $r = 91.7 \times 10^{-12}$ m. Substituting in the equation above gives

$$q = \frac{6.11 \times 10^{-30}\ \text{C m}}{91.7 \times 10^{-12}\ \text{m}} = 6.66 \times 10^{-20}\ \text{C}$$

The value of q in electronic charge units is therefore

$$q = 6.66 \times 10^{-20}\ \text{C} \times \left(\frac{1\ e^-}{1.602 \times 10^{-19}\ \text{C}}\right) = 0.416\ e^-$$

As in HCl, the hydrogen carries the positive charge, so the charge on the hydrogen end of the molecule is $+0.416\ e^-$ and the charge on the fluorine end is $-0.416\ e^-$.

Is the Answer Reasonable? If we look at the units, we see that they cancel correctly, so that gives us confidence that we've done the calculation correctly. The fact that our answer is between zero and one electronic charge unit, and therefore a partial electrical charge, further suggests that we've solved the problem correctly.

Practice Exercise 8.11

The chlorine end of the chlorine monoxide molecule carries a charge of $+0.167\ e^-$. The bond length is 154.6 pm. Calculate the dipole moment of the molecule in debye units. (*Hint:* Be sure to convert the charge to coulombs.)

Practice Exercise 8.12

Although isolated Na^+ and Cl^- ions are unstable, these ions can exist in the gaseous state as *ion pairs*. An ion pair consists of an NaCl unit in which the bond length is 236 pm. The dipole moment of the ion pair is 9.00 D. What are the actual amounts of charge on the sodium and chlorine atoms in this NaCl pair? What percentage of full $1+$ and $1-$ charges are these? (This is the *percentage ionic character* in the NaCl pair.)

Electronegativity

The degree to which a covalent bond is polar depends on the difference in the abilities of the bonded atoms to attract electrons. The greater the difference, the more polar the bond, and the more the electron density is shifted toward the atom that attracts electrons more.

The term that we use to describe the attraction an atom has for the electrons in a bond is called **electronegativity**. In HCl, for example, chlorine is *more electronegative* than hydrogen. This causes the electron pair of the covalent bond to spend more of its time around the more electronegative atom, which is why the Cl end of the bond acquires a partial negative charge.

The first scientist to develop numerical values for electronegativity was Linus Pauling (1901–1994). He observed that polar bonds have a bond energy larger than would be expected if the opposite ends of the bonds were electrically neutral. Pauling reasoned that the extra bond energy is caused by the attraction between the partial charges on opposite ends of the bond. By estimating the extra bond energy, he was able to develop a scale of electronegativities for the elements. Other scientists have used different approaches to measuring electronegativities, with similar results.

A set of numerical values for the electronegativities of the elements is shown in Figure 8.9. These data are useful because the *difference* in electronegativity provides an estimate of the degree of polarity of a bond. For instance, the data tell us fluorine is more electronegative than chlorine, so we expect HF to be more polar than HCl. (This is confirmed by the larger dipole moment of the HF molecule.) In addition, the relative magnitudes of the electronegativities indicate which ends of a bond carry the partial positive and negative charges. Thus, hydrogen is less electronegative than either fluorine or chlorine, so in both of these molecules the hydrogen bears the partial positive charge.

H—F̈:   H—C̈l:
δ⁺ δ⁻   δ⁺ δ⁻

Linus Pauling (1901–1994) contributed greatly to our understanding of chemical bonding. He was the winner of two Nobel Prizes, in 1954 for chemistry and in 1962 for peace.

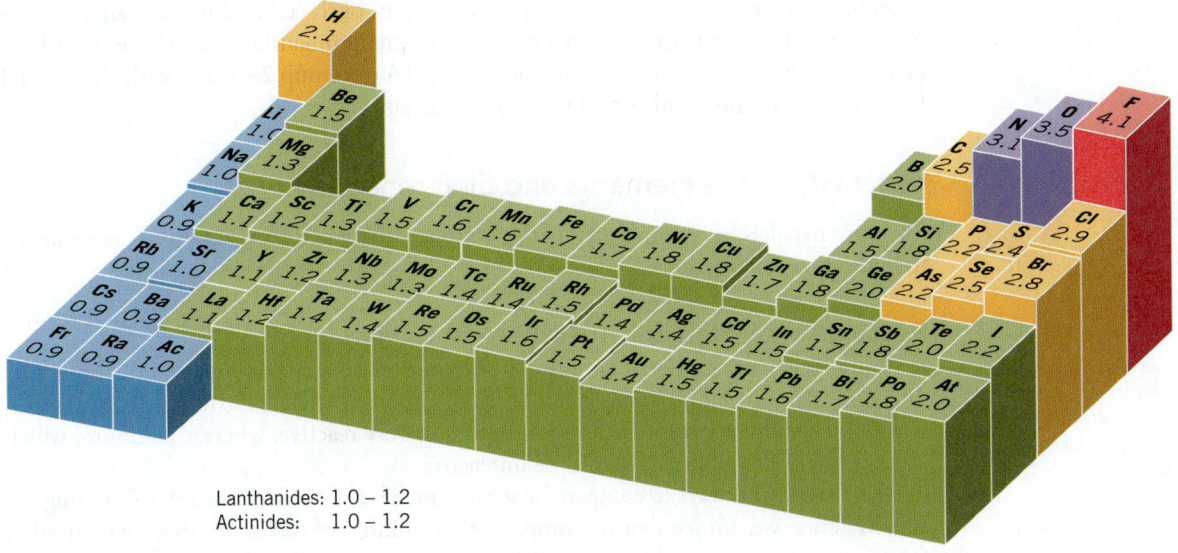

Figure 8.9 | The electronegativities of the elements. The noble gases are assigned electronegativities of zero and are omitted from the table.

By studying electronegativity values and their differences we find that there is no sharp dividing line between ionic and covalent bonding. Ionic bonding and *nonpolar covalent bonding* simply represent the two extremes. A bond is mostly ionic when the difference in electronegativity between two atoms is very large; the more electronegative atom acquires essentially complete control of the bonding electrons. In a **nonpolar covalent bond**, there is no difference in electronegativity, so the pair of bonding electrons is shared equally.

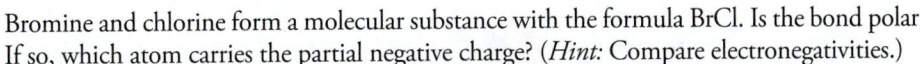

Cs^+ "bonding pair" held exclusively by fluorine

"bonding pair" shared equally by fluorine

The degree to which the bond is polar, which we might think of as the amount of **ionic character** of the bond, varies in a continuous way with changes in the electronegativity difference (Figure 8.10). The bond becomes more than 50% ionic when the electronegativity difference exceeds approximately 1.7.

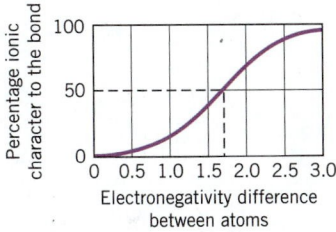

Figure 8.10 | Variation in the percentage ionic character of a bond with electronegativity difference. The bond becomes about 50% ionic when the electronegativity difference equals 1.7, which means that the atoms in the bond carry a partial charge of approximately $\pm 0.5\ e^-$.

Bromine and chlorine form a molecular substance with the formula BrCl. Is the bond polar? If so, which atom carries the partial negative charge? (*Hint:* Compare electronegativities.)

For each of the following bonds, choose the atom that carries the partial negative charge. Arrange them in order of increasing bond polarity: (a) P—Br, (b) Si—Cl, (c) S—Cl.

Practice Exercise 8.13

Practice Exercise 8.14

Periodic Trends in Electronegativity

An examination of Figure 8.9 reveals that within the periodic table, *electronegativity increases from bottom to top in a group, and from left to right in a period.* These trends follow those for ionization energy (IE); an atom that has a large IE will gain an electron more easily than an atom with a small IE, just as an atom with a large electronegativity will attract an electron pair more readily than an atom with a small electronegativity.

Elements located in the same region of the table (for example, the nonmetals) have similar electronegativities, which means that if they form bonds with each other, the electronegativity differences will be small and the bonds will be more covalent than ionic. On the

TOOLS

Periodic trends in electronegativity

other hand, if elements from widely separated regions of the table combine, large electronegativity differences occur and the bonds will be predominantly ionic. This is what happens, for example, when an element from Group 1A or Group 2A reacts with a nonmetal from the upper right-hand corner of the periodic table.

Reactivity of the Elements and Electronegativity

There are parallels between an element's electronegativity and its **reactivity**—its tendency to undergo redox reactions.

Reactivity of Metals and Their Ease of Oxidation

■ In general, *reactivity* refers to the tendency of a substance to react with something. The *reactivity of a metal* refers to its tendency specifically to undergo *oxidation*.

In nearly every compound containing a metal, the metal exists in a positive oxidation state. Therefore, for a metal, *reactivity* relates to how easily the metal is oxidized. For example, sodium, which is very easily oxidized, is said to be very reactive, whereas platinum, which is very difficult to oxidize, is said to be unreactive.

There are several ways to compare how easily metals are oxidized. In Table 5.3 using the activity series, we can see that by comparing the abilities of metals to displace each other from compounds we are able to establish their relative ease of oxidation.

Figure 8.11 illustrates how the ease of oxidation (reactivity) of metals varies in the periodic table. In general, these trends roughly follow the variations in electronegativity, with the metal being less easily oxidized as its electronegativity increases. You might expect this, because electronegativity is a measure of how strongly the atom of an element attracts electrons when combining with an atom of a different element. The more strongly the atom attracts electrons, the more difficult it is to oxidize. This relationship between reactivity and electronegativity is only approximate, however, because many other factors affect the stability of the compounds that are formed.

Elements of Groups 1A and 2A

In Figure 8.11, we see that the metals that are most easily oxidized are found at the far left in the periodic table. These are elements with very low electronegativities. The metals in Group 1A, for example, are so easily oxidized that all of them react with water to liberate hydrogen. Because of their reactivity toward moisture and oxygen, they have no useful applications that require exposure to the atmosphere, so we rarely encounter them as free metals. The same is true of the heavier metals in Group 2A, calcium through barium. These elements also react with water to liberate hydrogen. In Figure 8.9 we see that electronegativity decreases going down a group, which explains why the heavier elements in Group 2A are more reactive than those at the top of the group.

Reactivity of metals and the periodic table

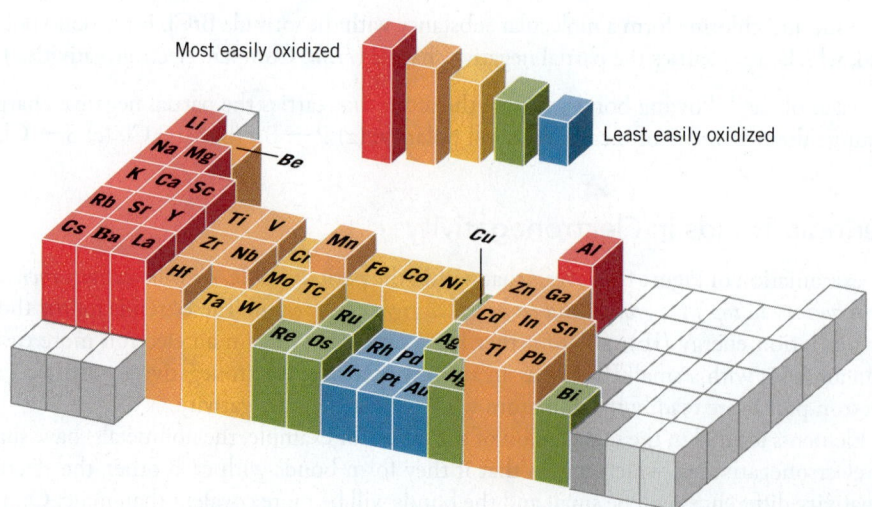

Figure 8.11 | Variations in the ease of oxidation of metals.

Noble Metals

In Figure 8.11 we can also locate the metals that are the most difficult to oxidize. They occur for the most part among the heavier transition elements in the center of the periodic table. Here we find the elements platinum and gold, which are sometimes called **noble metals** because of their extremely low degree of reactivity. Their bright luster and lack of any tendency to corrode in air or water combine to make them particularly attractive for use in fine jewelry. Their lack of reactivity also is responsible for their industrial uses. Gold, for example, is used to coat the electrical contacts in low-voltage circuits found in microcomputers, because even a small amount of corrosion would impede the flow of electricity so much as to make the devices unreliable.

The Oxidizing Power of Nonmetals

The reactivity of a nonmetal is determined by its ease of reduction, and therefore its ability to serve as an oxidizing agent. This ability also varies according to the element's electronegativity. Nonmetals with high electronegativities have strong tendencies to acquire electrons and are therefore strong oxidizing agents. In parallel with changes in electronegativities in the periodic table, *the oxidizing abilities of nonmetals increase from left to right across a period and from bottom to top in a group.* Thus, the most powerful oxidizing agent is fluorine, followed closely by oxygen, both of which appear in the upper right-hand corner of the periodic table.

Single replacement reactions (also called **displacement reactions**) occur among the nonmetals, just as with the metals (which you studied in Chapter 5). For example, heating a metal sulfide in oxygen causes the sulfur to be replaced by oxygen. The displaced sulfur then combines with additional oxygen to give sulfur dioxide. The equation for a typical reaction is

$$CuS(s) + \tfrac{3}{2}O_2(g) \longrightarrow CuO(s) + SO_2(g)$$

Displacement reactions are especially evident among the halogens, where a particular halogen in its elemental form will oxidize the *anion* of any halogen below it in Group 7A, as illustrated in the margin. Thus, F_2 will oxidize Cl^-, Br^-, and I^-; Cl_2 will oxidize Br^- and I^-, but not F^-; and Br_2 will oxidize I^-, but not F^- or Cl^-.

8.7 | Lewis Structures

In Section 8.5 we introduced you to Lewis structures and used them to describe various molecules. Examples included CO_2, Cl_2, N_2, as well as a variety of organic compounds.

Lewis structures are very useful in chemistry because they give us a relatively simple way to describe the structures of molecules. As a result, much chemical reasoning is based on them. In fact, in Chapter 9 you will learn how to use Lewis structures to make reasonably accurate predictions about the shapes of molecules. In this section we will develop a simple method for drawing Lewis structures for both molecules and polyatomic ions (which are also held together by covalent bonds).

Although the octet rule is important in covalent bonding, it is not always obeyed. For instance, there are some molecules in which one or more atoms must have more than an octet in the valence shell. Examples are PCl_5 and SF_6, whose Lewis structures are

In these molecules the formation of more than four bonds to the central atom requires that the central atom have a share of more than eight electrons.

This statue of Prometheus overlooking the skating rink in Rockefeller Center in New York City is covered in a thin layer of gold, providing both beauty and weather resistance.

TOOLS

Reactivity of nonmetals and the periodic trends

■ Fluorine:
$F_2 + 2Cl^- \longrightarrow 2F^- + Cl_2$
$F_2 + 2Br^- \longrightarrow 2F^- + Br_2$
$F_2 + 2I^- \longrightarrow 2F^- + I_2$

Chlorine:
$Cl_2 + 2Br^- \longrightarrow 2Cl^- + Br_2$
$Cl_2 + 2I^- \longrightarrow 2Cl^- + I_2$

Bromine:
$Br_2 + 2I^- \longrightarrow 2Br^- + I_2$

There are also some molecules (but not many) in which the central atom behaves as though it has less than an octet. The most common examples involve compounds of beryllium and boron.

$$\cdot Be \cdot + 2 \cdot \ddot{\underset{\cdot\cdot}{Cl}} : \longrightarrow \quad : \ddot{\underset{\cdot\cdot}{Cl}} - Be - \ddot{\underset{\cdot\cdot}{Cl}} :$$

four electrons around Be

$$\cdot \dot{B} \cdot + 3 \cdot \ddot{\underset{\cdot\cdot}{Cl}} : \longrightarrow \quad : \ddot{\underset{\cdot\cdot}{Cl}} - B - \ddot{\underset{\cdot\cdot}{Cl}} :$$

six electrons around B

Although Be and B sometimes have less than an octet, *the elements in Period 2 never exceed an octet.* The reason is because their valence shells, having $n = 2$, can hold a maximum of only 8 electrons. (This explains why the octet rule works so well for atoms of carbon, nitrogen, and oxygen.) However, elements in periods below Period 2, such as phosphorus and sulfur, sometimes do exceed an octet, because their valence shells can hold more than 8 electrons. For example, the valence shell for elements in Period 3, for which $n = 3$, could hold a maximum of 18 electrons, and the valence shell for Period 4 elements, which have *s, p, d,* and *f* subshells, could theoretically hold as many as 32 electrons.

TOOLS

Drawing Lewis structures

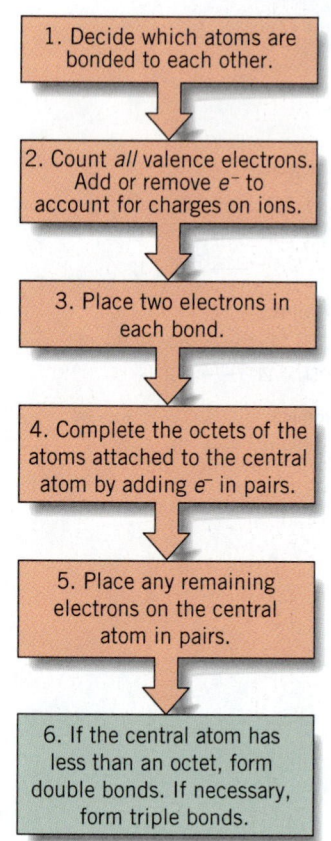

1. Decide which atoms are bonded to each other.

2. Count *all* valence electrons. Add or remove e^- to account for charges on ions.

3. Place two electrons in each bond.

4. Complete the octets of the atoms attached to the central atom by adding e^- in pairs.

5. Place any remaining electrons on the central atom in pairs.

6. If the central atom has less than an octet, form double bonds. If necessary, form triple bonds.

Figure 8.12 | Summary of the steps in drawing a Lewis structure. If you follow these steps, you will obtain a Lewis structure in which the octet rule is obeyed by the maximum number of atoms.

A Procedure for Drawing Lewis Structures

Figure 8.12 outlines a series of steps that provides a systematic method for drawing Lewis structures. The first step is to decide which atoms are bonded to each other. This is not always a simple matter. The formula can suggest the way the atoms are arranged because the central atom, which is usually the least electronegative one, is usually written first. For example, CO_2 and ClO_4^- have the following **skeletal structures** (i.e., arrangements of atoms):

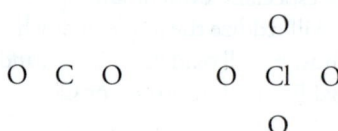

Sometimes, obtaining the skeletal structure is not quite so simple, especially when more than two elements are present. Some generalizations are possible, however. For example, the skeletal structure of nitric acid, HNO_3, is

$$\begin{array}{c} O \\ H \quad O \quad N \quad O \end{array} \quad \text{(correct)}$$

rather than one of the following:

$$\begin{array}{c} O \\ O \quad N \quad O \\ H \end{array} \qquad H \quad O \quad O \quad N \quad O$$

Nitric acid is an oxoacid (Section 4.5), and the hydrogen atoms that can be released from molecules of oxoacids are always bonded to oxygen atoms, which are in turn bonded to the third nonmetal atom. Therefore, recognizing HNO_3 as the formula of an oxoacid allows us to predict that the three oxygen atoms are bonded to the nitrogen, and the hydrogen is bonded to one of the oxygens. (It is also useful to remember that hydrogen forms only one bond, so we should not choose it to be a central atom.)

There are times when no reasonable basis can be found for choosing a particular skeletal structure. If you must make a guess, choose the most symmetrical arrangement of atoms, because it has the greatest chance of being correct.

After you've decided on the skeletal structure, the next step is to count all of the *valence electrons* to find out how many dots must appear in the final formula. Using the periodic table, locate the groups in which the elements in the formula occur to determine the number of valence electrons contributed by each atom. If the structure you wish to draw is that of an ion, *add one additional valence electron for each negative charge or remove a valence electron for each positive charge.* Some examples are as follows:

H_2SO_4	Sulfur (Group 6A) contributes $6e^-$	$1 \times 6 = 6e^-$
	Each oxygen (Group 6A) contributes $6e^-$	$4 \times 6 = 24e^-$
	Hydrogen (Group 1A) contributes $1e^-$	$\underline{2 \times 1 = 2e^-}$
		Total $= 32e^-$
ClO_4^-	Chlorine (Group 7A) contributes $7e^-$	$1 \times 7 = 7e^-$
	Each oxygen (Group 6A) contributes $6e^-$	$4 \times 6 = 24e^-$
	Add $1e^-$ for the $1-$ charge	$\underline{ = +1e^-}$
		Total $= 32e^-$
NH_4^+	Nitrogen (Group 5A) contributes $5e^-$	$1 \times 5 = 5e^-$
	Each hydrogen (Group 1A) contributes $1e^-$	$4 \times 1 = 4e^-$
	Subtract $1e^-$ for the $1+$ charge	$\underline{ = -1e^-}$
		Total $= 8e^-$

After we have determined the number of valence electrons, we place them into the skeletal structure in pairs following the steps outlined in Figure 8.12.

Using the first example, H_2SO_4, let's work our way through the steps to develop the Lewis structure of H_2SO_4.

We need to decide which atoms are bonded to each other in the skeletal structure. Because H_2SO_4 is an oxoacid, the hydrogen atoms are bonded to two oxygens, which in turn are bonded to the sulfur. The other oxygens are also bonded to the sulfur:

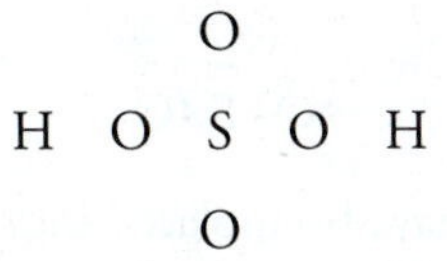

$$\begin{array}{ccccc} & & O & & \\ H & O & S & O & H \\ & & O & & \end{array}$$

We already determined that H_2SO_4 has 32 valence electrons, and we can start to distribute them by placing two electrons between the atoms to forms the bonds.

$$\begin{array}{c} O \\ H\!:\!O\!:\!S\!:\!O\!:\!H \\ O \end{array}$$

This uses 12 e^-, which leaves 20 electrons to go. We will now complete the octets of the atoms surrounding the sulfur atom. Hydrogen does not need any more electrons since it only needs two electrons to fill its valence shell. We will complete the valence shells of the oxygen atoms, which will use all the remaining 20 e^-:

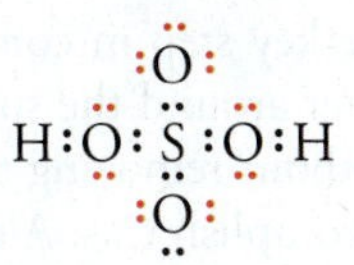

$$\begin{array}{c} :\!\ddot{O}\!: \\ H\!:\!\ddot{O}\!:\!\ddot{S}\!:\!\ddot{O}\!:\!H \\ :\!\ddot{O}\!: \end{array}$$

We may also write the structure of H_2SO_4 using dashes instead of electron pairs in the bonds:

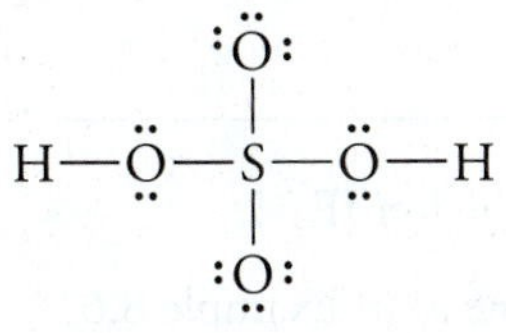

$$\begin{array}{c} :\!\ddot{O}\!: \\ | \\ H\!-\!\ddot{O}\!-\!S\!-\!\ddot{O}\!-\!H \\ | \\ :\!\ddot{O}\!: \end{array}$$

The sulfur and three oxygen atoms have octets, and the valence shell of hydrogen is complete with $2e^-$, so we are finished.

Let us go through two other examples to demonstrate how to draw Lewis structures with double bonds and structures with nonbonding pairs of electrons on the central atom.

Example 8.6
Drawing Molecular Lewis Structures

Draw the Lewis structure for the SO_3 molecule.

Analysis: We need to determine the skeletal structure of SO_3, then determine the total number of valence electrons, and finally, distribute these electrons in the molecule.

Assembling the Tools: We follow the procedure in Figure 8.12.

Solution: Sulfur is less electronegative than oxygen and it is written first in the formula, so we will place it as the central atom, surrounded by the three O atoms:

$$O$$
$$O \quad S \quad O$$

The total number of electrons in the formula is 24 ($6e^-$ from the sulfur, plus $6e^-$ from each oxygen). We begin to distribute the electrons by placing a pair in each bond. This gives

$$O$$
$$O : S : O$$

We have used $6e^-$, so there are $18e^-$ left. We next complete the octets around the oxygens, which uses the remaining electrons.

$$:\ddot{O}:$$
$$:\ddot{O}:S:\ddot{O}:$$

At this point all of the electrons have been placed into the structure, but we see that the sulfur has only six electrons around it. We cannot simply add more electrons because the total must be 24. Therefore, according to the last step of the procedure in Figure 8.12, we have to create a multiple bond. *To do this we move a pair of electrons that we have placed on an oxygen into a sulfur–oxygen bond so that it can be counted as belonging to both the oxygen and the sulfur.* In other words, we place a double bond between sulfur and one of the oxygens. It doesn't matter which oxygen we choose for this honor.

$$:\ddot{O}: \qquad :\ddot{O}: \qquad :\ddot{O}:$$
$$:\ddot{O}:S:\ddot{O}: \text{ gives } :\ddot{O}::S:\ddot{O}: \text{ or } :\ddot{O}=S—\ddot{O}:$$

Notice that each atom has an octet.

Is the Answer Reasonable? The key step in completing the structure is recognizing what we have to do to obtain an octet around the sulfur. We have to add more electrons to the valence shell of sulfur, but without removing them from any of the oxygen atoms. By forming the double bond, we accomplish this. A quick check also confirms that we've placed exactly the correct number of valence electrons into the structure.

Example 8.7
Drawing Ionic Lewis Structures

What is the Lewis structure for the ion IF_4^-?

Analysis: The analysis is the same as in Example 8.6.

- **Assembling the Tools:** As before, we follow the procedure in Figure 8.12. We begin by counting valence electrons, remembering to add an extra electron to account for the negative charge. Then we distribute the electrons in pairs.

- **Solution:** Iodine is less electronegative than fluorine and is first in the formula, so we can anticipate that iodine will be the central atom. Our skeletal structure is

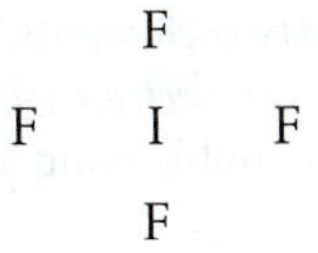

$$
\begin{array}{ccc}
 & F & \\
F & I & F \\
 & F &
\end{array}
$$

The iodine and fluorine atoms are in Group 7A and each contributes seven electrons, for a total of $35e^-$. The negative charge requires one additional electron to give a total of $36e^-$.

First we place $2e^-$ into each bond, and then we complete the octets of the fluorine atoms. This uses 32 electrons.

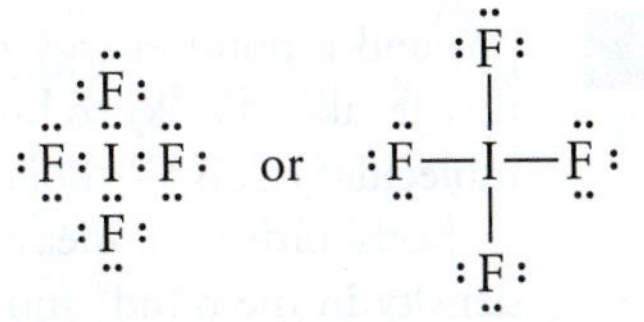

$$
\begin{array}{ccc}
\ddot{\text{F}} & & \ddot{\text{F}} \\
\ddot{\text{F}}\!:\!\ddot{\text{I}}\!:\!\ddot{\text{F}} & \text{or} & \ddot{\text{F}}\!-\!\text{I}\!-\!\ddot{\text{F}} \\
\ddot{\text{F}} & & \ddot{\text{F}}
\end{array}
$$

There are four electrons left, and according to Step 5 in Figure 8.12 they are placed on the central atom as *pairs* of electrons. This gives

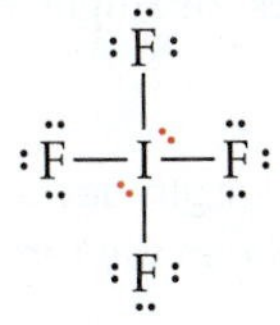

$$
\begin{array}{c}
:\ddot{\text{F}}: \\
| \\
:\ddot{\text{F}}\!-\!\text{I}\!-\!\ddot{\text{F}}: \\
| \\
:\ddot{\text{F}}:
\end{array}
$$

The last step is to add brackets around the formula and write the charge outside as a superscript.

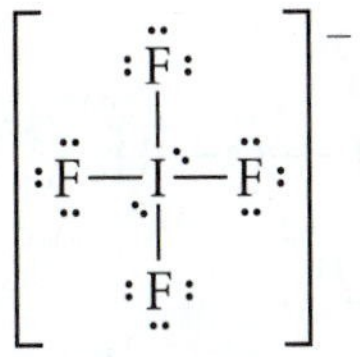

$$
\left[
\begin{array}{c}
:\ddot{\text{F}}: \\
| \\
:\ddot{\text{F}}\!-\!\text{I}\!-\!\ddot{\text{F}}: \\
| \\
:\ddot{\text{F}}:
\end{array}
\right]^{-}
$$

- **Is the Answer Reasonable?** We can recount the valence electrons, which tells us we have the right number of them, and all are in the Lewis structure. Each fluorine atom has an octet, which is proper. Notice that we have placed the "left-over" electrons onto the central atom. This gives iodine more than an octet, but that's okay because iodine is not a Period 2 element.

Determine the Lewis structure of $H_2PO_4^-$ by predicting a reasonable skeletal structure for $H_2PO_4^-$, determining the number of valence electrons that should be in its Lewis structure, and distributing the electrons to fill the valence shells of the atoms. (*Hint:* It's an ion derived from an oxoacid.)

Practice Exercise 8.15

Draw Lewis structures for OF_2, NH_4^+, SO_2, NO_3^-, ClF_3, and $HClO_4$.

Practice Exercise 8.16

Using the structures drawn in Practice Exercise 8.16, show the polarity of the bonds.

Practice Exercise 8.17

Formal Charges and Lewis Structures

Lewis structures are meant to describe how atoms share electrons in chemical bonds. Such descriptions are theoretical explanations or predictions that relate to the forces that hold molecules and polyatomic ions together. But, as you learned in Section 1.1, a theory is only as good as the observations on which it is based, so to have confidence in a theory about chemical bonding, we need to have a way to check it. We need experimental observations that relate to the description of bonding.

Bond Properties That Depend on Bond Order

To compare bonds between the same two elements, it's useful to define a quantity called the **bond order**, *which is the number of electron pairs shared between two atoms.* Thus, a single bond has a bond order of one, a double bond a bond order of two, and a triple bond a bond order of three.

Two properties that are related to the bond order are *bond length*, the distance between the nuclei of the bonded atoms, and *bond energy*, the energy required to separate the bonded atoms to give neutral particles. For example, we mentioned in Section 8.5 that measurements have shown that the H_2 molecule has a bond length of 75 pm and a bond energy of 435 kJ/mol, which means that it takes 435 kJ to break the bonds of 1 mol of H_2 molecules to give 2 mol of hydrogen atoms.

Bond order is a measure of the amount of electron density in the bond, and the greater the electron density, the more tightly the nuclei are held and the more closely they are drawn together. This is illustrated by the data in Table 8.3, which gives typical bond lengths and bond energies for single, double, and triple bonds between carbon atoms. In summary:

TABLE 8.3	Average Bond Lengths and Bond Energies Measured for Carbon–Carbon Bonds	
Bond	**Bond Length (pm)**	**Bond Energy (kJ/mol)**
C—C	154	348
C=C	134	615
C≡C	120	812

TOOLS

Correlation between bond properties and bond order

As the bond order increases, the bond length decreases and the bond energy increases, provided we are comparing bonds between the same elements.

With this as background, let's examine the Lewis structure of sulfuric acid that we drew earlier:

$$\text{H}-\overset{\displaystyle :\overset{..}{\text{O}}:}{\underset{\displaystyle :\overset{..}{\text{O}}:}{\overset{..}{\underset{..}{\text{O}}}}-\text{S}-\overset{..}{\underset{..}{\text{O}}}-\text{H}\qquad\text{Structure I}$$

It obeys the octet rule, and there doesn't seem to be any need to attempt to write any other structures for it. But a problem arises if we compare the predicted bond lengths with those found experimentally. In our Lewis structure, all four sulfur–oxygen bonds are shown as single bonds, which means they should have about the same bond lengths. However, experimentally it has been found that the bonds are not of equal length, as illustrated in Figure 8.13. The S—O bonds are shorter than the S—OH bonds, which means they must have a larger bond order. Therefore, we need to modify our Lewis structure to make it conform to reality.

Because sulfur is in Period 3, its valence shell has 3s, 3p, and 3d subshells, which together can accommodate more than eight electrons. Therefore, sulfur is able to form more than four bonds, so we are allowed to increase the bond order in the S—O bonds by moving electron pairs to create sulfur–oxygen double bonds as follows:

$$\text{H}-\overset{\displaystyle :\overset{..}{\text{O}}:}{\underset{\displaystyle :\overset{..}{\text{O}}:}{\overset{..}{\underset{..}{\text{O}}}}-\text{S}-\overset{..}{\underset{..}{\text{O}}}-\text{H}\quad\text{gives}\quad\text{H}-\overset{\displaystyle :\overset{..}{\text{O}}}{\underset{\displaystyle \overset{..}{\text{O}}:}{\overset{\parallel}{\underset{\parallel}{\text{S}}}}}-\overset{..}{\underset{..}{\text{O}}}-\text{H}\quad\text{Structure II}$$

Now we have a Lewis structure that better fits experimental observations because the sulfur–oxygen double bonds are expected to be shorter than the sulfur–oxygen single bonds. Because this second Lewis structure agrees better with the actual structure of the molecule, it is the *preferred* Lewis structure, even though it violates the octet rule.

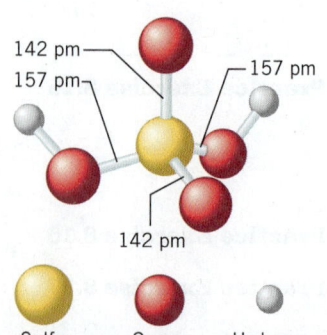

142 pm
157 pm
157 pm
142 pm

Sulfur   Oxygen   Hydrogen

Figure 8.13 | **The structure of sulfuric acid in the vapor state.** Notice the difference in the sulfur–oxygen bond lengths.

Assigning Formal Charges to Atoms

Are there any criteria that we could have applied that would have allowed us to predict that the second Lewis structure for H_2SO_4 is better than the one with only single bonds, even though it seems to violate the octet rule unnecessarily? To answer this question, let's take a closer look at the two Lewis structures we've drawn.

In Structure I, there are only single bonds between the sulfur and oxygen atoms. If the electrons in the bonds are shared equally by S and O, then each atom "owns" half of the electron pair, or the equivalent of one electron. In other words, the four single bonds place the equivalent of four electrons in the valence shell of the sulfur. An isolated single atom of sulfur, however, has six valence electrons, so in Structure I the sulfur appears to have two electrons *less* than it does as just an isolated atom. Thus, at least in a *bookkeeping* sense, it would appear that if sulfur obeyed the octet rule in H_2SO_4, it would have a charge of 2+. This *apparent* charge on the sulfur atom is called its **formal charge**.

■ The actual charges on the atoms in a molecule are determined by the relative electronegativities of the atoms.

Notice that in defining formal charge, we've stressed the word "apparent." *The formal charge arises because of the bookkeeping we've done and should not be confused with whatever the actual charge is on an atom in the molecule.* (The situation is somewhat similar to the oxidation numbers you learned to assign in Chapter 5, which are artificial charges assigned according to a set of rules.) Here's how formal charges are assigned.

Calculating the Formal Charge on Atoms in a Lewis Structure

1. For each atom, write down the number of valence electrons in an isolated atom of the element.
2. Using the Lewis structure, add up the valence electrons that "belong to" the atom in the molecule or ion, and then subtract this total from the value in Step 1. The result is the formal charge on the atom.
3. The sum of the formal charges in a structure must equal the charge on the particle.

Calculating formal charges

In performing the calculation in Step 2, electrons in bonds are divided equally between the two atoms, while unshared electrons are assigned exclusively to the atom on which they reside. For example, for Structure I above, we have

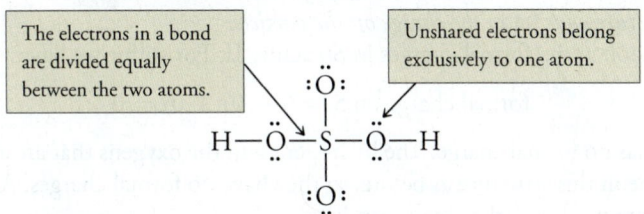

Therefore, the calculation of formal charge is summarized by the following equation:

Calculated number of electrons in the valence
shell of the atom in the Lewis structure

$$\text{formal charge} = \left[\begin{array}{c}\text{number of } e^- \text{ in valence} \\ \text{shell of the isolated atom}\end{array}\right] - \left[\begin{array}{c}\text{number of bonds} \\ \text{to the atom}\end{array} + \begin{array}{c}\text{number of} \\ \text{unshared } e^-\end{array}\right] \quad (8.3)$$

For example, for the sulfur in Structure I, we get

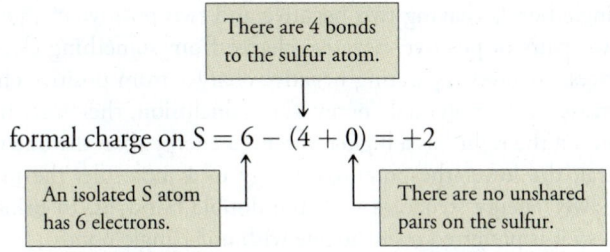

Let's also calculate the formal charges on the hydrogen and oxygen atoms in Structure I. An isolated H atom has one electron. In Structure I each H has one bond and no unshared electrons. Therefore,

$$\text{formal charge on H} = 1 - (1 + 0) = 0$$

In Structure I, we also see that there are two kinds of oxygen to consider. An isolated oxygen atom has six electrons, so we have, for the oxygens also bonded to hydrogen,

$$\text{formal charge} = 6 - (2 + 4) = 0$$

and for the oxygens not bonded to hydrogen,

$$\text{formal charge} = 6 - (1 + 6) = -1$$

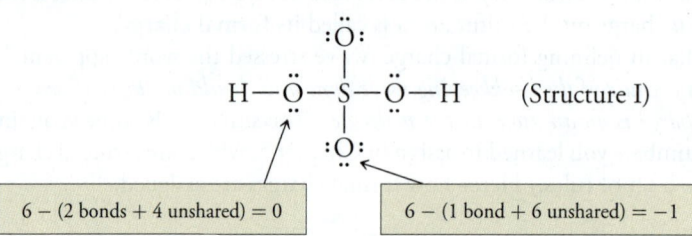

Nonzero formal charges are indicated in a Lewis structure by placing them in circles alongside the atoms, as follows:

Notice that the sum of the formal charges in the molecule adds up to zero. *After you've assigned formal charges, it is important to always check that the sum of the formal charges in the Lewis structure adds up to the charge on the particle.*

Now let's look at the formal charges in Structure II. For sulfur we have

$$\text{formal charge on S} = 6 - (6 + 0) = 0$$

so the sulfur has no formal charge. The hydrogens and the oxygens that are also bonded to H are the same in this structure as before, so they have no formal charges. And finally, the oxygens that are not bonded to hydrogen have

$$\text{formal charge} = 6 - (2 + 4) = 0$$

These oxygens also have no formal charges.

Now let's compare the two structures side by side.

Imagine changing the one with the double bonds (no formal charges on any atoms) to the one with the single bonds (having two negative and two positive charges). In doing so, we've created two pairs of positive–negative charge from something electrically neutral. Because the process involves separating negative charges from positive charges, it would produce an increase in the potential energy. Our conclusion, therefore, is that the singly bonded structure on the right has a higher potential energy than the one with the double bonds. In general, the lower the potential energy of a molecule, the more stable it is. Therefore, the lower energy structure with the double bonds is, in principle, the more stable structure, so it is preferred over the one with only single bonds.

This now gives us a rule that we can use in selecting the best Lewis structures for a molecule or ion:

When several Lewis structures are possible, the one with formal charges closest to zero is the most stable and is preferred.

TO⚗LS

Selecting the best Lewis structure

There are two other considerations we have to take into account. When we draw Lewis structures for polyatomic ions, it is not possible to have a formal charge of zero on all atoms. That is because the charge of the ion is the sum of the formal charges of all of the atoms. In addition, the most electronegative element should have the negative formal charge, and the least electronegative element should have the positive formal charge.

Example 8.8
Selecting Lewis Structures Based on Formal Charges

A student drew the following three Lewis structures for the nitric acid molecule:

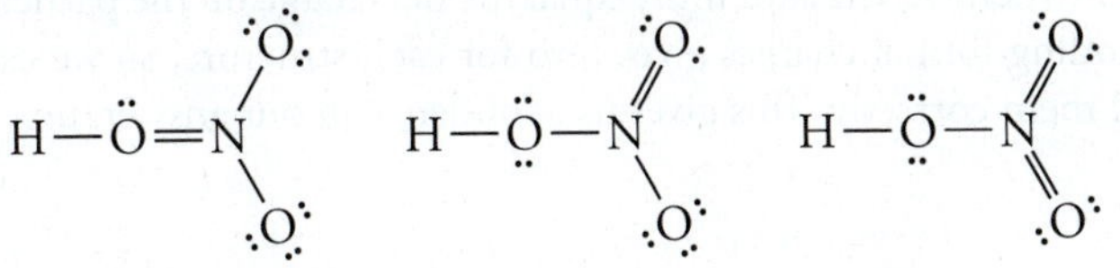

Which one is preferred?

Analysis: When we have to select among several Lewis structures to find the best one, we first have to assign formal charges to each of the atoms. As a rule, the structure with the formal charges closest to zero will be the best structure. We have to be careful, however, that we don't select a structure in which an atom is assigned more electrons than its valence shell can actually hold.

Assembling the Tools: To assign the formal charges, the tool we use is Equation 8.3.

$$\text{Formal charge} = \left[\begin{array}{c}\text{number of } e^- \text{ in valence} \\ \text{shell of the isolated atom}\end{array}\right] - \left[\begin{array}{c}\text{number of bonds} \\ \text{to the atom}\end{array} + \begin{array}{c}\text{number of} \\ \text{unshared } e^-\end{array}\right]$$

Solution: Except for hydrogen, all of the atoms in the molecule are from Period 2, and therefore can have a maximum of eight electrons in their valence shells. (Period 2 elements *never* exceed an octet because their valence shells have only *s* and *p* subshells and can accommodate a maximum of eight electrons.) Scanning the structures, we see that I and II show octets around both N and O. However, the nitrogen in Structure III has 5 bonds to it, which require 10 electrons. Therefore, this structure is not acceptable and can be eliminated immediately. Our choice is then between Structures I and II. Let's calculate formal charges on the atoms in each of them.

Structure I:

■ In Structure III, the formal charges are zero on each of the atoms, but this cannot be the "preferred structure" because the nitrogen atom has too many electrons in its valence shell.

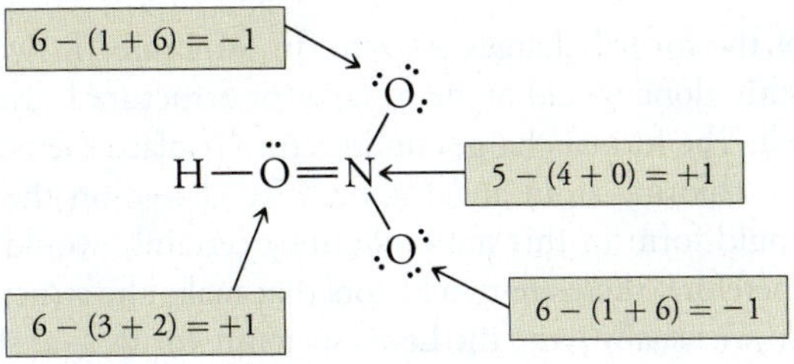

Structure II:

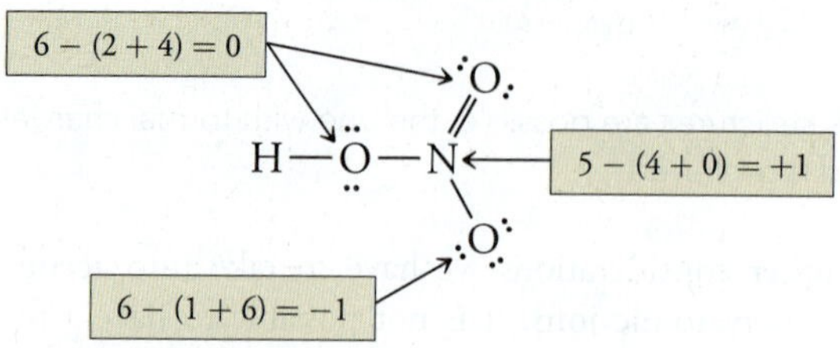

Next, we place the formal charges on the atoms in the structures.

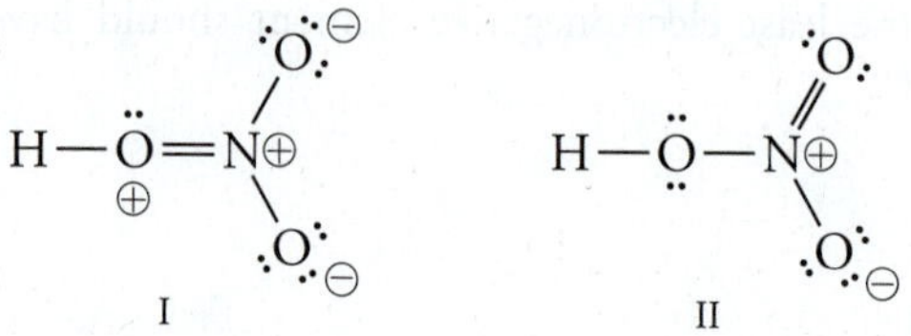

Because Structure II has fewer formal charges than Structure I, it is the lower-energy, preferred Lewis structure for HNO_3.

Is the Answer Reasonable? One simple check we can do is to add up the formal charges in each structure. The sum must equal the net charge on the particle, which is zero for HNO_3. Adding formal charges gives zero for each structure, so we can be confident we've assigned them correctly. This gives us confidence in our answer, too.

Example 8.9
Selecting Lewis Structures Based on Formal Charges

The following two structures can be drawn for BCl_3:

Why is the one that violates the octet rule preferred?

Analysis: We're asked to select between Lewis structures, which tells us that we have to consider formal charges. We'll assign them and then see whether we can answer the question.

Assembling the Tools: To assign formal charges we use Equation 8.3.

$$\text{Formal charge} = \begin{bmatrix} \text{number of } e^- \text{ in valence} \\ \text{shell of the isolated atom} \end{bmatrix} - \begin{bmatrix} \text{number of bonds} \\ \text{to the atom} \end{bmatrix} + \begin{bmatrix} \text{number of} \\ \text{unshared } e^- \end{bmatrix}$$

Solution: Assigning formal charges gives

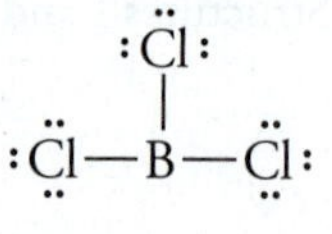

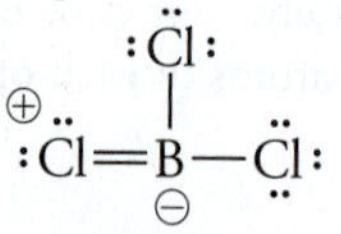

In Structure I all of the formal charges are zero. In Structure II, two of the atoms have formal charges, so this alone would argue in favor of Structure I. There is another argument in favor as well. The formal charges in Structure II place the positive charge on the more electronegative chlorine atom and the negative charge on the less electronegative boron. If charges could form in this molecule, they certainly would not be expected to form in this way. Therefore, there are two factors that make the structure with the double bond unfavorable, so we usually write the Lewis structure for BCl_3 as shown in Structure I.

● **Is the Answer Reasonable?** We've assigned the formal charges correctly, and our reasoning seems sound, so we appear to have answered the question adequately.

A student drew the following Lewis structure for the sulfite ion, SO_3^{2-}:

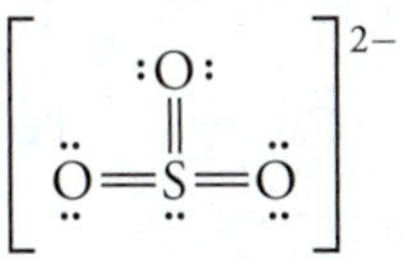

$$\left[\begin{array}{c} :\ddot{O}: \\ \| \\ \ddot{O}\!=\!\overset{}{S}\!=\!\ddot{O} \end{array}\right]^{2-}$$

Is this the best Lewis structure for the ion? (*Hint:* Negative formal charges should be on the more electronegative atoms.)

Assign formal charges to the atoms in the following Lewis structures:

(a) $:\ddot{N}\!-\!N\!\equiv\!O:$ (b) $[\ddot{S}\!=\!C\!=\!\ddot{N}]^-$

Draw the preferred Lewis structure for (a) SO_2, (b) $HClO_3$, and (c) H_3PO_4.

Practice Exercise 8.18

Practice Exercise 8.19

Practice Exercise 8.20

Coordinate Covalent Bonds

Often we use Lewis structures to follow the course of chemical reactions. For example, we can diagram how a hydrogen ion combines with a water molecule to form the hydronium ion, a process that occurs in aqueous solutions of acids.

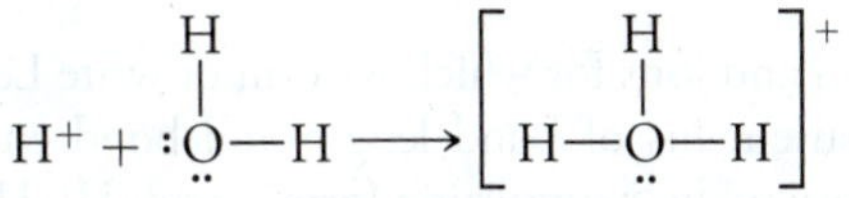

$$H^+ + :\overset{H}{\underset{}{\overset{|}{O}}}\!-\!H \longrightarrow \left[\overset{H}{\underset{}{\overset{|}{H}\!-\!\overset{|}{O}\!-\!H}}\right]^+$$

The formation of the bond between H^+ and H_2O follows a different path than the covalent bonds we discussed earlier in this chapter. For instance, when two H atoms combine to form H_2, each atom brings one electron to the bond.

$$H\cdot + \cdot H \longrightarrow H\!-\!H$$

But in the formation of H_3O^+, both of the electrons that become shared between the H^+ and the O originate on the oxygen atom of the water molecule. This type of bond, in which both electrons of the shared pair come from just one of the two atoms, is called a **coordinate covalent bond**.

Although we can make a distinction about the origin of the electrons shared in the bond, once the bond is formed a coordinate covalent bond is the same as any other covalent bond, and we can't tell where the electrons in the bond came from *after* the bond has been formed. In the H_3O^+ ion, for example, all three O—H bonds are identical once they've been formed.

The concept of a coordinate covalent bond is helpful in explaining what happens to atoms in a chemical reaction. For example, when ammonia is mixed with boron trichloride, an exothermic reaction takes place and the compound NH_3BCl_3 is formed in which there is a boron–nitrogen bond. Using Lewis structures, we can diagram this reaction:

■ All electrons are alike. We are using different colors for them so we can see where the electrons in the bond came from.

$$\overset{H}{\underset{\overset{|}{H}}{\overset{|}{H\!-\!N}}}: + \overset{:\ddot{Cl}:}{\underset{\overset{|}{:\ddot{Cl}:}}{\overset{|}{B\!-\!\ddot{Cl}:}}} \longrightarrow \overset{H}{\underset{\overset{|}{H}}{\overset{|}{H\!-\!N}}}:\overset{:\ddot{Cl}:}{\underset{\overset{|}{:\ddot{Cl}:}}{\overset{|}{B\!-\!\ddot{Cl}:}}}$$

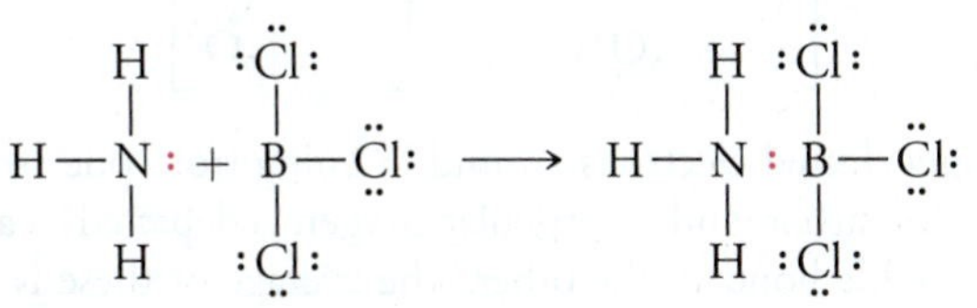

In the reaction, we might say that "the boron forms a coordinate covalent bond with the nitrogen of the ammonia molecule."

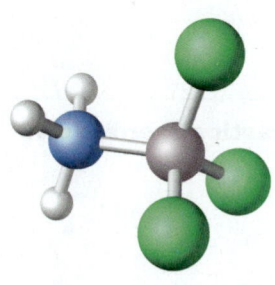

The addition compound formed from ammonia and boron trichloride. The light blue–colored atom is boron.

A curved arrow is used to represent the donation of a pair of electrons in a coordinate covalent bond. The direction of the arrow indicates the direction in which the electron pair is donated, in this case from the nitrogen to the boron.

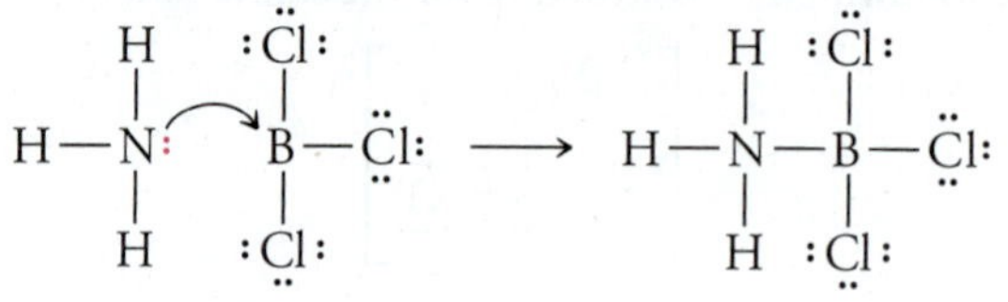

Compounds like BCl_3NH_3, which are formed by simply joining two smaller molecules, are sometimes called **addition compounds**.

Practice Exercise 8.21

Use Lewis structures to show how the formation of NH_4^+ from NH_3 and H^+ involves formation of a coordinate covalent bond. How does this bond differ from the other N—H bonds in NH_4^+? (*Hint:* Keep in mind the definition of a coordinate covalent bond.)

Practice Exercise 8.22

Use Lewis structures to explain how the reaction between hydroxide ion and hydrogen ion involves the formation of a coordinate covalent bond.

8.8 | Resonance Structures

There are some molecules and ions for which we cannot write Lewis structures that agree with experimental measurements of bond length and bond energy. An example is the formate ion, CHO_2^-, formed by neutralizing formic acid, $HCHO_2$. Following the usual steps, we would write the Lewis structure of the formate ion as shown on the right below.

Formic acid is the substance that causes the stinging sensation in bites from fire ants.

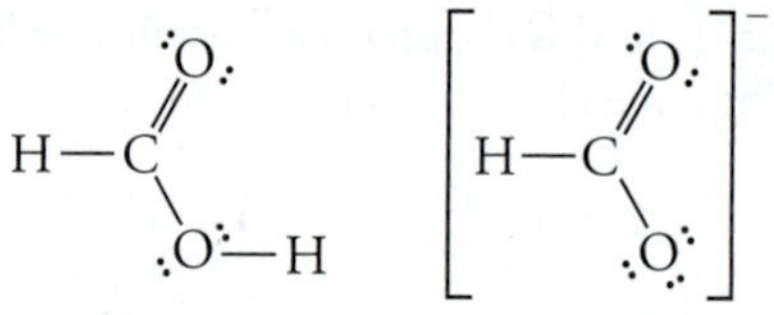

formic acid   formate ion

This structure suggests that in the CHO^- ion, one carbon–oxygen bond should be longer than the other, but experiments show that the C—O bond lengths are the same, with lengths that are about halfway between the expected values for a single bond and a double bond. The Lewis structure doesn't match the experimental evidence, and there's no way to write one that does. It would require showing all of the electrons in pairs and, at the same time, showing 1.5 pairs of electrons in each carbon–oxygen bond.

The way we get around problems like this is through the use of a concept called **resonance**. We view the structure of the molecule or ion as a composite, or average, of a number of Lewis structures that we can draw. For example, for formate ion we write

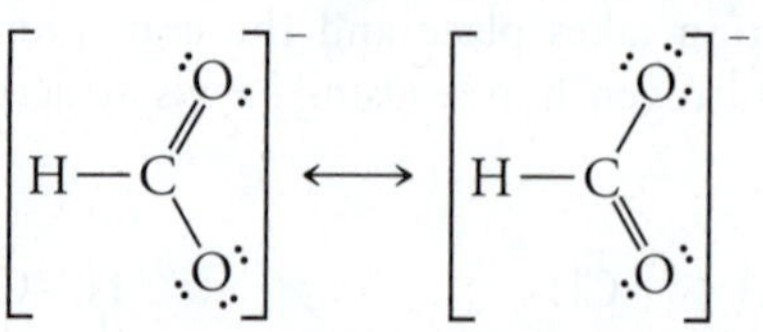

where we have simply shifted electrons around in going from one structure to the other. The bond between the carbon and a particular oxygen is depicted as a single bond in one structure and as a double bond in the other. The average of these is 1.5 bonds (halfway between a single and a double bond), which is in agreement with the experimental bond lengths. These two Lewis structures are called **resonance structures** or **contributing structures**. The actual structure of the ion, which we can't draw, is called a **resonance**

TOOLS

Resonance structures

hybrid of these two resonance structures. The double-ended arrow is used to show that we are drawing resonance structures and implies that the true hybrid structure is a composite of the two resonance structures.

Some students find the term *resonance* somewhat misleading. The word itself suggests that the actual structure flip-flops back and forth between the two structures shown. This is *not* the case! We are simply using structures that we are able to draw to describe an actual structure that we cannot draw. It's a bit like trying to imagine what a *zonkey* looks like by viewing photos of a zebra and a donkey. Although a zonkey may have characteristics of both parents, a zonkey is a zonkey; it doesn't flip back and forth between being a zebra one minute and a donkey the next. Similarly, a *resonance hybrid* also has characteristics of its "parents," but it doesn't flip back and forth between the contributing structures.

JIJI PRESS/AFP/Getty Images, Inc.

A zonkey is the hybrid offspring of a zebra and a donkey. It has characteristics of both of its parents, but is never exactly like either one of them. It certainly isn't a zebra one minute and a donkey the next!

When We Draw Resonance Structures

There is a simple way to determine when resonance should be applied to Lewis structures. If you find that you must move electrons to create one or more double bonds while following the procedures developed earlier, the number of resonance structures is equal to the number of equivalent choices for the locations of the double bonds between a central atom and two or more chemically equivalent atoms. For example, in drawing the Lewis structure for the NO_3^- ion, we reach the stage

$$\ddot{O}$$
$$|$$
$$\ddot{O} - N - \ddot{O}$$

A double bond must be created to give the nitrogen an octet. Since it can be placed in any one of three locations, there are three resonance structures for this ion.

$$
\left[\begin{array}{c} :\ddot{O}: \\ | \\ \ddot{O} \quad N \quad \ddot{O} \end{array}\right]^- \longleftrightarrow \left[\begin{array}{c} :O: \\ \| \\ \ddot{O} \quad N \quad \ddot{O} \end{array}\right]^- \longleftrightarrow \left[\begin{array}{c} :\ddot{O}: \\ | \\ \ddot{O} \quad N \quad O \end{array}\right]^-
$$

Notice that each structure is the same, except for the location of the double bond.

In the nitrate ion, the extra bond that appears to "move around" from one structure to another is actually divided among all three bond locations. Therefore, the **average bond order** in the N—O bonds is expected to be 1⅓, or 1.33. In general, we can calculate the bond order by adding up the total number of bonds and dividing by the number of equivalent positions. In the NO_3^- ion, we have a total of *four* bonds (two single bonds and a double bond) distributed over *three* equivalent positions, so the bond order is ⁴⁄₃ = 1⅓.

Average bond order

Example 8.10
Drawing Resonance Structures

Use formal charges to show that resonance applies to the preferred Lewis structure for the sulfite ion, SO_3^{2-}. Draw the resonance structures and determine the average bond order of the S—O bonds.

Analysis: Following our usual procedure we obtain the following Lewis structure:

$$
\left[\begin{array}{c} :\ddot{O}: \\ | \\ :\ddot{O} - S - \ddot{O}: \end{array}\right]^{2-}
$$

All of the valence electrons have been placed into the structure and we have octets around all of the atoms, so it doesn't seem that we need the concept of resonance. However, the question refers to the "preferred" structure, which suggests that we are going to have to assign formal charges and determine what the preferred structure is. Then we can decide whether the concept of resonance will apply.

Assembling the Tools: To assign formal charges, our tool is Equation 8.3.

$$\text{Formal charge} = \left[\begin{array}{c} \text{number of } e^- \text{ in valence} \\ \text{shell of the isolated atom} \end{array}\right] - \left[\begin{array}{c} \text{number of bonds} \\ \text{to the atom} \end{array} + \begin{array}{c} \text{number of} \\ \text{unshared } e^- \end{array}\right]$$

If a double bond occurs in the preferred structure, the number of equivalent positions equals the number of resonance structures. The average bond order will be calculated from the number of bonds distributed over the three equivalent bond locations.

Solution: When we assign formal charges, we get

$$\left[\begin{array}{c} \overset{\ominus}{:}\overset{\cdot\cdot}{O}: \\ | \\ :\overset{\cdot\cdot}{O} - \underset{\oplus}{S} - \overset{\cdot\cdot}{O}:\overset{\ominus}{} \\ \cdot\cdot \end{array}\right]^{2-}$$

We can obtain a better Lewis structure if we can reduce the number of formal charges. This can be accomplished by moving an unshared electron pair from one of the oxygens into an S—O bond, thereby forming a double bond. Let's do this using the oxygen at the left.

$$\left[\begin{array}{c} \overset{\ominus}{:}\overset{\cdot\cdot}{O}: \\ | \\ :\overset{\frown}{O} - S - \overset{\cdot\cdot}{O}:\overset{\ominus}{} \\ \cdot\cdot \end{array}\right]^{2-} \xrightarrow{\text{gives}} \left[\begin{array}{c} \overset{\ominus}{:}\overset{\cdot\cdot}{O}: \\ | \\ \overset{\cdot\cdot}{O} = S - \overset{\cdot\cdot}{O}:\overset{\ominus}{} \\ \cdot\cdot \end{array}\right]^{2-}$$

However, we could have done this with any of the three S—O bonds, so there are three equivalent choices for the location of the double bond. Therefore, there are three resonance structures.

$$\left[\begin{array}{c} :\overset{\cdot\cdot}{O}: \\ || \\ \overset{\cdot\cdot}{O} = S - \overset{\cdot\cdot}{O}: \\ \cdot\cdot \end{array}\right]^{2-} \longleftrightarrow \left[\begin{array}{c} :O: \\ || \\ :\overset{\cdot\cdot}{O} - S - \overset{\cdot\cdot}{O}: \\ \cdot\cdot \end{array}\right]^{2-} \longleftrightarrow \left[\begin{array}{c} :\overset{\cdot\cdot}{O}: \\ | \\ :\overset{\cdot\cdot}{O} - S = \overset{\cdot\cdot}{O} \\ \cdot\cdot \end{array}\right]^{2-}$$

As with the nitrate ion, we expect an average bond order of 1.33.

Is the Answer Reasonable? If we can answer "yes" to the following questions, the problem is solved correctly: Have we counted valence electrons correctly? Have we properly placed the electrons into the skeletal structure? Do the formal charges we've calculated add up to the charge on the SO_3^{2-} ion? Have we correctly determined the number of equivalent positions for the double bond? Have we computed the average bond order correctly? All the answers are "yes," so we have solved the problem correctly.

Practice Exercise 8.23

The phosphate ion has the following Lewis structure, where we've used formal charges to obtain the best structure.

$$\left[\begin{array}{c} :O: \\ || \\ :\overset{\cdot\cdot}{O} - P - \overset{\cdot\cdot}{O}: \\ | \\ :\overset{\cdot\cdot}{O}: \end{array}\right]^{3-}$$

How many resonance structures are there for this ion? (*Hint:* How many equivalent positions are there for the double bond?)

Practice Exercise 8.24

Draw the resonance structures for HCO_3^-.

Practice Exercise 8.25

Determine the preferred Lewis structure for the bromate ion, BrO_3^-, and, if appropriate, draw resonance structures.

Stability of Molecules with Resonance Structures

One of the benefits that a molecule or ion derives from existing as a resonance hybrid is that its total energy is lower than that of any one of its contributing structures. A particularly important example of this occurs with the compound benzene, C_6H_6. This is a flat hexagonal ring-shaped molecule (Figure 8.14) with a basic structure that appears in many important organic molecules, ranging from plastics to amino acids.

Two resonance structures are usually drawn for benzene.

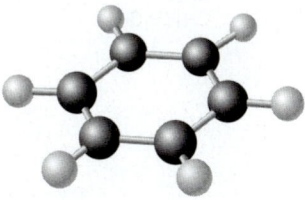

Figure 8.14 | **Benzene.** The molecule has a planar hexagonal structure.

These are generally represented as hexagons with dashes showing the locations of the double bonds. It is assumed that at each apex of the hexagon there is a carbon atom bonded to a hydrogen atom, which is not shown, as well as to the adjacent carbon atoms.

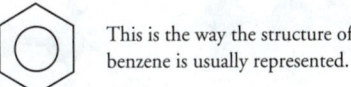

Usually, the actual structure of benzene (that of its resonance hybrid) is represented as a hexagon with a circle in the center. This is intended to show that the electron density of the three extra bonds is evenly distributed around the ring.

This is the way the structure of benzene is usually represented.

Although the individual resonance structures for benzene show double bonds, the molecule does not react like other organic molecules that have true carbon–carbon double bonds. The reason appears to be that the resonance hybrid is considerably more stable than either of the resonance forms. In fact, it has been calculated that the actual structure of the benzene molecule is more stable than either of its resonance structures by approximately 146 kJ/mol. This extra stability achieved through resonance is called the **resonance energy**.

8.9 | Covalent Compounds of Carbon

Covalent bonds are found in many of the substances we encounter on a daily basis. Most of them are classified as **organic compounds** in which carbon atoms are covalently bonded to other carbon atoms and to a variety of other nonmetals. They include the foods we eat, the fabrics we wear, the medicines that cure us, the fuels that power vehicles, and the fibers in the rope supporting a mountain climber. Because they are so common, organic compounds will be used frequently as examples in our discussions later in the book. For this reason, you will find it helpful to learn something now about their makeup.

Organic compounds fall into different classes according to the elements that are bonded to carbon and how atoms of those elements are arranged in the molecules. The kinds of compounds we will study in this section will include those in which carbon is bonded to hydrogen, oxygen, and nitrogen. As you learned in Section 2.7, such substances can be considered to be derived from hydrocarbons—compounds of carbon and hydrogen in which the basic molecular "backbones" are composed of carbon atoms linked to one another in a chainlike fashion.

■ A more comprehensive discussion of organic compounds is found in Chapter 22. In this section we look at some simple ways carbon atoms combine with other atoms to form certain important classes of organic substances that we encounter frequently.

One of the chief features of organic compounds is the tendency of carbon to complete its octet by forming four covalent bonds. For example, in the alkane series of hydrocarbons (which we described briefly in Section 2.7) all of the bonds are single bonds. The structures of the first three alkanes (methane, ethane, and propane) are

■ These structures can be written in a condensed form as

CH_4
CH_3CH_3
$CH_3CH_2CH_3$

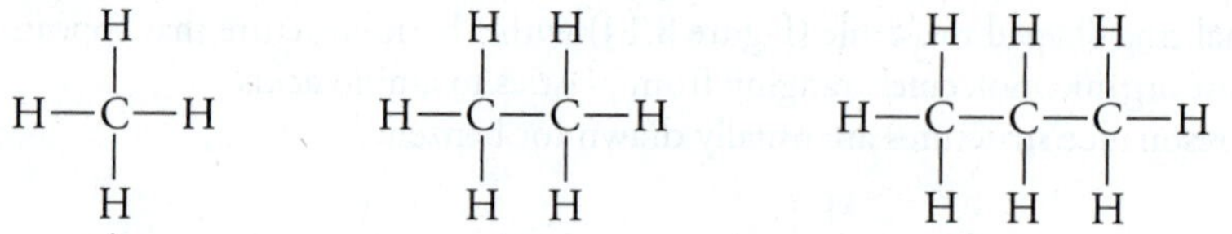

methane   ethane   propane

The shapes of their molecules are illustrated as space-filling models in Figure 2.30 in Section 2.7.

When more than four carbon atoms are present, matters become more complex because there is more than one way to arrange the atoms. For example, butane has the formula C_4H_{10}, but there are two ways to arrange the carbon atoms. These two arrangements occur in compounds commonly called butane and isobutane.

■ In condensed form, we can write their structures as

$CH_3CH_2CH_2CH_3$

$\underset{\displaystyle CH_3CHCH_3}{\overset{\displaystyle CH_3}{|}}$

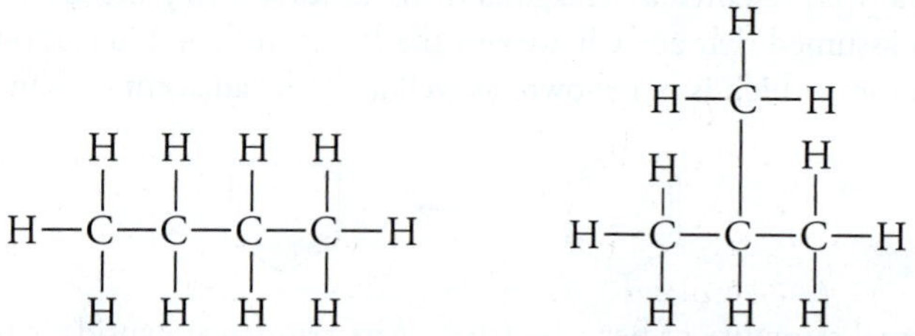

Butane

Methylpropane
(isobutane)

C_4H_{10}
bp = −0.5 °C

C_4H_{10}
bp = −11.7 °C

Even though they have the same molecular formula, these are actually different compounds with different properties, as you can see from the boiling points listed below their structures. The ability of atoms to arrange themselves in more than one way to give different compounds that have the same molecular formula is called **isomerism** and is discussed more fully in Chapters 21 and 22. The existence of *isomers* is one of the reasons there are so many organic compounds. For example, there are 366,319 different compounds, or isomers, that have the formula $C_{20}H_{42}$; they differ only in the way the carbon atoms are attached to each other.

Carbon can also complete its octet by forming double or triple bonds. The Lewis structures of ethene, C_2H_4, and ethyne, C_2H_2 (commonly called ethylene and acetylene, respectively) are as follows:[5]

ethene
(ethylene)

ethyne
(acetylene)

---

[5] In the IUPAC system for naming organic compounds, *meth-*, *eth-*, *prop-*, and *but-* indicate carbon chains of 1, 2, 3, and 4 carbon atoms, respectively. Organic nomenclature is discussed more fully in Chapter 22.

Compounds That Also Contain Oxygen and Nitrogen

Most organic compounds contain elements in addition to carbon and hydrogen. As we mentioned in Section 2.7, it is convenient to consider such compounds to be derived from hydrocarbons by replacing one or more hydrogens by other groups of atoms. Such compounds can be divided into various families according to the nature of the groups, called **functional groups**, attached to the parent hydrocarbon fragment. Some of these families are summarized in Table 8.4, in which the hydrocarbon fragment to which the functional group is attached is symbolized by the letter **R**.

Alcohol

In Chapter 2 we noted that alcohols are organic compounds in which one of the hydrogen atoms of a hydrocarbon is replaced by OH. The family name for these compounds is **alcohol**. Examples are methanol (methyl alcohol) and ethanol (ethyl alcohol), which have the following structures:

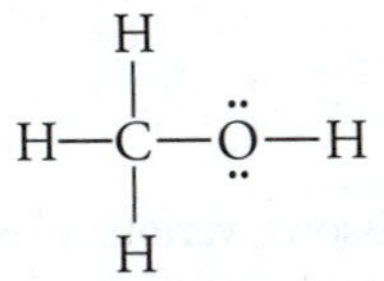

methanol
(methyl alcohol)

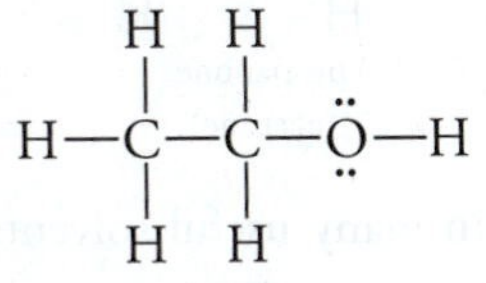

ethanol
(ethyl alcohol)

This container of "Canned Heat" contains methanol as the fuel. It is commonly used to heat food at buffets.

Andy Washnik

TABLE 8.4	Some Families of Oxygen- and Nitrogen-Containing Organic Compounds[a]	
Family Name	**General Formula**	**Example**
Alcohols	R—Ö—H	CH$_3$—Ö—H methanol
Aldehydes	R—C(=O)—H	CH$_3$—C(=O)—H ethanal (acetaldehyde)
Ketones	R—C(=O)—R	H$_3$C—C(=O)—CH$_3$ propanone (acetone)
Acids	R—C(=O)—Ö—H	H$_3$C—C(=O)—Ö—H ethanoic acid (acetic acid)
Amines	R—N̈H$_2$	H$_3$C—N̈H$_2$ methylamine
	R—N̈H—R	H$_3$C—N—CH$_3$ (with H above N) dimethylamine
	R—N—R (with R above N)	H$_3$C—N—CH$_3$ (with CH$_3$ above N) trimethylamine

[a] R stands for a hydrocarbon fragment such as CH$_3$— or CH$_3$CH$_2$—.

Some **condensed formulas** that we might write for these are CH_3OH and CH_3CH_2OH, or CH_3—OH and CH_3CH_2—OH. Methanol is used as a solvent and a fuel; ethanol is found in alcoholic beverages and is also blended with gasoline to yield a fuel called E85, containing 85% ethanol.

Ketones

In alcohols, the oxygen forms two single bonds to complete its octet, just as in water. But oxygen can also form double bonds, as you saw for CO_2. One family of compounds in which a doubly bonded oxygen replaces a pair of hydrogen atoms is called **ketones**. The simplest example is propanone, better known as acetone, a solvent often used in nail polish remover.

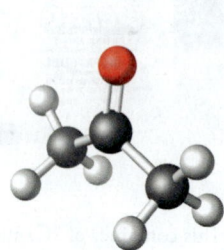

Acetone

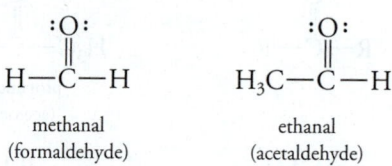

propanone
(acetone)

Ketones are found in many useful solvents that dissolve various plastics. An example is methyl ethyl ketone.

butanone
(methyl ethyl ketone)

Aldehydes

Notice that in ketones the carbon bonded to the oxygen is also attached to *two* other carbon atoms. If at least one of the atoms attached to the C=O group (called a **carbonyl group**, pronounced *car-bon-EEL*) is a hydrogen, a different family of compounds is formed called **aldehydes**. Examples are formaldehyde (used to preserve biological specimens, for embalming, and to make plastics) and acetaldehyde (used in the manufacture of perfumes, dyes, plastics, and other products).

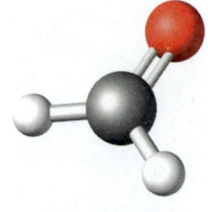

Formaldehyde

methanal
(formaldehyde)

ethanal
(acetaldehyde)

Organic Acids

Organic acids, also called **carboxylic acids**, constitute another very important family of oxygen-containing organic compounds. An example is acetic acid, which we described in Chapter 4. The shape of the molecule was illustrated in Figure 4.11, showing the single hydrogen atom that is capable of ionizing in the formation of H_3O^+. The Lewis structures of acetic acid and the acetate ion are

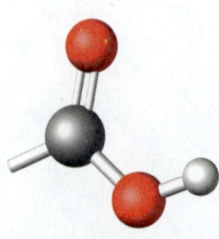

The carboxyl group

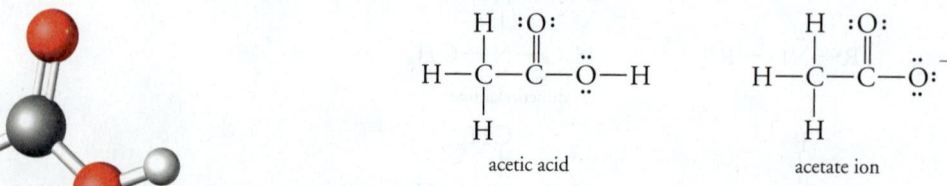

acetic acid

acetate ion

In general, the structures of organic acids are characterized by the presence of the **carboxyl group**, —CO_2H.

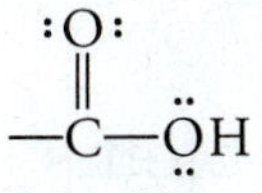

$$
\begin{array}{c}
:\!O\!: \\
\parallel \\
-C-\ddot{O}H
\end{array}
$$

carboxyl group

Notice that organic acids have both a doubly bonded oxygen and an OH group attached to the end carbon atom.

Amines

Nitrogen atoms need three electrons to complete an octet, and in most of its compounds, nitrogen forms three bonds. The common nitrogen-containing organic compounds can be imagined as being derived from ammonia by replacing one or more of the hydrogens of NH_3 with hydrocarbon groups. They're called **amines**, and an example is methylamine, CH_3NH_2.

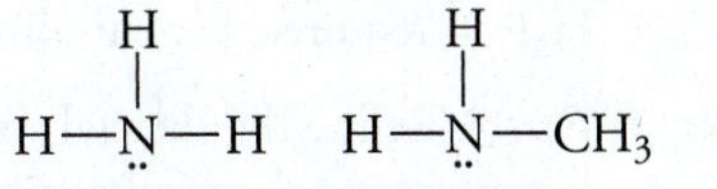

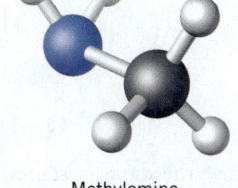

Methylamine

$$
\begin{array}{cc}
\quad H \quad & \quad H \quad \\
\quad | \quad & \quad | \quad \\
H-\underset{\cdot\cdot}{N}-H & H-\underset{\cdot\cdot}{N}-CH_3 \\
\text{ammonia} & \text{methylamine}
\end{array}
$$

Amines are strong-smelling compounds and often have a "fishy" odor. Like ammonia, they're weakly basic.[6]

$$CH_3NH_2(aq) + H_2O \rightleftharpoons CH_3NH_3{}^+(aq) + OH^-(aq)$$

As we noted in Chapter 5, the H^+ that is added to an amine becomes attached to the nitrogen atom.

Match the structural formulas on the left with the correct names of the families of organic compounds to which they belong. (*Hint:* How are the atoms connected to each other?)

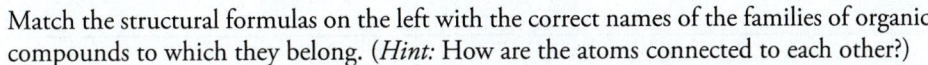

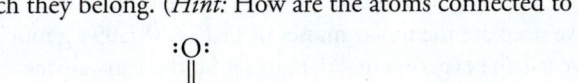

$$
\begin{array}{ll}
\begin{array}{c}
:\!O\!: \\
\parallel \\
CH_3-CH_2-C-H
\end{array} & \text{amine} \\[2ex]
\begin{array}{c}
H \\
| \\
CH_3-\underset{\cdot\cdot}{N}-CH_3
\end{array} & \text{alcohol} \\[2ex]
\begin{array}{c}
:\!O\!: \\
\parallel \\
H-C-\ddot{O}-H
\end{array} & \text{ketone} \\[2ex]
\begin{array}{c}
:\!O\!: \\
\parallel \\
CH_3-CH_2-C-CH_2-CH_3
\end{array} & \text{aldehyde} \\[2ex]
CH_3-CH_2-CH_2-\ddot{O}-H & \text{acid}
\end{array}
$$

Practice Exercise 8.26

The following questions apply to the compounds in Practice Exercise 8.26. (a) Which produces a basic solution in water? (b) Which produces an acidic aqueous solution? (c) For the acid, what is the Lewis structure of the anion formed when it is neutralized? (d) For the base, what is the Lewis structure of the cation formed when it reacts with H^+?

Practice Exercise 8.27

---

[6] Amino acids, which are essential building blocks of proteins in our bodies, contain both an amine group ($-NH_2$) and a carboxyl group ($-CO_2H$). The simplest of these is the amino acid glycine,

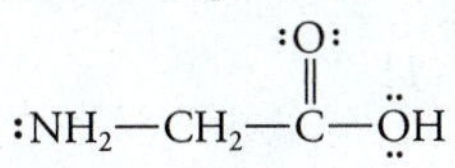

$$
\begin{array}{c}
:\!O\!: \\
\parallel \\
:NH_2-CH_2-C-\ddot{O}H
\end{array}
$$

glycine

Analyzing and Solving Multi-Concept Problems

Phosphorous acid has the formula H_3PO_3. It was found that a sample of the acid weighing 0.3066 g was neutralized completely by 0.4196 g of KOH. Use these data to write the preferred Lewis structure for H_3PO_3.

Analysis: Let's start by following the guidelines developed earlier in this chapter to arrive at the Lewis structure of oxoacids and place the hydrogen atoms on the oxygens to draw a triprotic acid.

H

O

H   O   P   O   H

We can use the data presented for the neutralization of the acid by KOH to confirm whether or not this is the correct arrangement of atoms.

From the neutralization data we can calculate the number of moles of KOH required to neutralize one mole of the acid.

This will tell us the number of "acidic" hydrogens in a molecule of the acid. If the acid is indeed a triprotic acid, then three moles of KOH should be needed to neutralize one mole of acid.

We can use the information from the stoichiometry of the neutralization reaction to tell us how to complete the Lewis structure either using the given skeletal structure or refining it, and then completing the Lewis structure.

Part 1: Use the data to determine the mole ratio KOH to H_3PO_3 required to neutralize the acid.

Part 2: Refine the skeletal structure, if necessary, to make it conform to the results of the computation in Part 1.

Part 3: Complete the Lewis structure, and then use formal charges to obtain the preferred structure.

PART 1

Assembling the Tools: The data we need are the molar masses of H_3PO_3 (81.994 g mol^{-1}) and KOH (56.108 g mol^{-1}). Then we use the experimental data to set up the equivalence

$$0.3066 \text{ g } H_3PO_3 \Leftrightarrow 0.4196 \text{ g of KOH}$$

Solution: This part of the problem can be expressed as

$$1 \text{ mol } H_3PO_3 \Leftrightarrow ? \text{ mol KOH}$$

Using dimensional analysis gives us

$$1 \text{ mol } H_3PO_3 \times \frac{81.994 \text{ g } H_3PO_3}{1 \text{ mol } H_3PO_3} \times \frac{0.4196 \text{ g KOH}}{0.3066 \text{ g } H_3PO_3} \times \frac{1 \text{ mol KOH}}{56.108 \text{ g KOH}} = 2.00 \text{ mol KOH}$$

Because only 2 moles of KOH are needed to neutralize 1 mole of the acid, the acid must be diprotic. We are going to have to make changes in the skeletal structure to reflect this.

PART 2

Assembling the Tools: We need to use the guidelines for constructing skeletal structures, modified to make the structure conform to the results of Part 1.

Solution: For oxoacid, acidic hydrogens are bonded to oxygen atoms, which are bonded in turn to the central atom. Since H_3PO_3 is diprotic, two such hydrogens are present, and the third one must be bonded to the phosphorus atom:

O

H   O   P   O   H

H

PART 3

Assembling the Tools: Our tool is the procedure for drawing Lewis structures, and we also use formal charges described in Section 8.8 to determine the preferred structure.

Solution: Phosphorus (Group 5A) contributes five electrons; each oxygen (Group 6A) contributes six electrons, and each hydrogen contributes one electron. The total is $26e^-$. Following the usual procedure, the Lewis structure is

$$
\begin{array}{ccc}
& :\!\ddot{\text{O}}\!: & \\
\text{H} :\!\ddot{\text{O}}\!:\!\text{P}\!:\!\ddot{\text{O}}\!:\!\text{H} & & \text{H}\!-\!\ddot{\text{O}}\!-\!\underset{\underset{\text{H}}{|}}{\overset{\overset{:\ddot{\text{O}}:}{|}}{\text{P}}}\!-\!\ddot{\text{O}}\!-\!\text{H} \\
\text{H} & &
\end{array}
$$

Assigning formal charges gives us

$$
\text{H}\!-\!\ddot{\text{O}}\!-\!\underset{\underset{\text{H}}{|}}{\overset{\overset{\overset{\ominus}{:\ddot{\text{O}}:}}{|}}{\overset{\oplus}{\text{P}}}}\!-\!\ddot{\text{O}}\!-\!\text{H}
$$

We can reduce the formal charges by moving a nonbonding electron pair from the oxygen at the top into the phosphorus–oxygen bond, giving a double bond.

$$
\text{H}\!-\!\ddot{\text{O}}\!-\!\underset{\underset{\text{H}}{|}}{\overset{\overset{\ddot{\text{O}}:}{\|}}{\text{P}}}\!-\!\ddot{\text{O}}\!-\!\text{H}
$$

This structure has no formal charges and is the preferred structure of the molecule.

Is the Answer Reasonable? The fact that the stoichiometry calculation gives a whole number of moles of KOH per mole of H_3PO_3 suggests this part of the problem has been solved correctly. Dimensional analysis also gives the correct units, so we can conclude the calculation is correct. The alternative skeletal structure agrees with the stoichiometry, so it is reasonable. Finally, we can check to be sure we've supplied the correct number of electrons, which we have, so our Lewis structure appears to be reasonable.

| Summary

Organized by Learning Objective

Describe the necessary condition for bond formation that produces a stable compound

In order for a compound to form from its elements, the potential energy of the compound must be less than the potential energy of the elements.

Explain the factors involved in ionic bonding

In ionic compounds, the electrostatic attractive forces between positive and negative ions are called **ionic bonds**. The formation of ionic compounds by electron transfer occurs between atoms of low ionization energy and atoms of high electron affinity. The chief stabilizing influence in the formation of ionic compounds is the **lattice energy** that is released.

Write electron configurations of ions

When atoms of the elements in Groups 1A and 2A as well as the nonmetals form ions, they usually gain or lose enough electrons to achieve a noble gas electron configuration. Transition elements lose their outer *s* electrons first, followed by loss of *d* electrons from the shell below the outer shell. Post-transition metals lose electrons from their outer *p* subshell first, followed by electrons from the outer *s* subshell.

Write Lewis symbols for atoms and ions

Lewis symbols are a bookkeeping device used to keep track of valence electrons in ionic and covalent bonds. The **Lewis symbol** of an element consists of the element's chemical symbol surrounded by a number of dots equal to the number of valence electrons.

Describe covalent bonds through the energetics of bond formation, octet rule, and multiple bonds

Shared electrons attract the positive nuclei, and this leads to a lowering of the potential energy of the atoms as a **covalent bond** forms. An atom tends to share enough electrons to complete its valence shell, which, except for hydrogen, usually holds eight electrons, which forms the basis of the **octet rule**; that is, atoms of the representative elements tend to acquire eight electrons in their outer shells when they form bonds. **Single**, **double**, and **triple bonds** involve the sharing of one, two, and three pairs of electrons, respectively, between two atoms. Boron and beryllium often have less than an octet in their compounds. Elements in Periods 3, 4, 5, and 6 can exceed an octet if they form more than four bonds.

Bond energy (the energy needed to separate the bonded atoms) and **bond length** (the distance between the nuclei of the atoms connected by the bond) are two experimentally measurable quantities that can be related to the number of pairs of electrons in the bond. For bonds between atoms of the same elements, bond energy increases and bond length decreases as the **bond order** increases.

Explain how electronegativity affects the polarity of covalent bonds and the reactivity of the elements

The attraction an atom has for the electrons in a bond is called the atom's **electronegativity**. When atoms of different electronegativities form a bond, the electrons are shared unequally and the bond is **polar**, with **partial positive** and **partial negative** charges at opposite ends. This causes the bond to be an electric **dipole**. In a **polar molecule**, such as HCl, the product of the charge at either end multiplied by the distance between the charges gives the **dipole moment, μ**. When the two atoms have the same electronegativity, the bond is **nonpolar**. When the electronegativity difference between the two bonded atoms is very large, ionic bonding results. A bond is approximately 50% ionic when the electronegativity difference is 1.7. In the periodic table, electronegativity increases from left to right across a period and from bottom to top in a group.

The **reactivity** of metals is related to the ease with which they are oxidized (lose electrons); for nonmetals it is related to the ease with which they are reduced (gain electrons). Metals with low electronegativities lose electrons easily, are good reducing agents, and tend to be very reactive. For nonmetals, the higher the electronegativity, the stronger is their ability to serve as oxidizing agents. In **displacement reactions**, a more electronegative nonmetal is often able to displace one of lower electronegativity from its compounds.

Draw Lewis structures for covalent molecules and calculate formal charges for individual atoms in a molecule

In the **Lewis structure** for an ionic compound, the Lewis symbol for the anion is enclosed in brackets (with the charge written outside) to show that all the electrons belong entirely to the ion. The Lewis structure for a molecule or polyatomic ion uses pairs of dots between chemical symbols to represent shared pairs of electrons. The electron pairs in covalent bonds usually are represented by dashes; one dash equals two electrons. The following procedure is used to draw the Lewis structure: (1) decide on the **skeletal structure** (remember that the least electronegative atom is usually the central atom and is usually first in the formula); (2) count all the valence electrons, taking into account the charge, if any; (3) place a pair of electrons in each bond; (4) complete the octets of atoms other than the central atom (but remember that hydrogen can only have two electrons); (5) place any left-over electrons on the central atom in pairs; (6) if the central atom still has less than an octet, move electron pairs to make double or triple bonds.

Formal Charges. The **formal charge** assigned to an atom in a Lewis structure (which usually differs from the actual charge on the atom) is calculated as the difference between the number of valence electrons of an isolated atom of the element and the number of electrons that "belong" to the atom because of its bonds to other atoms and its unshared valence electrons. The sum of the formal charges always equals the net charge on the molecule or ion. The most stable (lowest energy) Lewis structure for a molecule or ion is the one with formal charges closest to zero.

Coordinate Covalent Bonding. For bookkeeping purposes, we sometimes single out a covalent bond whose electron pair originated from one of the two bonded atoms. Once formed, a coordinate covalent bond is no different from any other covalent bond.

Draw and explain resonance structures

Bonds to chemically equivalent atoms must be the same; they must have the same bond length and the same bond energy, which means they must involve the sharing of the same number of electron pairs. Typically, this occurs when it is necessary to form multiple bonds during the drawing of a Lewis structure. When alternatives exist for the location of multiple bond among two or more equivalent atoms, then each possible Lewis structure is actually a **resonance structure** or **contributing structure**, and we draw them all. In drawing resonance structures, the relative locations of the nuclei must be identical in all. The **average bond order** is calculated from the total number of bonds distributed over the equivalent bond locations. Remember that none of the resonance structures corresponds to a real molecule, but their composite—the **resonance hybrid**—does approximate the actual structure of the molecule or ion.

Classify organic compounds and identify functional groups

Organic compounds are compounds with carbon atoms that are covalently bonded to other carbon atoms, hydrogen, oxygen, nitrogen, and other nonmetals. The functional groups discussed in this chapter are alcohols, ketones, aldehydes, organic acids, and amines.

TOOLS

Tools for Problem Solving The following tools were introduced in this chapter. Study them carefully so that you can select the appropriate tool when needed.

Octet rule (Section 8.3)
Atoms tend to gain or lose electrons until they have achieved an outer shell that contains an octet of electrons.

Order in which electrons are lost from an atom (Section 8.3)
Electrons are lost first from the shell with the largest n. Within a given shell, electrons are lost from the p subshell before the s subshell. If additional electrons are lost, they come from an available $(n - 1)$ d orbital.

Lewis symbols (Section 8.4)
Lewis symbols are a bookkeeping device that we use to keep track of valence electrons (the outermost s and p electrons) in atoms and ions. For a neutral atom of the representative elements, the Lewis symbol consists of the atomic symbol surrounded by dots equal in number to number of valence electrons.

Octet rule and covalent bonding (Section 8.5)
The octet rule helps us construct Lewis structures for covalently bonded molecules. Elements in Period 2 never exceed an octet in their valence shells.

Dipole moment (Section 8.6)
The dipole moment (μ), in debye units, of a diatomic molecule is calculated from the charge on an end of the molecule, q, and the bond length, r.

$$\mu = q \times r$$

Periodic trends in electronegativity (Section 8.6)
Electronegativity increases across a period and decreases down a group in the periodic table and can be used to estimate the degree of polarity of bonds and to estimate which atoms in a bond are the most electronegative.

Reactivity of metals and the periodic table (Section 8.6)
A knowledge of where the most reactive and least reactive metals are located in the periodic table gives a qualitative feel for how reactive a metal is.

Reactivity of nonmetals and the periodic trends (Section 8.6)
Oxidizing ability of nonmetals increases from left to right across a period and from bottom to top in a group.

Drawing Lewis structures (Section 8.7)
The method described in Figure 8.12 yields Lewis structures in which the maximum number of atoms obey the octet rule.

Correlation between bond properties and bond order (Section 8.7)
The correlations between increasing bond energy and decreasing bond length with bond order allow us to compare experimental covalent bond properties with those predicted by theory.

Calculating formal charges (Section 8.7)

$$\text{Formal charge} = \begin{bmatrix} \text{number of } e^- \text{ in valence} \\ \text{shell of the isolated atom} \end{bmatrix} - \begin{bmatrix} \text{number of bonds} \\ \text{to the atoms} \end{bmatrix} + \begin{bmatrix} \text{number of} \\ \text{unshared } e^- \end{bmatrix}$$

Selecting the best Lewis structure (Section 8.7)
A valid Lewis structure having the smallest formal charges is preferred.

Resonance structures (Section 8.8)
Expect resonance structures when there is more than one equivalent option for assigning the location of a double bond. The number of equivalent positions for the double bond is the number of resonance structures that can be drawn.

Average bond order (or bond order) (Section 8.8)

To calculate the bond order for molecular structures, add up the total number of bonds to a central atom and divide by the number of atoms bonded to that central atom.

WileyPLUS = *WileyPLUS*, an online teaching and learning solution. *Note to instructors*: Many of the end-of-chapter problems are available for assignment via the WileyPLUS system. **www.wileyplus.com**. Review Problems are presented in pairs separated by blue rules. Answers to problems whose numbers appear in blue are given in Appendix B. More challenging problems are marked with an asterisk ✳.

| Review Questions

Energy Requirements for Bond Formation

8.1 What must be true about the change in the total potential energy of a collection of atoms for a stable compound to be formed from the elements?

8.2 Under what conditions could a compound form from its elements if the potential energy for the compound is higher than the potential energy of the elements?

Ionic Bonding

8.3 What is an *ionic bond*?

8.4 Define the term *lattice energy*. In what ways does the lattice energy contribute to the stability of ionic compounds?

8.5 How is the tendency to form ionic bonds related to the IE and EA of the atoms involved?

8.6 What influence do ion size and charge have on lattice energies of ionic compounds?

Octet Rule and Electron Configurations of Ions

8.7 What is the *octet rule*? What is responsible for it?

8.8 Why doesn't hydrogen obey the octet rule?

8.9 Magnesium forms compounds containing the ion Mg^{2+} but not the ion Mg^{3+}. Why?

8.10 Why doesn't chlorine form the Cl^{2-} ion in compounds?

8.11 Why do many of the transition elements in Period 4 form ions with a 2+ charge? Why do some have 1+ ions?

8.12 If we were to compare the first, second, third, and fourth ionization energies of aluminum, between which pair of successive ionization energies would there be the largest difference?

Lewis Symbols: Keeping Track of Valence Electrons

8.13 The Lewis symbol for an atom only accounts for electrons in the valence shell of the atom. Why are we not concerned with the other electrons?

8.14 Which of these Lewis symbols is incorrect?

(a) $\cdot \ddot{O} \cdot$ (b) $\cdot \ddot{\underset{..}{Cl}} \cdot$ (c) $: \ddot{N} :$ (d) $: \ddot{\underset{..}{Sb}} :$

Covalent Bonds

8.15 Define *bond length* and *bond energy*.

8.16 Define *bond order*. How are bond energy and bond length related to bond order? Why are there these relationships?

8.17 The energy required to break the H—Cl bond to give H^+ and Cl^- ions would not be called the H—Cl bond energy. Why?

8.18 In terms of the potential energy change, why doesn't ionic bonding occur when two nonmetals react with each other?

8.19 When acetonitrile, $CH_3—C \equiv N :$, a molecular compound, is evaporated what type of particle is in the gas phase? How does this compare to the salt sodium chloride?

8.20 Describe what happens to the electron density around two hydrogen atoms as they come together to form an H_2. What happens to the energy of two hydrogen atoms as they approach each other? What happens to the spins of the electrons?

8.21 Is the formation of a covalent bond endothermic or exothermic?

8.22 What factors control the bond length in a covalent bond?

8.23 How many covalent bonds are normally formed by (a) hydrogen, (b) carbon, (c) oxygen, (d) nitrogen, and (e) chlorine?

Bond Polarity and Electronegativity

8.24 What is a polar covalent bond?

8.25 Define *dipole moment* in the form of an equation. What is the value of the *debye* (with appropriate units)?

8.26 Define electronegativity. On what basis did Pauling develop his scale of electronegativities?

8.27 Which element has the highest electronegativity? Which is the second most electronegative element? What are the horizontal and vertical periodic trends associated with electronegativity?

8.28 Among the following bonds, which are more ionic than covalent?

(a) Si—O (b) Ba—O (c) Se—Cl (d) K—Br

8.29 If an element has a low electronegativity, is it likely to be classified as a metal or a nonmetal? Explain your answer.

8.30 In what groups in the periodic table are the most reactive metals found? Where do we find the least reactive metals?

8.31 How is the electronegativity of a metal related to its reactivity?

8.32 When we say that aluminum is more *reactive* than iron, which kind of reaction of these elements are we describing?

8.33 Arrange the following metals in their approximate order of reactivity (most reactive first, least reactive last) based on their locations in the periodic table:

(a) iridium (b) silver (c) calcium (d) iron

8.34 Complete and balance the following equations. If no reaction occurs, write "N.R."

(a) $KCl + Br_2 \longrightarrow$

(b) $NaI + Cl_2 \longrightarrow$

(c) $KCl + F_2 \longrightarrow$

(d) $CaBr_2 + Cl_2 \longrightarrow$

(e) $AlBr_3 + F_2 \longrightarrow$

(f) $ZnBr_2 + I_2 \longrightarrow$

8.35 In each pair, choose the better oxidizing agent.

(a) O_2 or F_2

(b) As_4 or P_4

(c) Br_2 or I_2

(d) P_4 or S_8

(e) Se_8 or Cl_2

(f) As_4 or S_8

Lewis Structures

8.36 Without looking at the text, describe the steps for drawing Lewis structures.

8.37 Why do we usually place the least electronegative element in the center of a Lewis structure?

8.38 Why do Period 2 elements never form more than four covalent bonds? Why are Period 3 elements able to exceed an octet?

8.39 Define (a) single bond, (b) double bond, and (c) triple bond.

8.40 The Lewis structure for hydrogen cyanide is $H—C \equiv N:$. Draw circles enclosing electrons to show that carbon and nitrogen obey the octet rule.

8.41 How many electrons are in the valence shells of (a) Be in $BeCl_2$, (b) B in BCl_3, and (c) H in H_2O?

8.42 What is the minimum number of electrons that would be expected to be in the valence shell of As, in $AsCl_5$?

8.43 Nitrogen and arsenic are in the same group in the periodic table. Arsenic forms both $AsCl_3$ and $AsCl_5$, but with chlorine, nitrogen only forms NCl_3. On the basis of the electronic structures of N and As, explain why this is so.

Formal Charge

8.44 What is the definition of *formal charge*? How are formal charges indicated on a structural formula?

8.45 How are formal charges for atoms in a molecule determined? The sum of the formal charges of a molecule must add up to what?

8.46 How are formal charges used to select the best Lewis structure for a molecule? What is the basis for this method of selection?

8.47 What are the formal charges on the atoms in the HCl molecule? What are the actual charges on the atoms in this molecule? Are formal charges the same as actual charges?

Coordinate Covalent Bonds

8.48 What is a *coordinate covalent bond*?

8.49 Once formed, how (if at all) does a coordinate covalent bond differ from an ordinary covalent bond?

8.50 BCl_3 has an incomplete valence shell. Use Lewis structures to show how it could form a coordinate covalent bond with a water molecule.

Resonance Structures

8.51 Why is the concept of resonance needed?

8.52 What is a resonance hybrid? How does it differ from the resonance structures drawn for a molecule?

8.53 Draw the resonance structures of the benzene molecule. Why is benzene more stable than one would expect if the ring contained three carbon–carbon double bonds?

8.54 Polystyrene plastic is a hydrocarbon that consists of a long chain of carbon atoms joined by single bonds in which every other carbon is attached to a benzene ring. The ring is attached by replacing a hydrogen atom of benzene with a single bond to the carbon chain. Sketch a portion of a polystyrene molecule that contains five benzene rings.

Covalent Compounds of Carbon

8.55 Sketch the structures for (a) methane, (b) ethane, and (c) propane.

8.56 Draw the structure for a hydrocarbon that has a chain of six carbon atoms linked by single bonds. How many hydrogen atoms does the molecule have? What is the molecular formula for the compound?

8.57 How many *different* molecules have the formula C_5H_{12}? Sketch their structures.

8.58 What is a carbonyl group? In which classes of organic molecules that we've studied do we find a carbonyl group?

8.59 Match the compounds on the left with the family types on the right.

$$H_3C—CH_2—\overset{\overset{\displaystyle :O:}{\|}}{C}—\ddot{O}—H$$ hydrocarbon

$$H_3C—\overset{\overset{\displaystyle :O:}{\|}}{C}—CH_2—CH_3$$ aldehyde

$$H_3C—CH_2—\overset{\overset{\displaystyle :\ddot{O}H}{|}}{CH}—CH_3$$ amine

$$H_3C—CH_2—\overset{\overset{\displaystyle :O:}{\|}}{C}—H$$ acid

$$H_3C—\overset{\overset{\displaystyle H}{|}}{\underset{\displaystyle ..}{N}}—CH_2—CH_3$$ alcohol

$$H_3C—CH_2—\overset{\displaystyle CH}{\underset{\overset{\displaystyle |}{CH_3}}{}}—CH_3$$ ketone

8.60 Write a chemical equation for the ionization in water of the acid in Question 8.59. Is the acid strong or weak? Draw the Lewis structure for the ion formed by the acid when it ionizes.

8.61 Write a chemical equation for the reaction of the amine in Question 8.59 with water. Draw the Lewis structure for the ion formed by the amine.

8.62 Write the net ionic equation for the aqueous reaction of the acid and amine in Question 8.59 with each other. (Use molecular formulas.)

Review Problems

Ionic Bonding

8.63 In each of the following pairs of compounds, which would have the larger lattice energy: **(a)** CaO or Al_2O_3, **(b)** BeO or SrO, **(c)** NaCl or NaBr?

8.64 In each of the following pairs of compounds, which would have the larger lattice energy: **(a)** MgO or CaO, **(b)** Na_2O or MgO, **(c)** Ca_3P_2 or CaO?

*__8.65__ Use an enthalpy diagram and the following data to calculate the electron affinity of bromine. The standard heat of formation of NaBr is -360 kJ mol^{-1}. The energy needed to vaporize one mole of $Br_2(l)$ to give $Br_2(g)$ is 31 kJ mol^{-1}. The first ionization energy of Na is 495.4 kJ mol^{-1}. The bond energy of Br_2 is 192 kJ per mole of Br—Br bonds. The lattice energy of NaBr is -743.3 kJ mol^{-1}.

*__8.66__ Use an enthalpy diagram to calculate the lattice energy of $CaCl_2$ from the following information. Energy needed to vaporize one mole of Ca(s) is 192 kJ. For calcium, the first IE = 589.5 kJ mol^{-1}, the second IE = 1146 kJ mol^{-1}. The electron affinity of Cl is 349 kJ mol^{-1}. The bond energy of Cl_2 is 242.6 kJ per mole of Cl—Cl bonds. The standard heat of formation of $CaCl_2$ is -795 kJ mol^{-1}.

Octet Rule and Electron Configurations

8.67 Explain what happens to the electron configurations of Mg and Br when they react to form magnesium bromide.

8.68 Describe what happens to the electron configurations of lithium and nitrogen when they react to form the lithium nitride.

8.69 What are the electron configurations of the Pb^{2+} and Pb^{4+} ions?

8.70 What are the electron configurations of the Bi^{3+} and Bi^{5+} ions?

8.71 Write the abbreviated electron configuration of the Mn^{3+} ion. How many unpaired electrons does the ion contain?

8.72 Write the abbreviated electron configuration of the Co^{3+} ion. How many unpaired electrons does the ion contain?

Lewis Symbols: Keeping Track of Valence Electrons

8.73 Write Lewis symbols for the following atoms: **(a)** Si, **(b)** Sb, **(c)** Ba, **(d)** Al, **(e)** S.

8.74 Write Lewis symbols for the following atoms: **(a)** K, **(b)** Ge, **(c)** As, **(d)** Br, **(e)** Se.

8.75 Use Lewis symbols to diagram the reactions between **(a)** Mg and S, **(b)** Mg and Cl, and **(c)** Mg and N.

8.76 Use Lewis symbols to diagram the reactions between **(a)** Ca and Br, **(b)** Al and O, and **(c)** K and S.

Covalent Bonds

8.77 How much energy, in joules, is required to break the bond in *one* chlorine molecule? The bond energy of Cl_2 is 242.6 kJ/mol.

8.78 How much energy is released in the formation of one molecule of HCl by the following reaction?

$$H^+(g) + Cl^-(g) \longrightarrow HCl(g)$$

The bond energy of HCl is 431 kJ mol^{-1}. Additional data can be found in tables in Chapter 7.

8.79 The reason there is danger in exposure to high-energy radiation (e.g., ultraviolet and X rays) is that the radiation can rupture chemical bonds. In some cases, cancer can be caused by it. A carbon–carbon single bond has a bond energy of approximately 348 kJ per mole. What wavelength of light is required to provide sufficient energy to break the C—C bond? In which region of the electromagnetic spectrum is this wavelength located?

8.80 A mixture of H_2 and Cl_2 is stable, but a bright flash of light passing through it can cause the mixture to explode. The light causes Cl_2 molecules to split into Cl atoms, which are highly reactive. What wavelength of light is necessary to cause the Cl_2 molecules to split? The bond energy of Cl_2 is 242.6 kJ per mole.

8.81 Use Lewis structures to diagram the formation of (a) Br_2, (b) H_2O, and (c) NH_3 from neutral atoms.

8.82 Use Lewis structures to diagram the formation of (a) CH_4, (b) PH_3, (c) F_2 from neutral atoms.

8.83 Chlorine tends to form only one covalent bond because it needs just one electron to complete its octet. What are the Lewis structures for the simplest compound formed by chlorine with (a) nitrogen, (b) carbon, (c) sulfur, and (d) bromine?

8.84 Use the octet rule to predict the formula of the simplest compound formed from hydrogen and (a) selenium, (b) arsenic, and (c) silicon.

8.85 What would be the formula for the simplest compound formed from (a) phosphorus and chlorine, (b) carbon and fluorine, and (c) iodine and chlorine?

8.86 What would be the formula for the simplest compound formed from (a) arsenic and bromine, (b) silicon and chlorine, (c) bromine and fluorine?

Bond Polarity and Electronegativity

8.87 Use the data in Table 8.3 to calculate the amount of charge on the oxygen and nitrogen in the nitrogen monoxide molecule, expressed in electronic charge units ($e^- = 1.60 \times 10^{-19}$ C). Which atom carries the positive charge?

8.88 The molecule bromine monofluoride has a dipole moment of 1.42 D and a bond length of 176 pm. Calculate the charge on the ends of the molecule, expressed in electronic charge units ($e^- = 1.60 \times 10^{-19}$ C). Which atom carries the positive charge?

8.89 In the vapor state, cesium and fluoride ions pair to give CsF formula units that have a bond length of 0.255 nm and a dipole moment of 7.88 D. What is the actual charge on the cesium and fluorine atoms in CsF? What percentage of full 1+ and 1− charges is this?

8.90 The dipole moment of HF is 1.83 D and the bond length is 917 pm. Calculate the amount of charge (in electronic charge units) on the hydrogen and the fluorine atoms in the HF molecule.

8.91 Use Figure 8.9 to choose the atom in each of the following bonds that carries the partial positive charge: (a) N—S, (b) Si—I, (c) N—Br, (d) C—Cl.

8.92 Use Figure 8.9 to choose the atom that carries the partial negative charge in each of the following bonds: (a) Hg—I, (b) P—I, (c) Si—F, (d) Mg—N.

8.93 Which of the bonds in Problem 8.91 is the most polar?

8.94 Which of the bonds in the Problem 8.92 is the least polar?

Lewis Structures

8.95 Draw Lewis structures for (a) $SiCl_4$, (b) PF_3, (c) ClF_5, and (d) SCl_2.

8.96 Draw Lewis structures for (a) HIO_3, (b) H_2CO_3, (c) TeF_6, and (d) PH_3.

8.97 Draw Lewis structures for (a) $AsCl_4^+$, (b) ClO_2^-, (c) HNO_2, and (d) XeF_2.

8.98 Draw Lewis structures for (a) HCO_3^-, (b) PCl_4^+, (c) PF_6^-, and (d) XeF_4.

8.99 Draw Lewis structures for (a) carbon disulfide, (b) the cyanide ion, (c) germanium tetrachloride, and (d) the phosphate ion.

8.100 Draw Lewis structures for (a) selenium trioxide, (b) the carbonate ion, (c) hydrogen carbonate ion, and (d) selenium dioxide.

8.101 Draw Lewis structures for (a) AsH_3, (b) NO_2, (c) $SbCl_6$, and (d) IO_3^-.

8.102 Draw Lewis structures for (a) NO^+, (b) $HClO_2$, (c) H_2SeO_3, and (d) H_3AsO_4.

8.103 Draw the Lewis structure for (a) CH_2O (the central atom is carbon, which is attached to two hydrogens and an oxygen), and (b) SeO_2F_2 (the central atom is selenium, which is attached to two oxygens and two fluorines).

8.104 Draw Lewis structures for (a) the peroxide ion, O_2^{2-}, and (b) $POCl_3$ (the central atom is phosphorus, which is attached to the oxygen and the three chlorines).

8.105 Assign formal charges to each atom in the following structures:

(a) H—Ö—Cl—Ö: (b) Ö=S—Ö: (c) Ö=S—Ö:

8.106 Assign formal charges to each atom in the following structures:

(a) :Cl—N—Ö: (b) :F—N—Ö: (c) :F—S—F:

8.107 Draw the Lewis structure for $HClO_4$. Assign formal charges to each atom in the formula. Determine the preferred Lewis structure for this compound.

8.108 Draw the Lewis structure for SO_2Cl (sulfur bonded to two Cl and one O). Assign formal charges to each atom. Determine the preferred Lewis structure for this molecule.

8.109 Below are two structures for $BeCl_2$. Give two reasons why the one on the left is the preferred structure.

$$:\ddot{C}l—Be—\ddot{C}l: \qquad :\ddot{C}l—Be=\ddot{C}l:$$

8.110 The following are two Lewis structures that can be drawn for phosgene, a substance that has been used as a war gas.

$$\begin{matrix} :\ddot{O}: \\ | \\ :\ddot{C}l—C=\ddot{C}l: \end{matrix} \qquad \begin{matrix} \ddot{O}: \\ \| \\ :\ddot{C}l—C—\ddot{C}l: \end{matrix}$$

Which is the better Lewis structure? Why?

8.111 Use Lewis structures to show that the hydronium ion, H_3O^+, can be considered to be formed by the creation of a coordinate covalent bond between H_2O and H^+.

8.112 Use Lewis structures to show that the reaction

$$BF_3 + F \longrightarrow BF_4$$

involves the formation of a coordinate covalent bond.

| Additional Exercises

8.119 Give the formula and name of four different ionic compounds. What are the charges of the ions?

8.120 Use data from the tables of ionization energies and electron affinities in Chapter 7 to calculate the energy changes for the following reactions.

$$Na(g) + Cl(g) \longrightarrow Na^+(g) + Cl^-(g)$$
$$Na(g) + 2Cl(g) \longrightarrow Na^{2+}(g) + 2Cl^-(g)$$

Approximately how many times larger would the lattice energy of $NaCl_2$ have to be compared to the lattice energy of NaCl for $NaCl_2$ to be more stable than NaCl?

8.121 Changing 1 mol of Mg(s) and ½ mol of $O_2(g)$ to gaseous atoms requires a total of approximately 150 kJ of energy. The first and second ionization energies of magnesium are 737 and 1450 kJ/mol, respectively; the first and second electron affinities of oxygen are −141 and +844 kJ/mol, respectively; and the standard heat of formation of MgO(s) is −602 kJ/mol. Construct an enthalpy diagram similar to the one in Figure 8.4 and use it to calculate the lattice energy of magnesium oxide. How does the lattice energy of MgO compare with that of NaCl? What might account for the difference?

Resonance Structures

8.113 Draw all of the resonance structures for the N_2O_4 molecule and determine the average N—O bond order. The skeletal structure of the molecule is

$$\begin{matrix} O & & O \\ & N—N & \\ O & & O \end{matrix}$$

8.114 Draw the resonance structures for $CO_3{}^{2-}$. Calculate the average C—O bond order.

8.115 How should the N—O bond lengths compare in the $NO_3{}^-$ and $NO_2{}^-$ ions?

8.116 Arrange the following in order of increasing C—O bond length: CO, $CO_3{}^{2-}$, CO_2, $HCO_2{}^-$ (formate ion).

*__8.117__ The Lewis structure of CO_2 was given as

$$\ddot{O}=C=\ddot{O}$$

but two other resonance structures can also be drawn for it. What are they? On the basis of formal charges, why are they not preferred structures?

*__8.118__ Use formal charges to establish the preferred Lewis structures for the $ClO_3{}^-$ and $ClO_4{}^-$ ions. Draw resonance structures for both ions and determine the average Cl—O bond order in each. Which of these ions would be expected to have the shorter Cl—O bond length?

8.122 In many ways, tin(IV) chloride behaves more like a covalent molecular species than like a typical ionic chloride. Draw the Lewis structure for the tin(IV) chloride molecule.

8.123 In each pair, choose the one with the more polar bonds. (Use the periodic table to answer the question.)
(a) PCl_3 or $AsCl_3$
(b) SF_2 or GeF_4
(c) $SiCl_4$ or SCl_2
(d) SrO or SnO

8.124 How many electrons are in the outer shell of the Zn^{2+} ion?

8.125 The Lewis structure for carbonic acid (formed when CO_2 dissolves in water) is usually given as

$$\begin{matrix} \ddot{O}: \\ \| \\ H—\ddot{O}—C—\ddot{O}—H \end{matrix}$$

What is wrong with the following structures?

$$\begin{matrix} :\ddot{O}: \\ | \\ \text{(a) } H—\ddot{O}—C—\ddot{O}—H \end{matrix}$$

(b) H—O=C—O—H with :O: above C

(c) H—O=C=O—H with :O: above C

8.126 Are the following Lewis structures considered to be resonance structures? Explain. Which is the more likely structure for POCl₃?

:Cl—P=O with Cl atoms   :Cl—O=P with Cl atoms

8.127 Assign formal charges to all the atoms in the following Lewis structure of hydrazoic acid, HN₃.

H—N≡N—N:

Suggest a lower-energy resonance structure for this molecule.

8.128 Assign formal charges to all the atoms in the Lewis structure

:O—H, :O—As—O—H, H—O:

Suggest a lower-energy Lewis structure for this molecule.

8.129 The inflation of an "air bag" when a car experiences a collision occurs by the explosive decomposition of sodium azide, NaN₃, which yields nitrogen gas that inflates the bag. The following resonance structures can

be drawn for the azide ion, N₃⁻. Identify the best and worst of them.

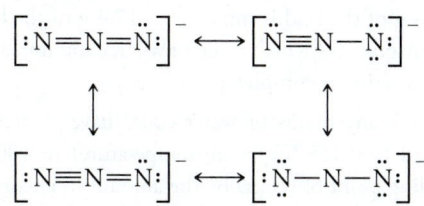

8.130 How should the sulfur–oxygen bond lengths compare for the species SO_3, SO_2, SO_3^{2-}, and SO_4^{2-}?

8.131 What is the most reasonable Lewis structure for S_2Cl_2?

*8.132 There are two acids that have the formula HCNO. Which of the following skeletal structures are most likely for them? Justify your answer.

H C O N   H N O C   H O C N
H C N O   H N C O   H O N C

8.133 In the vapor state, ion pairs of KF can be identified. The dipole moment of such a pair is measured to be 8.59 D and the K—F bond length is found to be 217 pm. Is the K—F bond 100% ionic? If not, what percent of full 1+ and 1− charges do the K and F atoms carry, respectively?

8.134 Below is a ball-and-stick model of a type of alcohol derived from a hydrocarbon. What is the formula for the hydrocarbon and what is its name?

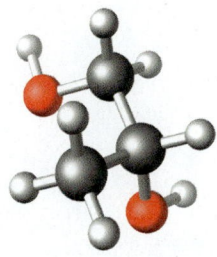

8.135 Explain why ions of the representative elements rarely have charges greater than +3 or smaller than −3.

|Multi-Concept Problems

8.136 Use Lewis structures to show the ionization of the following organic acid (a weak acid) in water.

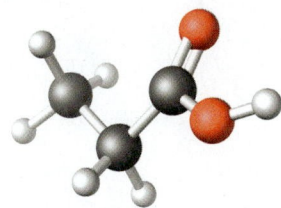

8.137 The compound below, an amine, is a weak base and undergoes ionization in water following a path similar to that of ammonia. Use Lewis structures to diagram the reaction of this amine with water.

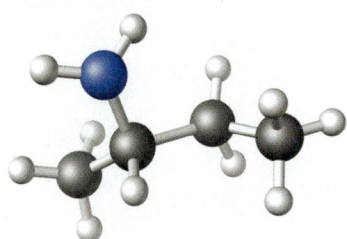

8.138 Use Lewis structures to diagram the reaction between the acid in Problem 8.136 with the base in Problem 8.137. If 12.5 g of the acid is mixed with 17.4 g of the base, how many grams of which reactant will remain unreacted after the reaction is complete?

*8.139 How many grams of water could have its temperature raised from 25 °C (room temperature) to 100 °C (the boiling point of water) by the amount of energy released in the formation of 1 mol of H_2 from hydrogen atoms? The bond energy of H_2 is 435 kJ/mol.

*8.140 What wavelength of light, if absorbed by a hydrogen molecule, could cause the molecule to split into the ions H^+ and H^-? (The data required are available in this and previous chapters.)

*8.141 A 38.40 mg sample of an organic acid composed of just carbon, hydrogen, and oxygen was burned in pure oxygen to give 37.54 mg CO_2 and 7.684 mg H_2O in a combustion analysis. In a separate experiment, the molecular mass was determined to be 90. In a titration, 40.2 mg of the acid dissolved in 50 mL of water required 14.28 mL of 0.0625 M NaOH for complete neutralization. Use these data to draw a reasonable Lewis structure for the compound.

|Exercises in Critical Thinking

8.142 What is the average bond energy of a C—C covalent bond? What wavelength of light provides enough energy to break such a bond? Using this information, explain why unfiltered sunlight is damaging to the skin.

8.143 One way of estimating the electronegativity of an atom is to use an average of its ionization energy and electron affinity. Why would these two quantities, taken together, be related to electronegativity?

8.144 The attractions between molecules of a substance can be associated with the size of the molecule's dipole moment. Explain why this is so.

8.145 The positive end of the dipole in a water molecule is not located on an atom. Explain why this happens and suggest other simple molecules that show the same effect.

8.146 In describing the structures of molecules, we use Lewis structures, formal charges, and experimental evidence. Rank these in terms of importance in deciding on the true structure of a molecule, and defend your choice.

9

Theories of Bonding and Structure

Chapter Outline

The honey bee is attracted to the flower by its scent. The honey bee collects the nectar and the pollen, which is transferred from one flower to another. Each different flower has a different scent, and it is the structure of the molecules that determines the scent. In this chapter, we will look at the structure of molecules and how bonding determines the structure.

© Kesipun/Shutterstock

This Chapter in Context

Molecules have three-dimensional shapes that are determined by the relative orientations of their covalent bonds, and this structure is maintained regardless of whether the substance is a solid, a liquid, or a gas. You've seen some of these shapes in previous chapters. Ionic substances are quite different. The structure of a solid ionic compound, such as NaCl, is controlled primarily by the sizes of the ions and their charges. The attractions between the ions have no preferred directions, so if an ionic compound is melted, this structure is lost and the ordered array of ions collapses into a jumbled liquid state.

Many of the properties of a molecule depend on the three-dimensional arrangement of its atoms. Just like the bee in the chapter opening photograph, we detect various aromas because of a precise fit between "odor molecules" and odor receptors in our olfactory system. In addition, shapes of active sites on enzyme molecules promote chemical reactions, while the unique shape of antibody binding sites toward pathogens provide protection against diseases. Similarly, the man-made structures of polymer molecules in plastics have a strong influence on the properties of materials made from them.

In this chapter we will explore the topic of molecular geometry and study theoretical models that allow us to explain, and in some cases predict, the shapes of small molecules. We will also examine theories that explain, in terms of wave mechanics and the electronic structures of atoms, how covalent bonds form and why they are so highly directional in nature. The knowledge gained here will be helpful in later chapters in this book when we examine the molecular basis of physical properties such as melting points and boiling points. These same principles will be extended to larger and larger molecules in your future studies of organic chemistry and biochemistry.

LEARNING OBJECTIVES

After reading this chapter, you should be able to:

- draw diagrams of the five basic molecular geometries, including bond angles
- predict the shape of a molecule or ion using the VSEPR model
- explain how the geometry of a molecule affects its polarity
- describe how the valence bond theory views bond formation
- explain how hybridization refines the valence bond theory of bonding
- describe the nature of multiple bonds using orbital diagrams and hybridization
- use molecular orbitals to explain the bonding of simple diatomic molecules
- compare and contrast resonance structures to delocalized molecular orbitals
- use band theory to explain bonding in solids and physical properties
- use hybridization and valence bond theory to explain allotropes of nonmetallic elements

9.1 | Five Basic Molecular Geometries

Although we live in a three-dimensional world made up of three-dimensional molecules, the Lewis structures we've been using to describe the bonding in molecules do not convey any information about shape. They are flat and two-dimensional, simply describing which atoms are bonded to each other. Our goal now is to examine theories that predict molecular shapes and explain covalent structures in terms of quantum theory. We begin by describing some of the kinds of shapes molecules have.

Molecular shape only becomes a question when there are at least three atoms present. If there are only two, there is no doubt as to how they are arranged; one is just alongside the other. But when there are three or more atoms in a molecule we find that its shape is often built from just one of five basic geometrical structures.

TOOLS

Basic molecular shapes

Linear Molecules

In a **linear molecule** the atoms lie in a straight line. When the molecule has three atoms, the angle formed by the covalent bonds, which we call the **bond angle**, equals 180° as illustrated below.

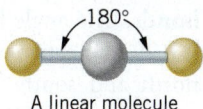

A linear molecule

Planar Triangular Molecules

A **planar triangular molecule** is one in which three atoms are located at the corners of a triangle and are bonded to a fourth atom that lies in the center of the triangle.

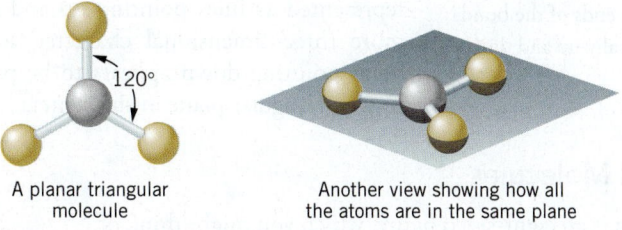

A planar triangular molecule

Another view showing how all the atoms are in the same plane

In this molecule, all four atoms lie in the same plane and the bond angles are all equal to 120°.

Tetrahedral Molecules

A **tetrahedron** is a four-sided geometric figure shaped like a pyramid with triangular faces. A **tetrahedral molecule** is one in which four atoms, located at the vertices of a tetrahedron, are bonded to a fifth atom in the center of the structure.

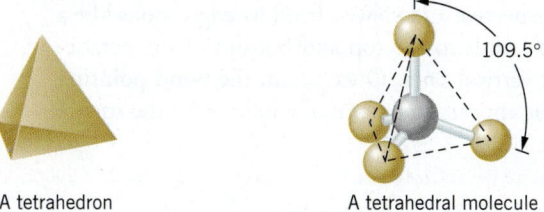

A tetrahedron

A tetrahedral molecule

All of the bond angles in a tetrahedral molecule are the same and are equal to 109.5°.

Trigonal Bipyramidal Molecules

A **trigonal bipyramid** consists of two trigonal pyramids (pyramids with triangular faces) that share a common base. In a **trigonal bipyramidal molecule**, the central atom is located in the middle of the triangular plane shared by the upper and lower trigonal pyramids and is bonded to five atoms that are at the vertices of the figure.

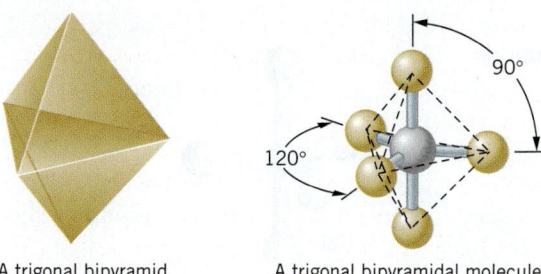

A trigonal bipyramid

A trigonal bipyramidal molecule

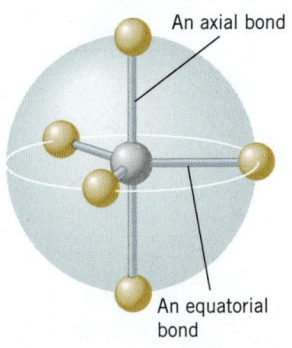

Figure 9.1 | **Axial and equatorial bonds** in a trigonal bipyramidal molecule.

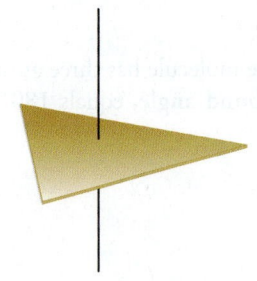

Figure 9.2 | **A simplified way of drawing a trigonal bipyramid.** To form a trigonal bipyramidal molecule, atoms would be attached at the corners of the triangle in the center and at the ends of the bonds that extend vertically up and down.

Figure 9.3 | **A simplified way of drawing an octahedron.** To form an octahedral molecule, atoms would be attached at the corners of the square in the center and at the ends of the bonds that extend vertically up and down.

Practice Exercise 9.1

In this molecule, not all the bonds are equivalent. If we imagine the trigonal bipyramid centered inside a sphere similar to earth, as illustrated in Figure 9.1, the atoms in the triangular plane are located around the equator. The bonds to these atoms are called **equatorial bonds**. The angle between any two equatorial bonds is 120°. The two vertical bonds pointing along the north and south axis of the sphere are 180° apart and are called **axial bonds**. The bond angle between an axial bond and an equatorial bond is 90°.

A simplified representation of a trigonal bipyramid is illustrated in Figure 9.2. The equatorial triangular plane is sketched as it would look tilted backward, so we're looking at it from its edge. The axial bonds are represented as lines pointing up and down. To add more three-dimensional character, notice that the bond pointing down appears to be partially hidden by the triangular plane in the center.

Octahedral Molecules

An **octahedron** is an eight-sided figure, which you might think of as two square pyramids (a pyramid with a square base and triangular sides) sharing a common square base. The octahedron has only six vertices, and in an **octahedral molecule** we find an atom in the center of the octahedron bonded to six other atoms at the vertices.

All of the bonds in an octahedral molecule are equivalent, with angles between adjacent bonds equal to 90°.

A simplified representation of an octahedron is shown in Figure 9.3. The square plane in the center of the octahedron, when drawn in perspective and viewed from its edge, looks like a parallelogram. The bonds to the top and bottom of the octahedron are shown as vertical lines. Once again, the bond pointing down is drawn so it appears to be partially hidden by the square plane in the center.

An octahedron

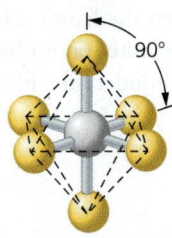

An octahedral molecule

Label the shapes of the following molecules:

(a) Carbon tetrabromide

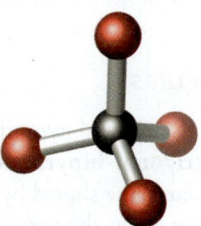

(b) Arsenic pentafluoride

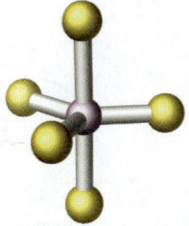

9.2 | Molecular Shapes and the VSEPR Model

In general, a useful theoretical model should explain known facts, and it should be capable of making accurate predictions. The **valence shell electron pair repulsion model** (called the **VSEPR model**, for short) is remarkably successful at both and is also conceptually simple. The model is based on the following idea:

Electron pairs (or groups of electron pairs) in the valence shell of an atom repel each other and will position themselves so that they are as far apart as possible, thereby minimizing the repulsions.

TOOLS
VSEPR model

The term we will use to describe the space occupied by a group of electrons in the valence shell of an atom is **electron domain**, so the preceding statement can be rephrased as: Electron domains stay as far apart as possible so as to minimize their mutual repulsions.

As you will see, in describing the shapes of molecules it is useful to divide electron domains into two categories—domains that contain electrons in bonds and domains that contain unshared electrons. They are called bonding domains and nonbonding domains, respectively.

- A **bonding domain** contains the electrons that are shared between two atoms. *Therefore, all of the electrons within a given single, double, or triple bond are considered to be in the same bonding domain.* A double bond (containing 4 electrons) will occupy more space than a single bond (with only 2 electrons) but all electrons shared by the atoms occupy the same general region in space, so they all belong to the same bonding domain.
- A **nonbonding domain** contains valence electrons that are associated with a *single atom.* A nonbonding domain is either an unshared pair of valence electrons (called a **lone pair**) or, in some cases, a single unpaired electron (found in molecules with an odd number of valence electrons).

■ An *electron domain* can be a bond, a lone pair, or an unpaired electron. Some prefer to call the VSEPR model (or VSEPR theory) the electron domain model.

Figure 9.4 shows the orientations assumed by different numbers of electron domains, which permit them to minimize repulsions by remaining as far apart as possible. These same orientations are achieved whether the domains are bonding or nonbonding. If all of them are bonding domains, molecules are formed having the shapes described in the preceding section, also shown in Figure 9.4.

Lewis Structures and the VSEPR Model

To apply the VSEPR model in predicting shape, we have to know how many electron domains are in the valence shell of the central atom. This is where Lewis structures are especially helpful.

Consider, for example, the $BeCl_2$ molecule. Using the tools from Chapter 8, we determine its Lewis structure as

$$:\ddot{C}l—Be—\ddot{C}l:$$

■ The bonds in $BeCl_2$ are covalent, rather than ionic.

Because there are just three atoms, only two shapes are possible. The molecule must be either linear or nonlinear; that is, the atoms lie in a straight line, or they form some angle less than 180°.

$$Cl—Be—Cl \qquad or \qquad \begin{array}{c} Be \\ Cl \quad \quad Cl \end{array}$$
$$180° \qquad\qquad\qquad <180°$$

To predict the structure, we begin by counting the number of electron domains in the valence shell of the Be atom. In this molecule Be forms two single bonds to Cl atoms, so

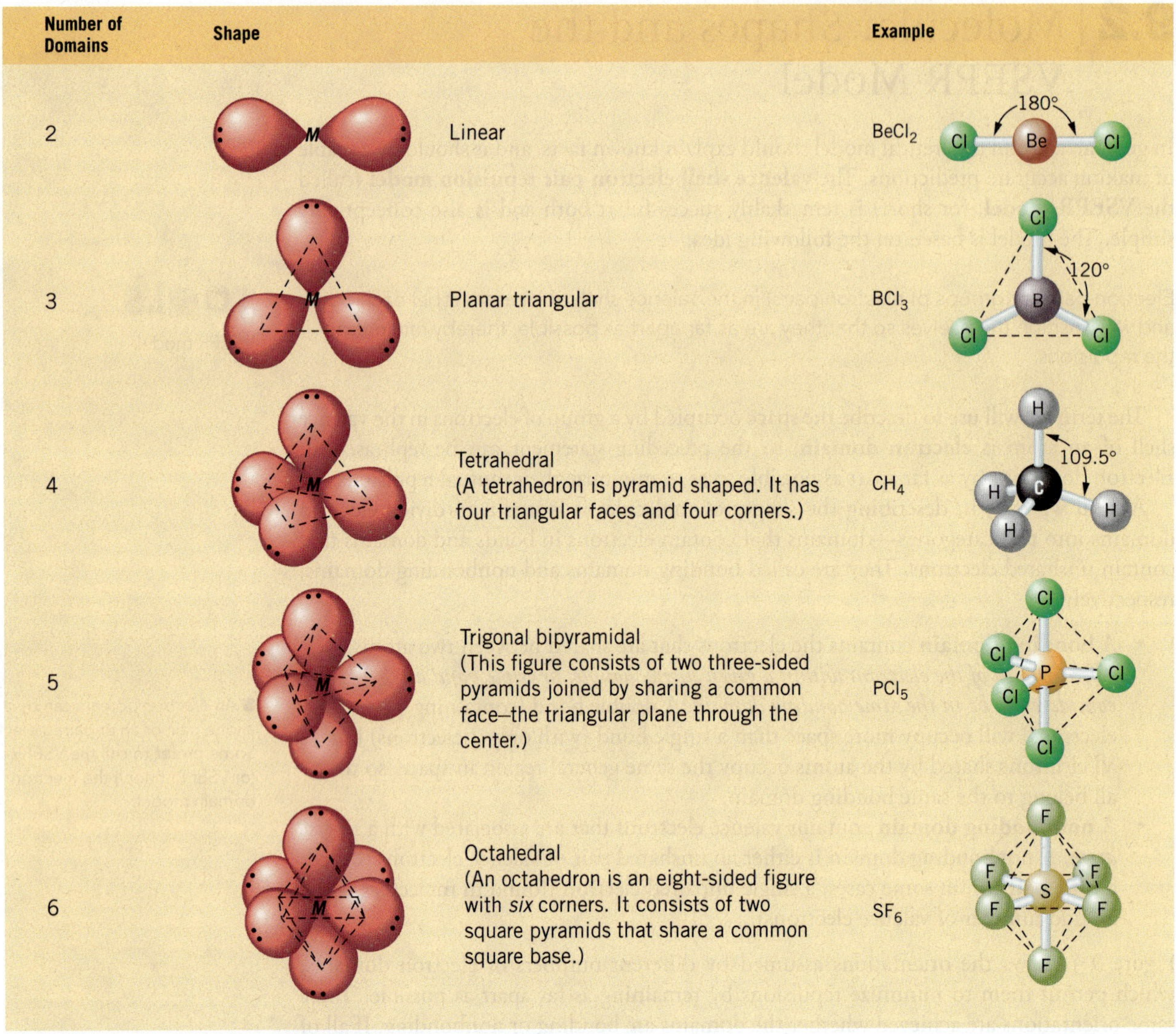

Number of Domains	Shape		Example
2	Linear		$BeCl_2$
3	Planar triangular		BCl_3
4	Tetrahedral (A tetrahedron is pyramid shaped. It has four triangular faces and four corners.)		CH_4
5	Trigonal bipyramidal (This figure consists of two three-sided pyramids joined by sharing a common face—the triangular plane through the center.)		PCl_5
6	Octahedral (An octahedron is an eight-sided figure with *six* corners. It consists of two square pyramids that share a common square base.)		SF_6

Figure 9.4 | **Shapes expected for different numbers of electron domains around a central atom, *M*.** Each lobe represents an electron domain.

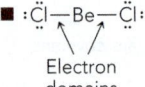

:Cl—Be—Cl:
Electron domains

Be has two bonding domains in its valence shell. In Figure 9.4, we see that when there are two electron domains in the valence shell of an atom, minimum repulsion occurs if they are on opposite sides of the nucleus, pointing in opposite directions. We can represent this as

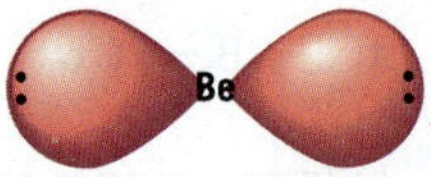

to suggest the approximate locations of the electron clouds of the valence shell electron pairs. In order for the electrons to be in the Be—Cl bonds, the Cl atoms must be placed where the electrons are; the result is that we predict that a $BeCl_2$ molecule should be linear.

$$Cl—Be—Cl$$

Experimentally it has been shown that this is the actual shape of $BeCl_2$ molecules in the vapor state.

Example 9.1
Predicting Molecular Shapes

Carbon tetrachloride was once used as a cleaning fluid until it was discovered that it causes liver damage if absorbed by the body. What is the shape of the molecule?

Analysis: To find the shape of the molecule, we need its Lewis structure so that we can count electron domains and apply the VSEPR model. To write the Lewis structure, we need the formula for the compound, and that requires the application of the rules of chemical nomenclature.

Assembling the Tools: First, we will need to use the rules of nomenclature in Chapter 2 to determine the chemical formula. Then we will use the main tool for solving this kind of problem, the Lewis structure, which we will draw following the procedure in Chapter 8. Finally, we'll use the VSEPR model as a tool to deduce the structure of the molecule.

Solution: From the name, carbon tetrachloride, we obtain the formula CCl_4. Following the procedure in Figure 8.12, the Lewis structure of CCl_4 is

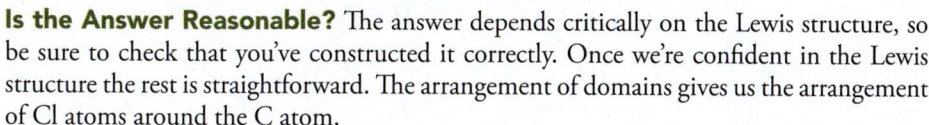

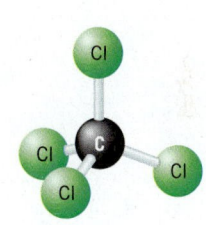

There are four bonds, each corresponding to a bonding domain around the carbon. According to Figure 9.4, the domains can be farthest apart when arranged tetrahedrally, so the molecule is expected to be tetrahedral. (This is, in fact, its structure.)

Is the Answer Reasonable? The answer depends critically on the Lewis structure, so be sure to check that you've constructed it correctly. Once we're confident in the Lewis structure the rest is straightforward. The arrangement of domains gives us the arrangement of Cl atoms around the C atom.

What is the shape of the SeF_6 molecule? (*Hint:* If necessary, refer to Figure 8.12.)

What shape is expected for the $SbCl_5$ molecule?

Practice Exercise 9.2

Practice Exercise 9.3

Nonbonding Domains and Molecular Shapes

Some molecules have a central atom with one or more nonbonding domains, each of which consists of an unshared electron pair (lone pair) or a single unpaired valence electron. These nonbonding domains affect the geometry of the molecule.

Molecules with Three Electron Domains around the Central Atom
An example of nonbonding domains affecting the geometry of a molecule is $SnCl_2$.

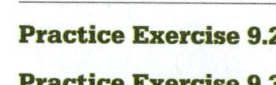

There are *three* domains around the tin atom: two bonding domains plus a nonbonding domain (the lone pair). According to Figure 9.4, the domains are farthest apart when at the corners of a triangle. For the moment, let's ignore the chlorine atoms and concentrate on how the electron domains are arranged.

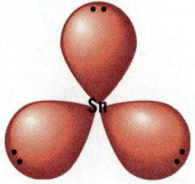

■ $SnCl_2$ is another example of a molecule that has properties indicating it has less than an octet around the tin atom. The electronegativity difference between tin and chlorine is only 1.1, which means the bonds have significant covalent character.

We can now find the shape of the molecule by placing the two Cl atoms where two of the domains are (below, left).

We can't describe this molecule as triangular, even though that is how the domains are arranged. *Molecular shape describes the arrangement of atoms, not the arrangement of electron domains.* Therefore, to describe the shape we ignore the lone pair, as shown above at the right, and say that the $SnCl_2$ molecule has a structure that is **nonlinear** or **bent** or **V-shaped**.

Notice that when there are three domains around the central atom, *two* different molecular shapes are possible. If all three are bonding domains, a molecule with a planar triangular shape is formed, as shown for BCl_3 in Figure 9.4. If one of the domains is nonbonding, as in $SnCl_2$, the arrangement of the atoms in the molecule is said to be nonlinear. The predicted shapes of both however, are *derived* by first noting the triangular arrangement of domains around the central atom and *then* adding the necessary number of atoms. We will use this method for determining all molecular shapes.

Molecules with Four Electron Domains around the Central Atom

There are many molecules with four electron pairs (an octet) in the valence shell of the central atom. When these electron pairs are used to form four single bonds, as in methane, CH_4, the resulting molecule is tetrahedral (Figure 9.4). There are many examples, however, where nonbonding domains are also present. For instance, two molecules you've encountered before are ammonia and water:

$$H - \overset{..}{N} - H \qquad H - \overset{..}{\underset{..}{O}} - H$$
$$\underset{}{\overset{|}{H}}$$

one lone pair　　　　two lone pairs

Figure 9.5 shows how the nonbonding domains affect the shapes of molecules of this type.

Number of Bonding Domains	Number of Nonbonding Domains	Structure	
4	0	109.5°	**Tetrahedral** (example, CH_4) All bond angles are 109.5°.
3	1		**Trigonal pyramidal** (pyramid shaped) (example, NH_3)
2	2		**Nonlinear, bent** (example, H_2O)

Figure 9.5 | Molecular shapes with four domains around the central atom. The molecules MX_4, MX_3, and MX_2 shown here all have four domains arranged tetrahedrally around the central atom.

With one nonbonding domain, as in NH_3, the central atom is at the top of a pyramid with three atoms at the corners of the triangular base. The resulting **trigonal pyramidal** structure is also called a **triangular pyramid**. When there are two nonbonding domains in the tetrahedron, as in H_2O, the three atoms of the molecule (the central atom plus the two atoms bonded to it) do not lie in a straight line, so the structure is described as nonlinear or bent.

Molecules with Five Electron Domains around the Central Atom

When five domains are present around the central atom, they are directed toward the vertices of a trigonal bipyramid. Molecules such as PCl_5 have this geometry, as shown in Figure 9.4, and are said to have a trigonal bipyramidal shape.

In the trigonal bipyramid, nonbonding domains always occupy positions in the *equatorial plane* (the triangular plane through the center of the molecule). This is because nonbonding domains, which have a positive nucleus only at one end, are larger than bonding domains, as illustrated in Figure 9.6. The larger nonbonding domains are less crowded in the equatorial plane, where they have just two closest neighbors at 90°, than they would be in an axial position, where they would have three closest neighbors at 90°.

Figure 9.7 shows the kinds of geometries that we find for different numbers of nonbonding domains in the trigonal bipyramid. When there is only one nonbonding domain, as in SF_4, the structure is described as a **distorted tetrahedron** or **seesaw** in which the

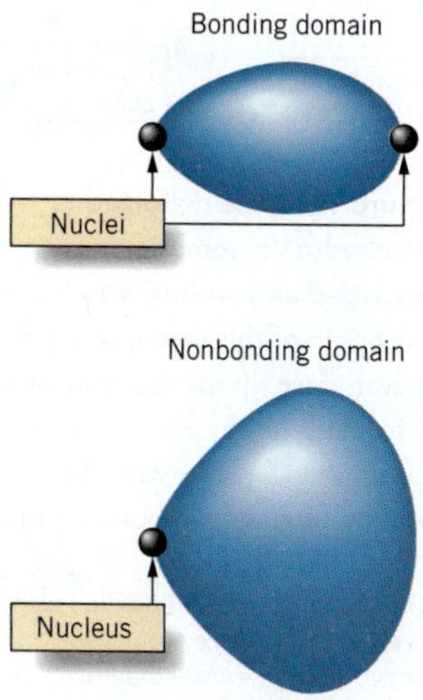

■ Once again, the arrangement of atoms describes the molecular shape, not the electron domains.

Figure 9.6 | Relative sizes of bonding and nonbonding domains.

Number of Bonding Domains	Number of Nonbonding Domains	Structure	
5	0	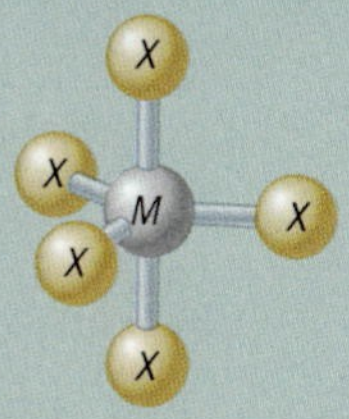	**Trigonal bipyramidal** (example, PCl_5)
4	1	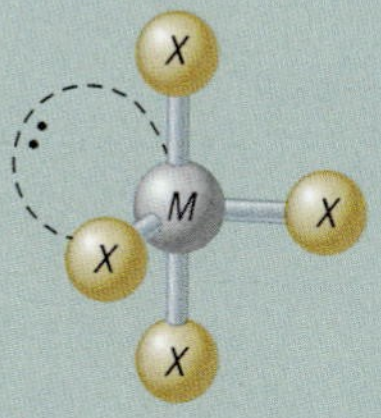	**Distorted tetrahedral** 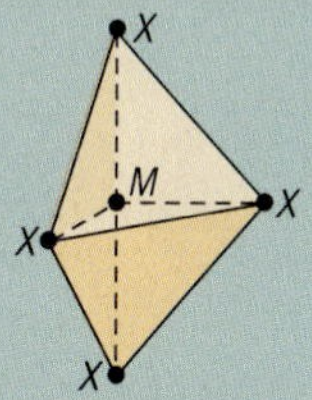 (example, SF_4)
3	2	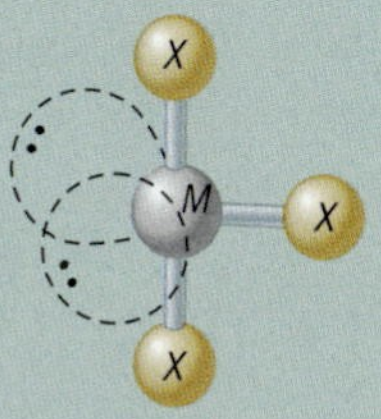	**T-shaped** 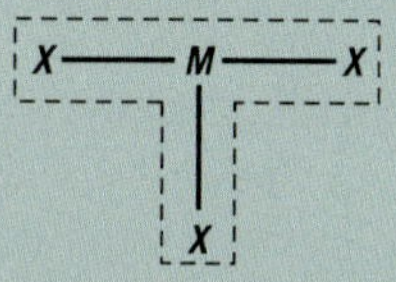 (example, ClF_3)
2	3	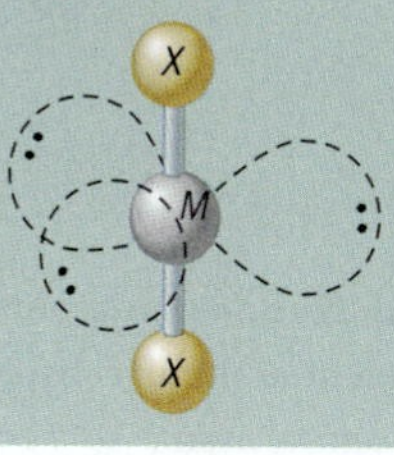	**Linear** (example, I_3^-)

Figure 9.7 | Molecular shapes with five domains around the central atom. Four different molecular structures are possible, depending on the number of nonbonding domains around the central atom M.

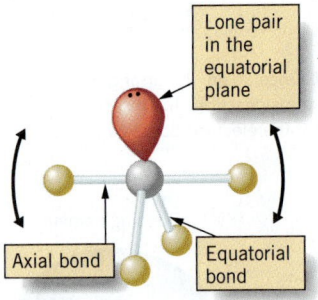

Figure 9.8 | **The distorted tetrahedron is sometimes described as a *seesaw structure*.** The origin of this description can be seen if we tip the structure over so it stands on the two atoms in the equatorial plane, with the nonbonding domain pointing up.

■ No common molecule or ion with six electron domains around the central atom has more than two nonbonding domains.

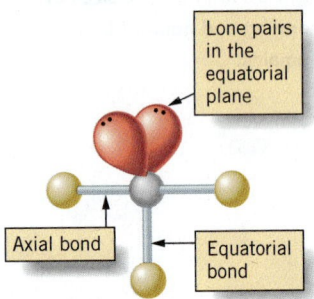

Figure 9.9 | **A molecule with two nonbonding domains in the equatorial plane of the trigonal bipyramid.** When tipped over, the molecule looks like a T, so it is called *T-shaped*.

central atom lies along one edge of a four-sided figure (Figures 9.7 and 9.8). With two nonbonding domains in the equatorial plane, the molecule is **T-shaped** (shaped like the letter T, Figure 9.9), and when there are three nonbonding domains in the equatorial plane, the molecule is linear.

Molecules with Six Electron Domains around the Central Atom

Finally, we come to molecules or ions that have six domains around the central atom. When all electron domains are involved in bonds, as in SF_6, the molecule is octahedral (Figure 9.4). When one nonbonding domain is present, the molecule or ion has the shape of a **square pyramid**, and when two nonbonding domains are present, they take positions on opposite sides of the nucleus and the molecule or ion has a **square planar** structure. These shapes are shown in Figure 9.10.

Steps in Using the VSEPR Model to Determine Molecular Shape

Now that we've seen how nonbonding and bonding domains influence molecular shape we can develop a strategy for applying the VSEPR model. In following this strategy it will be to your advantage to learn to sketch the five basic geometries (linear, planar triangular, tetrahedral, trigonal bipyramidal, and octahedral).

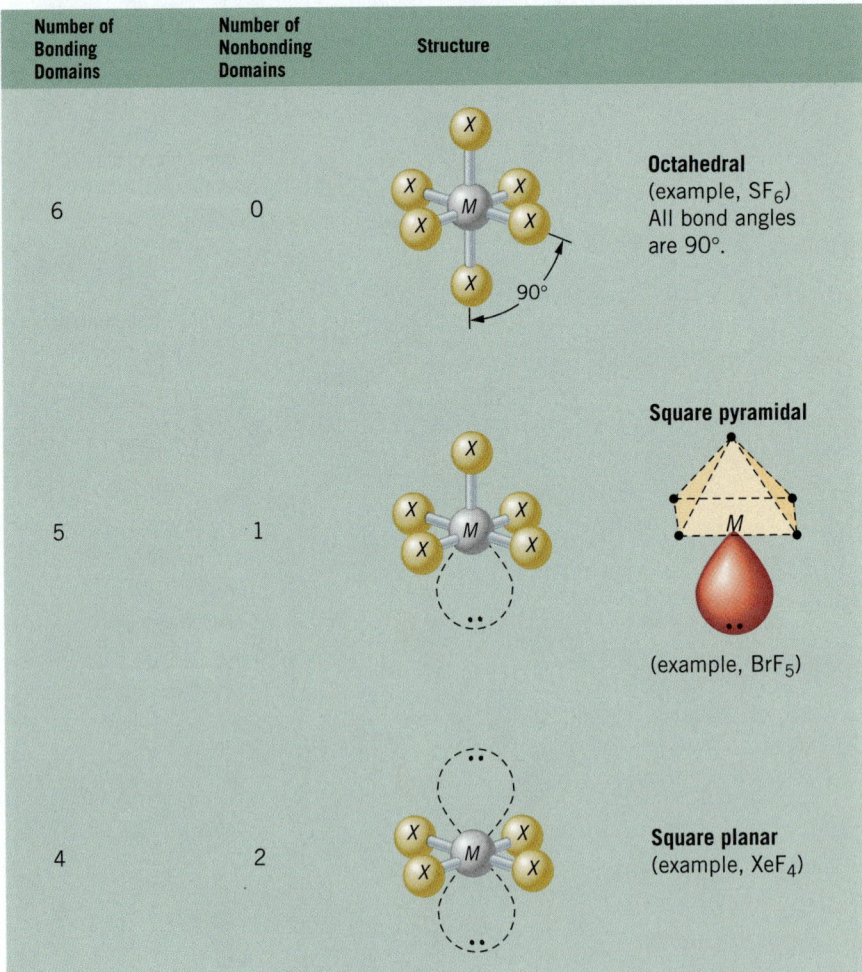

Figure 9.10 | **Molecular shapes with six domains around the central atom.** Although more are theoretically possible, only three different molecular shapes are observed, depending on the number of nonbonding domains around the central atom.

Determining Molecular Shape Using the VSEPR Model

Step 1. Use the procedure in Figure 8.12 to construct the Lewis formula for the molecule or ion.

Step 2. Count the total number of electron domains (bonding plus nonbonding).

Step 3. Using the result from Step 2, select the basic geometry upon which the shape of the molecule is based. [Make a drawing that shows the location of the central atom and the directions in which the domains (both bonding and nonbonding) are oriented.]

Step 4. Add the appropriate number of atoms to the bonding domains.

Step 5. If there are any nonbonding domains, use just the arrangement of atoms around the central atom to obtain the description of the molecular shape, as shown in Figures 9.5, 9.7, and 9.10.

This method works for determining the shapes of ions as well as molecules. The key is to keep track of the number of electrons that have to be added to make anions or removed resulting in cations.

TOOLS

Using the VSEPR model to predict molecular geometry

Example 9.2
Predicting the Shapes of Molecules

Xenon is one of the noble gases, and is generally quite unreactive. In fact, it was long believed that all the noble gases were totally unable to form compounds. It came as quite a surprise, therefore, when it was discovered that some compounds could be made. One of these is xenon difluoride. What would you expect the geometry of xenon difluoride to be—linear or nonlinear?

Analysis: This problem is similar to previous examples we've worked out. The first step will be to correctly write the formula of the compound. Then construct the Lewis formula, from which we can derive the shape of the molecule by applying the VSEPR model.

Assembling the Tools: First, we need the rules of nomenclature from Chapter 2 to obtain the formula. We then follow the procedure in Figure 8.12 to construct the Lewis formula. From here, we follow the strategy for applying the VSEPR model.

Solution: The formula for xenon difluoride is XeF_2. The outer shell of xenon, of course, has a noble gas configuration, which contains 8 electrons. Each fluorine has 7 valence electrons. Using this information we obtain the following Lewis structure for XeF_2:

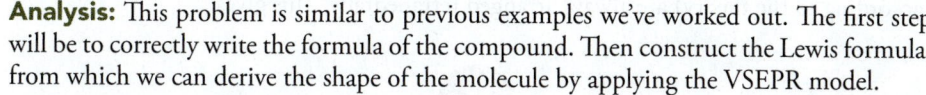

Next we count domains around xenon; there are five of them, three nonbonding and two bonding. When there are five domains, they are arranged as a trigonal bipyramid.

Now we must add the fluorine atoms. In a trigonal bipyramid, the nonbonding domains always occur in the equatorial plane through the center, so the fluorines go on the top and bottom. This gives

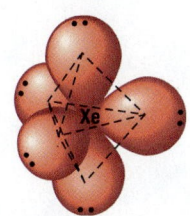

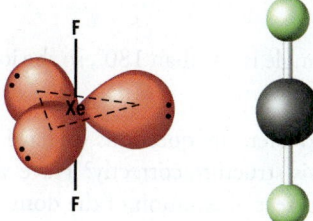

The three atoms, F—Xe—F, are arranged in a straight line, so the molecule is linear.

Is the Answer Reasonable? Is the Lewis structure correct? Yes. Have we selected the correct basic geometry? Yes. Have we attached the F atoms to the Xe correctly? Yes. All is in order, so the answer is correct.

Example 9.3
Predicting the Shapes of Ions

Do we expect the ClO_2^- ion to be linear?

Analysis: The VSEPR model is the only concept available for deducing molecular shapes, so we must use it.

Assembling the Tools: The strategy described above for applying the VSEPR model as a tool first requires construction of the Lewis structure, which will show us the bonding and nonbonding electron domains. The tool for this is Figure 8.12.

Solution: Following the usual procedure described in Figure 8.12, the Lewis structure for ClO_2^- is

$$\left[:\ddot{O}—\ddot{Cl}—\ddot{O}: \right]^-$$

Next, we count bonding and nonbonding domains around the central atom. There are four domains around the chlorine: two bonding and two nonbonding. Four domains (according to the theory) are always arranged tetrahedrally. This gives

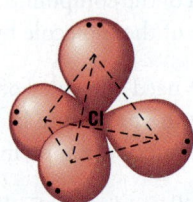

Now we add the two oxygens, as shown below on the left. It doesn't matter which locations in the tetrahedron we choose because all the bond angles are equal. Below, on the right, we see how the structure of the ion looks when we ignore the lone pairs.

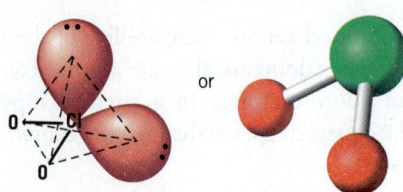

We see that the O—Cl—O angle is less than 180°, so the ion is predicted to be nonlinear (or bent or V-shaped).

Is the Answer Reasonable? Here are questions we have to answer to check our work. First, have we drawn the Lewis structure correctly? Have we counted the domains correctly? Have we selected the correct orientation of the domains? And finally, have we correctly described the structure obtained by adding the two oxygen atoms? Our answer to each of these questions is "Yes," so we can be confident our answer is right.

Example 9.4
Predicting the Shapes of Molecules with Multiple Bonds

The Lewis structure for the very poisonous gas hydrogen cyanide, HCN, is

$$H—C≡N:$$

Is the HCN molecule linear or nonlinear?

Analysis: We already have the Lewis structure, so we count electron domains and proceed as before. The critical link in this problem is remembering that a bonding domain connects one atom to the central atom and that *all the electron pairs in a given bond belong to the same bonding domain.*

Assembling the Tools: As before, the tool required is the strategy for applying the VSEPR model.

Solution: There are two bonding domains around the carbon: one for the triple bond with nitrogen and one for the single bond with hydrogen.

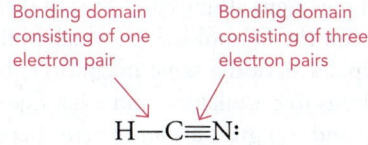

Bonding domain consisting of one electron pair

Bonding domain consisting of three electron pairs

$$H—C≡N:$$

Therefore, we expect the two bonds to locate themselves 180° apart, yielding a linear HCN molecule.

Is the Answer Reasonable? Have we correctly counted bonding domains? Yes. We can expect our answer to be correct.

The first known compound of the noble gas argon is HArF. What shape is expected for the HArF molecule? (*Hint:* Remember that argon is a noble gas with 8 electrons in its valence shell.)

Practice Exercise 9.4

What shape is expected for the IBr_2^- ion?

Practice Exercise 9.5

What shape is expected for the SO_3 molecule?

Practice Exercise 9.6

9.3 | Molecular Structure and Dipole Moments

The electrons are not always uniformly distributed in a molecule, and when that happens, a molecular **dipole moment** forms. The overall dipole moment of a molecule will affect its chemical and physical properties; therefore, we will need to be able to determine if a molecule has a dipole.

For a diatomic molecule, the dipole moment is determined solely by the polarity of the bond between the two atoms, as discussed in detail in Section 8.6. Thus, the HCl molecule has a nonzero dipole moment because the H—Cl bond is polar, whereas H_2 has a dipole moment equal to zero because the H—H bond is nonpolar.

For a molecule containing three or more atoms, each polar bond has its own **bond dipole**, which contributes to the overall dipole moment of the molecule. This happens because the bond dipoles have *vector properties*, with both direction and magnitude. In a molecule, the bond dipoles add as vectors do.

■ Any molecule composed of just two atoms that differ in electronegativity must be polar because the bond is polar.

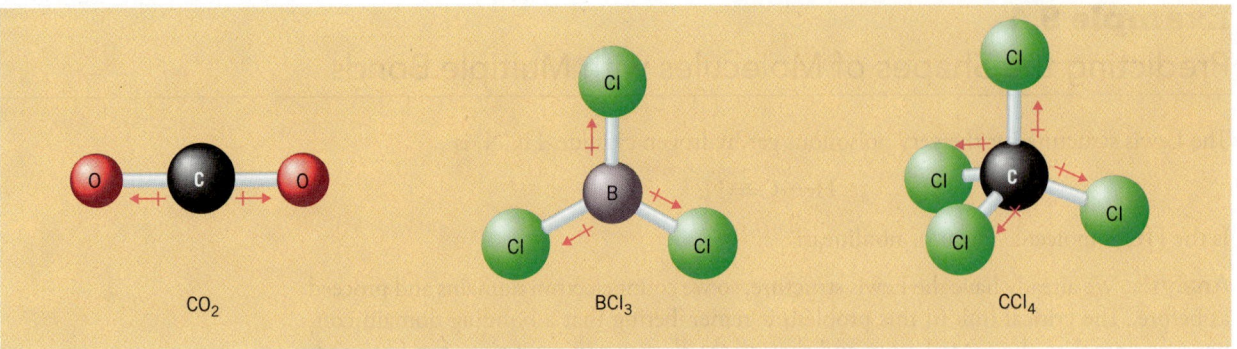

Figure 9.11 | **Molecular shape and its effect on the dipole moment.** Even though these molecules have polar bonds, vector addition of their bond dipoles leads to dipole moments of zero, which means the molecules are nonpolar.

Figure 9.11 shows three molecules in which vector addition of the bond dipoles leads to complete cancellation and dipole moments equal to zero. In the figure, the bond dipoles are indicated by arrows with plus signs at one end, $\longmapsto$, to show which end of the dipole is positive. On the left is the linear CO_2 molecule in which both C—O bonds are identical. The individual bond dipoles have the same magnitude, but they point in opposite directions. Vector addition leads to cancellation and a net dipole moment of zero, just as adding numbers such as +5 and −5 gives a sum of zero. Because the dipole moment is zero, the CO_2 molecule is nonpolar. Although it isn't so easy to visualize, the same thing also happens in BCl_3 and CCl_4. In each of these molecules, the influence of one bond dipole is canceled by the effects of the others, which causes the molecules to have net dipole moments of zero. As a result, BCl_3 and CCl_4 are **nonpolar molecules**.

Although the molecules in Figure 9.11 are nonpolar, the fact that they contain nonzero bond dipoles means that the atoms carry some partial positive or negative charge. In CO_2, for example, the carbon carries a partial positive charge and an equal negative charge is divided between the two oxygen atoms. Similarly, in BCl_3 and CCl_4, the central atoms have partial positive charges and the chlorine atoms have partial negative charges.

Perhaps you've noticed that the structures of the molecules in Figure 9.11 correspond to three of the basic shapes that we used to derive the shapes of molecules. Molecules with the remaining two structures, trigonal bipyramidal and octahedral, also are nonpolar if all the atoms attached to the central atom are the same. All of the basic shapes are "balanced," or **symmetric**,[1] if all of the domains and groups attached to them are identical. Examples are shown in Figure 9.12. The trigonal bipyramidal structure can be viewed as a planar triangular set of atoms (shown in blue) plus a pair of atoms arranged linearly (shown in red). All the bond dipoles in the planar triangle cancel, as do the two dipoles of the bonds arranged linearly, so the molecule is nonpolar overall. Similarly, we can look at the octahedral molecule as consisting of three linear sets of bond dipoles. Cancellation of bond dipoles occurs in each set, so overall the octahedral molecule is also nonpolar.

If all the atoms attached to the central atom are not the same, or if there are lone pairs in the valence shell of the central atom, the molecule is usually polar. For example, in $CHCl_3$, one of the atoms in the tetrahedral structure is different from the others. The C—H bond is less

---

[1] Symmetry is a more complex subject than we present here. When we describe the symmetry properties of a molecule, we are specifying the various ways the molecule can be turned and otherwise manipulated while leaving the molecule looking exactly as it appeared before the manipulation. For example, imagine the BCl_3 molecule in Figure 9.11 being rotated 120° around an axis perpendicular to the page, and that passes through the B atom. Performing this rotation leaves the molecule looking just as it did before the rotation. This rotation axis is a symmetry property of BCl_3.

Intuitively, we can recognize when an object possesses symmetry elements such as rotation axes (as well as other symmetry properties we haven't mentioned). Comparing objects, we can usually tell when one is more symmetric than another, and in our discussions in this chapter we rely on this qualitative sense of symmetry.

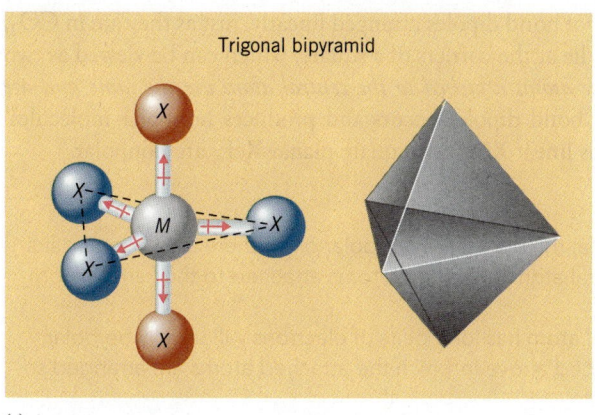

Trigonal bipyramid

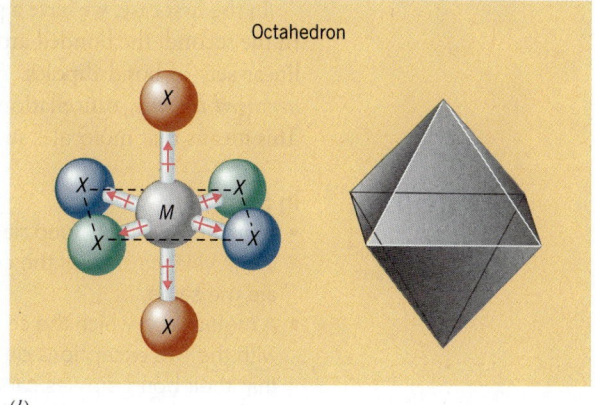

Octahedron

(a)

(b)

Figure 9.12 | Cancellation of bond dipoles in symmetric trigonal bipyramidal and octahedral molecules. In this figure, the colors of atoms are used to identify different sets of atoms within the molecule. (a) A trigonal bipyramidal molecule, MX_5, in which the central atom M is bonded to five identical atoms X. The set of three bond dipoles in the triangular plane in the center (in blue) cancel, as do the linear set of dipoles (red). Overall, the molecule is nonpolar. (b) An octahedral molecule MX_6 in which the central atom is bonded to six identical atoms. This molecule contains three linear sets of bond dipoles (red, blue, and green). Cancellation occurs for each set, so the molecule is nonpolar overall.

polar than the C—Cl bonds, and the bond dipoles do not cancel (Figure 9.13). An "unbalanced" structure such as this is said to be **asymmetric.**

Two familiar molecules that have lone pairs in the valence shells of their central atoms are shown in Figure 9.14. In these molecules the bond dipoles are oriented in such a way that their effects do not cancel. In water, for example, each bond dipole points partially in the same direction, toward the oxygen atom. As a result, the bond dipoles partially add to give a net dipole moment for the molecule. The same thing happens in ammonia, where three bond dipoles point partially in the same direction and add to give a polar NH_3 molecule.

Not every structure that contains nonbonding domains on the central atom produces polar molecules. The following are two exceptions.

■ The lone pairs can also influence the polarity of a molecule, but we will not explore this any further here.

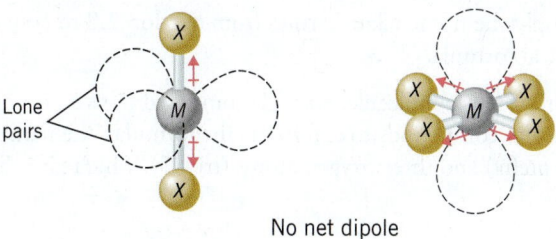

Lone pairs

No net dipole

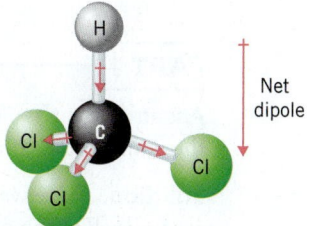

Net dipole

Figure 9.13 | Bond dipoles in the chloroform molecule, CHCL₃. Because C is slightly more electronegative than H, the C—H bond dipole points toward the carbon. The small C—H bond dipole actually adds to the effects of the C—C bond dipoles. All bond dipoles are additive, and this causes $CHCl_3$ to be a polar molecule.

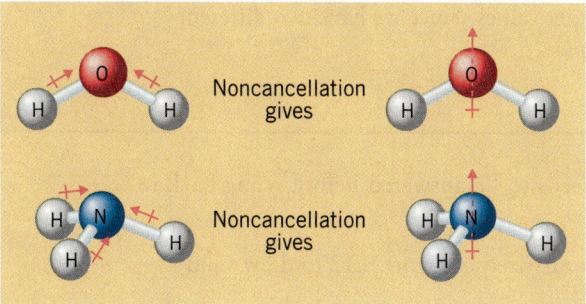

Noncancellation gives

Noncancellation gives

Figure 9.14 | Dipole moments in water and ammonia. Each of these molecules contains nonbonding domains in the valence shell of the central atoms and has a net dipole moment. When nonbonding domains occur on the central atom, the bond dipoles usually do not cancel, and polar molecules result.

In the first case, we have a pair of bond dipoles arranged linearly, just as they are in CO_2. In the second, the bonded atoms lie at the corners of a square, which can be viewed as two linear sets of bond dipoles. *If the atoms attached to the central atom are the same and are arranged linearly,* cancellation of bond dipoles occurs and produces nonpolar molecules. This means that molecules such as linear XeF_2 and square planar XeF_4 are nonpolar.

TOOLS

Molecular shape and molecular polarity

In Summary

- A molecule will be nonpolar if (a) the bonds are nonpolar or (b) there are no lone pairs in the valence shell of the central atom and all the atoms attached to the central atom are the same.
- A molecule in which the central atom has lone pairs of electrons will usually be polar, with the two exceptions described above in which the attached atoms are arranged so that their bond dipoles cancel.

On the basis of the preceding discussions, let's see how we apply these ideas to predict whether molecules are expected to be polar or nonpolar.

⚛ Analyzing and Solving Multi-Concept Problems

Do we expect the sulfur trioxide molecule to be polar or nonpolar?

Analysis: Answering this seemingly simple question actually requires answering a number of questions. First of all we need the formula for sulfur trioxide. Second, we ask ourselves whether the bonds in the molecule are polar or nonpolar. Third, if they are polar, do the bond dipoles cancel in the three-dimensional molecule? To answer that question, we need to know the shape of the molecule and whether there are lone pairs around the central atom. Fourth, to find the shape of the molecule we have to draw the Lewis structure and fifth we apply the VSEPR model. We see that answering the third part above requires that we answer parts four and five first. *Notice that after we define what needs to be done, the solution often does not follow the same sequence.*

PART 1

Assembling the Tools: We need to apply the nomenclature rules from Section 2.8 to convert the name of the compound into a chemical formula.

Solution: Since we are told that sulfur trioxide is a molecule, and it is composed of two nonmetals, we look to the nomenclature of molecular compounds to determine the formula. The name itself indicates that we have one sulfur (no prefix) and three oxygen atoms (tri – 3), which gives SO_3.

PART 2

Assembling the Tools: To determine if the bonds are polar all we need is a *sense* of the table of electronegativities (Figure 8.9). We do not need the values for the electronegativities.

Solution: We recall that most of the elements, particularly the nonmetals, have different electronegativities, and those closest to fluorine have the larger values. Since we have two different elements, sulfur and oxygen, *we expect the bonds to be polar.*

PART 4

As noted in the analysis, Part 4, along with Part 5, needs to be answered before going back to Part 3.

Assembling the Tools: We will need the method summarized in Figure 8.12 to draw valid Lewis structures.

Solution: We select sulfur as the central atom and arrange the available electrons in octets around the oxygen atoms. This results in sulfur having fewer than eight electrons. This is solved by creating a double bond (two electron pairs) between one oxygen atom and the sulfur. This structure we should end up with is shown in two equivalent ways, first with all the electrons shown and second with bonding pairs of electrons replaced by a solid line.

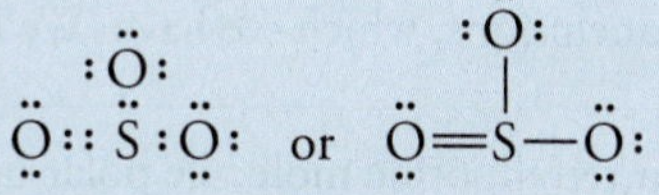

$$\overset{\ddot{\text{O}}\!:}{\underset{\ddot{\text{O}}\!:}{\ddot{\text{O}}::\ddot{\text{S}}:\ddot{\text{O}}\!:}} \quad \text{or} \quad \overset{:\ddot{\text{O}}\!:}{\underset{\ddot{\text{O}}\!:}{\ddot{\text{O}}\!=\!\text{S}\!-\!\ddot{\text{O}}\!:}}$$

PART 5

Assembling the Tools: To determine the geometric structure we need to determine how many electron domains are present and how many of those domains involve a bond to an atom. Then we can use Figures 9.4, 9.5, 9.7, and 9.10 to determine the atom geometry.

Solution: From the Lewis structure in Part 4 we see that there are three electron domains around the sulfur. All of the electron domains involve bonds between sulfur and an oxygen atom. We expect this to give us a triangular planar structure.

PART 3

Now we have the information from parts 4 and 5 to proceed with this part.

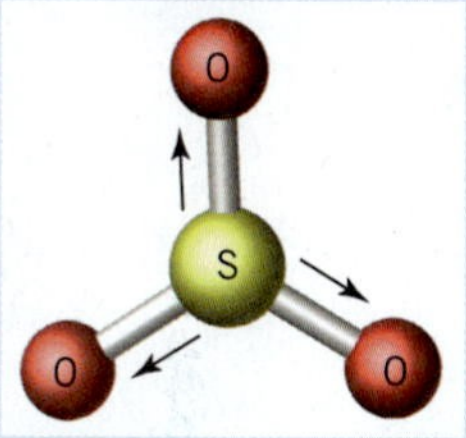

Assembling the Tools: Now we have the structure and the fact that all the bonds are equal (recall resonance structures) and must have the same polarity.

Solution: Since the molecule appears to be a symmetrical triangle with equal attraction for the electrons in all directions, we conclude that the molecule is nonpolar.

Is the Answer Reasonable?: There's not much to do here except to carefully check the Lewis structure to be sure it's correct and to check that we've applied the VSEPR theory correctly.

Example 9.5
Predicting Molecular Polarity

Would you expect the molecule HCN to be polar or nonpolar?

Analysis and Assembling the Tools: Our analysis proceeds as in the *Analyzing and Solving Multi-Concept Problems* we just discussed, so we will also apply the same tools.

Solution: To begin, we have polar bonds, because carbon is slightly more electronegative than hydrogen and nitrogen is slightly more electronegative than carbon. The Lewis structure of HCN is

$$\text{H}-\text{C}\equiv\text{N}\!:$$

There are two domains around the central carbon atom, so we expect a linear shape. However, the two bond dipoles do not cancel. One reason is that they are not of equal magnitude, which we know because the difference in electronegativity between C and H is 0.4, whereas the difference in electronegativity between C and N is 0.6. The other reason is that both bond dipoles point in the same direction, from the atom of low electronegativity

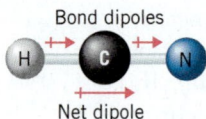

Bond dipoles

Net dipole

to the one of high electronegativity. This is illustrated in the margin. Notice that the bond dipoles combine by vector addition to give a net dipole moment for the molecule that is larger than either of the individual bond dipoles. As a result, HCN is a polar molecule.

Is the Answer Reasonable? As in the preceding example, we can check that we've obtained the correct Lewis structure and applied the VSEPR theory correctly, which we have. We can also check to see whether we've used the correct electronegativities to reach the right conclusions, which we have. We can be confident in our conclusions.

Practice Exercise 9.7

Is the sulfur tetrafluoride molecule polar or nonpolar? Explain your reasoning. (*Hint:* Use the VSEPR model to sketch the shape of the molecule.)

Practice Exercise 9.8

Explain how you decided which of the following molecules are expected to be polar: (a) TeF_6, (b) SeO_2, (c) $BrCl$, (d) AsH_3, (e) CF_2Cl_2.

9.4 | Valence Bond Theory

So far we have described the bonding in molecules using Lewis structures and the shapes using the VSEPR model. Lewis structures, however, tell us nothing about *why* covalent bonds are formed or how electrons manage to be shared between atoms. Nor does the VSEPR model explain why electrons group themselves into domains as they do. Thus, we must look beyond these simple models to understand more fully the covalent bond and the factors that determine molecular geometry.

There are two theories of covalent bonding that have evolved based on quantum theory: the **valence bond theory** (or **VB theory**, for short) and the **molecular orbital theory (MO theory)**. They differ principally in the way they construct a theoretical model of the bonding in a molecule. The valence bond theory imagines individual atoms, each with its own orbitals and electrons, coming together to form the covalent bonds of the molecule. The molecular orbital theory doesn't concern itself with how the molecule is formed. It just views a molecule as a collection of positively charged nuclei surrounded in some way by electrons that occupy a set of *molecular orbitals*, in much the same way that the electrons in an atom occupy *atomic orbitals*. (In a sense, MO theory would look at an atom orbital as if it were a special case—a molecule having only one positive center, instead of many.)

Bond Formation by Orbital Overlap

TOOLS

VB criteria for bond formation

According to VB theory, *a bond between two atoms is formed when two electrons with their spins paired are shared by two overlapping atomic orbitals, one orbital from each of the atoms joined by the bond.* By **overlap of orbitals** we mean that portions of two atomic orbitals from different atoms share the same space.[2]

An important part of the theory, as suggested by the italic type above, is that only one pair of electrons, with paired spins, can be shared by two overlapping orbitals. This electron pair becomes concentrated in the region of overlap and helps "cement" the nuclei together, so the amount that the potential energy is lowered when the bond is formed is determined in part by the extent to which the orbitals overlap. Therefore, *atoms tend to position themselves so that the maximum amount of orbital overlap occurs because this yields the minimum potential energy and therefore the strongest bonds.*

The way VB theory views the formation of a hydrogen molecule is shown in Figure 9.15. As the two atoms approach each other, their 1*s* orbitals begin to overlap and merge as the electron pair spreads out over both orbitals, thereby giving the H—H bond. The description of the bond in H_2 provided by VB theory is essentially the same as that discussed in Chapter 8.

---

[2] The concept of orbital overlap is actually more complicated than this and requires the application of quantum theory. For our purposes, the current definition will suffice.

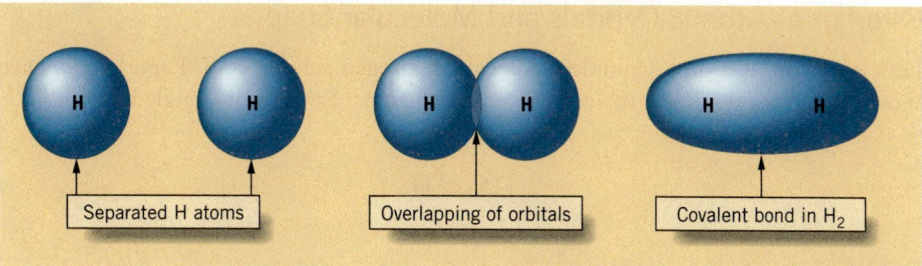

Now let's look at the HF molecule, which is a bit more complex than H_2. Following the usual rules we can write its Lewis structure as

$$H\text{—}\ddot{\underset{\cdot\cdot}{F}}\text{:}$$

and we can diagram the formation of the bond as

$$H\cdot + \cdot\ddot{\underset{\cdot\cdot}{F}}\text{:} \longrightarrow H\text{—}\ddot{\underset{\cdot\cdot}{F}}\text{:}$$

Our Lewis symbols suggest that the H—F bond is formed by the pairing of electrons, one from hydrogen and one from fluorine. To explain this according to VB theory, we must have two half-filled orbitals, one from each atom, that can be joined by overlap. (They must be half-filled, because we cannot place more than two electrons into the bond.) To see clearly what must happen, it is best to look at the orbital diagrams of the valence shells of hydrogen and fluorine.

$$
\begin{array}{cc}
H & \textcircled{\small$\uparrow$} \\
 & 1s
\end{array}
$$

$$
\begin{array}{cc}
F & \textcircled{\small$\uparrow\downarrow$}\quad \textcircled{\small$\uparrow\downarrow$}\textcircled{\small$\uparrow\downarrow$}\textcircled{\small$\uparrow$} \\
 & 2s\qquad\quad 2p
\end{array}
$$

The requirements for bond formation are met by overlapping the half-filled $1s$ orbital of hydrogen with the half-filled $2p$ orbital of fluorine; there are then two orbitals plus two electrons whose spins can adjust so they are paired. The formation of the bond is illustrated in Figure 9.16.

The overlap of orbitals provides a means for sharing electrons, thereby allowing each atom to complete its valence shell. It is sometimes convenient to indicate this using orbital diagrams. For example, the diagram below shows how the fluorine atom completes its $2p$ subshell by acquiring a share of an electron from hydrogen.

$$
\begin{array}{cc}
F \ (\text{in HF}) & \textcircled{\small$\uparrow\downarrow$}\quad \textcircled{\small$\uparrow\downarrow$}\textcircled{\small$\uparrow\downarrow$}\textcircled{\small$\uparrow\downarrow$} \\
 & 2s\qquad\quad 2p
\end{array}
\qquad (\text{Colored arrow is the H electron.})
$$

Notice that in both the Lewis and VB descriptions of the formation of the H—F bond, the atoms' valence shells are completed.

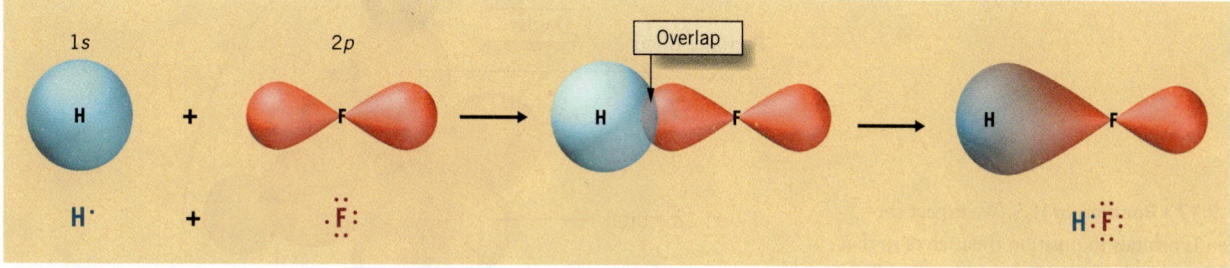

Figure 9.16 | The formation of the hydrogen fluoride molecule according to valence bond theory. For clarity, only the half-filled $2p$ orbital of fluorine is shown. The other $2p$ orbitals of fluorine are filled and cannot participate in bonding.

Overlap of Atomic Orbitals and Molecular Shapes

■ H_2S is the compound that gives rotten eggs their foul odor.

Let's look now at a more complex molecule, hydrogen sulfide, H_2S. Experiments have shown that it is a nonlinear molecule in which the H—S—H bond angle is about 92°.

$$\underset{92°}{H \overset{\overset{\displaystyle S}{\frown}}{} H}$$

Using Lewis symbols, we would diagram the formation of H_2S:

$$2H\cdot + \cdot\ddot{\underset{\cdot\cdot}{S}}\cdot \longrightarrow H{-}\ddot{S}{-}H$$

Our Lewis symbols suggest that each H—S bond is formed by the pairing of two electrons, one from H and one from S. Applying this to VB theory, each bond requires the overlap of two half-filled orbitals, one on H and one on S. Therefore, forming two H—S bonds in H_2S will require two half-filled orbitals on sulfur to form bonds to two separate H atoms. To clearly see what happens, let's look at the orbital diagrams of the valence shells of hydrogen and sulfur.

■ If necessary, review the procedure for drawing orbital diagrams in Section 7.7.

H   ①
      1s

S   ⑴⥮ ⑴⥮ ① ①
      3s   3p

Sulfur has two 3p orbitals that each contain only one electron. Each of these can overlap with the 1s orbital of a hydrogen atom, as shown in Figure 9.17. This overlap completes the 3p subshell of sulfur because each hydrogen provides one electron.

■ In valence bond theory, two orbitals from different atoms never overlap simultaneously with opposite ends of the same p orbital.

⑴⥮ ⑴⥮ ⑴⥮ ⑴⥮   (Colored arrows are H electrons.)
 3s   3p

In Figure 9.17, notice that when the 1s orbital of a hydrogen atom overlaps with a p orbital of sulfur, the best overlap occurs when the hydrogen atom lies along the axis of the p orbital. Because p orbitals are oriented at 90° to each other, the H—S bonds are expected to be at this angle, too. Therefore, the predicted bond angle is 90°. This is very close to the actual bond angle of 92° found by experiment. Thus, the VB theory requirement for maximum overlap quite nicely explains the geometry of the hydrogen sulfide molecule. Also, notice that both the Lewis and VB descriptions of the formation of the H—S bonds account for the completion of the atoms' valence shells. Therefore, *a Lewis structure can be viewed, in a very qualitative sense, as a shorthand notation for the valence bond description of a molecule.*

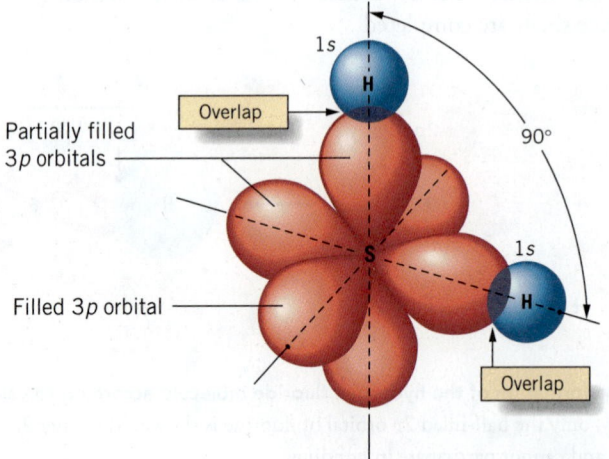

Figure 9.17 | **Bonding in H_2S.** We expect the hydrogen 1s orbitals to position themselves so that they can best overlap with the two partially filled 3p orbitals of sulfur, which gives a predicted bond angle of 90°. The experimentally measured bond angle of 92° is very close to the predicted angle.

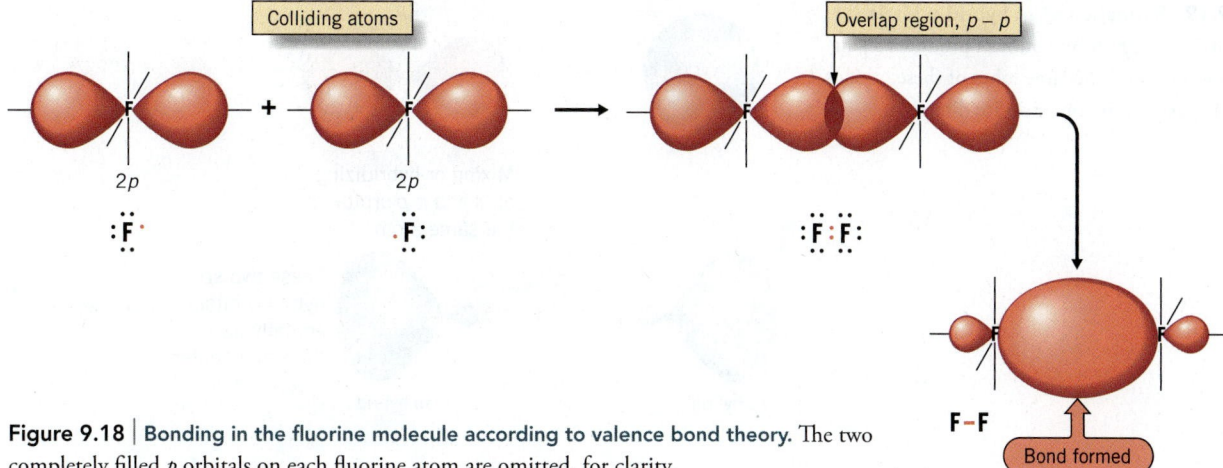

Figure 9.18 | **Bonding in the fluorine molecule according to valence bond theory.** The two completely filled *p* orbitals on each fluorine atom are omitted, for clarity.

Other kinds of orbital overlaps are also possible. For example, according to VB theory the bonding in the fluorine molecule, F_2, occurs by the overlap of two $2p$ orbitals, as shown in Figure 9.18. The formation of the other diatomic molecules of the halogens, all of which are held together by single bonds, could be similarly described.

Use the principles of VB theory to explain the bonding in HCl. Give the orbital diagram for chlorine in the HCl molecule and indicate the orbital that shares the electron with one from hydrogen. Sketch the orbital overlap that gives rise to the H—Cl bond. (*Hint:* Remember that a half-filled orbital on each atom is required to form the covalent bond.)

Practice Exercise 9.9

The phosphine molecule, PH_3, has a trigonal pyramidal shape with H—P—H bond angles equal to 93.7°, as shown in the margin. Give the orbital diagram for phosphorus in the PH_3 molecule and indicate the orbitals that share electrons with those from hydrogen. On a set of *xyz* coordinate axes, sketch the orbital overlaps that give rise to the P—H bonds.

Practice Exercise 9.10

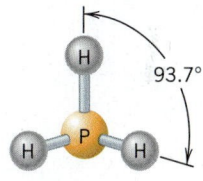

9.5 | Hybrid Orbitals and Molecular Geometry

There are many molecules for which the simple VB theory described in Section 9.4 fails to account for the correct shape. For example, there are no simple atomic orbitals that are oriented so they point toward the corners of a tetrahedron, yet there are many tetrahedral molecules. Therefore, to explain the bonds in molecules such as CH_4 we must study the way atomic orbitals of the same atom mix with each other to produce new orbitals, called **hybrid atomic orbitals**.[3] The new orbitals have new shapes and new directional properties, and they can overlap to give structures with bond angles that match those found by experiment. It's important to realize that hybrid orbitals are part of the valence bond theory. They can't be directly observed in an experiment. We use them to describe molecular structures that have been determined experimentally.

Hybrid Orbitals Formed from *s* and *p* Atomic Orbitals

Let's begin by studying what happens when we mix a $2s$ orbital with a $2p$ orbital to form a new set of two orbitals that we designate as two ***sp* hybrid orbitals**, illustrated in Figure 9.19.

---

[3] Mathematically, hybrid orbitals are formed by the addition and subtraction of the wave functions for the basic atomic orbitals. This process produces a new set of wave functions corresponding to the hybrid orbitals. The hybrid orbital wave functions describe the shape and directional properties of the hybrid orbitals.

Figure 9.19 | Formation of *sp* hybrid orbitals.
Mixing of 2*s* and 2*p* atomic orbitals produces a pair of *sp* hybrid orbitals. The large lobes of these orbitals have their axes pointing in opposite directions.

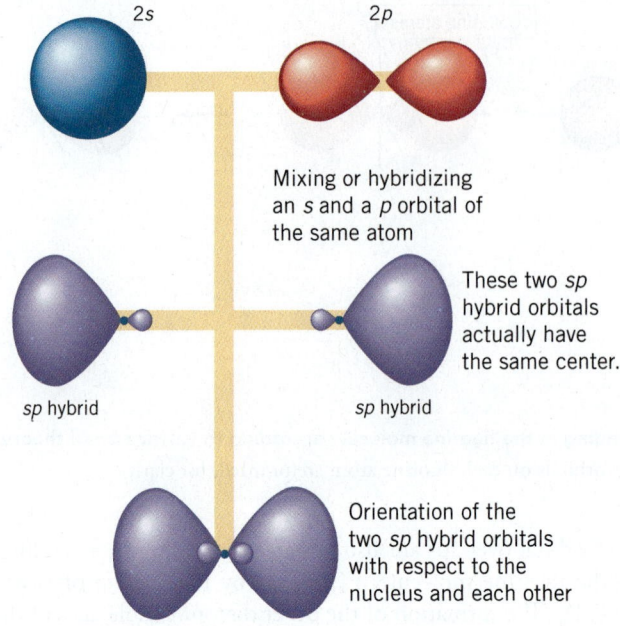

■ In general, the greater the overlap of two orbitals, the stronger the bond. At a given distance between nuclei, the greater size of a hybrid orbital gives better overlap than either an *s* or a *p* orbital.

First, notice that each has the same shape; each has one large lobe and another much smaller one. The large lobe extends farther from the nucleus than either the *s* or *p* orbital from which the hybrid was formed. This allows the hybrid orbital to overlap more effectively with an orbital on another atom when a bond is formed. As a result, hybrid orbitals form stronger, more stable, bonds than would be possible if just simple atomic orbitals were used.

Another point to notice in Figure 9.19 is that the large lobes of the two *sp* hybrid orbitals point in opposite directions; that is, they are 180° apart. If bonds are formed by the overlap of these hybrids with orbitals of other atoms, the other atoms will occupy positions on opposite sides of this central atom. Let's look at a specific example—the linear beryllium hydride molecule, BeH_2, as it would be formed in the gas phase.[4] The orbital diagram for the valence shell of beryllium is

Notice that the 2*s* orbital is filled and the three 2*p* orbitals are empty. For bonds to form at a 180° angle between beryllium and the two hydrogen atoms, two conditions must be met: (1) the two orbitals that beryllium uses to form the Be—H bonds must point in opposite directions, and (2) each of the beryllium orbitals must contain only one electron. In satisfying these requirements, the electrons of the beryllium atom become unpaired and the resulting half-filled *s* and *p* atomic orbitals become hybridized.

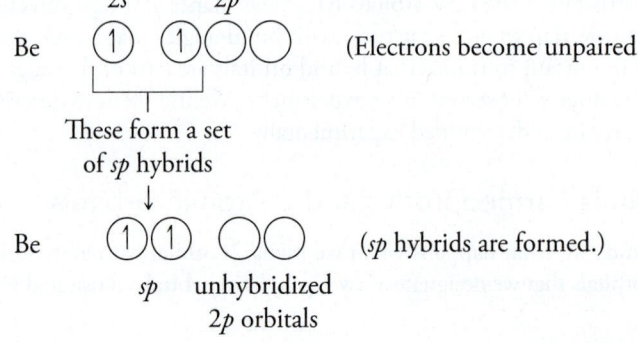

---

[4] In the solid state, BeH_2 has a complex structure not consisting of simple BeH_2 molecules.

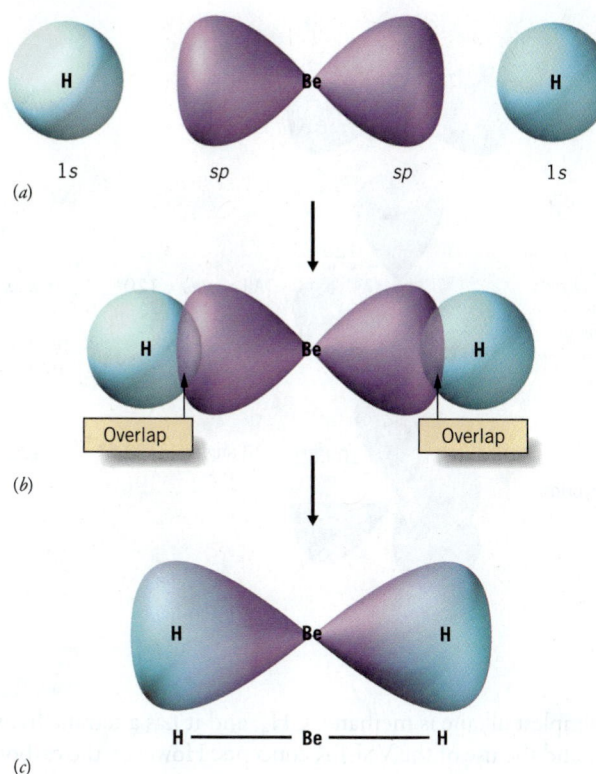

Figure 9.20 | **Bonding in BeH₂ according to valence bond theory.** **Figure 9.20** | **Bonding in BeH₂ according to valence bond theory.** Only the larger lobe of each *sp* hybrid orbital is shown. (*a*) The two hydrogen 1*s* orbitals approach the pair of *sp* hybrid orbitals of beryllium. (*b*) Overlap of the hydrogen 1*s* orbitals with the *sp* hybrid orbitals. (*c*) A representation of the distribution of electron density in the two Be—H bonds after they have been formed.

Now the 1*s* orbitals of the hydrogen atoms can overlap with the *sp* hybrids of beryllium to form the bonds, as shown in Figure 9.20. Because the two *sp* hybrid orbitals of beryllium are identical in shape and energy, the two Be—H bonds are alike except for the directions in which they point, and we say that the bonds are *equivalent*. Since the bonds point in opposite directions, the linear geometry of the molecule is also explained. The orbital diagram for beryllium in this molecule is

Be (in BeH₂)   ⟨↑↓⟩⟨↑↓⟩ ◯◯   (Colored arrows are H electrons.)

  sp   unhybridized
          2*p* orbitals

Even if we had not known the shape of the BeH₂ molecule, we could have obtained the same bonding picture by applying the VSEPR model first. The Lewis structure for BeH₂ is H:Be:H, with two bonding domains around the central atom, so the molecule is linear. Once the shape is known, we can apply the VB theory to explain the bonding in terms of orbital overlaps. Thus, the VB theory and VSEPR model complement each other well. The VSEPR model allows us to predict geometry in a simple way, and once the geometry is known, it is relatively easy to analyze the bonding in terms of VB theory.

In addition to *sp* hybrids, atomic *s* and *p* orbitals form two other kinds of hybrid orbitals. These are shown in Figure 9.21. In identifying hybrid orbitals, we specify which kinds of pure atomic orbitals, as well as the number of each, that are mixed to form the hybrids. Thus hybrid orbitals labeled *sp³* are formed by blending one *s* orbital and three *p* orbitals. *The total number of hybrid orbitals in a set is equal to the number of basic atomic orbitals used to form them.* Therefore, a set of **sp³ hybrids** consists of four orbitals, whereas a set of **sp² hybrid orbitals** consists of three orbitals.

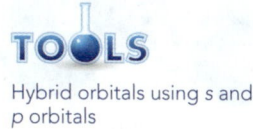

TOOLS

Hybrid orbitals using *s* and *p* orbitals

sp³ Hybrid Orbitals

The *sp³* hybrid is very important in organic chemistry, so it is worthwhile to take a deep look at how it forms and how it is very useful in understanding organic molecules, particularly

Figure 9.21 | Directional properties of hybrid orbitals formed from *s* and *p* atomic orbitals. (*a*) *sp* hybrid orbitals oriented at 180° to each other. (*b*) *sp²* hybrid orbitals formed from an *s* orbital and two *p* orbitals. The angle between them is 120°. (*c*) *sp³* hybrid orbitals formed from an *s* orbital and three *p* orbitals. The angle between any two of them is 109.5°.

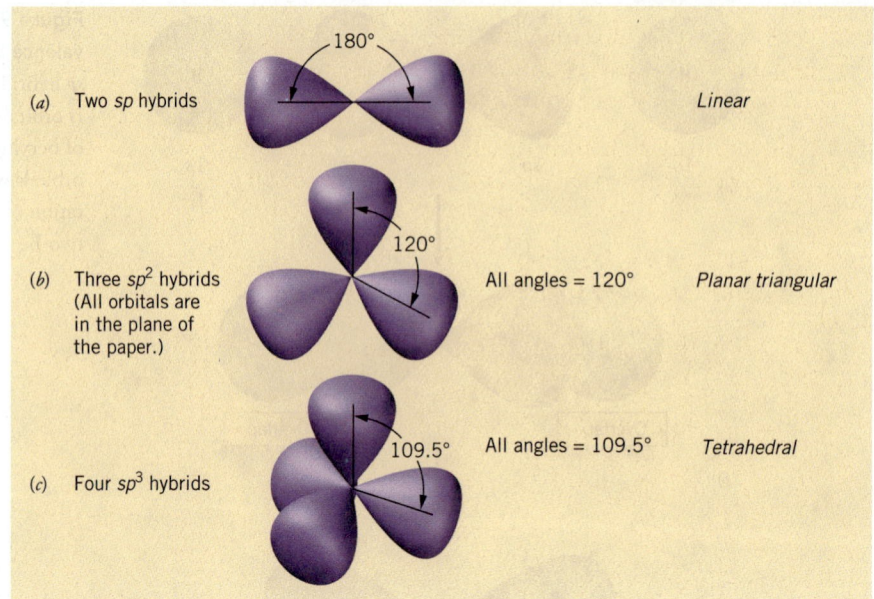

the alkanes. The simplest alkane is methane, CH_4, and it has a tetrahedral shape based on its Lewis structure and the use of the VSEPR concepts. However, the carbon atom consists of two paired 2*s* and two unpaired 2*p* electrons in its valence shell. Since the *s* orbital is nondirectional and the *p* obitals are arranged at 90° angles, they cannot produce the tetrahedral structure we predict and actually measure in the lab. By using the *sp³* hybrid we can explain the structure of CH_4.

Let's start by drawing the orbital diagram of the valence shell of carbon.

C (1↓) (1)(1)( )

      2*s*   2*p*

To form four C—H bonds, we need four half-filled orbitals. Unpairing the electrons in the 2*s* and moving one to the vacant 2*p* orbital satisfies this requirement.

    2*s*   2*p*

C (1) (1)(1)(1) (Electrons become unpaired.)

However, the *s* and *p* electrons have different properties. When we can hybridize all the orbitals to give the desired *sp³* set where each electron has the same properties, particularly the tetrahedral orientation,

C (1)(1)(1)(1)

     sp³ hybrids

Now the four bonds to hydrogen 1*s* orbitals can be formed by overlapping with the hybrid orbitals.

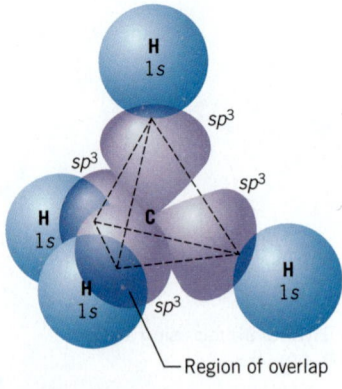

Figure 9.22 | Formation of the bonds in methane. Each bond results from the overlap of a hydrogen 1*s* orbital with an *sp³* hybrid orbital on the carbon atom.

C (in CH_4) (1↓)(1↓)(1↓)(1↓) (Colored arrows are H electrons.)

        sp³

This is illustrated in Figure 9.22. The following exercises allow an opportunity to apply this reasoning.

The BF$_3$ molecule has a planar triangular shape. What kind of hybrid orbitals does boron use in this molecule? Use orbital diagrams to explain how the bonds are formed. (*Hint:* What hybrid orbital is oriented correctly to give this molecular shape?)

Practice Exercise 9.11

In the gas phase, beryllium fluoride exists as linear molecules. Which kind of hybrid orbital does Be use in this compound? Use orbital diagrams to explain how the bonds are formed.

Practice Exercise 9.12

In methane, carbon forms four single bonds with hydrogen atoms by using sp^3 hybrid orbitals. In fact, carbon uses these same kinds of orbitals in all of its compounds in which it is bonded to four other atoms by single bonds. This makes the tetrahedral orientation of atoms around carbon one of the primary structural features of organic compounds, and organic chemists routinely think in terms of "tetrahedral carbon."

In the alkane series of hydrocarbons (compounds with the general formula $C_nH_{(2n+2)}$), carbon atoms are bonded to other carbon atoms. An example is ethane, C_2H_6.

$$H-\overset{\overset{\displaystyle H}{|}}{\underset{\underset{\displaystyle H}{|}}{C}}-\overset{\overset{\displaystyle H}{|}}{\underset{\underset{\displaystyle H}{|}}{C}}-H$$

In this molecule, the carbons are bonded together by the overlap of sp^3 hybrid orbitals (Figure 9.23). One of the most important characteristics of this bond is that the overlap of the orbitals in the C—C bond is hardly affected at all if one portion of the molecule rotates relative to the other around the bond axis. Such rotation, therefore, is said to occur freely and permits different possible relative orientations of the atoms in the molecule. These different relative orientations are called **conformations**. With complex molecules, the number of possible conformations is enormous. For example, Figure 9.24 illustrates three of the large number of possible conformations of the pentane molecule, C_5H_{12}, one of the low-molecular-weight organic compounds in gasoline.

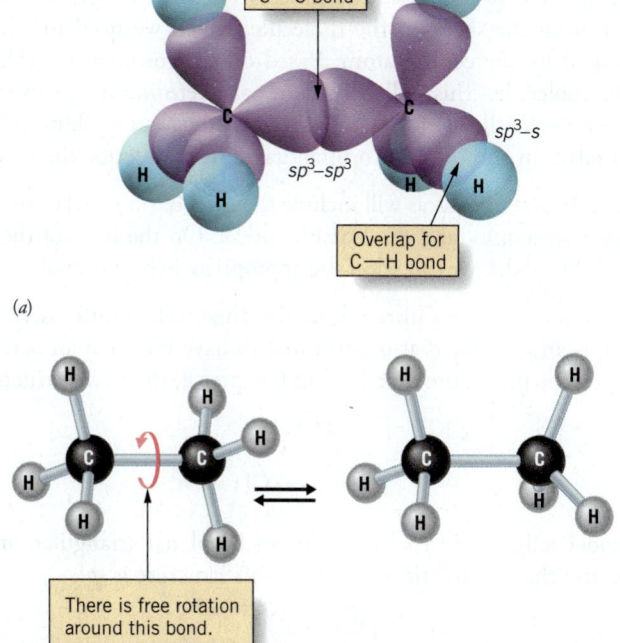

Figure 9.23 | **The bonds in the ethane molecule.** (*a*) Overlap of orbitals. (*b*) The degree of overlap of the sp^3 orbitals in the carbon–carbon bond is not appreciably affected by the rotation of the two CH$_3$— groups relative to each other around the bond.

Figure 9.24 | Three of the many conformations of the atoms in the pentane molecule, C_5H_{12}. Free rotation around single bonds makes these different conformations possible.

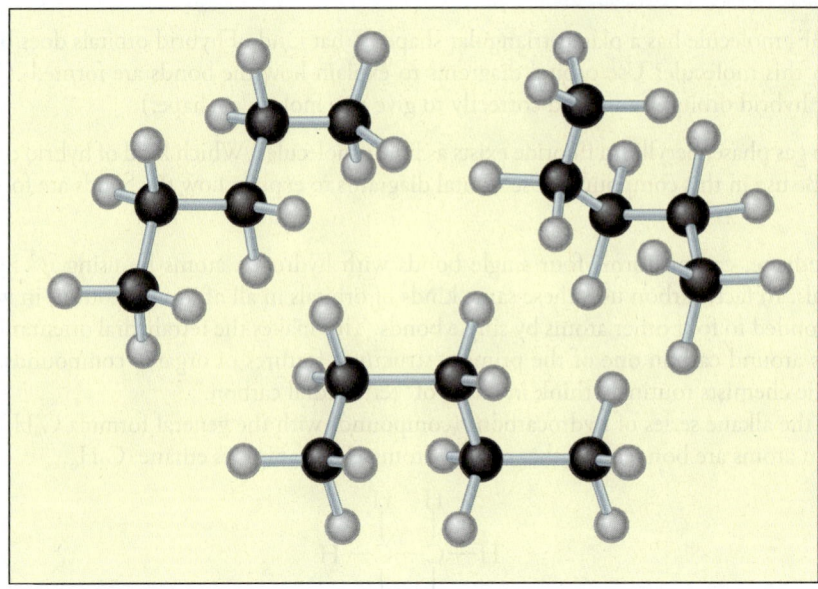

Using the VSEPR Model to Predict Hybridization

TOOLS

VSEPR model and hybridization

We've seen that if we know the structure of a molecule, we can make a reasonable guess as to the kind of hybrid orbitals that the central atom uses to form its bonds. Because the VSEPR model works so well in predicting geometry, we can use it to help us obtain VB descriptions of bonding. This is illustrated in the following example.

Example 9.6
Using the VSEPR Model to Predict Hybridization

Predict the shape of the boron trichloride molecule and describe the bonding in the molecule in terms of valence bond theory.

Analysis: This problem is similar to the process we used to describe BeH_2 and CH_4, except we're not given the shape of the molecule, which we need to select the kind of hybrid orbitals used by the central atom. Based on the formula, we will write a Lewis structure for the molecule. This will permit us to determine its geometry by applying VSEPR theory. From the shape of the molecule, we can select the kind of hybrid orbitals used by the central atom and then use orbital diagrams to describe the bonding.

Assembling the Tools: Our tools will include the rules of nomenclature, the method for constructing Lewis structures, and the VSEPR model. On the basis of the geometry predicted by the VSEPR model, we can select the appropriate hybrid orbitals from Figure 9.21.

Solution: The rules of nomenclature tell us the chemical formula is BCl_3. Keeping in mind that boron is an element that is permitted to have less than an octet in its valence shell and following the procedure described in Chapter 8, the Lewis structure for BCl_3 is

$$:\overset{\displaystyle ..}{\underset{\displaystyle ..}{Cl}}:$$
$$|$$
$$:\overset{\displaystyle ..}{\underset{\displaystyle ..}{Cl}}-B-\overset{\displaystyle ..}{\underset{\displaystyle ..}{Cl}}:$$

The VSEPR model tells us the molecule should be planar triangular, and referring to Figure 9.21, we find that the hybrid set that fits this structure is sp^2.

B   ⟨↑↓⟩ ⟨↑⟩◯◯

   $2s$ $2p$

To form the three bonds to chlorine, boron needs three half-filled orbitals. These can be obtained by unpairing the electrons in the 2s orbital and placing one of them in the 2p. Then the 2s orbital and two of the 2p orbitals can be combined to give the set of three sp^2 hybrids.

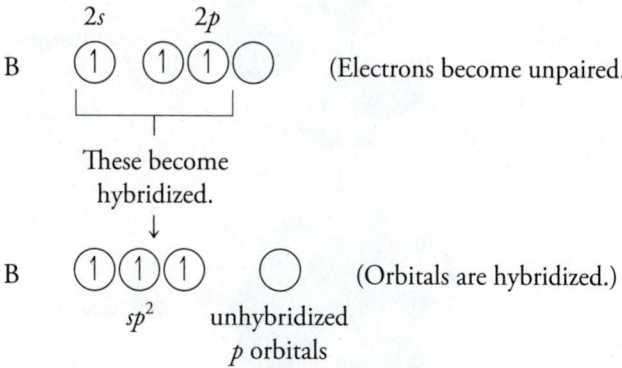

B   (Electrons become unpaired.)

These become hybridized.

B   (Orbitals are hybridized.)
    sp^2 unhybridized
         p orbitals

Now let's look at the valence shell of chlorine:

Cl   3s   3p

The half-filled 3p orbital of each chlorine atom can overlap with a hybrid sp^2 orbital of boron to give three B—Cl bonds.

B (in BCl_3)   (Colored arrows are Cl electrons.)
       sp^2 unhybridized
            p orbitals

Figure 9.25 illustrates the overlap of the orbitals to give the bonding in the molecule.

● **Is the Answer Reasonable?** Notice that the arrangement of atoms yields a planar triangular molecule, in agreement with the structure predicted by the VSEPR model. This internal consistency is convincing evidence that we've answered the question correctly.

What kind of hybrid orbitals are expected to be used by the central atom in SiH_4? (*Hint:* Use orbital diagrams to describe the bonding in the molecule.)

Practice Exercise 9.13

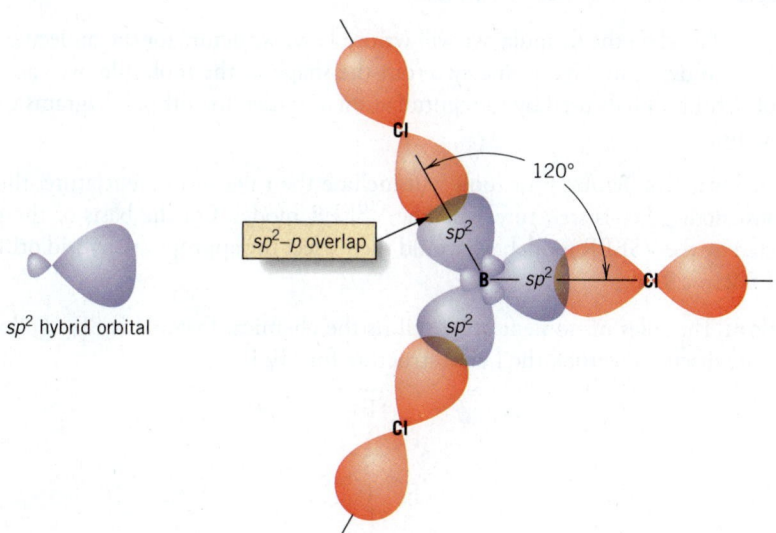

sp^2 hybrid orbital

sp^2–p overlap

120°

Figure 9.25 | **Description of the bonding in BCl_3.** Each B—Cl bond is formed by the overlap of a half-filled 3p orbital of á Cl atom with one of the sp^2 hybrid orbitals of boron. (For simplicity, only the half-filled 3p orbital of each Cl atom is shown.)

Figure 9.26 | Orientations of hybrid orbitals that involve *d* orbitals. (*a*) *sp³d* hybrid orbitals formed from an *s* orbital, three *p* orbitals, and a *d* orbital. The orbitals point toward the vertices of a trigonal bipyramid. (*b*) *sp³d²* hybrid orbitals formed from an *s* orbital, three *p* orbitals, and two *d* orbitals. The orbitals point toward the vertices of an octahedron.

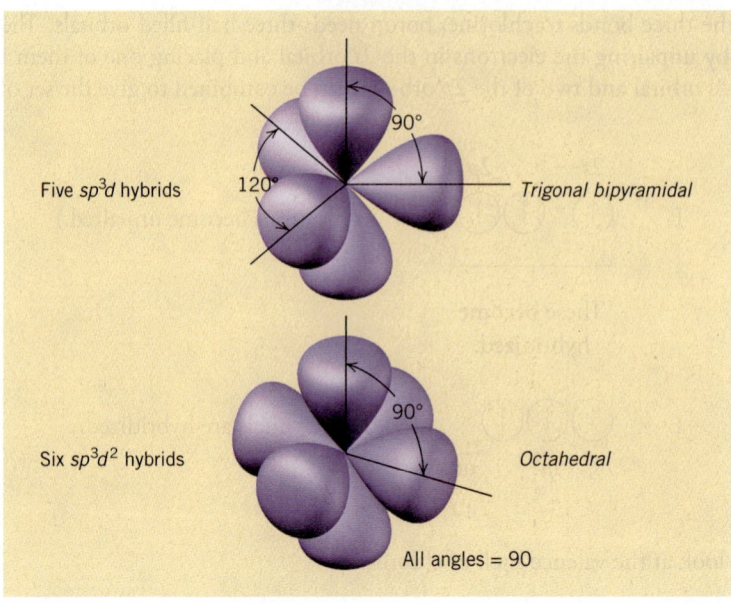

Hybrid Orbitals Formed from *s*, *p*, and *d* Orbitals

Hybrid orbitals using *s*, *p*, and *d* atomic orbitals

Earlier we saw that certain molecules have atoms that must violate the octet rule because they form more than four bonds. In these cases, the atom must reach beyond its *s* and *p* valence shell orbitals to form sufficient half-filled orbitals for bonding. This is because the *s* and *p* orbitals can be mixed to form a maximum of only four hybrid orbitals. When five or more hybrid orbitals are needed, *d* orbitals are brought into the mix. The two most common kinds of hybrid orbitals involving *d* orbitals are ***sp³d*** and ***sp³d²* hybrid orbitals**. Their directional properties are illustrated in Figure 9.26. Notice that the *sp³d* hybrids point toward the corners of a trigonal bipyramid and the *sp³d²* hybrids point toward the corners of an octahedron.

Example 9.7
Explaining Bonding with Hybrid Orbitals

Predict the shape of the sulfur hexafluoride molecule and describe the bonding in the molecule in terms of valence bond theory.

Analysis: Based on the formula, we will write a Lewis structure for the molecule. This will permit us to determine its geometry. From the shape of the molecule, we can select the kind of hybrid orbitals used by the central atom and then use orbital diagrams to describe the bonding.

Assembling the Tools: Our tools will include the rules of nomenclature, the method for constructing Lewis structures and the VSEPR model. On the basis of the geometry predicted by the VSEPR model, we would then select the appropriate hybrid orbitals from Figure 9.21 or 9.26.

Solution: The rules of nomenclature tell us the chemical formula is SF_6. Following the procedure discussed earlier, the Lewis structure for SF_6 is

The VSEPR model tells us the molecule should be octahedral, and referring to Figure 9.26, we find that the hybrid set that fits this structure is sp^3d^2.

Now let's examine the valence shell of sulfur.

S (⇅) (⇅)(1)(1)
   $3s$ $3p$

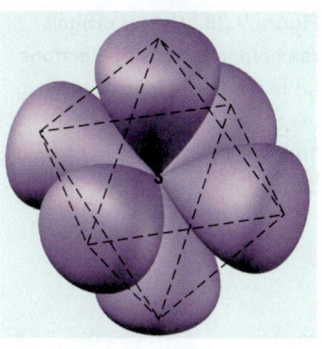

Figure 9.27 | The sp^3d^2 hybrid orbitals of sulfur in SF₆.

■ Sulfur is able to exceed an octet because of the availability of d orbitals in its valence shell.

To form six bonds to fluorine atoms we need six half-filled orbitals, but we show only four orbitals altogether. However, sp^3d^2 orbitals tell us we need to look for d orbitals to include in the set of hybrids.

An isolated sulfur atom has electrons only in its $3s$ and $3p$ subshells, so these are the only subshells we usually show in the orbital diagram. But the third shell also has a d subshell, which is empty in a sulfur atom. Therefore, let's rewrite the orbital diagram to show the vacant $3d$ subshell.

S (⇅) (⇅)(1)(1) ( )( )( )( )( )
   $3s$ $3p$ $3d$

Unpairing all of the electrons to give six half-filled orbitals, followed by hybridization, gives the required set of half-filled sp^3d^2 orbitals (Figure 9.27).

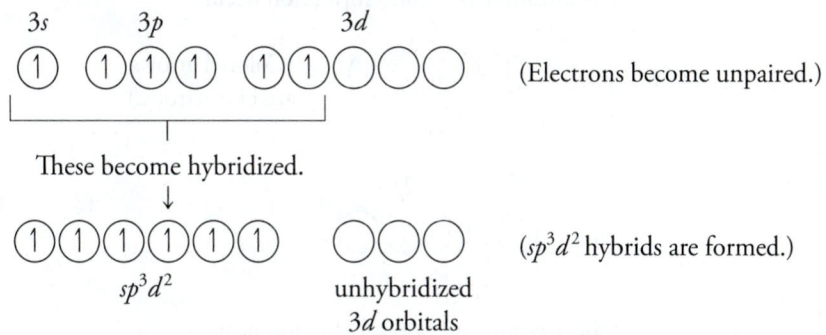

$3s$ $3p$ $3d$

(1) (1)(1)(1) (1)(1)( )( )( )   (Electrons become unpaired.)

These become hybridized.

↓

(1)(1)(1)(1)(1)(1) ( )( )( )   (sp^3d^2 hybrids are formed.)
   sp^3d^2 unhybridized
                     $3d$ orbitals

Finally, the six S—F bonds are formed by overlap of the half-filled $2p$ orbitals of the fluorine atoms with these half-filled sp^3d^2 hybrids.

S (in SF₆) (⇅)(⇅)(⇅)(⇅)(⇅)(⇅) ( )( )( )   (Colored arrows are F electrons.)
              sp^3d^2 unhybridized
                                  $3d$ orbitals

Is the Answer Reasonable? Once again, the fact that all the parts fit together so well to account for the structure of SF₆ makes us feel that our explanation of the bonding is reasonable.

What kind of hybrid orbitals are expected to be used by the central atom in PCl₅? (*Hint:* Which hybrid orbitals have the same geometry as the molecule?)

Practice Exercise 9.14

Use the VSEPR model to predict the shape of the AsCl₅ molecule and then describe the bonding in the molecules using valence bond theory.

Practice Exercise 9.15

Molecules with Nonbonding Domains

Methane is a tetrahedral molecule with sp^3 hybridization of the orbitals of carbon and H—C—H bond angles that are each equal to 109.5°. In ammonia, NH₃, the H—N—H bond angles are 107°, and in water the H—O—H bond angle is 104.5°. Both NH₃ and H₂O have H—X—H bond angles that are close to the bond angles expected for a

Figure 9.28 | Hybrid orbitals can hold lone pairs of electrons. (*a*) In water, two lone pairs on oxygen are held in sp^3 hybrid orbitals. (*b*) Ammonia has one lone pair in an sp^3 hybrid orbital.

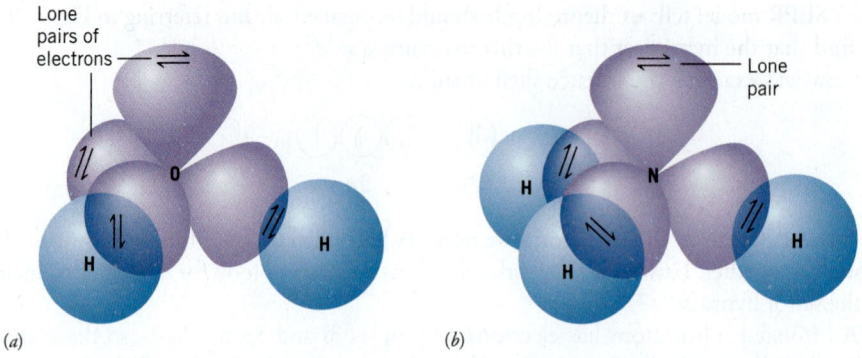

molecule whose central atom has sp^3 hybrids. The use of sp^3 hybrids by oxygen and nitrogen, therefore, is often used to explain the geometry of H_2O and NH_3.

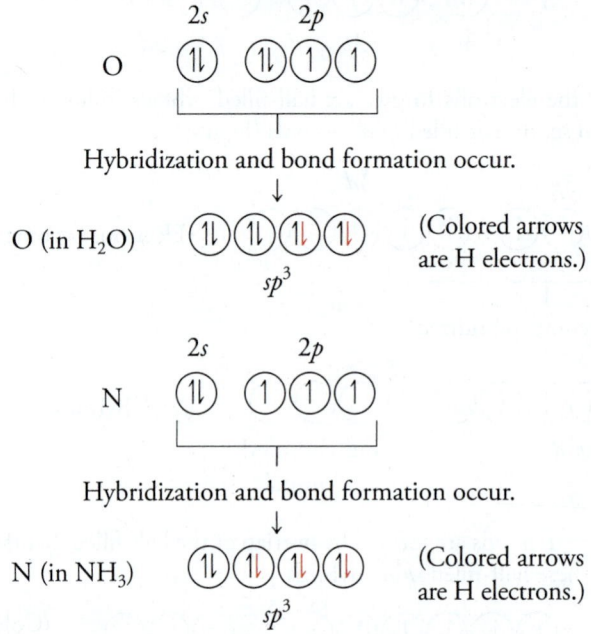

According to these descriptions, not all of the hybrid orbitals of the central atom must be used for bonding. Lone pairs of electrons can be accommodated in them too, as illustrated in Figure 9.28. In fact, putting the lone pair on the nitrogen in an sp^3 hybrid orbital gives a geometry that agrees well with the experimentally determined structure of the ammonia molecule.

Example 9.8
Explaining Bonding with Hybrid Orbitals

Use valence bond theory to explain the bonding in the SF_4 molecule.

Analysis: In solving this problem we will follow the same steps as in Example 9.7: (1) Draw the Lewis structure, (2) predict the geometry of the molecule, (3) select the hybrid orbital set that fits the structure, and (4) construct orbital diagrams to explain the bonding.

Assembling the Tools: Our tools are the same as before. We apply our method for drawing Lewis structures from Chapter 8. We apply the VSEPR theory. We use Figure 9.21 or 9.26 to select the hybrid orbital set.

Solution: We begin by constructing the Lewis structure for the molecule.

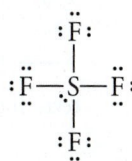

The VSEPR model predicts that the electron pairs around the sulfur should be in a trigonal bipyramidal arrangement, and the only hybrid set that fits this geometry is sp^3d (according to Figure 9.26). To see how the hybrid orbitals are formed, we look at the valence shell of sulfur, including the vacant $3d$ subshell.

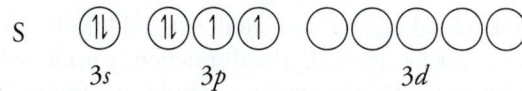

To form the four bonds to fluorine atoms, we need four half-filled orbitals, so we unpair the electrons in one of the filled orbitals. This gives

Next, we form the hybrid orbitals. In doing this, we use all the valence shell orbitals that have electrons in them.

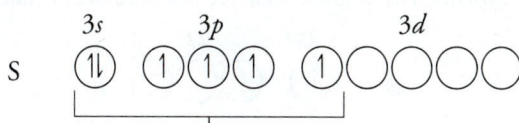

Now, four S—F bonds can be formed by overlap of half-filled $2p$ orbitals of fluorine with the sp^3d hybrid orbitals of sulfur to give the structure in the margin.

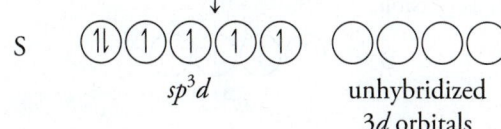

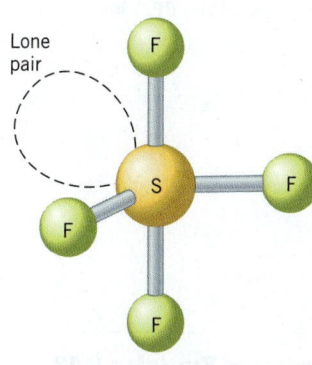

Lone pair

The structure of SF₄

Is the Answer Reasonable? Counting electrons in the Lewis structure gives a total of 34, which is how many electrons are in the valence shells of one sulfur and four chlorine atoms, so the Lewis structure appears to be correct. The rest of the answer flows smoothly, so the bonding description appears to be reasonable. Notice that the VSEPR model even shows us which of the hybrid orbitals contains the lone pair.

What kind of orbitals are used by Xe in the XeF₄ molecule? Give a step-by-step description of how you arrived at your answer. (*Hint:* An Xe atom has 8 valence electrons.)

Practice Exercise 9.16

Explain how to decide what kind of hybrid orbitals we would expect the central atom to use for bonding in (a) PCl₃ and (b) BrCl₃.

Practice Exercise 9.17

Formation of Coordinate Covalent Bonds

In Section 8.7, we defined a coordinate covalent bond as one in which both of the shared electrons are provided by just one of the joined atoms. Such a bond is formed when boron trifluoride, BF_3, combines with an additional fluoride ion to form the tetrafluoroborate ion BF_4^-

$$BF_3 + F^- \longrightarrow BF_4^-$$

tetrafluoroborate ion

We can diagram the reaction as follows:

$$\ddot{:}\underset{\displaystyle :\ddot{F}:}{\overset{\displaystyle :\ddot{F}:}{:\ddot{F}:\!B}} + \left[:\ddot{F}:\right]^- \longrightarrow \left[\underset{\displaystyle :\ddot{F}:}{\overset{\displaystyle :\ddot{F}:}{:\ddot{F}:\!B\!:\!\ddot{F}:}}\right]^-$$

As we mentioned previously, the coordinate covalent bond is really no different from any other covalent bond once it has formed. The distinction is made only for bookkeeping purposes. One place where such bookkeeping is useful is in keeping track of the orbitals and electrons used when atoms bond together.

The VB theory requirements for bond formation—two overlapping orbitals sharing two paired electrons—can be satisfied in two ways. One, as we have already seen, is by the overlapping of two half-filled orbitals. This gives an "ordinary" covalent bond. The other is by overlapping one filled orbital with one empty orbital. The atom with the filled orbital donates the shared pair of electrons, and a coordinate covalent bond is formed.

The structure of the BF_4^- ion, which the VSEPR model predicts to be tetrahedral (Figure 9.29), can be explained as follows. First, we examine the orbital diagram for boron.

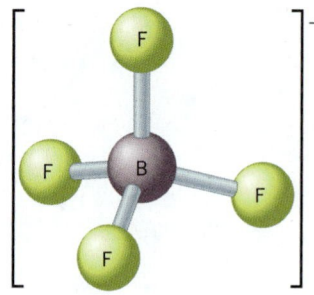

Figure 9.29 | The tetrahedral structure of the BF_4^- ion.

To form four bonds, we need four hybrid orbitals arranged tetrahedrally around the boron, so we expect boron to use sp^3 hybrids. Notice that we spread the electrons out over the hybrid orbitals as much as possible.

Boron forms three ordinary covalent bonds with fluorine atoms plus one coordinate covalent bond with a fluoride ion.

Coordinate covalent bond—both electrons are from F^- (Colored arrows are F electrons.)

Practice Exercise 9.18

If we assume that nitrogen uses sp^3 hybrid orbitals in NH_3, use valence bond theory to account for the formation of NH_4^+ from NH_3 and H^+. (*Hint:* Which atom donates the pair of electrons in the formation of the bond between H^+ and NH_3?)

Practice Exercise 9.19

What is the shape of the PCl_6^- ion? What hybrid orbitals are used by phosphorus in PCl_6^-? Draw the orbital diagram for phosphorus in PCl_6^-.

9.6 | Hybrid Orbitals and Multiple Bonds

The types of orbital overlap that we have described so far produce bonds in which the electron density is concentrated most heavily between the nuclei of the two atoms along an imaginary line that joins their centers. Any bond of this kind, whether formed from the

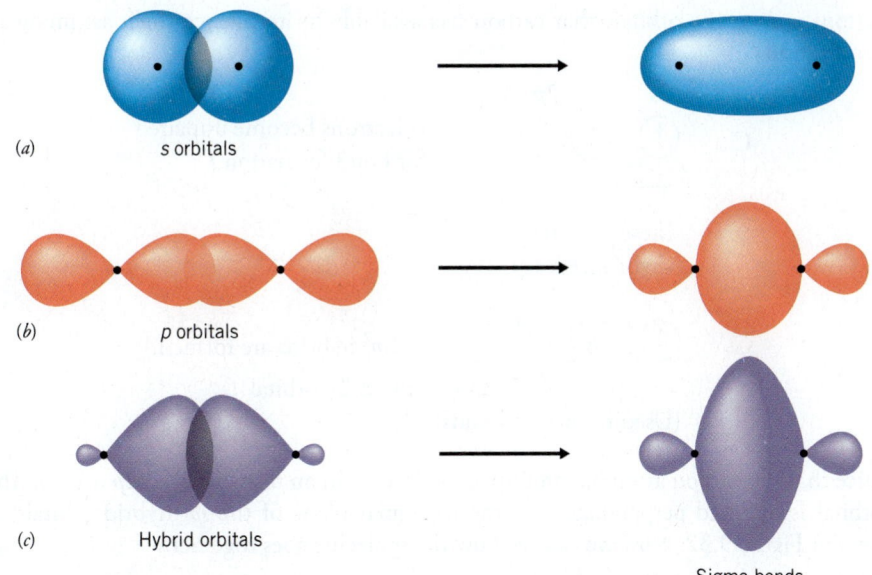

Figure 9.30 | Formation of σ bonds. Sigma bonds concentrate electron density along the line between the two atoms joined by the bond. (*a*) From the overlap of *s* orbitals. (*b*) From the end-to-end overlap of *p* orbitals. (*c*) From the overlap of hybrid orbitals.

overlap of *s* orbitals, *p* orbitals, or hybrid orbitals (Figure 9.30), is called a **sigma bond** (or **σ bond**).

Another way that *p* orbitals can overlap is shown in Figure 9.31. This produces a bond in which the electron density is divided between two separate regions that lie on opposite sides of an imaginary line joining the two nuclei. This kind of bond is called a **pi bond** (or **π bond**). Notice that a π bond, like a *p* orbital, consists of two parts, and each part makes up just half of the π bond; it takes both of them to equal one π bond. The formation of π bonds allows atoms to form double and triple bonds.

Double Bonds

A hydrocarbon that contains a double bond is ethene (also called ethylene), C_2H_4. It has the Lewis structure

$$
\begin{array}{ccc}
H & & H \\
\diagdown & & \diagup \\
& C = C & \\
\diagup & & \diagdown \\
H & & H
\end{array}
$$

ethene
(ethylene)

The molecule is planar, and each carbon atom lies in the center of a triangle surrounded by three other atoms (two H and one C atom). A planar triangular arrangement of bonds suggests that carbon uses sp^2 hybrid orbitals. Therefore, let's look at the distribution of

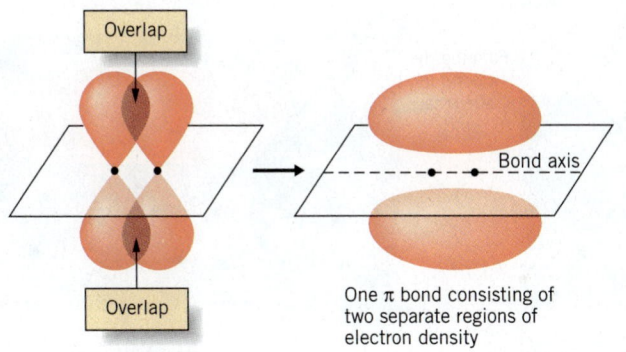

One π bond consisting of two separate regions of electron density

Figure 9.31 | Formation of a π bond. Two *p* orbitals overlap sideways instead of end-to-end. The electron density is concentrated in *two regions* on opposite sides of the bond axis, and taken together they constitute *one π bond*.

electrons among the orbitals that carbon has available in its valence shell, assuming sp^2 hybridization.

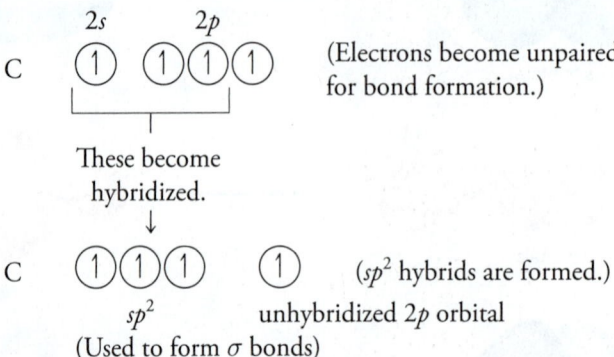

sp^2

unhybridized $2p$ orbital

(Used to form σ bonds)

■ The double bond is a little like a hot dog on a bun. The sigma bond is like the hot dog, and the π bond is like the two parts of the bun.

Notice that the carbon atom has an unpaired electron in an unhybridized $2p$ orbital. This p orbital is oriented perpendicular to the triangular plane of the sp^2 hybrid orbitals, as shown in Figure 9.32. Now we can see how the molecule goes together.

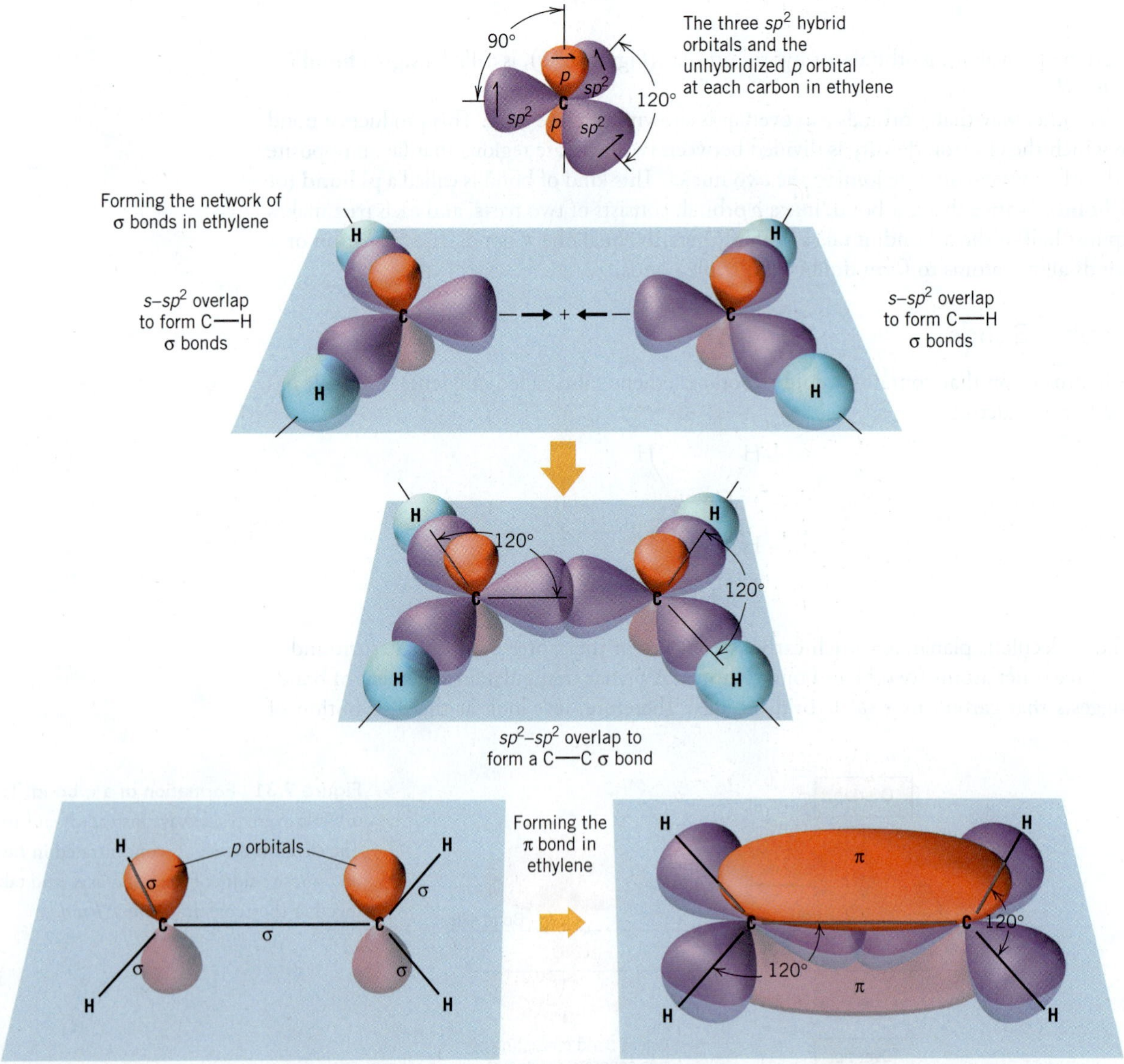

The three sp^2 hybrid orbitals and the unhybridized p orbital at each carbon in ethylene

90°

120°

Forming the network of σ bonds in ethylene

s–sp^2 overlap to form C—H σ bonds

s–sp^2 overlap to form C—H σ bonds

120°

120°

sp^2–sp^2 overlap to form a C—C σ bond

p orbitals

σ

σ

σ

σ

σ

Forming the π bond in ethylene

π

120°

120°

π

Figure 9.32 | The carbon–carbon double bond.

The basic framework of the molecule is determined by the formation of σ bonds. Each carbon uses two of its sp^2 hybrids to form σ bonds to hydrogen atoms. The third sp^2 hybrid on each carbon is used to form a σ bond between the two carbon atoms, thereby accounting for one of the two bonds of the double bond. Finally, the remaining unhybridized $2p$ orbitals, one from each carbon atom, overlap to produce a π bond, which accounts for the second bond of the double bond.

This description of the bonding in C_2H_4 accounts for one of the most important properties of double bonds: rotation of one portion of the molecule relative to the rest around the axis of the double bond occurs only with great difficulty. The reason for this is illustrated in Figure 9.33. We see that as one CH_2 group rotates relative to the other around the carbon–carbon bond, the unhybridized p orbitals become misaligned and can no longer overlap effectively. This destroys the π bond. In effect, then, rotation around a double bond involves bond breaking, which requires more energy than is normally available to molecules at room temperature. As a result, rotation around the axis of a double bond usually doesn't take place.

In almost every instance, a double bond consists of a σ bond and a π bond. Another example is the compound methanal (better known as formaldehyde, the substance used as a preservative for biological specimens and as an embalming fluid). The Lewis structure of this compound is

$$
\begin{array}{c}
\text{H} \\
\diagdown \\
\text{C}=\ddot{\text{O}}\!\!:\; \\
\diagup \\
\text{H}
\end{array}
$$

formaldehyde

As with ethene, the carbon forms sp^2 hybrids, leaving an unpaired electron in an unhybridized p orbital.

C   ① ① ①   ①

     sp^2 unhybridized
            $2p$ orbital

The oxygen can also form sp^2 hybrids, with electron pairs in two of them and an unpaired electron in the third. This means that the remaining unhybridized p orbital also has an unpaired electron.

      $2s$ $2p$

O   ⑴⥮   ⑴⥮ ① ①   (Ground state of oxygen)

These form sp^2 hybrids.
↓

O   ⑴⥮⑴⥮①   ①

     sp^2 unhybridized
            $2p$ orbital

Figure 9.34 shows how the carbon, hydrogen, and oxygen atoms form the molecule. As before, the basic framework of the molecule is formed by the σ bonds. These determine the molecular shape. The carbon–oxygen double bond also contains a π bond formed by the overlap of the unhybridized p orbitals.

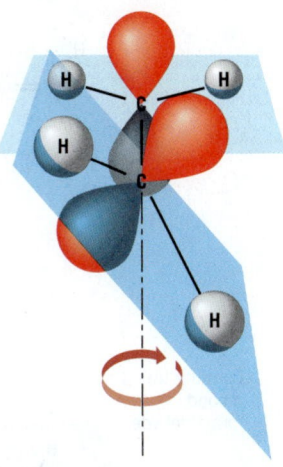

Figure 9.33 | Restricted rotation around a double bond. As the CH_2 group closest to us rotates relative to the one at the rear, the unhybridized p orbitals become misaligned, as shown here. This destroys the overlap and breaks the π bond. Bond breaking requires a lot of energy, more than is available to the molecule through the normal bending and stretching of its bonds at room temperature. Because of this, the rotation around the double bond axis is hindered or "restricted."

■ Restricted rotation around double bonds affects the properties of organic and biochemical molecules.

■ The two filled sp^2 hybrid orbitals on the oxygen become lone pairs on the oxygen atom in the molecule.

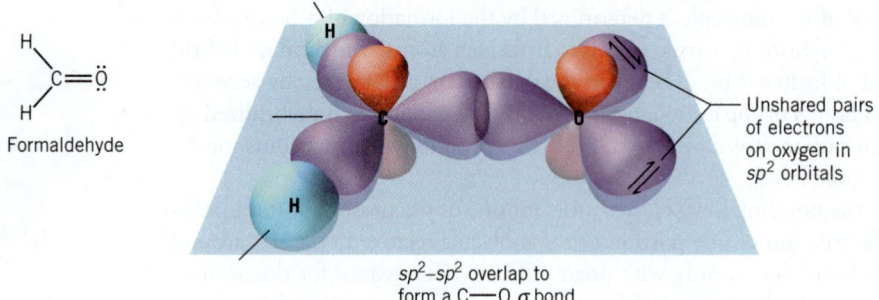

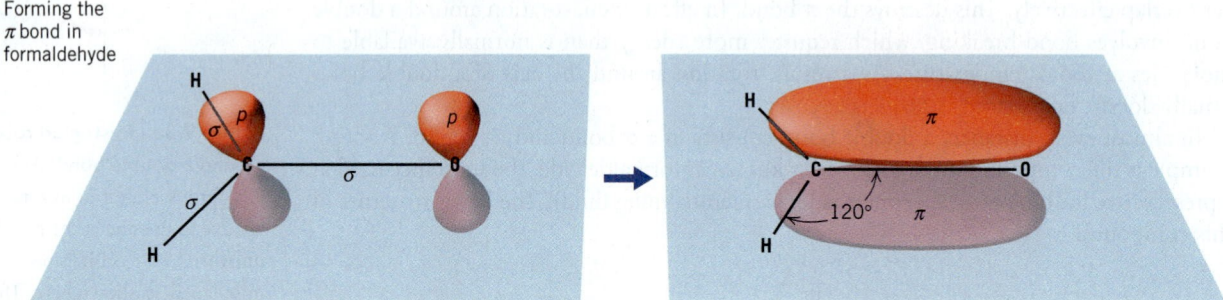

Figure 9.34 | **Bonding in formaldehyde.** The carbon–oxygen double bond consists of a σ bond and a π bond. The σ bond is formed by the overlap of sp^2 hybrid orbitals. The overlap of unhybridized p orbitals on the two atoms gives the π bond.

■ We label the orbitals p_x, p_y, and p_z for identification only; they are really equivalent.

Triple Bonds

An example of a molecule containing a triple bond is ethyne, also known as acetylene, C_2H_2 (a gas used as a fuel for welding torches).

$$H\text{—}C\equiv C\text{—}H$$
ethyne
(acetylene)

In the linear ethyne molecule, each carbon needs two hybrid orbitals to form two σ bonds—one to a hydrogen atom and one to the other carbon atom. These can be provided by mixing the $2s$ and one of the $2p$ orbitals to form sp hybrids. To help us visualize the bonding, we will imagine that there is an xyz coordinate system centered at each carbon atom and that it is the $2p_z$ orbital that becomes mixed in the hybrid orbitals.

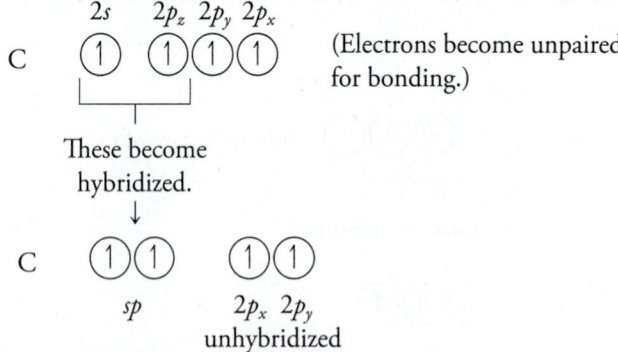

Figure 9.35 shows how the bonds in the molecule are formed. The sp orbitals point in opposite directions and are used to form the σ bonds. The unhybridized $2p_x$ and $2p_y$ orbitals are perpendicular to the C—C bond axis and overlap sideways to form two separate π bonds that surround the C—C sigma bond. Notice that we now have three pairs of electrons in three bonds—one σ bond and two π bonds—whose electron densities are concentrated in different places. Also notice that the use of sp hybrid orbitals for the σ bonds allows us to explain the linear arrangement of atoms in the molecule.

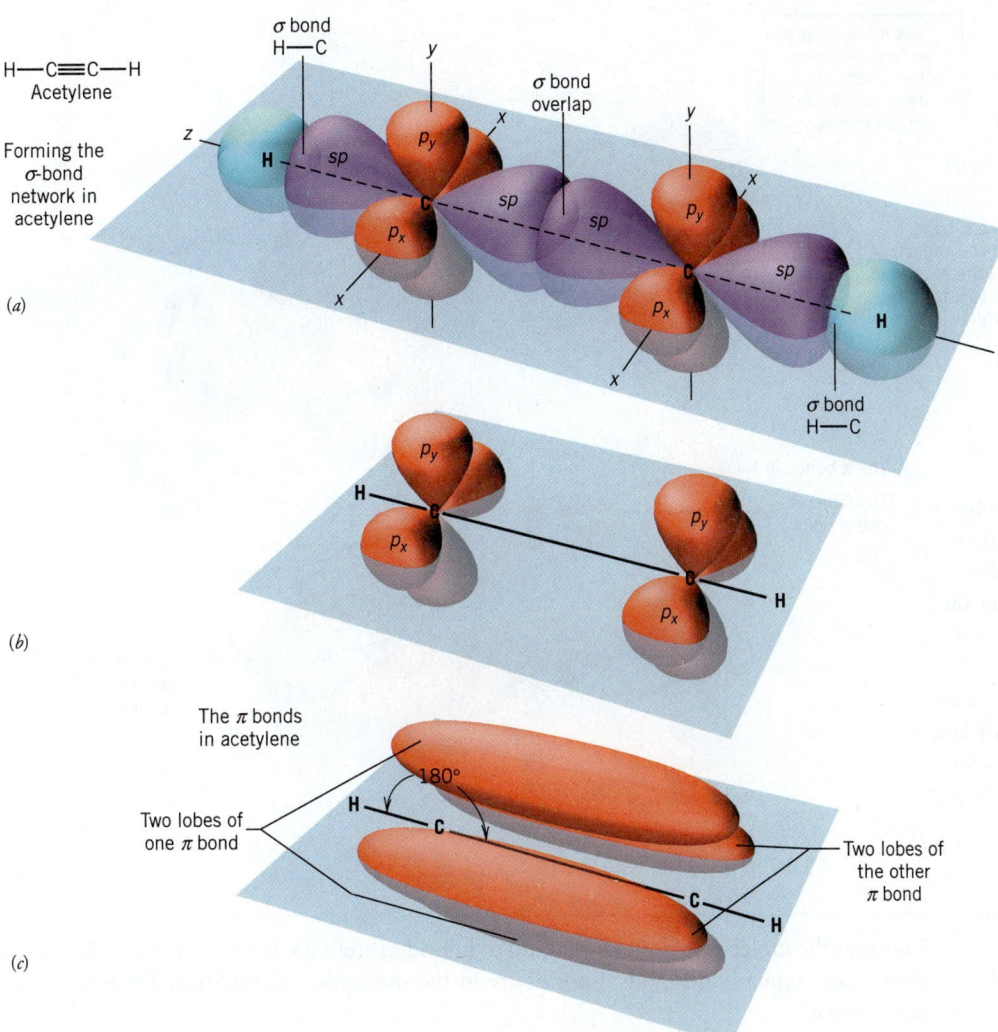

(a)

(b)

(c)

Figure 9.35 | The carbon–carbon triple bond in ethyne. (*a*) The *sp* hybrid orbitals on the carbon atoms are used to form sigma bonds to the hydrogen atoms and to each other. This accounts for one of the three bonds between the carbon atoms. (*b*) Sideways overlap of unhybridized $2p_x$ and $2p_y$ orbitals of the carbon atoms produces two π bonds. (*c*) The two π bonds in acetylene after they've formed surround the σ bond.

Similar descriptions can be used to explain the bonding in other molecules that have triple bonds. Figure 9.36, for example, shows how the nitrogen molecule, N_2, is formed. In it, too, the triple bond is composed of one σ bond and two π bonds.

Sigma Bonds and Molecular Structure

In the preceding discussions we examined several polyatomic molecules that contain both single and multiple bonds. The following observations will be helpful in applying the valence bond theory to a variety of similar molecules.

A Brief Summary

1. The basic molecular framework of a molecule is determined by the arrangement of its sigma (σ) bonds.
2. Hybrid orbitals are used by an atom to form its sigma bonds and to hold lone pairs of electrons.
3. The number of hybrid orbitals needed by an atom in a structure equals the number of atoms to which it is bonded plus the number of lone pairs of electrons in its valence shell.
4. When there is a double bond in a molecule, it consists of one σ bond and one π bond.
5. When there is a triple bond in a molecule, it consists of one σ bond and two π bonds.

Sigma and pi bonds and hybridization

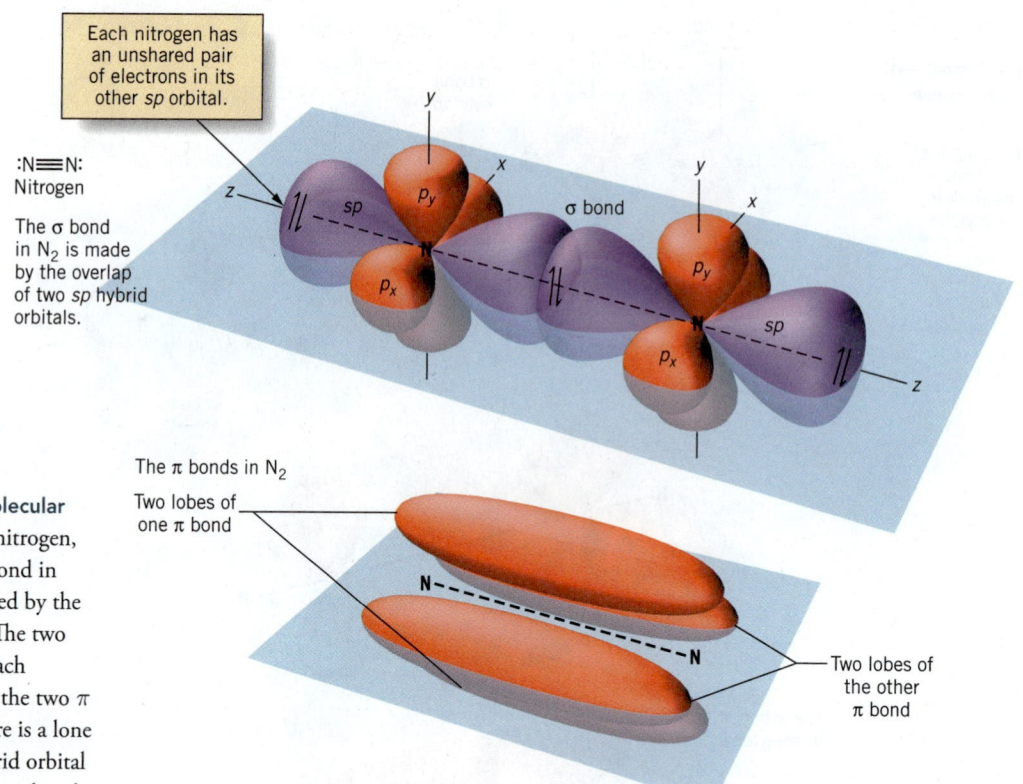

Each nitrogen has an unshared pair of electrons in its other *sp* orbital.

:N≡N:
Nitrogen

The σ bond in N₂ is made by the overlap of two *sp* hybrid orbitals.

The π bonds in N₂
Two lobes of one π bond

Two lobes of the other π bond

Figure 9.36 | **Bonding in molecular nitrogen.** The triple bond in nitrogen, N₂, is formed like the triple bond in ethyne. A sigma bond is formed by the overlap of *sp* hybrid orbitals. The two unhybridized 2*p* orbitals on each nitrogen atom overlap to give the two π bonds. On each nitrogen, there is a lone pair of electrons in the *sp* hybrid orbital that's not used to form the sigma bond.

Practice Exercise 9.20

Consider the molecule below. What kind of hybrid orbitals are used by atoms 1, 2, and 3? How many sigma bonds and pi bonds are in the molecule? (*Hint:* Study the brief summary above.)

Practice Exercise 9.21

Consider the molecule below. What kind of hybrid orbitals are used by atoms 1, 2, and 3? How many σ bonds and π bonds are in the molecule?

9.7 | Molecular Orbital Theory Basics

Molecular orbital theory takes the view that molecules and atoms are alike in one important respect: Both have energy levels that correspond to various orbitals that can be populated by electrons. In atoms, these orbitals are called atomic orbitals; in molecules, they are called **molecular orbitals**. (We shall frequently call them MOs.)

In most cases, the actual shapes and energies of molecular orbitals cannot be determined exactly. Nevertheless, reasonably good estimates of their shapes and energies can be obtained by combining the electron waves corresponding to the atomic orbitals of the atoms that make up the molecule. In forming molecular orbitals, these waves interact by constructive and destructive interference just like other waves we've seen in Chapter 7. Their intensities are either added or subtracted when the atomic orbitals overlap.

Formation of Molecular Orbitals from Atomic Orbitals

In Chapter 7 you learned that an atomic orbital is represented mathematically by a wave function, ψ, and that the square of the wave function, ψ^2, describes the distribution of electron density around the nucleus. In MO theory, molecular orbitals are also represented by wave functions, ψ_{MO}, that when squared describe the distribution of electron density around the entire set of nuclei that make up the molecule.

One of the properties of wave functions in general is they can have positive or negative algebraic signs in different regions of space. Let's look at two examples: the wave functions for a $1s$ orbital and for a $2p$ orbital. The $1s$ orbital has a positive sign in all regions, and we can indicate this as shown in Figure 9.37. The wave function for a $2p$ orbital, on the other hand, has a positive sign for one of its two lobes and a negative sign for the other. The signs of the wave functions become important when the orbitals combine to form molecular orbitals.

Theoreticians have found that they can obtain reasonable estimates of MO wave functions by mathematically adding and subtracting the wave functions for a pair of overlapping atomic orbitals. Adding the wave function of the atomic orbitals gives one ψ_{MO}, and subtracting them gives another ψ_{MO}. This gives two molecular orbitals with special properties, as illustrated in Figure 9.38 for two overlapping $1s$ orbitals centered on different nuclei. (For subtraction, the sign of one of the wave functions of the orbitals is changed before being added to the other, which is equivalent to subtracting one from the other.) The vertical scale in Figure 9.38 is energy and illustrates that when the wave functions of atomic orbitals are combined by addition, the resulting MO has a lower energy than the two $1s$ atomic orbitals. This is because by adding the wave function of the atomic orbitals, ψ_{MO} (and ψ^2) becomes large in regions between the nuclei where the atomic orbitals overlap. Stated another way, adding the wave function of atomic orbitals produces a molecular orbital wave function that shows that electron density is concentrated between the nuclei and leads to a lowering of the energy. This type of molecular orbital helps to stabilize a molecule when occupied by electrons and is called a **bonding molecular orbital**.

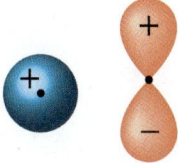

1s orbital   2p orbital

Figure 9.37 | **Algebraic signs of the wave functions for 1s and 2p atomic orbitals.** The wave function for a $1s$ orbital is positive everywhere. The wave function for a $2p$ orbital is positive for one lobe and negative for the other.

■ The number of MOs formed is always equal to the number of atomic orbitals that are combined.

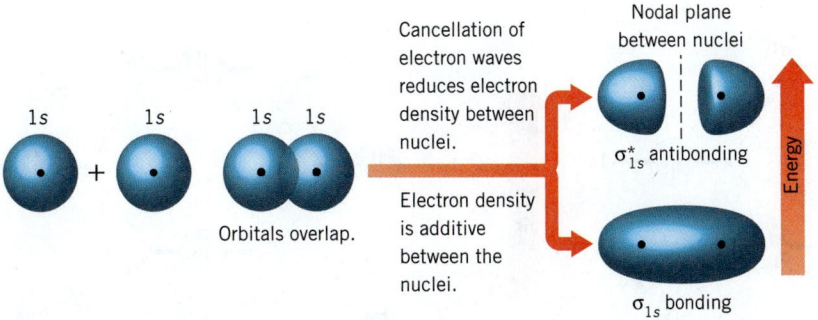

Figure 9.38 | **Combining 1s atomic orbitals to produce bonding and antibonding molecular orbitals.** These are σ-type orbitals because the electron density is concentrated along the imaginary line that passes through both nuclei. The antibonding orbital has a nodal plane between the nuclei where ψ_{MO}, ψ_{MO}^2, and the electron density drop to zero.

In Figure 9.38 we also see what happens when the wave function of one $1s$ orbital is subtracted from the other $1s$ orbital. This also yields a molecular orbital wave function, but the magnitude of ψ_{MO} becomes smaller between the nuclei where the atomic orbitals overlap. In fact, halfway between the nuclei the atomic orbital wave functions have the same magnitude but opposite sign, so when they are combined the resulting ψ_{MO} has a value of zero. As a result, between the two nuclei there is a nodal plane on which ψ_{MO} is equal to zero. On this plane ψ^2 is zero, and the electron density drops to zero. Notice that this MO, which is called an **antibonding molecular orbital**, has a higher energy than either of the two $1s$ atomic orbitals. This is because the electron density is reduced between the nuclei, causing the internuclear repulsions to outweigh their attractions for the electron density that remains between the nuclei. Antibonding molecular orbitals tend to destabilize a molecule when occupied by electrons.

Another thing to notice in Figure 9.38 is that both MOs have their maximum electron density on an imaginary line that passes through the two nuclei, giving them properties of sigma bonds. MOs like this are also designated as sigma (σ), with a subscript showing which atomic orbitals make up the MO and an asterisk indicating which is anti-bonding. Thus, the bonding and antibonding MOs formed by overlap of $1s$ orbitals are symbolized as σ_{1s} and σ^*_{1s}, respectively.

It is always true that bonding MOs are lower in energy than antibonding MOs formed from the same atomic orbitals, as shown in Figure 9.38. When electrons populate molecular orbitals, they fill the lower energy, bonding MOs first. The rules that apply to filling MOs are the same as those for filling atomic orbitals: *Electrons spread out over molecular orbitals of equal energy (Hund's rule), and two electrons can occupy the same orbital only if their spins are paired.* When filling the MOs, we also have to be sure we've accounted for all of the valence electrons of the separate atoms.

Why H₂ Exists but He₂ Does Not

Figure 9.39*a* is an MO energy level diagram for H_2. The energies of the separate $1s$ atomic orbitals are indicated at the left and right; those of the molecular orbitals are shown in the center. The H_2 molecule has two electrons, and both can be placed in the σ_{1s} orbital. The shape of this bonding orbital, shown in Figure 9.38, should be familiar. It's the same as the shape of the electron cloud that we described using the valence bond theory.

Next, let's consider what happens when two helium atoms come together. Why can't a stable molecule of He_2 be formed? Figure 9.39*b* is the energy diagram for He_2. Notice that both bonding and antibonding orbitals are filled. In situations such as this there is a net destabilization because the antibonding MO is raised in energy more than the bonding MO is lowered, relative to the orbitals of the separated atoms. This means the total energy of He_2 is larger than that of two separate He atoms, so the "molecule" is unstable and immediately comes apart.

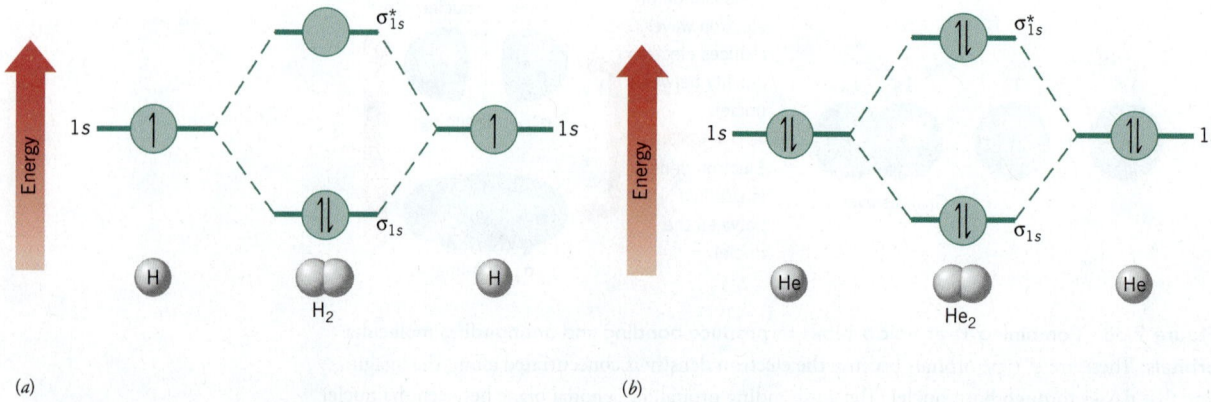

Figure 9.39 | Molecular orbital descriptions of H₂ and He₂. (*a*) Molecular orbital energy level diagram for H_2. (*b*) Molecular orbital energy level diagram for He_2.

In general, the effects of **antibonding electrons** (those in antibonding MOs) cancel the effects of an equal number of **bonding electrons**, and molecules with equal numbers of bonding and antibonding electrons are unstable. If we remove an antibonding electron from He_2 to give He_2^+, there is a net excess of bonding electrons, and the ion should be capable of existence. In fact, the emission spectrum of He_2^+ can be observed when an electric discharge is passed through a helium-filled tube, which shows that He_2^+ is present during the electric discharge. However, the ion is not very stable and cannot be isolated.

Bond Order

The concept of bond order was introduced in Section 8.7, where it was defined as the number of pairs of electrons shared between two atoms. To translate the MO description into these terms, we compute the bond order as follows:

$$\text{bond order} = \frac{(\text{number of bonding } e^-) - (\text{number of antibonding } e^-)}{2 \text{ electrons/bond}}$$

TOOLS

Calculating bond order in MO theory

For the H_2 molecule, we have

$$\text{bond order} = \frac{2 - 0}{2} = 1$$

A bond order of 1 corresponds to a single bond. For He_2 we have

$$\text{bond order} = \frac{2 - 2}{2} = 0$$

A bond order of zero means there is no bond, so the He_2 molecule is unable to exist. For the He_2^+ ion, which is able to form, the calculated bond order is

$$\text{bond order} = \frac{2 - 1}{2} = 0.5$$

This shows us that the bond order does not have to be a whole number. In this case, it indicates a bond character equivalent to about half a bond.

MO Description of Homonuclear Diatomic Molecules of Period 2

A **homonuclear diatomic molecule** is one in which both atoms are of the same element. Examples are N_2 and O_2, in which both elements are found in Period 2. As you have learned, the outer shell of a Period 2 element consists of $2s$ and $2p$ subshells. When atoms of this period bond to each other, the atomic orbitals of these subshells interact strongly to produce molecular orbitals. The $2s$ orbitals, for example, overlap to form σ_{2s} and σ^*_{2s} molecular orbitals having essentially the same shapes as the σ_{1s} and σ^*_{1s} MOs, respectively. Figure 9.40 shows the shapes of the bonding and antibonding MOs produced when the $2p$ orbitals overlap. If we label those that point toward each other as $2p_z$, a set of bonding and antibonding MOs are formed that we can label as σ_{2p_z} and $\sigma^*_{2p_z}$. The $2p_x$ and $2p_y$ orbitals, which are perpendicular to the $2p_z$ orbitals, overlap sideways to give π-type molecular orbitals. They are labeled π_{2p_x} and $\pi^*_{2p_x}$ and π_{2p_y} and $\pi^*_{2p_y}$, respectively. Notice in Figure 9.40 that we have indicated the algebraic signs of the $2p$ orbital wave functions. When the $2p$ orbitals overlap with the same sign, bonding MOs are formed, and when the signs are reversed, antibonding MOs are formed. In the bonding MOs, electron density is increased in the region between the two nuclei, whereas in antibonding MOs, electron density is reduced between the nuclei.

The approximate relative energies of the MOs formed from the second shell atomic orbitals are shown in Figure 9.41. Notice that from Li to N, the energies of the π_{2p_x} and π_{2p_y} orbitals are lower than the energy of the σ_{2p_z}. Then from O to Ne, the energies of the two levels are reversed. The reasons for this are complex and beyond the scope of this book.

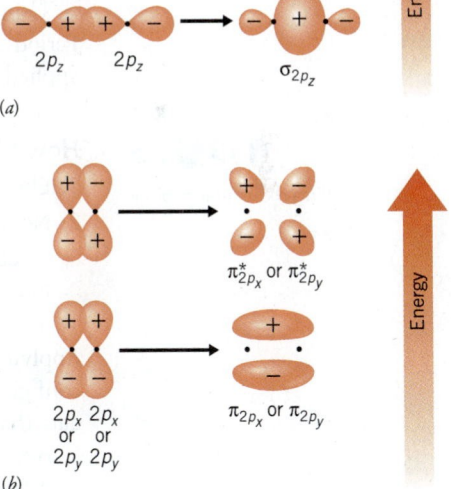

Figure 9.40 | **Formation of molecular orbitals by the overlap of p orbitals.** Where overlapping parts of the $2p$ orbitals have the same algebraic sign, a bonding MO is formed; where the signs are opposite, an antibonding MO is formed. (a) Two $2p_z$ orbitals that point at each other combine to give bonding and antibonding σ-type MOs. (b) Perpendicular to the $2p_z$ orbitals are $2p_x$ and $2p_y$ orbitals that overlap to give two sets of bonding and antibonding π-type MOs.

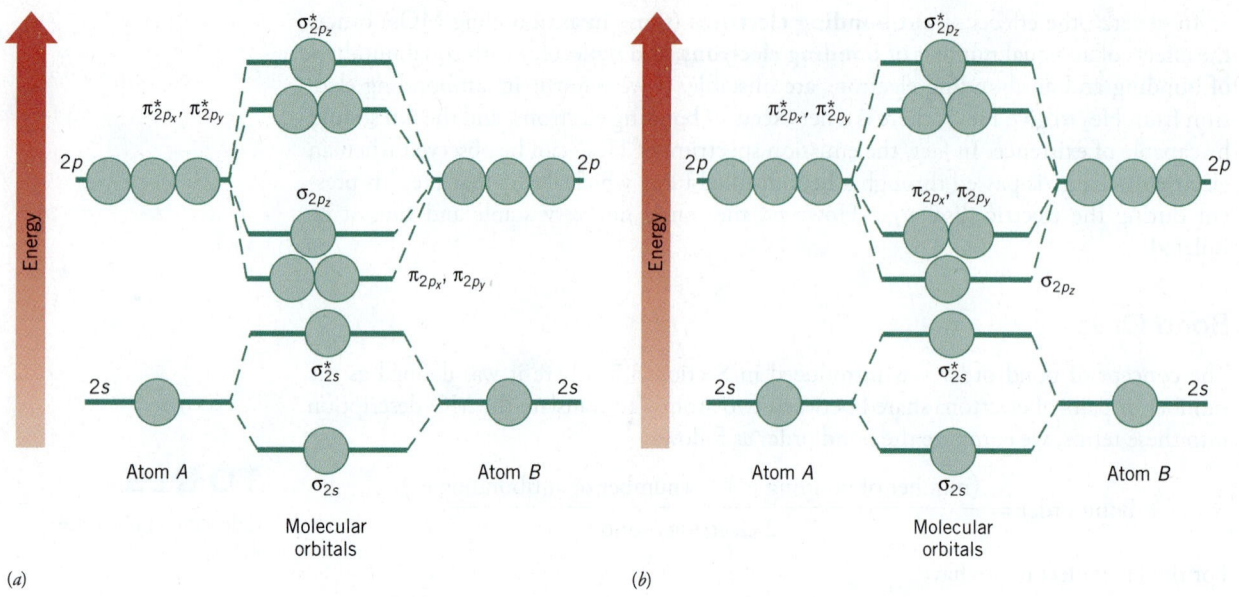

(a)

(b)

Figure 9.41 | **Approximate relative energies of molecular orbitals in Period 2 diatomic molecules.** (*a*) Li_2 through N_2. (*b*) O_2 through Ne_2.

Using Figure 9.41, we can predict the electronic structures of diatomic molecules of Period 2. These MO *electron configurations* are obtained using the same rules that are applied to the filling of atomic orbitals in atoms, as noted earlier.

Filling molecular orbitals

How Electrons Fill Molecular Orbitals

1. Electrons fill the lowest-energy orbitals that are available.
2. No more than two electrons, with spins paired, can occupy any orbital.
3. Electrons spread out as much as possible, with spins unpaired, over orbitals that have the same energy.

Applying these rules to the valence electrons of Period 2 atoms gives the MO electron configurations shown in Table 9.1. Let's see how well MO theory performs by examining data that are available for these molecules.

According to Table 9.1, MO theory predicts that molecules of Be_2 and Ne_2 should not exist at all because they have bond orders of zero. In beryllium vapor and in gaseous neon, no evidence of Be_2 or Ne_2 has ever been found. MO theory also predicts that diatomic molecules of the other Period 2 elements should exist because they all have bond orders greater than zero. These molecules have, in fact, been observed. Although lithium, boron, and carbon are complex solids under ordinary conditions, they can be vaporized. In the vapor, molecules of Li_2, B_2, and C_2 can be detected. Nitrogen, oxygen, and fluorine, as you know, are gaseous elements that exist as N_2, O_2, and F_2.

In Table 9.1, we also see that the predicted bond order increases from boron to carbon to nitrogen and then decreases from nitrogen to oxygen to fluorine. As the bond order increases, the *net* number of bonding electrons increases, so the bonds should become stronger and the bond lengths shorter. The *experimentally measured* bond energies and bond lengths given in Table 9.1 agree with these predictions quite nicely.

Molecular orbital theory is particularly successful in explaining the electronic structure of the oxygen molecule. Experiments show that O_2 is paramagnetic (it's weakly attracted to a magnet) and that the molecule contains two unpaired electrons. In addition, the bond length in O_2 is about what is expected for an oxygen–oxygen double bond. These data cannot be explained by valence bond theory. For example, if we write a

TABLE 9.1 Molecular Orbital Populations and Bond Orders for Period Two Diatomic Molecules[a]

Left panel (Energy increasing upward):

Orbital	Li₂	Be₂	B₂	C₂	N₂
$\sigma^*_{2p_z}$	○	○	○	○	○
$\pi^*_{2p_x}, \pi^*_{2p_y}$	○○	○○	○○	○○	○○
σ_{2p_z}	○	○	○	○	↑↓
π_{2p_x}, π_{2p_y}	○○	○○	↑ ↑	↑↓ ↑↓	↑↓ ↑↓
σ^*_{2s}	○	↑↓	↑↓	↑↓	↑↓
σ_{2s}	↑↓	↑↓	↑↓	↑↓	↑↓

Right panel (Energy increasing upward):

Orbital	O₂	F₂	Ne₂
$\sigma^*_{2p_z}$	○	○	↑↓
$\pi^*_{2p_x}, \pi^*_{2p_y}$	↑ ↑	↑↓ ↑↓	↑↓ ↑↓
π_{2p_x}, π_{2p_y}	↑↓ ↑↓	↑↓ ↑↓	↑↓ ↑↓
σ_{2p_z}	↑↓	↑↓	↑↓
σ^*_{2s}	↑↓	↑↓	↑↓
σ_{2s}	↑↓	↑↓	↑↓

	Li₂	Be₂	B₂	C₂	N₂	O₂	F₂	Ne₂
Number of Bonding Electrons	2	2	4	6	8	8	8	8
Number of Antibonding Electrons	0	2	2	2	2	4	6	8
Bond Order	1	0	1	2	3	2	1	0
Bond Energy (kJ/mol)	110	—	300	612	953	501	129	—
Bond Length (pm)	267	—	158	124	109	121	144	—

[a]Although the order of the energy levels corresponding to the σ_{2p_z}, and the π bonding MOs become reversed at oxygen, either sequence would yield the same result—a triple bond for N_2, a double bond for O_2, and a single bond for F_2.

Lewis structure for O_2 that shows a double bond and also obeys the octet rule, all the electrons appear in pairs.

$$:\ddot{O}::\ddot{O}: \qquad \text{(not acceptable based on experimental evidence because all electrons are paired)}$$

On the other hand, if we show the unpaired electrons, the structure has only a single bond and doesn't obey the octet rule.

$$:\ddot{\overset{\cdot}{O}}:\ddot{\overset{\cdot}{O}}: \qquad \text{(not acceptable based on experimental evidence because of the O—O single bond)}$$

With MO theory, we don't have any of these difficulties. By applying Hund's rule, the two electrons in the π^* orbitals of O_2 spread out over these orbitals with their spins unpaired because both orbitals have the same energy (see Table 9.1). The electrons in the two antibonding π^* orbitals cancel the effects of two electrons in the two bonding π orbitals, so the net bond order is 2 and the bond is effectively a double bond.

Some Simple Heteronuclear Diatomic Molecules

As molecules become more complex, the simple application of MO theory becomes much more difficult. This is because it is necessary to consider the relative energies of the individual atomic orbitals as well as the orientations of the orbitals relative to those on other atoms. Nevertheless, we can take a brief look at the MO descriptions of a couple of diatomic molecules to see what happens when both atoms in the molecule are not the same. Such molecules are said to be **heteronuclear**.

Hydrogen Fluoride

When we consider the possible interaction of the orbitals of different atoms to form molecular orbitals, the first factor we have to consider is the relative energies of the orbitals. This is because orbitals interact most effectively when they are of about equal energy; the greater the difference in energy between the orbitals, the less the orbitals interact, and the more the orbitals behave like simple atomic orbitals.

■ Although MO theory easily handles the bonding situations that VB theory has trouble with; MO theory loses the simplicity of VB theory. For even quite simple molecules, MO theory uses extensive calculations to make predictions.

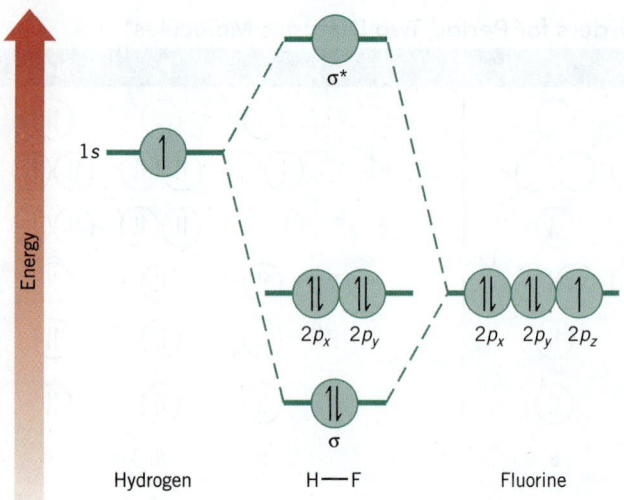

Figure 9.42 | **Molecular orbital energy diagram for HF.** Only the 1s orbital of hydrogen and the 2p orbitals of fluorine are shown.

In HF, the 1s orbital of hydrogen is higher in energy than either the 2s or 2p subshell of fluorine since the fluorine is more electronegative, but it is closest in energy to the 2p subshell (Figure 9.42). Taking the z axis as the internuclear axis, the hydrogen 1s orbital overlaps with the $2p_z$ orbital of fluorine to give bonding and antibonding σ-type orbitals, as illustrated in Figure 9.43. The $2p_x$ and $2p_y$ orbitals of fluorine, however, have no corresponding orbitals on hydrogen with which to interact, so they are unchanged when the molecule is formed. (This is shown in Figure 9.42 by representing the $2p_x$ and $2p_y$ orbitals as unattached circles at the same energy level they had before bonding occurred.) These two orbitals are said to be **nonbonding molecular orbitals** because they are neither bonding nor antibonding; they have no effect on the stability of the molecule, and do not contribute to the bond order calculation. The 1s orbital on the fluorine is not shown in the energy diagram because it is so low in energy, it does not participate in the bonding.

In the MO description of HF, we have a pair of electrons in the bonding MO formed by the overlap of the hydrogen 1s orbital with the fluorine $2p_z$ orbital. Earlier we saw that valence bond theory explains the bond in HF in the same way—as a pair of electrons shared between the hydrogen 1s orbital and a fluorine 2p orbital.

Carbon Monoxide

Carbon monoxide is a heteronuclear molecule in which both atoms are from Period 2, so we expect the orbitals of the second shell to be the ones used to form the MOs. The orbital overlaps are similar to those of the homonuclear diatomics of Period 2, and since it has the number and distribution of electrons as N_2, we will use the energy diagram resembling the one shown in Figure 9.41a for N_2.

Because the outer shell electrons of oxygen experience a larger effective nuclear charge than those of carbon, the oxygen orbitals will be somewhat lower in energy. This is shown in Figure 9.44. There are a total of 10 valence electrons (4 from carbon and 6 from oxygen) to distribute among the MOs of the molecule. When we do this, there are 8 bonding electrons and 2 antibonding electrons, so the net bond order is 3, corresponding to a triple bond. As expected, it consists of one σ bond and two π bonds.

Figure 9.43 | **Formation of σ and σ* orbitals in HF.** Once again, where overlapping atomic orbitals have the same sign, a bonding MO is formed; where the signs are opposite, an antibonding MO is formed. Notice that the antibonding σ* orbital has a nodal plane between the nuclei, which effectively removes electron density from the region between the nuclei and, if occupied, leads to destabilization of the molecule.

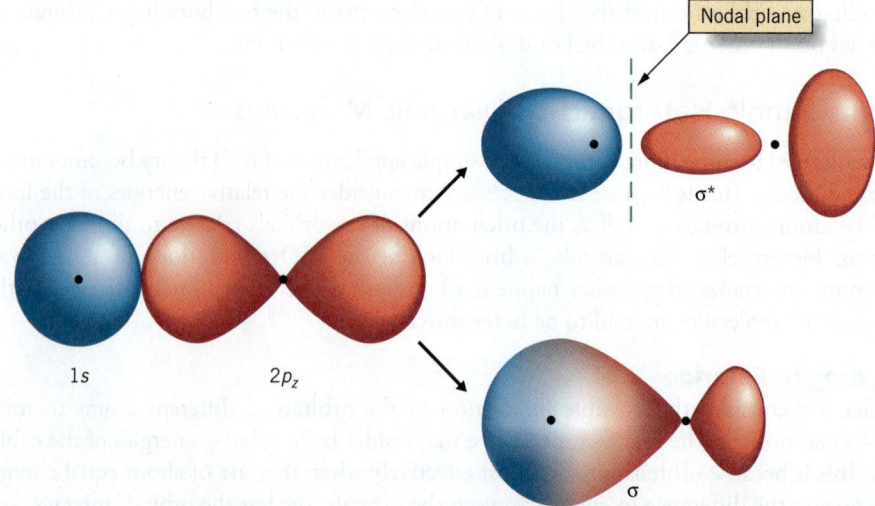

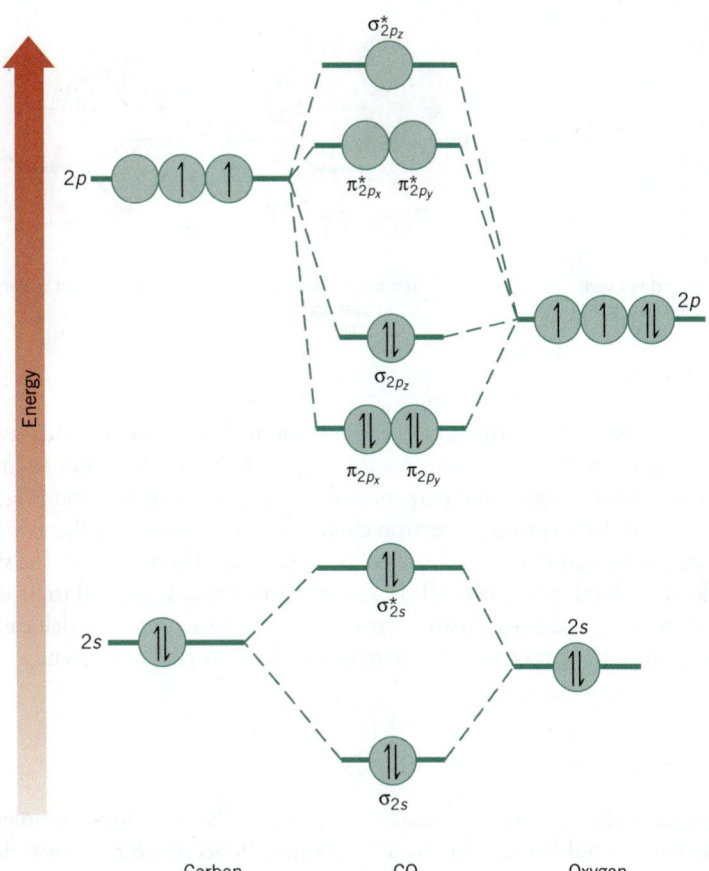

Figure 9.44 | **Molecular orbital energy diagram for carbon monoxide.** The oxygen orbitals are lower in energy than the corresponding carbon orbitals. The net bond order is 3.

The molecular orbital energy level diagram for the cyanide ion, CN^-, is similar to that of the earlier Period 2 homonuclear diatomics. Sketch the energy diagram for the ion and indicate the electron population of the MOs. What is the bond order in the ion? How does this agree with the bond order predicted from the Lewis structure of the ion? (*Hint:* How many valence electrons are there in the ion?)

Practice Exercise 9.22

The MO energy level diagram for the nitrogen monoxide molecule is essentially the same as that shown in Table 9.1 for O_2, except the oxygen orbitals are slightly lower in energy than the corresponding nitrogen orbitals. Sketch the energy diagram for nitrogen monoxide and indicate which MOs are populated. Calculate the bond order for the molecule. (*Hint:* Make adjustments to Figure 9.41*b*.)

Practice Exercise 9.23

9.8 | Delocalized Molecular Orbitals

One of the least satisfying aspects of the way valence bond theory explains chemical bonding is the need to write resonance structures for certain molecules and ions. For example, consider benzene, C_6H_6. As you learned earlier, this molecule has the shape of a ring whose resonance structures can be written as

The MO description of bonding in this molecule is as follows: The basic structure of the molecule is determined by the sigma-bond framework, which requires that the carbon

Figure 9.45 | Benzene. (*a*) The σ-bond framework. All atoms lie in the same plane. (*b*) The unhybridized *p* orbitals at each carbon prior to side-to-side overlap. (*c*) The double doughnut-shaped electron cloud formed by the delocalized π electrons.

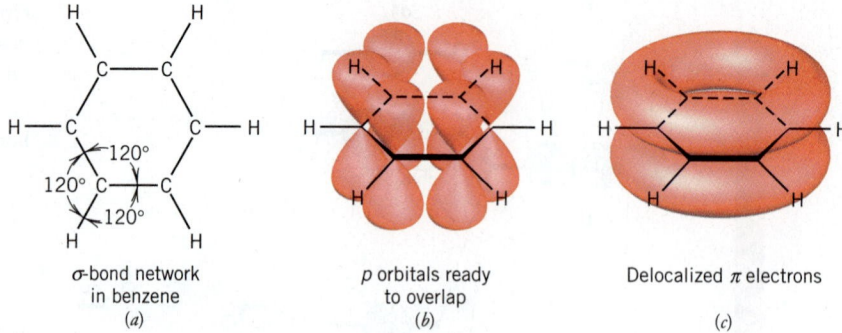

σ-bond network in benzene
(*a*)

p orbitals ready to overlap
(*b*)

Delocalized π electrons
(*c*)

atoms use sp^2 hybrid orbitals. This allows each carbon to form three σ bonds (two to other C atoms and one to an H atom) as illustrated in Figure 9.45*a*. Each carbon atom is left with a half-filled unhybridized *p* orbital perpendicular to the plane of the ring (Figure 9.45*b*). These *p* orbitals overlap to give a π-electron cloud that looks something like two doughnuts with the sigma-bond framework sandwiched between them (Figure 9.45*c*). The six electrons in this π-electron cloud spread over all six atoms in the ring and are said to be **delocalized** (i.e., they are not localized between any two atoms in the structure). The delocalized nature of the π electrons is the reason we usually represent the structure of benzene as

■ The terms **resonance energy**, **stabilization energy,** and **delocalization energy** describe the same physical property; they just come from different approaches to bonding theory.

One of the special characteristics of delocalized bonds is that they make a molecule or ion more stable than it would be if it had localized bonds. In Section 8.8 this was described in terms of resonance energy. In the molecular orbital theory, we no longer speak of resonance; instead, we refer to the electrons as **delocalized electrons**. The extra stability that is associated with this delocalization is therefore described, in the language of MO theory, as the **delocalization energy**.

Practice Exercise 9.24

The nitrate ion, NO_3^-, has three resonance structures. The electrons are delocalized over the bonds. What atomic orbitals are used to make up the delocalized molecular orbital?

9.9 | Bonding in Solids

Solids have some unique electrical properties that are familiar to everyone. For example, metals are good conductors of electricity, whereas nonmetallic substances are insulators; they are extremely poor electrical conductors. Between these extremes we find metalloids such as silicon and germanium, which are semiconductors; they are weak conductors of electricity. The theory developed to explain these widely differing properties is an extension of the principles of bonding discussed above and is called **band theory**.

According to band theory, an **energy band** in a solid is composed of a very large number of closely spaced energy levels that are formed by combining atomic orbitals of similar energy from each of the atoms within the substance. For example, in sodium the 1*s* atomic orbitals, one from each atom, combine to form a single 1*s* band. The number of energy levels in the band equals the number of 1*s* orbitals supplied by the entire collection of sodium atoms. The same thing occurs with the 2*s*, 2*p*, etc. orbitals, so that we also have 2*s*, 2*p*, etc. bands within the solid.

Figure 9.46 illustrates the energy bands in solid sodium. Notice that the electron density in the 1*s*, 2*s*, and 2*p* bands does not extend far from each individual nucleus, so these bands produce effectively localized energy levels in the solid. However, the 3*s* band (which is formed by overlap of the valence shell orbitals of the sodium atoms) is delocalized and

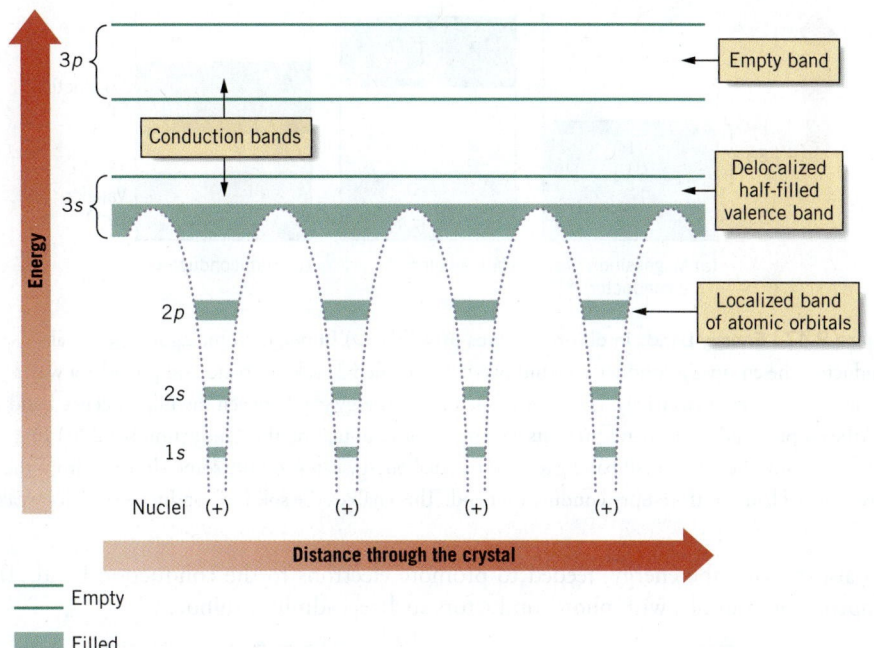

Figure 9.46 | **Energy bands in solid sodium.** The 1s, 2s, and 2p bands do not extend far in either direction from the nucleus, so the electrons in them are localized around each nucleus and can't move through the crystal. The delocalized 3s valence band extends throughout the entire solid, as do bands such as the 3p band formed from higher-energy orbitals of the sodium atoms.

extends continuously through the solid. The same applies to energy bands formed by higher-energy orbitals.

Sodium atoms have filled 1s, 2s, and 2p orbitals, so the corresponding bands in the solid are also filled. The 3s orbital of sodium, however, is only half-filled, which leads to a half-filled 3s band. The 3p and higher-energy bands in sodium are completely empty. When a voltage is applied across a piece of solid sodium, electrons in the half-filled 3s band can move from atom to atom with ease, and this allows sodium to conduct electricity well. However, electrons in the lower-energy filled bands are unable to move through the solid because orbitals on neighboring atoms are already filled and cannot accept an additional electron. Such electrons do not contribute to the conductivity of the solid.

We refer to the band containing the outer shell (valence shell) electrons as the **valence band**. Any band that is either vacant or partially filled and uninterrupted throughout the solid is called a **conduction band**, because electrons in it are able to move through the solid and thereby serve to carry electricity.

In metallic sodium the valence band and conduction band are the same, so sodium is a good conductor. In magnesium the 3s valence band is filled and therefore cannot be used to transport electrons. However, the vacant 3p conduction band actually overlaps the valence band and can easily be populated by electrons when voltage is applied (Figure 9.47a). This permits magnesium to be a conductor.

In an insulator such as glass, diamond, or rubber, all the valence electrons are used to form covalent bonds, so all the orbitals of the valence band are filled and cannot contribute to electrical conductivity. In addition, the energy separation, or **band gap**, between the filled valence band and the nearest conduction band (empty band) is large. As a result, electrons cannot populate the conduction band, so these substances are unable to conduct electricity (Figure 9.47b).

In a semiconductor such as silicon or germanium, the valence band is also filled, but the band gap between the filled valence band and the nearest conduction band is small (Figure 9.47c). At room temperature, thermal energy (kinetic energy associated with the temperature of the substance) possessed by the electrons is sufficient to promote some electrons to the conduction band and a small degree of electrical conductivity is observed. One of the interesting properties of semiconductors is that their electrical conductivity increases with increasing temperature. This is because as the temperature rises, the number of electrons with enough energy to populate the conduction band also increases. Photons

■ Recall that at any given temperature there is a distribution of kinetic energies among the particles of the substance. Thermal energy is kinetic energy a particle possesses because of its temperature. The higher the temperature, the higher the thermal energy.

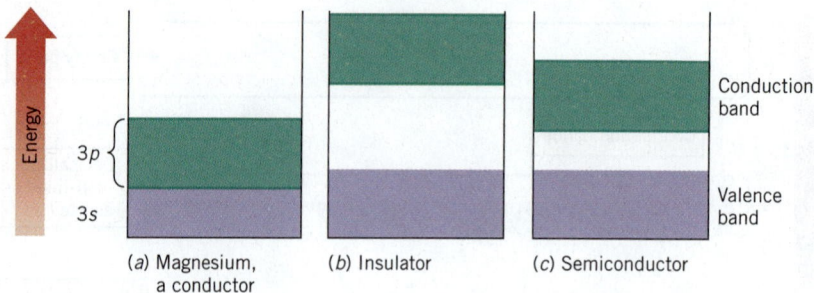

Figure 9.47 | **Energy bands in different types of solids.** (*a*) In magnesium, a good electrical conductor, the empty 3*p* conduction band overlaps the filled 3*s* valence band and provides a way for this metal to conduct electricity. (*b*) In an insulator, the energy gap between the filled valence band and the empty conduction band prevents electrons from populating the conduction band. (*c*) In a semiconductor, there is a small band gap, and thermal energy can promote some electrons from the filled valence band to the empty conduction band. This enables the solid to conduct electricity weakly.

can also provide the energy needed to promote electrons to the conduction band. This happens, for example, with photoconductors such as cadmium sulfide.

Practice Exercise 9.25

For the metal calcium, which orbitals on the metal form the valence band and which orbitals form the conduction band? (*Hint:* What are the highest energy filled orbitals?)

Practice Exercise 9.26

Arrange the following elements in order of increasing band gap: germanium, sulfur, and magnesium.

9.10 | Bonding of the Allotropes of the Elements

You learned earlier that the formation of σ bonds involves the overlap of *s* orbitals, or an end-to-end overlap of *p* orbitals or hybrid orbitals (Figure 9.30). Formation of a π bond normally requires the sideways overlap of unhybridized *p* orbitals (Figure 9.31).[5]

The strengths of *s* and *p* bonds depend on the sizes of the atoms involved—as atoms become larger, the bond strength generally decreases. In part, this is because the shared electrons are farther from the nuclei and are less effective at attracting them. Pi bonds are especially affected by atomic size. Small atoms, such as those in Period 2 of the periodic table, are able to approach each other closely. As a result, effective sideways overlap of their *p* orbitals can occur, and these atoms form strong π bonds. Atoms from Periods 3 through 6 are much larger and π-type overlap between their *p* orbitals is relatively ineffective, so π bonds formed by large atoms are relatively weak compared to σ bonds.

The effect of atomic size on the relative strengths of σ and π bonds leads to the following generalization:

TOOLS

Tendency to form multiple bonds

Elements in Period 2 are able to form multiple bonds fairly readily, while elements below them in Periods 3, 4, 5, and 6 have a tendency to prefer single bonds.

The consequences of this generalization are particularly evident when we examine the complexity of the structures that occur for the elemental nonmetals.

Among the nonmetals, only the noble gases exist in nature as single atoms. All the others are found in more complex forms in their free states—some as diatomic molecules and the rest in more complex molecular structures.

---

[5] Pi bonds can also be formed by *d* orbitals, but we do not discuss them in this book.

Nonmetals in Period 2

Fluorine (as well as the other halogens in Group 7A) has seven electrons in its outer shell and needs just one more electron to complete its valence shell. Fluorine is able to form the simple diatomic molecule F_2 with a single σ bond. The other halogens have similar formulas in their elemental states (Cl_2, Br_2, and I_2). Because only single bonds are involved, atomic size doesn't affect the complexity of the molecular structures of the halogens.

Moving to the left in the periodic table, the situation becomes more interesting for the elements in Groups 6A, 5A, and 4A. Oxygen and nitrogen have six and five electrons, respectively, in their valence shells. This means that an oxygen atom needs two electrons to complete its valence shell, and a nitrogen atom needs three. Oxygen and nitrogen, because of their small size, are capable of multiple bonding because they are able to form strong π bonds. This allows them to form a sufficient number of bonds with just a single neighbor to complete their valence shells, so they are able to form diatomic molecules.

The nitrogen molecule, which we discussed earlier, is able to complete its valence shell by forming a triple bond. Although a perfectly satisfactory Lewis structure for O_2 can't be drawn, experimental evidence suggests that the oxygen molecule does possess a double bond. Molecular orbital theory, which provides an excellent explanation of the bonding in O_2, also tells us that there is a double bond in the O_2 molecule.

Oxygen, in addition to forming the stable species O_2 (properly named dioxygen), also can exist in another very reactive molecular form called **ozone**, which has the formula O_3. The structure of ozone can be represented as a resonance hybrid,

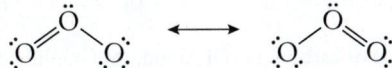

This unstable molecule can be generated by the passage of an electric discharge, or spark, through ordinary O_2. The pungent odor of ozone can often be detected in the vicinity of sparks from high-voltage wires, electrical motors, and electric trains. Near the ground, it is a pollutant and is harmful to living tissue. Ozone is also formed in limited amounts in the upper atmosphere by the action of ultraviolet radiation from the sun on O_2. The presence of ozone in the upper atmosphere shields life on Earth from exposure to intense and harmful ultraviolet light from the sun.

■ Ozone is also a component of photochemical smog formed by the interaction of sunlight with nitrogen oxides released in the exhaust of motor vehicles.

Existence of an element in more than one form, either as the result of differences in molecular structure, as with O_2 and O_3, or as the result of differences in the packing of molecules in the solid, is a phenomenon called **allotropy**. The different forms of the element are called **allotropes**. Thus, O_2 is one allotrope of oxygen and O_3 is another. Allotropy is not limited to oxygen, as you will soon see.

Allotropes of Carbon

An atom of carbon, also a Period 2 element, has four electrons in its valence shell, so it must share four electrons to complete its octet. There is no way for carbon to form a quadruple bond, so a simple C_2 species is not stable under ordinary conditions. Instead, carbon completes its octet in other ways, leading to four allotropic forms of the element. One of these is **diamond**, in which each carbon atom uses sp^3 hybrid orbitals to form covalent bonds to four other carbon atoms at the corners of a tetrahedron (Figure 9.48a).

In its other allotropes, carbon employs sp^2 hybrid orbitals to form ring structures with delocalized π systems covering their surfaces. The most stable form of carbon is **graphite**, which consists of layers of carbon atoms, each composed of many hexagonal "benzene-like" rings fused together in a structure reminiscent of chicken wire. A single such layer is called **graphene** (see On the Cutting Edge 9.1). In graphene, a $2p$ orbital on each carbon atom forms a p-type of interaction with identical $2p$ orbitals on its neighbors.

■ The 2010 Nobel Prize in physics was awarded to Andre Geim and Konstantin Novoselov, both of the University of Manchester, England, for their discovery that individual layers of graphene could be isolated from graphite.

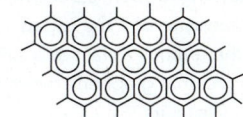

A fragment of a graphene layer in graphite.

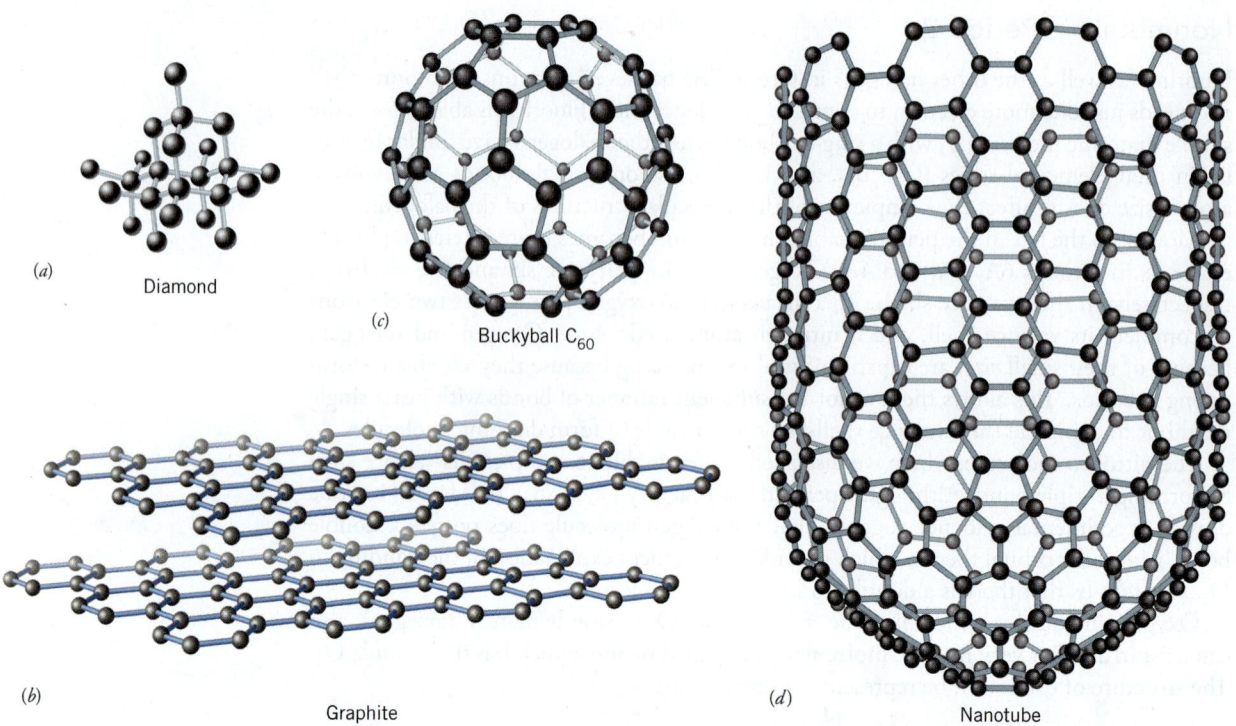

Figure 9.48 | **Molecular forms of carbon.** (*a*) Diamond. (*b*) Graphite. (*c*) Buckminsterfullerene, or "buckyball," C_{60}. (*d*) A portion of a carbon nanotube showing one closed end.

ON THE CUTTING EDGE (9.1)

Graphene and the Future of Electronics

It has long been suspected that single layers of graphite, given the name graphene (Figure 1), would have surprising and potentially useful properties. However, difficulties in preparing graphene sheets

Figure 1. An illustration depicting a single graphene layer.

by peeling them off crystals of graphite have prevented scientists from obtaining useful samples of the material, making accurate measurements impossible. These techniques yielded only small fragments of graphene.

Recently, scientists at IBM have developed a low-cost method of growing high-quality graphene on the surface of commercially available silicon carbide (SiC) wafers. Heating the wafers causes silicon to evaporate from the surface, and the carbon atoms left behind combine to form a graphene sheet.

Single sheets of graphene can transport electrons more quickly than other semiconductors, a property called electron mobility. The electron mobility of graphene, for example, is about 100 times greater than that of silicon, the semiconductor used in most currently available devices. For this reason, graphene is ideally suited for high-speed atomic scale operation. Furthermore, graphene's electrical properties can be controlled by switching it among conducting, nonconducting, and semiconducting states. This makes graphene a potential candidate for a variety of electronic devices.

According to scientists at IBM, they have already created a graphene-based transistor with the capability of operating at speeds of 100 GHz, about three times as fast as a silicon-based processor. Such discoveries suggest that graphene may someday replace silicon in high-speed electronic circuits.

In graphite, graphene layers are stacked one on top of another, as shown in Figure 9.48*b*. Graphite is an electrical conductor because of the delocalized π electron system that extends *along* the layers. Electrons can be pumped in at one end of a layer and removed from the other. Interestingly, graphite is a nonconductor if you try to pump electrons *through* the layers of graphite (from top to bottom of Figure 9.48*b*). Scientists have been able to isolate separate single layers of graphene, and there is much current interest in the physical and electrical properties of this carbon structure. For example, the breaking strength of graphene is more than 200 times that of steel!

In 1985, a new form of carbon was discovered that consists of tiny balls of carbon atoms, the simplest of which has the formula C_{60} (Figure 9.48*c*). They were named **fullerenes** and the C_{60} molecule itself was named **buckminsterfullerene** (nickname **buckyball**) in honor of R. Buckminster Fuller, the designer of a type of structure called a geodesic dome. The bonds between carbon atoms in the buckyball are arranged in a pattern of five- and six-membered rings arranged like the seams in a classic soccer ball, as well as the structural elements of the geodesic dome.

Carbon nanotubes, discovered in 1991, are another form of carbon that is related to the fullerenes. They are formed, along with fullerenes, when an electric arc is passed between carbon electrodes. The nanotubes consist of tubular carbon molecules that we can visualize as rolled-up sheets of graphene (with hexagonal rings of carbon atoms). The tubes are capped at each end with half of a spherical fullerene molecule, so a short tube would have a shape like a hot dog. A portion of a carbon nanotube is illustrated in Figure 9.48*d*. Carbon nanotubes have unusual properties that have made them the focus of much research in recent years.

■ Weight for weight, carbon nanotubes are about 100 times stronger than stainless steel and about 40 times stronger than the carbon fibers used to make tennis rackets and shafts for golf clubs.

Nonmetallic Elements below Period 2

In graphite, carbon exhibits multiple bonding, as do nitrogen and oxygen in their molecular forms. As we noted earlier, their ability to do this reflects their ability to form strong π bonds—a requirement for the formation of a double or triple bond. When we move to the third and successive periods, a different state of affairs exists. There, we have much larger atoms that are able to form relatively strong σ bonds but much weaker π bonds. Because their π bonds are so weak, these elements prefer single bonds (σ bonds), and the molecular structures of the free elements reflect this.

Elements of Group 6A

Below oxygen in Group 6A is sulfur, which has the Lewis symbol

$$: \overset{\displaystyle .}{\underset{\displaystyle ..}{S}} \cdot$$

A sulfur atom requires two electrons to complete its valence shell, so it must form two covalent bonds. However, sulfur doesn't form π bonds to other sulfur atoms; instead, it prefers to form two stronger single bonds to different sulfur atoms. Each of these also prefers to bond to two different sulfur atoms, and this gives rise to a

$$-\overset{\displaystyle ..}{\underset{\displaystyle ..}{S}}-\overset{\displaystyle ..}{\underset{\displaystyle ..}{S}}-\overset{\displaystyle ..}{\underset{\displaystyle ..}{S}}-\overset{\displaystyle ..}{\underset{\displaystyle ..}{S}}-\overset{\displaystyle ..}{\underset{\displaystyle ..}{S}}-$$

sequence. Actually, in sulfur's most stable form, called **orthorhombic sulfur**, the sulfur atoms are arranged in an eight-member ring to give a molecule with the formula S_8 (properly named cyclooctasulfur). The S_8 ring has a puckered crown-like shape, which is illustrated in Figure 9.49. Another allotrope is **monoclinic sulfur**, which also contains S_8 rings that are arranged in a slightly different crystal structure.

When solid sulfur is heated, it undergoes some interesting changes (Figure 9.50). When the solid first melts, the liquid that forms consists of S_8 molecules. As the liquid is heated further, the rings begin to open into chains of sulfur atoms that join to form long strands. The strands of sulfur atoms become intertwined, causing the liquid to become very viscous. Further heating causes the strands of sulfur to break into smaller fragments, and when the liquid boils it is no longer viscous.

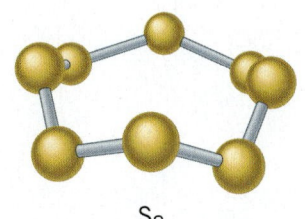

S_8

Figure 9.49 | The structure of the puckered S_8 ring.

Figure 9.50 | **Changes in sulfur as it is heated.** (*a*) Crystalline sulfur. (*b*) Molten sulfur just above its melting point. (*c*) Molten sulfur, just below 200 °C, is dark and viscous. (*d*) Boiling sulfur is dark red and no longer viscous.

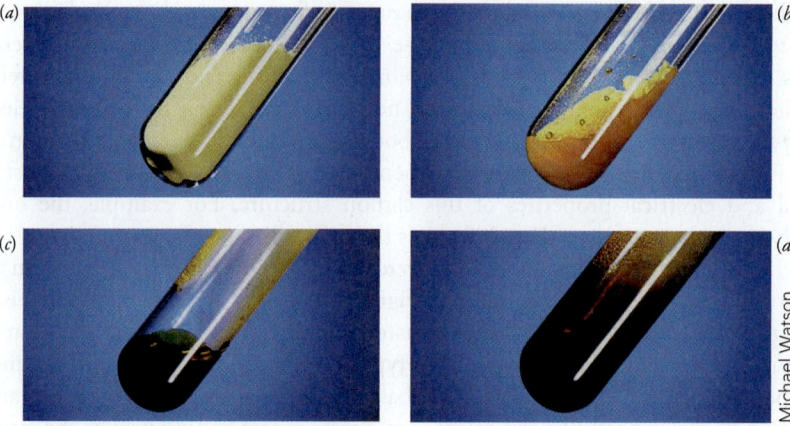

Selenium, below sulfur in Group 6A, also forms Se_8 rings in one of its allotropic forms. Both selenium and tellurium also can exist in a gray form in which there are long Se_x and Te_x chains (where the subscript x is a large number).

Elements of Group 5A

Like nitrogen, the other elements in Group 5A all have five valence electrons. Phosphorus is an example:

To achieve a noble gas structure, the phosphorus atom must acquire three more electrons. Because there is little tendency for phosphorus to form multiple bonds, as nitrogen does when it forms N_2, the octet is completed by the formation of three single bonds to three different phosphorus atoms.

The simplest elemental form of phosphorus is a wax-like solid called white phosphorus because of its appearance (Figure 9.51). It consists of P_4 molecules in which each phosphorus atom lies at a corner of a tetrahedron, as illustrated in Figure 9.52. Notice that in this structure each phosphorus is bound to three others. This allotrope of phosphorus is very reactive, partly because of the very small P—P—P bond angle of 60°. At this small angle, the *p* orbitals of the phosphorus atoms don't overlap very well, so the bonds are weak. As a result, breaking a P—P bond occurs easily. When a P_4 molecule reacts, this bond breaking is the first step, so P_4 molecules are readily attacked by other chemicals, especially oxygen. White phosphorus is so reactive toward oxygen that it ignites and burns spontaneously in air. For this reason, white phosphorus is used in military incendiary devices, and you've probably seen movies in which exploding phosphorus shells produce arching showers of smoking particles.

A second allotrope of phosphorus that is much less reactive is called **red phosphorus**. At the present time, its structure is unknown, although it has been suggested that it contains P_4 tetrahedra joined at the corners as shown in Figure 9.53. Red phosphorus is also

Figure 9.51 | **The two main allotropes of phosphorus.** (*a*) White phosphorus. (*b*) Red phosphorus.

■ The preferred angle between bonds formed by *p* orbitals is 90°. Each face of the P_4 tetrahedron is a triangle with 60° angles. This produces less overlap between the *p* orbitals in the bonds.

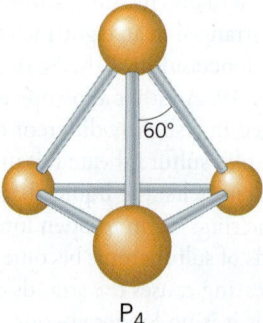

Figure 9.52 | **The molecular structure of white phosphorus, P_4.** The bond angles of 60° makes the phosphorus–phosphorus bonds quite weak and causes the molecule to be very reactive.

P_4

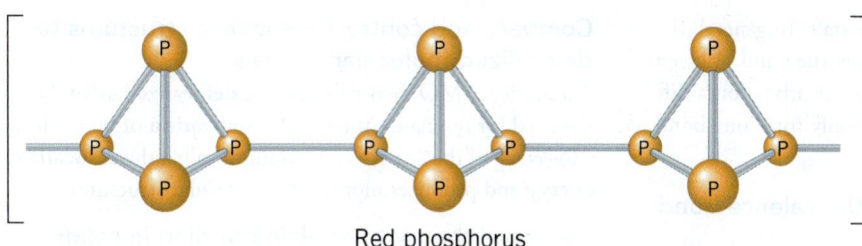

Red phosphorus

Figure 9.53 | Proposed molecular structure of red phosphorus. Red phosphorus is believed to be composed of long chains of P_4 tetrahedra connected at their corners.

used in explosives and fireworks, and it is mixed with fine sand and used on the striking surfaces of matchbooks. As a match is drawn across the surface, friction ignites the phosphorus, which then ignites the ingredients in the tip of the match.

A third allotrope of phosphorus is called **black phosphorus**, which is formed by heating white phosphorus at very high pressures. This variety has a layered structure in which each phosphorus atom in a layer is covalently bonded to three others in the same layer. As in graphite, these layers are stacked one atop another, with only weak forces between the layers. As you might expect, black phosphorus has many similarities to graphite.

The elements arsenic and antimony, which are just below phosphorus in Group 5A, are also able to form somewhat unstable yellow allotropic forms containing As_4 and Sb_4 molecules, but their most stable forms have a metallic appearance with structures similar to black phosphorus.

Elements of Group 4A

Finally, we look at silicon and germanium, the heavier nonmetallic elements in Group 4A. To complete their octets, each must form four covalent bonds. Unlike carbon, however, they have very little tendency to form multiple bonds, so they don't form allotropes that have a graphite structure. Instead, each of them forms a solid with a structure similar to diamond.

What is the hybridization of the carbon atom in (a) diamond, (b) graphite, and (c) buckyballs? (*Hint:* How many bonding domains does the carbon atom have in each substance?)

Practice Exercise 9.27

|Summary

Organized by Learning Objective

Draw diagrams of the five basic molecular geometries, including bond angles

The structures of most molecules can be described in terms of one or another of five basic geometries: **linear, planar triangular, tetrahedral, trigonal bipyramidal,** and **octahedral**.

Predict the shape of a molecule or ion using the VSEPR model

The **VSEPR theory** predicts molecular geometry by assuming that **electron domains**—regions of space that contain bonding electrons, unpaired valence electrons, or lone pairs—stay as far apart as possible from each other, while staying as close as possible to the central atom. Figures 9.4, 9.5, 9.7, and 9.10 illustrate the structures obtained with different numbers of groups of electrons in the valence shell of the central atom in a molecule or ion and with different numbers of lone pairs and attached atoms. The correct shape of a molecule or polyatomic

ion can usually be predicted by applying the VSEPR model to the Lewis structure.

Explain how the geometry of a molecule affects its polarity

A molecule that contains identical atoms attached to a central atom will be nonpolar if there are no lone pairs of electrons in the central atom's valence shell. It will be polar if lone pairs are present, except in two cases: (1) when there are three lone pairs and two attached atoms, and (2) when there are two lone pairs and four attached atoms. If all the atoms attached to the central atom are not alike, the molecule will usually be polar.

Describe how the valence bond theory views bond formation

According to VB theory, a covalent bond is formed between two atoms when an atomic orbital on one atom **overlaps** with an atomic orbital on the other and a pair of electrons with paired spins is shared between the overlapping orbitals. In general, the better the overlap of the orbitals, the stronger the bond. A given atomic orbital can overlap only with one other orbital on a different atom, so a given atomic orbital can only form one bond with an orbital on one other atom.

Explain how hybridization refines the valence bond theory of bonding

Hybrid orbitals are formed by mixing pure *s*, *p*, and *d* orbitals. Hybrid orbitals overlap better with other orbitals than the pure atomic orbitals from which they are formed, so bonds formed by hybrid orbitals are stronger than those formed by ordinary atomic orbitals. Sigma bonds (σ bonds) are formed by the following kinds of orbital overlap: *s–s*, *s–p*, end-to-end *p–p*, and overlap of hybrid orbitals. **Sigma bonds** allow free rotation around the bond axis. The side-by-side overlap of *p* orbitals produces a **pi bond** (π bond). Pi bonds do not permit free rotation around the bond axis because such a rotation involves bond breaking. In complex molecules, the basic molecular framework is built with σ bonds.

Describe the nature of multiple bonds using orbital diagrams and hybridization

A double bond consists of one σ bond and one π bond. A triple bond consists of one σ bond and two π bonds.

Use molecular orbitals to explain the bonding of simple diatomic molecules

MO theory begins with the supposition that molecules are similar to atoms, except that they have more than one positive center. They are treated as collections of nuclei and electrons, with the electrons of the molecule distributed among **molecular orbitals** of different energies. Molecular orbitals can spread over two or more nuclei, and can be considered to be formed by the constructive and destructive interference of the overlapping electron waves corresponding to the atomic orbitals of the atoms in the molecule. **Bonding MOs** concentrate electron density between nuclei; **antibonding MOs** remove electron density from between nuclei. **Nonbonding MOs** do not affect the energy of the molecule. The rules for the filling of MOs are the same as those for atomic orbitals.

Compare and contrast resonance structures to delocalized molecular orbitals

The ability of MO theory to describe **delocalized orbitals** avoids the need for resonance theory. **Delocalization of bonds** leads to a lowering of the energy by an amount called the **delocalization energy** and produces more stable molecular structures.

Use band theory to explain bonding in solids and physical properties

In solids, atomic orbitals of the atoms combine to yield **energy bands** that consist of many energy levels. The **valence band** is formed by orbitals of the valence shells of the atoms. A **conduction band** is a partially filled or empty band. In an **electrical conductor** the conduction band is either partially filled or is empty and overlaps a filled band. In an insulator, the **band gap** between the filled valence band and the empty conduction band is large, so no electrons populate the conduction band. In a **semiconductor**, the band gap between the filled valence band and the conduction band is small and thermal energy can promote some electrons to the conduction band.

Use hybridization and valence bond theory to explain allotropes of nonmetallic elements

Different forms of the same element are called **allotropes**. Oxygen exists in two allotropic forms: dioxygen (O_2) and **ozone** (O_3). Carbon forms several allotropes, including **diamond**, **graphite** (composed of layers of graphene), C_{60} molecules called **buckminsterfullerene** (one member of the fullerene family of structures), and **carbon nanotubes**. Sulfur forms S_8 molecules that can be arranged in two different allotropic forms. Phosphorus occurs as **white phosphorus** (P_4), **red phosphorus**, and **black phosphorus**. Silicon only forms a diamond-like structure.

Elements of Period 2, because of their small size, form strong π bonds. As a result, these elements easily participate in multiple bonding between like atoms, which accounts for diatomic molecules of O_2 and N_2, and the π-bonded structure of **graphene**. Elements of Periods 3, 4, and 5 are large and their *p* orbitals do not overlap well to form strong π bonds, so these elements prefer single σ bonds between like atoms, which leads to more complex molecular structures.

TOOLS

Tools for Problem Solving
The following tools were introduced in this chapter. Study them carefully so that you can select the appropriate tool when needed.

Basic molecular shapes (Section 9.1)
The five basic geometries are linear, planar triangular, tetrahedral, trigonal bipyramid, and octahedral.

VSEPR model (Section 9.2)
Electron groups repel each other and arrange themselves in the valence shell of an atom to yield minimum repulsions, which is what determines the shape of the molecule.

Using the VSEPR model to predict molecular geometry (Section 9.2)
To obtain a molecular structure, follow these steps: (1) draw the Lewis structure, (2) count electron domains, (3) select the basic geometry, (4) add atoms to bonding domains, and (5) ignore nonbonding domains to describe the shape.

Molecular shape and molecular polarity (Section 9.3)
We can use molecular shape to determine whether a molecule will be polar or nonpolar.

Valence bond theory criteria for bond formation (Section 9.4)
A bond requires overlap of two orbitals sharing two electrons with paired spins. Both orbitals can be half-filled, or one can be filled and the other empty. We use these criteria to establish which orbitals atoms use when bonds are formed.

Hybrid orbitals formed by *s* and *p* atomic orbitals (Section 9.5)
Two electron domains are formed by sp hybrid orbitals.
Three electron domains are formed by sp^2 hybrid orbitals.
Four electron domains are formed by sp^3 hybrid orbitals.
Five electron domains are formed by sp^3d hybrid orbitals.
Six electron domains are formed by sp^3d^2 hybrid orbitals.

The VSEPR model and hybrid orbitals involving *s* and *p* electrons (Section 9.5)
Lewis structures permit us to use the VSEPR model to predict molecular shape, which then allows us to select the correct hybrid orbitals for the valence bond description of bonding. We represent the central atom by *M*, the atoms attached to the central atom by *X*, and lone pairs by *E*. We can then signify the number of bonding and nonbonding domains around *M* as a formula MX_nE_m, where *n* is the number of bonding domains and *m* is the number of nonbonding domains. The following structures are obtained for two, three, and four domains.

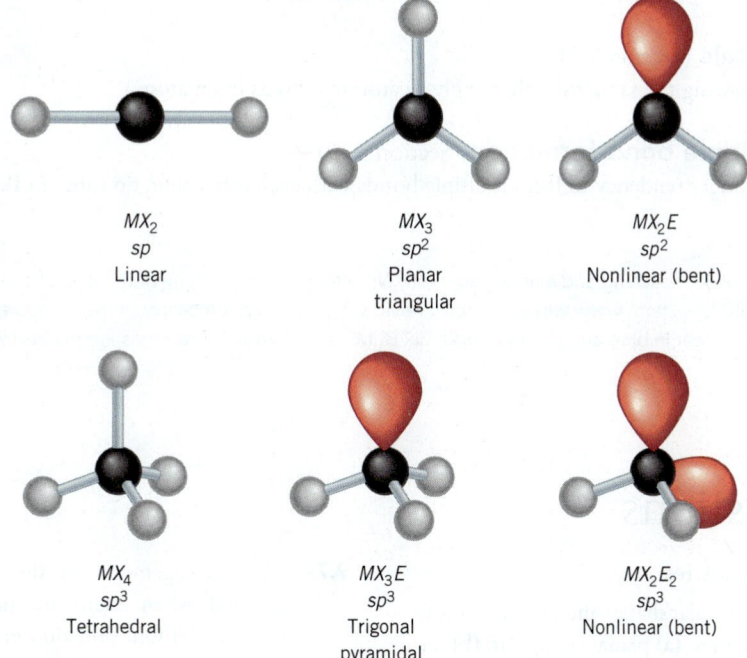

MX_2	MX_3	MX_2E
sp	sp^2	sp^2
Linear	Planar triangular	Nonlinear (bent)

MX_4	MX_3E	MX_2E_2
sp^3	sp^3	sp^3
Tetrahedral	Trigonal pyramidal	Nonlinear (bent)

The VSEPR model and hybrid orbitals involving *d* electrons (Section 9.5)
The following VSEPR structures are obtained for five and six domains using the generalized formula MX_nE_m.

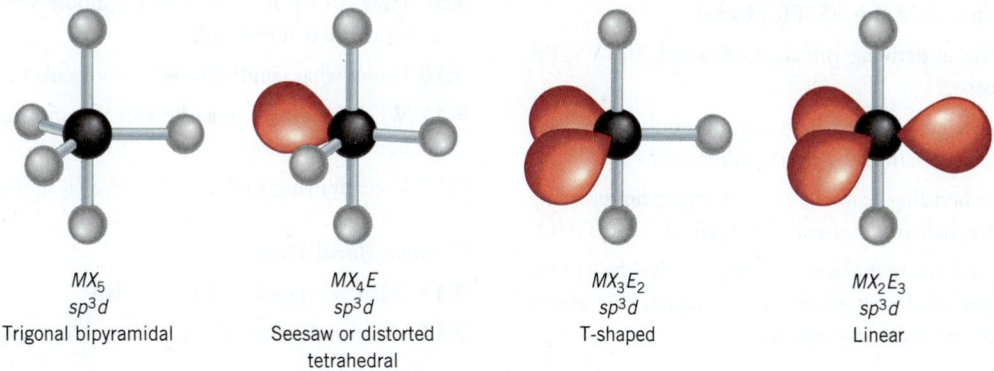

MX_5	MX_4E	MX_3E_2	MX_2E_3
sp^3d	sp^3d	sp^3d	sp^3d
Trigonal bipyramidal	Seesaw or distorted tetrahedral	T-shaped	Linear

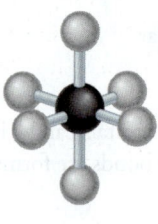

MX_6
sp^3d^2
Octahedral

MX_5E
sp^3d^2
Square pyramidal

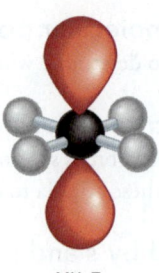

MX_4E_2
sp^3d^2
Square planar

Sigma (σ) and pi (π) bonds and hybridization (Section 9.6)

The Lewis structure for a polyatomic molecule allows us to apply these criteria to determine how many σ and π bonds are between atoms and the kind of hybrid orbitals each atom uses. The shape of the molecule is determined by the framework of σ bonds, with π bonds used for the double and triple bonds.

Calculating bond order in MO theory (Section 9.7)

$$\text{Bond order} = \frac{(\text{number of bonding } e^-) - (\text{number of antibonding } e^-)}{2 \text{ electrons/bond}}$$

Filling molecular orbitals (Section 9.7)

Electrons populate MOs following the same rules that apply to atomic orbitals in an atom.

Tendency toward multiple bond formation (Section 9.10)

Atoms of Period 2 have a stronger tendency to form multiple bonds with each other than do those in Periods 3 and below.

WileyPLUS = *WileyPLUS*, an online teaching and learning solution. *Note to instructors*: Many of the end-of-chapter problems are available for assignment via the WileyPLUS system. **www.wileyplus.com.** Review Problems are presented in pairs separated by blue rules. Answers to problems whose numbers appear in blue are given in Appendix B. More challenging problems are marked with an asterisk *.

| Review Questions

Five Basic Molecular Geometries

9.1 Sketch the following molecular shapes and give the various bond angles in the structures: **(a)** planar triangular, **(b)** tetrahedral, **(c)** octahedral.

9.2 Sketch the following molecular shapes and give the bond angles in the structures: **(a)** linear, **(b)** trigonal bipyramidal.

Molecular Shapes and the VSEPR Model

9.3 What is the underlying principle on which the VSEPR model is based?

9.4 What is an electron domain? How are nonbonding and double bonds described by electron domains?

9.5 How many bonding domains and how many nonbonding domains are there in a molecule of formaldehyde, HCHO?

9.6 Sketch the following molecular shapes and give the various bond angles in the structure: **(a)** T-shaped, **(b)** seesaw shaped, and **(c)** square pyramidal.

9.7 What arrangements of domains around an atom are expected when there are **(a)** three domains, **(b)** six domains, **(c)** four domains, or **(d)** five domains?

Molecular Structure and Dipole Moments

9.8 Why is it useful to know the polarities of molecules?

9.9 How do we indicate a bond dipole when we draw the structure of a molecule?

9.10 Under what conditions will a molecule be polar?

9.11 What condition must be met if a molecule having polar bonds is to be nonpolar?

9.12 Use a drawing to show why the SO_2 molecule is polar.

Valence Bond Theory

9.13 What is meant by orbital overlap?

9.14 How is orbital overlap related to bond energy?

9.15 Use sketches of orbitals to describe how VB theory would explain the formation of the H—Br bond in hydrogen bromide. How many lone pairs of electrons would be found on each?

Hybrid Orbitals and Molecular Geometry

9.16 Why do atoms usually use hybrid orbitals for bonding rather than pure atomic orbitals?

9.17 Sketch figures that illustrate the directional properties of the following hybrid orbitals: **(a)** sp, **(b)** sp^2, **(c)** sp^3.

9.18 Sketch figures that illustrate the directional properties of **(a)** sp^3d, **(b)** sp^3d^2.

9.19 Why do Period 2 elements never use sp^3d or sp^3d^2 hybrid orbitals for bond formation?

9.20 What relationship is there, if any, between Lewis structures and the valence bond descriptions of molecules?

9.21 How can the VSEPR model be used to predict the hybridization of an atom in a molecule?

9.22 If the central oxygen in the water molecule did not use sp^3 hybridized orbitals (or orbitals of any other kind of hybridization), what would be the expected bond angle in H_2O?

9.23 Using orbital diagrams, describe how sp^3 hybridization occurs in each atom: **(a)** carbon, **(b)** nitrogen, **(c)** oxygen. If these elements use sp^3 hybrid orbitals to form bonds, how many lone pairs of electrons would be found on each?

9.24 Sketch the way the orbitals overlap to form the bonds in each of the following: **(a)** CH_4, **(b)** NH_3, **(c)** H_2O. (Assume the central atom uses hybrid orbitals.)

9.25 We explained the bond angles of 107° in NH_3 by using sp^3 hybridization of the central nitrogen atom. If unhybridized p orbitals of nitrogen were used to overlap with $1s$ orbitals of each hydrogen, what would the H—N—H bond angles be? Explain.

9.26 Using sketches of orbitals and orbital diagrams, describe sp^2 hybridization of **(a)** boron and **(b)** carbon.

9.27 What two basic shapes have hybridizations that include d orbitals?

9.28 The ammonia molecule, NH_3, can combine with a hydrogen ion, H^+ (which has an empty $1s$ orbital), to form the ammonium ion, NH_4^+. (This is how ammonia can neutralize acid and therefore function as a base.) Sketch the geometry of the ammonium ion, indicating the bond angles.

9.29 How does the geometry around B and O change in the following reaction? How does the hybridization of each atom change?

$$\begin{array}{ccc} & H \quad\quad :\!\ddot{C}l\!: & \\ & | \quad\quad\quad | & \\ H\!-\!\ddot{O}\!: \;+\; & B\!-\!\ddot{C}l\!: & \longrightarrow \\ & | & \\ & :\!\ddot{C}l\!: & \end{array}$$

$$\begin{array}{c} H \;:\!\ddot{C}l\!: \\ |\quad\quad | \\ H\!-\!\ddot{O}\!-\!B\!-\!\ddot{C}l\!: \\ |\quad\quad| \\ :\!\ddot{C}l\!: \end{array}$$

Hybrid Orbitals and Multiple Bonds

9.30 How do σ and π bonds differ?

9.31 Why can free rotation occur easily around a σ-bond axis but not around a π-bond axis?

9.32 Using sketches, describe the bonds and bond angles in ethene, C_2H_4.

9.33 Sketch the way the bonds form in acetylene, C_2H_2.

9.34 How does VB theory treat the benzene molecule? (Draw sketches describing the orbital overlaps and the bond angles.)

Molecular Orbital Theory Basics

9.35 Why is the higher-energy MO in H_2 called an *antibonding* orbital? Make a sketch of the bonding and antibonding orbitals in H_2.

9.36 Below is an illustration showing two $3d$ orbitals about to overlap. The drawings also show the algebraic signs of the wave functions for both orbitals in this combination. Will this combination of orbitals produce a bonding or an antibonding MO? Sketch the shape of the MO.

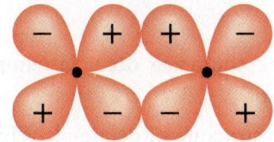

9.37 Will the combination of $3d$ orbitals in Question 9.36 yield a σ or π type of MO? Explain.

9.38 Explain why He_2 does not exist but H_2 does.

9.39 How does MO theory account for the paramagnetism of O_2?

9.40 On the basis of MO theory, explain why Li_2 molecules can exist but Be_2 molecules cannot. Could the ion Be_2^+ exist?

9.41 What relationship is there between bond order and bond energy?

9.42 Sketch the shapes of the π_{2p_y} and $\pi^*_{2p_y}$ MOs.

Valence Bond and Molecular Orbital Theories

9.43 What is the theoretical basis of both valence bond (VB) theory and molecular orbital (MO) theory?

9.44 What shortcomings of Lewis structures and VSEPR theory do VB and MO theories attempt to overcome?

9.45 What is the main difference in the way VB and MO theories view the bonds in a molecule?

Delocalized Molecular Orbitals

9.46 What is a delocalized MO? Explain, in terms of orbital overlap, why delocalized MOs are able to form in the benzene molecule.

9.47 What effect does delocalization have on the stability of the electronic structure of a molecule?

9.48 What is delocalization energy? How is it related to resonance energy?

Bonding in Solids

9.49 What is a conduction band? What is a valence band?

9.50 What is required to form a conduction band?

9.51 On the basis of the band theory of solids, how do conductors, insulators, and semiconductors differ?

9.52 Why does electrical conductivity in semiconductors increase with increasing temperature?

9.53 In calcium, why can't electrical conduction take place by movement of electrons through the $2s$ energy band? How does calcium conduct electricity?

Bonding in the Allotropes and of the Elements

9.54 What are *allotropes*? How do they differ from *isotopes*?

9.55 Why are the Period 2 elements able to form much stronger π bonds than the nonmetals of Period 3? Why does a Period 3 nonmetal prefer to form all σ bonds instead of one σ bond and several π bonds?

9.56 Even though the nonmetals of Periods 3, 4, and 5 do not tend to form π bonds between like atoms, each of the halogens is able to form diatomic molecules (Cl_2, Br_2, I_2). Why?

9.57 Which of the nonmetals occur in nature in the form of isolated atoms?

9.58 Describe the structure of diamond. What kind of hybrid orbitals does carbon use to form bonds in diamond? What is the geometry around carbon in this structure?

9.59 Describe the structure of graphene. What kind of hybrid orbitals does carbon use in the formation of the molecular framework of graphene?

9.60 How is the structure of graphite related to the structure of graphene?

9.61 Describe the C_{60} molecule. What is it called? What name is given to the series of similar substances?

9.62 How is the structure of a carbon nanotube related to the structure of graphene?

9.63 What is the molecular structure of silicon? Suggest a reason why silicon doesn't form an allotrope that's similar in structure to graphite.

9.64 Make a sketch that describes the molecular structure of white phosphorus.

9.65 What are the different allotropes of phosphorus?

9.66 What are the P—P—P bond angles in the P_4 molecule? If phosphorus uses p orbitals to form the phosphorus–phosphorus bonds, what bond angle would give the best orbital overlap? On the basis of your answers to these two questions, explain why P_4 is so chemically reactive.

9.67 What structure has been proposed for red phosphorus? How do the reactivities of red and white phosphorus compare?

9.68 What is the molecular structure of black phosphorus? In what way does the structure of black phosphorus resemble that of graphite?

9.69 What are the two allotropes of oxygen?

9.70 Draw the Lewis structure for O_3. Is the molecule linear, based on the VSEPR model? Assign formal charges to the atoms in the Lewis structure. Does this suggest the molecule is polar or nonpolar?

9.71 What beneficial function does ozone serve in earth's upper atmosphere?

9.72 What is the molecular structure of sulfur in its most stable allotropic form?

|Review Problems

Molecular Shapes and the VSEPR Model

9.73 Predict the shapes of (a) NH_2^-, (b) CO_3^{2-}, (c) IF_3, (d) Br_3^-, and (e) GaH_3.

9.74 Predict the shapes of (a) SF_3^+, (b) GeF_4, (c) SO_4^{2-}, (d) O_3, and (e) N_2O.

9.75 Predict the shapes of (a) FCl_2^+, (b) AsF_5, (c) AsF_3, (d) SbH_3, and (e) SeO_2.

9.76 Predict the shapes of (a) TeF_4, (b) $SbCl_6$, (c) NO_2, (d) PCl_4, and (e) PO_4^{3-}.

9.77 Predict the shapes of (a) IO_4^-, (b) IF_4^-, (c) TeF_6, (d) SiO_4 and (e) ICl_2.

9.78 Predict the shapes of (a) CS_2, (b) BrF_4^-, (c) ICl_3, (d) ClO_3^-, and (e) SeO_3.

9.79 Which of the following has a shape described by the figure below: (a) IO_4^-, (b) ICl_4^-, (c) $SnCl_4$, or (d) BrF_4^+?

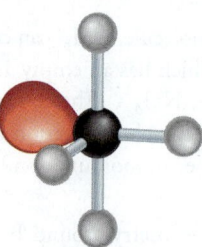

9.80 Which of the following has a shape described by the figure below: (a) BrF_3, (b) PF_3, (c) NO_3^-, or (d) SCl_3^-?

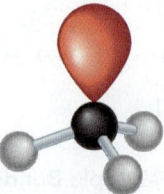

9.81 Ethene, also called ethylene, is a gas used to ripen tomatoes artificially. It has the Lewis structure:

$$\begin{array}{ccc} & H & H \\ & | & | \\ H-C&=&C-H \end{array}$$

What would you expect the H—C—H and H—C=C bond angles to be in this molecule? (*Caution:* Don't be fooled by the way the structure is drawn here.)

9.82 Ethyne, more commonly called acetylene, is a gas used in welding torches. It has the Lewis structure H—C≡C—H. What would you expect the H—C—C bond angle to be in this molecule?

9.83 Predict the bond angle for each of the following molecules: (a) Cl_2O, (b) H_2O, (c) SO_2, (d) HOCl, (e) SnF_2.

9.84 Predict the bond angle for each of the following ions: (a) I_3^-, (b) PH_2^-, (c) OCN^-, (d) PO_4^{3-}, (e) NH_2^-.

Molecular Structure and Dipole Moments

9.85 Which of the following molecules would be expected to be polar? (a) HBr, (b) $POCl_3$, (c) CH_2O, (d) $SnCl_4$, (e) $SbCl_5$.

9.86 Which of the following molecules would be expected to be polar? (a) PBr_3, (b) SO_3, (c) $AsCl_3$, (d) ClF_3, (e) BCl_3

9.87 Which of the following molecules or ions would be expected to have a net dipole moment? (a) ClNO, (b) XeF^+, (c) SeBr, (d) NO, (e) NO_2

9.88 Which of the following molecules or ions would be expected to have a net dipole moment? (a) H_2S, (b) BeH_2, (c) SCN^-, (d) CN^-, (e) $BrCl_3$

9.89 Explain why SF_6 is nonpolar, but SF_5Br is polar.

9.90 Explain why CH_3Cl is polar, but CCl_4 is not.

Valence Bond Theory

9.91 Use sketches of orbitals to show how VB theory explains the bonding in the Cl_2 molecule. Illustrate with appropriate orbital diagrams as well.

9.92 Hydrogen selenide is one of nature's most foul-smelling substances. Molecules of H_2Se have H—Se—H bond angles very close to 90°. How would VB theory explain the bonding in H_2Se? Use sketches of orbitals to show how the bonds are formed. Illustrate with appropriate orbital diagrams as well.

Hybrid Orbitals and Molecular Geometry

9.93 Use orbital diagrams to explain how the beryllium chloride molecule is formed. What kind of hybrid orbitals does beryllium use in this molecule?

9.94 Use orbital diagrams to describe the bonding in tin tetrachloride? What kind of hybrid orbitals does beryllium use in this molecule?

9.95 Use orbital diagrams to describe the bonding in antimony pentachloride, a substance used to add chlorine to carbon compounds. Be sure to indicate hybrid orbital formation.

9.96 Describe the bonding in tellurium hexafluoride, a toxic gas with an extremely unpleasant smell. Indicate the hybrid orbitals used by the tellurium.

9.97 Draw Lewis structures for the following and use the geometry predicted by the VSEPR model to determine what kind of hybrid orbitals the central atom uses in bond formation: (a) ClO_3^-, (b) SO_3, and (c) OF_2.

9.98 Draw Lewis structures for the following and use the geometry predicted by the VSEPR model to determine what kind of hybrid orbitals the central atom uses in bond formation: (a) $SbCl_6^-$, (b) PF_3, and (c) XeF_4.

9.99 Use the VSEPR model to help you describe the bonding in the following molecules according to VB theory: (a) arsenic trichloride and (b) chlorine trifluoride. Use orbital diagrams for the central atom to show how hybridization occurs.

9.100 Use the VSEPR model to help you describe the bonding in the following molecules according to VB theory: (a) antimony trichloride and (b) selenium dichloride. Use orbital diagrams for the central atom to show how hybridization occurs.

9.101 Use orbital diagrams to show that the bonding in SbF_6^- involves the formation of a coordinate covalent bond.

9.102 What kind of hybrid orbitals are used by tin in $SnCl_6^{2-}$? Draw the orbital diagram for Sn in $SnCl_6^{2-}$. What is the geometry of $SnCl_6^{2-}$?

Hybrid Orbitals and Multiple Bonds

*9.103 A nitrogen atom can undergo sp^2 hybridization when it becomes part of a carbon–nitrogen double bond, as in $H_2C=NH$.

    (a) Using a sketch, show the electron configuration of sp^2 hybridized nitrogen just before the overlapping occurs to make this double bond.

    (b) Using sketches (and the analogy to the double bond in C_2H_4), describe the two bonds of the carbon–nitrogen double bond.

    (c) Describe the geometry of $H_2C=NH$ (use a sketch to show all the expected bond angles.)

*9.104 A nitrogen atom can undergo sp hybridization and then become joined to carbon by a triple bond.

    (a) How many sigma and pi bonds make up the triple bond?

    (b) Draw the orbital diagram for the sp hybridization as it would look before any bonds are formed.

    (c) Draw sketches to show where the electron density of each bond in the triple bond is located.

    (d) What do you expect the bond angle for the HCN molecule to be?

9.105 Tetrachloroethylene, a common dry-cleaning solvent, has the formula C_2Cl_4. Its structure is

Use the VSEPR and VB theories to describe the bonding in this molecule. What are the expected bond angles in the molecule?

9.106 Phosgene, $COCl_2$, was used as a war gas during World War I. It reacts with moisture in the lungs of its victims to form CO_2 and gaseous HCl, which cause the lungs to fill with fluid. Phosgene is a simple molecule having the structure

Describe the bonding in this molecule using VSEPR and VB theory.

9.107 What kind of hybrid orbitals do the numbered atoms use in the following molecule?

9.108 What kind of hybrid orbitals do the numbered atoms use in the following molecule?

9.109 What kinds of bonds (σ, and π) are found in the numbered bonds in the following molecule?

|Additional Exercises

*****9.119** Construct the MO energy level diagram for the OH molecule assuming it is similar to that for HF. How many electrons are in **(a)** bonding MOs and **(b)** nonbonding MOs? What is the net bond order in the molecule?

*****9.120** If boron and nitrogen were to form a molecule with the formula BN, what would its MO energy level diagram look like, given that the energies of the $2p$ orbitals of nitrogen are lower than those of boron? If Figure 9.41*a*

9.110 What kinds of bonds (σ, π) are found in the numbered bonds in the following molecule?

Molecular Orbital Theory Basics

*****9.111** Construct the molecular orbital diagram for O_2. What is the net bond order in O_2?

*****9.112** Construct the molecular orbital diagram for N_2. What is the new bond order in N_2?

9.113 Use the MO energy diagram to predict **(a)** the bond orders for each molecule or ion, **(b)** which one has the longer bond length, and **(c)** which in each pair has the greater bond energy: (i) O_2 or O_2^+, (ii) O_2 or O^-.

9.114 Use the MO energy diagram to predict **(a)** the bond orders for each molecule or ion, **(b)** which one has the longer bond length, and **(c)** which in each pair has the greater bond energy: (i) C_2 or C_2^+, (ii) N_2 or N_2^+.

9.115 Assume that in the NO molecule the molecular orbital energy level sequence is similar to that for O_2. What happens to the NO bond length when an electron is removed from NO to give NO^+? How would the bond energy of NO compare to that of NO^+?

9.116 Assume that in the NO molecule the molecular orbital energy level sequence is similar to that for O_2. What happens to the NO bond length when an electron is added to NO to give NO^-? How would the bond energy of NO compare to that of NO^-?

9.117 Which of the following molecules or ions are paramagnetic? **(a)** O_2^+, **(b)** O_2, **(c)** O_2^-, **(d)** NO, **(e)** N_2

9.118 Which of the following molecules or ions are paramagnetic? **(a)** B_2, **(b)** C_2, **(c)** C_2^-, **(d)** C_2^{2-}, **(e)** F_2

applies, would the molecule be paramagnetic or diamagnetic? What is the net bond order in the molecule?

9.121 Formaldehyde has the Lewis structure

What would you predict its shape to be?

*9.122 The molecule XCl_3 is pyramidal. In which group in the periodic table is element X found? If the molecule were planar triangular, in which group would X be found? If the molecule were T-shaped, in which group would X be found? Why is it unlikely that element X is in Group 6A?

*9.123 Antimony forms a compound with hydrogen that is called stibine. Its formula is SbH_3 and the H—Sb—H bond angles are 91.3°. Which kinds of orbitals does Sb most likely use to form the Sb—H bonds: pure p orbitals or hybrid orbitals? Explain your reasoning.

9.124 Describe the changes in molecular geometry and hybridization that take place during the following reactions:

(a) $BF_3 + F^- \longrightarrow BF_4^-$

(b) $PCl_5 + Cl^- \longrightarrow PCl_4^-$

(c) $ICl_3 + Cl^- \longrightarrow ICl_4^-$

(d) $PCl_3 + Cl_2 \longrightarrow PCl_5$

(e) $C_2H_2 + H_2 \longrightarrow C_2H_4$

9.125 Which one of the following five diagrams best represents the structure of $BrCl_4^+$?

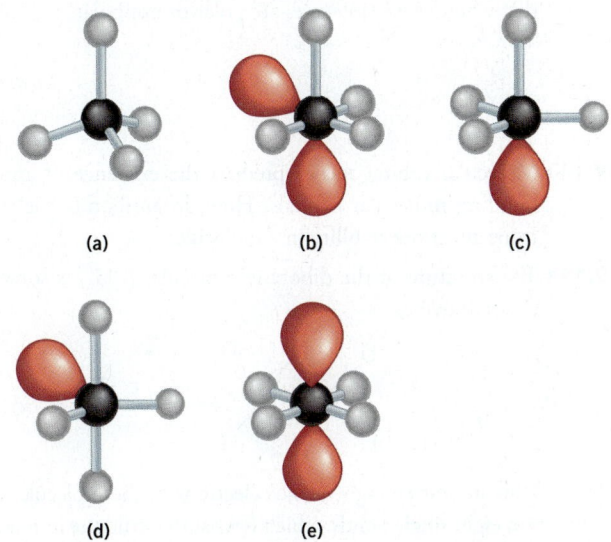

(a)   (b)   (c)

(d)   (e)

*9.126 Cyclopropane is a triangular molecule with C—C—C bond angles of 60°. Explain why the σ bonds joining carbon atoms in cyclopropane are weaker than the carbon–carbon σ bonds in the noncyclic propane.

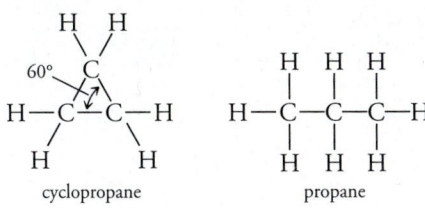

cyclopropane   propane

9.127 Phosphorus trifluoride, PF_3, has F—P—F bond angles of 97.8°. (a) How would VB theory use hybrid orbitals to explain these data? (b) How would VB theory use unhybridized orbitals to account for these data? (c) Do either of these models work very well?

9.128 A six-membered ring of carbons can hold a double bond but not a triple bond. Explain.

cyclohexene (exists)   cyclohexyne (unknown)

*9.129 *The more electronegative are the atoms bonded to the central atom, the less are the repulsions between the electron pairs in the bonds.* On the basis of this statement, predict the most probable structure for the molecule PCl_3F_2. Do we expect the molecule to be polar or nonpolar?

*9.130 *A lone pair of electrons in the valence shell of an atom has a larger effective volume than a bonding electron pair. Lone pairs therefore repel other electron pairs more strongly than do bonding pairs.* On the basis of these statements, describe how the bond angles in TeF_4 and BrF_5 deviate from those found in a trigonal bipyramid and an octahedron, respectively. Sketch the molecular shapes of TeF_4 and BrF_5 and indicate these deviations on your drawing.

*9.131 The two electron pairs in a double bond repel other electron pairs more than the single pair of electrons in a single bond. On the basis of this statement, which bond angles should be larger in SO_2Cl_2, the O—S—O bond angles or the Cl—S—Cl bond angles? In the molecule, sulfur is bonded to two oxygen atoms and two chlorine atoms. (*Hint:* Assign formal charges and work with the best Lewis structure for the molecule.).

9.132 In a certain molecule, a p orbital overlaps with a d orbital as shown.

The algebraic signs of the lobes of the d-orbital wave function are also indicated in the drawing. Which kind of bond is formed, σ or π? Explain your choice. Repeat the drawing twice on a separate sheet of paper. In one of them, indicate the signs of the lobes of the p-orbital wave function that would lead to a bonding MO. In the other, indicate the signs of the lobes of the p-orbital wave function that would lead to an antibonding MO.

*9.133 If we assign the internuclear axis in a diatomic molecule to be the z axis, what kind of p orbital (p_x, p_y, or p_z) on one atom would have to overlap with a d_{xz} orbital on the other atom to give a π bond?

Multi-Concept Problems

*9.134 The peroxynitrite ion, OONO⁻, is a potent toxin formed in cells affected by diseases such as diabetes and atherosclerosis. Peroxynitrite ion can oxidize and destroy biomolecules crucial for the survival of the cell. **(a)** Give the O—O—N and O—N—O bond angles in the peroxynitrite ion. **(b)** What is the hybridization of the N atom in the peroxynitrite ion? **(c)** Suggest why the peroxynitrite ion is expected to be much less stable than the nitrate ion, NO_3^-.

*9.135 An ammonia molecule, NH_3, is very polar, whereas NF_3 is almost nonpolar. Use this observation along with the valence bond description of bonding in these molecules to justify the following statement: *Lone pairs in hybrid orbitals contribute to the overall dipole moment of a molecule.*

*9.136 There exists a hydrocarbon called butadiene, which has the molecular formula C_4H_6 and the structure

The C=C bond lengths are 134 pm (about what is expected for a carbon–carbon double bond), but the C—C bond length in this molecule is 147 pm, which is shorter than a normal C—C single bond. The molecule is planar (i.e., all the atoms lie in the same plane).

(a) What kind of hybrid orbitals do the carbon atoms use in this molecule to form the C—C bonds?

(b) Between which pairs of carbon atoms do we expect to find sideways overlap of *p* orbitals (i.e., π-type *p-p* overlap)?

(c) Based on your answer to parts **(a)** and **(b)**, explain why the center carbon–carbon bond is shorter than a carbon–carbon single bond.

*9.137 A 0.244 g sample of a compound of phosphorus and bromine, when dissolved in water, reacted to give a solution containing phosphorus acid and bromide ion. Addition of excess $AgNO_3$ solution to the mixture led to precipitation of AgBr. When the precipitate of AgBr was collected and dried, it weighed 0.508 g. Determine the chemical formula of the phosphorus–bromine compound and predict whether its molecules are polar or nonpolar.

Exercises in Critical Thinking

9.138 Five basic molecular shapes were described for simple molecular structures containing a central atom bonded to various numbers of surrounding atoms. Can you suggest additional possible structures? Provide arguments about the likelihood that these other structures might actually exist.

9.139 Compare and contrast the concepts of delocalization and resonance.

9.140 Why doesn't a carbon–carbon quadruple bond exist?

9.141 What might the structure of the iodine heptafluoride molecule be? If you can think of more than one possible structure, which is likely to be of lowest energy based on the VSEPR model?

9.142 The F—F bond in F_2 is weaker than the Cl—Cl bond in Cl_2. How might the lone pairs on the atoms in the molecules be responsible for this?

9.143 Molecular orbital theory predicts the existence of antibonding molecular orbitals. How do antibonding electrons affect the stability in a molecule?

9.144 The structure of the diborane molecule, B_2H_6, is sometimes drawn as

There are not enough valence electrons in the molecule to form eight single bonds, which is what the structure implies. Assuming that the boron atoms use sp^3 hybrid orbitals, suggest a way that hydrogen 1*s* orbitals can be involved in forming delocalized molecular orbitals that bridge the two boron atoms. Use diagrams to illustrate your answer. What would be the average bond order in the bridging bonds?

Properties of Gases

10

Chapter Outline

The planet Saturn is one of the gas giants in our solar system. It is classified as a gas giant since its exterior is composed of gases. On the other hand, the atmosphere of the earth surrounds the planet with a thin layer of gases that is held close to the surface by gravitational forces. Our atmosphere is the only source of the oxygen that we, and other animals, breathe, and it is the repository for our exhaled carbon dioxide and water vapor. The carbon dioxide in our atmosphere is essential for plants that derive their energy from photosynthesis. However, no matter where we find gases, the behavior of all gases is remarkably alike.

©NASA

This Chapter in Context

In the preceding chapters we've discussed the chemical properties of a variety of different substances. We've also studied the kinds of chemical bonds that hold molecular and ionic substances together. In fact, it is the nature of the chemical bonds that dictates chemical properties. With this chapter we begin a systematic study of the *physical properties* of materials, including the factors that govern the behavior of gases, liquids, and solids. We study gases first because they are the easiest to understand and their behavior will help explain some of the properties of liquids and solids in Chapter 11.

Through everyday experience, you have become familiar with many of the properties that gases have. Our goal in this chapter is to refine this understanding in terms of the physical laws that govern the way gases behave. You will also learn how this behavior is interpreted in terms of the way we view gases at a molecular level. In our discussions we will describe how the energy concepts, first introduced to you in Chapter 6, provide an explanation of the gas laws. Finally, you will learn how a close examination of gas properties furnishes clues about molecular size and the attractions that exist between molecules.

LEARNING OBJECTIVES

After reading this chapter, you should be able to:

- describe the properties of gases at the macroscopic and molecular levels
- explain the measurement of pressure using barometers and manometers
- describe and use the gas laws of Dalton, Charles, Gay-Lussac, and the combined gas law
- perform stoichiometric calculations using the gas laws and Avogadro's principle
- explain and apply the ideal gas law, and explain how it incorporates the other gas laws
- describe Dalton's law of partial pressures, and apply it to collecting a gas over water
- state the main postulates of the kinetic theory of gases, and show how it explains the gas laws on a molecular level
- explain the physical significance of the corrections involved in the van der Waals equation for the differences between real and ideal gases

10.1 | A Molecular Look at Gases

Gas is one of the three common states of matter. Consider water, for example. We can see and feel it as a solid or a liquid, but it seems to disappear when it evaporates and surrounds us as water vapor. With this in mind, let's examine some of the properties of gaseous substances that explain the difference between gases and liquids or solids on the molecular level.

Since you have been surrounded by air all of your life, you already are aware of many properties that gases have. Let's look at two of them.

- You can wave your hand through air with little resistance. (Compare that with waving your hand through a tub filled with water.)
- The air in a bottle has little weight to it, so if a bottle of air is submerged under water and released, it quickly bobs to the surface.

Both of these observations suggest that a given volume of air doesn't have much matter in it. (We can express this by saying that air has a low density.) What else do you know about gases?

- Gases can be compressed. Inflating a tire involves pushing more and more air into the same container (the tire). This behavior is a lot different from that of liquids; you can't squeeze more water into an already filled bottle.
- Gases exert a pressure. Whenever you inflate a balloon you have an experience with gas pressure, and squeezing a balloon suggests that the pressure acts equally in all directions.
- The pressure of a gas depends on how much gas is confined. The more air you pump into a tire, the greater the pressure.
- Gases fill completely any container into which they're placed—you've never heard of half a bottle of air. If you put air in a container, it expands and fills the container's entire volume.

■ Dry air is roughly 21% O_2 and 78% N_2, but it has traces of several other gases.

©Elemental Imaging Getty Image, Inc.

Because air has so little weight for a given volume, it makes things filled with air float, much to the pleasure of these balloonists at a hot-air balloon festival.

- Gases mix freely and quickly with each other. You've experienced this when you've smelled the perfume of someone passing by. The vapors of the person's perfume mix with and spread through the air.
- The pressure of a gas rises when its temperature is increased. That's why there's the warning "Do Not Incinerate" printed on aerosol cans. A sealed can, if made too hot, is in danger of exploding from the increased pressure.

Other observations that have been made that can be explained by our knowledge of gases as another form of matter were a mystery when they were first observed.

- Rotting apples lost mass.
- Nails gained mass when they rusted.

Molecular Model of Gases

The simple qualitative observations about gases described above suggest what gases must be like when viewed at the molecular level (Figure 10.1). The fact that there's so little matter in a given volume suggests that there is a lot of space between the individual molecules, especially when compared to liquids or solids. Gases can be so easily compressed because squeezing a gas simply removes some of the empty space.

It also seems reasonable to believe that the molecules of a gas are moving around fairly rapidly. How else could we explain how molecules of a perfume move so quickly through the air? Furthermore, if gas molecules didn't move, gravity would cause them to settle to the bottom of a container (which they don't do). And if gas molecules are moving, some must be colliding with the walls of the container, and the force of these tiny collisions would explain the pressure a gas exerts.

Finally, the fact that gas pressure rises with increasing temperature suggests that the molecules move faster with increasing temperature, because faster molecules would exert greater forces when they collide with the walls.

Explain why adding more gas increases the pressure. (*Hint:* What causes the pressure of a gas?)

What chemical or physical processes would result in the fact that apples lose mass when rotting and that nails rust?

Lisa Passmore

Aerosol cans carry a warning about subjecting them to high temperatures because the internal pressure can become large enough to cause them to explode.

Practice Exercise 10.1

Practice Exercise 10.2

10.2 | Measurement of Pressure

As we discussed in Chapter 6, **pressure** *is force per unit area*, calculated by dividing the force by the area over which the force acts.

$$\text{Pressure} = \frac{\text{force}}{\text{area}}$$

For example, the earth exerts a gravitational force on everything with mass that is on it or near it. The *weight* of an object, like this book, is simply our measure of the force it exerts because gravity acts on it. Since gases fill a container and exert an equal force throughout the container, we will use the term pressure, so we do not have to take into account the area of the container.

The Barometer

Earth's gravity pulls on the air mass of the atmosphere, causing it to cover the Earth's surface like an invisible blanket. The molecules in the air collide with every object the air contacts, and by doing so, produce a pressure we call the **atmospheric pressure**.

At any particular location on Earth, the atmospheric pressure acts equally in all directions—up, down, and sideways. In fact, it presses against our bodies with a surprising amount of force, but we don't really feel it because the fluids in our bodies push back with

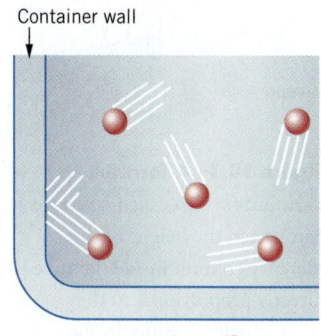

Container wall

Gas molecule = ●

Figure 10.1 | **A gas viewed at the molecular level.** Simple qualitative observations of the properties of gases lead us to conclude that a gas is composed of widely spaced molecules that are in constant motion. Collisions of molecules with the walls produce tiny forces that, when taken all together, are responsible for the gas pressure.

Figure 10.2 | **The effect of an unbalanced pressure.** (*a*) The pressure inside the can, P_{inside}, is the same as the atmospheric pressure outside, P_{atm}. The pressures are balanced; $P_{inside} = P_{atm}$. (*b*) When a vacuum pump reduces the pressure inside the can, P_{inside} becomes less than P_{atm}, and the unbalanced outside pressure quickly and violently makes the can collapse.

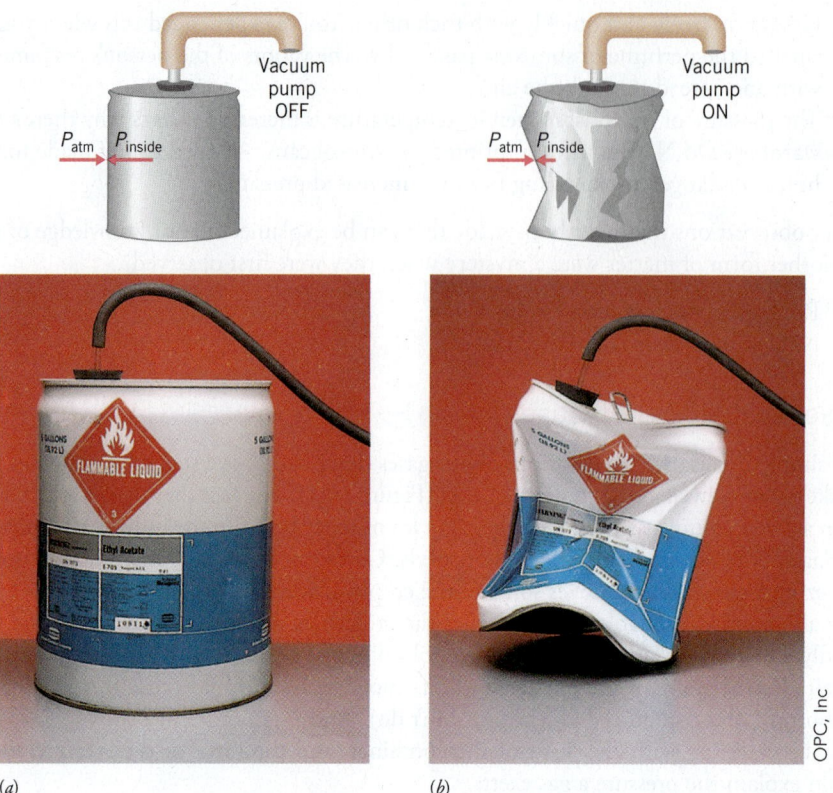

(*a*)   (*b*)

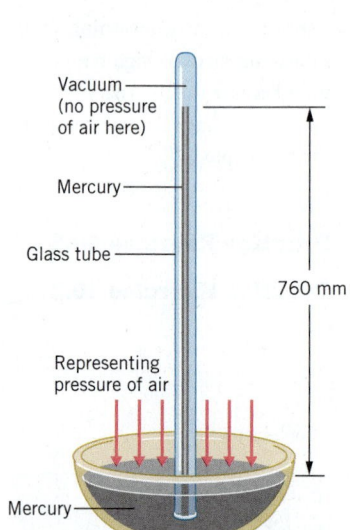

Figure 10.3 | **A Torricelli barometer.** Also called a mercury barometer, the height of the mercury column inside the tube is directly proportional to the atmospheric pressure. In the United States, weather reports often give the height of the mercury column in inches.

equal pressure. We can observe atmospheric pressure, however, if we pump the air from a collapsible container such as the can in Figure 10.2. Before air is removed, the walls of the can experience atmospheric pressure equally inside and out. When some air is pumped out, however, the pressure inside decreases, making the atmospheric pressure outside of the can greater than the pressure inside. The net inward pressure is sufficiently great to make the can crumple.

To measure atmospheric pressure we use a device called a **barometer**. The simplest type is the *Torricelli barometer*[1] (Figure 10.3), which consists of a glass tube sealed at one end, 80 cm or more in length. To set up the apparatus, the tube is filled with mercury, capped, inverted, and then its capped end is immersed in a dish of mercury. When the cap is removed, some, but not all, of the mercury runs out.[2] When, due to its pressure, mercury flows from the tube, the level of the mercury in the dish must rise. Atmospheric pressure, pushing on the surface of the mercury in the dish, opposes this rise in mercury level. Rapidly, the two pressures become equal, and no more mercury can run out. However, a space inside the tube above the mercury level has been created having essentially no atmosphere; it's a **vacuum**.

The height of the mercury column, measured from the surface of the mercury in the dish, is directly proportional to atmospheric pressure. On days when the atmospheric pressure is high, more mercury is forced from the dish into the tube and the height of the column increases. When the atmospheric pressure drops—during an approaching storm, for example—some mercury flows out of the tube and the height of the column decreases. Most people live where this height fluctuates between 730 and 760 mm.

---

[1] In 1643 Evangelista Torricelli, an Italian mathematician, suggested an experiment, later performed by a colleague, that demonstrated that atmospheric pressure determines the height to which a fluid will rise in a tube inverted over the same liquid. This concept led to the development of the Torricelli barometer, which is named in his honor.

[2] Today, mercury is kept as much as possible in closed containers. Although atoms of mercury do not readily escape into the gaseous state, mercury vapor is a dangerous poison.

Units of Pressure

At sea level, the height of the mercury column in a barometer fluctuates around a value of 760 mm. Some days it's a little higher, some a little lower, depending on the weather. The average pressure at sea level has long been used by scientists as a standard unit of pressure. The **standard atmosphere (atm)** was originally defined as the pressure needed to support a column of mercury 760 mm high measured at 0 °C.[3]

In the SI, the unit of pressure is the **pascal**, symbolized **Pa**. In SI units, the pascal is the ratio of force in *newtons* (N, the SI unit of force) to area in meters squared,

$$1 \text{ Pa} = \frac{1 \text{ N}}{1 \text{ m}^2} = 1 \text{ N m}^{-2}$$

It's a very small pressure; 1 Pa is approximately the pressure exerted by the weight of a lemon spread over an area of 1 m^2.

To bring the standard atmosphere unit in line with other SI units, it has been redefined in terms of the pascal as follows.

$$1 \text{ atm} = 101{,}325 \text{ Pa} \quad \text{(exactly)}$$

A unit of pressure related to the pascal is the **bar**, which is defined as 100 kPa. Consequently, one bar is slightly smaller than one standard atmosphere (1 bar = 0.9868 atm). You may have heard the **millibar** unit (1 mb = 10^{-3} bar) used in weather reports describing pressures inside storms such as hurricanes. For example, the lowest atmospheric pressure at sea level ever observed in the Atlantic basin was 882 mb during Hurricane Wilma on October 19, 2005. The storm later weakened but still caused extensive damage as it crossed Florida.

For ordinary laboratory work, the pascal (or kilopascal) is not a conveniently measured unit. Usually we use a unit of pressure called the **torr** (named after Torricelli). The torr is defined as 1/760th of 1 atm.

$$1 \text{ torr} = \frac{1}{760} \text{ atm}$$

$$1 \text{ atm} = 760 \text{ torr} \quad \text{(exactly)}$$

The torr is very close to the pressure that is able to support a column of mercury 1 mm high. In fact, the **millimeter of mercury** (abbreviated **mm Hg**) is often itself used as a pressure unit. Except when the most exacting measurements are being made, it is safe to use the relationship

$$1 \text{ torr} = 1 \text{ mm Hg}$$

■ In English units, one atmosphere of pressure is 14.7 lb/in.^{2} This means that at sea level, each square inch of your body is experiencing a force of nearly 15 pounds.

Terra MODIS data acquired by direct broadcast at the University of South Florida (Judd Taylor). Image processed at the University of Wisconsin-Madison (Lian Gumley)

A satellite photo of Hurricane Wilma when it reached Category 5 strength, with sustained winds of 175 miles per hour in the eye wall.

Standard atmosphere (atm) expressed in different pressure units.
760 torr
101,325 Pa
101.325 kPa
1.013 bar
1013 mb
14.7 lb in.$^{-2}$
1.034 kg cm^{-2}

[3] Because any metal, including mercury, expands or contracts as the temperature increases or decreases, the height of the mercury column varies with temperature (just as in a thermometer). Therefore, the definition of the standard atmosphere required that the temperature at which the mercury height is measured be specified.

Practice Exercise 10.3

Using the table above, determine the atmospheric pressure in pounds per square inch and inches of mercury when the barometer reads 730 mm Hg. (*Hint:* Recall your tools for converting units.)

Practice Exercise 10.4

The second lowest barometric pressure ever recorded at sea level in the Western Hemisphere was 888 mb during Hurricane Gilbert in 1988. What was the pressure in pascals and torr?

Manometers

Gases used as reactants or formed as products in chemical reactions are kept from escaping by using closed glassware. To measure pressures inside such vessels, a **manometer** is used. Two types are common: open-end and closed-end manometers.

Open-End Manometers

■ Advantages of mercury over other liquids are its low reactivity, its low melting point, and particularly its very high density, which permits short manometer tubes.

An **open-end manometer** consists of a U-tube partly filled with a liquid, usually mercury (see Figure 10.4). One arm of the U-tube is open to the atmosphere; the other is exposed to a container of some trapped gas. By comparing the levels of mercury in the two arms, we can determine the pressure of the gas in the flask. In part (*a*) of this diagram, the mercury levels are equal and the pressure inside the flask is equal to the atmospheric pressure. The second part of the figure, 10.4*b*, shows the mercury level as being higher in the arm exposed to the atmosphere. We conclude that the pressure in the flask is greater than atmospheric pressure. In Figure 10.4*c*, the Hg is higher in the arm connected to the container of gas, indicating that the pressure of the atmosphere must be *higher* than the gas pressure. With mercury in the tube, the difference in the heights in the two arms, represented here as P_{Hg}, is equal to the difference between the pressure of the gas and the pressure of the atmosphere. By measuring P_{Hg} in millimeters, the value equals the pressure difference in torr. For the situation shown in Figure 10.4*b*, we would calculate the pressure of the trapped gas as

$$P_{gas} = P_{atm} + P_{Hg} \quad (\text{when } P_{gas} > P_{atm})$$

In Figure 10.4*c* the calculation of the pressure is

$$P_{gas} = P_{atm} - P_{Hg} \quad (\text{when } P_{gas} < P_{atm})$$

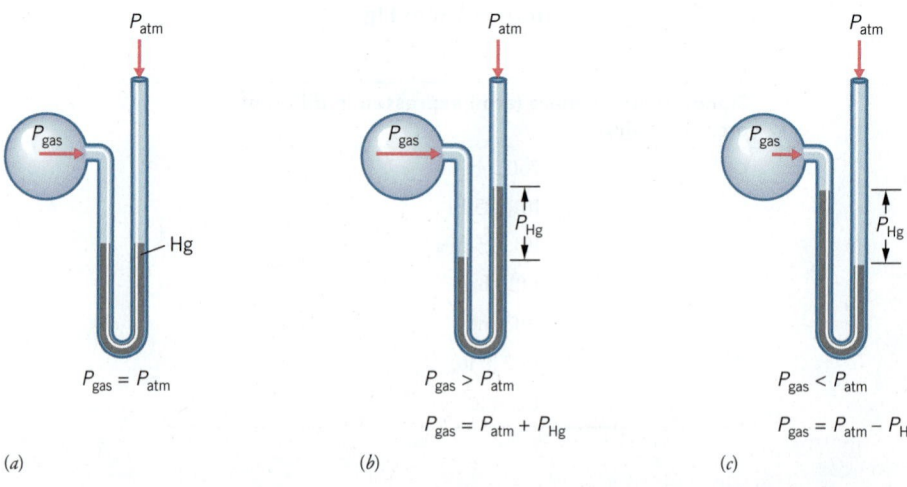

(*a*) (*b*) (*c*)

Figure 10.4 | An open-end manometer. The difference in the heights of the mercury in the two arms equals the pressure difference, in torr, between the atmospheric pressure, P_{atm}, and the pressure of the trapped gas, P_{gas}.

Example 10.1
Measuring the Pressure of a Gas Using a Manometer

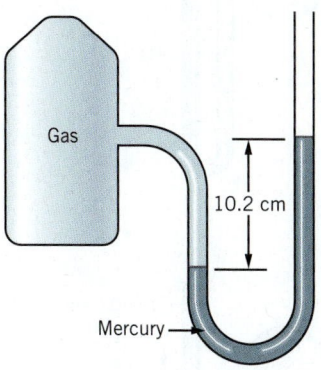

Gas

10.2 cm

Mercury

A student collected a gas in a flask connected to an open-end manometer, as illustrated in the figure in the margin. The difference in the heights of the mercury in the two columns was 10.2 cm and the atmospheric pressure was measured to be 756 torr. What was the pressure of the gas in the apparatus?

Analysis: From the preceding discussion, we know that when using an open-end manometer we will use the atmospheric pressure and either add to it or subtract from it. Let's use some common sense (something that will help a lot in working problems involving gases) to make the correct decision.

When we look at the diagram of the apparatus, we see that the mercury is pushed up into the arm of the manometer that's open to the air. Common sense tells us that the pressure of the gas inside must be larger than the pressure of the air outside. Therefore, we will add the pressure difference to the atmospheric pressure.

Finally, before we can do the arithmetic, we have to make our units the same, in this case, torr.

Assembling the Tools: We need the equation for determining the pressure in an open-end manometer,

$$P_{gas} = P_{atm} \pm P_{Hg}$$

And we have already decided to add P_{Hg} to P_{atm}. First, we use the usual conversion factors for SI prefixes from Table 2.4 to perform the conversion needed.

$$10^{-2}\ m = 1\ cm \quad \text{and} \quad 10^{-3}\ m = 1\ mm$$

The other conversion we need is the equality between 1 mm Hg and 1 torr.

Solution: We first need to convert cm Hg to mm Hg, so we use our conversion factors to write

$$10.2\ \text{cm} \times \frac{10^{-2}\ \text{m}}{1\ \text{cm}} \times \frac{1\ \text{mm}}{10^{-3}\ \text{m}} = 102\ \text{mm Hg}\left(\frac{1\ \text{torr}}{1\ \text{mm Hg}}\right) = 102\ \text{torr}$$

Based on our analysis, to find the gas pressure, we add 102 torr to the atmospheric pressure.

$$P_{gas} = P_{atm} + P_{Hg}$$
$$= 756\ \text{torr} + 102\ \text{torr}$$
$$= 858\ \text{torr}$$

The gas pressure is 858 torr.

Is the Answer Reasonable? It almost looks like the gas is trying to push the mercury out of the manometer, so we conclude that the gas pressure must be higher than atmospheric pressure. That's exactly what we've found in our calculation, so we can feel confident we've solved the problem correctly.

An open-end manometer is filled with mercury so that each side has a height equal to 15 cm. If the atmospheric pressure is 770 torr on a given day, what are the approximate maximum and minimum pressures that this manometer can measure? (*Hint:* What is the maximum difference in mercury height for this manometer?)

Practice Exercise 10.5

In another experiment that same day, it was found that the mercury level in the arm of the manometer attached to the container of gas was 11.7 cm higher than in the arm open to the air. What was the pressure of the gas?

Practice Exercise 10.6

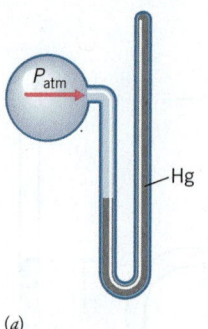

(a)

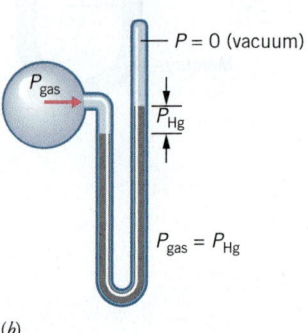

(b)

Figure 10.5 | A closed-end manometer for measuring gas pressures less than 1 atm or 760 torr. (a) When constructed, the tube is fully evacuated and then mercury is allowed to enter the tube to completely fill the closed arm. (b) When the tube is connected to a bulb containing a gas, the difference in the mercury heights (P_{Hg}) equals the pressure, in torr, of the trapped gas, P_{gas}.

■ Ten meters is about 33 feet or the height of a three-story building.

Closed-End Manometers

A **closed-end manometer** (Figure 10.5) is made by sealing the arm that will be farthest from the gas sample and then filling the closed arm completely with mercury. In this design, the mercury is pushed to the top of the closed arm when the open arm of the manometer is exposed to the atmosphere. When connected to a flask containing a gas, however, the mercury level in the sealed arm will drop, leaving a vacuum above it. The pressure of the gas can then be measured just by reading the difference in heights of the mercury in the two arms, P_{Hg}. No separate measurement of the atmospheric pressure is required.

Manometers with Liquids Other than Mercury

Scientists often use manometers with liquids other than mercury. One compelling reason is to reduce the possibility of a hazardous mercury spill. Another reason is that more precise measurements may be made using other liquids that have lower densities since larger changes in height will be observed. Afterwards, measurements using liquids such as water, ethyl alcohol, or mineral oil can be converted to the equivalent in mercury.

In order to convert the difference in heights of a liquid to the equivalent height in mercury, we need to know the density of the substitute liquid. We can reason that the height of different liquids, measuring the same pressure, should be inversely proportional to their densities:

$$\frac{\text{height liquid 1}}{\text{height liquid 2}} = \frac{\text{density liquid 2}}{\text{density liquid 1}}$$

If liquid 1 is mercury, we can rearrange the equation to read

$$\text{mm Hg} = \text{mm liquid} \times \frac{\text{density of liquid}}{\text{density Hg}}$$

This will allow us to use any liquid with a known density in a manometer and then convert the readings to the equivalent in mm Hg.

Let's consider what will happen if we use water instead of mercury in a manometer. Because of the low density of water compared to the high density of mercury, we will observe a height of water that is 13.6 times the height of mercury. The practical result of using a water manometer is that measurements will be 13.6 times more precise than those using a mercury manometer. The drawback is that a water manometer used to measure barometric pressure would have to be more than 10 meters high.

10.3 | Gas Laws

Earlier, we examined some properties of gases that are familiar to you. Our discussion was only qualitative, however, and now that we've discussed pressure and its units, we are ready to examine gas behavior quantitatively.

There are four variables that affect the properties of a gas—pressure, volume, temperature, and the amount of gas. In this section, we will study situations in which the amount of gas (measured either by grams or moles) remains constant and observe how gas samples respond to changes in pressure, volume, and temperature.

Pressure–Volume Law

When you inflate a bicycle tire with a hand pump, you squeeze air into a smaller volume and increase its pressure. Packing molecules into a smaller space causes an increased number of collisions with the walls, and because these collisions are responsible for the pressure, the pressure increases (Figure 10.6).

Robert Boyle, an Irish scientist (1627–1691), performed experiments to determine quantitatively how the volume of a fixed amount of gas varies with pressure. Because the volume is also affected by the temperature, he held that variable constant. A graph of typical data collected in his experiments is shown in Figure 10.7a and demonstrates that the volume of a given amount of gas held at constant temperature varies inversely with the applied pressure. Mathematically, this can be expressed as

$$V \propto \frac{1}{P} \quad \text{(temperature and amount of gas held constant)}$$

where V is the volume and P is the pressure. This relationship between pressure and volume is now called **Boyle's law** or the **pressure–volume law**.

In the expression above, the proportionality sign, $\propto$, can be removed by introducing a proportionality constant, C.

$$V = \frac{1}{P} \times C$$

Rearranging gives

$$PV = C$$

This equation tells us, for example, that if at constant temperature the gas pressure is doubled, the gas volume must be cut in half so that the product of $P \times V$ doesn't change.

What is remarkable about Boyle's discovery is that this relationship is essentially the same for all gases at temperatures and pressures usually found in the laboratory.

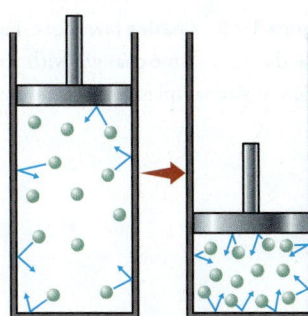

Figure 10.6 | **Compressing a gas increases its pressure.** A molecular view of what happens when a gas is squeezed into a smaller volume. By crowding the molecules together, the number of collisions within a given area of the walls increases, which causes the pressure to rise.

Ideal Gases

When very precise measurements are made, it's found that Boyle's law doesn't quite work. This is especially a problem when the pressure of the gas is very high or when the gas is at a low temperature where it's on the verge of changing to a liquid. Although real gases do not *exactly* obey Boyle's law or any of the other gas laws that we'll study, it is often useful to imagine a hypothetical gas that would. We call such a hypothetical gas an ideal gas. An **ideal gas** *obeys the gas laws exactly over all temperatures and pressures.* A real gas behaves more and more like an ideal gas as its pressure decreases and its temperature increases. Most gases we work with in the lab can be treated as ideal gases unless we're dealing with extremely precise measurements.

Temperature–Volume Law

In 1787 a French chemist and mathematician named Jacques Alexander Charles became interested in hot-air ballooning, which at the time was becoming popular in France. His new interest led him to study what happens to the volume of a sample of gas when the temperature is varied, keeping the pressure constant.

■ Jacques Alexandre César Charles (1746–1823), a French scientist, had a keen interest in hot-air balloons. He was the first to inflate a balloon with hydrogen.

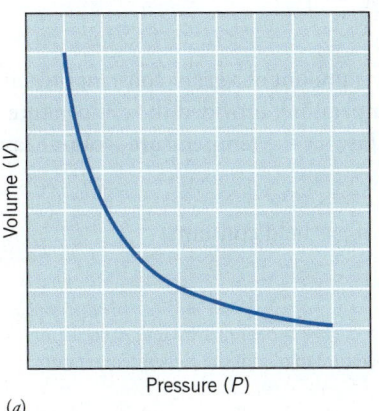

(a)
Pressure (P)

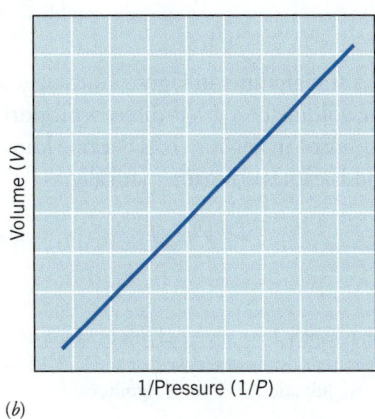

(b)
1/Pressure (1/P)

Figure 10.7 | **The variation of volume with pressure at constant temperature for a fixed amount of gas.** (a) A typical graph of volume versus pressure, showing that as the pressure increases, the volume decreases. (b) A straight line is obtained when volume is plotted against $1/P$, which shows that $V \propto 1/P$.

Figure 10.8 | **Charles law plots.** Each line shows how the gas volume changes with temperature for different size samples of the same gas.

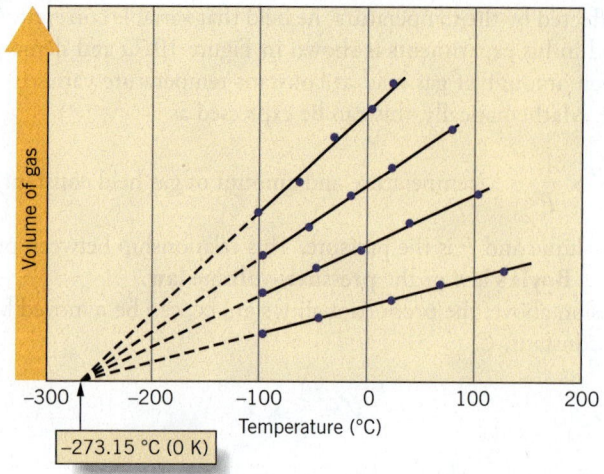

When data from experiments such as Charles' are plotted, a graph like that shown in Figure 10.8 is obtained. Here the volume of the gas is plotted against the temperature in degrees Celsius. The colored points correspond to typical data, and the lines are drawn to most closely fit the data. Each line represents data collected for a different sample. Because all gases eventually become liquids if cooled sufficiently, the solid portions of the lines correspond to temperatures at which measurements are possible; at lower temperatures the gas liquefies. However, if the lines are extrapolated (i.e., reasonably extended) back to a point where the volume of the gas would become zero if it didn't condense, all the lines meet at the same temperature, $-273.15\ ^\circ$C. Especially significant is the fact that this exact same behavior is exhibited by all gases; when plots of volume versus temperature are extrapolated to zero volume, the temperature axis is always crossed at $-273.15\ ^\circ$C. This point represents the temperature at which all gases, if they did not condense, would have a volume of zero, and below which they would have a negative volume. Negative volumes are impossible, of course, so it was reasoned that $-273.15\ ^\circ$C must be nature's coldest temperature, and it was called **absolute zero**.

As you learned earlier, absolute zero corresponds to the zero point on the Kelvin temperature scale, and to obtain a Kelvin temperature, we add 273.15 °C to the Celsius temperature.[4]

$$T_K = (t_C + 273.15\ ^\circ\text{C})\left(\frac{T_K}{t_C}\right)$$

■ Remember that we use the capital T for kelvin temperatures and from here forward we will not use the subscript K.

For most purposes, we will need only three significant figures, so we can use the following approximate relationship.

$$T_K = (t_C + 273\ ^\circ\text{C})\left(\frac{T_K}{t_C}\right)$$

The straight lines in Figure 10.8 suggest that for a given amount of a gas at constant pressure, the volume of a gas is directly proportional to its temperature, provided the temperature is expressed in kelvins. This became known as **Charles' law** (or the **temperature–volume law**) and is expressed mathematically as

$$V \propto T \qquad \text{(pressure and amount of gas held constant)}$$

---

[4] In Chapter 1 we presented this equation as $T_K = (t_C + 273.15\ ^\circ\text{C})(\frac{1\,\text{K}}{1\,^\circ\text{C}})$ to emphasize unit cancellation. Operationally, however, we just add 273.15 to the Celsius temperature to obtain the Kelvin temperature (or we just add 273 if three significant figures are sufficient).

Using a different proportionality constant, C', we can write

$$V = C'T \qquad \text{(pressure and amount of gas held constant)}$$

■ The value of C' depends on the size and pressure of the gas sample.

Pressure–Temperature Law

The French scientist Joseph Louis Gay-Lussac studied how the pressure and temperature of a fixed amount of gas at constant volume are related. (Such conditions exist, for example, when a gas is confined in a vessel with rigid walls, like an aerosol can.) The relationship that he established, called **Gay-Lussac's law** or the **pressure–temperature law**, states that the pressure of a fixed amount of gas held at constant volume is directly proportional to the Kelvin temperature. Thus,

$$P \propto T \qquad \text{(volume and amount of gas held constant)}$$

Using still another constant of proportionality, C'', Gay-Lussac's law becomes

$$P = C''T \qquad \text{(volume and amount of gas held constant)}$$

■ Joseph Louis Gay-Lussac (1778–1850), a French scientist, was a co-discoverer of the element boron.

■ We're using different symbols for the various gas law constants because they are different for each law.

Combined Gas Law

The three gas laws we've just examined can be brought together into a single equation known as the **combined gas law**, which states that the ratio PV/T is a constant for a fixed amount of gas.

$$\frac{PV}{T} = \text{constant} \qquad \text{(for a fixed amount of gas)}$$

Usually, we use the combined gas law in problems where we know some given set of conditions of temperature, pressure, and volume (for a fixed amount of gas), and wish to find out how one of these variables will change when the others are changed. If we label the initial conditions of P, V, and T with the subscript 1 and the final conditions with the subscript 2, the combined gas law can be written in the following useful form.

$$\frac{P_1 V_1}{T_1} = \frac{P_2 V_2}{T_2} \tag{10.1}$$

 **TOOLS**

Combined gas law

In applying this equation, T must always be in kelvins. The pressure and volume can have any units, but whatever the units are on one side of the equation, they must be the same on the other side.

It is simple to show that Equation 10.1 contains each of the other gas laws as special cases. Boyle's law, for example, applies when the temperature is constant. Under these conditions, T_1 equals T_2 and the temperature cancels from the equation. This leaves us with

$$P_1 V_1 = P_2 V_2 \qquad \text{(when } T_1 \text{ equals } T_2\text{)}$$

which is one way to write Boyle's law. Similarly, under conditions of constant pressure, P_1 equals P_2 and the pressure cancels, so Equation 10.1 reduces to

$$\frac{V_1}{T_1} = \frac{V_2}{T_2} \qquad \text{(when } P_1 \text{ equals } P_2\text{)}$$

This, of course, is another way of writing Charles' law. Under the constant volume conditions required by Gay-Lussac's law, V_1 equals V_2, and Equation 10.1 reduces to

$$\frac{P_1}{T_1} = \frac{P_2}{T_2} \qquad \text{(when } V_1 \text{ equals } V_2\text{)}$$

CHEMISTRY OUTSIDE THE CLASSROOM 10.1

Whipped Cream

The dessert in Figure 1 looks tempting, but up until 1934 everyone had to whip their own cream by hand. Food scientists at the time were in the midst of using chemistry to produce a wide array of "convenience foods," foods that needed little or no preparation or cooking time.

Many foods can be canned—just look at the rows of vegetables, meats, and fruits lined up in the supermarket. Not only are these handy, but they also have exceptional shelf life compared to fresh vegetables, meats, and fruits. Fresh foods could be kept longer in refrigerators, which became commonly available in the 1930s. Freezers for frozen foods and ice cream came shortly thereafter. Now the consumer could have ice cream at home, but the sundae still needed the whipped cream.

Around 1934, Charles Getz was a graduate student in chemistry at the University of Illinois when he developed and sought a patent for producing a fluffy aerosol of gas bubbles entrained in cream. The concept was simple. Put some cream in a can, pressurize it with a gas, and when the gas is released along with the cream, it should foam up into whipped cream. Hundreds of gases were tried as propellants and they were all failures. Getz found that carbon dioxide did create a foam with milk, but it also gave the food a bad taste because of the sour taste of the acidic CO_2. He switched to nitrous oxide because it had no taste and provided a good product. It is now believed that the nonpolar CO_2 and N_2O work by dissolving in the nonpolar fats in liquid cream.

When the pressure is released the gas expands to form tiny, relatively uniform bubbles to whip the cream. In 1935 Charles Getz applied for a process patent for his method of preparing whipped cream with nitrous oxide. He was issued Patent 2,294,172 in 1942. Aaron "Bunny" Lapin, an inventor from St. Louis, created Reddi-wip® real whipped cream in 1948.

Figure 1 Whipped cream dispensed from an aerosol container.

Envision/©Corbis

Example 10.2
Using the Combined Gas Law

Fuse /Getty Images

A compact fluorescent light bulb gives off light by exciting mercury atoms. In order to increase the chance of exciting the mercury atoms, argon gas is used to carry the electrons through the light bulb. Suppose a 12.0 L cylinder containing compressed argon at a pressure of 57.8 atm measured at 24 °C is to be used to fill light bulbs, each with a volume of 53.4 mL, to a pressure of 2.28 torr at 21 °C. How many of these light bulbs could be filled by the argon in the cylinder?

Analysis: What do we need to know to figure out how many light bulbs can be filled? If we knew the total volume of gas with a pressure of 2.28 torr at 21 °C, we could just divide by the volume of one light bulb; the result is the number of light bulbs that can be filled. So our main problem is determining what volume the argon in the cylinder will occupy when its pressure is reduced to 2.28 torr and its temperature is lowered to 21 °C.

Assembling the Tools: Because the total amount of argon isn't changing, we can use the combined gas law for the calculation.

$$\frac{P_1 V_1}{T_1} = \frac{P_2 V_2}{T_2}$$

Solution: First, we will also make sure that the temperature is expressed in kelvins and that the values for P and V have the same units so that they cancel. To do this we need to convert 57.8 atm to torr,

$$57.8 \text{ atm} \times \frac{760 \text{ torr}}{1 \text{ atm}} = 4.39 \times 10^4 \text{ torr}$$

To use the combined gas law, let's construct a table of data being sure to assign the proper values to either the initial or final conditions.

Initial (1)	Final (2)
P_1 4.39×10^4 torr	P_2 2.28 torr
V_1 12.0 L	V_2 ?
T_1 297 K (24 + 273)	T_2 294 K (21 + 273)

To obtain the final volume, we solve for V_2.

Ratio of temperatures

$$V_2 = V_1 \times \frac{P_1}{P_2} \times \frac{T_2}{T_1}$$

Ratio of pressures

Let's now substitute values from our table of data.

$$V_1 = 12.0 \text{ L} \times \frac{4.39 \times 10^4 \text{ torr}}{2.28 \text{ torr}} \times \frac{294 \text{ K}}{297 \text{ K}}$$

$$= 2.29 \times 10^5 \text{ L}$$

Before we can divide by the volume of one light bulb (53.4 mL), we have to convert the total volume to milliliters.

$$2.29 \times 10^5 \text{ L} \times \frac{1000 \text{ mL}}{1 \text{ L}} = 2.29 \times 10^8 \text{ mL}$$

The volume per light bulb gives us the relationship

$$1 \text{ light bulb} \Leftrightarrow 53.4 \text{ mL argon}$$

which we can use to construct a conversion factor to find the number of light bulbs that can be filled.

$$2.29 \times 10^8 \text{ mL argon} \times \frac{1 \text{ light bulb}}{53.4 \text{ mL argon}} = 4.28 \times 10^6 \text{ light bulbs}$$

That's over four million light bulbs! As you can see, there's not much argon in each one.

Is the Answer Reasonable? To determine whether we've set up the combined gas law properly, we check to see if the pressure and temperature ratios move the volume in the correct direction. Going from the initial to final conditions, the pressure *decreases* from 4.39×10^4 torr to 2.28 torr, so the volume should *increase* a lot. The pressure ratio we used is much larger than 1, so multiplying by it should increase the volume, which agrees with what we expect. Next, look at the temperature change; a *drop in temperature* should tend to *decrease* the volume, so we should be multiplying by a ratio that's less than 1. The ratio 294/298 is less than 1, so that ratio is correct, too.

Now that our setup is okay, we check the math by rounding our numbers. One way we can round the numbers to estimate the answer is:

$$V_2 = 12.0 \text{ L} \times \frac{4.5 \times 10^4 \text{ torr}}{2 \text{ torr}} \times \frac{300 \text{ K}}{300 \text{ K}}$$

The estimated answer for $V_2 = 27 \times 10^4$ L, which is very close to our calculator answer. We can multiply the liters by 10^3 to get 27×10^7 mL and then we will "round" the

53.4 mL/bulb to 50 mL/bulb. Dividing the two gives us 5×10^6 bulbs. This again agrees quite well with the calculated answer.

(An observant student might note that in the end 12 L of argon must remain in the cylinder. However, we can divide 12,000 mL by 53.4 to get an answer of approximately 224 bulbs. Subtracting 225 from 4.28×10^6 still leaves us with 4.28×10^6 light bulbs.)

Practice Exercise 10.7

Use the combined gas law to determine by what factor the pressure of an ideal gas must change if the Kelvin temperature is doubled and the volume is tripled. (*Hint:* Sometimes it is easier to assume a starting set of temperature, volume, and pressure readings and then apply the conditions of the problem.)

Practice Exercise 10.8

What will be the final pressure of a sample of nitrogen with a volume of 950 cm^3 at 745 torr and 25.0 °C if it is heated to 60.0 °C and given a final volume of 1150 cm^3?

10.4 | Stoichiometry Using Gas Volumes

Reactions at Constant *T* and *P*

When scientists studied reactions between gases quantitatively, they made an interesting discovery. If the volumes of the reacting gases, as well as the volumes of gaseous products, are measured under the same conditions of temperature and pressure, the volumes are simple whole-number ratios. For example, hydrogen gas reacts with chlorine gas to give gaseous hydrogen chloride. Beneath the names in the following equation are the relative volumes with which these gases interact (at the same *T* and *P*).

hydrogen + chlorine $\longrightarrow$ hydrogen chloride

1 volume   1 volume   2 volumes

What this means is that if we were to use 1.0 L of hydrogen, it would react with 1.0 L of chlorine and produce 2.0 L of hydrogen chloride. If we were to use 10.0 L of hydrogen, all the other volumes would be multiplied by 10 as well.

Similar simple, whole-number ratios by volume are observed when hydrogen combines with oxygen to give water, which is a gas above 100 °C.

hydrogen + oxygen $\longrightarrow$ water (gaseous)

2 volumes   1 volume   2 volumes

Notice that the reacting volumes, measured under identical temperatures and pressures, are in ratios of simple, whole numbers.[5]

Observations such as those above led Gay-Lussac to formulate his **law of combining volumes**, which states that when gases react at the same temperature and pressure, their combining volumes are in ratios of simple whole numbers. Much later, it was learned that these "simple whole numbers" are the coefficients of the equations for the reactions.

Avogadro's Principle

■ Amedeo Avogadro (1776–1856), an Italian scientist, helped to put chemistry on a quantitative basis.

The observation that the gases react in whole-number volume ratios led Amedeo Avogadro to conclude that, at the same *T* and *P*, equal volumes of gases must have identical numbers of molecules. Today, we know that "equal numbers of *molecules*" is the same as "equal numbers of *moles*," so Avogadro's insight, now called **Avogadro's principle**, is expressed as

---

[5] The great French chemist Antoine Laurent Lavoisier (1743–1794) was the first to observe the volume relationships of this particular reaction. In his 1789 textbook, *Elements of Chemistry*, he wrote that the formation of water from hydrogen and oxygen requires that two volumes of hydrogen be used for every volume of oxygen. Lavoisier was unable to extend the study of this behavior of hydrogen and oxygen to other gas reactions because he was beheaded during the French Revolution. (See Michael Laing, *The Journal of Chemical Education*, February 1998, page 177.)

TABLE 10.1	Molar Volumes of Some Gases at STP	
Gas	Formula	Molar Volume (L)
Helium	He	22.398
Argon	Ar	22.401
Hydrogen	H_2	22.410
Nitrogen	N_2	22.413
Oxygen	O_2	22.414
Carbon dioxide	CO_2	22.414

follows: When measured at the same temperature and pressure, equal volumes of gases contain equal numbers of moles. A corollary to Avogadro's principle is that the volume of a gas is directly proportional to its number of moles, n.

$$V \propto n \qquad \text{(at constant } T \text{ and } P\text{)}$$

Standard Molar Volume

Avogadro's principle implies that the volume occupied by one mole of *any* gas—its *molar volume*—must be identical for all gases under the same conditions of pressure and temperature. To compare the molar volumes of different gases, scientists agreed to use 1 atm and 273.15 K (0 °C) as the **standard conditions of temperature and pressure**,[6] or **STP**, for short. If we measure the molar volumes for a variety of gases at STP, we find that the values fluctuate somewhat because the gases are not "ideal." Some typical values are listed in Table 10.1, and if we were to examine the data for many gases we would find an average of around 22.4 L per mole. This value is taken to be the molar volume of an *ideal gas* at STP and is now called the **standard molar volume** of a gas.

For an ideal gas at STP:

$$1 \text{ mol gas} \Leftrightarrow 22.4 \text{ L gas}$$

Avogadro's principle was a remarkable advance in our understanding of gases. His insight enabled chemists for the first time to determine the formulas of gaseous elements.[7]

Stoichiometry Problems

For reactions involving gases, Avogadro's principle lets us use a new kind of stoichiometric equivalency, one between *volumes* of gases. Earlier, for example, we noted the following reaction and its gas volume relationships.

$$2H_2(g) + O_2(g) \longrightarrow 2H_2O(g)$$

2 volumes   1 volume   2 volumes

Provided we are dealing with gas volumes measured at the same temperature and pressure, we can write the following stoichiometric equivalencies.

■ The recognition that equivalencies in gas volumes are numerically the same as those for numbers of moles of gas in reactions involving gases simplifies many calculations.

---

[6] Today there are more than a dozen definitions of "standard temperature and pressure." Our selection of STP is predicated on tradition and the fact that pressure, commonly measured in mm Hg from a manometer, is easily converted to atmospheres.

[7] Suppose that hydrogen chloride, for example, is correctly formulated as HCl, not as H_2Cl_2 or H_3Cl_3 or higher, and certainly not as $H_{0.5}Cl_{0.5}$. Then the only way that *two* volumes of hydrogen chloride could come from just *one* volume of hydrogen and *one* of chlorine is if each particle of hydrogen and each of chlorine were to consist of *two* atoms of H and Cl, respectively, H_2 and Cl_2. If these particles were single-atom particles, H and Cl, then one volume of H and one volume of Cl could give only *one* volume of HCl, not two. Of course, if the initial assumption were incorrect so that hydrogen chloride is, say, H_2Cl_2 instead of HCl, then hydrogen would be H_4 and chlorine would be Cl_4. The extension to larger subscripts works in the same way.

$$2 \text{ volumes } H_2(g) \Leftrightarrow 1 \text{ volume } O_2(g) \qquad \text{just as} \qquad 2 \text{ mol } H_2 \Leftrightarrow 1 \text{ mol } O_2$$
$$2 \text{ volumes } H_2(g) \Leftrightarrow 2 \text{ volumes } H_2O(g) \qquad \text{just as} \qquad 2 \text{ mol } H_2 \Leftrightarrow 2 \text{ mol } H_2O$$
$$1 \text{ volume } O_2(g) \Leftrightarrow 2 \text{ volumes } H_2O(g) \qquad \text{just as} \qquad 1 \text{ mol } O_2 \Leftrightarrow 2 \text{ mol } H_2O$$

Relationships such as these can greatly simplify stoichiometry problems, as we see in Example 10.3.

Example 10.3
Stoichiometry of Reactions of Gases

How many liters of hydrogen, $H_2(g)$, measured at STP, are needed to combine exactly with 1.50 L of nitrogen, also measured at STP, to form ammonia?

Analysis: The problem states that the reactants are at STP, standard temperature (273 K) and pressure (760 torr). Therefore, all the gases are at the same temperature and pressure. That makes the ratios by volume the same as the ratio by moles, so we can then substitute the volume ratio for the mole ratio in our calculations.

Assembling the Tools: First, we need to balance the chemical equation as we learned to do in Section 3.5.

$$3H_2(g) + N_2(g) \longrightarrow 2NH_3(g)$$

Avogadro's principle then gives us the equivalency we need to perform the calculation.

$$3 \text{ volumes } H_2 \Leftrightarrow 1 \text{ volume } N_2$$

The stoichiometric procedures learned in Chapter 3 can be used to solve the problem.

Solution: We now restate the problem as

$$1.50 \text{ L } N_2 \Leftrightarrow ? \text{ L } H_2$$

We use the volume ratio, now in L units, to express the equivalence

$$3 \text{ L } H_2 \Leftrightarrow 1 \text{ L } N_2$$

This volume equivalence is used to construct the conversion factor, and our solution is

$$1.50 \text{ L } N_2 \times \frac{3 \text{ L } H_2}{1 \text{ L } N_2} = 4.50 \text{ L } H_2$$

Is the Answer Reasonable? The volume of H_2 needed is three times the volume of N_2, and 3×1.5 equals 4.5, so the answer is correct. Remember, however, that the simplicity of this problem arises because the volumes are at the same temperature and pressure.

Example 10.4
Stoichiometry Calculations when Gases Are Not at the Same *T* and *P*

Nitrogen monoxide, a pollutant released by automobile engines, is oxidized by molecular oxygen to give the reddish-brown gas nitrogen dioxide, which gives smog its characteristic color. The equation is

$$2NO(g) + O_2(g) \longrightarrow 2NO_2(g)$$

How many milliliters of O_2, measured at 22 °C and 755 torr, are needed to react with 184 mL of NO, measured at 45 °C and 723 torr?

Analysis: Once again, we have a stoichiometry problem involving gas volumes, but this one is more complicated than the one in Example 10.3 because the gases are not at the same temperature and pressure. We need to make the temperature and pressure the same for both gases. Since we are asked for the volume of O_2 at 22 °C and 755 torr, let's determine the volume of the NO under those conditions, then we can use the gas volumes for our stoichiometry calculation as diagrammed below.

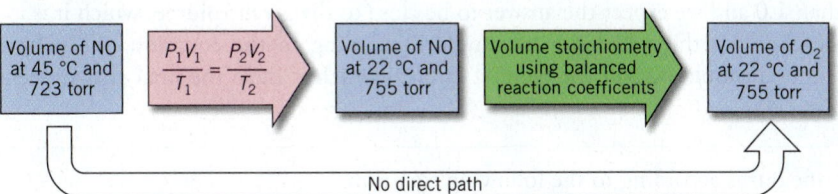

With that analysis, this problem is seen to contain two concepts: the *combined gas law* and a *stoichiometry calculation* using gas volumes.

Assembling the Tools: We can use the combined gas law to find what volume the NO would occupy if it were at the same temperature and pressure as the O_2.

$$\frac{P_1 V_1}{T_1} = \frac{P_2 V_2}{T_2}$$

We will again use Avogadro's principle once the volumes at the same temperature and pressure and the coefficients of the equation are determined to find the volume of O_2.

Solution: We use the combined gas law applied to the given volume of NO, so let's set up the data as usual.

Initial (1)		Final (2)	
P_1	723 torr	P_2	755 torr
V_1	184 mL	V_2	?
T_1	318 K (45 °C + 273)	T_2	295 K (22 °C + 273)

Solving the combined gas law for V_2 gives

$$V_2 = V_1 \times \left(\frac{P_1}{P_2}\right) \times \left(\frac{T_2}{T_1}\right)$$

Next we substitute values:

$$V_2 = 184 \text{ mL} \times \left(\frac{723 \text{ torr}}{755 \text{ torr}}\right) \times \left(\frac{295 \text{ K}}{318 \text{ K}}\right)$$

$$= 163.5 \text{ mL NO}$$

We can now use the coefficients of the equation to establish the equivalency

$$2 \text{ mL NO} \Leftrightarrow 1 \text{ mL O}_2$$

and apply it to find the volume of O_2 required for the reaction.

$$163.5 \text{ mL NO} \times \frac{1 \text{ mL O}_2}{2 \text{ mL NO}} = 81.7 \text{ mL O}_2$$

Note how we kept an extra significant figure in V_2 until we calculated the final volume of O_2.

Is the Answer Reasonable? For the first calculation, we can check to see whether the pressure and temperature ratios move the volume in the right direction. The pressure is increasing (723 torr ⟶ 755 torr), so that should decrease the volume. The pressure ratio is smaller than 1, so it is having the proper effect. The temperature is dropping (318 K ⟶ 295 K), so this change should also reduce the volume. The temperature ratio is smaller than 1, so it is also having the correct effect. The volume of NO at 22 °C and 755 torr is probably correct. We can also observe that the two ratios are only slightly less than 1.0 and we expect the answer to be close to the given volume, which it is.

The check of the calculation is simple. According to the equation, the volume of O_2 required should be half the volume of NO, which it is, so the final answer seems to be okay.

Practice Exercise 10.9

Methane burns according to the following equation.

$$CH_4(g) + 2O_2(g) \longrightarrow CO_2(g) + 2H_2O(g)$$

The combustion of 4.50 L of CH_4 consumes how many liters of O_2, both volumes measured at 25 °C and 740 torr? (*Hint:* Recall Avogadro's principle concerning the number of molecules in a fixed volume of gas at a given temperature and pressure.)

Practice Exercise 10.10

How many liters of air (air is 20.9% oxygen) are required for the combustion of 6.75 L of CH_4?

10.5 | Ideal Gas Law

In our discussion of the combined gas law, we noted that the ratio PV/T equals a constant for a fixed amount of gas. However, the value of this "constant" is actually proportional to the number of moles of gas, n, in the sample.[8]

To create an equation even more general than the combined gas law, therefore, we can write

$$\frac{P_1 V_1}{T_1} \propto n$$

We can replace the proportionality symbol with an equals sign by including another proportionality constant.

$$\frac{P_1 V_1}{T_1} = n \times \text{constant}$$

This new constant is given the symbol R and is called the **universal gas constant**. We can now write the combined gas law in a still more general form called the **ideal gas law**.

$$\frac{P_1 V_1}{T_1} = nR$$

■ Sometimes this equation is called the universal gas law.

An ideal gas would obey this law exactly over all ranges of the gas variables. The equation, sometimes called the **equation of state for an ideal gas**, is usually rearranged and written as

TOOLS
Ideal gas law

Ideal Gas Law (equation of state for an ideal gas)
$PV = nRT$

(10.2)

---

[8] In the problems we worked earlier, we were able to use the combined gas law expressed as

$$\frac{P_1 V_1}{T_1} = \frac{P_2 V_2}{T_2}$$

because the amount of gas remained fixed.

Equation 10.2 tells us how the four important variables for a gas, *P*, *V*, *n*, and *T*, are related. If we know the values of three, we can calculate the fourth. In fact, Equation 10.2 tells us that if values for three of the four variables are fixed for a gas, *the fourth can only have one value.* We can define the *state* of a gas simply by specifying any three of the four variables *P*, *V*, *T*, and *n*.

To use the ideal gas law, we have to know the value of the universal gas constant, *R*, which is equal to *PV/nT*. The value of *R* will depend on the values and units we choose for *P*, *V*, *T*, and *n*. We will use units of liters for *V*, kelvins for *T*, atmospheres for *P*, and moles for *n*. Earlier, in Section 10.4, we defined the standard temperature and pressure, STP, as 273 K and one atmosphere of pressure. We also noted that one mole of an ideal gas has a volume of 22.4 liters. Using these values we can calculate *R* as follows.

$$R = \frac{PV}{nT} = \frac{(1.00 \text{ atm})(22.4 \text{ L})}{(1.00 \text{ mol})(273 \text{ K})}$$

$$= 0.0821 \frac{\text{atm L}}{\text{mol K}}$$

or, arranging the units in a commonly used order,

$$R = 0.0821 \text{ L atm mol}^{-1} \text{ K}^{-1}$$

In later chapters we will find *R* defined using different units, and, of course, it will have a different numerical value.

■ More precise measurements give $R = 0.082057$ L atm mol^{-1} K^{-1}. If we use the bar as the standard pressure, then $R = 0.083144$ L bar mol^{-1} K^{-1}.

Example 10.5
Using the Ideal Gas Law

In Example 10.2, we described filling a 53.4 mL light bulb with argon at a temperature of 21 °C and a pressure of 2.28 torr. How many grams of argon are in the light bulb under these conditions?

Analysis: The question asks for the grams of Ar, and our stoichiometry experience (see Figure 3.6) tells us that the only way to calculate the mass of Ar is first to know the moles of Ar (or some way to calculate the moles of Ar). The first step will be to determine the moles of Ar in the light bulb; then we will convert moles to mass.

Assembling the Tools: The only equation that allows us to calculate the moles of a gas is the ideal gas law equation, *PV = nRT*. We will solve it for *n* by substituting values for *P*, *V*, *R*, and *T*. Once we have the number of moles of Ar, the tool to use is the relationship of atomic mass to moles. For Ar we write

1 mol Ar = 39.95 g Ar

A conversion factor 39.95 g Ar/1 mol Ar made from this relationship lets us convert moles to grams.

Solution: When using the ideal gas law we need to be sure that *P*, *V*, and *T* have the correct units. One helpful clue is that the units for *R* must cancel and therefore they prescribe the units we need for our variables. When using $R = 0.0821$ L atm mol^{-1} K^{-1}, we must have *V* in liters, *P* in atmospheres, and *T* in kelvins. Gathering the data and making the necessary unit conversions as we go, we have

$P = 3.00 \times 10^{-3}$ atm   from 2.28 torr $\times \dfrac{1 \text{ atm}}{760 \text{ torr}}$

$V = 0.158$ L   from 158 mL

$T = 294$ K   from (21 °C + 273)

Rearranging the ideal gas law for n gives us

$$n = \frac{PV}{RT}$$

Substituting the proper values of P, V, R, and T into this equation gives

$$n = \frac{(3.00 \times 10^{-3}\ \text{atm})(0.158\ \text{L})}{(0.0821\ \text{L atm mol}^{-1}\ \text{K}^{-1})(294\ \text{K})}$$

$$= 1.96 \times 10^{-5}\ \text{mol of Ar}$$

Now we convert moles of Ar to grams of Ar. Let's set up the question as an equation:

$$1.96 \times 10^{-5}\ \text{mol of Ar} = ?\ \text{g Ar}$$

Then we use the conversion factor above to complete the calculation.

$$1.96 \times 10^{-5}\ \text{mol Ar} \times \frac{39.95\ \text{g Ar}}{1\ \text{mol Ar}} = 7.85 \times 10^{-4}\ \text{g Ar}$$

Thus, the light bulb contains only about one milligram of argon.

Is the Answer Reasonable? As in other problems we approximate the answer by rounding all values to one significant figure to get:

$$n = \frac{(3 \times 10^{-3}\ \text{atm})(0.2\ \text{L})}{(0.1\ \text{L atm mol}^{-1}\ \text{K}^{-1})(300\ \text{K})} = \frac{0.6 \times 10^{-3}\ \text{mol}}{30} \approx 2 \times 10^{-5}\ \text{mol}$$

This is close to our answer for the moles of argon.

We multiply the estimate above by 40 g mol^{-1} Ar to get 8×10^{-4} grams. The result is very close to our calculated value. (Whenever you're working a problem where there is unit cancellation, be sure to check to be sure that the units do cancel as they are supposed to by adding your own cancellations here.)

Practice Exercise 10.11   How many grams of argon were in the 12.0 L cylinder of argon used to fill the light bulbs described in Example 10.2? The pressure of the argon was 57.8 atm and the temperature was 25 °C. (*Hint:* Are the units correct for the value of R, 0.0821 L atm mol^{-1} K^{-1}?)

Practice Exercise 10.12   Dry ice, solid $CO_2(s)$, can be made by allowing pressurized $CO_2(g)$ to expand rapidly. If 35% of the expanding $CO_2(g)$ ends up as $CO_2(s)$, how many grams of dry ice can be made from a tank of $CO_2(g)$ that has a volume of 6.0 cubic feet with a gauge pressure of 2.00×10^3 pounds per square inch (PSIG) at 22 °C? (*Note:* PSIG is the pressure above the prevailing atmospheric pressure that we can assume is 1.00 atm.)

Calculating Molar Mass of a Gas

When a chemist makes a new compound, one step in establishing its chemical identity is to determine its molar mass. If the compound is a gas, its molar mass can be found using experimental values of pressure, volume, temperature, and sample mass. The P, V, T data allow us to calculate the number of moles using the ideal gas law (as in Example 10.5).

$$n = \frac{PV}{RT}$$

Once we know the number of moles of gas and the mass of the gas sample, the molar mass is obtained by taking the ratio of grams to moles and then using the ideal gas law to substitute for moles.

TOOLS

Determination of molar mass from the ideal gas law

$$\text{molar mass} = \frac{\text{mass of sample}}{\text{mol of sample}} = \frac{\text{mass}}{n} = \frac{\text{mass}}{\frac{PV}{RT}} = \text{mass} \times \frac{RT}{PV}$$

We can use this equation to calculate the "average molar mass" of a gas mixture such as dry air. For example, if we took 22.4 liters of air at STP, we would have one mole of the mixture of gases that make up air. If we weighed the air, we would find that it weighed 28.96 g, and the "equivalent molar mass" would be 28.96 g mol^{-1}.

Example 10.6
Determining the Molar Mass of a Gas

As part of a rock analysis, a student added hydrochloric acid to a rock sample and observed a fizzing action, indicating that a gas was being evolved (see the figure in the margin). The student collected a sample of the gas in a 0.220 L gas bulb until its pressure reached 0.757 atm at a temperature of 25.0 °C. The sample weighed 0.299 g. What is the molar mass of the gas? What kind of compound was the likely source of the gas?

● **Analysis:** The strategy for finding the molar mass was described previously. We use the P, V, and T data to calculate the number of moles of gas in the sample. Then we divide the mass of the sample by the number of moles to find the molar mass.

● **Assembling the Tools:** The tool is the definition of molecular weight using the ideal gas law to determine the moles of the gas in the sample:

$$\text{molar mass} = \text{mass} \times \frac{RT}{PV}$$

● **Solution:** Any use of the ideal gas law requires the correct units to match $R = 0.0821$ L atm mol^{-1} K^{-1}. The pressure is in atmospheres, the volume is in liters, and the temperature is in kelvins. Gathering our data, we have

$$P = 0.757 \text{ atm} \qquad V = 0.220 \text{ L} \qquad T = 298 \text{ K} \quad (25.0\,°\text{C} + 273)$$
$$\text{mass} = 0.299 \text{ g}$$

Now we can substitute the data for P, V, and T, along with the value of R. This gives

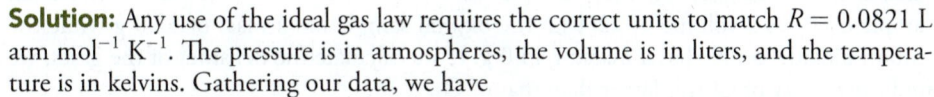

$$\text{molar mass} = 0.299 \text{ g} \times \frac{(0.0821 \text{ L atm mol}^{-1} \text{ K}^{-1})(298 \text{ K})}{(0.757 \text{ atm})(0.220 \text{ L})}$$
$$= 43.9 \text{ g mol}^{-1}$$

We now know the measured molar mass is 43.9, but what gas could this be? Looking back on our discussions in Chapter 4, what gases do we know are given off when a substance reacts with acids? In Table 4.2 we find some options; the gas might be H_2S, HCN, CO_2, or SO_2. Using atomic masses to calculate their molar masses, we get

H_2S	34 g mol^{-1}	CO_2	44 g mol^{-1}
HCN	27 g mol^{-1}	SO_2	64 g mol^{-1}

The only gas with a molar mass close to 43.9 is CO_2, and that gas would be evolved if we treat a carbonate with an acid. The rock probably contains a carbonate compound. (Limestone and marble are examples of such minerals.)

● **Are the Answers Reasonable?** We don't have to do much mental arithmetic to gain some confidence in our answer. The volume of the gas sample (0.220 L) is a bit less than $\frac{1}{100}$ of 22.4 L, the volume of one mole at STP. If the gas were at STP, then 0.299 g would be $\frac{1}{100}$ of a mole, so an entire mole would weigh about 30 g (rounded from 29.9 g). Although 30 g per mole is not very close to 43.9 g per mole, we do appear to have the

Hydrochloric acid reacting with a rock sample.

Andy Washnik

decimal in the right place. We can refine our estimate a bit. We see that the temperature is within 10% of the standard temperature and has a minor effect, but the pressure is only 75% of atmospheric pressure and is the major reason why our first estimate was off. Dividing the estimated molar mass by 0.75 (or multiplying by 1.33) gives an estimated answer of 40, very close to the one we calculated. In addition, the fact that the answer (43.9 g mol^{-1}) agrees so well with the molar mass of one of the gases formed when substances react with acids makes all the puzzle pieces fit. Viewed in total, therefore, we can feel confident in our answers.

Practice Exercise 10.13

The label on a cylinder of a noble gas became illegible, so a student allowed some of the gas to flow into an evacuated gas bulb with a volume of 300.0 mL until the pressure was 685 torr. The mass of the glass bulb increased by 1.45 g; its temperature was 27.0 °C. What is the molar mass of this gas? Which of the Group 8A gases was it? (*Hint:* What units are needed?)

Practice Exercise 10.14

A glass bulb is found to have a volume of 544.23 mL. The mass of the glass bulb filled with argon is 735.6898 g. The bulb is then flushed with a gaseous organic compound. The bulb, now filled with the organic gas, weighs 736.1310 g. The measurements were made at STP. What is the molar mass of the organic gas?

Gas Densities

■ Recall from Chapter 1 that the units for the density of liquids and solids we usually use are units of g mL^{-1} (or g cm^{-3}), but because gases have such low densities, units of g L^{-1} give numbers that are easier to comprehend.

Because one mole of any gas occupies the same volume at a particular pressure and temperature, the mass contained in that volume depends on the molar mass of the gas. Consider, for example, one-mole samples of O_2 and CO_2 at STP (Figure 10.9). Each sample occupies a volume of 22.4 L. The oxygen sample has a mass of 32.0 g while the carbon dioxide sample has a mass of 44.0 g. If we calculate the densities of the gases, we find the density of CO_2 is larger than that of O_2.

$$d_{O_2} = \frac{32.0 \text{ g}}{22.4 \text{ L}} = 1.43 \text{ g L}^{-1} \qquad d_{CO_2} = \frac{44.0 \text{ g}}{22.4 \text{ L}} = 1.96 \text{ g L}^{-1}$$

Because the volume of a gas is affected by temperature and pressure, the density of a gas changes as these variables change. Gases become less dense as their temperatures rise, which is why hot-air balloons are able to float; the less dense hot air inside the balloon floats in the more dense cool air that surrounds it. Gases also become more dense as their pressures increase because increasing the pressure packs more molecules into the same space.

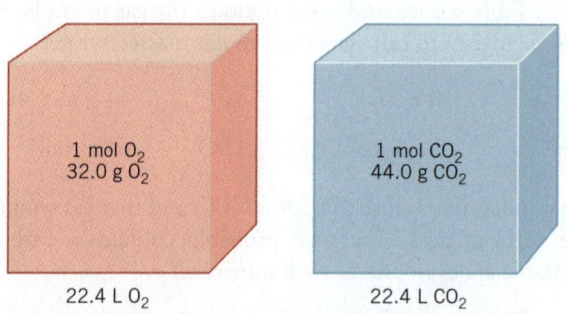

Figure 10.9 | **One-mole samples of O_2 and CO_2 at STP.** Each sample occupies 22.4 L, but the O_2 weighs 32.0 g whereas the CO_2 weighs 44.0 g. The CO_2 has more mass per unit volume than the O_2 and has the higher density.

To calculate the density of a gas at conditions other than STP, we can look at the equation we used to calculate the molar mass of a gas and the equation for density

$$\text{molar mass} = \text{mass} \times \frac{RT}{PV} \quad \text{and} \quad \text{density} = \frac{\text{mass}}{\text{volume}}$$

Putting these two equations together and rearranging gives

$$\text{density} = \frac{\text{mass}}{V} = \frac{\text{molar mass} \times P}{RT}$$

Example 10.7
Calculating the Density of a Gas

One procedure used to separate the isotopes of uranium to obtain material to construct a nuclear weapon employs a uranium compound with the formula UF_6. The compound boils at about 56 °C, so at 95 °C it is a gas. What is the density of UF_6 at 95 °C if the pressure of the gas is 746 torr? (Assume the gas contains the mix of uranium isotopes commonly found in nature.)

Analysis: To determine the density of a substance, we will need its mass and volume. We are given the volume, and the ideal gas law has the information we can use to determine the moles from which we can find the mass using the molar mass as shown above.

Assembling the Tools: We will need the ideal gas law and the equations for calculating the number of moles, n, and the density, d as shown above,

$$\text{density} = \frac{\text{molar mass} \times P}{RT}$$

For all gas law problems we need to be sure the units are correct, so we should be prepared to convert between torr and atmospheres as well as from °C to K.

Solution: After converting 746 torr to 0.982 atm and 95 °C to 368 K we enter the data into the equation along with the molar mass of 352.0 g/mol for UF_6 as

$$\frac{(352.0 \text{ g } UF_6/\text{mol } UF_6)(0.982 \text{ atm})}{(0.0821 \text{ L atm mol}^{-1} \text{ K}^{-1})(368 \text{ K})} = d = 11.4 \text{ g/L}$$

The density of gaseous UF_6 is 11.4 g/L.

Is the Answer Reasonable? At STP, the density would be equal to 352 g ÷ 22.4 L = 15.7 g L^{-1}. The given pressure is almost 1 atm, but the temperature is quite a bit higher than 273 K. Gases expand when heated, so a liter of the gas at the higher temperature will have less UF_6 in it. That means the density will be lower at the higher temperature, and our answer agrees with this analysis. Our answer is reasonable.

Practice Exercise 10.15

Sulfur dioxide is a gas that has been used in commercial refrigeration, but not in residential refrigeration because it is toxic, and you could be injured if the gas leaked. What is the density of SO_2 gas measured at −5 °C and a pressure of 96.5 kPa? (*Hint:* How does the number of moles relate to density?)

Practice Exercise 10.16

Radon, a radioactive gas, is formed in one step of the natural radioactive decay sequence of U-235 to Pb-207. Radon usually escapes harmlessly through the soil to the atmosphere. When the soil is frozen or saturated with water the only escape route is through cracks in the basements of houses and other buildings. In order to detect radon in a residence, would you place the sensor in the attic, the ground floor living area, or the basement? Justify your answer.

Calculating Molar Mass from Gas Density

One of the ways we can use the density of a gas is to determine the molar mass. To do this, we also need to know the temperature and pressure at which the density was measured. Example 10.8 illustrates the reasoning and calculation involved.[9]

Example 10.8
Calculating the Molecular Mass from Gas Density

A liquid sold under the trade name Perclene is used as a dry cleaning solvent. It has an empirical formula CCl_2 and a boiling point of 121 °C. When vaporized, the gaseous compound has a density of 4.93 g L^{-1} at 785 torr and 155 °C. What is the molar mass of the compound and what is its molecular formula?

Analysis: In Example 10.7 we saw how the density could be calculated using the ideal gas law if we knew P, T, and the molar mass. Thus, if we know the density then we can calculate the molar mass. When we know the empirical formula and the molar mass we can determine the molecular formula.

Assembling the Tools: Again, let's combine two concepts, the ideal gas law and the density, as we did in Example 10.7:

$$d = \frac{P \times \text{molar mass}}{RT}$$

As usual, we will be prepared to convert torr to atmospheres and °C to K temperature units. Finally, we add in the third concept involved in this question, the determination of molecular formulas from empirical formula and molecular masses. For this we use the tool found in Section 3.4.

Solution: Rearranging the equation above to solve for molar mass gives us

$$\text{molar mass} = \frac{dRT}{P}$$

To perform this calculation and cancel the units of R correctly we need to have the pressure in atmosphere units and the temperature in kelvins.

$$T = 428 \text{ K} \quad \text{from } (155 \text{ °C} + 273)$$

$$P = 1.03 \text{ atm} \quad \text{from } 785 \text{ torr} \times \frac{1 \text{ atm}}{760 \text{ torr}}$$

$$V = 1.00 \text{ L}$$

---

[9] From the ideal gas law we can derive an equation from which we could calculate the molar mass directly from the density. If we let the mass of gas equal g, we could calculate the number of moles, n, by the ratio

$$n = \frac{g}{\text{molar mass}}$$

Substituting into the ideal gas law gives

$$PV = nRT = \frac{gRT}{\text{molar mass}}$$

Solving for molar mass, we have

$$\text{molar mass} = \frac{gRT}{PV} = \left(\frac{g}{V}\right) \times \frac{RT}{P}$$

The quantity g/V is the ratio of mass to volume, which is the density d, so making this substitution gives

$$\text{molar mass} = \frac{dRT}{P}$$

This equation could also be used to solve the problem in Example 10.8 by substituting values for d, R, T, and P.

Then

$$\text{molar mass} = \frac{(4.93 \text{ g L}^{-1})(0.0821 \text{ L atm mol}^{-1} \text{ K}^{-1})(428 \text{ K})}{(1.03 \text{ atm})} = 168 \text{ g mol}^{-1}$$

To calculate the molecular formula, we divide the empirical formula mass, 82.9 g mol^{-1}, into the molecular mass, 168 g mol^{-1}, to see how many times CCl$_2$ occurs in the molecular formula:

$$\frac{168 \text{ g mol}^{-1}}{82.9 \text{ g mol}^{-1}} = 2.03$$

The result is close enough to 2.00, so to find the molecular formula, we multiply all of the subscripts of the empirical formula by 2:

$$\text{molecular formula} = C_{1 \times 2}Cl_{2 \times 2} = C_2Cl_4$$

(This is the formula for a compound commonly called tetrachloroethylene, which is indeed used as a dry cleaning fluid.)

● **Is the Answer Reasonable?** Sometimes we don't have to do any arithmetic to see that an answer is almost surely correct. The fact that the molar mass we calculated from the gas density is evenly divisible by the empirical formula mass suggests we've worked the problem correctly.

A gaseous compound of phosphorus and fluorine with an empirical formula of PF$_2$ has a density of 5.60 g L^{-1} at 23.0 °C and 750 torr. Determine the molecular formula of this compound. (*Hint:* Calculate the molar mass from the density.)

Practice Exercise 10.17

A compound composed of only carbon and hydrogen has a density of 5.55 g/L at 40.0 °C and 1.25 atm. What is the molar mass of the compound? What are the possible combinations of C and H that add up to that molar mass? Using the information in Section 2.7, determine which of your formulas is most likely the correct one.

Practice Exercise 10.18

Stoichiometry Using the Ideal Gas Law

Many chemical reactions either consume or give off gases. The ideal gas law can be used to relate the volumes of such gases to the amounts of other substances involved in the reaction, as illustrated by the following example.

Example 10.9
Calculating the Volume of a Gaseous Product Using the Ideal Gas Law

An important chemical reaction in the manufacture of Portland cement is the high temperature decomposition of calcium carbonate to give calcium oxide and carbon dioxide.

$$CaCO_3(s) \longrightarrow CO_2(g) + CaO(s)$$

Suppose a 1.25 g sample of CaCO$_3$ is decomposed by heating. How many milliliters of CO$_2$ gas will be evolved if the volume will be measured at 745 torr and 25 °C?

● **Analysis:** This appears to be a stoichiometry problem where we must convert the mass of one substance into the volume of a gas that would be expected from the reaction.

● **Assembling the Tools:** We need to assemble all of the techniques we used for stoichiometry calculations in Section 3.5. In addition we will need the ideal gas law to convert the moles of gaseous product into the expected volume of gas. The specific relationships we need are

$$1 \text{ mol CaCO}_3 \Leftrightarrow 1 \text{ mol CO}_2$$
$$1 \text{ mol CaCO}_3 = 100.1 \text{ g CaCO}_3$$

Our sequence of conversions will be

$$g\ CaCO_3 \longrightarrow mol\ CaCO_3 \longrightarrow mol\ CO_2 \longrightarrow mL\ CO_2$$

Solution: First, we will convert the 1.25 g of $CaCO_3$ to moles of $CaCO_3$, which is then converted to the number of moles of CO_2. We will use this value for n in the ideal gas law equation to find the volume of CO_2.

The formula mass of $CaCO_3$ is 100.1, so

$$\text{moles of CaCO} = 1.25\ \overline{g\ CaCO_3} \times \frac{1\ \text{mole } CaCO_3}{100.1\ \overline{g\ CaCO_3}}$$

$$= 1.25 \times 10^{-2}\ mol\ CaCO_3$$

Because 1 mol $CaCO_3 \Leftrightarrow$ 1 mol CO_2 we must have 1.25×10^{-2} mol CO_2, which we will use for n in the ideal gas law.

Before we use n in the ideal gas law equation, we must convert the given pressure and temperature into the units required by R.

$$P = 745\ \overline{torr}\ \frac{1\ atm}{760\ torr} = 0.980\ atm \qquad T = (25.0\ °C + 273) = 298\ K$$

By rearranging the ideal gas law equation we obtain

$$V = \frac{nRT}{P}$$

$$= \frac{(1.25 \times 10^{-2}\ mol)(0.0821\ L\ atm\ mol^{-1}\ K^{-1})(298\ K)}{0.980\ atm}$$

$$= 0.312\ L = 312\ mL$$

The reaction will yield 312 mL of CO_2 at the conditions specified.

Is the Answer Reasonable? We round all of the numbers in our calculation to one significant figure to get (cancel the units in the setup below)

$$V = \frac{(1 \times 10^{-2}\ mol)(0.1\ L\ atm\ mol^{-1}\ K^{-1})(300\ K)}{1\ atm} = 0.3\ L\ or\ 300\ mL$$

This answer is close to what we calculated above, and we may assume the calculation is correct. We also note that all units cancel to leave the desired liter units for volume.

Practice Exercise 10.19

Carbon disulfide is an extremely flammable liquid. It can be ignited by any small spark or even a very hot surface such as a steam pipe. The combustion reaction is

$$CS_2 + 3O_2 \longrightarrow CO_2 + 2SO_2$$

When 11.0 g of CS_2 are burned in excess oxygen, what total volume CO_2 and SO_2 is formed at 28 °C and 883 torr? (*Hint:* Treat this as an ordinary stoichiometry problem.)

Practice Exercise 10.20

In one lab, the gas-collecting apparatus used a gas bulb with a volume of 257 mL. How many grams of $CaCO_3(s)$ need to be heated to prepare enough $CO_2(g)$ to fill this bulb to a pressure of 738 torr at a temperature of 23 °C? The equation is

$$CaCO_3(s) \longrightarrow CaO(s) + CO_2(g)$$

10.6 | Dalton's Law of Partial Pressures

So far in our discussions we've dealt with only pure gases. However, gas mixtures, such as the air we breathe, are quite common. In general, gas mixtures obey the same laws as pure gases, so Boyle's law applies equally to both pure oxygen and to air. There are times, though, when we must be concerned with the composition of a gas mixture, such as when we are studying a pollutant in the atmosphere. In these cases, the variables affected by the composition of a gas mixture are the numbers of moles of each component and the contribution each component makes to the total observed pressure. Because gases mix completely, all the components of the mixture occupy the same volume—that of the container holding them. Furthermore, the temperature of each gaseous component is the same as the temperature of the entire mixture. Therefore, in a gas mixture, each of the components has the same volume and the same temperature.

Partial Pressures

In a mixture of nonreacting gases such as air, each gas contributes to the total pressure in proportion to the fraction (by moles) in which it is present (see Figure 10.10). This contribution to the total pressure is called the **partial pressure** of the gas. It is the pressure the gas would exert if it were the only gas in a container of the same size at the same temperature.

The general symbol we will use for the partial pressure of a gas A is P_A. For a particular gas, the formula of the gas may be put into the subscript, as in P_{O_2}. What John Dalton discovered about partial pressures is now called **Dalton's law of partial pressures**: The total pressure of a mixture of gases is the sum of their individual partial pressures. In equation form, the law is

$$P_{total} = P_A + P_B + P_C + \ldots \qquad (10.3)$$

In dry CO_2-free air at STP, for example, P_{O_2} is 159.12 torr, P_{N_2} is 593.44 torr, and P_{Ar} is 7.10 torr. These partial pressures add up to 759.66 torr, just 0.34 torr less than 760 torr or 1.00 atm. The remaining 0.34 torr is contributed by several trace gases, including other noble gases.

The explosive PETN, pentaerythritol tetranitrate, is one of the most powerful explosives known. It reacts with the following balanced equation

$$C(CH_2ONO_2)_4(s) \longrightarrow 2CO(g) + 4H_2O(g) + 3CO_2(g) + 2N_2(g)$$

If 0.0250 mol of PETN reacts to fill a 30.0 L sphere at a temperature of 25 °C, what is the partial pressure of each gas and the total pressure in the flask? (*Hint:* Start by calculating the moles of each gas present.)

Suppose a tank of oxygen-enriched air prepared for scuba diving has a volume of 17.00 L and a pressure of 237.0 atm at 25 °C. How many grams of oxygen are present if all the other gases in the tank exert a partial pressure of 115.0 atm?

Collecting Gases over Water

When gases that do not react with water are prepared in the laboratory, they can be trapped over water by an apparatus like that shown in Figure 10.11. Because of the way the gas is

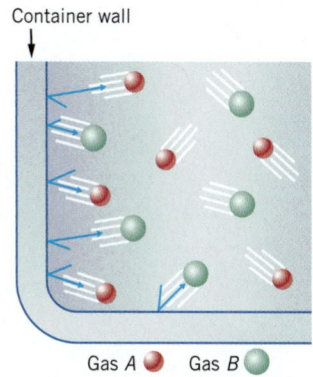

Gas A ● Gas B ○

Figure 10.10 | Partial pressures viewed at the molecular level. In a mixture of two gases, A and B, both collide with the walls of the container and thereby contribute their partial pressures to the total pressure.

TOOLS

Dalton's law of partial pressure

Practice Exercise 10.21

Practice Exercise 10.22

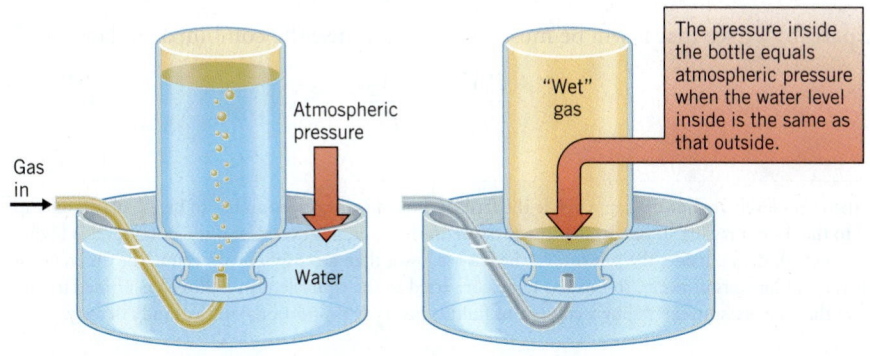

Atmospheric pressure

Gas in

Water

"Wet" gas

The pressure inside the bottle equals atmospheric pressure when the water level inside is the same as that outside.

Figure 10.11 | Collecting a gas over water. As the gas bubbles through the water, water vapor goes into the gas, so the total pressure inside the bottle includes the partial pressure of the water vapor at the temperature of the water.

TABLE 10.2	Vapor Pressure of Water at Various Temperatures[a]		
Temperature (°C)	Vapor Pressure (torr)	Temperature (°C)	Vapor Pressure (torr)
0	4.579	50	92.51
5	6.543	55	118.04
10	9.209	60	149.38
15	12.788	65	187.54
20	17.535	70	233.7
25	23.756	75	289.1
30	31.824	80	355.1
35	42.175	85	433.6
37[b]	47.067	90	525.8
40	55.324	95	633.9
45	71.88	100	760.0

[a] A more complete table is in Appendix C, Table 5.

[b] Human body temperature.

■ Even the mercury in a barometer has a tiny vapor pressure—0.0012 torr at 20 °C—which is much too small to affect readings of barometers and the manometers studied in this chapter.

collected, it is saturated with water vapor. (We say the gas is "wet.") Water vapor in a mixture of gases has a partial pressure like that of any other gas.

The vapor present in the space above *any* liquid always contains some of the liquid's vapor, which exerts its own pressure called the liquid's **vapor pressure**. Its value for any given substance depends only on the temperature. The vapor pressures of water at different temperatures, for example, are listed in Table 10.2. A more complete table is in Appendix C.5.

If we have adjusted the height of the collecting jar so the water level inside matches that outside, the total pressure of the trapped gas equals the atmospheric pressure, so the value for P_{total} is obtained from the laboratory barometer. We can now calculate P_{gas}, which is the pressure that the gas would exert if it were dry (i.e., without water vapor in it) using Dalton's law of partial pressures, $P_{total} = P_{gas} + P_{water\ vapor}$[10]

Example 10.10
Collecting a Gas over Water

A sample of oxygen is collected over water at 15 °C and a pressure of 738 torr. Its volume is 316 mL. (a) What is the partial pressure, in torr, of the oxygen? (b) What will be its volume, in mL, at STP when the water is removed?

Analysis: There are two parts to this problem. Part (a) concerns the collection of a gas over water, and the determination of the partial pressure of oxygen. Part (b) of the question requires that we perform a gas law calculation to determine the volume of oxygen at STP.

Assembling the Tools: For part (a) we will use Dalton's partial pressure equation,

$$P_{total} = P_{water\ vapor} + P_{O_2}$$

For part (b) we see that it will be most convenient to use the combined gas law.

$$\frac{P_1 V_1}{T_1} = \frac{P_2 V_2}{T_2}$$

---

[10] If the water levels are not the same inside the flask and outside, a correction has to be calculated and applied to the room pressure to obtain the true pressure in the flask. For example, if the water level is higher inside the flask than outside, the pressure in the flask is lower than atmospheric pressure. The difference in levels is in millimeters of *water*, so this has to be converted to the equivalent in millimeters of mercury using the fact that the pressure exerted by a column of fluid is inversely proportional to the fluid's density.

We might expect that we have to convert units as in previous problems. However, we note that R does not appear in any of our equations and we might not have to convert units; let's see.

Solution: Part (a): To calculate the partial pressure of the oxygen, we use Dalton's law. We will need the vapor pressure of water at 15 °C, which we find in Table 10.2 to be 12.788 torr, rounded to 12.8 torr. We rearrange the equation above to calculate P_{O_2}.

$$P_{O_2} = P_{total} - P_{water\ vapor}$$

$$= 738\ torr - 12.8\ torr = 725\ torr$$

The answer to part (a) is that the partial pressure of O_2 is 725 torr.

Part (b): We'll begin by making a table of our data.

Initial (1)		Final (2)	
P_1	725 torr (which is P_{O_2})	P_2	760 torr (standard pressure)
V_1	316 mL	V_2	the unknown is V_2
T_1	288 K (15.0 °C + 273)	T_2	273 K (standard temperature)

We use these in the combined gas law equation; solving for V_2 and rearranging the equation a bit we have

$$V_2 = V_1 \times \left(\frac{P_1}{P_2}\right) \times \left(\frac{T_2}{T_1}\right)$$

Now we can substitute values and calculate V_2.

$$V_2 = 316\ mL \times \left(\frac{725\ torr}{760\ torr}\right) \times \left(\frac{273\ K}{288\ K}\right)$$

$$= 286\ mL$$

Thus, when the water vapor is removed from the gas sample, the dry oxygen will occupy a volume of 286 mL at STP.

Are the Answers Reasonable? We know the pressure of the dry O_2 will be less than that of the wet gas, so the answer to part (a) seems reasonable. To check part (b), we can see if the pressure and temperature ratios move the volume in the correct direction. The pressure is increasing (720 torr ⟶ 760 torr), which should tend to lower the volume. The pressure ratio above will do that. The temperature change (293 K ⟶ 273 K) should also lower the volume, and once again, the temperature ratio above will have that effect. Our answer to part (b) is probably okay.

Suppose you prepared a sample of nitrogen and collected it over water at 15 °C at a total pressure of 745 torr and a volume of 317 mL. Find the partial pressure of the nitrogen, in torr, and the volume, in mL, it would occupy at STP. (*Hint:* Find the partial pressure of water at 15 °C.)

Practice Exercise 10.23

A 2.50 L sample of methane was collected over water at 28 °C until the pressure in the flask was 775 torr. A small amount of $CaSO_4(s)$ was then added to the flask to absorb the water vapor (forming $CaSO_4 \cdot 2H_2O(s)$). What is the pressure inside the flask once all the water is absorbed? Assume that the addition of $CaSO_4(s)$ absorbed all the water and did not change the volume of the flask. How many moles of $CH_4(g)$ have been collected?

Practice Exercise 10.24

Mole Fractions and Mole Percents

One of the useful ways of describing the composition of a mixture is in terms of the *mole fractions* of the components. The **mole fraction** *is the ratio of the number of moles of a given*

■ The concept of mole fraction applies to any uniform mixture in any physical state—gas, liquid, or solid.

TOOLS

Mole fractions

component to the total number of moles of all components. Expressed mathematically, the mole fraction of substance A in a mixture of $A, B, C, D, \ldots, Z$ substances is

$$X_A = \frac{n_A}{n_A + n_B + n_C + n_D + \ldots + n_Z} \tag{10.4}$$

where X_A is the mole fraction of component A, and $n_A, n_B, n_C, n_D, \ldots, n_Z$ are the numbers of moles of each component, $A, B, C, D, \ldots, Z$, respectively. The sum of all mole fractions for a mixture must always equal 1.

You can see in Equation 10.4 that both numerator and denominator have the same units (moles), so they cancel. As a result, a mole fraction has no units. Nevertheless, always remember the definition: a mole fraction stands for the ratio of *moles* of one component to the total number of *moles* of all components.

Sometimes the mole fraction composition of a mixture is expressed on a percentage basis; we call it a **mole percent (mol%)**. The mole percent is obtained by multiplying the mole fraction by 100 mol%.

$$\text{mole percent}_A = \frac{n_A}{n_A + n_B + n_C + n_D + \ldots + n_Z} \times 100\%$$

Mole Fractions and Partial Pressures

Partial pressure data can be used to calculate the mole fractions of individual gases in a gas mixture because the number of *moles* of each gas is directly proportional to its partial pressure. We can demonstrate this as follows. The partial pressure, P_A, for any one gas, A, in a gas mixture with a total volume V at a temperature T is found by the ideal gas law equation, $PV = nRT$. So to calculate the number of moles of A present, we have

$$n_A = \frac{P_A V}{RT}$$

For any particular gas mixture at a given temperature, the values of V, R, and T are all constants, making the ratio V/RT a constant, too. We can therefore simplify the previous equation by using C to stand for V/RT. In other words, we can write

$$n_A = P_A C$$

The result is the same as saying that *the number of moles of a gas in a mixture of gases is directly proportional to the partial pressure of the gas.* The constant C is the same for all gases in the mixture. So by using different letters to identify individual gases, we can let $P_B C$ stand for n_B, $P_C C$ stand for n_C, and so on in Equation 10.4. Thus,

$$X_A = \frac{P_A C}{P_A C + P_B C + P_C C + \ldots + P_Z C}$$

The constant, C, can be factored out and canceled, so

$$X_A = \frac{P_A}{P_A + P_B + P_C + \ldots + P_Z}$$

The denominator is the sum of the partial pressures of all the gases in the mixture, but this sum equals the total pressure of the mixture (Dalton's law of partial pressures). Therefore, the previous equation simplifies to

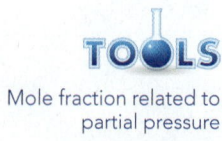

TOOLS

Mole fraction related to partial pressure

$$X_A = \frac{P_A}{P_{\text{total}}} \tag{10.5}$$

Thus, the mole fraction of a gas in a gas mixture is simply the ratio of its partial pressure to the total pressure. Equation 10.5 also gives us a simple way to calculate the partial pressure of a gas in a gas mixture when we know its mole fraction.

Example 10.11
Using Mole Fractions to Calculate Partial Pressures

Suppose a mixture of oxygen and nitrogen is prepared in which there are 0.200 mol O_2 and 0.500 mol N_2. If the total pressure of the mixture is 745 torr, what are the partial pressures of the two gases in the mixture?

Analysis: This problem asks us to determine the partial pressures of two gases, and we have been given the moles of each gas as well as the total pressure. We can use the mole fraction of each gas to determine the partial pressure of each gas.

Assembling the Tools: We need to first determine the mole fraction of each gas, Equation 10.4,

$$X_{O_2} = \frac{\text{moles of } O_2}{\text{mole of } O_2 + \text{moles of } N_2} \quad \text{and} \quad X_{N_2} = \frac{\text{moles of } N_2}{\text{mole of } O_2 + \text{moles of } N_2}$$

and to find the partial pressure, we rearrange Equation 10.5,

$$P_{O_2} = X_{O_2} P_{\text{total}} \quad \text{and} \quad P_{N_2} = X_{N_2} P_{\text{total}}$$

Solution: The mole fractions are calculated as follows:

$$X_{O_2} = \frac{\text{moles of } O_2}{\text{moles of } O_2 + \text{moles of } N_2}$$

$$= \frac{0.200 \text{ mol}}{0.200 \text{ mol} + 0.500 \text{ mol}}$$

$$= \frac{0.200 \text{ mol}}{0.700 \text{ mol}} = 0.286$$

Similarly, for N_2 we have[11]

$$= \frac{0.500 \text{ mol}}{0.200 \text{ mol} + 0.500 \text{ mol}} = 0.714$$

We can now use Equation 10.5 to calculate the partial pressure. Solving the equation for partial pressure, we have

$$P_{O_2} = X_{O_2} P_{\text{total}}$$

$$= 0.286 \times 745 \text{ torr}$$

$$= 213 \text{ torr}$$

$$P_{N_2} = X_{N_2} P_{\text{total}}$$

$$= 0.714 \times 745 \text{ torr}$$

$$= 532 \text{ torr}$$

Thus, the partial pressure of O_2 is 213 torr and the partial pressure of N_2 is 532 torr.

Is the Answer Reasonable? There are three things we can check here. First, the mole fractions add up to 1.000, which they must. Second, the partial pressures add up to 745 torr, which equals the given total pressure. Third, the mole fraction of N_2 is somewhat more

---

[11] Notice that the sum of the mole fractions (0.286 + 0.714) equals 1.000. In fact, we could have obtained the mole fraction of nitrogen with less calculation by subtracting the mole fraction of oxygen from 1.000.

$$X_{O_2} + X_{N_2} = 1.000$$

$$X_{N_2} = 1.000 - X_{O_2}$$

$$= 1.000 - 0.286 = 0.714$$

than twice that for O_2, so its partial pressure should be somewhat more than twice that of O_2. Examining the answers, we see this is true, so our answers should be correct.

Practice Exercise 10.25   Suppose a mixture containing 2.15 g H_2 and 34.0 g NO has a total pressure of 2.05 atm. What are the partial pressures of both gases in the mixture? (*Hint:* How many moles of H_2 and NO are in the flask?)

Practice Exercise 10.26   Sulfur dioxide and oxygen react according to the equation

$$2SO_2(g) + O_2(g) \longrightarrow 2SO_3(g)$$

If 50.0 g of $SO_2(g)$ is added to a flask resulting in a pressure of 0.750 atm, what will be the total pressure in the flask before and after the reaction when a stoichiometric amount of oxygen is added?

Graham's Law of Effusion

If you've ever walked past a restaurant and found your mouth watering after smelling the aroma of food, you've learned firsthand about diffusion! **Diffusion** is the spontaneous intermingling of the molecules of one gas, like those of the food aromas, with molecules of another gas, like the air outside the restaurant. (See Figure 10.12*a*). **Effusion**, on the other hand, is the gradual movement of gas molecules through a very tiny hole into a vacuum (Figure 10.12*b*). The rates at which both of these processes occur depends on the speeds of gas molecules; the faster the molecules move, the more rapidly diffusion and effusion occur.

The Scottish chemist Thomas Graham (1805–1869) studied the rates of diffusion and effusion of a variety of gases through porous clay pots and through small apertures. Comparing different gases at the same temperature and pressure, Graham found that their

CHEMISTRY AND CURRENT AFFAIRS (10.2)

Effusion and Nuclear Energy

The fuel used in almost all nuclear reactors is uranium, but only one of its naturally occurring isotopes, ^{235}U, can be easily split to yield energy. Unfortunately, this isotope is present in a very low concentration (about 0.72%) in naturally occurring uranium. Most of the element as it is mined consists of the more abundant isotope ^{238}U. Therefore, before uranium can be fabricated into fuel elements, it must be enriched to a ^{235}U concentration of about 2 to 5 percent. Enrichment requires that the isotopes be separated, at least to some degree.

Separating the uranium isotopes is not feasible by chemical means because the chemical properties of both isotopes are essentially identical. Instead, a method is required that is based on the very small difference in the masses of the isotopes. As it happens, uranium forms a compound with fluorine, UF_6, that is easily vaporized at a relatively low temperature. The UF_6 gas thus formed consists of two kinds of molecules, $^{235}UF_6$ and $^{238}UF_6$, with molecular masses of 349 and 352, respectively. Because of their different masses, their rates of effusion are slightly different; $^{235}UF_6$ effuses 1.0043 times faster than $^{238}UF_6$. Although the difference is small, it is sufficient to enable enrichment, provided the effusion is carried out over and over again enough times. In fact, it takes over 1400 separate effusion chambers arranged one after another to achieve the necessary level of enrichment.

Modern enrichment plants separate the $^{235}UF_6$ and $^{238}UF_6$ in a process using gas centrifuges. Inside the centrifuge the gases rotate at high speeds imparted by an impeller. The heavier $^{238}UF_6$ concentrates slightly toward the outer part of the centrifuge while the lighter

$^{235}UF_6$ has slightly higher concentrations toward the center as shown in Figure 1. These are continuously separated, and repeated centrifugation steps lead to the desired purity.

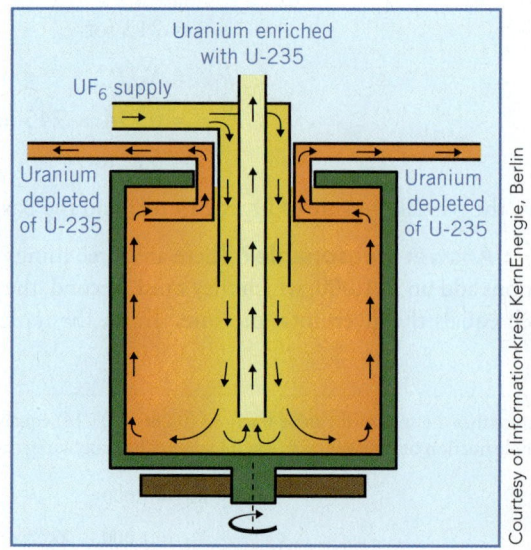

Figure 1 Isotopic separation by centrifugation

Courtesy of Informationkreis KernEnergie, Berlin

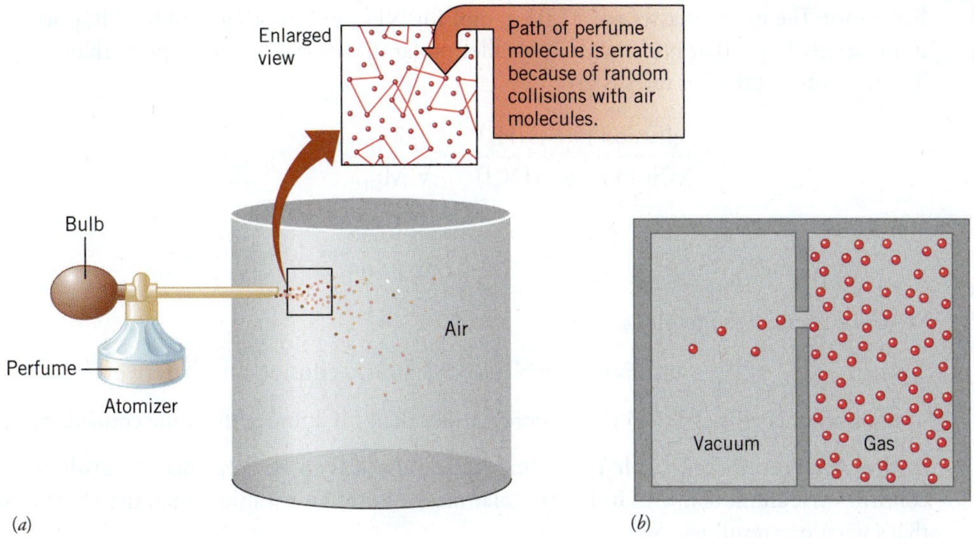

Enlarged view

Path of perfume molecule is erratic because of random collisions with air molecules.

Bulb

Perfume

Atomizer

Air

Vacuum   Gas

(a)   (b)

Figure 10.12 | Spontaneous movements of gases. (*a*) Diffusion. (*b*) Effusion.

rates of effusion were inversely proportional to the square roots of their densities. This relationship is now called **Graham's law**.

$$\text{Effusion rate} \propto \frac{1}{\sqrt{d}} \qquad \left(\begin{array}{c}\text{when compared at the}\\ \text{same } T \text{ and } P\end{array}\right)$$

Graham's law is usually used to compare the rates of effusion of different gases, so the proportionality constant can be eliminated and an equation can be formed by writing the ratio of effusion rates:

$$\frac{\text{effusion rate } (A)}{\text{effustion rate } (B)} = \frac{\dfrac{1}{\sqrt{d_A}}}{\dfrac{1}{\sqrt{d_B}}} = \frac{\sqrt{d_B}}{\sqrt{d_A}} = \sqrt{\frac{d_B}{d_A}} \qquad (10.6)$$

■ By taking the ratio, the proportionality constant cancels from numerator and denominator.

Earlier you saw that the density of a gas is directly proportional to its molar mass. Therefore, we can re-express Equation 10.6 as follows:

$$\frac{\text{effusion rate } (A)}{\text{effustion rate } (B)} = \sqrt{\frac{d_B}{d_A}} = \sqrt{\frac{\text{molar mass}_B}{\text{molar mass}_A}} \qquad (10.7)$$

TOOLS

Graham's law of effusion

where molar mass$_A$ and molar mass$_B$ are the molar masses of gases *A* and *B*, respectively.

Molecular masses also affect the rates at which gases undergo diffusion. Gases with low molar masses diffuse (and effuse) more rapidly than gases with high molar masses. Thus, hydrogen, H_2, with a molar mass of 2.02 g/mol will diffuse more rapidly than methane, CH_4, with a molar mass of 16.05 g/mol.

Example 10.12
Using Graham's Law

At a given temperature and pressure, which effuses more rapidly and by what factor: ammonia or hydrogen chloride?

Analysis: This is obviously a gas effusion problem that will require the use of Graham's law. Determining which effuses more rapidly, and obtaining a factor to describe how much more rapidly that one effuses, will require that we set up the correct ratio of effusion rates.

Assembling the Tools: We are going to need Graham's law of effusion (Equation 10.7), which we can write as

$$\frac{\text{effusion rate } (NH_3)}{\text{effustion rate } (HCl)} = \sqrt{\frac{M_{HCl}}{M_{NH_3}}}$$

Solution: The molar masses are 17.03 g/mol for NH_3 and 36.46 g/mol for HCl, so we immediately know that NH_3, with its smaller molar mass, effuses more rapidly than HCl. The ratio of the effusion rates is given by

$$\frac{\text{effusion rate } (NH_3)}{\text{effusion rate } (HCl)} = \sqrt{\frac{M_{HCl}}{M_{NH_3}}}$$

$$= \sqrt{\frac{36.46}{17.03}} = 1.463$$

We can rearrange the result as

$$\text{effusion rate } (NH_3) = 1.463 \times \text{effusion rate } (HCl)$$

Thus, ammonia effuses 1.463 times more rapidly than HCl under the same conditions.

Is the Answer Reasonable? The only quick check is to be sure that the arithmetic confirms that ammonia, with its lower molar mass, effuses more rapidly than the HCl, and that's what our result tells us.

Practice Exercise 10.27   Bromine has two isotopes with masses of 78.9 and 80.9 (to three significant figures), respectively. Bromine boils at 59 °C. At 75 °C, what is the expected ratio of the rate of effusion of $^{81}Br\text{—}^{81}Br$ compared to $^{81}Br\text{—}^{79}Br$ and $^{79}Br\text{—}^{79}Br$? (*Hint:* How do the rates of effusion relate to the molecular masses of gases?)

Practice Exercise 10.28   The hydrogen halide gases all have the same general formula, HX, where X can be F, Cl, Br, or I. If HCl(g) effuses 1.88 times more rapidly than one of the others, which hydrogen halide is the other: HF, HBr, or HI?

✹ Analyzing and Solving Multi-Concept Problems

A long time ago, in a poorly equipped lab that had only a balance and an oven, a chemist was asked to determine the formula for a substance that by its crystalline nature appeared to be a single pure substance. The first thing he did was to take a sample of this unknown compound, with a mass of 2.121 g, and heat it at 250 °C for six hours. When cooled, the mass of the substance was now 1.020 g. While waiting for the sample to dry, the chemist tested the substance's solubility and found it was insoluble in water but did dissolve with the release of a gas when a strong acid was added. With that knowledge the chemist set up an apparatus to react the dry sample with acid and collect the gas evolved by bubbling it through water into a 265 mL collection flask. At that time the temperature was 24 °C and the atmospheric pressure was 738 torr and the gas completely filled the flask. Finally, the chemist weighed the flask with the gas and found it to be 182.503 g. The flask weighed 182.346 g when it contained only air. What is the formula for this compound?

Analysis: Let's write out a plan. (1) Determine the moles of gas produced when the dry compound is dissolved in acid. (2) Determine the mass of the gas, its molar mass, and thus its identity, and then the anion that produced it. (3) Assuming that the cation could be M^+, M^{2+}, or M^{3+}, we use the mole ratio with the anion and determine the atomic mass of the cation. We identify the cation if possible. (4) We determine the number of water molecules in the hydrate and complete the chemical formula.

PART 1

Assembling the Tools: We will need the pressure of the dry gas:

$$P_{dry} = P_{total} - P_{water}$$

Next, we need to solve the ideal gas law for the moles of gas:

$$n = \frac{PV}{RT}$$

Solution: The tabulated data in Appendix C tell us that the vapor pressure of water at 24 °C is 22.4 torr and therefore our pressure of the dry gas is 738 torr − 24 torr = 714 torr. Converting torr to atmospheres, we get 0.939 atm. We convert the volume to 0.265 L, and the temperature in kelvins is 297 K. Calculating n gives us

$$n = \frac{(0.939 \text{ atm})(0.265 \text{ L})}{(0.0821 \text{ L atm mol}^{-1} \text{ K}^{-1})(297 \text{ K})} = 0.0102 \text{ mole}$$

PART 2

Assembling the Tools: We need to calculate the mass of air inside the flask so we can get the mass with absolutely nothing inside the flask. The average molar mass of dry air is 28.96 g mol^{-1}. The basic equation is the ideal gas law rearranged to calculate mass:

$$\text{mass} = \frac{PV \times \text{molar mass}}{RT}$$

Next we subtract the mass of the totally empty flask from the flask with our compound in it, to obtain the mass of the gas; then we can calculate the molar mass of the gas.

Solution: The mass of the air in the flask is

$$\text{mass of air} = \frac{(0.939 \text{ atm})(0.265 \text{ L})(28.96 \text{ g mol}^{-1})}{(0.0821 \text{ L atm mol}^{-1} \text{ K}^{-1})(297 \text{ K})} = 0.296 \text{ g air}$$

The mass of the totally empty flask is

$$182.346 \text{ g} - 0.296 \text{ g} = 182.050 \text{ g}$$

The mass of the unknown gas is then

$$182.503 \text{ g} - 182.050 \text{ g} = 0.453 \text{ g}$$

Dividing the mass of the unknown gas by the number of moles from Part 1 yields the molar mass,

$$\text{molar mass} = \frac{0.453 \text{ g gas}}{0.01020 \text{ moles gas}} = 44.4 \text{ g mol}^{-1}$$

Reviewing the gases in Table 4.2, we find that CO_2, MM = 44.01 g mol^{-1}, is the only gas that has a molar mass close to 44.4 g mol^{-1}. We also know that CO_2 is released when carbonates are treated with acid. Therefore, the anion must be CO_3^{2-}.

PART 3

Assembling the Tools: We have 0.102 moles of CO_3^{2-} in our compound, so we can use stoichiometry tools to find the mass of CO_3^{2-}, and the remaining mass in the sample will be the cation. The formula of the compound will be M_2CO_3, MCO_3, or $M_2(CO_3)_3$, and using the stoichiometry tools from Chapter 3, we can calculate the moles of the cation. As in Part 2, dividing the mass of the cation by the moles will give us the molar mass of the cation.

Solution: We have 0.0102 moles of CO_3^{2-} ions and the mass of carbonate is

$$0.0102 \text{ moles CO}_3^{2-} \times \frac{60.01 \text{ g CO}_3^{2-}}{1 \text{ mol CO}_3^{2-}} = 0.612 \text{ g CO}_3^{2-}$$

Subtracting this from the 1.020 g sample, we have 0.408 g of cation.

The moles of cation in our 0.102-mole sample will depend on the charge of the anion. If the cation is M^+ (M_2CO_3):

$$\text{moles of } M^+ = 0.0102 \ \overline{\text{mol CO}_3^{2-}} \times \frac{2 \text{ mol } M^+}{1 \ \overline{\text{mol CO}_3^{2-}}} = 0.0204 \text{ moles of } M^+$$

If the cation is M^{2+} (MCO_3):

$$\text{moles of } M^{2+} = 0.0102 \ \overline{\text{mol CO}_3^{2-}} \times \frac{2 \text{ mol } M^{2+}}{1 \ \overline{\text{mol CO}_3^{2-}}} = 0.0102 \text{ moles of } M^{2+}$$

If the cation is M^{3+} ($M_2(CO_3)_3$):

$$\text{moles of } M^{3+} = 0.0102 \ \overline{\text{mol CO}_3^{2-}} \times \frac{2 \text{ mol } M^{3+}}{3 \ \overline{\text{mol CO}_3^{2-}}} = 0.0068 \text{ moles of } M^{3+}$$

Finally, dividing 0.408 g by the moles of each cation gives us the molar masses of the possible cations. We calculate

$$M^+ = \frac{0.408 \text{ g } M}{0.0204 \text{ mol } M} = 20.0 \text{ g mol}^{-1}$$

$$M^{2+} = \frac{0.408 \text{ g } M}{0.0102 \text{ mol } M} = 40.0 \text{ g mol}^{-1}$$

$$M^{3+} = \frac{0.408 \text{ g } M}{0.00680 \text{ mol } M} = 50.0 \text{ g mol}^{-1}$$

Consulting the periodic table, we find that calcium with an atomic mass of 40 is the closest of all possibilities. Sodium, a +1 ion, has an atomic mass of 23 that is close to the calculated 20. However, our compound is insoluble and we know that sodium compounds are generally soluble, as shown in Table 4.1. Therefore, our compound appears to be $CaCO_3$.

PART 4

Assembling the Tools: The last step is to determine the number of water molecules in the hydrate. We have another stoichiometry step in which we calculate the moles of water per mole of compound. We know the moles of compound, so we can convert the mass loss, which is water, to moles of water using the stoichiometry concepts in Chapter 3:

$$\text{moles H}_2\text{O} = \frac{\text{g H}_2\text{O}}{\text{molar mass H}_2\text{O}}$$

Solution: The mass of water is the difference between the original mass and the dried mass.

$$\text{moles H}_2\text{O} = \frac{(2.121 \text{ g} - 1.020 \text{ g}) \text{ H}_2\text{O}}{18.0 \text{ g H}_2\text{O mol}^{-1}} = 0.0611 \text{ mol H}_2\text{O}$$

$$\text{moles H}_2\text{O in hydrate} = \frac{\text{mol H}_2\text{O}}{\text{mol CaCO}_3} = \frac{0.0611 \text{ mol H}_2\text{O}}{0.0102 \text{ mol CaCO}_3} = 5.99 \text{ mol H}_2\text{O}$$

This properly rounds to 6 moles of water, and we then write $CaCO_3 \cdot 6H_2O$ as the final answer.

Are the Answers Reasonable?: First, we recheck all of our calculations to be sure that all units cancel and that the math is correctly done. Finally, the fact that all the calculations result so precisely in a formula, $CaCO_3 \cdot 6H_2O$, with a whole number of waters of hydration, is a strong indication that the problem was solved correctly.

10.7 | Kinetic Molecular Theory of Gases

Scientists in the 19th century, who already knew the gas laws, wondered what had to be true, at the molecular level, about all gases to account for their conformity to a common set of gas laws. The **kinetic molecular theory of gases** provided an answer. We introduced some of its ideas in Chapter 6, and in Section 10.1 we described a number of observations that suggest what gases must be like when viewed at the molecular level. Let's look more closely now at the kinetic molecular theory to see how well it explains the behavior of gases.

The theory, often called simply the kinetic theory of gases, consisted of a set of postulates that describe the makeup of an ideal gas. Then the laws of physics and statistics were applied to see whether the observed gas laws could be predicted from the model. The results were splendidly successful.

Postulates of the Kinetic Theory of Gases

1. A gas consists of an extremely large number of very tiny particles that are in constant, random, and rapid motion.
2. The gas particles themselves occupy a net volume so small in relation to the volume of their container that their contribution to the total volume can be ignored.
3. The particles often collide in perfectly elastic collisions[12] with themselves and with the walls of the container, and they move in straight lines between collisions, neither attracting nor repelling each other.

In summary, the model pictures an ideal gas as a collection of very small, constantly moving billiard balls that continually bounce off each other and the walls of their container, and so exert a net pressure on the walls (as described in Figure 10.1). The gas particles are assumed to be so small that their individual volumes can be ignored, so an ideal gas is effectively all empty space.

■ The particles are assumed to be so small that they have no dimensions at all. They are essentially points in space.

Kinetic Theory and the Gas Laws

According to the model, gases are mostly empty space. As we noted earlier, this explains why gases, unlike liquids and solids, can be compressed so much (squeezed to smaller volumes). It also explains why we have gas laws for gases, and *the same laws for all gases*, but not comparable laws for liquids or solids. The chemical identity of the gas does not matter, because gas molecules do not touch each other except when they collide, and there are extremely weak interactions, if any, between them.

We cannot go over the mathematical details, but we can describe some of the ways in which the laws of physics and the model of an ideal gas account for the gas laws and other properties of matter.

Definition of Temperature

The greatest triumph of the kinetic theory came with its explanation of gas temperature, which we discussed in Section 6.2. What the calculations showed was that the product of gas pressure and volume, PV, is proportional to the average kinetic energy of the gas molecules.

$$PV \propto \text{average molecular KE}$$

But from the experimental study of gases, culminating in the equation of state for an ideal gas, we have another term to which PV is proportional—namely, the Kelvin temperature of the gas.

$$PV \propto T$$

(We know what the proportionality constant here is—namely, nR—because by the ideal gas law, PV equals nRT.) With PV proportional *both* to T and to the "average molecular

---

[12] In *perfectly elastic* collisions, no energy is lost by friction as the colliding objects deform momentarily.

Lower pressure

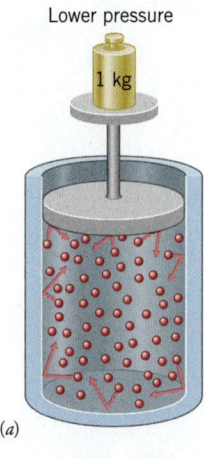

(a)

Higher pressure

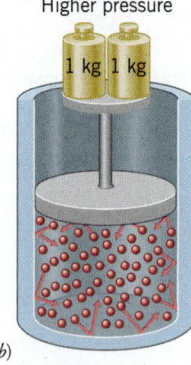

(b)

Figure 10.13 | The kinetic theory and the pressure–volume law (Boyle's law). When the gas volume is made smaller in going from (a) to (b), the number of collisions per second with each unit area of the container's walls increases. Therefore, the pressure increases.

KE," then it must be true that the temperature of a gas is proportional to the average molecular KE.

$$T \propto \text{average molecular KE} \qquad (10.8)$$

$$PV \propto T$$

Boyle's Law (Pressure–Volume Law)

Using the model of an ideal gas, physicists were able to demonstrate that gas pressure is the net effect of innumerable collisions made by gas particles with the walls of the container. Let's imagine that one wall of a gas container is a movable piston that we can push in (or pull out) and so change the volume (see Figure 10.13). If we reduce the volume by one-half, we double the number of molecules per unit volume. This would double the number of collisions per second with each unit area of the wall and therefore double the pressure. Thus, cutting the volume in half forces the pressure to double, which is exactly what Boyle discovered:

$$P \propto \frac{1}{V} \quad \text{or} \quad V \propto \frac{1}{P}$$

Gay-Lussac's Law (Pressure–Temperature Law)

As you learned earlier, the kinetic theory tells us that increasing the temperature increases the average velocity of gas particles. At higher velocities, the particles strike the container's walls more frequently and with greater force. If we don't change the volume, the *area* being struck remains the same, so the force per unit area (the pressure) must increase. Thus, the kinetic theory explains how the pressure of a fixed amount of gas is proportional to temperature (at constant volume), which is the pressure–temperature law of Gay-Lussac.

Charles' Law (Temperature–Volume Law)

We've just seen that the kinetic theory predicts that increasing the temperature should increase the pressure if the volume doesn't change. But suppose we wanted to keep the pressure constant when we raised the temperature. We could only do this if the volume of the container expanded (and therefore the surface area of the container walls increased) as the temperature increased. Therefore, a gas expands with increasing T in order to keep P constant, which is another way of saying that V is proportional to T at constant P. Thus, the kinetic theory explains Charles' law.

Dalton's Law of Partial Pressures

The law of partial pressures is actually evidence for that part of the third postulate in the kinetic theory that pictures particles of an ideal gas moving in straight lines between collisions, neither attracting nor repelling each other (see Figure 10.14). By not interacting with each other, the molecules act *independently*, so each gas behaves as though it were alone in the container. Only if the particles of each gas do act independently can the partial pressures of the gases add up in a simple way to give the total pressure.

Graham's Law of Effusion

The key conditions of Graham's law are that the rates of effusion of two gases with different molecular masses must be compared at the same pressure and temperature and under conditions where the gas molecules do not hinder each other. When two gases have the same temperature, their particles have identical average molecular kinetic energies. Using subscripts 1 and 2 to identify two gases with molecules having different masses m_1 and m_2, we can write that at a given temperature,

$$\overline{KE_1} = \overline{KE_2}$$

where the bar over KE signifies "average."

For a single molecule, its kinetic energy is $KE = \frac{1}{2}mv^2$. For a large collection of molecules of the same substance, the average kinetic energy is $\overline{KE} = \frac{1}{2}m\overline{v^2}$, where $\overline{v^2}$ is the *average of the velocities squared* (called the *mean square* velocity). We have not extended the "average" notation (the bar) over the mass because all the molecules of a given substance have the same mass (the average of their masses is just the mass).

Once again comparing two gases, 1 and 2, we take $\overline{v_1^2}$ and $\overline{v_2^2}$ to be the average of the velocities squared of their molecules. If both gases are at the same temperature, we have

$$\overline{KE_1} = \frac{1}{2}m_1\overline{v_1^2} = \frac{1}{2}m_2\overline{v_2^2} = \overline{KE_2}$$

Now let's rearrange the previous equation to get the ratio of $\overline{v^2}$ terms.

$$\frac{\overline{v_1^2}}{\overline{v_2^2}} = \frac{m_2}{m_1}$$

Next, we'll take the square root of both sides. When we do this, we obtain a ratio of quantities called the **root mean square** (abbreviated **rms**) **speeds**, which we will represent as $(\overline{v_1})_{rms}$ and $(\overline{v_2})_{rms}$.

$$\frac{(\overline{v_1})_{rms}}{(\overline{v_2})_{rms}} = \sqrt{\frac{m_2}{m_1}}$$

The rms speed, $\overline{v_{rms}}$, is not actually the same as the average speed of the gas molecules, but instead represents the speed of a molecule that would have the average kinetic energy. (The difference is subtle, and the two averages do not differ by much, as noted in the margin.)

For any substance, the mass of an individual molecule is proportional to the molar mass. Representing a molar mass of a gas by MM, we can restate this as $m \propto MM$. The proportionality constant is the same for all gases. (It's in grams per atomic mass unit when we express m in atomic mass units.) When we take a ratio of two molar masses, the constant cancels anyway, so we can write

$$\frac{(\overline{v_1})_{rms}}{(\overline{v_2})_{rms}} = \sqrt{\frac{m_2}{m_1}} = \sqrt{\frac{MM_2}{MM_1}}$$

According to the preceding equation, the rms speed of the molecules is inversely proportional to the square root of the molecular mass. This means that at a given temperature, molecules of a gas with a high molecular mass move more slowly, on average, than molecules of a gas with a low molecular mass.

As you might expect, fast-moving molecules will find an opening in the wall of a container more often than slow-moving molecules, so they will effuse faster. Therefore, the rate of effusion of a gas is proportional to the average speed of its molecules, and therefore, it is also proportional to $1/\sqrt{MM}$:

$$\text{effusion rate} \propto \overline{v_{rms}} \propto \frac{1}{\sqrt{MM}}$$

Let's use k as the proportionality constant. This gives

$$\text{effusion rate} = \frac{k}{\sqrt{MM}}$$

Comparing two gases, 1 and 2, and taking a ratio of effusion rates to cause k to cancel, we have

$$\frac{\text{effusion rate (gas 1)}}{\text{effusion rate (gas 2)}} = \sqrt{\frac{MM_2}{MM_1}}$$

This is the way we expressed Graham's law in Equation 10.7. Thus, still another gas law supports the model of an ideal gas.

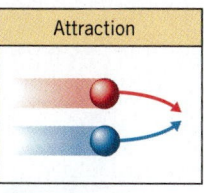

(a)

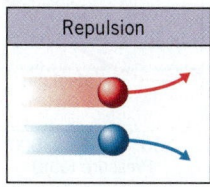

(b)

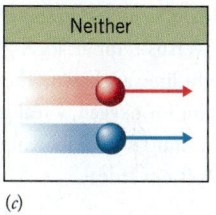

(c)

Figure 10.14 | Gas molecules act independently when they neither attract nor repel each other. Gas molecules would not travel in straight lines if they attracted each other as in (*a*) or repelled each other as in (*b*). They would have to travel farther between collisions with the walls and, therefore, would collide with the walls less frequently. This would affect the pressure. Only if the molecules traveled in straight lines with no attractions or repulsions, as in (*c*), would their individual pressures not be influenced by near misses or by collisions between the molecules.

■ Suppose we have two molecules with speeds of 6 and 10 m s^{-1}. The average speed is $\frac{1}{2}(6 + 10) = 8$ m s^{-1}. The rms speed is obtained by squaring each speed, averaging the squared values, and then taking the square root of the result. Thus,

$$\overline{v_{rms}} = \sqrt{\frac{1}{2}(6^2 + 10^2)} = 8.2 \text{ ms}^{-1}$$

Absolute Zero Defined

The kinetic theory found that the temperature is proportional to the average kinetic energy of the molecules:

$$T \propto \text{average molecular } KE \propto \frac{1}{2}m\overline{v^2}$$

If the average molecular KE becomes zero, the temperature must also become zero. However, mass (m) cannot become zero, so the only way that the average molecular KE can be zero is if v goes to zero. A particle cannot move any slower than it does at a dead standstill, so if the particles stop moving entirely, the substance is as cold as anything can get. It's at absolute zero.[13]

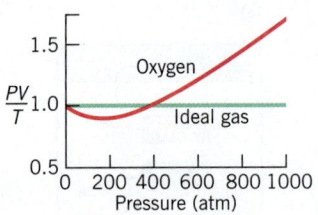

Figure 10.15 | **Deviation from the ideal gas law.** A graph of PV/T versus P for an ideal gas is a straight line, as shown. The same plot for oxygen, a real gas, is not a straight line, showing that O_2 is not "ideal."

10.8 | Real Gases

According to the ideal gas law, the ratio PV/T for a given gas sample equals a product of two terms, nR. Experimentally, however, for real gases PV/T is actually not quite a constant. When we use experimental values of P, V, and T for a real gas, such as O_2, to plot actual values of PV/T as a function of P, we get the curve shown in Figure 10.15. The *horizontal* line at $PV/T = 1$ in Figure 10.15 is what we should see if PV/T were truly constant over all values of P, as it would be for an ideal gas.

A real gas, like oxygen, deviates from ideal behavior for two important reasons. First, in an ideal gas there would be no attractions between molecules, but in a real gas molecules do experience weak attractions toward each other. Second, the model of an ideal gas assumes that gas molecules are infinitesimally small—that the individual molecules have no volume. But real molecules do take up some space. (If all of the kinetic motions of the gas molecules ceased and the molecules settled, you could imagine the net space that the molecules would occupy in and of themselves.)

At room temperature and atmospheric pressure, most gases behave nearly like an ideal gas, also for two reasons. First, the molecules are moving so rapidly and are so far apart that the attractions between them are hardly felt. As a result, the gas behaves almost as though there are no attractions. Second, the space between the molecules is so large that the volume occupied by the molecules themselves is insignificant. By doubling the pressure, we are able to squeeze the gas into very nearly half the volume.

Let's examine Figure 10.15 more closely. Starting at low pressures, close to $P = 1$, gases often behave as ideal gases. As the pressure is increased along the x-axis, the gas particles must be closer together. The first effect noticed as gas particles get closer is attractive forces. These forces between molecules reveal themselves by causing the pressure of a real gas to be slightly lower than that expected for an ideal gas. The attractions cause the paths of the molecules to bend whenever they pass near each other (Figure 10.16). Because the molecules are not traveling in straight lines, as they would in an ideal gas, they have to travel farther between collisions with the walls. As a result, the molecules of a real gas don't strike the walls as frequently as they would if the gas were ideal, and this reduced frequency of collision translates to a reduced pressure. Thus, the ratio PV/T is *less* than that for an ideal gas. The curve for O_2 in Figure 10.15, therefore, dips when we are just starting to increase the pressure.

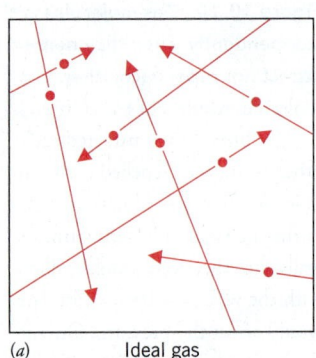

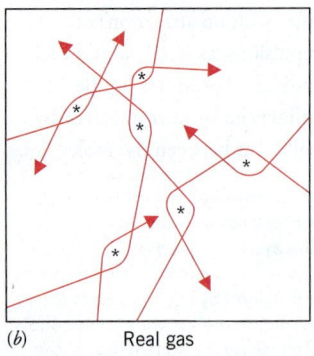

Figure 10.16 | **The effect of attractive forces on the pressure of a real gas.** (*a*) In an ideal gas, the molecules travel in straight lines. (*b*) In a real gas, the paths curve as one molecule passes close to another because the molecules attract each other. Asterisks indicate the points at which molecules came close to each other.

---

[13] Actually, even at 0 K, there must be some slight motion. It's required by the Heisenberg uncertainty principle, which says (in one form) that it's impossible to know precisely both the speed and the location of a particle simultaneously. (If one knows the speed, then there's uncertainty in the location, for example.) If the molecules were actually dead still at absolute zero, there would be no uncertainty in their speed—but then the uncertainty in their *position* would be infinitely great. We would not know where they were! But we do know; they're in this or that container. Thus some uncertainty in speed must exist to have less uncertainty in position and so locate the sample!

As we increase the pressure further, the fact that gas molecules must occupy some space becomes more important than the attractive forces. Continued increases in pressure can reduce only the empty space between the molecules, not the volume of the individual particles themselves. So, at very high pressure, the space occupied by the molecules themselves is a significant part of the total volume, and doubling the pressure cannot halve the total volume. As a result, the actual volume of a real gas is larger than expected for an ideal gas, and the ratio PV/T becomes larger as the pressure increases. We see this for O_2 at the right side of the graph in Figure 10.15.

The van der Waals Equation

Many attempts have been made to modify the equation of state of an ideal gas to get an equation that better fits the experimental data for individual real gases. One of the more successful efforts was that of J. D. van der Waals. He found ways to correct the measured values of P and V to give better fits of the data to the general gas law equation. The result of his derivation is called the **van der Waals equation** *of state for a real gas*. Let's take a brief look at how van der Waals made corrections to measured values of P and V to obtain expressions that fit the ideal gas law.

As you know, if a gas were ideal, it would obey the equation

$$P_{ideal}V_{ideal} = nRT$$

However, for a real gas, using the measured pressure, P_{meas}, and measured volume, V_{meas},

$$P_{meas}V_{meas} \neq nRT$$

The reason is because P_{meas} is smaller than P_{ideal} (as a result of attractive forces between real gas molecules), and because V_{meas} is larger than V_{ideal} (because real molecules do take up some space). Therefore, to get the pressure and volume to obey the ideal gas law, we have to *add* something to the measured pressure and *subtract* something from the measured volume. That's exactly what van der Waals did. Here's his equation.

■ J. D. van der Waals (1837–1923), a Dutch scientist, won the 1910 Nobel Prize in physics.

The measured pressure   The measured volume

$$\left(P_{meas} + \frac{n^2 a}{V^2}\right)(V_{meas} - nb) = nRT \qquad (10.9)$$

Correction to bring measured *P* up to the pressure an ideal gas would exert   Correction to reduce measured *V* to the volume an ideal gas would have

TOOLS

van der Waals equation of state for real gases

The constants a and b are called **van der Waals constants** (see Table 10.3). They are determined for each real gas by carefully measuring P, V, and T under varying conditions. Then trial calculations are made to figure out what values of the constants give the best matches between the observed data and the van der Waals equation.

Notice that the constant a involves a correction to the pressure term of the ideal gas law, so the size of a would indicate something about attractions between molecules. Larger values of a mean stronger attractive forces between molecules. Thus, the most easily liquefied substances, like water and ethyl alcohol, have the largest values of the van der Waals constant a, suggesting relatively strong attractive forces between their molecules.

The constant b helps to correct for the volume occupied by the molecules themselves, so the size of b indicates something about the sizes of particles in the gas. Larger values of b mean larger molecular sizes. Looking at data for the noble gases in Table 10.3, we see that as the atoms become larger from helium through xenon, the values of b become larger. In the next chapter we'll continue the study of factors that control the physical state of a substance, particularly attractive forces and their origins.

TABLE 10.3	Van der Waals Constants	
Substance	a (L^2 atm mol^{-2})	b ($L\ mol^{-1}$)
Noble Gases		
Helium, He	0.03421	0.02370
Neon, Ne	0.2107	0.01709
Argon, Ar	1.345	0.03219
Krypton, Kr	2.318	0.03978
Xenon, Xe	4.194	0.05105
Other Gases		
Hydrogen, H_2	0.02444	0.02661
Oxygen, O_2	1.360	0.03183
Nitrogen, N_2	1.390	0.03913
Methane, CH_4	2.253	0.04278
Carbon dioxide, CO_2	3.592	0.04267
Ammonia, NH_3	4.170	0.03707
Water, H_2O	5.464	0.03049
Ethyl alcohol, C_2H_5OH	12.02	0.08407

Summary

Organized by Learning Objective

Describe the properties of gases at the microscopic and molecular levels

Gases expand to fill any container, are easily compressed, have low densities, and are affected by temperature and pressure.

Explain the measurement of pressure using barometers and manometers

Atmospheric pressure is measured with a **barometer**, in which a pressure of one **standard atmosphere** (**1 atm**) will support a column of mercury 760 mm high. This is a pressure of 760 **torr**. By definition, 1 atm = 101,325 **pascals** (**Pa**) and 1 **bar** = 100 kPa. **Manometers**, both open-end and closed-end, are used to measure the pressure of trapped gases.

Describe and use the gas laws of Dalton, Charles, Gay-Lussac, and the combined gas law

Boyle's Law (Pressure–Volume Law). Volume varies inversely with pressure. $V \propto 1/P$, or $P_1V_1 = P_2V_2$.

Charles' Law (Temperature–Volume Law). Volume varies directly with the Kelvin temperature. $V \propto T$, or $V_1/V_2 = T_1/T_2$.

Gay-Lussac's Law (Temperature–Pressure Law). Pressure varies directly with Kelvin temperature. $P \propto T$, or $P_1/P_2 = T_1/T_2$.

Graham's Law of Effusion. The rate of effusion of a gas varies inversely with the square root of its density (or the square root of its molecular mass).

Combined Gas Law. PV divided by T for a given gas sample is a constant. $PV/T = C$, or $P_1V_1/T_1 = P_2V_2/T_2$.

Perform stoichiometric calculations using the gas laws and Avogadro's principle

Stoichiometric calculations are made using the gas laws and Avogadro's principle and the previous stoichiometric methods.

Whether we start with grams of some compound in a reaction, or the molarity of its solution plus a volume, or *P-V-T* data for a gas in the reaction, *we must get the essential calculation into moles*. After applying the coefficients, we can then move back to any other kind of unit we wish. The flowchart in Figure 10.17 summarizes what we have been doing.

Avogadro's Principle. Equal volumes of gases contain equal numbers of moles when compared at the same temperature and pressure. At **STP**, 273.15 K and 1.00 atm, 1 mol of an ideal gas occupies a volume of 22.4 L.

Gay-Lussac's Law of Combining Volumes. When measured at the same temperature and pressure, the volumes of gases consumed and produced in chemical reactions are in the same ratios as their coefficients.

Explain and apply the ideal gas law, and explain how it incorporates the other gas laws

For the ideal gas law: $PV = nRT$, when P is in atm and V is in L, the value of R is 0.0821 L atm mol^{-1} K^{-1} (T is, as usual, in kelvins). An **ideal gas** is a hypothetical gas that obeys the gas laws exactly over all ranges of pressure and temperature. Real gases exhibit ideal gas behavior most closely at low pressures and high temperatures, which are conditions remote from those that liquefy a gas.

Describe Dalton's law of partial pressures, and apply it to collecting a gas over water

Dalton's Law of Partial Pressures. The total pressure of a mixture of gases is the sum of the partial pressures of the individual gases. In terms of mole fractions, Dalton's law is $P_A = X_A P_{total}$ and $X_A = P_A/P_{total}$.

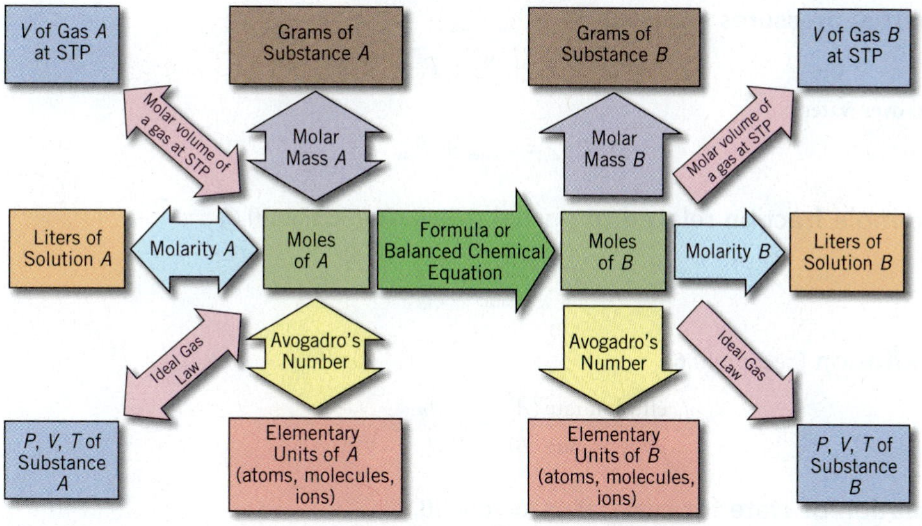

Figure 10.17 | Pathways for working stoichiometry problems. This diagram adds two additional ways to convert laboratory units to moles. As in Figures 3.6 and 4.27, the boxed items are given or calculated quantities while the arrows indicate where to find the conversion factors.

Mole Fraction. The mole fraction X_A of a substance A equals the ratio of the number of moles of A, n_A, to the total number of moles n_{total} of all the components of a mixture.

$$X_A = \frac{n_A}{n_{total}}$$

State the main postulates of the kinetic theory of gases, and show how it explains the gas laws on a molecular level

Kinetic Theory of Gases. An ideal gas consists of a large number of particles, each having essentially zero volume, that are in constant, chaotic, random motion, traveling in straight lines with no attractions or repulsions between them.

Explain the physical significance of the corrections involved in the van der Waals equation for the differences between real and ideal gases

Real Gases. The van der Waals constant a provides a measure of the strengths of the attractive forces between molecules, whereas the constant b gives a measure of the relative size of the gas molecules.

$$\left(P_{meas} + \frac{n^2 a}{V^2}\right)(V_{meas} - nb) = nRT$$

TOOLS

Tools for Problem Solving The following tools were introduced in this chapter. Study them carefully so you can select the appropriate tool when needed.

Combined gas law (Section 10.3)

$$\frac{P_1 V_1}{T_1} = \frac{P_2 V_2}{T_2}$$

Ideal gas law (Section 10.5)

$$PV = nRT$$

Determination of molar mass from the ideal gas law (Section 10.5)

$$\text{molar mass} = \frac{g}{V}\frac{RT}{P} = d\frac{RT}{P}$$

Dalton's law of partial pressures (Section 10.6)

$$P_{total} = P_A + P_B + P_C + \cdots$$

When a gas is collected over water,

$$P_{total} = P_{water} + P_{gas}$$

Mole fraction and mole fraction related to partial pressure (Section 10.6)

$$X_A = \frac{n_A}{n_{total}} = \frac{P_A}{P_{total}}$$

Graham's law of effusion (Section 10.6)

$$\frac{\text{effusion rate } (A)}{\text{effusion rate } (B)} = \sqrt{\frac{d_B}{d_A}} = \sqrt{\frac{M_B}{M_A}}$$

van der Walls equation of state for real gases (Section 10.8)

$$\left(P_{meas} + \frac{n^2 a}{V^2}\right)(V_{meas} - nb) = nRT$$

WileyPLUS = *WileyPLUS*, an online teaching and learning solution. *Note to instructors*: Many of the end-of-chapter problems are available for assignment via the WileyPLUS system. **www.wileyplus.com.** Review Problems are presented in pairs separated by blue rules. Answers to problems whose numbers appear in blue are given in Appendix B. More challenging problems are marked with an asterisk ⋆.

| Review Questions

Measurement of Pressure

10.1 If you get jabbed by a pencil, why does it hurt so much more if it's with the sharp point rather than the eraser? Explain in terms of the concepts of force and pressure.

10.2 Write expressions that could be used to form conversion factors to convert between: **(a)** kilopascal and atm, **(b)** torr and mm Hg, **(c)** bar and pascal, **(d)** torr and atm, **(e)** torr and pascal, **(f)** bar and atm.

10.3 At 20 °C the density of mercury is 13.6 g mL^{-1} and that of water is 1.00 g mL^{-1}. At 20 °C, the vapor pressure of mercury is 0.0012 torr and that of water is 18 torr. Give and explain two reasons why water would be an inconvenient fluid to use in a Torricelli barometer.

10.4 What is the advantage of using a closed-end manometer, rather than an open-end one, when measuring the pressure of a trapped gas? What is the disadvantage?

Gas Laws

10.5 Express the following gas laws in equation form: **(a)** temperature–volume law (Charles' law), **(b)** temperature–pressure law (Gay-Lussac's law), **(c)** pressure–volume law (Boyle's law), **(d)** combined gas law.

10.6 Which of the four important variables in the study of the physical properties of gases are assumed to be held constant in each of the following laws? **(a)** Boyle's law, **(b)** Charles' law, **(c)** Gay-Lussac's law, **(d)** combined gas law.

10.7 Determine how to plot the two properties of gases listed with each of the following gas laws so that the graph will be a straight line: **(a)** temperature–volume law (Charles' law), **(b)** temperature–pressure law (Gay-Lussac's law), **(c)** pressure–volume law (Boyle's law).

10.8 What is meant by an *ideal gas*? Under what conditions does a real gas behave most like an ideal gas?

Stoichiometry Using Gas Volumes

10.9 Explain Avogadro's principle in your own words. Explain why the volumes need to be measured at a constant temperature and pressure.

10.10 Explain why two liters of hydrogen gas and one liter of oxygen gas react to form two liters of water, when the pressures and temperatures are held constant.

Ideal Gas Law

10.11 State the ideal gas law in the form of an equation. What is the value of the gas constant in units of L atm mol^{-1} K^{-1}?

10.12 Using the ideal gas law, show that at STP, the molar volume of an ideal gas is 22.4 L.

⋆**10.13** The molar mass of a gas can be determined from its mass, volume, pressure, and temperature. Derive the equation from the ideal gas law and the definition of molar mass.

*10.14 The density of a gas can vary with pressure, volume, and temperature. Develop an equation for density from the ideal gas law.

Dalton's Law of Partial Pressures

10.15 What is partial pressure?

10.16 State Dalton's law of partial pressures in the form of an equation.

10.17 Define mole fraction. How is the partial pressure of a gas related to its mole fraction and the total pressure?

10.18 Consider the diagrams below that illustrate three mixtures of gases *A* and *B*. If the total pressure of the mixture is 1.00 atm, which of the drawings corresponds to a mixture in which the partial pressure of *A* equals 0.600 atm? What are the partial pressures of *A* in the other mixtures? What are the partial pressures of *B*?

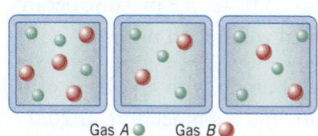

Gas *A* ● Gas *B* ●

10.19 Explain how a gas is collected over water. Why does the temperature of the gas need to be known to determine the pressure of the gas collected over water?

10.20 What is the difference between *diffusion* and *effusion*? State Graham's law in the form of an equation.

10.21 Why does the rate of effusion scale with the square root of the density or the molecular mass of the gas?

Kinetic Molecular Theory of Gases

10.22 Describe the model of a gas proposed by the kinetic theory of gases.

10.23 Use the kinetic molecular theory of gases to explain (a) Charles' law and (b) Boyle's law.

10.24 If the molecules of a gas at constant volume are somehow given a lower average kinetic energy, what two

measurable properties of the gas will change and in what direction?

10.25 Explain *how* raising the temperature of a gas causes it to expand at constant pressure. (*Hint:* Describe how the model of an ideal gas connects the increase in temperature to the gas expansion.)

10.26 Explain in terms of the kinetic theory how raising the temperature of a confined gas makes its pressure increase.

10.27 How does the kinetic theory explain the existence of an absolute zero, 0 K?

10.28 Which of the following gases has the largest value of v_{rms} 25 °C: N_2, CO_2, NH_3, or HBr?

10.29 How would you expect the rate of effusion of a gas to depend on (a) the pressure of the gas, and (b) the temperature of the gas?

Real Gases

10.30 Which postulates of the kinetic theory are not strictly true, and why?

10.31 What does a small value for the van der Waals constant *a* suggest about the molecules of the gas?

10.32 Which of the molecules below has the larger value of the van der Waals constant *b*? Explain your choice.

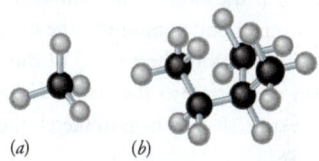

(*a*)   (*b*)

10.33 Under the same conditions of *T* and *V*, why is the pressure of a real gas less than the pressure the gas would exert if it were ideal? At a given *T* and *P*, why is the volume of a real gas larger than it would be if the gas were ideal?

10.34 Suppose we have a mixture of helium and argon. On average, which atoms are moving faster at 25 °C, and why?

| Review Problems

Measurement of Pressure

10.35 Carry out the following unit conversions: (a) 1.26 atm to torr, (b) 740 torr to atm, (c) 738 torr to mm Hg, (d) 1.45×10^3 Pa to torr.

10.36 Carry out the following unit conversions: (a) 0.625 atm to torr, (b) 825 torr to atm, (c) 62 mm Hg to torr, (d) 1.22 kPa to bar.

10.37 What is the pressure in torr of the following?

(a) 0.329 atm (summit of Mt. Everest, the world's highest mountain)

(b) 0.460 atm (summit of Mt. Denali, the highest mountain in the United States)

10.38 What is the pressure in atm of each of the following? (These are the values of the pressures exerted individually by N_2, O_2, and CO_2, respectively, in typical inhaled air.) (a) 595 torr, (b) 160 torr, (c) 0.300 torr

10.39 An open-end manometer containing mercury was connected to a vessel holding a gas at a pressure of 720 torr. The atmospheric pressure was 765 torr. Sketch a diagram of the apparatus showing the relative heights of the mercury in the two arms of the manometer. What is the difference in the heights of the mercury expressed in centimeters?

10.40 An open-end manometer containing mercury was connected to a vessel holding a gas at a pressure of 820 torr.

The atmospheric pressure was 750 torr. Sketch a diagram of the apparatus showing the relative heights of the mercury in the two arms of the manometer. What is the difference in the heights of the mercury expressed in centimeters?

10.41 An open-end mercury manometer was connected to a flask containing a gas at an unknown pressure. The mercury in the arm open to the atmosphere was 65 mm higher than the mercury in the arm connected to the flask. The atmospheric pressure was 748 torr. What was the pressure of the gas in the flask (in torr)?

10.42 An open-end mercury manometer was connected to a flask containing a gas at an unknown pressure. The mercury in the arm open to the atmosphere was 82 mm lower than the mercury in the arm connected to the flask. The atmospheric pressure was 752 torr. What was the pressure of the gas in the flask (in torr)?

10.43 Suppose that in a closed-end manometer the mercury in the closed arm was 12.5 cm higher than the mercury in the arm connected to a vessel containing a gas. What is the pressure of the gas expressed in torr?

10.44 Suppose a gas is in a vessel connected to both an open-end and a closed-end manometer. The difference in heights of the mercury in the closed-end manometer was 236 mm, while in the open-end manometer the mercury level in the arm open to the atmosphere was 512 mm below the level in the arm connected to the vessel. Calculate the atmospheric pressure. (It may help to sketch the apparatus.)

Gas Laws

10.45 A sample of nitrogen has a volume of 880 mL and a pressure of 740 torr. What pressure will change the volume to 550 mL at the same temperature?

10.46 A gas has a volume of 255 mL at 725 torr. What volume will the gas occupy at 365 torr if the temperature of the gas doesn't change?

10.47 A balloon has a volume of 2.50 L indoors at 22 °C. If the balloon is taken outdoors on a cold day when the air temperature is −15 °C (5 °F), what will its volume be in liters? Assume constant air pressure within the balloon.

10.48 A gas has a volume of 3.86 L at 45 °C. What will the volume of the gas be if its temperature is raised to 87 °C while its pressure is kept constant?

10.49 A sample of a gas has a pressure of 854 torr at 285 °C. To what Celsius temperature must the gas be heated to double its pressure if there is no change in the volume of the gas?

10.50 Before taking a trip, you check the air in a tire of your automobile and find it has a pressure of 45 lb in.$^{-2}$ on a day when the air temperature is 12 °C (54 °F). After traveling some distance, you find that the temperature of the air in the tire has risen to 43 °C (approximately 109 °F).

What is the air pressure in the tire at this higher temperature, expressed in units of lb in.$^{-2}$?

10.51 A sample of helium at a pressure of 745 torr and in a volume of 2.58 L was heated from 24.0 to 75.0 °C. The volume of the container expanded to 2.81 L. What was the final pressure (in torr) of the helium?

10.52 When a sample of neon with a volume of 648 mL and a pressure of 0.985 atm was heated from 16.0 to 63.0 °C, its volume became 689 mL. What was its final pressure (in atm)?

10.53 What must be the new volume of a sample of nitrogen (in L) if 2.68 L at 745 torr and 24.0 °C is heated to 375.0 °C under conditions that let the pressure change to 765 torr?

10.54 When 286 mL of oxygen at 741 torr and 18.0 °C was warmed to 33.0 °C, the pressure became 765 torr. What was the final volume (in mL)?

10.55 A sample of argon with a volume of 6.18 L, a pressure of 761 torr, and a temperature of 20.0 °C expanded to a volume of 9.45 L and a pressure of 373 torr. What was its final temperature in °C?

10.56 A sample of a refrigeration gas in a volume of 455 mL, at a pressure of 1.51 atm and at a temperature of 25.0 °C, was compressed into a volume of 222 mL with a pressure of 2.00 atm. To what temperature (in °C) did it have to change?

Stoichiometry Using Gas Volumes

10.57 How many milliliters of O_2 are consumed in the complete combustion of a sample of hexane, C_6H_{14}, if the reaction produces 855 mL of CO_2? Assume all gas volumes are measured at the same temperature and pressure. The reaction is

$$2C_6H_{14}(g) + 19O_2(g) \longrightarrow 12CO_2(g) + 14H_2O(g)$$

10.58 How many milliliters of oxygen are required to react completely with 175 mL of C_4H_{10} if the volumes of both gases are measured at the same temperature and pressure? The reaction is

$$2C_4H_{10}(g) + 13O_2(g) \longrightarrow 8CO_2(g) + 10H_2O(g)$$

*__10.59__ How many milliliters of O_2 measured at 27 °C and 654 torr are needed to react completely with 16.8 mL of CH_4 measured at 35 °C and 725 torr?

*__10.60__ How many milliliters of H_2O vapor, measured at 318 °C and 735 torr, are formed when 33.6 mL of NH_3 at 825 torr and 127 °C react with oxygen according to the following equation?

$$4NH_3(g) + 3O_2(g) \longrightarrow 2N_2(g) + 6H_2O(g)$$

*__10.61__ Calculate the maximum number of milliliters of CO_2, at 745 torr and 27 °C, that could be formed in the combustion of carbon monoxide if 0.300 L of CO at 683 torr and 25 °C is mixed with 155 mL of O_2 at 715 torr and 125 °C.

*10.62 A mixture of ammonia and oxygen is prepared by combining 0.300 L of NH_3 (measured at 0.750 atm and 28 °C) with 0.220 L of O_2 (measured at 0.780 atm and 50 °C). How many milliliters of N_2 (measured at 0.740 atm and 100.0 °C) could be formed if the following reaction occurs?

$$4NH_3(g) + 3O_2(g) \longrightarrow 2N_2(g) + 6H_2O(g)$$

Ideal Gas Law

10.63 What would be the value of the gas constant R in units of mL torr mol^{-1} K^{-1}?

10.64 The SI generally uses its base units to compute constants involving derived units. The SI unit for volume, for example, is the cubic meter, called the *stere*, because the meter is the base unit of length. We learned about the SI unit of pressure, the pascal, in this chapter. The temperature unit is the kelvin. Calculate the value of the gas constant in the SI units, m^3 Pa mol^{-1} K^{-1}.

10.65 What volume in liters does 0.136 g of O_2 occupy at 20.0 °C and 748 torr?

10.66 What volume in liters does 1.67 g of N_2 occupy at 22.0 °C and 756 torr?

10.67 What pressure (in torr) is exerted by 10.0 g of O_2 in a 2.50 L container at a temperature of 27 °C?

10.68 If 12.0 g of water is converted to steam in a 3.60 L pressure cooker held at a temperature of 108 °C, what pressure will be produced?

10.69 A sample of carbon dioxide has a volume of 26.5 mL at 20.0 °C and 624 torr. How many grams of CO_2 are in the sample?

10.70 Methane is formed in landfills by the action of certain bacteria on buried organic matter. If a sample of methane collected from a landfill has a volume of 255 mL at 758 torr and 27 °C, how many grams of methane are in the sample?

10.71 A chemist isolated a gas in a glass bulb with a volume of 255 mL at a temperature of 25.0 °C and a pressure (in the bulb) of 10.0 torr. The gas weighed 12.1 mg. What is the molar mass of this gas?

10.72 At 22.0 °C and a pressure of 755 torr, a gas was found to have a density of 1.13 g L^{-1}. Calculate its molar mass.

10.73 To three significant figures, calculate the density in g L^{-1} of the following gases at STP: (a) C_2H_6 (ethane), (b) N_2, (c) Cl_2, (d) Ar.

10.74 To three significant figures, calculate the density in g L^{-1} of the following gases at STP: (a) Ne, (b) O_2, (c) CH_4 (methane), (d) CF_4.

10.75 What density (in g L^{-1}) does oxygen have at 24.0 °C and 742 torr?

10.76 At 748.0 torr and 20.65 °C, what is the density of argon (in g L^{-1})?

*10.77 The explosive PETN, pentaerythritol tetranitrate, is one of the most powerful explosives known. It reacts with the following balanced equation

$$C(CH_2ONO_2)_4(s) \longrightarrow 2CO(g) + 4H_2O(g)$$
$$+ 3CO_2(g) + 2N_2(g)$$

How many grams of PETN will react to fill a 30.0 L sphere to a pressure of 720 torr at a temperature of 25 °C?

*10.78 TNT, trinitrotoluene, is an explosive that can react with the following balanced equation:

$$2C_7H_5N_3O_6(s) \longrightarrow 3N_2(g) + 5H_2O(g) + 7CO(g) + 7C(s)$$

What is the total volume of the products at 760 torr and 25 °C if 50 g of TNT did explode?

10.79 Propylene, C_3H_6, reacts with hydrogen under pressure to give propane, C_3H_8:

$$C_3H_6(g) + H_2(g) \longrightarrow C_3H_8(g)$$

How many liters of hydrogen (at 740 torr and 24 °C) react with 18.0 g of propylene?

*10.80 Nitric acid is formed when NO_2 is dissolved in water.

$$3NO_2(g) + H_2O(l) \longrightarrow 2HNO_3(aq) + NO(g)$$

How many milliliters of NO_2 at 25 °C and 752 torr are needed to form 12.0 g of HNO_3?

Dalton's Law of Partial Pressures

10.81 A mixture of gases contains 315 torr N_2, 275 torr O_2, and 285 torr Ar. What is the total pressure of the mixture? What is the mole fraction of O_2 in this mixture?

10.82 A certain biological incubator gas mixture contains 29 torr CO_2, 313 torr H_2, 383 torr N_2. What is the total pressure of the mixture? What is the mole fraction of H_2 in this mixture?

*10.83 A 1.00 L container was filled by pumping into it 1.00 L of N_2 at 20.0 cm Hg, 1.00 L of O_2 at 155 torr, and 1.00 L of He at 0.450 atm. All volumes and pressures were measured at the same temperature. What was the total pressure inside the container after the mixture was made?

*10.84 A special gas mixture, BAR 97 High without NO, is used in engine emission testing and contains 16.3 atm CO_2, 8270 torr CO, and 331 mm Hg propane with the 108 atm nitrogen. What was the total pressure in the gas cylinder?

10.85 What are the mole fraction and the mole percent of oxygen in exhaled air if P_{O_2} is 116 torr and P_{total} is 788 torr?

10.86 A mixture of 26,000 torr CO_2 and 104,000 torr N_2 is sold for packaging food. What are the mole fraction and mole percent of CO_2 in this mixture?

*10.87 A 22.4 L container at 0 °C contains 0.30 mol N_2, 0.20 mol O_2, 0.40 mol He, and 0.10 mol CO_2. What are the partial pressures of each of the gases in torr, atm, and bar?

10.88 A mixture of N_2, O_2, and CO_2 has a total pressure of 740 torr. In this mixture the partial pressure of N_2 is 12.0 cm Hg and the partial pressure of O_2 is 4.00 dm Hg. What is the partial pressure of the CO_2?

*10.89 A 0.200 mol sample of a mixture of N_2 and CO_2 with a total pressure of 845 torr was exposed to an excess of solid CaO, which reacts with CO_2 according to the equation

$$CaO(s) + CO_2(g) \longrightarrow CaCO_3(s)$$

After the reaction was complete, the pressure of the gas had dropped to 322 torr. How many moles of CO_2 were in the original mixture? (Assume no change in volume or temperature.)

*10.90 A sample of carbon monoxide was prepared and collected over water at a temperature of 22 °C and a total pressure of 754 torr. It occupied a volume of 268 mL. Calculate the partial pressure of the CO in torr as well as its dry volume (in mL) under a pressure of 1.00 atm and 25 °C.

10.91 The methane from the landfills, Problem 10.70, is saturated with water. What volume of "wet" methane would you have to collect at 20.0 °C and 742 torr to be sure that the sample contains 244 mL of dry methane (also at 742 torr)?

10.92 What volume of "wet" oxygen would you have to collect if you need the equivalent of 275 mL of dry oxygen at 1.00 atm? (The atmospheric pressure in the lab is 746 torr.) The oxygen is to be collected over water at 15.0 °C.

Graham's Law

10.93 Under conditions in which the density of CO_2 is 1.96 g L^{-1} and that of N_2 is 1.25 g L^{-1}, which gas will effuse more rapidly? What will be the ratio of the rates of effusion of N_2 to CO_2?

10.94 A mixture of He and Ne needs to be separated. What will be the ratio of the rates of effusion of He and Ne?

10.95 Arrange the following gases in order of increasing rate of diffusion at 25 °C: Cl_2, C_2H_4, SO_2.

10.96 For the gases CO, CO_2, NO, and NO_2, which gas will effuse the fastest? Which gas the slowest? Which gases will be the most difficult to separate by effusion?

10.97 An unknown gas X effuses 1.65 times faster than C_3H_8. What is the molecular mass of gas X?

10.98 To determine the molecular mass of an unknown gas, it was determined that it effuses 1.58 times slower than argon. What is the molecular mass of the gas, and what could the gas be?

| Additional Problems

10.99 Uranium hexafluoride is a white solid that readily passes directly into the vapor state. (Its vapor pressure at 20.0 °C is 120 torr.) A trace of the uranium in this compound—about 0.7%—is uranium-235, which can be used in a nuclear power plant. The rest of the uranium is essentially uranium-238, and its presence interferes with these applications for uranium-235. Gas effusion of UF_6 can be used to separate the fluoride made from uranium-235 and the fluoride made from uranium-238. Which hexafluoride effuses more rapidly? By how much? (You can check your answer by reading *Chemistry and Current Affairs* 10.1 in Section 10.6.)

*10.100 A bicycle pump has a barrel that is 75.0 cm long (about 30 in.). If air is drawn into the pump at a pressure of 1.00 atm during the upstroke, how long must the downstroke be, in centimeters, to raise the pressure of the air to 5.50 atm (approximately the pressure in the tire of a 10-speed bike)? Assume no change in the temperature of the air.

*10.101 One of the oldest units for atmospheric pressure is lb in.$^{-2}$ (pounds per square inch, or psi). Calculate the numerical value of the standard atmosphere in these units to three significant figures. Calculate the mass in pounds of a uniform column of water 33.9 ft high having an area of 1.00 in.^{2} at its base. (Use the following data: density of mercury = 13.6 g mL^{-1}; density of water = 1.00 g mL^{-1}; 1 mL = 1 cm^3; 1 lb = 454 g; 1 in. = 2.54 cm.)

*10.102 A typical automobile has a weight of approximately 3500 lb. If the vehicle is to be equipped with tires, each of which will contact the pavement with a "footprint" that is 6.0 in. wide by 3.2 in. long, what must the gauge pressure of the air be in each tire? (Gauge pressure is the amount that the gas pressure exceeds atmospheric pressure. Assume that atmospheric pressure is 14.7 lb in.$^{-2}$.)

*10.103 Suppose you were planning to move a house by transporting it on a large trailer. The house has an estimated weight of 45.6 tons (1 ton = 2000 lb). The trailer is expected to weigh 8.3 tons. Each wheel of the trailer will have tires inflated to a gauge pressure of 85 psi (which is actually 85 psi above atmospheric pressure). If the area of contact between a tire and the pavement can be no larger than 100.0 in.^{2} (10.0 in. × 10.0 in.), what is the minimum number of wheels the trailer must have? (Remember, tires are mounted in multiples of two on a trailer. Assume that atmospheric pressure is 14.7 psi.)

*10.104 The motion picture *Titanic* described the tragedy of the collision of the ocean liner of the same name with an iceberg in the North Atlantic. The ship sank soon after the collision on April 14, 1912, and now rests on the sea floor at a depth of 12,468 ft. Recently, the wreck was explored by the research vessel *Nautile*, which has successfully recovered a variety of items from the debris field surrounding the sunken ship. Calculate the pressure in atmospheres and pounds per square inch exerted on the hull of the *Nautile* as it explores the sea bed surrounding the *Titanic*. (*Hint:* The height of a column of liquid required to exert a given pressure is inversely proportional to the liquid's density. Seawater has a density of approximately 1.025 g mL^{-1}; mercury has a density of 13.6 g mL^{-1}; $1 \text{ atm} = 14.7 \text{ lb in.}^{-2}$.)

*10.105 Two flasks (which we will refer to as flask 1 and flask 2) are connected to each other by a U-shaped tube filled with an oil having a density of 0.826 g mL^{-1}. The oil level in the arm connected to flask 2 is 16.24 cm higher than in the arm connected to flask 1. Flask 1 is also connected to an open-end mercury manometer. The mercury level in the arm open to the atmosphere is 12.26 cm higher than the level in the arm connected to flask 1. The atmospheric pressure is 0.827 atm. What is the pressure of the gas in flask 2 expressed in torr? (See the hint given in Problem 10.104.)

*10.106 A bubble of air escaping from a diver's mask rises from a depth of 100 ft to the surface where the pressure is 1.00 atm. Initially, the bubble has a volume of 10.0 mL. Assuming none of the air dissolves in the water, how many times larger is the bubble just as it reaches the surface? Use your answer to explain why scuba divers constantly exhale as they slowly rise from a deep dive. (The density of seawater is approximately 1.025 g mL^{-1}; the density of mercury is 13.6 g mL^{-1}.)

*10.107 In a diesel engine, the fuel is ignited when it is injected into hot compressed air, heated by the compression itself. In a typical high-speed diesel engine, the chamber in the cylinder has a diameter of 10.7 cm and a length of 13.4 cm. On compression, the length of the chamber is shortened by 12.7 cm (a "5-inch stroke"). The compression of the air changes its pressure from 1.00 to 34.0 atm. The temperature of the air before compression is 364 K. As a result of the compression, what will be the final air temperature (in K and °C) just before the fuel injection?

*10.108 Early one cool (60.0 °F) morning you start on a bike ride with the atmospheric pressure at 14.7 lb in.^{-2} and the tire gauge pressure at 50.0 lb in.^{-2}. (Gauge pressure is the amount that the pressure exceeds atmospheric pressure.) By late afternoon, the air had warmed up considerably, and this plus the heat generated by tire friction sent the temperature inside the tire to 104 °F. What will the tire gauge now read, assuming that the volume of the air in the tire and the atmospheric pressure have not changed?

*10.109 The range of temperatures over which an automobile tire must be able to withstand pressure changes is roughly −50 to 120 °F. If a tire is filled to 35 lb in.^{-2} at −35 °F (on a cold day in Alaska, for example), what will be the pressure in the tire (in the same pressure units) on a hot day in Death Valley when the temperature is 115 °F? (Assume that the volume of the tire does not change.)

*10.110 A mixture was prepared in a 0.500 L reaction vessel from 0.300 L of O_2 (measured at 25 °C and 74.0 cm Hg) and 0.400 L of H_2 (measured at 45 °C and 1250 torr). The mixture was ignited and the H_2 and O_2 reacted to form water. What was the final pressure inside the reaction vessel after the reaction was over if the temperature was held at 122 °C?

*10.111 A student collected 18.45 mL of H_2 over water at 24 °C. The water level inside the collection apparatus was 8.5 cm higher than the water level outside. The barometric pressure was 746 torr. How many grams of zinc had to react with HCl(*aq*) to produce the H_2 that was collected?

*10.112 A mixture of gases is prepared from 87.5 g of O_2 and 12.7 g of H_2. After the reaction of O_2 and H_2 is complete, what is the total pressure of the mixture if its temperature is 162 °C and its volume is 12.0 L? What are the partial pressures of the gases remaining in the mixture?

10.113 A gas was found to have a density of $0.08747 \text{ mg mL}^{-1}$ at 17.0 °C and a pressure of 1.00 atm. What is its molecular mass? Can you tell what the gas most likely is?

*10.114 In one analytical procedure for determining the percentage of nitrogen in unknown compounds, weighed samples are made to decompose to N_2, which is collected over water at known temperatures and pressures. The volumes of N_2 are then translated into grams and then into percentages.

(a) Show that the following equation can be used to calculate the percentage of nitrogen in a sample having a mass of W grams when the N_2 has a volume of V in mL and is collected over water at t_C in °C at a total pressure of P in torr. The vapor pressure of water occurs in the equation as $P^\circ_{H_2O}$.

$$\text{Percentage N} = 0.04489 \times \frac{V(P - P^\circ_{H_2O})}{W(273 + t_C)}$$

(b) Use this equation to calculate the percentage of nitrogen in the sample described in Problem 10.117.

| Multi-Concept Problems

*10.115 A common laboratory preparation of hydrogen on a small scale uses the reaction of zinc with hydrochloric acid. Zinc chloride is the other product.

    (a) If 12.0 L of H_2 at 765 torr and 20.0 °C is wanted, how many grams of zinc are needed, in theory?

    (b) If the acid is available as 8.00 M HCl, what is the minimum volume of this solution (in milliliters) required to produce the amount of H_2 described in part (a)?

*10.116 A sample of an unknown gas with a mass of 3.620 g was made to decompose into 2.172 g of O_2 and 1.448 g of S. Before the decomposition, this sample occupied a volume of 1120 mL at 75.0 cm Hg and 25.0 °C. What is the molecular formula of this gas?

*10.117 A sample of a new anti-malarial drug with a mass of 0.2394 g was made to undergo a series of reactions that changed all of the nitrogen in the compound into N_2. This gas had a volume of 18.90 mL when collected over water at 23.80 °C and a pressure of 746.0 torr. At 23.80 °C, the vapor pressure of water is 22.110 torr. When 6.478 mg of the compound was burned in pure oxygen, 17.57 mg of CO_2 and 4.319 mg of H_2O were obtained. What are the percentages of C and H in this compound?

    (a) Assuming that any undetermined element is oxygen, write an empirical formula for the compound.

    (b) The molecular mass of the compound was found to be 324. What is its molecular formula?

10.118 The odor of a rotten egg is caused by hydrogen sulfide, H_2S. Most people can detect it at a concentration of 0.15 ppb (parts per billion), meaning 0.15 L of H_2S in 10^9 L of space. A typical student lab is 42 × 24 × 8.6 ft.

    (a) At STP, how many liters of H_2S could be present in a typical lab to have a concentration of 0.15 ppb?

    (b) How many milliliters of 0.100 M Na_2S would be needed to generate the amount of H_2S in part **(a)** by the reaction of hydrochloric acid with sodium sulfide? (Assume that all of the H_2S generated enters the atmosphere.)

10.119 Chlorine reacts with sulfite ion to give sulfate ion and chloride ion. How many milliliters of Cl_2 gas measured at 25 °C and 734 torr are required to react with all the SO_3^{2-} in 50.0 mL of 0.200 M Na_2SO_3 solution?

*10.120 In an experiment designed to prepare a small amount of hydrogen by the method described in Problem 10.115, a student was limited to using a gas-collecting bottle with a maximum capacity of 335 mL. The method involved collecting the hydrogen over water. What are the minimum number of grams of Zn and the minimum number of milliliters of 6.00 M HCl needed to produce the *wet* hydrogen that can exactly fit this collecting bottle at 0.932 atm and 25.0 °C?

*10.121 Carbon dioxide can be made in the lab by the reaction of hydrochloric acid with calcium carbonate. How many milliliters of dry CO_2 at 20.0 °C and 745 torr can be prepared from a mixture of 12.3 g of $CaCO_3$ and 185 mL of 0.250 M HCl?

10.122 Boron forms a variety of unusual compounds with hydrogen. A chemist isolated 6.3 mg of one of the boron hydrides in a glass bulb with a volume of 385 mL at 25.0 °C and a bulb pressure of 11 torr. Which of the following is likely to be its molecular formula: BH_3, B_2H_6, or B_4H_{10}?

| Exercises in Critical Thinking

10.123 Firefighters advise that you get out of a burning building by keeping close to the floor. We learned that carbon dioxide and most other hazardous compounds are more dense than air and they should settle to the floor. What other facts do we know that make the firefighter's advice correct?

10.124 Carbon dioxide is implicated in global warming. Propose ways to control or perhaps decrease the carbon dioxide content in our air. Rank each proposal based on its feasibility.

10.125 Methane is another gas implicated in the global warming problem. It has been proposed that much of the methane in the atmosphere is from ruminating cows. Evaluate that suggestion and come up with other possible sources of methane.

*10.126 One of the CFCs that is implicated in decreasing the ozone layer and formation of the ozone hole over Antarctica is Freon-12 or CCl_2F_2. Calculate the density of Freon-12 at STP and compare it to the average density of air. Suggest how Freon-12 can get into the stratosphere to threaten the ozone layer.

11

Intermolecular Attractions and the Properties of Liquids and Solids

Chapter Outline

Devil's Tower in Wyoming dramatically rises 1267 feet above the surrounding land. The rock is considered to be formed from lava cooling slowly in the earth's crust. This slow cooling allowed the crystalline columns to develop. While most liquid-to-solid phase transitions do not produce national monuments, they are important in our study of the physical properties of substances, which we will cover in this chapter.

© Blaine Harrington / Age Fotostock

This Chapter in Context

In Chapter 10 we studied the physical properties of gases, and we learned that all gases behave pretty much alike, regardless of their chemical composition. However, when we compare substances in their liquid or solid states (their *condensed states*), the situation is quite different. When a substance is a liquid or a solid, its particles are packed closely together and the forces between them, which we call *intermolecular forces*, are quite strong. Chemical composition and molecular structure play an important role in determining the strengths of such forces, and this causes one substance to behave differently from another when they are liquids or solids.

We begin our study by looking at the basic differences among the states of matter in terms of both common observable properties and the way the states of matter differ at the molecular level. For example, one observable property of matter is the shape of crystals. The striations running up the side of Devil's Tower in the chapter opening photo are actually huge crystals that have a six-sided (hexagonal) shape. This macromolecular shape gives us insights to the microscopic, molecular structure of these rocks. Other, everyday, observations can be used to explain fundamental properties of matter.

We will also examine how the intermolecular forces holding molecules together affect the physical properties of liquids and solids, and influence the energy changes associated with changes of state. Finally, we will discuss the structure and properties of crystalline solids.

LEARNING OBJECTIVES

After reading this chapter, you should be able to:

- describe the major intermolecular forces including dipole attractions, hydrogen bonding, and London forces
- explain how the molecular attractive forces affect macroscopic properties
- use the concept of dynamic equilibrium to understand changes of state
- explain what is meant by vapor pressure and how it reveals the relative strengths of intermolecular attractions
- understand the concepts of boiling points and normal boiling points
- calculate energy changes involved in changes of state
- understand and effectively use phase diagrams
- use Le Châtelier's principle to understand changes of state
- understand how heats of vaporization are determined experimentally
- describe how crystal structures are based on repeating molecular geometries
- illustrate the basics of X-ray crystallography
- understand that intermolecular forces determine the properties of crystal types

11.1 | Intermolecular Forces

There are differences among gases, liquids, and solids that are immediately obvious and familiar to everyone. For example, any gas will expand to fill whatever volume is available, even if it has to mix with other gases to do so. Liquids and solids, however, retain a constant volume when transferred from one container to another. A solid, such as an ice cube, also keeps its shape, but a liquid such as soda conforms to the shape of whatever bottle or glass we put it in. Properties such as the ones we've described can be understood in terms of the way the particles are distributed in the three states of matter, which is summarized in Figure 11.1.

Distance and Intermolecular Forces

If you've ever played with magnets, you know that their mutual attraction weakens rapidly as the distance between them increases. Intermolecular attractions are similarly affected by the distance between molecules. In gases, the molecules are far apart and intermolecular

■ The closer two molecules are, the more strongly they are attracted to each other.

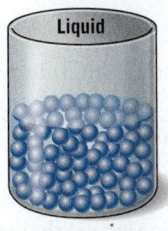

Observable Properties	Molecular Properties
Gas	
Easily compressed	Widely spaced molecules with much empty space between them
Expand spontaneously to fill container	Random motion
	Very weak attractions between the molecules
Liquid	
Retain volume	Molecules tightly packed but little order
Conform to shape of the container	Able to move past each other with little difficulty
Able to flow	Intermolecular attractive forces relatively strong
Nearly incompressible	
Solid	
Retain volume	Molecules tightly packed and highly ordered
Retain shape	Molecules locked in place
Virtually incompressible	Very strong molecular forces
Often have crystalline shape	

Figure 11.1 | **General properties of gases, liquids, and solids.** Properties can be understood in terms of the tightness of molecular packing and the strengths of the intermolecular attractions.

attractions negligible; as a result, chemical composition has little effect on the properties of a gas. However, in a liquid or a solid, the molecules are closer together and the attractions are much stronger. Differences among these attractions, caused by differences in chemical makeup, are greatly amplified, so the physical properties of liquids and solids depend quite heavily on chemical composition.

Intermolecular forces (the attractions *between* molecules) are always much weaker than the bonds between atoms *within* molecules. Bonds are considered to be **intramolecular** forces. In a molecule of NO, for example, the N and O atoms are held very tightly to each other by a covalent bond, and it is the strength of this bond that affects the *chemical properties* of NO. The strength of the chemical bond also keeps the molecule intact as it moves about. When a particular nitrogen atom moves, the oxygen atom bonded to it is forced to follow along (see Figure 11.2). Attractions between neighboring NO molecules, in contrast, are much weaker. In fact, they are only about 4% as strong as the covalent bond in NO. These weaker attractions are what determine the *physical properties* of liquid and solid NO.

There are several kinds of intermolecular attractions, which are discussed in this section. They all have something in common—namely, they arise from attractions between partial positive and negative electrical charges. Collectively, they are called **van der Waals forces**, after J. D. van der Waals, who studied the nonideal behavior of real gases.

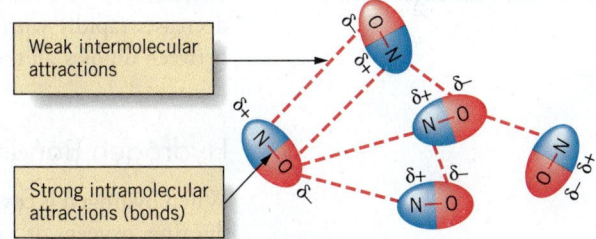

Figure 11.2 | **Attractions within and between nitrogen monoxide molecules.** Strong intramolecular attractions (chemical bonds) exist between N and O atoms *within* NO molecules. These attractions control the chemical properties of NO. Weaker intermolecular attractions exist *between* neighboring NO molecules. The intermolecular attractions control the physical properties of this substance.

Dipole–Dipole Attractions

In Sections 8.7 and 9.3 we learned how to determine if a molecule would be polar or nonpolar. We also learned to use the electronegativity of the atoms to help estimate the magnitude of the polarity of molecules. The polarities that we determined in Chapters 8

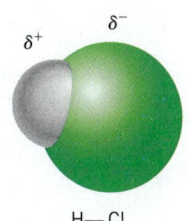

H—Cl

Figure 11.3 | **Dipole–dipole attractions.** Attractions between polar molecules occur because the molecules tend to align themselves so that opposite charges are near each other and like charges are as far apart as possible. The alignment is imperfect because the molecules are constantly moving and colliding.

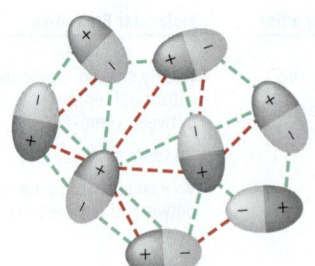

Attractions (– –) are greater than repulsions (– –), so the molecules feel a net attraction to each other.

and 9 are due to *permanent dipoles*. For instance, a molecule such as HCl always has a partial negative charge on the chlorine because electrons are drawn to the Cl because of its high electronegativity. Consequently HCl always has a partial positive charge on the hydrogen atom since the electrons from H are drawn away toward the chlorine. Because these partial charges are always present, the HCl molecule is considered to be a permanent dipole.

Since opposite charges attract, polar molecules tend to line up so the positive end of one dipole is near the negative end of another. However, molecules are in constant motion due to their thermal energy—that is, molecular kinetic energy—so they collide and at times two negative (or two positive) ends will also approach, but they repel each other. Overall the attractions last longer and are more effective than the repulsions, and as a result there is a net attraction between dipoles as shown in Figure 11.3. We classify this kind of intermolecular force a **dipole–dipole attraction**.

Because collisions lead to substantial misalignment of the dipoles and because the attractions are only between partial charges, dipole–dipole forces are much weaker than covalent bonds, being only about 1–4% as strong. Dipole–dipole attractions fall off rapidly with distance, with the energy required to separate a pair of dipoles being proportional to $1/d^3$, where d is the distance between the dipoles.

Practice Exercise 11.1	Sketch a plot of attractive force for two dipoles to qualitatively illustrate how such a force decreases rapidly with distance. (*Hint:* Plot the attractions on the *y*-axis and the distance between dipoles on the *x*-axis.)

Hydrogen Bonds

When hydrogen is covalently bonded to a very small, highly electronegative atom (usually fluorine, oxygen, or nitrogen), the result is a molecule with a large dipole moment. This leads to a particularly strong type of dipole–dipole attraction traditionally called **hydrogen bonding**. The electronegative atom pulls the electron density toward itself and gains a relatively large partial negative charge. In turn, the hydrogen carries a similarly large partial positive charge that strongly attracts the partial negative charge of the next atom. The structures below show some ways in which hydrogen bonds (dashed lines) can form.

$$H_3CH_2C \diagdown \atop H_3CH_2C \diagup N—H ---NH \diagup CH_2CH_3 \atop \diagdown CH_2CH_3$$

$$F—H ---NH \diagup CH_2CH_3 \atop \diagdown CH_2CH_3 \qquad F—H ---OH \diagdown H$$

$$H_3CH_2C \diagdown O—H --- NH \diagup CH_3 \atop \diagdown CH_3$$

$$H \diagdown O—H --- OH \diagdown CH_3 \qquad F—H --- F—H$$

■ A hydrogen bond is not a covalent bond. In water there are oxygen–hydrogen covalent bonds within H_2O molecules and hydrogen bonds between H_2O molecules.

Hydrogen bonds are exceptionally strong because F—H, O—H, and N—H bonds are very polar, and because the partial charges can get quite close since they are concentrated

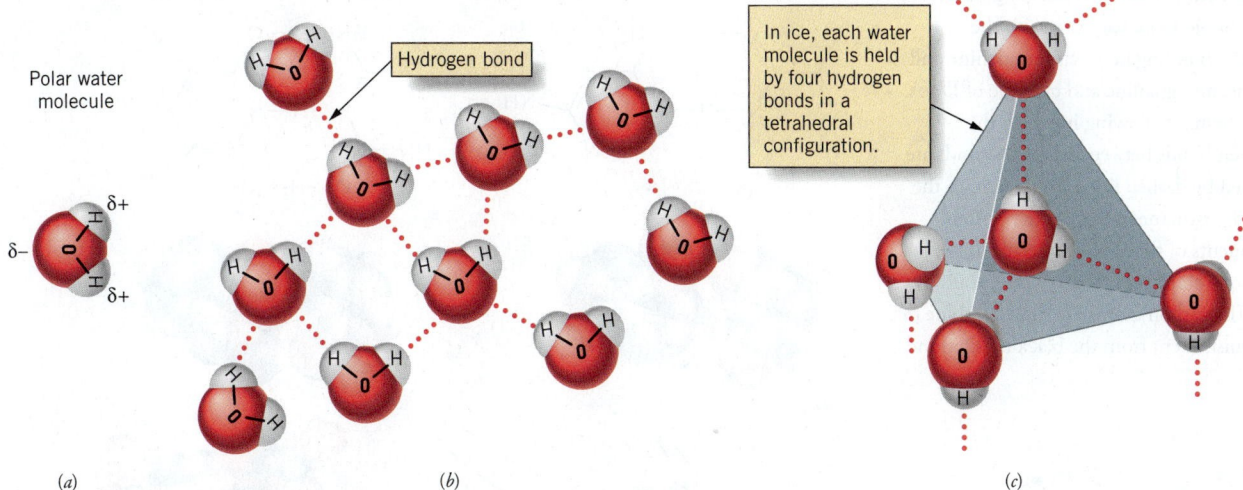

Figure 11.4 | Hydrogen bonding in water. (*a*) The polar water molecule. (*b*) Hydrogen bonding produces strong attractions between water molecules in the liquid. (*c*) Hydrogen bonding (dotted lines) between water molecules in ice, where each water molecule is held by four hydrogen bonds in a tetrahedral configuration.

on very small atoms. Typically, a hydrogen bond is about five to ten times stronger than other dipole–dipole attractions.

Hydrogen Bonds in Water and Biological Systems

Most substances become more dense when they change from a liquid to a solid. Water, however, is different (Figure 11.4). In liquid water, the molecules experience hydrogen bonds that continually break and re-form as the molecules move around (Figure 11.4*b*). As water begins to freeze, however, the molecules become locked in place, and each water molecule participates in four hydrogen bonds (Figure 11.4*c*). The resulting structure occupies a larger volume than the same amount of liquid water, so ice is less dense than the liquid. Because of this, ice cubes and icebergs float in the more dense liquid. The expansion of freezing water is capable of cracking a glass bottle of juice, or exploding a can of soda that you might try to cool quickly in a freezer. Ice formation is also responsible for erosion, causing rocks to split when water freezes after seeping into cracks. In northern cities, freezing water breaks up pavement, creating potholes in streets and sidewalks.

Hydrogen bonding is especially important in biological systems because many molecules in our bodies contain N—H and O—H bonds. Examples are proteins and DNA. Proteins are made up mostly (in some cases, entirely) of long chains of amino acids, linked head to tail to form polypeptides. Shown below is part of a polypeptide chain:

■ Amino acids are discussed briefly in Chapter 8. Polypeptides are examples of polymers, which are large molecules made by linking together many smaller units called monomers (in this case, amino acids).

$$\begin{array}{ccccccc}
\text{H} & \text{O} & \text{H} & \text{O} & \text{H} & \text{O} & \text{H} \\
| & \| & | & \| & | & \| & | \\
\text{---N—CHC} & \text{—N—CHC} & \text{—N—CHC—N—CHC---} \\
| & & | & & | & & | \\
\text{H} & & \text{CH}_3 & & \text{H} & & \text{CH}_2\text{C}_6\text{H}_5
\end{array}$$

One amino acid segment of a polypeptide

Hydrogen bonding between N—H units in one part of the chain and the C=O groups in another help determine the shape of the protein, which greatly influences its biological function. Hydrogen bonding is also responsible for the double helix structure of DNA, as illustrated in Figure 11.5, which carries our genetic information.

Figure 11.5 | **Hydrogen bonding holds the DNA double helix together.** (*a*) The hydrogen bonding between the adenine and thymine, and guanine and cytosine of DNA. (*b*) A schematic drawing in which the hydrogen bonds between the two strands are indicated by dashed lines. The legend to the left of the structure describes the various components of the DNA molecule. (*c*) A model of a short section of a DNA double helix. The carbon atoms are shown in blue to distinguish them from the black background.

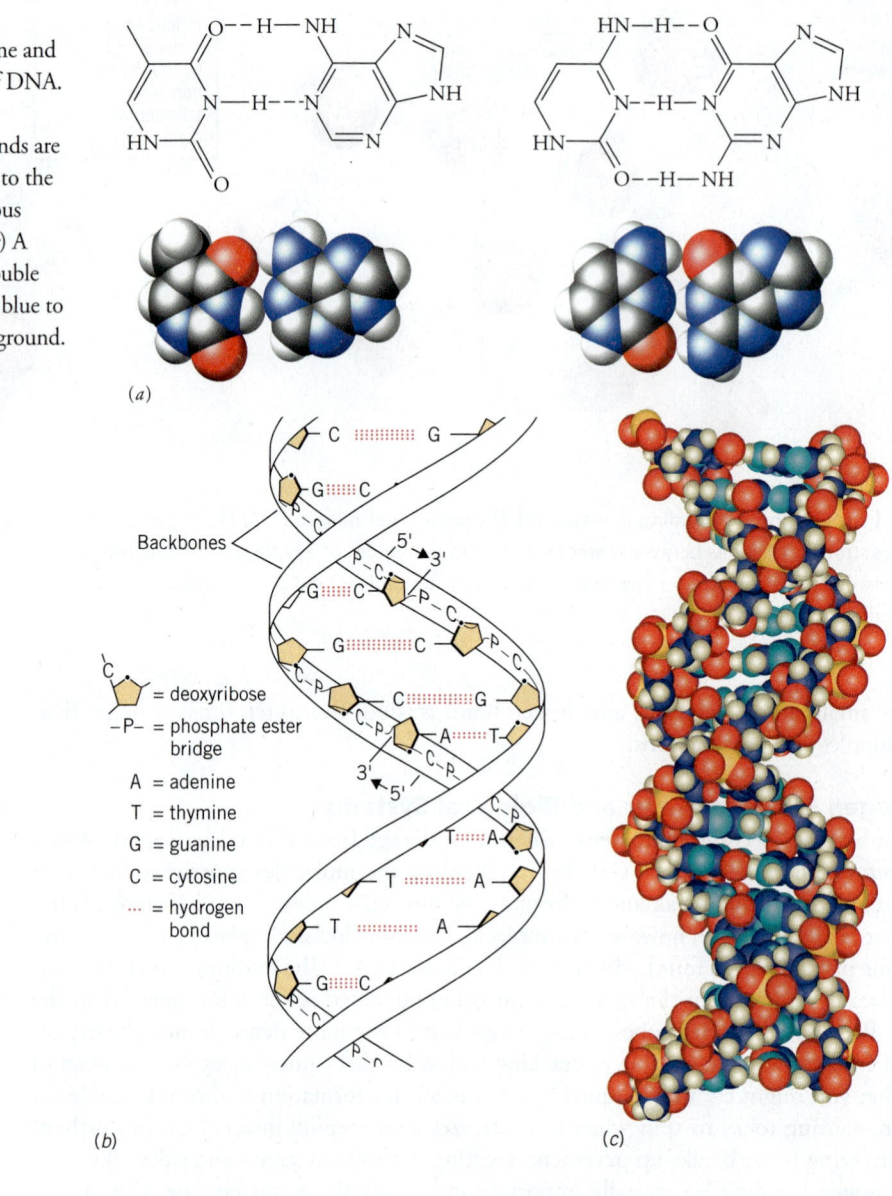

(a)

(b)

(c)

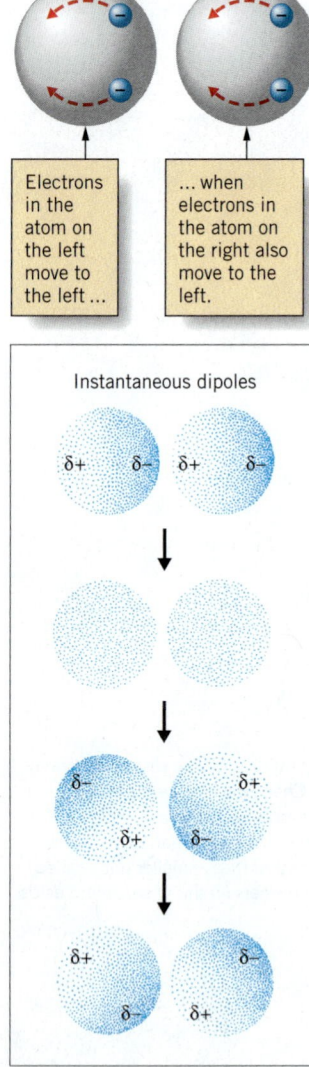

Figure 11.6 | **Instantaneous "frozen" views of the electron density** in two neighboring particles. Attractions exist between the instantaneous dipoles while they exist.

London Forces

Dipole–dipole and hydrogen bonding attractive forces also provide an explanation of why polar compounds can be condensed into liquids rather easily. Nonpolar substances must also experience intermolecular attractions, as evidenced by the ability of the noble gases and nonpolar molecules such as Cl_2 and CH_4 to condense to liquids, and even to solids, when cooled or compressed.

In 1930, Fritz London, a German physicist, explained how the particles in apparently nonpolar substances can experience intermolecular attractions. He noted that in any atom or molecule the electrons are constantly moving. If we could examine such motions in two neighboring particles, we would find that the movement of electrons in one influences the movement of electrons in the other. Because electrons repel each other, as an electron of one particle gets near the other particle, electrons on the second particle are pushed away. This happens continually as the electrons move around, so to some extent, the electron density in both particles flickers back and forth in a synchronous fashion. This is illustrated in Figure 11.6, which depicts a series of instantaneous "frozen" views of the electron density. Notice that *at any given moment the electron density of a particle can be unsymmetrical,* with more negative charge on one side than on the other. For that particular instant, the particle is a **dipole**, and we call it an **instantaneous dipole**.

As an instantaneous dipole forms in one particle, it causes the electron density in its neighbor to become unsymmetrical, too. As a result, this second particle also becomes a dipole. We call this an **induced dipole** because it is caused by, or *induced* by, the formation of the first dipole. Because of the way the dipoles are formed, the positive end of one is always near the negative end of the other, so there is an intermolecular attraction between the molecules. It is a very short-lived attraction, however, because the electrons keep moving; the dipoles vanish as quickly as they form. In another moment, however, the dipoles will reappear in a different orientation and there will be another brief *dipole–dipole attraction*. In this way the short-lived dipoles cause momentary tugs between the particles. When averaged over a period of time, there is a net overall attraction. It tends to be relatively weak, however, because the attractive forces are only "turned on" part of the time.

The momentary dipole–dipole attractions that we've just discussed are called *instantaneous dipole-induced dipole attractions.* They are also called **London dispersion forces** (or simply **London forces** or **dispersion forces**).

London forces exist between *all* molecules and ions. Although they are the only kind of attraction possible between nonpolar molecules, London forces even occur between oppositely charged ions, but their effects are relatively weak compared to ionic attractions.

The Strengths of London Forces

We can use boiling points to compare the strengths of intermolecular attractions. As we will explain in more detail later in this chapter, the higher the boiling point, the stronger the attractions between molecules in the liquid.

The strengths of London forces depend chiefly on three factors. One is the **polarizability** of the electron cloud of a particle. This is a measure of how easily the electron cloud can be distorted, and the ease of forming instantaneous and induced dipoles. In general, *as the volume of the electron cloud increases, its polarizability also increases.* When an electron cloud is large, the outer electrons are generally not held very tightly by the nucleus (or nuclei, if the particle is a molecule). This causes the electron cloud to be "mushy" and rather easily deformed, so instantaneous dipoles and induced dipoles form without much difficulty (see Figure 11.7). As a result, particles with large electron clouds experience stronger London forces than do similar particles with small electron clouds.

The effects of size can be seen if we compare the boiling points of the halogens or the noble gases (see Table 11.1). As the atoms become larger, the boiling points increase, reflecting increasingly stronger intermolecular attractions (stronger London forces). Since the size, or radius, of an atom generally increases with its mass, you may be misled into thinking that the mass determines the polarizability of an atom. However, it is the size and ease of distorting the electron cloud that determines the polarizability.

A second factor that affects the strengths of London forces is the number of atoms in a molecule. For molecules containing the same elements, London forces increase with the

■ London forces decrease very rapidly as the distance between particles increases. The energy required to separate particles held by London forces varies as $1/d^6$, where d is the distance between the particles.

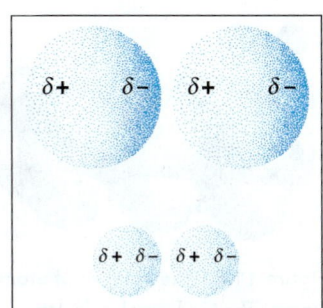

Figure 11.7 | Effect of molecular size on the strengths of London dispersion forces. A large electron cloud is more easily deformed than a small one, so in a large molecule the charges on opposite ends of an instantaneous dipole are larger than in a small molecule. Large molecules, therefore, experience stronger London forces than small molecules.

TABLE 11.1	Boiling Points of the Halogens and Noble Gases		
Group 7A	**Boiling Point (°C)**	**Group 8A**	**Boiling Point (°C)**
F_2	−188.1	He	−268.9
Cl_2	−34.6	Ne	−245.9
Br_2	58.8	Ar	−185.7
I_2	184.4	Kr	−152.3
		Xe	−107.1
		Rn	−61.8

TABLE 11.2	Boiling Points of Some Hydrocarbons[a]
Molecular Formula	Boiling Point at 1 atm (°C)
CH_4	164
C_2H_6	−88.6
C_3H_8	−42.1
C_4H_{10}	−0.5
C_5H_{12}	36.1
C_6H_{14}	68.7
⋮	⋮
$C_{10}H_{22}$	174.1
⋮	⋮
$C_{22}H_{46}$	368.6

[a]The molecules of each hydrocarbon in this table are often called "normal" hydrocarbons, where each end of the molecule is a —CH_3 group and the carbon atoms between the two end groups are all —CH_2— groups. For example, butane is $CH_3CH_2CH_2CH_3$.

number of atoms, as illustrated by the hydrocarbons listed in Table 11.2. As the number of atoms increases, there are more places along their lengths where instantaneous dipoles can develop and lead to London attractions (Figure 11.8). Even if the strength of attraction at each location is about the same, the *total* attraction experienced between the longer molecules is greater.[1]

The third factor that affects the strengths of London forces is molecular shape. Even with molecules that have the same number of the same kinds of atoms, those that have compact shapes experience weaker London forces than long chain-like molecules (Figure 11.9).

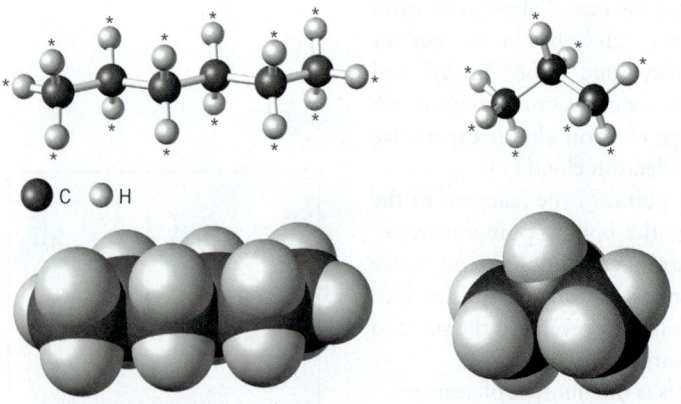

● C ○ H

Figure 11.8 | **The number of atoms in a molecule affects London forces.** The C_6H_{14} molecule, left, shown as both a ball-and-stick model and a space-filling model, has more sites (indicated by asterisks, *) along its chain where it can be attracted to other molecules nearby than does the shorter C_3H_8 molecule, right. As a result, the boiling point of C_6H_{14} (hexane, 68.7 °C) is higher than that of C_3H_8 (propane, −42.1 °C).

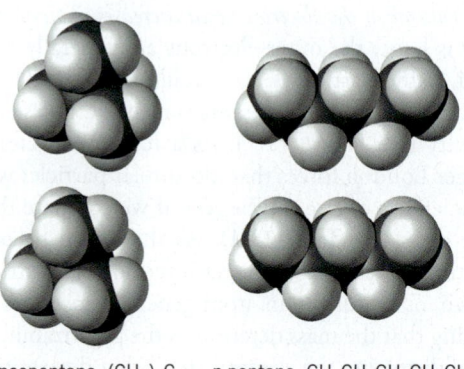

neopentane, $(CH_3)_4C$
bp = 9.5 °C

n-pentane, $CH_3CH_2CH_2CH_2CH_3$
bp = 36.1 °C

Figure 11.9 | **Molecular shape affects the strengths of London forces.** Shown at right are two compounds with the formula C_5H_{12}. Not all hydrogen atoms can be seen in these space-filling models. The compact neopentane molecule, $(CH_3)_4C$, has less area to interact with a neighboring molecule than the linear *n*-pentane molecule, $CH_3CH_2CH_2CH_2CH_3$, so overall the intermolecular attractions are weaker between the more compact molecules.

[1] The effect of large numbers of atoms on the total strengths of London forces can be compared to the bond between loop-and-hook layers of Velcro. Each loop-to-hook attachment is not very strong, but when large numbers of them are involved, the overall bond between Velcro layers is quite strong.

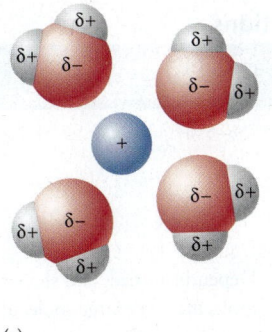

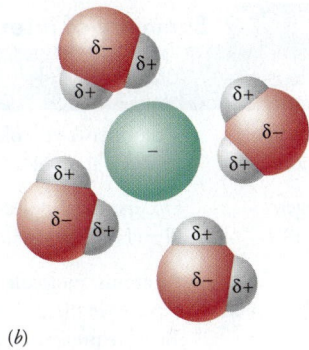

Figure 11.10 | Ion–dipole attractions between water molecules and ions. (*a*) The negative ends of water dipoles surround a cation and are attracted to the ion. (*b*) The positive ends of water molecules surround an anion, which gives a net attraction.

(*a*)   (*b*)

Presumably, because of the compact shape of the $(CH_3)_4C$ molecule, the area that can interact with a neighboring molecule is smaller than that of the chain-like $CH_3(CH_2)_3CH_3$ molecule.

In summary, all substances experience London attractive forces. Those molecules that are polar *also* experience dipole–dipole attractions, which includes the special case of hydrogen bonding.

Ion–Dipole and Ion–Induced Dipole Forces of Attraction

This chapter deals mainly with attractions between identical molecules in pure substances. We find that the same attractions that exist between neutral molecules in pure substances also arise in mixtures and will be discussed in the next chapter. However, we want to complete our list of attractive forces by considering ions interacting with molecules. Ions create two additional attractive forces. For example, ions are able to attract the charged ends of polar molecules to give **ion–dipole attractions**. This occurs in water, for example, when ionic compounds dissolve to give hydrated ions. Cations become surrounded by water molecules that are oriented with the negative ends of their dipoles pointing toward the cation. Similarly, anions attract the positive ends of water dipoles. This is illustrated in Figure 11.10. These same interactions can persist into the solid state as well. For example, aluminum chloride crystallizes from water as a hydrate with formula $AlCl_3 \cdot 6H_2O$. In it the Al^{3+} ion is surrounded by water molecules at the vertices of an octahedron, as illustrated in Figure 11.11. They are held there by ion–dipole attractions.

Ions are also capable of distorting nearby electron clouds, thereby creating dipoles in neighboring particles (like molecules of a solvent, or even other ions). This leads to **ion–induced dipole attractions**, which can be quite strong because the charge on the ion doesn't flicker on and off like the instantaneous charges responsible for ordinary London dispersion forces. The result is that the dipole that is induced by an ion lasts for a longer time and the overall attractive force is greater than the typical London force.

Figure 11.11 | Ion–dipole attractions hold water molecules in a hydrate. Water molecules are arranged at the vertices of an octahedron around an aluminum ion in $AlCl_3 \cdot 6H_2O$.

Estimating the Effects of Intermolecular Forces

In this section we have described a number of different types of intermolecular attractive forces and the kinds of substances in which they occur (see the summary in Table 11.3). With this knowledge, you should now be able to make some estimate of the nature and relative strengths of intermolecular attractions if you know the molecular structure of a substance. This will enable you to understand and sometimes predict how the physical properties of different substances compare. For example, we've already mentioned that boiling point is a property that depends on the strengths of intermolecular attractions. By being able to compare intermolecular forces in different substances, we can sometimes predict how their boiling points compare. This is illustrated in Example 11.1.

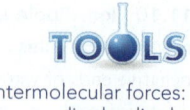

Intermolecular forces:
dipole–dipole
hydrogen bonds
London forces

Estimating intermolecular
attractive forces

TABLE 11.3	Summary of Intermolecular Attractions	
Intermolecular Attraction	Types of Substances that Exhibit Attraction	Strength Relative to a Covalent Bond
Dipole–dipole attractions	Occur between molecules that have permanent dipoles (i.e., polar molecules)	1%–5%
Hydrogen bonding	Occurs when molecules contain N—H, O—H, and F—H bonds	5%–10%
London dispersion forces	All atoms, molecules, and ions experience these kinds of attractions. They are present in all substances	Depends on sizes and shapes of molecules. For large molecules, the cumulative effect of many weak attractions can lead to a large net attraction
Ion–dipole attractions	Occur when ions interact with polar molecules	About 10%; depends on ion charge and polarity of molecule
Ion–induced dipole attractions	Occur when an ion creates a dipole in a neighboring particle, which may be a molecule or another ion	Variable, depending on the charge on the ion and the polarizability of its neighbor

Example 11.1
Using Relative Attractive Forces to Predict Properties

Below are structural formulas of ethanol (ethyl alcohol) and propylene glycol (a compound used as a nontoxic antifreeze). Which of these compounds would be expected to have the higher boiling point?

$$\begin{array}{cc} \text{ethanol} & \text{propylene glycol} \end{array}$$

Analysis: We know that boiling points are related to the strengths of intermolecular attractions—the stronger the attractions, the higher the boiling point. Therefore, if we can determine which kinds of intermolecular forces are present, and which compound has the stronger intermolecular attractions, we can answer the question.

Assembling the Tools: The tools we will use are the kinds of intermolecular attractions and their relative strengths and how these interactions affect the boiling points of the liquids.

Solution: We know that both substances will experience London forces, because they are present between *all* molecules. London forces become stronger as molecules become larger, so the London forces should be stronger in propylene glycol.

Looking at the structures, we see that both contain —OH groups (one in ethanol and two in propylene glycol). This means we can expect that there will be hydrogen bonding in both liquids. Because there are more —OH groups per molecule in propylene glycol than in ethanol, we might reasonably expect that there are more opportunities for the ethylene glycol molecules to participate in hydrogen bonding. This would make the hydrogen bonding forces greater in propylene glycol.

Our analysis tells us that both kinds of attractions are stronger in propylene glycol than in ethanol, so propylene glycol should have the higher boiling point.

Is the Answer Reasonable? There's not much we can do to check our answer other than to review the reasoning, which is sound. (We could also check a reference book, where we would find that the boiling point of ethanol is 78.5 °C and the boiling point of propylene glycol is 188.2 °C!)

List the following in order of their boiling points from lowest to highest.
(a) KBr, $CH_3CH_2CH_2CH_2CH_3$, CH_3CH_2OH
(b) $CH_3CH_2NH_2$, $CH_3CH_2–O–CH_2CH_3$, $HOCH_2CH_2CH_2CH_2OH$
(*Hint:* Determine the type of intermolecular attractive force that is important for each molecule.)

Practice Exercise 11.2

Propylamine and trimethylamine have the same molecular formula, C_3H_9N, but quite different structures, as shown below. Which of these substances is expected to have the higher boiling point? Why?

Practice Exercise 11.3

$$CH_3—CH_2—CH_2—NH_2 \qquad H_3C—\overset{\overset{\displaystyle CH_3}{|}}{N}—CH_3$$
propylamine $\qquad\qquad$ trimethylamine

11.2 | Intermolecular Forces and Physical Properties

In introducing this chapter we briefly described some properties of liquids and solids. We continue here with a more in-depth discussion, and we'll start by examining two properties that depend mostly on how tightly packed the molecules are—namely, *compressibility* and *diffusion*. Other properties depend much more on the strengths of intermolecular attractive forces, properties such as *retention of volume or shape*, *surface tension*, the ability of a liquid to *wet* a surface, the *viscosity* of a liquid, and a solid's or liquid's *tendency to evaporate*.

Properties that Depend on Tightness of Packing

Compressibility
The **compressibility** of a substance is a measure of its ability to be forced into a smaller volume. Gases are highly compressible because the molecules are far apart (Figure 11.12*a*). In a liquid or solid, however, most of the space is taken up by the molecules, and there is very little empty space into which to crowd other molecules (Figure 11.12*b*). As a result, it is very difficult to compress liquids or solids to a smaller volume by applying pressure, so we say that these states of matter are nearly **incompressible**. This is a useful property. When you "step on the brakes" of a car, for example, you rely on the incompressibility of the brake fluid to transmit the pressure you apply with your foot to the brake shoes on the wheels. The incompressibility of liquids is also the foundation of the engineering science of hydraulics, which uses fluids to transmit forces that lift or move heavy objects.

Diffusion
Diffusion occurs much more rapidly in gases than in liquids, and hardly at all in solids. In gases, molecules diffuse rapidly because they travel relatively long distances between

(*a*) Gases $\qquad\qquad$ (*b*) Liquids

Figure 11.12 | The compressibility of a gas and a liquid viewed at the molecular level; the arrows indicate added pressure.
(*a*) Gases compress easily because the molecules are far apart.
(*b*) Liquids are incompressible because the molecules are packed together.

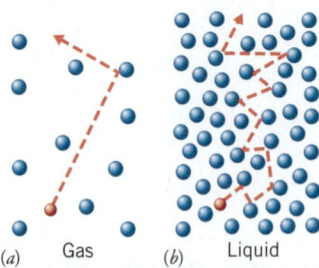

Figure 11.13 | Diffusion in a gas and a liquid viewed at the molecular level. (*a*) Diffusion in a gas is rapid because relatively few collisions occur between widely spaced molecules. (*b*) Diffusion in a liquid is slow because of the many collisions between closely spaced particles.

■ Because of strong hydrogen bonding, the surface tension of water is roughly two to three times larger than the surface tension of any common organic solvent.

Pat O'Hara/Stone/Getty Images

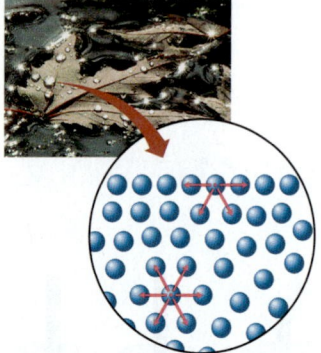

Figure 11.14 | Surface tension and intermolecular attractions. In water, as in other liquids, molecules at the surface are surrounded by fewer molecules than those below the surface. As a result, surface molecules experience fewer attractions than molecules within the liquid.

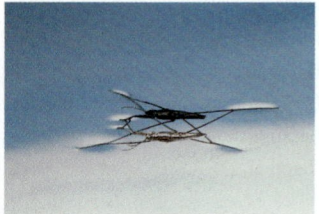

Hermann Eisenbeiss/Photo Researchers

collisions, as illustrated in Figure 11.13. In liquids, however, a given molecule collides many times as it moves about, so it takes longer to move from place to place, making diffusion much slower. Diffusion in solids is almost nonexistent at room temperature because the particles of a solid are held tightly in place. At high temperatures, though, the particles of a solid sometimes have enough kinetic energy to jiggle their way past each other, and diffusion can occur slowly. Such high-temperature, solid-state diffusion is used to produce modern electronic circuits for computers and smart phones.

Properties that Depend on Strengths of Intermolecular Attractions

Retention of Volume and Shape

In gases, intermolecular attractions are too weak to prevent the molecules from moving apart to fill an entire vessel, so a gas will conform to the shape and volume of its container, as shown in Figure 11.1. In liquids and solids, however, the attractions are much stronger and are able to hold the particles closely together. As a result, liquids and solids keep the same volume regardless of the size of their container. In a solid, the attractions are even stronger than in a liquid. They hold the particles more or less rigidly in place, so a solid retains its shape when moved from one container to another.

Surface Tension

A property that is especially evident for liquids is *surface tension*, which is related to the tendency of a liquid to seek a shape that yields the minimum surface area. For a given volume, the shape with the minimum surface area is a sphere—it's a principle of solid geometry. This is why liquid droplets tend to be little spheres.

To understand surface tension, we need to recall that, in general, a system becomes more stable when its potential energy decreases. Based on that principle, lets examine why molecules would prefer to be within the bulk of a liquid rather than at its surface.

In Figure 11.14, we see that a molecule *within* the liquid is surrounded by molecules on all sides, whereas one at the *surface* has neighbors beside and below it, but none above. As a result, a surface molecule is attracted to fewer neighbors than one within the liquid. Now, let's imagine how we might move an interior molecule to one at the surface. To do this, we would have to pull away some of the surrounding molecules. Because of intermolecular attractions, removing neighbors requires work, which means that the potential energy of the surface molecules increases. To stabilize, surface molecules can lose some of the excess potential energy by moving closer together and forming stronger attractions to neighboring molecules. The decrease in potential energy due to moving molecules closer together can also be viewed as an increase in the energy of attraction between surface molecules. The result is that it will require more energy to move two surface molecules apart than two interior molecules. More precisely, the **surface tension** of a liquid is proportional to the energy needed to expand its surface area.

In the previous paragraph we saw that surface molecules arranged closer together reduced the potential energy and made the system more stable. Another way to reduce the potential energy of a liquid is to reduce the total number of molecules at its surface. The lowest energy is achieved when the liquid has the smallest surface area possible and for a small amount of liquid, a droplet with a spherical shape is common.

The tendency of a liquid to spontaneously acquire a minimum surface area explains many common observations. For example, surface tension causes the sharp edges of glass tubing to become rounded when the glass is softened in a flame, an operation called "fire polishing." Surface tension is also what allows us to fill a water glass above the rim, giving the surface a rounded appearance. Figure 11.15 shows how the surface behaves as if it has a thin, invisible "skin" that lets the water in the glass pile up, trying to assume a spherical shape. If you push on the surface of a liquid, it resists expansion and pushes back, so the surface "skin" appears to resist penetration. This is what enables certain insects to "walk on water," as illustrated in the photo in the margin.

Surface tension is a property that varies with the strengths of intermolecular attractions. Liquids with strong intermolecular attractive forces have large differences in potential energy between their interior and surface molecules, and have large surface tensions. Not surprisingly, water's surface tension is among the highest known (with comparisons made at the same temperature); its intermolecular forces are hydrogen bonds, the strongest kind of dipole–dipole attraction. In fact, the surface tension of water is roughly three times that of gasoline, which consists of relatively nonpolar hydrocarbon molecules able to experience only London forces.

Wetting of a Surface by a Liquid

A property we associate with liquids, especially water, is their ability to wet things. **Wetting** is the spreading of a liquid across a surface to form a thin film. Water wets clean glass, such as the windshield of a car, by forming a thin film over the surface of the glass (see Figure 11.16*a*). Water won't wet a greasy windshield, however. On greasy glass, water forms tiny droplets or beads (see Figure 11.16*b*).

For wetting to occur, the intermolecular attractions between the liquid and the surface must be of about the same strength as the attractions within the liquid itself. Such a rough equality exists when water touches clean glass. This is because the glass surface contains lots of oxygen atoms to which water molecules can form hydrogen bonds. As a result, part of the energy needed to expand the water's surface area when wetting occurs is recovered by the formation of hydrogen bonds to the glass surface.

When the glass is coated by a film of oil or grease, the surface exposed to the water drop is now composed of relatively nonpolar oil and grease and molecules (Figure 11.16*b*).

Martin Shields / Science Source

Figure 11.15 | **Surface tension in a liquid.** Surface tension allows a glass to be filled with water above the rim.

■ Glass is a vast network of silicon–oxygen bonds.

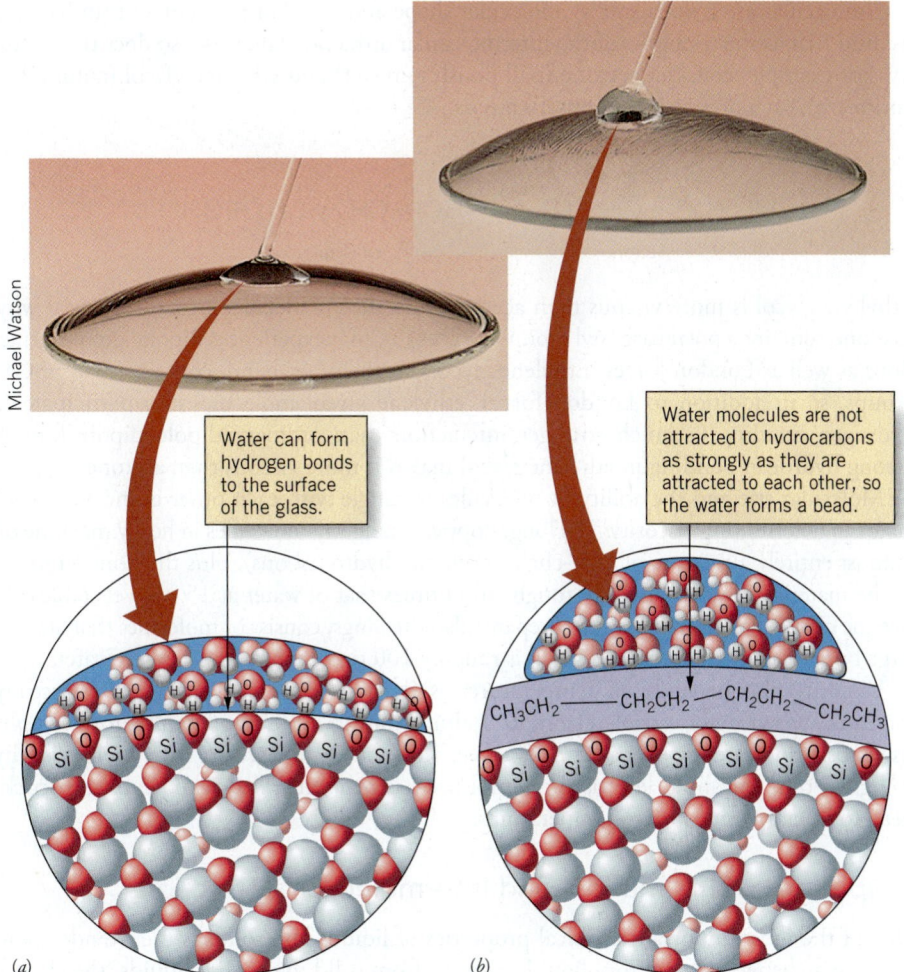

Michael Watson

Water can form hydrogen bonds to the surface of the glass.

Water molecules are not attracted to hydrocarbons as strongly as they are attracted to each other, so the water forms a bead.

CH_3CH_2 — CH_2CH_2 — CH_2CH_2 — CH_2CH_3

(a)   (b)

Figure 11.16 | **Intermolecular attractions affect the ability of water to wet a surface.** (*a*) Water wets a clean glass surface because the surface contains many oxygen atoms to which water molecules can form hydrogen bonds. (*b*) If the surface has a layer of grease, to which water molecules are only weakly attracted, the water doesn't wet it. The water resists spreading and forms a bead instead.

These attract other molecules (including water) largely by London forces, which are weak compared with hydrogen bonds. Therefore, the attractions *within* liquid water are much stronger than the attractions *between* water molecules and the greasy surface. The weak water-to-grease London forces can't overcome the hydrogen bonding within liquid water, so the water doesn't spread out; it forms beads, instead.

One of the reasons that detergents are used for such chores as doing laundry or washing floors is that detergents contain chemicals called **surfactants**, which drastically lower the surface tension of water. The water, with the added surfactant, has intermolecular forces similar to the grease and oil and the solution can spread more easily across the surface to be cleaned.

When a liquid has a low surface tension, like gasoline, we know that it has weak intermolecular attractions, and such a liquid easily wets solid surfaces. The weak attractions between the molecules in gasoline, for example, are readily replaced by attractions to almost any surface, so gasoline easily spreads to a thin film. If you've ever spilled a little gasoline, you have experienced firsthand that it doesn't bead.

Viscosity

As everybody knows, syrup flows less readily, or is more resistant to flow, than water when both are at the same temperature. Flowing is a change in the *form* of the liquid, and such resistance to a change in form is called the liquid's **viscosity**. We say that syrup is more *viscous* than water. The concept of viscosity is not confined to liquids. Solid things, even rock, also yield to forces acting to change their shapes, but normally do so only gradually and imperceptibly. Gases also have viscosity, but they respond almost instantly to form-changing forces.

Viscosity has been called the "internal friction" of a material. It is influenced both by intermolecular attractions and by molecular shape and size. For molecules of similar size, we find that as the strengths of the intermolecular attractions increase, so does the viscosity. For example, consider acetone (nail polish remover) and ethylene glycol (automotive antifreeze), each of which contains 10 atoms.

$$\underset{\text{acetone}}{H_3C-\overset{\overset{\textstyle O}{\|}}{C}-CH_3} \qquad \underset{\text{ethylene glycol}}{HO-CH_2-CH_2-OH}$$

■ If you've ever spilled a little acetone, or you know that it flows very easily. In contrast, ethylene glycol has an "oily" thickness to it and flows more slowly.

Ethylene glycol is more viscous than acetone, and their molecular structures reveal why. Acetone contains a polar carbonyl group ($>C=O$), so it experiences dipole–dipole attractions as well as London forces. Ethylene glycol, on the other hand, contains two —OH groups, so in addition to London forces, ethylene glycol molecules also participate in hydrogen bonding (a much stronger interaction than ordinary dipole–dipole forces). Strong hydrogen bonding in ethylene glycol makes it more viscous than acetone.

Molecular size and the ability of molecules to tangle with each other is another major factor in determining viscosity. The long, floppy, entangling molecules in heavy machine oil (almost entirely a mixture of long-chain, nonpolar hydrocarbons), plus the London forces in the material, give it a viscosity roughly 600 times that of water at 15 °C. Vegetable oils, like the olive oil or corn oil used to prepare salad dressings, consist of molecules that are also large but generally nonpolar. Olive oil is roughly 100 times more viscous than water.

■ As the temperature drops, molecules move more slowly and intermolecular forces become more effective at restraining flow.

Viscosity also depends on temperature; as the temperature decreases, the viscosity increases. When water is cooled from its boiling point to room temperature, for example, its viscosity increases by over a factor of three. The increase in viscosity with cooling is why a "light," thin (meaning less viscous), motor oil is recommended when cars and trucks need to be started in subzero temperatures.

Evaporation, Sublimation, and Intermolecular Attractions

One of the most important physical properties of liquids and solids is their tendency to undergo a change of state from liquid to gas or from solid to gas. For liquids, the change is called **evaporation**. For solids, the change directly to the gaseous state without going through the liquid state is called **sublimation**. Solid carbon dioxide is commonly called **dry ice** because it doesn't melt. Instead, at atmospheric pressure it *sublimes,* changing directly to gaseous CO_2. Naphthalene, the ingredient in some brands of moth repellant, is another substance that can sublime and seemingly disappear.

To understand evaporation and sublimation, we have to examine the motions of molecules. In a solid or liquid, molecules are not stationary; they bounce around, colliding with their neighbors. At a given temperature, there is *exactly the same* distribution of kinetic energies in a liquid or a solid as there is in a gas, which means that Figure 6.4 in Section 6.2 applies to liquids and solids, too. This figure tells us that at a given temperature a small fraction of the molecules have very large kinetic energies and therefore very high velocities. When one of these high-velocity molecules is at the surface and is moving outward fast enough, it can escape the attractions of its neighbors and enter the vapor state. We say the molecule has left by evaporation (or sublimation, if the substance is a solid).

Michael Watson

Cooling by Evaporation

One of the things we notice about the evaporation of a liquid is that it produces a cooling effect. You've experienced this if you've stepped out of a shower and been chilled by the air. The evaporation of water from your body produced this effect. In fact, our bodies use the evaporation of perspiration to maintain a constant body temperature.

We can see why liquids become cool during evaporation by examining Figure 11.17, which illustrates the kinetic energy distribution in a liquid at a particular temperature. An arrow along the horizontal axis points to the minimum kinetic energy needed by a molecule to escape the attractions of its neighbors. Only molecules with kinetic energies equal to or greater than the minimum can leave the liquid. Others may begin to leave, but before they can escape they slow to a stop and then fall back. Notice in Figure 11.17 that the minimum kinetic energy needed to escape is much larger than the average, which means that when molecules evaporate they carry with them large amounts of kinetic energy. As a result, the average kinetic energy of the molecules left behind decreases. (You might think of this as being similar to removing people taller than 6 feet from a large class of students. When this is done, the average height of those who are left is less.) Because the Kelvin temperature is directly proportional to the average kinetic energy, the temperature is also lower. In other words, evaporation causes the liquid that remains to be cooler.

Rate of Evaporation

Later in this chapter we will be concerned about the *rate of evaporation* of a liquid. There are several factors that control this. You are probably already aware of one of them—the surface area of the liquid. Because evaporation occurs from the liquid's surface and not from within, it makes sense that as the surface area is increased, more molecules are able

■ To understand how temperature and intermolecular forces affect the rate of evaporation, we must compare evaporation rates from the same-size surface area. In this discussion, therefore, "rate of evaporation" means "rate of evaporation per unit surface area."

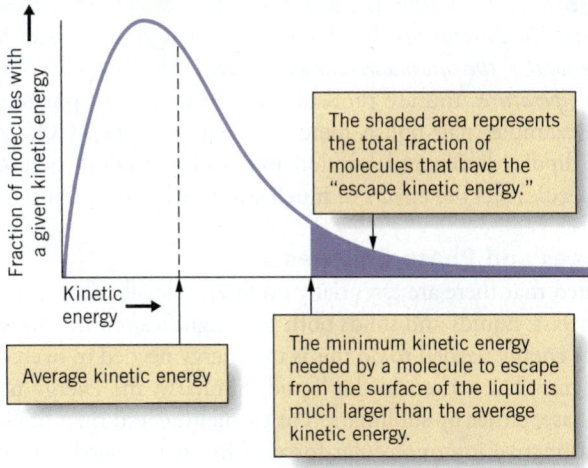

Figure 11.17 | Cooling of a liquid by evaporation. Molecules that are able to escape from the liquid have kinetic energies larger than the average. When they leave, the average kinetic energy of the molecules left behind is less, so the temperature is lower.

Figure 11.18 | **The effect of increasing the temperature on the rate of evaporation of a liquid.** At the higher temperature, the total fraction of molecules with enough kinetic energy to escape is larger, so the rate of evaporation is larger.

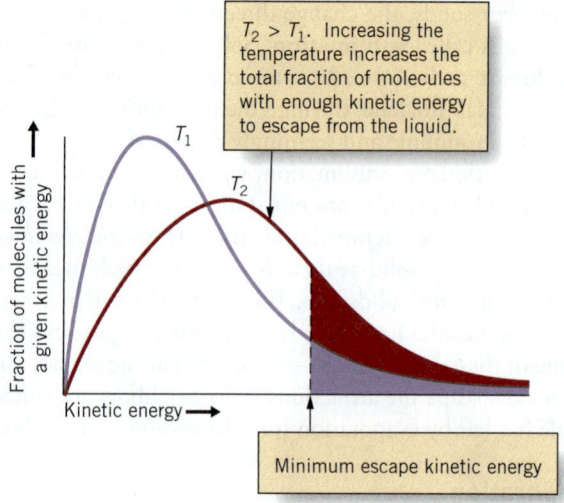

$T_2 > T_1$. Increasing the temperature increases the total fraction of molecules with enough kinetic energy to escape from the liquid.

Minimum escape kinetic energy

TOOLS

Factors that affect the rate of evaporation: temperature and intermolecular forces

to escape and the liquid evaporates more quickly. For liquids having the same surface area, the rate of evaporation depends on two additional factors—namely, temperature and the strengths of intermolecular attractions. Let's examine each of them separately.

As the temperature increases, so does the evaporation rate. Hot water, as you already know, evaporates faster than cold water. The reason can be seen by studying Figure 11.18, which shows kinetic energy distributions for the *same* liquid at two temperatures. Notice two important features of the figure. First, the same minimum kinetic energy is needed for the escape of molecules at both temperatures. This minimum is determined by the kinds of attractive forces between the molecules, and is independent of temperature. Second, the shaded area of the curve represents the fraction of molecules having kinetic energies equal to or greater than the minimum. At the higher temperature, this fraction is larger, which means that at the higher temperature a greater fraction of the molecules have the ability to evaporate. As you might expect, when more molecules have the needed energy, more evaporate in a unit of time. Therefore, the rate of evaporation per unit surface area of a given liquid is greater at a higher temperature.

The effect of intermolecular attractions on evaporation rate can be seen by studying Figure 11.19. Here we have kinetic energy distributions for two *different* liquids—call them *A* and *B*—both at the same temperature. In liquid *A*, the attractive forces are weak; they might be of the London type, for example. As we see, the minimum kinetic energy needed by *A* molecules to escape is not very large because they are not attracted very strongly to each other. In liquid *B*, the intermolecular attractive forces are much stronger; they might be hydrogen bonds, for instance. Molecules of *B*, therefore, are held more tightly to each other at the liquid's surface and must have a higher kinetic energy to evaporate. As you can see from the figure, the total fraction of molecules with enough energy to evaporate is greater for *A* than for *B*, which means that *A* evaporates faster than *B*. In general, then, *the weaker the intermolecular attractive forces, the faster is the rate of evaporation at a given temperature.* You are probably also aware of this phenomenon. At room temperature, for example, nail polish remover [acetone, $(CH_3)_2CO$], whose molecules experience weak dipole–dipole and London forces of attraction, evaporates faster than water, whose molecules feel the effects of much stronger hydrogen bonds.

Attractive Forces and Phase Changes

Previously we noted that there are essentially no intermolecular forces of attraction in the gas phase. By contrast, liquids and solids both have significant attractions. If we convert a liquid to a gas, the energy needed to do this is the energy needed to overcome all attractive intermolecular forces that existed in the liquid. Similarly, the energy needed to bring a solid to the gas phase, either by sublimation or by melting and then boiling, is the energy to overcome all attractive intermolecular forces. The energy needed to melt a substance

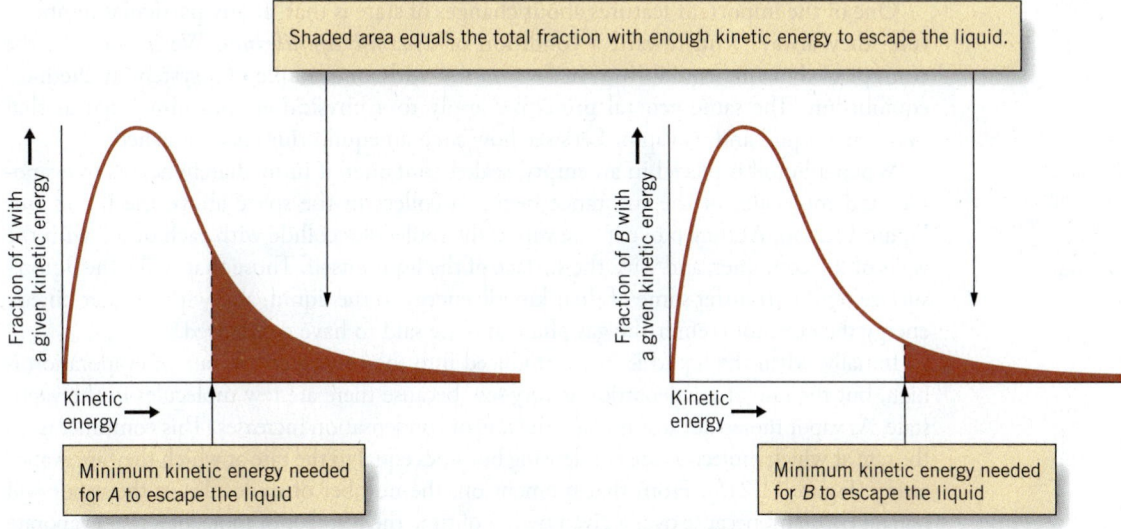

Shaded area equals the total fraction with enough kinetic energy to escape the liquid.

Minimum kinetic energy needed for *A* to escape the liquid

Minimum kinetic energy needed for *B* to escape the liquid

Figure 11.19 | **The kinetic energy distribution in two different liquids,** *A* and *B*, at the same temperature. The minimum kinetic energy required by molecules of *A* to escape is less than for *B* because the intermolecular attractions in *A* are weaker than in *B*. This causes *A* to evaporate faster than *B*.

only overcomes some of the intermolecular forces and as a result, we can only use vaporization of a liquid or sublimation of a solid as a measure of intermolecular forces.

People living in arid, dry, regions can cool their house using evaporative cooling. Use the kinetic molecular theory to explain what factors make this method of cooling less useful in more temperate climates. (*Hint:* Recall the explanation of heat transfer in Section 6.2.)

Practice Exercise 11.4

11.3 | Changes of State and Dynamic Equilibria

A **change of state** occurs when a substance is transformed from one physical state to another (Figure 11.20). The evaporation of a liquid into a gas and the sublimation of a solid into a gas are two examples. Others are the melting, or **fusion**, of a solid such as ice and the **crystallization**, or freezing, of a liquid such as water. Finally, a gas can become a solid in a process called **deposition** or a liquid through **condensation**.

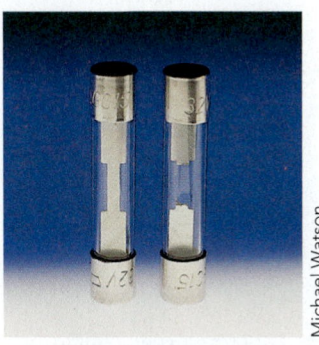

Michael Watson

Fusion means melting. The thin metal band in a fuse becomes hot as electricity passes through it. It protects valuable electrical devices by melting if too much current is drawn. On the left is a new fuse while on the right we see a fuse that has done its job.

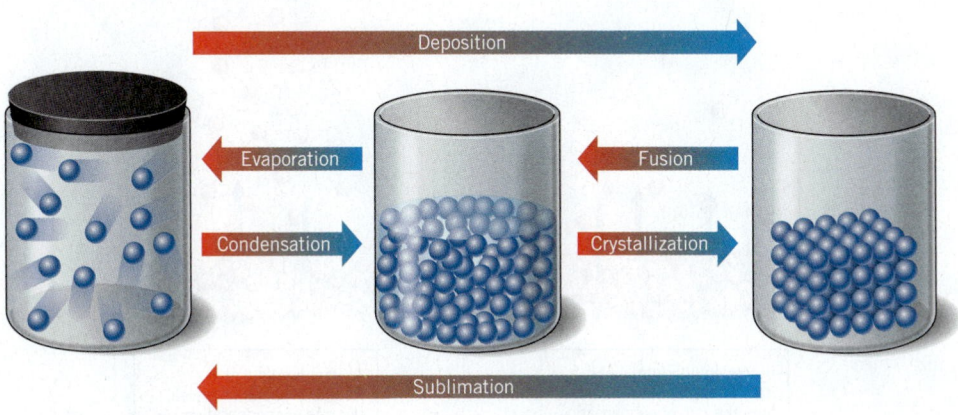

Deposition

Evaporation

Condensation

Fusion

Crystallization

Sublimation

Figure 11.20 | **Changes of state between gases, solids, and liquids.**

One of the important features about changes of state is that, at any particular temperature, they always tend toward a condition of *dynamic equilibrium*. We introduced the concept of dynamic equilibrium in Section 4.4 with an example of a system at chemical equilibrium. The same general principles apply to a physical equilibrium, such as that between a liquid and its vapor. Let's see how such an equilibrium is established.

When a liquid is placed in an empty, sealed container, it immediately begins to evaporate and molecules of the substance begin to collect in the space above the liquid (see Figure 11.21a). As they move in the vapor, the molecules collide with each other, with the walls of the container, and with the surface of the liquid itself. Those that strike the liquid's surface tend to transfer some of their kinetic energy to the liquid, and without that kinetic energy they cannot reenter the gas phase and are said to have condensed.

Initially, when the liquid is first introduced into the container, the rate of evaporation is high, but the rate of condensation is very low because there are few molecules in the vapor state. As vapor molecules accumulate, the rate of condensation increases. This continues until the rate at which molecules are condensing becomes equal to the rate at which they are evaporating (Figure 11.21b). From that moment on, the number of molecules in the vapor will remain constant, because over a given period of time the number of molecules that evaporate will equal those that condense. At this point we have a condition of *dynamic equilibrium,* one in which two opposing effects, evaporation and condensation, are occurring at equal rates.

Similar equilibria are also reached in melting and sublimation. At a temperature called the **melting point**, a solid begins to change to a liquid as heat is added. At this temperature a dynamic equilibrium can exist between molecules in the solid and those in the liquid. Molecules leave the solid and enter the liquid at the same rate as molecules leave the liquid and join the solid (Figure 11.22a). As long as no heat is added or removed from such a solid–liquid equilibrium mixture, melting and freezing occur at equal rates. For sublimation, the situation is exactly the same as in the evaporation of a liquid into a sealed container (see Figure 11.22b). After a few moments, the rates of sublimation and deposition become the same and equilibrium is established.

> ■ A vapor–liquid equilibrium is possible only in a closed container. When the container is open, vapor molecules drift away and the liquid might completely evaporate.

Practice Exercise 11.5

Use the kinetic molecular theory to explain why molecules in the gas phase will condense when collisions with the liquid phase occur. (*Hint:* Recall the explanation of heat transfer in Section 6.3.)

Figure 11.21 | **The evaporation of a liquid into a sealed container.** (*a*) The liquid has just begun to evaporate into the container. The rate of evaporation is greater than the rate of condensation. (*b*) A dynamic equilibrium is reached when the rate of evaporation equals the rate of condensation. In a given time period, the number of molecules entering the vapor equals the number that leave, so there is no net change in the number of gaseous molecules.

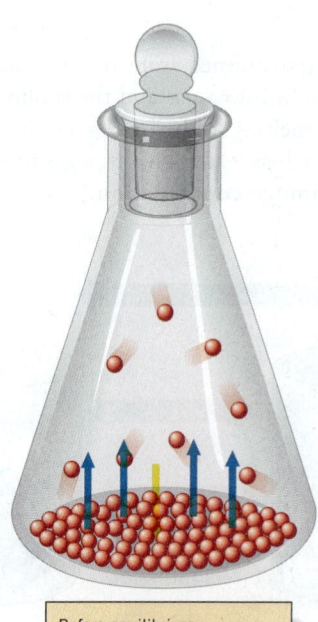

Before equilibrium:
Rate of evaporation is greater than the rate of condensation.

(a)

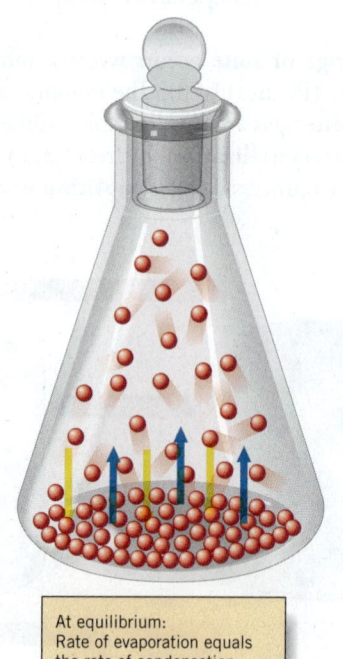

At equilibrium:
Rate of evaporation equals the rate of condensation.

(b)

Figure 11.22 | **Solid–liquid and solid–vapor equilibria.** (*a*) As long as no heat is added or removed, melting (red arrows) and freezing (black arrows) occur at equal rates and the number of particles in the solid remains constant. (*b*) Equilibrium is established when molecules sublime from the solid at the same rate as they deposit on the solid from the vapor.

11.4 | Vapor Pressures of Liquids and Solids

When a liquid evaporates, the molecules that enter the vapor exert a pressure called the **vapor pressure**. From the very moment a liquid begins to evaporate into the vapor space above it, there is a vapor pressure. If the evaporation is taking place inside a sealed container, this pressure grows until finally equilibrium is reached. Once the rates of evaporation and condensation become equal, the concentration of molecules in the vapor remains constant and the vapor exerts a constant pressure. This final pressure is called the **equilibrium vapor pressure of a liquid**. In general, when we refer to the *vapor pressure*, we really mean the equilibrium vapor pressure.

■ A liquid with a high vapor pressure at a given temperature is said to be volatile.

Factors that Determine the Equilibrium Vapor Pressure

Figure 11.23 shows plots of equilibrium vapor pressure versus temperature for a few liquids. From these graphs we see that both a liquid's temperature and its chemical composition are the major factors affecting its vapor pressure. Once we have selected a particular liquid, however, only the temperature matters. The reason is that the vapor pressure of a given liquid is a function solely of its rate of evaporation *per unit area of the liquid's surface.* When this rate is large, a large concentration of molecules in the vapor state is necessary to establish equilibrium, which is another way of saying that the vapor pressure is relatively high when the evaporation rate is high. As the temperature of a given liquid increases, so does its rate of evaporation and so does its equilibrium vapor pressure.

If we compare two liquids, the one that has the weaker intermolecular attractive forces will have a higher rate of evaporation and a higher vapor pressure. These relationships tell us that, of the four liquids in Figure 11.23, intermolecular attractions are strongest in

Figure 11.23 | The variation of vapor pressure with temperature for some common liquids.

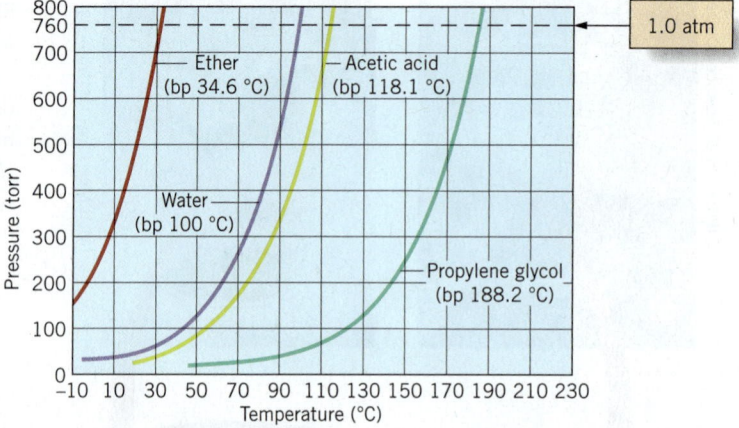

TOOLS

Factors that affect vapor pressure

propylene glycol, next strongest in acetic acid, third strongest in water, and weakest in ether. Generalizing this observation, vapor pressures are indicators of the relative strengths of the attractive forces in liquids.

In summary, two factors affect the equilibrium vapor pressure. The first is temperature: for a given substance, as the temperature increases, the vapor pressure increases. The second is the intermolecular forces: when comparing two substances, the weaker the intermolecular forces, the higher the vapor pressure.

Factors that Do Not Affect the Equilibrium Vapor Pressure

An important fact about vapor pressure is that its magnitude doesn't depend on the total surface area of the liquid, or on the volume of the liquid in the container, or on the volume of the container itself, just as long as some liquid remains when equilibrium is reached. The reason is that none of these factors affects the rate of evaporation *per unit surface area*.

Increasing the *total* surface area does increase the *total* rate of evaporation, but the larger area is also available for condensation; as a result, the rate at which molecules return to the liquid also increases. The rates of both evaporation and condensation are thus affected equally, and no change occurs to the equilibrium vapor pressure. Adding more liquid to the container can't affect the equilibrium either because evaporation occurs from the *surface*. Having more molecules in the bulk of the liquid does not change what is going on at the surface.

To understand why the vapor pressure doesn't depend on the *size* of the vapor space, consider a liquid in equilibrium with its vapor in a cylinder with a movable piston, as illustrated in Figure 11.24*a*. Withdrawing the piston (Figure 11.24*b*) increases the volume of the vapor space; as the vapor expands, the pressure it exerts becomes less, so there's a momentary drop in the pressure. The molecules of the vapor, being more spread out now,

Figure 11.24 | The effect of a volume change on the vapor pressure of a liquid. (*a*) Equilibrium exists between liquid and vapor. (*b*) The volume is increased, which upsets the equilibrium and causes the pressure to drop. The rate of condensation is now less than the rate of evaporation, which hasn't changed. (*c*) After more liquid has evaporated, equilibrium is restored and the vapor pressure has returned to its initial value.

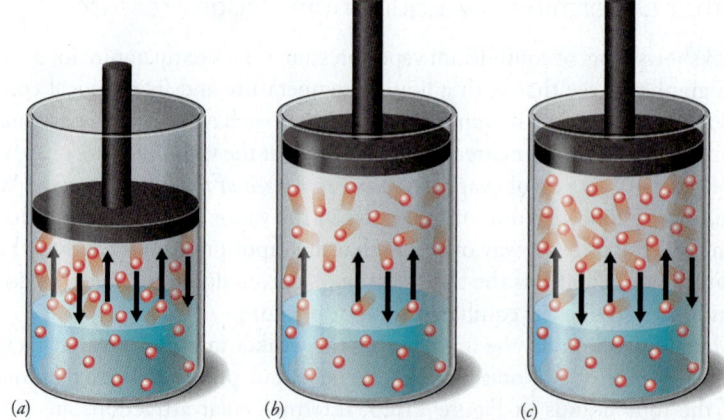

no longer strike the surface as frequently, so the rate of condensation has also decreased. The rate of evaporation hasn't changed, however, so for a moment the system is not at equilibrium and the substance is evaporating faster than it is condensing (Figure 11.24*b*). This condition prevails, changing more liquid into vapor, until the concentration of molecules in the vapor has risen enough to make the condensation rate equal again to the evaporation rate (Figure 11.24*c*). At this point, the vapor pressure has returned to its original value. Therefore, the net result of expanding the space above the liquid is to change more liquid into vapor, but it does not affect the equilibrium vapor pressure. Similarly, we expect that reducing the volume of the vapor space above the liquid will also not affect the equilibrium vapor pressure, either.

Considering Figure 11.24, in which direction should the piston be moved to decrease the number of molecules in the gas phase? (*Hint:* Consider what must happen to reestablish equilibrium after the piston is moved.)

Suppose a liquid is in equilibrium with its vapor in a piston–cylinder apparatus like that in Figure 11.24. If the piston is pushed in a short way and the system is allowed to return to equilibrium, what will have happened to the *total number* of molecules in both the liquid and the vapor?

Practice Exercise 11.6

Practice Exercise 11.7

Vapor Pressures of Solids

Solids have vapor pressures just as liquids do. In a crystal, the particles are constantly jiggling around, bumping into their neighbors. At a given temperature there is a distribution of kinetic energies, so some particles at the surface have large enough kinetic energies to break away from their neighbors and enter the vapor state. When particles in the vapor collide with the crystal, they can be recaptured, so deposition can occur, too. Eventually, the concentration of particles in the vapor reaches a point where the rate of sublimation equals the rate of deposition, and a dynamic equilibrium is established. The pressure of the vapor that is in equilibrium with a solid is called the **equilibrium vapor pressure of a solid**. As with liquids, this equilibrium vapor pressure is usually referred to simply as the vapor pressure. Like that of a liquid, the vapor pressure of a solid is determined by the strengths of the attractive forces between the particles and by the temperature.

■ In many solids, such as NaCl, the attractive forces are so strong that virtually no particles have enough kinetic energy to escape at room temperature, so essentially no evaporation occurs. Their vapor pressures at room temperature are virtually zero.

11.5 | Boiling Points of Liquids

If you were asked to check whether a pot of water was boiling, what would you look for? The answer, of course, is *bubbles*. When a liquid boils, large bubbles usually form at many places on the inner surface of the container and rise to the top. If you were to place a thermometer into the boiling water, you would find that the temperature remains constant, regardless of how you adjust the flame under the pot. A hotter flame just makes the water bubble faster, but it doesn't raise the temperature. Any pure liquid remains at a constant temperature while it is boiling, a temperature that's called the liquid's *boiling point*.

If you measure the boiling point of water in Los Angeles, New York, or any place else that is nearly at sea level, your thermometer will read 100 °C or very close to it. However, if you try this experiment in Denver, Colorado, you will find that water boils at about 95 °C. Denver, at a mile above sea level, has a lower atmospheric pressure. Thus, we find that the boiling point depends on the atmospheric pressure.

These observations raise some interesting questions. Why do liquids boil? And why does the boiling point depend on the pressure of the atmosphere? The answers become apparent when we realize that inside the bubbles of a boiling liquid is the *liquid's vapor*, not air. When water boils, the bubbles contain water vapor (steam); when alcohol boils, the bubbles contain alcohol vapor. As a bubble grows, liquid evaporates into it, and the pressure of the vapor pushes the liquid aside, making the size of the bubble increase

■ On the top of Mt. Everest, the world's tallest peak, water boils at only 69 °C.

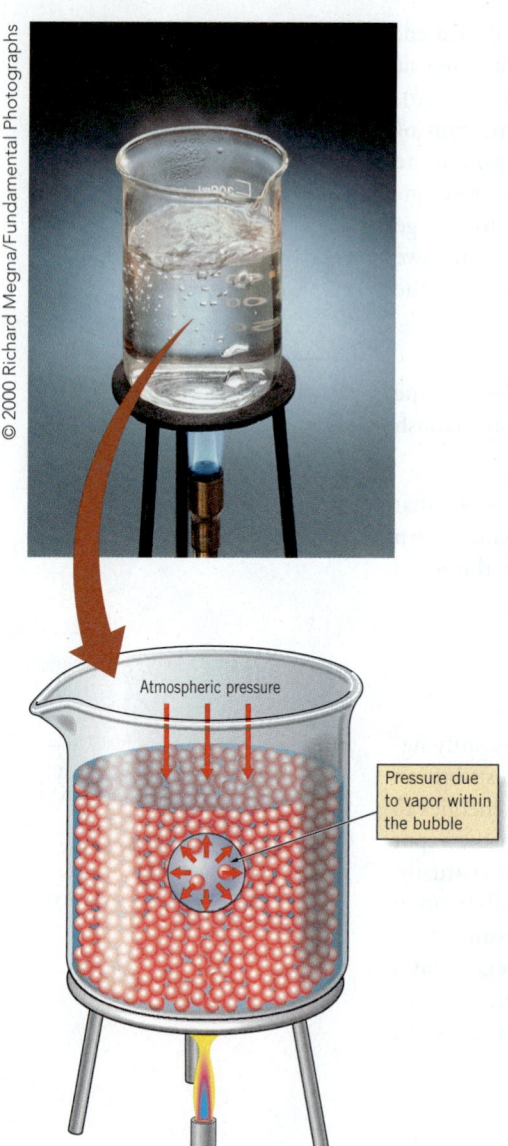

Atmospheric pressure

Pressure due to vapor within the bubble

Figure 11.25 | **A liquid at its boiling point.** The pressure of the vapor within a bubble in a boiling liquid pushes the liquid aside against the opposing pressure of the atmosphere. Bubbles can't form unless the vapor pressure of the liquid is at least equal to the pressure of the atmosphere.

TOOLS

Boiling points and intermolecular forces

(see Figure 11.25). Opposing the bubble's internal vapor pressure, however, is the pressure of the atmosphere pushing down on the top of the liquid, attempting to collapse the bubble. The only way the bubble can exist and grow is for the vapor pressure within it to equal (maybe just slightly exceed) the pressure exerted by the atmosphere. In other words, bubbles of vapor cannot even form until the temperature of the liquid rises to a point at which the liquid's vapor pressure equals the atmospheric pressure. Thus, in scientific terms, the **boiling point** is defined as the temperature at which the vapor pressure of the liquid is equal to the prevailing atmospheric pressure.

Now we can understand why water boils at a lower temperature in Denver than it does in New York City. Because the atmospheric pressure is lower in Denver, the water there doesn't have to be heated to as high a temperature to make its vapor pressure equal to the atmospheric pressure. The lower temperature of boiling water at places with high altitudes, like Denver, makes it necessary to cook foods longer. At the other extreme, a pressure cooker is a device that increases the pressure over the boiling water and thereby raises the boiling point. At the higher temperature, foods cook more quickly.

To make it possible to compare the boiling points of different liquids, chemists have chosen 1 atm as the reference pressure. The boiling point of a liquid at 1 atm is called its **normal boiling point**.[2] (If a boiling point is reported without also mentioning the pressure at which it was measured, we assume it to be the normal boiling point.) Notice in Figure 11.23 that we can find the normal boiling points of ether, water, acetic acid, and propylene glycol by noting the temperatures at which their vapor pressure curves cross the 1-atm pressure line.

Boiling Points and Intermolecular Attractions

Earlier we mentioned that the boiling point is a property whose value depends on the strengths of the intermolecular attractions in a liquid. When the attractive forces are strong, the liquid has a low vapor pressure at a given temperature, so it must be heated to a high temperature to bring its vapor pressure up to atmospheric pressure. High boiling points, therefore, result from strong intermolecular attractions, so we often use normal boiling point data to assess relative intermolecular attractions among different liquids. (In fact, we did this in solving Example 11.1.)

The effects of intermolecular attractions on boiling point are easily seen by examining Figure 11.26, which gives the plots of the boiling points versus period numbers for some families of binary hydrogen compounds. Notice, first, the gradual increase in boiling point for the hydrogen compounds of the Group 4A elements (CH_4 through GeH_4). These compounds are composed of nonpolar tetrahedral molecules. The boiling points increase from CH_4 to GeH_4 simply because the molecules become larger and their electron clouds become more polarizable, which leads to an increase in the strengths of the London forces.

When we look at the hydrogen compounds of the other nonmetals, we find the same trend from Period 3 through Period 5. Thus, for three compounds of the Group 5A series, PH_3, AsH_3, and SbH_3, there is a gradual increase in boiling point, corresponding again to the increasing strengths of London forces. Similar increases occur for the three Group 6A compounds (H_2S, H_2Se, and H_2Te) and for the three Group 7A compounds (HCl, HBr, and HI). Significantly, however, the Period 2 members of each of these series (NH_3, H_2O, and HF) have much higher boiling points than might otherwise be expected. The reason is that each is involved in hydrogen bonding, which is a much stronger attraction than dipole–dipole forces.

[2]The *standard boiling point* is the boiling point at 1 bar; the difference between the standard boiling point and the normal boiling point is negligible unless extremely precise measurements are being made.

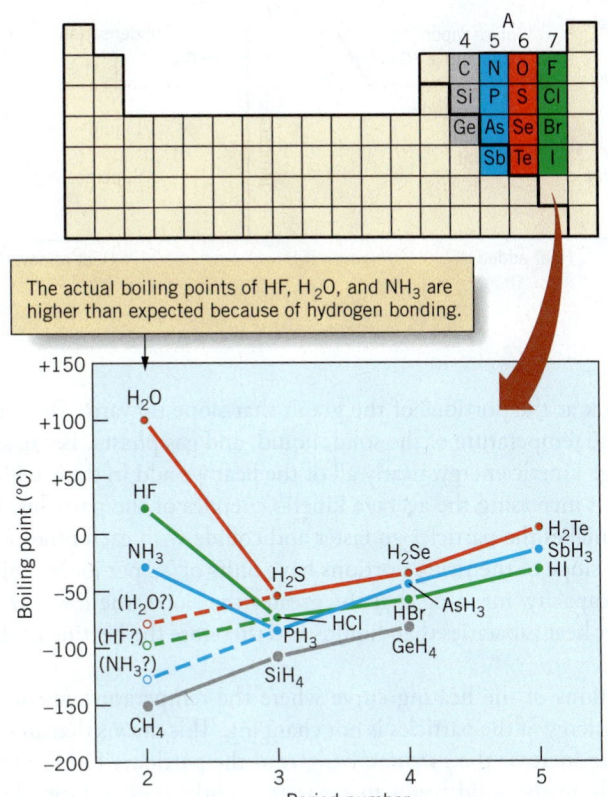

Figure 11.26 | The effects of intermolecular attractions on the boiling point. The boiling points of the hydrogen compounds of the elements of Groups 4A, 5A, 6A, and 7A of the periodic table.

The actual boiling points of HF, H_2O, and NH_3 are higher than expected because of hydrogen bonding.

One of the most interesting and far-reaching consequences of hydrogen bonding is that it causes water to be a liquid, rather than a gas, at temperatures near 25 °C. If it were not for hydrogen bonding, water would have a boiling point somewhere near −80 °C and could not exist as a liquid except at still lower temperatures. At such low temperatures it is unlikely that life as we know it could have developed.

The Dead Sea is approximately 1300 ft below sea level and its barometric pressure is approximately 830 torr. Will the boiling point of water be elevated from 100 °C by (a) less than 10 °C, (b) 10 °C to 25 °C, (c) 25 °C to 50 °C, or (d) above 50 °C? (*Hint:* Extrapolate from Figure 11.23.)

Practice Exercise 11.8

The atmospheric pressure at the summit of Mt. McKinley in Denali National Park in Alaska, 3.85 miles above sea level, is about 330 torr. Use Figure 11.23 to estimate the boiling point of water at the top of this mountain.

Practice Exercise 11.9

11.6 | Energy and Changes of State

When a liquid or solid evaporates or a solid melts, there are increases in the distances between the particles of the substance. Particles that normally attract each other are forced apart, increasing their potential energies. Such energy changes affect our daily lives in many ways, especially the energy changes associated with the changes in state of water, changes that even control the weather on our planet. To study these energy changes, let's begin by examining how the temperature of a substance varies as it is heated.

Heating Curves and Cooling Curves

Figure 11.27*a* illustrates the way the temperature of a substance changes as we add heat to it *at a constant rate*, starting with the solid and finishing with the gaseous state of the substance. The graph is sometimes called a **heating curve** for the substance.

Figure 11.27 | **Heating and cooling curves.** (*a*) A heating curve observed when heat is added to a substance at a constant rate. Superheating is shown as continued heating past the boiling point. (*b*) A cooling curve observed when heat is removed from a substance at a constant rate. Supercooling is seen here as the temperature of the liquid dips below its freezing point. Once a tiny crystal forms, the temperature rises to the freezing point.

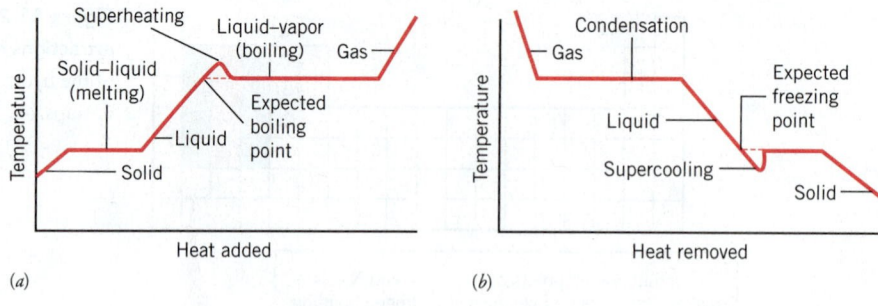

(*a*) (*b*)

■ When a solid or liquid is heated, the volume expands only slightly, so there are only small changes in the average distance between the particles. This means that very small changes in potential energy take place, so almost all of the heat added goes to increasing the kinetic energy.

First, let's look at the portions of the graph that slope upward. These occur where we are increasing the temperature of the solid, liquid, and gas phases. Because temperature is related to average kinetic energy, nearly all of the heat we add in these regions of the heating curve goes to increasing the average kinetic energies of the particles. In other words, the added heat makes the particles go faster and collide with each other with more force. In addition, the slope of the rising portions have units of °C per joule. This is the reciprocal of the heat capacity, meaning that the greater the slope, the lower the heat capacity. Gases have lower heat capacities than liquids and therefore the heating of the gas phase has the largest slope.

In those portions of the heating curve where the temperature remains constant, the average kinetic energy of the particles is not changing. This means that all of the heat being added must go to increase the *potential energies* of the particles. During melting, the particles held rigidly in the solid begin to separate slightly as they form the mobile liquid phase. The potential energy increase accompanying this process equals the amount of heat input during the melting process. During boiling, there is an even greater increase in the distance between the molecules. Here they go from the relatively tight packing in the liquid to the widely spaced distribution of molecules in the gas. This gives rise to an even larger increase in the potential energy, which we see as a longer flat region on the heating curve during the boiling of the liquid.

The opposite of a heating curve is a **cooling curve** (see Figure 11.27*b*). Here we start with a gas and gradually cool it—that is, remove heat from it at a constant rate—until we have reached a solid.

Superheating and Supercooling

Looking at Figure 11.27 again we notice two unusual features, one on each curve. There is a small "blip" on the heating curve near the transition from the liquid to a gas. A similar feature occurs when a liquid is cooled to a solid. These "blips" represent the phenomena of superheating and supercooling. **Superheating** occurs when the liquid is heated above the boiling point without boiling. If disturbed, a superheated liquid will erupt with a shower of vapor and liquid. Many people have discovered this effect when heating their favorite beverage in a microwave oven. When cooling a liquid it is possible to decrease the temperature below the freezing point without solidification occurring, resulting in a supercooled liquid. Once again, if the supercooled liquid is disturbed, very rapid crystallization occurs.

Molar Heats of Fusion, Vaporization, and Sublimation

Enthalpy changes during
phase changes

■ These are also called enthalpies of fusion, vaporization, and sublimation.

Because phase changes occur at constant temperature and pressure, the potential energy changes associated with melting and vaporization can be expressed as enthalpy changes. Usually, enthalpy changes are expressed on a "per mole" basis and are given special names to identify the kind of change involved. For example, using the word fusion instead of "melting," the **molar heat of fusion,** ΔH_{fusion}, is the heat absorbed by one mole of a solid when it melts to give a liquid at the same temperature and pressure. Similarly, the **molar heat of vaporization,** $\Delta H_{\text{vaporization}}$, is the heat absorbed when one mole of a liquid is changed to one mole of vapor at a constant temperature and pressure. Finally, the **molar**

heat of sublimation, $\Delta H_{\text{sublimation}}$, is the heat absorbed by one mole of a solid when it sublimes to give one mole of vapor, once again at a constant temperature and pressure. The values of ΔH for fusion, vaporization, and sublimation are all positive because the phase change in each case is endothermic, being accompanied by a net increase in potential energy. In order to calculate the amount of heat required to melt a specific amount of a substance, we can multiply the moles of the substance by the molar heat of fusion:

$$q = n \times \Delta H_{\text{fusion}} \qquad \text{(11.1)}$$

in which q is the heat required and n is the number of moles. This equation can also be used to determine the heat required for the process of sublimation or vaporization by substituting $\Delta H_{\text{sublimation}}$ or $\Delta H_{\text{vaporization}}$ for ΔH_{fusion}, respectively.

Examples of the influence of these energy changes on our daily lives abound. For example, you've added ice to a drink to keep it cool, because as the ice melts, it absorbs heat (its heat of fusion). Your body uses the heat of vaporization of water to cool itself through the evaporation of perspiration. During the summer, ice cream trucks carry dry ice because the sublimation of CO_2 absorbs heat and keeps the ice cream cold. And perhaps most important, weather on our planet is driven by the heat of vaporization of water, which converts solar energy into the energy of winds and storms. For example, large storms over oceans, such as hurricanes, rely on a continual supply of warm, moist air produced by the rapid evaporation of water from tropical waters. Continual condensation in the high clouds forms rain and supplies the energy needed to feed the storm's winds.

The solidification of a liquid to a crystal, the condensation of a gas to a liquid, or the deposition of a gas as a solid are simply the reverse of fusion, vaporization, and sublimation processes, respectively. Therefore, the heat of crystallization is equal to the heat of fusion but it has the opposite numerical sign. Similarly, the heats of condensation and deposition have the opposite signs of their counterparts.

Since heat is released when liquids solidify or when gases become solids or liquids, that heat can be put to practical use. For example, refrigerators work by condensing the refrigerant gas; this process releases heat, outside the refrigerator. When the liquid vaporizes in the reverse process, it absorbs heat, the temperature drops, and the inside of the refrigerator is cooled. Similarly, meteorologists can often use the "dew point" of the atmosphere to predict overnight low temperatures. The dew point is the temperature at which moisture in the air begins to condense. If the nighttime temperature decreases to the dew point, the heat released by the condensation of water usually keeps the temperature from falling much further.

■ $\Delta H_{\text{fusion}} = -\Delta H_{\text{crystallization}}$
$\Delta H_{\text{vaporization}} = -\Delta H_{\text{condensation}}$
$\Delta H_{\text{sublimation}} = -\Delta H_{\text{deposition}}$

TOOLS

Heat of fusion

■ The stronger the attractions, the more the potential energy will increase when the molecules become separated, and the larger will be the value of ΔH.

Example 11.2
Calculating the Amount of Heat Required during Changes of State

Liquid sodium metal is used as heat transfer material to cool nuclear reactors. How much heat is required to heat 75.0 g of sodium from 25.0 °C to 515.0 °C? The melting point of sodium is 97.8 °C; the specific heat of solid sodium is 1.23 J/g °C; the molar heat of fusion of sodium is 2.60 kJ/mol; and the specific heat of liquid sodium is 1.38 J/g °C. We will assume that the specific heats do not change with temperature.

Analysis: We are being asked to calculate the total amount of heat for the process of (1) heating the solid to the melting point, (2) melting the solid, and (3) heating the resulting liquid. We are given the amount of sodium and the specific heats of the solid and liquid sodium, in addition to the heat of fusion of sodium.

Assembling the Tools: Using Equation 6.7, $q = ms\,\Delta t$, the amount of heat, q, required to heat the solid and the liquid can be determined from the mass, m, the specific heat of the substance, s, and the change in temperature, Δt. For the process of melting the solid, we will use Equation 11.1, $q = n \times \Delta H_{\text{fusion}}$, in which ΔH_{fusion} is the heat of fusion of sodium.

Solution: Let us start at 25.0 °C to calculate the amount of heat required to heat the sodium to its melting point of 97.8 °C.

$$q_1 = ms\,\Delta t = 75.0\text{ g} \times 1.23\text{ J g}^{-1}\,°\text{C}^{-1} \times (97.8\,°\text{C} - 25.0\,°\text{C})$$

$$= 6716\text{ J}$$

The next step is melting the sodium at 97.8 °C, and we will use Equation 11.1 to calculate this amount of heat.

$$q_2 = n \times \Delta H_{fusion}$$

Before proceeding, we need to look at the units for ΔH_{fusion}, which are kJ/mol. The first thing to note is that we are given a mass of sodium, so we need to convert that to moles. Second, the energy units are in kJ but the energy units for heating the solid sodium are in J, so we have to convert them to J.

$$n = 75.0\text{ g Na} \times \frac{1\text{ mol Na}}{22.99\text{ g Na}} = 3.262\text{ mol Na}$$

$$q_2 = n \times \Delta H_{fusion} = 3.262\text{ mol Na} \times \frac{2.60\text{ kJ}}{1\text{ mol Na}} \times \frac{1000\text{ J}}{1\text{ kJ}}$$

$$q_2 = 8481\text{ J}$$

The final step in the process is heating the liquid sodium from 97.8 °C to 515.0 °C, so we can calculate the heat absorbed using the same equation as we did in the first step:

$$q_3 = ms\,\Delta t = 75.0\text{ g} \times 1.38\text{ J g}^{-1}\,°\text{C}^{-1} \times (515.0\,°\text{C} - 97.8\,°\text{C})$$

$$= 43{,}180\text{ J}$$

We add the three heats together to get the total heat:

$$q_{total} = q_1 + q_2 + q_3$$

$$q_{total} = 6716\text{ J} + 8481\text{ J} + 43{,}180\text{ J} = 58{,}400\text{ J}$$

Is the Answer Reasonable? Since this question has a number of calculations, it would be a good idea to estimate the answer and see whether we are in the ball park. The specific heats for the solid and liquid sodium are approximately 1 J/g, the mass is approximately 100 g, and the change in temperature is about 500 °C, so the heat, q, would be about $(1\text{ J/g}) \times (100\text{ g}) \times (500\,°\text{C}) = 50{,}000\text{ J}$. We have about 3 mol of sodium, 75 g Na/(1 mol Na/23 g), and the ΔH_{fusion} is about 3000 J/mol, so the heat required to melt the sodium is $(3\text{ mol Na}) \times (3000\text{ J/mol}) = 9000\text{ J}$. Adding these two heats together we get 59,000 J, which is about what we calculated above, indicating that our answer is correct.

Practice Exercise 11.10

Benzene has a boiling point of 80.1 °C, and a specific heat capacity of 1.8 J/g °C in the liquid phase and 1.92 J/g °C in the gas phase. The $\Delta H_{vaporization}$ of benzene is 30.77 kJ/mol. How much heat is released when 55 g of benzene in the gas phase is cooled from 105.0 °C to 25.0 °C? (*Hint:* The benzene is cooling from the gas to the liquid, so three steps are needed to solve this problem.)

Practice Exercise 11.11

Steam can cause more severe burns than water, even if both are at the same temperature. Calculate the amount of heat released from 10 g of steam at 100.0 °C as it cools to 37 °C (body temperature), and the amount of heat released from 10 g of water at 100.0 °C as it cools to 37 °C. The $\Delta H_{vaporization}$ of water is 43.9 kJ mol^{-1} and the specific heat of water is 4.184 J g^{-1} °C^{-1}.

Energy Changes and Intermolecular Attractions

As mentioned in Section 11.2, when a liquid evaporates or a solid sublimes, the particles go from a liquid (or solid) in which the attractive forces are very strong to the gas phase in which the attractive forces are so small they can almost be ignored. Therefore, the values of $\Delta H_{\text{vaporization}}$ and $\Delta H_{\text{sublimation}}$ give us directly the energy needed to separate molecules from each other. We can examine such values to obtain reliable comparisons of the strengths of intermolecular attractions.

In Table 11.4, notice that the heats of vaporization of water and ammonia are very large, which is just what we would expect for hydrogen-bonded substances. By comparison, CH_4, a nonpolar substance composed of atoms of a size similar to those of H_2O and NH_3, has a very small heat of vaporization. Note also that polar substances such as HCl and SO_2 have fairly large heats of vaporization compared with nonpolar substances. For example, compare HCl with Cl_2. Even though Cl_2 contains two relatively large atoms and therefore would be expected to have larger London forces than HCl, HCl has the larger $\Delta H_{\text{vaporization}}$. This must be due to dipole–dipole attractions between polar HCl molecules—attractions that are absent in nonpolar Cl_2.

Heats of vaporization also reflect the factors that control the strengths of London forces. For example, the data in Table 11.4 show the effect of chain length on the intermolecular attractions between hydrocarbons; as the chain length increases from one carbon in CH_4 to six carbons in C_6H_{14}, the heat of vaporization also increases, showing that the London forces also increase. Similarly, the heats of vaporization of the halogens in Table 11.4 show that the strengths of London forces increase as the electron clouds of the particles become larger.

TABLE 11.4	Some Typical Heats of Vaporization	
Substance	$\Delta H_{\text{vaporization}}$ (kJ/mol)	Type of Attractive Force
H_2O	+43.9	Hydrogen bonding and London
NH_3	+21.7	Hydrogen bonding and London
HCl	+15.6	Dipole–dipole and London
SO_2	+24.3	Dipole–dipole and London
F_2	+3.27	London
Cl_2	+10.2	London
Br_2	+15.0	London
I_2	+20.9	London
CH_4	+8.16	London
C_2H_6	+14.7	London
C_3H_8	+19.0	London
C_6H_{14}	+31.9	London

11.7 | Phase Diagrams

Sometimes it is useful to know under what combinations of temperature and pressure a substance will be a liquid, a solid, or a gas, or the conditions of temperature and pressure at which two phases are in equilibrium. A simple way to determine this is to use a **phase diagram**—a graphical representation of the pressure–temperature relationships that apply to the equilibria between the phases of the substance.

Figure 11.28 is the phase diagram for water. On it, there are three lines that intersect at a common point. Equilibrium between phases exists everywhere along the lines. For

TOOLS

Phase diagrams

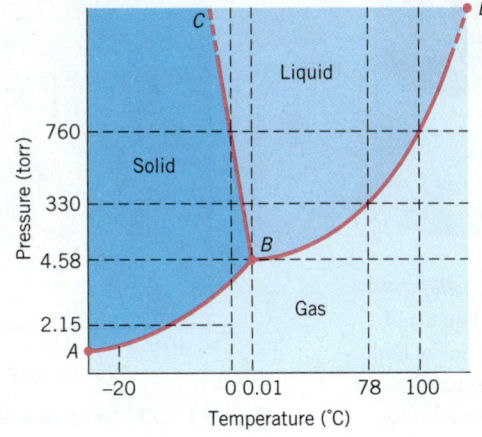

Figure 11.28 | **The phase diagram for water**, axis scales are not linear to emphasize certain features. Temperatures and pressures corresponding to the dashed lines on the diagram are referred to in the text discussion.

example, line *BD* is the vapor pressure curve for liquid water. It gives the temperatures and pressures at which the liquid and vapor are able to coexist in equilibrium. Notice that when the temperature is 100 °C, the vapor pressure is 760 torr. Therefore, this diagram also tells us that water boils at 100 °C when the pressure is 1 atm (760 torr).

The solid–vapor equilibrium line, *AB*, and the liquid–vapor line, *BD*, intersect at a common point, *B*. Because this point is on both lines, there is equilibrium between all three phases at the same time.

■ The melting point and boiling point can be read directly from the phase diagram.

$$\text{liquid} \underset{\nwarrow}{\overset{\nearrow}{}} \overset{\text{vapor}}{} \underset{\nearrow}{\overset{\nwarrow}{}} \text{solid}$$

vapor

liquid ⇌ solid

The temperature and pressure at which this triple equilibrium occurs define the **triple point** of the substance. For water, the triple point occurs at 0.01 °C and 4.58 torr. Every known chemical substance except helium has its own characteristic triple point, which is controlled by the balance of intermolecular forces in the solid, liquid, and vapor. The triple point of water is used in the SI to define our temperature scales.

Line *BC*, which extends upward from the triple point, is the *solid–liquid equilibrium line*, which shows how the melting point temperature changes as pressure changes. It gives temperatures and pressures at which the solid and the liquid are in equilibrium. At the triple point, the melting of ice occurs at +0.01 °C (and 4.58 torr); at 760 torr, melting occurs very slightly lower, at 0 °C. From the graph we can tell that *increasing the pressure on ice lowers its melting point* (Figure 11.29).

Now suppose we have ice at a pressure just below the solid–liquid line, *BC*. If, *at constant temperature,* we raise the pressure to a point just above the line, the ice will melt and become a liquid. This could only happen if the melting point decreases as the pressure increases.

Water is very unusual. Almost all other substances have melting points that increase with increasing pressure, as illustrated by the phase diagram for carbon dioxide (Figure 11.30). For CO_2 the solid–liquid line slants to the right (it slanted to the left for water). Also notice that carbon dioxide has a triple point that's above 1 atm. At atmospheric pressure, the only equilibrium that can be established is between solid carbon dioxide and its vapor. At a pressure of 1 atm, this equilibrium occurs at a temperature of −78 °C. This is the temperature of dry ice, which sublimes at atmospheric pressure at −78 °C.

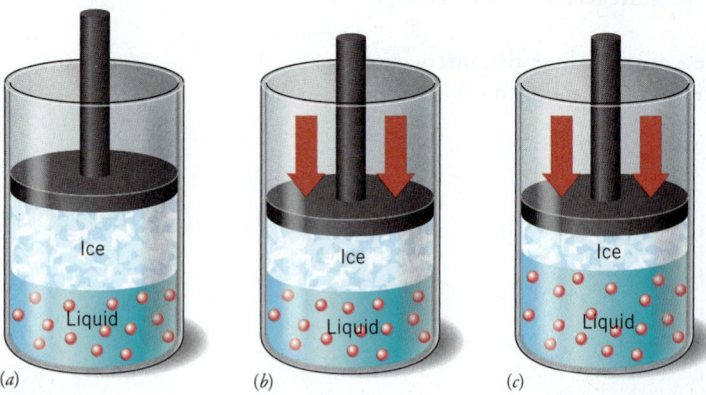

Figure 11.29 | **The effect of pressure on the equilibrium,** H_2O(*solid*) ⇌ H_2O(*liquid*). (*a*) The system is at equilibrium between water as a liquid and water as a solid. (*b*) Pushing down on the piston decreases the volume of both the ice and liquid water by a small amount and increases the pressure. (*c*) Some of the ice melts, producing the more dense liquid. As the total volume of ice and liquid water decreases, the pressure drops and equilibrium is restored.

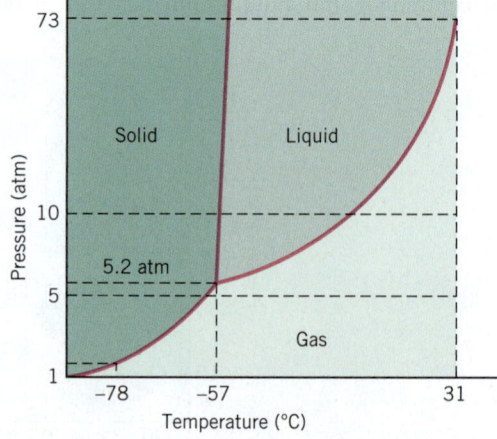

Figure 11.30 | **The phase diagram for carbon dioxide,** critical point is at 31.1 °C and 73 atm.

Interpreting a Phase Diagram

Besides specifying phase equilibria, the three intersecting lines on a phase diagram serve to define regions of temperature and pressure at which only a single phase can exist. For example, between lines *BC* and *BD* in Figure 11.28 are temperatures and pressures at which water exists as a liquid without being in equilibrium with either vapor or ice. At 760 torr, water is a liquid anywhere between 0 °C and 100 °C. According to the diagram, we can't have ice with a temperature of 25 °C if the pressure is 760 torr (which, of course, you already knew; ice never has a temperature of 25 °C.) The diagram also says that we can't have water vapor with a pressure of 760 torr when the temperature is 25 °C (which, again, you already knew; the temperature has to be taken to 100 °C for the vapor pressure to reach 760 torr). Instead, the phase diagram indicates that the *only* phase for pure water at 25 °C and 1 atm is the liquid. Below 0 °C at 760 torr, water is a solid; above 100 °C at 760 torr, water is a vapor. On the phase diagram for water, the phases that can exist in the different temperature–pressure regions are marked.

Example 11.3
Interpreting a Phase Diagram

What phase would we expect for water at 0 °C and 4.58 torr?

Analysis: The words "What phase . . .," as well as the specified temperature and pressure, suggest that we refer to the phase diagram of water (Figure 11.28).

Assembling the Tools: The tool we will use is the phase diagram of water.

Solution: First, we find 0 °C on the temperature axis of the phase diagram of water. Then, we move upward until we intersect a line corresponding to 4.58 torr. This intersection occurs in the "solid" region of the diagram. At 0 °C and 4.58 torr, then, water exists as a solid.

Is the Answer Reasonable? We've seen that the freezing point of water increases slightly when we lower the pressure, so below 1 atm, water should still be a solid at 0 °C. That agrees with the answer we obtained from the phase diagram.

Example 11.4
Interpreting a Phase Diagram

What phase changes occur if water at 0 °C is gradually compressed from a pressure of 2.15 torr to 800 torr?

Analysis: Asking "what phase changes occur" suggests once again that we use the phase diagram of water (Figure 11.28).

Assembling the Tools: Just as in Example 11.3, the tool we will use is the phase diagram of water.

Solution: According to the phase diagram, at 0 °C and 2.15 torr, water exists as a gas (water vapor). As the vapor is compressed, we move upward along the 0 °C line until we encounter the solid–vapor line. There, an equilibrium will exist as compression gradually transforms the gas into solid ice. Once all of the vapor has frozen, further compression raises the pressure and we continue the climb along the 0 °C line until we next encounter the solid–liquid line at 760 torr. As further compression takes place, the solid will melt. After all of the ice has melted, the pressure will continue to climb while the water remains a liquid. At 800 torr and 0 °C, the water will be liquid. The phase changes are gas to solid to liquid.

Is the Answer Reasonable? There's not too much we can do to check all this except to take a fresh look at the phase diagram. We do expect that above 760 torr, the melting point of ice will be less than 0 °C, so at 0 °C and 800 torr we can anticipate that water will be a liquid.

Practice Exercise 11.12 | The equilibrium line from point *B* to *D* in Figure 11.28 is present in another figure in this chapter. Identify what that line represents. (*Hint:* A review of the other figures will reveal the nature of the line.)

Practice Exercise 11.13 | What phase changes will occur if water at −20 °C and 2.15 torr is heated to 50 °C under constant pressure?

Practice Exercise 11.14 | What phase will water be in if it is at a pressure of 330 torr and a temperature of 50 °C?

Supercritical Fluids

For water (Figure 11.28), the vapor pressure line for the liquid, which begins at point *B*, terminates at point *D*, which is known as the **critical point**. The temperature and pressure at *D* are called the **critical temperature, T_c**, and the **critical pressure, P_c**. Above the critical temperature, a distinct liquid phase cannot exist, *regardless of the pressure.*

Figure 11.31 illustrates what happens to a substance as it approaches its critical point. In Figure 11.31*a*, we see a liquid in a container with some vapor above it. We can distinguish between the two phases because they have different densities, which causes them to bend light differently. This allows us to see the interface, or surface, between the more dense liquid and the less dense vapor. If this liquid is now heated, two things happen. First, more liquid evaporates. This causes an increase in the number of molecules per cubic centimeter of vapor, which, in turn, causes the density of the vapor to increase. Second, the liquid expands. This means that a given mass of liquid occupies more volume, so its density decreases. As the temperature of the liquid and vapor continue to increase, the vapor density rises and the liquid density falls; they approach each other. Eventually the densities become equal, and separate liquid and gas phases no longer exist; everything is the same (see Figure 11.31*b*). The highest temperature at which a liquid phase still exists is the critical temperature, and the pressure of the vapor at this temperature is the critical pressure. A substance that has a temperature above its critical temperature and a density near its liquid density is described as a **supercritical fluid**. Supercritical fluids have some unique properties that make them excellent solvents, and one that is particularly useful is supercritical carbon dioxide, which is used as a solvent to decaffeinate coffee and tea.

The values of the critical temperature and critical pressure are unique for every chemical substance and are controlled by the intermolecular attractions (see Table 11.5). Notice that liquids with strong intermolecular attractions, like water, tend to have high critical temperatures. Under pressure, the strong attractions between the molecules are able to

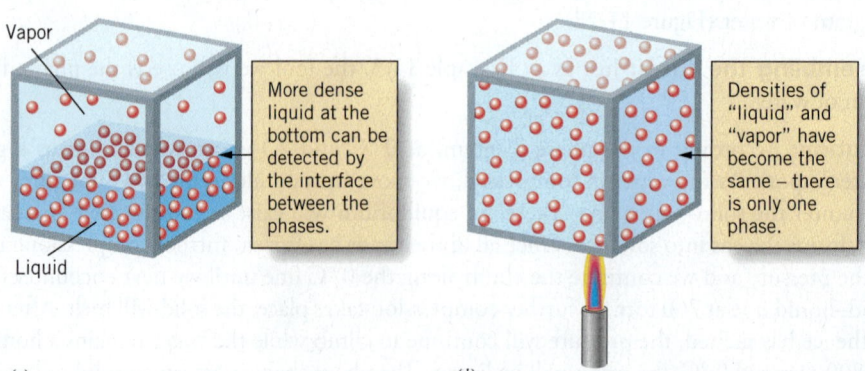

Figure 11.31 | **Changes that are observed when a liquid is heated in a sealed container.** (*a*) Below the critical temperature. (*b*) Above the critical temperature.

Vapor

More dense liquid at the bottom can be detected by the interface between the phases.

Liquid

Densities of "liquid" and "vapor" have become the same – there is only one phase.

(*a*)   (*b*)

hold them together in a liquid state even when the molecules are jiggling about violently at an elevated temperature. In contrast, substances with weak intermolecular attractions, such as methane and helium, have low critical temperatures. For these substances, even the small amounts of kinetic energy possessed by the molecules at low temperatures is sufficient to overcome the intermolecular attractions and prevent the molecules from sticking together as a liquid, despite being held close together under high pressure.

TABLE 11.5	Some Critical Temperatures and Pressures	
Compound	T_c (°C)	P_c (atm)
Water	374.1	217.7
Ammonia	132.5	112.5
Carbon dioxide	31.1	72.9
Ethane (C_2H_6)	32.2	48.2
Methane (CH_4)	−82.1	45.8
Helium	−267.8	2.3

CHEMISTRY OUTSIDE THE CLASSROOM 11.1

Decaffeinated Coffee and Supercritical Carbon Dioxide

Many people prefer to avoid caffeine, yet still enjoy their cup of coffee. For them, decaffeinated coffee is just the thing. To satisfy this demand, coffee producers remove caffeine from the coffee beans before roasting them.

Several methods have been used, some of which use solvents such as methylene chloride (CH_2Cl_2) or ethyl acetate ($CH_3CO_2C_2H_5$) to dissolve the caffeine. Even though only trace amounts of these solvents remain after the coffee beans are dried, there are those who would prefer not to have any such chemicals in their coffee. And that's where carbon dioxide comes into the picture.

It turns out that supercritical carbon dioxide is an excellent solvent for many organic substances, including caffeine. To make supercritical carbon dioxide, gaseous CO_2 is heated to a temperature (typically around 80 °C) above its critical temperature of 31 °C. It is then compressed to about 200 atm. This gives it a density near that of a liquid, but with some properties of a gas. The fluid has a very low viscosity and readily penetrates coffee beans that have been softened with steam, drawing out the water and the caffeine. After several hours, the CO_2 has removed as much as 97% of the caffeine, and the fluid containing the water and caffeine is then drawn off. When the pressure of the supercritical CO_2 solution is reduced, the CO_2 turns to a gas and the water and caffeine separate. The caffeine is recovered and sold to beverage or pharmaceutical companies. Meanwhile, the pressure over the coffee beans is also reduced and the beans are warmed to about 120 °C, causing residual CO_2 to evaporate. Because CO_2 is not a toxic gas, any traces of CO_2 that remain are harmless.

Decaffeination of coffee is not the only use of supercritical CO_2. It is also used to extract the essential flavor ingredients in spices and herbs for use in a variety of products. As with coffee, using supercritical CO_2 as a solvent completely avoids any potential harm that might be caused by small residual amounts of other solvents.

Andy Washnik

Liquefaction of Gases

When a gaseous substance is at a temperature below its critical temperature, it is capable of being liquefied by compressing it. For example, carbon dioxide is a gas at room temperature (approximately 25 °C). This is below its critical temperature of 31 °C. If the $CO_2(g)$ is gradually compressed, a pressure will eventually be reached that lies on the liquid–vapor curve for CO_2, and further compression will cause the CO_2 to liquefy. In fact, that's what happens when a CO_2 fire extinguisher is filled; the CO_2 that's pumped in is a liquid under high pressure. If you shake a filled CO_2 fire extinguisher, you can feel the liquid sloshing around inside, provided the temperature of the fire extinguisher is below 31 °C (88 °F). When the fire extinguisher is used, a valve releases the pressurized CO_2, which rushes out to extinguish the fire.

Gases such as O_2 and N_2, which have critical temperatures far below 0 °C, can never be liquids at room temperature. When they are compressed, they simply become high-pressure gases. To make liquid N_2 or O_2, the gases must be made very cold as well as be compressed to high pressures. Liquid nitrogen (B.P. −196 °C or 77 K) is often used to cool experiments to low temperatures in the lab. Liquid oxygen (B.P. −183 °C or 90 K) is used as an oxidizing agent and also as a source of $O_2(g)$ in hospitals.

■ On a very hot day, when the temperature is in the 90s, a filled CO_2 fire extinguisher won't give the sensation that it's filled with a liquid. At such temperatures, the CO_2 is in a supercritical state and no separate liquid phase exists.

11.8 | Le Châtelier's Principle and Changes of State

Throughout this chapter, we have studied various dynamic equilibria. One example was the equilibrium that exists between a liquid and its vapor in a closed container. You learned that when the temperature of the liquid is increased in this system, its vapor pressure also increases. Let's briefly review why this occurs.

Initially, the liquid is in equilibrium with its vapor, which exerts a certain pressure. When the temperature is increased, equilibrium no longer exists because evaporation occurs more rapidly than condensation. Eventually, as the concentration of molecules in the vapor increases, the system reaches a new equilibrium in which there is more vapor and a little less liquid. The greater concentration of molecules in the vapor causes a larger pressure.

The way a liquid's equilibrium vapor pressure responds to a temperature change is an example of a general phenomenon. Whenever a dynamic equilibrium is upset by some disturbance, the system changes in a way that will, if possible, bring the system back to equilibrium again. It's also important to understand that in the process of regaining equilibrium, the system undergoes a net change. Thus, when the temperature of a liquid is raised, there is some net conversion of liquid into vapor as the system returns to equilibrium. When the new equilibrium is reached, the amounts of liquid and vapor are not the same as they were before.

Throughout the remainder of this book, we will deal with many kinds of equilibria, both chemical and physical. It would be very time consuming and sometimes very difficult to carry out a detailed analysis each time we wish to know the effects of some disturbance on an equilibrium system. Fortunately, there is a relatively simple and fast method for predicting the effect of a disturbance, one based on a principle proposed in 1888 by a brilliant French chemist, Henry Le Châtelier (1850–1936).

TOOLS
Le Châtelier's principle

Le Châtelier's Principle

When a dynamic equilibrium in a system is upset by a disturbance, the system responds in a direction that tends to counteract the disturbance and, if possible, restore equilibrium.

Let's see how we can apply Le Châtelier's principle to a liquid–vapor equilibrium that is subjected to a temperature increase. We cannot increase a temperature, of course, without adding heat. Thus, the addition of heat is really the disturbing influence when a temperature is increased. So, let's incorporate "heat" as a member of the equation used to represent the liquid–vapor equilibrium.

$$\text{heat} + \text{liquid} \rightleftharpoons \text{vapor} \qquad (11.2)$$

Recall that we use double arrows, $\rightleftharpoons$, to indicate a dynamic equilibrium in an equation. They imply opposing changes happening at equal rates. Evaporation is endothermic, so the heat is placed on the left side of Equation 11.2 to show that heat is absorbed by the liquid when it changes to the vapor, and that heat is released when the vapor condenses to a liquid.

Le Châtelier's principle tells us that when we add heat to raise the temperature of the equilibrium system, the system will try to adjust in a way that absorbs some of the added heat. This can happen if some liquid evaporates, because vaporization is endothermic. When liquid evaporates, the amount of vapor increases and causes the pressure to rise. Thus, we have reached the correct conclusion in a very simple way—namely, that heating a liquid must increase its vapor pressure.

We often use the term **position of equilibrium** to refer to the relative amounts of the substances on opposite sides of the double arrows in an equilibrium expression such as Equation 11.2. Thus, we can think of how a disturbance affects the position of equilibrium. For example, increasing the temperature increases the amount of vapor and decreases the amount of liquid, and we say *the position of equilibrium has shifted*; in this case, it has shifted in the direction of the vapor, or it has *shifted to the right*. In using Le Châtelier's principle, it is often convenient to think of a disturbance as "shifting the position of equilibrium" in one direction or another in the equilibrium equation.

Practice Exercise 11.15

Use Le Châtelier's principle to predict how a temperature increase will affect the vapor pressure of a solid. (*Hint:* Solid + heat $\rightleftharpoons$ vapor.)

Practice Exercise 11.16

Designate whether each of the following physical processes is exothermic or endothermic: boiling, melting, condensing, subliming, and freezing. Can any of them be exothermic for some substances and endothermic for others?

11.9 | Determining Heats of Vaporization

The way the vapor pressure varies with temperature, which was described in Section 11.4 and is shown in Figure 11.32, depends on the *heat of vaporization* of a substance. The relationship between vapor pressure and temperature, however, is not a simple proportionality. Instead, it involves *natural logarithms*, which are logarithms to the base e, as compared to the more familiar base-10 logarithms. (If you are unfamiliar with natural logarithms, please refer to Appendix A.)

The Clausius–Clapeyron Equation

Rudolf Clausius (1822–1888), a German physicist, and Benoit Clapeyron (1799–1864), a French engineer, used the principles of thermodynamics (a subject that's discussed in Chapter 18) to derive the following equation that relates the vapor pressure, heat of vaporization (ΔH_{vap}), and temperature:

$$\ln P = \frac{-\Delta H_{vap}}{RT} + C \qquad (11.3)$$

Clausius–Clapeyron equation

The quantity $\ln P$ is the natural logarithm of the vapor pressure, R is the gas constant expressed in energy units ($R = 8.314\ \text{J mol}^{-1}\ \text{K}^{-1}$), T is the absolute temperature, and C is a constant. Scientists call this the **Clausius–Clapeyron equation**.

The Clausius–Clapeyron equation provides a convenient graphical method for determining heats of vaporization from experimentally measured vapor pressure–temperature data. To see this, let's rewrite the equation as follows:

$$\ln P = \left(\frac{-\Delta H_{vap}}{R}\right)\frac{1}{T} + C$$

Recall from algebra that a straight line is represented by the general equation

$$y = mx + b$$

where x and y are variables, m is the slope, and b is the intercept of the line with the y-axis. In this case, we can make the substitutions

$$y = \ln P \qquad x = \frac{1}{T} \qquad m = \left(\frac{-\Delta H_{vap}}{R}\right) \qquad b = C$$

Therefore, we have

$$
\begin{array}{ccccccc}
\ln P & = & \left(\dfrac{-\Delta H_{vap}}{R}\right) & \dfrac{1}{T} & + & C \\[2ex]
\updownarrow & & \updownarrow & \updownarrow & & \updownarrow \\[1ex]
y & = & m & x & + & b
\end{array}
$$

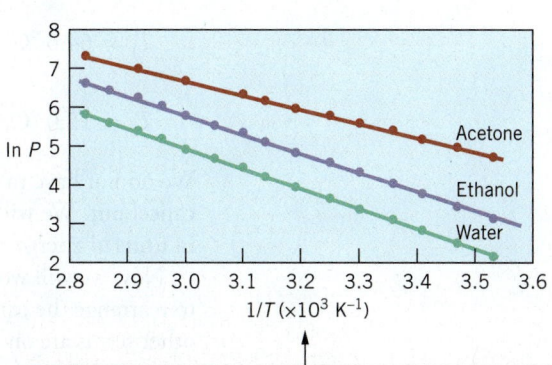

The numbers along the horizontal axis are equal to $1/T$ values multiplied by 1000 to make the axis easier to label.

Figure 11.32 | Plotting the Clausius–Clapeyron equation. A graph showing plots of $\ln P$ versus $\frac{1}{T}$ for acetone, ethanol, and water.

Thus, a graph of $\ln P$ versus $1/T$ should give a straight line that has a slope equal to $-\Delta H_{vap}/R$. Such straight-line relationships are illustrated in Figure 11.32, in which experimental data are plotted for water, acetone, and ethanol. From the graphs in Figure 11.32, the calculated values of ΔH_{vap} are as follows: for water, 43.9 kJ mol^{-1}; for acetone, 32.0 kJ mol^{-1}; and for ethanol, 40.5 kJ mol^{-1}.

Using Equation 11.2, a "two-point" form of the Clausius–Clapeyron equation can be derived that can be used to calculate ΔH_{vap} if the vapor pressure is known at two different temperatures. This equation is

$$\ln \frac{P_1}{P_2} = \frac{\Delta H_{vap}}{R}\left(\frac{1}{T_2} - \frac{1}{T_1}\right) \tag{11.4}$$

If we know the value of the heat of vaporization, Equation 11.4 can also be used to calculate the vapor pressure at some particular temperature (say, P_2 at a temperature T_2) if we already know the vapor pressure P_1 at a temperature T_1.

Example 11.5
Calculating the Heat of Vaporization from Temperatures and Vapor Pressures

Methanol, CH_3OH, experiences hydrogen bonding, dipole–dipole interactions, and London forces. At 64.6 °C, it has a vapor pressure of 1.00 atm, and at 12.0 °C, it has a vapor pressure of 0.0992 atm. What is the heat of vaporization for methanol?

Analysis: We are given two temperatures and two vapor pressures and we are asked to find the heat of vaporization. We will need to use the Clausius–Clapeyron equation to solve this problem.

Assembling the Tools: The main tool that we will use will be the application of the equation

$$\ln \frac{P_1}{P_2} = \frac{\Delta H_{vap}}{R}\left(\frac{1}{T_2} - \frac{1}{T_1}\right)$$

Solution: We are given the following data:

$$P_1 = 1.00 \text{ atm} \qquad t_1 = 64.6\,°C$$
$$P_2 = 0.0992 \text{ atm} \qquad t_2 = 12.0\,°C$$

In order to solve the equation we will have to convert the temperatures in °C into kelvins, K, and let's calculate their reciprocals too since they will be needed soon.

$$T_1 = 64.6\,°C + 273.15 = 337.8 \text{ K} \qquad \frac{1}{T_1} = 1/337.8 \text{ K} = 0.002960 \text{ K}^{-1}$$

$$T_2 = 12.0\,°C + 273.15 = 285.2 \text{ K} \qquad \frac{1}{T_2} = 1/285.2 \text{ K} = 0.003506 \text{ K}^{-1}$$

We do not have to convert the pressure units because they are in a ratio and the units will cancel out. We will use 8.314 J mol^{-1} K^{-1} for the value for R—namely, the gas constant in units of energy.

Now we will work with the Clausius–Clapeyron equation itself. One way to do this is to rearrange the equation so the item we need to solve for, ΔH_{vap}, is on the left side and all other terms are on the right side. We do this by multiplying both sides of the equation by R and divide by $\left(\dfrac{1}{T_2} - \dfrac{1}{T_1}\right)$. The result is

$$\Delta H_{vap} = \frac{R \times \ln \dfrac{P_1}{P_2}}{\left(\dfrac{1}{T_2} - \dfrac{1}{T_1}\right)}$$

Substituting the values for P_1, P_2, T_1, T_2, and R, we get

$$\Delta H_{vap} = \frac{8.314 \text{ J mol}^{-1} \text{ K}^{-1} \times \ln\dfrac{1.00 \text{ atm}}{0.0992 \text{ atm}}}{\left(\dfrac{1}{285.2 \text{ K}} - \dfrac{1}{337.8 \text{ K}}\right)}$$

We have already calculated the reciprocals of T, and we can calculate the natural log term to simplify the expression.

$$\Delta H_{vap} = \frac{8.314 \text{ J mol}^{-1} \text{ K}^{-1} \times 2.311}{(0.003506 \text{ K}^{-1} - 0.00296 \text{ K}^{-1})}$$

Now we perform the simple algebra operations, first on the numerator then on the denominator.

$$\Delta H_{vap} = \frac{19.21 \text{ J mol}^{-1} \text{ K}^{-1}}{0.000546 \text{ K}^{-1}}$$

$$\Delta H_{vap} = 35,180 \text{ J mol}^{-1} \quad \text{or} \quad 35.2 \text{ kJ mol}^{-1} \text{ when correctly rounded.}$$

● **Is the Answer Reasonable?** The first thing to check is the sign for the ΔH_{vap}, which is *always* positive for this endothermic process. Second, referring to Table 11.4, the magnitude of our result is similar to other compounds. Then we can review the math, and make sure that we did not make any math errors.

At 0.00 °C, hexane, C_6H_{14}, has a vapor pressure of 45.37 mm Hg. Its ΔH_{vap} is 30.1 kJ mol^{-1}. What is the vapor pressure of hexane at 62.2 °C? (*Hint:* For logarithms recall that $\ln(P_1/P_2) = \ln P_1 - \ln P_2$.)

What is the normal boiling point of ethanol, which has a $\Delta H_{vap} = 40.5$ kJ mol^{-1}, if its vapor pressure is 0.0992 atm at 27.3 °C?

Practice Exercise 11.17

Practice Exercise 11.18

■ A high degree of regularity is the principal feature that makes solids different from liquids. A liquid lacks this long-range repetition of structure because the particles in a liquid are jumbled and disorganized as they move about.

Notice that Example 11.5 and Practice Exercises 11.17 and 11.18 are all manipulations of the two-temperature, two-pressure form of the Clausius–Clapeyron equation. We just rearrange the equation mathematically to solve for different variables.

11.10 | Structures of Crystalline Solids

When many substances freeze, or when they separate as a solid from a solution, they tend to form crystals that have highly regular features. For example, Figure 11.33 is a photograph of crystals of sodium chloride—ordinary table salt. Notice that each particle is very nearly a perfect little cube. Whenever a solution of NaCl is evaporated, the crystals that form have edges that intersect at 90° angles. Thus, cubes are the norm for NaCl.

Crystals in general tend to have flat surfaces that meet at angles that are characteristic of the substance. The regularity of these surface features reflects the high degree of order among the particles that lie within the crystal. This is true whether the particles are atoms, molecules, or ions.

Lattices and Unit Cells

Any repetitive pattern has a symmetrical aspect about it, whether it is a wallpaper design or the orderly packing of particles in a crystal (Figure 11.34). For example, we can recognize certain repeating distances between the elements of the pattern, and we can see that the lines along which the elements of the pattern repeat are at certain angles to each other.

Figure 11.33 | **Crystals of table salt.** The size of the tiny cubic sodium chloride crystals can be seen in comparison.

The Photo Works

Figure 11.34 | Symmetry among repetitive patterns. A wallpaper design and particles arranged in a crystal each show a repeating pattern of structural units. The pattern can be described by the distances between the repeating units and the angles along which the repetition of structure occurs.

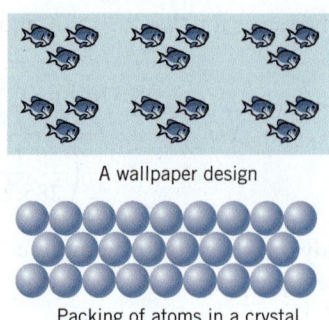

A wallpaper design

Packing of atoms in a crystal

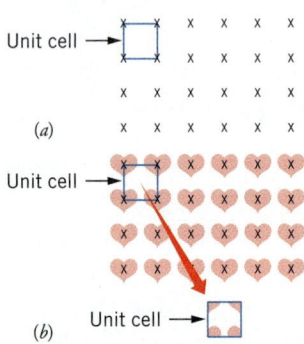

Figure 11.35 | A two-dimensional lattice. (*a*) A simple square lattice, for which the unit cell is a square with lattice points at the corners. (*b*) A wallpaper pattern formed by associating a design element (pink heart) with each lattice point. The x centered on each heart corresponds to a lattice point. The unit cell contains portions of a heart at each corner.

To concentrate on the symmetrical features of a repeating structure, it is convenient to describe it in terms of a set of points that have the same repeat distances as the structure, arranged along lines oriented at the same angles. Such a pattern of points is called a **lattice**, and when we apply it to describe the packing of particles in a solid, we often call it a **crystal lattice**.

In a crystal, the number of particles is enormous. If you could imagine being at the center of even the tiniest crystal, you would find that the particles go on as far as you can see in every direction. Describing the positions of all these particles or their lattice points is impossible and, fortunately, unnecessary. All we need to do is describe the repeating unit of the lattice, which we call the *unit cell*. To see this, and to gain an insight into the usefulness of the lattice concept, let's begin in two dimensions.

In Figure 11.35, we see a two-dimensional *square lattice*, which means the lattice points lie at the corners of squares. The repeating unit of the lattice, its **unit cell**, is indicated in the drawing. If we began with this unit cell, we could produce the entire lattice by moving it repeatedly left and right and up and down by distances equal to its edge length. In this sense, all of the properties of a lattice are contained in the properties of its unit cell.

An important fact about lattices is that the same lattice can be used to describe many different designs or structures. For example, in Figure 11.35*b*, we see a design formed by associating a pink heart with each lattice point. Using a square lattice, we could form any number of designs just by using different design elements (for example, a rose or a diamond) or by changing the lengths of the edges of the unit cell. *The only requirement is that the same design element must be associated with each lattice point.* In other words, if there is a rose at one lattice point, then there must be a rose at all the other lattice points.

Extending the lattice concept to three dimensions is relatively straightforward. Illustrated in Figure 11.36 is a simple cubic (also called a **primitive cubic**) lattice, the

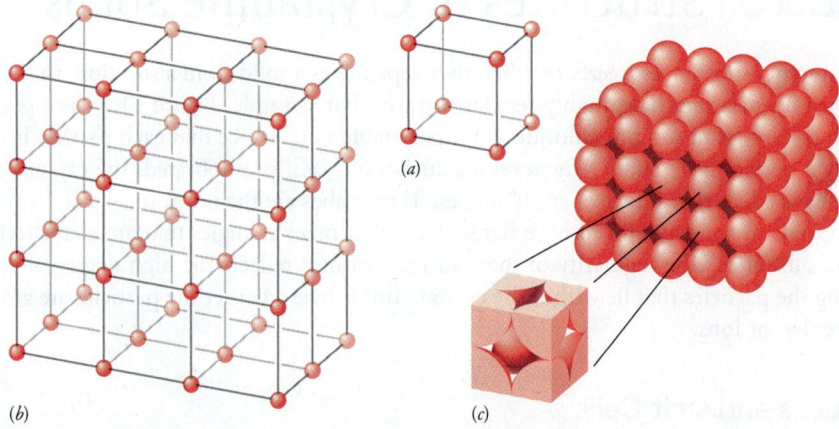

(*a*)

(*b*)

(*c*)

Figure 11.36 | A three-dimensional simple cubic lattice. (*a*) A simple cubic unit cell showing the locations of the lattice points. (*b*) A portion of a simple cubic lattice built by stacking simple cubic unit cells. (*c*) A hypothetical substance that forms crystals having a simple cubic lattice with identical atoms at the lattice points. Only a portion of each atom lies within this particular unit cell.

simplest and most symmetrical three-dimensional lattice. Its unit cell, the **simple cubic unit cell**, is a cube with lattice points only at its eight corners. Figure 11.36c shows the packing of atoms in a substance that might crystallize in a simple cubic lattice, as well as the unit cell for that substance.[3]

As with the two-dimensional lattice, we could use the same simple cubic lattice to describe the structures of many different substances. The *sizes* of the unit cells would vary because the sizes of atoms vary, but the essential symmetry of the stacking would be the same in them all. This fact about lattices makes it possible to describe limitless numbers of different compounds with just a limited set of three-dimensional lattices. In fact, it has been shown mathematically that there are only 14 different three-dimensional lattices possible, which means that all of the chemical substances that can exist must form crystals with one or another of these 14 lattice types.

The 14 lattice types are shown in Table 11.6. There are seven basic lattice shapes, which depend on the lengths of the sides, a_1, a_2, and a_3, and the angles between the edges of the shapes, α_{12}, α_{23}, and α_{13}. The variations on the basic shapes give the lattice types. The simple shape only has lattice points in the corners of the lattice. The body-centered structures have lattice points in the middle of the cells. The base-centered structures have a lattice point on two opposing faces, and the face-centered structures have lattice points on all of the faces.

Cubic Lattices

Cubic lattices are the simplest lattices because all of the sides have the same length and all of the angles are 90°. In addition to simple cubic, two other cubic lattices are possible: face-centered cubic and body-centered cubic. The **face-centered cubic** (abbreviated **fcc**) **unit cell** has lattice points (and, therefore, identical particles) at each of its eight corners plus another in the center of each face, as shown in Figure 11.37. Many common metals—copper, silver, gold, aluminum, and lead, for example—form crystals that have face-centered cubic lattices. Each of these metals has the same *kind* of lattice, but the *sizes* of their unit cells differ because the sizes of the atoms differ (see Figure 11.38).

The **body-centered cubic (bcc) unit cell** has lattice points at each corner plus one in the center of the cell, as illustrated in Figure 11.39. The body-centered cubic lattice is also common among a number of metals; examples include chromium, iron, and platinum. Again, these are substances with the same *kind* of lattice, but the dimensions of the lattices vary because of the *different sizes* of the particular atoms.

TOOLS

Cubic unit cells

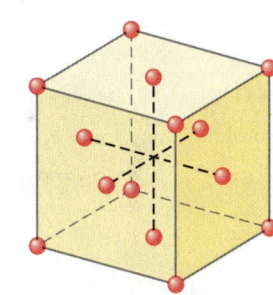

Figure 11.37 | A face-centered cubic unit cell. Lattice points are found at each of the eight corners and in the center of each face.

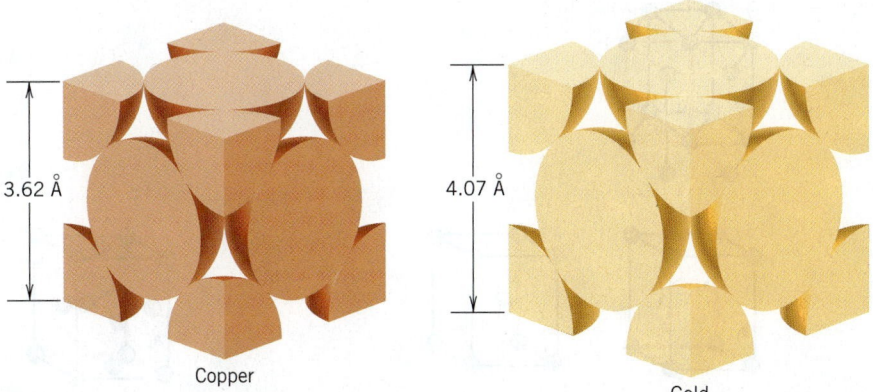

3.62 Å

Copper

4.07 Å

Gold

Figure 11.38 | Unit cells for copper and gold. These metals both crystallize in a face-centered cubic structure with similar face-centered cubic unit cells. The atoms are arranged in the same way, but their unit cells have edges of different lengths because the atoms are of different sizes. (1 Å = 1 × 10⁻¹⁰ m)

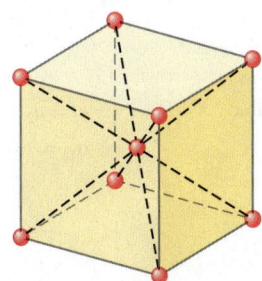

Figure 11.39 | A body-centered cubic unit cell. Lattice points are located at each of the eight corners and in the center of the unit cell.

[3] Polonium is the only metal that has an allotrope that crystallizes in a simple cubic lattice. Some compounds, however, do form simple cubic lattices.

TABLE 11.6	Crystal Lattice Types

The 7 Lattices	The 14 Lattice Types

The 7 Lattices	Simple	Body-Centered	Base-Centered	Face-Centered
Triclinic $a_1 \neq a_2 \neq a_3$ $\alpha_{12} \neq \alpha_{23} \neq \alpha_{31} \neq 90°$	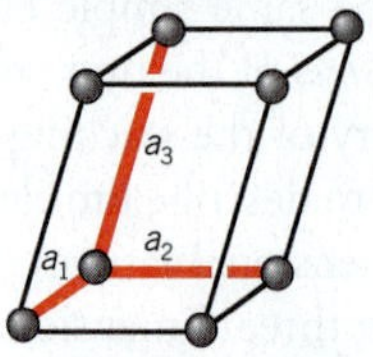			
Monoclinic $a_1 \neq a_2 \neq a_3$ $\alpha_{23} = \alpha_{31} = 90°$ $\alpha_{12} \neq 90°$	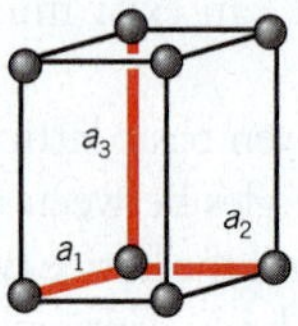	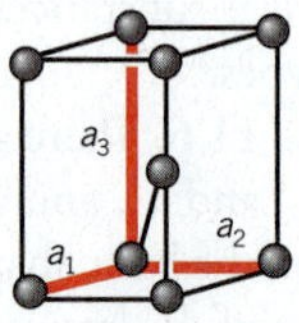		
Orthorhombic $a_1 \neq a_2 \neq a_3$ $\alpha_{12} = \alpha_{23} = \alpha_{31} = 90°$	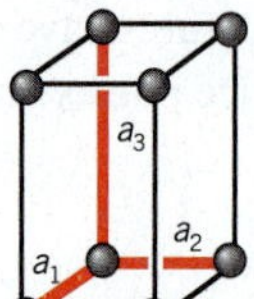	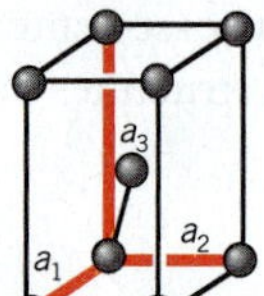	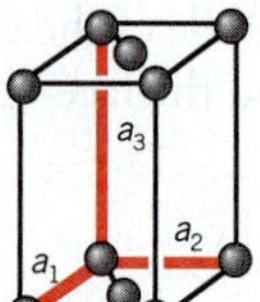	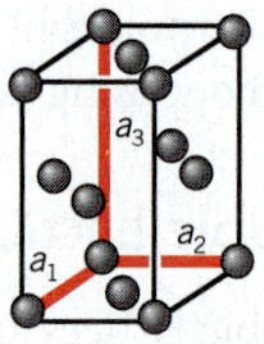
Tetragonal $a_1 = a_2 \neq a_3$ $\alpha_{12} = \alpha_{23} = \alpha_{31} = 90°$	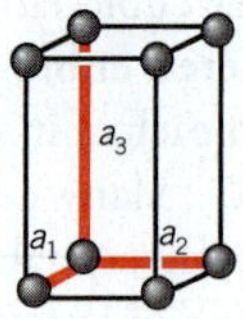	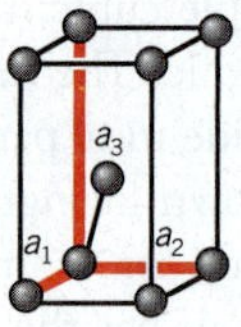		
Trigonal $a_1 = a_2 = a_3$ $\alpha_{12} = \alpha_{23} = \alpha_{31} < 120° \neq 90°$	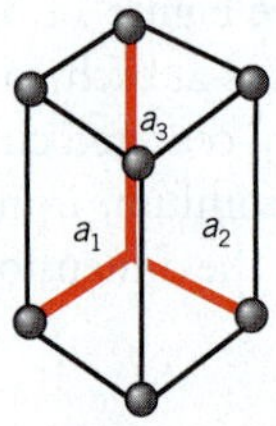			
Hexagonal $a_1 = a_2 \neq a_3$ $\alpha_{12} = 120°$ $\alpha_{23} = \alpha_{31} = 90°$	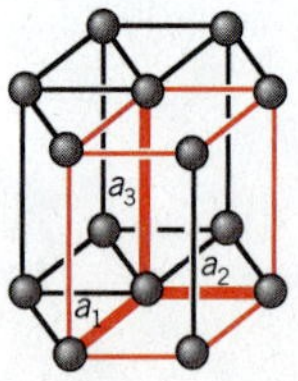			
Cubic $a_1 = a_2 = a_3$ $\alpha_{12} = \alpha_{23} = \alpha_{31} = 90°$	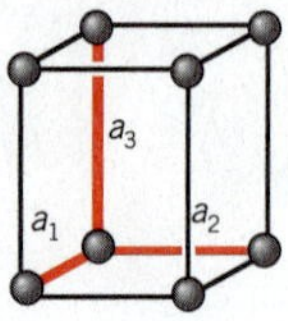	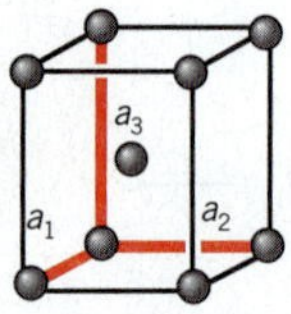		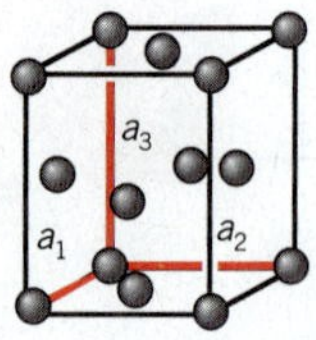

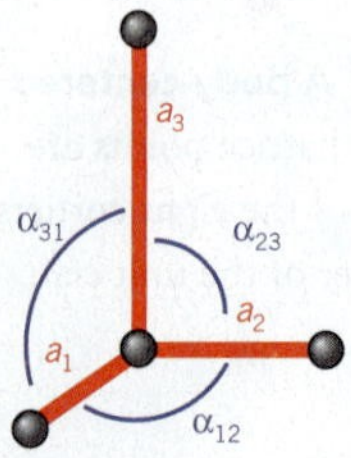

Not all unit cells are cubic. Some have edges of different lengths or edges that intersect at angles other than 90°, as shown in Table 11.6. Although you should be aware of the existence of other unit cells and lattices, we will limit the remainder of our discussion to cubic lattices and their unit cells.

If we look at the unit cell of the face-centered cubic structure of copper in Figure 11.38, we can see that the atoms are cut up into parts. We can show that the parts add up to whole atoms. Notice that when the unit cell is "carved out" of the crystal, we find only part of an atom (one-eighth of an atom, actually) at each corner. The rest of each atom resides in adjacent unit cells. Because the unit cell has eight corners, if we put all the corner pieces together we would obtain one complete atom.

$$8 \text{ corners} \times \frac{\frac{1}{8} \text{ atom}}{\text{corner}} = 1 \text{ atom}$$

Now, if we look at the atoms in the face, only half of each atom lies inside of the unit cell, but there are six faces and therefore six half atoms, for a total of three atoms.

$$6 \text{ faces} \times \frac{\frac{1}{2} \text{ atom}}{\text{face}} = 3 \text{ atoms}$$

The total number of atoms in the unit cell is four atoms: three from the faces and one from the corners. In addition, if an atom is on the edge of a cube, then it contributes ¼ of an atom to the unit cell for each edge, and if an atom is completely within the cell, it, of course, contributes one atom to the unit cell. All crystal lattice types have 12 edges except for the hexagonal lattice, which has 18.

Compounds that Crystallize with Cubic Lattices

We have seen that a number of metals have cubic lattices. The same is true for many compounds. Figure 11.40, for example, is a view of a portion of a sodium chloride crystal. If we look at the locations of identical particles (e.g., Cl^-, in green), we find them at lattice points that describe a face-centered cubic structure. The smaller gray spheres, the Na^+ ions, fill the spaces between the Cl^- ions. Thus, sodium chloride is said to have a face-centered cubic lattice, and the cubic shape of this lattice on the microscopic scale is what accounts for the cubic shape of a sodium chloride crystal on the macroscopic scale.

Many of the alkali halides (Group 1A–7A compounds), such as NaBr and KCl, crystallize with fcc lattices that have the same arrangement of ions as is found in NaCl. In fact, this arrangement of ions is so common that it's called the **rock salt structure** (rock salt is

■ If we focus only on the sodium ion lattice we also find a face-centered cubic structure.

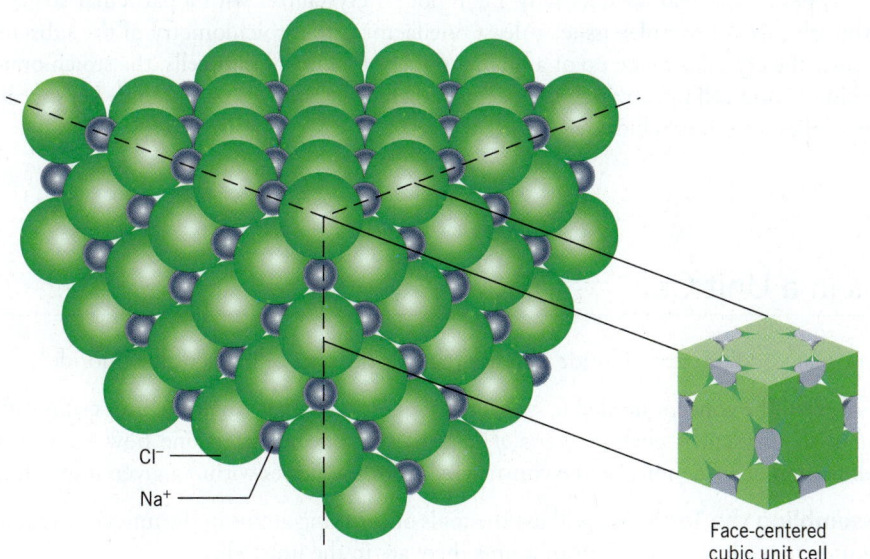

Cl⁻
Na⁺

Face-centered
cubic unit cell

Figure 11.40 | **The packing of ions in a sodium chloride crystal.** Chloride ions are shown here to be associated with the lattice points of a face-centered cubic unit cell, with the sodium ions placed between the chloride ions.

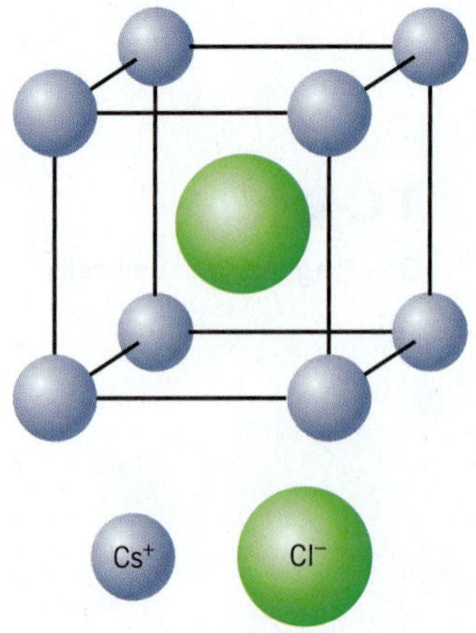

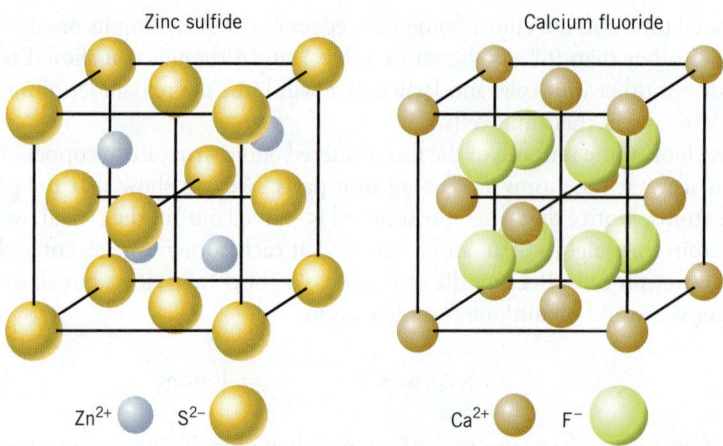

Figure 11.41 | **The unit cell for cesium chloride, CsCl.** The chloride ion is located in the center of the unit cell. *Note:* The ions are not shown full-size to make it easier to see their locations in the unit cell.

Figure 11.42 | **Crystal structures based on the face-centered cubic lattice.** Both zinc sulfide, ZnS, and calcium fluoride, CaF_2, have crystal structures that fit a face-centered cubic lattice. In ZnS, the sulfide ions are shown at the fcc lattice sites with the four zinc ions entirely within the unit cell. In CaF_2, the calcium ions are at the lattice points with the eight fluorides entirely within the unit cell. *Note:* The ions are not shown full size to make it easier to see their locations in the unit cells.

the mineral name of NaCl). Because sodium bromide and potassium chloride both have the same kind of lattice as sodium chloride, Figure 11.40 also could be used to describe their unit cells. The *sizes* of their unit cells are different, however, because K^+ is larger than Na^+ and Br^- is larger than Cl^-.

Other examples of cubic unit cells are shown in Figures 11.41 and 11.42. The structure of cesium chloride in Figure 11.41 is simple cubic, although at first glance it may appear to be body-centered. This is because in a crystal lattice, identical chemical units must be at each lattice point. In CsCl, Cs^+ ions are found at the corners, but not in the center, so the Cs^+ ions describe a simple cubic unit cell.

Both zinc sulfide and calcium fluoride in Figure 11.42 have face-centered cubic unit cells that differ from that for sodium chloride, which illustrates once again how the same basic kind of lattice can be used to describe a variety of chemical structures.

Effects of Stoichiometry on Crystal Structure

At this point, you may wonder why a compound crystallizes with a particular structure. Although this is a complex issue, at least one factor is the stoichiometry of the substance. Because the crystal is made up of a huge number of identical unit cells, the stoichiometry within the unit cell must match the overall stoichiometry of the compound. Let's see how this applies to sodium chloride.

Example 11.6
Counting Atoms or Ions in a Unit Cell

How many sodium and chloride ions are there in the unit cell of sodium chloride?

Analysis: The concept needed to answer this question is that when the unit cell is carved out of the crystal, it encloses *parts of ions*, so we have to determine how many *whole* sodium and chloride ions can be constructed from the pieces within a given unit cell.

Assembling the Tools: We will use the tools of counting atoms in the unit cell of sodium chloride to see how many parts of atoms there are in the unit cell.

Solution: Let's look at the "exploded" view of the NaCl unit cell shown in Figure 11.43. We have parts of chloride ions at the corners and in the center of each face. Let's add the parts.

$$8 \text{ corners} \times \tfrac{1}{8}Cl^- \text{ per corner} = 1\ Cl^-$$

$$6 \text{ faces} \times \tfrac{1}{2}Cl^- \text{ per face} = 3\ Cl^-$$

$$\text{Total} = 4\ Cl^-$$

For the sodium ions, we have parts along each of the 12 edges plus one whole Na^+ ion in the center of the unit cell. Let's add them.

$$12 \text{ edges} \times \tfrac{1}{4}Na^+ \text{ per edge} = 3\ Na^+$$

$$1\ Na^+ \text{ in the center} = 1\ Na^+$$

$$\text{Total} = 4\ Na^+$$

Thus, in one unit cell, there are four chloride ions and four sodium ions.

Is the Answer Reasonable? The ratio of the ions is 4 to 4, which is the same as 1 to 1. That's the ratio of the ions in NaCl, so the answer is reasonable.

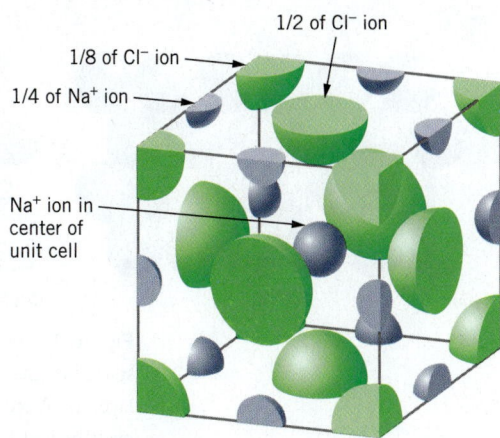

1/2 of Cl⁻ ion
1/8 of Cl⁻ ion
1/4 of Na⁺ ion
Na⁺ ion in center of unit cell

Figure 11.43 | **An exploded view of the unit cell of sodium chloride.**

Chromium crystallizes in a body-centered cubic structure. How many chromium atoms are in its unit cell? (*Hint:* Use Figure 11.39 as a guide.)

Practice Exercise 11.19

What is the ratio of the ions in the unit cell of cesium chloride, which is shown in Figure 11.41? Does this match the formula of cesium chloride? Why is this important?

Practice Exercise 11.20

The calculation in Example 11.6 shows why NaCl can have the crystal structure it does; the unit cell has the proper ratio of cations to anions. It also shows why a compound such as $CaCl_2$ could *not* crystallize with the same kind of unit cell as NaCl. The sodium chloride structure demands a 1-to-1 ratio of cation to anion, so it could not be used by $CaCl_2$ (which has a 1-to-2 cation-to-anion ratio).

Closest-Packed Solids

For many solids, particularly metals, the type of crystal structure formed is controlled by maximizing the number of neighbors that surround a given atom. The more neighbors an atom has, the greater are the number of interatomic attractions and the greater is the energy lowering when the solid forms. Structures that achieve the maximum density of packing are known as **closest-packed structures**, and there are two of them that have the same density, but are only slightly different in their packing. To visualize how these are produced, let's look at ways to pack spheres of identical size.

Figure 11.44*a* illustrates a layer of blue spheres packed as tightly as possible. Notice that each sphere is surrounded by six others that it touches in this layer. When we add a second layer, each sphere (red) rests in a depression formed by three spheres in the first layer, as illustrated in Figures 11.44*b* and 11.44*c*.

The difference between the two closest-packed structures lies in the relative orientations of the spheres in the first layer and those that form the third layer. In Figure 11.45*a*, the green spheres in the third layer each lie in a depression between red spheres that is directly above a depression between blue spheres in the first layer—that is, none of these three layers is directly above each other. This kind of packing is called **cubic closest packing**, abbreviated **ccp**, because when viewed from a different perspective, the atoms are located at positions corresponding to a face-centered cubic lattice (Figure 11.46*a*). Figure 11.45*b* describes the other closest-packed structure in which a green sphere in the third layer

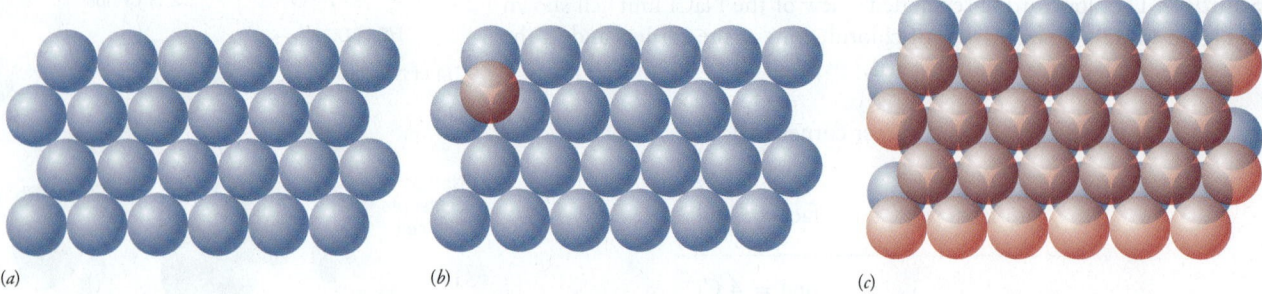

(a)   (b)   (c)

Figure 11.44 | **Packing of spheres.** (*a*) One layer of closely packed spheres. (*b*) A second layer is started by placing a sphere (colored red) in a depression formed between three spheres in the first layer. (*c*) A second layer of sphere shown slightly transparent so we can see how the atoms are stacked over the first layer.

rests in a depression between red spheres and directly above a blue sphere in the first layer. This arrangement of spheres is called **hexagonal closest packing,** abbreviated **hcp** (Figure 11.46*b*). This corresponds to the hexagonal structure shown in Table 11.6.

In the hcp structure, the layers alternate in an A-B-A-B . . . pattern, where A stands for the orientations of the first, third, fifth, etc., layers, and B stands for the orientations of the second, fourth, sixth, etc., layers. Thus, the spheres in the third, fifth, seventh, etc., layers are directly above those in the first, while spheres in the fourth, sixth, eighth, etc., layers are directly above those in the second. In the ccp structure, there is an A-B-C-A-B-C . . . pattern. The first layer is oriented like the fourth, the second like the fifth, and the third like the sixth.

Both the ccp and hcp structures yield very efficient packing of identically sized atoms. In both structures, each atom is in contact with 12 neighboring atoms—six atoms in its own

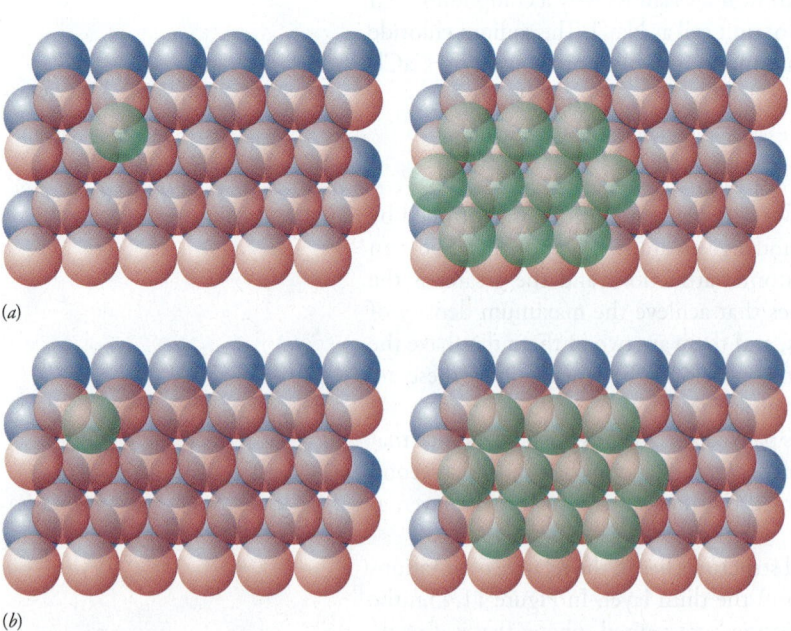

(a)

(b)

Figure 11.45 | **Closest-packed structures.** (*a*) Cubic closest packing of spheres. (*b*) Hexagonal closest packing of spheres. In both (*a*) and (*b*) the left diagram illustrates the position of one atom on the third layer and the right diagram shows the third layer partially complete. Notice that there are subtle differences between the two modes of packing.

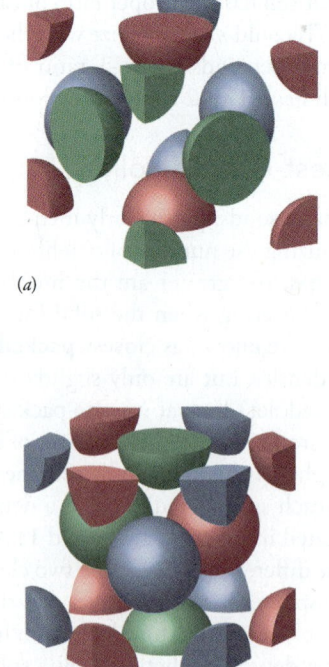

(a)

(b)

Figure 11.46 | **Unit cells of closest-packed structures.** (*a*) Cubic closest packing of spheres. (*b*) Hexagonal closest packing of spheres.

layer, three atoms in the layer below, and three atoms in the layer above. Metals that crystallize with the ccp structure include copper, silver, gold, aluminum, and lead. Metals with the hcp structure include titanium, zinc, cadmium, and magnesium.

Noncrystalline Solids

If a cubic salt crystal is broken, the pieces still have flat faces that intersect at 90° angles. If you shatter a piece of glass, on the other hand, the pieces often have surfaces that are not flat. Instead, they tend to be smooth and curved (see Figure 11.47). This behavior illustrates a major difference between crystalline solids, such as NaCl, and noncrystalline solids, also called **amorphous solids**, such as glass.

The word *amorphous* is derived from the Greek word *amorphos,* which means "without form." Amorphous solids do not have the kinds of long-range repetitive internal structures that are found in crystals. In some ways their structures, being jumbled, are more like liquids than solids. Examples of amorphous solids are ordinary glass and many plastics. In fact, the word **glass** is often used as a general term to refer to any amorphous solid.

As suggested in Figure 11.47, substances that form amorphous solids often consist of long, chain-like molecules that are intertwined in the liquid state somewhat like long strands of cooked spaghetti. To form a crystal from the melted material, these long molecules would have to become untangled and line up in specific patterns. However, as the liquid cools, the molecules slow down. Unless the liquid is cooled extremely slowly, the molecular motion decreases too rapidly for the untangling to take place, and the substance solidifies with the molecules still intertwined. As a result, amorphous solids are sometimes described as **supercooled liquids**, a term suggesting the kind of structural disorder found in liquids.

Robert Capece

> Within this solid, long molecules become tangled. Solid lacks the long-range order found in crystals.

Figure 11.47 | Glass is a noncrystalline solid. When glass breaks, the pieces have sharp edges, but their surfaces are not flat planes. This is because in an amorphous solid like glass, long molecules (much simplified here for clarity) are tangled and disorganized, so there is no long-range order characteristic of a crystal.

■ Some scientists prefer to reserve the term solid for crystalline substances, so they refer to amorphous solids as supercooled liquids.

11.11 | Determining the Structure of Solids

When atoms in a crystal are a beam of X rays, they absorb some of the radiation and then emit it again in all directions. In effect, each atom becomes a tiny X-ray source. If we look at radiation from two such atoms (Figure 11.48), we find that the X rays emitted are in phase in some directions but out of phase in others. In Chapter 7 you learned that constructive (in-phase) and destructive (out-of-phase) interferences create a phenomenon called *diffraction*. X-ray diffraction by crystals has enabled many scientists to win Nobel prizes by determining the structures of extremely complex compounds in a particularly elegant way.

In a crystal, there are enormous numbers of atoms, evenly spaced throughout the lattice. When a beam of X rays strikes a crystal, the radiation is diffracted because of constructive interference, and appears only in specific directions. In other directions, no X rays appear because of destructive interference. When the X rays coming from the crystal fall on photographic film, the diffracted beams form a **diffraction pattern** (see Figure 11.49). The film is darkened only where the X rays strike.[4]

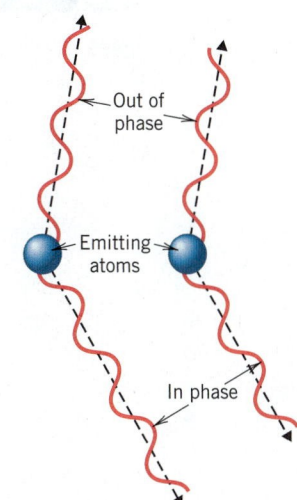

Figure 11.48 | The diffraction of X rays from atoms in a crystal. X rays emitted from atoms are in phase in some directions and out of phase in other directions.

■ The 2009 Nobel Prize in chemistry was awarded for the X-ray structure of the ribosome, the part of a cell where proteins are formed.

---

[4] Modern X-ray diffraction instruments use electronic devices to detect and measure the angles and intensities of the diffracted X rays. Computers have largely eliminated the tedious task of analyzing diffraction patterns to determine structures.

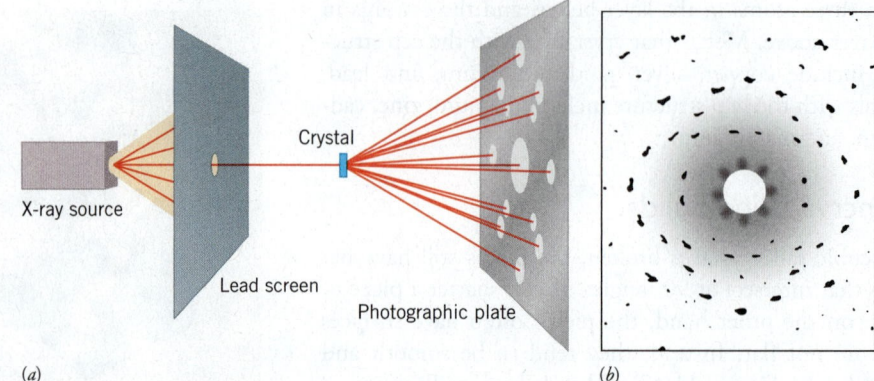

Figure 11.49 | X-ray diffraction. (*a*) The production of an X-ray diffraction pattern. (*b*) An X-ray diffraction pattern produced by sodium chloride recorded on photographic film.

In 1913, the British physicist William Henry Bragg and his son William Lawrence Bragg discovered that just a few variables control the appearance of an X-ray diffraction pattern. These are shown in Figure 11.50, which illustrates the conditions necessary to obtain constructive interference of the X rays from successive layers of atoms (planes of atoms) in a crystal. A beam of X rays having a wavelength λ strikes the layers at an angle θ. Constructive interference causes an intense diffracted beam to emerge at the same angle θ. The Braggs derived an equation, now called the **Bragg equation**, relating λ, θ, and the distance between the planes of atoms, *d*,

TOOLS

Bragg equation

$$n\lambda = 2d \sin \theta \tag{11.5}$$

where *n* is a whole number. The Bragg equation is the basic tool used by scientists in the study of solid structures.

To determine the structure of a crystal, the angles θ at which diffracted X-ray beams emerge from a crystal are measured. These angles are used to calculate the distances between the various planes of atoms in the crystal. The calculated interplanar distances are then used to work backward to deduce where the atoms in the crystal must be located so that layers of atoms are indeed separated by these distances. For complicated molecules and proteins, this is not a simple task, and some sophisticated mathematics as well as computers are needed to accomplish it. The efforts, however, are well rewarded because the calculations give the locations of atoms within the unit cell and the distances between them. This information, plus a lot of chemical "common sense," is used by chemists to arrive at the shapes and sizes of the molecules in the crystal. Example 11.7 illustrates how such data are used.

Figure 11.50 | The diffraction of X rays from successive layers of atoms in a crystal. The layers of atoms are separated by a distance *d*. The X rays of wavelength λ enter and emerge at an angle θ relative to the layers of the atoms. For the emerging beam of X rays to have any intensity, the condition $n\lambda = 2d \sin \theta$ must be fulfilled, where *n* is a whole number.

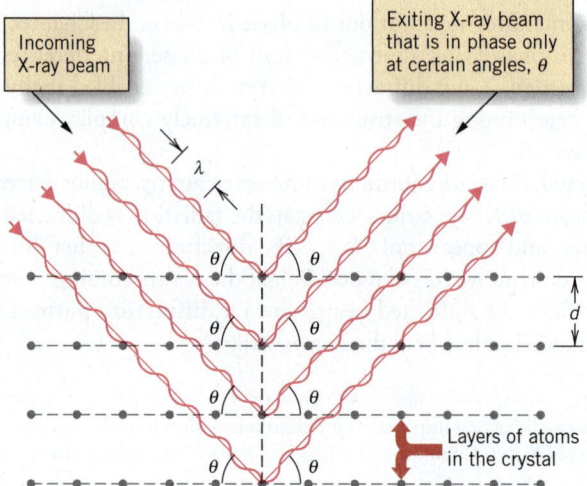

Example 11.7
Using Crystal Structure Data to Calculate Atomic Sizes

X-ray diffraction measurements reveal that copper crystallizes with a face-centered cubic lattice in which the unit cell length is 3.62 Å (see Figure 11.38). What is the radius of a copper atom expressed in angstroms and in picometers?

Analysis: We are given the length of the unit cell, 3.62 Å, and the shape of the unit cell, face-centered cubic. In the figure below, we see that copper atoms are in contact along a diagonal (the dashed line) that runs from one corner of a face to another corner.

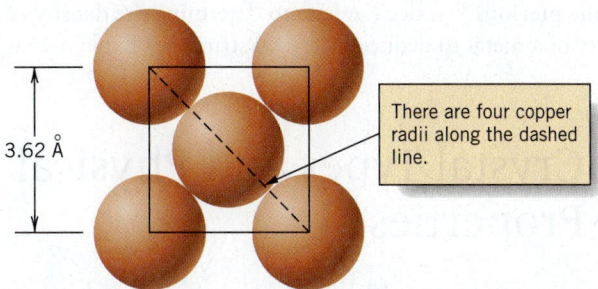

3.62 Å

There are four copper radii along the dashed line.

By geometry, we can calculate the length of this diagonal, which equals four times the radius of a copper atom. Once we calculate the radius in angstrom units we can convert to picometers.

Assembling the Tools: The two tools we will use are the arrangement of the atoms in the shape of the unit cell and the equation for determining the lengths of the sides of a right triangle, a and b, and the hypotenuse, d (the dashed line in the figure above, which is also called the diagonal):

$$a^2 + b^2 = d^2$$

Also, we will use the conversion factors for converting angstroms to picometers:

$$1\ \text{Å} = 1 \times 10^{-10}\ \text{m}$$
$$1\ \text{pm} = 1 \times 10^{-12}\ \text{m}$$

Solution: The unit cell is a cube, so the two sides of the triangle are the same length, l. Using the equation $a^2 + b^2 = d^2$, we can substitute l for a and b and rearrange the equation

$$l^2 + l^2 = 2l^2 = d^2$$

Taking the square root we get

$$\sqrt{2l^2} = \sqrt{2} \times l = d$$

We can now substitute in the actual length of the side of the unit cell and solve for the diagonal, d.

$$d = \sqrt{2} \times 3.62\ \text{Å} = 5.12\ \text{Å}$$

If we call the radius of the copper atom r_{Cu}, then the diagonal equals $4 \times r_{Cu}$. Therefore,

$$4 \times r_{Cu} = 5.12\ \text{Å}$$
$$r_{Cu} = 1.28\ \text{Å}$$

The calculated radius of the copper atom is 1.28 Å.

Next, we convert this to picometers. Knowing that the angstrom, Å, is defined as exactly 10^{-10} meters we can conveniently substitute 10^{-10} meters for the Å symbol.

$$1.28\ \text{Å} = 1.28 \times 10^{-10}\ \text{m} \times \frac{1\ \text{pm}}{1 \times 10^{-12}\ \text{m}} = 128\ \text{pm}$$

Is the Answer Reasonable? It's difficult to get an intuitive feel for the sizes of atoms, so we should be careful to check the calculation. The length of the diagonal seems about right; it is longer than the edge of the unit cell. If we look again at the diagram above, we can see that along the diagonal there are four copper radii. The rest of the arithmetic is okay, so our answer is correct.

Practice Exercise 11.21

Polonium is the only metal known to crystallize in a simple cubic structure. If the X-ray experiment shows that the side of a polonium crystal is 335 pm, what is the atomic radius for polonium? (*Hint:* Do the polonium atoms "touch" along a side, diagonal of a side, or diagonal through the center of the crystal?)

Practice Exercise 11.22

Use the data in the previous Practice Exercise to determine the density of polonium. Can we use the density of a metal to deduce its crystal structure? Explain your answer.

11.12 | Crystal Types and Physical Properties

Solids exhibit a wide range of properties. Some, such as diamond, are very hard, whereas others, such as dry ice and naphthalene (moth flakes), are relatively soft. Some, such as salt crystals, have high melting points whereas others, such as candle wax, melt at low temperatures. Also, some conduct electricity, but others are nonconducting. Physical properties such as these depend on the kinds of particles in the solid as well as on the strengths of the attractive forces holding the solid together. Even though we can't make exact predictions about such properties, some generalizations do exist. In discussing them, it is convenient to divide crystals into four types: ionic, molecular, covalent, and metallic.

Properties of crystal types

Ionic Crystals

Ionic crystals have ions at the lattice sites and the binding between them is mainly electrostatic, which is essentially nondirectional. As a result, the kind of lattice formed is determined mostly by the relative sizes of the ions and their charges. When the crystal forms, the ions arrange themselves to maximize attractions and minimize repulsions.

Because electrostatic forces are strong, ionic crystals tend to be hard. They also tend to have high melting points because the ions have to be given a lot of kinetic energy to enable them to break free of the lattice and enter the liquid state. The forces between ions can also be used to explain the brittle nature of many ionic compounds. For example, when struck by a hammer, a salt crystal shatters into many small pieces. A view at the atomic level reveals how this could occur (Figure 11.51). The slight movement of a layer of ions within

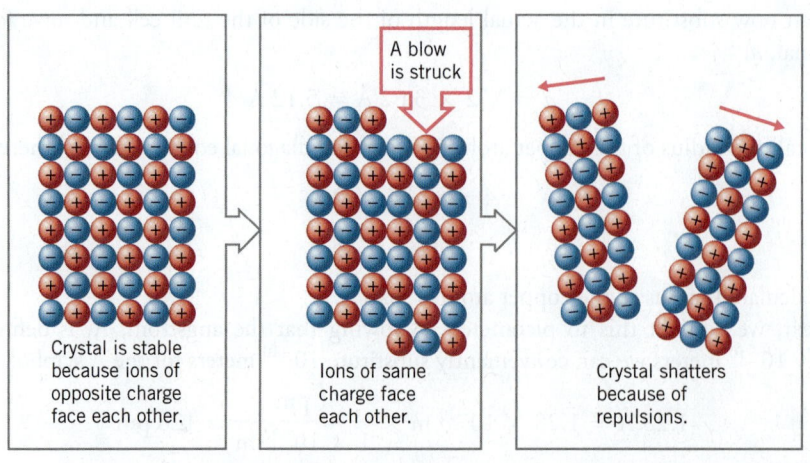

Figure 11.51 | **An ionic crystal shatters when struck.** In this microscopic view we see that striking an ionic crystal causes some of the layers to shift. This can bring ions of like charge face to face. The repulsions between the ions can then force parts of the crystal apart, causing the crystal to shatter.

an ionic crystal suddenly places ions of the *same* charge next to each another, and for that instant there are large repulsive forces that split the solid.

In the solid state, ionic compounds do not conduct electricity because the charges present are not able to move. However, when melted, ionic compounds are good conductors of electricity. Melting frees the electrically charged ions to move.

Molecular Crystals

Molecular crystals are solids in which the lattice sites are occupied either by atoms (as in solid argon or krypton) or by molecules (as in solid sulfur, PF_5, SO_2, or H_2O). If the molecules of such solids are relatively small, the crystals tend to be soft and have low melting points because the particles in the solid experience relatively weak intermolecular attractions. In crystals of argon, for example, the attractive forces are exclusively London forces. In SO_2, which is composed of polar molecules, there are dipole–dipole attractions as well as London forces. And, in water crystals (ice), the molecules are held in place primarily by strong hydrogen bonds. Molecular compounds do not conduct electricity in either the solid or liquid state because they are unable to transport electrical charges.

■ Molecular crystals are soft because little energy is needed to separate the particles or cause them to move past each other.

Covalent Crystals

Covalent crystals are solids in which lattice positions are occupied by atoms that are covalently bonded to other atoms at neighboring lattice sites. The result is a crystal that is essentially one gigantic molecule. These solids are sometimes called **network solids** because of the interlocking network of covalent bonds extending throughout the crystal in all directions. A typical example is diamond (see Figure 11.52). Covalent crystals tend to be very hard and to have very high melting points because of the strong attractions between covalently bonded atoms. Other examples of covalent crystals are quartz (SiO_2, found in some types of sand) and silicon carbide (SiC, a common abrasive used in sandpaper). Covalent crystals are poor conductors of electricity, although some, such as silicon, are semiconductors.

Metallic Crystals

Metallic crystals have properties that are quite different from those of the other three types. Metallic crystals conduct heat and electricity well, and they have the luster characteristically associated with metals. A number of different models have been developed to explain the properties of metals. One of the simplest models views the lattice positions of a metal lattice structure as being occupied by *positive ions* (nuclei plus core electrons). Surrounding them is a "cloud" of electrons formed by the valence electrons, which extends throughout the entire solid (see Figure 11.53). The electrons in this cloud belong to no

■ The band theory of solids discussed in Chapter 9 takes a close look at electronic structures of different kinds of solids.

Figure 11.52 | **The structure of diamond.** Each carbon atom is covalently bonded to four others at the corners of a tetrahedron. This is just a tiny portion of a diamond; the structure extends throughout the entire diamond crystal.

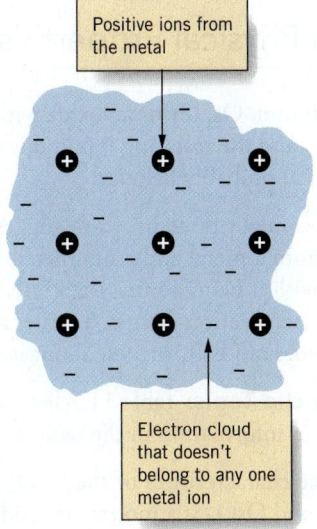

Positive ions from the metal

Electron cloud that doesn't belong to any one metal ion

Figure 11.53 | **The "electron sea" model of a metallic crystal.** In this highly simplified view of a metallic solid, metal atoms lose valence electrons to the solid as a whole and exist as positive ions surrounded by a mobile "sea" of electrons.

TABLE 11.7	Types of Crystals			
Crystal Type	Particles Occupying Lattice Sites	Type of Attractive Force	Typical Examples	Typical Properties
Ionic	Positive and negative ions	Attractions between ions of opposite charge	$NaCl$, $KMnO_4$, $NaNO_3$, $NaHCO_3$	Relatively hard; brittle; high melting points; nonconductors of electricity as solids, but conduct when melted
Molecular	Atoms or molecules	Dipole–dipole attractions, London forces, hydrogen bonding	HCl, SO_2, N_2, Ar, CH_4, H_2O	Soft; low melting points; nonconductors of electricity in both solid and liquid states
Covalent (network)	Atoms	Covalent bonds between atoms	Diamond, SiC (silicon carbide), SiO_2 (silicon dioxide, sand)	Very hard; very high melting points; nonconductors of electricity
Metallic	Positive ions	Attractions between positive ions and an electron cloud that extends throughout the crystal	Mg, Au, Cr, K, Bi	Range from very hard to very soft; melting points range from high to low; conduct electricity in both solid and liquid states; have characteristic luster

single positive ion, but rather to the metal as a whole. Because the electrons aren't localized on any one atom, they are free to move easily, which accounts for the high electrical conductivity of metals. The electrons can also carry kinetic energy rapidly through the solid, so metals are also good conductors of heat. This model explains the luster of metals, too. When light shines on the metal, the loosely held electrons vibrate easily and readily re-emit the light with essentially the same frequency and intensity.

Some metals, like tungsten, have very high melting points. Others, such as sodium (mp = 97.8 °C) and mercury (mp = −38.83 °C), have quite low melting points. To some extent, the melting point depends on the charge of the positive ions in the metallic crystal. The Group 1A metals have just one valence electron, so their cores are cations with a 1+ charge, which are only weakly attracted to the "electron sea" that surrounds them. Atoms of the Group 2A metals, however, form ions with a 2+ charge. These are attracted more strongly to the surrounding electron sea, so the Group 2A metals have higher melting points than their neighbors in Group 1A. Metals with very high melting points, like tungsten, must have very strong attractions between their atoms and between the atoms and the electron sea, which suggests that there probably is some covalent bonding between them as well.

The different ways of classifying crystals and a summary of their general properties are given in Table 11.7.

Example 11.8
Identifying Crystal Types from Physical Properties

The metal osmium, Os, forms an oxide with the formula OsO_4. The soft crystals of OsO_4 melt at 40 °C, and the resulting liquid does not conduct electricity. To which crystal type does solid OsO_4 probably belong?

Analysis: You might be tempted to suggest that the compound is ionic simply because it is formed from a metal and a nonmetal. However, the properties of the compound are inconsistent with it being ionic. Therefore, we have to consider that there may be exceptions to the generalization discussed earlier about metal–nonmetal compounds. If so, what do the properties of OsO_4 suggest about its crystal type?

Assembling the Tools: Table 11.7 lists the properties of the different types of crystals. We can use this information in the table as a tool to solve this problem.

Solution: The characteristics of the OsO_4 crystals—softness and a low melting point—suggest that solid OsO_4 is a molecular solid and that it contains molecules of OsO_4. This is further supported by the fact that liquid OsO_4 does not conduct electricity, which is evidence for the lack of ions in the liquid.

Is the Answer Reasonable? There's not much we can do to check ourselves here except to review our analysis.

Stearic acid is an organic acid that has a chain of 18 carbon atoms. It is a soft solid with a melting point of 70 °C. What crystal type best describes this compound? (*Hint:* Determine the dominant attractive forces that cause stearic acid to be a solid at room temperature.)

Practice Exercise 11.23

Boron nitride, which has the empirical formula BN, melts at 2730 °C and is almost as hard as diamond. What is the probable crystal type for this compound?

Practice Exercise 11.24

Crystals of elemental sulfur are easily crushed and melt at 113 °C to give a clear yellow liquid that does not conduct electricity. What is the probable crystal type for solid sulfur?

Practice Exercise 11.25

CHEMISTRY OUTSIDE THE CLASSROOM 11.2

Giant Crystals

The most common crystals that we use are salt crystals in our salt shakers. These crystals are less than a millimeter on an edge. However, in 2000, giant gypsum crystals were discovered by miners 300 meters below the Naica mountain in Mexico. These crystals measure up to 11 meters in length and can weigh up to 55 tons. Originally, the crystals were submerged, undisturbed, in an aqueous solution of calcium sulfate at 58 °C, which allowed them to slowly grow into these gigantic forms over many thousands of years. Gypsum is unusual in that it becomes less soluble as the temperature increases. If the temperature had been cooler, the crystals would have been smaller, because more calcium sulfate would have been in solution. Since the solution was kept at such a warm temperature, however, the crystals were able to grow larger and larger as more calcium sulfate precipitated.

Another interesting aspect of this cave is that while the temperature is 58 °C, the air is between 90 and 100% humidity. Under these conditions, a person would be one of the coolest objects in the cave, since body temperature is 37 °C. Breathing this air would cause the water in the air to condense in the lungs as it cooled inside the body. A person could only stay in the cave for 10 minutes at a time without drowning.

Carsten Peter/Speleoresearch & Films/NGS / Getty Images

| Summary

Organized by Learning Objective

Describe the major intermolecular forces including dipole attractions, hydrogen bonding, and London forces

Gases expand to fill the entire volume of a container. Liquids and solids retain a constant volume if transferred from one container to another. Solids also retain a constant shape. These characteristics are related to how tightly packed the particles are and to the relative strengths of the intermolecular attractions in the different states of matter. Polar molecules attract each other by **dipole–dipole attractions**, which arise because the positive end of one dipole attracts the negative end of another. **Hydrogen bonding**, a special case of dipole–dipole attractions, occurs between molecules in which hydrogen is covalently bonded to a small, very electronegative atom—principally, nitrogen, oxygen, or fluorine. Hydrogen bonding is much stronger than the other types of intermolecular attractions. Nonpolar molecules are attracted to each other by **London dispersion forces**, which are **instantaneous dipole–induced dipole attractions**. London forces are present between all particles, including atoms, polar and nonpolar molecules, and ions. Among different substances, London forces increase with an increase in the size of a particle's electron cloud; they also increase with increasing chain length among molecules such as the hydrocarbons. For large molecules, the cumulative effect of large numbers of weak London force interactions can be quite strong and outweigh other intermolecular attractions. **Ion–dipole attractions** occur when ions interact with polar substances. **Ion–induced dipole attractions** result when an ion creates a dipole in a neighboring nonpolar molecule or ion.

Explain how the molecular attractive forces affect macroscopic properties

Properties that depend mostly on the closeness of the packing of particles are **compressibility** (or the opposite, incompressibility) and **diffusion**. Diffusion is slow in liquids and almost nonexistent in solids at room temperature. Properties that depend mostly on the strengths of intermolecular attractions are the **retention of volume and shape**, **surface tension**, and the **ease of evaporation. Surface tension** is related to the energy needed to expand a liquid's surface area. A liquid can **wet** a surface if its molecules are attracted to the surface of an object about as strongly as they are attracted to each other. The **evaporation** of liquids and **sublimation** of solids is always endothermic. The overall rate of evaporation increases with increasing surface area, while the rate of evaporation per unit area of a liquid increases only with increasing temperature and with decreasing intermolecular attractions. The evaporation of a solid is called sublimation.

Use the concept of dynamic equilibrium to understand changes of state

Changes from one physical state to another—namely, **fusion**, freezing, **vaporization**, **condensation**, **deposition**, or **sublimation**—can occur as dynamic equilibria. In a **dynamic equilibrium**, opposing processes occur continually at equal rates, so over time there is no apparent change in the composition of the system. For liquids and solids, equilibria are established when vaporization occurs in a sealed container. A solid is in equilibrium with its liquid at the melting point.

Explain what is meant by vapor pressure and how it reveals the relative strengths of intermolecular attractions

When the rates of evaporation and condensation of a liquid are equal in a closed container, the vapor above the liquid exerts a pressure called the **equilibrium vapor pressure** (or more commonly, just the **vapor pressure**). The vapor pressure is controlled by the rate of evaporation per unit surface area. When the intermolecular attractive forces are large, the rate of evaporation is small and the vapor pressure is small. Vapor pressure increases with increasing temperature because the rate of evaporation increases as the temperature rises. Solids have vapor pressures just as liquids do.

Understand the concepts of boiling points and normal boiling points

A substance boils when its vapor pressure equals the prevailing atmospheric pressure. The **normal boiling point** of a liquid is the temperature at which its vapor pressure equals 1 atm. Substances with high boiling points have strong intermolecular attractions.

Calculate energy changes involved in changes of state

On a **heating curve**, flat portions correspond to phase changes in which the heat added increases the potential energies of the

particles without changing their average kinetic energy. **Superheating** may occur when the temperature of a liquid rises above its boiling point without boiling. On a **cooling curve**, flat portions correspond to phase changes in which the heat removed decreases the potential energies of the particles without changing their average kinetic energy. **Supercooling** sometimes occurs when the temperature of the liquid drops below the freezing point of the substance without crystallization. Both superheating and supercooling are unstable states. The enthalpy changes for melting, the vaporization of a liquid, and sublimation are the **molar heat of fusion**, ΔH_{fusion}, the **molar heat of vaporization**, $\Delta H_{vaporization}$, and the **molar heat of sublimation**, $\Delta H_{sublimation}$, respectively. They are all endothermic and are related in size as follows: $\Delta H_{fusion} < \Delta H_{vaporization} < \Delta H_{sublimation}$. The sizes of these enthalpy changes are large for substances with strong intermolecular attractive forces. The **molar heat of crystallization**, $\Delta H_{crystallization}$, the **molar heat of condensation**, $\Delta H_{condensation}$, and the **molar heat of deposition**, $\Delta H_{deposition}$, are all exothermic and are related in size as follows: $\Delta H_{crystallization} < \Delta H_{condensation} < \Delta H_{deposition}$.

Understand and effectively use phase diagrams

Temperatures and pressures at which equilibria can exist between phases are depicted graphically in a **phase diagram**. The three equilibrium lines intersect at the **triple point**. The liquid–vapor line terminates at the **critical point**. At the **critical temperature**, a liquid has a vapor pressure equal to its **critical pressure**. Above the critical temperature a liquid phase cannot be formed; the single phase that exists is called a **supercritical fluid**. The equilibrium lines also divide a phase diagram into temperature–pressure regions in which a substance can exist in just a single phase. Water is different from most substances in that its melting point decreases with increasing pressure.

Use Le Châtelier's principle to understand changes of state

When the equilibrium in a system is upset by a disturbance, the system changes in a direction that minimizes the disturbance and, if possible, brings the system back to equilibrium. By this principle, we find that raising the temperature favors an endothermic change. Decreasing the volume favors a change toward a more dense phase.

Understand how heats of vaporization are determined experimentally

The heat of vaporization of a substance is related to the variation of the vapor pressure with changes in temperature.

Describe how crystal structures are based on repeating molecular geometries

Crystalline solids have highly ordered arrangements of particles within them, which can be described in terms of repeating three-dimensional arrays of points called **lattices**. The simplest portion of a lattice is its **unit cell**. Many structures can be described with the same lattice by associating different units (i.e., atoms, molecules, or ions) to lattice points and by changing the dimensions of the unit cell. All crystals can be classified into one of the 14 different lattice types shown in Table 11.6. Three cubic unit cells are possible—**simple cubic**, **face-centered cubic**, and **body-centered cubic**. Sodium chloride and many other alkali metal halides crystallize in the **rock salt structure**, which contains four formula units per unit cell. Two modes of closest packing of atoms are **cubic closest packing (ccp)** and **hexagonal closest packing (hcp)**. The ccp structure has an A-B-C-A-B-C . . . alternating stacking of layers of spheres; the hcp structure has an A-B-A-B . . . stacking of layers. **Amorphous** solids lack the internal structure of crystalline solids. Glass is an amorphous solid and is sometimes called a **supercooled liquid**.

Illustrate the basics of X-ray crystallography

Information about crystal structures is obtained experimentally from **X-ray diffraction patterns** produced by the constructive and destructive interference of X rays scattered by atoms. Distances between planes of atoms in a crystal can be calculated by the Bragg equation, $n\lambda = 2d \sin \theta$, where n is a whole number, λ is the wavelength of the X rays, d is the distance between planes of atoms producing the diffracted beam, and θ is the angle at which the diffracted X-ray beam emerges relative to the planes of atoms producing the diffracted beam.

Understand that intermolecular forces determine the broad classes of crystals

Crystals can be divided into four general types: **ionic, molecular, covalent,** and **metallic**. Their properties depend on the kinds of particles within the lattice and on the attractions between the particles, as summarized in Table 11.7.

Tools for Problem Solving

The following tools were introduced in this chapter. Study them carefully so you can select the appropriate tool when needed.

Intermolecular forces: dipole–dipole, hydrogen bonds, and London forces (Section 11.1)

Dipole–dipole attractions, hydrogen bonding, and London forces are the three intermolecular forces that are responsible for the intermolecular attractions between molecules in the liquid and solid states. Ion–dipole and ion–induced dipole attractions are the interactions of ions with polar and nonpolar substances.

Estimating intermolecular attractive forces (Section 11.1)

From molecular structure (determined from VSEPR theory), we can determine whether a molecule is polar or not and whether it has O—H, N—H, or F—H bonds that result in hydrogen bonding. This lets us predict and compare the strengths of intermolecular attractions.

Factors that affect the rate of evaporation: temperature and intermolecular forces (Section 11.2)

As the temperature of a substance increases, the average kinetic energy of the molecules in the substance increases, and the rate of evaporation increases. The stronger the intermolecular forces, the slower the rate of evaporation.

Factors that affect vapor pressure (Section 11.4)

The two factors that affect the vapor pressure of a substance are the temperature of the system and the intermolecular forces.

Boiling points and intermolecular forces (Section 11.5)

Liquids boil when their vapor pressure equals the atmospheric pressure. The normal boiling point of a liquid is the boiling point of the liquid at 1 atm. Since the boiling point is related to vapor pressure, it too is a measure of intermolecular forces.

Enthalpy changes during phase changes (Section 11.6)

Each phase change has an associated enthalpy; fusion, vaporization, and sublimation are endothermic and condensation, deposition, and crystallization are exothermic. Two of these changes, $\Delta H_{\text{vaporization}}$ and $\Delta H_{\text{sublimation}}$, allow us to compare the strengths of intermolecular forces in substances.

Heat of fusion (Section 11.6)

The amount of heat required to melt a given amount of a substance is given by

$$q = n \times \Delta H_{\text{fusion}}$$

Similar equations are used for the other phase changes.

Phase diagrams (Section 11.7)

We use a phase diagram to identify temperatures and pressures at which equilibrium can exist between phases of a substance, and to identify conditions under which only a single phase can exist. The triple point and critical point can also be determined.

Le Châtelier's principle (Section 11.8)

Le Châtelier's principle enables us to predict the direction in which the position of equilibrium is shifted when a dynamic equilibrium is upset by a disturbance.

Clausius–Clapeyron equation (Section 11.9)

The relationship, Equation 11.3, between vapor pressure, heats of vaporization, and temperature is

$$\ln P = \frac{-\Delta H_{vap}}{RT}$$

This may be written as a "two-point" relationship as in Equation 11.4,

$$\ln \frac{P_1}{P_2} = \frac{\Delta H_{vap}}{R}\left(\frac{1}{T_2} - \frac{1}{T_1}\right)$$

Cubic unit cells (Section 11.10)

The three basic cubic lattice structures are simple cubic, face-centered cubic, and body-centered cubic. By knowing the arrangements of atoms in these unit cells, we can use the dimensions of the unit cell to calculate atomic radii and other properties.

Counting atoms in unit cells (Section 11.10)

The number of atoms in the unit cell can be counted by adding up the parts of the atoms that are contained in the unit cell: $\frac{1}{8}$ for an atom in a corner, $\frac{1}{2}$ for an atom on an edge, $\frac{1}{4}$ for an atom in a face, and 1 for an atom wholly in the unit cell.

Bragg equation (Section 11.11)

$$n\lambda = 2d \sin \theta$$

Properties of crystal types (Section 11.12)

By examining certain physical properties of a solid (i.e., the hardness, melting point, and electrical conductivity in the solid and liquid states), we can often predict the nature of the particles that occupy lattice sites in the solid and the kinds of attractive forces between them.

WileyPLUS = *WileyPLUS,* an online teaching and learning solution. *Note to instructors:* Many of the end-of-chapter problems are available for assignment via the WileyPLUS system. **www.wileyplus.com.** Review Problems are presented in pairs separated by blue rules. Answers to problems whose numbers appear in blue are given in Appendix B. More challenging problems are marked with an asterisk ✱.

| Review Questions

Intermolecular Forces

11.1 Why are the intermolecular attractive forces stronger in liquids and solids than they are in gases?

11.2 Compare the behavior of gases, liquids, and solids when they are transferred from one container to another.

11.3 For a given substance, how do the intermolecular attractive forces compare in its gaseous, liquid, and solid states?

11.4 Why do intermolecular attractions weaken as the distances between the molecules increase?

11.5 What are London forces? How are they affected by the sizes of the atoms in a molecule? How are they affected by the number of atoms in a molecule? How are they affected by the shape of a molecule?

11.6 Define polarizability. How does this property affect the strengths of London forces?

11.7 Describe dipole–dipole attractions.

11.8 Which nonmetals, besides hydrogen, are involved in hydrogen bonding? Why these elements and not others?

11.9 How do the strengths of covalent bonds and dipole–dipole attractions compare? How do the strengths of ordinary dipole–dipole attractions compare with the strengths of hydrogen bonds?

11.10 Which would give a stronger ion–dipole interaction with water molecules: Al^{3+} or Na^+?

11.11 For each pair, which ion produces the stronger ion-induced dipole attractions with oxygen molecules: **(a)** O^{2-} or S^{2-}, **(b)** Mg^{2+} or Al^{3+}?

Intermolecular Forces and Physical Properties

11.12 Which kinds of attractive forces, intermolecular or intramolecular, are responsible for chemical properties? Which kinds are responsible for physical properties?

11.13 Which is expected to have the higher boiling point, C_8H_{18} or C_4H_{10}? Explain your choice.

11.14 Ethanol and dimethyl ether have the same molecular formula, C_2H_6O. Ethanol boils at 78.4 °C, whereas dimethyl ether boils at −23.7 °C. Their structural formulas are

$$CH_3CH_2OH \qquad\qquad CH_3OCH_3$$
$$\text{ethanol} \qquad\qquad\qquad \text{dimethyl ether}$$

Explain why the boiling point of the ether is so much lower than the boiling point of ethanol.

11.15 Name two physical properties of liquids and solids that are controlled primarily by how tightly packed the particles are. Name three that are controlled mostly by the strengths of the intermolecular attractions.

11.16 What is compressibility? Which is the easiest to compress: solids, liquids, or gases? Why?

11.17 What is diffusion? Give an example of diffusion in everyday life.

11.18 Why does diffusion occur more slowly in liquids than in gases? Why does diffusion occur extremely slowly in solids?

11.19 On the basis of kinetic theory, would you expect the rate of diffusion in a liquid to increase or decrease as the temperature is increased? Explain your answer.

11.20 What is surface tension? Why do molecules at the surface of a liquid behave differently from those within the interior?

11.21 Which liquid is expected to have the larger surface tension at a given temperature, CCl_4 or H_2O? Explain your answer.

11.22 What does wetting of a surface mean? What is a surfactant? What is its purpose and how does it function?

11.23 Polyethylene plastic consists of long chains of carbon atoms, each of which is also bonded to hydrogens, as shown in the following structure:

```
     H  H  H  H  H  H  H  H
     |  |  |  |  |  |  |  |
···—C——C——C——C——C——C——C——C—···
     |  |  |  |  |  |  |  |
     H  H  H  H  H  H  H  H
```

Water forms beads when placed on a polyethylene surface. Why?

11.24 The structural formula for glycerol is

```
      H   H   H
      |   |   |
  H——C———C———C——H
      |   |   |
      OH  OH  OH
```

Would you expect this liquid to wet glass surfaces? Explain your answer.

11.25 Acetone (CH_3COCH_3), water (H_2O), and ethylene glycol (CH_2OHCH_2OH) are listed here, and shown below in ball-and-stick figures, in the order of increasing

viscosity for these three compounds. Why is this the correct order?

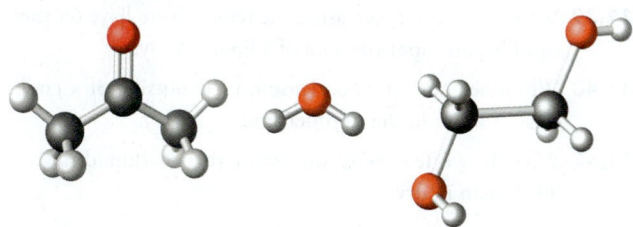

11.26 On the basis of what happens on the molecular level, why does evaporation lower the temperature of a liquid?

11.27 On the basis of the distribution of kinetic energies of the molecules of a liquid, explain why increasing the liquid's temperature increases the rate of evaporation.

11.28 How is the rate of evaporation of a liquid affected by increasing the surface area of the liquid? How is the rate of evaporation affected by the strengths of intermolecular attractive forces?

11.29 During the cold winter months, snow often disappears gradually without melting. How is this possible? What is the name of the process responsible for this phenomenon?

Changes of State and Dynamic Equilibrium

11.30 What terms do we use to describe the following changes of state?
(a) solid → gas, (b) gas → solid, (c) liquid → gas, (d) gas → liquid, (e) solid → liquid, (f) liquid → solid

11.31 When a molecule escapes from the surface of a liquid by evaporation, it has a kinetic energy that's much larger than the average KE. Why is it likely that after being in the vapor for a while its kinetic energy will be much less? If this molecule collides with the surface of the liquid, is it likely to bounce out again?

11.32 Why does a molecule of a vapor that collides with the surface of a liquid tend to be captured by the liquid, even if the incoming molecule has a large kinetic energy?

11.33 When equilibrium is established in the evaporation of a liquid into a sealed container, we refer to it as a dynamic equilibrium. Why?

11.34 Viewed at the molecular level, what is happening when a dynamic equilibrium is established between the liquid and solid forms of a substance? What is the temperature called when a liquid and a solid are in equilibrium?

11.35 Is it possible to establish a state of equilibrium between a solid and its vapor? Explain.

Vapor Pressures of Liquids and Solids

11.36 Define equilibrium vapor pressure. Why do we call the equilibrium involved a dynamic equilibrium?

11.37 Explain why changing the volume of a container in which there is a liquid–vapor equilibrium has no effect on the equilibrium vapor pressure.

11.38 Why doesn't a change in the surface area of a liquid cause a change in the equilibrium vapor pressure?

11.39 What effect does increasing the temperature have on the equilibrium vapor pressure of a liquid? Why?

11.40 Why does moisture condense on the outside of a cool glass of water in the summertime?

11.41 Why do we feel more uncomfortable in humid air at 90 °F than in dry air at 90 °F?

Boiling Points of Liquids

11.42 Define boiling point and normal boiling point.

11.43 Why does the boiling point vary with atmospheric pressure?

11.44 Mt. Kilimanjaro in Tanzania is the tallest peak in Africa (19,340 ft). The normal barometric pressure at the top of this mountain is about 345 torr. At what Celsius temperature would water be expected to boil there? (See Figure 11.23.)

11.45 When liquid ethanol begins to boil, what is present inside the bubbles that form?

11.46 The radiator cap of an automobile engine is designed to maintain a pressure of approximately 15 lb/in.^{2} above normal atmospheric pressure. How does this help prevent the engine from "boiling over" in hot weather?

11.47 Butane, C_4H_{10}, has a boiling point of -0.5 °C (which is 31 °F). Despite this, liquid butane can be seen sloshing about inside a typical butane lighter, even at room temperature. Why isn't the butane boiling inside the lighter at room temperature?

11.48 Why does H_2S have a lower boiling point than H_2Se? Why does H_2O have a much higher boiling point than H_2S?

11.49 An H—F bond is more polar than an O—H bond, so HF forms stronger hydrogen bonds than H_2O. Nevertheless, HF has a lower boiling point than H_2O. Explain why this is so.

Energy and Changes of State

11.50 The following is a cooling curve for one mole of a substance:

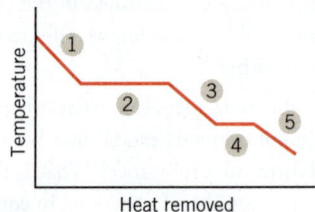

(a) On which portions of this graph do we find the average kinetic energy of the molecules of the substance changing?

(b) On which portions of this graph is the amount of heat removed related primarily to a lowering of the potential energy of the molecules?

(c) Which portion of the graph corresponds to the release of the heat of vaporization?

(d) Which portion of the graph corresponds to the release of the heat of fusion?

(e) Which is larger: the heat of fusion or the heat of vaporization?

(f) On the graph, indicate the melting point of the solid.

(g) On the graph, indicate the boiling point of the liquid.

(h) On the drawing, indicate how supercooling of the liquid would affect the graph.

11.51 Why is $\Delta H_{vaporization}$ larger than ΔH_{fusion}? How does $\Delta H_{sublimation}$ compare with $\Delta H_{vaporization}$? Explain your answer.

11.52 Would the "heat of condensation," $\Delta H_{condensation}$, be exothermic or endothermic?

11.53 Hurricanes can travel for thousands of miles over warm water, but they rapidly lose their strength when they move over a large land mass or over cold water. Why?

11.54 Ethanol (grain alcohol) has a molar heat of vaporization of 39.3 kJ/mol. Ethyl acetate, a common solvent, has a molar heat of vaporization of 32.5 kJ/mol. Which of these substances has the stronger intermolecular attractions?

11.55 A burn caused by steam is much more serious than one caused by the same amount of boiling water. Why?

11.56 Arrange the following substances in order of their increasing values of $\Delta H_{vaporization}$: HF, CH_4, CF_4, HCl.

Phase Diagrams

11.57 For most substances, the solid is more dense than the liquid. Use Le Châtelier's principle to explain why the melting point of such substances should *increase* with increasing pressure. Sketch the phase diagram for such a substance, being sure to have the solid–liquid equilibrium line slope in the correct direction.

11.58 Define critical temperature and critical pressure.

11.59 What is a supercritical fluid? Why is supercritical CO_2 used to decaffeinate coffee?

11.60 What phases of a substance are in equilibrium at the triple point?

11.61 Why doesn't CO_2 have a normal boiling point?

11.62 At room temperature, hydrogen can be compressed to very high pressures without liquefying. On the other hand, butane becomes a liquid at high pressure (at room temperature). What does this tell us about the critical temperatures of hydrogen and butane?

11.63 Sketch a generic phase diagram that shows the regions where liquids, solids, and gases exist, the lines indicating equilibria between different phases, and the triple point.

11.64 What is the significance of the triple point?

11.65 Explain how a unique physical property of water is reflected in its phase diagram.

Le Châtelier's Principle and Changes of State

11.66 State Le Châtelier's principle in your own words.

11.67 What do we mean by the position of equilibrium?

11.68 Use Le Châtelier's principle to predict the effect of adding heat in the equilibrium: solid + heat $\rightleftharpoons$ liquid.

11.69 Use Le Châtelier's principle to explain why lowering the temperature lowers the vapor pressure of a solid.

11.70 Most solids and liquids expand when heated. When a liquid is in equilibrium with its solid form, as the applied pressure is increased will the position of equilibrium move toward the liquid or the solid form? Explain your answer.

Determining Heats of Vaporization

11.71 According to the Clausius–Clapeyron equation, how does the vapor pressure of a substance change as the temperature increases?

11.72 Why can't Celsius temperatures be used in the Clausius–Clapeyron equation?

11.73 Why can any units of pressure be used, as long as you are solving for a temperature or the heat of vaporization?

11.74 What value of R must be used for the Clausius–Clapeyron equation?

Determining the Structure of Solids

11.75 What is the difference between a crystalline solid and an amorphous solid?

11.76 What is a lattice? What is a unit cell?

11.77 What relationship is there between a crystal lattice and its unit cell?

11.78 The diagrams that follow illustrate two typical arrangements of paving bricks in a patio or driveway. Sketch the unit cell that corresponds to each pattern of bricks.

Review Problems

Intermolecular Forces and Physical Properties

11.91 What kinds of intermolecular attractive forces (i.e., dipole–dipole, London, and hydrogen bonding) are present in the following substances?

    (a) HF   (b) PCl_3 (c) SF_6 (d) SO_2

11.79 The figure below illustrates the way the atoms of two different elements are packed in a certain solid. The nuclei occupy positions at the corners of a cube. Is this cube the unit cell for this substance? Explain your answer.

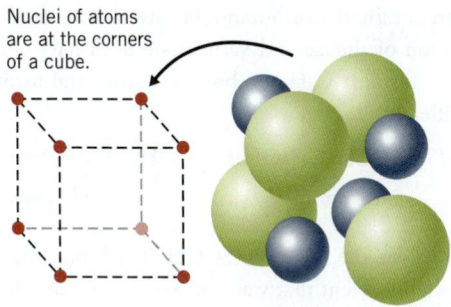

Nuclei of atoms are at the corners of a cube.

11.80 Make a sketch of a layer of sodium ions and chloride ions in a NaCl crystal. Indicate how the ions are arranged in a face-centered cubic pattern, regardless of whether we place lattice points at the Cl⁻ ions or the Na⁺ ions.

11.81 How do the crystal structures of copper and gold differ? In what way are they similar? On the basis of the locations of the elements in the periodic table, what kind of crystal structure would you expect for silver?

11.82 What kind of lattice does zinc sulfide have? What kind of lattice does calcium fluoride have?

11.83 Only 14 different kinds of crystal lattices are possible. How can this be true, considering the fact that there are millions of different chemical compounds that are able to form crystals?

11.84 Write the Bragg equation and define the symbols.

11.85 Why can't $CaCl_2$ or $AlCl_3$ form crystals with the same structure as NaCl?

Crystal Types and Physical Properties

11.86 What kinds of particles are located at the lattice sites in a metallic crystal?

11.87 What kinds of attractive forces exist between particles in **(a)** molecular crystals, **(b)** ionic crystals, and **(c)** covalent crystals?

11.88 Why are covalent crystals sometimes called network solids?

11.89 What does the word amorphous mean?

11.90 What is an amorphous solid? Compare what happens when crystalline and amorphous solids are broken into pieces.

11.92 What kinds of intermolecular attractive forces are present in the following substances?

    (a) $CH_3-\overset{\overset{\textstyle O}{\|}}{C}-OH$ **(b)** H_2S

    (c) SO_3 **(d)** CH_3NH_2

11.93 Which compound should have the higher vapor pressure at 25 °C, butanol or diethyl ether? Which should have the higher boiling point?

11.94 Which liquid evaporates faster at 25 °C, diethyl ether (an anesthetic) or butanol (a solvent used in the preparation of shellac and varnishes)? Both have the molecular formula $C_4H_{10}O$, but their structural formulas are different:

$CH_3CH_2CH_2CH_2OH$ $CH_3CH_2-O-CH_2CH_3$

butanol   diethyl ether

11.95 Consider the compounds $CHCl_3$ (chloroform, an important solvent that was once used as an anesthetic) and $CHBr_3$ (bromoform, which has been used as a sedative). Compare the strengths of their dipole–dipole attractions and the strengths of their London forces. Their boiling points are 61 °C and 149 °C, respectively. For these compounds, which kinds of attractive forces (i.e., dipole–dipole or London) are more important in determining their boiling points? Justify your answer.

11.96 Carbon dioxide does not liquefy at atmospheric pressure, but instead forms a solid that sublimes at −78 °C. Nitrogen dioxide forms a liquid that boils at 21 °C at atmospheric pressure. How do these data support the statement that CO_2 is a linear molecule whereas NO_2 is nonlinear?

11.97 Which of the following isomers of octane should have the higher viscosity?

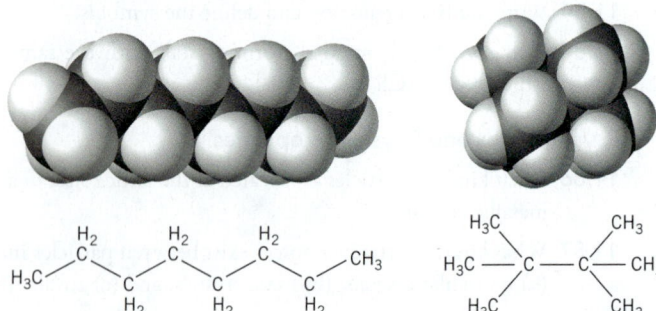

11.98 Which molecule below will wet a greasy glass surface more effectively, propylene glycol (left) or diethyl ether (right)?

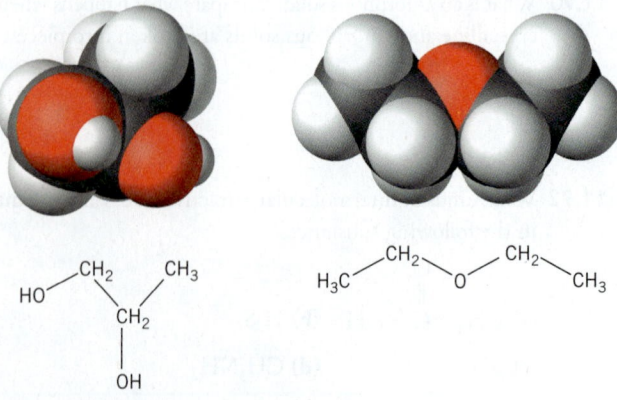

11.99 Which should have the higher boiling point, ethanol (CH_3CH_2OH, found in alcoholic beverages and fermented berries that are favored by wild bears) or ethanethiol (CH_3CH_2SH, a foul-smelling liquid found in the urine of rabbits that have feasted on cabbage)?

11.100 How do the strengths of London forces compare in $CO_2(l)$ and $CS_2(l)$? Which of these is expected to have the higher boiling point? How would you check your answer?

11.101 The following are the vapor pressures of some relatively common chemicals measured at 20 °C:

Benzene, C_6H_6	80 torr
Acetic acid, $HC_2H_3O_2$	11.7 torr
Acetone, C_3H_6O	184.8 torr
Diethyl ether, $C_4H_{10}O$	442.2 torr
Water, H_2O	17.5 torr

Arrange these substances in order of increasing intermolecular attractive forces.

11.102 The boiling points of some common substances are as follows:

Ethanol, C_2H_5OH	78.4 °C
Ethylene glycol, $C_2H_4(OH)_2$	197.2 °C
Water	100 °C
Diethyl ether, $C_4H_{10}O$	34.5 °C

Arrange these substances in order of increasing strengths of intermolecular attractions.

11.103 Using the information in Problem 11.101, list the compounds in order from lowest boiling point to highest boiling point.

11.104 Using the information in Problem 11.102, list the compounds in order from highest vapor pressure to lowest vapor pressure at 25 °C.

11.105 What intermolecular forces must the following substances overcome in the process of vaporization? (a) ethanol (C_2H_5OH), (b) acetonitrile (H_3CCN), (c) phosphorus pentachloride (PCl_5), and (d) sodium chloride (NaCl)

11.106 What intermolecular attractions will be formed when the following substances condense from the gas phase? (a) hexane (C_6H_{14}), (b) acetone (H_3CCOCH_3), (c) iodine (I_2), and (d) argon

Energy and Changes of State

11.107 The molar heat of vaporization of water at 25 °C is +43.9 kJ/mol. How many kilojoules of heat would be required to vaporize 125 mL (0.125 kg) of water?

11.108 The molar heat of vaporization of acetone, C_3H_6O, is 30.3 kJ/mol at its boiling point. How many kilojoules of heat would be liberated by the condensation of 5.00 g of acetone?

*__11.109__ Suppose 45.0 g of water at 85 °C is added to 105.0 g of ice at 0.0 °C. The molar heat of fusion of water is

6.01 kJ/mol, and the specific heat of water is 4.18 J/g °C. On the basis of these data, **(a)** what will be the final temperature of the mixture and **(b)** how many grams of ice will melt?

*11.110 A cube of solid benzene (C_6H_6) at its melting point and weighing 10.0 g is placed in 10.0 g of water at 35 °C. Given that the heat of fusion of benzene is 9.92 kJ/mol, to what temperature will the water have cooled by the time all of the benzene has melted?

Phase Diagrams

11.111 Sketch the phase diagram for a substance that has a triple point at −15.0 °C and 0.30 atm, melts at −10.0 °C at 1 atm, and has a normal boiling point of 90 °C.

11.112 Sketch the phase diagram for a substance that has a triple point at −57.0 °C and 5.10 atm. It has a vapor pressure of 1 atm at −78 °C and has a critical point at 31 °C and 73 atm. Its solid expands as temperature increases.

11.113 Based on the phase diagram sketched in Problem 11.111, below what pressure will the substance undergo sublimation? How does the density of the liquid compare with the density of the solid?

11.114 Based on the phase diagram drawn in Problem 11.112, will a solid sample of this substance melt and then vaporize as the temperature increases under laboratory conditions, or will the substance just sublime as the temperature increases? How does the density of the liquid compare to that of the solid at equilibrium?

11.115 Looking at the phase diagram for CO_2 (Figure 11.30), how can we tell that solid CO_2 is more dense than liquid CO_2?

11.116 According to Figure 11.30, what phase(s) should exist for CO_2 at **(a)** −60 °C and 6 atm, **(b)** −60 °C and 2 atm, **(c)** −40 °C and 10 atm, and **(d)** −57 °C and 5.2 atm?

Determining Heats of Vaporization

*11.117 Mercury is used in barometers since it has such a low vapor pressure. The heat of vaporization of mercury is 59.2 kJ mol^{-1}, and its normal boiling point is 629.9 K. What is the vapor pressure of mercury at 298.2 K?

*11.118 The heat of vaporization of water is 43.9 kJ mol^{-1}, and its normal boiling point is 373 K. Estimate the vapor pressure of water at 351 K.

*11.119 We can determine the heat of vaporization by using the Clausius–Clapeyron equation if we know the vapor pressures of a substance at two different temperatures. Determine the heat of vaporization of diethyl ether if the vapor pressure is 1.0 mm Hg at −74.3 °C and 425 mm Hg at 18.7 °C.

*11.120 If the vapor pressure of ethylene glycol is 7.23 mm Hg at 95.1 °C and 755 mm Hg at 197.1 °C, what is the heat of vaporization of ethylene glycol?

Determining the Structure of Solids

11.121 How many zinc and sulfide ions are present in the unit cell of zinc sulfide? (*Hint:* See Figure 11.42 and add up all of the parts of the atoms in the fcc unit cell.)

11.122 How many copper atoms are within the face-centered cubic unit cell of copper? (See Figure 11.39.)

11.123 The atomic radius of nickel is 1.24 Å. Nickel crystallizes in a face-centered cubic lattice. What is the length of the edge of the unit cell expressed in angstroms and in picometers?

11.124 Silver forms face-centered cubic crystals. The atomic radius of a silver atom is 144 pm. Draw the face of a unit cell with the nuclei of the silver atoms at the lattice points. The atoms are in contact along the diagonal. Calculate the length of an edge of this unit cell.

11.125 Potassium ions have a radius of 133 pm, and bromide ions have a radius of 195 pm. The crystal structure of potassium bromide is the same as for sodium chloride. Estimate the length of the edge of the unit cell in potassium bromide.

11.126 The unit cell edge in sodium chloride has a length of 564.0 pm. The sodium ion has a radius of 95 pm. What is the diameter of a chloride ion?

*11.127 Calculate the angles at which X rays of wavelength 229 pm will be observed to be diffracted from crystal planes spaced **(a)** 1.0×10^3 pm apart and **(b)** 2.5×10^2 pm apart. Assume $n = 1$ for both calculations.

*11.128 Calculate the interplanar spacings (in picometers) that correspond to diffracted beams of X rays at $\theta = 20.0°$, 27.4°, and 35.8°, if the X rays have a wavelength of 141 pm. Assume that $n = 1$.

*11.129 Cesium chloride forms a simple cubic lattice in which Cs^+ ions are at the corners and a Cl^- ion is in the center (see Figure 11.41). The cation–anion contact occurs along the body diagonal of the unit cell. (The body diagonal starts at one corner and then runs through the center of the cell to the opposite corner.) The length of the edge of the unit cell is 412.3 pm. The Cl^- ion has a radius of 181 pm. Calculate the radius of the Cs^+ ion.

11.130 Rubidium chloride has the rock salt structure. Cations and anions are in contact along the edge of the unit cell, which is 658 pm long. The radius of the chloride ion is 181 pm. What is the radius of the Rb^+ ion?

Crystal Types and Physical Properties

11.131 Tin(IV) chloride, $SnCl_4$, has soft crystals with a melting point of −30.2 °C. The liquid is nonconducting. What type of crystal is formed by $SnCl_4$?

11.132 Elemental boron is a semiconductor, is very hard, and has a melting point of about 2250 °C. What type of crystal is formed by boron?

11.133 Columbium is another name for one of the elements. This element is shiny, soft, and ductile. It melts at 2468 °C, and the solid conducts electricity. What kind of solid does columbium form?

11.134 Elemental phosphorus consists of soft white "waxy" crystals that are easily crushed and melt at 44 °C. The solid does not conduct electricity. What type of crystal does phosphorus form?

11.135 Indicate which type of crystal (i.e., ionic, molecular, covalent, or metallic) each of the following would form when it solidifies: (a) Br_2, (b) LiF, (c) MgO, (d) Mo, (e) Si, (f) PH_3, and (g) NaOH.

11.136 Indicate which type of crystal (i.e., ionic, molecular, covalent, or metallic) each of the following would form when it solidifies: (a) O_2, (b) H_2S, (c) Pt, (d) KCl, (e) Ge, (f) $Al_2(SO_4)_3$, and (g) Ne.

Additional Exercises

11.137 List all of the attractive forces that exist in solid Na_2SO_3. How many bonds are ionic and how many are covalent in this substance?

11.138 Calculate the mass of water vapor present in 10.0 L of air at 20.0 °C if the relative humidity is 75%.

11.139 Should acetone molecules be attracted to water molecules more strongly than to other acetone molecules? Explain your answer. The structure of acetone is as follows:

$$\begin{array}{ccccc} & H & O & H & \\ & | & \| & | & \\ H-&C-&C-&C-&H \\ & | & & | & \\ & H & & H & \end{array}$$

11.140 The following thermochemical equations apply to acetic acid.

$$HC_2H_3O_2(s) \longrightarrow HC_2H_3O_2(l) \quad \Delta H_{fusion} = 10.8 \text{ kJ/mol}$$

$$HC_2H_3O_2(l) \longrightarrow HC_2H_3O_2(g) \quad \Delta H_{vaporization} = 24.3 \text{ kJ/mol}$$

Use Hess's law to estimate the value for the heat of sublimation of acetic acid in kilojoules per mole.

11.141 Melting point is sometimes used as an indication of the extent of ionic bonding in a compound—the higher the melting point, the more ionic the substance. On this basis, oxides of metals seem to become less ionic as the charge on the metal ion increases. Thus, Cr_2O_3 has a melting point of 2266 °C while CrO_3 has a melting point of only 196 °C. The explanation often given is similar in some respects to explanations of the variations in the strengths of certain intermolecular attractions given in this chapter. Provide an explanation for the greater degree of electron sharing in CrO_3 as compared with Cr_2O_3.

11.142 When warm, moist air sweeps in from the ocean and rises over a mountain range, it expands and cools. Explain how this cooling is related to the attractive forces between gas molecules. Why does this cause rain to form? When the air drops down the far side of the range, its pressure rises as it is compressed. Explain why this causes the air temperature to rise. How does the humidity of this air compare with the air that originally came in off the ocean? Now explain why the coast of California is lush farmland, whereas valleys (such as Death Valley) that lie to the east of the tall Sierra Nevada mountains are arid and dry.

*****11.143** Gold crystallizes in a face-centered cubic lattice. The edge of the unit cell has a length of 407.86 pm. The density of gold is 19.31 g/cm^3. Use these data and the atomic mass of gold to calculate the value of Avogadro's number.

*****11.144** Gold crystallizes with a face-centered cubic unit cell with an edge length of 407.86 pm. Calculate the atomic radius of gold in units of picometers.

11.145 Identify the type of unit cell belonging to the following structures:

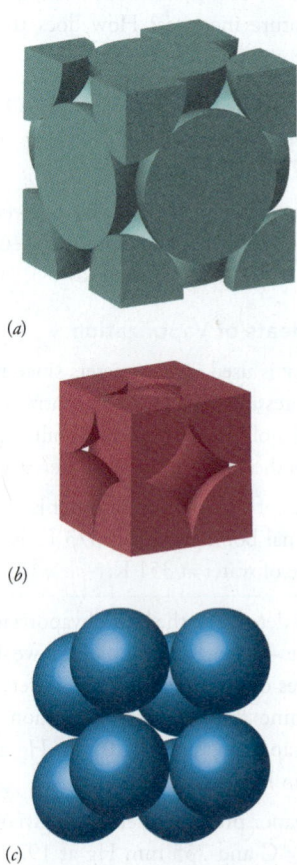

(a)

(b)

(c)

*11.146 Calculate the amount of empty space (in pm^3) in simple cubic, body-centered cubic, and face-centered cubic unit cells if the lattice points are occupied by identical atoms with a diameter of 1.00 pm. Which of these structures gives the most efficient packing of atoms?

*11.147 Silver has an atomic radius of 144 pm. What would be the density of silver in g cm^{-3} if it were to crystallize in **(a)** a simple cubic lattice, **(b)** a body-centered cubic lattice, and **(c)** a face-centered cubic lattice? The actual density of silver is 10.6 g cm^{-3}. Which cubic lattice does silver have?

*11.148 Potassium chloride crystallizes with the rock salt structure. When bathed in X rays, the layers of atoms corresponding to the surfaces of the unit cell produce a diffracted beam of X rays at an angle of 12.8°. Calculate the density of KCl. The wavelength of the X ray used was 154 pm.

11.149 Why do clouds form when the humid air of a weather system called a *warm front* encounters the cool, relatively dry air of a *cold front*?

|Multi-Concept Problems

*11.150 There are 270 Calories in a Hershey's® Milk Chocolate bar. If all of the energy in the bar were used to heat up 75 mL of ice at −25 °C, how hot would the water get? What state would the water be in? The following data might be useful: ΔH_{vap} = 43.9 kJ/mol, ΔH_{fusion} = 6.02 kJ/mol, $\Delta H_{sublimation}$ = 49.9 kJ/mol, specific heat of ice = 2.05 J g^{-1} °C^{-1}, specific heat of water = 4.18 J g^{-1} °C^{-1}, and specific heat of steam = 2.08 J g^{-1} °C^{-1}. Assume that the specific heat of water does not change with temperature.

*11.151 Seawater can be purified by distillation into potable water. If it costs 15 cents per kilowatt-hour for energy, what is the cost of energy to produce 100 gallons of drinking water at 25 °C if the seawater starts at 25 °C? (1 watt × 1 second = 1 joule)

*11.152 Freeze-drying is a process used to preserve food. If strawberries are to be freeze-dried, then they would be frozen to −80 °C and subjected to a vacuum of 0.001 torr. As the water is removed, it is frozen to a cold surface in the freeze-drier. If 150 mL of water has to be removed from a batch of strawberries, how much energy would be required? What changes of state are taking place? At any given time, how much water is in the atmosphere if the volume of the freeze-drier is 5 liters? If it takes one hour to remove one mL of water, how long will it take to freeze-dry the strawberries?

|Exercises in Critical Thinking

11.153 When reporting the vapor pressure for a substance the temperature at which the vapor pressure was measured must also be specified. However, we do not specify a temperature when reporting the enthalpy of a phase change. Why is that? What evidence, perhaps a figure in this chapter, can you provide to support your answer?

11.154 Supercritical CO_2 is used to decaffeinate coffee. Propose other uses for supercritical fluids.

11.155 Freshly precipitated crystals are usually very small. Over time the crystals tend to grow larger in a process called digestion. How can we use the concept of dynamic equilibrium to explain this phenomenon?

11.156 What are three "everyday" applications of Le Châtelier's principle? For example, we turn up the heat in an oven to cook a meal faster.

11.157 Lubricants, oils, greases, and so on are very important in everyday life. Explain how a lubricant works in terms of intermolecular forces.

11.158 Galileo's thermometer is a tube of liquid that has brightly colored glass spheres that float or sink depending on the temperature. Using your knowledge from the past two chapters, explain all of the processes that are involved in this thermometer.

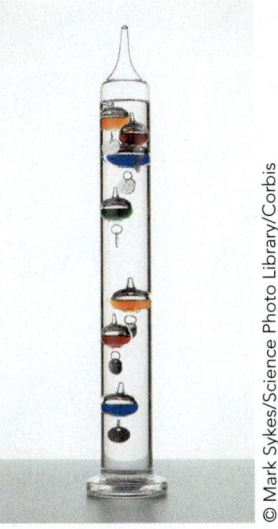

© Mark Sykes/Science Photo Library/Corbis

11.159 Use the Clausius–Clapeyron equation to plot the vapor pressure curve of a gas that has a heat of vaporization of 21.7 kJ mol^{-1} and a boiling point of 48 °C. Compare your results to the other vapor pressure curves in this chapter.

11.160 Will the near-weightless environment of the international space station have any effect on intermolecular forces and chemical reactions? What types of chemical reactions may benefit from a weightless environment?

11.161 Earlier in this chapter it was noted that the standard for pressure was changed to the bar from the atm. It was also stated that "this change would result in a negligible change in the normal boiling points." Do you believe that? Perform the appropriate calculation to see if the assertion is reasonable.

11.162 In looking up values for boiling points, heats of vaporization, heats of fusion, and melting points, you may find different values in different sources. **(a)** What source of data is considered the best? **(b)** How would you determine if a difference between two sources is significant?

Mixtures at the Molecular Level: Properties of Solutions

Chapter Outline

The value of gemstones lies in their beauty and rarity. However, as chemists, we can think of them as solutions: solid solutions with the solute that gives them their colors. Chromium in aluminum oxide for rubies, chromium in beryllium aluminosilicate for emeralds, iron and titanium in aluminum oxide for sapphires, and copper in copper phosphoaluminate for turquoise—these metals are dissolved in the solids.

Lawrence Lawry/Photo Researchers

This Chapter in Context

In Chapter 4 we discussed solutions as a medium for carrying out chemical reactions. Our focus at that time was on the kinds of reactions that take place in solution, and in particular, those that occur when water is the solvent. In this chapter we will examine how the addition of a solute affects the physical properties of the mixture. In our daily lives we take advantage of many of these effects. For example, we add ethylene glycol (antifreeze) to the water in a car's radiator to protect it against freezing and overheating because water has a lower freezing point and higher boiling point when solutes are dissolved in it.

Here we will study not only aqueous solutions, but also solutions involving other solvents. We will concentrate on liquid solutions, although solutions can also be gaseous or solid. In fact, all gases mix completely at the molecular level, so the gas mixtures we examined in Chapter 10 were gaseous solutions. Solid solutions, called alloys, include such materials as brass and bronze. Other solid solutions include paraffin wax, which is a mixture of long-chain hydrocarbons, and gemstones such as ruby, garnet, and jade, which are solid solutions where metal ions contribute to their distinctive colors, as shown in the opening photograph.

Gemstones Are Examples of Solid Solutions

Gem Name	Mineral	Colorant Atom(s)
Ruby	Al_2O_3	Cr, Fe
Sapphire	Al_2O_3	Cr, Ti, Fe
Turquoise	$CuAl_6(PO_4)_4(OH)_8 \cdot 5H_2O$	Cu
Emerald	$Be_3Al_2Si_6O_{18}$	Cr, V
Lapis lazuli	$(Na,Ca)_8(AlSiO_4)_6(SO_4,S,Cl)_2$	S
Jade	$NaAlSi_2O_6$	Cr
Amethyst	SiO_2	Mn
Tanzanite	$Ca_2Al_3(SiO_4)(Si_2O_7)O(OH)$	Cr
Malachite	$Cu_2CO_3(OH)_2$	Cu

LEARNING OBJECTIVES

After reading this chapter, you should be able to:

- describe how intermolecular forces affect the solution process for solids, liquids, and gases
- detail the energy changes that take place when two substances form a solution
- explain how temperature affects the solubility of a solute in a solvent
- using Henry's law, perform calculations involving the pressure of a gas in solution
- define solution concentrations: percent concentration, molal concentration, molar concentration, mole fraction, and mole percent, and convert between the different units
- describe the colligative properties, Raoult's law, freezing point depression, boiling point depression, and osmosis of a solution and use them in calculations for solutions with molecular and ionic solutes
- illustrate the characteristics of heterogeneous mixtures such as suspensions and colloids

12.1 | Intermolecular Forces and the Formation of Solutions

There is a wide variation in the ability of liquids to dissolve solutes. For example, water and gasoline do not dissolve in each other, but water and ethanol can be mixed in all proportions. Similarly, a salt such as potassium bromide will dissolve in water but not in liquid hydrocarbons such as paint thinner. To understand the reasons for these differences, we must examine the factors that drive solution formation as well as those that inhibit it.

Gas Molecules and Spontaneous Mixing

In Chapter 10 you learned that all gases mix spontaneously to form homogeneous mixtures (i.e., solutions). If two gases are placed in separate compartments of a container such as that in Figure 12.1 and the movable partition between them is removed, the molecules will begin to intermingle. The random motions of the molecules will cause them to diffuse one into the other until a uniform mixture is achieved.

The spontaneous mixing of gases illustrates one of nature's strong "driving forces" for change. A system, left to itself, will tend toward the most probable state.[1] At the instant we remove the partition, the container holds two separate gas samples, in contact but unmixed. This represents a highly improbable state because of the natural motions of the molecules. The vastly more probable distribution is one in which the molecules are thoroughly mixed, so formation of the gaseous solution involves a transition from a highly improbable state to a highly probable one.

The drive to attain the most probable state favors the formation of *any* solution. What limits the ability of most substances to mix completely, however, are intermolecular forces of attraction. Such attractions are negligible in gases, so regardless of the chemical makeup of the molecules, the forces are unable to prevent them from mixing. That's why *all gases* spontaneously mix to form solutions with each other. In liquids and solids, however, the situation is very different because of the intermolecular attractions.

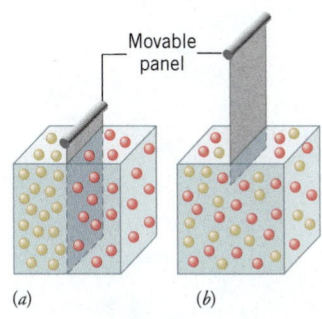

Figure 12.1 | **Mixing of gases.** When two gases, initially in separate compartments (*a*), suddenly find themselves in the same container (*b*), they mix spontaneously.

Liquids Dissolving in Liquids

For a liquid solution to form, the attractive forces between the solvent molecules, and between solute molecules, must be about as strong as attractions between solute and solvent molecules. Let's look at two examples: mixtures of water with benzene and water with ethanol.

benzene
(nonpolar)

ethanol
(contains a polar OH group)

Water and benzene (C_6H_6) are insoluble in each other. In water, there are strong hydrogen bonds between the molecules; in benzene the molecules attract each other by relatively weak London forces and are not able to form hydrogen bonds. Suppose that we did manage to disperse water molecules in benzene. As they move about, the water molecules would occasionally encounter each other. Because they attract each other so much more strongly than they attract benzene molecules, hydrogen bonds would cause water molecules to stick together at each such encounter. This would continue to happen until all the water was in a separate phase. Thus, a solution of water in benzene would not be stable and would gradually separate into two phases. We say water and benzene are **immiscible**, meaning they are mutually insoluble.

Water and ethanol (C_2H_5OH), on the other hand, are **miscible**, meaning that they are soluble in all proportions. This is because water and ethanol molecules can form hydrogen bonds with each other in mixtures that are nearly equivalent to those in the separate pure liquids (Figure 12.2). Mixing these molecules offers little resistance, so they are able to mingle relatively freely.

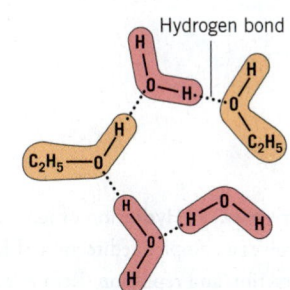

Figure 12.2 | **Hydrogen bonds in aqueous ethanol.** Ethanol molecules form hydrogen bonds (· · · · · ·) to water molecules.

---

[1] In Chapter 18 this driving force will be called *entropy.*

"Like dissolves like" rule

Like Dissolves Like Rule

Observations like those for benzene–water and ethanol–water mixtures led to a generalization often called the **"like dissolves like" rule**: when solute and solvent have molecules "like" each other in polarity, they tend to form a solution. When solute and solvent molecules are quite different in polarity, solutions of any appreciable concentration do not form. The rule has long enabled chemists to use chemical composition and molecular structure to predict the likelihood of two substances dissolving in each other.

The Solubility of Solids in Liquids

The "like dissolves like" principle also applies to the solubility of solids in liquid solvents. Polar solvents tend to dissolve polar and ionic compounds, whereas nonpolar solvents tend to dissolve nonpolar compounds.

Figure 12.3 depicts a section of a crystal of NaCl in contact with water. The dipoles of water molecules orient themselves so that the negative ends of some point toward Na^+ ions and the positive ends of others point at Cl^- ions. In other words, *ion–dipole* attractions occur that tend to tug and pull ions from the crystal. At the corners and edges of the crystal, ions are held by fewer neighbors within the solid and so are more readily dislodged than those elsewhere on the crystal's surface. As water molecules dislodge these ions, new corners and edges are exposed, and the crystal continues to dissolve.

As they become free, the ions become completely surrounded by water molecules (also shown in Figure 12.3). The phenomenon is called the **hydration** of ions. The *general* term for the surrounding of a solute particle by solvent molecules is **solvation**, so hydration is just a special case of solvation. Ionic compounds are able to dissolve in water when the attractions between water dipoles and ions overcome the attractions of the ions for each other within the crystal.

■ Water molecules collide everywhere along the crystal surface, but *successful* collisions—those that dislodge ions—are more likely to occur at corners and edges.

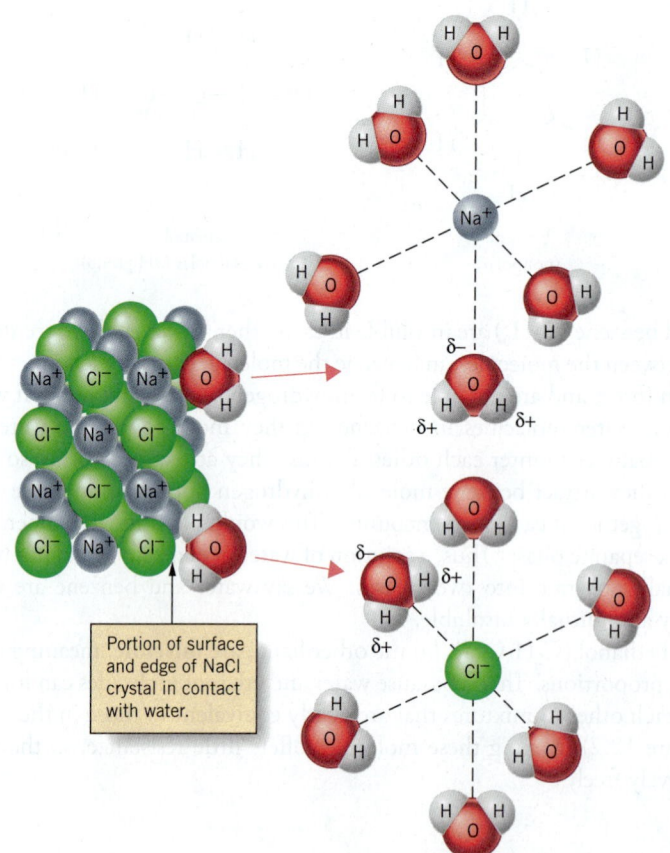

Figure 12.3 | **Hydration of ions.** Hydration involves a complex redirection of forces of attraction and repulsion. Before this solution forms, water molecules are attracted only to each other, and Na^+ and Cl^- ions have only each other to be attracted to in the crystal. In the solution, the ions have water molecules to take the places of their oppositely charged counterparts; in addition, water molecules are attracted to ions even more than they are to other water molecules.

Portion of surface and edge of NaCl crystal in contact with water.

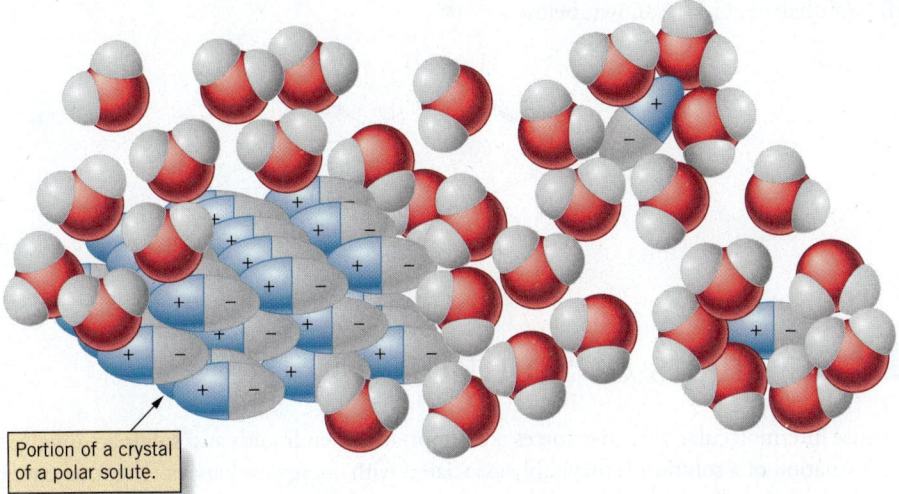

Figure 12.4 | **Hydration of a polar molecule.** A polar molecule of a molecular compound (such as the sugar glucose) can trade the forces of attraction it experiences for other molecules of its own kind for forces of attraction to molecules of water in an aqueous solution.

Portion of a crystal of a polar solute.

Similar events explain why solids composed of polar molecules, like those of sugar, dissolve in water (see Figure 12.4). Attractions between the solvent and solute dipoles help to dislodge molecules from the crystal and bring them into solution. Again we see that "like dissolves like," because a polar solute dissolves in a polar solvent.

The same reasoning explains why nonpolar solids like wax are soluble in nonpolar solvents such as benzene. Wax is a solid mixture of long-chain hydrocarbons, held together by London forces. The attractions between benzene molecules are also London forces, of comparable strength, so molecules of the wax can easily be dispersed among those of the solvent. However, since the intermolecular attractions are relatively weak, the driving force to attain the most probable state also pushes the formation of the solution.

When intermolecular attractive forces within solute and solvent are sufficiently different, the two do not form a solution. For example, ionic solids or very polar molecular solids (like sugar) have strong attractions between their particles that cannot be overcome by attractions to molecules of a nonpolar solvent such as benzene.

Sugar molecules such as this have polar OH groups that can form hydrogen bonds with water.

Which substances will be soluble in water? (a) sodium chloride, (b) sugar, (c) calcium carbonate, (d) methanol, (e) biphenyl:

Practice Exercise 12.1

(*Hint:* What makes a substance soluble in water?)

Which substances will be soluble in nonpolar benzene? (a) hexane, (b) potassium bromide, (c) a trans-fat, *trans*-oleic acid, shown below,

Practice Exercise 12.2

(d) naphthalene, C_8H_{10}, shown below,

12.2 | Heats of Solution

Because intermolecular attractive forces are important when liquids and solids are involved, the formation of a solution is inevitably associated with energy exchanges. The total energy absorbed or released when a solute dissolves in a solvent at constant pressure to make a solution is called the molar enthalpy of solution,[2] or usually just the **heat of solution, ΔH_{soln}**.

Energy is required to separate the particles of solute and also those of the solvent and make them spread out to make room for each other. This step is endothermic, because we must overcome the attractions between molecules to spread the particles out. However, once the particles come back together as a solution, energy is released as the attractive forces between approaching solute and solvent particles decrease the system's potential energy. This is an exothermic change. The enthalpy of solution, ΔH_{soln}, is simply the net result of these two opposing enthalpy contributions. Because enthalpy, H, is a state function, the magnitude of ΔH_{soln} doesn't depend on the path we take to form the solution from the separated solute and solvent.

Solutions of Solids in Liquids

For a solid dissolving in a liquid, it is convenient to imagine a two-step path.

Step 1. *Vaporize the solid to form individual solute particles.* The solute particles are molecules for molecular substances and ions for ionic compounds. The energy absorbed is equal in magnitude to the lattice energy of the solid, but opposite in sign.

Step 2. *Bring the separated gaseous solute particles into the solvent to form the solution.* This step is exothermic, and the enthalpy change when the particles from one mole of solute are dissolved in the solvent is called the **solvation energy**. If the solvent is water, the solvation energy can also be called the **hydration energy**.

The enthalpy diagram showing these steps for potassium iodide is given in Figure 12.5. Step 1 corresponds to the negative of the lattice energy of KI, which is represented by the thermochemical equation.

$$KI(s) \longrightarrow K^+(g) + I^-(g) \qquad \Delta H = +632 \text{ kJ}$$

Step 2 corresponds to the hydration energy of gaseous K^+ and I^- ions.

$$K^+(g) + I^-(g) \longrightarrow K^+(aq) + I^-(aq) \qquad \Delta H = -619 \text{ kJ}$$

The *enthalpy of solution* is obtained from the sum of the equations for Steps 1 and 2 and is the enthalpy change when one mole of crystalline KI dissolves in water (corresponding to the direct path in Figure 12.5).

$$KI(s) \longrightarrow K^+(aq) + I^-(aq) \qquad \Delta H_{soln} = +13 \text{ kJ}$$

---

[2] As a solute is dissolved, the properties of the solvent change because of the presence of previously dissolved solute. Therefore, even if we know the molar heat of solution, we cannot precisely calculate the heat evolved when a different amount than one mole of solute is dissolved in a liter of solvent.

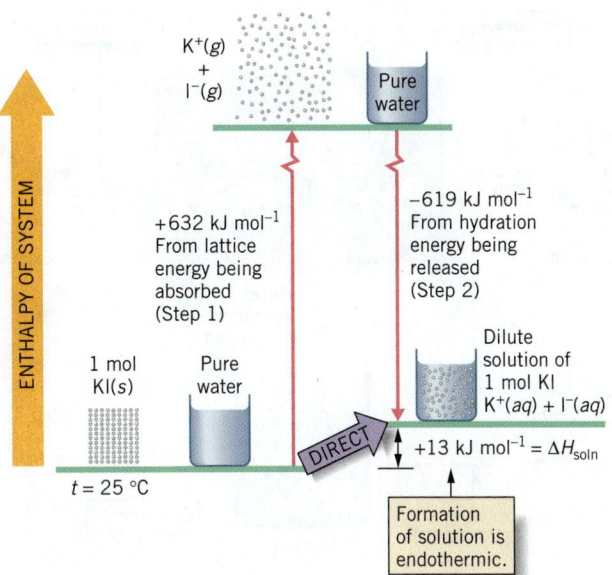

Figure 12.5 | Enthalpy diagram for the heat of solution of one mole of potassium iodide. Adding the negative of the lattice energy to the hydration energy gives a positive value for ΔH_{soln}, indicating that the solution process is endothermic.

The value of ΔH_{soln} indicates that the solution process is endothermic for KI, in agreement with the observation that when KI is added to water and the mixture is stirred, it becomes cool as the KI dissolves.

Table 12.1 provides a comparison between values of ΔH_{soln} obtained by the method described above and values obtained by direct measurements. The agreement between calculated and measured values may not seem particularly impressive, but this is partly because lattice and hydration energies are not precisely known and partly because the model used in our analysis is too simple. Notice, however, that when "theory" predicts relatively large heats of solution, the experimental values are also relatively large, and that both values have the same sign (except for NaBr). Notice also that the variations among the values follow the same trends when we compare the three chloride salts—LiCl, NaCl, and KCl—or the three bromide salts—LiBr, NaBr, and KBr.

■ Small percentage errors in very large numbers can cause huge uncertainties in the *differences* between such numbers.

Solutions of Liquids in Liquids

To consider heats of solution when liquids dissolve in liquids, it's useful to imagine both liquids expanding and then mixing to form the solution (see Figure 12.6). We will designate one liquid as the solute and the other as the solvent.

TABLE 12.1	Lattice Energy, Hydration Energies, and Heats of Solution for Some Group 1A Metal Halides			
			ΔH_{soln} [a]	
Compound	Lattice Energy (kJ mol^{-1})	Hydration Energy (kJ mol^{-1})	Calculated[b] ΔH_{soln} (kJ mol^{-1})	Measured ΔH_{soln} (kJ mol^{-1})
LiCl	−853	−883	−30	−37.0
NaCl	−786	−770	+16	+3.9
KCl	−715	−686	+29	+17.2
LiBr	−807	−854	−47	−49.0
NaBr	−747	−741	+6	−0.602
KBr	−682	−657	+25	+19.9
KI	−649	−619	+30	+20.33

[a]Heats of solution refer to the formation of extremely dilute solutions.
[b]Calculated ΔH_{soln} = hydration energy − lattice energy

Figure 12.6 | Enthalpy of solution for the mixing of two liquids. To analyze the enthalpy change for the formation of a solution of two liquids, we can imagine the hypothetical steps shown here. **Step 1:** The molecules of the liquid designated as the solvent move apart slightly to make room for the solute molecules, an endothermic process. **Step 2:** The molecules of the solute are made to take up a large volume to make room for the solvent molecules, which is also an endothermic change. **Step 3:** The expanded samples of solute and solvent spontaneously intermingle, their molecules also attracting each other, making the step exothermic.

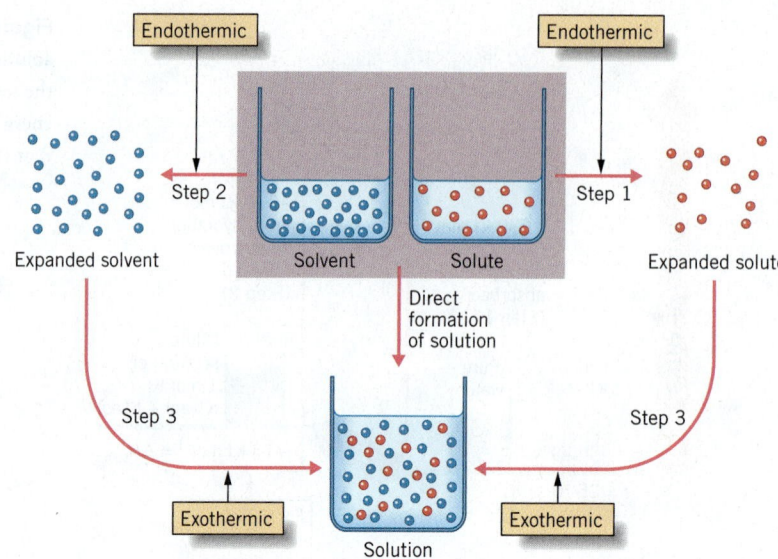

Step 1. *Expand the solute liquid.* First, we imagine that the molecules of one liquid are moved apart just far enough to make room for molecules of the other liquid. Because we have to overcome forces of attraction, this step increases the system's potential energy and so is *endothermic.*

Step 2. *Expand the solvent liquid.* The second step is like the first, but is done to the other liquid (the solvent). On an enthalpy diagram (Figure 12.7) we have climbed two energy steps and have both the solvent and the solute in their slightly "expanded" conditions.

Step 3. *Mix the expanded liquids.* The third step brings the molecules of the expanded solvent and solute together to form the solution. Because the molecules of the two liquids experience mutual forces of attraction, bringing them together lowers the system's potential energy, so Step 3 is exothermic. The value of ΔH_{soln} will, again, be the net energy change for these steps.

Figure 12.7 | Enthalpy changes in the formation of an ideal solution. The three-step and the direct-formation paths both start and end at the same place with the same enthalpy outcome. The sum of the positive ΔH values for the two endothermic steps, 1 and 2, numerically equals the negative ΔH value for the exothermic step, 3. The net ΔH for the formation of an ideal solution is therefore zero.

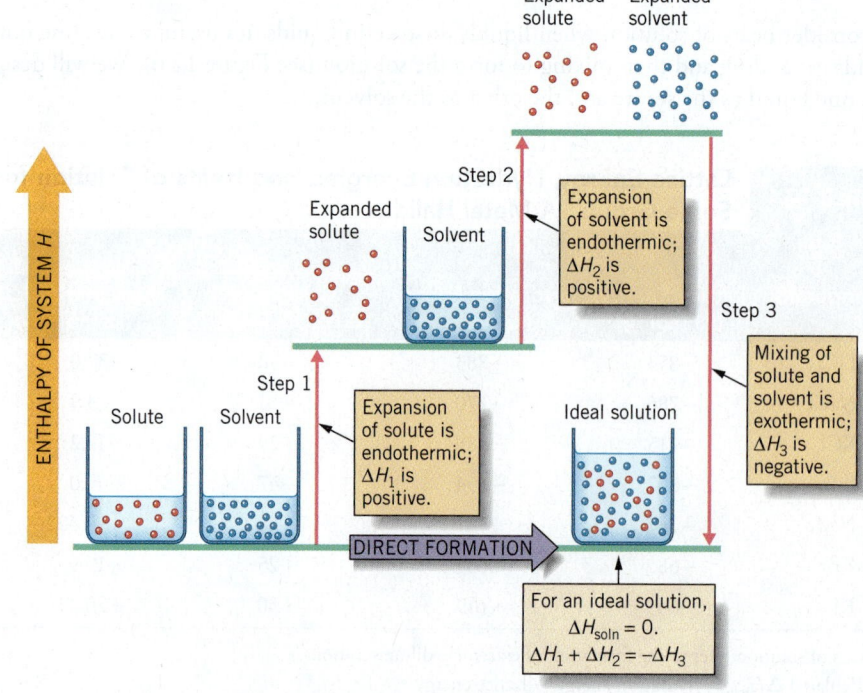

Ideal Solutions

The enthalpy diagram in Figure 12.7 shows the special case when the sum of the energy inputs for Steps 1 and 2 is equal to the energy released in Step 3, so the overall value of ΔH_{soln} is zero. This is very nearly the case when we make a solution of benzene and carbon tetrachloride. Attractive forces between molecules of benzene are almost exactly the same as those between molecules of CCl_4, or between molecules of benzene and CCl_4. If all such intermolecular forces were identical, the net ΔH_{soln} would be exactly zero, and the resulting solution would be called an **ideal solution**. Be sure to notice the difference between an *ideal solution* and an *ideal gas*. In an ideal gas, there are *no* attractive forces. In an ideal solution, there are attractive forces, but they are all the *same*.

For most liquids that are mutually soluble, ΔH_{soln} is not zero. Instead, heat is either given off or absorbed. For example, acetone and water are liquids that form a solution exothermically (ΔH_{soln} is negative). With these liquids, the third step releases more energy than the sum of the first two, chiefly because molecules of water and acetone attract each other more strongly than acetone molecules attract each other. This is because water molecules can form hydrogen bonds to acetone molecules in the solution, but acetone molecules cannot form hydrogen bonds to other acetone molecules in the pure liquid.

acetone

Another pair of liquids, ethanol and hexane, form a solution endothermically. In this case, the release of energy in the third step is not enough to compensate for the energy demands of Steps 1 and 2, so the solution cools as it forms. Since hydrogen-bonded ethanol molecules attract each other more strongly than they can attract hexane molecules, hexane molecules cannot push their way into ethanol without added energy to disrupt some of the hydrogen bonding between ethanol molecules.

■ $CH_3CH_2CH_2CH_2CH_2CH_3$
hexane

Gas Solubility

Unlike solid and liquid solutes, only very weak attractions exist between gas molecules, so the energy required to "expand the solute" is negligible. The heat absorbed or released when a gas dissolves in a liquid has essentially two contributions, as shown in Figure 12.8:

1. *Energy is absorbed when "pockets" in the solvent are opened to hold gas molecules.* The solvent must be expanded slightly to accommodate the molecules of the gas. This requires a small energy input, since attractions between solvent molecules must be overcome. Water is a special case; it already contains open holes in its network of loose hydrogen bonds around room temperature, so very little energy is required to create pockets.
2. *Energy is released when gas molecules occupy these pockets.* Intermolecular attractions between the gas molecules and the surrounding solvent molecules lower the total energy, and energy is released as heat. The stronger the attractions are, the more heat is released.
3. A larger amount of heat is released when a gas molecule is placed in the pocket in water than when it is placed in organic solvents.

These factors lead to two generalizations. *First, heats of solution for gases in organic solvents are often endothermic* because the energy required to open up pockets is greater than the energy released by attractions formed between the gas and solvent molecules. *Second, heats of solution for gases in water are often exothermic* because water already contains pockets to hold the gas molecules, and energy is released when water and gas molecules attract each other.

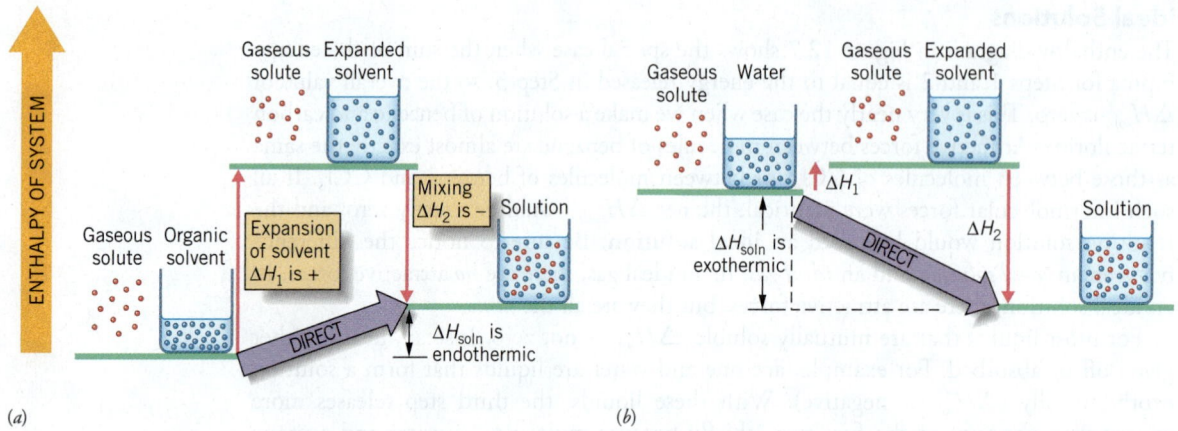

Figure 12.8 | **A molecular model of gas solubility.** (*a*) A gas dissolves in an organic solvent. Energy is absorbed to open "pockets" in the solvent that can hold the gas molecules. In the second step, energy is released when the gas molecules enter the pockets, where they are attracted to the solvent molecules. Here the solution process is shown to be endothermic. (*b*) At room temperature, water's loose network of hydrogen bonds already contains pockets that can accommodate gas molecules, so little energy is needed to prepare the solvent to accept the gas. In the second step, energy is released as the gas molecules take their places in the pockets, where they experience attractions to the water molecules. In this case, the solution process is exothermic.

Practice Exercise 12.3

Draw an enthalpy diagram for the endothermic process of dissolving ammonium chloride in water. (*Hint:* Which steps are exothermic and which steps are endothermic?)

Practice Exercise 12.4

When potassium hydroxide is dissolved in water, heat is evolved. What does this mean in terms of the enthalpy of the process of dissolving the solid?

12.3 | Solubility as a Function of Temperature

By "solubility" we mean the mass of solute that forms a *saturated* solution with a given mass of solvent at a specified temperature. The units often are grams of solute per 100 g of solvent. In such a solution there is a dynamic equilibrium between the undissolved solute and the solute dissolved in the solution, as shown in the equation below and in Figure 12.9.

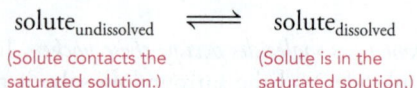

solute_undissolved ⇌ solute_dissolved
(Solute contacts the   (Solute is in the
saturated solution.)   saturated solution.)

As long as the temperature is held constant, the concentration of the solute in the solution remains the same. However, if the temperature of the mixture changes, this equilibrium tends to be upset and either more solute will dissolve or some will precipitate. To analyze how temperature affects solubility we can use Le Châtelier's principle, introduced in Chapter 11, which tells us that if a system at equilibrium is disturbed, the system will change in a direction that counteracts the disturbance and returns the system to equilibrium (if it can). If the formation of a solution is endothermic, adding heat will allow more solute to dissolve. On the other hand, if the formation of a solution is exothermic, adding heat will force some of the solute to precipitate.

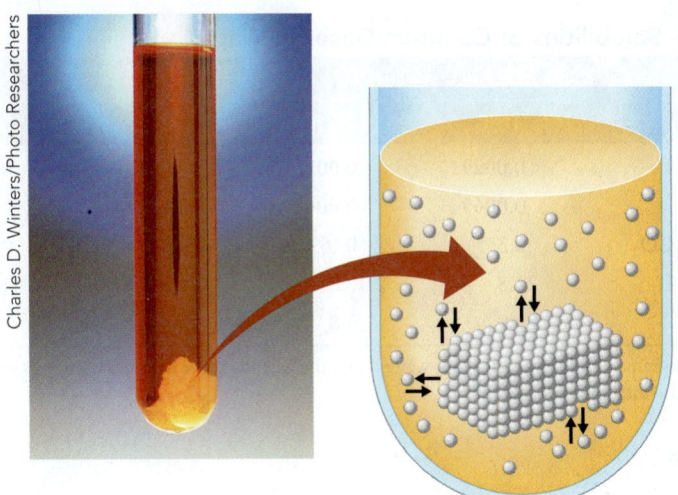

Charles D. Winters/Photo Researchers

Figure 12.9 | Molecular view of a saturated solution. A saturated solution involves a dynamic equilibrium between the solid crystal and the dissolved particles. The arrows indicate that dissolving is occurring as rapidly as crystallization.

Let's consider the endothermic process when the last few solute molecules are dissolved in preparing a saturated solution.[3]

$$\text{solute}_{\text{undissolved}} + \text{energy} \rightleftharpoons \text{solute}_{\text{dissolved}} \qquad \textbf{(12.1)}$$

According to Le Châtelier's principle, when we add heat energy to raise the temperature, the system responds by consuming some of the energy we've added. This causes the equilibrium in Equation 12.1 to "shift toward the products" and more solute is dissolved. Thus, when the solution process is endothermic (and it usually is endothermic near the saturation point), raising the temperature will increase the solubility of the solute. This is a common situation for solids dissolving in liquid solvents. How much the solubility is affected by temperature varies widely, as seen in Figure 12.10.

Some solutes, such as cerium(III) sulfate, $Ce_2(SO_4)_3$, become *less* soluble with increasing temperature (Figure 12.10). Energy must be *released* from a saturated solution of $Ce_2(SO_4)_3$ to make more solute dissolve. For its equilibrium equation, "energy" is on the right side because heat is released.

$$\text{solute}_{\text{undissolved}} \rightleftharpoons \text{solute}_{\text{dissolved}} + \text{energy} \qquad \textbf{(12.2)}$$

Raising the temperature by adding heat causes some of the dissolved solute to precipitate.

Temperature and Gas Solubility

The solubility of gases, like other solubilities, can increase or decrease with temperature, depending on the gas and the solvent. Table 12.2 lists the solubilities of several common gases in water at different temperatures, but all under 1 atm of pressure of the gas. In water, gases such as oxygen are usually more soluble at colder temperatures. Dissolved

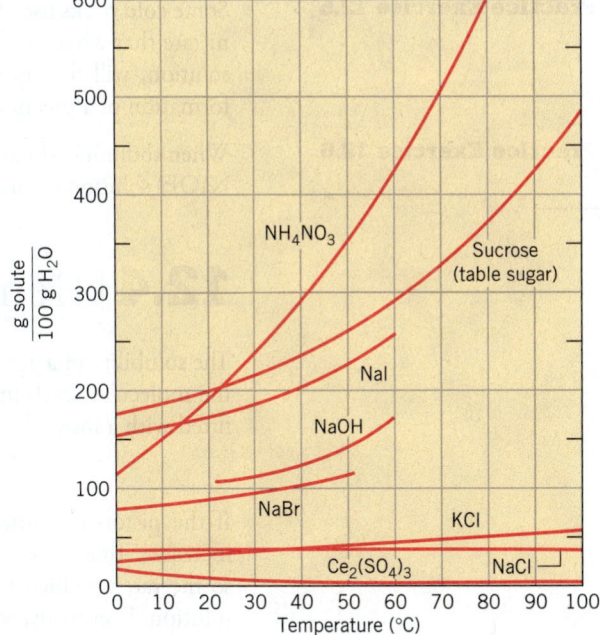

Figure 12.10 | Solubility in water versus temperature for several substances. Most substances become more soluble when the temperature of the solution is increased, but the amount of this increased solubility varies considerably.

---

[3] The heat of solution depends on the solute and the solvent. When preparing a solution, the added solute can be considered to be unchanging as it is added to the solution; however, the solvent does change. At the start, we add solute to pure solvent. Later in the process, we are adding solute to a mixture of solvent and solute. The solution process is one where solute is added to a continuously changing solvent. We find that the incremental heat of solution (i.e., heat evolved per gram of solute added) changes as the solvent changes and can even change from exothermic to endothermic. Interestingly, the incremental heats of solution of solutes added to a nearly saturated solution are almost all endothermic.

TABLE 12.2	Solubilities of Common Gases in Water[a]			
	Temperature			
Gas	0 °C	20 °C	50 °C	100 °C
Nitrogen, N_2	0.0029	0.0019	0.0012	0
Oxygen, O_2	0.0069	0.0043	0.0027	0
Carbon dioxide, CO_2	0.335	0.169	0.076	0
Sulfur dioxide, SO_2	22.8	10.6	4.3	1.8[b]
Ammonia, NH_3	89.9	51.8	28.4	7.4[c]

[a]Solubilities are in grams of solute per 100 g of water when the gaseous space over the liquid is saturated with the gas and the total pressure is 1 atm.
[b]Solubility at 90 °C
[c]Solubility at 96 °C

oxygen in lakes, streams, and oceans is extracted through the gills of marine animals. Fish generally seek colder waters because of the higher concentration of oxygen. **Hypoxia**, a condition in which oxygen concentrations are critically low, has been known to cause massive fish kills. However, in common organic solvents like carbon tetrachloride, toluene, and acetone the solubilities of H_2, N_2, CO, He, and Ne increase with rising temperature.

Practice Exercise 12.5

Some cold packs used for muscle strains consist of separate packs of water and ammonium nitrate that when combined form a solution endothermically. When heat is added to the solution, will the equilibrium shift toward the formation of the solution or toward the formation of a precipitate? (*Hint:* What side of this endothermic equation is heat?)

Practice Exercise 12.6

When sodium hydroxide is dissolved in water, the solution becomes hot. When a saturated NaOH solution is formed, and cooled, will sodium hydroxide precipitate?

12.4 | Henry's Law

The solubility of a gas in a liquid increases with increasing pressure. To understand this at the molecular level, imagine the following equilibrium established in a closed container fitted with a movable piston (Figure 12.11*a*).

$$\text{gas} + \text{solvent} \rightleftharpoons \text{solution} \qquad (12.3)$$

If the piston is pushed down (Figure 12.11*b*), the gas is compressed and its pressure increases. This causes the concentration of the gas molecules over the solution to increase, so the rate at which the gas dissolves is now greater than the rate at which it leaves the solution. Eventually, equilibrium is reestablished when the concentration of the gas in the solution has increased enough to make the rate of escape equal to the rate at which the gas dissolves (Figure 12.11*c*). At this point, the concentration of the gas in the solution is higher than before.

Figure 12.11 | **How pressure increases the solubility of a gas in a liquid.** (*a*) At some specific pressure, equilibrium exists between the vapor phase and the solution. (*b*) An increase in pressure upsets the equilibrium. More gas molecules are dissolving than are leaving the solution. (*c*) More gas has dissolved and equilibrium is restored.

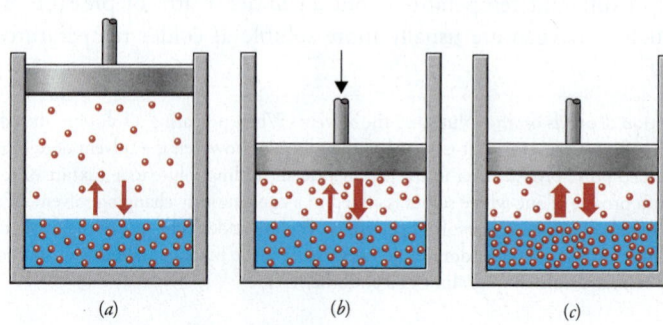

(*a*)   (*b*)   (*c*)

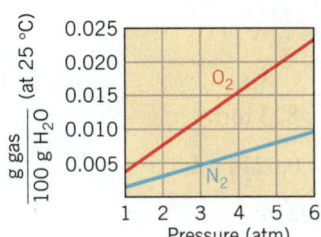

Figure 12.12 | Solubility in water versus pressure for two gases. The amount of gas that dissolves increases as the pressure is raised.

Bottled carbonated beverages fizz when the bottle is opened because the sudden drop in pressure causes a sudden drop in gas solubility.

The effect of pressure on the solubility of a gas can also be explained by Le Châtelier's principle and Equation 12.3. In this case, the disturbance is an increase in the pressure of the gas above the solution. How could the system counteract the pressure increase? The answer is: by having more gas dissolve in the solution. In this way, the pressure of the gas is reduced and the concentration of the gas in the solution is increased.

In Figure 12.12 the solubility in water of oxygen and nitrogen is plotted versus the change in pressure to give a straight line. This is expressed quantitatively by **Henry's law**, which states that the concentration of a gas in a liquid at any given temperature is directly proportional to the partial pressure of the gas over the solution.

$$C_{gas} = k_H P_{gas} \qquad (T \text{ is constant}) \tag{12.4}$$

where C_{gas} is the concentration of the gas and P_{gas} is the partial pressure of the gas above the solution. The proportionality constant, k_H, called the Henry's law constant, is unique to each gas. The equation is an approximation that works best at low concentrations and pressures and for gases that do not react with the solvent.

An alternate (and commonly used) form of Henry's law is

$$\frac{C_1}{P_1} = \frac{C_2}{P_2} \tag{12.5}$$

where the subscripts 1 and 2 refer to initial and final conditions, respectively. By taking the ratio, the Henry's law constant cancels.

TOOLS

Henry's law

■ William Henry (1774–1836), an English chemist, first reported the relationship between gas solubility and pressure.

Example 12.1
Solubility of Nitrogen Using Henry's Law

At 20.0 °C the solubility of N_2 in water is 0.0152 g L^{-1} when the partial pressure of nitrogen is 585 torr. What will be the solubility of N_2 in water at 20.0 °C when its partial pressure is 823 torr?

Analysis: This problem deals with the effect of gas pressure on gas solubility, so Henry's law applies.

Assembling the Tools: We use Henry's law in its form given by Equation 12.5,

$$\frac{C_1}{P_1} = \frac{C_2}{P_2}$$

to avoid having to know or calculate the Henry's law constant.

Solution: Let's gather the data first, keeping track of the initial (C_1 and P_1) and final conditions (C_2 and P_2).

$$C_1 = 0.0152 \text{ g L}^{-1} \qquad C_2 = \text{unknown}$$
$$P_1 = 585 \text{ torr} \qquad P_2 = 823 \text{ torr}$$

Using Equation 12.5, we have

$$\frac{0.152 \text{ g L}^{-1}}{585 \text{ torr}} = \frac{C_2}{823 \text{ torr}}$$

$$C_2 = 0.0214 \text{ g L}^{-1}$$

The solubility under the higher pressure is 0.0214 g L^{-1}.

Is the Answer Reasonable? In relationship to the initial concentration, the size of the answer makes sense because Henry's law tells us to expect a greater solubility at the higher pressure.

Practice Exercise 12.7   At 25 °C and standard pressure, a hydrogen sulfide solution can be made by bubbling $H_2S(g)$ into water; the result is a 0.11 molar solution. Since $H_2S(g)$ is more dense than air a layer of pure $H_2S(g)$ at atmospheric pressure will cover the solution. What is the value of Henry's law constant, and is the solubility of $H_2S(g)$ much greater or smaller than other gases mentioned in this section? If so, suggest a reason for the difference. (*Hint:* What is the pressure of the $H_2S(g)$?)

Practice Exercise 12.8   How many grams of nitrogen and oxygen are dissolved in 125 g of water at 20.0 °C when the water is saturated with air, in which P_{N_2} equals 593 torr and P_{O_2} equals 159 torr? At 1.00 atm pressure, the solubility of pure oxygen in water is 0.00430 g $O_2/100.0$ g H_2O, and the solubility of pure nitrogen in water is 0.00190 g $N_2/100.0$ g H_2O.

Solutions of Gases that React with Water

The gases sulfur dioxide, ammonia, and, to a lesser extent, carbon dioxide are far more soluble in water than are oxygen or nitrogen (see Table 12.2). Part of the reason is that SO_2, NH_3, and CO_2 molecules can form hydrogen bonds with water to help hold the gases in solution. More importantly, the more soluble gases also react with water to some extent, as the following chemical equilibria show.

■ We write these as the gas already dissolved in water. There is another equilibrium that is the gas dissolving in water.

$$CO_2(aq) + H_2O \rightleftharpoons H_2CO_3(aq) \rightleftharpoons H^+(aq) + HCO_3^-(aq)$$

$$SO_2(aq) + H_2O \rightleftharpoons H^+(aq) + HSO_3^-(aq)$$

$$NH_3(aq) + H_2O \rightleftharpoons NH_4^+(aq) + OH^-(aq)$$

The forward reactions contribute to the higher concentrations of the gases in solution, as compared to gases such as O_2 and N_2 that do not react with water at all. Gaseous sulfur trioxide is very soluble in water because it reacts *quantitatively* (i.e., completely) with water to form sulfuric acid.[4]

$$SO_3(g) + H_2O \longrightarrow H^+(aq) + HSO_4^-(aq)$$

12.5 | Concentration Units

In Section 4.7 you learned that for stoichiometry, *molar concentration* or *molarity*, mol/L, is a convenient unit of concentration because it lets us measure out moles of a solute simply by measuring volumes of solution. There are other concentration units—for example,

---

[4] The "concentrated sulfuric acid" of commerce has a concentration of 93 to 98% H_2SO_4, or roughly 18 M. This solution takes up water avidly and very exothermically, removing moisture even out of humid air itself (and so must be kept in stoppered bottles). Because it is dense, oily, sticky, and highly corrosive, it must be handled with extreme care. Safety goggles and gloves must be worn when dispensing concentrated sulfuric acid. When making a more dilute solution, *always pour (slowly with stirring) the concentrated acid into the water.* If water is poured onto concentrated sulfuric acid, it can layer on the acid's surface, because the density of the acid is so much higher than water's (1.8 g mL^{-1} vs. 1.0 g mL^{-1} for water). At the interface, such intense heat can be generated that the steam thereby created could explode from the container, spattering acid around.

percentage by mass, molality, mole fraction (or *mole percent*), and *mass–volume percentage.* For studying physical and chemical properties at different temperatures, units that depend on volume, such as molarity, are not preferred because volumes vary slightly with temperature. Most liquids expand slightly when heated, so a given solution will have a larger volume, which will change the molarity of the solution. We will first discuss the temperature-independent concentration units: percentage by mass, molality, and mole fraction (or mole percent). Then we will relate them to the temperature-dependent concentration units, molarity and percentage by mass–volume.

Percent Concentration

Concentrations of solutions are often expressed as a **percentage by mass** (sometimes called a *percent by weight*), which gives grams of solute per 100 grams of solution. Percentage by mass is indicated by % (w/w), where the "w" stands for "weight." To calculate a percentage by mass from solute and solution masses, we can use the **mass fraction** (the mass of solute divided by the mass of the solution) and multiply it by 100% as shown in the following formula:

$$\text{Percentage by mass} = \frac{\text{mass of solute}}{\text{mass of solution}} \times 100\% \qquad (12.6)$$

TOOLS

Percentage by mass

For example, a solution labeled "0.85% (w/w) NaCl" is one in which the ratio of solute to solution is 0.85 g of NaCl to 100.00 g of NaCl solution. This suggests that the percent concentration can be used as a conversion factor as illustrated in the following example. Also illustrated in this example is that often, the "(w/w)" symbol is omitted when writing mass percents.[5]

Example 12.2
Using Percent Concentrations

Seawater is typically 3.5% sea salt and has a density of 1.03 g mL^{-1}. How many grams of sea salt would be needed to prepare enough seawater solution to completely fill a 62.5 L aquarium?

Analysis: We need the number of grams of sea salt in 62.5 L of seawater solution. To solve the problem, we'll need the relationship that links seawater and sea salt: the percent concentration in Equation 12.6, assuming that the percent is a percentage by mass.

Assembling the Tools: We can write "3.5% sea salt" as the following conversion equality:

$$3.5 \text{ g sea salt} \Leftrightarrow 100 \text{ g soln}$$

Density (see Section 1.7) provides the next conversion equality:

$$1.03 \text{ g soln} \Leftrightarrow 1.00 \text{ mL soln}$$

Finally, we will use the SI prefixes in Table 1.5 to convert mL to liters.

---

[5] Clinical laboratories sometimes report concentrations as **percent by mass/volume**, using the symbol "%(w/v)":

$$\text{percent by mass/volume} = \frac{\text{mass of solute (g)}}{\text{volume of solution (mL)}} \times 100\%$$

For example, a solution that is 4%(w/v) SrCl$_2$ contains 4 g of SrCl$_2$ dissolved in 100 mL of solution. Notice that with a percent by mass, we can use any units for the mass of the solute and mass of the solution, as long as they are the same for both. With percent by mass/volume, we *must* use grams for the mass of solute and milliliters for the volume of solution. If a percent concentration does not include a (w/v) or (w/w) designation, we assume that it is a percentage by mass. The units used are often g dL^{-1}.

Solution: We will start by writing the question in equation form:

$$62.5 \text{ L soln} \Leftrightarrow \text{? g sea salt}$$

Now, let's use the dimensional analysis tools to arrive at the answer:

$$62.5 \text{ L soln} \times \frac{1000 \text{ mL soln}}{1 \text{ L soln}} \times \frac{1.03 \text{ g soln}}{1 \text{ mL soln}} \times \frac{3.5 \text{ g sea salt}}{100 \text{ g soln}} = 2.3 \times 10^3 \text{ g sea salt}$$

Is the Answer Reasonable? If seawater is about 4% sea salt, 100 g of seawater should contain about 4 g of salt. A liter of seawater weighs about 1000 g, so it would contain about 40 g of salt. We have about 60 L of seawater, which would contain 60 × 40 g of salt, or 2400 g of salt. This is not too far from our answer, so the answer seems reasonable.

Practice Exercise 12.9

What volume of water at 20.0 °C ($d = 0.9982 \text{ g mL}^{-1}$) is needed to dissolve 45.0 g of sucrose to make a 10.0 %(w/w) solution? (*Hint:* You need the mass of water to solve this problem.)

Practice Exercise 12.10

How many grams of NaBr are needed to prepare 25.0 g of 1.00%(w/w) NaBr solution in water? How many grams of water are needed? How many milliliters are needed, given that the density of water at room temperature is 0.998 g mL^{-1}?

■ In very dilute aqueous solutions

1 ppm = 1 mg/L
1 ppb = 1 μg/L

Percentages could also be called *parts per hundred*. Other similar expressions of concentration are *parts per million* (ppm) and *parts per billion* (ppb), where 1 ppm equals 1 g of component in 10^6 g of the mixture and 1 ppb equals 1 g of component in 10^9 g of the mixture. By analogy, we can see that ppm is simply the mass fraction multiplied by 10^6 ppm and ppb is the mass fraction multiplied by 10^9 ppb.

Molal Concentration

The number of moles of solute per kilogram of solvent is called the **molal concentration** or the **molality** of a solution. The usual symbol for molality is **m**.

Molal concentration

$$\text{molality} = m = \frac{\text{mol of solute}}{\text{kg of solvent}} \qquad (12.7)$$

For example, if we dissolve 0.500 mol of sugar in 1.00 kg of water, we have a 0.500 *m* solution of sugar. We would not need a volumetric flask to prepare the solution, because we weigh the solvent. Some important physical properties of a solution are related in a simple way to its molality, as we'll soon see.

It is important not to confuse molarity with molality.

■ Be sure to notice that molality is defined per kilogram of solvent, not kilogram of solution.

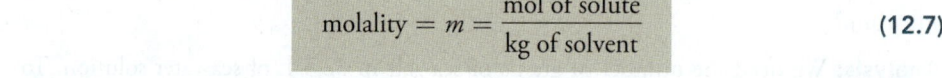

$$\text{molality} = m = \frac{\text{mol of solute}}{\text{kg of } \textit{solvent}} \qquad \text{molarity} = M = \frac{\text{mol of solute}}{\text{L of } \textit{solution}}$$

As we pointed out at the beginning of this section, the molarity of a solution changes slightly with temperature, but molality does not. Therefore, molality is more convenient in experiments involving temperature changes. As we saw with molarity, molality can be used as a conversion factor. For example, the sugar solution above gives us the following equivalence:

$$0.500 \text{ mol sugar} \Leftrightarrow 1.00 \text{ kg H}_2\text{O}$$

■ This only applies when the solvent is water. For other solvents, molarity and molality have quite different values for the same solution.

When water is the solvent, a solution's molarity approaches its molality as the solution becomes more dilute. In very dilute solutions, 1.00 L of solution is nearly 1.00 L of pure water, which has a mass close to 1.00 kg. Under these conditions, the ratio of moles/liter (molarity) is very nearly the same as the ratio of moles/kilogram (molality).

Example 12.3
Preparation of a Solution of a Given Molality

An experiment calls for a 0.150 m solution of sodium chloride in water. How many grams of NaCl would have to be dissolved in 500.0 g of water to prepare a solution of this molality?

Analysis: In this problem, we are given two numerical values: the molality and the mass of water. Since we just saw that the molality can be used as a conversion factor, we will use it to determine the mass of NaCl we need to dissolve in 500.0 g of water.

Assembling the Tools: First, we can use the given molality to write

$$0.150 \text{ mol NaCl} \Leftrightarrow 1.00 \text{ kg } H_2O$$

Next we need the molar mass of NaCl (22.99 + 35.45 = 58.44 g NaCl mol^{-1}) for the conversion equality between moles and mass,

$$1 \text{ mol NaCl} = 58.44 \text{ g NaCl}$$

Solution: We now write the question in equation form as

$$500.0 \text{ g } H_2O = ? \text{ g NaCl}$$

and then we use the appropriate conversion factors

$$500.0 \text{ g } H_2O \times \frac{0.150 \text{ mol NaCl}}{1.00 \times 10^3 \text{ g } H_2O} \times \frac{58.44 \text{ g NaCl}}{1 \text{ mol NaCl}} = 4.38 \text{ g NaCl}$$

When 4.38 g of NaCl is dissolved in 500.0 g of H_2O, the concentration is 0.150 m NaCl.

Is the Answer Reasonable? We'll round the formula mass of NaCl to 60, so 0.15 mol would weigh about 9 g. A 0.15 m solution would contain about 9 g NaCl in 1.00 kg of water. If we used only 500 g of water, we would need half as much salt, or about 4.5 g. Our answer is close to this, so it seems reasonable.

If you prepare a solution by dissolving 44.00 g of sodium sulfate in 250.0 g of water, what is the molality of the solution? Is the molarity of this solution numerically larger or smaller than its molality? (*Hint:* Recall the definition of molality and molarity.)

Practice Exercise 12.11

Water freezes at a lower temperature when it contains solutes. To study the effect of methanol on the freezing point of water, we might begin by preparing a series of solutions of known molalities. Calculate the number of grams of methanol (CH_3OH) needed to prepare a series of solutions from 0.050 to 0.250 m in 0.050 m increments. For each solution use 200.0 g of water.

Practice Exercise 12.12

Mole Fraction and Mole Percent

Mole fraction and mole percent are important tools that were discussed in Chapter 10 (Section 10.6), so for the sake of completeness we will just review the definitions. The mole fraction, X_A, of a substance A is given by

$$X_A = \frac{n_A}{n_A + n_B + n_C + n_D + \ldots + n_Z} \tag{12.8}$$

where $n_A, n_B, n_C, n_D, \ldots, n_Z$ are the numbers of moles of each component, $A, B, C, D \ldots, Z$, respectively. The sum of all mole fractions for a mixture must always equal 1. The **mole percent** is obtained by multiplying the mole fraction by 100 mol%.

$$\text{mol percent } A = \frac{n_A}{n_A + n_B + n_C + n_D + \ldots + n_Z} \times 100 \text{ mol%} \tag{12.9}$$

The mole fractions of all components of a mixture add up to 1.00, so the mole percents must add up to 100 mol%.

Molarity and Weight/Volume Concentration Units

The molarity has been shown to be useful as a conversion factor; however, it is temperature dependent because the denominator will change with changes in temperature.

$$M = \frac{\text{moles of solute}}{\text{liters of solution}} \tag{12.10}$$

Another temperature-dependent concentration unit is the **percentage by mass–volume**, defined as

■ Volume/volume percentages (%v/v) are also sometimes used for liquid solutions. This is the number of mL of solute in 100 mL of solution.

$$\%(\text{w/v}) = \frac{\text{grams solute}}{\text{mL solution}} \times 100\% \tag{12.11}$$

Conversions among Concentration Units

There are times when we need to change from one concentration unit to another. The following example shows that, when converting from one temperature-independent concentration unit to another, all the information we need is in the definitions of the units. As you perform more complex conversions, it may be helpful to construct a table of data similar to the one below.

Solvent	Solute	Solution
Mass	Mass	Total mass
Moles	Moles	Total moles
Volume	Volume	Total volume
Solvent molar mass	Solute molar mass	Density

Use a table such as this to keep track of solution data. Items in green are often found in tables if not given in a problem. Red items are replaced with known data. Perform conversions, as needed, to fill in the remaining spaces.

Example 12.4
Finding Molality from Mass Percent

Brine is a fairly concentrated solution of sodium chloride in water. One use for brine is in the curing process for certain cheeses, where the concentration of brine affects the quality of the cheese. A supplier offers a cheese manufacturer a good price on 1.90 *m* brine. The cheese maker needs a 10.0% (w/w) aqueous NaCl solution. Is the 1.90 *m* brine solution okay to use?

Analysis: We need to compare 1.90 *m* and 10.0% (w/w) brine to see whether they are equivalent. We can convert molality to %(w/w), or %(w/w) to molality. Let's do the latter conversion of %(w/w) to molality:

$$\frac{10.0 \text{ g NaCl}}{100.0 \text{ g NaCl soln}} \quad \text{to} \quad \frac{? \text{ mol NaCl}}{1.00 \text{ kg water}}$$

This will involve conversion of mass to moles for NaCl, and g NaCl solution to kg water.

Assembling the Tools: The equality for converting mass to moles is

$$58.44 \text{ g NaCl} = 1.00 \text{ mol NaCl}$$

We need to convert 100.0 g of the NaCl solution to the mass of water. Since the mass of the solution is the mass of solute plus the mass of water, we write

$$100.0 \text{ g NaCl solution} = 10.0 \text{ g NaCl} + 90.0 \text{ g H}_2\text{O}$$

From this we can write the equivalence

$$100.0 \text{ g NaCl solution} \Leftrightarrow 90.0 \text{ g H}_2\text{O}$$

Then we will use the SI prefixes for the conversion of grams to kilograms.

Solution: First, we write the question as an equation,

$$\frac{10.0 \text{ g NaCl}}{100.0 \text{ g NaCl soln}} = \frac{? \text{ mol NaCl}}{1.00 \text{ kg water}}$$

Let's use our conversion tools to convert mass of NaCl to moles NaCl, and the mass of solution to kg of water

$$\frac{10.0 \text{ g NaCl}}{100.0 \text{ g NaCl soln}} \times \frac{1 \text{ mol NaCl}}{58.44 \text{ g NaCl}} \times \frac{100.0 \text{ g NaCl soln}}{90.0 \text{ g water}} \times \frac{1000 \text{ g water}}{1 \text{ kg water}} = 1.90 \text{ } m \text{ NaCl}$$

We conclude that 10.0% NaCl solution would also be 1.90 molal and the supplier's product is acceptable.

Is the Answer Reasonable? A convenient rounding of the numbers in our calculation to one significant figure might be

$$\frac{10 \text{ g NaCl}}{100 \text{ g NaCl soln}} \times \frac{1 \text{ mol NaCl}}{50 \text{ g NaCl}} \times \frac{100 \text{ g NaCl soln}}{100 \text{ g water}} \times \frac{1000 \text{ g water}}{1 \text{ kg water}} = 2 \text{ } m \text{ water}$$

This is close to our calculated value, and we conclude we were correct. Notice that rounding the 58.6 down to 50 compensates for rounding 90 up to 100, so our estimate is closer to the correct number. We do not need a calculator for this estimate. First, cancel five zeros from the numerator numbers and five zeros from the denominator numbers. This leaves us with $^{10}\!/_5 = 2$.

A bottle on the stockroom shelf reads 52% sodium hydroxide solution. What is the molality of this solution? (*Hint:* Recall that the % means (w/w) percentage.)

Practice Exercise 12.13

Concentrated hydrochloric acid is about 37.0% HCl. Calculate the molality of this solution.

Practice Exercise 12.14

Conversions between the two temperature-dependent units are done in two steps: First, the numerators can be converted between moles and mass, and second, the denominators are then a simple conversion between mL and L units. The concentration units are summarized in Table 12.3.

TABLE 12.3	A Comparison of the Units Used to Calculate Concentration	
Unit	**Calculation**	**Temperature Dependence**
Molarity	$M = \dfrac{\text{moles of solute}}{\text{liters of solution}}$	Yes
Percentage by mass	$\text{Percentage by mass} = \dfrac{\text{mass of solute}}{\text{mass of solution}} \times 100\%$	No
Mass fraction	$\text{Mass fraction} = \dfrac{\text{mass of solute}}{\text{mass of solution}}$	No
Molality	$m = \dfrac{\text{mol of solute}}{\text{kg of solvent}}$	No
Mole fraction	$X_A = \dfrac{n_A}{n_A + n_B + n_C + n_D + \cdots + n_Z}$	No
Mole percent	$\text{mol percent } A = \dfrac{n_A}{n_A + n_B + n_C + n_D + \cdots + n_Z} \times 100 \text{ mol}\%$	No
Percentage by mass–volume	$\%(\text{w/v}) = \dfrac{\text{grams solute}}{\text{mL solution}} \times 100\%$	Yes

When we want to convert between a temperature-independent concentration unit such as molality or percentage concentration and a temperature-dependent unit such as molarity, we need to know the density of the solution. Example 12.5 illustrates this method.

Example 12.5
Finding Molarity from Mass Percent

A certain supply of concentrated hydrochloric acid has a concentration of 36.0% HCl. The density of the solution is 1.19 g mL^{-1}. Calculate the molar concentration of HCl.

Analysis: Recall the definitions of percentage by mass and molarity. We must perform the following conversion:

$$\frac{36.0 \text{ g HCl}}{100 \text{ g HCl soln}} \quad \text{to} \quad \frac{? \text{ mol HCl}}{1 \text{ L HCl soln}}$$

We have two conversions to perform. In the numerator we need to convert the mass of HCl to moles HCl, and in the denominator we convert the mass of solution to liters of solution.

Assembling the Tools: We need the mass-to-moles equality for HCl. The molar mass of HCl is 36.46, and we write the equality as

$$1.000 \text{ mol HCl} = 36.46 \text{ g HCl}$$

We use the density to write the equality:

$$1.19 \text{ g solution} = 1.00 \text{ mL solution}$$

Finally, we use the SI prefixes to obtain the correct volume units.

Solution: We first write the question in equation form,

$$\frac{36.0 \text{ g HCl}}{100 \text{ g HCl soln}} \quad \text{to} \quad \frac{? \text{ mol HCl}}{1 \text{ L HCl soln}}$$

We can do the conversions in either order. Here, we'll convert the numerator first:

$$\left(\frac{36.0 \text{ g HCl}}{100 \text{ g HCl soln}}\right)\left(\frac{1 \text{ mol HCl}}{36.46 \text{ g HCl}}\right)\left(\frac{1.19 \text{ g HCl soln}}{1.00 \text{ mL HCl soln}}\right)\left(\frac{1000 \text{ mL HCl soln}}{1 \text{ L HCl soln}}\right) = 11.7 \, M \text{ HCl}$$

So 36.0% HCl is also 11.7 M HCl.

Is the Answer Reasonable? We can estimate the answer in our head. First, cancel the 36.0 g HCl in the numerator and the 36.46 g HCl in the denominator to give ~1.0 (shown in red below). Now we divide 1000 in the numerator by 100 in the denominator to get 10 (shown in blue). All that is left is 1.19 × 10 = 11.9 M as shown below.

$$\left(\frac{36.0}{100}\right)\left(\frac{1 \text{ mol HCl}}{36.46}\right)\left(\frac{1.19}{1.00}\right)\left(\frac{1000}{1 \text{ L HCl soln}}\right) = 11.9 \, M \text{ HCl}$$

This is very close to our calculated answer, and we can be confident in our result.

Practice Exercise 12.15

Hydrobromic acid can be purchased as 40.0% solution. The density of this solution is 1.38 g mL^{-1}. What is the molar concentration of hydrobromic acid in this solution? (*Hints:* What are the units for percent composition and molarity?)

Practice Exercise 12.16

One gram of aluminum nitrate is dissolved in 1.00 liter of water at 20.0 °C. The density of water at this temperature is 0.9982 g cm^{-3} and the density of the resulting solution is 0.9989 g cm^{-3}. Calculate the molarity and molality of this solution.

Practice Exercise 12.16 illustrates that when the concentration of an aqueous solution is very low—one gram per liter, for example—the molality and the molarity are very close to the same numerical value. As a result, we can conveniently interchange molality and molarity when aqueous solutions are very dilute.

12.6 | Colligative Properties

The physical properties of solutions to be studied in this section are called **colligative properties**, because they depend mostly on the concentrations of particles in mixtures, not on their chemical identities. We start with an examination of the effects of solutes on the vapor pressures of solvents in liquid solutions.

■ After the Greek *kolligativ*, depending on number and not on nature.

Raoult's Law

All liquid solutions of nonvolatile solutes (solutes that have no tendency to evaporate) have lower vapor pressures than their pure solvents. The vapor pressure of such a solution is proportional to how much of the solution actually consists of the solvent. This proportionality is given by **Raoult's law**, which says that the vapor pressure of the solution, $P_{solution}$, equals the mole fraction of the solvent, $X_{solvent}$, multiplied by its vapor pressure when pure, $P^\circ_{solvent}$. In equation form, Raoult's law is expressed as follows.

■ Francois Marie Raoult (1830–1901) was a French scientist.

> **Raoult's Law Equation**
> $$P_{solution} = X_{solvent} P^\circ_{solvent}$$ (12.12)

Raoult's law

Because of the form of this equation, a plot of $P_{solution}$ versus $X_{solvent}$ should be linear at all concentrations when the system obeys Raoult's law (see Figure 12.13).

Notice that the mole fraction in Raoult's law refers to the solvent, not the solute. Usually we're more interested in the effect of the solute's mole fraction concentration, X_{solute}, on the vapor pressure. We can show that the change in vapor pressure, ΔP, is directly proportional to the mole fraction of solute, X_{solute}, as follows.[6]

$$\Delta P = X_{solute} P^\circ_{solvent}$$ (12.13)

The change in vapor pressure equals the mole fraction of the solute times the solvent's vapor pressure when pure.

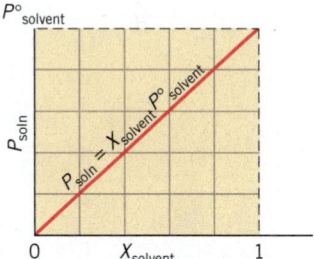

Figure 12.13 | A Raoult's law plot.

---

[6] To derive Equation 12.13, note that the value of ΔP is simply the following difference.

$$\Delta P = (P^\circ_{solvent} - P_{solution})$$

The mole fractions for our two-component system, $X_{solvent}$ and X_{solute}, must add up to 1; it's the nature of mole fractions.

$$X_{solvent} = 1 - X_{solute}$$

We now insert this expression for $X_{solvent}$ into the Raoult's law equation.

$$P_{solution} = X_{solvent} P^\circ_{solvent}$$

$$P_{solution} = (1 - X_{solute})P^\circ_{solvent} = P^\circ_{solvent} - X_{solute} P^\circ_{solvent}$$

So, by rearranging terms,

$$X_{solute} P^\circ_{solvent} = P^\circ_{solvent} - P_{solution} = \Delta P$$

This result gives us Equation 12.13.

Example 12.6
Determining Vapor Pressures with Raoult's Law

Carbon tetrachloride has a vapor pressure of 155 torr at 20 °C. This solvent can dissolve candle wax, which is essentially nonvolatile. Although candle wax is a mixture, we can take its molecular formula to be $C_{22}H_{46}$ (molar mass = 311 g mol^{-1}). Without any change in temperature, by how much will the vapor pressure of CCl_4 be lowered when a solution is prepared by dissolving 10.0 g of wax in 40.0 g of CCl_4 (molar mass = 154 g mol^{-1})? What is the final vapor pressure of this solution?

Analysis: This question uses Raoult's law, which, in turn, uses mole fractions. We have to work backward through this sequence to determine the change in vapor pressure of the solution. We then use that change to determine the actual vapor pressure.

Assembling the Tools: We will use the version of Raoult's law as given in Equation 12.13.

$$\Delta P = X_{solute} P^\circ_{solvent}$$

Since we know the vapor pressure of the pure solvent, we will have to use Equation 12.8 to determine the mole fraction of the solute to determine the change in vapor pressure.

$$X_{solute} = \frac{n_{solute}}{n_{solvent} + n_{solute}}$$

We are given the masses of the solvent and solute and can calculate the moles of each in order to use this equation using the mass-to-moles equalities of CCl_4 and $C_{22}H_{46}$.

$$1 \text{ mol } CCl_4 = 153.8 \text{ g } CCl_4 \quad \text{and} \quad 1 \text{ mol } C_{22}H_{46} = 310.7 \text{ g } C_{22}H_{46}$$

For the second part of the problem, we can use Raoult's law, Equation 12.12, or simply subtract the decrease in vapor pressure from the vapor pressure of pure carbon tetrachloride. The subtraction seems simpler, so we will use it.

Solution: Working backward in our analysis, we first have to determine the moles of each substance in the mixture. Our calculations are

$$\text{For } CCl_4 \qquad 40.0 \text{ g } CCl_4 \times \frac{1 \text{ mol } CCl_4}{153.8 \text{ g } CCl_4} = 0.260 \text{ mol } CCl_4$$

$$\text{For } C_{22}H_{46} \qquad 10.0 \text{ g } C_{22}H_{46} \times \frac{1 \text{ mol } C_{22}H_{46}}{310.7 \text{ g } C_{22}H_{46}} = 0.0322 \text{ mol } C_{22}Cl_{46}$$

The total number of moles is 0.292 mol. Now we can calculate the mole fraction of the solute, $C_{22}H_{46}$,

$$X_{C_{22}H_{46}} = \frac{0.0322 \text{ mol}}{0.292 \text{ mol}} = 0.110$$

With this mole fraction and the vapor pressure of pure CCl_4 (155 torr at 20 °C) we can calculate the amount that the vapor pressure is lowered, ΔP, by Equation 12.13.

$$\Delta P = 0.110 \times 155 \text{ torr} = 17.0 \text{ torr}$$

The presence of the wax in the CCl_4 lowers the vapor pressure of the CCl_4 by 17.0 torr. The actual vapor pressure will be 155 torr − 17 torr = 138 torr.

Is the Answer Reasonable? Our quickest check is that the vapor pressure is lowered and the amount it is lowered is approximately 10% of the original vapor pressure. Ten percent of 155 is close to our answer, 17.0 torr, and we are satisfied that it is reasonable. The last step to get the total vapor pressure is an easy subtraction.

Dibutyl phthalate, $C_{16}H_{22}O_4$ (molar mass 278 g mol^{-1}), is an oil sometimes used to soften plastic articles. Its vapor pressure is negligible around room temperature. What is the vapor pressure, at 20.0 °C, of a solution of 20.0 g of dibutyl phthalate in 50.0 g of pentane, C_5H_{12} (molar mass 72.2 g mol^{-1})? The vapor pressure of pure pentane at 20.0 °C is 541 torr. (*Hint:* Calculate mole fractions first.)

Practice Exercise 12.17

Acetone (molar mass = 58.0 g mol^{-1}) has a vapor pressure at a given temperature of 162 torr. At the same temperature, how many grams of nonvolatile stearic acid (molar mass = 284.5 g mol^{-1}) must be added to 156 g of acetone to decrease its vapor pressure to 155 torr?

Practice Exercise 12.18

Raoult's Law and Nonvolatile Solutes

In Chapter 11 we saw that the vapor pressure of a volatile liquid changes with temperature. In particular, if we lower the temperature, the vapor pressure decreases. Figure 12.14*a* illustrates that the lowered vapor pressure was due to a smaller fraction of all molecules having enough energy to leave the liquid state. Figure 12.14*b* shows how a similar decrease in the fraction of energetic molecules results from the presence of a nonvolatile solute.

In Figure 12.14*b* the lower curve represents the KE distribution for the volatile part of the mixture, substance *A*, and the KE distribution for the nonvolatile substance, *B*, is added to it to obtain the upper curve that represents the KE distribution for the entire solution. We can see that the height of the lower curve is consistently ⅔ that of the upper curve. This is because we assumed a mole fraction of 0.667 for *A*. Also, the fraction of *all* molecules that exceed the kinetic energy needed by *A* to escape the solution has not changed. What has changed is that only ⅔ of those molecules are *A*—the rest are *B*, and *B* can't escape the solution. The molecular view of this effect is shown in Figure 12.15.

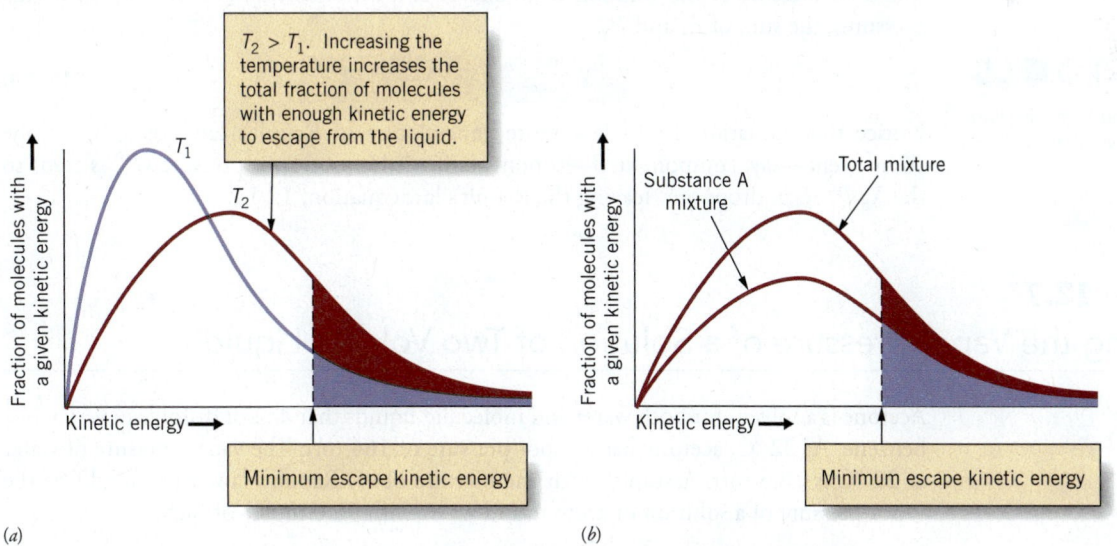

(*a*)   (*b*)

Figure 12.14 | **Kinetic energy distributions explain decreases in vapor pressure.** (*a*) Two KE distribution curves, one at a high temperature T_2 and one at a low temperature T_1. The purple-shaded portion for the low temperature shows a decrease in the fraction of molecules able to escape the liquid phase compared to the purple and red areas combined for the high temperature curve. (*b*) A KE distribution curve for a mixture with a 0.333 mole fraction of nonvolatile solute mixed with the same volatile solvent as in (*a*). The purple-shaded area represents the fraction of volatile molecules able to leave the liquid phase. The purple and red areas combined represent the fraction of all molecules with that energy, but the molecules represented by the red area cannot vaporize until they achieve a higher minimum escape KE for the nonvolatile solute.

Figure 12.15 | Molecular view of a nonvolatile solute reducing the vapor pressure of a solvent. (*a*) Equilibrium between a pure solvent and its vapor. With a high number of solvent molecules in the liquid phase, the rate of evaporation and condensation is relatively high. (*b*) In the solution, some of the solvent molecules have been replaced with solute molecules. There are fewer solvent molecules available to evaporate from solution. The evaporation rate is lower. When equilibrium is established, there are fewer molecules in the vapor. The vapor pressure of the solution is less than that of the pure solvent.

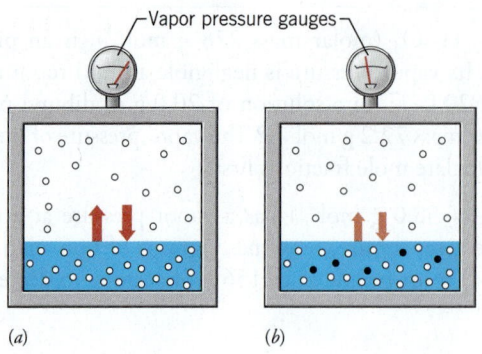

(*a*)　　　　(*b*)

Raoult's Law and Volatile Solutes

When two (or more) components of a liquid solution can evaporate, the vapor contains molecules of each. Each volatile substance contributes its own partial pressure to the total pressure. By Raoult's law, the partial pressure of a particular component is directly proportional to the component's mole fraction in the solution. By Dalton's law of partial pressures, the total vapor pressure will be the sum of all the partial pressures. To calculate these partial pressures, we use the Raoult's law equation for each component.

When component A is present in a mole fraction of X_A, its partial pressure (P_A) is this fraction of its vapor pressure when pure—namely, $P°$.

$$P_A = X_A P_A°$$

By the same argument, the partial pressure of component B, is

$$P_B = X_B P_B°$$

■ Remember, P_A and P_B here are the partial pressures as calculated by Raoult's law.

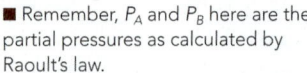

Raoult's law for two
volatile solvents

The total pressure of the solution of liquids A and B is then, by Dalton's law of partial pressures, the sum of P_A and P_B.

$$P_{\text{solution}} = X_A P_A° - X_B P_B° \qquad (12.14)$$

Notice that Equation 12.14 is a more general case of Raoult's law equation. If one component—say, component B—is nonvolatile, it has no vapor pressure ($P°$ is zero) so the $X_B P°$ term drops out, leaving the Raoult's law equation, 12.13.

Example 12.7
Calculating the Vapor Pressure of a Solution of Two Volatile Liquids

Acetone is a solvent for both water and molecular liquids that do not dissolve in water, like benzene. At 22 °C, acetone has a vapor pressure of 164 torr. The vapor pressure of water at 22 °C is 18.5 torr. Assuming that the mixture obeys Raoult's law, what would be the vapor pressure of a solution of acetone and water with 50.0 mol% of each?

Analysis: To find P_{total} we need to calculate the individual partial pressures and then add them together.

Tool: Our tool is Equation 12.14:

$$P_{\text{solution}} = X_A P_A° - X_B P_B°$$

Solution: A concentration of 50.0 mol% corresponds to a mole fraction of 0.500, so

$$P_{\text{acetone}} = 0.500 \times 164 \text{ torr} = 82.0 \text{ torr}$$

$$P_{\text{water}} = 0.500 \times 18.5 \text{ torr} = \underline{9.25 \text{ torr}}$$

$$P_{\text{total}} = 91.2 \text{ torr}$$

● **Is the Answer Reasonable?** The vapor pressure of the solution (91.2 torr) has to be much higher than that of pure water (18.5 torr) because of the volatile acetone, but much less than that of pure acetone (164 torr) because of the high mole fraction of water. The answer seems reasonable.

At 20.0 °C, the vapor pressure of cyclohexane, a nonpolar hydrocarbon, is 66.9 torr and that of toluene, a hydrocarbon related to benzene, is 21.1 torr. What is the vapor pressure of a solution of the two at 20.0 °C when toluene is present at a mole fraction of 0.250? (*Hint:* Recall that all mole fractions of the components of a mixture must add up to 1.00.)

Practice Exercise 12.19

Using the information from Practice Exercise 12.19, calculate the expected vapor pressure of a mixture of cyclohexane and toluene that consists of 122 grams of each liquid.

Practice Exercise 12.20

Only ideal solutions obey Raoult's law exactly. The vapor pressures of real solutions are sometimes higher or sometimes lower than Raoult's law would predict. These differences between ideal and real solution behavior are quite useful. They can tell us whether the attractive forces in the pure solvents, before mixing, are stronger or weaker than the attractive forces in the mixture. Comparing experimental vapor pressures with the predictions of Raoult's law provides a simple way to compare the relative strengths of attractions between molecules in solution. When the attractions between unlike molecules in the mixture are stronger than those in the pure solvents, the experimental vapor pressure will be lower than calculated with Raoult's law. Conversely, when the attractions between unlike molecules in the mixture are weaker than those in the pure solvents, the experimental vapor pressure will be larger than Raoult's law predicts.

Referring back to Figure 12.7, we can see that when attractions between the unlike molecules in the mixture are stronger than those in the pure solvents, we should expect Step 3, the mixing process, to be more exothermic than Steps 1 and 2 combined. Therefore, a lower vapor pressure than expected indicates an exothermic heat of mixing. Similarly, if the vapor pressure of the mixture is greater than expected by Raoult's law, the heat of mixing will be endothermic. Figure 12.16 illustrates the shapes of the vapor pressure versus mole fraction curves for ideal solutions and two types of nonideal solutions.

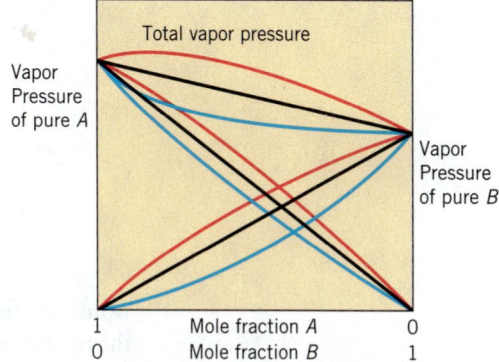

Figure 12.16 | Vapor pressure versus mole fraction of ideal and nonideal solutions. Solutions are made with two volatile solvents. (*a*) The black lines represent an ideal solution, where the attractive forces in the mixture and pure liquids are the same and $\Delta H_{soln} = 0$. (*b*) Blue lines represent solutions that have an exothermic ΔH_{soln} and have stronger attractions in the mixture. (*c*) Red lines represent solutions that have an endothermic ΔH_{soln} and have weaker attractions in the mixture.

Freezing Point Depression and Boiling Point Elevation

The lowering of the vapor pressure produced by the presence of a nonvolatile solute affects both the boiling and freezing points of a solution. This is illustrated for water in Figure 12.17. In this phase diagram, the solid blue lines correspond to the three equilibrium lines in the phase diagram for pure water, which we discussed in Section 11.7. Adding a nonvolatile solute lowers the vapor pressure of the solution, giving a new liquid–vapor equilibrium line for the solution, which is shown as the red line connecting points *A* and *B*.

When the solution freezes, the solid that forms is pure ice; there is no solute within the ice crystals. This is because the highly ordered structure of the solid doesn't allow solute molecules to take the place of water molecules. As a result, both pure water and the solution have the same solid–vapor equilibrium line on the phase diagram. Point *A* on the diagram is at the intersection of the liquid–vapor and solid–vapor equilibrium lines for the solution and represents the new triple point for the solution. Rising from this triple point is the solid–liquid equilibrium line for the solution, shown in red.

If we look at where the solid–liquid and liquid–vapor lines cross the 1 atm (760 torr) pressure line, we can find the normal freezing and boiling points. For water (the blue lines), the freezing point is 0 °C and the boiling point is 100 °C. Notice that the solid–liquid line for the solution meets the 760 torr line at temperatures *below* 0 °C. In other

■ Recall that the definition of the normal boiling and freezing points specifies that the pressure is 1 atm.

Figure 12.17 | **Phase diagrams for water and an aqueous solution.** Phase diagram for pure water (blue lines) and an aqueous solution of a nonvolatile solute (red lines).

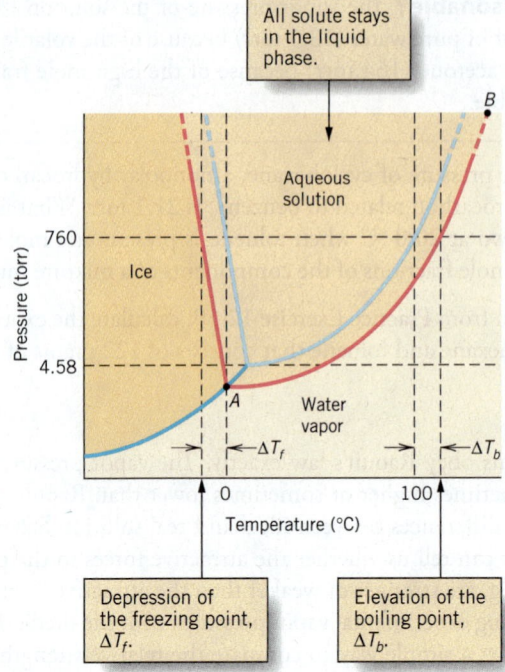

All solute stays in the liquid phase.

Aqueous solution

Ice

760

4.58

A

Water vapor

$\leftarrow \Delta T_f$

$\leftarrow \Delta T_b$

B

0

100

Temperature (°C)

Pressure (torr)

Depression of the freezing point, ΔT_f.

Elevation of the boiling point, ΔT_b.

words, the freezing point of the solution is below that of pure water. The amount by which the freezing point is lowered is called the **freezing point depression** and is given the symbol ΔT_f.

$$\Delta T_f = T_{f\,\mathrm{pure}} - T_{f\,\mathrm{observed}}$$

Similarly, we can see that the liquid–vapor line for the solution crosses the 760 torr line at a temperature *above* 100 °C, so the solution boils at a higher temperature than pure water. The amount by which the boiling point is raised is called the **boiling point elevation**. It is given the symbol ΔT_b.

$$\Delta T_b = T_{b\,\mathrm{observed}} - T_{b\,\mathrm{pure}}$$

Freezing point depression and boiling point elevation are both colligative properties. The magnitudes of ΔT_f and ΔT_b are proportional to the concentration of solute and solvent molecules. Molality (*m*) is the preferred concentration expression (rather than mole fraction or percent by mass) because of the resulting simplicity of the equations relating ΔT to concentration. The following equations work reasonably well only for dilute solutions, however.

TOOLS

Freezing point depression; boiling point elevation

$$\Delta T_f = K_f m \qquad \text{(12.15)}$$

$$\Delta T_b = K_b m \qquad \text{(12.16)}$$

K_f and K_b are proportionality constants and are called, respectively, the **molal freezing point depression constant** and the **molal boiling point elevation constant**. The values of both K_f and K_b are characteristic of each solvent (see Table 12.4). The units of each constant are °C m^{-1}. Thus, the value of K_f for a given solvent corresponds to the number of degrees of freezing point lowering for each molal unit of concentration. The K_f for water is 1.86 °C m^{-1}. A 1.00 *m* solution in water freezes at 1.86 °C *below* the normal freezing point of 0.00 °C, or at −1.86 °C. A 2.00 *m* solution should freeze at −3.72 °C. (We say "should" because systems are seldom this ideal.) Similarly, because K_b for water is 0.51 °C m^{-1}, a 1.00 *m* aqueous solution at 1 atm pressure boils at (100.00 + 0.51) °C or 100.51 °C, and a 2.00 *m* solution should boil at 101.02 °C.

TABLE 12.4	Molal Boiling Point Elevation and Freezing Point Depression Constants			
Solvent	BP (°C)	K_b (°C m^{-1})	MP (°C)	K_f (°C m^{-1})
Water	100	0.512	0	1.86
Acetic acid	118.3	3.07	16.6	3.90
Benzene	80.2	2.53	5.45	5.07
Chloroform	61.2	3.63	−63.5	4.68
Camphor	—	—	178.4	39.7
Cyclohexane	80.7	2.79	6.5	20.2

Example 12.8
Estimating a Freezing Point Using a Colligative Property

Estimate the freezing point, at 1.0 atm, of a solution made from 10.0 g of urea, $CO(NH_2)_2$ (molar mass 60.06 g mol^{-1}), and 125 g of water.

Analysis: We need to use the appropriate equation for the freezing point depression. To use that equation we also need to determine the molality of the solution.

Assembling the Tools: The main tool we need is expressed in Equation 12.15, the freezing point depression. We can look up the value for K_f and write the expression as

$$\Delta T_f = (1.86\ °C\ m^{-1})m$$

The equation for molality is

$$m = \frac{\text{mol urea}}{\text{kg water}}$$

and we use the molar mass of urea to convert the mass to the moles we need.

Solution: We write the equation that needs to be solved.

$$\Delta T_f = (1.86\ °C\ \text{molal}^{-1})m$$

Now to start the process for determining the molality, we'll convert grams of urea to moles.

$$10.0\ \overline{g\ CO(NH_2)_2} \times \frac{1\ \text{mol}\ CO(NH_2)_2}{60.06\ \overline{g\ CO(NH_2)_2}} = 0.166\ \text{mol}\ CO(NH_2)_2$$

and the 125 g to 0.125 kg of water. The molality is calculated as

$$\text{molality} = \frac{0.166\ \text{mol}}{0.125\ \text{kg}} = 1.33\ m$$

Now we solve the freezing point depression problem using the molality we just determined.

$$\Delta T_f = (1.86\ °C\ m^{-1})(1.33\ m)$$
$$= 2.47\ °C$$

The solution should freeze at 2.47 °C below 0 °C, or at −2.47 °C.

Is the Answer Reasonable? For every unit of molality, the freezing point must be depressed by about 2 °C. The molality of this solution is between 1 m and 2 m, so we expect the freezing point depression to be between about 2 °C and 4 °C.

Practice Exercise 12.21

In making candy, a certain recipe calls for heating an aqueous sucrose solution to the "soft-ball" stage, which has a boiling point of 235–240 °F. What is the range of mass percentages of the solutions of sugar ($C_{12}H_{22}O_{11}$) that boil at those two temperatures? (*Hint:* Remember that molal boiling point elevation constants are in °C m^{-1}.)

Practice Exercise 12.22

How many grams of glucose (molar mass = 180.9 g mol^{-1}) must be dissolved in 255 g of water to raise the boiling point to 102.36 °C?

Determining Molar Masses

We have described freezing point depression and boiling point elevation as *colligative* properties; they depend on the relative *numbers* of particles, not on their kinds. Because the effects are proportional to molal concentrations, experimentally measured values of ΔT_f or ΔT_b can be useful for calculating the molar masses of unknown solutes. To do this, let's examine the molality part of the temperature change in Equations 12.15 and 12.16 more closely. If we expand the molality equation we get

$$m = \frac{\text{mol solute}}{\text{kg solvent}} = \frac{\left(\dfrac{\text{g solute}}{\text{molar mass solute}}\right)}{\text{kg solvent}}$$

This indicates that if we measure ΔT_f or ΔT_b and know the mass of solute in each kg of solvent, we can estimate the last remaining unknown, the molar mass. Example 12.9 shows how to estimate molar mass.

Example 12.9
Calculating a Molar Mass from Freezing Point Depression Data

A solution made by dissolving 5.65 g of an unknown molecular compound in 110.0 g of benzene froze at 4.39 °C. What is the molar mass of the solute?

Analysis: To calculate the molar mass (i.e., g mol^{-1}), we need to know two things about the same sample: the number of moles and the number of grams. In this example, we've been given the number of grams, and we have to use the remaining data to calculate the number of moles.

Assembling the Tools: Our basic tool is the freezing point depression equation 12.15, expanding ΔT_f.

$$\Delta T_f = T_{f\,\text{pure}} - T_{f\,\text{observed}} = K_f m$$

We will also use the definition of molality

$$m = \frac{\text{mole solute}}{\text{kg solvent}} \quad \text{and} \quad \text{mol solute} = \frac{\text{g solute}}{\text{molar mass solute}}$$

Solution: Table 12.4 tells us that the melting point of pure benzene is 5.45 °C and that the value of K_f for benzene is 5.07 °C m^{-1}. The amount of freezing point depression is

$$\Delta T_f = 5.45 \,°C - 4.39 \,°C = 1.06 \,°C$$

We now use Equation 12.15 to find the molality of the solution,

$$\Delta T_f = K_f m$$

$$\text{molality} = \frac{\Delta T_f}{\Delta K_f} = \frac{1.06 \,°C}{5.07 \,°C \, m^{-1}} = 0.209 \, m$$

This means that for every kilogram of benzene in the solution, there are 0.209 mol of solute. However, we have only 110.0 g or 0.1100 kg of benzene, so the actual number of moles of solute in the given solution is

$$0.1100 \text{ kg benzene} \times \frac{0.209 \text{ mol solute}}{1 \text{ kg solvent}} = 0.0230 \text{ mol solute}$$

We can now obtain the molar mass. There are 5.65 g of solute and 0.0230 mol of solute.

$$\text{molar mass solute} = \frac{5.65 \text{ g solute}}{0.0230 \text{ mol solute}} = 246 \text{ g mol}^{-1}$$

The mass of one mole of the solute is 246 g.

Is the Answer Reasonable? A common mistake to avoid is using the given value of the freezing point, 4.39 °C, in Equation 12.15, instead of calculating ΔT_f. A check shows that we *have* calculated ΔT_f correctly. The ratio of ΔT_f to K_f is about ⅕, or 0.2 *m*, which corresponds to a ratio of 0.2 mol solute per 1 kg of benzene. The solution prepared has only 0.1 kg of solvent, so to have the same ratio, the amount of solute must be about 0.02 mol (2×10^{-2} mol). The sample weighs about 5 g, and dividing 5 g by 2×10^{-2} mol gives 2.5×10^2 g/mol, or 250 g/mol. This is very close to the answer we obtained, so we can feel confident it's correct.

A solution made by dissolving 3.46 g of an unknown compound in 85.0 g of benzene froze at 4.13 °C. What is the molar mass of the compound? (*Hint*: Recall the equation for calculating moles of a substance to find the key relationship.)

Practice Exercise 12.23

A mixture is prepared that is 5.0%(w/w) of an unknown substance mixed with naphthalene (molar mass = 128.2 g mol^{-1}). The freezing point of this mixture is found to be 77.3 °C. What is the molar mass of the unknown substance? (The melting point of naphthalene is 80.2 °C and $K_f = 6.9$ °C m^{-1}.)

Practice Exercise 12.24

Osmosis

In living things, membranes of various kinds keep mixtures and solutions organized and separated. Yet some substances have to be able to pass through membranes so that nutrients and products of chemical work can be distributed correctly. These membranes, in other words, must have a selective *permeability*. They must keep some substances from going through while letting others pass. Such membranes are said to be *semipermeable*.

The degree of permeability varies with the kind of membrane. Cellophane, for example, is permeable to water and small solute particles—ions or molecules—but impermeable to very large molecules, like those of starch or proteins. Special membranes can even be prepared that are permeable only to water, not to any solutes.

Depending on the kind of membrane separating solutions of different concentrations, two similar phenomena, *dialysis* and *osmosis*, can be observed. Both are functions of the relative concentrations of the particles of the dissolved materials on each side of the membrane. Therefore, physical properties of these systems are also classified as colligative properties.

When a membrane is able to let both water and *small* solute particles through, such as the membranes in living systems, the process is called **dialysis**, and the membrane is called a **dialyzing membrane**. It does not permit huge molecules through, like those of proteins and starch. Artificial kidney machines use dialyzing membranes to help remove the smaller molecules of wastes from the blood while letting the blood retain its large protein molecules.

When a semipermeable membrane will let only solvent molecules get through, this movement is called *osmosis*, and the special membrane needed to observe it is called an **osmotic membrane**.

When **osmosis** occurs, there's a net shift of solvent across the membrane from the more dilute solution (or pure solvent) into the more concentrated solution. This happens because there is a tendency toward the equalization of concentrations between the two solutions in contact with one another across the membrane. The rate of passage of solvent molecules through the membrane into the more concentrated solution is greater than their rate of passage in the opposite direction, presumably because at the surface of the membrane the solvent concentration is greater in the more dilute solution (Figure 12.18).

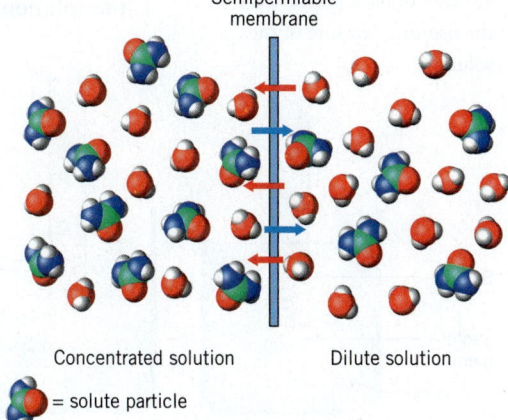

Semipermiable membrane

Concentrated solution   Dilute solution

= solute particle

Figure 12.18 | Osmosis. Solvent molecules pass more frequently from the more dilute solution into the more concentrated one, as indicated by the arrows. This leads to a gradual transfer of solvent from the less concentrated solution into the more concentrated one.

Figure 12.19 | Vapor pressure
and the transfer of the solvent.
(*a*) Because the two solutions have
unequal rates of evaporation, but
the same rate of condensation,
there is a gradual net transfer of
solvent from the more dilute
solution into the more concentrated one. (*b*) After time, the
concentration of the two solutions
is the same as the system reaches
equilibrium.

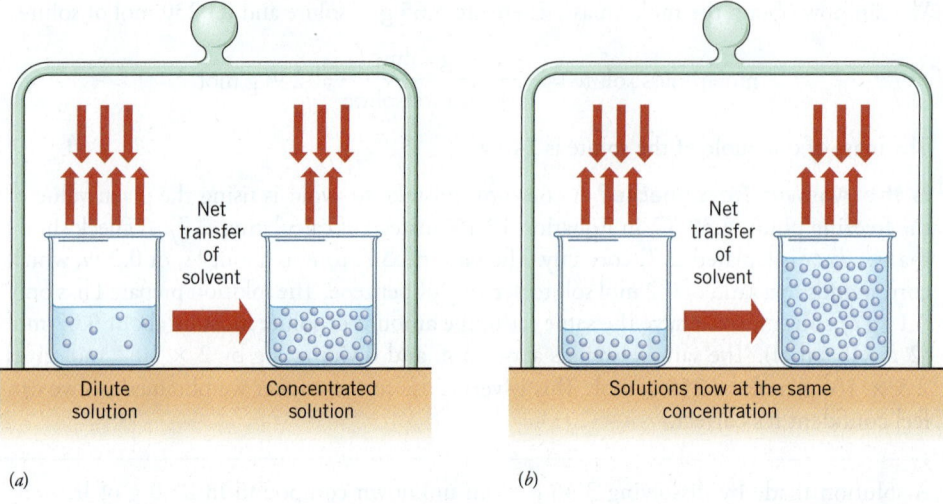

(*a*)   (*b*)

This leads to a gradual net flow of water through the membrane into the more concentrated solution.

We can observe an effect similar to osmosis due to vapor pressure, if two solutions with
unequal concentrations of a nonvolatile solute are placed in a sealed container (Figure
12.19). The rate of evaporation from the more dilute solution is greater than that of the
more concentrated solution, but the rate of return to each is the same because both solutions are in contact with the same gas phase. As a result, neither solution is in equilibrium
with the vapor. In the dilute solution, molecules of solvent are evaporating faster than
they're condensing. In the concentrated solution, however, just the opposite occurs; more
water molecules return to the solution than leave it. Therefore, over time there is a gradual
net transfer of solvent from the dilute solution into the more concentrated one until they
both achieve the same concentration and vapor pressure (Figure 12.19*b*).

Osmotic Pressure

Figure 12.20 | Osmosis and
osmotic pressure. (*a*) Initial
conditions. A solution, *B*, is
separated from pure water, *A*, by
an osmotic membrane; no osmosis
has yet occurred. (*b*) After a while,
the volume of fluid in the tube has
increased visibly. Osmosis has
taken place. (*c*) A back pressure is
needed to prevent osmosis. The
amount of back pressure is called
the osmotic pressure of the
solution.

An osmosis experiment is illustrated in Figure 12.20. Initially, we have a solution (*B*) in a
tube fitted with an osmotic membrane that dips into a container of pure water (*A*). As
time passes, the volume of liquid in the tube increases as solvent molecules transfer into
the solution. In Figure 12.20*b*, the net transport of water into the solution has visibly
increased the volume.

The weight of the rising fluid column in Figure
12.20*b* provides a push or opposing pressure that
makes it increasingly more difficult for molecules
of water to enter the solution inside the tube.
Eventually, this pressure becomes sufficient to
stop the osmosis. The exact opposing pressure
needed to prevent any osmotic flow *when one of
the liquids is pure solvent* is called the **osmotic
pressure** of the solution. If we apply further pressure, as illustrated in Figure 12.20*c*, we can force
enough water back through the membrane to
restore the system to its original condition.

Notice that the term "osmotic pressure" uses
the word *pressure* in a novel way. Apart from osmosis, a solution does not "have" a special pressure called osmotic pressure. What the solution has is a *concentration* that can
generate the occurrence of osmosis and the associated osmotic pressure under the right
circumstances. Then, in proportion to the solution's concentration, a specific back pressure is required to prevent osmosis. By *exceeding* this back pressure, osmosis can be
reversed. **Reverse osmosis** is widely used to purify seawater, both on ocean-going ships
and in arid locations where fresh water is scarce, polluted, or unavailable. (See more
details in Chemistry and Current Affairs.)

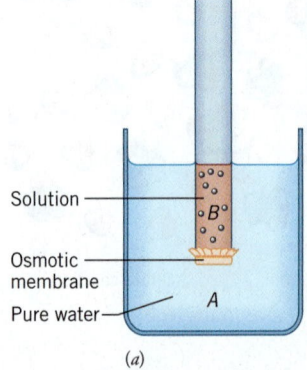

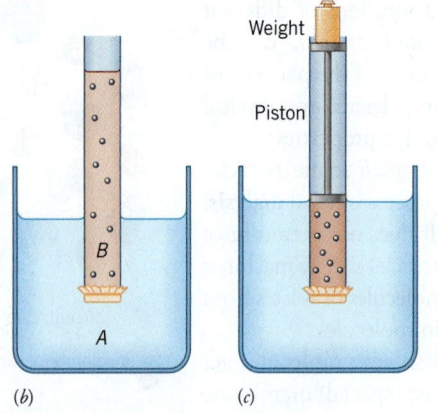

Weight

Piston

Solution ——
Osmotic ——
membrane
Pure water——

(*a*)   (*b*)   (*c*)

■ Reverse osmosis systems are
available at hardware stores for home
use. They can be installed to remove
impurities and foul tastes from
drinking water. Some bottled water
available on supermarket shelves
contains water that's been purified
by reverse osmosis.

CHEMISTRY AND CURRENT AFFAIRS 12.1

Pure Water by Reverse Osmosis

Many important issues face the world today, but none is more important than the availability of potable (pure and drinkable) water. As Table 1 shows, almost 97% of all water is saltwater found in the oceans and seas. Most of the freshwater is tied up in polar and glacial ice.

TABLE 1	Water Resources	
Water Source	Amount in the World	Amount per Person*
All water	1.4×10^9 km^3	0.2 km^3 = 2×10^8 m^3
Freshwater total	3.9×10^7 km^3	5.6×10^6 m^3
Freshwater, icebound	3.3×10^7 km^3	4.7×10^6 m^3
Freshwater, liquid	6.0×10^6 km^3	8.6×10^5 m^3

*For a world population of 7 billion people

The average person in a developed country uses approximately 54 m^3 of water per year. While it seems that there may be plenty of water—the problem is availability. Temperate climates get plenty of rain and usually have adequate water in lakes, streams, and aquifers. However, temperate climates often have high population densities, resulting in high water usage along with a greater possibility of water pollution. In arid regions, water is scarce even though the populations may be small. In all cases, natural water is a valuable resource that may have to be augmented using available technology.

There are many technologies that can be used to obtain freshwater. Boiling and condensing water uses too much energy to be practical. However, very large scale, multi-stage vacuum distillation produces about 85% of the desalinized water in the world but requires significant energy input. Reverse osmosis uses less energy and can be designed for smaller applications. Other methods such as ion-exchange, electro-dialysis, freezing, and solar dehumidification are also used.

Osmosis, as described in this chapter, involves spontaneous movement of pure solvent (usually water) through a membrane into a solution containing a solute. The tendency of the solvent to move through the membrane is measured as the osmotic pressure. To reverse the flow of water through an osmotic membrane, a pressure that exceeds the osmotic pressure needs to be applied to the solution side of the membrane. The result will be pure water extracted from a solution such as seawater. Figure 1 shows a schematic diagram of a reverse osmosis process. Figure 2 is a photo of a bank of reverse osmosis separation tubes in operation.

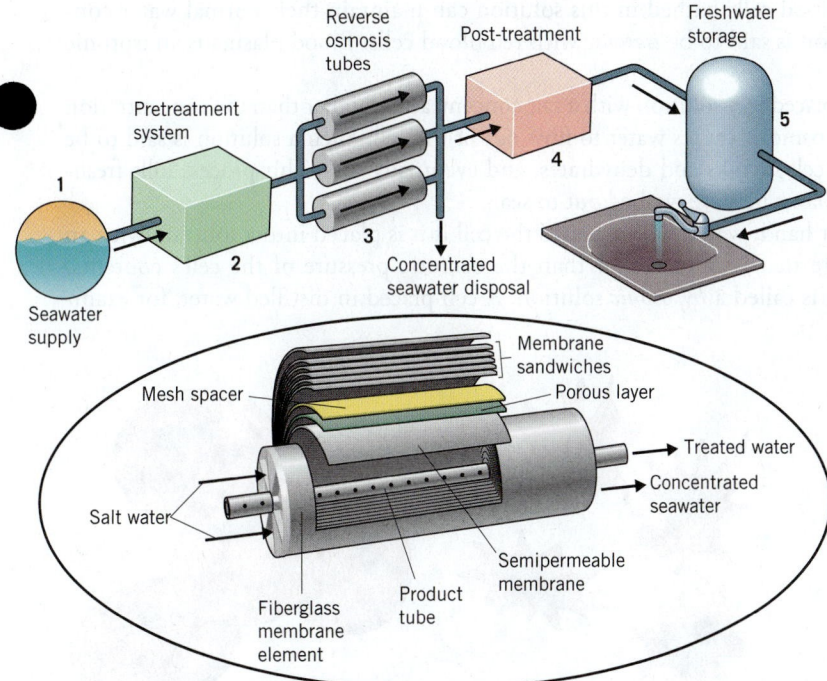

Figure 1 A schematic diagram of a reverse osmosis process. (1) The seawater is filtered (2) to remove virtually all particulate matter. (3) Pressure is applied to perform the separation. (4) Purified water is treated to adjust the acidity and add antibacterial agents before (5) storage. *Inset:* A diagram of the construction of the reverse osmosis tube.

© DenGuy/iStockphoto

Figure 2 A commercial reverse osmosis unit.

The symbol for osmotic pressure is the Greek capital pi, Π. In a dilute aqueous solution, Π is proportional both to temperature, T, and molar concentration of the solute in the solution, M:

$$\Pi \propto MT$$

The proportionality constant turns out to be the gas law constant, R, so for a *dilute* aqueous solution we can write

$$\Pi = MRT \tag{12.17}$$

TOOLS
Osmotic pressure

Of course, M is the ratio mol/L, which we can write as n/V, where n is the number of moles and V is the volume in liters. If we replace M with n/V in Equation 12.17 and rearrange terms, we have an equation for osmotic pressure identical in form to the ideal gas law.

$$\Pi V = nRT \tag{12.18}$$

Equation 12.18 is the *van't Hoff equation for osmotic pressure.*

Osmotic pressure is of tremendous importance in biology and medicine. Cells are surrounded with membranes that restrict the flow of salts but allow water to pass through freely. To maintain a constant amount of water, the osmotic pressure of solutions on either side of the cell membrane must be identical. For an example, a solution that is 0.85% NaCl by mass has the same osmotic pressure as the contents of red blood cells, and red blood cells bathed in this solution can maintain their normal water content. The solution is said to be *isotonic* with red blood cells. Blood plasma is an isotonic solution.

If the cell is placed in a solution with a salt concentration higher than the concentration within the cell, osmosis causes water to flow out of the cell. Such a solution is said to be *hypertonic*. The cell shrinks and dehydrates, and eventually dies. This process kills freshwater fish and plants that are washed out to sea.

On the other hand, water will flow into the cell if it is placed into a solution with an osmotic pressure that is much lower than the osmotic pressure of the cell's contents. Such a solution is called a *hypotonic* solution. A cell placed in distilled water, for exam-

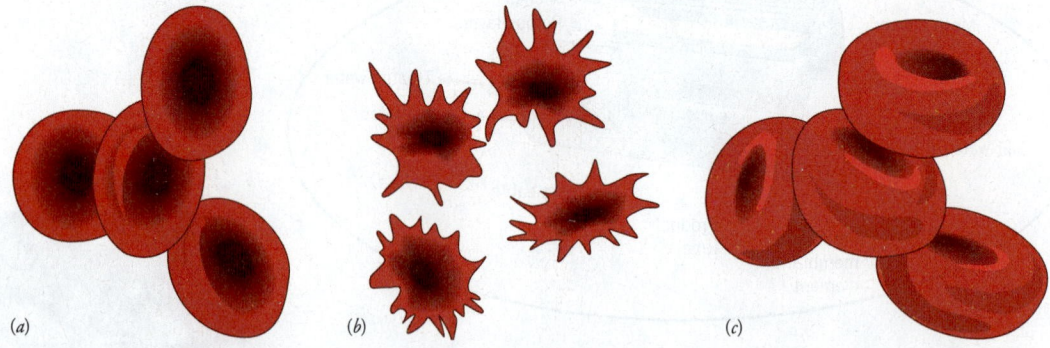

(a)   *(b)*   *(c)*

Figure 12.21 | Effects of isotonic, hypertonic, and hypotonic solutions on red blood cells. (*a*) In an isotonic solution (0.85% NaCl by mass), solutions on either side of the membrane have the same osmotic pressure, and there is no flow of water across the cell membrane. (*b*) In a hypertonic solution (5.0% NaCl by mass), water flows from areas of lower salt concentration (inside the cell) to areas of higher concentration (the hypertonic solution), causing the cell to dehydrate. (*c*) In a hypotonic solution (0.1% NaCl by mass), water flows from areas of lower salt concentration (the hypotonic solution) to areas of higher concentration (inside the cell). The cell swells and bursts.

ple, will swell and burst. If you've ever tried to soak a pair of contact lenses with tap water instead of an isotonic saline solution, you've experienced cell damage from a hypotonic solution. The effects of isotonic, hypotonic, and hypertonic solutions on cells are shown in Figure 12.21.

Obviously, the measurement of osmotic pressure can be very important in preparing solutions that will be used to culture tissues or to administer medicines intravenously. Osmotic pressures can be measured by an instrument called an *osmometer*, illustrated and explained in Figure 12.22. Osmotic pressures can be very high, even in dilute solutions, as Example 12.10 shows.

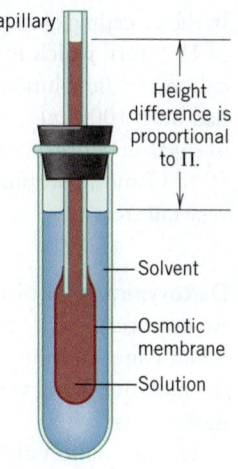

Figure 12.22 | Simple osmometer. When solvent moves into the solution by osmosis, the level of the solution in the capillary rises. The height reached can be related to the osmotic pressure of the solution.

Example 12.10
Calculating Osmotic Pressure

A very dilute solution, 0.00100 *M* sucrose in water, is separated from pure water by an osmotic membrane. What osmotic pressure in torr develops at 298 K?

Analysis: This question involves the application of the appropriate form of the osmotic pressure equation. Of course, we must be sure to use the correct value of *R*, so that the units work out correctly.

Assembling the Tools: We need to use Equation 12.17,

$$\Pi = MRT$$

Since we are given the temperature in kelvins and molarity has units of moles and liters, it is appropriate to choose $R = 0.0821$ L atm mol^{-1} K^{-1}.

Solution: Substituting into Equation 12.17,

$$\Pi = (0.00100 \text{ mol L}^{-1})(0.0821 \text{ L atm mol}^{-1}\text{K}^{-1})(298 \text{ K})$$

$$= 0.0245 \text{ atm}$$

In torr we have

$$\Pi = 0.0245 \text{ atm} \times \frac{760 \text{ torr}}{1 \text{ atm}} = 18.6 \text{ torr}$$

The osmotic pressure of 0.00100 *M* sugar in water is 18.6 torr.

Is the Answer Reasonable? A 1 *M* solution should have an osmotic pressure of *RT*. Rounding the gas law constant to 0.08 and the temperature to 300, the osmotic pressure of a 1 *M* solution should be about 24 atm. The osmotic pressure of an 0.001 *M* solution should be 1/1000 of this, or about 0.024 atm, which is very close to what we got before we converted to torr.

What is the osmotic pressure, in mm Hg and mm H_2O, of a protein solution when 5.00 g of the protein (molar mass = 235,000 g mol^{-1}) is used to prepare 100.0 mL of an aqueous solution at 4 °C? (*Hint:* The units needed for this calculation are specified by the value of *R* that is used.)

Practice Exercise 12.25

What is the osmotic pressure in torr of a 0.0115 *M* glucose solution at body temperature (37 °C)? What are the boiling and freezing points for the same solution?

Practice Exercise 12.26

In the preceding example, you saw that a 0.0010 M solution of sugar has an osmotic pressure of 18.6 torr, which is equivalent to 18.6 mm Hg. This pressure is sufficient to support a column of the solution (which is mostly water) roughly 25 cm or 10 in high. If the solution had been 100 times as concentrated, 0.100 M sugar—about the same as ¼ of a teaspoon of sugar in your tea—the height of the column supported would be roughly 25 m or over 80 ft! Osmotic pressure is one of several mechanisms by which water can reach the top of very tall trees.

Determining Molar Mass from Osmotic Pressure

An osmotic pressure measurement taken of a dilute solution can be used to determine the molar concentration of the solute, regardless of its chemical composition. Knowledge of the molarity, along with the mass of the solute in the solution, permits us to calculate the molar mass.

Determination of molar mass by osmotic pressure is much more sensitive than determination by the freezing point depression or boiling point elevation methods. The following example illustrates how osmotic pressure experiments can be used to determine molar mass.

Example 12.11
Determining Molar Mass from Osmotic Pressure

An aqueous solution with a volume of 100.0 mL and containing 0.122 g of an unknown molecular compound has an osmotic pressure of 16.0 torr at 20.0 °C. What is the molar mass of the solute?

Analysis: As we noted in Example 12.9, to determine a molar mass we need to measure two quantities for the same sample: its mass and the number of moles. Then, the molar mass is the ratio of grams to moles. We're given the number of *grams* of the solute, so we need to find the number of *moles* equivalent to this mass in order to compute the ratio.

Assembling the Tools: We can use the osmotic pressure Equation 12.18,

$$\Pi V = nRT$$

■ $16.0 \text{ torr} \times \dfrac{1 \text{ atm}}{760 \text{ torr}} = 0.0211 \text{ atm}$

to calculate the number of moles, n. We need to convert the given pressure to 0.0211 atm. The temperature, 20.0 °C, corresponds to 293 K (273 + 20.0) and the volume of 100.0 mL corresponds to 0.1000 L. Since we are given the mass, we just need the definition of molar mass as the tool to complete the calculation.

Solution: First, the number of moles, n, is calculated using the Equation 12.18 (with the data we just converted to the correct units),

$$(0.0211 \text{ atm})(0.1000 \text{ L}) = (n)(0.0821 \text{ L atm mol}^{-1} \text{ K}^{-1})(293 \text{ K})$$

$$n = 8.77 \times 10^{-5} \text{ mol}$$

The molar mass of the solute is the number of grams of solute per mole of solute:

$$\text{molar mass} = \frac{0.122 \text{ g}}{8.77 \times 10^{-5} \text{ mol}} = 1.39 \times 10^{3} \text{ g mol}^{-1}$$

Is the Answer Reasonable? In using Equation 12.18, it is essential that the pressure be in atmospheres when using $R = 0.0821$ L atm mol^{-1} K^{-1}. If all the units cancel correctly (and they do), we've computed the number of moles correctly. The moles of solute is approximately 10×10^{-5}, or 1×10^{-4}. Dividing the mass, roughly 0.12 g, by 1×10^{-4} mol gives 0.12×10^4 g per mol or 1.2×10^3 g per mol, which is close to the value we found.

Estimate the molecular mass of a protein when 137.2 mg of the protein, dissolved in 100.0 mL of water at 4 °C, supports a column of water that is 6.45 cm high. (*Hint:* The answer should be very large. Don't forget to convert cm of water to atmospheres of pressure.)

Practice Exercise 12.27

A solution of a carbohydrate prepared by dissolving 72.4 mg in 100 mL of solution has an osmotic pressure of 25.0 torr at 25.0 °C. What is the molecular mass of the compound?

Practice Exercise 12.28

Colligative Properties of Ionic Solutes

The molal freezing point depression constant for water is 1.86 °C m^{-1}. So you might think that a 1.00 m solution of NaCl would freeze at -1.86 °C. Instead, it freezes at -3.37 °C. This greater depression of the freezing point by the salt, which is almost twice -1.86 °C, is not hard to understand if we remember that colligative properties depend on the concentrations of *particles*. We know that NaCl(s) dissociates into ions in water.

$$NaCl(s) \longrightarrow Na^+(aq) + Cl^-(aq)$$

If the ions are truly separated, 1.00 m NaCl actually has a concentration of dissolved solute particles that is 2.00 m, twice the given molal concentration. Theoretically, 1.00 m NaCl should freeze at 2×1.00 $m \times (-1.86$ °C $m^{-1})$ or -3.72 °C. (Why it actually freezes a little higher than this, at -3.37 °C, will be discussed shortly.)

If we made up a solution of 1.00 m $(NH_4)_2SO_4$, we would have to consider the following dissociation.

$$(NH_4)_2SO_4(s) \longrightarrow 2NH_4^+(aq) + SO_4^{2-}(aq)$$

One mole of $(NH_4)_2SO_4$ can give a total of 3 mol of ions (2 mol of NH_4^+ and 1 mol of SO_4^{2-}). We would expect the freezing point of a 1.00 m solution of $(NH_4)_2SO_4$ to be 3×1.00 $m \times (-1.86$ °C $m^{-1}) = -5.58$ °C.

When we want to roughly *estimate* a colligative property of a solution of an electrolyte, we recalculate the solution's molality using an assumption about the way the solute dissociates or ionizes. For example, with ionic substances we may assume 100% dissociation, although we will see shortly that this assumption is only true for very dilute solutions.

Example 12.12
Estimating the Freezing Point of a Salt Solution

Estimate the freezing point of aqueous 0.106 m $MgCl_2$, assuming that it dissociates completely.

Analysis: Equation 12.15 relates a change in freezing temperature to molality, but we now have to include the fact that our solute dissociates into three particles, one Mg^{2+} and two Cl^- ions. Including this additional information allows us to make a better estimate of the freezing point.

Assembling the Tools: We will use the basic freezing point equation, with the K_f for water from Table 12.4.

$$\Delta T_f = (1.86 \,°C \, m^{-1})m$$

However, our molality must be the molality of all ions, not the molality of the salt.

$$MgCl_2(s) \longrightarrow Mg^{2+}(aq) + 2Cl^-(aq)$$

Solution: Because 1 mol of $MgCl_2$ gives 3 mol of ions, the effective (assumed) molality of ions in the solution is three times the molality of the salt, 0.106 m.

$$\text{Effective molality} = (3)(0.106 \, m) = 0.318 \, m$$

■ The effective molarity is often called the osmolarity in the biological sciences.

Now we can use Equation 12.15.

$$\Delta T_f = (1.86 \,°C \, m^{-1})(0.318 \, m)$$
$$= 0.591 \,°C$$

The freezing point is depressed below 0.000 °C by 0.591 °C, so we calculate that this solution freezes at −0.591 °C.

Is the Answer Reasonable? The molality of particles, as recalculated, is roughly 0.3, so 3/10 of 1.86 (call it 2) is about 0.6, which, after we add the unit, °C, and subtract from 0 °C gives −0.6 °C, which is close to the answer.

Practice Exercise 12.29 | Calculate the freezing point of aqueous 0.237 m LiCl on the assumption that it is 100% dissociated. Calculate the freezing point if the percent dissociation is 0%. (*Hint:* What is the concentration of particles in solution?)

Practice Exercise 12.30 | Determine the freezing point of aqueous solutions of $MgSO_4$ that are (a) 0.100 m, (b) 0.0100 m, and (c) 0.00100 m. Which of these could be measured with your laboratory thermometer that is graduated in one °C intervals?

 ## Analyzing and Solving Multi-Concept Problems

A recently synthesized compound was analyzed by a technician, but the analysis of the data is left to you. The technician took 16.59 g of the substance, and burned it in an excess of oxygen to give 28.63 g of carbon dioxide and 11.71 g of water. In another experiment, 1.395 g of the compound was decomposed by heating strongly in the presence of metallic sodium to destroy the organic material and release any chlorine as chloride ions, which were precipitated with silver nitrate as 2.541 g of AgCl. Finally, 5.41 g of this compound were mixed with 85.0 g of benzene and the freezing point was found to be 3.33 °C. What is the formula for the compound?

Analysis: We need to determine the empirical formula for the compound and then the molecular formula. We will (1) calculate the percent composition of the compound from the analysis reactions, (2) determine the empirical formula from the percentages, (3) use the freezing point depression to determine the molar mass, and (4) find the actual formula.

PART 1

Assembling the Tools: From the data we can determine the percentage of carbon, hydrogen, and chlorine in the compound. The percentages are calculated as follows:

$$\% \text{ composition} = \frac{\text{mass of element}}{\text{mass of sample}} \times 100\%$$

We can use the stoichiometric conversions introduced in Chapter 3.

Solution: Let's start by converting the mass of CO_2 to the mass of C,

$$28.63 \text{ g } CO_2 \times \frac{12.011 \text{ g C}}{44.01 \text{ g } CO_2} = 7.814 \text{ g C}$$

and the mass of H_2O to the mass of H,

$$11.71 \text{ g } H_2O\left(\frac{1 \text{ mol } H_2O}{18.015 \text{ g } H_2O}\right)\left(\frac{2 \text{ mol H}}{1 \text{ mol } H_2O}\right)\left(\frac{1.008 \text{ g H}}{1 \text{ mol H}}\right) = 1.310 \text{ g H}$$

We then calculate the percentage of C and H:

$$\% \text{ C} = \frac{7.817 \text{ g C}}{16.59 \text{ g sample}} \times 100\% = 47.10\% \text{ C}$$

$$\% \text{ H} = \frac{1.310 \text{ g H}}{16.59 \text{ g sample}} \times 100\% = 7.900\% \text{ H}$$

The mass of chlorine is calculated as

$$2.541 \text{ g } AgCl\left(\frac{1 \text{ mol } AgCl}{143.32 \text{ g } AgCl}\right)\left(\frac{1 \text{ mol } Cl}{1 \text{ mol } AgCl}\right)\left(\frac{35.45 \text{ g } Cl}{1 \text{ mol } Cl}\right) = 0.6285 \text{ g Cl}$$

The percentage of chloride in the sample is

$$\% Cl = \frac{0.6285 \text{ g Cl}}{1.395 \text{ g sample}} \times 100\% = 45.05\% \text{ Cl}$$

The percentages add up to about 100%, and we can conclude that there are no other elements in the compound except C, H, and Cl.

PART 2

Assembling the Tools: If we assume a 100 g sample, we can easily convert the percentages to mass units and use the stoichiometric conversions from mass to moles for the elements in the compound. Last, we use the procedures in Section 3.4 to determine the empirical formula.

Solution: We will convert the percentages to mass and then to moles for the three elements,

$$47.10 \text{ g C} \times \frac{1 \text{ mol C}}{12.011 \text{ g C}} = 3.921 \text{ mol C}$$

$$7.900 \text{ g H} \times \frac{1 \text{ mol H}}{1.008 \text{ g H}} = 7.837 \text{ mol H}$$

$$45.05 \text{ g Cl} \times \frac{1 \text{ mol Cl}}{35.45 \text{ g Cl}} = 1.271 \text{ mol Cl}$$

We divide each of the quantities above by the smallest value, 1.271, to get

$$3.08 \text{ mol C}, \quad 6.17 \text{ mol H}, \quad \text{and} \quad 1.00 \text{ mol Cl}$$

These numbers are close enough to integers to write

$$3 \text{ mol C}, \quad 6 \text{ mol H}, \quad \text{and} \quad 1 \text{ mol Cl}$$

The empirical formula is C_3H_6Cl.

PART 3

Assembling the Tools: We need to determine the molar mass from the freezing point depression, Equation 12.15. The freezing point of pure benzene from Table 12.4 is 5.45 °C and the freezing point depression constant is 5.07 °C m^{-1}. We can calculate the molality, then the molar mass.

Solution: The measured freezing point is 3.33 °C and ΔT_f is 2.12 °C. Using Equation 12.15 we write

$$\Delta T_f = 2.12\ °C = (5.07\ °C\ m^{-1})\ m$$

$$m = 0.418\ m$$

The molality is 0.418.

We can expand the definition of molality as

$$\text{molality} = \frac{\text{moles of solute}}{\text{kg solvent}} = \frac{\text{g solute/molar mass solute}}{\text{kg solvent}}$$

$$0.418\ m = \frac{5.41\ \text{g solute/molar mass of solute}}{0.085\ \text{kg solvent}}$$

$$\text{molar mass of solute} = \frac{5.41\ \text{g solute}}{0.085\ \text{kg solvent} \times 0.418\ m} = 152.3\ \text{g mol}^{-1}$$

PART 4

Assembling the Tools: Now we need to determine the molecular formula using the method described in Section 3.4 from the molar mass of the solute and the molar mass of the empirical formula.

Solution: We divide the molar mass by the empirical formula mass, to obtain the multiplication factor for the formula's subscripts,

$$\frac{152.3}{77.5} = 1.96 = 2$$

Therefore, we multiply all of the subscripts in the empirical formula by 2 and get the formula $C_6H_{12}Cl_2$.

Are the Answers Reasonable? In a four-part problem, if we end up with a simple, whole-number ratio after many calculations, it gives us confidence that the preceding parts were correct. In Part 1, the percentage of hydrogen is the smallest, as we expect for the lightest element. In Part 2 we get numbers that are very close to whole-number ratios for the formula. In Part 3 we find a molar mass that is neither extremely small nor large, as expected for a freezing point depression experiment. Each part seems to have reasonable results, and we are confident that the final result is correct.

Ion Pairing

Experiments show that neither the 1.00 m NaCl nor the 1.00 m $(NH_4)_2SO_4$ solution described earlier in this section freeze at temperatures quite as low as calculated. Our assumption that an electrolyte separates *completely* into its ions is incorrect. Some oppositely charged ions exist as very closely associated pairs, called **ion pairs**, which behave as single "molecules" (Figure 12.23). Clusters larger than two ions probably also exist. The formation of ion pairs and clusters makes the actual *particle* concentration in a 1.00 m NaCl solution somewhat less than 2.00 m. As a result, the freezing point depression of 1.00 m NaCl is not quite as large as calculated on the basis of 100% dissociation.

As solutions of electrolytes are made more and more *dilute*, the observed and calculated freezing points come closer and closer together. At greater dilutions, the **association** (coming together) of ions is less and less a complication because the ions collide less frequently, resulting in relatively fewer ion pairs.

■ "Association" is the opposite of dissociation. It is the coming together of particles to form larger particles.

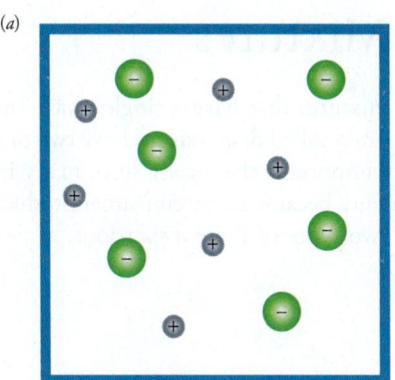

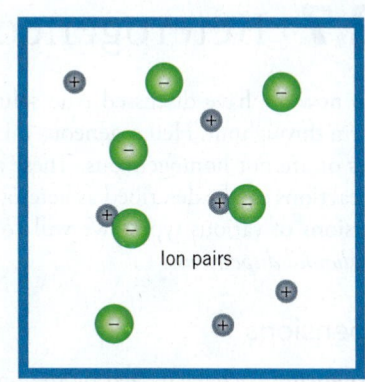

Ion pairs

Figure 12.23 | Ion pairs in a solution of NaCl. (*a*) If NaCl were completely dissociated in water, the Na⁺ and Cl⁻ ions would be totally independent. (*b*) Interionic attractions cause some ions to group together as ion pairs, which reduces the total number of independent particles in the solution. In this diagram, two ion pairs are shown. (Water molecules are not shown for clarity.)

Chemists compare the degrees of dissociation of electrolytes at different dilutions by a quantity called the **van't Hoff factor**, *i*. It is the ratio of the observed freezing point depression to the value calculated on the assumption that the solute dissolves as a nonelectrolyte.

$$i = \frac{(\Delta T_f)_{\text{measured}}}{(\Delta T_f)_{\text{calculated as a nonelectrolyte}}}$$

The predicted van't Hoff factor, *i*, is 2 for NaCl, KCl, and MgSO₄, which break up into two ions on 100% dissociation. For K₂SO₄, the theoretical value of *i* is 3 because one K₂SO₄ unit gives 3 ions. The actual van't Hoff factors for several electrolytes at different dilutions are given in Table 12.5. Notice that with decreasing concentration (that is, at higher dilutions) the experimental van't Hoff factors agree better with their corresponding hypothetical van't Hoff factors.

The increase in the percentage dissociation that comes with greater dilution is not the same for all salts. In going from concentrations of 0.1 to 0.001 *m*, the increase in percentage dissociation of KCl, as measured by the change in *i*, is only about 7%. However, for K₂SO₄ the increase for the same dilution is about 22%, a difference caused by the anion SO₄²⁻. It has twice the charge as Cl⁻ in KCl, and so the SO₄²⁻ ion attracts K⁺ more strongly than can Cl⁻. Hence, letting an ion of 2− charge and an ion of 1+ charge get farther apart by dilution has a greater effect on their acting independently than giving ions of 1− and 1+ charge more room.

TABLE 12.5 Van't Hoff Factors versus Concentration

	Van't Hoff Factor, *i*			
	Molal Concentration (mol salt/kg water)			Value of *i* if 100% Dissociation Occurred
Salt	0.1	0.01	0.001	
NaCl	1.87	1.94	1.97	2.00
KCl	1.85	1.94	1.98	2.00
K₂SO₄	2.32	2.70	2.84	3.00
MgSO₄	1.21	1.53	1.82	2.00

Molecular Association

Some molecular solutes produce *weaker* colligative effects than their molal concentrations would lead us to predict. These unexpectedly weak colligative properties are often evidence that solute molecules are clustering or associating in solution. For example, when dissolved in benzene, benzoic acid molecules associate as **dimers**. They are held together by hydrogen bonds, indicated by the dotted lines in the following equation.

■ *di*- signifies two, so a dimer is the result of the combination of two single molecules.

■ C₆H₅—, called the phenyl group in organic chemistry, has the structure

$$2C_6H_5-\overset{\overset{\displaystyle O}{\|}}{C}-O-H \;\rightleftharpoons\; C_6H_5-C\overset{O\cdots H-O}{\underset{O-H\cdots O}{<}}C-C_6H_5$$

benzoic acid   benzoic acid dimer

Because of association, the depression of the freezing point of a 1.00 *m* solution of benzoic acid in benzene is only about one-half the calculated value. By forming a dimer, benzoic acid has an effective molecular mass that is twice as much as normally calculated. The larger effective molecular mass reduces the molal concentration of particles by half, and the effect on the freezing point depression is reduced by one-half.

12.7 | Heterogeneous Mixtures

Up to now we have discussed true solutions, mixtures that have a single phase and are uniform throughout. Heterogeneous mixtures, often called **dispersions**, have two or more phases or are not homogeneous. These are very important chemically since many industrial reactions can be described as heterogeneous and because many consumer products are dispersions of various types. We will consider two types of these dispersions, *suspensions* and *colloidal dispersions*.

Suspensions

A **suspension** is characterized by the presence of relatively large particles mixed in a solvent. To prepare such a suspension, a substance can be mixed uniformly in a solvent, or dispersing medium, by mechanical means such as mixing or shaking. However, once the mixing process is stopped, the particles in a suspension will start to coalesce, due to attractive forces, into large particles that will separate into two phases. In other suspensions, such as fine sand dispersed in water, the two phases separate due to gravitational forces.

You may be most familiar with suspensions as common drugs. Solid magnesium hydroxide is suspended in water in a preparation called milk of magnesia. Many antibiotics are aqueous suspensions that you must "shake well before dispensing." Some salad dressings fall into the same category and will separate quickly into oil and vinegar layers soon after shaking.

Suspensions involving the gas phase are also familiar. Soot from fires is one example.[7] The result often is a layer of black grime that covers everything outdoors. Dust that settles out of the air and fine droplets of water within a fog are examples of a solid and a liquid suspended in the gas phase, respectively. Air bubbles within whipped cream or whipped egg whites are examples of a suspension of a gas in a liquid that will eventually separate, if not eaten first. In whipped cream and egg whites, the suspensions are relatively long lasting; the high viscosity hinders the ability of the gas bubbles to coalesce and form separate phases.

Colloidal Dispersions

When the particles in a suspension are very small, Brownian motion (see Section 2.7) keeps them from settling. These mixtures that do not settle, or separate into different phases, are called **colloidal dispersions**. Generally a colloid particle must have at least one dimension in the range of 1 to 1000 nm. Some of the particles of water in fog and soot in smoke, mentioned previously, qualify as true colloidal substances and do not separate. A small amount of starch dissolved in water is an example of a colloid in an aqueous system, also called a sol. Table 12.6 lists examples of colloidal mixtures and classifies them based on the dispersed phase (i.e., the colloid) and the dispersing phase (the solvent). In many mixtures the sizes of colloidal particles makes them invisible to the naked eye. Although a colloidal mixture may look very much like a true solution, the colloidal particles, with dimensions similar to the wavelengths of visible light, will scatter light. This scattering is called the **Tyndall effect** and can be observed by shining a narrow beam of light through the colloidal mixture. A dramatic illustration of the Tyndall effect is shown in Figure 12.24, where red laser light shines through a series of test tubes.

Figure 12.24 shows us some colloidal mixtures that look like true solutions. However, other colloidal dispersions, called **emulsions,** may look more like milk or whipped cream. In an emulsion the dispersing phase is a liquid and, often, the concentration of colloidal particles is very high. This means that colloidal particles may have a greater probability of colliding and coalescing, eventually into two phases.

■ John Tyndall (1820–1893), a prominent physicist, succeeded Michael Faraday as the Fullerian Professor of Chemistry at the Royal Institution of Great Britain.

---

[7] Coal- and oil-fired power plants used to be the major sources of soot and other particulates. Modern electrostatic precipitators and scrubbers remove a large majority of these particulates.

TABLE 12.6	Colloidal Systems		
Type	Dispersed Phase[a]	Dispersing Medium	Common Examples
Foam	Gas	Liquid	Soapsuds, whipped cream
Solid foam	Gas	Solid	Pumice, marshmallow
Liquid aerosol	Liquid	Gas	Mist, fog, cloud, air pollutants
Emulsion	Liquid	Liquid	Cream, mayonnaise, milk
Solid emulsion	Liquid	Solid	Butter, cheese
Smoke	Solid	Gas	Dust, particulates, in smog, aerogels
Sol (Colloidal dispersion)	Solid	Liquid	Starch in water, paint, jellies[b]
Solid sol	Solid	Solid	Alloys, pearls, opals

[a]Colloidal particles are in the dispersed phase.

[b]Semisolid or semirigid sols, such as gelatins and jellies, are called gels.

Chemists have devised a variety of strategies to slow or even stop the coalescing of colloidal particles in emulsions. In recent chapters we have learned that we can reduce the frequency of collisions in the liquid phase by (a) lowering the temperature, (b) raising the viscosity, or (c) decreasing attractive forces and increasing repulsive forces. Let's look at an example of each.[8] Old-time ice cream is an emulsion of milk, eggs, cream, sugar, and flavorings. In the case of ice cream, it is kept frozen and the colloidal particles rarely collide. In the whipped cream example, the usually fluid cream has colloidal-size air bubbles incorporated into it, resulting in a very viscous product. Once again, the limited mobility of the colloidal particles keeps them from coalescing. Figure 12.25 illustrates the microscopic view of two types of emulsion: mayonnaise, a liquid–liquid emulsion, and margarine, a liquid–solid emulsion.

The major method for stabilizing most commercial emulsions is to incorporate into one or both phases an electrical charge that keeps the colloidal particles apart. Surfactants such as sodium stearate are long molecules with a charged end and an uncharged hydrocarbon end. The charged end will dissolve in water and is said to be **hydrophilic**, while the hydrocarbon end dissolves more readily in oils and is called **hydrophobic.** When sodium stearate is added to an oil–water emulsion the hydrophobic end dissolves in the oil while the hydrophilic end extends into the aqueous phase. The result is often a spherical, negatively charged particle called a **micelle.** The negative charge of one micelle repels the negative charges of other micelles. The result is that micelles will not coalesce into separate oil and water phases. Proteins can also adsorb to oil particles and provide a charged surface. Figure 12.26 illustrates how adsorbed layers can produce charged colloidal particles. An emulsion stabilized in this fashion can be separated into two phases by adding an acid or base to neutralize the charged groups. This is called "breaking" an emulsion.

In crystals of small inorganic molecules a stabilizing charged layer can be established by an excess of one of the crystal's ions. For example, if lead(II) ions are precipitated with a large excess of chloride ions, the excess chloride ions will adsorb to the surface of the lead(II) chloride crystals in what is called the primary adsorbed layer. Since the primary adsorbed layer

Figure 12.24 | The Tyndall effect. Colloidal solutions may appear as true solutions. The first sample is a true solution and the light of the red laser beam is not scattered. The particles of the second sample scatter the light because the sample is a colloidal dispersion.

Charles D. Winters/ScienceSource

---

[8] Each of these examples actually uses a combination of all three strategies, but we focus on the main concept for each one.

Figure 12.25 | **Two common emulsions.** (*a*) The dispersion of oil in water occurs when mayonnaise is prepared. (*b*) Low-calorie and low-fat butter is often made by dispersing water in the solid butterfat.

(*a*)

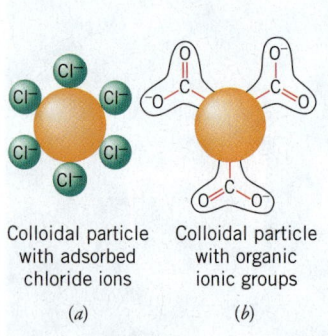

Colloidal particle with adsorbed chloride ions

(*a*)

Colloidal particle with organic ionic groups

(*b*)

Figure 12.26 Stabilizing colloidal dispersions. Colloidal particles that may tend to coalesce can be stabilized by acquiring an electrical charge. (*a*) Inorganic precipitates can have an adsorbed layer of ions that prevents them from coalescing into particles that can be filtered. (*b*) Colloidal particles can have charged groups that give it an overall electrical charge, or charged particles can be adsorbed to the surface of a colloidal particle with the same effect.

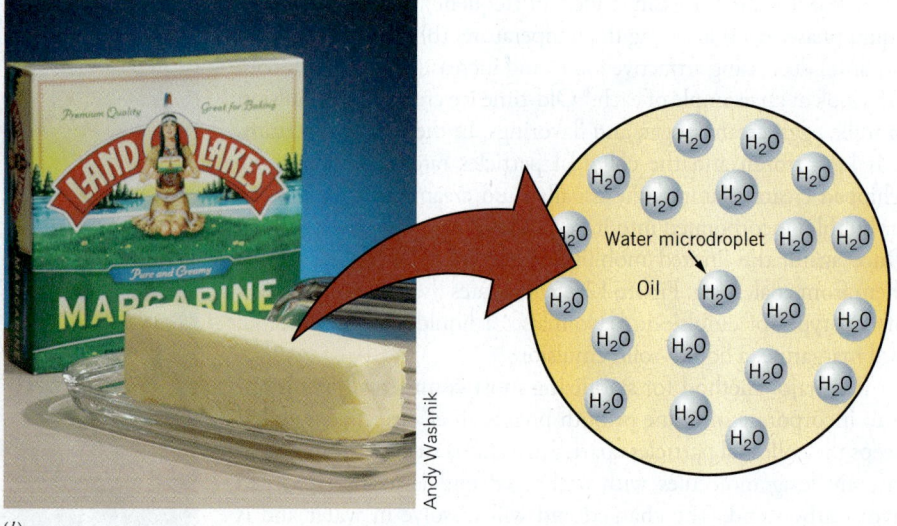

(*b*)

consists of chloride ions that have a negative charge, the tiny colloidal particles will not coalesce into larger crystals that can be collected by filtration. Figure 12.26 can also be used to visualize this system. Although stabilizing an emulsion is often a desired result in commercial preparations, here we see that it may also be a problem. The effect of the adsorbed layer can be minimized by having only a small excess of chloride ions during precipitation.

Modern nanoscience is very interested in nanometer-size particles that have unusual properties. Colloidal gold is a nanoparticle that may be useful for drug delivery to tumors. Interestingly, colloidal gold was first used by medieval artists who transcribed elegantly "illuminated" bibles, where this nanoparticle exhibits many different colors. Colloidal, magnetic, iron particles have many interesting properties, as do the fullerenes, substances related to the C_{60} allotrope of carbon that we discuss in Section 9.10.

Practice Exercise 12.31

Classify each of these as a colloid dispersion or a suspension: (**a**) milk, (**b**) jelly, (**c**) flour in water (**d**) air in cream. (*Hint:* What keeps particles suspended?)

Practice Exercise 12.32

Identify the solvent and the suspended particles of these colloids: (**a**) mayonnaise, (**b**) cheese, (**c**) marshmallows, (**d**) smog, (**e**) soapsuds.

Summary

Organized by Learning Objective

Describe how intermolecular forces affect the solution process for solids, liquids, and gases

Polar or ionic solutes generally dissolve well in polar solvents such as water. Nonpolar, molecular solutes dissolve well in nonpolar solvents. These observations are behind the **"like dissolves like" rule**. Nature's driving force for establishing the more statistically probable mixed state and intermolecular attractions are the two major factors in the formation of a solution. When both solute and solvent are nonpolar, nature's tendency toward the more probable mixed state dominates because intermolecular attractive forces are weak. Ion–dipole attractions and the **solvation** or **hydration** of dissolved species are major factors in forming solutions of ionic or polar solutes in something polar, like water.

Detail the energy changes that take place when two substances form a solution

The **molar enthalpy of solution** of a solid dissolved in a liquid, also called the **heat of solution,** is calculated as the **lattice energy** subtracted from the **solvation energy** (or, when water is the solvent, the **hydration energy**). The energy required to separate the ions or molecules into the gas phase is the same magnitude, but opposite in sign, to the lattice energy. The solvation energy corresponds to the potential energy lowering that occurs in the system from the subsequent attractions of these particles to the solvent molecules as intermingling occurs.

When liquids dissolve in liquids, an **ideal solution** forms if the potential energy increase needed to separate the molecules equals the energy lowering as the separated molecules come together.

When a gas dissolves in a gas, the energy cost to separate the particles is virtually zero, and they remain separated in the gas mixture, so the net enthalpy of solution is very small. When a gas dissolves in an organic solvent, the solution process is often endothermic. When a gas dissolves in water, the solution process is often exothermic.

Explain how temperature affects the solubility of a solute in a solvent

If the formation of a solution is endothermic, increasing the temperature will force more solute to dissolve, while for exothermic reactions, increasing the temperature will force the solute to precipitate.

Using Henry's law, perform calculations involving the pressure of a gas in solution

At pressures not too much different from atmospheric pressure, the solubility of a gas in a liquid is directly proportional to the partial pressure of the gas above the liquid—**Henry's law**.

Define solution concentrations: percent concentration, molal concentration, molar concentration, mole fraction, and mole percent, and convert between the different units

To study and use colligative properties, a solution's concentration ideally is expressed either in mole fractions or **molal concentration,** or **molality**. The **molality** of a solution, m, is the ratio of the number of moles of solute to the kilograms of *solvent* (not solution, but solvent). **Mass fractions** and **mass percents** are concentration expressions used often for chemicals when direct information about number of moles of solutes is not considered important. Mole fraction, molality, and mass percent are temperature-independent concentration expressions.

Describe the colligative properties, Raoult's law, freezing point depression, boiling point depression, and osmosis of a solution and use them in calculations for solutions with molecular and ionic solutes

Colligative properties are those that depend on the ratio of the particles of the solute to the molecules of the solvent. These properties include the **lowering of the vapor pressure**, the **depression of a freezing point**, the **elevation of a boiling point**, and the **osmotic pressure** of a solution.

According to **Raoult's law**, the vapor pressure of a solution of a nonvolatile (molecular) solute is the vapor pressure of the pure solvent times the mole fraction of the solvent. An alternative expression of this law is that the *change* in vapor pressure caused by the solute equals the mole fraction of the solute times the pure solvent's vapor pressure. When the components of a solution are volatile liquids, then Raoult's law calculations find the partial pressures of the vapors of the individual liquids. The sum of the partial pressures equals the total vapor pressure of the solution.

An ideal solution is one that would obey Raoult's law exactly and in which all attractions between molecules are equal. An ideal solution has $\Delta H_{soln} = 0$. Heats of solution of nonideal mixtures may be exothermic or endothermic and these correspond to vapor pressures that are less or greater than expected from Raoult's law, respectively.

In proportion to its molal concentration, a solute causes a **freezing point depression** and a **boiling point elevation**. The proportionality constants are the **molal freezing point depression constant** and the **molal boiling point elevation constant**, and they differ from solvent to solvent.

When a solution is separated from the pure solvent (or from a less concentrated solution) by an osmotic membrane, **osmosis** occurs, which is the net flow of solvent into the more concentrated solution. The back pressure required to prevent osmosis is called the solution's **osmotic pressure**.

When the membrane separating two different solutions is a dialyzing membrane, it permits **dialysis**, the passage not only of solvent molecules but also of small solute molecules or ions.

Because an electrolyte releases more ions in solution than indicated by the molal (or molar) concentration, a solution of an electrolyte has more pronounced colligative properties than a solution of a molecular compound at the same concentration. The dissociation of a strong electrolyte approaches 100% in very dilute solutions, particularly when both ions are singly charged. Weak electrolytes provide solutions more like those of nonelectrolytes.

The ratio of the value of a particular colligative property (e.g., ΔT_f) to the value expected if there were no dissociation is the **van't Hoff factor, i.** A solute whose formula unit breaks into two ions, like NaCl or $MgSO_4$, would have a van't Hoff factor of 2 if it were 100% dissociated. Observed van't Hoff factors are less than the theoretical values because of the formation of **ion pairs**, brought about by **association** of ions in solution.

Illustrate the characteristics of heterogeneous mixtures such as suspensions and colloids

Suspensions are mixtures where the particles are so large that they begin precipitating or coalescing as soon as the mixing process is stopped. The result is the formation of two or more phases. **Colloidal mixtures** are composed of particles with the largest dimension being 1000 nm. Colloidal mixtures do not settle and can be identified using the **Tyndall effect**. A special colloidal mixture called an **emulsion** is a mixture that can be stabilized by electrical charges or **hydrophobic** and **hydrophilic repulsions**. These factors keep the phases from coalescing. **Hydrophobic** and **hydrophilic** properties of surfactants often create **micelles** that stabilize emulsions.

TOOLS

Tools for Problem Solving
The following tools were introduced in this chapter. Study them carefully so you can select the appropriate tool when needed.

"Like dissolves like" rule (Section 12.1)
This rule uses the polarity and strengths of attractive forces, along with chemical composition and structure, to predict whether two substances can form a solution.

Henry's law (Section 12.4)

$$C_{gas} = k_H P_{gas} \quad \text{or} \quad \frac{C_1}{P_1} = \frac{C_2}{P_2}$$

Mass fraction; percentage by mass (Section 12.5)

$$\text{mass fraction} = \frac{\text{mass of solute}}{\text{mass of solution}}$$

$$\text{percentage by mass} = \%(\text{w/w}) = \frac{\text{mass of solute}}{\text{mass of solution}} \times 100\%$$

Molal concentration (Section 12.5)

$$\text{molality} = \frac{\text{moles of solute}}{\text{mass of solution}}$$

Raoult's law (Section 12.6)

$$P_{solution} = X_{solvent} P^\circ_{solvent}$$

Raoult's law for two volatile solvents (Section 12.6)

$$P_{solution} = X_A P^\circ_A + X_B P^\circ_B$$

Freezing point depression and boiling point elevation (Section 12.6)

$$\text{Freezing point depression:} \quad \Delta T_f = K_f m$$

$$\text{Boiling point elevation:} \quad \Delta T_b = K_b m$$

Osmotic pressure (Section 12.6)

$$\Pi V = nRT \quad \text{or} \quad \Pi = MRT$$

WileyPLUS = *WileyPLUS*, an online teaching and learning solution. *Note to instructors*: Many of the end-of-chapter problems are available for assignment via the WileyPLUS system. **www.wileyplus.com.** Review Problems are presented in pairs separated by blue rules. Answers to problems whose numbers appear in blue are given in Appendix B. More challenging problems are marked with an asterisk ★.

Review Questions

Intermolecular Forces and the Formation of Solutions

12.1 Why do two gases spontaneously mix when they are brought into contact?

12.2 When substances form liquid solutions, what two factors are involved in determining the solubility of the solute in the solvent?

12.3 Silver chloride is not soluble in water. What are the relative strengths of the intermolecular forces between Ag^+ and Cl^- ions and water molecules as compared to the intermolecular forces between water molecules and the forces that hold the silver chloride as a solid?

12.4 Methanol, CH_3OH, and water are miscible in all proportions. What does this mean? Explain how the OH unit in methanol contributes to this.

12.5 Hexane (C_6H_{12}) and water are immiscible. What does this mean? Explain why they are immiscible in terms of structural features of their molecules and the forces of attraction between them.

12.6 Explain how ion–dipole forces help to bring potassium chloride into solution in water.

12.7 Explain why potassium chloride will not dissolve in carbon tetrachloride, CCl_4.

12.8 Water and dichloromethane, CH_2Cl_2, are immiscible; however, when enough methanol is added, the three are soluble in each other. What properties of methanol allows the three liquids to form a homogeneous mixture?

Heats of Solution

12.9 The value of ΔH_{soln} for a soluble compound is, say, $+26$ kJ mol^{-1}, and a nearly saturated solution is prepared in an insulated container (e.g., a coffee cup calorimeter). Will the system's temperature increase or decrease as the solute dissolves?

12.10 Referring to Question 12.9, which value for this compound would be numerically larger, its lattice energy or its hydration energy?

12.11 Which would be expected to have the larger hydration energy, Al^{3+} or Li^+? Why? (Both ions are about the same size.)

12.12 Suggest a reason why the value of ΔH_{soln} for a gas such as CO_2, dissolving in water, is negative.

12.13 The value of ΔH_{soln} for the formation of an acetone–water solution is negative. Explain this in general terms using intermolecular forces of attraction.

12.14 The value of ΔH_{soln} for the formation of an ethanol–hexane solution is positive. Explain this in general terms that involve intermolecular forces of attraction.

12.15 When a certain solid dissolves in water, the solution becomes cool. Is ΔH_{soln} for this solute positive or negative? Explain your reasoning. Is the solubility of this substance likely to increase or decrease with increasing temperature? Explain your answer using Le Châtelier's principle.

12.16 If the value of ΔH_{soln} for the formation of a mixture of two liquids A and B is zero, what does this imply about the relative strengths of A—A, B—B, and A—B intermolecular attractions?

12.17 What kinds of data would have to be obtained to find out whether a solution of two miscible liquids is almost exactly an ideal solution?

Solubility as a Function of Temperature

12.18 If a saturated solution of NH_4NO_3 at 75 °C is cooled to 15 °C, how many grams of solute will separate if the quantity of the solvent is 125 g? (Use the data in Figure 12.10.)

12.19 Anglers know that on hot summer days, the largest fish will be found in deep sinks in lake bottoms. Use the temperature dependence of the density of water and the temperature dependence of the oxygen solubility in water to explain why.

Henry's Law

12.20 What is Henry's law?

12.21 Mountain streams often contain fewer living things than equivalent streams at sea level. Give one reason why this might be true in terms of oxygen solubilities at different pressures.

12.22 Why is ammonia so much more soluble in water than nitrogen? Would you expect hydrogen chloride gas to have a high or low solubility in water? Explain your answers to both questions.

12.23 Why does a bottled carbonated beverage fizz when you take the cap off?

12.24 Some chemical reactions need to be carried out in an atmosphere without oxygen present. One way to remove oxygen is to freeze the solvent, and remove the gas above the solid, then thaw the solvent—the freeze–thaw method. This is repeated three times. Why does this remove the oxygen?

Concentration Units

12.25 Write the definition for each of the following concentration units: **(a)** mole fraction, **(b)** mole percent, **(c)** molality, **(d)** percent by mass. What are the maximum possible values for the units in **(a)**, **(b)**, and **(d)**?

12.26 How does the molality of a solution vary with increasing temperature? How does the molarity of a solution vary with increasing temperature?

12.27 Suppose a 1.0 *m* solution of a solute is made using a solvent with a density of 1.15 g/mL. Will the molarity of this solution be numerically larger or smaller than 1.0? Justify your conclusion mathematically.

Colligative Properties

12.28 What specific fact about a physical property of a solution must be true to call it a colligative property?

12.29 What is Raoult's law?

12.30 Why does a nonvolatile solute decrease the vapor pressure of a solvent?

12.31 When octane is mixed with methanol, the vapor pressure of the octane over the solution is higher than what we would calculate using Raoult's law. Why? Explain the discrepancy in terms of intermolecular attractions.

12.32 Will a solution of pentane and hexane have an ideal Raoult's law vapor pressure curve? Explain your answer in terms of intermolecular attractions.

12.33 Explain why a nonvolatile solute dissolved in water makes the system have **(a)** a higher boiling point than water, and **(b)** a lower freezing point than water.

12.34 Why do we call dialyzing and osmotic membranes *semipermeable*? What is the opposite of *permeable*?

12.35 What is the key difference between dialyzing and osmotic membranes?

12.36 At a molecular level, explain why, in osmosis, there is a net migration of solvent from the side of the membrane less concentrated in solute to the side more concentrated in solute.

12.37 Two glucose solutions of unequal molarity are separated by an osmotic membrane. Which solution will lose water, the one with the higher molarity or the one with the lower molarity?

12.38 Which aqueous solution has the higher osmotic pressure, 10% glucose, $C_6H_{12}O_6$, or 10% sucrose, $C_{12}H_{22}O_{11}$? (Both are molecular compounds.)

12.39 When a solid is *associated* in a solution, what does this mean? What difference does it make to expected colligative properties?

12.40 What is the difference between a *hypertonic* solution and a *hypotonic* solution?

12.41 Why are colligative properties of solutions of ionic compounds usually more pronounced than those of solutions of molecular compounds of the same molalities?

12.42 What is the van't Hoff factor? What is its expected value for all nondissociating molecular solutes? If its measured value is slightly larger than 1.0, what does this suggest about the solute? What is suggested by a van't Hoff factor of approximately 0.5?

12.43 Which aqueous solution, if either, is likely to have the higher boiling point, 0.50 *m* NaI or 0.50 *m* Na_2CO_3?

Heterogeneous Mixtures

12.44 Determine whether each of the following is likely to be a true solution, a suspension, or a colloidal mixture: **(a)** orange juice, **(b)** apple juice, **(c)** perfume, **(d)** tea, **(e)** tea with lemon juice, **(f)** canned gravy. If you have a laser pointer, how can you confirm your answers?

12.45 Can sodium stearate stabilize a water-in-oil emulsion? Explain your conclusion.

12.46 Addition of a neutral salt to a colloidal precipitate such as the $BaSO_4(s)$ described in the text allows the crystals to grow in size. Develop a hypothesis on why this occurs.

12.47 What is the Tyndall effect?

12.48 What is a micelle, and why does it form?

| Review Problems

Heat of Solution

12.49 For an ionic compound dissolving in water, $\Delta H_{soln} = -56$ kJ mol^{-1} and the hydration energy is -894 kJ mol^{-1}. Estimate the lattice energy of the ionic compound.

12.50 Consider the formation of a solution of aqueous potassium chloride. Write the thermochemical equations for **(a)** the conversion of solid KCl into its gaseous ions and **(b)** the subsequent formation of the solution by hydration of the ions. The lattice energy of KCl is -715 kJ mol^{-1}, and the hydration energy of the ions is -686 kJ mol^{-1}. Calculate the enthalpy of solution of KCl in kJ mol^{-1}.

Henry's Law

12.51 The solubility of methane, the chief component of natural gas, in water at 20.0 °C and 1.0 atm pressure is 0.025 g L^{-1}. What is its solubility in water at 1.5 atm and 20.0 °C?

12.52 If the solubility of a gas in water is 0.010 g L^{-1} at 25 °C with the partial pressure of the gas over the solution at 1.0 atm, predict the solubility of the gas at the same temperature but at double the pressure.

12.53 At 740 torr and 20.0 °C, nitrogen has a solubility in water of 0.018 g L^{-1}. At 620 torr and 20.0 °C, its solubility is 0.015 g L^{-1}. Show that nitrogen obeys Henry's law.

12.54 Hydrogen gas has a solubility in water of 0.00157 g L^{-1} under 1.00 atm of H_2 pressure at 25 °C. At 0.85 atm and 25 °C, its solubility is 0.00133 g L^{-1}. Does hydrogen gas obey Henry's law?

12.55 If 100.0 mL of water is shaken with oxygen gas at 1.0 atm it will dissolve 0.0039 g O_2. Estimate the Henry's law constant for oxygen gas in water.

12.56 Helium gas can be used to displace other gases from a solvent by bubbling He through the solvent, a process called sparging, and leaving an atmosphere of helium above the solvent. At 760 torr of He, the concentration of He in water is 0.00148 g L^{-1} at 298 K. What is Henry's law constant for He at 298 K?

Concentration Units

12.57 Muriatic acid is the commercial name for hydrochloric acid that can be purchased from hardware stores as a solution that is 30%(w/w) HCl. What mass of this solution contains 7.5 g of HCl?

12.58 Hydrofluoric acid dissolves glass and must be stored in plastic containers. What mass of a 45%(w/w) solution of HF(*aq*) contains 1.5 g of HF?

12.59 What mass of a 0.853 molal solution of iron(III) nitrate is needed to obtain (**a**) 0.0200 moles of iron(III) nitrate, (**b**) 0.0500 moles of Fe^{3+} ions, (**c**) 0.00300 moles of nitrate ions?

12.60 In order to conduct three experiments that required different amounts of chloride ions, what mass of a 0.150 *m* NaCl solution is needed to obtain (**a**) 0.00100 mol Cl^-, (**b**) 0.00500 mol Cl, (**c**) 0.0200 mol Cl^-?

****12.61** What is the molality of NaCl in a solution that is 3.000 *M* NaCl, with a density of 1.07 g mL^{-1}?

****12.62** A solution of acetic acid, CH_3COOH, has a concentration of 0.143 *M* and a density of 1.00 g mL^{-1}. What is the molality of this solution?

12.63 The mole fraction of neon in the air is about 2.6×10^{-5}. If the average molar mass of air is 28.96 g/mol, what is the concentration of neon in air in ppm?

12.64 Botulinum toxin is one of the most acutely toxic substances known, with a lethal dose at about 140 ng for a 150 lb person. What would this concentration be in ppb?

12.65 A solution of fructose, $C_6H_{12}O_6$, a sugar found in many fruits, is made by dissolving 24.0 g of fructose in 1.00 kg of water. For this solution, what are (**a**) the molal concentration, (**b**) the mole fraction, (**c**) the mass percent, and (**d**) the molarity of fructose if the density of the solution is 1.0078 g/mL?

12.66 If you dissolved 11.5 g of NaCl in 1.00 kg of water, (**a**) what would be its molal concentration? (**b**) What are the mass percent NaCl and the mole percent NaCl in the solution? The volume of this solution is virtually identical to the original volume of the 1.00 kg of water. (**c**) What is the molar concentration of NaCl in this solution? (**d**) What would have to be true about any solvent for one of its dilute solutions to have essentially the same molar and molal concentrations?

12.67 A solution of ethanol, CH_3CH_2OH, in water has a concentration of 1.25 *m*. Calculate the mass percent of ethanol.

12.68 A solution of NaCl in water has a concentration of 19.5%. Calculate the molality of the solution.

12.69 A solution of NH_3 in water has a concentration of 7.50% by mass. Calculate the mole percent NH_3 in the solution. What is the molal concentration of the NH_3?

12.70 An aqueous solution of isopropyl alcohol, C_3H_8O, rubbing alcohol, has a mole fraction of alcohol equal to 0.250. What is the percent by mass of alcohol in the solution? What is the molality of the alcohol?

12.71 Sodium nitrate, $NaNO_3$, is sometimes added to tobacco to improve its burning characteristics. An aqueous solution of $NaNO_3$ has a concentration of 0.363 *m*. Its density is 1.0185 g mL^{-1}. Calculate the molar concentration of $NaNO_3$ and the mass percent of $NaNO_3$ in the solution. What is the mole fraction of $NaNO_3$ in the solution?

12.72 In an aqueous solution of sulfuric acid, the concentration is 1.89 mol% of acid. The density of the solution is 1.0645 g mL^{-1}. Calculate the following: (**a**) the molal concentration of H_2SO_4, (**b**) the mass percent of the acid, and (**c**) the molarity of the solution.

Colligative Properties

12.73 At 25 °C, the vapor pressure of water is 23.8 torr. What is the vapor pressure of a solution prepared by dissolving 65.0 g of $C_6H_{12}O_6$ (a nonvolatile solute) in 150 g of water? (Assume the solution is ideal.)

12.74 The vapor pressure of water at 20 °C is 17.5 torr. A 35% solution of the nonvolatile solute ethylene glycol, $C_2H_4(OH)_2$, in water is prepared. Estimate the vapor pressure of the solution.

12.75 At 25 °C the vapor pressures of benzene (C_6H_6) and toluene (C_7H_8) are 93.4 and 26.9 torr, respectively. A solution made by mixing 35.0 g of benzene and 65.0 g of toluene is prepared. What is the vapor pressure of this solution?

12.76 Pentane (C_5H_{12}) and heptane (C_7H_{16}) are two hydrocarbon liquids present in gasoline. At 20.0 °C, the vapor pressure of pentane is 422 torr and the vapor pressure of heptane is 36.0 torr. What will be the total vapor pressure (in torr) of a solution prepared by mixing equal masses of the two liquids?

****12.77** Benzene and toluene help achieve good engine performance from lead-free gasoline. At 40 °C, the vapor

pressure of benzene is 184 torr and that of toluene is 58 torr. Suppose you wished to prepare a solution of these liquids that will have a total vapor pressure of 96 torr at 40 °C. What must be the mole percent concentrations of each in the solution?

*12.78 The vapor pressure of pure methanol, CH_3OH, at 33 °C is 164 torr. How many grams of the nonvolatile solute glycerol, $C_3H_5(OH)_3$, must be added to 105 g of methanol to obtain a solution with a vapor pressure of 145 torr?

*12.79 A solution containing 8.3 g of a nonvolatile, nondissociating substance dissolved in 1.00 mol of chloroform, $CHCl_3$, has a vapor pressure of 511 torr. The vapor pressure of pure $CHCl_3$ at the same temperature is 526 torr. Calculate (a) the mole fraction of the solute, (b) the number of moles of solute in the solution, and (c) the molecular mass of the solute.

*12.80 At 21.0 °C, a solution of 18.26 g of a nonvolatile, nonpolar compound in 33.25 g of ethyl bromide, C_2H_5Br, had a vapor pressure of 336.0 torr. The vapor pressure of pure ethyl bromide at this temperature is 400.0 torr. Assuming an ideal solution, what is the molecular mass of the compound?

12.81 How many grams of sucrose ($C_{12}H_{22}O_{11}$) are needed to lower the freezing point of 125 g of water by 3.00 °C?

12.82 To make sugar candy, a concentrated sucrose solution is boiled until the temperature reaches 272 °F. What are the molality and the mole fraction of sucrose in this mixture?

12.83 A solution of 12.00 g of an unknown nondissociating compound dissolved in 200.0 g of benzene freezes at 3.45 °C. Calculate the molecular mass of the unknown.

12.84 A solution of 14 g of a nonvolatile, nondissociating compound in 0.10 kg of benzene boils at 81.7 °C. Calculate the molecular mass of the unknown.

*12.85 What are the molecular mass and molecular formula of a nondissociating molecular compound whose empirical formula is C_4H_2N if 3.84 g of the compound in 0.500 kg of benzene gives a freezing point depression of 0.307 °C?

*12.86 Benzene reacts with hot concentrated nitric acid dissolved in sulfuric acid to give chiefly nitrobenzene, $C_6H_5NO_2$. A by-product is often obtained, which consists of 42.86% C, 2.40% H, and 16.67% N (by mass). The boiling point of a solution of 5.5 g of the by-product in 45 g of benzene was 1.84 °C higher than that of benzene. (a) Calculate the empirical formula of the by-product. (b) Calculate a molecular mass of the by-product and determine its molecular formula.

12.87 (a) Show that the following equation is true.

$$\text{Molar mass of solute} = \frac{(\text{grams of solute})RT}{\Pi V}$$

(b) An aqueous solution of a compound with a very high molecular mass was prepared in a concentration of 2.0 g L^{-1} at 25 °C. Its osmotic pressure was 0.021 torr. Calculate the molecular mass of the compound.

12.88 A saturated solution is made by dissolving 0.400 g of a polypeptide (a substance formed by joining together in a chainlike fashion a number of amino acids) in water to give 1.00 L of solution. The solution has an osmotic pressure of 3.74 torr at 27 °C. What is the approximate molecular mass of the polypeptide?

12.89 The vapor pressure of water at 20.0 °C is 17.5 torr. At that temperature, what would be the vapor pressure of a solution made by dissolving 23.0 g of NaCl in 0.100 kg of water? (Assume complete dissociation of the solute and an ideal solution.)

12.90 How many grams of $AlCl_3$ would have to be dissolved in 0.150 L of water to give a solution that has a vapor pressure of 38.7 torr at 35 °C? Assume complete dissociation of the solute and ideal solution behavior. (At 35 °C, the vapor pressure of pure water is 42.2 torr.)

12.91 What is the osmotic pressure, in torr, of a 2.0% solution of NaCl in water when the temperature of the solution is 25 °C?

12.92 Below are the concentrations of the most abundant ions in seawater.

Ion	Molality
Chloride	0.566
Sodium	0.486
Magnesium	0.055
Sulfate	0.029
Calcium	0.011
Potassium	0.011
Bicarbonate	0.002

Use these data to estimate the osmotic pressure of seawater at 25 °C in units of atm. What is the minimum pressure in atm needed to desalinate seawater by reverse osmosis?

12.93 What is the expected freezing point of a 0.20 m solution of $CaCl_2$? (Assume complete dissociation.)

12.94 The freezing point of a 0.10 m solution of mercury(I) nitrate is approximately −0.27 °C. Show that these data suggest that the formula of the mercury(I) ion is Hg_2^{2+}.

12.95 The van't Hoff factor for the solute in 0.100 m $NiSO_4$ is 1.19. What would this factor be if the solution behaved as if it were 100% dissociated?

12.96 What is the expected van't Hoff factor for K_2SO_4 in an aqueous solution, assuming 100% dissociation?

Additional Problems

12.97 Vodka from the European Union must be at least 37.5% alcohol by volume or 75 proof. If the density of water is 1.00 g mL^{-1} and the density of ethanol is 0.789 g mL^{-1}, what are the molarity, mole fraction, and mass percent of alcohol in vodka?

*__12.98__ The "bends" is a medical emergency caused by the formation of tiny bubbles in the blood vessels of divers who rise too quickly to the surface from a deep dive. At 37 °C (normal body temperature), the solubility of N$_2$ in water is 0.015 g L^{-1} when its pressure over the solution is 1 atm. Air is approximately 78 mol% N$_2$. How many moles of N$_2$ are dissolved per liter of blood (essentially an aqueous solution) when a diver inhales air at a pressure of 1 atm? How many moles of N$_2$ dissolve per liter of blood when the diver is submerged to a depth of approximately 100 ft, where the total pressure of the air being breathed is 4 atm? If the diver suddenly surfaces, how many milliliters of N$_2$ gas, in the form of tiny bubbles, are released into the bloodstream from each liter of blood (at 37 °C and 1 atm)?

12.99 In order for mosquitos to survive the cold winter, they produce glycerol, C$_3$H$_8$O$_3$. What molality and mass fraction of glycerol is needed to keep water from freezing at 0 °F?

12.100 The vapor pressure of a mixture of 0.400 kg of carbon tetrachloride and 43.3 g of an unknown compound is 137 torr at 30 °C. At the same temperature, the vapor pressure of pure carbon tetrachloride is 143 torr, while that of the pure unknown is 85 torr. What is the approximate molecular mass of the unknown? Will our estimate of the molecular mass be too high or too low if the heat of solution is exothermic?

*__12.101__ Ethylene glycol, C$_2$H$_6$O$_2$, is used in many antifreeze mixtures. Protection against freezing to as low as −45 °F is sought.

(a) How many moles of solute are needed per kilogram of water to ensure this protection?

(b) The density of ethylene glycol is 1.11 g mL^{-1}. To how many milliliters of solute does your answer to part (a) correspond?

(c) Calculate the number of quarts of ethylene glycol that should be mixed with each quart of water to get the desired protection.

12.102 What is the osmotic pressure in torr of a 0.010 M aqueous solution of a molecular compound at 25 °C?

*__12.103__ The osmotic pressure of a dilute solution of a slightly soluble *polymer* (a compound composed of large molecules formed by linking many smaller molecules together) in water was measured using the osmometer in Figure 12.22. The difference in the heights of the liquid levels was determined to be 1.26 cm at 25 °C. Assume the solution has a density of 1.00 g mL^{-1}. (a) What is the osmotic pressure of the solution in torr? (b) What is the molarity of the solution? (c) At what temperature would the solution be expected to freeze? (d) On the basis of the results of these calculations, explain why freezing point depression cannot be used to determine the molecular masses of compounds composed of very large molecules.

12.104 A solution of ethanol, C$_2$H$_5$OH, in water has a concentration of 4.613 mol L^{-1}. At 20.0 °C, its density is 0.9677 g mL^{-1}. Calculate the following: (a) the molality of the solution and (b) the mass percentage of the alcohol in the mixture.

12.105 Consider an aqueous 1.00 m solution of Na$_3$PO$_4$, a compound with useful detergent properties.

(a) Calculate the boiling point of the solution on the assumption that it does not ionize at all in solution.

(b) Do the same calculation, assuming that the van't Hoff factor for Na$_3$PO$_4$ reflects 100% dissociation into its ions.

(c) The 1.00 m solution boils at 100.183 °C at 1 atm. Calculate the van't Hoff factor for the solute in this solution.

*__12.106__ A 2.50 g sample of aluminum chloride and sodium sulfate has a freezing point depression of 2.65 °C. Assuming complete dissociation of all ions, what are the molarities of aluminum chloride and sodium sulfate?

*__12.107__ A 0.118 m solution of LiCl has a freezing point of −0.415 °C. What is the van't Hoff factor for this solute at this concentration? What is the approximate osmotic pressure of a 0.118 m solution of LiCl at 10 °C? Express the answer in torr.

Multi-Concept Problems

*__12.108__ A sample containing only iron(II) nitrate and potassium chloride is dissolved in water and the boiling point of the mixture is 104.6 °C. That same solution is reacted with 36.3 mL of 0.220 M K$_2$Cr$_2$O$_7$ solution that completely converts the iron(II) to iron(III). What are the molarities of iron(II) nitrate and potassium chloride in the original solution? The density of the mixture is 1.032 g/mL, and we assume all soluble species are completely dissociated.

*12.109 A 25.00 mL sample of a 0.200 M solution of barium nitrate is mixed with 14.00 mL of a 0.250 M solution of potassium sulfate. Assuming that all ionic species are completely dissociated, what is the osmotic pressure of the mixture in torr?

*12.110 A compound is found to have a molecular formula of C_2H_6O. Write all of the possible structural formulas for this compound. With a relevant explanation, determine whether each of your structures will

 (a) be expected to be a solid, liquid, or gas at 25 °C.

 (b) be soluble in water or pentane.

 (c) form hydrogen bonds.

 (d) be a strong electrolyte, a weak electrolyte, or a non-electrolyte.

*12.111 How many mL of 0.223 M $K_2Cr_2O_7$ are needed to completely oxidize 155 mL of 0.650 M tin(II) chloride to tin(IV) in an acidic solution at 25 °C? When the reaction is complete, what will the osmotic pressure be, in torr?

*12.112 An experiment calls for the use of the dichromate ion, $Cr_2O_7^{2-}$, in sulfuric acid as an oxidizing agent for isopropyl alcohol, C_3H_8O. The chief product is acetone,

C_3H_6O, which forms according to the following equation.

$$3C_3H_8O + Na_2Cr_2O_7 + 4H_2SO_4 \longrightarrow$$
$$3C_3H_6O + Cr_2(SO_4)_3 + Na_2SO_4 + 7H_2O$$

The reaction has a by-product and you would like to determine what it is. In a typical reaction 21.4 g of isopropyl alcohol was reacted and 12.4 g of acetone was isolated. The oxidizing agent is available only as sodium dichromate dihydrate. What is the minimum number of grams of sodium dichromate dihydrate needed to oxidize 21.4 g of isopropyl alcohol? The reaction also produced 10.9 g of a volatile by-product. When a sample of it with a mass of 8.654 mg was burned in oxygen, it was converted into 22.368 mg of carbon dioxide and 10.655 mg of water, the sole products. (Assume that any unaccounted for element is oxygen.) A solution prepared by dissolving 1.338 g of the by-product in 115.0 g of benzene had a freezing point of 4.87 °C. Calculate the percentage composition of the by-product, determine its empirical formula, and calculate the molecular mass of the by-product and write its molecular formula. Is there another by-product of this reaction?

Exercises in Critical Thinking

*12.113 A certain organic substance is soluble in solvent A but it is insoluble in solvent B. If solvents A and B are miscible, will the organic compound be soluble in a mixture of A and B? What additional information is needed to answer this question?

*12.114 The situation described in Exercise 12.113 is actually quite common. How might it be used to purify the organic compound?

12.115 Compile and review all of the methods discussed for the determination of molecular masses. Assess which methods are the most reliable, which are the most sensitive, and which are the most convenient to use.

12.116 Having had some laboratory experience by now, evaluate whether preparation of a 0.25 *molar* solution is easier or more difficult than preparing a 0.25 *molal* solution. What experiments require the use of *molal* concentrations?

12.117 This chapter focused on the physical description of osmosis and its use in determining molar masses. What other uses are there for osmosis?

12.118 Using the principles developed in this chapter, explain why all solids will eventually result in a saturated solution if enough solid is added to the solvent.

12.119 It is observed that when 10.0 g of a substance is dissolved in 1.00 L of a certain solvent, 2.43 kJ of heat are

produced. When the next 10.0 g sample is added and dissolved, 2.16 kJ of heat are produced. Use the concepts of state functions and the principles in Section 12.2 to explain this observation.

*12.120 What would be the result if molarity rather than molality was used in freezing point depression (and boiling point elevation) experiments? What changes would be needed to make the molarity system work?

12.121 Compare the reasoning used to explain surface tension of liquids and the reasoning suggesting that edges and corners of crystals dissolve more rapidly than flat surfaces.

12.122 Consider a solution that has the maximum amount of $CaCl_2$ dissolved in it. Now add 5.0 mL of pure water to 100 mL of that solution. On the molecular level describe the difference between dissolving more $CaCl_2$ in this mixture and dissolving $CaCl_2$ in pure water.

12.123 When a 10.0 molar solution of sodium hydroxide in water is prepared, the beaker becomes very hot. Once the solution cools to room temperature, it is diluted with an equal volume of water and again the solution becomes very warm. Noting that a reaction that is exothermic in one direction must be endothermic in the opposite direction, provide an explanation that makes sense of the observations above.

Chemical Kinetics

Chapter Outline

The lit match in our chapter opening photo exemplifies many of the aspects of chemical kinetics we will study in this chapter. Chemical reactions run at many different speeds. Some, such as the rusting of iron or the breakdown of plastics in the environment, take place very slowly. Others, like the burning of a match, the combustion of gasoline, or the explosion of gunpowder, occur very quickly.

Image Source/Getty Images

This Chapter in Context

This Chapter in Context

Chemical kinetics is the study of the speeds (or rates) of chemical reactions. On a practical level, chemical kinetics is concerned with factors that affect the speeds of reactions and how reactions can be controlled. This is essential in industry, where synthesis reactions must take place at controlled speeds. If a reaction takes weeks or months to occur, it may not be economically feasible to carry out; if it occurs too quickly or uncontrollably, it may result in an explosion. For the consumer, studies on rates of decomposition allow a manufacturer to reliably determine the shelf life, or expiration date, of a product or drug. At a more fundamental level, a study of the speed of a reaction often gives clues that lead to an understanding about *how*, at the molecular level, reactants change into products. Understanding the reaction at this level of detail often allows more precise control of the reaction's speed, and suggests ways to modify the reaction to produce new types of products, or to improve the reaction's yield by preventing undesirable side reactions from occurring.

LEARNING OBJECTIVES

After reading this chapter, you should be able to:

- understand and use the five conditions that affect how rapidly chemicals react
- determine, from experimental data, the relative rates at which reactants disappear and products appear, and the rate of reaction, which is independent of the substance monitored
- use experimental initial rate data to determine rate laws
- use the basic results of integrated rate laws to determine the order of a reaction and calculate the time dependence of concentration for zero-, first-, and second-order reactions
- explain the rate of chemical reactions based on a molecular view of collisions that includes frequency, energy, and orientation that make up collision theory
- describe the basics of transition state theory including activated complexes and potential energy diagrams
- use the Arrhenius equation to determine the activation energy of a reaction
- use the concepts of reaction mechanisms to recognize reasonable mechanisms and suggest plausible mechanisms given experimental data
- relate the properties of homogeneous and heterogeneous catalysts and how they act to increase reaction rates

13.1 | Factors that Affect the Rate of Chemical Change

The **rate** of a given chemical change is the speed with which the reactants disappear and the products form. When a reaction is fast, more product is formed in a given period of time than in a slow reaction. This rate is measured by the amount of products produced or reactants consumed per unit time. Usually this is done by monitoring the concentrations of the reactants or products over time, as the reaction proceeds (see Figure 13.1).

Before we take up the quantitative aspects of these rates, let's look qualitatively at factors that can make a reaction run faster or slower. There are five principal factors that influence the rate of chemical change.

TOOLS
Factors that affect reaction rates

1. Chemical nature of the reactants themselves
2. Ability of the reactants to come in contact with each other
3. Concentrations of the reactants
4. Temperature
5. Availability of rate-accelerating agents called *catalysts*

Chemical Nature of the Reactants

Bonds break and new bonds form during reactions. The most fundamental differences among reaction rates, therefore, lie in the reactants themselves, in the inherent tendencies

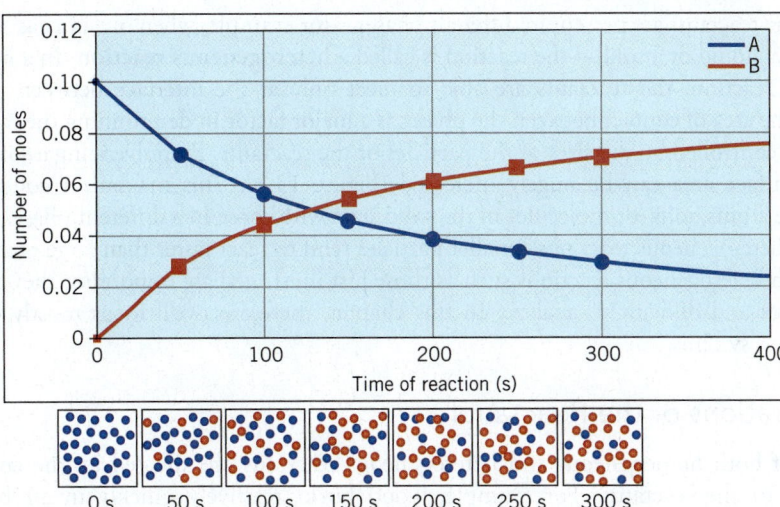

Figure 13.1 | Reaction rates are measured by monitoring concentration changes over time. For the reaction, $A \longrightarrow B$, the number of A molecules (in blue) decreases with time as the number of B molecules (in red) increases. The steeper the curves, the faster the rate of reaction. The filmstrip represents the relative numbers of molecules of A and B at each time.

for their atoms, molecules, or ions to undergo changes in chemical bonds or oxidation states. Some reactions are fast by nature and others are slow (Figure 13.2). For example, because sodium atoms lose electrons easily, a freshly exposed surface of metallic sodium tarnishes when exposed to air and moisture. Under identical conditions, potassium also reacts with air and moisture, but the reaction is much faster because potassium atoms lose electrons more easily than sodium atoms. The burning of the match in the chapter opening photograph is rapid while the rusting of iron in air is slow.

Ability of Reactants to Meet

Most reactions involve two or more reactants whose particles (i.e., atoms, ions, or molecules) must collide with each other for the reaction to occur. This is why reactions are often carried out in liquid solutions or in the gas phase. In these states, the particles are able to intermingle at a molecular level and collide with each other easily.

Reactions in which all of the reactants are in the same phase are called **homogeneous reactions**. Examples include the neutralization of sodium hydroxide by hydrochloric acid when both are dissolved in water, and the explosive reaction in the gas phase of gasoline vapor with oxygen that can occur when the two are mixed in the right proportions in a gasoline engine. (An *explosion* is an extremely rapid reaction that generates hot expanding gases.)

© 1994 Richard Megna - Fundamental Photographs

© 1991 Richard Megna - Fundamental Photographs

Figure 13.2 | The chemical nature of reactants affects reaction rates. (*a*) Sodium loses electrons easily, so it reacts quickly with water. (*b*) Potassium loses electrons even more easily than sodium, so its reaction with water is explosively fast.

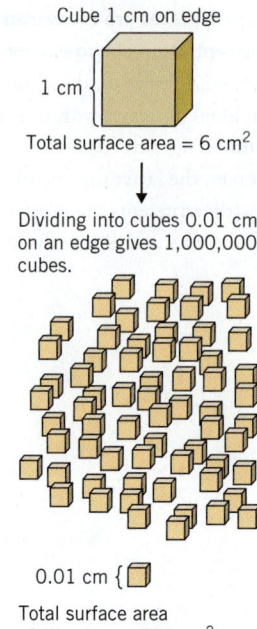

Cube 1 cm on edge

1 cm {

Total surface area = 6 cm²

↓

Dividing into cubes 0.01 cm on an edge gives 1,000,000 cubes.

0.01 cm {

Total surface area of all cubes = 600 cm²

Figure 13.3 | Effect of crushing a solid on surface area. When a single solid is subdivided into much smaller pieces, the total surface area on all of the pieces becomes very large.

When the reactants are present in different phases—for example, when one is a gas and the other is a liquid or a solid—the reaction is called a **heterogeneous reaction**. In a heterogeneous reaction, the reactants are able to meet only at the interface between the phases, so the area of contact between the phases is a major factor in determining the rate. This area is controlled by the sizes of the particles of the reactants. By pulverizing a solid, the total surface area can be hugely increased (Figure 13.3). This maximizes contact between the atoms, ions, or molecules in the solid state with those in a different phase. As a result, in heterogeneous reactions, smaller particles tend to react faster than large ones.

Although heterogeneous reactions such as those just illustrated are important, they are very complex and difficult to analyze. In this chapter, therefore, we'll focus mostly on homogeneous systems.

Concentrations of the Reactants

The rates of both homogeneous and heterogeneous reactions are affected by the concentrations of the reactants. For example, wood burns relatively quickly in air but extremely rapidly in pure oxygen. It has been estimated that if air were 30% oxygen instead of 21%, it would be impossible to put out forest fires. Even red-hot steel wool, which only sputters and glows in air, bursts into flame when thrust into pure oxygen (see Figure 13.4).

Temperature of the System

Chemical reactions occur faster at higher temperatures than they do at lower temperatures. You may have noticed, for example, that insects move more slowly when the air is cool. An insect is a cold-blooded creature, which means that its body temperature is determined by the temperature of its surroundings. As the air cools, insects cool off too, and so the rates of their metabolic reactions slow down, making the insects sluggish.

Presence of Catalysts

Catalysts are substances that increase the rates of chemical reactions without being used up. They affect every moment of our lives. In our bodies, substances called enzymes serve as catalysts for biochemical reactions. By making enzymes available or not, a cell is able to direct our body chemistry by controlling which chemical reactions can occur rapidly. Catalysts are used in the chemical industry to make gasoline, plastics, fertilizers, and other products that have become virtual necessities in our lives. We will discuss catalysts more in Section 13.9.

13.2 | Measuring Reaction Rates

The qualitative factors discussed in Section 13.1 can also be described quantitatively. To do this, we need to express reaction rates in mathematical terms. Let's start with the concept of **rate**, a change in quantity per unit time. This always implies a ratio in which a unit of time is in the denominator. Suppose, for example, that you have a job with a pay rate of ten dollars per hour. Because *per* can be interpreted as *divided by*, your rate of pay can be written as a fraction (abbreviating hour as hr):

$$\text{rate of pay} = 10 \text{ dollars per hour} = \frac{10 \text{ dollars}}{1 \text{ hr}} = 10 \text{ dollar hr}^{-1}$$

Here we've expressed the fraction 1/hr as hr^{-1}.

When chemical reactions occur, the *concentrations of reactants decrease* as they are used up, while the *concentrations of the products increase* as they form. So, one way to start describing a reaction's rate is to pick one substance in the reaction's chemical equation and measure its change in concentration per unit of time. Remembering that for change, Δ, we always take

Figure 13.4 | Effect of concentration on rate. Using pure oxygen instead of air, steel wool, after being heated to redness in a flame, burns spectacularly.

"final minus initial," the change in concentration of substance X with respect to time is given in Equation 13.1.

Change in concentration of X with respect to time

$$= \frac{(\text{conc. of } X \text{ at time } t_{final} - \text{conc. of } X \text{ at time } t_{initial})}{(t_{final} - t_{initial})}$$

$$= \frac{\Delta(\text{conc. of } X)}{\Delta t} \tag{13.1}$$

TOOLS

Rate with respect to one reactant or product

By convention, rates are reported as positive values, whether the rate was measured as the concentration of one of the products increasing or the concentration of one of the reactants decreasing. For instance, if the concentration of one product of a reaction increases by 0.50 mol/L each second, the rate of its formation is 0.50 mol L^{-1} s^{-1}. Similarly, if the concentration of a reactant changes by -0.20 mol/L per second, its rate with respect to the reactant is 0.20 mol L^{-1} s^{-1}; there is no minus sign.

When measuring rates, molarity (mol/L) is normally the concentration unit, and the second (s) is the most often used unit of time. Therefore, the units for rates of concentration change are most frequently

$$\frac{\text{mol}}{\text{L s}} \quad \text{or} \quad \text{mol } L^{-1} s^{-1} \quad \text{or} \quad M s^{-1}$$

Relative Rates and Reaction Stoichiometry

When we know the value of a rate of concentration change with respect to one substance, the coefficients of the reaction's balanced chemical equation can be used to find the rates with respect to the other substances. For example, in the combustion of propane,

$$C_3H_8(g) + 5O_2(g) \longrightarrow 3CO_2(g) + 4H_2O(g)$$

five moles of O_2 *must* be consumed per unit of time for each mole of C_3H_8 used in the same time. In this reaction, therefore, oxygen *must* react five times faster than propane in units of mol L^{-1} s^{-1}. Similarly, CO_2 forms three times faster than C_3H_8 reacts and H_2O forms four times faster. The magnitudes of the rates relative to each other are thus in the same relationship as the coefficients in the balanced chemical equation.

Example 13.1
Relationships of Rates within a Reaction

Butane, the fuel in cigarette lighters, burns in oxygen to give carbon dioxide and water. If, in a certain experiment, the butane concentration is decreasing at a rate of 0.20 mol L^{-1} s^{-1}, what is the rate at which the oxygen concentration is decreasing, and what are the rates at which the product concentrations are increasing?

Analysis: As always, we need a balanced chemical equation. Butane, C_4H_{10}, burns in oxygen to give CO_2 and H_2O according to the following balanced chemical equation:

$$2C_4H_{10}(g) + 13O_2(g) \longrightarrow 8CO_2(g) + 10H_2O(g)$$

We now need to relate the rate with respect to oxygen and the products to the given rate with respect to butane. The magnitudes of the rates relative to each other are in the same relationship as the coefficients in the balanced chemical equation.

Assembling the Tools: The chemical equation is the tool that links the stoichiometric amounts of these substances to the amount of butane. The other tool will be Equation 13.1 for calculating the rate of a reaction.

Solution: For oxygen,

$$\frac{0.20 \ \overline{\text{mol } C_4H_{10}}}{\text{L s}} \times \frac{13 \ \text{mol } O_2}{2 \ \overline{\text{mol } C_4H_{10}}} = \frac{1.3 \ \text{mol } O_2}{\text{L s}}$$

Oxygen is reacting at a rate of 1.3 mol L^{-1} s^{-1}. For CO_2 and H_2O, we have similar calculations:

$$\frac{0.20 \ \overline{\text{mol } C_4H_{10}}}{\text{L s}} \times \frac{8 \ \text{mol } CO_2}{2 \ \overline{\text{mol } C_4H_{10}}} = \frac{0.8 \ \text{mol } CO_2}{\text{L s}}$$

$$\frac{0.20 \ \overline{\text{mol } C_4H_{10}}}{\text{L s}} \times \frac{10 \ \text{mol } H_2O}{2 \ \overline{\text{mol } C_4H_{10}}} = \frac{1.0 \ \text{mol } H_2O}{\text{L s}}$$

Therefore,

Rate with respect to the formation of $CO_2 = 0.80$ mol L^{-1} s^{-1}
Rate with respect to the formation of $H_2O = 1.0$ mol L^{-1} s^{-1}

Are the Answers Reasonable? If what we've calculated is correct, then the ratio of the numerical values of the last two rates—namely, 0.80 to 1.0—should be the same as the ratio of the corresponding coefficients in the chemical equation—namely, 8 to 10 (the same ratio as 0.8 to 1.0).

Practice Exercise 13.1

The iodate ion reacts with sulfite ions in the reaction

$$IO_3^- + 3SO_3^{2-} \longrightarrow I^- + 3SO_4^{2-}$$

At what rate are the iodide and sulfate ions being produced if the sulfite ion is disappearing at a rate of 2.4×10^{-4} mol L^{-1} s^{-1}? (*Hint:* Pay attention to the stoichiometry.)

Practice Exercise 13.2

Hydrogen sulfide burns in oxygen to form sulfur dioxide and water. If sulfur dioxide is being formed at a rate of 0.30 mol L^{-1} s^{-1}, what are the rates of disappearance of hydrogen sulfide and oxygen?

Because the rates of disappearance of reactants and appearance of products are all related, it doesn't matter which species we pick to follow concentration changes over time. For example, to study the decomposition of hydrogen iodide, HI, into H_2 and I_2,

$$2HI(g) \longrightarrow H_2(g) + I_2(g)$$

it is easiest to monitor the I_2 concentration because it is the only colored substance in the reaction. As the reaction proceeds, purple iodine vapor forms, and there are instruments that allow us to relate the intensity of the color to the iodine concentration. Then, once we know the rate of formation of iodine, we also know the rate of formation of hydrogen. It is the same because the coefficients of H_2 and I_2 are the same in the balanced chemical equation. And the rate of disappearance of HI, which has a coefficient of 2 in the equation, is twice as fast as the rate of formation of I_2.

Because reactants and products are consumed and formed at different rates, we would like to find a way to express the speed of a reaction in a way that is independent of which reactant or product is monitored. To accomplish this, we divide the relative rate with respect to a given substance by the coefficient of the substance in the balanced chemical equation. The result is a quantity we will call the **rate of reaction**. Consider, for example, the combustion of propane discussed previously:

■ Notice that we have avoided the term *rate of reaction* until now to be sure it is not confused with the rate of change in concentration, or the rate with respect to a specific substance, which can be very different values.

$$C_3H_8(g) + 5O_2(g) \longrightarrow 3CO_2(g) + 4H_2O(g)$$

The rate at which O_2 is changing can be expressed as

$$\text{rate } O_2 = \frac{-\Delta[O_2]}{\Delta t}$$

where we have used square brackets around the chemical formula to denote molar concentration. Because the O_2 concentration is decreasing, we need a negative sign to make the rate a positive quantity. The *rate of reaction* would then be calculated by dividing the relative rate for O_2 by its coefficient, 5.

■ Square brackets are the standard way to denote molar concentration of the bracketed species.

$$\text{Rate of reaction} = -\frac{1}{5}\frac{\Delta[O_2]}{\Delta t}$$

Each of the other relative rates is treated in the same way to give the *same* value for the *rate of reaction*:

For a generalized reaction

$$aA + bB \longrightarrow cC + dD$$

the rate of reaction is given by Equation 13.2,

$$\text{rate of reaction} = -\frac{1}{a}\frac{\Delta[A]}{\Delta t} = -\frac{1}{b}\frac{\Delta[B]}{\Delta t} = \frac{1}{c}\frac{\Delta[C]}{\Delta t} = \frac{1}{d}\frac{\Delta[D]}{\Delta t} \quad (13.2)$$

TOOLS

Rate of reaction

Example 13.2
Rates of a Reaction

In Example 13.1, butane burned in oxygen to give carbon dioxide and water at a rate of 0.20 mol L^{-1} s^{-1} with respect to butane. What is the rate for the reaction independent of any of the reactants or products?

● **Analysis:** We are given the rate of the reaction with respect to butane. We can use the balanced chemical equation from Example 13.1 and the stoichiometric coefficients,

$$2C_4H_{10}(g) + 13O_2(g) \longrightarrow 8CO_2(g) + 10H_2O(g)$$

We will relate the rate of the reaction with respect to butane to the rate of the reaction independent of the reactants or products.

● **Assembling the Tools:** Just as in Example 13.1, we will use the balanced chemical equation as a tool because it gives the stoichiometric ratio of the reactants and products. We will then use Equation 13.2 to calculate the rate of the reaction.

● **Solution:** We are given the relative rate of the reaction of butane as 0.20 mol L^{-1} s^{-1}, since C_4H_{10} is a reactant, the change in concentration with respect to time is -0.20 mol L^{-1} s^{-1}. We also have the balanced chemical equation for the combustion of butane with the coefficient of 2 for butane.

$$\frac{\Delta[C_4H_{10}]}{\Delta t} = -\frac{0.20 \text{ mol}}{L\,s}$$

$$\text{rate}_{\text{reaction}} = -\frac{1}{2} \times \left(-\frac{0.20 \text{ mol}}{L\,s}\right)$$

$$= 0.10\frac{\text{mol}}{L\,s}$$

● **Are the Answers Reasonable?** Since the coefficient for butane is greater than one, the rate for the reaction should be less than the rate with respect to butane. Our value, 0.1 mol L^{-1} s^{-1}, is half the value of 0.2 mol L^{-1} s^{-1}, which makes sense.

Reaction Rates versus Time

A reaction rate is rarely constant throughout the reaction but commonly changes as the reactants are used up. This is because the rate usually depends on the concentrations of the

TABLE 13.1	Data, at 508 °C, for the Reaction $2HI(g) \longrightarrow H_2(g) + I_2(g)$	
Concentration of HI (mol/L)		**Time (s) ($\pm$1 s)**
0.100		0
0.0716		50
0.0558		100
0.0457		150
0.0387		200
0.0336		250
0.0296		300
0.0265		350

reactants, and these change as the reaction proceeds. For example, Table 13.1 lists data for the decomposition of hydrogen iodide at a temperature of 508 °C. The data, which show the changes in molar HI concentration over time, are plotted in Figure 13.5. Notice that the molar HI concentration decreases fairly rapidly during the first 50 seconds of the reaction, which means that the initial rate is relatively fast. However, later, in the interval between 300 s and 350 s, the concentration changes by only a small amount, so the rate has slowed considerably. Thus, the steepness of the curve at any moment reflects the rate of the reaction; the steeper the curve, the higher is the rate.

The rate at which the HI is being consumed at any particular moment is called the **instantaneous rate**. When we use the term "rate" we mean the instantaneous rate, unless we state otherwise. The instantaneous rate can be determined from the slope of the curve measured at the time we have chosen. The slope, which can be read off the graph, is the ratio (expressed positively) of the change in concentration to the change in time. In Figure 13.5, for example, the relative rate of the decomposition of hydrogen iodide is determined for a time 100 seconds from the start of the reaction. After the tangent to the curve is drawn, we measure a concentration change (a decrease of 0.029 mol/L) and the time change (110 s) from the graph of the tangent line. Because the rate is based on a *decreasing* concentration, we use a minus sign for the *equation* that describes this rate so that the rate itself will be a positive quantity.

■ The **average rate** is the slope of a line connecting two points on a concentration versus time graph. The instantaneous rate is the slope of a tangent line at a single point. The average and instantaneous rates are quite different from each other.

$$\text{rate}_{\text{with respect to HI}} = -\left\{ \frac{[\text{HI}]_{\text{final}} - [\text{HI}]_{\text{initial}}}{t_{\text{final}} - t_{\text{initial}}} \right\} = -\left\{ \frac{-0.029 \text{ mol/L}}{110 \text{ s}} \right\}$$

$$\text{rate}_{\text{with respect to HI}} = 2.6 \times 10^{-4} \text{mol L}^{-1} \text{ s}^{-1}$$

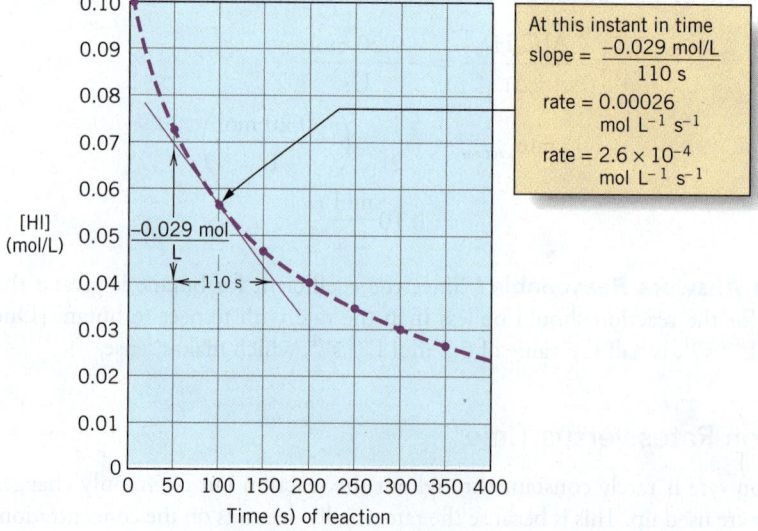

Figure 13.5 | **Effect of time on concentration.** The data for this plot of the change in the concentration of HI with time for the reaction, $2HI(g) \longrightarrow H_2(g) + I_2(g)$ at 508 °C are taken from Table 13.1. The slope is negative because we're measuring the *disappearance* of HI, but when its value is used as a rate of reaction, we express the rate as positive.

So, 110 s after the start of the reaction, the rate with respect to HI is 2.6×10^{-4} mol $L^{-1} s^{-1}$. In Example 13.3, we'll use this technique to obtain the *initial instantaneous rate of the reaction*—that is, the instantaneous rate of reaction at time zero.

Example 13.3
Estimating Initial Rates of Reaction

Use the experimental data shown in Figure 13.5 for the reaction

$$2HI(g) \longrightarrow H_2(g) + I_2(g)$$

to determine the initial rate of reaction at 508 °C.

Analysis: The problem asks us to determine the initial rate, which is the instantaneous rate of change in concentration at time zero. However, we have to be careful here to distinguish between the *rate with respect to HI* and the quantity called the *rate of reaction*, which is the rate independent of the coefficients in the balanced chemical equation. Using the slope of the tangent line at $t = 0$ will allow us to determine the instantaneous rate of reaction *with respect to HI*. To determine the *rate of reaction*, we then have to divide the rate with respect to HI by the coefficient of HI in the balanced equation.

Assembling the Tools: We will find the initial rate of reaction with respect to HI by determining the slope of the tangent line at time zero of the curve showing the HI concentration as a function of time in Figure 13.5. Remember that the slope of a straight line can be calculated from the coordinates of any two points (x_1, y_1) and (x_2, y_2) using the following equation:

$$\text{slope} = \frac{y_2 - y_1}{x_2 - x_1}$$

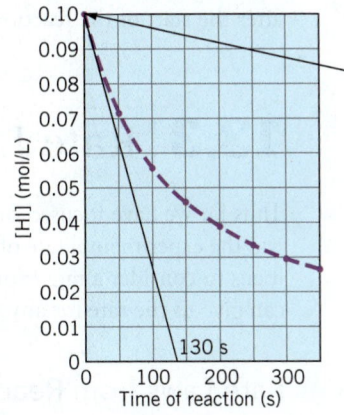

At time zero, slope of the tangent line =
$-(0.10 \text{ mol/L}^{-1}) / 130 \text{ s}$
rate = 7.7×10^{-4} mol L^{-1} s^{-1}

Solution: We can draw the line as shown in the graph in the margin. To precisely determine the slope of the tangent line, we choose two points that are as far apart as possible. The point on the curve (0 s, 0.10 mol/L) and the intersection of the tangent line with the time axis (130 s, 0 mol/L) are widely separated:

$$\text{slope} = \frac{0.0 \text{ mol/L} - 0.10 \text{ mol/L}}{130 \text{ s} - 0 \text{ s}} = -7.7 \times 10^{-4} \text{ mol L}^{-1} s^{-1}$$

The slope is negative because the concentration of HI is decreasing as time increases. Rates are positive quantities, so we can report the initial rate of reaction with respect to HI as 7.7×10^{-4} mol $L^{-1} s^{-1}$.

The initial rate of reaction (which is what we're asked to determine) is the rate with respect to HI divided by the coefficient of HI in the balanced equation:

$$\text{rate of reaction} = \frac{7.7 \times 10^{-4} \text{ mol L}^{-1} s^{-1}}{2} = 3.85 \times 10^{-4} \text{ mol L}^{-1} s^{-1}$$

When properly rounded, the answer is 3.8×10^{-4} mol $L^{-1} s^{-1}$.

Is the Answer Reasonable? The instantaneous rate for HI at time zero ought to be slightly larger than the *average rate* with respect to HI between zero and 50 seconds. We can compute the average rate directly from data in Table 13.1 by selecting a pair of concentrations of HI at two different times.

$$\text{slope} = \frac{0.0716 \text{ mol/L} - 0.100 \text{ mol/L}}{50 \text{ s} - 0 \text{ s}} = -5.7 \times 10^{-4} \text{ mol L}^{-1} s^{-1}$$

Thus, the average rate from 0 to 50 s is 5.7×10^{-4} mol L^{-1} s^{-1}. As expected, this is slightly less than 7.7×10^{-4} mol L^{-1} s^{-1} the instantaneous rate with respect to HI at time zero.

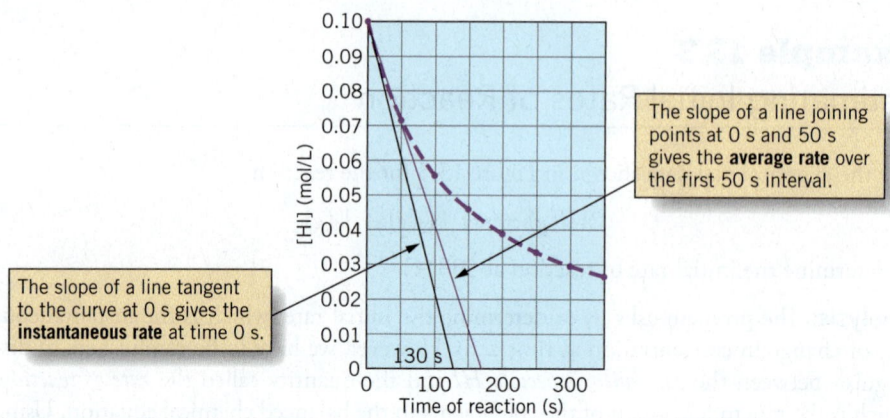

The slope of a line joining points at 0 s and 50 s gives the **average rate** over the first 50 s interval.

The slope of a line tangent to the curve at 0 s gives the **instantaneous rate** at time 0 s.

Practice Exercise 13.3

Use the graph in Figure 13.5 to estimate the rate of reaction with respect to HI at 2.00 minutes after the start of the reaction. (*Hint:* On the graph, the time is given in seconds.)

Practice Exercise 13.4

Use the graph in Figure 13.5 to estimate the rate of reaction with respect to HI at 250 seconds after the start of the reaction.

13.3 | Rate Laws

Thus far we have focused on a rate with respect to *one* component of a reaction. This rate was the experimental rate of a process for that particular reaction. We want to broaden our focus to consider a rate expression that includes all reactants, and have an expression that can give us the rate for any set of concentrations.

Rate Laws from Reaction Rates and Concentrations

The rate of a homogeneous reaction at any instant is proportional to the product of the molar concentrations of the reactants, each molarity raised to some power or exponent that has to be found experimentally. Let's consider a chemical reaction with an equation of the following form:

$$A + B \longrightarrow \text{products}$$

Its rate of reaction can be expressed as follows:

$$\text{rate} \propto [A]^m [B]^n \tag{13.3}$$

where the values of the exponents n and m are found by experiment.

The proportionality symbol, $\propto$, in Equation 13.3 can be replaced by an equals sign if we introduce a proportionality constant, **k**, which is called the **rate constant** for the reaction. This gives Equation 13.4:

$$\text{rate} = k[A]^m [B]^n \tag{13.4}$$

TOOLS

Rate law of a reaction

■ The value of k depends on the particular reaction being studied, as well as the temperature at which the reaction occurs.

Equation 13.4 is called the **rate law** for the reaction of A with B. Once we have found values for k, n, and m, the rate law allows us to calculate the rate of the reaction at any set of known values of concentrations. Consider, for example, the following reaction.

$$H_2SeO_3 + 6I^- + 4H^+ \longrightarrow Se + 2I_3^- + 3H_2O$$

Its rate law is of the form

$$\text{rate} = k[H_2SeO_3]^x[I^-]^y[H^+]^z$$

The exponents have been found experimentally to be the following for the initial rate of this reaction (i.e., the rate when the reactants are first combined).

$$x = 1, \quad y = 3, \quad \text{and} \quad z = 2$$

At 0 °C, k equals 5.0×10^5 L^5 mol^{-5} s^{-1}. (We have to specify the temperature because k varies with T.) Substituting the exponents and the value of k into the rate law equation gives the rate law for the reaction:

$$\text{rate} = (5.0 \times 10^5 \, L^5 \, mol^{-5} \, s^{-1})[H_2SeO_3][I^-]^3[H^+]^2 \quad \text{(at 0 °C)}$$

■ The units of the rate constant are such that the calculated rate will *always* have the units mol L^{-1} s^{-1}.

We can calculate the initial rate of the reaction at 0 °C for any set of concentrations of H_2SeO_3, I^-, and H^+ using this rate law.

Example 13.4
Calculating Reaction Rate from the Rate Law

In the stratosphere, molecular oxygen (O_2) can be broken into two oxygen atoms by ultraviolet radiation from the sun. If one of these oxygen atoms has enough energy and strikes an ozone (O_3) molecule in the stratosphere, the ozone molecule is destroyed, and two oxygen molecules are created:

$$O(g) + O_3(g) \longrightarrow 2O_2(g)$$

This reaction is part of the natural cycle of ozone destruction and creation in the stratosphere. What is the rate of ozone destruction *for this reaction alone* at an altitude of 25 km, if the rate law for the reaction is

$$\text{rate} = 4.15 \times 10^5 \, L \, mol^{-1} \, s^{-1}[O_3][O]$$

and the reactant concentrations at 25 km are the following: $[O_3] = 1.2 \times 10^{-8}$ M and $[O] = 1.7 \times 10^{-14}$ M?

Analysis: We are given the rate law, so the answer to this question is merely a matter of substituting the given molar concentrations into this law.

Assembling the Tools: We will use the given rate law for the reaction along with the concentrations to calculate the rate of the reaction.

Solution: To see how the units work out, let's write all of the concentration values as well as the rate constant's units in fraction form

$$\text{rate} = \left(\frac{4.15 \times 10^5 \, L}{mol \, s}\right) \times \left(\frac{1.2 \times 10^{-8} \, mol}{L}\right) \times \left(\frac{1.7 \times 10^{-14} \, mol}{L}\right)$$

Performing the arithmetic and canceling units yields

$$\text{rate} = \frac{8.5 \times 10^{-17} \, mol}{L \, s} = 8.5 \times 10^{-17} \, mol \, L^{-1} \, s^{-1}$$

Is the Answer Reasonable? There's no simple check. Multiplying the powers of ten for the rate constant and the concentrations together reassures us that the rate is of the correct order of magnitude, and we can see that at least the answer has the correct units for a reaction rate.

The rate law for the reaction

$$2NO(g) + 2H_2(g) \longrightarrow N_2(g) + 2H_2O(g) \text{ is}$$

$$\text{rate} = k[NO]^2[H_2]$$

Practice Exercise 13.5

If the rate of reaction is 7.86×10^{-3} mol L^{-1} s^{-1} when the concentrations of NO and H$_2$ are both 2.0×10^{-6} mol L^{-1}, (a) what is the value of the rate constant and (b) what are the units for the rate constant? (*Hint:* Note the units of the reaction rate.)

Practice Exercise 13.6

The rate law for the decomposition of HI to I$_2$ and H$_2$ is

$$\text{rate} = k[\text{HI}]^2$$

At 508 °C, the rate of the reaction of HI was found to be 2.5×10^{-4} mol L^{-1} s^{-1} when the HI concentration was 0.0558 M (see Figure 13.5). (a) What is the value of k? (b) What are the units of k?

Orders of Reaction

Although a rate law's exponents are generally unrelated to the chemical equation's coefficients, they sometimes are the same by coincidence, as is the case in the decomposition of hydrogen iodide:

$$2\text{HI}(g) \longrightarrow \text{H}_2(g) + \text{I}_2(g)$$

The rate law, as we've said, is

$$\text{rate} = k[\text{HI}]^2$$

The exponent of [HI] in the rate law—namely, 2—happens to match the coefficient of HI in the overall chemical equation, but *there is no way we could have predicted this match without experimental data.* Therefore, *never* simply assume the exponents and the coefficients are the same; it's a trap that many fall into.

An exponent in a rate law is called the **order of the reaction**[1] with respect to the corresponding reactant. For instance, the decomposition of gaseous N$_2$O$_5$ into NO$_2$ and O$_2$,

$$2\text{N}_2\text{O}_5(g) \longrightarrow 4\text{NO}_2(g) + \text{O}_2(g)$$

has the rate law

$$\text{rate} = k[\text{N}_2\text{O}_5]$$

■ When an exponent in an equation is found to be 1, it is usually omitted.

The exponent of [N$_2$O$_5$] is 1, so the reaction rate is said to be *first order* with respect to N$_2$O$_5$. The rate law for the decomposition of HI has an exponent of 2 for the HI concentration, so its reaction rate is *second order* in HI. The rate law

$$\text{rate} = k[\text{H}_2\text{SeO}_3][\text{I}^-]^3[\text{H}^+]^2$$

describes a reaction rate that is first order with respect to H$_2$SeO$_3$, third order with respect to I$^-$, and second order with respect to H$^+$.

The **overall order of a reaction** is the sum of the orders with respect to each reactant in the rate law. The decomposition of N$_2$O$_5$ is a first-order reaction. The overall order for the reaction involving H$_2$SeO$_3$ is $1 + 3 + 2 = 6$.

The exponents in a rate law are usually small whole numbers, but fractional and negative exponents are occasionally found. A negative exponent means that the concentration term really belongs in the denominator, which means that as the concentration of the species increases, the rate of reaction decreases.

There are even **zero-order reactions**. They have reaction rates that are independent of the concentration of any reactant. Zero-order reactions usually involve a small amount of a catalyst that is saturated with reactants. This is rather like the situation in a crowded supermarket with only a single checkout lane open. It doesn't matter how many people join the

---

[1] The reason for describing the *order* of a reaction is to take advantage of a great convenience—namely, the mathematics involved in the treatment of the data is the same for all reactions having the same order. We will not go into this very deeply, but you should be familiar with this terminology; it's often used to describe the effects of concentration on reaction rates.

line; the line will move at the same rate no matter how many people are standing in it. An example of a zero-order chemical reaction is the elimination of ethyl alcohol in the body, by the liver. Regardless of the blood alcohol level, the rate of alcohol removal by the body is constant, because the number of available catalyst molecules present in the liver is constant. Another zero-order reaction is the decomposition of gaseous ammonia into H_2 and N_2 on a hot platinum surface. The rate at which ammonia decomposes is the same, regardless of its concentration in the gas.

The rate law for a zero-order reaction is simply

$$rate = k$$

where the rate constant k has units of mol L^{-1} s^{-1}. The rate constant depends on the amount, quality, and available surface area of the catalyst. For example, forcing the ammonia through hot platinum powder (with a high surface area) would cause it to decompose faster than simply passing it over a hot platinum surface.

In all of the rate laws, the rate constant, k, indicates how fast a reaction proceeds. If the value for k is large, the reaction proceeds rapidly, and if k is small, the reaction is slow. The units for k must be such that the rate calculated from the rate law has units of mol L^{-1} s^{-1}. A list of units for k, as it depends on the overall order of the reaction, is given in Table 13.2.

TABLE 13.2	Units for Rate Constants Depend on the Order of the Rate Law
Order of Rate Law	Units for k
0	mol L^{-1} s^{-1}
1	s^{-1}
2	L mol^{-1} s^{-1}
3	L^2 mol^{-2} s^{-1}
4	L^3 mol^{-3} s^{-1}

Practice Exercise 13.7

The reaction,

$$BrO_3^- + 3SO_3^{2-} \longrightarrow Br^- + 3SO_4^{2-}$$

has the rate law

$$rate = k[BrO_3^-][SO_3^{2-}]$$

What is the order of the reaction with respect to each reactant? What is the overall order of the reaction? (*Hint:* Recall that a concentration with no exponent has, in effect, an exponent of 1.)

Practice Exercise 13.8

A certain reaction has an experimental rate law that is found to be second order in Cl_2 and first order in NO. Write the rate law for this reaction.

Practice Exercise 13.9

For the reaction of HCO_2H with bromine, the balanced chemical equation is

$$HCO_2H(aq) + Br_2(aq) \longrightarrow 2H^+(aq) + 2Br^-(aq) + CO_2(aq)$$

and the rate law for this reaction is

$$rate = k[Br_2]$$

What is the order of the reaction with respect to bromine and with respect to HCO_2H? What is the overall order of the reaction?

Obtaining Rate Laws from Experimental Data

We've mentioned several times that the exponents in the rate law of an overall reaction must be determined experimentally. This is the only way to know for sure what the exponents are. To determine the exponents, we study how changes in concentration affect the rate of the reaction. For example, consider again the following hypothetical reaction:

$$A + B \longrightarrow products$$

Suppose, further, that the data in Table 13.3 have been obtained in a series of five experiments. The form of the rate law for the reaction will be

$$rate = k[A]^n[B]^m$$

TOOLS

Determining rate laws

TABLE 13.3	Concentration Rate Data for the Hypothetical Reaction $A + B \longrightarrow$ Products		
	Initial Concentrations		Initial Rate of Formation
Experiment	[A] (mol L^{-1})	[B] (mol L^{-1})	of Products (mol L^{-1} s^{-1})
1	0.10	0.10	0.20
2	0.20	0.10	0.40
3	0.30	0.10	0.60
4	0.30	0.20	2.40
5	0.30	0.30	5.40

The values of n and m can be determined by looking for patterns in the rate data given in the table. *One of the easiest ways to reveal patterns in data is to form ratios of results using different sets of conditions.* Because this technique is quite generally useful, let's look in some detail at how it is applied to the problem of finding the rate law exponents. For experiments 1, 2, and 3 in Table 13.3, the concentration of B has been held constant at 0.10 M. Any change in the rate for these first three experiments must be due to the change in [A]. The rate law tells us that when the concentration of B is held constant, the rate must be proportional to $[A]^n$, so if we take the ratio of rate laws for experiments 2 and 1, we obtain

■ Subscripts attached to the concentrations and rates indicate the experiment number we use for the data.

$$\frac{\text{rate}_2}{\text{rate}_1} = \frac{k[A]_2^n[B]_2^m}{k[A]_1^n[B]_1^m} = \frac{k}{k}\left(\frac{[A]_2}{[A]_1}\right)^n\left(\frac{[B]_2}{[B]_1}\right)^m$$

For experiments 1 and 2, the ratio of the rates on the left side of this equation is

$$\frac{\text{rate}_2}{\text{rate}_1} = \frac{0.40 \text{ mol L}^{-1}\text{ s}^{-1}}{0.20 \text{ mol L}^{-1}\text{ s}^{-1}} = 2.0$$

and on the right side of the equation we enter all the concentrations and see that the identical values of $[B]_2^m$ and $[B]_1^m$ and the rate constants k in the numerator and denominator will cancel, as shown, to give

$$\frac{k[0.20 \text{ mol L}^{-1}]_2^n[0.10 \text{ mol L}^{-1}]_2^m}{k[0.10 \text{ mol L}^{-1}]_1^n[0.10 \text{ mol L}^{-1}]_1^m} = \frac{[0.20 \text{ mol L}^{-1}]_2^n}{[0.10 \text{ mol L}^{-1}]_1^n} = 2.0^n$$

Since we have simplified the left and the right side of our main equation, we can equate their results to read $2.0 = 2.0^n$. This equality is true only if $n = 1$. For each unique combination of our experiments 1, 2, and 3, we use the same procedure to obtain

$$2.0 = 2.0^n \qquad \text{(for experiments 2 and 1)}$$

$$3.0 = 3.0^n \qquad \text{(for experiments 3 and 1)}$$

$$1.5 = 1.5^n \qquad \text{(for experiments 3 and 2)}$$

The only value of n that makes all of these equations true is $n = 1$. Therefore, the reaction must be first order with respect to A.

Using the same procedure, while choosing experiments where the concentration of A is the same, will give us the exponent on [B]. In the final three experiments, the concentration of B is changed while the concentration of A is held constant. This time it is the concentration of B that affects the rate. Taking the ratio of rate laws for experiments 4 and 3, we have

$$\frac{\text{rate}_4}{\text{rate}_3} = \frac{k[A]_4^n[B]_4^m}{k[A]_3^n[B]_3^m}$$

The concentrations of A cancel, since $[A]_4 = [A]_3$, and both k's cancel, because the rate constant is constant, so

$$\frac{\text{rate}_4}{\text{rate}_3} = \frac{[B]_4^m}{[B]_3^m} = \left(\frac{[B]_4}{[B]_3}\right)^m$$

For each unique combination of experiments 3, 4, and 5, we have

$$4.0 = 2.0^m \qquad \text{(for experiments 4 and 3)}$$

$$9.0 = 3.0^m \qquad \text{(for experiments 5 and 3)}$$

$$2.25 = 1.5^m \qquad \text{(for experiments 5 and 4)}$$

The only value of m that makes these equations true is $m = 2$, so the reaction must be second order with respect to B.

If the value for the exponent is not obvious, we can also determine it by taking advantage of one of the mathematical properties of logarithms. If we take the preceding equation,

$$\frac{\text{rate}_4}{\text{rate}_3} = \left(\frac{[B]_4}{[B]_3}\right)^m$$

and take the logarithm of both sides,

$$\log\left(\frac{\text{rate}_4}{\text{rate}_3}\right) = \log\left(\frac{[B]_4}{[B]_3}\right)^m$$

■ There is more information on using logarithms in Appendix A.

then we can move the exponent, m, outside the log term so that it multiplies the log of the ratio of the concentrations by m:

$$\log\left(\frac{\text{rate}_4}{\text{rate}_3}\right) = m \times \log\left(\frac{[B]_4}{[B]_3}\right)$$

For example, using the data from experiments 5 and 4 from Table 13.3,

$$2.25 = 1.5^m$$

$$\log(2.25) = m\log(1.5)$$

$$0.352 = m(0.176)$$

$$m = \frac{0.352}{0.176} = 2$$

Having determined the exponents for the concentration terms, we now know that the rate law for the reaction must be

$$\text{rate} = k[A]^1[B]^2 = k[A][B]^2$$

To calculate the value of k, we substitute rate and concentration data into the rate law for any one of the sets of data.

$$k = \frac{\text{rate}}{[A][B]^2}$$

Using the data from the first set in Table 13.3,

$$k = \frac{0.20 \text{ mol L}^{-1}\text{ s}^{-1}}{(0.10 \text{ mol L}^{-1})(0.10 \text{ mol L}^{-1})^2}$$

$$= \frac{0.20 \text{ mol L}^{-1}\text{ s}^{-1}}{0.0010 \text{ mol}^3\text{ L}^{-3}}$$

After simplifying the units by cancellation, the value of k with the correct units is

$$k = 2.0 \times 10^2 \text{ L}^2\text{ mol}^{-2}\text{ s}^{-1}$$

Table 13.4 summarizes the reasoning used to determine the order with respect to each reactant from experimental data.

TABLE 13.4	Relationship Between the Order of a Reaction and Changes in Concentration and Rate		
Factor by Which Concentration is Changed	**Factor by Which the Rate Changes**	**Exponent on the Concentration Term in the Rate Law**	
2	Rate is unchanged.	0	
3	Rate is unchanged.	0	
4	Rate is unchanged.	0	
2	$2 = 2^1$	1	
3	$3 = 3^1$	1	
4	$4 = 4^1$	1	
2	$4 = 2^2$	2	
3	$9 = 3^2$	2	
4	$16 = 4^2$	2	
2	$8 = 2^3$	3	
3	$27 = 3^3$	3	
4	$64 = 4^3$	3	

Practice Exercise 13.10

Use the data from the other four experiments in Table 13.3 to calculate k for this reaction. What do you notice about the values of k? What are the units for k? (*Hint:* Don't forget the exponents in the rate law.)

Practice Exercise 13.11

Use the rate law determined in Practice Exercise 13.10 to describe what will happen to the reaction rate under the following conditions: (a) the concentration of B is tripled, (b) the concentration of A is tripled, and (c) the concentration of A is tripled and the concentration of B is halved.

Example 13.5
Determining the Exponents of a Rate Law

Sulfuryl chloride, SO_2Cl_2, is used to manufacture the antiseptic chlorophenol. The following data were collected on the decomposition of SO_2Cl_2 at a certain temperature.

$$SO_2Cl_2(g) \longrightarrow SO_2(g) + Cl_2(g)$$

Initial Concentration of SO_2Cl_2 (mol L^{-1})	Initial Rate of Formation of SO_2 (mol L^{-1} s^{-1})
0.100	2.2×10^{-6}
0.200	4.4×10^{-6}
0.300	6.6×10^{-6}

What is the rate law, and what is the value of the rate constant for this reaction?

Analysis: We are asked to write the rate law for the decomposition of SO_2Cl_2, and we can only do this by using the experimental data. After we determine the exponents for the rate law, we also have to determine the rate constant, k, for the rate law, including the units. Finally, we will use the rate law we determined and data from one of the experiments to calculate k.

Assembling the Tools: The first tool we will use is the general form of the expected rate law so we can see which exponents have to be determined. Then we use the tool for determining the order of reactions from the data. Then we can use the rate law with the orders of the reaction to determine k with the proper units.

Solution We expect the rate law to have the form rate $= k [SO_2Cl_2]^x$. If we write the ratio of rate laws for two experiments we get:

$$\frac{rate_2}{rate_1} = \frac{k[SO_2Cl_2]_2^x}{k[SO_2Cl_2]_1^x}$$

We cancel the rate constant, k, and notice that when we double the concentration from 0.100 M to 0.200 M, the initial rate doubles (from 2.2×10^{-6} mol L^{-1} s^{-1} to 4.4×10^{-6} mol L^{-1} s^{-1}). If we look at the first and third experiments, we see that when the concentration triples (from 0.100 M to 0.300 M), the rate also triples (from 2.2×10^{-6} mol L^{-1} s^{-1} to 6.6×10^{-6} mol L^{-1} s^{-1}). We can summarize these observations mathematically as $2 = 2^x$ and $3 = 3^x$. This behavior tells us that x must be one and the reaction must be first order in the SO_2Cl_2 concentration. The rate law is therefore

■ We can also use Experiments 2 and 3. From the second to the third, the rate increases by the same factor, 1.5, as the concentration, so by these data, too, the reaction must be first order.

$$rate = k[SO_2Cl_2]^1 = k[SO_2Cl_2]$$

To evaluate k, we can use any of the three sets of data. Choosing the first,

$$k = \frac{rate}{[SO_2Cl_2]}$$

$$k = \frac{2.2 \times 10^{-6} \text{ mol L}^{-1} \text{ s}^{-1}}{0.100 \text{ mol L}^{-1}} = 2.2 \times 10^{-5} \text{ s}^{-1}$$

Is the Answer Reasonable? We should get the same value of k by picking any other pair of rates and concentrations. Using the data from both the second and third experiments we get 2.2×10^{-5} s^{-1} each time. The fact that we get the same result each time confirms that both our rate law and calculations are reasonable.

Example 13.6
Determining the Exponents of a Rate Law

The following data were measured for the reduction of nitric oxide with hydrogen:

$$2H_2(g) + 2NO(g) \longrightarrow N_2(g) + 2H_2O(g)$$

Initial Concentrations (mol L^{-1})		Initial Rate of Formation of
[H$_2$]	[NO]	H$_2$O (mol L^{-1}s^{-1})
0.10	0.10	1.23×10^{-3}
0.20	0.10	2.46×10^{-3}
0.10	0.27	8.97×10^{-3}

What is the rate law for the reaction?

Analysis: This time we have two reactants. To see how their concentrations affect the rate, we must vary only one concentration at a time. Therefore, we choose two experiments in which the concentration of one reactant doesn't change and examine the effect of a change in the concentration of the other reactant. Then, we repeat the procedure for the second reactant.

Assembling the Tools: We will use the same tools we used in Example 13.5—namely, the general form of the expected rate law and the tool for determining the order of reactions from the experimental data.

Solution: We expect the rate law to have the form

$$rate = k[H_2]^m[NO]^n$$

Let's look at the first two experiments. Here the concentration of NO remains the same, so the rate is being affected by the change in the H_2 concentration. When we double the H_2 concentration, the rate doubles, so the reaction is first order with respect to H_2. This means $m = 1$.

Next, we need to pick two experiments in which the H_2 concentration doesn't change; that occurs in the first and third rows where $[H_2]_1 = [H_2]_2 = 0.01$ mol L^{-1}. In the first and third rows $[NO]_1 = 0.10$ mol L^{-1} and $[NO]_3 = 0.27$ mol L^{-1}. The initial rate of reaction for the first reaction is 1.23×10^{-3} mol L^{-1} s^{-1} and for the third reaction it is 8.97×10^{-3} mol L^{-1} s^{-1}.

These are not simple numbers to work with, so we will have to use the ratio of the two reaction rates,

$$\frac{\text{rate}_1}{\text{rate}_3} = \frac{\cancel{k}[\cancel{H_2}]_1^1[NO]_1^n}{\cancel{k}[\cancel{H_2}]_3^1[NO]_3^n}$$

We can eliminate k and $[H_2]$ as shown, since the values are equal, leaving us with

$$\frac{\text{rate}_1}{\text{rate}_3} = \frac{[NO]_1^n}{[NO]_3^n}$$

$$\log\left(\frac{\text{rate}_1}{\text{rate}_3}\right) = \log\left(\frac{[NO]_1}{[NO]_3}\right)^n = n\log\left(\frac{[NO]_1}{[NO]_3}\right)$$

Now we can use the experimental data to find the value for n:

$$\log\left(\frac{1.23 \times 10^{-3}}{8.97 \times 10^{-3}}\right) = n\log\left(\frac{0.10}{0.27}\right)$$

$$\log(0.137) = n\log(0.37)$$

$$-0.863 = n(-0.431)$$

$$n = 2$$

Therefore, the rate law for the reaction is

$$\text{rate} = k[H_2][NO]^2$$

Is the Answer Reasonable? If we have the correct rate law, the value of the rate constant should be the same, within the experimental error, for each row of data. The rate constant for the first row of data is calculated as

$$k_1 = \frac{\text{rate}_1}{[H_2]_1^1[NO]_1^2} = \frac{1.23 \times 10^{-3} \cancel{\text{mol}} \cancel{L^{-1}}}{(0.10 \cancel{\text{mol}} \cancel{L^{-1}})(0.10 \text{ mol } L^{-1})^2} = 1.23 \text{ } L^2 \text{ mol}^{-2} \text{ s}^{-1}$$

For the rate constants from the data in the second and third rows we get precisely the same value for the rate constant. Our rate law must be correct.

Practice Exercise 13.12

The following reaction is investigated to determine its rate law:

$$2H_2(g) + 2NO(g) \longrightarrow N_2(g) + 2H_2O(g)$$

Experiments yielded the following results:

Initial Concentrations (mol L^{-1})		Initial Rate of Formation of N$_2$ (mol L^{-1} s^{-1})
[H$_2$]	[NO]	
0.40×10^{-4}	0.30×10^{-4}	1.0×10^{-8}
0.80×10^{-4}	0.30×10^{-4}	4.0×10^{-8}
0.80×10^{-4}	0.60×10^{-4}	8.0×10^{-8}

(a) What is the rate law for the reaction? (b) What is the value of the rate constant? (c) What are the units for the rate constant? (*Hint:* Identify the two experiments where only [NO] changes and the two experiments where only the [H_2] varies.)

Practice Exercise 13.13

Ordinary sucrose (table sugar) reacts with water in an acidic solution to produce two simpler sugars, glucose and fructose, that have the same molecular formulas.

$$C_{12}H_{22}O_{11} + H_2O \longrightarrow C_6H_{12}O_6 + C_6H_{12}O_6$$

sucrose   glucose   fructose

In a series of experiments, the following data were obtained:

Initial Sucrose Concentration (mol L^{-1})	Rate of Formation of Glucose (mol L^{-1} s^{-1})
0.10	6.17×10^{-5}
0.36	2.22×10^{-4}
0.58	3.58×10^{-3}

(a) What is the order of the reaction with respect to sucrose? (b) What is the value of the rate constant, and what are its units?

Practice Exercise 13.14

A certain reaction has the following equation: $A + B \longrightarrow C + D$. Experiments yielded the following results:

Initial Concentrations (mol L^{-1})		Initial Rate of Formation of C (mol L^{-1} s^{-1})
[A]	[B]	
0.40	0.30	1.00×10^{-4}
0.60	0.30	2.25×10^{-4}
0.80	0.60	1.60×10^{-3}

(a) What is the rate law for the reaction? (b) What is the value of the rate constant? (c) What are the units for the rate constant? (d) What is the overall order of this reaction? (*Hint:* Solve for the exponent of [A], then use it to solve for the exponent of [B].)

13.4 | Integrated Rate Laws

The rate law tells us how the speed of a reaction varies with the concentrations of the reactants. Often, however, we are more interested in how the concentrations change over time. For instance, if we were preparing some compound, we might want to know how long it will take for the reactant concentrations to drop to some particular value, so we can decide when to isolate the products.

The relationship between the concentration of a reactant and time can be derived from a rate law using calculus. By summing or "integrating" the instantaneous rates of a reaction from the start of the reaction until some specified time, *t*, we can obtain *integrated rate laws* that quantitatively give concentration as a function of time. The form of the integrated rate law depends on the order of the reaction. The mathematical expressions that relate concentration and time in complex reactions can be complicated, so we will concentrate on using integrated rate laws for a few simple first- and second-order reactions with only one reactant.

First-Order Reactions

A **first-order reaction** is a reaction that has a rate law of the type

$$\text{rate} = k[A]$$

TOOLS

Integrated first-order rate law

Using calculus,[2] the following equation can be derived that relates the concentration of A and time:

$$\ln \frac{[A]_0}{[A]_t} = kt \qquad (13.5)$$

The symbol "ln" means natural logarithm. The expression to the left of the equals sign is the natural logarithm of the ratio of $[A]_0$ (the initial concentration of A when the time, $t = 0$) to $[A]_t$ (the concentration of A at a time t after the start of the reaction).

We can take the antilogarithm of both sides of Equation 13.5 and rearrange it algebraically to obtain the concentration at time t directly as a function of time:[3]

$$[A]_t = [A]_0\, e^{-kt} \qquad (13.6)$$

■ The value of e^{-kt} decreases as t increases because the complete exponent, $-kt$, becomes increasingly negative.

The e in Equation 13.6 is the base of the system of natural logarithms ($e = 2.718\ldots$). Equation 13.6 shows that the concentration of A decreases exponentially with time. Calculations can be done using the exponential form, Equation 13.5, or the logarithmic form, Equation 13.6. When both of the concentrations are known, it is easiest to use Equation 13.5; when we wish to calculate either $[A]_0$ or $[A]_t$, it may be easier to use Equation 13.6.

Example 13.7
Concentration–Time Calculations for First-Order Reactions

Dinitrogen pentoxide is not very stable. In the gas phase or dissolved in a nonaqueous solvent, such as carbon tetrachloride, it decomposes by a first-order reaction into dinitrogen tetroxide and molecular oxygen:

$$2N_2O_5 \longrightarrow 2N_2O_4 + O_2$$

The rate law is

$$\text{rate} = k[N_2O_5]$$

At 45 °C, the rate constant for the reaction in carbon tetrachloride is $6.22 \times 10^{-4}\ \text{s}^{-1}$. If the initial concentration of N_2O_5 in a carbon tetrachloride solution at 45 °C is 0.500 M, what will its concentration be after exactly one hour (i.e., 1.00 hr)?

---

[2] For a first-order reaction, the integrated rate law is obtained by calculus as follows: The instantaneous rate of change of the reactant A is given as

$$\text{rate} = \frac{-d[A]}{dt} = kt$$

which can be rearranged to give

$$\frac{-d[A]}{[A]} = -k\, dt$$

Next, we integrate between $t = 0$ and $t = t$ as the concentration of A changes from $[A]_0$ to $[A]_t$

$$\int_{[A]_0}^{[A]_t} \frac{d[A]}{[A]} = \int_0^t -k\, dt$$

$$\ln[A]_t - \ln[A]_0 = -kt$$

Using the properties of logarithms this is rearranged to give

$$\ln \frac{[A]_0}{[A]_t} = kt$$

[3] Because of the nature of logarithms, if $\ln x = a$, then $e^{\ln x} = e^a$. But $e^{\ln x} = x$, so $x = e^a$. A similar relationship exists for common (i.e., base 10) logarithms. If $\log x = a$, then $10^{\log x} = x = 10^a$.

● **Analysis:** We're dealing with a first-order reaction and the relationship between concentration and time, and we are looking for a concentration after one hour. We are given an initial concentration, 0.500 M, and the rate constant for the reaction, $6.22 \times 10^{-4}\,s^{-1}$.

● **Assembling the Tools:** The tool we have to apply is either Equation 13.5 or 13.6, the integrated rate law for first-order reactions. Specifically, we have to solve for an unknown concentration. The easiest form of the equation to use when one of the unknowns is a concentration term is Equation 13.6. Also, remember that the unit of k involves seconds, not hours, so we must convert the given time, 1.00 hr, into seconds (1.00 hr $= 3.60 \times 10^3$ s).

● **Solution:** Let's begin by listing the data.

$$[N_2O_5]_0 = 0.500\ M \qquad\qquad [N_2O_5]_t = ?\ M$$

$$k = 6.22 \times 10^{-4}\ s^{-1} \qquad\qquad t = 3.60 \times 10^3\ s$$

Using Equation 13.6, $[N_2O_5]_t = [N_2O_5]_0 e^{-kt}$, we substitute the values listed above

$$[N_2O_5]_t = [0.500\ M]_0 e^{-(6.22\times10^{-4}\,s^{-1})(3.60\times10^3\,s)}$$

$$= (0.500\ M) \times e^{-2.239}$$

$$= (0.500\ M) \times 0.1065$$

$$= 0.05327\ M \quad \text{(correctly rounded}^4 \text{ to } 0.053\ M)$$

■ Calculating $e^{-2.239}$ is a simple operation using a scientific calculator. In most cases it is the inverse of the ln function.

After one hour, the concentration of N_2O_5 will have decreased to 0.053 M.

The calculation could also have been done using Equation 13.5. We would begin by solving for the concentration ratio, substituting values for k and t and keeping an extra significant figure:

$$\ln\frac{[N_2O_5]_0}{[N_2O_5]_t} = (6.22 \times 10^{-4}\ s^{-1}) \times (3.60 \times 10^3\ s)$$

$$\ln\frac{[N_2O_5]_0}{[N_2O_5]_t} = 2.239$$

We now need to take the antilogarithm, antiln, of both sides of the last equation. The antiln of the left side is

$$\text{antiln}\left(\ln\frac{[N_2O_5]_0}{[N_2O_5]_t}\right) = \frac{[N_2O_5]_0}{[N_2O_5]_t}$$

To take the antilogarithm of 2.239 we raise e to the 2.239 power:

$$\text{antiln}\,(2.239) = e^{2.239} = 9.384 \quad \text{(keeping a few extra significant figures)}$$

Reuniting the two halves of the equation yields

$$\frac{[N_2O_5]_0}{[N_2O_5]_t} = 9.384$$

Now, we can substitute the known concentration, $[N_2O_5]_0 = 0.500\ M$.

$$\frac{0.500\ M}{[N_2O_5]_t} = 9.384$$

---

^{4} There are special rules for significant figures for logarithms and antilogarithms. In writing the logarithm of a quantity, the number of digits written *after the decimal point* equals the number of significant figures in the quantity. Raising e to the -2.24 power is the same as taking the antilogarithm of -2.24. Because the quantity -2.24 has two digits after the decimal, the antilogarithm, 0.1064 . . ., must be rounded to 0.11 to show just two significant figures.

Solving for $[N_2O_5]_t$ gives

$$[N_2O_5]_t = \frac{0.500\ M}{9.384} = 0.053\ M \quad \text{(when correctly rounded)}$$

Our rounding is based on the two decimal places in the time and rate constant. You can see that using Equation 13.6, the exponential form, is much easier for working this particular problem.

Is the Answer Reasonable? Notice that the final concentration of N_2O_5, the reactant in the reaction, is *less* than its initial concentration. You'd know that you made a huge mistake if the calculated final concentration was greater than the initial 0.500 *M* because reactants are consumed in chemical reactions. Also, we obtained the same answer using both Equations 13.5 and 13.6, so we can be confident it's correct. (If you're faced with a problem like this, you don't have to do it both ways; we just wanted to show that either equation could be used.)

Practice Exercise 13.15

When designing a consumer product, it is desirable that it have a two-year shelf life. Often this means that the active ingredient in the product should not decrease by more than 5% in two years. If the reaction is first order, what rate constant must the decomposition reaction of the active ingredient have? (*Hint:* What are the initial and final percentages of active ingredient?)

Practice Exercise 13.16

In Practice Exercise 13.13, the reaction of sucrose with water in an acidic solution was described.

$$\underset{\text{sucrose}}{C_{12}H_{22}O_{11}} + H_2O \longrightarrow \underset{\text{glucose}}{C_6H_{12}O_6} + \underset{\text{fructose}}{C_6H_{12}O_6}$$

The reaction is first order with a rate constant of $6.17 \times 10^{-4}\ s^{-1}$ at 35 °C, when $[H^+] = 0.10\ M$. Suppose that in an experiment the initial sucrose concentration was 0.40 *M*. (a) What will the concentration be after exactly 2 hours (i.e., 2.00 h)? (b) How many minutes will it take for the concentration of sucrose to drop to 0.30 *M*?

Determining Rate Constants of First-Order Reactions

Using the properties of logarithms,[5] Equation 13.5 can be rewritten as the following difference in logarithms:

$$\ln[A]_0 - \ln[A]_t = kt$$

This difference can then be rearranged into a form that corresponds to the equation for a straight line.

$$
\begin{array}{ccccc}
\ln[A]_t & = & -kt & + & \ln[A]_0 \\
\updownarrow & & \Updownarrow & & \updownarrow \\
y & = & mx & + & b
\end{array}
$$

■ The equation for a straight line is usually written

$$y = mx + b$$

where x and y are variables, m is the slope, and b is the intercept of the line with the y axis.

A plot of the values of $\ln[A]_t$ (vertical axis) versus values of t (horizontal axis) should give a straight line that has a slope equal to $-k$, the negative of the rate constant. Such a plot is illustrated in Figure 13.6 for the decomposition of N_2O_5 into N_2O_4 and O_2 in carbon tetrachloride solvent.

One method of determining whether a reaction is first order with respect to a reactant is to plot the natural logarithm of the concentration as it changes over time and see whether a straight line is obtained.

---

[5] The logarithm of a quotient or ratio, $\ln \frac{a}{b}$, can be written as the difference $\ln a - \ln b$.

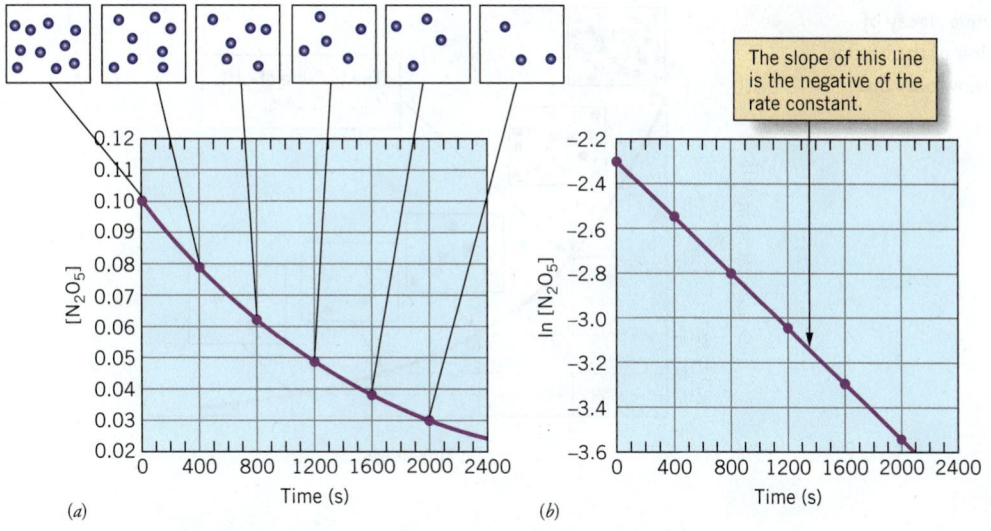

The slope of this line is the negative of the rate constant.

(a)

(b)

Figure 13.6 | The decomposition of N_2O_5. (a) A graph of concentration versus time for the decomposition at 45 °C. (b) A straight line is obtained if the logarithm of the concentration is plotted versus time. The slope of this line equals the negative of the rate constant for the reaction. Pictograms illustrate the relative amount of N_2O_5 molecules left.

Half-Life of a Reaction

The *half-life* of a reactant is a convenient way to describe how fast it reacts, particularly for a first-order process. A reactant's **half-life, $t_{1/2}$**, is the amount of time required for half of the reactant to disappear. A rapid reaction has a short half-life because half of the reactant disappears quickly. The equations for half-lives depend on the order of the reaction.

When we state the order of a reaction, we are referring to the overall reaction. Otherwise we state "with respect to" a specific reactant. When a reaction is first order, the half-life of the reactant can be obtained by setting $[A]_t$ equal to one-half of $[A]_0$ and using Equation 13.5.

$$[A]_t = \tfrac{1}{2}[A]_0$$

Substituting $\tfrac{1}{2}[A]_0$ for $[A]_t$ and $t_{1/2}$ for t in Equation 13.5, we have

$$\ln \frac{[A]_0}{\tfrac{1}{2}[A]_0} = kt_{1/2}$$

Note that the left-hand side of the equation simplifies to ln 2, so solving the equation for $t_{1/2}$ yields Equation 13.7:

$$t_{1/2} = \frac{\ln 2}{k} \tag{13.7}$$

Because k is a constant for a given reaction, the half-life is also a constant for any particular first-order reaction (at any given temperature). Remarkably, in other words, *the half-life of a first-order reaction is not affected by the initial concentration of the reactant*. This can be illustrated by one of the most common first-order events in nature, the change that radioactive isotopes undergo during radioactive "decay." In fact, you may have heard the term *half-life* used in reference to the life spans of radioactive substances.

Iodine-131 is an unstable, radioactive isotope of iodine and undergoes a nuclear reaction, emitting radiation[6] (more information about nuclear chemistry is presented in Chapter 21, Nuclear Chemistry). The intensity of the radiation emitted decreases, or *decays*, with time, as shown in Figure 13.7. Notice that the time it takes for the first half of the ^{131}I to disappear is eight days. During the next eight days, half of the remaining ^{131}I disappears, and so on. Regardless of the initial amount, it takes eight days for half of that amount of ^{131}I to disappear, which means that the half-life of ^{131}I is a constant, and the reaction is first-order.

TOOLS

Half-lives of first-order reactions

■ Because ln 2 equals 0.693, Equation 13.7 is sometimes written as

$$t_{1/2} = \frac{0.693}{k}$$

---

[6] Iodine-131 is used in the diagnosis of thyroid disorders. The thyroid gland is a small organ located just below the "Adam's apple" and astride the windpipe. It uses iodide ion to make a hormone, so when a patient is given a dose of ^{131}I mixed with nonradioactive I^-, both ions are taken up by the thyroid gland. The change in (temporary) radioactivity of the gland is a measure of thyroid activity.

Figure 13.7 | **First-order radioactive decay of iodine-131.** The initial concentration of the isotope is represented by $[I]_0$. Pictograms show only the relative amounts of ^{131}I left.

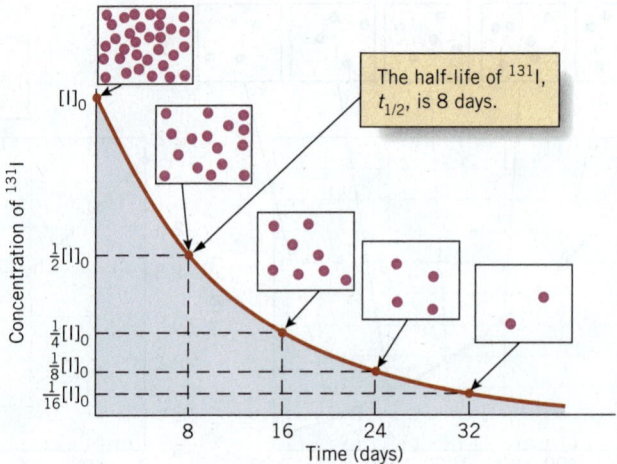

The half-life of ^{131}I, $t_{1/2}$, is 8 days.

Example 13.8
Half-Life Calculations

Suppose a patient is given a certain amount of iodine-131 as part of a diagnostic procedure for a thyroid disorder. Given that the half-life of radioactive iodine-131 is 8.02 days, what fraction of the initial iodine-131 would be present in a patient after 25.0 days if none of it were eliminated through natural body processes?

Analysis: We've learned that radioactive iodine-131 decays by a first-order process with a constant half-life. We can use the half-life to calculate the rate constant, k, for the reaction and then use the rate constant to calculate the amount of material left.

Assembling the Tools: We will need to use Equation 13.7 to determine the rate constant from the half-life,

$$t_{1/2} = \frac{\ln 2}{k}$$

and then Equation 13.5 to find the amount of material remaining after 25.0 days:

$$\ln \frac{[A]_0}{[A]_t} = kt$$

Solution: We will start by finding the first-order rate constant, k, which we can obtain from the half-life by rearranging Equation 13.7 (note that we keep one extra significant figure until the end):

$$k = \frac{\ln 2}{t_{1/2}} = \frac{0.693}{8.02 \text{ days}} = 0.08641 \text{ day}^{-1}$$

Then, we can use Equation 13.5 to compute the fraction, $\dfrac{[A]_0}{[A]_t}$,

$$\ln \frac{[A]_0}{[A]_t} = kt = (0.08641 \text{ day}^{-1})(25.0 \text{ days}) = 2.160$$

■ A logarithm must have three decimal places to give its anti-logarithm three significant figures.

Taking the antilogarithm of both sides, we have

$$\frac{[A]_0}{[A]_t} = e^{2.160} = 8.67$$

The initial concentration, $[A]_0$, is 8.67 times as large as the concentration after 25.0 days, so the fraction remaining after 25 days is

$$\frac{1}{8.67} = 0.115 = 0.12 \quad \text{when properly rounded}$$

Is the Answer Reasonable? A period of 24 days is exactly three half-lives, and 25 is very close to 24. Therefore, we can use the value of 24 days and apply the half-life concept three times. If we take the fraction initially present to be 1, we can set up the following table:

■ The fraction remaining after n half-lives is $(\frac{1}{2})^n$, or simply $\frac{1}{2^n}$.

Half-life	0	1	2	3
Fraction	1	$\frac{1}{2}$	$\frac{1}{4}$	$\frac{1}{8}$

Half of the iodine-131 is lost in the first half-life, half of that disappears in the second half-life, and so on. Therefore, the fraction remaining after three half-lives is $\frac{1}{8}$, which is very close to our value of $\frac{1}{8.67}$.

In Practice Exercise 13.13, the reaction of sucrose with water was found to be first order with respect to sucrose. The rate constant under the conditions of the experiments was $6.17 \times 10^{-4} \, s^{-1}$. Calculate the value of $t_{1/2}$ for this reaction in minutes. How many minutes would it take for three-quarters of the sucrose to react? (*Hint:* What fraction of the sucrose remains?)

Practice Exercise 13.17

From the answer to Practice Exercise 13.15, determine the half-life of an active ingredient that has a shelf life of 2.00 years.

Practice Exercise 13.18

The radioactive isotope, phosphorus-32, has a half-life of 14.26 days. What percent of phosphorus-32 will remain after 60 days?

Practice Exercise 13.19

Carbon-14 Dating

Carbon-14 is a radioactive isotope with a half-life of 5730 years and a rate constant, k, of $1.21 \times 10^{-4} \, yr^{-1}$. It is formed in small amounts in the upper atmosphere by the action of cosmic rays on nitrogen atoms. Once formed, the carbon-14 diffuses into the lower atmosphere. It becomes oxidized to carbon dioxide and enters earth's biosphere by means of photosynthesis. Carbon-14 thus becomes incorporated into plant substances and into the materials of animals that eat plants. Even as the carbon-14 decays, more is ingested by the living thing. The net effect is an overall steady-state availability of carbon-14 in the global system. As long as the plant or animal is alive, its ratio of carbon-14 atoms to carbon-12 atoms is constant. At death, an organism's remains have as much carbon-14 as they can ever have, and they now slowly lose this carbon-14 by radioactive decay. The decay is a first-order process with a rate independent of the *number* of original carbon atoms. The ratio of carbon-14 to carbon-12, therefore, can be related to the years that have elapsed between the time of death and the time of the measurement. The critical assumption in carbon-14 dating is that the steady-state availability of carbon-14 from the atmosphere has remained largely unchanged since the specimen stopped living.[7]

■ Willard F. Libby won the Nobel Prize in chemistry in 1960 for developing the carbon-14 method for dating ancient objects.

In contemporary biological samples, the ratio $^{14}C/^{12}C$ is about 1.2×10^{-12}. Thus, each fresh 1.0 g sample of biological carbon in equilibrium with the $^{14}CO_2$ of the atmosphere has a ratio of 5.8×10^{10} atoms of carbon-14 to 4.8×10^{22} atoms of carbon-12. The ratio decreases by a factor of 2 for each half-life period of ^{14}C.

■ In all dating experiments, the amounts of sample are extremely small, and extraordinary precautions must be taken to avoid contamination by "modern" materials.

The dating of an object makes use of the fact that radioactive decay is a first-order process. If we let r_0 stand for the $^{14}C/^{12}C$ ratio at the time of death of the carbon containing

---

[7] The available atmospheric pool of carbon-14 atoms fluctuates somewhat with the intensities of cosmic ray showers, with slow, long-term changes in earth's magnetic field, and with the huge injections of carbon-12 into the atmosphere from the large-scale burning of coal and petroleum in the 1900s. To reduce the uncertainties in carbon-14 dating, results of the method have been correlated with dates made by tree-ring counting. For example, an uncorrected carbon-14 dating of a Viking site at L'Anse aux Meadows, Newfoundland, gave a date of AD 895 ± 30. When corrected, the date of the settlement became AD 997, almost exactly the time indicated in Icelandic sagas for Leif Eriksson's landing at "Vinland," now believed to be the L'Anse aux Meadows site.

species and r_t stand for the $^{14}C/^{12}C$ ratio now, after the elapse of t years, we can substitute into Equation 13.5 to obtain

$$\ln \frac{r_0}{r_t} = kt \qquad (13.8)$$

where k is the rate constant for the decay (the *decay constant* for ^{14}C) and t is the elapsed time. We can now substitute $1.21 \times 10^{-4} \text{ yr}^{-1}$ for k into Equation 13.8 to give

TOOLS

Carbon-14 dating

$$\ln \frac{r_0}{r_t} = (1.21 \times 10^{-4} \text{ yr}^{-1})t \qquad (13.9)$$

Equation 13.9 can be used to calculate the age of a once-living object if its current $^{14}C/^{12}C$ ratio can be measured.

We have to realize here that we are not using the number of disintegrating nuclei, but the ratio of $^{14}C/^{12}C$, which is an intensive property. We need to use intensive properties to calculate the age of the object, since the size of the sample cannot be a determining factor.

Example 13.9
Calculating the Age of an Object by ^{14}C Dating

Using a mass spectrometer, a sample of an ancient wooden object was found to have a ratio of ^{14}C to ^{12}C equal to 3.3×10^{-13} as compared to a contemporary biological sample, which has a ratio of ^{14}C to ^{12}C of 1.2×10^{-12}. What is the age of the object?

Analysis: We are given the ratio of ^{14}C to ^{12}C and asked to find the age of the object.

Assembling the Tools: The tool we will use is Equation 13.9,

$$\ln \frac{r_0}{r_t} = (1.21 \times 10^{-4} \text{ yr}^{-1})t$$

and substitute the values given with $r_0 = 1.2 \times 10^{-12}$ and $r_t = 3.3 \times 10^{-13}$.

Solution: The contemporary ratio of ^{14}C to ^{12}C was given as 1.2×10^{-12}. This corresponds to r_0 in Equation 13.9. Substituting into Equation 13.9 gives

$$\ln \frac{1.2 \times 10^{-12}}{3.3 \times 10^{-13}} = (1.21 \times 10^{-4} \text{ yr}^{-1})t$$

$$\ln (3.6) = 1.281 = (1.21 \times 10^{-4} \text{ yr}^{-1})t$$

Solving for t ($1.281/1.21 \times 10^{-4} \text{ yr}^{-1} = t$) gives an age of 1.1×10^4 years (11,000 years).

Is the Answer Reasonable? We've been told that the object is ancient, so 11,000 years old seems to make sense. (Also, if you had substituted incorrectly into Equation 13.9, the answer would have been negative, and that certainly doesn't make sense!)

Practice Exercise 13.20

An ancient piece of wood was found to have a ^{14}C to ^{12}C ratio one-tenth of that in living trees. How many years old is this piece of wood ($t_{1/2} = 5730$ for ^{14}C)? (*Hint:* Recall the relationship between the integrated rate equation and half-life.)

Practice Exercise 13.21

The calibration curves for carbon-14 dating are constantly being updated. Before 2010, when using carbon-14 dating, samples that had decayed less than 5% and those that had decayed more than 95% may have had unacceptably large uncertainties. How many years before the present could carbon-14 be used for dating an object using the 5% and 95% limits?

Second-Order Reactions

For simplicity, we will only consider a **second-order reaction** with a rate law with the form

$$\text{rate} = k[B]^2$$

The relationship between concentration and time for a reaction with such a rate law is given by Equation 13.10, an equation that is quite different from that for a first-order reaction.

$$\frac{1}{[B]_t} - \frac{1}{[B]_0} = kt \qquad \text{(13.10)}$$

TOOLS

Integrated second-order rate law

$[B]_0$ is the initial concentration of B and $[B]_t$ is the concentration at time t.

Example 13.10
Concentration–Time Calculations for Second-Order Reactions

Nitrosyl chloride, NOCl, decomposes slowly to NO and Cl_2:

$$2NOCl \longrightarrow 2NO + Cl_2$$

The rate law for this reaction is second order in NOCl:

$$\text{rate} = k[NOCl]^2$$

The rate constant k equals 0.020 L mol^{-1} s^{-1} at a certain temperature. If the initial concentration of NOCl in a closed reaction vessel is 0.050 M, what will the concentration be after 35 minutes?

Analysis: We're given a rate law that is for a second-order reaction. We must calculate $[NOCl]_t$, the molar concentration of NOCl, after 35 minutes (i.e., 2100 s).

Assembling the Tools: Our tool for doing this is Equation 13.10, the integrated rate law for second-order reactions.

Solution: Let's begin by tabulating the data:

$$[NOCl]_0 = 0.050 \ M \qquad\qquad [NOCl]_t = ? \ M$$

$$k = 0.020 \text{ L mol}^{-1} \text{s}^{-1} \qquad\qquad t = 2100 \text{ s}$$

The equation we wish to substitute into is

$$\frac{1}{[NOCl]_t} - \frac{1}{[NOCl]_0} = kt$$

Making the substitutions gives

$$\frac{1}{[NOCl]_t} - \frac{1}{(0.050 \text{ mol L}^{-1})} = (0.020 \text{ L mol}^{-1} \text{s}^{-1}) \times (2100 \text{ s})$$

Solving for $1/[NOCl]_t$ gives

$$\frac{1}{[NOCl]_t} - 20 \text{ L mol}^{-1} = 42 \text{ L mol}^{-1}$$

$$\frac{1}{[NOCl]_t} = 62 \text{ L mol}^{-1}$$

Taking the reciprocals of both sides gives us the value of $[NOCl]_t$:

$$[NOCl]_t = \frac{1}{62 \text{ L mol}^{-1}} = 0.016 \text{ mol L}^{-1} = 0.016 \ M$$

The molar concentration of NOCl has decreased from 0.050 M to 0.016 M after 35 minutes.

Is the Answer Reasonable? The concentration of NOCl has decreased, so the answer appears to be reasonable.

Practice Exercise 13.22

For the reaction in Example 13.10, determine how many minutes it would take for the NOCl concentration to drop from 0.040 M to 0.010 M. (*Hint:* In solving Equation 13.10, time must be a positive value.)

Practice Exercise 13.23

A sample of nitrosyl chloride was collected for analysis at 10:35 am. At 3:15 pm the same day, the sample was analyzed and was found to contain 0.00035 M NOCl. What was the concentration of NOCl at the time the sample was collected?

Second-Order Rate Constants

The rate constant k for a second-order reaction, one with a rate following Equation 13.10, can be determined graphically by a method similar to that used for a first-order reaction. We can rearrange Equation 13.10 so that it corresponds to an equation for a straight line:

$$\frac{1}{[B]_t} = kt + \frac{1}{[B]_0}$$

$$\updownarrow \qquad \upuparrows \qquad \updownarrow$$

$$y = mx + b$$

When a reaction is second order, a plot of $1/[B]_t$ versus t should yield a straight line having a slope k. This is illustrated in Figure 13.8 for the decomposition of HI, using data in Table 13.1.

Half-Lives of Second-Order Reactions

The half-life of a second-order reaction *does* depend on initial reactant concentrations. We can see this by examining Figure 13.8, which follows the decomposition of gaseous HI, a second-order reaction. The reaction begins with a hydrogen iodide concentration of 0.10 M. After 125 seconds, the concentration of HI drops to 0.050 M, so 125 s is the observed half-life when the initial concentration of HI is 0.10 M. If we then take 0.050 M as the next "initial" concentration, we find that it takes 250 seconds (at a *total* elapsed time of 375 seconds) to drop to 0.025 M. If we cut the initial concentration in half, from 0.10 M to 0.050 M, the half-life doubles, from 125 to 250 seconds.

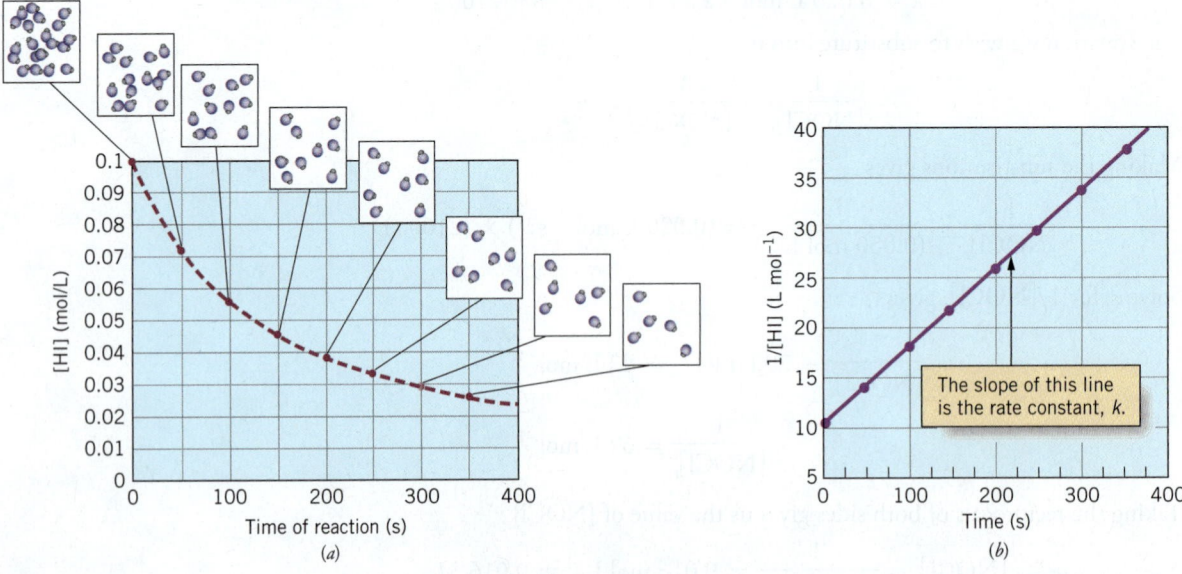

Figure 13.8 | **Second-order kinetics.** (*a*) A graph of concentration versus time for the decomposition of HI for the data in Table 13.1. (*b*) A straight line is obtained if $1/[HI]$ is plotted versus time. Pictogram illustrates relative number of HI molecules left.

It can be shown that for a second-order reaction of the type we're studying, the half-life is inversely proportional to the initial concentration of the reactant. The half-life is related to the rate constant by Equation 13.11.

$$t_{1/2} = \frac{1}{k \times (\text{initial concentration of reactant})} = \frac{1}{k \times [B]_0} \qquad \textbf{(13.11)}$$

Zero-Order Reactions

Zero-order reactions have an exponent of zero on the concentration term, so they do not depend on the concentration of any reactant:

$$\text{rate} = k[A]^0 = k$$

When the concentration of the reactant is plotted versus time, it will give a straight line if the reaction is a zero-order reaction, since the integrated rate law equation is

$$[A]_t = -kt + [A]_0 \qquad \textbf{(13.12)}$$

TOOLS

Integrated zero-order rate law

This can also be plotted in a similar fashion to the first-order and second-order reactions, which will give the rate constant, k, as the negative of the slope. The half-life expression for a zero-order reaction can be determined by substituting $[A]_t = \frac{1}{2}[A]_0$ into Equation 13.12 and solving the equation to get

$$t_{1/2} = \frac{1}{k}\frac{[A]_0}{2} \qquad \textbf{(13.13)}$$

We summarize the important points concerning rate laws, orders of reaction, integrated rate laws and half lives in Table 13.5.

TABLE 13.5	Summary of Important Rate Equations		
Order of Reaction	Rate Law	Integrated Rate Law	Equation for Half-life
0	Rate = k	$[A]_t = -kt + [A]_0$	$t_{1/2} = \dfrac{1}{k}\dfrac{[A]_0}{2}$
1	Rate = $k[A]$	$[A]_t = [A]_0\,e^{-kt}$	$t_{1/2} = \dfrac{\ln 2}{k}$
2	Rate = $k[B]^2$	$\dfrac{1}{[B]_t} - \dfrac{1}{[B]_0} = kt$	$t_{1/2} = \dfrac{1}{k \times [B]_0}$

Graphical Interpretation of Orders of Reactions

In the special case where a reaction has only one reactant that has a measurable change in concentration with time, the integrated rate laws and, in particular, their graphical representations are very useful. Concentration versus time data can be utilized to determine the order of the reaction using plots suggested by the integrated rate laws. Zero-order reactions yield a straight line when the concentration of the reactant is plotted versus time and the negative of the slope is the rate constant. First-order reactions have a linear graph when the natural log of the concentration of the reactant is plotted versus time and the negative slope of the line is the rate constant. Second-order reactions require the inverse of the concentration of the reactant plotted versus time to give a straight line, and the slope of the line is the rate constant. All of this information is summarized in Table 13.6.

TABLE 13.6	Graphical Interpretation of Orders of Reaction		
Order of Reaction	Straight-Line Plot	Rate Constant, k	Half-Life Dependence
0	$[X]$ vs. time	Negative slope of line	Depends on initial concentration
1	ln $[X]$ vs. time	Negative slope of line	No dependence on concentration
2	$1/[X]$ vs. time	Positive slope of line	Depends on initial concentration

Example 13.11
Half-Life Calculations

The reaction $2HI(g) \longrightarrow H_2(g) + I_2(g)$ has the rate law, rate $= k\,[HI]^2$, with $k = 0.079$ L mol^{-1} s^{-1} at 508 °C. What is the half-life for this reaction at this temperature when the initial HI concentration is 0.10 M?

Analysis: The rate law tells us that the reaction is second order, and we are asked for the half-life for this reaction. Keep in mind that for second-order reactions, the half-life depends on the initial concentration.

Assembling the Tools: To calculate the half-life, we need to use Equation 13.11 as the tool to solve this problem.

Solution: The initial concentration is 0.10 mol L^{-1}; $k = 0.079$ L mol^{-1} s^{-1}. Substituting these values into Equation 13.11 gives

$$t_{1/2} = \frac{1}{(0.079 \text{ L mol}^{-1} \text{ s}^{-1})(0.10 \text{ mol L}^{-1})}$$

$$= 1.3 \times 10^2 \text{ s}$$

Is the Answer Reasonable? To estimate the answer, we round the 0.079 to 0.1, so the estimated answer is $\dfrac{1}{0.1 \times 0.1} = 100$. This is close to our calculated value. In addition, we check that the units cancel to leave only the seconds units. Both of these checks support our answer.

Practice Exercise 13.24

The reaction $2NO_2 \longrightarrow 2NO + O_2$ is second order with respect to NO_2. If the initial concentration of $NO_2(g)$ is 6.54×10^{-4} mol L^{-1}, what is the rate constant if the initial reaction rate is 4.42×10^{-7} mol L^{-1} s^{-1}? What is the half-life of this system? (*Hint:* Start by setting up and solving the rate law before the integrated equation.)

Practice Exercise 13.25

Suppose that the value of $t_{1/2}$ for a certain reaction was found to be independent of the initial concentration of the reactants. What could you say about the order of the reaction? Justify your answer.

13.5 | Molecular Basis of Collision Theory

As a rule, the rate of a chemical reaction increases by a factor of approximately two for each ten degrees Celsius increase in temperature, although the actual amount of increase differs from one reaction to another. Temperature evidently has a strong effect on reaction rate. To understand why, we need to develop theoretical models that explain our observations. One of the simplest models is called *collision theory*.

Collision Theory

The basic postulate of **collision theory** is that the rate of a reaction is proportional to the number of *effective* collisions per second among the reactant molecules. An **effective collision** is one that actually gives product molecules. Anything that can increase the frequency of effective collisions should, therefore, increase the rate.

One of the several factors that influences the number of effective collisions per second is *concentration*. As reactant concentrations increase, the number of collisions per second of all types, including effective collisions, cannot help but increase. We'll return to the significance of concentration in Section 13.8.

Not *every* collision between reactant molecules actually results in a chemical change. We know this because the reactant atoms or molecules in a gas or a liquid undergo an enormous number of collisions per second with each other. If every collision were effective, all reactions would be over in an instant. *Only a very small fraction of all of the collisions can really lead to a net change.* Lets examine why so few collisions are effective.

Molecular Orientation

In most reactions, when two reactant molecules collide, they must be oriented correctly for a reaction to occur. For example, the reaction represented by the following balanced chemical equation,

$$2NO_2Cl \longrightarrow 2NO_2 + Cl_2$$

appears to proceed by a two-step sequence of reactions. One step involves the collision of an NO_2Cl molecule with a chlorine atom:

$$NO_2Cl + Cl \longrightarrow NO_2 + Cl_2$$

The orientation of the NO_2Cl molecule, when hit by the Cl atom, is important (see Figure 13.9). The poor orientation shown in Figure 13.9*a* cannot result in the formation of Cl_2 because the two Cl atoms are not being brought close enough together for a new Cl—Cl bond to form as an N—Cl bond breaks. Figure 13.9*b* shows the necessary orientation if the collision of NO_2Cl and Cl is to effectively lead to products.

■ When 0.10 *M* HI decomposes as in Figure 13.5, only about one in every billion billion (10^{18}) collisions leads to a net chemical reaction. In all the other collisions the HI molecules just bounce off each other.

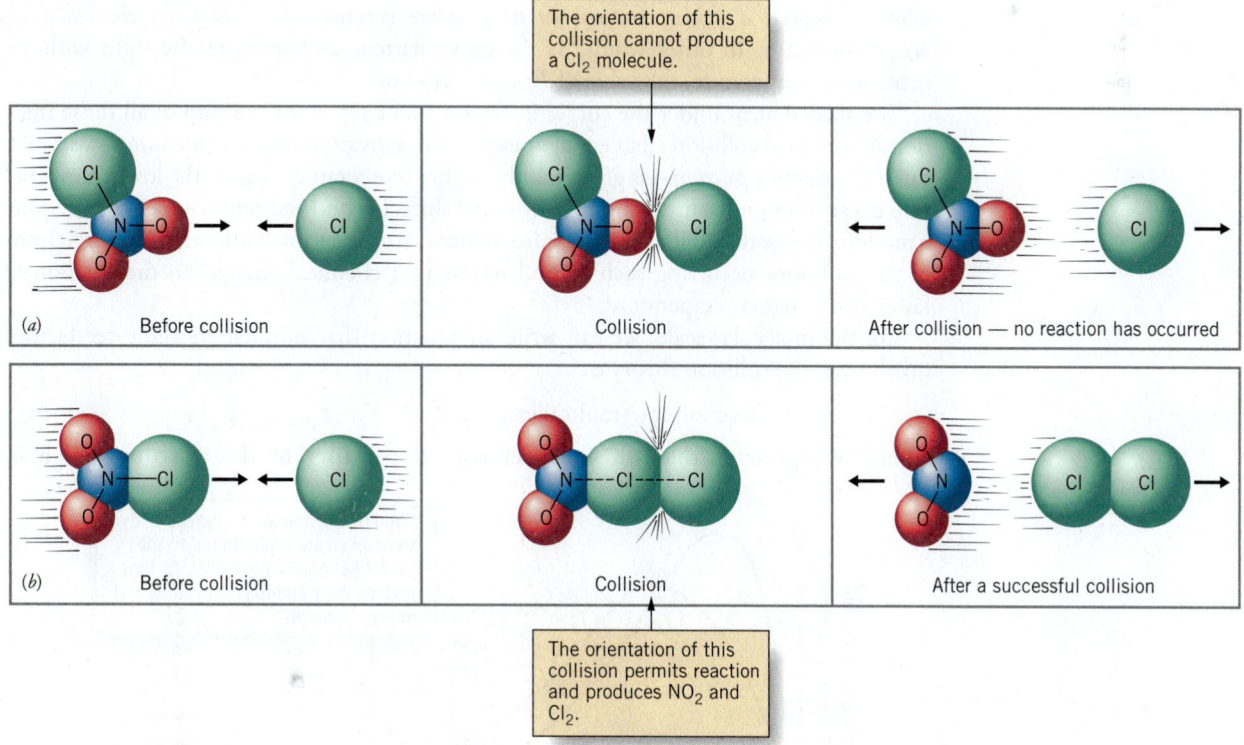

The orientation of this collision cannot produce a Cl_2 molecule.

(*a*) Before collision   Collision   After collision — no reaction has occurred

(*b*) Before collision   Collision   After a successful collision

The orientation of this collision permits reaction and produces NO_2 and Cl_2.

Figure 13.9 | **The importance of molecular orientation during a collision in a reaction.** The key step in the decomposition of NO_2Cl to NO_2 and Cl_2 is the collision of a Cl atom with a NO_2Cl molecule. (*a*) A poorly oriented collision. (*b*) An effectively oriented collision.

Molecular Kinetic Energy

Not all collisions, even those correctly oriented, are energetic enough to result in products, and this is the major reason that only a small percentage of all collisions actually lead to chemical change. The colliding particles must carry into the collision a certain minimum combined molecular kinetic energy, called the **activation energy, E_a**. In a successful collision, activation energy changes over to potential energy as the particles hit each other and chemical bonds in the reactants reorganize into those of the products. For most chemical reactions, the activation energy is quite large, and only a small fraction of all well-oriented, colliding molecules have it.

We can understand the existence of activation energy by studying in detail what actually takes place during a collision. For old bonds to break and new bonds to form, the atomic nuclei within the colliding particles must get close enough together. The molecules on a collision course must, therefore, be moving with a combined kinetic energy great enough to overcome the natural repulsions between electron clouds. Otherwise, the molecules simply veer away or bounce apart. Only fast molecules with large kinetic energies can collide with enough collision energy to enable their nuclei and electrons to overcome repulsions and thereby reach the positions required for the bond breaking and bond making that the chemical change demands, as shown in Figure 13.10.

■ The fraction of molecules, f_{KE}, having or exceeding the activation energy, E_a, is given by the expression

$$\ln f_{KE} = Ae^{-E_a/RT}$$

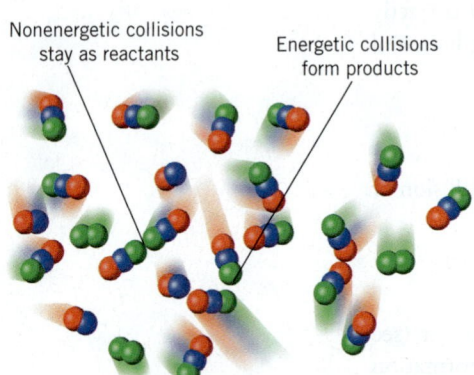

Nonenergetic collisions stay as reactants

Energetic collisions form products

Figure 13.10 | The energetics of collisions that lead to products.

Effect of Temperature

With the concept of activation energy, we can now explain why the rate of a reaction increases with increasing temperature. We'll use the two kinetic energy curves in Figure 13.11, each corresponding to a different temperature for the same mixture of reactants. Just like the graphs in Figure 11.18, each curve is a plot of the different *fractions* of all collisions (vertical axis) that have particular values of kinetic energy (horizontal axis). Notice what happens to the plots when the temperature is increased; the maximum point shifts to the right and the curve flattens somewhat. However, *a modest increase in temperature generally does not affect the reaction's activation energy*. In other words, as the curve flattens and shifts to the right with an increase in temperature, the value of E_a stays the same.

The shaded areas under the curves in Figure 13.11 represent the sum of all those fractions of the total collisions that equal or exceed the activation energy. This sum—we could call it the *reacting fraction*—is greater at the higher temperature than at the lower temperature because a significant fraction of the curve shifts beyond the activation energy in even a modest change to a higher temperature. In other words, at the higher temperature, more of the collisions occurring each second results in a chemical change, so the reaction is faster at the higher temperature.

On the molecular scale, we can write an equation that summarizes the three factors involved in the collision theory as

$$\text{reaction rate(molecules L}^{-1}\text{ s}^{-1}) = N \times f_{\text{orientation}} \times f_{KE}$$

where N represents the collisions per second per liter of the mixture, which is

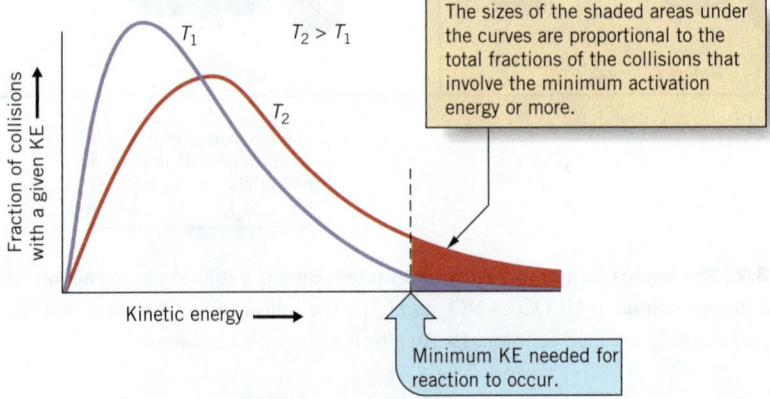

T_1 $T_2 > T_1$

T_2

The sizes of the shaded areas under the curves are proportional to the total fractions of the collisions that involve the minimum activation energy or more.

Fraction of collisions with a given KE

Kinetic energy

Minimum KE needed for reaction to occur.

Figure 13.11 | Kinetic energy distributions for a reaction mixture at two different temperatures.

approximately 10^{27} s^{-1}. The two other terms represent the fraction of collisions with the correct orientation, $f_{orientation}$, and the fraction of collisions with the required total kinetic energy, f_{KE}. To convert this to the laboratory scale where the rate of the reaction has units of mol L^{-1} s^{-1}, we divide the equation by Avogadro's number:

$$\text{reaction rate (mol L}^{-1}\text{ s}^{-1}) = \frac{\text{reaction rate (molecules L}^{-1}\text{ s}^{-1})}{6.02 \times 10^{23}(\text{molecules mol}^{-1})}$$

13.6 | Molecular Basis of Transition State Theory

Transition state theory is used to explain in detail what happens when reactant molecules come together in a collision. Most often, those in a head-on collision slow down, stop, and then emerge from the collision unchanged. When a collision does cause a reaction, the particles that separate are those of the products. Regardless of what happens to them, however, as the molecules on a collision course slow down, their total kinetic energy decreases as it changes into potential energy (PE). It's like the momentary disappearance of the kinetic energy of a tennis ball when it hits the racket. In the deformed racket and ball, this energy becomes potential energy, which soon changes back to kinetic energy as the ball takes off in a new direction.

Potential Energy Diagrams

To visualize the relationship between activation energy and the development of total potential energy, we sometimes use a *potential energy diagram* (Figure 13.12). The vertical axis represents changes in *potential* energy as the *kinetic* energy of the colliding particles changes over to potential energy. The horizontal axis is called the **reaction coordinate**, and it represents the extent to which the reactants have changed to the products. It helps us follow the path taken by the reaction as reactant molecules come together and change into product molecules. Activation energy appears as a potential energy "hill" or barrier between the reactants and products. Only colliding molecules, properly oriented, which can deliver sufficient kinetic energy that can be converted into potential energy at least as large as E_a, are able to climb over the hill and produce products.

We can use a potential energy diagram to follow the progress of both an unsuccessful and a successful collision (see Figure 13.13). As two reactant molecules collide, we say that they begin to climb the potential energy barrier as they slow down and experience the conversion of their kinetic energy into potential energy. However, if their combined initial kinetic energies are less than E_a, the molecules are unable to reach the top of the hill

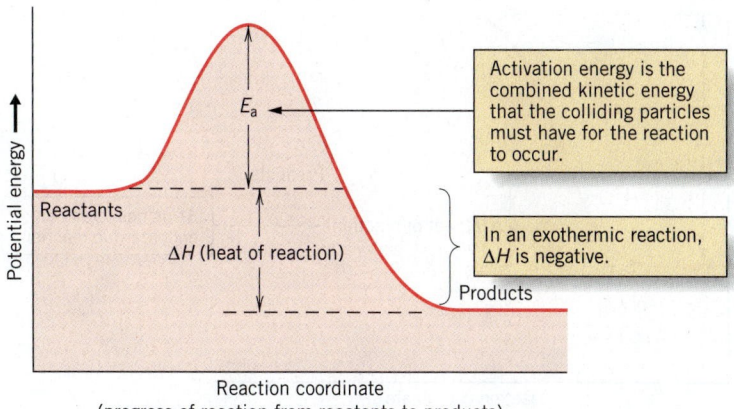

Reaction coordinate
(progress of reaction from reactants to products)

Figure 13.12 | Potential energy diagram for an exothermic reaction.

Figure 13.13 | The difference between an unsuccessful and a successful collision.

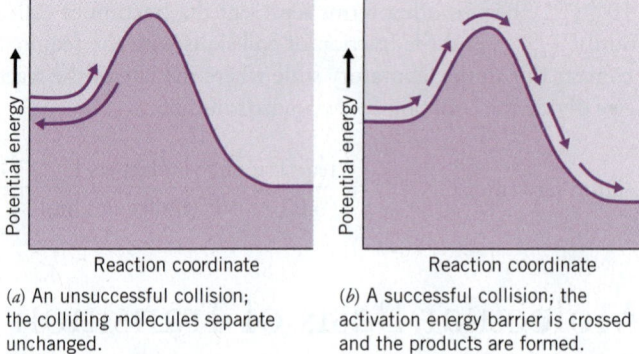

(a) An unsuccessful collision; the colliding molecules separate unchanged.

(b) A successful collision; the activation energy barrier is crossed and the products are formed.

(Figure 13.13a). Instead, they fall back toward the reactants. They bounce apart, chemically unchanged, with their original total kinetic energy; no net reaction has occurred. On the other hand, if the combined kinetic energy of the colliding molecules equals or exceeds E_a, and if the molecules are oriented properly, they are able to pass over the activation energy barrier and form product molecules (Figure 13.13b).

Potential Energy Diagrams and Heat of Reaction

A reaction's potential energy diagram, such as that of Figure 13.12, helps us to visualize the *heat of reaction*, ΔH, a concept introduced in Chapter 6. It's the difference between the potential energy of the products and the potential energy of the reactants. Figure 13.12 shows the potential energy diagram for an *exothermic* reaction because the products have a *lower* potential energy than the reactants. In such a system, the net decrease in potential energy appears as an increase in the molecular kinetic energy of the emerging product molecules. The temperature of the system increases during an exothermic reaction because the average molecular kinetic energy of the system increases.

A potential energy diagram for an endothermic reaction is shown in Figure 13.14. Now, the products have a *higher* potential energy than the reactants and, in terms of the heat of reaction, a net input of energy is needed to form the products. Endothermic reactions produce a cooling effect as they proceed because there is a net conversion of molecular kinetic energy to potential energy. As the *total* molecular kinetic energy decreases, the *average* molecular kinetic energy decreases as well, and the temperature drops.

Notice that E_a for an endothermic process is greater than the heat of reaction. If ΔH is both positive and *high*, E_a must also be high, making such reactions very slow. However, for an exothermic reaction (ΔH is negative), we cannot tell from ΔH how large E_a is. It could be high, making for a slow reaction, despite its being exothermic. If E_a is low, the reaction would be rapid and all its heat would appear quickly.

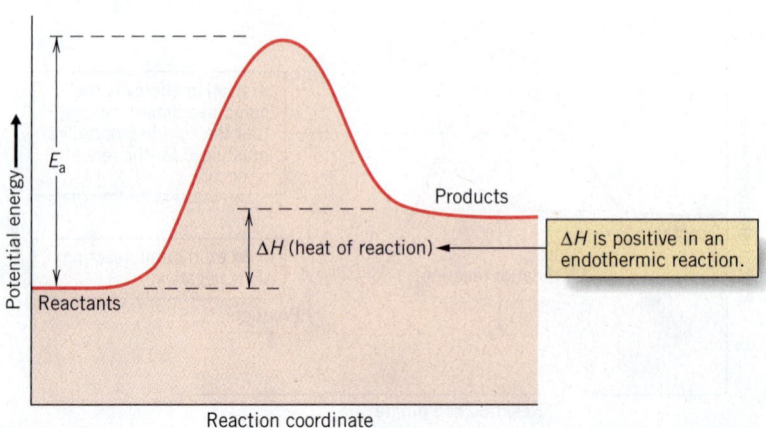

Figure 13.14 | A potential energy diagram for an endothermic reaction.

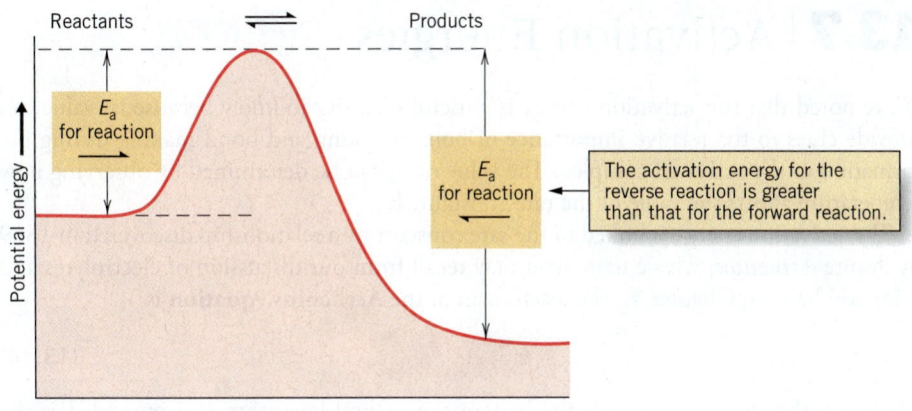

Figure 13.15 | The activation energy barrier for the forward and reverse reactions.

In Chapter 6, we saw that when the direction of a reaction is reversed, the sign given to the enthalpy change, ΔH, is reversed. In other words, a reaction that is exothermic in the forward direction *must* be endothermic in the reverse direction, and vice versa. This might seem to suggest that reactions are generally reversible. Many are, but if we look again at the energy diagram for a reaction that is exothermic in the forward direction (Figure 13.12), we see that the reaction is endothermic in the opposite direction *and must have a significantly higher activation energy* than the forward reaction. What differs most for the forward and reverse directions is the relative height of the activation energy barrier (Figure 13.15).

One of the main reasons for studying activation energies is that they provide information about what actually occurs during an effective collision. For example, in Figure 13.9*b*, we described a way that NO_2Cl could react successfully with a Cl atom during a collision. During this collision, there is a moment when the N—Cl bond is partially broken and the new Cl—Cl bond is partially formed. This brief moment during a successful collision is called the reaction's **transition state**. The potential energy of the transition state corresponds to the high point on the potential energy diagram (see Figure 13.16). The unstable chemical species that momentarily exists at this instant, $O_2N\cdots Cl\cdots Cl$, with its partially formed and partially broken bonds, is called the **activated complex**.

The size of the activation energy tells us about the relative importance of bond breaking and bond making during the formation of the activated complex. A very high activation energy suggests, for instance, that bond *breaking* contributes very heavily to the formation of the activated complex because bond breaking is an energy-absorbing process. On the other hand, a low activation energy may mean that bonds of about equal strength are being both broken and formed simultaneously.

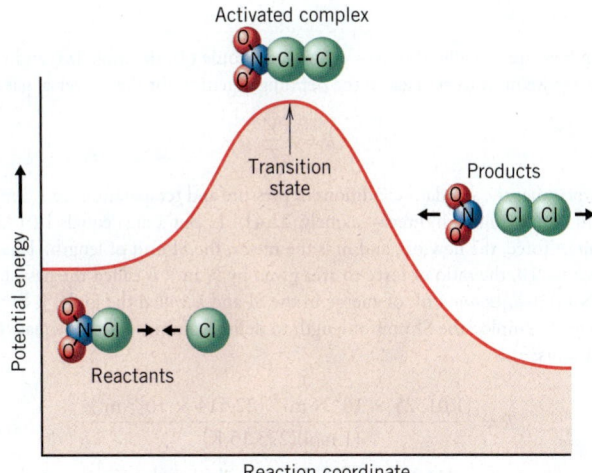

Figure 13.16 | The transition state and the activated complex. Formation of an activated complex in the reaction between NO_2Cl and Cl. $NO_2Cl + Cl \longrightarrow NO_2 + Cl_2$.

13.7 | Activation Energies

We've noted that the activation energy is a useful quantity to know because its value can provide clues to the relative importance of bond breaking and bond making during the formation of the activated complex. The value of E_a can be determined by observing how temperature affects the value of the rate constant, k.

The activation energy is linked to the rate constant by a relationship discovered in 1889 by Svante Arrhenius, whose name you may recall from our discussion of electrolytes and acids and bases in Chapter 4. The usual form of the **Arrhenius equation** is

TOOLS
Arrhenius equation

$$k = Ae^{-E_a/RT} \qquad (13.14)$$

where k is the rate constant, e is the base of the natural logarithm system, and T is the Kelvin temperature. A is a proportionality constant sometimes called the **frequency factor** or the **pre-exponential factor**. R is the gas constant, which we'll express in our study of kinetics in energy units—namely, R equals 8.314 J mol^{-1} K^{-1}.[8]

Graphical Determination of Activation Energy

Equation 13.14 is normally used in its logarithmic form. If we take the natural logarithm of both sides, we obtain

$$\ln k = \ln A - E_a/RT$$

Let's rewrite the equation as

$$\ln k = \ln A (E_a/R) \times (1/T) \qquad (13.15)$$

We know that the rate constant k varies with the temperature T, which also means that the quantity $\ln k$ varies with the quantity $(1/T)$. These two quantities—namely, $\ln k$ and $1/T$—are variables, so Equation 13.15 is in the form of an equation for a straight line:

$$\ln k = \ln A + (-E_a/R) \times (1/T)$$
$$\updownarrow \qquad \updownarrow \qquad \updownarrow \qquad \updownarrow$$
$$y = b + m \qquad x$$

To determine the activation energy, we can make a graph of $\ln k$ versus $1/T$, measure the slope of the line, and then use the relationship,

$$\text{slope} = -E_a/R$$

to calculate E_a. Example 13.12 illustrates how this is done.

---

[8] The units of R given here are actually SI units—namely, the joule (J), the mole (n), and the kelvin (K). To calculate R in these units we need to go back to the defining equation for the universal gas law:

$$R = \frac{PV}{nT}$$

In Chapter 10 we learned that the standard conditions of pressure and temperature are 1 atm and 273.15 K; we expressed the standard molar volume in liters—namely, 22.414 L. But 1 atm equals 1.01325×10^5 N m^{-2}, where N is the SI unit of force, the newton, and m is the meter, the SI unit of length. Thus, m^2 is area given in SI units. From Chapter 10, the ratio of force to area given by N m^{-2} is called the pascal, Pa, and force times distance (i.e., N m) defines one unit of energy in the SI and is called the joule, J. In the SI, volume must be expressed as m^3 to employ the SI unit of length to define volume, and 1 L equals 10^{-3} m^3. We can now calculate R in SI units:

$$R = \frac{(1.01325 \times 10^5 \text{ N m}^{-2})(22.414 \times 10^{-3} \text{ m}^3)}{(1 \text{ mol})(273.15 \text{ K})}$$

$$= 8.314 \text{ N m mol}^{-1} \text{ K}^{-1} = 8.314 \text{ J mol}^{-1} \text{ K}^{-1}$$

Example 13.12
Determining Energy of Activation Graphically

Consider again the decomposition of NO_2 into NO and O_2. The balanced chemical equation is

$$2NO_2(g) \longrightarrow 2NO(g) + O_2(g)$$

The following data were collected for the reaction

Rate Constant, k (L mol^{-1} s^{-1})	Temperature (°C)
7.8	405
9.9	415
14	425
18	435
24	445

Use the graphical method to determine the activation energy for the reaction in kilojoules per mole.

Analysis: We are asked to plot the given rate constants at different temperatures to determine the activation energy for the reaction from the slope of the graph, which is $-E_a/R$.

Assembling the Tools: Equation 13.15 is the tool that applies, but the use of rate data to determine the activation energy graphically requires that we plot ln k, not k, versus the *reciprocal* of the *Kelvin* temperature, so we have to convert the given data into ln k and $1/T$ before we can construct the graph.

Solution: To illustrate, using the first set of data, the conversions are

$$\ln k = \ln(7.8) = 2.05$$

$$\frac{1}{T} = \frac{1}{(405 + 273)K} = \frac{1}{678K}$$

$$= 1.475 \times 10^{-3} \text{ K}^{-1}$$

We are carrying extra "significant figures" for the purpose of graphing the data. The remaining conversions give the table in the margin. Then we plot ln k versus $1/T$ as shown in the following figure:

ln k	1/T (K^{-1})
2.05	1.475×10^{-3}
2.29	1.453×10^{-3}
2.64	1.433×10^{-3}
2.89	1.412×10^{-3}
3.18	1.393×10^{-3}

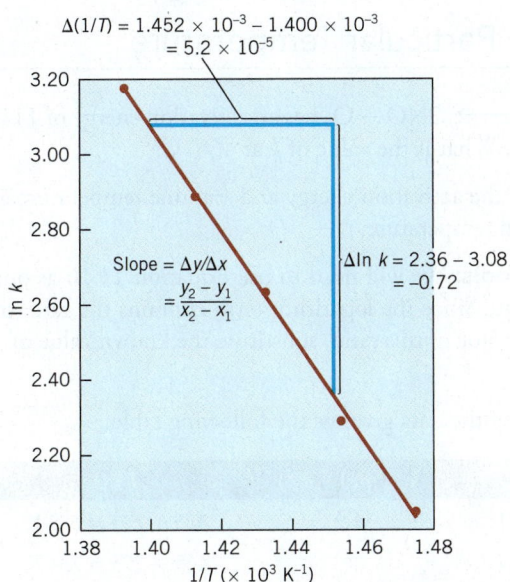

$\Delta(1/T) = 1.452 \times 10^{-3} - 1.400 \times 10^{-3}$
$= 5.2 \times 10^{-5}$

Slope $= \Delta y/\Delta x$
$= \dfrac{y_2 - y_1}{x_2 - x_1}$

$\Delta\ln k = 2.36 - 3.08$
$= -0.72$

■ The label on the x-axis may be confusing. However, what it means, for instance, is that the original value was multiplied by 10^3 to get the simpler number 1.38. Therefore, the original value was 1.38×10^{-3}.

The slope of the curve is obtained as the ratio

$$\text{slope} = \frac{\Delta(\ln k)}{\Delta(1/T)}$$

$$= \frac{-0.72}{5.2 \times 10^{-5} \text{ K}^{-1}}$$

$$= -1.4 \times 10^4 \text{ K} = -E_a/R$$

After changing signs and solving for E_a we have

$$E_a = (8.314 \text{ J mol}^{-1} \text{ K}^{-1})(1.4 \times 10^4 \text{ K})$$

$$= 1.2 \times 10^5 \text{ J mol}^{-1}$$

$$= 1.2 \times 10^2 \text{ kJ mol}^{-1}$$

Is the Answer Reasonable? Activation energies must always have a positive sign, and our result is positive. In addition, a check of the units shows that they cancel to give us the correct kJ mol^{-1}. We could try a different pair of points on the same graph to check our work.

Calculating Activation Energies from Rate Constants at Two Temperatures

If the activation energy and the rate constant at a particular temperature are known, the rate constant at another temperature can be calculated using the following relationship, which can be derived from Equation 13.14:

TOOLS

Arrhenius equation, alternate form

$$\ln\left(\frac{k_2}{k_1}\right) = \frac{-E_a}{R}\left(\frac{1}{T_2} - \frac{1}{T_1}\right) \tag{13.16}$$

This equation can also be used to calculate the activation energy from rate constants measured at two different temperatures. However, the graphical method discussed previously gives more precise values of E_a since more data points are used to plot the graph.

Example 13.13
Calculating the Rate Constant at a Particular Temperature

The reaction $2NO_2 \longrightarrow 2NO + O_2$ has an activation energy of 111 kJ/mol. At 385 °C, $k = 4.9$ L mol^{-1} s^{-1}. What is the value of k at 465 °C?

Analysis: We know the activation energy and k at one temperature and are asked for the value of k at a second temperature.

Assembling the Tools: We will need to use Equation 13.16 as our tool to obtain k at the other temperature. Since the logarithm term contains the ratio of the rate constants, we will solve for the value of this ratio, substitute the known value of k, and then solve for the unknown k.

Solution: Organizing the data gives us the following table:

	k (L mol^{-1} s^{-1})	T (K)
1	4.9	$385 + 273 = 658$ K
2	?	$465 + 273 = 738$ K

We must use $R = 8.314$ J mol^{-1} K^{-1} and express E_a in joules ($E_a = 1.11 \times 10^5$ J mol^{-1}). Next, we substitute values into the right side of the equation and solve for $\ln(k_2/k_1)$:

$$\ln\left(\frac{k_2}{k_1}\right) = \frac{-1.11 \times 10^5 \text{ J mol}^{-1}}{8.314 \text{ J mol}^{-1} \text{ K}^{-1}}\left(\frac{1}{738 \text{ K}} - \frac{1}{658 \text{ K}}\right)$$

$$= (-1.335 \times 10^4 \text{ K}) \times (-1.647 \times 10^{-4} \text{ K}^{-1})$$

Therefore,

$$\ln\left(\frac{k_2}{k_1}\right) = 2.20$$

Taking the antiln gives the ratio of k_2 to k_1:

$$\frac{k_2}{k_1} = e^{2.20} = 9.03$$

Solving for k_2,

$$k_2 = 9.03 \, k_1$$

Substituting the value of k_1 from the data table gives

$$k_2 = 9.03(4.9 \text{ L mol}^{-1} \text{ s}^{-1})$$

$$= 44 \text{ L mol}^{-1} \text{ s}^{-1}$$

● **Is the Answer Reasonable?** Although no simple check exists, we have at least found that the value of k for the higher temperature is greater than it is for the lower temperature, as it should be.

The reaction $CH_3I + HI \longrightarrow CH_4 + I_2$ was observed to have rate constants $k = 3.2$ L mol^{-1} s^{-1} at 355 °C and $k = 23$ L mol^{-1} s^{-1} at 405 °C. (a) What is the value of E_a expressed in kJ/mol? (b) What would be the rate constant at 310 °C? (*Hint:* Remember that the Arrhenius equation uses Kelvin temperatures.)

Practice Exercise 13.26

Ozone decomposes to form oxygen molecules and oxygen atoms, $O_3(g) \longrightarrow O_2(g) + O(g)$, in the upper atmosphere. The energy of activation for this reaction is 93.1 kJ/mol. At 600 K, the rate constant for this reaction is 3.37×10^3 M^{-1} s^{-1}. At what temperature will the rate constant be ten times larger, or 3.37×10^4 M^{-1} s^{-1}?

Practice Exercise 13.27

The rate constant is directly proportional to the reaction rate if the same reactant concentrations are used. When determining the stability of a consumer product, less than 5% should decompose in two years at 25 °C. What temperature should we set our oven to if we want to see that same 5% decomposition in one week? Assume the activation energy was previously determined to be 154 kJ mol^{-1}. (*Hint:* All of the information to solve the Arrhenius equation is here.)

Practice Exercise 13.28

Analyzing and Solving Multi-Concept Problems

A pressure cooker decreases the amount of time it takes to cook food. It does so by sealing the pot and using the water vapor to increase the pressure in the pot; this, in turn, increases the temperature at which water boils (see the vapor pressure graph of water, Figure 11.23). Because cooking involves chemical processes, food cooks faster at higher temperatures. Often, the pressure inside a pressure cooker increases to 2.0 atm. How long will it take to cook 1.0 pound of beans in a pressure cooker if it takes 55 minutes to cook them in boiling water at 1 atm? The pressure in the unsealed pot is 1.0 atm and the boiling point is 100 °C. We will assume that we are using a 4.0 liter pot and that the number of moles of gas in the pot increases from 0.13 mol to 0.25 mol, and that $E_a = 1.77 \times 10^5$ J mol^{-1} K^{-1}. In addition, since the cooking occurs in the liquid and the beans are covered in water, the reactant concentrations will not change, so the rate constant is directly proportional to the reaction rate.

Analysis: We are asked for the amount of time it takes to cook 1.0 pound of beans in a pressure cooker at a higher temperature than at the normal boiling point of water. Since we are looking for a rate that is related to temperature, we can use the Arrhenius equation. We need to decide which form of the Arrhenius equation to use. We are given a time for cooking, 55 min/lb, but that is not a rate or a rate constant, so we will have to convert it into a rate and use the fact that the reaction rates are directly proportional to the rate constants, since the concentrations of the reactions are not changing. We also have the temperature, 100 °C for the unpressurized reaction, as well as the activation energy

for the process, so we can use Equation 13.16 to find the rate at the higher temperature. In place of the ratio of the rate constants, we can use the rates, because they are directly proportional to each other.

Unfortunately, we are not given the higher temperature, so we will need to calculate it first. We are given a volume and the number of moles of gas, which leads us to the ideal gas law. Using the ideal gas law, $PV = nRT$, we can calculate the temperature of the gas at the higher pressure.

Now, we have one rate, the two temperatures, and the energy of activation. We can find the other rate. The three steps that we will have to carry out are (1) finding the temperature of the pressure cooker, (2) determining the rate of the reaction under atmospheric pressure, and (3) finding the rate of cooking at the higher temperature using the Arrhenius equation and then determining how long it will take to cook the beans.

PART 1

Assembling the Tools: We will use the ideal gas law ($PV = nRT$) as our first tool to find the temperature of the pressure cooker.

Solution: Let us start by assembling the data we will need:

$$\text{Pressure: } P = 2.0 \text{ atm} \qquad \text{Volume:: } V = 4.0 \text{ L}$$

$$\text{Number of moles: } n = 0.25 \text{ mol} \qquad \text{Gas constant: } R = 0.0821 \text{ L atm mol}^{-1} \text{ K}^{-1}$$

Now we can rearrange the equation to see how temperature is related to the rest of the equation:

$$PV = nRT$$

$$T = \frac{PV}{nR}$$

We can substitute these values into the equation to arrive at the higher temperature:

$$T = \frac{(2.0 \text{ atm})(4.0 \text{ L})}{(0.25 \text{ mol})(0.0821 \text{ L atm mol}^{-1} \text{ K}^{-1})}$$

$$T = 390 \text{ K}$$

PART 2

Assembling the Tools: We have the amount of time to cook the beans at 100 °C, 55 min/lb and we need to have a rate for that. If we take the inverse of the time, we will get the amount cooked per minute.

Solution:

$$\text{rate} = \frac{1.0 \text{ lb}}{55 \text{ min}}$$

$$\text{rate} = 0.0182 \text{ lb min}^{-1}$$

PART 3

Assembling the Tools: We have the two temperatures, the rate for the slow reaction, and the energy of activation. For this step, the tool we can use is the alternate form of the Arrhenius equation, Equation 13.16:

$$\ln\left(\frac{k_2}{k_1}\right) = \frac{-E_a}{R}\left(\frac{1}{T_2} - \frac{1}{T_1}\right)$$

Solution We will let k_2 represent the rate we are looking for, and set up the rest of the data as follows:

$$E_a = 1.77 \times 10^5 \text{ J mol}^{-1}$$

$$R = 0.0821 \text{ L atm mol}^{-1} \text{ K}^{-1}$$

$$T_1 = 373 \text{ K} \qquad \text{rate}_1 = 0.0182 \text{ lb/min}$$

$$T_2 = 390 \text{ K} \qquad \text{rate}_2 = ?$$

Substituting these values into Equation 13.16 gives us

$$\ln\left(\frac{k_2}{0.0182 \text{ lb/min}}\right) = \frac{-1.77 \times 10^5 \text{ J mol}^{-1}}{8.314 \text{ J mol}^{-1} \text{ K}^{-1}}\left(\frac{1}{390 \text{ K}} - \frac{1}{373 \text{ K}}\right)$$

$$\ln\left(\frac{k_2}{0.0182 \text{ lb/min}}\right) = -(2.13 \times 10^4 \text{ K})(2.56 \times 10^{-3} \text{ K}^{-1} - 2.68 \times 10^{-3} \text{ K}^{-1})$$

$$\ln\left(\frac{k_2}{0.0182 \text{ lb/min}}\right) = -(2.13 \times 10^4 \text{ K})(-1.17 \times 10^{-4} \text{ K}^{-1})$$

$$\ln\left(\frac{k_2}{0.0182 \text{ lb/min}}\right) = 2.49$$

$$\frac{k_2}{0.0182 \text{ lb/min}} = e^{2.49}$$

$$\frac{k_2}{0.0182 \text{ lb/min}} = 12$$

$$k_2 = 0.0182 \text{ lb/min} \times 12 = 0.22 \text{ lb/min}$$

Finally, we want to know how many minutes it takes to cook 1.0 lb of beans in the pressure cooker, so we need to take the inverse of the value for k_2 and multiply by 1.0 lb:

$$\text{time} = \frac{1 \text{ min}}{0.22 \text{ lb}} \times 1.0 \text{ lb}$$

$$\text{time} = 4.5 \text{ min, or about 5 minutes}$$

Are the Answers Reasonable?: One check is to look at the reasoning. Does the rate of the reaction go up if the temperature goes up? Another way to ask the question is: Does the time it takes to cook the beans go down? Both answers are "yes," so we are on the right track. Next, is the amount of time reasonable? Five minutes to cook something is reasonable, as long as the temperature is high enough.

13.8 | Mechanisms of Reactions

A balanced equation generally describes only a net overall change. Usually, however, the net change is the result of a series of simple reactions that are not at all evident from the equation. Consider, for example, the combustion of propane, C_3H_8:

$$C_3H_8(g) + 5O_2(g) \longrightarrow 3CO_2(g) + 4H_2O(g)$$

Anyone who has ever played billiards knows that this reaction simply cannot occur in a single, simultaneous collision between one propane molecule and five oxygen molecules. Just getting only three balls to come together with but one "click" on a flat, two-dimensional surface is extremely improbable. How unlikely it must be, then, for the *simultaneous* collision in three-dimensional space of six reactant molecules, one of which must be C_3H_8 and the other five O_2. Instead, the combustion of propane proceeds very rapidly by a series of much more probable steps. The series of individual steps that add up to the overall observed reaction is called the **reaction mechanism**. Information about reaction mechanisms is one of the dividends paid by the study of reaction rates.

Each individual step in a reaction mechanism is a simple chemical reaction called an *elementary process*. An **elementary process** is a reaction involving collisions between molecules. As you will soon see, a rate law can be written from the equation for an elementary process, using coefficients as exponents for the concentration terms without requiring experiments to determine the exponents. For most reactions, the individual elementary processes cannot actually be observed because they involve substances of fleeting existence; instead, we only see the net reaction. Therefore, the mechanism a chemist writes is really a *model* about what occurs step-by-step as the reactants are changed to the products.

Because the individual steps in a mechanism usually cannot be observed directly, devising a mechanism for a reaction requires some ingenuity. However, we can immediately tell whether a proposed mechanism is feasible, because *the rate law derived from the mechanism must agree with the experimental rate law for the reaction and give us the overall reaction stoichiometry.*

Elementary Processes

Consider the following elementary process that involves collisions between two identical molecules, leading directly to the products shown:

$$2NO_2 \longrightarrow NO_3 + NO \qquad \text{(13.17)}$$

$$\text{Rate} = k[NO_2]^x$$

How can we predict the value of the exponent, x? Suppose the NO_2 concentration were doubled. There would now be *twice* as many NO_2 molecules and *each* would have *twice* as many neighbors with which to collide (Figure 13.17). The number of NO_2-to-NO_2 collisions per second would doubly double—in other words, increase by a factor of 4. This would cause the rate to increase by a factor of 4, which is 2^2. Earlier we saw that when doubling a concentration leads to a four-fold increase in the rate, the concentration of that reactant is raised to the second power in the rate law. Thus, if Equation 13.17 represents an elementary process, its rate law should be

$$\text{rate} = k[NO_2]^2$$

Notice that the exponent in the rate law for this elementary process is the same as the coefficient in the chemical equation. Similar analyses for other types of elementary processes lead to similar observations and the following statement:

The exponents in the rate law for an elementary process are equal to the coefficients of the reactants in the chemical equation for that elementary process.

Remember that this rule applies only to elementary processes. If all we know is the balanced chemical equation for the overall reaction, the only way we can find the exponents of the rate law is by doing experiments.

Rate Laws and Rate-Determining Steps

How does the ability to predict the rate law of an elementary process help chemists predict reaction mechanisms? To answer this question, let's look at two reactions and what are believed to be their mechanisms. (There are many other, more complicated systems, and Chemistry Outside the Classroom 13.1 describes one type, the free radical chain reaction, that is particularly important.) Let's consider two ways in which mechanisms and experimental rate laws are used.

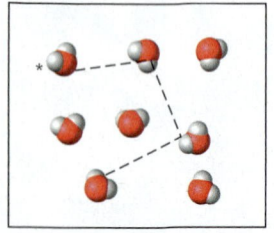

$t = 1s$

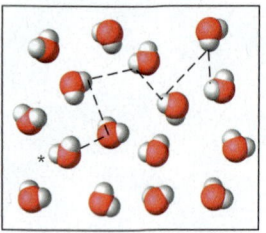

$t = 1s$

Figure 13.17 | The relative number of collisions in a given amount of time. For the molecule marked with an asterisk, *, the number of collisions in a given amount of time doubles when the concentration doubles. Since there are now twice as many molecules, each one has twice as many collisions.

Rate laws for elementary processes

CHEMISTRY OUTSIDE THE CLASSROOM 13.1

Free Radicals, Octane Ratings, Explosions, and Aging

A **free radical** is a very reactive species that contains one or more unpaired electrons. Examples are chlorine atoms formed when a Cl_2 molecule absorbs a photon (light) of the appropriate energy:

$$Cl_2 + \text{light energy}\,(h\nu) \longrightarrow 2Cl\cdot$$

(A dot placed next to the symbol of an atom or molecule represents an unpaired electron and indicates that the particle is a free radical.) The reason free radicals are so reactive is because of the tendency of electrons to become paired through the formation of either ions or covalent bonds.

Free radicals are important in many gaseous reactions, including those responsible for the production of photochemical smog in urban areas. Reactions involving free radicals have useful applications, too. For example, many plastics are made by reactions that take place by mechanisms that involve free radicals. In addition, free radicals play a part in *thermal cracking*, one of the most important processes in the petroleum industry. This reaction is used to break C—C and C—H bonds in long-chain hydrocarbons to produce the smaller molecules that give gasoline a higher octane rating. An example is the formation of free radicals in the thermal cracking reaction of butane. When butane is heated to 700–800 °C, one of the major reactions that occurs is

$$CH_3\!-\!CH_2 : CH_2\!-\!CH_3 \xrightarrow{\text{heat}} CH_3\!-\!CH_2\cdot + CH_3\!-\!CH_2\cdot$$

The central C—C bond of butane is shown here as a pair of dots, :, which represent the pair of electrons that make up the sigma bond between these carbon atoms. (Recall that we replaced bonding electron pairs with dashes to represent bonds when we made Lewis structures.) When the bond is broken, the electron pair is divided between the two free radicals that are formed. This reaction produces two ethyl radicals, $CH_3CH_2\cdot$.

Free radical reactions tend to have high initial activation energies because chemical bonds must be broken to form the radicals. Once the free radicals are formed, however, reactions in which they are involved tend to be very rapid.

Free Radical Chain Reactions

In many cases, a free radical reacts with a reactant molecule to give a product molecule plus another free radical. Reactions that involve such a step are called **chain reactions**.

Many explosive reactions are chain reactions involving free radical mechanisms. One of the most studied reactions of this type is the formation of water from hydrogen and oxygen. The elementary processes involved can be described according to their roles in the mechanism.

The reaction begins with an **initiation step** that gives free radicals:

$$H_2 + O_2 \xrightarrow{\text{hot surface}} 2OH\cdot \quad \text{(initiation)}$$

The chain continues with a **propagation step**, which produces the product plus another free radical.

$$OH\cdot + H_2 \longrightarrow H_2O + H\cdot \quad \text{(propagation)}$$

The reaction of H_2 and O_2 is explosive because the mechanism also contains **branching steps**:

$$\left.\begin{array}{l} H\cdot + O_2 \longrightarrow OH\cdot + O \\ O\cdot + H_2 \longrightarrow OH\cdot + H\cdot \end{array}\right\} \text{(branching)}$$

Thus, the reaction of one $H\cdot$ with O_2 leads to the net production of two $OH\cdot$ plus an $O\cdot$. Every time an $H\cdot$ reacts with oxygen, then, there is an increase in the number of free radicals in the system. The free radical concentration grows rapidly, and the reaction rate becomes explosively fast.

Chain mechanisms also contain **termination steps**, which remove free radicals from the system. In the reaction of H_2 and O_2, the wall of the reaction vessel serves to remove $H\cdot$, which tends to halt the chain process.

$$2H\cdot \xrightarrow{\text{vessel wall}} H_2$$

Free Radicals and Aging

Direct experimental evidence also exists for the presence of free radicals in functioning biological systems. These highly reactive species play many roles, but one of the most interesting is their apparent involvement in the aging process. One theory suggests that free radicals attack protein molecules in collagen. Collagen is composed of long strands of proteins and is found throughout the body, especially in the flexible tissues of the lungs, skin, muscles, and blood vessels. Attack by free radicals seems to lead to cross-linking between these fibers, which makes them less flexible. The most readily observable result of this is the stiffening and hardening of the skin that accompanies aging or too much sunbathing (see Figure 1).

Free radicals also seem to affect fats (lipids) within the body by promoting the oxidation and deactivation of enzymes that are normally involved in the metabolism of lipids. Over a long period, the accumulated damage gradually reduces the efficiency with which the cell carries out its activities. Interestingly, vitamin E appears to be a natural free radical inhibitor. Diets that are deficient in vitamin E produce effects resembling radiation damage and aging.

Still another theory of aging suggests that free radicals attack the DNA in the nuclei of cells. DNA is the substance that is responsible for directing the chemical activities required for the successful functioning of the cell. Reactions with free radicals cause a gradual accumulation of errors in the DNA, which reduce the efficiency of the cell and can lead, ultimately, to malfunction and cell death.

Franck Guiziou/Hemis/© Corbis

Figure 1 People exposed to sunlight over long periods, like this woman from Nepal, tend to develop wrinkles because ultraviolet radiation causes changes in their skin.

First Case: First Step as the Rate-Determining Step

First, consider the gaseous reaction,

$$2NO_2Cl \longrightarrow 2NO_2 + Cl_2 \qquad (13.18)$$

Experimentally, the rate is first order in NO_2Cl, so the rate law is

$$rate = k[NO_2Cl] \qquad (\textit{experimental})$$

The first question we might ask is, "Could the overall reaction (Equation 13.18) occur in a single step by the collision of two NO_2Cl molecules?" The answer is "no," because then it would be an elementary process and the rate law predicted for it would include a squared term, $[NO_2Cl]^2$. Instead, the experimental rate law is first order in NO_2Cl. Thus, the predicted and experimental rate laws don't agree, so we must look further to find the mechanism of the reaction.

On the basis of chemical intuition and other information that we won't discuss here, chemists believe the actual mechanism of the reaction in Equation 13.18 is the following two-step sequence of elementary processes:

$$NO_2Cl \longrightarrow NO_2 + Cl$$

$$NO_2Cl + Cl \longrightarrow NO_2 + Cl_2$$

■ The Cl atom formed here is called a *reactive intermediate*. It is often difficult to detect reactive intermediates due to their low concentrations and because they react so quickly.

Notice that when the two reactions are added, the *intermediate*, Cl, cancels out and we obtain the net overall reaction given in Equation 13.18. *Being able to add the elementary processes and thus to obtain the overall reaction is another major test of a mechanism.*

In any multi-step mechanism, one step is usually much slower than the others. In this mechanism, for example, it is believed that the first step is slow and that once a Cl atom forms, it reacts very rapidly with another NO_2Cl molecule to give the final products.

The final products of a multi-step reaction cannot appear faster than the products of the slow step, so the slow step in a mechanism is called the **rate-determining step** or the **rate-limiting step**. In the preceding two-step mechanism, then, the first reaction is the rate-determining step because the final products can't be formed faster than the rate at which Cl atoms form.

The rate-determining step is similar to a slow worker on an assembly line. The production rate depends on how quickly the slow worker works, regardless of how fast the other workers are. The factors that control the speed of the rate-determining step, therefore, also control the overall rate of the reaction. This means that *the rate law for the rate-determining step is directly related to the rate law for the overall reaction.*

Because the rate-determining step is an elementary process, we can predict its rate law from the coefficients of its reactants. The coefficient of NO_2Cl in its relatively slow breakdown to NO_2 and Cl is 1. Therefore, the rate law predicted for the first step is

$$rate = k[NO_2Cl] \qquad (\textit{predicted})$$

Thus, the predicted rate law derived for the two-step mechanism agrees with the experimentally measured rate law. Although this doesn't *prove* that the mechanism is correct, it does provide considerable support for it. From the standpoint of kinetics, therefore, the mechanism is reasonable.

Second Case: Second Step as the Rate-Determining Step

The second reaction mechanism that we will study is that of the following gas phase reaction:

$$2NO + 2H_2 \longrightarrow N_2 + 2H_2O \qquad (13.19)$$

The experimentally determined rate law is

$$\text{rate} = k[NO]^2[H_2] \qquad (\textit{experimental})$$

Based on this rate law, Equation 13.19 could *not* itself be an elementary process. If it were, the exponent for $[H_2]$ would have to be 2. Instead, a mechanism involving two or more steps must be involved.

A chemically reasonable mechanism that yields the correct form for the rate law consists of the following two steps:

$$2NO + H_2 \longrightarrow N_2O + H_2O \qquad (\text{slow})$$
$$N_2O + H_2 \longrightarrow N_2 + H_2O \qquad (\text{fast})$$

One test of the mechanism, as we said, is that the two equations must add to give the correct overall equation; and they do. Furthermore, the chemistry of the second step has actually been observed in separate experiments. N_2O is a known compound, and it does react with H_2 to give N_2 and H_2O. Another test of the mechanism involves the coefficients of NO and H_2 in the predicted rate law for the first step, the supposed rate-determining step:

$$\text{rate} = k[NO]^2[H_2] \qquad (\textit{predicted})$$

This rate equation does match the experimental rate law, but there is still a serious flaw in the proposed mechanism. If the postulated slow step actually described an elementary process, it would involve the simultaneous collision between three molecules, two NO and one H_2. A three-way collision is so unlikely that if it were really involved in the mechanism, the overall reaction would be extremely slow. Reaction mechanisms seldom include elementary processes that involve more than two-body or **bimolecular collisions**.

Chemists believe the reaction in Equation 13.19 proceeds by the following three-step sequence of bimolecular elementary processes:

$$2NO \rightleftharpoons N_2O_2 \qquad (\text{fast})$$
$$N_2O_2 + H_2 \longrightarrow N_2O + H_2O \qquad (\text{slow})$$
$$N_2O + H_2 \longrightarrow N_2 + H_2O \qquad (\text{fast})$$

In this mechanism, the first step is proposed to be a rapidly established equilibrium in which the unstable intermediate N_2O_2 forms in the forward reaction and then quickly decomposes into NO by the reverse reaction. The rate-determining step is the reaction of N_2O_2 with H_2 to give N_2O and a water molecule. The third step is the reaction mentioned above. Once again, notice that the three steps add to give the net overall change.

Since the second step is rate determining, the rate law for the reaction should match the rate law for this step. We predict this to be

$$\text{rate} = k[N_2O_2][H_2] \qquad \textbf{(13.20)}$$

However, the experimental rate law does not contain the species N_2O_2. Therefore, we must find a way to express the concentration of N_2O_2 in terms of the reactants in the overall reaction. To do this, let's look closely at the first step of the mechanism, which we view as a reversible reaction.

The rate in the forward direction, in which NO is the reactant, is

$$\text{rate (forward)} = k_f[NO]^2$$

The rate of the reverse reaction, in which N_2O_2 is the reactant, is

$$\text{rate (reverse)} = k_r[N_2O_2]$$

■ The detection of N_2O_2 as an intermediate, by very sensitive analytical methods, supports this proposed mechanism over others.

■ Recall that in a dynamic equilibrium, forward and reverse reactions occur at equal rates.

If we view this as a dynamic equilibrium, then the rate of the forward and reverse reactions are equal, which means that

$$k_f[NO]^2 = k_r[N_2O_2] \qquad (13.21)$$

Since we would like to eliminate N_2O_2 from the rate law in Equation 13.20, let's solve Equation 13.21 for $[N_2O_2]$:

$$[N_2O_2] = \frac{k_f}{k_r}[NO]^2$$

Substituting into the rate law in Equation 13.20 yields

$$\text{rate} = k\frac{k_f}{k_r}[NO]^2[H_2]$$

Combining all of the constants into one (k') gives

$$\text{rate} = k'[NO]^2[H_2] \qquad (\textit{predicted})$$

■ There are many reactions that do not follow simple first- or second-order rate laws and have mechanisms far more complex than those studied in this section. Even so, their more complex kinetics still serve as clues to their complex set of elementary processes.

Now, the rate law derived from the mechanism matches the rate law obtained experimentally. The three-step mechanism does appear to be reasonable on the basis of kinetics.

The procedure we have worked through here applies to many reactions that proceed by mechanisms involving sequential steps. Steps that precede the rate-determining step are considered to be rapidly established equilibria involving unstable intermediates.

Relating the Proposed Mechanism to the Experimental Rate Law

Although chemists may devise other experiments to help prove or disprove the correctness of a mechanism, one of the strongest pieces of evidence is the experimentally measured rate law for the overall reaction. No matter how reasonable a particular mechanism may appear, if its elementary processes cannot yield a predicted rate law that matches the experimental one, the mechanism is wrong and must be discarded.

In some cases the experimental rate law may suggest two different but equally probable mechanisms. One method for testing mechanisms is to find evidence of unique intermediates. In the previous mechanisms, both had N_2O as an intermediate, but the second mechanism had N_2O_2 as a unique intermediate. If experiments could show the existence of N_2O_2 during the reaction, the second mechanism would have added evidence that it is the correct mechanism.

Practice Exercise 13.29

Select the reactions below that may be elementary processes. For those not selected, explain why they are not elementary processes.

(a) $2N_2O_5 \longrightarrow 2N_2O_4 + O_2$

(b) $NO + O_3 \longrightarrow NO_2 + O_2$

(c) $2NO + H_2 \longrightarrow N_2O + H_2O$

(d) $C_3H_8(g) + 5O_2(g) \longrightarrow 3CO_2(g) + 4H_2O(g)$

(e) $C_{12}H_{22}O_{11} + H_2O \longrightarrow C_6H_{12}O_6 + C_6H_{12}O_6$

(f) $3H_2 + N_2 \longrightarrow 2NH_3$

(*Hint:* How many molecules are likely to collide at exactly the same time?)

Practice Exercise 13.30

Ozone, O_3, reacts with nitric oxide, NO, to form nitrogen dioxide and oxygen:

$$NO + O_3 \longrightarrow NO_2 + O_2$$

This is one of the reactions involved in the formation of photochemical smog. If this reaction occurs in a single step, what is the expected rate law for the reaction?

The mechanism for the decomposition of NO_2Cl is

$$NO_2Cl \rightleftharpoons NO_2 + Cl$$
$$NO_2Cl + Cl \longrightarrow NO_2 + Cl_2$$

What would the predicted rate law be if the second step in the mechanism were the rate-determining step?

13.9 | Catalysts

A **catalyst** is a substance that changes the rate of a chemical reaction without itself being used up. In other words, all of the catalyst added at the start of a reaction is present chemically unchanged after the reaction has gone to completion. The action caused by a catalyst is called **catalysis.** Broadly speaking, there are two kinds of catalysts. Positive catalysts speed up reactions, whereas negative catalysts, usually called inhibitors, slow reactions down. After this, when we use "catalyst" we'll mean positive catalyst, the usual connotation.

Although the catalyst is not part of the overall reaction, it does participate by changing the mechanism of the reaction. The catalyst provides a path to the products that has a rate-determining step with a lower activation energy than that of the uncatalyzed reaction (see Figure 13.18). Because the activation energy along this new route is lower, a greater fraction of the collisions of the reactant molecules have the minimum energy needed to react, so the reaction proceeds faster.

Catalysts can be divided into two groups—**homogeneous catalysts**, which exist in the same phase as the reactants, and **heterogeneous catalysts**, which exist in a separate phase from the reactants.

Homogeneous Catalysts

An example of homogeneous catalysis is found in the now outdated *lead chamber process* for manufacturing sulfuric acid. To make sulfuric acid by this process, sulfur is burned to

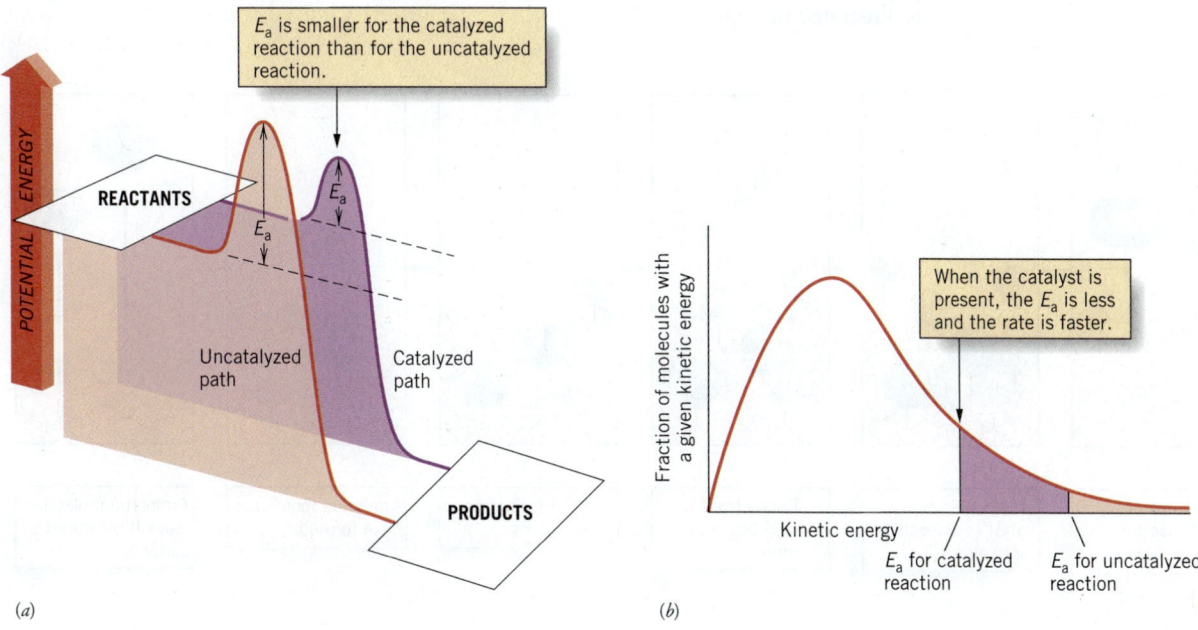

(a)　　　　　　　　　　　　　　　　(b)

Figure 13.18 | The effect of a catalyst on a reaction. (*a*) The catalyst provides an alternative, low-energy path from the reactants to the products. (*b*) A larger fraction of molecules have sufficient energy to react when the catalyzed path is available.

■ In the modern process for making sulfuric acid, the contact process, vanadium(V) oxide, V_2O_5, is a heterogeneous catalyst that promotes the oxidation of sulfur dioxide to sulfur trioxide.

give SO_2, which is then oxidized to SO_3. The SO_3 is dissolved in water as it forms to give H_2SO_4.

$$S + O_2 \longrightarrow SO_2$$

$$SO_2 + \tfrac{1}{2}O_2 \longrightarrow SO_3$$

$$SO_3 + H_2O \longrightarrow H_2SO_4$$

Unassisted, the second reaction, the oxidation of SO_2 to SO_3, occurs slowly. In the lead chamber process, the SO_2 is combined with a mixture of NO, NO_2, air, and steam in large, lead-lined reaction chambers. The NO_2 readily oxidizes the SO_2 to give NO and SO_3. The NO is then reoxidized to NO_2 by oxygen.

■ The NO_2 is regenerated in the second reaction and so is recycled over and over. Thus, only small amounts of it are needed in the reaction mixture to do an effective catalytic job.

$$NO_2 + SO_2 \longrightarrow NO + SO_3$$

$$NO + \tfrac{1}{2}O_2 \longrightarrow NO_2$$

The NO_2 serves as a catalyst by being an oxygen carrier and by providing a low-energy path for the oxidation of SO_2 to SO_3. Notice that the NO_2 is regenerated; it has not been permanently changed, a necessary condition for any catalyst.

Heterogeneous Catalysts

■ **Adsorption** means that molecules bind to a surface; **absorption** means that molecules enter the structure, as water soaked up by a sponge.

A heterogeneous catalyst is commonly a solid, and it usually functions by promoting a reaction on its surface. One or more of the reactant molecules are **adsorbed** onto the surface of the catalyst, where an interaction with the surface increases their reactivity. An example is the synthesis of ammonia from hydrogen and nitrogen by the Haber process.

$$3H_2 + N_2 \longrightarrow 2NH_3$$

The reaction takes place on the surface of an iron catalyst that contains traces of aluminum and potassium oxides. It is thought that hydrogen molecules and nitrogen molecules dissociate while being held on the catalytic surface. The hydrogen atoms then combine with the nitrogen atoms to form ammonia. Finally, the completed ammonia molecule breaks away, freeing the surface of the catalyst for further reaction. This sequence of steps is illustrated in Figure 13.19.

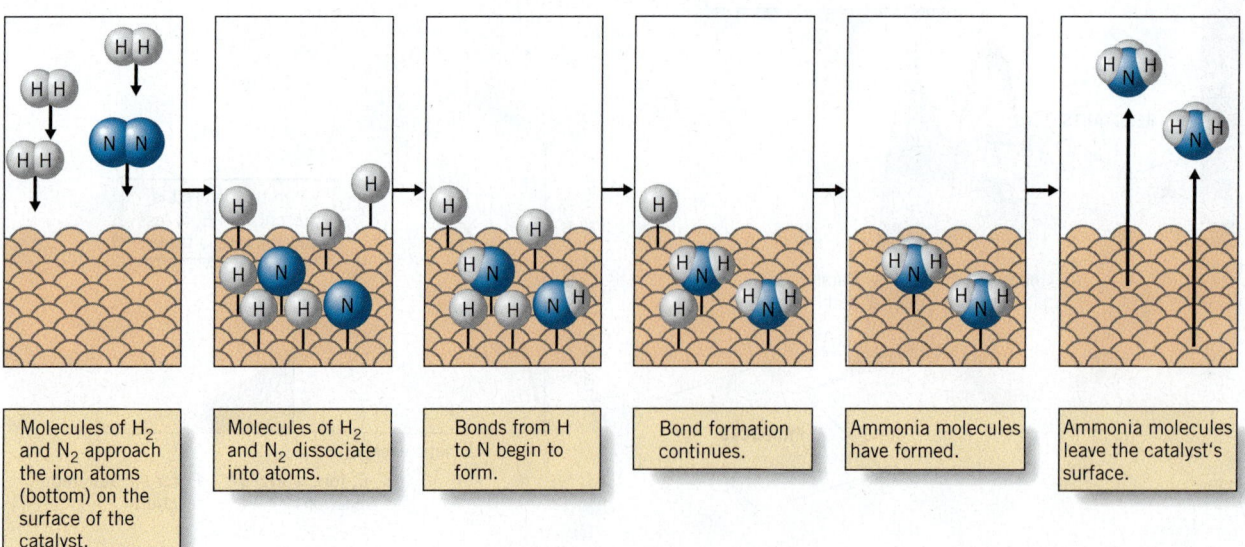

Figure 13.19 | **The Haber process.** Catalytic formation of ammonia molecules from hydrogen and nitrogen on the surface of a catalyst.

(a)   (b)

Figure 13.20 | Catalysts are very important in the petroleum industry. (*a*) Catalytic cracking towers at an oil refinery. (*b*) A variety of catalysts are available as beads, powders, or in other forms for various refinery operations.

Heterogeneous catalysts are used in many important commercial processes. The petroleum industry uses heterogeneous catalysts to crack hydrocarbons into smaller fragments and then re-form them into the useful components of gasoline (see Figure 13.20). The availability of such catalysts allows refineries to produce gasoline, jet fuel, or heating oil from crude oil in any ratio necessary to meet the demands of the marketplace.

A vehicle that uses unleaded gasoline is equipped with a catalytic converter (Figure 13.21) designed to lower the concentrations of exhaust gas pollutants, such as carbon monoxide, unburned hydrocarbons, and nitrogen oxides. The catalysts are nanometer-sized particles of platinum, ruthenium, and rhodium dispersed in a honeycomb of a high-temperature ceramic. The large ratio of surface area to mass enables the catalytic converter to react with large volumes of exhaust efficiently. Air is introduced into the exhaust stream that then passes over a catalyst that adsorbs CO, NO, and O_2. The NO dissociates into N and O atoms, and the O_2 also dissociates into atoms. Pairing of nitrogen atoms then produces N_2, and the oxidation of CO by oxygen atoms produces CO_2. Unburned hydrocarbons are also oxidized to CO_2 and H_2O. The catalysts in catalytic converters are deactivated or "poisoned" by lead-based octane boosters like tetraethyl lead [$Pb(C_2H_5)_4$]. Leaded gasoline was finally banned in 1995. Leaded gasoline also posed an environmental hazard from the lead emitted in automobile exhaust.

The poisoning of catalysts is also a major problem in many industrial processes. Methyl alcohol (methanol, CH_3OH), for example, is a promising fuel that can be made from coal and steam by the reaction

$$C \text{ (from coal)} + H_2O \longrightarrow CO + H_2$$

followed by

$$CO + 2H_2 \longrightarrow CH_3OH$$

A catalyst for the second step is copper(I) ion held in solid solution with zinc oxide. However, traces of sulfur, a contaminant in coal, must be avoided because sulfur reacts with the catalyst and destroys its catalytic activity.

Figure 13.21 | A modern catalytic converter of the type used in about 80% of new cars. Part of the converter has been cut away to reveal the porous ceramic material that serves as the support for the catalyst.

Enzymes

In living systems, complex protein-based molecules called *enzymes* catalyze almost every reaction that occurs in living cells. Enzymes contain a specially shaped area called an "active site" that lowers the energy of the transition state of the reaction being catalyzed. This causes the reaction rate to increase significantly. Many poisons have been shown to work by blocking important enzyme systems. Heavy metals bind to sulfur-containing groups and distort the active site. Many modern drugs work by inhibiting the catalytic activity of specific enzymes in the human body. In the pharmaceutical industry, chemists use molecular modeling to design drug molecules that have optimum shapes to fit enzymatic active sites. The better the fit, the more potent the drug will be.

Summary

Organized by Learning Objective

Understand and use the five conditions that affect how rapidly chemicals react

The speeds or **rates** of reactions are controlled by five factors: (1) the nature of the reactants, (2) the ability of the reactants to meet, (3) the concentrations of the reactants, (4) the temperature, and (5) the presence of catalysts. The rates of **heterogeneous reactions** are determined largely by the area of contact between the phases; the rates of **homogeneous reactions** are determined by the concentrations of the reactants.

Determine, from experimental data, the relative rates at which reactants disappear and products appear, and the rate of reaction, which is independent of the substance monitored

The rate is measured by monitoring the change in reactant or product concentrations with time:

$$\text{rate} = \Delta(\text{concentration})/\Delta(\text{time})$$

In any chemical reaction, the rates of formation of products and the rates of disappearance of reactants are related by the coefficients of the balanced overall chemical equation.

Use experimental initial rate data to determine rate laws

The **rate law** for a reaction relates the reaction rate to the molar concentrations of the reactants. The rate is proportional to the product of the molar concentrations of the reactants, each raised to an appropriate power. These exponents must be determined by experiments. The proportionality constant, **k**, is called the **rate constant**. The sum of the exponents in the rate law is the **order** (or overall order) of the reaction.

Use the basic results of integrated rate laws to determine the order of a reaction and calculate the time dependence of concentration for zero-, first-, and second-order reactions

Equations exist that relate the concentration of a reactant at a given time t to the initial concentration and the rate constant. The time required for half of a reactant to disappear is the **half-life, $t_{1/2}$**. For a first-order reaction, the half-life is a constant that depends only on the rate constant for the reaction; it is independent of the initial concentration. The half-life for a second-order reaction is inversely proportional both to the initial concentration of the reactant and to the rate constant. The half-life for a zero-order reaction is inversely proportional to the rate constant and proportional to the initial concentration.

Explain the rate of chemical reactions based on a molecular view of collisions that includes frequency, energy, and orientation that make up collision theory

According to **collision theory**, the rate of a reaction depends on the number of **effective collisions** per second of the reactant particles, which is only an extremely small fraction of the total number of collisions per second. This fraction is so small partly because the reactant molecules must be suitably oriented, but mostly because the colliding molecules must jointly possess a minimum molecular kinetic energy called the **activation energy**, E_a. As the temperature increases, a larger fraction of the collisions has this necessary energy, making more collisions effective each second and the reaction faster.

Describe the basics of transition state theory including activated complexes and potential energy diagrams

Transition state theory visualizes how the energies of molecules and the orientations of their nuclei interact as they collide. In this theory, the energy of activation is viewed as an energy barrier on the reaction's potential energy diagram. The *heat of reaction* is the net potential energy difference between the reactants and the products. In reversible reactions, the values of E_a for both the forward and reverse reactions can be identified on an energy diagram. The species at the high point on an energy diagram is the **activated complex** and is said to be in the **transition state**.

Use the Arrhenius equation to determine the activation energy of a reaction

The **Arrhenius equation** lets us see how changes in activation energy and temperature affect a rate constant. The Arrhenius equation also lets us determine E_a either graphically or by a calculation using the appropriate form of the equation. The calculation requires two rate constants determined at two temperatures. The graphical method uses more values of rate constants at more temperatures and thus usually yields more accurate results. The activation energy and the rate constant at one temperature can be used to calculate the rate constant at another temperature.

Use the concepts of reaction mechanisms to recognize reasonable mechanisms and suggest plausible mechanisms given experimental data

The detailed sequence of elementary processes that lead to the net chemical change is the **mechanism** of the reaction. Since intermediates usually cannot be detected, the mechanism is a theory. Support for a mechanism comes from matching the predicted rate law for the mechanism with the rate law obtained from experimental data. For the **rate-determining step** or for any **elementary process**, the corresponding rate law has exponents equal to the coefficients in the balanced chemical equation for the elementary process.

Relate the properties of homogeneous and heterogeneous catalysts and how they act to increase reaction rates

Catalysts are substances that change a reaction rate but are not used up by the reaction. Negative catalysts inhibit reactions.

Positive catalysts provide alternative paths for reactions for which at least one step has a smaller activation energy than the uncatalyzed reaction. **Homogeneous catalysts** are in the same phase as the reactants. **Heterogeneous catalysts** provide a path of lower activation energy by having a surface on which the reactants are adsorbed and react. Catalysts in living systems are called enzymes.

TO⚗LS

Tools for Problem Solving
The following tools were introduced in this chapter. Study them carefully so you can select the appropriate tool when needed.

Factors that affect reaction rates (Section 13.1)
The five factors that affect the rate of reactions are (1) the chemical nature of the reactants; (2) the ability of the reactants to come in contact with each other; (3) the concentrations of the reactants; (4) the temperature of the reaction; and (5) the availability of catalysts.

Rate with respect to one reactant of product (Section 13.2)

$$\text{rate} = \frac{\Delta(\text{conc. of } X)}{\Delta t}$$

The rates are always expressed as a positive value, so if the concentration is decreasing for a reactant, the sign is changed to the positive.

Rate of a reaction (Section 13.2)
For the reaction,

$$aA + bB \longrightarrow cC + dD$$

the rate expression is

$$rate = -\frac{1}{a}\frac{\Delta[A]}{\Delta t} = -\frac{1}{b}\frac{\Delta[B]}{\Delta t} = \frac{1}{c}\frac{\Delta[C]}{\Delta t} = \frac{1}{d}\frac{\Delta[D]}{\Delta t}$$

Rate law of a reaction (Section 13.3)

$$\text{rate} = k[A]^n[B]^m$$

Determining rate laws (Section 13.3)

$$\frac{\text{rate}_2}{\text{rate}_1} = \frac{k[A]_2^n[B]_2^m}{k[A]_1^n[B]_1^m} = \frac{k}{k}\left(\frac{[A]_2}{[A]_1}\right)^n\left(\frac{[B]_2}{[B]_1}\right)^m$$

Integrated first-order rate law (Section 13.4)

$$\ln\frac{[A]_0}{[A]_t} = kt$$

Half-lives of first-order reactions (Section 13.4)

$$\ln 2 = kt_{1/2}$$

Carbon-14 dating (Section 13.4)

The half-life for carbon-14 is 5730 years and the rate constant is 1.21×10^{-4} yr^{-1}. The age of an object can be found by comparing the ratio of $^{14}C/^{12}C$, r_t, from an object in question to the ratio of $^{14}C/^{12}C$ in a contemporary sample, r_0.

$$\ln\frac{r_0}{r_t} = (1.21 \times 10^{-4}\ yr^{-1})t$$

Integrated second-order rate law (Section 13.4)

For a second-order reaction of the form, rate $= k[B]^2$, with known k, this equation is used to calculate the concentration of a reactant at some specified time after the start of the reaction, or the time required for the concentration to decrease to some specified value.

$$\frac{1}{[B]_t} - \frac{1}{[B]_0} = kt$$

Integrated zero-order rate law (Section 13.4)

In a zero-order rate law with the form, rate $= k$, the rate is independent of the concentration of the reactants. The integrated rate law is

$$[A]_t = -kt + [A]_0$$

Arrhenius equation (Section 13.7)

$$k = Ae^{-E_a/RT}$$

Arrhenius equation, alternate form (Section 13.7)

$$\ln\left(\frac{k_2}{k_1}\right) = \frac{-E_a}{R}\left(\frac{1}{T_2} - \frac{1}{T_1}\right)$$

Rate laws for elementary processes (Section 13.8)

The exponents in the rate laws are equal to the coefficients of the reactants in the chemical equations for that elementary process.

WileyPLUS = *WileyPLUS*, an online teaching and learning solution. *Note to instructors*: Many of the end-of-chapter problems are available for assignment via the WileyPLUS system. **www.wileyplus.com.** Review Problems are presented in pairs separated by blue rules. Answers to problems whose numbers appear in blue are given in Appendix B. More challenging problems are marked with an asterisk ★.

|Review Questions

Factors that Affect the Rate of Chemical Change

13.1 Why are chemical reactions usually carried out in solution?

13.2 Give an example from everyday experience of **(a)** a very fast reaction, **(b)** a moderately fast reaction, and **(c)** a slow reaction.

13.3 What is a *homogeneous reaction*? What is a *heterogeneous reaction*? Give examples.

13.4 How does particle size affect the rate of a heterogeneous reaction? Why?

13.5 What is the major factor that affects the rate of a heterogeneous reaction?

13.6 The rate of hardening of epoxy glue depends on the amount of hardener that is mixed into the glue. What factor affecting reaction rates does this illustrate?

13.7 A Polaroid™ instant photograph develops faster if it's kept warm than if it is exposed to cold. Why?

13.8 Insects have no way of controlling their body temperatures like mammals do. In cool weather, they become sluggish and move less quickly. How can this be explained using the principles developed in this chapter?

13.9 Persons who have been submerged in very cold water and who are believed to have drowned sometimes can be revived. On the other hand, persons who have been submerged in warmer water for the same length of time have died. Explain this in terms of factors that affect the rates of chemical reactions.

Measuring Reaction Rates

13.10 How does an instantaneous rate of reaction differ from an average rate of reaction?

13.11 What is the difference between the rate of reaction and the reaction rate with respect to one component of the reaction?

13.12 Explain how the initial instantaneous rate of reaction can be determined from experimental concentration versus time data.

13.13 What are the units of reaction rate? What is the sign of a reaction rate?

13.14 Describe how to determine the instantaneous rate of chemical change. What data do you need and how will you use it?

Rate Laws

13.15 What are the units of the rate constant for **(a)** a first-order reaction, **(b)** a second-order reaction, and **(c)** a zero-order reaction?

13.16 How does the dependence of reaction rate on concentration differ between a zero-order and a first-order reaction? Between a first-order and second-order reaction?

13.17 Is there any way of using the coefficients in the balanced overall equation for a reaction to predict with certainty what the exponents are in the rate law?

13.18 If the concentration of a reactant is doubled and the reaction rate is unchanged, what must be the order of the reaction with respect to that reactant?

13.19 If the concentration of a reactant is doubled and the reaction rate doubles, what must be the order of the reaction with respect to that reactant?

13.20 If the concentration of a reactant is doubled, by what factor will the rate increase if the reaction is second order with respect to that reactant?

13.21 In an experiment, the concentration of a reactant was tripled. The rate increased by a factor of 27. What is the order of the reaction with respect to that reactant?

13.22 Biological reactions usually involve the interaction of an enzyme with a *substrate*, the substance that actually undergoes the chemical change. In many cases, the rate of reaction depends on the concentration of the enzyme but is independent of the substrate concentration. What is the order of the reaction with respect to the substrate in such instances?

Integrated Rate Laws

13.23 Rearrange the integrated rate equations for **(a)** a first-order reaction, **(b)** a second-order reaction, and **(c)** a zero-order reaction to calculate $[A]_t$. Use the symbol $[A]_0$ to represent the initial concentration if needed.

13.24 How is the half-life of a first-order reaction affected by the initial concentration of the reactant?

13.25 How is the half-life of a second-order reaction affected by the initial reactant concentration?

13.26 How is the half-life of a zero-order reaction affected by the initial reactant concentration?

13.27 Derive the equations for $t_{1/2}$ for first- and second-order reactions from Equations 13.7 and 13.11 respectively.

13.28 The integrated rate law for a zero-order reaction is

$$[A]_t = -kt + [A]_0$$

Derive an equation for the half-life of a zero-order reaction.

13.29 Which of the following graphs represents the data collected for a first-order reaction? A second-order reaction? A zero-order reaction?

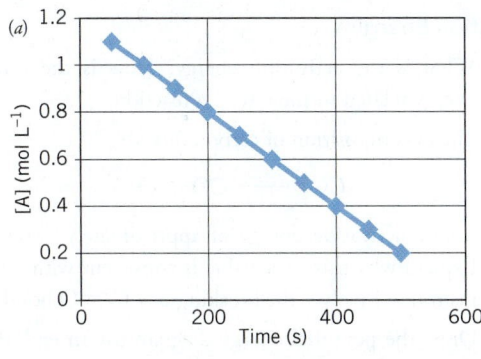

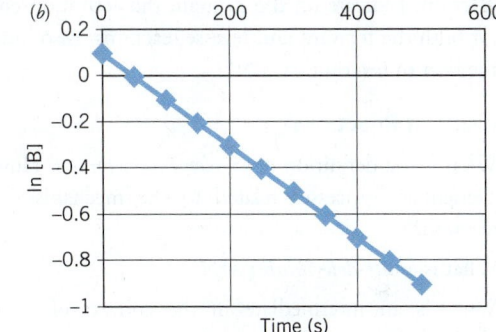

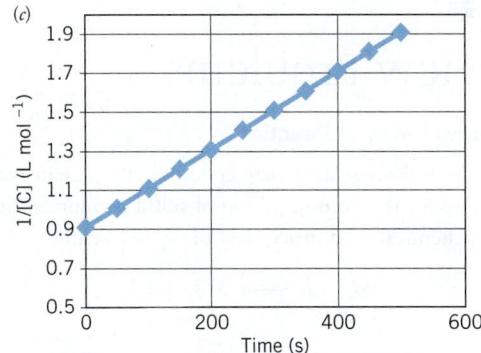

Molecular Basis of Collision Theory

13.30 What is the basic postulate of collision theory?

13.31 What two factors influence the effectiveness of molecular collisions in producing chemical change?

13.32 In terms of the kinetic theory, why does an increase in temperature increase the reaction rate?

13.33 Some might say that the collision theory "treats reactions as collisions between hard spheres." Critique this statement and compare to the transition state theory as needed.

Molecular Basis of Transition State Theory

13.34 What does the transition state theory attempt to describe about chemical reactions?

13.35 Draw a potential energy diagram for an exothermic reaction and indicate on the diagram the location of the transition state.

13.36 Some might say that the "transition state theory tries to describe what happens from the moment molecules start to collide until they finally separate." Critique this statement, comparing to the collision theory as needed.

Activation Energies

13.37 What is the activation energy? How is the activation energy related to the rate of a reaction?

13.38 The decomposition of carbon dioxide,

$$CO_2 \longrightarrow CO + O$$

has an activation energy of approximately 460 kJ/mol. Explain why this large value is consistent with a mechanism that involves the breaking of a $C\!\!=\!\!O$ bond.

13.39 Draw the potential energy diagram for an endothermic reaction. Indicate on the diagram the activation energy for both the forward and reverse reactions. Also indicate the heat of reaction.

Mechanisms of Reactions

13.40 What is the definition of an *elementary process?* How are elementary processes related to the mechanism of a reaction?

13.41 What is a *rate-determining step?*

13.42 What is an intermediate in the context of reaction mechanisms?

13.43 Free radicals are discussed in Chemistry Outside the Classroom 13.1. What is a free radical? Why are free radical mechanisms important in many reactions?

13.44 Suppose we compared two reactions, one requiring the simultaneous collision of three molecules and the other requiring a collision between two molecules. From the standpoint of statistics, and all other factors being equal, which reaction should be faster? Explain your answer.

13.45 In what way is the rate law for a reaction related to the rate-determining step?

13.46 How does an elementary process relate to **(a)** the rate law for that process, **(b)** the rate law for a reaction, and **(c)** the coefficients of the chemical reaction?

Catalysts

13.47 How does a catalyst increase the rate of a chemical reaction?

13.48 What is a *homogeneous catalyst*? How does it function, in general terms?

13.49 What is the purpose of the catalytic converter that most automobiles use today? Is the catalyst heterogeneous or homogeneous?

13.50 Tell how you would recognize a catalyst in a description of a chemical process or experimental procedure.

13.51 What is the difference in meaning between the terms *adsorption* and *absorption*? Which one applies to heterogeneous catalysts?

13.52 Why should leaded gasoline not be used in cars equipped with catalytic converters?

| Review Problems

Measuring Rates of Reaction

13.53 The following data were collected at a certain temperature for the decomposition of sulfuryl chloride, SO_2Cl_2, a chemical used in a variety of organic syntheses.

$$SO_2Cl_2 \longrightarrow SO_2 + Cl_2$$

Time (min)	$[SO_2Cl_2]$ (mol L^{-1})
0	0.1000
1.00×10^2	0.0876
2.00×10^2	0.0768
3.00×10^2	0.0673
4.00×10^2	0.0590
5.00×10^2	0.0517
6.00×10^2	0.0453
7.00×10^2	0.0397
8.00×10^2	0.0348
9.00×10^2	0.0305
1.000×10^3	0.0267
1.100×10^3	0.0234

Make a graph of concentration versus time and determine the instantaneous rate of formation of SO_2 using the tangent to the curve at $t = 200$ minutes and $t = 600$ minutes.

13.54 The following data were collected for the decomposition of acetaldehyde. CH_3CHO, (used in the manufacture of a variety of chemicals including perfumes, dyes, and plastics), into methane and carbon monoxide. The data were collected at 535 °C.

$$CH_3CHO \longrightarrow CH_4 + CO$$

$[CH_3CHO]$ (mol L^{-1})	Time (s)
0.200	0
0.153	0.20×10^2
0.124	0.40×10^2
0.104	0.60×10^2
0.090	0.80×10^2
0.079	1.00×10^2
0.070	1.20×10^2
0.063	1.40×10^2
0.058	1.60×10^2
0.053	1.80×10^2
0 049	2.00×10^2

Make a graph of concentration versus time and determine, using the tangent to the curve, the instantaneous rate of reaction of CH_3CHO after 60 seconds and after 120 seconds.

13.55 For the reaction, $2A + B \longrightarrow 3C$, it was found that the rate of disappearance of B was 0.30 mol L^{-1} s^{-1}. What were the rates of disappearance of A and the rate of appearance of C?

13.56 In the reaction, $3H_2 + N_2 \longrightarrow 2NH_3$, how does the rate of disappearance of hydrogen compare to the rate of disappearance of nitrogen? How does the rate of appearance of NH_3 compare to the rate of disappearance of nitrogen?

13.57 In the combustion of hexane (a low-boiling component of gasoline),

$$2C_6H_{14}(g) + 19O_2(g) \longrightarrow 12CO_2(g) + 14H_2O(g)$$

it was found that the rate of decrease of C_6H_{14} was 1.20 mol L^{-1} s^{-1}.

(a) What was the rate of reaction with respect to O_2?

(b) What was the rate of formation of CO_2?

(c) What was the rate of formation of H_2O?

13.58 At a certain moment in the reaction,

$$2 N_2O_5 \longrightarrow 4NO_2 + O_2$$

N_2O_5 is decomposing at a rate of 2.5×10^{-6} mol L^{-1} s^{-1}. What are the rates of formation of NO_2 and O_2?

13.59 Consider the reaction,

$$CH_3Cl(g) + 3Cl_2(g) \longrightarrow CCl_4(g) + 3HCl(g)$$

(a) Express the rate of the reaction with respect to each of the reactants and products.

(b) If the instantaneous rate of the reaction with respect to HCl is 0.029 $M s^{-1}$, what is the instantaneous rate of the reaction?

13.60 The decomposition of phosphine, a very toxic gas, forms phosphorus and hydrogen in the following reaction:

$$4PH_3(g) \longrightarrow P_4(g) + 6H_2(g)$$

(a) Express the rate with respect to each of the reactants and products.

(b) If the instantaneous rate with respect to PH_3 is 0.34 $M s^{-1}$, what is the instantaneous rate of the reaction?

Rate Laws

13.61 Estimate the rate of the reaction,

$$H_2SeO_3 + 6I^- + 4H^+ \longrightarrow Se + 2I_3^- + 3H_2O$$

given that the rate law for the reaction at 0 °C is

$$\text{rate} = (5.0 \times 10^5 \text{ L}^5 \text{ mol}^{-5} \text{ s}^{-1})[H_2SeO_3][I^-]^3[H^+]^2$$

The reactant concentrations are $[H_2SeO_3] = 2.0 \times 10^{-2}$ M, $[I^-] = 2.0 \times 10^{-3}$ M, and $[H^+] = 1.0 \times 10^{-3}$ M.

13.62 Estimate the rate of the reaction,

$$H^+(aq) + OH^-(aq) \longrightarrow H_2O$$

given the rate law for the reaction is

$$\text{rate} = (1.3 \times 10^{11} \text{ L mol}^{-1} \text{ s}^{-1})[OH^-][H^+]$$

for neutral water, where $[H^+] = 1.0 \times 10^{-7}$ M and $[OH^-] = 1.0 \times 10^{-7}$ M.

13.63 The oxidation of NO (released in small amounts in the exhaust of automobiles) produces the brownish-red gas NO_2, which is a component of urban air pollution.

$$2NO(g) + O_2(g) \longrightarrow 2NO_2(g)$$

The rate law for the reaction is rate $= k [NO]^2[O_2]$. At 25 °C, $k = 7.1 \times 10^9$ L^2 mol^{-2} s^{-1}. What would be the rate of the reaction if $[NO] = 0.0010$ mol L^{-1} and $[O_2] = 0.034$ mol L^{-1}?

13.64 The rate law for the decomposition of N_2O_5 is rate $= k[N_2O_5]$. If $k = 1.0 \times 10^{-5}$ s^{-1}, what is the reaction rate when the N_2O_5 concentration is 0.0010 mol L^{-1}?

13.65 The rate law for a certain enzymatic reaction is zero order with respect to the substrate. The rate constant for the reaction is 6.4×10^2 $M s^{-1}$. If the initial concentration of the substrate is 0.275 mol L^{-1}, what is the initial rate of the reaction?

13.66 Radon-220 is radioactive, and decays into polonium-216 by emitting an alpha particle. This is a first-order process with a rate constant of 0.0125 s^{-1}. When the concentration of radon-220 is 1.0×10^{-9} mol L^{-1}, what is the rate of the reaction?

13.67 The following data were collected for the reaction $M + N \longrightarrow P + Q$:

Initial Concentrations (mol L^{-1})		Initial Rate of Reaction (mol L^{-1} s^{-1})
[M]	[N]	
0.010	0.010	2.5×10^{-3}
0.020	0.010	5.0×10^{-3}
0.020	0.030	4.5×10^{-2}

What is the rate law for the reaction? What is the value of the rate constant (with correct units)?

13.68 Cyclopropane, C_3H_6, is a gas used as a general anesthetic. It undergoes a slow molecular rearrangement to propylene.

cyclopropane   propylene

At a certain temperature, the following data were obtained relating concentration and rate:

Initial Concentration of Cyclopropane (mol L^{-1})	Initial Rate of Formation of Propylene (mol L^{-1} s^{-1})
0.050	2.95×10^{-5}
0.100	5.90×10^{-5}
0.150	8.85×10^{-5}

What is the rate law for the reaction? What is the value of the rate constant, with correct units?

13.69 The reaction of iodide ion with hypochlorite ion, OCl^- (the active ingredient in a "chlorine bleach" such as Clorox), follows the equation $OCl^- + I^- \longrightarrow OI^- + Cl^-$. It is a rapid reaction that gives the following rate data:

Initial Concentrations (mol L^{-1})		Initial Rate of Formation of Cl$^-$ (mol L^{-1} s^{-1})
[OCl$^-$]	[I$^-$]	
1.6×10^{-3}	1.6×10^{-3}	1.75×10^4
7.6×10^{-3}	1.6×10^{-3}	8.31×10^4
1.6×10^{-3}	9.6×10^{-3}	1.05×10^5

What is the rate law for the reaction? Determine the value of the rate constant with its correct units.

13.70 The formation of small amounts of nitrogen oxide, NO, in automobile engines is the first step in the formation of smog. As noted in Problem 13.63, nitrogen oxide is readily oxidized to nitrogen dioxide by the reaction $2NO(g) + O_2(g) \longrightarrow 2NO_2(g)$. The following data were collected in a study of the rate of this reaction:

Initial Concentrations (mol L^{-1})		Initial Rate of Formation of NO$_2$ (mol L^{-1} s^{-1})
[O$_2$]	[NO]	
0.0010	0.0010	7.10
0.0037	0.0010	26.3
0.0037	0.0062	1011

What is the rate law for the reaction? Determine the value of the rate constant with its correct units.

13.71 At a certain temperature, the following data were collected for the reaction, $2ICl + H_2 \longrightarrow I_2 + 2HCl$:

Initial Concentrations (mol L^{-1})		Initial Rate of Formation of I$_2$ (mol L^{-1} s^{-1})
[ICl]	[H$_2$]	
0.12	0.12	0.0015
0.78	0.12	0.0098
0.12	0.089	0.0011

Determine the rate law and the rate constant (with correct units) for the reaction.

13.72 The following data were obtained for the reaction of $(CH_3)_3CBr$ with hydroxide ion at 55 °C.

$$(CH_3)_3CBr + OH^- \longrightarrow (CH_3)_3COH + Br^-$$

Initial Concentrations (mol L^{-1})		Initial Rate of Formation of (CH$_3$)$_3$ COH (mol L^{-1} s^{-1})
[(CH$_3$)$_3$CBr]	[OH$^-$]	
0.19	0.10	1.0×10^{-3}
0.46	0.10	2.4×10^{-3}
0.73	0.10	3.8×10^{-3}
0.19	0.25	1.0×10^{-3}
0.19	0.37	1.0×10^{-3}

What is the rate law for the reaction? What is the value of the rate constant (with correct units) at this temperature?

Integrated Rate Laws

13.73 Data for the decomposition of SO_2Cl_2 according to the equation $SO_2Cl_2(g) \longrightarrow SO_2(g) + Cl_2(g)$, were given in Problem 13.53. Show graphically that these data fit a first-order rate law. Graphically determine the rate constant for the reaction.

13.74 For the data in Problem 13.54, decide graphically whether the reaction is first or second order. Graphically determine the rate constant for the reaction described in that problem.

13.75 The decomposition of SO_2Cl_2 described in Problem 13.53 has a first-order rate constant, $k = 2.2 \times 10^{-5}$ s^{-1} at 320 °C. If the initial SO_2Cl_2 concentration in a container is 0.0040 M, what will its concentration be (a) after 1.00 hour and (b) after 1.00 day?

13.76 The decomposition of acetaldehyde, CH_3CHO, was described in Problem 13.54, and the order of the reaction and the rate constant for the reaction at 530 °C were determined in Problem 13.74. If the initial concentration of acetaldehyde is 0.300 M, what will the concentration be (a) after 30 minutes, (b) after 180 minutes?

13.77 If it takes 75.0 min for the concentration of a reactant to drop to 25.0% of its initial value in a first-order reaction, what is the rate constant for the reaction in the units min^{-1}?

13.78 It takes 15.4 minutes for the concentration of a reactant to drop to 5.0% of its initial value in a second-order reaction. What is the rate constant for the reaction in the units of L mol^{-1} min^{-1}?

13.79 The concentration of a drug in the body is often expressed in units of milligrams per kilogram of body weight. The initial dose of a drug in an animal was 25.0 mg/kg body weight. After 2.00 hours, this concentration had dropped to 15.0 mg/kg body weight. If the drug is eliminated metabolically by a first-order process, what is the rate constant for the process in units of min^{-1}?

13.80 Phosphine, PH_3, decomposes into phosphorus, P_4, and hydrogen at 680 °C, as described in Problem 13.60. If the initial concentration of phosphine is 1.1×10^{-6} g L^{-1}, and after 180 s the concentration has dropped to 0.30×10^{-6} g L^{-1}, what is the rate constant for this process in s^{-1}?

13.81 Hydrogen iodide decomposes according to the equation,

$$2HI(g) \longrightarrow H_2(g) + I_2(g)$$

The reaction is second order and has a rate constant equal to 1.6×10^{-3} L mol^{-1} s^{-1} at 750 °C. If the initial concentration of HI in a container is 3.4×10^{-2} M, how many minutes will it take for the concentration to be reduced to 8.0×10^{-4} M?

13.82 The reaction of $NOBr(g)$ to form $NO(g)$ and $Br_2(g)$ is second order:

$$2NOBr(g) \longrightarrow 2NO(g) + Br_2(g)$$

The rate constant is 0.556 L mol^{-1} s^{-1} at some temperature. If the initial concentration of NOBr in the container is 0.25 M, how long will it take for the concentration to decrease to 0.025 M?

13.83 Using the information determined in Problem 13.79, calculate what the initial dose of the drug must be in order for the drug concentration 3.00 hours afterward to be 5.0 mg/kg body weight.

13.84 The second-order rate constant for the decomposition of HI at 750 °C was given in Problem 13.81. At 2.5×10^3 minutes after a particular experiment had begun, the HI concentration was equal to 4.5×10^{-4} mol L^{-1}. What was the initial molar concentration of HI in the reaction vessel?

13.85 The half-life of a certain first-order reaction is 15 minutes. What fraction of the original reactant concentration will remain after 2.0 hours?

13.86 Strontium-90 has a half-life of 28 years. How long will it take for all of the strontium-90 presently on earth to be reduced to $\frac{1}{32}$ of its present amount?

13.87 Using the graph from Problem 13.53, determine the time required for the SO_2Cl_2 concentration to drop from 0.100 mol L^{-1} to 0.050 mol L^{-1}. How long does it take for the concentration to drop from 0.050 mol L^{-1} to 0.025 mol L^{-1}? What is the order of this reaction? (*Hint:* How is the half-life related to concentration?)

13.88 Using the graph from Problem 13.54, determine how long it takes for the CH_3CHO concentration to decrease from 0.200 mol L^{-1} to 0.100 mol L^{-1}. How long does it take the concentration to drop from 0.100 mol L^{-1} to 0.050 mol L^{-1}? What is the order of this reaction? (*Hint:* How is the half-life related to concentration?)

13.89 Hydrogen peroxide, which decomposes in a first-order reaction, has a half-life of 10 hours in air. How long will it take for hydrogen peroxide to decompose to 10% of its original concentration?

13.90 SO_2Cl_2 decomposes in a first-order process with a half life of 4.88×10^3 s. If the original concentration of

SO_2Cl_2 is 0.012 M, how many seconds will it take for the SO_2Cl_2 to reach 0.0020 M?

13.91 A 500 mg sample of rock was found to have 2.45×10^{-6} mol of potassium-40 ($t_{1/2} = 1.3 \times 10^9$ yr) and 2.45×10^{-6} mol of argon-40. How old was the rock? (*Hint:* What assumption is made about the origin of the argon-40?)

13.92 A tree killed by being buried under volcanic ash was found to have a ratio of carbon-14 atoms to carbon-12 atoms of 4.8×10^{-14}. How long ago did the eruption occur?

13.93 A wooden door lintel from an excavated site in Mexico would be expected to have what ratio of carbon-14 to carbon-12 atoms if the lintel were 9.0×10^3 years old?

13.94 If a rock sample was found to contain 1.16×10^{-7} mol of argon-40, how much potassium-40 ($t_{1/2} = 1.3 \times 10^9$ yr) would also have to be present for the rock to be 1.3×10^9 years old? See assumption in Problem 13.91.

Activation Energies

13.95 The following data were collected for a reaction:

Rate Constant (L mol^{-1} s^{-1})	Temperature (°C)
2.88×10^{-4}	3.20×10^2
4.87×10^{-4}	3.40×10^2
7.96×10^{-4}	3.60×10^2
1.26×10^{-3}	3.80×10^2
1.94×10^{-3}	4.00×10^2

Determine the activation energy for the reaction in kJ/mol both graphically and by calculation using Equation 13.16. For the calculation of E_a, use the first and last sets of data in the table.

13.96 Rate constants were measured at various temperatures for the reaction, $HI(g) + CH_3I(g) \longrightarrow CH_4(g) + I_2(g)$. The following data were obtained:

Rate Constant (L mol^{-1} s^{-1})	Temperature (°C)
1.91×10^{-2}	2.05×10^2
2.74×10^{-2}	2.10×10^2
3.90×10^{-2}	2.15×10^2
5.51×10^{-2}	2.20×10^2
7.73×10^{-2}	2.25×10^2
1.08×10^{-1}	2.30×10^2

Determine the activation energy in kJ/mol both graphically and by calculation using Equation 13.16. For the calculation of E_a, use the first and last sets of data in the table.

13.97 NOCl decomposes as: $2NOCl \longrightarrow 2NO + Cl_2$, and has $k = 9.3 \times 10^{-5}$ L mol^{-1} s^{-1} at 100 °C and $k = 1.0 \times 10^{-3}$ L mol^{-1} s^{-1} at 130 °C. What is E_a for this reaction in kJ mol^{-1}? Use the data at 373 K to calculate the frequency factor, A. What is the rate constant at 473 K?

13.98 The conversion of cyclopropane, an anesthetic, to propylene (see Problem 13.68) has a rate constant $k = 1.3 \times 10^{-6} s^{-1}$ at 673 K and $k = 1.1 \times 10^{-5} s^{-1}$ at 703 K. What is the activation energy in kJ/mol? Use the data given at 703 K to calculate the frequency factor, A, for this reaction. What is the rate constant for the reaction at 350 °C?

13.99 The decomposition of N_2O_5 has an activation energy of 103 kJ/mol and a frequency factor of $4.3 \times 10^{13} s^{-1}$. What is the rate constant for this decomposition at **(a)** 25 °C and **(b)** 373 K?

13.100 At 35 °C, the rate constant for the reaction

$$C_{12}H_{22}O_{11} + H_2O \longrightarrow C_6H_{12}O_6 + C_6H_{12}O_6$$
$$\underset{\text{sucrose}}{} \qquad\qquad \underset{\text{glucose}}{} \quad \underset{\text{fructose}}{}$$

is $k = 6.2 \times 10^{-5} s^{-1}$. The activation energy for the reaction is 108 kJ mol^{-1}. What is the rate constant for the reaction at 45 °C?

Mechanisms of Reactions

13.101 The oxidation of NO to NO_2, one of the reactions in the production of smog, appears to involve carbon monoxide. A possible mechanism is

$$CO + \cdot OH \longrightarrow CO_2 + H\cdot$$

$$H\cdot + O_2 \longrightarrow HOO\cdot$$

$$HOO\cdot + NO \longrightarrow \cdot OH + NO_2$$

(The formulas with dots represent extremely reactive species with unpaired electrons and are called *free radicals*.) Write the net chemical equation for the reaction.

13.102 A reaction has the following mechanism:

$$2NO \longrightarrow N_2O_2$$

$$N_2O_2 + H_2 \longrightarrow N_2O + H_2O$$

$$N_2O + H_2 \longrightarrow N_2 + H_2O$$

What is the net overall change that occurs in this reaction? Identify any intermediates in the reaction.

13.103 If the reaction

$$NO_2 + CO \longrightarrow NO + CO_2$$

occured by a one-step collision process, what would be the expected rate law for the reaction? The actual rate law is rate $= k [NO_2]^2$. Could the reaction actually occur by a one-step collision between NO_2 and CO? Explain.

13.104 If the reaction

$$2NO_2(g) + F_2(g) \longrightarrow 2NO_2 F(g)$$

occurred by a one-step process, what would be the expected rate law for the reaction? The actual rate law is rate $= k [NO_2][F_2]$, why is this a better rate law?

13.105 Consider the general reaction

$$AB + C \longrightarrow AC + B$$

If this reaction occurs in one step, what would be the expected rate law for the reaction?

13.106 Nitrogen dioxide reacts with carbon monoxide to produce nitrogen monoxide and carbon dioxide in the reaction, $NO_2(g) + CO(g) \longrightarrow NO(g) + CO_2(g)$. The first step in the reaction is proposed to be two NO_2 molecules colliding to form NO_3 and NO. The NO_3 intermediate then reacts with CO to form CO_2. What is the rate law for this proposed mechanism?

13.107 The oxidation of nitrogen monoxide with oxygen to nitrogen dioxide has a possible mechanism of

$$2NO(g) \rightleftharpoons N_2O_2(g) \qquad \text{(fast)}$$
$$N_2O_2(g) + O_2(g) \longrightarrow 2NO_2(g) \qquad \text{(slow)}$$

What are the intermediates in the proposed mechanism? What is the balanced equation for the overall reaction? What is the rate law for the reaction?

13.108 The reaction of chloroform and chlorine forms carbon tetrachloride and hydrogen chloride in the following proposed mechanism:

$$Cl_2(g) \rightleftharpoons 2Cl(g) \qquad\qquad \text{(fast)}$$
$$Cl(g) + CHCl_3(g) \longrightarrow HCl(g) + CCl_3(g) \quad \text{(slow)}$$
$$CCl_3(g) + Cl(g) \longrightarrow CCl_4(g) \qquad\qquad \text{(fast)}$$

What are the intermediates in the proposed mechanism? What is the balanced equation for the overall reaction? What is the rate law for the reaction?

| Additional Exercises

13.109 The following data were collected for the reaction of cyclobutane, $C_4H_8(g)$, to form ethylene, $C_2H_4(g)$.

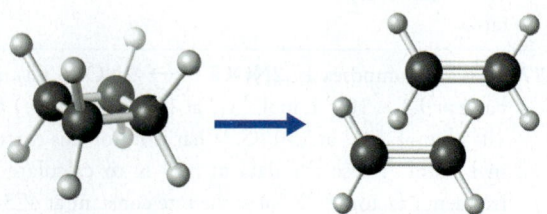

$$C_4H_8(g) \longrightarrow 2C_2H_4(g)$$

Time (min)	$[C_4H_8]$ (mol L^{-1})
0	0.00912
2.00×10^3	0.00720
4.00×10^3	0.00565
6.00×10^3	0.00447
8.00×10^3	0.00353
1.00×10^4	0.00278

Make a graph of concentration versus time for the *formation* of C_2H_4 and the decomposition of C_4H_8 on the same graph. What are the rates of formation of C_2H_4 at $t = 1.00 \times 10^3$ s and $t = 9.00 \times 10^3$ s, and the rates of decomposition of C_4H_8 at $t = 1.00 \times 10^3$ s and $t = 9.00 \times 10^3$ s? What can be said about the relationship between the values of the rates of formation and decomposition?

13.110 The age of wine can be determined by measuring the trace amount of radioactive tritium, ^3H, present in a sample. Tritium is formed from hydrogen in water vapor in the upper atmosphere by cosmic bombardment, so all naturally occurring water contains a small amount of this isotope. Once the water is in a bottle of wine, however, the formation of additional tritium from the water is negligible, so the tritium initially present gradually diminishes by a first-order radioactive decay with a half-life of 12.5 years. If a bottle of wine is found to have a tritium concentration that is 10.0% of freshly bottled wine, what is the age of the wine?

13.111 On the following graph, label the products, reactant, $\Delta H_{reaction}$, the energy of activation (E_a) and the transition state. Is the reaction exothermic or endothermic? On the graph, draw a path for a catalyzed reaction.

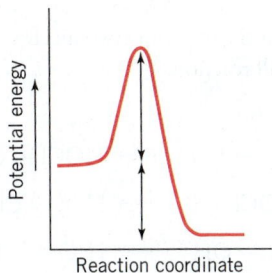

13.112 Carbon-14 dating can be used to estimate the age of formerly living materials because the uptake of carbon-14 from carbon dioxide in the atmosphere stops once the organism dies. If tissue samples from a mummy contain about 81.0% of the carbon-14 expected in living tissue, how old is the mummy? The half-life for decay of carbon-14 is 5730 years.

***13.113** What percentage of cesium chloride made from cesium-137 ($t_{1/2} = 30$ yr) remains after 150 years? What *chemical* product forms?

***13.114** For the following reactions, predict how the rate of the reaction will change as the concentration of the reactants triple.

(a) $SO_2Cl_2 \longrightarrow SO_2 + Cl_2$ rate = $k[SO_2Cl_2]$
(b) $2HI \longrightarrow H_2 + I_2$ rate = $k[HI]^2$
(c) $ClOO \longrightarrow Cl + O_2$ rate = k
(d) $NH_4^+(aq) + NO_2^-(aq) \rightarrow N_2(g) + 2H_2O$
      rate = $k[NH_4^+][NO_2^-]$
(e) $2H_2(g) + 2NO(g) \longrightarrow N_2(g) + 2H_2O(g)$
      rate = $k[H_2][NO]^2$

***13.115** One of the reactions that occurs in polluted air in urban areas is $2NO_2(g) + O_3(g) \longrightarrow N_2O_5(g) + O_2(g)$. It is believed that a species with the formula NO_3 is involved in the mechanism, and the observed rate law for the overall reaction is rate = $k[NO_2][O_3]$. Propose a mechanism for this reaction that includes the species NO_3 and is consistent with the observed rate law.

***13.116** Suppose a reaction occurs with the following mechanism:

(1) $2A \rightleftharpoons A_2$ (fast)
(2) $A_2 + E \longrightarrow B + C$ (slow)

in which the first step is a very rapid reversible reaction that can be considered to be essentially an equilibrium (forward and reverse reactions occurring at the same rate) and the second is a slow step.

(a) Write the rate law for the forward reaction in step (1).
(b) Write the rate law for the reverse reaction in step (1).
(c) Write the rate law for the rate-determining step.
(d) What is the chemical equation for the net reaction that occurs in this chemical change?

Use the results of parts (a) and (b) to rewrite the rate law of the rate-determining step in terms of the concentrations of the reactants in the overall balanced chemical equation for the reaction.

13.117 The decomposition of urea, $(NH_2)_2CO$, in 0.10 M HCl follows the equation

$(NH_2)_2CO(aq) + 2H^+(aq) + H_2O \longrightarrow 2NH_4^+(aq) + CO_2(g)$

At 65 °C, $k = 5.84 \times 10^{-6}$ min^{-1}, and at 75 °C, $k = 2.25 \times 10^{-5}$ min^{-1}. If this reaction is run at 85 °C starting with a urea concentration of 0.0020 M, how many minutes will it take for the urea concentration to drop to 0.0012 M?

13.118 Show that for a reaction that obeys the general rate law, rate = $k[A]^n$ a graph of log(rate) versus log [A] should yield a straight line with a slope equal to the order of the reaction. For the reaction in Problem 13.57, measure the rate of the reaction at $t = 150, 300, 450,$ and 600 s. Then graph log(rate) versus log[SO_2Cl_2] and determine the order of the reaction with respect to SO_2Cl_2.

13.119 The rates of many reactions approximately double for each 10 °C rise in temperature. Assuming a starting temperature of 25 °C, what would the activation energy be, in kJ mol^{-1}, if the rate of a reaction were to be twice as large at 35 °C?

13.120 If the rate constant for a first-order reaction is doubled by heating the reaction, what happens to the rate of the reaction if the concentration is kept the same?

***13.121** For the following potential energy diagram, which path represents a catalyzed reaction? How many steps would be proposed in the mechanism for the catalyzed reaction, and which step would be rate-determining?

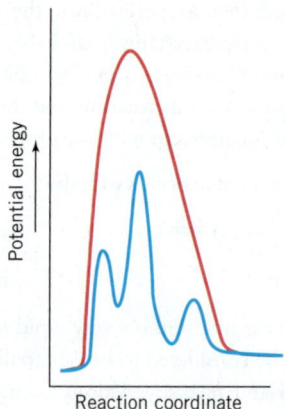

Reaction coordinate

*13.122 The development of a photographic image on film is a process controlled by the kinetics of the reduction of silver halide by a developer. The time required for development at a particular temperature is inversely proportional to the rate constant for the process. Below are published data on development times for Kodak's TriX film using Kodak D-76 developer. From these data, estimate the activation energy (in units of kJ mol^{-1}) for the development process. Also estimate the development time at 15 °C.

Temperature (°C)	Development Time (minutes)
18	10
20	9
21	8
22	7
24	6

*13.123 The rate at which crickets chirp depends on the ambient temperature, because crickets are cold-blooded insects whose body temperature follows the temperature of their environment. It has been found that the Celsius temperature can be estimated by counting the number of chirps in 8 seconds and then adding 4. In other words, t_c = (number of chirps in 8 seconds) + 4.

(a) Calculate the number of chirps in 8 seconds for temperatures of 20, 25, 30, and 35 °C.

(b) The number of chirps per unit of time is directly proportional to the rate constant for a biochemical reaction involved in the cricket's chirp.

(c) On the basis of this assumption, make a graph of ln(chirps in 8 s) versus (1/T).

(d) Calculate the activation energy for the biochemical reaction involved.

(e) How many chirps would a cricket make in 8 seconds at a temperature of 40 °C?

*13.124 The cooking of an egg involves the denaturation of a protein called albumen. The time required to achieve a

particular degree of denaturation is inversely proportional to the rate constant for the process. This reaction has a high activation energy, $E_a = 418$ kJ mol^{-1}. Calculate how long it would take to cook a traditional three-minute egg on top of Mt. McKinley in Alaska on a day when the atmospheric pressure there is 355 torr.

*13.125 The following question is based on Chemistry Outside the Classroom 13.1. The reaction of hydrogen and bromine appears to follow the mechanism shown,

$$Br_2 \longrightarrow 2Br\cdot$$

$$Br\cdot + H_2 \longrightarrow HBr + H\cdot$$

$$H\cdot + Br_2 \longrightarrow HBr + Br\cdot$$

$$2Br\cdot \longrightarrow Br_2$$

(a) Identify the initiation step in the mechanism.

(b) Identify any propagation steps.

(c) Identify the termination step.

(d) The mechanism also contains the reaction

$$H\cdot + HBr \longrightarrow H_2 + Br\cdot$$

How does this reaction affect the rate of formation of HBr?

13.126 Show that the following two mechanisms give the same net overall reaction.

Mechanism 1

$$OCl^- + H_2O \longrightarrow HOCl + OH^-$$

$$HOCl + I^- \longrightarrow HOI + Cl^-$$

$$HOI + OH^- \longrightarrow H_2O + OI^-$$

Mechanism 2

$$OCl^- + H_2O \longrightarrow HOCl + OH^-$$

$$I^- + HOCl \longrightarrow ICl + OH^-$$

$$ICl + 2OH^- \longrightarrow OI^- + Cl^- + H_2O$$

13.127 The experimental rate law for the reaction

$$NO_2 + CO \longrightarrow CO_2 + NO,$$

is rate = $k [NO_2]^2$. If the mechanism is

$$2NO_2 \longrightarrow NO_3 + NO \quad (slow)$$

$$NO_3 + CO \longrightarrow NO_2 + CO_2 \quad (fast)$$

show that the predicted rate law is the same as the experimental rate law.

13.128 Radioactive samples are considered to become nonhazardous after 10 half-lives. If the half-life is less than 88 days, the radioactive sample can be stored through a decay-in-storage program in which the material is kept in a lead-lined cabinet for at least 10 half-lives. What percent of the initial material will remain after 10 half-lives?

13.129 Use a spreadsheet to generate a graph for the data in (a) Problem 13.95 and (b) Problem 13.96. Obtain the equation for the straight line and then determine the slope. Use this slope to calculate the activation energy in kJ mol^{-1}.

13.130 Use a spreadsheet to generate separate graphs for the data given in problems (a) 13.53, (b) 13.54, and (c) 13.109. Fit each curve to an appropriate polynomial equation trying several exponents to determine the best fit. Next, take the first derivative of the best equation to determine the slope at the desired time and using this slope, calculate the answer. If you did these calculations by hand, how do the results compare?

13.131 Radon-220 is radioactive, and decays into polonium-216 by emitting an alpha particle. This is a first-order process with a rate constant of 0.0125 s^{-1}. How many alpha particles per second will be emitted by one picogram of ^{220}Rn inhaled on a particle of dust?

Multi-Concept Problems

*__13.132__ The catalyzed decomposition of ethanol at 327 °C has a rate constant of 4.00×10^{-5} mol L^{-1} s^{-1}. The plot of concentration of ethanol versus time gives a straight line. The balanced equation for this reaction is

$$C_2H_5OH(g) \longrightarrow C_2H_4(g) + H_2O(g)$$

If the initial concentration of ethanol is 0.020 mol L^{-1}, how long will it take for the pressure to reach 1.4 atm at 327 °C?

*__13.133__ On December 19, 2007, the T2 Laboratories, Inc., reactor exploded in a runaway reaction. The reaction of methyl cyclopentadienyl dimer and sodium produces sodium methyl cyclopentadiene and hydrogen:

The reactor has to be cooled when its temperature reaches 182 °C. On the day of the explosion, the cooling system malfunctioned and the reactor reached 199 °C, at which point a second, more exothermic reaction began to occur. The typical energy of activation for an organic reaction is about 50 kJ/mol. The heat of this reaction is 40 kJ mol^{-1}, and the heat capacity for an industrial scale reaction (i.e., 10,000 mol) can be 7.5×10^6 J °C^{-1}. If the rate constant for a typical reaction is 1×10^{-5} s^{-1} at 25 °C, what will be the rate constant, if all the heat is supplied to the reaction all at once, without any external cooling? If the reaction were started at 182 °C, would the reaction reach 199 °C if all the heat were released at once?

Exercises in Critical Thinking

*__13.134__ Primary explosives are extremely sensitive to impact, friction, or static electricity, and only a small amount of energy is required to initiate the explosion. Secondary and tertiary explosives require more energy to start a reaction, and often primary explosives are used to initiate their reaction. Using a potential energy diagram, show how a primary explosive would start the reaction of a secondary explosive.

13.135 Provide three examples of ordinary occurrences that mimic a reaction mechanism and have a rate-limiting step.

13.136 Can a reaction have a negative activation energy? Explain your response.

*__13.137__ Assume you have a three-step mechanism. Would the potential energy diagram have three peaks? If so, how would you distinguish the rate-limiting step?

13.138 What range of ages can ^{14}C dating reliably determine?

13.139 Why are initial reaction rates used to determine rate laws?

13.140 If a reaction is reversible (i.e., the products can react to re-form the reactants), what would the rate law look like?

13.141 The ozone layer protects us from high-energy radiation. The description of how the ozone layer works is a kinetics problem concerning the formation and destruction of ozone to produce a steady state. How can we write this mathematically, assuming that the processes are elementary reactions?

*__13.142__ How would you measure the rate of an extremely fast reaction?

*__13.143__ For a reaction done on the ton scale, would a heterogeneous catalyst or a homogeneous catalyst be more effective to use?

13.144 Can we use molality instead of molarity in constructing rate laws? Can mole fractions be used?

Chemical Equilibrium

14 •

Chapter Outline

14.1 | Dynamic Equilibrium in Chemical Systems

14.2 | Equilibrium Laws

14.3 | Equilibrium Laws Based on Pressures or Concentrations

14.4 | Equilibrium Laws for Heterogeneous Reactions

14.5 | Position of Equilibrium and the Equilibrium Constant

14.6 | Equilibrium and Le Châtelier's Principle

14.7 | Calculating Equilibrium Constants

14.8 | Using Equilibrium Constants to Calculate Concentrations

In a high wire act, the artists must maintain their equilibrium, otherwise disaster may ensue. If a tightrope walker is pushed in one direction, he or she will bend in the other direction to counteract the motion. Chemical reactions in equilibrium also respond to reestablish equilibrium when disturbed. In the pages ahead we will take a close look at chemical equilibria and examine the factors that affect the relative amounts of reactants and products present when a dynamic chemical equilibrium is achieved.

© Liang Xu/Xinhua Press/Corbis

In Chapter 13 you learned that most chemical reactions slow down as the reactants are consumed and the products form. The focus there was learning about the factors that affect how fast the reactant and product concentrations change. In this and the following three chapters we will turn our attention to the ultimate fate of chemical systems, dynamic chemical equilibrium—the situation that exists when the concentrations cease to change.

You have already encountered the concept of a dynamic equilibrium several times in this book. Recall that such an equilibrium is established when two opposing processes occur at equal rates. In a liquid–vapor equilibrium, for example, the processes are evaporation and condensation. In a chemical equilibrium, such as the ionization of a weak acid, they are the forward and reverse reactions represented by the chemical equation. Understanding the factors that affect equilibrium systems is important not only in chemistry but in other sciences as well. They apply to biology because living cells must control the concentrations of substances within them in order to survive. And they are fundamental in understanding chemical reactions that affect many current environmental problems, including global climate change, acid rain, and stratospheric ozone depletion.

This Chapter in Context

LEARNING OBJECTIVES

After reading this chapter, you should be able to:

- distinguish, describe, and explain the nature of a dynamic equilibrium
- explain the basics of equilibrium laws
- write and convert between equilibrium laws based on molar concentration and gas pressures
- understand and explain why solids and pure liquids are not included in equilibrium laws
- interpret the value of the equilibrium constant as an indicator of the position of equilibrium
- use Le Châtelier's principle and the reaction quotient, Q, to interpret where a system lies in relation to equilibrium
- describe experiments to determine the equilibrium constant and how to use the data from these experiments
- use tabulated values of equilibrium constants and experimental data to calculate the equilibrium concentrations (or pressures) of all species in an equilibrium mixture

14.1 | Dynamic Equilibrium in Chemical Systems

We begin our discussion of equilibrium by examining principles that apply to chemical systems in general. In Chapters 15 through 17 we will focus on equilibria in aqueous solutions. Then in Chapter 18 we will tie together the concept of chemical equilibrium with the energetics of chemical change.

A Molecular Interpretation of Equilibrium

Let's review the process by which a system reaches equilibrium. For this purpose, we'll examine the reaction for the decomposition of gaseous N_2O_4 into NO_2.

$$N_2O_4(g) \longrightarrow 2NO_2(g) \qquad (14.1)$$

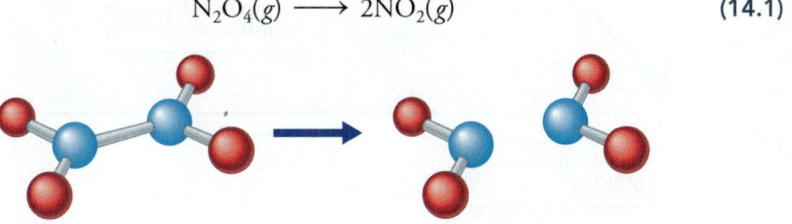

If this reaction were irreversible, over time the concentration of N_2O_4 would continue to drop until it reached zero and all of the N_2O_4 had reacted. However, if we look at what actually happens in the reaction (Figure 14.1), we see that the N_2O_4 concentration decreases until it reaches some final, steady nonzero value. Meanwhile, the NO_2 concentration rises until it also reaches a final steady value.

The explanation for these observations is that the reaction above is actually able to proceed in both directions. As NO_2 is formed and starts to accumulate in the reaction mixture, collisions between NO_2 molecules are able to produce N_2O_4 molecules, so the following reaction begins to occur.

$$2NO_2(g) \longrightarrow N_2O_4(g) \tag{14.2}$$

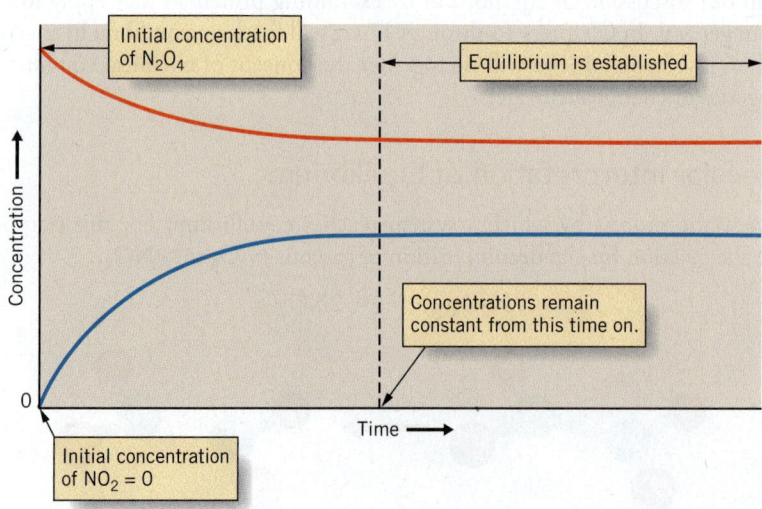

- For the equilibrium $A \rightleftharpoons B$, the reaction read from left to right is the **forward reaction**; the reaction read from right to left is the **reverse reaction**.

$A \longrightarrow B$ (forward reaction)

$A \longleftarrow B$ (reverse reaction)

Even after equilibrium has been reached, both the forward and reverse reactions continue to occur.

As the NO_2 concentration rises because of Equation 14.1, the rate of the reaction in Equation 14.2 increases. Because N_2O_4 is being remade by Equation 14.2, the drop in N_2O_4 concentration starts to slow down. As you see in Figure 14.1, eventually a state is reached in which the concentrations cease to change. At this point the rate of the **forward reaction** (Equation 14.1) is equal to the rate of the **reverse reaction** (Equation 14.2); N_2O_4 is being formed as fast as it's decomposing, and we've reached *dynamic equilibrium*. As you learned in Section 4.4, when we write the equation to describe the equilibrium, we use double arrows ($\rightleftharpoons$)

$$N_2O_4(g) \rightleftharpoons 2NO_2(g) \tag{14.3}$$

Most chemical systems eventually reach a state of dynamic equilibrium, given enough time. Sometimes, though, this equilibrium is extremely difficult (or even impossible) to detect, because in some reactions the amounts of either the reactants or products present at equilibrium are virtually zero. For instance, in water vapor at room temperature there are no detectable amounts of either H_2 or O_2 from the equilibrium

$$2H_2O(g) \rightleftharpoons 2H_2(g) + O_2(g)$$

Water molecules are so stable that we can't detect whether any of them decompose. Even in cases such as this, however, it is often *convenient* to assume that an equilibrium does exist.

Figure 14.1 | **The approach to equilibrium.** In the decomposition of $N_2O_4(g)$ into $NO_2(g)$, given by the equation $N_2O_4(g) \longrightarrow 2NO_2(g)$, the concentrations of the N_2O_4 and NO_2 change relatively fast at first. As time passes, the concentrations change more and more slowly. When equilibrium is reached, the concentrations of N_2O_4 and NO_2 no longer change with time; they remain constant.

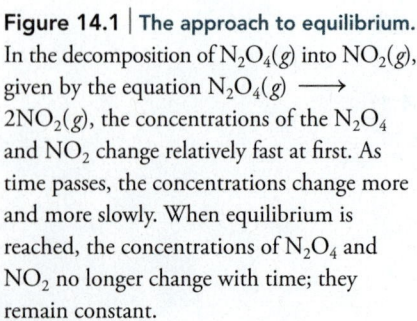

The Meaning of Reactants and Products in a Chemical Equilibrium

For the forward reaction in the equilibrium described by Equation 14.3, N_2O_4 is the reactant and NO_2 is the product. But the reverse reaction is also occurring, where NO_2 is the reactant and N_2O_4 is the product. When the system is at equilibrium, the usual definitions of reactant and product don't make a lot of sense. *In these situations, when we use the term **reactants**, we simply mean those substances written to the left of the double arrows. Similarly, we use the term **products** to mean the substances to the right of the double arrows.*

Approaching Equilibrium from Reactants or Products

One of the interesting (and useful) things about a chemical equilibrium is that the composition of an equilibrium mixture is independent of whether we begin the reaction from the "reactant side" or the "product side." For example, suppose we set up the two experiments shown in Figure 14.2. In the first 1.00 L flask we place 0.0350 mol N_2O_4. Because no NO_2 is present, some N_2O_4 must decompose for the mixture to reach equilibrium, so the reaction in Equation 14.3 will proceed in the forward direction (i.e., from left to right). When equilibrium is reached, we find the concentration of N_2O_4 has dropped to 0.0292 mol L^{-1} and the concentration of NO_2 has increased from zero to 0.0116 mol L^{-1}.

In the second 1.00 L flask we place 0.0700 mol of NO_2 (*precisely* the amount of NO_2 that would form if 0.0350 mol of N_2O_4—the amount placed in the first flask—decomposed completely). In this second flask there is no N_2O_4 present initially, so NO_2 molecules must combine, following the reverse reaction (right to left) in Equation 14.3, to give enough N_2O_4 to achieve equilibrium. When we measure the concentrations at equilibrium in the second flask, we find, once again, 0.0292 mol L^{-1} of N_2O_4 and 0.0116 mol L^{-1} of NO_2.

We see here that the same equilibrium composition is reached whether we begin with pure NO_2 or pure N_2O_4, as long as the *total* amount of nitrogen and oxygen to be divided between these two substances is the same. Similar observations apply to other chemical systems as well, which leads to the following generalization.

For a given *overall* system composition, we always reach the same equilibrium concentrations whether equilibrium is approached from the forward or reverse direction.

There are times when it is very convenient to be able to imagine the approach to equilibrium starting from the product side of the equation.

In what way does the chapter-opening photo represent a dynamic equilibrium and in what ways does it differ from a dynamic equilibrium?

■ The units mol L^{-1} (or mol/L) combined represent molarity, *M*.

TOOLS

Approach to equilibrium

Practice Exercise 14.1

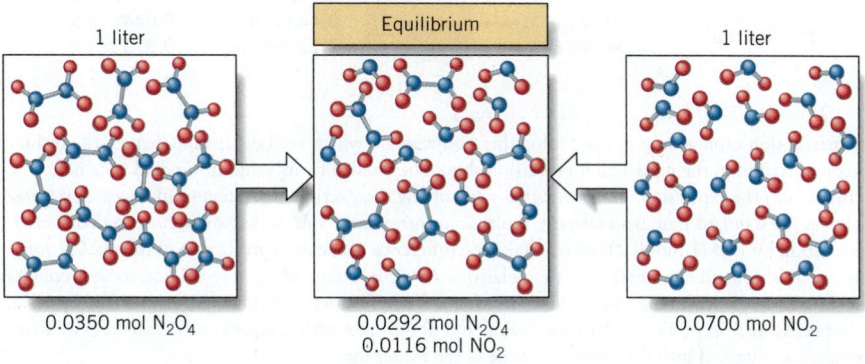

| 1 liter | Equilibrium | 1 liter |

0.0350 mol N_2O_4

0.0292 mol N_2O_4
0.0116 mol NO_2

0.0700 mol NO_2

Figure 14.2 | **Reaction reversibility for the equilibrium** $N_2O_4(g) \rightleftharpoons 2NO_2(g)$

The same equilibrium composition is reached from either the forward or reverse direction, provided the *overall* system composition is the same.

14.2 | Equilibrium Laws

For any chemical system at equilibrium, there exists a simple, predictable relationship among the molar concentrations of the reactants and products. To study this, let's consider the gaseous reaction of hydrogen with iodine to form hydrogen iodide.

$$H_2(g) + I_2(g) \rightleftharpoons 2HI(g)$$

Figure 14.3 shows the results of several experiments measuring equilibrium amounts of each gas, starting with different amounts of the reactants and product in a 10.0 L reaction vessel. When equilibrium is reached, the amounts of H_2, I_2, and HI are different for each experiment, as are their molar concentrations. This isn't particularly surprising, but what is amazing is that the relationship among the concentrations is very simple.

For each experiment in Figure 14.3, if we square the molar concentration of HI at equilibrium and then divide this by the product of the equilibrium molar concentrations of H_2 and I_2, we obtain the same numerical value. This is shown in Table 14.1, where we have once again used square brackets around formulas as symbols for molar concentrations.

The fraction used to calculate the values in the last column of Table 14.1,

$$\frac{[HI]^2}{[H_2][I_2]}$$

is called the **mass action expression**.[1] The origin of this term isn't important; just consider it a name we use to refer to this fraction. The numerical value of the mass action expression

◼ The molar concentrations are obtained by dividing the number of moles of each substance by the volume, 10.0 L.

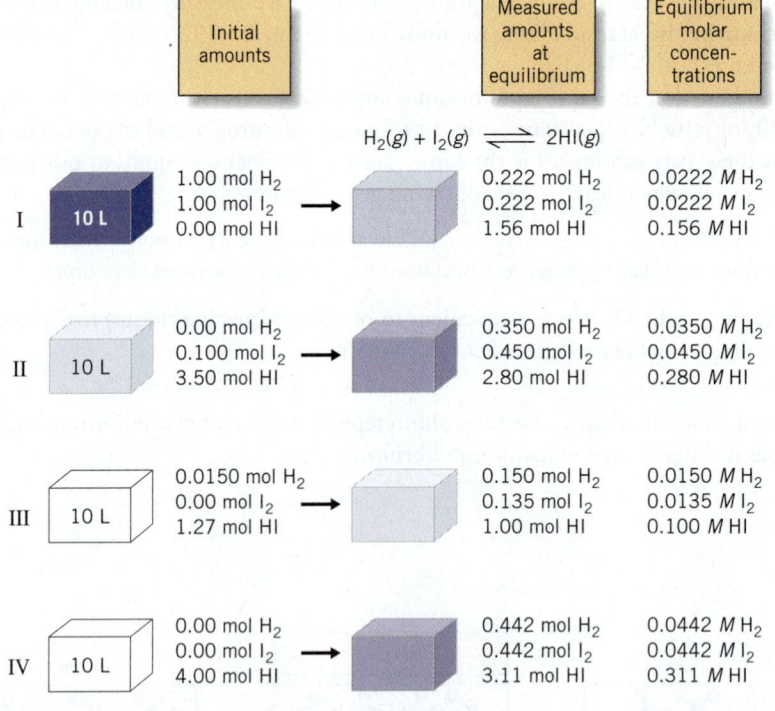

Figure 14.3 | **Four experiments to study the equilibrium among H_2, I_2, and HI gases.** Different amounts of the reactants and product are placed in a 10.0 L reaction vessel at 440 °C where the gases establish the equilibrium

$$H_2(g) + I_2(g) \rightleftharpoons 2HI(g)$$

When equilibrium is reached, different amounts of reactants and products remain in each experiment, which give different equilibrium concentrations.

---

[1] The mass action expression is derived using thermodynamics, which we'll discuss in Chapter 18. Technically, each concentration in the mass action expression should be divided by its value at standard state before inserting it into the expression. This makes the value of any mass action expression unitless. *For substances in solution*, the standard state is an effective molar concentration of 1 *M*, so we would have to divide each concentration by 1 *M* (1 mol L^{-1}), which makes no difference in the numerical value of the mass action expression as long as all concentrations are molarities. *For gases*, the standard state is 1 bar, so we would have to divide each gas concentration by its concentration at that pressure, or each gas partial pressure by 1 bar. For simplicity, we'll leave the standard state values out of our mass action expressions. Later we'll see that these values are lumped into the numerical value of the expression.

TABLE 14.1	Equilibrium Concentrations and the Value of the Mass Action Expression				
	Equilibrium Concentrations (mol L^{-1})				$\dfrac{[HI]^2}{[H_2][I_2]}$
Experiment	$[H_2]$	$[I_2]$	$\rightleftharpoons$	$[HI]$	
I	0.0222	0.0222		0.156	$(0.156)^2/(0.0222)(0.0222) = 49.4$
II	0.0350	0.0450		0.280	$(0.280)^2/(0.0350)(0.0450) = 49.8$
III	0.0150	0.0135		0.100	$(0.100)^2/(0.0150)(0.0135) = 49.4$
IV	0.0442	0.0442		0.311	$(0.311)^2/(0.0442)(0.0442) = 49.5$
					Average $= 49.5$

is called the **reaction quotient**, and is often symbolized by the letter **Q**. For this reaction, we can write

$$Q = \frac{[HI]^2}{[H_2][I_2]}$$

This equation applies for any set of H_2, I_2, and HI concentrations, whether we are at equilibrium or not.

In Table 14.1, notice that when H_2, I_2, and HI are in dynamic equilibrium at 440 °C, Q is equal to essentially the same constant value of 49.5. In fact, if we repeated the experiments in Figure 14.3 over and over again starting with different amounts of H_2, I_2, and HI, we would still obtain the same reaction quotient, provided the systems had reached equilibrium and the temperature was 440 °C. Therefore, for this reaction at equilibrium we can write

$$\frac{[HI]^2}{[H_2][I_2]} = 49.5 \qquad \text{(at equilibrium at 440 °C)} \qquad \textbf{(14.4)}$$

This relationship is called the **equilibrium law** for the system. Significantly, it tells us that for a mixture of these three gases to be at equilibrium at 440 °C, the value of the mass action expression (the reaction quotient) must equal 49.5. If the reaction quotient has any other value, then the gases are not in equilibrium at this temperature. The constant 49.5, which characterizes this equilibrium system, is called the **equilibrium constant**. The equilibrium constant is usually symbolized by K_c (we use the subscript **c** because we write the mass action expression using molar concentrations). Thus, we can state the equilibrium law as follows:

■ Sometimes Equation 14.4 is called an equilibrium expression, or the law of mass action.

$$\frac{[HI]^2}{[H_2][I_2]} = K_c = 49.5 \qquad \text{(at 440 °C)} \qquad \textbf{(14.5)}$$

It is often useful to think of an equilibrium law such as Equation 14.5 as a *condition* that must be met for equilibrium to exist.

For chemical equilibrium to exist in a reaction mixture, the value of the reaction quotient Q must be equal to the equilibrium constant, K_c.

As you may have noticed, we have repeatedly mentioned the temperature when referring to the value of K_c. This is because the value of the equilibrium constant changes when the temperature changes. Thus, if we had performed the experiments in Figure 14.3 at a temperature other than 440 °C, we would have obtained a different value for K_c.

Homogeneous Equilibria

A **homogeneous reaction** (or **homogeneous equilibrium**) is one in which all of the reactants and products are in the same gas or liquid phase. Equilibria among gases, such as the one described above, are homogeneous because all gases mix freely with each other, so a

TOOLS

Requirement for equilibrium

■ In general, it is necessary to specify the temperature when giving a value of K_c, because K_c changes when the temperature changes. For example, for the reaction

$CH_4(g) + H_2O(g) \rightleftharpoons CO(g) + 3H_2(g)$

$K_c = 1.78 \times 10^{-3}$	at 800 °C
$K_c = 4.68 \times 10^{-2}$	at 1000 °C
$K_c = 5.68$	at 1500 °C

single phase exists. There are also many equilibria in which reactants and products are dissolved in the same liquid phase, such as an aqueous solution. You will study equilibria in aqueous solutions in Chapters 16 and 17.

An important fact about homogeneous reactions is that the mass action expression and the equilibrium law can *always* be predicted from the coefficients of the balanced chemical equation for the reaction. For example, for the general chemical equation

$$dD + eE \rightleftharpoons fF + gG \tag{14.6}$$

where D, E, F, and G represent chemical formulas and d, e, f, and g are their coefficients, the mass action expression is

$$\frac{[F]^f[G]^g}{[D]^d[E]^e}$$

The exponents in the mass action expression are the same as the stoichiometric coefficients in the balanced equation.

The condition for equilibrium in this reaction is given by the equation

$$\frac{[F]^f[G]^g}{[D]^d[E]^e} = K_c \tag{14.7}$$

TOOLS

The equilibrium law

■ We *cannot* predict a *rate law* from the balanced overall equation, but we *can* predict the *equilibrium law*.

where the only concentrations that satisfy the equation are *equilibrium concentrations*.

Notice that in writing the mass action expression the molar concentrations of the products are always placed in the numerator and those of the reactants are always placed in the denominator. Also note that after being raised to appropriate powers the concentration terms are multiplied, not added.

Example 14.1
Writing the Equilibrium Law

Most of the hydrogen produced in the United States is derived from methane in natural gas, using the forward reaction of the equilibrium,

$$CH_4(g) + H_2O(g) \rightleftharpoons CO(g) + 3H_2(g)$$

What is the equilibrium law for this reaction?

Analysis: This is a straightforward question, and finding the answer is a task you will repeat many times as you work problems dealing with chemical equilibria.

Assembling the Tools: The pattern we follow is represented by Equations 14.6 and 14.7, simply substituting formulas and coefficients.

Solution: The equilibrium law sets the mass action expression equal to the equilibrium constant. To form the mass action expression, we place the concentrations of the products in the numerator and the concentrations of the reactants in the denominator. The coefficients in the balanced chemical equation become exponents on the concentrations. The equilibrium law is

$$\frac{[CO][H_2]^3}{[CH_4][H_2O]} = K_c$$

Is the Answer Reasonable? Check to see that the products are on top of the fraction and that the reactants are on the bottom. Also check the exponents and be sure they are the same as the coefficients in the balanced chemical equation. Notice that we omit writing the exponent when it is equal to 1.

Practice Exercise 14.2

Write the equilibrium law for each of the following:
(a) $2H_2(g) + O_2(g) \rightleftharpoons 2H_2O(g)$
(b) $CH_4(g) + 2O_2(g) \rightleftharpoons CO_2(g) + 2H_2O(g)$
(*Hint:* Recall that it is always products divided by reactants.)

The equilibrium law for a reaction is

Practice Exercise 14.3

$$\frac{[NO_2]^4}{[N_2O_3]^2[O_2]} = K_c$$

Write the chemical equation for this equilibrium law.

The rule that we always write the concentrations of the products in the numerator of the mass action expression and the concentrations of the reactants in the denominator is not required by nature. Rather, it is a convention chemists have agreed to follow. Certainly, if the mass action expression is equal to a constant, its reciprocal is also equal to a constant (let's call it K_c'):

$$\frac{[HI]^2}{[H_2][I_2]} = K_c \qquad \frac{[H_2][I_2]}{[HI]^2} = \frac{1}{K_c} = K_c'$$

Having a set rule for constructing the mass action expression is very useful; it means we don't have to specify the mass action expression when we give the equilibrium constant for a reaction. For example, suppose we're told that at a particular temperature $K_c = 10.0$ for the reaction

$$2NO_2(g) \rightleftharpoons N_2O_4(g)$$

From the chemical equation, we can write the correct mass action expression and the correct equilibrium law.

$$K_c = \frac{[N_2O_4]}{[NO_2]^2} = 10.0$$

The balanced chemical equation contains all the information we need to write the equilibrium law.

Manipulating Equilibrium Laws

Sometimes it is useful to be able to combine chemical equilibria to obtain the equation for some other reaction of interest. In doing this, we perform various operations such as reversing an equation, multiplying the coefficients by some factor, and adding the equations to give the desired equation. In our discussion of thermochemistry, you learned how such manipulations affect ΔH values. Some different rules apply to changes in the mass action expressions and equilibrium constants.

TOOLS

Manipulating equilibrium equations

Changing the Direction of an Equilibrium

When the direction of an equation is reversed, the new equilibrium constant is the reciprocal of the original. You have just seen this in the preceding discussion. As another example, when we reverse the equilibrium

$$PCl_3 + Cl_2 \rightleftharpoons PCl_5 \qquad K_c = \frac{[PCl_5]}{[PCl_3][Cl_2]}$$

we obtain

$$PCl_5 \rightleftharpoons PCl_3 + Cl_2 \qquad K_c' = \frac{[PCl_3][Cl_2]}{[PCl_5]}$$

The mass action expression for the second reaction is the reciprocal of that for the first, so K_c' equals $1/K_c$.

Multiplying the Coefficients by a Factor

When the coefficients in an equation are multiplied by a factor, the equilibrium constant is raised to a power equal to that factor. For example, suppose we multiply the coefficients of the equation

$$PCl_3 + Cl_2 \rightleftharpoons PCl_5 \qquad K_c = \frac{[PCl_5]}{[PCl_3][Cl_2]}$$

by 2. This gives

$$2PCl_3 + 2Cl_2 \rightleftharpoons 2PCl_5 \qquad K_c'' = \frac{[PCl_5]^2}{[PCl_3]^2[Cl_2]^2}$$

Comparing mass action expressions, we see that $K_c'' = K^2$.

Adding Chemical Equilibria

When chemical equilibria are added, their equilibrium constants are multiplied. For example, suppose we add the following two equations.

$$2N_2 + O_2 \rightleftharpoons 2N_2O \qquad K_{c_1} = \frac{[N_2O]^2}{[N_2]^2[O_2]}$$

$$2N_2O + 3O_2 \rightleftharpoons 4NO_2 \qquad K_{c_2} = \frac{[NO_2]^4}{[N_2O]^2[O_2]^3}$$

$$2N_2 + 4O_2 \rightleftharpoons 4NO_2 \qquad K_{c_3} = \frac{[NO_2]^4}{[N_2]^2[O_2]^4}$$

■ We have numbered the equilibrium constants just to distinguish one from the other.

If we multiply the mass action expression for K_{c_1} by that for K_{c_2}, we obtain the mass action expression for K_{c_3}.

$$\frac{[N_2O]^2}{[N_2]^2[O_2]} \times \frac{[NO_2]^4}{[N_2O]^2[O_2]^3} = \frac{[NO_2]^4}{[N_2]^2[O_2]^4}$$

Therefore, $K_{c_1} \times K_{c_2} = K_{c_3}$.

Practice Exercise 14.4

At 25 °C, $K_c = 7.0 \times 10^{25}$ for the reaction

$$2SO_2(g) + O_2(g) \rightleftharpoons 2SO_3(g)$$

What is the value of K_c for the reaction: $SO_3(g) \rightleftharpoons SO_2(g) + \frac{1}{2}O_2(g)$? (*Hint:* What do you have to do with the original equation to obtain the new equation?)

Practice Exercise 14.5

At 25 °C, the following reactions have the equilibrium constants noted to the right of their equations.

$$2CO(g) + O_2(g) \rightleftharpoons 2CO_2(g) \qquad K_c = 3.3 \times 10^{91}$$
$$2H_2(g) + O_2(g) \rightleftharpoons 2H_2O(g) \qquad K_c = 9.1 \times 10^{80}$$

Use these data to calculate the numerical value of K_c for the reaction

$$H_2O(g) + CO(g) \rightleftharpoons CO_2(g) + H_2(g)$$

14.3 | Equilibrium Laws Based on Pressures or Concentrations

When all of the reactants and products are gases, we can formulate mass action expressions in terms of partial pressures as well as molar concentrations. This is possible because the molar concentration of a gas is proportional to its partial pressure (Figure 14.4). If we start with the ideal gas law,

$$PV = nRT$$

and solve for P, we can write the result as

$$P = \left(\frac{n}{V}\right)RT$$

The quantity n/V has units of mol/L and is simply the molar concentration. Therefore, we can write

$$P = (\text{molar concentration}) \times RT \qquad \textbf{(14.8)}$$

This equation applies whether the gas is by itself in a container or part of a mixture. In the case of a gas mixture, P is the partial pressure of the gas.

The relationship expressed in Equation 14.8 lets us rewrite the mass action expression, for reactions between gases, in terms of partial pressures. For the generalized equation of a completely gas phase equilibrium that obeys Equation 14.6,

$$dD + eE \rightleftharpoons fF + gG$$

the general form of the mass action expression using partial pressures is

$$K_P = \frac{(P_F)^f (P_G)^g}{(P_D)^d (P_E)^e} \qquad \textbf{(14.9)}$$

However, when we make a switch from concentrations to pressures, we can't expect the numerical values of the equilibrium constants to be the same, so we use two different symbols for K. When molar concentrations are used, we use the symbol K_c. When partial pressures are used, then $\textbf{\textit{K}}_\textbf{P}$ is the symbol. For example, the equilibrium law for the reaction of nitrogen with hydrogen to form ammonia,

$$N_2(g) + 3H_2(g) \rightleftharpoons 2NH_3(g)$$

can be written in either of the following two ways:

$$\frac{[NH_3]^2}{[N_2][H_2]^3} = K_c \qquad \left(\begin{array}{l}\text{because molar concentrations are}\\\text{used in the mass action expression}\end{array}\right)$$

$$\frac{P_{NH_3}^2}{P_{N_2} P_{H_2}^3} = K_P \qquad \left(\begin{array}{l}\text{because partial pressures are used}\\\text{in the mass action expression}\end{array}\right)$$

The equilibrium molar concentrations can be used to calculate K_c, whereas the equilibrium partial pressures can be used to calculate K_P.

$V = 1.00$ L $V = 0.500$ L
(a)   (b)

Figure 14.4 | **Molar concentration in a gas is proportional to the pressure.** Going from (a) to (b), we double the pressure, which halves the volume.

Equilibrium law using partial pressures

■ If you double the number of molecules of a gas per liter without changing the temperature, you double the pressure.

Example 14.2
Writing Expressions for K_P

Most of the world's supply of methanol, CH_3OH, is produced by the following reaction:

$$CO(g) + 2H_2(g) \rightleftharpoons CH_3OH(g)$$

What is the expression for K_P for this equilibrium?

Analysis: This is very similar to Example 14.1, except we're writing the equilibrium law for K_P instead of K_c.

Assembling the Tools: Our tools are the templates given by Equations 14.6 and 14.9.

$$dD + eE \rightleftharpoons fF + gG$$

$$K_P = \frac{(P_F)^f (P_G)^g}{(P_D)^d (P_E)^e}$$

Solution: In the mass action expression, it's always products over reactants. In this case we use partial pressures raised to powers equal to the coefficients in the balanced equation:

$$K_P = \frac{P_{CH_3OH}}{P_{H_2}^2 P_{CO}}$$

Is the Answer Reasonable? Check to see that the mass action expression gives products over reactants, and not the other way around. Check each exponent against the coefficients in the balanced chemical equation.

Practice Exercise 14.6

Write the equilibrium law in terms of partial pressures for the following reaction. (*Hint:* Keep in mind how the mass action expression is written.)

$$2N_2(g) + O_2(g) \rightleftharpoons 2N_2O(g)$$

Practice Exercise 14.7

Using partial pressures, write the equilibrium law for the reaction

$$H_2(g) + I_2(g) \rightleftharpoons 2HI(g)$$

Relating K_P to K_c

If all of the reactants and products are gases or pure substances K_c can be written as K_P, but if the reaction contains substances in solution, then we can only write K_c equilibrium laws. However, for the reactions that can have K_P equations, some K_P values are equal to K_c, but for many others the two constants have different values. It is therefore desirable to have a way to calculate one from the other. Converting between K_P and K_c uses the relationship between partial pressure and molarity given in Equation 14.8. We can substitute

$$(\text{molar concentration}) \times RT$$

for the partial pressure of each gas in the mass action expression for K_P. Similarly, K_c can be changed to K_P by solving Equation 14.8 for the molar concentrations, and then substituting the result, P/RT, into the appropriate expression for K_c. This sounds like a lot of work, and it is. Fortunately, there is a general equation (Equation 14.10), which can be derived from these relationships, that we can use to make the conversions simple.

TOOLS

Converting between K_P and K_c

$$K_P = K_c(RT)^{\Delta n_g} \tag{14.10}$$

In this equation, the value of Δn_g is equal to the change in the *number of moles of gas* in going from the reactants to the products.

$$\Delta n_g = (\text{moles of gaseous products}) - (\text{moles of gaseous reactants}) \tag{14.11}$$

We use the coefficients of the balanced equation for the reaction to calculate the numerical value of Δn_g. For example, the equation

$$2N_2(g) + O_2(g) \rightleftharpoons 2N_2O(g)$$

tells us that two moles of N_2O are formed when two moles of N_2 and one mole of O_2 react. In other words, two moles of gaseous product are formed from a total of three moles of gaseous reactants. That's a decrease of one mole of gas, so Δn_g for this reaction equals -1.

For some reactions, the value of Δn_g is equal to zero. An example is the decomposition of HI.

$$2HI(g) \rightleftharpoons H_2(g) + I_2(g)$$

If we take the coefficients to mean moles, there are two moles of gas on each side of the equation, so $\Delta n_g = 0$. Because (RT) raised to the zero power is equal to 1, $K_P = K_c$.

Example 14.3
Converting between K_P and K_c

At 500 °C the reaction between nitrogen and hydrogen to form ammonia,

$$N_2(g) + 3H_2(g) \rightleftharpoons 2NH_3(g)$$

has $K_c = 6.0 \times 10^{-2}$. What is the numerical value of K_P for this reaction?

Analysis: Quite clearly, this conversion will require Equation 14.10. However, before we can begin we have to be sure about the units of R and T. When used to represent temperature, a capital letter T in an equation always means the Kelvin temperature. Therefore we will have to convert the Celsius temperature to kelvins. Next we must choose an appropriate value for R. Referring back to Equation 14.8, if the partial pressures are expressed in atm and the concentration in mol L^{-1}, the only value of R that includes all of these units (L, mol, atm, and K) is $R = 0.0821$ L atm mol^{-1} K^{-1}, and this is the *only* value of R that can be used in Equation 14.10.

Assembling the Tools: As we've noted, we are going to have to substitute into Equation 14.10,

$$K_P = K_c(RT)^{\Delta n_g}$$

To three significant figures, the temperature in kelvins is calculated by the equation: $T_K = t_C + 273$. We also have to calculate Δn_g using Equation 14.11,

$$\Delta n_g = \text{(moles of gaseous products)} - \text{(moles of gaseous reactants)}$$

Solution: Let's start by calculating Δn_g. From the coefficients in the equation, we have two moles of gaseous products and four moles of gaseous reactants [1 mol $N_2(g)$ and 3 mol $H_2(g)$]. Therefore, $\Delta n_g = (2) - (1 + 3) = -2$.

Assembling the data, then, we have

$$K_c = 6.0 \times 10^{-2} \qquad T = (500 + 273) \text{ K} = 773 \text{ K}$$

$$\Delta n_g = -2 \qquad R = 0.0821 \text{ L atm mol}^{-1} \text{ K}^{-1}$$

Substituting these into the equation for K_P gives

$$K_P = (6.0 \times 10^{-2}) \times [(0.0821) \times (773)]^{-2}$$

$$= (6.0 \times 10^{-2}) \times (63.5)^{-2}$$

$$= 1.5 \times 10^{-5}$$

In this case, K_P has a numerical value quite different from that of K_c.

Is the Answer Reasonable? In working these problems, check to be sure you have used the correct value of R and that the temperature is expressed in kelvins. Notice that if Δn_g is negative, as it is in this reaction, the expression $(RT)^{\Delta n_g}$ will be less than one. That should make K_P smaller than K_c, which is consistent with our results.

Nitrous oxide, N_2O, is a gas used as an anesthetic; it is sometimes called "laughing gas." This compound has a strong tendency to decompose into nitrogen and oxygen following the equation

Practice Exercise 14.8

$$2N_2O(g) \rightleftharpoons 2N_2(g) + O_2(g)$$

but the reaction is so slow that the gas appears to be stable at room temperature (25 °C). The decomposition reaction has $K_c = 7.3 \times 10^{34}$. What is the value of K_P for this reaction at 25 °C? (*Hint:* Be careful in calculating Δn_g.)

Practice Exercise 14.9

Methanol, CH_3OH, is a promising fuel that can be synthesized from carbon monoxide and hydrogen according to the equation

$$CO(g) + 2H_2(g) \rightleftharpoons CH_3OH(g)$$

For this reaction at 200 °C, $K_P = 3.8 \times 10^{-2}$. Do you expect K_P to be larger or smaller than K_c? Calculate the value of K_c at this temperature.

14.4 | Equilibrium Laws for Heterogeneous Reactions

When more than one phase exists in a reaction mixture, we call it a **heterogeneous reaction**. A common example is the combustion of charcoal, in which a solid fuel reacts with gaseous oxygen. Another is the thermal decomposition of sodium bicarbonate (baking soda), which occurs when the compound is spread on a fire.

$$2NaHCO_3(s) \longrightarrow Na_2CO_3(s) + H_2O(g) + CO_2(g)$$

If $NaHCO_3$ is placed in a sealed container so that no CO_2 or H_2O can escape, the gases and solids come to a heterogeneous equilibrium.

$$2NaHCO_3(s) \rightleftharpoons Na_2CO_3(s) + H_2O(g) + CO_2(g)$$

Following the procedure we developed earlier, we would write the equilibrium law for this reaction as

$$\frac{[Na_2CO_3(s)][H_2O(g)][CO_2(g)]}{[NaHCO_3(s)]^2} = K$$

However, the equilibrium law for reactions involving pure liquids and solids can be written in an even simpler form. This is because the concentration of a pure liquid or solid is a constant at a given temperature. *For any pure liquid or solid, the ratio of amount of substance to volume of substance is a constant.* For example, if we had a 1 mole crystal of $NaHCO_3$, it would occupy a volume of 38.9 cm³. Two moles of $NaHCO_3$ would occupy twice this volume, 77.8 cm³ (Figure 14.5), but the *ratio* of moles to liters (i.e., the molar concentration) remains the same.

Similar reasoning shows that the concentration of Na_2CO_3 in pure solid Na_2CO_3 is a constant, too. This means that the equilibrium law now has three constants, K plus two of the concentration terms. It makes sense to combine all of the numerical constants together.

$$[H_2O(g)][CO_2(g)] = \frac{[NaHCO_3(s)]^2}{[Na_2CO_3(s)]}K = K_c$$

The equilibrium law for a heterogeneous reaction is written without concentration terms for pure solids or pure liquids.

Equilibrium constants that are given in tables represent all of the constants combined.[2]

So far all of the reactions we have discussed are reactions that include solids, liquids, and gases. When we talk about reactions in solution, the reactants and products are not pure solids or liquids, but have variable concentrations. Therefore, we can use their molarities in the equilibrium law and use K_c values.

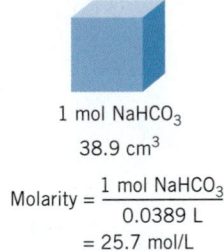

1 mol $NaHCO_3$
38.9 cm³

Molarity $= \dfrac{1 \text{ mol } NaHCO_3}{0.0389 \text{ L}}$
$= 25.7$ mol/L

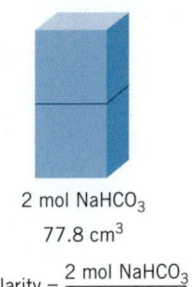

2 mol $NaHCO_3$
77.8 cm³

Molarity $= \dfrac{2 \text{ mol } NaHCO_3}{0.0778 \text{ L}}$
$= 25.7$ mol/L

Figure 14.5 | The concentration of a substance in the solid state is a constant. Doubling the number of moles also doubles the volume, but the *ratio* of moles to volume remains the same.

TOOLS

Equilibrium laws for heterogeneous reactions

---

[2] Thermodynamics handles heterogeneous equilibria in a more elegant way by expressing the mass action expression in terms of "effective concentrations," or **activities**. In doing this, thermodynamics *defines* the *activity* of any pure liquid or solid as equal to 1, which means terms involving such substances drop out of the mass action expression.

Example 14.4
Writing the Equilibrium Law for a Heterogeneous Reaction

The air pollutant sulfur dioxide can be removed from a gas mixture by passing the gases over calcium oxide. The equation is

$$CaO(s) + SO_2(g) \rightleftharpoons CaSO_3(s)$$

Write the equilibrium law for this reaction.

Analysis: For a heterogeneous equilibrium, we do not include pure solids or pure liquids in the mass action expression. Therefore, we exclude CaO and $CaSO_3$, and include only SO_2.

Assembling the Tools: Our tool is the method for constructing the equilibrium law.

Solution: Since everything in the balanced chemical equation is a solid except for $SO_2(g)$, the equilibrium law is simply

$$\frac{1}{[SO_2(g)]} = K_c$$

Is the Answer Reasonable? There's not much to check here except to review the reasoning, which appears to be okay.

Write the equilibrium law for the following reaction. (*Hint:* Follow the reasoning above.)

$$NH_3(g) + HCl(g) \rightleftharpoons NH_4Cl(s)$$

Practice Exercise 14.10

Write the equilibrium law for each of the following heterogeneous reactions.

Practice Exercise 14.11

(a) $2Hg(l) + Cl_2(g) \rightleftharpoons Hg_2Cl_2(s)$
(b) Dissolving solid Ag_2CrO_4 in water: $Ag_2CrO_4(s) \rightleftharpoons 2Ag^+(aq) + CrO_4^{2-}(aq)$
(c) $CaCO_3(s) + H_2O(l) + CO_2(aq) \rightleftharpoons Ca^{2+}(aq) + 2HCO_3^-(aq)$

14.5 | Position of Equilibrium and the Equilibrium Constant

Whether we work with K_P or K_c, a bonus of always writing the mass action expression with the product concentrations in the numerator is that the size of the equilibrium constant gives us a measure of the **position of equilibrium** (how far the reaction proceeds toward completion when equilibrium is reached). For example, the reaction

$$2H_2(g) + O_2(g) \rightleftharpoons 2H_2O(g)$$

has $K_c = 9.1 \times 10^{80}$ at 25 °C. For an equilibrium between these gases,

$$K_c = \frac{[H_2O]^2}{[H_2]^2[O_2]} = \frac{9.1 \times 10^{80}}{1}$$

By writing K_c in this way, we can emphasize the fact that the numerator of this mass action expression is enormous compared with the denominator, which means that the concentration of H_2O is huge compared with the concentrations of H_2 and O_2. At equilibrium, therefore, almost all of the hydrogen and oxygen atoms in the system are found in H_2O molecules and very few are present in H_2 and O_2. Thus, the enormous value of K_c tells us that *the position of equilibrium in this reaction lies far toward the right* and that the reaction of H_2 with O_2 goes essentially to completion.

■ Actually, you would need about 200,000 L of water vapor at 25 °C just to find one molecule of O_2 and two molecules of H_2.

● Reactant ● Product

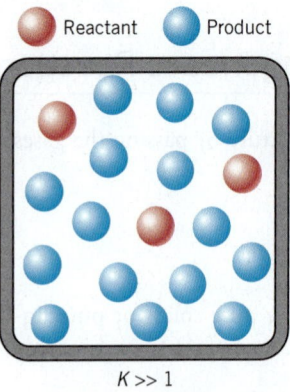
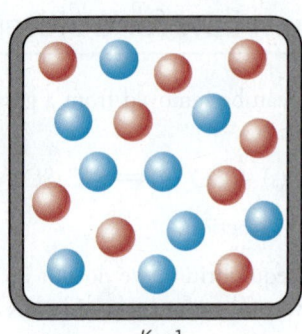
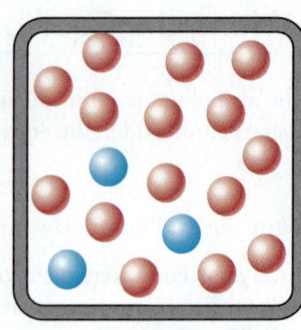

$K \gg 1$

The reaction proceeds far toward completion.

$K = 1$

Equilibrium is reached with equal concentrations of reactant and product.

$K \ll 1$

Hardly any reaction has occurred when equilibrium is reached.

Figure 14.6 | **The position of equilibrium and the magnitude of K.** Shown are three views of an equilibrium mixture for reactions of the type

reactant $\rightleftharpoons$ product

If $K \gg 1$, the concentration of product is much larger than the concentration of reactant. The two concentrations are equal if $K = 1$, and the concentration of the product is much smaller than that of the reactant if $K \ll 1$.

The reaction between N_2 and O_2 to give NO,

$$N_2(g) + O_2(g) \rightleftharpoons 2NO(g)$$

has a very small equilibrium constant; $K_c = 4.8 \times 10^{-31}$ at 25 °C. The equilibrium law for the reaction is

$$\frac{[NO]^2}{[N_2][O_2]} = 4.8 \times 10^{-31} = \frac{4.8}{10^{31}}$$

■ In air at 25 °C, the equilibrium concentration of NO should be about 10^{-17} mol/L. It is usually higher because NO is formed in various reactions, such as those responsible for air pollution caused by automobiles.

Here the denominator is huge compared with the numerator, so the concentrations of N_2 and O_2 must be very much larger than the concentration of NO. This means that in a mixture of N_2 and O_2 at this temperature, the amount of NO that is formed is negligible. The reaction hardly proceeds at all toward completion before equilibrium is reached, and *the position of equilibrium lies far toward the left.*

The relationship between the equilibrium constant and the position of equilibrium can thus be summarized as follows (also see Figure 14.6):

When K is very large	The reaction proceeds far toward completion. The position of equilibrium lies far to the right, toward the products.
When $K \approx 1$	The concentrations of reactants and products are nearly the same at equilibrium. The position of equilibrium lies approximately midway between reactants and products.
When K is very small	Extremely small amounts of products are formed. The position of equilibrium lies far to the left, toward the reactants.

TOOLS

Significance of the magnitude of K

Notice that we have omitted the subscript for K in this summary. The same qualitative predictions about the extent of reaction apply whether we use K_P or K_c.

One of the ways that we can use equilibrium constants is to compare the extents to which two or more reactions proceed to completion. Take care in making such comparisons, however, because unless the values of K are greatly different, the comparison is valid only if both reactions have the same number of reactant and product molecules appearing in their balanced chemical equations.

Practice Exercise 14.12

Suppose a mixture contained equal concentrations of H_2, Br_2, and HBr. Given that the reaction $H_2(g) + Br_2(g) \rightleftharpoons 2HBr(g)$ has $K_c = 1.4 \times 10^{-21}$, will the reaction proceed to the left or right in order to reach equilibrium? (*Hint:* In the mass action expression for this reaction, how should the size of the numerator compare with that of the denominator when the system reaches equilibrium?)

Practice Exercise 14.13

Arrange the following reactions in order of their increasing tendency to proceed toward completion.

(a) $H_2(g) + Br_2(g) \rightleftharpoons 2HBr(g)$ $K_c = 1.4 \times 10^{-21}$

(b) $2NO(g) \rightleftharpoons N_2(g) + O_2(g)$ $K_c = 2.1 \times 10^{30}$

(c) $2BrCl \rightleftharpoons Br_2 + Cl_2$ (in CCl_4 solution) $K_c = 0.14$

14.6 | Equilibrium and Le Châtelier's Principle

In Section 14.8 you will see that it is possible to perform calculations that tell us what the composition of an equilibrium system is. However, many times we really don't need to know exactly what the equilibrium concentrations are. Instead, we may want to know what actions we should take to control the relative amounts of the reactants or products at equilibrium. For instance, if we were designing gasoline engines, we would like to know what could be done to minimize the formation of nitrogen oxide pollutants. Or, if we were preparing ammonia, NH_3, by the reaction of N_2 with H_2, we might want to know how to maximize the yield of NH_3.

Le Châtelier's principle, introduced in Chapter 11, provides us with the means for making qualitative predictions about changes in chemical equilibria. It does this in much the same way that it allows us to predict the effects of outside influences on equilibria that involve physical changes, such as liquid–vapor equilibria. Recall that **Le Châtelier's principle** states that *if an outside influence upsets an equilibrium, the system undergoes a change in a direction that counteracts the disturbing influence and, if possible, returns the system to equilibrium.* Let's examine the kinds of "outside influences" that affect chemical equilibria.

TOOLS

Le Châtelier's principle

Adding or Removing a Reactant or Product

When a system is at equilibrium, the reaction quotient, Q, is equal to K. For a homogeneous system, if we add or remove some of a reactant or product, the value of Q changes, which means it no longer equals K and the system is no longer at equilibrium. In other words, the equilibrium has been upset, and according to Le Châtelier's principle the system should change in a direction that opposes the disturbance we've introduced. If we've added a substance, the reaction should proceed in a direction to remove some of it; if we've removed a substance, the reaction should proceed in a direction to replace it.

By comparing the value of Q to K the direction of a reaction can be deduced. If Q is larger than K, the reaction will move in the reverse direction, toward the reactants, to decrease Q. On the other hand, if Q is smaller than K, then the reaction will have to move toward products to increase Q and reach equilibrium. This is summarized in Table 14.2.

As an example, let's study the following equilibrium between two **complex ions** of copper.

$$Cu(H_2O)_4^{2+}(aq) + 4Cl^-(aq) \rightleftharpoons CuCl_4^{2-}(aq) + 4H_2O$$

blue   yellow

As noted, $Cu(H_2O)_4^{2+}(aq)$ is blue and $CuCl_4^{2-}(aq)$ is yellow. Mixtures of the two have an intermediate color and therefore appear blue-green, as illustrated in Figure 14.7, center.

Suppose we add chloride ion to an equilibrium mixture of these copper ions. The system can remove some Cl^- by reacting it with $Cu(H_2O)_4^{2+}(aq)$. This gives more $CuCl_4^{2-}(aq)$ (Figure 14.7, right), and we say that the equilibrium has "shifted to the right" or "shifted toward the products." In the new position of equilibrium, there is less $Cu(H_2O)_4^{2+}(aq)$ and more $CuCl_4^{2-}$ and uncombined H_2O. There is also more Cl^-, because not all that we add reacts. In other words, in this new position of equilibrium, *all* the concentrations have changed in a way that causes Q to become equal to K_c. Similarly,

■ Recall that the reaction quotient, Q, is the numerical value of the mass action expression.

TABLE 14.2	Comparison of Q and K
	Reaction Direction
$Q > K$	To reactants
$Q < K$	To products
$Q = K$	Stays at equilibrium

■ $Cu(H_2O)_4^{2+}$ and $CuCl_4^{2-}$ are called **complex ions**. Some of the interesting properties of complex ions of metals are discussed in Chapter 21.

Figure 14.7 | **The effect of concentration changes on the position of equilibrium.** The solution in the center contains a mixture of blue $Cu(H_2O)_4^{2+}$ and $CuCl_4^{2-}$, so it has a blue-green color. At the right, some of the same solution after the addition of concentrated HCl. It has a more pronounced green color because the equilibrium is shifted toward $CuCl_4^{2-}$. At the left is some of the original solution after the addition of water. It is blue because the equilibrium has shifted toward $Cu(H_2O)_4^{2+}$.

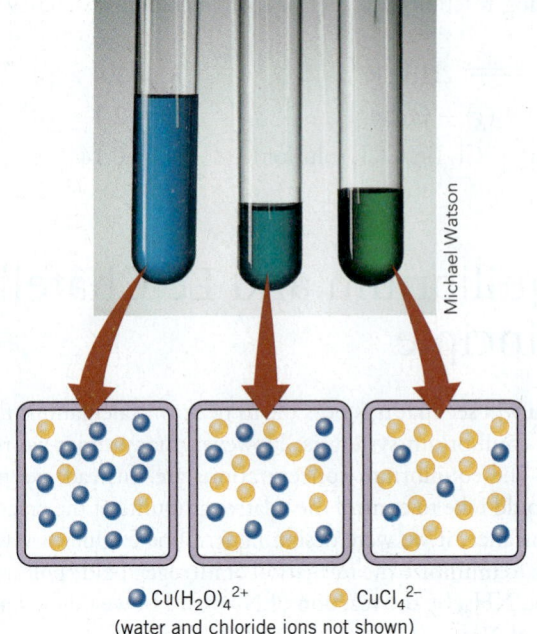

● $Cu(H_2O)_4^{2+}$ ● $CuCl_4^{2-}$
(water and chloride ions not shown)

■ The reaction shifts in a direction that will remove a substance that has been added or replace a substance that has been removed.

the position of equilibrium is shifted to the left when we add water to the mixture (Figure 14.7, left). The system is able to get rid of some of the H_2O by reaction with $CuCl_4^{2-}$, so more of the blue $Cu(H_2O)_4^{2+}$ is formed.

If we were able to remove a reactant or product, the position of equilibrium would also be changed. For example, if we add Ag^+ to a solution that contains both copper ions in equilibrium, we see an enhancement of the blue color.

$$Cu(H_2O)_4^{2+}(aq) + 4Cl^-(aq) \rightleftharpoons CuCl_4^{2-}(aq) + 4H_2O$$
blue   yellow

$$Ag^+(aq) + Cl^-(aq) \rightleftharpoons AgCl(s)$$

As the Ag^+ reacts with Cl^- to form insoluble AgCl in the second reaction, the equilibrium in the first reaction shifts to the left to replace some of the Cl^-, which also produces more blue $Cu(H_2O)_4^{2+}$. The position of equilibrium shifts to replace the substance that is removed.

The position of equilibrium shifts in a way to remove reactants or products that have been added, or to replace reactants or products that have been removed.

Changing the Volume of a Gaseous Equilibrium

Changing the volume of a mixture of reacting gases changes molar concentrations and partial pressures, so we would expect volume changes to have some effect on the position of equilibrium. Let's consider the equilibrium involved in the industrial synthesis of ammonia, which is the subject of Chemistry Outside the Classroom 14.1.

$$3H_2(g) + N_2(g) \rightleftharpoons 2NH_3(g)$$

If we reduce the volume of the reaction mixture, we cause the pressure to increase. The system can counteract the pressure increase if it is able to produce fewer molecules of gas, because fewer molecules exert a lower pressure. For the reaction above, forming more ammonia does just that. When the reaction proceeds to the right, two NH_3 molecules appear when four molecules (one N_2 and three H_2) disappear. Therefore, this equilibrium is shifted to the right when the volume of the reaction mixture is reduced.

Now let's look at the equilibrium

$$H_2(g) + I_2(g) \rightleftharpoons 2HI(g)$$

CHEMISTRY OUTSIDE THE CLASSROOM 14.1

The Haber Process: Helping to Feed the World's Population

In the popular press, chemistry is often given a bloody nose. Rarely is it mentioned that without the chemical industry, widespread famine would exist today across the world. This is because nearly one-third of the earth's human population is sustained by fertilizers generated from ammonia produced by the reaction of H_2 with N_2 in an industrial process known as the Haber process (also called the Haber–Bosch process).

At room temperature and pressure, nitrogen has little tendency to react with anything. The strong triple bond in the N_2 molecule means that any reaction that involves the breaking of the $N \equiv N$ bond will have a very high activation energy, which makes the reaction extremely slow. Making the reaction between H_2 and N_2 industrially viable requires extreme conditions of pressure and temperature. In the summer of 1909, Fritz Haber demonstrated how to overcome these obstacles, and Carl Bosch later scaled up the process to industrial size. Both Haber and Bosch were later awarded Nobel Prizes for their work.

In a modern ammonia synthesis plant, hydrogen is obtained from natural gas, which is composed chiefly of methane, CH_4. Methane is reacted with steam over a nickel oxide catalyst.

$$CH_4(g) + H_2O(g) \xrightarrow{\text{nickle oxide catalyst}} CO(g) + 3H_2(g)$$

$$CO(g) + H_2O(g) \xrightarrow{\text{nickle oxide catalyst}} CO_2(g) + H_2(g)$$

The net result is that methane and steam are converted to hydrogen gas and carbon dioxide. Nitrogen for the Haber process is obtained from air, where it is present at a level of 78%. The final stage of the process is to combine hydrogen and nitrogen over an iron oxide catalyst.

$$3H_2(g) + N_2(g) \rightleftharpoons 2NH_3(g) \quad \Delta H° = -92.4 \text{ kJ mol}^{-1}$$

High pressures in the range of 150–250 atm are used to help force the position of equilibrium to the right. High temperatures (300–500 °C) are also used, even though they tend to shift the position of equilibrium

away from ammonia. This is necessary to make the catalyst effective and make the reaction proceed at an acceptable rate. Under these conditions, one pass over the catalyst bed produces a yield of only about 15%, but the mixture is passed over multiple catalyst beds, with cooling between passes to remove NH_3. Any unreacted gases are recycled, so eventually a total yield of nearly 98% is achieved.

Ammonia is used as a fertilizer by injecting it directly into the ground, where it is captured by moisture in the soil (Figure 1). The gas also serves as a raw material for the production of nitrates and nitrites, which can be isolated as potassium or ammonium salts. The latter are solids that can be incorporated with other necessary plant nutrients to produce more complete fertilizers.

Figure 1 A farmer operates machinery that injects ammonia gas directly into the soil, where it will act as a fertilizer.

Doug Martin/Photo Researchers

If this reaction proceeds in either direction, there is no change in the number of molecules of gas. This reaction, therefore, cannot respond to pressure changes, so changing the volume of the reaction vessel has virtually no effect on the equilibrium.

The simplest way to analyze the effects of a volume change on an equilibrium system involving gases is to count the number of molecules of *gaseous* substances on both sides of the equation.

Changing the Temperature

To change the temperature of an equilibrium system, heat is either added or removed. For example, the decomposition of $N_2O_4(g)$ into $NO_2(g)$ discussed in Section 14.1 is endothermic in the forward direction, with $\Delta H° = +57.9$ kJ.

■ Temperature is the *only* factor that can change K for a given reaction.

$$N_2O_4(g) \rightleftharpoons 2NO_2(g) \qquad \Delta H° = +57.9 \text{ kJ}$$

If we add heat to raise the temperature of the equilibrium mixture, the position of equilibrium will shift in the direction that absorbs heat. Because the reaction from left to right is endothermic, the position of equilibrium will shift to the right. This increases the concentration of NO_2 and decreases the concentration of N_2O_4.

Sometimes it's easier to analyze the effect if we include energy in the equation. For this reaction, heat is absorbed in the forward direction, so we include heat as a reactant.

$$\text{Heat} + N_2O_4(g) \rightleftharpoons 2NO_2(g)$$

Adding a reactant drives the reaction in the forward direction, with heat behaving like any other reactant in this case.

Lowering the temperature by removing heat from an equilibrium system has the opposite effect from adding heat. Such a disturbance to the equilibrium promotes an exothermic change, which partially replaces the heat removed. In the preceding reaction, the reaction from right to left is exothermic, so removing heat from the system shifts the equilibrium position to the left. This causes the concentration of N_2O_4 to increase and the concentration of NO_2 to decrease.

Increasing the temperature increases the concentration of products in an endothermic reaction.
Decreasing the temperature increases the concentration of products in an exothermic reaction.

This gives us a way to experimentally determine whether a reaction is exothermic or endothermic, without using a thermometer. Earlier we described the equilibrium involving complex ions of copper. The effect of temperature on this equilibrium is demonstrated in Figure 14.8. When the reaction mixture is heated, we see from the color change that the equilibrium shifts toward the products. Therefore, the reaction must be endothermic.

Effect of Temperature Changes on K for a Reaction

Changes in concentrations or volume can shift the position of equilibrium, but they do not change the equilibrium constant. Changing the temperature, however, causes the position of equilibrium to shift because it changes the value of K. The enthalpy change of the reaction is the critical factor. Consider, for example, the equilibrium for the formation of ammonia.

■ We'll see how to quantitatively relate K and temperature in Chapter 18.

$$3H_2(g) + N_2(g) \rightleftharpoons 2NH_3(g) \qquad \Delta H° = -46.19 \text{ kJ}$$

$$\frac{[NH_3]^2}{[N_2][H_2]^3} = K_c$$

Because the reaction is exothermic, increasing the temperature shifts the equilibrium to the left. When equilibrium is restored, the concentration of NH_3 has decreased and the concentrations of N_2 and H_2 have increased. Therefore, the numerator of the mass action expression has become smaller and the denominator larger. This gives a smaller reaction quotient and therefore a smaller value of K_c.

Increasing the temperature of an exothermic reaction makes its equilibrium constant smaller.
Increasing the temperature of an endothermic reaction makes its equilibrium constant larger.

Figure 14.8 | The effect of temperature on the equilibrium

$$Cu(H_2O)_4^{2+}(aq) + 4Cl^-(aq) \rightleftharpoons$$
$$CuCl_4^{2-}(aq) + 4H_2O$$

In the center is an equilibrium mixture of the two complex ions. When the solution is cooled in ice (left), the equilibrium shifts toward the blue $Cu(H_2O)_4^{2+}$. When heated in boiling water (right), the equilibrium shifts toward $CuCl_4^{2-}$. This behavior indicates that the reaction is endothermic in the forward direction.

Notice floating ice cubes

Notice steam condensate from hot water

Michael Watson

Catalysts and the Position of Equilibrium

Recall that catalysts are substances that affect the speeds of chemical reactions without actually being used up. However, *catalysts do not affect the position of equilibrium in a system*. The catalyst's only effect is to bring the system to equilibrium faster.

■ Catalysts don't appear in the chemical equation for a reaction, so they don't appear in the mass action expression, either.

Pressure Changes Caused by Adding an Inert Gas at Constant Volume

A change in volume is not the only way to change the pressure in an equilibrium system of gaseous reactants and products. The pressure can also be changed by keeping the volume the same and adding another gas. If this gas cannot react with any of the gases already present (i.e., if the added gas is inert toward the substances in equilibrium), the concentrations and partial pressures of the reactants and products won't change. They will continue to satisfy the equilibrium law and the reaction quotient will continue to equal K_c, so there will be no change in the position of equilibrium.

Example 14.5
Application of Le Châtelier's Principle

The reaction $CH_4(g) + H_2O(g) \rightleftharpoons CO(g) + 3H_2(g)$ is endothermic with $K_c = 5.67$ at 1500 °C. How will the amount of $CO(g)$ at equilibrium be affected by (a) adding more $H_2O(g)$, (b) lowering the pressure by increasing the volume of the container, (c) raising the temperature of the reaction mixture, and (d) adding a catalyst to the system? (e) Which of these will change the equilibrium constant and in which direction (increase or decrease) will K_c change?

Analysis: The problem asks us to predict the effects of various changes on the position of equilibrium, which is the clue that the problem requires application of Le Châtelier's principle.

Assembling the Tools: Each of these kinds of disturbances and the tools to deal with them was described in the preceding discussion. We simply select the tool according to the type of disturbance involved.

Solution: (a) Adding H_2O will cause the equilibrium to shift to the right, in a direction that will consume some of the added H_2O. The amount of CO will increase.

(b) When the pressure in the system drops, the system responds by producing more molecules of gas, which will tend to raise the pressure and partially offset the change. If the reaction goes from left to right, we use up two molecules (one CH_4 and one H_2O) and form four molecules (one CO and three H_2). This yields a net increase in the number of molecules, so some CH_4 and H_2O react and the amount of CO at equilibrium will increase.

(c) Because the reaction is endothermic, we write the equation showing heat as a reactant:

$$\text{Heat} + CH_4(g) + H_2O(g) \rightleftharpoons CO(g) + 3H_2(g)$$

Raising the temperature is accomplished by adding heat, so the system will respond by absorbing heat. This means that the equilibrium will shift to the right and the amount of CO at equilibrium will increase.

(d) A catalyst causes a reaction to reach equilibrium more quickly, but it has no effect on the position of chemical equilibrium. Therefore, the amount of CO at equilibrium will not be affected.

(e) Finally, the *only* change that alters K is the temperature change. Raising the temperature (adding heat) will increase K_c for this endothermic reaction.

Are the Answers Reasonable? There's not much we can do here other than review our reasoning. Doing so will reveal we've answered the questions correctly.

Practice Exercise 14.14

Consider the equilibrium

$$PCl_5(g) + 4H_2O(g) \rightleftharpoons H_3PO_4(l) + 5HCl(g)$$

How will the amount of H_3PO_4 at equilibrium be affected by decreasing the volume of the container holding the reaction mixture? (*Hint:* Notice that this is a heterogeneous equilibrium.)

Practice Exercise 14.15

Consider the equilibrium

$$PCl_3(g) + Cl_2(g) \rightleftharpoons PCl_5(g)$$

for which $\Delta H° = -88$ kJ. How will the amount of Cl_2 at equilibrium be affected by (a) adding PCl_3, (b) adding PCl_5, (c) raising the temperature, and (d) decreasing the volume of the container? How (if at all) will each of these changes affect K_P for the reaction?

14.7 | Calculating Equilibrium Constants

In Section 14.5 you learned that the magnitude of an equilibrium constant gives us some feel for the extent to which the reaction proceeds at equilibrium. Sometimes, however, it is necessary to have more than merely a qualitative knowledge of equilibrium concentrations. This requires that we be able to use the equilibrium law for purposes of calculation.

Equilibrium calculations for gaseous reactions can be performed using either K_P or K_c, but for reactions in solution we must use K_c. Whether we deal with concentrations or partial pressures, however, the same basic principles apply.

Overall, we can divide equilibrium calculations into two main categories:

1. Calculating equilibrium constants from known equilibrium concentrations or partial pressures.
2. Calculating one or more equilibrium concentrations or partial pressures using the known value of K_c or K_P.

In this section we will study how to calculate equilibrium constants. To keep things as simple as possible, we will stick to working problems involving K_c and molar concentrations.

One way to determine the value of K_c is to carry out the reaction, measure the concentrations of reactants and products after equilibrium has been reached, and then use the equilibrium values in the equilibrium law to compute K_c. As an example, let's look again at the decomposition of N_2O_4.

$$N_2O_4(g) \rightleftharpoons 2NO_2(g)$$

In Section 14.1, we saw that if 0.0350 mol of N_2O_4 is placed into a 1.00 liter flask at 25 °C, the concentrations of N_2O_4 and NO_2 at equilibrium are

$$[N_2O_4] = 0.0292 \text{ mol/L} \qquad [NO_2] = 0.0116 \text{ mol/L}$$

To calculate K_c for the reaction, we substitute the equilibrium concentrations into the mass action expression of the equilibrium law.

$$\frac{[NO_2]^2}{[N_2O_4]} = K_c$$

$$\frac{[0.0116]^2}{[0.0292]} = K_c$$

Performing the arithmetic gives

$$K_c = 4.61 \times 10^{-3}$$

Concentration Tables

Only rarely are we given all the equilibrium concentrations required to calculate the equilibrium constant. Usually, we have information about the composition of a reaction mixture when it is first prepared, as well as some additional data that we can use to figure out what the equilibrium concentrations are. To help organize our thinking, our tool will be a **concentration table** that we will set up beneath the chemical equation. In this table we keep track of the concentrations of the substances involved in the reaction as it proceeds toward equilibrium. Normally the table has the same number of data columns as there are substances in the equilibrium equation. The table has three rows, which we label *Initial Concentrations*, *Changes in Concentrations*, and *Equilibrium Concentrations*. Entries in the table are either numbers with units of molar concentration (mol L^{-1}) or variables (e.g., x) that represent molar concentrations. Let's look at how the data for each row are obtained.

TOOLS

Concentration table

Initial Concentrations

The initial concentrations are those present in the reaction mixture when it's prepared; we imagine that no reaction occurs until everything is mixed. Often, the statement of a problem will give the initial molar concentrations, and these are then entered into the table under the appropriate formulas. In other cases, the mass or number of moles of reactants or products dissolved in a specified volume are given and we must then calculate the molar concentrations. If the initial amount or concentration of a reactant or product is not mentioned in the problem, we will assume none was added to the reaction mixture initially, so its initial concentration is entered as zero.

■ Remember that when using K_c, all entries in the concentration table must be molarities for equilibrium calculations.

Changes in Concentrations

If the reaction mixture is not at equilibrium when it's prepared, the chemical reaction must continue (either to the left or right) to bring the system to equilibrium. When this happens, the concentrations change. We will use a positive sign to indicate that a concentration increases and a negative sign to show a decrease in concentration.

The changes in concentrations always *occur in the same ratio as the coefficients in the balanced equation.* For example, if we were dealing with the equilibrium

$$3H_2(g) + N_2(g) \rightleftharpoons 2NH_3(g)$$

and found that the $N_2(g)$ concentration decreases by 0.10 M during the approach to equilibrium, the entries in the "change" row would be as follows:

	$3H_2(g)$	+	$N_2(g)$	$\rightleftharpoons$	$2NH_3(g)$
Changes in Concentrations:	$-3 \times (0.10\ M)$		$-1 \times (0.10\ M)$		$+2 \times (0.10\ M)$
	$\downarrow$		$\downarrow$		$\downarrow$
	$-0.30\ M$		$-0.10\ M$		$+0.20\ M$

In constructing the "change" row, *be sure that the reactant concentrations all change in the same direction and that the product concentrations all change in the opposite direction.* If the concentrations of the reactants decrease, all the entries for the reactants in the "change" row should have a minus sign, and all the entries for the products should be positive. In this example, N_2 reacts with H_2 (so their concentrations both decrease) and form NH_3 (so its concentration increases).

In Section 14.8, where we will be trying to find equilibrium concentrations from K_c and initial concentrations, our goal will be to find numerical values for the quantities in the change row. These quantities will be unknowns and will be represented symbolically by variables.

■ The changes in concentrations are controlled by the stoichiometry of the reaction. If we can find one of the changes, we can calculate the others from it.

Equilibrium Concentrations

These are the concentrations of the reactants and products when the system finally reaches equilibrium. We obtain them by adding the changes in concentrations to the initial concentrations.

$$\begin{pmatrix} \text{initial} \\ \text{concentration} \end{pmatrix} + \begin{pmatrix} \text{change in} \\ \text{concentration} \end{pmatrix} = \begin{pmatrix} \text{equilibrium} \\ \text{concentration} \end{pmatrix}$$

The equilibrium concentrations are the only quantities that satisfy the equilibrium law. In this section we will use the numerical values in the last row to calculate K_c.

The concentration table is one of our most useful tools for analyzing and setting up equilibrium problems. As you will see, we will use it in almost every problem. The following example illustrates how we can use it in calculating an equilibrium constant.

Example 14.6
Calculating K_c from Equilibrium Concentrations

At a certain temperature, a mixture of H_2 and I_2 was prepared by placing 0.200 mol of H_2 and 0.200 mol of I_2 into a 2.00 liter flask. After a period of time the following equilibrium was established:

$$H_2(g) + I_2(g) \rightleftharpoons 2HI(g)$$

The purple color of the I_2 vapor was used to monitor the reaction, and from the decreased intensity of the purple color it was determined that, at equilibrium, the I_2 concentration had dropped to 0.020 mol L^{-1}. What is the value of K_c for this reaction at this temperature?

Analysis: The first step in *any* equilibrium problem is to write the balanced chemical equation and the related equilibrium law. The equation is already given, and the equilibrium law corresponding to it is

$$\frac{[HI]^2}{[H_2][I_2]} = K_c$$

To calculate the value of K_c we must substitute the equilibrium concentrations of H_2, I_2, and HI into the mass action expression. But what are they? We have been given only one directly, the value of $[I_2]$. To obtain the others, we have to do some chemical reasoning.

Assembling the Tools: The principal tool to solve this problem will be a concentration table, constructed following the procedure described previously. Once we've filled in the last row of the table, we can substitute the values for the equilibrium concentrations into the next tool, the equilibrium law, and calculate K_c. Because the table must contain molar concentrations, another tool we will apply is the definition of molarity.

Solution: The first step is to enter the data given in the statement of the problem. As noted above, the entries must be molar concentrations, but we've been given moles and the volume of the container. Therefore, for the initial concentrations we have to calculate the ratio of moles to liters for both the H_2 and I_2, (0.200 mol/2.00 L) = 0.100 M. Because no HI was placed into the reaction mixture, its initial concentration is set to zero. These quantities are shown in red type in the first row. The problem statement also gives us the equilibrium concentration of I_2, so we enter this in the table in the last row (also shown in red type). The values in regular type are then derived as described below.

	$H_2(g)$	+	$I_2(g)$	$\rightleftharpoons$	$2HI(g)$
Initial concentrations (M)	0.100		0.100		0
Changes in concentrations (M)	−0.080		−0.080		+2(0.080)
Equilibrium concentrations (M)	0.020		0.020		0.160

Changes in Concentrations We have been given both the initial and equilibrium concentrations of I_2, so by difference we can calculate the change for I_2 (−0.080 M). The other changes are then calculated from the mole ratios specified by the coefficients of the balanced equation. Because H_2 and I_2 have the same coefficients (i.e., 1), their changes are equal. The coefficient of HI is 2, so its change must be twice that of I_2. Reactant concentrations are decreasing, so their changes are negative; the product concentration increases, so its change is positive.

Equilibrium Concentrations For H_2 and HI, we just add the change to the initial value.

The hard work has now been done. We substitute the equilibrium concentrations from the last row of the table into the mass action expression and calculate K_c.

$$K_c = \frac{(0.160)^2}{(0.020)(0.020)}$$

$$= 64$$

● **Is the Answer Reasonable?** First, carefully examine the equilibrium law. As always, products must be placed in the numerator and reactants in the denominator. Be sure that each exponent corresponds to the correct coefficient in the balanced chemical equation.

Next, check the concentration table to be sure all entries are molar concentrations. Note that the initial concentration of HI is zero. The change for HI must be positive, because the HI concentration cannot become smaller than zero. The positive value in the table agrees with this assessment. Notice, also, that the changes for both reactants have the same sign, which is opposite that of the product. This relationship among the signs of the changes is always true and serves as a useful check when you construct a concentration table.

Finally, we can check to be sure we've entered each of the equilibrium concentrations into the mass action expression correctly. At this point we can be confident we've set up the problem correctly. To be sure of the answer, redo the calculation on your calculator.

In a particular experiment, it was found that when $O_2(g)$ and $CO(g)$ were mixed and allowed to react according to the equation

$$2CO(g) + O_2(g) \rightleftharpoons 2CO_2(g)$$

The O_2 concentration had decreased by 0.030 mol L^{-1} when the reaction reached equilibrium. How had the concentrations of CO and CO_2 changed? (*Hint:* How are the changes in concentrations related to one another as the reaction approaches equilibrium?)

Practice Exercise 14.16

An equilibrium was established for the reaction

$$CO(g) + H_2O(g) \rightleftharpoons CO_2(g) + H_2(g)$$

at 500 °C. (This is an industrially important reaction for the preparation of hydrogen.) At equilibrium, the following concentrations were found in the reaction vessel: $[CO] = 0.180\ M$, $[H_2O] = 0.0411\ M$, $[CO_2] = 0.150\ M$, and $[H_2] = 0.200\ M$. What is the value of K_c for this reaction?

Practice Exercise 14.17

A student placed 0.200 mol of $PCl_3(g)$ and 0.100 mol of $Cl_2(g)$ into a 1.00 liter container at 250 °C. After the reaction

$$PCl_3(g) + Cl_2(g) \rightleftharpoons PCl_5(g)$$

came to equilibrium it was found that the flask contained 0.120 mol of PCl_3.
(a) What were the initial concentrations of the reactants and product?
(b) By how much had the concentrations changed when the reaction reached equilibrium?
(c) What were the equilibrium concentrations?
(d) What is the value of K_c for this reaction at this temperature?

Practice Exercise 14.18

14.8 | Using Equilibrium Constants to Calculate Concentrations

In the simplest calculation of this type, all but one of the equilibrium concentrations are known and a few algebraic steps produce the answer, as illustrated in Example 14.7.

Example 14.7
Using K_c to Calculate Concentrations at Equilibrium

The reversible reaction

$$CH_4(g) + H_2O(g) \rightleftharpoons CO(g) + 3H_2(g)$$

has been used as a commercial source of hydrogen. At 1500 °C, $K_c = 5.67$ and an equilibrium mixture of these gases was found to have the following concentrations: $[CO] = 0.300\ M$, $[H_2] = 0.800\ M$, and $[CH_4] = 0.400\ M$. At 1500 °C, what was the equilibrium concentration of $H_2O(g)$ in the mixture?

Analysis: We have all of the equilibrium concentrations except one. Because the equilibrium concentrations are related to each other by the equilibrium law, we should be able to substitute values and solve for $[H_2O]$.

Assembling the Tools: In solving any equilibrium problem, the tools we need are the balanced chemical equation and the equilibrium law. We already have the equation, so we use it to write

$$K_c = \frac{[CO][H_2]^3}{[CH_4][H_2O]}$$

(By now you should be familiar with the procedure for doing this. If not, review Section 14.2.)

Solution: We substitute the values given in the statement of the problem into the equilibrium law and solve for the unknown quantity.

$$5.67 = \frac{[0.300][0.800]^3}{[0.400][H_2O]}$$

Solving for $[H_2O]$ gives

$$[H_2O] = \frac{[0.300][0.800]^3}{[0.400](5.67)} = \frac{0.154}{2.27} = 0.0678$$

Is the Answer Reasonable? *We can always check problems of this kind by substituting all of the equilibrium concentrations into the mass action expression to see if the value of Q we get equals K_c. Let's estimate the answer by rounding the water concentration to 0.06 M.*

$$Q = \frac{[CO][H_2]^3}{[CH_4][H_2O]} = \frac{(0.300)(0.800)^3}{(0.400)(0.0600)} = 6.4$$

Try doing the math in your head by cancelling the numbers before reaching for a calculator. We see that rounding off the H_2O concentration caused a slight error, but Q is reasonably close to K_c so that we can be sure we've solved the problem correctly.

Practice Exercise 14.19

The decomposition of N_2O_4 at 25 °C,

$$N_2O_4(g) \rightleftharpoons 2NO_2(g)$$

has $K_c = 4.61 \times 10^{-3}$. A 2.00 L vessel contained 0.0466 mol N_2O_4 at equilibrium. What was the concentration of NO_2 in the vessel? (*Hint:* What was the concentration of N_2O_4 in the container at equilibrium?)

Practice Exercise 14.20

Ethyl ethanoate (commonly called ethyl acetate), $CH_3CO_2C_2H_5$, is an important solvent used in nail polish remover, lacquers, adhesives, the manufacture of plastics, and even as a food flavoring. It is produced from acetic acid and ethanol by the reaction

$$CH_3CO_2H + C_2H_5OH \rightleftharpoons CH_3CO_2C_2H_5 + H_2O$$

ethanoic acid   ethanol   ethyl ethanoate
(acetic acid)   (ethyl acetate)

At 25 °C, $K_c = 4.10$ for this reaction. In a reaction mixture, the following equilibrium concentrations were observed: $[CH_3CO_2H] = 0.210\ M$, $[H_2O] = 0.00850\ M$, and $[CH_3CO_2C_2H_5] = 0.910\ M$. What was the concentration of C_2H_5OH in the mixture?

■ Recall that acetic acid has the structure

where the hydrogen released by ionization is indicated in red. The formula of the acid is written either as CH_3CO_2H (which emphasizes its molecular structure) or as $HC_2H_3O_2$ (which emphasizes that the acid is monoprotic).

Using K_c and Initial Concentrations

A more complex type of calculation involves the use of initial concentrations and K_c to compute equilibrium concentrations. Although some of these problems can be so complicated that a computer is needed to solve them, we can learn the general principles involved by working on simple calculations. Even these, however, require a little applied algebra. This is where the concentration table can be especially helpful.

For example, we can calculate the equilibrium concentrations of each of the four molecules in the reaction

$$CO(g) + H_2O(g) \rightleftharpoons CO_2(g) + H_2(g)$$

if we know
  (a) the equilibrium law
  (b) the value of the equilibrium constant K_c
  (c) the initial concentrations
  (d) how much the initial concentrations have changed when equilibrium is reached

We can write the equilibrium law from the chemical equation, $K_c = 4.06$ at 500 °C, but it can be looked up in tables if it is not given, and the initial concentrations must be given as part of the question. In this case, let's say that 0.100 mol of CO and 0.100 mol of $H_2O(g)$ are placed in a 1.00 liter reaction vessel at 500 °C. All we need to do is determine how much the concentrations change in reaching equilibrium. The concentration table comes in handy here.

To build the table, we need quantities to enter into the "initial concentrations," "changes in concentrations," and "equilibrium concentrations" rows.

Initial Concentrations
The initial concentrations of CO and H_2O are each 0.100 mol/1.00 L = 0.100 M. Since neither CO_2 nor H_2 is initially placed into the reaction vessel, their initial concentrations both are zero. These are shown in red in the table.

Changes in Concentration
Some CO_2 and H_2 must form for the reaction to reach equilibrium. This also means that some CO and H_2O must react. But how much? If we knew the answer, we could calculate the equilibrium concentrations. Therefore, the changes in concentration are our unknown quantities.

Let us allow x to be equal to the number of moles per liter of CO that react. The *change* in the concentration of CO *must be* $-x$ (it is negative because the change decreases the CO concentration). Because CO and H_2O react in a 1:1 mole ratio, the change in the H_2O concentration is also $-x$. Since one mole each of CO_2 and H_2 are formed from one mole of CO, the CO_2 and H_2O concentrations each increase by x (their changes are $+x$). These quantities are shown in blue.

■ We could just as easily have chosen x to be the number of mol/L of H_2O that reacts or the number of mol/L of CO_2 or H_2 that forms. There's nothing special about having chosen CO to define x.

Equilibrium Concentrations
We obtain the equilibrium concentrations as

$$\left(\begin{array}{c}\text{initial}\\\text{concentration}\end{array}\right) + \left(\begin{array}{c}\text{change in}\\\text{concentration}\end{array}\right) = \left(\begin{array}{c}\text{equilibrium}\\\text{concentration}\end{array}\right)$$

■ The quantities representing the equilibrium concentrations must satisfy the equation given by the equilibrium law.

Here is the completed concentration table.

	CO(g)	+	H$_2$O(g)	⇌	CO$_2$(g)	+	H$_2$(g)
Initial concentrations (M)	0.100		0.100		0		0
Changes in concentrations (M)	$-x$		$-x$		$+x$		$+x$
Equilibrium concentrations (M)	$0.100 - x$		$0.100 - x$		x		x

We know that the quantities in the last row must satisfy the equilibrium law, so we substitute them into the mass action expression and solve for x.

$$\frac{(x)(x)}{(0.100 - x)(0.100 - x)} = 4.06$$

which we can write as

$$\frac{(x)^2}{(0.100 - x)^2} = 4.06$$

In this problem we can solve the equation for x most easily by taking the square root of both sides.

$$\sqrt{\frac{(x)^2}{(0.100 - x)^2}} = \frac{(x)}{(0.100 - x)} = \sqrt{4.06} = 2.01$$

Multiplying both sides by $(0.100 - x)$ gives

$$x = 2.01(0.100 - x)$$
$$x = 0.201 - 2.01x$$

Adding $2.01x$ to both sides gives

$$x + 2.01x = 0.201$$
$$3.01x = 0.201$$
$$x = 0.0668$$

Now that we know the value of x, we can calculate the equilibrium concentrations from the last row of the table.

$$[CO] = 0.100 - x = 0.100 - 0.0668 = 0.033 \ M$$
$$[H_2O] = 0.100 - x = 0.100 - 0.0668 = 0.033 \ M$$
$$[CO_2] = x = 0.0668 \ M$$
$$[H_2] = x = 0.0668 \ M$$

We can check this answer by substituting the equilibrium concentrations we've found into the mass action expression and evaluate the reaction quotient.

$$Q = \frac{(0.0668)(0.0668)}{(0.033)(0.033)} = 4.1$$

Since Q is quite close to K_c the answers are reasonable. In the following examples we will see variations of this method that build on our chemical and mathematical knowledge.

Example 14.8
Using K_c to Calculate Equilibrium Concentrations

In the preceding discussion it was stated that the reaction

$$CO(g) + H_2O(g) \rightleftharpoons CO_2(g) + H_2(g)$$

has $K_c = 4.06$ at 500 °C. Suppose 0.0600 mol each of CO and H$_2$O are mixed with 0.100 mol each of CO$_2$ and H$_2$ in a 1.00 L reaction vessel. What will the concentrations of all of the substances be when the mixture reaches equilibrium at this temperature?

Analysis: We will proceed in much the same way as before. However, this time determining the algebraic signs of x will not be quite so simple because none of the initial concentrations is zero. The best way to determine the algebraic signs is to use the initial concentrations to calculate the initial reaction quotient. We can then compare Q to K_c, and by reasoning, figure out which way the reaction must proceed to make Q equal to K_c.

Assembling the Tools: The equilibrium law for the reaction is

$$\frac{[CO_2][H_2]}{[CO][H_2O]} = K_c$$

One of our tools will be the mass action expression, from which we can calculate the initial value of Q. Once we know in which direction the reaction must proceed to reach equilibrium, we can construct the concentration table.

Solution: First we use the initial concentrations, shown in the first row of the concentration table below, to determine the initial value of the reaction quotient.

$$Q_{initial} = \frac{(0.100)(0.100)}{(0.0600)(0.0600)} = 2.78 < K_c$$

■ All we need to do is estimate the value of Q to see if it is larger or smaller than K_c. In this case, $0.1/0.06 = 1.66$ and we know that 1.66×1.66 must be less than 4.

As indicated, $Q_{initial}$ is less than K_c, so the system is not at equilibrium. To reach equilibrium Q must become larger, which requires an increase in the concentrations of CO_2 and H_2 as the reaction proceeds. This means that for CO_2 and H_2, the change must be positive, and, for CO and H_2O, the change must be negative. Here is the completed concentration table.

	CO(g)	+	H₂O(g)	⇌	CO₂(g)	H₂(g)
Initial concentrations (M)	0.0600		0.0600		0.100	0.100
Changes in concentrations (M)	$-x$		$-x$		$+x$	$+x$
Equilibrium concentrations (M)	$0.0600 - x$		$0.0600 - x$		$0.100 + x$	$0.100 + x$

As before, we substitute equilibrium quantities into the mass action expression in the equilibrium law, which gives us

$$\frac{(0.100 + x)(0.100 + x)}{(0.0600 - x)(0.0600 - x)} = \frac{(0.100 + x)^2}{(0.0600 - x)^2} = 4.06$$

Taking the square root of both sides yields

$$\sqrt{\frac{(0.100 + x)^2}{(0.0600 - x)^2}} = \sqrt{4.06} \Rightarrow \text{which is} \Rightarrow \frac{(0.100 + x)}{(0.0600 - x)} = 2.01$$

To solve for x we first multiply each side by $(0.0600 - x)$ to obtain

$$0.100 + x = 2.01(0.0600 - x)$$

$$0.100 + x = 0.121 - 2.01x$$

Collecting terms in x on one side and the constants on the other gives

$$x + 2.01x = 0.121 - 0.100$$

$$3.01x = 0.021$$

$$x = 0.0070$$

Now we can calculate the equilibrium concentrations:

$$[CO] = [H_2O] = (0.0600 - x) = 0.0600 - 0.0070 = 0.0530 \ M$$

$$[CO_2] = [H_2] = (0.100 + x) = 0.100 + 0.0070 = 0.107 \ M$$

Is the Answer Reasonable? As a check, let's evaluate the reaction quotient using the calculated equilibrium concentrations.

$$\frac{(0.107)(0.107)}{(0.0530)(0.0530)} = 4.08$$

This is acceptably close to the value of K_c. (It is not *exactly* equal to K_c because some answers were rounded during the calculations.)

In each of the preceding examples, we were able to simplify the solution by taking the square root of both sides of the algebraic equation obtained by substituting equilibrium concentrations into the mass action expression. Such simplifications are not always possible, however, as illustrated in the next example.

Example 14.9
Using K_c to Calculate Equilibrium Concentrations

At a certain temperature, $K_c = 4.50$ for the reaction

$$N_2O_4(g) \rightleftharpoons 2NO_2(g)$$

If 0.300 mol of N_2O_4 is placed into a 2.00 L container at this temperature, what will be the equilibrium concentrations of both gases?

Analysis: As in Example 14.8, we will use the fact that at equilibrium the mass action expression must be equal to K_c. Because neither of the equilibrium concentrations is given, we will have to find algebraic expressions for the equilibrium concentrations and substitute them into the mass action expression.

Assembling the Tools: As usual, our principal tools will be the equilibrium law,

$$\frac{[NO_2]^2}{[N_2O_4]} = 4.50$$

and the concentration table for the reaction. We also have to use the definition of molarity to calculate the initial molar concentration of N_2O_4 from the data given in the problem. Finally, we should be prepared to use the quadratic equation, if needed.

Solution: The first step is constructing the concentration table.

Initial Concentrations The initial concentration of N_2O_4 is 0.300 mol/2.00 L = 0.150 M. Since no NO_2 was placed in the reaction vessel, its initial concentration is 0.000 M.

■ Any time an initial concentration equals zero, we know its "change" must have a + sign since it cannot become less than zero.

Changes in Concentrations There is no NO_2 in the reaction mixture, so we know its concentration must increase. This means the N_2O_4 concentration must decrease as some of the NO_2 is formed. Let's allow x to be the number of moles per liter of N_2O_4 that reacts, so the change in the N_2O_4 concentration is $-x$. Because of the stoichiometry of the reaction, the NO_2 concentration must increase by $2x$, so its change in concentration is $+2x$.

■ Setting all coefficients of x so that they are the same as the coefficients in the balanced equation ensures they are in the correct ratios.

Equilibrium Concentrations As before, we add the change to the initial concentration in each column to obtain expressions for the equilibrium concentrations. Here is the completed concentration table.

	$N_2O_4(g)$ $\rightleftharpoons$	$2NO_2(g)$
Initial concentrations (M)	0.150	0
Changes in concentrations (M)	$-x$	$+2x$
Equilibrium concentrations (M)	$0.150 - x$	$2x$

To obtain the solution, we substitute the equilibrium quantities into the mass action expression.

$$\frac{(2x)^2}{(0.150 - x)} = \frac{4x^2}{(0.150 - x)} = 4.50 \qquad (14.12)$$

This time the left side of the equation is not a perfect square, so we cannot just take the square root of both sides as in Example 14.8. However, because the equation involves terms in x^2, x, and a constant, we can use the quadratic formula. Recall that for a *quadratic equation* of the form

$$ax^2 + bx + c = 0$$

we can solve for x using the *quadratic formula*

$$x = \frac{-b \pm \sqrt{b^2 - 4ac}}{2a} \qquad (14.13)$$

 **TOOLS**

Quadratic formula

Returning to Equation 14.12 we multiply both sides by $(0.150 - x)$

$$4x^2 = 4.50(0.150 - x)$$
$$4x^2 = 0.675 - 4.50x$$

Arranging terms in the standard order for a quadratic equation gives

$$4x^2 + 4.50x - 0.675 = 0$$

Therefore, the quantities we will substitute into the quadratic formula are as follows: $a = 4$, $b = 4.50$, and $c = -0.675$. Making these substitutions gives

$$x = \frac{-4.50 \pm \sqrt{(4.50)^2 - 4(4)(-0.675)}}{2(4)}$$

$$x = \frac{-4.50 \pm \sqrt{31.05}}{8}$$

$$x = \frac{-4.50 \pm 5.57}{8}$$

Because of the $\pm$ term, there are always two values of x that satisfy the quadratic equation from a mathematical point of view. In this case they are $x = 0.134$ and $x = -1.26$. Chemically however, only the positive, $x = 0.134$, makes any sense since the negative value results in a negative concentration, which is impossible. Using $x = 0.134$, the equilibrium concentrations are

$$[N_2O_4] = 0.150 - 0.134 = 0.016 \ M$$
$$[NO_2] = 2(0.134) = 0.268 \ M$$

In general, whenever you use the quadratic equation in a chemical calculation, one root will be satisfactory and the other will lead to answers that are nonsense in a chemical sense.

Is the Answer Reasonable? Once again, we can evaluate the reaction quotient using the calculated equilibrium values. When we do this, we obtain $Q = 4.49$, which is acceptably close to the value of K_c given.

During an experiment, 0.200 mol of H_2 and 0.200 mol of I_2 were placed into a 1.00 liter vessel where the reaction

Practice Exercise 14.21

$$H_2(g) + I_2(g) \rightleftharpoons 2HI(g)$$

came to equilibrium. For this reaction, $K_c = 49.5$ at the temperature of the experiment. What were the equilibrium concentrations of H_2, I_2, and HI? (*Hint:* See if taking a square root can simplify the mathematics.)

Practice Exercise 14.22

In an experiment, 0.200 mol H_2 and 0.100 mol I_2 were placed in a 1.00 L vessel where the following equilibrium was established:

$$H_2(g) + I_2(g) \rightleftharpoons 2HI(g)$$

For this reaction, $K_c = 49.5$ at the temperature of the experiment. What were the equilibrium concentrations of H_2, I_2, and HI? (*Hint:* Use the quadratic equation.)

Equilibrium problems can be much more complex than the ones we have just discussed, sometimes resulting in the third, fourth, or even higher powers of x. However, sometimes you can make assumptions that simplify the problem so that a solution can be easily obtained.

Calculations when K_c Is Very Small

When the K_c for a reaction is very small, the position of equilibrium lies far to the left and very small amounts of product are formed. In these cases, the initial amounts of the reactants will decrease by such a small amount that the change may not be noticed, or even be measurable! In those cases we can assume that the initial and equilibrium concentrations are the same and this permits us to simplify the calculations considerably. For example, if we start with a 0.352 M solution of hydrofluoric acid and only 0.0002 M reacts, the remaining hydrofluoric acid is

$$0.325\ M_{initial} - 0.0002\ M_{reacted} = 0.3248\ M_{left} \quad \text{(when properly rounded} = 0.325\ M)$$

We can see that the amount that reacted made no difference when we used significant figures properly. Let's see how this can be applied to equilibrium calculations as shown in Example 14.10.

Example 14.10
Simplifying Equilibrium Calculations for Reactions with a Small K_c

Hydrogen, a potential fuel, is found in great abundance in water. Before the hydrogen can be used as a fuel, however, it must be separated from the oxygen; the water must be split into H_2 and O_2. One possibility is thermal decomposition, but this requires very high temperatures. Even at 1000 °C, $K_c = 7.3 \times 10^{-18}$ for the reaction

$$2H_2O(g) \rightleftharpoons 2H_2(g) + O_2(g)$$

If at 1000 °C the H_2O concentration in a reaction vessel is set initially at 0.100 M, what will the H_2 concentration be when the reaction reaches equilibrium?

Analysis: As in earlier problems of this type, we can anticipate that we will have to solve an algebraic equation to obtain the equilibrium concentrations, none of which are given in the problem. We will follow that approach and see where it leads us.

Assembling the Tools: One tool we will definitely need is the equilibrium law:

$$\frac{[H_2]^2[O_2]}{[H_2O]^2} = 7.3 \times 10^{-18}$$

Another will be the concentration table.

Solution: The first step will be to construct the concentration table. The changes in concentrations are unknown and will be represented by expressions that contain the unknown quantity x.

Initial Concentrations The initial concentration of H_2O is 0.100 M; those of H_2 and O_2 are both 0 M.

Changes in Concentrations We know the changes must be in the same ratio as the coefficients in the balanced equation, so we place x's in this row with coefficients equal to those in the chemical equation. Because there are no products present initially, their changes must be positive and the change for the water must be negative. The complete concentration table is shown here.

	2H$_2$O(g) $\rightleftharpoons$	2H$_2$(g) +	O$_2$(g)
Initial concentrations (*M*)	0.100	0	0
Changes in concentrations (*M*)	$-2x$	$+2x$	$+x$
Equilibrium concentrations (*M*)	$0.100 - 2x$	$2x$	x

Next, we substitute the equilibrium quantities from the last row into the mass action expression.

$$\frac{(2x)^2(x)}{(0.100 - 2x)^2} = 7.3 \times 10^{-18}$$

or

$$\frac{(4x^3)}{(0.100 - 2x)^2} = 7.3 \times 10^{-18}$$

This is a *cubic equation* (one term involves x^3), which is difficult to solve. In this instance the very small value of K_c tells us that very little of the H$_2$O will decompose, so we can assume that

$$0.100 - 2x \approx 0.100$$

This assumption greatly simplifies the math since the $(0.100 - 2x)^2$ in the denominator is replaced by $(0.100)^2$. We now have

$$\frac{(4x^3)}{(0.100)^2} = 7.3 \times 10^{-18}$$

$$4x^3 = (7.3 \times 10^{-18})(0.100)^2 = 7.3 \times 10^{-20}$$

$$x^3 = (7.3 \times 10^{-20})/4 = 1.8 \times 10^{-20}$$

$$x = \sqrt[3]{1.8 \times 10^{-20}}$$

$$x = 2.6 \times 10^{-7}$$

■ Even before we solve the problem, we know that hardly any H$_2$ and O$_2$ will be formed, because K_c is so small.

Notice that the value of x that we've obtained is indeed very small. If we double it to 0.00000052 and subtract the answer from 0.100, we still get 0.100 when we round to the correct number of significant figures (the third decimal place) as shown.

0.100
−0.00000052 (red digits are rounded)
0.100   (rounded correctly)

■ We take advantage of the significant figure rules for adding and subtracting. For example,

0.100
−0.10000052
0.100

The digits in red are rounded off.

This check verifies that our assumption was valid. Finally, we have to obtain the H$_2$ concentration. Our table gives

$$[H_2] = 2x$$

Therefore,

$$[H_2] = 2(2.6 \times 10^{-7}) = 5.2 \times 10^{-7} \, M$$

Is the Answer Reasonable? The equilibrium constant is very small (7.3×10^{-18}) for this reaction, so we expect the amount of product at equilibrium to be small as well. The very low concentration of H$_2$ seems reasonable. The value of x was positive, which means that there was a decrease in reactant concentrations and an increase in product as we would expect if only reactants were originally present. Finally, we can check the calculation by seeing if the reaction quotient is equal to the equilibrium constant:

$$Q = \frac{[H_2]^2[O_2]}{[H_2O]^2} = \frac{(5.2 \times 10^{-7})^2(2.6 \times 10^{-7})}{(0.100)^2} = 7.0 \times 10^{-18}$$

which differs from the original equilibrium constant only because the value of x was rounded off when we took the cube root.

Situations that Suggest Simplification of the Math Will Work

When the value of K_c or K_P is very small, the position of equilibrium lies far to the left, very near the reactants. This means that if we start with just reactants, the concentrations will hardly change. Under these conditions, the simplifying assumption made in Example 14.10 is valid because the small change in the reactant concentration, x, is subtracted from a much larger value. We could also have neglected x (or $2x$) if it were a very small number that was being added to a much larger one. Remember that you can only neglect an x that's *added* or *subtracted*; you can never drop an x that occurs as a multiplying or dividing factor. Some examples are

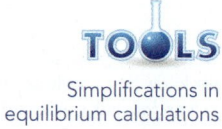

TOOLS

Simplifications in equilibrium calculations

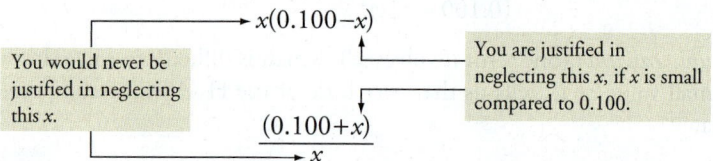

As a rule of thumb, you can expect that these simplifying assumptions will be valid if the concentration from which x is subtracted, or to which x is added, is at least 1000 times greater than K. For instance, in the preceding example, $2x$ was subtracted from 0.100. Since 0.100 is much larger than $1000 \times (7.3 \times 10^{-18})$ we expect the assumption $0.100 - 2x \approx 0.100$ to be valid. However, even though the simplifying assumption is *expected* to be valid, always check to see if it really is after finishing the calculation. If the assumption proves invalid, then some other way to solve the algebra must be found.

Practice Exercise 14.23

At 25 °C, the reaction $2NH_3(g) \rightleftharpoons N_2(g) + 3H_2(g)$ has $K_c = 2.3 \times 10^{-9}$. If 0.041 mol NH_3 is placed in a 1.00 L container, what will the concentrations of N_2 and H_2 be when equilibrium is established? (*Hint:* Review the math below the concentration table in Example 14.10.)

Practice Exercise 14.24

In air at 25 °C and 1.00 atm, the N_2 concentration is 0.033 M and the O_2 concentration is 0.00810 M. The reaction

$$N_2(g) + O_2(g) \rightleftharpoons 2NO(g)$$

has $K_c = 4.8 \times 10^{-31}$ at 25 °C. Taking the N_2 and O_2 concentrations given above as initial values, calculate the equilibrium NO concentration that should exist in our atmosphere from this reaction at 25 °C.

Analyzing and Solving Multi-Concept Problems

The reaction

$$2NO(g) + O_2(g) \rightleftharpoons 2NO_2(g)$$

has $K_P = 1.65 \times 10^{11}$ at 25 °C. Suppose a mixture of the reactants is prepared at 25 °C by transferring 1.15 g of NO and 645 mL of O_2 measured at 33 °C and 800 torr into a 1.00 L

vessel. When the mixture comes to equilibrium, what will be the concentrations of each of the three gases?

Analysis: As with all equilibrium problems, we are going to work with the equilibrium law. For this gaseous reaction, we can write this as

$$\frac{[NO_2]^2}{[NO]^2[O_2]} = K_c \quad \text{or} \quad \frac{P_{NO_2}^2}{P_{NO}^2 P_{O_2}} = K_P$$

We've been given K_P, but we know we can convert this to K_c. This was covered in Section 14.3. (Although we could work with K_P, this problem will be easier if we use K_c instead, because we're looking for equilibrium concentrations.)

Before we can construct the concentration table, we need the initial concentrations of the reactants. For NO, we have to convert grams to moles, which you learned to do in Chapter 4. For O_2, we must use the P, V, T data to find moles of O_2.

Assembling the Tools: To find K_c, we have to use Equation 14.10, $K_P = K_c(RT)^{\Delta n_g}$. Solving for K_c gives

$$K_c = \frac{K_P}{(RT)^{\Delta n_g}} = K_P(RT)^{-\Delta n_g} \tag{14.14}$$

Remember that $R = 0.0821$ L atm mol^{-1} K^{-1} and T is the Kelvin temperature.

To calculate the moles of NO placed in the reaction vessel, our tool is the molar mass of NO, which is 30.01 g mol^{-1}. To find moles of O_2, the tool is the ideal gas law, which we solve for n_{O_2},

$$n_{O_2} = \frac{PV}{RT}$$

The definition of molarity provides the tool to calculate the molarity of each reactant.

Solution: First we calculate K_c. Taking coefficients to represent moles, the reaction has two moles of gaseous products and three moles of gaseous reactants. Therefore, $\Delta n_g = -1$. Substituting values into Equation 14.14 gives

$$K_c = 1.65 \times 10^{11}[(0.0821 \text{ L atm mol}^{-1} \text{ K}^{-1}) \times (298 \text{ K})]^{-(-1)}$$

$$= 4.04 \times 10^{12}$$

Next, we calculate initial moles of each reactant.

Moles of NO

$$1.15 \text{ g NO} \times \frac{1 \text{ mol NO}}{30.01 \text{ g NO}} = 0.0383 \text{ mol NO}$$

Moles of O_2

$$n_{O_2} = \frac{PV}{RT} = \frac{\left(\dfrac{800 \text{ torr}}{760 \text{ torr/atm}}\right) \times 0.645 \text{ L}}{(0.0821 \text{ L atm mol}^{-1} \text{ K}^{-1}) \times (298 \text{ K})}$$

$$= 0.0278 \text{ mol } O_2$$

Both reactants are in the 1.00 L vessel, so their initial concentrations are $[NO] = 0.0383$ M and $[O_2] = 0.0278$ M. No NO_2 was placed in the vessel, so its initial concentration is zero. Here's the concentration table, assembled in the usual way.

	2NO(g)	+	O₂(g)	⇌	2NO₂(g)
Initial concentrations (M)	0.0383		0.0278		0
Changes in concentrations (M)	$-2x$		$-x$		$+2x$
Equilibrium concentrations (M)	$0.0383 - 2x$		$0.0278 - x$		$2x$

Substituting equilibrium quantities into the equilibrium law gives

$$K_c = 4.04 \times 10^{12} = \frac{(2x)^2}{(0.0383 - 2x)^2(0.0278 - x)}$$

We might be tempted to assume that x (or $2x$) is negligibly small and can be dropped, which would make the algebra manageable. However, this isn't going to work here. The very large value of K_c tells us this reaction proceeds far toward completion, so that almost all of at least one of the reactants will be consumed. *This means the changes in concentration are not small and can't be ignored.* So, the question is, can we reshape the problem to make it possible to solve it?

Recasting the Problem

Analysis: The algebra will be manageable if we can start the calculation with the system very close to equilibrium. In this reaction the position of equilibrium lies far to the right, as indicated by the value of K_c. Because the same equilibrium concentrations are reached regardless of whether we approach equilibrium from the left or the right, we are going to assume equilibrium in this system is reached in two steps. Step 1 will involve the complete reaction of the initial NO and O_2 to give NO_2. This is actually a limiting reactant problem. Step 2 will involve approaching equilibrium from the right. This will require that we rework the concentration table starting from the product side.

Assembling the Tools: The tool is the method of working limiting reactant problems, which you learned in Chapter 3.

Solution: We start with 0.0383 mol NO and 0.0278 mol O_2. If all the NO were to react, it would require only 0.0192 mol O_2,

$$0.0383 \; \cancel{\text{mol NO}} \times \frac{1 \; \text{mol} \; O_2}{2 \; \cancel{\text{mol NO}}} = 0.0192 \; \text{mol} \; O_2$$

Therefore, O_2 is in excess and NO is limiting. Based on the coefficients in the equation, if all the NO reacts, it will form 0.0383 mol NO_2 and the amount of O_2 remaining will be (0.0278 mol − 0.0192 mol) = 0.0086 mol.

Assembling the Tools: The tools will be the same as those we tried initially when we attempted to solve the problem.

Solution: Here are our new initial starting conditions:

$$[NO] = 0 \; M \quad \text{(We've assumed all initially present reacted.)}$$
$$[O_2] = 0.0086 \; M$$
$$[NO_2] = 0.0383 \; M$$

The new concentration table is shown below. Notice that the signs have changed for the "changes in concentration" row. This reflects that we are now starting at the product side and that the reaction that's occurring is the one from right to left.

	2NO(g)	+	O_2(g)	⇌	2NO$_2$(g)
Initial concentrations (M)	0		0.0086		0.0383
Changes in concentrations (M)	$+2x$		$+x$		$-2x$
Equilibrium concentrations (M)	$2x$		$0.0086 + x$		$0.0383 - 2x$

Substituting values from the last row into the mass action expression gives

$$K_c = 4.04 \times 10^{12} = \frac{(0.0383 - 2x)^2}{(2x)^2(0.0086 + x)}$$

We know that, by starting at the product side, we are very close to equilibrium and therefore x and $2x$ are going to be small compared to the initial concentrations, so we feel safe in neglecting them. This gives

$$K_c = 4.04 \times 10^{12} = \frac{(0.0383)^2}{(2x)^2(0.0086)}$$

This can be solved for x to give $x = 1.03 \times 10^{-7}$. Using the quantities in the last row of the concentration table above, the equilibrium concentrations are as follows:

$$[NO] = 2 \times (1.03 \times 10^{-7}) = 2.06 \times 10^{-7} \, M$$

$$[O_2] = 0.0086 + (1.03 \times 10^{-7}) = 0.0086 \, M \qquad \text{(correctly rounded)}$$

$$[NO_2] = [0.0383 - 2 \times (1.03 \times 10^{-7})] = 0.0383 \, M \qquad \text{(correctly rounded)}$$

Are the Answers Reasonable? First, notice that after recasting the problem, we are justified in making the simplifying assumptions that made the mathematical equation easily solvable. We can also check the answers by substituting them into the mass action expression.

$$Q = \frac{(0.0383)^2}{(2.06 \times 10^{-7})^2(0.0086)} = 4.02 \times 10^{12}$$

Q compares favorably to K_c (which is 4.04×10^{12}), so we can be confident they are correct.

Summary

Organized by Learning Objective

Distinguish, describe, and explain the nature of a dynamic equilibrium

When the forward and reverse reactions in a chemical system occur at equal rates, a dynamic equilibrium exists and the concentrations of the reactants and products remain constant. For a given overall chemical composition, the amounts of reactants and products that are present at equilibrium are the same regardless of whether the equilibrium is approached from the direction of pure "reactants," pure "products," or any mixture of them.

Explain the basics of equilibrium laws

The **mass action expression** is a fraction. The concentrations of the products, raised to powers equal to their coefficients in the chemical equation for a **homogeneous equilibrium**, are multiplied together in the numerator. The denominator is constructed in the same way from the concentrations of the reactants raised to powers equal to their coefficients. The numerical value of the mass action expression is the **reaction quotient, Q**. At equilibrium, the reaction quotient is equal to the **equilibrium constant, K_c**. If partial pressures of gases are used in the mass action expression, is represented as K_P. The magnitude of the equilibrium constant is roughly proportional to the extent to which the reaction proceeds to completion when equilibrium is reached. Equilibrium equations can be manipulated by multiplying the coefficients by a common factor, changing the direction of the reaction, and by adding two or more equilibria. The rules given in the description of the *Tools for Problem Solving* below apply.

Write and convert between equilibrium laws based on molar concentration and gas pressures

The values of K_P and K_c are only equal if the same number of moles of gas are represented on both sides of the chemical equation. When the numbers of moles of gas are different, K_P is related to K_c by the equation $K_P = K_c(RT)^{\Delta n_g}$. Remember to use $R = 0.0821$ L atm mol^{-1} K^{-1} and $T = $ absolute temperature. Also, be careful to calculate Δn_g as the difference between the number of moles of *gaseous* products and the number of moles of *gaseous* reactants in the balanced equation.

Understand and explain why solids and pure liquids are not included in equilibrium laws

An equilibrium involving substances in more than one phase is a **heterogeneous equilibrium**. The mass action expression for a heterogeneous equilibrium omits concentration terms for pure liquids and/or pure solids because their values are constants that contribute to the value of K.

Interpret the value of the equilibrium constant as an indicator of the position of equilibrium

When the value of K_c is very large, at equilibrium the reaction is mostly products. When the value of K_c is very small, at equilibrium, the reaction is mostly reactants. When the value of K_c is close to 1, then a mixture of products and reactions is formed at equilibrium.

Use Le Châtelier's principle and the reaction quotient, Q, to interpret where a system lies in relation to equilibrium

By comparing Q, the value for the mass action expression, to K, we can determine in which direction the reaction will proceed. If Q is smaller than K, then the reaction will move to products. If Q is larger than K, then the reaction will move to reactants. If Q is equal to K, then the reaction is at equilibrium.

Le Châtelier's principle states that *when an equilibrium is upset by some disturbance, a chemical change occurs in a direction that counteracts the disturbing influence and brings the system to equilibrium again, if possible.* Adding a reactant or a product causes a reaction to occur that uses up part of what has been added. Removing a reactant or a product causes a reaction that replaces part of what has been removed. Increasing the pressure (by reducing the volume) drives a reaction in the direction of the fewer number of moles of gas. Altering the pressure on an equilibrium system involving only solids and liquids has virtually no effect on the position of equilibrium. Raising the temperature causes an equilibrium to shift in an endothermic direction. The value of K increases with increasing temperature for reactions that are endothermic in the forward direction. A *change in temperature is the only factor that changes K*. Addition of an inert gas to increase pressure or a catalyst has no effect on an equilibrium.

Describe experiments to determine the equilibrium constant and how to use the data from these experiments

Only equilibrium concentrations satisfy the equilibrium law. When these are used, the value of the mass action expression, Q, is equal to K_c or K_P.

Use tabulated values of equilibrium constants and experimental data to calculate the equilibrium concentrations (or pressures) of all species in an equilibrium mixture

The initial concentrations in a chemical system are controlled by the person who combines the chemicals at the start of the reaction. The relative changes in concentration are determined by the stoichiometry of the reaction. When a change in concentration is very small compared to the initial concentration, the change may be neglected and the algebraic equation derived from the equilibrium law can be simplified. In general, this simplification is valid if the initial concentration is at least 1000 times larger than K. Since equilibrium can be approached from either direction, a calculation can be recast in terms of initial amounts of product rather than reactants in order to take advantage of possible simplifications. In general the initial amounts should be in terms of reactants if K is small and in terms of products if K is large.

Tools for Problem Solving
The following tools were introduced in this chapter. Study them carefully so you can select the appropriate tool when needed.

The approach to equilibrium (Section 14.1)
Remember that the same equilibrium composition is reached regardless of whether it is approached from the reactants or from the products. When K_c or K_P is very large, equilibrium problems can be greatly simplified if we first imagine the reaction going to completion (converting all the reactants to products) and then approaching the equilibrium from the direction of the products.

Requirement for equilibrium (Section 14.2)
For a system to be at equilibrium, the numerical value of Q must equal the equilibrium constant. Use this tool to determine whether a system is at equilibrium when you have a way to calculate Q. By comparing Q with K, you can determine the direction the reaction must proceed to reach equilibrium.

The equilibrium law (Section 14.2)
For a homogeneous reaction, an equation of the form

$$dD + eE \rightleftharpoons fF + gG$$

has the equilibrium law

$$K_c = \frac{[F]^f[G]^g}{[D]^d[E]^e}$$

Manipulating equilibrium equations (Section 14.2)
There are occasions when it is necessary to modify an equation for an equilibrium, or to combine two or more equilibrium equations. These tools are used to obtain the new equilibrium constants for the new equations.

- When two equations are added, we multiply their Ks to obtain the new K.
- When an equation is multiplied by a factor n to obtain a new equation, we raise its K to the power n to obtain the K for the new equation.
- When an equation is reversed, we take the reciprocal of its K to obtain the new K.

Equilibrium law using partial pressures for gas phase reactions (Section 14.3)

$$dD + eE \rightleftharpoons fF + gG$$

$$K_P = \frac{(P_F)^f(P_G)^g}{(P_D)^d(P_E)^e}$$

Converting between K_P and K_c (Section 14.3)

To relate K_P to K_c, we use the equation

$$K_P = K_c(RT)^{\Delta n_g}$$

Remember to use $R = 0.0821$ L atm mol^{-1} K^{-1}. Also, remember that Δn_g ($= n_{products} - n_{reactants}$) is the change in moles of gas on going from reactants to products.

Equilibrium laws for heterogeneous reactions (Section 14.4)

Remember that pure solids and pure liquids do not appear in the mass action expression.

Significance of the magnitude of K (Section 14.5)

Use this tool to make a rough estimate of the position of equilibrium.

- When K is very large, the position of equilibrium lies far to the right (toward the products).
- When K is very small, the position of equilibrium lies far to the left (toward the reactants).

Le Châtelier's principle (Section 14.6)

Factors that influence the shift of the position of equilibrium are:

- Adding or removing a reactant or product
- Changing the volume of a gaseous equilibrium
- Raising or lowering the temperature
- Changing temperature changes K
- Catalysts or inert gases have no effect on the position of equilibrium.

Concentration table (Section 14.7)

Remember the following points when constructing the table:

- All entries in the table must have units of molarity (mol L^{-1}).
- Any reactant or product for which an initial concentration or amount is not given in the statement of the problem is assigned an initial concentration of zero.
- Any substance with an initial concentration of zero *must* have a positive change in concentration when the reaction proceeds to equilibrium.
- The changes in concentration are in the same ratio as the coefficients in the balanced equation. When the changes are unknown, the coefficients of x can be the same as the coefficients in the balanced equation.
- Only quantities in the last row (equilibrium concentrations) satisfy the equilibrium law.

Quadratic Formula (Section 14.8)

This equation allows us to analytically solve equations of the form $ax^2 + bx + c = 0$:

$$x = \frac{-b \pm \sqrt{b^2 - 4ac}}{2a}$$

Simplifications in equilibrium calculations (Section 14.8)

When the initial reactant concentrations are larger than $1000 \times K$, they will change only slightly as the reaction approaches equilibrium. You can therefore expect to be able to neglect the change when it is being added to or subtracted from an initial concentration. *If you use simplifying approximations, be sure to check their validity after obtaining an answer.* This tool is especially useful when working problems in which K is very small and the initial concentrations of the reactants are known, *or* if K is very large and initial concentrations of products are known.

Review Questions

Dynamic Equilibrium in Chemical Systems

14.1 Sketch a graph showing how the concentrations of the reactant A and product P of a typical chemical reaction ($A \rightleftharpoons P$) vary with time during the course of the reaction. Assume that no products are present at the start of the reaction. Indicate on the graph where the system has reached equilibrium.

14.2 Using black circles to represent A and open circles to represent P, show the relative amounts of A and P in beakers at two points before equilibrium is reached and two points after equilibrium is reached in your answer to Question 14.1. The original beaker looks like this.

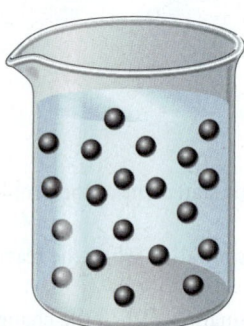

14.3 Repeat the exercise in Question 14.1 but this time start with a mixture that is 25% product and 75% reactant.

14.4 Using black circles to represent A and open circles to represent P, show the relative amounts of A and P in beakers at two points before equilibrium is reached and two points after equilibrium is reached in your answer to Question 14.3. The original beaker looks like this.

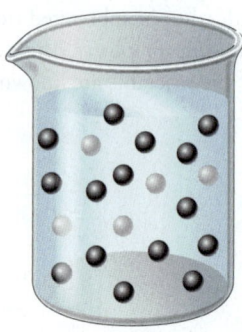

14.5 What meanings do the terms *reactants* and *products* have when describing a chemical equilibrium?

Equilibrium Laws

14.6 What is an *equilibrium law*? How does it differ from the *mass action expression*?

14.7 How is the term *reaction quotient* defined? What symbol is it given?

14.8 When a chemical equation and its equilibrium constant are given, why is it not necessary to also specify the form of the mass action expression?

14.9 Under what conditions does the reaction quotient equal K_c?

14.10 Describe in words how the mass action expression is written.

Equilibrium Laws Based on Pressures or Concentrations

14.11 State in words how K_P is written.

14.12 State the equation relating K_P to K_c and define all terms. Which is the only value of R that can be properly used in this equation?

14.13 Consider the following equilibrium.

$$2\text{NaHCO}_3(s) \rightleftharpoons \text{Na}_2\text{CO}_3(s) + \text{CO}_2(g) + \text{H}_2\text{O}(g)$$

If you were converting between K_P and K_c, what value of Δn_g would you use?

14.14 Use the ideal gas law to show that the partial pressure of a gas is directly proportional to its molar concentration. What is the proportionality constant?

Equilibrium Laws for Heterogeneous Reactions

14.15 What is the difference between a *heterogeneous equilibrium* and a *homogeneous equilibrium*?

14.16 Why do we omit the concentrations of pure liquids and pure solids from the mass action expression of heterogeneous reactions?

Position of Equilibrium and the Equilibrium Constant

14.17 Suppose for the reaction $A \longrightarrow B$ the value of Q is less than K_c. Which way does the reaction have to proceed to reach equilibrium—in the forward or reverse direction?

14.18 For the reaction in Question 14.17, sketch a graph of the concentrations of A and B with time and indicate where on the graph equilibrium is reached. Sketch graphs for **(a)** K is large, **(b)** K is small, and **(c)** $K = 1$.

14.19 At 225 °C, $K_P = 6.3 \times 10^{-3}$ for the reaction

$$CO(g) + 2H_2(g) \rightleftharpoons CH_3OH(g)$$

Would we expect this reaction to go nearly to completion?

14.20 Here are some reactions and their equilibrium constants.

(a) $2CH_4(g) \rightleftharpoons C_2H_6(g) + H_2(g)$

$$K_c = 9.5 \times 10^{-13}$$

(b) $CH_3OH(g) + H_2(g) \rightleftharpoons CH_4(g) + H_2O(g)$

$$K_c = 3.6 \times 10^{20}$$

(c) $H_2(g) + Br_2(g) \rightleftharpoons 2HBr(g)$ $K_c = 2.0 \times 10^9$

Arrange these reactions in order of their increasing tendency to go to completion.

Equilibrium and Le Châtelier's Principle

14.21 State Le Châtelier's principle in your own words.

14.22 Explain, using its effect on the reaction quotient, why adding a reactant to the following equilibrium shifts the position of equilibrium to the right.

$$PCl_3(g) + Cl_2(g) \rightleftharpoons PCl_5(g)$$

Review Problems

Equilibrium Laws

14.27 Write the equilibrium law for each of the following gas phase reactions in terms of molar concentrations:

(a) $2PCl_3(g) + O_2(g) \rightleftharpoons 2POCl_3(g)$

(b) $2SO_3(g) \rightleftharpoons 2SO_2(g) + O_2(g)$

(c) $N_2H_4(g) + 2O_2(g) \rightleftharpoons 2NO(g) + 2H_2O(g)$

(d) $N_2H_4(g) + 6H_2O_2(g) \rightleftharpoons 2NO_2(g) + 8H_2O(g)$

(e) $SOCl_2(g) + H_2O(g) \rightleftharpoons SO_2(g) + 2HCl(g)$

14.28 Write the equilibrium law for each of the following gas phase reactions in terms of molar concentrations.

(a) $3Cl_2(g) + NH_3(g) \rightleftharpoons NCl_3(g) + 3HCl(g)$

(b) $PCl_3(g) + PBr_3(g) \rightleftharpoons PCl_2Br(g) + PClBr_2(g)$

(c) $NO(g) + NO_2(g) + H_2O(g) \rightleftharpoons 2HNO_2(g)$

(d) $H_2O(g) + Cl_2O(g) \rightleftharpoons 2HOCl(g)$

(e) $Br_2(g) + 5F_2(g) \rightleftharpoons 2BrF_5(g)$

14.29 Write the equilibrium law for the following reactions in aqueous solution.

(a) $Ag^+(aq) + 2NH_3(aq) \rightleftharpoons Ag(NH_3)_2^+(aq)$

(b) $Cd^{2+}(aq) + 4SCN^-(aq) \rightleftharpoons Cd(SCN)_4^{2-}(aq)$

14.30 Write the equilibrium law for the following reactions in aqueous solution.

(a) $HClO(aq) + H_2O \rightleftharpoons H_3O+(aq)^+ ClO^-(aq)$

(b) $CO_3^{2-}(aq) + HSO^-(aq) \rightleftharpoons$

$$HCO_3^-(aq) + SO_4^{2-}(aq)$$

14.31 At 25 °C, $K_c = 1 \times 10^{-85}$ for the reaction

$$7IO_3^-(aq) + 9H_2O + 7H^+(aq) \rightleftharpoons I_2(aq) + 5H_5IO_6(aq)$$

14.23 Halving the volume of a gas doubles its pressure. Using the reaction quotient corresponding to K_P, explain why halving the volume shifts the following equilibrium to the left.

$$N_2O_4(g) \rightleftharpoons 2NO_2(g)$$

14.24 How will the value of K_P for the following reactions be affected by an increase in temperature?

(a) $CO(g) + 2H_2(g) \rightleftharpoons CH_3OH(g)$

$$\Delta H° = -18 \text{ kJ}$$

(b) $N_2O(g) + NO_2(g) \rightleftharpoons 3NO(g)$

$$\Delta H° = +155.7 \text{ kJ}$$

(c) $2NO(g) + Cl_2(g) \rightleftharpoons 2NOCl(g)$

$$\Delta H° = -77.07 \text{ kJ}$$

14.25 Why doesn't a catalyst affect the position of equilibrium in a chemical reaction?

14.26 Why doesn't adding an inert gas to increase the pressure, while keeping the volume constant, have any effect on the position of equilibrium?

What is the value of K_c for the following reaction?

$$I_2(aq) + 5H_5IO_6(aq) \rightleftharpoons 7IO_3^-(aq) + 9H_2O + 7H^+(aq)$$

14.32 Use the following equilibria

$$2CH_4(g) \rightleftharpoons C_2H_6(g) + H_2(g) \quad K_c = 9.5 \times 10^{-13}$$

$$CH_4(g) + H_2O(g) \rightleftharpoons CH_3OH(g) + H_2(g) \ K_c = 2.8 \times 10^{-21}$$

to calculate K_c for the reaction

$$2CH_3OH(g) + H_2(g) \rightleftharpoons C_2H_6(g) + 2H_2O(g)$$

14.33 Write the equilibrium law for each of the following reactions in terms of molar concentrations:

(a) $H_2(g) + Cl_2(g) \rightleftharpoons 2HCl(g)$

(b) $\frac{1}{2}H_2(g) + \frac{1}{2}Cl_2(g) \rightleftharpoons HCl(g)$

How does K_c for reaction (a) compare with K_c for reaction (b)?

14.34 Write the equilibrium law for the reaction

$$2HCl(g) \rightleftharpoons H_2(g) + Cl_2(g)$$

How does K_c for this reaction compare with K_c for reaction (a) in Problem 14.33?

Equilibrium Laws Based on Pressures and Concentrations

14.35 Write the equilibrium law for the reactions in Problem 14.27 in terms of partial pressures.

14.36 Write the equilibrium law for the reactions in Problem 14.28 in terms of partial pressures.

14.37 A 345 mL vessel contains NH_3 at a pressure of 745 torr and a temperature of 45 °C. What is the molar concentration of ammonia in the container?

14.38 In a certain container at 145 °C the concentration of water vapor is 0.0200 M. What is the partial pressure of H_2O in the container?

14.39 For which of the following reactions does $K_P = K_c$?
(a) $CO_2(g) + H_2(g) \rightleftharpoons CO(g) + H_2O(g)$
(b) $PCl_3(g) + Cl_2(g) \rightleftharpoons PCl_5(g)$
(c) $N_2O_4(g) \rightleftharpoons 2NO_2(g)$

14.40 For which of the following reactions does $K_P = K_c$?
(a) $2H_2(g) + C_2H_2(g) \rightleftharpoons C_2H_6(g)$
(b) $N_2(g) + O_2(g) \rightleftharpoons 2NO(g)$
(c) $2NO(g) + O_2(g) \rightleftharpoons 2NO_2(g)$

14.41 The reaction $CO(g) + 2H_2(g) \rightleftharpoons CH_3OH(g)$ has a $K_P = 6.3 \times 10^{-3}$ at 225 °C. What is the value of K_c at this temperature?

14.42 The reaction $HCO_2H(g) \rightleftharpoons CO(g) + H_2O(g)$ has a $K_P = 1.6 \times 10^6$ at 400 °C. What is the value of K_c for this reaction at this temperature?

14.43 The reaction $N_2O(g) + NO_2(g) \rightleftharpoons 3NO(g)$ has a $K_c = 4.2 \times 10^{-4}$ at 500 °C. What is the value of K_P at this temperature?

14.44 One possible way of removing NO from the exhaust of a gasoline engine is to cause it to react with CO in the presence of a suitable catalyst.

$$2NO(g) + 2CO(g) \rightleftharpoons N_2(g) + 2CO_2(g)$$

At 300 °C, this reaction has $K_c = 2.2 \times 10^{59}$. What is K_P at 300 °C?

14.45 At 773 °C the reaction $CO(g) + 2H_2(g) \rightleftharpoons CH_3OH(g)$ has $K_c = 0.40$. What is K_P at this temperature?

14.46 The reaction $COCl_2(g) \rightleftharpoons CO(g) + Cl_2(g)$ has $K_P = 4.6 \times 10^{-2}$ at 395 °C. What is K_c at this temperature?

Equilibrium Laws for Heterogeneous Reactions

14.47 Calculate the molar concentration of water in (a) 18.0 mL of $H_2O(l)$, (b) 100.0 mL of $H_2O(l)$, and (c) 1.00 L of $H_2O(l)$. Assume that the density of water is 1.00 g/mL.

14.48 The density of sodium chloride is 2.164 g cm^{-3}. What is the molar concentration of NaCl in a 12.0 cm^3 sample of pure NaCl? What is the molar concentration of NaCl in a 25.0 g sample of pure NaCl?

14.49 Write the equilibrium law corresponding to K_c for each of the following heterogeneous reactions.
(a) $2C(s) + O_2(g) \rightleftharpoons 2CO(g)$
(b) $2NaHSO_3(s) \rightleftharpoons Na_2SO_3(s) + H_2O(g) + SO_2(g)$

(c) $2C(s) + 2H_2O(g) \rightleftharpoons CH_4(g) + CO_2(g)$
(d) $CaCO_3(s) + 2HF(g) \rightleftharpoons$

$$CO_2(g) + CaF_2(s) + H_2O(g)$$

(e) $CuSO_4 \cdot 5H_2O(s) \rightleftharpoons CuSO_4(s) + 5H_2O(g)$

14.50 Write the equilibrium law corresponding to K_c for each of the following heterogeneous reactions.
(a) $CaCO_3(s) + SO_2(g) \rightleftharpoons CaSO_3(s) + CO_2(g)$
(b) $AgCl(s) + Br^-(aq) \rightleftharpoons AgBr(s) + Cl^-(aq)$
(c) $Cu(OH)_2(s) \rightleftharpoons Cu^{2+}(aq) + 2OH^-(aq)$
(d) $Mg(OH)_2(s) \rightleftharpoons MgO(s) + H_2O(g)$
(e) $3CuO(s) + 2NH_3(g) \rightleftharpoons$

$$3Cu(s) + N_2(g) + 3H_2O(g)$$

14.51 The heterogeneous reaction $2HCl(g) + I_2(s) \rightleftharpoons 2HI(g) + Cl_2(g)$ has $K_c = 1.6 \times 10^{-34}$ at 25 °C. Suppose 0.100 mol of HCl and solid I_2 are placed in a 1.00 L container. What will be the equilibrium concentrations of HI and Cl_2 in the container?

14.52 At 25 °C, $K_c = 360$ for the reaction

$$AgCl(s) + Br^-(aq) \rightleftharpoons AgBr(s) + Cl^-(aq)$$

If solid AgCl is added to a solution containing 0.10 M Br$^-$, what will be the equilibrium concentrations of Br$^-$ and Cl$^-$?

Equilibrium and Le Châtelier's Principle

14.53 The reaction

$$CO(g) + 2H_2(g) \rightleftharpoons CH_3OH(g)$$

has $\Delta H° = -18$ kJ. How will the amount of CH_3OH present at equilibrium be affected by the following changes?
(a) Adding $CO(g)$
(b) Removing $H_2(g)$
(c) Decreasing the volume of the container
(d) Adding a catalyst
(e) Increasing the temperature

14.54 How will the position of equilibrium in the reaction

$$heat + CH_4(g) + 2H_2S(g) \rightleftharpoons CS_2(g) + 4H_2(g)$$

be affected by the following changes?
(a) Adding $CH_4(g)$
(b) Adding $H_2(g)$
(c) Removing $CS_2(g)$
(d) Decreasing in the volume of the container
(e) Increasing the temperature

14.55 Consider the equilibrium

$$N_2O(g) + NO_2(g) \rightleftharpoons 3NO(g) \qquad \Delta H° = +155.7 \text{ kJ}$$

In which direction will this equilibrium be shifted by the following changes?
(a) Adding N_2O

(b) Removing NO_2

(c) Adding NO

(d) Increasing the temperature of the reaction mixture

(e) Adding helium gas to the reaction mixture at constant volume

(f) Decreasing the volume of the container at constant temperature

14.56 Consider the equilibrium

$$2NO(g) + Cl_2(g) \rightleftharpoons 2NOCl(g)$$

for which $\Delta H° = -77.07$ kJ. How will the amount of Cl_2 at equilibrium be affected by the following changes?

(a) Removing $NO(g)$

(b) Adding $NOCl(g)$

(c) Raising the temperature

(d) Decreasing the volume of the container at constant temperature

Calculating Equilibrium Constants

14.57 At 773 °C, a mixture of $CO(g)$, $H_2(g)$, and $CH_3OH(g)$ was allowed to come to equilibrium. The concentrations were determined to be: $[CO] = 0.105$ M, $[H_2] = 0.250$ M, $[CH_3OH] = 0.00261$ M. Calculate K_c for the reaction

$$CO(g) + 2H_2(g) \rightleftharpoons CH_3OH(g)$$

14.58 Ethylene, C_2H_4, and water react under appropriate conditions to give ethanol. The reaction is:

$$C_2H_4(g) + H_2O(g) \rightleftharpoons C_2H_5OH(g)$$

An equilibrium mixture of these gases at a certain temperature had the following partial pressures: $P_{C_2H_4} = 0.575$ atm, $P_{H_2O} = 1.30$ atm, and $P_{C_2H_5OH} = 6.99$ atm. What is the value of K_P?

14.59 At high temperature, 2.00 mol of HBr was placed in a 4.00 L container where it decomposed in the reaction:

$$2HBr(g) \rightleftharpoons H_2(g) + Br_2(g)$$

At equilibrium the concentration of Br_2 was measured to be 0.0955 M. What is K_c for this reaction at this temperature?

14.60 A 0.050 mol sample of formaldehyde vapor, CH_2O, was placed in a heated 500 mL vessel and some of it decomposed. The reaction is

$$CH_2O(g) \rightleftharpoons H_2(g) + CO(g)$$

At equilibrium, the $CH_2O(g)$ concentration was 0.066 mol L^{-1}. Calculate the value of K_c for this reaction.

14.61 The reaction $NO_2(g) + NO(g) \rightleftharpoons N_2O(g) + O_2(g)$ reached equilibrium at a certain high temperature. Originally, the reaction vessel contained the following initial concentrations: $[N_2O] = 0.184$ M, $[O_2] = 0.377$ M, $[NO_2] = 0.0560$ M, and $[NO] = 0.294$ M. The concen-

tration of the NO_2, the only colored gas in the mixture, was monitored by following the intensity of the color. At equilibrium, the NO_2 concentration had become 0.118 M. What is the value of K_c for this reaction at this temperature?

14.62 At 25 °C, 0.0560 mol O_2 and 0.020 mol N_2O were placed in a 1.00 L container where the following equilibrium was then established.

$$2N_2O(g) + 3O_2(g) \rightleftharpoons 4NO_2(g)$$

At equilibrium, the NO_2 concentration was 0.020 M. What is the value of K_c for this reaction?

Using Equilibrium Constants to Calculate Concentrations

14.63 At a certain temperature, $K_c = 0.18$ for the equilibrium

$$PCl_3(g) + Cl_2(g) \rightleftharpoons PCl_5(g)$$

Suppose a reaction vessel at this temperature contained these three gases at the following concentrations: $[PCl_3] = 0.0420$ M, $[Cl_2] = 0.0240$ M, $[PCl_5] = 0.00500$ M.

(a) Compute the reaction quotient and use it to determine whether the system is in a state of equilibrium.

(b) If the system is not at equilibrium, in which direction will the reaction proceed to get to equilibrium?

14.64 At 460 °C, the reaction

$$SO_2(g) + NO_2(g) \rightleftharpoons NO(g) + SO_3(g)$$

has $K_c = 85.0$. A reaction flask at 460 °C contains these gases at the following concentrations: $[SO_2] = 0.00250$ M, $[NO_2] = 0.00350$ M, $[NO] = 0.0250$ M, and $[SO_3] = 0.0400$ M.

(a) Is the reaction at equilibrium?

(b) If not, in which direction will the reaction proceed to arrive at equilibrium?

14.65 At a certain temperature, the reaction

$$CO(g) + 2H_2(g) \rightleftharpoons CH_3OH(g)$$

has $K_c = 0.500$. If a reaction mixture at equilibrium contains 0.180 M CO and 0.220 M H_2, what is the concentration of $CH_3OH(g)$?

14.66 $K_P = 0.018$ for the reaction

$$N_2(g) + 3H_2(g) \rightleftharpoons 2NH_3(g)$$

at a certain temperature. Suppose it was found that in an equilibrium mixture of these gases the partial pressure of $NH_3(g)$ is 22.2 atm and the partial pressure of N_2 is 1.18 atm. What is the partial pressure of $H_2(g)$?

14.67 At 25 °C, $K_c = 0.145$ for the following reaction in the solvent CCl_4.

$$2BrCl \rightleftharpoons Br_2 + Cl_2$$

If the initial concentration of BrCl in the solution is 0.050 M, what will the equilibrium concentrations of Br_2 and Cl_2 be?

14.68 At 25 °C, $K_c = 0.145$ for the following reaction in the solvent CCl_4.

$$2BrCl \rightleftharpoons Br_2 + Cl_2$$

If the initial concentrations of Br_2 and Cl_2 are each 0.0250 M, what will their equilibrium concentrations be?

14.69 The equilibrium constant, K_c, for the reaction

$$SO_3(g) + NO(g) \rightleftharpoons NO_2(g) + SO_2(g)$$

was found to be 0.500 at a certain temperature. If 0.240 mol of SO_3 and 0.240 mol of NO are placed in a 2.00 L container and allowed to react, what will be the equilibrium concentration of each gas?

14.70 For the reaction in Problem 14.69, a reaction mixture is prepared in which 0.120 mol NO_2 and 0.120 mol of SO_2 are placed in a 1.00 L vessel. After the system reaches equilibrium, what will be the equilibrium concentrations of all four gases? How do these equilibrium values compare to those calculated in Problem 14.69? Account for your observation.

14.71 At a certain temperature the reaction

$$CO(g) + H_2O(g) \rightleftharpoons CO_2(g) + H_2(g)$$

has $K_c = 0.400$. Exactly 1.00 mol of each gas was placed in a 100.0 L vessel and the mixture underwent reaction. What was the equilibrium concentration of each gas?

14.72 At 25 °C, $K_c = 0.145$ for the following reaction in the solvent CCl_4.

$$2BrCl \rightleftharpoons Br_2 + Cl_2$$

If the initial concentrations of each substance in a solution are 0.0400 M, what will their equilibrium concentrations be?

14.73 The reaction $2HCl(g) \rightleftharpoons H_2(g) + Cl_2(g)$ has $K_c = 3.2 \times 10^{-34}$ at 25 °C. If a reaction vessel contains initially 0.0500 mol L^{-1} of HCl and then reacts to reach equilibrium, what will be the concentrations of H_2 and Cl_2?

14.74 At 200 °C, $K_c = 1.4 \times 10^{-10}$ for the reaction

$$N_2O(g) + NO_2(g) \rightleftharpoons 3NO(g)$$

If 0.200 mol of N_2O and 0.400 mol NO_2 are placed in a 4.00 L container, what would the NO concentration be if this equilibrium were established?

14.75 At 2000 °C, the decomposition of CO_2,

$$2CO_2(g) \rightleftharpoons 2CO(g) + O_2(g)$$

has $K_c = 6.4 \times 10^{-7}$. If a 1.00 L container holding 1.00×10^{-2} mol of CO_2 is heated to 2000 °C, what will be the concentration of CO at equilibrium?

14.76 At 500 °C, the decomposition of water into hydrogen and oxygen,

$$2H_2O(g) \rightleftharpoons 2H_2(g) + O_2(g)$$

has $K_c = 6.0 \times 10^{-28}$. How many moles of H_2 and O_2 are present at equilibrium in a 5.00 L reaction vessel at this temperature if the container originally held 0.015 mol H_2O?

14.77 At a certain temperature, $K_c = 0.18$ for the equilibrium

$$PCl_3(g) + Cl_2(g) \rightleftharpoons PCl_5(g)$$

If 0.026 mol of PCl_5 is placed in a 2.00 L vessel at this temperature, what will the concentration of PCl_3 be at equilibrium?

14.78 At 460 °C, the reaction

$$SO_2(g) + NO_2(g) \rightleftharpoons NO(g) + SO_3(g)$$

has $K_c = 85.0$. Suppose 0.100 mol of SO_2, 0.0600 mol of NO_2, 0.0800 mol of NO, and 0.120 mol of SO_3 are placed in a 10.0 L container at this temperature. What will be the concentrations of all the gases be when the system reaches equilibrium?

14.79 At a certain temperature, $K_c = 0.500$ for the reaction

$$SO_3(g) + NO(g) \rightleftharpoons NO_2(g) + SO_2(g)$$

If 0.100 mol SO_3 and 0.200 mol NO are placed in a 2.00 L container and allowed to come to equilibrium, what will the NO_2 and SO_2 concentrations be?

14.80 At 25 °C, $K_c = 0.145$ for the following reaction in the solvent CCl_4.

$$2BrCl \rightleftharpoons Br_2 + Cl_2$$

A solution was prepared with the following initial concentrations: $[BrCl] = 0.0400$ M, $[Br_2] = 0.0300$ M, and $[Cl_2] = 0.0200$ M. What will their equilibrium concentrations be?

***14.81** At a certain temperature, $K_c = 4.3 \times 10^5$ for the reaction

$$HCO_2H(g) \rightleftharpoons CO(g) + H_2O(g)$$

If 0.200 mol of HCO_2H is placed in a 1.00 L vessel, what will be the concentrations of CO and H_2O when the system reaches equilibrium? (*Hint:* Where does the position of equilibrium lie when K is very large?)

***14.82** The reaction $H_2(g) + Br_2(g) \rightleftharpoons 2HBr(g)$ has a $K_c = 2.0 \times 10^9$ at 25 °C. If 0.100 mol of H_2 and 0.200 mol of Br_2 are placed in a 10.0 L container, what will all the equilibrium concentrations be at 25 °C? (*Hint:* Where does the position of equilibrium lie when K is very large?)

| Additional Exercises

14.83 The reaction $N_2O_4(g) \rightleftharpoons 2NO_2(g)$ has $K_P = 0.140$ at 25 °C. In a reaction vessel containing these gases in equilibrium at this temperature, the partial pressure of N_2O_4 was 0.250 atm.

(a) What was the partial pressure of the NO_2 in the reaction mixture?

(b) What was the total pressure of the mixture of gases?

14.84 At 25 °C, the following concentrations were found for the gases in an equilibrium mixture for the reaction

$$N_2O(g) + NO_2(g) \rightleftharpoons 3NO(g)$$

$[NO_2] = 0.24\ M$, $[NO] = 4.8 \times 10^{-8}\ M$, and $[N_2O] = 0.023\ M$. What is the value of K_P for this reaction?

14.85 The following reaction in aqueous solution has $K_c = 1 \times 10^{-85}$ at a temperature of 25 °C.

$$7IO_3^-(aq) + 9H_2O + 7H^+(aq) \rightleftharpoons I_2(aq) + 5H_5IO_6(aq)$$

What is the equilibrium law for this reaction?

*__14.86__ At a certain temperature, $K_c = 0.914$ for the reaction

$$NO_2(g) + NO(g) \rightleftharpoons N_2O(g) + O_2(g)$$

A mixture was prepared containing 0.200 mol NO_2, 0.300 mol NO, 0.150 mol N_2O, and 0.250 mol O_2 in a 4.00 L container. What will be the equilibrium concentrations of each gas?

*__14.87__ At 27 °C, $K_P = 1.5 \times 10^{18}$ for the reaction

$$3NO(g) \rightleftharpoons N_2O(g) + NO_2(g)$$

If 0.030 mol of NO were placed in a 1.00 L vessel and this equilibrium were established, what would be the equilibrium concentrations of NO, N_2O, and NO_2?

14.88 Consider the equilibrium

$$2NaHSO_3(s) \rightleftharpoons Na_2SO_3(s) + H_2O(g) + SO_2(g)$$

How will the position of equilibrium be affected by the following changes?

(a) Adding $NaHSO_3$ to the reaction vessel

(b) Removing Na_2SO_3 from the reaction vessel

(c) Adding H_2O to the reaction vessel

(d) Increasing the volume of the reaction vessel at constant temperature

14.89 At a certain temperature, $K_c = 0.914$ for the reaction

$$NO_2(g) + NO(g) \rightleftharpoons N_2O(g) + O_2(g)$$

A mixture was prepared containing 0.200 mol of each gas in a 5.00 L container. What will be the equilibrium concentration of each gas? How will the concentrations then change if 0.050 mol of NO_2 is added to the equilibrium mixture?

*__14.90__ At a certain temperature, $K_c = 0.914$ for the reaction

$$NO_2(g) + NO(g) \rightleftharpoons N_2O(g) + O_2(g)$$

Equal amounts of NO and NO_2 are to be placed in a 5.00 L container until the N_2O concentration at equilibrium is 0.0500 M. How many moles of NO and NO_2 must be placed in the container?

14.91 Two 1.00 L bulbs are filled with 0.500 atm of $F_2(g)$ and $PF_3(g)$, respectively, as illustrated in the accompanying figure. At a particular temperature, $K_P = 4.0$ for the reaction of these gases to form

$$F_2(g) + PF_3(g) \rightleftharpoons PF_5(g)$$

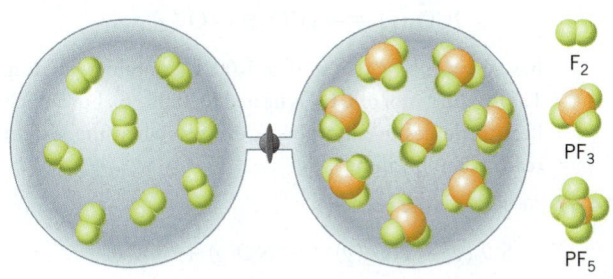

(a) The stopcock between the bulbs is opened and the pressure falls. Make a sketch of this apparatus showing the gas composition once the pressure is stable.

(b) List any chemical reactions that continue to occur once the pressure is stable.

(c) Suppose all the gas in the left bulb is forced into the bulb on the right and then the stopcock is closed. Make a second sketch to show the composition of the gas mixture after equilibrium is reached.

(d) Comment on how the composition of the one-bulb system differs from that for the two-bulb system.

14.92 For the equilibrium

$$3NO_2(g) \rightleftharpoons N_2O_5(g) + NO(g)$$

$K_c = 1.0 \times 10^{-11}$. If a 4.00 L container initially holds 0.20 mol of NO_2, how many moles of N_2O_5 will be present when this system reaches equilibrium?

14.93 To study the following reaction at 20 °C,

$$NO(g) + NO_2(g) + H_2O(g) \rightleftharpoons 2HNO_2(g)$$

a mixture of $NO(g)$, $NO_2(g)$, and $H_2O(g)$ was prepared in a 10.0 L glass bulb. For NO, NO_2, and HNO_2, the initial concentrations were as follows: $[NO] = [NO_2] = 2.59 \times 10^{-3}\ M$ and $[HNO_2] = 0\ M$. The initial partial pressure of $H_2O(g)$ was 17.5 torr. When equilibrium was reached, the HNO_2 concentration was $4.0 \times 10^{-4}\ M$. Calculate the equilibrium constant, K_c, for this reaction.

14.94 At 100.0 °C, $K_c = 0.135$ for the reaction

$$3H_2(g) + N_2(g) \rightleftharpoons 2NH_3(g)$$

In a reaction mixture at equilibrium at this temperature, $[NH_3] = 0.030\ M$ and $[N_2] = 0.50\ M$. What are the partial pressures of all of the gases in the mixture? Use them to calculate K_P and compare the result with K_P calculated using Equation 14.10.

Multi-Concept Problems

*14.95 For the reaction below, $K_P = 1.6 \times 10^6$ at 400.0 °C.

$$HCHO_2(g) \rightleftharpoons CO(g) + H_2O(g)$$

A mixture of $CO(g)$ and $H_2O(l)$ was prepared in a 2.00 L reaction vessel at 25 °C in which the pressure of CO was 0.177 atm. The mixture also contained 0.391 g of H_2O. The vessel was sealed and heated to 400 °C. When equilibrium was reached, what was the partial pressure of the $HCHO_2$ at 400 °C?

14.96 At 2000 °C, the decomposition of CO_2,

$$2CO_2(g) \rightleftharpoons 2CO(g) + O_2(g)$$

has $K_c = 6.4 \times 10^{-7}$. If a 1.00 L container holding 1.00×10^{-2} mol of CO_2 is heated to 2000 °C, how much heat, in joules, is absorbed or evolved when the system reaches equilibrium?

*14.97 At 460 °C, the reaction

$$SO_2(g) + NO_2(g) \rightleftharpoons NO(g) + SO_3(g)$$

has $K_c = 85.0$. If 0.0255 mol of SO_2, 0.0600 mol of NO_2, 0.0800 mol of NO, and 0.0446 mol of SO_3 are placed in a 10.0 L container at this temperature, what is the change in pressure of the system when equilibrium is achieved?

*14.98 At 200.0 °C, $K_c = 1.4 \times 10^{-10}$ for the reaction

$$N_2O(g) + NO_2(g) \rightleftharpoons 3NO(g)$$

If 300 mL of NO measured at 800 torr and 25 °C is placed in a 4.00 L container, what will be the N_2O and NO molar concentrations at equilibrium? What will be the total pressure of the mixture (in torr) at equilibrium at 25 °C?

*14.99 At 400 °C, $K_c = 2.9 \times 10^4$ for the reaction:

$$HCHO_2(g) \rightleftharpoons CO(g) + H_2O(g)$$

A mixture was prepared with the following initial concentrations: $[CO] = 0.20$ M, $[H_2O] = 0.30$ M. No formic acid, $HCHO_2$, was initially present. What was the equilibrium concentration of $HCHO_2$? What is the density of the equilibrium mixture at this temperature?

*14.100 The reaction $N_2O_4(g) \rightleftharpoons 2NO_2(g)$ has $K_P = 0.140$ at 25.0 °C. Into a 4.50 L reaction vessel at this temperature was placed 0.0245 mol N_2O_4 and 0.0116 mol NO_2. What will be the change in density, at this temperature, when equilibrium is established?

Exercises in Critical Thinking

14.101 In an equilibrium law, coefficients in the balanced equation appear as exponents on concentrations. Why, in general, does this not also apply for rate laws?

14.102 Why are equilibrium concentrations useful to know?

14.103 Suppose we set up a system in which water is poured into a vessel having a hole in the bottom. If the rate of water inflow is adjusted so that it matches the rate at which water drains through the hole, the amount of water in the vessel remains constant over time. Is this an equilibrium system? Explain.

14.104 Do equilibrium laws apply to other systems outside of chemistry? Give examples.

14.105 What might prevent a system from reaching dynamic equilibrium? Illustrate your answer with examples.

14.106 After many centuries the earth's atmosphere still has not come to equilibrium with the oceans and land. What factors are responsible for this?

14.107 If a mixture consisting of many small crystals in contact with a saturated solution is studied for a period of time, the smallest of the crystals are observed to dissolve while the larger ones grow even larger. Explain this phenomenon in terms of a dynamic equilibrium.

14.108 Why doesn't Le Châtelier's principle apply to the removal of some of a solid or pure liquid from a heterogeneous reaction mixture?

14.109 Le Châtelier's principle qualitatively describes what will occur if a reactant or product is added or removed from a reaction mixture. Describe as many ways as possible to remove a reactant or product from a mixture at equilibrium.

Acids and Bases, A Molecular Look

Lisa Passmore

The weather is hot, you are thirsty, and you would like a drink of cold water with a slice of lime. You pull out your ceramic knife, slice the fruit, pour the water, and add ice and the lime. After drinking, you feel much better. The lime added a slight acidity to the water, and helped quench your thirst.

This Chapter in Context

In Chapter 4 we introduced you to compounds that we call acids and bases. Many common substances, from foods such as the lime in the opening photograph, to household products such as vinegar and ammonia, to biologically important compounds such as amino acids, are conveniently classified as either an acid or a base. The acid–base reaction is such a useful relationship that the acid–base concept has been expanded far beyond the limited Arrhenius definition we discussed in Chapter 4. We now return to acid–base chemistry to learn about these broader classifications, study the periodic trends of the strengths of acids and bases, and examine the acid–base chemistry in the production of useful high-tech materials, including ceramics such as the ceramic knife that is cutting the lime.

LEARNING OBJECTIVES

After reading this chapter, you should be able to:

- identify Brønsted–Lowry acids and bases and explain the terms conjugate acid–base pair and amphoteric substances
- compare the strengths of Brønsted–Lowry acids and bases, how they are classified as strong or weak, and the strengths of conjugate acid–base pairs
- using the periodic table, describe the trends in the strengths of binary acids and oxoacids
- define Lewis acids and bases and compare them to the Arrhenius and Brønsted–Lowry definitions
- using the periodic table, describe which elements are most likely to form acids or bases
- describe how the production of advanced ceramics depends on acid–base chemistry, including the sol-gel process

Andy Washnik

Figure 15.1 | The reaction of gaseous HCl with gaseous NH₃. As each gas escapes from its concentrated aqueous solution and mingles with the other, a cloud of microcrystals of NH₄Cl forms above the bottles.

15.1 | Brønsted–Lowry Acids and Bases

In Chapter 4, an acid was described as a substance that produces H_3O^+ in water, whereas a base gives OH^-. An acid–base neutralization reaction, according to Arrhenius, occurs when an acid and a base combine to produce water and a salt. However, many reactions resemble neutralizations without involving H_3O^+, OH^-, or even H_2O. For example, when open bottles of concentrated hydrochloric acid and concentrated aqueous ammonia are placed side by side, a white cloud forms when the vapors from the two bottles mix (see Figure 15.1). The cloud consists of tiny crystals of ammonium chloride, which form when ammonia and hydrogen chloride gases, escaping from the open bottles, mix in air and react.

$$NH_3(g) + HCl(g) \longrightarrow NH_4Cl(s)$$

What's interesting is that this is the same net reaction that occurs when an aqueous solution of ammonia (a base) is neutralized by an aqueous solution of hydrogen chloride (an acid). Yet, the gaseous reaction doesn't fit the description of an acid–base neutralization according to the Arrhenius definition because there's no water involved.

If we look at both the aqueous and gaseous reactions, they do have something in common. Both involve the transfer of a *proton* (a hydrogen ion, H^+) from one particle to another.[1] In water, where HCl is completely ionized, the transfer is from H_3O^+ to NH_3, as we discussed in Section 4.6. The ionic equation is shown below.

$$NH_3(aq) + H_3O^+(aq) + Cl^-(aq) \longrightarrow \underbrace{NH_4^+(aq) + Cl^-(aq)}_{\text{The ions of NH}_4\text{Cl}} + H_2O$$

---

[1] When the single electron is removed from a hydrogen atom, what remains is just the nucleus of the atom, which is a proton. Therefore, a hydrogen ion, H^+, consists of a proton, and the terms *proton* and *hydrogen ion* are often used interchangeably.

In the gas phase, the proton is transferred directly from the HCl molecule to the NH_3 molecule.

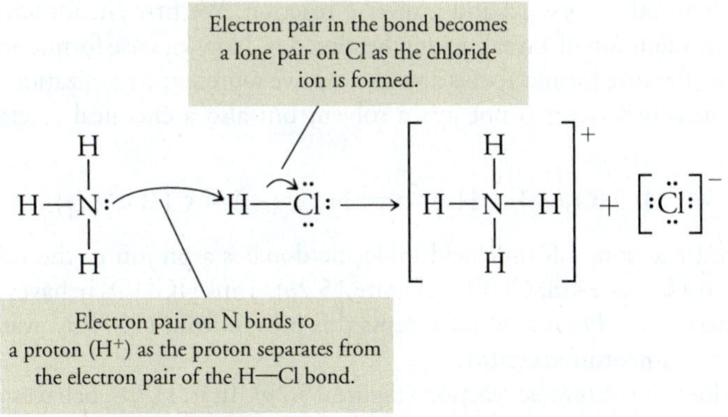

Electron pair in the bond becomes a lone pair on Cl as the chloride ion is formed.

Electron pair on N binds to a proton (H^+) as the proton separates from the electron pair of the H—Cl bond.

- We are using "curved" arrows to show how electrons shift and rearrange during the formation of the ions. The electron pair on the nitrogen atom of the NH_3 molecule binds to H^+ as it is removed from the electron pair in the H—Cl bond. The electron pair in the H—Cl bond shifts entirely to the Cl as the Cl^- ion is formed.

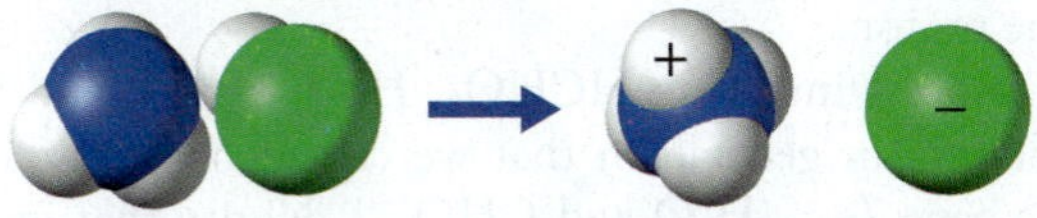

- Although the positive charge is shown on just one of the hydrogen atoms in NH_4^+, it is spread equally over all four hydrogen atoms.

As ammonium ions and chloride ions form, they gather, and settle as crystals of ammonium chloride.

Proton Transfer Reactions

Johannes Brønsted (1879–1947), a Danish chemist, and Thomas Lowry (1874–1936), a British scientist, realized that the important event in most acid–base reactions is simply the transfer of a proton from one particle to another. Therefore, they redefined *acids* and *bases*:

TOOLS

Brønsted–Lowry definitions

Brønsted–Lowry Definitions of Acids and Bases[2]
A **Brønsted–Lowry acid** is a proton donor.
A **Brønsted–Lowry base** is a proton acceptor.

The heart of the *Brønsted–Lowry concept of acids and bases* is that *acid–base reactions are proton transfer reactions*. Accordingly, hydrogen chloride is an acid because when it reacts with ammonia, HCl molecules donate protons to NH_3 molecules. Similarly, ammonia is a base because NH_3 molecules accept protons.

Even when water is the solvent, chemists use the Brønsted–Lowry definitions more often than those of Arrhenius. Thus, the reaction between hydrogen chloride and water to form hydronium ion (H_3O^+) and chloride ion (Cl^-), another proton transfer reaction, is clearly a Brønsted–Lowry acid–base reaction. Molecules of HCl are the acid in this reaction, and water molecules are the base. HCl molecules collide with water molecules and protons transfer during the collisions.

| Water | Hydrogen chloride | Collision of molecules permits proton transfer | Hydronium ion | Chloride ion |

- Again, in H_3O^+, the positive charge is spread equally over all three hydrogen atoms.

---

[2] Although Brønsted and Lowry are both credited with defining acids and bases in terms of proton transfer, Brønsted carried the concepts further. For the sake of brevity, we will often use the terms *Brønsted acid* and *Brønsted base* when referring to the substances involved in proton transfer reactions.

H—O—C—H (with O double-bonded to C)
formic acid (HCHO$_2$)

Only the H in red is available in an acid–base reaction.

O—C—H (with O double-bonded to C, and ⁻O)
formate ion (CHO$_2^-$)

TOOLS

Conjugate acid–base pairs

Conjugate Acids and Bases

Let's use the Brønsted–Lowry view to consider an acid–base reaction as a chemical equilibrium, having both a forward and a reverse reaction. We first encountered a chemical equilibrium in a solution of a weak acid in Section 4.4. Now let's use formic acid, HCHO$_2$, as an example. Because formic acid is a weak acid, we represent its ionization as a chemical equilibrium in which water is not just a solvent but also a chemical reactant, a proton acceptor.[3]

$$HCHO_2(aq) + H_2O \rightleftharpoons H_3O^+(aq) + CHO_2^-(aq)$$

In the forward reaction, a formic acid molecule donates a proton to the water molecule and changes to a formate ion, CHO$_2^-$ (Figure 15.2a). Thus HCHO$_2$ behaves as a Brønsted acid, a **proton donor**. Because water accepts this proton from HCHO$_2$, water behaves as a Brønsted base, a **proton acceptor**.

Now let's look at the reverse reaction (Figure 15.2b). In it, H$_3$O$^+$ behaves as a Brønsted acid because it donates a proton to the CHO$_2^-$. The CHO$_2^-$ ion behaves as a Brønsted base by accepting the proton.

The equilibrium process involving HCHO$_2$, H$_2$O, H$_3$O$^+$, and CHO$_2$ is typical of proton transfer equilibria in general, in that we can identify *two* acids (e.g., HCHO$_2$ and H$_3$O$^+$) and *two* bases (e.g., H$_2$O and CHO$_2^-$). Notice that in the aqueous formic acid equilibrium, the acid on the right of the arrows (H$_3$O$^+$) is formed from the base on the left (H$_2$O), and the base on the right (CHO$_2^-$) is formed from the acid on the left (HCHO$_2$).

Two substances that differ from each other only by one proton are referred to as a **conjugate acid–base pair**. Thus, H$_3$O$^+$ and H$_2$O are such a pair; they are alike except that H$_3$O$^+$, the acid, has one more proton than H$_2$O. One member of the pair is called the **conjugate acid** because it is the proton donor. The other member is the **conjugate base** because it is the pair's proton acceptor. We say that H$_3$O$^+$ is the conjugate acid of H$_2$O, and H$_2$O is the conjugate base of H$_3$O$^+$. Notice that *the acid member of the pair always has one more H$^+$ than the base member*.

The pair HCHO$_2$ and CHO$_2^-$ is the other conjugate acid–base pair in the aqueous formic acid equilibrium. HCHO$_2$ has one more H$^+$ than CHO$_2^-$, so the conjugate acid of CHO$_2^-$ is HCHO$_2$; the conjugate base of HCHO$_2$ is CHO$_2^-$. One way to highlight

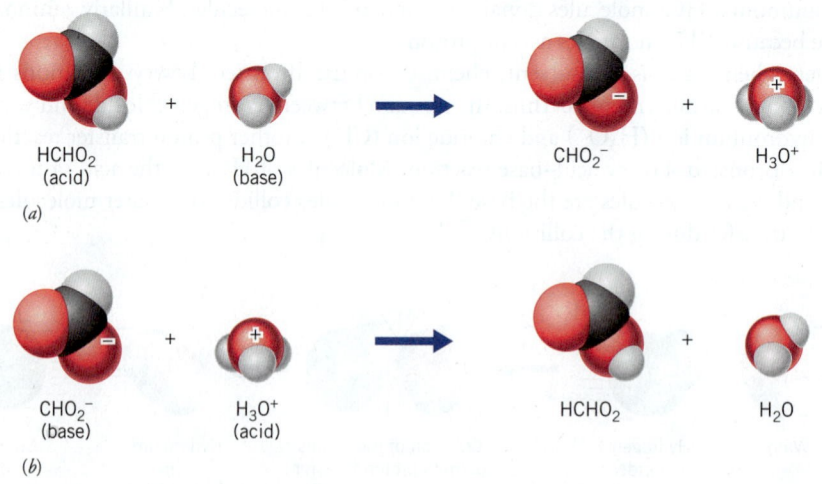

Figure 15.2 | Brønsted–Lowry acids and bases in aqueous formic acid. (*a*) Formic acid transfers a proton to a water molecule. HCHO$_2$ is the acid and H$_2$O is the base. (*b*) When hydronium ion transfers a proton to the CHO$^-$ ion, H$_3$O$^+$ is the acid and CHO$_2^-$ is the base.

HCHO$_2$ (acid) + H$_2$O (base) → CHO$_2^-$ + H$_3$O$^+$

(*a*)

CHO$_2^-$ (base) + H$_3$O$^+$ (acid) → HCHO$_2$ + H$_2$O

(*b*)

---

[3] Just as we have represented acetic acid as HC$_2$H$_3$O$_2$, placing the ionizable or acidic hydrogen first (as we do in HCl or HNO$_3$), so we give the formula of formic acid as HCHO$_2$ and the formula of the formate ion as CHO$_2^-$. As the chemical structure shows, however, the acidic hydrogen in formic acid is bonded to O, not C. Many chemists prefer to write formic acid as HCO$_2$H.

the two members of a conjugate acid–base pair in an equilibrium equation is to connect them by a line.

conjugate pair

$$HCHO_2 \ + \ H_2O \ \rightleftharpoons \ H_3O^+ \ + \ CHO_2^- \qquad (15.1)$$

acid   base   acid   base

conjugate pair

TOOLS

Conjugate acid–base equilibria

In any Brønsted–Lowry acid–base equilibrium, there are invariably *two* conjugate acid–base pairs, and you will need to learn how to pick them out of an equation by inspection and to write them from formulas.

Example 15.1
Determining the Formulas of Conjugate Acids and Bases

What is the conjugate base of nitric acid, HNO_3, and what is the conjugate acid of the hydrogen sulfate ion, HSO_4^-?

Analysis: Any reference to a conjugate acid or a conjugate base signals that the problem involves the Brønsted–Lowry acid–base concept.

Assembling the Tools: The conjugate acid-base pair tool states that members of any conjugate acid–base pair differ by one (and *only* one) H^+, and the member that has the extra hydrogen ion is the acid.

Solution: We are asked to find the conjugate base of HNO_3, which means that HNO_3 must be the acid member of the pair. To find the formula of the base, we remove one H^+ (both the atom and the charge) from the acid, HNO_3, leaving NO_3^-, the nitrate ion, as the conjugate base.

We are also asked to find the conjugate acid of HSO_4^-, making HSO_4^- the base member of an acid–base pair. We add one H^+ to the base, which is adding one hydrogen to the formula of the base and increasing its positive charge (or decreasing its negative charge) by one unit. Thus, adding an H^+ to HSO_4^- gives its conjugate acid, H_2SO_4.

Are the Answers Reasonable? As a check, we can quickly compare the two formulas in each pair.

$$HNO_3 \qquad NO_3^-$$

$$H_2SO_4 \qquad HSO_4^-$$

In each case, the formula on the right has one less H^+ than the one on the left, so it is the conjugate base.

Which of the following are conjugate acid–base pairs? Describe why the others are not true conjugate acid–base pairs: (a) H_3PO_4 and $H_2PO_4^-$, (b) HI and H^+, (c) NH_2^- and NH_3, (d) HNO_2 and NH_4^+, (e) CO_3^{2-} and CN^-, (f) HPO_4^{2-} and $H_2PO_4^-$. (*Hint:* Recall that conjugate acid–base pairs must differ by one H^+.)

Practice Exercise 15.1

Write the formula of the conjugate *base* for each of the following Brønsted–Lowry acids: (a) H_2O, (b) HI, (c) H_3PO_4, (d) $H_2PO_4^-$, (e) NH_4^+.

Write the formula of the conjugate *acid* for each of the following Brønsted–Lowry bases: (f) HO_2^-, (g) CO_3^{2-}, (h) NH_2^-, (i) NH_3, (j) HPO_4^{2-}.

Practice Exercise 15.2

Example 15.2
Identifying Conjugate Acid–Base Pairs in a Brønsted–Lowry Acid–Base Reaction

■ Sodium hydrogen sulfate (also called sodium bisulfate) is used to manufacture certain kinds of cement and to clean oxide coatings from metals.

The anion of sodium hydrogen sulfate, HSO_4^-, reacts as follows with the phosphate ion, PO_4^{3-}:

$$HSO_4^-(aq) + PO_4^{3-}(aq) \longrightarrow SO_4^{2-}(aq) + HPO_4^{2-}(aq)$$

Identify the two conjugate acid–base pairs.

Analysis: Two things are needed to identify the conjugate acid–base pairs in an equation. (1) The members of a conjugate pair differ by one hydrogen and one positive charge. (2) The members of each pair must be on opposite sides of the arrow of the equation, as in Equation 15.1. In each pair, of course, the acid is the one with the greater number of hydrogens.

Assembling the Tools: We will use Equation 15.1, the conjugate acid-base equilibrium, as a template to establish the relationships among the acid–base members of the two pairs.

Solution: Two of the formulas in the equation contain "PO_4^{3-}" unit so they must belong to the same conjugate pair. The one with the greater number of hydrogens, HPO_4^{2-}, must be the Brønsted–Lowry acid, and the other, PO_4^{3-}, must be the base. Therefore, one conjugate acid–base pair is HPO_4^{2-} and PO_4^{3-}. The other two ions, HSO_4^- and SO_4^{2-}, belong to the second conjugate acid–base pair; HSO_4^{2-} is the conjugate acid and SO_4^{2-} is the conjugate base.

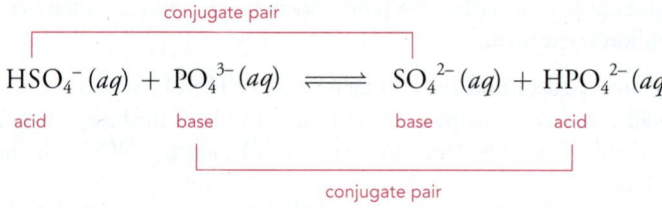

Is the Answer Reasonable? A check assures us that we have fulfilled the requirements that each conjugate pair has one member on one side of the arrow and the other member on the opposite side of the arrow and that the members of each pair differ from each other by one (and *only* one) H^+.

Practice Exercise 15.3

Sodium cyanide solution, when poured into excess hydrochloric acid, releases hydrogen cyanide as a gas. The reaction is

$$NaCN(aq) + HCl(aq) \longrightarrow HCN(g) + NaCl(aq)$$

Identify the conjugate acid–base pairs in this reaction. (*Hint:* It may be more obvious if the spectator ions are removed.)

Practice Exercise 15.4

One kind of baking powder contains sodium bicarbonate and calcium dihydrogen phosphate. When water is added, a reaction occurs by the following net ionic equation.

$$HCO_3^-(aq) + H_2PO_4^-(aq) \longrightarrow H_2CO_3(aq) + HPO_4^{2-}(aq)$$

Identify the two Brønsted acids and the two Brønsted bases in this reaction. (The H_2CO_3 decomposes to release CO_2, which causes cake batter to rise.)

Amphoteric Substances

Some molecules or ions are able to function either as an acid or as a base, depending on the kind of substance mixed with them. For example, in its reaction with hydrogen chloride, water behaves as a *base* because it *accepts* a proton from the HCl molecule.

$$H_2O + HCl(g) \longrightarrow H_3O^+(aq) + Cl^-(aq)$$
$$\text{base} \qquad\quad \text{acid}$$

On the other hand, water behaves as an *acid* when it reacts with the weak base ammonia.

$$H_2O + NH_3(aq) \longrightarrow NH_4^+(aq) + OH^-(aq)$$
$$\text{acid} \qquad\quad \text{base}$$

Here, H_2O *donates* a proton to NH_3 in the forward reaction.

A substance that can react as either an acid or a base depending on the other substance present is said to be **amphoteric**. Another term is **amphiprotic**, which describes substances that can donate and accept protons.

Amphoteric or amphiprotic substances may be either molecules or ions. For example, anions of acid salts, such as the bicarbonate ion of baking soda, are amphoteric. The HCO_3^- ion can either donate a proton to a base or accept a proton from an acid. Thus, toward the hydroxide ion, the bicarbonate ion is an acid; it donates its proton to OH^-.

$$HCO_3^-(aq) + OH^-(aq) \longrightarrow CO_3^{2-}(aq) + H_2O$$
$$\text{acid} \qquad\qquad \text{base}$$

Toward hydronium ion, however, HCO_3^- is a base; it accepts a proton from H_3O^+.

$$HCO_3^-(aq) + H_3O^+(aq) \longrightarrow H_2CO_3(aq) + H_2O$$
$$\text{base} \qquad\qquad \text{acid}$$

■ All amphiprotic substances are amphoteric, but not all amphoteric substances are amphiprotic.

Which of the following are amphoteric and which are not? Provide reasons for your decisions: (a) $H_2PO_4^-$, (b) H_2S, (c) H_3PO_4, (d) NH_4^+, (e) H_2O, (f) HI, (g) HNO_2. (*Hint:* Amphoteric substances must be able to provide an H^+ and also react with an H^+.)

The anion of sodium monohydrogen phosphate, Na_2HPO_4, is amphoteric. Using H_3O^+ and OH^-, write net ionic equations that illustrate this property.

Practice Exercise 15.5

Practice Exercise 15.6

15.2 | Strengths of Brønsted–Lowry Acids and Bases

Brønsted–Lowry acids and bases have differing abilities to lose or gain protons. In this section we examine how these abilities can be compared and how we can anticipate differences according to the locations of key elements in the periodic table.

Comparing Acids and Bases to a Relative Standard

When we speak of the strength of a Brønsted–Lowry acid, we are referring to its ability to donate a proton to a base. We measure this by determining the extent to which the reaction of the acid with the base proceeds toward completion—the more complete the reaction, the stronger the acid. To compare the strengths of a series of acids, we have to select some reference base, and because we are interested most in reactions in aqueous media, that base is usually water (although other reference bases could also be chosen).

In Chapter 4 we discussed strong and weak acids from the Arrhenius point of view, and much of what we said there applies when the same acids are studied using the Brønsted–Lowry concept. Thus, acids such as HCl and HNO_3 react completely with water to give H_3O^+ because they are strong proton donors. Hence, we classify them as strong Brønsted–Lowry acids. On the other hand, acids such as HNO_2 (nitrous acid) and $HC_2H_3O_2$ (acetic acid) are much weaker proton donors. Their reactions with water are far from complete, and we classify them as weak acids.

In a similar manner, the relative strengths of Brønsted bases are assigned according to their abilities to accept and bind protons. Once again, to compare strengths, we have to choose a standard acid. Because water is amphiprotic, it can serve as the standard acid as well. Substances that are powerful proton acceptors, such as the oxide ion, react completely with water and are considered to be strong Brønsted–Lowry bases.

$$O^{2-} + H_2O \xrightarrow{100\%} 2OH^-$$

Weaker proton acceptors, such as ammonia, undergo incomplete reactions with water; we classify them as weak bases.

Hydronium Ion and Hydroxide Ion in Water

Both HCl and HNO_3 are very powerful proton donors. When placed in water they react completely, losing their protons to water molecules to yield H_3O^+ ions. Representing them by the general formula HA, we have

$$\underset{\text{acid}}{HA} + \underset{\text{base}}{H_2O} \xrightarrow{100\%} \underset{\text{acid}}{H_3O^+} + \underset{\text{base}}{A^-}$$

Because both reactions go to completion, we really can't tell which of the two, HCl or HNO_3, is actually the better proton donor (stronger acid). This would require a reference base less willing than water to accept protons. In water, both HCl and HNO_3 converted quantitatively to another acid, H_3O^+. The conclusion, therefore, is that *H_3O^+ is the strongest acid we will ever find in an aqueous solution*, because stronger acids react completely with water to give H_3O^+.

A similar conclusion is reached regarding hydroxide ion. We noted that the strong Brønsted base O^{2-} reacts completely with water to give OH^-. Another very powerful proton acceptor is the amide ion, NH_2^-, which also reacts completely with water.

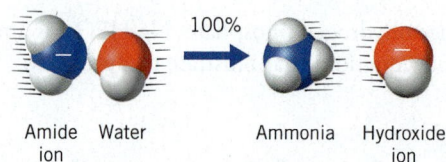

| Amide | Water | Ammonia | Hydroxide |
| ion | | | ion |

Of course, we can also express this reaction in the form of a regular chemical equation,

$$\underset{\text{base}}{NH_2^-} + \underset{\text{acid}}{H_2O} \xrightarrow{100\%} \underset{\text{acid}}{NH_3} + \underset{\text{base}}{OH^-}$$

Using water as the reference acid, we can't tell which is the better proton acceptor, O^{2-} or NH_2^-, because both react completely, being replaced by another base, OH^-. Therefore, we can say that *OH^- is the strongest base we will ever find in an aqueous solution*, because stronger bases react completely with water to give OH^-.

Comparing Acid–Base Strengths of Conjugate Pairs

The chemical equation for the ionization of an acid in water actually shows two Brønsted–Lowry acids. One is the acid itself and the other is hydronium ion. In almost every case, one of the acids is stronger than the other, and the position of equilibrium tells us which of the two acids is stronger. Let's see how this works using the acetic acid equilibrium, in which the position of equilibrium lies to the left.

$$\underset{\text{acid}}{HC_2H_3O_2(aq)} + \underset{\text{base}}{H_2O} \rightleftharpoons \underset{\text{acid}}{H_3O^+(aq)} + \underset{\text{base}}{C_2H_3O_2^-(aq)}$$

The two Brønsted–Lowry acids in this equilibrium are $HC_2H_3O_2$ and H_3O^+, and it's helpful to think of them as competing with each other in donating protons to acceptors. The fact that nearly all potential protons stay on the $HC_2H_3O_2$ molecules, and only a relative few spend their time on the H_3O^+ ions, means that the hydronium ion is a better proton donor than the acetic acid molecule. Thus, the hydronium ion is a stronger Brønsted–Lowry acid than acetic acid, and we inferred this relative acidity from the position of equilibrium.

The acetic acid equilibrium also has two bases, $C_2H_3O_2^-$ and H_2O. Both compete for any available protons. But at equilibrium, most of the protons originally carried by acetic acid are still found on $HC_2H_3O_2$ molecules; relatively few are joined to H_2O in the form of H_3O^+ ions. This means that acetate ions must be more effective than water molecules at obtaining and holding protons from proton donors. This is the same as saying that the acetate ion is a stronger base than the water molecule, a fact we are able to infer, once again, from the position of the acetic acid equilibrium.

The preceding analysis of the acetic acid equilibrium brings out an important point. Notice that *both* the weaker of the two acids and the weaker of the two bases are found on the same side of the equation, which is the side favored by the position of equilibrium.

The position of an acid–base equilibrium favors the weaker acid and base.

TOOLS

Acid–base strength and the position of equilibrium

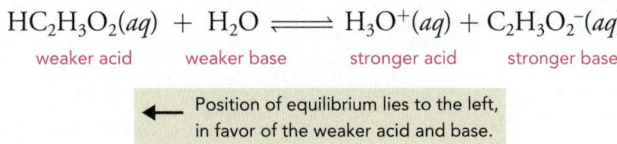

$$HC_2H_3O_2(aq) + H_2O \rightleftharpoons H_3O^+(aq) + C_2H_3O_2^-(aq)$$

weaker acid   weaker base   stronger acid   stronger base

← Position of equilibrium lies to the left, in favor of the weaker acid and base.

Reciprocal Relationship in Acid–Base Strength

One aid in predicting the relative strengths of conjugate acids and bases is the existence of a reciprocal relationship.

The stronger the Brønsted acid is, the weaker its conjugate base.

TOOLS

Reciprocal relationship in acid–base strengths

To illustrate, recall that $HCl(g)$ is a very strong Brønsted acid; it's 100% ionized in a dilute aqueous solution.

$$HCl(g) + H_2O \xrightarrow{100\%} H_3O^+(aq) + Cl^-(aq)$$

As we explained in Chapter 4, we don't write double equilibrium arrows for the ionization of a strong acid. Not doing so with HCl is another way of saying that the chloride ion, the conjugate base of HCl, must be a very weak Brønsted–Lowry base. Even in the presence of H_3O^+, a very strong proton donor, chloride ions don't recombine with protons. So HCl, the strong acid, has a particularly weak conjugate base, Cl^-.

There's a matching **reciprocal relationship**.

The weaker the Brønsted acid is, the stronger its conjugate base.

Consider, for example, the conjugate pair, OH^- and O^{2-}. The hydroxide ion is the conjugate acid and the oxide ion is the conjugate base. But the hydroxide ion must be an *extremely* weak Brønsted–Lowry acid; in fact, we've known it so far only as a base. Given the extraordinary weakness of OH^- as an acid, its conjugate base, the oxide ion, must be an exceptionally strong base. And as you've already learned, oxide ion is such a strong base that its reaction with water is 100% complete. That's why we don't write double equilibrium arrows in the equation for the reaction.

$$O_2^- + H_2O \xrightarrow{100\%} OH^- + OH^-$$

base   acid   acid   base

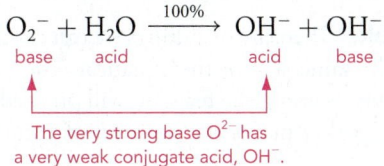

The very strong base O^{2-} has a very weak conjugate acid, OH^-.

An amphoteric substance will act as a base if mixed with an acid, but it will act as an acid if mixed with a base. You might ask: If two amphoteric substances are mixed together, which will act as the acid and which as the base? The obvious answer is that the stronger acid will act as the acid and the other will be the base. In the reaction below, we find that the position of equilibrium lies to the left (with the reactants).

$$H_2S(aq) + HCO_3^-(aq) \rightleftharpoons HS^-(aq) + H_2CO_3(aq)$$

acid   base   base   acid

> ←
> Position of equilibrium lies to the left.
> HCO_3^- must be a weaker base than HS^-.

This can be interpreted as meaning that $H_2CO_3(aq)$ is a stronger acid than $H_2S(aq)$, which in turn makes $HCO_3^-(aq)$ a weaker base than $HS^-(aq)$. Therefore, if we mix a solution containing $HS^-(aq)$ with a solution containing $HCO_3^-(aq)$, the reaction will be

$$HS^-(aq) + HCO_3^-(aq) \rightleftharpoons H_2S(aq) + CO_3^{2-}(aq)$$

base   acid   acid   base

because the stronger base, HS^-, will remove the H^+ from the weaker base, HCO_3^-, causing the latter to actually behave as an acid.

Example 15.3
Using Reciprocal Relationships to Predict Equilibrium Positions

In the reaction below, will the position of equilibrium lie to the left or to the right, given the fact that acetic acid is known to be a stronger acid than the hydrogen sulfite ion?

$$HSO_3^-(aq) + C_2H_3O_2^-(aq) \rightleftharpoons HC_2H_3O_2(aq) + SO_3^{2-}(aq)$$

Analysis: In the equation above, the two acids are $HC_2H_3O_2$ and HSO_3. To find the answer to the question posed in the problem, we have to ask: How are the relative strengths of the acids related the position of equilibrium?

Assembling the Tools: The position of an acid–base equilibrium favors the weaker acid and base. That's one of the tools we need to solve this problem. The other is the reciprocal relationship between the strengths of the members of a conjugate acid–base pair.

Solution: We'll write the equilibrium equation, using the given fact about the relative strengths of acetic acid and the hydrogen sulfite ion to start writing labels.

$$HSO_3^-(aq) + C_2H_3O_2^-(aq) \rightleftharpoons HC_2H_3O_2(aq) + SO_3^{2-}(aq)$$

weaker acid   stronger acid

Now we'll use our tool about the reciprocal relationships to label the two bases. The stronger acid must have the weaker conjugate base; the weaker acid must have the stronger conjugate base.

$$HSO_3^-(aq) + C_2H_3O_2^-(aq) \rightleftharpoons HC_2H_3O_2(aq) + SO_3^{2-}(aq)$$

weaker acid   weaker base   stronger acid   stronger base

Finally, because the position of equilibrium favors the weaker acid and base, the position of equilibrium lies to the left.

Is the Answer Reasonable? There are two things we can check. First, both of the weaker conjugates should be on the same side of the equation. They are, so that suggests we've made the correct assignments. Second, the reaction will proceed farther in the direction of the weaker acid and base, so that places the position of equilibrium on the left, which agrees with our answer.

Practice Exercise 15.7

Given that HSO_4^- is a stronger acid than HPO_4^{2-}, what is the chemical reaction if solutions containing these ions are mixed together? (*Hint:* One of these must act as an acid and the other as a base.)

Practice Exercise 15.8

Given that HClO is a weaker acid than HNO_2, determine whether the substances on the left of the arrows or those on the right are favored in the following equilibrium.

$$HClO(aq) + NO_2^-(aq) \rightleftharpoons ClO^-(aq) + HNO_2(aq)$$

15.3 | Periodic Trends in the Strengths of Acids

In most cases, acids are formed by nonmetals, either in the form of solutions of their compounds with hydrogen [such as HCl(*aq*)] or as solutions of their oxides, which react with water to form oxoacids (such as H_2SO_4). The strengths of these acids vary in a systematic way with the location of the nonmetal in the periodic table.

Trends in the Strengths of Binary Acids

Many (but not all) of the binary compounds between hydrogen and nonmetals, which we may represent by HX and H_2X, are acidic and are called **binary acids**. Table 15.1 lists those that are acids in water. The strong acids are marked by asterisks.

The relative strengths of binary acids correlate with the periodic table in two ways.

> The strengths of the binary acids increase from left to right within the same period.
> The strengths of binary acids increase from top to bottom within the same group.

The strength of a binary acid is inversely related to the strength of the H—X bond. Breaking this bond is essential for the hydrogen to separate as an H^+ ion, so anything that contributes to decreasing the bond strength will increase acid strength and vice versa. The two factors that account for bond strength are the electronegativity of the nonmetal X and the length of the H—X bond.

The electronegativity of the atom X affects the polarity of the H—X bond, so as electronegativity increases, the partial positive charge ($\delta+$) on H becomes greater. This makes the bonded hydrogen more like the free hydrogen ion and so it is easier for the hydrogen to separate as H^+. The result is that the molecule becomes a better proton donor.

Going from left to right across a period, atomic size varies little, so the lengths of the H—X bonds are nearly the same and have little effect on acid strength. The major influence is the variation in the electronegativity of X, which increases from left to right. For example, as we go from left to right within Period 3, from S to Cl, the electronegativity increases and we find that HCl is a stronger acid than H_2S. A similar increase in electronegativity occurs going left to right in Period 2, from O to F, so HF is a stronger acid than H_2O.

In going down a group, there are two opposing factors. The electronegativity decreases, so the H—X bonds become less polar. At the same time, the atoms X become much larger, and this leads to a large increase in bond length and a drop in the H—X bond strength.[4] The decrease in polarity tends to make the acids weaker, but the increase in

TABLE 15.1	Acidic Binary Compounds of Hydrogen and Nonmetals[a]	
Period Number	Group 6A	
2	(H_2O)	
3	H_2S	Hydrosulfuric acid
4	H_2Se	Hydroselenic acid
5	H_2Te	Hydrotelluric acid
	Group 7A	
2	HF	Hydrofluoric acid
3	*HCl	Hydrochloric acid
4	*HBr	Hydrobromic acid
5	*HI	Hydroioidic acid

[a]The *names* are for the aqueous solutions of these compounds.
*Strong acids.

TOOLS

Periodic trends in strengths of binary acids

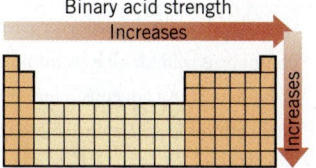

Binary acid strength
Increases

increases

---

[4] Small atoms tend to form stronger bonds than large atoms, so as we go down a group there is a rapid decrease in the H—X bond strength as the atoms X become larger.

bond length tends to make them stronger. Changes in bond length seem to win out, because the acids H_nX become stronger as we go down a group. For example, among the binary acids of the halogens the following order of relative acidity is observed.[5]

$$HF < HCl < HBr < HI$$

Thus, HF is the weakest acid in the series; in fact it is a weak acid, and HI is the strongest. The identical trend occurs in the series of the binary acids of Group 6A elements, having formulas of the general type H_2X.

Practice Exercise 15.9

Order the following groups of acids from the weakest to the strongest: (a) HI, HF, HBr; (b) HCl, PH_3, H_2S; (c) H_2Te, H_2O, H_2Se; (d) AsH_3, HBr, H_2Se; (e) HI, PH_3, H_2Se. (*Hint:* These are all binary acids.)

Practice Exercise 15.10

Using *only* the periodic table, choose the stronger acid of each pair: (a) H_2Se or HBr, (b) H_2Se or H_2Te, (c) H_2O or H_2S.

Trends in the Strengths of Oxoacids

In Chapter 4 you learned that acids composed of hydrogen, oxygen, and some other element are called **oxoacids** (see Table 15.2). Those that are strong acids in water are identified in the table by asterisks.

TABLE 15.2 **Some Oxoacids of Nonmetals and Metalloids**[a]

Period Number	Group 4A	Group 5A	Group 6A	Group 7A
2	H_2CO_3 Carbonic acid	*HNO_3 Nitric acid HNO_2 Nitrous acid		HFO Hypofluorous acid
3		H_3PO_4 Phosphoric acid H_3PO_3 Phosphorous acid[b]	*H_2SO_4 Sulfuric acid H_2SO_3 Sulfurous acid[c]	*$HClO_4$ Perchloric acid *$HClO_3$ Chloric acid $HClO_2$ Chlorous acid $HClO$ Hypochlorous acid
4		H_3AsO_4 Arsenic acid H_3AsO_3 Arsenous acid	*H_2SeO_4 Selenic acid H_2SeO_3 Selenous acid	*$HBrO_4$ Perbromic acid[d] *$HBrO_3$ Bromic acid
5			$Te(OH)_6$ Telluric acid[e] H_2TeO_3 Tellurous acid	HIO_4 Periodic acid (H_5IO_6)[f] HIO_3 Iodic acid

[a]Strong acids are indicated with asterisks. Only the first proton of strong diprotic acids completely ionize.

[b]Phosphorus acid, despite its formula, is only a diprotic acid.

[c]Hypothetical. An aqueous solution actually contains just dissolved sulfur dioxide, $SO_2(aq)$.

[d]Pure perbromic acid is unstable; a dihydrate is known.

[e]$Te(OH)_6$ is a diprotic acid.

[f]H_5IO_6 is formed from $HIO_4 + 2H_2O$.

---

[5] We compare acid strengths by the acid's ability to donate a proton to a particular base. As we noted earlier, for strong acids such as HCl, HBr, and HI, water is too strong a proton acceptor to permit us to see differences among their proton-donating abilities. All three of these acids are completely ionized in water and so appear to be of equal strength, a phenomenon called the *leveling effect* (the differences are obscured or *leveled* out). To compare the acidities of these acids, a solvent that is a weaker proton acceptor than water (liquid HF or $HC_2H_3O_2$, for example) has to be used.

   A feature common to the structures of all oxoacids is the presence of O—H groups bonded to some central atom. For example, the structures of two oxoacids of the Group 6A elements are

$$
\begin{array}{cc}
\ddot{\text{O}}: & \ddot{\text{O}}: \\
\parallel & \parallel \\
\text{H}-\ddot{\text{O}}-\text{S}-\ddot{\text{O}}-\text{H} & \text{H}-\ddot{\text{O}}-\text{Se}-\ddot{\text{O}}-\text{H} \\
\parallel & \parallel \\
:\ddot{\text{O}}: & :\ddot{\text{O}}: \\
\text{H}_2\text{SO}_4 & \text{H}_2\text{SeO}_4 \\
\text{sulfuric acid} & \text{selenic acid}
\end{array}
$$

When an oxoacid ionizes, the hydrogen that is lost as an H^+ *always* comes from an O—H bond. The strength of an oxoacid is determined by how the group of atoms attached to the oxygen affects the polarity, and thus the strength, of the O—H bond. If this group of atoms makes the O—H bond more polar, the bond strength will be smaller. This allows the H to come off more easily as H^+, increasing the strength of the acid.

$$
\overset{\delta-}{}\overset{\delta+}{} \\
G-\text{O}-\text{H} \\
\longleftarrow
$$

> If the group of atoms, G, attached to the O—H group is able to draw electron density from the O atom, the O will pull electron density from the O—H bond, thereby making the bond more polar.

It turns out that there are two principal factors that determine how the polarity of the O—H bond is affected. One is the electronegativity of the central atom in the oxoacid and the other is the number of oxygen atoms attached to the central atom.

Effect of the Electronegativity of the Central Atom

To study the effects of the electronegativity of the central atom, we must compare oxoacids having the same number of oxygens. When we do this, we find that as the electronegativity of the central atom increases, the oxoacid becomes a better proton donor (i.e., a stronger acid). The following diagram illustrates the effect.

$$
\overset{|}{\underset{|}{-\text{X}}}\overset{\delta-}{-\text{O}}\overset{\delta+}{-\text{H}} \\
\longleftarrow
$$

> As the electronegativity of X increases, electron density is drawn away from O, which draws electron density away from the O—H bond. This makes the bond more polar and makes the molecule a better proton donor.

Because electronegativity increases from bottom to top within a group and from left to right within a period, we can make the following generalization.

When the central atoms of oxoacids are bonded to the same number of oxygen atoms, the acid strength increases from bottom to top within a group and from left to right within a period.

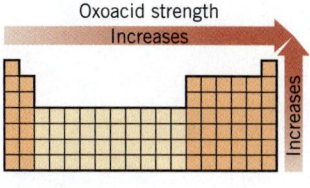

Oxoacid strength
Increases
Increases

Trends in the strengths of oxoacids

In Group 6A, for example, H_2SO_4 is a stronger acid than H_2SeO_4 because sulfur is more electronegative than selenium. Similarly, among the halogens, acid strength increases for acids with the formula HXO_4 as follows:

$$HIO_4 < HBrO_4 < HClO_4$$

Going from left to right within Period 3, we can compare the acids H_3PO_4, H_2SO_4, and $HClO_4$, where we find the following order of acidities.

$$H_3PO_4 < H_2SO_4 < HClO_4$$

Practice Exercise 15.11

Explain why one acid is stronger than the other in the following pairs: (a) $HClO_3$ and $HBrO_3$, (b) H_3PO_4 and H_2SO_4. (*Hint:* Note that each pair has the same number of oxygen atoms.)

Practice Exercise 15.12

Explain why one acid is weaker than the other in the following pairs: (a) H_3PO_4 and H_3AsO_4, (b) HIO_4 and H_2TeO_4.

Effect of the Number of Oxygen Atoms Bound to the Central Atom

Comparing oxoacids with the same central atom, as the number of *lone oxygen atoms* increases, the oxoacid becomes a better proton donor. (A *lone oxygen* is one that is bonded only to the central atom and not to a hydrogen.) Thus, comparing HNO_3 with HNO_2, we find that HNO_3 is the stronger acid. To understand why, let's look at their molecular structures.

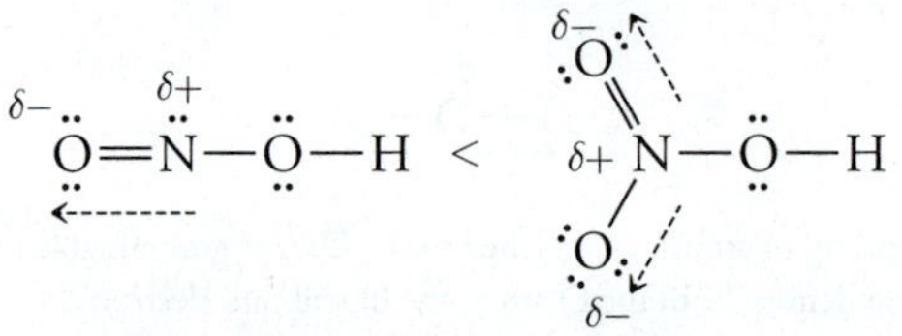

nitrous acid   nitric acid

■ Oxygen is the second most electronegative element and has a strong tendency to pull electron density away from any atom to which it is attached.

In an oxoacid, lone oxygens pull electron density away from the central atom, which increases the central atom's ability to draw electron density away from the O—H bond. It's as though the lone oxygens make the central atom more electronegative. Therefore, more lone oxygens result in more polar O—H bonds and a stronger acid. Thus, the two lone oxygens in HNO_3 produce a greater effect than the one lone oxygen in HNO_2, so HNO_3 is the stronger acid.

Similar effects are seen among other oxoacids, as well. For the oxoacids of chlorine, for instance, we find the following trend:

$$HClO < HClO_2 < HClO_3 < HClO_4$$

Comparing their structures, we have

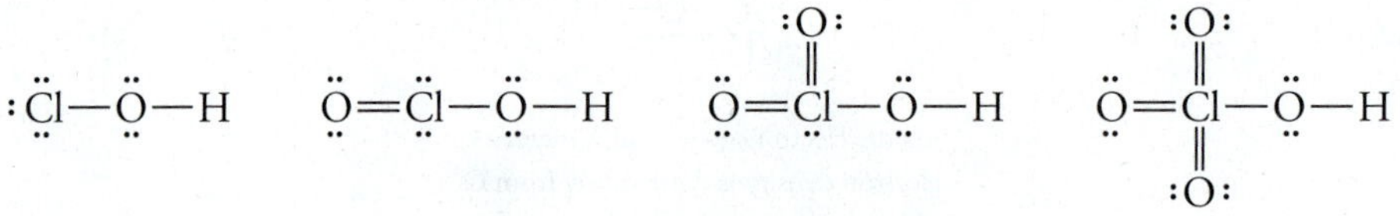

■ Usually, the formula for hypochlorous acid is written HOCl to reflect its molecular structure. We've written it HClO here to make it easier to follow the trend in acid strengths among the oxoacids of chlorine.

HClO   HClO_2   HClO_3   HClO_4

These structures all have only one O—H bond, so we can easily count the lone oxygens by subtracting 1 from the total number of oxygen atoms. This leads to another generalization.

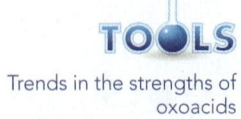

TOOLS

Trends in the strengths of oxoacids

For a given central atom, the acid strength of an oxoacid increases with the number of lone oxygen atoms bonded to the central atom.

The ability of lone oxygens to affect acid strength extends to organic compounds as well. For example, compare the molecules below.

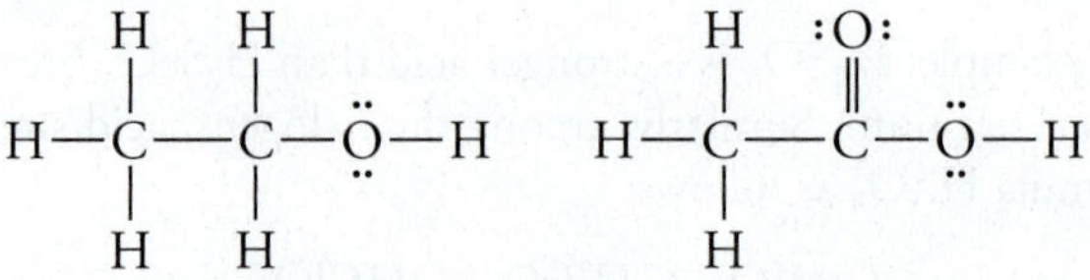

ethanol   acetic acid

In water, ethanol (ethyl alcohol) is not acidic at all. Replacing the two hydrogens on the carbon adjacent to the OH group with an oxygen, however, yields acetic acid. The greater ability of oxygen to pull electron density from the carbon produces a greater polarity of the O—H bond, which is one factor that causes acetic acid to be a better proton donor than ethanol.

The Reciprocal Relationship—A Molecular Interpretation

In Section 15.2 we noted the reciprocal relationship between the strength of an acid and that of its conjugate base. For oxoacids, the lone oxygens play a part in determining the basicity of the anion formed in the ionization reaction. Consider the acids $HClO_3$ and $HClO_4$.

$$
\begin{array}{cc}
\overset{\displaystyle :O:}{\underset{}{H-\ddot{O}-\overset{\|}{Cl}=\ddot{O}}} & \overset{\displaystyle :O:}{\underset{:O:}{H-\ddot{O}-\overset{\|}{\underset{\|}{Cl}}=\ddot{O}}} \\
HClO_3 & HClO_4
\end{array}
$$

From our earlier discussion, we expect $HClO_4$ to be a stronger acid than $HClO_3$, which it is. Ionization of their protons yields the anions ClO_3^- and ClO_4^-.

$$
\begin{array}{cc}
\left[\overset{\displaystyle :O:}{\underset{}{:\ddot{O}-\overset{\|}{Cl}=\ddot{O}}} \right]^- & \left[\overset{\displaystyle :O:}{\underset{:O:}{:\ddot{O}-\overset{\|}{\underset{\|}{Cl}}=\ddot{O}}} \right]^- \\
ClO_3^- & ClO_4^-
\end{array}
$$

In an **oxoanion** (an anion formed from an oxoacid) the *lone* oxygens carry most of the negative charge, which is delocalized over the lone oxygens. In the ClO_3^- ion, the single negative charge is spread over three oxygens, so each carries an effective charge of about $\frac{1}{3}-$. This is not a *formal* charge obtained by the rules in Section 8.7. We're saying that the $1-$ charge on the ion is delocalized over three oxygens to give each an effective charge of $\frac{1}{3}-$. By the same reasoning, in the ClO_4^- ion each oxygen carries a charge of about $\frac{1}{4}-$. The smaller negative charge on the oxygens in ClO_4^- makes this ion less able than ClO_3^- to attract H^+ ions from H_3O^+, so ClO_4^- is the weaker base than ClO_3^-. Thus, the anion of the stronger acid is the weaker base.

■ The concept of delocalization of electrons to provide stability was presented in Section 9.8.

In each pair, explain why one is a stronger acid than the other: (a) HIO_3 and HIO_4, (b) H_2TeO_3 and H_2TeO_4, (c) H_3AsO_3 and H_3AsO_4. (*Hint:* This problem focuses on the effect of oxygen atoms in acids.)

In each pair, explain why one is a weaker acid than the other: (a) H_2SO_4 and $HClO_4$, (b) H_3AsO_4 and H_2SO_4.

Practice Exercise 15.13

Practice Exercise 15.14

Strengths of Organic Acids

We just noted that replacing two H atoms in an alcohol by one lone oxygen causes the resulting molecule to be acidic. The acidity of the molecule can be increased further if other electronegative groups such as halogens are bonded to carbon atoms near the —CO_2H group of the acid. Such a group, labeled X below, pulls electron density from the carbon atom of the carboxyl group, further increasing the polarity of the O—H bond. This causes an increase in the acidity of the molecule.

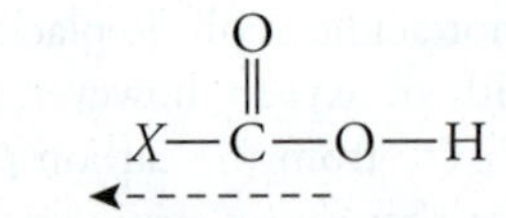

$$
\begin{array}{c}
O \\
\parallel \\
X-C-O-H
\end{array}
$$

As the ability of X to draw electrons away from C increases, the polarity of the O—H bond increases and the molecule becomes more acidic.

Thus, chloroacetic acid is a stronger acid than acetic acid, and dichloroacetic acid and trichloroacetic acid are increasingly stronger acids.

$$CH_3CO_2H < CH_2ClCO_2H < CHCl_2CO_2H < CCl_3CO_2H$$

Addition of each chlorine effectively withdraws more electron density from the O—H bond, resulting in a weaker bond and a stronger acid.

Practice Exercise 15.15

How would you expect the acidities of the following molecules to compare? (*Hint*: How does the electronegativity of the group attached to the acid group change?)

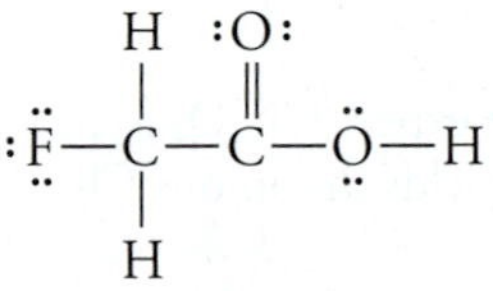

(a) 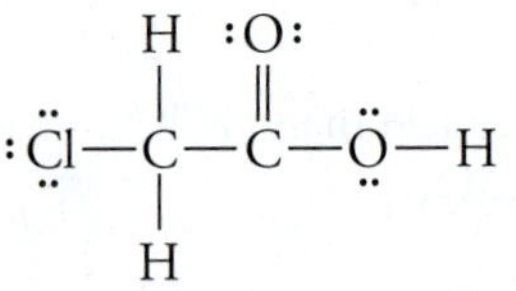 Cl–C–C–O–H (b) F–C–C–O–H (c) Br–C–C–O–H

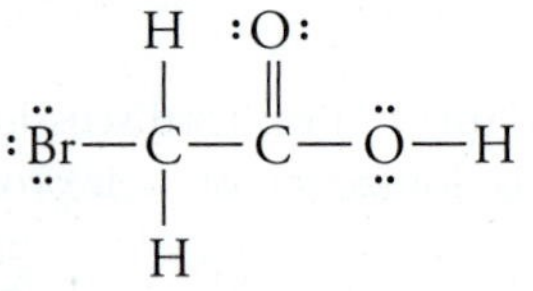

Practice Exercise 15.16

List these acids in terms of increasing acidity: CF_3CO_2H, CH_3CO_2H, CHF_2CO_2H, CH_2FCO_2H.

15.4 | Lewis Acids and Bases

In our preceding discussions, acids and bases have been characterized by their tendency to lose or gain protons. However, there are many reactions not involving proton transfer that have properties we associate with acid–base reactions. For example, if gaseous SO_3 is passed over solid CaO, a reaction occurs in which $CaSO_4$ forms.

$$CaO(s) + SO_3(g) \longrightarrow CaSO_4(s) \tag{15.2}$$

If these reactants are dissolved in water first, they react to form $Ca(OH)_2$ and H_2SO_4, and when their solutions are mixed the following reaction takes place.

$$Ca(OH)_2(aq) + H_2SO_4(aq) \longrightarrow CaSO_4(s) + 2H_2O \tag{15.3}$$

The same two initial reactants, CaO and SO_3, form the same ultimate product, $CaSO_4$. It certainly seems that if Reaction 15.3 is an acid–base reaction, we should be able to consider Reaction 15.2 to be an acid–base reaction, too. But there are no protons being transferred, so our definitions require further generalizations. These were provided by G. N. Lewis, after whom Lewis symbols are named.

Lewis Definitions of Acids and Bases

TOOLS

Lewis acid–base definitions

1. A **Lewis acid** is any ionic or molecular species that can accept a pair of electrons in the formation of a coordinate covalent bond.
2. A **Lewis base** is any ionic or molecular species that can donate a pair of electrons in the formation of a coordinate covalent bond.
3. **Neutralization** is the formation of a coordinate covalent bond between the donor (base) and the acceptor (acid).

Examples of Lewis Acid–Base Reactions

The reaction between BF_3 and NH_3 illustrates a Lewis **acid–base neutralization**. The reaction is exothermic because a bond is formed between N and B, with the nitrogen donating an electron pair and the boron accepting it.

■ A similar reaction was described in Chapter 8 between BCl_3 and ammonia. Compounds like BF_3NH_3, which are formed by simply joining two smaller molecules, are called **addition compounds**.

New
coordinate
covalent bond

Lewis
base

Lewis
acid

NH_3 BF_3 NH_3BF_3

Note that as the bond forms from ammonia to boron trifluoride, the geometry around the boron changes from planar triangular to tetrahedral, which is what is expected with four bonds to the boron.

The ammonia molecule thus acts as a *Lewis base*. The boron atom in BF_3, having only six electrons in its valence shell and needing two more to achieve an octet, accepts the pair of electrons from the ammonia molecule. Hence, BF_3 is functioning as a *Lewis acid*.

As this example illustrates, Lewis bases are substances that have completed valence shells *and unshared pairs of electrons* (e.g., NH_3, H_2O, and O^{2-}). A Lewis acid, on the other hand, can be a substance with an incomplete valence shell, such as BF_3 or H^+.

A substance can also be a Lewis acid even when it has a central atom with a complete valence shell. This works when the central atom has a double bond that, by the shifting of an electron pair to an adjacent atom, can make room for an incoming pair of electrons from a Lewis base. Carbon dioxide is an example. When carbon dioxide is bubbled into aqueous sodium hydroxide, the gas is instantly trapped as the bicarbonate ion.

$$CO_2(g) + OH^-(aq) \longrightarrow HCO_3^-(aq)$$

Lewis acid–base theory represents the movement of electrons in this reaction as follows.

One of the electron pairs of the double bond shifts to the oxygen atom as a covalent bond forms from the O atom of the OH^- ion to the C atom.

New covalent bond (coordinate covalent)

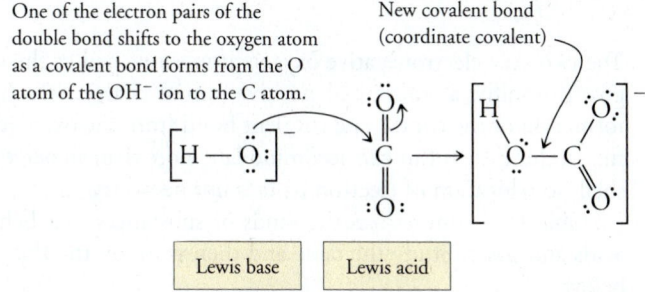

Lewis base   Lewis acid

■ Remember, a coordinate covalent bond is just like any other covalent bond once it has formed. By using this term, we are following the origin of the electron pair that forms the bond.

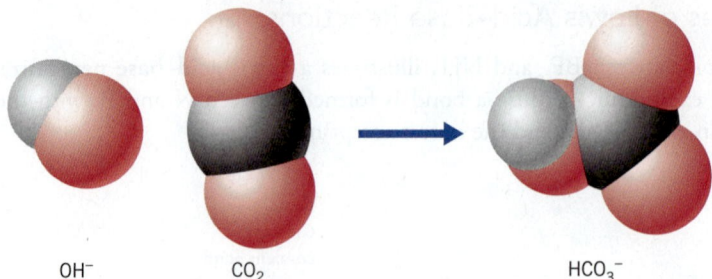

OH⁻   CO₂   HCO₃⁻

The donation of an electron pair *from* the oxygen of the OH^- ion produces a bond, so the OH^- ion is the Lewis base. The carbon atom of the CO_2 accepts the electron pair, so CO_2 is the Lewis acid.

Lewis acids can also be substances that have valence shells capable of holding more electrons. For example, consider the reaction of sulfur dioxide as a Lewis acid with oxide ion as a Lewis base to make the sulfite ion. This reaction occurs when gaseous sulfur dioxide, an acidic anhydride, mingles with solid calcium oxide, a basic anhydride, to give calcium sulfite, $CaSO_3$.

$$SO_2(g) + CaO(s) \longrightarrow CaSO_3(s)$$

Let's see how electrons relocate as the sulfite ion forms. We use one of the two resonance structures of SO_2 and one of the three such structures for the sulfite ion.

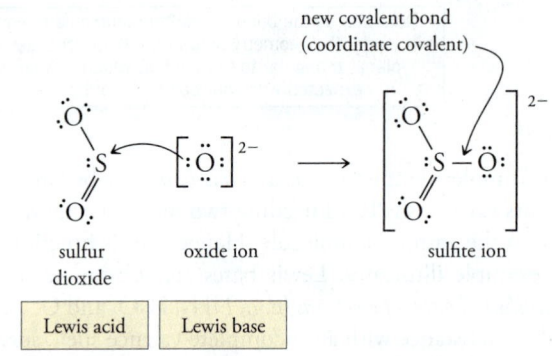

new covalent bond
(coordinate covalent)

sulfur dioxide   oxide ion   sulfite ion

| Lewis acid | Lewis base |

| TABLE 15.3 | Types of Substances that Are Lewis Acids and Lewis Bases |

Lewis Acids

Molecules or ions with incomplete valence shells (e.g., BF_3, H^+)

Molecules or ions with complete valence shells, but with multiple bonds that can be shifted to make room for more electrons (e.g., CO_2)

Molecules or ions that have central atoms capable of holding additional electrons (usually, atoms of elements in Periods 3 through 7) (e.g., SO_2)

Metals or metal ions that can accept Lewis bases (e.g., Fe, Zn^{2+})

Lewis Bases

Molecules or ions that have unshared pairs of electrons and that have complete valence shells (e.g., O^{2-}, NH_3)

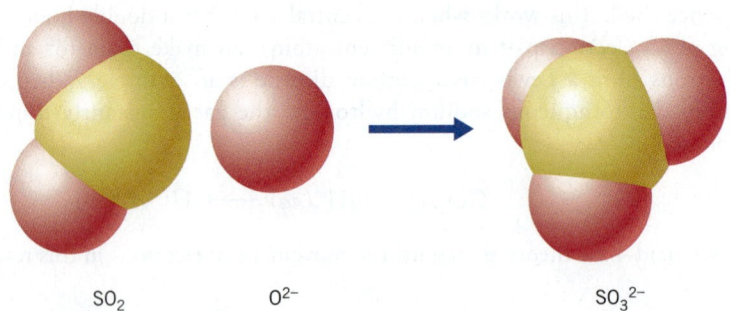

SO_2 O^{2-} SO_3^{2-}

The two very electronegative oxygen atoms attached to the sulfur in SO_2 give the sulfur a substantial positive partial charge, which induces the formation of the coordinate covalent bond from the oxide ion to the sulfur. In this case, sulfur can accommodate more than an octet in its valence shell, so relocation of electron pairs is not necessary.

Table 15.3 summarizes the kinds of substances that behave as Lewis acids and bases. Study the table and then work on the Practice Exercises below.

Identify the Lewis acid and Lewis base in each reaction.

Practice Exercise 15.17

(a) $NH_3 + H^+ \rightleftharpoons NH_4^+$

(b) $SeO_3 + Na_2O \rightleftharpoons Na_2SeO_4$

(c) $Ag^+ + 2NH_3 \rightleftharpoons Ag(NH_3)_2^+$

(*Hint:* Draw Lewis structures of the reactants.)

(a) Is the fluoride ion more likely to behave as a Lewis acid or a Lewis base? Explain.

Practice Exercise 15.18

(b) Is the $BeCl_2$ molecule more likely to behave as a Lewis acid or a Lewis base? Explain.

(c) Is the SO_3 molecule more likely to behave as a Lewis acid or a Lewis base? Explain.

Interpreting Brønsted–Lowry Acid–Base Reactions Using the Lewis Acid–Base Concept

Reactions involving the transfer of a proton can be analyzed by either the Brønsted–Lowry or Lewis definitions of acids and bases. Consider, for example, the reaction between hydronium ion and ammonia in aqueous solution.

$$H_3O^+ + NH_3 \longrightarrow H_2O + NH_4^+$$

Applying the Brønsted–Lowry definitions, we have an acid, H_3O^+, reacting with a base, NH_3, to give the corresponding conjugates: H_2O being the conjugate base of H_3O^+ and NH_4^+ being the conjugate acid of NH_3.

To interpret this reaction using the Lewis definitions, we view it as the stronger Lewis base (NH_3) donates a pair of electrons to the Lewis acid (the proton, H^+) and takes it from the weaker Lewis base (H_2O). To emphasize that it is the proton that is transferring, we might write the equation as

$$H_2O{-}H^+ + NH_3 \longrightarrow H_2O + H^+{-}NH_3$$

We can diagram this as follows using Lewis structures:

Electron pair in the OH bond shifts to the oxygen as the water molecule is formed.

The electron pair of the nitrogen captures H^+ (a Lewis acid) from the oxygen of the hydronium ion.

The molecular view of the reaction is

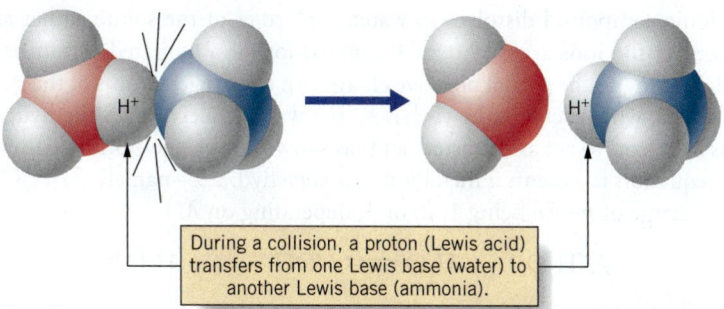

During a collision, a proton (Lewis acid) transfers from one Lewis base (water) to another Lewis base (ammonia).

In general, all Brønsted–Lowry acid–base reactions can be analyzed from the Lewis point of view following this same pattern, with the position of equilibrium favoring the proton being attached to the stronger Lewis base.

A word of caution is in order here. In analyzing proton transfer reactions, stick to one interpretation or the other (Brønsted–Lowry or Lewis). Usually we will use the Brønsted–Lowry approach because it is useful to think in terms of conjugate acid–base pairs. However, if you switch to the Lewis interpretation, don't try to apply Brønsted–Lowry terminology at the same time; it won't work and you'll only get confused.

15.5 | Acid–Base Properties of Elements and Their Oxides

The elements most likely to form acids are the nonmetals in the upper right-hand corner of the periodic table. Those most likely to form basic hydroxides are similarly grouped in one general location in the table, among the metals, particularly those in Groups 1A (alkali metals) and 2A (alkaline earth metals).

In general, the experimental basis for classifying an element according to its abilities to form an acid or base depends on how its *oxide* behaves toward water. Earlier you learned that metal oxides like Na_2O and CaO are called *base anhydrides* ("anhydride" means "without water") because they react with water to form hydroxides.

$$Na_2O + H_2O \longrightarrow 2NaOH \qquad \text{sodium hydroxide}$$

$$CaO + H_2O \longrightarrow Ca(OH)_2 \qquad \text{calcium hydroxide}$$

The reactions of metal oxides with water are really reactions of oxide ions, which take H^+ from molecules of H_2O, leaving ions of OH^-.

Many metal oxides are insoluble in water, so their oxide ions are unable to take H^+ ions from H_2O molecules. Many are, however, able to react with acids, which attack the oxide ions in the solid. Iron(III) oxide, for example, reacts with acid as follows.

$$Fe_2O_3(s) + 6H^+(aq) \longrightarrow 2Fe^{3+}(aq) + 3H_2O$$

This is a common method for removing rust from iron in industrial processes.

Nonmetal oxides are usually *acid anhydrides*; those that react with water give acidic solutions. Typical examples of the formation of acids from nonmetal oxides are the following reactions.

$$SO_3(g) + H_2O \longrightarrow H_2SO_4(aq) \qquad \text{sulfuric acid}$$

$$N_2O_5(g) + H_2O \longrightarrow 2HNO_3(aq) \qquad \text{nitric acid}$$

$$CO_2(g) + H_2O \longrightarrow H_2CO_3(aq) \qquad \text{carbonic acid}$$

Acidity of Hydrated Metal Ions

■ The force of attraction between an ion and water molecules is often strong enough to persist when the solvent water evaporates. Solid hydrates crystallize from solution— for example, $CuSO_4 \cdot 5H_2O$, shown in Figure 2.15.

When an ionic compound dissolves in water, molecules of the solute gather around the ions and we say the ions are *hydrated*. The metal ion of a hydrated cation behaves as a Lewis acid, binding to the partial negative charges on the oxygen atoms of the surrounding water molecules, which serve as Lewis bases. The water molecules on the hydrated metal ions themselves tend to act as Brønsted acids as shown in the equilibrium below. For simplicity, the equation represents a metal ion as a *monohydrate*—namely, $M(H_2O)^{n+}$ with a net positive charge of $n+$ (n being 1, 2, or 3, depending on M).

$$M(H_2O)^{n+} + H_2O \rightleftharpoons MOH^{(n-1)+} + H_3O^+$$

In other words, hydrated metal ions tend to be proton donors in water. Let's see why.

The positive charge on the metal ion attracts the water molecule and draws electron density from the O—H bonds, causing them to become more polar (Figure 15.3). This increases the partial positive charge on H and weakens the O—H bond with respect to the transfer of H$^+$ to another nearby water molecule in the formation of a hydronium ion. The entire process can be illustrated as follows.

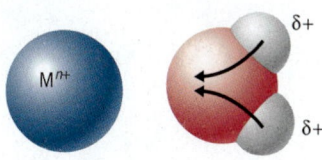

Figure 15.3 | Polarization of a water molecule by a metal cation. The positive charge on the metal ion pulls electron density away from the H atoms of the water molecule. This decreases the H—O bond strength making it easier for an H$^+$ to transfer to a nearby water molecule.

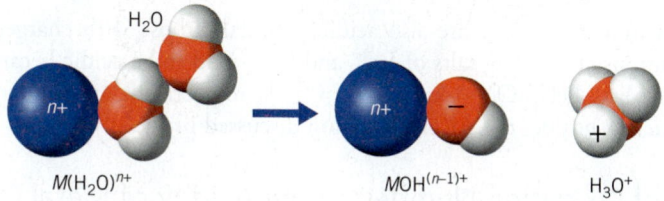

Electron density is reduced in the O—H bonds of water (indicated by the curved arrows) by the positive charge of the metal ion, thereby increasing the partial positive charge on the hydrogens. This promotes the transfer of H$^+$ to a water molecule.

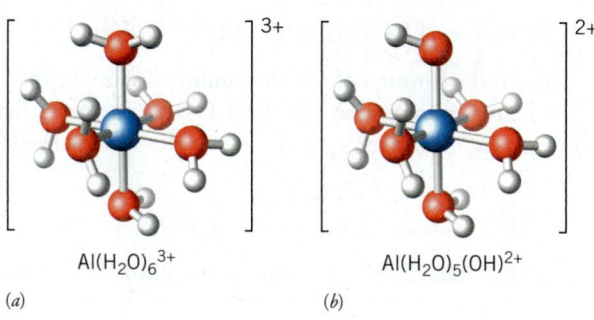

The degree to which metal ions produce acidic solutions depends primarily on two things: the amount of charge on the metal ion and the metal ion's size. As the metal ion's charge increases, the polarizing effect is increased, which thereby favors the release of H$^+$. This means that highly charged metal ions ought to produce solutions that are more acidic than ions of lower charge, and that is generally the case.

The size of the cation also affects its acidity because when the cation is small, the positive charge is highly concentrated. A highly concentrated positive charge is better able to pull electrons from an O—H bond than a positive charge that is more spread out. Therefore, for a given positive charge, the smaller the cation, the more acidic are its solutions.

Both size and amount of charge can be considered simultaneously by referring to a metal ion's *positive charge density*, the ratio of the positive charge to the volume of the cation (its ionic volume):

$$\text{charge density} = \frac{\text{ionic charge}}{\text{ionic volume}}$$

The higher the positive charge density, the more effective the metal ion is at drawing electron density from the O—H bond and the more acidic is the hydrated cation.

Very small cations with large positive charges have large positive charge densities and tend to be quite acidic. An example is the hydrated aluminum ion, the hexahydrate, $Al(H_2O)_6{}^{3+}$ (see Figure 15.4). The ion is acidic in water because of the equilibrium

$$Al(H_2O)_6{}^{3+}(aq) + H_2O \rightleftharpoons Al(H_2O)_5(OH)^{2+}(aq) + H_3O^+(aq)$$

TOOLS

Acidity of metal cations

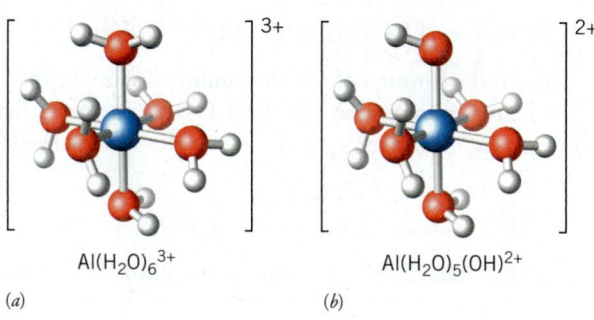

$Al(H_2O)_6{}^{3+}$ $Al(H_2O)_5(OH)^{2+}$

(a)   (b)

Figure 15.4 | Hydrated aluminum ion is a weak proton donor in water. (*a*) The highly charged Al^{3+} ion (blue) binds six water molecules in an aqueous solution to give an octahedral structure with the formula $Al(H_2O)_6{}^{3+}$. (*b*) When it donates a proton, the ion loses one unit of positive charge as a water molecule is transformed into a hydroxide ion, shown at the top of structure (*b*). The resulting ion now has the formula $Al(H_2O)_5(OH)^{2+}$.

The equilibrium, while not actually *strongly* favoring the products, does produce enough hydronium ion so that a 0.1 M solution of $AlCl_3$ in water has about the same concentration of hydronium ions as a 0.1 M solution of acetic acid, roughly $1 \times 10^{-3} M$.

Periodic Trends in the Acidity of Metal Ions

Within the periodic table, atomic size increases down a group and decreases from left to right in a period. Cation sizes follow these same trends, so within a given group, the cation of the metal at the top of the group has the smallest volume and the largest charge density. Therefore, hydrated metal ions at the top of a group in the periodic table are the most acidic within the group.

The cations of the Group 1A metals (Li^+, Na^+, K^+, Rb^+, and Cs^+), with charges of just 1+, have little tendency to increase the H_3O^+ concentration in an aqueous solution.

Within Group 2A, the Be^{2+} cation is very small and has sufficient charge density to cause the hydrated ion to be a weak acid. The other cations of Group 2A (Mg^{2+}, Ca^{2+}, Sr^{2+}, Ba^{2+}) have charge densities that become progressively smaller as we go down the group. Although their hydrated ions all generate some hydronium ion in water, the amount is negligible.

Some transition metal ions are also acidic, especially those with charges of 3+. For example, solutions containing salts of Fe^{3+} and Cr^{3+} tend to be acidic because their ions in solution exist as $Fe(H_2O)_6^{3+}$ and $Cr(H_2O)_6^{3+}$, respectively, and undergo the same ionization reaction as does the $Al(H_2O)_6^{3+}$ ion discussed previously.

Influence of Oxidation Number on the Acidity of Metal Oxides

Not all metal oxides are basic. As the oxidation number (or charge) on a metal ion increases, the metal ion becomes more *acidic*; it becomes a better electron pair acceptor. For metal hydrates, we've seen that this causes electron density to be pulled from the OH bonds of water molecules, causing the hydrate itself to become a weak proton donor. The increasing acidity of metal ions with increasing charge also affects the basicity of their oxides.

When the metal is in a very high oxidation state, the oxide itself becomes acidic. Chromium(VI)oxide, CrO_3, is an example. When dissolved in water, the resulting solution is quite acidic and is called chromic acid. One of the principal species in the solution is H_2CrO_4, which is a strong acid that is more than 95% ionized. The acid forms salts containing the chromate ion, CrO_4^{2-}.

When the positive charge on a metal is small, the oxide tends to be basic, as we've seen for oxides such as Na_2O and CaO. With ions having a 3+ charge, the oxides are less basic and begin to take on acidic properties as well; they become amphoteric. (Recall from Section 15.1 that a substance is *amphoteric* if it is capable of reacting as either an acid or a base.) Aluminum oxide is an example; it can react with both acids and bases. It has basic properties when it dissolves in acid.

$$Al_2O_3(s) + 6H^+(aq) \longrightarrow 2Al^{3+}(aq) + 3H_2O$$

As noted earlier, the hydrated aluminum ion has six water molecules surrounding it, so in an acidic solution the aluminum exists primarily as $Al(H_2O)_6^{3+}$.

On the other hand, aluminum oxide exhibits acidic properties when it dissolves in a base. One way to write the equation for the reaction is

$$Al_2O_3(s) + 2OH^-(aq) \longrightarrow 2AlO_2^-(aq) + H_2O$$

Actually, in basic solution the formula of the aluminum-containing species is more complex than this and is better approximated by $Al(H_2O)_2(OH)_4$. Note that the difference between the two formulas is just the number of water molecules involved in the formation of the ion.

$$Al(H_2O)_2(OH)_4^- \text{ is equivalent to } AlO_2^- + 4H_2O$$

However we write the formula for the aluminum-containing ion, it is an anion, not a cation.

15.6 | Advanced Ceramics and Acid–Base Chemistry

Ceramic materials have a long history, dating to prehistoric times. Examples of pottery about 13,000 years old have been found in several parts of the world. Today, manufactured **ceramics** include common inorganic building materials such as brick, cement, and glass. We find ceramics around the home as porcelain dinnerware, tiles, sinks, toilets, and artistic pottery and figurines. These materials are made from inorganic minerals such as clay, silica (sand), and other silicates (compounds containing anions composed of silicon and oxygen), which are taken from the earth's crust.

In recent times, an entirely new set of materials, generally referred to as *advanced ceramics*, have been prepared by chemists in laboratories and have high-tech applications. We find such ceramics in places we wouldn't expect, such as in cell phones, knives, and diesel engines.

Several methods are used to form ceramics from their raw materials. In many cases, the components of the ceramic are first pulverized to give a very fine powder. The powder is then mixed with water to form a slurry that can be poured into a mold where it sets up into a solid form that has little structural strength. (Alternatively, the fine powder is mixed with a binder and pressed into the desired shape.) The newly formed object is next placed in a kiln where it is heated to a high temperature, often over 1000 °C. At these high temperatures the fine particles stick together by a process called **sintering**, which produces the finished ceramic. Although sintering is sometimes accompanied by partial melting of the fine particles, often the particles remain entirely solid during the process.

There are some problems with making ceramics by the traditional methods just described. Because it is difficult to produce uniform and very small particle sizes by grinding, ceramics made from materials prepared this way often contain small cracks and voids, which adversely affect physical properties such as strength. In addition, the chemical composition of the ceramic cannot be easily and reproducibly controlled by mixing various powdered components.

CHEMISTRY OUTSIDE THE CLASSROOM 15.1

Applications of Advanced Ceramic Materials

There are some properties that all ceramics have in common, and there are others that can be tailored by controlling the ceramic's composition and method of preparation. For example, virtually all ceramic materials have extremely high melting points and are much harder than steel. Some, such as boron nitride, BN, are nearly as hard as diamond, the hardest substance known. Silicon carbide, SiC, is a hard and relatively inexpensive ceramic to make in bulk and has long been used as an abrasive in sandpaper and grinding wheels.

The high strength and relatively low density of ceramic materials make them useful for applications in space. They are used to line the surfaces of rocket engines because they transmit heat very slowly and are able to protect the metals used to construct the engines from melting. The exterior of the space shuttle (Figure 1) is coated with protective ceramic tiles that serve as thermal "heat" shields during re-entry from space. The tiles are made of a low-density porous silica ceramic derived from common sand, SiO_2. They are able to withstand very high temperatures and have a very low thermal conductivity, preventing heat transfer to the body of the shuttle.

High-tech ceramics are relatively new materials, and new applications for them are continually being found. It is interesting to note how varied their uses are. Here are a few examples.

Figure 1 Space shuttle heat shield. Individual heat-shielding ceramic tiles are visible on the surface of the space shuttle *Challenger*. The photo was taken shortly after the spacecraft was completed. In 1986, the *Challenger* was lost in a tragic accident shortly after lift-off. Damage to the tiles during liftoff is believed to be responsible for the loss of the *Columbia* and its crew in February 2003 when the spacecraft broke apart during reentry into the earth's atmosphere.

James L. Amos/Science Source

Figure 2 Titanium nitride. A drill bit with a thin gold-colored coating of titanium nitride, TiN, will retain its sharpness longer than steel drill bits.

Thin ceramic films, deposited by the sol-gel process or other methods, are found on optical surfaces where they serve as anti-reflective coatings and as filters for lighting. Tools such as drill bits are given thin coatings of titanium nitride (TiN) to make them more wear resistant (Figure 2).

Zirconia (ZrO_2) is used to make ceramic golf spikes for golf shoes, as well as portions of hip-joint replacement parts for medicine and knives that stay sharp much longer than steel like the ones in the chapter-opening photograph. All these applications are possible because of the hardness of the ZrO_2 ceramic.

Boron nitride (BN) powder is composed of flat, plate-like crystals that can easily slide over one another, and one of its uses is in cosmetics, where it gives a silky texture to the product (Figure 3). Another boron-containing ceramic, boron carbide, is used along with Kevlar polymer to make bulletproof vests. When a bullet strikes the vest, it is shattered by the ceramic, which absorbs most of the kinetic energy, with residual energy being absorbed by the Kevlar backing.

Silicon nitride, Si_3N_4, is used to make engine components for diesel engines because it is extremely hard and wear resistant, has a high stiffness with low density, and can withstand extremely high temperatures and harsh chemical environments.

Piezoelectric ceramics have the property that they produce an electric potential when their shape is deformed. Conversely, they deform when an electric potential is applied to them. A company makes "smart skis" that incorporate piezoelectric devices that use both properties. Vibrations in the skis are detected by the potential developed when they deform. A potential is then applied to cancel the vibration. This "smart materials" technology was first developed to dampen vibrations in optical components of the "Star Wars" weapons system designed to protect the United States from missile attack.

Figure 3 Ceramics in cosmetics. A variety of cosmetics use boron nitride, BN, as an ingredient.

The Sol-Gel Process

For some types of ceramics, the problems of particle size and uniformity can be avoided by using a method called the **sol-gel process**. The chemistry involved is based on acid–base reactions similar to those discussed earlier in this chapter. The starting materials are metal salts or compounds in which a metal or metalloid (e.g., Si) is bonded to some number of *alkoxide* groups. An **alkoxide** is an anion formed by removing a hydrogen ion from an alcohol.[6] For instance, removal of H^+ from ethanol yields the ethoxide ion.

$$C_2H_5OH \xrightarrow{-H^+} C_2H_5O^-$$

ethanol   ethoxide ion

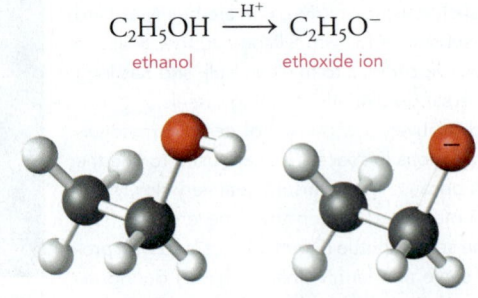

[6] An **alkyl** group is a hydrocarbon fragment derived from an alkane by removal of a hydrogen atom. For example, the methyl group, CH_3—, is derived from methane, CH_4, and is an alkyl group. If we represent an alcohol by the general formula R—OH, where R represents an alkyl group, then the general formula of an alkoxide ion is R—O$^-$.

The ethoxide ion forms salts with metal ions that generally are soluble in ethanol, which should not be surprising considering the similarities between the solute and solvent.

Alcohols are extremely weak acids with very little tendency to lose H^+ ions, so alkoxide ions are very strong bases with a *very strong* affinity for H^+. When placed in water they react immediately and completely by removing H^+ from H_2O to give the alcohol and hydroxide ion.

$$C_2H_5O^- + H_2O \xrightarrow{100\%} C_2H_5OH + OH^-$$

This reaction forms the basis for the start of a sequence of reactions that ultimately yields the ceramic material. Let's use zirconium(IV) ethoxide, $Zr(C_2H_5O)_4$, to illustrate the process. The sequence of reactions begins with the gradual addition of water to an alcohol solution of $Zr(C_2H_5O)_4$, which causes ethoxide ions to react to form ethanol and be replaced by hydroxide ions. This reaction is called **hydrolysis** because it involves a reaction with water. The first step in the process is

■ Typical metals used in the sol-gel process are Si, Ti, Zr, Al, Sn, and Ce, all with high oxidation states.

$$Zr(C_2H_5O)_4 + H_2O \longrightarrow Zr(C_2H_5O)_3OH + C_2H_5OH$$

The large positive charge on the zirconium ion binds the OH^- tightly and polarizes the O—H bond, causing it to weaken. When two $Zr(C_2H_5O)_3OH$ units encounter each other they undergo an acid–base reaction. A proton is transferred from the OH attached to one zirconium ion to the OH attached to another zirconium ion. The result is the simultaneous formation of a water molecule and an oxygen bridge between the two zirconium ions.

As more and more water is added, this process continues, with ethoxide ions changing to ethanol molecules and more and more oxygen bridges being formed between zirconium ions. The result is a network of zirconium atoms bridged by oxygens, which produces extremely small insoluble particles that are essentially oxides with many residual hydroxide ions. These particles are suspended in the alcohol solvent and have a gel-like quality.[7]

Once the sol-gel suspension is formed, it can be used in a number of ways, as indicated in Figure 15.5. It can be deposited on a surface by dip coating to yield very thin ceramic coatings. It can be cast into a mold to produce a semisolid gelatin-like material (wet gel). The wet gel can be dried by evaporation of the solvent to give a porous solid called a **xerogel**. Further heating causes the porous structure of the xerogel to collapse and form a dense ceramic or glass with a uniform structure. If the solvent is removed from the wet gel under supercritical conditions (at a temperature above the critical temperature of the solvent), a very porous and extremely low density solid called an **aerogel** is formed (see the photo in Review Question 1.92). By adjusting the viscosity of the gel suspension, ceramic fibers can be formed. And by precipitation, ultrafine and uniform ceramic powders are formed. What is amazing is that all these different forms can be made from the same material depending on how the suspension is handled.

■ Xerogels have very large internal surface areas because of their fine porous structure. Some have been found to be useful catalysts.

---

[7] The particle size is larger than one would find in a true solution, but smaller than in a typical precipitate. This type of mixture is called a *colloid*, and a colloid composed of tiny solid particles suspended in a liquid medium is called a *sol*. That's where part of the name *sol-gel* comes from.

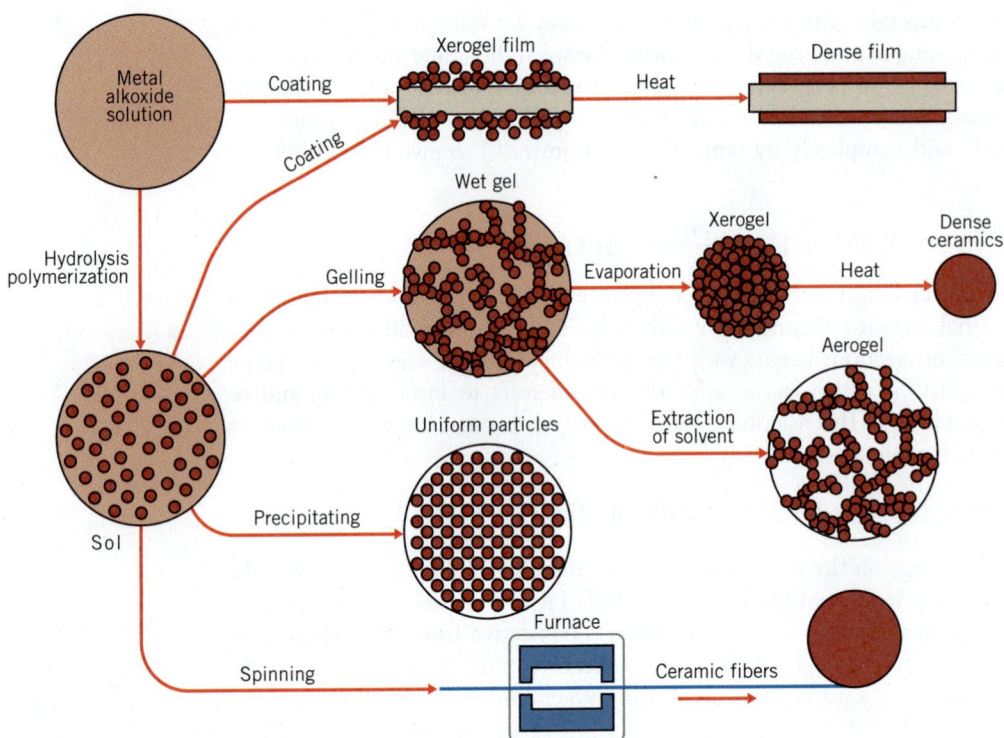

Figure 15.5 | Sol-gel technologies and their products.

Summary

Organized by Learning Objective

Identify Brønsted–Lowry acids and bases and explain the terms conjugate acid–base pair and amphoteric substances

A **Brønsted–Lowry acid** (often simply referred to as a **Brønsted acid**) is a proton donor; a **Brønsted–Lowry base** (or simply **Brønsted base**) is a proton acceptor. According to the Brønsted–Lowry approach, an acid–base reaction is a proton transfer event. In an equilibrium involving a Brønsted acid and base, there are two **conjugate acid–base pairs**. The members of any given acid–base pair differ from each other by only one H^+, with the acid having one more H^+ than the base. A substance that can be either an acid or a base, depending on the nature of the other reactant, is **amphoteric** or, with emphasis on proton-transfer reactions, **amphiprotic**.

Compare the strengths of Brønsted–Lowry acids and bases, how they are classified as strong or weak, and the strengths of conjugate acid–base pairs

A Brønsted–Lowry acid–base equation is composed of two sets of conjugate acid–base pairs. The acid of one acid–base pair is a reactant and the base of the other acid–base pair is the second reactant, and their counterparts are products. If the position of equilibrium lies with the reactants, the reactants are the weak members of the two conjugate acid–base pairs. The reverse is true if the position of equilibrium lies with the product. This information also tells us which of the two conjugate acids is stronger. The conjugate acid that is a reactant is weaker than the

conjugate acid that is a product if the position lies with the reactants. The reverse is true if the position of equilibrium lies with the reactants. We use the same reasoning to predict that the reactant base is the stronger of the two bases if the position of equilibrium lies with the products.

Using the periodic table, describe the trends in the strengths of binary acids and oxoacids

Binary acids contain only hydrogen and another nonmetal. Their strengths increase from top to bottom within a group and left to right across a period. **Oxoacids**, which contain oxygen atoms in addition to hydrogen and another element, increase in strength as the number of oxygen atoms on the same central atom increases. Delocalization of negative charge enhances the stability of oxoacid anions, making them weaker bases and, as a result, their conjugate acids are correspondingly stronger. Oxoacids having the same number of oxygens generally increase in strength as the central atom moves from bottom to top within a group and from left to right across a period.

Define Lewis acids and bases and compare them to the Arrhenius and Brønsted–Lowry definitions

A **Lewis acid** accepts a pair of electrons from a **Lewis base** in the formation of a coordinate covalent bond. Lewis bases often have filled valence shells and must have at least one unshared electron pair. Lewis acids have an incomplete valence shell that can accept an electron pair, have double bonds that allow electron pairs to be moved to make room for an incoming electron pair from a Lewis base, or have valence shells that can accept more than an octet of electrons.

Using the periodic table, describe which elements are most likely to form acids or bases

Oxides of metals are basic anhydrides when the charge on the ion is small. Those of the Group 1A and 2A metals neutralize acids and tend to react with water to form soluble metal hydroxides. The hydrates of metal ions tend to be proton donors when the positive charge density of the metal ion is itself sufficiently high, as it is when the ion has a 3+ charge. The small beryllium ion, Be^{2+}, forms a weakly acidic hydrated ion. Metal oxides become more acidic as the oxidation number of the metal

becomes larger. Aluminum oxide is amphoteric, dissolving in both acids and bases. Chromium(VI) oxide is acidic, forming chromic acid when it dissolves in water.

Describe how the production of advanced ceramics depends on acid–base chemistry, including the sol-gel process

Common ceramic materials are made from inorganic minerals obtained from the earth's crust. They are ground to fine powders, formed into shapes, and fired at high temperatures where **sintering** occurs, causing the fine particles to stick together to give a rigid solid. The **sol-gel process** uses the reaction of water with metal **alkoxides** to form extremely small, gel-like particles suspended in alcohol that can be used in a variety of ways to give coatings, **xerogels**, **aerogels**, fibers, and ultrafine powders for making ceramics.

Tools for Problem Solving
The following tools were introduced in this chapter. Study them carefully so you can select the appropriate tool when needed.

Brønsted–Lowry definitions (Section 15.1)
Acids are proton donors and bases are proton acceptors.

Conjugate acid–base pairs (Section 15.1)
Every conjugate acid–base pair consists of an acid plus a base with one less proton (H^+).

Conjugate acid–base equilibria (Section 15.1)
The conjugate acid is always on the opposite side of an equation from its conjugate base, and the Brønsted–Lowry acid–base equilibrium is typically written with two sets of conjugate acid–base pairs.

Acid–base strength and the position of equilibrium (Section 15.2)
In an acid–base equilibrium, the equilibrium favors the weaker acid and base. You can also use the position of equilibrium to establish the relative strengths of the acids or bases in an equilibrium.

Reciprocal relationship in acid–base strengths (Section 15.2)
The stronger an acid, the weaker is its conjugate base. This tool can help establish the position of equilibrium when dealing with more than one acid–base pair.

Periodic trends in strengths of binary acids (Section 15.3)
The acidities of H—X bonds increases from left to right across a period and from top to bottom in a group.

Trends in the strengths of oxoacids (Section 15.3)
As the electronegativity of the central atom increases, the acidity of the oxoacid increases, as long as the number of oxygen atoms on the atom remains the same. For the same central atom, as the number of oxygen atoms on the central atom increases, the strength of the oxoacid increases.

Lewis acid–base definitions (Section 15.4)
A Lewis acid accepts a pair of electrons and a Lewis base donates a pair of electrons. The bond formed in the acid–base neutralization reaction is a coordinate covalent bond.

Acidity of metal cations (Section 15.5)
Metal ions become better at polarizing H_2O molecules as their charge increases and their size decreases. Use this to compare the relative abilities of different metal ions to produce acidic solutions in water.

Review Questions

Brønsted–Lowry Acids and Bases

15.1 How is a Brønsted–Lowry acid defined? How is a Brønsted–Lowry base defined? How do these definitions differ from the Arrhenius definition of acids and bases?

15.2 How are the formulas of the members of a conjugate acid–base pair related to each other? Within the pair, how can you tell which is the acid?

15.3 Is H_2SO_4 the conjugate acid of SO_4^{2-}? Explain your answer.

15.4 What is meant by the term *amphoteric*? Give two chemical equations that illustrate the amphoteric nature of water.

15.5 Define the term amphiprotic.

Strengths of Brønsted–Lowry Acids and Bases

15.6 Explain why H_3O^+ is the strongest acid found in water. In pure sulfuric acid, H_2SO_4, what is the strongest acid found? Explain why OH^- is the strongest base found in water.

15.7 In an acid–base equilibrium, can the stronger acid be on the same side of the arrow as the weaker base? Why or why not?

15.8 The position of equilibrium in the equation below lies far to the left. Identify the conjugate acid–base pairs. Which of the two acids is stronger?

$$HOCl(aq) + H_2O \rightleftharpoons H_3O^+(aq) + OCl^-(aq)$$

15.9 Consider the following: CO_3^{2-} is a weaker base than hydroxide ion, and HCO_3^- is a stronger acid than water. In the equation below, would the position of equilibrium lie to the left or to the right? Justify your answer.

$$CO_3^{2-}(aq) + H_2O \rightleftharpoons HCO_3^-(aq) + OH^-(aq)$$

15.10 Acetic acid, $HC_2H_3O_2$, is a weaker acid than nitrous acid, HNO_2. How do the strengths of the bases $C_2H_3O^-$ and NO^- compare?

15.11 Nitric acid, HNO_3, is a very strong acid. It is 100% ionized in water. In the reaction below, would the position of equilibrium lie to the left or to the right?

$$NO_3^-(aq) + H_2O \rightleftharpoons HNO_3(aq) + OH^-(aq)$$

15.12 $HClO_4$ is a stronger proton donor than HNO_3, but in water both acids appear to be of equal strength; they are both 100% ionized. Why is this so? What solvent

property would be necessary in order to distinguish between the acidities of these two Brønsted–Lowry acids?

★**15.13** Formic acid, $HCHO_2$, and acetic acid, $HC_2H_3O_2$, are classified as weak acids, but in water $HCHO_2$ is more fully ionized than $HC_2H_3O_2$. However, if we use liquid ammonia as a solvent for these acids, they both appear to be of equal strengths; both are 100% ionized in liquid ammonia. Explain why this is so.

Periodic Trends in the Strength of Acids

15.14 Explain how the strength of the bond between the molecule and a hydrogen atom defines the strength of an acid.

15.15 What are the two properties that have an effect on the strength of the bond between the molecule and the hydrogen atom.

15.16 Within the periodic table, how do the strengths of the binary acids vary from left to right across a period? How do they vary from top to bottom within a group?

15.17 Explain why H_2S is a stronger acid than H_2O.

15.18 Within the periodic table, how do the strengths of the oxoacids vary from left to right across a period? How do they vary from top to bottom within a group?

15.19 Explain why nitric acid is a stronger acid than nitrous acid.

$$HO-NO_2 \qquad HO-NO$$
$$\text{nitric acid} \qquad \text{nitrous acid}$$

15.20 Astatine, atomic number 85, is radioactive and does not occur in appreciable amounts in nature. On the basis of what you have learned in this chapter, answer the following.

  (a) How would the acid strength of HAt compare with that of HI?

  (b) How would the acid strength of $HAtO_3$ compare with that of $HBrO_3$?

15.21 Which is the stronger Brønsted–Lowry base, $CH_3CH_2O^-$ or $CH_3CH_2S^-$? What is the basis for your selection?

15.22 Explain why $HClO_4$ is a stronger acid than H_2SeO_4.

15.23 Which of the molecules below is expected to be the stronger Brønsted–Lowry acid? Why?

15.24 Which of the molecules below has the stronger conjugate base? Explain your choice.

Lewis Acids and Bases

15.25 Define Lewis acid and Lewis base.

15.26 In terms of atomic orbitals, what is required for Lewis acids and Lewis bases?

15.27 Explain why the addition of a proton to a water molecule to give H_3O^+ can be considered a Lewis acid–base reaction.

15.28 Methylamine has the formula CH_3NH_2 and the structure

Use Lewis structures to illustrate the reaction of methylamine with boron trifluoride.

15.29 Use Lewis structures to show the Lewis acid–base reaction between SO_3 and H_2O to give H_2SO_4. Identify the Lewis acid and the Lewis base in the reaction.

15.30 Explain why the oxide ion, O^{2-}, can function as a Lewis base but not as a Lewis acid.

15.31 The molecule SbF_5 is able to function as a Lewis acid. Explain why it is able to be a Lewis acid.

15.32 In the reaction of calcium with oxygen to form calcium oxide, each calcium atom gives a pair of electrons to an oxygen atom. Why isn't this viewed as a Lewis acid–base reaction?

Acid–Base Properties of the Elements and Their Oxides

15.33 Why do nonmetals oxides tend to form acids and metal oxides tend to form bases?

15.34 Suppose that a new element was discovered. Based on the discussions in this chapter, what properties (both physical and chemical) might be used to classify the element as a metal or a nonmetal?

15.35 If the oxide of an element dissolves in water to give an acidic solution, is the element more likely to be a metal or a nonmetal?

15.36 Boric acid is very poisonous and is used in ant bait (to kill ant colonies) and to poison cockroaches. It is a weak acid with a formula often written as H_3BO_3, although it is better written as $B(OH)_3$. It functions not as a Brønsted–Lowry acid, but as a Lewis acid. Using Lewis structures, show how $B(OH)_3$ can bind to a water molecule and cause the resulting product to be a weak Brønsted acid.

15.37 Why do small highly charged hydrated metal ions form acidic solutions?

15.38 Many chromium salts crystallize as hydrates containing the ion $Cr(H_2O)_6{}^{3+}$. Solutions of these salts tend to be acidic. Explain why.

15.39 Which ion is expected to give the more acidic solution, Fe^{2+} or Fe^{3+}? Why?

15.40 Ions of the alkali metals have little effect on the acidity of a solution. Why?

15.41 What acid is formed when the following oxides react with water? **(a)** SO_3 **(b)** CO_2 **(c)** P_4O_{10}

15.42 Consider the following oxides: CrO, Cr_2O_3, CrO_3.

(a) Which is most acidic?

(b) Which is most basic?

(c) Which is most likely to be amphoteric?

15.43 Write equations for the reaction of Al_2O_3 with **(a)** a strong acid, and **(b)** a strong base.

Advanced Ceramics and Acid–Base Chemistry

15.44 What is a ceramic? How are ceramics formed from their raw materials?

15.45 What is sintering?

15.46 What type of reaction is used in the sol-gel process? What small molecule is formed in the process for making the sol-gel suspension?

15.47 How does the sol-gel process work and how does it make uniform particle sizes?

15.48 How can a xerogel be converted into an aerogel? Why does an aerogel have such a low density?

Review Problems

Brønsted–Lowry Acids and Bases

15.49 Write the formula for the conjugate acid of each of the following.

(a) F^- **(b)** C_5H_5N **(c)** $HCrO_4{}^-$ **(d)** CN^-

15.50 Write the formula for the conjugate acid of each of the following.

(a) N_2H_4 **(b)** O^{2-} **(c)** $SO_4{}^{2-}$ **(d)** $H_2PO_4{}^-$

15.51 Write the formula for the conjugate base of each of the following.

(a) NH_2OH **(b)** HCN **(c)** HNO_2 **(d)** $HPO_4{}^{2-}$

15.52 Write the formula for the conjugate base of each of the following.

(a) $HSO_3{}^-$ **(b)** H_5IO_6 **(c)** HNO_3 **(d)** H_2S

15.53 Identify the conjugate acid–base pairs in the following reactions.

(a) $HNO_3 + N_2H_4 \rightleftharpoons NO_3^- + N_2H_5^+$

(b) $NH_3 + N_2H_5^+ \rightleftharpoons NH_4^+ + N_2H_4$

(c) $H_2PO_4^- + CO_3^{2-} \rightleftharpoons HPO_4^{2-} + HCO_3^-$

(d) $HIO_3 + HC_2O_4^- \rightleftharpoons IO_3^- + H_2C_2O_4$

15.54 Identify the conjugate acid–base pairs in the following reactions.

(a) $HSO_4^- + SO_3^{2-} \rightleftharpoons HSO_3^- + SO_4^{2-}$

(b) $S^{2-} + H_2O \rightleftharpoons HS^- + OH^-$

(c) $CN^- + H_3O^+ \rightleftharpoons HCN + H_2O$

(d) $H_2Se + H_2O \rightleftharpoons HSe^- + H_3O^+$

Periodic Trends in the Strengths of Acids

15.55 Choose the stronger acid: (a) HBr or HCl, (b) H_2O or HF, (c) H_2S or HBr. Give your reasons.

15.56 Choose the stronger acid: (a) H_2S or H_2Se, (b) H_2Te or HI, (c) PH_3 or NH_3. Give your reasons.

15.57 Choose the stronger acid and give your reason: (a) HOCl or $HClO_2$, (b) H_2SeO_4 or H_2SeO_3.　•

15.58 Choose the stronger acid and give your reason: (a) HIO_3 or HIO_4, (b) H_3AsO_4 or H_3AsO_3.

15.59 Choose the stronger acid: (a) $HClO_3$ or HIO_3, (b) HIO_2 or $HClO_3$, (c) H_2SeO_3 or $HBrO_4$. Give your reasons.

15.60 Choose the stronger acid: (a) H_3AsO_4 or H_3PO_4, (b) H_2CO_3 or HNO_3, (c) H_2SeO_4 or $HClO_4$. Give your reasons.

Lewis Acids and Bases

15.61 Use Lewis symbols to diagram the reaction

$$NH_2^- + H^+ \longrightarrow NH_3$$

Identify the Lewis acid and Lewis base in the reaction.

15.62 Use Lewis symbols to diagram the reaction

$$BF_3 + F^- \longrightarrow BF_4^-$$

15.63 Beryllium chloride, $BeCl_2$, exists in the solid as a polymer composed of long chains of $BeCl_2$ units arranged as follows.

The formula of the chain can be represented as $(BeCl_2)_n$, where n is a large number. Use Lewis structures to show how the reaction $nBeCl_2 \longrightarrow (BeCl_2)_n$ is a Lewis acid–base reaction.

15.64 Aluminum chloride, $AlCl_3$, forms molecules with itself with the formula Al_2Cl_6. Its structure is

Use Lewis structures to show how the reaction $2AlCl_3 \longrightarrow Al_2Cl_6$ is a Lewis acid–base reaction.

15.65 Use Lewis structures to diagram the reaction

$$CO_2 + H_2O \longrightarrow H_2CO_3$$

Identify the Lewis acid and Lewis base in the reaction.

15.66 Use Lewis structures to diagram the reaction

$$CO_2 + O^{2-} \longrightarrow CO_3^{2-}$$

Identify the Lewis acid and Lewis base in the reaction.

15.67 Use Lewis structures to show how the following reaction involves the transfer of a Lewis base from one Lewis acid to another. Identify the Lewis acid and Lewis base.

$$CO_3^{2-} + SO_2 \longrightarrow CO_2 + SO_3^{2-}$$

15.68 Use Lewis structures to show how the following reaction can be viewed as the displacement of one Lewis base by another Lewis base from a Lewis acid. Identify the two Lewis bases and the Lewis acid.

$$NH_2^- + H_2O \longrightarrow NH_3 + OH^-$$

Acid–Base Properties of Elements and Their Oxides

15.69 The ion $Cr(H_2O)_6^{3+}$ is weakly acidic. Write an equation showing its behavior as a Brønsted–Lowry acid in water.

15.70 The ion $Hg(H_2O)_6^{2+}$ acts as an acid. Write an equation showing its behavior as a Brønsted–Lowry acid in water.

15.71 The compound $Mg(OH)_2$ is basic, but $Si(OH)_4$ is an acid (silicic acid). Why?

15.72 Explain why CaO is an anhydrous base and SO_2 is an anhydrous acid.

| Additional Exercises

15.73 What is the formula of the conjugate acid of dimethyl-amine, $(CH_3)_2NH$? What is the formula of its conjugate base? Trimethylamine has a similar structure to dimethyl-amine; why does it not have a conjugate base?

15.74 Using liquid ammonia as a solvent, sodium amide reacts with ammonium chloride in an acid–base neutralization

reaction. Assuming that these compounds are completely dissociated in liquid ammonia, write molecular, ionic, and net ionic equations for the reaction. Which substance is the acid and which is the base?

15.75 In liquid SO_2 as a solvent, $SOCl_2$ reacts with Na_2SO_3 in a reaction that can be classified as neutralization (acid

plus base yields solvent plus a salt). Write an equation for the reaction. Which solute is the acid and which is the base? Describe what is happening in terms of the Lewis definition of acids and bases.

*15.76 The following space-filling model depicts the structure of a compound called ethanamide.

How would you expect its base strength to compare with that of ammonia? Justify your answer.

15.77 Which of the following compounds is the stronger base? Explain.

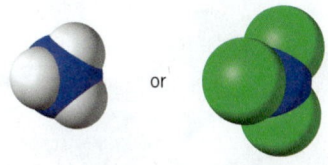

15.78 Which of the two molecules below is the stronger Brønsted–Lowry acid? Why?

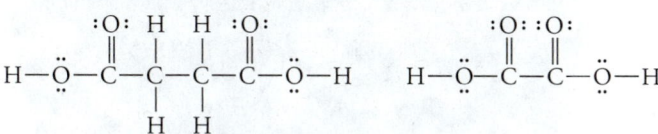

15.79 Write equations that illustrate the amphiprotic nature of the bicarbonate ion.

Multi-Concept Problems

*15.86 On the basis of the VSEPR theory, sketch the structure of the $Al(H_2O)_6^{3+}$ ion. Sketch the two possible structures of the ion formed when a proton is removed from four of the six water molecules. Which is the more likely of the two structures?

Exercises in Critical Thinking

15.88 Are all Arrhenius acids Brønsted–Lowry acids? Are they all Lewis acids? Give examples if they are not. Give a reasoned explanation if they are.

15.89 How could you determine whether HBr is a stronger acid than HI?

15.90 Alcohols are organic compounds that have an —OH group. Are alcohols acids or bases? Sugars have an —OH group on almost every carbon atom; are they acids or bases? Phenol is a benzene ring with an —OH

15.80 Hydrogen peroxide is a stronger Brønsted–Lowry acid than water.

(a) Explain why this is so.

(b) Is an aqueous solution of hydrogen peroxide acidic or basic?

*15.81 Sodium hydroxide, NaOH, is basic. Aluminum hydroxide, $Al(H_2O)_3(OH)_3$, is amphoteric. The compound O_3ClOH (usually written $HClO_4$) is acidic. Considering that each compound contains one or more OH groups, why are their acid–base properties so different?

15.82 Hydrazine, N_2H_4, is a weaker Brønsted–Lowry base than ammonia. In the following reaction, would the position of equilibrium lie to the left or to the right? Justify your answer.

$$N_2H_5^+ + NH_3 \rightleftharpoons N_2H_4 + NH_4^+$$

15.83 Identify the two Brønsted–Lowry acids and two bases in the reaction

$$NH_2OH + CH_3N^+ \rightleftharpoons NH_3OH^+ + CH_3NH_2$$

15.84 In the reaction in the preceding exercise, the position of equilibrium lies to left. Identify the stronger acid in each of the conjugate pairs in the reaction.

*15.85 How would you expect the degree of ionization of $HClO_3$ to compare in the solvents $H_2O(l)$ and $HF(l)$? The reactions are

$$HClO_3 + H_2O \rightleftharpoons H_3O^+ + ClO_3^-$$
$$HClO_3 + HF \rightleftharpoons H_2F^+ + ClO_3^-$$

Justify your answer.

15.87 A mixture is prepared containing 0.10 M of each of the following: arsenic acid, sodium arsenate, arsenous acid, and sodium arsenite. What are the formulas and chemical structures of the predominant arsenic-containing species present in the solution at equilibrium? Write the equation for the equilibrium.

group in place of one hydrogen atom. Is phenol an acid or base?

15.91 Acid rain, acid mine runoff, and acid leaching of metals from soils are important environmental considerations. What do these topics refer to and how do they affect you as a person?

15.92 Using just Figure 7.30, find the five most acidic cations. Explain your methods. Where would you look to confirm your conclusions?

16

Acid–Base Equilibria in Aqueous Solutions

"Have a Coke and a smile" was the advertisement for Coca-Cola in the 1980s. One of the ironies of that advertisement was that Coca-Cola contains phosphoric acid, which in turn is used by dentists to etch teeth to allow dental appliances such as braces to adhere to the surface.

Paul Johnson/Getty Images

Foods tend to be acidic while cleaners tend to be basic. Coca-Cola contains two acids, carbonic acid and phosphoric acid, which contribute to the tartness and taste of the soda. In this chapter, we will discuss acids and bases, and expand the concepts of the chemical properties of acids and bases that we introduced in Chapter 15 and apply the principles of chemical equilibrium from Chapter 14 to acids and bases. Using the principles of molecular polarity and the concept of delocalization of electrons, we were able to qualitatively compare the strengths of acids and bases. Our goal in this chapter is to bring these concepts together as we examine the quantitative aspects of acid–base chemistry. In particular we are interested in acids, bases, and their mixtures as they affect the concentration of hydronium ions in solution through the concept of pH.

The principles we develop here have applications in laboratories that investigate environmental, forensic, and biochemical problems. High-tech laboratories interested in materials science and nanotechnology often use the principles of acid–base equilibria. The consumer-oriented industries that make products such as cosmetics, foods, beverages, and cleaning chemicals all employ chemists who know that control of pH is very important in safe and effective consumer products.

This Chapter in Context

LEARNING OBJECTIVES

After reading this chapter, you should be able to:

- define pH and explain the use of "p" notation
- explain how to determine the pH of strong acids or bases in aqueous solution
- write expressions for the acid ionization constant, K_a, and base ionization constant, K_b, and explain how they are related to each other
- describe how to determine acid and base ionization constants from experimental data
- determine equilibrium concentrations and pH for weak acid or base solutions
- explain how a salt solution can be acidic or basic
- describe what a buffer solution is, how it works, and how to calculate pH changes in a buffer
- define polyprotic acids and describe how to calculate concentrations of all species in a solution of a polyprotic acid
- draw and explain titration curves for reactions of strong or weak acids and bases

16.1 | Water, pH, and "p" Notation

There are literally thousands of weak acids and bases, and their strengths vary widely. Both the acetic acid in vinegar and the carbonic acid in Coca-Cola, for example, are classified as weak. Yet carbonic acid is only about 3% as strong an acid as acetic acid. To study such differences quantitatively, we need to explore acid–base equilibria in greater depth. This requires that we first discuss a particularly important equilibrium that exists in *all* aqueous solutions—namely, the ionization of water itself.

Using sensitive instruments, pure water is observed to weakly conduct electricity, indicating the presence of very small concentrations of ions. They arise from the very slight self-ionization, or *autoionization,* of water itself, represented by the following equilibrium equation.

$$H_2O + H_2O \rightleftharpoons H_3O^+ + OH^-$$

The forward reaction requires a collision of two H_2O molecules.

■ We've omitted the usual (*aq*) following the symbols for ions in water for the sake of clarity.

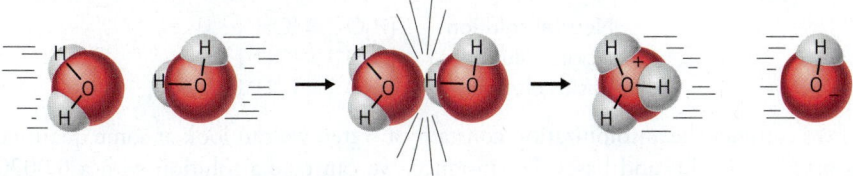

TOOLS

Ion product constant of water, K_w

Its equilibrium law, following the procedures developed in Section 14.4, is

$$[H_3O^+][OH^-] = K_w \qquad (16.1)$$

Since water is a pure liquid, with a constant 55.6 molar concentration, it does not appear in this equilibrium law. Because of the importance of the autoionization equilibrium, its equilibrium constant is given the special symbol, K_w, that is called the **ion product constant of water**.

Often, for convenience, we omit the water molecule that carries the hydrogen ion and write H^+ in place of H_3O^+. The equilibrium equation for the autoionization of water then simplifies to

$$H_2O \rightleftharpoons H^+ + OH^-$$

The equation for K_w based on this is likewise simplified.

$$[H^+][OH^-] = K_w \qquad (16.2)$$

In pure water, the concentrations of H^+ and OH^- produced by the autoionization are equal because the ions are formed in equal numbers. It's been found that the concentrations have the following values at 25 °C:

$$[H^+] = [OH^-] = 1.0 \times 10^{-7} \text{ mol L}^{-1}$$

Therefore, at 25 °C,

$$K_w = (1.0 \times 10^{-7}) \times (1.0 \times 10^{-7})$$
$$K_w = 1.0 \times 10^{-14} \qquad (16.3)$$

TABLE 16.1	K_w at Various Temperatures
Temperature (°C)	K_w
0	1.1×10^{-15}
10	2.9×10^{-15}
20	6.8×10^{-15}
25	1.0×10^{-14}
30	1.5×10^{-14}
37[a]	2.5×10^{-14}
40	2.9×10^{-14}
50	5.5×10^{-14}
60	9.6×10^{-14}

[a]Normal body temperature

As with other equilibrium constants, the value of K_w varies with temperature (see Table 16.1). Unless stated otherwise, we will deal with systems at 25 °C.

Effect of Solutes on [H⁺] and [OH⁻]

Water's autoionization takes place in *any* aqueous solution, and because of the effects of other solutes, the molar concentrations of H^+ and OH^- may not be equal. Nevertheless, their product, K_w, is always the same. Although Equations 16.1–16.3 were derived for pure water, they also apply to dilute aqueous solutions. The significance of this must be emphasized. *In any aqueous solution, the product of* [H⁺] *and* [OH⁻] *equals* K_w, although the two molar concentrations may not actually equal each other.

Criteria for Acidic, Basic, and Neutral Solutions

One of the consequences of the autoionization of water is that *in any aqueous solution, there are always both H_3O^+ and OH^- ions, regardless of what solutes are present.* This means that in a solution of the acid HCl there is some OH^-, and in a solution of the base NaOH, there is some H_3O^+. We call a solution acidic or basic depending on which ion has the largest concentration.

A **neutral solution** is one in which the molar concentrations of H_3O^+ and OH^- are equal. An **acidic solution** is one in which some solute has made the molar concentration of H_3O^+ greater than that of OH^-. On the other hand, a **basic solution** exists when the molar concentration of OH^- exceeds that of H_3O^+.

Neutral solution	$[H_3O^+] = [OH^-]$
Acidic solution	$[H_3O^+] > [OH^-]$
Basic solution	$[H_3O^+] < [OH^-]$

If we consider the autoionization constant of water, we can look at some quantitative relationships of acids and bases. For instance, we can take a solution with a 0.0020 *M* concentration of H_3O^+, and calculate the extremely small OH^- concentration that exists in the same solution. Rearranging Equation 16.3, we get

$$[OH^-] = \frac{K_w}{[H_3O^+]} = \frac{1.0 \times 10^{-14}}{2.0 \times 10^{-3}} = 5.0 \times 10^{-12}$$

A similar calculation allows us to determine the hydronium ion concentration if we know the hydroxide ion concentration.

Practice Exercise 16.1

Commercial, concentrated, hydrochloric acid has a concentration of 12 moles per liter. If the molarity of hydronium ions is assumed to also be 12 M, what is the concentration of hydroxide ions? (*Hint:* Use the relationships just discussed.)

Practice Exercise 16.2

An aqueous solution of sodium bicarbonate, $NaHCO_3$, has a molar concentration of hydroxide ions of 7.8×10^{-6} M. What is the molar concentration of hydrogen ions? Is the solution acidic, basic, or neutral?

The pH Concept

In most solutions of weak acids and bases, the molar concentrations of H^+ and OH^- are very small, like those we just discussed. Writing and comparing two exponential values is not always easy. A Danish chemist, S. P. L. Sørenson (1868–1939), suggested an easier approach.

To make comparisons of small values of $[H^+]$ easier, Sørenson defined a quantity that he called the **pH** of the solution as follows.

$$pH = -\log[H^+] \tag{16.4}[1]$$

The properties of logarithms let us rearrange Equation 16.4 as follows.

$$[H^+] = 10^{-pH} \tag{16.5}$$

Equation 16.4 can be used to calculate the pH of a solution if its molar concentration of H^+ is known. On the other hand, if the pH is known and we wish to calculate the molar concentration of hydrogen ions, we apply Equation 16.5. For example, if the hydrogen ion concentration, $[H^+]$, is 3.6×10^{-4}, we take the logarithm of this value (which is -3.44) and change its sign so that the result is a pH of 3.44. With digital calculators, it's as simple as pressing the log key and *then* changing the sign. Determining the hydrogen ion concentration using Equation 16.5 requires that we change the sign of the pH and then take the antilogarithm. For instance, if we are given a pH of 12.58 we first change it to -12.58. We then take the antilogarithm according to the instructions that came with your calculator. In this case the result is $2.6 \times 10^{-13} = [H^+]$.

Acidic, Basic, and Neutral Solutions

One meaning attached to pH is that it is a measure of the acidity of a solution. Hence, we may define acidic, basic, and neutral in terms of pH values. At 25 °C, in pure water, or in any solution that is neutral,

$$[H^+] = [OH^-] = 1.0 \times 10^{-7} \, M$$

Therefore, by Equation 16.4, the pH of a neutral solution at 25 °C is 7.00.[2]

An acidic solution is one in which $[H^+]$ is larger than 10^{-7} M and so has a pH less than 7.00. Thus, as a solution's acidity increases, its pH decreases.

TOOLS

Definition of pH

■ Calculating pH:

$$\begin{aligned} pH &= -\log[H^+] \\ &= -\log(3.6 \times 10^{-4}) \\ &= -(-3.44) \\ &= 3.44 \end{aligned}$$

Calculating $[H^+]$:

$$\begin{aligned} -\log[H^+] &= pH \\ -\log[H^+] &= 12.58 \\ \log[H^+] &= -12.58 \\ [H^+] &= \text{antilog}(-12.58) \\ [H^+] &= 10^{-12.58} \\ [H^+] &= 2.6 \times 10^{-13} \, M \end{aligned}$$

■ If you still have questions about these calculations, see the mathematics review in Appendix A.

■ Neutral pH is 7.00 only for 25 °C. At other temperatures a neutral solution still has $[H^+] = [OH^-]$, but because K_w changes with temperature, so do the values of $[H^+]$ and $[OH^-]$. This causes the pH of the neutral solution to differ from 7.00. For example, at normal body temperature (37 °C), a neutral solution has $[H^+] = [OH^-] = 1.6 \times 10^{-7}$, so the pH = 6.80.

[1] In this equation and similar ones, like Equation 16.6, the logarithm of only the numerical part of the bracketed term is taken. The physical units must be mol L^{-1}, but they are set aside for the calculation.
[2] Recall that the rule for significant figures in logarithms is that the number of decimal places in the logarithm of a number equals the number of significant figures in the number. For example, 3.2×10^{-5} has just two significant figures. The logarithm of this number, displayed on a pocket calculator, is -4.494850022. We write the logarithm, when correctly rounded, as -4.49.

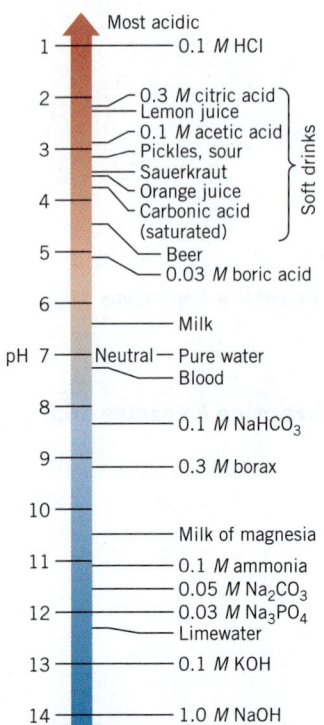

Figure 16.1 | The pH scale.

Defining the p-function, pX

Relating pH and pOH to pK_w

■ Calculating pH and pOH:

$$pH = -\log(6.3 \times 10^{-5})$$
$$= -(-4.20)$$
$$= 4.20$$
$$pK_w = 14.00 = 4.20 + pOH$$
$$pOH = 14.00 - 4.20$$
$$= 9.80$$

A basic solution is one in which the value of $[H^+]$ is less than 10^{-7} M and so has a pH that is greater than 7.00. As a solution's acidity decreases, its pH increases. These simple relationships may be summarized as follows. At 25 °C:

pH = 7.00	Neutral solution
pH < 7.00	Acidic solution
pH > 7.00	Basic solution

The pH values for some common substances are given on a pH scale in Figure 16.1.

One of the deceptive features of pH is how much the hydrogen ion concentration changes with a relatively small change in pH. Because of the logarithmic scale, a change of one pH unit corresponds to a ten-fold change in the hydrogen ion concentration. For example, at pH 6 the hydrogen ion concentration is 1×10^{-6} mol L^{-1}. At pH 5, it's 1×10^{-5} mol L^{-1}, or ten times greater. That's a major difference.

Solutions having hydrogen ion concentrations greater than 1 M can be prepared and they have negative pH values. However, at such high concentrations it is usually easier to simply write the molarity of the hydrogen ions. For example, in a solution where the value of $[H^+]$ is 2.00 M, the "2.00" is a simple enough number; we can't improve it further by using its corresponding pH value. When $[H^+]$ is 2.00 M, the pH, calculated by Equation 16.4, would be $-\log(2.00)$ or -0.301. There is nothing wrong with negative pH values, but they offer no advantage over just writing the actual molarity.

"p" Notation

The logarithmic definition of pH, Equation 16.4, has proved to be so useful that it has been adapted to quantities other than $[H^+]$. Thus, for any quantity X, we may define a term, pX, in the following way

$$pX = -\log X \qquad (16.6)$$

For example, to express small concentrations of hydroxide ion, we can define the **pOH** of a solution as

$$pOH = -\log[OH^-]$$

Similarly, for K_w we can define **pK_w** as follows.

$$pK_w = -\log K_w$$

The numerical value of pK_w at 25 °C equals $-\log(1.0 \times 10^{-14})$ or $-(-14.00)$. Thus,

$$pK_w = 14.00 \qquad \text{(at 25 °C)}$$

A useful relationship among pH, pOH, and pK_w can be derived by taking the logarithm of all terms in Equation 16.1 or 16.2 and changing the sign to obtain Equation 16.7:

$$pH + pOH = pK_w = 14.00 \qquad \text{(at 25 °C)} \qquad (16.7)$$

This tells us that in an aqueous solution of any solute at 25 °C, the sum of pH and pOH is 14.00.

pH Calculations

Let's now look at some pH calculations using Equations 16.4 and 16.7. These are calculations you will be performing often in the rest of this chapter.

If we know the $[H^+]$ of a solution—let's use 6.3×10^{-5} M as an example—we can calculate the pH by taking the logarithm and changing the sign. The pH we calculate should be 4.20. We now have two ways to determine the pOH. Using Equation 16.2 allows us to determine $[OH^-]$, and then we can take the logarithm and change the sign to calculate the pOH. An easier method is to use Equation 16.7 and simply subtract the pH from 14.00. This subtraction tells us that the pOH must be 9.80. Equations 16.4 and 16.7 tell us that if we know one of the four terms, $[H^+]$, $[OH^-]$, pH, or pOH, we can calculate the other three. For instance, if you know the pOH but a problem requires the $[H^+]$, we know how to easily calculate the needed value. The following practice exercises make use of these calculations.

Water draining from old coal and mineral mines often has pH values of 4.0 or lower. The cause is thought to stem from reactions of ground water and oxygen with iron pyrite, FeS. What are the pOH, $[H^+]$, and $[OH^-]$ in an acid mine drainage sample that has a pH of 3.75? (*Hint:* Recall that pH, pOH, $[H^+]$, and $[OH^-]$ are all interrelated using K_w and pK_w.)

Practice Exercise 16.3

Because rain washes pollutants out of the air, the lakes in many parts of the world have undergone pH changes. In a New England state, the water in one lake was found to have a hydrogen ion concentration of 3.2×10^{-5} mol L^{-1}. What are the calculated pH and pOH values of the lake's water? Is the water acidic or basic?

Practice Exercise 16.4

Find the values of $[H^+]$ and $[OH^-]$ that correspond to each of the following values of pH. State whether each solution is acidic or basic.

Practice Exercise 16.5

(a) 2.90 (the approximate pH of lemon juice)

(b) 3.85 (the approximate pH of sauerkraut)

(c) 10.81 (the pH of milk of magnesia, an antacid)

(d) 4.11 (the pH of orange juice, on the average)

(e) 11.61 (the pH of dilute, household ammonia)

pH Meters

One of the remarkable things about pH is that it can be easily measured using an instrument called a pH meter (Figure 16.2). An electrode system sensitive to the hydrogen ion concentration in a solution is first dipped into a standard solution of known pH to calibrate the instrument. Once calibrated, the apparatus can then be used to measure the pH of any other solution simply by immersing the electrodes into it. The accuracy of standard pH calibrating solutions, and therefore pH measurements, is usually ±0.02 pH units.

16.2 | pH of Strong Acid and Base Solutions

Many solutes affect the pH of an aqueous solution. In this section we examine how strong acids and bases behave and how to calculate the pH of their solutions. Weak acids and bases have similar effects, which we will discuss in the next section.

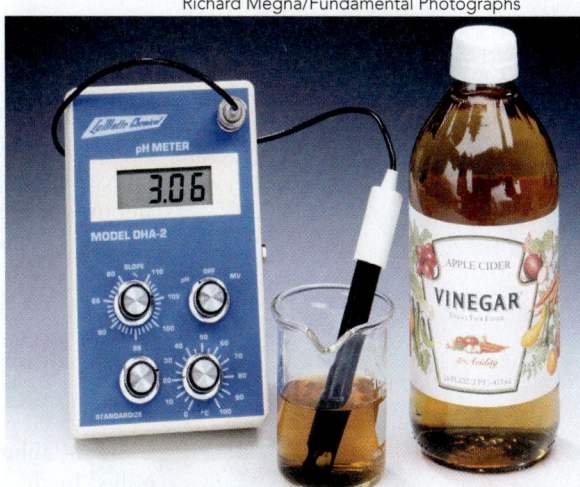

Richard Megna/Fundamental Photographs

Figure 16.2 | **A pH meter.** After the instrument has been calibrated with a solution of a known pH, the combination electrode is dipped into the solution to be tested and the pH is read from the meter.

Strong Acids and Bases

In Section 15.2 we saw that in an aqueous solution strong acids are considered to be completely ionized and bases are considered to be completely dissociated. This makes calculating the concentrations of H^+ and OH^- in their solutions a relatively simple task.

When the solute is a strong monoprotic acid, such as HCl or HNO_3, the molarity of H^+ will be the same as the molarity of the strong acid. Thus, a 0.010 M solution of HCl contains 0.010 mol L^{-1} of H^+ and a 0.0020 M solution of HNO_3 contains 0.0020 mol L^{-1} of H^+.

To calculate the pH of a solution of a strong monoprotic acid, we use the molarity of the H^+ obtained from the stated molar concentration of the acid. Thus, the 0.010 M HCl solution mentioned above has a pH of 2.00.

■ If necessary, review the list of strong acids in Section 4.4.

For strong bases, calculating the OH^- concentration is similarly straightforward. A 0.050 M solution of NaOH contains 0.050 mol L^{-1} of OH^- because the base is fully dissociated and each mole of NaOH releases one mole of OH^- when it dissociates. For bases such as $Ba(OH)_2$ we have to recognize that two moles of OH^- are released by each mole of the base.

$$Ba(OH)_2(s) \longrightarrow Ba^{2+}(aq) + 2OH^-(aq)$$

Therefore, if a solution contained 0.010 mol $Ba(OH)_2$ per liter, the concentration of OH^- would be 0.020 M. Of course, once we know the OH^- concentration we can calculate pOH, from which we can calculate the pH. Working through the following practice exercises will help to solidify these methods.

Practice Exercise 16.6

Calculate the $[H^+]$, pH, and pOH in 0.0050 M HNO_3. (*Hint:* HNO_3 is a strong acid.)

Practice Exercise 16.7

Calculate the pOH, $[H^+]$, and pH in a solution made by weighing 1.20 g of KOH and dissolving it in sufficient water to make 250.0 mL of solution.

Effect of Solute on the Ionization of Water

In the preceding calculations, we have made a critical and correct assumption—namely, that the autoionization of water contributes negligibly to the total $[H^+]$ in a solution of an acid and to the total $[OH^-]$ in a solution of a base.[3] Let's take a closer look at this.

In a solution of an acid, there are actually *two* sources of H^+. One is from the ionization of the acid itself and the other is from the autoionization of water. Thus,

■ A similar equation applies to the equilibrium OH^- concentration in a solution of a base.

$[OH^-]_{total} =$

$[OH^-]_{from\ solute} + [OH^-]_{from\ H_2O}$

$$[H^+]_{total} = [H^+]_{from\ solute} + [H^+]_{from\ H_2O}$$

Except in very dilute solutions of acids, the amount of H^+ contributed by the water ($[H^+]_{from\ H_2O}$) is small compared to the amount of H^+ contributed by the acid ($[H^+]_{from\ solute}$). For instance, in 0.020 M HCl the molarity of OH^- is 5.0×10^{-13} M. The only source of OH^- in this acidic solution is from the autoionization of water, and the amounts of OH^- and H^+ formed by the autoionization of water must be equal. Therefore, $[H^+]_{from\ H_2O}$ also equals 5.0×10^{-13} M. If we now look at the total $[H^+]$ for this solution, we have

$$[H^+]_{total} = 0.020\ M + 5.0 \times 10^{-13}\ M$$

(from HCl)   (from H₂O)

$$= 0.020\ M \qquad \text{(rounded correctly)}$$

In any solution of an acid, the autoionization of water is suppressed by the H^+ furnished by the solute. It's simply an example of Le Châtelier's principle. If we look at the autoionization reaction, we can see that if some H^+ is provided by an external source (an acidic solute, for example), the position of equilibrium will be shifted to the left.

$$H_2O \rightleftharpoons H^+ + OH^-$$

Adding H^+ from a solute causes the position of equilibrium to shift to the left.

As the results of our calculation have shown, the concentrations of H^+ and OH^- from the autoionization reaction are reduced well below their values in a neutral solution (1.0×10^{-7} M). Therefore, except for *very* dilute solutions (10^{-6} M or less), we will assume that all of the H^+ in the solution of an acid comes from the solute. Similarly, we'll assume that in a solution of a base, all the OH^- comes from the dissociation of the solute.

---

[3] The autoionization of water cannot be neglected when working with very dilute solutions of acids and bases.

16.3 | Ionization Constants, K_a and K_b

As you've learned in Chapter 15, weak acids and bases are incompletely ionized in water and exist in solution as molecules that are in equilibria with the ions formed by their reactions with water. To deal quantitatively with these equilibria it is essential that you be able to write correct chemical equations for the equilibrium reactions, from which you can then obtain the corresponding correct equilibrium laws. Fortunately, there is a pattern that applies to the way these substances react. Once you've learned how to write the correct equation for one acid or base, you can write the correct equation for any other.

Reaction of a Weak Acid with Water

In aqueous solutions, all weak acids behave the same way. They are Brønsted acids and, therefore, proton donors. Some examples are $HC_2H_3O_2$, HSO_4^- and NH_4^+. In water, these participate in the following equilibria.

$$HC_2H_3O_2 + H_2O \rightleftharpoons H_3O^+ + C_2H_3O_2^-$$

$$HSO_4^- + H_2O \rightleftharpoons H_3O^+ + SO_4^{2-}$$

$$NH_4^+ + H_2O \rightleftharpoons H_3O^+ + NH_3$$

Notice that in each case, the acid reacts with water to give H_3O^+ and the corresponding conjugate base. We can represent these reactions in a general way using HA to represent the formula of the weak acid and A^- to represent the anion.

$$HA + H_2O \rightleftharpoons H_3O^+ + A^- \tag{16.8}$$

As you can see from the equations above, HA does not have to be electrically neutral; it can be a molecule such as $HC_2H_3O_2$, a negative ion such as HSO_4^{2-}, or a positive ion such as NH_4^+. (Of course, the actual charge on the conjugate base will then depend on the charge on the parent acid.)

Following the procedure developed in Chapter 14, we can also write a general equation for the equilibrium law, omitting the concentration of liquid water as usual, for Equation 16.8,

$$\frac{[H_3O^+][A^-]}{[HA]} = K_a \tag{16.9}$$

The new constant K_a is called an **acid ionization constant**. Abbreviating H_3O^+ as H^+, the equation for the ionization of the acid can be simplified as

$$HA \rightleftharpoons H^+ + A^-$$

from which the expression for K_a is obtained directly.

$$K_a = \frac{[H^+][A^-]}{[HA]} \tag{16.10}$$

You should learn how to write the chemical equation for the ionization of a weak acid and be able to write the equilibrium law corresponding to its K_a. For example, if we have a weak acid such as HNO_2 we can write the ionization reaction as

$$HNO_2 \rightleftharpoons H^+ + NO_2^-$$

and the corresponding equilibrium law is

$$K_a = \frac{[H^+][NO_3^-]}{[HNO_3]}$$

TOOLS

General equation for the ionization of a weak acid

■ Some call K_a the *acid dissociation constant*.

The meat products shown here contain nitrite ion as a preservative.

Andy Washnik

Practice Exercise 16.8

For each of the following acids, write the equation for its ionization in water and the appropriate expression for K_a: (a) $HC_2H_3O_2$, (b) $(CH_3)_3NH^+$, (c) H_3PO_4. (*Hint:* Determine the conjugate base for each of these acids.)

Practice Exercise 16.9

For each of the following acids, write the equation for its ionization in water and the appropriate expression for K_a: (a) $HCHO_2$, (b) $(CH_3)_2NH_2$, (c) $H_2PO_4^-$.

For weak acids, values of K_a are usually quite small and can be conveniently represented in a logarithmic form similar to pH. Thus, we can define the **pK_a** of an acid as

■ Also, K_a = antilog $(-pK_a)$ or $K_a = 10^{-pK_a}$

$$pK_a = -\log K_a$$

■ The values of K_a for strong acids are very large and are not tabulated.

The strength of a weak acid is determined by its value of K_a; the larger the K_a, the stronger and more fully ionized the acid. Because of the negative sign in the defining equation for pK_a, the stronger the acid, the *smaller* is its value of pK_a. The values of K_a and pK_a for some typical weak acids are given in Table 16.2. A more complete list is located in Appendix C.

Practice Exercise 16.10

Use Table 16.2 to find all the acids that are stronger than acetic acid and weaker than formic acid. (*Hint:* It may be easier to focus on the K_a values since they are directly related to acid strength.)

Practice Exercise 16.11

Two acids, HA and HB, have pK_a values of 3.16 and 4.14, respectively. Which is the stronger acid? What are the K_a values for these acids?

Reaction of a Weak Base with Water

As with weak acids, all weak bases behave in a similar manner in water. They are weak Brønsted bases and are therefore proton acceptors. Examples are ammonia, NH_3, and acetate ion, $C_2H_3O_2^-$. Their reactions with water are

$$NH_3 + H_2O \rightleftharpoons NH_4^+ + OH^-$$

$$C_2H_3O_2^- + H_2O \rightleftharpoons HC_2H_3O_2 + OH^-$$

TABLE 16.2	K_a and pK_a Values for Weak Monoprotic Acids at 25 °C		
Name of Acid	Formula	K_a	pK_a
Iodic acid	HIO_3	1.7×10^{-1}	0.77
Chloroacetic acid	$HC_2H_2O_2Cl$	1.4×10^{-3}	2.85
Nitrous acid	HNO_2	4.6×10^{-4}	3.34
Hydrofluoric acid	HF	3.5×10^{-4}	3.46
Cyanic acid	HOCN	2×10^{-4}	3.7
Formic acid	$HCHO_2$	1.8×10^{-4}	3.74
Barbituric acid	$HC_4H_3N_2O_3$	9.8×10^{-5}	4.01
Hydrazoic acid	HN_3	2.5×10^{-5}	4.60
Acetic acid	$HC_2H_3O_2$	1.8×10^{-5}	4.74
Butanoic acid	$HC_4H_7O_2$	1.5×10^{-5}	4.82
Propanoic acid	$HC_3H_5O_2$	1.3×10^{-5}	4.89
Hypochlorous acid	HOCl	3.0×10^{-8}	7.52
Hydrocyanic acid	HCN	4.9×10^{-10}	9.31
Phenol	HC_6H_5O	1.3×10^{-10}	9.89
Hydrogen peroxide	H_2O_2	2.4×10^{-12}	11.62

Notice that in each instance, the base reacts with water to give OH^- and the corresponding conjugate acid. We can also represent these reactions by a general equation. If we represent the base by the symbol B, the reaction is

$$B + H_2O \rightleftharpoons BH^+ + OH^- \tag{16.11}$$

(As with weak acids, B does not have to be electrically neutral.) This yields the equilibrium law, where water is omitted as usual. We have a new equilibrium constant that is called the **base ionization constant**, with the symbol K_b.

$$K_b = \frac{[BH^+][OH^-]}{[B]} \tag{16.12}$$

To illustrate, let's write the ionization reaction for the weak base, propylamine, $CH_3CH_2CH_2NH_2$.

$$CH_3CH_2CH_2NH_2 + H_2O \rightleftharpoons CH_3CH_2CH_2NH_3^+ + OH^-$$

The corresponding equilibrium law is

$$K_b = \frac{[CH_3CH_2CH_2NH_3^+][OH^-]}{[CH_3CH_2CH_2NH_2]}$$

For each of the following bases, write the equation for its ionization in water and the appropriate expression for K_b. (*Hint:* See Equation 16.12.) (a) $(CH_3)_3N$ (trimethylamine), (b) SO_3^{2-} (sulfite ion), (c) NH_2OH (hydroxylamine).

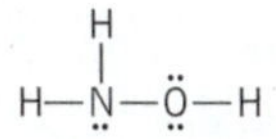

Write the ionization reaction, and the K_b expression, when each of the following amphiprotic substances acts as a base: (a) HS^-, (b) $H_2PO_4^-$, (c) HPO_4^{2-}, (d) HCO_3^-, (e) HSO_3^-.

Because the K_b values for weak bases are usually small numbers, the same kind of logarithmic notation is often used to represent their equilibrium constants. Thus, **pK_b** is defined as

$$pK_b = -\log K_b$$

Table 16.3 lists some molecular bases and their corresponding values of K_b and pK_b. A more complete list is located in Appendix C.

TOOLS

General equation for the ionization of a weak base

■ K_b is also called the *base dissociation constant*.

Practice Exercise 16.12
■
$$H-\overset{\displaystyle H}{\underset{\displaystyle |}{N}}-\overset{..}{\underset{..}{O}}-H$$
hydroxylamine, NH_2OH

Practice Exercise 16.13

■ Also, K_b = antilog ($-pK_b$) or $K_b = 10^{-pK_b}$

■ Tables of ionization constants usually give values for only the molecular member of an acid–base pair.

TABLE 16.3 K_b and pK_b Values for Weak Molecular Bases at 25 °C

Name of Acid	Formula	K_b	pK_b
Butylamine	$C_4H_9NH_2$	5.9×10^{-4}	3.23
Methylamine	CH_3NH_2	4.5×10^{-4}	3.35
Ammonia	NH_3	1.8×10^{-5}	4.74
Strychnine	$C_{21}H_{22}N_2O_2$	1.8×10^{-6}	5.74
Morphine	$C_{17}H_{19}NO_3$	1.6×10^{-6}	5.80
Hydrazine	N_2H_4	1.3×10^{-6}	5.89
Hydroxylamine	$HONH_2$	1.1×10^{-8}	7.96
Pyridine	C_5H_5N	1.8×10^{-9}	8.75
Aniline	$C_6H_5NH_2$	4.3×10^{-10}	9.37

■ Formic acid and formate ion have the structures

$$H-\overset{\overset{\displaystyle O}{\|}}{C}-O-H$$

formic acid, $HCHO_2$
(Acidic hydrogen shown in red.)

$$H-\overset{\overset{\displaystyle O}{\|}}{C}-O^-$$

formate ion, CHO_2^-

Often the formula for formic acid is written HCOOH to emphasize its molecular structure. In this chapter we will write the formulas for acids with the acidic hydrogens first in the formula, which is why we've given the formula for formic acid as $HCHO_2$.

The Product of K_a and K_b

Formic acid, $HCHO_2$ (a substance partly responsible for the sting of a fire ant), is a typical weak acid that ionizes according to the equation

$$HCHO_2 + H_2O \rightleftharpoons H_3O^+ + CHO_2^-$$

As you've seen, we write its K_a expression as

$$K_a = \frac{[H^+][CHO_2^-]}{[HCHO_2]}$$

The conjugate base of formic acid is the formate ion, CHO_2^-, and when a solute that formate ion, CHO_2^- contains this ion (e.g., $NaCHO_2$) is dissolved in water, the solution is slightly basic. In other words, the formate ion acts as a weak base since it is the conjugate base of formic acid. The reaction of formate ions with water is

$$CHO_2^- + H_2O \rightleftharpoons HCHO_2 + OH^-$$

The K_b expression for this base is

$$K_b = \frac{[HCHO_2][OH^-]}{[CHO_2^-]}$$

There is an important relationship between the equilibrium constants for this conjugate acid–base pair: the product of K_a times K_b equals K_w. We can see this by multiplying the mass action expressions above.

$$K_a \times K_b = \frac{[H^+][\cancel{CHO_2^-}]}{[\cancel{HCHO_2}]} \times \frac{[\cancel{HCHO_2}][OH^-]}{[\cancel{CHO_2^-}]} = [H^+][OH^-] = K_w$$

In fact, this same relationship exists for *any* acid–base conjugate pair.

For *any* acid–base conjugate pair:

Inverse relationship between K_a and K_b

$$\boxed{K_a \times K_b = K_w} \tag{16.13}$$

Another useful relationship, which can be derived by taking the negative logarithm of each side of Equation 16.13, is

Relationship between pK_a and pK_b

■ $-\log(K_a \times K_b) = -\log K_w$
$(-\log K_a) + (-\log K_b) = -\log K_w$
$pK_a + pK_b = pK_w$

$$\boxed{pK_a + pK_b = pK_w = 14.00 \qquad \text{(at 25 °C)}} \tag{16.14}$$

There are some important consequences of the relationship expressed in Equation 16.13. One is that it is not necessary to tabulate both K_a and K_b for the members of an acid–base pair; if one K is known, the other can be calculated. For example, the K_a for $HCHO_2$ and the K_b for NH_3 will be found in most tables, and those tables usually will not contain the K_b for CHO_2^- or the K_a for NH_4^+. If they are needed, we can calculate them using Equation 16.13.

Practice Exercise 16.14

The base methylamine CH_3NH_2 has a K_b of 4.5×10^{-4}. What is the K_a for the methylammonium ion, $CH_3NH_3^+$? (*Hint:* Recall that K_a and K_b are inversely proportional to each other.)

Practice Exercise 16.15

The value of K_a for the $HCHO_2$ ion is 1.8×10^{-4}. What is the value of K_b for CHO_2^-?

Another interesting and useful observation is that *there is an inverse relationship between the strengths of the acid and base members of a conjugate pair.* This is illustrated graphically in Figure 16.3. Because the product of K_a and K_b is a constant, the larger the value of K_a the smaller is the value of K_b. In other words, *the stronger the acid, the weaker is its conjugate base* (a fact that we noted in Chapter 15 in our discussion of the strengths of Brønsted acids and bases). We will say more about this relationship in Section 16.6.

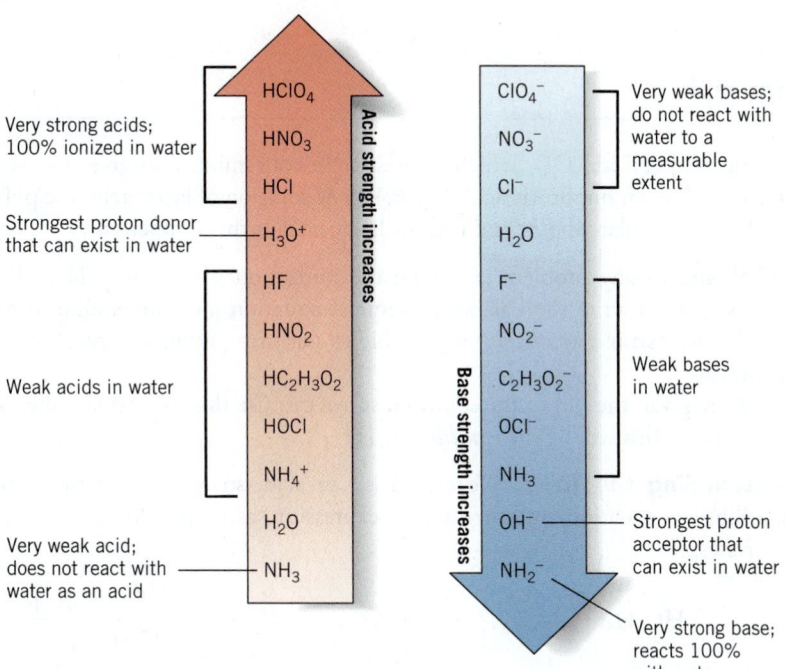

Figure 16.3 | **The relative strengths of conjugate acid–base pairs.** The stronger the acid is, the weaker is its conjugate base. The weaker the acid is, the stronger is its conjugate base. Very strong acids are 100% ionized, and their conjugate bases do not react with water to any measurable extent.

16.4 | Determining K_a and K_b Values

Our goal in this section is to develop a general strategy for dealing quantitatively with the equilibria of weak acids and bases in water. Generally, these calculations fall into two categories. The first involves calculating the value of K_a or K_b from the initial concentration of the acid or base and the measured pH of the solution (or some other data about the equilibrium composition of the solution). The second involves calculating equilibrium concentrations given K_a or K_b and initial concentrations, which we consider in Section 16.5.

Methods Using Initial Concentrations and Equilibrium Data

In problems of this type, our *first* goal is to obtain numerical values for *all* the equilibrium concentrations that are needed to evaluate the mass action expression in the definition of K_a or K_b. (The reason, of course, is that at equilibrium the reaction quotient, Q, equals the equilibrium constant.) We are usually given the molar concentration of the acid or base as it would appear on the label of a bottle containing the solution. Also provided is information from which we can obtain directly at least one of the equilibrium concentrations. Thus, we might be given the measured pH of the solution, which provides the equilibrium concentration of H^+. (Equilibrium is achieved rapidly in these solutions, so when we measure the pH, the value obtained can be used to calculate the equilibrium concentration of H^+.) Alternatively, we might be given the **percentage ionization** of the acid or base, which we define as follows:

$$\text{percentage ionization} = \frac{\text{moles ionized per liter}}{\text{moles available per liter}} \times 100\% \qquad (16.15)$$

Percentage ionization

Let's look at some examples that illustrate how to determine K_a and K_b from the kind of data mentioned.

Example 16.1

Calculating K_a and pK_a from pH

Susan Kaprov

For some, nothing goes better on a hot dog than sauerkraut and mustard. Lactic acid gives sauerkraut its sour taste.

Lactic acid ($HC_3H_5O_3$), which is present in sour milk, also gives sauerkraut its tartness. It is a monoprotic acid. In a 0.100 M solution of lactic acid, the pH is 2.44 at 25 °C. Calculate the K_a and pK_a for lactic acid at this temperature.

Analysis: In any problem involving the ionization of a weak acid or base, the first step is to write the balanced chemical equation and the equilibrium law. If needed we can use a concentration table to perform any necessary stoichiometric reasoning.

We're given the pH of the solution, so we can use this to calculate the H^+ concentration. This will be the *equilibrium* $[H^+]$.

Assembling the Tools: We're told it's an acid, so we'll begin by writing the equilibrium equation, and then the K_a expression, as we have done in the practice exercises,

$$HC_3H_5O_3 \rightleftharpoons H^+ + C_3H_5O_3^- \qquad K_a = \frac{[H^+][C_3H_5O_3^-]}{[HC_3H_5O_3]}$$

Equation 16.5 is needed to calculate $[H^+]$, and we will construct and use a concentration table using the outline given in Section 14.8 as our guide.

Solution: Looking at our tools, we see that we need the equilibrium values for $[H^+]$, $[C_3H_5O_3^-]$, and $[HC_3H_5O_3]$ to calculate K_a. At this point it will help to set up a concentration table. To obtain the initial concentrations we see that the only solute is $HC_3H_5O_3$, (0.100 M), so the only source of the ions, H^+ and $C_3H_5O_3^-$, is the ionization of the acid. Their concentrations are listed initially as zero. The change in concentrations will be represented as the unknown x. Adding the initial concentration and change rows gives us the terms for the equilibrium concentrations. These entries are shown in black.

We now find the $[H^+]$ from pH. This gives us the *equilibrium* value of $[H^+]$, shown in red.

$$[H^+] = 10^{-2.44}$$

$$= 0.0036 \ M$$

From this we see that x must be 0.0036. Knowing x, we can calculate the equilibrium concentrations of $HC_3H_5O_3$ and $C_3H_5O_3^-$ as shown in green in the table.

	$HC_3H_5O_3$	$\rightleftharpoons$	H^+	+	$HC_3H_5O_3^-$
Initial concentrations (M)	0.100		0		0
Changes in concentrations (M)	$-x$		$+x$		$+x$
Equilibrium concentrations (M)	$(0.100 - x)$		x		x
Equilibrium concentration values based on pH	$(0.100 - 0.0036) = 0.096$ (correctly rounded)		0.0036		0.0036

The last row of data now contains the equilibrium concentrations that we now use to calculate K_a, by substituting them into the K_a expression.

$$K_a = \frac{(3.6 \times 10^{-3})(3.6 \times 10^{-3})}{0.096} = 1.4 \times 10^{-4}$$

Thus the acid ionization constant for lactic acid is 1.4×10^{-4}. To find pK_a, we take the negative logarithm of K_a.

$$pK_a = -\log K_a = -\log(1.4 \times 10^{-4}) = 3.85$$

Is the Answer Reasonable? Weak acids have small ionization constants, so the value we obtained for K_a seems to be reasonable. We should also check the entries in the concentration table to be sure they are reasonable. For example, the "changes" for the ions are both positive, meaning both of their concentrations are increasing. This is the way it *must* be because neither ion can have a concentration less than zero. The changes are caused by the ionization reaction, so they both have to change in the same direction. Also, we have the concentration of the molecular acid decreasing, as it should if the ions are being formed by the ionization.

Sometimes we have the opportunity to determine the percentage ionization of a substance. Obvious examples are osmotic pressure measurements and measurements of freezing point depression and boiling point elevations that we discussed in Section 12.6. The following example illustrates how we can determine the value of K_b of a base such as methylamine when we know its percentage ionization.

Example 16.2
Calculating K_b and pK_b from the Percentage Ionization

Methylamine, CH_3NH_2, is a weak base and one of several substances that give herring brine its pungent odor. In 0.100 M CH_3NH_2, only 6.4% of the base is ionized. What are K_b and pK_b of methylamine?

Analysis: In this problem we've been given the percentage ionization of the base. We will use this to calculate the equilibrium concentrations of the ions in the solution. After we know these, solving the problem follows the same path as in the preceding example.

Assembling the Tools: As in the previous example, we need the balanced chemical equation (see the format for a weak base in Equation 16.11) so we can write the equilibrium law (see the format for the equilibrium law of a weak base in Equation 16.12). These are written as

Methylamine is responsible in part for the fishy aroma of pickled herring.

$$CH_3NH_2 + H_2O \rightleftharpoons CH_3NH_3^+ + OH^-$$

$$K_b = \frac{[CH_3NH_3^+][OH^-]}{[CH_3NH_2]}$$

We will also use the percentage ionization, Equation 16.15, to calculate at least one of the equilibrium concentrations.

Solution: We can start by constructing the equilibrium table with the chemical equation and the appropriate rows and columns. The initial concentrations are those concentrations before any reaction takes place. Therefore the concentrations of the ions are listed as zeros. Usually, we use the variable x to indicate the changes expected, in the change row, and the equilibrium concentrations in the last row, which we recall is the sum of the first two rows. However, the percentage composition allows us to complete the table directly.

The percentage ionization tells us that 6.4% of the CH_3NH_2 has reacted. Entering the information into Equation 16.15 gives us

■ To find 6.4% of 0.100 M, Equation 16.15 has been rearranged.

$$\frac{\text{moles per liter of } CH_2NH_2 \text{ ionized}}{0.100\ M} \times 100\% = 6.4\%$$

Therefore, the number of moles per liter of the base that has ionized at equilibrium in this solution is:

$$\text{moles per liter of } CH_3NH_2 \text{ ionized} = \frac{6.4\% \times 0.100\ M}{100\%} = 0.0064\ M$$

This value represents the *decrease* in the concentration of CH_3NH_2, which is -0.0064 moles L^{-1} of CH_3NH_2. We can now use this change in $[CH_3NH_2]$ to determine the concentrations of the other species. Since both OH^- and $CH_3NH_3^+$ form when CH_3NH_2 reacts, $[CH_3NH_3^+]$ and $[OH^-]$ are both 0.0064 mol L^{-1}.

	H_2O +	CH_3NH_2	$\rightleftharpoons$	$CH_3NH_3^+$ +	OH^-
Initial concentrations (M)		0.100		0	0
Changes in concentrations (M)		-0.0064		0.0064	0.0064
Equilibrium concentrations (M)		$(0.100 - 0.0064) = 0.094$		0.0064	0.0064

Now we can calculate K_b by substituting equilibrium quantities into the mass action expression.

$$K_b = \frac{(0.0064)(0.0064)}{(0.096)} = 4.3 \times 10^{-4}$$

and

$$pK_b = -\log(4.3 \times 10^{-4}) = 3.37$$

Is the Answer Reasonable? First, the value of K_b is small, so that suggests we've probably worked the problem correctly. We can also check the "Change" row to be sure the algebraic signs are correct. Both initial concentrations start out at zero, so they must both increase, which they do. The change for the CH_3NH_2 is opposite in sign to that for the ions, and that's as it should be. Therefore, we appear to have set up the problem correctly.

Practice Exercise 16.16

Salicylic acid reacts with acetic acid to form aspirin, acetyl salicylic acid. A 0.200 molar solution of salicylic acid has a pH of 1.83. Calculate the K_a and pK_a of salicylic acid. (*Hint:* Convert the pH to $[H^+]$ first.)

Practice Exercise 16.17

■

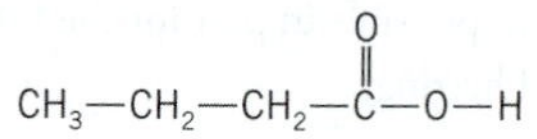

$$CH_3-CH_2-CH_2-\overset{\overset{\displaystyle O}{\|}}{C}-O-H$$

butanoic acid

$HC_4H_7O_2$

When butter turns rancid, its foul odor is mostly that of butanoic acid, $HC_4C_7O_2$, a weak monoprotic acid similar to acetic acid in structure. In a 0.0100 M solution of butanoic acid at 20 °C, the acid is 3.8% ionized. Calculate the K_a and pK_a of butanoic acid at this temperature.

Practice Exercise 16.18

Few substances are more effective in relieving intense pain than morphine. Morphine is an alkaloid—an alkali-like compound obtained from plants—and alkaloids are all weak bases. In 0.010 M morphine, the pOH is 3.90. Calculate the K_b and pK_b for morphine. (You don't need to know the formula for morphine, just that it's a base. Use whatever symbol you want to write the equation.)

16.5 | pH of Weak Acid and Weak Base Solutions

Someday you may synthesize an acid or base and need to characterize its properties by determining its K_a or K_b value. However, it is more common to use those values for everyday laboratory calculations of pH or $[H^+]$.

Calculating Equilibrium Concentrations

The key to calculating equilibrium concentrations is examining the statement of the problem carefully so you can construct the correct chemical equation, which will then lead you to write the equilibrium law that's required to obtain the solution. With this in mind, let's

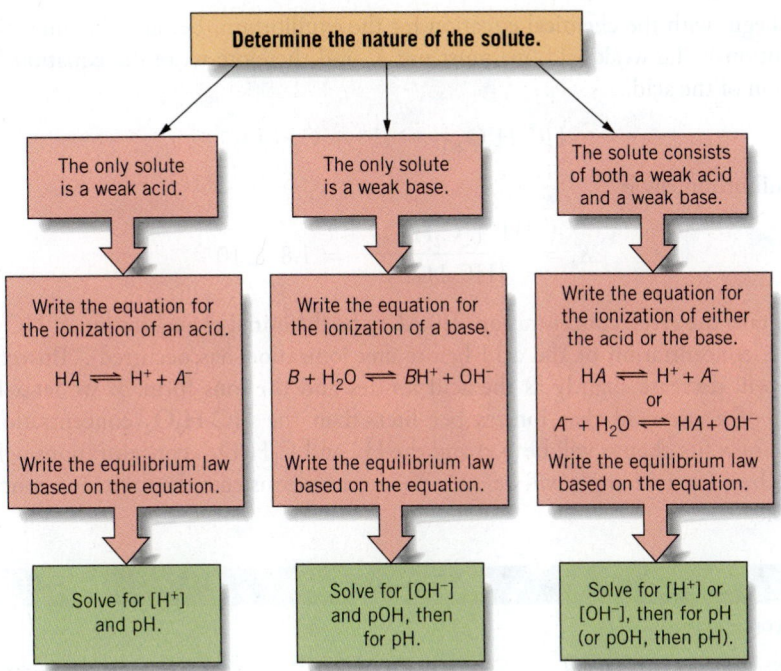

Figure 16.4 | Determining how to proceed in acid–base equilibrium problems. The nature of the solute species determines how the problem is approached. Following this diagram will get you started in the right direction.

take an overview of the "landscape" to develop a method for selecting the correct approach to a problem.

Almost any problem in which you are given a value of K_a or K_b falls into one of three categories: (1) the aqueous solution contains a weak acid as its *only* solute, (2) the solution contains a weak base as its *only* solute, or (3) the solution contains *both* a weak acid and its conjugate base. The approach we take for each of these conditions is described below and is summarized in Figure 16.4.

1. If the solution contains only a weak acid as the solute, then the problem *must* be solved using K_a. This means that the correct chemical equation for the problem is the ionization of the weak acid. It also means that if you are given the K_b for the acid's conjugate base, you will have to calculate the K_a using Equation 16.13 in order to solve the problem.

2. If the solution contains only a weak base, the problem *must* be solved using K_b. The correct chemical equation is for the ionization of the weak base. If the problem gives you K_a for the base's conjugate acid, then you must use Equation 16.13 to calculate K_b to solve the problem.

3. Solutions that contain two solutes, one a weak acid and the other its conjugate base, have some special properties that we will discuss later. (Such solutions are called *buffers.*) To work problems for these kinds of mixtures, we can use *either* K_a or K_b—it doesn't matter which we use, because we will obtain the same answers either way. However, if we elect to use K_a, then the chemical equation we use must be for the ionization of the acid. If we decide to use K_b, then the correct chemical equation is for the ionization of the base. Usually, the choice of whether to use K_a or K_b is made on the basis of which constant is most readily available.

■ Conditions 1 and 2 apply to solutions of weak molecular acids and bases and to solutions of salts that contain an ion that is an acid or a base. Condition 3 applies to solutions called *buffers*, which are discussed in Section 16.7.

Simplifications in Calculations

In Chapter 14 you learned that when the equilibrium constant is small, it is frequently possible to make simplifying assumptions that greatly reduce the algebraic effort in obtaining equilibrium concentrations. For many acid–base equilibrium problems such simplifications are particularly useful.

Let's consider a solution of 1.0 M acetic acid, $HC_2H_3O_2$, for which $K_a = 1.8 \times 10^{-5}$, and see how to determine the equilibrium concentrations in the solution.

■
$$H_3C-\overset{\overset{\displaystyle O}{\|}}{C}-O-H$$
acetic acid
$HC_2H_3O_2$

We begin with the chemical equation for the equilibrium. Because the only solute in the solution is the weak acid, we must use K_a and therefore write the equation for the ionization of the acid.

$$HC_2H_3O_2 \rightleftharpoons H^+ + C_2H_3O_2^-$$

The equilibrium law is

$$K_a = \frac{[H^+][C_2H_3O_2^-]}{[HC_2H_3O_2]} = 1.8 \times 10^{-5}$$

We will take the given concentration, 1.0 M, to be the initial concentration of $HC_2H_3O_2$ (i.e., the concentration of the acid before any ionization has occurred). This concentration will decrease slightly as the acid ionizes and the ions form. If we let x be the amount of acetic acid that ionizes per liter, then the $HC_2H_3O_2$ concentration will decrease by x (its change will be $-x$) and the H^+ and $C_2H_3O_2^-$ concentrations will each increase by x (their changes will be $+x$). We can now construct the following concentration table.

■ The initial concentrations of the ions are set equal to zero because none of them have yet been supplied by a solute.

	$HC_2H_3O_2$	$\rightleftharpoons$	H^+	$+$	$C_2H_3O_2^-$
Initial concentrations (M)	1.0		0		0
Changes in concentrations (M)	$-x$		$+x$		$+x$
Equilibrium concentrations (M)	$1.0 - x$		x		x

The values in the last row of the table should satisfy the equilibrium law. When they are substituted into the mass action expression we obtain

$$\frac{(x)(x)}{1.0 - x} = 1.8 \times 10^{-5}$$

This equation involves a term in x^2, and it could be solved using the quadratic formula as we did in Chapter 14. However, this and many other similar calculations involving weak acids and bases can be simplified. Let's look at the reasoning.

The equilibrium constant, 1.8×10^{-5}, is quite small, so we can anticipate that very little of the acetic acid will be ionized at equilibrium. This means x will be very small, so we make the approximation $(1.0 - x) \approx 1.0$. Replacing $1.0 - x$ with 1.0 yields the equation

■ We are assuming that when x is subtracted from 1.0 and *the result is rounded to the correct number of significant figures*, the answer will round to 1.0.

$$\frac{(x)(x)}{1.0 - x} = \frac{x^2}{1.0} = 1.8 \times 10^{-5}$$

Solving for x gives $x = 0.0042$ M. We see that the value of x is indeed negligible compared to 1.0 M (i.e., if we subtract 0.0042 M from 1.0 M and round correctly, we obtain 1.0 M). This gives us $[H^+] = [C_2H_3O_2^-] = 0.0042$ M, and $[HC_2H_3O_2] = 1.0$ M. We can also calculate the pH of the solution

$$pH = -\log[H^+]$$

$$= -\log(0.0042)$$

$$= 2.38$$

Notice that when we make this approximation, *the initial concentration of the acid is used as if it were the equilibrium concentration.* The approximation is valid when the equilibrium constant is small and the concentrations of the solutes are reasonably high—conditions that will apply to most situations you will encounter in this chapter. After a few examples we will discuss the conditions under which these approximations are not valid.

Let's try some typical acid–base equilibrium problems.

A student planned an experiment that would use 0.10 M propionic acid, $HC_3H_5O_2$. Calculate the value of $[H^+]$ and the pH for this solution. For propionic acid, $K_a = 1.3 \times 10^{-5}$. (*Hint:* Consider using an assumption to solve this rather than the quadratic formula.)

Benzoic acid, $HC_6H_5CO_2$, is a monoprotic acid (only one H^+ ionizes) with a $K_a = 6.3 \times 10^{-5}$. Calculate $[H^+]$ and the pH of a 0.023 M solution of benzoic acid.

Nicotinic acid, $HC_6H_4NO_2$, is a B vitamin. It is also a weak acid with $K_a = 1.4 \times 10^{-5}$. Calculate $[H^+]$ and the pH of a 0.050 M solution of $HC_2H_4NO_2$.

The calcium salt of propionic acid, calcium propionate, is used as a preservative in baked products.

Andy Washnik

Practice Exercise 16.19

propionic acid
$HC_3H_5O_2$

Practice Exercise 16.20

Practice Exercise 16.21

When we have the initial concentration of a molecular base, the determination of pH involves the steps we just described. However, we need to be careful to remember that the calculation yields the hydroxide ion concentration. That requires an additional conversion step from pOH to pH as shown in the following example.

Example 16.3
Calculating the pH of a Solution and the Percentage Ionization of the Solute

A solution of hydrazine, N_2H_4, has an initial concentration of 0.25 M. What is the pH of the solution, and what is the percentage ionization of the hydrazine? Hydrazine has $K_b = 1.3 \times 10^{-6}$.

Analysis: Hydrazine must be a weak base, because we have its value of K_b (the "b" in K_b tells us this is a *base* ionization constant). Since hydrazine is the only solute in the solution, we will have to write the equation for the ionization of a weak base and then set up the K_b expression.

Assembling the Tools: To write the chemical equation we need the formula for the conjugate acid of N_2H_4. It has one additional H^+, so its formula is $N_2H_5^+$. Now we write the balanced chemical equation and the equilibrium law as

$$N_2H_4 + H_2O \rightleftharpoons N_2H_5^+ + OH^-$$

$$K_b = \frac{[N_2H_5^+][OH^-]}{[N_2H_4]}$$

We will use a concentration table and appropriate assumptions.

To calculate the percentage ionization, we need to use Equation 16.15:

$$\text{percentage ionization} = \frac{\text{moles ionized per liter}}{\text{moles available per liter}} \times 100\%$$

This requires that we determine the number of moles per liter of the base that has ionized. The concentration table will help us perform this calculation correctly.

Solution: Let's once again construct the concentration table. The only sources of $N_2H_5^+$ and OH^- are the ionization of the N_2H_4, so their initial concentrations are both zero. They will form in equal amounts, so we let their changes in concentration equal $+x$. The concentration of N_2H_4 will decrease by x, so its change is $-x$.

	H_2O	+	N_2H_4	$\rightleftharpoons$	$N_2H_5^+$	+	OH^-
Initial concentrations (M)			0.25		0		0
Changes in concentrations (M)			$-x$		$+x$		$+x$
Equilibrium concentrations (M)			$(0.25 - x) \approx 0.25$		x		x

Because K_b is so small, $[N_2H_4] \approx 0.25\ M$. (As before, we assume the initial concentration will be effectively the same as the equilibrium concentration, an approximation that we expect to be valid.) Substituting into the K_b expression,

$$\frac{(x)(x)}{0.25 - x} \approx \frac{(x)(x)}{0.25} = 1.3 \times 10^{-6}$$

Solving for x gives: $x = 5.7 \times 10^{-4}$. This value represents the hydroxide ion concentration, from which we can calculate the pOH.

$$pOH = -\log(5.7 \times 10^{-4}) = 3.24$$

The pH of the solution can then be obtained from the relationship

$$pH + pOH = 14.00$$

Thus,

$$pH = 14.00 - 3.24 = 10.76$$

For the percent ionization we look to the concentration table and see that the value of x represents the amount of N_2H_4 that ionizes per liter (i.e., the change in the N_2H_4 concentration). Therefore,

$$\text{percentage ionization} = \frac{5.7 \times 10^{-4}\ M}{0.25\ M} \times 100\%$$

$$= 0.23\%$$

The base is 0.23% ionized.

Are the Answers Reasonable? First, we can quickly check to see whether the value of x satisfies our assumption and it does ($0.25 - 0.00057 = 0.25$). The value of the hydroxide ion concentration is small and the pH indicates a basic solution. Additionally, the small concentration of hydroxide ions agrees with the small percentage ionization we calculated. We could use our equilibrium concentrations to calculate K_b as in the previous example.

Practice Exercise 16.22 Aniline, $C_6H_5NH_2$, is a precursor for many dyes used to color fabrics, known as aniline dyes. Aniline has a K_b of 4.3×10^{-10}. What is the pOH of a 0.025 molar solution of aniline? (*Hint:* Calculate the OH^- concentration.)

Practice Exercise 16.23 Pyridine, C_5H_5N, is a bad-smelling liquid for which $K_b = 1.8 \times 10^{-9}$. What is the pH of a 0.010 M aqueous solution of pyridine?

Practice Exercise 16.24 Phenol is an acidic organic compound for which $K_a = 1.3 \times 10^{-10}$. What is the pH of a 0.15 M solution of phenol in water?

When Simplifications Fail

In the examples we studied above, we made a simplification by assuming that we could use the initial concentration as the equilibrium concentration of the weak acid or weak base. We confirmed this assumption by observing that, if we correctly used the significant figure rules for subtraction, the initial concentration did not change when we accounted for the amount that ionized.

In some instances, we will calculate an answer where the initial concentration will change slightly when we subtract the amount that ionizes. In those cases we may want to use the quadratic formula or the method of successive approximations described in Appendix A. Since these two methods are tedious and have more steps where mistakes might be made, we'll consider relaxing the rules for using our simplifications.

When we ask ourselves how precise our pH calculations need to be, we find two factors to consider. First, most ionization constants only have two significant figures, therefore the hydrogen ion concentrations we calculate should have two significant figures. We recall that the uncertainty is usually taken as ± 1 in the last digit of a measurement. The uncertainty of a two-digit number ranges from 1% to 10%. Second, the experimental determination of pH usually has an uncertainty of ± 0.02 pH units or $\pm 5\%$ in terms of hydrogen ion concentration. In general, there is usually little needed to calculate a pH more precisely than we can measure it in the laboratory. These two sources of error tell us that any simplification that results in an error less than approximately $\pm 5\%$ will give us acceptable results. Without going into the detailed explanation, we find a simple rule for testing whether a simplification will work. The rule states that a simplifying assumption will work if the initial concentration of a weak acid, C_{HA}, or a weak base, C_B, is more than 100 times the value of the ionization constant.[4]

■ If the simplifications do not work, we can solve quadratic equations by using the quadratic formula. A detailed description of how to use the quadratic formula is given in Section 14.8, Example 14.9.

$$C_{HA} > 100 \times K_a \quad \text{or} \quad C_B > 100 \times K_b$$

Since we can easily multiply a K_a or K_b by 100, we can test to see if our assumptions will work *before* we do the calculation.

TOOLS

Criteria for predicting when simplifying assumptions will work

16.6 | Acid–Base Properties of Salt Solutions

When we study solutions of salts, we find that many have a neutral pH. Examples are NaCl and KNO_3. However, not all salt solutions behave this way. Some, such as $NaC_2H_3O_2$, are slightly basic while others, such as NH_4Cl, are slightly acidic. To understand this behavior, we have to examine how the ions of a salt can affect the pH of a solution.

In Section 16.3 you saw that weak acids and bases are not limited to molecular substances. For instance, NH_4^+ was given as an example of a weak acid, and $C_2H_3O_2^-$ was cited as an example of a weak base. Ions such as these always come to us in compounds in which there are both a cation *and* an anion. Therefore, to place NH_4^+ in water, we need a salt such as NH_4Cl, and to place $C_2H_3O_2^-$ in water, we need a salt such as $NaC_2H_3O_2$.

Because a salt contains two ions, the pH of its solution can potentially be affected by either the cation or the anion, or perhaps even both. Therefore, we have to consider *both* ions if we wish to predict the effect of a salt on the pH of a solution.

Acidic Cations

If the cation of a salt is able to influence the pH of a solution, it does so by behaving as a weak acid. Not all cations are acidic, however, so let's look at the possibilities.

■ The reaction of a salt with water to give either an acidic or basic solution is sometimes called **hydrolysis**. Hydrolysis means *reaction with water*.

---

[4] If we calculate the hydrogen ion concentration analytically, using the quadratic formula, we obtain the "true" value. We can compare that value to the concentration obtained by using a simplification and calculate the error as

$$\frac{[H^+]_{\text{simplification}} - [H^+]_{\text{quadratic}}}{[H^+]_{\text{quadratic}}} \times 100\% = \% \text{ error}$$

When we relate this percent error to the ratio of C_a/K_a it is found that when this ratio is greater than 100, the error is less than approximately 5%.

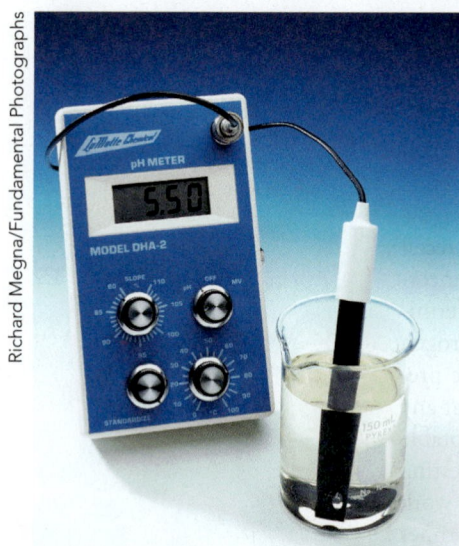

Figure 16.5 | **Ammonium ion as a weak acid.** The measured pH of a solution of the salt NH_4NO_3 is less than 7, which indicates the solution is acidic. Bromothymol blue, a pH indicator discussed in Section 16.9, has been added and has a faint yellow color.

TOOLS

Identifying acidic cations

Conjugate Acids of Molecular Bases

The ammonium ion, NH_4^+, is the conjugate acid of the molecular base, NH_3. We have already learned that NH_4^+, supplied for example by NH_4Cl, is a weak acid. As seen in Figure 16.5, solutions of this salt have a pH lower than 7. The equation for the reaction as an acid is

$$NH_4^+(aq) + H_2O \rightleftharpoons NH_3(aq) + H_3O^+(aq)$$

Writing this in a simplified form, along with the K_a expression, we have

$$NH_4^+(aq) \rightleftharpoons NH_3(aq) + H^+(aq) \qquad K_a = \frac{[NH_3][H^+]}{[NH_4^+]}$$

Because the K_a values for ions are seldom tabulated, we would usually expect to calculate the K_a value using the relationship: $K_a \times K_b = K_w$. In Table 16.3 the K_b for NH_3 is given as 1.8×10^{-5}. Therefore,

$$K_a = \frac{K_w}{K_b} = \frac{1.0 \times 10^{-14}}{1.8 \times 10^{-5}} = 5.6 \times 10^{-10}$$

Another example is the hydrazinium ion, $N_2H_5^+$, which is also a weak acid.

$$N_2H_5^+(aq) \rightleftharpoons N_2H_4(aq) + H^+(aq) \qquad K_a = \frac{[N_2H_4][H^+]}{[N_2H_5^+]}$$

The tabulated K_b for N_2H_4 is 1.3×10^{-6}. From this, the calculated value of K_a equals 7.7×10^{-9}.

These examples illustrate a general phenomenon:

Cations that are the conjugate acids of molecular bases tend to be weakly acidic.

Cations that are not conjugate acids of molecular bases are generally metal ions such as Na^+, K^+, and Ca^{2+}. In Chapter 15 we learned that the cations of the Group 1A metals are extremely weak acids and are unable to affect the pH of a solution. Except for Be^{2+}, the cations of Group 2A also do not affect pH.

Basic Anions

When a Brønsted acid loses a proton, its conjugate base is formed. Thus Cl^- is the conjugate base of HCl, and $C_2H_3O_2^-$ is the conjugate base of $HC_2H_3O_2$.

$$HCl + H_2O \longrightarrow Cl^- + H_3O^+$$

$$HC_2H_3O_2 + H_2O \rightleftharpoons C_2H_3O_2^- + H_3O^+$$

Although both Cl^- and $C_2H_3O^-$ are bases, only the latter affects the pH of an aqueous solution. Our knowledge of conjugate acid–base pairs from Chapter 15 helps us understand why.

Earlier you learned that there is an inverse relationship between the strength of an acid and its conjugate base—the stronger the acid, the weaker is the conjugate base. Therefore, when an acid is *extremely strong*, as in the case of HCl or other "strong" acids that are 100% ionized (such as HNO_3), the conjugate base is *extremely weak*—too weak to affect in a measurable way the pH of a solution. Consequently, we have the following generalization:

The anion of a strong acid is too weak a base to influence the pH of a solution.

Acetic acid is much weaker than HCl, as evidenced by its value of K_a (1.8×10^{-5}). Because acetic acid is a weak acid, its conjugate base is much stronger than Cl^- and we can expect it to affect the pH (Figure 16.6). We can calculate the value of K_b for $C_2H_3O^-$ from the K_a for $HC_2H_3O_2$, also with the equation $K_a \times K_b = K_w$.

$$K_b = \frac{1.0 \times 10^{-14}}{1.8 \times 10^{-5}} = 5.6 \times 10^{-10}$$

This leads to another conclusion:

The anion of a weak acid is a weak base and can influence the pH of a solution. It will tend to make the solution basic.

TOOLS

Identifying basic anions

Acid–Base Properties of Salts

To decide whether any given salt will affect the pH of an aqueous solution, we must examine each of its ions and see what it alone might do. There are four possibilities:

1. If neither the cation nor the anion can affect the pH, the solution should be neutral.
2. If only the cation of the salt is acidic, the solution will be acidic.
3. If only the anion of the salt is basic, the solution will be basic.
4. If a salt has a cation that is acidic and an anion that is basic, the pH of the solution is determined by the *relative* strengths of the acid and base based on the K_a and K_b of the ions.

Let's consider two salts, sodium hypochlorite, NaOCl, and calcium nitrate, $Ca(NO_3)_2$, and see how we predict if their solutions will be acidic, basic, or neutral. We can write the equations for dissolving both salts as

$$NaOCl(s) \xrightarrow{H_2O} Na^+(aq) + OCl^-(aq)$$

$$Ca(NO_3)_2(s) \xrightarrow{H_2O} Ca^{2+}(aq) + 2NO_3^-(aq)$$

Now we consider whether or not each ion is related to a weak acid (for anions) or to a weak base (for cations). For the two anions, if we add a hydrogen ion to each, we can identify the acid that they are related to, HOCl and HNO_3. HNO_3 is one of our strong acids so the NO_3^- ion will not affect the pH of a solution. The HOCl on the other hand is a weak acid and will produce the conjugate base OCl^-. Turning our attention to the cations, we can see that the two cations are in Groups 1A and 2A, respectively, and are extremely weak acids that do not affect the pH of a solution. Finally, of all four ions, only the OCl^- will affect the pH. We identified the hypochlorite ion as a conjugate base and therefore its solution will be basic. Calcium nitrate solutions will be neutral since neither ion affects the pH.

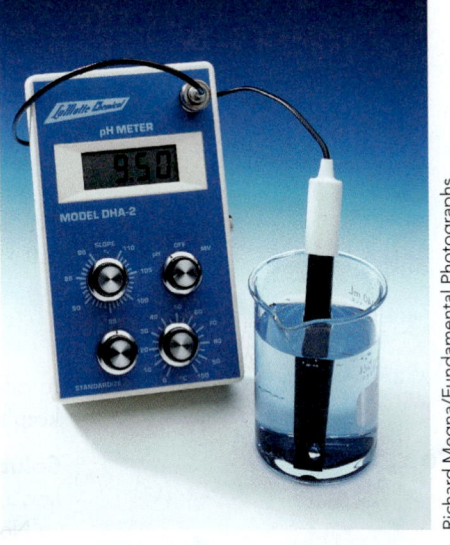

Figure 16.6 | Acetate ion is a weak base in water. The pH of a solution of $KC_2H_3O_2$ is greater than 7. Bromothymol blue has been added and has a blue color indicating a basic solution.

■ The only time an anion might be acidic is if it is from a partially neutralized polyprotic acid—for example, HSO_4^-.

Richard Megna/Fundamental Photographs

Are solutions of (a) $NaNO_2$, (b) KCl, and (c) NH_4Br acidic, basic, or neutral? (*Hint:* Write the ions for these compounds. What are the conjugate acids and bases of these ions?)

Practice Exercise 16.25

Are solutions of (a) $NaNO_3$, (b) KF, and (c) NH_4NO_3 acidic, basic, or neutral?

Practice Exercise 16.26

Now that we have reinforced the concept that anions are conjugate bases and cations are conjugate acids, we can calculate the pH of their solutions. This will be done in exactly the same way in which we calculated the pH of solutions containing weak molecular acids or weak molecular bases.

Example 16.4
Calculating the pH of a Salt Solution

Household bleach used for washing clothes, and removing mildew and moss from buildings and outdoor furniture, is a solution of sodium hypochlorite, NaOCl. What is the pH of a 0.10 *M* solution of NaOCl? For HOCl, $K_a = 3.0 \times 10^{-8}$.

Analysis: Problems such as this are just like the other acid–base equilibrium problems you have learned to solve. (1) We need to determine the nature of the solute: is it a weak acid, a weak base, or are both a weak acid and weak base present? (2) We then write the appropriate chemical equation and equilibrium law. (3) We proceed to the solution.

Assembling the Tools: Since sodium is in Group 1A, that tells us that the Na^+ does not affect the pH. However the OCl^- is the conjugate base of the weak acid, HOCl. Therefore we have a solution of a weak base and we need to use K_b and the appropriate chemical reaction. The reaction we need describes how OCl^- reacts with water.

$$OCl^- + H_2O \rightleftharpoons HOCl + OH^-$$

Then we write the appropriate equilibrium law as

$$K_b = \frac{[HOCl][OH^-]}{[OCl^-]}$$

We do not know the value of K_b for OCl^- but we do know that the K_a for HOCl is 3.0×10^{-8}. We will use the equation

$$K_b = \frac{K_w}{K_a} = \frac{1.0 \times 10^{-14}}{3.0 \times 10^{-8}} = 3.3 \times 10^{-7}$$

to calculate the needed value for K_b. Now we can set up a concentration table to help us keep track of the numbers in our calculation.

Solution: We've written the balanced chemical equation above and also the equilibrium law, and we calculated K_b from the given value of K_a.

Now let's set up the concentration table. The only sources of HOCl and OH^- are the reaction of the OCl^-, so their concentrations will each increase by x and the concentration of OCl^- will decrease by x.

	H_2O +	OCl^- $\rightleftharpoons$	HOCl +	OH^-
Initial concentrations (M)		0.10	0	0
Changes in concentrations (M)		$-x$	$+x$	$+x$
Equilibrium concentrations (M)		$(0.10 - x) \approx 0.10$	x	x

At equilibrium, the concentrations of HOCl and OH^- are the same, x.

$$[HOCl] = [OH^-] = x$$

Because K_b is so small, $[OCl^-] \approx 0.10\ M$. (Once again, we expect x to be so small that we are able to use the initial concentration of the base as if it were the equilibrium concentration.) Substituting into the K_b expression gives

$$\frac{(x)(x)}{0.10 - x} \approx \frac{(x)(x)}{0.10} = 3.3 \times 10^{-7}$$

$$x = 1.8 \times 10^{-4}\ M$$

At this point we need to remember that the value of x represents the OH^- concentration, from which we can calculate the pOH and then the pH.

$$[OH^-] = 1.8 \times 10^{-4} M$$

$$pOH = -\log(1.8 \times 10^{-4}) = 3.74$$

$$pH = 14.00 - pOH$$

$$= 14.00 - 3.74$$

$$= 10.26$$

The pH of this solution is 10.26.

■ If you have trouble writing the correct chemical equation for this equilibrium, you should review Section 16.3.

• **Is the Answer Reasonable?** First, we check our assumption and find it is correct $(0.10 - 0.00018 = 0.10)$. Also, from the nature of the salt, we expect the solution to be basic. The calculated pH corresponds to a basic solution, so the answer seems to be reasonable. You've also seen that we can check the accuracy of the answer by substituting the calculated equilibrium concentrations into the mass action expression.

$$\frac{[HOCl][OH^-]}{[OCl^-]} = \frac{(x)(x)}{0.10} = \frac{(1.8 \times 10^{-4})^2}{0.10} = 3.2 \times 10^{-7}$$

The result is quite close to our value of K_b, so our answers are reasonable.

What is the pH of a solution when 25.0 g of methylammonium chloride, CH_3NH_3Cl, is dissolved in enough water to make 500.0 mL of solution? (*Hint:* What is the molarity of the CH_3NH_3Cl solution?)

Practice Exercise 16.27

What is the pH of a 0.10 M solution of $NaNO_2$?

Practice Exercise 16.28

If 500.0 mL of a 0.20 M solution of ammonia is mixed with 500.0 mL of a 0.20 M solution of HBr, what is the pH of the resulting solution of NH_4Br?

Practice Exercise 16.29

Salts of a Weak Acid and a Weak Base

There are many salts whose ions are both able to affect the pH of the salt solution. Whether or not the salt has a net effect on the pH now depends on the relative strengths of both ions. If the cation and anion are identical in their respective strengths, the salt has no net effect on pH. In ammonium acetate, for example, the ammonium ion is an acidic cation and the acetate ion is a basic anion. However, the K_a of NH_4^+ is 5.6×10^{-10} and the K_b of $C_2H_3O_2^-$ just happen to be the same, 5.6×10^{-10}. The cation tends to produce H^+ ions to the same extent that the anion tends to produce OH^-. So in aqueous ammonium acetate, $[H^+] = [OH^-]$, and the solution has a pH of 7.

Now consider ammonium formate, NH_4CHO_2. The formate ion, CHO_2^-, is the conjugate base of the weak acid, formic acid, so it is a Brønsted base. Its K_b is 5.6×10^{-11}. Comparing this value to the (slightly larger) K_a of the ammonium ion, 5.6×10^{-10}, we see that NH_4^+ is slightly stronger as an acid than the formate ion is as a base. As a result, a solution of ammonium formate is slightly acidic. We are not concerned here about calculating a pH, only in predicting if the solution is acidic, basic, or neutral.

■ If *neither* the cation nor the anion is able to affect the pH, the salt solution will be neutral (provided no other acidic or basic solutes are present).

Will an aqueous solution of ammonium cyanide, NH_4CN, be acidic, basic, or neutral? (*Hint:* These anions and cations are conjugates of an acid and a base listed in Tables 16.2 and 16.3.)

Practice Exercise 16.30

Will an aqueous solution of ammonium nitrite, NH_4NO_2, be acidic, basic, or neutral?

Practice Exercise 16.31

16.7 | Buffer Solutions

Many chemical and biological systems are quite sensitive to pH. For example, if the pH of your blood were to change from what it should be, within the range of 7.35 to 7.42, either to 7.00 or to 8.00, you would die. Lakes and streams with a pH less than 5 often cannot support fish life. Thus, a change in pH can produce unwanted effects, and systems that are sensitive to pH must be protected from the H^+ or OH^- that might be formed or consumed by some reaction. Buffers are mixtures of solutes that accomplish this. Such a solution is said to be buffered or it is described as a buffer solution.

Composition of a Buffer

A **buffer** contains solutes that enable it to resist large changes in pH when small amounts of either strong acid or strong base are added to it. Ordinarily, the buffer consists of two solutes, one a weak Brønsted acid and the other its conjugate base. If the acid is molecular, then the conjugate base is supplied by a soluble salt of the acid. For example, a common buffer system consists of acetic acid plus sodium acetate, with the salt's acetate ion serving as the Brønsted base. In your blood, carbonic acid (H_2CO_3, a weak diprotic acid) and the bicarbonate ion (HCO_3^-, its conjugate base) serve as one of the buffer systems used to maintain a remarkably constant pH in the face of the body's production of organic acids by metabolism. Another common buffer consists of the weakly acidic cation, NH_4^+, supplied by a salt like NH_4Cl, and its conjugate base, NH_3.

One important point about buffers is the distinction between keeping a solution at a particular pH and keeping it neutral—at a pH of 7. Although it is certainly possible to buffer a solution at pH 7, buffers can be made that will work at any pH value throughout the pH scale.

How a Buffer Works

TOOLS
How buffers work

To work, a buffer must be able to neutralize either a strong acid or strong base that is added. This is precisely what the conjugate acid and base components of the buffer do. Consider, for example, a buffer composed of acetic acid, $HC_2H_3O_2$, and acetate ion, $C_2H_3O_2^-$, supplied by a salt such as $NaC_2H_3O_2$. If we add extra H^+ to the buffer (from a strong acid) the acetate ion (the weak conjugate base) can react with it as follows.

$$H^+(aq) + C_2H_3O_2^-(aq) \longrightarrow HC_2H_3O_2(aq)$$

■ Not all the added H^+ is neutralized, so the pH is lowered a little. Soon we'll see how much it changes.

Thus, the added H^+ changes some of the buffer's Brønsted base, $C_2H_3O_2^-$, to its conjugate (weak) acid, $HC_2H_3O_2$. This reaction prevents a large buildup of H^+ and a corresponding decrease in pH that would otherwise be caused by the addition of the strong acid.

A similar response occurs when a strong base is added to the buffer. The OH^- from the strong base will react with some $HC_2H_3O_2$.

$$HC_2H_3O_2(aq) + OH^-(aq) \longrightarrow C_2H_3O_2^-(aq) + H_2O$$

Here the added OH^- changes some of the buffer's Brønsted acid, $HC_2H_3O_2$, into its conjugate base, $C_2H_3O_2^-$. This prevents a buildup of OH^-, which would otherwise cause a large increase in the pH. Thus, one member of a buffer team neutralizes H^+ that might get into the solution, and the other member neutralizes OH^-. Understanding buffers is an important tool for chemists to use in applications ranging from the protocols of a research project to designing a consumer product.

Practice Exercise 16.32

Acetic acid, $HC_2H_3O_2$, and sodium acetate, $NaC_2H_3O_2$ (this provides the acetate ion $C_2H_3O_2^-$), can be used to make an "acetate" buffer. Does the acetate ion or the acetic acid increase when a strong acid is added to the buffer? Is it the acetate ion or acetic acid that decreases when a strong base is added to the buffer? Explain your answers. (*Hint:* Which of the buffer components will react with HCl? Which will react with NaOH?)

Practice Exercise 16.33

For a buffer composed of NH_3 and NH_4^+ (from NH_4Cl), write chemical equations that show what happens when (a) a small amount of strong acid is added, and (b) a small amount of strong base is added.

Calculating the pH of a Buffer Solution

Calculating the pH of a buffer mixture follows the same procedures we employed in Section 16.5 with a few small changes. The following example illustrates the principles involved.

Example 16.5
Calculating the pH of a Buffer

To study the effects of a weakly acidic medium on the rate of corrosion of a metal alloy, a student prepared a buffer solution containing both 0.110 M $NaC_2H_3O_2$ and 0.090 M $HC_2H_3O_2$. What is the pH of this solution?

Analysis: We recognize that this is a buffer solution because it contains *both* the weak acid $HC_2H_3O_2$ and its conjugate base $C_2H_3O_2^-$. Also, earlier we noted that when both solute species are present we can use *either* K_a or K_b to perform calculations, whichever is handy.

Assembling the Tools: In our tables we find $K_a = 1.8 \times 10^{-5}$ for $HC_2H_3O_2$, so the simplest approach is to use K_a and the corresponding balanced chemical equation for the ionization of the acid.

$$HC_2H_3O_2 \rightleftharpoons H^+ + C_2H_3O_2^-$$

$$K_a = \frac{[H^+][C_2H_3O_2^-]}{[HC_2H_3O_2]}$$

We will also be able to use the simplifying approximations developed earlier; these almost always work for buffers.

Solution: As usual, we begin with the chemical equation and the expression for K_a that we wrote above. Next, let's set up the concentration table so we can proceed logically. We will take the initial concentrations of $HC_2H_3O_2$ and $C_2H_3O_2^-$ to be the values given in the problem. There's no H^+ present from a strong acid, so we set this concentration equal to zero. If the initial concentration of H^+ is zero, its concentration must increase on the way to equilibrium, so under H^+ in the change row we enter $+x$. The other changes follow from that. For buffer solutions the quantity x will be very small, so it is safe to make the simplifying approximations in the last row. Here's the completed table.

	$HC_2H_3O_2$	$\rightleftharpoons$	H^+	$+$	$C_2H_3O_2^-$
Initial concentrations (M)	0.090		0		0.110
Changes in concentrations (M)	$-x$		$+x$		$+x$
Equilibrium concentrations (M)	$(0.090 - x) \approx 0.090$		x		$(0.110 + x) \approx 0.110$

What remains, then, is to substitute the quantities from the last row of the table into the K_a expression.

$$\frac{(x)(0.110 + x)}{(0.090 - x)} \approx \frac{(x)(0.110)}{(0.090)} = 1.8 \times 10^{-5}$$

Solving for x gives us

$$x = \frac{(0.090) \times (1.8 \times 10^{-5})}{(0.110)} = 1.47 \times 10^{-5}$$

Because x equals $[H^+]$, we now have $[H^+] = 1.47 \times 10^{-5}$ M. Then we calculate pH:

$$pH = -\log(1.47 \times 10^{-5}) = 4.833$$

Thus, the pH of the buffer is 4.83, when properly rounded.

Is the Answer Reasonable? Again, we check our assumptions and find they worked ($0.090 - 0.0000147 = 0.090$ and $0.110 + 0.0000147 = 0.110$). Also, we can check the answer in the usual way by substituting our calculated equilibrium values into the mass action expression. Let's do it.

$$\frac{[H^+][C_2H_3O_2^-]}{[HC_2H_3O_2]} = \frac{(1.5 \times 10^{-5})(0.110)}{(0.0.90)} = 1.8 \times 10^{-5}$$

■ Notice how small x is compared to the initial concentrations. The simplification was valid.

The reaction quotient equals K_a, so the values we've obtained are correct equilibrium concentrations.

Practice Exercise 16.34

Calculate the pH of the buffer solution in the preceding example by using the K_b for $C_2H_3O_2^-$. Be sure to write the chemical equation for the equilibrium as the reaction of $C_2H_3O_2^-$ with water. Then use the chemical equation as a guide in setting up the equilibrium expression for K_b. (*Hint:* If you work the problem correctly, you should obtain the same answer as above.)

Practice Exercise 16.35

One liter of buffer is made by dissolving 100.0 grams of acetic acid, $HC_2H_3O_2$, and 100.0 grams of sodium acetate, $NaC_2H_3O_2$, in enough water to make one liter. What is the pH of this solution?

The "Common Ion Effect"

If the solution in the preceding example had contained only acetic acid with a concentration of 0.090 M, the calculated $[H^+]$ would have been $1.3 \times 10^{-3} M$, considerably higher than that of the buffer, which also contains 0.110 M $C_2H_3O_2^-$. The effect of adding sodium acetate, a substance containing $C_2H_3O_2^-$ ions, to a solution of acetic acid is to suppress the ionization of the acid—it's an example of Le Châtelier's principle. Suppose, for example, we had established the equilibrium

$$HC_2H_3O_2 \rightleftharpoons H^+ + C_2H_3O_2^-$$

According to Le Châtelier's principle, if we add $C_2H_3O_2^-$, the reaction will proceed in a direction to remove some of it. This will shift the equilibrium to the left, thereby reducing the concentration of H^+.

In this example, acetate ion is said to be a **common ion**, in the sense that it is *common* to both the acetic acid equilibrium and to the salt we added, sodium acetate. The suppression of the ionization of acetic acid by addition of the common ion is referred to as the **common ion effect**. We will encounter this phenomenon again in Chapter 17 when we consider the solubilities of salts.

Simplifications in Buffer Calculations

There are two useful simplifications that we can use in working buffer calculations. The first is the one we described in Section 16.5:

■ There are some buffer systems for which these simplifications might not work. However, you will not encounter them in this text.

Because the initial concentrations are so close to the equilibrium concentrations in the buffer mixture, we can use *initial* concentrations of both the weak acid and its conjugate base as though they were equilibrium values.

A further simplification can be made because the mass action expression contains the ratio of the molar concentrations (in units of moles per liter) of the acid and conjugate base. Let's enter these units for the acid and its conjugate base into the mass action expression. For an acid HA,

$$K_a = \frac{[H^+][A^-]}{[HA]} = \frac{[H^+](\text{mol } A^- \text{ L}^{-1})}{(\text{mol } HA^- \text{ L}^{-1})} = \frac{[H^+](\text{mol } A^-)}{(\text{mol } HA^-)} \quad \textbf{(16.16)}$$

Notice that the units L^{-1} can be canceled from the numerator and denominator. This means that for a given acid–base pair, $[H^+]$ is determined by the *mole* ratio of conjugate base to conjugate acid; we don't *have* to use molar concentrations.

For buffer solutions *only*, we can use either molar concentrations or moles in the K_a (or K_b) expression to express the amounts of the acid and its conjugate base. However, we must use the same units for each member of the pair.

A further consequence of the relationship derived in Equation 16.16 is that the pH of a buffer should not change if the buffer is diluted. Dilution changes the volume of a solution but it does not change the number of moles of the solutes, so their mole *ratio* remains constant and so does [H$^+$].

Preparing Buffers with a Desired pH

The hydrogen ion concentration of a buffer is controlled by both K_a and the ratio of concentrations (or ratio of moles) of the members of the acid–base pair. This can be seen by rearranging Equation 16.16 to solve for [H$^+$]:

$$[H^+] = K_a \frac{[HA]}{[A^-]} \qquad (16.17)$$

or

$$[H^+] = K_a \frac{\text{mol } HA}{\text{mol } A^-} \qquad (16.18)$$

Solutions of conjugate acid–base pairs will generally act as buffers when the mole (or molarity) ratio of acid to base is between 0.1 and 10. When this ratio is less than 0.1, added base will cause large pH changes. When the ratio is greater than 10, small additions of acid will cause large pH changes. When the ratio is close to 1.0, the solution will be effective if either strong acid or strong base is added. Therefore, if we want to prepare a buffer that works well near some specified pH, we look for an acid with a K_a as close to the desired pH as possible. In general, the pK_a should be within ±1 pH unit of the desired pH:

$$pH = pK_a \pm 1$$

Also, a high concentration of both the conjugate acid and base will increase the buffer's ability to resist pH change. This is called the **buffer capacity**, and it is equal to the moles of strong acid, or strong base, needed to change the pH of one liter of buffer by 1.0 pH unit. For experiments in biology, the toxicity of the members of the acid–base pair must also be considered, and that often narrows the choices considerably.

Example 16.6
Preparing a Buffer Solution to Have a Predetermined pH

A solution buffered at a pH of 5.00 is needed in an experiment. Can we use acetic acid and sodium acetate to make it? If so, what is the mole ratio of acetic acid and sodium acetate that will result in a pH = 5.00 solution? How many moles of $NaC_2H_3O_2$ must be added to 1.0 L of a solution that contains 1.0 mol $HC_2H_3O_2$ to prepare the buffer?

Analysis: We need to do three things in this problem. First, we need to decide if, in fact, acetic acid can be used for a pH = 5.00 buffer. Second, we need to use the appropriate equations to determine the mole ratio and then, third, the amount of sodium acetate to use.

Assembling the Tools: The first tool describes the necessary relationship between pK_a and pH to successfully prepare a buffer solution:

$$pH = pK_a \pm 1$$

If we only have the K_a we also need Equation 16.6 to calculate the pK_a. Next, we need the equations to convert pH to [H$^+$] and then use it to solve Equation 16.16 to calculate the necessary mole ratio. Once we have this, we can proceed to calculate the number of moles of $C_2H_3O_2^-$ needed, and then the number of moles of $NaC_2H_3O_2$.

Solution: Looking up the K_a for acetic acid we find it is 1.8×10^{-5}. Because we want the pH to be 5.00, the pK_a of the selected acid should be 5.00 ± 1. We use Equation 16.6 to calculate the pK_a.

$$pK_a = -\log(1.8 \times 10^{-5}) = 4.74$$

From this we determine that acetic acid is appropriate for this buffer.

Next, we use Equation 16.18 to find the mole ratio of solutes.

$$[\text{H}^+] = K_a \times \frac{(\text{mol HC}_2\text{H}_3\text{O}_2)}{(\text{mol C}_2\text{H}_3\text{O}_2{}^-)}$$

First, let's solve for the mole ratio.

$$\frac{(\text{mol HC}_2\text{H}_3\text{O}_2)}{(\text{mol C}_2\text{H}_3\text{O}_2{}^-)} = \frac{[\text{H}^+]}{K_a}$$

The desired pH = 5.00, so $[\text{H}^+] = 1.0 \times 10^{-5}$; also $K_a = 1.8 \times 10^{-5}$. Substituting gives

$$\frac{(\text{mol HC}_2\text{H}_3\text{O}_2)}{(\text{mol C}_2\text{H}_3\text{O}_2{}^-)} = \frac{1.0 \times 10^{-5}}{1.8 \times 10^{-5}} = 0.56$$

This is the *mole* ratio of the buffer components that is needed.

For the final step, the solution we are preparing contains 1.0 mol $\text{HC}_2\text{H}_3\text{O}_2$, so the number of moles of acetate ion required is

$$\text{moles C}_2\text{H}_3\text{O}_2{}^- = \frac{1.0 \text{ mol HC}_2\text{H}_3\text{O}_2}{0.56}$$

$$= 1.8 \text{ mol C}_2\text{H}_3\text{O}_2{}^-$$

For each mole of $\text{NaC}_2\text{H}_3\text{O}_2$ there is one mole of $\text{C}_2\text{H}_3\text{O}_2{}^-$, so to prepare the solution we need 1.8 mol $\text{NaC}_2\text{H}_3\text{O}_2$.

Are the Answers Reasonable? A 1-to-1 mole ratio of $\text{C}_2\text{H}_3\text{O}_2{}^-$ to $\text{HC}_2\text{H}_3\text{O}_2$ would give pH = pK_a = 4.74. The desired pH of 5.00 is slightly more basic than 4.74, so can expect that the amount of conjugate base needed should be larger than the amount of conjugate acid. Our answer of 1.8 mol of $\text{NaC}_2\text{H}_3\text{O}_2$ appears to be reasonable.

Practice Exercise 16.36

From Table 16.2 select an acid that, along with its sodium salt, can be used to make a buffer that has a pH of 5.25. If you have 500.0 mL of a 0.200 M solution of that acid, how many grams of the corresponding sodium salt do you have to dissolve to obtain the desired pH? (*Hint:* There is more than one correct answer to this problem. The first step is to determine the ratio of molarities of the conjugate acid and conjugate base.)

Practice Exercise 16.37

A chemist needed an aqueous buffer with a pH of 3.90. Would formic acid and its salt, sodium formate, make a good pair for this purpose? If so, what mole ratio of the acid, HCHO_2, to the anion of this salt, $\text{CHO}_2{}^-$, is needed? How many grams of NaCHO_2 would have to be added to a solution that contains 0.10 mol HCHO_2?

If you take a biology course, you're likely to run into a logarithmic form of Equation 16.17 called the **Henderson–Hasselbalch equation**. This is obtained by taking the negative logarithm of both sides of Equation 16.9 and rearranging to get

TOOLS
Henderson–Hasselbalch equation

$$\text{pH} = pK_a + \log\frac{[A^-]}{[HA]} \qquad (16.19)$$

In most buffers used in the life sciences, the anion A^- comes from a salt in which the cation has a charge of 1+, such as NaA, and the acid is monoprotic. With these as conditions, the equation is sometimes written

$$\text{pH} = pK_a + \log\frac{[\text{salt}]}{[\text{acid}]} \qquad (16.20)$$

For practice, you may wish to apply this equation to buffer problems at the end of the chapter.

Calculating pH Change for a Buffer

Earlier we described how a buffer is able to neutralize small amounts of either strong acid or base to keep the pH from changing by large amounts. We can use this knowledge now to calculate how much the pH will change.

Example 16.7
Calculating How a Buffer Resists Changes in pH

How much will the pH change if 0.020 mol of HCl is added to a buffer solution that was made by dissolving 0.12 mol of NH_3 and 0.095 mol of NH_4Cl in 250 mL of water?

Analysis: This problem will require two calculations. The first is to determine the pH of the original buffer. The second is to determine the pH of the mixture after the HCl has been added. For the second calculation, we have to determine how the acid changes the amounts of NH_3 and NH_4^+.

Assembling the Tools: For the pH of the initial solution we need the chemical reaction and the K_b for NH_3.

$$NH_3 + H_2O \rightleftharpoons NH_4^+ + OH^- \qquad K_b = \frac{[NH_4^+][OH^-]}{[NH_3]} = 1.8 \times 10^{-5}$$

To determine the pH of the solution after the HCl was added we need the chemical reaction for HCl in this mixture. Since HCl is a strong acid, it will react quantitatively (completely) with the conjugate base of the buffer—in this case, NH_3. The balanced molecular equation is

$$NH_3 + HCl \longrightarrow NH_4Cl$$

In this case it is better to write the net ionic equation as

$$NH_3 + H^+ \longrightarrow NH_4^+$$

We then do a limiting reactant calculation to determine how much of the NH_3 will be converted to NH_4^+. We then use the new amounts of NH_3 and NH_4^+ to determine the pH.

Solution: We begin by calculating the pH of the buffer before the HCl is added. The solution contains 0.12 mol of NH_3 and 0.095 mol of NH_4^+ from the complete dissociation of the salt NH_4Cl. We can enter these quantities into the mass action expression in place of concentrations and solve for $[OH^-]$.

> ■ The solution also contains 0.095 mol of Cl^-, of course, but this ion is not involved in the equilibrium.

$$1.8 \times 10^{-5} = \frac{(mol\ NH_4^+) \times [OH^-]}{(mol\ NH_3)} = \frac{(0.095) \times [OH^-]}{(0.12)}$$

solving for $[OH^-]$ gives

$$[OH^-] = 2.3 \times 10^{-5}$$

To calculate the pH, we obtain pOH and subtract it from 14.00.

$$pOH = -\log(2.3 \times 10^{-5}) = 4.64$$

$$pH = 14.00 - 4.64 = 9.36$$

This is the pH before we add any HCl.

Next, we consider the reaction that takes place when we add the HCl to the buffer. The 0.020 mol of HCl is completely ionized, so we're adding 0.020 mol H^+. The 0.020 mol of acid will react with 0.020 mol of NH_3 to form 0.020 mol of NH_4^+. This causes the

■ The equilibrium constant for this reaction is 1.8×10^{10}, so the reaction goes almost to completion. Nearly all the added H^+ reacts with NH_3.

number of moles of NH_3 to *decrease* by 0.020 mol and the number of moles of NH_4^+ to *increase* by 0.020 mol. After addition of the acid, we have

Number of moles in the original buffer mixture | Change in the number of moles because of reaction

$$\text{moles of } NH_3 = 0.12 \text{ mol} - 0.020 \text{ mol} = 0.10 \text{ mol } NH_3$$

$$\text{moles of } NH_4^+ = 0.095 \text{ mol} + 0.020 \text{ mol} = 0.115 \text{ mol } NH_4^+$$

We now use these new amounts of NH_3 and NH_4^+ to calculate the new pH of the buffer solution. First, we calculate

$$1.8 \times 10^{-5} = \frac{(\text{mol } NH_4^+) \times [OH^-]}{(\text{mol } NH_3)} = \frac{(0.115)[OH^-]}{(0.10)}$$

Solving for $[OH^-]$ gives

$$[OH^-] = 1.6 \times 10^{-5}$$

To calculate the pH, we obtain pOH and subtract it from 14.00.

$$\text{pOH} = -\log(1.6 \times 10^{-5}) = 4.80$$

$$\text{pH} = 14.00 - 4.80 = 9.20$$

This is the pH after addition of the acid. We're asked for the change in pH, so we take the difference between the two values:

$$\text{change in pH} = 9.20 - 9.36 = -0.16$$

Thus, the pH has decreased by 0.16 pH units.

Is the Answer Reasonable? In general, we can assume that changes in pH should be small (less than 1.0 pH unit), otherwise there's no simple way to estimate how *much* the pH will change when H^+ or OH^- is added to the buffer. However, we can anticipate the *direction* of the change. Adding H^+ to a buffer will lower the pH somewhat and adding OH^- will raise it. In this example H^+ is added to the buffer; so its pH should decrease.

Practice Exercise 16.38 How much will the pH change if we add 0.15 mol NaOH to 1.00 L of a buffer that contains 1.00 mol $HC_2H_3O_2$ and 1.00 mol $NaC_2H_3O_2$? (*Hint:* Will the NaOH react with $HC_2H_3O_2$ or $NaC_2H_3O_2$? Write the equation for the reaction.)

Practice Exercise 16.39 A buffer is prepared by mixing 50.0 g of NH_3 and 50.0 g of NH_4Cl in 0.500 L of solution. What is the pH of this buffer, and what will the pH change to if 5.00 g of HCl is then added to the mixture?

If you look back over the calculations in the preceding example, you can get some feel for how effective buffers can be at preventing large swings in pH. Notice that we've added enough HCl to react with approximately 20% of the NH_3 in the solution, but the pH changed by only 0.16 pH units. If we were to add this same amount of HCl to 0.250 L of pure water, it would give a solution with $[H^+] = 0.080\ M$. Such a solution would have a pH of 1.10, which is quite acidic. However, the buffer mixture stayed basic, with a pH of 9.20, even after addition of the acid. That's pretty remarkable!

16.8 | Polyprotic Acids

Until now our discussion of weak acids has focused entirely on equilibria involving monoprotic acids. There are, of course, many acids capable of supplying more than one H^+ per molecule. Recall that these are called polyprotic acids. Examples include sulfuric acid,

H_2SO_4, carbonic acid, H_2CO_3, and phosphoric acid, H_3PO_4. These acids undergo ionization in a series of steps, each of which releases one proton. For weak polyprotic acids, such as H_2CO_3 and H_3PO_4, each step is an equilibrium. Even sulfuric acid, which we consider a strong acid, is not completely ionized. Loss of the first proton to yield the HSO_4^- ion is complete, but the loss of the second proton is incomplete and involves an equilibrium. In this section, we will focus our attention on weak polyprotic acids as well as solutions of their salts.

Let's begin with the weak diprotic acid, H_2CO_3. In water, the acid ionizes in two steps, each of which is an equilibrium that transfers an H^+ ion to a water molecule.

$$H_2CO_3 + H_2O \rightleftharpoons H_3O^+ + HCO_3^-$$

$$HCO_3^- + H_2O \rightleftharpoons H_3O^+ + CO_3^{2-}$$

As usual, we can use H^+ in place of H_3O^+ and simplify these equations to give

$$H_2CO_3 \rightleftharpoons H^+ + HCO_3^-$$

$$HCO_3^- \rightleftharpoons H^+ + CO_3^{2-}$$

Each step has its own ionization constant, K_a, which we identify as K_{a_1} for the first step and K_{a_2} for the second. For carbonic acid,

$$K_{a_1} = \frac{[H^+][HCO_3^-]}{[H_2CO_3]} = 4.3 \times 10^{-7}$$

$$K_{a_2} = \frac{[H^+][CO_3^{2-}]}{[HCO_3^-]} = 5.6 \times 10^{-11}$$

Notice that each ionization makes a contribution to the total molar concentration of H^+, and one of our goals here is to relate the K_a values and the concentration of the acid to $[H^+]$. At first glance, this seems to be a formidable task, but certain simplifications are justified that make the problem relatively simple to solve.

Simplifications in Calculations

The principal factor that simplifies calculations involving many polyprotic acids is the large differences between successive ionization constants. Notice that for H_2CO_3, K_{a_1} is much larger than K_{a_2} (they differ by a factor of nearly 10,000). Similar differences between K_{a_1} and K_{a_2} are observed for many diprotic acids. One reason is that an H^+ is lost much more easily from the neutral H_2A molecule than from the HA^- ion. The stronger attraction of the opposite charges inhibits the second ionization. Typically, K_{a_1} is greater than K_{a_2} by a factor of between 10^4 and 10^5, as the data in Table 16.4 show. For a triprotic acid, such as phosphoric acid, H_3PO_4, the second acid ionization constant is similarly greater than the third.

Because K_{a_1} is so much larger than K_{a_2}, at equilibrium virtually all the H^+ in a solution of the acid comes from the first step in the ionization. In other words,

$$[H^+]_{equilibrium} = [H^+]_{first\ step} + [H^+]_{second\ step}$$

$$[H^+]_{first\ step} \gg [H^+]_{second\ step}$$

Therefore, we make the approximation that

$$[H^+]_{equilibrium} \approx [H^+]_{first\ step}$$

This means that as far as calculating the H^+ concentration is concerned, we can treat the acid as though it were a monoprotic acid and ignore the second step in the ionization. In addition we have

$$[HCO_3^-]_{equilibrium} = [HCO_3^-]_{first\ step} - [CO_3^{2-}]_{second\ step}$$

However, since $[CO_3^{2-}]_{second\ step} = [H^+]_{second\ step}$ and $[H^+]_{second\ step}$ is very small we can say that

$$[HCO_3^-]_{equilibrium} \approx [HCO_3^-]_{first\ step}$$

Vitamin C. Many fruits and vegetables contain ascorbic acid (vitamin C), a weak diprotic acid with the formula $H_2C_6H_6O_6$.

Polyprotic acids ionize stepwise

■ Additional K_a values of polyprotic acids are located in Appendix C.

TABLE 16.4	**Acid Ionization Constants for Polyprotic Acids**				
			Acid Ionization Constant for Successive Ionization (25 °C)		
Name	Formula	1st	2nd	3rd
Arsenic acid	H_3AsO_4	5.5×10^{-3}	1.7×10^{-7}	5.1×10^{-12}
Ascorbic acid (vitamin C)	$H_2C_6H_6O_6$	8.0×10^{-5}	1.6×10^{-12}	
Carbonic acid	H_2CO_3	4.3×10^{-7}	5.6×10^{-11}	
Citric acid (18 °C)	$H_3C_6H_5O_7$	7.4×10^{-4}	1.7×10^{-5}	4.0×10^{-7}
Hydrosulfuric acid	$H_2S(aq)$	8.9×10^{-8}	1×10^{-19}	
Oxalic acid	$H_2C_2O_4$	5.9×10^{-2}	6.4×10^{-5}	
Phosphoric acid	H_3PO_4	7.5×10^{-3}	6.2×10^{-8}	4.2×10^{-13}
Selenic acid	H_2SeO_4	Large	1.2×10^{-2}	
Selenous acid	H_2SeO_3	3.5×10^{-3}	1.5×10^{-9}	
Sulfuric acid	H_2SO_4	Large	1.2×10^{-2}	
Sulfurous acid	H_2SO_3	1.5×10^{-2}	6.3×10^{-8}	
Telluric acid	H_6TeO_6	2×10^{-8}	1×10^{-11}	
Tellurous acid	H_2TeO_3	3.5×10^{-3}	2.0×10^{-8}	

It is therefore reasonable to conclude that

$$[H^+]_{equilibrium} = [HCO_3^-]_{equilibrium}$$

This last equality leads to the interesting result

$$K_{a_2} = \frac{[H^+][CO_3^{2-}]}{[HCO_3^-]} = 5.6 \times 10^{-11} = [CO_3^{2-}]$$

meaning the molar concentration of the carbonate ion is simply K_{a_2} when the only solute is carbonic acid. This will be true for other diprotic acids that have widely different K_{a_1} and K_{a_2} values.

Example 16.8 illustrates how these relationships are applied.

Example 16.8
Calculating the Concentrations of Solute Species in a Solution of a Polyprotic Acid

Calculate the concentrations of all the species in a solution of 0.040 M H_2CO_3 as well as the pH of the solution.

Analysis: We will use the procedures developed above to determine $[H^+]$, pH, $[H_2CO_3]$, $[HCO_3^-]$, and $[CO_3^{2-}]$.

Assembling the Tools: As in any equilibrium problem, we begin with the chemical equations and the K_a expressions

$$H_2CO_3 \rightleftharpoons H^+ + HCO_3^- \qquad K_{a_1} = \frac{[H^+][HCO_3^-]}{[H_2CO_3]} = 4.3 \times 10^{-7}$$

$$HCO_3^- \rightleftharpoons H^+ + CO_3^{2-} \qquad K_{a_2} = \frac{[H^+][CO_3^{2-}]}{[HCO_3^-]} = 5.6 \times 10^{-11}$$

Once we have these equations, it is a matter of applying them, with our chemical insights, in the appropriate manner. The two approximations that will be very useful are

$$[H^+]_{equilibrium} \approx [H^+]_{first\ step}$$

and

$$[HCO_3^-]_{equilibrium} \approx [HCO_3^-]_{formed\ in\ first\ step}$$

Solution: To start with, we can treat H_2CO_3 as though it were a monoprotic acid as we did in Section 16.5. We know some of the H_2CO_3 ionizes, let's call this amount x. If x moles per liter of H_2CO_3 ionizes, then x moles per liter of H^+ and H_2CO_3 are formed, and the concentration of H_2CO_3 is reduced by x.

	H_2CO_3	$\rightleftharpoons$	H^+	$+$	HCO_3^-
Initial concentrations (M)	0.040		0		0
Changes in concentrations (M)	$-x$		$+x$		$+x$
Equilibrium concentrations (M)	$0.040 - x$		x		x

Substituting into the expression for K_{a_1} gives

$$\frac{x^2}{(0.040 - x)} = \frac{x^2}{0.040} = 4.3 \times 10^{-7}$$

Solving for x gives $x = 1.3 \times 10^{-4}$, so $[H^+] = [CO_3^-] = 1.3 \times 10^{-4}$. Notice that our simplifying assumption is true and $[H_2CO_3]$ remains at 0.040 M. From this we can now calculate the pH.

$$pH = -\log(1.3 \times 10^{-4}) = 3.89$$

Now let's calculate the concentration of CO_3^{2-}. We obtain this from the expression for K_{a_2}.

$$K_{a_2} = \frac{[H^+][CO_3^{2-}]}{[HCO_3^-]} = 5.6 \times 10^{-11}$$

In our analysis, we concluded that we can use the approximations

$$[H^+]_{equilibrium} \approx [H^+]_{first\ step}$$

$$[HCO_3^-]_{equilibrium} \approx [HCO_3^-]_{formed\ in\ first\ step}$$

Substituting the values obtained above gives us

$$\frac{(1.3 \times 10^{-4})[CO_3^{2-}]}{(1.3 \times 10^{-4})} = 5.6 \times 10^{-11}$$

$$[CO_3^{2-}] = 5.6 \times 10^{-11} = K_{a_2}$$

Let's summarize the results. At equilibrium, we have

$$[H_2CO_3] = 0.040\ M$$

$$[H^+] = [HCO_3^{2-}] = 1.3 \times 10^{-4}\ M \quad and \quad pH = 3.89$$

$$[CO_3^{2-}] = 5.6 \times 10^{-11}\ M$$

■ Keep in mind that for a given solution, the values of H^+ used in the expressions for K_{a_1} and K_{a_2} are identical. At equilibrium there is only *one* equilibrium H^+ concentration.

Are the Answers Reasonable? To start, we check our assumptions. First, the initial concentration was assumed to be much larger than the hydrogen ion concentration, and it is $(0.040 - 0.00013 = 0.040)$. The second approximation was that the amount of CO_3^{2-} formed would be very small in relation to the HCO_3^- (and also H^+) concentration calculated in the first step. This assumption is true too $(1.3 \times 10^{-4} - 5.6 \times 10^{-11} = 1.3 \times 10^{-4})$. We can also check that the hydrogen ions added in the second step are inconsequential $(1.3 \times 10^{-4} + 5.6 \times 10^{-11} = 1.3 \times 10^{-4})$.

We've checked our assumptions and they are okay. The pH is less than 7; this is what we expect from an acid. If more verification is needed, you can always perform a quick check of the answers by substituting them into the appropriate mass action expressions:

$$\frac{(1.3 \times 10^{-4})(1.3 \times 10^{-4})}{0.040} = 4.2 \times 10^{-7} \qquad \text{which is very close to } K_{a_1}$$

$$\frac{(1.3 \times 10^{-4})(5.6 \times 10^{-11})}{(1.3 \times 10^{-4})} = 5.6 \times 10^{-11} \qquad \text{which equals } K_{a_2}$$

One last thing: the problem asked for *all species* in this solution of H_2CO_3. Two other substances are in the reaction mixture, water and hydroxide ions, and a rigorous solution would also include these species.

Practice Exercise 16.40

Write the three ionization steps for phosphoric acid, H_3PO_4, and write the K_a expressions for each step. (*Hint:* Recall that one proton is ionized in each step.)

Practice Exercise 16.41

Ascorbic acid (vitamin C) is a diprotic acid, $H_2C_6H_6O_6$. (See Table 16.4.) Calculate $[H^+]$, pH, and $[C_6H_6O_6^{2-}]$ in a 0.10 M solution of ascorbic acid.

One of the most interesting observations we can derive from the preceding example is that *in a solution that contains a polyprotic acid as the only solute, the concentration of the ion formed in the second step of the ionization equals K_{a_2}.*

Salts of Polyprotic Acids

You learned earlier that the pH of a salt solution is controlled by whether the salt's cation, anion, or both are able to react with water. For simplicity, the salts of polyprotic acids that we will discuss here will be limited to those containing *nonacidic* cations, such as sodium or potassium ion. In other words, we will study salts in which the anion alone is basic and thereby affects the pH of a solution.

A typical example of a salt of a polyprotic acid is sodium carbonate, Na_2CO_3. It is a salt of H_2CO_3, and the salt's carbonate ion is a Brønsted base that is responsible for *two* equilibria that furnish OH^- ion and thereby affect the pH of the solution. These equilibria and their corresponding expressions for K_b are as follows:

$$CO_3^{2-}(aq) + H_2O \rightleftharpoons HCO_3^-(aq) + OH^-(aq) \qquad \textbf{(16.21)}$$

$$K_{b_1} = \frac{[HCO_3^-][OH^-]}{[CO_3^{2-}]}$$

$$HCO_3^-(aq) + H_2O \rightleftharpoons H_2CO_3(aq) + OH^-(aq) \qquad \textbf{(16.22)}$$

$$K_{b_2} = \frac{[H_2CO_3][OH^-]}{[HCO_3^-]}$$

The calculations required to determine the concentrations of the species involved in these equilibria are very much like those involving weak diprotic acids. The difference is that the chemical reactions here are those of bases instead of acids. In fact, *the simplifying assumptions in our calculations will be almost identical to those for weak polyprotic acids.*

First, let's determine the value of K_{b_1} and K_{b_2} for the successive equilibria in Equations 16.21 and 16.22. To obtain these constants, we will use the relationship that $K_a \times K_b = K_w$ along with the tabulated values of K_{a_1} and K_{a_2} for H_2CO_3.

At the start of this section we saw that $HCO_3^-(aq)$ and $CO_3^{2-}(aq)$ are associated with K_{a_2}. When a salt containing CO_3^{2-} reacts with water, the CO_3^{2-} is the base (Equation 16.21),

■ In general, K_b values for anions of polyprotic acids are not tabulated. When needed, they are calculated from the appropriate K_a values for the acids.

the conjugate acid is HCO_3^-. Therefore, K_{b_1} contains the same conjugate acid–base pair as K_{a_2}. Therefore, we calculate K_{b_1} from K_{a_2} as

$$K_{b_1} = \frac{K_w}{K_{a_2}} = \frac{1.0 \times 10^{-14}}{5.6 \times 10^{-11}} = 1.8 \times 10^{-4}$$

Similarly, we see that $H_2CO_3(aq)$ and $HCO_3^-(aq)$ are associated with K_a. When HCO_3^- is the base (Equation 16.22), the conjugate acid is H_2CO_3. Therefore, to calculate K_{b_2} we must use K_{a_1} because both have the same conjugate acid–base pair in the equilibrium expression

$$K_{b_2} = \frac{K_w}{K_{a_1}} = \frac{1.0 \times 10^{-14}}{4.3 \times 10^{-7}} = 2.3 \times 10^{-8}$$

■ Remember: K_{b_1} comes from K_{a_2}, and K_{b_2} comes from K_{a_1}.

Now we can compare the two K_b values. For CO_3^{2-}, $K_{b_1} = 1.8 \times 10^{-4}$, and for the (much) weaker base, HCO_3^-, K_{b_2} is 2.3×10^{-8}. Thus CO_3^{2-} has a K_b nearly 10,000 times that of HCO_3^-. This means that the reaction of CO_3^{2-} with water (Equation 16.21) generates far more OH^- than the reaction of HCO_3^- (Equation 16.22). The contribution of the latter to the total pool of OH^- is relatively so small that we can safely ignore it. The simplification here is very much the same as in our treatment of the ionization of weak polyprotic acids.

$$[OH^-]_{equilibrium} = [OH^-]_{formed\ in\ first\ step} + [OH^-]_{formed\ in\ second\ step}$$

$$[OH^-]_{formed\ in\ first\ step} \gg [OH^-]_{formed\ in\ second\ step}$$

$$[OH^-]_{equilibrium} \approx [OH^-]_{formed\ in\ first\ step}$$

Thus, if we wish to calculate the pH of a solution of a basic anion of a polyprotic acid, *we may work exclusively with K_{b_1} and ignore any further reactions.* Let's study an example of how this works.

Example 16.9
Calculating the pH of a Solution of a Salt of a Weak Diprotic Acid

What is the pH of a 0.11 M solution of Na_2CO_3?

Analysis: Based on the discussion above we will approach this question in the same way that we solved Example 16.4 previously for a salt of a monoprotic acid.

Assembling the Tools: We can ignore the reaction of HCO_3^- with water, so the only relevant equilibrium is

$$CO_3^{2-} + H_2O \rightleftharpoons HCO_3^- + OH^- \qquad K_{b_1} = \frac{[HCO_3^-][OH^-]}{[CO_3^{2-}]} = 1.8 \times 10^{-4}$$

(The value of K_{b_1} was calculated from K_{a_2} in the discussion preceding this example.)

Solution: Some of the CO_3^{2-} will react; we'll let this be x mol L^{-1}. The concentration table, with our simplifying assumption, is as follows.

	H_2O +	CO_3^{2-}	$\rightleftharpoons$	HCO_3^- +	OH^-
Initial concentrations (M)		0.11		0	0
Changes in concentrations (M)		$-x$		$+x$	$+x$
Equilibrium concentrations (M)		$(0.11 - x) \approx 0.11$		x	x

Now we're ready to insert the values in the last row of the table into the equilibrium expression for the carbonate ion.

$$\frac{[HCO_3^-][OH^-]}{[CO_3^{2-}]} = \frac{(x)(x)}{0.11 - x} \approx \frac{(x)(x)}{0.11} = 1.8 \times 10^{-4}$$

Some detergents that use sodium carbonate as their caustic ingredient. Detergents are best able to dissolve grease and grime if their solutions are basic. In these products, the basic ingredient is sodium carbonate. The salt Na_2CO_3 is relatively nontoxic and the carbonate ion it contains is a relatively strong base. The carbonate ion also serves as a water softener by precipitating Ca^{2+} ions as insoluble $CaCO_3$.

$$x = [OH^-] = 4.5 \times 10^{-3} \text{mol L}^{-1}$$

and

$$pOH = -\log(4.5 \times 10^{-3}) = 2.35$$

Finally, we calculate the pH by subtracting pOH from 14.00.

$$pH = 14.00 - 2.35 = 11.65$$

Thus, the pH of 0.11 M Na$_2$CO$_3$ is calculated to be 11.65. Once again we see that an aqueous solution of a salt composed of a basic anion and a neutral cation is basic.

Is the Answer Reasonable? Our result was reasonable since our assumption was true: $(0.11 - 0.0048 = 0.11)$ when properly rounded. Also, the pH we obtained corresponds to a basic solution, which is what we expect when the anion of the salt reacts as a weak base. The answer seems reasonable.

Practice Exercise 16.42

Sodium bicarbonate gives solutions that are slightly alkaline (approximately pH 8.3), and similar solutions of sodium hydroxide (lye) are very basic with a pH around 12 or more. Lye will cause major damage, if not death, if swallowed, while sodium bicarbonate is a common antacid. Sodium carbonate forms solutions that are more basic than solutions of sodium bicarbonate. Calculate the pH of a 0.10 M sodium carbonate solution and decide if it is an acceptable substitute for sodium bicarbonate. (*Hint:* The conjugate base in this problem is the carbonate ion.)

Practice Exercise 16.43

What is the pH of a 0.20 M solution of Na$_2$SO$_3$ at 25 °C? For the diprotic acid, H$_2$SO$_3$, $K_{a_1} = 1.5 \times 10^{-2}$ and $K_{a_2} = 6.3 \times 10^{-8}$.

Practice Exercise 16.44

Reasoning by analogy from what you learned about solutions of weak polyprotic acids, what is the molar concentration of H$_2$SO$_3$ in a 0.010 M solution of Na$_2$SO$_3$?

16.9 | Acid–Base Titrations

In Chapter 4 we studied the overall procedure for an acid–base titration, and we saw how titration data can be used in various stoichiometric calculations. In performing this procedure, the titration is halted at the **end point** when a change in color of an **indicator** occurs. Ideally, this end point should coincide with the theoretical **equivalence point**, where the calculated, stoichiometrically equivalent, amounts of acid and base have combined. To obtain this ideal result, selecting an appropriate indicator requires foresight. We will understand this better by studying how the pH of the solution being titrated changes with the addition of titrant.

When the pH of a solution at different stages of a titration is plotted against the volume of titrant added, we obtain a **titration curve**. The values of pH in these plots can be experimentally measured using a pH meter during the titration, or they can be calculated by the procedures we will describe below.

We will use the calculation method, and also demonstrate that all titration curves can be described by four calculations. Two calculations describe single points on the curve, the starting point and the equivalence point. The other two calculations are simple limiting reactant calculations for (a) mixtures between the start and the equivalence point and for (b) after the equivalence point.

■ The *titrant* is the solution being slowly added from a buret to a solution in the receiving flask.

TOOLS

Titration curves have four parts

■ Not all acid–base reactions react the acid and base in a 1:1 ratio. We must always base our calculations on a balanced chemical equation.

Strong Acid–Strong Base Titrations

The titration of HCl(aq) with a standardized NaOH solution illustrates the titration of a strong acid by a strong base. The molecular and net ionic equations are

$$HCl(aq) + NaOH(aq) \longrightarrow NaCl(aq) + H_2O$$

$$H^+(aq) + OH^-(aq) \longrightarrow H_2O$$

Let's consider what happens to the pH of a solution, initially 25.00 mL of 0.2000 M HCl, as small amounts of the titrant, 0.2000 M NaOH, are added. We will calculate the pH of the resulting solution at various stages of the titration, retaining only two decimal places, and plot the values against the volume of titrant.

At the start, before any titrant has been added, the receiving flask contains only 0.2000 M HCl. Because this is a strong acid, we know that

$$[H^+] = [HCl] = 0.2000\ M$$

So the initial pH is

$$pH = -\log(0.2000) = 0.70$$

After the start but before the equivalence point we need to determine the concentration of the excess reactant in a simple limiting reactant problem. Let's first calculate the amount of HCl initially present in 25.00 mL of 0.2000 M HCl.

$$25.00\ \text{mL HCl soln} \times \frac{0.2000\ \text{mol HCl}}{1000\ \text{mL HCl soln}} = 5.000 \times 10^{-3}\ \text{mol HCl}$$

Now suppose we add 10.00 mL of 0.2000 M NaOH from the buret. The amount of NaOH added is

$$10.00\ \text{mL NaOH soln} \times \frac{0.2000\ \text{mol NaOH}}{1000\ \text{mL NaOH soln}} = 2.000 \times 10^{-3}\ \text{mol NaOH}$$

We can see that we have more moles of HCl than NaOH (and the mole ratio in the chemical equation is 1:1) so that the base neutralizes 2.000×10^{-3} mol of HCl, and the amount of HCl remaining is

$$(5.000 \times 10^{-3} - 2.000 \times 10^{-3})\ \text{mol HCl} = 3.000 \times 10^{-3}\ \text{mol HCl remaining}$$

To obtain the concentration we divide by the total volume of the solution that is now 25.00 mL + 10.00 mL = 35.00 mL = 0.03500 L. The $[H^+]$ is

$$[H^+] = \frac{3.000 \times 10^{-3}\ \text{mol}}{0.03500\ \text{L}} = 8.571 \times 10^{-2}\ M$$

The corresponding pH is 1.07.

We can repeat this calculation for as many different volumes as we care to choose as long as the volume of NaOH is less than the equivalence point volume. A spreadsheet will allow us to quickly calculate the pH of large numbers of data points, but only a few are needed to see the nature of the curve.

At the equivalence point we can calculate the volume of NaOH added as

$$M_{\text{HCl}} \times V_{\text{HCl}} = M_{\text{NaOH}} \times V_{\text{NaOH}}$$

$$V_{\text{NaOH}}(\text{at eq pt}) = \frac{(0.2000\ M_{\text{HCl}})(25.00\ \text{mL HCl})}{(0.2000\ M\ \text{NaOH})}$$

$$= 25.00\ \text{mL}$$

At this point we have exactly neutralized all of the acid with base and neither HCl nor NaOH is in excess. The solution contains only NaCl, and we have already observed that NaCl solutions are neutral with a pH = 7.00. The pH at the equivalence point of all strong acid–strong base titrations is 7.00.

After the equivalence point we will have an excess of base. As we did before, we calculate the moles of acid and the moles of base. Now the moles of base will be larger, and we subtract the moles of acid to find the excess moles of base. This is divided by the total volume to obtain the $[OH^-]$ from which the pH is calculated.

Table 16.5 shows the results of calculating the pH of the solution in the receiving flask after the addition of further small volumes of the titrant. Figure 16.7 shows a plot of the pH of the solution in the receiving flask versus the volume of the 0.2000 M NaOH solution

An acid–base titration.
Phenolphthalein is often used as an indicator to detect the end point.

■ Despite the high precision assumed for the molarity and the volume of the titrant, we will retain only two significant figures in the calculated pH. Precision higher than this is hard to obtain and seldom sought in actual lab work.

■ $[H^+] = \dfrac{M_A V_A - M_B V_B}{V_A + V_B}$

■ The pH, at the equivalence point, of all strong acid–strong base titrations is 7.00.

■ $[OH^-] = \dfrac{M_B V_B - M_A V_A}{V_A + V_B}$

TABLE 16.5	Titration of a Strong Acid with a Strong Base. The sample is 25.00 mL of 0.2000 M HCl and the titrant is 0.2000 M NaOH.						
Initial Volume of HCl (mL)	Initial Amount of HCl (mol)	Volume of NaOH Added (mL)	Amount of NaOH (mol)	Amount of Excess Reagent (mol)	Total Volume of Solution (mL)	Molarity of Ion in Excess (mol L^{-1})	pH
25.00	5.000×10^{-3}	0	0	5.000×10^{-3} (H$^+$)	25.00	0.2000 (H$^+$)	0.70
25.00	5.000×10^{-3}	10.00	2.000×10^{-3}	3.000×10^{-3} (H$^+$)	35.00	8.571×10^{-2} (H$^+$)	1.07
25.00	5.000×10^{-3}	20.00	4.000×10^{-3}	1.000×10^{-3} (H$^+$)	45.00	2.222×10^{-2} (H$^+$)	1.65
25.00	5.000×10^{-3}	24.00	4.800×10^{-3}	2.000×10^{-4} (H$^+$)	49.00	4.082×10^{-3} (H$^+$)	2.39
25.00	5.000×10^{-3}	24.90	4.980×10^{-3}	2.000×10^{-5} (H$^+$)	49.90	4.000×10^{-4} (H$^+$)	3.40
25.00	5.000×10^{-3}	24.99	4.998×10^{-3}	2.000×10^{-6} (H$^+$)	49.99	4.000×10^{-5} (H$^+$)	4.40
25.00	5.000×10^{-3}	25.00	5.000×10^{-3}	0	50.00	0	7.00
25.00	5.000×10^{-3}	25.01	5.002×10^{-3}	2.000×10^{-6} (OH$^-$)	50.01	3.999×10^{-5} (OH$^-$)	9.60
25.00	5.000×10^{-3}	25.10	5.020×10^{-3}	2.000×10^{-5} (OH$^-$)	50.10	3.992×10^{-4} (OH$^-$)	10.60
25.00	5.000×10^{-3}	26.00	5.200×10^{-3}	2.000×10^{-4} (OH$^-$)	51.00	3.922×10^{-3} (OH$^-$)	11.59
25.00	5.000×10^{-3}	50.00	1.000×10^{-2}	5.000×10^{-3} (OH$^-$)	75.00	6.667×10^{-2} (OH$^-$)	12.82

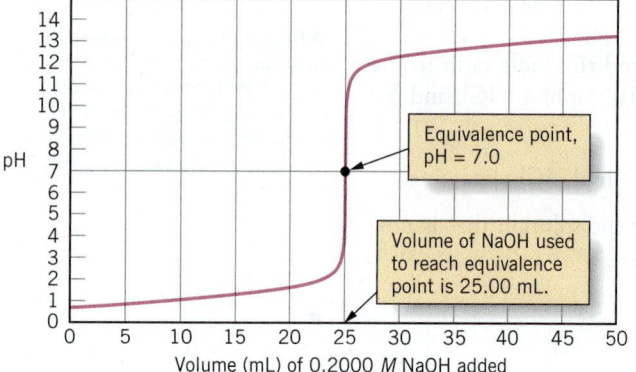

Figure 16.7 | Titration curve for titrating a strong acid with a strong base. This graph shows how the pH changes during the titration of 0.2000 M HCl with 0.2000 M NaOH.

added—the *titration curve* for the titration of a strong acid with a strong base. Notice how the pH of the solution increases slowly in a linear fashion until we are very close to the equivalence point, pH = 7.00. With the addition of a very small amount of base, the curve rises very sharply and then, almost as suddenly, bends to reflect just a gradual increase in pH.

Weak Acid–Strong Base Titrations

The calculations for the titration of a weak acid by a strong base are a bit more complex than when both the acid and base are strong. This is because we have to consider the equilibria involving the weak acid and its conjugate base. As an example, let's do the calculations and draw a titration curve for the titration of 25.0 mL of 0.2000 M HC$_2$H$_3$O$_2$ with 0.2000 M NaOH. The same four parts are calculated but they use different equations.

The molecular and net ionic equations we will use are

$$HC_2H_3O_2 + NaOH \longrightarrow NaC_2H_3O_2 + H_2O$$

$$HC_2H_3O_2 + OH^- \longrightarrow C_2H_3O_2^- + H_2O$$

Because the concentrations of both solutions are the same, and because the acid and base react in a one-to-one mole ratio, we know that we will need exactly 25.0 mL of the base to reach the equivalence point. With this as background, let's look at what is involved in calculating the pH at various points along the titration curve.

At the start the solution is simply a solution of the weak acid, HC$_2$H$_3$O$_2$. We must use K_a to calculate the pH. We have seen that in many situations the assumptions hold and the calculations are summarized as

$$K_a = \frac{[H^+][C_2H_3O_2^-]}{[HC_2H_3O_2]} = \frac{(x)(x)}{0.200 - x} \approx \frac{(x)(x)}{0.200} = 1.8 \times 10^{-5}$$

$$x^2 = 3.6 \times 10^{-6} \quad \text{(rounded)}$$

$$x = [H^+] = 1.9 \times 10^{-3}$$

This result gives us a pH of 2.72. This is the pH before any NaOH is added.

Between the start and the equivalence point. As we add NaOH to the $HC_2H_3O_2$, the chemical reaction produces $C_2H_3O_2^-$, so the solution contains both $HC_2H_3O_2$ and $C_2H_3O_2^-$ (it is a buffer solution). Our equilibrium law can be rearranged to read

$$[H^+] = \frac{K_a[HC_2H_3O_2]}{[C_2H_3O_2^-]} = \frac{K_a(\text{mol } HC_2H_3O_2)}{(\text{mol } C_2H_3O_2^-)}$$

Now all we need to do is solve a limiting reactant calculation to determine the moles of acetic acid, $HC_2H_3O_2$, left after each addition of NaOH and at the same time calculate the moles of acetate ion, $C_2H_3O_2^-$, formed.

Let's see what happens when we add 10.00 mL of NaOH solution. The initial moles of acetic acid are

$$0.2000 \, M \, HC_2H_3O_2 \times 0.02500 \, L = 5.000 \times 10^{-3} \, \text{mol } HC_2H_3O_2$$

the moles of NaOH added are

$$0.20000 \, M \, NaOH \times 0.01000 \, L = 2.000 \times 10^{-3} \, \text{mol NaOH}$$

The moles of acetic acid left are

$$5.000 \times 10^{-3} \, \text{mol} - 2.000 \times 10^{-3} \, \text{mol} = 3.000 \times 10^{-3} \, \text{mol } HC_2H_3O_2$$

From the chemical equations above, the moles of NaOH added are stoichiometrically equal to the moles of acetate ion formed. We therefore have 2.000×10^{-3} mol $C_2H_3O_2^-$. Entering the values for the K_a and the moles of acetic acid and acetate ions we get

$$[H^+] = \frac{(1.8 \times 10^{-5})(3.000 \times 10^{-3} \, \text{mol } HC_2H_3O_2)}{(2.000 \times 10^{-3} \, \text{mol } C_2H_3O_2^-)} = 2.7 \times 10^{-5} \, M$$

and the pH is 4.57. We can obtain more data points by selecting volumes of added NaOH between zero and 25 mL and repeating the calculation methods we just used. To generate a very large number of data points, repetitive calculations can be performed using a spreadsheet and the equations used above.

At the equivalence point. Here all the $HC_2H_3O_2$ has reacted and the solution contains the salt $NaC_2H_3O_2$ (i.e., we have a mixture of the ions Na^+ and $C_2H_3O_2^-$).

We began with 5.00×10^{-3} mol of $HC_2H_3O_2$, so we now have 5.00×10^{-3} mol of $C_2H_3O_2^-$ (in 0.0500 L). The concentration of $C_2H_3O_2^-$ is

$$[C_2H_3O_2^-] = \frac{5.00 \times 10^{-3} \, \text{mol}}{0.0500 \, L} = 0.100 \, M$$

Because $C_2H_3O_2^-$ is a base, we have to use K_b and write the chemical equation for the reaction of a base with water.

$$C_2H_3O_2^-(aq) + H_2O \rightleftharpoons HC_2H_3O_2(aq) + OH^-(aq)$$

$$K_b = \frac{[OH^-][HC_2H_3O_2]}{[C_2H_3O_2^-]} = \frac{K_w}{K_a} = 5.6 \times 10^{-10}$$

Substituting into the mass action expression gives

$$\frac{[OH^-][HC_2H_3O_2]}{[C_2H_3O_2^-]} = \frac{(x)(x)}{0.100 - x} \approx \frac{(x)(x)}{0.100} = 5.6 \times 10^{-10}$$

$$x^2 = 5.6 \times 10^{-11}$$

$$x = [OH^-] = 7.5 \times 10^{-6} \, M$$

The pOH calculates to be 5.12. Finally, the pH is $(14.00 - 5.12)$ or 8.88. The pH at the equivalence point in this titration is 8.88, which is somewhat basic because the solution contains the Brønsted base, $C_2H_3O_2^-$. *In the titration of any weak acid with a strong base, the pH at the equivalence point will be greater than 7.*

■ The end point in a titration can be found using a pH meter by plotting pH versus volume of base added and noting the sharp rise in pH. If a pH meter is used, we don't have to use an indicator.

■ $[H^+] = \sqrt{K_a M_{HA}}$

■ Recall from Chapter 4 that molarity multiplied by volume (in liters) gives moles of solute.

■ $[H^+] = K_a \dfrac{M_A V_A - M_B V_B}{V_A + V_B}$

■ $[OH^-] = \sqrt{K_b \dfrac{M_A V_A}{V_A + V_B}}$

■ It is interesting that although we describe the overall reaction in the titration as *neutralization*, at the equivalence point the solution is slightly basic, not neutral.

After the equivalence point. Now the additional OH^- beyond the equivalence point shifts the following equilibrium to the left:

$$C_2H_3O_2^-(aq) + H_2O \rightleftharpoons HC_2H_3O_2(aq) + OH^-(aq)$$

The production of OH^- *by this route* is thus suppressed as we add more and more NaOH solution. The only source of OH^- that affects the pH is from the base added after the equivalence point. From here on, therefore, the pH calculations become identical to the post-equivalence point calculations of the titration curve for HCl and NaOH shown above. Each point is just a matter of calculating the additional number of moles of NaOH, calculating the new final volume, taking their ratio (to obtain the molar concentration), and then calculating pOH and pH.

Table 16.6 summarizes the titration data, and the titration curve is given in Figure 16.8. Once again, notice how the change in pH occurs slowly at first, then shoots up through the equivalence point, and finally looks just like the last half of the curve in Figure 16.7.

■ $$[OH^-] = \frac{M_BV_B - M_AV_A}{V_A + V_B}$$

Practice Exercise 16.45

When a weak acid such as acetic acid, $HC_2H_3O_2$, is titrated with KOH what are ALL of the species in solution at (a) the equivalence point, (b) at the start of the titration, (c) after the equivalence point, and (d) before the equivalence point. For each part, list the species in order from highest concentration to lowest. (*Hint:* H_2O should be the first item in all your lists and either $[H^+]$ or $[OH^-]$ should be the lowest.)

Practice Exercise 16.46

Suppose we titrate 20.0 mL of 0.100 *M* $HCHO_2$ with 0.100 *M* NaOH. Calculate the pH (a) before any base is added, (b) when half of the $HCHO_2$ has been neutralized, (c) after a total of 15.0 mL of base has been added, and (d) at the equivalence point.

Practice Exercise 16.47

Suppose 30.0 mL of 0.15 *M* NaOH is added to 50.0 mL of 0.20 *M* $HCHO_2$. What will be the pH of the resulting solution?

Weak Base–Strong Acid Titrations

The calculations involved here are nearly identical to those of the weak acid–strong base titration. We will review what is required but will not actually perform the calculations.

If we titrate 25.00 mL of 0.2000 *M* NH_3 with 0.2000 *M* HCl, the molecular and net ionic equations are

$$NH_3(aq) + HCl(aq) \longrightarrow NH_4Cl(aq)$$

$$NH_3(aq) + H^+(aq) \longrightarrow NH_4^+(aq)$$

TABLE 16.6	Titration of a Weak Acid with a Strong Base. The sample is 25.00 mL of 0.2000 *M* HC_2H_3O and the titrant is 0.2000 *M* NaOH.		
Volume of Base Added (mL)	**Molar Concentration of Species in Parentheses**		**pH**
None	1.9×10^{-3} (H^+)		2.72
10.00	2.7×10^{-5} (H^+)		4.57
24.90	7.2×10^{-8} (H^+)		7.14
24.99	7.1×10^{-9} (H^+)		8.14
25.00	7.5×10^{-6} (OH^-)		8.88
25.01	4.0×10^{-5} (OH^-)		9.60
25.10	4.0×10^{-4} (OH^-)		10.60
26.00	3.9×10^{-3} (OH^-)		11.59
35.00	3.3×10^{-2} (OH^-)		12.52

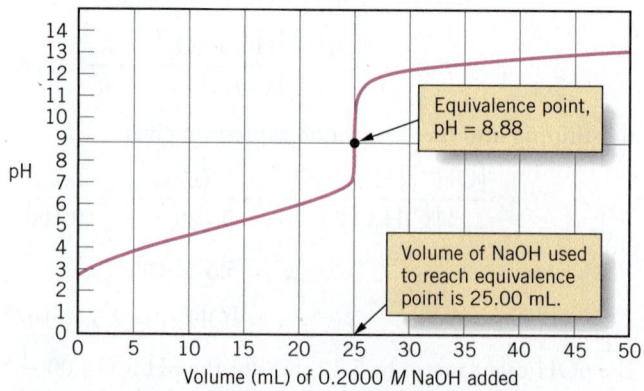

Figure 16.8 | **Titration curve for titrating a weak acid with a strong base.** In this titration, we follow the pH as 0.2000 *M* acetic acid is titrated with 0.2000 *M* NaOH.

Before the titration begins. The solution at this point is simply a solution of the weak base, NH_3. Since the only solute is a base, we must use K_b to calculate the pH.

Between the start and the equivalence point. As we add HCl to the NH_3, the neutralization reaction produces NH_4^+, so the solution contains both NH_3 and NH_4^+ (it is a buffer solution). We do a limiting reactant calculation to determine the moles of ammonia left and the moles of ammonium ions produced by the addition of HCl. We use the equilibrium law to perform the calculation as we did with the weak acid.

At the equivalence point. Here all the NH_3 has reacted and the solution contains the salt NH_4Cl. We calculate the concentration of ammonium ions and then use the techniques we developed to determine the pH of the conjugate acid of a weak base.

After the equivalence point. The H^+ introduced by further addition of HCl has nothing with which to react, so it causes the solution to become more and more acidic. The concentration of H^+ calculated from the excess of HCl is used to calculate the pH.

Figure 16.9 illustrates the titration curve for this system.

Titration Curves for Diprotic Acids

When a weak diprotic acid such as ascorbic acid (vitamin C) is titrated with a strong base, there are two protons to be neutralized and there are two equivalence points. Provided that the values of K_{a_1} and K_{a_2} differ by at least four powers of 10, the neutralization takes place stepwise and the resulting titration curve shows two sharp increases in pH. We won't perform the calculations here because they're complex, but it is important to recognize the general shape of the titration curve, which is shown in Figure 16.10.

Acid–Base Indicators

In Chapter 5 you learned that the end point of a titration could be detected when the color of the solution changed dramatically from one color to another. The substance that changed color was an acid–base indicator. Now we can understand how such indicators work.

Most dyes used as acid–base indicators are also weak acids. Therefore, let's represent an indicator molecule by the formula HIn. In its unionized state, HIn has one color. Its conjugate base, In^-, has a different color, the more strikingly different the better. In solution, the indicator is involved in a typical acid–base equilibrium:

$$H\mathit{In}(aq) \rightleftharpoons H^+(aq) + \mathit{In}^-(aq)$$

Acid form   base form
(one color)   (another color)

■ Indicators and other molecules can absorb part of the visible spectrum of light. What is not absorbed is observed as the color. If a molecule absorbs the red and yellow wavelengths of light, it will look blue.

TOOLS

How acid–base indicators work

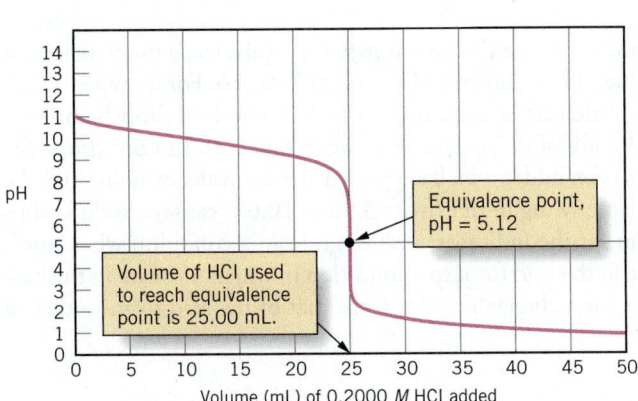

Figure 16.9 | Titration curve for the titration of a weak base with a strong acid. This graph shows how the pH changes as 0.2000 *M* NH_3 is titrated with 0.2000 *M* HCl.

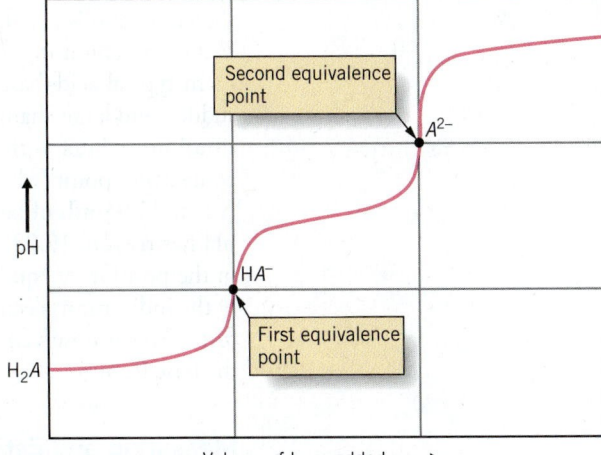

Figure 16.10 | Titration of a diprotic acid, H_2A, by a strong base. As each equivalence point is reached, there is a sharp rise in the pH.

TABLE 16.7	Common Acid–Base Indicators	
Indicator	Approximate pH Range over which the Color Changes	Color Change (lower to higher pH)
Methyl green	0.2–1.8	Yellow to blue
Thymol blue	1.2–2.8	Yellow to blue
Methyl orange	3.2–4.4	Red to yellow
Ethyl red	4.0–5.8	Colorless to red
Methyl purple	4.8–5.4	Purple to green
Bromocresol purple	5.2–6.8	Yellow to purple
Bromothymol blue	6.0–7.6	Yellow to blue
Phenol red	6.4–8.2	Yellow to red/violet
Litmus	4.7–8.3	Red to blue
Cresol red	7.0–8.8	Yellow to red
Thymol blue	8.0–9.6	Yellow to blue
Phenolphthalein	8.2–10.0	Colorless to pink
Thymolphthalein	9.4–10.6	Colorless to blue
Alizarin yellow R	10.1–12.0	Yellow to red
Clayton yellow	12.2–13.2	Yellow to amber

The corresponding acid ionization constant, K_{In}, is given by

$$K_{In} = \frac{[H^+][In^-]}{[HIn]}$$

In a strongly acidic solution, when the H^+ concentration is high, the equilibrium is shifted to the left and most of the indicator molecules exist in its protonated or "acid form." Under these conditions, the color we observe is that of HIn. If the solution is made basic, the H^+ concentration decreases and the equilibrium shifts to the right, toward In^-, and the color we observe is that of the unprotonated "base form" of the indicator. Table 16.7 provides approximate pH ranges for the color changes of some common pH indicators.

It is generally agreed that if the ratio of $[HIn]/[In^-]$ is greater than 10 (i.e., the pH is 1 pH unit below the pK_{In}) the eye will see the color associated with the HIn molecule. If the same ratio is less than 0.1 (i.e., the pH is 1 pH unit above the pK_{In}), the color seen is that of the anion In^-.

In typical acid–base titrations, we see that as we pass the equivalence point there is a sudden and large change in the pH, as shown in Figures 16.7–16.10. For example, in the titration of HCl with NaOH described earlier, the pH just one-half drop before the equivalence point (when 24.97 mL of the base has been added) is 3.92. Just one drop later (when 25.03 mL of base has been added) we have passed the equivalence point and the pH has risen to 10.08. This large swing in pH (from 3.92 to 10.08) causes a sudden shift in the position of equilibrium for the indicator, and we go from a condition where most of the indicator molecules are in the acid form to a condition in which most are in the base form. This is observed visually as a change in color from that of the acid form to that of the base form.

Measuring and Estimating pH

Earlier we saw that pH is usually measured with an instrument called a pH meter (Figure 16.2). We can now understand that a buffer solution is used to calibrate the pH meter because the pH of a buffer is reasonably stable. Typically, the accuracy of a pH meter is limited by the ±0.02 pH unit accuracy of the standardizing buffer. Modern pH meters can give much more precise readings if we are interested in changes in pH rather than the absolute pH value.

Another less precise method of obtaining the pH uses acid–base indicators. Table 16.7 gives several examples. Indicators change color over a narrow range of pH values, and Figure 16.11 shows the colors of some indicators at opposite ends of their color-change ranges. In addition, pH test papers are available that are impregnated with one or more indicator dyes. To obtain a rough idea of the pH, a drop of the solution to be tested is touched to a strip of the test paper and the resulting color is compared with a color code. Some commercial test papers (e.g., Hydrion) are impregnated with several dyes, with their containers carrying the color code (see Figure 16.12). Litmus paper was one of the first pH test papers ever made, and it is still commonly found in chemistry labs. It consists of porous paper impregnated with litmus dye, made either red or blue by exposure to acid or base and then dried.

Below pH 4.7 litmus is red, and above pH 8.3 it is blue. The transition for the color change occurs over a pH range of 4.7 to 8.3 with the center of the change at about 6.5, very nearly a neutral pH. To test whether a solution is acidic, a drop of it is placed on blue litmus paper. If the dye turns pink, the solution is acidic. Similarly, to test if the solution is basic, a drop is touched to red litmus paper. If the dye turns blue, the solution is basic. If a solution is close to neutral, neither red nor blue litmus will change color.

Figure 16.12 | Using a pH test paper. The color of this Hydrion test strip changed to orange when a drop of the lemon juice solution in the beaker was placed on it. According to the color code, the pH of the solution is closer to 3 than to the color for pH 5.

■ Litmus is extracted from a type of moss called lichens.

pH 8.2   pH 10.0
(a)   Phenolphthalein

pH 3.2   pH 4.4
(b)   Methyl orange

pH 6.0   pH 7.6
(c)   Bromothymol blue

pH 4.8   pH 5.4
(d)   Methyl purple

Figure 16.11 | Colors of some common acid–base indicators.

Summary

Organized by Learning Objective

Define pH and explain the use of "p" notation

Water reacts with itself to produce small amounts of H_3O^+ (often abbreviated H^+) and OH^- ions. The concentrations of these ions, both in pure water and dilute aqueous solutions, are related by the expression

$$[H^+][OH^-] = K_w = 1.0 \times 10^{-14} \quad \text{(at 25 °C)}$$

K_w is the **ion product constant of water**. In pure water

$$[H^+] = [OH^-] = 1.0 \times 10^{-7}$$

The **pH** of a solution is defined by the equation, $pH = -\log[H^+]$. As the pH decreases, the acidity, or $[H^+]$, increases. The comparable term, **pOH** $(= -\log[OH^-])$, is used to describe a solution that is basic. A solution is acidic if the hydrogen ion concentration exceeds 1.0×10^{-7} or the pH is less than 7.00. Similarly, a solution is basic if the hydroxide ion concentration exceeds 1.0×10^{-7} or if the pH is greater than 7.00.

Explain how to determine the pH of strong acids or bases in aqueous solution

When calculating the pH of strong acids or strong bases, we assume that they are 100% ionized.

Write expressions for the acid ionization constant, K_a, and base ionization constant, K_b, and explain how they are related to each other

A weak acid HA ionizes according to the general equation

$$HA + H_2O \rightleftharpoons H_3O^+ + A^-$$

or more simply,

$$HA \rightleftharpoons H^+ + A^-$$

The equilibrium constant is called the **acid ionization constant, K_a**:

$$K_a = \frac{[H_3O^+][A^-]}{[HA]}$$

A weak base B ionizes by the general equation

$$B + H_2O \rightleftharpoons BH^+ + OH^-$$

The equilibrium constant is called the **base ionization constant, K_b**:

$$K_b = \frac{[BH^+][OH^-]}{[B]}$$

The smaller the values of K_a (or K_b), the weaker are the substances as Brønsted acids (or bases).

Another way to compare the relative strengths of acids or bases is to use the negative logarithms of K_a and K_b, called **pK_a** and **pK_b**, respectively.

$$pK_a = -\log K_a \qquad pK_b = -\log K_b$$

Stronger acids or bases have *smaller* pK_a or pK_b values, respectively.

Describe how to determine acid and base ionization constants from experimental data

The values of K_a and K_b can be obtained from initial concentrations of the acid or base and either the pH of the solution or the percentage ionization of the acid or base. The measured pH gives the equilibrium value for $[H^+]$.

Determine equilibrium concentrations and pH for weak acid or base solutions

Problems fall into one of three categories: (1) the only solute is a weak acid (we must use the K_a expression), (2) the only solute is a weak base (we must use the K_b expression), and (3) the solution contains both a weak acid and its conjugate base (we can use either K_a or K_b).

When the initial concentration of the acid (or base) is larger than 100 times the value of K_a (or K_b), it is safe to use initial concentrations of acid or base as though they were equilibrium values in the mass action expression. When this approximation cannot be used, we can use the quadratic formula.

Explain how a salt solution can be acidic or basic

A solution of a salt is acidic if the cation is acidic but the anion is neutral. Metal ions with high charge densities generally are acidic, but those of Groups 1A and 2A (except Be^{2+}) are not. Cations such as NH_4^+, which are the conjugate acids of weak *molecular* bases, are themselves proton donors and are acidic.

When the anion of a salt is the conjugate base of a *weak* acid, the anion is a weak base. Anions of strong acids, such as Cl^- and NO_3^-, are such weak Brønsted bases that they cannot affect the pH of a solution.

If the salt is derived from a weak acid *and* a weak base, its net effect on pH has to be determined on a case-by-case basis by determining which of the two ions is the stronger.

Describe what a buffer solution is, how it works, and how to calculate pH changes in a buffer

A solution that contains both a weak acid and a weak base (usually an acid–base conjugate pair) is called a **buffer**, because the anion, A^-, is able to absorb $[H^+]$ from a strong acid

$$A^- + H^+ \longrightarrow HA$$

or the acid, HA, is able to absorb $[OH^-]$ from a strong base without large changes in pH.

$$HA + OH^- \longrightarrow A^- + H_2O$$

The pH of a buffer is controlled by the ratio of weak acid to weak base, expressed either in terms of molarities or moles.

$$[H^+] = K_a\frac{[HA]}{[A^-]} \qquad [H^+] = K_a\frac{\text{mol } HA}{\text{mol } A^-}$$

Because the $[H^+]$ is determined by the mole ratio of HA to A^-, dilution does not change the pH of a buffer. Buffers are most effective when the pK_a of the acid member of the buffer pair lies within ± 1 unit of the desired pH.

Define polyprotic acids and describe how to calculate concentrations of all species in a solution of a polyprotic acid

A polyprotic acid has a K_a value for the ionization of each of its hydrogen ions. Successive values of K_a often differ by a factor of 10^4 to 10^5. This allows us to calculate the pH of a solution of a polyprotic acid by using just the value of K_{a_1}. If the polyprotic acid is the only solute, the anion formed in the second step of the ionization, A^{2-}, has a concentration equal to K_{a_2}.

The anions of weak polyprotic acids are bases that react with water in successive steps, the last of which has the molecular polyprotic acid as a product. For a diprotic acid, $K_{b_1} = K_w/K_{a_2}$ and $K_{b_2} = K_w/K_{a_1}$. Usually, $K_{b_1} \gg K_{b_2}$, so virtually all the OH^- produced in the solution comes from the first step. The pH of the solution can be calculated using just K_{b_1} and the reaction

$$A^{2-} + H_2O \rightleftharpoons HA^- + OH^-$$

Draw and explain titration curves for reactions of strong or weak acids and bases

A graph of the pH values versus the volume of titrant gives us a **titration curve**. The titration curve contains important information about the chemical system including the equivalence point and equilibrium constant data.

Each calculated titration curve has four distinct calculations:

(a) The starting pH of a solution that contains only an acid or base
(b) The pH values for titrant volumes between the start and the equivalence points can be used to calculate the
(c) The pH at the equivalence point
(d) The pH values for titrant volumes past the equivalence point

Acid–base indicators are weak acids in which the conjugate acid and base have different colors. The sudden change in pH at the equivalence point causes a rapid shift from one color of the indicator to the other. If matched to the equivalence point, a pH indicator will change color abruptly when a titration reaches the equivalence point.

TOOLS

Tools for Problem Solving
The following tools were introduced in this chapter. Study them carefully so you can select the appropriate tool when needed.

Ion product constant of water, K_w (Section 16.1)

$$K_w = [H^+][OH^-] \quad \text{or} \quad K_w = [H_3O^+][OH^-]$$

At 25 °C the value of K_w is 1.0×10^{-14}.

Definition of pH (Section 16.1)

$$pH = -\log[H^+]$$

Defining the p-function, pX (Section 16.1)

$$pX = -\log X$$

Relating pH and pOH to pK_w (Section 16.1)

$$pH + pOH = pK_w \qquad pK_w = 14.00 \text{ at } 25\,°C$$

General equation for the ionization of a weak acid (Section 16.3)

$$HA + H_2O \rightleftharpoons H_3O^+ + A^- \quad \text{or} \quad HA \rightleftharpoons H^+ + A^-$$

General equation for the ionization of a weak base (Section 16.3)

$$B + H_2O \rightleftharpoons BH^+ + OH^-$$

Inverse relationship between K_a and K_b (Section 16.3)

$$K_a \times K_b = K_w$$

Relationship between pK_a and pK_b (Section 16.3)

$$pK_a + pK_b = pK_w \qquad pK_w = 14.00 \quad \text{at } 25\,°C$$

Percentage ionization (Section 16.4)

$$\text{percentage ionization} = \frac{\text{moles ionized per liter}}{\text{moles available per liter}} \times 100\%$$

Criterion for predicting if simplifying assumptions will work (Section 16.5)

$$C_{HA} > 100 \times K_a \quad \text{or} \quad C_B > 100 \times K_b$$

Identification of acidic cations and basic anions (Section 16.6)

We use the concepts developed here to determine whether a salt solution is acidic, basic, or neutral. This is also the first step in calculating the pH of a salt solution.

How buffers work (Section 16.7)

These reactions illustrate how the concentrations of conjugate acid and base change when a strong acid or strong base is added to a buffer. Adding H^+ decreases $[A^-]$ and increases $[HA]$; adding OH^- decreases $[HA]$ and increases $[A^-]$. These reactions keep the pH of the solution from changing drastically.

Henderson–Hasselbalch equation (Section 16.7)

$$pH = pK_a + \log\frac{[A^-]}{[HA]}$$

Polyprotic acids ionize stepwise (Section 16.8)

Polyprotic acids have more than one ionizable proton in their formulas. Each proton will ionize sequentially, and the chemical equation will have a form similar to the general equation for the ionization of a weak acid tool above.

Titration curves have four parts (Section 16.9)

To plot the graph of pH as a function of volume of titrant added, that is, a titration curve, there are four different stages of calculation. They are (a) the starting point, (b) the stage from the start to the equivalence point, (c) the equivalence point, and (d) the stage after the equivalence point.

How acid–base indicators work (Section 16.9)

Titration indicators are weak acids that have one color for the conjugate acid and a different color for the conjugate base, and the change in color depends on the pH of the solution.

WileyPLUS = *WileyPLUS*, an online teaching and learning solution. *Note to instructors*: Many of the end-of-chapter problems are available for assignment via the WileyPLUS system. **www.wileyplus.com.** Review Problems are presented in pairs separated by blue rules. Answers to problems whose numbers appear in blue are given in Appendix B. More challenging problems are marked with an asterisk ★.

| Review Questions

Water, pH and "p" Notation

16.1 Write the chemical equation for **(a)** the autoionization of water and **(b)** the equilibrium law for K_w.

16.2 How are acidic, basic, and neutral solutions in water defined **(a)** in terms of $[H^+]$ and $[OH^-]$ and **(b)** in terms of pH and pOH?

16.3 At 25 °C, how are the pH and pOH of a solution related to each other?

16.4 Why do chemists use pH notation instead of the concentration of H^+ ions?

16.5 Explain how acids and bases suppress the ionization of water, often called the common ion effect.

16.6 Explain the leveling effect of water.

16.7 Could you use the p-notation for the concentration of a very dilute solution of chloride ion for a solution made when a tablespoon of water is added to a gallon of water? How would it be defined?

pH of Strong Acid and Base Solutions

16.8 List the strong acids.

16.9 What chemical property is central to our classifying an acid as a strong acid?

16.10 Explain the difference between strength and concentration of an acid.

16.11 How are strong bases identified?

16.12 Some strong bases can be used to safely reduce stomach acidity (heartburn). What differentiates strong bases that are safe medicines from those that could cause irreparable harm?

16.13 Explain why we can ignore the autoionization of water in a 1.0 M solution of a strong acid.

Determining K_a and K_b Values

16.14 Write the general equation for the ionization of a weak acid, HA, in water. Give the equilibrium law corresponding to K_a.

16.15 Which diagram best represents a weak acid? A strong acid? A nonelectrolyte?

16.16 Why do we use equilibrium constants, K_a and K_b, for weak acids and bases, but not for the strong acids and bases?

16.17 Write the chemical equation for the ionization of each of the following weak acids in water. (For polyprotic acids, write only the equation for the first step in the ionization.)

(a) HNO_2 (c) $HAsO_4^{2-}$

(b) H_3PO_4 (d) $(CH_3)_3NH^+$

16.18 For each of the acids in Review Question 16.17 write the appropriate K_a expression.

16.19 Write the general equation for the ionization of a weak base, B, in water. Give the equilibrium law corresponding to K_b.

16.20 Write the chemical equation for the ionization of each of the following weak bases in water.

(a) $(CH_3)_3N$ (c) NO_2^-

(b) AsO_4^{3-} (d) $(CH_3)_2N_2H_2$

16.21 For each of the bases in Review Question 16.20, write the appropriate K_b expression.

16.22 The pK_a of HCN is 9.31 and that of HF is 3.46. Which is the stronger Brønsted base, CN^- or F^-?

16.23 Write the structural formulas for the conjugate acids of the following:

(a) $CH_3-CH_2-\overset{\overset{\displaystyle CH_3}{|}}{N}-H$ (b) [pyridine structure] N:

(c) $H-\overset{\overset{\displaystyle H}{|}}{\underset{\cdot\cdot}{O}}-N-H$

16.24 Write the structural formulas for the conjugate bases of the following:

(a) [benzene] NH_3^+ (b) $[CH_3-\overset{\overset{\displaystyle CH_3}{|}}{\underset{\displaystyle CH_3}{|}}{N}-H]^+$

(c) $[H-\overset{\overset{\displaystyle H}{|}}{N}-\overset{\overset{\displaystyle H}{|}}{\underset{\displaystyle H}{N}}-H]^+$

16.25 How is percentage ionization defined? Write the equation.

pH of Weak Acid and Weak Base Solutions

16.26 What criterion do we use to determine whether or not the equilibrium concentration of an acid or base will be effectively the same as its initial concentration when we calculate the pH of the solution?

16.27 For which of the following are we permitted to make the assumption that the equilibrium concentration of the acid or base is the same as the initial concentration when we calculate the pH of the solution specified?

(a) 0.020 M $HC_2H_3O_2$ (c) 0.002 M N_2H_4

(b) 0.10 M CH_3NH_2 (d) 0.050 M $HCHO_2$

16.28 What is the quadratic formula? When is it appropriate to use it finding equilibrium concentrations?

Acid–Base Properties of Salt Solutions

16.29 Aspirin is acetylsalicylic acid, a monoprotic acid whose K_a value is 3.3×10^{-4}. Does a solution of the sodium salt of aspirin in water test acidic, basic, or neutral? Explain.

16.30 The K_b value of the oxalate ion, $C_2O_4^{2-}$, is 1.6×10^{-10}. Is a solution of $K_2C_2O_4$ acidic, basic, or neutral? Explain.

16.31 Consider the following compounds and suppose that 0.5 M solutions are prepared of each: NaI, NH_4Br, KF, KCN, $KC_2H_3O_2$, $CsNO_3$, and KBr. Write the *formulas* of those that have solutions that are (a) acidic, (b) basic, and (c) neutral.

16.32 Will an aqueous solution of $AlCl_3$ turn litmus red or blue? Explain.

16.33 A solution of hydrazinium acetate is slightly acidic. Without looking at the tables of equilibrium constants, is K_a for acetic acid larger or smaller than K_b for hydrazine? Justify your answer.

16.34 When ammonium nitrate is added to a suspension of magnesium hydroxide in water, the $Mg(OH)_2$ dissolves. Write a net ionic equation to show how this occurs.

Buffer Solutions

16.35 To form a buffer, two substances are needed. How are these substances related?

16.36 Write ionic equations that illustrate how each pair of compounds can serve as a buffer pair.

(a) H_2CO_3 and NaHCO (the "carbonate" buffer in blood)

(b) NaH_2PO_4 and Na_2HPO_4 (the "phosphate" buffer inside body cells)

(c) NH_4Cl and NH_3

(d) Phenol and sodium phenolate

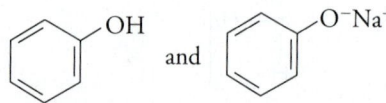

16.37 The hydrogen phosphate ion is able to act as a buffer all by itself. Write chemical equations that show how this ion reacts with **(a)** H^+ and **(b)** OH^-.

Polyprotic Acids

16.38 When sulfur dioxide, an air pollutant from the burning of sulfur-containing coal or oil, dissolves in water, an acidic solution is formed that can be viewed as containing sulfurous acid, H_2SO_3.

$$H_2O + SO_2(g) \rightleftharpoons H_2SO_3(aq)$$

Write the expression for K_{a_1} and K_{a_2} for sulfurous acid.

16.39 Which diagram best describes a diprotic acid with the first proton a strong acid and the second proton a weak acid?

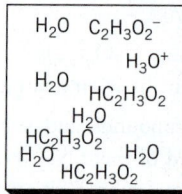

Review Problems

Water, pH and "p" Notation

16.49 Calculate the $[H^+]$, pH, and pOH in each of the following solutions in which the hydroxide ion concentrations are

(a) $0.0068\ M$ **(c)** $1.6 \times 10^{-8}\ M$

(b) $6.4 \times 10^{-5}\ M$ **(d)** $8.2 \times 10^{-12}\ M$

16.50 Calculate the $[OH^-]$, pH, and pOH for each of the following solutions in which the H^+ concentrations are

(a) $3.5 \times 10^{-7}\ M$ **(c)** $2.5 \times 10^{-11}\ M$

(b) $0.0017\ M$ **(d)** $7.9 \times 10^{-2}\ M$

16.40 Citric acid, found in citrus fruits, is a triprotic acid, $H_3C_6H_5O_7$. Write chemical equations for the three-step ionization of this acid in water and write the three K_a expressions (equilibrium laws).

16.41 What simplifying assumptions do we usually make in working problems involving the ionization of polyprotic acids? Why are they usually valid? Under what conditions do they fail?

16.42 Write the equations for the chemical equilibria that exist in solutions of **(a)** Na_2SO_3, **(b)** Na_3PO_4, and **(c)** $K_2C_4H_4O_6$.

16.43 What simplifying assumptions do we usually make in working problems involving equilibria of salts of polyprotic acids? Why are they usually valid? Under what conditions do they fail?

Acid–Base Titrations

16.44 Define the terms equivalence point and end point as they apply to an acid–base titration.

16.45 Will the solution be acidic, neutral, or basic at the equivalence point for

(a) a formic acid solution that is titrated with sodium hydroxide?

(b) a solution of hydrazine that is titrated with hydrochloric acid?

(c) a solution of hydrochloric acid that is titrated with sodium hydroxide?

16.46 Qualitatively, describe how an acid–base indicator works. Why do we want to use a minimum amount of indicator in a titration?

16.47 If you use methyl orange in the titration of $HC_2H_3O_2$ with NaOH, will the end point of the titration correspond to the equivalence point? If not, suggest a better indicator for this titration. Justify your answer.

16.48 Using the list of indicators in Table 16.7, choose the appropriate indicators for the titrations depicted in Figures 16.8 and 16.9.

16.51 Calculate the molar concentrations of H^+ and OH^- in solutions that have the following pH values.

(a) 8.14 **(b)** 2.56 **(c)** 11.25 **(d)** 13.28 **(e)** 6.70

16.52 Calculate the molar concentrations of H^+ and OH^- in solutions that have the following pH values.

(a) 12.67 **(b)** 5.18 **(c)** 11.55 **(d)** 4.22 **(e)** 6.06

16.53 Calculate the molar concentrations of H^+ and OH^- in solutions that have the following pOH values.

(a) 7.19 **(b)** 1.26 **(c)** 10.85 **(d)** 13.15 **(e)** 5.24

16.54 Calculate the molar concentrations of H^+ and OH^- in solutions that have the following pOH values.

(a) 12.27 **(b)** 6.14 **(c)** 10.65 **(d)** 4.28 **(e)** 3.76

16.55 A certain brand of beer had a H^+ concentration equal to 1.9×10^{-5} mol L^{-1}. What is the pH of the beer?

16.56 A soft drink was put on the market with $[H^+] = 1.4 \times 10^{-5}$ mol L^{-1}. What is its pH?

16.57 A sample of Windex had a $[OH^-] = 6.3 \times 10^{-5}$ mol L^{-1}. What is the pH of the sample?

16.58 Bases tend to be used for cleaning. A solution of Tide laundry detergent had a $[OH^-] = 1.6 \times 10^{-4}$ mol L^{-1}. What is the pH of the sample?

16.59 Deuterium oxide, D_2O, ionizes like water. At 20 °C its K_w or ion product constant, analogous to that of water, is 8.9×10^{-16}. Calculate $[D^+]$ and $[OD^-]$ in deuterium oxide at 20 °C. Calculate also the pD and the pOD. What would be the neutral pD?

16.60 At the temperature of the human body, 37 °C, the value of K_w is 2.5×10^{-14}. Calculate $[H^+]$, $[OH^-]$, pH, and pOH of pure water at this temperature. What is the relationship between pH, pOH, and pK_w at this temperature? Is pH 7.00 water neutral at this temperature?

16.61 The interaction of water droplets in rain with carbon dioxide that is naturally present in the atmosphere causes rain water to be slightly acidic because CO_2 is an acid anhydride. As a result, pure clean rain has a pH of about 5.7. What are the hydrogen ion and hydroxide ion concentrations in this rain water?

16.62 "Acid rain" forms when rain falls through air polluted by oxides of sulfur and nitrogen. Trees and plants are affected if the acid rain has a pH of 3.5 or lower. What is the hydrogen ion concentration in acid rain that has a pH of 3.16? What is the pH of a solution having twice your calculated hydrogen ion concentration?

pH of Strong Acid and Base Solutions

16.63 What is the concentration of H^+ in 0.00065 M HNO_3? What is the pH of this solution? What is the OH^- concentration in this solution?

16.64 What is the concentration of H^+ in 0.031 M $HClO_4$? What is the pH of the solution? What is the OH^- concentration in the solution? By how much does the pH change if the concentration of H^+ is doubled?

16.65 A sodium hydroxide solution is prepared by dissolving 6.0 g NaOH in 1.00 L of solution. What is the molar concentration of OH^- in the solution? What are the pOH and the pH of the solution? What is the hydrogen ion concentration in the solution?

16.66 A solution was made by dissolving 0.837 g $Ba(OH)_2$ in 100 mL final volume. What is the molar concentration of OH^- in the solution? What are the pOH and the pH? What is the hydrogen ion concentration in the solution?

16.67 A solution of $Ca(OH)_2$ has a measured pH of 11.60. What is the molar concentration of the $Ca(OH)_2$ in the solution? What is the molar concentration of $Ca(OH)_2$ if the solution is diluted so that the pH is 10.60?

16.68 A solution of HCl has a pH of 2.50. How many grams of HCl are there in 0.250 L of this solution? How many grams of HCl are in 250 mL of an HCl solution that has twice this pH?

16.69 In a 0.0020 M solution of NaOH, how many moles per liter of OH^- come from the ionization of water?

16.70 In a certain solution of HCl, the ionization of water contributes 3.4×10^{-11} moles per liter to the H^+ concentration. What is the total H^+ concentration in the solution?

****16.71** A solution was prepared with 2.64 micrograms of $Ba(OH)_2$ in 1.00 liter of water. What is the pH of the solution, and what concentration of hydrogen ions is provided by the ionization of water?

****16.72** What is the pH of a 3.0×10^{-7} M solution of HCl? What concentration of hydrogen ions is provided by the ionization of water?

16.73 Rhododendrons are shrubs that produce beautiful flowers in the springtime. They only grow well in soil that has a pH that is 5.5 or slightly lower. What is the hydrogen ion concentration in the soil moisture if the pH is 5.5?

16.74 As eggs age, the pH of the egg white increases from about 7.9 to 9.3 as the carbon dioxide diffuses out of the egg. What is the hydrogen ion concentration in an egg, if the pH is 8.3?

Acid and Base Ionization Constants: K_a and K_b

16.75 The K_a for HF is 3.5×10^{-4}. What is the K_b for F^-?

16.76 The barbiturate ion, $C_4H_3N_2O_3$, has $K_b = 1.0 \times 10^{-10}$. What is the K_a for barbituric acid?

16.77 Iodic acid, HIO_3, has a pK_a of 0.77. (a) What are the formula and the K_b of its conjugate base? (b) Is its conjugate base a stronger or a weaker base than the acetate ion?

16.78 Lactic acid, $HC_3H_5O_3$, is responsible for the sour taste of old milk. At 25 °C, its $K_a = 1.4 \times 10^{-4}$. (a) What is the K_b of its conjugate base, the lactate ion, $C_3H_5O_3^-$? (b) Is its conjugate base a stronger or a weaker base than the acetate ion?

pH of Weak Acid and Weak Bases

****16.79** A 1.0 M solution of acetic acid has a pH of 2.37. What percentage of the acetic acid is ionized in the solution?

****16.80** A 0.250 M solution of NH_3 has a pH of 11.32. What percentage of the ammonia is ionized in this solution?

16.81 A 0.20 M solution of a weak acid, HA, has a pH of 3.22. What is the percentage ionization of the acid? What is the value of K_a for the acid?

16.82 A 0.15 M solution of an acid found in milk has a pH of 2.80. What is the percentage ionization of the acid? What is the K_a for the acid?

16.83 A 0.12 M solution of a weak base is 0.012% ionized. What is the pH of the solution? What is the value of K_b for this base?

16.84 If a weak base is 0.030% ionized in 0.030 M solution, what is the pH of the solution? What is the value of K_b for the base?

16.85 Periodic acid, HIO_4, is an important oxidizing agent and a moderately strong acid. In a 0.10 M solution, $[H^+] = 3.8 \times 10^{-2}$ mol L^{-1}. Calculate the K_a and pK_a for periodic acid.

16.86 Chloroacetic acid, $HC_2H_2O_2Cl$, is a stronger monoprotic acid than acetic acid. In a 0.10 M solution, the pH is 1.96. Calculate the K_a and pK_a for chloroacetic acid.

16.87 Ethylamine, $CH_3CH_2NH_2$, has a strong, pungent odor similar to that of ammonia. Like ammonia, it is a Brønsted base. A 0.10 M solution has a pH of 11.87. Calculate the K_b and pK_b for ethylamine. What is the percentage ionization of ethylamine in this solution?

16.88 Hydroxylamine, $HONH_2$, like ammonia, is a Brønsted base. A 0.15 M solution has a pH of 10.11. What are the K_b and pK_b values for hydroxylamine? What is the percentage ionization of the $HONH_2$?

*16.89 What are the concentrations of all the solute species in 0.150 M lactic acid, $HC_3H_5O_2$? What is the pH of the solution? This acid has a $K_a = 1.4 \times 10^{-4}$.

*16.90 What are the concentrations of all the solute species in a 1.0 M solution of hydrogen peroxide, H_2O_2? What is the pH of the solution? For H_2O_2, $K_a = 2.4 \times 10^{-12}$.

16.91 Codeine, a cough suppressant extracted from crude opium, is a weak base with a pK_b of 5.80. What will be the pH of a 0.020 M solution of codeine? (Use *Cod* as a symbol for codeine.)

16.92 Pyridine, C_5H_5N, is a bad-smelling liquid that is a weak base in water. Its pK_b is 8.77. What is the pH of a 0.20 M aqueous solution of this compound?

16.93 A solution of acetic acid has a pH of 2.54. What is the concentration of acetic acid in this solution?

16.94 How many moles of NH_3 must be dissolved in water to give 500.0 mL of solution with a pH of 11.22?

*16.95 What is the pH of a 0.0050 M solution of sodium cyanide?

*16.96 What is the pH of a 0.020 M solution of chloroacetic acid, for which $K_a = 1.4 \times 10^{-3}$?

16.97 The compound *para*-aminobenzoic acid (PABA) is a powerful sun-screening agent whose salts were once used widely in sun tanning and screening lotions. The parent acid, which we may symbolize as H-*Paba*, is a weak acid with a pK_a of 4.92 (at 25 °C). What are the $[H^+]$ and pH of a 0.030 M solution of this acid?

16.98 Barbituric acid, $HC_4H_3N_2O_3$ (which we will abbreviate H-*Bar*), was discovered by the Nobel Prize–winning organic chemist Adolph von Bayer and named after his friend, Barbara. It is the parent compound of widely used sleeping drugs, the barbiturates. Its pK_a is 4.01. What are the $[H^+]$ and pH of a 0.020 M solution of H-*Bar*?

Acid–Base Properties of Salt Solutions

16.99 Calculate the pH of 0.20 M NaCN. What is the concentration of HCN in the solution?

16.100 Calculate the pH of 0.40 M KNO_2. What is the concentration of HNO_2 in the solution?

16.101 Calculate the pH of 0.15 M CH_3NH_3Cl. For methylamine, CH_3NH_2, $K_b = 4.5 \times 10^{-4}$.

16.102 Calculate the pH of 0.10 M hydrazinium chloride, N_2H_5Cl.

16.103 A 0.18 M solution of the sodium salt of nicotinic acid (also known pharmaceutically as niacin) has a pH of 9.05. What is the value of K_a for nicotinic acid?

16.104 A weak base B forms the salt $BHCl$, composed of the ions BH^+ and Cl^-. A 0.15 M solution of the salt has a pH of 4.28. What is the value of K_b for the base B?

*16.105 Liquid chlorine bleach is really nothing more than a solution of sodium hypochlorite, NaOCl, in water. Usually, the concentration is approximately 5.1% NaOCl by weight. Use this information to calculate the approximate pH of a bleach solution, assuming no other solutes are in the solution except NaOCl. (Assume the bleach has a density of 1.0 g/mL.)

*16.106 The conjugate acid of a molecular base has the general formula BH^+ and a pK_a of 5.00. A solution of a salt of this cation, BHY, tests slightly basic. Will the conjugate acid of Y^-, HY, have a pK_a greater than 5.00 or less than 5.00? Explain.

Buffer Solutions

16.107 What is the pH of a solution that contains 0.15 M $HC_2H_3O_2$ and 0.25 M $C_2H_3O_2^-$? Use $K_a = 1.8 \times 10^{-5}$ for $HC_2H_3O_2$.

16.108 A buffer is prepared containing 0.25 M NH_3 and 0.45 M NH_4^+. Calculate the pH of the buffer using the K_b for NH_3.

16.109 Rework Review Problem 16.107 using the K_b for the acetate ion. (Be sure to write the proper chemical equation and equilibrium law.)

16.110 Calculate the pH of the buffer in Review Problem 16.108 using the K_a for NH_4^+.

*16.111 By how much will the pH change if 0.025 mol of HCl is added to 1.00 L of the buffer in Review Problem 16.107?

16.112 By how much will the pH change if 25.0 mL of 0.20 M NaOH is added to 0.500 L of the buffer in Review Problem 16.108?

*16.113 Certain proteolytic enzymes react in alkaline solutions. One of these enzymes produces 1.8 micromoles of hydrogen ions per second in a 2.50 mL portion of a buffer composed of 0.25 M NH_3 and 0.20 M NH_4Cl.

By how much will the concentrations of the NH_3 and NH_4^+ ions change after the reaction has run for 35 seconds?

*16.114 A hydrolytic enzyme consumes 1.8×10^{-7} moles of hydrogen ions per minute in 2.50 L of a buffer that contained 0.15 M $HC_2H_3O_2$ and 0.25 M $C_2H_3O_2^-$. By how much did the concentrations of $HC_2H_3O_2$ and $C_2H_3O_2^-$ change after the enzyme reacted for 3.45 minutes?

*16.115 What are the initial and final pH values for the buffer in Review Problem 16.113? What is the change in pH?

*16.116 What are the initial and final pH values for the buffer in Review Problem 16.114? What is the change in pH?

16.117 How many grams of sodium acetate, $NaC_2H_3O_2$, would have to be added to 1.0 L of 0.15 M acetic acid (pK_a 4.74) to make the solution a buffer for pH 4.00?

16.118 How many grams of sodium formate, $NaCHO_2$, would have to be dissolved in 1.0 L of 0.12 M formic acid (pK_a 3.74) to make the solution a buffer for pH 3.50?

16.119 Suppose 30.00 mL of 0.100 M HCl is added to an acetate buffer prepared by dissolving 0.100 mol of acetic acid and 0.110 mol of sodium acetate in 0.100 L of solution. What are the initial and final pH values? What would be the pH if the same amount of HCl solution were added to 125 mL of pure water?

16.120 How many milliliters of 0.15 M HCl would have to be added to the original 0.100 L of the buffer described in Review Problem 16.108 to make the pH decrease by 0.05 pH unit? How many milliliters of the same HCl solution would, if added to 0.100 L of pure water, make the pH decrease by 0.05 pH unit?

Polyprotic Acids

*16.121 Calculate the concentrations of all the solute species in a 0.15 M solution of ascorbic acid (vitamin C). What is the pH of the solution?

*16.122 Tellurium, in the same family as sulfur, forms an acid analogous to sulfuric acid and called telluric acid. It exists, however, as H_6TeO_6 (which can be understood as the formula $H_2TeO_4 + 2H_2O$). It is a diprotic acid with $K_{a_1} = 2 \times 10^{-8}$ and $K_{a_2} = 1 \times 10^{-11}$. Calculate the concentrations of H^+, $H_5TeO_6^-$, and $H_4TeO_6^{2-}$ in a 0.25 M solution of H_6TeO_6. What is the pH of the solution?

*16.123 Calculate the concentrations of all of the solute species involved in the equilibria in a 2.0 M solution of H_3PO_4. Calculate the pH of the solution.

*16.124 What is the pH of a 0.25 M solution of arsenic acid, H_3AsO_4? In this solution, what are the concentrations of $H_2AsO_4^-$ and $HAsO_4^{2-}$?

*16.125 Phosphorous acid, H_3PO_3, is actually a diprotic acid for which $K_{a_1} = 5.0 \times 10^{-2}$ and $K_{a_2} = 2.0 \times 10^{-7}$. What are the values of $[H^+]$, $[H_2PO_3^-]$, and $[HPO_3^{2-}]$ in a 1.0 M solution of H_3PO_3? What is the pH of the solution?

16.126 What is the pH of a 0.20 M solution of oxalic acid, $H_2C_2O_4$?

*16.127 Calculate the pH of 0.24 M Na_2SO_3. What are the concentrations of all ions and H_2SO_3 in the solution?

*16.128 Calculate the pH of 0.33 M K_2CO_3. What are the concentrations of all ions and H_2CO_3 in the solution?

16.129 Sodium citrate, $Na_3C_6H_5O_7$, is used as an anticoagulant in the collection of blood. What is the pH of a 0.10 M solution of this salt?

16.130 What is the pH of a 0.25 M solution of sodium oxalate, $Na_2C_2O_4$?

16.131 What is the pH of a 0.50 M solution of Na_3PO_4? In this solution, what are the concentrations of HPO_4^{2-}, $H_2PO_4^-$, and H_3PO_4?

*16.132 The pH of a 0.10 M Na_2CO_3 solution is adjusted to 12.00 using a strong base. What is the concentration of HCO_3^- in this solution?

Acid–Base Titrations

16.133 What is the pH of a solution prepared by mixing 25.0 mL of 0.180 M $HC_2H_3O_2$ with 40.0 mL of 0.250 M NaOH?

*16.134 What is the pH of a solution prepared by mixing exactly 25.0 mL of 0.200 M $HC_2H_3O_2$ with 15.0 mL of 0.400 M KOH?

16.135 When 50.0 mL of 0.050 M formic acid, $HCHO_2$, is titrated with 0.050 M sodium hydroxide, what is the pH at the equivalence point? (Be sure to take into account the change in volume during the titration.) Select a good indicator for this titration from Table 16.7.

16.136 When 25 mL of 0.12 M aqueous ammonia is titrated with 0.12 M hydrobromic acid, what is the pH at the equivalence point? Select a good indicator for this titration from Table 16.7.

*16.137 For the titration of 75.00 mL of 0.1000 M acetic acid with 0.1000 M NaOH, calculate the pH **(a)** before the addition of any NaOH solution, **(b)** after 25.00 mL of the base has been added, **(c)** after half of the $HC_2H_3O_2$ has been neutralized, and **(d)** at the equivalence point.

*16.138 For the titration of 50.00 mL of 0.1000 M ammonia with 0.1000 M HCl, calculate the pH **(a)** before the addition of any HCl solution, **(b)** after 20.00 mL of the acid has been added, **(c)** after half of the NH_3 has been neutralized, and **(d)** at the equivalence point.

| Additional Exercises

*16.139 Calculate the percentage ionization of acetic acid in solutions having concentrations of 1.0 M, 0.10 M, and 0.010 M. How does the percentage ionization of a weak acid change as the acid becomes more dilute?

*16.140 What is the pH of a solution that is 0.100 M in HCl and also 0.125 M in $HC_2H_3O_2$? What is the concentration of acetate ion in this solution?

16.141 A solution is prepared by mixing 325 mL of 0.500 M NH_3 and 175 mL of 0.500 M HCl. Assuming that the volumes are additive, what is the pH of the resulting mixture?

*16.142 A solution is prepared by dissolving 15.0 g of pure $HC_2H_3O_2$ and 25.0 g of $NaC_2H_3O_2$ in 775 mL of solution (the final volume). (a) What is the pH of the solution? (b) What would the pH of the solution be if 25.0 mL of 0.250 M NaOH were added? (c) What would the pH be if 25.0 mL of 0.40 M HCl were added to the original 775 mL of buffer solution?

*16.143 For an experiment involving what happens to the growth of a particular fungus in a slightly acidic medium, a biochemist needs 255 mL of an acetate buffer with a pH of 5.12. The buffer solution has to be able to hold the pH to within ±0.10 pH unit of 5.12 even if 0.0100 mol of NaOH or 0.0100 mol of HCl enters the solution.

(a) What is the minimum number of grams of acetic acid and of sodium acetate dihydrate that must be used to prepare the buffer?

(b) Describe the buffer by giving its molarity in acetic acid and its molarity in sodium acetate.

(c) What is the pH of an unbuffered solution made by adding 0.0100 mol of NaOH to 255 mL of pure water?

(d) What is the pH of an unbuffered solution made by adding 0.0100 mol of HCl to 255 mL of pure water?

*16.144 Predict whether the pH of 0.120 M NH_4CN is greater than, less than, or equal to 7.00. Give your reasons.

*16.145 What is the pH of a 4.5×10^{-2} M solution of ammonium acetate, $NH_4C_2H_3O_2$?

*16.146 How many milliliters of ammonia gas measured at 25.2 °C and 745 torr must be dissolved in 0.250 L of 0.050 M HNO_3 to give a solution with a pH of 9.26?

16.147 $HClO_4$ is a stronger proton donor than HNO_3, but in water both acids appear to be of equal strength; they are both 100% ionized. Why is this so? What solvent property would be necessary in order to distinguish between the acidities of these two Brønsted acids?

16.148 The hydrogen sulfate ion, HSO_4^-, is a moderately strong Brønsted acid with a K_a of 1.2×10^{-2}.

(a) Write the chemical equation for the ionization of the acid and give the appropriate K_a expression.

(b) What is the value of $[H^+]$ in 0.010 M HSO_4^- (furnished by the salt, $NaHSO_4$)? Do NOT make simplifying assumptions; use the quadratic equation.

(c) What is the calculated $[H^+]$ in 0.010 M HSO_4^-, obtained by using the usual simplifying assumption?

(d) How much error is introduced by incorrectly using the simplifying assumption?

16.149 Some people who take megadoses of ascorbic acid will drink a solution containing as much as 6.0 g of ascorbic acid dissolved in a glass of water. Assuming the volume to be 0.250 L, calculate the pH of this solution.

16.150 For the titration of 25.00 mL of 0.1000 M HCl with 0.1000 M NaOH, calculate the pH of the reaction mixture after each of the following total volumes of base have been added to the original solution. (Remember to take into account the change in total volume.) Construct a graph showing the titration curve for this experiment.

(a) 0.00 mL   (d) 24.99 mL   (g) 25.10 mL
(b) 10.00 mL   (e) 25.00 mL   (h) 26.00 mL
(c) 24.90 mL   (f) 25.01 mL   (i) 50.00 mL

16.151 Below is a diagram illustrating a mixture HF and F^- in an aqueous solution. For this mixture, does pH equal pK_a for HF? Explain. Describe how the number of HF molecules and F^- ions will change after three OH^- ions are added. How will the number of HF and F^- change if two H^+ ions are added? Explain your answers by using chemical equations.

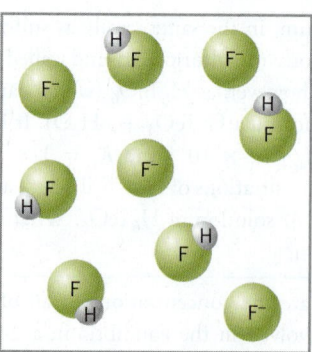

Multi-Concept Problems

*16.152 How many milliliters of 0.10 M NaOH must be added to 0.200 L of 0.010 M HCl to give a mixture with a pH of 3.00?

16.153 Milk of magnesia is a suspension of magnesium hydroxide in water. Although $Mg(OH)_2$ is relatively insoluble, a small amount does dissolve in the water, which makes the mixture slightly basic and gives it a pH of 10.08. How many grams of $Mg(OH)_2$ are actually dissolved in 2.0 tablespoons of milk of magnesia?

16.154 How many milliliters of 0.0100 M KOH are needed to completely neutralize the HCl in 325 mL of a hydrochloric acid solution that has a pH of 2.25?

16.155 It was found that 25.20 mL of an HNO_3 solution are needed to react completely with 30.0 mL of a LiOH solution that has a pH of 12.05. What is the molarity of the HNO_3 solution?

16.156 Suppose 38.0 mL of 0.000200 M HCl is added to 40.0 mL of 0.000180 M NaOH. What will be the pH of the final mixture?

*16.157 Suppose 10.0 mL of HCl gas at 25 °C and 734 torr is bubbled into 257 mL of pure water. What will be the pH of the resulting solution, assuming all the HCl dissolves in the water?

*16.158 Suppose the HCl described in the preceding problem is bubbled into 20.00 L of a solution of NaOH that has a pH of 10.50. What will the pH of the resulting solution be?

*16.159 What is the approximate freezing point of a 0.50 M solution of dichloroacetic acid, $HC_2HO_2Cl_2$ ($K_a = 5.0 \times 10^{-2}$)? Assume the density of the solution is 1.0 g/mL.

Exercises in Critical Thinking

16.160 What happens to the pH of a solution as it is heated? Does that mean that the autoionization of water is exothermic or endothermic?

16.161 Can the pH of a solution ever have a negative value? Give an example of a situation where the pH has a negative value.

16.162 In simplifying our calculations, we were satisfied with a maximum error of 5% in the concentration of hydronium ions when we compare the value calculated from the quadratic formula to the value obtained using simplifying assumptions. What does an error of 5% in hydrogen ion concentration mean in terms of pH? Is the pH error the same for a $1.0 \times 10^{-3}\,M\,H^+$ solution as it is for a $5.0 \times 10^{-6}\,M$ solution?

16.163 In the 1950s it was discovered that lakes in the northeastern United States had declining fish populations due to increased acidity. In order to address the problem of lake acidification, make a list of what you need to know in order to start addressing the problem.

16.164 Where are buffers found in everyday consumer products? Propose reasons why a manufacturer would use a buffer in a given product of your choice.

16.165 Why must the acid used for a buffer have a pK_a within one pH unit of the pH of the buffer? What happens if that condition is not met?

16.166 What conjugate acid–base pairs are used to buffer **(a)** over-the-counter drugs, **(b)** foods, **(c)** cosmetics, and **(d)** shampoos?

16.167 Your blood at 37 °C needs to be maintained within a narrow pH range of 7.35 to 7.45 to maintain optimal health. What possible conjugate acid–base pairs are present in blood that can buffer the blood and keep it within this range?

16.168 Develop a list of the uses of phosphoric acid in various consumer products.

16.169 What would our pH scale look like if Arrhenius decided that $pH = -\ln [H^+]$? What would the pH of a neutral solution be?

Solubility and Simultaneous Equilibria

In the middle of the Atlantic Ocean sits the Lost City, a field of hydrothermal vents. Discovered in December 2000, these vents made of calcium carbonate rise 30 to 60 meters above the seafloor. The very low solubility of calcium carbonate allows these towers to grow and form these fanciful shapes. In this chapter, we will discuss the solubility of these insoluble salts, and how their solubility can be used as the driving force of a reaction, or how the solubility of insoluble salts can be changed.

NSF, NOAA, University of Washington

In the preceding chapters you learned how the principles of equilibrium can be applied to aqueous solutions of acids and bases. In this chapter we extend these principles to aqueous reactions that involve the formation and dissolving of precipitates (a topic we introduced in Chapter 4). Many such reactions are common in the world around us. For example, water seeping into the seafloor becomes superheated and dissolves calcium carbonate; as the water emerges and cools the calcium carbonate precipitates into the majestic thermal vent shown in the chapter-opening photo. Closer to us, groundwater rich in carbon dioxide dissolves deposits of calcium carbonate, producing vast underground caverns, and as the remaining calcium-containing solution gradually evaporates, stalactites and stalagmites form. Within living organisms, precipitation reactions form the hard calcium carbonate shells of clams, oysters, and coral, as well as the unwanted calcium oxalate and calcium phosphate deposits we call kidney stones (Chemistry Outside the Classroom 4.1). Also, dilute acids in our mouths promote the dissolving of tooth enamel, which is composed of a mineral made up of calcium phosphate and calcium hydroxide.

This Chapter in Context

LEARNING OBJECTIVES

After reading this chapter, you should be able to:

- perform solubility calculations using neutral salts
- understand how hydrogen ion concentration affects the solubility of basic salts
- describe why oxides and sulfides are special cases and need unique methods to determine solubilities
- use selective precipitation to separate ions for qualitative and quantitative analysis
- describe and use equilibria related to reactions of complexes
- enhance solubility by using complexation reactions

17.1 | Equilibria in Solutions of Slightly Soluble Salts

Solubility Product Constant, K_{sp}

None of the salts we described in Chapter 4 as being insoluble are *totally* insoluble. For example, the solubility rules tell us that AgCl is "insoluble," but if some solid silver chloride is placed in water, a very small amount does dissolve. Once the solution has become saturated, the following equilibrium is established between the undissolved AgCl and its ions in the solution:

$$AgCl(s) \rightleftharpoons Ag^+(aq) + Cl^-(aq)$$

This is a heterogeneous equilibrium because it involves solid silver chloride in equilibrium with its ions in aqueous solution. Using the procedure developed in Section 14.4, we write the equilibrium law as shown in Equation 17.1, omitting the solid from the mass action expression.

$$[Ag^+][Cl^-] = K_{sp} \qquad (17.1)$$

The equilibrium constant, K_{sp}, is called the **solubility product constant** (because the system is a solubility equilibrium and the *constant* equals a *product* of ion concentrations).

It's important that you understand the distinction between solubility and solubility product. The solubility of a salt is the amount of the salt that dissolves in a given amount of solvent to give a **saturated solution**. The **solubility product** is the mathematical expression that is the product of the molar concentrations of the ions in the saturated solution, raised to appropriate powers (see below).

The solubilities of salts change with temperature, so a value of K_{sp} applies only at the temperature at which it was determined. Some typical K_{sp} values are listed in Table 17.1 and in Appendix C.

■ Recall that when salts go into solution, they dissociate essentially completely, so the equilibrium is between the solid and the ions that are in the solution.

TOOLS

Solubility product constant, K_{sp}

■ In this case we use the mathematical meaning of "product" rather than the chemical meaning.

TABLE 17.1		Solubility Product Constants		
Type	**Salt**		**Ions of Salt**	**K_{sp} (25 °C)**
Halides	CaF_2	$\rightleftharpoons$	$Ca^{2+} + 2F^-$	3.4×10^{-11}
	PbF_2	$\rightleftharpoons$	$Pb^{2+} + 2F^-$	3.3×10^{-8}
	$AgCl$	$\rightleftharpoons$	$Ag^+ + Cl^-$	1.8×10^{-10}
	$AgBr$	$\rightleftharpoons$	$Ag^+ + Br^-$	5.4×10^{-13}
	AgI	$\rightleftharpoons$	$Ag^+ + I^-$	8.5×10^{-17}
	$PbCl_2$	$\rightleftharpoons$	$Pb^{2+} + 2Cl^-$	1.7×10^{-5}
	$PbBr_2$	$\rightleftharpoons$	$Pb^{2+} + 2Br^-$	6.6×10^{-6}
	PbI_2	$\rightleftharpoons$	$Pb^{2+} + 2I^-$	9.8×10^{-9}
Hydroxides	$Al(OH)_3$	$\rightleftharpoons$	$Al^{3+} + 3OH^-$	3×10^{-34} [a]
	$Ca(OH)_2$	$\rightleftharpoons$	$Ca^{2+} + 2OH^-$	5.0×10^{-6}
	$Fe(OH)_2$	$\rightleftharpoons$	$Fe^{2+} + 2OH^-$	4.9×10^{-17}
	$Fe(OH)_3$	$\rightleftharpoons$	$Fe^{3+} + 3OH^-$	2.8×10^{-39}
	$Mg(OH)_2$	$\rightleftharpoons$	$Mg^{2+} + 2OH^-$	5.6×10^{-12}
	$Zn(OH)_2$	$\rightleftharpoons$	$Zn^{2+} + 2OH^-$	3×10^{-17} [b]
Carbonates	Ag_2CO_3	$\rightleftharpoons$	$2Ag^+ + CO_3^{2-}$	8.5×10^{-12}
	$MgCO_3$	$\rightleftharpoons$	$Mg^{2+} + CO_3^{2-}$	6.8×10^{-8}
	$CaCO_3$	$\rightleftharpoons$	$Ca^{2+} + CO_3^{2-}$	3.4×10^{-9} [c]
	$SrCO_3$	$\rightleftharpoons$	$Sr^{2+} + CO_3^{2-}$	5.6×10^{-10}
	$BaCO_3$	$\rightleftharpoons$	$Ba^{2+} + CO_3^{2-}$	2.6×10^{-9}
	$CoCO_3$	$\rightleftharpoons$	$Co^{2+} + CO_3^{2-}$	1.0×10^{-10}
	$NiCO_3$	$\rightleftharpoons$	$Ni^{2+} + CO_3^{2-}$	1.4×10^{-7}
	$ZnCO_3$	$\rightleftharpoons$	$Zn^{2+} + CO_3^{2-}$	1.5×10^{-10}
Chromates	Ag_2CrO_4	$\rightleftharpoons$	$2Ag^+ + CrO_4^{2-}$	1.1×10^{-12}
	$PbCrO_4$	$\rightleftharpoons$	$Pb^{2+} + CrO_4^{2-}$	1.8×10^{-14} [d]
Sulfates	$CaSO_4$	$\rightleftharpoons$	$Ca^{2+} + SO_4^{2-}$	4.9×10^{-5}
	$SrSO_4$	$\rightleftharpoons$	$Sr^{2+} + SO_4^{2-}$	3.4×10^{-7}
	$BaSO_4$	$\rightleftharpoons$	$Ba^{2+} + SO_4^{2-}$	1.1×10^{-10}
	$PbSO_4$	$\rightleftharpoons$	$Pb^{2+} + SO_4^{2-}$	2.5×10^{-8}
Oxalates	CaC_2O_4	$\rightleftharpoons$	$Ca^{2+} + C_2O_4^{2-}$	2.3×10^{-9}
	MgC_2O_4	$\rightleftharpoons$	$Mg^{2+} + C_2O_4^{2-}$	4.8×10^{-6}
	BaC_2O_4	$\rightleftharpoons$	$Ba^{2+} + C_2O_4^{2-}$	1.2×10^{-7}
	FeC_2O_4	$\rightleftharpoons$	$Fe^{2+} + C_2O_4^{2-}$	2.1×10^{-7}
	PbC_2O_4	$\rightleftharpoons$	$Pb^{2+} + C_2O_4^{2-}$	2.7×10^{-11}

[a]Alpha form. [b]Amorphous form. [c]Calcite form. [d]At 10 °C

Ion Product, the Reaction Quotient for Slightly Soluble Salts

In preceding chapters we described the value of the mass action expression as the **reaction quotient**, Q. For simple solubility equilibria like those discussed in this section, the mass action expression is a product of ion concentrations raised to appropriate powers, so Q is often called the **ion product** of the salt. Thus, for AgCl,

$$\text{ion product} = [Ag^+][Cl^-] = Q$$

At any dilution of a salt throughout the range of possibilities for an *unsaturated* solution, there will be varying values for the ion concentrations and, therefore, for Q. However, Q acquires a constant value, K_{sp}, in a *saturated* solution. When a solution is less than saturated, the value of Q is less than K_{sp}. Thus, we can use the numerical value of Q for a given solution as a test for saturation by comparing it to the value of K_{sp}.

Another aspect of solubility products is that many salts produce more than one of a given ion per formula unit when they dissociate, and this introduces exponents into the ion product expression. For example, when mercury(II) iodide, HgI_2, precipitates (Figure 17.1), it enters into the following solubility equilibrium:

$$HgI_2(s) \rightleftharpoons Hg^{2+}(aq) + 2I^-(aq)$$

The equilibrium law is obtained following the procedure we developed in Chapter 14, using coefficients as exponents in the mass action expression:

$$[Hg^{2+}][I^-]^2 = K_{sp}$$

Thus, the ion product contains the ion concentrations raised to powers equal to the number of ions released per formula unit. This means that to obtain the correct ion product expression you have to know the formulas of the ions that make up the salt. In other words, you have to realize that BiI_3 is composed of one Bi^{3+} ion and three iodide, I^-, anions.

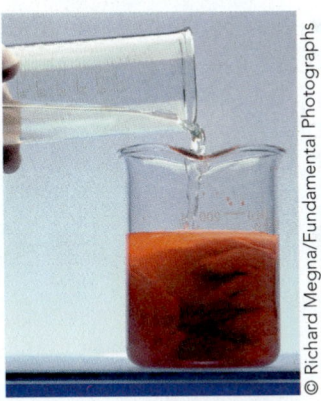

Figure 17.1 | **Mercury(II) iodide.** When mercury(II) acetate chromate is added to a solution of sodium iodide, bright red "insoluble" mercury(II) iodide, HgI_2, precipitates.

© Richard Megna/Fundamental Photographs

What are the ion product expressions for the following salts: (a) lead(II) bromide, (b) aluminum hydroxide? (*Hint:* Remember that they are heterogeneous equilibria.)

Practice Exercise 17.1

Calomel, mercury(I) chloride, is extremely poisonous and at the same time extremely insoluble, so much so that it was routinely prescribed as a medicine in the early 1900s. Remembering that mercury(I) forms a polyatomic cation, what is the K_{sp} equation for calomel?

Practice Exercise 17.2

Determining K_{sp} from Molar Solubilities

One way to determine the value of K_{sp} for a salt is to measure its solubility—that is, how much of the salt is required to give a saturated solution in a specified amount of solution. It is useful to express this as the **molar solubility**, which equals the number of moles of salt dissolved in one liter of its *saturated solution*. The molar solubility can be used to calculate the K_{sp} under the assumption that all of the salt that dissolves is 100% dissociated into the ions implied in the salt's formula.[1]

Molar solubility

One note here about calculations involving solubility products. Many of the calculations we will carry out include changes in concentration, and the use of concentration tables, as introduced in Chapter 14, will help organize our thoughts and make the calculations easier. If used properly, the concentration tables lay out all of the values for the concentrations and changes that occur during the solubility process. This is illustrated in Example 17.3.

Example 17.1
Calculating K_{sp} from Solubility Data

Silver bromide, AgBr, is the light-sensitive compound used in nearly all photographic film. The solubility of AgBr in water was measured to be 1.4×10^{-4} g L^{-1} at 25 °C. Calculate K_{sp} for AgBr at this temperature.

---

[1] This assumption works reasonably well for slightly soluble salts made up of singly charged ions, like silver bromide. For simplicity, and to illustrate the nature of calculations involving solubility equilibria, we will work on the assumption that all salts behave as though they are 100% dissociated. This is not entirely true, especially for the more soluble salts of multiply charged ions, so the accuracy of our calculations is limited. We discussed the reasons for the incomplete dissociation of salts in Section 12.6.

Analysis: For any question about the solubility of a slightly soluble salt, we start with the balanced chemical equation and then construct the equation for the solubility product constant, K_{sp}.

To determine the value of the K_{sp}, we will need the molar concentrations of the ions in solution. We are given a solubility of AgBr in g L^{-1}, so we will have to convert that to mol L^{-1} of AgBr and then the concentrations of the individual ions.

Assembling the Tools: The first tool that we need is the equation for the K_{sp}, which we derive from the balanced chemical equation:

$$AgBr(s) \rightleftharpoons Ag^+(aq) + Br^-(aq) \qquad K_{sp} = [Ag^+][Br^-]$$

For the concentrations of the ions we will use the tools from Section 3.2 to convert the mass to moles in order to convert grams of AgBr L^{-1} to moles of AgBr L^{-1}.

Solution: We need to start with the balanced chemical equation and the expression for the solubility product constant that we wrote in the tools section.

Then we calculate the number of moles of AgBr dissolved per liter starting with the given data and using the molar mass as our conversion factor:

$$\text{molar solubility} = \frac{1.4 \times 10^{-4} \text{ g AgBr}}{1.00 \text{ L soln}} \times \frac{1.00 \text{ mol AgBr}}{187.77 \text{ g AgBr}} = 7.5 \times 10^{-7} \text{ mol AgBr L}^{-1}$$

We can now use the formula, AgBr, and stoichiometry to get the concentrations we need:

$$\text{molarity of Ag}^+ = \frac{7.5 \times 10^{-7} \text{ mol AgBr}}{1.00 \text{ L soln}} \times \frac{1 \text{ mol Ag}^+}{1 \text{ mol AgBr}} = 7.5 \times 10^{-7} \text{ mol Ag}^+ \text{ L}^{-1}$$

$$\text{molarity of Br}^- = \frac{7.5 \times 10^{-7} \text{ mol AgBr}}{1.00 \text{ L soln}} \times \frac{1 \text{ mol Br}^-}{1 \text{ mol AgBr}} = 7.5 \times 10^{-7} \text{ mol Br}^- \text{ L}^{-1}$$

Finally we substitute these concentrations into the K_{sp} equation,

$$K_{sp} = [Ag^+][Br^-] = (7.5 \times 10^{-7} M \text{ Ag}^+)(7.5 \times 10^{-7} M \text{ Br}^-) = 5.6 \times 10^{-13}$$

The K_{sp} of AgBr is thus calculated to be 5.6×10^{-13} at 25 °C.

Is the Answer Reasonable? The math is easy to check. For the molarity of AgBr, try entering the mass as 0.00014 g/L into your calculator rather than as an exponent, and for the solubility product just multiply 7.5×7.5 and work out the exponents by hand. Both come out to our original results and we have confidence in the answer.

■ The K_{sp} we calculated from the solubility data here differs by only 4% from the value in Table 17.1.

Example 17.2
Calculating K_{sp} from Molar Solubility Data

Mercury(II) bromide, HgI_2, is a bright red solid that has a solubility of 1.9×10^{-10} mol L^{-1} at 25 °C in water. What is the K_{sp} for HgI_2?

Analysis: This problem is very similar to Example 17.1, except that the ions are not in a one-to-one ratio. In this case, when $HgBr_2$ dissolves, it produces two Br^- ions for each $HgBr_2$.

Assembling the Tools: The tools we will need for this problem are the same as the tools in Example 17.1, except we are given the solubility in mol L^{-1}, so we do not have to convert from grams to moles. We will need the balanced equilibrium equation and the

K_{sp} expression. As in Example 17.1, to determine the value for K_{sp}, we will use the molar solubility of the salt, and the equilibrium equation to calculate the concentrations of the mercury and bromide ions.

● **Solution:** For the balanced equilibrium equation and K_{sp} expression, we have

$$HgI_2(s) \rightleftharpoons Hg^{2+}(aq) + 2I^-(aq) \qquad K_{sp} = [Hg^{2+}][I^-]^2$$

Now we need $[Hg^{2+}]$ and $[I^-]$ to solve for the K_{sp} value. We start with the molar solubility given at the start and use the chemical equation to generate the conversion factors:

$$\text{molarity of } Hg^{2+} = \frac{1.9 \times 10^{-10} \text{ mol } HgI_2}{1.00 \text{ L soln}} \times \frac{1 \text{ mol } Hg^{2+}}{1 \text{ mol } HgI_2} = 1.9 \times 10^{-10} \text{ mol } Hg^{2+} \text{ L}^{-1}$$

$$\text{molarity of } I^- = \frac{1.9 \times 10^{-10} \text{ mol } HgI_2}{1.00 \text{ L soln}} \times \frac{2 \text{ mol } I^-}{1 \text{ mol } HgI_2} = 3.8 \times 10^{-10} \text{ mol } I^- \text{ L}^{-1}$$

Substituting these equilibrium concentrations into the mass action expression for K_{sp} gives

$$K_{sp} = (1.9 \times 10^{-10})(3.8 \times 10^{-10})^2$$

This time let's multiply *just the numerical part*, $1.9 \times 3.8 \times 3.8 = 27.4$, and then add the exponents, $10 + 10 + 10 = 30$, to get the answer:

$$= 27.4 \times 10^{-30}$$

$$= 2.7 \times 10^{-29} \quad \text{(when properly rounded)}$$

So the K_{sp} of HgI_2 at 25 °C is calculated to be 2.7×10^{-29}.

Notice that we avoided entering the entire number into our calculator; we just needed it for the numerical part since it is quite easy to add the exponents separately. Also, in this calculation all the significant figures were kept until finally rounding to the two significant figures that are justified.

● **Is the Answer Reasonable?** The math was straightforward. We could repeat the calculation by entering the entire number into our calculators to be sure we get the same answer. Obtaining the molarities of the ions required the correct mole ratios for the conversions and we should check them. Since one term is squared, *it is very important to place the molar concentrations in the correct places.* If they were accidentally reversed, the solubility product would be a factor of four too small.

The solubility of thallium(I) iodide, TlI, in water at 20 °C is 5.9×10^{-3} g L^{-1}. Using this fact, calculate K_{sp} for TlI on the assumption that it is 100% dissociated in the solution. (*Hint:* What is the molar solubility of TlI?)

Practice Exercise 17.3

One liter of water will dissolve 4.3×10^{-5} mol of Ag_3PO_4 at 25 °C. Calculate the value of K_{sp} for Ag_3PO_4.

Practice Exercise 17.4

Determining Molar Solubility from K_{sp}

Besides calculating K_{sp} from solubility information, we can also compute solubilities from tabulated values of K_{sp}. Examples 17.3 and 17.4 illustrate the calculations.[2]

---

[2] As noted earlier, these calculations ignore the fact that the salt that dissolves is not truly 100% dissociated into the ions implied in the salt's formula. This is particularly a problem with the salts of multiply charged ions, so the concentrations of the ions of such salts, when calculated from their K_{sp} values, must be taken as rough estimates. In fact, many calculations involving K_{sp} give values that are merely estimates.

Example 17.3
Calculating Molar Solubility from K_{sp}

What is the molar solubility of AgCl in pure water at 25 °C?

Analysis: To solve the problem, we're going to need a balanced chemical equation for the solubility equilibrium, the equilibrium law (K_{sp} expression), and the value of K_{sp}, which we can obtain from Table 17.1. We will also need to set up a concentration table, as in Section 14.7, to keep track of the information.

Assembling the Tools: The relevant equations to solve this problem are

$$AgCl(s) \rightleftharpoons Ag^+(aq) + Cl^-(aq) \qquad K_{sp} = [Ag^+][Cl^-] = 1.8 \times 10^{-10}$$

With these equations, we can build a concentration table. First, we see that the solvent is pure water, so neither Ag^+ nor Cl^- is present in solution at the start; their initial concentrations are zero. Next, we turn to the "change" row. The molar amount that dissolves is unknown so *we will define our unknown, s, as the molar solubility of the salt*—that is, the number of moles of AgCl that dissolves in one liter. Because 1 mol of AgCl yields 1 mol Ag^+ and 1 mol Cl^-, the concentration of each of these ions increases by $+s$ and is entered as such in the concentration table.

> ■ We're getting tired of always using x for the unknown. Let's use s instead; it also seems like a better symbol for solubility.

Solution: Our concentration table is as follows:

	AgCl(s)	$\rightleftharpoons$	Ag⁺(aq)	+	Cl⁻(aq)
Initial Concentrations (*M*)	NE*		0		0
Concentration Change (*M*)	NE		+s		+s
Equilibrium Concentration (*M*)	NE		s		s

> ■ By defining s as the molar solubility, the coefficients of s in the "change" row are the same as the coefficients of Ag^+ and Cl^- in the equation for the equilibrium.

*NE = No entry. No entries are made in this column since AgCl(s) is not part of the mass action expression. However, it is understood that a saturated solution usually has some solid present to be certain it is saturated.

We make the substitutions into the K_{sp} expression, using equilibrium quantities from the last row of the table:

$$[Ag^+][Cl^-] = 1.8 \times 10^{-10}$$
$$(s)(s) = 1.8 \times 10^{-10}$$
$$s = 1.3 \times 10^{-5}$$

The calculated molar solubility of AgCl in water at 25 °C is 1.3×10^{-5} mol L^{-1}.

Is the Answer Reasonable? The solvent is water, so the initial concentrations are zero. If *s* is the molar solubility, then the changes in the concentrations of Ag^+ and Cl^- when the salt dissolves must also be equal to *s*. We can also check the algebra, which we've done correctly, so the answer seems to be okay.

Example 17.4
Calculating Molar Solubility from K_{sp}

Calculate the molar solubility of lead(II) iodide in water at 25 °C from its K_{sp}.

Analysis: The key to this problem is realizing that there are two iodide ions released for each lead(II) iodide formula unit. For this problem, we will need the formula for lead(II) iodide, so that we can set up the balanced chemical equation and the K_{sp} expression. Then we will need to use a concentration table to find the molar solubilities of the lead(II) iodide.

Assembling the Tools: The tools are the same tools we used in Example 17.3, except that the first thing we need to do is determine the formula for lead(II) iodide, which is obtained from the rules of nomenclature in Chapter 2. We then use the equation for the equilibrium and the K_{sp} expression.

Solution: We determine that the formula for lead(II) iodide is, PbI_2. We then write the chemical equation and the K_{sp} for this process. Next, we obtain K_{sp} from Table 17.1.

$$PbI_2(s) \rightleftharpoons Pb^{2+}(aq) + 2I^-(aq) \qquad K_{sp} = [Pb^{2+}][I^-]^2 = 9.8 \times 10^{-9}$$

For the concentration table, we can start with the solvent, water, and the initial concentrations of the ions, which are zero. As before, we will define s as the molar solubility of the salt. By doing this, the coefficients of s in the "change" row will be identical to the coefficients of the ions in the chemical equation. This assures us that the changes in the concentrations are in the correct mole ratio.

Here is the concentration table.

	$PbI_2(s)$	$\rightleftharpoons$	$Pb^{2+}(aq)$	+	$2I^-(aq)$
Initial Concentrations (M)	NE		0		0
Concentration Change (M)	NE		+s		+2s
Equilibrium Concentration (M)	NE		s		s

Substituting equilibrium quantities into the K_{sp} expression gives

$$[Pb^{2+}][I^-]^2 = 9.8 \times 10^{-9}$$

$$(s)(2s)^2 = 4s^3 = 9.8 \times 10^{-9}$$

$$s^3 = 2.45 \times 10^{-9}$$

$$s = 1.3 \times 10^{-3} \quad \text{(when correctly rounded)}$$

Thus, the molar solubility of PbI_2 in pure water is calculated to be 1.3×10^{-3} mol L^{-1}.

Is the Answer Reasonable? We check the entries in the table. The solvent is water, so the initial concentrations are equal to zero. By setting s equal to the moles per liter of PbI_2 that dissolves (i.e., the molar solubility), the coefficients s in the "change" row have to be the same as the coefficients of Pb^{2+} and Cl^- in the balanced chemical equation. The entries in the "change" row are okay. In performing the algebra, notice that when we square $2s$, we get $4s^2$. Multiplying $4s^2$ by s gives $4s^3$. After that, if our math is correct, the answer is correct.

A precipitate of yellow PbI_2 forms when the two colorless solutions, one with sodium iodide and the other of lead(II) nitrate, are mixed.

Lawrence Migdale/Photo Researchers

What is the calculated molar solubility in water at 25 °C of (a) AgBr and (b) PbBr$_2$? (*Hint:* Review the algebra in Example 17.4.)

Practice Exercise 17.5

Calculate the molar solubility of Hg_2Cl_2 in water. Its K_{sp} is 1.4×10^{-18}.

Practice Exercise 17.6

The Common Ion Effect

Suppose that we stir some lead(II) chloride (a compound having a low solubility) with water long enough to establish the following equilibrium:

$$PbCl_2(s) \rightleftharpoons Pb^{2+}(aq) + 2Cl^-(aq)$$

If we now add a concentrated solution of a soluble lead compound, such as $Pb(NO_3)_2$, the increased concentration of Pb^{2+} in the $PbCl_2$ solution will drive the position of equilibrium to the left, causing some $PbCl_2$ to precipitate. The phenomenon is simply an application of Le Châtelier's principle, the net result being that $PbCl_2$ is less soluble in a solution that

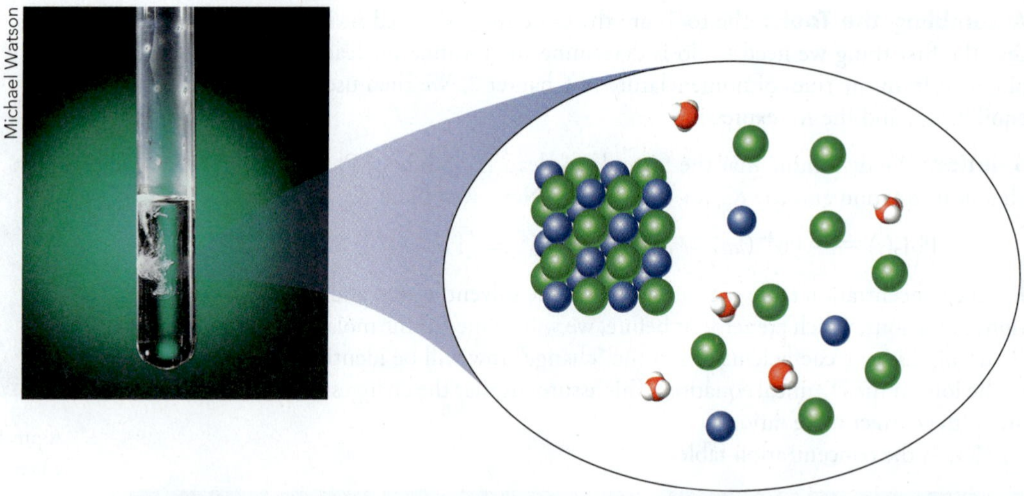

Figure 17.2 | **The common ion effect.** The test tube shown here initially held a saturated solution of NaCl, where the equilibrium, $NaCl(s) \rightleftharpoons Na^+(aq) + Cl^-(aq)$, had been established. Addition of a few drops of concentrated HCl, containing a high concentration of the common ion Cl^-, in green, forced the equilibrium to shift to the left. This caused some white crystals of solid NaCl to precipitate.

contains Pb^{2+} from another source than it is in pure water. The same effect is produced if a concentrated solution of a soluble chloride salt such as NaCl is added to the saturated $PbCl_2$ solution. The added Cl^- will drive the equilibrium to the left, reducing the amount of dissolved $PbCl_2$.

The phenomenon we just described is an example of the **common ion effect**, which we also described in Chapter 16. In this case, Pb^{2+} is the **common ion** when we add $Pb(NO_3)_2$, and Cl^- is the common ion when we add NaCl. Figure 17.2 shows how even a relatively soluble salt, NaCl, can be forced out of its saturated solution simply by adding concentrated hydrochloric acid, which serves as a source of the common ion, Cl^-. The common ion effect can dramatically lower the solubility of a salt. Example 17.5 demonstrates the typical wording and solution of a common ion problem. It also demonstrates how we can simplify the math when the concentrations of the ions of the insoluble salt are significantly larger than the value for K_{sp}.

Example 17.5
Calculations Involving the Common Ion Effect

What is the molar solubility of PbI_2 in a 0.10 M NaI solution?

Analysis: Because PbI_2 and NaI have the same ion in common, I^-, this is an example of a calculation involving the common ion effect. As usual, to solve the problem we're going to need a balanced chemical equation for the solubility equilibrium, the equilibrium law (i.e., the K_{sp} expression), and the value of K_{sp}, which is listed in Table 17.1. We will also need to set up a concentration table as in Example 17.4, but this time we have to take into account the solute already present in the solution.

Assembling the Tools: We begin with the balanced chemical equation for the equilibrium, the appropriate K_{sp} expression, and the value of K_{sp} obtained from Table 17.1:

$$PbI_2(s) \rightleftharpoons Pb^{2+}(aq) + 2I^-(aq) \qquad K_{sp} = [Pb^{2+}][I^-]^2 = 9.8 \times 10^{-9}$$

As before, we imagine that we are adding the solid PbI_2 to a solvent into which it dissolves. This time, however, the solvent isn't water; it's a solution of NaI, which contains one of the ions of the salt PbI_2. The NaI completely dissociates and yields $0.10\,M\,Na^+$ and $0.10\,M\,I^-$. The initial concentration of I^- is therefore $0.10\,M$. Next, we let s be the molar solubility of PbI_2. When s mol of PbI_2 dissolves per liter, the concentration of Pb^{2+} changes by $+s$ and that of I^- by twice as much, or $+2s$. Finally, the equilibrium concentrations are obtained by summing the initial concentrations and the changes.

Solution: Here is the concentration table.

	$PbI_2(s)$	$\rightleftharpoons$	$Pb^{2+}(aq)$	$+$	$2I^-(aq)$
Initial Concentrations (M)	NE		0		0.10
Concentration Change (M)	NE		$+s$		$+2s$
Equilibrium Concentration (M)	NE		s		$0.10 + 2s$

Substituting equilibrium values into the K_{sp} expression gives

$$K_{sp} = [Pb^{2+}][I^-]^2$$
$$(s)(0.10 + 2s)^2 = 9.8 \times 10^{-9}$$

Just a brief inspection reveals that solving this expression for s will be difficult if we cannot simplify the math. Fortunately, a simplification is possible because the small value of K_{sp} for PbI_2 indicates that the salt has a very low solubility. This means very little of the salt will dissolve, so s (or even $2s$) will be quite small. Let's assume that $2s$ will be much smaller than 0.10. If this is so, then $0.10 + 2s \approx 0.10$. Substituting $0.10\,M$ for the I^- concentration gives

$$(s)(0.10)^2 = 9.8 \times 10^{-9}$$
$$(s) = \frac{9.8 \times 10^{-9}}{(0.10)^2}$$
$$(s) = 9.8 \times 10^{-7}$$

Thus, the molar solubility of PbI_2 in $0.10\,M\,NaI$ solution is calculated to be $9.8 \times 10^{-7}\,M$. We should always check to see if our simplifying assumption is valid. Notice that $2s$, which equals 2.0×10^{-6}, is indeed vastly smaller than 0.10, just as we anticipated. (If we add 2.0×10^{-6} to 0.10 and round correctly, we obtain 0.10.)

Is the Answer Reasonable? First, we can check the entries in the concentration table. The initial concentrations come from the solvent, which contains no Pb^{2+} but does contain $0.10\,M\,I^-$. By letting s equal the molar solubility, the coefficients of s in the "change" row equal the coefficients in the balanced equation. The equilibrium row is the sum of the first two rows. At this point, we have checked the concentration table. To check the answer, we can recalculate the K_{sp}.

Calculate the molar solubility of AgI in a $0.20\,M\,CaI_2$ solution. Compare the answer to the calculated molar solubility of AgI in pure water. (*Hint:* What is the I^- concentration in $0.20\,M\,CaI_2$?)

Practice Exercise 17.7

Calculate the molar solubility of $Ni_3(PO_4)_2$ in seawater where the nominal concentration of phosphate ions is $2.0 \times 10^{-6}\,M$. Assume the dissociation of $Ni_3(PO_4)_2$ is 100%. The K_{sp} for $Ni_3(PO_4^{3-})_2$ is 4.7×10^{-32}. (*Optional:* How many gallons of seawater will yield one ounce (28.3 g) of nickel assuming that seawater is a saturated solution of $Ni_3(PO_4)_2$?)

Practice Exercise 17.8

In Example 17.4, we found that the molar solubility of PbI_2 in *pure* water was $1.3 \times 10^{-3}\,M$. In water that contains $0.10\,M\,NaI$ (Example 17.5), the solubility of PbI_2 is $9.8 \times 10^{-7}\,M$, well over a thousand times smaller. As we said, the common ion effect often causes huge reductions in the solubilities of sparingly soluble compounds by shifting

■ A Mistake to Avoid: Only use coefficients to multiply unknowns, such as s, in the concentration table. Never multiply numerical data, such as the initial concentration, by a coefficient.

the equilibrium toward the formation of the solid. The common ion effect *never increases the solubility* of a salt.

Determining Whether a Precipitate Will Form

If we know the anticipated concentrations of the ions of a salt in a solution, we can use the value of K_{sp} for the salt to predict whether or not a precipitate should form. This is because the computed value for Q (the ion product) can tell us whether a solution is unsaturated, saturated, or supersaturated. *For a precipitate of a salt to form, the solution must be supersaturated.*

If the solution is unsaturated, its ion concentrations are less than required for saturation, Q is less than K_{sp}, and no precipitate will form. For a saturated solution, we have Q equal to K_{sp}, and no precipitate will form. (If a precipitate were to form, the solution would become unsaturated and the precipitate would re-dissolve.) If the solution is supersaturated, the ion concentrations exceed those required for saturation and Q is larger than K_{sp}. Only in this last instance should we expect a precipitate to form so that Q will be reduced to the value of the K_{sp}. This can be summarized as follows:

TOOLS

Predicting precipitation

Precipitate will form $\quad Q > K_{sp}$ $\quad$ (supersaturated)

No precipitate will form $\begin{cases} Q = K_{sp} & \text{(saturated)} \\ Q < K_{sp} & \text{(unsaturated} \end{cases}$

For example, will a precipitate of $PbCl_2$ form if we prepare a solution containing 0.015 M in NaCl and 0.15 M $Pb(NO_3)_2$? Since there is one chloride ion and one lead ion in each formula, the concentrations of these ions will be the same as the salts. We can then use those values to calculate Q for the solubility product of $PbCl_2$ as:

$$Q = [Pb^{2+}][Cl^-]^2 = (0.15)(0.015)^2 = 3.4 \times 10^{-5}$$

■ It is usually difficult to prevent the extra salt from precipitating out of a supersaturated solution. (Sodium acetate is a notable exception. Supersaturated solutions of this salt are easily made.)

From Table 17.1, we find that $K_{sp} = 1.7 \times 10^{-5}$. Since $Q > K_{sp}$ we conclude that a precipitate *will* form.

Try some sample calculations to predict whether precipitation will occur.

Practice Exercise 17.9 | Calculate the ion product for the solution in which the Hg^{2+} concentration is 0.0025 M and the I^- concentration is 0.030 M. Will a precipitate of HgI_2 form in this solution? (*Hint:* What is the form of the ion product for HgI_2?)

Practice Exercise 17.10 | Calculate the ion product for a solution containing 3.4×10^{-4} M Br^- and 4.8×10^{-5} M Ag^+. Will a precipitate form?

The practice exercises were straightforward since the concentrations of the ions were given and we simply had to compare Q to the value of the K_{sp}. We can use the same principles when mixing two solutions with known concentrations. The difference is that we must calculate the concentrations of the ions when the solutions are mixed and also take dilution into account. The following example illustrates the process.

Example 17.6
Predicting whether a Precipitate Will Form when Solutions Are Mixed

What possible precipitate might form by mixing 50.0 mL of 1.0×10^{-4} M NaCl with 50.0 mL of 1.0×10^{-6} M $AgNO_3$? Will it form? (Assume the volumes are additive.)

Analysis: In this problem we are being asked, in effect, whether a metathesis reaction will occur between NaCl and $AgNO_3$. We should be able to use the solubility rules in Chapter 4 to predict whether this might occur. If the solubility rules suggest a precipitate, we can then calculate the ion product for the compound using the concentrations of the ions in the final solution. If this ion product exceeds K_{sp} for the salt, then a precipitate is expected.

To calculate Q correctly requires that we use the concentrations of the ions after the solutions have been mixed. Therefore, before computing the ion product, we must first take into account that mixing the solutions dilutes each of the solutes. Dilution problems were covered in Section 4.7.

■ To obtain the solution to this problem we must bring together tools from two different chapters.

Assembling the Tools: Our first tool is the solubility rules (Section 4.6), which helps us determine whether each product is soluble or insoluble. Then, we need to use the dilution equation (Equation 4.6, Section 4.7)

$$V_{dil} \times M_{dil} = V_{conc} \times M_{conc}$$

to find the concentrations of the ions for the ion product. Finally, we will use the ion product and compare Q to K_{sp} to determine whether a precipitate forms.

Solution: Let's begin by writing the equation for the potential metathesis reaction between $NaCl$ and $AgNO_3$, using the solubility rules to help us determine whether each product is soluble or insoluble.

$$NaCl(aq) + AgNO_3(aq) \longrightarrow AgCl(s) + NaNO_3(aq)$$

The solubility rules indicate that we expect a precipitate of AgCl. But are the concentrations of Ag^+ and Cl^- actually high enough?

In the original solutions, the 1.0×10^{-6} M $AgNO_3$ contains 1.0×10^{-6} M Ag^+ and the 1.0×10^{-4} M $NaCl$ contains 1.0×10^{-4} M Cl^-. What are the concentrations of these ions after dilution? To determine these, we use Equation 4.6:

$$V_{dil} \times M_{dil} = V_{conc} \times M_{conc}$$

Solving for M_{dil} gives

$$M_{dil} = \frac{M_{conc} \times V_{conc}}{V_{dil}}$$

The initial volumes of the more concentrated solutions are 50.0 mL, and when the two solutions are combined, the final total volume is 100.0 mL. Therefore,

$$[Ag^+]_{dil} = \frac{(50.0 \text{ mL})(1.0 \times 10^{-6} \text{ } M)}{100.0 \text{ mL}} = 5.0 \times 10^{-7} \text{ } M$$

$$[Cl^-]_{dil} = \frac{(50.0 \text{ mL})(1.0 \times 10^{-4} \text{ } M)}{100.0 \text{ mL}} = 5.0 \times 10^{-5} \text{ } M$$

Now we use these to calculate Q for AgCl, which we can obtain from the dissociation reaction of the salt:

$$AgCl(s) \rightleftharpoons Ag^+(aq) + Cl^-(aq)$$

$$Q = [Ag^+][Cl^-]$$

Substituting the concentrations computed above gives

$$Q = (5.0 \times 10^{-7})(5.0 \times 10^{-5}) = 2.5 \times 10^{-11}$$

In Table 17.1, the K_{sp} for AgCl is given as 1.8×10^{-10}. Notice that Q is smaller than K_{sp}, which means that the final solution is unsaturated in AgCl and a precipitate will not form.

Is the Answer Reasonable? There are several things we should check here. They include writing the equation for the metathesis reaction, calculating the concentrations of the ions after dilution (we've doubled the volume, so the concentrations are halved), and setting up the correct ion product. All appear to be correct, so the answer should be okay.

Practice Exercise 17.11   What precipitate might be expected if we mix 100.0 mL of 1.0×10^{-3} M Pb(NO$_3$)$_2$ and 100.0 mL of 2.0×10^{-3} M MgBr$_2$? Will some precipitate form? Assume that the volumes are additive. (*Hint:* What tools will you need to solve the problem?)

Practice Exercise 17.12   What precipitate might possibly be formed if we mix 50.0 mL of 0.10 M Pb(NO$_3$)$_2$ and 20.0 mL of 0.040 M NaCl? Will some of that precipitate form? Assume that the volumes are additive.

17.2 | Solubility of Basic Salts Is Influenced by Acids

In the previous section we considered "neutral salts" or those salts that are formed by reacting a strong acid with a strong base. Basic salts are formed in the reaction of a weak acid with a strong base and in water they are slightly basic.

When a basic salt is slightly soluble, very few ions are formed, and even fewer react with water. The result is that in many cases we can do the same calculations with basic salts as we did in Section 17.1 with neutral salts. The results will be most accurate with salts that have low molar solubilities and if the weak acid used to prepare the salt is relatively strong.

Solubilities of basic salts are strongly affected by acids. If we add solid calcium fluoride to an acidic solution, the fluoride ions will react with available hydrogen ions to make hydrogen fluoride. This removes F$^-$ and, using Le Châtelier's principle, we can reason that removing fluoride ions will result in more calcium fluoride dissolving.

$$CaF_2(s) \rightleftharpoons Ca^{2+}(aq) + 2F^-(aq) \quad \text{then} \quad 2F^-(aq) + 2H^+ \rightleftharpoons 2HF(aq)$$

The result is that basic salts such as CaF$_2$ will be more soluble in acid solution. This two-step process actually occurs as a single reaction made up of two simultaneous processes. Let's see how that works in the case of FePO$_4$.

Consider the case of iron(III) phosphate, FePO$_4$. The dissolution equation is:

$$FePO_4(s) \rightleftharpoons Fe^{3+}(aq) + PO_4^{3-}(aq) \tag{17.2}$$

The phosphate ion is actually a rather strong base and it will react with available hydrogen ions as in Equation 17.3.

$$PO_4^{3-}(aq) + H^+(aq) \rightleftharpoons HPO_4^{2-}(aq) \tag{17.3}$$

When PO$_4^{3-}$ and H$^+$ react the phosphate ion concentration decreases. Using Le Châtelier's principle, more FePO$_4$ will then dissolve to reestablish equilibrium.

One important conclusion from the discussion above is that addition of an acid to a basic salt will increase the amount of salt that dissolves because the anion is removed. If we add the two reactions together, which we can do if we add acid to a solution with FePO$_4$, we get

$$FePO_4(s) + \cancel{PO_4^{3-}(aq)} + H^+(aq) \rightleftharpoons Fe^{3+}(aq) + \cancel{PO_4^{3-}(aq)} + HPO_4^{2-}(aq)$$

and we can see that if we add acid, the reaction is pushed in the forward direction. Using the same argument, but for acid salts, adding a base will increase the amount that dissolves.

Simultaneous Equilibria

Simultaneous equilibrium describes a situation where two or more equilibrium reactions are occurring at the same time in a given mixture. In those cases, the mathematical relationships of all equilibrium equations must be satisfied at the same time. The way to do this is to combine the chemical equations to obtain an expression of the overall reaction and then use our concepts for manipulating equilibrium constants (Section 14.2) to determine the value of the equilibrium constant.

We can treat the simultaneous equilibria of a slightly soluble salt and a weak acid by adding equations 17.2 and 17.3 to get the overall reaction

$$FePO_4(s) + H^+(aq) \rightleftharpoons Fe^{3+}(aq) + HPO_4^{2-}(aq) \tag{17.4}$$

■ We are adding Equation 17.2 and 17.3 so we multiply the K_{sp} and K_{a_3}.

The equilibrium constant for Equation 17.2 is the K_{sp} and for Equation 17.3 it is $1/K_a$ for the weak acid HPO_4^{2-}, which is the third ionization reaction for H_3PO_4, or K_{a_3}. The overall equilibrium constant and mass action expression for Equation 17.4 is

$$K_{overall} = K_{sp}\frac{1}{K_{a_3}} = \frac{[Fe^{3+}][HPO_4^{2-}]}{[H^+]} = 9.9 \times 10^{-16}\left(\frac{1}{4.2 \times 10^{-13}}\right) = 2.3 \times 10^{-3}$$

Solving this equation for the solubility of $FePO_4$ is beyond the scope of this book. However, the multi-concept problem that follows shows that solutions for acidic solvents can be obtained if we know the concentration of the acid beforehand.

Analyzing and Solving Multi-Concept Problems

What is the solubility of CaF_2 in a 1.50 M solution of nitric acid?

Analysis: Even a quite simply stated question can be a multi-concept problem as we'll see. In this situation (1) the salt has to be classified as neutral, acidic, or basic. (2) We need to identify all the equilibrium reactions that occur and combine them appropriately. (3) We will use the appropriate equilibrium calculations to determine the answer.

PART 1

Assembling the Tools: We need to recall the concepts related to the salts of acids and bases from Section 16.6.

Solution: To classify CaF_2 we write the two ions from the formula: Ca^{2+} and F^-. Next we add the OH^- to the cation and H^+ to the anion to determine which acid and base were used to make the salt. We get

$$Ca^{2+} + 2OH^- \rightleftharpoons Ca(OH)_2 \quad \text{and} \quad F^- + H^+ \rightleftharpoons HF$$

Our salt was formed from the strong base $Ca(OH)_2$ and the weak acid HF. Now we conclude that Ca^{2+} must be a *very weak* conjugate acid while F^- is a *weak* conjugate base. So, because a weak conjugate base is *stronger than* a very weak conjugate acid, we have a basic salt and we should expect to solve a simultaneous equilibrium problem.

PART 2

Assembling the Tools: In this part we need to assemble the different equilibria involved in this problem. These are the balanced chemical equations and equilibrium equations that go along with the chemical equations. Finally we will need to look up the values for the K_{sp} and K_a equations.

Solution: The solubility equation for CaF_2 is:

$$CaF_2(s) \rightleftharpoons Ca^{2+}(aq) + 2F^-(aq)$$

and the ionization equation for the weak acid HF is:

$$HF(aq) \rightleftharpoons F^-(aq) + H^+(aq)$$

We look up the values and find $K_{sp}(CaF_2) = 3.4 \times 10^{-11}$ and $K_a(HF) = 3.5 \times 10^{-4}$. To combine these equations (so that $2F^-$ cancel) we need to double the coefficients and reverse the ionization for HF so that we get

$$CaF_2(s) \rightleftharpoons Ca^{2+}(aq) + 2F^-(aq)$$

$$\underline{2F^-(aq) + 2H^+(aq) \rightleftharpoons 2HF(aq)}$$

$$CaF_2(s) + 2H^+(aq) \rightleftharpoons Ca^{2+}(aq) + 2HF(aq)$$

The overall equilibrium constant is

$$K_{overall} = \frac{K_{sp}}{K_a K_a} = \frac{3.4 \times 10^{-11}}{(3.5 \times 10^{-4})(3.5 \times 10^{-4})} = 2.8 \times 10^{-4}$$

Noting that nitric acid is a strong acid that is completely ionized, we can take 1.50 M as the initial $[H^+]$. The concentration table is

	CaF$_2$	+	2H$^+$	$\rightleftharpoons$	2HF	+	Ca^{2+}
Initial Concentrations (M)	NE		1.50		0		0
Concentration Changes (M)	NE		$-2s$		$+2s$		$+s$
Equilibrium Concentrations (M)	NE		$1.50-2s$		$2s$		s

Now we can enter the terms from the equilibrium row of the concentration table and solve for the molar solubility of CaF$_2$.

$$K_{overall} = 2.8 \times 10^{-4} = \frac{[Ca^{2+}][HF]^2}{[H^+]^2}$$

$$2.8 \times 10^{-4} = \frac{(s)(2s)^2}{(1.50-2s)^2}$$

As we've done before for complex equations we will attempt to simplify this equation by assuming that $2s \ll 1.50\ M$.

$$2.8 \times 10^{-4} = \frac{(s)(2s)^2}{(1.50)^2}$$

$$6.3 \times 10^{-4} = s(4s^2) = 4s^3$$

$$1.58 \times 10^{-4} = s^3$$

$$s = 5.4 \times 10^{-2}$$

Our assumption seems to be true since $2s = 0.11\ M$ and that is less than one-tenth of the original 1.50 M. The molar solubility of CaF$_2$ is 5.4×10^{-2} mol L^{-1} when the solvent is a 1.5 M acid solution.

Is the Answer Reasonable? We've checked that the assumption is true. A check of the math also reveals no apparent errors.

Practice Exercise 17.13   Determine whether adding an acid will increase the solubility of the following salts: (a) AgBr, (b) CaCO$_3$, (c) Ag$_2$CrO$_4$, and (d) Zn(CN)$_2$. (*Hint:* Classify the salts as a neutral, acidic, or basic.)

Practice Exercise 17.14   Determine the molar solubility of Ag$_2$CrO$_4$ in 1.0 M H$^+$ solution.

In the multi-concept problem we found the solubility of CaF_2 when we took into account the reaction of fluoride ions with acid. Our initial analysis indicated that the simultaneous reaction would reduce the fluoride ion concentration and Le Châtelier's principle would predict an increase in solubility to restore equilibrium. If we calculate the solubility of CaF_2 and ignore the reaction of fluoride with water the result we get is 2.0×10^{-4} mol L^{-1}. Using the simultaneous equilibria gave us a larger solubility that also agrees with Le Châtelier's principle.

Later, in Section 17.4, we will use the effect of $[H^+]$ on molar solubilities to determine experimental conditions to selectively precipitate one cation in the presence of others.

17.3 | Equilibria in Solutions of Metal Oxides and Sulfides

Aqueous equilibria involving insoluble oxides and sulfides are more complex than those we've considered so far because of reactions of the anions with the solvent, water. When a metal oxide dissolves in water, it does so by reacting with water instead of by a simple dissociation of ions that remain otherwise unchanged. Sodium oxide, for example, consists of Na^+ and O^{2-} ions, and it readily dissolves in water. The solution, however, does not contain the oxide ion, O^{2-}. Instead, the hydroxide ion forms. The equation for the reaction is

$$Na_2O(s) + H_2O \longrightarrow 2NaOH(aq)$$

This actually involves the reaction of oxide ions with water as the crystals of Na_2O break up:

$$O^{2-}(s) + H_2O \longrightarrow 2OH^-(aq)$$

The oxide ion is simply too powerful a base to exist in water at any concentration worthy of experimental note. We can understand why from the extraordinarily high (estimated) value of K_b for O^{2-}, 1×10^{22}. Thus, there is no way to supply oxide ions to an aqueous solution in order to form an insoluble metal oxide directly. Oxide ions react with water, instead, to generate the hydroxide ion. When an insoluble metal oxide instead of an insoluble metal hydroxide does precipitate from a solution, it forms because the specific metal ion is able to react with OH^-, extract O^{2-}, and leave H^+ (or H_2O) in the solution. The silver ion, for example, precipitates as brown silver oxide, Ag_2O, when OH^- is added to aqueous silver salts (Figure 17.3):

$$2Ag^+(aq) + 2OH^-(aq) \longrightarrow Ag_2O(s) + H_2O$$

The reverse of this reaction, written as an equilibrium, corresponds to the solubility equilibrium for Ag_2O:

$$Ag_2O(s) + H_2O \rightleftharpoons 2Ag^+(aq) + 2OH^-(aq)$$

Solubility Equilibria for Metal Sulfides and Oxides

When we shift from oxygen to sulfur in Group 6A and consider metal sulfides, we find many similarities to oxides. One is that the sulfide ion, S^{2-}, like the oxide ion, O^{2-}, is such a strong Brønsted–Lowry base that it does not exist in any ordinary aqueous solution. The sulfide ion has not been detected in an aqueous solution, even in the presence of $8\ M$ NaOH, where one might think that the reaction

$$OH^- + HS^- \longrightarrow H_2O + S^{2-}$$

could generate some detectable S^{2-}. An $8\ M$ NaOH solution is at a concentration well outside the bounds of the "ordinary." Thus, Na_2S, like Na_2O, dissolves in water by reacting with it, not by releasing an otherwise unchanged divalent anion, S^{2-}:

$$Na_2S(s) + H_2O \longrightarrow 2Na^+(aq) + HS^-(aq) + OH^-(aq)$$

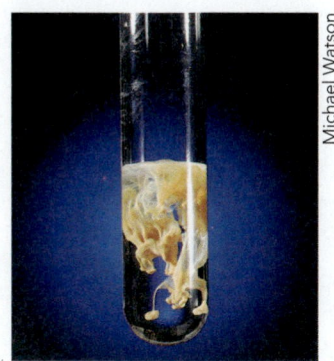

Figure 17.3 | Silver oxide. A brown, mud-like precipitate of silver oxide forms as a sodium hydroxide solution is added to a solution of silver nitrate.

Michael Watson

Figure 17.4 | The colors of some metal sulfides.

As with Na_2O, water reacts with the sulfide ion as the crystals break apart and the ions enter the solution:

$$S^{2-}(s) + H_2O \longrightarrow HS^-(aq) + OH^-(aq)$$

■ In the laboratory, the qualitative analysis of metal ions uses the precipitation of metal sulfides to separate some metal ions from others.

The sulfides of most metals have quite low solubilities in water. Simply bubbling hydrogen sulfide gas into an aqueous solution of any one of a number of metal ions—Cu^{2+}, Pb^{2+}, and Ni^{2+}, for example—causes their sulfides to precipitate. Many of these have distinctive colors (Figure 17.4) that can be used to help identify which metal ion is in solution. To write the solubility equilibrium for an insoluble metal sulfide, we have to take into account the reaction of the sulfide ion with water. For example, for CuS the equilibrium is best written as

$$CuS(s) + H_2O \rightleftharpoons Cu^{2+}(aq) + HS^-(aq) + OH^-(aq) \tag{17.5}$$

This equation yields the ion product $[Cu^{2+}][HS^-][OH^-]$, and the solubility product constant for CuS is expressed by the equation,

$$K_{sp} = [Cu^{2+}][HS^-][OH^-]$$

Values of K_{sp} of this form for a number of metal sulfides are listed in the last column in Table 17.2. Notice particularly how much the K_{sp} values vary—from 2×10^{-53} to 3×10^{-11}, a spread of a factor of 10^{42}.

TABLE 17.2 Metal Ions Separable by Selective Precipitation of Sulfides[a]

Metal Ion	Sulfide	K_{spa}	K_{sp}
Acid-Insoluble Sulfides			
Hg^{2+}	HgS (black form)	2×10^{-32}	2×10^{-53}
Ag^+	Ag_2S	6×10^{-30}	6×10^{-51}
Cu^{2+}	CuS	6×10^{-16}	6×10^{-37}
Cd^{2+}	CdS	8×10^{-7}	8×10^{-28}
Pb^{2+}	PbS	3×10^{-7}	3×10^{-28}
Sn^{2+}	SnS	1×10^{-5}	1×10^{-26}
Base-Insoluble Sulfides (Acid-Soluble Sulfides)			
Zn^{2+}	α-ZnS	2×10^{-4}	2×10^{-25}
	β-ZnS	3×10^{-2}	3×10^{-23}
Co^{2+}	CoS	5×10^{-1}	5×10^{-22}
Ni^{2+}	NiS	4×10^{1}	4×10^{-20}
Fe^{2+}	FeS	6×10^{2}	6×10^{-19}
Mn^{2+}	MnS (pink form)	3×10^{10}	3×10^{-11}
	MnS (green form)	3×10^{7}	3×10^{-14}

[a]Data are for 25 °C. See R. J. Meyers, *J. Chem. Ed.*, vol. 63, 1986, p. 687.

Acid-Insoluble Sulfides

Many metal sulfides are able to react with acid and thereby dissolve. An example is zinc sulfide:

$$ZnS(s) + 2H^+(aq) \longrightarrow Zn^{2+}(aq) + H_2S$$

However, some metal sulfides, referred to as the acid-insoluble sulfides, have K_{sp} values so low that they do not dissolve in acid. The cations in this group can be precipitated from the other cations simply by bubbling hydrogen sulfide into a sufficiently acidic solution that contains several metal ions. When the solution is acidic, we have to treat the solubility equilibria differently. In acid, HS^- and OH^- would be neutralized, leaving their conjugate acids, H_2S and H_2O. Under these conditions, the equation for the solubility equilibrium becomes

$$CuS(s) + 2H^+(aq) \rightleftharpoons Cu^{2+}(aq) + H_2S(aq)$$

This changes the mass action expression for the solubility product equilibrium, which we will now call the **acid solubility product, K_{spa}**. The "a" in the subscript indicates that the medium is acidic.

$$K_{spa} = \frac{[Cu^{2+}][H_2S]}{[H^+]^2}$$

TOOLS

Acid solubility product constant, K_{spa}

Table 17.2 also lists K_{spa} values for the metal sulfides. Notice that all K_{spa} values are 10^{21} larger than the K_{sp} values. Metal sulfides are clearly vastly more soluble in dilute acid than in water. Yet several—the acid-insoluble sulfides—are so insoluble that even the most soluble of them, SnS, barely dissolves, even in moderately concentrated acid. So, there are two families of sulfides, the acid-insoluble sulfides and the acid-soluble ones, otherwise known as the base-insoluble sulfides. As we will discuss in Section 17.4, the differing solubilities of their sulfides in acid provide a means of separating the two classes of metal ions from each other.

Example 17.7
Molar Solubility of Iron(II) Sulfide

What pH is needed to achieve a molar solubility of FeS equal to 0.075 M?

Analysis: We will use the desired concentration of FeS to determine the concentrations of the Fe^{2+} and H_2S and then solve for the hydrogen ion concentration. From there the pH can be easily calculated.

Assembling the Tools: We need the appropriate equilibrium expressions and the K_{spa} value for FeS from Table 17.2, from which we can write the equilibrium law needed to work the problem:

$$FeS(s) + 2H^+(aq) \rightleftharpoons Fe^{2+}(aq) + H_2S(aq) \qquad K_{spa} = \frac{[Fe^{2+}][H_2S]}{[H^+]^2} = 6 \times 10^2$$

The appropriate stoichiometry calculations will also be needed to convert the molarity of FeS to the $[Fe^{2+}]$ and $[H_2S]$ values needed.

Solution: Starting with 0.075 mol FeS L^{-1}, we see that one mol of FeS provides one mol of Fe^{2+} so $[Fe^{2+}]$ must be 0.075 M. One mol of FeS also provides one mole of S^{2-} that is converted into one mole of H_2S and so $[H_2S]$ is also 0.075 M.

$$\frac{[Fe^{2+}][H_2S]}{[H^+]^2} = \frac{(0.075)(0.075)}{[H^+]^2} = 6 \times 10^2$$

Now we solve for $[H^+]$:

$$\frac{(0.075)(0.075)}{6 \times 10^2} = 9.3 \times 10^{-6} = [H^+]^2$$

$$[H^+] = 3.1 \times 10^{-3}$$

$$pH = 2.5 \quad \text{(rounded correctly)}$$

We will need a pH of 2.5 to achieve a 0.075 M solution of FeS.

Is the Answer Reasonable? The K_{spa} values tell us that FeS is soluble in acid and that we can achieve moderate concentrations as set in this question. There's not much more we can do to check the answer, other than to review the entire set of calculations.

Practice Exercise 17.15   What pH is needed to achieve a concentration of 0.022 M NiS? (*Hint:* Is NiS soluble in acid according to Table 17.2?)

Practice Exercise 17.16   What pH is needed to dissolve 2.00 g PbS in 1.5×10^6 L of water?

17.4 | Selective Precipitation

Selective precipitation means causing one ion to precipitate while another remains in solution. Often this is possible because of large differences in the solubilities of salts that we would generally consider to be insoluble. For example, the K_{sp} values for AgCl and PbCl$_2$ are 1.8×10^{-10} and 1.7×10^{-5}, respectively, which gives them molar solubilities in water of 1.3×10^{-5} M for AgCl and 1.6×10^{-2} M for PbCl$_2$. In terms of molar solubilities, lead(II) chloride is approximately 1200 times more soluble in water than AgCl. If we had a solution containing both 0.10 M Pb^{2+} and 0.10 M Ag$^+$ and began adding Cl$^-$, AgCl would precipitate first (Figure 17.5). In fact, we can calculate that before any PbCl$_2$ starts to precipitate, the concentration of Ag$^+$ will have been reduced to 1.6×10^{-8} M. Thus, nearly all, 99.99998%, of the silver is removed from the solution without precipitating any of the lead, and the ions are effectively separated. All we need to do is find a way of adjusting the concentration of the Cl$^-$ to achieve the separation. In Chapter 19 we will discuss electrochemical methods that will enable us to know when we have achieved the necessary chloride concentration.

Figure 17.5 | **Selective precipitation.** When dilute sodium chloride is added to a solution containing both Ag$^+$ ions and Pb^{2+} ions (both dissolved as their nitrate salts), the less soluble AgCl precipitates before the more soluble PbCl$_2$. Precipitation of the AgCl is nearly complete before any PbCl$_2$ begins to form.

Example 17.8
Selective Precipitation of Silver Halides

If a solution contains $0.10\ M\ Cl^-$ and $0.10\ M\ I^-$, then what concentration of $AgNO_3$ will allow the separation of these halides by selective precipitation?

Analysis: We are being asked to find the concentration of Ag^+ ions that will remove one of the ions from the solution while leaving the other behind. We will have to decide which solid will precipitate first, and then determine the concentration of the silver ions that will precipitate the first solid and leave the second solid dissolved in the solution.

Assembling the Tools: From Table 17.1, we can find the ion product constants

$$AgCl(s) \rightleftharpoons Ag^+(aq) + Cl^-(aq) \qquad K_{sp} = 1.8 \times 10^{-10}$$

$$AgI(s) \rightleftharpoons Ag^+(aq) + I^-(aq) \qquad K_{sp} = 8.5 \times 10^{-17}$$

Silver iodide is much less soluble than silver chloride because it has a much smaller K_{sp}. So silver iodide will precipitate first, and the silver iodide solid can be filtered off to leave the chloride ions in solution. Next, we have to determine the concentration of the silver ions that will precipitate the iodide. To do this, we can imagine that we start with no silver in solution, and that we slowly add the silver ions. As we increase the concentration of the silver, the iodide begins to precipitate; this is the lowest concentration of silver that needs to be added. Then we add more silver, and more iodide precipitates, until at a certain concentration, the solution becomes saturated in AgCl. This is the maximum concentration of Ag^+ that we can have without causing AgCl to precipitate. This is what we need to calculate.

Solution: We can start by determining what concentration of Ag^+ will precipitate AgCl. We know that the concentration of Cl^- is $0.10\ M$, so we can use the solubility product expression to solve for the concentration of Ag^+ required to produce saturation:

$$AgCl(s) \rightleftharpoons Ag^+(aq) + Cl^-(aq)$$

$$K_{sp} = [Ag^+][Cl^-]$$

$$1.8 \times 10^{-10} = [Ag^+](0.10\ M)$$

$$1.8 \times 10^{-9}\ M = [Ag^+]$$

As long as the concentration of the silver ion is kept at or below $1.8 \times 10^{-9}\ M$, then the chloride ion will not precipitate. Now the question is, will the I^- concentration be reduced enough to consider the ions to be effectively separated? Let's use the concentration of silver ions to calculate how much iodide will be in solution:

$$AgI(s) \rightleftharpoons Ag^+(aq) + I^-(aq)$$

$$K_{sp} = [Ag^+][I^-]$$

$$8.5 \times 10^{-17} = (1.8 \times 10^{-9}\ M)[I^-]$$

$$4.7 \times 10^{-8}\ M = [I^-]$$

The iodide was $0.10\ M$ before the addition of the silver ions, and now it is $4.7 \times 10^{-8}\ M$. We can consider the iodide completely precipitated. So the maximum concentration of silver ion needed to achieve effective separation is $1.8 \times 10^{-9}\ M$.

Is the Answer Reasonable? From our work we showed that the silver iodide precipitated well before the silver chloride. Considering the differences in the K_{sp} values, this is to be expected. We can go back and check the math, and make sure that our reasoning was correct.

Practice Exercise 17.17

If a solution contains calcium ions (0.25 *M*) and barium ions (0.05 *M*), what concentration of SO_4^{2-} would be required to selectively precipitate the barium ions from solution? (*Hint:* First find the concentration of sulfate ions that will cause the calcium ions to precipitate.)

Practice Exercise 17.18

A solution contains 0.20 *M* $CaCl_2$ and 0.10 *M* $MgCl_2$. What concentration of hydroxide ions would be required to separate the magnesium as $Mg(OH)_2$ without precipitating the calcium as $Ca(OH)_2$? What pH is needed for this separation?

Metal Sulfides

The large differences in K_{spa} values between the acid-insoluble and the base-insoluble metal sulfides make it possible to separate the corresponding cations from each other by adjusting the pH of the solution. The sulfides of the acid-insoluble cations are selectively precipitated by hydrogen sulfide from a solution by adjusting the pH so that the other cations remain in solution. A solution saturated in H_2S is used, for which the molarity of H_2S is 0.1 *M*. Example 17.9 shows how we can calculate the pH needed to selectively precipitate two metal cations as their sulfides. We will use Cu^{2+} and Ni^{2+} ions to represent a cation from each class.

Example 17.9
Selective Precipitation of the Sulfides of Acid-Insoluble Cations from Base-Insoluble Cations

Over what range of hydrogen ion concentrations (and pH) is it possible to separate Cu^{2+} from Ni^{2+} when both metal ions are present in a solution at a concentration of 0.010 *M* and the solution is made saturated in H_2S (where $[H_2S] = 0.1$ *M*)?

Analysis: Both the concentration of the H_2S and the Cu^{2+} and Ni^{2+} ions are known, so we are only changing the concentration of the hydrogen ion, and we are being asked for a range of hydrogen ion concentrations. The low end of the hydrogen ion concentration will be the more basic solution, and the high end will be the more acidic solution. Sulfides become more soluble as the hydrogen ion concentration increases. This problem is reduced to two questions. The first question is: What is the most basic solution that will keep the more soluble sulfide in solution? The answer to this will be the lower limit on $[H^+]$. At any lower H^+ concentration, this compound will precipitate, so we want an H^+ concentration equal to or greater than this value. The second question is: What hydrogen ion concentration would be needed to keep the less soluble sulfide in solution? (The answer to this question will give us the upper limit on $[H^+]$; we would really want a lower H^+ concentration, so the less soluble sulfide will precipitate.) Once we know these limits, we know that any hydrogen ion concentration in between them will permit the less soluble sulfide to precipitate but retain the more soluble sulfide in solution.

Assembling the Tools: We need the appropriate equilibrium expressions and the K_{spa} values from Table 17.2, from which we can write the equilibrium laws needed to work the problem:

$$CuS(s) + 2H^+(aq) \rightleftharpoons Cu^{2+}(aq) + H_2S(aq) \qquad K_{spa} = \frac{[Cu^{2+}][H_2S]}{[H^+]^2} = 6 \times 10^{-16}$$

$$NiS(s) + 2H^+(aq) \rightleftharpoons Ni^{2+}(aq) + H_2S(aq) \qquad K_{spa} = \frac{[Ni^{2+}][H_2S]}{[H^+]^2} = 4 \times 10^1$$

NiS is much more soluble in an acidic solution than CuS. As the concentration of H^+ increases, the more soluble the NiS will be. (Actually, both will become more soluble, but

NiS is much more soluble than CuS.) Therefore, we need an H^+ concentration that will be large enough to prevent NiS from precipitating, but small enough so that CuS does precipitate. We will use the equilibrium constant expressions to calculate the H^+ concentrations for the limits of the solubility of CuS and NiS.

Solution: We will obtain the lower limit first, by calculating the $[H^+]$ required based on a concentration of Ni^{2+} equal to 0.010 M and an H_2S concentration of 0.1 M. If we keep the value of $[H^+]$ equal to or larger than this value, then NiS will be prevented from precipitating. First, we substitute values into the K_{spa} expression:

$$K_{spa} = \frac{[Ni^{2+}][H_2S]}{[H^+]^2} = \frac{(0.010)(0.10)}{[H^+]^2} = 4 \times 10^1$$

Now we solve for $[H^+]$:

$$[H^+]^2 = \frac{(0.010)(0.10)}{4 \times 10^1}$$

$$[H^+] = 5 \times 10^{-3} \, M$$

At any $[H^+]$ lower than this, the $[H^+]$ will not be great enough to keep the NiS in solution. This is a pH of 2.3.

Next, we will find the upper limit. If CuS does not precipitate, the Cu^{2+} concentration will be the given value, 0.010 M, so we substitute this along with the H_2S concentration (0.1 M) into the expression for K_{spa} of Cu^{2+}:

$$K_{spa} = \frac{[Cu^{2+}][H_2S]}{[H^+]^2} = \frac{(0.010)(0.10)}{[H^+]^2} = 6 \times 10^{-16}$$

Now we solve for $[H^+]$:

$$[H^+]^2 = \frac{(0.010)(0.10)}{6 \times 10^{-16}} = 2 \times 10^{12}$$

$$[H^+] = 1 \times 10^6 \, M$$

If we could make $[H^+] = 1 \times 10^6 \, M$, we could prevent CuS from forming. However, it isn't possible to have 10^6 or a million moles of H^+ per liter! What the calculated $[H^+]$ tells us, therefore, is that no matter how acidic the solution is, we cannot prevent CuS from precipitating when we saturate the solution with H_2S. (You can now see why CuS is classed as an "acid-insoluble sulfide.")

Now we have our limits: the lower limit for the concentration of H^+ is $5 \times 10^{-3} \, M$ and there is no actual higher limit, since the theoretical higher limit is beyond any possible physical means.

If we maintain the pH of the solution of 0.010 M Cu^{2+} and 0.010 M Ni^{2+} at 2.3 or lower (more acidic), as we make the solution saturated in H_2S, then virtually all of the Cu^{2+} will precipitate as CuS, but all of the Ni^{2+} will remain in solution.

Is the Answer Reasonable? The K_{spa} values certainly tell us that CuS is quite insoluble in acid and that NiS is a lot more soluble, so from that standpoint the results seem reasonable. There's not much more we can do to check the answer, other than to review the entire set of calculations.

Suppose a solution contains 0.0050 M Co^{2+} and is saturated with H_2S (with $[H_2S]$ = 0.1 M). Would CoS precipitate if the solution has a pH of 3.5? (*Hint:* What relationship do we use to determine whether a precipitate is going to form in a solution?)

Practice Exercise 17.19

Consider a solution containing Hg^{2+} and Fe^{2+}, both at molarities of 0.010 M. It is to be saturated with H_2S. Calculate the highest pH that this solution could have that would keep Fe^{2+} in solution while causing Hg^{2+} to precipitate as HgS.

Practice Exercise 17.20

In actual lab work, separation of the acid-insoluble sulfides from the base-insoluble ones is performed with $[H^+] \approx 0.3 \, M$, which corresponds to a pH of about 0.5. This ensures that the base-insoluble sulfides will not precipitate. The calculation in Example 17.9 also demonstrated why NiS can be classified as a "base-insoluble sulfide." If the solution is basic when it is made saturated in H_2S, the pH will surely be larger than 2.3 and NiS will precipitate.

Metal Carbonates

The principles of selective precipitation by the control of pH apply to any system where the anion is that of a weak acid. The metal carbonates can also be selectively precipitated, with many being quite insoluble in water (see Table 17.1). The dissociation of magnesium carbonate in water, for example, involves the following equilibrium and K_{sp} equations:

$$MgCO_3(s) \rightleftharpoons Mg^{2+}(aq) + CO_3^{2-}(aq) \qquad K_{sp} = 6.8 \times 10^{-8}$$

For strontium carbonate, the equations are

$$SrCO_3(s) \rightleftharpoons Sr^{2+}(aq) + CO_3^{2-}(aq) \qquad K_{sp} = 5.6 \times 10^{-10}$$

Would it be possible to separate the magnesium ion from the strontium ion by taking advantage of the difference in K_{sp} of their carbonates? We can do so if we can control the carbonate ion concentration. Because the carbonate ion is a relatively strong Brønsted–Lowry base, its control is available, indirectly, by adjusting the pH of the solution. This is because the hydrogen ion is one of the species in each of the following equilibria.[3]

$$H_2CO_3(aq) \rightleftharpoons H^+(aq) + HCO_3^-(aq) \qquad K_{a_1} = 4.3 \times 10^{-7}$$

$$HCO_3^-(aq) \rightleftharpoons H^+(aq) + CO_3^{2-}(aq) \qquad K_{a_2} = 5.6 \times 10^{-11}$$

$$CO_2(aq) + H_2O \rightleftharpoons H_2CO_3(aq) \rightleftharpoons H^+(aq) + HCO_3^-(aq)$$

You can see that if we increase the hydrogen ion concentration, both equilibria will shift to the left, in accordance with Le Châtelier's principle, and this will reduce the concentration of the carbonate ion. The value of $[CO_3^{2-}]$ thus decreases with decreasing pH. On the other hand, if we decrease the hydrogen ion concentration, making the solution more basic, we will cause the two equilibria to shift to the right. The concentration of the carbonate ion thus increases with increasing pH. For the calculations ahead, it will be useful to combine the two hydrogen carbonate equilibria into one overall equation that relates the molar concentrations of carbonic acid, carbonate ion, and hydrogen ion. So, we first add the two:

$$H_2CO_3(aq) \rightleftharpoons H^+(aq) + HCO_3^-(aq)$$

$$\underline{HCO_3^-(aq) \rightleftharpoons H^+(aq) + CO_3^{2-}(aq)}$$

$$H_2CO_3(aq) \rightleftharpoons 2H^+(aq) + CO_3^{2-}(aq) \qquad \text{(17.6)}$$

Recall from Chapter 14 that when we add two equilibria to get a third, the equilibrium constant of the latter is the product of those of the two equilibria that are combined. Thus,

---

[3] The situation involving aqueous carbonic acid is complicated by the presence of dissolved CO_2, which we could represent in an equation as $CO_2(aq)$. In fact, this is how most of the dissolved CO_2 exists—namely, as $CO_2(aq)$, not as $H_2CO_3(aq)$. However, we may use $H_2CO_3(aq)$ as a surrogate or stand-in for $CO_2(aq)$, because the latter changes smoothly to the former on demand. The following two successive equilibria involving carbonic acid exist in a solution of aqueous CO_2:

$$CO_2(aq) + H_2O \rightleftharpoons H_2CO_3(aq) \rightleftharpoons H^+(aq) + HCO_3^-(aq)$$

The value of K_{a_1} cited here for $H_2CO_3(aq)$ is really the product of the equilibrium constants of these two equilibria.

for Equation 17.6, the equilibrium constant, which we'll symbolize as K_a, is obtained as follows:

$$K_a = K_{a_1} K_{a_2} = \frac{[H^+]^2[CO_3^{2-}]}{[H_2CO_3]} \qquad (17.7)$$

$$K_a = (4.3 \times 10^{-7})(5.6 \times 10^{-11})$$

$$= 2.4 \times 10^{-17}$$

TOOLS

Combined K_a expression for a diprotic acid

■ We can only use Equation 17.7 only when two of the three concentrations are known.

With this K_a value we may now study how to find the pH range within which Mg^{2+} and Sr^{2+} can be separated by taking advantage of their difference in K_{sp} values. To perform such a separation we would saturate a solution with CO_2, which provides H_2CO_3 through the rapidly established equilibrium,

$$CO_2(aq) + H_2O \rightleftharpoons H_2CO_3(aq)$$

Thus, a solution that contains 0.030 M CO_2 has an effective H_2CO_3 concentration of 0.030 M.

Example 17.10
Separation of Metal Ions by the Selective Precipitation of Their Carbonates

A solution contains magnesium nitrate and strontium nitrate, each at a concentration of 0.10 M. Carbon dioxide is to be bubbled in to make the solution saturated in $CO_2(aq)$—approximately 0.030 M. What pH range would make it possible for the carbonate of one cation to precipitate but not that of the other?

Analysis: There are really two parts to this. First, what is the range in values of $[CO_3^{2-}]$ that allow one carbonate to precipitate but not the other? Second, given this range, what are the values of the solution's pH that produce this range in $[CO_3^{2-}]$ values? To answer the first question, we will use the K_{sp} values of the two carbonate salts and their molar concentrations to find the CO_3^{2-} concentrations in their saturated solutions. To keep the more soluble carbonate from precipitating, the CO_3^{2-} concentration must be at or below that required for saturation. For the less soluble of the two, we must have a CO_3^{2-} concentration larger than in a saturated solution so that it will precipitate.

To answer the second question, we will use the combined K_a expression for H_2CO_3 to find the $[H^+]$ that gives the necessary CO_3^{2-} concentrations obtained in answering the first question. Having these $[H^+]$, it is then a simple matter to convert them to pH values.

Assembling the Tools: The tools we need to solve the problem are the K_{sp} values and equilibria for $MgCO_3$ and $SrCO_3$:

$$K_{sp} = [Mg^{2+}][CO_3^{2-}] = 6.8 \times 10^{-8}$$

$$K_{sp} = [Sr^{2+}][CO_3^{2-}] = 5.6 \times 10^{-10}$$

We also need the combined K_a expression for carbonic acid:

$$K_a = \frac{[H^+]^2[CO_3^{2-}]}{[H_2CO_3]} = 2.4 \times 10^{-17}$$

Solution: Let's find the $[CO_3^{2-}]$ values required to give saturated solutions of these two salts.

For magnesium carbonate:

$$[CO_3^{2-}] = \frac{K_{sp}}{[Mg^{2+}]} = \frac{6.8 \times 10^{-8}}{(0.10)} = 6.8 \times 10^{-7}\ M$$

For strontium carbonate:

$$[CO_3{}^{2-}] = \frac{K_{sp}}{[Sr^{2+}]} = \frac{5.6 \times 10^{-10}}{(0.10)} = 5.6 \times 10^{-9} \, M$$

On the basis of the K_{sp} values, we see that $MgCO_3$ is the more soluble of the two. For separation, we have to keep the $[CO_3{}^{2-}]$ larger than $5.6 \times 10^{-9} \, M$ and less than or equal to $6.8 \times 10^{-7} \, M$ to keep the $MgCO_3$ in solution and to precipitate the $SrCO_3$.

The second phase of our calculation now asks what values of $[H^+]$ correspond to the calculated limits on $[CO_3{}^{2-}]$. For this we will use the K_a expression given by Equation 17.7. First, we solve for the square of $[H^+]$, using the molarity of the dissolved CO_2, 0.030 M, as the value of $[H_2CO_3]$. This gives us

$$[H^+]^2 = K_a \times \frac{[H_2CO_3]}{[CO_3{}^{2-}]}$$

$$[H^+]^2 = 2.4 \times 10^{-17} \times \frac{(0.030)}{[CO_3{}^{2-}]}$$

We will use this equation for each of the two boundary values of $[CO_3{}^{2-}]$ to calculate the corresponding two values of $[H^+]^2$. Once we have them, we can readily determine $[H^+]$ and pH. For magnesium carbonate, to prevent the precipitation of $MgCO_3$, $[CO_3{}^{2-}]$ must be no higher than $6.8 \times 10^{-7} \, M$, so $[H^+]^2$ must not be less than

$$[H^+]^2 = 2.4 \times 10^{-17} \times \frac{(0.030)}{(6.8 \times 10^{-7})} = 1.1 \times 10^{-12}$$

$$[H^+] = 1.0 \times 10^{-6} \, M$$

This corresponds to a pH of 6.00. At a higher (more basic) pH, magnesium carbonate precipitates. Thus, pH $\leq$ 6.00 prevents the precipitation of $MgCO_3$. For strontium carbonate, to have a saturated solution of $SrCO_3$ with $[Sr^{2+}] = 0.10 \, M$, the value of $[CO_3{}^{2-}]$ would be $5.6 \times 10^{-9} \, M$, as we calculated previously. This corresponds to a value of $[H^+]^2$ found as follows:

$$[H^+]^2 = 2.4 \times 10^{-17} \times \frac{(0.030)}{(5.6 \times 10^{-9})} = 1.3 \times 10^{-10}$$

$$[H^+] = 1.1 \times 10^{-5} \, M$$

$$pH = 4.95$$

To cause $SrCO_3$ to precipitate, the $[CO_3{}^{2-}]$ would have to be higher than $5.6 \times 10^{-9} \, M$, and that would require that $[H^+]$ be less than $1.1 \times 10^{-5} \, M$. If $[H^+]$ were less than $1.1 \times 10^{-5} \, M$, the pH would have to be higher than 4.95. Thus, to cause $SrCO_3$ to precipitate from the given solution, pH > 4.95.

In summary, when the pH of the given solution is kept above 4.95 and less than or equal to 6.00, Sr^{2+} will precipitate as $SrCO_3$, but Mg^{2+} will remain dissolved.

Is the Answer Reasonable? The only way to be sure of the answer is to go back over the reasoning and the calculations. However, there is a sign that the answer is probably correct. Notice that the two K_{sp} values are not vastly different—they differ by just a factor of about 40. Therefore, it's not surprising that to achieve separation we would have to keep the pH within a rather narrow range (from 4.95 to 6.00).

Practice Exercise 17.21

The K_{sp} for barium oxalate, BaC_2O_4, is 1.2×10^{-7} and for oxalic acid, $H_2C_2O_4$, $K_{a_1} = 5.9 \times 10^{-2}$ and $K_{a_2} = 6.4 \times 10^{-5}$. If a solution containing 0.10 M $H_2C_2O_4$ and 0.050 M $BaCl_2$ is prepared, what must the minimum H^+ concentration be in the solution to prevent the formation of a BaC_2O_4 precipitate? (*Hint:* Work with a combined K_a expression for $H_2C_2O_4$.)

A solution contains calcium nitrate and nickel nitrate, each at a concentration of 0.10 M. Carbon dioxide is to be bubbled in to make its concentration equal 0.030 M. What pH range would make it possible for the carbonate of one cation to precipitate but not that of the other?

Practice Exercise 17.22

Separating Metal Ions: Qualitative Analysis

The method for determining what elements are present in a sample is called qualitative analysis. Metal ions can be separated and isolated by selective precipitation with different anions at different pH values. The metal cations have been divided into five groups, depending on the solubilities of their compounds. The first group is the chloride group, which will precipitate Pb^{2+}, Hg_2^{2+}, and Ag^+. This precipitation reaction is carried out by adding HCl to the solution. To precipitate the next group, the concentration of HCl is brought to 0.3 M and H_2S is added. All of the acid-insoluble sulfides will precipitate (Hg^{2+}, Ag^+, Cu^{2+}, Cd^{2+}, Pb^{2+}, and Sn^{2+} from Table 17.2, as well as Bi^{3+}, As^{3+}, and Sb^{3+}). The third group is precipitated by the addition of NaOH, which will make the solution basic. The compounds MnS, FeS, NiS, CoS, and ZnS will precipitate as base-insoluble sulfides along with the hydroxides $Fe(OH)_3$, $Al(OH)_3$, and $Cr(OH)_3$. For the fourth group, CO_3^{2-} is added to the solution and the following carbonates are precipitated: $MgCO_3$, $CaCO_3$, $SrCO_3$, and $BaCO_3$. The final group is the metal cations that are completely soluble— namely, Na^+, K^+, and NH_4^+. This scheme is summarized in Figure 17.6. Many other "qual schemes" have been developed over the years and if a certain ion is not found in Figure 17.6, a literature search can often find a scheme that does.

17.5 | Equilibria Involving Complex Ions

Formation of Complex Ions

In our previous discussions of metal-containing compounds, we left you with the impression that the only kinds of bonds in which metals are ever involved are ionic bonds. For some metals, like the alkali metals of Group 1A, this is close enough to the truth to warrant no modifications. However, for many other metal ions, especially those of the transition metals and the post-transition metals, it is not. This is because the ions of many of these metals

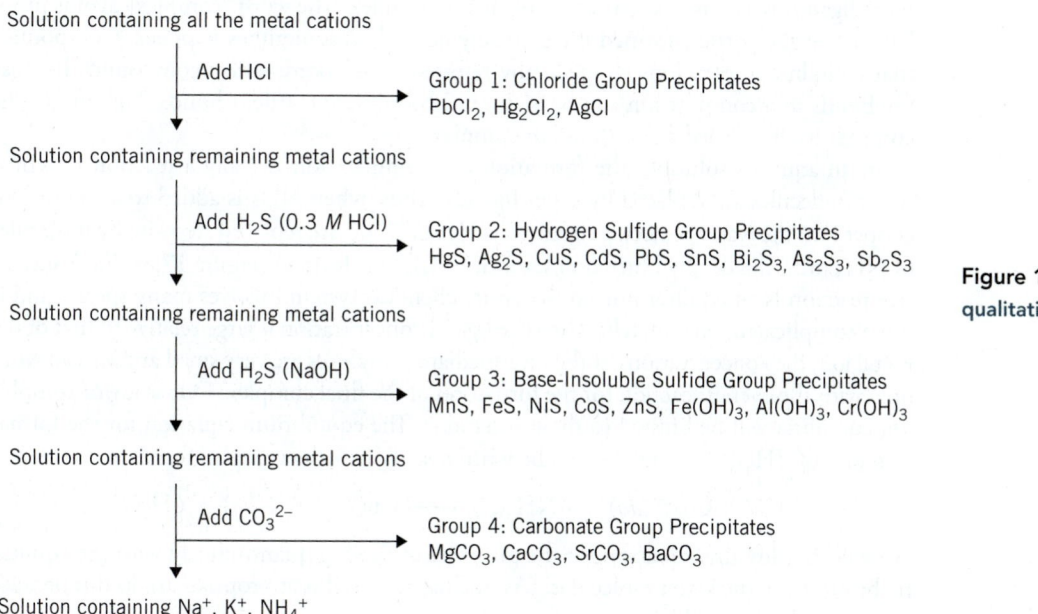

Solution containing all the metal cations

Add HCl → Group 1: Chloride Group Precipitates
$PbCl_2$, Hg_2Cl_2, AgCl

Solution containing remaining metal cations

Add H_2S (0.3 M HCl) → Group 2: Hydrogen Sulfide Group Precipitates
HgS, Ag_2S, CuS, CdS, PbS, SnS, Bi_2S_3, As_2S_3, Sb_2S_3

Solution containing remaining metal cations

Add H_2S (NaOH) → Group 3: Base-Insoluble Sulfide Group Precipitates
MnS, FeS, NiS, CoS, ZnS, $Fe(OH)_3$, $Al(OH)_3$, $Cr(OH)_3$

Solution containing remaining metal cations

Add CO_3^{2-} → Group 4: Carbonate Group Precipitates
$MgCO_3$, $CaCO_3$, $SrCO_3$, $BaCO_3$

Solution containing Na^+, K^+, NH_4^+

Figure 17.6 | Outline of a qualitative analysis scheme.

Figure 17.7 | The complex ion of Cu^{2+} and water. A solution containing copper sulfate has a blue color because it contains the complex ion $Cu(H_2O)_6^{2+}$.

are able to behave as Lewis acids (electron pair **acceptors** in the formation of coordinate covalent bonds). Thus, by participating in Lewis acid–base reactions they become covalently bonded to other atoms. The copper(II) ion is a typical example. In aqueous solutions of copper(II) salts, such as $CuSO_4$ or $Cu(NO_3)_2$, the copper is not present as simple Cu^{2+} ions. Instead, each Cu^{2+} ion becomes bonded to six water molecules to give a pale blue ion with the formula $Cu(H_2O)_6^{2+}$ (see Figure 17.7). We call this species a **complex ion** because it is composed of a number of simpler species (i.e., it is complex, not simple). The chemical equation for the formation of the $Cu(H_2O)_6^{2+}$ ion is

$$Cu^{2+} + 6H_2O \longrightarrow Cu(H_2O)_6^{2+}$$

which can be diagrammed using Lewis structures as follows:

$$Cu(H_2O)_6^{2+}$$

- Recall that a Lewis base is an electron pair donor in the formation of a coordinate covalent bond.

As you can see in this analysis, the Cu^{2+} ion accepts pairs of electrons from the water molecules, so Cu^{2+} is a Lewis acid and the water molecules are each Lewis bases.

The number of complex ions formed by metals, especially the transition metals, is enormous, and the study of the properties, reactions, structures, and bonding in complex ions like $Cu(H_2O)_6^{2+}$ has become an important specialty within chemistry. We will provide a more complete discussion of them in Chapter 21. For now, we will introduce you to some of the terminology that we use in describing these substances.

- **Ligand**: from the Latin *ligare*, meaning "to bind."

A Lewis base that attaches itself to a metal ion is called a **ligand**. Ligands can be neutral molecules with unshared pairs of electrons (e.g., H_2O), or they can be anions (e.g., Cl^- or OH^-). The atom in the ligand that actually provides the electron pair is called the **donor atom**, and the metal ion is the acceptor. The result of combining a metal ion with one or more ligands is a complex ion, or simply just a complex. The word "complex" avoids problems when the particle formed is electrically neutral, as sometimes happens. Compounds that contain complex ions are generally referred to as coordination compounds because the bonds in a complex ion can be viewed as coordinate-covalent bonds. Sometimes the complex itself is called a coordination complex.

In an aqueous solution, the formation of a complex ion is really a reaction in which water molecules are replaced by other ligands. Thus, when NH_3 is added to a solution of copper ion, the water molecules in the $Cu(H_2O)_6^{2+}$ ion are replaced stepwise by molecules of NH_3 until the deeply blue complex $Cu(NH_3)_4^{2+}$ is formed (Figure 17.8). Each successive reaction is an equilibrium, so the entire chemical system involves many species and is quite complicated. Fortunately, when the ligand concentration is large relative to that of the metal ion, the concentrations of the intermediate complexes are very small and we can work only with the overall reaction for the formation of the final complex. Our study of complex ion equilibria will be limited to these situations. The equilibrium equation for the formation of $Cu(NH_3)_4^{2+}$, therefore, can be written as though the complex forms in one step.

$$Cu(H_2O)_6^{2+}(aq) + 4NH_3(aq) \rightleftharpoons Cu(NH_3)_4^{2+}(aq) + 6H_2O$$

We will simplify this equation for the purposes of dealing quantitatively with the equilibria by omitting the water molecules. (As in Chapter 14, it is appropriate to do this because the concentration of H_2O in aqueous solutions is taken to be effectively a constant and

Figure 17.8 | **The complex ion of Cu²⁺ and ammonia.** Ammonia molecules displace water molecules from $Cu(H_2O)_6^{2+}$ (left) to give the deep blue $Cu(NH_3)_4^{2+}$ ion (right). In the molecular view we show two water molecules that are also weakly attached to the copper ion in the $Cu(NH_3)_4^{2+}$ cation.

need not be included in mass action expressions.) In simplified form, we write the preceding equilibrium as follows:

$$Cu^{2+}(aq) + 4NH_3(aq) \rightleftharpoons Cu(NH_3)_4^{2+}(aq)$$

We have two goals here—namely, to study such equilibria themselves and to learn how they can be used to influence the solubilities of metal ion salts.

■ According to Le Châtelier's principle, the high concentration of ammonia shifts this reaction to completion, so effectively all Cu²⁺ ions are complexed with ammonia molecules.

Formation Constants

When the chemical equation for the equilibrium is written so that the complex ion is the product, the equilibrium constant for the reaction is called the **formation constant, K_{form}**. The equilibrium law for the formation of $Cu(NH_3)_4^{2+}$ in the presence of excess NH_3, for example, is

$$\frac{\left[Cu(NH_3)_4^{2+}\right]}{\left[Cu^{2+}\right]\left[NH_3\right]^4} = K_{formation} = K_{form}$$

Sometimes this equilibrium constant is called the **stability constant.** The larger its value, the greater is the concentration of the complex at equilibrium, and so the more stable is the complex.

Table 17.3 provides several more examples of complex ion equilibria and their associated equilibrium constants. (Additional examples are in Appendix C.) The most stable complex in the table, $Co(NH_3)_6^{3+}$, has the largest value of K_{form}.

TOOLS

Formation constants of complex ions

Andy Washnik

TABLE 17.3	Formation Constants and Instability Constants for Some Complex Ions		
Ligand	Equilibrium	$K_{formation}$	$K_{instability}$
NH_3	$Ag^+ + 2NH_3 \rightleftharpoons Ag(NH_3)_2^+$	1.6×10^7	6.3×10^{-8}
	$Co^{2+} + 6NH_3 \rightleftharpoons Co(NH_3)_6^{2+}$	5.0×10^4	2.0×10^{-5}
	$Co^{3+} + 6NH_3 \rightleftharpoons Co(NH_3)_6^{3+}$	4.6×10^{33}	2.2×10^{-34}
	$Cu^{2+} + 4NH_3 \rightleftharpoons Cu(NH_3)_4^{2+}$	1.1×10^{13}	9.1×10^{-14}
	$Hg^{2+} + 4NH_3 \rightleftharpoons Hg(NH_3)_4^{2+}$	1.8×10^{19}	5.6×10^{-20}
F^-	$Al^{3+} + 6F^- \rightleftharpoons AlF_6^{3-}$	1×10^{20}	1×10^{-20}
	$Sn^{4+} + 6F^- \rightleftharpoons SnF_6^{2-}$	1×10^{25}	1×10^{-25}
Cl^-	$Hg^{2+} + 4Cl^- \rightleftharpoons HgCl_4^{2-}$	5.0×10^{15}	2.0×10^{-16}
Br^-	$Hg^{2+} + 4Br^- \rightleftharpoons HgBr_4^{2-}$	1.0×10^{21}	1.0×10^{-21}
I^-	$Hg^{2+} + 4I^- \rightleftharpoons HgI_4^{2-}$	1.9×10^{30}	5.3×10^{-31}
CN^-	$Fe^{2+} + 6CN^- \rightleftharpoons Fe(CN)_6^{4-}$	1.0×10^{24}	1.0×10^{-24}
	$Fe^{3+} + 6CN^- \rightleftharpoons Fe(CN)_6^{3-}$	1.0×10^{31}	1.0×10^{-31}

Instability Constants

Some chemists prefer to describe the relative stabilities of complex ions differently. Their approach focuses attention on the breakdown of the complex, not its formation. Therefore, the associated equilibrium equation is written as the reverse of the formation of the complex. The equilibrium for the copper–ammonia complex would be written as follows, for example:

$$Cu(NH_3)_4^{2+}(aq) \rightleftharpoons Cu^{2+}(aq) + 4NH_3(aq)$$

The equilibrium constant for this equilibrium is called the **instability constant**,

$$K_{instability} = K_{inst} = \frac{[Cu^{2+}][NH_3]^4}{[Cu(NH_3)_4^{2+}]} = \frac{1}{K_{form}}$$

Notice that K_{inst} is the reciprocal of K_{form}. The K_{inst} is called an instability constant because the larger its value is, the more unstable the complex is. The data in the last column of Table 17.3 show this. The least stable complex in the table, $Co(NH_3)_6^{2+}$, has the largest value of K_{inst}.

CHEMISTRY OUTSIDE THE CLASSROOM 17.1

No More Soap Scum—Complex Ions and Solubility

A problem that has plagued homeowners with hard water—water that contains low concentrations of divalent cations, especially Ca^{2+}—is the formation of insoluble deposits of "soap scum" as well as "hard water spots" on surfaces such as shower tiles, shower curtains, and bathtubs. These deposits form when calcium ions interact with large anions in soap to form precipitates and also when hard water that contains bicarbonate ion evaporates, causing the precipitation of calcium carbonate, $CaCO_3$.

$$Ca^{2+}(aq) + 2HCO_3^-(aq) \longrightarrow CaCO_3(s) + CO_2(g) + H_2O$$

A variety of consumer products are sold that contain ingredients intended to prevent these precipitates from forming, and they accomplish this by forming complex ions with calcium ions, which has the effect of increasing the solubilities of the soap scum and calcium carbonate deposits. A principal ingredient in these products is an organic compound called **ethylenediaminetetraacetic acid** (mercifully abbreviated **EDTA**). The structure of the compound, which is also abbreviated as H_4EDTA to emphasize that it contains four acidic hydrogens that are part of carboxyl groups, is

H_4EDTA

Acidic hydrogens are shown in blue,
donor atoms are shown in red.

Acidic hydrogens are shown in blue, donor atoms are shown in red. This molecule is an excellent complex-forming ligand; it contains a total of six donor atoms (in red) that can bind to a metal ion, enabling the ligand to wrap itself around the metal ion as illustrated above, right. (The usual colors are used to identify the various elements in the ligand.) A metal ion can be surrounded octahedrally by all six donor atoms if H_4EDTA loses all of the acidic hydrogens.

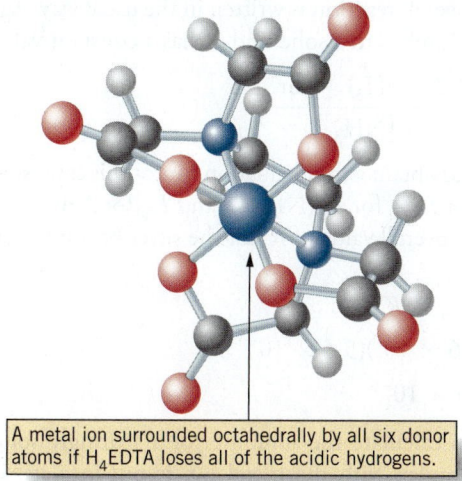

A metal ion surrounded octahedrally by all six donor atoms if H_4EDTA loses all of the acidic hydrogens.

One consumer product, called Clean Shower (Figure 1), contains H_4EDTA along with substances called surfactants. When the EDTA binds with calcium ions, it releases just two H^+ ions,

$$H_4EDTA(aq) + Ca^{2+}(aq) \longrightarrow CaH_2EDTA(aq) + 2H^+(aq)$$

The H^+ combine with anions of the soap to form neutral organic compounds called fatty acids that are usually not water soluble. The surfactants in the product, however, enable the fatty acids to dissolve, preventing the formation of precipitates. The Clean Shower product is sprayed on the wet walls after taking a shower, forming the products just described. The next time you take a shower, the water washes away the soluble products, keeping soap scum from building up and keeping the shower walls clean.

Andy Washnik

Figure 1 The product Clean Shower contains EDTA, which prevents the formation of soap scum and hard water spots in showers and on bathtubs.

17.6 | Complexation and Solubility

The silver halides are extremely insoluble salts. The K_{sp} of AgBr at 25 °C, for example, is only 5.4×10^{-13}. In a saturated aqueous solution, the concentration of each ion of AgBr is only 7.3×10^{-7} mol L^{-1}. Suppose that we start with a saturated solution in which undissolved AgBr is present and equilibrium exists. Now, suppose that we begin to add aqueous ammonia to the system. Because NH_3 molecules bind strongly to silver ions, they begin to form $Ag(NH_3)_2^+$ with the trace amount of Ag^+ ions initially in solution. The reaction is complex ion equilibrium:

$$Ag^+(aq) + 2NH_3(aq) \rightleftharpoons Ag(NH_3)_2^+(aq) \qquad K_{form} = 1.6 \times 10^7$$

Because the forward reaction removes uncomplexed Ag^+ ions from solution, it upsets the solubility equilibrium:

$$AgBr(s) \rightleftharpoons Ag^+(aq) + Br^-(aq) \qquad K_{sp} = 5.4 \times 10^{-13}$$

By withdrawing uncomplexed Ag^+ ions, the ammonia causes the solubility equilibrium to shift to the right to generate more Ag^+ ions from AgBr(s). The way the two equilibria are related can be viewed as follows:

$$AgBr(s) \rightleftharpoons Ag^+(aq) + Br^-(aq)$$
$$+$$

Addition of ammonia removes free silver ion from the solution, causing the solubility equilibrium to shift to the right

$$2NH_3(aq)$$
$$\updownarrow$$
$$Ag(NH_3)_2^+(aq)$$

Thus, adding ammonia causes more AgBr(s) to dissolve—an example of Le Châtelier's principle at work. Our example illustrates the following general phenomenon: The solubility of a slightly soluble salt increases when one of its ions can be changed to a soluble complex ion. To analyze what happens, let's put the two equilibria together:

Complex ion equilibrium: $\quad Ag^+(aq) + 2NH_3(aq) \rightleftharpoons Ag(NH_3)_2^+(aq)$

Solubility equilibrium: $\qquad\qquad\qquad AgBr(s) \rightleftharpoons Ag^+(aq) + Br^-(aq)$

Sum of equilibria: $\qquad AgBr(s) + 2NH_3(aq) \rightleftharpoons Ag(NH_3)_2^+(aq) + Br^-(aq)$

The equilibrium constant for the net overall reaction is written in the usual way. Again, the term for [AgBr(s)] is omitted because it refers to a solid and so has a constant value:

$$K_c = \frac{[Ag(NH_3)_2^+][Br^-]}{[NH_3]^2}$$

Because we have added two equations to obtain a third equation, K_c for this expression is the product of K_{form} and K_{sp}. The values for K_{form} for $Ag(NH_3)_2^+$ and K_{sp} for AgBr are known, so by multiplying the two, we find the overall value of K_c for the silver bromide–ammonia system.

$$K_c = K_{form} \times K_{sp}$$
$$K_c = (1.6 \times 10^7)(5.4 \times 10^{-13})$$
$$= 8.6 \times 10^{-6}$$

This approach thus gives us a way to calculate the solubility of a sparingly soluble salt when a substance able to form a complex with the metal ion is put into its solution. Example 17.11 shows how this works.

Example 17.11
Calculating the Solubility of a Slightly Soluble Salt in the Presence of a Ligand

How many moles of AgBr can dissolve in 1.0 L of 1.0 M NH$_3$?

Analysis: Silver bromide is insoluble in water, but with the addition of NH$_3$, more AgBr can be dissolved. This question is asking us to use the K_{form} and the K_{sp} together to determine the amount of AgBr dissolved in water.

Assembling the Tools: A few preliminaries have to be done before we can take advantage of a concentration table. We need the overall balanced chemical equation and its associated equation for the equilibrium constant, which serve as our tools for dealing with these kinds of problems. The overall equilibrium is

$$AgBr(s) + 2NH_3(aq) \rightleftharpoons Ag(NH_3)_2^+(aq) + Br^-(aq)$$

The equation for K_c is

$$K_c = \frac{[Ag(NH_3)_2^+][Br^-]}{[NH_3]^2} = 8.6 \times 10^{-6} \qquad \text{(as calculated earlier)}$$

Next, we will need to use the concentration table. To do this, we imagine we are adding solid AgBr to the ammonia solution. Before any reaction takes place, the concentration of NH$_3$ is 1.0 M and the concentration of Ag(NH$_3$)$_2^+$ is zero. The concentration of Br$^-$ is also zero, because there is none of it in the ammonia solution. If we define s as the molar solubility of the AgBr in the solution, then the concentrations of Ag(NH$_3$)$_2^+$ and Br$^-$ both increase by s and the concentration of ammonia decreases by $2s$ (because of the coefficient of NH$_3$ in the balanced chemical equation). The equilibrium values are obtained as usual by adding the initial and change rows. However, we need to note that letting the concentration of Br$^-$ equal that of Ag(NH$_3$)$_2^+$ is valid because and only because K_{form} is such a large number. Essentially all Ag$^+$ ions that do dissolve from the insoluble AgBr are changed into the complex ion. There are relatively few uncomplexed Ag$^+$ ions in the solution, so the number of Ag(NH$_3$)$_2^+$ and Br$^-$ ions in the solution are very nearly the same.

- **Solution:** The concentration table is as follows.

	AgBr	+	2NH$_3$	$\rightleftharpoons$	Ag(NH$_3$)$_2$$^+$	+	Br$^-$
Initial Concentrations (*M*)	NE		1.0		0		0
Concentration Change (*M*)	NE		$-2s$		$+s$		$+s$
Equilibrium Concentrations (*M*)	NE		$1.0-2s$		s		s

We've done all the hard work. Now all we need to do is substitute the values in the last row of the concentration table into the equation for K_c,

$$K_c = \frac{[\text{Ag(NH}_3)_2{}^+][\text{Br}^-]}{[\text{NH}_3]^2} = \frac{(s)(s)}{(1.0-2s)^2} = 8.6 \times 10^{-6}$$

This can be solved analytically by taking the square root of both sides, which gives

$$\frac{(s)}{(1.0-2s)} = \sqrt{8.6 \times 10^{-6}} = 2.9 \times 10^{-3}$$

Solving for s and rounding to the correct number of significant figures gives

$$s = 2.9 \times 10^{-3}$$

Because we've defined s as the molar solubility of AgBr in the ammonia solution, we can say that 2.9×10^{-3} mol of AgBr dissolves in 1.0 L of 1.0 M NH$_3$. (This is not very much, but in contrast, only 7.1×10^{-7} mol of AgBr dissolves in 1.0 L of pure water. Thus, AgBr is nearly 4000 times more soluble in the 1.0 M NH$_3$ than in pure water.)

- **Is the Answer Reasonable?** Everything depends on the reasoning used in preparing the concentration table, particularly letting $-2s$ represent the change in the concentration of NH$_3$ as a result of the presence of AgBr and the formation of the complex. If you are comfortable with this, then the calculation should be straightforward.

Calculate the solubility of silver chloride in 0.10 M NH$_3$ and compare it with its solubility in pure water. Refer to Table 17.1 for the K_{sp} for AgCl. (*Hint:* Assume that all of the Ag$^+$ from the AgCl becomes incorporated into Ag(NH$_3$)$_2$$^+$.)

Practice Exercise 17.23

How many moles of NH$_3$ have to be added to 1.0 L of water to dissolve 0.20 mol of AgCl? The complex ion Ag(NH$_3$)$_2$$^+$ forms.

Practice Exercise 17.24

Summary

Organized by Learning Objective

Perform solubility calculations using neutral salts

The mass action expression for the solubility equilibrium of a salt is called the **ion product**. It equals the product of the molar concentrations of its ions, each raised to a power equal to the subscript of the ion in the salt's formula. At a given temperature, the value of the ion product, Q, in a saturated solution of the salt equals a constant called the **solubility product constant** or K_{sp}. If the ion product, Q, exceeds the solubility product constant of the salt a precipitate is expected. For solubility equilibria, the common ion effect is the ability of an ion of a soluble salt to suppress solubility.

Understand how hydrogen ion concentration affects the solubility of basic salts

Basic salts have anions that are conjugate bases of weak acids. The anions of basic salts will react with available hydrogen ions to reform the weak acid, thus removing some of the

anion from solution. Acid solutions increase the solubility of basic salts, and basic solutions increase the solubility of acid salts.

Describe why oxides and sulfides are special cases and need unique methods to determine solubilities

Metal sulfides, when they dissolve, never form the sulfide ion, S^{2-}; and the oxide ion, O^{2-}, does not occur in aqueous solution. Many sulfides and oxides are vastly more soluble in an acidic solution than in water. To express the solubility of sulfides in acid, the **acid solubility products**, K_{spa}, is used.

Use selective precipitation to separate ions for qualitative and quantitative analysis

Acid-insoluble sulfides can be selectively separated from the base-insoluble sulfides by making the solution both quite acidic as well as saturated in H_2S. By adjusting the pH, salts of other weak acids, like the carbonate salts, can also be selectively precipitated. Chloride is used to selectively precipitate Ag^+, Pb^{2+}, and Hg_2^{2+} ions. A sequential use of selective precipitation allows for the qualitative identification of ions.

Describe and use equilibria related to reactions of complexes

Coordination compounds contain complex ions (also called complexes or coordination complexes) formed from a metal ion and a number of ligands. **Ligands** are Lewis bases (electron pair donors) and may be electrically neutral or negatively charged.

Water and ammonia are common neutral ligands. Metal ions act as electron pair acceptors (i.e., as Lewis acids). The equilibrium constant for the formation of a metal–ligand complex in the presence of an excess of ligand is called the **formation constant, K_{form},** of the complex.

Enhance solubility by using complexation reactions

Salts whose cations form stable complexes, like Ag^+ reacting with ammonia to form the complex $Ag(NH_3)_2^+$, are made more soluble when the ligand is present. This increase in solubility can be explained chemically as simultaneous equilibria involving a dissolution process and a complexation reaction.

TOOLS

Tools for Problem Solving
The following tools were introduced in this chapter. Study them carefully so you can select the appropriate tool when needed.

Solubility product constant, K_{sp} (Section 17.1)
The K_{sp} is the name of an equilibrium constant that applies when the equilibrium law describes a saturated solution.

Molar solubility (Section 17.1)
This is defined as the moles of solute dissolved in a liter of a saturated solution.

Predicting precipitation (Section 17.1) $Q > K_{sp}$, a precipitate forms
$Q \leq K_{sp}$, a precipitate does not form

Acid solubility product constant, K_{spa} (Section 17.3)
The K_{spa} data are used to calculate the solubility of a metal sulfide at a given the pH.

Combined K_a expression for a diprotic acid (Section 17.4)
Combining expressions for K_{a_1} and K_{a_2} for a diprotic acid, H_2A, yields the equation, for example,

$$K_a = K_{a_1}K_{a_2} = \frac{[H^+]^2[CO_3^{2-}]}{[H_2CO_3]}$$

Formation constants of complex ions (Section 17.5)
We can use them to make judgments concerning the relative stabilities of complexes and the solubilities of salts in solutions containing a ligand.

| Review Questions

Equilibria in Solutions of Slightly Soluble Salts

17.1   What is the difference between an ion product and an ion product constant?

17.2   Use the following equilibrium to demonstrate why the K_{sp} expression does not include the concentration of $Ba_3(PO_4)_2$ in the denominator:

$$Ba_3(PO_4)_2(s) \rightleftharpoons 3Ba^{2+}(aq) + 2PO_4^{3-}(aq)$$

17.3   What is the common ion effect? How does Le Châtelier's principle explain it? Use the solubility equilibrium for AgCl and the addition of NaCl to the solution to illustrate the common ion effect.

17.4   With respect to K_{sp}, what conditions must be met if a precipitate is going to form in a solution?

17.5   What limits the accuracy and reliability of solubility calculations based on K_{sp} values?

17.6 Why do we not use K_{sp} values for soluble salts such as NaCl?

17.7 Using Figure 12.10, how does the K_{sp} change with temperature for KCl and NH_4NO_3?

Solubility of Basic Salts Is Influenced by Acids

17.8 What makes the equilibrium for the solubility of a neutral salt different from the solubility of a basic salt?

17.9 Is $Ba_3(PO_4)_2$ a basic salt? Justify your answer.

$$Ba_3(PO_4)_2(s) \rightleftharpoons 3Ba^{2+}(aq) + 2PO_4{}^{3-}(aq)$$

17.10 Water can provide ions to a solution. What type of salt would have a common ion with water?

17.11 How does Le Châtelier's principle explain the solubility of a basic salt?

17.12 A solution of $MgBr_2$ can be changed to a solution of $MgCl_2$ by adding $AgCl(s)$ and stirring the mixture well. In terms of the equilibria involved, explain how this happens.

17.13 On the basis of Le Châtelier's principle, explain how the addition of solid NH_4Cl to a beaker containing solid $Mg(OH)_2$ in contact with water is able to cause the $Mg(OH)_2$ to dissolve. Write equations for all of the chemical equilibria that exist in the solution after the addition of the NH_4Cl.

17.14 Which of the following will be more soluble if acid is added to the mixture? Will adding base increase the solubility of any of these? Are any of their solubilities independent of the pH of the solution? **(a)** ZnS, **(b)** $Ca(OH)_2$, **(c)** $MgCO_3$, **(d)** AgCl, **(e)** PbF_2

Equilibria in Solutions of Metal Oxides and Sulfides

17.15 Potassium oxide is readily soluble in water, but the resulting solution contains essentially no oxide ion. Explain, using an equation, what happens to the oxide ion.

17.16 What chemical reaction takes place when solid sodium sulfide is dissolved in water? Write the chemical equation.

17.17 Consider cobalt(II) sulfide. **(a)** Write its solubility equilibrium and K_{sp} equation for a saturated solution in water. (*Remember:* There is no free sulfide ion in the solution.) **(b)** Write its solubility equilibrium and K_{spa} equation for a saturated solution in aqueous acid.

Selective Precipitation

17.18 Use Le Châtelier's principle to explain how adjusting the pH makes it possible to control the concentration of $C_2O_4{}^{2-}$ in a solution of oxalic acid, $H_2C_2O_4$.

17.19 Suppose you wanted to control the $PO_4{}^{3-}$ concentration in a solution of phosphoric acid by controlling the pH of the solution. If you assume you know the H_3PO_4 concentration, what combined equation would be useful for this purpose?

17.20 If you had a solution with Ag^+, Co^{2+}, and Cu^{2+}, how could you separate the metal ions by selective precipitation?

Equilibria Involving Complex Ions

17.21 Explain the interaction between a metal ion and ammonia in a metal complex. Could the metal be neutral, or does it have to be a cation? Explain.

17.22 What is a formation constant and what is an instability constant?

Complexation and Solubility

17.23 Using Le Châtelier's principle, explain how the addition of aqueous ammonia dissolves silver chloride. If HNO_3 is added after the AgCl has dissolved in the NH_3 solution, it causes AgCl to re-precipitate. Explain why.

17.24 For $PbCl_3{}^-$, $K_{form} = 2.5 \times 10^1$. If a solution containing this complex ion is diluted with water, $PbCl_2$ precipitates. Write the equations for the equilibria involved and use them together with Le Châtelier's principle to explain how this happens.

Review Problems

Equilibria in Solutions of Slightly Soluble Salts

17.25 Write the K_{sp} expressions for each of the following compounds: **(a)** Hg_2Cl_2, **(b)** AgBr, **(c)** $PbBr_2$, **(d)** CuCl, **(e)** HgI_2.

17.26 Write the K_{sp} expressions for each of the following compounds: **(a)** AgI, **(b)** CuI, **(c)** $AuCl_3$, **(d)** Hg_2I_2, **(e)** PbI_2.

17.27 Write the K_{sp} expressions for each of the following compounds: **(a)** CaF_2, **(b)** Ag_2CO_3, **(c)** $PbSO_4$, **(d)** $Fe(OH)_3$, **(e)** PbF_2, **(f)** $Cu(OH)_2$.

17.28 Write the K_{sp} expressions for each of the following compounds: **(a)** $Fe_3(PO_4)_2$, **(b)** Ag_3PO_4, **(c)** $PbCrO_4$, **(d)** $Al(OH)_3$, **(e)** $ZnCO_3$, **(f)** $Zn(OH)_2$.

17.29 Using Problems 17.25 and 17.27, group the compounds so that the K_{sp} values can directly inform you about their relative solubilities.

17.30 Using Problems 17.26 and 17.28, group the compounds so that the K_{sp} values can directly inform you about their relative solubilities.

17.31 Which compound is more soluble in water, $ZnCO_3$ or $MgCO_3$?

17.32 Which compound is more soluble in water, $BaSO_4$ or $SrCO_3$?

17.33 In water, the solubility of lead(II) chloride is 0.016 *M*. Use this information to calculate the value of K_{sp} for $PbCl_2$.

17.34 The solubility of zinc oxalate is 7.9×10^{-3} *M*. Calculate the K_{sp} for ZnC_2O_4.

17.35 Barium sulfate is so insoluble that it can be swallowed without significant danger even though Ba^{2+} is toxic. At 25 °C, 1.00 L of water dissolves only 0.00245 g of $BaSO_4$. Calculate the molar solubility and K_{sp} for $BaSO_4$.

17.36 A student found that a maximum of 0.800 g silver acetate is able to dissolve in 100.0 mL of water. What are the molar solubility and the K_{sp} for this salt?

17.37 A student evaporated 100.0 mL of a saturated BaF_2 solution and found the solid BaF_2 she recovered weighed 0.132 g. From these data, calculate K_{sp} for BaF_2.

17.38 A student prepared a saturated solution of $CaCrO_4$ and found that when 156 mL of the solution was evaporated, 0.649 g of $CaCrO_4$ was left behind. What is the value of K_{sp} for this salt?

17.39 At 25 °C, the molar solubility of silver phosphate is 1.8×10^{-5} mol L^{-1}. Calculate K_{sp} for this salt.

17.40 The molar solubility of barium phosphate in water at 25 °C is 1.4×10^{-8} mol L^{-1}. What is the value of K_{sp} for this salt?

17.41 What is the molar solubility of $PbBr_2$ in water?

17.42 What is the molar solubility of Ag_2CrO_4 in water?

17.43 Calculate the molar solubility of $Zn(CN)_2$ in water. (Ignore the reaction of the CN^- ion with water.)

17.44 Calculate the molar solubility of PbF_2 in water. (Ignore the reaction of the F^- ion with water.)

17.45 At 25 °C, the value of K_{sp} for LiF is 1.8×10^{-3}, and that for BaF_2 is 1.7×10^{-6}. Which salt, LiF or BaF_2, has the larger molar solubility in water? Calculate the solubility of each in units of mol L^{-1}.

17.46 At 25 °C, the value of K_{sp} for AgCN is 6.0×10^{-17} and that for $Zn(CN)_2$ is 3×10^{-16}. In terms of grams per 100 mL of solution, which salt is more soluble in water?

17.47 A salt whose formula is MX has a K_{sp} equal to 3.2×10^{-10}. Another sparingly soluble salt, MX_3, must have what value of K_{sp} if the molar solubilities of the two salts are to be identical?

17.48 A salt having a formula of the type M_2X_3 has $K_{sp} = 2.2 \times 10^{-20}$. Another salt, M_2X, has to have what K_{sp} value if M_2X has twice the molar solubility of M_2X_3?

17.49 Calcium sulfate is found in plaster. At 25 °C the value of K_{sp} for $CaSO_4$ is 4.9×10^{-5}. What is the calculated molar solubility of $CaSO_4$ in water?

17.50 Chalk is $CaCO_3$, and at 25 °C its $K_{sp} = 3.4 \times 10^{-9}$. What is the molar solubility of $CaCO_3$? How many grams of $CaCO_3$ dissolve in 0.100 L of aqueous solution? (Ignore the reaction of CO_3^{2-} with water.)

17.51 It was found that the molar solubility of $BaSO_3$ in 0.10 M $BaCl_2$ is 8.0×10^{-6} M. What is the value of K_{sp} for $BaSO_3$?

17.52 The molar solubility of Ag_2CrO_4 in 0.10 M Na_2CrO_4 is 1.7×10^{-6} M. What is the value of K_{sp} for Ag_2CrO_4?

17.53 Copper(I) chloride has $K_{sp} = 1.7 \times 10^{-7}$. Calculate the molar solubility of copper(I) chloride in **(a)** pure water, **(b)** 0.0200 M HCl solution, **(c)** 0.200 M HCl solution, and **(d)** 0.150 M $CaCl_2$ solution.

17.54 Mercury(I) chloride has $K_{sp} = 1.4 \times 10^{-18}$. Calculate the molar solubility of mercury(I) chloride in **(a)** pure water, **(b)** 0.010 M HCl solution, **(c)** 0.010 M $MgCl_2$ solution, and **(d)** 0.010 M $Hg_2(NO_3)_2$ solution.

17.55 Calculate the molar solubility of $Mg(OH)_2$ in a solution that is basic with a pH of 12.50.

17.56 Calculate the molar solubility of $Al(OH)_3$ in a solution that is slightly basic with a pH of 9.50.

17.57 What is the highest concentration of Pb^{2+} that can exist in a solution of 0.10 M HCl?

17.58 Will lead(II) bromide be less soluble in 0.10 M $Pb(C_2H_3O_2)_2$ or in 0.10 M NaBr?

17.59 Calculate the molar solubility of Ag_2CrO_4 at 25 °C in **(a)** 0.200 M $AgNO_3$ and **(b)** 0.200 M Na_2CrO_4. For Ag_2CrO_4 at 25 °C, $K_{sp} = 1.1 \times 10^{-12}$.

17.60 What is the molar solubility of $Mg(OH)_2$ in **(a)** 0.20 M NaOH and **(b)** 0.20 M $MgSO_4$? For $Mg(OH)_2$, $K_{sp} = 5.6 \times 10^{-12}$.

17.61 What is the molar solubility of $Fe(OH)_2$ in a buffer that has a pH of 9.50?

17.62 What is the molar solubility of $Ca(OH)_2$ in **(a)** 0.10 M $CaCl_2$ and **(b)** 0.10 M NaOH?

*17.63 In an experiment, 2.20 g of NaOH(s) is added to 250 mL of 0.10 M $FeCl_2$ solution. What mass of $Fe(OH)_2$ will be formed? What will the molar concentration of Fe^{2+} be in the final solution?

*17.64 Suppose that 1.75 g of NaOH(s) is added to 250 mL of 0.10 M $NiCl_2$ solution. What mass, in grams, of $Ni(OH)_2$ will be formed? What will be the pH of the final solution? For $Ni(OH)_2$, $K_{sp} = 5.5 \times 10^{-16}$.

17.65 Does a precipitate of $PbCl_2$ form when 0.0150 mol of $Pb(NO_3)_2$ and 0.0120 mol of NaCl are dissolved in 1.00 L of solution?

17.66 Silver acetate, $AgC_2H_3O_2$, has $K_{sp} = 2.3 \times 10^{-3}$. Does a precipitate form when 0.015 mol of $AgNO_3$ and 0.25 mol of $Ca(C_2H_3O_2)_2$ are dissolved in a total volume of 1.00 L of solution?

17.67 Does a precipitate of $PbBr_2$ form if 50.0 mL of 0.0100 M $Pb(NO_3)_2$ is mixed with **(a)** 50.0 mL of 0.0100 M KBr and **(b)** 50.0 mL of 0.100 M NaBr?

17.68 Would a precipitate of silver acetate form if 22.0 mL of 0.100 M $AgNO_3$ were added to 45.0 mL of 0.0260 M $NaC_2H_3O_2$? For $AgC_2H_3O_2$, $K_{sp} = 2.3 \times 10^{-3}$.

17.69 Suppose that 50.0 mL each of 0.0100 M solutions of NaBr and $Pb(NO_3)_2$ are poured together. Does a precipitate form? If so, calculate the molar concentrations of the ions at equilibrium.

17.70 If a solution of 0.10 M Mn^{2+} and 0.10 M Cd^{2+} is gradually made basic, what will the concentration of Cd^{2+} be when $Mn(OH)_2$ just begins to precipitate? Assume no change in the volume of the solution.

*17.71 In an aqueous suspension of Ca(OH)$_2$, the only other component is water. Use K_{sp} to estimate the pH of the suspension.

*17.72 Suppose that 25.0 mL of 0.10 M HCl is added to 1.000 L of saturated Mg(OH)$_2$ in contact with more than enough Mg(OH)$_2$(s) to react with all of the HCl. After reaction has ceased, what will the molar concentration of Mg^{2+} be? What will the pH of the solution be?

*17.73 When solid NH$_4$Cl is added to a suspension of Mg(OH)$_2$(s), some of the Mg(OH)$_2$ dissolves. (a) Write equations for all of the chemical equilibria that exist in the solution after the addition of the NH$_4$Cl. (b) How many moles of NH$_4$Cl must be added to 1.0 L of a suspension of Mg(OH)$_2$ to dissolve 0.10 mol of Mg(OH)$_2$? (c) What is the pH of the solution after the 0.10 mol of Mg(OH)$_2$ has dissolved in the solution containing the NH$_4$Cl?

*17.74 After solid CaCO$_3$ was added to a slightly basic solution, the pH was measured to be 8.50. What was the molar solubility of CaCO$_3$ in this solution?

*17.75 Will a precipitate form in a solution made by dissolving 1.0 mol of AgNO$_3$ and 1.0 mol HC$_2$H$_3$O$_2$ in 1.0 L of solution? For AgC$_2$H$_3$O$_2$, K_{sp} = 2.3 × 10^{-3}, and for HC$_2$H$_3$O$_2$, K_a = 1.8 × 10^{-5}.

17.76 What is the molar solubility of Fe(OH)$_3$ in water? (*Hint:* You have to take into account the self-ionization of water.)

Solubility of Basic Salts Is Influenced by Acids

*17.77 Magnesium hydroxide, Mg(OH)$_2$, found in milk of magnesia, has a solubility of 7.05 × 10^{-3} g L^{-1} at 25 °C. Calculate K_{sp} for Mg(OH)$_2$.

*17.78 What is the molar solubility of vanadium(III) hydroxide in pure water? What concentration of hydroxide ions does this compound produce?

*17.79 What is the molar solubility of BaC$_2$O$_4$ in pure water, and in 1.0 M HCl? (*Hint:* In this solution, assume the oxalate ion is only mono-protonated.)

17.80 What is the molar solubility of Mg$_3$(PO$_4$)$_2$ in pure water, and in 2.0 M HCl? (*Hint:* In this solution, the phosphate ion can only be mono-protonated.)

*17.81 How many grams of solid sodium acetate must be added to 0.200 L of a solution containing 0.200 M AgNO$_3$ and 0.10 M nitric acid to cause silver acetate to begin to precipitate? For HC$_2$H$_3$O$_2$, K_a = 1.8 × 10^{-5}, and for AgC$_2$H$_3$O$_2$, K_{sp} = 2.3 × 10^{-3}.

*17.82 How many grams of solid potassium fluoride must be added to 200 mL of a solution that contains 0.20 M AgNO$_3$ and 0.10 M acetic acid to cause silver acetate to begin to precipitate? For HF, K_a = 3.5 × 10^{-4}; for HC$_2$H$_3$O$_2$, K_a = 1.8 × 10^{-5}; and for AgC$_2$H$_3$O$_2$, K_{sp} = 2.3 × 10^{-3}.

*17.83 What is the molar solubility of Mg(OH)$_2$ in 0.1 M NH$_3$ solution? Remember that NH$_3$ is a weak base.

*17.84 If 100 mL of 2.0 M NH$_3$ is added to 0.400 L of a solution containing 0.10 M Mn^{2+} and 0.10 M Sn^{2+}, what min-

imum number of grams of HCl would have to be added to the mixture to prevent Mn(OH)$_2$ from precipitating? For Mn(OH)$_2$, K_{sp} = 1.6 × 10^{-13}. Assume that virtually all of the tin is precipitated as Sn(OH)$_2$ by the following reaction:

$$Sn^{2+}(aq) + 2NH_3(aq) + 2H_2O \longrightarrow Sn(OH)_2(s) + 2NH_4^+(aq)$$

Equilibria in Solutions of Metal Oxides and Sulfides

*17.85 Calculate the molar solubility of CoS in water and in 6.0 M HNO$_3$.

*17.86 Calculate the molar solubility of β–ZnS in water and in 2.0 M HNO$_3$.

*17.87 Does nickel(II) sulfide dissolve in 4 M HCl? Perform the calculations that prove your answer?

*17.88 Does iron(II) sulfide dissolve in 8 M HCl? Perform the calculations that prove your answer.

Selective Precipitation

17.89 Both AgCl and AgI are very sparingly soluble salts, but the solubility of AgI is much less than that of AgCl, as can be seen by their K_{sp} values. Suppose that a solution contains both Cl$^-$ and I$^-$ with [Cl$^-$] = 0.050 M and [I$^-$] = 0.050 M. If solid AgNO$_3$ is added to 1.00 L of this mixture (so that no appreciable change in volume occurs), what is the value of [I$^-$] when AgCl first begins to precipitate?

17.90 Suppose that Na$_2$SO$_4$ is added gradually to 100.0 mL of a solution that contains both Ca^{2+} ion (0.15 M) and Sr^{2+} ion (0.15 M). (a) What will the Sr^{2+} concentration be (in mol L^{-1}) when CaSO$_4$ just begins to precipitate? (b) What percentage of the strontium ion has precipitated when CaSO$_4$ just begins to precipitate?

17.91 What value of [H$^+$] and what pH permit the selective precipitation of the sulfide of just one of the two metal ions in a solution that has a concentration of 0.010 M Pb^{2+} and 0.010 M Co^{2+}?

17.92 What pH would yield the maximum separation of Mn^{2+} from Sn^{2+} in a solution that is 0.010 M in Mn^{2+}, 0.010 M in Sn^{2+}, and saturated in H$_2$S? (Assume the green form of MnS in Table 17.2.)

17.93 What range of pH would permit the selective precipitation of Cu^{2+} as Cu(OH)$_2$ from a solution that contains 0.10 M Cu^{2+} and 0.10 M Mn^{2+}? For Mn(OH)$_2$, K_{sp} = 1.6 × 10^{-13}, and for Cu(OH)$_2$, K_{sp} = 4.8 × 10^{-20}.

17.94 Kidney stones often contain insoluble calcium oxalate, CaC$_2$O$_4$, which has a K_{sp} = 2.3 × 10^{-9}. Calcium oxalate is considerably less soluble than magnesium oxalate, MgC$_2$O$_4$, which has a K_{sp} = 4.8 × 10^{-6}. Suppose a solution contained both Ca^{2+} and Mg^{2+} at a concentration of 0.10 M. What pH would be required to achieve maximum separation of these ions by precipitation of CaC$_2$O$_4$ if the solution also contains oxalic acid, H$_2$C$_2$O$_4$ at a concentration of 0.10 M? For H$_2$C$_2$O$_4$, K_{a_1} = 5.9 × 10^{-2} and K_{a_2} = 6.4 × 10^{-5}.

17.95 Solid Mn(OH)$_2$ is added to a solution of 0.100 M FeCl$_2$. After reaction, what will be the molar concentrations of

Mn^{2+} and Fe^{2+} in the solution? What will be the pH of the solution? For $Mn(OH)_2$, $K_{sp} = 1.6 \times 10^{-13}$.

17.96 What value of $[H^+]$ and what pH would allow the selective separation of the carbonate of just one of the two metal ions in a solution that has a concentration of 0.010 M La^{3+} and 0.010 M Pb^{2+}? For $La_2(CO_3)_3$, $K_{sp} = 4.0 \times 10^{-34}$; for $PbCO_3$, $K_{sp} = 7.4 \times 10^{-14}$. A saturated solution of CO_2 in water has a concentration of H_2CO_3 equal to 3.3×10^{-2} M.

17.97 A chemist has a solution containing Ag^+ and Ni^{2+} and would like to separate them by precipitating one of the metals by forming the carbonate of that metal. If carbon dioxide is bubbled into the solution to make a saturated $CO_2(aq)$ solution, approximately 0.030 M, in what pH range would it be possible to precipitate the silver ion as silver(I) carbonate and leave the nickel(II) ions in solution?

*17.98 In the metal plating industry, the waste is often contaminated with Zn^{2+} and Ni^{2+}. In order to save the metals, they are recovered by precipitating them as the carbonate salts. In what pH range would it be possible to separate and collect the zinc(II) ions and then the nickel(II) ions? The concentration of CO_2 is 0.030 M.

Equilibria Involving Complex Ions

17.99 Write the chemical equilibria and equilibrium laws that correspond to K_{form} for the following complexes: (a) $CuCl_4^{2-}$, (b) AgI_2^-, (c) $Cr(NH_3)_6^{3+}$.

17.100 Write the chemical equilibria and equilibrium laws that correspond to K_{form} for the following complexes: (a) $Ag(S_2O_3)_2^{3-}$, (b) $Zn(NH_3)_4^{2+}$, (c) SnS_3^{2-}.

17.101 Write equilibria that correspond to K_{form} for each of the following complex ions and write the equations for K_{form}: (a) $Co(NH_3)_6^{3+}$, (b) HgI_4^{2-}, (c) $Fe(CN)_6^{4-}$.

17.102 Write the equilibria that are associated with the equations for K_{form} for each of the following complex ions. Write also the equations for the K_{form} of each: (a) $Hg(NH_3)_4^{2+}$, (b) SnF_6^{2-}, (c) $Fe(CN)_6^{3-}$.

17.103 Write equilibria that correspond to K_{inst} for each of the following complex ions and write the equations for K_{inst}: (a) $Co(NH_3)_6^{3+}$, (b) HgI_4^{2-}, (c) $Fe(CN)_6^{4-}$.

17.104 Write the equilibria that are associated with the equations for K_{inst} for each of the following complex ions. Write also the equations for the K_{inst} of each: (a) $Hg(NH_3)_4^{2+}$, (b) SnF_6^{2-}, (c) $Fe(CN)_6^{3-}$.

17.105 For $PbCl_3^-$, $K_{form} = 2.5 \times 10^1$. Use this information plus the K_{sp} for $PbCl_2$ to calculate K_c for the following reaction: $PbCl_2(s) + Cl^-(aq) \rightleftharpoons PbCl_3^-(aq)$.

17.106 The overall formation constant for $Ag(CN)_2^-$ equals 5.3×10^{18}, and the K_{sp} for $AgCN$ equals 6.0×10^{-17}. Calculate K_c for the following reaction: $AgCN(s) + CN^-(aq) \rightleftharpoons Ag(CN)_2^-(aq)$.

17.107 How many grams of solid $NaCN$ have to be added to 1.2 L of water to dissolve 0.11 mol of $Fe(OH)_3$ in the form of $Fe(CN)_6^{3-}$? Use data as needed from Tables 17.1 and 17.3. (For simplicity, ignore the reaction of CN^- ion with water.)

17.108 In photography, unexposed silver bromide is removed from film by soaking the film in a solution of sodium thiosulfate, $Na_2S_2O_3$. Silver ion forms a soluble complex with thiosulfate ion, $S_2O_3^{2-}$, which has the formula $Ag(S_2O_3)_2^{3-}$, and formation of the complex causes the $AgBr$ in the film to dissolve. The $Ag(S_2O_3)_2^{3-}$ complex has $K_{form} = 2.0 \times 10^{13}$. How many grams of $AgBr$ ($K_{sp} = 5.4 \times 10^{-13}$) will dissolve in 125 mL of 1.20 M $Na_2S_2O_3$ solution?

17.109 Silver iodide is very insoluble and can be difficult to remove from glass apparatus, but it forms a relatively stable complex ion, AgI_2^- ($K_{form} = 1 \times 10^{11}$), that makes AgI fairly soluble in a solution containing I^-. When a solution containing the AgI_2^- ion is diluted with water, AgI precipitates. Explain why this happens in terms of the equilibria that are involved. How many grams of AgI will dissolve in 0.100 L of 1.0 M KI solution?

17.110 Silver forms a sparingly soluble iodide salt, AgI, and it also forms a soluble iodide complex, AgI_2^- ($K_{form} = 1 \times 10^{11}$). How many grams of potassium iodide must be added to 0.100 L of water to dissolve 0.020 mol of AgI? (*Hint:* Be sure to include all of the iodide that's added to the solution.)

17.111 The formation constant for $Ag(CN)_2^-$ equals 5.3×10^{18}. Use the data in Table 17.1 to determine the molar solubility of AgI in 0.010 M KCN solution.

17.112 Suppose that some dipositive cation, M^{2+}, is able to form a complex ion with a ligand, L, by the following balanced equation: $M^{2+} + 2L \rightleftharpoons M(L)_2^{2+}$. The cation also forms a sparingly soluble salt, MCl_2. In which of the following circumstances would a given quantity of ligand be more able to bring larger quantities of the salt into solution? Explain and justify the calculation involved: (a) $K_{form} = 1 \times 10^2$ and $K_{sp} = 1 \times 10^{-15}$, (b) $K_{form} = 1 \times 10^{10}$ and $K_{sp} = 1 \times 10^{-20}$.

17.113 Calculate the molar solubility of $Cu(OH)_2$ in 2.0 M NH_3. (For simplicity, ignore the reaction of NH_3 as a base.)

17.114 The molar solubility of $Zn(OH)_2$ in 1.0 M NH_3 is 5.7×10^{-3} mol L^{-1}. Determine the value of the instability constant of the complex ion, $Zn(NH_3)_4^{2+}$. Ignore the reaction, $NH_3 + H_2O \rightleftharpoons NH_4^+ + OH^-$.

*17.115 What ratio of K_{sp} values will allow for the use of selective precipitation to separate two ionic compounds in which the less soluble one has less than 0.01 mol% remaining when the more soluble one begins to precipitate? To make the calculation easier, assume that both compounds have one-to-one ratios of cation to anion.

*17.116 A student had a solution that contained Pb^{2+}, Ni^{2+}, and Cd^{2+} and wanted to separate the ions. The student added H_2S and acidified the solution to pH 0.5. The precipitate was collected and then the student raised the pH to 8.0 and another solid precipitated. Finally, the student added HCl, expecting to collect lead(II) chloride. Was the student able to collect the $PbCl_2$? What was in the first and second precipitates?

*17.117 Suppose that 50.0 mL of 0.12 M $AgNO_3$ is added to 50.0 mL of 0.048 M NaCl solution. **(a)** What mass of AgCl will form? **(b)** Calculate the final concentrations of all of the ions in the solution that is in contact with the precipitate. **(c)** What percentage of the Ag^+ ions have precipitated?

*17.118 A sample of hard water was found to have 278 ppm Ca^{2+} ion. Into 1.00 L of this water, 1.00 g of Na_2CO_3 was dissolved. What is the new concentration of Ca^{2+} in parts per million? (Assume that the addition of Na_2CO_3 does not change the volume, and assume that the density of the aqueous solutions involved are all 1.00 g mL^{-1}.)

Complexation and Solubility

*17.119 What are the concentrations of Pb^{2+}, Br^-, and I^- in an aqueous solution that's in contact with both PbI_2 and $PbBr_2$?

*17.120 In Example 17.11 we say, "there are relatively few uncomplexed Ag^+ ions in the solution." Calculate the molar concentration of Ag^+ ion actually left after the complex forms as described in Example 17.11.

*17.121 What is the molar solubility of $PbCl_2$ in 6.0 M HCl? What is the formula of the complex formed?

*17.122 An old method for mining gold was to wash the ore with a cyanide solution and oxidizing agent to create Au^+. What is the solubility of Au^+ in 0.50 M NaCN solutions?

| Additional Exercises

*17.123 What is the molar concentration of Cu^{2+} ion in a solution prepared by mixing 0.50 mol of NH_3 and 0.050 mol of $CuSO_4$ in 1.00 L of solution? For NH_3, $K_b = 1.8 \times 10^{-5}$; for $Cu(OH)_2$, $K_{sp} = 4.8 \times 10^{-20}$; and for $Cu(NH_3)_4^{2+}$, $K_{form} = 1.1 \times 10^{13}$.

*17.124 On the basis of the K_{sp} of $Al(OH)_3$, what would be the pH of a mixture consisting of solid $Al(OH)_3$ mixed with pure water? (Assume 100% dissociation of the aluminum hydroxide.)

*17.125 The compound EDTA is used to remove soap scum, as it was described in the *Chemistry Outside the Classroom* 17.1. It has also been used to treat heavy metal poisoning, especially lead and mercury poisoning. The reaction of lead with EDTA is

$$Pb^{2+}(aq) + EDTA^{4-}(aq) \rightleftharpoons PbEDTA^{2-}(aq)$$

$$K = 1.1 \times 10^{18}$$

If a 1.0 L solution of 0.15 M $Pb(NO_3)_2$ containing 0.10 M $EDTA^{-4}$ is at a pH of 12.50, will $Pb(OH)_2$ precipitate?

17.126 A chemist has a solution that contains a mixture of the following metal ions: Cu^{2+}, Co^{2+}, Ba^{2+}, Pb^{2+}, Ag^+, Mn^{2+}, Ca^{2+}, and Bi^{3+}. How can these ions be separated? Can all of them be separated? Which ones cannot be separated from each other?

17.127 You are given a sample containing NaCl(*aq*) and NaBr(*aq*), both with concentrations of 0.020 *M*. Some of the figures shown on the next page represent a series of snapshots of what molecular-level views of the sample would show as a 0.200 *M* $Pb(NO_3)_2(aq)$ solution is slowly added. In these figures, Br^- is red-brown, and Cl^- is green. Also, spectator ions are not shown, nor are ions whose concentrations are much less than the other ions. The K_{sp} constants for $PbCl_2$ and $PbBr_2$ are 1.7×10^{-5} and 6.6×10^{-6}, respectively. (1) Arrange the figures in a time sequence to show what happens as the lead(II) nitrate solution is added, excluding any that do not "make sense." (2) Explain why you excluded any figures that did not belong in the observed time sequence.

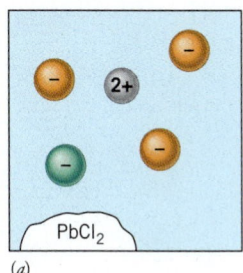

(a)

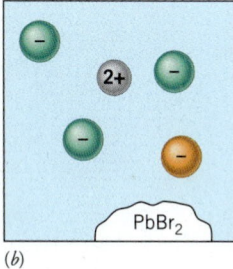

(b)

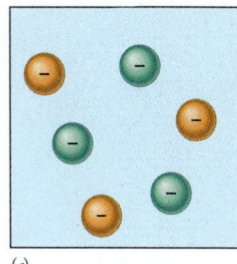

(c)

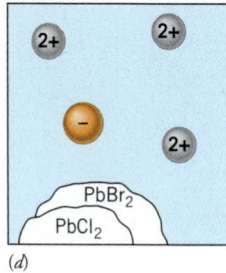

(d)

(e)

| Multi-Concept Problems

*17.128 What is the age of a stalagmite, if it has accumulated from a saturated $CaCO_3$ solution dropping at a rate of 1 drop per second? The stalagmite is a cone 10 ft in diameter and 16 ft high. The density of $CaCO_3$ is 2.71 g cm^{-3}. Assume that there are 20 drops of water in one milliliter of water and that the water evaporates when it hits the stalagmite.

*17.129 In modern construction, walls and ceilings are constructed of "drywall," which consists of plaster sandwiched between sheets of heavy paper. Plaster is composed of calcium sulfate, $CaSO_4 \cdot 2H_2O$. Suppose you had a leak in a water pipe that was dripping water on a drywall ceiling ½ in. thick at a rate of 2.00 L per day. Use the K_{sp} of calcium sulfate to estimate how many days it would take to dissolve a hole 1.0 cm in diameter. Assume the density of the plaster is 0.97 g cm^{-3}.

*17.130 How many milliliters of 0.10 M HCl would have to be added to 100 mL of a saturated solution of $PbCl_2$ in contact with 50.0 g of solid $PbCl_2$ to reduce the Pb^{2+} concentration to 0.0050 M?

*17.131 Marble is composed primarily of $CaCO_3$, a slightly soluble salt. Unfortunately, it is attacked by the acid in acid rain. What is the solubility of $CaCO_3$ in rain water (pH = 5.60) and acid rain (pH = 4.20)?

*17.132 The pH of a saturated solution of $Mg(OH)_2$ is 9.8. Determine the K_{sp} for $Mg(OH)_2$.

*17.133 The osmotic pressure of a saturated solution of $NiCO_3$ is 13 torr at 25 °C. From this information, calculate the K_{sp} for $NiCO_3$. Ignore the reaction of the carbonate ion with water.

*17.134 A saturated solution of $PbCl_2$ has a freezing point of −0.089 °C. What is the K_{sp} of $PbCl_2$ as determined by the freezing point depression value? Assume that lead(II) chloride dissociates completely and that the density of the solution is 1.00 g mL^{-1}.

| Exercises in Critical Thinking

17.135 Consider mercury(II) sulfide, HgS, which has a solubility product of 2×10^{-53}. Suppose some solid HgS was added to 1.0 L of water. How many ions of Hg^{2+} and S^{2-} are present in the water when equilibrium is reached? If your answer is accurate, is there a true equilibrium between HgS(s) and the ions in the solution? Explain your answer.

17.136 If aqueous ammonia is added gradually to a solution of copper sulfate, a pale blue precipitate forms that then dissolves to give a deep blue solution. Describe the chemical reactions that take place during these changes.

17.137 From a practical standpoint, can you effectively separate Pb^{2+} and Sr^{2+} ions by selective precipitation of their sulfates? Support your conclusions using calculations.

17.138 A salt with the formula of MX_2, and a molar mass of approximately 200, is slightly soluble in water. Estimate the minimum reliable mass you can determine on your laboratory balance. Use that estimate to determine the smallest value of the solubility product of MX_2 that could be determined by evaporating a liter of a solution that is saturated in MX_2.

17.139 In older textbooks, the solubility equilibrium for lead(II) sulfide was written as $PbS(s) \rightleftharpoons Pb^{2+}(aq) + S^{2-}(aq)$, with $K_{sp} = 3 \times 10^{-28}$. Calculate the solubility of lead(II) sulfide using K_{sp} (assuming no reaction of S^{2-} with water) and with K_{spa}. Is there a difference between the two answers? Discuss the results of your calculations.

17.140 Suppose two silver wires, one coated with silver chloride and the other coated with silver bromide, are placed in a beaker containing pure water. Over time, what if anything will happen to the compositions of the coatings on the two wires? Justify your answer.

18

Thermodynamics

Chapter Outline

As the dye moves through the water, as you know, it is more likely that the dye will spread out and color all of the water instead of forming pockets of color in the colorless water. This mixing is spontaneous, and in the chapter, we will study what drives this mixing.

Photo Researchers, Inc./Science Source

This Chapter in Context

Our experience leads us to expect the complete mixing of the dye with water. The chemistry we observe in our world is controlled both by what *can* happen and what *cannot*. As another example, hydrocarbon fuels such as octane (C_8H_{18}) *can* burn, forming CO_2 and H_2O and releasing heat. When combustion reactions are started, they proceed *spontaneously* (i.e., on their own, without further assistance), and we use them to provide energy to move vehicles, generate electricity, and heat homes and offices. On the other hand, if we mix CO_2 and H_2O, there is nothing we can do to entice them to react spontaneously to form hydrocarbons. These reactions *cannot* take place on their own. If they could, our problems with fossil fuel supplies and greenhouse gas production could easily be solved.

Observations like those described above raise the fundamental question: What determines whether or not a chemical reaction is possible when substances are combined? The answer to this question is found in the study of thermodynamics, which expands on topics introduced in Chapter 6—energy and chemical change. As you will see, not only will we be able to use thermodynamics to determine the possibility of reaction, but also we will find another explanation of chemical equilibrium and another way of determining numerical values for equilibrium constants.

LEARNING OBJECTIVES

After reading this chapter, you should be able to:

- state the first law of thermodynamics and explain how the change in energy differs from the change in enthalpy
- explain what is meant by the direction of spontaneous change
- describe the factors that affect the magnitude of entropy and the sign of an entropy change in a chemical reaction
- state the second law of thermodynamics in terms of entropy, and explain the significance of Gibbs free energy
- explain how the third law of thermodynamics leads to a standard entropy of formation
- calculate standard free energy changes
- explain the connection between the Gibbs free energy change and the maximum energy produced by a reaction
- explain the connection between Gibbs free energy change and the position of equilibrium
- describe how to calculate a Gibbs free energy change from an equilibrium constant
- define bond energy in terms of enthalpy

18.1 | First Law of Thermodynamics

■ *Thermo* implies heat, *dynamics* implies movement.

Chemical thermodynamics is the study of the role of energy in chemical changes and the spontaneity of chemical processes. It is based on a few laws that summarize centuries of experimental observation. Each law is a statement about relationships between energy, heat, work, and temperature. Because the laws manifest themselves in so many different ways and underlie so many different phenomena, there are many alternative but equivalent ways to state them. For ease of reference, the laws are identified by number and are called the first law, the second law, and the third law of thermodynamics.

The *first law of thermodynamics*, which was discussed in Chapter 6, states that internal energy may be transferred as heat or work but it cannot be created or destroyed. This law, you recall, serves as the foundation for Hess's law, which we used in our computations involving enthalpy changes. Let's review the first law in more detail.

Recall that the **internal energy (E)** of a system, which is given the symbol E, is the system's total energy—the sum of all the kinetic and potential energies of its particles. For a chemical reaction, a change in the internal energy, ΔE, is defined as

$$\Delta E = E_{products} - E_{reactants}$$

Thus, ΔE is positive if energy flows into a system and negative if energy flows out.

The first law of thermodynamics considers two ways by which energy can be exchanged between a system and its surroundings. One is by the absorption or release of heat, which is given the symbol q. The other involves work, w. If *work is done on a system*, as in the compression of a gas, the system gains and stores energy. Conversely, if *the system does work on the surroundings*, as when a gas expands and pushes a piston, the system loses some energy by changing part of its potential energy to kinetic energy, which is transferred to the surroundings. The **first law of thermodynamics** expresses the net change in energy mathematically by the equation

$$\Delta E = q + w$$

In Chapter 6 you learned that a positive sign on an energy change indicates energy gained by the system, whereas a negative sign indicates energy lost by the system as in the locomotive in Figure 18.1. Thus, when . . .

q is $(+)$ Heat is absorbed by the system.
q is $(-)$ Heat is released by the system.
w is $(+)$ Work is done on the system.
w is $(-)$ Work is done by the system.

In the lab you may observe the reaction beaker becoming cool if q for the reaction has a positive sign and it absorbs the heat from the solvent and flask, and, conversely, you may feel it warm when heat is released. Similarly, you might observe the system expanding, especially when a gas is evolved, when work has a negative sign.

In Chapter 6 you also learned that ΔE is the difference between two state functions, $E_{final} - E_{initial}$, which means that its value does not depend on how a change from one state to another is carried out. The values of q and w strongly depend on how the system changes, as we illustrated in Figure 6.8 in Section 6.5, and are not state functions. Engineers often try to maximize the work obtained for a given change while minimizing the loss of energy in the form of heat. The ratio of $w/\Delta E$ is called the thermodynamic efficiency of a given process.

Pressure–Volume Work

There are two kinds of work that chemical systems can do (or have done on them) that are of concern to us. One is electrical work, which is examined in Chapter 19 and will be discussed there. The other is work associated with the expansion or contraction of a system under the influence of an external pressure. An example is the work you perform on a gas when you compress it to fill a tire (Figure 18.2). Such "pressure–volume" or P–V work was discussed in Chapter 6, where it was shown that this kind of work is given by the equation

$$w = -P\,\Delta V$$

where P is the *external pressure* on the system.

If P–V work, similar to the work that powers the diesel engine in Figure 18.3, is the only kind of work involved in a chemical change, the equation for ΔE takes the form

$$\Delta E = q + (-P\,\Delta V) = q - P\,\Delta V$$

In Chapter 6, we also showed that when a reaction takes place in a container whose volume cannot change, the entire energy change must appear as heat. Therefore, ΔE is described as the heat at constant volume (q_v),

$$\Delta E = q_v$$

■ $\Delta E =$ (heat input) + (work input)

Figure 18.1 | Expanding gases can do work. In a steam engine, such as the one powering this locomotive, heat is absorbed by water to turn it into high temperature, high pressure steam (q is positive). The steam then expands, pushing pistons and causing the locomotive to move. In the expansion, the steam does work and w is negative.

Figure 18.2 | Work being done on a gas. When a gas is *compressed* by an external pressure, w is positive because work is done on the gas. Because V decreases when the gas is compressed, the negative sign assures that w will be positive for the compression of a gas.

Figure 18.3 | **Diesel engines are highly efficient.** In the cylinders of a diesel engine, air becomes very hot when compressed to a small volume because the work energy of compression is converted into heat energy. The temperature is so high that the fuel ignites when injected into the cylinder. The small explosion that occurs pushes the pistons to power large trucks and heavy equipment such as this diesel locomotive.

John Griffin/The Image Works

Practice Exercise 18.1

Molecules of an ideal gas have no intermolecular attractions and, therefore, undergo no change in potential energy on expansion of the gas. If the expansion is also at a constant temperature, there is no change in the kinetic energy, so the **isothermal** (constant temperature) expansion of an ideal gas has $\Delta E = 0$. Suppose such a gas expands at constant temperature from a volume of 1.0 L to 12.0 L against a constant opposing pressure of 14.0 atm. In units of *L atm*, what are q and w for this change? (*Hints:* Is the system doing work, or is work done on the system? What must be the sum of q and w in this case?)

Practice Exercise 18.2

If a gas is compressed under **adiabatic conditions** (allowing no transfer of heat to or from the surroundings) by application of an external pressure, the temperature of the gas increases. Why?

Enthalpy

Rarely do we carry out reactions in containers of fixed volume. Usually, reactions take place in containers open to the atmosphere where they are exposed to a constant pressure. To study heats of reactions under constant pressure conditions, the term enthalpy is used. Recall that the **enthalpy**, *H*, is defined by the equation

$$H = E + PV$$

In Section 6.6 we showed that at constant pressure, the **enthalpy change**, ΔH, is equal to q_p

$$\Delta H = q_p$$

where q_p is the **heat at constant pressure**.

The Difference between ΔE and ΔH

In Chapter 6 we noted that ΔE and ΔH are not equal. They differ by the pressure–volume work, $-P\Delta V$.

$$\Delta E - \Delta H = -P\Delta V$$

■ The ΔV values for reactions involving only solids and liquids are very tiny, so ΔE and ΔH for these reactions are nearly the same size.

The only time ΔE and ΔH differ by a significant amount is when gases are formed or consumed in a reaction, and even then they do not differ by much. To calculate ΔE from ΔH (or ΔH from ΔE), we must have a way of calculating the pressure–volume work.

If we assume the gases in the reaction behave as ideal gases, then we can use the ideal gas law, rearranged to solve for *V*:

$$V = \frac{nRT}{P}$$

A volume change can therefore be expressed as

$$\Delta V = \Delta\left(\frac{nRT}{P}\right)$$

If pressure and temperature are held constant, only n can vary. We show this as Δn and write

$$\Delta V = \Delta n \left(\frac{RT}{P} \right)$$

Thus, when the reaction occurs at constant pressure, the volume change is caused by a change in the number of moles of *gas*. Of course, in chemical reactions not all reactants and products need to be gases, so to be sure we compute Δn correctly, let's express the change in the number of moles of gas as Δn_{gas}. It's defined as

$$\Delta n_{gas} = (n_{gas})_{products} - (n_{gas})_{reactants}$$

The $P \Delta V$ product is therefore,

$$P \Delta V = P \times \Delta n \left(\frac{RT}{P} \right) = \Delta n_{gas} RT$$

Substituting into the equation for ΔH gives

$$\Delta H = \Delta E + \Delta n_{gas} RT \qquad (18.1)$$

The following example illustrates how small the difference is between ΔE and ΔH.

TOOLS

Converting between ΔE and ΔH

Example 18.1
Conversion between ΔE and ΔH

The decomposition of calcium carbonate in limestone is used industrially to make lime, CaO, for manufacturing cement.

$$CaCO_3(s) \longrightarrow CaO(s) + CO_2(g)$$

The reaction is endothermic and has $\Delta H° = +571$ kJ. What is the value of $\Delta E°$ for this reaction, and what is the difference between $\Delta H°$ and $\Delta E°$?

Analysis: The difference between the internal energy and the enthalpy is $-P \Delta V$, and when temperature and pressure are held constant, $-P \Delta V$ is equal to $-\Delta n_{gas} RT$. To determine T, we have to realize that the system is at standard state, as indicated by the superscript °, so the temperature is 25 °C, or 298 K; the number of moles, n, can be obtained from the coefficients in the balanced chemical equation.

Assembling the Tools: We need to use the balanced chemical equation and Equation 18.1.

$$\Delta H = \Delta E + \Delta n_{gas} RT$$

We add the superscript ° to ΔH and ΔE, and that defines a temperature of 298 K (standard temperature for measuring heats of reaction). It is also important to remember that in calculating Δn_{gas}, we count just the moles of gas. Also, because we wish to calculate the term nRT in units of kilojoules, we will use $R = 8.314$ J mol^{-1} K^{-1}, being sure to convert J to kJ at the appropriate time.

Solution: Solving for ΔE and applying the superscript ° gives

$$\Delta E° = \Delta H° - \Delta n_{gas} RT$$

To calculate Δn_{gas} our chemical equation shows that there is one mole of gas among the products and no moles of gas among the reactants, so

$$\Delta n_{gas} = 1 \text{ mol} - 0 \text{ mol} = 1 \text{ mol}$$

The temperature is 298 K, and $R = 8.314$ J mol^{-1} K^{-1}. Substituting,

$$\Delta E° = +571 \text{ kJ} - (1 \text{ mol})(8.314 \text{ J mol}^{-1} \text{ K}^{-1})(298 \text{ K})$$

$$= +571 \text{ kJ} - 2480 \text{ J}$$

We cannot add units of kJ to units of J, so we convert 2480 J to 2.48 kJ:

$$\Delta E° = +571 \text{ kJ} - 2.48 \text{ kJ}$$

$$= +569 \text{ kJ}$$

The difference between $\Delta H°$ and $\Delta E°$ is $\Delta n_{gas}RT$, or 2.48 kJ.

Is the Answer Reasonable? The equation we used predicts that $\Delta E°$ should be smaller than $\Delta H°$ if there is an increase in the moles of gas. Our result agrees with that assessment. We found the difference to be relatively small, which also agrees with the discussion. Finally, the energy needed to decompose $CaCO_3$ at constant volume ($\Delta E°$) should be less than the energy needed at constant pressure ($\Delta H°$), because the enthalpy includes additional energy to do the work of pushing back the atmosphere as the system expands.

Practice Exercise 18.3

Calculate the difference, in kilojoules, between ΔE and ΔH for the following exothermic reaction at 45 °C. Which is more exothermic, ΔE or ΔH? (*Hint:* Note the temperature.)

$$2N_2O(g) + 3O_2(g) \longrightarrow 4NO_2(g)$$

Practice Exercise 18.4

The reaction

$$CaO(s) + 2HCl(g) \longrightarrow CaCl_2(s) + H_2O(g)$$

has $\Delta H° = -217.1$ kJ. Calculate $\Delta E°$ for this reaction. What is the percentage difference between $\Delta E°$ and $\Delta H°$? (Use one significant figure for your percentage.)

When we examine the result of Example 18.1 we find that the difference between $\Delta E°$ and $\Delta H°$ is very small. In this case it works out to be approximately 1%.

18.2 | Spontaneous Change

Now that we've reviewed the way thermodynamics treats energy changes, we turn our attention to one of our main goals in studying this subject—finding relationships among the factors that determine whether or not a change in a system will occur spontaneously. By **spontaneous change** we mean one that occurs by itself, without continuous outside assistance. Examples are water flowing over a waterfall, the burning of natural gas on a stove, and the melting of ice cubes in a cold drink on a warm day. These are events that proceed on their own.

Some spontaneous changes occur very rapidly. An example is the detonation of a stick of dynamite, or the exposure of photographic film. Other spontaneous events, such as the rusting of iron or the erosion of stone, occur slowly, and many years may pass before a change is noticed. Still others occur at such an extremely slow rate under ordinary conditions that they appear not to be spontaneous at all. Gasoline–oxygen mixtures appear perfectly stable indefinitely at room temperature because under these conditions they react very slowly. However, if exposed to a spark, their rate of reaction increases tremendously and they react explosively.

Each day we also witness events that are obviously *not* spontaneous. We may pass by a pile of bricks in the morning and later in the day find that they have become a brick wall. We know from experience that the wall didn't get there by itself. A pile of bricks becoming a brick wall is *not* spontaneous; it requires the intervention of a bricklayer. Similarly, the decomposition of water into hydrogen and oxygen is not spontaneous. We see water all the time and we know that it's stable. Nevertheless, we can cause water to decompose by passing an electric current through it in a process called *electrolysis* (Figure 18.4).

$$2H_2O(l) \xrightarrow{\text{electrolysis}} 2H_2(g) + O_2(g)$$

This decomposition will continue, however, only as long as the electric current is maintained. As soon as the supply of electricity is cut off, the decomposition ceases. This example demonstrates the difference between spontaneous and nonspontaneous changes. Once a spontaneous event begins, it has a tendency to continue until it is finished. A nonspontaneous event, on the other hand, can continue only as long as it receives some sort of outside assistance.

Nonspontaneous changes have another common characteristic. They are able to occur only when accompanied by some spontaneous change. For example, a bricklayer consumes food, and a series of spontaneous biochemical reactions then occur that supply the necessary muscle power to build the brick wall. Similarly, the nonspontaneous electrolysis of water requires some sort of spontaneous mechanical or chemical change to generate the needed electricity. In short, *all nonspontaneous events occur at the expense of spontaneous ones.* Everything that happens can be traced, either directly or indirectly, to spontaneous changes.

The reaction of gasoline vapor with oxygen, mentioned previously, illustrates an important observation about spontaneous changes. Even though the reaction between these substances has a strong tendency to occur, and is therefore spontaneous, the *rate of the reaction* at room temperature is so slow that the mixture appears to be stable. In other words, the reaction *appears* to be nonspontaneous because its rate is so slow. There are many reactions in nature that are spontaneous but occur at such a slow rate that they aren't observed. Biochemical reactions often fall into this category. Without the presence of a catalyst, an enzyme for instance, these reactions are so slow that effectively they do not occur. Living systems control their chemical reactions by selectively making enzymes available when spontaneous reactions or their products are needed.

■ The driving of nonspontaneous reactions to completion by linking them to spontaneous ones is an important principle in biochemistry.

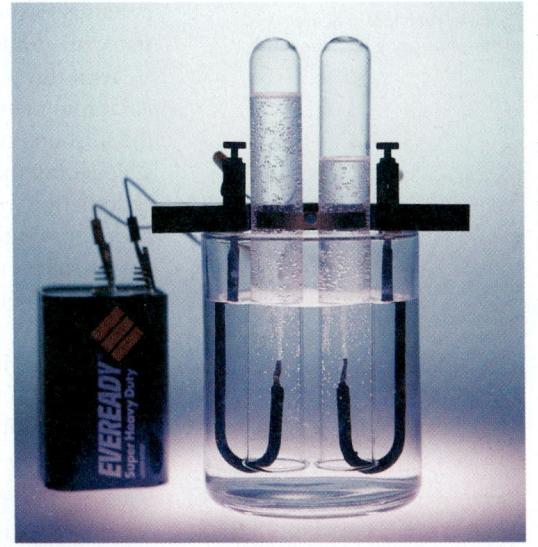

Figure 18.4 | **The electrolysis of water produces H$_2$ and O$_2$ gases.** It is a nonspontaneous change that only continues as long as electricity is supplied.

Direction of Spontaneous Change

What determines the direction of spontaneous change? Let's begin by examining some everyday events such as those depicted in Figure 18.5. When iron rusts, heat is released, so the reaction lowers the internal energy of the system. Similarly, the chemical substances in the gasoline–oxygen mixture lose chemical energy by evolving heat and light as the gasoline burns to produce CO$_2$ and H$_2$O.

Figure 18.5 | **Three common spontaneous events.** Iron rusts, fuel burns, and an ice cube melts at room temperature.

Since many spontaneous reactions are obviously exothermic, we might be tempted to conclude (as some 19th century chemists did) that spontaneous events occur because heat energy flows out of a system. This energy-lowering process was labeled a "driving force" behind spontaneous change. In addition to the loss of heat energy in chemical reactions a change in potential energy can occur. For instance, when we drop a book, it spontaneously falls to the floor because that lowers its potential energy. Since a change that lowers the potential energy of a system is also exothermic, the old concept of spontaneity could be stated as: *Exothermic changes have a tendency to proceed spontaneously.*

■ However, most, but not all, chemical reactions that are exothermic occur spontaneously.

Yet if this is so, how do we explain the third photograph in Figure 18.5? The melting of ice at room temperature is clearly a spontaneous process. But ice absorbs heat energy from the surroundings as it melts. This absorbed heat gives the water from the melted ice cube a higher internal energy than the original ice had. This is an example of a spontaneous but endothermic process. There are many other examples of spontaneous endothermic processes: the evaporation of acetone in nail polish remover, the expansion of carbon dioxide gas into a vacuum, the reaction of acetic acid with sodium carbonate, and the operation of a chemical "cold pack."

In Chapter 13 we discussed the solution process, and we said that one of the principal driving forces in the formation of a solution is the fact that the mixed state is much more probable than the unmixed state. It turns out that arguments like this can be used to explain the direction of any spontaneous process.

Practice Exercise 18.5

Are the following processes spontaneous?

(a) Ice cream melting on a hot day
(b) Silver chloride dissolving in water that contains sodium nitrate
(c) Isolating argon from the air

(*Hint:* What is needed for a reaction to be spontaneous?)

Practice Exercise 18.6

The following processes are not spontaneous as written, what can be done to make them spontaneous activities?

(a) A stone rolling uphill
(b) A glass of iced tea freezing completely as it sits on the counter
(c) The formation of sodium hydroxide and hydrochloric acid when sodium chloride dissolves in water

18.3 | Entropy

In Chapter 6 you learned that when a hot object is placed in contact with a colder one, heat will flow spontaneously from the hot object to the colder one. But why? The law of conservation of energy doesn't help, since energy is conserved no matter in which direction the heat flows.

To analyze what happens in the spontaneous flow of heat, let's imagine a situation in which we have two objects in contact with one another, one with a high temperature and the other with a low temperature. Because of the relationship between average kinetic energy and temperature, we expect the high-temperature object to have many rapidly moving molecules, whereas the low-temperature object will have more slow-speed molecules. Where the objects touch, many of the molecular collisions will involve fast-moving "hot" molecules and slow-moving "cold" ones. In such a collision, it is very unlikely that the fast molecule will gain kinetic energy at the expense of the slow one. Instead, the fast molecule will lose kinetic energy and slow down, while the slow molecule will gain kinetic energy and speed up. Over time, the result of many such collisions is that the hot object cools and the cool object warms. While this happens, the combined kinetic energy of the two objects becomes distributed over all the molecules in the system.

■ Imagine a game of billiards where a moving ball, the cue ball, hits a stationary ball. In the collision, we observe that the cue ball loses some, but not all, of its speed and the other ball is now in motion. The result is the transfer of energy from the "hot" cue ball to the "cold" stationary ball.

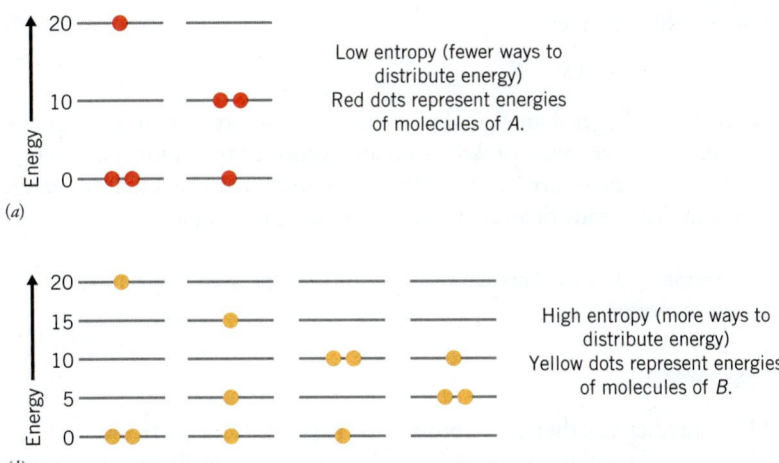

(a)

(b)

Figure 18.6 | A positive value for ΔS means an increase in the number of ways energy can be distributed among a system's molecules. Consider the reaction $A \longrightarrow B$, where A can take on energies that are multiples of 10 energy units, and B can take on energies that are multiples of 5 units. Suppose that the total energy of the reacting mixture of three molecules is 20 units. (*a*) There are two ways to distribute 20 units of energy among the three molecules of *A*. (*b*) There are four ways to distribute 20 units of energy among the three molecules of *B*. The entropy of *B* is higher than the entropy of *A* because there are more ways to distribute the same amount of energy in *B* than in *A*.

What we learn here is that heat flows because of the probable outcome of intermolecular collisions. Viewed on a larger scale, the uniform distribution of kinetic energy between the objects, caused by the flow of heat, is much more probable than a situation in which no heat flow occurs, which illustrates the role of probability in determining the direction of a spontaneous change. *Spontaneous processes tend to proceed from states of low probability to states of higher probability.* The higher probability states are those that allow more options for distributing energy among the molecules, so we can also say that spontaneous processes tend to disperse energy.

Distributing Energy in a System

Because statistical probability is so important in determining the outcome of chemical and physical events, thermodynamics defines a state function, called **entropy** (symbol *S*), which is related to the possible number of equivalent ways that energy can be distributed in a system. The greater the number of ways to distribute energy, the larger is its statistical probability, and therefore the larger the value of the entropy. This is shown in Figure 18.6 for two systems of molecules that can each have a different way to distribute its energy. Figure 18.7 uses money to represent another way to visualize the spreading of energy into equivalent states.

Because entropy is a state function, an **entropy change**, Δ*S*, is independent of the path taken between initial and final states. Therefore, we can write

■ The greater the statistical probability of a particular state, the greater is the entropy.

$$\Delta S = S_{final} - S_{initial}$$

(a)　　　Two ways to count out $2 with paper money

(b)　　　Five ways to count out $2 with 50¢ and 25¢ coins

Figure 18.7 | Entropy. If energy were money, entropy would describe the number of different ways of counting it out. For example, (*a*) there are only two ways of counting out $2 using American paper money, but (*b*) there are five ways of counting out $2 using 50 cent and 25 cent coins. One could consider the system using coins as having more entropy than the system using just paper money.

For a chemical reaction this becomes

$$\Delta S = S_{\text{products}} - S_{\text{reactants}}$$

As you can see, when S_{final} is larger than S_{initial} (or when S_{products} is larger than $S_{\text{reactants}}$), the value of ΔS is positive. A positive value for ΔS means an increase in the number of energy-equivalent ways the system can be arranged, and we have seen that this kind of change tends to be spontaneous. This leads to a general statement about entropy:

Any event that is accompanied by an increase in the entropy of the system will have a *tendency* to occur spontaneously.

Factors that Affect ΔS

It is often possible to predict whether ΔS is positive or negative for a particular change. This is because several factors influence the magnitude of the entropy change in predictable ways.

Volume

For gases, the entropy increases with increasing volume, as illustrated in Figure 18.8. At the left we see a gas confined to one side of a container, separated from a vacuum by a removable partition. Let's suppose the partition could be pulled away in an instant, as shown in Figure 18.8*b*. Immediately, we have a situation in which all the molecules of the gas are at one end of a larger container. There are many more possible ways that the total kinetic energy can be distributed among the molecules in the larger volume. That makes the configuration in Figure 18.8*b* extremely unlikely. The gas expands spontaneously to achieve a more probable (higher entropy) particle distribution, Figure 18.8*c*.

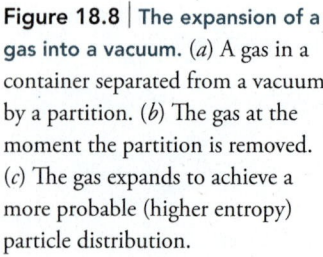

Figure 18.8 | **The expansion of a gas into a vacuum.** (*a*) A gas in a container separated from a vacuum by a partition. (*b*) The gas at the moment the partition is removed. (*c*) The gas expands to achieve a more probable (higher entropy) particle distribution.

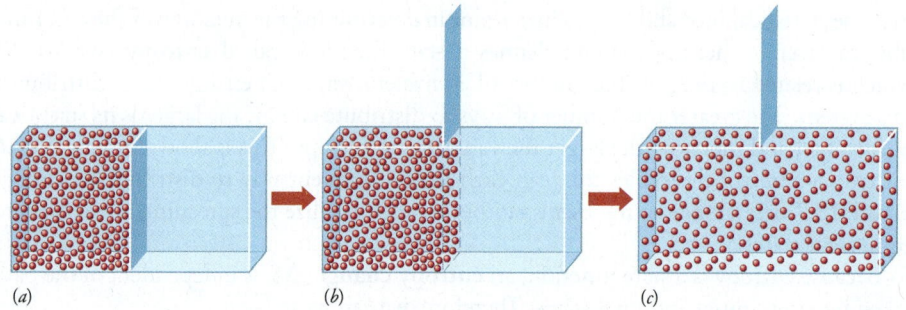

(*a*)   (*b*)   (*c*)

Temperature

The entropy is also affected by temperature; the higher the temperature, the larger is the entropy. For example, when a substance is a solid at absolute zero, its particles are essentially motionless. There is relatively little kinetic energy, and so there are few ways to distribute kinetic energy among the particles, so the entropy of the solid is relatively low (Figure 18.9*a*). If some heat is added to the solid, the kinetic energy of the particles increases along with the temperature. This causes the particles to move and vibrate within the crystal, so at a particular moment (Figure 18.9*b*) the particles are not found exactly at their lattice sites. There is more kinetic energy than at the lower temperature, and there are more ways to distribute it among the molecules, so the entropy is larger. If the temperature is raised further, the particles are given even more kinetic energy with an even larger number of possible ways to distribute it, causing the solid to have a still higher entropy (Figure 18.9*c*).

Physical State

One of the major factors that affects the entropy of a system is its physical state, which is demonstrated in Figure 18.10. Suppose that the diagrams represent ice, liquid water, and water vapor all at the same temperature. There is greater freedom of molecular movement

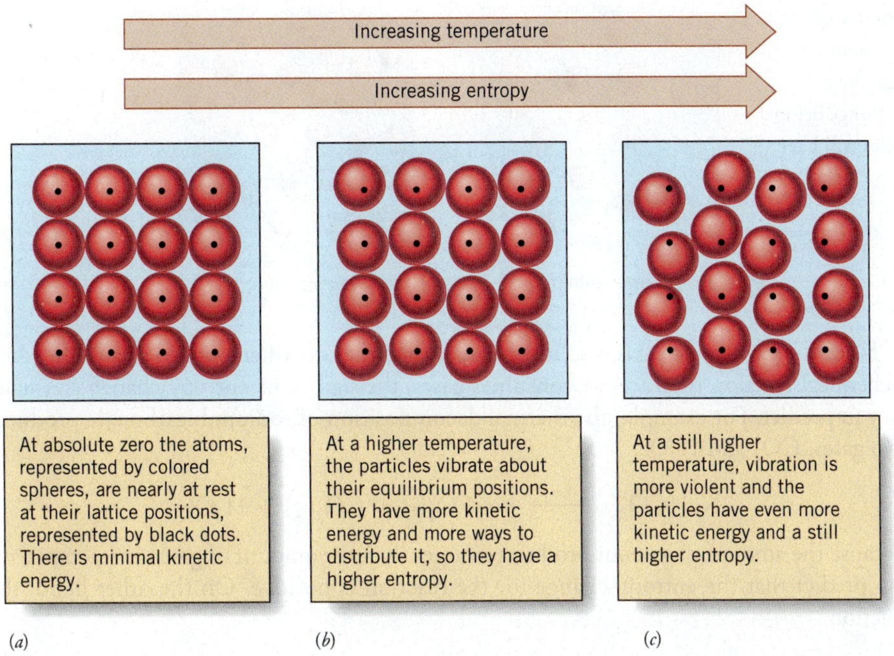

Increasing temperature

Increasing entropy

At absolute zero the atoms, represented by colored spheres, are nearly at rest at their lattice positions, represented by black dots. There is minimal kinetic energy.

(a)

At a higher temperature, the particles vibrate about their equilibrium positions. They have more kinetic energy and more ways to distribute it, so they have a higher entropy.

(b)

At a still higher temperature, vibration is more violent and the particles have even more kinetic energy and a still higher entropy.

(c)

Figure 18.9 | Variation of entropy with temperature. (*a*) At absolute zero the atoms, represented by colored circles, rest at their equilibrium lattice positions, represented by black dots. Entropy is essentially zero. (*b*) At a higher temperature, the particles vibrate about their equilibrium positions and there are more different ways to distribute kinetic energy among the molecules. Entropy is higher than in (*a*). (*c*) At a still higher temperature, vibration is more violent and at any instant the particles are found in even more arrangements. Entropy is higher than in (*b*).

in water than in ice at the same temperature, and so there are more ways to distribute kinetic energy among the molecules of liquid water than there are in ice. The water molecules in the vapor are free to move through the entire container. They are able to distribute their kinetic energies in a very large number of ways. In general, therefore, there are many more possible ways to distribute kinetic energy among gas molecules than there are in liquids and solids. In fact, a gas has such a large entropy compared with a liquid or solid that changes which produce gases from liquids or solids are almost always accompanied by increases in entropy.

Number of Particles

Entropy, S, is an extensive property of a system just as H, the enthalpy, is an extensive property that depends on the amount of substance in the system. If we add more molecules to a system, there will be more ways to distribute the total energy of the system, and therefore the entropy is larger, as shown in Figure 18.11.

Predicting the Sign and Magnitude of ΔS for a Chemical Reaction

Two factors can help in predicting the sign of the entropy change in a reaction: the production or consumption of a gas, and the number of molecules in the products as compared to the number in the reactants.

TOOLS

Factors affecting entropy: number of particles

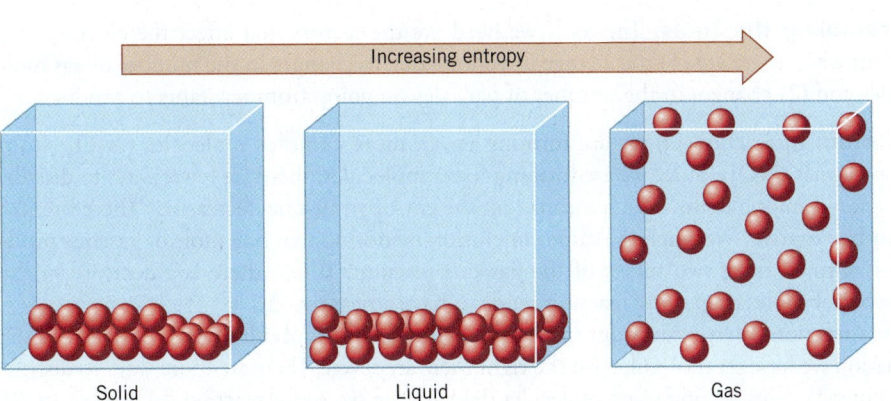

Increasing entropy

Solid   Liquid   Gas

Figure 18.10 | Comparison of the entropies of the solid, liquid, and gaseous states of a substance. The crystalline solid has a very low entropy. The liquid has a higher entropy because its molecules can move more freely and there are more ways to distribute kinetic energy among them. All the particles are still found at one end of the container. The gas has the highest entropy because the particles are randomly distributed throughout the entire container, so there are many, many ways to distribute kinetic energy among the molecules.

Figure 18.11 | **Entropy is affected by number of particles.** Adding additional particles to a system increases the number of ways that energy can be distributed in the system, so with all other things being equal, a reaction that produces more particles will have a positive value of ΔS.

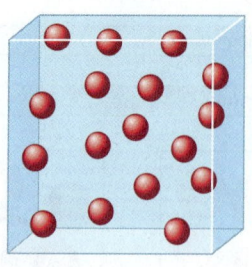
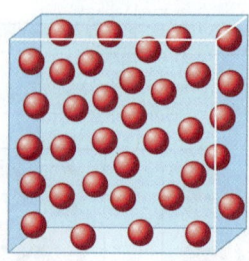

Lower entropy   Higher entropy

Since the entropy of a gas is so much larger than that of either a liquid or a solid, when a chemical reaction produces or consumes gases, the sign of its entropy change is usually easy to predict. For example, the thermal decomposition of sodium bicarbonate produces two gases, CO_2 and H_2O.

$$2NaHCO_3(s) \xrightarrow{\text{heat}} Na_2CO_3(s) + CO_2(g) + H_2O(g)$$

Because the amount of gaseous products is larger than the amount of gaseous reactants, we can predict that the entropy change for the reaction is positive. On the other hand, the reaction

$$CaO(s) + SO_2(g) \longrightarrow CaSO_3(s)$$

(which can be used to remove sulfur dioxide from a gas mixture) has a negative entropy change because the number of gas molecules decreases as the reaction progresses.

For chemical reactions, another major factor that affects the sign of ΔS is a change in the total number of molecules as the reaction proceeds. When more molecules are produced during a reaction, more ways of distributing the energy among the molecules are possible. *When all other things are equal, reactions that increase the number of particles in the system tend to have a positive entropy change.*

Example 18.2
Predicting the Sign of ΔS

Predict the algebraic sign of ΔS for the reactions:
(a) $2NO_2(g) \longrightarrow N_2O_4(g)$
(b) $C_3H_8(g) + 5O_2(g) \longrightarrow 3CO_2(g) + 4H_2O(g)$

Analysis: We will need to consider all of the factors that can have an effect on the entropy change of a particular system: volume, temperature, physical state, and number of particles. We note that all substances are in the gas phase, and no temperature change is specified. Therefore, phase change and temperature changes do not need to be considered.

Assembling the Tools: The tools we need are the factors that affect the entropy of a system when there are (1) the change in volume due to changes in the number of gas molecules and (2) changes in the number of particles on going from reactants to products.

Solution: In reaction (a) we are forming fewer, more complex molecules (N_2O_4) from simpler ones (NO_2). Since we are forming fewer molecules, there are fewer ways to distribute energy among them, which means that the entropy must be decreasing. Therefore, ΔS must be negative. We reach the same conclusion by noting that one mole of gaseous product is formed from two moles of the gaseous reactant. When there is a decrease in the number of moles of gas, the reaction tends to have a negative ΔS.

For reaction (b), we can count the number of molecules on both sides. On the left of the equation we have six molecules; on the right there are seven. There are more ways to distribute kinetic energy among seven molecules than among six, so for reaction (b), we expect ΔS to be positive. We reach the same conclusion by counting the number of moles of gas on both sides of the equation. On the left there are 6 moles of gas; on the right there are 7 moles of gas. Because the number of moles of gas is increasing, we expect ΔS to be positive.

Is the Answer Reasonable? The only check we can perform here is to carefully review our reasoning. It appears sound, so the answers appear to be correct.

Would you expect the ΔS to be positive or negative for the following:

$$Ag^+(aq) + Br^-(aq) \longrightarrow AgBr(s)$$

(*Hint:* How is the freedom of movement of the ions affected?)

Practice Exercise 18.7

Predict the sign of the entropy change for (a) the condensation of steam to liquid water and (b) the sublimation of solid iodine.

Practice Exercise 18.8

Predict the sign of ΔS for the following reactions:

(a) $2SO_2(g) + O_2(g) \longrightarrow 2SO_3(g)$
(b) $2H_2(g) + O_2(g) \longrightarrow 2H_2O(l)$
(c) $N_2(g) + 3H_2(g) \longrightarrow 2NH_3(g)$
(d) $Ca(OH)_2(s) \xrightarrow{H_2O} Ca^{2+}(aq) + 2OH^-(aq)$

Practice Exercise 18.9

18.4 | Second Law of Thermodynamics

So far, you have learned that enthalpy and entropy are two factors that affect the spontaneity of a physical or chemical event. Sometimes they work together to favor a change, as in the combustion of gasoline where heat is given off (an exothermic change) and large volumes of gases are formed (an entropy increase).

In many situations, the enthalpy and entropy changes oppose each other, as in the melting of ice. The process of melting absorbs heat and is endothermic, which tends to make the process nonspontaneous. However, the greater freedom of motion of the molecules that accompanies melting has the opposite effect and tends to make the change spontaneous.

When the enthalpy and entropy changes conflict, temperature becomes a critical factor that can influence the direction in which the change is spontaneous. For example, consider a mixture of solid ice and liquid water. If we raise the temperature of the mixture to 25 °C, all of the solid will melt. At 25 °C the change *solid* $\longrightarrow$ *liquid* is spontaneous. If we cool the mixture to −25 °C, freezing occurs, so at −25 °C the opposite change, *liquid* $\longrightarrow$ *solid*, is spontaneous. *Thus, there are actually three factors that can influence spontaneity: the enthalpy change, the entropy change, and the temperature.*

■ Knowledge of ΔS, ΔH, and T will allow us to determine whether a system will undergo a spontaneous change.

Second Law of Thermodynamics

One of the most far-reaching observations in science is incorporated into the **second law of thermodynamics**, which states, in effect, that *whenever a spontaneous event takes place in our universe, the total entropy of the universe increases* ($\Delta S_{universe} > 0$). Notice that the increase in entropy that's referred to here is for the *total* entropy of the *universe* (system *plus* surroundings), not just the system alone. This means that a system's entropy can decrease, just as long as there is a larger increase in the entropy of the surroundings so that the *overall* entropy change is positive. Because everything that happens relies on spontaneous changes of some sort, the entropy of the universe is constantly rising.

Let's examine the entropy change for the universe. As we've suggested, this quantity equals the sum of the entropy change for the system plus the entropy change for the surroundings:

$$\Delta S_{universe} = \Delta S_{system} + \Delta S_{surroundings}$$

It can be shown that the entropy change for the surroundings is equal to the heat transferred *to* the surroundings *from* the system, $q_{surroundings}$, divided by the Kelvin temperature, T, at which it is transferred.

$$\Delta S_{surroundings} = \frac{q_{surroundings}}{T}$$

The law of conservation of energy requires that the heat transferred to the surroundings equals the negative of the heat added to the system, so we can write

$$q_{surroundings} = -q_{system}$$

In our study of the first law of thermodynamics we saw that for changes at constant temperature and pressure, $q_{system} = \Delta H$ for the system. By substitutions, therefore, we arrive at the relationship

$$\Delta S_{surroundings} = \frac{-\Delta H_{system}}{T}$$

and the entropy change for the entire universe becomes

$$\Delta S_{universe} = \Delta S_{system} - \frac{\Delta H_{system}}{T}$$

By rearranging the right side of this equation, we obtain

$$\Delta S_{universe} = \frac{T\Delta S_{system} - \Delta H_{system}}{T}$$

Now let's multiply both sides of the equation by T to give

$$T\Delta S_{universe} = T\Delta S_{system} - \Delta H_{system}$$

or

$$T\Delta S_{universe} = -(\Delta H_{system} - T\Delta S_{system})$$

■ The origin of *free* in *free energy* is discussed in Section 18.7.

Because $\Delta S_{universe}$ must be positive for a spontaneous change, the quantity in parentheses, $(\Delta H_{system} - T\Delta S_{system})$, must be negative. Therefore, we can state that for a change to be spontaneous,

$$\Delta H_{system} - T\Delta S_{system} < 0 \tag{18.2}$$

The Gibbs Free Energy

Equation 18.2 gives us a way of examining the balance between ΔH, ΔS, and temperature in determining the spontaneity of an event, and it becomes convenient at this point to introduce another thermodynamic state function. It is called the **Gibbs free energy**, *G*, named to honor one of America's greatest scientists, Josiah Willard Gibbs (1839–1903). (It's called *free energy* because it is related, as we will see later, to the maximum energy in a change that is "free" or "available" to do useful work.) The Gibbs free energy is defined as

$$G = H - TS \tag{18.3}$$

so for the **Gibbs free energy change**, **ΔG**, at constant T and P, this equation becomes

$$\Delta G = \Delta H - T\Delta S \tag{18.4}$$

Because G is defined entirely in terms of state functions, it is also a state function. This means that

$$\Delta G = G_{final} - G_{initial} \tag{18.5}$$

By comparing Equations 18.2 and 18.4, we arrive at the special importance of the free energy change:

Gibbs free energy

TOOLS

ΔG as a predictor of spontaneity

At constant temperature and pressure, a change can only be spontaneous if it is accompanied by a decrease in the free energy of the system.

In other words, for a change to be spontaneous, G_{final} must be less than $G_{initial}$ and ΔG must be negative. With this in mind, we can now examine how ΔH, ΔS, and T can be used to determine spontaneity.

■ Reactions that occur with a free energy decrease are sometimes said to be **exergonic**. Those that occur with a free energy increase are sometimes said to be **endergonic**.

When ΔH Is Negative and ΔS Is Positive

The combustion of octane (a component of gasoline),

$$2C_8H_{18}(l) + 25\,O_2(g) \longrightarrow 16CO_2(g) + 18H_2O(g)$$

is a very exothermic reaction. There is also a large increase in entropy because the number of particles in the system increases and large volumes of gases are formed. For this change, ΔH is negative and ΔS is positive, both of which favor spontaneity. Let's analyze how this affects the sign of ΔG.

$$\Delta H \text{ is negative } (-)$$
$$\Delta S \text{ is positive } (+)$$
$$\Delta G = \Delta H - T\Delta S$$
$$= (-) - [T(+)]$$

Notice that regardless of the absolute temperature, which must be a positive number, ΔG will be negative. This means that regardless of the temperature, such a change must be spontaneous. In fact, once started, fires will continue to consume all available fuel or oxygen at nearly any temperature because combustion reactions are always spontaneous (see Figure 18.12a).

When ΔH Is Positive and ΔS Is Negative

When a change is endothermic and is accompanied by a lowering of the entropy, both factors work against spontaneity.

$$\Delta H \text{ is positive } (+)$$
$$\Delta S \text{ is negative } (-)$$
$$\Delta G = \Delta H - T\,\Delta S$$
$$= (+) - [T(-)]$$

(a) (b) (c)

Figure 18.12 | **Process spontaneity can be predicted if ΔH, ΔS, and T are known.** (a) When ΔH is negative and ΔS is positive, as in any combustion reaction, the reaction is spontaneous at any temperature. (b) When ΔH is positive and ΔS is negative, the process is not spontaneous. Left to themselves, ash, carbon dioxide, and water will not spontaneously combine to form wood. (c) When ΔH and ΔS have the same sign, temperature determines whether the process is spontaneous or not. Water spontaneously becomes ice below 0 °C, and ice spontaneously melts into liquid water above 0 °C.

Now, no matter what the temperature is, ΔG will be positive and the change must be nonspontaneous. An example would be carbon dioxide and water coming back together to form wood and oxygen again in a fire (Figure 18.12*b*).[1] If you saw such a thing happen in a video, experience would tell you that the video was being played backwards.

When ΔH and ΔS Have the Same Sign

When ΔH and ΔS have the same algebraic sign, the temperature becomes the determining factor in controlling spontaneity. If ΔH and ΔS are both positive, then

$$\Delta G = (+) - [T(+)]$$

Thus, ΔG is the difference between two quantities, ΔH and $T\Delta S$. This difference will only be negative if the term $T\Delta S$ is larger in magnitude than ΔH, and this will only be true when the temperature is high. In other words, *when ΔH and ΔS are both positive, the change will be spontaneous at high temperature but not at low temperature.* A change noted above is the melting of ice.

$$H_2O(s) \longrightarrow H_2O(l)$$

This is a change that is endothermic and also accompanied by an increase in entropy. We know that at high temperature (above 0 °C) melting is spontaneous, but at low temperature (below 0 °C) it is not.

For similar reasons, when ΔH and ΔS are both negative, ΔG will be negative (and the change spontaneous) only when the temperature is low.

$$\Delta G = (-) - [T(-)]$$

Only when the negative value of ΔH is larger in magnitude than the negative value of $T\Delta S$ will ΔG be negative. Such a change is only spontaneous at low temperature. An example is the freezing of water (see Figure 18.12*c*).

$$H_2O(l) \longrightarrow H_2O(s)$$

This is an exothermic change that is accompanied by a decrease in entropy; it is only spontaneous at low temperatures (i.e., below 0 °C).

Figure 18.13 summarizes the effects of the signs of ΔH and ΔS on ΔG, and hence on the spontaneity of physical and chemical events.

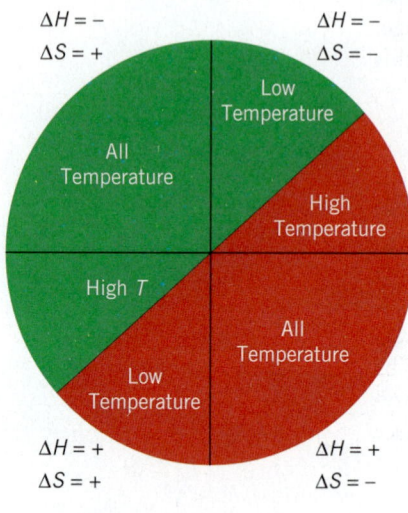

$\Delta H = -$
$\Delta S = +$

$\Delta H = -$
$\Delta S = -$

Low Temperature

All Temperature

High Temperature

High *T*

All Temperature

Low Temperature

$\Delta H = +$
$\Delta S = +$

$\Delta H = +$
$\Delta S = -$

■ Spontaneous

■ Nonspontaneous

Figure 18.13 | Summary of the effects of the signs of ΔH and ΔS on spontaneity as a function of temperature. When ΔH and ΔS have the same sign, spontaneity is determined by the temperature.

Practice Exercise 18.10

Would you expect the following reactions to be spontaneous?

(a) $2N_2(g) + O_2(g) \longrightarrow 2N_2O(g)$ $\quad\quad\quad\quad\quad$ $\Delta H = +163.2$ kJ

(b) $C_3H_8(g) + 5O_2(g) \longrightarrow 3CO_2(g) + 4H_2O(g)$ $\quad$ $\Delta H = -2044$ kJ

(c) $2H_2O_2(l) \longrightarrow 2H_2O(l) + O_2(g)$ $\quad\quad\quad\quad\quad$ $\Delta H = -98.0$ kJ

(*Hint:* Remember how entropy and enthalpy affect spontaneity.)

Practice Exercise 18.11

What change in temperature would make the process spontaneous?

(a) $H_2O(l) \longrightarrow H_2O(g)$

(b) $Ag^+(aq) + Cl^-(aq) \longrightarrow AgCl(s)$

---

[1] Plants can manufacture wood from carbon dioxide and water that they take in, but the reaction by itself is *not* spontaneous. Plant cells manufacture wood by coupling this nonspontaneous reaction to a complex series of reactions with negative values of ΔG, so that the entire chain of reactions *taken together* is spontaneous overall. This same trick is used to drive nonspontaneous processes forward in human cells. The high negative free energy change from the breakdown of sugar and other nutrients is coupled with the nonspontaneous synthesis of complex proteins from simple starting materials, driving cell growth and making life possible.

18.5 | Third Law of Thermodynamics

Earlier we described how the entropy of a substance depends on temperature, and we noted that at absolute zero the order within a crystal is at a maximum and the entropy is at a minimum. The **third law of thermodynamics** goes one step farther by stating: *At absolute zero the entropy of a perfectly ordered pure crystalline substance is zero.*

$$S = 0 \quad \text{at} \quad T = 0\,\text{K}$$

Because we know the point at which entropy has a value of zero, it is possible by *experimental measurement* and calculation to determine the total amount of entropy that a substance has at temperatures above 0 K.

If the entropy of one mole of a substance is determined at a temperature of 298 K (25 °C) and a pressure of 1 atm, we call it the **standard entropy**, $S°$. Table 18.1 lists the standard entropies for a number of substances.[2] Notice that entropy has the dimensions of energy/temperature (i.e., joules per kelvin); this is explained in Chemistry Outside the Classroom 18.1.

■ 25 °C and 1 atm are the same standard conditions we used in our discussion of ΔH in Chapter 6.

TABLE 18.1	Standard Entropies of Some Typical Substances at 298.15 K				
Substance	$S°$ $(J\,mol^{-1}\,K^{-1})$	**Substance**	$S°$ $(J\,mol^{-1}\,K^{-1})$	**Substance**	$S°$ $(J\,mol^{-1}\,K^{-1})$
$Ag(s)$	42.55	$CaO(s)$	40	$N_2(g)$	191.5
$AgCl(s)$	96.2	$Ca(OH)_2(s)$	76.1	$NH_3(g)$	192.5
$Al(s)$	28.3	$CaSO_4(s)$	107	$NH_4Cl(s)$	94.6
$Al_2O_3(s)$	51.0	$CaSO_4 \cdot \frac{1}{2}H_2O(s)$	131	$NO(g)$	210.6
$C(s)$ (graphite)	5.69	$CaSO_4 \cdot 2H_2O(s)$	194.0	$NO_2(g)$	240.5
$CO(g)$	197.9	$Cl_2(g)$	223.0	$N_2O(g)$	220.0
$CO_2(g)$	213.6	$Fe(s)$	27	$N_2O_4(g)$	304
$CH_4(g)$	186.2	$Fe_2O_3(s)$	90.0	$Na(s)$	51.0
$CH_3Cl(g)$	234.2	$H_2(g)$	130.6	$Na_2CO_3(s)$	136
$CH_3OH(l)$	126.8	$H_2O(g)$	188.7	$NaHCO_3(s)$	102
$CO(NH_2)_2(s)$	104.6	$H_2O(l)$	69.96	$NaCl(s)$	72.38
$CO(NH_2)_2(aq)$	173.8	$HCl(g)$	186.7	$NaOH(s)$	64.18
$C_2H_2(g)$	200.8	$HNO_3(l)$	155.6	$Na_2SO_4(s)$	149.4
$C_2H_4(g)$	219.8	$H_2SO_4(l)$	157	$O_2(g)$	205.0
$C_2H_6(g)$	229.5	$HC_2H_3O_2(l)$	160	$PbO(s)$	67.8
$C_8H_{18}(l)$	466.9	$Hg(l)$	76.1	$S(s)$	31.9
$C_2H_5OH(l)$	161	$Hg(g)$	175	$SO_2(g)$	248.5
$Ca(s)$	41.4	$K(s)$	64.18	$SO_3(g)$	256.2
$CaCO_3(s)$	92.9	$KCl(s)$	82.59		
$CaCl_2(s)$	114	$K_2SO_4(s)$	176		

[2] In our earlier discussions of standard states (Chapter 6) we defined the *standard pressure* as 1 atm. This was the original pressure unit. However, in the SI, the recognized unit of pressure is the pascal (Pa), not the atmosphere. After considerable discussion, the SI adopted the **bar** as the standard pressure for thermodynamic quantities: 1 bar = 10^5 Pa. One bar differs from one atmosphere by only 1.3%, and for thermodynamic quantities that we deal with in this text, their values at 1 atm and at 1 bar differ by an insignificant amount. Since most available thermodynamic data are still specified at 1 atm rather than 1 bar, we shall continue to use 1 atm for the standard pressure.

CHEMISTRY OUTSIDE THE CLASSROOM (18.1

Carved in Stone

In 1870 Ludwig von Boltzmann (Figure 1) started developing the ideas and concepts that led to the modern field of science called statistical mechanics. Statistical mechanics describes the properties of large groups of particles based on the laws of probability and how the particles interact.

tells us this is not quite true.) However, the result is that a perfect crystal at 0 K has only one possible microstate, a totally ordered arrangement of all the atoms. Solving the Boltzmann equation when $W = 1$ tells us the entropy must be zero for a perfect crystal at absolute zero. This is the third law of thermodynamics.

Figure 1 Ludwig von Boltzmann (February 20, 1844 to September 5, 1906).

Figure 2 A photo of the gravesite of Boltzmann with his equation engraved above his bust.

In developing this field, Boltzmann introduced his famous equation for entropy,

$$S = k \ln W$$

which is indeed carved in stone at his gravesite in Vienna, Austria (Figure 2). The equation states that the entropy of a particle is equal to the Boltzmann constant, $k = 1.38 \times 10^{-23}$ J mol^{-1} K^{-1}, times the natural logarithm of W, the number of different microstates that the particle can visit. The combination of the properties of all the microstates defines the macrostate, or state, of the system, which we observe on the basis of pressure, volume, temperature, and number of particles. Describing and counting microstates is beyond the scope of our discussion. However, two interesting results of Boltzmann's equation can be described easily.

First, let's consider a simple system, a perfect crystal of one of the elements. In such a crystal, there is only one arrangement of the atoms, often face-centered or body-centered cubic as shown in Chapter 11. Now let's lower the temperature to absolute zero. Since the temperature is directly proportional to the kinetic energy, this suggests that the kinetic energy, or movement, of the atoms is zero. (Quantum mechanics

A second case of interest involves entropy change. If we consider an ideal gas in a container, it will have some number of microstates, $W_{initial}$, that define its pressure, volume, temperature, and number of gas molecules. After a change in the system, we will have a different number of microstates, W_{final}. We can express the *change* as

$$\Delta S = k \ln W_{final} - k \ln W_{initial}$$
$$= k \ln (W_{final}/W_{initial})$$

For an ideal gas, the microstates are defined on the basis of the position and momentum of the particles. If we expand the gas isothermally, the temperature, and therefore momentum, mv, of the particles does not change. However, the number of positions that a particle can visit will change in direct proportion to the volume change. For example, if we take one mole of an ideal gas and triple its volume the entropy change will be

$$\Delta S = (1.38 \times 10^{-23} \text{ J molecule}^{-1} \text{ K}^{-1}) \times \ln (3)$$
$$= 1.52 \times 10^{-23} \text{ J molecule}^{-1} \text{ K}^{-1}$$

For a mole of substance we multiply by Avogadro's number

$$\Delta S = \left(\frac{1.52 \times 10^{-23} \text{ J}}{\text{molecule K}} \right)\left(\frac{6.02 \times 10^{-23} \text{ molecules}}{\text{mole}} \right)$$
$$\Delta S = 9.13 \text{ J mol}^{-1} \text{ K}^{-1}$$

Since we expanded one mole of gas, the entropy change is 9.13 J K^{-1}.

Calculating $\Delta S°$ for a Reaction

Once we have the standard entropies of a variety of substances, we can calculate the **standard entropy change**, $\Delta S°$, for chemical reactions in much the same way as we calculated $\Delta H°$ in Chapter 6.

$$\Delta S° = (\text{sum of } S° \text{ of the products}) - (\text{sum of } S° \text{ of the reactants}) \qquad (18.6)$$

If the reaction we are working with happens to correspond to the formation of 1 mol of a compound from its elements, then the $\Delta S°$ that we calculate can be referred to as the **standard entropy of formation**, $\Delta S_f°$. Values of $\Delta S_f°$ are not tabulated, however; if we need them for some purpose, we must calculate them from tabulated values of $S°$.

TOOLS

Standard entropy changes

■ This is simply a Hess's law type of calculation. Note, however, that elements have nonzero $S°$ values, which must be included in the bookkeeping.

Example 18.3
Calculating $\Delta S°$ from Standard Entropies

Urea (a compound found in urine) is manufactured commercially from CO_2 and NH_3. One of its uses is as a fertilizer, where it reacts slowly with water in the soil to produce ammonia and carbon dioxide. The ammonia provides a source of nitrogen for growing plants.

$$CO(NH_2)_2(aq) + H_2O(l) \longrightarrow CO_2(g) + 2NH_3(g)$$

urea

What is the standard entropy change when one mole of urea reacts with water?

Analysis: We need to combine standard entropies in the way that represents what occurs in the chemical reaction. Notice the similarity to the methods we learned for the determination of standard heats of reaction in Chapter 6.

Assembling the Tools: We need to use Equation 18.6 to compute the standard entropy change for the reaction. To do this we need the balanced chemical equation, given above, and the data found in Table 18.1 that are listed in the table to the right.

Substance	$S°$ (J mol^{-1} K^{-1})
$CO(NH_2)_2(aq)$	173.8
$H_2O(l)$	69.96
$CO_2(g)$	213.6
$NH_3(g)$	192.5

Solution: Applying Equation 18.6, we have

$$\Delta S° = [S°_{CO_2(g)} + 2\,S°_{NH_3(g)}] - [S°_{CO(NH_2)_2(aq)} + S°_{H_2O(l)}]$$

$$= \left[1 \text{ mol } CO_2 \times \left(\frac{213.6 \text{ J}}{(\text{mol } CO_2) \text{ K}}\right) + 2 \text{ mol } NH_3 \times \left(\frac{192.5 \text{ J}}{(\text{mol } NH_3) \text{ K}}\right)\right]$$

$$- \left[1 \text{ mol } CO(NH_2)_2 \times \left(\frac{173.8 \text{ J}}{(\text{mol } CO(NH_2)_2) \text{ K}}\right) + 1 \text{ mol } H_2O \times \left(\frac{69.96 \text{ J}}{(\text{mol } H_2O) \text{ K}}\right)\right]$$

$$= (598.6 \text{ J/K}) - (243.8 \text{ J/K})$$

$$= 354.8 \text{ J/K}$$

Thus, the standard entropy change for this reaction is $+354.8$ J/K (which we can also write as $+354.8$ J K^{-1}).

Is the Answer Reasonable? In the reaction, gases are formed from liquid reactants. Since gases have much larger entropies than liquids, we expect $\Delta S°$ to be positive, which agrees with our answer.

■ Notice that the unit *mol* cancels in each term, so the units of $\Delta S°$ are joules per kelvin.

Calculate $\Delta S_f°$ for $NH_3(g)$. (*Hint:* Write the equation for the reaction; see Section 6.8.)

Calculate the standard entropy change, $\Delta S°$, in J K^{-1} for the following reactions:

(a) $CaO(s) + 2HCl(g) \longrightarrow CaCl_2(s) + H_2O(l)$
(b) $C_2H_4(g) + H_2(g) \longrightarrow C_2H_6(g)$

Practice Exercise 18.12

Practice Exercise 18.13

18.6 | Standard Free Energy Change, ΔG°

TOOLS

Calculating ΔG° from ΔH° and ΔS°

When ΔG is determined at 25 °C (298 K) and 1 atm, we call it the **standard free energy change**,[3] ΔG°. There are several ways of obtaining ΔG° for a reaction. One of them is to compute ΔG° from ΔH° and ΔS° using an expression analogous to Equation 18.4.

$$\Delta G^\circ = \Delta H^\circ - (298.15 \text{ K}) \Delta S^\circ$$

The value of ΔG° can also be calculated from experimentally determined equilibrium constants, and we will discuss how this is done later.

Example 18.4
Calculating ΔG° from ΔH° and ΔS°

Calculate ΔG° for the reaction of urea with water from values of ΔH° and ΔS°.

$$CO(NH_2)_2(aq) + H_2O(l) \longrightarrow CO_2(g) + 2NH_3(g)$$

Analysis: This problem has two parts. From Chapter 6 we calculated the standard heat of reaction, ΔH°, and in this chapter we found out how to determine ΔS°. Once we have ΔH° and ΔS°, we need to combine them in the correct way to obtain ΔG°.

Assembling the Tools: We can calculate ΔG° with the equation

$$\Delta G^\circ = \Delta H^\circ - 298.15 \text{ K} \Delta S^\circ$$

Before calculating any values we need the balanced chemical equation, which is given in the statement of the problem. To calculate ΔH°, we use Hess's law as a tool with the data in Table 6.2. To obtain ΔS°, we would do a similar calculation using Equation 18.6 as a tool with data from Table 18.1. However, we already performed this calculation in Example 18.3.

Solution: First we calculate ΔH° from data in Table 6.2.

$$\Delta H^\circ = \left[\Delta H^\circ_{f\,CO_2(g)} + 2\Delta H^\circ_{f\,NH_3(g)}\right] - \left[\Delta H^\circ_{f\,CO(NH_2)_2(aq)} + 2\Delta H^\circ_{f\,H_2O(l)}\right]$$

$$= \left[1 \text{ mol} \times \left(\frac{-393.5 \text{ kJ}}{\text{mol}}\right) + 2 \text{ mol} \times \left(\frac{-46.19 \text{ kJ}}{\text{mol}}\right)\right]$$

$$- \left[1 \text{ mol} \times \left(\frac{-319.2 \text{ kJ}}{\text{mol}}\right) + 1 \text{ mol} \times \left(\frac{-285.9 \text{ kJ}}{\text{mol}}\right)\right]$$

$$= (-485.9 \text{ kJ}) - (-605.1 \text{ kJ})$$

$$= +119.2 \text{ kJ}$$

In Example 18.3 we found ΔS° to be +354.8 J K^{-1}. To calculate ΔG° we also need the Kelvin temperature, which we must express to at least four significant figures to match the number of significant figures in ΔS°. Since standard temperature is *exactly* 25 °C, $T_K = (25.00 + 273.15) \text{ K} = 298.15 \text{ K}$. Also, we must be careful to express the energy units of ΔH° and $T \Delta S^\circ$ with the same SI prefix so they combine correctly. We'll change ΔS° from +354.8 J K^{-1} to +0.3548 kJ K^{-1}. Substituting into the equation for ΔG°,

---

[3] Sometimes, the temperature is specified as a subscript in writing the symbol for the standard free energy change. For example, ΔG° can also be written ΔG°_{298}. As you will see later, there are times when it is desirable to indicate the temperature explicitly.

$$\Delta G° = +119.2 \text{ kJ} - (298.15 \text{ K})(0.3548 \text{ kJ K}^{-1})$$

$$= +119.2 \text{ kJ} - 105.8 \text{ kJ}$$

$$= +13.4 \text{ kJ}$$

Therefore, for this reaction, $\Delta G° = +13.4$ kJ.

● **Is the Answer Reasonable?** There's not much we can do to check the reasonableness of the answer. We just need to check to be sure we've done the calculations correctly.

Calculate $\Delta G°$ for the formation of one mole of N_2O_4 from its elements using $\Delta H_f°$ and $\Delta S_f°$ for N_2O_4. (*Hint:* Write the chemical equation.)

Practice Exercise 18.14

Use the data in Table 6.2 and Table 18.1 to calculate $\Delta G°$ for the formation of iron(III) oxide (the iron oxide in rust). The equation for the reaction is

Practice Exercise 18.15

$$4Fe(s) + 3O_2(g) \longrightarrow 2Fe_2O_3(s)$$

In Section 6.8 you learned that it is useful to have tabulated standard heats of formation, $\Delta H_f°$, because they can be used with Hess's law to calculate $\Delta H°$ for many different reactions. **Standard free energy of formation**, $\Delta G_f°$, can be used in similar calculations to obtain $\Delta G°$ for a reaction or physical change.

$$\Delta G° = (\text{sum of } \Delta G_f° \text{ of products}) - (\text{sum of } \Delta G_f° \text{ of reactants}) \quad \textbf{(18.7)}$$

TO☐LS

Calculating $\Delta G°$ using $\Delta G_f°$ values

The $\Delta G_f°$ values for some typical substances are listed in Table 18.2. Example 18.5 shows how we can use them to calculate $\Delta G°$ for a reaction.

TABLE 18.2	Standard Free Energies of Formation of Typical Substances at 298.15 K				
Substance	**$\Delta G_f°$ (kJ mol^{-1})**	**Substance**	**$\Delta G_f°$ (kJ mol^{-1})**	**Substance**	**$\Delta G_f°$ (kJ mol^{-1})**
Ag(s)	0	CaO(s)	−604.2	K$_2$SO$_4$(s)	−1316.4
AgCl(s)	−109.7	Ca(OH)$_2$(s)	−896.76	N$_2$(g)	0
Al(s)	0.00	CaSO$_4$(s)	−1320.3	NH$_3$(g)	−16.7
Al$_2$O$_3$(s)	−1576.4	CaSO$_4 \cdot \frac{1}{2}$H$_2$O(s)	−1435.2	NH$_4$Cl(s)	−203.9
C(s) (graphite)	0	CaSO$_4$·2H$_2$O(s)	−1795.7	NO(g)	+ 86.69
CO(g)	−137.3	Cl$_2$(g)	0	NO$_2$(g)	+51.84
CO$_2$(g)	−394.4	Fe(s)	0	N$_2$O(g)	+103.6
CH$_4$(g)	−50.79	Fe$_2$O$_3$(s)	−741.0	N$_2$O$_4$(g)	+98.28
CH$_3$Cl(g)	−58.6	H$_2$(g)	0	Na(s)	0
CH$_3$OH(l)	−166.2	H$_2$O(g)	−228.6	Na$_2$CO$_3$(s)	−1048
CO(NH$_2$)$_2$(s)	−197.2	H$_2$O(l)	−237.2	NaHCO$_3$(s)	−851.9
CO(NH$_2$)$_2$(aq)	−203.8	HCl(g)	−95.27	NaCl(s)	−384.0
C$_2$H$_2$(g)	+209	HNO$_3$(l)	−79.91	NaOH(s)	−382
C$_2$H$_4$(g)	+68.12	H$_2$SO$_4$(l)	−689.9	Na$_2$SO$_4$(s)	−1266.8
C$_2$H$_6$(g)	−32.9	HC$_2$H$_3$O$_2$(l)	−392.5	O$_2$(g)	0
C$_2$H$_5$OH(l)	−174.8	Hg(l)	0	PbO(s)	−189.3
C$_8$H$_{18}$(l)	+17.3	Hg(g)	+31.8	S(s)	0
Ca(s)	0	K(s)	0	SO$_2$(g)	−300.4
CaCO$_3$(s)	−1128.8	KCl(s)	−408.3	SO$_3$(g)	−370.4
CaCl$_2$(s)	−750.2				

Example 18.5
Calculating $\Delta G°$ from $\Delta G_f°$

© Daniel Acker/Bloomberg/ Getty Images

Ethanol, C_2H_5OH, is made from grain by fermentation and is used as an additive to gasoline to produce a fuel mix called E85 (85% ethanol, 15% gasoline). What is $\Delta G°$ for the combustion of liquid ethanol to give $CO_2(g)$ and $H_2O(g)$?

Analysis: The process for determining $\Delta G°$ is the same as we used for determining $\Delta H°$ and $\Delta S°$.

Assembling the Tools: Our tool is Equation 18.7.

$$\Delta G° = (\text{sum of } \Delta G_f° \text{ of products}) - (\text{sum of } \Delta G_f° \text{ of reactants})$$

To use it we will need the balanced chemical equation for the reaction and $\Delta G_f°$ for each reactant and product. The free energy data are available in Table 18.2.

Solution: First, we write the balanced equation for the reaction.

$$C_2H_5OH(l) + 3O_2(g) \longrightarrow 2CO_2(g) + 3H_2O(g)$$

Applying Equation 18.7, we have

$$\Delta G° = [2\Delta G_{f\,CO_2(g)}° + 3\Delta G_{f\,H_2O(g)}°] - [\Delta G_{f\,C_2H_5O(l)}° + 3\Delta G_{f\,O_2(g)}°]$$

As with $\Delta H_f°$, the $\Delta G_f°$ for any element in its standard state is zero. Therefore, using the data from Table 18.2,

$$\Delta G° = \left[2 \text{ mol} \times \left(\frac{-394.4 \text{ kJ}}{\text{mol}} \right) + 3 \text{ mol} \times \left(\frac{-228.6 \text{ kJ}}{\text{mol}} \right) \right]$$
$$- \left[1 \text{ mol} \times \left(\frac{-174.8 \text{ kJ}}{\text{mol}} \right) + 3 \text{ mol} \times \left(\frac{0 \text{ kJ}}{\text{mol}} \right) \right]$$
$$= (-1474.6 \text{ kJ}) - (-174.8 \text{ kJ})$$
$$= -1299.8 \text{ kJ}$$

The standard free energy change for the reaction equals -1299.8 kJ.

Is the Answer Reasonable? This is a combustion reaction for a compound containing only C, H, and O. Very often, such combustion reactions are spontaneous with negative values for $\Delta G°$. Our answer is obviously in the right ballpark.

Practice Exercise 18.16

Calculate $\Delta G°$ for the reaction of iron(III) oxide with carbon monoxide to give elemental iron and carbon dioxide. (*Hint:* Be careful writing the balanced chemical equation.)

Practice Exercise 18.17

Calculate $\Delta G_{\text{reaction}}°$ in kilojoules for the following reactions using the data in Table 18.2.
(a) $2NO(g) + O_2(g) \longrightarrow 2NO_2(g)$
(b) $Ca(OH)_2(s) + 2HCl(g) \longrightarrow CaCl_2(s) + 2H_2O(g)$

18.7 | Maximum Work and ΔG

One of the chief uses of spontaneous chemical reactions is the production of useful work. For example, fuels are burned in gasoline or diesel engines to power automobiles and heavy machinery, and chemical reactions in batteries start our autos and run all sorts of modern electronic gadgets, including cell phones, iPods, and laptop computers.

When chemical reactions occur, however, their energy is not always harnessed to do work. For instance, if gasoline is burned in an open dish, the energy evolved is lost entirely as heat and no useful work is accomplished. Engineers, therefore, seek ways to capture as much energy as possible in the form of work. One of their primary goals is to maximize the efficiency with which chemical energy is converted to work and to minimize the amount of energy transferred unproductively to the environment as heat.

Scientists have discovered that the maximum conversion of chemical energy to work occurs if a reaction is carried out under conditions that are said to be thermodynamically reversible. A process is defined as **thermodynamically reversible** if its driving force is opposed by another force that is just the slightest bit weaker, so that the slightest increase in the opposing force will cause the direction of the change to be reversed. An example of a nearly **reversible process** is illustrated in Figure 18.14, where we have a compressed gas in a cylinder pushing against a piston that's held in place by liquid water above it. If a water molecule evaporates, the external pressure drops slightly and the gas can expand just a bit, performing a small amount of work. Gradually, as one water molecule after another evaporates, the gas inside the cylinder slowly expands and performs work on the surroundings. At any time, however, the process can be reversed by the condensation of a water molecule.[4]

Although we could obtain the maximum work by carrying out a change reversibly, a thermodynamically reversible process requires so many steps that it proceeds at an extremely slow speed. If the work cannot be done at a reasonable rate, it is of little value to us. Our goal, then, is to approach thermodynamic reversibility for maximum efficiency, but to carry out the change at a pace that will deliver work at acceptable rates.

The relationship of useful work to thermodynamic reversibility was illustrated earlier (Section 6.5) in our discussion of the discharge of an automobile battery. Recall that when the battery is short-circuited with a heavy wrench, no work is done and all the energy appears as heat. In this case there is nothing opposing the discharge, and it occurs in the most thermodynamically irreversible manner possible. However, when the current is passed through a small electric motor, the motor itself offers resistance to the passage of the electricity and as the battery is discharged, some of the energy is released as work. In this instance, the discharge occurs in a more nearly thermodynamically reversible manner because of the opposition provided by the motor, and a relatively large amount of the available energy appears in the form of the work accomplished by the motor.

The preceding discussion leads naturally to the question: Is there a limit to the amount of the available energy in a reaction that can be harnessed as useful work? The answer to this question is to be found in the Gibbs free energy.

■ Gas and diesel engines are not very efficient, so most of the energy produced in the combustion of fuel appears as heat, not work. That's why these engines require cooling systems.

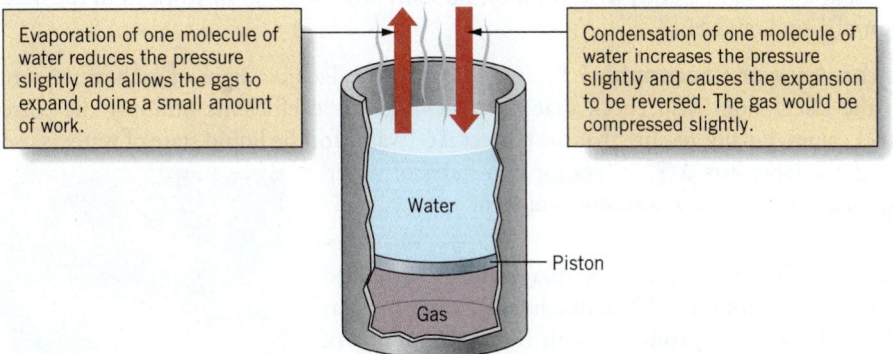

Figure 18.14 | **A reversible expansion of a gas.** As water molecules evaporate one at a time, the external pressure gradually decreases and the gas slowly expands, performing a small amount of work in each step. The process would be reversed if a molecule of water were to condense into the liquid. The ability of the expansion to be reversed by the slightest increase in the opposing pressure is what makes this a reversible process.

---

[4] Although we sometimes say that a chemical reaction is "reversible" because it can run in both the forward and reverse directions, we cannot say the reaction is *thermodynamically reversible* unless the concentrations are only infinitesimally different from their equilibrium values as the reaction occurs.

The maximum amount of energy produced by a reaction that can be *theoretically* harnessed as work is equal to ΔG.

This is the energy that need not be lost to the surroundings as heat and is therefore *free* to be used for work. Thus, by determining the value of ΔG, we can find out whether or not a given reaction will be an effective source of useful energy. Also, by comparing the actual amount of work derived from a given system with the ΔG values for the reactions involved, we can measure the efficiency of the system.

Example 18.6
Calculating Maximum Work

Calculate the maximum work available, expressed in kilojoules, from the oxidation of 1 mol of octane, $C_8H_{18}(l)$, by oxygen to give $CO_2(g)$ and $H_2O(l)$ at 25 °C and 1 atm.

Analysis: The maximum work is equal to ΔG for the reaction. Standard thermodynamic conditions are specified, so we need to determine $\Delta G°$.

Assembling the Tools: The tool for this calculation is Equation 18.7.

$$\Delta G° = (\text{sum of } \Delta G_f° \text{ of products}) - (\text{sum of } \Delta G_f° \text{ of reactants})$$

Solution: First, we need a balanced equation for the reaction. For the complete oxidation (combustion) of one mole of C_8H_{18} we have

$$C_8H_{18}(l) + 12\tfrac{1}{2}O_2(g) \longrightarrow 8CO_2(g) + 9H_2O(l)$$

Then we apply Equation 18.7:

$$\Delta G° = [8\Delta G_f°{}_{CO_2\,(g)} + 9\Delta G_f°{}_{H_2O(l)}] - [\Delta G_f°{}_{C_8H_{18}(l)} + 12.5\Delta G_f°{}_{O_2(g)}]$$

Referring to Table 18.2 and dropping the canceled mol units,

$$\Delta G° = [8 \times (-394.4)\text{ kJ} + 9 \times (-237.2)\text{ kJ}] - [1 \times (+17.3)\text{ kJ} + 12.5 \times (0)\text{ kJ}]$$

$$= (-5290\text{ kJ}) - (+17.3\text{ kJ})$$

$$= -5307\text{ kJ}$$

Thus, at 25 °C and 1 atm, we can expect no more than 5307 kJ of work from the oxidation of 1 mol of C_8H_{18}.

Is the Answer Reasonable? Be sure to check the algebraic signs of each of the terms in the calculation. Also check that the stoichiometric coefficients multiply the correct $\Delta G°$ values. Finally, be sure that the correct $\Delta G°$ value for the liquid state of water is used. Since the table lists $\Delta G°$ values for two states of water, liquid and gas, this is a common source of error.

Practice Exercise 18.18

Calculate the maximum work that could be obtained from the combustion of 125 g of ethanol, $C_2H_5OH(l)$ at 25 °C. How does it compare with the maximum work from the combustion of 125 g of $C_8H_{18}(l)$? Which is the better fuel? (*Hint:* The work available *per mole* has already been calculated in the worked examples.)

Practice Exercise 18.19

Calculate the maximum work that could be obtained at 25 °C and 1 atm from the oxidation of 1.00 mol of aluminum by $O_2(g)$ to give $Al_2O_3(s)$. (The combustion of aluminum in booster rockets provides part of the energy that lifts the space shuttle off its launching pad, Figure 18.15.)

Figure 18.15 | **Blastoff of the space shuttle.** The large negative heat of formation of Al_2O_3 provides power to the solid-state booster rockets that lift the space shuttle from its launch pad.

John Raoux/AP Photo

ON THE CUTTING EDGE 18.2

Thermodynamic Efficiency and Sustainability

The concept of sustainability incorporates the idea that the human race must consider the cumulative effects of our everyday activities and reinvent these activities so that we will not drive ourselves into extinction. It has been stated that 99.9% of all species that ever lived are now extinct. Humans have the possibility, because of our ability to think and plan, to avoid this fate. Chemists, physicists, and biologists will be very important in developing the tools that will make this effort successful.

Thermodynamics enters this discussion because it sets limits on our ability to use resources. For instance, the first law of thermodynamics suggests that a change in internal energy is the sum of the work and heat changes in the system. If work is the useful part of an energy change and heat is wasted energy, we can say that $w/(q + w)$ is the efficiency of the system. The efficiencies of some devices are shown in Table 1. There are other measures of efficiency. For example, Nicolas Léonard Sadi Carnot suggested that the efficiency of a heat engine is $(T_{high} - T_{low})/T_{high}$, where the high temperature exists inside the engine and the low temperature is the temperature of the exhaust. The best Carnot efficiency is realized in engines that have the greatest possible difference between the operating temperature and the waste heat temperature. Table 2 lists the efficiencies of some heat engines.

TABLE 1	Efficiencies of Various Devices
Tungsten light bulb	2–3%
Candle	0.04%
Gas mantle	0.2%
Light-emitting diode	Up to 22%
Fluorescent tube	Up to 15%
Compact fluorescent bulb	Up to 12%
Electric motors	75 to 90%

Engineers and research scientists realize that one part of achieving a sustainable energy economy is to maximize efficiency.

Figure 1 Combustion produces energy with a low efficiency and increased amounts of pollutants.

Figure 2 Wind turbines convert wind energy to electricity with little on-site pollution.

TABLE 2	Approximate Heat Engine Efficiencies
Gasoline engines	25–30%
Steam engines	8%
Diesel engines	40%
Steam turbines	60%

Modern sustainability calculations of efficiency include, along with the thermodynamic efficiency, the costs of all components of the process. For an electric generating facility, it includes the cost of the land that the facility uses, the cost of constructing the facility, the cost of waste disposal, and the cost of environmental impacts such as CO_2, NO_x, SO_2, and heat emissions to the environment (Figures 1 and 2). Employee, management, and regulatory costs are also included. The difficulty in sustainability calculations is finding an appropriate method for adding all these factors into a final, meaningful value.

With the modern realization of the many factors that comprise a simple activity such as the generation of electricity, there are many opportunities for chemists and other scientists to make significant contributions. Advanced lubricants can help increase the efficiency of many mechanical devices. Chemists and materials scientists design and fabricate materials that can withstand extremely high temperatures for long periods of time. Durable materials increase efficiency by reducing the "down-time" for frequent repairs. Although tungsten has a melting point of 3683 K, it is not used in electric turbines because it is too brittle. However, tungsten nitride and silicon nitride are used successfully as very durable coatings in high-temperature generators. These same materials are used in the turbines of modern jet engines in commercial aircraft (Figure 3).

Figure 3 Cutaway of a jet engine showing high-speed, high-temperature turbine blades.

18.8 | Free Energy and Equilibrium

We have seen that when the value of ΔG for a given change is negative, the change occurs spontaneously, and have also seen that a change is nonspontaneous when ΔG is positive. However, when ΔG is neither positive nor negative, the change is neither spontaneous nor nonspontaneous—the system is in a state of equilibrium. This occurs when ΔG is equal to zero.

When a system is in a state of dynamic equilibrium,

$$G_{products} = G_{reactants} \quad \text{and} \quad \Delta G = 0$$

Let's again consider the freezing of water:

$$H_2O(l) \rightleftharpoons H_2O(s)$$

Below 0 °C, ΔG for this change is negative and the freezing is spontaneous. On the other hand, above 0 °C we find that ΔG is positive and freezing is nonspontaneous. When the temperature is exactly 0 °C, $\Delta G = 0$ and an ice–water mixture exists in a condition of equilibrium. As long as heat isn't added or removed from the system, neither freezing nor melting is spontaneous and the ice and liquid water can exist together indefinitely.

Equilibrium and Work

We have identified ΔG as a quantity that specifies the amount of work that is available from a system. Since ΔG is zero at equilibrium, the amount of work available is zero also. Therefore, when a system is at equilibrium, no work can be extracted from it. As an example, consider again the common lead storage battery that we use to start our cars.

When the battery is fully charged, there are virtually no products of the discharge reaction present. The chemical reactants, however, are present in large amounts. Therefore, the total free energy of the reactants far exceeds the total free energy of products and, since $\Delta G = G_{products} - G_{reactants}$, the ΔG of the system has a large negative value. This means that a lot of energy is available to do work. As the battery discharges, the reactants are converted to products and $G_{products}$ gets larger while $G_{reactants}$ gets smaller. The result is that ΔG becomes less negative, and less energy is available to do work. Finally, the battery reaches equilibrium. The total free energies of the reactants and the products have become equal, so $G_{products} - G_{reactants} = 0$ and $\Delta G = 0$. No further work can be extracted and we say the battery is "dead."

Estimating Melting and Boiling Points

For a phase change such as $H_2O(l) \longrightarrow H_2O(s)$, equilibrium can only exist at one particular temperature at atmospheric pressure. For water, that temperature is 0 °C. Above 0 °C, only liquid water can exist, and below 0 °C all the liquid will freeze to give ice. This yields an interesting relationship between ΔH and ΔS for a phase change. Since $\Delta G = 0$,

$$\Delta G = 0 = \Delta H - T\Delta S$$

Therefore,

$$\Delta H = T\Delta S$$

and

$$\Delta S = \frac{\Delta H}{T} \tag{18.8}$$

Thus, if we know ΔH for the phase change and the temperature at which the two phases coexist, we can calculate ΔS for the phase change. Another interesting relationship that we can obtain is

$$T = \frac{\Delta H}{\Delta S} \tag{18.9}$$

Thus, if we know ΔH and ΔS, we can calculate the temperature at which equilibrium will occur.

Example 18.7
Calculating the Equilibrium Temperature for a Phase Change

For the phase change $Br_2(l) \longrightarrow Br_2(g)$, $\Delta H° = +31.0$ kJ mol^{-1} and $\Delta S° = 92.9$ J mol^{-1} K^{-1}. Assuming that ΔH and ΔS are nearly temperature independent, calculate the approximate Celsius temperature at which $Br_2(l)$ will be in equilibrium with $Br_2(g)$ at 1 atm (i.e., the normal boiling point of liquid Br_2).

Analysis: The normal boiling point of a substance is the temperature at which the liquid and gas phases are in equilibrium at one atmosphere of pressure; see Section 11.5. The wording of the problem strongly suggests that we can substitute $\Delta H°$ and $\Delta S°$ for ΔH and ΔS, respectively. If we do that we can use the equations we just discussed.

■ ΔH and ΔS do not change much with changes in temperature. This is because temperature changes affect the enthalpies and entropies of both the reactants and products by about the same amount, so the differences between reactants and products stays fairly constant.

Assembling the Tools: The temperature at which equilibrium exists is given by Equation 18.9. When we assume that ΔH and ΔS do not depend much on temperature, then we can use $\Delta H°$ and $\Delta S°$ in this equation to approximate the boiling point. That is,

$$T \approx \frac{\Delta H°}{\Delta S°}$$

Solution: Substituting the data given in the problem,

$$T \approx \frac{3.10 \times 10^4 \text{ J mol}^{-1}}{92.9 \text{ J mol}^{-1} \text{ K}^{-1}}$$

$$= 334 \text{ K}$$

The Celsius temperature is $334 - 273 = 61$ °C. Notice that we were careful to express $\Delta H°$ in joules, not kilojoules, so the units would cancel correctly. It is also interesting that the boiling point we calculated is quite close to the measured normal boiling point of 58.8 °C.

Is the Answer Reasonable? The $\Delta H°$ value equals 31,000 and the $\Delta S°$ value equals approximately 100, which means the temperature should be about 310 K. Our value, 334 K, is not far from that, so the answer is reasonable.

The heat of vaporization of ammonia is 21.7 kJ mol^{-1} and the boiling point of ammonia is -33.3 °C. Estimate the entropy change for the vaporization of liquid ammonia. (*Hint:* What algebraic sign do we expect for this entropy change?)

Practice Exercise 18.20

The heat of vaporization of mercury is 60.7 kJ/mol. For $Hg(l)$, $S° = 76.1$ J mol^{-1} K^{-1}, and for $Hg(g)$, $S° = 175$ J mol^{-1} K^{-1}. Estimate the normal boiling point of liquid mercury.

Practice Exercise 18.21

Free Energy Diagrams

One way to gain a better understanding of how the free energy changes during a reaction is by studying **free energy diagrams**. As an example, let's study a reaction you've seen before—the decomposition of N_2O_4 into NO_2.

$$N_2O_4(g) \longrightarrow 2NO_2(g)$$

In our discussion of chemical equilibrium in Section 14.1, we noted that equilibrium in this system can be approached from *either* direction, with the same equilibrium concentrations being achieved provided we begin with the same overall system composition.

Figure 18.16 shows the free energy diagram for the reaction, which depicts how the free energy changes as we proceed from the reactant to the product. On the left of the diagram

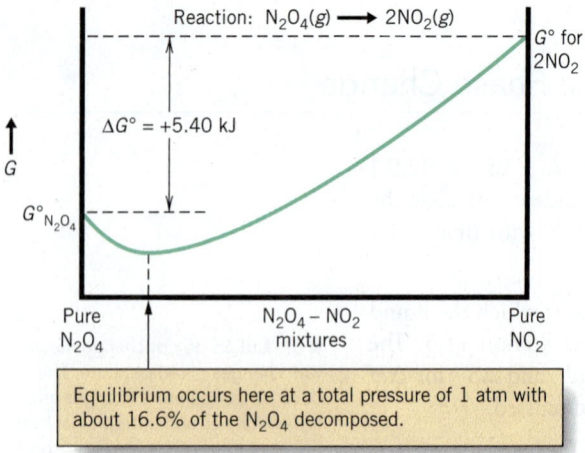

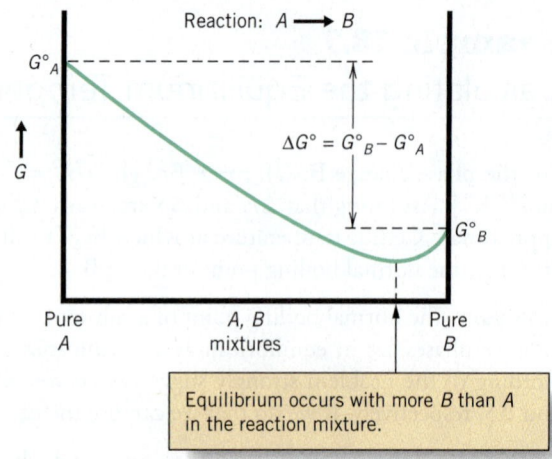

Figure 18.16 | **Free energy diagram for the decomposition of $N_2O_4(g)$.** The minimum on the curve indicates the composition of the reaction mixture at equilibrium. Because $\Delta G°$ is positive, the position of equilibrium lies close to the reactants. Not much product will form by the time the system reaches equilibrium.

Figure 18.17 | **Free energy curve for a reaction having a negative $\Delta G°$.** Because $G_B°$ is lower than $G_A°$, $\Delta G°$ is negative. This causes the position of equilibrium to lie far to the right, near the products. When the system reaches equilibrium, there will be a large amount of products present and few reactants.

we have the free energy of one mole of pure $N_2O_4(g)$, and on the right the free energy of two moles of pure $NO_2(g)$. Points along the horizontal axis represent mixtures of both substances. Notice that in going from reactant (N_2O_4) to product ($2NO_2$), the free energy has a minimum. It drops below that of either pure N_2O_4 or pure NO_2.

Any system will spontaneously seek the lowest point on its free energy curve. If we begin with pure $N_2O_4(g)$, the reaction will proceed from left to right and some $NO_2(g)$ will be formed, because proceeding in the direction of NO_2 leads to a lowering of the free energy. If we begin with pure $NO_2(g)$, a change also will occur. Going downhill on the free energy curve now takes place as the reverse reaction occurs [i.e., $2NO_2(g) \longrightarrow N_2O_4(g)$]. Once the bottom of the "valley" is reached, the system has come to equilibrium. As you learned in Sections 11.8 and 14.6, if the system isn't disturbed, the composition of the equilibrium mixture will remain constant. Now we see that the reason is because any change (moving either to the left or right) would require an uphill climb. Free energy increases are not spontaneous, so this doesn't happen.

■ When a system moves "downhill" on its free energy curve, $G_{final} < G_{initial}$ and ΔG is negative. Changes with negative ΔG are spontaneous.

An important thing to notice in Figure 18.16 is that some reaction takes place spontaneously in the forward direction even though $\Delta G°$ is positive. However, the reaction doesn't proceed far before equilibrium is reached. For comparison, Figure 18.17 shows the shape of the free energy curve for a reaction with a negative $\Delta G°$. We see here that at equilibrium there has been a much greater conversion of reactants to products. Thus, $\Delta G°$ *informs us where the position of equilibrium lies between pure reactants and pure products.*

$\Delta G°$ and the Position of Equilibrium

Using $\Delta G°$ to assess the position of equilibrium

- When $\Delta G°$ is positive, the position of equilibrium lies close to the reactants and little reaction occurs by the time equilibrium is reached. The reaction will appear to be non-spontaneous.

- When $\Delta G°$ is negative, the position of equilibrium lies close to the products and a large amount of products will have formed by the time equilibrium is reached. The reaction will appear to be spontaneous.

- When $\Delta G° = 0$, the position of equilibrium will lie about midway between reactants and products. Substantial amounts of both reactants and products will be present when equilibrium is reached. The reaction will appear to be spontaneous whether we begin with pure reactants or pure products.

$\Delta G°$ and the Position of Equilibrium

In general, the value of $\Delta G°$ for most reactions is much larger numerically than the $\Delta G°$ for the $N_2O_4 \longrightarrow NO_2$ reaction. In addition, the extent to which a reaction proceeds is very sensitive to the size of $\Delta G°$. If the $\Delta G°$ value for a reaction is reasonably large—about 20 kJ or more—almost no observable reaction will occur when $\Delta G°$ is positive. On the other hand, the reaction will go almost to completion if $\Delta G°$ is both large and negative.[5] From a practical standpoint, then, *the size and sign of $\Delta G°$ serve as indicators of whether an observable spontaneous reaction will occur.*

Example 18.8
Using $\Delta G°$ as a Predictor of the Outcome of a Reaction

Would we expect to be able to observe the following reaction at 25 °C?

$$NH_4Cl(s) \longrightarrow NH_3(g) + HCl(g)$$

Analysis: To answer this question we need to know the magnitude and algebraic sign of $\Delta G°$; that will allow us to estimate the position of equilibrium.

Assembling the Tools: The relationship between $\Delta G°$ and the position of equilibrium is one tool needed to answer this question. The other tool is Equation 18.7, which is needed to obtain the standard free energy change, $\Delta G°$.

Solution: First, let's calculate $\Delta G°$ for the reaction using the data in Table 18.2. The procedure is the same as that discussed earlier.

$$\Delta G° = [\Delta G°_{f\,NH_3(g)} + [\Delta G°_{f\,HCl(g)}] - [\Delta G°_{f\,NH_4Cl(s)}]$$
$$= [1 \text{ mol} \times (-16.7 \text{ kJ/mol}) + 1 \text{ mol} \times (-95.27 \text{ kJ/mol})]$$
$$- [1 \text{ mol} \times (-203.9 \text{ kJ/mol})]$$
$$= +91.9 \text{ kJ}$$

Because $\Delta G°$ is large and positive, only extremely small amounts of products can form at this temperature. Therefore, we would not expect to observe any decomposition of the NH_4Cl.

Is the Answer Reasonable? Be sure to check the algebraic signs of each term in the calculation. Also check that the free energies of the reactants are subtracted from those of the products. This problem involves adding (and subtracting) numbers. If we add all the raw numbers, without regard to sign, that should be our maximum possible positive answer. Similarly, the negative of this sum will be the largest possible negative value we could have. Our answer easily passes that test.

Use the data in Table 18.2 to determine whether the reaction

$$SO_2(g) + \tfrac{1}{2}O_2(g) \longrightarrow SO_3(g)$$

should "occur spontaneously" at 25 °C. (*Hint:* Be careful with algebraic signs.)

Practice Exercise 18.22

Use the data in Table 18.2 to determine whether we should expect to see the formation of $CaCO_3(s)$ in the following reaction at 25 °C.

$$CaCl_2(s) + H_2O(g) + CO_2(g) \longrightarrow CaCO_3(s) + 2HCl(g)$$

Practice Exercise 18.23

[5] As we discussed earlier, to actually see a change take place, the speed of a spontaneous reaction must be reasonably fast. For example, the decomposition of the nitrogen oxides into N_2 and O_2 is thermodynamically spontaneous ($\Delta G°$ is negative), but their rates of decomposition are so slow that these substances appear to be stable.

$\Delta G°$ Varies with Temperature

So far, we have confined our discussion of the relationship of free energy and equilibrium to a special case, 25 °C. But what about other temperatures? Equilibria certainly can exist at temperatures other than 25 °C, and in Chapter 14 you learned how to apply Le Châtelier's principle to predict the way temperature affects the position of equilibrium. Now let's see how thermodynamics deals with this.

You've learned that at 25 °C the position of equilibrium is determined by the difference between the free energy of pure products and the free energy of pure reactants. This difference is given by $\Delta G°_{298}$, where we have now used the subscript "298" to indicate the temperature, 298 K. We define $\Delta G°_{298}$ as

$$\Delta G°_{298} = (\Delta G°_{298})_{product} - (\Delta G°_{298})_{reactant}$$

At temperatures other than 25 °C, it is still the difference between the free energies of the products and reactants that determines the position of equilibrium. We write this as $\Delta G°_T$. Thus, at a temperature other than 25 °C (298 K), we have

$$\Delta G°_T = (\Delta G°_T)_{product} - (\Delta G°_T)_{reactant}$$

where $(\Delta G°_T)_{product}$ and $(\Delta G°_T)_{reactant}$ are the total free energies of the pure products and reactants, respectively, at this other temperature.

Next, we must find a way to compute $\Delta G°_T$. Earlier we saw that $\Delta G°$ can be obtained from the equation

$$\Delta G° = \Delta H° - (298 \text{ K}) \Delta S°$$

At a different temperature, T, the equation becomes

$$\Delta G°_T = \Delta H°_T - T \Delta S°_T$$

Now we seem to be getting closer to our goal. If we can compute or estimate the values of $\Delta H°_T$ and $\Delta S°_T$ for a reaction, we have solved our problem.

The size of $\Delta G°_T$ obviously depends very strongly on the temperature—the equation above has temperature as one of its variables. However, as noted in the margin comment in Example 18.7, the magnitudes of the ΔH and ΔS for a reaction are relatively insensitive to changes in temperature. Therefore, we can use $\Delta H°_{298}$ and $\Delta S°_{298}$ as reasonable approximations of $\Delta H°_T$ and $\Delta S°_T$. This allows us to rewrite the equation for $\Delta G°_T$ as

TO**O**LS

Calculating $\Delta G°$ at temperatures other than 25 °C

$$\boxed{\Delta G°_T \approx \Delta H°_{298} - T \Delta S°_{298}} \tag{18.10}$$

The following example illustrates how this equation is useful.

Example 18.9
$\Delta G°$ at Temperatures Other than 25 °C

Earlier we saw that at 25 °C the value of $\Delta G°$ for the reaction

$$N_2O_4(g) \longrightarrow 2NO_2(g)$$

has a value of +5.40 kJ. What is the approximate value of $\Delta G°_T$ for this reaction at 105 °C?

Analysis: In this problem we can't calculate $\Delta G°$ from standard free energies of formation because we only have such values for 25 °C. Therefore, we're going to have to use the alternative approach described above, which requires Equation 18.10 and values of $\Delta H°$ and $\Delta S°$.

Assembling the Tools: Equation 18.10,

$$\Delta G_T° \approx \Delta H_{298}° - T\Delta S_{298}°$$

is the tool needed to solve the problem. To use it, we need values of $\Delta H°$ and $\Delta S°$. The $\Delta H°$ is calculated using Equation 6.13, Hess's law. The $\Delta S°$ is determined using Equation 18.6,

$$\Delta S° = (\text{sum of } S° \text{ of the products}) - (\text{sum of } S° \text{ of the reactants})$$

Using the data in Table 6.2 we find the following standard heats of formation for use in Equation 6.13.

$$N_2O_4(g) \qquad \Delta H_f° = +9.67 \text{ kJ/mol}$$
$$NO_2(g) \qquad \Delta H_f° = +33.8 \text{ kJ/mol}$$

Next, we compute $\Delta S°$ for the reaction using absolute entropy data from Table 18.1.

$$N_2O_4(g) \qquad S° = 304 \text{ J/mol K}$$
$$NO_2(g) \qquad S° = 240.5 \text{ J/mol K}$$

Solution: First, we compute $\Delta H°$ using Hess's law and the data we just gathered:

$$\Delta H° = [2\Delta H_{f\,NO_2(g)}°] - [\Delta H_{f\,N_2O_4(g)}°]$$

$$= \left[2 \text{ mol} \times \left(\frac{33.8 \text{ kJ}}{\text{mol}}\right)\right] - \left[1 \text{ mol} \times \left[\frac{9.67 \text{ kJ}}{\text{mol}}\right]\right]$$

$$= +57.9 \text{ kJ}$$

Next, we use Equation 18.6 to calculate $\Delta S°$.

$$\Delta S° = \left[2 \text{ mol} \times \left(\frac{240.5 \text{ J}}{\text{mol K}}\right)\right] - \left[1 \text{ mol} \times \left(\frac{304 \text{ J}}{\text{mol K}}\right)\right]$$

$$= +177 \text{ J K}^{-1} \text{ or } 0.177 \text{ kJ K}^{-1}$$

The temperature is 105 °C, which is 378 K, so we can write the free energy change as $\Delta G_{378}°$. Substituting into Equation 18.10 using $T = 378$ K, we have

$$\Delta G_{378}° \approx (+57.9 \text{ kJ}) - (378 \text{ K})(0.177 \text{ kJ K}^{-1})$$

$$\approx -9.0 \text{ kJ}$$

Notice that the $\Delta G_T°$ has changed from +5.40 kJ at 25 °C to −9.0 at 105 °C.

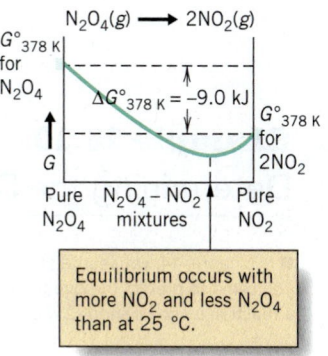

Is the Answer Reasonable? As usual, we can double-check the algebraic signs to be sure we've performed the calculations correctly. For this problem, however, we can check that our answer makes sense based on Le Châtelier's principle.

Since the reaction is endothermic, heat may be written as a reactant:

$$\text{heat} + N_2O_4(g) \rightleftharpoons 2NO_2(g)$$

When heat is added, we expect the position of equilibrium to shift toward the products. The positive value of $\Delta G_T°$ at 25 °C means that the position of equilibrium lies nearest the reactants. When $\Delta G_T°$ has a negative value at 105 °C, it means that the position of equilibrium now lies closer to the products. The result is a shift in the position of equilibrium toward the products, in agreement with our expectations.

Practice Exercise 18.24

In Examples 18.3 and 18.4 we computed $\Delta S°$ and $\Delta H°$ for the reaction

$$CO(NH_2)_2(aq) + H_2O(l) \longrightarrow CO_2(g) + 2NH_3(g)$$

What is $\Delta G_T°$ for this reaction at 75 °C? (*Hint:* Be careful with the units.)

Practice Exercise 18.25

Use the data in Table 18.2 to determine $\Delta G_{298}°$ for the reaction

$$2NaHCO_3(s) \longrightarrow Na_2CO_3(s) + CO_2(g) + H_2O(g)$$

Then calculate the approximate value for $\Delta G_{490}°$ for the reaction at 217 °C using the data in Tables 6.2 and 18.1. How does the position of equilibrium for this reaction change as the temperature is increased?

18.9 | Equilibrium Constants and $\Delta G°$

In the preceding discussion, you learned in a qualitative way that the position of equilibrium in a reaction is determined by the sign and magnitude of $\Delta G°$. You also learned that the direction in which a reaction proceeds depends on where the system composition stands relative to the minimum on the free energy curve. Thus, the reaction will proceed spontaneously in the forward direction only if it will lead to a lowering of the free energy (i.e., if ΔG is negative).

Quantitatively, the relationship between ΔG and $\Delta G°$ is expressed by the following equation, which we will not attempt to justify.

TOOLS

Relating the reaction quotient to ΔG

■ In Chapter 14 we defined *Q* as the *reaction quotient,* the numerical value of the mass action expression.

$$\Delta G = \Delta G° + RT \ln Q \qquad (18.11)$$

Here R is the gas constant in appropriate energy units (i.e., 8.314 J mol^{-1} K^{-1}), T is the Kelvin temperature, and ln Q is the natural logarithm of the reaction quotient that was introduced in Section 14.2. For gaseous reactions, Q is calculated using partial pressures expressed in atm;[6] for reactions in solution, Q is calculated from molar concentrations. Equation 18.11 allows us to predict the direction of the spontaneous change in a reaction mixture if we know $\Delta G°$ and the composition of the mixture, as illustrated in Example 18.10.

Example 18.10
Determining the Direction of a Spontaneous Reaction

■ In Chapter 14, you learned that you can also predict the direction of a reaction by comparing *Q* with *K*.

The reaction $2NO_2(g) \rightleftharpoons N_2O_4(g)$ has $\Delta G_{298}° = -5.40$ kJ per mole of N_2O_4. In a reaction mixture, the partial pressure of NO_2 is 0.25 atm and the partial pressure of N_2O_4 is 0.60 atm. In which direction must this reaction proceed, at 25 °C, to reach equilibrium?

Analysis: Since we know that reactions proceed spontaneously *toward* equilibrium, we are really being asked to determine whether the reaction will proceed spontaneously in the forward or reverse direction. In Chapter 14 we learned that when we compare Q to K we can determine in which direction a reaction will proceed. Determining the sign of ΔG will also allow us to predict the direction of a chemical reaction.

---

[6] Strictly speaking, we should express pressures in atm or bar depending on which was used to obtain the $\Delta G°$ data. For simplicity, we will use the more familiar pressure unit atm in all our calculations. Because the atm and bar differ by only about 1 percent, any errors that might be introduced will be very small.

● **Assembling the Tools:** We can use Equation 18.11 to calculate ΔG for the forward reaction. If ΔG is negative, then the forward reaction is spontaneous. However, if the calculated ΔG is positive, the forward reaction is nonspontaneous, and it is really the reverse reaction that is spontaneous. To solve Equation 18.11 we need the correct form for the mass action expression so we can calculate Q. Expressed in terms of partial pressures, the mass action expression defines Q for this reaction:

$$Q = \frac{P_{N_2O_4}}{(P_{NO_2})^2}$$

● **Solution:** Using the mass action expression for Q, the equation we will use is

$$\Delta G = \Delta G° + RT \ln\left(\frac{P_{N_2O_4}}{(P_{NO_2})^2}\right)$$

Next, let's assemble the data:

$$\Delta G°_{298} = -5.40 \text{ kJ mol}^{-1} \qquad T = 298 \text{ K}$$
$$= -5.40 \times 10^3 \text{ J mol}^{-1} \qquad P_{N_2O_4} = 0.60 \text{ atm}$$
$$R = 8.314 \text{ J mol}^{-1} \text{ K}^{-1} \qquad P_{NO_2} = 0.25 \text{ atm}$$

Notice that we have changed the energy units of $\Delta G°$ to joules so they will be the same as the energy units in the R we are using. The next step is to substitute the data into the equation for ΔG. In doing this, we leave off the units for the partial pressures because reaction quotients are always dimensionless quantities.[7] Here is the calculation.[8]

$$\Delta G = -5.40 \times 10^3 \text{ J mol}^{-1} + (8.314 \text{ J mol}^{-1} \text{ K}^{-1})(298 \text{ K}) \ln\left[\frac{0.60}{(0.25)^2}\right]$$

$$= -5.40 \times 10^3 \text{ J mol}^{-1} + (8.314 \text{ J mol}^{-1} \text{ K}^{-1})(298 \text{ K}) \ln(9.6)$$

$$= -5.40 \times 10^3 \text{ J mol}^{-1} + (8.314 \text{ J mol}^{-1} \text{ K}^{-1})(298 \text{ K})(2.26)$$

$$= -5.40 \times 10^3 \text{ J mol}^{-1} + 5.60 \times 10^3 \text{ J mol}^{-1}$$

$$= +2.0 \times 10^2 \text{ J mol}^{-1}$$

Since ΔG is positive, the forward reaction is nonspontaneous. The reverse reaction is the one that will occur, and to reach equilibrium, some N_2O_4 will have to decompose.

● **Is the Answer Reasonable?** There is no simple check. However, we can check to be sure the energy units in both terms on the right are the same. Notice that here we have changed the units for $\Delta G°$ to joules to match those of R. Also notice that the temperature is expressed in kelvins, to match the temperature units in R.

Calculate ΔG for the reaction described in the preceding example if the partial pressure of NO_2 is 0.260 atm and the partial pressure of N_2O_4 is 0.598 atm. Where does the system stand relative to equilibrium? (*Hint:* What is ΔG for a system at equilibrium?)

Practice Exercise 18.26

In which direction will the reaction described in Example 18.10 proceed to reach equilibrium if the partial pressure of NO_2 is 0.60 atm and the partial pressure of N_2O_4 is 0.25 atm?

Practice Exercise 18.27

---

[7] For reasons beyond the scope of this text, quantities in mass action expressions used to compute reaction quotients and equilibrium constants are actually ratios. For partial pressures, it's the pressure in atm divided by 1 atm; for concentrations, it's the molarity divided by 1 M. This causes the units to cancel to give a dimensionless value for Q or K.

[8] As we noted in the footnote in Section 13.4, when taking the logarithm of a quantity, the number of digits *after the decimal point* should equal the number of significant figures in the quantity. Since 0.60 atm and 0.25 atm both have two significant figures, the logarithm of the quantity in square brackets (2.26) is rounded to give two digits after the decimal point.

Computing Thermodynamic Equilibrium Constants

Earlier you learned that when a system reaches equilibrium the free energy of the products equals the free energy of the reactants and ΔG equals zero. We also know that at equilibrium the reaction quotient equals the equilibrium constant.

$$\text{At equilibrium} \quad \begin{cases} \Delta G = 0 \\ Q = K \end{cases}$$

If we substitute these into Equation 18.11, we obtain

$$0 = \Delta G^{\circ} + RT \ln K$$

TOOLS

Determining thermodynamic equilibrium constants

which can be rearranged to give

$$\Delta G^{\circ} = -RT \ln K \qquad \text{(18.12)}$$

The equilibrium constant **K** calculated from this equation is called the **thermodynamic equilibrium constant** and corresponds to K_{p} for reactions involving gases (with partial pressures expressed in atm) and to K_{c} for reactions in solution (with concentrations expressed in mol L^{-1}).[9]

Equation 18.12 is useful because it permits us to determine equilibrium constants from either measured or calculated values of ΔG°. As you know, ΔG° can be determined by a Hess's law type of calculation from tabulated values of $\Delta G_{\text{f}}^{\circ}$. It also allows us to obtain values of ΔG° from measured equilibrium constants.

Example 18.11
Calculating the Gibbs Free Energy Change

The brownish haze often associated with air pollution is caused by nitrogen dioxide, NO_2, a red-brown gas. Nitric oxide, NO, is formed in automobile engines, and some of it escapes into the air where it is oxidized to NO_2 by oxygen.

$$2NO(g) + O_2(g) \rightleftharpoons 2NO_2(g)$$

The value of K_{p} for this reaction is 1.7×10^{12} at 25.00 °C. What is ΔG° for the reaction, (a) expressed in joules per mole and (b) in kilojoules per mole?

Analysis: We have an equation that relates the equilibrium constant to the standard free energy. All we have to do is collect the appropriate information to perform the calculation.

Assembling the Tools: The relationship between the standard free energy change, ΔG°, and an equilibrium constant is expressed in Equation 18.12.

$$\Delta G^{\circ} = -RT \ln K_{\text{p}}$$

We'll need the data below to calculate ΔG°. Note that the temperature is given to the nearest hundredth of a degree, so we'll use $T_{\text{K}} = t_{\text{C}} + 273.15$ to compute the Kelvin temperature.

$$R = 8.314 \text{ J mol}^{-1} \text{ K}^{-1}$$
$$T = 298.15 \text{ K}$$
$$K_{\text{p}} = 1.7 \times 10^{12}$$

[9] The thermodynamic equilibrium constant requires that gases *always* be included as partial pressures in mass action expressions. The pressure units must be standard pressure units (atmospheres or bars, depending on which convention was used for collecting the ΔG° data). For a heterogeneous equilibrium involving a gas and a substance in solution, the mass action expression will mix partial pressures (for the gas) with molarities (for the dissolved substance).

Solution: (a) Substituting these values into the equation, we have

$$\Delta G° = -(8.314 \text{ J mol}^{-1} \text{ K}^{-1} \times 298.15 \text{ K}) \ln(1.7 \times 10^{12})$$

$$= -(8.314 \text{ J mol}^{-1} \text{ K}^{-1} \times 298.15 \text{ K}) \times (28.16)$$

$$= -6.980 \times 10^4 \text{ J mol}^{-1} \quad \text{(to four significant figures)}$$

(b) Changing J to kJ,

$$\Delta G° = -69.80 \text{ kJ mol}^{-1}$$

Is the Answer Reasonable? The value of K_p tells us that the position of equilibrium lies far to the right, which means $\Delta G°$ must be large and negative. Therefore, the answer, -69.80 kJ mol^{-1}, seems reasonable.

■ The units in this calculation give $\Delta G°$ on a per mole (mol^{-1}) basis. This reminds us that we are viewing the coefficients of the reactants and products as representing moles, rather than some other sized quantity.

Example 18.12
Thermodynamic Equilibrium Constants

Sulfur dioxide, which is sometimes present in polluted air, reacts with oxygen when it passes over the catalyst in automobile catalytic converters. The product is the very acidic oxide, sulfur trioxide.

$$2SO_2(g) + O_2(g) \rightleftharpoons 2SO_3(g)$$

For this reaction, $\Delta G° = -1.40 \times 10^2$ kJ mol^{-1} at 25 °C. What is the value of K_p?

Analysis: Once again we have an equation that we can solve once we collect all of the data for the equation's variables.

Assembling the Tools: We will use Equation 18.12.

$$\Delta G° = -RT \ln K_p$$

The chemical equation, in this case, is extra information. We see from the equation that we need to find T and $\Delta G°$ in the statement of the problem, and if necessary look up the value for R. We can list our data as

$$R = 8.314 \text{ J mol}^{-1} \text{ K}^{-1}$$

$$T = 298 \text{ K}$$

$$\Delta G° = -1.40 \times 10^2 \text{ kJ mol}^{-1}$$

$$= -1.40 \times 10^5 \text{ J mol}^{-1}$$

To make the units cancel correctly, we have added the unit mol^{-1} to the value of $\Delta G°$. This just emphasizes that the amounts of reactants and products are specified in mole-sized quantities.

Solution: To calculate K_p, let's first solve for $\ln K_p$.

$$\ln K_p = \frac{\Delta G°}{RT}$$

Substituting values gives

$$\ln K_p = \frac{-(-1.40 \times 10^5 \text{ J mol}^{-1})}{(8.314 \text{ J mol}^{-1} \text{ K}^{-1})(298 \text{ K})}$$

To calculate K_p, we take the antilogarithm,

$$K_p = e^{56.5}$$

$$= 3 \times 10^{24}$$

Notice that we have expressed the answer to only one significant figure. As discussed earlier, when taking a logarithm, the number of digits written after the decimal place equals the number of significant figures in the number. Conversely, the number of significant figures in the antilogarithm equals the number of digits after the decimal in the logarithm.

Is the Answer Reasonable? The value of $\Delta G°$ is large and negative, so the position of equilibrium should favor the products. The large value of K_p also indicates that the products are favored and, therefore, the result is reasonable.

Practice Exercise 18.28   The reaction $N_2(g) + 3H_2(g) \rightleftharpoons 2NH_3(g)$ has $K_p = 6.9 \times 10^5$ at 25.0 °C. Calculate $\Delta G°$ for this reaction in units of kilojoules. (*Hint:* Be careful with units and significant figures.)

Practice Exercise 18.29   The reaction $H_2(g) + I_2(g) \rightleftharpoons 2HI(g)$ has $\Delta G° = +3.3$ kJ mol^{-1} at 25.0 °C. What is the value of K_p at this temperature?

18.10 | Bond Energies

You have seen how thermodynamic data allow us to predict the spontaneity of chemical reactions as well as the nature of chemical systems at equilibrium. In addition to these useful and important benefits of the study of thermodynamics, there is a bonus. By studying heats of reaction, and heats of formation in particular, we can obtain fundamental information about the chemical bonds in the substances that react. The reasoning is that a chemical reaction results in products that have different bonds than the reactants did. The difference in the total energy of the bonds in the products compared to the total energy of the bonds in the reactants shows up as heat that is lost or gained.

Recall that the **bond energy** is the *amount of energy needed to break a chemical bond to give electrically neutral fragments.* It is a useful quantity to know in the study of chemical properties, because during chemical reactions, bonds within the reactants are broken and new ones are formed as the products appear. The first step—bond breaking—is one of the factors that controls the reactivity of substances. Elemental nitrogen, for example, has a very low degree of reactivity, which is generally attributed to the very strong triple bond in N_2. Reactions that involve the breaking of this bond in a single step simply do not occur. When N_2 does react, it is by a stepwise involvement of its three bonds, one at a time.

Determining Bond Energies

■ The heats of formation of the nitrogen oxides are endothermic because of the large bond energy of the N_2 molecule.

The bond energies of simple diatomic molecules such as H_2, O_2, and Cl_2 are usually measured *spectroscopically*. A flame or an electric spark is used to excite (energize) the molecules, causing them to emit light. An analysis of the spectrum of emitted light allows scientists to compute the amount of energy needed to break the bond.

■ The kind of bond breaking described here divides the electrons of the bond equally between the two atoms. It could be symbolized as

$$A{:}B \longrightarrow A{\cdot} + {\cdot}B$$

For more complex molecules, thermochemical data can be used to calculate bond energies in a Hess's law kind of calculation. We will use the standard heat of formation of methane to illustrate how this is accomplished. However, before we can attempt such a calculation, we must first define a thermochemical quantity that we will call the **atomization energy**, symbolized ΔH_{atom}. This is the amount of energy needed to

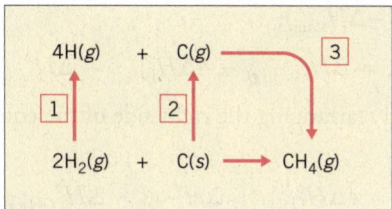

Figure 18.18 | **Two paths for the formation of methane from its elements in their standard states.** As described in the text, steps 1, 2, and 3 of the upper path involve the formation of gaseous atoms of the elements and the formation of bonds in CH_4. The lower path corresponds to the direct combination of the elements in their standard states to give CH_4. Because H is a state function, the sum of the enthalpy changes along the upper path must equal the enthalpy change for the lower path (ΔH_f°).

rupture all the chemical bonds in one mole of gaseous molecules to give gaseous atoms as products. For example, the atomization of methane is

$$CH_4(g) \longrightarrow C(g) + 4H(g)$$

and the enthalpy change for the process is ΔH_{atom}. For this particular molecule, ΔH_{atom} corresponds to the total amount of energy needed to break all the C—H bonds in one mole of CH_4; therefore, division of ΔH_{atom} by 4 would give the average C—H bond energy in methane, expressed in kJ mol^{-1}.

Figure 18.18 shows how we can use the standard heat of formation ΔH_f°, to calculate the atomization energy. Across the bottom we have the chemical equation for the formation of CH_4 from its elements. The enthalpy change for this reaction, of course, is ΔH_f°. In this figure we also can see an alternative three-step path that leads to $CH_4(g)$. One step is the breaking of H—H bonds in the H_2 molecules to give gaseous hydrogen atoms, another is the vaporization of carbon to give gaseous carbon atoms, and the third is the combination of the gaseous atoms to form CH_4 molecules. These changes are labeled 1, 2, and 3 in the figure.

Since $\Delta H = H_{final} - H_{initial}$, a difference in state functions, the net enthalpy change from one state to another is the same regardless of the path that we follow. This means that the sum of the enthalpy changes along the upper path must be the same as the enthalpy change along the lower path, ΔH_f°. Perhaps this can be more easily seen in Hess's law terms if we write the changes along the upper path in the form of thermochemical equations.

Steps 1 and 2 have enthalpy changes that are called *standard heats of formation of gaseous atoms*. Values for these quantities have been measured for many of the elements, and some are given in Table 18.3. Step 3 is the opposite of atomization, and its enthalpy change will therefore be the negative of ΔH_{atom} (recall that if we reverse a reaction, we change the sign of its ΔH).

■ A more complete table of standard heats of formation of gaseous atoms is located in Appendix C.

TABLE 18.3	Standard Heats of Formation of Some Gaseous Atoms from the Elements in Their Standard States
Atom	**ΔH_f° per mole of atoms (kJ mol^{-1})[a]**
H	217.89
Li	161.5
Be	324.3
B	560
C	716.67
N	472.68
O	249.17
S	276.98
F	79.14
Si	450
P	332.2
Cl	121.47
Br	112.38
I	107.48

(Step 1) $2H_2(g) \longrightarrow 4H(g)$ $\Delta H_1^\circ = 4\Delta H_{f\,H(g)}^\circ$

(Step 2) $C(s) \longrightarrow C(g)$ $\Delta H_2^\circ = \Delta H_{f\,C(g)}^\circ$

(Step 3) $\dfrac{4H(g) + C(g) \longrightarrow CH_4(g)}{2H_2(g) + C(s) \longrightarrow CH_4(g)}$ $\dfrac{\Delta H_3^\circ = -\Delta H_{atom}}{\Delta H^\circ = 4\Delta H_{f\,CH_4(g)}^\circ}$

Notice that by adding the first three equations, $C(g)$ and $4H(g)$ cancel and we get the equation for the formation of CH_4 from its elements in their standard states. This means that adding the ΔH° values of the first three equations should give ΔH_f° for CH_4:

$$\Delta H_1^\circ + \Delta H_2^\circ + \Delta H_3^\circ = \Delta H_{f\,CH_4(g)}^\circ$$

Let's substitute for ΔH_1°, ΔH_2°, and ΔH_3°, and then solve for ΔH_{atom}. First, we substitute for the ΔH° quantities:

$$4\Delta H_{f\,H(g)}^\circ + \Delta H_{f\,C(g)}^\circ + (-\Delta H_{atom}) = \Delta H_{f\,CH_4(g)}^\circ$$

[a]All values are positive because forming the gaseous atoms from the elements involves bond breaking and is endothermic.

TABLE 18.4	Some Average Bond Energies		
Bond	Bond Energy (kJ mol^{-1})	Bond	Bond Energy (kJ mol^{-1})
C—C	348	C—Br	276
C=C	612	C—I	238
C≡C	960	H—H	436
C—H	412	H—F	565
C—N	305	H—Cl	431
C=N	613	H—Br	366
C≡N	890	H—I	299
C—O	360	H—N	388
C=O	743	H—O	463
C—F	484	H—S	338
C—Cl	338	H—Si	376

Next, we solve for $(-\Delta H_{atom})$:

$$-\Delta H_{atom} = \Delta H^{\circ}_{f\,CH_4(g)} - 4\Delta H^{\circ}_{f\,H(g)} - \Delta H^{\circ}_{f\,C(g)}$$

Changing signs and rearranging the right side of the equation just a bit gives

$$\Delta H_{atom} = 4\Delta H^{\circ}_{f\,H(g)} + \Delta H^{\circ}_{f\,C(g)} - \Delta H^{\circ}_{f\,CH_4(g)}$$

Now all we need are the ΔH°_f values for the substances indicated on the right side. From Table 18.3 we obtain $\Delta H^{\circ}_{f\,H(g)}$ and $\Delta H^{\circ}_{f\,C(g)}$, and the value of $\Delta H^{\circ}_{f\,CH_4(g)}$ is obtained from Table 6.2. We will round these to the nearest 0.1 kJ/mol.

$$\Delta H^{\circ}_{f\,H(g)} = +217.9 \text{ kJ/mol}$$

$$\Delta H^{\circ}_{f\,C(g)} = +716.7 \text{ kJ/mol}$$

$$\Delta H^{\circ}_{f\,CH_4(g)} = -74.8 \text{ kJ/mol}$$

Substituting these values gives

$$\Delta H_{atom} = 1663.1 \text{ kJ/mol}$$

and division by 4 gives an estimate of the average C—H bond energy in this molecule.

$$\text{Bond energy} = \frac{1663.1 \text{ kJ/mol}}{4}$$

$$= 415.8 \text{ kJ/mol of C—H bonds}$$

This value is quite close to the one in Table 18.4, which is an average of the C—H bond energies in many different compounds. The other bond energies in Table 18.4 are also based on thermochemical data and were obtained by similar calculations.

Estimating Heats of Formation

An amazing thing about many covalent bond energies is that they are very nearly the same in many different compounds. This suggests, for example, that a C—H bond is very nearly the same in CH_4 as it is in C_3H_6 or $C_{15}H_{32}$ or a large number of other compounds that contain this kind of bond.

Because the bond energy doesn't vary much from compound to compound, we can use tabulated bond energies to estimate the heats of formation of substances. For example, let's calculate the standard heat of formation of methyl alcohol vapor, $CH_3OH(g)$. The structural formula for methanol is

$$\begin{array}{c} H \\ | \\ H-C-O-H \\ | \\ H \end{array}$$

To perform this calculation, we set up two paths from the elements to the compound, as shown in Figure 18.19. The lower path has an enthalpy change corresponding to $\Delta H^{\circ}_{f\,CH_3OH(g)}$, while the upper path takes us to the gaseous elements and then through the energy released when the bonds in the molecule are formed. This latter energy can be computed from the bond energies in Table 18.4. As before, the sum of the energy changes along the upper path must be the same as the energy change along the lower path, and this permits us to compute $\Delta H^{\circ}_{f\,CH_3OH(g)}$.

Steps 1, 2, and 3 involve the formation of the gaseous atoms from the elements, and their enthalpy changes are obtained from Table 18.3.

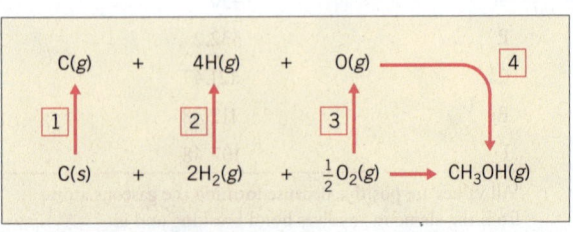

Figure 18.19 | Two paths for the formation of methyl alcohol vapor from its elements in their standard states. The numbered paths are discussed in the text.

$$\Delta H_1^\circ = \Delta H_{fC(g)}^\circ \quad = 1 \text{ mol} \times 716.7 \text{ kJ/mol} \quad = 716.7 \text{ kJ}$$

$$\Delta H_2^\circ = 4\Delta H_{fH(g)}^\circ = 4 \text{ mol} \times (217.9 \text{ kJ/mol}) = 871.6 \text{ kJ}$$

$$\Delta H_3^\circ = \Delta H_{fO(g)}^\circ \quad = 1 \text{ mol} \times (249.2 \text{ kJ/mol}) = 249.2 \text{ kJ}$$

The sum of these values, $+1837.5$ kJ, is the total energy-input (the net ΔH°) for the first three steps.

The formation of the CH_3OH molecule from the gaseous atoms is exothermic because *energy is always released* when atoms become joined by covalent bonds. In this molecule we can count three C—H bonds, one C—O bond, and one O—H bond. Their formation releases energy equal to their bond energies, which we obtain from Table 18.4.

Bond	Energy (kJ)
3(C—H)	$3 \times (412) = 1236$
C—O	360
O—H	463

Adding these together gives a total of 2059 kJ. Therefore, for Step 4 in Figure 18.19 the enthalpy, ΔH_4°, is -2059 kJ (because bond formation is exothermic). Now we can compute the total enthalpy change for the upper path.

$$\Delta H^\circ = (+1837.5 \text{ kJ}) + (-2059 \text{ kJ})$$

$$= -222 \text{ kJ}$$

The value just calculated should be equal to ΔH_f° for $CH_3OH(g)$. For comparison, it has been found experimentally that ΔH_f° for this molecule (in the vapor state) is -201 kJ/mol. At first glance, the agreement doesn't seem very good, but on a relative basis the calculated value (-222 kJ) differs from the experimental one by only about 10%.

Determine the heat of formation of gaseous 1-propanol ($CH_3CH_2CH_2OH$) and 2-bromobutane ($CH_3CHBrCH_2CH_3$). (*Hint:* Remember that bond formation is exothermic.)

Practice Exercise 18.30

Determine the heat of formation of gaseous cyclohexane, C_6H_{12}. Cyclohexane is a six-member carbon ring with two hydrogen atoms attached to each carbon atom. Do the same for benzene, C_6H_6. (See Section 8.8, for the structure.)

Practice Exercise 18.31

| Summary

Organized by Learning Objective

State the first law of thermodynamics and explain how the change in energy differs from the change in enthalpy

The **first law of thermodynamics** states that the change in the **internal energy** of a system, ΔE, equals the sum of the heat absorbed by the system, q, and the work done on the system, w. For pressure–volume work, $w = -P\Delta V$, where P is the external pressure and $\Delta V = V_{final} - V_{initial}$. The heat at constant volume, q_v, is equal to ΔE, whereas the heat at constant pressure, q_p, is equal to ΔH. The value of ΔH differs from ΔE by the work expended in pushing back the atmosphere when the change occurs at constant atmospheric pressure. For a chemical reaction, $\Delta H = \Delta E + \Delta n_{gas} RT$, where Δn_{gas} is the change in the number of moles of gas when going from reactants to products.

Explain what is meant by the direction of spontaneous change

A **spontaneous change** occurs without outside assistance, whereas a nonspontaneous change requires continuous help and can occur only if it is accompanied by and linked to some other spontaneous event.

Describe the factors that affect the magnitude of entropy and the sign of an entropy change in a chemical reaction

Spontaneity is associated with statistical probability—spontaneous processes tend to proceed from lower probability states to those of higher probability. The thermodynamic quantity associated with the probability of a state is **entropy, S.** Entropy is a state function that measures the number of energetically equivalent ways—microstates, W—a state can be realized. In general, gases have much higher entropies than liquids, which have somewhat higher entropies than solids. Entropy increases with volume for a gas and with the temperature. During a chemical reaction, the entropy tends to increase if the number of molecules increases.

State the second law of thermodynamics in terms of entropy, and explain the significance of Gibbs free energy

The **second law of thermodynamics** states that the entropy of the universe (system plus surroundings) increases whenever a spontaneous change occurs.

The **Gibbs free energy**, G, is a state function and its change, ΔG, allows us to determine the combined effects of enthalpy and entropy changes on the spontaneity of a chemical or physical change. A change is spontaneous only if the free energy of the system decreases (ΔG is negative). When ΔH and ΔS have the same algebraic sign, the temperature becomes the critical factor in determining spontaneity.

Explain how the third law of thermodynamics leads to a standard entropy of formation

Third Law of Thermodynamics: The entropy of a pure crystalline substance is equal to zero at absolute zero (0 K). Standard entropies, $S°$, are calculated for 25 °C and 1 atm (Table 18.1) and can be used to calculate $\Delta S°$ for chemical reactions.

Calculate standard free energy changes

When ΔG is measured at 25 °C and 1 atm, it is the **standard free energy change, $\Delta G°$**. The **standard free energies of formation, $\Delta G_f°$**, can be used to obtain $\Delta G°$ for chemical reactions by a Hess's law type of calculation.

Explain the connection between the Gibbs free energy change and the maximum energy produced by a reaction

For any system, the value of ΔG is equal to the maximum amount of energy that can be obtained in the form of useful work, which can be obtained only if the change takes place by a **reversible process.** All real changes are irreversible and we always obtain less work than is theoretically available; the rest is lost as heat.

Explain the connection between Gibbs free energy change and the position of equilibrium

When a system reaches equilibrium, $\Delta G = 0$ and no useful work can be obtained from it. For a phase change (e.g., solid $\rightleftharpoons$ liquid) ΔS and ΔH are related by the equation $\Delta S = \Delta H/T$, where T is the temperature at which the two phases are in equilibrium.

In chemical reactions, equilibrium occurs at a minimum on the free energy curve part way between pure reactants and pure products. The composition of the equilibrium mixture is determined by where the minimum lies along the reactant $\longrightarrow$ product axis; when it lies close to the products, the proportion of product to reactants is large and the reaction goes far toward completion.

When a reaction has a value of $\Delta G°$ that is both large and negative, it will appear to occur spontaneously because a lot of products will be formed by the time equilibrium is reached. If $\Delta G°$ is large and positive, it may be difficult to observe any reaction at all because only tiny amounts of products will be formed.

Describe how to calculate a Gibbs free energy change from an equilibrium constant

The spontaneity of a reaction is determined by ΔG (how the free energy changes with a change in concentration). This is related to the standard free energy change, $\Delta G°$, by the equation $\Delta G = \Delta G° + RT \ln Q$, where Q is the reaction quotient for the system. At equilibrium, $\Delta G° = -RT \ln K$, where $K = K_p$ for gaseous reactions and $K = K_c$ for reactions in solution.

Define bond energy in terms of enthalpy

The **bond energy** equals the amount of energy needed to break a bond to give neutral fragments. The sum of all the bond energies in a molecule is the **atomization energy, ΔH_{atom}**, and, on a mole basis, it corresponds to the energy needed to break one mole of molecules into individual atoms. The heat of formation of a gaseous compound equals the sum of the energies needed to form atoms of the elements that are found in the substance plus the negative of the atomization energy.

Tools for Problem Solving

The following tools were introduced in this chapter. Study them carefully so you can select the appropriate tool when needed.

Converting between ΔE and ΔH (Section 18.1)

$$\Delta H = \Delta E + \Delta n_{gas} RT$$

Factors that affect the entropy (Section 18.3)

- *Volume*: Entropy increases with increasing volume.
- *Temperature*: Entropy increases with increasing temperature.
- *Physical state*: $S_{gas} \gg S_{liquid} > S_{solid}$. When gases are formed in a reaction, ΔS is almost always positive.
- *Number of particles*: Entropy increases when the number of particles increases.

Gibbs free energy (Section 18.4)

$$\Delta G = \Delta H - T\Delta S$$

ΔG as a predictor of spontaneity (Section 18.4)

- If ΔG is less than zero, the reaction is spontaneous.
- If ΔG is greater than zero, the reaction is nonspontaneous.

Standard entropy change (Section 18.5)

$$\Delta S° = (\text{sum of } S° \text{ of products}) - (\text{sum of } S° \text{ of reactants})$$

Calculating $\Delta G°$ from $\Delta H°$ and $\Delta S°$ (Section 18.6)

$$\Delta G° = \Delta H° - (298.15 \text{ K})\Delta S°$$

Calculating $\Delta G°$ using $\Delta G_f°$ values (Section 18.6)

$$\Delta G° = (\text{sum of } \Delta G_f° \text{ of products}) - (\text{sum of } \Delta G_f° \text{ of reactants})$$

Using $\Delta G°$ to assess the position of equilibrium (Section 18.8)

- $\Delta G°$ is large and positive: position of equilibrium is close to reactants.
- $\Delta G°$ is large and negative: position of equilibrium is close to products.
- $\Delta G° = 0$: position of equilibrium lies about midway between reactants and products.

Calculating $\Delta G°$ at temperatures other than 25 °C (Section 18.8)

$$\Delta G_T° \approx \Delta H_{298}° - T\Delta S_{298}°$$

Relating the reaction quotient to ΔG (Section 18.9)

$$\Delta G = \Delta G° + RT \ln Q$$

- If ΔG is negative, the reaction is spontaneous in the forward direction.
- If ΔG is positive, the reaction is spontaneous in the reverse direction.
- If ΔG is zero, the reaction is at equilibrium.

Determining thermodynamic equilibrium constants (Section 18.9)

$$\Delta G° = -RT \ln K$$

WileyPLUS = *WileyPLUS*, an online teaching and learning solution. *Note to instructors*: Many of the end-of-chapter problems are available for assignment via the WileyPLUS system. **www.wileyplus.com.** Review Problems are presented in pairs separated by blue rules. Answers to problems whose numbers appear in blue are given in Appendix B. More challenging problems are marked with an asterisk ＊.

| Review Questions

First Law of Thermodynamics

18.1 What is the origin of the name *thermodynamics*?

18.2 State the first law of thermodynamics in your own words. What equation defines the change in the internal energy in terms of heat and work? Define the meaning of the symbols, including the significance of their algebraic signs.

18.3 How is a change in the internal energy defined in terms of the initial and final internal energies?

18.4 What is the algebraic sign of ΔE for an endothermic change? Why?

18.5 Which quantities in the statement of the first law are state functions and which are not?

18.6 Which thermodynamic quantity corresponds to the heat at constant volume? Which corresponds to the heat at constant pressure?

18.7 What are the units of $P\Delta V$ if pressure is expressed in pascals and volume is expressed in cubic meters?

18.8 If there is a decrease in the number of moles of gas during an exothermic chemical reaction, which is numerically larger, ΔE or ΔH? Why?

18.9 Which of the following changes is accompanied by the most negative value of ΔE? **(a)** A spring is compressed and heated. **(b)** A compressed spring expands and is cooled. **(c)** A spring is compressed and cooled. **(d)** A compressed spring expands and is heated.

Spontaneous Change

18.10 What is a *spontaneous change*? What role does kinetics play in determining the apparent spontaneity of a chemical reaction?

18.11 List five changes that you have encountered recently that occurred spontaneously. List five changes that are nonspontaneous that you have caused to occur.

18.12 Which of the items that you listed in Question 18.11 are exothermic (leading to a lowering of the potential energy) and which are endothermic (accompanied by an increase in potential energy)?

18.13 At constant pressure, what role does the enthalpy change play in determining the spontaneity of an event?

18.14 How do the probabilities of the initial and final states in a process affect the spontaneity of the process?

Entropy

18.15 An instant cold pack purchased in a pharmacy contains a packet of solid ammonium nitrate surrounded by a pouch of water. When the packet of NH_4NO_3 is broken, the solid dissolves in water and a cooling of the mixture occurs because the solution process for NH_4NO_3 in water is endothermic. Explain, in terms of what happens to the molecules and ions, why this mixing occurs spontaneously.

18.16 What is *entropy*?

18.17 How is the entropy of a substance affected by **(a)** an increase in temperature, **(b)** a decrease in volume, **(c)** changing from a liquid to a solid, and **(d)** dissociating into individual atoms?

18.18 Will the entropy change for each of the following be positive or negative?

(a) Moisture condenses on the outside of a cold glass.

(b) Raindrops form in a cloud.

(c) Gasoline vaporizes in the carburetor of an automobile engine.

(d) Air is pumped into a tire.

(e) Frost forms on the windshield of your car.

(f) Sugar dissolves in coffee.

18.19 On the basis of our definition of entropy, suggest why entropy is a state function.

Second Law of Thermodynamics

18.20 State the second law of thermodynamics.

18.21 How can a process have a negative entropy change for the system, and yet still be spontaneous?

18.22 Explain the terms $\Delta S_{universe}$, ΔS_{system}, and $\Delta S_{surroundings}$, and how they relate to each other.

18.23 Explain how the entropy of the surroundings of a reaction is related to the enthalpy of the reaction.

18.24 What is the driving force for a reaction that is endothermic?

18.25 Define Gibbs free energy in your own words.

18.26 In terms of the algebraic signs of ΔH and ΔS, under which of the following circumstances will a change be spontaneous:

(a) At all temperatures?

(b) At low temperatures but not at high temperatures?

(c) At high temperatures but not at low temperatures?

18.27 Under what circumstances will a change be nonspontaneous regardless of the temperature?

Third Law of Thermodynamics

18.28 State the third law of thermodynamics in your own words.

18.29 Explain why the units for entropy have a dependence on temperature—for example, why are the units $J\ mol^{-1}\ K^{-1}$?

18.30 Explain why the values of ΔH for elements in their standard state are $0\ kJ\ mol^{-1}$, but the values for the standard entropy, S, for elements are not $0\ J\ mol^{-1}\ K^{-1}$.

18.31 Would you expect the entropy of an alloy (a solution of two metals) to be zero at 0 K? Explain your answer.

18.32 Why does entropy increase with increasing temperature?

18.33 Does glass have $S = 0$ at 0 K? Explain.

Standard Free Energy Change, $\Delta G°$

18.34 What is the equation expressing the change in the Gibbs free energy for a reaction occurring at constant temperature and pressure?

18.35 Why can $\Delta G°$ be calculated at different temperatures using $\Delta H°$ and $\Delta S°$ at 298.15 K?

Maximum Work and ΔG

18.36 How is free energy related to useful work?

18.37 What is a thermodynamically reversible process? How is the amount of work obtained from a change related to thermodynamic reversibility?

18.38 How is the *rate* at which energy is withdrawn from a system related to the amount of that energy that can appear as useful work?

18.39 When glucose is oxidized by the body to generate energy, part of the energy is used to make molecules of ATP (adenosine triphosphate). However, of the total energy released in the oxidation of glucose, only 38% actually goes to making ATP. What happens to the rest of the energy?

18.40 Why are real, observable changes not considered to be thermodynamically reversible processes?

Free Energy and Equilibrium

18.41 In what way is free energy related to equilibrium?

18.42 How can boiling points or melting points be estimated from the $\Delta S°$ and $\Delta H°$ values for the phase change?

18.43 Considering the fact that the formation of a bond between two atoms is exothermic and is accompanied by an entropy decrease, explain why all chemical compounds decompose into individual atoms if heated to a high enough temperature.

18.44 When a warm object is placed in contact with a cold one, they both gradually come to the same temperature. On a molecular level, explain how this is related to entropy and spontaneity.

18.45 Sketch the shape of the free energy curve for a chemical reaction that has a positive $\Delta G°$. Indicate the composition of the reaction mixture corresponding to equilibrium.

18.46 Many reactions that have large, negative values of $\Delta G°$ are not actually observed to happen at 25 °C and 1 atm. Why?

Equilibrium Constants and $\Delta G°$

18.47 Suppose a reaction has a negative $\Delta H°$ and a negative $\Delta S°$. Will more or less product be present at equilibrium as the temperature is raised?

18.48 Write the equation that relates the free energy change to the value of the reaction quotient for a reaction.

18.49 How is the equilibrium constant related to the standard free energy change for a reaction? (Write the equation.)

18.50 What is the value of $\Delta G°$ for a reaction for which $K = 1$?

18.51 How does the thermodynamic equilibrium constant differ from an equilibrium constant determined experimentally?

Bond Energies

18.52 Define the term atomization energy.

18.53 Why are the heats of formation of gaseous atoms from their elements endothermic quantities?

18.54 The gaseous C_2 molecule has a bond energy of 602 kJ mol^{-1}. Why isn't the standard heat of formation of $C(g)$ equal to half this value?

| Review Problems

First Law of Thermodynamics

18.55 A certain system absorbs 0.300 kJ of heat and has 0.700 kJ of work performed on it. What is the value of ΔE for the change? Is the overall change exothermic or endothermic?

18.56 The value of ΔE for a certain change is -1455 J. During the change, the system absorbs 812 J of heat. Did the system do work, or was work done on the system? How much work, expressed in joules, was involved?

18.57 Suppose that you were pumping an automobile tire with a hand pump that pushed 24.0 in.^{3} of air into the tire on each stroke, and that during one such stroke the opposing pressure in the tire was 30.0 lb/in.^{2} above the normal atmospheric pressure of 14.7 lb/in.^{2}. Calculate the number of joules of work accomplished during each stroke. (1 L atm = 101.325 J)

18.58 Consider the reaction between aqueous solutions of baking soda, $NaHCO_3$, and vinegar, $HC_2H_3O_2$.

$$NaHCO_3(aq) + HC_2H_3O_2(aq) \longrightarrow$$
$$NaC_2H_3O_2(aq) + H_2O(l) + CO_2(g)$$

If this reaction occurs at 1.00 atmospheres of pressure, how much work, expressed in *L atm*, is done by the system in pushing back the atmosphere when 1.00 mol $NaHCO_3$ reacts at a temperature of 25 °C? (*Hint:* Review the gas laws.)

18.59 Calculate $\Delta H°$ and $\Delta E°$ for the following reactions at 25 °C. (If necessary, refer to the data in Table C.2 in Appendix C.)

(a) $3PbO(s) + 2NH_3(g) \longrightarrow$
$$3Pb(s) + N_2(g) + 3H_2O(g)$$
(b) $NaOH(s) + HCl(g) \longrightarrow NaCl(s) + H_2O(l)$
(c) $Al_2O_3(s) + 2Fe(s) \longrightarrow Fe_2O_3(s) + 2Al(s)$
(d) $2CH_4(g) \longrightarrow C_2H_6(g) + H_2(g)$

18.60 Calculate $\Delta H°$ and $\Delta E°$ for the following reactions at 25 °C. (If necessary, refer to the data in Table C.2 in Appendix C.)

(a) $2C_2H_2(g) + 5O_2(g) \longrightarrow 4CO_2(g) + 2H_2O(g)$
(b) $C_2H_2(g) + 5N_2O(g) \longrightarrow$
$$2CO_2(g) + H_2O(g) + 5N_2(g)$$
(c) $NH_4Cl(s) \longrightarrow NH_3(g) + HCl(g)$
(d) $(CH_3)_2CO(l) + 4O_2(g) \longrightarrow$
$$3CO_2(g) + 3H_2O(g)$$

18.61 The reaction

$$2N_2O(g) \longrightarrow 2N_2(g) + O_2(g)$$

has $\Delta H° = -163.14$ kJ. What is the value of ΔE for the decomposition of 186 g of N_2O at 25 °C? If we assume that ΔH doesn't change appreciably with temperature, what is ΔE for this same reaction at 217 °C?

18.62 A 10.0 L vessel at 22 °C contains butane, $C_4H_{10}(g)$, at a pressure of 2.00 atm. What is the maximum amount of work that can be obtained by the combustion of this butane if the gas is first brought to a pressure of 1 atm and the temperature is brought to 28 °C? Assume the products are also returned to this same temperature and pressure.

Spontaneous Change

18.63 Predict the sign of ΔS for the following reactions.

(a) The sublimation of $I_2(s)$

(b) $CO(g) + 2H_2(g) \longrightarrow CH_3OH(g)$

(c) $2Na(s) + Cl_2(g) \longrightarrow 2NaCl(s)$

18.64 Predict the sign of ΔS for the following reactions.

(a) $Ba(s) + Cl_2(g) \longrightarrow BaCl_2(s)$

(b) $NaCl(s) \longrightarrow Na^+(aq) + Cl^-(aq)$

(c) $2NaHCO_3(s) \xrightarrow{\text{heat}}$

$$2Na_2CO_3(s) + H_2O(g) + CO_2(g)$$

18.65 Use the data from Table 6.2 to calculate $\Delta H°$ for the following reactions. On the basis of their values of $\Delta H°$, which are favored to occur spontaneously?

(a) $CaO(s) + CO_2(g) \longrightarrow CaCO_3(s)$

(b) $C_2H_2(g) + 2H_2(g) \longrightarrow C_2H_6(g)$

(c) $3CaO(s) + 2Fe(s) \longrightarrow 3Ca(s) + Fe_2O_3(s)$

18.66 Use the data from Table 6.2 to calculate $\Delta H°$ for the following reactions. On the basis of their values of $\Delta H°$, which are favored to occur spontaneously?

(a) $NH_4Cl(s) \longrightarrow NH_3(g) + HCl(g)$

(b) $2C_2H_2(g) + 5O_2(g) \longrightarrow 4CO_2(g) + 2H_2O(g)$

(c) $C_2H_2(g) + 5N_2O(g) \longrightarrow$

$$2CO_2(g) + H_2O(g) + 5N_2(g)$$

Entropy

18.67 There are two chemical systems, A and B, each with two particles that have a total energy of 20 J. For the first system, the two particles each can have energies of 0, 5, 10, 15, or 20 J, and for the second system, the particles can have energies of 0, 10, or 20 J. Which configuration will have the least entropy?

18.68 A chemical system has three particles that can have energies of 0, 5, 10, 15, or 20 J. If the total energy of the system is 30 J, how many different ways can the particles be organized?

18.69 Which system has a higher entropy? Explain your reasoning.

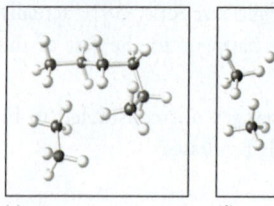 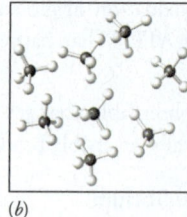 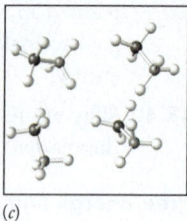

(a)   (b)   (c)

18.70 Which system has a higher entropy?

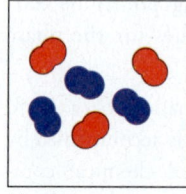 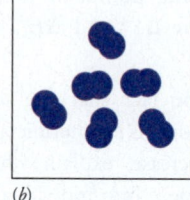 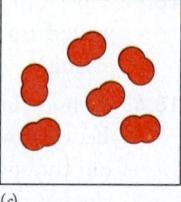

(a)   (b)   (c)

18.71 What factors must you consider to determine the sign of ΔS for the reaction $2N_2O(g) \longrightarrow 2N_2(g) + O_2(g)$ if it occurs at constant temperature?

18.72 What factors must you consider to determine the sign of ΔS for the reaction: $2HI(g) \longrightarrow H_2(g) + I_2(g)$?

18.73 Predict the algebraic sign of the entropy change for the following reactions.

(a) $PCl_3(g) + Cl_2(g) \longrightarrow PCl_5(g)$

(b) $SO_2(g) + CaO(s) \longrightarrow CaSO_3(s)$

(c) $CO_2(g) + H_2O(l) \longrightarrow H_2CO_3(aq)$

(d) $Ni(s) + 2HCl(aq) \longrightarrow H_2(g) + NiCl_2(aq)$

18.74 Predict the algebraic sign of the entropy change for the following reactions.

(a) $I_2(s) \longrightarrow I_2(g)$

(b) $Br_2(g) + 3Cl_2(g) \longrightarrow 2BrCl_3(g)$

(c) $NH_3(g) + HCl(g) \longrightarrow NH_4Cl(s)$

(d) $CaO(s) + H_2O(l) \longrightarrow Ca(OH)_2(s)$

Second Law of Thermodynamics

18.75 Under what conditions will the reaction be spontaneous?

(a) $CH_4(g) + 2O_2(g) \longrightarrow CO_2(g) + 2H_2O(g)$

$$\Delta S < 0 \text{ and } \Delta H < 0$$

(b) $Ba(NO_3)_2(s) \longrightarrow Ba^{2+}(aq) + 2NO_3^-(aq)$

$$\Delta S > 0 \text{ and } \Delta H > 0$$

18.76 Under what conditions will the reaction be spontaneous?

(a) $Al_2O_3(s) + 2Fe(s) \longrightarrow Fe_2O_3(s) + 2Al(s)$

$$\Delta S > 0 \text{ and } \Delta H > 0$$

(b) $CS_2(g) \longrightarrow CS_2(l)$ $\Delta S < 0 \text{ and } \Delta H < 0$

Third Law of Thermodynamics

18.77 Calculate $\Delta S°$ for the following reactions in J K^{-1} from the data in Table 18.1. Considering only the values of $\Delta S°$, which of these reactions are favored to occur spontaneously?

(a) $N_2(g) + 3H_2(g) \longrightarrow 2NH_3(g)$

(b) $CO(g) + 2H_2(g) \longrightarrow CH_3OH(l)$

(c) $2C_2H_6(g) + 7O_2(g) \longrightarrow 4CO_2(g) + 6H_2O(g)$

(d) $Ca(OH)_2(s) + H_2SO_4(l) \longrightarrow$
$$CaSO_4(s) + 2H_2O(l)$$

(e) $S(s) + 2N_2O(g) \longrightarrow SO_2(g) + 2N_2(g)$

18.78 Calculate $\Delta S°$ for the following reactions in J K^{-1}, using the data in Table 18.1.

(a) $2Ag(s) + Cl_2(g) \longrightarrow 2AgCl(s)$

(b) $H_2(g) + \frac{1}{2}O_2(g) \longrightarrow H_2O(g)$

(c) $H_2(g) + \frac{1}{2}O_2(g) \longrightarrow H_2O(l)$

(d) $CaCO_3(s) + H_2SO_4(l) \longrightarrow$
$$CaSO_4(s) + H_2O(g) + CO_2(g)$$

(e) $NH_3(g) + HCl(g) \longrightarrow NH_4Cl(s)$

18.79 Calculate $\Delta S_f°$ for these compounds in J mol^{-1} K^{-1}.

(a) $C_2H_4(g)$

(b) $HC_2H_3O_2(l)$

(c) $CaSO_4 \cdot 2H_2O(s)$

18.80 Calculate $\Delta S_f°$ for these compounds in J mol^{-1} K^{-1}.

(a) $Al_2O_3(s)$

(b) $CaSO_4 \cdot \frac{1}{2}H_2O(s)$

(c) $NH_4Cl(s)$

18.81 Nitrogen dioxide, NO_2, an air pollutant, dissolves in rainwater to form a dilute solution of nitric acid. The equation for the reaction is

$$3NO_2(g) + H_2O(l) \longrightarrow 2HNO_3(l) + NO(g)$$

Calculate $\Delta S°$ for this reaction in J K^{-1}.

18.82 Good wine will turn to vinegar if it is left exposed to air, because the alcohol is oxidized to acetic acid. The equation for the reaction is

$$C_2H_5OH(l) + O_2(g) \longrightarrow HC_2H_3O_2(l) + H_2O(l)$$

Calculate $\Delta S°$ for this reaction in J mol^{-1} K^{-1}.

Standard Free Energy Change, $\Delta G°$

18.83 Phosgene, $COCl_2$, was used as a war gas during World War I. It reacts with the moisture in the lungs to produce HCl, which causes the lungs to fill with fluid, and CO, which asphyxiates the victim. Both lead ultimately to death. For $COCl_2(g)$, $S° = 284$ J/mol K and $\Delta H_f° = -223$ kJ/mol. Use this information and the data in Table 18.1 to calculate $\Delta G_f°$ for $COCl_2(g)$ in kJ mol^{-1}.

18.84 Aluminum oxidizes rather easily, but forms a thin protective coating of Al_2O_3 that prevents further oxidation of the aluminum beneath. Use the data for $\Delta H_f°$ (Table 6.2)

and $S°$ (Table 18.1) to calculate $\Delta G_f°$ for $Al_2O_3(s)$ in kJ mol^{-1}.

18.85 Compute $\Delta G°$ in kJ for the following reactions, using the data in Table 18.2.

(a) $SO_3(g) + H_2O(l) \longrightarrow H_2SO_4(l)$

(b) $2NH_4Cl(s) + CaO(s) \longrightarrow$
$$CaCl_2(s) + H_2O(l) + 2NH_3(g)$$

(c) $CaSO_4(s) + 2HCl(g) \longrightarrow CaCl_2(s) + H_2SO_4(l)$

18.86 Compute $\Delta G°$ in kJ for the following reactions, using the data in Table 18.2.

(a) $2HCl(g) + CaO(s) \longrightarrow CaCl_2(s) + H_2O(g)$

(b) $2AgCl(s) + Ca(s) \longrightarrow CaCl_2(s) + 2Ag(s)$

(c) $3NO_2(g) + H_2O(l) \longrightarrow 2HNO_3(l) + NO(g)$

*18.87 Given the following,

$$4NO(g) \longrightarrow 2N_2O(g) + O_2(g)$$
$$\Delta G° = -139.56 \text{ kJ}$$
$$2NO(g) + O_2(g) \longrightarrow 2NO_2(g)$$
$$\Delta G° = -69.70 \text{ kJ}$$

calculate $\Delta G°$ for the reaction

$$2N_2O(g) + 3O_2(g) \longrightarrow 4NO_2(g)$$

*18.88 Given the following reactions and their $\Delta G°$ values,

$$COCl_2(g) + 4NH_3(g) \longrightarrow$$
$$CO(NH_2)_2(s) + 2NH_4Cl(s) \quad \Delta G° = -332.0 \text{ kJ}$$
$$COCl_2(g) + H_2O(l) \longrightarrow CO_2(g) + 2HCl(g)$$
$$\Delta G° = -141.8 \text{ kJ}$$
$$NH_3(g) + HCl(g) \longrightarrow NH_4Cl(s)$$
$$\Delta G° = -91.96 \text{ kJ}$$

calculate the value of $\Delta G°$ for the reaction

$$CO(NH_2)_2(s) + H_2O(l) \longrightarrow CO_2(g) + 2NH_3(g)$$

Maximum Work and ΔG

18.89 Gasohol is a mixture of gasoline and ethanol (grain alcohol), C_2H_5OH. Calculate the maximum work that could be obtained at 25 °C and 1 atm by burning 1 mol of C_2H_5OH.

$$C_2H_5OH(l) + 3O_2(g) \longrightarrow 2CO_2(g) + 3H_2O(g)$$

18.90 What is the maximum amount of useful work that could possibly be obtained at 25 °C and 1 atm from the combustion of 48.0 g of natural gas, $CH_4(g)$ to give $CO_2(g)$ and $H_2O(g)$?

Free Energy and Equilibrium

18.91 Chloroform, formerly used as an anesthetic and now believed to be a carcinogen, has a heat of vaporization $\Delta H_{vaporization} = 31.4$ kJ mol^{-1}. The change, $CHCl_3(l) \longrightarrow CHCl_3(g)$, has $\Delta S° = 94.2$ J mol^{-1} K^{-1}. At what temperature do we expect $CHCl_3$ to boil (i.e., at what temperature will liquid and vapor be in equilibrium at 1 atm pressure)?

18.92 For the melting of aluminum, $Al(s) \longrightarrow Al(l)$, $\Delta H^\circ = 10.0$ kJ mol^{-1}, and $\Delta S^\circ = 9.50$ J/mol K. Calculate the melting point of Al. (The actual melting point is 660 °C.)

18.93 Isooctane, a minor constituent of gasoline, has a boiling point of 99.3 °C and a heat of vaporization of 37.7 kJ mol^{-1}. What is ΔS (in J mol^{-1} K^{-1}) for the vaporization of 1 mol of isooctane?

18.94 Acetone (nail polish remover) has a boiling point of 56.2 °C. The change, $(CH_3)_2CO(l) \longrightarrow (CH_3)_2CO(g)$, has $\Delta H^\circ = 31.9$ kJ mol^{-1}. What is ΔS° for this change?

18.95 Determine whether the following reaction (equation unbalanced) will be spontaneous at 25 °C. (Do we expect appreciable amounts of products to form?)

$$C_2H_4(g) + HNO_3(l) \longrightarrow$$
$$HC_2H_3O_2(l) + H_2O(l) + NO(g) + NO_2(g)$$

18.96 Which of the following reactions (equations unbalanced) would be expected to be spontaneous at 25 °C and 1 atm?

(a) $PbO(s) + NH_3(g) \longrightarrow Pb(s) + N_2(g) + H_2O(g)$

(b) $NaOH(s) + HCl(g) \longrightarrow NaCl(s) + H_2O(l)$

(c) $Al_2O_3(s) + Fe(s) \longrightarrow Fe_2O_3(s) + Al(s)$

(d) $2CH_4(g) \longrightarrow C_2H_6(g) + H_2(g)$

Equilibrium Constants and ΔG°

18.97 Calculate the value of the thermodynamic equilibrium constant for the following reactions at 25 °C. (Refer to the data in Appendix C.)

(a) $2PCl_3(g) + O_2(g) \rightleftharpoons 2POCl_3(g)$

(b) $2SO_3(g) \rightleftharpoons 2SO_2(g) + O_2(g)$

18.98 Calculate the value of the thermodynamic equilibrium constant for the following reactions at 25 °C. (Refer to the data in Appendix C.) $\Delta G^\circ_{f,H_2O_2} = -105.6$ kJ mol^{-1}

(a) $N_2H_4(g) + 2O_2(g) \rightleftharpoons 2NO(g) + 2H_2O(g)$

(b) $N_2H_4(g) + 6H_2O_2(g) \rightleftharpoons 2NO_2(g) + 8H_2O(g)$

18.99 The reaction $NO_2(g) + NO(g) \rightleftharpoons N_2O(g) + O_2(g)$ has $\Delta G^\circ_{1273} = -9.67$ kJ. A 1.00 L reaction vessel at 1000.0 °C contains 0.0200 mol NO_2, 0.040 mol NO, 0.015 mol N_2O, and 0.0350 mol O_2. Is the reaction at equilibrium? If not, in which direction will the reaction proceed to reach equilibrium?

18.100 The reaction $CO(g) + H_2O(g) \rightleftharpoons HCHO_2(g)$ has $\Delta G^\circ_{673} = +79.8$ kJ mol^{-1}. If a mixture at 673 K contains 0.040 mol CO, 0.022 mol H_2O, and 3.8×10^{-3} mol $HCHO_2$ in a 2.50 L container, is the reaction at equilibrium? If not, in which direction will the reaction proceed spontaneously?

18.101 A reaction that can convert coal to methane (the chief component of natural gas) is

$$C(s) + 2H_2(g) \rightleftharpoons CH_4(g)$$

for which $\Delta G^\circ = -50.79$ kJ mol^{-1}. What is the value of K_p for this reaction at 25 °C? Does this value of K_p suggest that studying this reaction as a means of methane production is worthwhile pursuing?

18.102 One of the important reactions in living cells from which the organism draws energy is the reaction of adenosine triphosphate (ATP) with water to give adenosine diphosphate (ADP) and free phosphate ion.

$$ATP + H_2O \rightleftharpoons ADP + PO_4^{3-}$$

The value of ΔG°_{310} for this reaction at 37 °C (normal human body temperature) is -33 kJ mol^{-1}. Calculate the value of the equilibrium constant for the reaction at this temperature.

18.103 What is the value of the equilibrium constant for a reaction for which $\Delta G^\circ = 0$? What will happen to the composition of the system if we begin the reaction with the pure products?

18.104 Methanol, a potential replacement for gasoline as an automotive fuel, can be made from H_2 and CO by the reaction

$$CO(g) + 2H_2(g) \rightleftharpoons CH_3OH(g)$$

At 500.0 K, this reaction has $K_p = 6.25 \times 10^{-3}$. Calculate ΔG°_{500} for this reaction in units of kilojoules.

Bond Energies

18.105 Use the data in Table 18.4 to compute the approximate atomization energy of NH_3.

18.106 Approximately how much energy would be released during the formation of the bonds in one mole of acetone molecules? Acetone, the solvent usually found in nail polish remover, has the structural formula

18.107 The standard heat of formation of ethanol vapor, $C_2H_5OH(g)$, is -235.3 kJ mol^{-1}. Use the data in Table 18.3 and the average bond energies for C—C, C—H, and O—H bonds to estimate the C—O bond energy in this molecule. The structure of the molecule is

18.108 The standard heat of formation of ethylene, $C_2H_4(g)$, is $+52.284$ kJ mol^{-1}. Calculate the C=C bond energy in this molecule.

18.109 Carbon disulfide, CS_2, has the Lewis structure $\ddot{S}=C=\ddot{S}$, and for $CS_2(g)$, $\Delta H_f^\circ = +115.3$ kJ mol^{-1}. Use the data in Table 18.3 to calculate the average $C=S$ bond energy in this molecule.

18.110 Gaseous hydrogen sulfide, H_2S, has $\Delta H_f^\circ = -20.15$ kJ mol^{-1}. Use the data in Table 18.3 to calculate the average $S-H$ bond energy in this molecule.

18.111 For $SF_6(g)$, $\Delta H^\circ = -1096$ kJ mol^{-1}. Use the data in Table 18.3 to calculate the average $S-F$ bond energy in SF_6.

18.112 Use the results of the preceding problem and the data in Table C.3 of Appendix C to calculate the standard heat of formation of $SF_4(g)$. The measured value of ΔH_f° for $SF_4(g)$ is -718.4 kJ mol^{-1}. What is the per-centage difference between your calculated value of ΔH_f° and the experimentally determined value?

18.113 Use the data in Tables 18.3 and 18.4 to estimate the standard heat of formation of acetylene, $H-C\equiv C-H$, in the gaseous state.

18.114 What would be the approximate heat of formation of CCl_4 vapor at 25 °C and 1 atm?

★ 18.115 Which substance should have the more exothermic heat of formation, CF_4 or CCl_4?

★ 18.116 Would you expect the value of ΔH_f° for benzene, C_6H_6, computed from tabulated bond energies, to be very close to the experimentally measured value of ΔH_f°? Justify your answer.

Additional Exercises

18.117 Look at Table C.2 in Appendix C. Some of the ions when dissolved in water have entropies that are negative—for example, $Ca^{2+}(aq)$ $S^\circ = -52.1$ J mol^{-1} K^{-1}. Explain how a solvated ion can have a negative entropy.

18.118 Calculate the ΔG° for the dissolution of calcium chloride and calcium carbonate. Using your values, explain the solubility of calcium chloride and calcium carbonate as expressed in the solubility rules.

18.119 If pressure is expressed in atmospheres and volume is expressed in liters, $P\,\Delta V$ has units of L atm (liters × atmospheres). In Chapter 10 you learned that 1 atm = 101,325 Pa, and in Chapter 1 you learned that 1 L = 1 dm^3. Use this information to determine the number of joules corresponding to 1 L atm.

18.120 Calculate the work, in joules, done by a gas as it expands at constant temperature from a volume of 3.00 L and a pressure of 5.00 atm to a volume of 8.00 L. The external pressure against which the gas expands is 1.00 atm. (1 atm = 101,325 Pa)

★ 18.121 When an ideal gas expands at a constant temperature, $\Delta E = 0$ for the change. Why?

★ 18.122 When a real gas expands at a constant temperature, $\Delta E > 0$ for the change. Why?

18.123 An ideal gas in a cylinder fitted with a piston expands at constant temperature from a pressure of 5 atm and a volume of 6.0 L to a final volume of 12 L against a constant opposing pressure of 2.5 atm. How much heat does the gas absorb, expressed in units of L atm (liter × atm)? (*Hint:* See Exercise 18.121.)

★ 18.124 A cylinder fitted with a piston contains 5.00 L of a gas at a pressure of 4.00 atm. The entire apparatus is contained in a water bath to maintain a constant temperature of 25 °C. The piston is released and the gas expands until the pressure inside the cylinder equals the atmospheric pressure outside, which is 1 atm. Assume ideal gas behavior and calculate the amount of work done by the gas as it expands at constant temperature.

★ 18.125 The experiment described in Exercise 18.124 is repeated, but this time a weight, which exerts a pressure of 2 atm, is placed on the piston. When the gas expands, its pressure drops to this 2 atm pressure. Then the weight is removed and the gas is allowed to expand again to a final pressure of 1 atm. Throughout both expansions the temperature of the apparatus was held at a constant 25 °C. Calculate the amount of work performed by the gas in each step. How does the combined total amount of work in this two-step expansion compare with the amount of work done by the gas in the one-step expansion described in Exercise 18.124? How can even more work be obtained by the expansion of the gas?

18.126 When potassium iodide dissolves in water, the mixture becomes cool. For this change, which is of a larger magnitude, $T\,\Delta S$ or ΔH?

18.127 The enthalpy of combustion $\Delta H_{combustion}^\circ$, of oxalic acid, $H_2C_2O_4(s)$, is -246.05 kJ mol^{-1}. Consider the following data:

Substance	ΔH_f° J mol^{-1}	S° (J mol^{-1} K^{-1})
$C(s)$	0	5.69
$CO_2(g)$	-393.5	213.6
$H_2(g)$	0	130.6
$H_2O(l)$	-285.8	69.96
$O_2(g)$	0	205.0
$H_2C_2O_4(s)$	?	120.1

(a) Write the balanced thermochemical equation that describes the combustion of one mole of oxalic acid.

(b) Write the balanced thermochemical equation that describes the formation of one mole of oxalic acid.

(c) Use the information in the table above and the equations in parts (a) and (b) to calculate ΔH_f° for oxalic acid.

(d) Calculate ΔS_f° for oxalic acid and ΔS° for the combustion of one mole of oxalic acid.

(e) Calculate ΔG_f° for oxalic acid and ΔG° for the combustion of one mole of oxalic acid.

18.128 Many biochemical reactions have positive values for ΔG° and seemingly should not be expected to be spontaneous. They occur, however, because they are chemically coupled with other reactions that have negative values of ΔG°. An example is the set of reactions that forms the beginning part of the sequence of reactions involved in the metabolism of glucose, a sugar. Given these reactions and their corresponding ΔG° values,

$$\text{glucose} + PO_4^{3-} \longrightarrow \text{glucose 6-phosphate} + H_2O$$
$$\Delta G^\circ = +13.13 \text{ kJ}$$

$$ATP + H_2O \longrightarrow ADP + PO_4^{3-}$$
$$\Delta G^\circ = -32.22 \text{ kJ}$$

calculate ΔG° for the coupled reaction

$$\text{glucose} + ATP \longrightarrow \text{glucose 6-phosphate} + ADP$$

*18.129 The reaction

$$2C_4H_{10}(g) + 13O_2(g) \longrightarrow 8CO_2(g) + 10H_2O(g)$$

has $\Delta G^\circ = -5407$ kJ. Determine the value of ΔG_f° for $C_4H_{10}(g)$. Calculate the value of K_c for the reaction at 25 °C.

*18.130 At 1500 °C, $K_c = 5.67$ for the reaction

$$CH_4(g) + H_2O(g) \rightleftharpoons CO(g) + 3H_2(g)$$

Calculate the value of ΔG_{1773}° for the reaction at this temperature.

18.131 Given the following reactions and their values of ΔG°, calculate the value of ΔG_f° for $N_2O_5(g)$.

$$2H_2(g) + O_2(g) \longrightarrow 2H_2O(l)$$
$$\Delta G^\circ = -474.4 \text{ kJ}$$

$$N_2O_5(g) + H_2O(l) \longrightarrow 2HNO_3(l)$$
$$\Delta G^\circ = -37.6 \text{ kJ}$$

$$\tfrac{1}{2}N_2(g) + \tfrac{3}{2}O_2(g) + 2H_2(g) \longrightarrow HNO_3(l)$$
$$\Delta G^\circ = -79.91 \text{ kJ}$$

*18.132 At room temperature (25 °C), the gas ClNO is impure because it decomposes slightly according to the equation

$$2ClNO(g) \rightleftharpoons Cl_2(g) + 2NO(g)$$

The extent of decomposition is about 5%. What is the approximate value of ΔG_{298}° for this reaction at this temperature?

*18.133 The reaction

$$N_2O(g) + O_2(g) \rightleftharpoons NO_2(g) + NO(g)$$

has $\Delta H^\circ = -42.9$ kJ and $\Delta S^\circ = -26.1$ J/K. Suppose 0.100 mol of N_2O and 0.100 mol of O_2 are placed in a 2.00 L container at 475 °C and this equilibrium is established. What percentage of the N_2O has reacted? (*Note:* Assume that ΔH and ΔS are relatively insensitive to temperature, so ΔH_{298}° and ΔS_{298}° are about the same as ΔH_{773}° and ΔS_{773}°, respectively.)

18.134 Use the data in Table 18.3 to calculate the bond energy in the nitrogen molecule and the oxygen molecule.

18.135 The heat of vaporization of carbon tetrachloride, CCl_4, is 29.9 kJ mol^{-1}. Using this information and data in Tables 18.3 and 18.4, estimate the standard heat of formation of liquid CCl_4.

18.136 At 25 °C, 0.0560 mol O_2 and 0.020 mol N_2O were placed in a 1.00 L container where the following equilibrium was then established.

$$2N_2O(g) + 3O_2(g) \rightleftharpoons 4NO_2(g)$$

At equilibrium, the NO_2 concentration was 0.020 M. Calculate the value of ΔG° for the reaction.

18.137 For the substance $SO_2F_2(g)$, $\Delta H_f^\circ = -858$ kJ mol^{-1}. The structure of the SO_2F_2 molecule is

Use the value of the S—F bond energy calculated in Problem 18.111 and the data in Appendix C.3 to determine the average S=O bond energy in SO_2F_2 in units of kJ mol^{-1}.

Multi-Concept Questions

*18.138 Ethyl alcohol, C_2H_5OH, has been suggested as an alternative to gasoline as a fuel. In Example 18.5 we calculated ΔG° for combustion of 1 mol of C_2H_5OH; in Example 18.6 we calculated ΔG° for combustion of 1 mol of octane. Let's assume that gasoline has the same properties as octane (one of its constituents). The density of C_2H_5OH is 0.7893 g/mL; the density of octane,

C_8H_{18}, is 0.7025 g/mL. Calculate the maximum work that could be obtained by burning 1 gallon (3.78 liters) each of C_2H_5OH and C_8H_{18}. On a *volume* basis, which is the better fuel?

18.139 When solutions of sodium hydroxide are used to neutralize hydrochloric acid, the standard heat of reaction is found to be -55.86 kJ. When propionic acid $HC_3H_5O_2$ is neutralized with sodium hydroxide the standard heat of the reaction is -49.23 kJ. What is the standard heat and entropy change for the ionization of propionic acid?

*18.140 You've recently inherited a fortune and immediately after that an inventor approaches you with the investment deal of a lifetime. The inventor has found a "system" to react water with methane to make methanol and hydrogen gas, two potentially valuable fuels. In your interview with the inventor, it is claimed that a revolutionary new catalyst shifts the equilibrium position of the reaction to favor products and that 134 kg of methanol are pro-

duced for each 117 cubic meters of methane used (at STP). Use your knowledge of chemical principles to find all of the flaws in this project before deciding to risk your new fortune on it.

*18.141 A certain weak acid has a pK_a of 5.83. When 100.0 mL of a 0.00525 M solution of this weak acid at 21.26 °C is reacted with 45.6 mL of 0.00634 M sodium hydroxide ($HA + OH^- \longrightarrow A^- + H_2O$) at 22.18 °C, the temperature of the mixture rises to 24.88 °C. What is the entropy change associated with the ionization ($HA \longrightarrow A^- + H^+$) of this weak acid? (Assume that the volumes are additive and the densities of the two solutions are 1.000 g cm^{-3}, with a specific heat that is the same as pure water.)

|Exercises in Critical Thinking

18.142 The average C—H bond energy calculated using the procedure in Section 18.10 is not quite equal to the energy needed to cause the reaction $CH_4(g) \longrightarrow CH_3(g) + H(g)$. Suggest reasons why this is so.

18.143 Discuss this statement: A world near absolute zero would be controlled almost entirely by potential energy. NASA states that the average temperature of the universe is 2.73 K. What does this mean about the universe?

18.144 If a catalyst were able to affect the position of equilibrium in a reaction, it would be possible to construct a perpetual motion machine (a machine from which

energy could be extracted without having to put energy into it). Imagine how such a machine could be made. Why would it violate the first law of thermodynamics?

18.145 At the beginning of this chapter we noted that the reaction of CO_2 with H_2O to form a hydrocarbon fuel is nonspontaneous. According to thermodynamics, where would the position of equilibrium lie for such a reaction. Why?

Electrochemistry

Chapter Outline

Erik Isakson/Getty Images

People around the world are coming to the conclusion that we must develop "green" economic and consumption policies that are sustainable far into the future. This means that the use of resources should be minimized, and those resources that are used for a wide variety of human activities must be replenished rapidly. One way to achieve a sustainable energy economy is to convert solar and wind energy directly into electricity, using technology such as the windmills shown in the chapter opening photograph, bypassing petroleum and all of its production, refining, and pollution problems. Since solar energy is only available for a part of a day and wind energy is variable, economical storage of electrical energy in high-tech batteries is essential.

In this chapter we will study how it is possible to separate the processes of oxidation (electron loss) and reduction (electron gain) and cause them to occur in different physical locations. When we are able to do this in an arrangement called a galvanic cell, we can use spontaneous redox reactions to produce electricity. And by reversing the process, we can use electricity to make nonspontaneous redox reactions happen to produce important products by a process called *electrolysis*.

This Chapter in Context

LEARNING OBJECTIVES

After reading this chapter, you should be able to:

- set up and use galvanic cells
- predict galvanic cell potentials
- use standard reduction potentials to predict spontaneous reactions
- relate standard cell potentials and standard free energies
- relate the concentration of solutes to cell potentials
- understand how electricity is produced
- explain the nature of electrolysis
- use electrolysis information quantitatively
- describe some practical applications of electrolysis

19.1 | Galvanic (Voltaic) Cells

Batteries have become common sources of portable power for a wide range of consumer products, from cell phones to iPods to laptops and hybrid cars. The energy from a battery comes from a spontaneous redox reaction in which the electron transfer is forced to take place through a wire. The apparatus that provides electricity in this way is called a **galvanic cell**, after Luigi Galvani (1737–1798), an Italian anatomist who discovered that electricity can cause the contraction of muscles. (It is also called a **voltaic cell**, after another Italian scientist, Allesandro Volta (1745–1827), whose inventions led ultimately to the development of modern batteries.)

Construction of a Galvanic Cell

If a shiny piece of metallic copper is placed into an aqueous solution of silver nitrate, a spontaneous reaction occurs. Gradually, a grayish white deposit forms on the copper and the liquid itself becomes pale blue as hydrated Cu^{2+} ions enter the solution (see Figure 19.1). The equation for this reaction is

$$2Ag^+(aq) + Cu(s) \longrightarrow Cu^{2+}(aq) + 2Ag(s)$$

Although the reaction is exothermic, no usable energy can be harnessed from it because all the energy is dispersed as heat.

To produce useful *electrical* energy, the two half-reactions involved in the net reaction must be made to occur in separate containers or compartments called **half-cells**. When this is done, electrons must flow through external wires to complete the circuit. The flow of electrons can then be used to power devices such as a laser pointer, a laptop computer, or your iPod. An apparatus to accomplish this—a galvanic cell—is made up of two half-cells, as illustrated in Figure 19.2. On the left, a silver electrode dips into a solution of $AgNO_3$, and, on the right, a copper electrode dips into a solution of $Cu(NO_3)_2$. The two electrodes are connected by an external electrical circuit and the two solutions are connected by a *salt bridge*, the function of which will be described shortly. When the circuit is completed by closing the switch, the reduction of Ag^+ to Ag occurs spontaneously in the half-cell on the left and oxidation of Cu to Cu^{2+} occurs spontaneously in the half-cell on the right. The reaction that takes place in each half-cell is a *half-reaction* of the type you

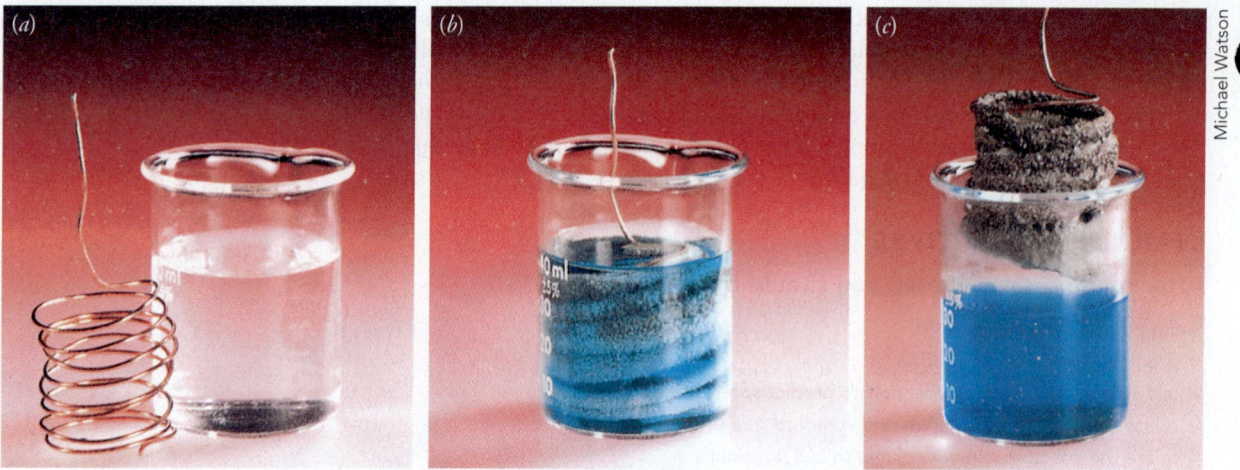

Figure 19.1 | **Reaction of copper with a solution of silver nitrate.** (*a*) A coil of copper wire stands next to a beaker containing a silver nitrate solution. (*b*) When the copper wire is placed in the solution, copper dissolves, giving the solution its blue color, and metallic silver deposits as glittering crystals on the wire. (*c*) After a while, much of the copper has dissolved and nearly all of the silver has deposited as the free metal.

learned to balance by the ion–electron method in Chapter 5. In the silver half-cell, the following half-reaction occurs.

$$Ag^+(aq) + e^- \longrightarrow Ag(s) \qquad \text{(reduction)}$$

In the copper half-cell, the half-reaction is

$$Cu(s) \longrightarrow Cu^{2+}(aq) + 2e^- \qquad \text{(oxidation)}$$

When these reactions take place, electrons left behind by oxidation of the copper travel through the external circuit to the other electrode where they are transferred to the silver ions, as Ag^+ is reduced to the familiar shiny silver metal.

Cell Reactions

The overall reaction that takes place in the galvanic cell is called the **cell reaction**. To obtain it, we combine the individual electrode half-reactions, making sure that the number of electrons gained in one half-reaction equals the number lost in the other. Thus, to obtain

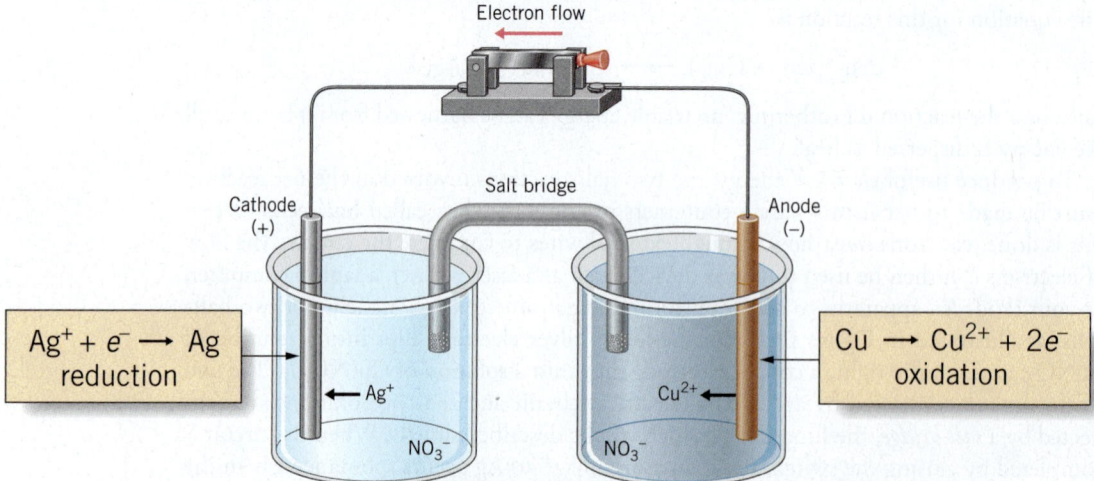

Figure 19.2 | **A galvanic cell.** The cell consists of two half-cells. Oxidation takes place in one half-cell and reduction in the other, as indicated by the half-reactions.

the cell reaction we multiply the half-reaction for the reduction of silver by 2 and then add the two half-reactions to obtain the net reaction. (Notice that $2e^-$ appears on each side, and so they cancel.) This is exactly the same process we used to balance redox reactions by the ion–electron method described in Section 5.2.

$$2Ag^+(aq) + 2e^- \longrightarrow 2Ag(s) \qquad \text{(reducation)}$$
$$Cu(s) \longrightarrow Cu^{2+}(aq) + 2e^- \qquad \text{(oxidation)}$$
$$\overline{2Ag^+(aq) + Cu(s) + 2e^- \longrightarrow 2Ag(s) + Cu^{2+}(aq) + 2e^-} \qquad \text{(cell reaction)}$$

Naming Electrodes in a Galvanic Cell

The electrodes in electrochemical systems are identified by the names *cathode* and *anode*. The names are *always* assigned according to the nature of the chemical changes that occur at the electrodes. In *all electrochemical systems:*

The **cathode** is the electrode at which reduction (electron gain) occurs.
The **anode** is the electrode at which oxidation (electron loss) occurs.

Thus, in the galvanic cell we've been discussing, the silver electrode is the cathode and the copper electrode is the anode.

Conduction of Charge

In the external circuit of a galvanic cell, electrical charge is transported from one electrode to the other by the movement of *electrons* through the wires. This is called **metallic conduction**, and it is how metals in general conduct electricity. In this external circuit, electrons always travel from the anode, where they are a product of the oxidation process, to the cathode, where they are picked up by the substance being reduced.

In electrochemical cells another kind of electrical conduction also takes place. In a solution that contains ions (or in a molten ionic compound), *electrical charge is carried through the liquid by the movement of ions, not electrons.* The transport of electrical charge by ions is called **electrolytic conduction**.

When the reactions take place in the copper–silver galvanic cell, positive copper ions *enter* the liquid that surrounds the anode while positive silver ions *leave* the liquid that surrounds the cathode (Figure 19.3). For the galvanic cell to work, the solutions in both

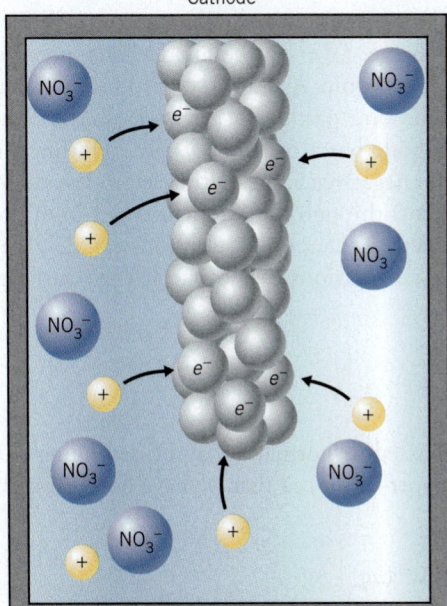

Cathode

Reduction of silver ions at the cathode extracts electrons from the electrode, so the electrode becomes positively charged.

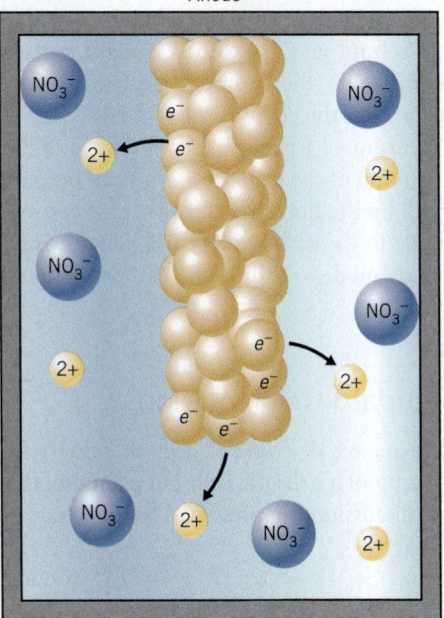

Anode

Oxidation of copper atoms at the anode leaves electrons behind on the electrode, which becomes negatively charged.

■ When electrons appear as a reactant, the process is reduction; when they appear as a product, it is oxidation.

■ The two electrons that are canceled in this reaction are also the number of electrons transferred. These will be important later when we discuss the Nernst equation.

TOOLS

Electrode names and reactions

Figure 19.3 | **Expanded view of Figure 19.2 to show changes that take place at the anode and cathode in the copper–silver galvanic cell.** (Not drawn to scale.) At the anode, Cu^{2+} ions enter the solution when copper atoms are oxidized, leaving electrons behind on the electrode. Unless Cu^{2+} ions move away from the electrode or NO_3^- ions move toward it, the electrode will become positively charged. At the cathode Ag^+ ions leave the solution and become silver atoms by acquiring electrons from the electrode surface. Unless more silver ions move toward the cathode or negative ions move away, the solution around the electrode will become negatively charged.

half-cells must remain electrically neutral. This requires that ions be permitted to enter or leave the solutions. For example, when copper is oxidized, the solution surrounding the anode becomes filled with Cu^{2+} ions, so negative ions are needed to balance their charge. Similarly, when Ag^+ ions are reduced, NO_3^- ions are left behind in the solution and positive ions are needed to maintain neutrality. The **salt bridge** shown in Figure 19.2 allows the movement of ions required to keep the solutions electrically neutral. The salt bridge is also essential to complete the electrical circuit.

A salt bridge is a tube filled with a solution of a salt composed of ions not involved in the cell reaction. Often KNO_3 or KCl is used. The tube is fitted with porous plugs at each end that prevent the solution from pouring out, but at the same time enabling the solution in the salt bridge to exchange ions with the solutions in the half-cells.

During operation of the cell, negative ions can diffuse from the salt bridge into the copper half-cell, or, to a much smaller extent, Cu^{2+} ions can leave the solution and enter the salt bridge. Both processes help keep the copper half-cell electrically neutral. In the silver half-cell, positive ions from the salt bridge can enter or negative NO_3^- ions can, once again to a much smaller extent, leave the half-cell by entering the salt bridge, to keep it electrically neutral too.

Without the salt bridge, electrical neutrality could not be maintained and no electrical current could be produced by the cell. Therefore, *electrolytic contact*—contact by means of a solution containing ions—must be maintained for the cell to function.

If we look closely at the overall movement of ions during the operation of the galvanic cell, we find that negative ions (*anions*) move away from the cathode, where they are present in excess, and *toward the anode*, where they are needed to balance the charge of the positive ions being formed. Similarly, we find that positive ions (*cations*) move away from the anode, where they are in excess, and *toward the cathode*, where they can balance the charge of the anions left in excess. In fact, the reason positive ions are called cations and negative ions are called anions is because of the nature of the electrodes toward which they move. In summary:

Cations move in the general direction of the cathode.
Anions move in the general direction of the anode.

Charges of the Electrodes

At the anode of the galvanic cell described in Figures 19.2 and 19.3, copper atoms spontaneously leave the electrode and enter the solution as Cu^{2+} ions. The electrons that are left behind give the anode a slight negative charge. (We say the anode has a *negative polarity*.) At the cathode, electrons spontaneously join Ag^+ ions to produce neutral atoms, but the effect is the same as if Ag^+ ions become part of the electrode, so the cathode acquires a slight positive charge. (The cathode has a *positive polarity*.) During the operation of the cell, the amount of positive and negative charge on the electrodes is kept small by the flow of electrons (an electric current) through the external circuit from the anode to the cathode when the circuit is complete. In fact, unless electrons can flow out of the anode and into the cathode, the chemical reactions that occur at their surfaces will cease.

Standard Cell Notation

As a matter of convenience, chemists have devised a shorthand way of describing the makeup of a galvanic cell. For example, the copper–silver cell that we have been using in our discussion is represented as follows:

$$Cu(s) \mid Cu^{2+}(aq) \parallel Ag^+(aq) \mid Ag(s)$$

By convention, in **standard cell notation**, the anode half-cell is specified on the left, with the electrode material of the anode given first. In this case, the anode is copper metal. The

■ Often the salt bridge is prepared by saturating a hot agar–agar solution with KNO_3 or KCl. After it's poured into a U-shaped tube the agar–agar solidifies on cooling. The salt ions can move but the agar–agar stays in place.

■ The small difference in charge between the electrodes is formed by the spontaneity of the overall reaction—that is, by the favorable free energy change. Nature's tendency toward electrical neutrality prevents a large buildup of charge on the electrodes and promotes the spontaneous flow of electricity through the external circuit.

■ Standard cell notation is also called a cell diagram.

single vertical bar represents the *phase boundary* between the copper electrode and the solution that surrounds it. The double vertical bars represent the two-phase boundaries, one at each end of the salt bridge, which connects the solutions in the two half-cells. On the right, the cathode half-cell is described, with the material of the cathode given last. Thus, the electrodes themselves (copper and silver) are specified at opposite ends of the cell description.

■ In the anode compartment, Cu is the reactant and is oxidized to Cu^{2+}, whereas in the cathode compartment, Ag^+ is the reactant and is reduced to Ag.

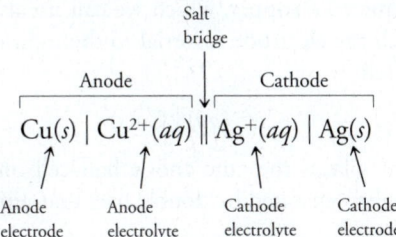

Sometimes, both the oxidized and reduced forms of the reactants in a half-cell are soluble and cannot be used as an electrode. In these cases, an inert electrode composed of platinum, graphite, or gold is used to provide a site for electron transfer. For example, a galvanic cell can be made using an anode composed of a zinc electrode dipping into a solution containing Zn^{2+} and a cathode composed of a platinum electrode dipping into a solution containing both Fe^{2+} and Fe^{3+} ions. The cell reaction is

$$2Fe^{3+}(aq) + Zn(s) \longrightarrow 2Fe^{2+}(aq) + Zn^{2+}(aq)$$

The cell notation for this galvanic cell is written as follows:

$$Zn(s) \mid Zn^{2+}(aq) \parallel Fe^{2+}(aq), Fe^{3+}(aq) \mid Pt(s)$$

■ Since they are thoroughly mixed in solution, it does not matter if Fe^{2+} or Fe^{3+} is written first.

where we have separated the formulas for the two iron ions by a comma. In this cell, the reduction of the Fe^{3+} to Fe^{2+} takes place at the surface of the inert platinum electrode.

Example 19.1
Describing Galvanic Cells

The following spontaneous reaction occurs when metallic zinc is dipped into a solution of copper sulfate.

$$Zn(s) + Cu^{2+}(aq) \longrightarrow Zn^{2+}(aq) + Cu(s)$$

Describe a galvanic cell that could take advantage of this reaction. What are the half-cell reactions? What is the standard cell notation? Make a sketch of the cell and label the cathode and anode, the charges on each electrode, the direction of ion flow, and the direction of electron flow.

Analysis: Answering all these questions relies on identifying the anode and cathode in the equation for the cell reaction; this is often the key to solving problems of this type. By definition, the anode is the electrode at which oxidation happens, and the cathode is where reduction occurs.

Assembling the Tools: We can use the principles of the ion–electron method for balancing redox reactions that were learned in Section 5.2. When the half-reactions are balanced, the oxidation reaction will occur at the anode and the reduction reaction will occur at the cathode. Finally, we will use the procedures just discussed to write the standard cell notation and draw our diagram.

Solution: The two half-reactions are

$$Zn(s) \longrightarrow Zn^{2+}(aq) + 2e^- \qquad \text{(anode reaction)}$$
$$Cu^{2+}(aq) + 2e^- \longrightarrow Cu(s) \qquad \text{(cathode reaction)}$$

We start writing the standard cell notation with the oxidation reaction that occurs at the anode. The anode half-cell is therefore the zinc metal as the electrode dipping into a solution that contains Zn^{2+} [e.g., from dissolved $Zn(NO_3)_2$ or $ZnSO_4$]. Symbolically, the anode

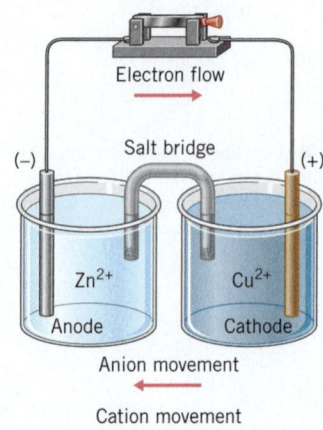

Electron flow

Salt bridge

(−)   (+)

Zn²⁺   Cu²⁺

Anode   Cathode

Anion movement

Cation movement

The zinc–copper cell. In the cell notation described in this example, we indicate the anode on the left and the cathode on the right. In this drawing of the apparatus, the anode half-cell is also shown on the left, but it could just as easily be shown on the right, as in Figure 19.2. Be sure you understand that where we place the apparatus on the lab bench doesn't affect which half-cell is the anode and which is the cathode.

half-cell is written with the electrode material at the left of the vertical bar and the oxidation product at the right.

$$Zn(s) \mid Zn^{2+}(aq)$$

Copper ions in solution, $Cu^{2+}(aq)$ [e.g., from dissolved $Cu(NO_3)_2$ or $CuSO_4$], gain electrons and are reduced to metallic copper, which we can use as the cathode. The copper half-cell is represented with the electrode material to the right of the vertical bar and the substance reduced on the left.

$$Cu^{2+}(aq) \mid Cu(s)$$

The standard cell notation places the zinc anode half-cell on the left and the copper cathode half-cell on the right, separated by double bars that represent the salt bridge.

$$Zn(s) \mid Zn^{2+}(aq) \parallel Cu^{2+}(aq) \mid Cu(s)$$
$$\phantom{Zn(s) \mid Zn^{2}}\text{anode}\phantom{(aq) \parallel Cu^{2+}}\text{cathode}$$

A sketch of the cell is shown in the margin. The anode always carries a negative charge in a galvanic cell, so the zinc electrode is negative and the copper electrode is positive. Electrons in the external circuit travel from the negative electrode to the positive electrode (i.e., from the Zn anode to the Cu cathode). Anions move toward the anode and cations move toward the cathode. In the salt bridge we also find that anions will flow toward the anode and cations toward the cathode as shown.

Are the Answers Reasonable? All of the answers depend on determining which substance is oxidized and which is reduced, so we check that first. Oxidation is electron loss, and Zn must lose electrons to become Zn^{2+}, so zinc is oxidized and must be the anode. If zinc is the anode, then copper must be the cathode. We can then reason that the oxidation of zinc produces electrons that flow from the anode to the cathode.

Practice Exercise 19.1

Sketch and label a galvanic cell that makes use of the following spontaneous redox reaction.

$$Mg(s) + Fe^{2+}(aq) \longrightarrow Mg^{2+}(aq) + Fe(s)$$

Write the half-reactions for the anode and cathode. Give the standard cell notation. (*Hint:* Determine which half-reaction represents oxidation and which represents reduction.)

Practice Exercise 19.2

Write the anode and cathode half-reactions for the following galvanic cell. Write the equation for the overall cell reaction.

$$Al(s) \mid Al^{3+}(aq) \parallel Ni^{2+}(aq) \mid Ni(s)$$

Sketch and label a galvanic cell that corresponds to this standard cell notation.

19.2 | Cell Potentials

A galvanic cell has an ability to push electrons through the external circuit. The magnitude of this ability is expressed as a **potential**. Potential is expressed in an electrical unit called the **volt (V)**, which is a measure of the amount of energy, in joules, that can be delivered per **coulomb** (the SI unit of charge) as the charges move through the circuit. Thus, a charge flowing under a potential of 1 volt can deliver 1 joule of energy per coulomb.

$$1\,V = 1\,J/C \tag{19.1}$$

■ The potential generated by a galvanic cell has also been called the **electromotive force** (emf). Modern electrochemistry uses the preferred E_{cell} or E°_{cell} to note cell potentials and standard cell potentials, respectively.

The voltage or potential of a galvanic cell varies with the amount of charge flowing through the circuit. The *maximum* potential that a given cell can generate is called its **cell potential, E_{cell}**, and it depends on the composition of the electrodes, the concentrations of the ions in the half-cells, and the temperature. The **standard state** for electrochemistry is defined as a system where the temperature is 25 °C, all concentrations are 1.00 M, and any gases are at 1.00 atm pressure. When the system is at standard state, the potential of a galvanic cell is the **standard cell potential**, symbolized by E°_{cell}.

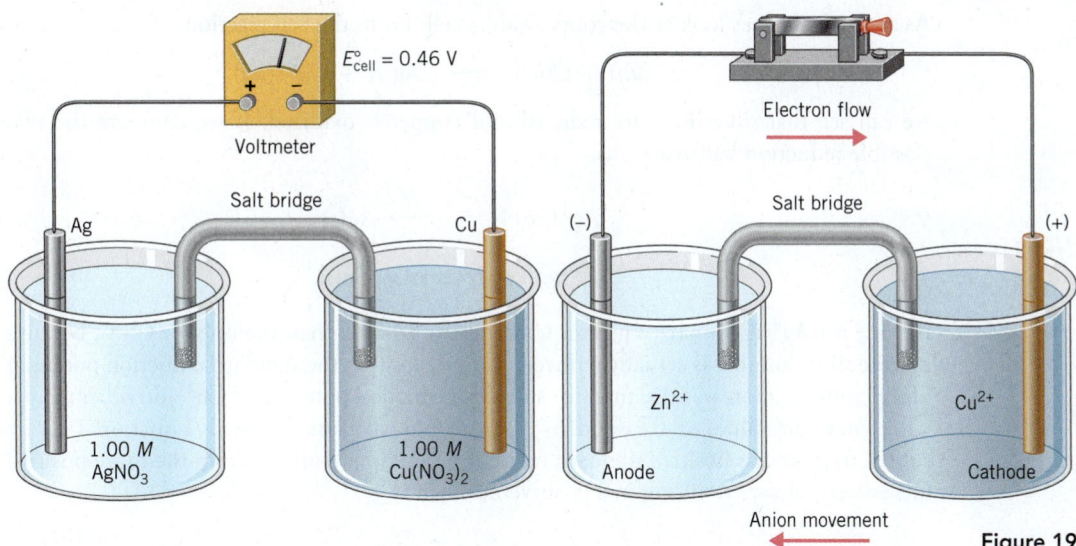

Figure 19.4 | A cell designed to generate the standard cell potential. The concentrations of the Cu^{2+} and Ag^+ ions in the half-cells are 1.00 *M*. It is very important to always connect the negative terminal of the voltmeter to the anode for correct readings.

Cell potentials are rarely larger than a few volts. For example, the standard cell potential for the galvanic cell constructed from silver and copper electrodes shown in Figure 19.4 is only 0.46 V, and one cell in an automobile battery produces only about 2 V. Batteries that generate higher voltages, such as the automobile battery, contain a number of cells connected one after the other (i.e., in series) so that their potentials are additive.

In Section 19.5 we will see that we can calculate the cell potential, E_{cell}, for systems that are not at standard state. As long as we know the temperature, concentration of solutes and pressures of gases, we will be able to calculate how much the cell potential differs from the standard cell potential. Without getting into the details that are presented later, it is important to note that in most cases the standard cell potential and the calculated cell potential for a reaction usually differ by less than 0.5 volt and usually have the same algebraic sign. Therefore we will make generalizations and demonstrate them using standard reduction potentials. These generalizations will usually apply to nonstandard state cell potentials.

Reduction Potentials

It is useful to think of each half-cell as having a certain natural tendency to acquire electrons and proceed as a *reduction*. The magnitude of this tendency is expressed by the half-reaction's **reduction potential**. When measured under standard conditions the reduction potential is called the **standard reduction potential**. To represent a standard reduction potential, we will add a subscript to the symbol $E°$ that identifies the substance undergoing reduction. Thus, the standard reduction potential for the half-reaction

$$Cu^{2+}(aq) + 2e^- \longrightarrow Cu(s)$$

is specified as $E°_{Cu^{2+}}$. Sometimes both the reactant and product are specified as part of the symbol, for example, $E°_{Cu^{2+}/Cu}$.

When two half-cells are connected to make a galvanic cell, the one with the larger standard reduction potential (the one with the greater tendency to undergo reduction) acquires electrons from the half-cell with the lower standard reduction potential, which is therefore forced to undergo oxidation. The standard cell potential, which is always taken to be a positive number, represents the *difference* between the standard reduction potential of one half-cell and the standard reduction potential of the other. In general, therefore,

■ If charge flows from a cell, some of the cell's voltage is lost overcoming its own internal resistance, and the measured voltage is less than the original E_{cell}.

■ Standard reduction potentials are also called **standard electrode potentials**.

■ In a galvanic cell, the measured cell potential is always taken to be a positive value. This is important to remember.

$$E°_{cell} = \begin{bmatrix} \text{standard reduction} \\ \text{potential of the} \\ \text{substance reduced} \end{bmatrix} - \begin{bmatrix} \text{standard reduction} \\ \text{potential of the} \\ \text{substance oxidized} \end{bmatrix}$$ (19.2)

TOOLS

$E°_{cell} = E°_{reduction} - E°_{oxidation}$

As an example, let's look at the copper–silver cell. From the cell reaction,

$$2Ag^+(aq) + Cu(s) \longrightarrow 2Ag(s) + Cu^{2+}(aq)$$

we can see that silver ions are reduced and copper is oxidized. If we compare the two possible reduction half-reactions,

$$Ag^+(aq) + e^- \longrightarrow Ag(s)$$

$$Cu^{2+}(aq) + 2e^- \longrightarrow Cu(s)$$

■ Notice that we use the standard cell potentials without any adjustment for the number of electrons in the half-reaction.

the one for Ag^+ must have a greater tendency to proceed than the one for Cu^{2+}, because it is the silver ion that is actually reduced. This means that the standard reduction potential of Ag^+ must be more positive than the standard reduction potential of Cu^{2+}. In other words, if we knew the values of $E^\circ_{Ag^+}$ and $E^\circ_{Cu^{2+}}$, we could calculate E°_{cell} with Equation 19.2 by subtracting the less positive standard reduction potential (copper) from the more positive one (silver), always resulting in a positive number.

$$E^\circ_{cell} = E^\circ_{Ag^+} - E^\circ_{Cu^{2+}}$$

The Hydrogen Electrode

Unfortunately there is no way to measure the standard reduction potential of an isolated half-cell. All we can measure is the difference in potential produced when two half-cells are connected. Therefore, to assign numerical values for standard reduction potentials, a reference electrode has been arbitrarily chosen and its standard reduction potential has been defined as *exactly zero volts*. This reference electrode is called the **standard hydrogen electrode** (see Figure 19.5). Gaseous hydrogen at a pressure of 1.00 atm is bubbled over a platinum electrode coated with very finely divided platinum, which provides a large catalytic surface area on which the electrode reaction can occur. This electrode is surrounded by a solution whose temperature is 25 °C and in which the hydrogen ion concentration is 1.00 M. The half-cell reaction at the platinum surface, written as a reduction, is

$$2H^+(aq, 1.00\ M) + 2e^- \rightleftharpoons H_2(g, 1.00\ atm) \qquad E^\circ_{H^+} \equiv 0\ V \quad (exactly)$$

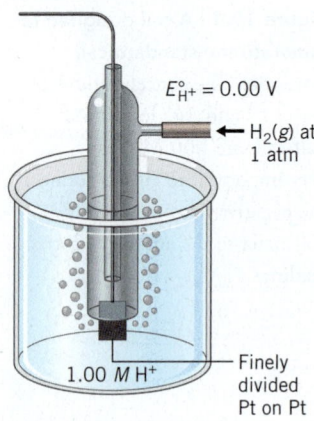

$E^\circ_{H^+} = 0.00\ V$

H$_2$(g) at 1 atm

Finely divided Pt on Pt

1.00 M H$^+$

Figure 19.5 | The hydrogen electrode. The half-reaction is $2H^+(aq) + 2e^- \rightleftharpoons H_2(g)$.

The double arrows indicate only that the reaction is reversible, not that there is true equilibrium. Whether the half-reaction occurs as reduction or oxidation depends on the standard reduction potential of the half-cell with which it is paired.

Figure 19.6 illustrates the hydrogen electrode connected to a copper half-cell to form a galvanic cell. When we use a voltmeter to measure the potential of the cell, we find that the copper electrode has a positive charge and the hydrogen electrode has a negative charge. Therefore, copper must be the cathode, and Cu^{2+} is reduced to Cu when the cell operates. Similarly, the hydrogen electrode must be the anode, and H_2 is oxidized to H^+. The half reactions and cell reaction, therefore, are

$$Cu^{2+}(aq) + 2e^- \longrightarrow Cu(s) \qquad (cathode\ reaction)$$

$$H_2(g) \longrightarrow 2H^+(aq) + 2e^- \qquad (anode\ reaction)$$

$$\overline{Cu^{2+}(aq) + H_2(g) \longrightarrow Cu(s) + 2H^+(aq)} \qquad (cell\ reaction)[1]$$

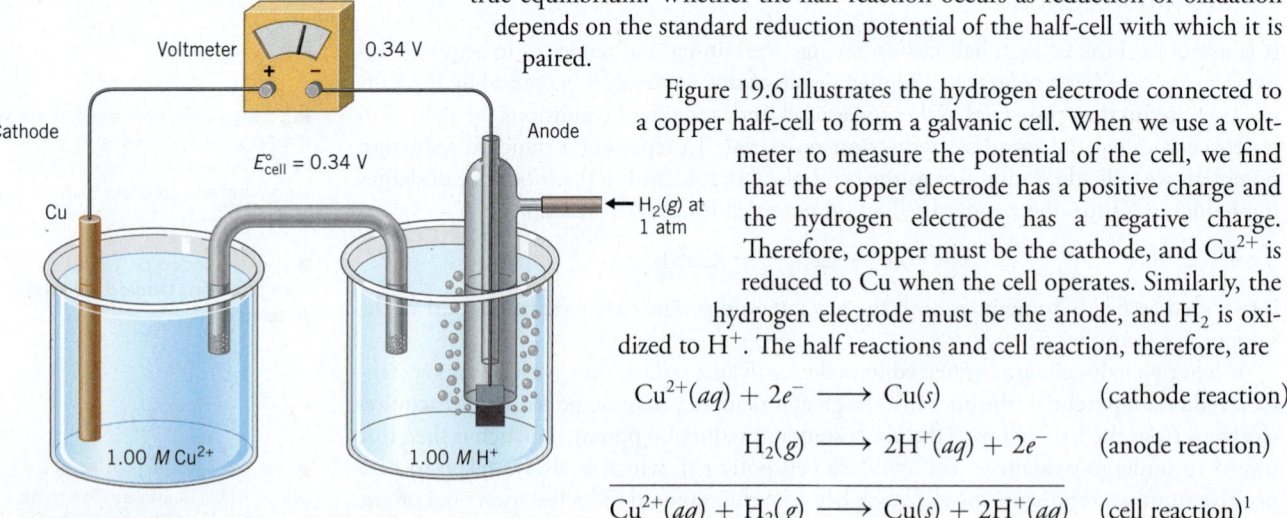

Voltmeter   0.34 V

Cathode   Anode

$E^\circ_{cell} = 0.34\ V$

Cu   H$_2$(g) at 1 atm

1.00 M Cu^{2+} 1.00 M H$^+$

Figure 19.6 | A galvanic cell composed of copper and hydrogen half-cells. The cell reaction is $Cu^{2+}(aq) + H_2(g) \longrightarrow$ $Cu(s) + 2H^+(aq)$.

---

[1] The cell notation for this cell is written as

$$Pt(s)\ |\ H_2(g)\ |\ H^+(aq)\ ||\ Cu^{2+}(aq)\ |\ Cu(s)$$

The notation for the hydrogen electrode (the anode in this case) is shown at the left of the double vertical bars.

Using Equation 19.2, we can express $E°_{cell}$ in terms of $E°_{Cu^{2+}}$ and $E°_{H^+}$.

$$E°_{cell} = E°_{Cu^{2+}} - E°_{H^+}$$

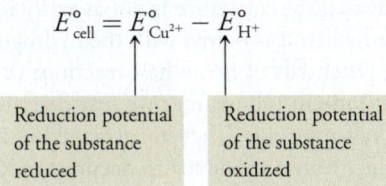

Reduction potential of the substance reduced

Reduction potential of the substance oxidized

The measured standard cell potential is 0.34 V, and $E°_{H^+}$ equals 0.00 V. Therefore,

$$0.34\ \text{V} = E°_{Cu^{2+}} - 0.00\ \text{V}$$

Relative to the hydrogen electrode, then, the standard reduction potential of Cu^{2+} is +0.34 V. (We have written the value with a plus sign because some standard reduction potentials are negative, as we will see.)

Now let's look at a galvanic cell set up between a zinc electrode and a hydrogen electrode (see Figure 19.7). This time we find that the hydrogen electrode is positive and the zinc electrode is negative, which tells us that the hydrogen electrode is the cathode and the zinc electrode is the anode. This means that the hydrogen ion is being reduced and zinc is being oxidized. Based on this reasoning the half-reactions and cell reaction are

$$2H^+(aq) + 2e \longrightarrow H_2(g) \qquad \text{(cathode reaction)}$$

$$Zn(s) \longrightarrow Zn^{2+}(aq) + 2e^- \qquad \text{(anode reaction)}$$

$$\overline{2H^+(aq) + Zn(s) \longrightarrow H_2(g) + Zn^{2+}(aq)} \quad \text{(cell reaction)}^2$$

From Equation 19.2, the standard cell potential is given by

$$E°_{cell} = E°_{H^+} + E°_{Zn^{2+}}$$

Substituting into this the measured standard cell potential of 0.76 V and $E°_{H^+} = 0.00$ V, we have

$$0.76\ \text{V} = 0.00\ \text{V} - E°_{Zn^{2+}}$$

which gives

$$E°_{Zn^{2+}} = -0.76\ \text{V}$$

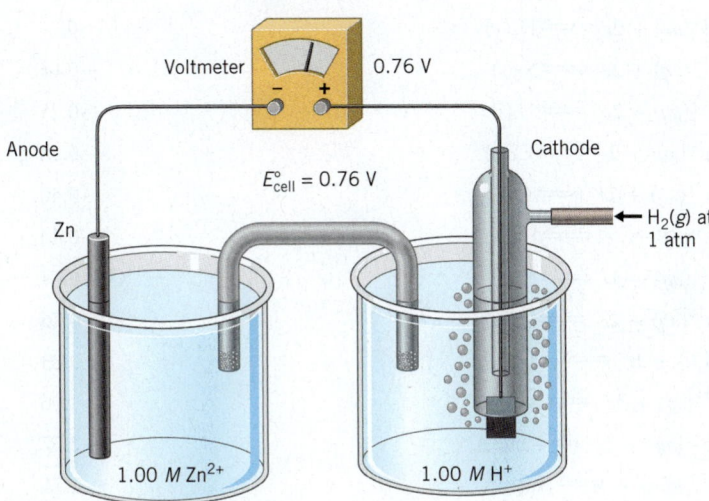

Figure 19.7 | A galvanic cell composed of zinc and hydrogen half-cells. The cell reaction is $Zn(s) + 2H^+(aq) \longrightarrow Zn^{2+}(aq) + H_2(g)$.

---

2 This cell is represented as

$$Zn(s)\ |\ Zn^{2+}(aq)\ ||\ H^+(aq)\ |\ H_2(g)\ |\ Pt(s)$$

This time the hydrogen electrode is the cathode and appears at the right of the double vertical bars.

Notice that the standard reduction potential of zinc is negative. A negative standard reduction potential simply means that the substance is not as easily reduced as H^+. In this case, it tells us that Zn is oxidized when it is paired with the hydrogen electrode.

The standard reduction potentials of many half-reactions can be compared to that for the standard hydrogen electrode in the manner we just described. Table 19.1 lists values obtained for some typical half-reactions. They are arranged in decreasing order—the half reactions at the top have the greatest tendency to occur as reductions, while those at the bottom have the least tendency to occur as reduction reactions.

■ Substances located to the left of the double arrows are oxidizing agents, because they become reduced when the reactions proceed in the forward direction. The best oxidizing agents are those found at the top of the table (e.g., F_2).

TABLE 19.1 Standard Reduction Potentials at 25 °C[a]	
Half-Reaction	**$E°$ (volts)**
$F_2(g) + 2e^- \rightleftharpoons 2F^-(aq)$	+2.87
$S_2O_8^{2-}(aq) + 2e^- \rightleftharpoons 2SO_4^{2-}(aq)$	+2.01
$PbO_2(s) + HSO_4^- + 3H^+ + 2e^- \rightleftharpoons PbSO_4(s) + 2H_2O$	+1.69
$MnO_4^-(aq) + 8H^+(aq) + 5e^- \rightleftharpoons Mn^{2+}(aq) + 4H_2O$	+1.51
$PbO_2(s) + 4H^+(aq) + 2e^- \rightleftharpoons Pb^{2+}(aq) + 2H_2O$	+1.46
$BrO_3^-(aq) + 6H^+(aq) + 6e^- \rightleftharpoons Br^-(aq) + 3H_2O$	+1.44
$Au^{3+}(aq) + 3e^- \rightleftharpoons Au(s)$	+1.42
$Cl_2(g) + 2e^- \rightleftharpoons 2Cl^-(aq)$	+1.36
$O_2(g) + 4H^+(aq) + 4e^- \rightleftharpoons 2H_2O$	+1.23
$Br_2(aq) + 2e^- \rightleftharpoons 2Br^-(aq)$	+1.07
$NO_3^- + 4H^+(aq) + 3e^- \rightleftharpoons NO(g) + 2H_2O$	+0.96
$Ag^+(aq) + e^- \rightleftharpoons Ag(s)$	+0.80
$Fe^{3+}(aq) + e^- \rightleftharpoons Fe^{2+}(aq)$	+0.77
$I_2(s) + 2e^- \rightleftharpoons 2I^-(aq)$	+0.54
$NiO_2(s) + 2H_2O + 2e^- \rightleftharpoons Ni(OH)_2(s) + 2OH^-(aq)$	+0.49
$Cu^{2+}(aq) + 2e^- \rightleftharpoons Cu(s)$	+0.34
$SO_4^{2-}(aq) + 4H^+(aq) + 2e^- \rightleftharpoons H_2SO_3(aq) + H_2O$	+0.17
$AgBr(s) + e^- \rightleftharpoons Ag(s) + Br^-(aq)$	+0.07
$2H^+(aq) + 2e^- \rightleftharpoons H_2(g)$	0
$Sn^{2+}(aq) + 2e^- \rightleftharpoons Sn(s)$	−0.14
$Ni^{2+}(aq) + 2e^- \rightleftharpoons Ni(s)$	−0.25
$Co^{2+}(aq) + 2e^- \rightleftharpoons Co(s)$	−0.28
$Cd^{2+}(aq) + 2e^- \rightleftharpoons Cd(s)$	−0.40
$Fe^{2+}(aq) + 2e^- \rightleftharpoons Fe(s)$	−0.44
$Cr^{3+}(aq) + 3e^- \rightleftharpoons Cr(s)$	−0.74
$Zn^{2+}(aq) + 2e^- \rightleftharpoons Zn(s)$	−0.76
$2H_2O + 2e^- \rightleftharpoons H_2(g) + 2OH^-(aq)$	−0.83
$Al^{3+}(aq) + 3e^- \rightleftharpoons Al(s)$	−1.66
$Mg^{2+}(aq) + 2e^- \rightleftharpoons Mg(s)$	−2.37
$Na^+(aq) + e^- \rightleftharpoons Na(s)$	−2.71
$Ca^{2+}(aq) + 2e^- \rightleftharpoons Ca(s)$	−2.87
$K^+(aq) + e^- \rightleftharpoons K(s)$	−2.92
$Li^+(aq) + e^- \rightleftharpoons Li(s)$	−3.05

■ Substances located to the right of the double arrows are reducing agents; they become oxidized when the reactions proceed from right to left. The best reducing agents are those found at the bottom of the table (e.g., Li).

[a]See Appendix C, Table C.9 for a more complete listing.

CHEMISTRY OUTSIDE THE CLASSROOM ⟮19.1⟯

Corrosion of Iron and Cathodic Protection

A problem that has plagued humanity ever since the discovery of methods for obtaining iron and other metals from their ores has been corrosion—the reaction of a metal with substances in the environment. The rusting of iron in particular is a serious problem because iron and steel have so many uses.

The rusting of iron is a complex chemical reaction that involves both oxygen and moisture (see Figure 1). Iron won't rust in pure water that's oxygen free, and it won't rust in pure oxygen in the absence of moisture. The corrosion process is apparently electrochemical in nature, as shown in the accompanying diagram. At one place on the surface, iron becomes oxidized in the presence of water and enters solution as Fe^{2+}.

$$Fe(s) \longrightarrow Fe^{2+}(aq) + 2e^-$$

At this location the iron is acting as an anode.

The electrons that are released when the iron is oxidized travel through the metal to some other place where the iron is exposed to oxygen. This is where reduction takes place (it's a cathodic region on the metal surface), and oxygen is reduced to give hydroxide ion.

$$\tfrac{1}{2}O_2(aq) + H_2O + 2e^- \longrightarrow 2OH^-(aq)$$

The iron(II) ions that are formed at the anodic regions gradually diffuse through the water and eventually contact the hydroxide ions. This causes a precipitate of $Fe(OH)_2$ to form, which is very easily oxidized by O_2 to give $Fe(OH)_3$. This hydroxide readily loses water. In fact, complete dehydration gives the oxide,

$$2Fe(OH)_3 \longrightarrow Fe_2O_3 + 3H_2O$$

When partial dehydration of the $Fe(OH)_3$ occurs, *rust* is formed. It has a composition that lies between that of the hydroxide and that of the oxide, Fe_2O_3, and is usually referred to as a *hydrated oxide*. Its formula is generally represented as $Fe_2O_3 \cdot xH_2O$.

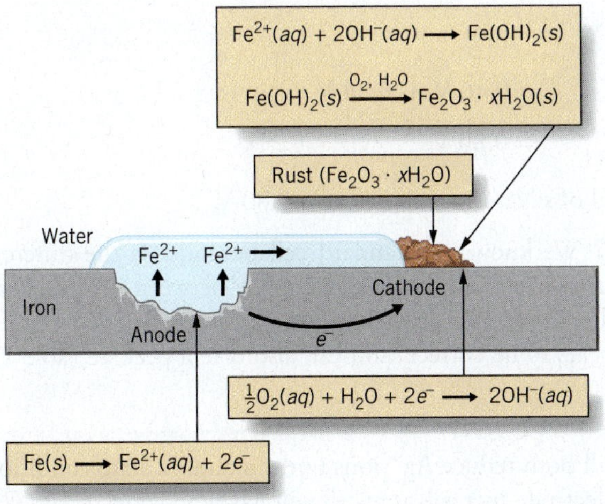

This mechanism for the rusting of iron explains one of the more interesting aspects of this damaging process. Perhaps you've noticed that when rusting occurs on the body of a car, the rust appears at and around a break (or a scratch) in the surface of the paint, but the damage extends under the painted surface for some distance. Apparently, the Fe^{2+} ions that are formed at the anode sites are able to diffuse rather long distances to the hole in the paint, where they finally react with air to form the rust.

Cathodic Protection

One way to prevent the rusting of iron is to coat it with another metal. This is done with "tin" cans, which are actually steel cans that have been coated with a thin layer of tin. However, if the layer of tin is scratched and the iron beneath is exposed, the corrosion is accelerated because iron has a lower reduction potential than tin; the iron becomes the anode in an electrochemical cell and is easily oxidized.

Another way to prevent corrosion is called *cathodic protection*. It involves placing the iron in contact with a metal that is *more easily* oxidized. This causes iron to be a cathode and the other metal to be the anode. If corrosion occurs, the iron is protected from oxidation because it is cathodic and the other metal reacts instead.

Zinc is most often used to provide cathodic protection to other metals. For example, zinc sacrificial anodes can be attached to the rudder of a boat (see Figure 2). When the rudder is submerged, the zinc will gradually corrode but the metal of the rudder will not. Periodically, the anodes are replaced to provide continued protection.

Steel objects that must withstand the weather are often coated with a layer of zinc, a process called galvanizing. You've probably seen this on chain-link fences and metal garbage pails. Even if the steel is exposed through a scratch, it is not oxidized because it is in contact with a metal that is more easily oxidized.

$$Fe^{2+}(aq) + 2OH^-(aq) \longrightarrow Fe(OH)_2(s)$$

$$Fe(OH)_2(s) \xrightarrow{O_2,\ H_2O} Fe_2O_3 \cdot xH_2O(s)$$

Rust ($Fe_2O_3 \cdot xH_2O$)

Water Fe^{2+} Fe^{2+}

Iron   Anode   Cathode e^-

$$\tfrac{1}{2}O_2(aq) + H_2O + 2e^- \longrightarrow 2OH^-(aq)$$

$$Fe(s) \longrightarrow Fe^{2+}(aq) + 2e^-$$

Figure 1 Corrosion of iron. Iron dissolves in anodic regions to give Fe^{2+}. Electrons travel through the metal to cathodic sites where oxygen is reduced, forming OH^-. The combination of the Fe^{2+} and OH^-, followed by air oxidation, gives rust.

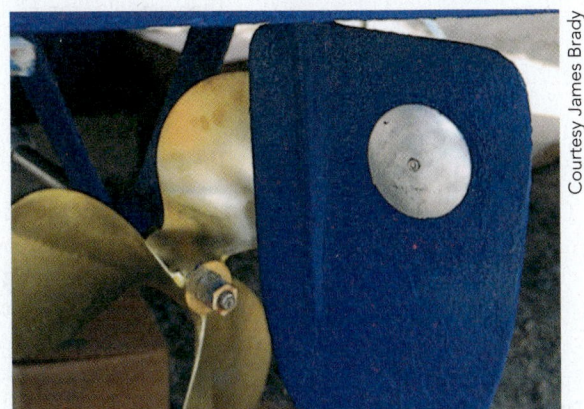

Courtesy James Brady

Figure 2 Cathodic protection. Before launching, a shiny new zinc anode disk is attached to the bronze rudder of this boat to provide cathodic protection. Over time, the zinc will corrode instead of the less reactive bronze. (The rudder is painted with a special blue paint to inhibit the growth of barnacles.)

Example 19.2
Calculating Standard Cell Potentials

We mentioned earlier that the standard cell potential of the silver–copper galvanic cell has a value of $+0.46$ V. The cell reaction is

$$2Ag^+(aq) + Cu(s) \longrightarrow 2Ag(s) + Cu^{2+}(aq)$$

and we have seen that the standard reduction potential of Cu^{2+}, $E°_{Cu^{2+}}$, is $+0.34$ V. What is the value of the standard reduction potential of Ag^+, $E°_{Ag^+}$?

Analysis: The standard cell potential is the difference between two standard reduction potentials. Since we know two of these three variables, we just have to be sure our math is correct.

Assembling the Tools: We know the standard potential of the cell and one of the two standard reduction potentials, so Equation 19.2 will be our tool to calculate the unknown standard reduction potential.

$$E°_{cell} = E°_{reduction} - E°_{oxidation}$$

The principles of oxidation and reduction will be used to determine which reactant is reduced and which is oxidized, allowing us to set up the equation correctly.

Solution: Silver changes from Ag^+ to Ag; its oxidation number decreases from $+1$ to 0, so we conclude that Ag^+ is reduced. Similar reasoning tells us that copper is oxidized from Cu to Cu^{2+}. Therefore, according to Equation 19.2,

$$E°_{cell} = E°_{Ag^+} - E°_{Cu^{2+}}$$

| Standard reduction potential of substance reduced | Standard reduction potential of substance oxidized |

Substituting values for $E°_{cell}$ and $E°_{Cu^{2+}}$,

$$0.46 \text{ V} = E°_{Ag^+} - 0.34 \text{ V}$$

Then we solve for $E°_{Ag^+}$.

$$E°_{Ag^+} = 0.46 \text{ V} + 0.34 \text{ V}$$
$$= 0.80 \text{ V}$$

The standard reduction potential of silver ion is therefore $+0.80$ V.

Is the Answer Reasonable? We know the standard cell potential is the difference between the two standard reduction potentials. The difference between $+0.80$ V and $+0.34$ V (subtracting the smaller from the larger) is 0.46 V. Our calculated standard reduction potential for Ag^+ appears to be correct. You can also take a peek at Table 19.1 now for a final check.

Practice Exercise 19.3

Copper metal and zinc metal will both reduce Ag^+ ions under standard state conditions. Which metal when used as an electrode in a galvanic cell will have the larger $E°_{cell}$? (*Hint:* Write the two possible chemical reactions.)

Practice Exercise 19.4

A galvanic cell has a standard cell potential of 1.93 V. One of the half-cells contains Mg^{2+} and $Mg(s)$ as the solute and electrode, respectively. Calculate the standard reduction potential of the other half cell and use Table 19.1 to identify the corresponding half reaction.

At this point you may have wondered why the term *cell potential* is used in some places and *standard cell potential* is used in others. Based on our definitions, the term standard cell potential is used in places where the system is at standard state. Cell potential is used for *any* set of concentrations, pressures, and temperatures, including the standard conditions. For convenience, our calculations in the next two sections will use standard cell potentials.

19.3 | Utilizing Standard Reduction Potentials

In the previous section we saw how to define standard reduction potentials and standard cell potentials. In addition, by defining the standard reduction potential of the hydrogen electrode, we saw that standard reduction potentials could be determined. Now we will begin investigating some of the uses of standard cell potentials.

Predicting Spontaneous Reactions

It's easy to predict the spontaneous reaction when the substances in two half-reactions, at standard state, are mixed together. This is because we know that *the half-reaction with the more positive reduction potential always takes place as written (namely, as a reduction), while the other half-reaction is forced to run in reverse (as an oxidation).* Table 19.1 is arranged with the largest positive reduction potential at the top, to the smallest (most negative) at the bottom. The activity series shown in Table 5.3 is also arranged in the same way. We can extend the principles developed in Section 5.6 to predict spontaneous reactions based on the *position of the half-reactions* in Table 19.1.

For *any* pair of half-reactions, the one higher up in the table has the more positive standard reduction potential and occurs as a reduction. The other half-reaction is reversed and occurs as an oxidation. This concept is useful if we just want to know if a reaction is spontaneous without calculating E°_{cell}, or as a quick check after solving a problem. Additionally, *for a spontaneous reaction, the reactants are found on the left side of the higher half-reaction and on the right side of the lower half-reaction.* (This is usually, but not always, true of systems that are not at standard state.)

■ Strictly speaking, the E° values only tell us what to expect under standard conditions. However, only when E°_{cell} is small can changes in concentration change the direction of the spontaneous reaction.

Example 19.3
Predicting a Spontaneous Reaction

What spontaneous reaction occurs if Cl_2 and Br_2 are added to a solution that contains both Cl^- and Br^-? Assume standard conditions.

Analysis: We know that in the spontaneous redox reaction, the more easily reduced substance will be the one that undergoes reduction. We need the data to decide which substance will be reduced, Cl_2 or Br_2; we already know that neither Cl^- nor Br^- can be reduced since they cannot be in lower oxidation states.

Assembling the Tools: We can use the positions of the half-reactions for the reduction of Cl_2 and Br_2 in Table 19.1 to compare their E° values.

Solution: There are two possible reduction half-reactions:

$$Cl_2(g) + 2e^- \longrightarrow 2Cl^-(aq)$$
$$Br_2(aq) + 2e^- \longrightarrow 2Br^-(aq)$$

In Table 19.1 we observe that the reduction of Cl_2 is found *above* Br_2 in the table so the reduction of Cl_2 must have a *more positive* reduction potential. That leads us to conclude that Cl_2 *must be reduced* and that Br^- *must be oxidized.*

$$Cl_2(g) + 2e^- \longrightarrow 2Cl^-(aq) \qquad \text{(a reduction)}$$

$$2Br^-(aq) \longrightarrow Br_2(aq) + 2e^- \quad \text{(an oxidation)}$$

The net reaction is obtained by combining the half-reactions.

$$Cl_2(g) + 2Br^-(aq) \longrightarrow Br_2(aq) + 2Cl^-(aq)$$

Is the Answer Reasonable? We can calculate the standard cell potential (E°_{cell} = 1.36 V − 1.07 V = +0.29 V) for our proposed reaction. Seeing that E°_{cell} is positive confirms we are correct.

Practice Exercise 19.5

Using the positions of the respective half-reactions in Table 19.1, predict whether the following reaction is spontaneous if all the ions are 1.0 M at 25 °C. If it is not, write the equation that represents a spontaneous reaction. (*Hint:* Find the half-reactions for this equation.)

$$Ni^{2+}(aq) + 2Fe^{2+}(aq) \longrightarrow Ni(s) + 2Fe^{3+}(aq)$$

Practice Exercise 19.6

Use the positions of the half-reactions in Table 19.1 to predict the spontaneous reaction when Br^-, SO_4^{2-}, H_2SO_3, and Br_2 are mixed in an acidic solution at standard state.

Calculating Standard Cell Potentials

We've just seen that we can use standard reduction potentials, or the positions of two half-reactions in Table 19.1, to predict spontaneous redox reactions. If we intend to use these reactions in a galvanic cell, we can also determine the numerical value of the standard cell potential, as illustrated in the next example.

Example 19.4
Predicting the Cell Reaction and Standard Cell Potential of a Galvanic Cell

A typical cell of a lead storage battery of the type used to start automobiles is constructed using electrodes made of lead and lead(IV) oxide (PbO_2) and with sulfuric acid as the electrolyte. The half-reactions and their standard reduction potentials in this system are

$$PbO_2(s) + 3H^+(aq) + HSO_4^-(aq) + 2e^- \rightleftharpoons PbSO_4(s) + 2H_2O \qquad E^\circ_{PbO} = 1.69 \text{ V}$$
$$PbSO_4(s) + H^+(aq) + 2e^- \rightleftharpoons Pb(s) + HSO_4^{2-}(aq) \quad E^\circ_{PbSO_4} = -0.36 \text{ V}$$

What is the cell reaction and what is the standard potential of this cell?

Analysis: We can easily predict spontaneous reactions if the system is at standard state. In the spontaneous cell reaction, the half-reaction with the larger (more positive) standard reduction potential will take place as reduction while the other half-reaction will be reversed and occur as oxidation since this is the only way to end up with a positive value for the standard cell potential.

Assembling the Tools: The standard cell potential is calculated using Equation 19.2.

$$E^\circ_{cell} = E^\circ_{reduction} - E^\circ_{oxidation}$$

We also use the methods in Section 5.2 to combine the two half-reactions, once we determine which should be the oxidation and which is the reduction.

● **Solution:** PbO_2 has a larger, more positive standard reduction potential than $PbSO_4$, so the first half-reaction will occur in the direction written. The second must be reversed to occur as an oxidation. In the cell, therefore, the half-reactions are

$$PbO_2(s) + 3H^+(aq) + HSO_4^-(aq) + 2e^- \longrightarrow PbSO_4(s) + 2H_2O$$

$$Pb(s) + HSO_4^-(aq) \longrightarrow PbSO_4(s) + H^+(aq) + 2e^-$$

■ Remember that half-reactions are combined following the same procedure used in the ion–electron method of balancing redox reactions (Section 5.2).

Since both half reactions have two electrons each, we can add the two half-reactions and cancel the electrons and one $H^+(aq)$ to obtain the cell reaction,

$$PbO_2(s) + Pb(s) + 2H^+(aq) + 2HSO_4^-(aq) \longrightarrow 2PbSO_4(s) + 2H_2O$$

The standard potential of the cell is

$$E^\circ_{cell} = (E^\circ \text{ of substance reduced}) - (E^\circ \text{ of substance oxidized})$$

Since the first half-reaction occurs as a reduction and the second as an oxidation,

$$E^\circ_{cell} = E^\circ_{PbO} - E^\circ_{PbSO_4}$$
$$= (1.69 \text{ V}) - (-0.36 \text{ V})$$
$$= 2.05 \text{ V}$$

● **Are the Answers Reasonable?** The half-reactions involved in the problem are located in Table 19.1, and their relative positions tell us that PbO_2 will be reduced and that lead will be oxidized. Therefore, we've combined the half-reactions correctly. Since E°_{cell} is positive, we feel confident it was calculated correctly.

What are the overall cell reaction and the standard cell potential of a galvanic cell employing the following half-reactions?

$$NiO_2(s) + 2H_2O + 2e^- \rightleftharpoons Ni(OH)_2(s) + 2OH^-(aq) \qquad E^\circ_{NiO} = 0.49 \text{ V}$$

$$Fe(OH)_2(s) + 2e^- \rightleftharpoons Fe(s) + 2OH^-(aq) \qquad E^\circ_{Fe(OH)_2} = -0.89 \text{ V}$$

(*Hint:* Recall that standard cell potentials are based on a spontaneous reaction.)

Practice Exercise 19.7

■ These are the reactions in an Edison cell, a type of rechargeable storage battery.

What are the overall cell reaction and the standard cell potential of a galvanic cell employing the following half-reactions?

$$Cr^{3+}(aq) + 3e^- \rightleftharpoons Cr(s)$$

$$MnO_4^-(aq) + 8H^+(aq) + 5e^- \rightleftharpoons Mn^{2+}(aq) + 4H_2O$$

Practice Exercise 19.8

 Analyzing and Solving Multi-Concept Problems

A galvanic cell is constructed by placing 1.0 *M* aluminum nitrate in one beaker with an aluminum electrode and a second beaker with 1.0 *M* copper(II) nitrate and a copper electrode. When a salt bridge is inserted to connect the two beakers, what cell potential do we expect to measure between the copper and aluminum electrodes? Which metal is the anode and which is the cathode? What is the spontaneous chemical reaction?

Analysis: This problem is very similar to the preceding one; however, it involves many parts that were included as given data in the previous example. So we will proceed in the same way, but we have a bit more work to do. *Part 1* will involve writing the correct formulas from their names, using our nomenclature skills. Then we will write ions for substances, as needed, using the solubility rules. For *Part 2*, we will take the substances in Part 1 and write half reactions, eliminating spectator ions. Then we combine these half-reactions into a net ionic equation. *Part 3* involves looking up and using the standard reduction potentials to make appropriate decisions to arrive at the spontaneous reaction and its standard cell potential. *Part 4* involves using the concept that the reaction that occurs at an electrode will define it as either an anode or a cathode.

PART 1

Assembling the Tools: We will review Section 2.6 for nomenclature, if necessary, to write formulas for the compounds and their ions as needed. We need the solubility rules from Table 4.1 to determine if these compounds are soluble in water, and if they are soluble, the correct formulas for the ions are needed.

Solution: For aluminum nitrate we can write $Al(NO_3)$ based on the fact that aluminum will always produce the Al^{3+} ion and the formula for the polyatomic nitrate ion. The Roman numeral (II) tells us we have Cu^{2+} and its chloride must be $CuCl_2$. Both compounds should be soluble, and their ions are Al^{3+}, Cu^{2+}, NO_3^-, and Cl^-. Remember that we also have aluminum and copper in their metallic or elemental forms.

PART 2

Assembling the Tools: We need the concepts of oxidation numbers (Section 5.1) and reduction half-reactions (Section 5.2) to write an equation where the electrons and the substance in the oxidized form are on the left of the arrow and the oxidized form of the substance is on the right side. We will need two half-reactions.

Solution: We see that Al^{3+} and Cu^{2+} are the oxidized forms of Al and Cu, respectively, so we can write:

$$Al^{3+} + 3e^- \longrightarrow Al$$

$$Cu^{2+} + 2e^- \longrightarrow Cu$$

In the process of writing these half-reactions, we did not have to use the NO_3^- and the Cl^- ions, and since their oxidation numbers do not change, they are, by definition, spectator ions.

Without looking at Table 19.1, we have two ways to combine these half-reactions. After equalizing the numbers of electrons (by multiplying the first half-reaction by two and the second half-reaction by three), the two half-reactions are subtracted to get either

$$2Al^{3+} + 3Cu \longrightarrow 3Cu^{2+} + 2Al$$

or

$$3Cu^{2+} + 2Al \longrightarrow 2Al^{3+} + 3Cu$$

PART 3

Assembling the Tools: We need Equation 19.2 to determine the standard cell potential. Using the data in Table 19.1 we find values for $E^\circ_{Al^{3+}}$ and $E^\circ_{Cu^{2+}}$.

Solution: The values for the standard reduction potentials are $E^\circ_{Al^{3+}} = -1.66$ V and $E^\circ_{Cu^{2+}} = +0.34$ V. For the two different reactions in Part 2 we get

$$2Al^{3+} + 3Cu \longrightarrow 3Cu^{2+} + 2Al \qquad E^\circ_{cell} = -1.66 \text{ V} - (+0.34 \text{ V}) = -2.00 \text{ V}$$

and

$$3Cu^{2+} + 2Al \longrightarrow 2Al^{3+} + 3Cu \qquad E^\circ_{cell} = +0.34 \text{ V} - (-1.66 \text{ V}) = +2.00 \text{ V}$$

Since E°_{cell} is a positive value for the second reaction, it must be the spontaneous reaction.

PART 4

Assembling the Tools: We need the definitions of anode and cathode to determine the identity of the electrodes.

Solution: Oxidation always occurs at the anode and in our spontaneous equation Al is oxidized to Al^{3+}, so the aluminum electrode is the anode. Similarly, reduction always occurs at the cathode. Copper ions are reduced to copper metal at the copper electrode, so it must be the cathode.

Are the Answers Reasonable? If we locate these half-reactions in Table 19.1, their relative positions tell us we've chosen the correct equation for the spontaneous reaction. It also means we've identified correctly the substances reduced and oxidized, so we've correctly applied Equation 19.2 and correctly identified the anode and cathode.

A 1.0 M solution of copper(II) perchlorate and 1.0 M chromium(III) nitrate along with some copper wire and a chunk of chromium are placed in the same beaker, at 25 °C. Write the balanced equation for the spontaneous reaction, and determine the standard cell potential if the same materials are used in a galvanic cell. (*Hint:* Write the half-reactions first.)

Practice Exercise 19.9

A galvanic cell is constructed with two platinum electrodes and one beaker containing 1.0 M iron(II) sulfate and 1.0 M iron(III) sulfate. The second beaker has 1.0 M solutions of tin(II) nitrate and tin(IV) nitrate. What are the spontaneous reaction and the standard cell voltage?

Practice Exercise 19.10

Cell Potentials of Spontaneous Reactions

Because we can predict the spontaneous redox reaction that will take place among a mixture of reactants, it is also possible to predict whether or not a particular reaction, as written, can occur spontaneously.

If the standard cell potential we calculate is positive, the reaction is spontaneous, and if the potential is negative, the reverse reaction is spontaneous.

$E°_{cell}$ and spontaneous reactions

For example, to obtain the standard cell potential for a spontaneous reaction in our previous examples, we subtracted the standard reduction potentials in a way that gave a positive answer. Therefore, if we compute the standard cell potential for a particular reaction *based on the way the equation is written* and the standard potential comes out positive, we know the reaction is spontaneous. If the calculated standard cell potential comes out negative, however, the reaction is *nonspontaneous*. In fact, it is really *spontaneous in the opposite direction*.

Example 19.5
Determining If a Reaction Is Spontaneous by Using the Calculated Standard Cell Potential

Determine whether the following reactions, at standard state, are spontaneous as written. If a reaction is not spontaneous, write the equation for the reaction that is.

(a) $Cu(s) + 2H^+(aq) \longrightarrow Cu^{2+}(aq) + H_2(g)$

(b) $3Cu(s) + 2NO_3^-(aq) + 8H^+(aq) \longrightarrow 3Cu^{2+}(aq) + 2NO(g) + 4H_2O$

Analysis: Our goal is to calculate the standard cell potential based on the reaction as written. To do this, for each reaction we need to write and identify its oxidation half-reaction and its reduction half-reaction. Then we can calculate the standard cell potential, which, if positive, indicates a spontaneous reaction. Otherwise the reverse reaction is spontaneous.

Assembling the Tools: Since the algebraic sign of $E°_{cell}$ informs us about spontaneity, we will use Equation 19.2. We also need half-reactions that are described in Section 5.2 and their corresponding standard reduction potentials found in Table 19.1.

Solution: We will determine $E°_{cell}$ for the two reactions as written.
(a) The half-reactions involved in this equation are

$$Cu(s) \longrightarrow Cu^{2+}(aq) + 2e^- \quad \text{(oxidation)}$$
$$2H^+(aq) + 2e^- \longrightarrow H_2(g) \quad \text{(reduction)}$$

The H^+ is reduced and Cu is oxidized, so Equation 19.2 will take the form

$$E°_{cell} = E°_{H^+} - E°_{Cu^{2+}}$$

| Standard reduction potential of substance reduced | Standard reduction potential of substance oxidized |

Substituting values from Table 19.1 gives

$$E°_{cell} = (0.00 \text{ V}) - (0.34 \text{ V})$$
$$= -0.34 \text{ V}$$

The calculated standard cell potential is negative, so reaction (a) is not spontaneous in the forward direction. The spontaneous reaction is actually the reverse of (a).

$$Cu^{2+}(aq) + H_2(g) \longrightarrow Cu(s) + 2H^+(aq)$$
Reaction (a) reversed

(b) The half-reactions involved in this equation are

$$Cu(s) \longrightarrow Cu^{2+}(aq) + 2e^-$$
$$NO_3^-(aq) + 4H^+(aq) + 3e^- \longrightarrow NO(g) + 2H_2O$$

The Cu is oxidized while the NO_3^- is reduced. According to Equation 19.2,

$$E°_{cell} = E°_{NO_3^-} - E°_{Cu^{2+}}$$

Substituting values from Table 19.1 gives

$$E°_{cell} = (0.96 \text{ V}) - (0.34 \text{ V})$$
$$= +0.62 \text{ V}$$

Because the calculated standard cell potential is positive, reaction (b) is spontaneous in the forward direction, as written.

■ Copper dissolves in HNO_3 because it contains the oxidizing agent NO_3^-.

Are the Answers Reasonable? By noting the relative positions of the half-reactions in Table 19.1, you can confirm that we've answered the questions correctly.

Practice Exercise 19.11

Write each of the following equations so that they represent spontaneous reactions at standard state.

(a) $2Br^-(aq) + I_2(s) \longrightarrow Br_2(aq) + 2I^-(aq)$

(b) $Mn^{2+}(aq) + 5Ag^+(aq) + 4H_2O \longrightarrow MnO_4^-(aq) + 5Ag(s) + 8H^+(aq)$

(*Hint:* Determine the sign of the calculated standard cell potential.)

Practice Exercise 19.12

Under standard state conditions, which of the following reactions occur spontaneously? Justify your answer.

(a) $2Br^-(aq) + 2HOCl(aq) + 2H^+(aq) \longrightarrow Br_2(aq) + Cl_2(g) + 2H_2O$

(b) $3Zn^{2+}(aq) + 2Cr(s) \longrightarrow 3Zn(s) + 2Cr^{3+}(aq)$

19.4 | E°_{cell} and ΔG°

The fact that cell potentials allow us to predict the spontaneity of redox reactions is no coincidence. There is a relationship between the cell potential and the free energy change for a reaction. In Chapter 18, we saw that ΔG for a reaction is a measure of the maximum useful work that can be obtained from a chemical reaction. Specifically, the relationship is

$$-\Delta G = \text{maximum work} \qquad (19.3)$$

In an electrical system, work is supplied by the flow of electric charge created by the potential of the cell. It can be calculated from the equation

$$\text{maximum work} = n\,\mathscr{F}\,E_{cell} \qquad (19.4)$$

where n is the number of moles of electrons transferred in the equation, $\mathscr{F}$ is a constant called the **Faraday constant**, which is equal to the number of coulombs of charge equivalent to 1 mol of electrons (9.65×10^4 coulombs per mole of electrons), and E_{cell} is the potential of the cell in volts. We can show that Equation 19.4 has the same units as work (which has the units of energy) by analyzing the units. In Equation 19.1 you saw that 1 volt = 1 joule/coulomb. Therefore,

$$\text{maximum work} = \underbrace{\cancel{\text{mol } e^-}}_{n} \times \underbrace{\frac{\cancel{\text{coulombs}}}{\cancel{\text{mol } e^-}}}_{\mathscr{F}} \times \underbrace{\frac{\text{joule}}{\cancel{\text{coulomb}}}}_{E_{cell}} = \text{joule}$$

Combining Equations 19.3 and 19.4 gives us

$$\Delta G_{rxn} = -n\,\mathscr{F}\,E_{cell} \qquad (19.5)$$

At standard state we are dealing with the *standard* cell potential, so we can calculate the *standard* free energy change.

$$\Delta G^{\circ}_{rxn} = -n\,\mathscr{F}\,E^{\circ}_{cell} \qquad (19.6)$$

Referring back to Chapter 18, if ΔG has a negative value, a reaction will be spontaneous, and this corresponds to a positive value of E_{cell}.

Up to now we have been careful to predict spontaneity for systems at standard state where E_{cell} is equal to E°_{cell}. Later in Section 19.5, we will see how to calculate E_{cell} under nonstandard state conditions and precisely predict if a reaction is spontaneous. In addition we must consider how units cancel. When we write ΔH°, ΔG°, or any other standard thermodynamic parameter, they all refer to a specific chemical equation. Usually it is the balanced equation with the lowest possible whole-number coefficients. But, as explained in Section 6.6, the amount of heat energy or free energy refers to reacting the stoichiometric number of moles indicated by the coefficients of the reaction. We often write the chemical equation along with the term ΔH°_{rxn} or ΔG°_{rxn} and its value, to avoid any confusion.

TOOLS

Faraday constant

■ More precisely, 1 **faraday** ($\mathscr{F}$) = 96,485 C.

TOOLS

ΔG is related to E_{cell}

TOOLS

ΔG° is related to E°_{cell}

Example 19.6
Calculating the Standard Free Energy Change

Calculate ΔG°_{rxn} for the following reaction, given that its standard cell potential is 0.320 V at 25 °C.

$$NiO_2(s) + 2Cl^-(aq) + 4H^+(aq) \longrightarrow Cl_2(g) + Ni^{2+}(aq) + 2H_2O$$

Analysis: We will need to use the relationship between E°_{cell} and ΔG°_{rxn}. The given chemical equation allows us to determine n, which is used in the equation. We also need to convert volts into joules per coulomb.

Assembling the Tools: Our tool for solving this problem is Equation 19.6.

$$\Delta G^\circ_{rxn} = -n\,\mathscr{F}\,E^\circ_{cell}$$

We know $\mathscr{F}$ and E°_{cell}, and the value of n, the total number of electrons transferred in the balanced chemical equation, can be computed with the aid of principles developed in Section 5.2. We will also use the Faraday constant (the SI abbreviation for coulomb is C).

$$1\,\mathscr{F} = \left(\frac{9.65 \times 10^4\ C}{1\ mol\ e^-}\right)$$

Equation 19.1 is used to convert volts to joules per coulomb.

Solution: To solve Equation 19.6 we determine n by taking the coefficients in the chemical equation to stand for *moles*; two moles of Cl are oxidized to Cl_2 and two moles of electrons are transferred to the NiO_2. Therefore we conclude that $n = 2$ mol e^- per mol of reaction. Finally, we write 0.320 V as 3.20 J/C. Using Equation 19.6, we have

$$\Delta G^\circ_{rxn} = \frac{-2\ \cancel{mol\ e^-}}{mol\ rxn} \times \left(\frac{9.65 \times 10^4\ \cancel{C}}{1\ \cancel{mol\ e^-}}\right) \times \left(\frac{0.320\ J}{\cancel{C}}\right)$$

$$= -6.18 \times 10^4\ J/mol\ rxn \quad or\ simply \quad -6.18 \times 10^4\ J/mol$$

$$= -61.8\ kJ/mol$$

This is the energy that is produced when one mole of $NiO_2(s)$ reacts with two moles of chloride ions and four moles of hydrogen ions.

Is the Answer Reasonable? Let's do some approximate arithmetic. The Faraday constant equals approximately 100,000, or 10^5. The product $2 \times 0.32 = 0.64$, so ΔG° should be about 0.64×10^5 or 6.4×10^4. The answer seems to be okay. Importantly, the algebraic signs tell us the same thing: that is, the positive E°_{cell} and the negative ΔG° both predict a spontaneous reaction.

Practice Exercise 19.13

A certain reaction has an E°_{cell} of 0.107 volts and has a ΔG° of -30.9 kJ. How many electrons are transferred in the reaction? (*Hint:* See Equation 19.6.)

Practice Exercise 19.14

Calculate ΔG° for the reactions that take place in the galvanic cells as written in Practice Exercises 19.11 and 19.12.

E°_{cell} and Equilibrium Constants

One of the useful applications of electrochemistry is the determination of equilibrium constants. In Section 18.9 you saw that ΔG°_{rxn} is related to the equilibrium constant by the expression

$$\Delta G^\circ_{rxn} = -RT \ln K_c$$

where we used K_c for the equilibrium constant because electrochemical reactions occur in solution. Earlier in this section we saw that ΔG°_{rxn} is also related to E°_{cell},

$$\Delta G^\circ_{rxn} = -n\,\mathscr{F}\,E^\circ_{cell}$$

Therefore, E°_{cell} and the equilibrium constant are also related. Equating the right sides of the two equations, we have

$$-n\,\mathscr{F}\,E^\circ_{cell} = -RT \ln K_c$$

Solving for $E°_{cell}$ gives[3]

$$E°_{cell} = \frac{RT}{n\mathscr{F}} \ln K_c \qquad (19.7)$$

$E°_{cell}$ is related to K_c

For the units to work out correctly, the value of R must be 8.314 J mol^{-1} K^{-1}, T stands for the temperature in kelvins, $\mathscr{F}$ equals 9.65×10^4 C per mole of e^-, and n equals the number of moles of electrons transferred per mol of reaction.

Example 19.7
Calculating Equilibrium Constants from $E°_{cell}$

Calculate K_c for the reaction in Example 19.6.

Analysis: We now have an equation that relates the standard cell potential and the equilibrium constant. We need to choose the appropriate values for each variable to be sure that units cancel properly.

Assembling the Tools: Equation 19.7 is our tool for solving this problem.

$$E°_{cell} = \frac{RT}{n\mathscr{F}} \ln K_c$$

We need to collect the terms to insert in this equation to solve it. We need to find $E°_{cell}$, n, R, T, and $\mathscr{F}$. Remember that T is in kelvins and R must have the appropriate units of J (mol rxn)$^{-1}$ K^{-1}.

Solution: First, lets collect all the values needed:

$$T = 25\,°C = 298\ K \qquad\qquad E°_{cell} = 0.320\ V = 0.320\ J\ C^{-1}$$

$$R = 8.314\ J\ (mol\ rxn)^{-1}\ K^{-1} \qquad \mathscr{F} = 9.65 \times 10^4\ C\ per\ mole\ of\ e^-$$

$$n = 2\ mole\ of\ e^-\ per\ mol\ rxn$$

With all of the data, let's rearrange Equation 19.7 for $\ln K_c$ and then substitute values.

$$\ln K = E°_{cell}\frac{n\mathscr{F}}{RT}$$

$$\ln K = (0.320\ J\ C^{-1})\ \dfrac{\dfrac{2\ \cancel{mol\ e^-}}{\cancel{mol\ rxn}} \times \dfrac{9.65 \times 10^4\ C}{\cancel{mol\ e^-}}}{\dfrac{8.314\ J}{(\cancel{mol\ rxn})\ K} \times 298\ K}$$

$$= \frac{0.320\ \cancel{J}}{\cancel{C}} \times \frac{77.90\ \cancel{C}}{\cancel{J}} = 24.9$$

---

[3] For historical reasons, Equation 19.7 is sometimes expressed in terms of common logs (base 10 logarithms). Natural and common logarithms are related by the equation

$$\ln x = 2.303 \log x$$

For reactions at 25 °C (298 K), all of the constants (R, T, and $\mathscr{F}$) can be combined with the factor 2.303 to give 0.0592 joules/coulomb. Because joules/coulomb equals volts, Equation 19.7 reduces to

$$E°_{cell} = \frac{0.0592\ V}{n} \log K_c$$

where n is the number of moles of electrons transferred in the cell reaction as it is written.

Taking the antilogarithm,

$$K_c = e^{24.9} = 7 \times 10^{10}$$

Is the Answer Reasonable? As a rough check, we can look at the magnitude of E°_{cell} and apply some simple reasoning. When E°_{cell} is positive, ΔG° is negative, and in Chapter 18 you learned that when ΔG° is negative, the position of equilibrium lies toward the product side of the reaction. Therefore, we expect that K_c will be large, and that agrees with our answer.

Practice Exercise 19.15

The calculated standard cell potential for the reaction

$$Cu^{2+}(aq) + 2Ag(s) \rightleftharpoons Cu(s) + 2Ag^+(aq)$$

is $E^\circ_{cell} = -0.46$ V. Calculate K_c for this reaction as written. Is this reaction spontaneous? If not, what is K_c for the spontaneous reaction? (*Hint:* The tool described by Equation 19.7 is important here.)

Practice Exercise 19.16

Use the following half-reactions and the data in Table 19.1 to write the spontaneous reaction. Write the equilibrium law for this reaction and use the standard cell voltage to determine the value of the equilibrium constant.

$$Ag^+(aq) + e^- \longrightarrow Ag(s)$$

$$AgBr(s) + e^- \longrightarrow Ag(s) + Br^-(aq)$$

Walther Nernst (1864–1941) was a professor of physical chemistry who developed the equation named after him.

TOOLS

Nernst equation

19.5 | Cell Potentials and Concentrations

At 25 °C when all of the ion concentrations in a cell are 1.00 *M* and when the partial pressures of any gases involved in the cell reaction are 1.00 atm, the cell potential is equal to the standard potential. When the concentrations or pressures change, however, so does the potential. For example, in an operating cell or battery, the potential gradually drops as the reactants are used up and as the cell reaction approaches its natural equilibrium status. When it reaches equilibrium, the potential has dropped to zero—the battery is "dead."

The Nernst Equation

The effect of concentration on the cell potential can be obtained from thermodynamics. In Section 18.9, you learned that the free energy change is related to the reaction quotient Q by the equation

$$\Delta G = \Delta G^\circ + RT \ln Q$$

Substituting for ΔG and ΔG° from Equations 19.5 and 19.6 gives

$$-n \mathscr{F} E_{cell} = -n \mathscr{F} E^\circ_{cell} + RT \ln Q$$

Dividing both sides by $-n \mathscr{F}$ gives

$$E_{cell} = E^\circ_{cell} - \frac{RT}{n \mathscr{F}} \ln Q \qquad (19.8)$$

This equation is commonly known as the **Nernst equation,**[4] named after Walther Nernst, a German chemist and physicist. Notice that if $Q = 1$, then $\ln Q = 0$ and $E_{cell} = E^\circ_{cell}$.

---

[4] Using common logarithms instead of natural logarithms and calculating the constants for 25 °C gives another form of the Nernst equation that is sometimes used:

$$E_{cell} = E^\circ_{cell} - \frac{0.0592 \text{ V}}{n} \log Q$$

In writing the Nernst equation for a galvanic cell, we will construct the mass action expression (from which we calculate the value of Q) using molar concentrations for ions and partial pressures in atmospheres for gases.[5] Thus, for the following cell using a hydrogen electrode (with the partial pressure of H_2 not necessarily equal to 1 atm) and having the reaction

$$Cu^{2+}(aq) + H_2(g) \longrightarrow Cu(s) + 2H^+(aq)$$

the Nernst equation would be written

$$E_{cell} = E_{cell}^\circ - \frac{RT}{n\mathscr{F}} \ln \frac{[H^+]^2}{[Cu^{2+}]P_{H_2}}$$

■ This is a heterogeneous reaction, so we have not included the concentration of the solid, Cu(s), in the mass action expression.

Example 19.8
Calculating the Effect of Concentration on E_{cell}

Suppose a galvanic cell employs the following half-reactions.

$$Ni^{2+}(aq) + 2e^- \rightleftharpoons Ni(s) \qquad E_{Ni^{2+}}^\circ = -0.25\text{ V}$$
$$Cr^{3+}(aq) + 3e^- \rightleftharpoons Cr(s) \qquad E_{Cr^{3+}}^\circ = -0.74\text{ V}$$

Calculate the cell potential when $[Ni^{2+}] = 4.87 \times 10^{-4}\ M$ and $[Cr^{3+}] = 2.48 \times 10^{-3}\ M$.

Analysis: We will be calculating the cell potential for a system that does not have standard state concentrations by using the Nernst equation. As with previous problems, we will need to write the appropriate chemical equations so that the variables can be identified and used properly.

Assembling the Tools: We start with the Nernst equation, Equation 19.8, to determine E_{cell},

$$E_{cell} = E_{cell}^\circ - \frac{RT}{n\mathscr{F}} \ln Q$$

The balanced chemical equation is needed (Chapter 5) to determine the number of electrons transferred, n, and the correct form of the mass action expression (Chapter 14) from which we calculate the value of Q.

Solution: Nickel has the more positive (less negative) standard reduction potential, so its half-reaction will occur as a reduction. This means that chromium will be oxidized. Making electron gain ($6e^-$) equal to electron loss ($6e^-$), the cell reaction is found as follows.

$$3[Ni^{2+}(aq) + 2e^- \longrightarrow Ni(s)] \qquad \text{(reduction)}$$
$$2[Cr(s) \longrightarrow Cr^{3+}(aq) + 3e^-] \qquad \text{(oxidation)}$$

$$\overline{3Ni^{2+}(aq) + 2Cr(s) \longrightarrow 3Ni(s) + 2Cr^{3+}(aq)} \quad \text{(cell reaction)}$$

The total number of electrons transferred is six, which means $n = 6$. Now we can write the Nernst equation for the system.

$$E_{cell} = E_{cell}^\circ - \frac{RT}{n\mathscr{F}} \ln \frac{[Cr^{3+}]^2}{[Ni^{2+}]^3}$$

Notice that we've constructed the mass action expression, from which we will calculate the reaction quotient, using the concentrations of the ions raised to powers equal to their

[5] Because of interionic attractions, ions do not always behave as though their concentrations are equal to their molarities. Strictly speaking, therefore, we should use effective concentrations (called activities) in the mass action expression. Effective concentrations are difficult to calculate, so for simplicity we will use molarities and accept the fact that our calculations are not entirely accurate.

coefficients in the net cell reaction, and that we have not included concentration terms for the two solids. This is the procedure we followed for heterogeneous equilibria in Section 15.4.

Next we need $E°_{cell}$. Since Ni^{2+} is reduced,

$$E°_{cell} = E°_{Ni^{2+}} - E°_{Cr^{3+}}$$

$$= (-0.25 \text{ V}) - (-0.74 \text{ V})$$

$$= 0.49 \text{ V}$$

Now we can substitute this value for $E°_{cell}$ along with $R = 8.314$ J (mol rxn)$^{-1}$ K^{-1}, $T = 298$ K, $n = 6$ mol e^- per mol rxn, $\mathscr{F} = 9.65 \times 10^4$ C/mol e^-, $[Ni^{2+}] = 4.87 \times 10^{-4}$ M, and $[Cr^{3+}] = 2.48 \times 10^{-3}$ M into the Nernst equation. Notice that we have changed mol to mol rxn in the units for R to reflect the fact that we are working with a balanced equation. Using these terms we have

$$E_{cell} = 0.49 \text{ V} - \frac{\dfrac{8.314 \text{ J}}{(\text{mol rxn}) \text{ K}} \times 298 \text{ K}}{\dfrac{6 \text{ mol } e^-}{\text{mol rxn}} \times \dfrac{9.65 \times 10^4 \text{ C}}{\text{mol } e^-}} \ln \frac{(2.48 \times 10^{-3})^2}{(4.87 \times 10^{-4})^3}$$

Since the volt, V, is equal to one J/C we make that substitution to obtain

$$E_{cell} = 0.49 \text{ V} - (0.00428 \text{ V}) \ln (5.32 \times 10^4)$$

$$= 0.49 \text{ V} - (0.00428 \text{ V})(10.882)$$

$$= 0.49 \text{ V} - 0.0466 \text{ V}$$

$$= 0.44 \text{ V}$$

The potential of the cell is expected to be 0.44 V.

Is the Answer Reasonable? There's no simple way to check the answer, but the small numerical value of 0.44 V is reasonable. In addition, there are certain important points to consider. First, check that you've combined the half-reactions correctly to (a) calculate $E°_{cell}$ and (b) give the balanced cell reaction. Now check that we wrote the correct mass action expression from the balanced equation and that we have the correct number of electrons transferred, n. Finally, check that $R = 8.314$ J mol^{-1} K^{-1}, and the Kelvin temperature were used.

Example 19.9
Spontaneity of a Reaction May Be Concentration Dependent

The reaction of tin metal with acid can be written as

$$Sn(s) + 2H^+(aq) \longrightarrow Sn^{2+}(aq) + H_2(g)$$

Calculate the cell potential at 25 °C when (a) the system is at standard state, (b) the pH is 2.00, and (c) the pH is 5.00. Assume that $[Sn^{2+}] = 0.010$ M and the partial pressure of H_2 is 0.965 atm for parts (b) and (c).

Analysis: We will be using the equations developed in this chapter for computing the values for the standard cell potential and various cell potentials.

Assembling the Tools: Part (a) is at standard state, and we have used Equation 19.2 to determine the standard cell potential $E°_{cell}$, which is also the cell potential E_{cell}.

For parts (b) and (c) we must use the Nernst equation as our tool, Equation 19.8. We determine the number of electrons transferred, n, from the chemical reaction. We can also set up Q, (see Section 14.2), for the Nernst equation and it will have the form:

$$Q = \frac{[Sn^{2+}] P_{H_2}}{[H^+]^2}$$

Solutions: For part (a) we find the difference between standard reduction potentials as

(a) $E^\circ_{cell} = 0.00 - (-0.14) = +0.14$ V

For parts (b) and (c) we substitute into the Nernst equation using $R = 8.314$ J mol^{-1} K^{-1}, $T = 298$ K, and F $= 9.65 \times 10^4$ C/mol e^-.

$$E_{cell} = E^\circ_{cell} - \frac{RT}{n\mathscr{F}} \ln \frac{[Sn^{2+}] P_{H_2}}{[H^+]^2}$$

From the reaction we determine that $n = 2$ mol e^- since Sn^{2+} is reduced by two electrons to Sn, and we can also calculate the hydrogen ion concentrations for the pH 2.00 and pH 5.00 solutions as 1.0×10^{-2} M and 1.0×10^{-5} M, respectively. For parts (b) and (c) we make the following substitutions:

(b) $E_{cell} = 0.14$ V $- \dfrac{\dfrac{8.314\ \text{J}}{(\text{mol rxn})K} \times 298\ \text{K}}{\dfrac{2\ \text{mol } e^-}{\text{mol rxn}} \times \dfrac{9.65 \times 10^4\ \text{C}}{\text{mol } e^-}} \ln \dfrac{(0.010)(0.965)}{(1.0 \times 10^{-2})^2}$

$E_{cell} = 0.14$ V $- 0.06$ V $= +0.08$ V

(c) $E_{cell} = 0.14$ V $- \dfrac{\dfrac{8.314\ \text{J}}{(\text{mol rxn})\ K} \times 298\ \text{K}}{\dfrac{2\ \text{mol } e^-}{\text{mol rxn}} \times \dfrac{9.65 \times 10^4\ \text{C}}{\text{mol } e^-}} \ln \dfrac{(0.010)(0.965)}{(1.0 \times 10^{-5})^2}$

$E_{cell} = 0.14$ V $- 0.24$ V $= -0.10$ V

At standard state the reaction is spontaneous, and at pH $= 2.00$ it is spontaneous, but the potential is barely positive. At pH 5.00 the reaction is not spontaneous, but the reverse reaction will be.

Is the Answer Reasonable? The first question to answer is "does this make sense?" and indeed it does. That the results vary uniformly with pH is an indication the results are correct. We cannot easily estimate natural logarithms, so a check of calculations may be easiest if the values are entered into your calculator in a different sequence from your first calculation.

A galvanic cell is constructed with a copper electrode dipping into a 0.015 M solution of Cu^{2+} ions and an electrode made of magnesium is immersed in a 2.2×10^{-6} M solution of magnesium ions. What is the balanced chemical equation and the cell potential at 25 °C? (*Hint:* Set up the Nernst equation for this reaction.)

Practice Exercise 19.17

In Example 19.9, assume all conditions are the same except for the pH. At what pH will the cell potential be zero?

Practice Exercise 19.18

Concentration from E_{cell} Measurements

One of the principal uses of the relationship between concentration and cell potential is for the measurement of concentrations of redox reactants and products in a galvanic cell. Experimental determination of cell potentials combined with modern developments in electronics has provided a means of monitoring and analyzing the concentrations of all

■ The ease of such operations and the fact that they lend themselves well to automation and computer analysis make electrochemical analyses especially attractive to scientists.

sorts of substances in solution, even some that are not themselves ionic and that are not involved directly in electrochemical changes. In fact, the operation of a pH meter relies on the logarithmic relationship between hydrogen ion concentration and the potential of a special kind of electrode, the glass electrode, shown in Figure 19.8.

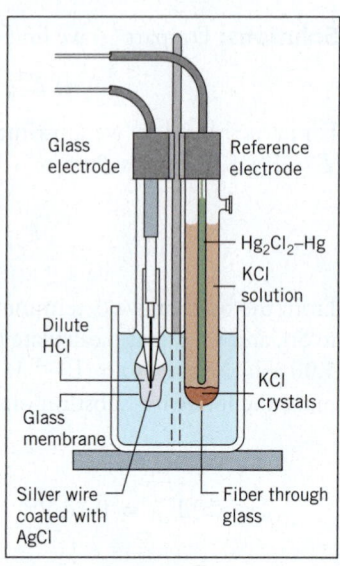

Figure 19.8 | **Electrodes used with a pH meter.** The electrode on the left is called a *glass electrode*. It contains a silver wire, coated with AgCl, dipping into a dilute solution of HCl. This half-cell has a potential that depends on the difference between the [H$^+$] inside and outside a thin glass membrane at the bottom of the electrode. On the right is a reference electrode that forms the other half-cell. The galvanic cell formed by the two electrodes produces a potential that is proportional to the pH of the solution into which they are dipped.

Example 19.10
Using the Nernst Equation to Determine Concentrations

A laboratory was assigned the job of determining the copper(II) ion concentration in thousands of water samples. To make these measurements an electrochemical cell was assembled that consists of a silver electrode, dipping into a 0.225 M solution of AgNO$_3$, connected by a salt bridge to a second half-cell containing a copper electrode. The copper half-cell was then filled with one water sample after another, with the cell potential measured for each sample. In the analysis of one sample, the cell potential at 25 °C was measured to be 0.62 V. The copper electrode was observed to carry a negative charge, so it served as the anode. What was the concentration of copper ion in this sample?

Analysis: In this problem, we've been given the cell potential, E_{cell}, and we can calculate $E°_{cell}$. The unknown quantity is one of the concentration terms in Q that is part of the Nernst equation. Once we assemble all of the variables for the Nernst equation, an algebraic solution is possible.

Assembling the Tools: We will need Equation 19.8, the Nernst equation,

$$E_{cell} = E°_{cell} - \frac{RT}{n\mathscr{F}} \ln Q$$

and the chemical reaction represented by the description of the galvanic cell. Because copper is the anode, it is being oxidized. This also means that Ag$^+$ is being reduced. Therefore, the equation for the cell reaction is

$$Cu(s) + 2Ag^+(aq) \longrightarrow Cu^{2+}(aq) + 2Ag(s)$$

From the balanced equation we can determine the mass action expression, n, and $E°_{cell}$. Since R and Faraday's constant are known, along with the concentration of silver ions, the remaining variable is [Cu^{2+}] that the question asks us to compute.

• **Solution:** Two electrons are transferred, so $n = 2$ and the Nernst equation is

$$E_{cell} = E_{cell}^\circ - \frac{RT}{2\mathcal{F}} \ln \frac{[Cu^{2+}]}{[Ag^+]^2}$$

The value of E_{cell}° can be obtained from the tabulated standard reduction potentials in Table 19.1. Following our usual procedure and recognizing that silver ion is reduced,

$$E_{cell}^\circ = E_{Ag^+}^\circ - E_{Cu^{2+}}^\circ$$
$$= (0.80\ V) - (0.34\ V)$$
$$= 0.46\ V$$

Now we can substitute values into the Nernst equation and solve for the concentration ratio in the mass action expression.

$$0.62\ V = 0.46\ V - \frac{\dfrac{8.314\ J}{(mol\ rxn)\ K} \times 298\ K}{\dfrac{2\ mol\ e^-}{mol\ rxn} \times \dfrac{9.65 \times 10^4\ C}{mol\ e^-}} \ln \frac{[Cu^{2+}]}{[Ag^+]^2}$$

Solving for $\ln([Cu^{2+}]/[Ag^+]^2)$ gives (retaining one extra significant figure)

$$\ln \frac{[Cu^{2+}]}{[Ag^+]^2} = -12.5$$

Taking the antilog gives us the value of the mass action expression.

$$\frac{[Cu^{2+}]}{[Ag^+]^2} = 4 \times 10^{-6}$$

Since we know that the concentration of Ag^+ is 0.225 M, we can now solve for the Cu^{2+} concentration.

$$[Cu^{2+}] = 4 \times 10^{-6} \times [Ag^+]^2 = 4 \times 10^{-6} \times (0.225)^2$$
$$[Cu^{2+}] = 2 \times 10^{-7} M \quad \text{(correctly rounded)}$$

• **Is the Answer Reasonable?** All we can do easily is check to be sure we've written the correct chemical equation, on which all the rest of the solution to the problem depends. Be careful about algebraic signs and that you select the proper value for R and the temperature is expressed in kelvins. Also, notice that we first solved for the logarithm of the ratio of concentration terms. Then, after taking the (natural) antilogarithm, we substitute the known value for $[Ag^+]$ and solve for $[Cu^{2+}]$.

As a final point, notice that the Cu^{2+} concentration is indeed very small, and that it can be obtained very easily by simply measuring the potential generated by the electrochemical cell. Determining the concentrations in many samples is also very simple—just change the water sample and measure the potential again.

In the analysis of two other water samples by the procedure described in Example 19.10, cell potentials (E_{cell}) of 0.57 V and 0.82 V were obtained. Calculate the Cu^{2+} ion concentration in each of these samples. (*Hint:* Your answers should be on either side of the answer in the example.)

Practice Exercise 19.19

A galvanic cell is constructed with a copper electrode dipping into a 0.015 M solution of Cu^{2+} ions, and a magnesium electrode immersed in a solution with an unknown concentration of magnesium ions. The cell potential is measured as 2.79 volts at 25 °C. What is the concentration of magnesium ions?

Practice Exercise 19.20

$$Mg(s) + Cu^{2+}(aq) \longrightarrow Mg^{2+}(aq) + Cu(s)$$

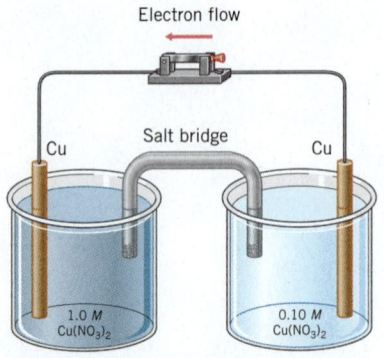

Electron flow

Cu

Salt bridge

Cu

1.0 *M*
Cu(NO₃)₂

0.10 *M*
Cu(NO₃)₂

Figure 19.9 | **A concentration cell.** When the circuit is completed, reactions occur that tend to make the concentrations of Cu^{2+} the same in the two half-cells. Oxidation occurs in the more dilute half-cell, and reduction occurs in the more concentrated one.

Concentration Cells

The dependence of cell potential on concentration allows us to construct a galvanic cell from two half-cells composed of the same substances but having different concentrations of the solute species. An example would be a pair of copper electrodes dipping into solutions that have different concentrations of Cu^{2+}, say 0.10 M Cu^{2+} in one and 1.0 M in the other (Figure 19.9). When this cell operates, reactions take place that tend to bring the two Cu^{2+} concentrations toward the same value. Thus, in the half-cell containing 0.10 M Cu^{2+}, copper is oxidized, which adds Cu^{2+} to the more dilute solution. In the other cell, Cu^{2+} is reduced, removing Cu^{2+} from the more concentrated solution. This makes the more dilute half-cell the anode and the more concentrated half-cell the cathode.

$$Cu(s) \mid Cu^{2+}(0.10\ M) \parallel Cu^{2+}(1.0\ M) \mid Cu(s)$$
anode   cathode

The half-reactions in the spontaneous cell reaction are

$$Cu(s) \longrightarrow Cu^{2+}(0.10\ M) + 2e^-$$

$$Cu^{2+}(1.0\ M) + 2e^- \longrightarrow Cu(s)$$

$$\overline{Cu^{2+}(1.0\ M) \longrightarrow Cu^{2+}(0.10\ M)}$$

The Nernst equation for this cell is

$$E_{cell} = E_{cell}^{\circ} - \frac{RT}{n\mathscr{F}} \ln \frac{[Cu^{2+}]_{dilute}}{[Cu^{2+}]_{conc.}}$$

Because we're dealing with the same substances in the cell, $E_{cell}^{\circ} = 0$ V. When the cell operates, $n = 2$, and we'll take $T = 298$ K. Substituting values,

$$E_{cell} = 0.0\ V - \frac{\dfrac{8.314\ J}{(mol\ rxn)K} \times 298K}{\dfrac{2\ mol\ e^-}{mol\ rxn} \times \dfrac{9.65 \times 10^4\ C}{mol\ e^-}} \ln \frac{(0.10)}{(1.0)}$$

$$= 0.030\ V$$

In this concentration cell, one solution is ten times more concentrated than the other, yet the potential generated is only 0.03 V. In general, the potential generated by concentration differences are quite small. Yet they are significant in biological systems, where electrical potentials are generated across biological membranes by differences in ion concentrations (e.g., K^+). Membrane potentials are important in processes such as the transmission of nerve impulses.

These small differences in potential also illustrate that if Q, in a system that is not at standard state, is between 0.10 and 10 we can conclude that $E_{cell} \approx E_{cell}^{\circ}$ and we can generalize the results in Sections 19.2 and 19.3 when predicting spontaneous reactions.

19.6 | Electricity

Large amounts of electricity are generated mechanically by rotating coils of wire in a magnetic field in a turbine. To rotate the coils, superheated steam, flowing water, and the wind are used in a variety of turbine designs, such as the windmills in the chapter-opening photograph. Superheated steam is produced reliably by burning coal, oil, natural gas, and even municipal wastes. Such turbines are used mainly in large central electrical generators near major population centers. Since the energy of the fuel is not completely converted into electricity, the efficiency of these generators ranges from 35 to 45% and they require a significant investment in pollution control.

Hydroelectric power plants are claimed to have efficiencies around 90%. They are clean and nonpolluting. However, hydroelectric power often requires building large dams to maintain a reservoir of water. Additionally they are often located long distances from their customers, and transmission of electricity over these distances incurs significant losses. Wind-powered turbines and windmills are also considered to be clean and nonpolluting. They are approximately 50% efficient and can be located very close to the end-user. However, daily, even hourly, fluctuations in wind speed may introduce inconvenient variations in power production.

Chemical research is important in producing new turbine materials and combustion processes to increase efficiencies of major electric generators. In this section we will concentrate on how chemical potential and intrinsic electrical properties of materials can be used to generate smaller, but important, amounts of electricity.

Batteries

One of the most familiar uses of a galvanic cell, popularly called a **battery**, is the generation of portable electrical energy.[6] These devices are classified as either **primary cells** (cells not designed to be recharged; they are discarded after their energy is depleted) or **secondary cells** (cells designed for repeated use; they are able to be recharged).

Lead Storage Batteries

The common **lead storage battery** used to start an automobile is composed of a number of secondary cells, each having a potential of about 2 V, that are connected in series so that their voltages are additive. Most automobile batteries contain six such cells and give about 12 V, but 6, 24, and 32 V batteries are also available.

A typical lead storage battery is illustrated in Figure 19.10. The anode of each cell is composed of a set of lead plates, the cathode consists of another set of plates that hold a coating of PbO_2, and the electrolyte is sulfuric acid. When the battery is discharging, the electrode reactions are

$$PbO_2(s) + 3H^+(aq) + HSO_4^-(aq) + 2e^- \longrightarrow PbSO_4(s) + 2H_2O \qquad \text{(cathode)}$$

$$Pb(s) + HSO_4^-(aq) \longrightarrow PbSO_4(s) + H^+(aq) + 2e^- \quad \text{(anode)}$$

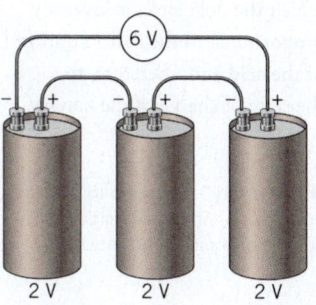

The oldest known electric battery in existence, discovered in 1938 in Baghdad, Iraq, consists of a copper tube surrounding an iron rod. If filled with an acidic liquid such as vinegar, the cell could produce a small electric current.

If three 2 volt cells are connected in series, their voltages are additive to provide a total of 6 volts. In today's autos, 12 volt batteries containing six cells are the norm.

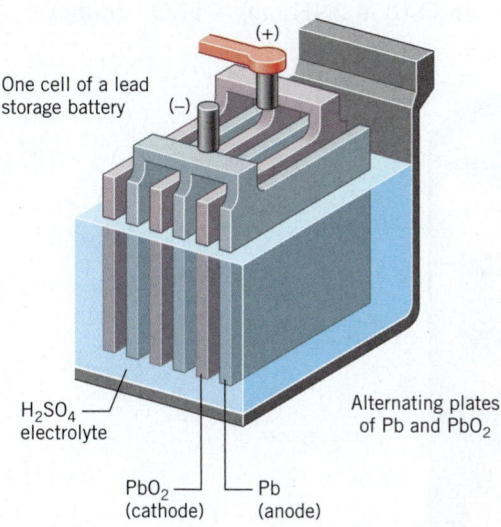

One cell of a lead storage battery

(+)
(−)

H_2SO_4 electrolyte

PbO_2 (cathode)   Pb (anode)

Alternating plates of Pb and PbO_2

Figure 19.10 | Lead storage battery. A 12 volt lead storage battery, such as those used in most automobiles, consists of six cells like the one shown here. Notice that the anode and cathode each consist of several plates connected together. This allows the cell to produce the large currents necessary to start a car.

[6] Strictly speaking, a cell is a single electrochemical unit consisting of a cathode and an anode. A battery is a collection of cells connected in series.

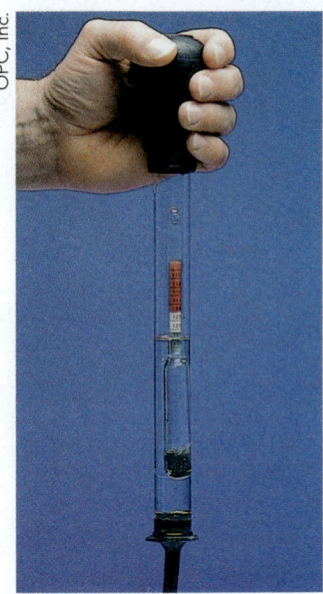

Figure 19.11 | A battery hydrometer. Battery acid is drawn into the glass tube. The depth to which the float sinks is inversely proportional to the concentration of the acid and, therefore, to the state of charge of the battery.

■ This is "dry" compared to the lead-acid battery, which has a significant volume of liquid sulfuric acid.

The net reaction taking place in each cell is

$$PbO_2(s) + Pb(s) + \underbrace{2H^+(aq) + 2HSO_4^-(aq)}_{2H_2SO_4} \longrightarrow 2PbSO_4(s) + 2H_2O$$

As the cell discharges, the sulfuric acid concentration decreases, which in turn, causes the density of the electrolyte to drop. The state of charge of the battery can be monitored with a **hydrometer**, which consists of a rubber bulb that is used to draw the battery fluid into a glass tube containing a float (see Figure 19.11). The depth to which the float sinks is inversely proportional to the density of the liquid—the deeper the float sinks, the lower is the density of the acid and the weaker is the charge on the battery. The narrow neck of the float is usually marked to indicate the state of charge of the battery.

The principal advantage of the lead storage battery is that the cell reactions that occur spontaneously during discharge can be reversed by the application of a voltage from an external source. In other words, the battery can be recharged by electrolysis. The reaction for battery recharge is

$$2PbSO_4(s) + 2H_2O \xrightarrow{\text{electrolysis}} PbO_2(s) + Pb(s) + 2H^+(aq) + 2HSO_4^-(aq)$$

Improper recharging of lead-acid batteries can produce potentially explosive H_2 gas. Most modern lead storage batteries use a lead–calcium alloy as the anode. That reduces the need to have the individual cells vented, and the battery can be sealed, thereby preventing spillage of the corrosive electrolyte.

Zinc–Manganese Dioxide Cells

The original, relatively inexpensive 1.5 V dry cell is the **zinc–manganese dioxide cell**, or **Leclanché cell** (named after its inventor George Leclanché). It is a primary cell used to power flashlights, toys, and the like, but it is not really dry (see Figure 19.12). Its outer shell is made of zinc, which serves as the anode. The cathode—the positive terminal of the battery—consists of a carbon (graphite) rod surrounded by a moist paste of graphite powder, manganese dioxide, and ammonium chloride.

The anode reaction is simply the oxidation of zinc.

$$Zn(s) \longrightarrow Zn^{2+}(aq) + 2e^- \qquad \text{(anode)}$$

The cathode reaction is complex, and a mixture of products is formed. One of the major reactions is

$$2MnO_2(s) + 2NH_4^+(aq) + 2e^- \longrightarrow Mn_2O_3(s) + 2NH_3(aq) + H_2O \quad \text{(cathode)}$$

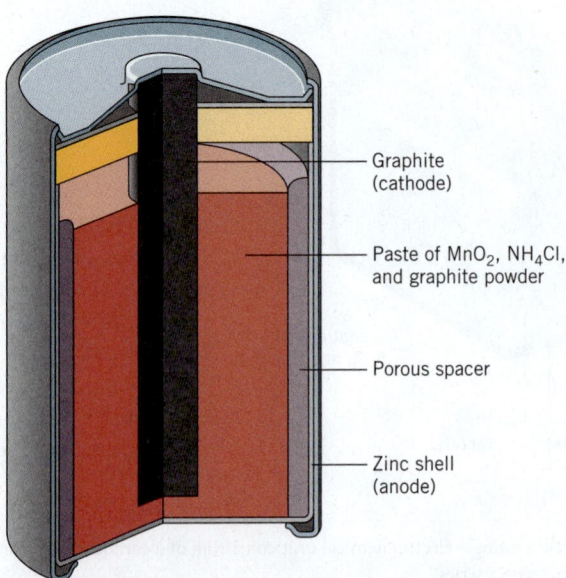

Graphite (cathode)

Paste of MnO_2, NH_4Cl, and graphite powder

Porous spacer

Zinc shell (anode)

Figure 19.12 | A cut-away view of a zinc–carbon dry cell (Leclanché cell).

The ammonia that forms at the cathode reacts with some of the Zn^{2+} produced from the anode to form a complex ion, $Zn(NH_3)_4$. Because of the complexity of the cathode half-cell reaction, no simple overall cell reaction can be written.

A more popular version of the Leclanché battery uses a basic, or *alkaline*, electrolyte and is called an **alkaline battery** or **alkaline dry cell**. It too uses Zn and MnO_2 as reactants, but under basic conditions (Figure 19.13). The half-cell reactions are

$$Zn(s) + 2OH^-(aq) \longrightarrow ZnO(s) + H_2O + 2e^- \qquad \text{(anode)}$$

$$2MnO_2(s) + H_2O + 2e^- \longrightarrow Mn_2O_3(s) + 2OH^-(aq) \qquad \text{(cathode)}$$

$$Zn(s) + 2MnO_2(s) \longrightarrow ZnO(s) + Mn_2O_3(s) \qquad \text{(net cell reaction)}$$

and the voltage is about 1.54 V. It has a longer shelf life and is able to deliver higher currents for longer periods than the less expensive zinc–carbon cell.

Nickel–Metal Hydride Batteries

These secondary cells, which are often referred to as Ni-MH batteries, have largely supplanted the nickel–cadmium, or "Nicad," battery to power devices such as cell phones, digital cameras, and even electric vehicles.

The cathode in the Ni-MH cell is NiO(OH), a compound of nickel in the +3 oxidation state, and the electrolyte is a solution of KOH. Using the symbol MH to stand for the metal hydride, the reactions in the cell during discharge are:

$$MH(s) + OH^-(aq) \longrightarrow M(s) + H_2O + e^- \qquad \text{(anode)}$$

$$NiO(OH)(s) + H_2O + e^- \longrightarrow Ni(OH)_2(s) + OH^-(aq) \qquad \text{(cathode)}$$

The overall cell reaction is

$$MH(s) + NiO(OH)(s) \longrightarrow Ni(OH)_2(s) + M(s) \qquad E^\circ_{cell} = 1.35 \text{ V}$$

When the cell is recharged, these reactions are reversed.

The anode reactant, written as MH, is actually hydrogen that is oxidized to the +1 state in water. This is possible because certain metal alloys [e.g., $LaNi_5$ (an alloy of lanthanum and nickel) and Mg_2Ni (an alloy of magnesium and nickel)] have the ability to absorb and hold substantial amounts of hydrogen, and the hydrogen can be made to participate in reversible electrochemical reactions. Here, the term *metal hydride* is used to describe the hydrogen-absorbing alloy.

The principal advantage of nickel–metal hydride cells is that they can store about 50% more energy in the same volume as older rechargeable Nicad batteries. This means, for example, that comparing cells of the same size and weight, a nickel–metal hydride cell can last about 50% longer than a nickel–cadmium cell.

Lithium Batteries

Lithium has the most negative standard reduction potential of any metal (Table 19.1), so it has a lot of appeal as an anode material. Furthermore, lithium is a very lightweight metal, so a cell employing lithium as a reactant would also be lightweight. The major problem with using lithium in a galvanic cell is that the metal reacts vigorously with water to produce hydrogen gas and lithium hydroxide.

$$2Li(s) + 2H_2O \longrightarrow 2LiOH(aq) + H_2(g)$$

Therefore, to employ lithium in a galvanic cell, scientists had to find a way to avoid aqueous electrolytes. This became possible in the 1970s with the introduction of organic solvents and solvent mixtures that were able to dissolve certain lithium salts and thereby serve as electrolytes.

Today's lithium batteries fall into two categories, primary batteries that can be used once and then discarded when fully discharged, and rechargeable cells.

One of the most common nonrechargeable cells is the **lithium–manganese dioxide battery**, which accounts for about 80% of all primary lithium cells. This cell uses a solid

■ The alkaline battery is also a primary cell.

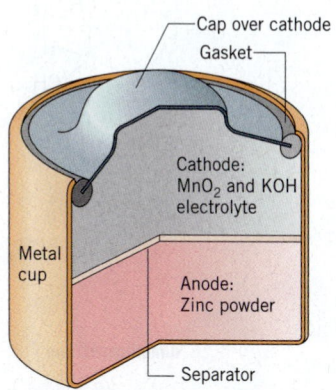

Figure 19.13 | A simplified diagram of an alkaline dry cell.

■ If a battery can supply large amounts of energy and is contained in a small package, it will have a desirably high **energy density** (available energy per unit volume.)

■ Ni-MH batteries have the advantage of being lightweight and having a high energy density—that is, more electrons per cm^3 of battery.

lithium anode and a cathode made of heat-treated MnO_2. The electrolyte is a mixture of propylene carbonate and dimethoxyethane (see the structures in the margin) containing a dissolved lithium salt such as $LiClO_4$. The cell reactions are as follows (superscripts are the oxidation numbers of the manganese):

$$Li \longrightarrow Li^+ + e^- \qquad \text{(anode)}$$
$$\underline{Mn^{IV}O_2 + Li^+ + e^- \longrightarrow Mn^{III}O_2(Li^+) \qquad \text{(cathode)}}$$
$$Li + Mn^{IV}O_2 \longrightarrow Mn^{III}O_2(Li^+) \qquad \text{(net cell reaction)}$$

This cell produces a voltage of about 3.4 V, which is more than twice that of an alkaline dry cell, and because of the light weight of the lithium, it produces more than twice as much energy for a given weight. These cells are used in applications that require a higher current drain or energy pulses (e.g., a photoflash).

Lithium Ion Cells

Rechargeable lithium batteries found in many cell phones, digital cameras, and laptop computers do not contain metallic lithium. They are called **lithium ion cells** and use lithium ions instead. In fact, the cell's operation doesn't actually involve true oxidation and reduction. Instead, it uses the transport of Li^+ ions through the electrolyte from one electrode to the other accompanied by the transport of electrons through the external circuit to maintain charge balance. Here's how it works.

It was discovered that Li^+ ions are able to slip between layers of atoms in certain crystals such as graphite[7] and $LiCoO_2$ (a process called **intercalation**). When the cell is constructed, it is in its "uncharged" state, with no Li^+ ions between the layers of carbon atoms in the graphite. When the cell is charged (Figure 19.14a), Li^+ ions leave $LiCoO_2$ and travel through the electrolyte to the graphite (represented below by the formula C_6).

Initial charging:

$$LiCoO_2 + C_6 \longrightarrow Li_{1-x}CoO_2 + Li_xC_6$$

When the cell spontaneously discharges to provide electrical power (Figure 19.14b), Li^+ ions move back through the electrolyte to the cobalt oxide while electrons move through

Margin notes

propylene carbonate

dimethoxyethane

■ Pronounced in-ter-ka-lay-shun. Rhymes with percolation.

■ In these equations, the subscript x indicates that some, but not all, of the Li^+ ions migrate to the graphite in the charging process. Similarly, the subscript y represents the fraction of Li^+ ions that migrate back as electron flow is generated.

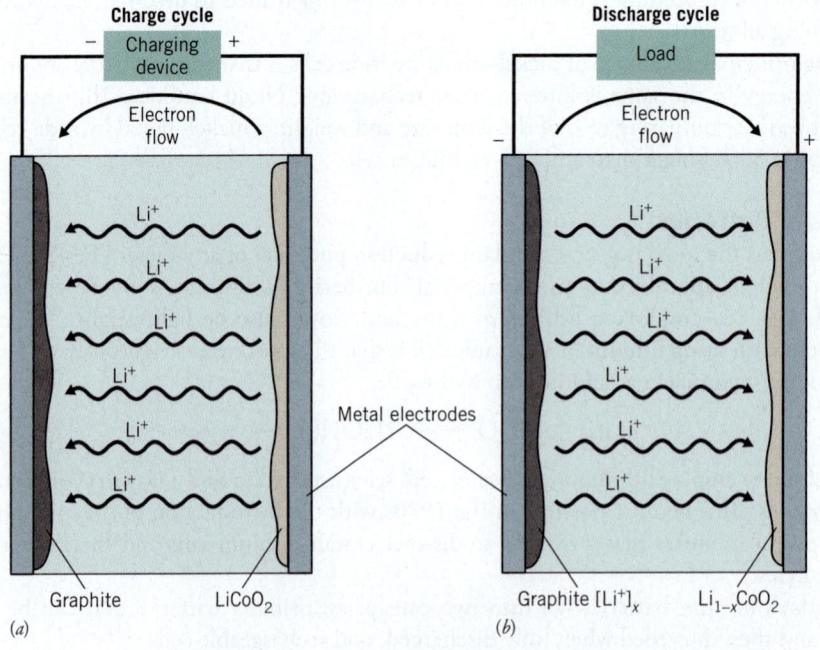

Figure 19.14 | **Lithium ion cell.** (a) During the charging cycle, an external voltage forces electrons through the external circuit and causes lithium ions to travel from the $LiCoO_2$ electrode to the graphite electrode. (b) During discharge, the lithium ions spontaneously migrate back to the $LiCoO_2$ electrode, and electrons flow through the external circuit to balance the charge.

---

[7] Graphite is one of the forms of elemental carbon and consists of layers of joined benzene-like rings.

the external circuit from the graphite electrode to the cobalt oxide electrode. If we represent the amount of Li$^+$ transferring by y, the discharge "reaction" is

Discharge:

$$Li_{1-x}CoO_2 + Li_xC_6 \longrightarrow Li_{1-x+y}CoO_2 + Li_{x-y}C_6$$

Thus, the charging and discharging cycles simply sweep Li$^+$ ions back and forth between the two electrodes, with the electrons flowing through the external circuit to keep the charge in balance.

Fuel Cells

The galvanic cells we've discussed until now can only produce power for a limited time because the electrode reactants are eventually depleted. Fuel cells are different; they are electrochemical cells in which the electrode reactants are supplied continuously and are able to operate without theoretical limit as long as the supply of reactants is maintained. This makes fuel cells an attractive source of power where long-term generation of electrical energy is required.

Figure 19.15 illustrates an early design of a hydrogen–oxygen **fuel cell**. The electrolyte, a hot (~200 °C) concentrated solution of potassium hydroxide in the center compartment, is in contact with two porous electrodes that contain a catalyst (usually platinum) to facilitate the electrode reactions. Gaseous hydrogen and oxygen under pressure are circulated so as to come in contact with the electrodes. At the cathode, oxygen is reduced.

$$O_2(g) + 2H_2O + 4e^- \longrightarrow 4OH^-(aq) \quad \text{(cathode)}$$

At the anode, hydrogen is oxidized to water.

$$H_2(g) + 2OH^-(aq) \longrightarrow 2H_2O + 2e^- \quad \text{(anode)}$$

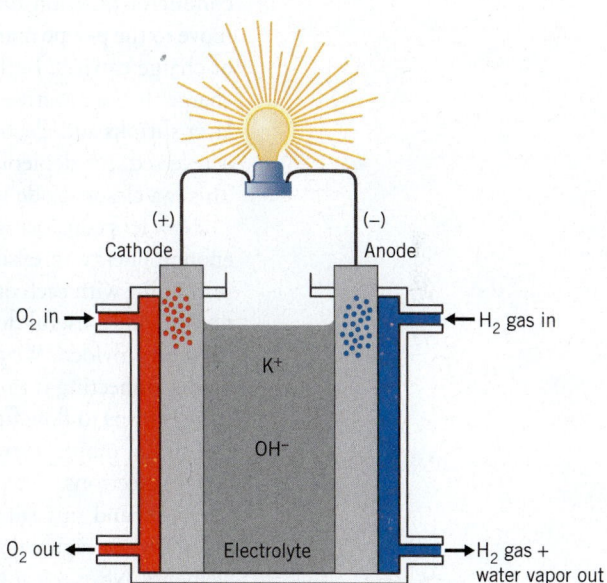

Figure 19.15 | A hydrogen–oxygen fuel cell.

Part of the water formed at the anode leaves as steam mixed with the circulating hydrogen gas. The net cell reaction, after making electron loss equal to electron gain, is

$$2H_2(g) + O_2(g) \longrightarrow 2H_2O \quad \text{(net cell reaction)}$$

Hydrogen–oxygen fuel cells are an attractive alternative to fossil fuel-powered engines in part because they are essentially pollution-free—the only product of the reaction is harmless water. Fuel cells are also quite thermodynamically efficient, converting as much as 75% of the available energy to useful work, compared to approximately 25 to 30% for gasoline and diesel engines. Among the major obstacles are the energy costs of generating the hydrogen fuel and problems in providing storage and distribution of the highly flammable H$_2$.

Photovoltaic Cells

There exists a wide variety of devices where the electrical properties of materials, combined with the energy of photons, can be used to generate a flow of electrons. These devices are called *photovoltaics*.

A photovoltaic device must have a means to (a) separate an electron from a substance and (b) only allow that electron to return to its original site after moving through an external electrical circuit. In the process of traveling through the electrical circuit, the energy of that electron can be used to perform useful work.

Although modern photovoltaic devices are quite sophisticated, we can investigate a simple model to understand the basics. We start with ultrapure silicon (or germanium). These are semiconductors and are not useful until a controlled amount of impurity, in the form of phosphorous or boron, is added (this process is called doping). We know that silicon should form a tetrahedral network crystal similar to diamond. If boron is added, it will replace a silicon atom in the crystal with one important difference: it can only form three bonds with the four neighboring silicon atoms. The absent bond is called a **hole** or a **positive charge carrier**. If phosphorus is added, it has five valence electrons. Four of these bond with four neighboring silicon atoms, leaving one extra electron that can move through the crystal as a **negative charge carrier**. For simplicity we call the boron doped silicon a **p-type semiconductor** and the phosphorus doped silicon an **n-type semiconductor**.

When n-type and p-type semiconductor materials are joined together we form a semiconductor junction that has unique properties. First, electrons from the n-type material move to the p-type material and combine with available holes. The result is a layer depleted of charge carriers, both electrons and holes. This depletion layer inhibits further flow of charge. If the negative end of a battery is connected to the n-type material, the depletion layer shrinks and electrons can flow from the n-type to the p-type material. If the battery is reversed, the depletion layer expands and electrons cannot flow in the reverse direction. This is a classic diode where electrons can flow only one way.

Now let's consider what happens if we hit these semiconductors with photons that have enough energy to excite some electrons, creating electron–hole pairs. These pairs could recombine with each other, or with other holes and electrons within the diode. That would not provide a useful flow of electrons. To be useful, another, easier, path for the electrons must be provided. We provide this pathway by adding wires (electrodes) to each side of the diode connecting it to an external circuit. This external circuit provides an easier route for the electrons to flow from the n-type back to the p-type material. Of course, we can add a light bulb, motor, meter, computer, or any other electric device to take advantage of this flow of electrons.

To expand our simple photocell to a useful device, we need to add a few extras as illustrated in Figure 19.16. First, we add a glass covering to protect our device from the elements. Next, to capture all the sunlight possible, an anti-reflective coating is added, and then we have a metallic mesh cathode added to the n-type material so that light can get through. Next is the junction with the p-type semiconductor and, finally, the metal that forms the anode. Since each photovoltaic cell produces about 0.5 volt, 24 cells may be connected in series to produce a 12 volt device. Different groupings of the photocells may be designed to improve either the voltage or current rating of the device. Figure 19.17 illustrates part of a large array of photovoltaics that could be used to power a small group of homes or stores.

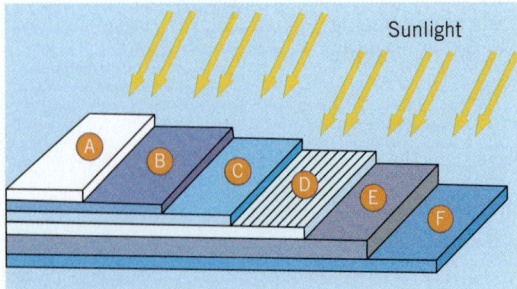

Figure 19.16 | Diagram of the internal parts of a photovoltaic cell. (*a*) Glass protective covering, (*b*) antireflective coating, (*c*) electrode mesh, (*d*) n-type semiconductor material, (*e*) p-type semiconductor material, (*f*) electrode. The thickness of this cell is approximately 0.3 mm, and the n-type semiconductor is approximately 0.002 mm thick.

Figure 19.17 | Acres of photovoltaic cells can produce large amounts of energy.

© Fotosearch/Purestock/SuperStock

19.7 | Electrolytic Cells

In our preceding discussions, we've examined how spontaneous redox reactions can be used to generate electrical energy. We now turn our attention to the opposite process, the use of electrical energy to force nonspontaneous redox reactions to occur.

When electricity is passed through a molten (melted) ionic compound or through a solution of an electrolyte, a chemical reaction occurs that we call **electrolysis**. A typical electrolysis apparatus, called an **electrolysis cell** or **electrolytic cell**, is shown in Figure 19.18. This particular cell contains molten sodium chloride. (A substance undergoing electrolysis must be molten or in solution so its ions can move freely and conduction can occur.) *Inert electrodes*—electrodes that won't react with the molten NaCl or the electrolysis products—are dipped into the cell and then connected to a source of direct current (DC) electricity.

The DC source serves as an "electron pump," pulling electrons away from one electrode and pushing them through the external wiring onto the other electrode. The electrode from which electrons are removed becomes positively charged, while the other electrode becomes negatively charged. When electricity starts to flow, chemical changes begin to happen. At the positive electrode, oxidation occurs as electrons are pulled away from negatively charged chloride ions. Because of the nature of the chemical change, therefore, *the positive electrode becomes the anode.* The DC source pumps the electrons through the external electrical circuit to the negative electrode. Here reduction takes place as the electrons are forced onto positively charged sodium ions, so *the negative electrode is the cathode.*

The chemical changes that occur at the electrodes can be described by chemical equations.

$$Na^+(l) + e^- \longrightarrow Na(l) \qquad \text{(cathode)}$$

$$2Cl^-(l) \longrightarrow Cl_2(g) + 2e^- \qquad \text{(anode)}$$

As in a galvanic cell, the overall reaction that takes place in the electrolysis cell is called the *cell reaction.* To obtain it, we add the individual electrode half-reactions together, making sure that the number of electrons gained in one half-reaction equals the number lost in the other.

$$2Na^+(l) + 2e^- \longrightarrow 2Na(l) \qquad \text{(cathode)}$$

$$\underline{2Cl^-(l) \longrightarrow Cl_2(g) + 2e^- \qquad \text{(anode)}}$$

$$2Na^+(l) + 2Cl^-(l) + 2e^- \longrightarrow 2Na(l) + Cl_2(g) + 2e^- \qquad \text{(cell reaction)}$$

■ To perform electrolysis, we must use direct current in which electrons move in only one direction, not in the oscillating, back and forth pattern of alternating current.

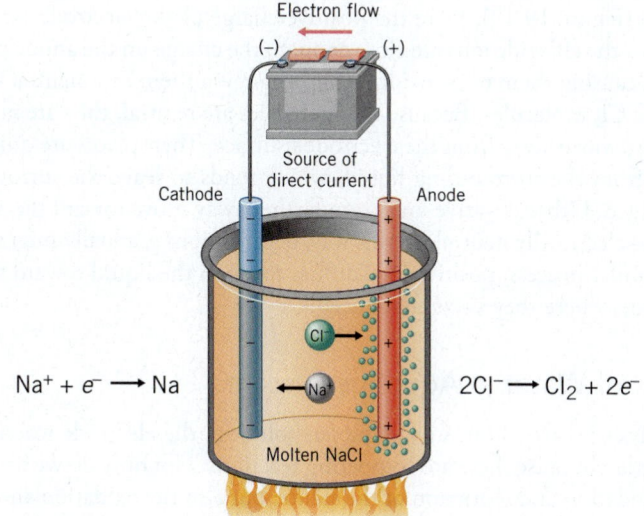

$$Na^+ + e^- \longrightarrow Na \qquad\qquad 2Cl^- \longrightarrow Cl_2 + 2e^-$$

Figure 19.18 | Electrolysis of molten sodium chloride. In this electrolysis cell, the passage of an electric current decomposes molten sodium chloride into metallic sodium and gaseous chlorine. Unless the products are kept apart, they react on contact to re-form NaCl.

As you know, table salt is quite stable. It doesn't normally decompose because the reverse reaction, the reaction of sodium and chlorine to form sodium chloride, is highly spontaneous. Therefore, we often write the word *electrolysis* above the arrow in the equation to show that electricity is the driving force for this otherwise nonspontaneous reaction.

$$2Na^+(l) + 2Cl^-(l) \xrightarrow{\text{electrolysis}} 2Na(l) + Cl_2(g)$$

Comparing Electrolytic and Galvanic Cells

In a galvanic cell, the spontaneous cell reaction deposits electrons on the anode and removes them from the cathode. As a result, the anode carries a slight negative charge and the cathode a slight positive charge. In most galvanic cells the reactants must be placed in separate compartments. In an *electrolysis cell*, the situation is reversed. Often the two electrodes are immersed in the same liquid. Also, the oxidation at the anode must be forced to occur, which requires that the anode be positive so it can remove electrons from the reactant at that electrode. On the other hand, the cathode must be made negative so it can force the reactant at the electrode to accept electrons.

■ By agreement among scientists, the names anode and cathode are assigned according to the nature of the reaction taking place at the electrode. If the reaction is oxidation, the electrode is called the anode; if it's reduction, the electrode is called the cathode.

Electrolytic Cell	Galvanic Cell
Cathode is negative (reduction).	Cathode is positive (reduction).
Anode is positive (oxidation).	Anode is negative (oxidation).
Anode and cathode are often in the same compartment.	Anode and cathode are usually in separate compartments.

Even though the charges on the cathode and anode differ between electrolytic cells and galvanic cells, the ions in solution always move in the same direction. In both types of cells, positive ions (cations) move toward the cathode. They are attracted there by the negative charge on the cathode in an electrolysis cell; they diffuse toward the cathode in a galvanic cell to balance the charge of negative ions left behind when ions are reduced. Similarly, negative ions (anions) move toward the anode. They are attracted to the positive anode in an electrolysis cell, and they diffuse toward the anode in our galvanic cell to balance the charge of the positive ions entering the solution.

Electrolysis at the Molecular Level

The electrical conductivity of a molten salt or a solution of an electrolyte is possible only because of the reactions that take place at the surface of the electrodes. For example, when charged electrodes are dipped into molten NaCl they become surrounded by a layer of ions of the opposite charge. Let's look closely at what happens at one of the electrodes—say, the anode (Figure 19.19). Here the positive charge of the electrode attracts negative Cl^- ions. When the chloride ions are close enough, the charge on the anode pulls electrons from the ions, causing them to be oxidized and changing them into neutral Cl atoms that join to become Cl_2 molecules. Because the molecules are neutral, they are not held by the electrode and so move away from the electrode's surface. Their places are quickly taken by negative ions from the surrounding liquid, which tends to leave the surrounding liquid positively charged. Other negative ions from farther away move toward the anode to keep the liquid there electrically neutral. In this way, negative ions gradually migrate toward the anode. By a similar process, positive ions diffuse through the liquid toward the negatively charged cathode, where they become reduced.

■ Cations (positive ions) move toward the cathode and anions (negative ions) migrate toward the anode. This happens in both electrolytic and galvanic cells.

Electrolysis of Water in Aqueous Systems

When electrolysis is carried out in an aqueous solution, the electrode reactions are more difficult to predict because there are competing reactions. Not only do we have to consider the possible oxidation and reduction of the solute, but also the oxidation and reduction of the solvent, water. For example, consider what happens when electrolysis is performed on

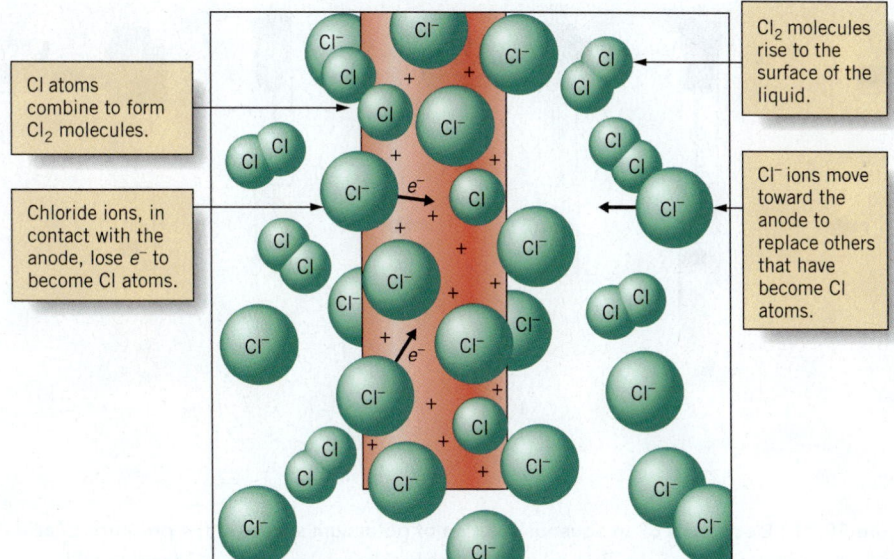

The positive charge of the electrode attracts a coating of Cl⁻ ions. At the surface of the electrode, electrons are pulled from the ions, yielding neutral Cl atoms, which combine to form Cl₂ molecules that move away from the electrode and eventually rise to become Cl atoms.

Cl₂ molecules rise to the surface of the liquid.

Cl atoms combine to form Cl₂ molecules.

Chloride ions, in contact with the anode, lose e^- to become Cl atoms.

Cl⁻ ions move toward the anode to replace others that have become Cl atoms.

a solution of potassium sulfate (Figure 19.20). The products are hydrogen and oxygen. At the cathode, water is reduced, not K^+.

$$2H_2O(l) + 2e^- \longrightarrow H_2(g) + 2OH^-(aq) \qquad \text{(cathode)}$$

At the anode, water is oxidized, not the sulfate ion.

$$2H_2O(l) \longrightarrow O_2(g) + 4H^+(aq) + 4e^- \qquad \text{(anode)}$$

Color changes of an acid–base indicator dissolved in the solution confirm that the solution becomes basic around the cathode, where OH is formed, and acidic around the anode, where H^+ is formed (see Figure 19.21). In addition, the gases H_2 and O_2 can be separately collected.

We can understand why these redox reactions happen if we examine the standard reduction potential data from Table 19.1. For example, at the cathode we have the following competing reactions.

$$K^+(aq) + e^- \longrightarrow K(s) \qquad\qquad E^\circ_{K^+} = -2.92 \text{ V}$$

$$2H_2O(l) + 2e^- \longrightarrow H_2(g) + 2OH^-(aq) \qquad E^\circ_{H_2O} = -0.83 \text{ V}$$

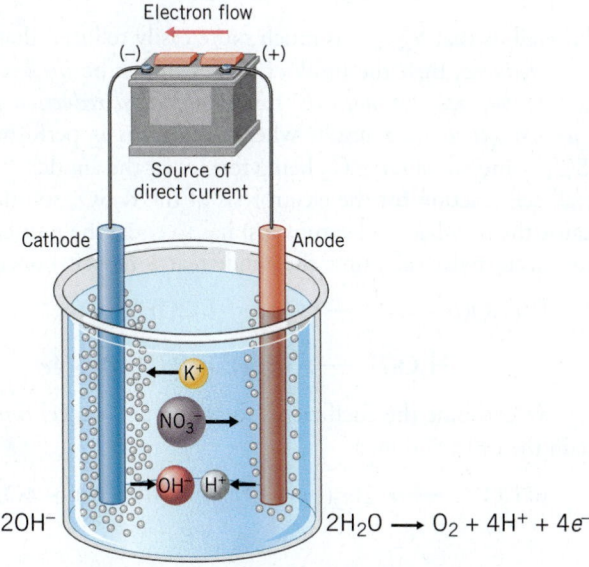

Electron flow

(−)   (+)

Source of direct current

Cathode   Anode

K^+

NO_3^-

OH^- H^+

$2H_2O + 2e^- \longrightarrow H_2 + 2OH^-$ $2H_2O \longrightarrow O_2 + 4H^+ + 4e^-$

Figure 19.20 | Molecular view of the electrolysis of an aqueous solution of potassium sulfate. The products of the electrolysis are the gases hydrogen and oxygen.

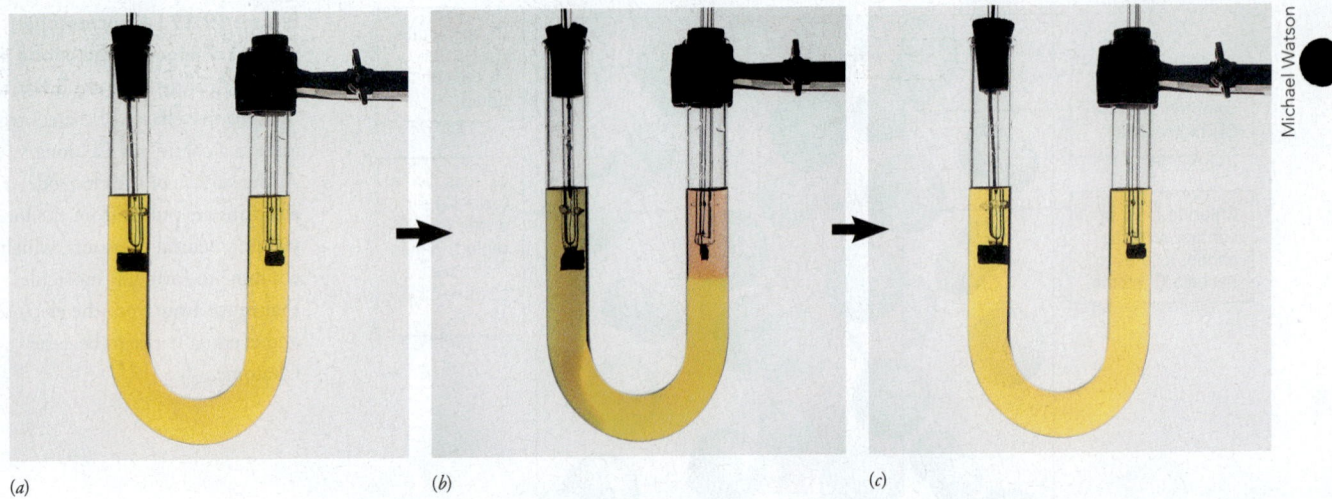

(a)　　　　　(b)　　　　　(c)

Figure 19.21 | **Electrolysis of an aqueous solution of potassium sulfate in the presence of acid–base indicators.** (*a*) The initial yellow color indicates that the solution is neutral (neither acidic nor basic). (*b*) As the electrolysis proceeds, H^+ is produced at the anode (along with O_2) and causes the solution there to become pink. At the cathode, H_2 is evolved and OH^- ions are formed, which turns the solution around that electrode a bluish violet. (*c*) After the electrolysis is stopped and the solution is stirred, the color becomes yellow again as the H^+ and OH^- ions formed by the electrolysis neutralize each other.

Water has a much less negative (and therefore more positive) standard reduction potential than K^+, which means H_2O is much easier to reduce than K^+. As a result, when the electrolysis is performed the more easily reduced substance is reduced and we observe H_2 being formed at the cathode.

At the anode we have the following possible oxidation half-reactions.

$$2SO_4^{2-}(aq) \longrightarrow S_2O_8^{2-}(aq) + 2e^-$$

$$2H_2O \longrightarrow 4H^+(aq) + O_2(g) + 2e^-$$

In Table 19.1, we find them written in the opposite direction:

$$S_2O_8^{2-}(aq) + 2e^- \longrightarrow 2SO_4^{2-}(aq) \qquad E^\circ_{S_2O_8^{2-}} = +2.01 \text{ V}$$

$$O_2(g) + 4H^+(aq) + 4e^- \longrightarrow 2H_2O \qquad E^\circ_{O_2} = +1.23 \text{ V}$$

The E° values tell us that $S_2O_8^{2-}$ is much more easily reduced than O_2. But if $S_2O_8^{2-}$ is the *more easily reduced*, then the product, SO_4^{2-}, must be *the less easily oxidized*. Stated another way, *the half-reaction with the smaller standard reduction potential is more easily reversed as an oxidation.* As a result, when electrolysis is performed, water is oxidized instead of SO_4^{2-} and we observe O_2 being formed at the anode.

The overall cell reaction for the electrolysis of the K_2SO_4 solution can be obtained as before. Because the number of electrons lost has to equal the number gained, the cathode reaction must occur twice each time the anode reaction occurs once.

$$2 \times [2H_2O(l) + 2e^- \longrightarrow H_2(g) + 2OH^-(aq)] \qquad \text{(cathode)}$$

$$2H_2O(l) \longrightarrow O_2(g) + 4H^+(aq) + 4e^- \qquad \text{(anode)}$$

After adding, we combine the coefficients for water and cancel four electrons from both sides to obtain the cell reaction.

$$6H_2O(l) \longrightarrow 2H_2(g) + O_2(g) + 4H^+(aq) + 4OH^-(aq)$$

Notice that hydrogen ions and hydroxide ions are produced in equal numbers. In Figure 19.21, we see that when the solution is stirred, they combine to form water.

$$6H_2O(l) \longrightarrow 2H_2(g) + O_2(g) + \underbrace{4H^+(aq) + 4OH^-(aq)}_{4H_2O}$$

The net change, then, is

$$2H_2O \xrightarrow{\text{electrolysis}} 2H_2(g) + O_2(g)$$

Function of Spectator Ions in Electrolysis

Although neither K^+ nor SO_4^{2-} is changed by the reaction, K_2SO_4 or some other electrolyte is needed for the electrolysis to proceed. Its function is to maintain electrical neutrality at the electrodes. At the anode, H^+ ions are formed and their charge can be balanced by mixing with SO_4^{2-} ions of the solute. Similarly, at the cathode the K^+ ions are able to mix with OH^- ions as they are formed, thereby balancing the charge and keeping the solution electrically neutral. In this way, at any moment, each small region of the solution is able to contain the same number of positive and negative charges and thereby remain neutral.

Standard Reduction Potentials and Electrolysis Products

Suppose we wish to know what products are expected in the electrolysis of an aqueous solution of copper(II) bromide, $CuBr_2$. Let's examine the possible electrode half-reactions and their respective standard reduction potentials. At the cathode, possible reactions are the reduction of copper ion and the reduction of water. From Table 19.1,

$$Cu^{2+}(aq) + 2e^- \longrightarrow Cu(s) \qquad\qquad E^\circ_{Cu^{2+}} = +0.34 \text{ V}$$

$$2H_2O(l) + 2e^- \longrightarrow H_2(g) + 2OH^-(aq) \qquad E^\circ_{H_2O} = -0.83 \text{ V}$$

The much more positive reduction standard potential for Cu^{2+} tells us to anticipate that Cu^{2+} will be reduced at the cathode.

At the anode, possible reactions are the oxidation of Br and the oxidation of water. The half-reactions are

$$2Br^-(aq) \longrightarrow Br_2(aq) + 2e^-$$

$$2H_2O(l) \longrightarrow O_2(g) + 4H^+(aq) + 4e^-$$

In Table 19.1, they are written as reductions with the following E° values.

$$Br_2(aq) + 2e^- \longrightarrow 2Br^-(aq) \qquad E^\circ_{Br_2} = +1.07 \text{ V}$$

$$O_2(g) + 4H^+(aq) + 4e^- \longrightarrow 2H_2O(l) \qquad E^\circ_{O_2} = +1.23 \text{ V}$$

The data tell us that O_2 is more easily reduced than Br_2, which means that Br is more easily oxidized than H_2O. Therefore, we expect that Br^- will be oxidized at the anode.

In fact, our predictions are confirmed when we perform the electrolysis. The cathode, anode, and net cell reactions are

$$Cu^{2+}(aq) + 2e^- \longrightarrow Cu(s) \qquad\qquad \text{(cathode)}$$

$$2Br^-(aq) \longrightarrow Br_2(aq) + 2e^- \qquad\qquad \text{(anode)}$$

$$\overline{\qquad\qquad\qquad\qquad\qquad\qquad\qquad\qquad\qquad}$$

$$Cu^{2+}(aq) + 2Br^-(aq) \xrightarrow{\text{electrolysis}} Cu(s) + Br_2(aq) \qquad \text{(net reaction)}$$

Example 19.11
Predicting the Products in an Electrolysis Reaction

Electrolysis is planned for an aqueous solution that contains a mixture of 0.50 M $ZnSO_4$ and 0.50 M $NiSO_4$. On the basis of standard reduction potentials, what products are expected to be observed at the electrodes? What is the expected net cell reaction?

Analysis: We need to consider the competing reactions at the cathode and the anode. At the cathode, the half-reaction with the *most positive* standard reduction potential will be the one most expected to occur. At the anode, the half-reaction with the *least positive* standard reduction potential is the one that will most readily occur in the reverse, or oxidation, direction.

Assembling the Tools: Our main tool will be a table of standard reduction potentials. We can use Table 19.1, the expanded table in Appendix Table C.9, or even more extensive tables in reference works.

Solution: At the cathode, the competing reduction reactions involve the two cations and water. The reactions and their standard reduction potentials are

$$Ni^{2+}(aq) + 2e^- \rightleftharpoons Ni(s) \qquad\qquad E° = -0.25 \text{ V}$$

$$Zn^{2+}(aq) + 2e^- \rightleftharpoons Zn(s) \qquad\qquad E° = -0.76 \text{ V}$$

$$2H_2O + 2e^- \rightleftharpoons H_2(g) + 2OH^-(aq) \qquad E° = -0.83 \text{ V}$$

The least negative standard reduction potential is that of Ni^{2+}, so we expect this ion to be reduced at the cathode, resulting in solid nickel deposited onto the cathode.

At the anode, the competing oxidation reactions are for water and SO_4^{2-} ion. In Table 19.1, substances oxidized are found on the right side of the half-reactions. The two half-reactions having these as products are

$$S_2O_8^{2-}(aq) + 2e^- \rightleftharpoons 2SO_4^{2-}(aq) \qquad E° = +2.01 \text{ V}$$

$$O_2(g) + 4H^+(aq) + 4e^- \rightleftharpoons 2H_2O \qquad E° = +1.23 \text{ V}$$

The half-reaction with the least positive $E°$ (the second one here) is most easily reversed as an oxidation, so we expect the oxidation half-reaction to be

$$2H_2O \rightleftharpoons O_2(g) + 4H^+(aq) + 4e^-$$

At the anode, we expect O_2 to be formed.

The predicted net cell reaction is obtained by combining the two expected electrode half-reactions, making the electron loss equal to the electron gain.

$$2H_2O \longrightarrow O_2(g) + 4H^+(aq) + 4e^- \qquad \text{(anode)}$$

$$2 \times [Ni^{2+}(aq) + 2e^- \longrightarrow Ni(s)] \qquad \text{(cathode)}$$

$$\overline{2H_2O + 2Ni^{2+}(aq) \longrightarrow O_2(g) + 4H^+(aq) + 2Ni(s)} \qquad \text{(net cell reaction)}$$

Are the Answers Reasonable? We can check the locations of the half-reactions in Table 19.1 to confirm our conclusions. For the reduction step, the higher up in the table a half-reaction is, the greater its tendency to occur as reduction. Among the competing half-reactions at the cathode, the one for Ni^{2+} is highest, so we expect that Ni^{2+} is the easiest to reduce and $Ni(s)$ should be formed at the cathode.

For the oxidation step, the lower down in the table a half-reaction is, the easier it is to reverse and cause to occur as oxidation. On this basis, the oxidation of water is easier than the oxidation of SO_4^{2-}, so we expect H_2O to be oxidized and O_2 to be formed at the anode.

Of course, we could also test our prediction by carrying out the electrolysis experimentally.

In the electrolysis of an aqueous solution containing Fe^{2+} and I^-, what product do we expect at the anode? (*Hint:* Write the three oxidation half-reactions possible at the anode for Fe^{2+}, I^-, and H_2O.)

Practice Exercise 19.21

In the electrolysis of an aqueous solution containing both Cd^{2+} and Sn^{2+}, what product do we expect at the cathode?

Practice Exercise 19.22

Sometimes Predictions Fail

Although we can use standard reduction potentials to predict electrolysis reactions most of the time, there are occasions when standard reduction potentials do not successfully predict electrolysis products. Sometimes concentrations, far from standard conditions, will change the sign of the cell potential. The formation of complex ions can also interfere and produce unexpected results. And sometimes the electrodes themselves are the culprits. For example, in the electrolysis of aqueous NaCl using inert platinum electrodes, we find experimentally that Cl_2 is formed at the anode. Is this what we would have expected? Let's examine the standard reduction potentials of O_2 and Cl_2 to find out.

$$Cl_2(g) + 2e^- \rightleftharpoons 2Cl^-(aq) \qquad E^\circ = +1.36 \text{ V}$$

$$O_2(g) + 4H^+(aq) + 4e^- \rightleftharpoons 2H_2O \qquad E^\circ = +1.23 \text{ V}$$

Because of its less positive standard reduction potential, we would expect the oxygen half-reaction to be the easier to reverse (with water being oxidized to O_2). Thus, standard reduction potentials predict that O_2 should be formed, but experiment shows that Cl_2 is produced. The nature of the electrode surface and how it interacts with oxygen is part of the answer. Further explanation for why this happens is beyond the scope of this text, but the unexpected result does teach us that we must be cautious in predicting products in electrolysis reactions solely on the basis of standard reduction potentials.

19.8 | Electrolysis Stoichiometry

In about 1833, Michael Faraday discovered that the amount of chemical change that occurs during electrolysis is directly proportional to the amount of electrical charge that is passed through an electrolysis cell. For example, the reduction of copper ion at a cathode is given by the equation

$$Cu^{2+}(aq) + 2e^- \longrightarrow Cu(s)$$

The equation tells us that to deposit one mole of metallic copper requires two moles of electrons. Thus, the half-reaction for an oxidation or reduction relates the amount of chemical substance consumed or produced to the amount of electrons that the electric current must supply. To use this information, however, we must be able to relate it to electrical measurements that can be made in the laboratory.

The SI unit of electric current is the **ampere (A)** and the SI unit of charge is the *coulomb* (C). A coulomb is the amount of charge that passes by a given point in a wire when an electric current of one ampere flows for one second. This means that coulombs are the product of amperes of current multiplied by seconds. Thus

$$1 \text{ coulomb} = 1 \text{ ampere} \times 1 \text{ second}$$

$$1 \text{ C} = 1 \text{ A s} \tag{19.9}$$

For example, if a current of 4 A flows through a wire for 10 s, 40 C pass by a given point in the wire.

$$(4 \text{ A}) \times (10 \text{ s}) = 40 \text{ A s} = 40 \text{ C}$$

© Classic Image/Alamy

Michael Faraday (1791–1867), a British scientist and both a chemist and a physicist, made key discoveries leading to electric motors, generators, and transformers.

TOOLS

Coulombs = (amperes) × (time in seconds)

Figure 19.22 | Flowchart including electrolysis stoichiometry.

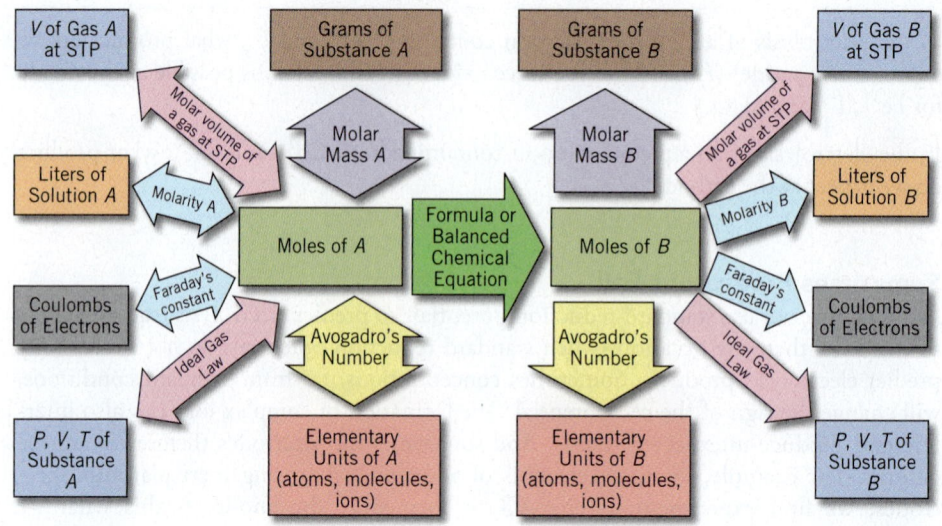

As we noted previously, it has been determined that 1 mol of electron carries a charge of 9.65×10^4 C, which, in honor of Michael Faraday, is often called the Faraday constant, $\mathscr{F}$.

$$1 \text{ mol } e^- \Leftrightarrow 9.65 \times 10^4 \text{ C} \quad \text{(to three significant figures)}$$

$$1 \,\mathscr{F} = 9.65 \times 10^4 \text{ C/mol } e^-$$

Now we have a way to relate laboratory measurements to the amount of chemical change that occurs during an electrolysis. Measuring the current in amperes and the time in seconds allows us to calculate the charge sent through the system in coulombs. From this we can get the amount of electrons (in moles), which we can then use to calculate the amount of chemical change produced. The following examples demonstrate the principles involved for electrolysis, but similar calculations also apply to reactions in galvanic cells. Figure 19.22 illustrates how this fits in with our previous stoichiometric calculations. We used similar pathways in Figures 3.6, 4.24, and 10.17.

Example 19.12
Calculations Related to Electrolysis

How many grams of copper are deposited on the cathode of an electrolytic cell if an electric current of 2.00 A passes through a solution of $CuSO_4$ for a period of 19.0 min?

Analysis: The wording of this question suggests that our solution will be related to a stoichiometry type of calculation in which we will convert the time and electric current to grams of copper.

Assembling the Tools: Figure 19.22 helps us assemble the needed tools. Equation 19.9,

$$1 \text{ C} = 1 \text{ A s}$$

is our tool for determining the coulombs of electrons used. Faraday's constant is used as a conversion factor to determine the moles of electrons. Next, the balanced half-reaction serves as our tool for relating moles of electrons to moles of Cu. The ion being reduced is Cu^{2+}, so the half-reaction is

$$Cu^{2+} + 2e^- \longrightarrow Cu$$

Therefore,

$$1 \text{ mol Cu} \Leftrightarrow 2 \text{ mol } e^-$$

Finally, we convert the moles of Cu to mass of Cu using the relationships from Chapter 3. Here's a diagram of the path to the solution.

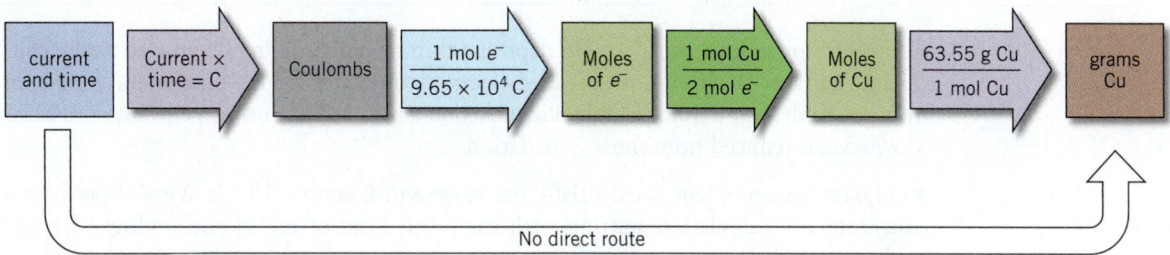

Since the calculation will be done stepwise we will keep an extra significant figure that will be rounded at the end.

Solution: First we convert minutes to seconds

$$19.0 \text{ min} = 1.14 \times 10^3 \text{ s}$$

Then we multiply the current by the time to obtain the number of coulombs (1 A s = 1 C).

$$(1.14 \times 10^3 \text{ s}) \times (2.00 \text{ A}) = 2.28 \times 10^3 \text{ A s}$$

$$= 2.28 \times 10^3 \text{ C}$$

Because 1 mol $e^- \Leftrightarrow 9.65 \times 10^4$ C,

$$2.28 \times 10^3 \text{ C} \times \left(\frac{1 \text{ mol } e^-}{9.65 \times 10^4 \text{ C}} \right) = 0.02363 \text{ mol } e^-$$

Next, we use the relationship between mol e^- and mol Cu from the balanced half-reaction along with the atomic mass of copper.

$$0.02363 \text{ mol } e^- \left(\frac{1 \text{ mol Cu}}{2 \text{ mol } e^-} \right) \left(\frac{63.55 \text{ g Cu}}{1 \text{ mol Cu}} \right) = 0.7508 \text{ g Cu}$$

With proper rounding the electrolysis will deposit 0.751 g of copper on the cathode.

   We could have combined all these steps in a single calculation by stringing together the various conversion factors and using the factor-label method to cancel units. In the equation below we are writing 9.65×10^4 C as the equivalent 9.65×10^4 A s.

$$2.00 \text{ A} \times 19.0 \text{ min} \times \frac{60 \text{ s}}{1 \text{ min}} \times \frac{1 \text{ mol } e^-}{9.65 \times 10^4 \text{ A s}} \times \frac{1 \text{ mol Cu}}{2 \text{ mol } e^-} \times \frac{63.55 \text{ g Cu}}{1 \text{ mol Cu}} = 0.751 \text{ g Cu}$$

Is the Answer Reasonable? As before, we first ascertain that the problem is set up correctly by checking that the units cancel correctly. Then, round all numbers to one significant figure to estimate the answer.

$$g \text{ Cu} = 2\text{A} \times 20 \text{ min} \times \frac{60 \text{ s}}{1 \text{ min}} \times \frac{1 \text{ mol } e^-}{10 \times 10^4 \text{ A s}} \times \frac{1 \text{ mol Cu}}{2 \text{ mol } e^-} \times \frac{60 \text{ g Cu}}{1 \text{ mol Cu}}$$

Since the units cancel properly, we will cancel one set of 2s and write all numbers greater than ten in exponential notation to get

$$\frac{2 \times 2 \times 10^1 \times 6 \times 10^1 \times 6 \times 10^1}{10 \times 10^4 \times 2} = \frac{2 \times 6 \times 6 \times 10^3}{10^5} = 72 \times 10^{-2} = 0.72 \text{ g Cu}$$

This is very close to our answer, and we are even more confident that the calculation was correct when we supply the cancellation of units to our estimate.

Example 19.13
Calculations Related to Electrolysis

Electrolysis provides a useful way to deposit a thin metallic coating on an electrically conducting surface. This technique is called electroplating. How much time would it take, in minutes, to deposit 0.500 g of metallic nickel on a metal object using a current of 3.00 A? The nickel is reduced from the +2 oxidation state.

Analysis: This problem is essentially the reverse of Example 19.12. We will perform a stoichiometric calculation starting with the given mass of nickel and ending with the coulombs needed.

Assembling the Tools: We should be very familiar with converting mass to moles; the next step requires a balanced chemical equation. Because the nickel is reduced to the free metal from the +2 state, we can write

$$Ni^{2+} + 2e^- \longrightarrow Ni(s)$$

This gives the relationship

$$1 \text{ mol Ni} \Leftrightarrow 2 \text{ mol } e^-$$

which allows us to calculate the number of moles of electrons required. This, in turn, is used with the tool for the Faraday constant to determine the number of coulombs required. Equation 19.9 tells us that this is the product of amperes and seconds; so we can calculate the time needed to deposit the metal. The diagram below isolates our sequence of steps that can also be deduced from Figure 19.22.

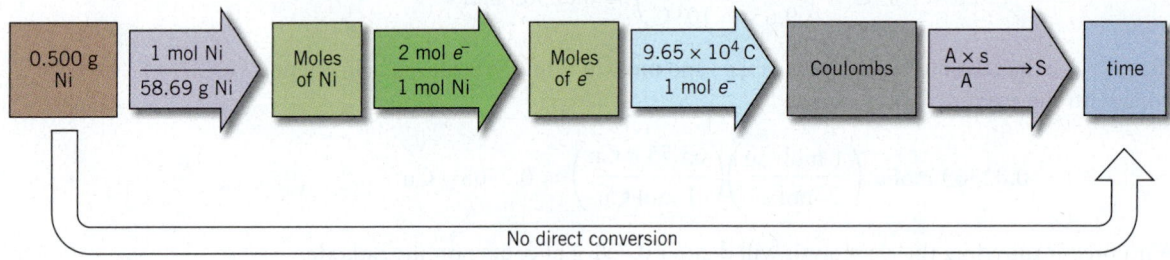

■ In stepwise calculations we usually carry one or more extra significant figures to minimize rounding errors.

Solution: First, we calculate the number of moles of electrons required (keeping at least one extra significant figure).

$$0.500 \text{ g Ni}\left(\frac{1 \text{ mol Ni}}{58.69 \text{ g Ni}}\right)\left(\frac{2 \text{ mol } e^-}{1 \text{ mol Ni}}\right) = 0.01704 \text{ mol } e^-$$

Then we calculate the number of coulombs needed.

$$0.01704 \text{ mol } e^-\left(\frac{9.65 \times 10^4 \text{ C}}{1 \text{ mol } e^-}\right) = 1.644 \times 10^3 \text{ C} = 1.644 \times 10^3 \text{ A s}$$

■ Once we've calculated the number of coulombs required, we can calculate the time required if we know the current, or the current needed to perform the electrolysis in a given time.

This tells us that the product of current multiplied by time equals 1.644×10^3 A s. The current is 3.00 A. Dividing 1.644×10^3 A s by 3.00 A gives the time required in seconds, which we then convert to minutes.

$$\left(\frac{1.644 \times 10^3 \text{ A s}}{3.00 \text{ A}}\right)\left(\frac{1 \text{ min}}{60 \text{ s}}\right) = 9.133 \text{ min}$$

Properly rounded, this becomes 9.13 min. We could also have combined these calculations in a single string of conversion factors.

$$0.500 \text{ g Ni}\left(\frac{1 \text{ mol Ni}}{58.69 \text{ g Ni}}\right)\left(\frac{2 \text{ mol } e^-}{1 \text{ mol Ni}}\right)\left(\frac{9.65 \times 10^4 \text{ C}}{1 \text{ mol } e^-}\right)\left(\frac{1 \text{ A s}}{1 \text{ C}}\right)\left(\frac{1}{3.00 \text{ A}}\right)\left(\frac{1 \text{ min}}{60 \text{ s}}\right) = 9.133 \text{ min}$$

- **Is the Answer Reasonable?** As in the preceding example, let's check that the units cancel properly and then round all numbers to one digit and write them in scientific notation, without the units for clarity, to get:

$$0.5\left(\frac{1}{60}\right)\left(\frac{2}{1}\right)\left(\frac{10 \times 10^4}{1}\right)\left(\frac{1}{1}\right)\left(\frac{1}{3}\right)\left(\frac{1}{60}\right) = \frac{0.5 \times 2 \times 10 \times 10^4}{60 \times 3 \times 60}$$

$$= \frac{10^5}{6 \times 3 \times 6 \times 10^2}$$

$$= \frac{10^5}{108 \times 10^2} = \frac{10^5}{10^4} = 10$$

This result, along with the proper cancellation of units, indicates that our setup and calculations were reasonable.

How many moles of hydroxide ion will be produced at the cathode during the electrolysis of water with a current of 4.00 A for a period of 13.00 minutes? The cathode reaction is

$$2e^- + 2H_2O \longrightarrow H_2 + 2OH^-$$

(*Hint:* The stepwise diagram in Example 19.13 will be helpful.)

How many minutes will it take for a current of 14.0 A to deposit 3.00 g of gold from a solution of $AuCl_3$?

What current must be supplied to deposit 0.0500 g of gold from a solution of $AuCl_3$ in 20.0 min?

Suppose the solutions in the galvanic cell depicted in Figure 19.2 (Section 19.1) have a volume of 125 mL and suppose the cell is operated for a period of 3.46 hr with a constant current of 0.550 A flowing through the external circuit. By how much will the concentration of the copper ions increase during this time period?

Practice Exercise 19.23

Practice Exercise 19.24

Practice Exercise 19.25

Practice Exercise 19.26

19.9 | Practical Applications of Electrolysis

Electrochemistry has many applications both in science and in our everyday lives. In this limited space, we can only touch on some of the more common and important examples of the industrial uses of electrolysis.

Industrial Applications

Besides being a useful tool in the chemistry laboratory, electrolysis has many important industrial applications. In this section we will briefly examine the chemistry of electroplating and the production of some of our most common chemicals.

Electroplating

Electroplating, which was mentioned in Examples 19.12 and 19.13, is a procedure in which electrolysis is used to apply a thin (generally 0.03 to 0.05 mm thick) ornamental or protective coating of one metal over another. It is a common technique for improving the appearance and durability of metal objects. For instance, a thin, shiny coating of metallic chromium is applied over steel objects to make them attractive and to prevent rusting.

The exact composition of the electroplating bath varies, depending on the metal to be deposited, and it can affect the appearance and durability of the finished surface. For example, silver deposited from a solution of silver nitrate ($AgNO_3$) does not stick to other metal surfaces very well. However, if it is deposited from a solution of silver cyanide containing $Ag(CN)_2^-$, the coating adheres well and is bright and shiny. Other metals that are

This motorcycle sparkles with chrome plating that was deposited by electrolysis. The shiny, hard coating of chromium is both decorative and a barrier to corrosion.

Syracuse Newspapers/The Image Works

electroplated from a cyanide bath are gold and cadmium. Nickel, which can also be applied as a protective coating, is plated from a nickel sulfate solution, and chromium is plated from a chromic acid (H_2CrO_4) solution.

Production of Aluminum

Aluminum is a useful but highly reactive metal. It is so difficult to reduce that ordinary metallurgical methods for obtaining it do not work. Early efforts to produce aluminum by electrolysis failed because its anhydrous halide salts (those with no water of hydration) are difficult to prepare and are volatile, tending to evaporate rather than melt. On the other hand, its oxide, Al_2O_3, has such a high melting point (over 2000 °C) that no practical method of melting it could be found.

In 1886, Charles M. Hall discovered that Al_2O_3 dissolves in the molten form of a mineral called cryolite, Na_3AlF_6, to give a conducting mixture with a relatively low melting point from which aluminum could be produced electrolytically. The process was also discovered by Paul Héroult in France at nearly the same time, and today this method for producing aluminum is usually called the **Hall–Héroult process** (see Figure 19.23). Purified aluminum oxide, which is obtained from an ore called *bauxite,* is dissolved in molten cryolite in which the oxide dissociates to give Al^{3+} and O^{2-} ions. At the cathode, aluminum ions are reduced to produce the free metal, which forms as a layer of molten aluminum below the less dense solvent. At the carbon anodes, the oxide ion is oxidized to give free O_2.

$$4[Al^{3+} + 3e^- \longrightarrow Al(l)] \qquad \text{(cathode)}$$
$$\underline{3[2O^{2-} \longrightarrow O_2(g) + 4e^-] \qquad \text{(anode)}}$$
$$4Al^{3+} + 6O^{2-} \longrightarrow 4Al(l) + 3O_2(g) \qquad \text{(cell reaction)}$$

The oxygen formed at the anode reacts with the carbon electrodes (producing CO_2), so the electrodes must be replaced frequently.

■ Aluminum is used today as a structural metal, in alloys, and in such products as aluminum foil, electrical wire, window frames, and kitchen pots and pans.

The production of aluminum consumes enormous amounts of electrical energy and is therefore very costly, not only in terms of dollars but also in terms of energy resources. For this reason, recycling of aluminum has a high priority as we seek to minimize our use of energy.

Production of Sodium

Sodium is prepared by the electrolysis of molten sodium chloride (see Section 19.7). The metallic sodium and the chlorine gas that form must be kept apart or they will react

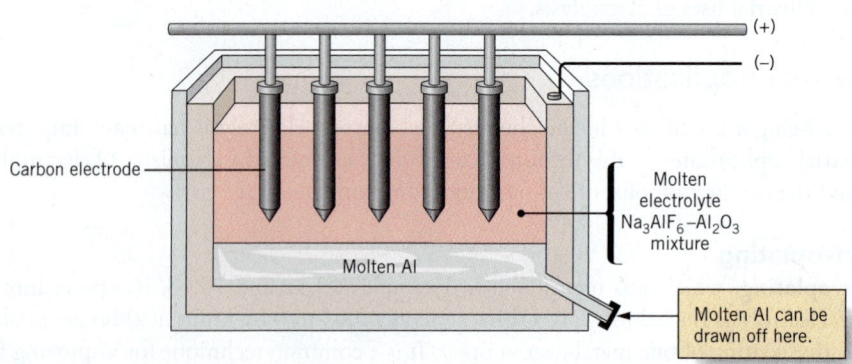

Figure 19.23 | **Production of aluminum by electrolysis.** In the apparatus used to produce aluminum electrolytically by the Hall–Héroult process, Al_2O_3 is dissolved in molten cryolite, Na_3AlF_6. The Al^{3+} is reduced to metallic Al, and O^{2-} is oxidized to O_2, which reacts with the carbon anodes to give CO_2. Periodically, molten aluminum is drawn off at the bottom of the cell and additional Al_2O_3 is added to the cryolite. The carbon anodes also must be replaced from time to time as they are consumed by their reaction with O_2.

violently and re-form NaCl. A specialized apparatus called a **Downs cell** accomplishes this separation.

Both sodium and chlorine are commercially important. Chlorine is used largely to manufacture plastics such as polyvinyl chloride (PVC), many solvents, and industrial chemicals. A small percentage of the annual chlorine production is used to chlorinate drinking water.

Sodium has been used in the manufacture of tetraethyl lead, an octane booster for gasoline that has been phased out in the United States but is still used in many other countries. Sodium is also used in the production of energy-efficient sodium vapor lamps, which give streetlights and other commercial lighting a bright yellow-orange color.

Refining Copper

When copper is first obtained from its ore, it is about 99% pure. The impurities—mostly silver, gold, platinum, iron, and zinc—decrease the electrical conductivity of the copper enough that even 99% pure copper must be further refined before it can be used in electrical wire.

The impure copper is used as the anode in an electrolysis cell that contains a solution of copper sulfate and sulfuric acid as the electrolyte (see Figure 19.24). The cathode is a thin sheet of very pure copper. When the cell is operated at the correct voltage, only copper and impurities more easily oxidized than copper (iron and zinc) dissolve at the anode. The less active metals simply fall off the electrode and settle to the bottom of the container. At the cathode, copper ions are reduced, but the zinc ions and iron ions remain in solution because they are more difficult to reduce than copper. Gradually, the impure copper anode dissolves and the copper cathode, about 99.96% pure, grows larger. The accumulating sludge—called anode mud—is removed periodically, and the value of the silver, gold, and platinum recovered from it virtually pays for the entire refining operation.

Electrolysis of Brine

One of the most important commercial electrolysis reactions is the electrolysis of concentrated aqueous sodium chloride solutions called **brine**. Water is much more easily reduced than sodium ions, so H_2 forms at the cathode.

$$2H_2O(l) + 2e^- \longrightarrow H_2(g) + 2OH^-(aq) \quad \text{(cathode)}$$

As we noted earlier, even though water is more easily oxidized than chloride ions, complicating factors at the electrode surface actually allow chloride ions to be oxidized more readily than water. At the anode, therefore, we observe the formation of Cl_2.

$$2Cl^-(aq) \longrightarrow Cl_2(g) + 2e^- \quad \text{(anode)}$$

■ PVC is used to make a large variety of products, from raincoats to wire insulation to pipes and conduits for water and sanitary systems. In the United States, the annual demand for PVC plastics is about 15 billion pounds.

Chris R Sharp/PhotoResearchers/Getty Images

Copper refining. Copper cathodes, 99.96% pure, are pulled from the electrolytic refining tanks at Kennecott's Utah copper refinery. It takes about 28 days for the impure copper anodes to dissolve and deposit the pure metal on the cathodes.

■ Copper refining provides one-fourth of the silver and one-eighth of the gold produced annually in the United States.

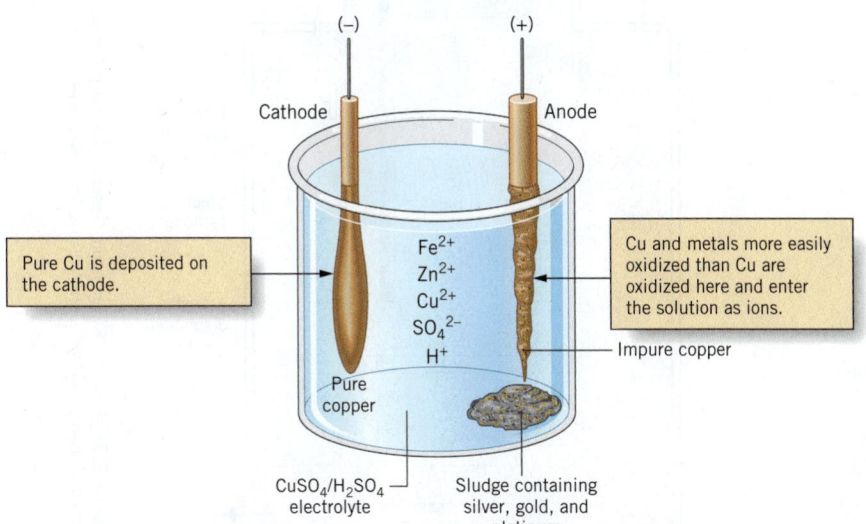

Pure Cu is deposited on the cathode.

Cu and metals more easily oxidized than Cu are oxidized here and enter the solution as ions.

Cathode (−)   Anode (+)

Fe^{2+}
Zn^{2+}
Cu^{2+}
SO_4^{2-}
H^+

Pure copper

Impure copper

$CuSO_4/H_2SO_4$ electrolyte

Sludge containing silver, gold, and platinum

Figure 19.24 | Purification of copper by electrolysis. Impure copper anodes dissolve and pure copper is deposited on the cathodes. Metals less easily oxidized than copper settle to the bottom of the apparatus as "anode mud," while metal ions less easily reduced than copper remain in solution.

The net cell reaction is therefore

$$2Cl^-(aq) + 2H_2O(l) \longrightarrow H_2(g) + Cl_2(g) + 2OH^-(aq)$$

If we include the sodium ion, already in the solution as a spectator ion and not involved in the electrolysis directly, we can see why this is such an important reaction.

$$\underbrace{2Na^+(aq) + 2Cl^-(aq)}_{2NaCl(aq)} + 2H_2O \longrightarrow H_2(g) + Cl_2(g) + \underbrace{2Na^+(aq) + 2OH^-(aq)}_{2NaOH(aq)}$$

■ Sodium hydroxide is commonly known as lye or caustic soda.

Thus, the electrolysis converts inexpensive salt to valuable chemicals: H_2, Cl_2, and NaOH. The hydrogen is used to make other chemicals, including hydrogenated vegetable oils. The chlorine is used for the purposes mentioned earlier. Among the uses of sodium hydroxide, one of industry's most important bases, are the manufacture of soap and paper, the neutralization of acids in industrial reactions, and the purification of aluminum ores.

In the industrial electrolysis of brine, it is necessary to capture the H_2 and Cl_2 separately to prevent them from mixing and reacting (explosively). Second, the NaOH from the reaction is contaminated with unreacted NaCl. Third, if Cl_2 is left in the presence of NaOH, the solution becomes contaminated by hypochlorite ions (OCl^-), which form when Cl_2 reacts with OH^-.

$$Cl_2(g) + 2OH^-(aq) \longrightarrow Cl^-(aq) + OCl^-(aq) + H_2O$$

In one manufacturing operation, however, the Cl_2 is not removed as it forms, and its reaction with hydroxide ion is used to manufacture aqueous sodium hypochlorite. For this purpose, the solution is stirred vigorously during the electrolysis so that very little Cl_2 escapes. As a result, a stirred solution of NaCl gradually changes during electrolysis to a solution of NaOCl, a dilute, 5% solution of which is sold as liquid laundry bleach (e.g., Clorox®).

Most of the pure NaOH manufactured today is made in an apparatus called a **diaphragm cell**. The design varies somewhat, but Figure 19.25 illustrates its basic features. The cell consists of an iron wire mesh cathode that encloses a porous asbestos shell—the diaphragm. The NaCl solution is added to the top of the cell and seeps slowly through the diaphragm.

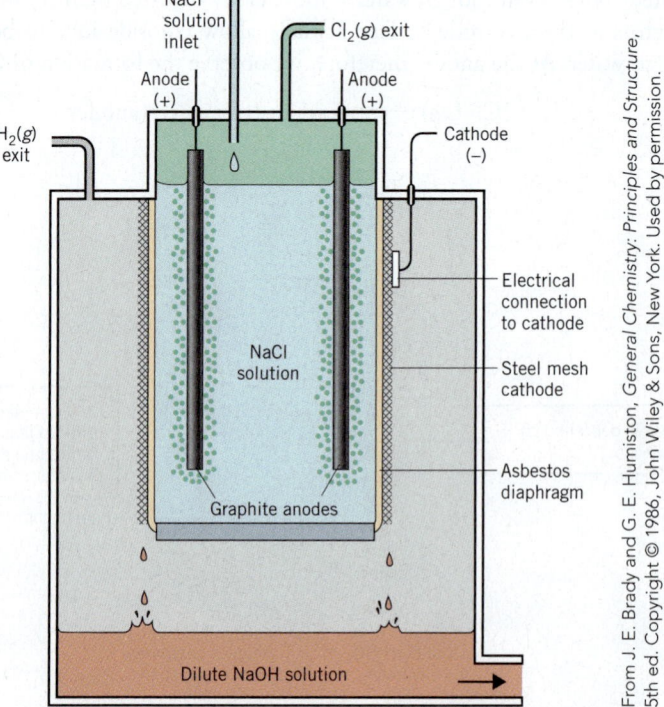

Figure 19.25 | **A diaphragm cell used in the commercial production of NaOH by the electrolysis of aqueous NaCl.** This is a cross section of a cylindrical cell in which the NaCl solution is surrounded by an asbestos diaphragm supported by a steel mesh cathode.

From J. E. Brady and G. E. Humiston, *General Chemistry: Principles and Structure*, 5th ed. Copyright © 1986, John Wiley & Sons, New York. Used by permission.

When it contacts the iron cathode, hydrogen is evolved and is pumped out of the surrounding space. The solution, now containing dilute NaOH, drips off the cell into the reservoir below. Meanwhile, within the cell, chlorine is generated at the anodes. Because there is no OH^- near the anodes, the Cl_2 can't react to form the OCl^- ion and simply bubbles out of the solution and is captured as illustrated in the figure.

Summary

Organized by Learning Objective

Set up and use galvanic cells

A **galvanic cell** has two **half-cells**, each containing an **electrode** in contact with an electrolyte reactant mixture representative of one of the reaction's half-reactions. The reaction is thus divided into separate oxidation and reduction half-reactions, with the electron transfer occurring through an external electrical circuit. Reduction occurs at the **cathode**; oxidation occurs at the **anode**. In a galvanic cell, the cathode is positively charged and the anode is negatively charged. The half-cells are connected by a **salt bridge** that allows cations to move toward the cathode and anions toward the anode to maintain electrical neutrality.

Predict Galvanic cell potentials

The **potential** (expressed in volts) produced by a cell is equal to the **standard cell potential** when all ion concentrations are $1.00\ M$ and the partial pressures of any gases involved equal 1.00 atm and the temperature is $25.0\ °C$.

Use standard reduction potentials to predict spontaneous reactions

The standard cell potential is the difference between the **standard reduction potentials** of the half-cells. In the spontaneous reaction, the half-cell with the higher standard reduction potential undergoes reduction and forces the other to undergo oxidation. The standard reduction potentials of isolated half-cells can't be measured, but values are assigned using the **hydrogen electrode** as a reference electrode; $E° = 0.00$ V. Species more easily reduced than H^+ have positive standard reduction potentials; those less easily reduced have negative standard reduction potentials. Standard reduction potentials are used to calculate $E°_{cell}$. Standard reduction potentials can be used to predict if a redox reaction will be spontaneous, whether or not the reaction is carried out in a galvanic cell.

Relate standard cell potentials and standard free energies

The values of $\Delta G°$ and K_c for a reaction can be calculated from $E°_{cell}$. They all involve the **faraday**, $\mathscr{F}$, a constant equal to the number of **coulombs (C)** of charge per mole of electrons ($1\mathscr{F} = 96,485$ C/mol e^-).

Relate the concentration of solutes to cell potentials

The **Nernst equation** relates the cell potential to the standard cell potential and the reaction quotient. It is used to calculate the cell potential under nonstandard conditions. If the cell potential is measured, the Nernst equation may be used to determine concentrations.

Understand how electricity is produced

The **zinc–manganese dioxide cell** (the **Leclanché cell** or common dry cell) and the common **alkaline battery** (which uses essentially the same reactions as the less expensive dry cell) are **primary cells** and are not rechargeable. The **lead storage battery** and the **nickel–metal hydride (NiMH) battery** are **secondary cells** and are rechargeable. The state of charge of the lead storage battery can be tested with a **hydrometer**, which measures the density of the sulfuric acid electrolyte. The rechargeable nickel–metal hydride battery uses hydrogen contained in a metal alloy as its anode reactant and has a high **energy density**. Primary **lithium–manganese dioxide cells** and rechargeable **lithium ion cells** produce a large cell potential and have a very large energy density. Lithium ion cells store and release energy by transferring lithium ions between electrodes where the Li^+ are **intercalated** between layers of atoms in the electrode materials. **Fuel cells**, which have high thermodynamic efficiencies, are able to provide continuous power because they consume fuel that can be fed continuously. **Photovoltaic cells** are based on semiconductor technology where the junction between a **p-type** and an **n-type** semiconductor produces a device called a **diode** that allows electrons to flow in only one direction. Once photons separate an electron from the semiconductor material, the properties of a diode make it easier for electrons to flow through an external circuit.

Explain the nature of electrolysis

In an **electrolytic cell,** a flow of electricity causes an otherwise nonspontaneous reaction to occur. A negatively charged **cathode** causes reduction of one reactant and a positively charged **anode** causes oxidation of another. Ion movement instead of electron transport occurs in the electrolyte. The electrode reactions are determined by which species is most easily reduced and which is most easily oxidized, but in aqueous solutions complex surface effects at the electrodes can alter the natural order. In the electrolysis of water, an electrolyte must be present to maintain electrical neutrality at the electrodes.

Use electrolysis information quantitatively

The product of current (**amperes**) and time (seconds) gives coulombs. This relationship and the half-reactions that occur at the anode or cathode permit us to relate the amount of chemical change to measurements of current and time.

Describe some practical applications of electrolysis

Electroplating, the production of aluminum, the **refining** of copper, and the electrolysis of molten and aqueous sodium chloride are examples of practical applications of electrolysis.

Tools for Problem Solving
The following tools were introduced in this chapter. Study them carefully so you can select the appropriate tool when needed.

Electrode names and reactions (Section 19.1)
The **cathode** is the electrode at which reduction (electron gain) occurs. The **anode** is the electrode at which oxidation (electron loss) occurs.

Standard reduction potentials used to calculate $E°_{cell}$ (Section 19.2)

$$E°_{cell} = E°_{reduction} - E°_{oxidation}$$

$E°_{cell}$ and spontaneous reactions (Section 19.3)
A reaction is spontaneous if $E°_{cell} > 0$ V.

Faraday constant (Section 19.4)

$$1\mathscr{F} = 96{,}485.340 \text{ C/mol } e^-$$

Standard cell potentials are related to thermodynamic quantities (Section 19.4)

$$\Delta G = -n\mathscr{F}E$$

$$E°_{cell} = \frac{RT}{n\mathscr{F}} \ln K_c$$

Nernst equation (Section 19.5)

$$E_{cell} = E°_{cell} - \frac{RT}{n\mathscr{F}} \ln K_c$$

Coulombs are related to current in amperes and time in seconds (Section 19.8)

$$C = A \times t$$

WileyPLUS = *WileyPLUS*, an online teaching and learning solution. *Note to instructors*: Many of the end-of-chapter problems are available for assignment via the WileyPLUS system. **www.wileyplus.com.** Review Problems are presented in pairs separated by blue rules. Answers to problems whose numbers appear in blue are given in Appendix B. More challenging problems are marked with an asterisk ✱.

| Review Questions

Galvanic Cells

19.1 What is a galvanic cell? What is a half-cell?

19.2 What is the function of a *salt bridge*?

19.3 In a copper–silver cell, why must the Cu^{2+} and Ag^+ solutions be kept in separate containers?

19.4 What is the general name we give to reactions that take place at the anode and those that take place at the cathode in a galvanic cell? What is the sign of the electrical charges on the anode and cathode in a galvanic cell?

19.5 In a galvanic cell, do electrons travel from anode to cathode, or from cathode to anode? Explain.

19.6 Explain how the movement of the ions relative to the electrodes is the same in both galvanic and electrolytic cells.

19.7 Aluminum will displace tin from solution according to the equation

$$2Al(s) + 3Sn^{2+}(aq) \longrightarrow 2Al^{3+}(aq) + 3Sn(s)$$

What would be the individual half-cell reactions if this were the cell reaction for a galvanic cell? Which metal would be the anode and which the cathode?

19.8 Make a sketch of the galvanic cell in Review Question 19.7.

19.9 Make a sketch of a galvanic cell for which the cell notation is

$$Fe(s) \mid Fe^{3+}(aq) \parallel Ag^+(aq) \mid Ag(s)$$

(a) Label the anode and the cathode.

(b) Indicate the charge on each electrode.

(c) Indicate the direction of electron flow in the external circuit.

(d) Write the equation for the net cell reaction.

19.10 Make a sketch of a galvanic cell represented by the following cell notation:

$$Pt(s) \mid Fe^{2+}(aq), Fe^{3+}(aq) \parallel Br_2(aq), Br^-(aq) \mid Pt(s)$$

Label the diagram and indicate the composition of the electrolytes in the two cell compartments. Show the signs of the electrodes and label the anode and cathode. Write the equation for the net cell reaction.

Cell Potentials

19.11 What is the difference between a *cell potential* and a *standard cell potential*?

19.12 How are standard reduction potentials combined to give the standard cell potential for a spontaneous reaction?

19.13 If you set up a galvanic cell using metals not found in Table 19.1, what experimental information will tell you which is the anode and which is the cathode in the cell?

19.14 Galvanic cells are set up so that the cell potential always has a positive sign. What does this signify about the chemical reaction occurring in the cell?

Utilizing Standard Reduction Potentials

19.15 Describe the hydrogen electrode. What is the value of its standard reduction potential?

19.16 What do the positive and negative signs of reduction potentials tell us?

19.17 If $E^{\circ}_{Cu^{2+}}$ had been chosen as the standard reference electrode and had been assigned a potential of 0.00 V, what would be the reduction potential of the hydrogen electrode relative to it?

19.18 Describe how to experimentally determine the standard reduction potential of a metal.

19.19 Compare Table 5.3 with Table 19.1. What can you say about the basis for the activity series for metals?

E°_{cell} and ΔG°

19.20 Write the equation that relates the standard cell potential to the standard free energy change for a reaction.

19.21 What is the equation that relates the equilibrium constant to the cell potential?

19.22 Show how the equation that relates the equilibrium constant to the cell potential (Equation 19.7) can be derived from the Nernst equation (Equation 19.8).

19.23 What is the cell potential of a galvanic cell when the cell reaction has reached equilibrium?

Cell Potentials and Concentration

19.24 The cell reaction during the discharge of a lead storage battery is

$$Pb(s) + PbO_2(s) + 2H^+(aq) + 2HSO_4^-(aq) \longrightarrow$$
$$2PbSO_4(s) + 2H_2O$$

The standard cell potential is 2.05 V. What is the correct form of the Nernst equation for this reaction at 25 °C?

19.25 What is a concentration cell? Why is the E°_{cell} equal to zero for such a cell?

19.26 Describe what happens if a galvanic cell is short circuited by connecting the anode directly to the cathode.

Electricity

19.27 What are the anode and cathode reactions during the discharge of a lead storage battery? How can a battery produce a potential of 12 V if the cell reaction has a standard potential of only 2 V?

19.28 What are the anode and cathode reactions during the charging of a lead storage battery?

19.29 How is a hydrometer constructed? How does it measure density? Why can a hydrometer be used to check the state of charge of a lead storage battery?

19.30 What reactions occur at the electrodes in the ordinary dry cell?

19.31 What chemical reactions take place at the electrodes in an alkaline dry cell?

19.32 How is hydrogen held as a reactant in a nickel–metal hydride battery? Write the chemical formula for a typical alloy used in this battery. What is the electrolyte?

19.33 What are the anode, cathode, and net cell reactions that take place in a nickel–metal hydride battery during discharge? What are the reactions when the battery is charged?

19.34 Give two reasons why lithium is such an attractive anode material for use in a battery. What are the problems associated with using lithium for this purpose?

19.35 What are the electrode materials in a typical primary lithium cell? Write the equations for the anode, cathode, and cell reactions.

19.36 What are the electrode materials in a typical lithium ion cell? Explain what happens when the cell is charged. Explain what happens when the cell is discharged.

19.37 Platinum and palladium intercalate large amounts of hydrogen in a manner similar to lithium. Why are these metals not used for batteries?

19.38 Write the cathode, anode, and net cell reaction in a hydrogen–oxygen fuel cell.

19.39 What advantages do fuel cells offer over conventional means of obtaining electrical power by the combustion of fuels?

Electrolytic Cells

19.40 What electrical charges do the anode and the cathode carry in an electrolytic cell? What does the term *inert electrode* mean?

19.41 Why must electrolysis reactions occur at the electrodes in order for electrolytic conduction to continue?

19.42 Why must NaCl be melted before it is electrolyzed to give Na and Cl_2? Write the anode, cathode, and overall cell reactions for the electrolysis of molten NaCl.

19.43 Write half-reactions for the oxidation and the reduction of water.

19.44 What happens to the pH of the solution near the cathode and anode during the electrolysis of a solution of K_2SO_4? Does K_2SO_4 get electrolyzed? What function does K_2SO_4 serve in the electrolysis of a K_2SO_4 solution?

Electrolysis Stoichiometry

19.45 What is a *faraday*? What relationships relate faradays to current and time measurements?

19.46 Using the same current, which will require the greater length of time: depositing 0.10 mol Cu from a Cu^{2+} solution, or depositing 0.10 mol of Cr from a Cr^{3+} solution? Explain your reasoning.

19.47 An electric current is passed through two electrolysis cells connected in series (so the same amount of current passes through each of them). One cell contains Cu^{2+} and the other contains Ag^+. In which cell will the larger number of moles of metal be deposited? Explain your answer.

19.48 An electric current is passed through two electrolysis cells connected in series (so the same amount of current passes through each of them). One cell contains Cu^{2+} and the other contains Fe^{2+}. In which cell will the greater mass of metal be deposited? Explain your answer.

Practical Applications of Electrolysis

19.49 What is *electroplating*? Sketch an apparatus to electroplate silver.

19.50 Describe the Hall–Héroult process for producing metallic aluminum. What half-reaction occurs at the anode? What half-reaction occurs at the cathode? What is the overall cell reaction?

19.51 In the Hall–Héroult process, why must the carbon anodes be replaced frequently?

19.52 How is metallic sodium produced? Write equations for the anode and cathode reactions. What are some uses of metallic sodium?

19.53 Describe the electrolytic refining of copper. What economic advantages offset the cost of electricity for this process? What chemical reactions occur at **(a)** the anode and **(b)** the cathode?

19.54 Describe the electrolysis of aqueous sodium chloride. How do the products of the electrolysis compare for stirred and unstirred reactions? Write chemical equations for the reactions that occur at the electrodes.

Review Problems

Galvanic Cells

19.55 Write the half-reactions and the balanced cell reaction for the following galvanic cells.

(a) $Zn(s) \mid Zn^{2+}(aq) \parallel Cr^{3+}(aq) \mid Cr(s)$

(b) $Pb(s), PbSO_4(s) \mid HSO_4^-(aq) \parallel$
$$H^+(aq), HSO_4^{2-}(aq) \mid PbO_2(s), PbSO_4(s)$$

(c) $Mg(s) \mid Mg^{2+}(aq) \parallel Sn^{2+}(aq) \mid Sn(s)$

19.56 Write the half-reactions and the balanced cell reaction for the following galvanic cells.

(a) $Cd(s) \mid Cd^{2+}(aq) \parallel Au^{3+}(aq) \mid Au(s)$

(b) $Fe(s) \mid Fe^{2+}(aq) \parallel Br_2(aq), Br^-(aq) \mid Pt(s)$

(c) $Cr(s) \mid Cr^{3+}(aq) \parallel Cu^{2+}(aq) \mid Cu(s)$

19.57 Write the cell notation for the following galvanic cells. For half-reactions in which all the reactants are in solution or are gases, assume the use of inert platinum electrodes.

(a) $NO_3^-(aq) + 4H^+(aq) + 3Fe^{2+}(aq) \longrightarrow$
$$3Fe^{3+}(aq) + NO(g) + 2H_2O$$

(b) $Cl_2(g) + 2Br^-(aq) \longrightarrow Br_2(aq) + 2Cl^-(aq)$

(c) $Au^{3+}(aq) + 3Ag(s) \longrightarrow Au(s) + 3Ag^+(aq)$

19.58 Write the cell notation for the following galvanic cells. For half-reactions in which all the reactants are in solution or are gases, assume the use of inert platinum electrodes.

(a) $Cd^{2+}(aq) + Fe(s) \longrightarrow Cd(s) + Fe^{2+}(aq)$

(b) $NiO_2(s) + 4H^+(aq) + 2Ag(s) \longrightarrow$
$$Ni^{2+}(aq) + 2H_2O + 2Ag^+(aq)$$

(c) $Mg(s) + Cd^{2+}(aq) \longrightarrow Mg^{2+}(aq) + Cd(s)$

Cell Potentials

19.59 For each pair of substances, use Table 19.1 to choose the better reducing agent.

(a) Sn(s) or Ag(s) **(c)** Co(s) or Zn(s)

(b) $Cl^-(aq)$ or $Br^-(aq)$ **(d)** $I^-(aq)$ or Au(s)

19.60 For each pair of substances, use Table 19.1 to choose the better oxidizing agent.

(a) $NO_3^-(aq)$ or $MnO_4^-(aq)$

(b) $Au^{3+}(aq)$ or $Co^{2+}(aq)$

(c) $PbO_2(s)$ or $Cl_2(g)$

(d) $NiO_2(s)$ or $HOCl(aq)$

19.61 Use the data in Table 19.1 to calculate the standard cell potential for each of the following reactions:

(a) $NO_3^-(aq) + 4H^+(aq) + 3Fe^{2+}(aq) \longrightarrow$
$$3Fe^{3+}(aq) + NO(g) + 2H_2O$$

(b) $Br_2(aq) \longrightarrow 2Cl^-(aq) \longrightarrow Cl_2(g) + 2Br^-(aq)$

(c) $Au^{3+}(aq) + 3Ag(s) \longrightarrow Au(s) + 3Ag^+(aq)$

19.62 Use the data in Table 19.1 to calculate the standard cell potential.

(a) $Cd^{2+}(aq) + Fe(s) \longrightarrow Cd(s) + Fe^{2+}(aq)$

(b) $NiO_2(s) + 4H^+(aq) + 2Ag(s) \longrightarrow$
$$Ni^{2+}(aq) + 2H_2O + 2Ag^+(aq)$$

(c) $Mg(s) + Cd^{2+}(aq) \longrightarrow Mg^{2+}(aq) + Cd(s)$

19.63 From the positions of the half-reactions in Table 19.1, determine whether the following reactions are spontaneous under standard state conditions.

(a) $2Au^{3+} + 6I^- \longrightarrow 3I_2 + 2Au$

(b) $H_2SO_3 + H_2O + Br_2 \longrightarrow 4H^+ + SO_4^{2-} + 2Br^-$

(c) $3Ca + 2Cr^{3+} \longrightarrow 2Cr + 3Ca^{2+}$

19.64 Use the data in Table 19.1 to determine which of the following reactions should occur spontaneously under standard state conditions.

(a) $Br_2 + 2Cl \longrightarrow Cl_2 + 2Br^-$

(b) $3Fe^{2+} + 2NO + 4H_2O \longrightarrow 3Fe + 2NO_3 + 8H^+$

(c) $Ni^{2+} + Fe \longrightarrow Fe^{2+} + Ni$

19.65 From the half-reactions below, determine the cell reaction and standard cell potential.

$$BrO_3^- + 6H^+ + 6e^- \rightleftharpoons Br^- + 3H_2O$$
$$E^\circ_{BrO_3^-} = 1.44 \text{ V}$$

$$I_2 + 2e^- \rightleftharpoons 2I^- \qquad E^\circ_{I_2} = 0.54 \text{ V}$$

19.66 What is the standard cell potential and the net reaction in a galvanic cell that has the following half reactions?

$$MnO_2 + 4H^+ + 2e^- \rightleftharpoons Mn^{2+} + 2H_2O$$
$$E^\circ_{MnO_2} = 1.23 \text{ V}$$

$$PbCl_2 + 2e^- \rightleftharpoons Pb + 2Cl^- \qquad E^\circ_{PbCl_2} = -0.27 \text{ V}$$

19.67 What will be the spontaneous reaction among H_2SO_3, $S_2O_3^{2-}$, HOCl, and Cl_2? The half-reactions involved are:

$$2H_2SO_3 + 2H^+ + 4e^- \rightleftharpoons S_2O_3^{2-} + 3H_2O$$
$$E^\circ_{H_2SO_3} = 0.40 \text{ V}$$

$$2HOCl + 2H^+ + 2e^- \rightleftharpoons Cl_2 + 2H_2O$$
$$E^\circ_{HOCl} = 1.63 \text{ V}$$

19.68 What will be the spontaneous reaction among Br_2, I_2, Br^-, and I^-?

19.69 Will the following reaction occur spontaneously under standard state conditions?

$$SO_4^{2-} + 4H^+ + 2I^- \longrightarrow H_2SO_3 + I_2 + H_2O$$

Use E°_{cell} calculated from data in Table 19.1 to answer this question.

19.70 Determine whether the reaction:

$$S_2O_8^{2-} + Ni(OH)_2 + 2OH^- \longrightarrow$$
$$2SO_4^{2-} + NiO_2 + 2H_2O$$

will occur spontaneously under standard state conditions. Use E°_{cell} calculated from the data below to answer the question.

$$NiO_2 + 2H_2O + 2e^- \rightleftharpoons Ni(OH)_2 + 2OH^-$$
$$E^\circ_{NiO_2} = 0.49 \text{ V}$$

$$S_2O_8^{2-} + 2e^- \rightleftharpoons 2SO_4^{2-} \qquad E^\circ_{SO_4^{2-}} = 2.01 \text{ V}$$

E°_{cell} and ΔG°

19.71 Calculate ΔG° for the following reaction *as written*:

$$2Br^- + I_2 \longrightarrow 2I^- + Br_2$$

19.72 Calculate ΔG° for the reaction

$$2MnO_4^- + 6H^+ + 5HCHO_2 \longrightarrow$$
$$2Mn^{2+} + 8H_2O + 5CO_2$$

for which $E^\circ_{cell} = 1.69 \text{ V}$.

19.73 Given the following half-reactions and their standard reduction potentials,

$$2ClO_3^- + 12H^+ + 10e^- \rightleftharpoons Cl_2 + 6H_2O$$
$$E^\circ_{ClO_3^-} = 1.47 \text{ V}$$

$$S_2O_8^{2-} + 2e^- \rightleftharpoons 2SO_4^{2-} \qquad E^\circ_{S_2O_8^{2-}} = 2.01 \text{ V}$$

calculate (a) E°_{cell}, (b) ΔG° for the cell reaction, and (c) the value of K_c for the cell reaction.

19.74 Calculate K_c for the system

$$Ni^{2+} + Co \rightleftharpoons Ni + Co^{2+}$$

Use the data in Table 19.1 and assume $T = 298 \text{ K}$.

19.75 The system

$$2AgI + Sn \rightleftharpoons Sn^{2+} + 2Ag + 2I^-$$

has a calculated $E^\circ_{cell} = -0.015 \text{ V}$. What is the value of K_c for this system?

19.76 Determine the value of K_c at 25 °C for the reaction

$$2H_2O + 2Cl_2 \rightleftharpoons 4H^+ + 4Cl^- + O_2$$

Cell Potentials and Concentrations

19.77 The cell reaction

$$NiO_2(s) + 4H^+(aq) + 2Ag(s) \longrightarrow$$
$$Ni^{2+}(aq) + 2H_2O + 2Ag^+(aq)$$

has $E°_{cell} = 2.48$ V. What will be the cell potential at a pH of 2.00 when the concentrations of Ni^{2+} and Ag^+ are each 0.030 M?

19.78 The $E°_{cell} = 0.135$ V for the reaction

$$3I_2(s) + 5Cr_2O_7{}^{2-}(aq) + 34H^+ \longrightarrow$$
$$6IO_3{}^-(aq) + 10Cr^{2+}(aq) + 17H_2O$$

What is E_{cell} if $[Cr_2O_7{}^{2-}] = 0.0010\ M$, $[H^+] = 0.10\ M$, $[IO_3{}^-] = 0.00010\ M$, and $[Cr] = 0.0010\ M$?

*19.79 A cell was set up having the following reaction.

$$Mg(s) + Cd^{2+}(aq) \longrightarrow Mg^{2+}(aq) + Cd(s)$$

$$E°_{cell} = 1.97\ V$$

The magnesium electrode was dipped into a 1.00 M solution of $MgSO_4$ and the cadmium electrode was dipped into a solution of unknown Cd^{2+} concentration. The potential of the cell was measured to be 1.54 V. What was the unknown Cd^{2+} concentration?

*19.80 A silver wire coated with AgCl is sensitive to the presence of chloride ion because of the half-cell reaction

$$AgCl(s) + e^- \rightleftharpoons Ag(s) + Cl^-\quad E°_{AgCl} = 0.2223\ V$$

A student, wishing to measure the chloride ion concentration in a number of water samples, constructed a galvanic cell using the AgCl electrode as one half-cell and a copper wire dipping into 1.00 M $CuSO_4$ solution as the other half-cell. In one analysis, the potential of the cell was measured to be 0.0895 V with the copper half-cell serving as the cathode. What was the chloride ion concentration in the water? (Take $E°_{Cu^{2+}} = 0.3419$ V.)

*19.81 At 25 °C, a galvanic cell was set up having the following half-reactions.

$$Fe^{2+}(aq) + 2e^- \rightleftharpoons Fe(s)\qquad E°_{Fe^{2+}} = -0.447\ V$$

$$Cu^{2+}(aq) + 2e^- \rightleftharpoons Cu(s)\qquad E°_{Cu^{2+}} = +0.3419\ V$$

The copper half-cell contained 100 mL of 1.00 M $CuSO_4$. The iron half-cell contained 50.0 mL of 0.100 M $FeSO_4$. To the iron half-cell was added 50.0 mL of 0.500 M NaOH solution. The mixture was stirred and the cell potential was measured to be 1.175 V. Calculate the concentration of Fe^{2+} in the solution.

*19.82 Suppose a galvanic cell was constructed at 25 °C using a Cu/Cu^{2+} half-cell (in which the molar concentration of Cu^{2+} was 1.00 M) and a hydrogen electrode having a partial pressure of H_2 equal to 1 atm. The hydrogen electrode dips into a solution of unknown hydrogen ion

concentration, and the two half-cells are connected by a salt bridge. The precise value of $E°_{cell}$ is +0.3419 V.

(a) Derive an equation for the pH of the solution with the unknown hydrogen ion concentration, expressed in terms of E_{cell} and $E°_{cell}$.

(b) If the pH of the solution were 5.15, what would be the observed potential of the cell?

(c) If the potential of the cell were 0.645 V, what would be the pH of the solution?

*19.83 What is the potential of a concentration cell at 25.0 °C if it consists of silver electrodes dipping into two different solutions of $AgNO_3$, one with a concentration of 0.015 M and the other with a concentration of 0.50 M? What would be the potential of the cell if the temperature of the cell were 75 °C?

*19.84 What is the potential of a concentration cell at 25.0 °C if it consists of copper electrodes dipping into two different solutions of $Cu(NO_3)_2$, one with a concentration of 0.015 M and the other with a concentration of 0.50 M? What would be the potential of the cell if the temperature of the cell were 75 °C?

Electrolytic Cells

19.85 If electrolysis is carried out on an aqueous solution of aluminum sulfate, what products are expected at the electrodes? Write the equation for the net cell reaction.

19.86 If electrolysis is carried out on an aqueous solution of cadmium iodide, what products are expected at the electrodes? Write the equation for the net cell reaction.

19.87 What products would we expect at the electrodes if a solution containing both KBr and $CuSO_4$ were electrolyzed?

19.88 What products would we expect at the electrodes if a solution containing both $BaCl_2$ and CuI_2 were electrolyzed? Write the equation for the net cell reaction.

19.89 Using Table 19.1, list the ions in aqueous solution that we would not expect to be reduced at the cathode.

19.90 Using Table 19.1, list the ions, in aqueous solution, that we would not expect to be oxidized at the anode.

Electrolysis Stoichiometry

19.91 How many moles of electrons are required to (a) reduce 0.20 mol Fe^{2+} to Fe? (b) Oxidize 0.70 mol Cl to Cl_2? (c) Reduce 1.50 mol Cr^{3+} to Cr? (d) Oxidize 1.0×10^{-2} mol Mn^{2+} to MnO_4 ?

19.92 How many moles of electrons are required to (a) produce 5.00 g Mg from molten $MgCl_2$? (b) Form 41.0 g Cu from a $CuSO_4$ solution?

19.93 How many grams of $Fe(OH)_2$ are produced at an iron anode when a basic solution undergoes electrolysis at a current of 8.00 A for 12.0 min?

19.94 How many grams of Cl_2 would be produced in the electrolysis of molten NaCl by a current of 4.25 A for 35.0 min?

19.95 How many hours would it take to produce 75.0 g of metallic chromium by the electrolytic reduction of Cr^{3+} with a current of 2.25 A?

19.96 How many hours would it take to generate 35.0 g of lead from $PbSO_4$ during the charging of a lead storage battery using a current of 1.50 A? The half-reaction is $Pb + HSO_4^- \longrightarrow PbSO_4 + H^+ + 2e^-$.

19.97 How many amperes would be needed to produce 60.0 g of magnesium during the electrolysis of molten $MgCl_2$ in 2.00 hr?

19.98 A large electrolysis cell that produces metallic aluminum from Al_2O_3 by the Hall–Héroult process is capable of yielding 900 lb (409 kg) of aluminum in 24 hr. What current is required?

****19.99** The electrolysis of 250 mL of a brine solution (NaCl) was carried out for a period of 20.00 min with a current of 2.00 A in an apparatus that prevented Cl_2 from reacting with other products of the electrolysis. What is the concentration of hydroxide ions in the solution?

****19.100** A 100.0 mL sample of 2.00 M NaCl was electrolyzed for a period of 25.0 min with a current of 1.45 amperes, with stirring. What is the concentration of sodium hypochlorite produced?

Additional Exercises

****19.101** A watt is a unit of electrical power and is equal to one joule per second (1 watt = 1 J s^{-1}). How many hours can a calculator drawing 2.0×10^{-3} watt be operated by a mercury battery having a cell potential equal to 1.34 V if a mass of 1.00 g of HgO is available at the cathode? The cell reaction is

$$HgO(s) + Zn(s) \longrightarrow ZnO(s) + Hg(l)$$

****19.102** Suppose that a galvanic cell were set up having the net cell reaction

$$Zn(s) + 2Ag^+(aq) \longrightarrow Zn^{2+}(aq) + 2Ag(s)$$

The Ag^+ and Zn^{2+} concentrations in their respective half-cells initially are 1.00 M, and each half-cell contains 100 mL of electrolyte solution. If this cell delivers current at a constant rate of 0.10 A, what will be the cell potential after 15.00 hr?

19.103 In a certain zinc–copper cell,

$$Zn(s) + Cu^{2+}(aq) \longrightarrow Zn^{2+}(aq) + Cu(s)$$

the ion concentrations are $[Cu^{2+}] = 0.0100$ M and $[Zn^{2+}] = 1.0$ M. What is the cell potential at 25 °C?

****19.104** The value of K_{sp} for AgBr is 5.4×10^{-13}. What will be the potential of a cell constructed of a standard hydrogen electrode as one half-cell and a silver wire coated with AgBr dipping into 0.10 M HBr as the other half-cell. For the Ag/AgBr electrode,

$$AgBr(s) + e^- \rightleftharpoons Ag(s) + Br^-(aq)$$
$$E^\circ_{AgBr} = +0.070 \text{ V}$$

19.105 Based only on the half-reactions in Table 19.1, determine what spontaneous reaction will occur if the following substances are mixed together: **(a)** I_2, I^- and Fe^{2+}, Fe^{3+}; **(b)** Mg, Mg^{2+} and Cr, Cr^{3+}; **(c)** Co, Co^{2+} and H_2SO_3, SO_4^{2-}. (*Hint:* Use Table 19.1 to write the possible half-reactions and, if necessary, determine E°_{cell}.)

****19.106** A student set up an electrolysis apparatus and passed a current of 1.22 A through a 3 M H_2SO_4 solution for 30.0 min. The H_2 formed at the cathode was collected and found to have a volume, over water at 27 °C, of 288 mL at a total pressure of 767 torr. Use these data to calculate the charge on the electron, expressed in coulombs.

****19.107** A hydrogen electrode is immersed in a 0.10 M solution of acetic acid at 25 °C. This electrode is connected to another consisting of an iron nail dipping into 0.10 M $FeCl_2$. What will be the measured potential of this cell? Assume $P_{H_2} = 1.00$ atm.

****19.108** What current would be required to deposit 1.00 m^2 of chrome plate having a thickness of 0.050 mm in 4.50 hr from a solution of H_2CrO_4? The density of chromium is 7.19 g cm^{-3}.

****19.109** A solution containing vanadium in an unknown oxidation state was electrolyzed with a current of 1.50 A for 30.0 min. It was found that 0.475 g of V was deposited on the cathode. What was the original oxidation state of the vanadium ion?

****19.110** Consider the reduction potentials of the following sets of substances: (a) F_2, Cl_2, Br_2, and I_2 (b) Li^+, Na^+, K^+, Rb^+, and Cs^+ (c) Mg^{2+}, Ca^{2+}, Sr^{2+}, and Ba^{2+}. How do the reduction potentials correlate with properties such as electron affinity, ionization energy, and electronegativity? If necessary, suggest why a correlation does not occur.

****19.111** An Ag/AgCl electrode dipping into 1.00 M HCl has a standard reduction potential of +0.2223 V. The half reaction is

$$AgCl(s) + e^- \rightleftharpoons Ag(s) + Cl^-(aq)$$

A second Ag/AgCl electrode is dipped into a solution containing Cl at an unknown concentration. The cell generates a potential of 0.0478 V, with the electrode in the solution of unknown concentration having a negative charge. What is the molar concentration of Cl in the unknown solution?

19.112 A galvanic cell was constructed with a nickel electrode that was dipped into 1.20 M $NiSO_4$ solution and a chromium electrode that was immersed into a solution containing Cr^{3+} at an unknown concentration. The potential of the cell was measured to be 0.552 V, with the chromium serving as the anode. The standard cell potential for this system was determined to be 0.487 V.

What was the concentration of Cr^{3+} in the solution of unknown concentration?

19.113 Consider the following galvanic cell:

$$Ag(s) \mid Ag^+(3.0 \times 10^{-4}M) \parallel$$

$$Fe^{3+}(1.1 \times 10^{-3} M), Fe^{2+}(0.040 M) \mid Pt(s)$$

Calculate the cell potential. Determine the sign of the electrodes in the cell. Write the equation for the spontaneous cell reaction.

Multi-Concept Problems

*19.114 The electrolysis of 0.250 L of a brine solution (NaCl) was carried out for a period of 20.00 min with a current of 2.00 A in an apparatus that prevented Cl_2 from reacting with other products of the electrolysis. The resulting solution was titrated with 0.620 M HCl. How many mL of the HCl solution was required for the titration?

*19.115 A solution of NaCl in water was electrolyzed with a current of 2.50 A for 15.0 min. How many milliliters of dry Cl_2 gas would be formed if it was collected over water at 25 °C and a total pressure of 750 torr?

*19.116 How many milliliters of dry gaseous H_2, measured at 20.0 °C and 735 torr, would be produced at the cathode in the electrolysis of dilute H_2SO_4 with a current of 0.750 A for 15.00 min?

*19.117 At 25 °C, a galvanic cell was set up having the following half-reactions.

$$Fe^{2+}(aq) + 2e^- \rightleftharpoons Fe(s) \qquad E^\circ_{Fe^{2+}} = -0.447 \text{ V}$$

$$Cu^{2+}(aq) + 2e^- \rightleftharpoons Cu(s) \qquad E^\circ_{Cu^{2+}} = +0.3419 \text{ V}$$

The copper half-cell contained 100 mL of 1.00 M $CuSO_4$. The iron half-cell contained 50.0 mL of 0.100 M $FeSO_4$. To the iron half-cell was added 50.0 mL of 0.500 M NaOH solution. The mixture was stirred and the cell potential was measured to be 1.175 V. Calculate the value of K_{sp} for $Fe(OH)_2$.

*19.118 Given the following reduction half-reactions and their standard potentials, what is the molar solubility of silver bromide?

$$Ag^+ + e^- \longrightarrow Ag \qquad E^\circ_{Ag^+} = +0.80 \text{ V}$$

$$AgBr + e^- \longrightarrow Ag + Br^- \qquad E^\circ_{AgBr} = +0.07 \text{ V}$$

*19.119 The normal range of chloride ions in blood serum (serum is the fluid left after the red blood cells clot) ranges from 0.096 to 0.106 M. An electrolysis apparatus can be set up to generate silver ions from a silver anode to precipitate the chloride ions. If the current used in the experiment is 0.500 ampere and if 3.00 mL of serum is transferred to the electrolysis cell, what range of time, in seconds, will a normal sample take to precipitate all of the chloride ions?

*19.120 An unstirred solution of 2.00 M NaCl was electrolyzed for a period of 25.0 min and then titrated with 0.250 M HCl. The titration required 15.5 mL of the acid. What was the average current in amperes during the electrolysis?

*19.121 What masses of H_2 and O_2 in grams would have to react each second in a fuel cell at 110 °C to provide 1.00 kilowatt (kW) of power if we assume a thermodynamic efficiency of 70%? (*Hint:* Use data in Chapters 6 and 18 to compute the value of ΔG° for the reaction $H_2(g) + \frac{1}{2}O_2(g) \longrightarrow H_2O(g)$ at 110 °C. 1 watt = 1 J s^{-1}.)

Exercises in Critical Thinking

*19.122 Draw an atomic-level diagram of the events that occur in a galvanic cell at a cathode made of magnesium metal dipping into a solution containing magnesium nitrate.

*19.123 In biochemical systems, the normal standard state that requires $[H^+] = 1.00$ M is not realistic. **(a)** Which half reactions in Table 19.1 will have different potentials if pH = 7.00 is defined as the standard state for hydronium ions? **(b)** What will be the new standard reduction potentials at pH = 7.00 for these reactions? These

are called E°_{cell} with the prime indicating the potential at pH = 7.00.

19.124 Calculate a new version of Table 19.1 using the lithium half-reaction to define zero. Does this change the results of any problems involving standard cell potentials?

19.125 In Problem 19.83, the potential at 75 °C was calculated. Does the change in molarity of the solutions, due to the change in density of water, have an effect on the potentials?

19.126 There are a variety of methods available for generating electricity. List as many methods as you can. Rank each of these methods based on your knowledge of **(a)** the efficiency of the method and **(b)** the environmental pollution caused by each method.

19.127 Using the cost of electricity in your area, how much will it cost to produce a case of 24 soda cans, each weighing 0.45 ounce? Assume that alternating current can be converted to direct current with 100% efficiency.

*__**19.128**__ Most flashlights use two or more batteries in series. Use the concepts of galvanic cells in this chapter to explain why a flashlight with two new batteries and one "dead" battery will give only a dim light if any light is obtained at all.

19.129 If two electrolytic cells are placed in series, the same number of electrons must pass through both cells. One student argues that you can get twice as much product if two cells are placed in series compared to a single cell and therefore the cost of production (i.e., the cost of electricity) will decrease greatly and profits will increase. Is the student correct? Explain your reasoning based on the principles of electrochemistry.

19.130 Current lithium ion batteries are controversial. What is the issue involved and what solutions have been proposed?

19.131 A new battery called the "redox flow battery" is being developed as a solution to the problem of variable production of electricity by solar and wind energy. Using the knowledge developed so far, investigate this technology and develop either a list of scientific reasons to invest, or not invest, your money in this technology.

Nuclear Reactions and Their Role in Chemistry

Chapter Outline

© Free Soul Production / Shutterstock

Daily Herald Archive/SSPL/Getty Images

Glow-in-the-dark watches are not a new invention. In the early 1900s, the numbers and watch hands were painted with radium, which glowed as the radioactive element decayed and emitted light.

Our opening photo shows the eerie glow of a watch face from the early 1900s where emissions from radioactive isotopes of radium cause the numbers and dial hands to glow in the dark. Unfortunately, the nature and effects of radioactivity were just becoming known and many workers, mainly women, who painted these watch dials suffered and died from radiation poisoning. Today, radium is actually used to cure cancer. It is no surprise that work with radioactive materials can be hazardous and is highly controlled by government agencies. However, radioactive materials are also very useful for modern science, as we explain in this chapter.

From the standpoint of chemistry, our interest in the atomic nucleus stems primarily from its role in determining the number and energies of an atom's electrons. This is because it is the electron distribution in an atom that controls chemical properties. Although the nuclei of many isotopes are exceptionally stable, many of the elements also have one or more isotopes with unstable nuclei that have unique properties which make them particularly useful and interesting. These unstable nuclei are able to undergo nuclear reactions, and tend to emit radiation consisting of particles and/or energy. Nuclear reactions and decay, and the particles they emit, have been applied to a wide variety of theoretical, medical, analytical, environmental, safety, and energy solutions in 21st century science and technology. We will investigate some of these properties in this chapter, and you are encouraged to research this rapidly changing field of science on your own.

This Chapter in Context

LEARNING OBJECTIVES

After reading this chapter, you should be able to:

- utilize the unified laws of mass and energy conservation
- grasp the importance of the nuclear binding energy
- understand the nature of radionuclides and their emissions
- explore the nature of the "band of stability"
- explain modern transmutation methods
- demonstrate a knowledge of how radioactivity is measured
- show an appreciation for practical applications of radionuclides
- develop an understanding of nuclear fission and fusion

20.1 | Conservation of Mass and Energy

Changes involving unstable atomic nuclei generally involve large amounts of energy, amounts that are considerably greater than in chemical reactions. To understand how these energy changes arise, we begin our study by reexamining two physical laws that, until this chapter, have been assumed to be separate and independent—namely, the laws of conservation of energy and conservation of mass. They may be treated as separate laws for chemical reactions but not for nuclear reactions. These two laws, however, are only different aspects of a deeper, more general law.

As atomic and nuclear physics developed in the early 1900s, physicists realized that the mass of a particle cannot be treated as a constant in all circumstances. The mass, m, of a particle depends on the particle's velocity, v, relative to the observer. A particle's mass is related to this velocity, and to the velocity of light, c, by Equation 20.1.

$$m = \frac{m_0}{\sqrt{1 - (v/c)^2}} \qquad (20.1)$$

■ $c = 2.99792458 \times 10^8$ m s^{-1}, the speed of light.

Notice what happens when v is zero and the particle has no velocity relative to the observer. The ratio v/c is then zero, the entire denominator reduces to a value of 1, and Equation 20.1 becomes

$$m = m_0$$

Thus the symbol m_0 stands for the particle's rest mass.

Rest mass is what we measure in all lab operations, because any object, like a chemical sample, is either at rest (from our viewpoint) or is not moving extraordinarily rapidly. Only as the particle's velocity approaches the speed of light, c, does the v/c term in Equation 20.1 become important. As v approaches c, the ratio v/c approaches 1, and so $[1 - (v/c)^2]$ gets closer and closer to 0. The whole denominator, in other words, approaches a value of zero. If it actually reached zero, then m, which would be $(m_0 \div 0)$, would become infinitely large. The mass, m, of the particle moving at the velocity of light would be infinitely heavy, a physical impossibility. Thus the speed of light is seen as the absolute upper limit of the speed that any particle can approach.

At the velocities of everyday experience, the mass of anything calculated by Equation 20.1 equals the rest mass to four or five significant figures. The difference cannot be detected by weighing devices. Thus, in all of our normal work, mass appears to be conserved, and the law of conservation of mass functions this way in chemistry.

Practice Exercise 20.1

What is the mass of a tennis ball traveling at 53.6 m s^{-1}? The rest mass of a tennis ball is 57.7 g, as regulated by the International Tennis Federation. (*Hint:* How does the velocity of the tennis ball affect the mass?)

Practice Exercise 20.2

What is the mass of the same tennis ball in Practice Exercise 20.1 if the velocity of the ball is increased to 5.36×10^7 m s^{-1}?

Photo Researchers

Albert Einstein (1879–1955) won the Nobel Prize for physics in 1921.

■ The Einstein equation is given as $E = mc^2$ in the popular press.

Einstein equation

Law of Conservation of Mass–Energy

We know that matter cannot appear from nothing, so the extra mass an object acquires as it goes faster must come from the energy supplied to increase the object's velocity. Physicists, therefore, realized that mass and energy are interconvertible and that in the world of high-energy physics, the laws of conservation of mass and conservation of energy are not separate and independent. What emerged was a single law now called the **law of conservation of mass–energy**.

Law of Conservation of Mass–Energy

The sum of all of the energy in the universe and of all of the mass (expressed as an equivalent in energy) is a constant.

The Einstein Equation

Albert Einstein was able to show that when mass converts to energy, the change in energy, ΔE, is related to the change in rest mass, Δm_0, by the following equation, now called the **Einstein equation**:

$$\Delta E = \Delta m_0 c^2 \qquad \text{(20.2)}$$

Again, c is the velocity of light, 3.00×10^8 m s^{-1}.

Because the velocity of light is very large, even if an energy change is enormous, the change in mass, Δm_0, is extremely small. For example, the combustion of methane releases considerable heat per mole:

$$CH_4(g) + 2O_2(g) \longrightarrow CO_2(g) + 2H_2O(l) \qquad \Delta H° = -890 \text{ kJ}$$

The release of 890 kJ of heat energy corresponds to a loss of mass, which by the Einstein equation equals a loss of 9.89 ng. This is about 1×10^{-7}% of the total mass of one mol of CH_4 and two mol of O_2. Such a tiny change in mass is not detectable by laboratory balances, so for all practical purposes, mass is conserved. Although the Einstein equation has no direct applications in chemistry involving the rearrangement of electrons in chemical reactions, its importance became clear when atomic **fission** (i.e., the breaking apart of heavy atoms to form lighter fragments) was first observed in 1939.

20.2 | Nuclear Binding Energy

As we will discuss further in Section 20.3, an atomic nucleus is held together by extremely powerful forces of attraction that are able to overcome the repulsions between protons. To break a nucleus into its individual **nucleons**—that is, protons and neutrons—therefore requires an enormous input of energy. This energy is called the **nuclear binding energy**.

Absorption of the binding energy that was released when the nucleus was formed would produce the individual nucleons by breaking up the nucleus. These nucleons would now carry extra mass corresponding to the mass-equivalent of the energy they had absorbed. If we add up their masses, the sum should be larger than the mass of the nucleus from which they had come. And this is exactly what is observed. For a given atomic nucleus, the sum of the rest masses of all of its nucleons is always a little larger than the actual mass of the nucleus. The mass difference is called the **mass defect**, and its energy equivalence is the nuclear binding energy.

Keep in mind that nuclear binding energy is not energy actually possessed by the nucleus but is, instead, the energy the nucleus would have to absorb to break apart. Thus, the higher the binding energy, the more stable is the nucleus.

Calculating Nuclear Binding Energies

We can calculate nuclear binding energy with the Einstein equation, using helium-4 as an example. The rest mass of an isolated proton is 1.0072764668 u and that of a neutron is 1.0086649160 u. Helium-4 has atomic number 2, so its nucleus consists of four nucleons (two protons and two neutrons). The rest mass of one helium-4 nucleus is known to be 4.0015061792 u. However, the sum of the rest masses of its four separated nucleons is slightly more, 4.0318827656 u, which we can show as follows:

$$\text{For 2 protons: } 2 \times 1.0072764668 \text{ u} = 2.0145529336 \text{ u}$$

$$\text{For 2 neutrons: } 2 \times 1.0086649160 \text{ u} = 2.0173298320 \text{ u}$$

$$\text{Total rest mass of nucleons in } {}^4\text{He} = 4.0318827656 \text{ u}$$

The mass defect, the difference between the calculated and measured rest masses for the helium-4 nucleus, is 0.0303765864 u:

$$\underbrace{4.0318827656 \text{ u}}_{\text{mass of the 4 nucleons}} - \underbrace{4.0015061792 \text{ u}}_{\text{mass of nucleus}} = \underbrace{0.0303765864 \text{ u}}_{\text{mass defect}}$$

Using Einstein's equation, let's calculate to four significant figures the nuclear binding energy that is equivalent to the mass defect for the ^{4}He nucleus. We will round the mass defect to 0.03038 u. To obtain the energy in joules, we have to remember that 1 J = 1 kg m² s⁻², so the mass in atomic mass units (u) must be converted to kilograms. The table of constants inside the rear cover of the book gives 1 u = $1.660538921 \times 10^{-24}$ g, which equals $1.660538921 \times 10^{-27}$ kg. Substituting into the Einstein equation gives $1.6606538921 \times 10^{-27}$ kg

$$\Delta E = \Delta mc^2 = (0.03038 \text{ u}) \times \underbrace{\frac{1.6606538921 \times 10^{-27} \text{ kg}}{1 \text{ u}}}_{\Delta m \text{ (in kg)}} \times \underbrace{(2.9979 \times 10^8 \text{ m s}^{-1})^2}_{c^2}$$

$$= 4.534 \times 10^{-12} \text{ kg m}^2 \text{ s}^{-2}$$

$$= 4.534 \times 10^{-12} \text{ J}$$

There are four nucleons in the helium-4 nucleus, so the binding energy per nucleon is $(4.534 \times 10^{-12}$ J/4 nucleons) or 1.134×10^{-12} J/nucleon.

The formation of just one nucleus of ^{4}He releases 4.534×10^{-12} J, which is a very small amount of energy. If we could make Avogadro's number or 1 mol of ^{4}He nuclei (with a total mass of only 4 g), the net release of energy would be

$$(6.022 \times 10^{23} \text{ nuclei mol}^{-1}) \times (4.534 \times 10^{-12} \text{ J/nucleus}) = 2.731 \times 10^{12} \text{ J mol}^{-1}$$

This is a huge amount of energy from forming only 4 g of helium. It could keep a 100 watt incandescent light bulb lit for nearly 900 years!

Practice Exercise 20.3

Calculate the mass defect for the iron-56 nucleus. The mass of iron-56 is 55.934939 u. (*Hint:* How many protons and neutrons does iron-56 have?)

Practice Exercise 20.4

Calculate the binding energy per nucleon for the carbon-12 nucleus, whose mass is 11.996708 u.

Binding Energy and Nuclear Stability

Figure 20.1 shows a plot of binding energies per nucleon versus mass numbers for most of the elements. We are interested in this because the larger the binding energy per nucleon, the more stable the nucleus. The curve passes through a maximum at iron-56, which means that the iron-56 nuclei are the most stable of all. The plot in Figure 20.1, however, does not have a sharp maximum. Thus, a large number of elements with intermediate mass numbers in the broad center of the periodic table include the most stable isotopes in nature.

Nuclei of low mass number have small binding energies per nucleon. Joining two such nuclei to form a heavier nucleus, a process called **nuclear fusion**, leads to a more stable nucleus and a large increase in binding energy per nucleon. This extra energy is released when the two lighter nuclei fuse (join) and is the origin of the energy released in the cores of stars as discussed in Chapter 0 and the detonation of a hydrogen bomb. Nuclear fusion is discussed further in Section 20.8.

■ Another term for fusion is nucleosynthesis, a word we used in Chapter 0.

As we follow the plot of Figure 20.1 to the higher mass numbers, the nuclei decrease in stability as the binding energies decrease. Among the heaviest atoms, therefore, we might expect to find isotopes that could change to more stable forms by breaking up into lighter

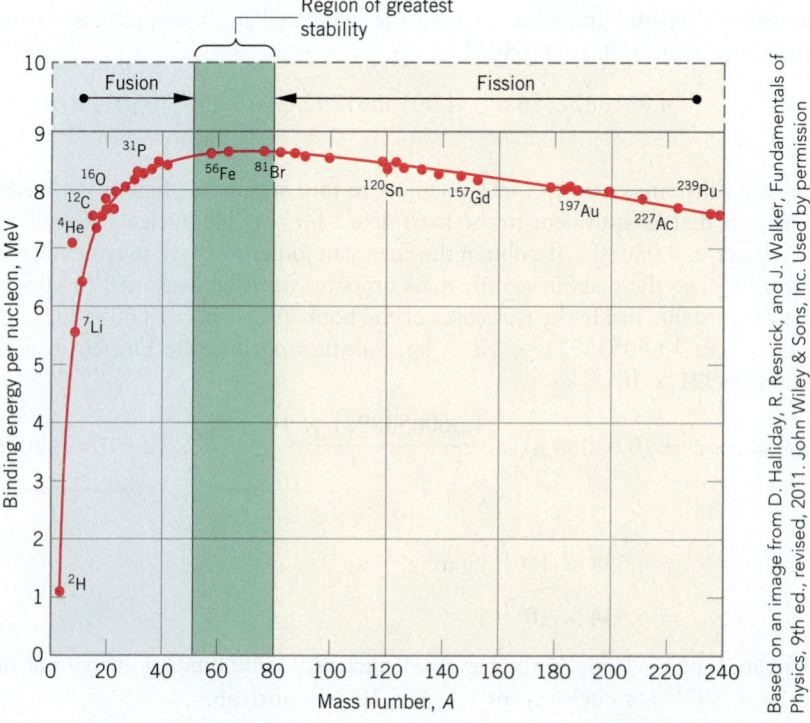

Figure 20.1 | **Binding energies per nucleon.**

Based on an image from D. Halliday, R. Resnick, and J. Walker, Fundamentals of Physics, 9th ed., revised, 2011. John Wiley & Sons, Inc. Used by permission

nuclei, by undergoing nuclear fission. **Nuclear fission**, the subject of Section 20.8, is the spontaneous breaking apart of a nucleus to form isotopes of intermediate mass number.

■ Quite often you'll see the terms "atomic fusion" and "atomic fission" used for nuclear fusion and nuclear fission, respectively.

20.3 | Radioactivity

Except for hydrogen, all atomic nuclei have more than one proton, each of which carries a positive charge. Because like charges repel, we might wonder how any nucleus could be stable. However, the electrostatic forces of attraction and repulsion, such as the kinds present among the ions in a crystal of sodium chloride, are not the only forces at work in the nucleus. Protons do, indeed, repel each other electrostatically, but another force, a force of attraction called the **nuclear strong force**, also acts in the nucleus. The nuclear strong force, effective only at very short distances, overcomes the electrostatic force of repulsion between protons, and it binds both protons and neutrons into a nuclear package. Moreover, the neutrons, by helping to keep the protons farther apart, also lessen repulsions between protons.

One consequence of the difference between the nuclear strong force and the electrostatic force occurs among nuclei that have large numbers of protons but too few intermingled neutrons to dilute the electrostatic repulsions between protons. Such nuclei are often unstable because their nuclei carry excess energy in order to hold the nucleus together. To achieve a lower energy and thus more stability, unstable nuclei have a tendency to eject small nuclear fragments, and many simultaneously release high-energy electromagnetic radiation. The stream of particles (or photons) coming from the sample is called **nuclear radiation** and the phenomenon is called **radioactivity**. Isotopes that exhibit this property are called **radionuclides**. About 60 of the approximately 350 naturally occurring isotopes are radioactive.

In a sample of a given radionuclide, not all of the atoms undergo change at once. The rate at which radiation is emitted (which translates into the intensity of the radiation) depends on the identity of the isotope in the sample. Over time, as radioactive nuclei change into stable ones, the number of atoms of the radionuclide remaining in the sample decreases, causing the intensity of the radiation to drop, or decay. The radionuclide is said to undergo **radioactive decay**.

Naturally occurring **atomic radiation** consists principally of three kinds: alpha, beta, and gamma radiation, as discussed below. Other types of radiation also occur. They are summarized in Table 20.1 and are described on the pages ahead.

■ Adjacent neutrons experience no electrostatic repulsion between each other, only the attractive strong force.

Alpha Radiation

Alpha radiation consists of a stream of helium nuclei called **alpha particles**, symbolized as ^4_2He or $^4_2\alpha$, where 4 is the mass number and 2 is the atomic number. The alpha particle bears a charge of 2+, but the charge is omitted from the symbol.

Alpha particles are the most massive of those commonly emitted by radionuclides. When ejected (Figure 20.2), alpha particles move through the atom's electron orbitals, emerging from the atom at speeds of up to one-tenth the speed of light. Their size, however, prevents them from going far. After traveling at most only a few centimeters in air,

■ In Chapter 0 you learned that an isotope is identified by writing its mass number, A, as a superscript and its atomic number, Z, as a subscript in front of the chemical symbol, X, as in $^A_Z X$. We use this same notation in representing particles involved in nuclear reactions. For particles that are not atomic nuclei, Z stands for the charge on the particle.

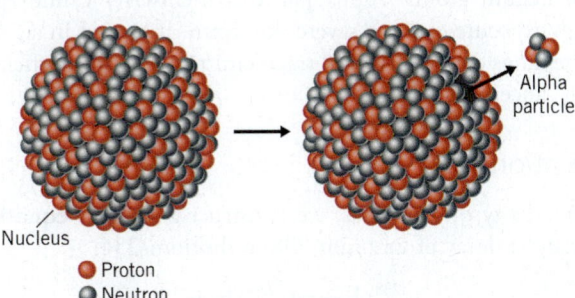

Nucleus

● Proton
◐ Neutron

Alpha particle

Figure 20.2 | Emission of an alpha particle from an atomic nucleus. Removal of ^4_2He from a nucleus decreases the atomic number by 2 and the mass number by 4.

TABLE 20.1 **Nuclear Reactions**

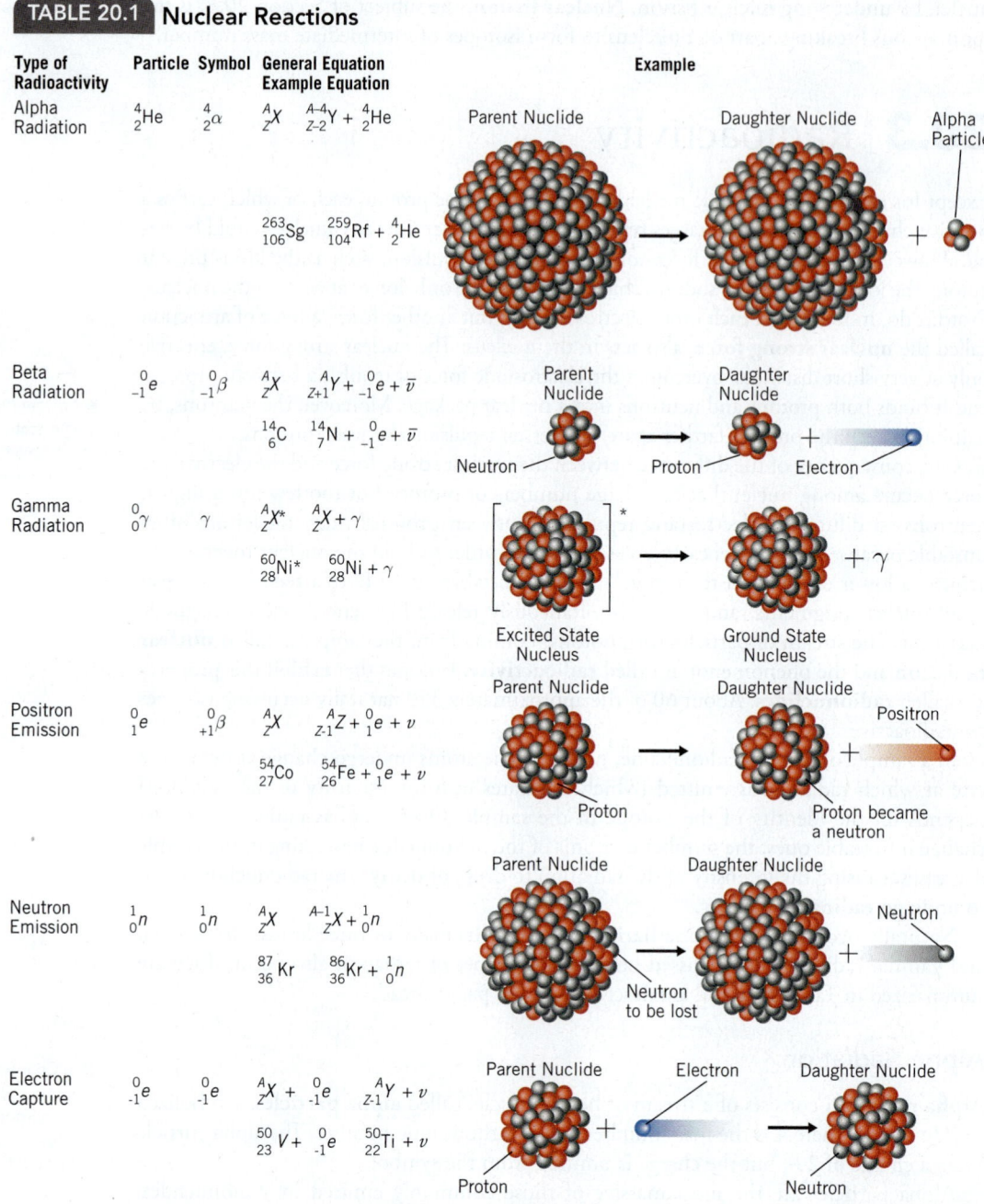

Type of Radioactivity	Particle	Symbol	General Equation Example Equation	Example
Alpha Radiation	^4_2He	$^4_2\alpha$	$^A_Z X \quad ^{A-4}_{Z-2}Y + ^4_2\text{He}$	Parent Nuclide → Daughter Nuclide + Alpha Particle
			$^{263}_{106}\text{Sg} \quad ^{259}_{104}\text{Rf} + ^4_2\text{He}$	
Beta Radiation	$^0_{-1}e$	$^0_{-1}\beta$	$^A_Z X \quad ^A_{Z+1}Y + ^0_{-1}e + \bar{v}$	Parent Nuclide → Daughter Nuclide
			$^{14}_6\text{C} \quad ^{14}_7\text{N} + ^0_{-1}e + \bar{v}$	Neutron → Proton + Electron
Gamma Radiation	$^0_0\gamma$	γ	$^A_Z X^* \quad ^A_Z X + \gamma$	Excited State Nucleus → Ground State Nucleus + γ
			$^{60}_{28}\text{Ni}^* \quad ^{60}_{28}\text{Ni} + \gamma$	
Positron Emission	$^0_1 e$	$^0_{+1}\beta$	$^A_Z X \quad ^A_{Z-1}Z + ^0_1 e + v$	Parent Nuclide → Daughter Nuclide + Positron
			$^{54}_{27}\text{Co} \quad ^{54}_{26}\text{Fe} + ^0_1 e + v$	Proton → Proton became a neutron
Neutron Emission	$^1_0 n$	$^1_0 n$	$^A_Z X \quad ^{A-1}_Z X + ^1_0 n$	Parent Nuclide → Daughter Nuclide + Neutron
			$^{87}_{36}\text{Kr} \quad ^{86}_{36}\text{Kr} + ^1_0 n$	Neutron to be lost
Electron Capture	$^0_{-1}e$	$^0_{-1}e$	$^A_Z X + ^0_{-1}e \quad ^A_{Z-1}Y + v$	Parent Nuclide + Electron → Daughter Nuclide
			$^{50}_{23}V + ^0_{-1}e \quad ^{50}_{22}\text{Ti} + v$	Proton → Neutron

alpha particles collide with air molecules, lose kinetic energy, pick up electrons, and become neutral helium atoms. Alpha particles themselves cannot penetrate the skin, although enough exposure causes a severe skin burn. If carried in air or on food into the soft tissues of the lungs or the intestinal tract, emitters of alpha particles can cause serious harm, including cancer.

Nuclear Equations

To symbolize the decay of a nucleus, we construct a **nuclear equation**, which we can illustrate by the alpha decay of uranium-238 to thorium-234:

$$^{238}_{92}\text{U} \longrightarrow ^{234}_{90}\text{Th} + ^4_2\text{He}$$

■ A nuclear transformation such as the alpha decay of ^{238}U does not depend on the chemical environment. The same nuclear equation applies whether the uranium is in the form of the free element or in a compound.

Unlike chemical reactions, a **nuclear reaction** produces new isotopes, so we need separate rules for balancing nuclear equations.

Rules for Balancing a Nuclear Equation:
1. The sum of the mass numbers on each side of the arrow must be the same.
2. The sum of the atomic numbers (nuclear charge) on each side of the arrow must be the same.

In the nuclear equation for the decay of uranium-238, the atomic numbers balance ($90 + 2 = 92$), and the mass numbers balance ($234 + 4 = 238$). Notice that electrical charges are not shown, even though they are there (initially). The alpha particle, for example, has a charge of $2+$. If it was emitted by a neutral uranium atom, then the thorium particle initially has a charge of $2-$. These charged particles, however, eventually pick up or lose electrons either from each other or from molecules in the matter through which they travel.

Beta Radiation

Naturally occurring **beta radiation** consists of a stream of electrons, which in this context are called **beta particles**. In a nuclear equation, the beta particle has the symbol $_{-1}^{0}e$ or $_{-1}^{0}\beta$ because the electron's mass number is 0 and its charge is $1-$. Hydrogen-3 (tritium) is a beta emitter that decays by the following equation:

$$_{1}^{3}\text{H} \longrightarrow {}_{2}^{3}\text{He} + {}_{-1}^{0}e + \bar{v}$$

tritium   helium-3   beta-particle   antineutrino
(electron)

■ Sometimes the beta particle is given the symbol $_{-1}^{0}\beta$, or simply β^{-}.

Both the antineutrino (to be described further shortly) and the beta particle come from the atom's nucleus, not its electron shells. We do not think of them as having a prior existence in the nucleus, any more than a photon exists before its emission from an excited atom (see Figure 20.3). Both the beta particle and the antineutrino are created during the decay process in which a neutron is transformed into a proton:

$$_{0}^{1}n \longrightarrow {}_{-1}^{0}e + {}_{1}^{1}p + \bar{v}$$

neutron   beta particle   proton   antineutrino
(in the   (emitted)   (remains in   (emitted)
nucleus)   the nucleus)

Unlike alpha particles, which are all emitted with the same discrete energy from a given radionuclide, beta particles emerge from a given beta emitter with a continuous spectrum of energies. Their energies vary from zero to some characteristic fixed upper limit for each radionuclide. This fact once gave nuclear physicists considerable trouble, partly because it was an apparent violation of energy conservation. To solve this problem, Wolfgang Pauli proposed in 1927 that beta emission is accompanied by emission of yet another decay particle, this one electrically neutral and almost massless. Enrico Fermi suggested the name neutrino ("little neutral one"), but eventually it was named the **antineutrino**, symbolized by $\bar{v}$.

An electron is extremely small, so a beta particle is less likely to collide with any atoms or molecules in the medium through which it travels. Depending on its initial kinetic

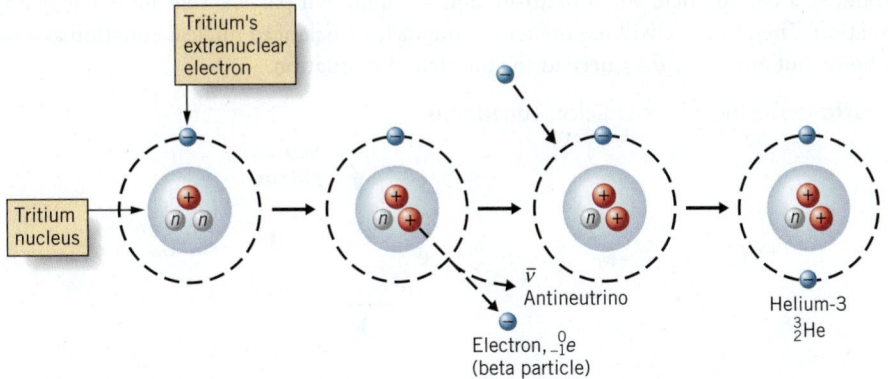

Tritium's extranuclear electron

Tritium nucleus

$\bar{v}$
Antineutrino

Electron, $_{-1}^{0}e$
(beta particle)

Helium-3
$_{2}^{3}$He

Figure 20.3 | **Emission of a beta particle from a tritium nucleus.** Emission of a beta particle changes a neutron into a proton. This results in a positively charged ion, which picks up an electron to become a neutral atom.

energy, a beta particle can travel up to 300 cm (about 10 ft) in dry air, much farther than alpha particles. However, only the highest-energy beta particles can penetrate the skin.

Gamma Radiation

■ Gamma photons have the symbol $^0_0\gamma$ because they have neither charge nor mass.

Gamma radiation, which often accompanies most nuclear decays, consists of high-energy photons given the symbol $^0_0\gamma$ or, often, simply γ in equations. Gamma radiation is extremely penetrating and is effectively blocked only by very dense materials, like lead.

The emission of gamma radiation involves transitions between energy levels within the nucleus. Nuclei have energy levels of their own, much as atoms have orbital energy levels. When a nucleus emits an alpha or beta particle, it sometimes is left in an excited energy state. By the emission of a gamma-ray photon, the nucleus relaxes into a more stable state. For example, cobalt-60 emits a beta particle and forms a nickel-60 isotope in a nuclear excited state, which then emits two γ particles.

$$^{60}_{27}\text{Co} \longrightarrow {}^{60}_{28}\text{Ni*} + {}^{0}_{-1}e \longrightarrow {}^{60}_{28}\text{Ni} + 2\gamma$$

The excited state is indicated by an asterisk.

The Electron Volt

The energy carried by a given radiation is usually described by an energy unit new to our study, the **electron volt**, abbreviated **eV**; 1 eV is the energy an electron receives when accelerated under the influence of 1 volt. It is related to the joule as follows:

$$1 \text{ eV} = 1.602 \times 10^{-19} \text{ J}$$

■ In the interconversion of mass and energy,

$1 \text{ eV} \Leftrightarrow 1.783 \times 10^{-36} \text{ kg}$

$1 \text{ MeV} \Leftrightarrow 1.783 \times 10^{-27} \text{ g}$

$1 \text{ GeV} \Leftrightarrow 1.783 \times 10^{-24} \text{ g}$

As you can see, the electron volt is an extremely small amount of energy, so multiples are commonly used, such as the kilo-, mega-, and gigaelectron volt. An alpha particle emitted by radium-224 has an energy of 5 MeV. Hydrogen-3 (tritium) emits beta radiation at an energy of 0.05 to 1 MeV. The gamma radiation from cobalt-60, the radiation currently used to kill bacteria and other pests in certain foods, consists of photons with energies of 1.173 MeV and 1.332 MeV.

Example 20.1
Writing a Balanced Nuclear Equation

■ Ions of cesium, which is in Group 1A with sodium, travel in the body to many of the same sites where sodium ions go.

Cesium-137, ^{137}Cs, one of the radioactive wastes from a nuclear power plant or an atomic bomb explosion, emits beta and gamma radiation. Write the nuclear equation for the decay of cesium-137.

Analysis: We are being asked to balance a nuclear equation, and when we do this, we need to keep track of the atomic numbers and the mass numbers so they are balanced.

Assembling the Tools: We will start with an incomplete equation using the given information. We have one reactant, the Cs that decays to give three particles: two known products, a beta particle and a neutron, and an unknown particle that must balance the equation. Therefore, we will use the requirements for a balanced nuclear equation as a tool to figure out any other data needed to complete the equation.

Solution: The incomplete nuclear equation is

$$^{137}_{55}\text{Cs} \longrightarrow {}^{0}_{-1}e + {}^{0}_{0}\gamma + \underline{\qquad}$$

Mass number goes here.

Atomic symbol goes here.

Atomic number goes here.

We need to determine Z and A and then we'll use X to identify the symbol for the element. For the superscripts we see that 137 will be the superscript, A, for X since all other superscripts are zero. For the subscripts we see that X must be 56 (since $55 = -1 + 56$). We can now use 56 to identify the element as barium, and use its symbol Ba. Now we fill in the last term as $^{137}_{56}\text{Ba}$ the balanced equation is:

$$^{137}_{55}\text{Cs} \longrightarrow {}^{0}_{-1}e + {}^{0}_{0}\gamma + {}^{137}_{56}\text{Ba}$$

Is the Answer Reasonable? Besides double-checking that element 56 is barium, the answer satisfies the requirements of a nuclear equation—namely, the sums of the mass numbers, 137, are the same on both sides, as are the sums of the atomic numbers.

Marie Curie earned one of her two Nobel Prizes for isolating the element radium, which soon became widely used to treat cancer. Radium-226, ^{226}Ra, emits a gamma photon plus an alpha particle to give radon-222. Write a balanced nuclear equation for its decay and identify the particle that's emitted. (*Hint:* Be sure to balance atomic numbers and mass numbers.)

Practice Exercise 20.5

Write the balanced nuclear equation for the decay of iodine-131, a beta emitter. (Iodine-131 is one of the many radionuclides present in the wastes of operating nuclear power plants.)

Practice Exercise 20.6

Positron and Neutron Emission

Many synthetic isotopes (isotopes not found in nature, but synthesized by methods discussed in Section 20.5) emit **positrons**, which are particles with the mass of an electron but a positive instead of a negative charge, and its symbol is $^{0}_{1}e$. It forms in the nucleus by the conversion of a proton to a neutron (Figure 20.4). Positron emission, like beta emission, is accompanied by a chargeless and virtually massless particle, a neutrino (v), the counterpart of the antineutrino ($\bar{v}$) in the realm of antimatter (defined below). Cobalt-54, for example, is a positron emitter and changes to a stable isotope of iron:

$$^{54}_{27}\text{Co} \longrightarrow {}^{54}_{26}\text{Fe} + \underset{\text{positron}}{{}^{0}_{1}e} + \underset{\text{neutrino}}{v}$$

■ The symbol $^{+}\beta$ is sometimes used for the positron.

A positron, when emitted, eventually collides with an electron, and the two annihilate each other (Figure 20.5). Their masses change entirely into the energy of two gamma-ray photons called annihilation radiation photons, each with an energy of 511 keV.

Because a positron destroys a particle of ordinary matter (an electron), it is called a particle of antimatter. To be classified as **antimatter**, a particle must have a counterpart among one of ordinary matter, and the two must annihilate each other when they collide. For example, a neutron and an antineutron represent such a pair and annihilate each other when they come in contact.

Neutron emission, another kind of nuclear reaction, leads to a different isotope of the same element. Neutron emission occurs when an isotope has an excess of neutrons. Krypton-87, for example, decays as follows to krypton-86:

$$^{87}_{36}\text{Kr} \longrightarrow {}^{86}_{36}\text{Kr} + \underset{\text{neutron}}{{}^{1}_{0}n}$$

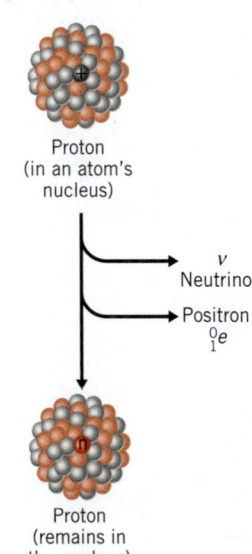

Proton
(in an atom's nucleus)

v
Neutrino

Positron
$^{0}_{1}e$

Proton
(remains in the nucleus)

Figure 20.4 | The emission of a positron replaces a proton by a neutron.

X Rays and Electron Capture

X rays, like gamma rays, consist of high-energy electromagnetic radiation, but their energies are usually less than those of gamma radiation. Although X rays are emitted by some synthetic radionuclides, when needed for medical diagnostic work they are generated by focusing a high-energy electron beam onto a metal target.

Electron capture, yet another kind of nuclear reaction, is very rare among natural isotopes but common among synthetic radionuclides. For example, an electron can be

■ X rays used in diagnosis typically have energies of 100 keV or less.

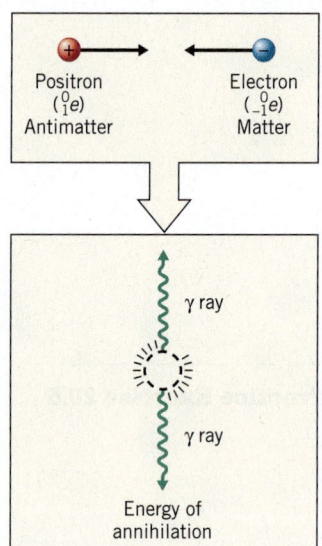

Figure 20.5 | **Gamma radiation is produced when a positron and an electron collide.** The annihilation of a positron and an electron leads to two gamma ray photons that go in exactly opposite directions.

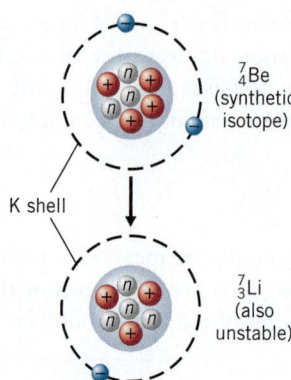

Figure 20.6 | **Electron capture.** Electron capture is the collapse of an orbital electron into the nucleus, and this changes a proton into a neutron. A gap is left in a low-energy electron orbital. When an electron drops from a higher orbital to fill the gap, an X-ray photon is emitted.

captured from the orbital electron shell having $n = 1$ or $n = 2$ by a vanadium-50 nucleus, causing it to change to a stable ^{50}Ti nucleus. The transformation is accompanied by the emission of X rays and a neutrino.

$$^{50}_{23}\text{V} \xrightarrow{\text{electron capture}} \,^{50}_{22}\text{Ti} + \text{X ray} + v$$

The net effect in the nucleus of electron capture is the conversion of a proton into a neutron (Figure 20.6):

$$\underset{\substack{\text{proton} \\ \text{(in the nucleus)}}}{^{1}_{1}p} \quad + \quad \underset{\substack{\text{electron (captured} \\ \text{from the 1s orbital)}}}{^{0}_{-1}e} \quad \longrightarrow \quad \underset{\substack{\text{neutron} \\ \text{(in the nucleus)}}}{^{1}_{0}n}$$

Electron capture does not change an atom's mass number, only its atomic number. It also leaves a hole in the first or second electron shell, and the atom emits photons of X rays as other orbital electrons drop down to fill the hole. Moreover, the nucleus that has just captured an orbital electron may be in an excited energy state and so can emit a gamma-ray photon.

Radioactive Disintegration Series

Sometimes a radionuclide does not decay directly to a stable isotope, but decays instead to another unstable radionuclide. The decay of one radionuclide after another will continue until a stable isotope forms. The sequence of such successive nuclear reactions is called a **radioactive disintegration series**. Four such series occur naturally. Uranium-238 is at the head of one of them (Figure 20.7).

The rates of decay of radionuclides vary and are usually described by specifying their half-lives, $t_{1/2}$, a topic we studied in Section 14.4. One *half-life* period in nuclear science is the time it takes for a given sample of a radionuclide to decay to one-half of its initial amount.

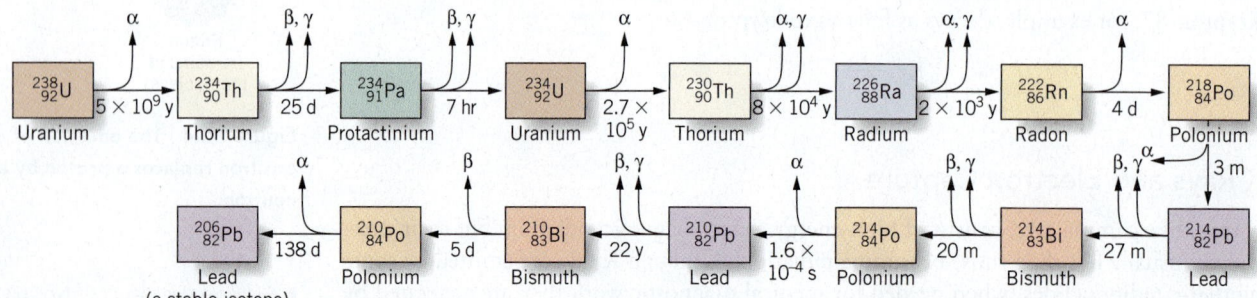

Figure 20.7 | **The uranium-238 radioactive disintegration series.** The time given beneath each arrow is the half-life period of the preceding isotope (y = year, m = month, d = day, hr = hour, mi = minute, and s = second).

TABLE 20.2	Typical Half-Life Periods		
Element	Isotope	Half-life	Radiations or Mode of Decay
Naturally occurring radionuclides			
Potassium	$^{40}_{19}K$	1.3×10^9 yr	beta, gamma
Tellurium	$^{123}_{52}Te$	1.2×10^{13} yr	electron capture
Neodymium	$^{144}_{60}Nd$	5×10^{15} yr	alpha
Samarium	$^{149}_{62}Sm$	4×10^{14} yr	alpha
Rhenium	$^{187}_{75}Re$	7×10^{10} yr	beta
Radon	$^{222}_{86}Rn$	3.82 day	alpha
Radium	$^{226}_{88}Ra$	1590 yr	alpha, gamma
Thorium	$^{230}_{90}Th$	8×10^4 yr	alpha, gamma
Uranium	$^{238}_{92}U$	4.51×10^9 yr	alpha
Synthetic radionuclides			
Tritium	$^3_1H, \, ^3_1T$	12.26 yr	beta
Oxygen	$^{15}_8O$	124 s	positron
Phosphorus	$^{32}_{15}P$	14.3 day	beta
Technetium	$^{99m}_{43}Tc$	6.02 hr	gamma
Iodine	$^{131}_{53}I$	8.07 day	beta
Cesium	$^{137}_{55}Cs$	30.1 yr	beta
Strontium	$^{90}_{38}Sr$	28.1 yr	beta
Plutonium	$^{238}_{94}Pu$	87.8 yr	alpha
Americium	$^{243}_{95}Am$	7.37×10^3 yr	alpha

Radioactive decay is a first-order process, so the period of time taken by one half-life is independent of the initial number of nuclei. The huge variations in the half-lives of several radionuclides are shown in Table 20.2.

Carbon-11 is an isotope with a neutron-poor nucleus and is a positron emitter. Write the balanced nuclear equation for the decay of carbon-11. (*Hint:* What nuclide is formed when a positron is emitted?)

Practice Exercise 20.7

Beryllium-13 is neutron rich and is a neutron emitter. Write the balanced nuclear equation for the decay of beryllium-13.

Practice Exercise 20.8

Selenium-72 decays to arsenic-72 by electron capture, which in turn is used for PET imaging. Write a balanced nuclear equation for the decay of selenium-72.

Practice Exercise 20.9

20.4 | Band of Stability

Figure 20.8 plots all of the known isotopes of all of the elements, both stable and unstable, according to their numbers of protons and neutrons. The two curved lines in Figure 20.8 enclose a zone, called the **band of stability**, within which lie all stable nuclei. (No isotope above element 83, bismuth, is included in Figure 20.8 because none has a stable isotope.) Within the band of stability are also some unstable isotopes, because smooth lines cannot be drawn to exclude them.

Any isotope not represented anywhere on the array, inside or outside the band of stability, probably has a half-life too short to permit its detection. For example, an isotope with 50 neutrons and 60 protons would lie well below the band of stability and would likely be extremely unstable. Any attempt to synthesize it would likely be a waste of time and money.

Figure 20.8 | **The band of stability.** Stable nuclei fall within a narrow band in a plot of the number of neutrons versus the number of protons. Nuclei far outside the band are too unstable to exist.

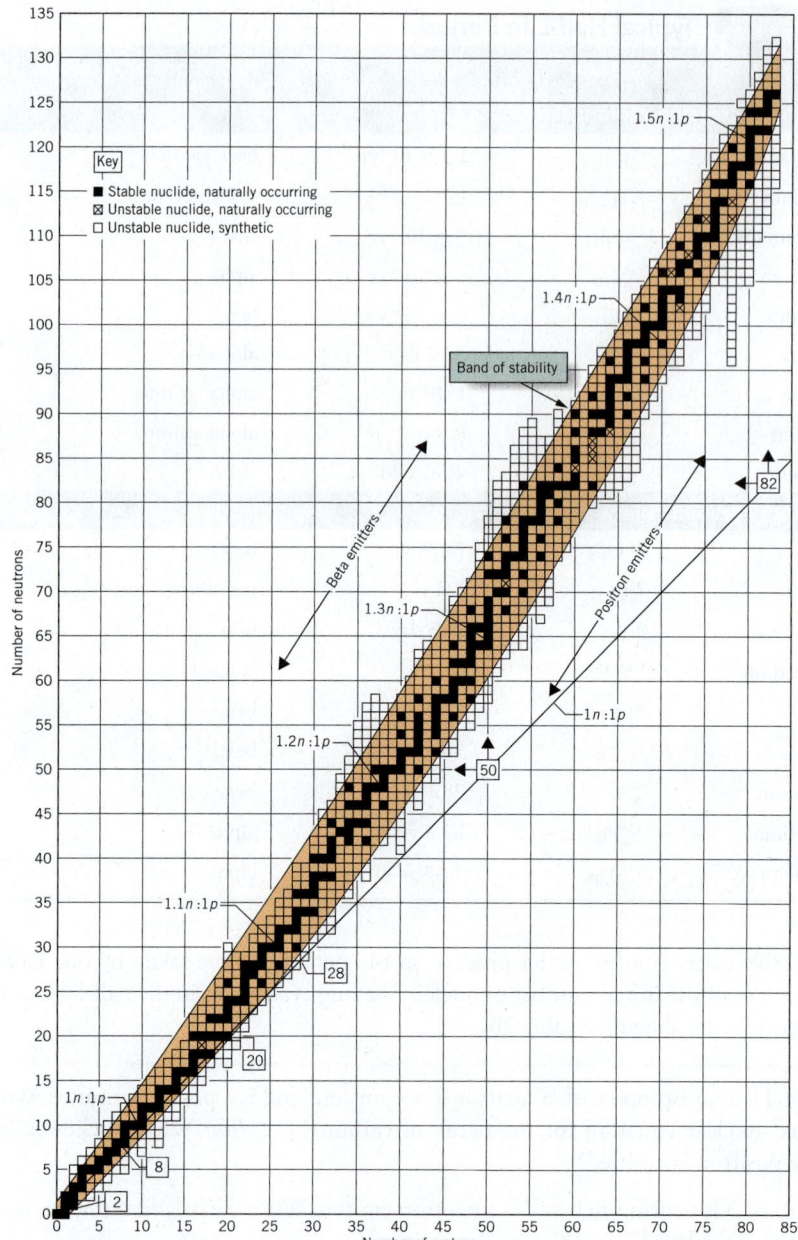

Notice that the band curves slightly upward as the number of protons increases. The curvature means that the ratio of neutrons to protons gradually increases from 1:1, a ratio indicated by the straight line in Figure 20.8. This occurs because more protons require more neutrons to provide a compensating nuclear strong force and to dilute electrostatic proton–proton repulsions.

Nuclear Reactions of Unstable Nuclei

The isotopes with atomic numbers above 83 tend to be alpha emitters. Isotopes occurring above and to the left of the band of stability tend to be beta emitters. Isotopes lying below and to the right of the band are positron emitters. There are some reasonable explanations for these tendencies.

Alpha Emitters

The alpha emitters, as we said, occur mostly among the radionuclides above atomic number 83. Their nuclei have too many protons, and the most efficient way to lose protons is by the loss of an alpha particle.

Beta Emitters

Beta emitters are generally above the band of stability and so have neutron-to-proton ratios that evidently are too high. By beta decay, a nucleus loses a neutron and gains a proton, thus decreasing the ratio:

$$\ _{0}^{1}n \longrightarrow \ _{1}^{1}p + \ _{-1}^{0}e + \bar{\nu}$$

For example, by beta decay, fluorine-20 decreases its neutron-to-proton ratio from 11/9 to 10/10.

$$\ _{9}^{20}F \longrightarrow \ _{10}^{20}Ne + \ _{-1}^{0}e + \bar{\nu}$$

$$\frac{neutron}{proton} = \frac{11}{9} \longrightarrow \frac{10}{10}$$

The surviving nucleus, neon-10, is closer to the center of the band of stability. Figure 20.9, an enlargement of the fluorine part of Figure 20.8, explains this change further. Figure 20.9 also shows how the beta decay of magnesium-27 to aluminum-27 lowers the neutron-to-proton ratio.

■ The proton can also be given the symbol $_{1}^{1}$H in nuclear equations.

Positron Emitters

In nuclei with too few neutrons to be stable, positron emission increases the neutron-to-proton ratio by changing a proton into a neutron. A fluorine-17 nucleus, for example, increases its neutron-to-proton ratio, improves its stability, and moves into the band of stability by emitting a positron and a neutrino and changing to oxygen-17 (see Figure 20.9):

$$\ _{9}^{17}F \longrightarrow \ _{8}^{17}O + \ _{1}^{0}e + \nu$$

$$\frac{neutron}{proton} = \frac{8}{9} \longrightarrow \frac{9}{8}$$

Positron decay by magnesium-23 to sodium-23 produces a similarly favorable shift (also shown in Figure 20.9).

Odd–Even Rule

Study of the compositions of stable nuclei reveals that nature favors even numbers for protons and neutrons. This is summarized by the **odd–even rule**.

Odd–Even Rule

When the numbers of neutrons and protons in a nucleus are both even, the isotope is far more likely to be stable than when both numbers are odd.

TOOLS

Odd–even rule

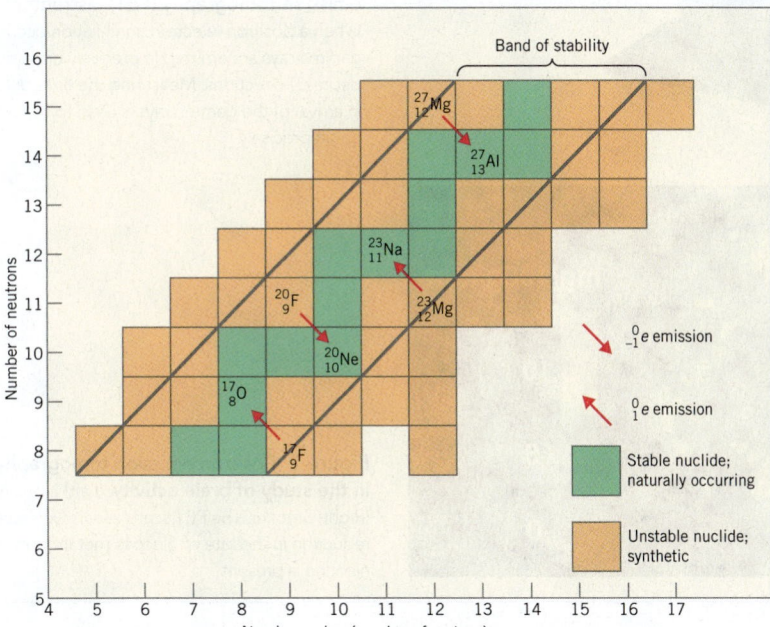

Figure 20.9 | **An enlarged section of the band of stability.** Beta decay from magnesium-27 and fluorine-20 reduces their neutron-to-proton ratio and moves them closer to the band of stability. Positron decay from magnesium-23 and fluorine-17 increases this ratio and moves these nuclides closer to the band of stability, too.

CHEMISTRY ON THE CUTTING EDGE 20.1

Positron Emission Tomography (PET)

Positron emitters are used in an important method for studying brain function called the PET scan, for positron emission tomography. The technique begins by chemically incorporating positron-emitting radionuclides into molecules, like glucose, that can be absorbed by the brain directly from the blood. It's like inserting radiation generators that act from within the brain rather than focusing X rays or gamma rays from the outside. Carbon-11, for example, is a positron emitter whose atoms can be used in place of carbon-12 atoms in glucose molecules, $C_6H_{12}O_6$. (One way to prepare such glucose is to let a leafy vegetable, Swiss chard, use $^{11}CO_2$ to make the glucose by photosynthesis.)

A PET scan using tiny amounts of carbon-11 glucose detects situations where glucose is not taken up normally—for example, in manic depression, schizophrenia, and Alzheimer's disease. After the carbon-11 glucose is ingested by the patient, radiation detectors outside the body pick up the annihilation radiation produced when electrons react with positrons emitted at specifically those brain sites that use glucose. These positrons emit gamma rays in opposite directions. When the pairs of gamma rays are detected 180 degrees apart, the acquisition time can be used to back calculate exactly where the radiation came from. By mapping the locations of the brain sites using the tagged glucose, a picture showing brain function can be formed. PET scan technology (Figure 1), for example, demonstrated that the uptake of glucose by the brains of smokers is less than that of non-smokers, as shown in Figure 2.

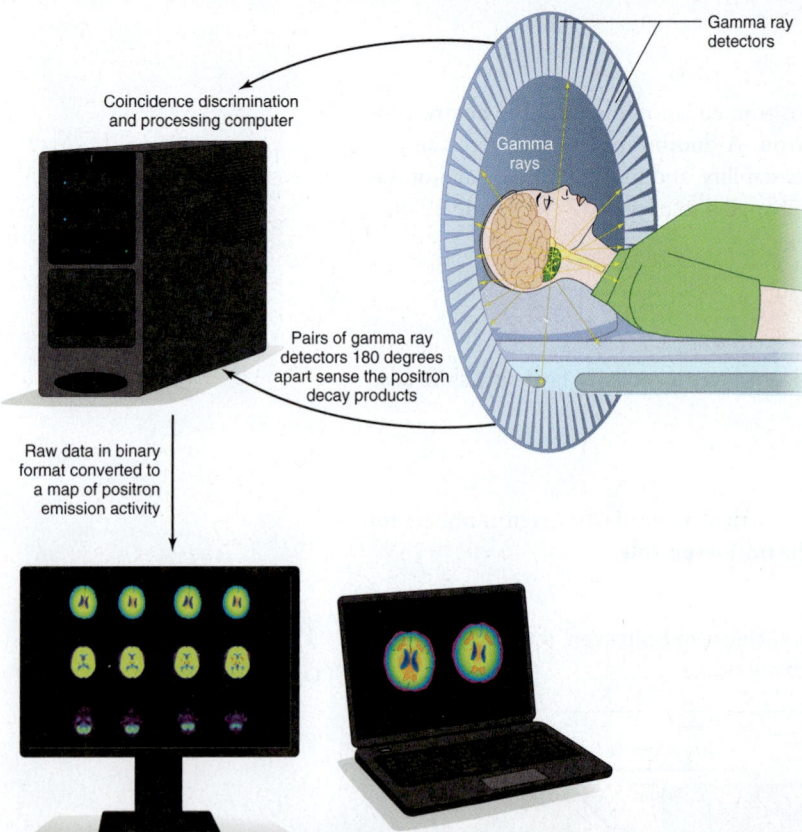

Figure 1 A diagram of how the positron emission tomography (PET) instrument works. When a positron–electron annihilation occurs two gamma rays are emitted in precisely opposite (180 degrees) directions. Measuring the time difference in arrival of the gamma rays is used to calculate their origin precisely.

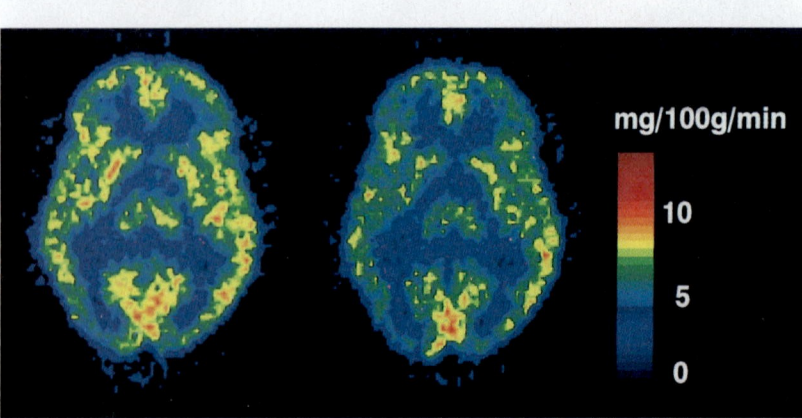

Figure 2 Positron emission tomography (PET) in the study of brain activity. (Left) Nonsmoker. (Right) Smoker. The PET scan reveals widespread reduction in the rate of glucose metabolism when nicotine is present.

Of the 264 stable isotopes, only 5 have odd numbers of both protons and neutrons, whereas 157 have even numbers of both. The rest have an odd number of one nucleon and an even number of the other. To see this, notice in Figure 20.8 how the horizontal lines with the largest numbers of dark squares (stable isotopes) most commonly correspond to even numbers of neutrons. Similarly, the vertical lines with the most dark squares most often correspond to even numbers of protons.

■ These five stable isotopes, ^1_1H, ^6_3Li, $^{10}_5B$, $^{14}_7N$, and $^{138}_{57}La$, all have odd numbers of both protons and neutrons.

The odd–even rule is related to the spins of nucleons. Both protons and neutrons behave as though they spin like orbital electrons. When two protons or two neutrons have paired spins, meaning the spins are opposite, their combined energy is less than when the spins are unpaired. Only when there are even numbers of protons and neutrons can all of the spins be paired and so give the nucleus a lower energy and greater stability. The least stable nuclei tend to be those with both an odd number of protons and an odd number of neutrons.

Nuclear Magic Numbers

Another rule of thumb for nuclear stability is based on magic numbers of nucleons. Isotopes with specific numbers of protons or neutrons, the magic numbers, are more stable than the rest. The **magic numbers** of nucleons are 2, 8, 20, 28, 50, 82, and 126, and where they fall is shown in Figure 20.8 (except for magic number 126).

When the numbers of both protons and neutrons are the same magic number, as they are in ^4_2He, $^{16}_8O$, and $^{40}_{20}Ca$, the isotope is very stable. $^{100}_{50}Sn$ also has two identical magic numbers. Although this isotope of tin lies well outside the band of stability and is unstable, having a half-life of only several seconds, it is much more stable than nearby radionuclides, whose half-lives are in milliseconds. It is stable enough to be observed. One stable isotope of lead, $^{208}_{82}Pb$, involves two different magic numbers—namely, 82 protons and 126 neutrons.

■ Magic numbers do not cancel the need for a favorable neutron/proton ratio. An atom with 82 protons and 82 neutrons lies far outside the band of stability, and yet 82 is a magic number.

The **nuclear shell model** is based on the idea that a nucleus has a shell structure with energy levels analogous to electron energy levels. Electron levels, as you already know, are associated with their own special numbers, those that equal the maximum number of electrons allowed in a principal energy level: 2, 8, 18, 32, 50, 72, and 98 (for principal levels with n equal to 1, 2, 3, 4, 5, 6, and 7, respectively). The total numbers of electrons in the atoms of the most chemically stable elements—the noble gases—also make up a special set: 2, 10, 18, 36, 54, and 86 electrons. Thus, special sets of numbers are not unique to nuclei.

Curium-242 has 96 protons and 146 neutrons. What is the most probable form of radioactive decay for Cm-242? (*Hint:* Where would Cm-242 be found relative to the band of stability in Figure 20.8?)

Practice Exercise 20.10

Glenn Seaborg was the discoverer of more synthetic elements than anyone and seaborgium (At. No. 106) is named for him. What is the most probable radioactive decay route for the seaborgium-262 isotope?

Practice Exercise 20.11

20.5 | Transmutation

The change of one isotope into another is called **transmutation**, and radioactive decay is only one cause. Transmutation can also be forced by the bombardment of nuclei with high-energy particles, such as alpha particles from natural emitters, neutrons from atomic reactors, or protons made by stripping electrons from hydrogen. To make them better bombarding missiles, protons and alpha particles can be accelerated in an electrical field by a cyclotron, synchrotron, or linear accelerator as shown in Figure 20.10. This gives them greater energy and enables them to sweep through the target atom's orbital electrons and become buried in its nucleus. Although beta particles can also be accelerated, they are repelled by a target atom's own electrons.

Figure 20.10 | **Linear accelerator.** In this linear accelerator at Brookhaven National Laboratory on Long Island, New York, protons can be accelerated to just under the speed of light and given up to 33 GeV of energy before they strike their target. (1 GeV $= 10^9$ eV)

Compound Nuclei

Both the energy and the mass of a bombarding particle enter the target nucleus at the moment of capture. The new nucleus, called a **compound nucleus**, has excess energy that quickly becomes distributed among all of the nucleons, but the nucleus is nevertheless rendered somewhat unstable. To get rid of the excess energy, a compound nucleus generally ejects something (e.g., a neutron, proton, or electron) and often emits gamma radiation as well. This leaves a new nucleus of an isotope different from the original target, so a transmutation has occurred overall.

■ Compound here refers only to the idea of combination, not to a chemical.

Ernest Rutherford observed the first example of artificial transmutation. When he let alpha particles pass through a chamber containing nitrogen atoms, an entirely new radiation was generated, one much more penetrating than alpha radiation. It proved to be a stream of protons (Figure 20.11). Rutherford was able to show that the protons came from the decay of the compound nuclei of fluorine-18, produced when nitrogen-14 nuclei captured bombarding alpha particles:

■ The asterisk, *, symbolizes a high-energy nucleus—a compound nucleus.

$$\underset{\text{alpha particle}}{^{4}_{2}\text{He}} \;+\; \underset{\text{nitrogen nucleus}}{^{14}_{7}\text{N}} \;\longrightarrow\; \underset{\text{flourine (compound nucleus)}}{^{18}_{9}\text{F}^{*}} \;\longrightarrow\; \underset{\text{oxygen (a rare but stable isotope)}}{^{17}_{8}\text{O}} \;+\; \underset{\text{proton (high energy)}}{^{1}_{1}p}$$

Another example is the synthesis of alpha particles from lithium-7, in which protons are used as bombarding particles. The resulting compound nucleus, beryllium-8, splits into two alpha particles.

$$\underset{\text{proton}}{^{1}_{1}p} \;+\; \underset{\text{lithium}}{^{7}_{3}\text{Li}} \;\longrightarrow\; \underset{\text{beryllium}}{^{8}_{4}\text{Be}^{*}} \;\longrightarrow\; \underset{\text{alpha particles}}{2\,^{4}_{2}\text{He}}$$

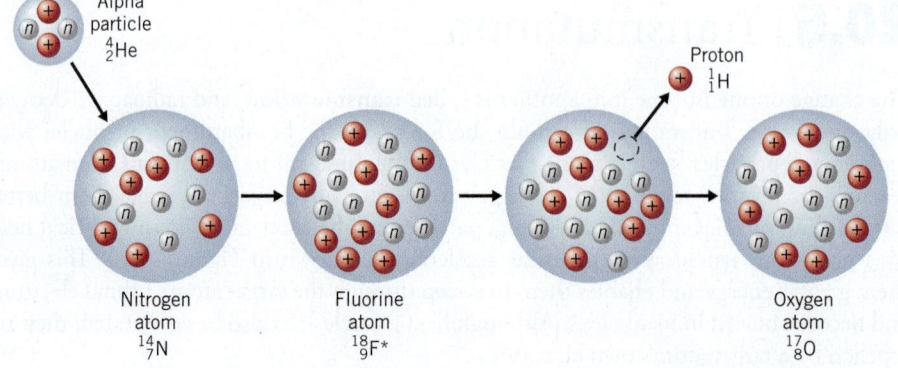

Figure 20.11 | **Transmutation of nitrogen into oxygen.** When the nucleus of nitrogen-14 captures an alpha particle, it becomes a compound nucleus of fluorine-18. This then expels a proton and becomes the nucleus of oxygen-17.

Modes of Decay

A given compound nucleus can be made in a variety of ways. Aluminum-27, for example, forms by any of the following routes:

$$\ce{^4_2He + ^{23}_{11}Na \longrightarrow ^{27}_{13}Al^*}$$

$$\ce{^1_1p + ^{26}_{12}Mg \longrightarrow ^{27}_{13}Al^*}$$

$$\ce{^2_1H + ^{25}_{12}Mg \longrightarrow ^{27}_{13}Al^*}$$

Each compound nucleus $^{27}_{13}\text{Al}^*$ has a different amount of energy (*) depending on the path used to create it. Depending on this energy, different paths of decay are available, and all of the following routes have been observed. They illustrate how the synthesis of such a large number of synthetic isotopes, some stable and others unstable, has been possible.

■ $^2_1$H is a deuteron. It is the nucleus of a deuterium atom, just as a proton is the nucleus of a hydrogen (protium) atom.

$$\begin{array}{c} ^{27}_{13}\text{Al} \quad + \quad ^0_0\gamma \\ \text{aluminum-27} \quad \text{gamma} \\ \text{(now stable)} \quad \text{radiation} \end{array}$$

$$\begin{array}{c} ^{26}_{12}\text{Mg} \quad + \quad ^1_1p \\ \text{magnesium-26} \quad \text{proton} \\ \text{(stable)} \quad \text{radiation} \end{array}$$

The mode of decay is a function of the nuclear energy of aluminum-27.

$^{27}_{13}\text{Al}^*$

$$\begin{array}{c} ^{26}_{13}\text{Al} \quad + \quad ^1_0n \\ \text{aluminum-26} \quad \text{neutron} \\ \text{(unstable;} \quad \text{radiation} \\ t_{1/2} = 7.4 \times 10^5 \text{ yr)} \end{array}$$

$$\begin{array}{c} ^{25}_{12}\text{Mg} \quad + \quad ^1_0n \quad + \quad ^1_1p \\ \text{magnesium-25} \quad \text{neutron} \quad \text{proton} \\ \text{(stable)} \quad \text{radiation} \quad \text{radiation} \end{array}$$

$$\begin{array}{c} ^{23}_{11}\text{Na} \quad + \quad ^4_2\text{He} \\ \text{sodium-23} \quad \text{alpha} \\ \text{(stable)} \quad \text{radiation} \end{array}$$

Synthetic Elements

Over a thousand isotopes have been made by transmutations. Most, nearly 900 in number, do not occur naturally; they appear in the band of stability (Figure 20.8) as open squares. The naturally occurring radioactive isotopes above atomic number 83 all have very long half-lives. Others might have existed, but their half-lives probably were too short to permit them to last into our era. All of the elements beyond neptunium (atomic number 93 and higher, known as the **transuranium elements**, are synthetic. Elements from atomic numbers 93 to 103 complete the actinide series of the periodic table, which starts with element 90 (thorium). Beyond this series, elements 104 through 118 have also been made. In 2013 the IUPAC proposed the name flerovium (Fl) for element 114 and livermorium (Lv) for element 116.

To make the heaviest elements, bombarding particles larger than neutrons are used, such as alpha particles or the nuclei of heavier atoms. For example, element 116— livermorium-293—was made from the compound nucleus formed by the fusion of calcium-48 and curium-248, which then ejected three neutrons. Flerovium-289, and three neutrons, were formed when plutonium-244 was bombarded with calcium-48.

■ Elements 114 and 116 were named after the places of their discovery: the Flerov Laboratory in Dubna, Russia, and the Lawrence Livermore Lab in Livermore, California.

$$\ce{^{48}_{20}Ca + ^{248}_{96}Cm \longrightarrow ^{296}_{116}Lv^* \longrightarrow ^{293}_{116}Lv + 3^1_0n}$$

$$\ce{^{48}_{20}Ca + ^{244}_{94}Pu \longrightarrow ^{292}_{114}Fl \longrightarrow ^{289}_{114}Fl + 3^1_0n}$$

Figure 20.12 | **Scintillation counter.** Energy received from radiations striking the phosphor at the end of this probe is processed by the photomultiplier unit and the counts are measured on the dial.

■ The becquerel is named after Henri Becquerel (1852–1908), the discoverer of radioactivity, who won a Nobel Prize in 1903.

Similarly, some atoms of element 111 (roentgenium, Rg) formed when a neutron was lost from the compound nucleus made by bombarding bismuth-209 with nickel-64. Most of these heavy atoms are extremely unstable, with half-lives measured in fractions of milliseconds. An exception is element 114; two isotopes have been detected with half-lives reported to be in the tens of seconds.

20.6 | Measuring Radioactivity

Atomic radiation is often described as **ionizing radiation** because it creates ions by knocking electrons from molecules in the matter through which the radiation travels. The generation of ions is behind some of the devices for detecting radiation.

The Geiger–Müller tube, one part of a **Geiger counter**, detects beta and gamma radiation having energy high enough to penetrate the tube's window. Inside the tube is a gas under low pressure in which ions form when radiation enters. The ions permit a pulse of electricity to flow, which activates a current amplifier, a pulse counter, and an audible click.

A **scintillation probe** (Figure 20.12) measures tiny flashes of light that are produced when a screen that contains a phosphor is struck by a particle of ionizing radiation. In biological work, experimental mixtures are added to a liquid phosphor, which produces a flash of light for each disintegration event. These flashes must be amplified by a photomultiplier to be automatically counted.

The darkening of a photographic film (a polymer impregnated with light-sensitive chemicals) exposed to radiation over a period of time is proportional to the total quantity of radiation received. **Film dosimeters** work on this principle (Figure 20.13), and people who work near radiation sources wear them. The dosimeter measures the total exposure to radiation, and if a predetermined limit is exceeded, as evidenced by darkening of the film, the worker must be reassigned to a workplace with lower radiation levels.

Units of Radiation

The **activity** of a radioactive material is the number of disintegrations per second. The SI unit of activity is the **becquerel (Bq)**, and it equals one disintegration per second. For example, a liter of air has an activity of about 0.04 Bq, due to the carbon-14 in its carbon dioxide. A gram of natural uranium has an activity of about 2.6×10^4 Bq.

The **curie (Ci)**, named after Marie Curie, who discovered radium, is an older unit, being equal to the activity of a 1.0 g sample of radium-226:

$$1 \text{ Ci} = 3.7 \times 10^{10} \text{ disintegrations s}^{-1} = 3.7 \times 10^{10} \text{ Bq} \qquad \textbf{(20.3)}$$

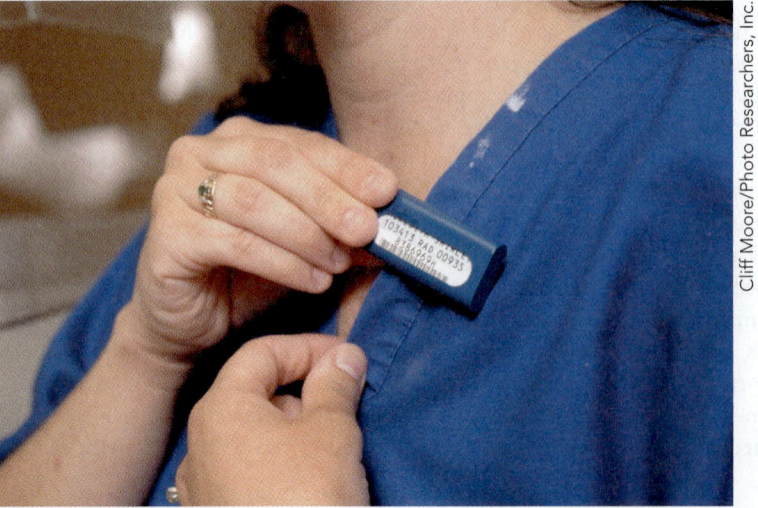

Figure 20.13 | **Badge dosimeter.**

For a sufficiently large sample of a radioactive material, the activity is experimentally found to be proportional to the number of radioactive nuclei, N:

$$activity = kN$$

The constant of proportionality, k, is called the **decay constant**. The decay constant is characteristic of the particular radioactive nuclide, and it gives the activity per nuclide in the sample. Since activity is the number of disintegrations per second, and therefore the change in the number of nuclei per second, we can write

$$activity = -\frac{\Delta N}{\Delta t} = kN \qquad (20.4)$$

which is called the **law of radioactive decay**.[1] The law shows that radioactive decay is a first-order kinetic process, and the decay constant is really just a first-order rate constant in terms of number of nuclei, rather than concentrations.

Recall from Chapter 13 that the half-life of a first-order reaction is given by Equation 13.7, which we repeat here.

$$t_{1/2} = \frac{\ln 2}{k} = \frac{0.693}{k}$$

If we know the half-life of a radioisotope, we can use this relationship to compute its decay constant and also the activity of a known mass of the radioisotope, as Example 20.2 demonstrates.

■ In this discussion, a *disintegration* refers to a single nuclear change that results in the emission of a burst of radiation, which could consist of an alpha particle, beta particle, or gamma photon.

TOOLS
Law of radioactive decay

TOOLS
Half-life of a radionuclide

Example 20.2
Using the Law of Radioactive Decay

Deep-space probes such as NASA's *Cassini* spacecraft need to keep their instruments warm enough to operate effectively. Since solar power is not available in the darkness of deep space, these craft generate heat from the radioactive decay of small pellets of plutonium dioxide, pictured in the margin. Each pellet is about the size of a pencil eraser and weighs about 2.7 g. If the pellets are pure PuO_2, with the plutonium being ^{238}Pu, what is the activity of one of the fuel pellets, in becquerels?

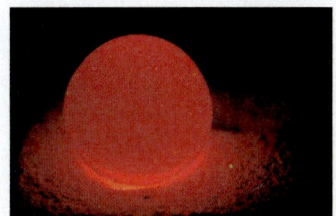
U.S. Department of Energy/Photo Researchers

Analysis: We are looking for the activity of a fuel pellet, and to do this, we will need to know how much material there is and the rate of decay. We can find the number of ^{238}Pu atoms in the sample and then we will need to find the rate constant, k, from the half-life of ^{238}Pu, since it is a first-order process.

Assembling the Tools: The tool for finding the activity is Equation 20.4. But to use the equation, we'll need the decay constant, k, and the number of plutonium-238 atoms, N, in the fuel pellet.

From Table 20.2, the half-life of ^{238}Pu is 87.8 years. We can rearrange Equation 13.7 to compute the decay constant (in units of s^{-1}), from the half-life.

We know that the pellet contains 2.7 g of PuO_2. To calculate N, we'll have to use our stoichiometry tools from Chapter 2 to calculate the number of atoms of Pu. The conversion sequence is:

$$2.7 \text{ g } PuO_2 \longrightarrow mol\ PuO_2 \longrightarrow mol\ Pu \longrightarrow atoms\ Pu$$

Solution: First, let's compute the decay constant from the half-life. We must convert the half-life into seconds:

$$87.8 \text{ years} \times \frac{365 \text{ days}}{1 \text{ year}} \times \frac{24 \text{ hours}}{1 \text{ day}} \times \frac{60 \text{ min}}{1 \text{ hour}} \times \frac{60 \text{ s}}{1 \text{ min}} = 2.77 \times 10^9 \text{ s}$$

---

[1] The minus sign is introduced to make the activity a positive number, since the change in the number of radioactive nuclei, ΔN, is negative.

Now, we can solve Equation 13.7 for the decay constant:

$$k = \frac{\ln 2}{t_{1/2}} = \frac{0.693}{2.77 \times 10^9 \text{ s}}$$

$$= 2.50 \times 10^{-10} \text{ s}^{-1}$$

Next, we need the number of ^{238}Pu atoms in the fuel pellet:

$$2.7 \text{ g PuO}_2 \times \frac{1 \text{ mol PuO}_2}{270 \text{ g PuO}_2} \times \frac{1 \text{ mol Pu}}{1 \text{ mol PuO}_2} \times \frac{6.02 \times 10^{23} \text{ atoms Pu}}{1 \text{ mol Pu}} = 6.0 \times 10^{21} \text{ atoms Pu}$$

From the law of radioactive decay, Equation 20.4, the activity of the fuel pellet is

$$\text{activity} = kN$$

$$= (2.50 \times 10^{-10} \text{ s}^{-1}) \times (6.0 \times 10^{21} \text{ atoms Pu})$$

$$= 1.5 \times 10^{12} \text{ atoms Pu/s}$$

Since the becquerel is defined as the number of disintegrations per second, and each plutonium atom corresponds to one disintegration, we can report the activity as 1.5×10^{12} Bq.

Is the Answer Reasonable? There is no easy way to check the size of the decay constant beyond checking the arithmetic. The number of Pu atoms in the pellet makes sense, because if we have 2.7 g PuO_2 and the formula mass is 270 g mol^{-1}, we have 1/100th of a mole of PuO_2 and so 1/100 a mole of Pu. We should have (1/100) of 6.02×10^{23} atoms, or 6.02×10^{21} atoms of Pu.

The fuel pellet has about 40 times the activity of a gram of radium-226, as indicated by Equation 20.3, so the activity of ^{238}Pu per gram is about 15 times the activity of ^{226}Ra per gram.

Practice Exercise 20.12

A 2.00 g sample of a mixture of plutonium-238 with a nonradioactive metal has an activity of 6.22×10^{11} Bq. What is the percentage by mass of plutonium in the sample? (*Hint:* How many atoms of Pu are in the sample?)

Practice Exercise 20.13

The EPA limit for radon-222 is 4 pCi (picocuries) per liter of air. How many atoms of ^{222}Rn per liter of air will produce this activity?

Nuclear radiation can have varying effects depending on the energy of the radiation and its ability to be absorbed. The **gray (Gy)** is the SI unit of absorbed dose, and 1 gray corresponds to 1 joule of energy absorbed per kilogram of absorbing material. The **rad** is an older unit of absorbed dose, where 1 rad is the absorption of 10^{-2} joule per kilogram of tissue. Thus, 1 Gy equals 100 rad. In terms of danger, if every individual in a large population received 4.5 Gy (450 rad), roughly half of the population would die in 60 days.

The gray is not a good basis for comparing the biological effects of radiation in tissue, because these effects depend not just on the energy absorbed but also on the kind of radiation and the tissue itself. The **sievert (Sv)**, the SI unit of dose equivalent, was invented to meet this problem. The **rem** is an older unit of dose equivalent, one still used in medicine. Its value is generally taken to equal 10^{-2} Sv. The U.S. government has set guidelines of 0.3 rem per week as the maximum exposure workers may receive. (For comparison, a chest X ray typically involves about 0.004 rem or 4 mrem.)

■ The gray is named after Harold Gray, a British radiologist. Rad stands for radiation absorbed dose.

■ *Rem* stands for roentgen equivalent for man, where the roentgen is a unit related to X ray and gamma radiation.

Radiation and Living Tissue

A whole-body dose of 25 rem (0.25 Sv) induces noticeable changes in human blood. A set of symptoms called radiation sickness develops at about 100 rem, becoming severe at 200 rem. Among the symptoms are nausea, vomiting, a drop in the white cell count, diarrhea, dehydration, prostration, hemorrhaging, and loss of hair. If each person in a large population absorbed 400 rem, half would die in 60 days. A 600 rem dose would kill everyone in the group in a week. Many workers received at least 400 rem in the moments following the steam explosion that tore apart one of the nuclear reactors at the Ukraine energy park near Chernobyl in 1986. Although not measured, the workers who painted radium watch dials, similar to our chapter-opening photo, also received an excess of radiation exposure.

Free Radical Formation

Even small absorbed doses of radiation can be biologically harmful. The danger does not lie in the heat energy associated with the dose, which is usually very small. Rather, the harm is in the ability of ionizing radiation to create unstable ions or neutral species with odd (unpaired) electrons, species that can set off other reactions. Water, for example, can interact as follows with ionizing radiation:

$$\text{H} - \ddot{\text{O}} - \text{H} \xrightarrow{\text{radiation}} \left[\text{H} - \dot{\text{O}} - \text{H} \right]^+ + {}_{-1}^{0}e$$

The new cation, $\left[\text{H} - \dot{\text{O}} - \text{H} \right]^+$, is unstable and one breakup path is

$$\left[\text{H} - \dot{\text{O}} - \text{H} \right]^+ \longrightarrow \underset{\text{proton}}{\text{H}^+} + \underset{\substack{\text{hydroxyl} \\ \text{radical}}}{:\dot{\text{O}} - \text{H}}$$

The proton might pick up a stray electron to become a hydrogen atom, H·.

Both the hydrogen atom and the hydroxyl radical are examples of **free radicals** (often simply called *radicals*), which are neutral or charged particles having one or more unpaired electrons. Chemically, they are very reactive. What they do once formed depends on the other chemical species nearby, but radicals can set off a series of totally undesirable chemical reactions inside a living cell. This is what makes the injury from absorbed radiation of far greater magnitude than what the energy alone could inflict. A dose of 600 rem is a lethal dose in a human, but the same dose absorbed by pure water causes the ionization of only one water molecule in every 36 million.

Background Radiation

The presence of radionuclides in nature makes it impossible for us to be free from all exposure to ionizing radiation. Cosmic rays composed of high-energy photons shower on us from the sun and interstellar space. They interact with the air's nitrogen molecules to produce carbon-14, a beta emitter, which enters the food chain via photosynthesis, which converts CO_2 to sugars and starch. From the earth's crust comes the radiation from naturally present radionuclides. The top 40 cm of earth's soil holds an average of 1 gram of radium, an alpha emitter, per square kilometer. Naturally occurring potassium-40, a beta emitter, adds its radiation wherever potassium ions are found in the body. The presence of carbon-14 and potassium-40 together produce about 5×10^5 nuclear disintegrations per minute inside an adult human. Radon gas, which is part of the ^{238}U decay series shown in Figure 20.7, seeps into basements from underground rock formations. The World Nuclear Organization estimates that a little over half of the radiation we experience, on average, is from radon-222 and its decay products as summarized in Figure 20.14.

Diagnostic X rays, both medical and dental, also expose us to ionizing radiation. All of these sources produce a combined **background radiation** that averages 360 mrem per person annually in the United States. The averages are roughly 82% from natural radiation and 18% from medical sources.

Radiation Protection

Gamma radiation and X rays are so powerful that they are effectively shielded only by very dense materials, like lead. Otherwise, one should stay as far from a source as possible, because the intensity of radiation diminishes with the square of the distance. This relationship is the **inverse square law**, which can be written mathematically as follows, where d is the distance from the source:

$$\text{radiation intensity} \propto \frac{1}{d^2}$$

Figure 20.14 | **Sources of radiation.** The estimated percentages of the 610 millirems of radiation that people living in the United States encounter. Thoron is the name for a radioactive isotope of thorium.

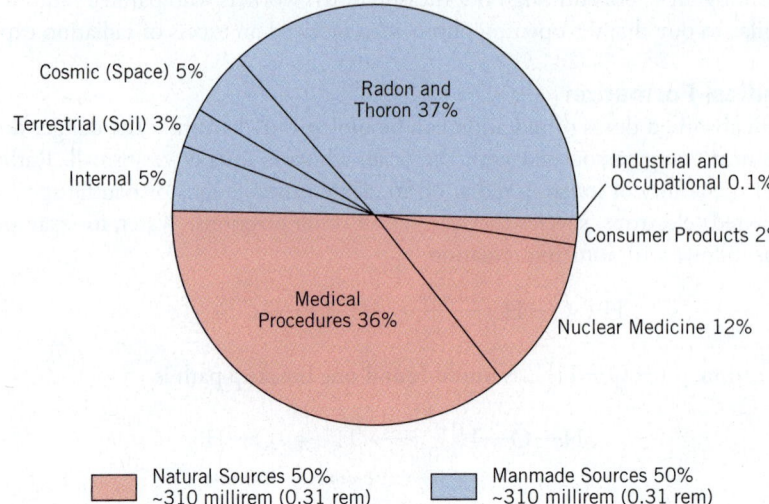

Source of Radiation Exposure in the United States

Cosmic (Space) 5%
Terrestrial (Soil) 3%
Internal 5%
Radon and Thoron 37%
Industrial and Occupational 0.1%
Consumer Products 2%
Medical Procedures 36%
Nuclear Medicine 12%

Natural Sources 50%
~310 millirem (0.31 rem)

Manmade Sources 50%
~310 millirem (0.31 rem)

Based on data from the National Council on Radiation Protection and Measurements, report number 160 (2009). www.ncrppublications.org.

When the intensity, I_1, is known at distance d_1, then the intensity I_2 at distance d_2 can be calculated by Equation 20.5:

TOOLS
Inverse square law

$$\frac{I_1}{I_2} = \frac{d_2^{\,2}}{d_1^{\,2}} \tag{20.5}$$

This law applies only to a small source that radiates equally in all directions, with no intervening shields.

Practice Exercise 20.14

If the intensity of radiation from a radioactive sources is 4.8 units at a distance of 5.0 m, how far from the source would you have to move to reduce the intensity to 0.30 unit? (*Hint:* How does intensity vary with distance?)

Practice Exercise 20.15

If an operator 10 m from a small source is exposed to 1.4 units of radiation, what will be the intensity of the radiation if he moves to 1.2 m from the source?

20.7 | Medical and Analytical Applications of Radionuclides

Because chemical properties depend on the number and arrangement of orbital electrons and not on the specific makeup of nuclei, both radioactive and stable isotopes of an element behave the same chemically. This fact forms the basis for some uses of radionuclides. The chemical and physical properties enable scientists to get radionuclides into place in systems of interest. Then the radiation is exploited for medical or analytical purposes. Tracer analysis, neutron activation analysis, and radiological dating are three examples we will discuss.

Tracer Analysis

In **tracer analysis**, the chemical form and properties of a radionuclide enable the system to distribute it to a particular location. The intensity of the radiation then tells something about how that site is working. For example, in the form of the iodide ion, iodine-131 is carried by the body to the thyroid gland, the only organ in the body that uses iodide ion.

The thyroid takes up the iodide ion to synthesize the hormone thyroxine. The activity of the thyroid can be monitored since an underactive thyroid gland is unable to concentrate iodide ion normally and will emit less intense radiation than a normal thyroid.

Isotopes can be used to correct gravimetric analysis to obtain more accurate results. For example, calcium sulfate is slightly soluble and gravimetric analysis using $CaSO_4$ as the precipitate has a large error due to the calcium that remains in solution. A known amount of radioactive ^{45}Ca, a beta emitter, can be added to the solution and will precipitate along with the unknown calcium. Then, the activity of the calcium-45 in the precipitate can be compared to the total activity that was added. Lets suppose that 0.500 g of $CaSO_4$ is collected. If the activity of the radioactive calcium is only 75% of the amount initially added, we can say that only 75% of *all* the calcium ions in the unknown was precipitated. Now, since we know that 0.500 g $CaSO_4$ is 75% of the true value, dividing 0.500 g/0.75 = 0.667 g $CaSO_4$ gives us the correct answer.

Tracer analyses are also used to pinpoint the locations of brain tumors, which are uniquely able to concentrate the pertechnetate ion, TcO_4^-, made from technetium-99*m*.[2] This strong gamma emitter, which resembles the chloride ion in some respects, is one of the most widely used radionuclides in medicine.

■ Technetium-99*m* is also used in bone scans to detect and locate bone cancer. Active cancer sites concentrate the Tc, which can be detected using external scanners.

Neutron Activation Analysis

A number of stable nuclei can be changed into emitters of gamma radiation by capturing neutrons, and this makes possible a procedure called **neutron activation analysis**. Neutron capture followed by gamma emission can be represented by the following equation, where A is a mass number and X stands for the atomic symbol of an element that captures neutrons:

$$^{A}X \;+\; {}_{0}^{1}n \;\longrightarrow\; ^{A+1}X* \;\longrightarrow\; ^{A+1}X \;+\; {}_{0}^{0}\gamma$$

isotope of element X being analyzed	neutron		compound nucleus (unstable)	more stable form of a new isotope of X	gamma radiation

A neutron-enriched compound nucleus emits gamma radiation at its own set of unique frequencies (or energies), and these sets of frequencies are now known for each isotope. (Not all isotopes, however, become gamma emitters by neutron capture.) The element can be identified by measuring the specific frequencies (energies) of gamma radiation emitted. The concentration of the element can be determined by measuring the intensity of the gamma radiation.

Neutron activation analysis is so sensitive that concentrations as low as 10^{-9}% of some elements can be determined. Many elements can be determined at the same time, so it is a multi-element method of analysis and is also nondestructive in that the sample is not consumed. This technique has found use in authenticating art and archaeological objects based on the chemical composition of the pigments or trace elements present. Trace amounts of heavy metals have been found in the hair of persons poisoned by lead, mercury, or arsenic. Since hair grows at a relatively constant rate, the time of poisoning can also be estimated by measuring how far the segment of hair containing the heavy metal is from the scalp.

Radiological Dating

The determination of the age of a geological deposit or an archaeological find by the use of the radionuclides that are naturally present is called **radiological dating**. It is based partly on the premise that the half-lives of radionuclides have been constant throughout the entire geological period. This premise is supported by the finding that half-lives are insensitive to all environmental forces such as heat, pressure, magnetism, or electrical stresses. We have already discussed radiological dating of archaeological objects by carbon-14 analyses in Section 13.4 where our focus was on kinetics.

---

[2] The *m* in technetium-99*m* means that the isotope is in a metastable form. Its nucleus is at a higher energy level than the nucleus in technetium-99, to which technetium-99*m* decays.

Dating on the Geological Time Scale

In geological dating, a pair of isotopes is sought that are related as a "parent" to a "daughter" in a radioactive disintegration series, like the uranium-238 series (Figure 20.7). Uranium-238 (as "parent") and lead-206 (as "daughter") thus have been used as a radiological dating pair of isotopes. The atoms of ^{206}Pb plus the atoms of ^{238}U in a given size sample represent the initial amount of ^{238}U present, N_0, when the rock solidified. The atoms of ^{238}U present today represent the amount of this isotope that remain or N_t. The half-life of uranium-238 is very long, a necessary criterion for geological dating. Put simply, after the number of atoms of uranium-238 and lead-206 are determined in a rock specimen, the integrated first-order rate equation is used to calculate the age of the rock. This works since many magmas are depleted of volatile lead so that newly solidified rock has essentially no lead to start with.

Probably the most widely used isotopes for dating rock are the potassium-40/argon-40 pair. Potassium-40 is a naturally occurring radionuclide with a half-life nearly as long as that of uranium-238. One of its modes of decay is electron capture in which argon-40 is formed:

$$^{40}_{19}\text{K} + ^{0}_{-1}e \longrightarrow ^{40}_{18}\text{Ar}$$

The argon produced by this reaction remains trapped within the crystal lattices of the rock and is freed only when the rock sample is melted. How much $^{40}_{18}$Ar has accumulated and the amount of $^{40}_{19}$K remaining is measured with a mass spectrometer (Section 0.5). The sum of argon-40 and potassium-40 represents the total $^{40}_{19}$K present when the rock solidified; the amount of $^{40}_{19}$K present today represents the amount left after time $= t$. The observed ratio of argon-40 to potassium-40, together with the half-life of the parent, permits the estimation of the age of the specimen.

- For ^{238}U, $t_{1/2} = 4.51 \times 10^9$ yr
  For ^{40}K, $t_{1/2} = 1.3 \times 10^9$ yr

Because the half-lives of uranium-238 and potassium-40 are so long, samples have to be at least 300,000 years old for either of the two parent–daughter isotope pairs to provide reliable results.

Carbon-14 for Dating Once-Living Material

As discussed in detail in Chapter 13, measurements of the ^{14}C to ^{12}C ratio in an ancient organic sample, such as an object made of wood or bone, permits calculation of the sample's age.

There are two approaches to carbon-14 dating; both require burning some of the sample to produce CO_2. The older method, introduced by its discoverer, Willard F. Libby (Nobel Prize in chemistry, 1960), measures the radioactivity of a sample, per gram of carbon, taken from the specimen, which represents N_t. The radioactivity of a modern sample, per gram of carbon, is used as N_0 in the integrated first-order rate equation.

- For carbon dating to be accurate, extraordinary precautions must be taken to ensure that specimens are not contaminated by more recent sources of carbon or carbon compounds.

The newer method of carbon-14 dating uses a mass spectrometer that is able to separate the atoms of carbon-14 from the other isotopes of carbon (as well as from nitrogen-14) and count all of them, not just the carbon-14 atoms that decay. This approach permits the burning of smaller samples (0.5–5 mg vs. 1–20 g for the Libby method) which is important for valuable artifacts. Objects of up to 70,000 years old can be dated, but the highest accuracy involves systems no older than 60,000 years.

20.8 | Nuclear Fission and Fusion

Nuclear fission is a process whereby a heavy atomic nucleus splits into two lighter fragments. Nuclear fusion, on the other hand, is a process whereby very light nuclei join to form a heavier nucleus. Both processes release large amounts of energy, as we will discuss shortly.

Nuclear Fission Reactions

Because of their electrical neutrality, neutrons penetrate an atom's electron cloud relatively easily and so are able to enter the nucleus. Enrico Fermi discovered in the early 1930s that even slow-moving, thermal neutrons can be captured. (Thermal neutrons are those whose average kinetic energy is the same as any other group of particles at room temperature.) When he directed them at a uranium target, he discovered that several different species of nuclei, all much lighter than uranium, were produced.

Without realizing it, what Fermi had observed was the nuclear fission of one particular isotope, uranium-235, present in small concentrations in naturally occurring uranium. The general reaction can be represented as follows.

$$^{235}_{92}U + ^{1}_{0}n \longrightarrow X + Y + b\,^{1}_{0}n$$

X and Y can be a large variety of nuclei with intermediate atomic numbers. Over 30 have been identified. The coefficient b has an average value of 2.47, the average number of neutrons produced by fission events. A typical specific fission is

$$^{235}_{92}U + ^{1}_{0}n \longrightarrow ^{236}_{92}U^{*} \longrightarrow ^{94}_{36}Kr + ^{135}_{56}Ba + 3\,^{1}_{0}n$$

What actually undergoes fission is the compound nucleus of uranium-236. It has 144 neutrons and 92 protons, giving it a neutron-to-proton ratio of roughly 1.6. Initially, the emerging krypton and barium isotopes have the same ratio, and this is much too high for them. The neutron-to-proton ratio for stable isotopes with 36 to 56 protons is nearer 1.2 to 1.3 (Figure 20.8). Therefore, the initially formed, neutron-rich krypton and barium nuclides promptly eject neutrons, called secondary neutrons that generally have much higher energies than thermal neutrons.

An isotope that can undergo fission after neutron capture is called a **fissile isotope**. The naturally occurring fissile isotope of uranium used in reactors is uranium-235, whose abundance among the uranium isotopes today is only 0.72%. Two other fissile isotopes, uranium-233 and plutonium-239, can be made in nuclear reactors.

Nuclear Chain Reactions

The secondary neutrons released by fission become thermal neutrons as they are slowed by collisions with surrounding materials. They can now be captured by unchanged uranium-235 atoms. Because each fission event produces, on the average, more than two new neutrons, the potential exists for a **nuclear chain reaction** (Figure 20.15). A chain reaction is a self-sustaining process whereby products from one event cause one or more repetitions of the process.

If the sample of uranium-235 is small enough, the loss of neutrons to the surroundings is sufficiently rapid to prevent a chain reaction. However, at a certain **critical mass** of uranium-235 (i.e., about 50 kilograms), this loss of neutrons is insufficient to prevent a sustained reaction. A virtually instantaneous fission of the sample ensues—in other words, an atomic bomb explosion. To trigger an atomic bomb, therefore, two subcritical masses of uranium-235 (or plutonium-239) are forced together to form a critical mass.

Nuclear Power

The binding energy per nucleon (Figure 20.1) in uranium-235 (about 7.6 MeV) is less than the binding energies of the new nuclides (about 8.5 MeV). The net change for a single fission event can be calculated as follows (where we intend only two significant figures in each result):

■ 1 MeV $= 1.602 \times 10^{-13}$ J

Binding energy in krypton-94:

$$(8.5 \text{ MeV/nucleon}) \times 94 \text{ nucleons} = 800 \text{ MeV}$$

Binding energy in barium-139:

$$(8.5 \text{ MeV/nucleon}) \times 139 \text{ nucleons} = \underline{1200 \text{ MeV}}$$

$$\text{Total binding energy of products: } 2000 \text{ MeV}$$

Binding energy in uranium-235:

$$(7.6 \text{ MeV/nucleon}) \times 235 \text{ nucleons} = 1800 \text{ MeV}$$

Figure 20.15 | **Nuclear chain reaction.** Whenever the concentration of a fissile isotope is high enough (i.e., at or above the critical mass), the neutrons released by one fission can be captured by enough unchanged nuclei to cause more than one additional fission event. In civilian nuclear reactors, the fissile isotope is too dilute for this to get out of control. In addition, control rods of nonfissile materials that are able to capture excess neutrons can be inserted or withdrawn from the reactor core to make sure that the heat generated can be removed as fast as it forms.

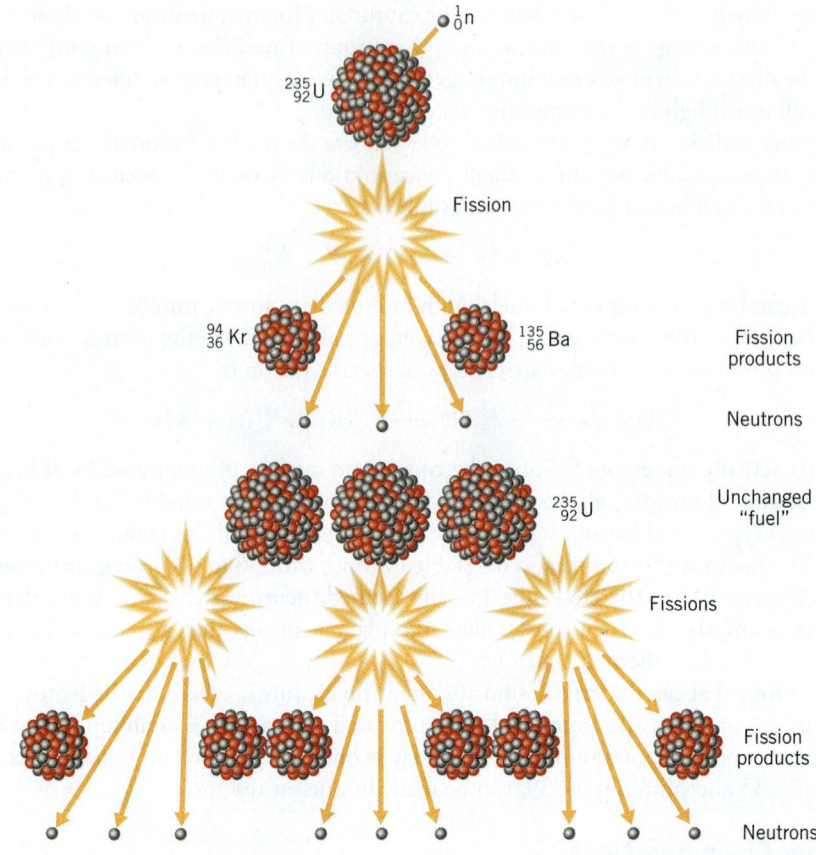

^1_0n

$^{235}_{92}U$

Fission

$^{94}_{36}Kr$ $^{135}_{56}Ba$ Fission products

Neutrons

$^{235}_{92}U$ Unchanged "fuel"

Fissions

Fission products

Neutrons

■ The energy available from 1 kg of uranium-235 is equivalent to the energy of combustion of 3000 tons of soft coal or 13,200 barrels of oil.

The difference in total binding energy is (2000 MeV − 1800 MeV), or 200 MeV (3.2×10^{-11} J), which has to be taken as just a rough calculation. This is the energy released by each fission event going by the equation given. The energy produced by the fission of 1 kg (4.25 mole) of uranium-235 is calculated to be roughly 8×10^{13} J, enough to keep a 100 watt light bulb in energy for 3000 years.

Nuclear Power Plants

Virtually all civilian nuclear power plants throughout the world operate on the same general principles. The energy of fission is used, as heat, either directly or indirectly to increase the pressure of some gas that then drives an electrical generator.

The heart of a nuclear power plant is the reactor, where fission takes place in the fuel core. The nuclear fuel is generally uranium oxide, enriched to 2–4% in uranium-235 and formed into glasslike pellets. These are housed in long, sealed metal tubes called cladding. Groups of tubes are assembled with spacers that permit a coolant to circulate around the tubes. A reactor has several such assemblies in its fuel core. The coolant carries away the heat of fission.

There is no danger of a nuclear power plant undergoing an atomic bomb–type explosion. An atomic bomb requires uranium-235 at a concentration of 85% or greater, or plutonium-239 at a concentration of at least 93%. The concentration of fissile isotopes in a reactor is in the range of 2 to 4%, and much of the remainder is the common, nonfissile uranium-238. However, if the coolant fails to carry away the heat of fission, the reactor core can melt, and the molten mass might even go through the thick-walled containment vessel in which the reactor is kept. Alternatively, the high heat of the fission might split molecules of coolant water into hydrogen and oxygen, which, on recombining, would produce an immense explosion. This type of reactor failure occurred in Japan on March 11, 2011, after a massive earthquake and tsunami disabled the reactor cooling systems.

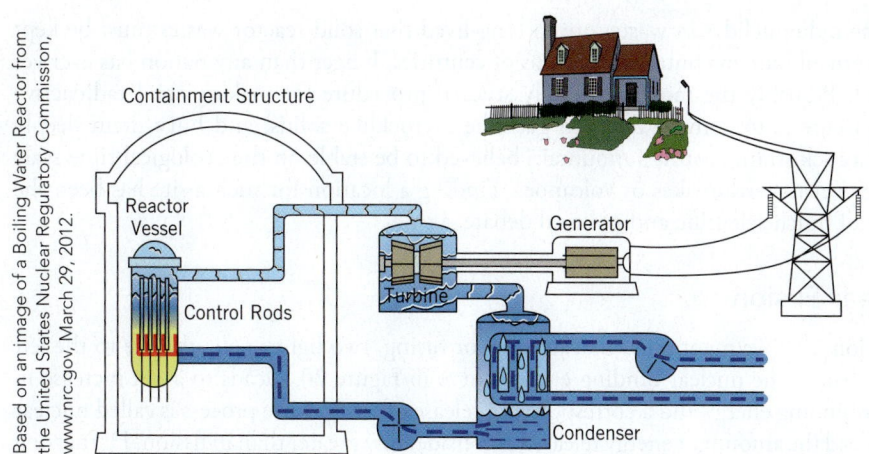

Based on an image of a Boiling Water Reactor from the United States Nuclear Regulatory Commission, www.nrc.gov, March 29, 2012.

Figure 20.16 | Pressurized water reactor, the type used in most of the nuclear power plants in the United States. Water in the primary coolant loop is pumped around and through the fuel elements in the core, and it carries away the heat of the nuclear chain reactions. The hot water delivers its heat to the cooler water in the secondary coolant loop, where steam is generated to drive the turbines.

To convert secondary neutrons to thermal neutrons, the fuel core has a moderator, which is the coolant water itself in nearly all civilian reactors. Collisions between secondary neutrons and moderator molecules heats up the moderator. This heat energy eventually generates steam that enables an electric turbine to run. Ordinary water is a good moderator, but so are heavy water (D_2O) and graphite.

Two main types of reactors dominate civilian nuclear power: the boiling water reactor and the pressurized water reactor. Both use ordinary water as the moderator and so are sometimes called light water reactors. Roughly two-thirds of the reactors in the United States are the pressurized-water type (Figure 20.16). Such a reactor has two loops, and water circulates in both. The primary loop moves water through the reactor core, where it picks up thermal energy from fission. The water is kept in the liquid state by being under high pressure (hence the name pressurized-water reactor).

The hot water in the primary loop transfers thermal energy to the secondary loop at the steam generator (Figure 20.16). This makes steam at high temperature and pressure, which is piped to the turbine. As the steam drives the turbine, the steam pressure drops. The condenser at the end of the turbine, cooled by water circulating from a river or lake or from huge cooling towers, forces a maximum pressure drop within the turbine by condensing the steam to liquid water. The returned water is then recycled to high-pressure steam. (In the boiling water reactor, there is only one coolant loop. The water heated in the reactor itself is changed to the steam that drives the turbine.)

■ D_2O is deuterium oxide. Deuterium is an isotope of hydrogen, $_1^2H$.

Radioactive Waste

Radioactive wastes from nuclear power plants occur as gases, liquids, and solids. The gases are mostly radionuclides of krypton and xenon, but with the exception of krypton-85 ($t_{1/2} = 10.4$ years), the gases have short half-lives and decay quickly. During decay, they must be contained, and this is one function of the cladding. Other dangerous radionuclides produced by fission include iodine-131, strontium-90, and cesium-137.

Iodine-131 must be contained because the human thyroid gland concentrates iodide ion to make the hormone thyroxine. Once the isotope is in the gland, beta radiation from iodine-131 could cause harm, possibly thyroid cancer or impaired thyroid function. An effective countermeasure to iodine-131 poisoning is to take doses of ordinary iodine (as sodium iodide). Statistically, this increases the likelihood that the thyroid will take up stable iodide ion rather than the unstable anion of iodine-131.

Cesium-137 and strontium-90 also pose problems to humans. Cesium is in Group 1A together with sodium, so radiating cesium-137 cations travel to some of the same places in the body where sodium ions go. Strontium is in Group 2A with calcium, so strontium-90 cations can replace calcium ions in bone tissue, sending radiation into bone marrow and possibly causing leukemia. The half-lives of cesium-137 and strontium-90, however, are relatively short.

■ Cesium-137 has a half-life of 30 years; that of strontium-90 is 28.1 years. Both are beta emitters.

Some radionuclides in wastes are so long-lived that solid reactor wastes must be kept away from all human contact for dozens of centuries, longer than any nation has ever yet endured. Probably the most intensively studied procedure for making solid radioactive wastes secure is to convert them to glasslike or rocklike solids, and bury them deeply within a rock stratum or in a mountain believed to be stable on the geological time scale with respect to earthquakes or volcanoes. Finding a location for such a site has been the subject of much scientific and political debate.

Nuclear Fusion

In Section 20.2 we mentioned that joining, or fusing, two light nuclei that lie to the left of the peak in the nuclear binding energy curve in Figure 20.1 leads to a net increase in nuclear binding energy and a corresponding release of energy. The process is called nuclear fusion, and the amount of energy released is considerably greater than in fission. Harnessing this energy for peaceful purposes is still a long way off, however, because many immensely difficult scientific and engineering problems still remain to be solved.

Deuterium, $_1^2H$, an isotope of hydrogen, is a key fuel in all approaches to fusion. It is naturally present as 0.015% of all hydrogen atoms, including those in water. Despite this low percentage, the earth has so much water that the supply of deuterium is virtually limitless.

The fusion reaction most likely to be used in a successful fusion reactor involves the fusion of deuterium and tritium, 3H, another isotope of hydrogen:

$$\underset{\text{deuterium}}{_1^2H} \quad + \quad \underset{\text{tritium}}{_1^3H} \quad \longrightarrow \quad \underset{\text{helium}}{_2^4He} \quad + \quad \underset{\text{neutron}}{_0^1n} \quad + \quad 17.6 \text{ MeV}$$

This corresponds to an energy yield of 2.82×10^{-12} J for each atom of helium formed, or 1.70×10^9 kJ per mole of He formed. One problem with this reaction is that tritium is radioactive with a relatively short half-life, so it doesn't occur naturally. It can be made in several ways, however, from lithium or even deuterium via other nuclear reactions.

Comparing fission and fusion on a mass basis, the fission of one kilogram of ^{235}U yields approximately 8×10^{13} J, whereas forming one kilogram of ^4He by the preceding fusion reaction yields 4.2×10^{14} J. Therefore, on a mass basis, fusion yields more than five times as much energy as fission. The potential energy yield from fusion is so great that the deuterium in just 0.005 cubic kilometers of ocean would supply the energy needs of the United States for one year.

Thermonuclear Fusion

The central scientific problem with fusion is to get the fusing nuclei close enough for a long enough time that the nuclear strong force (of attraction) can overcome the electrostatic force (of repulsion). As we learned in Section 20.3, the strong force acts over a much shorter range than the electrostatic force. Two nuclei on a collision course, therefore, repel each other virtually until they touch and get into the range of the strong force. The kinetic energies of two approaching nuclei must therefore be very substantial if they are to overcome this electrostatic barrier. Achieving such energies, moreover, must be accomplished by large numbers of nuclei all at once in batch after batch if there is to be any practical generation of electrical power by nuclear fusion. Relatively isolated fusion events achieved in huge accelerators will not do. The only practical way to give batch quantities of nuclei enough energy is to transfer thermal energy to them, and so the overall process is called thermonuclear fusion. Temperatures required to provide such thermal energy are very high—namely, more than 100 million degrees Celsius!

The atoms whose nuclei we want to fuse must first be stripped of their electrons. Thus, a high energy cost is exacted from the start but, overall, the energy yield will more than pay for it. The product is an electrically neutral, gaseous mixture of nuclei and unattached electrons called a **plasma**. The plasma must then be made so dense that like-charged nuclei are within 2 fm (2×10^{-15} m) of each other, meaning a plasma density of roughly 200 g cm^{-3} as compared with 200 mg cm^{-3} under ordinary conditions. To achieve this, the plasma must be confined at a pressure of several billion atmospheres long enough for the separate nuclei to fuse. The temperature needed is several times the temperature at the center of our sun.

Although practical peaceful uses for thermonuclear fusion are still in the distant future, military applications have been around for over 50 years. Thermonuclear fusion is the source of the energy released in the explosion of a hydrogen bomb. The energy needed to trigger the fusion is provided by the explosion of a fission bomb based on either uranium or plutonium.

Fusion in the Sun and Stars

Nature has used thermonuclear fusion since the origin of the universe as the source of energy in stars, where high temperatures (about 15 megakelvins) and huge gravity provide the kinetic energy and high density needed to initiate fusion reactions. The chief process in solar-mass stars is called the proton–proton cycle.

■ The interior of the sun is at a temperature of approximately 15 million kelvins (15 MK).

The Proton–Proton Cycle

$$2({}_1^1H + {}_1^1H) \longrightarrow 2{}_1^2H + 2{}_{1}^{0}e + 2\nu$$

$$2({}_1^1H + {}_1^2H) \longrightarrow 2{}_2^3He + 2\gamma$$

$$\underline{{}_2^3He + {}_2^3He \longrightarrow {}_2^4He + 2{}_1^1H}$$

$$\text{Net: } 4{}_1^1H \longrightarrow {}_2^4He + 2{}_{1}^{0}e + 2\nu + 2\gamma$$

The positrons produced combine with electrons in the plasma, annihilate each other, and generate additional energy and gamma radiation. Virtually all the neutrinos escape the sun and move into the solar system (and beyond), carrying with them a little less than 2% of the energy generated by the cycle. Not counting the energy of the neutrinos, each operation of one cycle generates 26.2 MeV or 4.20×10^{-12} J, which is equivalent to 2.53×10^{12} J mole^{-1} of alpha particles produced. This is the source of the solar energy radiated throughout our system, which can continue in this way for another 5 billion years.

⚛ Analyzing and Solving Multi-Concept Problems

Another element formed in the sun is beryllium-7. It has a mass of 7.01692983 u. Combustion of how much oil, in kg, would release the same amount of energy as the amount of energy released by the formation of 1.00 kg of ^{7}Be? Since oil is made up of many different compounds, assume that this oil is $C_{20}H_{42}$, which has a ΔH_f° of -2330 kJ/mol.

Analysis: We are being asked to compare two reactions and the energy that they produce. The first reaction is a nuclear reaction, and the amount of energy released in this reaction will be used to calculate the amount of material in the second, combustion reaction. The solution will be divided into four parts: (1) for the first reaction, we will determine the binding energy for ^{7}Be, and (2) then we will determine the amount of energy released in that process for 1.00 kg of ^{7}Be. We will then turn our attention to the oil and do the following: (3) find the heat of combustion for $C_{20}H_{42}$ and then (4) the amount of $C_{20}H_{42}$ that gives off the same amount of heat as the ^{7}Be.

PART 1

Assembling the Tools: We will find the nuclear binding energy of ^{7}Be from the mass defect using the rest masses of the protons, and neutrons in the ^{7}Be nucleus and the rest mass of the ^{7}Be nucleus itself. We will have to consult a reference source for the rest mass of ^{7}Be.

■ When searching, use recognized sources such as the Nuclear Regulatory Commission (NRC), the National Institutes of Standards and Technology (NIST), or the International Union of Pure and Applied Chemistry (IUPAC).

Solution: We start with the makeup of ^{7}Be and the masses of protons and neutrons. There are four protons and three neutrons in ^{7}Be, and from Section 20.2 the mass of a proton is 1.0072764668 u and the mass of a neutron is 1.0086649160 u are given.

$$\text{For 4 protons:}\quad 4 \times 1.0072764668 \text{ u} = 4.0291058672 \text{ u}$$

$$\text{For 3 neutrons:}\quad 3 \times 1.0086649160 \text{ u} = \underline{3.0259947480 \text{ u}}$$

$$\text{Total rest mass of nucleons in } ^7\text{Be} = 7.0551006152 \text{ u}$$

The mass defect is calculated from the difference between the mass of the nucleons and the mass of the ^{7}Be nucleus: 7.01692983

$$7.0551006152 \text{ u} - 7.01692983 \text{ u} = 0.03817079 \text{ u}$$

$$\underset{\text{mass of the 7 nucleons}}{} \qquad \underset{\text{mass of nucleus}}{} \qquad \underset{\text{mass defect}}{}$$

PART 2

Assembling the Tools: Now that we have the mass defect, we can apply the Einstein equation to calculate the amount of energy equivalent to the mass lost,

$$E = mc^2$$

in which the m is the mass lost, 0.038171 u (rounded to 5 significant figures), and c is 2.9979×10^8 m s^{-1}. After we calculate the energy lost for one atom, we can then calculate it for 1.00 kg of ^{7}Be by converting from one atom to one mole to mass using Avogadro's number and the molar mass of ^{7}Be, respectively.

Solution: For Part 2, the first step is using the Einstein equation to calculate the amount of energy released upon the formation of one ^{7}Be nucleus:

$$E = mc^2$$

$$E = (0.038171 \text{ u}) \times \frac{1.660654 \times 10^{-27} \text{ kg}}{1 \text{ u}} \times (2.9979 \times 10^8 \text{ m s}^{-1})^2$$

$$= 5.6970 \times 10^{-12} \text{ kg m}^2 \text{ s}^{-2}$$

Recalling that kg m^2 s^{-2} are the SI units for the joule, we can write the answer as 5.6970×10^{-12} J.

The energy we just calculated represents one ^{7}Be nucleus so now we need to calculate the amount of energy released for 1.00 kg of ^{7}Be nuclei:

$$E = 1.00 \text{ kg Be} \times \frac{1000 \text{ g}}{1 \text{ kg}} \times \frac{1 \text{ mol Be}}{7.0169298 \text{ g Be}} \times \frac{6.022 \times 10^{23} \text{ atoms Be}}{1 \text{ mol Be}} \times \frac{5.6970 \times 10^{-12} \text{ J}}{1 \text{ atom Be}}$$

$$= 4.83 \times 10^{14} \text{ J}$$

We can readily convert this to kilojoule units as 4.83×10^{11} kJ. This is the amount of energy released by the formation of 1.00 kg of ^{7}Be.

PART 3

Assembling the Tools: Now, we need to find out how much oil will release this much energy upon combustion. We need to balance the combustion equation as covered in Section 3.5. Then we can determine the heat released by burning a mole of $C_{20}H_{42}$ using Hess's law as we did in Section 6.8.

Solution: We write the skeleton equation since we know that O_2 is a reactant and that gaseous CO_2 and H_2O are the products:

$$C_{20}H_{42}(l) + O_2(g) \longrightarrow CO_2(g) + H_2O(g)$$

Balancing the equation yields

$$2C_{20}H_{42}(l) + 61O_2(g) \longrightarrow 40CO_2(g) + 42H_2O(g)$$

From Table C.2 in Appendix C, we can find the enthalpies of formation for $CO_2(g)$ and $H_2O(g)$, and use it in the equation:

$$\Delta H^\circ = \left[40 \text{ mol } CO_2(g) \times \left(\frac{-395 \text{ kJ}}{\text{mol } CO_2(g)} \right) + 42 \text{ mol } H_2O(g) \times \left(\frac{-241.8 \text{ kJ}}{\text{mol } H_2O(g)} \right) \right]$$

$$- \left[2 \text{ mol } C_{20}H_{42}(l) \times \left(\frac{-2330 \text{ kJ}}{\text{mol } C_{20}H_{42}(l)} \right) \right]$$

$$\Delta H^\circ = -21,240 \text{ kJ}$$

This is the amount of heat given off by the combustion of two moles of $C_{20}H_{42}$.

PART 4

Assembling the Tools: For the last step in this problem, we have to find out how many kilograms of $C_{20}H_{42}$ will give off 4.83×10^{11} kJ of energy. One conversion factor we have is 21,240 kJ of energy given off by the combustion of 2 mol of $C_{20}H_{42}$. The other conversion factor needed is the molar mass of $C_{20}H_{42}$.

Solution: For this step, we will convert 4.89×10^{11} kJ into kg of mol $C_{20}H_{42}$ using the heat of reaction and the molar mass as follows:

$$4.83 \times 10^{11} \text{ kJ} \times \frac{2 \text{ mol } C_{20}H_{42}}{2.124 \times 10^4 \text{ kJ}} \times \frac{282.6 \text{ g } C_{20}H_{42}}{1 \text{ mol } C_{20}H_{42}} = 1.28 \times 10^{10} \text{ g } C_{20}H_{42}$$

Now we convert g to kg to obtain:

$$1.28 \times 10^{10} \text{ g } C_{20}H_{42} \times \frac{1 \text{ kg}}{1000 \text{ g}} = 1.28 \times 10^7 \text{ kg } C_{20}H_{42}$$

Are the Answers Reasonable? This is an incredibly large number. In order to check that our answer is in the correct ballpark, check all exponents and then combine them to see if the exponent is correct.

|Summary

Organized by Learning Objective

Utilize the unified laws of mass and energy conservation

Mass and energy are interconvertible, and the **Einstein equation**, $\Delta E = \Delta mc^2$ (where c is the speed of light), lets us calculate one from the other. The total of the energy in the universe and all of the mass calculated as an equivalent of energy is a constant, which is a statement of the **law of conservation of mass–energy**.

Grasp the importance of the nuclear binding energy

When a nucleus forms from its nucleons, some mass changes into energy. This amount of energy, the **nuclear binding energy**, is required to break up the nucleus again. The higher the binding energy per nucleon, the more stable is the nucleus.

Understand the nature of radionuclides and their emissions

Radioactivity is the change of a nucleus into another nucleus. The electrostatic force by which protons repel each other is overcome in the nucleus by the nuclear strong force. The ratio of neutrons to protons is a factor in nuclear stability. By radioactivity, the naturally occurring **radionuclides** adjust their neutron/proton ratios, lower their energies. The loss of an **alpha particle** leaves a nucleus with four fewer units of mass number and two fewer of atomic number. Loss of a **beta particle** leaves a nucleus with the same mass number and an atomic number one unit higher. **Gamma radiation** lets a nucleus lose some energy without a change in mass or atomic number.

Nuclear equations are balanced when the mass numbers and atomic numbers on either side of the reaction arrow respectively balance. The energies of emission are usually described in **electron volts (eV)** or multiples thereof, where $1 \text{ eV} = 1.602 \times 10^{-19}$ J. A few very long-lived radionuclides in nature, like ^{238}U, are at the heads of **radioactive disintegration series**, which represent the successive decays of "daughter" radionuclides until a stable isotope forms.

Explore the nature of the "band of stability"

Stable nuclides generally fall within a curving band, called the **band of stability**, when all known nuclides are plotted according to their numbers of neutrons and protons. Radionuclides with neutron/proton ratios that are too high eject beta particles to adjust their ratios downward. Those with neutron/proton ratios that are too low generally emit **positrons** to change their ratios upward. Isotopes whose nuclei consist of even numbers of both neutrons and protons are generally much more stable than those with odd numbers of both; this is the **odd–even rule**. Having all neutrons paired and all protons paired is energetically better (i.e., more stable) than having any nucleon unpaired. Isotopes with specific numbers of protons or neutrons, the **magic numbers** of 2, 8, 20, 28, 50, 82, and 126, are generally more stable than others.

Explain modern transmutation methods

When a bombardment particle—a proton, deuteron, alpha particle, or neutron—is captured, the resulting **compound nucleus** contains the energy of both the captured particle and its nucleons. The mode of decay of the compound nucleus is a function of its extra energy, not its extra mass. Many **radionuclides** have been made by these nuclear reactions, including all of the elements from 93 and higher.

Depending on the specific isotope, synthetic radionuclides emit alpha, beta, and gamma radiation. Some emit **positrons** (positive electrons, a form of **antimatter**) that produce gamma radiation by annihilation collisions with electrons. Other synthetic radionuclides decay by **electron capture** and emit **X rays**. Some radionuclides emit **neutrons**.

Demonstrate a knowledge of how radioactivity is measured

The detection and measurement of **ionizing radiation** using **Geiger counters** or **scintillation counters**, takes advantage of the ability of radiation to generate ions in air or other matter. Such radiation can harm living tissue by producing **free radicals**. The **curie (Ci)** and the **becquerel (Bq)**, the SI unit, describe how active a source is; $1 \text{ Ci} = 3.7 \times 10^{10}$ Bq where 1 Bq = 1 disintegration/s. The normal **background radiation** causes millirem exposures per year. Naturally occurring radon, cosmic rays, radionuclides in soil and stone building materials, medical X rays, and releases from nuclear tests or from nuclear power plants all contribute to this background.

The SI unit of absorbed dose, the **gray (Gy)**, is used to describe how much energy is absorbed by a unit mass of absorber; 1 Gy = 1 J/kg. An older unit, the **rad**, is equal to 0.01 Gy. The **sievert**, an SI unit, and the **rem**, an older unit, are used to compare doses absorbed by different tissues and caused by different kinds of radiation. Protection against radiation can be achieved by using dense shields (e.g., lead or thick concrete), by avoiding the overuse of radionuclides or X rays in medicine, and by taking advantage of the **inverse square law**. This law tells us that the intensity of radiation decreases with the square of the distance from its source.

Show an appreciation for practical applications of radionuclides

Tracer analysis uses small amounts of a radionuclide, which can be detected using devices like the scintillation counter, to follow the path of chemical and biological processes. In **neutron activation analysis**, neutron bombardment causes some elements to become γ emitters. The radiation can be detected and measured, giving the identity and concentration of the activated elements. **Radiological dating** uses the known half-lives of naturally occurring radionuclides to date geological and archeological objects. Radionuclides can also tell the chemist about the efficiency of a reaction or the yield of a product.

Develop an understanding of nuclear fission and fusion

Uranium-235, which occurs naturally, and plutonium-239, which can be made from uranium-238, are **fissile isotopes** that serve as the fuel in present-day nuclear reactors. When either isotope captures a thermal neutron, the isotope splits in one of several ways to give two smaller isotopes plus energy and more neutrons. The neutrons can generate additional fission events, enabling a **nuclear chain reaction**. If a critical mass of a fissile isotope is allowed to form, the nuclear chain reaction proceeds out of control, and the material detonates as an atomic bomb explosion. Pressurized water reactors are the most commonly used fission reactors for power generation. One major problem with nuclear energy is the storage of radioactive wastes.

Thermonuclear fusion joins two light nuclei to form a heavier nucleus with the release of more energy than nuclear fission. A typical reaction combines $^2_1$H and $^3_1$H to give $^4_2$He and a neutron. High temperatures and pressures are necessary to initiate the fusion reaction. In stars, gravity is able to contain the high temperature **plasma** and allow fusion to occur. In a hydrogen bomb, the reaction is initiated by a fission bomb. Scientific and engineering hurdles must still be overcome before fusion can be a viable peaceful energy source.

Tools for Problem Solving
The following tools were introduced in this chapter. Study them carefully so that you can select the appropriate tool when needed.

The Einstein equation (Section 20.1)
$$\Delta E = \Delta m_0 c^2 \text{ (or often just } E = mc^2),$$

Balancing nuclear equations (Section 20.3)
When balancing a nuclear equation apply the following two criteria:
1. The sums of the mass numbers on each side of the arrow must be equal.
2. The sums of the atomic numbers on each side must be the same.

The odd–even rule (Section 20.4)
When the numbers of neutrons and protons in a nucleus are both even, the isotope is far more likely to be stable than when both numbers are odd.

Law of radioactive decay (Section 20.6)
$$\text{Activity} = \frac{\Delta N}{\Delta t} = kN$$

Activity has units of disintegrations per second, dps, or Bq. The decay constant, k, is a first-order rate constant with units of s^{-1}, and the number of atoms of the radionuclide in the sample is represented by N.

Half-life of a radionuclide (Section 20.6)
$$t_{1/2} = \frac{\ln 2}{k} = \frac{0.693}{k}$$

The half-life is represented by $t_{1/2}$ and the decay constant by k.

Inverse square law (Section 20.6)
$$\frac{I_1}{I_2} = \frac{d_2^{\,2}}{d_1^{\,2}}$$

I represents the intensity of radiation and d represents the distance from the source of the radiation.

> **WileyPLUS** = *WileyPLUS*, an online teaching and learning solution. *Note to instructors*: Many of the end-of-chapter problems are available for assignment via the WileyPLUS system. **www.wileyplus.com.** Review Problems are presented in pairs separated by blue rules. Answers to problems whose numbers appear in blue are given in Appendix B. More challenging problems are marked with an asterisk ✱.

| Review Questions

Conservation of Mass and Energy

20.1 In calculations involving chemical reactions, we can regard the law of conservation of mass as a law independent of the law of conservation of energy despite Einstein's union of the two. What fact(s) make(s) this possible?

20.2 How can we know that the speed of light is the absolute upper limit on the speed of any object?

20.3 State the following:
    (a) the law of conservation of mass–energy
    (b) the Einstein equation

20.4 What is the difference between the rest mass of a particle and the mass of a particle in motion? Why do we use the rest mass of the particle for most calculations?

Nuclear Binding Energy

20.5 Why isn't the sum of the masses of all nucleons in one nucleus equal to the mass of the actual nucleus?

20.6 Which element has the largest binding energy per nucleon? What happens to the nuclei when the binding energy is too small?

Radioactivity

20.7 When a substance is described as radioactive, what does that mean? Why is the term radioactive decay used to describe the phenomenon?

20.8 Three kinds of radiation make up nearly all of the radiation observed from naturally occurring radionuclides. What are they?

20.9 Give the composition of each of the following: **(a)** alpha particle, **(b)** beta particle, **(c)** positron, **(d)** deuteron.

20.10 Why is the penetrating ability of alpha radiation less than that of beta or gamma radiation?

20.11 With respect to their formation, how do gamma rays and X rays differ?

20.12 How does electron capture generate X rays?

20.13 For the process of beta emission, where does the electron originate? What particle emits the electron and what else is formed?

Band of Stability

20.14 What data are plotted and what criterion is used to identify the actual band in the band of stability?

20.15 Both barium-123 and barium-140 are radioactive, but which is more likely to have the longer half-life? Explain your answer.

20.16 Tin-112 is a stable nuclide but indium-112 is radioactive and has a very short half-life ($t_{1/2} = 14$ min). What does tin-112 have that indium-112 does not to account for this difference in stability?

20.17 Lanthanum-139 is a stable nuclide but lanthanum-140 is unstable ($t_{1/2} = 40$ hr). What rule of thumb concerning nuclear stability is involved?

20.18 As the atomic number increases, the neutron/proton ratio increases. What does this suggest is a factor in nuclear stability?

20.19 Radionuclides of high atomic number are far more likely to be alpha emitters than those of low atomic number. Offer an explanation for this phenomenon.

20.20 Although lead-164 has two magic numbers, 82 protons and 82 neutrons, this isotope is unknown. Lead-208, however, is known and stable. What problem accounts for the nonexistence of lead-164?

20.21 What decay particle is emitted from a nucleus of low to intermediate atomic number but a relatively high neutron/proton ratio? How does the emission of this particle benefit the nucleus?

20.22 What decay particle is emitted from a nucleus of low to intermediate atomic number but a relatively low neutron/proton ratio? How does the emission of this particle benefit the nucleus?

20.23 Does electron capture increase, decrease, or maintain the neutron/proton ratio in a nucleus? Are radionuclides above or below the band of stability more likely to undergo this change?

Transmutations

20.24 Compound nuclei form and then decay almost at once. What accounts for the instability of a compound nucleus?

20.25 What explains the existence of several decay modes for the compound nucleus aluminum-27?

20.26 Rutherford theorized that a compound nucleus forms when helium nuclei hit nitrogen-14 nuclei. If this compound nucleus decayed by the loss of a neutron instead of a proton, what would be the other product?

20.27 Why is it easier to use neutrons to form compound nuclei rather than alpha particles or other nuclei?

Measuring Radioactivity

20.28 What specific property of nuclear radiation is used by the Geiger counter?

20.29 Dangerous doses of radiation can actually involve very small quantities of energy. Explain.

20.30 What units, SI and common, are used to describe each of the following?
(a) the activity of a radioactive sample
(b) the energy of a particle or of a photon of radiation given off by a nucleus
(c) the amount of energy absorbed by a given mass by a dose of radiation
(d) dose equivalents for comparing biological effects

20.31 Explain the necessity in health sciences for the Sievert.

20.32 Which is more dangerous: ingesting a radioactive compound with a short half-life or a radioactive compound with a long half-life if the compound can be eliminated from the body within a day?

Medical and Analytical Applications of Radionuclides

20.33 Why should a radionuclide used in medical diagnostic procedures have a short half-life? If the half-life is too short, what problem arises?

20.34 An alpha emitter is not used in medical diagnostic procedures. Why?

20.35 In general terms, explain how neutron activation analysis is used and how it works.

20.36 What properties of isotopes make them useful for following chemical reactions?

20.37 What is one assumption in the use of the uranium/lead ratio for dating ancient geologic formations?

20.38 If a sample used for carbon-14 dating is contaminated by air, there is a potentially serious problem with the method. What is it?

20.39 List some of the kinds of radiation that make up our background radiation.

20.40 In your own words and diagrams, explain how the precise spot in the body where a positron and electron annihilate each other can be found.

Nuclear Fission and Fusion

20.41 What do each of the following terms mean? **(a)** thermal neutron, **(b)** nuclear fission, **(c)** fissile isotope, **(d)** nuclear fusion, **(e)** critical mass

20.42 Which fissile isotope occurs in nature?

20.43 What fact about the fission of uranium-235 makes it possible for a chain reaction to occur?

20.44 Explain in general terms why fission generates more neutrons than needed to initiate it.

20.45 Why is naturally occurring U-235 subcritical? What needs to be done to make U-235 critical?

20.46 What purpose is served by a moderator in a nuclear reactor?

|Review Problems

Conservation of Mass and Energy

*__20.51__ Calculate the mass in kilograms of a 1.00 kg object when its velocity, relative to us, is **(a)** 3.00×10^7 m s^{-1}, **(b)** 2.90×10^8 m s^{-1}, and **(c)** 2.99×10^8 m s^{-1}. (Notice the progression of these numbers toward the velocity of light, 3.00×10^8 m s^{-1}.)

*__20.52__ Calculate the velocity, relative to us, in m s^{-1} of an object with a rest mass, m_0, of 1.000 kg when its mass, m, is **(a)** 1.005 kg, **(b)** 1.1 kg, and **(c)** 5.0 kg.

20.53 Calculate the mass equivalent in grams of 1.00 kJ of energy.

20.54 Calculate the energy equivalent in kJ of 1.00 g of mass.

20.55 Calculate the amount of mass in nanograms that is changed into energy when one mole of liquid water forms by the combustion of hydrogen, all measurements being made at 1 atm and 25 °C. What percentage is this of the total mass of the reactants?

20.56 Show that the mass equivalent of the energy released by the complete combustion of 1 mol of methane (890 kJ) is 9.89 ng.

Nuclear Binding Energies

20.57 Calculate the binding energy in J/nucleon of the deuterium nucleus whose mass is 2.0141 u.

20.58 Calculate the binding energy in J/nucleon of the tritium nucleus whose mass is 3.0160 u.

20.59 What is the mass defect for copper-63 if the mass of the nucleus is 62.9295975 u? Calculate the binding energy for Cu-63.

20.60 What is the mass defect for americium-241 if the mass of the nucleus is 241.0568291 u? Calculate the binding energy for Am-241.

Radioactivity

20.61 Complete the following nuclear equations by writing the symbols of the missing particles:

(a) $^{211}_{82}\text{Pb} \longrightarrow {}^{0}_{-1}e +$ _____

(b) $^{177}_{73}\text{Ta} \xrightarrow{\text{electron capture}}$ _____

(c) $^{220}_{86}\text{Rn} \longrightarrow {}^{4}_{2}\text{He} +$ _____

(d) $^{19}_{10}\text{Ne} \longrightarrow {}^{0}_{1}e$ _____

20.47 Why is there no possibility of an atomic bomb–type explosion from a nuclear power plant?

20.48 Why are Cs, Sr, and I radioactive waste isotopes of particular concern to regulators and the public?

20.49 Write the nuclear equation for the fusion reaction between deuterium and tritium. Why must tritium be synthesized for this reaction?

20.50 What obstacles make constructing a reactor for controlled nuclear fusion especially difficult?

20.62 Write the symbols of the missing particles to complete the following nuclear equations:

(a) $^{245}_{56}\text{Cm} \longrightarrow {}^{4}_{2}\text{He} +$ _____

(b) $^{146}_{56}\text{Ba} \longrightarrow {}^{0}_{-1}e +$ _____

(c) $^{58}_{29}\text{Cu} \longrightarrow {}^{0}_{1}e +$ _____

(d) $^{68}_{32}\text{Ge} \xrightarrow{\text{electron capture}}$ _____

20.63 Write a balanced nuclear equation for each of the following changes: **(a)** alpha emission from plutonium-242, **(b)** beta emission from magnesium-28, **(c)** positron emission from silicon-26, **(d)** electron capture by argon-37.

20.64 Write the balanced nuclear equation for each of the following nuclear reactions: **(a)** electron capture by iron-55, **(b)** beta emission by potassium-42, **(c)** positron emission by ruthenium-93, **(d)** alpha emission by californium-251.

20.65 Write the symbols, including the atomic and mass numbers, for the radionuclides that would give each of the following products: **(a)** fermium-257 by alpha emission, **(b)** bismuth-211 by beta emission, **(c)** neodymium-141 by positron emission, **(d)** tantalum-179 by electron capture.

20.66 Each of the following nuclides forms by the decay mode described. Write the symbols of the parents, giving both atomic and mass numbers: **(a)** rubidium-80 formed by electron capture, **(b)** antimony-121 formed by beta emission, **(c)** chromium-50 formed by positron emission, **(d)** californium-253 formed by alpha emission.

20.67 Krypton-87 decays to krypton-86. What other particle forms?

20.68 Write the symbol of the nuclide that forms from cobalt-58 when it decays by electron capture.

Band of Stability

20.69 If an atom of potassium-38 had the option of decaying by positron emission or beta emission, which route would it likely take, and why? Write the balanced nuclear equation.

20.70 Suppose that an atom of argon-37 could decay by either beta emission or electron capture. Which route would it likely take, and why? Write the balanced nuclear equation.

20.71 What is the most probable decay process for polonium-209? Write the balanced nuclear equation.

20.72 What is the most probable decay process for silicon-32? Write the balanced nuclear equation.

20.73 If we begin with 3.00 mg of iodine-131 ($t_{1/2}$ = 8.07 hr), how much remains after six half-life periods?

20.74 A sample of technetium-99m with a mass of 9.00 ng will have decayed to how much of this radionuclide after four half-life periods (about 1 day)?

Transmutation

20.75 When vanadium-51 captures a deuteron ($^2_1$H), what compound nucleus forms? (Write its symbol.) This particle expels a proton (^1_1p). Write the balanced nuclear equation for the overall change from vanadium-51.

20.76 The alpha-particle bombardment of fluorine-19 generates sodium-22 and neutrons. Write the balanced nuclear equation, including the intermediate compound nucleus.

20.77 Gamma-ray bombardment of bromine-81 causes a transmutation in which a neutron is one product. Write the symbol of the other product.

20.78 Neutron bombardment of cadmium-115 results in neutron capture and the release of gamma radiation. Write the balanced nuclear equation.

20.79 When manganese-55 is bombarded by protons, each ^{55}Mn nucleus releases a neutron. What else forms? Write the balanced nuclear equation.

20.80 Which nuclide forms when sodium-23 is bombarded by alpha particles and the compound nucleus emits a gamma-ray photon?

20.81 The nuclei of which isotope of zinc would be needed as bombardment particles to make nuclei of element 112 from lead-208 if the intermediate compound nucleus loses a neutron?

20.82 Write the symbol of the nuclide whose nuclei would be the target for bombardment by nickel-64 nuclei to produce nuclei of $^{272}_{111}$Rg after the intermediate compound nucleus loses a neutron.

Measuring Radioactivity

20.83 Suppose that a radiologist who is 2.0 m from a small, unshielded source of radiation receives 2.8 units of radiation. To reduce the exposure to 0.28 units of radiation, to what distance from the source should the radiologist move?

20.84 By what percentage should a radiation specialist increase the distance from a small unshielded source to reduce the radiation intensity by 10.0%?

20.85 If exposure from a distance of 1.60 m gave a worker a dose of 8.4 rem, how far should the worker move away

from the source to reduce the dose to 0.50 rem for the same period?

20.86 During work with a radioactive source, a worker was told that he would receive 55 mrem at a distance of 4.0 m during 30 min of work. What would be the received dose if the worker moved closer, to 0.50 m, for the same period?

20.87 A sample giving 3.7×10^{10} disintegrations per second has what activity in Ci and in Bq?

20.88 A sample of a radioactive metal used in medicine has an activity of 3.5 mCi. What is its activity in Bq and how many disintegrations/s does it give?

20.89 Smoke detectors contain a small amount of americium-241, which has a half-life of 1.70×10^5 days. If the detector contains 0.20 mg of ^{241}Am, what is the activity, in becquerels? In microcuries?

20.90 Strontium-90 is a dangerous radioisotope present in fallout produced by nuclear weapons. ^{90}Sr has a half-life of 1.00×10^4 days. What is the activity of 1.00 g of ^{90}Sr, in becquerels? In microcuries?

20.91 Iodine-131 is a radioisotope present in radioactive fallout that targets the thyroid gland. If 1.00 mg of ^{131}I has an activity of 4.6×10^{12} Bq, what is the decay constant for ^{131}I? What is the half-life, in seconds?

20.92 A 10.0 mg sample of thallium-201 has an activity of 7.9×10^{13} Bq. What is the decay constant for ^{201}Tl? What is the half-life of ^{201}Tl, in seconds?

Medical and Analytical Applications of Radionuclides

*__20.93__ What percentage of cesium chloride made from cesium-137 (beta emitter, $t_{1/2}$ = 30.1 y) remains after 150 y? What chemical product forms?

*__20.94__ A sample of waste has a radioactivity, caused solely by strontium-90 (beta emitter, $t_{1/2}$ = 28.1 yr), of 0.245 Ci g^{-1}. How many years will it take for its activity to decrease to 1.00×10^{-6} Ci g^{-1}? 2.45×10^{-6} mol of argon-40. How old was the rock? (What assumption is made about the origin of the argon-40?)

20.95 A worker in a laboratory unknowingly became exposed to a sample of radio-labeled sodium iodide made from iodine-131 (beta emitter, $t_{1/2}$ = 8.07 day). The mistake was realized 28.0 days after the accidental exposure, at which time the activity of the sample was 25.6×10^{-5} Ci/g. The safety officer needed to know how active the sample was at the time of the exposure. Calculate this value in Ci/g.

20.96 Technetium-99m (gamma emitter, $t_{1/2}$ = 6.02 hr) is widely used for diagnosis in medicine. A sample prepared in the early morning for use that day had an activity of 4.52×10^{-6} Ci. What will its activity be at the end of the day—that is, after 8.00 hr?

*20.97 To an Erlenmeyer flask containing 100.0 mL of a seawater sample, a chemist added 2.25×10^{-6} Ci of ^{45}Ca. After adding some sodium sulfate solution, filtering, and drying the resulting precipitate the mass of $CaSO_4$ was found to be 65.6 mg. When the activity of the precipitate was measured, it was found to be 1.75×10^{-6} Ci. What is the concentration of calcium ions in this seawater sample?

*20.98 To an Erlenmeyer flask containing 21.0 mL of a milk sample mixed with approximately 50 mL of water; a chemist added 7.32×10^{-6} Ci of ^{45}Ca. After adding some sodium sulfate solution, filtering, and drying the resulting precipitate, the mass of $CaSO_4$ was found to be 12.8 mg. When the activity of the precipitate was measured, it was found to be 2.75×10^{-6} Ci. What is the concentration of calcium ions in the milk? Does this sample meet the concentrations stated on a milk carton? (The RDA for calcium is 1300 mg between the ages of 14 and 19.)

20.99 A 0.500 g sample of rock was found to have 2.45×10^{-6} mol of potassium-40 ($t_{1/2} = 1.3 \times 10^9$ yr) and 1.22×10^{-6} mol of argon-40. How old was the rock? (What assumption is made about the origin of the argon-40?)

20.100 If a rock sample was found to contain 1.16×10^{-7} mol of argon-40, how much potassium-40 would also have to be present for the rock to be 1.3×10^9 years old?

*20.101 A tree killed by being buried under volcanic ash was found to have a ratio of carbon-14 atoms to carbon-12 atoms of 4.8×10^{-14}. How long ago did the eruption occur?

*20.102 A wooden door lintel from an excavated site in Mexico would be expected to have what ratio of carbon-14 to carbon-12 atoms if the lintel is 9.0×10^3 y old?

Nuclear Fission and Fusion

20.103 Complete the following nuclear equation by supplying the symbol for the other product of the fission.

$$^{235}_{92}U + ^{1}_{0}n \longrightarrow ^{94}_{38}Sr + \underline{\quad} + 2^{1}_{0}n$$

20.104 Both products of the fission in Problem 20.103 are unstable. According to Figures 20.8 and 20.9, what is the most likely way for each of them to decay: by alpha emission, beta emission, or positron emission? Explain. What are some of the possible fates of the extra neutrons produced by the fission shown in Problem 20.103?

Additional Exercises

*20.105 The age of "young" groundwater can be determined by measuring the ratio of tritium to helium-3 in water. If "young" groundwater is no more than 40 years old, what percent of tritium will remain in water at 40 years? The amount of ^{3}He in the water is also measured at the same time as the amount of tritium. What would the sum of the moles of tritium and ^{3}He represent?

20.106 What is the balanced nuclear equation for each of the following changes? (a) beta emission from aluminum-30, (b) alpha emission from einsteinium-252, (c) electron capture by molybdenum-93, (d) positron emission by phosphorus-28

*20.107 Calculate to five significant figures the binding energy in J/nucleon of the nucleus of an atom of iron-56. The observed mass of one atom is 55.9349 u. Which figure in this chapter lets us know that no isotope has a larger binding energy per nucleon?

20.108 Give the balanced nuclear equation for each of the following changes: (a) positron emission by carbon-10, (b) alpha emission by curium-243, (c) electron capture by vanadium-49, (d) beta emission by oxygen-20.

*20.109 If a positron is to be emitted spontaneously, how much more mass (as a minimum) must an atom of the parent have than an atom of the daughter nuclide? Explain.

*20.110 If a proton and an antiproton were to collide and produce two annihilation photons, what would the wavelength of the photons be, in meters? Which of the

decimal multipliers in Table 2.4 would be most appropriate for expressing this wavelength?

20.111 There is a gain in binding energy per nucleon when light nuclei fuse to form heavier nuclei. However, a tritium atom and a deuterium atom, in a mixture of these isotopes, does not spontaneously fuse to give helium (and energy). Explain why not.

*20.112 ^{214}Bi decays to isotope A by alpha emission; A then decays to B by beta emission, which decays to C by another beta emission. Element C decays to D by still another beta emission, and D decays by alpha emission to a stable isotope, E. What is the proper symbol of element E? (Contributed by Prof. W. J. Wysochansky, Georgia Southern University.)

*20.113 ^{15}O decays by positron emission with a half-life of 124 s. (a) Give the proper symbol of the product of the decay. (b) How much of a 750 mg sample of ^{15}O remains after 5.0 min of decay? (Contributed by Prof. W. J. Wysochansky, Georgia Southern University.)

*20.114 Alpha decay of ^{238}U forms ^{234}Th. What kind of decay from ^{234}Th produces ^{234}Ac? (Contributed by Prof. Mark Benvenuto, University of Detroit–Mercy.)

20.115 A sample of rock was found to contain 2.07×10^{-5} mol of ^{40}K and 1.15×10^{-5} mol of ^{40}Ar. If we assume that all of the ^{40}Ar came from the decay of ^{40}K, what is the age of the rock in years? ($t_{1/2} = 1.3 \times 10^9$ years for ^{40}K.)

20.116 The ^{14}C content of an ancient piece of wood was found to be one-tenth of that in living trees. How many years old is this piece of wood? ($t_{1/2} = 5730$ years for ^{14}C.)

*__20.117__ Dinitrogen trioxide, N_2O_3, is largely dissociated into NO and NO_2 in the gas phase where there exists the equilibrium, $N_2O_3 \rightleftharpoons NO + NO_2$. In an effort to determine the structure of N_2O_3, a mixture of NO and *NO_2 was prepared containing isotopically labeled N in the NO_2. After a period of time the mixture was analyzed and found to contain substantial amounts of both *NO and *NO_2. Explain how this is consistent with the structure for N_2O_3 being ONONO.

*__20.118__ The reaction, $(CH_3)_2Hg + HgI_2 \longrightarrow 2CH_3HgI$, is believed to occur through a transition state with the structure,

If this is so, what should be observed if CH_3HgI and *HgI_2 are mixed, where the asterisk denotes a radioactive isotope of Hg? Explain your answer.

20.119 A large, complex piece of apparatus has built into it a cooling system containing an unknown volume of cooling liquid. The volume of the coolant needs to be measured without draining the lines. To the coolant was added 10.0 mL of methanol whose molecules included atoms of ^{14}C and that had a specific activity (activity per gram) of 580 counts per minute per gram (cpm/g), determined using a Geiger counter. The coolant was permitted to circulate to assure complete mixing before a sample was withdrawn that was found to have a specific activity of 29 cpm/g. Calculate the volume of coolant in the system in milliliters. The density of methanol is 0.792 g/mL, and the density of the coolant is 0.884 g/mL.

20.120 Iodine-131 is used to treat Graves disease, a disease of the thyroid gland. The amount of ^{131}I used depends on the size of the gland. If the dose is 86 microcuries per gram of thyroid gland, how many grams of ^{131}I should be administered to a patient with a thyroid gland weighing 20 g? Assume all the iodine administered accumulates in the thyroid gland.

*__20.121__ The fuel for a thermonuclear bomb (hydrogen bomb) is lithium deuteride, a salt composed of the ions $^6_3Li^+$ and $^2_1H^-$. Considering the nuclear reaction

$$^6_3Li^+ + {}^2_1H^- \longrightarrow {}^4_2He + {}^3_1H$$

explain how a ^{235}U fission bomb could serve as a trigger for a fusion bomb. Write appropriate balanced nuclear equations.

20.122 In 2006, the synthesis of ^{294}Uuo (an isotope of element 118) was reported to involve the bombardment of ^{245}Cf with ^{48}Ca. Write an equation for the nuclear reaction, being sure to include any other products of the reaction.

*__20.123__ Radon, a radioactive noble gas, is an environmental problem in some areas, where it can seep out of the ground and into homes. Exposure to radon-222, an α emitter with a half-life of 3.823 days, can increase the risk of lung cancer. At an exposure level of 4 pCi per liter (the level at which the EPA recommends action), the lifetime risk of death from lung cancer due to radon exposure is estimated to be 62 out of 1000 for current smokers, compared with 73 out of 10,000 for non-smokers. If the air in a home was analyzed and found to have an activity of 4.1 pCi L^{-1}, how many atoms of ^{222}Rn are there per liter of air?

*__20.124__ The isotope ^{145}Pr decays by emission of beta particles with an energy of 1.80 MeV each. Suppose a person swallowed, by accident, 1.0 mg of Pr having a specific activity (activity per gram) of 140 Bq/g. What would be the absorbed dose from ^{145}Pr in units of Gy and rad over a period of 10 minutes? Assume all of the beta particles are absorbed by the person's body.

20.125 Why is the isotope ^{36}Cl radioactive while ^{35}Cl and ^{37}Cl are not? The isotope decays by beta emission, what isotope is formed? Write the nuclear equation that shows the process.

|Multi-Concept Questions

*__20.126__ A complex ion of chromium(III) with oxalate ion was prepared from ^{51}Cr-labeled $K_2Cr_2O_7$, having a specific activity of 843 cpm/g (counts per minute per gram), and ^{14}C-labeled oxalic acid, $H_2C_2O_4$, having an specific activity of 345 cpm/g. Chromium-51 decays by electron capture with the emission of gamma radiation, whereas ^{14}C is a pure beta emitter. Because of the characteristics of the beta and gamma detectors, each of these isotopes may be counted independently. A sample of the complex ion was observed to give a gamma count of 165 cpm and a beta count of 83 cpm. From these data, determine the number of oxalate ions bound to each Cr(III) in the complex ion. (*Hint:* For the starting materials calculate the cpm per mole of Cr and oxalate, respectively.)

*__20.127__ Iridium-192, a beta emitter, is used in medicine to treat certain cancers. The ^{192}Ir is used as a wire and inserted near the tumor for a given amount of time and then removed. The ^{192}Ir wire is coated with pure platinum, and is sold by diameter and amount of radiation emitted.

One type of wire sold is 140 mm long, has a linear apparent activity of 30–130 MBq/cm, an active diameter (which is the iridium-192) of 0.1 mm, and an outer diameter (the platinum) of 0.3 mm. How much ^{192}Ir is in the wire if the apparent activity is 30 MBq/cm? If the wire is placed in the tumor for 50 hours, how much radiation is absorbed by the tumor? The wire is shipped in lead tubes—why?

Exercises in Critical Thinking

20.128 A silver wire coated with nonradioactive silver chloride is placed into a solution of sodium chloride that is labeled with radioactive ^{36}Cl. After a while, the AgCl was analyzed and found to contain some ^{36}Cl. How do you interpret the results of this experiment?

20.129 Suppose you were given a piece of cotton cloth and told that it was believed to be 2000 years old. You performed a carbon dating test on a tiny piece of the cloth and your data indicated that it was only 800 years old. If the cloth really was 2000 years old, what factors might account for your results?

20.130 In 2006, the former Soviet spy Alexander Litvinenko was poisoned by the polonium isotope ^{210}Po. He died 23 days after ingesting the isotope, which is an alpha emitter. Find data on the Internet to answer the following: Assuming Litvinenko was fed 1 mg of ^{210}Po and that it became uniformly distributed through the cells in his body, how many atoms of ^{210}Po made their way into each cell in his body? (Assume his body contained the average number of cells found in an adult human.) Also, calculate the number of cells affected by the radiation each second, being sure to take into account an estimate of the average number of cells affected by an alpha particle emitted by a ^{210}Po nucleus.

20.131 In 2012 Palestinians asserted that clothing worn by Yassar Arafat showed the presence of polonium-210 (see the previous question). On November 27, 2012, his body was exhumed for testing. Research the final results of this testing and make an argument either for or against a possible murder case.

20.132 What would be the formula of the simplest hydrogen compound of element 116? Would a solution of this compound in water be acidic, basic, or neutral? Explain your reasoning.

20.133 Astatine is a halogen. Its most stable isotope, ^{110}At, has a half-life of only 8.3 hours, and only very minute amounts of the element are ever available for study (<0.001 mg). This amount is so small as to be virtually invisible. Describe experiments you might perform that would tell you whether the silver salt of astatine, AgAt, is insoluble in water.

20.134 The International Union of Pure and Applied Chemistry has named the elements with atomic numbers of 114 and 116 flerovium and livermorium, respectively. Research the criteria that the IUPAC uses to determine that a new element has actually been discovered (or synthesized).

20.135 Using a graphing program, such as Excel, and Equation 20.1, plot m/m_0 on the y-axis and velocity on the x-axis. The velocity should range from 0 to the speed of light. Comment on how the mass changes as the velocity increases.

Metal Complexes

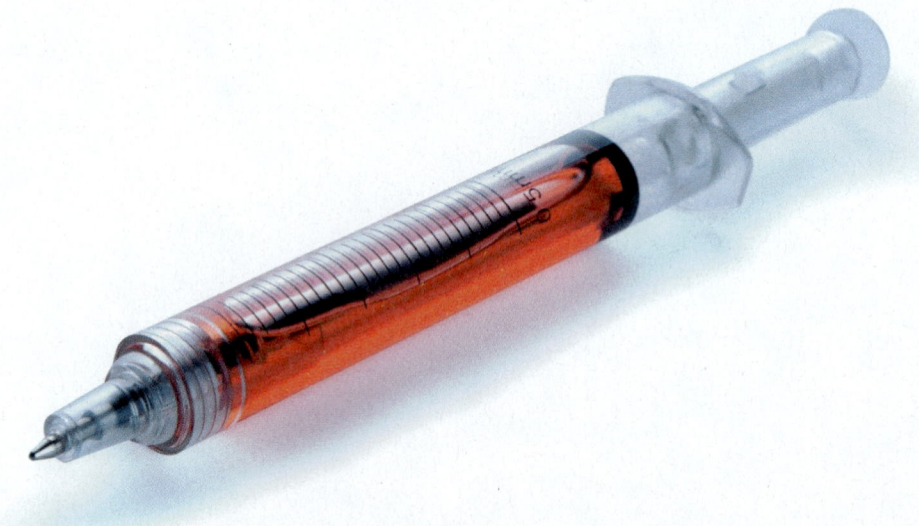

Vaccinations are a procedure that most children go through to inoculate them against childhood diseases. If you remember your shots, the liquid was most likely colorless. However, the vitamin B$_{12}$ in the syringe, which some people need to take for pernicious anemia, is a metal complex with cobalt bound in a molecule, which gives it the intense red color.

aperturesound/Shutterstock

I n previous chapters you learned many of the concepts that chemists have used to develop an understanding of how the elements react with each other and the kinds of compounds produced. For example, in our discussions of chemical kinetics, you learned how various factors affect the rates of reactions, and in our study of thermodynamics you learned how enthalpy and entropy changes affect the possibility of observing chemical changes. Our focus, however, was primarily on the concepts themselves, with examples of chemical behavior being used to reinforce and justify them. With all of these concepts available to us now, we will direct our emphasis in the opposite direction and apply the concepts to examine some of the physical and chemical properties of the compounds of metals.

This chapter is devoted to an in-depth look at complex ions of metals, a topic first introduced in Chapter 17. These substances—which have applications from food preservatives to catalyzing biochemical reactions, such as vitamin B_{12}, which is involved in metabolism—have a variety of structures and colors.

This Chapter in Context

LEARNING OBJECTIVES

After reading this chapter, you should be able to:

- describe the different kinds of ligands and the rules for writing formulas for metal complexes containing them
- use the rules for naming metal complexes
- show the different geometries associated with each of the common coordination numbers, and draw the structures
- write structural formulas for all the isomers of a given metal complex
- describe how crystal field theory explains the properties of metal complexes
- explain the biological function of several common metal complexes in the body

21.1 | Complex Ions

In Section 17.5, we introduced you to **complex ions** (or more simply, **complexes**) of metals. Recall that these are substances formed when molecules or anions become bonded covalently to metal ions to form more complex species. Two examples that were given were the pale blue $[Cu(H_2O)_6]^{2+}$ ion and the deep blue $[Cu(NH_3)_4]^{2+}$ ion. We also discussed how the formation of complex ions can affect the solubilities of salts, but the importance of these substances reaches far beyond solubility equilibria. The number of complex ions formed by metals, especially the transition metals, is enormous, and the study of the properties, reactions, structures, and bonding in complexes like $[Cu(NH_3)_4]^{2+}$ has become an important specialty within chemistry. The study of metal complexes extends into biochemistry—for example, the cobalt ions in vitamin B_{12} in the chapter-opening photograph. This is because, ultimately, nearly all of the metals in our bodies become bound in complexes in order to perform their biochemical functions.

Before we proceed further, let's consider some of the basic terminology we will use in our discussions. The molecules or ions that become bound, with a coordinate covalent bond, to a metal ion (e.g., the NH_3 molecules in $[Cu(NH_3)_4]^{2+}$) are called **ligands**. Ligands are either neutral molecules or anions that contain one or more atoms with at least one unshared electron pair that can be donated to the metal ion in the formation of a metal–ligand bond. The reaction of a ligand with a metal ion is therefore a Lewis acid–base reaction in which the ligand is the Lewis base and the metal ion is the Lewis acid.

■ Recall that a coordinate covalent bond is one in which both of the shared electrons originate on the same atom. Once formed, of course, a coordinate covalent bond is just like any other covalent bond.

Silver ion (Lewis acid)   Ammonia molecule (Lewis base)   Coordinate covalent bond between ammonia and the silver ion

A ligand atom that donates an electron pair to the metal is said to be a **donor atom** and the metal is the **acceptor**. Thus, in the preceding $AgNH_3^+$ example, the nitrogen of the ammonia molecule is the donor atom and the silver ion is the acceptor. Because of the way the metal–ligand bond is formed it is a coordinate covalent bond, and compounds that contain metal complexes are often called **coordination compounds**. The complexes themselves are sometimes referred to as **coordination complexes**.

Types of Ligands

Ligands, the Lewis bases in these metal complexes, can range from monatomic ions to complex ligands that can bind metals at more than one site, and they can be neutral or anionic.

Anions that serve as ligands include many simple monatomic ions, such as the halide ions (F^-, Cl^-, Br^-, and I^-) and the sulfide ion (S^{2-}). Common polyatomic anions that are ligands are nitrite ion (NO_2^-), cyanide ion (CN^-), hydroxide ion (OH^-), thiocyanate ion (SCN^-), and thiosulfate ion ($S_2O_3^{2-}$). (This is really only a small sampling, not a complete list.)

The most common neutral molecule that serves as a ligand is water, and most of the reactions of metal ions in aqueous solutions are actually reactions of their complex ions—ions in which the metal is attached to some number of water molecules. Copper(II) ions,[1] for example, form the complex ion $[Cu(H_2O)_6]^{2+}$, and cobalt(II) also combines with water molecules to form $[Co(H_2O)_6]^{2+}$. Another common neutral ligand is ammonia, NH_3, which has one lone pair of electrons on the nitrogen atom. If ammonia is added to an aqueous solution containing the $[Ni(H_2O)_6]^{2+}$ ion, for example, the color changes dramatically from green to blue as ammonia molecules displace water molecules (see Figure 21.1):

$$[Ni(H_2O)_6]^{2+}(aq) + 5NH_3(aq) \longrightarrow [Ni(NH_3)_5(H_2O)]^{2+}(aq) + 5H_2O$$

(green)　　　　　　　　　　　　　　　　　(blue)

Figure 21.1 | **Complex ions of nickel.** (*Left*) A solution of nickel chloride, which contains the green $[Ni(H_2O)_6]^{2+}$ ion. (*Right*) Adding ammonia to the nickel chloride solution forms the blue $[Ni(NH_3)_5(H_2O)]^{2+}$ ion.

Andy Washnik

---

[1] The formula $Cu(H_2O)_4^{2+}$ we used in previous chapters is actually an oversimplification. Copper(II) complexes in water usually have two water molecules loosely attached to the Cu^{2+} at a greater distance than the other four ligand atoms. Therefore, the copper(II) complex with water could also be written $Cu(H_2O)_6^{2+}$ to indicate the four strongly bound water molecules and the two weakly bound water molecules on the Cu^{2+} ion.

Each of the ligands that we have discussed so far is able to use just one atom to attach itself to a metal ion. Such ligands are called **monodentate ligands**, indicating that they have only "one tooth" with which to "bite" the metal ion.

There are also many ligands that have two or more donor atoms, and collectively they are referred to as **polydentate ligands**. The most common of these have two donor atoms, so they are called **bidentate ligands**. When they form complexes, *both* donor atoms become attached to the same metal ion. Oxalate ion and ethylenediamine (abbreviated *en* in writing the formula for a complex) are examples of bidentate ligands.

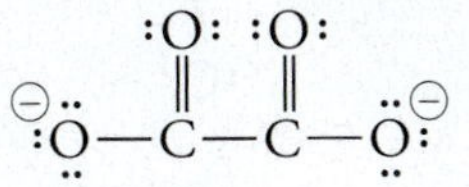

oxalate ion 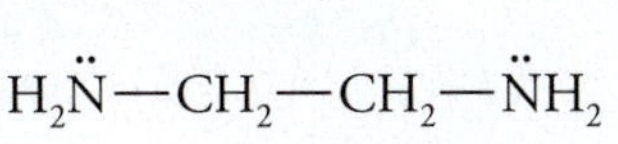   ethylenediamine, en

When these ligands become attached to a metal ion, the following ring structures are formed:

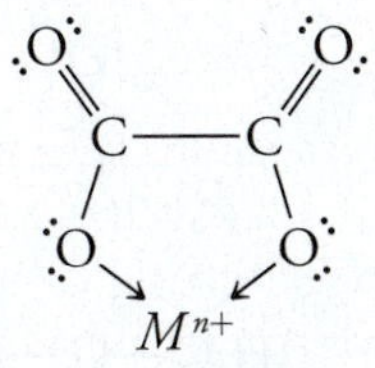

an oxalate complex 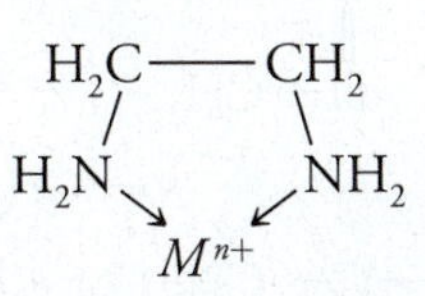   an ethylenediamine complex

Complexes that contain such ring structures are called **chelates**.[2] Structures like these are important in "complex ion chemistry," as we shall see later in this chapter.

A list of ligands is given in Table 21.1. The atom that binds to the metal is underlined. Some ligands can bind through different atoms; for example, the NO_2^- ion can bind either through the N or through the O.

TABLE 21.1	Some Common Ligands
Name as a Ligand	**Formula[a]**
acetato	CH_3COO^-
ammine	NH_3
aqua	H_2O
azido	N_3^-
bromo	Br^-
cyano	CN^-
chloro	Cl^-
ethylenediamine (en)	$NH_2CH_2CH_2NH_2$
ethylenediaminetetraacetate ($EDTA^{4-}$)	$(OOCCH_2)_2NCH_2CH_2N(CH_2COO)_2$
hydroxo	OH^-
iodo	I^-
nitrito	ONO^-
nitro	ONO^-
oxalato (ox^{2-})	$COOCOO^{2-}$
sulfido	S^{2-}
thiosulfato	$S_2OO_2^{2-}$

[a]The red colored elements are the ones that bond to the metal ions.

[2] The term *chelate* comes from the Greek *chele*, meaning claw. These bidentate ligands grasp the metal ions with two "claws" (donor atoms) somewhat like a crab holds its prey. (Who says scientists have no imagination?)

■ Sometimes the ligand EDTA⁴⁻ is abbreviated using small letters—that is, edta⁴⁻.

One of the most common polydentate ligands is a compound called ethylenediamine-tetraacetate, mercifully abbreviated $EDTA^{4-}$ (or Y^{4-} by analytical chemists).

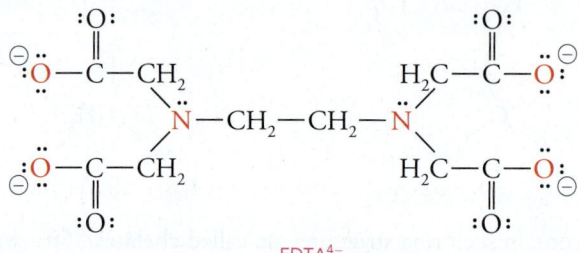

Ethylenediaminetetraacetic acid, H_4EDTA

The H atoms attached to the oxygen atoms are easily removed as protons, which gives an anion with a charge of 4−. The structure of the anion, $EDTA^{4-}$, is as follows with the donor atoms in red:

$EDTA^{4-}$

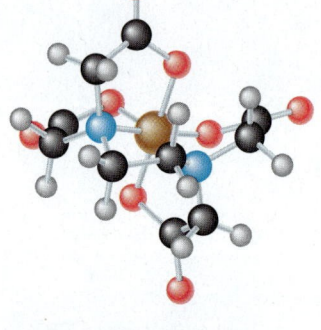

The structure of an EDTA complex. The metal ion is in the center of the complex bonded to the two nitrogens and four oxygens.

The $EDTA^{4-}$ ion has six donor atoms, and this permits it to wrap itself around a metal ion and form very stable complexes.

EDTA is a particularly useful and important ligand. It is relatively nontoxic, which allows it to be used in small amounts in foods to retard spoilage. If you look at the labels on bottles of salad dressings, for example, you often will find that one of the ingredients is $CaNa_2EDTA$ (calcium disodium EDTA). The $EDTA^{4-}$ available from this salt forms soluble complex ions with any traces of metal ions that might otherwise promote reactions of the salad oils with oxygen, and thereby lead to spoilage.

■ The *calcium* salt of EDTA is used because the EDTA⁴⁻ ion would otherwise extract Ca²⁺ ions from bones, and that would be harmful.

Many shampoos contain Na_4EDTA to soften water. The $EDTA^{4-}$ binds to Ca^{2+}, Mg^{2+}, and Fe^{3+} ions, which removes them from the water and prevents them from interfering with the action of detergents in the shampoo. A similar application was described in Chemistry Outside the Classroom 17.1.

EDTA stops the clotting of blood by complexing calcium ions that are required for the process. For that reason, EDTA is added to whole blood that is drawn for analysis by clinical chemists. EDTA has even been used as a treatment in cases of heavy metal poisoning because it can help remove the poisonous heavy metal ions, like Hg^{2+} and Pb^{2+}, from the body when they have been accidentally ingested.

Writing Formulas for Metal Complexes

TOOLS

Rules for writing formulas for complexes

When we write the formula for a complex, we utilize the following rules:

1. The symbol for the metal ion is always given first, followed by the ligands.
2. When more than one kind of ligand is present, anionic ligands are written first (in alphabetical order), followed by neutral ligands (also in alphabetical order).
3. The charge on the complex is the algebraic sum of the charge on the metal ion and the charges on the ligands.
4. The formula is placed inside of square brackets with the charge of the complex as a superscript outside the brackets. When the charge is zero the brackets and charge are omitted.

■ Square brackets here do not mean molar concentration. It is usually clear from the context of a discussion whether we intend the brackets to mean molarity or a chemical formula.

For example, the formula of the complex ion of Cu^{2+} and NH_3, which we mentioned earlier, was written $[Cu(NH_3)_4]^{2+}$ with the Cu first followed by the ligands. The charge on the complex is 2+ because the copper ion has a charge of 2+ and the ammonia molecules are neutral. Copper(II) ion also forms $[Cu(CN)_4]^{2-}$, a complex ion with four cyanide ions, CN^-. The metal ion contributes two (+) charges and the four ligands contribute a total of four (−) charges, one for each cyanide ion. The algebraic sum is −2, so the complex ion has a charge of 2−.

The brackets emphasize that the ligands are attached to the metal ion and are not free to roam about. One of the many complex ions formed by the chromium(III) ion contains five water molecules and one chloride ion as ligands. To indicate that all are attached to the Cr^{3+} ion, we use brackets and write the complex ion as $[CrCl(H_2O)_5]^{2+}$. When this complex is isolated as a chloride salt, the formula is written $[CrCl(H_2O)_5]Cl_2$, in which $[CrCl(H_2O)_5]^{2+}$ is the cation and so is written first. The formula $[CrCl(H_2O)_5]Cl_2$ shows that five water molecules and a chloride ion are bonded to the chromium ion, and the other two chloride ions are said to be **counter ions**, meaning they are present to provide electrical neutrality for the salt.

In Chapter 2, you learned about hydrates, and one was the beautiful blue hydrate of copper(II) sulfate, $CuSO_4·5H_2O$. It was much too early then to make the distinction, but the formula should have been written as $[Cu(H_2O)_4]SO_4·H_2O$ to show that four of the five water molecules are held in the crystal as part of the complex ion $[Cu(H_2O)_4]^{2+}$. The fifth water molecule is held in the crystal by being hydrogen bonded to the sulfate ion.

Many other hydrates of metal salts actually contain complex ions of the metals in which water is the ligand. Cobalt salts like cobalt(II) chloride, for example, crystallize from aqueous solutions as hexahydrates (meaning they contain six H_2O molecules per formula unit of the salt). The compound $CoCl_2·6H_2O$ (Figure 21.2) actually is $[Co(H_2O)_6]Cl_2$, and it contains the pink complex $[Co(H_2O)_6]^{2+}$. This ion also gives solutions of cobalt(II) salts a pink color as can be seen in Figure 21.2. Although most hydrates of metal salts contain complex ions, the distinction is seldom made, and it's acceptable to write the formula for these hydrates in the usual fashion—for example, $CuSO_4·5H_2O$ instead of $[Cu(H_2O)_4]SO_4·H_2O$.

Figure 21.2 | **Color of the cobalt(II) ion in hexahydrate salts.** The ion $[Co(H_2O)_6]^{2+}$ is pink and gives its color both to crystals and to an aqueous solution of $CoCl_2·6H_2O$.

Example 21.1
Writing the Formula for a Complex Ion

Write the formula for the complex ion formed by the metal ion Cr^{3+} and six NO_2^- ions as ligands. Decide whether the complex could be isolated as a chloride salt or a potassium salt, and write the formula for the appropriate salt.

Analysis: We are being asked to determine the formula, including charge, for the complex ion with one Cr^{3+} and six NO_2^-. If the net charge on the complex is positive, it would require a negative ion to form a salt, so it could be isolated as a chloride salt; if the complex is a negative ion, then a potassium salt could form.

Assembling the Tools: The rules for writing formulas for metal complexes are the tools we'll use.

Solution: Six NO_2^- ions contribute a total charge of 6−; the metal contributes a charge of 3+. The algebraic sum is (6−) + (3+) = 3−. The formula of the complex ion is therefore $[Cr(NO_2)_6]^{3-}$.

Because the complex is an anion, it requires a cation to form a neutral salt, so the complex could be isolated as a potassium salt, not as a chloride salt. For the salt to be electrically neutral, three K^+ ions are required for each complex ion, $[Cr(NO_2)_6]^{3-}$. The formula of the salt would therefore be $K_3[Cr(NO_2)_6]$. Notice that in writing the formula for the salt, we've specified the cation (K^+) first, followed by the anion.

Is the Answer Reasonable? Things to check: Have we written the formula with the metal ion first, followed by the ligands? Have we computed the charge on the complex correctly? Does the formula for the salt correspond to an electrically neutral substance? The answer to each question is yes, so all seems to be okay.

Practice Exercise 21.1

Write the formula of the complex ion formed by Ag^+ and two thiosulfate ions, $S_2O_3^{2-}$. If we were able to isolate this complex ion as its ammonium salt, what would be the formula for the salt? (*Hint:* Remember, salts are electrically neutral.)

Practice Exercise 21.2

Aluminum chloride crystallizes from aqueous solutions as a hexahydrate. Write the formula for the salt and suggest a formula for the complex ion formed by aluminum ion and water.

Practice Exercise 21.3

What is the formula of the complex ion that is formed from chromium(III), four water molecules, and two chloride ions? Is the counter ion more likely to be a halide or an alkali metal?

The Chelate Effect

An interesting aspect of the complexes formed by bidentate ligands such as ethylenediamine and oxalate ion is that their stabilities are higher as compared to similar complexes formed by monodentate ligands. For example, the complex $[Ni(en)_3]^{2+}$ is considerably more stable than $[Ni(NH_3)_6]^{2+}$, even though both complexes have six nitrogen atoms bound to a Ni^{2+} ion. We can compare them quantitatively by examining their formation constants:

■ Ammonia, a monodentate ligand, supplies one donor atom to a metal. Ethylenediamine (en), a bidentate ligand, has two nitrogen donor atoms. Therefore, six ammonia molecules and three ethylenediamine molecules supply the same number of donor atoms.

$$Ni^{2+}(aq) + 6NH_3(aq) \rightleftharpoons [Ni(NH_3)_6]^{2+}(aq) \qquad K_{form} = 2.0 \times 10^8$$

$$Ni^{2+}(aq) + 3en(aq) \rightleftharpoons [Ni(en)_3]^{2+}(aq) \qquad K_{form} = 4.1 \times 10^{17}$$

The ethylenediamine complex is more stable than the ammonia complex by a factor of 2×10^9 (2 billion)! This exceptional stability of complexes formed with polydentate ligands is called the **chelate effect**, so named because it occurs in compounds in which the ligands have two or more sites that bind to one metal center. A ring forms involving the metal and the ligand, and extra stability is gained if the ring has five or six atoms in it.

There are two related reasons for the chelate effect, which we can understand best if we examine the ease with which the complexes undergo dissociation once formed. One reason appears to be associated with the probability of the ligand leaving the vicinity of the metal ion when the donor atom becomes detached. If one end of a bidentate ligand comes loose from the metal, the donor atom cannot wander very far because the other end of the ligand is still attached. There is a high probability that the loose end will become reattached to the metal ion before the other end can let go, so overall the ligand appears to be bound tightly. With a monodentate ligand, however, there is nothing to hold the ligand near the metal ion if it becomes detached. The ligand can easily move off into the surrounding solution and be lost. As a result, a monodentate ligand is not as firmly attached to the metal ion as a polydentate ligand.

The second reason is related to the entropy change for the dissociation. In Chapter 18 you learned that when there is an increase in the number of particles in a chemical reaction, the entropy change is positive. Dissociation of both complexes produces more particles, so the entropy change is positive for both of them. However, comparing the two reactions,

$$[Ni(NH_3)_6]^{2+}(aq) \rightleftharpoons Ni^{2+}(aq) + 6NH_3(aq) \qquad K_{inst} = 5.0 \times 10^{-9}$$

$$[Ni(en)_3]^{2+}(aq) \rightleftharpoons Ni^{2+}(aq) + 3en(aq) \qquad K_{inst} = 2.4 \times 10^{-18}$$

■ Recall that

$$\Delta G° = -RT\ln K$$

A more favorable $\Delta G°$ for dissociation should yield a larger equilibrium constant, and, indeed, we find the instability constant is larger for the ammonia complex.

we see that dissociation of the ammonia complex gives a net increase of *six* in the number of particles, whereas dissociation of the ethylenediamine complex yields a net increase of *three*. This means the entropy change is more positive for the dissociation of the ammonia complex than for the ethylenediamine complex. The larger entropy change translates into a more favorable $\Delta G°$ for the decomposition, so at equilibrium the ammonia complex should be dissociated to a greater extent. This is, in fact, what is suggested by the values of their instability constants, K_{inst} (recall from Section 17.5 that $K_{inst} = 1/K_{form}$).

21.2 | Metal Complex Nomenclature

The naming of chemical compounds was introduced in Chapter 2, where we discussed the nomenclature system for simple inorganic compounds. This system, revised and kept up to date by the International Union of Pure and Applied Chemistry (IUPAC), has been extended to cover metal complexes. Below are some of the rules that have been developed to name coordination complexes. As you will see, some of the names arrived at following the rules are difficult to pronounce at first, and may even sound a little odd. However, the primary purpose of this and any other system of nomenclature is to provide a method that gives each compound its own unique name and permits us to write the formula of the compound given the name.

Rules of Nomenclature for Coordination Complexes

1. **Cationic species are named before anionic species.** This is the same rule that applies to other ionic compounds such as NaCl, where we name the cation first, followed by the anion (i.e., sodium chloride).

2. **The names of anionic ligands always end in the suffix –o.**
   (a) Ligands whose names end in –*ide* usually have this suffix changed to –*o*.

TOOLS

Rules for naming complexes

■ Names such as chloro- and bromo- in the table are the ones traditionally used and are acceptable. Strict IUPAC names simply drop the -e and replace it with -o to give chlorido, bromido, etc. Thus, S would be named sulfido as a ligand.

Anion		Ligand
chloride	Cl^-	chloro-
bromide	Br^-	bromo-
cyanide	CN^-	cyano-
oxide	O^{2-}	oxo-

   (b) Ligands whose names end in -*ite* or -*ate* become -*ito* and -*ato*, respectively.

Anion		Ligand
carbonate	CO_3^{2-}	carbonato-
thiosulfate	$S_2O_3^{2-}$	thiosulfato-
thiocyanate	SCN^-	thiocyanato- (when bonded through sulfur)
		isothiocyanato- (when bonded through nitrogen)
oxalate	$C_2O_4^-$	oxalato-
nitrite	NO_2^-	nitrito- (when bonded through oxygen; written ONO in the formula for the complex)
	NO_2^-	nitro- (when bonded through nitrogen; it is an exception)

3. **A neutral ligand is given the same name as the neutral molecule.** Thus, the molecule ethylenediamine, when serving as a ligand, is called ethylenediamine in the name of the complex. Two very important exceptions to this, however, are water and ammonia. These are named as follows when they serve as ligands:

H_2O aqua NH_3 ammine (note the double *m*)

4. **When there is more than one of a particular ligand, their number is specified by the prefixes di- = 2, tri- = 3, tetra- = 4, penta- = 5, hexa- = 6, and so forth. When confusion might result by using these prefixes, the following are used instead: bis- = 2, tris- = 3, and tetrakis- = 4.** Following this rule, the presence of two chloride ligands in a complex would be indicated as *dichloro*- (notice, too, the ending on the ligand name). However, if two ethylenediamine ligands are present, use of the prefix *di*- might cause confusion. Someone reading the name might wonder whether diethylenediamine means two ethylenediamine molecules or one molecule of a substance called diethylenediamine. To avoid this problem we place the ligand name in parentheses preceded by *bis*; that is, *bis(ethylenediamine)*.

■ *Bis*- is employed here so that if the name is used in verbal communication it is clear that the meaning is two ethylenediamine molecules.

■ The ligands are alphabetized according to the first letter of the name of the ligand, not the first letter of the numbering prefix.

5. In the *formula* of a complex, the symbol for the metal is written first, followed by those of the ligands. Among the ligands, anionic ligands are written first (in alphabetical order), followed by neutral ligands (also in alphabetical order). In the *name* of the complex, the ligands are named first, in alphabetical order *without regard to charge*, followed by the name of the metal. For example, suppose we had a complex composed of Co^{3+}, two Cl^- (chloro- ligands), one CN^- (cyano- ligand), and three NH_3 (ammine- ligands). The formula of this electrically neutral complex would be written $[CoCl_2(CN)(NH_3)_3]$. In the name of this complex, the ligands would be specified before the metal as *triamminedichlorocyano-* (*triammine-* for the three NH_3 ligands, *dichloro-* for the two Cl^- ligands, and *cyano-* for the CN^- ligand). When alphabetizing the names of the ligands, we ignore the prefixes *tri-* and *di-*. Thus, *triammine-* is written before *dichloro-* because *ammine-* precedes *chloro-* alphabetically. For the same reason, *dichloro-* is written before *cyano-*. The complete name for the complex is given below under Rule 7.

6. **Negative (anionic) complex ions always end in the suffix *-ate*.** This suffix is appended to the English name of the metal atom in most cases. However, if the name of the metal ends in *-ium*, *-um*, or *-ese*, the ending is dropped and replaced by *-ate*.

Metal	Metal as Named in an Anionic Complex
aluminum	aluminate
chromium	chormate
manganese	manganate
nickel	nickelate
cobalt	cobaltate
zinc	zincate
platinum	plantinate
vanadium	vanadate

For metals whose symbols are derived from their Latin names, the suffix *-ate* is appended to the Latin stem. (An exception, however, is mercury; in an anion it is named *mercurate*.)

Metal	Stem	Metal as Named in an Anionic Complex
iron	ferr-	ferrate
copper	cupr-	cuprate
lead	plumb-	plumbate
silver	argent-	argentite
gold	aur-	aurate
tin	stann-	stannate

For neutral or positively charged complexes, the metal is *always* specified with the English name for the element, *without any suffix*.

7. **The oxidation state of the metal in the complex is written in Roman numerals within parentheses following the name of the metal.**[3] For example,

$[CoCl_2(CN)(NH_3)_3]$ is triamminedichlorocyanocobalt(III)

$[Co(NH_3)_6]^{3+}$ is the hexaamminecobalt(III) ion

$[CuCl]^{2-}$ is the tetrachlorocuprate(II) ion

No spaces

---

[3] Alternatively, the charge on the complex as a whole can be used instead of the oxidation number—for example, tetrachloridocuprate(2−).

Notice that there are no spaces between the names of the ligands and the name of the metal, and that there is no space between the name of the metal and the parentheses that enclose the oxidation state expressed in Roman numerals.

The following are some additional examples. Study them carefully to see how the nomenclature rules apply. Then try the practice exercises that follow.

$[Ni(CN)_4]^{2-}$	tetracyanonickelate(II) ion
$K_3[CoCl_6]$	potassium hexachlorocobaltate(III)
$[CoCl_2(NH_3)_4]^+$	tetraamminedichlorocobalt(III) ion
$Na_3[Co(NO_2)_6]$	sodium hexanitrocobaltate(III)
$[Ag(NH_3)_2]^+$	diamminesilver(I) ion
$[Ag(S_2O_3)_2]^{3-}$	dithiosulfatoargentate(I) ion
$[Mn(en)_3]Cl_2$	tris(ethylenediamine)manganese(II) chloride
$[PtCl_2(NH_3)_2]$	diamminedichloroplatinum(II)

■ Notice, once again, that the alphabetical order of the ligands is determined by the first letter in the name of the ligand, not the first letter in the prefix.

Practice Exercise 21.4

Write the formula for each of the following: (a) hexachlorostannate(IV) ion, (b) ammonium diaquatetracyanoferrate(II), and (c) dibromobis(ethylenediamine)osmium(II). (*Hint:* Divide each name into its various parts.)

Practice Exercise 21.5

Name the following compounds: (a) $K_3[Fe(CN)_6]$, (b) $[CrCl_2(en)_2]_2SO_4$, and (c) $[Co(H_2O)_6]_2[CrF_6]$ (in which the cobalt is in the +2 oxidation state).

21.3 | Coordination Number and Structure

One of the most interesting aspects of the study of complexes is the different kinds of structures that they form. In many ways, this is related to the **coordination number** of the metal ion, which we define as *the number of coordinate covalent bonds to the metal ion*. For example, in the complex $[Ni(CN)_4]^{2-}$, four cyanide ions bind the nickel ion, so the coordination number of Ni^{2+} in this complex is 4. Similarly, the coordination number of Cr^{3+} in the $[Cr(H_2O)_6]^{3+}$ ion is 6, and the coordination number of Ag^+ in $[Ag(NH_3)_2]^+$ is 2.

Sometimes the coordination number isn't immediately obvious from the formula of the complex. For example, you learned that there are many polydentate ligands that contain more than one donor atom that can bind simultaneously to a metal ion. Often, a metal is able to accommodate two or more polydentate ligands to give complexes with formulas such as $[Cr(H_2O)_2(en)_2]^{3+}$ and $[Cr(en)_3]^{3+}$. In each of these examples, the coordination number of the Cr^{3+} is 6. In the $[Cr(en)_3]^{3+}$ ion, there are three ethylenediamine ligands that each supply two donor atoms, for a total of 6, and in $[Cr(H_2O)_2(en)_2]^{3+}$, the two ethylenediamine ligands supply a total of 4 donor atoms and the two H_2O molecules supply another 2, so once again the total is 6.

Coordination Number and Geometry

For metal complexes, there are certain geometries that are usually associated with particular coordination numbers.

■ Ordinarily, the VSEPR model isn't used to predict the structures of transition metal complexes because it can't be relied on to give correct results when the metal has a partially filled *d* subshell.

Coordination Number 2

Examples are complexes such as $[Ag(NH_3)_2]^+$ and $[Ag(CN)_2]^-$. Usually, these complexes have a linear structure such as

$$[H_3N\text{—}Ag\text{—}NH_3]^+ \quad \text{and} \quad [NC\text{—}Ag\text{—}CN]^-$$

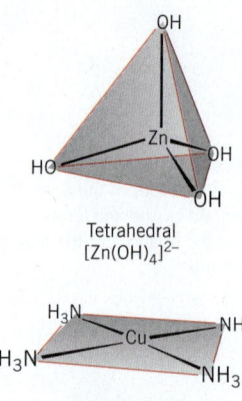

Tetrahedral
$[Zn(OH)_4]^{2-}$

Square planar
$[Cu(NH_3)_4]^{2+}$

Figure 21.3 | Tetrahedral and square planar geometries. Tetrahedral and square planar geometries are structures that occur for complexes in which the metal ion has a coordination number of 4. For the copper complex, we are viewing the square planar structure tilted back into the plane of the page.

Figure 21.4 | Octahedral complexes. Complexes with this geometry can be formed with either monodentate ligands, such as water, or with polydentate ligands, such as ethylenediamine (en). To simplify the drawing of the ethylenediamine complex, the atoms joining the donor nitrogen atoms in the ligand, —CH₂—CH₂—, are represented as the curved line between the N atoms. Also notice that the nitrogen atoms of the bidentate ligand span adjacent positions within the octahedron. This is the case for all polydentate ligands that you will encounter in this book.

Since the Ag⁺ ion has a filled *d* subshell, it behaves as any of the representative elements as far as predicting geometry by VSEPR theory, so these structures are exactly what we would expect based on that theoretical model.

Coordination Number 4

Two common geometries occur when four ligand atoms are bonded to a metal ion—tetrahedral and square planar. These are illustrated in Figure 21.3. The tetrahedral geometry is usually found with metal ions that have completely filled *d* subshells, such as Zn^{2+}. The complexes $[Zn(NH_3)_4]^{2+}$ and $[Zn(OH)_4]^{2-}$ are examples.

Square planar geometries are observed for complexes of Cu^{2+}, Ni^{2+}, Pd^{2+} and especially Pt^{2+}. Examples are $[Cu(NH_3)_4]^{2+}$, $[Ni(CN)_4]^{2-}$, and $[PtCl_4]^{2-}$. The most well-studied square planar complexes are those of Pt^{2+}, because they are considerably more stable than the others.

Coordination Number 6

The most common coordination number for complex ions is 6. Examples are $[Al(H_2O)_6]^{3+}$, $[Co(C_2O_4)_3]^{3-}$, $[Ni(en)_3]^{2+}$, and $[Co(EDTA)]^-$. With few exceptions, all complexes with a coordination number of 6 are octahedral. This holds true for those formed from both monodentate and bidentate ligands, as illustrated in Figure 21.4. In describing the shapes of octahedral complexes, most chemists use one of the drawings of the octahedron shown in Figure 21.5. Some common coordination numbers for the metal ions are summarized in Table 21.2.

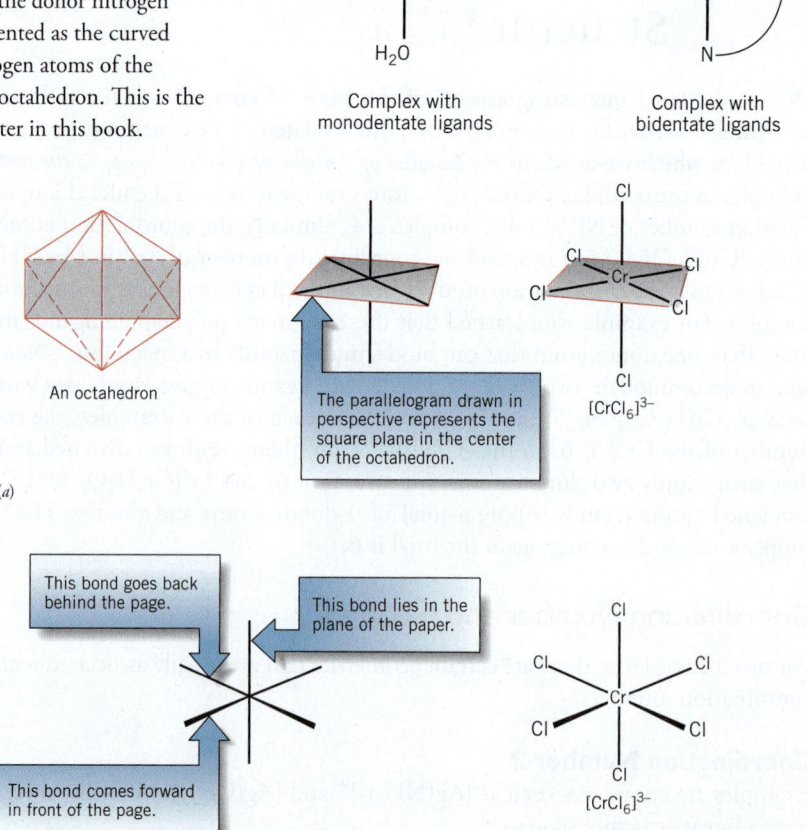

$[Ni(H_2O)_6]^{2+}$

Complex with monodentate ligands

$[Ni(en)_3]^{2+}$
(N⌒N is en.)

Complex with bidentate ligands

An octahedron

The parallelogram drawn in perspective represents the square plane in the center of the octahedron.

(a)

$[CrCl_6]^{3-}$

This bond goes back behind the page.

This bond lies in the plane of the paper.

This bond comes forward in front of the page.

$[CrCl_6]^{3-}$

Figure 21.5 | Representations of the octahedral complex, $[CrCl_6]^{3-}$.
(a) Drawings similar to those you learned to construct in Section 9.1. (b) An alternative method of representing the octahedron.

(b)

TABLE 21.2	Some Common Coordination Numbers of Metal Ions		
Metal Ion	Coordination Number	Metal Ion	Coordination Number
Al^{3+}	4, 6	Ni^{2+}	4, 6
Sc^{3+}	6	Cu^+	2, 4
Ti^{4+}	6	Cu^{2+}	4, 6
V^{3+}	6	Zn^{2+}	4
Cr^{3+}	6	Pd^{2+}	4
Mn^{2+}	6	Ag^+	2
Fe^{2+}	6	Pt^{2+}	4
Fe^{3+}	6	Pt^{4+}	6
Co^{2+}	4, 6	Au^+	2, 4
Co^{3+}	6	Au^{3+}	4

What is the coordination number of the metal ion in $[Cr(C_2O_4)_3]^{3-}$? (*Hint:* Identify the ligand and sketch its structure.)

Practice Exercise 21.6

What is the coordination number of the metal ion in (a) $[CoCl_2(C_2O_4)_2]^{3-}$, (b) $[Cr(C_2O_4)_2(en)]^-$, and (c) $[Co(EDTA)]^-$?

Practice Exercise 21.7

What is the coordination number of the metal ion in (a) $[PtCl_2(NH_3)_2]$ and (b) $[Ag(S_2O_3)_2]^{3-}$? What are the coordination geometries for the complexes?

Practice Exercise 21.8

21.4 | Isomers of Metal Complexes

From a formula, we can name the complex and determine its coordination number, but determining the exact structure of the molecule may be more difficult. Sometimes we can use simple rules to make reasonable structural guesses, as in the discussion of the drawing of Lewis structures in Chapter 8, but these rules apply only to simple molecules and ions. For more complex substances, there usually is no way of knowing for sure what the structure of the molecule or ion is without performing experiments to determine the structure.

One of the reasons that structures can't be predicted with certainty from chemical formulas alone is that there are usually many different ways that the atoms in the formula can be arranged. In fact, it is frequently possible to isolate two or more compounds that actually have the same chemical formula. For example, three different solids, each with its own characteristic color and other properties, can be isolated from a solution of chromium(III) chloride. All three have the same overall composition, and in the absence of other data, their formulas are written $CrCl_3 \cdot 6H_2O$. However, experiments have shown that these solids are actually the salts of three different complex ions. Their actual formulas (and colors) are

$$[Cr(H_2O)_6]Cl_3 \qquad \text{purple}$$

$$[CrCl(H_2O)_5]Cl_2 \cdot H_2O \qquad \text{blue-green}$$

$$[CrCl_2(H_2O)_4]Cl \cdot 2H_2O \qquad \text{green}$$

Even though their overall compositions are the same, each of these substances is a distinct chemical compound with its own unique set of properties.

The existence of two or more compounds, each having the same chemical formula, is known as **isomerism.** In the preceding example, each salt is said to be an **isomer** of $CrCl_3 \cdot 6H_2O$. For coordination compounds, isomerism can occur in a variety of ways. In the case of $CrCl_3 \cdot 6H_2O$, isomers exist because of the different possible ways that the water

Chromium(III) chloride hexahydrate. The hydrated chromium(III) chloride purchased from chemical supply companies is actually $[CrCl_2(H_2O)_4]Cl \cdot 2H_2O$. Its green color in both the solid state and solution is due to the complex ion, $[CrCl_2(H_2O)_4]^+$.

Michael Watson

molecules and chloride ions can be held in the crystals. In one instance, all of the water molecules serve as ligands, while in the other two, part of the water is present as water of hydration and some of the chloride is bonded to the metal ion. Another example, which is similar in some respects, is $Cr(NH_3)_5SO_4Br$. This "substance" can be isolated as two isomers:

$$[CrSO_4(NH_3)_5]Br \quad and \quad [CrBr(NH_3)_5]SO_4$$

They can be distinguished chemically by their differing abilities to react with Ag^+ and Ba^{2+}. The first isomer reacts in aqueous solution with Ag^+ to give a precipitate of AgBr, but it doesn't react with Ba^{2+}. This tells us that Br^- exists as a free ion in the solution. It also suggests that the SO_4^{2-} is bound to the chromium and is unavailable to react with Ba^{2+} to give insoluble $BaSO_4$.

Aqueous solutions of the second isomer, $[CrBr(NH_3)_5]SO_4$, reacts in solution with Ba^{2+} to give a precipitate of $BaSO_4$, which means there is free SO_4^{2-} in the solution. There is no reaction with Ag^+, however, because Br^- is bonded to the chromium and is unavailable as free Br^- in the solution. Thus, even though both isomers have the same overall composition, they behave chemically in quite different ways and are therefore distinctly different compounds.

■ Ag_2SO_4 is soluble but $BaSO_4$ is insoluble. $BaBr_2$ is soluble but AgBr is insoluble.

Stereoisomerism

One of the more interesting kinds of isomerism found among coordination compounds is called **stereoisomerism**, which is defined as *differences among isomers that arise as a result of the various possible orientations of their atoms in space.* In other words, when stereoisomerism exists, we have compounds in which the same atoms are attached to each other, but they differ in the way those atoms are arranged in space relative to one another.

One form of stereoisomerism is called **geometric isomerism**; it is best understood by referring to an example. Consider the square planar complexes having the formula $PtCl_2(NH_3)_2$. There are two ways to arrange the ligands around the platinum, as illustrated below. In one isomer, called the **cis isomer**, the chloride ions are *next to each other* and the ammonia molecules are also next to each other. In the other isomer, called the **trans isomer**, identical ligands are *opposite each other*. In identifying (and naming) isomers, *cis* means "on the same side," whereas *trans* means "on opposite sides."

■ The isomer on the left, *cis–*$PtCl_2(NH_3)_2$, is the anticancer drug known as *cisplatin*. It is interesting that only the *cis* isomer of this compound is clinically active against tumors. The *trans* isomer is totally ineffective.

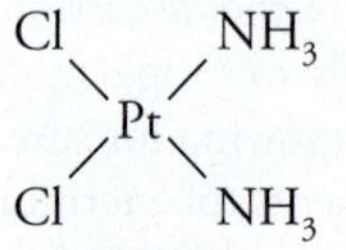

cis isomer
cis-diamminedichloroplatinum(II)

and

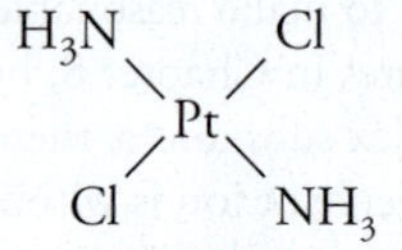

trans isomer
trans-diamminedichloroplatinum(II)

Geometric isomerism also occurs for octahedral complexes. For example, consider the ions $[CrCl_2(H_2O)_4]^+$ and $[CrCl_2(en)_2]^+$. Both can be isolated as *cis* and *trans* isomers.

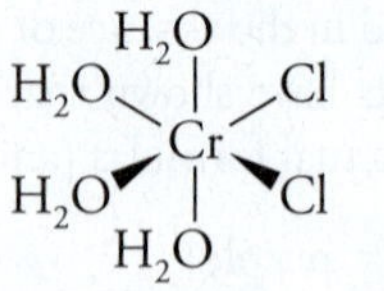

cis

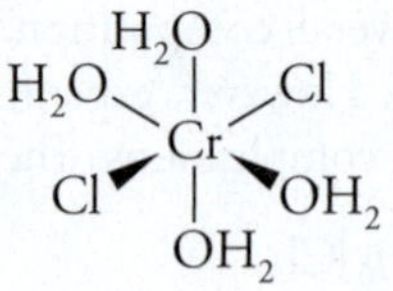

trans

$[CrCl_2(H_2O)_4]^+$

In the *cis* isomer, the chloride ligands are both on the same side of the metal ion; in the *trans* isomer, the chloride ligands are on opposite ends of a line that passes through the center of the metal ion.

Chirality

There is a second kind of stereoisomerism that is much more subtle than geometric isomerism. This occurs when molecules are exactly the same except for one small difference—they are *nonsuperimposable mirror images of each other.* They bear the same relationship to each other as do your left and right hands. (In fact this relationship between such isomers is sometimes referred to by the term *handedness*.)

Although similar in appearance, your two hands are not exactly alike. If you place one hand over the other, palms down, your thumbs point in opposite directions. **Superimposability** is the ability of two objects to fit "one within the other" with no mismatch of parts. Your left and right hands lack this ability, so we say they are *nonsuperimposable.* This is why your right hand doesn't fit into your left-hand glove (Figure 21.6); your right hand does not match *exactly* the space that corresponds to your left hand.

Your hands are also mirror images of each other. If you hold your left hand so it faces a mirror, and then look at its reflection as illustrated in Figure 21.7, you will see that it looks exactly like your right hand. If it were possible to reach "through the looking glass," your right-hand glove would fit the reflection of your left hand perfectly. Thus, your two hands are *nonsuperimposable mirror images* of each other. If two objects are nonsuperimposable mirror images of each other, they are not identical and are said to be **chiral** or have **chirality**. Your left and right hands have this property and are not exactly alike.

Chiral Complexes with Bidentate Ligands

The most common examples of chirality among coordination compounds occur with octahedral complexes that contain two or three bidentate ligands—for instance, $[CoCl_2(en)_2]^+$ and $[Co(en)_3]^{3+}$. For the complex $[Co(en)_3]^{3+}$, the two nonsuperimposable isomers, called **enantiomers**, are shown in Figure 21.8. For the complex $[CoCl_2(en)_2]^+$, only the *cis* isomer is chiral, as described in Figure 21.9.

As you can see, chiral isomers differ in only a very minor way from each other. This difference is so subtle that most of their properties are identical. They have identical melting points and boiling points, and in nearly all of their reactions, their behavior is exactly alike. The only way that the difference between chiral molecules or ions manifests itself is in the way they interact with physical or chemical "probes" that also have a handedness about them. For example, if two reactants are both chiral, then a given isomer of one of them will usually behave slightly differently toward each of the two isomers of the other reactant. This has very profound effects in biochemical reactions, in which many of the molecules involved, such as amino acids, are chiral.

Optical Isomers

One way that chiral isomers differ is in the way they affect polarized light. Light is *electromagnetic radiation* that possesses both electric and magnetic components that behave like vectors. These vectors oscillate in directions perpendicular to the direction in which the light wave is traveling (Figure 21.10*a*). In ordinary light, the oscillations of the electric and magnetic fields of the photons are oriented randomly around the direction of the

Figure 21.6 | One difference between left and right hands. A left-hand glove won't fit the right hand.

■ Chirality is the technical term for "handedness," meaning objects structurally related to each other as are our left and right hands.

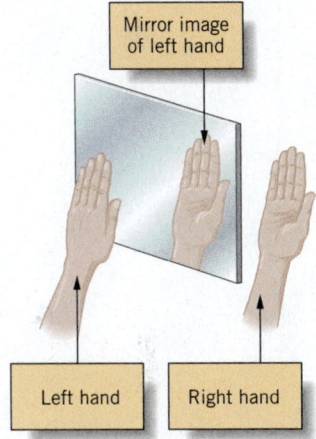

Figure 21.7 | Hands are mirror images of each other. The image of the left hand reflected in the mirror appears the same as the right hand.

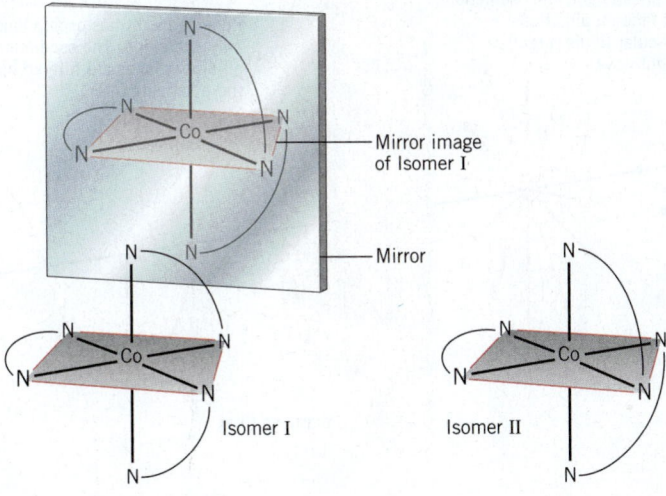

Figure 21.8 | The two isomers of $[Co(en)_3]^{3+}$. Isomer II is constructed as the mirror image of isomer I. No matter how Isomer II is turned about, it is not superimposable on Isomer I.

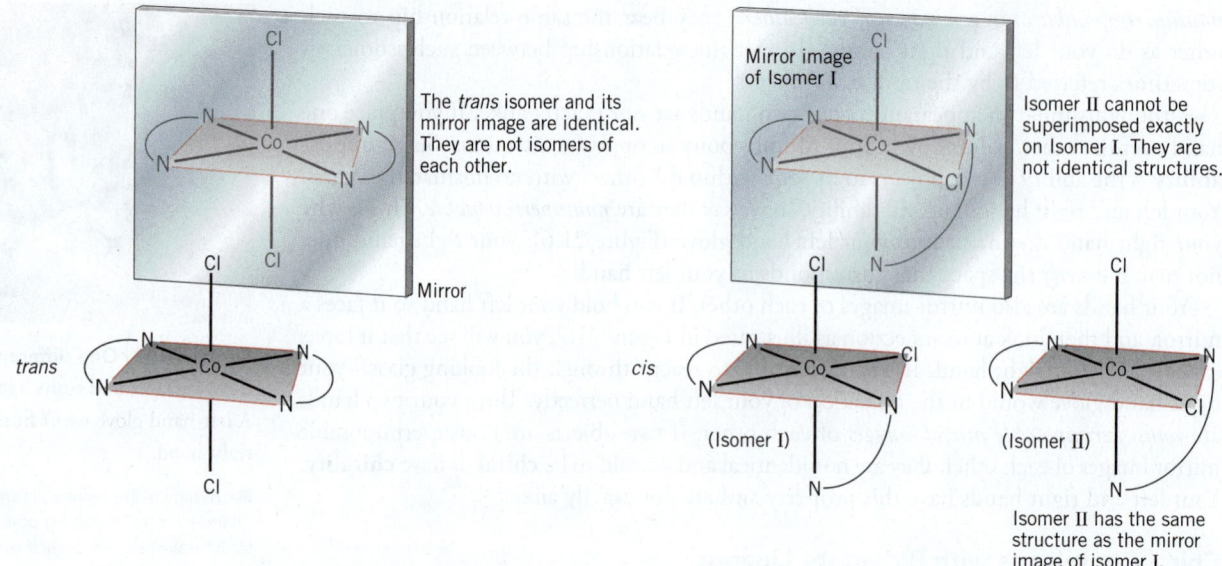

The *trans* isomer and its mirror image are identical. They are not isomers of each other.

Mirror image of Isomer I

Isomer II cannot be superimposed exactly on Isomer I. They are not identical structures.

trans

cis

(Isomer I)

(Isomer II)

Isomer II has the same structure as the mirror image of isomer I.

Figure 21.9 | **Isomers of the $[CoCl_2(en)_2]^+$ ion.** The mirror image of the *trans* isomer can be superimposed exactly on the original, so the *trans* isomer is not chiral. The *cis* isomer (Isomer I) is chiral, however, because its mirror image (Isomer II) cannot be superimposed on the original.

light beam. In **plane-polarized light**, all of the oscillations occur in the same plane (Figure 21.10*b*). Ordinary light can be polarized in several ways. One is to pass it through a special film of plastic, as in a pair of polarized sunglasses. This has the effect of filtering out all of the vibrations except those that are in one plane (Figure 21.10*b*).

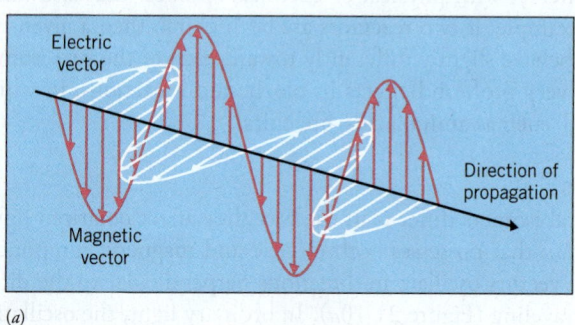

Electric vector

Direction of propagation

Magnetic vector

(a)

Figure 21.10 | **Unpolarized and polarized light.** (*a*) Light possesses electric and magnetic components that oscillate perpendicular to the direction of the propagation of the light wave. (*b*) In unpolarized light, the electromagnetic oscillations of the photons are oriented at random angles around the axis of the propagation of the light wave. The polarizing filter has the effect of preventing oscillations from passing through unless they are in one particular plane. The result is called plane-polarized light.

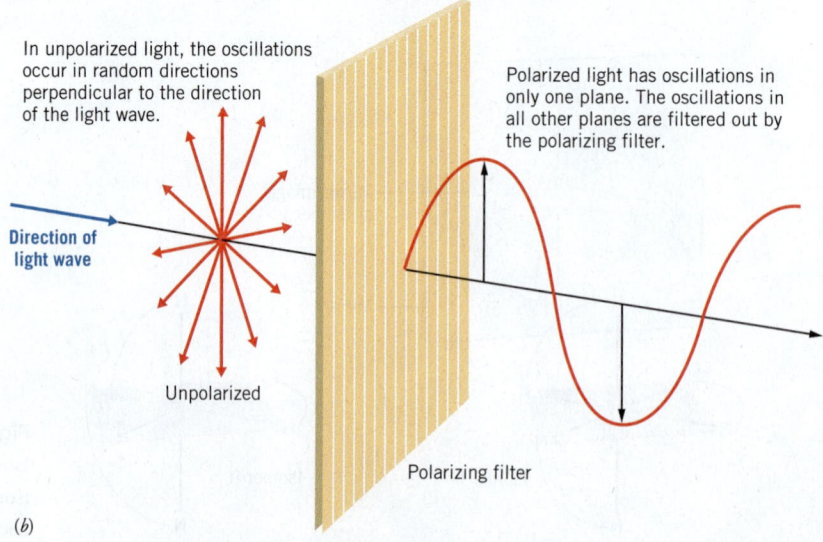

In unpolarized light, the oscillations occur in random directions perpendicular to the direction of the light wave.

Polarized light has oscillations in only one plane. The oscillations in all other planes are filtered out by the polarizing filter.

Direction of light wave

Unpolarized

Polarizing filter

(b)

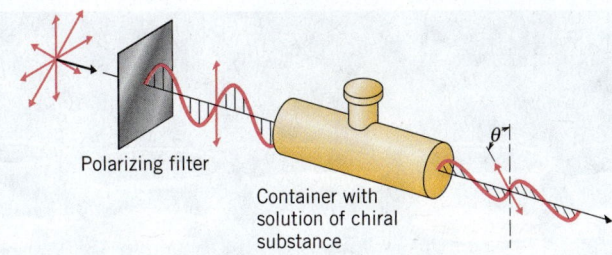

Figure 21.11 | **Rotation of polarized light by a chiral substance.** When plane polarized light passes through a solution of a chiral substance, the plane of polarization is rotated either to the left or to the right. In this illustration, the plane of polarization of the light is rotated to the left (as seen facing the light source).

Chiral isomers like those described in Figures 21.8 and 21.9 have the ability to rotate plane polarized light, as illustrated in Figure 21.11. Because of this phenomenon, chiral isomers are said to be **optical isomers**.

What type of isomers are these compounds? (*Hint:* How does the linkage of the atoms differ between the compounds?)

Practice Exercise 21.9

(a)

(b)

How many isomers can be drawn for $[Cr(H_2O)_6]Br_3$? Draw the coordination compounds.

Practice Exercise 21.10

21.5 | Bonding in Metal Complexes

Complexes of the transition metals differ from the complexes of other metals in two special ways: (1) they are usually colored, whereas complexes of the representative metals are usually (but not always) colorless, and (2) their magnetic properties are often affected by the ligands attached to the metal ion. For example, it is not unusual for a given metal ion to form complexes with different ligands to give a rainbow of colors, as illustrated in Figure 21.12 for a series of complexes of cobalt. Also, because transition metal ions often have incompletely filled d subshells, we expect to find many of them with unpaired d electrons, and compounds that contain them should be paramagnetic. However, for a given metal ion, the number of unpaired electrons is not always the same from one complex to another. For example, Fe^{2+} has four of its six $3d$ electrons unpaired in the $[Fe(H_2O)_6]^{2+}$ ion, but all of its electrons are paired in the $[Fe(CN)_6]^{4-}$ ion. As a result, the $[Fe(H_2O)_6]^{2+}$ ion is paramagnetic and the $[Fe(CN)_6]^{4-}$ ion is diamagnetic.

Figure 21.12 | **Colors of complex ions depend on the nature of the ligands.** Each of these brightly colored solutions contains a complex ion of Co^{3+}. The variety of colors arises because of the different ligands (molecules or anions) that are bonded to the cobalt ion in the complexes.

Crystal Field Theory

Any theory that attempts to explain the bonding in complex ions must also explain their colors and magnetic properties. One of the simplest theories that does this is the **crystal field theory**. This theory gets its name from its original use in explaining the behavior of transition metal ions in crystals. It was discovered later that the theory also works well for transition metal complexes.

Crystal field theory ignores covalent bonding in complexes. It assumes that the primary stability comes from the electrostatic attractions between the positively charged metal ion and the negative charges of the electrons being donated by the ligands. Crystal field theory's unique approach, though, is the way it examines how the negative charges on the ligands affect the energy of the complex by influencing the energies of the *d* orbitals of the metal ion, and this is what we will focus our attention on here. To understand the theory, therefore, it is essential that you know the shapes and orientation of the *d* orbitals. The *d* orbitals were described in Chapter 7, and they are illustrated again in Figure 21.13.

First, notice that four of the *d* orbitals have the same shape but point in different directions. These are the $d_{x^2-y^2}$, d_{xy}, d_{xz}, and d_{yz} orbitals. Each has four lobes of electron density. The fifth *d* orbital, labeled d_{z^2}, has two lobes that point in opposite directions along the *z* axis plus a small donut-shaped ring of electron density around the center that is concentrated in the *xy* plane.

Of prime importance to us are the *directions* in which the lobes of the *d* orbitals point. Notice that three of them—d_{xy}, d_{xz}, and d_{yz}—point *between* the *x*, *y*, and *z* axes. The other two—the d_{z^2} and $d_{x^2-y^2}$ orbitals—have their maximum electron densities *along* the *x*, *y*, and *z* axes.

Now let's consider constructing an octahedral complex within this coordinate system. We can do this by bringing ligands in along each of the axes as shown in Figure 21.14. The question we want to answer is: How do these ligands affect the energies of the *d* orbitals?

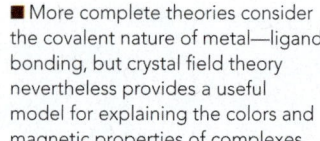

■ More complete theories consider the covalent nature of metal—ligand bonding, but crystal field theory nevertheless provides a useful model for explaining the colors and magnetic properties of complexes.

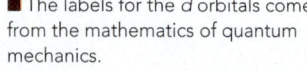

■ The labels for the *d* orbitals come from the mathematics of quantum mechanics.

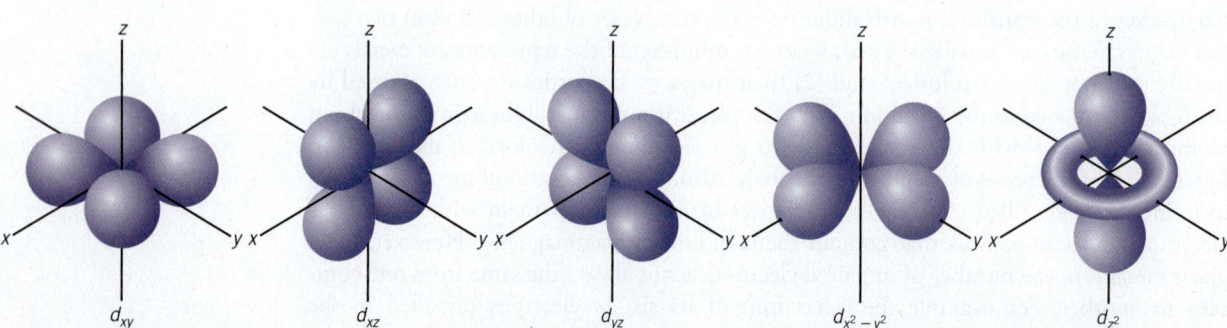

Figure 21.13 | The shapes and directional properties of the five *d* orbitals of a *d* subshell.

In an isolated atom or ion, all of the *d* orbitals of a given *d* subshell have the same energy. Therefore, an electron will have the same energy regardless of which *d* orbital it occupies. In an octahedral complex, however, this is no longer true. If the electron is in the $d_{x^2-y^2}$ or d_{z^2} orbital, it is forced to be nearer the negative charge of the ligands than if it is in a d_{xy}, d_{xz}, or d_{yz} orbital. Since the electron itself is negatively charged and is repelled by the charges of the ligands, the electron's potential energy will be higher in the d_{z^2} and $d_{x^2-y^2}$ orbitals than in a d_{xy}, d_{xz}, or d_{yz} orbital. Therefore, as the complex is formed, the *d* subshell actually splits into *two* new energy levels, as shown in Figure 21.15. Here we see that regardless of which orbital the electron occupies, its energy increases because it is repelled by the negative charges of the approaching ligands. However, the electron is repelled *more* (and has a higher energy) if it is in an orbital that points directly at the ligands than if it occupies an orbital that points between them.

In an octahedral complex, the energy difference between the two sets of *d*-orbital energy levels is called the **crystal field splitting**. It is usually given the symbol Δ (delta), and its magnitude depends on the following factors:

The nature of the ligand. Some ligands produce a larger splitting of the energies of the d orbitals than others. For a given metal ion, for example, cyanide always gives a large value of Δ and F^- always gives a small value. We will have more to say about this later.

The oxidation state of the metal. For a given metal and ligand, the size of Δ increases with an increase in the oxidation number of the metal. As electrons are removed from a metal and the charge on the ion becomes more positive, the ion becomes smaller. This means that the ligands are attracted to the metal more strongly and they can approach the center of the complex more closely. As a result, they also approach the d orbitals along the x, y, and z axes more closely, and thereby cause a greater repulsion. This causes a greater splitting of the two d-orbital energy levels and a larger value of Δ.

The row in which the metal occurs in the periodic table. For a given ligand and oxidation state, the size of Δ increases going down a group. In other words, for a given ligand, an ion of an element in the first row of transition elements has a smaller value of Δ than the ion of a heavier member of the same group. Thus, comparing complexes of Ni^{2+} and Pt^{2+} with the same ligand, we find that the platinum complex has the larger crystal field splitting. The explanation of this is that in the larger ion (e.g., Pt^{2+}), the d orbitals are larger and extend farther from the nucleus in the direction of the ligands. This provides more overlap with the ligand orbitals and produces a larger repulsion between the ligands and the orbitals that point at them.

The magnitude of Δ is very important in determining the properties of complexes, including the stabilities of oxidation states of the metal ions, the colors of complexes, and their magnetic properties. Let's look at some examples.

Stabilities of Oxidation States

Comparing the cations formed by chromium, it is found that the Cr^{2+} ion is very easily oxidized to Cr^{3+}. This can be explained by crystal field theory. In water, we expect these

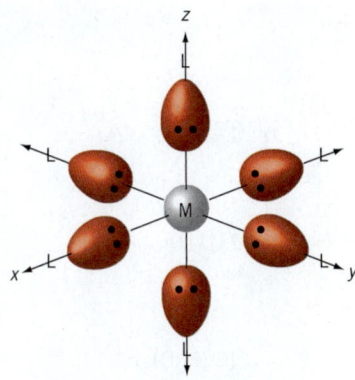

Figure 21.14 | **An octahedral complex ion with ligands that lie along the x, y, and z axes.** The dots represent the lone electron pairs on the ligands, which are oriented so that they point at the metal ion.

TOOLS

Crystal field splitting for octahedral complexes

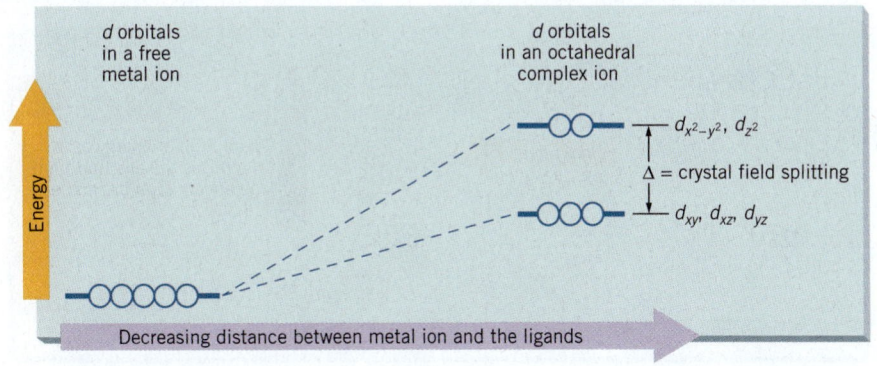

d orbitals in a free metal ion

d orbitals in an octahedral complex ion

$d_{x^2-y^2}$, d_{z^2}

Δ = crystal field splitting

d_{xy}, d_{xz}, d_{yz}

Energy

Decreasing distance between metal ion and the ligands

Figure 21.15 | **The changes in the energies of the d orbitals of a metal ion as an octahedral complex is formed.** As the ligands approach the metal ion, the *d* orbitals split into two new energy levels.

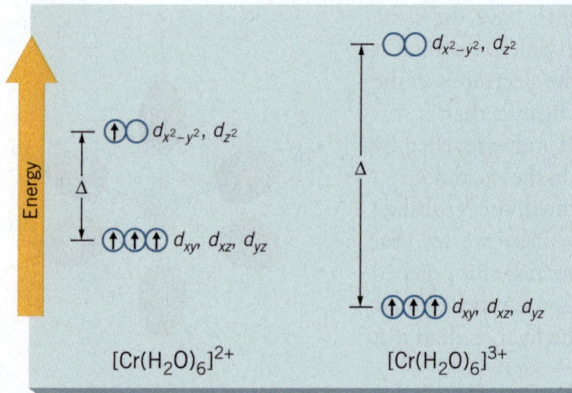

Figure 21.16 | Energy level diagrams for the [Cr(H₂O)₆]²⁺ and [Cr(H₂O)₆]³⁺ ions. The magnitude of Δ is larger for the chromium(III) complex because the Cr³⁺ ion is smaller than the Cr²⁺ ion and the ligands are drawn closer to the metal ion, thereby increasing repulsions felt by the $d_{x^2-y^2}$ and d_{z^2} orbitals.

ions to exist as the complexes $[Cr(H_2O)_6]^{2+}$ and $[Cr(H_2O)_6]^{3+}$, respectively. Let's examine the energies and electron populations of the *d*-orbital energy levels in each complex (Figure 21.16).

The element chromium has the electron configuration

$$Cr \qquad [Ar]3d^5 4s^1$$

Removing two electrons gives the Cr^{2+} ion, and removing three gives Cr^{3+}:

$$Cr^{2+} \qquad [Ar]3d^4$$

$$Cr^{3+} \qquad [Ar]3d^3$$

Next, we distribute the *d* electrons among the various *d* orbitals following Hund's rule, but for Cr^{2+} we have to make a choice. Should the electrons all be forced into the lower of the two energy levels, or should they be spread out? From the diagram we see that the fourth electron does not pair with one of the others in the lower energy *d*-orbital level. Instead, three electrons go into the lower level and the fourth is in the upper level. We will discuss *why* this happens later, but for now, let's use the two energy diagrams to explain why Cr^{2+} is so easy to oxidize.

There are actually two factors that favor the oxidation of Cr(II) to Cr(III). First, the electron that is removed from Cr^{2+} to give Cr^{3+} comes from the *higher* energy level, so oxidizing the chromium(II) *removes* a high-energy electron. The second factor is the effect caused by increasing the oxidation state of the chromium. As we've pointed out, this increases the magnitude of Δ, and as you can see, the energy of the three electrons that remain is lowered. Thus, both the removal of a high-energy electron and the lowering of the energy of the electrons that are left behind help make the oxidation occur, so the $[Cr(H_2O)_6]^{2+}$ ion is very easily oxidized to $[Cr(H_2O)_6]^{3+}$.

Colors of Metal Complexes

When light is absorbed by an atom, molecule, or ion, the energy of the photon raises an electron from one energy level to another. In many substances, such as sodium chloride, the energy difference between the highest energy populated level and the lowest energy unpopulated level is quite large, so the wavelength of a photon that has the necessary energy lies outside the visible region of the spectrum. The substance appears white because visible light is unaffected; it is reflected unchanged.

For complex ions of the transition metals, the energy difference between the *d*-orbital energy levels is not very large, and photons with wavelengths in the visible region of the spectrum are able to raise an electron from the lower-energy set of *d* orbitals to the higher-energy set. This is shown in Figure 21.17 for the $[Cr(H_2O)_6]^{3+}$ ion.

As you know, white light contains photons of all of the wavelengths and colors in the visible spectrum. If we shine white light through a solution of a complex, the light that

■ Remember, $E = h\nu$. The energy of the photon absorbed determines the frequency (and wavelength) of the absorbed light.

Figure 21.17 | Absorption of a photon by the [Cr(H₂O)₆]³⁺ complex. (*a*) The electron distribution in the ground state of the [Cr(H₂O)₆]³⁺ ion. (*b*) Light energy raises an electron from the lower energy set of *d* orbitals to the higher energy set.

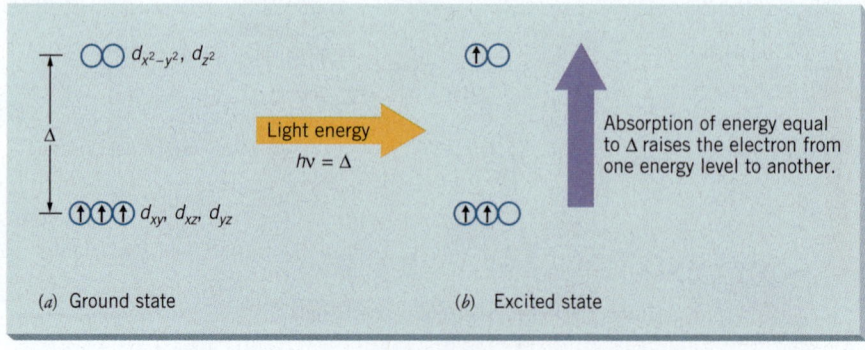

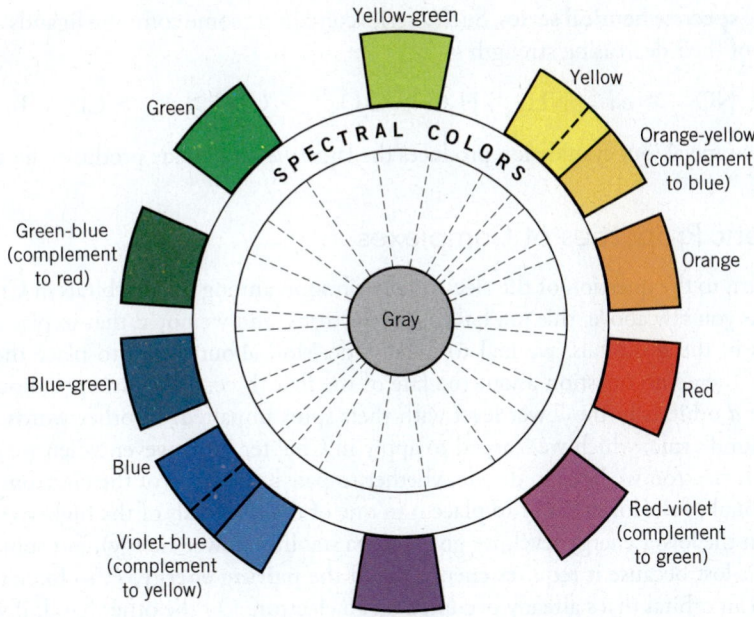

Figure 21.18 | A color wheel. Colors that are across from each other are called complementary colors. When a substance absorbs a particular color, light that is reflected or transmitted has the color of its complement. Thus, something that absorbs red light appears green-blue, and vice versa.

passes through has all of the colors *except* those that have been absorbed. It is not difficult to determine what will be seen if we know what colors are being absorbed. All we need is a color wheel like the one shown in Figure 21.18. Across from each other on the color wheel are **complementary colors**. Green-blue is the complementary color to red, and yellow is the complementary color to violet-blue. If a substance absorbs a particular color when bathed in white light, the perceived color of the reflected or transmitted light is the complementary color. In the case of the $[Cr(H_2O)_6]^{3+}$ ion, the light absorbed when the electron is raised from one set of *d* orbitals to the other has a wavelength of 575 nm, which is the color of yellow light. This is why a solution of this ion appears violet.[4] An instrument that measures the wavelength of light and the amount of light absorbed by a compound is a spectrophotometer (shown in Figure 21.19).

Because of the relationship between the energy and the frequency of light, we see that the color of the light absorbed by a complex depends on the magnitude of Δ; the larger the size of Δ, the more energy the photon must have and the higher will be the frequency of the absorbed light. For a given metal in a given oxidation state, the size of Δ depends on the ligand. Some ligands give a large crystal field splitting, while others give a small splitting. For example, ammonia produces a larger splitting than water, so the complex $[Cr(NH_3)_6]^{3+}$ absorbs light of higher energy and higher frequency than $[Cr(H_2O)_6]^{3+}$. (The $[Cr(NH_3)_6]^{3+}$ absorbs blue light and appears yellow.) Because changing the ligand changes Δ, the same metal ion is able to form a variety of complexes with a large range of colors.

A ligand that produces a large crystal field splitting with one metal ion also produces a large Δ in complexes with other metals. For example, cyanide ion is a very effective ligand and always gives a very large Δ, regardless of the metal to which it is bound. Ammonia is less effective than cyanide ion but more effective than water. Thus, ligands can be arranged in order of their effectiveness at producing a large crystal field splitting. This sequence is

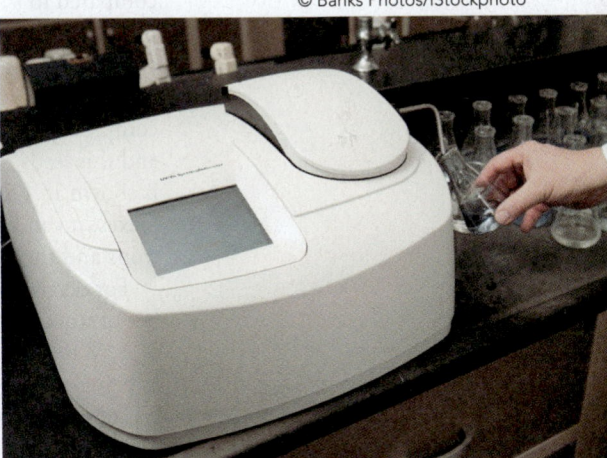

Figure 21.19 | A spectrophotometer typically found in a research laboratory.

---

[4] The perception of color is actually somewhat more complex than this because of the varying sensitivity of the human eye to various wavelengths. For example, the eye is much more sensitive to green than to red. If a compound reflects both of these colors with equal intensity, it will appear greenish simply because the eye sees green better than it sees red.

■ The order of the ligands can be determined by measuring the frequencies of the light absorbed by complexes.

called the **spectrochemical series**. Such a series containing some common ligands arranged in order of their decreasing strength is

$$CN^- > NO_2^- > en > NH_3 > H_2O > C_2O_4^{2-} > OH^- > F^- > Cl^- > Br^- > I^-$$

For a given metal ion, cyanide ion produces the largest Δ and iodide produces the smallest.

Magnetic Properties of Complexes

Let's return to the question of the electron distribution among the d orbitals in Cr^{2+} complexes. As you saw above, this ion has four d electrons, and we noted that in placing these electrons in the d orbitals, we had to make a decision about where to place the fourth electron. There's no question about the fate of the first three. They just spread out across the three d orbitals in the lower level with their spins unpaired. In other words, we just follow Hund's rule, which we learned to apply in Chapter 7. However, when we come to the fourth electron we have to decide whether to pair it with one of the electrons already in a d orbital of the lower set or to place it in one of the d orbitals of the higher set.[5] If we place it in the lower energy level, we give it extra stability (lower energy), but some of this stability is lost because it requires energy, called the **pairing energy**, P, to force the electron into an orbital that's already occupied by an electron. On the other hand, if we place it in the higher level, we are relieved of the burden of pairing the electron, but it also tends to give the electron a higher energy. Thus, for the fourth electron, "pairing" and "placing" affect the energy of the complex in opposite ways.

The critical factor in determining whether the fourth electron enters the lower level and becomes paired, or whether it enters the higher level with the same spin as the other d electrons, is the magnitude of Δ. If Δ is larger than the pairing energy P, then greater stability is achieved if the fourth electron is paired with one in the lower level. If Δ is small compared to P, then greater stability is obtained by spreading the electrons out as much as possible. The complexes $[Cr(H_2O)_6]^{2+}$ and $[Cr(CN)_6]^{4-}$ illustrate this well.

Water is a ligand that does not produce a large Δ, so $P > \Delta$, and minimum pairing of the electrons takes place. This explains the energy level diagram for the $[Cr(H_2O)_6]^{2+}$ complex in Figure 21.20a. When cyanide is the ligand, however, a very large Δ is obtained, and this leads to pairing of the fourth electron with one in the lower set of d orbitals. This is shown in Figure 21.20b. By measuring the degree of paramagnetism of the two complexes, it can be demonstrated experimentally that $[Cr(H_2O)_6]^{2+}$ has four unpaired electrons and the $[Cr(CN)_6]^{4-}$ ion has just two.

For octahedral chromium(II) complexes, there are two possibilities in terms of the number of unpaired electrons. They contain either four or two, depending on the magnitude of Δ. When there is the maximum number of unpaired electrons, the complex is described as being a **high-spin complex**; when there is the minimum number of unpaired electrons it is described as being a **low-spin complex**. High- and low-spin octahedral complexes are

Figure 21.20 | The effect of Δ on the electron distribution in a complex with four d electrons. (*a*) When Δ is small, the electrons remain unpaired. (*b*) When Δ is large, the lower energy level accepts all four electrons and two electrons become paired.

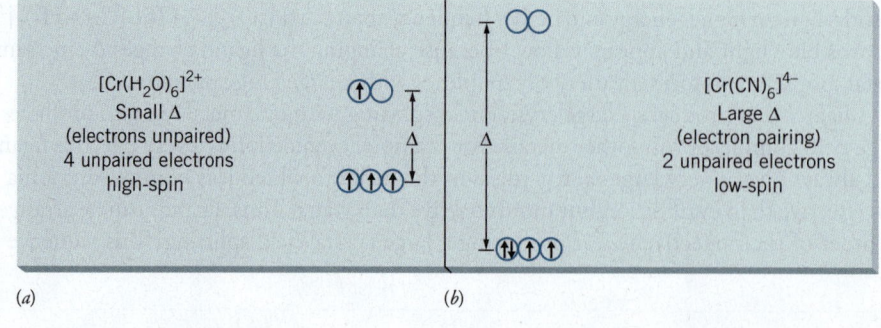

[5] We've never had to make this kind of decision before because the energy levels in atoms were always widely spaced. In complex ions, however, the spacing between the two d-orbital energy levels is fairly small.

possible when the metal has a d^4, d^5, d^6, or d^7 electron configuration. Let's look at another example—one containing the Fe^{2+} ion, which has the electron configuration,

$$Fe^{2+} \qquad [Ar]3d^6$$

At the beginning of this section we mentioned that the $[Fe(H_2O)_6]^{2+}$ ion is paramagnetic and has four unpaired electrons, while the $[Fe(CN)_6]^{4-}$ ion is diamagnetic, meaning it has no unpaired electrons. Now we can see why, by referring to Figure 21.21. Water produces a weak splitting and a minimum amount of pairing of electrons. When the six d electrons in the Fe^{2+} ion are distributed, one must be paired in the lower level after filling the upper level. The result is four unpaired d electrons, and a high-spin complex. Cyanide ion, however, produces a large splitting, so $\Delta > P$. This means that maximum pairing of electrons in the lower level takes place, and a low-spin complex is formed. Six electrons are just the right amount to completely fill all three of these d orbitals, and since all of the electrons are paired, the complex is diamagnetic.

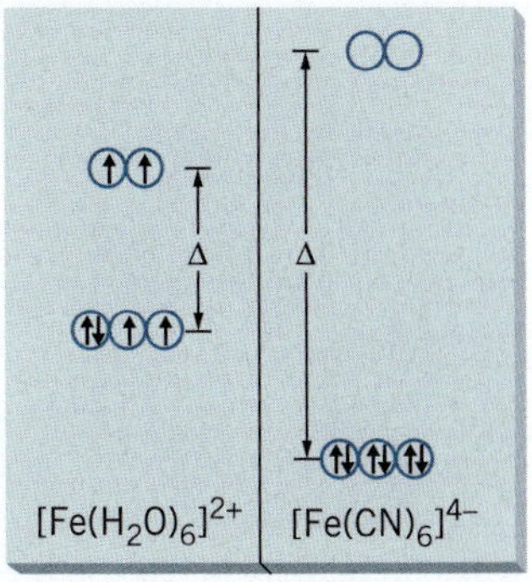

Crystal Field Theory and Other Geometries

The crystal field theory can be extended to geometries other than octahedral. The effect that changing the structure of the complex has on the energies of the d orbitals is to change the splitting pattern.

Square Planar Complexes
We can form a square planar complex from an octahedral one by removing the ligands that lie along the z axis. As this happens, the ligands in the xy plane are able to approach the metal a little closer because they are no longer being repelled by ligands along the z axis. The effect of these changes on the energies of the d orbitals is illustrated in Figure 21.22.

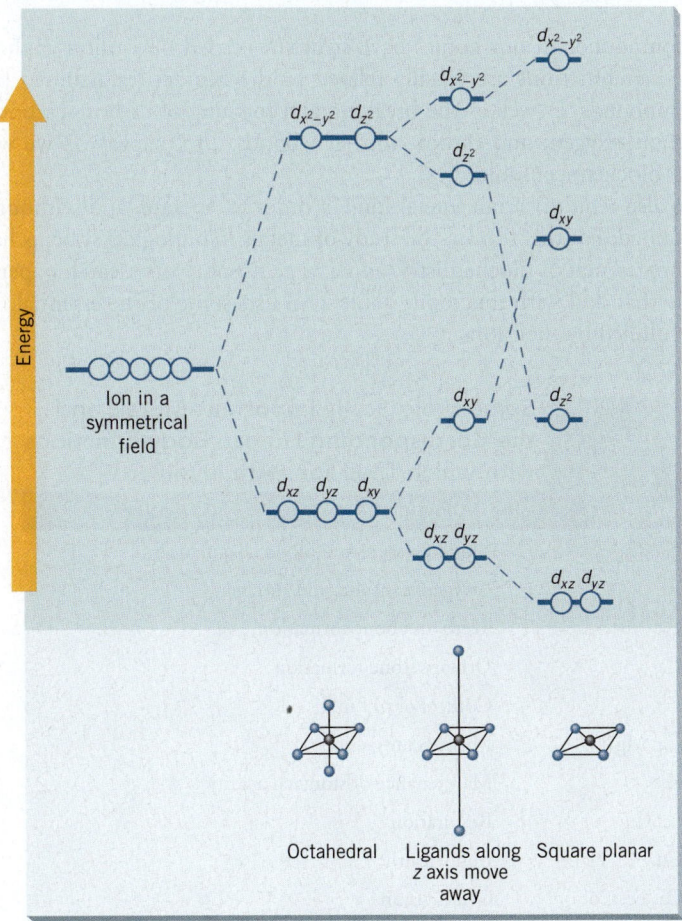

Figure 21.21 | The distribution of d electrons in $[Fe(H_2O)_6]^{2+}$ and $[Fe(CN)_6]^{4-}$. The magnitude of Δ for the cyanide complex is much larger than for the water complex. This produces a maximum pairing of electrons in the lower energy set of d orbitals in $[Fe(CN)_6]^{4-}$.

■ In Figure 21.22, as the energies of the d orbitals change, the d_{z^2} orbital drops in energy by the same amount as the $d_{x^2-y^2}$ rises. Similarly, the energies of the d_{xz} and d_{yz} orbitals drop only half as much as the energy of the d_{yz} rises. In this way, if all the d orbitals were filled, the changes in geometry would have no effect on the total energy of the complex.

Figure 21.22 | Energies of d orbitals in complexes with various structures. The splitting pattern of the d orbitals changes as the geometry of the complex changes.

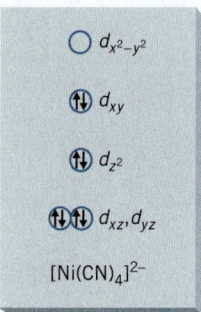

Figure 21.23 | Distribution of the electrons among the *d* orbitals of nickel in the diamagnetic [Ni(CN)₄]²⁻ ion.

The repulsions felt by the *d* orbitals that point in the *z* direction are reduced, so we find that the energies of the d_{z^2}, d_{xz}, and d_{yz} orbitals drop. At the same time, the energies of the orbitals in the *xy* plane feel greater repulsions, so the $d_{x^2-y^2}$ and d_{xy} rise in energy.

Nickel(II) ion (which has eight *d* electrons) forms a complex with cyanide ion that is square planar and diamagnetic. In this complex the strong field produced by the cyanide ions yields a large energy separation between the d_{xy} and $d_{x^2-y^2}$ orbitals, so that a low-spin complex results. The electron distribution in this complex is illustrated in Figure 21.23.

Tetrahedral Complexes

The splitting pattern for the *d* orbitals in a tetrahedral complex is illustrated in Figure 21.24. Now, the ligands are approaching the metal center between the axes, so the order of the energy levels is exactly opposite to that in an octahedral complex. In addition, the size of the crystal field splitting, Δ, is also much smaller for a tetrahedral complex than for an octahedral one. (Actually, $\Delta_{\text{tet}} \approx \frac{4}{9} \Delta_{\text{oct}}$ for the same metal ion with the same ligands.) This small Δ is always less than the pairing energy, so tetrahedral complexes are almost always high-spin complexes.

Practice Exercise 21.11

Octahedral cobalt(III) complexes can be either low spin or high spin. Draw an energy level diagram for a high-spin complex and a low-spin complex, and include the electrons. How many unpaired electrons does each complex have? Is either complex diamagnetic? Why or why not? (*Hint:* Determine the number of electrons in Co³⁺).

Practice Exercise 21.12

Construct an energy level diagram for a tetrahedral Ni(0) complex. How many unpaired electrons would this compound have?

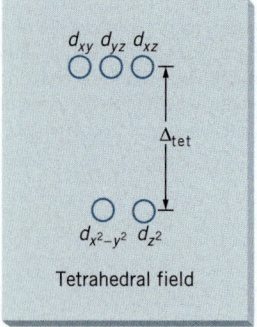

Figure 21.24 | Splitting pattern of the *d* orbitals for a tetrahedral complex. The crystal field splitting for the tetrahedral structure (Δ_{tet}) is smaller than that for the octahedral structure (Δ_{oct}). For complexes with the same ligands, $\Delta_{\text{tet}} \approx \frac{4}{9} \Delta_{\text{oct}}$.

21.6 | Biological Functions of Metal Ions

Most of the compounds in our bodies have structures based on carbon as the principal element, and their functions are usually related to the geometries assumed by carbon-containing compounds, as well as the breaking and forming of carbon–carbon, carbon–hydrogen, carbon–oxygen, and carbon–nitrogen bonds. In Chapter 22 we will discuss some classes of biochemical molecules.

Our bodies also require certain metal ions in order to operate, and without them life cannot be sustained. For this reason, the study of metals in biological systems has become a very important branch of biochemistry, and a large number of research papers are published annually that deal with this topic. Table 21.3 lists some of the essential metals and the functions fulfilled by their ions.

TABLE 21.3	Some Biologically Important Metals and the Corresponding Human Body Functions with which Their Ions Are Involved
Metal	**Body Function**
Na, Ca	Blood pressure and blood coagulation
Fe	Oxygen transport and storage
Ca	Teeth and bone formation
Ca	Urinary stone formation
Zn	Control of pH in blood
Ca, Mg	Muscle contraction
K	Maintenance of stomach acidity
Fe, Cu	Respiration
Cu	Bone health
Ca, Fe, Co	Cell division

A few metals, such as sodium and potassium, are found as simple hydrated monatomic ions in body fluids. Most metal ions, however, become bound by other ligands and do their work as part of metal complexes. As examples, we will look briefly at two metals—iron and cobalt.

Iron is one of the essential elements required by our bodies. We obtain it in a variety of ways in our diets. Iron is involved in oxygen transport in our blood and in retaining oxygen in muscle tissue so that it's available when needed. The iron is present as Fe^{2+} held in a complex in which the basic ligand structure is

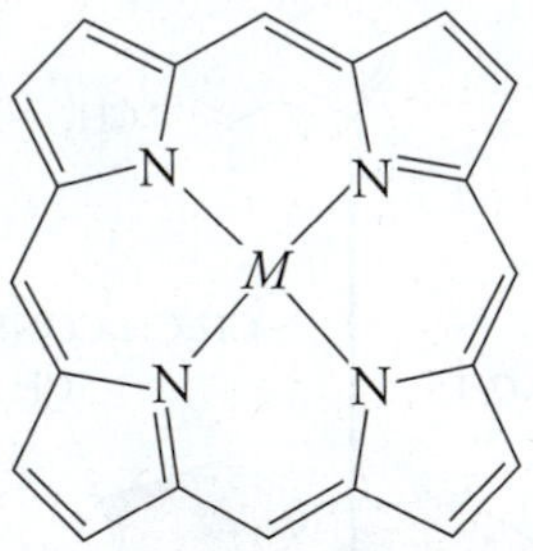

This ligand composition, with its square planar arrangement of nitrogen atoms that bind to a metal ion, is called a *porphyrin structure*. It is the ligand structure in a biologically active unit called heme. Heme is the oxygen-carrying component of the blood protein hemoglobin and in myoglobin, which is found in muscle tissue.

In the lungs, O_2 molecules are absorbed by the blood and become bound to Fe^{2+} ions in the heme units of hemoglobin (Figure 21.25). Blood circulation then carries the O_2 to tissues where it is needed, at which time it is released by the Fe^{2+}. One of the important functions of the porphyrin ligand in this process is to prevent the Fe^{2+} from being oxidized by the O_2. (In fact, if the iron is oxidized to Fe^{3+}, it is no longer able to carry O_2.) In muscle tissue, heme units in the protein myoglobin take O_2 from hemoglobin and hold it until it's needed. In this way, muscle tissue is able to store O_2 so that plenty of it is available when the muscle must work hard.

Heme units are also present in proteins called cytochromes, where the iron is involved in electron transfer reactions that employ the +2 and +3 oxidation states of iron.

■ The porphyrin structure is present in chlorophyll too, where the metal bound in the center is Mg^{2+}. Chlorophyll absorbs sunlight (solar energy) which is used by plants to convert carbon dioxide and water into glucose and oxygen.

■ The iron in hemoglobin also binds CO_2 and helps transport it back to the lungs to be exhaled.

■ The ability of transition metals to exist in different oxidation states is one reason they are used in biological systems. They can easily take part in oxidation–reduction reactions. Copper(I) and copper(II) ions constitute another pair involved in catalyzing biochemical redox reactions.

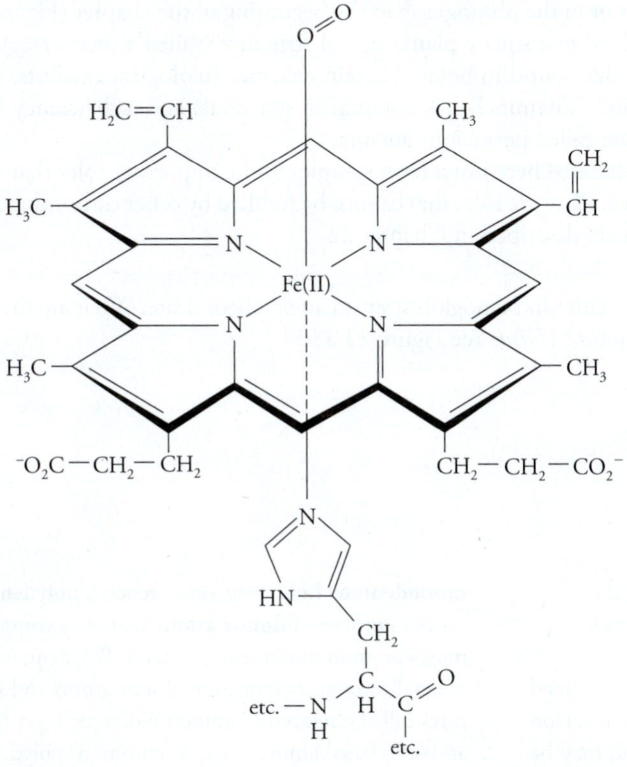

Figure 21.25 | Iron(II) bound to oxygen in heme.
The porphyrin ring ligand in heme surrounds an Fe^{2+} ion that binds an O_2 molecule. In its oxygenated form, the Fe^{2+} is octahedrally coordinated in the structure.

Figure 21.26 | **The structure of cyanocobalamin.**
Notice the cobalt ion in the center of the square
planar arrangement of nitrogen atoms that are part
of the ligand structure. Overall, the cobalt is
surrounded octahedrally by donor atoms.

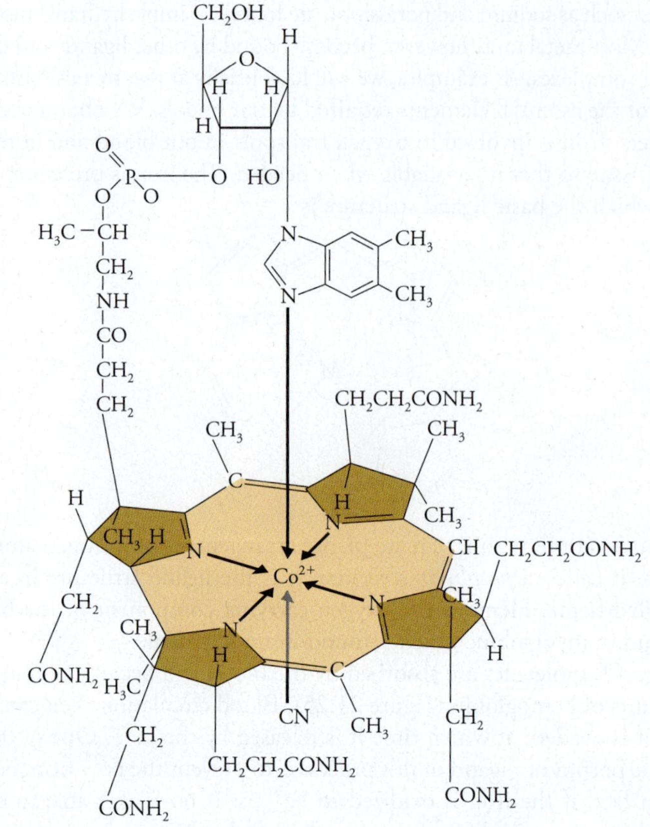

Vitamin B_{12}

A structure similar to heme is found in cyanocobalamin, the form of vitamin B_{12} found
in vitamin pills or in the photograph at the beginning of the chapter (Figure 21.26). Here,
a Co^{2+} ion is held in a square planar ligand structure (called a *corrin ring*) that is slightly
different from that found in heme. Certain enzymes (biological catalysts) require cobala-
mins to function. Vitamin B_{12} is essential in our diets, and a deficiency in this vitamin
leads to a disease called pernicious anemia.

We have illustrated here just a few examples of the important roles that metal ions play
in living systems. They are roles that cannot be fulfilled by other carbon-based compounds,
some of which are described in Chapter 22.

Practice Exercise 21.13   The iron metal center in hemoglobin sits in an octahedral site. What are the ligands for the
iron in hemoglobin? (*Hint:* See Figure 22.25.)

Summary

Organized by Learning Objective

Describe the different kinds of ligands and the rules for writing formulas for metal complexes containing them

Coordination compounds contain **complex ions** (also called
complexes or **coordination complexes**), formed from a metal ion
and a number of ligands. **Ligands** are Lewis bases and may be

monodentate, **bidentate**, or, in general, **polydentate**, depending
on the number of **donor atoms** that they contain. Water is the
most common monodentate ligand. Polydentate ligands bind to
a metal through two or more donor atoms and yield ring structures called **chelates**. Common bidentate ligands are oxalate ion
and ethylenediamine (en); a common polydentate ligand is

ethylenediaminetetraacetic acid (EDTA), which has six donor atoms.

Complexes of polydentate ligands are more stable than similar complexes formed with monodentate ligands. The phenomenon is called the **chelate effect**.

Use the rules for naming metal complexes
Complexes are named following a set of rules developed by the IUPAC. These are summarized in Section 21.2.

Show the different geometries associated with each of the common coordination numbers, and draw the structures
The **coordination number** of a metal ion in a complex is the number of donor atoms attached to the metal ion. Polydentate ligands supply two or more donor atoms, which must be taken into account when determining the coordination number from the formula of the complex. Geometries associated with common coordination numbers are as follows: for coordination number 2, linear; for coordination number 4, tetrahedral and square planar (especially for Pt^{2+} complexes); and for coordination number 6, octahedral.

Write structural formulas for all the isomers of a given metal complex
When two or more distinct compounds have the same chemical formula, they are **isomers** of each other. **Stereoisomers** have the same atoms attached to each other, but the atoms are arranged differently in space. In a *cis* **isomer**, attached groups of atoms are on the same side of some reference plane through the molecule. In a *trans* **isomer**, they are on opposite sides. *Cis* and *trans* isomerism is a form of **geometric isomerism. Chiral** isomers are exactly the same in every way but one—they are not superimposable on their mirror images. These kinds of isomers exist for complexes of the type $M(AA)_3$, where M is a metal ion and AA is a bidentate ligand, and also for complexes of the type *cis*-$M(AA)_2a_2$, where a is a monodentate ligand. Chiral isomers that are mirror images of each other are said to be **enantiomers**. Because they are able to rotate **plane-polarized light**, they are called **optical isomers**.

Describe how crystal field theory explains the properties of metal complexes
In an octahedral complex, the ligands influence the energies of the d orbitals by splitting the d subshell into two energy sublevels. The lower-energy one consists of the d_{xy}, d_{xz}, and d_{yz} orbitals; the higher-energy level consists of the d_{z^2} and $d_{x^2-y^2}$ orbitals. The energy difference between the two new d sublevels is the **crystal field splitting**, Δ, and for a given ligand it increases with an increase in the oxidation state of the metal and it depends on the period number in which the metal is found. For a given metal ion, Δ depends on the ligand.

In the **spectrochemical series**, ligands are arranged in order of their ability to cause a large Δ. Cyanide ion produces the largest crystal field splitting, whereas iodide ion produces the smallest. **Low-spin** complexes result when Δ is larger than the **pairing energy**—the energy needed to cause two electrons to become paired in the same orbital. **High-spin** complexes occur when Δ is smaller than the pairing energy. Light of energy equal to Δ is absorbed when an electron is raised from the lower-energy set of d orbitals to the higher set, and the color of the complex is determined by the colors that remain in the transmitted light. Crystal field theory can also explain the relative stabilities of oxidation states, in many cases.

Different splitting patterns of the d orbitals occur for other geometries. The patterns for square planar and tetrahedral geometries are described in Figures 21.22 and 21.24, respectively. The value of Δ for a tetrahedral complex is only about $\frac{4}{9}$ that of Δ for an octahedral complex.

Explain the biological function of several common metal complexes in the body
Most metals required by living organisms perform their actions when bound as complex ions. Heme contains Fe^{2+} held in a square planar porphyrin ligand and binds to O_2 in hemoglobin and myoglobin. Heme is also found in cytochromes, where it participates in redox reactions involving Fe^{2+} and Fe^{3+}. Vitamin B_{12}, required by the body to prevent the vitamin deficiency disease called pernicious anemia, contains Co^{2+} in a corrin ring structure, which is similar to the porphyrin ring.

Tools for Problem Solving
The following tools were introduced in this chapter. Study them carefully so you can select the appropriate tool when needed.

Rules for writing formulas for complexes (Section 21.1)
1. The symbol for the metal ion is always given first, followed by the ligands.
2. When more than one kind of ligand is present, anionic ligands are written first (in alphabetical order), followed by neutral ligands (also in alphabetical order).
3. The charge on the complex is the algebraic sum of the charge on the metal ion and the charges on the ligands.
4. The formula is placed inside of square brackets with the charge of the complex as a superscript outside the brackets, if it is not zero.

Rules for naming complexes (Section 21.2)
Naming complexes follows rules that are an extension of the rules you learned earlier.

Crystal field splitting pattern for octahedral complexes (Section 21.5)

Figure 21.15 forms the basis for applying the principles of crystal field theory to octahedral complexes. For a complex under consideration, set up the splitting diagram and place electrons into the d orbitals following Hund's rule. For d^4, d^5, d^6, and d^7 configurations, you may have to decide whether a high- or low-spin configuration is preferred.

Spectrochemical series (Section 21.5)

$$CN^- > NO^- > en > NH_3 > H_2O > C_2O_4^{2-} > OH^- > F^- > Cl^- > Br^- > I^-$$

WileyPLUS = *WileyPLUS*, an online teaching and learning solution. *Note to instructors*: Many of the end-of-chapter problems are available for assignment via the WileyPLUS system. **www.wileyplus.com.** Review Problems are presented in pairs separated by blue rules. Answers to problems whose numbers appear in blue are given in Appendix B. More challenging problems are marked with an asterisk ★.

|Review Questions

Complex Ions

21.1 The formation of the complex ion $[Cu(H_2O)_6]^{2+}$ is described as a Lewis acid–base reaction. Explain.

   (a) What are the formulas of the Lewis acid and the Lewis base in this reaction?

   (b) What is the formula of the ligand?

   (c) What is the name of the species that provides the donor atom?

   (d) What atom is the donor atom, and why is it so designated?

   (e) What is the name of the species that is the acceptor?

21.2 To be a ligand, a substance should also be a Lewis base. Explain.

21.3 Give two examples of a charged ligand and two examples of an uncharged ligand.

21.4 Why are substances that contain complex ions often called coordination compounds?

21.5 Use Lewis structures to diagram the formation of $Cu(NH_3)_4^{2+}$ and $CuCl_4^{2-}$ ions from their respective components.

21.6 What must be true about the structure of a ligand classified as *bidentate*?

21.7 Label the following as monodentate, bidentate, or polydentate.

   (a) Cl^-

   (b)

   (c) NH_3

   (d) $H_2\ddot{N}-CH_2-CH_2-\ddot{N}H_2$

   (e) CN^-

(f)

21.8 What is a *chelate*? Use Lewis structures to diagram the way that the oxalate ion, $C_2O_4^{2-}$, functions as a chelating agent.

21.9 How many donor atoms does $EDTA^{4-}$ have?

21.10 Explain how a salt of $EDTA^{4-}$ can retard the spoilage of salad dressing.

21.11 How does a salt of $EDTA^{4-}$ in shampoo make the shampoo work better in hard water?

21.12 The cobalt(III) ion, Co^{3+}, forms a 1:1 complex with $EDTA^{4-}$. What is the net charge, if any, on this complex, and what would be a suitable formula for it (using the symbol EDTA)?

21.13 Which complex is more stable, $[Cr(NH_3)_6]^{3+}$ or $[Cr(en)_3]^{3+}$? Why?

21.14 What is the chelate effect? How does the dissociation of a ligand influence the chelate effect? What role does entropy have in the chelate effect?

Metal Complex Nomenclature

21.15 When naming a complex, how are the metal, and ligands ordered?

21.16 How are the names of anionic ligands changed when they are part of the name of a coordination complex?

21.17 How are water and ammonia names in coordination complexes?

21.18 In a formula, what is the order of the metal and ligands?

21.19 What suffix is used for anionic complexes.

21.20 How are the oxidation states of a metal indicated in coordination complexes?

Coordination Number and Structure

21.21 What is a *coordination number*? What structures are generally observed for complexes in which the central metal ion has a coordination number of 4? What is the most common structure observed for coordination number 6?

21.22 Sketch the structure of an octahedral complex that contains only identical monodentate ligands. Use M for the metal and L for the ligand.

*21.23 Sketch the structure of the octahedral $[Co(EDTA)]^-$ ion. Remember that adjacent donor atoms in a polydentate ligand span adjacent positions in the octahedron.

21.24 Give the geometry or geometries for coordination number (a) 2, (b) 4, and (c) 6.

21.25 Draw (a) a tetrahedral structure, (b) a square planar structure, (c) a tetrahedral structure, and (d) an octahedral structure.

Isomers of Metal Complexes

21.26 What are *isomers*?

21.27 Define *stereoisomerism*, *geometric isomerism*, *chiral isomers*, and *enantiomers*.

21.28 What are *cis* and *trans* isomers?

21.29 What condition must be fulfilled in order for a molecule or ion to be chiral?

21.30 What are optical isomers?

21.31 What are the differences between optical and geometric isomers?

Bonding in Metal Complexes

21.32 On appropriate coordinate axes, sketch and label the five d orbitals.

21.33 Which d orbitals point *between* the x, y, and z axes? Which point *along* the coordinate axes?

21.34 Explain why an electron in a d_{z^2} or $d_{x^2-y^2}$ orbital in an octahedral complex will experience greater repulsions because of the presence of the ligands than an electron in a d_{xy}, d_{xz}, or d_{yz} orbital.

21.35 Sketch the d-orbital energy level diagram for a typical octahedral complex.

21.36 Explain the role of (a) the ligand, (b) the oxidation state of the metal, (c) and the position of the metal in the periodic table on the crystal field splitting of the d orbitals in a metal complex.

21.37 Explain why octahedral cobalt(II) complexes are easily oxidized to cobalt(III) complexes. Sketch the d-orbital energy diagram and assume a large value of Δ when placing electrons in the d orbitals.

21.38 Explain how the same metal in the same oxidation state is able to form complexes of different colors.

21.39 If a complex appears red, what color light does it absorb? What color light is absorbed if the complex appears yellow?

21.40 What does the term *spectrochemical series* mean? How can the order of the ligands in the series be determined?

21.41 What do the terms *low-spin complex* and *high-spin complex* mean?

21.42 For which d-orbital electron configurations are both high-spin and low-spin octahedral complexes possible?

21.43 What factors about the metal and ligands affect the magnitude of Δ?

21.44 Indicate by means of a sketch what happens to the d-orbital electron configuration of the $[Fe(CN)_6]^{4-}$ ion when it absorbs a photon of visible light.

21.45 The complex $[Co(C_2O_4)_3]^{3-}$ is diamagnetic. Sketch the d-orbital energy level diagram for this complex and indicate the electron populations of the orbitals.

21.46 Consider the complex $[M(H_2O)_2Cl_4]^-$ illustrated below on the left. Suppose the structure of this complex is distorted to give the structure on the right, where the water molecules along the z axis have moved away from the metal somewhat and the four chloride ions along the x and y axes have moved closer. What effect will this distortion have on the energy level splitting pattern of the d orbitals? Use a sketch of the splitting pattern to illustrate your answer.

Biological Functions of Metals Ions

21.47 In what ways is the porphyrin structure important in biological systems?

21.48 If a metal ion is held in the center of a porphyrin ring structure, what is its coordination number? (Assume the porphyrin is the only ligand.)

21.49 What function does heme serve in hemoglobin? What does it do in myoglobin?

21.50 How are the ligand ring structures similar in vitamin B_{12} and in heme? What metal is coordinated in cobalamin?

21.51 What are some of the roles played by calcium ion in the body?

21.52 List some biologically important metals. What role do they play in the body?

Review Problems

Complex Ions

21.53 The iron(III) ion forms a complex with six cyanide ions that is often called the ferricyanide ion. What is the net charge on this complex ion, and what is its formula? What is the IUPAC name for the complex?

21.54 The silver ion forms a complex ion with two ammonia molecules. What is the formula of this ion, and what is its IUPAC name? Can this complex ion exist as a salt with the sodium ion or with the chloride ion? Write the formula of the possible salt. (Use brackets and parentheses correctly.)

21.55 Write the formula, including its correct charge, for a complex that contains Cr^{3+}, two NH_3 ligands, and four NO_2^- ligands.

21.56 Write the formula, including its correct charge, for a complex that contains Co^{3+}, two Cl^-, and two ethylenediamine ligands.

Metal Complex Nomenclature

21.57 How would the following molecules or ions be named as ligands when writing the name of a complex ion?

(a) NH_3 (b) N_3^- (c) SO_4^{2-} (d) $C_2H_3O_2^-$

21.58 How would the following molecules or ions be named as ligands when writing the name of a complex ion?

(a) $C_2O_4^-$ (c) S^{2-}
(b) $(CH_3)_2NH$ (dimethylamine) (d) Cl^-

21.59 Give IUPAC names for each of the following:

(a) $[Ni(NH_3)_6]Cl_2$ (d) $[Mn(CN)_4(NH_3)_2]^{2-}$
(b) $[CrCl_3(NH_3)_3]^-$ (e) $K_3[Fe(C_2O_4)_3]$
(c) $[Co(NO_2)_6]^{3-}$

21.60 Give IUPAC names for each of the following:

(a) $[AgI_2]^-$ (d) $[CrCl(NH_3)_5]SO_4$
(b) $[SnS_3]^{2-}$ (e) $K_3[Co(C_2O_4)_3]$
(c) $[Co(en)_2(H_2O)_2]SO_4$

21.61 Write chemical formulas for each of the following:

(a) tetraaquadicyanoiron(III) ion
(b) tetraammineoxalatonickel(II)
(c) pentaaquahydroxoaluminum(III) chloride
(d) potassium hexathiocyanatomanganate(III)
(e) tetrachlorocuprate(II) ion

21.62 Write chemical formulas for each of the following:

(a) tetrachloroaurate(III) ion
(b) bis(ethylenediamine)dinitroiron(III) ion
(c) tetraamminedicarbonatocobalt(III) nitrate
(d) ethylenediaminetetraacetatoferrate(II) ion
(e) diamminedichloroplatinum(II)

Coordination Number and Structure

21.63 What is the coordination number of nickel in $[Ni(C_2O_4)_2(NO_2)_2]^{4-}$?

21.64 What is the coordination number of Fe in $FeCl_2(H_2O)_2(en)]$? What is the oxidation number of iron in this complex?

21.65 Draw a reasonable structure for (a) $[Zn(NH_3)_4]^{2+}$ and (b) trioxalatochromate(III) ion.

21.66 Draw a reasonable structure for (a) $[CoBr(NH_3)_3]^{2+}$ and (b) dichloroethylenediamineplatinum(II).

21.67 NTA is the abbreviation for nitrilotriacetic acid, a substance that was used at one time in detergents. Its structure is

The four donor atoms of this ligand are shown in red. Sketch the structure of an octahedral complex containing this ligand. Assume that two water molecules are also attached to the metal ion and that each oxygen donor atom in the NTA is bonded to a position in the octahedron that is adjacent to the nitrogen of the NTA.

21.68 The following compound is called diethylenetriamine and is abbreviated "dien":

$$\ddot{N}H_2-CH_2-CH_2-\ddot{N}H-CH_2-CH_2-\ddot{N}H_2$$

When it bonds to a metal, it is a ligand with three donor atoms.

(a) Which are most likely the donor atoms?
(b) What is the coordination number of cobalt in the complex $Co(dien)^{3+}$?
(c) Sketch the structure of the complex $[Co(dien)_2]^{3+}$.
(d) Which complex would be expected to be more stable in aqueous solution, $[Co(dien)_2]^{3+}$ or $[Co(NH_3)_6]^{3+}$?
(e) What would be the structure of triethylenetetraamine?

Isomers of Metal Complexes

*21.69 Below are two structures drawn for a complex. Are they actually different isomers, or are they identical? Explain your answer.

Structure I   Structure II

*21.70 Below is a structure for one of the isomers of the complex [Co(H₂O)₃(dien)]³⁺. Are isomers of this complex chiral? Justify your answer. See Problem 21.68.

(represents dien)

*21.71 Sketch and label the isomers of the square planar complex [PtBrCl(NH₃)₂].

*21.72 The complex [CoCl₃(NH₃)₃] can exist in two isomeric forms. Sketch them.

*21.73 Sketch the chiral isomers of [CrCl₂(en)₂]⁺. Is there a non-chiral isomer of this complex?

*21.74 Sketch the chiral isomers of [Co(C₂O₄)₃]³⁻.

Bonding in Metal Complexes

*21.75 In which complex do we expect to find the larger Δ?

(a) [Cr(H₂O)₆]²⁺ or [Cr(H₂O)₆]³⁺

(b) [Cr(en)₃]³⁺ or [CrCl₆]³⁻

*21.76 Arrange the following complexes in order of increasing wavelength of the light absorbed by them: [Cr(H₂O)₆]³⁺, [CrCl₆]³⁻, [Cr(en)₃]³⁺, [Cr(CN)₆]³⁻, [Cr(NO₂)₆]³⁻, [CrF₆]³⁻, and [Cr(NH₃)₆]³⁺.

*21.77 Which complex should be expected to absorb light of the highest frequency: [Cr(H₂O)₆]³⁺, [Cr(en)₃]³⁺, or [Cr(CN)₆]³⁻?

*21.78 Which complex should absorb light at the longer wavelength?

(a) [Fe(H₂O)₆]²⁺ or [Fe(CN)₆]⁴⁻

(b) [Mn(CN)₆]³⁻ or [Mn(CN)₆]⁴⁻

*21.79 In each pair below, which complex is expected to absorb light of the shorter wavelength? Justify your answers.

(a) [RuCl(NH₃)₅]²⁺ or [FeCl(NH₃)₅]²⁺

(b) [Ru(NH₃)₆]²⁺ or [Ru(NH₃)₆]³⁺

*21.80 A complex [CoA₆]³⁺ is red. The complex [CoB₆]³⁺ is green. Which ligand, A or B, produces the larger crystal field splitting, Δ? Explain your answer.

*21.81 Referring to the two ligands, A and B, described in Problem 21.80, which complex would be expected to be more easily oxidized, [CoA₆]²⁺ or [CoB₆]²⁺? Explain your answer.

*21.82 Referring to the complexes in Problems 21.80 and 21.81, would the color of [CoA₆]²⁺ more likely be red or blue?

*21.83 Would the complex [CoF₆]⁴⁻ more likely be low spin or high spin? Could it be diamagnetic?

*21.84 Would the complex [Cr(CN)₆]⁴⁻ more likely be low spin or high spin? Could it be diamagnetic?

*21.85 Sketch the d-orbital energy level diagrams for [Fe(H₂O)₆]³⁺ and [Fe(CN)₆]³⁻ and predict the number of unpaired electrons in each.

*21.86 Sketch the d-orbital energy level diagrams for [MnF₆]²⁻ and [Mn(CN)₆]³⁻ and predict the number of unpaired electrons in each.

|Additional Exercises

*21.87 Ni²⁺ ions can be either four coordinate or six coordinate; which one is more likely to be a red compound and which is likely to be a blue compound?

*21.88 For the complex (NH₄)₂[Pd(SCN)₄], draw the structure of the free ligand, draw the structure of the complex, indicate the shape of the complex, and give the oxidation state of the central metal ion.

*21.89 Most of the first row transition metals form 2+ ions, while other oxidation states are less common. Why do these metals form 2+ ions?

*21.90 Is the complex [Co(EDTA)]⁻ chiral? Illustrate your answer with sketches.

*21.91 The complex [PtCl₂(NH₃)₂] can be obtained as two distinct isomeric forms. Make a model of a tetrahedron and show that if the complex were tetrahedral, two isomers would be impossible.

*21.92 [Ag(NH₃)₂]⁺ and [Ag(CN)₂]⁻ are both linear complexes. Draw the crystal field energy level diagram for the d

orbitals for a linear complex, if the ligands lie along the z-axis.

*21.93 A solution was prepared by dissolving 0.500 g of CrCl₃ · 6H₂O in 100 mL of water. A silver nitrate solution was added and gave a precipitate of AgCl that was filtered from the mixture, washed, dried, and weighed. The AgCl had a mass of 0.538 g.

(a) What is the formula of the complex ion of chromium in this compound?

(b) What is the correct formula for the compound?

(c) Sketch the structure of the complex ion in this compound.

(d) How many different isomers of the complex can be drawn?

*21.94 The compound Cr₂(NH₃)₃(H₂O)₃Cl is a neutral salt in which the cation and anion are both octahedral complex ions. How many isomers (including possible structural and chiral isomers) are there that have this overall composition?

21.95 The complex $[Co(CN)_6]^{4-}$ is not expected to be perfectly octahedral. Instead, the ligands in the *xy* plane are pulled closer to the Co^{2+} ion while those along the *z* axis move slightly farther away. Using information available in this chapter, explain why the distortion of the octahedral geometry leads to a net lowering of the energy of this complex.

21.96 Sketch all of the isomers of $[CoCl_2(en)(NH_3)_2]^+$. Label the *cis* chloro and *trans* chloro isomers.

21.97 Copper(II) aqueous solutions are blue and zinc(II) aqueous solutions are colorless. Why?

Multi-Concept Problems

21.98 Silver plating processes often use $Ag(CN)_2^-$ as the source of Ag^+ in solution. To make the solution, a chemist used 4.0 L of 3.00 *M* NaCN and 50 L of 0.2 *M* $AgNO_3$. What is the concentration of free Ag^+ ions? Why is AgCl not used in this process?

21.99 Cobalt forms complexes in both the +2 and +3 oxidation states. The standard reduction potential for Co^{3+} to Co^{2+} is 1.82 V. In water, the species in solution is $[Co(H_2O)_6]^{3+}$. The equilibrium constants for the formation of the hexaammine complexes are as follows:

$$Co^{2+} + 6NH_3 \rightleftharpoons [Co(NH_3)_6]^{2+} \quad K_{form} = 1.3 \times 10^{14}$$

$$Co^{3+} + 6NH_3 \rightleftharpoons [Co(NH_3)_6]^{3+} \quad K_{form} = 4.8 \times 10^{48}$$

Why is the cobalt(III) complex more stable than the cobalt(II) species? What is the reduction potential for the reduction of $[Co(NH_3)_6]^{2+}$? Which is a stronger oxidizing agent, $[Co(NH_3)_6]^{3+}$ or $[Co(H_2O)_6]^{3+}$?

21.100 Platinum(IV) makes compounds with coordination numbers of 6. A Pt^{4+} compound was made with chloride and amine ligands. The empirical formula for the compound was determined to be $PtCl_4(NH_3)_6$. The isolated compound has an osmotic pressure of 0.734 atm for a 0.1 *M* solution, and when 0.126 g of the compound was reacted with $AgNO_3$, 0.0209 g of AgCl was formed. How many ions does the compound form in water? What is the structure of this compound? What other isomers could have been formed? How many ions would they have released in solution?

Exercises in Critical Thinking

21.101 It was mentioned in footnote 5 in Section 21.5 that *d* orbitals are capable of participating in the formation of π bonds. Make a sketch that illustrates how such a bond could be formed between two *d* orbitals and between a *d* and a *p* orbital.

21.102 The two chiral isomers of $[Co(C_2O_4)_3]^{3-}$ (shown at the right) can be viewed as propellers having either a right- or left-handed twist, respectively. They are identified by the labels Δ and Λ, as indicated. A 50–50 mixture of both isomers is said to be *racemic* and will not rotate plane-polarized light. Using various laboratory procedures, the two isomers present in a racemic mixture can be separated from each other. For this complex, however, a solution of a single isomer is unstable and gradually reverts to a mixture of the two isomers by a process called *racemization*. Racemization involves the

conversion of one isomer to the other until an equilibrium between the two isomers is achieved (i.e., $\Delta \rightleftharpoons \Lambda$). One mechanism proposed for the racemization of the isomers of $[Co(C_2O_4)_3]^{3-}$ involves the dissociation of an oxalate ion followed by the rearrangement and then the reattachment of $C_2O_4^{2-}$.

Δ isomer $C_2O_4^{2-}$ Λ isomer

What experiment could you perform to test whether this is really the mechanism responsible for the racemization of $[Co(C_2O_4)_3]^{3-}$?

22

Organic Compounds, Polymers, and Biochemicals

Chapter Outline

When we think of organic compounds, we might think of dyes, drugs, or gasoline compounds made of carbon. Other people might think of biological compounds such as proteins or DNA, or plastics such as polystyrene or these polymer clay beads made of polyvinylchloride, PVC. PVC is an organic polymer made from repeating units of CH_2CHCl linked together; it is also used for plastic pipes, bottles, and the vinyl siding for houses.

© Goruppa/iStockphoto

This Chapter in Context

In Section 8.9 we noted that many of the substances we encounter on a daily basis have molecular structures based on atoms of carbon linked to one another by covalent bonds. You've already seen a variety of them as examples used in discussions throughout this book. They include such substances as hydrocarbons, alcohols, ketones (such as acetone), and organic acids. The number of such compounds is enormous and their study constitutes the branch of chemistry called organic chemistry, so named because at one time it was believed that such substances could only be synthesized by living organisms. At the molecular level of life, nature uses compounds of carbon. The amazing variety of living systems, down to the uniqueness of each individual, is possible largely because of the properties of this element. In this chapter we will introduce you to organic chemistry.

LEARNING OBJECTIVES

After reading this chapter, you should be able to:

- write structural formulas for organic compounds, highlighting their functional groups
- describe the nomenclature rules and major reactions of alkanes, alkenes, alkynes, and aromatic hydrocarbons
- describe the names and typical reactions of common alcohols, ethers, aldehydes, ketones, carboxylic acids, and esters
- describe the names and typical reactions of amines and amides
- explain the structures, synthesis, and properties of polymers
- name common carbohydrates, lipids, and proteins and describe their properties
- explain how the structures of DNA and RNA enable the transmission of genetic information and the synthesis of proteins

22.1 | Organic Structures and Functional Groups

Organic chemistry is the study of the preparation, properties, and reactions of those compounds of carbon not classified as inorganic. The latter include the oxides of carbon, the bicarbonates and carbonates of metal ions, the metal cyanides, and a handful of other compounds. There are tens of millions of known carbon compounds, and all but a very few are classified as organic.

Uniqueness of the Element Carbon

What makes the existence of so many organic compounds possible is the ability of carbon atoms to form strong covalent bonds to each other while at the same time bonding strongly to atoms of other nonmetals. For example, molecules in the plastic polyvinylchloride, which is used to make the polymer beads in the chapter-opening photograph, have carbon chains that are thousands of carbon atoms long, with hydrogen and chlorine atoms attached to the carbon atoms.

■ Sulfur atoms can also form long chains, but they are unable to bond to the atoms of any other element strongly at the same time.

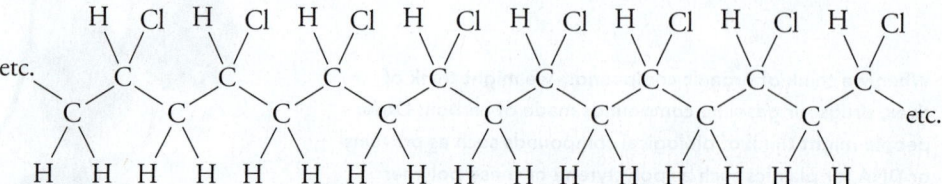

polyvinylchloride (small segment of one molecule)

The longest known sequence of atoms of the other members of Group 4A, each also bonded to hydrogen atoms, is eight for silicon, five for germanium, two for tin, and one for lead.

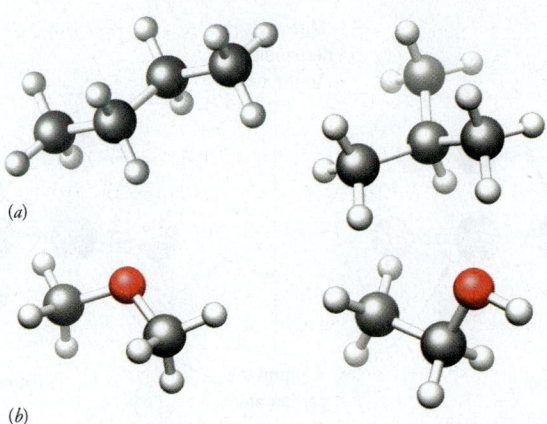

Figure 22.1 | **Isomers.** (*a*) The isomers of C_4H_{10}. Butane is on the left and methylpropane (isobutane) is on the right. (*b*) The isomers of C_2H_6O. Dimethyl ether is on the left and ethanol is on the right.

Another reason for the huge number of organic compounds is isomerism, which was mentioned briefly in Section 8.9. (Isomerism among complex ions is discussed in Chapter 21.) Recall that isomers are compounds with identical molecular formulas, but whose molecules have different structures. There are a number of different kinds of isomers, as we will discuss below. Two examples of isomers of C_4H_{10} and C_2H_6O are shown in Figure 22.1.

The structural formulas of the compounds in Figure 22.1 can be represented in different ways. Complete Lewis structures can be drawn showing all of the bonds between atoms.

■ The isomers of butane were also discussed in Section 8.9.

■ VSEPR theory permits us to predict the tetrahedral geometry around carbon observed in Figure 22.1.

$$
\begin{array}{cccc}
& & & \\
\underset{\substack{\text{butane}\\ \text{BP}-0.5\,°\text{C}}}{H-\overset{\overset{\displaystyle H}{|}}{\underset{\underset{\displaystyle H}{|}}{C}}-\overset{\overset{\displaystyle H}{|}}{\underset{\underset{\displaystyle H}{|}}{C}}-\overset{\overset{\displaystyle H}{|}}{\underset{\underset{\displaystyle H}{|}}{C}}-\overset{\overset{\displaystyle H}{|}}{\underset{\underset{\displaystyle H}{|}}{C}}-H} &
&
&
\end{array}
$$

butane
BP −0.5 °C

2-methyl propane
(isobutane)
BP −11.7 °C

ethanol
BP 78.5 °C

dimethyl ether
BP −23 °C

Notice that each isomer has a different boiling point. As we noted in Chapter 8, isomers are different chemical substances, each with its own set of properties.

To save space and time when writing structural formulas we can write **condensed structural formulas** (also called **condensed structures**) such as those below, in which the C—H, C—C, and C—O bonds are "understood." We can do this because in nearly all organic compounds carbon forms four bonds, nitrogen forms three, oxygen forms two, and hydrogen forms one.

TOOLS

Condensed structures

$$CH_3CH_2CH_2CH_3 \qquad CH_3\overset{\overset{\displaystyle CH_3}{|}}{C}HCH_3 \qquad CH_3CH_2OH \qquad CH_3OCH_3$$

butane

2-methyl propane
(isobutane)

ethanol

dimethyl ether

Take a moment to compare the condensed structural formulas with the full Lewis structures. Also note that there can be more than one way to write a condensed structure, as illustrated by 2-methylpropane.

$$CH_3\overset{\overset{\displaystyle CH_3}{|}}{C}HCH_3 \quad \text{or} \quad CH_3CH(CH_3)CH_3 \quad \text{or} \quad (CH_3)_3CH$$

Figure 22.2 | **Chiral isomers of butan-2-ol.** Isomer 2 is the mirror image of Isomer 1. No matter how you twist or turn Isomer 1, you cannot get it to fit exactly onto Isomer 2, which means the two isomers are nonsuperimposable mirror images of each other.

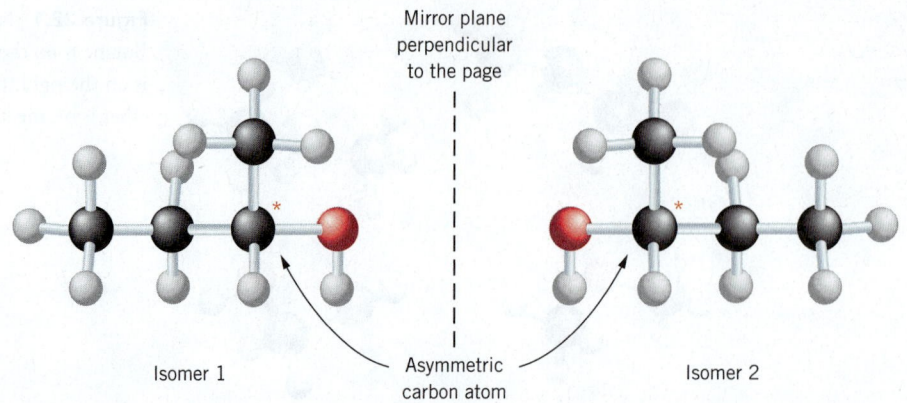

Mirror plane perpendicular to the page

Isomer 1   Asymmetric carbon atom   Isomer 2

■ Few of the possible isomers of the compounds with large numbers of carbon atoms have actually been made, but there is nothing except too much crowding within the molecules of some of them to prevent their existence.

Formula	Number of Isomers
C_8H_{18}	18
$C_{10}H_{22}$	75
$C_{20}H_{42}$	366,319
$C_{40}H_{82}$	6.25×10^{13} (estimated)

We will use condensed structures frequently throughout this chapter. In writing condensed structures, be sure there are sufficient numbers of hydrogens so that the total number of bonds to each carbon equals four.

In Chapter 21 we described the chiral properties of certain isomers. An organic compound exhibits **chirality** when it contains an **asymmetric carbon atom**, which is a carbon atom that is bonded to four different atoms or groups. An example is butan-2-ol (also called 2-butanol) in which the asymmetric carbon atom, in red, is bonded to H, OH, CH_3, and CH_2CH_3.

$$\underset{\text{butan-2-ol}}{\underset{\text{2-butanol}}{CH_3\overset{\overset{\displaystyle OH}{|}}{C}HCH_2CH_3}}$$

The two nonsuperimposable mirror image isomers of this compound, also called optical isomers or stereoisomers, are shown in Figure 22.2.

The table in the margin illustrates that as the number of carbons per molecule increases, the number of possible isomers for any given formula becomes enormous.

Open-Chain and Ring Compounds

The continuous sequence of carbon atoms in polyvinylchloride (depicted at the start of this Section) is called a **straight chain**. This means *only* that no carbon atom bonds to more than two other carbons. It does not refer to the *shape* of the molecule. If we made a molecular model of polyvinylchloride and coiled it into a spiral, we would still say that its carbon skeleton has a straight chain. It has no branches.

Branched chains are also very common. For example, isooctane, the standard for the octane rating of gasoline, has a main chain of five carbon atoms (in black) and three CH_3 branches (shown in red).

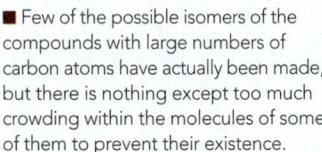

Straight chain

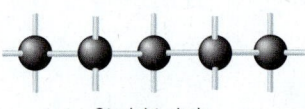

Branched chain

■ Isooctane, one of the many alkanes in gasoline, is the standard for the octane ratings of various types of gasoline. Pure isooctane is assigned an octane rating of 100.

$$\underset{\text{isooctane}}{CH_3\overset{\overset{\displaystyle CH_3}{|}}{\underset{\underset{\displaystyle CH_3}{|}}{C}}CH_2\overset{\overset{\displaystyle CH_3}{|}}{C}HCH_3}$$

Carbon rings are also common. Cyclohexane, for example, has a ring of six carbon atoms:

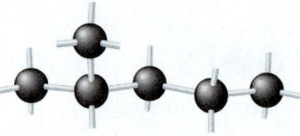

Carbon ring

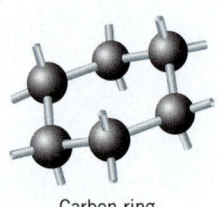

cyclohexane   cyclohexane (fully condensed)

Just about everything is "understood" in the very convenient, fully condensed structure of cyclohexane. When polygons, like the hexagon, are used to represent rings, the following conventions are used:

Using Polygons to Represent Rings
1. C occurs at each corner unless O or N (or another atom capable of forming more than one bond) is explicitly written at a corner.
2. A line connecting two corners is a covalent bond (single bond) between adjacent ring atoms.
3. Remaining bonds, as required by the carbon at a corner to have a total of four bonds, are understood to bond H atoms.
4. Double bonds are always explicitly shown.

We can illustrate these rules with the following cyclic compounds:

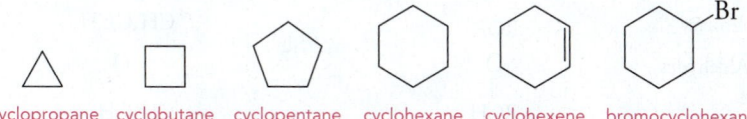

cyclopropane cyclobutane cyclopentane cyclohexane cyclohexene bromocyclohexane

There is no theoretical upper limit to the size of a ring.

Many compounds have **heterocyclic rings**. Their molecules carry an atom other than carbon in a ring position, as in pyrrole, piperidine, and tetrahydropyran.

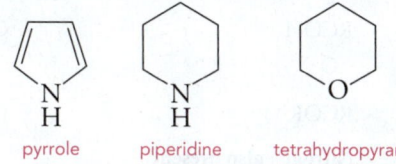

pyrrole   piperidine   tetrahydropyran

The atom other than carbon is called the heteroatom, and any atoms of H or other element bonded to the heteroatom are not "understood" but are always shown.

Notation similar to that for rings is often used for open-chain molecules as well. For example, the structure of pentane can be represented as follows:

$$CH_3-CH_2-CH_2-CH_2-CH_3$$

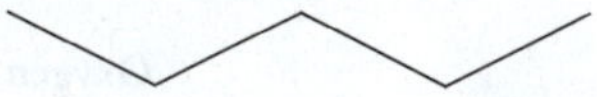

pentane

On the right, the four bonds between carbon atoms are represented by four connected line segments. At the end of each line segment (including where line segments join) there is a carbon atom plus enough hydrogen atoms to account for the four bonds that the carbon must have. If an atom is shown in a structure, it replaces the carbon atom that would have been in that position.

Practice Exercise 22.1

Give the molecular formula and the condensed structural formula for these organic compounds (*Hint:* How are carbons represented in these structures?)

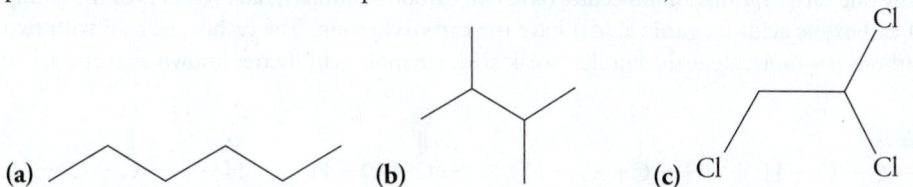

(a)   (b)   (c) Cl

Practice Exercise 22.2

Draw an organic compound with five carbons in a ring, and a CH_3 group on one of the carbons in the ring. What is the formula for this compound?

Organic Families and Their Functional Groups

The study of the huge number of organic compounds is manageable because they can be sorted into *organic families* defined by *functional groups*, a number of which were described in Section 8.9. A more complete list is found in Table 22.1.

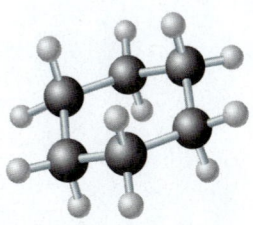

Cyclohexane

TOOLS

Functional groups

TABLE 22.1 Some Important Families of Organic Compounds

Family		
	Only C and H present	
Hydrocarbon	**Families of Hydrocarbons**	
	Alkanes: only single bonds	H_3C-CH_3
	Alkenes	$H_2C=CH_2$
	Alkynes	$HC\equiv CH$
	Aromatic: Benzene ring	 ⬡
	Oxygen also present	
Alcohols	ROH	CH_3CH_3OH
Ethers	ROR′	CH_3OCH_3
Aldehydes	$\overset{\overset{\displaystyle O}{\|\|}}{RCH}$	$\overset{\overset{\displaystyle O}{\|\|}}{CH_3CH}$
Ketones	$\overset{\overset{\displaystyle O}{\|\|}}{RCR'}$	$\overset{\overset{\displaystyle O}{\|\|}}{CH_3CCH_3}$
Carboxylic acids	$\overset{\overset{\displaystyle O}{\|\|}}{RCOH}$	$\overset{\overset{\displaystyle O}{\|\|}}{CH_3COH}$
Esters	$\overset{\overset{\displaystyle O}{\|\|}}{RCOR'}$	$\overset{\overset{\displaystyle O}{\|\|}}{CH_3COCH_3}$
	Nitrogen also present	
Amines	RNH_2, $RNHR'$, $RNR'R''$	CH_3NH_2
		CH_3NHCH_3
		$\overset{\overset{\displaystyle CH_3}{\|}}{CH_3NCH_3}$
	Oxygen and nitrogen also present	
Amides (peptides)	$\overset{\overset{\displaystyle O}{\|\|}}{RC}-\overset{\overset{\displaystyle R''(H)}{\|}}{NR'(H)}$	

Functional groups are small structural units within molecules at which most of the compound's chemical reactions occur. For example, all *alcohols* have the *alcohol group*, which imparts certain chemical properties that are common to them all. The simplest member of the alcohol family, methanol (also called methyl alcohol), has molecules with only one carbon. Ethanol molecules have two carbons. Similarly, all members of the family of carboxylic acids (**organic acids**) have the carboxyl group. The carboxylic acid with two carbons per molecule is the familiar weak acid, ethanoic acid (better known as acetic acid).

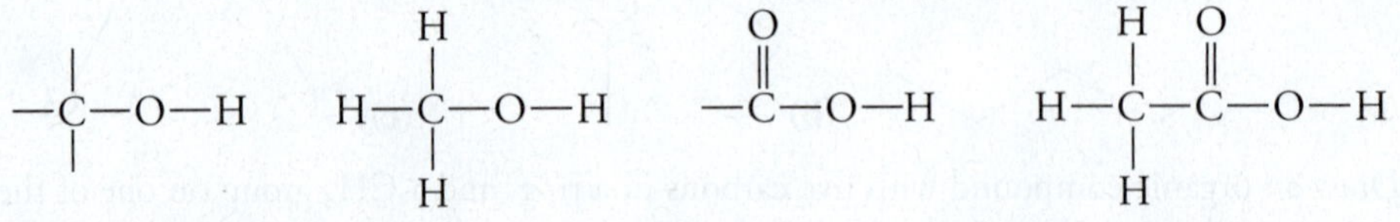

■ Recall from Chapter 8 that organic acids contain the carboxyl group.

$$-\overset{\overset{\displaystyle |}{}}{\underset{\underset{\displaystyle |}{}}{C}}-O-H \qquad H-\overset{\overset{\displaystyle H}{|}}{\underset{\underset{\displaystyle H}{|}}{C}}-O-H \qquad -\overset{\overset{\displaystyle O}{\|\|}}{C}-O-H \qquad H-\overset{\overset{\displaystyle H}{|}}{\underset{\underset{\displaystyle H}{|}}{C}}-\overset{\overset{\displaystyle O}{\|\|}}{C}-O-H$$

alcohol group   methanol   carboxyl group   ethanoic acid
                      (methyl alcohol)   (acetic acid)

One family in Table 22.1, the alkane family, has no functional group, just C—C and C—H single bonds. These bonds are virtually nonpolar, because C and H are so alike in electronegativity. Therefore, alkane molecules are the least able of all organic molecules to attract ions or polar molecules. As a result, they do not react, at least at room temperature, with polar or ionic reactants such as strong acids and bases and common oxidizing agents, like dichromate or permanganate ions.

The Symbol R in Structural Formulas

In Tables 8.4 and 22.1 the functional groups are shown attached to the symbol **R**. Chemists use the symbol R in a structure to represent purely alkane-like hydrocarbon groups, which are unreactive toward substances that are able to react with the functional group. For example, one type of **amine** is represented by the formula $R—NH_2$, where R could be CH_3, CH_3CH_2, $CH_3CH_2CH_2$, and so forth. Because the amines are compounds with an ammonia-like group, they are weak Brønsted bases and, like ammonia, react with hydronium ions. We can, therefore, summarize in just one equation the reaction of a strong acid with *any* amine of the $R—NH_2$ type, regardless of how large the hydrocarbon portion of the molecule is, by the following equation:

$$R—NH_2 + H_3O^+ \longrightarrow R—NH_3^+ + H_2O$$

22.2 | Hydrocarbons: Structure, Nomenclature, and Reactions

Hydrocarbons are compounds whose molecules consist of only C and H atoms. The fossil fuels—coal, petroleum, and natural gas—supply us with virtually all of these substances. One of the operations in petroleum refining is to boil crude oil (petroleum from which natural gas has been removed) and selectively condense the vapors between preselected temperature ranges. Gasoline, for example, is the fraction boiling roughly between 40 and 200 °C and is composed of molecules that have 5 to 12 carbon atoms. The kerosene and jet fuel fraction contains molecules with more carbon atoms and overlaps this range, going from 175 to 325 °C.

Petroleum refinery. The towers contain catalysts that break up large molecules in hot crude oil to smaller molecules suitable for vehicle engines.

Hydrocarbons can be divided into four types: alkanes, alkenes, alkynes, and aromatic hydrocarbons (see Table 22.1). All are insoluble in water. All burn, giving carbon dioxide and water if sufficient oxygen is available.

The **alkanes**, mentioned earlier, contain carbon atoms bonded to other atoms by only single bonds. In the alkanes all of the carbon atoms use sp^3 hybrid orbitals for bonding, so the orientation of atoms around each carbon is tetrahedral. Open-chain members of the alkane family have the general formula C_nH_{2n+2}.

Alkenes contain one or more carbon-carbon double bonds of the type described in Figure 9.32. **Alkynes** contain one or more triple bonds formed as shown in Figure 9.35. Alkenes and alkynes are said to be **unsaturated compounds** because under appropriate conditions their molecules can take up additional H atoms to give alkane molecules. We will study these reactions later. The alkanes are **saturated compounds**, hydrocarbons with only single bonds, and do not take on additional H atoms.

The **aromatic hydrocarbons**, typified by benzene, are also unsaturated because the carbon atoms of their rings, when represented by simple Lewis structures, also have double bonds.

■ Alkanes are saturated in the sense that they have as many hydrogen atoms as the carbon atoms can hold, just as a saturated solution has as much solute as it can hold.

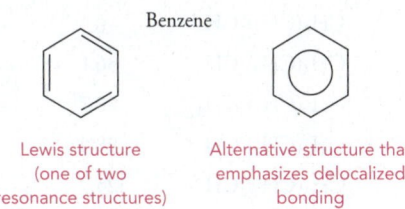

Benzene

Lewis structure
(one of two
resonance structures)

Alternative structure that
emphasizes delocalized
bonding

The delocalized nature of the π bond system in the ring causes them to behave quite differently from alkenes and alkynes in their chemical reactions, as we discuss later.

IUPAC Nomenclature of Alkanes

■ In earlier discussions we gave both the official IUPAC names as well as common names for organic compounds.

TOOLS

Naming alkanes

The **IUPAC rules** for naming organic compounds follow a regular pattern. The last sylla-ble in an IUPAC name designates the family of the compound. The names of all saturated hydrocarbons, for example, end in *-ane*. For each family there is a rule for picking out and naming the parent chain or parent ring within a specific molecule. The compound is then regarded as having substituents attached to its parent chain or ring. Let's see how these principles work with naming alkanes.

IUPAC Rules for Naming the Alkanes

1. The name ending for all alkanes (and cycloalkanes) is *-ane*.
2. The parent chain is the longest continuous chain of carbons in the structure. For example, the branched-chain alkane

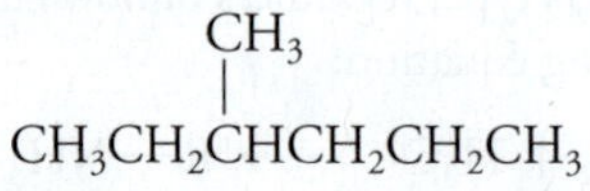

$$CH_3$$
$$|$$
$$CH_3CH_2CHCH_2CH_2CH_3$$

is regarded as being "made" from the following parent

$$CH_3CH_2CH_2CH_2CH_2CH_3$$

by replacing a hydrogen atom on the third carbon from the left with CH_3.

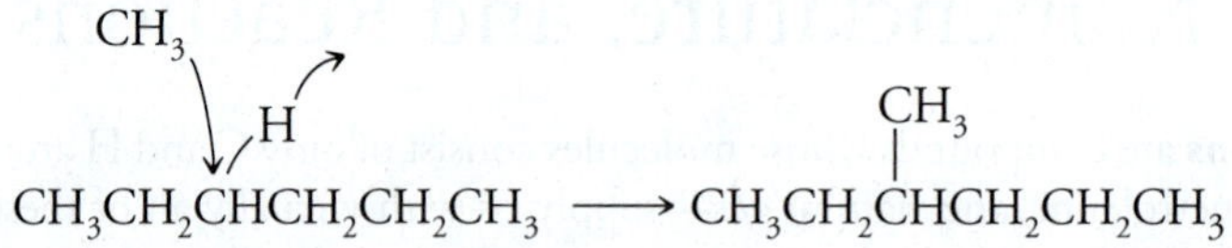

$$CH_3 \qquad\qquad\qquad\qquad CH_3$$
$$\searrow H \qquad\qquad\qquad\qquad |$$
$$CH_3CH_2CHCH_2CH_2CH_3 \longrightarrow CH_3CH_2CHCH_2CH_2CH_3$$

3. A prefix is attached to the name ending, *-ane*, to specify the number of carbon atoms in the parent chain. The prefixes through chain lengths of 10 carbons are as follows. The names in Table 22.2 show their use.

meth-	1 C	hex-	6 C
eth-	2 C	hept-	7 C
prop-	3 C	oct-	8 C
but-	4 C	non-	9 C
pent-	5 C	dec-	10 C

The parent chain of our example has six carbons, so the parent is named hexane—*hex-* for six carbons and *-ane* for being in the alkane family. Therefore, the alkane whose name we are devising is viewed as a derivative of this parent, hexane.

TABLE 22.2 **Straight-Chain Alkanes**

IUPAC Name	Molecular Formula	Structure	Boiling Point (°C)	Melting Point (°C)	Density (g mL^{-1}, 20 °C)
Methane	CH_4	CH_4	−161.5	−182.5	
Ethane	C_2H_6	CH_3CH_3	−88.6	−183.3	
Propane	C_3H_8	$CH_3CH_2CH_3$	−42.1	−189.7	
Butane	C_4H_{10}	$CH_3(CH_2)_2CH_3$	−0.5	−138.4	
Pentane	C_5H_{12}	$CH_3(CH_2)_3CH_3$	36.1	−129.7	0.626
Hexane	C_6H_{14}	$CH_3(CH_2)_4CH_3$	68.7	−95.3	0.659
Heptane	C_7H_{16}	$CH_3(CH_2)_5CH_3$	98.4	−90.6	0.684
Octane	C_8H_{18}	$CH_3(CH_2)_6CH_3$	125.7	−56.8	0.703
Nonane	C_9H_{20}	$CH_3(CH_2)_7CH_3$	150.8	−53.5	0.718
Decane	$C_{10}H_{22}$	$CH_3(CH_2)_8CH_3$	174.1	−29.7	0.730

4. The carbon atoms of the parent chain are numbered starting from whichever end of the chain gives the location of the first branch the lower of two possible numbers. Thus, the correct direction for numbering our example is from left to right, not right to left, because this locates the branch (CH_3) at position 3, not position 4.

$$CH_3 \atop CH_3CH_2CHCH_2CH_2CH_3$$
1   2   3   4   5   6
(correct direction of numbering)

$$CH_3 \atop CH_3CH_2CHCH_2CH_2CH_3$$
6   5   4   3   2   1
(incorrect direction of numbering)

5. Each branch attached to the parent chain is named, so we must now learn the names of some of the alkane-like branches.

Alkyl Groups

Any branch that consists only of carbon and hydrogen and has only single bonds is called an alkyl group, and the names of all **alkyl groups** end in -*yl*. Think of an alkyl group as an alkane minus one of its hydrogen atoms. For example, if we remove a hydrogen from methane and ethane we get the methyl and **ethyl groups**, respectively:

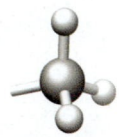

methyl group

$$H-\overset{\overset{\displaystyle H}{|}}{\underset{\underset{\displaystyle H}{|}}{C}}-H \xrightarrow{\text{remove one H}} H-\overset{\overset{\displaystyle H}{|}}{\underset{\underset{\displaystyle H}{|}}{C}}- \quad \text{or} \quad CH_3-$$

methane   methyl group

$$H-\overset{\overset{\displaystyle H}{|}}{\underset{\underset{\displaystyle H}{|}}{C}}-\overset{\overset{\displaystyle H}{|}}{\underset{\underset{\displaystyle H}{|}}{C}}-H \xrightarrow{\text{remove one H}} H-\overset{\overset{\displaystyle H}{|}}{\underset{\underset{\displaystyle H}{|}}{C}}-\overset{\overset{\displaystyle H}{|}}{\underset{\underset{\displaystyle H}{|}}{C}}- \quad \text{or} \quad CH_3CH_2-$$

ethane   ethyl group

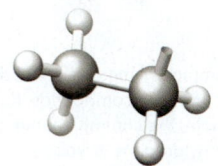

ethyl group

Two alkyl groups, propyl and 1-methylethyl (or **isopropyl**), can be obtained from propane because the middle position in its chain of three is not equivalent to either of the end positions.

$$H-\overset{\overset{\displaystyle H}{|}}{\underset{\underset{\displaystyle H}{|}}{C}}-\overset{\overset{\displaystyle H}{|}}{\underset{\underset{\displaystyle H}{|}}{C}}-\overset{\overset{\displaystyle H}{|}}{\underset{\underset{\displaystyle H}{|}}{C}}-H \xrightarrow{\text{remove one H}} H-\overset{\overset{\displaystyle H}{|}}{\underset{\underset{\displaystyle H}{|}}{C}}-\overset{\overset{\displaystyle H}{|}}{\underset{\underset{\displaystyle H}{|}}{C}}-\overset{\overset{\displaystyle H}{|}}{\underset{\underset{\displaystyle H}{|}}{C}}- \quad \text{or} \quad CH_3CH_2CH_2-$$

propane   propyl

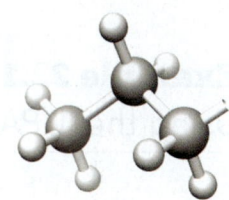

propyl group

$$H-\overset{\overset{\displaystyle H}{|}}{\underset{\underset{\displaystyle H}{|}}{C}}-\overset{\overset{\displaystyle H}{|}}{\underset{\underset{\displaystyle H}{|}}{C}}-\overset{\overset{\displaystyle H}{|}}{\underset{\underset{\displaystyle H}{|}}{C}}-H \xrightarrow{\text{remove one H}} H-\overset{\overset{\displaystyle H}{|}}{\underset{\underset{\displaystyle H}{|}}{C}}-\overset{\displaystyle H}{\underset{\underset{\displaystyle H}{|}}{C}}-\overset{\overset{\displaystyle H}{|}}{\underset{\underset{\displaystyle H}{|}}{C}}-H \quad \text{or} \quad CH_3CHCH_3$$

propane   1-methylethyl (isopropy)

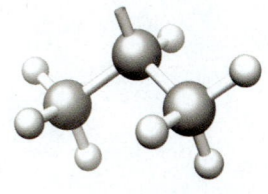

1-methylethyl group (isopropyl group)

■ Isopropyl is the common name; 1-methylethyl is the official IUPAC name.

We will not need to know the IUPAC names for any alkyl groups with four or more carbon atoms. Now let's continue with our IUPAC rules.

6. The name of each alkyl group is attached to the name of the parent as a prefix, placing its chain location number in front, separating the number from the name by a hyphen. Thus, the original example is named 3-methylhexane.

$$CH_3 \atop CH_3CH_2CHCH_2CH_2CH_3$$

7. When two or more groups are attached to the parent, each is named and located with a number. The names of alkyl substituents are assembled in their alphabetical order. Always use hyphens to separate numbers from words. The following molecule provides an illustration:

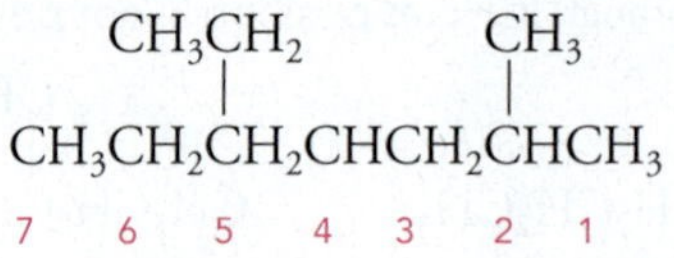

$$CH_3CH_2 \qquad CH_3$$
$$|\qquad\qquad |$$
$$CH_3CH_2CH_2CHCH_2CHCH_3$$

7   6   5   4   3   2   1

4-ethyl-2-methylheptane

8. When two or more substituents are identical, multiplier prefixes are used: di- (for 2), tri- (for 3), tetra- (for 4), and so forth. The location number of every group must occur in the final name. Always separate a number from another number in a name by a comma. For example,

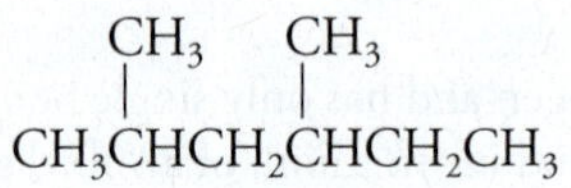

$$CH_3 \qquad CH_3$$
$$|\qquad\quad |$$
$$CH_3CHCH_2CHCH_2CH_3$$

Correct name: 2,4-dimethylhexane

Incorrect names: 2,4-methylhexane
3,5-dimethylhexane
2-methyl-4-methylhexane
2-4-dimethylhexane

9. When identical groups are on the same carbon, the number of this position is repeated in the name. For example,

$$CH_3$$
$$|$$
$$CH_3CCH_2CH_2CH_3$$
$$|$$
$$CH_3$$

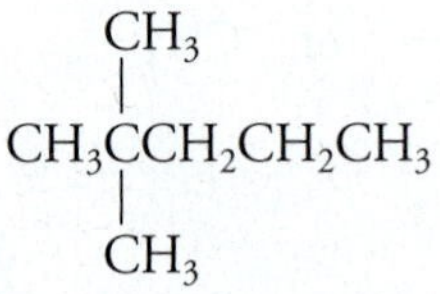

Correct name: 2,2-dimethylpentane

Incorrect names: 2-dimethylpentane
2,2-methylpentane
2,2-methylpentane

■ Common names are still widely used for many compounds. For example, the common name of 2-methylpropane is isobutane.

These are not all of the IUPAC rules for alkanes, but they will handle our needs.

Example 22.1
Using the IUPAC Rules to Name an Alkane

What is the IUPAC name for the following compound?

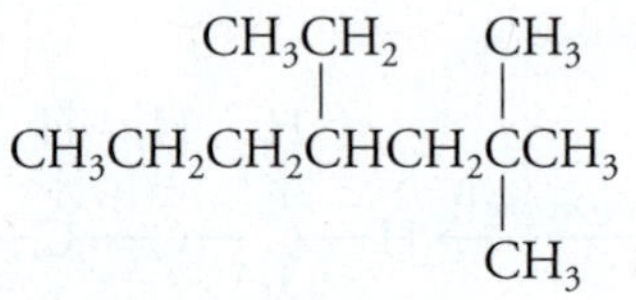

$$CH_3CH_2 \qquad CH_3$$
$$|\qquad\qquad |$$
$$CH_3CH_2CH_2CHCH_2CCH_3$$
$$|$$
$$CH_3$$

Analysis: In naming an organic compound, the first step is to identify which family it belongs to, because the rules differ slightly according to the functional group. Then the parent chain needs to be identified, followed by the branch chains.

Assembling the Tools: Our tool, of course, is the IUPAC rules for naming alkanes.

Solution: The ending to the name must be -*ane*. The longest chain is seven carbons long, so the name of the parent alkane is heptane. We have to number the chain from right to left in order to reach the first branch with the lower number.

At carbon 2, there are two one-carbon methyl groups. At carbon 4, there is a two-carbon ethyl group. Alphabetically, ethyl comes before methyl, so we must assemble these names as follows to make the final name. (Names of alkyl groups are alphabetized before any prefixes such as di- or tri- are affixed.)

4-ethyl-2,2-dimethylheptane

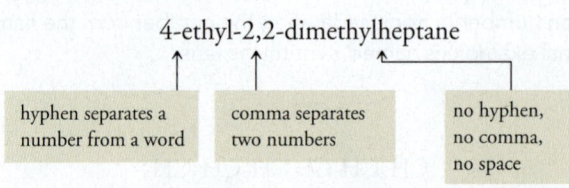

| hyphen separates a number from a word | comma separates two numbers | no hyphen, no comma, no space |

● **Is the Answer Reasonable?** The most common mistake is to pick a shorter chain than the true "parent," a mistake we did not make. Another common mistake is to number the chain incorrectly. One overall check that some people use is to count the number of carbons implied by the name (in our example, $2 + 1 + 1 + 7 = 11$) and compare it to the count obtained directly from the structure. If the counts don't match, you know you can't be right.

A student incorrectly named a compound 1,1,1-trimethylethane. What should the name be? (*Hint:* Write the structure of 1,1,1-trimethylethane.)

Practice Exercise 22.3

Write the IUPAC names of the following compounds. In searching for the parent chain, be sure to look for the longest continuous chain of carbons even if the chain twists and goes around corners.

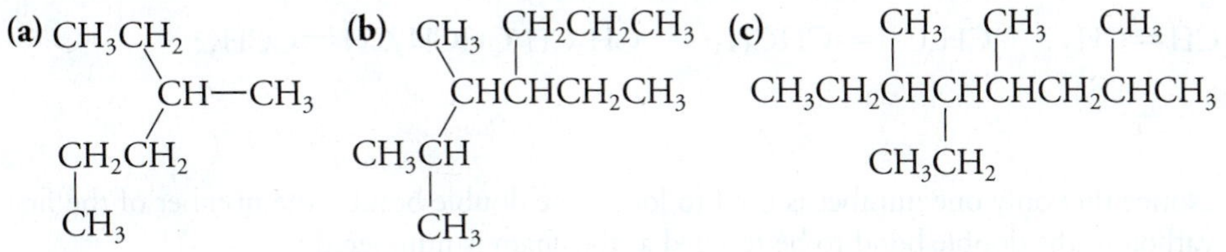

Practice Exercise 22.4

(a) CH_3CH_2
　　　　$\mid$
　　　$CH—CH_3$
　　　$\mid$
　　CH_2CH_2
　　　$\mid$
　　　CH_3

(b) CH_3　$CH_2CH_2CH_3$
　　　$\mid$　　$\mid$
　　　$CHCHCH_2CH_3$
　　　$\mid$
　　CH_3CH
　　　$\mid$
　　　CH_3

(c) 　　　CH_3　CH_3　CH_3
　　　　　$\mid$　　$\mid$　　$\mid$
　$CH_3CH_2CHCHCHCH_2CHCH_3$
　　　　　　$\mid$
　　　　CH_3CH_2

Chemical Properties of Alkanes

Alkanes are relatively unreactive at room temperature. They do not react with concentrated acids such as H_2SO_4 or concentrated aqueous bases such as NaOH. Like all hydrocarbons, alkanes burn in air to give water and oxides of carbon (CO_2 and CO). Hot nitric acid, chlorine, and bromine also react with alkanes. The chlorination of methane, for example, can result in the following compounds:

CH_3Cl　　　　CH_2Cl_2　　　　$CHCl_3$　　　CCl_4
methyl chloride　methylene chloride　chloroform　carbon tetrachloride

■ Halogens form single bonds to carbon atoms. These are the common names for the chlorinated derivatives of methane, not the IUPAC names.

When heated at high temperatures in the absence of air, alkanes "crack," meaning that they break up into smaller molecules. The cracking of methane, for example, yields finely powdered carbon and hydrogen.

$$CH_4 \xrightarrow{\text{high temperature}} C + 2H_2$$

The controlled cracking of ethane gives ethene, commonly called ethylene.

$$CH_3CH_3 \xrightarrow{\text{high temperature}} CH_2{=}CH_2 + H_2$$
ethane　　　　　　　　　　　ethene

Ethylene is one of the most important raw materials in the organic chemicals industry. It is used to make polyethylene plastic items as well as ethanol and ethylene glycol (an antifreeze).

■ $HOCH_2CH_2OH$
ethylene glycol

Alkenes and Alkynes

Hydrocarbons with one or more double bonds are members of the alkene family. Open-chain alkenes have the general formula, C_nH_{2n}. Hydrocarbons with triple bonds are in the alkyne family, and have the general formula, C_nH_{2n-2} (when open-chain).

Alkenes and alkynes, like all hydrocarbons, are virtually nonpolar and therefore insoluble in water, and they are flammable. The most familiar alkenes are ethylene and propene (commonly called propylene), the raw materials for polyethylene and polypropylene, respectively. Ethyne ("acetylene"), an important alkyne, is the fuel for oxyacetylene torches.

$CH_2{=}CH_2$　　　　$CH_3CH{=}CH_2$　　　$HC{\equiv}CH$
ethene (ethylene)　　propene (propylene)　　ethyne (acetylene)

Naming alkenes and alkynes

IUPAC Nomenclature of Alkenes and Alkynes

The IUPAC rules for the names of alkenes are adaptations of those for alkanes, but with two important differences. First, the parent chain must include the double bond even if this means that the parent chain is shorter than another. Second, the parent alkene chain must be numbered from whichever end gives the first carbon of the double bond the lower of two possible numbers. This (lower) number, preceded and followed by a hyphen, is inserted into the name of the parent chain just before the *-ene* or *-yne* ending, unless there is no ambiguity about where the double bond occurs. The numbers for the locations of branches are not considered in numbering the chain. Otherwise, alkyl groups are named and located as before. Some examples of correctly named alkenes are as follows:

$CH_3CH_2CH{=}CH_2$

but-1-ene
(not but-3-ene)

$CH_3CH{=}CHCH_3$

but-2-ene

$CH_3CH_2CHCH_2CH{=}CCH_3$
(with CH₃ groups at positions 2 and 5)

2,5-dimethylhept-2-ene
(not 3,6-dimethyl-5-heptene)

cyclohexene

Notice that only one number is used to locate the double bond—the number of the first carbon of the double bond to be reached as the chain is numbered.

Some alkenes have two double bonds and are called *dienes*. Some have three double bonds and are called *trienes*, and so forth. Each double bond has to be located by a number.

$CH_2{=}CHCH{=}CHCH_3$

penta-1,3-diene

$CH_2{=}CHCH_2CH{=}CH_2$

penta-1,4-diene

$CH_2{=}CHCH{=}CHCH{=}CH_2$

hexa-1,3,5-triene

Geometric Isomerism

As explained in Section 9.6, rotation of one part of a molecule relative to another around a carbon–carbon double bond cannot occur without bond breaking. As a result, the orientations of the atoms bonded to the carbons of the double bond are locked in place. This allows many alkenes to exhibit **geometric isomerism**. Thus, *cis*-but-2-ene and *trans*-but-2-ene (commonly called *cis*-2-butene and *trans*-2-butene, respectively) are **geometric isomers** of each other (see below). They not only have the same molecular formula, C_4H_8, but also the same skeletons and the same organization of atoms and bonds—namely, $CH_3CH{=}CHCH_3$. The two 2-butenes differ in the *directions* taken by the two CH_3 groups attached at the double bond.

■ Because ring structures also lock out free rotation, geometric isomers of ring compounds are also possible. These two isomers of 1,2-dimethylcyclopropane are examples. In the representations below, the ring is viewed tilting back into the page, with the front edge of the ring shown as a heavy line.

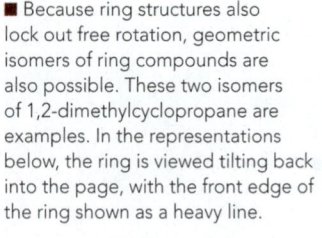

trans isomer
BP 28 °C

cis isomer
BP 37 °C

H_3C CH_3
   $C{=}C$
 H H

cis-but-2-ene
BP 3.72 °C

H CH_3
   $C{=}C$
H_3C H

trans-but-2-ene
BP 0.88 °C

cis means "on the same side;" ***trans*** means "on opposite sides." This difference in orientation gives the two geometric isomers of 2-butene measurable differences in physical properties, as their boiling points show. Because each has a double bond, however, the *chemical* properties of *cis*- and *trans*-2-butene are very similar.

Reactions of Alkenes

A double bond has more electron density than a single bond, so electron-seeking species such as protons are naturally attracted to the electron density at the π bond of the double bond. As a result, alkenes can react readily with protons provided by strong proton donors to give **addition reactions** in which two parts of a reactant become separately attached to the two carbons of a double bond. Ethylene, for example, readily reacts with hydrogen chloride as follows:

$$CH_2{=}CH_2 + H{-}Cl(g) \longrightarrow Cl{-}CH_2{-}CH_3$$

We say that the hydrogen chloride molecule adds across the double bond, the hydrogen of HCl going to the carbon at one end and the chlorine going to the other end. The pair of electrons of the π bond move out and take H^+ from HCl, which releases Cl^-. A positive $(+)$ charge is left at one end of the original carbon–carbon double bond as a bond to H forms at the other end.

$$CH_2{=}CH_2 + H{-}Cl \longrightarrow \overset{+}{C}H_2{-}CH_2 \longrightarrow Cl{-}CH_2{-}CH_3$$

ethyl carbocation   chloroethane

■ Recall that curved arrows indicate how electron pairs rearrange; they usually are not used to signify the movement of atoms.

The result of H^+ transfer is a very unstable cation called a carbocation, an ion with a positive charge on carbon. This charged site then quickly attracts the Cl^- ion to give the product, chloroethane.

Another example is 2-butene; like ethylene, it too adds HCl.

$$CH_3CH{=}CHCH_3 + HCl \longrightarrow CH_3CHCH_2CH_3$$
$$\underset{\text{but-2-ene}}{} \qquad\qquad \overset{|}{Cl}$$

2-chlorobutane

■ A carbocation, having a carbon with only six valence electrons, not an octet, generally has only a fleeting existence.

Alkene double bonds also add hydrogen bromide, hydrogen iodide, and sulfuric acid ("hydrogen sulfate"). Alkynes undergo similar addition reactions.

A water molecule adds to a double bond if an acid catalyst, like sulfuric acid, is present. The resulting product is an alcohol. For example, adding H_2O to the double bond of ethylene gives ethanol as the product.

$$H_2C{=}CH_2 + HOH \xrightarrow[\text{catalyst}]{H_2SO_4} H_2C{-}CH_2$$
$$\underset{\text{ethene}}{} \qquad\qquad\qquad \overset{|}{H}\;\; \overset{|}{OH}$$

ethanol

Other inorganic compounds that add to an alkene double bond are chlorine, bromine, and hydrogen. Chlorine and bromine react rapidly at room temperature. Ethylene, for example, reacts with bromine to give 1,2-dibromoethane.

$$CH_2{=}CH_2 + Br{-}Br \longrightarrow CH_2{-}CH_2$$
$$\underset{\text{ethene}}{} \qquad\qquad\qquad \overset{|}{Br}\;\; \overset{|}{Br}$$

1,2-dibromoethane

The product of the addition of hydrogen to an alkene is an alkane and the reaction is called hydrogenation. It requires a catalyst—powdered platinum, for example—and sometimes a higher temperature and pressure than available under an ordinary room temperature and pressure. The hydrogenation of 2-butene (*cis* or *trans*) gives butane.

$$CH_3CH{=}CHCH_3 + H{-}H \xrightarrow[\text{catalyst}]{\text{heat, pressure}} CH_3CH{-}CHCH_3 \text{ or } CH_3CH_2CH_2CH_3$$
$$\underset{\substack{\text{but-2-ene}\\\text{(cis or trans)}}}{} \qquad\qquad\qquad\qquad \overset{|}{H}\;\; \overset{|}{H} \qquad\qquad \underset{\text{butane}}{}$$

TOOLS

Reactions of alkenes and alkynes

Aromatic Hydrocarbons

■ Hydrocarbons and their oxygen or nitrogen derivatives that are not aromatic are called aliphatic compounds.

The most common **aromatic compounds** contain the benzene ring, a ring of six carbon atoms, each bonded to one H or one other atom or group. The benzene ring is represented as a hexagon either with alternating single and double bonds or with a circle. For example,

benzene   toluene   ethylbenzene
                 (methylbenzene)

The circle better suggests the delocalized bonds of the benzene ring, which are described in Section 9.8.

In the molecular orbital view of benzene the delocalization of the ring's π electrons strongly stabilizes the ring. This explains why benzene does not easily give addition reactions; they would interfere with the delocalization of electron density. Instead, the benzene ring most commonly undergoes **substitution reactions**, those in which one of the ring H atoms is replaced by another atom or group. For example, benzene reacts with chlorine in the presence of iron(III) chloride to give chlorobenzene, instead of a 1,2-dichloro compound. To dramatize this point, we must use a resonance structure for benzene.

■ In this reaction, FeCl$_3$ is a catalyst. The HCl is a product, indicated by the curved arrow.

chlorobenzene   (This would form if chlorine
                        added to the double bond.)

You can infer that substitution, but not addition, leaves intact the delocalized and very stable π electron network of the benzene ring.

Provided that a suitable catalyst is present, benzene reacts by substitution with chlorine, bromine, and nitric acid as well as with sulfuric acid. (Recall that Cl$_2$ and Br$_2$ readily add to alkene double bonds.)

$$C_6H_6 + Br_2 \xrightarrow{\text{FeBr}_3 \text{ catalyst}} C_6H_5{-}Br + HBr$$
$$\text{bromobenzene}$$

$$C_6H_6 + HNO_3 \xrightarrow{\text{H}_2\text{SO}_4 \text{ catalyst}} C_6H_5{-}NO_2 + H_2O$$
$$\text{nitrobenzene}$$

$$C_6H_6 + H_2SO_4 \longrightarrow C_6H_5{-}SO_3H + H_2O$$
$$\text{benzenesulfonic acid}$$

22.3 | Organic Compounds Containing Oxygen

Most of the organic compounds we encounter are not hydrocarbons, although they all contain fragments that are considered to be derived from hydrocarbons. In this section we study compounds that possess the oxygen-containing functional groups in Table 22.1.

Alcohols and Ethers

Alcohols were discussed in Chapter 8. In general, they include any compound with an OH group bonded to a carbon that is also attached to three other groups by single bonds.

Using the symbol R to represent any alkyl group, alcohols have ROH as their general structure.

The four structurally simplest alcohols are the following (their common names are in parentheses below their IUPAC names):

$$CH_3OH \qquad CH_3CH_2OH \qquad CH_3CH_2CH_2OH \qquad CH_3\underset{\underset{\displaystyle OH}{|}}{C}HCH_3$$

methanol   ethanol   propan-1-ol   propan-2-ol
(methyl alcohol)   (ethyl alcohol)   (propyl alcohol)   (isopropyl alcohol)
BP 65 °C   BP 78.5 °C   BP 97 °C   BP 82 °C

Ethanol is the alcohol in beverages and is also added to gasoline to make "gasohol" and makes up 85% of the fuel E-85 (the rest is gasoline).

Molecules of **ethers** contain two alkyl groups joined to one oxygen, the two R groups can be either the same or different. We give only the common names for the following examples:

$$CH_3OCH_3 \qquad CH_3CH_2OCH_2CH_3 \qquad CH_3OCH_2CH_3 \qquad R{-}O{-}R'$$

dimethyl ether   diethyl ether   methyl ethyl ether   ethers
BP −23 °C   BP 34.5 °C   BP 11 °C   (general structure)

Diethyl ether is the "ether" that was once widely used as an anesthetic during surgery.

The contrasting boiling points of alcohols and ethers illustrate the influence of hydrogen bonding. Hydrogen bonds cannot exist between molecules of ethers and the simple ethers have very low boiling points, being substantially lower than those of alcohols of comparable molecular sizes. For example, 1-butanol, $CH_3CH_2CH_2CH_2OH$ (BP 117 °C) boils 83 degrees higher than its structural isomer, diethyl ether (BP 34.5 °C).

Naming Alcohols

The name ending for an alcohol is *-ol*. It replaces the *-e* ending of the name of the hydrocarbon that corresponds to the parent. The parent chain of an alcohol must be the longest that includes the carbon bonded to the OH group. The chain is numbered starting at the end that gives the site of the OH group the lowest number regardless of where alkyl substituents occur.

Major Reactions of Alcohols and Ethers

Ethers are almost as chemically inert as alkanes. They burn (as do alkanes), and they are split apart when boiled in concentrated acids.

The alcohols, in contrast, have a rich chemistry. When the carbon atom of the alcohol system, the alcohol carbon atom, is also bonded to at least one H, this H can be removed by an oxidizing agent. The H atom of the OH group will also leave, and the oxygen will form a double bond with the carbon. The two H atoms become part of a water molecule, and we may think of the oxidizing agent as providing the O atom for H_2O. At this point, the number of H atoms that remain on the original alcohol carbon then determines the family of the product.

The oxidation of an alcohol of the RCH_2OH type, with two H atoms on the alcohol carbon, produces first an aldehyde, which is further oxidized to a carboxylic acid. (We will say more about aldehydes and carboxylic acids later in this section.)

$$RCH_2OH \xrightarrow{\text{oxidation}} \overset{\overset{\displaystyle O}{\|}}{RCH} \xrightarrow{\text{further oxidation}} \overset{\overset{\displaystyle O}{\|}}{RCOH}$$

aldehyde   carboxylic acid

The net ionic equation for the formation of the aldehyde when dichromate ion is used is

$$3RCH_2OH + Cr_2O_7{}^{2-} + 8H^+ \longrightarrow 3RCH{=}O + 2Cr^{3+} + 7H_2O$$

TOOLS

Naming alcohols

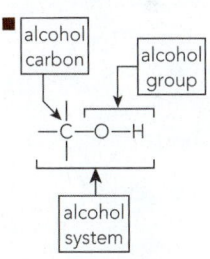

TOOLS

Oxidation of alcohols

■ The aldehyde group, CH=O, is one of the most easily oxidized, and the carboxyl group, CO_2H, is one of the most oxidation resistant of the functional groups.

Aldehydes are much more easily oxidized than alcohols, so unless the aldehyde is removed from the solution as it forms, it will consume any oxidizing agent that has not yet reacted and be changed to the corresponding carboxylic acid.

The oxidation of an alcohol of the R_2CHOH type produces a *ketone*. For example, the oxidation of 2-propanol gives propanone, commonly known as acetone.

$$3CH_3\overset{\overset{\displaystyle OH}{|}}{C}HCH_3 + Cr_2O_7{}^{2-} + 8H^+ \longrightarrow 3CH_3\overset{\overset{\displaystyle O}{\|}}{C}CH_3 + 2Cr^{3+} + 7H_2O$$

propan-2-ol   propanone
(acetone, a ketone)

Ketones strongly resist oxidation, so a ketone does not have to be removed from the oxidizing agent as it forms.

Alcohols of the type R_3COH have no removable H atom on the alcohol carbon, so they cannot be oxidized in a similar manner.

Other Reactions of Alcohols

Dehydration of alcohols

In the presence of a strong acid, like concentrated sulfuric acid, an alcohol molecule can undergo the loss of a water molecule, leaving behind a carbon–carbon double bond. This reaction, called **dehydration**, is one example of an **elimination reaction**:

$$\overset{\overset{\displaystyle CH_2—CH_2}{\underset{\underset{\displaystyle H \qquad OH}{|\qquad\;|}}{}}}{}\;\xrightarrow[\text{heat}]{\text{acid catalyst,}}\; CH_2{=}CH_2 + H_2O$$

ethanol   ethylene

$$\overset{\overset{\displaystyle CH_3CH—CH_2}{\underset{\underset{\displaystyle H \qquad OH}{|\qquad\;|}}{}}}{}\;\xrightarrow[\text{heat}]{\text{acid catalyst,}}\; CH_3CH{=}CH_2 + H_2O$$

propan-1-ol   propylene

What makes the elimination of water possible is the proton-accepting ability of the O atom of the OH group. Alcohols resemble water in that they react like Brønsted bases toward concentrated strong acids to give an equilibrium mixture involving a protonated form. Ethanol, for example, dissolves in and reacts with concentrated sulfuric acid by the following reaction (H_2SO_4 is written as $H{-}OSO_3H$):[1]

$$\overset{\overset{\displaystyle H}{|}}{CH_2}{-}CH_2 + H{-}OSO_3H \rightleftharpoons \overset{\overset{\displaystyle H}{|}}{H_2C}{-}CH_2 + {}^-OSO_3H$$

HO   (catalyst) $\overset{+}{O}$
                             H   H

This bond is now weak.

The organic cation is nothing more than the ethyl derivative of the hydronium ion, $CH_3CH_2OH_2{}^+$, and *all three bonds to oxygen in this cation are weak*, just like all three bonds to oxygen in H_3O^+.

A water molecule now leaves, taking with it the electron pair that held it to the CH_2 group. The remaining organic species is a carbocation.

---

[1] The first curved arrow illustrates how an atom attached to one molecule, like the O of the HO group, uses an unshared pair of electrons to pick up an atom of another molecule, like the H of the catalyst. The second curved arrow shows that both electrons of the bond from H to O in sulfuric acid remain in the hydrogen sulfate ion that forms. Thus H transfers as H^+.

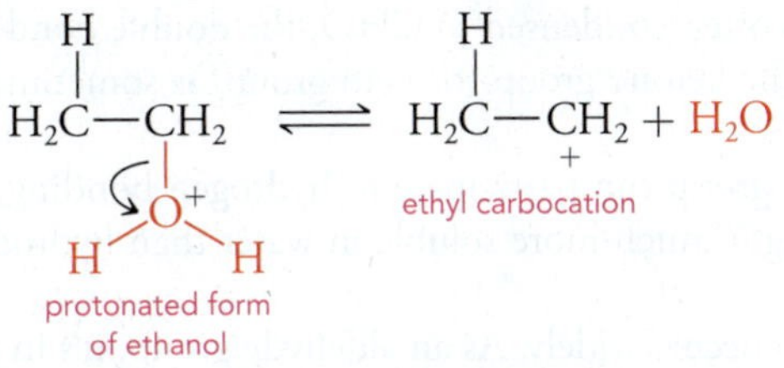

$$H_2\overset{\displaystyle H}{C}-CH_2 \;\rightleftharpoons\; H_2\overset{\displaystyle H}{C}-\overset{+}{C}H_2 + H_2O$$

protonated form of ethanol

ethyl carbocation

As we stated earlier, carbocations are unstable. The ethyl carbocation, $CH_3CH_2^+$, loses a proton to become more stable, donating it to a proton acceptor such as HSO_3O^- to reform the catalyst HSO_3H. The electron pair that was used in the bonding to the departing proton stays behind to become the second bond of the new double bond in the product. The next equation illustrates this step. All carbon atoms now have outer octets.

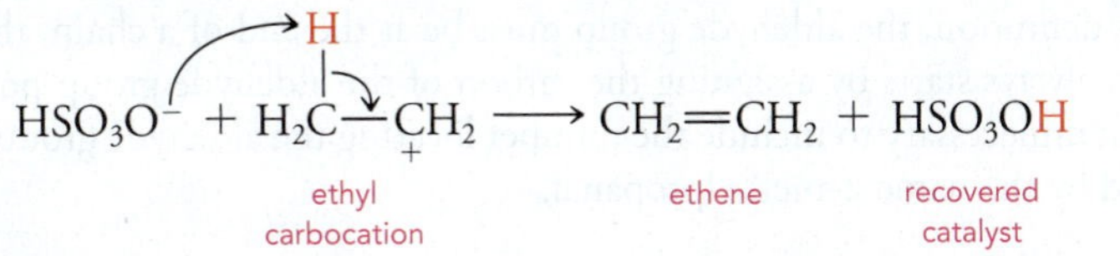

$$HSO_3O^- + H_2\overset{\displaystyle H}{C}-\overset{+}{C}H_2 \longrightarrow CH_2{=}CH_2 + HSO_3OH$$

ethyl carbocation

ethene

recovered catalyst

The last few steps in the dehydration of ethanol—the separation of H_2O, loss of the proton, formation of the double bond, and recovery of the catalyst—probably occur simultaneously. Notice two things about the catalyst, H_2SO_4. First, it works to convert a species with strong bonds, the alcohol, to one with weak bonds strategically located, the protonated alcohol. Second, the H_2SO_4 is recovered, as it must be if it is to be considered a catalyst.

Under acidic conditions, the OH group of an alcohol can be replaced by a halogen atom using a concentrated strong binary acid:

$$CH_3CH_2OH + HI(\text{conc.}) \xrightarrow{\text{heat}} CH_3CH_2I + H_2O$$

ethanol

iodoethane (ethyl iodide)

$$CH_3CH_2CH_2OH + HBr(\text{conc.}) \xrightarrow{\text{heat}} CH_3CH_2CH_2Br + H_2O$$

propan-1-ol

1-bromopropane (propyl bromide)

TOOLS

Substitution reactions of alcohols

These reactions, like that between chlorine and benzene in Section 22.2, are **substitution reactions**. The first step in each is the transfer of H^+ to the OH of the alcohol to give the protonated form of the OH group.

$$R-OH + H^+ \longrightarrow R-OH_2^+$$

The acid catalyst works to weaken the R—O bond in R—OH by adding H^+ to form $R-OH_2^+$. Given the high concentration of halide ion, X^-, it is this species that successfully reacts with $R-OH_2^+$ to displace OH_2 (H_2O) and yield R—X.

Aldehydes and Ketones

In Chapter 8 we described the **carbonyl group**, $>C{=}O$ and noted its presence in aldehydes and ketones, and we've seen how these compounds are formed in the oxidation of alcohols.

What differentiates aldehydes from ketones are the atoms attached to the carbon of the $C{=}O$ group. When the carbon atom of the carbonyl group binds an H atom plus a hydrocarbon group (or a second H), the compound is an aldehyde. When $C{=}O$ is bonded to two hydrocarbon groups at C, the compound is a ketone.

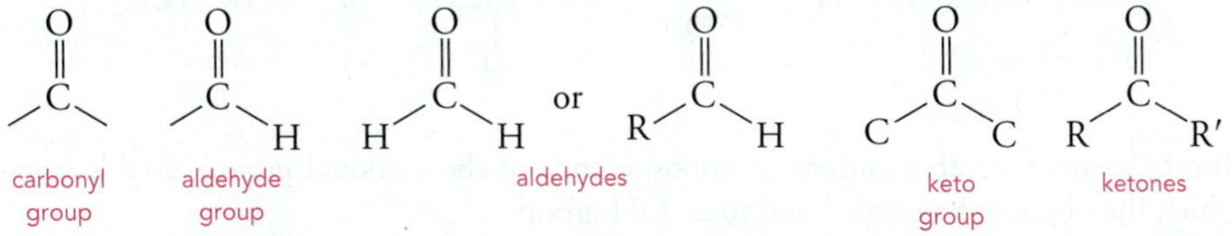

carbonyl group

aldehyde group

aldehydes

keto group

ketones

The aldehyde group is often condensed to CHO, the double bond of the carbonyl group being "understood." The ketone group (or keto group) is sometimes similarly condensed to CO.

The polar carbonyl group can participate in hydrogen bonding, which helps to make compounds containing it much more soluble in water than hydrocarbons of roughly the same molecular size.

The carbonyl group occurs widely. As an aldehyde group, it's in the molecules of most sugars, like glucose. Another common sugar, fructose, has the keto group.

Naming Aldehydes and Ketones

Naming aldehydes and ketones

The IUPAC name ending for an aldehyde is *-al*. The parent chain is the longest chain that includes the aldehyde group. Thus, the three-carbon aldehyde is named propanal, because "propane" is the name of the three-carbon alkane and the *-e* in propane is replaced by *-al*. Because, by definition, the aldehyde group must be at the end of a chain, the numbering of the chain always starts by assigning the carbon of the aldehyde group position 1. This rule makes it unnecessary to include the number locating the aldehyde group in the name, as illustrated by the name 2-methylpropanal.

$$
\begin{array}{cccc}
\underset{\substack{\text{methanal}\\ \text{(formaldehyde)}\\ \text{BP} -21\ ^\circ\text{C}}}{\overset{\overset{\displaystyle O}{\|}}{HCH}} & \underset{\substack{\text{ethanal}\\ \text{(acetaldehyde)}\\ \text{BP } 21\ ^\circ\text{C}}}{\overset{\overset{\displaystyle O}{\|}}{CH_3CH}} & \underset{\substack{\text{propanal}\\ \text{BP } 49\ ^\circ\text{C}}}{\overset{\overset{\displaystyle O}{\|}}{CH_3CH_2CH}} & \underset{\substack{\text{2-methylpropanal}\\ \text{(not 2-methyl-1-propanal)}\\ \text{BP } 64\ ^\circ\text{C}}}{\overset{\overset{\displaystyle CH_3}{|}}{CH_3CH}\!-\!\overset{\overset{\displaystyle O}{\|}}{CH}}
\end{array}
$$

Aldehydes cannot form hydrogen bonds between their own molecules, so they boil at lower temperatures than alcohols of comparable molecular size.

The name ending for the IUPAC names of ketones is *-one*. The parent chain must include the carbonyl group and be numbered from whichever end reaches the carbonyl carbon first. The number of the ketone group's location must be part of the name whenever there would otherwise be uncertainty.

■ We need not write "propan-2-one," because if the carbonyl carbon is anywhere else in a three-carbon chain, the compound could not be a ketone; it would be the *aldehyde*, propanal.

$$
\begin{array}{ccc}
\underset{\substack{\text{propanone}\\ \text{(acetone)}\\ \text{BP } 56.5\ ^\circ\text{C}}}{\overset{\overset{\displaystyle O}{\|}}{CH_3CCH_3}} & \underset{\substack{\text{pentan-3-one}\\ \text{BP } 101.5\ ^\circ\text{C}}}{\overset{\overset{\displaystyle O}{\|}}{CH_3CH_2CCH_2CH_3}} & \underset{\substack{\text{5-methylhexan-2-one}\\ \text{(not 2-methylhexan-5-one)}\\ \text{BP } 145\ ^\circ\text{C}}}{\overset{\overset{\displaystyle CH_3}{|}}{CH_3CHCH_2CH_2CCH_3}\ \overset{\overset{\displaystyle O}{\|}}{}}
\end{array}
$$

Reactions of Aldehydes and Ketones

Hydrogen is capable of adding across the double bond of the carbonyl group in both aldehydes and ketones. The reaction is just like the addition of hydrogen across the double bond of an alkene and takes place under roughly the same conditions—namely, with a metal catalyst, heat, and pressure. The reaction is called either hydrogenation or reduction. For example,

Reduction of aldehydes and ketones

$$
\underset{\text{ethanal}}{\overset{\overset{\displaystyle O}{\|}}{CH_3CH}} + H\!-\!H \xrightarrow[\text{catalyst}]{\text{heat, pressure}} \overset{\overset{\displaystyle O\!-\!H}{|}}{\underset{\displaystyle H}{CH_3CH}} \quad \text{or} \quad \underset{\text{ethanol}}{CH_3CH_2OH}
$$

$$
\underset{\substack{\text{propanone}\\ \text{(acetone)}}}{\overset{\overset{\displaystyle O}{\|}}{CH_3CCH_3}} + H\!-\!H \xrightarrow[\text{catalyst}]{\text{heat, pressure}} \overset{\overset{\displaystyle O\!-\!H}{|}}{\underset{\displaystyle H}{CH_3CCH_3}} \quad \text{or} \quad \underset{\text{propan-2-ol}}{\overset{\overset{\displaystyle OH}{|}}{CH_3CHCH_3}}
$$

The H atoms take up positions at opposite ends of the carbonyl group's double bond, which then becomes a single bond to an OH group.

Aldehydes and ketones are in separate families because of their remarkably different behavior toward oxidizing agents: aldehydes are easily oxidized, but ketones strongly resist oxidation. Even in storage, aldehydes are slowly oxidized by the oxygen in the air trapped in the bottle and the air that inevitably seeps in.

Carboxylic Acids and Esters

A **carboxylic acid** has an OH bonded to the carbon of the carbonyl group.

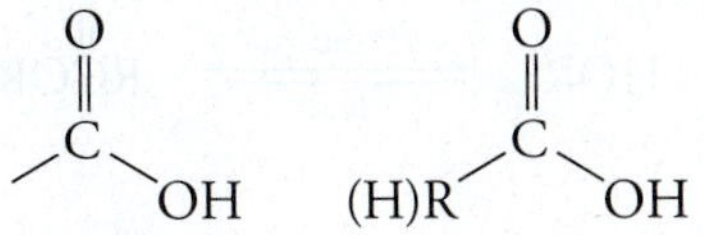

carboxyl group

carboxylic acids

The H in parentheses means the group attached to the C of the carboxyl group can either be R or H. In **condensed formulas** the **carboxyl group** is often written as CO_2H (or sometimes COOH).

The name ending of the IUPAC names of carboxylic acids is *-oic acid*. The parent chain must be the longest that includes the carboxyl carbon, which is numbered position 1. The name of the hydrocarbon with the same number of carbons as the parent is then changed by replacing the terminal *-e* with *-oic acid*. For example,

HCO_2H

methanoic acid
(formic acid)
BP 101 °C

CH_3CO_2H

ethanoic acid
(acetic acid)
BP 118 °C

$CH_3CHCH_2CO_2H$

3-methylbutanoic acid
BP 176 °C

Carboxylic acids are used to synthesize two important kinds of derivatives, esters and amides (which are discussed in the next section). In **esters**, the OH of the carboxyl group is replaced by OR.

$(H)RCOR'$ or $(H)RCO_2R'$,   for example, $CH_3CO(CH_2)_7CH_3$

esters

octyl ethanoate
(fragrance of oranges)

The IUPAC name of an ester begins with the name of the alkyl group attached to the O atom. This is followed by a separate word, which is taken from the name of the parent carboxylic acid by changing *-ic acid* to *-ate*. For example,

HCO_2CH_3

methyl methanoate
BP 31.5 °C

$CH_3CH_2CO_2CH_2CH_3$

ethyl propanoate
BP 99 °C

$CH_3CHCH_2CO_2CHCH_3$

isopropyl 3-methylbutanoate
BP 142 °C

Reactions of Carboxylic Acids and Esters

The carboxyl group is weakly acidic, causing compounds that contain it to be weak acids in water. In fact, we've used both formic acid and acetic acid in previous discussions of weak acids. All carboxylic acids, both water soluble and water insoluble, neutralize Brønsted bases such as the hydroxide, bicarbonate, and carbonate ions. The general equation for the reaction with OH⁻ is

$$RCO_2H + OH^- \xrightarrow{\text{H}_2\text{O}} RCO_2^- + H_2O$$

TOOLS

Naming organic acids

■ Because carboxylic acids have both a lone oxygen and an OH group, their molecules strongly hydrogen bond to each other. Their high boiling points, relative to alcohols of comparable molecular size, reflect this.

■ A derivative of a carboxylic acid is a compound that can be made from the acid, or can be changed to the acid by hydrolysis.

TOOLS

Naming esters

The carboxyl group is present in all of the building blocks of proteins, the amino acids, which are discussed further in Section 22.4.

One way to prepare an ester is to heat a solution of the parent carboxylic acid and the alcohol in the presence of an acid catalyst. (We'll not go into the details of how this happens.) The following kind of equilibrium forms, but a substantial stoichiometric excess of the alcohol (usually the less expensive reactant) is commonly used to drive the position of the equilibrium toward the ester.

TOOLS
Formation of esters

$$\underset{\substack{\text{carboxylic acid}}}{\overset{O}{\underset{||}{RCOH}}} + \underset{\substack{\text{alcohol}}}{HOR'} \underset{\text{heat}}{\overset{H^+ \text{ catalyst}}{\rightleftarrows}} \underset{\substack{\text{ester}}}{\overset{O}{\underset{||}{RCOR'}}} + H_2O$$

For example,

$$\underset{\substack{\text{butanoic acid} \\ \text{(BP 166 °C)}}}{\overset{O}{\underset{||}{CH_3CH_2CH_2COH}}} + \underset{\substack{\text{ethanol} \\ \text{(BP 78.5 °C)}}}{HOCH_2CH_3} \underset{\text{heat}}{\overset{H^+ \text{ catalyst}}{\rightleftarrows}} \underset{\substack{\text{ethyl butanoate (BP 120 °C)} \\ \text{(fragrance of pineapple)}}}{\overset{O}{\underset{||}{CH_3CH_2CH_2COCH_2CH_3}}} + H_2O$$

An ester is hydrolyzed to its parent acid and alcohol when the ester is heated together with a stoichiometric excess of water (plus an acid catalyst). The identical equilibrium as shown above forms, but in ester hydrolysis water is in excess, so the equilibrium shifts to the left to favor the carboxylic acid and alcohol, another illustration of Le Châtelier's principle.

Esters are also split apart by the action of aqueous base, only now the carboxylic acid emerges not as the free acid but as its anion. The reaction is called ester **saponification**. We can illustrate it by the action of aqueous sodium hydroxide on ethyl ethanoate, a simple ester:

TOOLS
Saponification of esters

$$\underset{\substack{\text{ethyl ethanoate} \\ \text{(ethyl acetate)}}}{\overset{O}{\underset{||}{CH_3COCH_2CH_3}}}(aq) + NaOH(aq) \overset{\text{heat}}{\longrightarrow} \underset{\substack{\text{ethanoate ion} \\ \text{(acetate ion)}}}{\overset{O}{\underset{||}{CH_3CO^-}}}(aq) + Na^+(aq) + \underset{\substack{\text{ethanol}}}{HOCH_2CH_3}(aq)$$

Ester groups abound among the molecules of the fats and oils in our diets. The hydrolysis of their ester groups occurs when we digest them, with an enzyme as the catalyst, not a strong acid. Because this digestion occurs in a region of the intestinal tract where the fluids are slightly basic, the anions of carboxylic acids form. Esters also tend to have pleasant aromas and are responsible for the fragrance of many fruits. For example, octyl acetate,

$$\underset{\substack{\text{octyl acetate}}}{\overset{O}{\underset{||}{CH_3COCH_2CH_2CH_2CH_2CH_2CH_2CH_2CH_3}}}$$

gives oranges their characteristic odor. See the table in the margin for other esters and their aromas.

Ester	Aroma
$HCO_2CH_2CH_3$	Rum
$HCO_2CH_2CH(CH_3)_2$	Raspberries
$CH_3CO_2(CH_2)_4CH_3$	Bananas
$CH_3CO_2(CH_2)_2CH(CH_3)_2$	Pears
$CH_3CO_2(CH_2)_7CH_3$	Oranges
$CH_3(CH_2)_2CO_2CH_2CH_3$	Pineapples
$CH_3(CH_2)_2CO_2(CH_2)_4CH_3$	Apricots

Example 22.2
Alcohol Oxidation Products

What is the structure of the organic product made by the oxidation of butan-2-ol with dichromate ion? If no oxidation can occur, so state.

Analysis: We first have to determine the structure of butan-2-ol. The oxidation product then depends on the number of hydrogens attached to the alcohol carbon atom.

Assembling the Tools: First, we have to use the nomenclature rules to draw the structure of butan-2-ol. If there are no H atoms on the alcohol carbon, oxidation by dichromate ion cannot occur. If there is one H atom on the alcohol carbon, the tool we use predicts that the product will be a ketone. If there are two H atoms on the alcohol carbon, the product will be an aldehyde which can be further oxidized to the carboxylic acid.

Solution: The nomenclature rules tell us that for butan-2-ol the OH group of the alcohol is on carbon number 2 of a four-carbon chain of singly bonded carbon atoms. In drawing the structure we have to remember that each carbon atom must have four bonds.

or

$$CH_3\overset{\overset{\Large OH}{|}}{C}HCH_2CH_3$$

butan-2-ol

Bond the C to the O in OH by moving the OH to the right

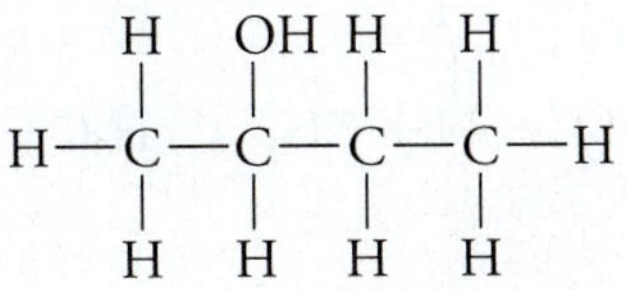

The alcohol carbon has one H atom, so the product will be a ketone. We simply erase the two H atoms shown in red and insert a double bond from C to O. (The product is called butanone.)

$$CH_3\overset{\overset{\Large O}{\|}}{C}CH_2CH_3$$

butanone (a ketone)

Is the Answer Reasonable? The starting material had two alkyl groups, CH_3 and CH_2CH_3. It doesn't matter what these alkyl groups are, however, because the reaction takes the same course in all cases. All alcohols of the R_2CHOH type can be oxidized to ketones in this way.

The "skeleton" of heavy atoms, C and O, does not change in this kind of oxidation, and butanone has the same skeleton as butan-2-ol. The oxidation also produces a double bond from C to O, as we showed.

Practice Exercise 22.5

Oxidation of an alcohol gave the following product. What was the formula of the original alcohol? (*Hint:* How many hydrogens are removed from the alcohol carbon by oxidation?)

$$CH_3CH_2-\overset{\overset{\Large O}{\|}}{C}-CH_3$$

Practice Exercise 22.6

What are the structures of the products (if any) that form from the oxidation of the following alcohols by dichromate ion? (a) ethanol, (b) pentan-3-ol, (c) 2-methylpropan-2-ol

Practice Exercise 22.7

Write the structural formula(s) for the principal organic product(s) in the following reactions:

$$\text{(a)}\ CH_3\overset{\overset{\Large OH}{|}}{C}HCH_2CH_3 + HCl(\text{conc.}) \xrightarrow{\text{heat}} \qquad \text{(b)}\ CH_3CH_2CH_2OH \xrightarrow[\text{Conc.}]{H_2SO_4}$$

Practice Exercise 22.8

Complete the following equation and write the structural formula for the principal organic product. What is the name of the product?

$$CH_3CH_2\overset{\overset{\Large O}{\|}}{C}-OH + CH_3OH \underset{\text{heat}}{\overset{H^+\ \text{catalyst}}{\rightleftharpoons}}$$

Practice Exercise 22.9

Write structural formulas for the following: **(a)** 3-methylbutanal, **(b)** ethyl propionate, and **(c)** butanone.

Practice Exercise 22.10

Write the structural formula(s) for the principal organic product(s) in the following reaction:

$$CH_3CH_2O\overset{\overset{\Large O}{\|}}{C}CH_3 + OH^- \underset{\text{heat}}{\overset{H_2O}{\longrightarrow}}$$

22.4 | Organic Derivatives of Ammonia

As described in Chapter 8, amines can be viewed as organic derivatives of ammonia in which one, two, or three hydrocarbon groups have replaced hydrogens. Examples, together with their common (not IUPAC) names, are

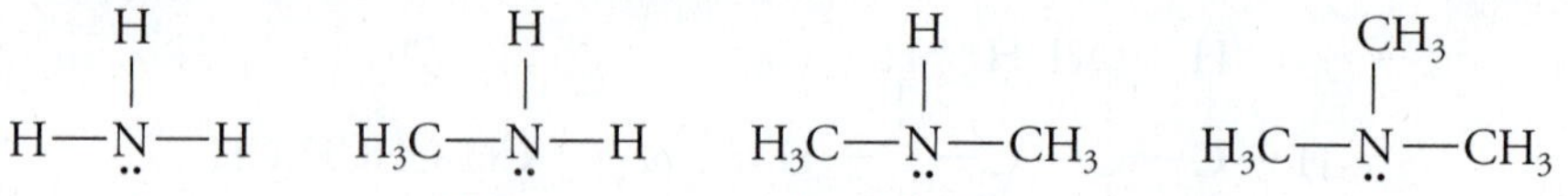

■ The following are in the order of decreasing strengths of intermolecular attractions.

CH_3CH_2OH, BP 78.5 °C

$CH_3CH_2NH_2$, BP 17 °C

$CH_3CH_2CH_3$, BP −42 °C

ammonia
BP −33.4 °C

methylamine
BP −8 °C

dimethylamine
BP 8 °C

trimethylamine
BP 3 °C

The N—H bond is not as polar as the O—H bond, so hydrogen bonding is not as strong and amines boil at lower temperatures than alcohols of comparable molecular size. Amines of low molecular mass are soluble in water due to hydrogen bonding between molecules of water and the amine.

Basicity and Reactions of Amines

In Section 8.9, amines were described as weak Brønsted bases. Behaving like ammonia, amines that are able to dissolve in water establish an equilibrium in which a low concentration of hydroxide ion exists. For example,

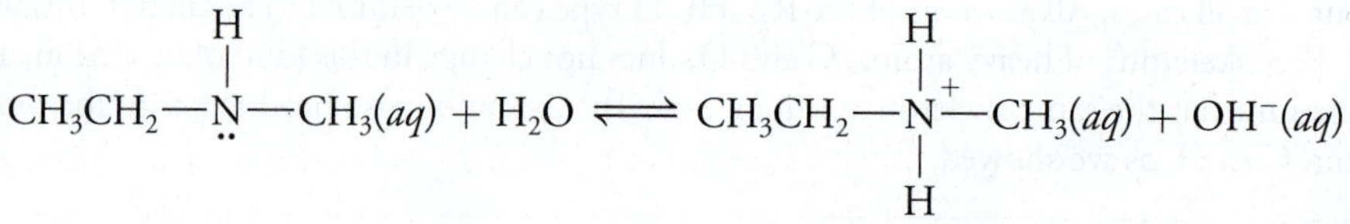

ethylmethylamine

ethylmethylammonium ion

As a result, aqueous solutions of amines test basic to litmus and have pH values above 7.

When an amine is mixed with a strong acid such as hydrochloric acid, the amine and acid react almost quantitatively. The amine accepts a proton and changes almost entirely into its protonated form. For example,

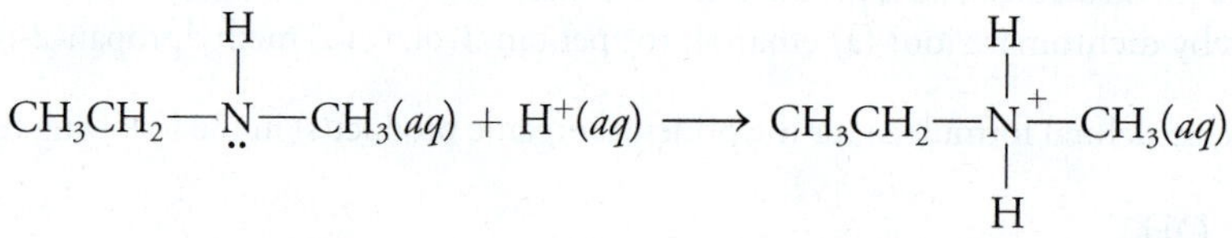

ethylmethylamine

ethylmethylammonium ion
(a protonated amine)

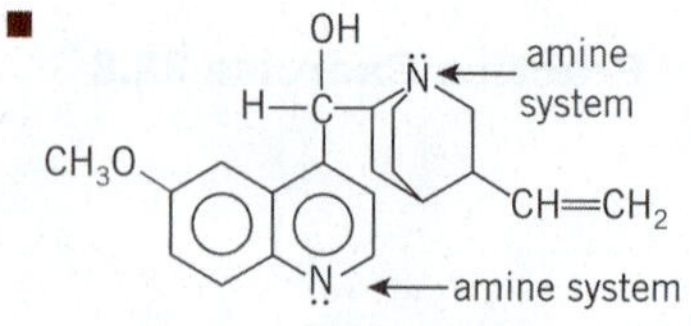

Even water-insoluble amines undergo this reaction, and the resulting salt is much more soluble in water than the original electrically neutral amine.

Many important medicinal chemicals, like quinine (which is used to treat malaria), are amines, but they are usually supplied to patients in protonated forms so that the drug can be administered as an aqueous solution, not as a solid. This strategy is particularly important for medicines that must be given by intravenous drip.

■
OH — amine system
CH_3O
CH=CH$_2$
amine system
quinine
(an antimalarial drug)

Protonated Amines as Weak Brønsted Acids

A protonated amine is a substituted ammonium ion. Like the ammonium ion itself, protonated amines are weak Brønsted acids. They can neutralize strong base:

$$CH_3NH_3{}^+(aq) + OH^-(aq) \longrightarrow CH_3NH_2(aq) + H_2O$$

methylammonium ion

methylamine

This reverses the protonation of an amine and releases the uncharged amine molecule.

Amides: Derivatives of Carboxylic Acids

Carboxylic acids can also be converted to amides, a functional group found in proteins. In **amides**, the OH of the carboxyl group is replaced by nitrogen, which may also be bonded to any combination of H atoms or hydrocarbon groups.

$$\text{(H)R}\overset{\overset{\displaystyle O}{\|}}{C}\text{NH}_2 \quad \text{or} \quad \text{(H)RCONH}_2, \qquad \text{for example,} \quad \text{CH}_3\overset{\overset{\displaystyle O}{\|}}{C}\text{NH}_2$$

simple amides   ethanamide
(acetamide)

The simple amides are those in which the nitrogen bears no hydrocarbon groups, only two H atoms. In the place of either or both of these H atoms, however, there can be a hydrocarbon group, and the resulting substance is still an amide.

The IUPAC names of the simple amides are devised by first writing the name of the parent carboxylic acid. Then its ending, *-oic acid*, is replaced by *-amide*. For example,

$$\text{CH}_3\text{CH}_2\text{CONH}_2 \qquad \text{CH}_3\text{CH}_2\text{CH}_2\text{CH}_2\text{CONH}_2 \qquad \overset{\overset{\displaystyle CH_3}{|}}{\text{CH}_3\text{CHCH}_2\text{CH}_2\text{CONH}_2}$$

propanamide   pentanamide   4-methylpentanamide

One of the ways to prepare simple amides parallels that of the synthesis of esters—that is, by heating a mixture of the carboxylic acid and an excess of ammonia.

In general: $\quad \text{R}\overset{\overset{\displaystyle O}{\|}}{C}\text{OH} \quad + \quad \text{H}-\text{NH}_2 \xrightarrow{\text{heat}} \text{R}\overset{\overset{\displaystyle O}{\|}}{C}\text{NH}_2 \quad + \quad \text{H}_2\text{O}$

carboxylic acid   ammonia   simple amide

TOOLS

Formation of amides

An example: $\quad \text{CH}_3\overset{\overset{\displaystyle O}{\|}}{C}\text{OH} \quad + \quad \text{H}-\text{NH}_2 \xrightarrow{\text{heat}} \text{CH}_3\overset{\overset{\displaystyle O}{\|}}{C}\text{NH}_2 \quad + \quad \text{H}_2\text{O}$

Amides, like esters, can be hydrolyzed. When simple amides are heated with water, they change back to their parent carboxylic acids and ammonia. Both strong acids and strong bases promote the reaction. As the following equations show, the reaction is the reverse of the formation of an amide.

In general: $\quad \text{R}\overset{\overset{\displaystyle O}{\|}}{C}\text{NH}_2 \quad + \quad \text{H}-\text{OH} \xrightarrow{\text{heat}} \text{R}\overset{\overset{\displaystyle O}{\|}}{C}\text{OH} \quad + \quad \text{NH}_3$

simple amide   carboxylic acid

TOOLS

Hydrolysis of an amide

An example: $\quad \text{CH}_3\overset{\overset{\displaystyle O}{\|}}{C}\text{NH}_2 \quad + \quad \text{H}-\text{OH} \xrightarrow{\text{heat}} \text{CH}_3\overset{\overset{\displaystyle O}{\|}}{C}\text{OH} \quad + \quad \text{NH}_3$

ethanamide   ethanoic acid
(acetic acid)

Urea, excreted in urine, is the amide of carbonic acid, H_2CO_3, and when it is hydrolyzed the products are ammonia and carbon dioxide (formed in the decomposition of H_2CO_3).

$$\text{H}_2\text{N}-\overset{\overset{\displaystyle O}{\|}}{C}-\text{NH}_2 + 2\text{H}_2\text{O} \longrightarrow \text{HO}-\overset{\overset{\displaystyle O}{\|}}{C}-\text{OH} + 2\text{NH}_3 \quad \nearrow \quad \text{CO}_2 + \text{H}_2\text{O}$$

urea   carbonic acid

This reaction explains why a baby's wet diaper gradually develops the odor of ammonia. Urea is used commercially as a fertilizer because its hydrolysis in the soil releases ammonia, a good source of nitrogen for plants.

Nonbasicity of Amides

Despite the presence of the NH_2 group in simple amides, the amides are not Brønsted bases like amines or ammonia. This can be understood by examining the following two resonance structures:

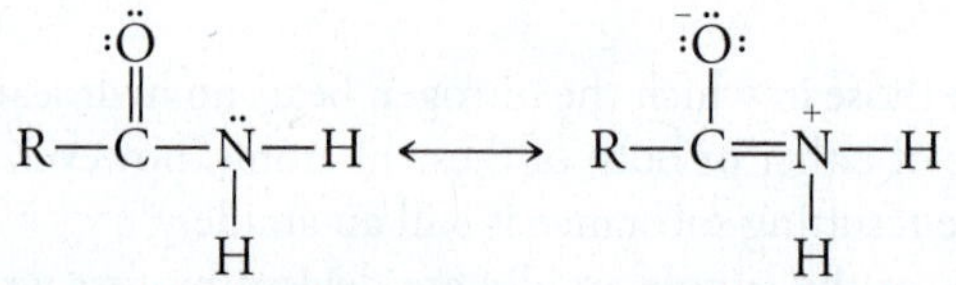

$$R-\overset{\overset{\displaystyle :\ddot{O}:}{\|}}{C}-\overset{\overset{\displaystyle H}{|}}{\underset{\underset{\displaystyle H}{|}}{\ddot{N}}}-H \longleftrightarrow R-\overset{\overset{\displaystyle :\ddot{O}:^-}{|}}{C}=\overset{\overset{\displaystyle +}{\underset{\underset{\displaystyle H}{|}}{N}}}-H$$

Structure 1   Structure 2

Effectively, the lone pair on the nitrogen in Structure 1 becomes partially delocalized onto the oxygen, as shown in Structure 2. This makes the lone pair on the amide nitrogen less available for donation to an H^+ than the corresponding lone pair on the nitrogen of an amine. As a result, the amide nitrogen has little tendency to acquire a proton, so amides are neutral compounds in an acid–base sense.

Practice Exercise 22.11

Complete the following equation by drawing structural formulas for the products.

$$H_2N-CH_2CH_2CH_3(aq) + CH_3CH_2-\overset{\overset{\displaystyle O}{\|}}{C}-OH(aq) \xrightarrow{\text{heat}}$$

(*Hint:* Identify the functional groups and their properties.)

Practice Exercise 22.12

Write the structural formula(s) for the principal organic product(s) in the following reactions:

(a) $CH_3CH_2\overset{\overset{\displaystyle O}{\|}}{C}-\overset{\overset{\displaystyle CH_2CH_3}{|}}{N}H + H_2O \xrightarrow{\text{heat}}$

(b) $CH_3CH_2-\overset{\overset{\displaystyle O}{\|}}{C}-OCH(CH_3)_2 \xrightarrow{OH^-(aq)}$

(c) $CH_3CH_2CH_2CH_2OH \xrightarrow[\text{Conc.}]{H_2SO_4}$

22.5 | Organic Polymers

Nearly all of the compounds that we've studied so far are composed of relatively small molecules. Many of the substances we encounter in the world around us, however, are composed of **macromolecules**—molecules made up of hundreds or even thousands of atoms. In nature, substances with macromolecules are almost everywhere you look. They include the fibers that give strength to trees and things made of wood, and they include proteins and the DNA found in all living creatures. Macromolecules of biological origin are discussed in Sections 22.6 and 22.7; in this section we look at synthetic macromolecules, such as those found in plastics and synthetic fibers.

Order within Polymer Molecules

A **polymer** is a macromolecular substance whose molecules all have a small characteristic structural feature that repeats itself over and over. An example is polypropylene, a polymer with many uses, including dishwasher-safe food containers, indoor-outdoor carpeting, and artificial turf. The molecules of polypropylene have the following general structure:

$$\text{etc.} - CH_2 - \underset{\underset{CH_3}{|}}{CH} - CH_2 - \underset{\underset{CH_3}{|}}{CH} - CH_2 - \underset{\underset{CH_3}{|}}{CH} - CH_2 - \underset{\underset{CH_3}{|}}{CH} - CH_2 - \underset{\underset{CH_3}{|}}{CH} - \text{etc.}$$

polypropylene

Notice that the polymer consists of a long carbon chain (the polymer's **backbone**) with CH_3 groups attached at periodic intervals.[2]

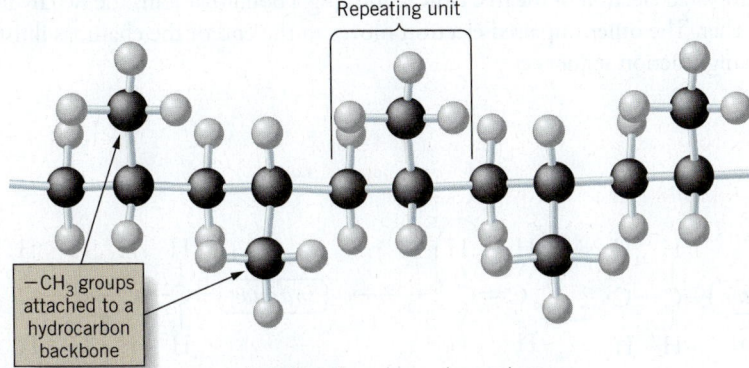

Repeating unit

$-CH_3$ groups attached to a hydrocarbon backbone

A portion of a polypropylene polymer

If you study this structure, you can see that one structural unit occurs repeatedly (actually thousands of times). In fact, the structure of a polymer is usually represented by the use of only its repeating unit, enclosed in parentheses, with a subscript n standing for several thousand units.

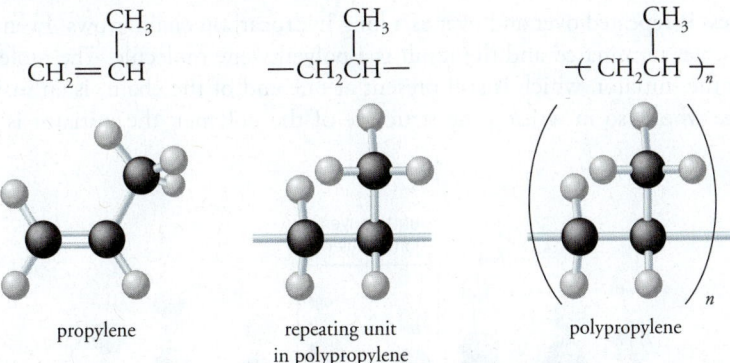

$$CH_2 = \underset{\underset{CH_3}{|}}{CH}$$

$$- CH_2 \underset{\underset{CH_3}{|}}{CH} -$$

$$\left(CH_2 \underset{\underset{CH_3}{|}}{CH} \right)_n$$

propylene

repeating unit in polypropylene

polypropylene

The value of n is not a constant for every molecule in a given polymer sample. A polymer, therefore, does not consist of molecules identical in size, just identical in kind; they have the same repeating unit. Notice that despite the *-ene* ending (which is used in naming alkenes), "polypropylene" has no double bonds. The polymer is named after its starting material, propylene (IUPAC name, propene).

The repeating unit of a polymer is contributed by a chemical raw material called a **monomer**. Thus propylene is the monomer for polypropylene. The reaction that makes a polymer out of a monomer is called **polymerization**, and the verb is "to polymerize." Most (but not all) useful polymers are formed from monomers that are considered to be organic compounds.

---

[2] The drawing of the molecule shows one way the CH_3 groups can be oriented relative to the backbone and each other. Other orientations are also possible, and they affect the physical properties of the polymer.

Chain-Growth Polymers

There are basically two ways that monomers become joined to form polymers—addition and condensation. In the simple addition of one monomer unit to another, the process continues over and over until a very long chain of monomer units is produced. Polymers formed by this process are called **chain-growth polymers** or **addition polymers**. Polypropylene, discussed previously, is an example of a chain-growth polymer, as is polyethylene, formed from ethylene (CH_2CH_2) monomer units. Under the right conditions and with the aid of a substance called an initiator, a pair of electrons in the carbon–carbon double bond of ethylene becomes unpaired. The initiator binds to one carbon, leaving an unpaired electron on the other carbon. The result is a very reactive substance called a free radical, which can attack the double bond of another ethylene. In the attack, one of the electron pairs of the double bond becomes unpaired. One of the electrons becomes shared with the unpaired electron of the free radical, forming a bond that joins the two hydrocarbon units together. The other unpaired electron moves to the end of the chain, as illustrated in the following reaction sequence:

TOOLS

Addition polymerization

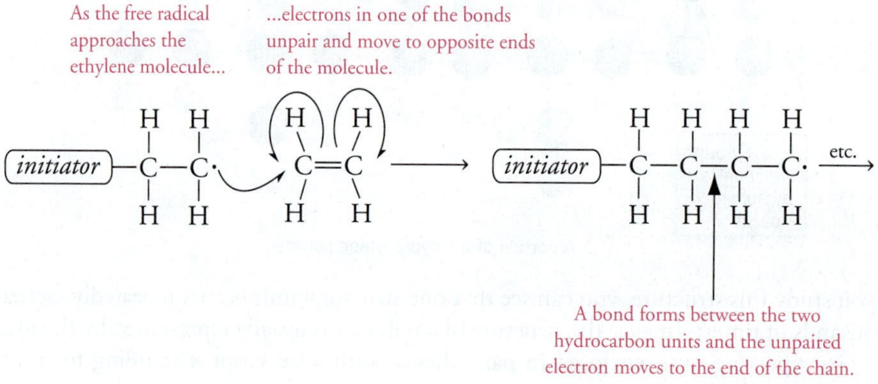

As the free radical approaches the ethylene molecule...

...electrons in one of the bonds unpair and move to opposite ends of the molecule.

A bond forms between the two hydrocarbon units and the unpaired electron moves to the end of the chain.

This process is repeated over and over as a long hydrocarbon chain grows. Eventually the chain becomes terminated and the result is a polyethylene molecule. The molecule is so large that the initiator, which is still present at one end of the chain, is an insignificant part of the whole, so in writing the structure of the polymer, the initiator is generally ignored.

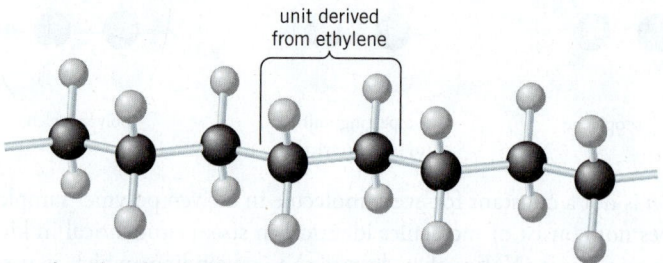

unit derived from ethylene

A portion of a polyethylene molecule. Although formed from a monomer with the formula C_2H_4, the actual repeating unit is CH_2.

The process by which the polymer forms has a significant influence on its ultimate structure. For example, the least expensive method for making polyethylene leads to **branching**, which means that polymer chains grow off the main backbone of the molecule as it grows longer. Other, more expensive procedures produce molecules without branching, which has a significant impact on the properties of the polymer.

In addition to polyethylene and polypropylene, another very common chain-growth polymer, called **polystyrene**, is formed by polymerizing styrene,

styrene (C₆H₅—CH=CH₂)   polystyrene

Figure 22.3 | Styrofoam. Made of polystyrene, Styrofoam is widely used as an insulation in construction.

Notice that styrene is similar to propylene, but with a benzene ring in place of the methyl group (—CH_3). Therefore, in the polymer we find C_6H_5— attached to every other atom in the hydrocarbon backbone. Polystyrene has almost as many applications as polyethylene. It's used to make clear plastic drinking glasses, molded car parts, and housings of things like computers and kitchen appliances. Sometimes a gas, like carbon dioxide, is blown through molten polystyrene as it is molded into articles. As the hot liquid congeals, tiny pockets of gas are trapped, and the product is a foamed plastic, like the familiar insulated foam cups or insulation materials (Figure 22.3).

Hundreds of substances similar to ethylene, propylene, and styrene, as well as their halogen derivatives, have been tested as monomers, and Table 22.3 lists some examples.

■ A "halogen derivative" is a compound in which one or more halogen atoms substitute for hydrogens in a parent molecule. Thus, CH_3Cl is halogen derivative of CH_4.

TABLE 22.3	Some Addition Polymers Formed from Compounds Related to Ethylene, $H_2C=CH_2$	
Polymer	**Monomer**	**Uses**
Polyethylene	$H_2C=CH_2$	Grocery bags, bottles, children's toys, bullet-proof vests
Polypropylene	$H_2C=CH$ with CH_3	Dishwasher-safe plastic kitchenware, indoor-outdoor carpeting, rope
Polystyrene	C₆H₅—CH=CH₂	Plastic cups, toys, housings for kitchen appliances, Sytrofoam insulation.
Polyvinyl chloride (PVC)	$H_2C=CH$ with Cl	Insulation, credit cards, vinyl siding for houses, bottles, plastic pipe
Polytetrafluoroethylene (Teflon)	$F_2C=CF_2$	Nonstick surfaces on cookware, valves
Polyvinyl acetate (PVA)	$H_2C=C$ with H and O—C(=O)—CH_3	Latex paint, coatings, glue, molded items
Polymethyl methacrylate (Lucite)	$H_2C=C$ with CH_3 and O—C(=O)—CH_3	Shatter-resistant windows, coatings, acrylic paints, molded items

Example 22.3
Writing the Formula for a Chain-Growth Polymer

Use the information in Table 22.3 to write the structure of the polymer polyvinyl chloride showing three repeat units. Write the general formula for this polymer.

Analysis: The polymer we're dealing with is a chain-growth polymer, so we can anticipate that the entire $CH_2 = CHCl$ molecule will repeat over and over. The polymer will be formed by opening the double bond, as we saw for ethylene.

Assembling the Tools: The tool we use to solve the problem is the method by which monomer units add to each other in the formation of the polymer.

Solution: When the double bond opens, bonds to two other monomer units will form.

$$\text{other monomer unit} -\overset{\displaystyle H}{\underset{\displaystyle H}{\overset{|}{\underset{|}{C}}}}-\overset{\displaystyle Cl}{\underset{\displaystyle H}{\overset{|}{\underset{|}{C}}}}- \text{other monomer unit}$$

We need to attach three repeating units to have the answer to the problem.

$$-\overset{\displaystyle H}{\underset{\displaystyle H}{\overset{|}{\underset{|}{C}}}}-\overset{\displaystyle Cl}{\underset{\displaystyle H}{\overset{|}{\underset{|}{C}}}}-\overset{\displaystyle H}{\underset{\displaystyle H}{\overset{|}{\underset{|}{C}}}}-\overset{\displaystyle Cl}{\underset{\displaystyle H}{\overset{|}{\underset{|}{C}}}}-\overset{\displaystyle H}{\underset{\displaystyle H}{\overset{|}{\underset{|}{C}}}}-\overset{\displaystyle Cl}{\underset{\displaystyle H}{\overset{|}{\underset{|}{C}}}}-$$

repeating unit

The general formula for the polymer should indicate the repeating unit occurring n times.

$$\left(\overset{\displaystyle H}{\underset{\displaystyle H}{\overset{|}{\underset{|}{C}}}}-\overset{\displaystyle Cl}{\underset{\displaystyle H}{\overset{|}{\underset{|}{C}}}}\right)_{n}$$

Is the Answer Reasonable? There's not much to check here, other than to be sure that the repeating unit has the same molecular formula as the monomer, which it does.

Practice Exercise 22.13

Suppose but-2-ene were polymerized. Make a sketch showing three repeating units of the monomer in the polymer. (*Hint:* What is the structure of but-2-ene?)

Practice Exercise 22.14

Write a formula showing three repeating units of the monomer in the polymer Teflon. (See Table 22.3.)

Step-Growth Polymers

Condensation polymerization

The second way that monomer units can combine to form a polymer is by a process called **condensation**, in which a small molecule is eliminated when the two monomer units become joined. In a simplified way, we can diagram this as follows.

$$A-A-A-A-A-A-\boxed{OH \quad H}-B-B-B-B-B-B$$
$$\downarrow \quad \rightarrow H_2O$$
$$A-A-A-A-A-A-B-B-B-B-B-B \ + \ H_2O$$

In this example, an OH group from one molecule combines with an H from another, forming a water molecule. At the same time, the two molecules A and B become joined by a covalent bond. If this can be made to happen at both ends of A and B, long chains are formed and a **condensation polymer** (also called a **step-growth polymer**) is the result.

The two most familiar condensation polymers are nylon and polyesters. Nylon is formed by combining two different compounds, so it's considered a copolymer. The first nylon to be manufactured is called **nylon 6,6** because it forms by combining two compounds each with six carbon atoms.

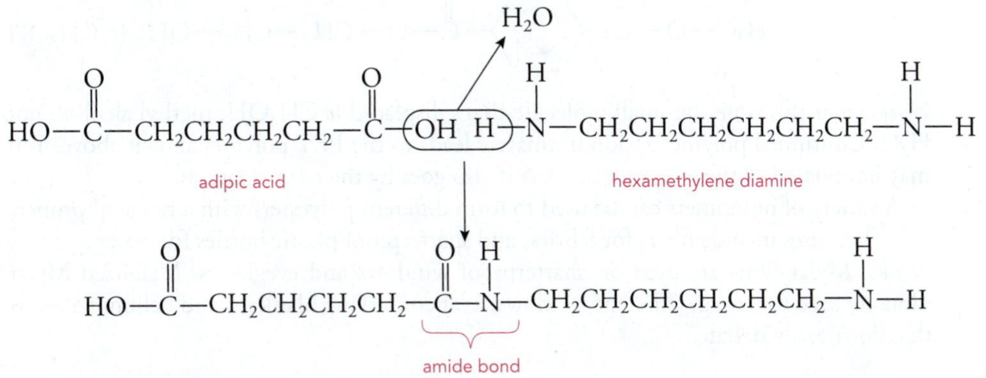

adipic acid   hexamethylene diamine

amide bond

Notice that adipic acid contains two carboxyl groups (it's called a dicarboxylic acid). The other compound is a diamine because it has two amine groups. By the elimination of water, the two molecules become joined by an amide bond, which we discussed in Section 22.4. This same linkage is found in proteins, including silk and the proteins found in our bodies.

The reaction product with the amide bond still has a carboxyl group on one end and an amine group on the other, so further condensation can occur, ultimately leading to the formation of nylon 6,6.

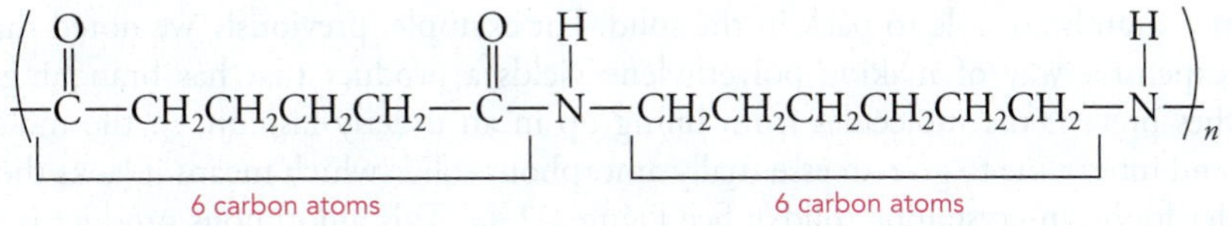

6 carbon atoms   6 carbon atoms

nylon 6,6

Nylon was invented in 1940 and became popular as a substitute for silk in women's stockings. It forms strong elastic fibers and is used to make fishing line as well as fibers found in all sorts of clothing and many other products.

Another type of condensation polymer uses the formation of an ester bond to join the monomers. An example of this is polyester, which is shown below.

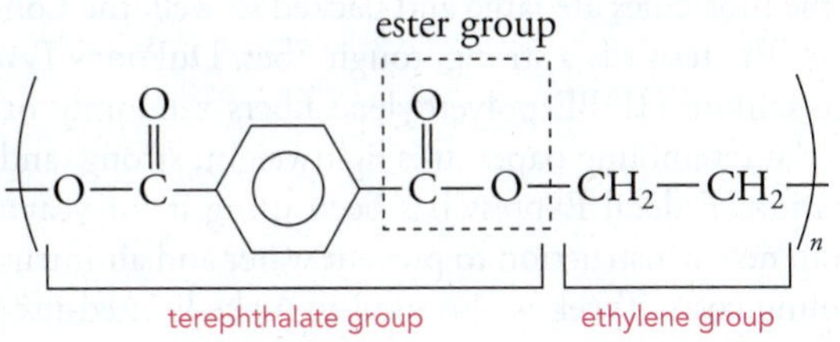

ester group

terephthalate group   ethylene group

poly(ethylene terephthalate), also known as PET

This copolymer is made by condensation of ethylene glycol, an alcohol, and dimethyl terephthalate, an ester. The first step is as follows:

$$H_3C-O-\overset{\overset{\displaystyle O}{\|}}{C}-\bigcirc-\overset{\overset{\displaystyle O}{\|}}{C}-\boxed{O-CH_3 \qquad HO}-CH_2-CH_2-OH$$

dimethyl terephthalate   ethylene glycol

$$\downarrow$$

$$H_3C-O-\overset{\overset{\displaystyle O}{\|}}{C}-\bigcirc-\overset{\overset{\displaystyle O}{\|}}{C}-O-CH_2-CH_2-OH \;+\; CH_3OH$$

Notice that this time the small molecule that's displaced is CH_3OH, methyl alcohol, not H_2O. Continued polymerization ultimately leads to the PET polymer shown above. You may have heard of this polymer because it also goes by the name Dacron.

A variety of monomers can be used to form different polyesters with a range of properties. Their uses include fibers for fabrics, and shatterproof plastic bottles for water and soft drinks. Mylar films are used on shatterproof windows and eyeglasses. Metalized Mylar films are used for energy-efficient windows, heat conserving blankets, and helium balloons that don't easily deflate.

Physical Properties and Polymer Crystallinity

Beyond chemical stability, physical properties are the features of polymers most sought after. Desirable properties of Teflon, for example, are its chemical inertness and its slipperiness toward just about anything. Nylon isn't eaten by moths—a chemical property, in fact—but its superior strength and its ability to be made into fibers and fabrics of great beauty are what make nylon valuable. Dacron resists mildewing and its fibers are not weakened by mildew like those of cotton. Dacron also has greater strength with lower mass than cotton, and Dacron fibers do not stretch as much, which are properties that account for Dacron's use for making sails.

In many ways, the physical properties of a polymer are related to how the individual polymer strands are able to pack in the solid. For example, previously we noted that the least expensive way of making polyethylene yields a product that has branching. The branches prevent the molecules from lining up in an orderly fashion, so the molecules twist and intertwine to give an essentially amorphous solid (which means it lacks the kind of order found in crystalline solids). See Figure 22.4a. This amorphous product is called low-density polyethylene (LDPE); the polymer molecules have a relatively low molecular mass and the solid has little structural strength. LDPE is the kind of polyethylene used to make the plastic bags from grocery stores.

Following different methods, polyethylene can be made to form without branching and with molecular masses ranging from 200,000 to 500,000 u. This polymer is called linear polyethylene or high-density polyethylene, HDPE. In HDPE, the polymer strands are able to line up alongside each other to produce a large degree of order (and therefore, crystallinity), as illustrated in Figure 22.4b. This enables the molecules to form fibers easily, and because the molecules are large and packed so well, the London forces between them are very strong. The result is a strong, tough fiber. DuPont's Tyvek®, for example, is made from thin, crystalline HDPE polyethylene fibers randomly oriented and pressed together into a material resembling paper. It is lightweight, strong, and resists water, tears, punctures, and abrasion. Federal Express has been using it for years for envelopes, and builders use it to wrap new construction to prevent water and air intrusion, thereby lowering heating and cooling costs. Tyvek is also used to make limited-use protective clothing for use in hazardous environments.

Under the right conditions, linear polyethylene molecules can be made extremely long, yielding ultrahigh-molecular-weight polyethylene (UHMWPE) with molecules having

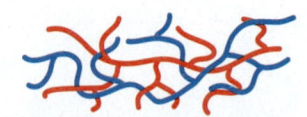

(a) Two branched polymer chains become twisted with little order.

(b) Polymer chains are able to align, producing a tightly packed structure with a large degree of crystallinity.

Figure 22.4 | Amorphous and crystalline polyethylene.
(a) When branching occurs in LDPE, the polymer strands are unable to become aligned and an amorphous product results.
(b) Linear HDPE has a high degree of crystallinity, which makes for excellent strong fibers.

■ HDPE molecules contain approximately 30,000 CH_2 units linked end to end!

■ UHMWPE molecules contain between 200,000 and 400,000 CH_2 units linked end to end!

molecular masses of three to six million. The fibers produced from this polymer are so strong that they are used to make bulletproof vests! Honeywell is producing an oriented polyethylene polymer it calls Spectra, which forms flexible fibers that can be woven into a strong, cut-resistant fabric. It is used to make thin, lightweight liners for surgical gloves that resist cuts by scalpels, industrial work gloves, and even sails for sailboats (see Figure 22.5). Mixed with other plastics, it can be molded into strong rigid forms such as helmets for military or sporting applications.

Other good fiber-forming polymers also have long molecules with shapes that permit strong interactions between the individual polymer strands. Nylon, for example, possesses polar carbonyl groups, $>C=O$, and N—H bonds that form strong hydrogen bonds between the individual molecules.

A carbon atom is at each vertex.

hydrogen bonds

Three strands of nylon 6,6 bound to each other by hydrogen bonds

Kevlar, another type of nylon, also forms strong hydrogen bonds between polymer molecules and is very crystalline. Its strong fibers are also used to make bulletproof vests. Because the fibers are so strong, they are also used to make thin yet strong hulls of racing boats. This lightweight construction improves speed and performance without sacrificing safety.

Figure 22.5 | Spectra fibers. Strong, lightweight sails made of Spectra fibers propelled Brad Van Liew to victory in the 2002–2003 "Around Alone" around-the-world yacht race. Van Liew sailed 31,094 miles in seven months to win the race.

High-speed racing boats often are built with hulls made of Kevlar. Because of the polymer's high strength, the hulls can be made thin, thereby reducing weight and increasing speed.

22.6 | Carbohydrates, Lipids, and Proteins

Biochemistry is the systematic study of the chemicals of living systems, their organization into cells, and their chemical interactions. Biochemicals have no life in themselves as isolated compounds, yet life has a molecular basis. Only when chemicals are organized into cells in tissues can interactions occur that enable tissue repair, cell reproduction, the generation of energy, the removal of wastes, and the management of a number of other functions. The world of living things is composed mostly of organic compounds, including many natural polymers such as proteins, starch, cellulose, and the chemicals of genes.

For its existence, a living system requires materials, energy, and information or "blueprints." Our focus in this section will be on the structures of substances that supply them—carbohydrates, lipids, proteins, and nucleic acids—the basic materials whose molecules, together with water and a few kinds of ions, make up cells and tissues.

Carbohydrates

Most **carbohydrates**, which serve as a major source of the chemical energy we need to function, are polymers of simpler units called **monosaccharides**. The most common monosaccharide is glucose, a pentahydroxyaldehyde, and probably the most widely occurring structural unit in the entire living world. Glucose is the chief carbohydrate in blood, and it provides the building unit for such important polysaccharides as cellulose and starch.

■ Starch, table sugar, and cotton are all carbohydrates.

Figure 22.6 | **Structures of glucose.** Three forms of glucose are in equilibrium in an aqueous solution. The curved arrows in the open form show how bonds become reoriented as it closes into a cyclic form. Depending on how the CH=O group is turned at the moment of ring closure, the new OH group at C-1 takes up one of two possible orientations, α or β. In these structures, the heavier bonds are used to indicate the part of the molecule that is closer to the viewer.

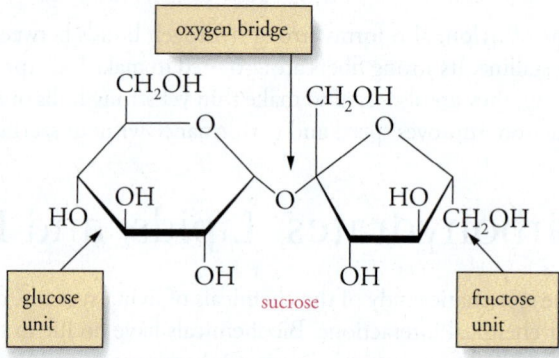

Ring opens here.

• Rotation at C-1 carbon
• Movement of H to form OH

Ring opens here.

α-glucose

Open form
(polyhydroxyaldehyde)

β-glucose

$$CH_2-CH-CH-CH-CH-CH$$
$$\quad|\quad\ |\quad\ |\quad\ |\quad\ |\quad\quad \overset{O}{\overset{\|}{}}$$
$$OH\ \ OH\ OH\ OH\ OH$$

glucose (open-chain form)
a polyhydroxyaldehyde

$$CH_2-CH-CH-CH-\overset{O}{\overset{\|}{C}}-CH_2$$
$$\ |\quad\ |\quad\ |\quad\ |\qquad\quad\ |$$
$$OH\ \ OH\ OH\ OH\qquad OH$$

fructose (open-chain form)
a polyhydroxyketone

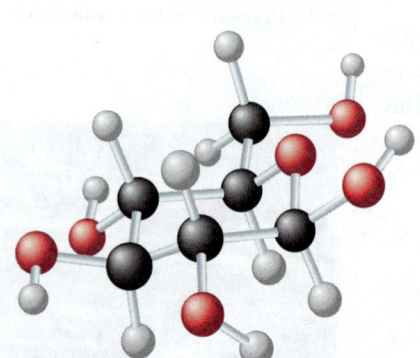

Because of the tetrahedral arrangement of atoms around each carbon, the six-membered rings of the cyclic forms of monosaccharides are not actually flat, as implied by the structural formulas we've drawn for them. The β-glucose molecule, for example, has the structure shown above.

Fructose, a pentahydroxyketone, is produced together with glucose when we digest table sugar. Honey is also rich in fructose.

When dissolved in water, the molecules of most monosaccharides exist in an equilibrium involving more than one structure. Glucose, for example, exists as two cyclic forms and one open-chain form in equilibrium in water (see Figure 22.6). The open-chain form, the only one with a free aldehyde group, is represented by less than 0.1% of the solute molecules. Yet, the solute in an aqueous glucose solution still gives the reactions of a polyhydroxyaldehyde. This is possible because the equilibrium between this form and the two cyclic forms shifts rapidly to supply more of any of its members when a specific reaction occurs to just one (in accordance with Le Châtelier's principle).

Disaccharides

Carbohydrates whose molecules are split into two monosaccharide molecules by reacting with water are called **disaccharides**. Sucrose (table sugar, cane sugar, or beet sugar) is an example, and its hydrolysis gives glucose and fructose.

oxygen bridge

CH_2OH CH_2OH

glucose unit

sucrose

fructose unit

To simplify, let's represent sucrose as Glu—O—Fru, where Glu is a glucose unit and Fru is a fructose unit, both joined by an oxygen bridge, —O—. The hydrolysis of sucrose, the chemical reaction by which we digest it, can thus be represented as follows.[3]

$$Glu-O-Fru + H_2O \xrightarrow[\text{hydrolysis}]{\text{digestions}} glucose + fructose$$

sucrose

Lactose (milk sugar) hydrolyzes to glucose and galactose (Gal), an isomer of glucose, and this is the reaction by which we digest lactose.

$$Gal-O-Glu + H_2O \xrightarrow[\text{hydrolysis}]{\text{digestions}} galactose + glucose$$

lactose

[3] Considerable detail is lost with the simplified structure of sucrose (and other carbohydrates like it to come). We leave the details, however, to other books because we seek a broader view.

Polysaccharides

Plants store glucose units for energy needs in molecules of **starch**, a type of large polymeric sugar molecule called a **polysaccharide**. Starch is found often in seeds and tubers (e.g., potatoes). It consists of two kinds of glucose polymers. The structurally simpler kind is amylose, which makes up roughly 20% of starch. We may represent its structure as follows, where O is the oxygen bridge linking glucose units.

$$Glu \left(O - Glu \right)_n OH$$

amylose; *n* is very large

The average amylose molecule has over 1000 glucose units linked together by oxygen bridges. These are the sites that are attacked and broken when water reacts with amylose during digestion. Molecules of glucose are released and are eventually delivered into circulation in the blood stream.

$$Amylose + nH_2O \xrightarrow[\text{hydrolysis}]{\text{digestion}} n \text{ glucose}$$

The bulk of starch is made up of amylopectin, whose molecules are even larger than those of amylose. The amylopectin molecule consists of several amylose molecules linked by oxygen bridges from the end of one amylose unit to a site somewhere along the "chain" of another amylose unit.

$$
\begin{array}{c}
\text{etc.} \\
| \\
Glu \left(O - Glu \right)_m O \\
| \\
Glu \left(O - Glu \right)_n O \\
| \\
Glu \left(O - Glu \right)_o O \\
| \\
\text{etc.}
\end{array}
$$

amylopectin (*m*, *n*, and *o* are large numbers)

Molecular masses ranging from 50,000 to several million are observed for amylopectin samples from the starches of different plant species. (A molecular mass of 1 million corresponds to about 6000 glucose units.)

Animals store glucose units for energy as **glycogen**, a polysaccharide with a molecular structure very similar to that of amylopectin. When we eat starchy foods and deliver glucose molecules into the bloodstream, any excess glucose not needed to maintain a healthy concentration in the blood is removed from circulation by particular tissues, like the liver and muscles. Liver and muscle cells convert glucose to glycogen. Later, during periods of high-energy demand or fasting, glucose units are released from the glycogen reserves so that the concentration of glucose in the blood stays high enough for the needs of the brain and other tissues.

Cellulose

Cellulose is a polymer of glucose, much like amylose, but with the oxygen bridges oriented with different geometries. We lack the enzyme needed to hydrolyze its oxygen bridges, so we are unable to use cellulose materials like lettuce for food, only for fiber. Animals, such as cows that eat grass and leaves, have bacteria living in their digestive tracts that convert cellulose into small molecules, which the host organism then appropriates for its own use.

■ Cellulose is the chief material in a plant cell wall, and it makes up about 100% of cotton.

Lipids

Lipids are natural products that are water insoluble, but tend to dissolve in nonpolar solvents such as diethyl ether or benzene. The lipid family is huge because the only structural requirement is that lipid molecules be relatively nonpolar with large segments that are entirely hydrocarbon-like. For example, the lipid family includes cholesterol as well as sex hormones, like estradiol and testosterone. You can see from their structures how largely hydrocarbon-like they are.

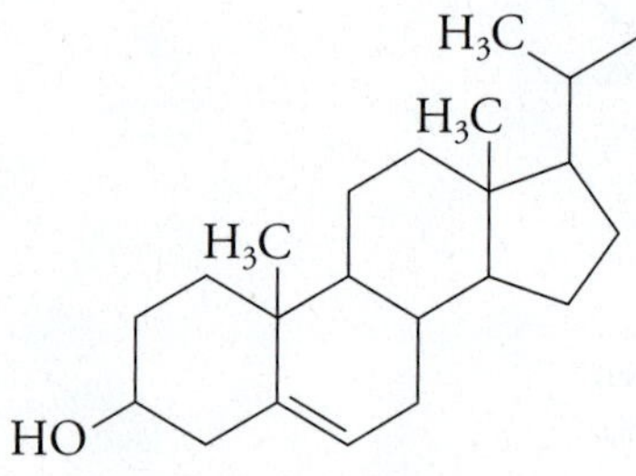

cholesterol

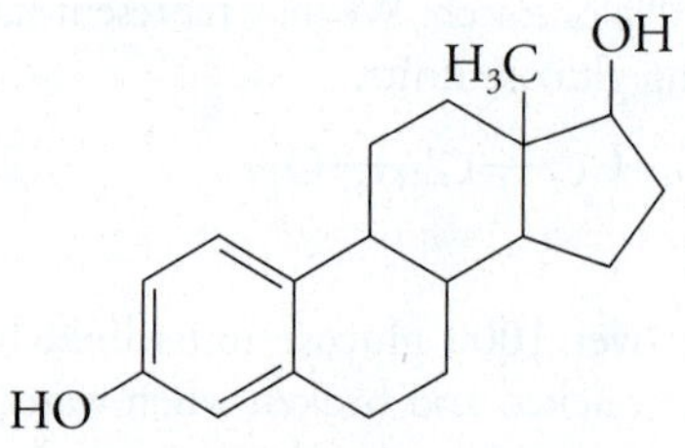

estradiol
(a female sex hormone)

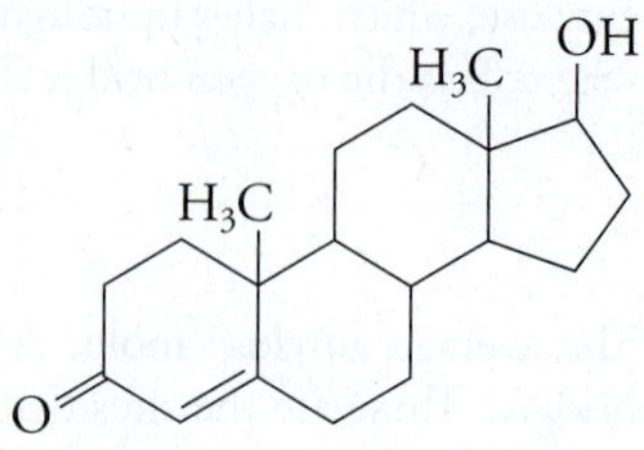

testosterone
(a male sex hormone)

Triacylglycerols

- CH₂OH
  |
  CHOH
  |
  CH₂OH
  glycerol

The lipid family also includes the edible fats and oils in our diets—substances such as olive oil, corn oil, butterfat, and lard. These are **triacylglycerols**, which are esters between glycerol, an alcohol with three OH groups, and any three of several long-chain carboxylic acids.

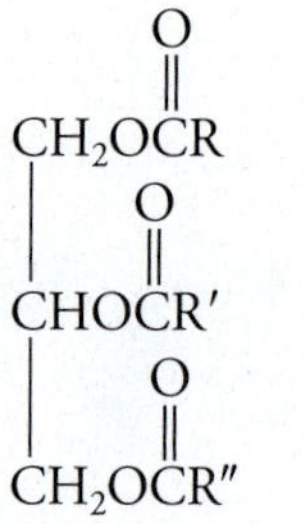

triacylglycerol
(general structural
features)

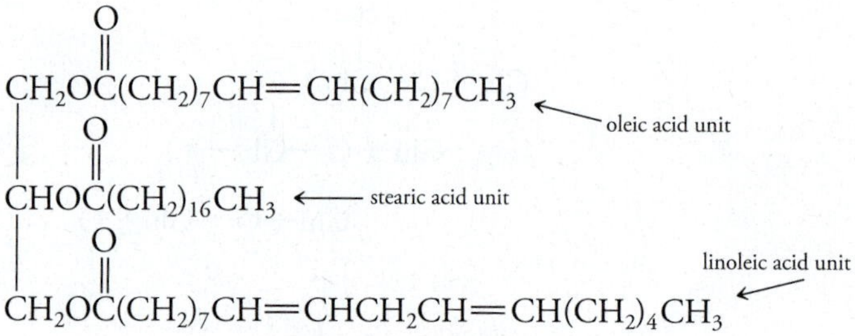

triacylglycerol—typical molecule present
in a vegetable oil

oleic acid unit

stearic acid unit

linoleic acid unit

The carboxylic acids used to make triacylglycerols are called **fatty acids** and generally have just one carboxyl group on an unbranched chain with an even number of carbon atoms (Table 22.4). Their long hydrocarbon chains make triacylglycerols mostly like hydrocarbons in physical properties, including insolubility in water. Many fatty acids have one or more carbon–carbon double bonds.

- Cooking oils are advertised as polyunsaturated because of their several alkene groups per molecule.

Triacylglycerols obtained from vegetable sources, like olive oil, corn oil, and peanut oil, are called vegetable oils and are liquids at room temperature. Triacylglycerols from animal sources, like lard and tallow, are called animal fats and are solids at room temperature. The vegetable oils generally have more double bonds per molecule than animal fats, and so are said to be polyunsaturated. The double bonds are usually *cis*, and so the molecules are

TABLE 22.4	Common Fatty Acids		
Fatty Acid	Number of Carbon Atoms	Structure	Melting Point (°C)
Myristic acid	14	$CH_3(CH_2)_{12}CO_2H$	54
Palmitic acid	16	$CH_3(CH_2)_{14}CO_2H$	63
Stearic acid	18	$CH_3(CH_2)_{16}CO_2H$	70
Oleic acid	18	$CH_3(CH_2)_7CH{=}CH(CH_2)_7CO_2H$	4
Linoleic acid	18	$CH_3(CH_2)_4CH{=}CHCH_2CH{=}CH(CH_2)_7CO_2H$	−5
Linolenic acid	18	$CH_3CH_2CH{=}CHCH_2CH{=}CHCH_2CH{=}CH(CH_2)_7CO_2H$	−11

kinked, making it more difficult for them to nestle close together and experience London forces. So they tend to be liquids rather than solids.

Digestion of Triacylglycerols

We digest the triacylglycerols by hydrolysis, meaning by their reaction with water. Our digestive juices in the upper intestinal tract have enzymes called lipases that catalyze these reactions. For example, the complete digestion of the triacylglycerol shown previously occurs by the following reaction:

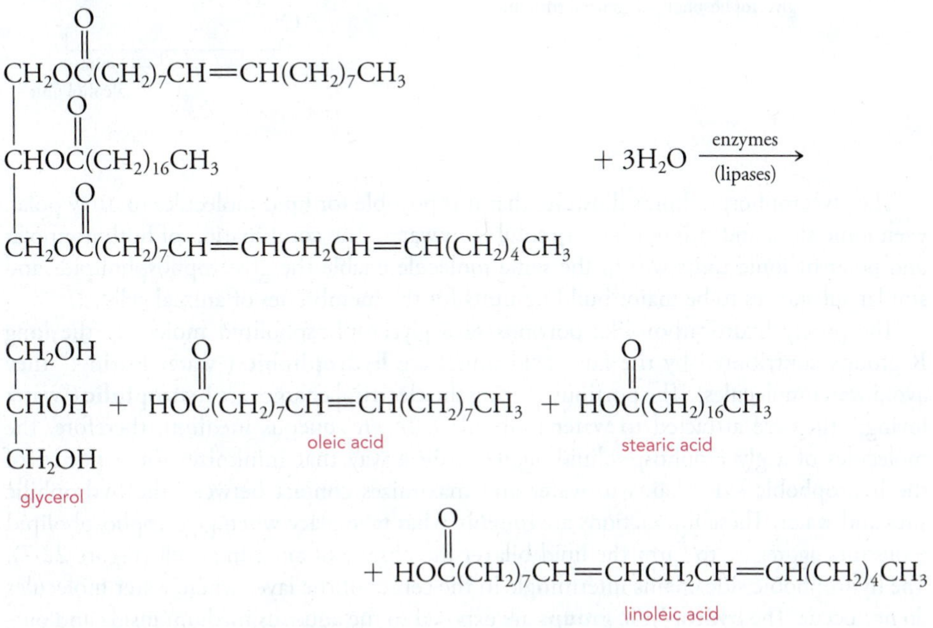

$$CH_2OC(CH_2)_7CH{=}CH(CH_2)_7CH_3$$
$$CHOC(CH_2)_{16}CH_3 \qquad + 3H_2O \xrightarrow[\text{(lipases)}]{\text{enzymes}}$$
$$CH_2OC(CH_2)_7CH{=}CHCH_2CH{=}CH(CH_2)_4CH_3$$

$$CH_2OH$$
$$CHOH \; + \; HOC(CH_2)_7CH{=}CH(CH_2)_7CH_3 \; + \; HOC(CH_2)_{16}CH_3$$
$$CH_2OH$$

glycerol   oleic acid   stearic acid

$$+ \; HOC(CH_2)_7CH{=}CHCH_2CH{=}CH(CH_2)_4CH_3$$

linoleic acid

Actually, the anions of the acids form, because the medium in which lipid digestion occurs is basic.

When the hydrolysis of a triacylglycerol is carried out in the presence of sufficient base so as to release the fatty acids as their anions, the reaction is called saponification. A mixture of the salts of long-chain fatty acids is what makes up ordinary soap.

Hydrogenation of Vegetable Oils

Vegetable oils are generally less expensive to produce than butterfat, but because the oils are liquids, few people care to use them as bread spreads. Animal fats, like butterfat, are solids at room temperature, and vegetable oils differ from animal fats only in the number of carbon–carbon double bonds per molecule. Simply adding hydrogen to the double bonds of a vegetable oil, therefore, changes the lipid from a liquid to a solid.

Partial hydrogenation of a vegetable oil can lead to rearrangement of the atoms around the double bonds from *cis* to *trans*. The resulting products are known as *trans fats*. Because ingestion of trans fats has been associated with coronary artery disease, food product labels are now required to show the amounts of these fats in the food. Fast-food chains such as McDonalds, Wendy's, KFC, and Taco Bell have switched from using trans fat oils for cooking their French fries. In New York City, trans fats are banned from all restaurants.

■ Hydrogenation of a vegetable oil will add hydrogen atoms to both *cis* and *trans* isomers and result in a straight-chain, saturated, molecule that maximizes the London attractive forces.

Cell Membranes in Animals

The lipids involved in the structures of cell membranes in animals are not triacylglycerols. Some are diacylglycerols with the third site on the glycerol unit taken up by an attachment to a phosphate unit. This, in turn, is joined to an amino alcohol unit by an ester-like network. The phosphate unit carries one negative charge, and the amino unit has a positive charge. These lipids are called glycerophospholipids. Lecithin is one example; it is used in cooking as an emulsifier.

■ The amino alcohol unit in lecithin is contributed by choline, a cation:

$$HOCH_2CH_2\overset{+}{N}(CH_3)_3$$

Ethanolamine (in its protonated form),

$$HOCH_2CH_2NH_3^+$$

is another amino alcohol that occurs in phospholipids.

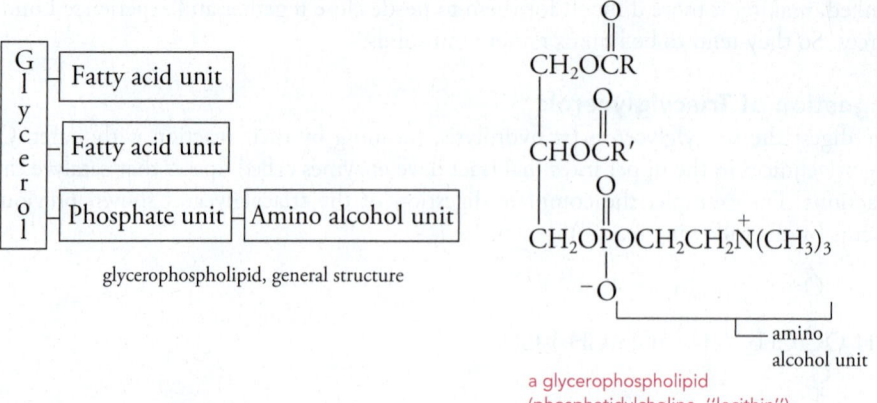

glycerophospholipid, general structure

a glycerophospholipid
(phosphatidylcholine, "lecithin")

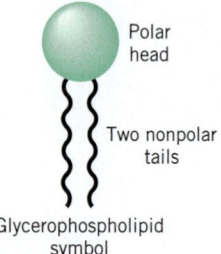

Polar head

Two nonpolar tails

Glycerophospholipid symbol

The glycerophospholipids illustrate that it is possible for lipid molecules to carry polar, even ionic sites, and still not be very soluble in water. The combination of both nonpolar and polar or ionic units within the same molecule enable the glycerophospholipids and similar substances to be major building units for the membranes of animal cells.

The purely hydrocarbon-like portions of a glycerophospholipid molecule (the long R groups contributed by the fatty acid units) are **hydrophobic** ("water fearing;" they avoid water molecules). The portions bearing the electrical charges are **hydrophilic** ("water loving;" they are attracted to water molecules). In an aqueous medium, therefore, the molecules of a glycerophospholipid aggregate in a way that minimizes the exposure of the hydrophobic side chains to water and maximizes contact between the hydrophilic sites and water. These interactions are roughly what take place when glycerophospholipid molecules aggregate to form the lipid bilayer membrane of an animal cell (Figure 22.7). The hydrophobic side chains intermingle in the center of the layer where water molecules do not occur. The **hydrophilic groups** are exposed to the aqueous medium inside and outside of the cell. Not shown in Figure 22.7 are cholesterol and cholesterol ester molecules, which help to stiffen the membranes. Thus cholesterol is essential to the cell membranes of animals.

Cell membranes also include protein units, which provide several services. Some are molecular recognition sites for molecules such as hormones and neurotransmitters. Others provide channels for the movements of ions, like Na^+, K^+, Ca^{2+}, Cl^-, HCO_3^-, and others into or out of the cell. Some are channels for the transfer of small organic molecules, like glucose.

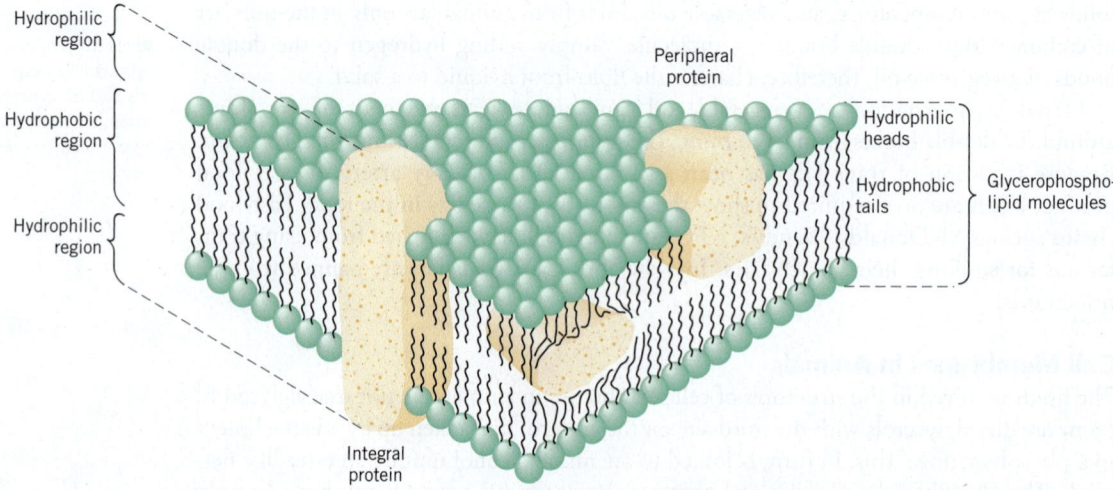

Figure 22.7 | **Animal cell membrane.** The lipid molecules of an animal cell membrane are organized as a bilayer.

Proteins

Proteins are a large family of substances that make up about half of the human body's dry weight. They are found in all cells and serve as enzymes, hormones, and neurotransmitters. They carry oxygen in the blood stream as well as some of the waste products of metabolism. No other group of compounds has such a variety of functions in living systems.

The dominant structural units of **proteins** are macromolecules called **polypeptides**, which are made from a set of monomers called **α-amino acids**. Most protein molecules include, besides their polypeptide units, small organic molecules or metal ions, and the whole protein lacks its characteristic biological function without these species.

The monomer units for polypeptides are a group of about 20 α-amino acids, all of which have structural features in common. Some examples of the set of 20 amino acids used to make proteins are given below. The symbol R stands for an amino acid side chain. All amino acids are known by their common names and each also has been given a three-letter symbol.

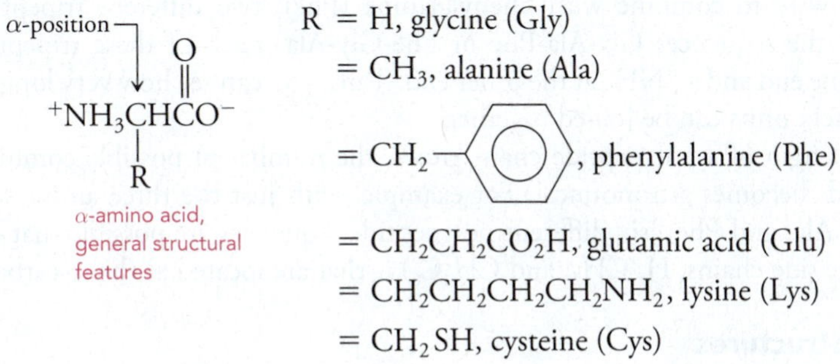

■ Neurotransmitters are small molecules that travel between the end of one nerve cell and the surface of the next to transmit the nerve impulse.

α-position

$$^+NH_3CHCO^-$$

|
R

α-amino acid, general structural features

$R = H$, glycine (Gly)

$= CH_3$, alanine (Ala)

$= CH_2 - \bigcirc$, phenylalanine (Phe)

$= CH_2CH_2CO_2H$, glutamic acid (Glu)

$= CH_2CH_2CH_2CH_2NH_2$, lysine (Lys)

$= CH_2SH$, cysteine (Cys)

The simplest amino acid is glycine, for which the "side chain" is H.

Glycine, like all of the amino acids in their pure states, exists as a dipolar ion, with the positive and negative charges separate but on the same molecule. Such an ion results from an internal self-neutralization, by the transfer of a proton from the proton-donating carboxyl group to the proton-accepting amino group.

$$NH_2CH_2CO_2H \longrightarrow {}^+NH_3CH_2CO_2{}^-$$

glycine

glycine, dipolar ionic form

■ The formal name for the dipolar ionic form is **zwitterion**.

Polypeptides

Polypeptides are copolymers of the amino acids. The carboxyl group of one amino acid becomes joined to the amino group of another by means of an amide bond—the same kind of carbonyl–nitrogen bond found in nylon, but here in a peptide it is called a **peptide bond**. Let's see how two amino acids, glycine and alanine, can become linked by a (multistep) splitting out of water.

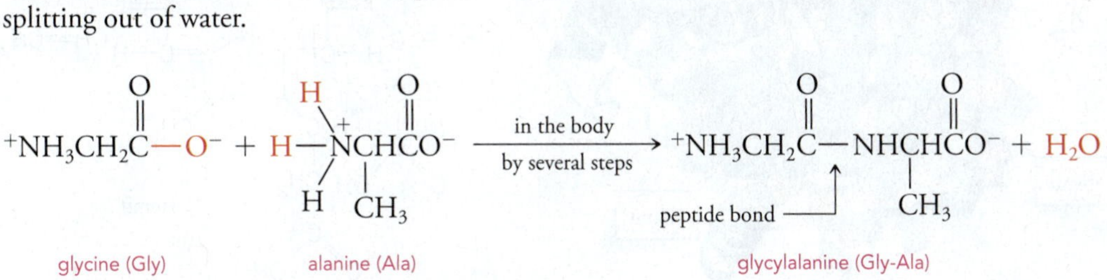

$$^+NH_3CH_2\overset{O}{\overset{\|}{C}}-O^- + H-\overset{+}{\underset{H}{\overset{H}{N}}}CHCO^- \xrightarrow[\text{by several steps}]{\text{in the body}} {}^+NH_3CH_2\overset{O}{\overset{\|}{C}}-NHCHCO^- + H_2O$$

glycine (Gly)   alanine (Ala)   peptide bond   glycylalanine (Gly-Ala)

The product of this reaction, glycylalanine, is an example of a dipeptide, two amino acids joined by a peptide bond. (Biochemists often use the three-letter symbols for the amino acids to describe the structural formulas of such products, in this case Gly-Ala.)

We could have taken glycine and alanine in different roles and written the equation for the formation of a different dipeptide, Ala-Gly.

$$\text{{}^+NH_3CHC{-}O^- + H{-}NCH_2C{-}O^-} \xrightarrow[\text{by several steps}]{\text{in the body,}} \text{{}^+NH_3CHC{-}NHCH_2CO^- + H_2O}$$

alanine (Ala)   glycine (Gly)   alanylglycine (Ala-Gly)

peptide bond

■ The artificial sweetener aspartame (NutraSweet®) is the methyl ester of a dipeptide.

methyl ester group

$$\begin{array}{l} O \\ \| \\ COCH_3 \\ | \\ CHCH_2C_6H_5 \\ | \\ NH \leftarrow \text{peptide bond} \\ | \\ C{=}O \\ | \\ CHCH_2CO_2^- \\ | \\ NH_3^+ \end{array}$$
aspartame

The formation of these two dipeptides is like the formation of two 2-letter words from the letters N and O. Taken in one order we get NO; in the other, ON. They have entirely different meanings, yet are made of the same pieces.

Notice that each dipeptide, Gly-Ala and Ala-Gly, has a CO_2^- at one end of the chain and a $^+NH_3$ group at the other end. Each end of a dipeptide molecule, therefore, could form yet another peptide bond involving any of the 20 amino acids. For example, if glycylalanine were to combine with phenylalanine (Phe), two different tripeptides could form with the sequences Gly-Ala-Phe or Phe-Gly-Ala. Each of these tripeptides has a CO_2^- at one end and a $^+NH_3$ at the other end. Thus, you can see how very long sequences of amino acid units can be joined together.

As the length of a polypeptide chain grows, the number of possible combinations of amino acids becomes astronomical. For example, with just the three amino acids we've used (Gly, Ala, and Phe), six different polypeptide sequences are possible that differ only in the three side chains, H, CH_3, and $CH_2C_6H_5$ that are located at the α-carbon atoms.

Protein Structures

Many proteins consist of a single polypeptide. Most proteins, however, involve assemblies of two or more polypeptides. These are identical in some proteins, but in others, the aggregating polypeptides are different. Moreover, a relatively small organic molecule may be included in the aggregation, and a metal ion is sometimes present, as well. Thus, the terms "protein" and "polypeptide" are not synonyms. Hemoglobin, for example, has all of the features just described (Figure 22.8). It is made of four polypeptides—two similar pairs—and four molecules of heme, the organic compound that causes the red color of blood. Heme, in turn, holds an iron(II) ion. The entire package is the protein, hemoglobin. If one piece is missing or altered in any way—for example, if iron occurs as Fe^{3+} instead of Fe^{2+}—the substance is not hemoglobin, and it does not transport oxygen in the blood.

■ Hemoglobin is the oxygen carrier in blood.

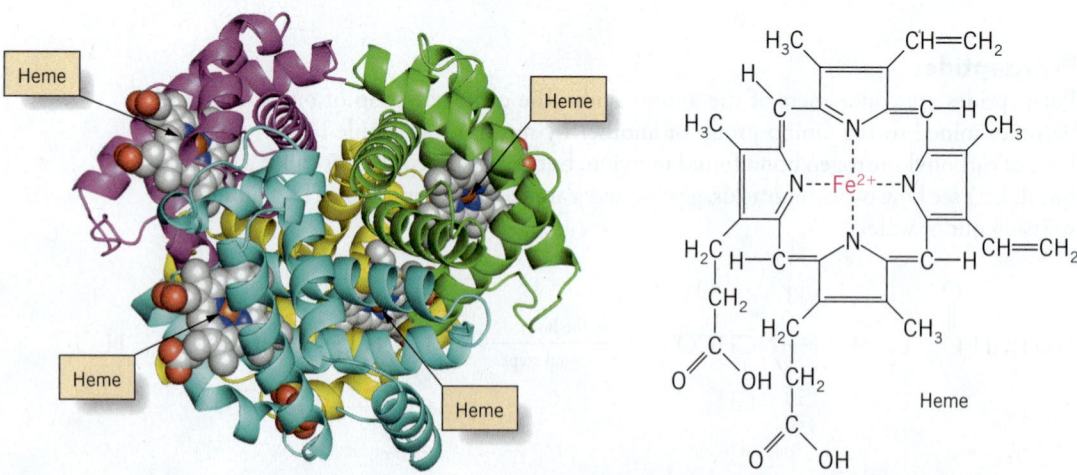

Figure 22.8 | **Hemoglobin.** Its four polypeptide chains, each shown as different colored ribbons, twist and bend around the four embedded heme units. Each heme unit contains an Fe^{2+} ion in its center that is able to bind to O_2.

Notice in Figure 22.8 how the strands of each polypeptide unit in hemoglobin are coiled and that the coils are kinked and twisted. Such shapes of polypeptides are determined by the amino acid sequence, because the side chains are of different sizes and some are hydrophilic and others are hydrophobic. Polypeptide molecules become twisted and coiled in whatever way minimizes the contact of hydrophobic groups with the surrounding water and maximizes the contacts of hydrophilic groups with water molecules. In addition, hydrogen bonding between N—H and C=O groups in neighboring loops helps hold the shapes of the helices.

The final shape of a protein, called its native form, is as critical to its ability to function as anything else about its molecular architecture. For example, just the exchange of one side-chain R group by another changes the shape of hemoglobin and causes a debilitating condition known as sickle-cell anemia.

■ In one of the subunits of the hemoglobin in those with sickle-cell anemia, an isopropyl group, —CH(CH₃)₂, is a side-chain where a —CH₂CH₂CO₂H side-chain should be.

Enzymes

The catalysts in living cells are called **enzymes**, and virtually all are proteins. Some enzymes require a metal ion, such as Mn^{2+}, Co^{2+}, Cu^{2+}, or Zn^{2+}, all of which are on the list of the trace elements that must be in a good diet. Some enzymes also require molecules of the B vitamins to be complete enzymes.

Some of our most dangerous poisons work by deactivating enzymes, often those needed for the transmission of nerve signals. For example, the botulinum toxin that causes botulism, a deadly form of food poisoning, deactivates an enzyme in the nervous system. Heavy metal ions, like Hg^{2+} or Pb^{2+}, are poisons because they also deactivate enzymes.

■ Heavy metal ions bond to the HS groups of cysteine side-chains in polypeptides.

Label these compounds as carbohydrates, lipids, or amino acids. (*Hint:* What are the structural features of each family of compounds?)

Practice Exercise 22.15

(a)

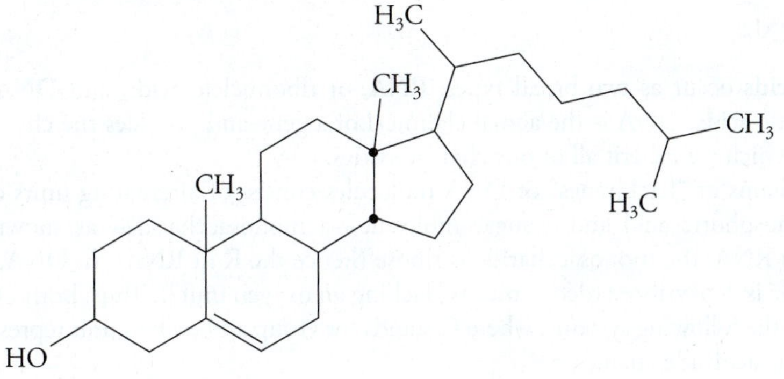

(b)

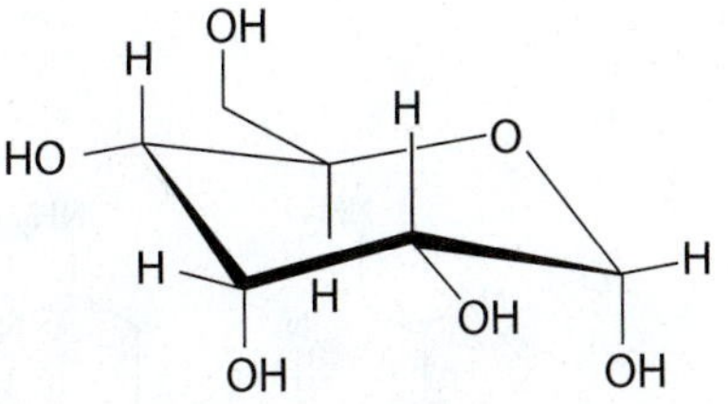

(c)

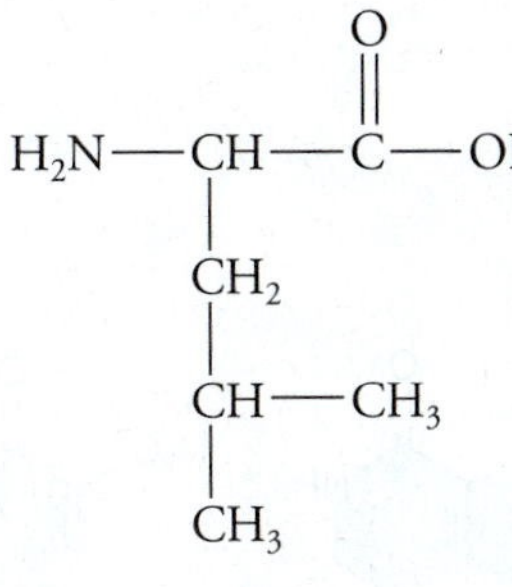

(d)

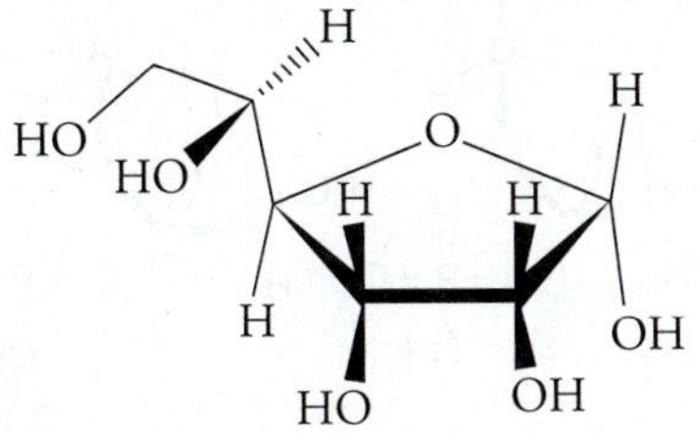

(e)

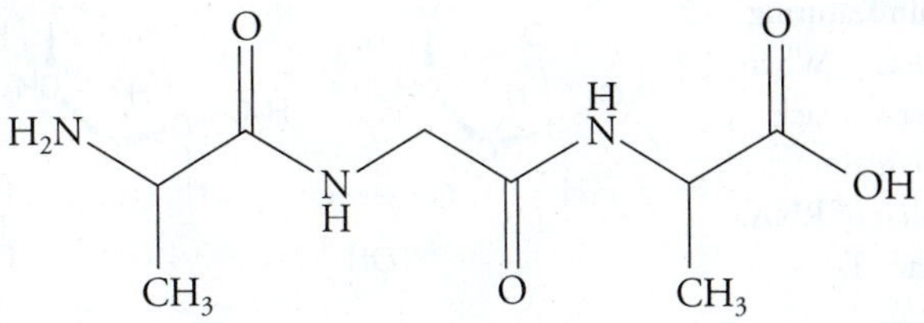

(f) CH$_2$OC(O)—(CH$_2$)$_{16}$CH$_3$

  CHOC(O)—(CH$_2$)$_{16}$CH$_3$

  CH$_2$OC(O)—(CH$_2$)$_{16}$CH$_3$

Practice Exercise 22.16

Label the hydrophobic and hydrophilic portions of the glycerophospholipid.

CH$_2$OC(O)—

CHOC(O)—

CH$_2$O—P—O—

22.7 | Nucleic Acids, DNA and RNA

The polypeptides of an organism are made under the chemical direction of a family of compounds called the nucleic acids. Both the similarities and the uniqueness of every species as well as every individual member of a species depend on structural features of these compounds.

DNA and RNA

The **nucleic acids** occur as two broad types: **RNA**, or ribonucleic acids, and **DNA**, or deoxyribonucleic acids. DNA is the actual chemical of a gene and provides the chemical basis through which we inherit all of our characteristics.

The main chains or "backbones" of DNA molecules consist of alternating units contributed by phosphoric acid and a sugar molecule—a monosaccharide—as shown in Figure 22.9. In RNA, the monosaccharide is ribose (hence the R in RNA). In DNA, the monosaccharide is deoxyribose (deoxy means "lacking an oxygen unit"). Thus, both DNA and RNA have the following systems, where G stands for group and each G unit represents a unique nucleic acid side chain.

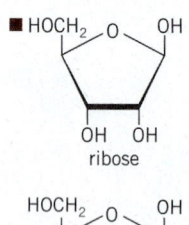

ribose

deoxyribose

Figure 22.9 | Nucleic acids. A segment of a DNA chain featuring each of the four DNA bases. When the sites marked by asterisks each carry an OH group, the main "backbone" would be that of RNA. In RNA, U would replace T.

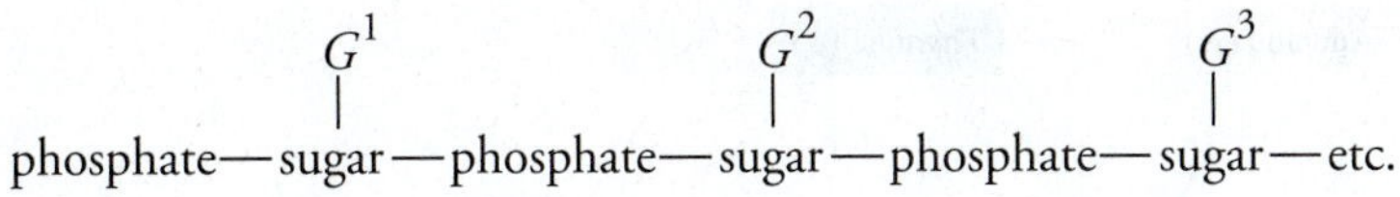

$$G^1 \qquad\qquad G^2 \qquad\qquad G^3$$
$$|\qquad\qquad\quad |\qquad\qquad\quad |$$

phosphate—sugar—phosphate—sugar—phosphate—sugar—etc.

Backbone system in all nucleic acids is many thousands of repeating units long.
In DNA, the sugar is deoxyribose.
In RNA, the sugar is ribose.

The side chains, G, are all heterocyclic amines whose molecular shapes have much to do with their function. Being amines, which are basic in water, they are referred to as the bases of the nucleic acids and are represented by single letters—A for adenine, T for thymine, U for uracil, G for guanine, and C for cytosine.

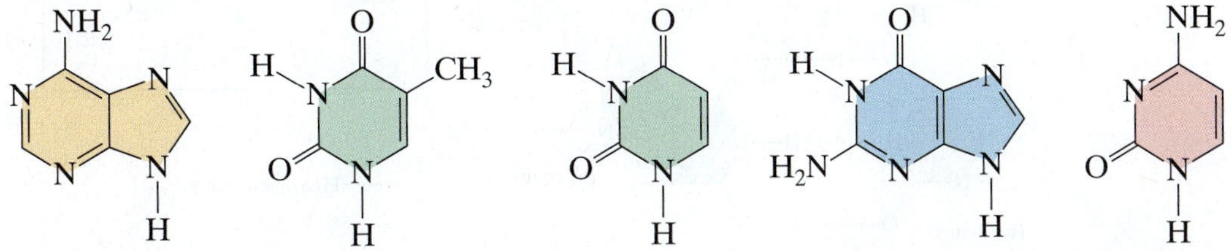

| adenine A | thymine T | uracil U | guanine G | cytosine C |

In DNA, the bases are A, T, G, and C; in RNA they are A, U, G, and C. These few bases are the "letters" of the genetic alphabet. Amazingly, the messages describing our entire genetic makeup are composed with just four "letters"—A, T, G, and C.

The DNA Double Helix

In 1953, F. H. C. Crick of England and J. D. Watson, an American, deduced that DNA occurs in cells as two intertwined, oppositely running strands of molecules coiled like a spiral staircase and called the **DNA double helix**. Hydrogen bonds help to hold the two strands side by side, but other factors are also involved.

The bases have N—H and O=C groups, which enable hydrogen bonds (· · ·) to form between them.

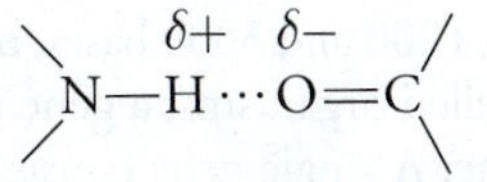

$$\begin{array}{ccc} & \delta+ & \delta- \\ \diagdown N{-}H & \cdots & O{=}C \diagup \\ \diagup & & \diagdown \end{array}$$

However, the bases with the best "matching" molecular geometries for maximum hydrogen bonding occur only as particular pairs of bases. The functional groups of each pair are in exactly the right locations in their molecules to allow hydrogen bonds between pair members. In DNA, A pairs only with T, never with G or C (see Figure 22.10). Similarly, C pairs only with G, never with A or T. Thus, opposite every G on one strand in a DNA double helix, a C occurs on the other. Opposite every A on one strand in DNA, a T is found on the other.

Adenine (A) can also pair with uracil (U), but U occurs in RNA. The A-to-U pairing is an important factor in the work of RNA.

DNA Replication

Prior to cell division, the cell produces duplicate copies of its DNA so that each daughter cell will have a complete set. Such reproductive duplication is called DNA **replication**.

The accuracy of DNA replication results from the limitations of the base pairings: A only with T, and C only with G (Figure 22.11). The unattached letters, A, T, G, and C, in Figure 22.11 here represent not just the bases but the whole monomer molecules of DNA.

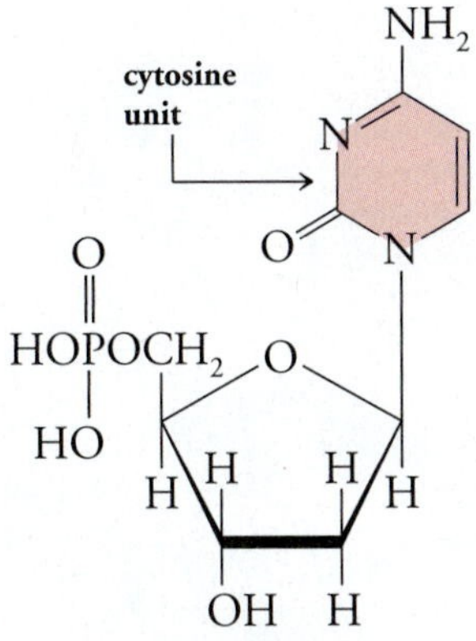

■ Crick and Watson shared the 1962 Nobel Prize in physiology or medicine with Maurice Wilkins, using X-ray data from Rosalind Franklin to deduce the helical structure of DNA. The structure of DNA is illustrated in Figure 11.5.

a typical nucleotide, one using cytosine as its side chain

Figure 22.10 | **Base pairing in DNA.** The hydrogen bonds are indicated by dotted lines.

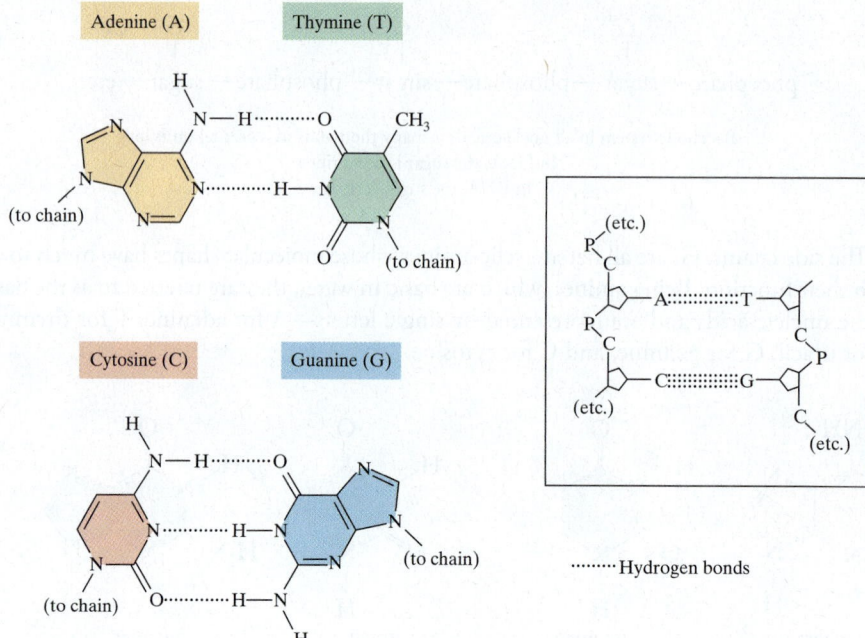

The monomers are called nucleotides and are molecules made of a phosphate-sugar unit bonded to one particular base (see the illustration in the margin). The nucleotides are made by the cell and are present in the cellular "soup."

An enzyme catalyzes each step of replication. As replication occurs, the two strands of the parent DNA double helix separate, and the monomers for new strands assemble along the exposed single strands. Their order of assembling is completely determined by the specificity of base pairing. For example, a base such as T on a parent strand can accept only a nucleotide with the base A. Two daughter double helixes result that are identical to the parent, and each carries one strand from the parent. One new double helix goes to one daughter cell and the second goes to the other.

Genes and Polypeptide Synthesis

A single human gene has between 1000 and 3000 bases, but they do not occur continuously on a DNA molecule. In multi-celled organisms, a gene is neither a whole DNA molecule nor a continuous sequence in one. A single gene consists of the totality of particular segments of a DNA chain that, taken together, carry the necessary genetic instruction for a particular polypeptide.

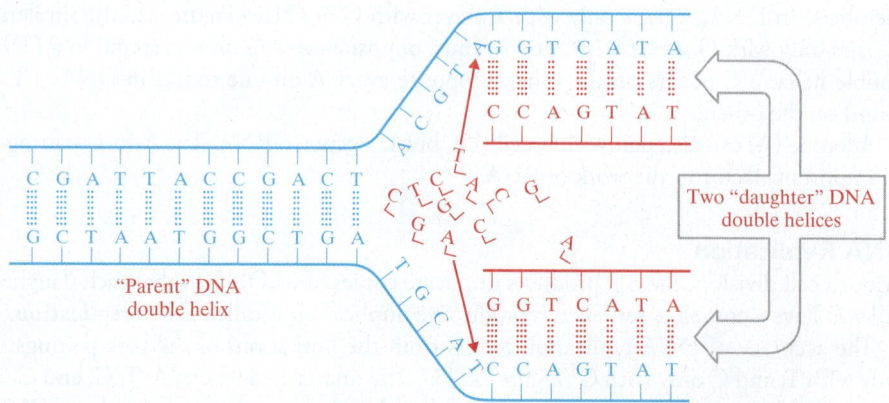

Figure 22.11 | Base pairing and replication of DNA.

Each polypeptide in a cell is made under the direction of its own gene. In broad outline, the steps between a gene and a polypeptide occur as follows.

DNA $\xrightarrow{\text{transcription}}$ RNA $\xrightarrow{\text{translation}}$ Polypeptide

(The genetic message is read off in the cell nucleus and transferred to RNA.)

(The genetic message, now on RNA outside the nucleus, is used to direct the synthesis of a polypeptide.)

The step labeled **transcription** accomplishes just that—the genetic message, represented by a sequence of bases on DNA, is transcribed into a complementary sequence of bases on RNA, but U, not T, is on the RNA. **Translation** means the conversion of the base sequence on RNA into a side-chain sequence on a new polypeptide. It is like translating from one language (the DNA/RNA base sequences) to another language (the polypeptide side-chain sequence).

The translation step in polypeptide synthesis involves several steps and several different kinds of RNA. In the process, the cell tells which amino acid unit must come next in a growing polypeptide chain by means of a code that relates unique groups of three nucleic acid bases, called **codons**, to specific amino acids.

Which amino acid goes with which codon constitutes the **genetic code**, which enables the cell to translate from the four-letter RNA alphabet (A, U, G, and C) to the 20-letter alphabet of amino acids (the amino acid side chains). One of the most remarkable features of the code is that it is essentially universal. The codon specifying alanine in humans, for example, also specifies alanine in the genetic machinery of bacteria, aardvarks, camels, rabbits, and stinkbugs. Our chemical kinship with the entire living world is profound (even humbling).

Draw a ribose ring and a deoxyribose ring. What is the difference between these two rings? (*Hint:* RNA uses the monosaccharide ribose and DNA uses the monosaccharide deoxyribose.)

Practice Exercise 22.17

Which base pairs match in a double helix? Which elements hydrogen bond to form these pairs?

Practice Exercise 22.18

| Summary

Organized by Learning Objective

Write structural formulas for organic compounds, highlighting their functional groups

Organic chemistry is the study of the covalent compounds of carbon, except for its oxides, carbonates, bicarbonates, and a few others. **Functional groups** attached to hydrocarbon groups are the basis for the classification of organic compounds. Members of a family have the same kinds of chemical properties, which are a function of the relative proportions of functional and nonfunctional groups. Opportunities for hydrogen bonding strongly influence the boiling points and water solubilities of compounds with OH or NH groups. Organic compounds exhibit isomerism because often there is more than one possible structure corresponding to a given molecular formula. Molecules containing an **asymmetric carbon atom** exhibit chirality.

The **alkanes** are the least polar and the least reactive of the organic families. Their carbon skeletons can exist as **straight chains**, **branched chains**, or **rings**. When they are open-chain, there is free rotation about single bonds. Rings can include heteroatoms, but **heterocyclic compounds** are not hydrocarbons.

Describe the nomenclature rules and major reactions of alkanes, alkenes, alkynes, and aromatic hydrocarbons

By the IUPAC rules, the names of alkanes end in -*ane*. A prefix denotes the number of carbon atoms in the parent chain, the longest continuous chain, which is numbered from whichever end is nearest a branch. The **alkyl groups** are alkanes substituents.

Alkanes in general do not react with strong acids or bases. At high temperatures, their molecules crack to give H_2 and unsaturated hydrocarbons.

For unsaturated hydrocarbons, the lack of free rotation at a carbon–carbon double bond makes **geometric isomerism** possible. The pair of electrons of a π bond makes alkenes act as Brønsted bases, enabling alkenes to undergo **addition reactions** with strong binary acids, or with water in the presence of an acid catalyst. Alkenes also add hydrogen (in the presence of a metal catalyst). Bromine and chlorine add to alkenes under mild, uncatalyzed conditions. Strong oxidizing agents, like ozone, readily attack alkenes and break their molecules apart.

Aromatic hydrocarbons, like benzene, do not give the addition reactions shown by alkenes; instead, they undergo **substitution reactions**. In the presence of suitable catalysts, benzene undergoes these reactions with chlorine, bromine, nitric acid, and sulfuric acid.

Describe the names and typical reactions of common alcohols, ethers, aldehydes, ketones, carboxylic acids, and esters

The **alcohols**, ROH, and the **ethers**, ROR′, are alkyl derivatives of water. Ethers have almost as few chemical properties as alkanes. Alcohols undergo an **elimination reaction** that splits out H_2O and changes them to alkenes. Concentrated strong binary acids, like HI, react with alcohols by a **substitution reaction**, which replaces the OH group by halogen. Oxidizing agents convert alcohols of the type RCH_2OH first into aldehydes and then into carboxylic acids. Oxidizing agents convert alcohols of the type R_2CHOH into ketones. Alcohols form esters with carboxylic acids.

Aldehydes, RCH=O, among the most easily oxidized organic compounds, are oxidized to carboxylic acids. **Ketones**, RCOR′, strongly resist oxidation. Aldehydes and ketones add hydrogen to make alcohols.

The **carboxylic acids**, RCO_2H, are weak acids but neutralize strong bases. Their anions are Brønsted bases. Carboxylic acids can also be changed to esters by heating them with alcohols.

Esters, $RCO_2R′$, react with water to give back their parent acids and alcohols. When aqueous base is used for hydrolysis, the process is called saponification and the products are the salts of the parent carboxylic acids as well as alcohols.

Describe the names and typical reactions of amines and amides

The simple **amines**, RNH_2, as well as their more substituted relatives, RNHR′ and RNR′R″ are organic derivatives of ammonia and thus are weak bases that can neutralize strong acids. The conjugate acids of amines are good proton donors and can neutralize strong bases.

The simple **amides**, $RCONH_2$, can be made from acids and ammonia, and they can be hydrolyzed back to their parent acids and trivalent nitrogen compounds. The amides are not basic.

Explain the structures, synthesis, and properties of polymers

Polymers, which are **macromolecules**, are made up of a very large number of atoms in which a small characteristic feature repeats over and over many times. **Polypropylene**, a **chain-growth** or **addition polymer** that consists of a hydrocarbon **backbone** with a methyl group, CH_3, attached to every other carbon, is formed by **polymerization** of the **monomer** propylene. Polyethylene and polystyrene, which are also addition polymers, are formed from ethylene and styrene, respectively. **Condensation polymers (step-growth polymers)** are formed by elimination of a small molecule such as H_2O or CH_3OH from two monomer units accompanied by the formation of a covalent bond between the monomers. **Nylons** and **polyesters** are copolymers because they are formed from two different monomers.

The polymerization of ethylene can lead to **branching**, which produces an amorphous polymer called low-density polyethylene (LDPE). High-density polyethylene (HDPE) and ultrahigh-molecular-weight polyethylene (UHMWPE) are not branched and are more crystalline, which makes them stronger. Nylon's properties are affected by hydrogen bonding between polymer strands.

Name common carbohydrates, lipids, and proteins and describe their properties

The glucose unit is present in **starch**, glycogen, cellulose, sucrose, lactose, and simply as a monosaccharide in honey. It's the chief "sugar" in blood. As a **monosaccharide**, glucose is a pentahydroxyaldehyde in one form but it also exists in two cyclic forms. The latter are joined by oxygen bridges in the **disaccharides** and **polysaccharides**. In lactose (milk sugar), a glucose and a galactose unit are joined. In sucrose (table sugar), glucose and fructose units are linked.

The major polysaccharides—starch (amylose and amylopectin), glycogen, and cellulose—are all polymers of glucose but with different patterns and geometries of branching. The digestion of the disaccharides and starch requires the hydrolyses of the oxygen bridges to give monosaccharides.

Natural products in living systems that are relatively soluble in nonpolar solvents, but not in water, are members of a large group of compounds called **lipids**. They include sex hormones and cholesterol, the triacylglycerols of nutritional importance, and the glycerophospholipids and other phospholipids needed for cell membranes.

The **triacylglycerols** are triesters between glycerol and three of a number of long-chain, often unsaturated fatty acids. In the digestion of the triacylglycerols, the hydrolysis of the ester groups occurs, and anions of **fatty acids** form (together with glycerol).

Molecules of glycerophospholipids have large segments that are hydrophobic and others that are hydrophilic. Glycerophospholipid molecules are the major components of cell membranes, where they are arranged in a lipid bilayer.

The **amino acids** in nature are mostly compounds with $-NH_3^+$ and $-CO_2^-$ groups joined to the same carbon atom (called the alpha position of the amino acid). About 20 amino acids are important to the structures of polypeptides and proteins.

When two amino acids are linked by a **peptide bond** (an amide bond), the compound is a dipeptide. In **polypeptides**, several (sometimes thousands) of regularly spaced peptide bonds occur, involving as many amino acid units. The uniqueness of a polypeptide lies in its chain length and in the sequence of the side-chain groups.

Some **proteins** consist of only one polypeptide, but most proteins involve two or more (sometimes different) polypeptides and often a nonpolypeptide (an organic molecule) and a metal ion.

Nearly all **enzymes** are proteins. Major poisons and some drugs work by deactivating enzymes.

Explain how the structures of DNA and RNA enable the transmission of genetic information and the synthesis of proteins

The formation of a polypeptide with the correct amino acid sequence needed for a given protein is under the control of **nucleic acids**. Nucleic acids are polymers whose backbones are made of a repeating sequence of pentose sugar units. On each sugar unit is a **base**—a heterocyclic amine such as adenine (A), thymine (T), guanine (G), cytosine (C), or uracil (U). In **DNA**, the sugar is deoxyribose, and the bases are A, T, G, and C. In **RNA**, the sugar is ribose and the bases are A, U, G, and C.

DNA exists in cell nuclei as **double helices**. The two strands are associated side by side, with hydrogen bonds occurring between particular pairs of bases. In DNA, A is always paired to T; C is always paired to G. The base U replaces T in RNA, and A can also pair with U. The bases, A, U, G, and C, are the four letters of the genetic alphabet, and the specific combination that codes for one amino acid in polypeptide synthesis are three bases, side by side, on a strand. The **replication** of DNA is the synthesis by the cell of exact copies of the original two strands of a DNA double helix.

Tools for Problem Solving The following tools were introduced in this chapter. Study them carefully so you can select the appropriate tool when needed.

Condensed structures (Section 22.1)
The number of covalent bonds formed by the nonmetals in achieving an octet:

Group 4A	4 bonds	Group 5A	3 bonds	Group 6A	2 bonds
Group 7A	1 bond	hydrogen	1 bond		

In writing a condensed structure, be sure the number of hydrogens with a given element is sufficient to make up the difference between the number of non-hydrogen atoms attached and the total number of bonds required by the element in question.

Polygons as condensed structures (Section 22.1)
A carbon atom is understood at each corner; other elements in the ring are explicitly written. Edges of the polygon represent covalent bonds; double bonds are explicitly shown. The remaining bonds, as required by the covalence of the atom at a corner, are understood to bond H atoms.

Functional groups (Section 22.1)
The functional groups, Table 22.1, in a molecule places the structure into an organic family and allows a certain set of reactions the compound can undergo.

IUPAC rules of nomenclature for organic compounds
For all compounds, the longest chain of carbon atoms or the longest chain with the functional group defines the parent chain for which the compound is named.

For alkanes (Section 22.2)
The chain is numbered from whichever end gives the lowest number to the first substituent. The name ending is *-ane*, and the prefixes that describe the number of carbon atoms in the chain are given in Table 22.2. The locations of alkyl substituents are specified by a number and a hyphen preceding the name of the alkyl group. Alkyl substituents are specified in alphabetical order.

Alkenes and alkynes (Section 22.2)
The name ending for alkenes is *-ene*, and the name ending for alkynes is *-yne*. The rules are similar to those for naming alkanes, except that the chain is numbered so that the first carbon of the double or triple bond has the lower of two possible numbers.

Alcohols (Section 22.3)
The name ending for alcohols is *-ol*. The parent chain is numbered from whichever end gives the lowest number to the —OH group.

Aldehydes and ketones (Section 22.3)

The name ending for aldehydes is *-al*. For ketones, the name ending is *-one*. The parent chains are numbered from whichever end gives the lowest number to the carbonyl carbon.

Carboxylic acids (Section 22.3)

The name ending is *-oic acid*.

Esters (Section 22.3)

The name begins with the name of the alkyl group attached to the O atom. This is followed by a separate word taken from the name of the parent carboxylic acid, but altered by changing *-ic acid* to *-ate*.

Reactions of alkenes and alkynes (Section 22.2)

Components of a molecule add across a double bond.

$$CH_2{=}CH_2 + A{-}B \longrightarrow \underset{\overset{|}{A} \quad \overset{|}{B}}{CH_2{-}CH_2}$$

Reactions of alcohols

Oxidation of alcohols (Section 22.3)

The product(s) depend on the number of hydrogens attached to the alcohol carbon.

$$RCH_2OH \xrightarrow{\text{oxidation}} \underset{\text{aldehyde}}{R\overset{\overset{\text{O}}{\|}}{C}H} \xrightarrow{\text{further oxidation}} \underset{\text{carboxylic acid}}{R\overset{\overset{\text{O}}{\|}}{C}OH}$$

$$\underset{}{R\overset{\overset{\text{OH}}{\|}}{C}HR'} \xrightarrow{\text{oxidation}} \underset{\text{ketone}}{R\overset{\overset{\text{O}}{\|}}{C}R'}$$

$$\underset{\overset{|}{R''}}{\overset{\overset{R'}{|}}{R}COH} \xrightarrow{\text{oxidation}} \text{no reaction}$$

Dehydration (Section 22.3)

$$\underset{\overset{|}{H} \quad \overset{|}{OH}}{RCH{-}CH_2} \xrightarrow[\text{heat}]{\text{acid catalyst}} RCH{=}CH_2 + H_2O$$

Substitution (Section 22.3)

$$ROH + HX(\text{conc.}) \xrightarrow{\text{heat}} RX + H_2O$$

Reactions of aldehydes and ketones: reduction (Section 22.3)

$$\underset{\text{aldehyde}}{R\overset{\overset{\text{O}}{\|}}{C}H} + H{-}H \xrightarrow[\text{catalyst}]{\text{heat, pressure}} R\overset{\overset{\text{OH}}{|}}{C}H_2$$

$$\underset{\text{ketone}}{R\overset{\overset{\text{O}}{\|}}{C}R'} + H{-}H \xrightarrow[\text{catalyst}]{\text{heat, pressure}} R\overset{\overset{\text{OH}}{|}}{C}HR'$$

Reactions of organic acids

Neutralization (Section 22.3)

Carboxylic acids react with bases to give salts.

$$RCO_2H + OH^- \longrightarrow RCO_2^- + H_2O$$

Formation of esters (Section 22.3)

Esters are formed in the reaction of a carboxylic acid with an alcohol.

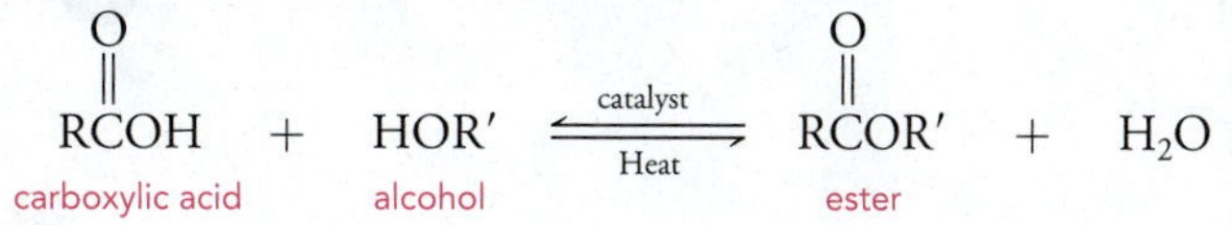

Saponification of esters (Section 22.3)

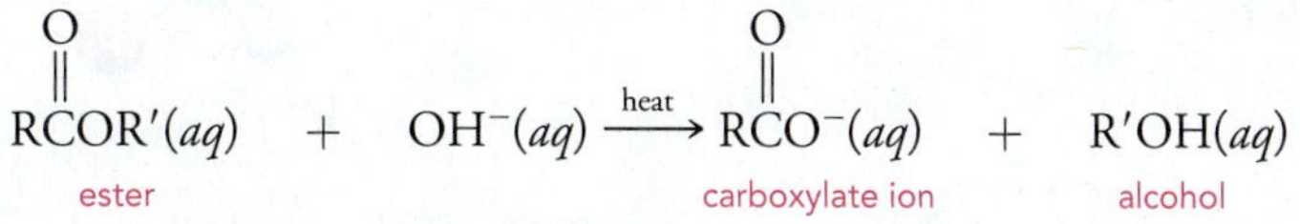

Formation of amides (Section 22.4)

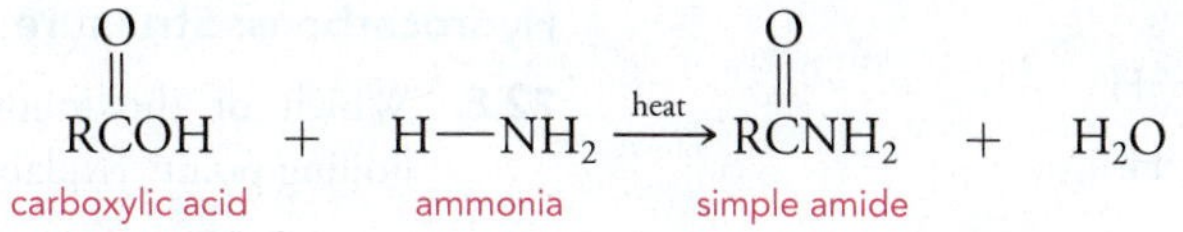

Hydrolysis of amides (Section 22.4)

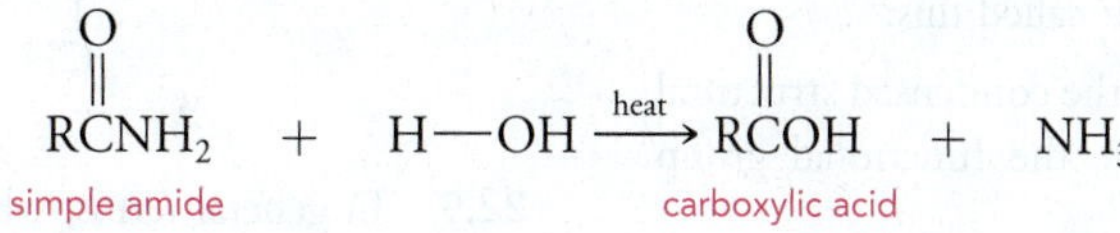

Addition polymerization (Section 22.5)

The addition of monomer units to a gradually growing chain by a free radical mechanism yields a long polymer chain. The repeating unit of the polymer has the formula of the monomer itself.

Condensation polymerization (Section 22.5)

Monomers are linked by the elimination of a small molecule such as H_2O or CH_3OH and the formation of a covalent bond between the remaining fragments. For copolymers, the repeating unit is that of the two monomers minus the small molecule that was eliminated.

WileyPLUS = WileyPLUS, an online teaching and learning solution. *Note to instructors*: Many of the end-of-chapter problems are available for assignment via the WileyPLUS system. **www.wileyplus.com.** Review Problems are presented in pairs separated by blue rules. Answers to problems whose numbers appear in blue are given in Appendix B. More challenging problems are marked with an asterisk *.

|Review Questions

Organic Structures and Functional Groups

22.1 In general terms, what makes possible the large number of organic compounds?

22.2 What does R represent in an organic structure? What is the condensed structure of R if R—CH_3 represents the compound ethane?

22.3 Which of the following structures are possible, given the numbers of bonds that various atoms normally form?

(a) $CH_2CH_2CH_3$

(b) $CH_3CHCH_2CH_3$

(c) $CH_3CH{=\!=}CHCH_2CH_3$

22.4 Write condensed structures of the following:

(a)

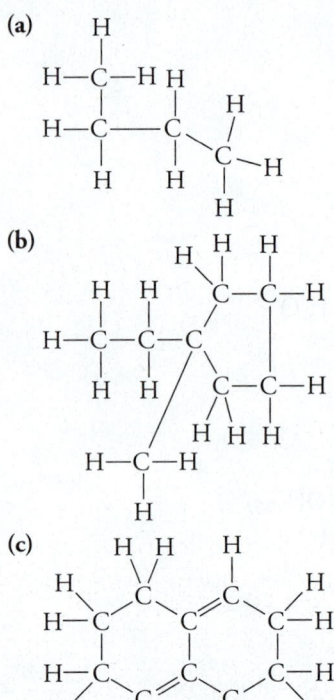

(b)

(c)

22.5 In $CH_3CH_2NH_2$, the NH_2 group is called the functional group. In general terms, why is it called this?

22.6 For each of the following, write the condensed structural formula and identify and name the functional group present in the molecule.

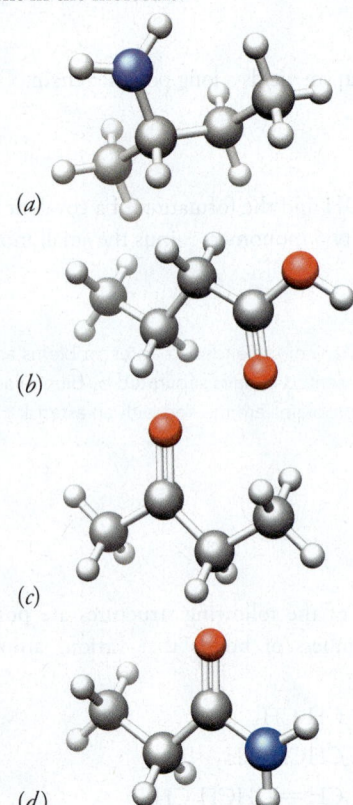

(a)

(b)

(c)

(d)

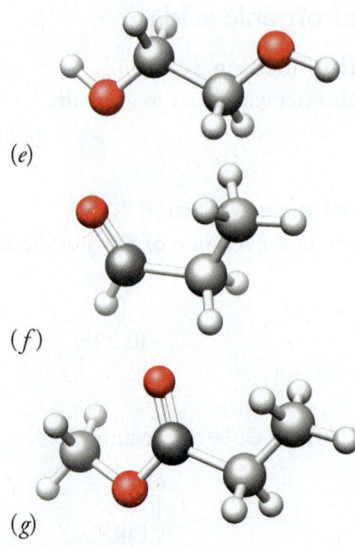

(e)

(f)

(g)

22.7 What must be true about two substances if they are to be called isomers of each other?

Hydrocarbons: Structure, Nomenclature, and Reactions

22.8 Which of the following compounds has the higher boiling point? Explain.

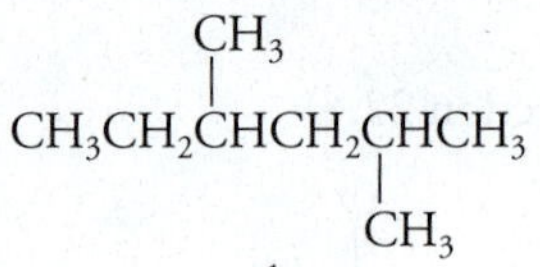

$$CH_3CH_2CHCH_2CHCH_3 \qquad CH_3CH_2CH_2CH_2CH_3$$

with CH_3 substituents

A 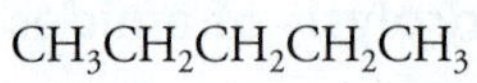   **B**

22.9 In general terms, why do functional groups impart more chemical reactivity to molecules that have them? Why don't the alkanes display as many reactions as, say, the amines?

22.10 Write the condensed structures of the following compounds:

(a) 2,2-dimethyloctane

(b) 1,3-dimethylcyclopentane

(c) 1,1-diethylcyclohexane

(d) 6-ethyl-5-isopropyl-7-methyloct-1-ene

(e) *cis*-pent-2-ene

22.11 What is the difference between geometric isomers and stereoisomers?

22.12 What prevents free rotation about a double bond and so makes geometric isomers possible?

22.13 No number is needed to identify the location of the double bond in $CH_3CH{=}CH_2$, propene. Why?

22.14 2-Methylpropene reacts with hydrogen chloride as follows:

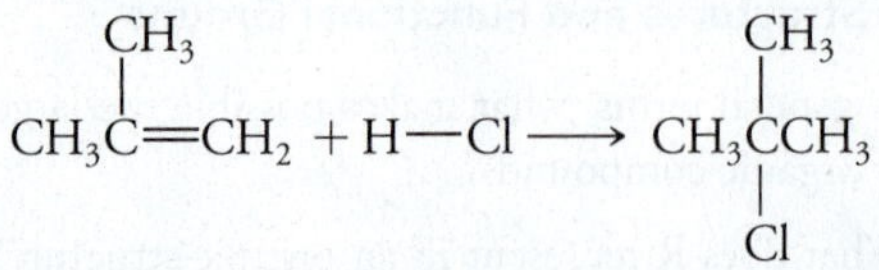

$$CH_3C{=}CH_2 + H{-}Cl \longrightarrow CH_3CCH_3$$

with CH_3 group on left carbon and Cl on product

The reaction occurs in two steps. Write the equation of each step.

22.15 Propene is known to react with concentrated sulfuric acid as follows:

$$CH_3CH=CH_2 + H_2SO_4 \longrightarrow CH_3CHCH_3$$
$$\underset{\displaystyle OSO_3H}{|}$$

The reaction occurs in two steps. Write the equation for these two steps.

22.16 The compound but-2-ene exists as two isomers, but when both isomers are hydrogenated, the products are identical. Explain.

22.17 In general terms, why doesn't benzene undergo the same kinds of addition reactions as cyclohexene?

22.18 What is the difference between a substitution reaction and an addition reaction?

Organic Compounds Containing Oxygen

22.19 Explain why $CH_3CH_2CH_2OH$ is more soluble in water than $CH_3CH_2CH_2CH_3$. Which one has the higher boiling point? Explain

22.20 Examine the structures of the following compounds:

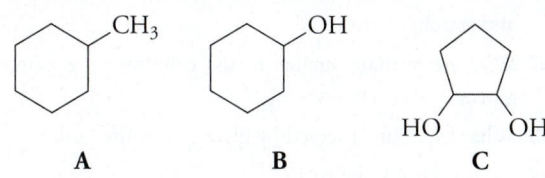

A   **B**   **C**

(a) Which has the highest boiling point? Explain.

(b) Which has the lowest boiling point? Explain.

22.21 Write the condensed structures for the following. Which is more soluble in water? Explain.

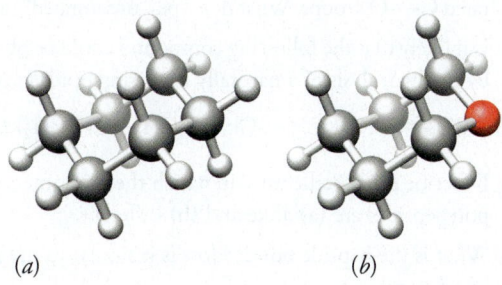

(a)   (b)

22.22 Why do aldehydes and ketones have boiling points that are lower than those of their corresponding alcohols?

22.23 Acetic acid boils at 118 °C, higher even than 1-propanol, which boils at 97 °C. How can the higher boiling point of acetic acid be explained?

22.24 Methyl ethanoate has many more atoms than its parent acid, ethanoic acid. Yet methyl ethanoate (BP 59 °C) boils at a much lower temperature than ethanoic acid (BP 118 °C). How can this be explained?

22.25 Write condensed structures of the following compounds:

(a) 3-methylbutanal   (c) 2-chloropropanoic acid

(b) 4-methyloctan-2-one   (d) 1-methylethylethanoate

22.26 Write condensed structures of the following compounds:

(a) butan-2,3-dione

(b) butanedicarboxylic acid

(c) 2-aminopropanal

(d) cyclohexyl 2-methylpropanoate

22.27 "3-Butanol" is not a proper name, but a structure could still be written for it. What is this structure, and what IUPAC name should be used?

22.28 Briefly explain how the C—O bond in isopropyl alcohol is weakened when a strong acid is present.

*__22.29__ Which isomer of butanol cannot be oxidized by dichromate ion? Write its structure and IUPAC name.

22.30 A monofunctional organic oxygen compound dissolves in aqueous base but not in aqueous acid. The compound is in which of the families of organic compounds that we studied? Explain.

22.31 Write the equation for the equilibrium that is present in a solution of propanoic acid and methanol with a trace of strong acid.

Organic Derivatives of Ammonia

22.32 Ethanamide is a solid at room temperature, while 1-aminobutane has about the same molecular mass and is a liquid. Explain.

22.33 Amines, RNH_2, do not have boiling points as high as alcohols with comparable numbers of atoms. Why?

22.34 A monofunctional organic nitrogen compound dissolves in aqueous hydrochloric acid but not in aqueous sodium hydroxide. What kind of organic compound is it?

22.35 Why are aqueous solutions of amides neutral while amines are basic?

*__22.36__ Hydrazine is a Brønsted base but urea does not exhibit basic properties. Offer an explanation.

$$NH_2NH_2 \qquad \qquad NH_2CNH_2$$

hydrazine   urea

22.37 Write the equilibrium equation for an aqueous solution of $CH_3CH_2\,CH_2NH_2$.

22.38 Write the products that can be expected to form in the following situations. If no reaction occurs, write "no reaction."

(a) $CH_3CH_2CH_2NH_2 + HBr(aq) \longrightarrow$

(b) $CH_3CH_2CH_2CH_3 + HI(aq) \longrightarrow$

(c) $CH_3CH_2CH_2NH_3^+ + H_3O^+ \longrightarrow$

(d) $CH_3CH_2CH_2NH_3^+ + OH^- \longrightarrow$

Organic Polymers

22.39 What is a macromolecule? Name two naturally occurring macromolecular substances.

22.40 What is a polymer? Are all macromolecules polymers?

22.41 What do we mean by the term polymer backbone?

22.42 What is a monomer? Draw structures for the monomers used to make **(a)** polypropylene, **(b)** polytetrafluroethylene, and **(c)** polyvinyl chloride.

22.43 How do propylene and the repeating unit in polypropylene differ?

22.44 What is the difference between an addition polymer and condensation polymer?

22.45 What is the repeating unit in polypropylene? Write the formula for the polypropylene polymer. Give three uses for polypropylene.

22.46 Write the structure of polystyrene showing three of the repeating units. What are three uses for polystyrene plastic?

22.47 What is a copolymer?

22.48 Write the structural formula for **(a)** nylon 6,6 and **(b)** poly(ethylene terephthalate).

22.49 The structural formula for Kevlar is

Identify the amide bond in this polymer.

22.50 Polycarbonate polymers are polyesters. An example is

Identify the structural feature that makes it a polyester. Identify the structural feature that makes it a polycarbonate.

22.51 What is meant by the term branching as applied to polymers? How does branching affect the properties of low-density polyethylene?

22.52 What is high-density polyethylene (HDPE)? Why does it form fibers easily?

22.53 What is ultrahigh-molecular-weight polyethylene (UHMWPE)? What range of molecular masses is found for this polymer?

22.54 How does polymer crystallinity affect the physical properties of the polymer?

22.55 Why doesn't low-density polyethylene form strong crystalline polymer fibers?

22.56 Why does nylon form strong fibers?

22.57 Show how hydrogen bonding can bind Kevlar polymer chains together. (See Question 23.49.)

*22.58 What are some applications of crystalline polymers such as HDPE, UHMWPE, and Kevlar?

Carbohydrates, Lipids, and Proteins

22.59 What are the three fundamental needs for sustaining life, and what are the general names for the substances that supply these needs?

22.60 How are carbohydrates defined?

22.61 What monosaccharide forms when the following polysaccharides are completely hydrolyzed? **(a)** starch, **(b)** glycogen, **(c)** cellulose

22.62 Glucose exists in three forms, only one of which is a polyhydroxyaldehyde. How, then, can an aqueous solution of any form of glucose undergo the reactions of an aldehyde?

22.63 Name the compounds that form when sucrose is digested.

22.64 The digestion of lactose gives what compounds? Name them.

22.65 The complete hydrolysis of the following compounds gives what product(s)? (Give the names only.)

(a) amylose

(b) amylopectin

22.66 Describe the relationships among amylose, amylopectin, and starch.

22.67 Why are humans unable to use cellulose as a source of glucose?

22.68 What function is served by glycogen in the body?

22.69 How are lipids defined?

22.70 Why are lipids more soluble than carbohydrates in nonpolar solvents?

22.71 Cholesterol is not an ester, yet it is defined as a lipid. Why?

22.72 A product such as corn oil is advertised as "polyunsaturated," but all nutritionally important lipids have unsaturated $C=O$ groups. What does "polyunsaturated" refer to?

22.73 Is it likely that the following compound could be obtained by the hydrolysis of a naturally occurring lipid? Explain.

$$CH_3(CH_2)_2CH=CHCH_2CH=CHCH_2CH=CH(CH_2)_7CO_2H$$

22.74 Describe the specific ways in which the monomers for all polypeptides are **(a)** alike and **(b)** different.

22.75 What is the peptide bond? How is it similar to the amide bond in nylon?

22.76 Describe the structural way in which two isomeric polypeptides would be different.

22.77 Describe the structural ways in which two different polypeptides can differ.

22.78 Why is a distinction made between the terms polypeptide and protein?

22.79 In general terms, what attractive forces are at work that determine the final shape of a polypeptide strand?

22.80 What kind of substance makes up most enzymes?

22.81 The most lethal chemical poisons act on what kinds of substances? Give examples.

22.82 In general terms only, how does the body solve the problem of getting a particular amino acid sequence, rather than a mixture of randomly organized sequences, into a polypeptide?

Nucleic Acids, DNA and RNA

22.83 What kind of attractive force occurs between the two DNA strands in a double helix?

22.84 How are the two DNA strands in a double helix structurally related?

22.85 In what ways do DNA and RNA differ structurally?

22.86 Which base pairs with **(a)** A in DNA, **(b)** A in RNA, and **(c)** C in DNA or RNA?

22.87 The process of transcription begins with which nucleic acid and ends with which one?

22.88 The process of translation begins with which nucleic acid? What is the end result of translation?

| Review Problems

Organic Structures and Functional Groups

22.89 Write full (expanded) structures for each of the following molecular formulas. Remember how many bonds the various kinds of atoms must have. In some you will have to use double or triple bonds. (*Hint:* A trial-and-error approach will have to be used.)

(a) CH_5N **(c)** $CHCl_3$ **(e)** C_2H_2

(b) CH_2Br_2 **(d)** NH_3O **(f)** N_2H_4

22.90 Write full (expanded) structures for each of the following molecular formulas. Follow the guidelines of Problem 22.89.

(a) C_2H_6 **(c)** CH_2O **(e)** C_3H_3N

(b) CH_2O_2 **(d)** HCN **(f)** CH_4O

22.91 Name the family to which each compound belongs.

(a) $CH_3CH{=}CH_2$ **(d)** $CH_3CH_2CH_2\overset{\overset{\displaystyle O}{\|}}{C}OH$

(b) CH_3CH_2OH **(e)** $CH_3CH_2CH_2NH_2$

(c) $CH_3CH_2\overset{\overset{\displaystyle O}{\|}}{C}OCH_3$ **(f)** $HOCH_2CH_2CH_3$

22.92 To what organic family does each compound belong?

(a) $CH_3C{\equiv}CH$ **(d)** $CH_3{-}O{-}CH_2CH_3$

(b) $CH_3CH_2\overset{\overset{\displaystyle O}{\|}}{C}H$ **(e)** $CH_3CH_2NH_2$

(c) $CH_3\overset{\overset{\displaystyle O}{\|}}{C}CH_2CH_2CH_3$ **(f)** $CH_3CH_2\overset{\overset{\displaystyle O}{\|}}{C}NH_2$

22.93 Identify the compounds in Problem 22.91 that are saturated compounds.

22.94 Identify the compounds in Problem 23.92 that are saturated compounds.

22.95 Classify the following compounds as amines or amides. If another functional group occurs, name it too.

(a) $CH_3CH_2NH_2$ **(c)** $CH_3CH_2\overset{\overset{\displaystyle O}{\|}}{C}NH_2$

(b) $CH_3CH_2NHCH_3$ **(d)** $CH_3\overset{\overset{\displaystyle O}{\|}}{C}CH_2NH_2$

22.96 Name the functional groups present in each of the following structures:

(a)

(b)

(c)

(d) $CH_3O\overset{\overset{\displaystyle O}{\|}}{C}CH_2CH_2\overset{\overset{\displaystyle O}{\|}}{C}OH$

22.97 Decide whether the members of each pair are identical, isomers, or unrelated.

(a) $CH_3{-}CH_3$ and $\overset{\overset{\displaystyle CH_3}{|}}{CH_3}$

(b)   and

(c) CH_3CH_2OH and $CH_3CH_2CH_2OH$

(d) $CH_3CH{=}CH_2$ and

(e) H—$\overset{\overset{\displaystyle O}{\|}}{C}$—CH$_3$ and CH$_3$—$\overset{\overset{\displaystyle O}{\|}}{C}$—H

(f) CH$_3$CHCH$_3$ and CH$_3$CH
$\quad\quad\ \ \overset{|}{CH_3}$ $\quad\quad\quad\quad\quad\ \overset{|}{CH_3}$
$\quad\quad\quad\quad\quad\quad\quad\quad\quad\quad\underset{|}{}$
$\quad\quad\quad\quad\quad\quad\quad\quad\quad\quad CH_3$

(g) CH$_3$CH$_2$CH$_2$NH$_2$ and CH$_3$CH$_2$—NH—CH$_3$

22.98 Examine each pair and decide if the two are identical, isomers, or unrelated.

(a) CH$_3$CH$_2$$\overset{\overset{\displaystyle O}{\|}}{C}$OH and HO$\overset{\overset{\displaystyle O}{\|}}{C}CH_2CH_3$

(b) H$\overset{\overset{\displaystyle O}{\|}}{C}OCH_2CH_3$ and CH$_3$CH$_2$$\overset{\overset{\displaystyle O}{\|}}{C}$OH

(c) H$\overset{\overset{\displaystyle O}{\|}}{C}OCH_2CH_2$OH and HOCH$_2CH_2$$\overset{\overset{\displaystyle O}{\|}}{C}$OH

(d) CH$_3$$\overset{\overset{\displaystyle O}{\|}}{C}CH_2CH_3$ and CH$_3$CH$_2$$\overset{\overset{\displaystyle O}{\|}}{C}CH_3$

(e) CH$_3$—CH—CH$_3$
$\quad\quad\quad\ \overset{|}{CH_2}$—CH$_2$ CH$_3$ CH$_3$
$\quad\quad\quad\quad\quad\ \ \overset{|}{CH_2}$—$\overset{|}{C}$—$\overset{|}{CH}$
$\quad\quad\quad\quad\quad\quad\quad\quad\ \overset{|}{CH_3}\quad\overset{\diagdown}{CH_3}$

and

$\quad\quad\ CH_3$ $\quad\quad\quad\quad\quad\quad CH_3\ \ CH_3$
$\quad\quad\ \overset{|}{}$ $\quad\quad\quad\quad\quad\quad\quad \overset{|}{}\quad\ \overset{|}{}$
CH$_3$—CH—CH$_2$—CH$_2$—CH$_2$—C——CH—CH$_3$
$\quad\quad\quad\quad\quad\quad\quad\quad\quad\quad\quad\overset{|}{CH_3}$

Hydrocarbons: Structure, Nomenclature, and Reactions

22.99 Write the IUPAC names of the following hydrocarbons:

(a) CH$_3$CH$_2$CH$_2$CH$_2$CH$_3$

(b) CH$_3$CH$_2$CH$_2$CHCH$_3$
$\quad\quad\quad\quad\quad\quad\ \overset{|}{CH_3}$

(c) $\quad\ CH_3$
$\quad\quad\ \overset{|}{}$
CH$_3$CHCH$_2$CHCH$_2$CH$_3$
$\quad\quad\quad\quad\quad\ \overset{|}{CH_3}$

22.100 Write the IUPAC names of the following compounds:

(a) $\quad\quad\quad CH_3$
$\quad\quad\quad\quad\ \overset{|}{}$
CH$_3$CH$_2$CHCH$_2$CHCH$_3$
$\quad\quad\quad\quad\quad\quad\quad\ \overset{|}{CH_3}$

(b) CH$_3$CH$_2$CH=CHCH$_2$CH$_3$

(c) $\quad\ CH_3$
$\quad\quad\ \overset{|}{}$
CH$_3$CHCH=CHCH$_3$

22.101 Write the structures of the *cis* and *trans* isomers, if any, for the following:

(a) CH$_2$=CHBr

(b) CH$_3$CH=CHCH$_2$CH$_3$

(c) CH$_3$C=CHBr
$\quad\quad\ \overset{|}{Br}$
$\quad\quad CH_3\overset{|}{C}\ =CHBr$

22.102 Write the structures of the *cis* and *trans* isomers, if any, for the following compounds:

(a) $\quad\quad CH_3$
$\quad\quad\quad\ \overset{|}{}$
CH$_3$C=CHCH$_3$

(b) $\quad\quad\quad\quad CH_3$
$\quad\quad\quad\quad\quad\ \overset{|}{}$
CH$_3$CH=CCH$_2$CH$_3$

(c) $\quad\quad\quad\quad\quad Cl\ \ Cl$
$\quad\quad\quad\quad\quad\quad\ \overset{|}{}\quad\overset{|}{}$
CH$_3$CH$_2$C=CCH$_2$CH$_3$

22.103 Write the structures of the products that form when ethylene reacts with each of the following substances by an addition reaction. (Assume that needed catalysts or other conditions are provided.)

(a) H$_2$ $\quad\quad$ (c) Br$_2$ $\quad\quad$ (e) HBr(g)

(b) Cl$_2$ $\quad\quad$ (d) HCl(g) $\quad\quad$ (f) H$_2$O (in acid)

22.104 The isopropyl cation is more stable than the propyl cation, so the former forms almost exclusively when propene undergoes addition reactions with proton-donating species. On the basis of this fact, predict the final products of the reaction of propene with each of the following reagents:

(a) hydrogen chloride

(b) hydrogen iodide

(c) water in the presence of an acid catalyst

22.105 Repeat Problem 22.103 using but-2-ene.

22.106 Repeat Problem 22.104 using cyclohexene.

22.107 If one mole of benzene were to add one mole of Br$_2$, what would form? What forms, instead, when Br$_2$, benzene, and a FeBr$_3$ catalyst are heated together? (Write the structures.)

22.108 Predict the products of the reaction of benzene and nitric acid with a sulfuric acid catalyst.

Organic Compounds Containing Oxygen

22.109 Write condensed structures and the IUPAC names for all of the saturated alcohols with three or fewer carbon atoms per molecule.

22.110 Write the condensed structures and the IUPAC names for all of the possible alcohols with the general formula C$_4$H$_{10}$O.

*22.111 Write the condensed structures of all of the possible ethers with the general formula $C_4H_{10}O$. Following the pattern for the common names of ethers given in the chapter, what are the likely common names of these ethers?

22.112 Write the structures of the isomeric alcohols of the formula $C_4H_{10}O$ that could be oxidized to aldehydes. Write the structure of the isomer that could be oxidized to a ketone.

*22.113 Write the structures of the products of the acid-catalyzed dehydrations of the following compounds:

(a) OH

(b) OH

—CHCH₃

(c)

—CH₂CH₂OH

*22.114 Write the structure of the product of the acid-catalyzed dehydration of propan-2-ol. Write the mechanism of the reaction.

*22.115 Write the structures of the substitution products that form when the alcohols of Problem 22.113 are heated with concentrated hydroiodic acid.

*22.116 Write the structures of the aldehydes or ketones that can be prepared by the oxidation of the compounds given in Problem 22.113.

*22.117 When ethanol is heated in the presence of an acid catalyst, ethene and water form; an elimination reaction occurs. When butan-2-ol is heated under similar conditions, two alkenes form (although not in equal quantities). Write the structures and IUPAC names of these two alkenes.

*22.118 If the formation of the two alkenes in Problem 23.117 were determined purely by statistics, then the mole ratio of the two alkenes should be in the ratio of what whole numbers? Why?

*22.119 Which of the following two compounds could be easily oxidized? Write the structure of the product of the oxidation.

$$\underset{\textbf{A}}{CH_3CH_2\overset{\overset{\displaystyle O}{\|}}{C}H} \qquad \underset{\textbf{B}}{CH_3CH_2\overset{\overset{\displaystyle O}{\|}}{C}\underset{\underset{\displaystyle CH_3}{|}}{C}HCH_3}$$

22.120 Which of the following compounds has the chemical property given below? Write the structure of the product of the reaction.

$$\underset{\textbf{A}}{CH_3CH_2CH_2OH} \qquad \underset{\textbf{B}}{CH_3CH_2\overset{\overset{\displaystyle O}{\|}}{C}OH} \qquad \underset{\textbf{C}}{CH_3CH_2\overset{\overset{\displaystyle O}{\|}}{C}H}$$

(a) Is easily neutralized

(b) Neutralizes NaOH

(c) Forms an ester with methyl alcohol

(d) Can be oxidized to an aldehyde

(e) Can be dehydrated to an alkene

Organic Derivatives of Ammonia

*22.121 Write the structures of the products that form in each of the following situations. If no reaction occurs, write "no reaction."

(a) $CH_3CH_2CO_2^- + HCl(aq) \longrightarrow$

(b) $CH_3CH_2CO_2CH_3 + H_2O \xrightarrow{heat}$

(c) $CH_3CH_2CH_2CO_2H + NaOH(aq) \longrightarrow$

*22.122 Write the structures of the products that form in each of the following situations. If no reaction occurs, write "no reaction."

(a) $CH_3CH_2CO_2H + NH_3 \xrightarrow{heat}$

(b) $CH_3CH_2CH_2CONH_2 + H_2O \xrightarrow{heat}$

(c) $CH_3CH_2CO_2CH_3 + NaOH(aq) \xrightarrow{heat}$

*22.123 What organic products with smaller molecules form when the following compound is heated with water?

$$CH_3CH_2\underset{\underset{\displaystyle CH_3CH_2}{|}}{N}\overset{\overset{\displaystyle O}{\|}}{C}CH_3$$

*22.124 Which of the following species undergo the reaction specified? Write the structures of the products that form.

$$CH_3CH_2\overset{\overset{\displaystyle O}{\|}}{C}NH_2 \qquad CH_3CH_2CH_2NH_2 \qquad CH_3CH_2NH_3^+$$

(a) Neutralizes dilute hydrochloric acid

(b) Hydrolysis

(c) Neutralizes dilute sodium hydroxide

Organic Polymers

22.125 The structure of vinyl acetate is shown below. This compound forms a chain-growth polymer in the same way as ethylene and propylene. Draw a segment of the poly(vinyl acetate) polymer that contains three of the repeating units.

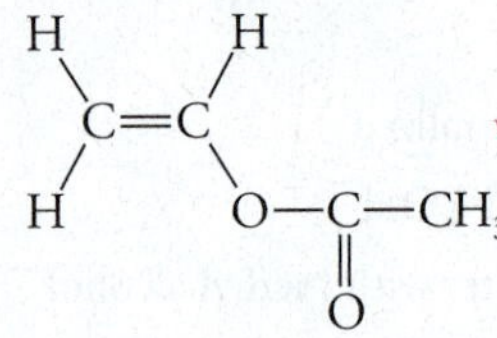

yinyl acetate

22.126 If the following two compounds polymerized in the same way that Dacron forms, what would be the repeating unit of their polymer?

$$HO\overset{\overset{O}{\|}}{C}CH_2CH_2\overset{\overset{O}{\|}}{C}OH \qquad HOCH_2CH_2OH$$

22.127 Kodel, a polyester made from the following monomers, is used to make fibers for weaving crease-resistant fabrics.

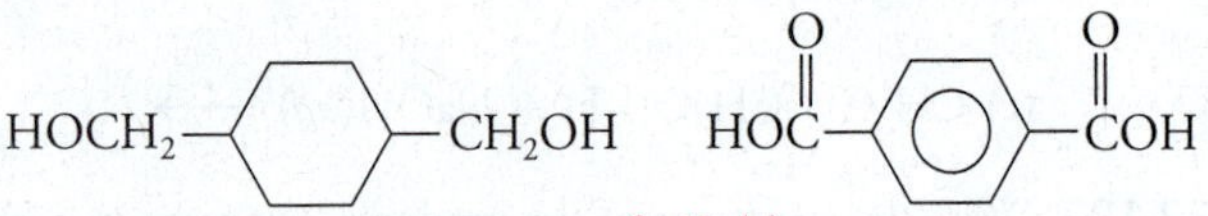

monomers for Kodel

What is the structure of the repeating unit in Kodel?

22.128 If the following two compounds polymerized in the same way that nylon forms, what would be the repeating unit in the polymer?

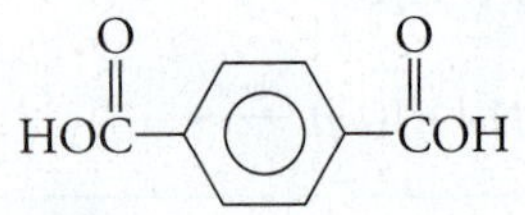 $NH_2CH_2CH_2CH_2CH_2NH_2$

Carbohydrates, Lipids, and Protiens

22.129 Write the structure of a triacylglycerol that could be made from palmitic acid, oleic acid, and linoleic acid.

22.130 Write the structures of the products of the complete hydrolysis of the following triacylglycerol:

$$\begin{array}{l} CH_2O\overset{\overset{O}{\|}}{C}(CH_2)_7CH{=}CHCH_2CH{=}CH(CH_2)_4CH_3 \\ | \\ CHO\overset{\overset{O}{\|}}{C}(CH_2)_{12}CH_3 \\ | \\ CH_2O\overset{\overset{O}{\|}}{C}(CH_2)_7CH{=}CH(CH_2)_7CH_3 \end{array}$$

*__22.131__ Write the structure of the triacylglycerol that would result from the complete hydrogenation of the lipid whose structure is given in Problem 22.130.

*__22.132__ If the compound in Problem 22.130 is saponified, what are the products? Give their names and structures.

*__22.133__ What parts of glycerophospholipid molecules provide hydrophobic sites? Which provide hydrophilic sites?

*__22.134__ In general terms, describe the structure of the membrane of an animal cell. What kinds of molecules are present? How are they arranged? What forces keep the membrane intact? What functions are served by the protein components?

22.135 Write the structure of the dipeptide that could be hydrolyzed to give two molecules of glycine.

22.136 What is the structure of the tripeptide that could be hydrolyzed to give three molecules of alanine?

22.137 What are the structures of the two dipeptides that could be hydrolyzed to give glycine and phenylalanine?

22.138 Write the structures of the dipolar ionic forms of the amino acids that are produced by the complete digestion of the following compound:

$$^{+}NH_3\underset{\underset{CH_3}{|}}{CH}\overset{\overset{O}{\|}}{C}NH\underset{\underset{CH_2SH}{|}}{CH}\overset{\overset{O}{\|}}{C}NH\underset{\underset{CH_2C_6H_5}{|}}{CH}\overset{\overset{O}{\|}}{C}O^{-}$$

| Additional Exercises

22.139 Predict which of the following species can neutralize dilute, aqueous HCl at room temperature.

(a) $CH_3\overset{\overset{O}{\|}}{C}O^{-}$

(b) CH_3NH_2

(c) HO^{-}

(d) $CH_3\overset{\overset{O}{\|}}{C}NH_2$

(e) $CH_3CH_2CH_2CH_3$

(f) 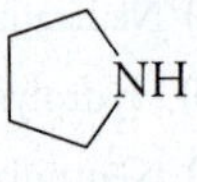

22.140 Write molecular formulas for the following:

(a) 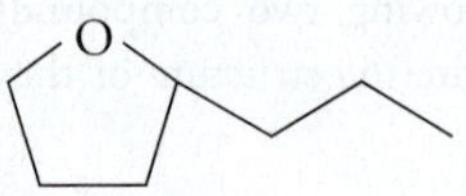

(b) 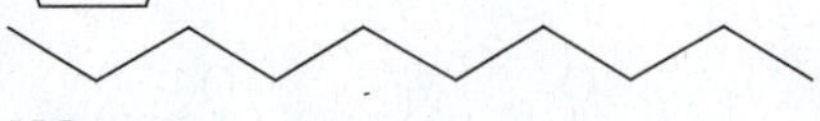

(c) 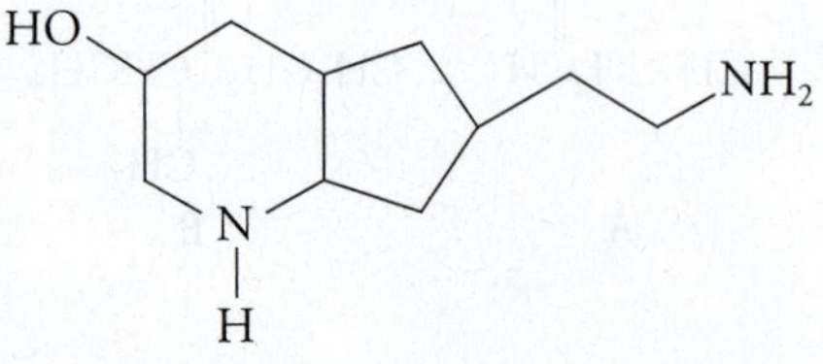

22.141 The H_2SO_4-catalyzed addition of water to 2-methyl-propene could give two isomeric alcohols.

    (a) Write their condensed structures.

    (b) Actually, only one forms. It is not possible to oxidize this alcohol to an aldehyde or to a ketone having four carbons. Which alcohol, therefore, is the product?

    (c) Write the structures of the two possible carbocations.

    (d) One of the two carbocations of part (c) is more stable than the other. Which one is it, and how can you tell?

22.142 Suggest a reason why trimethylamine, $(CH_3)_3N$, has a lower boiling point (BP 3 °C) than dimethylamine, $(CH_3)_2NH$ (BP 8 °C), despite having a larger number of atoms.

*__22.143__ Write the structures of the organic products that form in each situation. If no reaction occurs, write "no reaction."

    (a)

$$HOC\overset{\displaystyle O}{\|}\!\!-\!\!\bigcirc\!\!-\!\!COC\overset{\displaystyle O}{\|}H + NaOH(aq) \longrightarrow$$
excess

(b)
$$CH_3\overset{\displaystyle O}{\overset{\|}{C}}CH_2CH_2CH_3 + H_2 \xrightarrow[\text{catalyst}]{\text{heat, pressure,}}$$

(c) $CH_3\overset{+}{N}H_2CH_2CH_2CH_3 + OH^-(aq) \longrightarrow$

(d)
$$CH_3CH_2OCH_2CH_2\overset{\displaystyle O}{\overset{\|}{C}}OCH_3 + H_2O \xrightarrow{\text{heat}}$$

(e) ⬠ $+ H_2O \xrightarrow[\text{catalyst}]{H_2SO_4}$

(f) ◯—CH=CH—◯ $+ HBr \longrightarrow$

*__22.144__ How many tripeptides can be made from three different amino acids? If the three are glycine (Gly), alanine (Ala), and phenylalanine (Phe), what are the structures of the possible tripeptides. (*Hint:* Gly-Ala-Phe is one structure.)

22.145 Why doesn't PVC have chlorines on adjacent atoms?

| Multi-Concept Problems

*__22.146__ An unknown alcohol was either *A* or *B* below. When the unknown was oxidized with an excess of strong oxidizing agent, sodium dichromate in acid, the organic product isolated was able to neutralize sodium hydroxide.

    ⬠—CH₂OH   OH above ⬠—CH₃

        A   **B**

    (a) Which was the alcohol? Give the reasons for this choice.

    (b) What is the molecular formula of the product of the oxidation?

    (c) Draw the structural formula of the product.

    (d) What is the minimum number of grams of sodium dichromate needed to oxidize 36.8 g of the alcohol?

    (e) If 36.8 g of the alcohol was oxidized and all of the product was collected, how many grams of NaOH could it neutralize?

*__22.147__ Estimate the number of kilojoules of heat that would be liberated in the complete combustion of 11.35 g of gaseous compound *B* in the preceding problem. (*Hint:* Use bond energy data to estimate the standard heat of formation of the compound.)

*__22.148__ A 6.225 mg sample of a monoprotic organic acid, when burned in pure oxygen, gave 12.44 mg of CO_2 and 5.091 mg H_2O. What are the empirical and molecular formulas for this acid? How many milliliters of 0.224 *M* KOH are needed to react completely with 0.822 g of the acid? Draw the structure of the ester that this acid would form with ethanol.

22.149 When 0.2081 g of an organic acid was dissolved in 50.00 mL of 0.1016 *M* NaOH, it took 23.78 mL of 0.1182 *M* HCl to neutralize the NaOH that was not used up by the sample of the organic acid. Calculate the formula mass from the data given. Is the answer necessarily the molecular mass of the organic acid? Explain.

22.150 The compound that causes your eyes to water when you chop an onion has the following molecular structure:

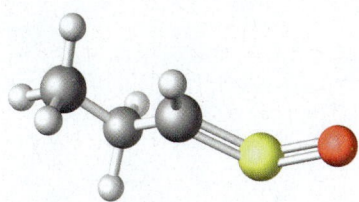

How many grams of oxygen would be required for the complete combustion of 1.225 g of this compound?

| Exercises in Critical Thinking

22.151 The compound CH_2Cl_2 only exists as one isomer. Why does this support the statement that the H and Cl atoms in the compound are arranged tetrahedrally around the carbon rather than in a square planar structure?

22.152 The α-carbon atom in an amino acid is a chiral center (except for glycine). Of the two possible enantiomers of these optically active substances, only one is produced naturally. What does this suggest about the mechanism whereby amino acids are synthesized in living creatures?

22.153 Use resonance structures to explain why urea, $(NH_2)_2C{=}O$, is not basic like typical amines. Without using resonance structures, use electronegativities to explain the same fact.

22.154 Below is the ring structure of a sugar molecule. How many chiral centers does this sugar molecule have? How many isomers are possible?

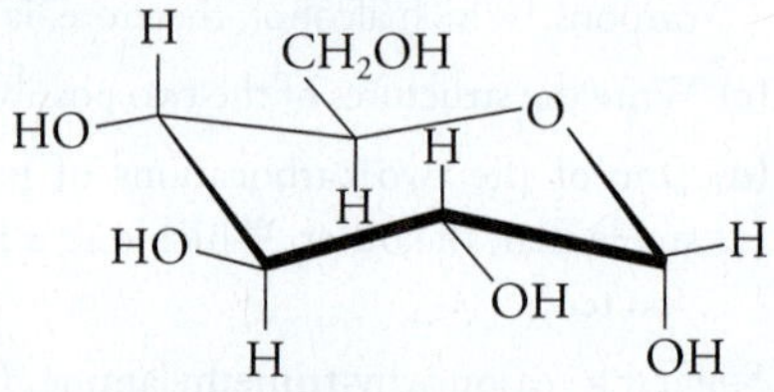

22.155 What would be the repeating unit of the polymer formed by condensation polymerization of the two monomers shown below?

$$CH_3O-\overset{\displaystyle O}{\overset{\|}{C}}-CH_2-CH_2-CH_2-\overset{\displaystyle O}{\overset{\|}{C}}-OCH_3$$

$$H_2N-CH_2-CH_2-NH_2$$

Review of Mathematics

This appendix provides a brief review of the math topics that you will use in solving the chemistry problems in this book. When you are having difficulty in following the math in worked examples, when you are unsure how to proceed with solving the math in an exercise or problem, or when the answers you obtain do not match those in Appendix B, refer to the topics discussed. A more complete review is available on WileyPlus.

A.1 | Exponential and Scientific Notation

Very large and very small numbers can be expressed as powers of ten in which a number between 1 and 10 is multiplied by 10 raised to a power (e.g., 3.2×10^5 or 1.66×10^{-6}). This format of writing numbers is called **exponential notation**, **standard scientific notation**, or **scientific notation**. Some examples are given in Table A.1.

The Meaning of Positive and Negative Exponents

Positive Exponents

Suppose you come across a number such as 6.4×10^4, which is just 64,000, or 6.4 multiplied by 10 four times.

$$6.4 \times 10^4 = 6.4 \times 10 \times 10 \times 10 \times 10 = 64,000$$

Changing 6.4×10^4 from exponential notation to a regular number involves changing the exponent on the 10 to zero (recall that $10^0 = 1$) and then moving the decimal point in the 6.4 so we don't change the *value of the number*. To do this, just increase the value of the 6.4 the same amount that we decreased the power of 10. In this case, move the decimal point four places to the right.

For changing numbers greater than 1.0 into scientific notation, count the number of places you would have to move the decimal point to the left to put it just after the first digit in the number. For example, for 60,530, you would have to move the decimal point four places to the left

$$\underset{4}{6} \; \underset{3}{0} \; \underset{2}{5} \; \underset{1}{3} \; 0$$

which decreased the value by ten for each move of the decimal point. To avoid changing the value of the number, the exponent on the 10 is increased by one for each move of the decimal point. We rewrite it as 6.0530×10^4.

We check our answer by being sure that if we decreased the exponential part, the numerical part is increased (or vice-versa).

Negative Exponents

When we have a small number (less than 1) in scientific notation, for example 1.4×10^{-4}, the power of ten is a negative number. Negative powers of ten are treated the same way as positive values, always being sure to increase the number while decreasing the exponent or vice-versa. We can convert 1.4×10^{-4} to a regular number as we did before, Since we will be *increasing* the exponential from 10^{-4} to 10^0, we need to *decrease* the numerical part from 1.4 to 0.00014.

TABLE A.1	
Number	Exponential Form
1	1×10^0
10	1×10^1
100	1×10^2
1000	1×10^3
10,000	1×10^4
100,000	1×10^5
1,000,000	1×10^6
0.1	1×10^{-1}
0.01	1×10^{-2}
0.001	1×10^{-3}
0.000 1	1×10^{-4}
0.000 01	1×10^{-5}
0.000 001	1×10^{-6}
0.000 000 1	1×10^{-7}

To express a number smaller than 1 in scientific notation, we count the number of places the decimal has to be moved to put it to the right of the first nonzero digit.

$$0.000068901$$

5 places

Since we have increased the value of the numerical part, we need to decrease the exponential by the same amount. So we must decrease the exponential from its starting value, 10^0, to 10^{-5}. This number, 0.000068901 can be re-written as 6.8901×10^{-5}.

Multiplying and Dividing Numbers Written in Exponential and Scientific Notation

Let's take an example: Solve the following equation for v.

$$v = \frac{(7.284 \times 10^4)(6.45 \times 10^{31})}{(1.238 \times 10^{28})(2.37 \times 10^{-2})}$$

Now we will group the numbers and then group the exponents as

$$v = \frac{(7.284 \times 6.45)(10^{31} \times 10^4)}{(1.238 \times 2.37)(10^{-2} \times 10^{28})}$$

We all know how to multiply and divide regular numbers and get 16.0 when properly rounded. Now for the exponents.

$$v = 16.0 \frac{(10^{31} \times 10^4)}{(10^{-2} \times 10^{28})}$$

Multiplication of powers of 10 involves adding the exponents. The numerator will be $10^{31+4} = 10^{35}$ and the denominator will be $10^{-2+28} = 10^{26}$. Division of powers of 10 involves subtracting the denominator exponent from the numerator exponent. We get $10^{35-26} = 10^9$. Our answer is

$$v = 16.0 \times 10^9 = 1.60 \times 10^{10}$$

To review, there are two steps in multiplying and/or dividing numbers that are written in scientific notation.

1. Multiply and divide the decimal parts of the numbers (the parts that precede the multiplication signs).
2. To obtain the exponent on the 10 in the answer, add the exponents of the 10s for multiplication and subtract the denominator exponent from the numerator exponent for division.

A.2 | Logarithms

A logarithm is simply an exponent. For example, if we have an expression

$$N = a^b$$

we see that to obtain N we must raise a to the b power. The exponent b is said to be the logarithm of N; it is the exponent to which we must raise a, which we call the base, to obtain the number N.

$$\log_a N = b$$

We can define a logarithm for any base we want. The two most frequently encountered logarithm systems in the sciences are *common logarithms* with a base of 10, or $\log_{10}$, and *natural logarithms*, in which the base is e, where $e = 2.7182818 \ldots$ However, for any system of logarithms there are some useful relationships that evolve from the behavior of exponents in arithmetic operations.

For Multiplication:

If $N = A \times B$
then $\log N = \log A + \log B$

For Division:

If $N = \dfrac{A}{B}$
then $\log N = \log A - \log B$

For Exponentials:

If $N = A^b$
then $\log N = b \log A$

Common Logarithms

The common logarithm or $\log_{10}$ of a number N is the exponent to which 10 must be raised to equal the number N. To determine the common logarithm, or log for short, all you need to do is enter the number in your calculator and press the "log" key. To reverse the process, take the antilogarithm or antilog. Enter the logarithm in your calculator and press the "10^x" key(s).

Natural Logarithms

In the sciences there are many phenomena in which observed quantities are related to each other logarithmically, and the exponential relationship involves powers of e, which is said to be the base of the natural logarithms. The quantity e is derived from calculus and is an irrational number—that is a nonrepeating decimal. To eight significant figures its value is

$$e = 2.7182818 \ldots$$

Natural logarithms use the symbol ln instead of log. Thus, if the number N equals e^x,

$$N = e^x$$

then

$$\log_e N = \ln e^x = x$$

As with common logarithms, the natural logarithm is obtained on your calculator with a press of a key, only now the key to use is the "ln" key. To reverse the process, enter the natural logarithm and press the "e^x" key(s). This gives the antinatural logarithm or antiln for short.

Relating Common and Natural Logarithms

There is a simple relationship between common and natural logarithms. To four significant figures,

$$\ln N = 2.303 \log N$$

A.3 | Method of Successive Approximations

The *method of successive approximations* is not only much faster than using the quadratic formula, particularly with a scientific calculator, but it is also just as accurate. This method can be used even when the exponents are greater than two. The procedure is outlined in Figure A.1, which we will apply to solving the same equation below for x.

Figure A.1 | Steps in the method of successive approximations.

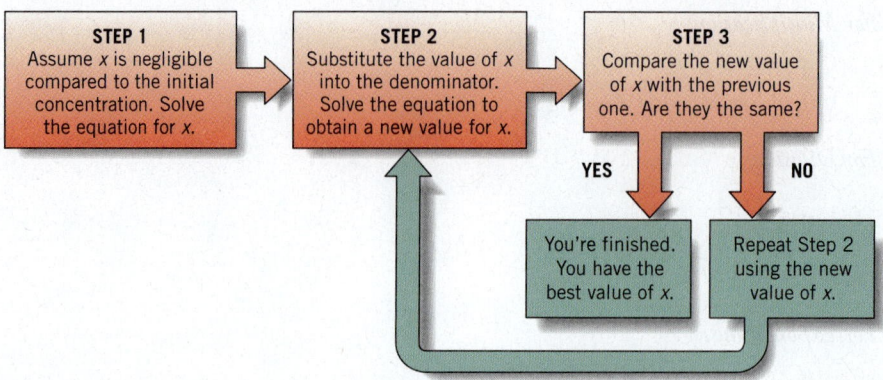

Step 1 We apply the simplifying approximation that $x \ll 0.010$ to the equation, even though we know it is not really applicable.

$$\frac{x^2}{(0.010 - x)} = 1.4 \times 10^{-3}$$

for which we obtain the solution $x = 3.7 \times 10^{-3}$. This is our *first approximation*.

Step 2 We substitute $x = 3.7 \times 10^{-3}$ (our *first approximation*) into the term in the denominator, $(0.100 - x)$, and then re-compute x. This gives us

$$\frac{x^2}{(0.010 - 0.0037)} = 1.4 \times 10^{-3}$$

or

$$x^2 = (0.006) \times 1.4 \times 10^{-3}$$

Taking only the positive root,

$$x = 2.9 \times 10^{-3} \ (second\ approximation)$$

We call $x = 2.9 \times 10^{-3}$ our *second approximation*.

Step 3 We compare the second approximation to the first and we see they are not the same, so we repeat Step 2.

Step 2 again We substitute $x = 2.9 \times 10^{-3}$ (our second approximation) into the term in the denominator, $(0.100 - x)$, and then re-compute x. This gives us a *third approximation*,

$$\frac{x^2}{(0.010 - 0.0029)} = 1.4 \times 10^{-3}$$

$$x^2 = (0.007) \times 1.4 \times 10^{-3}$$

$$x = 3.1 \times 10^{-3} \ (third\ approximation)$$

Step 3 again The third approximation is not the same as the second approximation, so we repeat Step 2 again.

Step 2 again We substitute $x = 3.1 \times 10^{-3}$ (our third approximation) into the term in the denominator, $(0.100 - x)$, and then re-compute x. This gives us

$$\frac{x^2}{(0.010 - 0.0031)} = 1.4 \times 10^{-3}$$

$$x^2 = (0.007) \times 1.4 \times 10^{-3}$$

$$x = 3.1 \times 10^{-3} \ (third\ approximation)$$

Step 3 again When we compare the third and fourth approximations, we see that they are identical, so we are finished. The method of successive approximations tells us that $x = 3.1 \times 10^{-3}$.

Each succeeding approximation differs from the preceding one by a smaller and smaller amount. We stop the calculation when the difference between two approximations is insignificant.

Answers to Practice Exercises and Selected Problems

Chapter 0
Practice Exercises

0.1 This answer is student dependent. **0.2** Nucleosynthesis would occur at temperatures around 1 billion degrees and a high density of nucleons. **0.3** Only light elements were synthesized during the big bang because the temperature was too high for the heavier elements to form. **0.4** Core and enriched layers or nuclei are needed for nucleosynthesis in stars because the enriched layers are the lighter layers, which react to form the heavier nuclei. The heavier nuclei are more dense and move to the center of the star. **0.5** Experimental evidence for the concept of nucleosynthesis lies in the layers of the stars and that stars have different compositions and that the most abundant elements are the lightest ones. **0.6** Iron is the heaviest element in stars since iron has the most stable nucleus. **0.7** Supernovas provide for the synthesis of heavier elements because the energy in the supernova makes the atomic collisions strong enough for the nuclei to combine. **0.8** $^{240}_{94}$Pu, 94 electrons. The bottom number is the atomic number, found on the periodic table (number of protons). The top number is the mass number (sum of the number of protons and the number of neutrons). Since it is a neutral atom, it has 94 electrons. **0.9** $^{35}_{17}$Cl contains 17 protons, 17 electrons, and 18 neutrons. We can discard the 17. **0.10** 26.9814 u **0.11** 5.2955 times heavier **0.12** Any other atom could have been used as the standard for atomic mass; for example, oxygen used to be the standard, and one atomic mass unit was 1/16 the mass of oxygen, but that was before they took into account the isotopes of oxygen. Scientists needed to use one isotope of an element. **0.13** 10.8 u **0.14** 20.18 u

Review Problems

0.31 27.00 u; aluminum **0.33 (a)** 88 protons, 88 electrons, and 138 neutrons **(b)** 82 protons, 82 electrons, and 124 neutrons **(c)** 6 protons, 6 electrons, and 8 neutrons **(d)** 11 protons, 11 electrons, and 12 neutrons **0.35** 53 protons, 53 electrons, and 78 neutrons **0.37** 1.008u **0.39** 2.01588 u **0.41** 63.55 u

Chapter 1
Practice Exercises

1.1 The scientific method is an iterative process of gathering information through making observations and collecting data and then formulating explanations that lead to a conclusion.
1.2 (a) element **(b)** mixture, homogeneous **(c)** compound **(d)** mixture, heterogeneous **(e)** element **1.3 (a)** chemical change **(b)** physical change **(c)** physical change **(d)** physical change **1.4 (a)** intensive **(b)** extensive **(c)** intensive **(d)** extensive **1.5** meter^3 or m^3 **1.6** $\text{kg}\left(\dfrac{m}{s^2}\right)$ or kg m s^{-2} **1.7** 671 °F **1.8** 13 °C; 293 K
1.9 (a) 42.0 g **(b)** 0.857 g/mL **(c)** 149 cm **1.10 (a)** 54.155 g

(b) 11.3 g **(c)** 3.62 ft **(d)** 0.36 m^2 **1.11** 11.5 m^2
1.12 (a) 1.40×10^5 in^3 **(b)** 1.25×10^5 cm **(c)** 92.7 g **(d)** 8.59 km/L
1.13 0.0163 g/mL **1.14** 2.19 g mL^{-1} **1.15** 16.5 g/cm^3. The object is not composed of pure gold. **1.16** 647 lb **1.17** 720 lb **1.18** The specific gravity of urine is below the normal range.

Review Problems

1.39 (a) Physical change **(b)** Physical change **(c)** Chemical change **(d)** Chemical change **(e)** Chemical change **1.41 (a)** gas **(b)** solid **(c)** gas **(d)** liquid **1.43 (a)** 0.01 m **(b)** 1000 m **(c)** 10^{12} pm **(d)** 0.1 m **(e)** 0.001 kg **(f)** 0.01 **1.45 (a)** 135 °F **(b)** -31.9 °C **(c)** 105 °C **(d)** 242 K **1.47 (a)** 15.7×10^6 °C **(b)** 2.83×10^7 °F **1.49** 39.7 °C. This dog has a fever; the temperature is out of normal canine range. **1.51** 26.2 cm, 3 sig. figs.; 26.25 cm, 4 sig. fig. **1.53 (a)** 4 sig. fig. **(b)** 5 sig. fig. **(c)** 4 sig. fig. **(d)** 2 sig. fig. **(e)** 4 sig. fig. **1.55 (a)** 0.72 m^2 **(b)** 84.24 kg **(c)** 4.19 g/cm^3 **(d)** 19.42 g/mL **(e)** 858.0 cm^2 **1.57 (a)** finite number of sig. fig. **(b)** exact number **(c)** finite number of sig. fig. **(d)** exact number **1.59 (a)** 11.5 km/hr **(b)** 8.2×10^6 µg/L **(c)** 7.53×10^{-5} kg **(d)** 0.1375 L **(e)** 25 mL **(f)** 3.42×10^{-20} dm **1.61 (a)** 91 cm **(b)** 2.3 kg **(c)** 2800 mL **(d)** 200 mL **(e)** 88 km/hr **(f)** 80.4 km **1.63 (a)** 7,800 cm^2 **(b)** 577 km^2 **(c)** 6.54×10^6 cm^3 **1.65** 4,000 pistachios **1.67** 90 m/s **1.69** 1520 mi/hr **1.71** 5.1×10^{13} mi **1.73** 0.798 g/mL **1.75** 31.6 mL **1.77** 276 g **1.79** 11 g/cm^3 **1.81** density $=$ 0.591 lb gal^{-1}; sp. gr. of liquid hydrogen $= 0.0708$; density $=$ 0.0709 g mL^{-1}

Chapter 2
Practice Exercises

2.1 (a) 1S, 6F **(b)** 4C, 12H, 2N **(c)** 3Ca, 2P, 8O **(d)** 1Co, 2N, 12O, 12H **2.2 (a)** 2N, 4H, 3O **(b)** 1Fe, 1N, 4H, 2S, 8O **(c)** 1Mo, 2N, 11O, 10H **(d)** 6C, 4H, 1Cl, 1N, 2O **2.3** Reactants: 4N, 12H, and 6O; Products: 4N, 12H, and 6O **2.4** Reactants: 6N, 42H, 2P, 20O, 3Ba, 12C; Products: 3Ba, 2P, 20O, 6N, 42H, 12C; The reaction is balanced. **2.5 (a)** 26 protons and 26 electrons **(b)** 26 protons and 23 electrons **(c)** 7 protons and 10 electrons **(d)** 7 protons and 7 electrons **2.6 (a)** 8 protons and 8 electrons **(b)** 8 protons and 10 electrons **(c)** 13 protons and 10 electrons **(d)** 13 protons and 13 electrons **2.7 (a)** NaF **(b)** Na$_2$O **(c)** MgF$_2$ **(d)** Al$_4$C$_3$ **2.8 (a)** Ca$_3$N$_2$ **(b)** AlBr$_3$ **(c)** K$_2$S **(d)** CsCl **2.9 (a)** CrCl$_3$ and CrCl$_2$, Cr$_2$O$_3$ and CrO **(b)** CuCl, CuCl$_2$, Cu$_2$O and CuO **2.10 (a)** Au$_2$S and Au$_2$S$_3$ Au$_3$N and AuN **(b)** TiS, Ti$_2$S$_3$, and TiS$_2$; Ti$_3$N$_2$, TiN, and Ti$_3$N$_4$ **2.11 (a)** Na$_2$CO$_3$ **(b)** (NH$_4$)$_2$SO$_4$ **2.12 (a)** KC$_2$H$_3$O$_2$ **(b)** Sr(NO$_3$)$_2$ **(c)** Fe(C$_2$H$_3$O$_2$)$_3$ **2.13 (a)** K$_2$S **(b)** BaBr$_2$ **(c)** NaCN **(d)** Al(OH)$_3$ **(e)** Ca$_3$P$_2$ **2.14 (a)** aluminum chloride **(b)** barium hydroxide **(c)** sodium bromide **(d)** calcium fluoride **(e)** potassium phosphide **2.15** lithium sulfide, magnesium phosphide, nickel(II) chloride, titanium(II) chloride,

iron(III) oxide **2.16 (a)** Al_2S_3 **(b)** SrF_2 **(c)** TiO_2 **(d)** CoO
(e) Au_2O_3 **2.17 (a)** lithium carbonate **(b)** potassium
permanganate **(c)** iron(III) hydroxide **2.18 (a)** $KClO_3$ **(b)** NaOCl
(c) $Ni_3(PO_4)_2$ **2.19** C_8H_{18}, $CH_3CH_2CH_2CH_2CH_2CH_2CH_2CH_3$
2.20 (a) C_3H_8O, $CH_3CH_2CH_2OH$ **(b)** $C_4H_{10}O$, $CH_3CH_2CH_2CH_2OH$
2.21 (a) phosphorus trichloride **(b)** sulfur dioxide **(c)** dichlorine
heptoxide **(d)** hydrogen sulfide **2.22 (a)** $AsCl_5$ **(b)** SCl_6 **(c)** S_2Cl_2
(d) H_2Te **2.23** diiodine pentoxide **2.24** chromium(III) acetate

Review Problems

2.67 1Cr, 6C, 9H, 6O **2.69** $MgSO_4$ **2.71** CH_3COOH or $C_2H_4O_2$

2.73 NH_3 **2.75 (b)**

$$H-\underset{\underset{H}{|}}{\overset{\overset{H}{|}}{N}}-H$$

2.77 (a) 2K potassium,
2C carbon, 4O oxygen **(b)** 2H hydrogen, 1S sulfur, 3O oxygen **(c)** 12C
carbon, 26H hydrogen **(d)** 4H hydrogen, 2C carbon, 2O oxygen
(e) 9H hydrogen, 2N nitrogen, 1P phosphorus, 4O
oxygen **2.79 (a)** 1Ni nickel, 2Cl chlorine, 8O oxygen **(b)** 1C carbon,
1O oxygen, 2Cl chlorine **(c)** 2K potassium, 2Cr chromium,
7O oxygen **(d)** 2C carbon, 4H hydrogen, 2O oxygen **(e)** 2N nitrogen,
9H hydrogen, 1P phosphorus, 4 O oxygen **2.81 (a)** 14C, 28H,
14O **(b)** 4N, 8H, 2C, 2O **(c)** 15C, 40H, 15O **2.83 (a)** 6Na **(b)** 3C
(c) 27O **(d)** 2Fe **2.85** No, $S(CH_3)_2$, O_2, SO_2, CO_2, H_2O
2.87 Reaction is not balanced as written. **2.89 (a)** K^+ **(b)** Br^-
(c) Mg^{2+} **(d)** S^{2-} **(e)** Al^{3+} **2.91 (a)** NaBr **(b)** KI **(c)** BaO **(d)** $MgBr_2$
(e) Al_2S_3 **(f)** BaF_2 **2.93 (a)** $CoCl_2$ and $CoCl_3$ **(b)** $TiCl_2$, $TiCl_3$, and
$TiCl_4$ **(c)** $MnCl_2$ and $MnCl_3$ **2.95 (a)** KNO_3 **(b)** $Ca(C_2H_3O_2)_2$
(c) NH_4Cl **(d)** $Fe_2(CO_3)_3$ **(e)** $Mg_3(PO_4)_2$ **2.97 (a)** PbO and
PbO_2 **(b)** SnO and SnO_2 **(c)** MnO and Mn_2O_3 **(d)** FeO
and Fe_2O_3 **(e)** Cu_2O and CuO **2.99 (a)** calcium sulfide
(b) aluminum bromide **(c)** sodium phosphide **(d)** barium
arsenide **(e)** rubidium sulfide **2.101 (a)** silicon dioxide
(b) xenon tetrafluoride **(c)** tetraphosphorus decoxide
(d) dichlorine heptoxide **2.103 (a)** iron(II) sulfide
(b) copper(II) oxide **(c)** tin(IV) oxide **(d)** cobalt(II) chloride
hexahydrate **2.105 (a)** sodium nitrite **(b)** potassium
permanganate **(c)** magnesium sulfate heptahydrate **(d)** potassium
thiocyanate **2.107 (a)** ionic, chromium(II) chloride **(b)** molecular,
disulfur dichloride **(c)** molecular, sulfur trioxide **(d)** ionic,
ammonium acetate **(e)** ionic, potassium iodate **(f)** molecular,
tetraphosphorus hexoxide **(g)** ionic, calcium sulfite **(h)** ionic,
silver cyanide **(i)** ionic, zinc(II) bromide **(j)** molecular, hydrogen
selenide **2.109 (a)** Na_2HPO_4 **(b)** Li_2Se **(c)** $Cr(C_2H_3O_2)_3$
(d) S_2F_{10} **(e)** $Ni(CN)_2$ **(f)** Fe_2O_3 **(g)** SbF_5 **2.111 (a)** $(NH_4)_2S$
(b) $Cr_2(SO_4)_3 \cdot 6H_2O$ **(c)** SiF_4 **(d)** MoS_2 **(e)** $SnCl_4$ **(f)** H_2Se
(g) P_4S_7 **2.113 (a)** Incorrect: Hg_2Cl_2 **(b)** Correct **(c)** Correct
(d) Incorrect: $Al_2(CO_3)_3$ **(e)** Incorrect: $KMnO_4$ **(f)** Incorrect:
CaF_2 **(g)** Incorrect: Na_2O **2.115 (a)** Incorrect: iron(II) sulfide,
iron(III) sulfide **(b)** Incorrect: sodium chloride **(c)** Incorrect:
calcium bromide **(d)** Correct: $Al_2(CO_3)_3$ **(e)** Correct: $Ca(OCl)_2$
(f) Incorrect: magnesium phosphate **(g)** Incorrect: potassium
sulfide **2.117** diselenium hexasulfide and diselenium tetrasulfide

Chapter 3

Practice Exercise

3.1 0.129 mol Al **3.2** 5.84 g I_2 **3.3** $\pm 1.15 \times 10^{-5}$ mol K_2SO_4
3.4 8.67×10^{22} atoms Au **3.5** 5.68×10^{-22} g **3.6** No
3.7 0.0516 mol Al^{3+} **3.8** 3.44 mol N atoms **3.9** 3.87 mol O atoms
3.10 59.6 g Fe **3.11** 0.116 g F **3.12** 36.84% N, 63.16% O
3.13 13.04% H, 52.17% C. It is likely that the compound contains

another element since the percentages do not add up to 100%.
3.14 63.65% N, 36.35% O **3.15** NO: 46.68% N, 53.32% O
NO_2: 30.45% N, 69.55% O N_2O_3: 36.86% N, 63.14% O
N_2O_4: 30.45% N, 69.55% O The compound N_2O_3 corresponds
to the data in Practice Exercise 3.12. **3.16** NH_2 **3.17** SO_2
3.18 NI_3 **3.19** Al_2O_3 **3.20** Na_2SO_4 **3.21** C_9H_8O **3.22** CS_2
3.23 CH_2O **3.24** N_2H_4 **3.25** $C_2H_4Cl_2$, $C_6H_6Cl_6$
3.26 $3Ba(NO_3)_2(aq) + 2(NH_4)_3PO_4(aq) \longrightarrow Ba_3(PO_4)_2(s) +$
$6NH_4NO_3(aq)$ **3.27** $CaCl_2(aq) + K_2CO_3(aq) \longrightarrow CaCO_3(s) +$
$2KCl(aq)$ **3.28** 3.38 mol O_2 **3.29** 0.183 mol H_2SO_4 **3.30** 78.5 g
Al_2O_3 **3.31** 1.18×10^2 g CO_2 **3.32** 55.0 g CO_2, 34 g HCl
remaining **3.33** 30.01 g NO **3.34** 36.78 g $HOOCC_6H_4O_2C_2H_3$,
83.5% **3.35** 30.9 g $HC_2H_3O_2$, 86.1% **3.36** Overall three-step yield
is 68.6%, and the two-step yield is 72.1%. The two-step process is
the preferred process.

Review Problems

3.31 (a) 75.4 g Fe **(b)** 392 g O **(c)** 35.1 g Ca **3.33 (a)** $NaHCO_3$:
84.0066 g/mol **(b)** $(NH_4)_2CO_3$: 96.0858 g/mol **(c)** $CuSO_4 \cdot 5H_2O$:
249.685 g/mole **(d)** potassium dichromate: 294.1846 g/mole
(e) aluminum sulfate: 342.151 g/mol **3.35 (a)** 388 g $Ca_3(PO_4)_2$
(b) 3.49×10^{-5} g C_4H_{10} **(c)** 0.151 g $Fe(NO_3)_3$ **(d)** 139 g $(NH_4)_2CO_3$
3.37 (a) 9.16×10^{-11} mol NH_3 **(b)** 4.31×10^{-8} mol Na_2CrO_4
(c) 0.215 mol $CaCO_3$ **(d)** 7.94×10^{-2} mol $Sr(NO_3)_2$ **3.39** 0.302
mol Ni **3.41** 2.59×10^{-3} mole Ta **3.43** 1.30×10^{-10} g K
3.45 3.01×10^{23} atoms C-12 **3.47 (a)** 6 atom C:11 atom H
(b) 12 mole C:11 mole O **(c)** 2 atom H:1 atom O **(d)** 2 mole H:
1 mole O **3.49** 1.05 mol Bi **3.51** 4.32 mol Cr

3.53 (a) $\left(\dfrac{2 \text{ mol Al}}{3 \text{ mol S}}\right)$ or $\left(\dfrac{3 \text{ mol S}}{2 \text{ mol Al}}\right)$

(b) $\left(\dfrac{3 \text{ mol S}}{1 \text{ mol Al}_2(SO_4)_3}\right)$ or $\left(\dfrac{1 \text{ mol Al}_2(SO_4)_3}{3 \text{ mol S}}\right)$ **(c)** 0.600 mol Al

(d) 3.48 mol S **3.55** 0.0725 mol N_2, 0.218 mol H_2 **3.57** 0.833 mol
UF_6 **3.59** 9.33×10^{23} atoms C **3.61** 3.76×10^{24} atoms
3.63 0.0750 mol Ca; 3.01 g Ca **3.65** 1.30 mol N, 62.5 g
$(NH_4)_2CO_3$ **3.67** 3.43 kg fertilizer **3.69 (a)** 12.2% N, 5.26% H,
26.9% P, 55.6% O **(b)** 62.0% C, 10.4% H, 27.6% O **(c)** 19.2% Na,
1.68% H, 25.8% P, 53.3% O **(d)** 23.3% Ca, 18.6% S, 55.7% O,
2.34% H **3.71** Heroin has a higher percentage oxygen. **3.73** Freon
141b has a higher percentage chlorine. **3.75** 22.9% P, 77.1% Cl
3.77 These data are consistent with the experimental values cited
in the problem. **3.79** 0.474 g O **3.81 (a)** SCl **(b)** CH_2O **(c)** NH_3
(d) AsO_3 **(e)** HO **3.83** $NaTcO_4$ **3.85** $C_9H_8O_2$ **3.87** CCl_2
3.89 C_2H_6O **3.91** $C_{19}H_{30}O_2$ **3.93 (a)** $Na_2S_4O_6$ **(b)** $C_6H_4Cl_2$
(c) $C_6H_3Cl_3$ **3.95** $C_{19}H_{30}O_2$ **3.97** HgBr, Hg_2Br_2 **3.99** CHNO,
$C_4H_4N_4O_4$ **3.101 (a)** $Mg(OH)_2 + 2HBr \longrightarrow MgBr_2 + 2H_2O$
(b) $2HCl + Ca(OH)_2 \longrightarrow CaCl_2 + 2H_2O$ **(c)** $Al_2O_3 + 3H_2SO_4$
$\longrightarrow Al_2(SO_4)_3 + 3H_2O$ **(d)** $2KHCO_3 + H_3PO_4 \longrightarrow K_2HPO_4 +$
$2H_2O + 2CO_2$ **(e)** $C_9H_{20} + 14O_2 \longrightarrow 9CO_2 + 10H_2O$
3.103 (a) $Ca(OH)_2 + 2HCl \longrightarrow CaCl_2 + 2H_2O$ **(b)** $2AgNO_3 +$
$CaCl_2 \longrightarrow Ca(NO_3)_2 + 2AgCl$ **(c)** $Pb(NO_3)_2 + Na_2SO_4 \longrightarrow$
$PbSO_4 + 2NaNO_3$ **(d)** $2Fe_2O_3 + 3C \longrightarrow 4Fe + 3CO_2$
3.105 $4NH_2CHO + 5O_2 \longrightarrow 4CO_2 + 6H_2O + 2N_2$
3.107

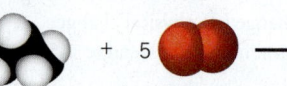 + 5  $\longrightarrow$ 3 + 4

3.109 $4Fe(s) + 3O_2(g) \longrightarrow 2Fe_2O_3(s)$ **3.111 (a)** 2 atoms of Ba,
20 atoms of O, and 36 atoms of H **(b)** 2 moles of Ba, 20 moles of O,

and 36 moles of H **3.113 (a)** 0.030 mol $Na_2S_2O_3$ **(b)** 0.24 mol HCl
(c) 0.15 mol H_2O **(d)** 0.15 mol H_2O **3.115 (a)** 552 g O_2
(b) 23.4 g CO_2 **(c)** 350. g H_2O **3.117 (a)** $4P + 5O_2 \longrightarrow P_4O_{10}$
(b) 8.85 g O_2 **(c)** 14.2 g P_4O_{10} **(d)** 3.26 g P **3.119** 30.28 g HNO_3
3.121 0.47 kg O_2 **3.123** 2 molecules of SO_2 **3.125 (a)** Fe_2O_3
(b) 195 g Fe **3.127** 26.7 g $FeCl_3$ **3.129** 0.91 mg HNO_3
3.131 66.98 g $BaSO_4$, 96.22% **3.133** 88.72% **3.135** 2.03 g MnF_3

Chapter 4

Practice Exercises

4.1 (a) $CaI_2(s) \longrightarrow Ca^{2+}(aq) + 2I^-(aq)$
  (b) $K_3PO_4(s) \longrightarrow 3K^+(aq) + PO_4^{3-}(aq)$
4.2 (a) $Al(NO_3)_3(s) \longrightarrow Al^{3+}(aq) + 3NO_3^-(aq)$
  (b) $Na_2CO_3(s) \longrightarrow 2Na^+(aq) + CO_3^{2-}(aq)$
4.3 Molecular: $(NH_4)_2SO_4(aq) + Ba(NO_3)_2(aq) \longrightarrow$
$$BaSO_4(s) + 2NH_4NO_3(aq)$$
  Ionic: $2NH_4^+(aq) + SO_4^{2-}(aq) + Ba^{2+}(aq) + 2NO_3^-(aq) \longrightarrow$
$$BaSO_4(s) + 2NH_4^+(aq) + 2NO_3^-(aq)$$
  Net ionic: $Ba^{2+}(aq) + SO_4^{2-}(aq) \longrightarrow BaSO_4(s)$
4.4 Molecular: $CdCl_2(aq) + Na_2S(aq) \longrightarrow CdS(s) + 2NaCl(aq)$
  Ionic: $Cd^{2+}(aq) + 2Cl^-(aq) + 2Na^+(aq) + S^{2-}(aq) \longrightarrow$
$$CdS(s) + 2Na^+(aq) + 2Cl^-(aq)$$
  Net ionic: $Cd^{2+}(aq) + S^{2-}(aq) \longrightarrow CdS(s)$
4.5 $HC_3H_7O_2(aq) \rightleftharpoons H_3O^+(aq) + C_3H_7O_2^-(aq)$
4.6 $HNO_3(aq) + H_2O \longrightarrow H_3O^+(aq) + NO_3^-(aq)$
  $HNO_2(aq) + H_2O \rightleftharpoons H_3O^+(aq) + NO_2^-(aq)$
4.7 $(C_2H_5)_3N(aq) + H_2O \rightleftharpoons (C_2H_5)_3NH^+(aq) + OH^-(aq)$
4.8

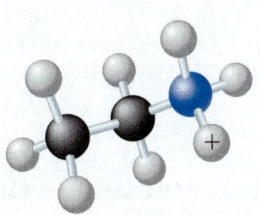

  $CH_3CH_2NH_2(aq) + H_2O \rightleftharpoons CH_3CH_2NH_3^+(aq) + OH^-(aq)$
4.9 $H_3C_6H_5O_7(s) + H_2O \rightleftharpoons H_3O^+(aq) + H_2C_6H_5O_7^-(aq)$
  $H_2C_6H_5O_7^-(aq) + H_2O \rightleftharpoons H_3O^+(aq) + HC_6H_5O_7^{2-}(aq)$
  $HC_6H_5O_7^{2-}(aq) + H_2O \rightleftharpoons H_3O^+(aq) + C_6H_5O_7^{3-}(aq)$
4.10 $H_2S(aq) + H_2O \rightleftharpoons H_3O^+(aq) + HS^-(aq)$
  $HS^-(aq) + H_2O \rightleftharpoons H_3O^+(aq) + S^{2-}(aq)$
4.11 HF: Hydrofluoric acid
  HI: Hydroiodic acid
4.12 Iodic acid
4.13 Molecular: $Zn(NO_3)_2(aq) + Ca(C_2H_3O_2)_2(aq) \longrightarrow$
$$Zn(C_2H_3O_2)_2(aq) + Ca(NO_3)_2(aq)$$
  Ionic: $Zn^{2+}(aq) + 2NO_3^-(aq) + Ca^{2+}(aq) + 2C_2H_3O_2^-(aq)$
  $\longrightarrow Zn^{2+}(aq) + 2C_2H_3O_2^-(aq) + Ca^{2+}(aq) + 2NO_3^-(aq)$
  Net ionic: No reaction
4.14 (a) Molecular: $AgNO_3(aq) + NH_4Cl(aq) \longrightarrow$
$$AgCl(s) + NH_4NO_3(aq)$$
  Ionic: $Ag^+(aq) + NO_3^-(aq) + NH_4^+(aq) + Cl^-(aq) \longrightarrow$
$$AgCl(s) + NH_4^+(aq) + NO_3^-(aq)$$
  Net ionic: $Ag^+(aq) + Cl^-(aq) \longrightarrow AgCl(s)$
  (b) Molecular: $Na_2S(aq) + Pb(C_2H_3O_2)_2(aq) \longrightarrow$
$$2NaC_2H_3O_2(aq) + PbS(s)$$
  Ionic: $2Na^+(aq) + S^{2-}(aq) + Pb^{2+}(aq) + 2C_2H_3O_2^-(aq)$
  $\longrightarrow 2Na^+(aq) + 2C_2H_3O_2^-(aq) + PbS(s)$
  Net ionic: $S^{2-}(aq) + Pb^{2+}(aq) \longrightarrow PbS(s)$

4.15 Molecular: $2HNO_3(aq) + Ca(OH)_2(aq) \longrightarrow$
$$Ca(NO_3)_2(aq) + 2H_2O$$
  Ionic: $2H^+(aq) + 2NO_3^-(aq) + Ca^{2+}(aq) + 2OH^-(aq) \longrightarrow$
$$Ca^{2+}(aq) + 2NO_3^-(aq) + 2H_2O$$
  Net ionic: $H^+(aq) + OH^-(aq) \longrightarrow H_2O$
4.16 (a) Molecular: $HCl(aq) + KOH(aq) \longrightarrow H_2O + KCl(aq)$
  Ionic: $H^+(aq) + Cl^-(aq) + K^+(aq) + OH^-(aq) \longrightarrow$
$$H_2O + K^+(aq) + Cl^-(aq)$$
  Net ionic: $H^+(aq) + OH^-(aq) \longrightarrow H_2O$
  (b) Molecular: $2HCl(aq) + Ba(OH)_2(aq) \longrightarrow$
$$2H_2O + BaCl_2(aq)$$
  Ionic: $2H^+(aq) + 2Cl^-(aq) + Ba^{2+}(aq) + 2OH^-(aq) \longrightarrow$
$$2H_2O + Ba^{2+}(aq) + 2Cl^-(aq)$$
  Net ionic: $H^+(aq) + OH^-(aq) \longrightarrow H_2O$
4.17 Molecular: $CH_3NH_2(aq) + HCHO_2(aq) \longrightarrow$
$$CH_3NH_3CHO_2(aq)$$
  Ionic: $CH_3NH_2(aq) + HCHO_2(aq) \longrightarrow$
$$CH_3NH_3^+(aq) + CHO_2^-(aq)$$
  Net ionic: $CH_3NH_2(aq) + HCHO_2(aq) \longrightarrow$
$$CH_3NH_3^+(aq) + CHO_2^-(aq)$$
4.18 $NaHSO_3$, sodium hydrogen sulfite
4.19 $H_3AsO_4(aq) + NaOH(aq) \longrightarrow NaH_2AsO_4(aq) + H_2O$
$$\text{sodium dihydrogen arsenate}$$
  $NaH_2AsO_4(aq) + NaOH(aq) \longrightarrow Na_2HAsO_4(aq) + H_2O$
$$\text{sodium hydrogen arsenate}$$
  $Na_2HAsO_4(aq) + NaOH(aq) \longrightarrow Na_3AsO_4(aq) + H_2O$
$$\text{sodium arsenate}$$
  The acid is completely neutralized in the last step.
4.20 Molecular: $2HCHO_2(aq) + Co(OH)_2(s) \longrightarrow$
$$Co(CHO_2)_2(aq) + 2H_2O$$
  Ionic: $2HCHO_2(aq) + Co(OH)_2(s) \longrightarrow$
$$2CHO_2^-(aq) + Co^{2+}(aq) + 2H_2O$$
  Net ionic: $2HCHO_2(aq) + Co(OH)_2(s) \longrightarrow$
$$2CHO_2^-(aq) + Co^{2+}(aq) + 2H_2O$$
4.21 Molecular: $MgCl_2(aq) + (NH_4)_2SO_4(aq) \longrightarrow$
$$MgSO_4(aq) + 2NH_4Cl(aq)$$
  Ionic: $Mg^{2+}(aq) + 2Cl^-(aq) + 2NH_4^+(aq) + SO_4^{2-}(aq) \longrightarrow$
$$Mg^{2+}(aq) + 2Cl^-(aq) + 2NH_4^+(aq) + SO_4^{2-}(aq)$$
  Net ionic: No reaction
4.22 $CuO(s) + 2HNO_3(aq) \longrightarrow Cu(NO_3)_2(aq) + H_2O$
  or
  $Cu(OH)_2(s) + 2HNO_3(aq) \longrightarrow Cu(NO_3)_2(aq) + 2H_2O$
4.23 You want to use a metathesis reaction that produces CoS, which is insoluble, and a second product that is soluble. You may want the reactants to be soluble.
  $CoCl_2(aq) + Na_2S(aq) \longrightarrow CoS(s) + 2NaCl(aq)$
4.24 0.500 M LiBr **4.25** 0.133 M KI **4.26** 284 mL $NaNO_3$ solution **4.27** 0.0438 mol HCl, 1.60 g HCl **4.28** 2.11 g $Sr(NO_3)_2$
4.29 0.531 g $AgNO_3$ **4.30** 1.065 g Na_2SO_4 **4.31** 250.0 mL
4.32 600 mL of water **4.33** 31.6 mL H_3PO_4 **4.34** 26.8 mL
NaOH **4.35** 0.40 M Fe^{3+}, 1.2 M Cl^- **4.36** 0.750 M Na^+
4.37 0.0449 M $CaCl_2$ **4.38** 60.0 mL KOH **4.39** 0.605 g
Na_2SO_4 **4.40** 0.385 g Na_2CO_3 **4.41** 0.178 M H_2SO_4
4.42 0.0220 M HCl, 0.0802% HCl **4.43** 30.0% $CaCl_2$, 70% $MgCl_2$

Review Problems

4.43 (a) strong electrolyte **(b)** nonelectrolyte **(c)** strong electrolyte
(d) nonelectrolyte
4.45 (a) $LiF(aq) \longrightarrow Li^+(aq) + F^-(aq)$
  (b) $KMnO_4(aq) \longrightarrow K^+(aq) + MnO_4^-(aq)$

(c) $K_2CO_3(aq) \longrightarrow 2K^+(aq) + CO_3{}^{2-}(aq)$
(d) $Co(NO_3)_2(aq) \longrightarrow Co^{2+}(aq) + 2NO_3{}^-(aq)$
4.47 (a) Ionic: $2NH_4{}^+(aq) + CO_3{}^{2-}(aq) + Ba^{2+}(aq) + 2Cl^-(aq)$
$\longrightarrow 2NH_4{}^+(aq) + 2Cl^-(aq) + BaCO_3(s)$
Net: $Ba^{2+}(aq) + CO_3{}^{2-}(aq) \longrightarrow BaCO_3(s)$
(b) Ionic: $Cu^{2+}(aq) + 2Cl^-(aq) + 2Na^+(aq) + 2OH^-(aq) \longrightarrow$
$Cu(OH)_2(s) + 2Na^+(aq) + 2Cl^-(aq)$
Net: $Cu^{2+}(aq) + 2OH^-(aq) \longrightarrow Cu(OH)_2(s)$
(c) Ionic: $3Fe^{2+}(aq) + 3SO_4{}^{2-}(aq) + 6Na^+(aq) + 2PO_4{}^{3-}(aq)$
$\longrightarrow Fe_3(PO_4)_2(s) + 6Na^+(aq) + 3SO_4{}^{2-}(aq)$
Net: $3Fe^{2+}(aq) + 2PO_4{}^{3-}(aq) \longrightarrow Fe_3(PO_4)_2(s)$
(d) Ionic: $2Ag^+(aq) + 2C_2H_3O_2{}^-(aq) + Ni^{2+}(aq) + 2Cl^-(aq)$
$\longrightarrow 2AgCl(s) + Ni^{2+}(aq) + 2C_2H_3O^-(aq)$
Net: $2Ag^+(aq) + 2Cl^-(aq) \longrightarrow 2AgCl(s)$
4.49 Na^+ and Cl^-, $Co^{2+}(aq) + 2OH^-(aq) \longrightarrow Co(OH)_2(s)$
4.51 $HClO_4(aq) + H_2O \longrightarrow H_3O^+(aq) + ClO_4{}^-(aq)$
4.53 $HI(g) + H_2O \longrightarrow H_3O^+(aq) + I^-(aq)$
4.55 $N_2H_4(aq) + H_2O \rightleftharpoons N_2H_5{}^+(aq) + OH^-(aq)$
4.57 $HNO_2(aq) + H_2O \rightleftharpoons H_3O^+(aq) + NO_2{}^-(aq)$
4.59 $H_2CO_3(aq) + H_2O \rightleftharpoons H_3O^+(aq) + HCO_3{}^-(aq)$
$HCO_3{}^-(aq) + H_2O \rightleftharpoons H_3O^+(aq) + CO_3{}^{2-}(aq)$
4.61 (a) $H_2S(g)$: hydrogen sulfide **(b)** $H_2S(aq)$: hydrosulfuric acid
4.63 (a) perbromic acid **(b)** bromic aicd **(c)** bromous acid
(d) hypobromous acid **(e)** hydrobromic acid
4.65 (a) perbromate **(b)** bromate **(c)** bromite **(d)** hypobromite
(e) bromide
4.67 (a) H_2CrO_4 **(b)** H_2CO_3 **(c)** $H_2C_2O_4$
4.69 (a) sodium bicarbonate or sodium hydrogen carbonate
(b) potassium dihydrogen phosphate **(c)** ammonium hydrogen
phosphate
4.71 (a) hypochlorous acid sodium hypochlorite $NaOCl$
(b) iodous acid sodium iodite $NaIO_2$
(c) perchloric acid sodium perchlorate $NaClO_4$
4.73 H_2SO_4 **4.75** sodium butyrate **4.77** formic acid **4.79 (a)**,
(b), and **(d)**
4.81 (a) Ionic: $3Fe^{2+}(aq) + 3SO_4{}^{2-}(aq) + 6K^+(aq) + 2PO_4{}^{3-}(aq)$
$\longrightarrow Fe_3(PO_4)_2(s) + 6K^+(aq) + 3SO_4{}^{2-}(aq)$
Net: $3Fe^{2+}(aq) + 2PO_4{}^{3-}(aq) \longrightarrow Fe_3(PO_4)_2(s)$
(b) Ionic: $3Ag^+(aq) + 3C_2H_3O_2{}^-(aq) + Al^{3+}(aq) + 3Cl^-(aq)$
$\longrightarrow 3AgCl(s) + Al^{3+}(aq) + 3C_2H_3O_2{}^-(aq)$
Net: $Ag^+(aq) + Cl^-(aq) \longrightarrow AgCl(s)$
(c) Ionic: $2Cr^{3+}(aq) + 6Cl^-(aq) + 3Ba^{2+}(aq) + 6OH^-(aq)$
$\longrightarrow 2Cr(OH)_3(s) + 3Ba^{2+}(aq) + 6Cl^-(aq)$
Net: $Cr^{3+}(aq) + 3OH^-(aq) \longrightarrow Cr(OH)_3(s)$
4.83 Molecular: $NaCl(aq) + AgNO_3(aq) \longrightarrow$
$AgCl(s) + NaNO_3(aq)$
Ionic: $Na^+(aq) + Cl^-(aq) + Ag^+(aq) + NO_3{}^-(aq) \longrightarrow$
$AgCl(s) + Na^+(aq) + NO_3{}^-(aq)$
Net: $Cl^-(aq) + Ag^+(aq) \longrightarrow AgCl(s)$
4.85 Molecular: $Na_2S(aq) + Cu(NO_3)_2(aq) \longrightarrow$
$CuS(s) + 2NaNO_3(aq)$
Ionic: $2Na^+(aq) + S^{2-}(aq) + Cu^{2+}(aq) + 2NO_3{}^-(aq) \longrightarrow$
$CuS(s) + 2Na^+(aq) + 2NO_3{}^-(aq)$
Net: $Cu^{2+}(aq) + S^{2-}(aq) \longrightarrow CuS(s)$
4.87 (a) Molecular: $Ca(OH)_2(aq) + 2HNO_3(aq) \longrightarrow$
$Ca(NO_3)_2(aq) + 2H_2O$
Ionic: $Ca^{2+}(aq) + 2OH^-(aq) + 2H^+(aq) + 2NO_3{}^-(aq)$
$\longrightarrow Ca^{2+}(aq) + 2NO_3{}^-(aq) + 2H_2O$
Net: $H^+(aq) + OH^-(aq) \longrightarrow H_2O$
(b) Molecular: $Al_2O_3(s) + 6HCl(aq) \longrightarrow$
$2AlCl_3(aq) + 3H_2O$

Ionic: $Al_2O_3(s) + 6H^+(aq) + 6Cl^-(aq) \longrightarrow$
$2Al^{3+}(aq) + 6Cl^-(aq) + 3H_2O$
Net: $Al_2O_3(s) + 6H^+(aq) \longrightarrow 2Al^{3+}(aq) + 3H_2O$
(c) Molecular: $Zn(OH)_2(aq) + H_2SO_4(aq) \longrightarrow$
$ZnSO_4(aq) + 2H_2O$
Ionic: $Zn(OH)_2(s) + 2H^+(aq) + SO_4{}^{2-}(aq) \longrightarrow$
$Zn^{2+}(aq) + SO_4{}^{2-}(aq) + 2H_2O$
Net: $Zn(OH)_2(s) + 2H^+(aq) \longrightarrow Zn^{2+}(aq) + 2H_2O$
4.89 The electrical conductivity would decrease regularly, until one
solution had neutralized the other, forming a nonelectrolyte:
$Ba^{2+}(aq) + 2OH^-(aq) + 2H^+(aq) + SO_4{}^{2-}(aq)$
$\longrightarrow BaSO_4(s) + 2H_2O$

After neutralization had been reached, excess sulfuric acid would
cause the conductivity to increase, because sulfuric acid is a strong
electrolyte itself.
4.91 (a) $2H^+(aq) + CO_3{}^{2-}(aq) \longrightarrow H_2O + CO_2(g)$
(b) $NH_4{}^+(aq) + OH^-(aq) \longrightarrow NH_3(g) + H_2O$
4.93 $Na_2S(aq) + 2HCl(aq) \longrightarrow 2NaCl(aq) + H_2S(g)$
Net: $S^{2-}(aq) + 2H^+(aq) \longrightarrow H_2S(g)$
4.95 (a) formation of insoluble $Cr(OH)_3$
(b) formation of water, a weak electrolyte
4.97 (a) Molecular: $3HNO_3(aq) + Cr(OH)_3(s) \longrightarrow$
$Cr(NO_3)_3(aq) + 3H_2O$
Ionic: $3H^+(aq) + 3NO_3{}^-(aq) + Cr(OH)_3(s) \longrightarrow$
$Cr^{3+}(aq) + 3NO_3{}^-(aq) + 3H_2O$
Net: $3H^+(aq) + Cr(OH)_3(s) \longrightarrow Cr^{3+}(aq) + 3H_2O$
(b) Molecular: $HClO_4(aq) + NaOH(aq) \longrightarrow$
$NaClO_4(aq) + H_2O$
Ionic: $H^+(aq) + ClO_4{}^-(aq) + Na^+(aq) + OH^-(aq) \longrightarrow$
$Na^+(aq) + ClO_4{}^-(aq) + H_2O$
Net: $H^+(aq) + OH^-(aq) \longrightarrow H_2O$
(c) Molecular: $Cu(OH)_2(s) + 2HC_2H_3O_2(aq) \longrightarrow$
$Cu(C_2H_3O_2)_2(aq) + 2H_2O$
Ionic: $Cu(OH)_2(s) + 2H^+(aq) + 2C_2H_3O_2{}^-(aq) \longrightarrow$
$Cu^{2+}(aq) + 2C_2H_3O_2{}^-(aq) + 2H_2O$
Net: $Cu(OH)_2(s) + 2H^+(aq) \longrightarrow Cu^{2+}(aq) + 2H_2O$
(d) Molecular: $ZnO(s) + H_2SO_4(aq) \longrightarrow ZnSO_4(aq) + H_2O$
Ionic: $ZnO(s) + 2H^+(aq) + SO_4{}^{2-}(aq) \longrightarrow$
$Zn^{2+}(aq) + SO_4{}^{2-}(aq) + H_2O$
Net: $ZnO(s) + 2H^+(aq) \longrightarrow Zn^{2+}(aq) + H_2O$
4.99 (a) Molecular: $Na_2SO_3(aq) + Ba(NO_3)_2(aq) \longrightarrow$
$BaSO_3(s) + 2NaNO_3(aq)$
Ionic: $2Na^+(aq) + SO_3{}^{2-}(aq) + Ba^{2+}(aq) + 2NO_3{}^-(aq)$
$\longrightarrow BaSO_3(s) + 2Na^+(aq) + 2NO_3{}^-(aq)$
Net: $Ba^{2+}(aq) + SO_3{}^{2-}(aq) \longrightarrow BaSO_3(s)$
(b) Molecular: $2HCHO_2(aq) + K_2CO_3(aq) \longrightarrow$
$CO_2(g) + H_2O + 2KCHO_2(aq)$
Ionic: $2HCHO_2(aq) + 2K^+(aq) + CO_3{}^{2-}(aq) \longrightarrow$
$CO_2(g) + H_2O + 2K^+(aq) + 2CHO_2{}^-(aq) + 2CHO_2{}^-(aq)$
Net: $2HCHO_2(aq) + CO_3{}^{2-}(aq) \longrightarrow$
$CO_2(g) + H_2O + 2CHO_2{}^-(aq)$
(c) Molecular: $2NH_4Br(aq) + Pb(C_2H_3O_2)_2(aq) \longrightarrow$
$2NH_4C_2H_3O_2(aq) + PbBr_2(s)$
Ionic: $2NH_4{}^+(aq) + 2Br^-(aq) + Pb^{2+}(aq) + 2C_2H_3O_2{}^-(aq)$
$\longrightarrow 2NH_4{}^+(aq) + 2C_2H_3O_2{}^-(aq) + PbBr_2(s)$
Net: $Pb^{2+}(aq) + 2Br^-(aq) \longrightarrow PbBr_2(s)$
(d) Molecular: $2NH_4ClO_4(aq) + Cu(NO_3)_2(aq) \longrightarrow$
$Cu(ClO_4)_2(aq) + 2NH_4NO_3(aq)$
Ionic: $2NH_4{}^+(aq) + 2ClO_4{}^-(aq) + Cu^{2+}(aq) + 2NO_3{}^-(aq)$
$\longrightarrow Cu^{2+}(aq) + 2ClO_4{}^-(aq) + 2NO_3{}^-(aq) + 2NH_4{}^+(aq)$
Net: No reaction

4.101 There are numerous possible answers. One of many possible sets of answers would be:
(a) $NaHCO_3(aq) + HCl(aq) \longrightarrow NaCl(aq) + CO_2(g) + H_2O$
(b) $FeCl_2(aq) + 2NaOH(aq) \longrightarrow Fe(OH)_2(s) + 2NaCl(aq)$
(c) $Ba(NO_3)_2(aq) + K_2SO_3(aq) \longrightarrow BaSO_3(s) + 2KNO_3(aq)$
(d) $2AgNO_3(aq) + Na_2S(aq) \longrightarrow Ag_2S(s) + 2NaNO_3(aq)$
(e) $ZnO(s) + 2HCl(aq) \longrightarrow ZnCl_2(aq) + H_2O$
4.103 (a) $2.0 \, M \, H_2SO_4$ **(b)** $0.56 \, M \, NaNO_3$
4.105 (a) $1.00 \, M \, NaOH$ **(b)** $0.577 \, M \, CaCl_2$ **4.107** 658 mL
$NaC_2H_3O_2$ **4.109 (a)** 1.46 g NaCl **(b)** 16.2 g $C_2H_{12}O_6$
(c) 6.13 g H_2SO_4 **4.111** $0.11 \, M \, H_2SO_4$ **4.113** 300 mL
4.115 225 mL water **4.117 (a)** 0.0147 mol Ca^{2+},
0.0294 mol Cl^- **(b)** 0.0204 mol Al^{3+}, 0.0612 mol Cl^-
4.119 (a) $0.25 \, M \, Cr^{2+}$, $0.50 \, M \, NO_3^-$ **(b)** $0.10 \, M \, Cu^{2+}$,
$0.10 \, M \, SO_4^{2-}$ **(c)** $0.48 \, M \, Na^+$, $0.16 \, M \, PO_4^{3-}$ **(d)** $0.15 \, Al^{3+}$,
$0.23 \, M \, SO_4^{2-}$ **4.121** 1.07 g $Al_2(SO_4)_3$ **4.123** 11.9 mL $NiCl_2$
soln, 0.39 g $NiCO_3$ **4.125** 0.113 M KOH, $KOH(aq) + HCl(aq)$
$\longrightarrow KCl(aq) + H_2O$ **4.127** 0.969 g $Al_2(SO_4)_3$
4.129 2.00 mL $FeCl_3$ soln, 0.129 g AgCl **4.131** 13.3 mL $AlCl_3$
4.133 $0.167 \, M \, Fe^{3+}$, 3.67 g **4.135** $0.114 \, M$ HCl
4.137 (a) 2.67×10^{-3} mol $HC_3H_5O_3$ **(b)** 0.240 g $HC_3H_5O_3$
4.139 (a) 3.56×10^{-3} mol Pb **(b)** 0.7386 g Pb **(c)** 48.40% Pb in
the sample **(d)** Yes

Chapter 5

Practice Exercises

5.1 Oxygen is reduced since it gains electrons. **5.2** Aluminum is oxidized and is, therefore, the reducing agent. Chlorine is reduced and is, therefore, the oxidizing agent. **5.3** Cl +3 **5.4 (a)** Ni +2; Cl −1 **(b)** Mg +2; Ti +4; O −2 **(c)** K +1; Cr +6; O −2 **(d)** H +1; P +5; O −2 **(e)** V +3; C 0; H +1; O −2 **(f)** N −3; H +1
5.5 $KClO_3 + 3HNO_2 \longrightarrow KCl + 3HNO_3$ is a redox reaction. $KClO_3$ is reduced and HNO_2 is oxidized. **5.6** Cl_2 is reduced and is the oxidizing agent. $NaClO_2$ is oxidized and is the reducing agent. **5.7** $2Al(s) + 3Cu^{2+}(aq) \longrightarrow 2Al^{3+}(aq) + 3Cu(s)$
5.8 $3Sn^{2+} + 16H^+ + 2TcO_4^- \longrightarrow 2Tc^{4+} + 8H_2O + 3Sn^{4+}$
5.9 $4OH^- + SO_2 \longrightarrow SO_4^{2-} + 2e^- + 2H_2O$ **5.10** $2MnO_4^- + 3C_2O_4^{2-} + 4OH^- \longrightarrow 2MnO_2 + 6CO_3^{2-} + 2H_2O$
5.11 $Zn \longrightarrow Zn^{2+} + 2e^-$; $2H^+ + 2e^- \longrightarrow H_2$ **5.12 (a)** molecular: $Mg(s) + 2HCl(aq) \longrightarrow MgCl_2(aq) + H_2(g)$; ionic: $Mg(s) + 2H^+(aq) + 2Cl^-(aq) \longrightarrow Mg^{2+}(aq) + 2Cl^-(aq) + H_2(g)$; net ionic: $Mg(s) + 2H^+(aq) \longrightarrow Mg^{2+}(aq) + H_2(g)$ **(b)** molecular: $2Al(s) + 6HCl(aq) \longrightarrow 2AlCl_3(aq) + 3H_2(g)$; ionic: $2Al(s) + 6H^+(aq) + 6Cl^-(aq) \longrightarrow 2Al^{3+}(aq) + 6Cl^-(aq) + 3H_2(g)$; net ionic: $2Al(s) + 6H^+(aq) \longrightarrow 2Al^{3+}(aq) + 3H_2(g)$ **5.13** $Cu^{2+}(aq) + Mg(s) \longrightarrow Cu(s) + Mg^{2+}(aq)$ **5.14 (a)** $2Al(s) + 3Cu^{2+}(aq) \longrightarrow 2Al^{3+}(aq) + 3Cu(s)$ **(b)** $Ag(s) + Mg^{2+}(aq) \longrightarrow$ No reaction **5.15** $C_4H_4S(s) + 6O_2(g) \longrightarrow 4CO_2(g) + 2H_2O + SO_2(g)$ **5.16** $C_3H_8O_2(l) + 4O_2(g) \longrightarrow 3CO_2(g) + 4H_2O$ **5.17** $2Sr(s) + O_2(g) \longrightarrow 2SrO(s)$ **5.18** $4Fe(s) + 3O_2(g) \longrightarrow 2Fe_2O_3(s)$

Review Problems

5.29 (a) ClO_4^-: Cl +7, O −2 **(b)** Cl^-: Cl −1 **(c)** SF_6: S +6, F −1 **(d)** $Au(NO_3)_3$: Au +3, N +5, O −2 **5.31 (a)** Na +1; O −2; Br +1 **(b)** Na +1; O −2; Br +3 **(c)** Na +1; O −2; Br +5 **(d)** Na +1; O −2; Br +7 **5.33 (a)** S −2; Bi +3 **(b)** Cl −1; Ce +4 **(c)** Cs +1; O −1/2 **(d)** F −1; O +1 **5.35** Ti +3; N −3 **5.37** Oxidation number of O is 0 **5.39 (a)** substance reduced (and oxidizing agent): HNO_3 substance oxidized (and reducing agent): H_3AsO_3

(b) substance reduced (and oxidizing agent): HOCl substance oxidized (and reducing agent): NaI **(c)** substance reduced (and oxidizing agent): $KMnO_4$ substance oxidized (and reducing agent): $H_2C_2O_4$ **(d)** substance reduced (and oxidizing agent): H_2SO_4 substance oxidized (and reducing agent): Al **5.41** In the forward direction: Cl_2 is reduced and oxidized. In the reverse direction: Cl^- is the reducing agent and HOCl is the oxidizing agent. **5.43 (a)** $OCl^- + 2S_2O_3^{2-} + 2H^+ \longrightarrow S_4O_6^{2-} + Cl^- + H_2O$ **(b)** $2NO_3^- + Cu + 4H^+ \longrightarrow 2NO_2 + Cu^{2+} + 2H_2O$ **(c)** $IO_3^- + 3H_3AsO_3 \longrightarrow I^- + 3H_3AsO_4$ **(d)** $Zn + SO_4^{2-} + 4H^+ \longrightarrow Zn^{2+} + SO_2 + 2H_2O$ **5.45 (a)** $3Sn + 4NO_3^- + 4H^+ \longrightarrow 3SnO_2 + 4NO + 2H_2O$ **(b)** $PbO_2 + 4Cl^- + 4H^+ \longrightarrow PbCl_2 + Cl_2 + 2H_2O$ **(c)** $Ag + 2H^+ + NO_3^- \longrightarrow Ag^+ + NO_2 + H_2O$ **(d)** $4Fe^{3+} + 2NH_3OH^+ \longrightarrow 4Fe^{2+} + N_2O + 6H^+ + H_2O$ **5.47 (a)** $2CrO_4^{2-} + 3S^{2-} + 4H_2O \longrightarrow 2CrO_2^- + 3S + 8OH^-$ **(b)** $3C_2O_4^{2-} + 2MnO_4^- + 4H_2O \longrightarrow 6CO_2 + 2MnO_2 + 8OH^-$ **(c)** $4ClO_3^- + 3N_2H_4 \longrightarrow 4Cl^- + 6NO + 6H_2O$ **(d)** $3SO_3^{2-} + 2MnO_4^- + H_2O \longrightarrow 3SO_4^{2-} + 2MnO_2 + 2OH^-$ **5.49** $10H^+(aq) + NO_3^-(aq) + 4Mg(s) \longrightarrow NH_4^+(aq) + 4Mg^{2+}(aq) + 3H_2O(aq)$ **5.51** $H_2O + C \longrightarrow CO + H_2$ **5.53** $3H_2C_2O_4 + K_2Cr_2O_7 + 8H^+ \longrightarrow 6CO_2 + 2K^+ + 2Cr^{3+} + 7H_2O$ **5.55** $O_3 + Br^- \longrightarrow BrO_3^-$ **5.57 (a)** Molecular: $Mn(s) + 2HCl(aq) \longrightarrow MnCl_2(aq) + H_2(g)$; Ionic: $Mn(s) + 2H^+(aq) + 2Cl^-(aq) \longrightarrow Mn^{2+}(aq) + 2Cl^-(aq) + H_2(g)$; Net Ionic: $Mn(s) + 2H^+(aq) \longrightarrow Mn^{2+}(aq) + H_2(g)$ **(b)** M: $Cd(s) + 2HCl(aq) \longrightarrow CdCl_2(aq) + H_2(g)$; I: $Cd(s) + 2H^+(aq) + 2Cl^-(aq) \longrightarrow Cd^{2+}(aq) + 2Cl^-(aq) + H_2(g)$; Net Ionic: $Cd(s) + 2H^+(aq) \longrightarrow Cd^{2+}(aq) + H_2(g)$ **(c)** M: $Sn(s) + 2HCl(aq) \longrightarrow SnCl_2(aq) + H_2(g)$; I: $Sn(s) + 2H^+(aq) + 2Cl^-(aq) \longrightarrow Sn^{2+}(aq) + 2Cl^-(aq) + H_2(g)$; Net Ionic: $Sn(s) + 2H^+(aq) \longrightarrow Sn^{2+}(aq) + H_2(g)$ **5.59 (a)** Molecular: $Ni(s) + H_2SO_4(aq) \longrightarrow NiSO_4(aq) + H_2(g)$; Ionic: $Ni(s) + 2H^+(aq) + SO_4^{2-}(aq) \longrightarrow Ni^{2+}(aq) + SO_4^{2-}(aq) + H_2(g)$; Net Ionic: $Ni(s) + 2H^+(aq) \longrightarrow Ni^{2+}(aq) + H_2(g)$ **(b)** Molecular: $2Cr(s) + 3H_2SO_4(aq) \longrightarrow Cr_2(SO_4)_3(aq) + 3H_2(g)$; Ionic: $2Cr(s) + 6H^+(aq) + 3SO_4^{2-}(aq) \longrightarrow 2Cr^{3+}(aq) + 3SO_4^{2-}(aq) + 3H_2(g)$; Net Ionic: $2Cr(s) + 6H^+(aq) \longrightarrow 2Cr^{3+}(aq) + 3H_2(g)$ **5.61 (a)** $3Ag(s) + 4HNO_3(aq) \longrightarrow 3AgNO_3(aq) + 2H_2O + NO(g)$ **(b)** $Ag(s) + 2HNO_3(aq) \longrightarrow AgNO_3(aq) + H_2O + NO_2(aq)$ **5.63** Net ionic equation: $Cu(s) + H_2SO_4(aq) + 2H^+(aq) \longrightarrow Cu^{2+}(aq) + SO_2(g) + 2H_2O$; Molecular equation: $Cu(s) + 2H_2SO_4(aq) \longrightarrow CuSO_4(aq) + SO_2(g) + 2H_2O$ **5.65 (a)** $Mn(s) + Fe^{2+}(aq) \longrightarrow Mn^{2+}(aq) + Fe(s)$ **(b)** N.R. **(c)** $Mg(s) + Co^{2+}(aq) \longrightarrow Mg^{2+}(aq) + Co(s)$ **(d)** $2Cr(s) + 3Sn^{2+}(aq) \longrightarrow 2Cr^{3+}(aq) + 3Sn(s)$ **5.67** Pt, Ru, Tl, Pu **5.69** $Cd(s) + 2TlCl(aq) \longrightarrow CdCl_2(aq) + 2Tl(s)$ **5.71** $Mg(s) + Zn^{2+}(aq) + 2Cl^-(aq) \longrightarrow Mg^{2+}(aq) + 2Cl^-(aq) + Zn(s)$ **5.73** $2Li(s) + 2H_2O \longrightarrow H_2(g) + 2OH^-(aq) + 2Li^-(aq)$ **5.75 (a)** $2C_6H_6(l) + 15O_2(g) \longrightarrow 12CO_2(g) + 6H_2O(g)$ **(b)** $2C_4H_{10}(g) + 13O_2(g) \longrightarrow 8CO_2(g) + 10H_2O(g)$ **(c)** $C_{21}H_{44}(s) + 32O_2(g) \longrightarrow 21CO_2(g) + 22H_2O(g)$ **5.77 (a)** $2C_6H_6(l) + 9O_2(g) \longrightarrow 12CO(g) + 6H_2O(g)$; $2C_4H_{10}(g) + 9O_2(g) \longrightarrow 8CO(g) + 10H_2O(g)$; $2C_{21}H_{44}(s) + 43O_2(g) \longrightarrow 42CO(g) + 44H_2O(g)$ **(b)** $2C_6H_6(l) + 3O_2(g) \longrightarrow 12C(s) + 6H_2O(g)$; $2C_4H_{10}(g) + 5O_2(g) \longrightarrow 8C(s) + 10H_2O(g)$; $C_{21}H_{44}(s) + 11O_2(g) \longrightarrow 21C(s) + 22H_2O(g)$ **5.79** $C_6H_{12}O_6(s) + 6O_2(g) \longrightarrow 6CO_2(g) + 6H_2O(g)$ **5.81** $2(CH_3)_2S(g) + 9O_2(g) \longrightarrow 4CO_2(g) + 6H_2O(g) + 2SO_2(g)$ **5.83 (a)** $2Zn(s) + O_2(g) \longrightarrow 2ZnO(s)$ **(b)** $4Al(s) + 3O_2(g) \longrightarrow 2Al_2O_3(s)$ **(c)** $2Mg(s) + O_2(g) \longrightarrow 2MgO(s)$ **(d)** $4Fe(s) + 3O_2(g) \longrightarrow 2Fe_2O_3(s)$ **5.85** 3.53 g Cu **5.87 (a)** $2MnO_4^- + 5Sn^{2+} + 16H^+ \longrightarrow 2Mn^{2+} + 5Sn^{4+} + 8H_2O$ **(b)** 17.4 mL $KMnO_4$ **5.89 (a)** $IO_3^- + 3SO_3^{2-} \longrightarrow I^- + 3SO_4^{2-}$ **(b)** 9.55 g Na_2SO_3

5.91 (a) $6.38 \times 10^{-3}\,M\,I_3^-$ **(b)** 1.01×10^{-3} g SO_2
(c) 2.10×10^{-3} % **(d)** 21 ppm **5.93 (a)** 9.463% **(b)** 18.40%
5.95 (a) 0.02994 g H_2O_2 **(b)** 2.994% H_2O_2 **5.97 (a)** $2CrO_4^{2-} + 3SO_3^{2-} + H_2O \longrightarrow 2CrO_2^- + 3SO_4^{2-} + 2OH^-$ **(b)** 0.875 g Cr
(c) 25.4% Cr **5.99 (a)** 5.405×10^{-3} mol $C_2O_4^{2-}$ **(b)** 0.5999 g
$CaCl_2$ **(c)** 24.35% $CaCl_2$

Chapter 6

Practice Exercises

6.1 5.9×10^5 J, 590 kJ **6.2** 160 kcal, 1.6×10^5 cal, 670 kJ, 6.7×10^5 J **6.3** 7 °C **6.4** 55.1 J °C^{-1} **6.5** 5330 J, 5.33 kJ, 1280 cal, 1.28 kcal
6.6 128 °C **6.7** exothermic **6.8** endothermic **6.9** second run
6.10 second run **6.11** 14.6 kJ °C^{-1} **6.12** -5640 kJ mol^{-1}, -3.94 Cal/g
6.13 -58 kJ mol^{-1} NaOH **6.14** $HC_2H_3O_2 + OH^- \longrightarrow H_2O + C_2H_3O_2^-$, $\Delta H = 57$ kJ/mol $HC_2H_3O_2$ **6.15** $3CH_4(g) + 6O_2(g) \longrightarrow 3CO_2(g) + 6H_2O(l)$ $\Delta H = -2671.5$ kJ
6.16 $^{2.5}N_2(g) + ^{7.5}H_2(g) \longrightarrow 2.5NH_3(g)$ $\Delta H = -115.5$ kJ
6.17

2Cu(s) + O$_2$(g)

−310 kJ

−169 kJ

Cu$_2$O(s) + ½O$_2$(g)

−141 kJ

2CuO(s)

Exothermic

6.18

NO(g) + ½O$_2$(g)

+90.4 kJ

−56.6 kJ

NO$_2$(g)

½N$_2$(g) + O$_2$(g)

+33.8 kJ

Endothermic

6.19 $N_2O(g) + \frac{1}{2}O_2(g) \longrightarrow 2NO(g)$ $\Delta H = +99.2$ kJ
6.20 $C_2H_4(g) + H_2O(l) \longrightarrow C_2H_5OH(l)$ $\Delta H° = -44.0$ kJ
6.21 385 kJ **6.22** 2.62×10^6 kJ **6.23** $K(s) + \frac{1}{2}Br_2(l) \longrightarrow KBr(s)$ $\Delta H_f° = -393.8$ kJ **6.24** $Na(s) + \frac{1}{2}H_2(g) + C(s) + \frac{3}{2}O_2(g) \longrightarrow NaHCO_3(s)$ $\Delta H_f° = -947.7$ kJ/mol
6.25 -11.0 kJ **6.26** $+98.3$ kJ **6.27 (a)** $= -113.1$ kJ
(b) -177.8 kJ

Review Problems

6.49 Four-fold increase. Decreases 9 fold **6.51** 3.4×10^{-21} J
6.53 -1320 J, -315 cal **6.55** 45.7 g H_2O **6.57 (a)** 1.67×10^3 J
(b) 1.67×10^3 J **(c)** 2.32 J °C^{-1} **(d)** 0.464 J g^{-1} °C^{-1}
6.59 25.12 J mol^{-1} °C^{-1} **6.61** -30.4 kJ **6.63** -53 kJ mol^{-1}
6.65 (a) $C_3H_8(g) + 5O_2(g) \longrightarrow 3CO_2(g) + 4H_2O(l)$ **(b)** 2.22×10^5 J
(c) -222 kJ/mol **6.67** -17 J **6.69** $+100$ J **6.71** 250 J
6.73 (a) -283 kJ **(b)** $\frac{2}{3}NH_3(g) + \frac{7}{6}O_2(g) \longrightarrow \frac{2}{3}NO_2(g) + H_2O(g)$ $\Delta H° = -189$ kJ **6.75** 162 kJ **6.77** 8.64 g
6.79

Ge(s) + O$_2$(g)

−534.7 kJ

−255 kJ

GeO(s) + ½O$_2$(g)

−280 kJ

GeO$_2$(s)

6.81 $2NO_2(g) \longrightarrow N_2O_4(g)$ $\Delta H° = -57.93$ kJ
$2NO(g) + O_2(g) \longrightarrow 2NO_2(g)$ $\Delta H° = -113.14$ kJ

Adding, we have:
$2NO(g) + O_2(g) \longrightarrow N_2O_4(g)$ $\Delta H° = -171.07$ kJ
6.83 $\Delta H° = -57.27$ kJ **6.85** $\Delta H° = -53$ kJ **6.87** $\Delta H° = -463$ kJ
6.89 Only **(c)** satisfies this requirement.
6.91 (a) $2C(s, graphite) + 2H_2(g) + O_2(g) \longrightarrow HC_2H_3O_2(l)$
$\Delta H_f° = -487.0$ kJ
(b) $2C(s, graphite) + \frac{1}{2}O_2(g) + 3H_2(g) \longrightarrow C_2H_5OH(l)$
$\Delta H_f° = -277.63$ kJ
(c) $Ca(s) + \frac{1}{8}S_8(s) + 3O_2(g) + 2H_2(g) \longrightarrow CaSO_4 \cdot 2H_2O(s)$
$\Delta H_f° = -2021.1$ kJ
(d) $2Na(s) + \frac{1}{8}S_8(s) + 2O_2(g) \longrightarrow Na_2SO_4(s)$
$\Delta H_f° = -1384.5$ kJ
6.93 (a) -196.6 kJ **(b)** -177.8 kJ **6.95** -2.21×10^3 kJ

Chapter 7

Practice Exercises

7.1 5.10×10^{14} Hz **7.2** 3.25 m **7.3** 2.63 μm, infrared region
7.4 656.5 nm, red **7.5** 7.57×10^{-20} J **7.6** 1st energy level
7.7 6.01×10^{-20} J, 1.51×10^{-20} J **7.8** 660 nm **7.9 (a)** 4, 2
(b) 5, 3 **(c)** 7, 0

7.10 When $n = 2$, $\ell = 0, 1$, the s, and p subshells
When $n = 5$, $\ell = 0, 1, 2, 3, 4$, the s, p, d, f and g subshells
The number of subshells is equal to the value of n.
7.11 10 electrons **7.12** 50 electrons

7.13 (a) Na

1s 2s 2p 3s

(b) S

1s 2s 2p 3s 3p

(c) Ar

1s 2s 2p 3s 3p

7.14 (a) Si $1s^2 2s^2 2p^6 3s^1 3p^2$ **(b)** Ti $1s^2 2s^2 p^6 3s^2 3p^6 4s^2 3d^2$
(c) Se $1s^2 2s^2 2p^6 3s^2 3p^6 4s^2 3d^{10} 4p^4$
7.15 Yes
7.16 (a) 0 unpaired electrons
(b) 2 unpaired electrons
(c) 0 unpaired electrons
(d) 8 unpaired electrons

7.17 (a) Mg $1s^2 2s^2 2p^6 3s^2$
(b) Ge $1s^2 2s^2 2p^6 3s^2 3p^6 3d^{10} 4s^2 4p^2$
(c) Cd $1s^2 2s^2 2p^6 3s^2 3p^6 3d^{10} 4s^2 4p^6 4d^{10} 5s^2$
(d) Gd $1s^2 2s^2 2p^6 3s^2 3p^6 3d^{10} 4s^2 4p^6 4d^{10} 4f^7 5s^2 5p^6 5d^1 6s^2$
7.18 (a) O $1s^2 2s^2 2p^4$
    S $1s^2 2s^2 2p^6 3s^2 3p^4$
    Se $1s^2 2s^2 2p^6 3s^2 3p^6 3d^{10} 4s^2 4p^4$
(b) P $1s^2 2s^2 2p^6 3s^2 3p^3$
    N $1s^2 2s^2 2p^3$
    Sb $1s^2 2s^2 2p^6 3s^2 3p^6 3d^{10} 4s^2 4p^6 4d^{10} 5s^2 5p^3$

The elements have the same number of electrons in the valence shell, and the only differences between the valence shells are the energy levels.

7.19 (a) Zr   [Kr] $4d^2 5s^2$

[Kr] (↑)(↑)(○)(○) (↑↓)
      4d   5s

(b) Po   [Xe] $4f^{14} 5d^{10} 6s^2 6p^4$

[Xe] (↑↓)(↑↓)(↑↓)(↑↓)(↑↓)(↑↓)(↑↓) (↑↓)(↑↓)(↑↓)(↑↓)(↑↓)
              4f   5d

(↑↓) (↑↓)(↑)(↑)
 6s   6p

7.20 (a) P   [Ne]$3s^2 3p^3$

[Ne] (↑↓) (↑)(↑)(↑)
      3s   3p

3 unpaired electrons

(b) Sn   [Kr]$4d^{10} 5s^2 5p^2$

[Kr] (↑↓)(↑↓)(↑↓)(↑↓)(↑↓) (↑↓) (↑)(↑)(○)
              4d   5s   5p

2 unpaired electrons

7.21 Based on the definition of valence, there are no examples where more than 8 electrons would occupy the valence shell. For representative elements the valence shell is defined as the occupied shell with the highest value of n. In the ground state atom, only s and p electrons fit this definition. The transition elements have outer electron configurations: $(n-1)d^a ns^b$, so the valence shell is the ns subshell.

7.22 (a) Se $4s^2 4p^4$ 6 valence electrons
(b) Sn $5s^2 5p^2$ 4 valence electrons
(c) I $5s^2 5p^5$ 7 valence electrons
7.23 (a) Sn **(b)** Ga **(c)** Cr **(d)** S^{2-} **7.24 (a)** P **(b)** Fe^{3+}
(c) Fe **(d)** Cl^-
7.25 (a) Be **(b)** C **7.26 (a)** C^{2+} **(b)** Mg^{2+}

Review Problems

7.79 6.88×10^{14} Hz **7.81** 1.02×10^{15} Hz **7.83** 2.965 m
7.85 2.65×10^{-19} J/photon, 1.6×10^5 J mol^{-1} **7.87 (a)** violet
(b) 7.307×10^{14} s^{-1} **(c)** 4.842×10^{-19} J **7.89** 1094 nm; infrared region; not visible **7.91** 1.74×10^{-6} m, 1.14×10^{-19} J; infrared region **7.93 (a)** p **(b)** f **7.95 (a)** 3 **(b)** 2 **7.97 (a)** $n = 3$, $\ell = 0$
(b) $n = 5$, $\ell = 2$ **7.99** 0, 1, 2, 3, 4, or 5 **7.101** 8 **7.103 (a)** 1, 0, or -1 **(b)** 3, 2, 1, 0, -1, -2, or -3 **7.105** When $m_\ell = -4$ the minimum value of ℓ is 4 and the minimum value of n is 5.
7.107

n	ℓ	m_ℓ	m_s
2	1	-1	$+1/2$
2	1	-1	$-1/2$
2	1	0	$+1/2$
2	1	0	$-1/2$
2	1	$+1$	$+1/2$
2	1	$+1$	$-1/2$

7.109 21 electrons have $\ell = 1$
4 electrons have $m_\ell = 2$
7.111 (a) S $1s^2 2s^2 2p^6 3s^2 3p^4$
(b) K $1s^2 2s^2 2p^6 3s^2 3p^6 4s^1$

(c) Ti $1s^2 2s^2 2p^6 3s^2 3p^6 3d^2 4s^2$
(d) Sn $1s^2 2s^2 2p^6 3s^2 3p^6 3d^{10} 4s^2 4p^6 4d^{10} 5s^2 5p^2$
7.113 (a) Mn: paramagnetic **(b)** As: paramagnetic **(c)** S: paramagnetic
(d) Sr: not paramagnetic **(e)** Ar: not paramagnetic
7.115 (a) Mg: 0 unpaired electrons **(b)** P: 3 unpaired electrons
(c) V: 3 unpaired electrons
7.117 (a) Ni   [Ar]$3d^8 4s^2$
(b) Cs   [Xe]$6s^1$
(c) Ge   [Ar]$3d^{10} 4s^2 4p^2$
(d) Br   [Ar]$3d^{10} 4s^2 4p^5$
(e) Bi   [Xe]$4f^{14} 5d^{10} 6s^2 6p^3$
7.119 (a) Mg

(↑↓) (↑↓) (↑↓)(↑↓)(↑↓) (↑↓)
 1s   2s   2p   3s

(b) Ti

(↑↓) (↑↓) (↑↓)(↑↓)(↑↓) (↑↓) (↑↓)(↑↓)(↑↓) (↑↓)
 1s   2s   2p   3s   3p   4s

(↑)(↑)(○)(○)(○)
      3d

7.121 (a) Ni

[Ar] (↑↓) (↑↓)(↑↓)(↑↓)(↑)(↑)
      4s   3d

(b) Cs

[Xe] (↑)
      6s

(c) Ge

[Ar] (↑↓) (↑↓)(↑↓)(↑↓)(↑↓)(↑↓) (↑)(↑)(○)
      4s   3d   4p

(d) Br

[Ar] (↑↓) (↑↓)(↑↓)(↑↓)(↑↓)(↑↓) (↑↓)(↑↓)(↑)
      4s   3d   4p

7.123 (a) 5 **(b)** 4 **(c)** 4 **(d)** 6 **7.125 (a)** Na $3s^1$ **(b)** Al $3s^2 3p^1$
(c) Ge $4s^2 4p^2$ **(d)** P $3s^2 3p^3$

7.127 (a) Na:

(↑)
3s

(b) Al

(↑↓) (↑)(○)(○)
 3s   3p

(c) Ge

(↑↓) (↑)(↑)(○)
 4s   4p

(d) P

(↑↓) (↑)(↑)(↑)
 3s   3p

7.129 (a) $+1$ **(b)** $+6$ **(c)** $+7$ **7.131 (a)** Mg **(b)** Bi
7.133 As < Ge < Sn < Sb **7.135 (a)** Na **(b)** Co^{2+} **(c)** Cl^-
7.137 (a) N **(b)** S **(c)** Cl **7.139 (a)** Br **(b)** As **7.141** Mg

Chapter 8

Practice Exercises

8.1 (a) CaBr$_2$ (b) LiF (c) CaO

8.2

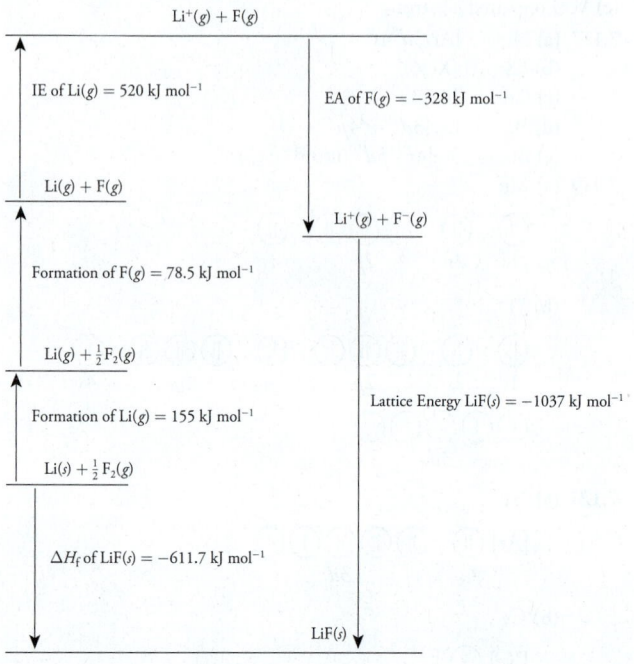

Li$^+$(g) + F(g)

IE of Li(g) = 520 kJ mol^{-1}

EA of F(g) = −328 kJ mol^{-1}

Li(g) + F(g)

Li$^+$(g) + F$^-$(g)

Formation of F(g) = 78.5 kJ mol^{-1}

Li(g) + $\frac{1}{2}$F$_2$(g)

Formation of Li(g) = 155 kJ mol^{-1}

Li(s) + $\frac{1}{2}$F$_2$(g)

Lattice Energy LiF(s) = −1037 kJ mol^{-1}

ΔH_f of LiF(s) = −611.7 kJ mol^{-1}

LiF(s)

8.3 Indium should lose its 5p^1 electron to have an electron configuration of 1$s^1$2$s^2$2$p^6$3$s^2$3$p^6$3d^{10}4$s^2$4$p^6$4d^{10}5s^2

8.4 Cr: [Ar]3$d^5$4s^1

  (a) Cr^{2+}: [Ar]3d^4 The 4s electron and one 3d electron are lost.

  (b) Cr^{3+}: [Ar]3d^3 The 4s electron and two 3d electrons are lost.

  (c) Cr^{6+}: [Ar]   The 4s electron and all of the 3d electrons are lost.

8.5 The electron configurations are identical, [Ar].

8.6

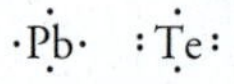

·P̈b· :T̈e:

They are in the fourth and sixth groups respectively

8.7

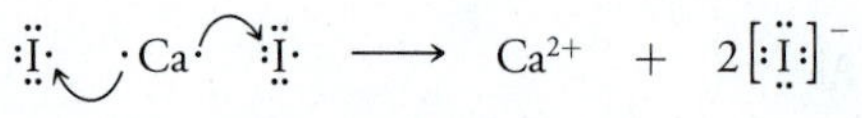

:Ï: ·Ca· :Ï· ⟶ Ca^{2+} + 2[:Ï:]$^-$

8.8

:Ö: ·Mg· ⟶ Mg^{2+} + [:Ö:]$^{2-}$

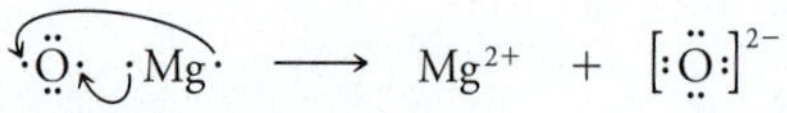

8.9 (a) 2 bonds (b) 3 bonds (c) 4 bonds

8.10 Both F atoms need an additional pair of electrons.
The O with a double bond needs two more electrons.
The C with a double bond needs to lose two electrons.

8.11 1.24 D

8.12 0.795 e^-
79.5% positive charge on the Na and 79.5% negative charge on the Cl.

8.13 The bond is polar and the Cl carries the negative charge.

8.14 (a) Br (b) Cl (c) Cl
S—Cl < P—Br < Si—Cl

8.15

O
H O P O H
O

Number of valence electrons:
O   6 each and 4 O atoms total 24 electrons
P   5 electrons
H   1 each and 2 H atoms total 2 electrons
Negative charge 1 electron
Total 32 valence electrons

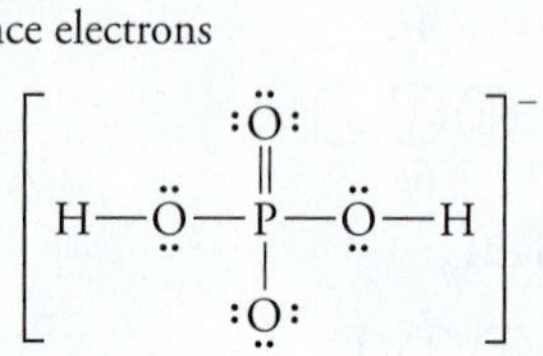

$$\left[\begin{array}{c} :\ddot{O}: \\ \| \\ H-\ddot{O}-P-\ddot{O}-H \\ | \\ :\ddot{O}: \end{array}\right]^-$$

8.16

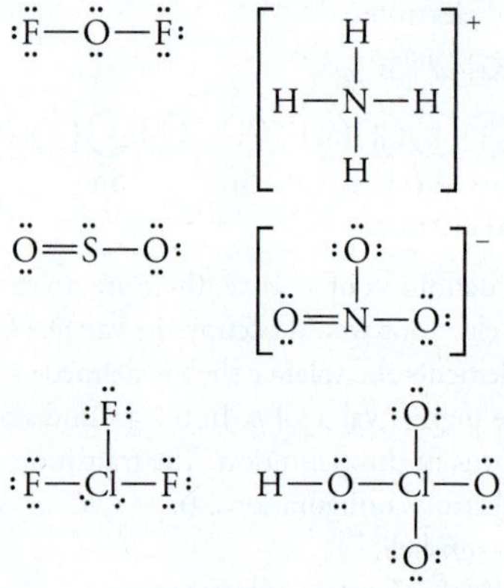

:F̈—Ö—F̈: $\left[\begin{array}{c} H \\ | \\ H-N-H \\ | \\ H \end{array}\right]^+$

Ö=S̈—Ö: $\left[\begin{array}{c} :\ddot{O}: \\ | \\ \ddot{O}=N-\ddot{O}: \end{array}\right]^-$

:F̈:   :Ö:
:F̈—Cl—F̈:   H—Ö—Cl—Ö:
                   :Ö:

8.17 F—Ö—F $\left[\begin{array}{c} H \\ | \\ H-N-H \\ | \\ H \end{array}\right]^+$

Ö=S̈—Ö: $\left[\begin{array}{c} :\ddot{O}: \\ | \\ \ddot{O}=N-\ddot{O}: \end{array}\right]^-$

:F̈:   :Ö:
:F̈—Cl—F̈:   H—Ö—Cl—Ö:
                   :Ö:

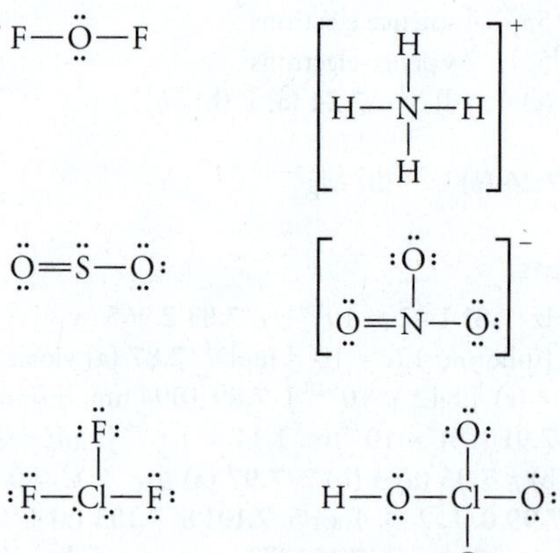

8.18 Two of the oxygen atoms should have a single bond and three lone pairs and the sulfur should have one double bond, two single bonds, and a lone pair. This is the best formal structure. The single bonded oxygens each have a 1− charge and the sulfur and double bonded oxygen have a zero formal charge.

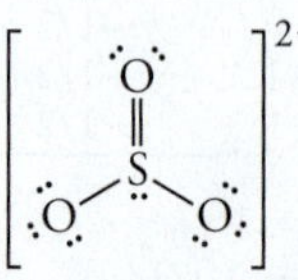

$$\left[\begin{array}{c} :\ddot{O}: \\ \ddot{S} \\ :\ddot{O}: \quad \ddot{O}: \end{array}\right]^{2-}$$

8.19 (a) :N̈—N≡O: (b) $\left[\ddot{S}=C=\ddot{N}\right]^-$
     (−2) (+1) (+1)   (0) (0) (−1)

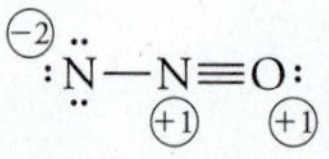

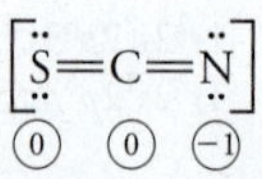

8.20 (a) SO_2

$$\ddot{O}=\ddot{S}=\ddot{O}$$

(b) $HClO_3$

$$H-\ddot{O}-\overset{\overset{\ddot{O}}{\|}}{Cl}=\ddot{O}:$$

(c) H_3PO_4

$$H-\overset{:\ddot{O}-H}{\underset{:\ddot{O}:}{\overset{|}{\underset{|}{P}}}}-\ddot{O}-H$$

8.21 $H-\overset{\overset{H}{|}}{\underset{H}{N}}: \; + \; H^+ \longrightarrow \left[H-\overset{\overset{H}{|}}{\underset{H}{N}}-H \right]^+$

There is no difference between the coordinate covalent bond and the other covalent bonds.

8.22 $H-\ddot{O}:^- \; + \; H^+ \longrightarrow H-\ddot{O}-H$

coordinate covalent bond

8.23 There are four resonance structures.

$$\left[:\overset{:\ddot{O}:}{\underset{:\ddot{O}:}{\overset{|}{\underset{|}{\ddot{O}-P-\ddot{O}:}}}} \right]^{3-} \longleftrightarrow \left[\overset{:\ddot{O}:}{\underset{:\ddot{O}:}{\overset{\|}{\underset{|}{\ddot{O}=P-\ddot{O}:}}}} \right]^{3-} \longleftrightarrow$$

$$\left[\overset{:\ddot{O}:}{\underset{:\ddot{O}:}{\overset{|}{\underset{\|}{\ddot{O}-P-\ddot{O}:}}}} \right]^{3-} \longleftrightarrow \left[\overset{:\ddot{O}:}{\underset{:\ddot{O}:}{\overset{|}{\underset{\|}{:\ddot{O}-P=\ddot{O}}}}} \right]^{3-}$$

8.24 $\left[\overset{:\ddot{O}:}{\underset{}{\overset{\|}{\ddot{O}=C-\ddot{O}:}}} \right]^- \longleftrightarrow \left[\overset{:\ddot{O}:}{\underset{}{\overset{|}{:\ddot{O}-C=\ddot{O}}}} \right]^-$

8.25

$$\left[\overset{:\ddot{O}:}{\underset{}{\overset{\|}{\ddot{O}=Br-\ddot{O}:}}} \right]^- \longleftrightarrow \left[:\ddot{O}=Br=\ddot{O} \right]^- \longleftrightarrow \left[\overset{:\ddot{O}:}{\underset{}{\overset{\|}{:\ddot{O}-Br=\ddot{O}}}} \right]^-$$

8.26

$CH_3-CH_2-\overset{\overset{:O:}{\|}}{C}-H$ aldehyde

$CH_3-\overset{\overset{H}{|}}{N}-CH_3$ amine

$H-\overset{\overset{:O:}{\|}}{C}-\ddot{O}-H$ acid

$CH_3-CH_2-\overset{\overset{:O:}{\|}}{C}-CH_2-CH_3$ ketone

$CH_3-CH_2-CH_2-\ddot{O}-H$ alcohol

8.27 (a) CH_3NHCH_3 **(b)** $HCOOH$

(c) $\overset{\overset{:\ddot{O}:}{\|}}{\underset{H}{C}}\overset{}{\underset{\ddot{O}:^\ominus}{}}$ **(d)** $\left[H_3C-\overset{\overset{H}{|}}{\underset{H}{N}}-CH_3 \right]^+$

Review Problems

8.63 (a) Al_2O_3 **(b)** BeO **(c)** $NaCl$

8.65

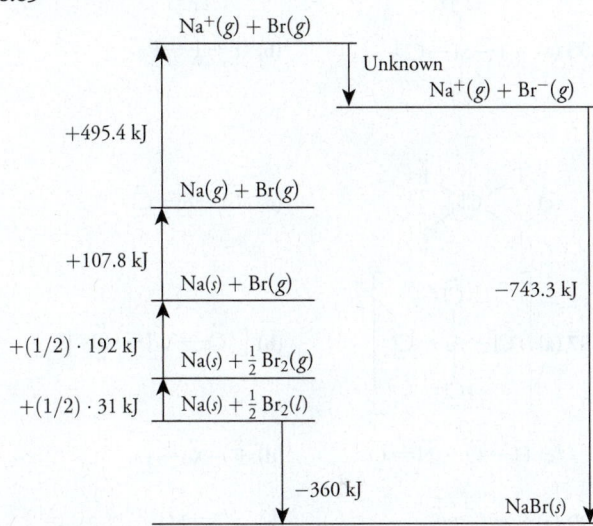

Electron affinity for $Br = -331.4$ kJ/mol

8.67 Magnesium loses two electrons.
Bromine gains an electron.
To keep the overall change of the formula unit neutral, two Br^- ions combine with one Mg^{2+} ion to form $MgBr_2$.

8.69 $[Xe]\,4f^{14}5d^{10}6s^2$, $[Xe]4f^{14}5d^{10}$

8.71 $[Ar]3d^4$, 4 unpaired electrons

8.73 (a) $\cdot\overset{\cdot}{Si}\cdot$ **(b)** $:\overset{\cdot}{Sb}\cdot$ **(c)** $Ba\cdot$ **(d)** $\cdot\overset{\cdot}{Al}\cdot$ **(e)** $:\overset{\cdot}{Sb}:$

8.75 (a) $\cdot Mg\overset{\frown}{\cdot}\overset{\cdot\cdot}{S}\overset{\cdot}{_{\cdot}} \longrightarrow Mg^{2+}\cdot \quad \left[:\overset{\cdot\cdot}{\underset{\cdot\cdot}{S}}:\right]^{2-}$

(b) $:\overset{\cdot}{Cl}\overset{\curvearrowleft}{\cdot}\; \cdot Mg\cdot \;\overset{\curvearrowright}{:}\overset{\cdot}{Cl}\overset{\cdot}{\cdot} \longrightarrow Mg^{2+} + 2\left[:\overset{\cdot\cdot}{\underset{\cdot\cdot}{Cl}}:\right]^-$

(c)

$\overset{\frown}{\cdot Mg}\overset{\frown}{\cdot}\overset{\cdot}{N}\overset{\frown}{_{\cdot}}\overset{\frown}{\cdot Mg}\overset{\frown}{\cdot}\overset{\cdot}{N}\overset{\frown}{\cdot Mg}\cdot \longrightarrow 3\,Mg^{2+} + 2\left[:\overset{\cdot\cdot}{\underset{\cdot\cdot}{N}}:\right]^{3-}$

8.77 4.029×10^{-19} J/molecule

8.79 344 nm; ultraviolet region

8.81 (a) $:\overset{\cdot\cdot}{\underset{\cdot\cdot}{Br}}\overset{\curvearrowleft\curvearrowright}{\cdot}\; \cdot\overset{\cdot\cdot}{\underset{\cdot\cdot}{Br}}: \longrightarrow :\overset{\cdot\cdot}{\underset{\cdot\cdot}{Br}}-\overset{\cdot\cdot}{\underset{\cdot\cdot}{Br}}:$

(b) $H\cdot\overset{\curvearrowright}{} \;\cdot\overset{\cdot\cdot}{\underset{\cdot\cdot}{O}}\cdot\; \overset{\curvearrowleft}{}\cdot H \longrightarrow H-\ddot{O}-H$

(c) $H\cdot\overset{\curvearrowright}{}\; \cdot\overset{\cdot\cdot}{\underset{}{N}}\cdot\; \overset{\curvearrowleft}{}\cdot H \longrightarrow H-\overset{\overset{\cdot\cdot}{}}{\underset{\underset{H}{|}}{N}}-H$

$H\cdot\overset{\curvearrowright}{_{\cdot}}$

8.83 (a) 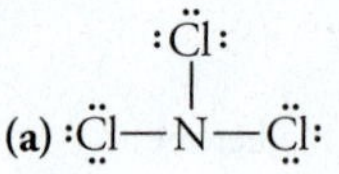 :Cl̈—N̈—Cl̈: (with Cl̈: above N) **(b)** 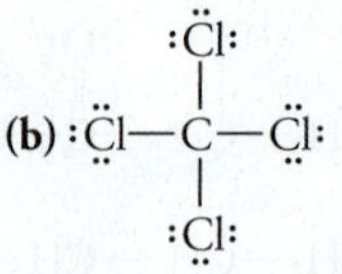 :Cl̈—C—Cl̈: (with :Cl̈: above and :Cl̈: below C)

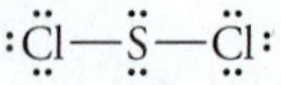

(c) :Cl̈—S̈—Cl̈: **(d)** 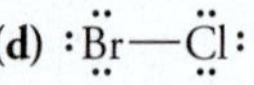 :B̈r—Cl̈:

8.85 (a) PCl_3 **(b)** CF_4 **(c)** ICl
8.87 $0.029\ e^-$
The nitrogen atom is positive.
8.89 $0.645\ e^-$
This is 64.5% of a positive charge on Cs and 64.5% of a negative charge on F.
8.91 (a) S **(b)** Si **(c)** Br **(d)** C
8.93 N—S

8.95 (a) 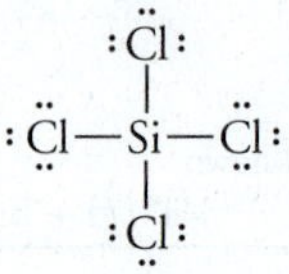 :Cl̈—Si—Cl̈: (with :Cl̈: above and :Cl̈: below Si) **(b)** 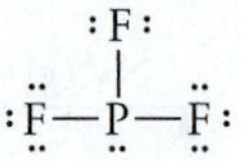 :F̈—P—F̈: (with :F̈: above P)

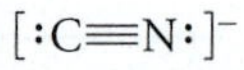

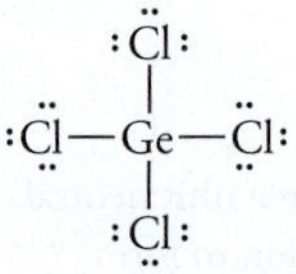

(c) 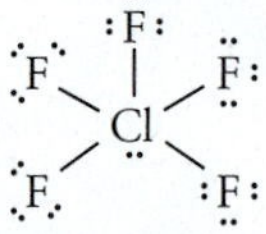 F Cl with six F's (ClF₆ octahedral arrangement) **(d)** 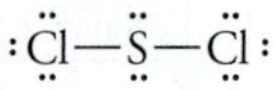 :Cl̈—S̈—Cl̈:

8.97 (a) 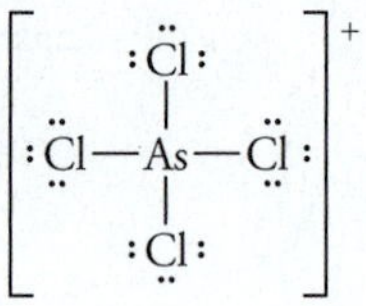 [:Cl̈—As—Cl̈: with :Cl̈: above and :Cl̈: below]⁺ **(b)** 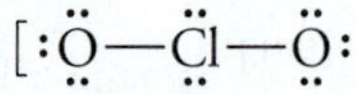 [:Ö—Cl̈—Ö:]⁻

(c) 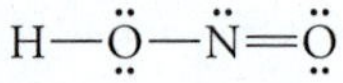 H—Ö—N̈=Ö **(d)** 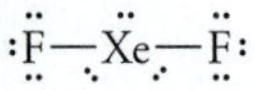 :F̈—Xe—F̈:

8.99 (a) 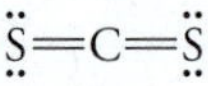 S̈=C=S̈ **(b)** [:C≡N:]⁻

(c) :Cl̈—Ge—Cl̈: (with :Cl̈: above and :Cl̈: below) **(d)** 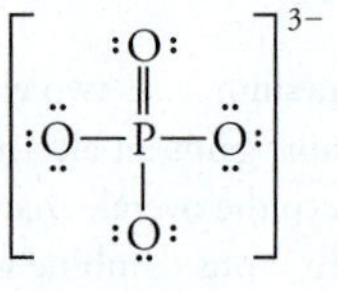 [:Ö—P—Ö: with :O: above (double bond) and :O: below]³⁻

8.101 (a) 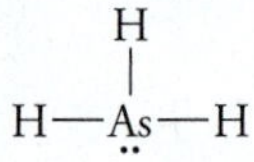 H—As—H (with H above) **(b)** 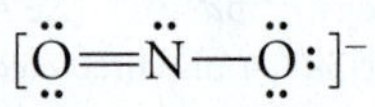 [Ö=N̈—Ö:]⁻

(c) 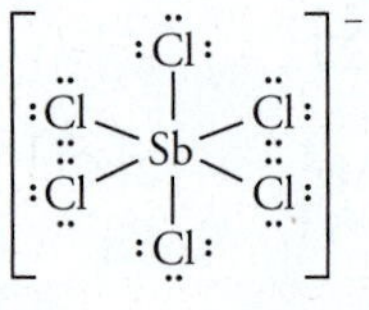 [:Cl̈—Sb—Cl̈: with :Cl̈ above, Cl̈: right, :Cl̈ left, :Cl̈: below]⁻ **(d)** 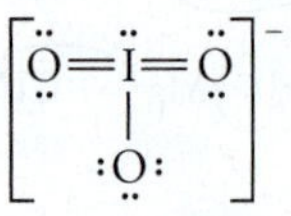 [Ö=I=Ö with :O: below]⁻

8.103 (a) 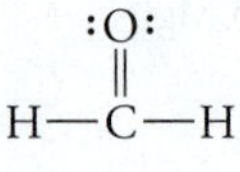 H—C—H (with :O: double bond above) **(b)** 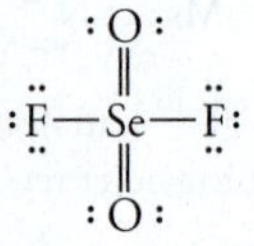 :F̈—Se—F̈: (with :O: above and :O: below)

8.105 (a) 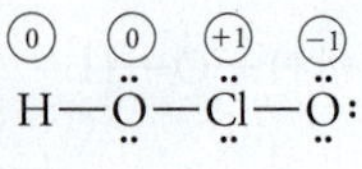 H—Ö—Cl̈—Ö: with formal charges ⓪ ⓪ (+1) (−1)

(b) 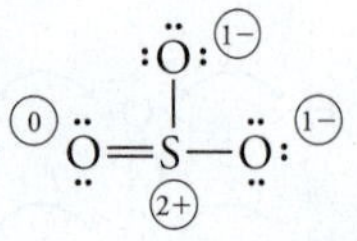 Ö=S—Ö: with formal charges ⓪ on O=, (2+) on S, (1−) on Ö: ; :Ö:(1−) above

(c) 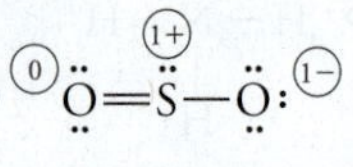 ⓪ Ö=S—Ö:(1−) with (1+) above S

8.107

:Ö:(1−) ... (1−):Ö—Cl—Ö—H(0) with (3+) on Cl, (0) on middle O, :Ö:(1−) below

:O: ... :Ö—Cl—Ö—H with :O: below (double bonds)

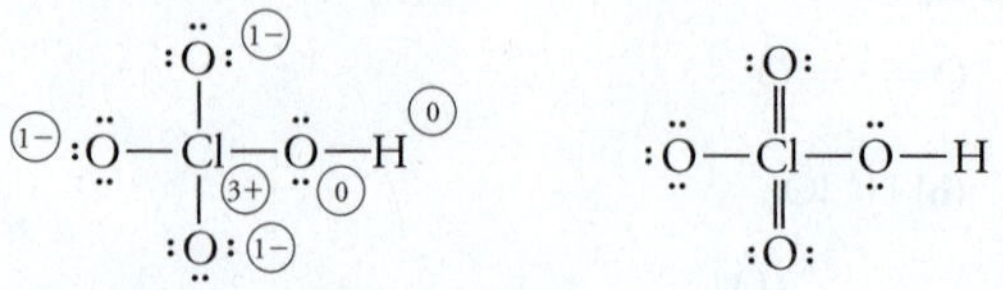

The preferred structure is the one on the right.
8.109 (1) The formal charges on all of the atoms of the left structure are zero; therefore, the potential energy of this molecule is lower and it is more stable. (2) Group 2 atoms normally have only two bonds.

8.111 H—Ö—H + H⁺ ⟶ [H—Ö—H with H below]⁺

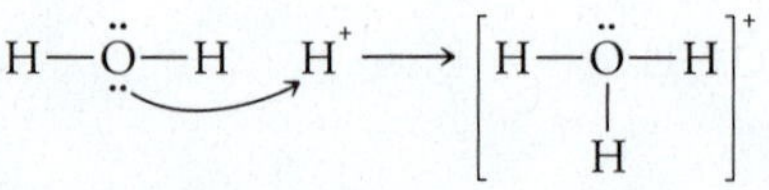

8.113

:O: ... N—N ... (four resonance structures with O's arranged around the N—N bond)

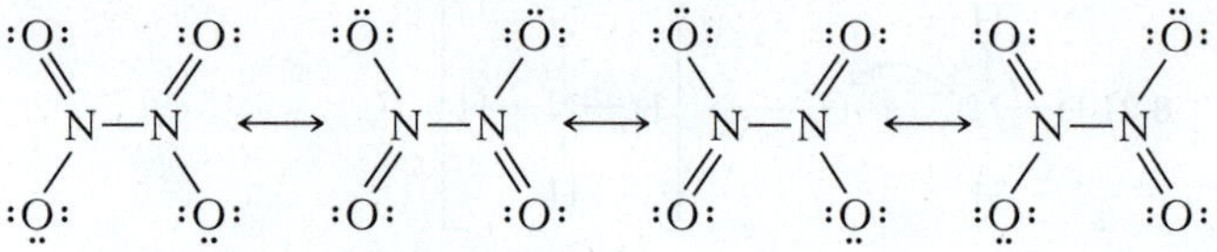

The average bond order is ³⁄₂.
8.115 N—O bond in NO_2^- should be shorter than that in NO_3^-.
8.117 :O≡C—Ö: ⟷ :Ö—C≡O:
These are not preferred structures, because in each Lewis diagram, one oxygen bears a formal charge of +1 whereas the other bears a formal charge of −1. The structure with the formal charges of zero has a lower potential energy and is more stable.

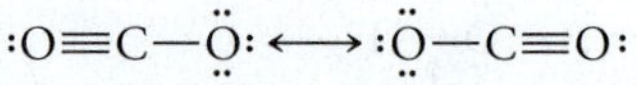

Chapter 9

Practice Exercises

9.1 (a) Tetrahedral **(b)** Trigonal bipyramidal **9.2** Octahedral
9.3 Trigonal bipyramidal **9.4** Linear **9.5** Linear **9.6** Trigonal planar **9.7** Polar **9.8 (a)** Nonpolar **(b)** Polar **(c)** Polar **(d)** Polar
(e) Polar **9.9** The H—Cl bond is formed by the overlap of the half-filled $1s$ atomic orbital of a H atom with the half-filled $3p$ valence orbital of a Cl atom:

Cl atom in HCl (x = H electron):

⇅ ⇅⇅(⇅x)
$3s$ $3p$

9.10 The half-filled $1s$ atomic orbital of each H atom overlaps with a half-filled $3p$ atomic orbital of the P atom, to give three P—H bonds. This should give a bond angle of 90°.

P atom in PH_3 (x = H electron):

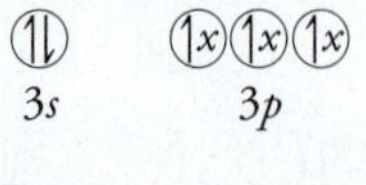

⇅ (↑x)(↑x)(↑x)
$3s$ $3p$

Note: Only half of each p orbital is shown:

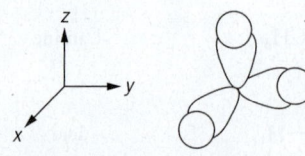

9.11 sp^2 hybrid orbitals on the B, x = F electron

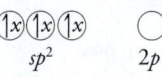

$$sp^2 \qquad 2p$$

9.12 sp hybrid orbitals on the Be, x = F electron

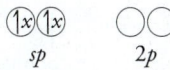

$$sp \qquad 2p$$

9.13 sp^3 **9.14** sp^3d **9.15** Trigonal bipyramidal, sp^3d—$3p$ overlap
9.16 sp^3d^2 **9.17 (a)** sp^3 **(b)** sp^3d **9.18** NH_3 is sp^3 hybridized. Three of the electron pairs are used for bonding. The fourth pair of electrons is a lone pair of electrons. This pair of electrons is used for the formation of the bond between the nitrogen and the hydrogen ion, H^+.
9.19 Octahedral, sp^3d^2
P atom in PCl_6^- (x = Cl electron):

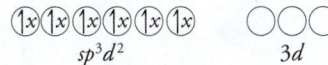

$$sp^3d^2 \qquad 3d$$

9.20 Atom 1: sp^2; Atom 2: sp^3; Atom 3: sp^2; 10 σ bonds and 2 π bonds
9.21 Atom 1: sp; Atom 2: sp^2; Atom 3: sp^3; 9 σ bonds and 3 π bonds
9.22 The bond order of the ion is 3. The MO diagram agrees with the Lewis structure.

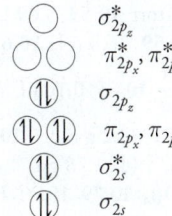

9.23

$$\begin{aligned}
&\bigcirc \qquad \sigma^*_{2p_z} \\
&\textcircled{1}\,\bigcirc \qquad \pi^*_{2p_x}, \pi^*_{2sp_y} \\
&\textcircled{1}\,\textcircled{1} \qquad \pi_{2p_x}, \pi_{2p_y} \\
&\textcircled{1} \qquad \sigma_{2p_z} \\
&\textcircled{1} \qquad \sigma^*_{2s} \\
&\textcircled{1} \qquad \sigma_{2s}
\end{aligned}$$

The bond order is calculated to be 5/2.
9.24 The p orbitals on the N and O that are not involved in forming the σ bond framework. **9.25** Valence band: $4s$ orbitals. Conduction band: $3d$ orbitals. **9.26** Magnesium < germanium < sulfur
9.27 (a) sp^3 **(b)** sp^2 **(c)** sp^2

Review Problems

9.73 (a) Bent **(b)** Planar triangular **(c)** T-shaped **(d)** Linear **(e)** Planar triangular **9.75 (a)** Bent **(b)** Trigonal bipyramidal **(c)** Trigonal pyramidal **(d)** Trigonal pyramidal **(e)** Bent **9.77 (a)** Tetrahedral **(b)** Square planar **(c)** Octahedral **(d)** Tetrahedral **(e)** Linear
9.79 BrF_4^+ **9.81** 120° **9.83 (a)** 109.5° **(b)** 109.5° **(c)** 120° **(d)** 109.5° **(e)** 120° **9.85 (a)**, **(b)**, and **(c)** **9.87** All are polar.
9.89 In SF_6, the individual bonds in this substance are polar bonds, the geometry of the bonds is symmetrical, so the individual dipole moments of the various bonds cancel one another.

In SF_5Br, one of the six bonds has a different polarity, so the individual dipole moments of the various bonds do not cancel one another.
9.91 One $3p$ orbital overlaps with a $3p$ orbital on the other Cl. Each Cl atom (x = an electron from the other Cl atom):

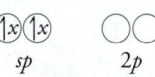

$$3s \qquad 3p$$

9.93 Hybridized Be: (x = a Cl electron)

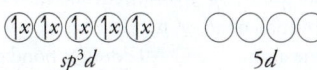

$$sp \qquad 2p$$

Beryllium uses sp hybrid orbitals.
9.95 Sb atom in $SbCl_5$ (x = a Cl electron)

$$\textcircled{1x}\textcircled{1x}\textcircled{1x}\textcircled{1x}\textcircled{1x} \qquad \bigcirc\bigcirc\bigcirc\bigcirc$$
$$sp^3d \qquad\qquad 5d$$

Antimony uses sp^3d hybrid orbitals.
9.97 (a) Tetrahedral: sp^3

$$\left[\begin{array}{c} :\ddot{O}: \\ | \\ :\ddot{O}-Cl: \\ | \\ :\ddot{O}: \end{array}\right]^-$$

(b) Planar triangle: sp^2

$$\begin{array}{c} :O: \\ \| \\ :\ddot{O} \quad S \quad \ddot{O}: \end{array}$$

(c) Tetrahedral: sp^3

$$\begin{array}{c} \ddot{O} \\ :\ddot{F} \qquad \ddot{F}: \end{array}$$

9.99 (a) sp^3
The hybrid orbital diagram for As: (x = a Cl electron)

$$\textcircled{1}\textcircled{1x}\textcircled{1x}\textcircled{1x}$$
$$sp^3$$

(b) sp^3d
The hybrid orbital diagram for Cl: (x = a F electron)

$$\textcircled{1}\textcircled{1}\textcircled{1x}\textcircled{1x}\textcircled{1x} \qquad \bigcirc\bigcirc\bigcirc\bigcirc$$
$$sp^3d \qquad\qquad 3d$$

9.101 Sb in SbF_6^-: (xx = an electron pair from the donor F^-)

$$\textcircled{1}\textcircled{1}\textcircled{1}\textcircled{1}\textcircled{1}\textcircled{xx}$$
$$sp^3d^2$$

9.103 (a) N in the C=N system:

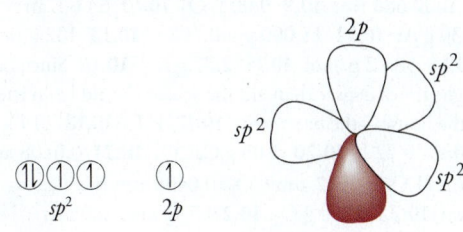

$$\textcircled{1}\textcircled{1}\textcircled{1} \qquad \textcircled{1}$$
$$sp^2 \qquad\qquad 2p$$

(b) sigma bond   pi bond

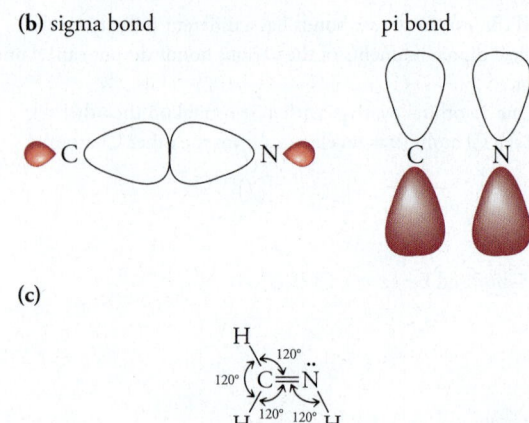

(c)

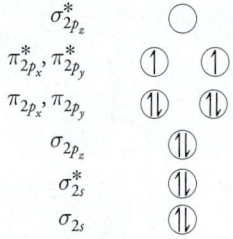

9.105 Each carbon atom is sp^2 hybridized, and each C—Cl bond is formed by the overlap of an sp^2 hybrid of carbon with a p atomic orbital of a chlorine atom. The C=C double bond consists first of a C—C σ bond formed by "head-on" overlap of sp^2 hybrids from each C atom. Second, the C=C double bond consists of a side-to-side overlap of unhybridized p orbitals of each C atom, to give one π bond. The molecule is planar, and the expected bond angles are all 120°. **9.107** 1. sp^3 2. sp 3. sp^2 4. sp^2 **9.109** 1. One σ bond 2. One σ bond and two π bond 3. One σ bond 4. One σ bond and one π bond

9.111

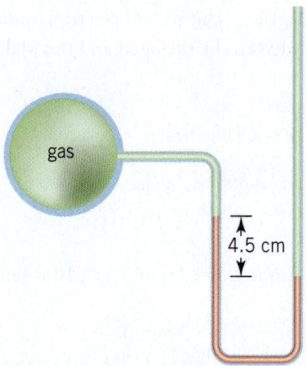

Bond order = 2.
9.113 (a) (i) O_2: BO = 2, O_2^+: BO = 2.5; (ii) O_2: BO = 2, O_2^-: BO = 1.5 **(b)** (i) O_2, (ii) O_2^- **(c)** (i) O_2^+, (ii) O_2 **9.115** The bond length of NO is longer than the bond length of NO^+. The bond energy of NO is lower than the bond energy of NO^+. **9.117** O_2^+, O_2, O_2^-, and NO

Chapter 10

Practice Exercises

10.1 Adding more gas increases the pressure because there are more particles to bounce against the walls of the container, increasing the force per unit area. **10.2** When apples rot, they lose mass because they are emitting a gas that would be the by-product of the chemical reaction. When nails rust, the iron combines with the oxygen in the air, thus gaining mass. **10.3** 14.1 psi, 28.7 in. Hg **10.4** 88,800 Pa, 666 torr **10.5** 1070 mm Hg, 470 mm Hg **10.6** 653 mm Hg **10.7** $\frac{2}{3}P$ **10.8** 688 torr **10.9** 9.00 L O_2 **10.10** 64.6 L air **10.11** 1130 g Ar **10.12** 15,000 g solid CO_2 **10.13** 132 g mol^{-1}, xenon **10.14** 58.12 g/mol **10.15** 2.77 g L^{-1} **10.16** Since radon is almost eight times denser than air, the sensor should be in the lowest point in the house—the basement. **10.17** P_2F_4 **10.18** 114 g mol^{-1}, C_8H_{18} **10.19** 9.22 L **10.20** 1.03 g $CaCO_3$ **10.21** 0.0408 atm CO, 0.0816 atm H_2O, 0.0612 atm CO_2, 0.0408 atm N_2, P_{total} = 0.2244 atm **10.22** 2710 g O_2 **10.23** 732 torr, 289 mL

10.24 747 torr, 0.0994 mol CH_4 **10.25** 0.996 atm H_2, 1.05 atm NO **10.26** 1.125 atm
10.27 $\dfrac{\text{effusion rate}_{^{81}Br^{81}Br}}{\text{effusion rate}_{^{81}Br^{79}Br}} = 0.994$

$\dfrac{\text{effusion rate}_{^{81}Br^{81}Br}}{\text{effusion rate}_{^{79}Br^{79}Br}} = 0.988$

10.28 HI

Review Problems

10.35 (a) 958 torr **(b)** 0.974 atm **(c)** 738 mm Hg **(d)** 10.9 torr
10.37 (a) 250 torr **(b)** 350 torr **10.39** 4.5 cm Hg

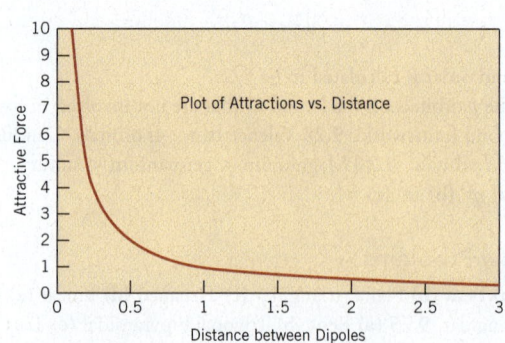

10.41 813 torr **10.43** 125 torr **10.45** 1180 torr **10.47** 2.18 L
10.49 843 °C **10.51** 801 torr **10.53** 5.69 L **10.55** −53 °C
10.57 1.35×10^3 mL **10.59** 36.3 mL **10.61** 224 mL
10.63 $6.24 \times 10^4 \dfrac{\text{mL torr}}{\text{mol K}}$ **10.65** 0.104 L **10.67** 2340 torr
10.69 0.0398 g CO_2 **10.71** 88.2 g mol$^-$ **10.73 (a)** 1.34 g L^{-1}
(b) 1.25 g L^{-1} **(c)** 3.17 g L^{-1} **(d)** 1.78 g L^{-1} **10.75** 1.28 g L^{-1}
10.77 33.2 g $C(CH_2ONO_2)_4$ **10.79** 10.7 L H_2 **10.81** 875 torr,
$X = 0.314$ **10.83** 697 torr **10.85** 0.147, 14.7% **10.87** P_{N_2}:
228 torr, 0.30 atm, 0.304 bar P_{O_2}: 152 torr, 0.20 atm, 0.203 bar
P_{He}: 304 torr, 0.40 atm, 0.405 bar P_{CO_2}: 76 torr, 0.10 atm, 0.101 bar
10.89 0.124 mol CO_2 **10.91** 250 total mL **10.93** Nitrogen, 1.25
10.95 $Cl_2 < SO_2 < C_2H_4$ **10.97** 16.2 g mol^{-1}

Chapter 11

Practice Exercises

11.1

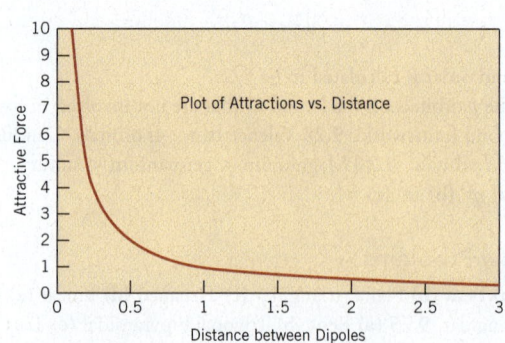

11.2 (a) $CH_3CH_2CH_2CH_2CH_3 < CH_3CH_2OH < KBr$
(b) $CH_3CH_2OCH_2CH_3 < CH_3CH_2NH_2 <$
$HOCH_2CH_2CH_2CH_2OH$ **11.3** Propylamine, because of its ability to form hydrogen bonds **11.4** Evaporative cooling works because as the water molecules evaporate, they require heat from the surroundings, and the surroundings include the air. **11.5** Molecules in the gas

phase will condense when they collide with a liquid phase because some of the kinetic energy that the gas phase molecules have is transferred to the liquid, leaving the gas molecule with too little kinetic energy to remain in the gas phase. **11.6** The piston should be pushed in. **11.7** The total number of the molecules in the vapor and the liquid remains the same. **11.8** **(a)** less than 10 °C above 100 °C **11.9** Approximately 75 °C **11.10** 3.0×10^1 kJ **11.11** 2.7×10^1 kJ, 2.6 kJ **11.12** The vapor pressure curve, see Figure 11.23. **11.13** Solid to gas **11.14** Liquid **11.15** The vapor pressure increases with increasing temperature.

11.16

Boiling	Endothermic
Melting	Endothermic
Condensing	Exothermic
Subliming	Endothermic
Freezing	Exothermic

No, each physical change is always exothermic, or always endothermic as shown. **11.17** 530 mm Hg **11.18** 77.4 °C **11.19** 2 Cr atoms per unit cell **11.20** The ratio is 1 to 1. **11.21** 168 pm **11.22** 9.23 g cm^{-3} **11.23** A molecular crystal **11.24** A covalent or network solid **11.25** A molecular solid

Review Problems

11.91 **(a)** London forces, dipole–dipole, hydrogen bonding **(b)** London forces, dipole–dipole **(c)** London forces **(d)** London forces, dipole–dipole **11.93** Diethyl ether should have a higher vapor pressure. Butanol has a higher boiling point. **11.95** Chloroform would be expected to display larger dipole–dipole attractions, and bromoform would be expected to show stronger London forces due to having larger electron clouds, which are more polarizable than those of chlorine. Since bromoform has a higher boiling point than chloroform, it experiences stronger intermolecular attractions than chloroform. Therefore, London forces are more important. **11.97** *n*-octane; the elongated structure **11.99** Ethanol **11.101** Diethyl ether < acetone < benzene < water < acetic acid **11.103** Diethyl ether < acetone < benzene < water < acetic acid

11.105

Compound	Intermolecular Forces Broken
(a) C_2H_5OH	London, dipole–dipole, hydrogen bonding
(b) H_3CCN	London, dipole–dipole
(c) PCl_5	London
(d) NaCl	London, ionic bonds

11.107 305 kJ
11.109 **(a)** 0.0 °C **(b)** 48.0 g of solid water
11.111

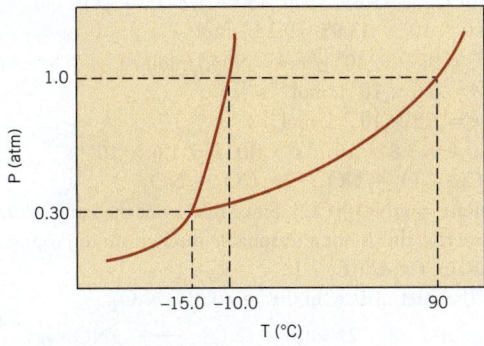

11.113 Below 0.30 atm. The density of the solid is higher than that of the liquid. **11.115** The solid–liquid line slants toward the right.

11.117 3.46×10^{-6} atm **11.119** 31,400 J mol⁻¹ **11.121** 4 Zn²⁺, 4S²⁻ **11.123** 3.51 Å, 351 pm **11.125** 656 pm **11.127** **(a)** 6.57° **(b)** 27.3° **11.129** 176 pm **11.131** Molecular solid **11.133** Metallic solid

11.135 **(a)** molecular   **(d)** metallic   **(g)** ionic
     (b) ionic   **(e)** covalent
     (c) ionic   **(f)** molecular

Chapter 12

Practice Exercises

12.1 a, b, and d **12.2** a, c, and d
12.3

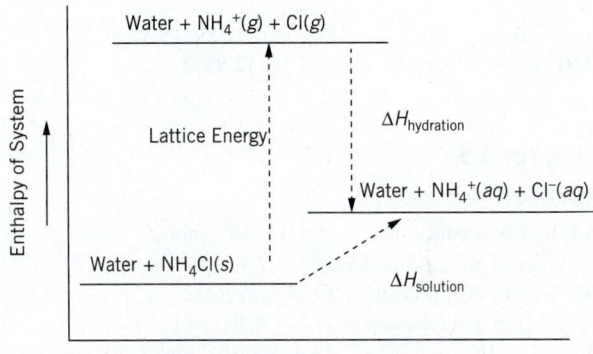

12.4 More heat is released when the ions are solvated than the amount of heat required to form the gaseous ions. **12.5** Equilibrium will shift toward solution. **12.6** NaOH will precipitate. **12.7** $k_H = 3.7 \text{ g L}^{-1}$ Hydrogen sulfide is more soluble in water than nitrogen and oxygen. Hydrogen sulfide reacts with the water to form hydronium ions and HS⁻. **12.8** 1.12 mg O_2, 1.85 mg N_2 **12.9** 406 mL water **12.10** 0.250 g NaBr, 24.75 g H_2O, 24.8 mL H_2O **12.11** 1.239 m, $M < m$

12.12

0.320 g	CH_3OH
0.640 g	CH_3OH
0.960 g	CH_3OH
1.28 g	CH_3OH
1.60 g	CH_3OH

12.13 27 m NaOH **12.14** 16.1 m **12.15** 6.82 M HBr **12.16** 0.00469 M Al(NO₃)₃, 0.00470 m Al(NO₃)₃ **12.17** 4.90×10^2 torr **12.18** 34.5 g stearic acid **12.19** 55.4 torr **12.20** 45.1 torr **12.21** 81.8% to 84.6% **12.22** 213 g glucose **12.23** 157 g mol⁻¹ **12.24** 125 g mol⁻¹ **12.25** 3.68 mm Hg, 50.0 mm H_2O **12.26** 222 torr, $T_f = -0.021$ °C, $T_{bp} = 100.006$ °C **12.27** 5.00×10^3 g mol⁻¹ **12.28** 5.38×10^2 g mol⁻¹ **12.29** -0.882 °C, -0.441 °C **12.30** **(a)** -0.372 °C **(b)** -0.0372 °C **(c)** -0.00372 °C The first freezing point depression **12.31** **(a)** colloid dispersion **(b)** colloid dispersion **(c)** suspension **(d)** colloid dispersion **12.32** **(a)** water and oil **(b)** water and fat **(c)** sugar and air **(d)** air and soot particles **(e)** soap and air

Review Problems

12.49 838 kJ mol⁻¹ **12.51** 0.038 g/L
12.53

$$C_2 = \frac{C_1 \times P_2}{P_1} = \frac{(0.018 \text{ g L}^{-1}) \times (620 \text{ torr})}{(740 \text{ torr})} = 0.015 \text{ g/L}$$

12.55 3.9×10^{-5} g mL^{-1} atm^{-1}
12.57 25 g **12.59** (a) 23.4 g (b) 58.6 g (c) 1.17 g **12.61** 3.35 m
12.63 18 ppm **12.65** (a) 0.133 m, (b) 2.39×10^{-3}, (c) 2.38%,
(d) 0.131 M **12.67** 5.45% **12.69** 7.89%, 4.76 m **12.71** 0.359 M,
3.00%, 6.49×10^{-3} **12.73** 22.8 torr **12.75** 52.7 torr
12.77 70 mol% toluene and 30 mol% benzene **12.79** (a) 0.029
(b) 2.99×10^{-2} moles (c) 278 g/mol **12.81** 68.9 g
12.83 152 g/mol **12.85** 127 g/mol, $C_8H_4N_2$ **12.87** (a) If the
equation is correct, the units on both sides of the equation should be
g/mol. The units on the right side of this equation are:

$$\frac{(g) \times (L\ atm\ mol^{-1}\ K^{-1}) \times (K)}{L \times atm} = g/mol$$

which is correct. (b) 1.8×10^6 g/mol **12.89** 15.3 torr
12.91 1.3×10^4 torr **12.93** $-1.1\ °C$ **12.95** 2

Chapter 13

Practice Exercises

13.1 Rate of production of I$^-$ = 8.0×10^{-5} mol L^{-1} s^{-1}
Rate of production of SO$_4^{2-}$ = 2.4×10^{-4} mol L^{-1} s^{-1}
13.2 Rate of disappearance of O$_2$ = 0.45 mol L^{-1} s^{-1}
Rate of disappearance of H$_2$S = 0.30 mol L^{-1} s^{-1}
13.3 2.1×10^{-4} mol L^{-1} s^{-1} **13.4** 1×10^{-4} mol L^{-1} s^{-1}
13.5 (a) $k = 9.8 \times 10^{14}$ L^2 mol^{-2} s^{-1} (b) L^2 mol^{-2} s^{-1}
13.6 (a) $k = 8.0 \times 10^{-2}$ L mol^{-1} s^{-1} (b) L mol^{-1} s^{-1}
13.7 Order of the reaction with respect to $[BrO_3^-] = 1$
Order of the reaction with respect to $[SO_3^{2-}] = 1$
Overall order of the reaction = 2
13.8 Rate = $k[Cl_2]^2[NO]$
13.9 Order with respect to Br$_2$ is 1.
Order with respect to HCO$_2$H is zero.
Overall order of the reaction is 1.
13.10 2.0×10^2 L^2 mol^{-2} s^{-1}. All sets of data give the same value
for k. **13.11** (a) The rate will increase nine-fold. (b) The rate will
increase three-fold. (c) The rate will decrease to three fourths of
the original rate. **13.12** (a) rate = $k[NO]^2[H_2]^1$ (b) 2.1×10^5
L^2 mol^{-2} s^{-1} (c) L^2 mol^{-2} s^{-1} **13.13** (a) first order (b) The rate
constant is 6.17×10^{-4} s^{-1} **13.14** (a) rate = $k[A]^2[B]^2$
(b) 6.9×10^{-3} L^3 mol^{-3} s^{-1} (c) L^3 mol^{-3} s^{-1} (d) Overall order = 4
13.15 2.56×10^{-2} yr^{-1} **13.16** (a) 4.71×10^{-3} M
(b) 7.8 min **13.17** 18.7 min, 37.4 min **13.18** 27.1 yr **13.19** 5.41%
13.20 1.90×10^4 yr **13.21** The upper limit of dates is 25,000 years
before present. The lower limit of dates is 420 years before present.
13.22 62 min **13.23** 2.0×10^{-2} M **13.24** $k = 1.03$ L mol^{-1} s^{-1},
$t_{1/2} = 1.48 \times 10^3$ s **13.25** The reaction is first order. **13.26** (a)
1.4×10^2 kJ/mol (b) 0.40 L mol^{-1} s^{-1} **13.27** 684 K **13.28** 322 K
or 49 °C **13.29** (a), (b), and (e) may be elementary processes.
Equations (c), (d), and (f) are not elementary processes because
they have more than two molecules colliding at one time.
13.30 Rate = $k[NO][O_3]$
13.31 Rate = $\dfrac{k[NO_2Cl]^2}{[NO_2]}$

Review Problems

13.53 At 200 min: 1×10^{-4} M/min. At 600 minutes:
7×10^{-5} M/min

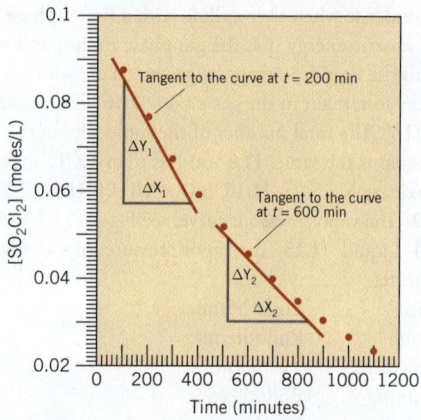

13.55 Rate of disappearance of A = 0.60 mol A L^{-1} s^{-1}
Rate of appearance of C = 0.90 mol C L^{-1} s^{-1}
13.57 (a) rate for O$_2$ = 11.4 mol O$_2$ L^{-1} s^{-1} (b) rate for
CO$_2$ = 7.20 mol CO$_2$ L^{-1} s^{-1} (c) rate for H$_2$O = 8.40 mol
H$_2$O L^{-1} s^{-1}
13.59 (a) $-\dfrac{\Delta[CH_3Cl]}{\Delta t} = -\dfrac{1}{3}\dfrac{\Delta[Cl_2]}{\Delta t} = \dfrac{\Delta[CCl_4]}{\Delta t} = \dfrac{1}{3}\dfrac{\Delta[HCl]}{\Delta t}$
(b) 9.7×10^{-3} M s^{-1}
13.61 8.0×10^{-11} mol L^{-1} s^{-1} **13.63** 2.4×10^2 mol L^{-1} s^{-1}
13.65 6.4×10^2 M s^{-1}. **13.67** Rate = $k[M][N]^2$ $k = 2.5 \times 10^3$ L^2
mol^{-2} s^{-1} **13.69** Rate = $k[OCl^-][I^-]$ $k = 6.8 \times 10^9$ L mol^{-1} s^{-1}
13.71 Rate = k[ICl][H$_2$] $k = 1.04 \times 10^{-1}$ L mol^{-1} s^{-1}
13.73

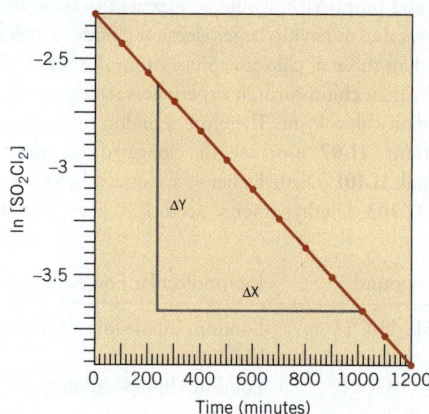

1.32×10^{-3} min^{-1}
13.75 (a) 3.7×10^{-3} M (b) 6.0×10^{-4} M **13.77** 0.0185 min^{-1}
13.79 4.26×10^{-3} min^{-1} **13.81** 1.3×10^4 min **13.83** 11 mg/kg
13.85 1/256 **13.87** Approximately 500 min. The reaction is first-
order in SO$_2$Cl$_2$. **13.89** 33 hr **13.91** 1.3×10^9 years old.
13.93 4.0×10^{-13} **13.95** 79.1 kJ/mol
13.97 E_a = 9.89×10^4 J/mol = 99 kJ/mol,
A = 6.6×10^9 L mol^{-1} s^{-1},
k = 7.9×10^{-2} L mol^{-1} s^{-1}
13.99 (a) k = 3.8×10^{-5} s^{-1} (b) k = 1.6×10^{-1} s^{-1}
13.101 CO + O$_2$ + NO $\longrightarrow$ CO$_2$ + NO$_2$
13.103 Rate = $k[NO_2][CO]$. Since this is not the same as the observed rate law, this is not a reasonable mechanism to propose.
13.105 Rate = $k[AB][C]$
13.107 The intermediate in this reaction is N$_2$O$_2$.

$$2NO(g) + O_2(g) \longrightarrow 2NO_2(g)$$

The rate law is determined from the slow step.

$$Rate = k[NO]^2[O_2]$$

Chapter 14

Practice Exercises

14.1 The chapter opening photograph of the tight-rope walker represents a dynamic equilibrium because the tightrope walker is constantly adjusting his balance to keep himself upright. It differs from a dynamic equilibrium because two processes that are in equilibrium are not occurring at the same time.

14.2 (a) $\dfrac{[H_2O]^2}{[H_2]^2[O_2]} = K_c$ (b) $\dfrac{[CO_2][H_2O]^2}{[CH_4][O_2]^2} = K_c$

14.3 $2N_2O_3(g) + O_2(g) \longrightarrow 4NO_2(g)$ **14.4** 1.2×10^{-13}

14.5 1.9×10^5

14.6 $K_P = \dfrac{(P_{N_2O})^2}{(P_{N_2})^2(P_{O_2})}$

14.7 $K_P = \dfrac{(P_{HI})^2}{(P_{H_2})(P_{I_2})}$

14.8 1.8×10^{36} **14.9** K_P is smaller than K_c. $K_c = 57$

14.10 $K_c = \dfrac{1}{[NH_3(g)][HCl(g)]}$

14.11 (a) $K_c = \dfrac{1}{[Cl_2(g)]}$ (b) $K_c = [Ag^+(aq)]^2[CrO_4^{2-}(aq)]$

(c) $K_c = \dfrac{[Ca^{+2}(aq)][HCO_3^-(aq)]^2}{[CO_2(aq)]}$

14.12 The reaction will move to the left. **14.13** (a) < (c) < (b)
14.14 There will be no change in the amount of H_3PO_4.
14.15 (a) Decrease $[Cl_2]$, K_P unchanged (b) Increase $[Cl_2]$, K_P unchanged (c) Increase $[Cl_2]$, K_P decreased (d) Decrease $[Cl_2]$, K_P unchanged **14.16** [CO] decreases by 0.060 mol/L and $[CO_2]$ increases by 0.060 mol/L. **14.17** $K_c = 4.06$
14.18 (a) $[PCl_3] = 0.200\ M$
  $[Cl_2] = 0.100\ M$
  $[PCl_5] = 0.000\ M$
  (b) $[PCl_3]$ and $[Cl_2]$ decreased by 0.080 M. $[PCl_5]$ increased by 0.080 M. (c) $[PCl_3] = 0.120\ M$, $[PCl_5] = 0.080\ M$, $[Cl_2] = 0.020\ M$ (d) $K_c = 33$
14.19 $[NO_2] = 1.04 \times 10^{-2}\ M$ **14.20** $[C_2H_5OH] = 8.98 \times 10^{-3}\ M$
14.21 $[H_2] = [I_2] = 0.044\ M$, $[HI] = 0.312\ M$
14.22 $[H_2] = 0.107\ M$ $[I_2] = 0.007\ M$ $[HI] = 0.187\ M$
14.23 $[N_2] = 6.2 \times 10^{-4}$ $[H_2] = 1.9 \times 10^{-3}$
14.24 $[NO] = 1.1 \times 10^{-17}\ M$

Review Problems

14.27 (a) $K_c = \dfrac{[POCl_3]^2}{[PCl_3]^2[O_2]}$ (b) $K_c = \dfrac{[SO_2]^2[O_2]}{[SO_3]^2}$

(c) $K_c = \dfrac{[NO]^2[H_2O]^2}{[N_2H_4][O_2]^2}$ (d) $K_c = \dfrac{[NO_2]^2[H_2O]^8}{[N_2H_4][H_2O_2]^6}$

(e) $K_c = \dfrac{[SO_2][HCl]^2}{[SOCl_2][H_2O]}$

14.29 (a) $K_c = \dfrac{[Ag(NH_3)_2^+]}{[Ag^+][NH_3]^2}$ (b) $K_c = \dfrac{[Cd(SCN)_4^{2-}]}{[Cd^{2+}][SCN^-]^4}$

14.31 $K_c = 1 \times 10^{85}$

14.33 (a) $K_c = \dfrac{[HCl]^2}{[H_2][Cl_2]}$ (b) $K_c = \dfrac{[HCl]}{[H_2]^{1/2}[Cl_2]^{1/2}}$

K_c for reaction (b) is the square root of K_c for reaction (a).

14.35 (a) $K_P = \dfrac{(P_{POCl_3})^2}{(P_{PCl_3})^2(P_{O_2})}$ (b) $K_P = \dfrac{(P_{SO_2})^2(P_{O_2})}{(P_{SO_3})^2}$

(c) $K_P = \dfrac{(P_{NO})^2(P_{H_2O})^2}{(P_{N_2H_4})(P_{O_2})^2}$ (d) $K_P = \dfrac{(P_{NO_2})^2(P_{H_2O})^8}{(P_{N_2H_4})(P_{H_2O_2})^6}$

(e) $K_P = \dfrac{(P_{SO_2})(P_{HCl})^2}{(P_{SOCl_2})(P_{H_2O})}$

14.37 $0.038\ M$ **14.39** (a) **14.41** 11 **14.43** 2.7×10^{-2}
14.45 5.4×10^{-5} **14.47** (a) $55.5\ M$ (b) $55.5\ M$ (c) $55.5\ M$

14.49 (a) $K_c = \dfrac{[CO]^2}{[O_2]}$ (b) $K_c = [H_2O][SO_2]$

(c) $K_c = \dfrac{[CH_4][CO_2]}{[H_2O]^2}$ (d) $K_c = \dfrac{[H_2O][CO_2]}{[HF]^2}$

(e) $K_c = [H_2O]^5$

14.51 $[HI] = 1.47 \times 10^{-12}\ M$ $[Cl_2] = 7.37 \times 10^{-13}\ M$
14.53 (a) Increase; we are adding a reactant. (b) Decrease; we are removing a reactant. (c) Increase; decreasing the volume favors the side with fewer gas molecules. (d) No change; a catalyst increases the rate but does not affect the concentration. (e) Decrease; this is an exothermic reaction so heat may be considered a product.
14.55 (a) right (b) left (c) left (d) right (e) no effect (f) left
14.57 0.398 **14.59** 0.0955 **14.61** 0.915
14.63 (a) $Q = 4.96$; the system is not at equilibrium.
  (b) The system must shift to the left to reach equilibrium.
14.65 $[CH_3OH] = 4.36 \times 10^{-3}\ M$ **14.67** $[Br_2] = [Cl_2] = 0.012\ M$
14.69 $[NO_2] = [SO_2] = 0.0497$ mol/L
  $[NO] = [SO_3] = 0.0703$ mol/L
14.71 $[H_2] = [CO_2] = 7.7 \times 10^{-3}\ M$ $[CO] = [H_2O] = 0.012\ M$
14.73 $[H_2] = [Cl_2] = 8.9 \times 10^{-19}\ M$
14.75 $[CO] = 5.0 \times 10^{-4}\ M$
14.77 $[PCl_3] = 0.013\ M$
14.79 $[NO_2] = [SO_2] = 0.0281\ M$
14.81 $[CO] = [H_2O] = 0.200\ M$

Chapter 15

Practice Exercises

15.1 The conjugate acid–base pairs are (a), (c), and (f). (b): The conjugate base of HI is I^-. (d): The conjugate base of HNO_2 is NO_2^- and the conjugate base of NH_4^+ is NH_3. (e): The conjugate acid of CO_3^{2-} is HCO_3^- and the conjugate acid of CN^- is HCN.
15.2 (a) OH^- (b) I^- (c) $H_2PO_4^-$ (d) HPO_4^{2-} (e) NH_3 (f) H_2O_2 (g) HCO_3^- (h) NH_3 (i) NH_4^+ (j) $H_2PO_4^-$
15.3 HCN and CN^-; HCl and Cl^-
15.4 Brønsted acids: $H_2PO_4^-(aq)$ and $H_2CO_3(aq)$
  Brønsted bases: $HCO_3^-(aq)$ and $HPO_4^{2-}(aq)$
15.5 (a) $H_2PO_4^-$ amphoteric, since it can both accept and donate a proton
  (b) H_2S amphoteric, since it can both accept and donate a proton
  (d) H_3PO_4 not amphoteric: it can only donate protons
  (e) NH_4^+ not amphoteric: it can only donate protons
  (f) H_2O amphoteric, since it can both accept and donate a proton
  (g) HI not amphoteric: it can only donate protons
  (h) HNO_2 not amphoteric: it can only donate protons

15.6 $HPO_4^{2-}(aq) + OH^-(aq) \longrightarrow PO_4^{3-}(aq) + H_2O$
$HPO_4^{2-}(aq) + H_3O^+(aq) \longrightarrow H_2PO_4^-(aq) + H_2O$
15.7 $HSO_4^-(aq) + HPO_4^{2-}(aq) \longrightarrow SO_4^{2-}(aq) + H_2PO_4^-(aq)$
15.8 The substances on the left are favored.
15.9 (a) $HF < HBr < HI$ (b) $PH_3 < H_2S < HCl$ (c) $H_2O <$
$H_2Se < H_2Te$ (d) $AsH_3 < H_2Se < HBr$ (e) $PH_3 < H_2Se < HI$
15.10 (a) HBr (b) H_2Te (c) H_2S **15.11** (a) $HClO_3$
(b) H_2SO_4 **15.12** (a) H_3AsO_4 (b) H_2TeO_4 **15.13** (a) HIO_4
(b) H_2TeO_4 (c) H_3AsO_4 **15.14** (a) H_2SO_4 (b) H_3AsO_4
15.15 The acid strength decreases as follows:
$FCH_2COOH > ClCH_2COOH > BrCH_2COOH$
15.16 The acid strength increases as
$CH_3CO_2H < CH_2FCO_2H < CHF_2CO_2H < CF_3CO_2H$
15.17 (a) NH_3, Lewis base H^+, Lewis acid (b) Na_2O, Lewis base
SeO_3, Lewis acid (c) Ag^+, Lewis acid NH_3, Lewis base
15.18 (a) Fluoride ions have a filled octet of electrons and are
likely to behave as Lewis bases. (b) $BeCl_2$ is a likely Lewis acid
since it has an incomplete shell. (c) SO_3 is a Lewis acid, since the
central sulfur bears a significant positive charge.

Review Problems

15.49 (a) HF (b) $C_5H_5NH^+$ (c) H_2CrO_4 (d) HCN
15.51 (a) NH_2O^- (b) CN^- (c) NO_2^- (d) PO_4^{3-}
15.53 (a) conjugate pair

15.55 (a) HBr (b) HF (c) HBr **15.57** (a) $HClO_2$ (b) H_2SeO_4
15.59 (a) $HClO_3$ (b) $HClO_3$ (c) $HBrO_4$
15.61

Lewis acid: H^+
Lewis base: NH_2^-

15.63

15.65

15.67

Lewis base: an O in CO_3^{2-} and SO_3^{2-}
Lewis acids: SO_2 and CO_2

15.69
$Cr(H_2O)_6^{3+}(aq) + H_2O \longrightarrow Cr(H_2O)_5OH^{2+}(aq) + H_3O^+(aq)$

15.71 $Mg(OH)_2$ dissociates in water to produce OH^- ions, so the
compound is a basic species. $Si(OH)_4$ is a molecular compound and
the hydrogen ion will dissociate when the compound is dissolved in
water. Thus, $Si(OH)_4$ is an acidic species.

Chapter 16

Practice Exercises

16.1 8.3×10^{-16} M **16.2** 1.3×10^{-9} M; the solution is basic.
16.3 $10.25, 1.8 \times 10^{-4}$ $M, 5.6 \times 10^{-11}$ M **16.4** $4.49, 9.51$; the
solution is acidic. **16.5** (a) 1.3×10^{-3} $M, 7.7 \times 10^{-12}$ M The
solution is acidic. (b) 1.4×10^{-4} $M, 7.1 \times 10^{-11}$ M The solution
is acidic. (c) 1.5×10^{-11} $M, 6.5 \times 10^{-4}$ M The solution is basic.
(d) 7.8×10^{-5} $M, 1.3 \times 10^{-10}$ M The solution is acidic.
(e) 2.5×10^{-12} $M, 4.1 \times 10^{-3}$ M The solution is basic.
16.6 0.0050 $M, 2.30, 11.70$ **16.7** $1.07, 1.2 \times 10^{-13}$ $M, 12.93$
16.8 (a) $HC_2H_3O_2 + H_2O \rightleftharpoons H_3O^+ + C_2H_3O_2^-$

$$K_a = \frac{[H_3O^+][C_2H_3O_2^-]}{[HC_2H_3O_2]}$$

(b) $(CH_3)_3NH^+ + H_2O \rightleftharpoons H_3O^+ + (CH_3)_3N$

$$K_a = \frac{[H_3O^+][(CH_3)_3N]}{[(CH_3)_3NH^+]}$$

(c) $H_3PO_4 + H_2O \rightleftharpoons H_3O^+ + H_2PO_4^-$

$$K_a = \frac{[H_3O^+][H_2PO_4^-]}{[H_3PO_4]}$$

16.9 (a) $HCHO_2 + H_2O \rightleftharpoons H_3O^+ + CHO_2^-$

$$K_a = \frac{[H_3O^+][CHO_2^-]}{[HCHO_2]}$$

(b) $(CH_3)_2NH_2^+ + H_2O \rightleftharpoons H_3O^+ + (CH_3)_2NH$

$$K_a = \frac{[H_3O^+][(CH_3)_2NH]}{[(CH_3)_2NH_2^+]}$$

(c) $H_2PO_4^- + H_2O \rightleftharpoons H_3O^+ + HPO_4^{2-}$

$$K_a = \frac{[H_3O^+][HPO_4^{2-}]}{[H_2PO_4^-]}$$

16.10 Barbituric acid and hydrazoic acid

16.11 The acid with the smaller pK_a (HA) is the strongest acid.
For HA: 6.9×10^{-4}
For HB: 7.2×10^{-5}

16.12 (a) $(CH_3)_3N + H_2O \rightleftharpoons (CH_3)_3NH^+ + OH^-$

$$K_b = \frac{[(CH_3)_3NH^+][OH^-]}{[(CH_3)_3N]}$$

(b) $SO_3^{2-} + H_2O \rightleftharpoons HSO_3^- + OH^-$

$$K_b = \frac{[HSO_3^-][OH^-]}{[SO_3^{2-}]}$$

(c) $NH_2OH + H_2O \rightleftharpoons NH_3OH^+ + OH^-$

$$K_b = \frac{[NH_3OH^+][OH^-]}{[NH_2OH]}$$

16.13 (a) $HS^- + H_2O \rightleftharpoons H_2S + OH^-$

$$K_b = \frac{[H_2S][OH^-]}{[HS^-]}$$

(b) $H_2PO_4^- + H_2O \rightleftharpoons H_3PO_4 + OH^-$

$$K_b = \frac{[H_3PO_4][OH^-]}{[H_2PO_4^-]}$$

(c) $HPO_4^{2-} + H_2O \rightleftharpoons H_2PO_4^- + OH^-$

$$K_b = \frac{[H_2PO_4^-][OH^-]}{[HPO_4^{2-}]}$$

(d) $HCO_3^- + H_2O \rightleftharpoons H_2CO_3 + OH^-$

$$K_b = \frac{[H_2CO_3][OH^-]}{[HCO_3^-]}$$

(e) $HSO_3^- + H_2O \rightleftharpoons H_2SO_3 + OH^-$

$$K_b = \frac{[H_2SO_3][OH^-]}{[HSO_3^-]}$$

16.14 2.2×10^{-11} **16.15** 5.6×10^{-11} **16.16** 1.2×10^{-3}, 2.92
16.17 1.5×10^{-5}, 4.82 **16.18** 1.6×10^{-6}, 5.79 **16.19** 1.1×10^{-3} M,
2.96 **16.20** 1.2×10^{-3} M, 2.92 **16.21** 8.4×10^{-4} M, 3.08
16.22 5.48 **16.23** 8.63 **16.24** 5.35 **16.25** (a) basic (b) neutral
(c) acidic **16.26** (a) neutral (b) basic (c) acidic **16.27** 5.39
16.28 8.17 **16.29** 5.13 **16.30** Basic **16.31** Acidic
16.32 Upon addition of a strong acid, the concentration of $HC_2H_3O_2$
will increase. When a strong base is added, the concentration of the
acetic acid will decrease. **16.33** (a) $H^+ + NH_3 \longrightarrow NH_4^+$
(b) $OH^- + NH_4^+ \longrightarrow H_2O + NH_3$ **16.34** 4.83 **16.35** 4.61
16.36 Hydrazoic acid, 29.1 NaN_3
Acetic acid, 26.2 g $NaC_2H_3O_2$
Butanoic acid, 29.6 g $NaC_4H_7O_2$
Propanoic acid, 22.1 g $NaC_3H_5O_2$
16.37 Yes, formic acid and sodium formate would make a good buffer solution. 0.70, 9.7 g $NaCHO_2$.
16.38 0.13 pH units **16.39** 9.75, 9.67
16.40

$$H_3PO_4 + H_2O \rightleftharpoons H_3O^+ + H_2PO_4^- \quad K_a = \frac{[H_3O^+][H_2PO_4^-]}{[H_3PO_4]}$$

$$H_2PO_4^- + H_2O \rightleftharpoons H_3O^+ + HPO_4^{2-} \quad K_a = \frac{[H_3O^+][HPO_4^{2-}]}{[H_2PO_4^-]}$$

$$HPO_4^{2-} + H_2O \rightleftharpoons H_3O^+ + PO_4^{3-} \quad K_a = \frac{[H_3O^+][PO_4^{3-}]}{[HPO_4^{2-}]}$$

16.41 2.8×10^{-3} M, 2.55, 1.6×10^{-12} **16.42** 11.63 It is too basic
to be a substitute for $NaHCO_3$. **16.43** 10.24 **16.44** 6.7×10^{-13}.
16.45 (a) H_2O, K^+, $HC_2H_3O_2$, H^+, $C_2H_3O_2^-$, and OH^-

$$[H_2O] > [K^+] > [C_2H_3O_2^-] > [OH^-] > [HC_2H_3O_2] > [H^+]$$

(b) H_2O, $HC_2H_3O_2$, H^+, $C_2H_3O_2^-$, and OH^-

$$[H_2O] > [HC_2H_3O_2] > [H^+] \approx [C_2H_3O_2^-] > [OH^-]$$

(c) H_2O, K^+, $HC_2H_3O_2$, H^+, $C_2H_3O_2^-$, and OH^-

$$[H_2O] > [K^+] > [OH^-] > [C_2H_3O_2^-] > [HC_2H_3O_2] > [H^+]$$

(d) H_2O, K^+, $HC_2H_3O_2$, H^+, $C_2H_3O_2^-$, and OH^-

$$[H_2O] > [HC_2H_3O_2] > [K^+] > [C_2H_3O_2^-] > [H^+] > [OH^-]$$

16.46 (a) 2.37 (b) 3.74 (c) 4.22 (d) 8.22 **16.47** 3.66

Review Problems

16.49 (a) 1.5×10^{-12} M H^+, 11.83, 2.17
(b) 1.6×10^{-10} M H^+, 9.81, 4.19
(c) 6.3×10^{-7} M H^+, 6.20, 7.80
(d) 1.2×10^{-3} M H^+, 2.91, 11.09
16.51 (a) 7.2×10^{-9} M H^+, 1.4×10^{-6} M OH^-
(b) 2.8×10^{-3} M H^+, 3.6×10^{-12} M OH^-
(c) 5.6×10^{-12} M H^+, 1.8×10^{-3} M OH^-
(d) 5.2×10^{-14} M H^+, 1.9×10^{-1} M OH^-
(e) 2.0×10^{-7} M H^+, 5.0×10^{-8} M OH^-
16.53 (a) 1.5×10^{-7} M H^+, 6.5×10^{-8} M OH^-
(b) 1.8×10^{-13} M H^+, 5.5×10^{-2} M OH^-
(c) 7.1×10^{-4} M H^+, 1.4×10^{-11} M OH^-
(d) 1.4×10^{-1} M H^+, 7.1×10^{-14} M OH^-
(e) 1.7×10^{-9} M H^+, 5.8×10^{-6} M OH^-
16.55 4.72 **16.57** 9.80 **16.59** 3.0×10^{-8} M, 7.52
16.61 $[H^+] = 2 \times 10^{-6}$ M, $[OH^-] = 5 \times 10^{-9}$ M
16.63 $[H^+] = 0.00065$ M, pH = 3.19, $[OH^-] = 1.54 \times 10^{-11}$ M
16.65 0.15 M OH^-, 0.82, 13.18, 6.67×10^{-14} M H^+
16.67 2.0×10^{-3} M $Ca(OH)_2$, 2.0×10^{-4} M $Ca(OH)_2$
16.69 5.0×10^{-12} M **16.71** 7.07, 8.6×10^{-8} M H^+
16.73 3.16×10^{-6} M H^+ **16.75** 2.9×10^{-11} **16.77** (a) The conjugate base is IO_3^-. 5.9×10^{-14} (b) IO_3^- is a weaker base than an
acetate anion. **16.79** 0.43% **16.81** 0.30%, 1.8×10^{-6}
16.83 9.15, 1.7×10^{-9} **16.85** 2.3×10^{-2}, 1.64 **16.87** 5.9×10^{-4},
3.23, 7.4% **16.89** pH = 2.34, $[HC_3H_5O_2] = 0.145$, $[H^+] =$
4.6×10^{-3}, $[C_3H_5O_2^-] = 4.6 \times 10^{-3}$ **16.91** 10.25
16.93 0.46 M **16.95** 10.50 **16.97** 6.0×10^{-4} M H^+, 3.22
16.99 11.30, 2.0×10^{-3} M HCN **16.101** 5.74 **16.103** 1.5×10^{-5}
16.105 10.68 **16.107** 4.97 **16.109** 4.97 **16.111** -0.12 pH units
16.113 2.5×10^{-2} M
16.115 Initial pH = 9.36, final pH = 9.26, pH decreases by
0.10 pH
16.117 2.2 g $NaC_2H_3O_2$ **16.119** Initial pH = 4.79, final pH = 4.76,
after addition of HCl pH = 1.71
16.121 $[H_2C_6H_6O_6] \cong 0.15$ M
$[H_3O^+] = [HC_6H_6O_6^-] = 3.5 \times 10^{-3}$ M
$[C_6H_6O_6^{2-}] = 1.6 \times 10^{-12}$ M
pH = 2.46

16.123 $0.12\,M = [H_3O^+] = [H_2PO_4^-]$
$1.9\,M = [H_3PO_4]$
$6.2 \times 10^{-8}\,M = [HPO_4^{2-}]$
$2.2 \times 10^{-19}\,M = [PO_4^{3-}]$
$pH = 0.92$

16.125 $0.20\,M = [H_3O^+] = [H_2PO_3^-]$
$[HPO_3^{2-}] = 2.0 \times 10^{-7}\,M$
$pH = 0.70$

16.127 $pH = 10.69$
$[SO_3^{2-}] = 0.24\,M$
$[HSO_3^-] = [OH^-] = 4.9 \times 10^{-4}\,M$
$[Na^+] = 0.24\,M$
$[H_3O^+] = 2.0 \times 10^{-11}$
$[H_2SO_3] = 6.7 \times 10^{-13}$

16.129 9.70

16.131 $pH = 12.99$
$[HPO_4^{2-}] = 9.8 \times 10^{-2}\,M$
$[H_2PO_4^-] = 1.6 \times 10^{-7}$
$[H_3PO_4] = 2.1 \times 10^{-18}\,M$

16.133 12.93 **16.135** 8.07, Cresol red **16.137** (a) 2.87 (b) 4.44
(c) 4.74 (d) 8.72

Chapter 17

Practice Exercises

17.1 (a) $K_{sp} = [Pb^{2+}][Br^-]^2$ (b) $K_{sp} = [Al^{3+}][OH^-]^3$
17.2 $K_{sp} = [Hg_2^{2+}][Cl^-]^2$ **17.3** 3.2×10^{-10} **17.4** 9.2×10^{-17}
17.5 (a) $7.3 \times 10^{-7}\,M$ (b) 0.012 M **17.6** $7.0 \times 10^{-7}\,M$
17.7 $2.1 \times 10^{-16}\,M$ in 0.20 M CaI_2; $9.2 \times 10^{-9}\,M$ in pure water
17.8 $7.6 \times 10^{-8}\,M$; 5.5×10^5 gallons seawater **17.9** 2.2×10^{-6}
A precipitate will form. **17.10** 1.6×10^{-8} A precipitate will form.
17.11 $PbBr_2(s)$ No precipitate is expected.
17.12 $PbCl_2$ No precipitate is expected.
17.13 (a) Adding acid will not increase the solubility of AgBr.
(b) Adding acid will increase the solubility of $CaCO_3$.
(c) Adding acid will increase the solubility of Ag_2CrO_4.
(d) Adding acid will increase the solubility of $Zn(CN)_2$.
17.14 $5.7 \times 10^{-3}\,M$ **17.15** 2.5 or lower **17.16** 5.0
17.17 $2.0 \times 10^{-4} > [SO_4^{2-}] > 2.2 \times 10^{-9}$
17.18 $5 \times 10^{-3}\,M\,OH^-$, 11.70 **17.19** CoS will precipitate. **17.20** 2.9
17.21 0.40 M **17.22** $5.34 < pH < 6.14$
17.23 The molar solubility of AgCl in 0.10 M NH_3 is $4.9 \times 10^{-3}\,M$.
The molar solubility of AgCl in water is $1.3 \times 10^{-5}\,M$.
17.24 4.1 mol NH_3

Review Problems

17.25 (a) $K_{sp} = [Hg_2^{2+}][Cl^-]^2$ (d) $K_{sp} = [Cu^+][Cl^-]$
(b) $K_{sp} = [Ag^+][Br^-]$ (e) $K_{sp} = [Hg^{2+}][I^-]^2$
(c) $K_{sp} = [Pb^{2+}][Br^-]^2$
17.27 (a) $K_{sp} = [Ca^{2+}][F^-]^2$ (d) $K_{sp} = [Fe^{3+}][OH^-]^3$
(b) $K_{sp} = [Ag^+]^2[CO_3^{2-}]$ (e) $K_{sp} = [Pb^{2+}][F^-]^2$
(c) $K_{sp} = [Pb^{2+}][SO_4^{2-}]$ (f) $K_{sp} = [Cu^{2+}][OH^-]^2$
17.29 1:1 ratio cation to anion
$AgBr(s)$ $CuCl(s)$ $PbSO_4(s)$
1:2 ratio cation to anion, or 2:1 ratio cation to anion
$Hg_2Cl_2(s)$ $PbBr_2(s)$ $CaF_2(s)$ $PbF_2(s)$ $Cu(OH)_2(s)$ $Ag_2CO_3(s)$
$HgI_2(s)$
1:3 ratio cation to anion
$Fe(OH)_3(s)$

17.31 $MgCO_3$ **17.33** 1.6×10^{-5}
17.35 $1.05 \times 10^{-5}\,M$; 1.10×10^{-10} **17.37** 1.7×10^{-6}
17.39 2.8×10^{-18} **17.41** 1.2×10^{-2} **17.43** $4.2 \times 10^{-6}\,M$
17.45 LiF: $4.2 \times 10^{-2}\,M$ BaF_2: $7.5 \times 10^{-3}\,M$
LiF has a greater than the molar solubility than BaF_2.
17.47 2.8×10^{-18} **17.49** $7.0 \times 10^{-3}\,M$ **17.51** 8.0×10^{-7}
17.53 (a) $4.1 \times 10^{-4}\,M$ (b) $8.5 \times 10^{-6}\,M$ (c) $8.5 \times 10^{-7}\,M$
(d) $5.7 \times 10^{-7}\,M$ **17.55** $5.6 \times 10^{-9}\,M$ **17.57** $1.7 \times 10^{-3}\,M$
17.59 (a) $2.8 \times 10^{-11}\,M$ (b) $1.2 \times 10^{-6}\,M$ **17.61** $4.9 \times 10^{-8}\,M$
17.63 2.2 g $Fe(OH)_2$; $1.2 \times 10^{-13}\,M$ **17.65** No precipitate will form.
17.67 (a) No precipitate will form. (b) A precipitate will form.
17.69 No precipitate will form. **17.71** 12.34
17.73 (a) $Mg(OH)_2(s) \rightleftharpoons Mg^{2+}(aq) + 2OH^-(aq)$
$NH_4^+(aq) + OH^-(aq) \rightleftharpoons NH_3(aq) + H_2O$
$Mg(OH)_2(s) + 2NH_4^+(aq) \rightleftharpoons Mg^{2+}(aq) + 2H_2O + 2NH_3(aq)$
(b) 0.68 mol NH_4Cl (c) 11.28
17.75 A precipitate will form. **17.77** 7.09×10^{-12}
17.79 $3.5 \times 10^{-4}\,M$; 0.13 M **17.81** 1.84 g $NaC_2H_3O_2$
17.83 $3.3 \times 10^{-6}\,M$ **17.85** 2×10^{-11}, 1.8 M. **17.87** NiS is very
soluble in 4 M acid. **17.89** $2.3 \times 10^{-8}\,M$ **17.91** 0.045 M H^+;
$pH = 1.35$ **17.93** $4.8 < pH < 8.1$
17.95 $[Fe^{2+}] = 3 \times 10^{-5}\,M$; $[Mn^{2+}] = 0.100\,M$; $pH = 8.11$
17.97 $4.54 < pH < 6.14$
17.99 (a)
$$Cu^{2+}(aq) + 4Cl^-(aq) \rightleftharpoons CuCl_4^{2-}(aq) \quad K_{form} = \frac{[CuCl_4^{2-}]}{[Cu^{2+}][Cl^-]^4}$$
(b)
$$Ag^+(aq) + 2I^-(aq) \rightleftharpoons AgI_2^-(aq) \quad K_{form} = \frac{[AgI_2^-]}{[Ag^+][I^-]^2}$$
(c)
$$Cr^{3+}(aq) + 6NH_3(aq) \rightleftharpoons Cr(NH_3)_6^{3+}(aq) \quad K_{form} = \frac{[Cr(NH_3)_6^{3+}]}{[Cr^{3+}][NH_3]^6}$$
17.101 (a)
$$Co^{3+}(aq) + 6NH_3(aq) \rightleftharpoons Co(NH_3)_6^{3+}(aq) \quad K_{form} = \frac{[Co(NH_3)_6^{3+}]}{[Co^{3+}][NH_3]^6}$$
(b)
$$Hg^{2+}(aq) + 4I^-(aq) \rightleftharpoons HgI_4^{2-}(aq) \quad K_{form} = \frac{[HgI_4^{2-}]}{[Hg^{2+}][I^-]^4}$$
(c)
$$Fe^{2+}(aq) + 6CN^-(aq) \rightleftharpoons Fe(CN)_6^{4-}(aq) \quad K_{form} = \frac{[Fe(CN)_6^{4-}]}{[Fe^{2+}][CN^-]^6}$$
17.103 (a)
$$Co(NH_3)_6^{3+}(aq) \rightleftharpoons Co^{3+}(aq) + 6NH_3(aq) \quad K_{inst} = \frac{[Co^{3+}][NH_3]^6}{[Co(NH_3)_6^{3+}]}$$
(b)
$$HgI_4^{2-}(aq) \rightleftharpoons Hg^{2+}(aq) + 4I^-(aq) \quad K_{inst} = \frac{[Hg^{2+}][I^-]^4}{[HgI_4^{2-}]}$$
(c)
$$Fe(CN)_6^{4-}(aq) \rightleftharpoons Fe^{2+}(aq) + 6CN^-(aq) \quad K_{inst} = \frac{[Fe^{2+}][CN^-]^6}{[Fe(CN)_6^{4-}]}$$
17.105 4.3×10^{-4} **17.107** 412 g NaCN **17.109** 2.0×10^{-4} g AgI
17.111 $4.9 \times 10^{-3}\,M$ **17.113** $1.2 \times 10^{-2}\,M$

17.115 $\dfrac{K_{sp(1)}}{K_{sp(2)}} = 1 \times 10^4$

17.117 (a) 0.34 g AgCl **(b)** $[Ag^+] = 3.6 \times 10^{-2} M$
$[NO_3^-] = 6.0 \times 10^{-2} M$
$[Na^+] = 2.4 \times 10^{-2} M$
$[Cl^-] = 5.0 \times 10^{-9} M$
(c) 40%

17.119 $[Pb^{2+}] = 8.40 \times 10^{-3} M$; $[I^-] = 9.70 \times 10^{-4} M$;
$[Br^-] = 1.58 \times 10^{-2} M$ **17.121** $1.2 \times 10^5 M$; $[PbCl_4]^{2-}$

Chapter 18

Practice Exercises

18.1 $w = -154$ L atm; $q = +154$ L atm
18.2 ΔV is negative for a compression so ΔE increases and T increases. Energy is added to the system in the form of work.
18.3 -2.64 kJ; ΔE is more exothermic. **18.4** -214.6 kJ; 1%
18.5 (a) yes, spontaneous **(b)** no, nonspontaneous
(c) no, nonspontaneous **18.6 (a)** Push the stone uphill, or let it roll downhill. **(b)** The ice in the iced tea can melt, or the glass of iced tea can be put in the freezer. **(c)** Mix sodium hydroxide and hydrochloric acid in water to form sodium chloride.
18.7 Negative **18.8 (a)** negative **(b)** positive **18.9 (a)** negative **(b)** negative **(c)** negative **(d)** positive **18.10 (a)** nonspontaneous **(b)** spontaneous **(c)** spontaneous
18.11 (a) high temperatures **(b)** low temperatures
18.12 -99.1 J K^{-1} **18.13 (a)** -229 J/K **(b)** -120.9 J/K
18.14 98.3 kJ mol^{-1} **18.15** -1482 kJ **18.16** -30.3 kJ
18.17 (a) -69.7 kJ/mol **(b)** -120.1 kJ/mol
18.18 3590 kJ; 5810 kJ; octane **18.19** 788 kJ **18.20** 90.5 J K^{-1}
18.21 341 °C **18.22** Spontaneous **18.23** Not spontaneous
18.24 -4.3 kJ
18.25 $+32.8$ kJ; -34.7 kJ
The equilibrium shifts to products.
18.26 0 kJ mol^{-1}
The system is at equilibrium, so that the reaction will not move.
18.27 The reaction will proceed in the forward direction.
18.28 -33 kJ **18.29** 0.26 **18.30** -260.5 kJ mol^{-1} for 1-propanol, -88 kJ mol^{-1} for 2-bromobutane
18.31 -117 kJ mol^{-1} for C_6H_{12}, $+255.6$ kJ mol^{-1} for C_6H_6

Review Problems

18.55 $+1.000$ kJ, endothermic
18.57 81 J
18.59 (a) $\Delta H° = +24.58$ kJ $\Delta E = +19.6$ kJ
  (b) $\Delta H° = -178$ kJ $\Delta E = -175$ kJ
  (c) $\Delta H° = +847.6$ kJ $\Delta E = \Delta H°$
  (d) $\Delta H° = +65.029$ kJ $\Delta E = \Delta H°$
18.61 -350 kJ, $\Delta E_{217\,°C} = -354$ kJ
18.63 (a) positive **(b)** negative **(c)** negative
18.65 (a) -178 kJ, favored **(b)** -311 kJ, favored **(c)** $+1084.3$ kJ, not favorable **18.67** The second system **18.69** System (*b*)
18.71 (1) the number of moles of products versus reactants
  (2) the state of the products versus reactants
  (3) the complexity of the molecules
18.73 (a) negative **(b)** negative **(c)** negative **(d)** positive
18.75 (a) low temperatures **(b)** high temperatures
18.77 (a) $\Delta S° = -198.3$ J/K, not spontaneous

(b) $\Delta S° = -332.3$ J/K, not favored
(c) $\Delta S° = +92.6$ J/K, favorable
(d) $\Delta S° = +14$ J/K, favorable
(e) $\Delta S° = +159$ J/K, favorable
18.79 (a) -52.8 J mol^{-1} K^{-1} **(b)** -318 J mol^{-1} K^{-1}
(c) -868.9 J mol^{-1} K^{-1} **18.81** -269.7 J/K **18.83** -209 kJ/mol
18.85 (a) -82.3 kJ **(b)** -8.8 kJ **(c)** $+70.7$ kJ
18.87 $\Delta G° = +0.16$ kJ **18.89** -1299.8 kJ **18.91** 333 K
18.93 101 J mol^{-1} K^{-1} **18.95** Spontaneous **18.97 (a)** 9.20×10^{42}
(b) 5.37×10^{-13} **18.99** The system is not at equilibrium and must shift to the right to reach equilibrium. **18.101** $K_p = 8.000 \times 10^8$
This reaction is worth studying as a method for methane production.
18.103 $K_c = 1$ The equilibrium will shift toward the reactants.
18.105 1.16×10^3 kJ/mol **18.107** 354 kJ/mol **18.109** 577.7 kJ/mol
18.111 308.0 kJ/mol **18.113** 85 kJ/mol **18.115** CF_4

Chapter 19

Practice Exercises

19.1 Anode: $Mg(s) \longrightarrow Mg^{2+}(aq) + 2e^-$
Cathode: $Fe^{2+}(aq) + 2e^- \longrightarrow Fe(s)$
Cell notation: $Mg(s) \mid Mg^{2+}(aq) \parallel Fe^{2+}(aq) \mid Fe(s)$

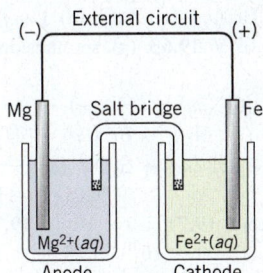

19.2 Anode: $Al(s) \longrightarrow Al^{3+}(aq) + 3e^-$
Cathode: $Ni^{2+}(aq) + 2e^- \longrightarrow Ni(s)$
Overall: $3Ni^{2+}(aq) + 2Al(s) \longrightarrow 2Al^{3+}(aq) + 3Ni(s)$
19.3 Zinc **19.4** -0.44 V, $Fe^{2+}(aq) + 2e^- \longrightarrow Fe(s)$
19.5 Not spontaneous, $Ni(s) + 2Fe^{3+}(aq) \longrightarrow$
$Ni^{2+}(aq) + 2Fe^{2+}(aq)$
19.6 $Br_2(aq) + H_2SO_3(aq) + H_2O \longrightarrow 2Br^-(aq) +$
$SO_4^{2-}(aq) + 4H^+(aq)$
19.7 $NiO_2(s) + Fe(s) + 2H_2O \longrightarrow Ni(OH)_2(s) + Fe(OH)_2(s)$
$E°_{cell} = 1.38$ V
19.8 $5Cr(s) + 3MnO_4^-(aq) + 24H^+(aq) \rightleftharpoons$
$3Mn^{2+}(aq) + 12H_2O(l) + 5Cr^{3+}(aq)$ $E°_{cell} = 2.25$ V
19.9
$3Cu^{2+}(aq) + 2Cr(s) \longrightarrow 3Cu(s) + 2Cr^{3+}(aq)$ $E°_{cell} = 1.08$ V
19.10
$2Fe^{3+}(aq) + Sn^{2+}(aq) \longrightarrow 2Fe^{2+}(aq) + Sn^{4+}(aq)$ $E°_{cell} = 0.62$ V
19.11 (a) $Br_2(aq) + 2I^-(aq) \longrightarrow 2Br^-(aq) + I_2(s)$
  (b) $MnO_4^-(aq) + 5Ag(s) + 4H_2O \longrightarrow$
$Mn^{2+}(aq) + 5Ag^+(aq) + 8H^+(aq)$
19.12 (a) spontaneous **(b)** nonspontaneous
19.13 3 mol e^- **19.14** From Practice Exercise 19.11, **(a)** 102 kJ,
(b) 343 kJ From Practice Exercise 19.12, **(a)** -108 kJ, **(b)** 11.6 kJ
19.15 2.7×10^{-16}, 3.7×10^{15}
19.16
$Ag^+(aq) + Br^-(aq) \longrightarrow AgBr(s)$ $K = \dfrac{1}{[Ag^+][Br^-]} = 4.5 \times 10^{-13}$

19.17 $Cu^{2+}(aq) + Mg(s) \longrightarrow Cu(s) + Mg^{2+}(aq)$ $E_{cell} = 2.82$ V
19.18 3.38 **19.19** $9.6 \times 10^{-6} M$, $3.4 \times 10^{-14} M$

19.20 2.9×10^{-5} M **19.21** I_2 **19.22** $Sn(s)$
19.23 3.23×10^{-2} mol OH^- **19.24** 5.24 min **19.25** 0.0612 A
19.26 $+0.284$ M

Review Problems

19.55 **(a)** anode: $Zn(s) \longrightarrow Zn^{2+}(aq) + 2e^-$
cathode: $Cr^{3+}(aq) + 3e^- \longrightarrow Cr(s)$
cell: $3Zn(s) + 2Cr^{3+}(aq) \longrightarrow 3Zn^{2+}(aq) + 2Cr(s)$
(b) anode: $Pb(s) + HSO_4^-(aq) \longrightarrow$
$$PbSO_4(s) + H^+(aq) + 2e^-$$
cathode: $PbO_2(s) + HSO_4^-(aq) + 3H^+(aq) + 2e^- \longrightarrow$
$$PbSO_4(s) + 2H_2O$$
cell: $Pb(s) + PbO_2(s) + 2HSO_4^{2-}(aq) + 2H^+(aq) \longrightarrow$
$$2PbSO(s) + 2H_2O$$
(c) anode: $Mg(s) \longrightarrow Mg^{2+}(aq) + 2e^-$
cathode: $Sn^{2+}(aq) + 2e^- \longrightarrow Sn(s)$
cell: $Mg(s) + Sn^{2+}(aq) \longrightarrow Mg^{2+}(aq) + Sn(s)$
19.57 **(a)** $Pt(s) \mid Fe^{2+}(aq), Fe^{3+}(aq) \parallel$
$$NO_3^-(aq), H^+(aq) \mid NO(g) \mid Pt(s)$$
(b) $Pt(s) \mid Br_2(aq), Br^-(aq) \parallel Cl^-(aq) \mid Cl_2(g) \mid Pt(s)$
(c) $Ag(s) \mid Ag^+(aq) \parallel Au^{3+}(aq) \mid Au(s)$
19.59 **(a)** $Sn(s)$ **(b)** $Br^-(aq)$ **(c)** $Zn(s)$ **(d)** $I^-(aq)$ **19.61** **(a)** 0.19 V
(b) -0.29 V **(c)** 0.62 V **19.63** **(a)** spontaneous **(b)** spontaneous
(c) spontaneous
19.65 $BrO_3^-(aq) + 6I^-(aq) + 6H^+(aq) \longrightarrow$
$$3I_2(s) + Br^-(aq) + 3H_2O \quad E^\circ_{cell} = 0.90 \text{ V}$$
19.67 $4HOCl(aq) + 2H^+(aq) + S_2O_3^{2-}(aq) \longrightarrow$
$$2Cl_2(g) + H_2O + 2H_2SO_3(aq)$$
19.69 Not spontaneous **19.71** 1.0×10^2 kJ **19.73** **(a)** 0.54 V
(b) -5.2×10^2 kJ **(c)** 2.1×10^{91} **19.75** 0.31 **19.77** 2.38 V
19.79 2.86×10^{-15} M **19.81** 8.66×10^{-14} M **19.83** 0.090 V, 0.105 V
19.85 Cathode: $H_2(g)$
Anode: $O_2(g)$
Net cell reaction: $2H_2O \rightleftharpoons 2H_2(g) + O_2(g)$
19.87 Br_2 and Cu **19.89** $Al^{3+}, Mg^{2+}, Na^+, Ca^{2+}, K^+,$ and Li^+
19.91 **(a)** 0.40 mol e^- **(b)** 0.70 mol e^- **(c)** 4.50 mol e^-
(d) 5.0×10^{-2} mol e^- **19.93** 2.68 g $Fe(OH)_2$ **19.95** 51.5 hr
19.97 66.2 amp **19.99** 0.0996 M OH^-

Chapter 20

Practice Exercises

20.1 57.7 g **20.2** 58.6 g **20.3** 0.514197 u
20.4 1.23057×10^{-12} J/nucleon
20.5 $^{226}_{88}Ra \longrightarrow ^{222}_{86}Rn + ^4_2He + ^0_0\gamma$ An alpha particle is emitted.
20.6 $^{131}_{53}I \longrightarrow ^{131}_{54}Xe + ^0_{-1}e + \bar{\nu}$ **20.7** $^{11}_6C \longrightarrow ^{11}_5B + ^0_1e + \nu$
20.8 $^{13}_4Be \longrightarrow ^{12}_4Be + ^1_0n$
20.9 $^{72}_{34}Se + ^0_{-1}e \longrightarrow ^{72}_{33}As + X \text{ ray} + \nu$
20.10 $^{242}_{96}Cm \longrightarrow ^{238}_{94}Pu + ^4_2He$ **20.11** $^{262}_{106}Sg \longrightarrow ^{258}_{104}Rf + ^4_2He$
20.12 49.2% Pu **20.13** 7.05×10^4 atoms Rn-222 **20.14** 20 m
20.15 100 units

Review Problems

20.51 **(a)** 1.01 kg **(b)** 3.91 kg **(c)** 12.3 kg **20.53** 1.11×10^{-11} g
20.55 $1.77 \times 10^{-8}\%$ **20.57** 1.8×10^{-13} J per nucleon
20.59 0.5760271 u, 8.5966×10^{-11} J
20.61 **(a)** $^{211}_{83}Bi$ **(b)** $^{177}_{72}Hf$ **(c)** $^{216}_{84}Po$ **(d)** $^{19}_9F$
20.63 **(a)** $^{242}_{94}Pu \longrightarrow ^4_2He + ^{238}_{92}U$ **(b)** $^{28}_{12}Mg \longrightarrow ^0_{-1}e + ^{28}_{13}Al + \nu$
(c) $^{26}_{14}Si \longrightarrow ^0_1e + \bar{\nu} + ^{26}_{13}Al$ **(d)** $^{37}_{18}Ar + ^0_{-1}e \longrightarrow ^{37}_{17}Cl$

20.65 **(a)** $^{261}_{102}No$ **(b)** $^{211}_{82}Pb$ **(c)** $^{141}_{61}Pm$ **(d)** $^{179}_{74}W$
20.67 ^1_0n **20.69** Positron emission, $^{38}_{19}K \longrightarrow ^0_1e + ^{38}_{18}Ar$
20.71 Alpha emission, $^{209}_{84}Po \longrightarrow ^{205}_{82}Pb + ^4_2He$
20.73 0.0469 mg I-131
20.75 $^{53}_{24}Cr^*, ^{51}_{23}V + ^2_1H \longrightarrow ^{53}_{24}Cr^* \longrightarrow ^1_1p + ^{52}_{23}V$
20.77 $^{80}_{35}Br$
20.79 $^{55}_{26}Fe; ^{55}_{25}Mn + ^1_1p \longrightarrow ^1_0n + ^{55}_{26}Fe$
20.81 $^{70}_{30}Zn + ^{208}_{82}Pb \longrightarrow ^{278}_{112}Uub \longrightarrow ^1_0n + ^{277}_{112}Uub$
20.83 6.3 m **20.85** 6.6 m
20.87 1 Ci; 3.7×10^{10} Bq.
20.89 2.4×10^7 s^{-1}, 2.4×10^7 Bq, 6.5×10^2 μCi
20.91 1.0×10^{-6} s^{-1}, 6.9×10^5 s
20.93 The chemical product is $BaCl_2$. 3.2%.
20.95 2.84×10^{-3} Ci g^{-1} **20.97** 6.20×10^{-3} M Ca^{2+}
20.99 7.6×10^8 years old **20.101** 2.7×10^4 years ago
20.103 $^{235}_{92}U + ^1_0n \longrightarrow ^{94}_{38}Sr + ^{140}_{54}Xe + 2^1_0n$

Chapter 21

Practice Exercises

21.1 $[Ag(S_2O_3)_2]^{3-}$, $(NH_4)_3Ag(S_2O_3)_2$ **21.2** $AlCl_3 \cdot 6H_2O$,
$[Al(H_2O)_6]^{3+}$ **21.3** $[CrCl_2(H_2O)_4]^+$, The counter ion would be a
halide. **21.4** **(a)** $[SnCl_6]^{2-}$ **(b)** $(NH_4)_2[Fe(CN)_4(H_2O)_2]$
(c) $OsBr_2(H_2NCH_2CH_2NH_2)_2$ or $OsBr_2(en)_2$
21.5 **(a)** potassium hexacyanoferrate(III)
(b) dichlorobis(ethylenediamine)chromium(III) sulfate
(c) hexaaquacobalt(II) hexafluorochromate(II) **21.6** six
21.7 **(a)** six **(b)** six **(c)** six **21.8** **(a)** four, square planar
(b) two, linear **21.9** **(a)** optical isomers, or enantiomers
(b) geometric isomers
21.10 Six isomers

21.11

Low spin has zero unpaired electrons and is diamagnetic.
High spin has four unpaired electrons and is paramagnetic.
Diamagnetic compounds have no unpaired electrons.

21.12

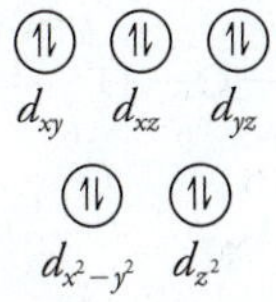

$$d_{xy} \quad d_{xz} \quad d_{yz}$$

$$d_{x^2-y^2} \quad d_{z^2}$$

The Ni(0) complex has no unpaired electrons.
21.13 Porphyrin, imidazole, and oxygen.

Review Problems

21.53 -3, $[Fe(CN)_6]^{3-}$, hexacyanoferrate(III) ion
21.55 $[Cr(NH_3)_2(NO_2)_4]^-$ **21.57** (a) ammine (b) azido
(c) sulfato (d) acetato **21.59** (a) hexaamminenickel(II) chloride
(b) triamminetrichlorochromate(II) ion (c) hexanitrocobaltate(III)
ion (d) diamminetetracyanomanganate(II) ion (e) potassium
trioxalatoferrate(III) or potassium trisoxalatoferrate(III)
21.61 (a) $[Fe(CN)_2(H_2O)_4]^+$ (b) $[Ni(C_2O_4)(NH_3)_4]$
(c) $[Al(H_2O)_5(OH)]Cl_2$ (d) $K_3[Mn(SCN)_6]$ (e) $[CuCl_4]^{2-}$
21.63 Six
21.65 (a)

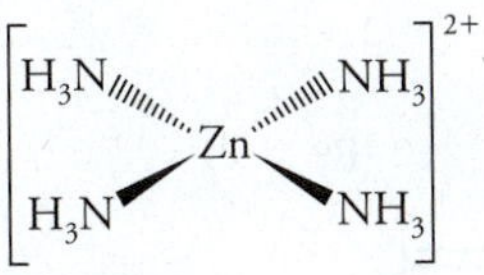

(b)

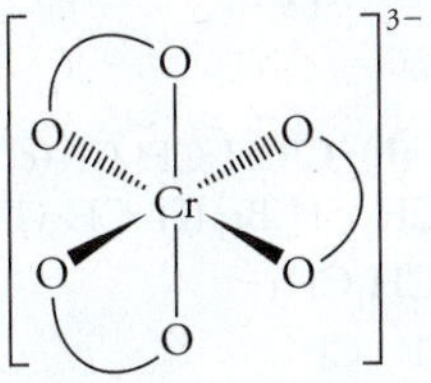

The curved lines represent the backbone of the oxalate ion.
21.67

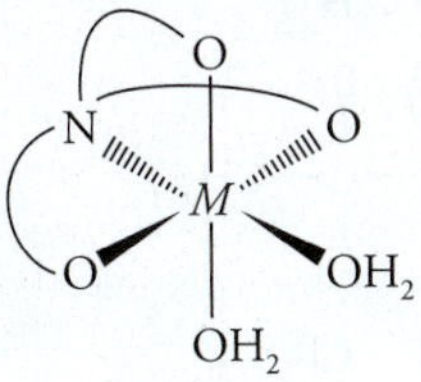

The curved lines represent $-CH_2-\overset{\overset{\displaystyle O}{\|}}{C}-$ groups.
21.69 Identical
21.71 cis: trans:

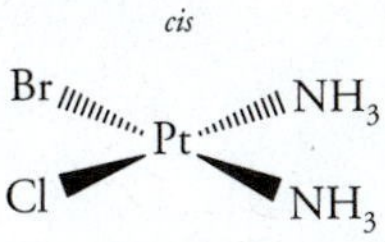

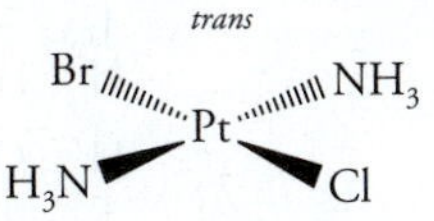

21.73 The cis isomer is chiral:

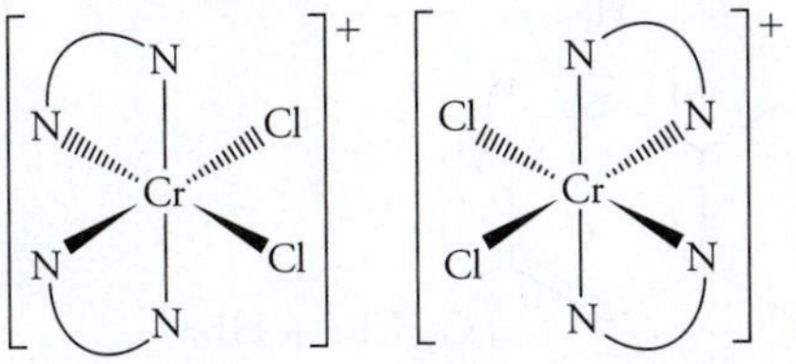

The trans isomer is nonchiral:

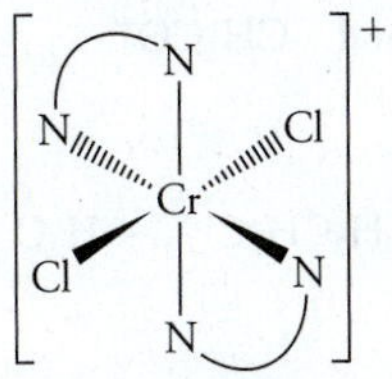

21.75 (a) $[Cr(H_2O)_6]^{3+}$ (b) $[Cr(en)_3]^{3+}$ **21.77** $[Cr(CN)_6]^{3-}$
21.79 (a) $[RuCl(NH_3)_5]^{2+}$ (b) $[Ru(NH_3)_6]^{3+}$ **21.81** CoA_6^{3+}
21.83 High-spin; it cannot be diamagnetic.
21.85

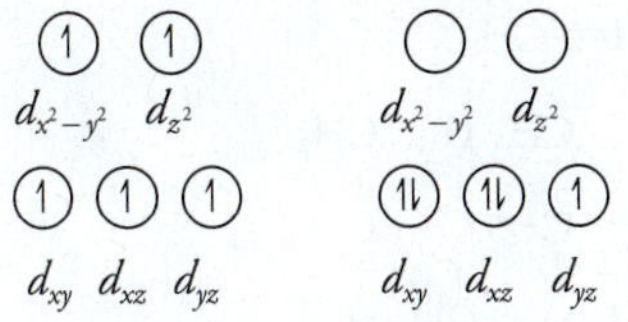

$$d_{x^2-y^2} \quad d_{z^2} \qquad\qquad d_{x^2-y^2} \quad d_{z^2}$$

$$d_{xy} \quad d_{xz} \quad d_{yz} \qquad d_{xy} \quad d_{xz} \quad d_{yz}$$

$$Fe(H_2O)_6^{3+} \qquad\qquad Fe(CN)_6^{3-}$$

Chapter 22

Practice Exercises

22.1 (a) C_6H_{14}, $CH_3CH_2CH_2CH_2CH_2CH_3$
(b) C_6H_{14}, $(CH_3)_2CH_2CH_2(CH_3)_2$
(c) $C_2H_3Cl_3$, $ClCH_2CHCl_2$

22.2

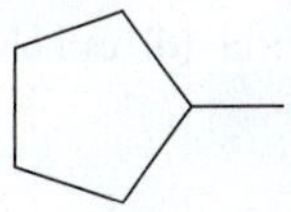

$-CH_3$,

C_6H_{12}

22.3 2,2–dimethylpropane **22.4** (a) 3-methylhexane
(b) 4-ethyl-2,3-dimethylheptane (c) 5-ethyl-2,4,6-trimethyloctane

22.5

$$CH_3CH_2 - \overset{\overset{\displaystyle OH}{|}}{\underset{\underset{\displaystyle H}{|}}{C}} - CH_3$$

22.6 (a) $CH_3\overset{\overset{\displaystyle O}{\|}}{C}H$ or $CH_3\overset{\overset{\displaystyle O}{\|}}{C}OH$ (b) $CH_3CH_2\overset{\overset{\displaystyle O}{\|}}{C}CH_2CH_3$

(c) No oxidation.

22.7 (a) $CH_3\overset{\overset{\displaystyle Cl}{|}}{C}HCH_2CH_3$ (b) $CH_3CH=CH_2$

22.8 $CH_3CH_2\overset{\overset{\displaystyle O}{\|}}{C}OH + CH_3OH \underset{\longleftarrow}{\overset{H^+ \text{ catalyst}}{\rightleftharpoons}}$

$$CH_3CH_2\overset{\overset{\displaystyle O}{\|}}{C}OCH_3 + H_2O$$

methyl propanoate

22.9 (a) $CH_3\overset{\overset{\displaystyle CH_3}{|}}{C}HCH_2\overset{\overset{\displaystyle O}{\|}}{C}H$ (b) $CH_3CH_2\overset{\overset{\displaystyle O}{\|}}{C}OCH_2CH_3$

(c) $CH_3\overset{\overset{\displaystyle O}{\|}}{C}CH_2CH_3$

22.10 CH_3CH_2OH and CH_3CO^- (with =O above C)

22.11 $CH_3CH_2CNHCH_2CH_2CH_3 + H_2O$ (with =O above the first C of CNH)

22.12 (a) $CH_3CH_2COH + CH_3CH_2NH_2$ (with =O above C of COH)

(b) $CH_3CH_2CO^- + HOCH(CH_3)_2$ (with =O above C)

(c) $CH_3CH_2CH=CH_2$

22.13

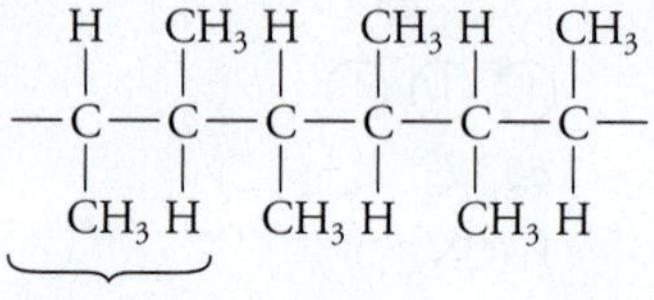

H CH₃ H CH₃ H CH₃
—C—C—C—C—C—C—
CH₃ H CH₃ H CH₃ H

Repeat unit

22.14

F F F F F F
—C—C—C—C—C—C—
F F F F F F

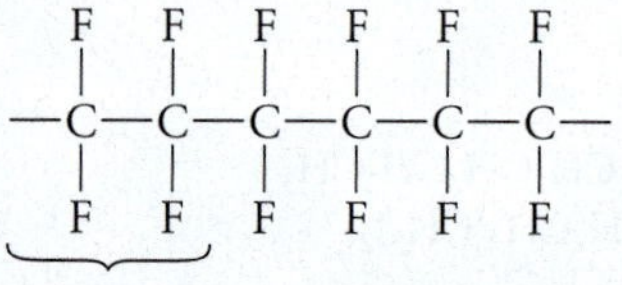

Repeat unit

22.15 (a) lipid **(b)** carbohydrate **(c)** amino acid **(d)** carbohydrate **(e)** amino acid **(f)** lipid

22.16

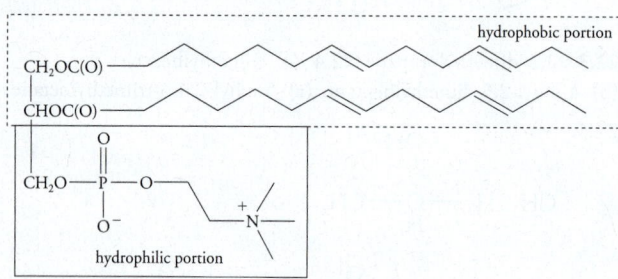

hydrophobic portion

$CH_2OC(O)$—

$CHOC(O)$—

CH_2O—P—O—CH_2CH_2—N^+(CH₃)₃

hydrophilic portion

22.17

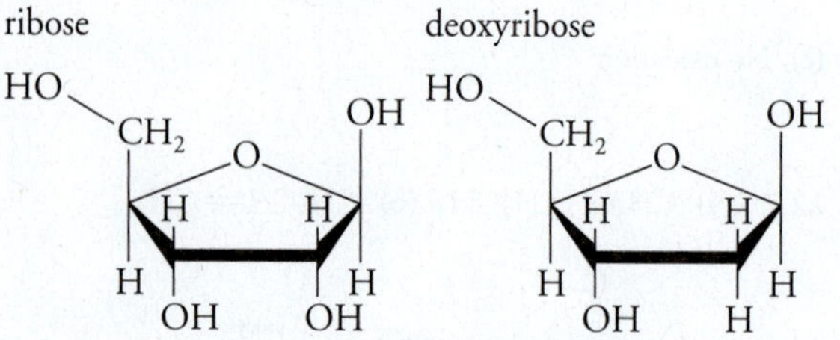

ribose   deoxyribose

The difference is the extra OH group on the ribose ring.

22.18 A hydrogen bonds to T using N—H—O and N—H—N hydrogen bonds. G hydrogen bonds to C using N—H—O, N—H—N, and O—H—N hydrogen bonds

Review Problems

22.89 (a)

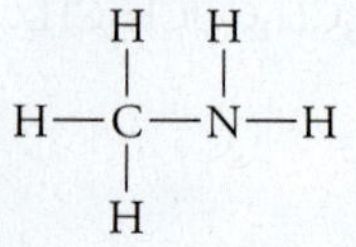

H H   H
H—C—N—H **(b)** Br—C—Br
H   H

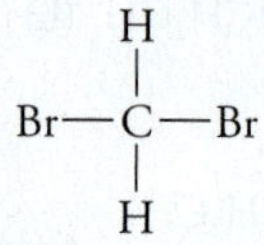

(c)

Cl
H—C—Cl
Cl

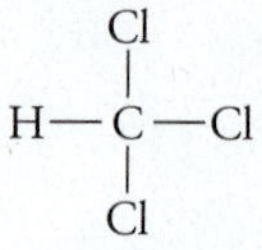

(d)

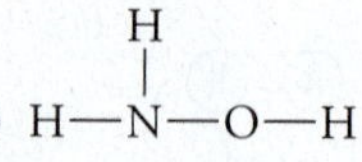

H
H—N—O—H

(e) 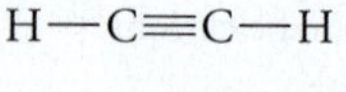 H—C≡C—H **(f)**

H H
H—N—N—H

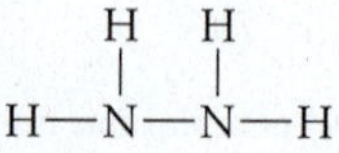

22.91 (a) alkene **(b)** alcohol **(c)** ester **(d)** carboxylic acid **(e)** amine **(f)** alcohol **22.93** b, c, d, e, f **22.95 (a)** amine **(b)** amine **(c)** amide **(d)** amine, ketone **22.97 (a)** identical **(b)** identical **(c)** unrelated **(d)** isomers **(e)** identical **(f)** identical **(g)** isomers **22.99 (a)** pentane **(b)** 2-methylpentane **(c)** 2,4-dimethylhexane

22.101 (a) no isomers

(b)

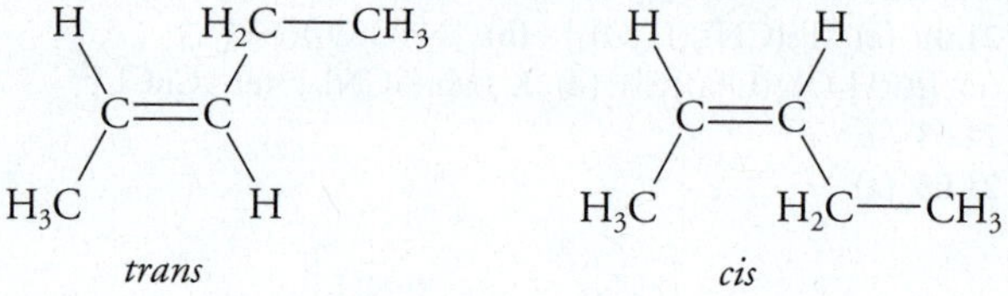

H H_2C—CH_3 $\backslash$C=C$/$ H_3C H *trans*

H H $\backslash$C=C$/$ H_3C H_2C—CH_3 *cis*

(c)

Br Br
C=C
H₃C H   *cis*

Br H
C=C
H₃C Br   *trans*

22.103 (a) CH_3CH_3 **(b)** $ClCH_2CH_2Cl$ **(c)** $BrCH_2CH_2Br$ **(d)** CH_3CH_2Cl **(e)** CH_3CH_2Br **(f)** CH_3CH_2OH

22.105 (a) $CH_3CH_2CH_2CH_3$

(b)

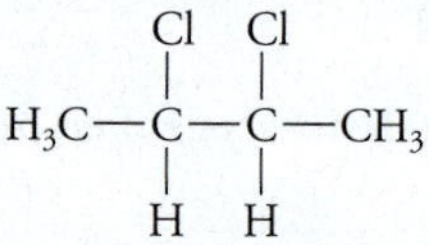

Cl Cl
H₃C—C—C—CH₃
H H

(c)

Br Br
H₃C—C—C—CH₃
H H

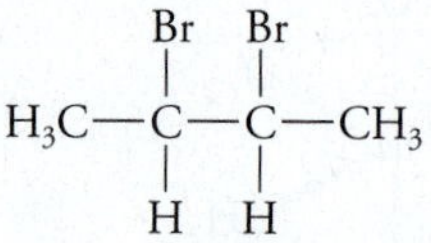

(d)

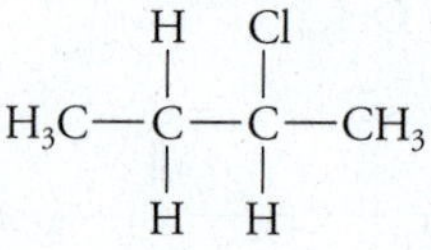

H Cl
H₃C—C—C—CH₃
H H

(e)

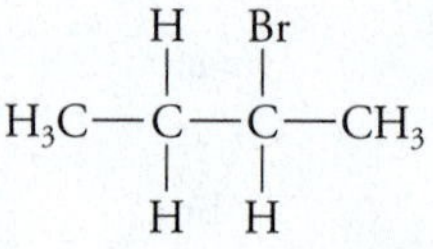

H Br
H₃C—C—C—CH₃
H H

(f)

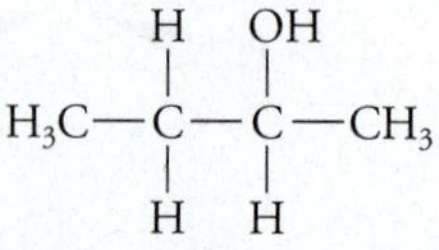

H OH
H₃C—C—C—CH₃
H H

22.107

Br ... Br , $C_6H_5Br + HBr$

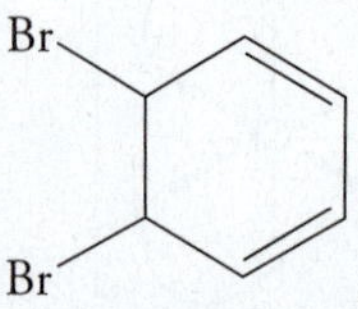

22.109 CH_3OH, methanol, methyl alcohol
CH_3CH_2OH, ethanol, ethyl alcohol
$CH_3CH_2CH_2OH$, 1–propanol, propyl alcohol
$(CH_3)_2COH$, 2–propanol, isopropyl alcohol

22.111 $CH_3CH_2CH_2—O—CH_3$, methyl propyl ether
$CH_3CH_2—O—CH_2CH_3$, diethyl ether or ether
$(CH_3)_2CH—O—CH_3$, methyl 2–propyl ether

22.113 (a)

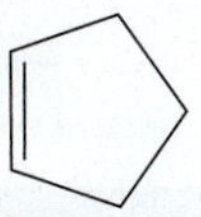

(b)

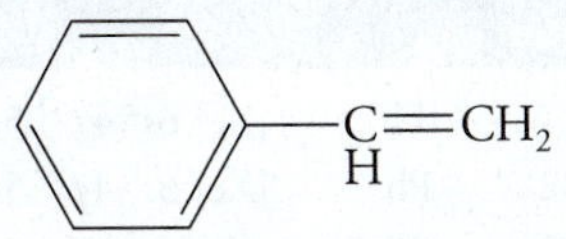

(c)

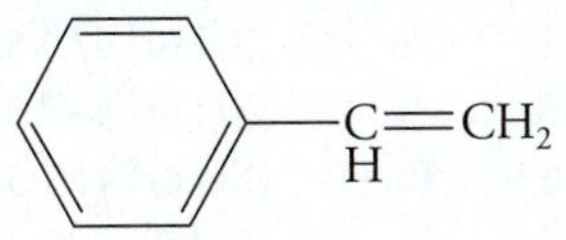

22.115 (a) 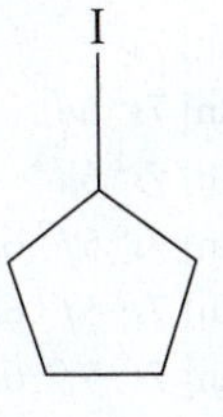 **(b)**

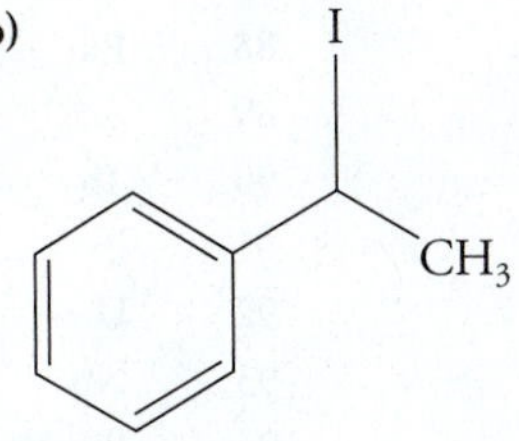

(c)

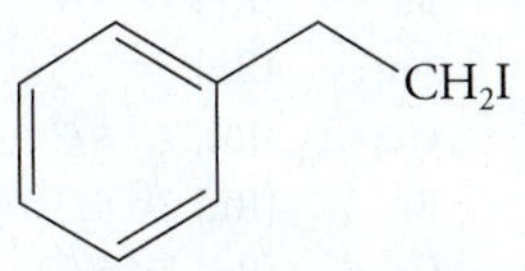

22.117 $CH_2=CHCH_2CH_3$, 1–butene
$CH_3CH=CHCH_3$, 2–butene

22.119 The aldehyde is more easily oxidized.

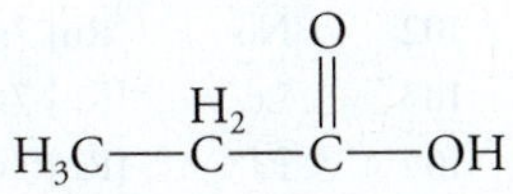

22.121 (a) $CH_3CH_2CO_2H$
(b) $CH_3CH_2CO_2H + CH_3OH$
(c) $Na^+ + CH_3CH_2CH_2CO_2^- + H_2O$

22.123 $CH_3CO_2H + CH_3CH_2NHCH_2CH_3$

22.125

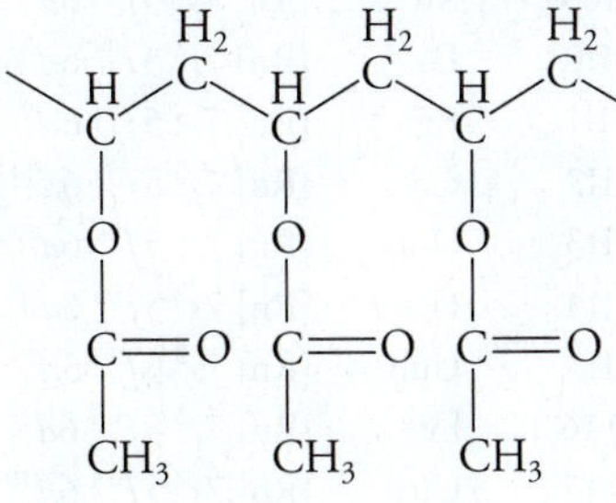

22.127

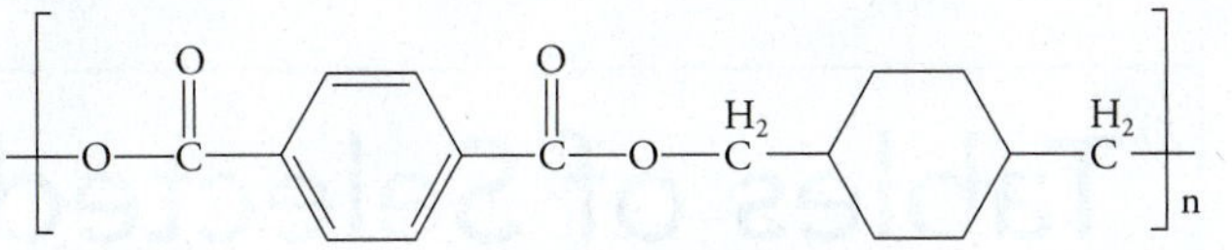

22.129

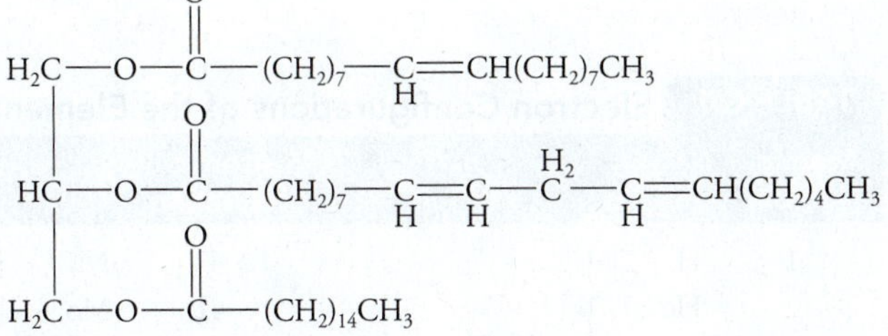

22.131

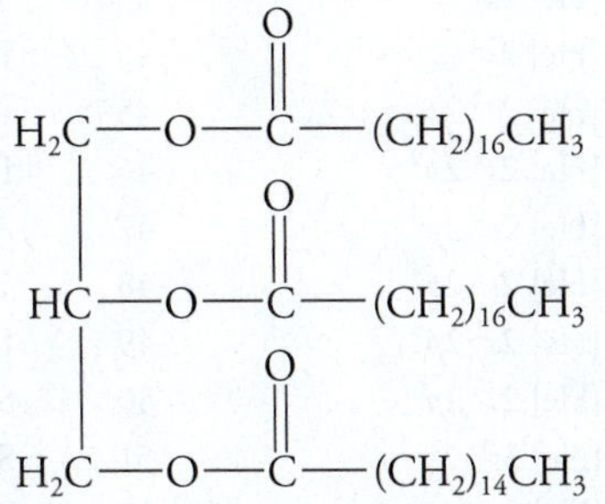

22.133 Hydrophobic sites are composed of fatty acid units. Hydrophilic sites are composed of charged units.

22.135

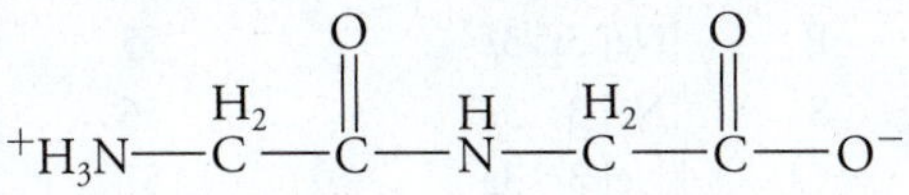

22.137

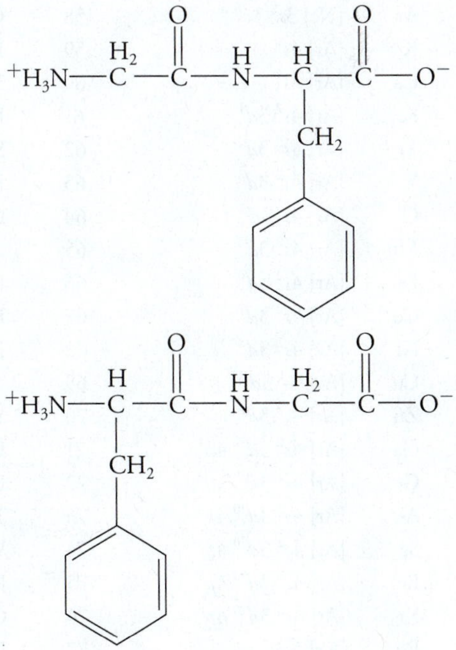

Tables of Selected Data

TABLE C.1 Electron Configurations of the Elements

Atomic Number			Atomic Number			Atomic Number		
1	H	$1s^1$	41	Nb	$[Kr]\,5s^1\,4d^4$	81	Tl	$[Xe]\,6s^2\,4f^{14}\,5d^{10}\,6p^1$
2	He	$1s^2$	42	Mo	$[Kr]\,5s^1\,4d^5$	82	Pb	$[Xe]\,6s^2\,4f^{14}\,5d^{10}\,6p^2$
3	Li	$[He]\,2s^1$	43	Tc	$[Kr]\,5s^2\,4d^5$	83	Bi	$[Xe]\,6s^2\,4f^{14}\,5d^{10}\,6p^3$
4	Be	$[He]\,2s^2$	44	Ru	$[Kr]\,5s^1\,4d^7$	84	Po	$[Xe]\,6s^2\,4f^{14}\,5d^{10}\,6p^4$
5	B	$[He]\,2s^2\,2p^1$	45	Rh	$[Kr]\,5s^1\,4d^8$	85	At	$[Xe]\,6s^2\,4f^{14}\,5d^{10}\,6p^5$
6	C	$[He]\,2s^2\,2p^2$	46	Pd	$[Kr]\,4d^{10}$	86	Rn	$[Xe]\,6s^2\,4f^{14}\,5d^{10}\,6p^6$
7	N	$[He]\,2s^2\,2p^3$	47	Ag	$[Kr]\,5s^1\,4d^{10}$	87	Fr	$[Rn]\,7s^1$
8	O	$[He]\,2s^2\,2p^4$	48	Cd	$[Kr]\,5s^2\,4d^{10}$	88	Ra	$[Rn]\,7s^2$
9	F	$[He]\,2s^2\,2p^5$	49	In	$[Kr]\,5s^2\,4d^{10}\,5p^1$	89	Ac	$[Rn]\,7s^2\,6d^1$
10	Ne	$[He]\,2s^2\,2p^6$	50	Sn	$[Kr]\,5s^2\,4d^{10}\,5p^2$	90	Th	$[Rn]\,7s^2\,6d^2$
11	Na	$[Ne]\,3s^1$	51	Sb	$[Kr]\,5s^2\,4d^{10}\,5p^3$	91	Pa	$[Rn]\,7s^2\,5f^3\,6d^1$
12	Mg	$[Ne]\,3s^2$	52	Te	$[Kr]\,5s^2\,4d^{10}\,5p^4$	92	U	$[Rn]\,7s^2\,5f^3\,6d^1$
13	Al	$[Ne]\,3s^2\,3p^1$	53	I	$[Kr]\,5s^2\,4d^{10}\,5p^5$	93	Np	$[Rn]\,7s^2\,5f^4\,6d^1$
14	Si	$[Ne]\,3s^2\,3p^2$	54	Xe	$[Kr]\,5s^2\,4d^{10}\,5p^6$	94	Pu	$[Rn]\,7s^2\,5f^6$
15	P	$[Ne]\,3s^2\,3p^3$	55	Cs	$[Xe]\,6s^1$	95	Am	$[Rn]\,7s^2\,5f^7$
16	S	$[Ne]\,3s^2\,3p^4$	56	Ba	$[Xe]\,6s^2$	96	Cm	$[Rn]\,7s^2\,5f^7\,6d^1$
17	Cl	$[Ne]\,3s^2\,3p^5$	57	La	$[Xe]\,6s^2\,5d^1$	97	Bk	$[Rn]\,7s^2\,5f^9$
18	Ar	$[Ne]\,3s^2\,3p^6$	58	Ce	$[Xe]\,6s^2\,4f^1\,5d^1$	98	Cf	$[Rn]\,7s^2\,5f^{10}$
19	K	$[Ar]\,4s^1$	59	Pr	$[Xe]\,6s^2\,4f^3$	99	Es	$[Rn]\,7s^2\,5f^{11}$
20	Ca	$[Ar]\,4s^2$	60	Nd	$[Xe]\,6s^2\,4f^4$	100	Fm	$[Rn]\,7s^2\,5f^{12}$
21	Sc	$[Ar]\,4s^2\,3d^1$	61	Pm	$[Xe]\,6s^2\,4f^5$	101	Md	$[Rn]\,7s^2\,5f^{13}$
22	Ti	$[Ar]\,4s^2\,3d^2$	62	Sm	$[Xe]\,6s^2\,4f^6$	102	No	$[Rn]\,7s^2\,5f^{14}$
23	V	$[Ar]\,4s^2\,3d^3$	63	Eu	$[Xe]\,6s^2\,4f^7$	103	Lr	$[Rn]\,7s^2\,5f^{14}\,6d^1$
24	Cr	$[Ar]\,4s^1\,3d^5$	64	Gd	$[Xe]\,6s^2\,4f^7\,5d^1$	104	Rf	$[Rn]\,7s^2\,5f^{14}\,6d^2$
25	Mn	$[Ar]\,4s^2\,3d^5$	65	Tb	$[Xe]\,6s^2\,4f^9$	105	Db	$[Rn]\,7s^2\,5f^{14}\,6d^3$
26	Fe	$[Ar]\,4s^2\,3d^6$	66	Dy	$[Xe]\,6s^2\,4f^{10}$	106	Sg	$[Rn]\,7s^2\,5f^{14}\,6d^4$
27	Co	$[Ar]\,4s^2\,3d^7$	67	Ho	$[Xe]\,6s^2\,4f^{11}$	107	Bh	$[Rn]\,7s^2\,5f^{14}\,6d^5$
28	Ni	$[Ar]\,4s^2\,3d^8$	68	Er	$[Xe]\,6s^2\,4f^{12}$	108	Hs	$[Rn]\,7s^2\,5f^{14}\,6d^6$
29	Cu	$[Ar]\,4s^1\,3d^{10}$	69	Tm	$[Xe]\,6s^2\,4f^{13}$	109	Mt	$[Rn]\,7s^2\,5f^{14}\,6d^7$
30	Zn	$[Ar]\,4s^2\,3d^{10}$	70	Yb	$[Xe]\,6s^2\,4f^{14}$	110	Ds	$[Rn]\,7s^2\,5f^{14}\,6d^8$
31	Ga	$[Ar]\,4s^2\,3d^{10}\,4p^1$	71	Lu	$[Xe]\,6s^2\,4f^{14}\,5d^1$	111	Rg	$[Rn]\,7s^2\,5f^{14}\,6d^9$
32	Ge	$[Ar]\,4s^2\,3d^{10}\,4p^2$	72	Hf	$[Xe]\,6s^2\,4f^{14}\,5d^2$	112	Cn	$[Rn]\,7s^2\,5f^{14}\,6d^{10}$
33	As	$[Ar]\,4s^2\,3d^{10}\,4p^3$	73	Ta	$[Xe]\,6s^2\,4f^{14}\,5d^3$	113	Uut	$[Rn]\,7s^2\,5f^{14}\,6d^{10}\,7p^1$
34	Se	$[Ar]\,4s^2\,3d^{10}\,4p^4$	74	W	$[Xe]\,6s^2\,4f^{14}\,5d^4$	114	Fl	$[Rn]\,7s^2\,5f^{14}\,6d^{10}\,7p^2$
35	Br	$[Ar]\,4s^2\,3d^{10}\,4p^5$	75	Re	$[Xe]\,6s^2\,4f^{14}\,5d^5$	115	Uup	$[Rn]\,7s^2\,5f^{14}\,6d^{10}\,7p^3$
36	Kr	$[Ar]\,4s^2\,3d^{10}\,4p^6$	76	Os	$[Xe]\,6s^2\,4f^{14}\,5d^6$	116	Lv	$[Rn]\,7s^2\,5f^{14}\,6d^{10}\,7p^4$
37	Rb	$[Kr]\,5s^1$	77	Ir	$[Xe]\,6s^2\,4f^{14}\,5d^7$	117	Uus	$[Rn]\,7s^2\,5f^{14}\,6d^{10}\,7p^5$
38	Sr	$[Kr]\,5s^2$	78	Pt	$[Xe]\,6s^1\,4f^{14}\,5d^9$	118	Uuo	$[Rn]\,7s^2\,5f^{14}\,6d^{10}\,7p^6$
39	Y	$[Kr]\,5s^2\,4d^1$	79	Au	$[Xe]\,6s^1\,4f^{14}\,5d^{10}$			
40	Zr	$[Kr]\,5s^2\,4d^2$	80	Hg	$[Xe]\,6s^2\,4f^{14}\,5d^{10}$			

TABLE C.2 Thermodynamic Data for Selected Elements, Compounds, and Ions (25 °C)

Substance	ΔH_f° (kJ mol^{-1})	S° (J mol^{-1} K^{-1})	ΔG_f° (kJ mol^{-1})	Substance	ΔH_f° (kJ mol^{-1})	S° (J mol^{-1} K^{-1})	ΔG_f° (kJ mol^{-1})
Aluminum				$HBr(g)$	−36	198.5	+53.1
$Al(s)$	0	28.3	0	$Br^-(aq)$	−121.55	82.4	−103.96
$Al^{3+}(aq)$	−524.7		−481.2	**Cadmium**			
$AlCl_3(s)$	−704	110.7	−629	$Cd(s)$	0	51.8	0
$Al_2O_3(s)$	−1669.8	51.0	−1576.4	$Cd^{2+}(aq)$	−75.90	−73.2	−77.61
$Al_2(SO_4)_3(s)$	−3441	239	−3100	$CdCl_2(s)$	−392	115	−344
				$CdO(s)$	−258.2	54.8	−228.4
Arsenic				$CdS(s)$	−162	64.9	−156
$As(s)$	0	35.1	0	$CdSO_4(s)$	−933.5	123	−822.6
$AsH_3(g)$	+66.4	223	+68.9	**Calcium**			
$As_4O_6(s)$	−1314	214	−1153	$Ca(s)$	0	41.4	0
$As_2O_5(s)$	−925	105	−782	$Ca^{2+}(aq)$	−542.83	−53.1	−553.58
$H_3AsO_3(aq)$	−742.2			$CaCO_3(s)$	−1207	92.9	−1128.8
$H_3AsO_4(aq)$	−902.5			$CaF_2(s)$	−741	80.3	−1166
Barium				$CaCl_2(s)$	−795.0	114	−750.2
$Ba(s)$	0	66.9	0	$CaBr_2(s)$	−682.8	130	−663.6
$Ba^{2+}(aq)$	−537.6	9.6	−560.8	$CaI_2(s)$	−535.9	143	
$BaCO_3(s)$	−1219	112	−1139	$CaO(s)$	−635.5	40	−604.2
$BaCrO_4(s)$	−1428.0			$Ca(OH)_2(s)$	−986.59	76.1	−896.76
$BaCl_2(s)$	−860.2	125	−810.8	$Ca_3(PO_4)_2(s)$	−4119	241	−3852
$BaO(s)$	−553.5	70.4	−525.1	$CaSO_3(s)$	−1156		
$Ba(OH)_2(s)$	−998.22	−8	−875.3	$CaSO_4(s)$	−1433	107	−1320.3
$Ba(NO_3)_2(s)$	−992	214	−795	$CaSO_4 \cdot \frac{1}{2}H_2O(s)$	−1575.2	131	−1435.2
$BaSO_4(s)$	−1465	132	−1353	$CaSO_4 \cdot \frac{1}{2}2H_2O(s)$	−2021.1	194.0	−1795.7
Beryllium				**Carbon**			
$Be(s)$	0	9.50	0	$C(s, \text{graphite})$	0	5.69	0
$BeCl_2(s)$	−468.6	89.9	−426.3	$C(s, \text{diamond})$	+1.88	2.4	+2.9
$BeO(s)$	−611	14	−582	$CCl_4(l)$	−134	214.4	−65.3
Bismuth				$CO(g)$	−110.5	197.9	−137.3
$Bi(s)$	0	56.9	0	$CO_2(g)$	−393.5	213.6	−394.4
$BiCl_3(s)$	−379	177	−315	$CO_2(aq)$	−413.8	117.6	−385.98
$Bi_2O_3(s)$	−576	151	−497	$H_2CO_3(aq)$	−699.65	187.4	−623.08
Boron				$HCO_3^-(aq)$	−691.99	91.2	−586.77
$B(s)$	0	5.87	0	$CO_3^{2-}(aq)$	−677.14	−56.9	−527.81
$BCl_3(g)$	−404	290	−389	$CS_2(l)$	+89.5	151.3	+65.3
$B_2H_6(g)$	+36	232	+87	$CS_2(g)$	+117	237.7	+67.2
$B_2O_3(s)$	−1273	53.8	−1194	$HCN(g)$	+135.1	201.7	+124.7
$B(OH)_3(s)$	−1094	88.8	−969	$CN^-(aq)$	+150.6	94.1	+172.4
Bromine				$CH_4(g)$	−74.848	186.2	−50.79
$Br_2(l)$	0	152.2	0	$C_2H_2(g)$	+226.75	200.8	+209
$Br_2(g)$	+30.9	245.4	+3.11	$C_2H_4(g)$	+52.284	219.8	+68.12

(Continued)

TABLE C.2　Thermodynamic Data for Selected Elements, Compounds, and Ions (25 °C) (Continued)

Substance	ΔH_f° (kJ mol^{-1})	S° (J mol^{-1} K^{-1})	ΔG_f° (kJ mol^{-1})	Substance	ΔH_f° (kJ mol^{-1})	S° (J mol^{-1} K^{-1})	ΔG_f° (kJ mol^{-1})
$C_2H_6(g)$	−84.667	229.5	−32.9	$CuCl_2(s)$	−172	119	−131
$C_3H_8(g)$	−104	269.9	−23	$Cu_2O(s)$	−168.6	93.1	−146.0
$C_4H_{10}(g)$	−126	310.2	−17.0	$CuO(s)$	−155	42.6	−127
$C_6H_6(l)$	+49.0	173.3	+124.3	$Cu_2S(s)$	−79.5	121	−86.2
$CH_3OH(l)$	−238.6	126.8	−166.2	$CuS(s)$	−53.1	66.5	−53.6
$C_2H_5OH(l)$	−277.63	161	−174.8	$CuSO_4(s)$	−771.4	109	−661.8
$HCHO_2(g)$	−363	251	+335	$CuSO_4 \cdot 5H_2O(s)$	−2279.7	300.4	−1879.7
$HC_2H_3O_2(l)$	−487.0	160	−392.5	**Fluorine**			
$HCHO(g)$	−108.6	218.8	−102.5	$F_2(g)$	0	202.7	0
$CH_3CHO(g)$	−167	250	−129	$F^-(aq)$	−332.6	−13.8	−278.8
$(CH_3)_2CO(l)$	−248.1	200.4	−155.4	$HF(g)$	−271	173.5	−273
$C_6H_5CO_2H(s)$	−385.1	167.6	−245.3	**Gold**			
$CO(NH_2)_2(s)$	−333.19	104.6	−197.2	$Au(s)$	0	47.7	0
$CO(NH_2)_2(aq)$	−391.2	173.8	−203.8	$Au_2O_3(s)$	+80.8	125	+163
$CH_2(NH_2)CO_2H(s)$	−532.9	103.5	−373.4	$AuCl_3(s)$	−118	148	−48.5
Chlorine				**Hydrogen**			
$Cl_2(g)$	0	223.0	0	$H_2(g)$	0	130.6	0
$Cl^-(aq)$	−167.2	56.5	−131.2	$H_2O(l)$	−285.9	69.96	−237.2
$HCl(g)$	−92.30	186.7	−95.27	$H_2O(g)$	−241.8	188.7	−228.6
$HCl(aq)$	−167.2	56.5	−131.2	$H_2O_2(l)$	−187.6	109.6	−120.3
$HClO(aq)$	−131.3	106.8	−80.21	$H_2Se(g)$	+76	219	+62.3
Chromium				$H_2Te(g)$	+154	234	+138
$Cr(s)$	0	23.8	0	**Iodine**			
$Cr^{3+}(aq)$	−232			$I_2(s)$	0	116.1	0
$CrCl_2(s)$	−326	115	−282	$I_2(g)$	+62.4	260.7	+19.3
$CrCl_3(s)$	−563.2	126	−493.7	$HI(g)$	+26.6	206	+1.30
$Cr_2O_3(s)$	−1141	81.2	−1059	**Iron**			
$CrO_3(s)$	−585.8	72.0	−506.2	$Fe(s)$	0	27	0
$(NH_4)_2Cr_2O_7(s)$	−1807			$Fe^{2+}(aq)$	−89.1	−137.7	−78.9
$K_2Cr_2O_7(s)$	−2033.01			$Fe^{3+}(aq)$	−48.5	−315.9	−4.7
Cobalt				$Fe_2O_3(s)$	−822.2	90.0	−741.0
$Co(s)$	0	30.0	0	$Fe_3O_4(s)$	−1118.4	146.4	−1015.4
$Co^{2+}(aq)$	−59.4	−110	−53.6	$FeS(s)$	−100.0	60.3	−100.4
$CoCl_2(s)$	−325.5	106	−282.4	$FeS_2(s)$	−178.2	52.9	−166.9
$Co(NO_3)_2(s)$	−422.2	192	−230.5	**Lead**			
$CoO(s)$	−237.9	53.0	−214.2	$Pb(s)$	0	64.8	0
$CoS(s)$	−80.8	67.4	−82.8	$Pb^{2+}(aq)$	−1.7	10.5	−24.4
Copper				$PbCl_2(s)$	−359.4	136	−314.1
$Cu(s)$	0	33.15	0	$PbO(s)$	−219.2	67.8	−189.3
$Cu^{2+}(aq)$	+64.77	−99.6	+65.49	$PbO_2(s)$	−277	68.6	−219
$CuCl(s)$	−137.2	86.2	−119.87				

TABLE C.2 **Thermodynamic Data for Selected Elements, Compounds, and Ions (25 °C)** *(Continued)*

Substance	ΔH_f° (kJ mol^{-1})	S° (J mol^{-1} K^{-1})	ΔG_f° (kJ mol^{-1})	Substance	ΔH_f° (kJ mol^{-1})	S° (J mol^{-1} K^{-1})	ΔG_f° (kJ mol^{-1})
Pb(OH)$_2$(s)	−515.9	88	−420.9	NiO(s)	−244	38	−216
PbS(s)	−100	91.2	−98.7	NiO$_2$(s)			−199
PbSO$_4$(s)	−920.1	149	−811.3	NiSO$_4$(s)	−891.2	77.8	−773.6
				NiCO$_3$(s)	−664.0	91.6	−615.0
Lithium				Ni(CO)$_4$(g)	−220	399	−567.4
Li(s)	0	28.4	0				
Li$^+$(aq)	−278.6	10.3		**Nitrogen**			
LiF(s)	−611.7	35.7	−583.3	N$_2$(g)	0	191.5	0
LiCl(s)	−408	59.29	−383.7	NH$_3$(g)	−46.19	192.5	−16.7
LiBr(s)	−350.3	66.9	−338.87	NH$_4^+$(aq)	−132.5	113	−79.37
Li$_2$O(s)	−596.5	37.9	−560.5	N$_2$H$_4$(g)	+95.40	238.4	+159.3
Li$_3$N(s)	−199	37.7	−155.4	N$_2$H$_4$(l)	+50.6	121.2	+149.4
				NH$_4$Cl(s)	−315.4	94.6	−203.9
Magnesium				NO(g)	+90.37	210.6	+86.69
Mg(s)	0	32.5	0	NO$_2$(g)	+33.8	240.5	+51.84
Mg^{2+}(aq)	−466.9	−138.1	−454.8	N$_2$O(g)	+81.57	220.0	+103.6
MgCO$_3$(s)	−1113	65.7	−1029	N$_2$O$_4$(g)	+9.67	304	+98.28
MgF$_2$(s)	−1124	79.9	−1056	N$_2$O$_5$(g)	+11	356	+115
MgCl$_2$(s)	−641.8	89.5	−592.5	HNO$_3$(l)	−173.2	155.6	−79.91
MgCl$_2 \cdot$ 2H$_2$O(s)	−1280	180	−1118	NO$_3^-$(aq)	−205.0	146.4	−108.74
Mg$_3$N$_2$(s)	−463.2	87.9	−411				
MgO(s)	−601.7	26.9	−569.4	**Oxygen**			
Mg(OH)$_2$(s)	−924.7	63.1	−833.9	O$_2$(g)	0	205.0	0
				O$_3$(g)	+143	238.8	+163
Manganese				OH$^-$(aq)	−230.0	−10.75	−157.24
Mn(s)	0	32.0	0				
Mn^{2+}(aq)	−223	−74.9	−228	**Phosphorus**			
MnO$_4^-$(aq)	−542.7	191	−449.4	P(s, white)	0	41.09	0
KMnO$_4$(s)	−813.4	171.71	−713.8	P$_4$(g)	+314.6	163.2	+278.3
MnO(s)	−385	60.2	−363	PCl$_3$(g)	−287.0	311.8	−267.8
Mn$_2$O$_3$(s)	−959.8	110	−882.0	PCl$_5$(g)	−374.9	364.6	−305.0
MnO$_2$(s)	−520.9	53.1	−466.1	PH$_3$(g)	+5.4	210.2	+12.9
Mn$_3$O$_4$(s)	−1387	149	−1280	P$_4$O$_6$(s)	−1640		
MnSO$_4$(s)	−1064	112	−956	POCl$_3$(g)	−558.5	325.5	−512.9
				POCl$_3$(l)	−597.1	222.5	−520.8
Mercury				P$_4$O$_{10}$(s)	−2984	228.9	−2698
Hg(l)	0	76.1	0	H$_3$PO$_4$(s)	−1279	110.5	−1119
Hg(g)	+61.38	175	+31.8				
Hg$_2$Cl$_2$(s)	−265.2	192.5	−210.8	**Potassium**			
HgCl$_2$(s)	−224.3	146.0	−178.6	K(s)	0	64.18	0
HgO(s)	−90.83	70.3	−58.54	K$^+$(aq)	−252.4	102.5	−283.3
HgS(s, red)	−58.2	82.4	−50.6	KF(s)	−567.3	66.6	−537.8
				KCl(s)	−435.89	82.59	−408.3
Nickel				KBr(s)	−393.8	95.9	−380.7
Ni(s)	0	30	0	KI(s)	−327.9	106.3	−324.9
NiCl$_2$(s)	−305	97.5	−259				

TABLE C.2	Thermodynamic Data for Selected Elements, Compounds, and Ions (25 °C) *(Continued)*						
Substance	ΔH_f° (kJ mol^{-1})	S° (J mol^{-1} K^{-1})	ΔG_f° (kJ mol^{-1})	Substance	ΔH_f° (kJ mol^{-1})	S° (J mol^{-1} K^{-1})	ΔG_f° (kJ mol^{-1})
KOH(s)	−424.8	78.9	−379.1	Na$_2$O(s)	−510	72.8	−376
K$_2$O(s)	−361	98.3	−322	NaOH(s)	−426.8	64.18	−382
K$_2$SO$_4$(s)	−1433.7	176	−1316.4	Na$_2$SO$_4$(s)	−1384.49	149.49	−1266.83
Silicon				**Sulfur**			
Si(s)	0	19	0	S(s, rhombic)	0	31.9	0
SiH$_4$(g)	+33	205	+52.3	SO$_2$(g)	−296.9	248.5	−300.4
SiO$_2$(s, alpha)	−910.0	41.8	−856	SO$_3$(g)	−395.2	256.2	−370.4
Silver				H$_2$S(g)	−20.6	206	−33.6
Ag(s)	0	42.55	0	H$_2$SO$_4$(l)	−811.32	157	−689.9
Ag$^+$(aq)	+105.58	72.68	+77.11	H$_2$SO$_4$(aq)	−909.3	20.1	−744.5
AgCl(s)	−127.0	96.2	−109.7	SF$_6$(g)	−1209	292	−1105
AgBr(s)	−100.4	107.1	−96.9	**Tin**			
AgNO$_3$(s)	−124	141	−32	Sn(s, white)	0	51.6	0
Ag$_2$O(s)	−31.1	121.3	−11.2	Sn^{2+}(aq)	−8.8	−17	−27.2
Sodium				SnCl$_4$(l)	−511.3	258.6	−440.2
Na(s)	0	51.0	0	SnO(s)	−285.8	56.5	−256.9
Na$^+$(aq)	−240.12	59.0	−261.91	SnO$_2$(s)	−580.7	52.3	−519.6
NaF(s)	−571	51.5	−545	**Zinc**			
NaCl(s)	−411.0	72.38	−384.0	Zn(s)	0	41.6	0
NaBr(s)	−360	83.7	−349	Zn^{2+}(aq)	−153.9	−112.1	−147.06
NaI(s)	−288	91.2	−286	ZnCl$_2$(s)	−415.1	111	−369.4
NaHCO$_3$(s)	−947.7	102	−851.9	ZnO(s)	−348.3	43.6	−318.3
Na$_2$CO$_3$(s)	−1131	136	−1048	ZnS(s)	−205.6	57.7	−201.3
Na$_2$O$_2$(s)	−510.9	94.6	−447.7	ZnSO$_4$(s)	−982.8	120	−874.5

TABLE C.3	Heats of Formation of Gaseous Atoms from Elements in Their Standard States				
Element	ΔH_f° (kJ mol^{-1})[a]	Element	ΔH_f° (kJ mol^{-1})[a]	Element	ΔH_f° (kJ mol^{-1})[a]
Group 1A		Sr	163.6	**Group 6A**	
H	217.89	Ba	177.8	O	249.17
Li	161.5	**Group 3A**		S	276.98
Na	107.8	B	560	**Group 7A**	
K	89.62	Al	329.7	F	79.14
Rb	82.0	**Group 4A**		Cl	121.47
Cs	78.2	C	716.67	Br	112.38
Group 2A		Si	450	I	107.48
Be	324.3	**Group 5A**			
Mg	146.4	N	472.68		
Ca	178.2	P	332.2		

[a]All values in this table are positive because forming the gaseous atoms from the elements is endothermic: it involves bond breaking.

TABLE C.4	Average Bond Energies		
Bond	Bond Energy (kJ mol^{-1})	Bond	Bond Energy (kJ mol^{-1})
C—C	348	C—Br	276
C=C	612	C—I	238
C≡C	960	H—H	436
C—H	412	H—F	565
C—N	305	H—Cl	431
C=N	613	H—Br	366
C≡N	890	H—I	299
C—O	360	H—N	388
C=O	743	H—O	463
C—F	484	H—S	338
C—Cl	338	H—Si	376

TABLE C.5	Vapor Pressure of Water as a Function of Temperatures						
Temp (°C)	Vapor Pressure (torr)	Temp (°C)	Vapor Pressure (torr)	Temp (°C)	Vapor Pressure (torr)	Temp (°C)	Vapor Pressure (torr)
0	4.58	26	25.2	52	102.1	78	327.3
1	4.93	27	26.7	53	107.2	79	341.0
2	5.29	28	28.3	54	112.5	80	355.1
3	5.68	29	30.0	55	118.0	81	369.7
4	6.10	30	31.8	56	123.8	82	384.9
5	6.54	31	33.7	57	129.8	83	400.6
6	7.01	32	35.7	58	136.1	84	416.8
7	7.51	33	37.7	59	142.6	85	433.6
8	8.04	34	39.9	60	149.4	86	450.9
9	8.61	35	42.2	61	156.4	87	468.7
10	9.21	36	44.6	62	163.8	88	487.1
11	9.84	37	47.1	63	171.4	89	506.1
12	10.5	38	49.7	64	179.3	90	525.8
13	11.2	39	52.4	65	187.5	91	546.0
14	12.0	40	55.3	66	196.1	92	567.0
15	12.8	41	58.3	67	205.0	93	588.6
16	13.6	42	61.5	68	214.2	94	610.9
17	14.5	43	64.8	69	223.7	95	633.9
18	15.5	44	68.3	70	233.7	96	657.6
19	16.5	45	71.9	71	243.9	97	682.1
20	17.5	46	75.6	72	254.6	98	707.3
21	18.7	47	79.6	73	265.7	99	733.2
22	19.8	48	83.7	74	277.2	100	760.0
23	21.1	49	88.0	75	289.1		
24	22.4	50	92.5	76	301.4		
25	23.8	51	97.2	77	314.1		

TABLE C.6	**Solubility Product Constants**					
Salt	K_{sp}	**Salt**	K_{sp}	**Salt**	K_{sp}
Fluorides		$Co(OH)_2$	5.9×10^{-15}	**Carbonates**	
MgF_2	5.2×10^{-11}	$Co(OH)_3$	3×10^{-45}	$MgCO_3$	6.8×10^{-8}
CaF_2	3.4×10^{-11}	$Ni(OH)_2$	5.5×10^{-16}	$CaCO_3$	3.4×10^{-9}
SrF_2	4.3×10^{-9}	$Cu(OH)_2$	4.8×10^{-20}	$SrCO_3$	5.6×10^{-10}
BaF_2	1.8×10^{-7}	$V(OH)_3$	4×10^{-35}	$BaCO_3$	2.6×10^{-9}
LiF	1.8×10^{-3}	$Cr(OH)_3$	2×10^{-30}	$MnCO_3$	2.2×10^{-11}
PbF_2	3.3×10^{-8}	Ag_2O	1.9×10^{-8}	$FeCO_3$	3.1×10^{-11}
Chlorides		$Zn(OH)_2$	3×10^{-17}	$CoCO_3$	1.0×10^{-10}
$CuCl$	1.7×10^{-7}	$Cd(OH)_2$	7.2×10^{-15}	$NiCO_3$	1.4×10^{-7}
$AgCl$	1.8×10^{-10}	$Al(OH)_3$ (alpha form)	3×10^{-34}	$CuCO_3$	2.5×10^{-10}
Hg_2Cl_2	1.4×10^{-18}	**Cyanides**		Ag_2CO_3	8.5×10^{-12}
$TlCl$	1.9×10^{-4}	$AgCN$	6.0×10^{-17}	Hg_2CO_3	3.6×10^{-17}
$PbCl_2$	1.7×10^{-5}	$Zn(CN)_2$	3×10^{-16}	$ZnCO_3$	1.5×10^{-10}
$AuCl_3$	3.2×10^{-25}	**Sulfites**		$CdCO_3$	1.0×10^{-12}
Bromides		$CaSO_3 \cdot \frac{1}{2}H_2O$	3.1×10^{-7}	$PbCO_3$	7.4×10^{-14}
$CuBr$	6.3×10^{-9}	Ag_2SO_3	1.5×10^{-14}	**Phosphates**	
$AgBr$	5.4×10^{-13}	$BaSO_3$	5.0×10^{-10}	$Ca_3(PO_4)_2$	2.1×10^{-33}
Hg_2Br_2	6.4×10^{-23}	**Sulfates**		$Mg_3(PO_4)_2$	1.0×10^{-24}
$HgBr_2$	6.2×10^{-20}	$CaSO_4$	4.9×10^{-5}	$SrHPO_4$	1.2×10^{-7}
$PbBr_2$	6.6×10^{-6}	$SrSO_4$	3.4×10^{-7}	$BaHPO_4$	4.0×10^{-8}
Iodides		$BaSO_4$	1.1×10^{-10}	$LaPO_4$	3.7×10^{-23}
CuI	1.3×10^{-12}	$RaSO_4$	3.7×10^{-11}	$Fe_3(PO_4)_2$	1×10^{-36}
AgI	8.5×10^{-17}	Ag_2SO_4	1.2×10^{-5}	Ag_3PO_4	8.9×10^{-17}
Hg_2I_2	5.2×10^{-29}	Hg_2SO_4	6.5×10^{-7}	$FePO_4$	9.9×10^{-16}
HgI_2	2.9×10^{-29}	$PbSO_4$	2.5×10^{-8}	$Zn_3(PO_4)_2$	5×10^{-36}
PbI_2	9.8×10^{-9}	**Chromates**		$Pb_3(PO_4)_2$	3.0×10^{-44}
Hydroxides		$BaCrO_4$	1.2×10^{-10}	$Ba_3(PO_4)_2$	5.8×10^{-38}
$Mg(OH)_2$	5.6×10^{-12}	$CuCrO_4$	3.6×10^{-6}	**Ferrocyanides**	
$Ca(OH)_2$	5.0×10^{-6}	Ag_2CrO_4	1.1×10^{-12}	$Zn_2[Fe(CN)_6]$	2.1×10^{-16}
$Mn(OH)_2$	1.6×10^{-13}	Hg_2CrO_4	2.0×10^{-9}	$Cd_2[Fe(CN)_6]$	4.2×10^{-18}
$Fe(OH)_2$	4.9×10^{-17}	$CaCrO_4$	7.1×10^{-4}	$Pb_2[Fe(CN)_6]$	9.5×10^{-19}
$Fe(OH)_3$	2.8×10^{-39}	$PbCrO_4$	1.8×10^{-14}		

TABLE C.7 Formation Constants of Complexes (25 °C)

Complex Ion Equilibrium	K_{form}	Complex Ion Equilibrium	K_{form}
Halide Complexes		**Complexes with Other Monodentate Ligands**	
$Al^{3+} + 6F^- \rightleftharpoons [AlF_6]^{3-}$	1×10^{20}	**Methylamine (CH_3NH_2)**	
$Al^{3+} + 4F^- \rightleftharpoons [AlF_4]^-$	2.0×10^8	$Ag^+ + 2CH_3NH_2 \rightleftharpoons [Ag(CH_3NH_2)_2]^+$	7.8×10^6
$Be^{2+} + 4F^- \rightleftharpoons [BeF_4]^{2-}$	1.3×10^{13}	**Thiocyanate Ion (SCN^-)**	
$Sn^{4+} + 6F^- \rightleftharpoons [SnF_6]^{2-}$	1×10^{25}	$Cd^{2+} + 4SCN^- \rightleftharpoons [Cd(SCN)_4]^{2-}$	1×10^3
$Cu^+ + 2Cl^- \rightleftharpoons [CuCl_2]^-$	3×10^5	$Cu^{2+} + 2SCN^- \rightleftharpoons [Cu(SCN)_2]$	5.6×10^3
$Ag^+ + 2Cl^- \rightleftharpoons [AgCl_2]^-$	1.8×10^5	$Fe^{3+} + 3SCN^- \rightleftharpoons [Fe(SCN)_3]$	2×10^6
$Pb^{2+} + 4Cl^- \rightleftharpoons [PbCl_4]^{2-}$	2.5×10^{15}	$Hg^{2+} + 4SCN^- \rightleftharpoons [Hg(SCN)_4]^{2-}$	5.0×10^{21}
$Zn^{2+} + 4Cl^- \rightleftharpoons [ZnCl_4]^{2-}$	1.6	**Hydroxide Ion (OH^-)**	
$Hg^{2+} + 4Cl^- \rightleftharpoons [HgCl_4]^{2-}$	5.0×10^{15}	$Cu^{2+} + 4OH^- \rightleftharpoons [Cu(OH)_4]^{2-}$	1.3×10^{16}
$Cu^+ + 2Br^- \rightleftharpoons [CuBr_2]^-$	8×10^5	$Zn^{2+} + 4OH^- \rightleftharpoons [Zn(OH)_4]^{2-}$	2×10^{20}
$Ag^+ + 2Br^- \rightleftharpoons [AgBr_2]^-$	1.7×10^7	**Complexes with Bidentate Ligands***	
$Hg^{2+} + 4Br^- \rightleftharpoons [HgBr_4]^{2-}$	1×10^{21}	$Mn^{2+} + 3\,en \rightleftharpoons [Mn(en)_3]^{2+}$	6.5×10^5
$Cu^+ + 2I^- \rightleftharpoons [CuI_2]^-$	8×10^8	$Fe^{2+} + 3\,en \rightleftharpoons [Fe(en)_3]^{2+}$	5.2×10^9
$Ag^+ + 2I^- \rightleftharpoons [AgI_2]^-$	1×10^{11}	$Co^{2+} + 3\,en \rightleftharpoons [Co(en)_3]^{2+}$	1.3×10^{14}
$Pb^{2+} + 4I^- \rightleftharpoons [PbI_4]^{2-}$	3×10^4	$Co^{3+} + 3\,en \rightleftharpoons [Co(en)_3]^{3+}$	4.8×10^{48}
$Hg^{2+} + 4I^- \rightleftharpoons [HgI_4]^{2-}$	1.9×10^{30}	$Ni^{2+} + 3\,en \rightleftharpoons [Ni(en)_3]^{2+}$	4.1×10^{17}
Ammonia Complexes		$Cu^{2+} + 2\,en \rightleftharpoons [Cu(en)_2]^{2+}$	3.5×10^{19}
$Ag^+ + 2NH_3 \rightleftharpoons [Ag(NH_3)_2]^+$	1.6×10^7	$Mn^{2+} + 3\,bipy \rightleftharpoons [Mn(bipy)_3]^{2+}$	1×10^6
$Zn^{2+} + 4NH_3 \rightleftharpoons [Zn(NH_3)_4]^{2+}$	7.8×10^8	$Fe^{2+} + 3\,bipy \rightleftharpoons [Fe(bipy)_3]^{2+}$	1.6×10^{17}
$Cu^{2+} + 4NH_3 \rightleftharpoons [Cu(NH_3)_4]^{2+}$	1.1×10^{13}	$Ni^{2+} + 3\,bipy \rightleftharpoons [Ni(bipy)_3]^{2+}$	3.0×10^{20}
$Hg^{2+} + 4NH_3 \rightleftharpoons [Hg(NH_3)_4]^{2+}$	1.8×10^{19}	$Co^{2+} + 3\,bipy \rightleftharpoons [Co(bipy)_3]^{2+}$	8×10^{15}
$Co^{2+} + 6NH_3 \rightleftharpoons [Co(NH_3)_6]^{2+}$	5.0×10^4	$Mn^{2+} + 3\,phen \rightleftharpoons [Mn(phen)_3]^{2+}$	2×10^{10}
$Co^{3+} + 6NH_3 \rightleftharpoons [Co(NH_3)_6]^{3+}$	4.6×10^{33}	$Fe^{2+} + 3\,phen \rightleftharpoons [Fe(phen)_3]^{2+}$	1×10^{21}
$Cd^{2+} + 6NH_3 \rightleftharpoons [Cd(NH_3)_6]^{2+}$	2.6×10^5	$Co^{2+} + 3\,phen \rightleftharpoons [Co(phen)_3]^{2+}$	6×10^{19}
$Ni^{2+} + 6NH_3 \rightleftharpoons [Ni(NH_3)_6]^{2+}$	2.0×10^8	$Ni^{2+} + 3\,phen \rightleftharpoons [Ni(phen)_3]^{2+}$	2×10^{24}
Cyanide Complexes		$Co^{2+} + 3C_2O_4^{2-} \rightleftharpoons [Co(C_2O_4)_3]^4$	4.5×10^6
$Fe^{2+} + 6CN^- \rightleftharpoons [Fe(CN)_6]^{4-}$	1.0×10^{24}	$Fe^{3+} + 3C_2O_4^{2-} \rightleftharpoons [Fe(C_2O_4)_3]^{3-}$	3.3×10^{20}
$Fe^{3+} + 6CN^- \rightleftharpoons [Fe(CN)_6]^{3-}$	1.0×10^{31}	**Complexes of Other Polydentate Ligands***	
$Ag^+ + 2CN^- \rightleftharpoons [Ag(CN)_2]^-$	5.3×10^{18}	$Zn^{2+} + EDTA^{4-} \rightleftharpoons [Zn(EDTA)]^{2-}$	3.8×10^{16}
$Cu^+ + 2CN^- \rightleftharpoons [Cu(CN)_2]^-$	1.0×10^{16}	$Mg^{2+} + 2NTA^{3-} \rightleftharpoons [Mg(NTA)_2]^{4-}$	1.6×10^{10}
$Cd^{2+} + 4CN^- \rightleftharpoons [Cd(CN)_4]^{2-}$	7.7×10^{16}	$Ca^{2+} + 2NTA^{3-} \rightleftharpoons [Ca(NTA)_2]^{4-}$	3.2×10^{11}
$Au^+ + 2CN^- \rightleftharpoons [Au(CN)_2]^-$	2×10^{38}		

*en = ethylenediamine

bipy = bipyridyl

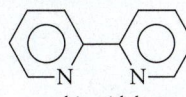

bipyridyl

phen = 1,10-phenanthroline

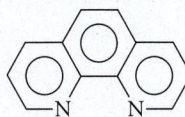

1,10-phenanthroline

$EDTA^{4-}$ = ethylenediaminetetraacetate ion

NTA^{3-} = nitrilotriacetate ion

TABLE C.8	Ionization Constants of Weak Acids and Bases (Alternative Formulas in Parentheses)	

Monoprotic Acid	Name	K_a
$HC_2O_2Cl_3$ (Cl_3CCO_2H)	trichloroacetic acid	2.2×10^{-1}
HIO_3	iodic acid	1.7×10^{-1}
$HC_2HO_2Cl_2$ (Cl_2CHCO_2H)	dichloroacetic acid	5.0×10^{-2}
$HC_2H_2O_2Cl$ (ClH_2CCO_2H)	chloroacetic acid	1.4×10^{-3}
HNO_2	nitrous acid	4.6×10^{-4}
HF	hydrofluoric acid	3.5×10^{-4}
$HOCN$	cyanic acid	2×10^{-4}
$HCHO_2$ (HCO_2H)	formic acid	1.8×10^{-4}
$HC_3H_5O_3$ [$CH_3CH(OH)CO_2H$]	lactic acid	1.4×10^{-4}
$HC_4H_3N_2O_3$	barbituric acid	9.8×10^{-5}
$HC_7H_5O_2$ ($C_6H_5CO_2H$)	benzoic acid	6.3×10^{-5}
$HC_4H_7O_2$ ($CH_3CH_2CH_2CO_2H$)	butanoic acid	1.5×10^{-5}
HN_3	hydrazoic acid	2.5×10^{-5}
$HC_2H_3O_2$ (CH_3CO_2H)	acetic acid	1.8×10^{-5}
$HC_3H_5O_2$ ($CH_3CH_2CO_2H$)	propanoic acid	1.3×10^{-5}
$HC_2H_4NO_2$	nicotinic acid (niacin)	1.4×10^{-5}
$HOCl$	hypochlorous acid	3.0×10^{-8}
$HOBr$	hypobromous acid	2.1×10^{-9}
HCN	hydrocyanic acid	4.9×10^{-10}
HC_6H_5O	phenol	1.3×10^{-10}
HOI	hypoiodous acid	2.3×10^{-11}
H_2O_2	hydrogen peroxide	2.4×10^{-12}

Polyprotic Acid	Name	K_{a_1}	K_{a_2}	K_{a_3}
H_2SO_4	sulfuric acid	large	1.2×10^{-2}	
H_2CrO_4	chromic acid	5.0	1.5×10^{-6}	
$H_2C_2O_4$	oxalic acid	5.9×10^{-2}	6.4×10^{-5}	
H_3PO_3	phosphorous acid	5.0×10^{-2}	2.0×10^{-7}	
H_2S	hydrosulfuric acid	8.9×10^{-8}	1×10^{-19}	
H_2SO_3	sulfurous acid	1.5×10^{-2}	6.3×10^{-8}	
H_2SeO_4	selenic acid	large	1.2×10^{-2}	
H_2SeO_3	selenous acid	3.5×10^{-3}	1.5×10^{-9}	
H_6TeO_6	telluric acid	2×10^{-8}	1×10^{-11}	
H_2TeO_3	tellurous acid	3.3×10^{-3}	2.0×10^{-8}	
$H_2C_3H_2O_4$ ($HO_2CCH_2CO_2H$)	malonic acid	1.4×10^{-3}	2.0×10^{-6}	
$H_2C_8H_4O_4$	phthalic acid	1.1×10^{-3}	3.9×10^{-6}	
$H_2C_4H_4O_6$	tartaric acid	9.2×10^{-4}	4.3×10^{-5}	
$H_2C_6H_6O_6$	ascorbic acid	8.0×10^{-5}	1.6×10^{-12}	
H_2CO_3	carbonic acid	4.3×10^{-7}	5.6×10^{-11}	
H_3PO_4	phosphoric acid	7.5×10^{-3}	6.2×10^{-8}	4.2×10^{-13}
H_3AsO_4	arsenic acid	5.5×10^{-3}	1.7×10^{-7}	5.1×10^{-12}
$H_3C_6H_5O_7$	citric acid	7.4×10^{-4}	1.7×10^{-5}	6.0×10^{-7}

(Continued)

TABLE C.8	Ionization Constants of Weak Acids and Bases *(Continued)*	
Weak Base	Name	K_b
$(CH_3)_2NH$	dimethylamine	9.6×10^{-4}
$C_4H_9NH_2$	butylamine	5.9×10^{-4}
CH_3NH_2	methylamine	4.5×10^{-4}
$CH_3CH_2NH_2$	ethylamine	4.3×10^{-4}
$(CH_3)_3N$	trimethylamine	7.4×10^{-5}
NH_3	ammonia	1.8×10^{-5}
$C_{21}H_{22}N_2O_2$	strychnine	1.8×10^{-6}
N_2H_4	hydrazine	1.3×10^{-6}
$C_{17}H_{19}NO_3$	morphine	1.6×10^{-6}
NH_2OH	hydroxylamine	1.1×10^{-8}
C_5H_5N	pyridine	1.8×10^{-9}
$C_6H_5NH_2$	aniline	4.3×10^{-10}
PH_3	phosphine	1×10^{-28}

TABLE C.9 Standard Reduction Potentials (25 °C)

$E°$ (Volts)	Half-Cell Reaction
+2.87	$F_2(g) + 2e^- \rightleftharpoons 2F^-(aq)$
+2.07	$O_3(g) + 2H^+(aq) + 2e^- \rightleftharpoons O_2(g) + H_2O$
+2.01	$S_2O_8^{2-}(aq) + 2e^- \rightleftharpoons 2SO_4^{2-}(aq)$
+1.84	$Co^{3+}(aq) + e^- \rightleftharpoons Co^{2+}(aq)$
+1.77	$H_2O_2(aq) + 2H^+(aq) + 2e^- \rightleftharpoons 2H_2O$
+1.68	$MnO_4^-(aq) + 4H^+(aq) + 3e^- \rightleftharpoons MnO_2(s) + 2H_2O$
+1.69	$PbO_2(s) + HSO_4^-(aq) + 3H^+(aq) + 2e^- \rightleftharpoons PbSO_4(s) + 2H_2O$
+1.63	$2HOCl(aq) + 2H^+(aq) + 2e^- \rightleftharpoons Cl_2(g) + 2H_2O$
+1.51	$Mn^{3+}(aq) + e^- \rightleftharpoons Mn^{2+}(aq)$
+1.51	$MnO_4^-(aq) + 8H^+(aq) + 5e^- \rightleftharpoons Mn^{2+}(aq) + 4H_2O$
+1.46	$PbO_2(s) + 4H^+(aq) + 2e^- \rightleftharpoons Pb^{2+}(aq) + 2H_2O$
+1.44	$BrO_3^-(aq) + 6H^+(aq) + 6e^- \rightleftharpoons Br^-(aq) + 3H_2O$
+1.42	$Au^{3+}(aq) + 3e^- \rightleftharpoons Au(s)$
+1.36	$Cl_2(g) + 2e^- \rightleftharpoons 2Cl^-(aq)$
+1.33	$Cr_2O_7^{2-}(aq) + 14H^+(aq) + 6e^- \rightleftharpoons 2Cr^{3+}(aq) + 7H_2O$
+1.24	$O_3(g) + H_2O + 2e^- \rightleftharpoons O_2(g) + 2OH^-(aq)$
+1.23	$MnO_2(s) + 4H^+(aq) + 2e^- \rightleftharpoons Mn^{2+}(aq) + 2H_2O$
+1.23	$O_2(g) + 4H^+(aq) + 4e^- \rightleftharpoons 2H_2O$
+1.20	$Pt^{2+}(aq) + 2e^- \rightleftharpoons Pt(s)$
+1.07	$Br_2(aq) + 2e^- \rightleftharpoons 2Br^-(aq)$
+0.96	$NO_3^-(aq) + 4H^+(aq) + 3e^- \rightleftharpoons NO(g) + 2H_2O$
+0.94	$NO_3^-(aq) + 3H^+(aq) + 2e^- \rightleftharpoons HNO_2(aq) + H_2O$
+0.91	$2Hg^{2+}(aq) + 2e^- \rightleftharpoons Hg_2^{2+}(aq)$
+0.87	$HO_2^-(aq) + H_2O + 2e^- \rightleftharpoons 3OH^-(aq)$
+0.81	$NO_3^-(aq) + 4H^+(aq) + 2e^- \rightleftharpoons 2NO_2(g) + 2H_2O$
+0.80	$Ag^+(aq) + e^- \rightleftharpoons Ag(s)$
+0.77	$Fe^{3+}(aq) + e^- \rightleftharpoons Fe^{2+}(aq)$
+0.68	$O_2(g) + 2H^+(aq) + 2e^- \rightleftharpoons H_2O_2(aq)$
+0.54	$I_2(s) + 2e^- \rightleftharpoons 2I^-(aq)$
+0.49	$NiO_2(s) + 2H_2O + 2e^- \rightleftharpoons Ni(OH)_2(s) + 2OH^-(aq)$
+0.45	$SO_2(aq) + 4H^+(aq) + 4e^- \rightleftharpoons S(s) + 2H_2O$
+0.401	$O_2(g) + 2H_2O + 4e^- \rightleftharpoons 4OH^-(aq)$
+0.34	$Cu^{2+}(aq) + 2e^- \rightleftharpoons Cu(s)$
+0.27	$Hg_2Cl_2(s) + 2e^- \rightleftharpoons 2Hg(l) + 2Cl^-(aq)$
+0.25	$PbO_2(s) + H_2O + 2e^- \rightleftharpoons PbO(s) + 2OH^-(aq)$
+0.2223	$AgCl(s) + e^- \rightleftharpoons Ag(s) + Cl^-(aq)$
+0.172	$SO_4^{2-}(aq) + 4H^+(aq) + 2e^- \rightleftharpoons H_2SO_3(aq) + H_2O$
+0.169	$S_4O_6^{2-}(aq) + 2e^- \rightleftharpoons 2S_2O_3^{2-}(aq)$
+0.16	$Cu^{2+}(aq) + e^- \rightleftharpoons Cu^+(aq)$
+0.15	$Sn^{4+}(aq) + 2e^- \rightleftharpoons Sn^{2+}(aq)$
+0.14	$S(s) + 2H^+(aq) + 2e^- \rightleftharpoons H_2S(g)$
+0.07	$AgBr(s) + e^- \rightleftharpoons Ag(s) + Br^-(aq)$
0 (exactly)	$2H^+(aq) + 2e^- \rightleftharpoons H_2(g)$
−0.12	$Pb^{2+}(aq) + 2e^- \rightleftharpoons Pb(s)$

(Continued)

TABLE C.9   Standard Reduction Potentials (25 °C) *(Continued)*

$E°$ (Volts)	Half-Cell Reaction
−0.14	$Sn^{2+}(aq) + 2e^- \rightleftharpoons Sn(s)$
−0.15	$AgI(s) + e^- \rightleftharpoons Ag(s) + I^-(aq)$
−0.25	$Ni^{2+}(aq) + 2e^- \rightleftharpoons Ni(s)$
−0.28	$Co^{2+}(aq) + 2e^- \rightleftharpoons Co(s)$
−0.34	$In^{3+}(aq) + 3e^- \rightleftharpoons In(s)$
−0.34	$Tl^+(aq) + e^- \rightleftharpoons Tl(s)$
−0.36	$PbSO_4(s) + H^+(aq) + 2e^- \rightleftharpoons Pb(s) + HSO_4^-(aq)$
−0.40	$Cd^{2+}(aq) + 2e^- \rightleftharpoons Cd(s)$
−0.44	$Fe^{2+}(aq) + 2e^- \rightleftharpoons Fe(s)$
−0.56	$Ga^{3+}(aq) + 3e^- \rightleftharpoons Ga(s)$
−0.58	$PbO(s) + H_2O + 2e^- \rightleftharpoons Pb(s) + 2OH^-(aq)$
−0.74	$Cr^{3+}(aq) + 3e^- \rightleftharpoons Cr(s)$
−0.76	$Zn^{2+}(aq) + 2e^- \rightleftharpoons Zn(s)$
−0.81	$Cd(OH)_2(s) + 2e^- \rightleftharpoons Cd(s) + 2OH^-(aq)$
−0.83	$2H_2O + 2e^- \rightleftharpoons H_2(g) + 2OH^-(aq)$
−0.89	$Fe(OH)_2(s) + 2e^- \rightleftharpoons Fe(s) + 2OH^-(aq)$
−0.91	$Cr^{2+}(aq) + 2e^- \rightleftharpoons Cr(s)$
−1.16	$N_2(g) + 4H_2O + 4e^- \rightleftharpoons N_2O_4(aq) + 4OH^-(aq)$
−1.18	$V^{2+}(aq) + 2e^- \rightleftharpoons V(s)$
−1.216	$ZnO_2^-(aq) + 2H_2O + 2e^- \rightleftharpoons Zn(s) + 4OH^-(aq)$
−1.63	$Ti^{2+}(aq) + 2e^- \rightleftharpoons Ti(s)$
−1.66	$Al^{3+}(aq) + 3e^- \rightleftharpoons Al(s)$
−1.79	$U^{3+}(aq) + 3e^- \rightleftharpoons U(s)$
−2.02	$Sc^{3+}(aq) + 3e^- \rightleftharpoons Sc(s)$
−2.36	$La^{3+}(aq) + 3e^- \rightleftharpoons La(s)$
−2.37	$Y^{3+}(aq) + 3e^- \rightleftharpoons Y(s)$
−2.37	$Mg^{2+}(aq) + 2e^- \rightleftharpoons Mg(s)$
−2.71	$Na^+(aq) + e^- \rightleftharpoons Na(s)$
−2.87	$Ca^{2+}(aq) + 2e^- \rightleftharpoons Ca(s)$
−2.89	$Sr^{2+}(aq) + 2e^- \rightleftharpoons Sr(s)$
−2.90	$Ba^{2+}(aq) + 2e^- \rightleftharpoons Ba(s)$
−2.92	$Cs^+(aq) + e^- \rightleftharpoons Cs(s)$
−2.92	$K^+(aq) + e^- \rightleftharpoons K(s)$
−2.93	$Rb^+(aq) + e^- \rightleftharpoons Rb(s)$
−3.05	$Li^+(aq) + e^- \rightleftharpoons Li(s)$

Glossary

This glossary has the definitions of the key terms that were marked in boldface throughout the chapters. The numbers in parentheses that follow the definitions are the numbers of the sections in which the glossary entries received their principal discussions.

A

α-Amino Acid: One of about 20 monomers found in naturally occurring polypeptides. (22.6)

Abbreviated Electron Configurations: A shorthand method to write electron configurations using the preceding noble gas to represent core electrons. (7.8)

Absolute Zero: 0 K, −273.15 °C. Nature's lowest temperature. (1.4, 10.3)

Absorption: The process of having one substance penetrate and enter the structure of another, usually a liquid or gas in a solid. (13.9)

Acceptor: A Lewis acid that accepts an electron pair from a Lewis base; the central metal ion in a complex ion. (17.5, 21.1)

Accuracy: The closeness of a measurement to the true value. (1.5)

Acid: *Arrhenius definition:* A substance that produces hydronium ions (hydrogen ions) in water. (4.4) *Brønsted definition:* A proton donor. (16.1) *Lewis definition:* An electron-pair acceptor. (16.4)

Acid Anhydride: An oxide that reacts with water to make the solution acidic. (4.4)

Acid–Base Indicator: A dye with one color in acid and another color in base. (4.9)

Acid–Base Neutralization: The reaction of an acid with a base. (4.6, 15.4)

Acidic Solution: An aqueous solution in which $[H^+] > [OH^-]$. (16.1)

Acid Ionization Constant (K_a):

$$K_a = \frac{[H^+][A^-]}{[HA]} \text{ for the equilibrium,}$$

$$[HA] \rightleftharpoons H^+ + A^- \qquad (16.3)$$

Acid Salt: A salt of a partially neutralized polyprotic acid, for example, $NaHSO_4$ or $NaHCO_3$. (4.6)

Acid Solubility Product, K_{spa}: The special solubility product expression for metal sulfides in dilute acid and related to the equation for their dissolving. (17.3)

Actinide Elements (Actinide Series): Elements 90–103. (2.1)

Activated Complex: The chemical species that exists with partly broken and partly formed bonds in the transition state. (13.6)

Activation Energy (E_a): The minimum kinetic energy that must be possessed by the reactants in order to result in an effective collision (one that produces products). (13.5)

Activity: (1) The effective concentrations of substances that take into account intermolecular and ionic attractive forces. Pure liquids and solids are assigned an activity of 1.00 (14.4). (2) For radioactive materials, the number of disintegrations per second. (20.6)

Activity Series: A list of metals in order of their reactivity as reducing agents. (5.4)

Actual Yield: See *Yield, Actual.*

Addition Compound: A molecule formed by the joining of two simpler molecules through formation of a covalent bond (usually a coordinate covalent bond). (8.7, 15.4)

Addition Polymer: A polymer formed by the simple addition of one monomer unit to another, a process that continues over and over until a very long chain of monomer units is produced. (22.5)

Addition Reaction: The addition of a molecule to a double or triple bond. (22.2)

Adiabatic Conditions: Conditions within a system during which no energy enters or leaves the system. (18.1)

Adsorption: The process of one substance adhering to the surface of a solid. (13.9)

Aerogel: A modern ceramic that is extremely porous and has a low density (15.6)

Alcohol: An organic compound whose molecules have the —OH group attached to carbon. (2.7)

Alcohol: An organic compound whose molecules have the —OH group attached to tetrahedral carbon. (8.8, 22.3)

Aldehyde: An organic compound whose molecules have a carbonyl group ($\diagdown C = O$) bonded to at least one hydrogen. (8.8)

Alkali Metals: The Group 1A elements (except hydrogen)—lithium, sodium, potassium, rubidium, cesium, and francium. (2.1)

Alkaline Battery (Alkaline Dry Cell): A zinc–manganese dioxide galvanic cell of 1.54 V used commonly in flashlight batteries. (19.6)

Alkaline Earth Metals: The Group 2A elements—beryllium, magnesium, calcium, strontium, barium, and radium. (2.1)

Alkane: A hydrocarbon whose molecules have only single bonds. (2.7, 22.2)

Alkene: A hydrocarbon whose molecules have one or more double bonds. (22.2)

Alkoxide: An anion formed by removing a hydrogen ion from an alcohol (15.6)

Alkyl Group: An organic group of carbon and hydrogen atoms related to an alkane but with one less hydrogen atom (e.g., CH_3—, methyl; CH_3CH_2—, ethyl). (15.6, 22.2)

Alkyne: A hydrocarbon whose molecules have one or more triple bonds. (22.2)

Allotrope: One of two or more forms of an element. (9.10)

Allotropy: The existence of an element in two or more molecular or crystalline forms called allotropes. (9.10)

Alpha Particle: The nucleus of a helium atom. (20.3)

Alpha Radiation: A high-velocity stream of alpha particles produced by radioactive decay. (20.3)

Amide: An organic compound whose molecules have any one of the following groups: (22.4)

$$\underset{-CNH_2}{\overset{\overset{\displaystyle O}{\|}}{}} \quad \underset{-CNHR}{\overset{\overset{\displaystyle O}{\|}}{}} \quad \underset{-CNR_2}{\overset{\overset{\displaystyle O}{\|}}{}}$$

Amine: An organic compound whose molecules contain the group NH_2, NHR, or NR_2. (8.8, 9.6, 22.4)

Amorphous Solid: A noncrystalline solid. A glass. (11.10)

Ampere (A): The SI unit for electric current; one coulomb per second. (19.8)

Amphiprotic: A compound that can act either as a proton donor or as a proton acceptor. (15.1)

Amphoteric: A compound that can react as either an acid or a base. (15.1)

Amplitude: The height of a wave, which is a measure of the wave's intensity. (7.1)

amu: See *Atomic Mass Unit.*

Angstrom (Å): 1 Å = 10^{-10} m = 100 pm = 0.1 nm. (7.10)

Anion: A negatively charged ion. (2.5)

Anode: The positive electrode in a gas discharge tube. The electrode at which oxidation occurs during an electrochemical change. (0.5, 19.1)

Antibonding Electrons: Electrons that occupy antibonding molecular orbitals. (9.7)

Antibonding Molecular Orbital: A molecular orbital that has no electron density between nuclei and destabilizes a molecule when occupied by electrons. (9.7)

Antimatter: Any particle annihilated by a particle of ordinary matter. (20.3)

Antineutrino: The antimatter form of the neutrino. (20.3)

Aromatic Compound: An organic compound whose molecules have the benzene ring system. (22.2)

Aromatic Hydrocarbon: Hydrocarbons that contain a benzene ring or similar ring systems with alternating single and double bonds. (22.2)

Arrhenius Equation: An equation that relates the rate constant of a reaction to the reaction's activation energy. (13.7)

Association: The joining together of molecules by hydrogen bonds. (12.6)

Asymmetric: Molecules in which the substituents around the central atom are not the same or there are lone pairs of electrons on the central atom. (9.3)

Asymmetric Carbon Atom: A carbon atom that is bonded to four different groups and which is a chiral center. (22.1)

Atmosphere, Standard (atm): 101,325 Pa. The pressure that supports a column of mercury 760 mm high at 0 °C; 760 torr. (6.5, 10.2)

Atmospheric Pressure: The pressure exerted by the mixture of gases in our atmosphere. (6.5, 7.5, 10.2)

Atom: A neutral particle having one nucleus; the smallest representative sample of an element. (1.1)

Atomic Force Microscope: A modern instrument that determines the topology of materials with a resolution on the order of atomic dimensions. Orientations of atoms and structures of molecular assemblies on these surfaces can be deduced from these measurements. (0.4)

Atomic Mass: The average mass (in u) of the atoms of the isotopes of a given element as they occur naturally. (0.5)

Atomic Mass Unit (u): $1.6605402 \times 10^{-24}$ g; 1/12th the mass of one atom of carbon-12. Sometimes given the symbol amu. (0.5)

Atomic Number: The number of protons in a nucleus. (0.5)

Atomic Radiation: Radiation consisting of particles or electromagnetic radiation given off by radioactive elements. (20.3)

Atomic Spectrum: The line spectrum produced when energized or excited atoms emit electromagnetic radiation. (7.2)

Atomic Symbol: See *Chemical Symbol.*

Atomic Weight: See *Atomic Mass.*

Atomization Energy (ΔH_{atom}): The energy needed to rupture all of the bonds in one mole of a substance in the gas state and produce its atoms, also in the gas state. (18.10)

Aufbau Principle: A set of rules enabling the construction of an electron structure of an atom from its atomic number. (7.7)

Average: See *Mean.*

Average Bond Order: The bond order of two or more equivalent bonds, determined by dividing the total number of sigma and pi bonds by the number of bonding electron domains. (8.8)

Average Kinetic Energy: Average of all kinetic energies of particles in a system. Estimated as the point slightly beyond the maximum of a kinetic energy distribution graph. (6.2)

Average Rate: Slope between two points on a concentration versus time plot. (13.2)

Avogadro's Number (Avogadro's Constant): 6.022×10^{23}; the number of particles or formula units in one mole. (3.1)

Avogadro's Principle: Equal volumes of gases contain equal numbers of molecules when they are at identical temperatures and pressures. (10.4)

Axial Bonds: Covalent bonds oriented parallel to the vertical axis in a trigonal bipyramidal molecule. (9.1)

Azimuthal Quantum Number (ℓ): See *Secondary Quantum Number.*

B

Backbone (Polymer): The long chain of atoms in a polymer to which other groups are attached. (22.5)

Background Radiation: The atomic radiation from the natural radionuclides in the environment and from cosmic radiation. (20.6)

Balance: An apparatus for measuring mass. (1.4)

Balanced Equation: A chemical equation that has on opposites sides of the arrow the same number of each atom and the same net charge. (2.4, 3.5)

Ball-and-Stick Model: The representation of molecules using colored balls for atoms and sticks representing bonds. (2.3)

Band Gap: The energy difference between the valence band and the conduction band. (9.9)

Band of Stability: The envelope that encloses just the stable nuclides in a plot of all nuclides constructed according to their numbers of neutrons versus their numbers of protons. (20.4)

Band Theory: A theory used to explain electron conduction in metals, semiconductors and insulators. (9.9)

Bar: The standard pressure for thermodynamic quantities; 1 bar = 10^5 pascals, 1 atm = 101,325 Pa. (6.5, 10.2, 18.5)

Barometer: An apparatus for measuring atmospheric pressure. (10.2)

Base: *Arrhenius theory:* A substance that releases OH^- ions in water. (4.4) *Brønsted theory:* A proton acceptor. (16.1) *Lewis theory:* An electron-pair acceptor. (16.3)

Base Anhydride: Oxides of metals that when dissolved in water produce a basic solution (4.4)

Base Ionization Constant, K_b:

$$K_b = \frac{[BH^+][OH^-]}{[B]} \text{ for the equilibrium,}$$

$$B + H_2O \rightleftharpoons BH^+ + OH^- \qquad (16.3)$$

Base Units: The units of the fundamental measurements of the SI. (1.4)

Basic Solution: An aqueous solution in which $[H^+] < [OH^-]$. (16.1)

Battery: One or more galvanic cells arranged in series to serve as a practical source of electricity. (19.6)

Becquerel (Bq): 1 disintegration per second, dps. The SI unit for the activity of a radioactive source. (20.6)

Bent Molecule (Nonlinear or V-Shaped Molecule): A three-atom molecule that is nonlinear. (9.2)

Beta Particle: An electron emitted by radioactive decay. (20.3)

Beta Radiation: A stream of electrons produced by radioactive decay. (20.3)

Bidentate Ligand: A ligand that has two atoms that can become simultaneously bonded to the same metal ion. (21.1)

Big-bang Theory: The cosmological theory that postulates that the universe started as an extremely small and dense entity that violently exploded into our currently known universe. (0.2)

Bimolecular Collision: A collision of two molecules. (13.8)

Binary Acid: An acid with the general formula H_nX, where X is a nonmetal. (4.5, 15.3)

Binary Compound: A compound composed of two different elements. (2.5)

Binding Energy, Nuclear: The energy equivalent of the difference in mass between an atomic nucleus and the sum of the masses of its nucleons. (20.2)

Biochemistry: The study of the organic substances in organisms. (2.7, 3.6, 22.6)

Black Phosphorus: An allotrope of phosphorus that has a layered structure. (9.10)

Body-Centered Cubic (bcc) Unit Cell: A unit cell having identical atoms, molecules, or ions at the corners of a cube plus one more particle in the center of the cube. (11.10)

Bohr Radius: The approximate distance between the nucleus and electron in a hydrogen atom in its ground state. (7.3)

Boiling Point: The temperature at which the vapor pressure of the liquid equals the atmospheric pressure. (12.5)

Boiling Point Elevation: A colligative property of a solution by which the solution's boiling point is higher than that of the pure solvent. (12.6)

Bomb Calorimeter: A calorimeter designed to maintain a constant volume. (6.6)

Bond Angle: The angle formed by two bonds that extend from the same atom. (9.1)

Bond Dipole: A dipole within a molecule associated with a specific bond. (9.3)

Bond Distance: See *Bond Length*.

Bond Energy: The energy needed to break one mole of a particular bond to give electrically neutral fragments. (8.5, 18.10)

Bonding Domain: A region between two atoms that contains one or more electron pairs in bonds and that influences molecular shape. (9.2)

Bonding Electrons: Electrons that occupy bonding molecular orbitals. (9.7)

Bonding Molecular Orbital: A molecular orbital that introduces a buildup of electron density between nuclei and stabilizes a molecule when occupied by electrons. (9.7)

Bond Length: The distance between two nuclei that are held together by a chemical bond. (8.5)

Bond Order: (a) The average number of bonds to a central atom in a molecule. (8.7) (b) The sum of the bonding electrons less the antibonding electrons divided by two. (9.7)

Born-Haber Cycle: A method for visualizing energy contributions to a physical or chemical change such as the dissolution of a compound in a solvent or the formation of a compound. (8.2)

Boundary: The interface between a system and its surroundings across which energy or matter might pass. (6.3)

Boyle's Law: See *Pressure–Volume Law*.

Bragg Equation: $n\lambda = 2d \sin \theta$. The equation used to convert X-ray diffraction data into a crystal structure. (11.11)

Branched Chain: An organic compound in whose molecules the carbon atoms do not all occur one after another in a continuous sequence. (22.1)

Branching (Polymer): The formation of side chains (branches) along the main backbone of a polymer. (22.5)

Branching Step: A step in a chain reaction that produces more chain- propagating species than it consumes. (Chemistry Outside the Classroom 13.1)

Brine: A concentrated aqueous solution of sodium chloride, often with other salts; usually concentrated seawater. (19.9)

Brønsted–Lowry Acid: A proton donor. (15.1)

Brønsted–Lowry Base: A proton acceptor. (15.1)

Brownian Motion: The random, erratic motions of colloidally dispersed particles in a fluid. (2.7)

Buckminsterfullerene: The C-60 molecule. Also called buckyball. (9.10)

Buckyball: See *Buckminsterfullerene*.

Buffer: (a) A pair of solutes, a conjugate weak acid–weak base pair, that can keep the pH of a solution almost constant if either acid or base is added. (b) A solution containing such a pair of solutes. (16.7)

Buffer Capacity: A measure of how much strong acid or strong base is needed to change the pH of a buffer by some specified amount. (16.7)

Buret: A long tube of glass usually marked in mL and 0.1 mL units and equipped with a stopcock for the controlled addition of a liquid to a receiving flask. (4.9)

By-product: The substances formed by side reactions. (3.7)

C

Calorie (cal): 4.184 J. The energy that will raise the temperature of 1.00 g of water from 14.5 to 15.5 °C. (The nutritional *Calorie*, with a capital C, means 1000 cal or 1 kcal.) (6.1)

Calorimeter: An apparatus used in the determination of the heat of a reaction. (6.5)

Calorimetry: The science of measuring the quantities of heat that are involved in a chemical or physical change. (6.5)

Carbohydrates: Polyhydroxyaldehydes or polyhydroxyketones or substances that yield these by hydrolysis and that are obtained from plants or animals. (22.6)

Carbon Nanotube: Tubular carbon molecules that can be visualized as rolled up sheets of graphite (with hexagonal rings of carbon atoms) capped at each end by half of a spherical fullerene molecule. (9.10)

Carbon Ring: A series of carbon atoms arranged in a ring. (22.1)

Carbonyl Group: An organic functional group consisting of a carbon atom joined to an oxygen atom by a double bond; $C{=}O$. (8.8, 22.3)

Carboxyl Group: The organic functional group of an acid consisting of a carbon joined to an oxygen atom with a double bond and an OH group; $-CO_2H$ or $-C{\overset{\displaystyle O}{\underset{\displaystyle OH}{\diagup}}}$. (8.8, 22.3)

Carboxylic Acid: An organic compound whose molecules have the carboxyl group (8.8, 22.3)

Catalysis: Rate enhancement caused by a catalyst. (13.9)

Catalyst: A substance that in relatively small proportion accelerates the rate of a reaction without being permanently chemically changed. (13.9)

Cathode Ray: A stream of electrons ejected from a hot metal and accelerated toward a positively charged site in a vacuum tube. (0.5)

Cathode: The negative electrode in a gas discharge tube. The electrode at which reduction occurs during an electrochemical change. (0.5, 19.1)

Cation: A positively charged ion. (2.5)

Cell Potential, E_{cell}: The potential (voltage) of a galvanic cell when no current is drawn from the cell. (19.2)

Cell Reaction: The overall chemical change that takes place in an electrolytic cell or a galvanic cell. (19.1)

Celsius Scale: A temperature scale on which water freezes at 0 °C and boils at 100 °C (at 1 atm) and that has 100 divisions called Celsius degrees between those two points. (1.4)

Centimeter (cm): 0.01 m. (1.4)

Ceramics: A solid, inorganic material made from clay, silica and other silicates. (15.6)

Chain-Growth Polymer: See *Addition Polymer*.

Chain Reaction: A self-sustaining change in which the products of one event cause one or more new events. (Chemistry Outside the Classroom. 13.1)

Change of State: Transformation of matter from one physical state to another. In thermochemistry, any change in a variable used to define the state of a particular system—a change in composition, pressure, volume, or temperature. (11.3)

Charles' Law: See *Temperature–Volume Law*.

Chemical Bond: The force of electrical attraction that holds atoms together in compounds. (2.7, 8, This Chapter in Context)

Chemical Change: A change that converts substances into other substances; a chemical reaction. (1.2)

Chemical Energy: The potential energy of chemicals that is transferred during chemical reactions. (6.1)

Chemical Equation: A before-and-after description that uses formulas and coefficients to represent a chemical reaction. (2.4)

Chemical Equilibrium: Dynamic equilibrium in a chemical system. (4.4)

Chemical Formula: A formula written using chemical symbols and subscripts that describes the composition of a chemical compound or element. (2.3)

Chemical Kinetics: The study of rates of reaction. (13 This Chapter in Context)

Chemical Property: The ability of a substance, either by itself or with other substances, to undergo a change into new substances. (1.3)

Chemical Reaction: A change in which new substances (products) form from starting materials (reactants). (1.2)

Chemical Symbol: A formula for an element. (0.5, 1.2)

Chemical Thermodynamics: See *Thermodynamics*.

Chirality: The "handedness" of an object; the property of an object (like a molecule) that makes it unable to be superimposed onto a model of its own mirror image. (22.1)

cis **Isomer:** A stereoisomer whose uniqueness is in having two groups on the same side of some reference plane. (22.2)

Clausius–Clapeyron Equation: The relationship between the vapor pressure, the temperature, and the molar heat of vaporization of a substance (where C is a constant). (11.9)

$$\ln P = \frac{\Delta H_{\text{vap}}}{RT} + C$$

Closed-End Manometer: See *Manometer*.

Closed System: A system that can absorb or release energy but not mass across the boundary between the system and its surroundings. (6.3)

Closest-Packed Structure: A crystal structure in which atoms or molecules are packed as efficiently as possible. (11.10)

Codon: An individual unit of hereditary instruction that consists of three, side by side, side chains on a molecule of mRNA. (22.7)

Coefficients: Numbers in front of formulas in chemical equations. (2.4)

Colligative Property: A property such as vapor pressure lowering, boiling point elevation, freezing point depression, and osmotic pressure whose physical value depends only on the ratio of the numbers of moles of solute and solvent particles and not on their chemical identities. (12.6)

Colloidal Dispersion: Particles in a suspension that do not settle or separate into different phases. (12.7)

Combined Gas Law: See *Gas Law, Combined*.

Combustion: A rapid reaction with oxygen accompanied by a flame and the evolution of heat and light. (5.5)

Combustion Analysis: A method for obtaining percentage composition data for organic compounds. (3.4)

Common Ion: The ion in a mixture of ionic substances that is common to the formulas of at least two. (16.7, 17.1)

Common Ion Effect: The solubility of one salt is reduced by the presence of another having a common ion. The addition of an ion, that is present in the net ionic equation, will alter the position of equilibrium. (16.7, 17.1)

Competing Reaction: A reaction that reduces the yield of the main product by forming by-products. (3.7)

Complex Ion (Complex): The Lewis acid-base product formed between one or more ligands and a metal ion. (14.6, 15.6, 17.5, 22.1)

Composition: The elements and their ratios that make up a substance. (1.2)

Compound Nucleus: An atomic nucleus carrying excess energy following its capture of some bombarding particle. (20.5)

Compressibility: Capable of undergoing a reduction in volume under increasing pressure. (11.2)

Concentrated Solution: A solution that has a large ratio of the amounts of solute to solvent. (4.1)

Concentration: The ratio of the quantity of solute to the quantity of solution (or the quantity of solvent). (4.1) (Also see *Molal Concentration, Molar Concentration, Mole Fraction, Percentage Concentration*.)

Concentration Table: A part of the strategy for organizing data needed to make certain calculations, particularly any involving equilibria. (14.7)

Conclusion: A statement that is based on what we think about a series of observations. (1.1)

Condensation: The change of a vapor to its liquid state. (11.3)

Condensation (Polymerization): The process of forming a condensation polymer. (22.5)

Condensation Polymer (Step-Growth Polymer): A polymer formed from monomers by splitting out a small molecule such as H_2O or CH_3OH. (22.5)

Condensed Formula: (Condensed Structural Formulas, or Condensed Structures): Generally a formula of an organic compound written to show the sequence of carbon atoms and functional groups without individual bonds. (8.8, 22.3)

Condensed Structural Formulas (Condensed Structures): See *Condensed Formula*.

Conduction Band: In band theory, this consists of any band of atomic energy levels that is continuous throughout the solid and empty or partially filled with electrons. (9.9)

Conformation: A particular relative orientation or geometric form of a flexible molecule. (9.5)

Conjugate Acid: The species in a conjugate acid–base pair that has the greater number of H-units. (15.1)

Conjugate Acid–Base Pair: Two substances (ions or molecules) whose formulas differ by only one H-unit. (15.1)

Conjugate Base: The species in a conjugate acid–base pair that has the fewer number of H-units. (15.1)

Continuous Spectrum: The electromagnetic spectrum corresponding to the mixture of frequencies present in white light. (7.2)

Contributing Structure: One of a set of two or more Lewis structures used in applying the theory of resonance to the structure of a compound. A resonance structure. (8.8)

Conversion Factor: A ratio constructed from the relationship between two units such as 2.54 cm/1 in., from 1 in. = 2.54 cm. (1.6)

Cooling Curve: A graph showing how the temperature and physical state of a substance changes as heat is removed from it at a constant rate. (11.6)

Coordinate Covalent Bond: A covalent bond in which both electrons originated from one of the joined atoms, but otherwise is the same as a covalent bond in all respects. (8.7)

Core Electrons: The inner electrons of an atom that are not exposed to the electrons of other atoms when chemical bonds form. (7.10)

Corrosion: The slow oxidation of metals exposed to air or water. (5.5)

Coulomb (C): The SI unit of electrical charge; the charge on 6.24×10^{18} electrons; the amount of charge that passes a fixed point of a wire conductor when a current of 1 A flows for 1 s. (19.2)

Coulomb's Law: This is the mathematical equation that describes the attractive and repulsive forces between charged particles (8.2)

Covalent Bond: A chemical bond that results when atoms share electron pairs. (8.5)

Covalent Crystal (Network Solid): A crystal in which the lattice positions are occupied by atoms that are covalently bonded to the atoms at adjacent lattice sites. (11.12)

Critical Mass: The mass of a fissile isotope above which a self-sustaining chain reaction occurs. (20.8)

Critical Point: The point at the end of a vapor pressure versus temperature curve for a liquid and that corresponds to the critical pressure and the critical temperature. (11.7)

Critical Pressure (P_c): The vapor pressure of a substance at its critical temperature. (11.7)

Critical Temperature (T_c): The temperature above which a substance cannot exist as a liquid regardless of the pressure. (11.7)

Crystal Lattice: The repeating symmetrical pattern of atoms, molecules, or ions that occurs in a crystal. (11.10)

Crystallization: The process in which a substance forms a crystal lattice from a liquid, gas, or solution. (11.3)

Cubic Closest Packing (ccp): Efficient packing of spheres with an A-B-C-A-B-C…

alternating stacking of layers of spheres. (11.10)

Cubic Meter (m^3): The SI derived unit of volume. (1.4)

Curie (Ci): A unit of activity for radioactive samples, equal to 3.7×10^{10} disintegrations per second. (20.6)

D

Dalton's Atomic Theory: Matter consists of tiny, indestructible particles called atoms. All atoms of one element are identical. The atoms of different elements have different masses. Atoms combine in definite ratios by atoms when they form compounds. (0.4)

Dalton's Law of Partial Pressures: See *Partial Pressures, Law of.*

Data: The information (often in the form of physical quantities) obtained in an experiment or other experience or from literature references. (1.1)

Debye: Unit used to express dipole moments. 1 D = 3.34×10^{-30} C m (coulomb meter). (8.6)

Decay Constant: The first-order rate constant for radioactive decay. (20.6)

Decimal Multipliers: Factors—exponentials of 10 or decimals—that are used to define larger or smaller SI units. (1.4)

Decomposition: A chemical reaction that changes one substance into two or more simpler substances. (1.2)

Dehydration: Formation of a carbon–carbon double bond by removal of the components of water from an alcohol. (22.3)

Delocalization Energy: The difference between the energy a substance would have if its molecules had no delocalized molecular orbitals and the energy it has because of such orbitals. (9.8)

Delocalized electrons: A π-electron cloud that is spread out over more than two atoms in a structure. (9.8)

Density: The ratio of an object's mass to its volume. (1.7)

Deposition: The change of a vapor to its solid state. The term describing the opposite of sublimation. (11.3)

Derived Unit: Any unit defined solely in terms of base units. (1.4)

Deuterium: The isotope of hydrogen with a mass number of 2. (20.8)

Diagonal Trend: Physical or chemical properties that generally vary diagonally from one corner to the other in the periodic table (e.g., ionization energy, electron affinity, and electronegativity). (7.10)

Dialysis: The passage of small molecules and ions, but not species of a larger

size, through a semipermeable membrane. (12.6)

Dialyzing Membrane: A membrane that allows the passage of solvent and small solutes through its pores, but large molecules cannot pass through the pores. (12.6)

Diamagnetic: A substance that contains no unpaired electrons, whereby the substance is repelled weakly by a magnet. (7.6)

Diamond: A crystalline form of carbon in which each carbon atom is bonded in a tetrahedral structure to four other carbon atoms. (9.10)

Diaphragm Cell: An electrolytic cell used to manufacture sodium hydroxide by the electrolysis of aqueous sodium chloride. (19.9)

Diatomic Molecule: A molecular substance made from two atoms. (2.3)

Diffraction: Constructive and destructive interference by waves. (7.4)

Diffraction Pattern: The image formed on a screen or a photographic film caused by the diffraction of electromagnetic radiation such as visible light or X rays. (11.11)

Diffusion: The spontaneous intermingling of one substance with another. (10.6)

Dilute Solution: A solution in which the ratio of the quantities of solute to solvent is small. (4.1)

Dilution: The process whereby a concentrated solution is made more dilute. (4.7)

Dimensional Analysis: A method for converting one set of units to another that relies on a logical sequence of steps and unit cancellation. (1.6)

Dimer: Two monomer units joined by chemical bonds or intramolecular forces. (12.6)

Dipole–Dipole Attraction: Attraction between molecules that are dipoles. (11.1)

Dipole (Electric): Partial positive and partial negative charges separated by a distance. (8.6, 11.1, 12.2)

Dipole Moment (μ): The product of the sizes of the partial charges in a dipole multiplied by the distance between them; a measure of the polarity of a molecule. (8.6, 9.3)

Diprotic Acid: An acid that can furnish two H^+ per molecule. (4.4)

Disaccharide: A carbohydrate whose molecules can be hydrolyzed into two monosaccharides. (22.6)

Dispersion Forces: Another term for London forces. (11.1)

Dispersions: Mixtures of very small particles of immiscible substances that do not readily separate into different phases. Often used when describing emulsions. (12.7)

Displacement Reactions: See *Single Replacement Reaction*

Dissociation: The separation of preexisting ions when an ionic compound dissolves or melts. (4.2)

Distorted Tetrahedron: A description of a molecule in which the central atom is surrounded by five electron pairs, one of which is a lone pair of electrons. The central atom is bonded to four other atoms. The structure is also said to have a seesaw shape. (9.2)

DNA Double Helix: Two oppositely running strands of DNA held in a helical configuration by inter-strand hydrogen bonds. (22.7)

DNA: Deoxyribonucleic acid; a nucleic acid that hydrolyzes to deoxyribose, phosphate ion, adenine, thymine, guanine, and cytosine, and that is the carrier of genes. (22.7)

Donor Atom: The atom on a ligand that makes an electron pair available in the formation of a complex. (17.5, 22.1)

Double Bond: (a) A covalent bond formed by sharing two pairs of electrons. (8.5) (b) A covalent bond consisting of one sigma bond and one pi bond.

Double Replacement Reaction (Metathesis Reaction): A reaction of two salts in which cations and anions exchange partners
(e.g., $AgNO_3 + NaCl \longrightarrow$
$\qquad AgCl + NaNO_3$). (4.6)

Downs Cell: An electrolytic cell for the industrial production of sodium. (19.9)

Dry Ice: Common name for solid carbon dioxide. (11.2)

Ductility: A metal's ability to be drawn (or stretched) into wire. (2.2)

Dynamic Equilibrium: A condition in which two opposing processes are occurring at equal rates. (4.4)

E

Effective Collision: A collision between molecules that is capable of leading to a net chemical change. (13.5)

Effective Nuclear Charge: The net positive charge an outer electron experiences as a result of the partial screening of the full nuclear charge by core electrons. (7.10)

Effusion: The movement of a gas through a very tiny opening into a vacuum. (10.6)

Effusion, Law of: See *Graham's Law.*

Einstein Equation: $\Delta E = m_0 c^2$ where ΔE is the energy obtained when a quantity of rest mass, Δm_0, is destroyed, or the energy lost when this quantity of mass is created. (20.1)

Electric Dipole: Two poles of electric charge separated by a distance. (8.6)

Electrolysis: The production of a chemical change by the passage of electricity through a solution that contains ions or through a molten ionic compound. (19.7)

Electrolysis Cell: An apparatus for electrolysis. (19.7)

Electrolyte: A compound that conducts electricity either in solution or in the molten state. (4.2)

Electrolytic Cell: See *Electrolysis Cell.*

Electrolytic Conduction: The transport of electrical charge by ions. (19.1)

Electromagnetic Radiation: See *Electromagnetic Wave*

Electromagnetic Spectrum: The distribution of frequencies of electromagnetic radiation among various types of such radiation—microwave, infrared, visible, ultraviolet, X rays, and gamma rays. (7.1)

Electromagnetic Wave (Electromagnetic Radiation): The successive series of oscillations in the strengths of electrical and magnetic fields associated with light, microwaves, gamma rays, ultraviolet rays, infrared rays, and the like. (7.1)

Electromotive Force (emf): An older term for the cell potential. (19.2)

Electron (e^- or $_{-1}^{0}e$): A subatomic particle with a charge of 1– and mass of 0.0005486 u (9.109383×10^{-28} g) that occurs outside an atomic nucleus. The particle that moves when an electric current flows. (0.5)

Electron Affinity (EA): The energy change (usually expressed in kJ mol^{-1}) that occurs when an electron adds to an isolated gaseous atom or ion. (7.10)

Electron Capture: The capture by a nucleus of an orbital electron and that changes a proton into a neutron in the nucleus. (20.3)

Electron Cloud: Because of its wave properties, an electron's influence spreads out like a cloud around the nucleus. (7.9)

Electron Density: The concentration of the electron's charge within a given volume. (7.9)

Electron Domain: A region around an atom where one or more electron pairs are concentrated and which influences the shape of a molecule. (9.2)

Electronegativity: The relative ability of an atom to attract electron density toward itself when joined to another atom by a covalent bond. (8.6)

Electronic Configuration: See *Electron Structure*

Electronic Structure: The distribution of electrons in an atom's orbitals. (7.7)

Electron Pair Bond: A covalent bond. (8.5)

Electron Spin: The spinning of an electron about its axis that is believed to occur because the electron behaves as a tiny magnet. (7.6)

Electron Volt (eV): The energy an electron receives when it is accelerated under the influence of 1 V and equal to 1.6×10^{-19} J. (20.3)

Electroplating: Depositing a thin metallic coating on an object by electrolysis. (19.9)

Elementary Process: One of the individual steps in the mechanism of a reaction. (13.8)

Element: A substance in which all of the atoms have the same atomic number. A substance that cannot be broken down by chemical reactions into anything that is both stable and simpler. (0.5, 1.2)

Elimination Reaction: The loss of a small molecule from a larger molecule as in the elimination of water from an alcohol. (22.3)

Emission Spectrum: See *Atomic Spectrum.*

Empirical Formula: A chemical formula that uses the smallest whole-number subscripts to give the proportions by atoms of the different elements present. (3.4)

Emulsion: A mixture of small globules of one immiscible liquid in another. (12.7)

Enantiomers: Stereoisomers whose molecular structures are related as an object to its mirror image but that cannot be superimposed. (21.4)

Endergonic: Descriptive of a change accompanied by an increase in free energy. (18.4)

Endothermic: Descriptive of a change in which a system's internal energy increases. (6.4)

End Point: The moment in a titration when the indicator changes color and the titration is ended. (4.9, 16.9)

Energy: Something that matter possesses by virtue of an ability to do work. (6.1)

Energy Band: Closely spaced energy levels in materials that can be described as a band of energy (9.9)

Energy Density: For a galvanic cell, the ratio of the energy available to the volume of the cell. (19.6)

Energy Level: A particular energy an electron can have in an atom or a molecule. (7.3)

English System: A system of weights and measures that includes such units as pounds, ounces, gallons, and miles. (1.4)

Enthalpy (H): The heat content of a system. (6.6, 18.1)

Enthalpy Change (ΔH): The difference in enthalpy between the initial state and the final state for some change. (6.6, 18.1)

Enthalpy Diagram: A graphical depiction of enthalpy changes following different paths from reactants to products. (6.8)

Entropy (S): A thermodynamic quantity related to the number of equivalent ways the energy of a system can be distributed. The greater this number, the more probable is the state and the higher is the entropy. (18.3)

Entropy Change (ΔS): The difference in entropy between the initial state and the final state for some change. (18.3)

Enzyme: A protein that acts as a catalyst. (22.6)

Equation of State for an Ideal Gas: See *Gas Law, Ideal*.

Equatorial Bond: A covalent bond located in the plane perpendicular to the long axis of a trigonal bipyramidal molecule. (9.1)

Equilibrium Constant, K_c: The value that the mass action expression has when the system is at equilibrium and quantities are expressed in mol L^{-1}. (14.2)

Equilibrium Constant, K_P: The value that the mass action expression has when the system is at equilibrium and quantities are expressed as partial pressures in atm or bar. (14.2)

Equilibrium Law: The mathematical equation for a particular equilibrium system that sets the mass action expression equal to the equilibrium constant. (14.2)

Equilibrium Vapor Pressure of a Liquid: The pressure exerted by a vapor in equilibrium with its liquid state. (11.4)

Equilibrium Vapor Pressure of a Solid: The pressure exerted by a vapor in equilibrium with its solid state. (11.4)

Equivalence Point: The point in a calculated titration curve where the equivalents of one reactant equals the number of equivalents of the other reactant.

Error (in a Measurement): The difference between a measurement and the "true" value we are trying to measure. (1.5)

Ester: An organic compound whose molecules have the ester group. (22.3)

$$\overset{\displaystyle O}{\overset{\displaystyle \|}{-C-O-C}}$$
easter group

Ether: An organic compound in whose molecules two hydrocarbon groups are joined by oxygen. (22.3)

Ethylenediaminetetraacetic acid (EDTA): A ligand with the formula. $CH_2(N(CH_3COOH)_2)_2 CH_2 (N(CH_3COOH)_2)_2$ and four acid groups and two amines, all of which can bind to a metal center. (Chemistry Outside the Classroom (17.1)

Ethyl Group: Alkyl group derived from ethane, $-CH_2CH_3$ (22.2)

Evaporation: Change of state from a liquid to a vapor. (11.2)

Excess Reactant: The reactant left over once the limiting reactant is used up. (3.6)

Excited State: A term describing an atom or molecule where all of the electrons are not in their lowest possible energy levels. (7.2)

Exergonic: Descriptive of a change accompanied by a decrease in free energy. (18.4)

Exothermic: Descriptive of a change in which energy leaves a system and enters the surroundings. (6.4)

Expansion Work: See *Pressure–Volume Work*.

Extensive Property: A property of an object that is described by a physical quantity whose magnitude is proportional to the size or amount of the object (e.g., mass or volume). (1.3)

F

Face-Centered Cubic (fcc) Unit Cell: A unit cell having identical atoms, molecules, or ions at the corners of a cube and also in the center of each face of the cube. (11.10)

Fahrenheit Scale: A temperature scale on which water freezes at 32 °F and boils at 212 °F (at 1 atm) and between which points there are 180 degree divisions called Fahrenheit degrees. (1.4)

Family of Elements: See *Group*.

Faraday ($\mathcal{F}$): One mole of electrons; 9.65×10^4 coulombs. (19.4)

Faraday Constant ($\mathcal{F}$): 9.65×10^4 coulombs/mol e^-. (19.4)

Fatty Acid: One of several long-chain carboxylic acids produced by the hydrolysis (digestion) of a lipid. (22.6)

Film Dosimeter: A device used by people working with radioactive isotopes that records doses of atomic radiation by the darkening of photographic film. (20.6)

First Law of Thermodynamics: A formal statement of the law of conservation of energy. $\Delta E = q + w$. (6.5, 18.1)

First-Order Reaction: A reaction with a rate law in which rate $= k[A]$, where A is a reactant. (13.4)

Fissile Isotope: An isotope capable of undergoing fission following neutron capture. (20.8)

Fission: The breaking apart of atomic nuclei into smaller nuclei accompanied by the release of energy, and the source of energy in nuclear reactors. (20.1)

Formal Charge: The apparent charge on an atom in a molecule or polyatomic ion as calculated by a set of rules. (8.7)

Formation Constant (K_{form}): The equilibrium constant for an equilibrium involving the formation of a complex ion. Also called the stability constant. (17.5)

Formula Mass: The sum of the atomic masses (in u) of all of the atoms represented in a chemical formula. Often used with units of g mol^{-1} to represent masses of ionic substances. (3.1) See also *Molar Mass*.

Formula Unit: A particle that has the composition given by the chemical formula. (2.5)

Forward Reaction: In a chemical equation, the reaction as read from left to right. (4.4, 14.1)

Free Element: An element that is not combined with another element in a compound. (2.3)

Free Energy Diagram: A plot of the changes in free energy for a multicomponent system versus the composition. (18.8)

Free Radical: An atom, molecule, or ion that has one or more unpaired electrons. (20.6, Chemistry Outside the Classroom 13.1)

Freezing Point Depression: A colligative property of a liquid solution by which the freezing point of the solution is lower than that of the pure solvent. (12.6)

Frequency Factor: The proportionality constant, A, in the Arrhenius equation. (13.7)

Fuel Cell: An electrochemical cell in which electricity is generated from the redox reactions of common fuels. (19.6)

Fuel: Commonly one component of a combustion reaction; specifically the reducing agent. (6.9)

Fullerene: An allotrope of carbon made of an extended joining together of five- and six-membered rings of carbon atoms. (9.10)

Functional Group: The group of atoms of an organic molecule that enters into a characteristic set of reactions that are independent of the rest of the molecule. (8.8)

Fusion: (a) Melting; the change of a solid to its liquid state. (11.3, 12.4) (b) The formation of atomic nuclei by the joining together of the nuclei of lighter atoms. (0.2, 20.2)

G

Galvanic Cell: An electrochemical cell in which a spontaneous redox reaction produces electricity. (19.1)

Gamma Radiation: Electromagnetic radiation with wavelengths in the range of 1 Å or less (the shortest wavelengths of the spectrum). (20.3)

Gas: One of the states of matter. A gas consists of rapidly moving widely spaced atomic or molecular sized particles. (1.3)

Gas Constant, Universal (R):
$R = 0.0821$ liter atm mol^{-1} K^{-1} or $R = 8.314$ J mol^{-1} K^{-1} (10.5)

Gas Law, Combined: For a given mass of gas, the product of its pressure and volume divided by its Kelvin temperature is a constant. (10.3)
$PV/T =$ a constant

Gas Law, Ideal: See *Ideal Gas Law.*

Gay-Lussac's Law: See *Pressure– Temperature Law.*

Geiger Counter: A device that detects beta and gamma radiation. (20.6)

Genetic Code: The correlation of codons with amino acids. (22.7)

Geometric Isomer: One of a set of isomers that differ only in geometry. (22.2)

Geometric Isomerism: The existence of isomers whose molecules have identical atomic organizations but different geometries; e.g., *cis-trans* isomers. (21.4, 22.2)

Gibbs Free Energy (G): A thermodynamic quantity that relates enthalpy (H), entropy (S), and temperature (T) by the equation: (18.4)

$$G = H - TS$$

Gibbs Free Energy Change (ΔG): The difference given by: (18.4)

$$\Delta G = \Delta H - T\Delta S$$

Glass: Any amorphous solid. (11.10)

Glycogen: A polysaccharide that animals use to store glucose units for energy. (22.6)

Graham's Law (Law of Effusion)**:** The rates of effusion of gases are inversely proportional to the square roots of their densities when compared at identical pressures and temperatures.

$$\text{Effusion rate} \propto \frac{1}{\sqrt{d}} \text{ (constant } P \text{ and } T)$$

where d is the gas density. (10.6)

Gram (g): 0.001 kg. (1.4)

Graphene: A sheet of graphite, one atom thick. (9.10)

Graphite: The most stable allotrope of carbon, consisting of layers of joined six-membered rings of carbon atoms. (9.10)

Gray (Gy): The SI unit of radiation absorbed dose. (20.6)

$$1 \text{ Gy} = 1 \text{ J kg}^{-1}$$

Ground State: The lowest energy state of an atom or molecule. (7.3)

Group: A vertical column of elements in the periodic table. (2.1)

H

Half-Cell: That part of a galvanic cell in which either oxidation or reduction takes place. (19.1)

Half-life ($t_{1/2}$): The time required for a reactant concentration or the mass of a radionuclide to be reduced by half. (13.4)

Half-Reaction: A hypothetical reaction that constitutes either the oxidation or the reduction half of a redox reaction. This balanced reaction is unique in that it uses electrons to balance the electrical charge. (5.2)

Hall–Héroult Process: A method for manufacturing aluminum by the electrolysis of aluminum oxide in molten cryolite. (19.9)

Halogens: Group 7A in the periodic table—fluorine, chlorine, bromine, iodine, and astatine. (2.1)

Hard Water: Water with dissolved Mg^{2+}, Ca^{2+}, Fe^{2+}, or Fe^{3+} ions at a concentration high enough (above 25 mg L^{-1}) to interfere with the use of soap. (Chemistry Outside the Classroom 4.2)

Heat: Energy that flows from a hot object to a cold object as a result of their difference in temperature. (6.2)

Heat at Constant Pressure (q_p): The heat of a reaction in an open system, ΔH. (18.1)

Heat Capacity: The quantity of heat needed to raise the temperature of an object by 1 °C. (6.3)

Heating Curve: A graph showing how the temperature and physical state of a substance changes as heat is added to it at a constant rate. (11.6)

Heat of Combustion (ΔH_c): The heat evolved in the combustion of a substance. (6.6)

Heat of Reaction: The heat exchanged between a system and its surroundings when a chemical change occurs in the system. (6.5)

Heat of Reaction at Constant Pressure (q_p): The heat of a reaction in an open system, ΔH. (6.6)

Heat of Reaction at Constant Volume (q_v): The heat of a reaction in a sealed vessel, such as a bomb calorimeter, ΔE. (6.6)

Heat of Solution (ΔH_{soln}): The energy exchanged between the system and its surroundings when one mole of a solute dissolves in a solvent to make a dilute solution. (12.2)

Henderson–Hasselbalch Equation:

$$pH = pK_a + \log \frac{[A^-]}{[HA]} \text{ or}$$

$$pH = pK_a + \log \frac{[\text{salt}]}{[\text{acid}]} \qquad (16.7)$$

Henry's Law: The concentration of a gas dissolved in a liquid at any given temperature is directly proportional to the partial pressure of the gas above the solution. $C_{gas} = k_H P_{gas}$. (12.4)

Hertz (Hz): 1 cycle s^{-1}; the SI unit of frequency. (7.1)

Hess's Law Equation: For the change, $\Delta H°_{\text{reaction}} =$
$$\left(\begin{array}{c} \text{sum of } \Delta H_f° \text{ of all} \\ \text{of the products} \end{array} \right) - \left(\begin{array}{c} \text{sum of } \Delta H_f° \text{ of all} \\ \text{of the reactants} \end{array} \right)$$
(6.9)

Hess's Law of Heat Summation (Hess's Law): For any reaction that can be written in steps, the standard heat of reaction is the same as the sum of the standard heats of reaction for the steps. (6.8)

Heterocyclic Rings: Molecular rings that include one or more atoms other than carbon within the ring. (22.1)

Heterogeneous Catalyst: A catalyst that is in a different phase than the reactants and onto whose surface the reactant molecules are adsorbed and where they react. (13.9)

Heterogeneous Mixture: A mixture that has two or more phases with different properties. (1.2)

Heterogeneous Reaction: A reaction in which not all of the chemical species are in the same phase. (13.1, 14.4)

Heteronuclear Molecule: A molecule in which not all atoms are of the same element. (9.7)

Hexagonal Closest Packing (hcp): Efficient packing of spheres with an A-B-A-B-... alternating stacking of layers of spheres. (11.10)

High-Spin Complex: A complex ion or coordination compound in which there is the maximum number of unpaired electrons. (21.5)

Hole: An electron-deficient site in a semiconductor. A positive charge carrier. (19.6)

Homogeneous Catalyst: A catalyst that is in the same phase as the reactants. (13.9)

Homogeneous Equilibrium: An equilibrium system in which all components are in the same phase. (14.2)

Homogeneous Mixture: A mixture that has only one phase and that has uniform properties throughout; a solution. (1.2)

Homogeneous Reaction: A reaction in which all of the chemical species are in the same phase. (13.1, 14.2)

Homonuclear Diatomic Molecule: A diatomic molecule in which both atoms are of the same element. (9.7)

Hund's Rule: Electrons that occupy orbitals of equal energy are distributed with unpaired spins as much as possible among all such orbitals. (7.7)

Hybrid Atomic Orbitals: Orbitals formed by mixing two or more of the basic atomic orbitals of an atom and that make possible more effective overlaps with the orbitals of adjacent atoms than do ordinary atomic orbitals. (9.5)

Hydrate: A compound that contains molecules of water in a definite ratio to other components. (2.3)

Hydrated Ion: An ion surrounded by water molecules that are attracted by the charge on the ion. (4.2)

Hydration: The development in an aqueous solution of a cage of water molecules about ions or polar molecules of the solute. (12.1)

Hydration Energy: The enthalpy change associated with the hydration of gaseous ions or molecules as they dissolve in water. (12.2)

Hydride: (a) A binary compound of hydrogen. (b) A compound containing the hydride ion (H^-). (5.1)

Hydrocarbon: An organic compound whose molecules consist entirely of carbon and hydrogen atoms. (2.7, 22.2)

Hydrogen Bonding: An extra strong dipole–dipole attraction between a hydrogen bound covalently to nitrogen, oxygen, or fluorine and another nitrogen, oxygen, or fluorine atom. (11.1)

Hydrogen Electrode: The standard of comparison for reduction potentials and for which E_H° has a value of 0.00 V at 25 °C, when $P_{H_2} = 1$ atm and $[H^+] = 1\ M$ in the reversible half-cell reaction: $2H^+(aq) + 2e^- \rightleftharpoons H_2(g)$ (19.2)

Hydrolysis: A reaction with water (15.6, 16.6)

Hydrometer: A device for measuring specific gravity. (1.7, 19.6)

Hydronium Ion: H_3O^+. (4.4)

Hydrophilic: Miscible (soluble in water), literally, "water loving." (12.7, 22.6)

Hydrophilic Group: A polar molecular unit capable of having dipole–dipole attractions or hydrogen bonds with water molecules. (22.6)

Hydrophobic. Not soluble in water, literally, "afraid of water." (12.7, 22.6)

Hypothesis: A tentative explanation of the results of experiments. (1.1)

Hypoxia: Condition of natural water where there is not dissolved oxygen to support aquatic life. (12.3)

I

Ideal Gas: A hypothetical gas that obeys the gas laws exactly. (10.3)

Ideal Gas Law: $PV = nRT$. (10.5)

Ideal Solution: A hypothetical solution that would obey the vapor pressure–concentration law (Raoult's law) exactly. (12.2)

Immiscible: Mutually insoluble. Usually used to describe liquids that are insoluble in each other. (12.1)

Incompressible: Incapable of losing volume under increasing pressure. (11.2)

Indicator: A chemical put in a solution being titrated and whose change in color signals the end point. (16.9)

Induced Dipole: A dipole created when the electron cloud of an atom or a molecule is distorted by a neighboring dipole or by an ion. (11.1)

Initiation Step: The step in a chain reaction that produces reactive species that can start chain propagation steps. (Chemistry Outside the Classroom 13.1)

Inner Transition Elements: Members of the two long rows of elements below the main body of the periodic table—elements 58–71 and elements 90–103. (2.1)

Inorganic Compound: A compound made from any elements except those compounds of carbon classified as organic compounds. (2.6)

Instability Constant (K_{inst}): The reciprocal of the formation constant for an equilibrium in which a complex ion forms. (17.5)

Instantaneous Dipole: A momentary dipole in an atom, ion, or molecule caused by the erratic movement of electrons. (11.1)

Instantaneous Rate: The rate of reaction at any particular moment during a reaction. (13.2)

Intensive Property: A property whose physical magnitude is independent of the size of the sample, such as density or temperature. (1.3)

Intercalation: The insertion of small atoms or ions between layers in a crystal such as graphite. (19.6)

Interference Fringes: Pattern of light produced by waves that undergo diffraction. (7.4)

Intermolecular Forces (Intermolecular Attractions): Attractions *between* neighboring molecules. (11.1)

Internal Energy (E, $E_{Internal}$): The sum of all of the kinetic energies and potential energies of the particles within a system. (18.1, 6.2)

International System of Units (SI): The successor to the metric system of measurements that retains most of the units of the metric system and their decimal relationships but employs new reference standards. (1.4)

Intramolecular Forces: Forces of attraction within molecules; chemical bonds. (11.1)

Inverse Square Law: The intensity of a radiation is inversely proportional to the square of the distance from its source. (20.6)

Ion: An electrically charged particle on the atomic or molecular scale of size. (2.5)

Ion–Dipole Attraction: The attraction between an ion and the charged end of a polar molecule. (11.1)

Ion–Electron Method: A method for balancing redox reactions that uses half-reactions. (5.2)

Ionic Bond: The attractions between ions that hold them together in ionic compounds. (8.2)

Ionic Character: The extent to which a covalent bond has a dipole moment and is polarized. (8.6)

Ionic Compound: A compound consisting of positive and negative ions. (2.5)

Ionic Crystal: A crystal that has ions located at the lattice points. (11.12)

Ionic Equation: A chemical equation in which soluble strong electrolytes are written in dissociated or ionized form. (4.3)

Ion–Induced Dipole Attraction: Attraction between an ion and a dipole induced in a neighboring molecule. (11.1)

Ionization Energy (IE): The energy needed to remove an electron from an isolated, gaseous atom, ion, or molecule (usually given in units of kJ mol^{-1}). (7.10)

Ionization Reaction: A reaction of chemical particles that produces ions. (4.4)

Ionizing Radiation: Any high-energy radiation—X rays, gamma rays, or radiations from radionuclides—that generates ions as it passes through matter. (20.6)

Ion Pair: A more or less loosely associated pair of ions in a solution. (12.6)

Ion Product: The mass action expression for the solubility equilibrium involving the ions of a salt and equal to the product of the molar concentrations of the ions, each concentration raised to a power that equals the number of ions obtained from one formula unit of the salt. (17.1)

Ion Product Constant of Water (K_w): $K_w = [H^+][OH^-]$ (16.1)

Isolated System: A system that cannot exchange matter or energy with its surroundings. (6.3)

Isomerism: The existence of sets of isomers. (8.8, 21.4)

Isomer: One of a set of compounds that have identical molecular formulas but different structures. (21.4)

Isopropyl Group: Alkyl group derived from propane, $-CH(CH_3)_2$ (22.2)

Isothermal: A system where the temperature remains unchanged. (18.1)

Isotopes: Atoms of the same element with different atomic masses. Atoms of the same element with different numbers of neutrons in their nuclei. (0.5)

IUPAC: The International Union of Pure and Applied Chemistry, which develops the formal rules for naming substances. (0.5)

IUPAC Rules: The formal rules for naming substances as developed by the International Union of Pure and Applied Chemistry. (22.2)

J

Joule (J): The SI unit of energy. (6.1) 1 J = 1 kg m^2 s^{-2} 4.184 J = 1 cal (exactly)

K

K_a: See *Acid Ionization Constant.*

K_b: See *Base Ionization Constant.*

K_c: See *Equilibrium Constant, K_c.*

Kelvin (K): One degree on the Kelvin scale of temperature and identical in size to the Celsius degree. (1.4)

Kelvin Scale: The temperature scale on which water freezes at 273.15 K and boils at 373.15 K and that has 100 degree divisions called kelvins between these points. K = °C + 273.15. (1.4)

Ketone: An organic compound that has a carbonyl group ($>C=O$) bonded to two hydrocarbon groups. (8.8)

K_{form}: See *Formation Constant.*

Kilocalorie (kcal): 1000 cal. (6.1)

Kilogram (kg): The base unit for mass in the SI and equal to the mass of a cylinder of platinum–iridium alloy kept by the International Bureau of Weights and Measures at Sevres, France. 1 kg = 1000 g. (1.4)

Kilojoule (kJ): 1000 J. (6.1)

Kinetic Energy (KE): Energy of motion. KE = (½)mv^2. (6.1)

Kinetic Molecular Theory: Molecules of a substance are in constant motion with a distribution of kinetic energies at a given temperature. The average kinetic energy of the molecules is proportional to the Kelvin temperature. (6.2)

Kinetic Molecular Theory of Gases: A set of postulates used to explain the gas laws. A gas consists of an extremely large number of very tiny, very hard particles in constant, random motion. They have negligible volume and, between collisions, experience no forces between themselves. (10.7)

K_{inst}: See *Instability Constant.*

K_P: See *Equilibrium Constant, K_P.*

K_{spa}: See *Acid Solubility Product.*

K_{sp}: See *Solubility Product Constant.*

K_w: See *Ion Product Constant of Water.*

L

Lampblack: Finely divided carbon that is formed from the incomplete combustion of organic compounds. (5.5)

Lanthanide Elements: Elements 58–71. (2.1)

Lattice: A symmetrical pattern of points arranged with constant repeat distances along lines oriented at constant angles. (11.10)

Lattice Energy: Energy released by the imaginary process in which isolated ions come together to form a crystal of an ionic compound. (8.2)

Law: A description of behavior (and not an *explanation* of behavior) based on the results of many experiments. (1.1)

Law of Combining Volumes: When gases react at the same temperature and pressure, their combining volumes are in ratios of simple whole numbers. (10.4)

Law of Conservation of Energy: The energy of the universe is constant; it can be neither created nor destroyed but only transferred and transformed. (6.1)

Law of Conservation of Mass–Energy: The sum of all the mass in the universe and of all of the energy, expressed as an equivalent in mass (calculated by the Einstein equation), is a constant. (20.1)

Law of Effusion (Graham's Law): See *Graham's Law.*

Law of Radioactive Decay:

$$\text{Activity} = -\frac{\Delta N}{\Delta t} = kN$$

where ΔN is the change in the number of radioactive nuclei during the time span Δt, and k is the decay constant. (20.6)

Lead Storage Battery: A galvanic cell of about 2 V involving lead and lead(IV) oxide in sulfuric acid. (19.6)

Le Châtelier's Principle: When a system that is in dynamic equilibrium is subjected to a disturbance that upsets the equilibrium, the system undergoes a change that counteracts the disturbance and, if possible, restores the equilibrium. (14.6)

Leclanché Cell: See *Zinc–Manganese Dioxide Cell.*

Lewis Acid: An electron-pair acceptor. (15.4)

Lewis Base: An electron-pair donor. (15.4)

Lewis Formula: See *Lewis structure.*

Lewis Structure: A structural formula drawn with Lewis symbols and that uses dots and dashes to show the valence electrons and shared pairs of electrons. (8.5)

Lewis Symbol: The symbol of an element that includes dots to represent the valence electrons of an atom of the element. (8.4)

Ligand: A molecule or an anion that can form a coordinate-covalent bond with a metal ion to form a complex. (17.5, 21.1)

Like Dissolves Like Rule: Strongly polar and ionic solutes tend to dissolve in polar solvents and nonpolar solutes tend to dissolve in nonpolar solvents. (12.1)

Limiting Reactant: The reactant that determines how much product can form when nonstoichiometric amounts of reactants are used. (3.6)

Linear Molecule: A molecule all of whose atoms lie on a straight line. (9.1)

Lipid: Any substance found in plants or animals that can be dissolved in nonpolar solvents. (22.6)

Liquid: One of the states of matter. A liquid consists of tightly packed atomic or molecular sized particles that can move past each other. (1.3)

Liter (L): 1 dm³. 1 L = 1000 mL = 1000 cm³. (1.4)

Lithium Ion Cell: A cell in which lithium ions are transferred between the electrodes through an electrolyte, while electrons travel through the external circuit. (19.6)

Lithium–Manganese Dioxide Battery: A battery that uses metallic lithium as the anode and manganese dioxide as the cathode. (19.6)

London Forces (Dispersion Forces): Weak attractive forces caused by instantaneous dipole–induced dipole attractions. (11.1)

Lone Pair: A pair of electrons in the valence shell of an atom that is not shared with another atom. An unshared pair of electrons. (9.2)

Low-Spin Complex: A coordination compound or a complex ion with electrons paired as much as possible in the lower energy set of d orbitals. (21.5)

M

Macromolecule: A molecule whose molecular mass is very large. (22.5)

Macroscopic: A scale that includes masses in the range of 1 to 1000 g or the equivalent. Also considered the laboratory scale. (1.1)

Magic Numbers: The numbers 2, 8, 20, 28, 50, 82, and 126, numbers whose significance in nuclear science is that a nuclide in which the number of protons or neutrons equals a magic number has nuclei that are relatively more stable than those of other nuclides nearby in the band of stability. (20.4)

Magnetic Quantum Number (m_ℓ): A quantum number that describes the directionality of an orbital and can have integral values from $-\ell$ to zero to $+\ell$, where ℓ is the secondary quantum number. (7.5)

Main Group Elements: Elements in any of the A groups in the periodic table. (2.1)

Main Reaction: The desired reaction between the reactants as opposed to competing reactions that give by-products. (3.7)

Malleability: A metal's ability to be hammered or rolled into thin sheets. (2.2)

Manometer: A device for measuring the pressure within a closed system. The two types—*closed end* and *open end*—differ according to whether the operating fluid (e.g., mercury) is exposed at one end to the atmosphere. (10.2)

Mass: A measure of the amount of matter that there is in a given sample. (1.2)

Mass Action Expression: A fraction in which the numerator is the product of the molar concentrations of the products, each raised to a power equal to its coefficient in the equilibrium equation, and the denominator is the product of the molar concentrations of the reactants, each also raised to the power that equals its coefficient in the equation. (For gaseous reactions, partial pressures can be used in place of molar concentrations.) (14.2)

Mass Defect: For a given isotope, it is the mass that changed into energy as the nucleons gathered to form the nucleus, this energy being released from the system. (20.2)

Mass Fraction: The ratio of the mass of one component of a mixture to the total mass of that same mixture. (12.5)

Mass Number: The numerical sum of the protons and neutrons in an atom of a given isotope. (0.5)

Matter: Anything that has mass and occupies space. (1.2)

Mean: The sum of N numerical values divided by N; the average. (1.5)

Measurement: A numerical observation. (1.4)

Melting Point: The temperature at which a substance melts; the temperature at which a solid is in equilibrium with its liquid state. (11.3)

Metal: An element or an alloy that is a good conductor of electricity, that has a shiny surface, and that is malleable and ductile; an element that normally forms positive ions and has an oxide that is basic. (2.2)

Metallic Conduction: Conduction of electrical charge by the movement of electrons. (19.1)

Metallic Crystal: A solid having positive ions at the lattice positions that are attracted to a "sea of electrons" that extends throughout the entire crystal. (11.12)

Metalloids: Elements with properties that lie between those of metals and nonmetals, and that are found in the periodic table around the diagonal line running from boron (B) to astatine (At). (2.2)

Metathesis Reaction: See *Double Replacement Reaction.*

Meter (m): The SI base unit for length (1.4)

Micelle: A structure formed from molecules that have a hydrophobic and hydrophilic end. The result is a circular or spherical structure with the molecules forming a rudimentary membrane. (12.7)

Millibar: 1 mb = 10^{-3} bar. (10.2)

Milliliter (mL): 0.001 L. 1000 mL = 1 L. (1.4)

Millimeter (mm): 0.001 m. 1000 mm = 1 m. (1.4)

Millimeter of Mercury (mm Hg): A unit of measurement that is proportional to pressure; equal to 1/760 atm. 760 mm Hg = 1 atm. 1 mm Hg = 1 torr. (10.2)

Miscible: Mutually soluble. (12.1)

Mixture: Any matter consisting of two or more substances physically combined in no particular proportion by mass. (1.2)

Model, Theoretical: A picture or a mental construction derived from a set of ideas and assumptions that are imagined to be true because they can be used to explain certain observations and measurements (e.g., the model of an ideal gas). (1.1)

Molal Boiling Point Elevation Constant (K_b): The number of degrees (°C) per unit of molal concentration that a boiling point of a solution is higher than that of the pure solvent. (12.6)

Molal Concentration (m): The number of moles of solute in 1000 g of solvent. (12.5)

Molal Freezing Point Depression Constant (K_f): The number of degrees (°C) per unit of molal concentration that a freezing point of a solution is lower than that of the pure solvent. (12.6)

Molality: The molal concentration. (12.5)

Molar Concentration (M): The number of moles of solute per liter of solution. The molarity of a solution. (4.7)

Molar Heat Capacity: The heat that can raise the temperature of 1 mol of a substance by 1 °C; the heat capacity per mole. (6.3)

Molar Heat of Fusion, ΔH_{fusion}: The heat absorbed when 1 mol of a solid melts to give 1 mol of the liquid at constant temperature and pressure. (11.6)

Molar Heat of Sublimation, $\Delta H_{sublimation}$: The heat absorbed when 1 mol of a solid sublimes to give 1 mol of its vapor at constant temperature and pressure. (11.6)

Molar Heat of Vaporization, $\Delta H_{vaporization}$: The heat absorbed when 1 mol of a liquid changes to 1 mol of its vapor at constanttemperature and pressure. (11.6)

Molarity: See *Molar Concentration.*

Molar Mass: The mass of one mole of a substance; the mass in grams equal to the sum of the atomic masses of the atoms in a substance, with units of g mol⁻¹. (3.1)

Molar Solubility: The number of moles of solute required to give 1 L of a saturated solution of the solute. (17.1)

Molar Volume, Standard: The volume of 1 mol of a gas at STP; 22.4 L mol⁻¹. (10.4)

Mole (mol): The SI unit for amount of substance; the formula mass in grams of an element or compound; an amount of a chemical substance that contains 6.022×10^{23} formula units. (3.1)

Molecular Compound: A compound consisting of electrically neutral molecules. (2.7)

Molecular Crystal: A crystal that has molecules or individual atoms at the lattice points. (11.12)

Molecular Equation: A chemical equation that gives the full formulas of all of the reactants and products and that is used to plan an actual experiment. (4.3)

Molecular Formula: A chemical formula that gives the actual composition of one molecule. (3.4)

Molecular Kinetic Energy: The energy associated with the motions of and within molecules as they fly about, spinning and vibrating. (6.2)

Molecular Mass: The sum of the atomic masses (in u) of all of the atoms represented in a molecular chemical substance; also called the *molecular weight.* May be used with units of g mol^{-1}. (3.1) See also *Molar Mass.*

Molecular Orbital (MO): An orbital that extends over two or more atomic nuclei. (9.7)

Molecular Orbital Theory (MO Theory): A theory about covalent bonds that views a molecule as a collection of positive nuclei surrounded by electrons distributed among a set of bonding, antibonding, and nonbonding orbitals of different energies. (9.4)

Molecular Weight: See *Molecular Mass.*

Molecule: A neutral particle composed of two or more atoms combined in a definite ratio of whole numbers. (2.3)

Mole Fraction: The ratio of the number of moles of one component of a mixture to the total number of moles of all components. (10.6)

Mole Percent (mol%): The mole fraction of a component expressed as a percent; mole fraction $\times 100\%$. (10.6, 12.5)

Monatomic: A particle consisting of just one atom. (2.6)

Monoclinic Sulfur: An allotrope of sulfur. (9.10)

Monodentate Ligand: A ligand that can form only one bond with a metal ion. (21.1)

Monomer: A substance of relatively low formula mass that is used to make a polymer. (22.5)

Monoprotic Acid: An acid that can furnish one H$^+$ per molecule. (4.4)

Monosaccharide: A carbohydrate that cannot be hydrolyzed. (22.6)

N

Natural Gas: A naturally occurring hydrocarbon gas mixture consisting primarily of methane. (2.3)

Negative Charge: A type of electrical charge possessed by certain particles such as the electron. A negative charge is attracted by a positive charge and is repelled by another negative charge. (0.5)

Negative Charge Carrier: In a semiconductor, the electrons are considered to be negative charge carriers. (19.6)

Nernst Equation: An equation relating cell potential and concentration. (19.5)

$$E_{\text{cell}} = E_{\text{cell}}^{\circ} - \frac{RT}{n\mathscr{F}} \ln Q$$

Net Ionic Equation: A chemical equation that shows only the reacting ions and from which spectator ions have been omitted. It is balanced when both atoms and electrical charge balance. (4.3)

Network Solid: See *Covalent Crystal.*

Neutralization, Acid–Base: See *Acid–Base Neutralization.*

Neutral Solution: A solution in which [H$^+$] = [OH$^-$]. (16.1)

Neutron (^1_0n): A subatomic particle with a charge of zero, a mass of 1.0086649 u $(1.674927 \times 10^{-24}$ g) and that exists in all atomic nuclei except those of hydrogen-1 isotope. (0.5)

Neutron Activation Analysis: A technique to analyze for trace impurities in a sample by studying the frequencies and intensities of the gamma radiations they emit after they have been rendered radioactive by neutron bombardment of the sample. (20.7)

Neutron Emission: A nuclear reaction in which a neutron is ejected. (20.3)

Nitrogen Family: Group 5A in the periodic table—nitrogen, phosphorus, arsenic, antimony, and bismuth. (2.1)

Noble Gases: Group 8A in the periodic table—helium, neon, argon, krypton, xenon, and radon. (2.1)

Noble Metals: The group of transition metals, including gold, silver and platinum, that have relatively low chemical reactivity. (8.6)

Nodal Plane: A plane that can be drawn to separate opposing lobes of *p*, *d*, and *f* orbitals. (7.9)

Node: A place where the amplitude or intensity of a wave is zero. (7.4)

Nomenclature: The names of substances and the rules for devising names. (2.6)

Nonbonding Domain: A region in the valence shell of an atom that holds an unshared pair of electrons and that influences the shape of a molecule. (9.2)

Nonbonding Molecular Orbitals: A molecular orbital that has no net effect on the stability of a molecule when populated with electrons and is localized on one atom in the molecule. (9.8)

Nonelectrolyte: A compound that in its molten state or in solution cannot conduct electricity. (4.2)

Nonionizing hydrogen: Hydrogen atoms attached to a molecule which do not ionize when the substance is added to water. (4.4)

Nonlinear (Molecule): A molecule in which the atoms do not lie in a straight line. (9.2)

Nonmetal: A nonductile, nonmalleable, nonconducting element that tends to form negative ions (if it forms them at all) far more readily than positive ions and whose oxide is likely to show acidic properties. (2.2)

Nonmetal Hydrides: Usually binary compounds that contain hydrogen and a nonmetal. (2.7)

Nonoxidizing Acid: An acid in which the anion is a poorer oxidizing agent than H$^+$ (e.g., HCl, H$_2$SO$_4$, H$_3$PO$_4$). (5.3)

Nonpolar Covalent Bond: A covalent bond in which the electron pair(s) are shared equally by the two atoms. (8.6)

Nonpolar Molecule: A molecule that has no net dipole moment. (9.3)

Normal Boiling Point: The temperature at which the vapor pressure of a liquid equals 1 atm. (11.5)

n-type Semiconductor: A semiconductor that carries electrical charge by means of electron movement. (19.6)

Nuclear Binding Energy: See *Binding Energy, Nuclear.*

Nuclear Chain Reaction: A self-sustaining nuclear reaction. (20.8)

Nuclear Equation: A description of a nuclear reaction that uses the special symbols of isotopes, that describes some kind of nuclear transformation or disintegration, and that is balanced when the sums of the atomic numbers on either side of the arrow are equal and the sums of the mass numbers are also equal. (20.3)

Nuclear Fission: See *Fission.*

Nuclear Radiation: Alpha, beta, or gamma radiation emitted by radioactive nuclei. (20.3)

Nuclear Reaction: A change in the composition or energy of the nuclei of isotopes accompanied by one or more events such as the radiation of nuclear particles or electromagnetic energy, transmutation, fission, or fusion. (13.4, 20.3)

Nuclear Shell Model: A model of the atomic nucleus showing shell-like structure similar to the electron energy shells. (20.4)

Nuclear Strong Force: The attractive force within the nucleus that overcomes the repulsive forces of positively charged protons. (20.3)

Nucleic Acids: Polymers in living cells that store and translate genetic information and whose molecules hydrolyze to give a sugar unit (ribose from ribonucleic acid, RNA, or deoxyribose from deoxyribonucleic acid, DNA), a phosphate, and a set of four of the five nitrogen-containing, heterocyclic bases (adenine, thymine, guanine, cytosine, and uracil). (22.7)

Nucleon: A proton or a neutron. (0.5, 20.2)

Nucleosynthesis: The synthesis of heavy nuclei from lighter nuclei within a star. (0.2)

Nylon 6,6: A polymer of a six-carbon dicarboxylic acid and a six-carbon diamine. (22.5)

O

Observation: A statement that accurately describes something we see, hear, taste, feel, or smell. (1.1)

Octahedral Molecule: A molecule in which a central atom is surrounded by six atoms located at the vertices of an imaginary octahedron. (9.1)

Octahedron: An eight-sided figure that can be envisioned as two square pyramids sharing the common square base. (9.1)

Octet of Electrons: Eight electrons in the valence shell of an atom. (8.3)

Octet Rule: An atom tends to gain or lose electrons until its outer shell has eight electrons. (8.3)

Odd–Even Rule: When the numbers of protons and neutrons in an atomic nucleus are both even, the isotope is more likely to be stable than when both numbers are odd. (20.4)

Open-End Manometer: See *Manometer.*

Open System: A system that can exchange both matter and energy with its surroundings. (6.3)

Optical Isomers: Stereoisomers other than geometric (*cis–trans*) isomers and that

include substances that can rotate the plane of plane-polarized light. (21.4)

Orbital: An electron waveform with a particular energy and a unique set of values for the quantum numbers n, ℓ, and m_ℓ. (7.5)

Orbital Angular Momentum Number: See *Secondary Quantum Number*

Orbital Diagram: A diagram in which the electrons in an atom's orbitals are represented by arrows to indicate paired and unpaired spins. (7.7)

Order (of the Reaction): The sum of the exponents in the rate law is the *overall* order. Each exponent gives the order of the reaction with respect to a specific reactant. (13.3)

Organic Acid: An acid that contains the carboxyl group $-\overset{\overset{\displaystyle O}{\|}}{C}-OH$ (8.8) *See also Carboxylic acid.*

Organic Chemistry: The study of the compounds of carbon that are not classified as inorganic. (2.7, 22.1)

Organic Compound: Any compound of carbon other than a carbonate, bicarbonate, cyanide, cyanate, carbide, or gaseous oxide. (8.8)

Orthorhombic Sulfur: The most stable allotrope of sulfur, composed of S_8 rings. (9.10)

Osmosis: The passage of solvent molecules, but not those of solutes, through a semipermeable membrane; the limiting case of dialysis. (12.6)

Osmotic Membrane: A membrane that allows passage of solvent, but not solute particles. (12.6)

Osmotic Pressure: The back pressure that would have to be applied to prevent osmosis; one of the colligative properties. (12.6)

Outer Electrons: The electrons in the occupied shell with the largest principal quantum number. An atom's electrons in its valence shell. (7.8)

Outer Shell: The occupied shell in an atom having the highest principal quantum number (n). (7.8)

Overall Order of a Reaction: The sum of the exponents on the concentration terms in a rate law. (13.3)

Overlap of Orbitals: A portion of two orbitals from different atoms that share the same space in a molecule. (9.4)

Oxidation: A change in which an oxidation number increases (becomes more positive). A loss of electrons. (5.1)

Oxidation Number: The charge that an atom in a molecule or ion would have if all of the electrons in its bonds belonged entirely to the more electronegative atoms; the oxidation state of an atom. (5.1)

Oxidation–Reduction Reaction: A chemical reaction in which changes in oxidation numbers occur. (5.1)

Oxidation State: See *Oxidation Number*.

Oxidizing Acid: An acid in which the anion is a stronger oxidizing agent than H^+ (e.g., $HClO_4$, HNO_3). (5.3)

Oxidizing Agent: The substance that causes oxidation and that is itself reduced. (5.1)

Oxoacid: An acid that contains oxygen as well as hydrogen and another element (e.g., HNO_3, H_3PO_4, H_2SO_4). (4.5, 15.3)

Oxoanion: The anion of an oxoacid (e.g., ClO_4^-, HSO_4^-). (15.3)

Ozone: A very reactive allotrope of oxygen with the formula O_3. (9.10)

P

Pairing Energy: The energy required to force two electrons to become paired and occupy the same orbital. (21.5)

Paramagnetism: The weak magnetism of a substance whose atoms, molecules, or ions have one or more unpaired electrons. (7.6)

Partial Charge: Charges at opposite ends of a dipole that are fractions of full 1+ or 1− charges. (8.6)

Partial Pressure: The pressure contributed by an individual gas to the total pressure of a gas mixture. (10.6)

Partial Pressure, Law of (Dalton's Law of Partial Pressures): The total pressure of a mixture of gases equals the sum of their partial pressures. (10.6)

Particulates: Small solid particles suspended in air or a solvent such as water. (5.5)

Pascal (Pa): The SI unit of pressure equal to 1 newton m^{-2}; 1 atm = 101,325 Pa. (10.2)

Pauli Exclusion Principle: No two electrons in an atom can have the same values for all four of their quantum numbers. (7.6)

Peptide Bond: The amide linkage in molecules of polypeptides. (22.6)

Percentage by Mass (Percentage by Weight): (a) The number of grams of an element combined in 100 g of a compound. (3.3) (b) The number of grams of a substance in 100 g of a mixture or solution. (12.5)

Percentage by Mass/Volume: The mass of a solute in a given volume of solvent. Most often g/100 mL. (12.5)

Percentage Composition: A list of the percentages by weight of the elements in a compound. (3.3)

Percentage Concentration: A ratio of the amount of solute to the amount of solution expressed as a percent. (4.1) *Weight/weight:* Grams of solute in 100 g of solution. *Weight/volume:* Grams of solute in 100 mL of solution. *Volume/volume:* Volumes of solute in 100 volumes of solution.

Percentage Ionization: An equation that quantifies the extent of ionization of a substance in solution. (16.4) Percentage ionization

$$= \frac{\text{amount of substance ionized}}{\text{initial amount of substance}} \times 100\%$$

Percentage Yield: See *Yield, Percentage*

Period: A horizontal row of elements in the periodic table. (2.1)

Periodic Table: A table in which symbols for the elements are displayed in order of increasing atomic number and arranged so that elements with similar properties lie in the same column (group). (Inside front cover, 0.5, 2.1)

Peroxide: A binary compound in which oxygen has a -1 oxidation state. (5.1)

Phase: A homogeneous region within a sample. (1.2)

Phase Diagram: A pressure–temperature graph on which are plotted the temperatures and the pressures at which equilibrium exists between the states of a substance. It defines regions of T and P in which the solid, liquid, and gaseous states of the substance can exist. (11.7)

Photoelectric Effect: The phenomenon whereby energetic photons can eject electrons from certain materials. (7.1)

Physical Change: A change that is not accompanied by a change in chemical makeup. (1.2)

Physical Property: A property that can be specified without reference to another substance and that can be measured without causing a chemical change. (1.3)

pH: $-\log [H^+]$ (16.1)

Pi Bond (π Bond): A bond formed by the sideways overlap of a pair of p orbitals and that concentrates electron density into two separate regions that lie on opposite sides of a plane that contains an imaginary line joining the nuclei. (9.6)

pK_a: $-\log K_a$ (16.3)
pK_b: $-\log K_b$ (16.3)
pK_w: $-\log K_w$ (16.1)

Planar Triangular Molecule: A molecule in which a central atom holds three other atoms located at the corners of an equilateral triangle and that includes the central atom at its center. (9.1)

Planck's Constant (h): The ratio of the energy of a photon to its frequency; $6.6260755 \times 10^{-34}$ J s. (7.1)

Plane-Polarized Light: Light in which all the oscillations occur in one plane. (21.4)

Plasma: An electrically neutral, very hot gaseous mixture of nuclei and unattached electrons. (20.8)

pOH: $-\log [OH^-]$ (16.1)

Polar Covalent Bond (Polar Bond): A covalent bond in which more than half of the bond's negative charge is concentrated around one of the two atoms. (8.6)

Polarizability: A term that describes the ease with which the electron cloud of a molecule or ion is distorted. (11.1)

Polar Molecule: A molecule in which individual bond polarities do not cancel and in which, therefore, the centers of density of negative and positive charges do not coincide. (8.6)

Polyatomic Ion: An ion composed of two or more atoms. (2.5)

Polydentate Ligand: A ligand that has two or more atoms with lone electron pairs that can become simultaneously bonded to a metal ion. (21.1)

Polymer: A substance consisting of macromolecules that have repeating structural units. (22.5)

Polymerization: A chemical reaction that converts a monomer into a polymer. (22.5)

Polypeptide: A polymer of α-amino acids that makes up all or most of a protein. (22.6)

Polyprotic Acid: An acid that can furnish more than one H^+ per molecule. (4.4)

Polysaccharide: A carbohydrate whose molecules can be hydrolyzed into many monosaccharide molecules. (22.6)

Polystyrene: An addition polymer of styrene with the following structure.

polystyrene (22.5)

Position of Equilibrium: The relative amounts of the substances on both sides of the double arrows in the equation for an equilibrium. (4.4, 11.8, 14.5)

Positive Charge: A type of electrical charge possessed by certain particles such as the proton. A positive charge is attracted by a negative charge and is repelled by another positive charge. (0.5)

Positive Charge Carrier: In a semiconductor material doped with an electron deficient element, the holes carry the charge. (19.6)

Positron: A positively charged particle with the mass of an electron. (20.3)

Post-transition Metal: A metal that occurs in the periodic table immediately to the right of a row of transition elements. (2.5)

Potential: See *Volt*.

Potential Energy (PE): Stored energy. (6.1)

Precipitate: A solid that separates from a solution usually as the result of a chemical reaction. (4.1)

Precipitation Reaction: A reaction in which a precipitate forms. (4.6)

Precision: How reproducible measurements are; the fineness of a measurement as indicated by the number of significant figures reported in the physical quantity. (1.5)

Pre-exponential Factor: A number or variable that precedes the exponential part of a number. (13.7)

Pressure: Force per unit area. (6.5, 10.2)

Pressure–Temperature Law (Gay-Lussac's Law): The pressure of a given mass of gas is directly proportional to its Kelvin temperature if the volume is kept constant. $P \propto T$. (10.3)

Pressure–Volume Law (Boyle's Law): The volume of a given mass of a gas is inversely proportional to its pressure if the temperature is kept constant. $V \propto 1/P$. (10.3)

Pressure–Volume Work (P–V Work): The energy transferred as work when a system expands or contracts against the pressure exerted by the surroundings. At constant pressure, $w = -P\,\Delta V$. (6.5)

Primary Cell: A galvanic cell (battery) not designed to be recharged; it is discarded after its energy is depleted. (19.6)

Primitive Cubic: See *Simple Cubic Unit Cell*.

Principal Quantum Number (n): The quantum number that defines the principal energy levels and that can have values of 1, 2, 3, ..., $\propto$. (7.5)

Products: The substances produced by a chemical reaction and whose formulas follow the arrows in chemical equations. (2.4, 14.1)

Propagation Step: A step in a chain reaction for which one product must serve in a succeeding propagation step as a reactant and for which another (final) product accumulates with each repetition of the step. (Chemistry Outside the Classroom 13.1)

Property: A characteristic of matter. (1.3, 2.1)

Protein: A macromolecular substance found in cells that consists wholly or mostly of one or more polypeptides that often are combined with an organic molecule or a metal ion. (22.6)

Proton (^1_1p or $^1_1H^+$): A subatomic particle, with a charge of 1+ and a mass of 1.0072765 u ($1.6726218 \times 10^{-24}$ g) and that is found in atomic nuclei. (0.5)

Proton Acceptor: A Brønsted base. (15.1)

Proton Donor: A Brønsted acid. (15.1)

p-type Semiconductor: A semiconductor material with a deficit of electrons causing charge to be carried by positive charge carriers or "holes." (19.6)

Pure Substance: An element or a compound. (1.2)

Q

Q: See *Reaction Quotient.*

Qualitative Analysis: The use of experimental procedures to determine what elements are present in a substance. (4.9)

Qualitative Observation: Observations that do not involve numerical information. (1.4)

Quanta: Packets of electromagnetic radiation now commonly called photons. (7.1)

Quantitative Analysis: The use of experimental procedures to determine the percentage composition of a compound or the percentage of a component of a mixture. (4.9)

Quantitative Observation: An observation involving a measurement and numerical information. (1.4)

Quantized: Descriptive of a discrete, definite amount as of *quantized energy.* (7.3)

Quantum: The energy of one photon. (7.1)

Quantum Number: A number related to the energy, shape, or orientation of an orbital, or to the spin of an electron. (7.3)

R

R: A symbol used in organic chemistry to represent a hydrocarbon fragment such as —CH_3 or —CH_2CH_3. (8.8, 8.9, 22.2)

R: See *Gas Constant, Universal.*

Rad (rd): A unit of radiation-absorbed dose and equal to 10^{-5} J g^{-1} or 10^{-2} Gy. (20.6)

Radioactive Decay: The change of a nucleus into another nucleus (or into a more stable form of the same nucleus) by the loss of a small particle or a gamma ray photon. (20.3)

Radioactive Disintegration Series: A sequence of nuclear reactions beginning with a very long-lived radionuclide and ending with a stable isotope of lower atomic number. (20.3)

Radioactivity: The emission of one or more kinds of radiation from an isotope with unstable nuclei. (20.3)

Radiological Dating: A technique for measuring the age of a geologic formation or an ancient artifact by determining the ratio of the concentrations of two isotopes, one radioactive and the other a stable decay product. (13.4, 20.7)

Radionuclide: A radioactive isotope. (20.3)

Raoult's Law: The vapor pressure of one component above a mixture of molecular compounds equals the product of its vapor pressure when pure and its mole fraction. (12.6)

Rate: A ratio in which a unit of time appears in the denominator, for example, 40 miles hr^{-1} or 3.0 mol L^{-1} s^{-1}. (13.2)

Rate Constant (k): The proportionality constant in the rate law; the rate of reaction when all reactant concentrations are 1 M. (13.3)

Rate-Determining Step (Rate-Limiting Step): The slowest step in a reaction mechanism. (13.8)

Rate Law: An equation that relates the rate of a reaction to the molar concentrations of the reactants raised to powers. (13.3)

Rate of Reaction: How quickly the reactants disappear and the products form and usually expressed in units of mol L^{-1} s^{-1}. (13.2)

Reactants: The substances brought together to react and whose formulas appear before the arrow in a chemical equation. (2.4, 14.1)

Reaction Coordinate: The horizontal axis of a potential energy diagram of a reaction. (13.6)

Reaction Mechanism: A sequence of elementary processes that sum to the overall reaction and possess one process that can derive the experimental rate law. (13.8)

Reaction Quotient (Q): The numerical value of the mass action expression regardless of whether or not the system is at equilibrium. (17.1, 14.2) See *Mass Action Expression.*

Reactivity: A description of the tendency for a substance to undergo reaction. For a metal, it is the tendency to undergo oxidation. (8.6)

Reciprocal Relationship: The stronger the Brønsted acid, the weaker its conjugate base. (15.2)

Red Giant: A star which is about to explode in a supernova. (0.2)

Redox Reaction: An oxidation–reduction reaction. (5.1)

Red Phosphorus: A relatively unreactive allotrope of phosphorus. (9.10)

Reducing Agent: A substance that causes reduction and is itself oxidized. (5.1)

Reduction: A change in which an oxidation number decreases (becomes less positive and more negative). A gain of electrons. (5.1)

Reduction Potential: A measure of the tendency of a given half-reaction to occur as a reduction. (19.2)

Relative Abundance: A fraction representing the mass or number of atoms of an isotope compared to the total mass or number. (0.5)

Rem: A dose in rads multiplied by a factor that takes into account the variations that different radiations have in their damage-causing abilities in tissue. (20.6)

Replication: In nucleic acid chemistry, the reproductive duplication of DNA double helices prior to cell division. (22.7)

Representative Element: An element in one of the A groups in the periodic table. (2.1)

Reproducible: A measurement or experiment that, when repeated, gives the same result within experimental uncertainty. (1.17)

Resonance: A concept in which the actual structure of a molecule or polyatomic ion is represented as a composite or average of two or more Lewis structures, which are called the resonance or contributing structures (and none of which has real existence). (8.8)

Resonance Energy: The difference in energy between a substance and its principal resonance (contributing) structure. (8.8, 9.8)

Resonance Hybrid: The actual structure of a molecule or polyatomic ion taken as a composite or average of the resonance or contributing structures. (8.8)

Resonance Structure: A Lewis structure that contributes to the hybrid structure in resonance-stabilized systems; a contributing structure. (8.8)

Reverse Osmosis: The procedure where a pressure, exceeding the osmotic pressure, is applied to an osmotic membrane to force pure solvent from a mixture. (12.6)

Reverse Reaction: In a chemical equation, the reaction as read from right to left. (4.4, 14.1)

Reversible Process: A process capable of proceeding in either the forward or reverse direction. (18.7)

Ring, Carbon: A closed-chain sequence of carbon atoms. (22.1)

RNA: Ribonucleic acid; a nucleic acid that gives ribose, phosphate ion, adenine, uracil, guanine, and cytosine when hydrolyzed. It occurs in several varieties. (22.7)

Rock Salt Structure: The face-centered cubic structure observed for sodium chloride, which is also possessed by crystals of many other compounds. (11.10)

Root Mean Square Speed (rms Speed): The square root of the average of the speeds-squared of the molecules in a substance. (10.7)

Rust: A complex mixture of iron oxides, often formed from the weathering of iron. (5.5)

Rydberg Equation: An equation used to calculate the wavelengths of all the spectral lines of hydrogen. (7.2)

S

Salt: An ionic compound in which the anion is not OH^- or O^{2-} and the cation is not H^+. (4.2)

Salt Bridge: A tube that contains an electrolyte that connects the two half-cells of a galvanic cell. (19.1)

Saponification: The reaction of an organic ester with a strong base to give an alcohol and the salt of the organic acid. (22.3)

Saturated Compound: A compound whose molecules have only single bonds. (22.2)

Saturated Solution: A solution that holds as much solute as it can at a given temperature. A solution in which there is an equilibrium between the dissolved and the undissolved states of the solute. (4.1, 5.1, 17.1)

Scanning Tunneling Microscope (STM): An instrument that enables the imaging of individual atoms on the surface of an electrically conducting specimen. (0.4)

Scientific Law: See *Law*.

Scientific Method: The observation, explanation, and testing of an explanation by additional experiments. (1.1)

Scientific Notation: The representation of a quantity as a decimal number between 1 and 10 multiplied by 10 raised to a power (e.g., 6.02×10^{23}). (1.5)

Scintillation Probe: A device for measuring nuclear radiation that contains a sensor composed of a substance called a *phosphor* that emits a tiny flash of light when struck by a particle of ionizing radiation. These flashes can be magnified electronically and automatically counted. (20.6)

Secondary Cell: A galvanic cell (battery) designed for repeated use; it is able to be recharged. (19.6)

Secondary Quantum Number (ℓ): The quantum number that indicates the shape of an orbital whose values can be 0, 1, 2, ..., $(n-1)$, where n is the principal quantum number. (7.5)

Second Law of Thermodynamics: Whenever a spontaneous event takes place, it is accompanied by an increase in the entropy of the universe. (18.4)

Second-Order Reaction: A reaction with a rate law of the type: rate = $k[A]^2$ or rate = $k[A][B]$, where A and B are reactants. (13.4)

Seesaw-Shaped Molecule: See *Distorted Tetrahedron*.

Selective Precipitation: A technique that uses differences in the solubilities of specific salts to separate ions from each other. (17.4)

Semiconductor: A substance that conducts electricity weakly. (2.2)

Shell: All of the orbitals associated with a given value of n (the principal quantum number). (7.5)

Sievert (Sv): The SI unit for dose equivalent. The dose equivalent H is calculated from D (the dose in grays), Q (a measure of the effectiveness of the radiation at causing harm), and N (a variable that accounts for other modifying factors). $H = DQN$. (20.6)

Sigma Bond (σ Bond): A bond formed by the head-to-head overlap of two atomic orbitals and in which electron density becomes concentrated along and around the imaginary line joining the two nuclei. (9.6)

Significant Figures (Significant Digits): The digits in a physical measurement that are known to be certain plus the first digit that contains uncertainty. (1.5)

Simple Cubic Unit Cell: A cubic unit cell that has atoms only at the corners of the cell. (11.10)

Single Bond: A covalent bond in which a single pair of electrons is shared. (8.5)

Single Replacement Reaction: A reaction in which one element replaces another in a compound; usually a redox reaction. (5.4)

Sintering: The process of forming a ceramic by heating to a high temperature its components which are pulverized. (15.6)

SI Units (International System of Units): The modified metric system adopted in 1960 by the General Conference on Weights and Measures. (1.4)

Skeletal Structure: A diagram of the arrangement of atoms in a molecule, which is the first step in constructing the Lewis structure. (8.7)

Skeleton Equation: An unbalanced equation showing only the formulas of reactants and products. (5.2)

Sol-Gel Process: A method for producing sol gels, very porous materials. (15.6)

Solid: One of the states of matter. A solid consists of tightly packed atomic or molecular sized particles held rigidly in place. (1.3)

Solubility: The ratio of the quantity of solute to the quantity of solvent in a saturated solution and that is usually expressed in units of (g solute)/(100 g solvent) at a specified temperature. (4.1)

Solubility Product: The mass action expression for the dissolution of a slightly soluble salt. (17.1)

Solubility Product Constant (K_{sp}): The equilibrium constant for the solubility of a salt and that, for a saturated solution, is equal to the product of the molar concentrations of the ions, each raised to a power equal to the number of its ions in one formula unit of the salt. (17.1) See also *Acid Solubility Product*.

Solubility Rules: A set of rules describing salts that are soluble and those that are insoluble. They enable the prediction of the formation of a precipitate in a metathesis reaction. (4.6)

Solute: Something dissolved in a solvent to make a solution. (4.1)

Solution: A homogeneous mixture in which all particles are of the size of atoms, small molecules, or small ions. (1.2, 4.1)

Solvation: The development of a cagelike network of a solution's solvent molecules about a molecule or ion of the solute. (12.1)

Solvation Energy: The enthalpy of the interaction of gaseous molecules or ions of solute with solvent molecules during the formation of a solution. (12.2)

Solvent: A medium, usually a liquid, into which something (a solute) is dissolved to make a solution. (4.1)

sp Hybrid Orbital: A hybrid orbital formed by mixing one s and one p atomic orbital. The angle between a pair of sp hybrid orbitals is 180°. (9.5)

sp^2 Hybrid Orbital: A hybrid orbital formed by mixing one s and two p atomic orbitals. sp^2 hybrids are planar triangular with the angle between two sp^2 hybrid orbitals being 120°. (9.5)

sp^3 Hybrid Orbital: A hybrid orbital formed by mixing one s and three p atomic orbitals. sp^3 hybrids point to the corners of a tetrahedron; the angle between two sp^3 hybrid orbitals is 109.5°. (9.5)

sp^3d Hybrid Orbital: A hybrid orbital formed by mixing one s, three p, and one d atomic orbital. sp^3d hybrids point to the corners of a trigonal bipyramid. (9.5)

sp^3d^2 Hybrid Orbital: A hybrid orbital formed by mixing one s, three p, and two d atomic orbitals. sp^3d^2 hybrids point to the corners of an octahedron. (9.5)

Space-Filling Model: A method of depicting molecules by the shape and volume that the molecule actually occupies. (2.3)

Specific Gravity: A dimensionless property of matter. Its value is determined by dividing the density by the density of water under the same conditions. (1.7)

Specific Heat Capacity (Specific Heat): The quantity of heat that will raise the temperature of 1 g of a substance by 1 °C, usually in units of cal g^{-1} °C^{-1} or J g^{-1} °C^{-1}. (6.3)

Spectator Ion: An ion whose formula appears in an ionic equation identically on both sides of the arrow, that does not participate in the reaction, and that is excluded from the net ionic equation. (4.3)

Spectrochemical Series: A listing of ligands in order of the wavelength of light absorbed by complexes containing those ligands. These wavelengths are related to the energy of the crystal field splitting. (21.5)

Spin Quantum Number (m_s): The quantum number associated with the spin of a subatomic particle and for the electron can have a value of $+\frac{1}{2}$ or $-\frac{1}{2}$. (7.6)

Spontaneous Change: A change that occurs by itself without outside assistance. (18.2)

Square Planar Molecule: A molecule with a central atom having four bonds that point to the corners of a square. (9.2)

Square Pyramid: A pyramid with four triangular sides and a square base. (9.2)

Stability Constant: See *Formation Constant*.

Stabilization Energy: See *Resonance Energy*. (9.8)

Standard Atmosphere: See *Atmosphere, Standard*

Standard Cell Notation: A way of describing the anode and cathode half-cells in a galvanic cell. The anode half-cell is specified on the left, with the electrode material of the anode given first and a vertical bar representing the phase boundary between the electrode and the solution. Double bars represent the salt bridge between the half-cells. The cathode half-cell is specified on the right, with the material of the cathode given last. Once again, a single vertical bar represents the phase boundary between the solution and the electrode. (19.1)

Standard Cell Potential (E°_{cell}): The potential of a galvanic cell at 25 °C and when all ionic concentrations are exactly 1 M and the partial pressures of all gases are 1 atm. (19.2)

Standard Conditions of Temperature and Pressure (STP): Standard reference conditions for gases. 273.15 K (0 °C) and 1 atm (760 torr). (10.4)

Standard Electrode Potential: The electrode potential measured versus the standard hydrogen electrode at standard state conditions. (19.2)

Standard Enthalpy of Formation (ΔH°_f): See *Standard Heat of Formation*.

Standard Entropy (S°): The entropy of 1 mol of a substance at 25 °C and 1 atm. (18.5)

Standard Entropy Change (ΔS°): The entropy change of a reaction when determined with reactants and products at 25 °C and 1 atm and on the scale of the mole quantities given by the coefficients of the balanced equation. (18.5)

Standard Entropy of Formation (ΔS°_f): The value of ΔS° for the formation of one mole of a substance from its elements in their standard states. (18.5)

Standard Free Energy Change (ΔG°): $\Delta G^\circ = \Delta H^\circ - T\,\Delta S^\circ$. (18.6)

Standard Free Energy of Formation (ΔG°_f): The value of ΔG° for the formation of *one* mole of a compound from its elements in their standard states. (18.6)

Standard Heat of Combustion (ΔH°_c): The enthalpy change for the combustion of one mole of a compound under standard conditions. (6.9)

Standard Heat of Formation (ΔH°_f): The amount of heat absorbed or evolved when one mole of the compound is formed from its elements in their standard states. (6.9)

Standard Heat of Reaction (ΔH°): The enthalpy change of a reaction when determined with reactants and products at 25 °C and 1 atm and on the scale of the mole quantities given by the coefficients of the balanced equation. (6.7)

Standard Hydrogen Electrode: See *Hydrogen Electrode*.

Standard Molar Volume: See *Molar Volume, Standard*.

Standard Reduction Potential, E°: The reduction potential of a half-reaction at 25 °C when all ion concentrations are 1 M and the partial pressures of all gases are 1 atm. Also called standard electrode potential. (19.2)

Standard Solution: Any solution whose concentration is accurately known. (4.9)

Standard State: The condition in which a substance is in its most stable form at 25 °C and 1 atm. (6.7, 6.9, 7.7, 7.9, 19.2)

Standing Wave: A wave whose peaks and nodes do not change position. (7.4)

Starch: A polymer of glucose used by plants to store energy. (22.6)

State: The set of specific values of the physical properties of a system—its composition, physical form, concentration, temperature, pressure, and volume. (6.2)

State Function: A quantity whose value depends only on the initial and final states of the system and not on the path taken by the system to get from the initial to the final state. (P, V, T, H, S, and G are all state functions.) (6.2)

Stereoisomerism: The existence of isomers whose structures differ only in spatial orientations (e.g., geometric isomers and optical isomers). (21.4)

Stock System: A system of nomenclature that uses Roman numerals to specify oxidation states. (2.6)

Stoichiometry: A description of the relative quantities by moles of the reactants and products in a reaction as given by the coefficients in the balanced equation. (3 This Chapter in Context)

Stopcock: A valve on a buret that is used to control the flow of titrant. (4.9)

Stored Energy: See *Potential Energy*.

Straight Chain: An organic compound that has carbon atoms joined in one continuous open-chain sequence. (22.1)

Strong Acid: An acid that is essentially 100% ionized in water. A good proton donor. An acid with a large value of K_a. (4.4)

Strong Base: Any powerful proton acceptor. A base with a large value of K_b. A metal hydroxide that dissociates essentially 100% in water. (4.4)

Strong Electrolyte: Any substance that ionizes or dissociates in water to essentially 100%. (4.2)

Structural Formula (Lewis Structure): A chemical formula that shows how the atoms of a molecule or polyatomic ion are arranged, to which other atoms they are bonded, and the kinds of bonds (single, double, or triple). (8.5)

Subatomic Particles: Electrons, protons, neutrons, and atomic nuclei. (0.5)

Sublimation: The conversion of a solid directly into a gas without passing through the liquid state. (11.2)

Subscript: In a chemical formula, a number after a chemical symbol, written below the line, and indicating the number of the preceding atoms in the formula (e.g., CH_4). Subscripts are also used to differentiate many variables such as the acid ionization constant (K_a) and the base ionization constant (K_b). (2.3)

Subshell: All of the orbitals of a given shell that have the same value of their secondary quantum number, ℓ. (7.5)

Substance: See *Pure Substance*.

Substitution Reaction: The replacement of an atom or group on a molecule by another atom or group. (22.2, 22.3)

Supercooled Liquid: A liquid at a temperature below its freezing point. An amorphous solid. (11.10)

Supercritical Fluid: A substance at a temperature above its critical temperature. (11.7)

Superheating: The process where a substance is heated to a temperature above its normal boiling point. (11.6)

Superimposability: A test of structural chirality in which a model of one structure and a model of its mirror image are compared to see if the two match perfectly, with every part of one coinciding simultaneously with the parts of the other. (21.4)

Supernova: Usually the explosion of a red giant star that has synthesized a significant amount of iron. (0.2)

Supersaturated Solution: A solution that contains more solute than it would hold if the solution were saturated. Supersaturated solutions are unstable and tend to produce precipitates. (4.1)

Surface Tension: A measure of the amount of energy needed to expand the surface area of a liquid. (11.2)

Surfactant: A substance that lowers the surface tension of a liquid and promotes wetting. (11.2)

Surroundings: That part of the universe other than the system being studied and separated from the system by a real or an imaginary boundary. (6.3)

Suspension: A heterogeneous mixture that will eventually separate into its component parts under the influence of gravity. (12.7)

Symmetric: An object is symmetric if it looks the same when rotated, reflected in a mirror, or reflected through a point. (9.3)

System: That part of the universe under study and separated from the surroundings by a real or an imaginary boundary. (6.3, 6.7)

T

Tarnishing: See *Corrosion*.

Temperature: A measure of the hotness or coldness of something. A property related to the average kinetic energy of the atoms and molecules in a sample. A property that determines the direction of

heat flow—from high temperature to low temperature. (6.2)

Temperature–Volume Law (Charles' Law): The volume of a given mass of a gas is directly proportional to its Kelvin temperature if the pressure is kept constant. $V \propto T$. (10.3)

Termination Step: A step in a chain reaction in which a reactive species needed for a chain propagation step disappears without helping to generate more of this species. (Chemistry Outside the Classroom 13.1)

Tetrahedral Molecule: A molecule with a central atom bonded to four other atoms located at the corners of an imaginary tetrahedron. (9.1)

Tetrahedron: A four-sided figure with four triangular faces and shaped like a pyramid. (9.1)

Theoretical Model: See *Model, Theoretical*.

Theoretical Yield: See *Yield, Theoretical*

Theory: A tested explanation of the results of many experiments. (1.1)

Thermal Energy: The molecular kinetic energy possessed by molecules as a result of the temperature of the sample. Energy that is transferred as heat. (6.2)

Thermal Equilibrium: A condition reached when two or more substances in contact with each other come to the same temperature. (6.2)

Thermochemical Equation: A balanced chemical equation accompanied by the value of $\Delta H°$ that corresponds to the mole quantities specified by the coefficients. (6.7)

Thermochemistry: The study of the energy changes of chemical reactions. (6 This Chapter in Content)

Thermodynamically Reversible: A process that occurs by an infinite number of steps during which the driving force for the change is just barely greater than the force that resists the change. (18.7)

Thermodynamic Equilibrium Constant (K): The equilibrium constant that is calculated from $\Delta G°$ (the standard free energy change) for a reaction at T, in kelvins, by the equation, $\Delta G° = RT \ln K$. (18.9)

Thermodynamics (Chemical Thermodynamics): The study of the role of energy in chemical change and in determining the behavior of materials. (6 This Chapter in Content, 18.1)

Third Law of Thermodynamics: For a pure crystalline substance at 0 K, $S = 0$. (18.5)

Titrant: The solution added from a buret during a titration. (4.9)

Titration: An analytical procedure in which a solution of unknown concentration

is combined slowly and carefully with a standard solution until a color change of some indicator or some other signal shows that equivalent quantities have reacted. Either solution can be the titrant in a buret with the other solution being in a receiving flask. (4.9)

Titration Curve: For an acid–base titration, a graph of pH versus the volume of titrant added. (16.9)

Torr: A unit of pressure equal to 1/760 atm. (10.2)

Tracer Analysis: The use of small amounts of a radioisotope to follow (trace) the course of a chemical or biological change. (20.7)

trans **(Isomer):** A stereoisomer whose uniqueness is having two groups on opposite sides of some reference plane. (22.2)

Transcription: The synthesis of mRNA at the direction of DNA. (22.7)

Trans **Isomer:** A stereoisomer whose uniqueness lies in having two groups that project on opposite sides of a reference plane. (21.4)

Transition Elements: The elements located between Groups 2A and 3A in the periodic table. (2.1)

Transition State: The structure and energy of the substances in an elementary process of a reaction mechanism when the species involved have acquired the minimum amount of potential energy and geometric configuration needed for a successful reaction. (13.6)

Transition State Theory: A theory about reaction rates that is based on the formation and breakup of activated complexes. (13.6)

Translation: The synthesis of a polypeptide at the direction of a molecule of mRNA. (22.7)

Transmutation: The conversion of one isotope into another. (20.5)

Transuranium Elements: Elements 93 and higher. (20.5)

Traveling Wave: A wave whose peaks and nodes move. (7.4)

Triacylglycerol: An ester of glycerol and three fatty acids. (22.6)

Triangular Pyramid: A three-sided pyramid. (9.2)

Trigonal Bipyramid: A six-sided figure made of two three-sided pyramids that share a common face. (9.1)

Trigonal Bipyramidal Molecule: A molecule with a central atom holding five other atoms that are located at the corners of a trigonal bipyramid. (9.1)

Trigonal Pyramidal Molecule: A molecule that consists of an atom, situated at the top of a three-sided pyramid, that is bonded to three other atoms located at the corners of the base of the pyramid. (9.2)

Triple Bond: A covalent bond comprised of one sigma bond and two pi bonds. Three pairs of electrons are shared between two atoms. (8.5)

Triple Point: The temperature and pressure at which the liquid, solid, and vapor states of a substance can coexist in equilibrium. (11.7)

T-Shaped Molecule: A molecule having five electron domains in its valence shell, two of which contain lone pairs. The other three are used in bonds to other atoms. The molecule has the shape of the letter T, with the central atom located at the intersection of the two crossing lines. (9.2)

Tyndall Effect: Colloidal dispersions scatter light while true solutions do not scatter light. This light scattering is known as the Tyndall effect. (12.7)

U

u: See *Atomic Mass Unit.*

Uncertainty: The amount by which a measured quantity deviates from the true or actual value. (1.5)

Uncertainty Principle: There is a limit to our ability to measure a particle's speed and position simultaneously. (7.9)

Unit Cell: The smallest portion of a crystal that can be repeated over and over in all directions to give the crystal lattice. (11.10)

Unit of Measurement: A reference quantity, such as the meter or kilogram, in terms of which the sizes of measurements can be expressed. (1.4)

Universal Gas Constant (R): See *Gas Constant, Universal.*

Universe: The system and surroundings taken together. (6.3)

Unsaturated Compound: A compound whose molecules have one or more double or triple bonds. (22.2)

Unsaturated Solution: Any solution with a concentration less than that of a saturated solution of the same solute and solvent. (4.1)

V

Vacuum: An enclosed space containing no matter whatsoever. A *partial vacuum* is an enclosed space containing a gas at a very low pressure. (10.2)

Valence Band: Very closely spaced energy levels of valence electrons, particularly in metals is called the valence band. (9.9)

Valence Bond Theory (VB Theory): A theory of covalent bonding that views a bond as being formed by the sharing of one pair of electrons between two overlapping atomic or hybrid orbitals. (9.4)

Valence Electrons: The electrons of an atom in its valence shell that participate in the formation of chemical bonds. (7.8)

Valence Shell: The electron shell with the highest principal quantum number, *n*, that is occupied by electrons. (7.8)

Valence Shell Electron Pair Repulsion Model (VSEPR Model): The bonding and nonbonding (lone pair) electron domains in the valence shell of an atom seek an arrangement that leads to minimum repulsions and thereby determine the geometry of a molecule. (9.2)

Van der Waals' Constants: Empirical constants that make the van der Waals' equation conform to the gas law behavior of a real gas. (10.8)

Van der Waals' Equation: An equation of state for a real gas that corrects *V* and *P* for the excluded volume and the effects of intermolecular attractions. (10.8)

Van der Waals' Forces: Attractive forces including dipole–dipole, ion–dipole, and induced dipole forces. (11.1)

Van't Hoff Factor (i): The ratio of the observed freezing point depression to the value calculated on the assumption that the solute dissolves as un-ionized molecules. (12.6)

Vapor Pressure: The pressure exerted by the vapor above a liquid (usually referring to the *equilibrium* vapor pressure when the vapor and liquid are in equilibrium with each other). (10.6, 11.4)

Viscosity: A liquid's resistance to flow. (11.2)

Visible Spectrum: That region of the electromagnetic spectrum whose frequencies can be detected by the human eye. (7.1)

Volt (V): The SI unit of electric potential or emf in joules per coulomb. (19.2)

$$1\,V = 1\,J\,C^{-1}$$

Voltaic Cell: See *Galvanic Cell.*

Volumetric Flask: A piece of glassware calibrated to hold an exact volume of liquid. Volumetric flasks come in a variety of sizes. (4.7)

VSEPR Model: See *Valence Shell Electron Pair Repulsion Model.*

V-Shaped Molecule: See *Bent Molecule* and *Nonlinear Molecule.*

W

Wave: An oscillation that moves outward from a disturbance. (7.1)

Wave Function (ψ): A mathematical function that describes the intensity of an electron wave at a specified location in an atom. The square of the wave function at a particular location specifies the probability of finding an electron there. (7.4, 7.9)

Wavelength (λ): The distance between crests in the wavelike oscillations of electromagnetic radiations. (7.1)

Weak Acid: An acid with a low percentage ionization in solution; a poor proton donor; an acid with a low value of K_a. (4.4)

Weak Base: A base with a low percentage ionization in solution; a poor proton acceptor; a base with a low value of K_b. (4.4)

Weak Electrolyte: A substance that has a low percentage ionization or dissociation in solution. (4.4)

Weight: The force with which something is attracted to the earth by gravity. (1.2)

Wetting: The spreading of a liquid across a solid surface. (11.2)

X

Xerogel: An extremely low density material that, among other things, can serve as a very low weight insulating material. (15.6)

X Ray: A stream of very high-energy photons emitted by substances when they are bombarded by high-energy beams of electrons. (20.3)

Y

Yield, Actual: The amount of a product obtained in a laboratory experiment. (3.7)

Yield, Percentage: The ratio, given as a percent, of the quantity of product actually obtained in a reaction to the theoretical yield. (3.7)

Z

Zero-Order Reaction: A reaction that occurs at a constant rate regardless of the concentration of the reactant. (13.4)

Zinc–Manganese Dioxide Cell (Leclanché Cell): A galvanic cell of about 1.5 V involving zinc and manganese dioxide under mildly acidic conditions. (19.6)

Zwitterion: A compound that contains both acid and a base functional groups, in which the acid protonates the base. (22.6)

Index

A

Abbreviated electron configurations, 329–330
Absolute zero, 39, 474, 504, 871
Absorption, 672
Acceptor atoms, in complex ions, 1004
Accuracy, 43, 54
Acetaldehyde, 390
Acetate ion(s):
   and ammonium ions, 785
   as base, 739
   in buffer solutions, 786
   structure of, 170, 390
   and water, 770
   as weak base, 782, 783
Acetic acid:
   acidity of, 744–746, 763
   basic anions from, 782
   in buffer solution, 786
   as carboxylic acid, 1038
   equilibrium of, 166–167, 170, 738–739
   from ethanamide, 1055
   ionization of, 170
   and sodium acetate, 788
   and strong base, 177, 179
   structure of, 390, 711
Acetone, 528, 583, 598
Acetylene (ethyne), 94, 388, 438–439, 1043
Acids. *See also specific acids*
   amino, 391n.6, 519, 1069
   Arrhenius, 164
   binary, 172–173, 741–742
   Brønsted–Lowry, *see* Brønsted–Lowry acids and bases
   carboxylic, *see* Carboxylic acids
   conjugate, *see* Conjugate acids and bases
   diprotic, 170, 803
   fatty, 181, 1066–1067
   hydronium ions from, 164–165
   Lewis, 746–750, 842, 1003
   monoprotic, 170, 770
   nomenclature of, 172–174
   nonoxidizing, 226, 227
   nucleic, 1072–1075
   organic, 390–391, 745–746, 1038
   oxidizing, 226, 227
   oxoacids, 173, 374–376, 742–746
   polyprotic, *see* Polyprotic acids
   redox reactions in, 225–228
   strength of, 165–167, 741–746
   strong, *see* Strong acids
   weak, *see* Weak acids
Acid anhydrides, 171–172, 750

Acid–base chemistry, 731–757
   in aqueous solutions, 164–172
   Brønsted–Lowry acids and bases, 732–741
   and ceramics, 753–756
   for elements and oxides, 750–752
   Lewis acids and bases, 746–750
   and metathesis reactions, 176–180
   strengths of acids and bases, 737–746
Acid–base equilibrium, 762–807
   and autoionization water, 763–765, 768
   buffer solutions, 785–792
   ionization constants, 769–776
   pH, 765–768, 776–781
   and "p" notation, 766
   polyprotic acids, 792–798
   properties of salt solutions, 781–785
   strong acids and bases, 767–768
   and titrations, 798–805
   weak acids and bases, 776–781
Acid–base indicators, 803–805
   defined, 198
   electrolysis in presence of, 942
   at end point, 798
   measuring and estimating pH with, 804–805
   phenolphthalein as, 799
Acid–base neutralizations, 176–180, 746–747
Acid–base titrations, 197–199, 798–805
Acid dissociation constant, *see* Acid ionization constant (K_a)
Acidic anions, 783
Acidic cations, 781–782
Acidic solutions:
   balancing redox reactions in, 220–224
   hydronium and hydroxide ions in, 764–765
   pH of, 765–766
Acid-insoluble sulfides, 833–834, 836–837, 841
Acid ionization constant (K_a):
   of acid–base indicators, 804
   of buffer solutions, 789
   from initial concentration and equilibrium data, 773–775
   of polyprotic acids, 793–796
   product of K_b and, 772–773
   for weak acid in water, 769–770
Acidity:
   of hydrated metal ions, 750–752
   of metal oxides, 752

of nonmetal oxides, 171–172
   and solubility of basic salts, 828–831
Acid salts, 179–180
Acid solubility product (K_{spa}), 833, 836
Actinide elements, 66
Activated complexes, 659
Activation energies (E_a), 656, 659–663
Activity:
   of radioactive material, 980–981
   in thermodynamics, 698n.2
Activity series, of metals, 230–233
Actual yield, 141
Addition, 44, 694, 717, 718
Addition compounds, 384, 747
Addition polymers (chain-growth polymers), 1058–1060
Addition reactions, of alkenes, 1045
Adiabatic (term), 258, 858
Adipic acid, 1061
Adsorption, 672
Advanced ceramics, 753–756
Aerogels, 755
Aerosols, 467
AFM (atomic force microscope), 8, 9
Aging, free radicals and, 667
Air, 69, 466
Alanine, 1069–1070
Alanylglycine, 1070
Alcohols, 389–390, 1046–1049
   and alkanes, 94
   major reactions of, 1047–1049
   nomenclature of, 1047
   oxidation of, 1052–1053
Alcohol group, 1038
Aldehydes, 390, 1047–1051
Aldehyde group, 1049
Aliphatic compounds, 1046
Alkali metals (alkalis), 66–67, 175, 841
Alkaline batteries (alkaline dry cells), 935
Alkaline earth metals, 67
Alkanes, 93–94, 1039
   bonding in, 426
   chemical properties of, 1043
   nomenclature of, 1040–1043
Alkenes, 1039, 1043–1045
Alkoxides, 754
Alkyl groups, 754n.6, 1041–1042
Alkynes, 1039, 1043–1044
Allotropes, 450–455
Allotropy, 451
α-amino acids, 1069
Alpha particles (alpha emitters):
   and alpha radiation, 967–968
   and transmutation, 978
   and unstable nuclei, 974
Alpha radiation, 967–968

Alpha rays, 12
Aluminum:
   combustion of, 878
   metallic character of, 70
   oxidation of, 232, 237
   production of, 950
Aluminum-27, 979
Aluminum ions, 82, 751–752
Aluminum oxide, 237, 752, 950
Amides, 1055–1056
Amide ions, 738
Amines, 167, 391, 1039, 1054
Amino acids, 391n.6, 519, 1069
Ammonia, 96
   and acid, 178–179
   and amines, 1054
   and BF_3, 747
   bond dipoles in, 417
   and boron trichloride, 383–384
   and copper, 842–844
   dynamic equilibrium, 167–168
   Haber process for, 672
   and hydrochloric acid, 732–733
   and hydronium ions, 749
   industrial synthesis of, 702–704
   Lewis structure of, 367, 391
   as ligand, 1004, 1006–1008
   in metal complexes, 1021
   molar heat of vaporization for, 541
   as molecular base, 167, 168
   molecular shape of, 410, 411
   organic derivatives of, 1054–1056
   production of, 275–276
   shape and hybridization of orbitals, 431, 432
   and silver halides, 845
   solubility of, 588
   with water, 770
Ammonium acetate, 785
Ammonium chloride, 732, 733
Ammonium dihydrogen phosphate, 86
Ammonium formate, 785
Ammonium hydroxide, 174n.6
Ammonium ions, 90, 175, 782, 785
Ammonium sulfate, 160
Amorphous solids, 557, 1062
Ampere (unit), 945
Amphiprotic substances, 737
Amphoteric substances, 737, 740, 752
Amplitude, of electromagnetic waves, 302
amu (atomic mass units), 16
Amylopectin, 1065
Amylose, 1065
Angstrom (unit), 36, 337
Ångström, Anders Jonas, 337
Anhydrides, 171–172, 750
Aniline, 780

RELATIONSHIPS AMONG UNITS

(Values in boldface are exact.)

Length

1 in. = **2.54** cm

1 ft = **30.48** cm

1 yd = **0.9144** m

1 mi = **5280** ft

1 ft = **12** in.

1 yd = **36** in.

Volume

1 liq. oz = **29.57353** mL

1 qt = **946.352946** mL

1 gallon = **3.785411784** L

1 gallon = **4** qt = **8** pt

1 qt = **2** pt = **32** liq. oz

Mass

1 oz = **28.349523125** g

1 lb = **453.59237** g

1 lb = **16** oz

Pressure

1 atm = **760** torr

1 atm = **101,325** Pa

1 atm = 14.696 psi (lb/in.^{2})

1 atm = 29.921 in. Hg

Energy

1 cal = **4.184** J

1 ev = 1.6022×10^{-19} J

1 ev/molecule = 96.49 kJ/mol

1 ev/molecule = 23.06 kcal/mol

1 J = 1 kg m^2 s^{-2} = 10^7 erg

PHYSICAL CONSTANTS

Rest mass of electron	$m_e = 5.485799094 \times 10^{-4}$ u ($9.10938291 \times 10^{-28}$ g)
Rest mass of proton	$m_p = 1.0072764668$ u ($1.672621777 \times 10^{-24}$ g)
Rest mass of neutron	$m_n = 1.0086649160$ u ($1.674927351 \times 10^{-24}$ g)
Electronic charge	$e = 1.602176586 \times 10^{-19}$ C
Atomic mass unit	u = $1.660538921 \times 10^{-24}$ g
Gas constant	$R = 0.0820575$ L atm mol^{-1} K^{-1}
	$= 8.3144621$ J mol^{-1} K^{-1}
	$= 1.98721$ cal mol^{-1} K^{-1}
Molar volume, ideal gas	$= 22.4140$ L (at STP)
Avogadro's number	$= 6.02214129 \times 10^{23}$ things/mol
Speed of light in a vacuum	$c = 2.99792458 \times 10^8$ m s^{-1} (Exactly)
Planck's constant	$h = 6.62606957 \times 10^{-34}$ J s
Faraday constant	$\mathscr{F} = 9.64853365 \times 10^4$ C mol^{-1}

LABORATORY REAGENTS

(Values are for the average concentrated reagents available commercially.)

Reagent	Percent (w/w)	Mole Solute Liter Solution	Gram Solute 100 mL Solution
NH_3	29	15	26
$HC_2H_3O_2$	99.7	17	105
HCl	37	12	44
HNO_3	71	16	101
H_3PO_4	85	15	144
H_2SO_4	96	18	177

SOURCES OF USEFUL INFORMATION

...vity series for metals	Table 5.3
Atomic and ionic radii	Figure 7.30
Bond energies	App C.4
Density of water as a function of temperature	Table 1.7
Electron affinities	Table 7.3
Electron configurations of the elements	App C.1
Electronegativities of the elements	Figure 8.9
Entropies (25 °C, 1 atm)	App C.2
Formation and instability constants of complexes	App C.7
Free energies of formation (25 °C, 1 atm)	App C.2
Gases formed in metathesis reactions	Table 4.2
Half-lives, radionuclides	Table 20.2
Heats of formation (25 °C, 1 atm)	App C.2
Heats of formation of gaseous atoms (25 °C, 1 atm)	App C.3
Indicators, acid–base	Table 16.7
Ionization energies of first 12 elements	Table 7.2
Ions formed by representative elements	Table 2.2
Ions formed by transition and post-transition elements	Table 2.3
K_a values for monoprotic acids (25 °C)	App C.8
K_a values for polyprotic acids	App C.8
...values for bases (25 °C)	App C.8
K_{sp} values for insoluble salts	App C.6
K_w values at various temperatures	Table 16.1
Lewis structures, rules for drawing	Figure 8.12
Non-SI units used in chemistry	Table 1.3
Oxidation numbers, rules for calculating	Section 5.1
Organic compounds, families of	Table 22.1
Polyatomic ions	Table 2.4
Reduction potentials, standard	App C.9
SI base units	Table 1.2
SI prefixes and multiplication factors	Table 1.5
Solubility rules	Table 4.1
Strong acids	Section 4.4
Vapor pressure of water as a fuction of temperature	App C.5

PERIODIC TRENDS

(Trends in the following properties that can be correlated with the periodic table are discussed in the given sections.)

Atomic sizes	7.10
Crystal field splitting	21.5
Electrical charges on ions	2.5
Electron affinities	7.10
Electron configurations of atoms	7.8
Electronegativities	8.6
Ionization energies	7.10
Metallic versus nonmetallic properties	2.2
Reactivities of metals	8.6
Strengths of binary acids	15.3
Strengths of oxoacids	15.3

Carbon	C
Hydrogen	H
Nitrogen	N
Oxygen	O
Phosphorus	P
Sulfur	S
Fluorine	F
Chlorine	Cl
Bromine	Br
Iodine	I
Silicon	Si
Boron	B